THE

Atlas OF North America's Migratory Birds

032
1825

THE Atlas OF North America's Migratory Birds

Tracking Movement Across Seasons and Continents

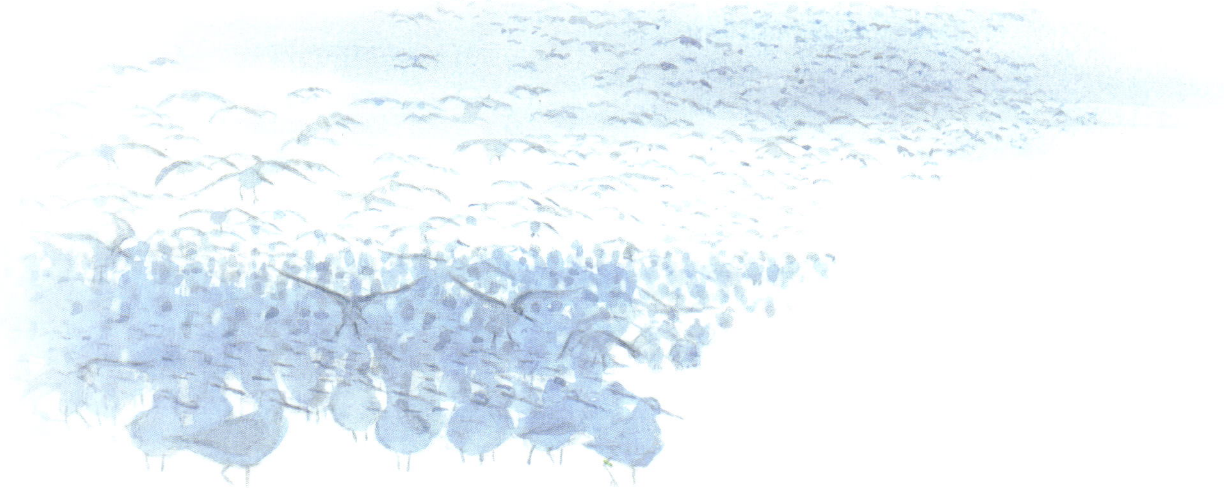

Peter P. Marra
Georgetown University

Amy L. Scarpignato
Smithsonian Institution

Illustrated by John C. Anderton

eBird map data provided by Cornell Lab of Ornithology

Published in association with
Princeton University Press
Princeton and Oxford

DEDICATION

This Atlas is dedicated to all the amateur and professional bird enthusiasts, catchers, mist netters, taggers and collectors from the past, the present, and those still to come in the future. It is their love and insatiable curiosity about birds and their desire to know just a little bit more about where birds go that makes this Atlas possible and necessary.

Published by Princeton University Press
41 William Street, Princeton, New Jersey 08540
99 Banbury Road, Oxford OX2 6JX

GPSR Authorized Representative: Easy Access System Europe - Mustamäe tee 50, 10621 Tallinn, Estonia, gpsr.requests@easproject.com

press.princeton.edu

Map data from Fink, D., T. Auer, A. Johnston, M. Strimas-Mackey, S. Ligocki, O. Robinson, W. Hochachka, L. Jaromczyk, C. Crowley, K. Dunham, A. Stillman, I. Davies, A. Rodewald, V. Ruiz-Gutierrez, and C. Wood. 2023. eBird Status and Trends, Data Version 2022. Cornell Lab of Ornithology, Ithaca, NY. https://doi.org/10.2173/ebirdst.2022.

Map data from BirdLife International and *Handbook of the Birds of the World* (2023). Bird species distribution maps of the world. Version 2023.1. Available at http://datazone.birdlife.org/species/requestdis.

ISBN 978-0-691-16318-5
ISBN (PDF) 978-0-691-27718-9
ISBN (epub) 978-0-691-29186-4

British Library Cataloging-in-Publication Data is available

Editorial: Robert Kirk, Laura Lassen, and Megan Mendonça
Production Editorial: Karen Carter
Text Design: D & N Publishing, Wiltshire, UK
Cover Design: Wanda España
Production: Steven Sears
Publicity: Matthew Taylor and Caitlyn Robson-Iszatt
Copyeditor: Susan Matheson
Cover art by John C. Anderton

All net proceeds from the sale of this book will be used to support Latin American and Caribbean ornithologists and the conservation of migratory birds.

This book has been composed in Helvetica Neue (headings) and Minion Pro (text)

Printed in Malaysia

10 9 8 7 6 5 4 3 2 1

MIX
Paper | Supporting responsible forestry
FSC™ C005748

Recommended Citation:
Peter P. Marra and Amy L. Scarpignato. *The Atlas of North America's Migratory Birds: Tracking Movement Across the Seasons.* With contributions by Autumn-Lynn Harrison, Valentine Herrmann, Kenneth V. Rosenberg, Anthony J. Roberts, Timothy Guida, Richard B. Lanctot, M. Clay Green, Tom Will, Laurie J. Goodrich, Emily B. Cohen, Allison E. Huysman, Camilla Gómez, Melanie A. Smith, Jason P. Loghry, Jean-François Therrien, Tom Auer, Stuart A. Mackenzie, and Kristen C. Ruegg. Princeton, NJ: Princeton University Press, 2026.

CONTENTS

ACKNOWLEDGMENTS

This Atlas was only possible because of the contributions of countless dedicated people and organizations—all committed to the study and protection of birds. We cannot overstate how much this book is built on and simply a culmination of all their efforts. In many ways we are only compilers. Millions of people have contributed to what is presented throughout this book. This includes both professional and amateur scientists, who, over the years, have spent tireless hours applying for grants and organizing projects; waking up at the crack of dawn; in many cases, volunteering time and money; catching, banding, and tagging; measuring, sampling, entering, and analyzing data; and then making the data available, either through peer-reviewed publications or through data repositories, simply because they want the data to see the light of day.

In the case of tracking information, all data contributors are acknowledged within each species account and in the combined Data Contributors section near the end of the book. All datasets are acknowledged in the Dataset Citations section. All data contributors had the opportunity to review the visualizations of their data and the text of their species accounts. In the case of banding data, it is impossible to credit the tens of thousands of individuals who have contributed tirelessly to the banding of birds out of their desire to protect birds and to contribute to a collective understanding of them. From the beginning of banding practices in the early 1900s until today, volunteers and professionals have together banded and marked over 80 million birds, with only approximately 6 million ever being recovered (dead or alive). The areas birds occupy are so vast, and the birds themselves so small, that banding by itself, with only passive means of recovery, is clearly insufficient for understanding migratory connectivity.

Many organizations have been instrumental in the development of this Atlas. Funding for the Migratory Connectivity Project (MCP) and the creation of this Atlas has been provided by ConocoPhillips. Without their support it would not have been possible. Specifically, Natalie Riley has been a tremendous supporter of this Atlas and our work and for this we are grateful. The Smithsonian Institution provided resources for many aspects of this work, including hosting the MCP. Georgetown University, through their support of Marra and his lab, has also contributed in many ways to the development of this Atlas.

The US Geological Survey's (USGS) Bird Banding Lab (BBL) has been invaluable, as have its dedicated chiefs and staff since its start in 1920 and their work in issuing federal permits and bands, supporting training and animal care, and, most importantly, maintaining the extensive North American Bird Banding Program's database of banding and recapture data.

The foresight, effort, and continued support from the BBL and the Canadian Wildlife Service's Bird Banding Office, with backing from the USGS and the US Fish and Wildlife Service, are truly commendable. We would like to extend special thanks to former BBL chief Bruce Peterjohn, current BBL chief Antonio Celis-Murillo, and BBL staff members Danny Bystrak, Jo Lutmerding, Matt Rogosky, and Rita Malarodova for their unwavering support of this project.

The Cornell Lab of Ornithology has changed the landscape of birding, ornithology, and conservation science by creating, supporting, and continuing to develop and expand eBird, their participatory science program. eBird has transformed our ability to visualize and understand the distribution and abundance, and even the status and trends, of birds. We would like to thank the entire eBird team at the Cornell Lab, including key contributors to the maps—Daniel Fink, Ali Johnston, Matt Strimas-Mackey, Shawn Ligocki, Orin Robinson, Wesley Hochachka, Lauren Jaromczyk Oldham, Amanda Rodewald, Chris Wood, Ian Davies, and Andrew Spencer, as well as Steve Kelling and John Fitzpatrick—for their enthusiasm and support for using eBird maps in this Atlas. Equally, these maps were made possible by more than a million data contributors, who eat, drink, and sleep eBird, contributing data from across the hemisphere. It has changed how we all view the biogeography of birds, and we are honored and thankful to be able to illustrate physical eBird maps for the birds of North America.

Thanks to the National Audubon Society, especially Miguel Jiminez and Chad Witko, for assistance with data outreach and collection; Chad Wilsey and Jill Deppe for general and some financial support of the Atlas project; and Nathaniel Seavey, Lotem Taylor, and Erika Knight.

We also appreciate all of our colleagues at the various tag and tracking companies around the world that continue to develop and refine tag and upload technology. The last few years represent a remarkable period of research and development and, importantly, advances in our ability to track smaller and smaller animals, at greater frequencies, and with more precision. These advances are essential for the science and the conservation of all birds.

Many interns and staff have contributed to the Atlas in various ways, including Smithsonian interns Danielle Aube, Paige Becker, Adrienne Dale, Stacey Hill, Justin Saunders, Julie Shieldcastle, Kevin Trujillo, Amanda Zeigler, and Rebecca Zurlo, and Smithsonian staff Leah Culp, Nick Garnhart, and Candace Stenzel. Birds Canada, specifically staff that support the Motus Wildlife Tracking System—Lucas Berrigan, Erica Geldart, Elizabeth Gow, Catherine Jardine, and Denis Lepage, and all collaborating individuals and organizations—and the Bird Genoscape Project—Jacob Job and Christen Bossu—were all instrumental in supporting data management and availability for the Atlas. The Marine Geospatial Ecology Lab at Duke University supporting the Migratory Connectivity in the Ocean initiative (an open-access system synthesizing knowledge of marine migratory connectivity) collaborated to jointly request tracking data from seabirds and to develop initial processing and modeling approaches; thanks to Corrie Curtice, Sara DeLand, Daniel Dunn, Pat Halpin, Ellie Heywood, and Brittany Pashkow. Tammy Davies and Ana Carneiro, staff for BirdLife International's Seabird Tracking Database, helped facilitate data requests and answered questions about dataset formatting. Kristen Orr provided help with design and layout. Jeffrey Hostetler provided help with data analysis.

AUTHOR CONTRIBUTIONS

This Atlas was conceived by Peter P. Marra in 2009 and was planned as a primary contribution of the Migratory Connectivity Project (MCP). The MCP is housed at the Smithsonian Migratory Bird Center, and its goal is to advance the conservation and understanding of animals throughout their full life cycle. Marra, while at the Smithsonian Institution, acquired the funding from ConocoPhillips to support the MCP and the production of this Atlas. Marra co-led the production with Amy L. Scarpignato, who also managed its development and outreach to all data contributors. Map creation was led by Scarpignato and Valentine Herrmann and assisted by Autumn-Lynn Harrison and Allison Huysman. Chapters 1 and 2 were led by Marra; chapters 3 and 4 were led by Scarpignato, and all authors had an opportunity to provide edits.

Writing of species accounts was led by Marra and Scarpignato, as well as by Anthony J. Roberts (waterfowl), waders and wetland birds (M. Clay Green), shorebirds (Richard B. Lanctot and Jason P. Loghry), passerines (Kenneth V. Rosenberg, Camila Gómez, Timothy Guida, Allison E. Huysman, and Melanie A. Smith), diurnal raptors (Laurie J. Goodrich), owls (Jean-François Therrien), larids (Autumn-Lynn Harrison and Guida), grebes and cormorants (Guida), seabirds, including alcids, frigatebirds, marine loons, procellariiformes, skuas, sulids, tropicbirds (Harrison), near-passerines and woodpeckers (Smith), and all species that only have data from the Cornell Lab of Ornithology's eBird online database (Tom Will). All accounts were edited by Marra and Scarpignato. Estimates of migratory connectivity were provided by Emily B. Cohen and Valentine Herrmann. Data support and information based on eBird, genoscape, and Motus programs were supported and provided by Tom Auer, Kristen C. Ruegg, and Stuart A. Mackenzie, respectively.

All pieces of art (minus a few technical illustrations) were painted by John Anderton. Anderton's paintings are not only about the bird; more importantly, they perfectly capture the essence of each species in its environment. He paints behavior, geographic location, and the nuances of each species, almost as if you are looking at the actual bird through binoculars. Anderton has placed his subjects in varied habitats across the year so that the reader appreciates how a species might interact with its environment throughout its full annual cycle. Each stroke of Anderton's brush makes the Atlas sing and adds to the depth, richness, and excitement that the connectivity data reveal for the first time but that the data couldn't do alone. He has also painted the occasional band, wing tag, and transmitter on individual birds.

Maps and data reveal the hidden journeys birds take throughout the year. The artwork helps us experience what it is like to truly see them. Together, we hope this integration brings these pages to life and fosters continued observation and conservation of these remarkable animals.

Peter P. Marra
The Earth Commons Institute; Department of Biology, Georgetown University
Washington, D.C., USA

Amy L. Scarpignato
Smithsonian's National Zoo and Conservation Biology Institute, Migratory Bird Center; Road to Recovery, The Earth Commons Institute, Georgetown University
Washington, D.C., USA

Contributors

Autumn-Lynn Harrison
Smithsonian's National Zoo and Conservation Biology Institute
Washington, D.C., USA

Valentine Herrmann
Smithsonian's National Zoo and Conservation Biology Institute
Front Royal, VA, USA

Kenneth V. Rosenberg
Cornell Lab of Ornithology, Cornell University
Ithaca, NY, USA

Anthony J. Roberts
US Fish and Wildlife Service, Migratory Bird Program
Rutland, VT, USA

Timothy Guida
Coastal Research and Extension Center, Mississippi State University
Mississippi State, MS, USA

Richard B. Lanctot
US Fish and Wildlife Service, Migratory Bird Program
Anchorage, Alaska, USA

M. Clay Green
Department of Biology, Texas State University
San Marcos, TX, USA

Tom Will
Road to Recovery & Partners in Flight

Laurie J. Goodrich
Hawk Mountain Sanctuary Association
Orwigsburg, PA, USA

Emily B. Cohen
Appalachian Laboratory, University of Maryland Center for Environmental Science
Frostburg, MD, USA

Allison E, Huysman
Department of Entomology, University of Maryland
College Park, MD, USA

Camila Gómez
SELVA
Bogotá, Colombia

Melanie A. Smith
National Audubon Society
New York, New York, USA

Jason P. Loghry
Coastal Bend Bays and Estuaries Bird Program
Corpus Christi, TX, USA

Jean-Francois Therrien
Hawk Mountain Sanctuary Association
Orwigsburg, PA, USA

Tom Auer
Cornell Lab of Ornithology, Cornell University
Ithaca, NY, USA

Stuart A. Mackenzie
Birds Canada
Ontario, Canada

Kristen C. Ruegg
Department of Biology, Colorado State University
Fort Collins, Colorado, USA

Illustrator

John C. Anderton

CHAPTER 1

DISCOVERING UNKNOWN MIGRATIONS

Sunrise and sunset, the ebb and flow of tides along coastal land masses, and the seasonal transitions between winter, spring, summer, and fall are all reflective of Earth's physical cycles. Cycles are events that repeat in time and over space. In addition to physical cycles, biological cycles are omnipresent: biogeochemical cycles move and transport chemicals (such as water, nitrogen, carbon dioxide) between systems and living organisms; plants and animals have seasonally changing physiologies for reproductive and nonreproductive periods; and species proceed through life cycle stages that are adapted to differing conditions and environments. Cycles are pervasive throughout all physical and biological systems, and in many cases, the biological cycles are driven by the physical cycles themselves, all in a chase for energy and food.

The rotation of the tilted Earth around the sun—arguably the most fundamental physical cycle of our planet—entrains enormous numbers of living things into these daily and annual biological cycles. Biological cycles occur at multiple levels of organization, from the biochemical and cellular to the individual, population, and species level. Daily physical cycles can involve transitions from periods of light to dark and can regulate critical behaviors and life histories, from flowering and feeding, to exercise and sleep, to driving the annual cycles of plants and animals through the seasonal changes in weather, climate, and primary productivity. Consequently, the annual cycles of most vertebrates are organized into a series of breeding and nonbreeding phases (often including a large-scale movement phase and a stationary phase), events that vary in duration and often result in the movement of individual animals across geographic locations at various spatial scales as they track critical resources. Nowhere is this more evident than with species that make repeated annual movements to take advantage of pulses of food and other resources and to avoid food scarcity, predators, disease, and harsh seasonal environments—all to maximize reproduction and survival. These repeated annual journeys, directed and intentional movements to-and-fro, from breeding to nonbreeding areas, characterize and define *migration* (Newton 2023).

Migration is one of the most spectacular behaviors we see in nature and can be an astounding phenomenon to observe, in part because of the sheer numbers of animals that can be seen migrating at a single place in time along their journey. The diversity of taxa exhibiting migratory behavior is equally impressive and likely underestimated. Migratory taxa include invertebrates (from moths to butterflies to dragonflies) and vertebrates including fish (from eels to salmon to sharks), amphibians (salamanders), reptiles (from snakes to sea turtles), birds (from hummingbirds to starlings to albatross), and mammals (from bats to wildebeest to walruses and whales).

Where and why animals move has fascinated humans for millennia. In some cases, the migrations are completed by individuals within a single generation, and in others, one annual cycle migration is completed over the course of multiple generations, as is the case for most species of migratory insects such as the Monarch butterfly (*Danaus plexippus*; Reppert and de Roode 2018) and the Green Darner dragonfly (*Anax junius*; Hallworth et al. 2018).

Despite the breadth of migratory behavior demonstrated across taxa in the animal kingdom, bird migration is the exemplar of the phenomenon. Birds are extremely observable, as they traverse human-dominated habitats, and their movements en masse are extraordinary. Almost everyone has witnessed a flock of geese, robins, or starlings fly across the sky—sometimes in a ballet-like murmuration. Birds move through the air, often in remarkable numbers, making their migrations highly observable. Birds are also migration champions—the Arctic Tern (*Sterna paradisaea*) undertakes the longest known migration of any animal in the world, moving almost from pole to pole (Egevang et al. 2010). Despite their observability, many migratory bird species have also been challenging to study because their movements occur under the dark of night and over vast areas. As easy as it is to see them at a single place in time, it was for many years impossible to follow individuals throughout a migratory journey, especially for smaller species, once out of sight.

Globally, there are eight recognized primary avian migration systems (Greenberg and Marra 2005; Newton 2023). Each of these migration systems can be segregated into fairly distinct global geographical areas, include several other distinct flyways, and can have migratory birds spending time within and between temperate and tropical zones. Seabird migration systems, six in total, largely fall outside of these eight systems (Morten et al. 2025). One largely terrestrial example, and a focus of this book, is the Nearctic-Neotropical migration of birds that breed in temperate North America. There is enormous variation of species and pathways within this migration system, including many species that migrate to the tropics for their nonbreeding period (tanagers, warblers, and vireos) and others whose migratory movements are restricted within the Nearctic (juncos, robins, and bluebirds).

Despite the importance of migratory movements in the life cycle of these species, the periods of migration themselves are often a relatively short component (one to eight weeks) of the annual cycle. There is typically a migration prior to breeding (termed the pre-breeding period; e.g., March–May in the northern spring) as individuals return to nesting areas, and another after breeding (termed the post-breeding period; e.g., August–October in the northern autumn) as birds return to their nonbreeding grounds (e.g., to spend the austral winter months for Nearctic breeders). Annual migrations are intense and critical bookends, if you will, around the breeding, molting, and stationary nonbreeding periods, which together make up the annual cycle for many migratory birds. The post-migration, nonbreeding period has indeed generally been considered stationary, but this is becoming less the norm and more the exception (Teitelbaum et al. 2023). We now know there is quite a bit of variation in the spacing behavior of migratory birds during the nonbreeding period. For example, species like the Wood Thrush (*Hylocichla mustelina*) and American Redstart (*Setophaga ruticilla*) use a territorial/floating system for several months (Rappole et al. 1989; Marra and Holmes 2001; Stanley et al. 2021a; Dossman et al. 2024). There are species with regular flights between foraging and roosting areas, like the Northern Waterthrush (*Seiurus noveboracensis*; J.A. Smith et al. 2008) and American Oystercatcher

(*Haematopus palliatus*; Sanders et al. 2013). Some species move between a series of sites separated by hundreds of kilometers, such as the Veery (*Catharus fuscescens*; Heckscher et al. 2011) and Bobolink (*Dolichonyx oryzivorus*; Renfrew et al. 2013); and several species of seabirds engage in wandering movements across vast oceans (Bernard et al. 2021).

The nonbreeding period has sometimes been referred to as the "wintering" period by Northern Hemisphere ornithologists, which has caused confusion because it is not the winter if the birds are below the equator in the tropics during these months. For this reason, in this Atlas, we avoid using the names of seasons to define periods of the annual cycle. Specifically, we avoid the word "wintering" and will instead use "nonbreeding" to refer to the nonbreeding nonmigration period of the annual cycle. We recognize that migration is also technically a nonbreeding activity, but we will separate the migratory phase of the nonbreeding period as pre- and post-breeding migration rather than use the phrases "spring" and "fall" migration, which similarly represents the seasons in the North and not necessarily the South (Albert and Siegel 2024).

A classic annual cycle of a migratory terrestrial bird comes from the well-studied species, the American Redstart (*Setophaga ruticilla*). The redstart is a small but tenacious and colorful songbird that weighs less than a US quarter. Depending on the breeding location, redstarts will embark on a pre-breeding migration, lasting approximately two weeks, from tropical nonbreeding areas from March to May and arrive onto their breeding grounds from April to June. Having arrived to breeding areas, males and females will pair; females will build a nest, lay eggs, and fledge a single brood of young by July. After fledging, both parents will continue to feed the brood outside of the nest for an additional two weeks. Afterward, adults and the newly independent young birds will molt and then undergo a post-breeding migration south to the nonbreeding grounds, where they will spend the majority of the annual cycle from approximately October to April. Redstarts have a broad range and distribution and can fly thousands of kilometers during their migrations, much of it at night, and sometimes across large bodies of water. Their nonbreeding range extends from northern Peru to Central America, Mexico, the Caribbean, and southern Florida, USA. Redstarts breed from the bayous of Louisiana, USA, east and north to the Atlantic provinces of Canada and west through Montana to northern Alberta and British Columbia. Individuals show high between-year site fidelity to both breeding and nonbreeding areas, but how individuals and regional populations are connected across these annual cycles has been a mystery for the American redstart and many other migratory species. In other words, seeing a map with a breeding and nonbreeding range for a given species fails to portray the movement or origins between these two important periods in the life history of these species. Needed is an understanding of how populations within a species are connected in space and time throughout the year between different stages of the annual cycle, i.e., their migratory connectivity (Webster et al. 2002).

We define a population as a group of individuals within a species that live in a specific area or region. Defining where a population starts and stops is not a simple undertaking and in fact is poorly understood for most species. In some cases, there are genetic and morphological distinctions (i.e., populations sometimes defined as subspecies), and in others there are simply spatial separations that create hypothetical populations. Regardless, understanding how individual birds and their populations move within a year (whether they stay together or separate) and throughout their lifetimes is essential for our understanding of a species' biology and also for its conservation.

This book provides the first comprehensive and systematic summaries of all tracking, movement, and observational data available for the migratory birds of North America. This includes data collected by observation, by banding and other extrinsic tracking tools, and by intrinsic markers based on genetics and stable isotopes for 564 species of birds, either breeding within the United States or Canada, or in the case of some species of seabirds, species that breed in the Southern Hemisphere and migrate to North American waters for their nonbreeding period. We use summary text and a collection of maps to illustrate their migratory connectivity. Only recently have we been able to discover the migratory connectivity for so many species (Scarpignato et al. 2023). This is due, in part, to recent advances in an assortment of tracking techniques, including both intrinsic (e.g., genetics, stable isotopes)

and extrinsic (e.g., bands and markers, light-level geolocators, satellite tags) tools, but also because of an almost century-old effort of banding and resighting of birds, managed by the USGS Bird Banding Lab, by thousands of passionate people who are dedicated to the study of birds. Each map may be compiled from a variety of these tools and provides insights into the patterns of migratory connectivity. We also use data from the Cornell Lab of Ornithology's eBird online database, which includes single observations of individual birds that are compiled and scaled up to include contributions by tens of thousands of dedicated bird-watchers. When these individual points in space and time feed into distribution models at larger spatial scales, they provide the most detailed patterns of distribution and abundance of bird species ever developed. Where available, these models form the base maps for species in the Atlas. In addition, eBird data can provide novel insights into movements of distinct populations. For the many species with little or no tracking data, we turn to these for possible insights into migratory connectivity. In this introductory chapter we define migratory connectivity and explain why understanding it is so important for advancing fundamental biology and for conservation. The specifics for how we can track individual birds and populations with extrinsic tags and intrinsic markers are detailed in chapter 2. In chapter 3, we outline the analyses and methodologies used to create this Atlas, and in Chapter 4 we define the components of the species accounts, which constitute the majority of this Atlas.

WHAT IS MIGRATORY CONNECTIVITY?

Globally, at least 2,500 species of birds, approximately one-quarter of all avian species, are considered migratory—and this is likely a gross underestimate. In part, because of the variation in migration strategies, we have yet to understand movement patterns within and between species. For example, of the approximately 75% of North American species considered migratory, some are considered *obligate migrants*—where all or most individuals within a species undergo a fixed, repeated, and predictable migration (virtually all Arctic and boreal breeders; e.g., Red Knot, *Calidris canutus*; Blackpoll Warbler, *Setophaga striata*)—and others are considered *facultative migrants*—where there appears to be flexibility, over years, within an individual bird in its tendency to migrate (e.g., American Robin, *Turdus migratorius*; Brown and Miller 2016). Some populations of species within facultative migrants also exhibit *partial migration*, where some individuals within a population may migrate while others remain resident year-round (e.g., Skylark, *Alauda arvensis*; Hegemann et al. 2015). In general, we know more about the biology of obligate migrant species than facultative or partial migrants, in part because the former is easier to document. Obligate migrants disappear entirely from one area and show up in another; whereas facultative migrants like American Robins and Eastern Bluebirds (*Sialia sialis*) will stay within geographical regions year-round. Until an individual tracking or banding study is conducted, one does not know if the robins present in your yard in December and January are the same birds that bred in that location in June, or if these robins bred farther north and are now spending the nonbreeding season in your yard.

Given the enormous complexity and fascinating variation to avian migration systems within and between species, it is important not to stereotype our thinking based on any one taxonomic group of birds. For example, in contrast to the American Redstart example above, pelagic seabirds spend most of their lives foraging over the open ocean. Seabirds provide an excellent case study of how variable the life history schedules of birds can be. Some species undertake classic long-distance migrations between two distinct ranges (e.g., Arctic Terns; Wong et al. 2021), moving from nest sites in the Arctic tundra to large oceanic, nonbreeding ranges in the Southern Hemisphere. Others may embark from their breeding period spent on islands to wander over ocean areas the size of continents during a single nomadic, nonbreeding period (e.g., Laysan Albatross, *Phoebastria immutabilis*; and Black-footed Albatross, *Phoebastria nigripes*; Harrison et al. 2018). The anatomical and physiological adaptations of seabirds enable them to track patchy and unpredictable ocean prey. Seabirds also can challenge our preconceived notions of when and how often migrations occur. Once individuals hit breeding maturity, which, in some species such as the Black-Browed Albatross (*Thalassarche melanophris*), may not start until 6–13 years of age, breeding can comprise much of the annual cycle (del Hoyo et al. 2020a). Once paired and on islands, both male and female can participate in incubation and chick-rearing for up to six months, during which time adults undertake commuting flights to marine foraging areas. These flights during the breeding season can be as long as some terrestrial bird migration distances. Following breeding, pelagic seabirds depart for a life completely at sea, wandering vast expanses of the ocean tracking resources. Clearly, the differences across avian taxa from terrestrial to pelagic can be extreme, but what remains consistent is the need to understand where individuals and populations go throughout their life histories and annual cycles.

Regardless of the avian system, or the time schedules making up the various components of the annual cycle, quantifying migratory connectivity—how individuals and populations within a species redistribute across space and time—is essential for both an understanding of fundamental biology and for species conservation. This is not a new realization. Early migration biologists, such as Finn Salomonsen (1955) emphasized the importance of annual cycles, as well as how migratory bird populations might be connected throughout the year. Our knowledge of migratory connectivity and its importance have advanced since that first influential paper, but these basic facts still have been poorly described for most species.

The spatial component of migratory connectivity includes both a geographic pattern (consider, for example, "leapfrog," "chain," or "parallel" migrations) and a strength component (the extent to which individuals and populations remain together between phases of the annual cycle in both space and time; Bauer et al. 2016; Knight et al. 2021a; Roberts et al. 2023). Regarding geographic patterns, individuals and populations could use a *leapfrog migration*, where the northernmost, nonbreeding populations of a species breed in the southernmost portions of the breeding range, while more southern nonbreeding populations of the same species leapfrog southern breeding populations and spend their breeding seasons farther north. This would result in a differential migration distance between populations. Alternatively, individuals may use *chain migration*, where the populations that spend the nonbreeding season in the north also spend the breeding season in the north, and the populations that spend the nonbreeding season in the south also spend the breeding season in the south. Therefore both sets of populations travel similar distances for migration, but the two patterns of migration occur across different sets of latitudes. *Parallel migration* involves different sets of populations moving north to south across the same sets of latitudes but positioned at different longitudes throughout the annual cycle and largely not crossing (Rose-breasted Grosbeak, *Pheucticus ludovicianus*; figure 1.1A).

The second spatial component refers to the strength of migratory connectivity and emphasizes the degree to which individuals from different populations stay together during migration and on their breeding and nonbreeding grounds. Migratory connectivity is often described categorically on a scale from strong to moderate to weak (figure 1.1) but can also be estimated quantitatively from 0 (weak) to –1 or 1 (strong). The strength of migratory connectivity is, to some degree, scale-dependent. How we define a population and how sampling or tag deployment is done can influence our measures of strength (see the "Quantifying the Strength of Migratory Connectivity" section in chapter 2). Regardless, considering both the pattern and the strength of the migratory connections can have profound implications for understanding many aspects of the ecology and evolution of migratory species and, most importantly, for our ability to conserve their populations.

WHY MIGRATORY CONNECTIVITY MATTERS

Today, we find ourselves in a race against time to reverse population declines and prevent possible extinctions of many bird species (K.V. Rosenberg et al. 2019; P.A. Smith et al. 2023). Almost half of the migratory bird species in North America are experiencing significant declines. Over the last fifty years, the overall North American avian population size has declined by approximately 3 billion birds. For the most part, we understand broad threats (e.g., habitat loss, invasive species, contaminants), but we often do not know why individual species or populations are declining. Knowing where individuals and populations go throughout their annual cycles is the first step to understanding the threats faced by individuals, helping us to decipher where, when, and why individuals are dying—all critical to understanding the causes of population decline. Thus, understanding migratory connectivity is an essential foundation to finding a solution and implementing effective measures to curb declines, and also to predict how species will respond to climate change, habitat loss, or pesticides.

For some range-restricted species, i.e., those that have small breeding, migratory, or nonbreeding areas, the migratory connectivity may

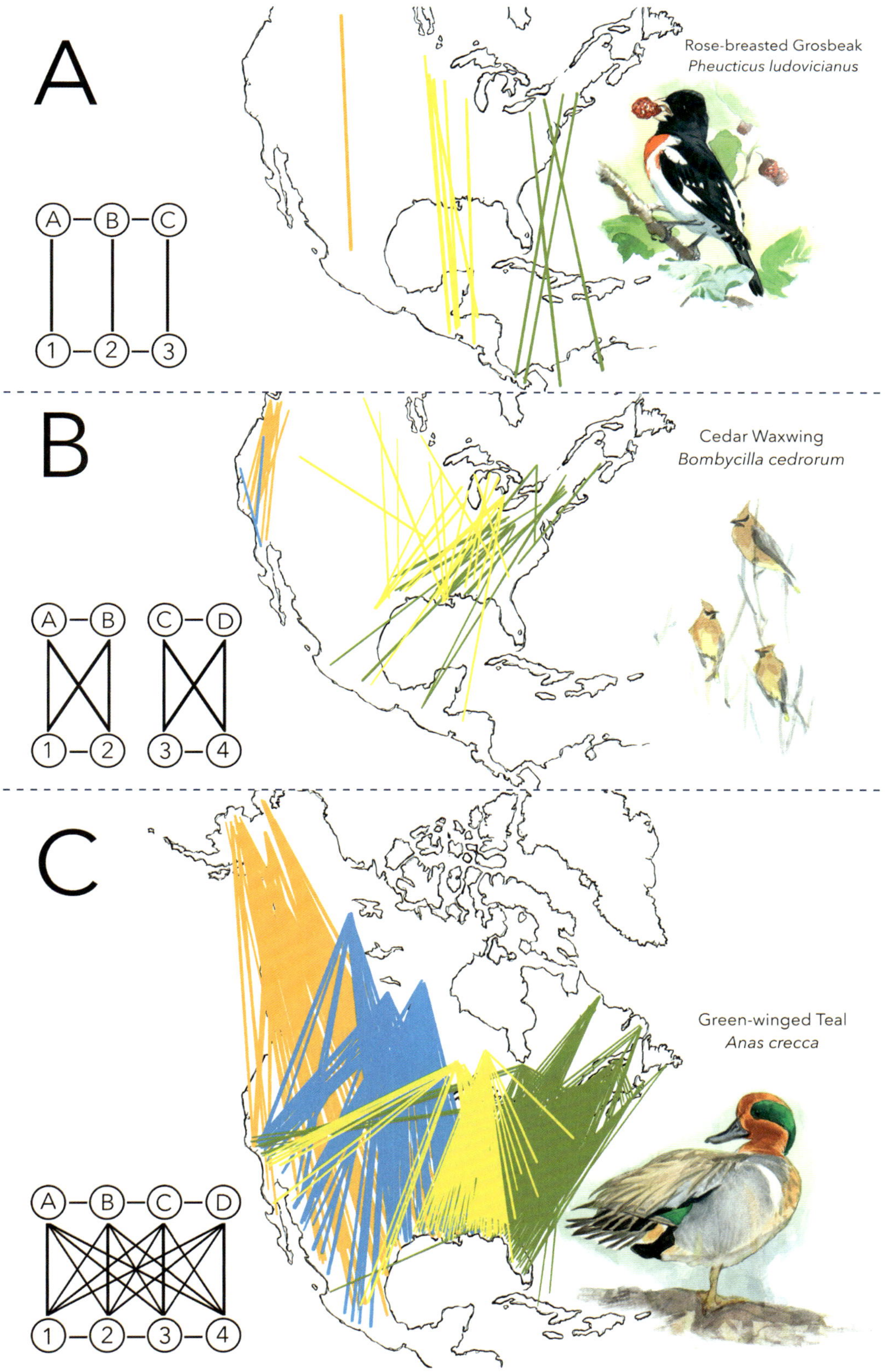

FIGURE 1.1 *Migratory connectivity can be considered on a gradient: (A) strong, (B) moderate, (C) weak. Figure adapted from Cohen et al. 2018.*

or may not be known, but conservation of the species will depend on the protection of one or all of those restricted geographic areas. Nowhere is this better illustrated than with the Kirtland's Warbler (*Setophaga kirtlandii*), a migratory warbler almost lost to extinction in the late 1970s (Mayfield 1983). The Kirtland's Warbler was one of North America's most endangered songbirds but has recently been removed from the US endangered species list. Over 95% of the population of Kirtland's Warblers breed in the Jack Pine forests of Michigan, and the locations of sizable nonbreeding populations, primarily on four islands in the Bahamas, were discovered only in the last twenty-five years (Wunderle et al. 2010). Prior to this, including when the species was reduced to just 167 singing males in the 1970s and 1980s, we were not confident where Kirtland's Warblers spent the nonbreeding season (Probst 1986). At the time, the entire nonbreeding area was known from only a few observations and historical specimen records, and almost nothing was known about the routes individuals took during migration. Only in the last few years have we identified the migration pathways of Kirtland's Warblers thanks to the advent of a variety of miniaturized tracking technologies (N.W. Cooper et al. 2017). Knowing the migratory connectivity provided critical insights to support the future protection of the entire species and their habitats throughout the annual cycle. Today, due in large part to conservation taking place throughout the annual cycle of this extremely range-restricted species, Kirtland's Warblers have a population of nearly 2,000 singing males.

For most migratory species in North America, we know the ranges occupied throughout the annual cycle. Indeed, species ranges have, for the most part, been documented and available in popular field guides for years. However, it is important to remember that each species' breeding range may be composed of multiple populations that can, for example, vary in demography and population trajectory, morphology, and/or strength and timing for how individuals from those populations move throughout their annual cycles. Individuals from each of these populations may migrate south to a nonbreeding area where they spend the majority of the annual cycle, and the obvious question becomes: Do these populations remain distinct?

Considering both the pattern and the strength of the migratory connections can have profound implications for our ability to conserve both populations and species (Cohen et al. 2018). Why is this important? For many reasons, although a primary reason is that knowing connectivity is critical to understanding the drivers of population change: Where, when, and what are the threats (e.g., habitat loss, pesticides, illegal hunting, and overharvesting) potentially impacting individuals and populations? This is important for conservation and also for simply understanding population dynamics. Several classic examples illustrate the critical nature of connectivity information for understanding population dynamics and conservation of species; one of these is the Black-crowned Night Heron (*Nycticorax nycticorax*).

The Black-crowned Night Heron is a migratory waterbird that is fairly widely distributed across the United States. Across their range in the late 1960s and early 1970s, Black-crowned Night Herons were declining, presumably due to DDT or DDE exposure. After DDT was banned in 1972, most populations rebounded. A breeding population in the northwest of the United States was still found to have high levels of DDE, low numbers of nestlings, and a high incidence of cracked eggs (Henny and Blus 1986). Interestingly, no DDE was found in prey species on the breeding grounds, so it was not understood where night herons obtained the DDE in the eggs. Despite knowing the entire nonbreeding range for the species, specific nonbreeding locations of individuals from this declining subpopulation were unknown. In 1979, Charles Henny and Lawrence Blus attached radio transmitters to 15 night herons at a breeding population in Ruby Lake, Nevada, known to have birds with heavily contaminated eggs, and attached transmitters to 14 night herons near Pocatello, Idaho, where the eggs had significantly lower levels of DDE. Using aerial surveys and some band recoveries, the researchers were able to determine that these two breeding populations used mostly different nonbreeding areas, consistent with differences in DDE prevalence in eggs. Pocatello birds that had no DDE exposure primarily spent the nonbreeding season on coastal zones of central Mexico, where there was minimal DDT in the environment, whereas contaminated night herons from Ruby Lake spent the nonbreeding season in the southwestern US and interior and northern Mexico (where DDT was still in the environment at high levels). In this study, knowing

Population-level seasonal interactions, like those at the individual level, can occur across any season or periods of the year.

Collectively, both individual and population-level seasonal interactions illustrate how events throughout the full annual cycle interact to drive the fundamental ecology of more and more species. Indeed, a growing body of research is demonstrating that, although the stages of the annual cycle are often temporary and/or geographically separate, they are inextricably linked. While the terminology and empirical evidence is relatively new, seasonal interactions have actually been in the literature for a long time, including from scholars as noteworthy as Charles Darwin, who is credited with noticing that events prior to the breeding period can influence female fecundity in migratory birds. Capital breeding versus income breeding, another important example of a carry-over effect, refers to where and when birds access nutrients for reproduction, either from stored capital or from current intake; the tradeoffs related to these methods are supported by a robust literature (Stephens et al. 2009). Other examples exist in evolutionary biology, including morphological traits such as color, as is the case for delayed plumage maturation (Rohwer and Butcher 1988).

the migratory connectivity was critical in confirming how a nonbreeding location was contributing to differing exposure rates and population declines within a migratory species.

Clearly understanding migratory connectivity is fundamental to our ability to identify when and where in the annual cycle vital rates (i.e., survival, reproduction) are being impacted. Nothing is as simple as it seems, and events that happen throughout the annual cycle are not independent simply because we characterize them into discrete periods. In fact, overwhelming evidence supports the idea that events in one period or season of the annual cycle can interact to affect individuals (i.e., carry-over effects) and populations (i.e., density-dependent interactions) in subsequent periods (e.g., Marra et al. 1998, X.A. Harrison et al. 2011). Collectively, we refer to these as seasonal interactions, and these complex ecological relationships can occur at the individual and population levels (Marra et al. 2015). These are more than likely the rule rather than the exception but have just been understudied. To study seasonal interactions, it is essential to understand migratory connectivity.

At the level of the individual, the effects of ecological events (e.g., low food availability, high individual density, high precipitation) occurring during one period can influence individuals and carry over, through nonfatal effects, to impact the same individuals in subsequent periods (for a review, see X.A. Harrison et al. 2011). Carry-over effects are types of seasonal interactions that impact individuals and occur between seasons. One example is when individuals in poor physical condition or with a delayed departure from nonbreeding areas, carry over these effects into migration or to the breeding grounds, by arriving late at key migration sites (after most resources have been depleted) and are delayed on breeding initiation, resulting in ultimately lower reproductive success in that later season. It is important to point out that carry-over effects (which are nonfatal) can originate at any point in the annual cycle and then influence subsequent events, condition, performance, or behaviors in a later season (O'Connor et al. 2014).

At the population level, high reproductive success on breeding areas can result in high recruitment of individuals to the nonbreeding grounds. Depending on the strength of migratory connectivity, more individuals per unit area theoretically may result in negative density-dependent feedback. That is, because there are more individuals, density goes up on the nonbreeding grounds, resulting in either fewer resources per individual or birds being forced into habitats of lower suitability, resulting in a decline in an individual bird's physical condition.

Despite long recognizing the importance of events throughout the annual cycle, our fundamental understanding of animal ecology and, likewise, our ability to implement effective conservation strategies to save species, particularly birds, may make significant advances as our understanding of migratory connectivity improves for many bird species. Currently we have some minimal level of connectivity information for 68% (383 out of 564) of North American bird species (this Atlas). The good news is that the technological playing field for tracking animals is improving rapidly. In the next chapter, we describe the emergence of animal tracking, including new scientific tools and techniques, along with powerful citizen science efforts, such as eBird, that allow for unprecedented advances in the pace and magnitude of data collection, analysis, and modeling of animal movements and migrations. The use of radio telemetry (e.g., the Motus Wildlife Tracking System), light-level and multisensor geolocators, satellite transmitters (e.g., Argos and GPS), stable isotope analysis, and genomics are rapidly advancing our ability to track individuals. Similarly, eBird provides a continental-scale perspective of not only where birds are but how entire species and populations move over the full annual cycle. When these data are combined with climate and land use data, and shared with conservation practitioners and policymakers (B.L. Sullivan et al. 2017; Davies et al. 2021; A.-L. Harrison et al. 2024), the emerging insights and advances into basic bird biology and conservation have been nothing short of magnificent.

Ultimately, the conservation of migratory species and the grand phenomenon of migration itself will depend not only on an improved biological understanding of how populations are linked throughout the annual cycle but will also require us to surround this migratory connectivity framework with social, legal, and enforcement support structures to better protect linked populations and the habitats they depend on (A. Farnsworth et al. 2024). Connecting individuals and populations of migrating species throughout the annual cycle will also, through targeted educational efforts and campaigns, improve stronger social connections between the people who are unknowingly bound through the wide-ranging species they share. It is essential that we continue to promote understanding and appreciation of these remarkable species, true global citizens, if we hope to protect them for future generations.

CHAPTER 2

A NEW AGE OF ANIMAL TRACKING

APPROACHES AND TOOLS FOR QUANTIFYING MIGRATORY CONNECTIVITY

The annual and seasonal appearance and disappearance of birds and other animals have piqued the curiosity of natural historians for well over 2,000 years. Some of the early and creative explanations of these disappearing acts include: species transformed into other species, birds hibernated underwater with fish, and some species even migrated to the moon. Thankfully, biologists eventually realized that the evidence supporting these wild theories had no basis. A similarly preposterous idea eventually emerged that the disappearing animals were actually moving around the globe every year. This was a more reasonable hypothesis and was supported by real data, including specimen collections, bird observations, and eventually a smattering of banding recovery data from throughout the annual cycle and from distant localities. These early types of data provided a more convincing interpretation of why birds were seasonally disappearing. Equally interesting was the early approximations of the annual and season-specific ranges of birds. Yet, it has been only recently, in the last twenty years or so, that our understanding has become somewhat detailed and more accurate of the annual ranges, including both the distribution and the abundance of many, but certainly not all, bird species, excluding, for example, seabirds and boreal species. This improved understanding emerged through a timely combination of the exponential growth of bird-watching and citizen science, a new ability to analyze massive amounts of data with almost a single keystroke, and a renewed interest in animal tracking. Almost simultaneously, several new tracking technologies were developed, along with new innovative quantitative approaches, which have allowed scientists to track individual animals over time and space to better understand migratory connectivity. Clearly, the last quarter century has been a renaissance for tracking animal movement.

Determining where individuals and populations of migratory animals move throughout their lives (both their annual and full life cycles) has challenged biologists for many years. In fact, it has been one of the grand challenges in not only ornithology but also more broadly in animal ecology (e.g., Webster et al. 2002; Kays et al. 2015). We glimpse a bird in the woods, in a grassland, or on the beach and it can be gone in seconds, never to be seen again. Even for a captured bird with a simple leg band, once released the probability is extremely low that the individual bird will ever be seen again, unless the bird is site faithful. Until recently, much of our research and understanding of birds and other animals has come from these brief observations of individuals and usually from during only one period in their annual cycle (Marra et al. 2015). We then build theories and hypotheses around these glimpses of observations or collections of data to explain that which we cannot see. Times have changed. This chapter describes new approaches and tools, from genetics to stable isotope analysis, to massive citizen-science endeavors, to tags attached to the backs of live birds that communicate to receivers on satellites circling the Earth. These tools are allowing us to understand how birds have populated the hemispheres over the past 100,000 years, to visualize their contemporary distributions, to follow almost in real time how, when, and where individuals move in space and time through the annual cycle. We are truly in a new age of animal tracking, with advancements in technology occurring at a tremendous pace, opening up new windows in our understanding of the biology of these remarkable species.

The marking of birds by amateurs and scientists has had a long and diverse history in human culture, going back as far as 200 BCE. Examples include crows found with thread attached to a leg, presumably by Roman soldiers, and in the Middle Ages, falcons fitted with metal plates embossed with family seals. Frederick Lincoln (1921), writing about the history of bird banding, describes other types of attempts, including marking flight and tail feathers with indelible ink or metal discs, or mutilating parts of the bill or feet in some way. None of these attempts were ever suitable for actually tracking birds. The first record of a bird being banded was a Gray Heron (*Ardea cinerea*) captured in Germany in 1710 with multiple metal bands attached to its leg, one of which apparently had been attached in Turkey years prior, but these data are poorly substantiated (Wood 1945). Although this record from a heron may be the first successful resighting of an individual banded bird, the jump from a resighting event to understanding the broader phenomenon of migration still took some time.

It was still a few hundred years of head scratching and of looking under rocks and toward the stars by our ornithological predecessors before the annual disappearance and appearance of birds, the phenomenon of migration, was first truly deciphered. Starting around 1822 in Germany, the observation of White Storks (*Ciconia ciconia*) that had arrows pierced through their necks (termed "Arrow Stork" or *Pfeilstorch* in German) made the birds traceable to African tribes (Kinzelbach 2005). Finally, long-distance migration was convincingly demonstrated.

Now, two hundred years later, we have seen the evolution of a fascinating panoply of intrinsic and extrinsic tracking technologies, all developed to understand various aspects of the migratory biology, behavior, and connectivity of birds and other organisms. Through all these efforts—observations from birders, banding and reencounter data, data from tracking birds with devices on their backs—we have learned an enormous amount about where and how birds move over space and time, and one clear lesson is that no single technique satisfies all needs. The diversity of birds and their varied habitats understandably requires a diversity of approaches, techniques, and technologies.

In this chapter, we provide an overview of the approaches, techniques, and technologies that scientists have been using to track the movements of birds. First, we describe how weather radar and bird-watching itself, combined with new massive data processing capabilities, have been essential to advancing our understanding and visualization of the distribution and abundance of bird species in ways never before possible. These data are extremely helpful for understanding emerging patterns of migratory connectivity. Second, we provide an overview of the history of bird banding and detail some of the issues of this hundred-year-old plus approach to tracking bird movements. Third, we describe tracking approaches that use *extrinsic* devices affixed to the body of a bird that archive data and that require recovery, and also devices that transmit locations to orbiting satellites and to stationary, terrestrial receivers. We also describe some of the ways extrinsic tags and devices have been attached to birds. Last, we describe how *intrinsic* markers can be used to determine the origin of an individual bird; for example, a genetic signature can determine if a population is isolated, and a stable isotopic ratio can diagnose what a bird has recently eaten. Collectively, the advances in tracking birds, and animals more broadly, have transformed our understanding and represent some of the most exciting research areas in ecology and evolutionary biology. Information that is also essential for their conservation. Even more importantly, they allow us to tell the stories of these migratory animals as they transcend the hemispheres and the globe, connecting us beyond our borders in ways we still do not fully comprehend. Moreover, once visualized, the scale and awe of the tracks themselves draws in professional ornithologists, amateurs, and even those who have no prior knowledge of birds. This is the inescapable lure of migratory birds.

METHODS FOR TRACKING BIRDS

— OBSERVATIONAL APPROACHES

Scientists have been able to describe the pre- and post-breeding migrations of hundreds of millions of birds, moving primarily at night, across the Western Hemisphere. During the course of these multiday journeys, migratory birds need to stop to rest and refuel during the day, and then they depart within an hour after sunset and fly through the night. Direct observation of nocturnally migrating birds can be extremely difficult, and a variety of approaches have been employed to study these nocturnal migrations, including: coordinating networks of birders to use telescopes at night to observe silhouettes of birds crossing in front of the moon, called a moon watch (e.g., Lowery and Newman 1966; Bridge et al. 2024); recovering and analyzing dead birds (individuals attracted by lights but then killed by collisions with buildings, lighthouses, and towers, e.g., Weir 1976; Avery 1979); and analyzing nocturnal flight calls (e.g., A.D. Smith et al. 2014). Another fascinating and accidental tool emerged almost simultaneously with moon watching as radar technology expanded across the country for aviation and climate forecasting. In addition to planes, and water droplets, radar observers also detected "angels" on nighttime screens (Harfer 1957), which turned out to be birds. Eventually, with advancing technology, and the widespread deployment of the Next Generation Weather Radar (NEXRAD) system in the United States in the late 1990s, along with new analytical tools, radar approaches have become powerful tools for tracking bird migration over broad spatial scales (e.g., Gauthreaux and Belser 1998).

Now, in addition to detecting and depicting meteorological phenomena, weather radar data can generate detailed information about continent-scale migratory movements of birds. Weather radar can also detect the movements of insects and bats. A major limitation is that, in most cases, it does not allow for species identification. Despite this limitation, radars have become extremely effective for mapping patterns of migration across broad extents because they are systematically collecting data—year-round, twenty-four hours a day—which are archived and freely available from the National Oceanic and Atmospheric Administration's National Centers for Environmental Information (Kelly et al. 2012). Reflectivity and radial velocity from weather radars can be used to measure where and when birds occur in the airspace, their flight altitudes, their direction and speed, and where they stop over on the ground. By observing the patterns of bird movement during a migration season, weather radars allow for a spatial and temporal assessment of the importance of stopover sites and airspace corridors across large geographic areas. No other tool offers such comprehensive and unbiased observations of bird migration over such a large area. Radar data have been applied comprehensively to map: the distributions of birds migrating through and stopping in entire regions (Guo et al. 2023), the proportion of migrating birds that stop relative to the numbers migrating overhead (Cohen et al. 2021), the right-siting of renewable energy development to minimize collision risk (Cohen et al. 2022), and the timing of migratory movements (K.G. Horton et al. 2020). Data from the US-wide network of weather radars are used to make real-time forecasts of nocturnal migration used by many birders in a project called BirdCast (https://birdcast.info/). These forecasts are also used to trigger lights-out alerts for many cities and states; the effects of artificial light pollution, especially in big cities, can cause skyglow, which can be disorienting to migrating birds and cause them to collide with buildings and windows, which is often deadly. This technology provides unprecedented insight into the broad-scale movements of birds at continental and regional scales. Although radar does not identify individual species or populations, radar data can be paired with migratory connectivity data to understand migratory movements at multiple scales. Finally, a recent powerful application aggregates data from the radar network to estimate the magnitude of migratory flights (seasonally and annually) to help quantify the loss of abundance of migratory birds across North America (Dokter et al. 2018; K.V. Rosenberg et al. 2019).

Not all birds migrate at night. In fact, most raptors migrate during the day, taking advantage of thermals to move north and south. Thermals are created as the sun rises, heating up the ground, creating wind currents that can be used to assist with lift and migrations. Raptor biologists have long known of unique places, in some cases geographic bottlenecks, where magnificent congregations of hawks can be observed, such as in Hawk Mountain, Pennsylvania, or the strip of land in eastern Mexico that borders the Gulf, known as the River of Raptors in Veracruz, Mexico. Opportunities for watching individual raptors migrate have led to organized monitoring efforts for counting raptors and to several key insights into the biology of these species (Hussell and Ruelas Inzunza 2008). HawkWatch International operates the largest, single-run, standardized network of migration sites in the world, and the Hawk Migration Association collects hawk count data from almost two hundred affiliated raptor monitoring sites throughout the United States, Canada, and Mexico, and coordinates HawkCount, a near real-time international database of hawk sitings across the continent. All these efforts for watching unmarked individuals, from moon watching to radar networks to raptor counting, along with existing citizen science efforts (e.g., the Breeding Bird Survey in Sauer et al. 2003) and the meteoric rise of bird-watching generally, contribute in some way to an entirely different opportunity to merge all of these various count data into one place, where the analysis and the interpretation possibilities create a new level of biological visualization and understanding (B.L. Sullivan et al. 2009).

According to a recent survey conducted by the US Fish and Wildlife Service (https://www.fws.gov/program/national-survey-fishing-hunting-and-wildlife-associated-recreation-fhwar), around 96 million, or one-third, of Americans observed, fed, photographed, or supported habitat for birds. People love to watch birds, and whether a professional or an amateur pursuit, this preoccupation has resulted in the collection of important data over time. Much of what we know about bird distributions, including seasonal changes, has come from observational data collected, in large part, by amateurs. Expert knowledge, combined with historical specimen data, resulted in basic range maps for most species presented in field guides and summarized by BirdLife International's DataZone (https://datazone.birdlife.org).

More recently, the bird-monitoring project eBird, initiated in 2002 (B.L. Sullivan et al. 2009), has harnessed the power of observational data to transform the traditional range map into a set of dynamic maps and models that depict the year-round distribution and abundance of birds (Fink et al. 2023). eBird has grown into a network of local, national, and international collaborators, with more than one million contributors, functioning as one of the largest on-the-ground biodiversity sensor networks in the world. With over a billion bird observations and data from every country in the world, this database provides a rich source of fine-scale, year-round avian data—all information to understand how bird populations move throughout the year across the hemispheres.

The goal of the eBird Status and Trends data products is to leverage eBird data to document the location, concentration, movement, and change of bird populations throughout the annual cycle. Translating these observations into reliable, high-resolution information requires statistical models to fill spatiotemporal gaps and control for the biases inherent in species observations collected by volunteers. Doing this at scale, both taxonomically, across a large and varied set of species, and geographically, across continental spatial extents, requires a scalable and automated modeling workflow. This Atlas uses eBird Status and Trends data products, which include estimates of abundance and range, to depict the seasonal geography of species populations (Fink et al. 2023). The range-wide, population-level information provided by citizen-derived data provides full-annual-cycle maps of distribution and abundance and forms the highly informative base maps of migratory connectivity for most of the species presented in this Atlas.

— EXTRINSIC MARKERS

The formal marking of birds for scientific purposes did not begin until the late 1800s. Bird "banding" (as it's known in the Western Hemisphere) or "ringing" (Eastern Hemisphere) is one of the oldest and frequently used methods of marking birds for scientific studies (Bairlein 2001). The process involves placing an aluminum band with a unique identification number, typically issued by a national government, on the bird's leg (Gosler 2004). Some studies may also add unique combinations of colored plastic bands, leg flags, neck collars, and wing tags to help identify individuals, eliminating the need to recapture birds to read the number printed on the band (Cleminson and Nebel 2012).

Organized, large-scale banding was started in 1890 by Danish ornithologist Christian C. Mortenson (see table 1). He banded a total of 1,550 European starlings with zinc bands at first, but realized these were too heavy and switched to aluminum, the same material used today (Preuss 2001). In North America, the first systematic use of bird banding was in 1902 by Smithsonian scientist Paul Bartsch when he banded Black-crowned Night Herons at the National Zoo in Washington, DC. In 1920, in response to the 1916 Migratory Bird Convention, the US Geological Survey (USGS) Bird Banding Laboratory (BBL) was founded (Tautin 2008; figure 2.1, table 1).

The BBL oversees the collection, archiving, management, and dissemination of all data from banded and marked birds in North America. The BBL has collaborated with Canada's Bird Banding Office to administer the North American Bird Banding Program and manage all banding and reencounter (live resightings/recaptures and dead recoveries) records for the birds in the United States and Canada. From 1960 to 2025, more than 79 million birds have been banded. Since 1914 there have been over 6.5 million band reencounters from almost 700 species. Despite 125 years of organized bird banding in North America, there has not been a thorough, systematic analysis of band reencounter data across all species to understand patterns of migratory connectivity. This book provides the first such synthesis of North American banding data.

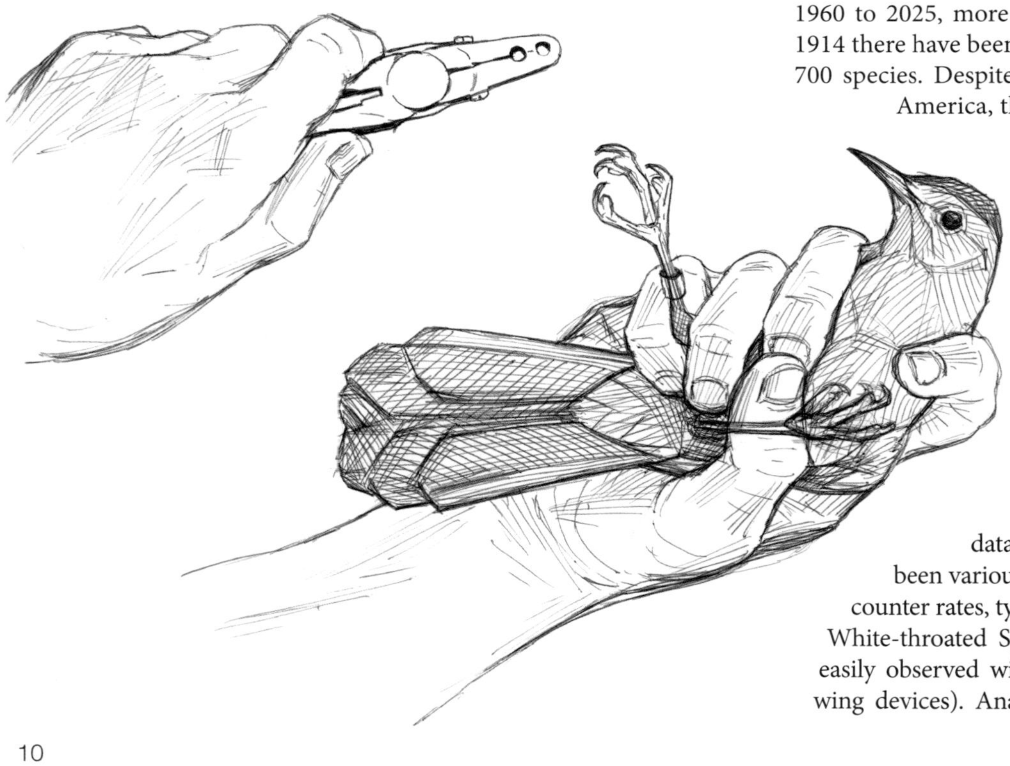

Among the challenges to analyzing banding data is that the BBL has not digitized banding records prior to 1954, only reencounters, so we do not know the actual number of birds banded from 1914 to 1954. From the 1960s forward, the BBL database maintains the original banding records necessary to build individual capture histories, data largely not available within other large banding datasets (Korner-Nievergelt et al. 2012). There have been various assessments of individual species with high reencounter rates, typically short-distance migrants (American Robins, White-throated Sparrows), large-bodied birds (Brants), or species easily observed with identifying attachment devices (e.g., leg flags, wing devices). Analyses from large-bodied birds (>100 g), such as

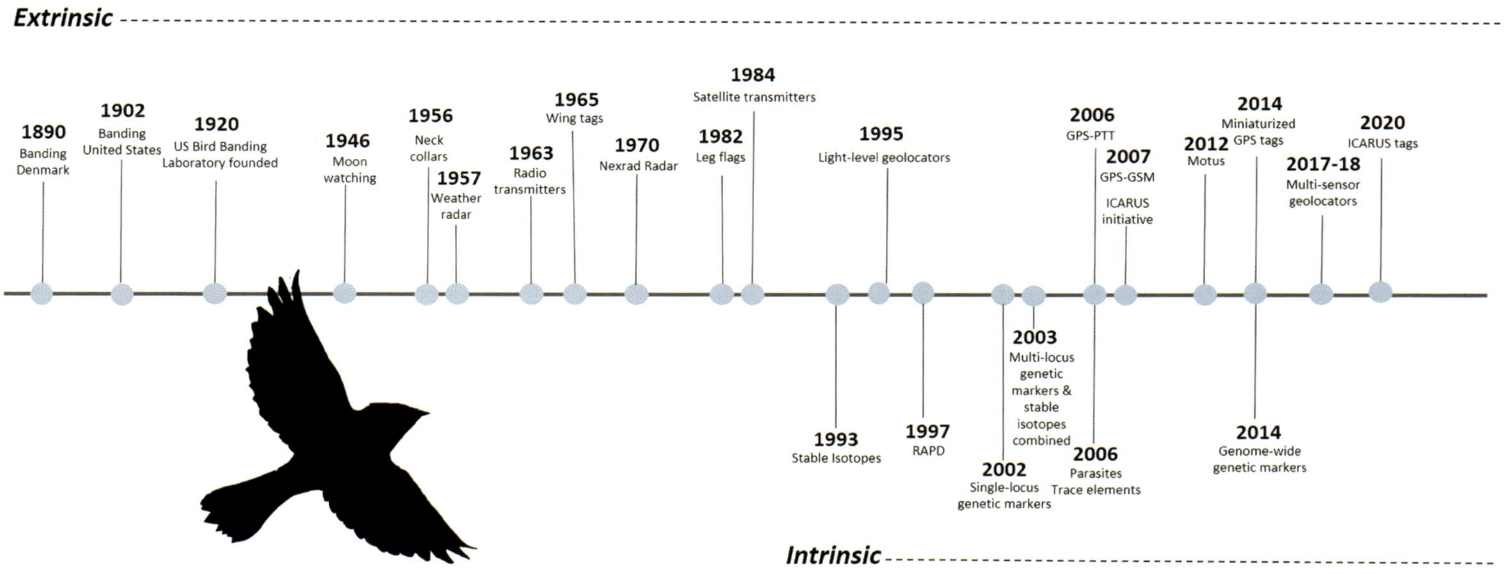

FIGURE 2.1 *Timeline of extrinsic approaches (above the timeline) and intrinsic approaches (below the timeline) for tracking birds in space and time. Detailed explanations for each can be found in table 1 (extrinsic approaches) and table 3 (intrinsic approaches).*

ducks and other waterfowl, confirm that these data can be useful for assessing migratory connectivity (Roberts et al. 2023). Waterfowl and game species have a large amount of reencounter data thanks to band returns from hunting. Smaller shorebirds, such as the Piping Plover, and temperate short-distance migrants (with ranges within the US and Canada), such as the American Robin, can often have large numbers of encounters (Brown and Miller 2016). Species such as these occur in places, such as along beaches or other human-occupied habitats, where the probability of being seen, recaptured, or found dead by humans is high. In general, for smaller-bodied migratory species, the frequency of band reencounters has not appeared to live up to expectations. For example, across all species of passerines and near-passerines, <2% of birds have been recaptured, despite more than 42 million banded (BBL).

Given the availability of banding data as a source of information about migratory connectivity and given its apparent lack of utility for many smaller-bodied species, banding data have too often been dismissed and underutilized. This is true at least in part because enormous biases can exist with the analysis of banding and reencounter data due to geographic variation in banding effort and reencounter and reporting probabilities. And the results provided in this book are unfortunately not immune from such issues. That does not mean we should avoid making use of this extensive dataset of animal movement; it only means we should consider the biases, quantitatively where possible, as we examine the biological patterns. Banding and reencounter data are the most spatially accurate source of information on migratory connectivity for many species and are often the only long-term dataset available for most Nearctic-Neotropical migratory bird species. Thus, interpretation of reencounter locations requires accounting for the variation in the data due to the unknown observer and reporting distribution (Korner-Nievergelt et al. 2010; Cohen et al. 2014). Better understanding of things such as human population density, time of year, weather, where, and when banding stations are operating, the size and color of the bird, the habitat its body settles in (e.g., beach versus forest), and how fast it might decay (e.g., tropics versus Arctic, summer versus winter) or be scavenged by a predator are all necessary to interpret reencounter data. For example, coastal birds tend to have higher reencounter rates in part because they are bigger so they can be seen, caught by fisherman, or found when their bodies wash up along beaches, where beachcombers might notice the band. The same is not true for a small warbler or a thrush that dies in a forest and quickly disappears. All these factors need to be considered when interpreting the species-specific migratory connectivity maps.

Also problematic with using banding data alone for estimating migratory connectivity is that even if a bird is found, the vast majority of reencounter data only represent two points in time, the initial banding and then a second point, making the actual "track of movement" impossible to discern. In addition, most reencounters of a banded bird away from breeding areas occur during migration, when many banding stations are in operation and also when the highest mortality occurs. For game species like waterfowl, reencounters occur primarily through hunting. Overall, for most birds, and certainly for smaller-bodied species, the problem remains that with banding efforts it is very rare to reencounter the same individual even once and all but impossible to reencounter it at multiple distant locations throughout the annual cycle. This is especially true for species that spend the nonbreeding season in areas with low human population density and with underdeveloped or nonexistent banding programs, such as in areas within the Southern Hemisphere where a large percentage of North American species spend the majority of the annual cycle. Despite these various biases, researchers have developed several approaches to quantify migratory connectivity (see the "Quantifying the Strength of Migratory Connectivity" section below). In the next chapter, we describe how we have tried to address the biases in the migratory connectivity data included in this volume.

Across the Western Hemisphere, only a small number of organized and systematic federal banding programs exist. In addition to the North American Bird Banding program of the United States and Canada, bird banding schemes exist in Brazil (Centro de Estudos de Migraciones de Aves; CEMAVE), Chile (Agricultural and Livestock Service; SAG), and Argentina (Centro Nacional de Anillado de Aves; CENAA). In Peru, although there is no federally organized banding scheme, there is an organized effort to manage banding (Centro de Ornitologia y Biodiversidad; CORBIDI). Similarly, Colombia does not have a banding scheme. A nonprofit group dedicated to the conservation and research of Colombian birds, called SELVA, deposits all its banding data in the SIB (Sistema de Información de Biodiversidad), a national biodiversity database administered by Instituto Humboldt. Although SIB is not specifically designed for banding data, it is centralized, public, and keeps data backed up and safe. No country-specific banding schemes exist in Central America. Several Caribbean islands that have connections to Europe (e.g., British Virgin Islands) cooperate with EURING (Europe's

TABLE 1 DETAILED HISTORY OF KEY ADVANCES WITH EXTRINSIC APPROACHES FOR TRACKING BIRDS

1890	Banding	Christian Mortenson rings starlings in Denmark.	Preuss 2001
1902	Banding	Paul Bartsch bands Black-Crowned Night Herons in Washington, DC, using serially numbered leg bands with the year and a "Return to Smithsonian Institution" address inscribed on them.	Bartsch 1904
1920	Banding	United States Bird Banding Laboratory is founded to oversee banding at a national level.	Tautin 2008
1946	Moon Watching	A technique for observing and quantifying nocturnal migration by scanning for bird silhouettes passing in front of the moon.	Rense 1946
1956	Neck Collars	Neck collars are used to study individual migrations of Tundra Swans, in accordance with the color and coding of the Continental protocol for bands; this results in a higher percentage of resightings when compared to resightings of swans banded with metal and plastic tarsus bands. Yellow plastic collars were first used on Tundra Swans in the US by Harvey K. Nelson in 1956 and subsequently by John Frye at Shiawassee National Wildlife Refuge, Saginaw, Michigan, from 1960 to 1969, after which they joined the Continental Program.	Sladen 1973
1957	Weather Radar	Weather radar detects odd angel-shaped images later determined to be birds flying through the atmosphere.	Harfer 1957
1963	Radio Telemetry	Transmitters are attached to migrating thrushes and tracked with a vehicle.	Cochran and Lord 1963
1965	Wing Tags	Common Eiders are resighted in nonbreeding and breeding areas using bands and wing tags; the wing tags provide greater recognizability while the birds are swimming or in flocks.	Milne 1965
1970	NEXRAD Radar	Weather radar deployed across the continental United States that detects and stores the movement of birds, bats, and insects through the atmosphere. Data collection is automated, stored, and available and has transformed large-scale analysis of pre- and post-migration of birds.	Gauthreaux 1970
1982	Leg Flags	Leg flags are used to follow shorebird migrations, particularly those of Sanderlings, Red Knots, and Black-bellied Plovers, throughout the Americas (Pan American Shorebird Program).	Myers 1983
1984	Satellite Telemetry	Satellite transmitters are used on Bald Eagles, Trumpeter Swans, and Tundra Swans; the method provides precise data about habitat and migration routes and may be used to study long-distance bird migrations on a global scale.	Strikwerda et al. 1985, 1986
1995	Light-level Geolocators	Light-level geolocators, which determine bird position using loggers that record light intensity, are used to track Wandering Albatrosses from breeding colonies to nonbreeding sites; as an alternative to satellite telemetry, light-level geolocators are less costly and less difficult to implement for yearlong tracking.	Weimerskirch and Wilson 2000
2006	GPS-PTT	Radar and satellite global positioning system–platform transmitter terminal (GPS-PTT) tags are used on Turkey Vultures, providing complementary information on the movements and behaviors of individual birds. The radar aspect allows for continuous monitoring, while the satellite tag component allows for the identification of individuals. Furthermore, the added GPS technology reports altitude, speed, and heading, as well as locations that are more accurate than satellite telemetry.	Beason et al. 2010
2007	GPS-GSM	GPS-GSM tags are used on Eastern Imperial Eagle in order to study the migration, nonbreeding locations, pair formation, and other stages of life of juvenile eagles. The devices record the geographic coordinates of the locations of tagged birds through a GPS system. This information is then sent through the network of a GSM operator, using the General Packet Radio Service. In addition to the location of the bird, the transmitter can provide data on the bird's activity—flying or resting, height, speed, and direction of flight.	Gradev et al. 2011
2012	Motus	The Motus Wildlife Tracking System—a collaborative network of researchers using coordinated automated radio-telemetry arrays to study movements of small flying organisms at local, regional, and hemispheric scale—is started. By 2023, over 1,800 automated stations are operating across 31 countries, and more than 40,000 animals of more than 300 species have been tagged and tracked by system collaborators.	P.D. Taylor et al. 2017
2014	Archival GPS	Miniaturized GPS tags are used on Ovenbirds to identify and characterize nonbreeding territories, quantify migratory connectivity, and identify population boundaries. The miniaturized technology allows for all the benefits of GPS tag tracking, except transmission of data, to be applied to small birds.	Hallworth and Marra 2015
2017–2018	Multisensor Geolocators	Multisensor geolocators use accelerometer, temperature, and pressure data to estimate locations. First used on large-bodied birds and then miniaturized for passerines.	Vansteelant et al. 2017; Liechti et al. 2018
2020	ICARUS	The ICARUS—International Cooperation for Animal Research Using Space—Initiative is launched, with the goal of implementing a small-animal satellite tracking system that would enable the global monitoring of animals. These miniature measuring devices have various sensors that continuously record the behavioral and health data of the animals. The transmitters can also record environmental conditions such as temperature, humidity, and air pressure. In 2020, ICARUS tags are deployed on blackbirds and thrushes in a cross-continental, satellite-based animal monitoring endeavor. ICARUS stopped transmitting data on March 3, 2022, due to the Ukraine–Russia war and strained relations with the International Space Station, Russia, and NATO block countries. As of this date, ICARUS is developing an independent satellite capability.	Wikelski et al. 2007

banding scheme; euring.org). In addition, there are coordinated efforts across the Western Hemisphere for individual species such as Red Knots, Purple Martins, and Brown Pelicans.

In addition to aluminum leg bands, other types of extrinsic markers have been developed and deployed to document a reencounter based on a visual observation. These involve leg bands with unique color combinations, neck collars, wing tags, and leg flags. Plastic leg bands come in eight to ten colors and can form an enormous number of unique combinations when placed on both legs (sometimes with readable alphanumeric codes), which allow observers to identify individuals. Such approaches are typically used for intense observations of individual birds for within-season studies, but sometimes there are observations of color-banded individuals across season, especially with shorebirds. Numbered neck collars were used to facilitate resighting of individual birds for species like geese and swans. One of the first of these studies was conducted by Harvey K. Nelson in 1956 on Tundra Swans using yellow plastic collars (unpublished study referenced in Sladen 1973). A later analysis of encounter rates of Tundra Swans using leg bands versus neck collars by Sladen (1973) clearly demonstrated how effective the neck collars were in increasing data reporting. Similarly, in 1965, H. Milne placed numbered wing tags on Common Eiders so they could be resighted when swimming or with flocks. Since then, wing tags have been used on a large diversity of species, from Magnificent Frigatebirds to penguins, gulls, and crows. While all have experienced higher reencounter rates, some researchers have reported negative impacts on individual birds (Trefry et al. 2013). Finally, leg bands with flags, either numbered or colored, have successfully been used to increase reencounter rates, especially with shorebirds that are easily observed in open habitats with spotting scopes and binoculars. In addition, to facilitate and amplify the number of resightings, an organization has been created to centralize reporting of leg flags (bandedbirds.org). Data from this effort were not accessible for the Atlas at the time of publication.

Electronic Tracking Approaches (Transmitting, Receiving, and Archiving)

Electronic tracking devices (tags) attached to migrating animals are now a standard tool for studying movement and migratory connectivity. To date, about half of all migratory bird species in North America have been tracked (Scarpignato et al. 2023; figure 2.2). Tracking tags can be classified by the type of data they collect (GPS data, Argos data, ambient light levels) and how the data is received by the user (via satellite, by retrieving the tag and its archived data, via cellular towers). At a minimum, tags provide the information needed to estimate a time series of an individual animal's locations, usually over a year but can be

FIGURE 2.2 *Proportions of migratory birds, by order, that have been tracked (Scarpignato et al. 2023).*

more frequent and for multiple years. At their most sophisticated level, combined with the location data, tags may provide additional parameters of interest to help infer ecological relationships, for example, altitude, acceleration, conductivity, and air or sea surface temperature (Cade et al. 2021).

The earliest and simplest versions of tracking tags were capable of transmitting a radio signal to a nearby receiver with an antenna (MacDonald 1978). This has been done most commonly with a person tracking individuals on foot. In fact, radio transmitters have been used extensively to track local-scale movements of many animal species, but this technology has had very limited success at larger-scale movements (Millspaugh and Marzluff 2001). A first attempt to develop tags to track individual birds over larger spatial scales was led by Cochran and Lord (1963; figure 2.1, table 1). Cochran, an electrical engineer, built and deployed the earliest version of a radio transmitter on thrushes. He was motivated to understand how migratory birds navigate as they move across landscapes, and he tracked these movements, sometimes with the use of planes and automobiles. One tracking session of a Swainson's Thrush lasted for nine consecutive days. Radio transmitter technology has since become extremely lightweight and more versatile. The limitation is in the range of detection and the need for a receiving unit and antennas to be within range to detect the transmitter. In recent years, advances in automated radio telemetry have allowed for a much larger geographic coverage using a coordinated array of receivers called the Motus Wildlife Tracking System (more below).

Thankfully, remote-transmission tracking technologies have since expanded considerably and now include tags attached to animals that communicate with a receiver on an orbiting satellite system. Examples of satellite systems used for tracking animals include Argos and Global Positioning System (GPS) and also the ICARUS antenna (International Cooperation for Animal Research Using Space; figure 2.1, table 1) attached not to a satellite but to the International Space Station (ISS). A modification to these systems includes the Groupe Special Mobile (GSM; figure 2.1, table 1) or "cell phone tags" that use GPS satellites to determine location and cell phone towers to receive the data when the bird flies in close proximity to a cell phone tower/receiver.

Among these remote tracking methods, tags that communicate with satellite systems provide some of the most unbiased estimates of location available. Satellite tags, also known as Platform Transmitting Terminal (PTT; figure 2.1, table 1) tags, are transmitters that send user-defined periodic messages to the Argos system, a global satellite-based location and data collection system dedicated to studying and protecting our environment (figure 2.3). These polar-orbiting Argos satellites fly 850 km above the Earth. They pick up the signals, store them on-board, and relay them back to Earth to receiving stations or processing centers. The processing centers collect all incoming data and calculate locations by measuring the Doppler effect on transmission frequency (401.650 MHz ± 30 kHz). The Doppler effect is the change in frequency of a sound wave or electromagnetic wave that occurs when the source of vibration and observer are moving relative to each other. The processing centers distribute the data directly to Argos users.

The GPS is a freely accessible global navigation satellite system, maintained by the

FIGURE 2.3 *One of the most commonly used satellite-based tracking systems is the Argos system, a global satellite-based location and data collection system that can be used for tracking animals, such as the albatross here, as they move long distances over remote areas.*

US government, used to provide location and time information. In recent years, the use of GPS technology to track species has increased (Scarpignato et al. 2023). If a GPS receiver on the ground has an unobstructed line of sight to four or more GPS satellites, it can use the information (time of transmission, precise orbital information of the satellite, and general system health / rough orbits of all GPS satellites) transmitted to calculate accurate time and location. GPS archival tags do not maintain continuous location data but instead calculate location at user-defined intervals, providing a "track" view of the path taken by the tag. To access GPS data, the device the animal is carrying must either transmit the data to an Argos satellite or a GSM cell tower; or the GPS archival tag (see below) must be retrieved directly from the animal and the data downloaded or data collected indirectly with a receiver.

Although satellite tags can generate some spectacular data, there are also several downsides. One of the most significant is that devices using satellites for their transmission process are by far the most expensive for studying migratory connectivity and movement, typically ranging from US$1,000 to US$4,000 per tag. In addition, there are annual data delivery fees to the satellite or cell phone companies, which can exceed US$1,000 per tag per year. This restricts deployments for most species and the ability to attain suitable sample sizes for many research questions. In addition to cost, the size of the tag has severely limited the species that can be tracked (figure 2.4). Until recently many satellite tags exceeded the recommended 3% of a bird's body weight largely due to battery size (Murray and Fuller 2000; Bridge et al. 2011). Improvements that have increased tag life expectancy and decreased tag weight include the change from battery to solar power and circuit board miniaturization. Currently the smallest satellite tags (Argos/GPS) are 1–5 g. The other important consideration is the area of resolution that each type of satellite tag provides for the actual geolocation of the bird. Argos PTTs are accurate from <250 m to 1.5 km or more (Douglas et al. 2012), whereas GPS transmitters are accurate to within several meters. Electronic transmitting tags will likely continue to shrink in size and cost, presenting an unparalleled opportunity to discover how distant breeding and nonbreeding areas connect and interact in space and time for even the smallest animals. Hopefully, tracking individuals for their entire life cycles will be possible with lightweight and inexpensive tags.

The ICARUS tracking system, which had a receiving antenna on the Russian side of the ISS, is now unfortunately offline because of political tensions related to the Ukraine–Russian war. The Russian space agency Roscosmos has terminated its cooperation with many partners, and the transmission of data from the ISS was stopped on March 3, 2022. As of this date, ICARUS is trying to develop an independent satellite capability.

ICARUS was considered a strong possibility for improving tracking by taking advantage of the relatively low-orbit ISS (400 km versus 850 km for Argos). This minimized the transmission power requirements of ICARUS tags. The smallest transmitting device was just under 5 g, transmitted at least once, if not multiple times, per day, and was fairly inexpensive (<US$750).

Automated radio telemetry is another form of remote-transmission tracking technology that has been growing widely, especially over the past decade in North America. Automated radio telemetry generally allows for tracking the smallest animals with relatively high geographic precision and the greatest temporal precision. This level of temporal precision allows for detailed examination of behavioral characteristics, such as daily/nightly activity levels, individual flight times/speeds, specific departure/arrival times, and habitat use patterns. Coded tags send out repeat identification signals on customized intervals, from 2 seconds to 60 seconds, which allows for very specific estimates of state or behavioral changes in animals.

Radio tags that weigh <0.2 g are now available, which allows tracking of almost any animal weighing <6 g. However, the life of the tag is directly related to the size of the battery and the frequency of radio transmissions (burst/pulse rates), which can be limiting for smaller tag sizes. The smallest tags with modest transmission rates may only last 2–3 weeks, whereas tags weighing 1 g can last up to 1 year, and tags larger than 1 g can last 2–3 years. Recent advances in solar technology mean that solar tags (<0.5 g) with no battery and "hybrid" tags with batteries charged by a solar panel have unlimited life. In both instances, tags can theoretically last for the life of the bird, but there are other physical limitations that may influence tag retention.

An automated radio telemetry station consists of a receiver, connected to one or more antennas, which automatically records and stores

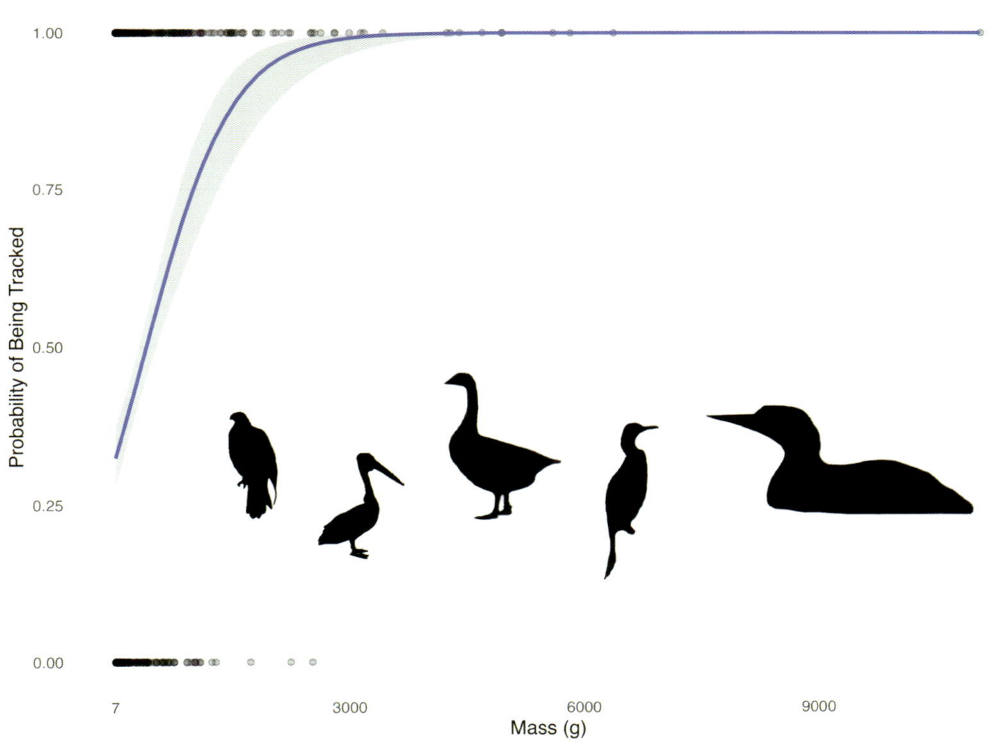

FIGURE 2.4 *The probability of a species being tracked, given its mass, for all North American migratory bird species that can feasibly be tracked with current technologies (Scarpignato et al. 2023).*

signals from coded radio tags carried by birds, along with detailed time-stamps recorded from GPS in the receiver (figure 2.5). Stations can be mobile but are most often at fixed locations. Stations can have many antennas, allowing for tags to be detected in any direction. Tags need to be within the range of an antenna to be detected, and therefore the geographic precision is limited to the detection distance, or radius of the antenna being used, which can vary from 100 m up to >20 km, depending on the antenna being used, the height and configuration of the station within the landscape, and the orientation of the bird/tag in relation to the station. When multiple automated stations are placed at strategic places on the landscape, birds can be tracked in detail at local, regional, and even hemispheric scales.

Networks of coordinated arrays of automated stations all monitoring the same frequency or channel help detect and track individual tagged animals over large spatial scales. The utility of automated telemetry to track birds over long distances is limited to the extent of the array of stations positioned on the landscape. However, coordination and cooperation between researchers and organizations can broaden the spatial and temporal scale over which animals can be tracked, as realized by the Motus Wildlife Tracking System (Motus; https://motus.org; Taylor et al. 2017). Motus is a program of Birds Canada, in partnership with a collaborative global network of researchers, organizations, educators, and citizens using coordinated automated radio telemetry to track small flying organisms (birds, bats, and insects).

The Motus community participates and contributes to the system through deploying and maintaining one or more Motus automated radio telemetry stations (Motus stations), and/or conducting projects tracking the behavior and movement of wildlife using Motus tags and infrastructure. Detection data from Motus stations are uploaded, either autonomously (networked receivers) or manually by collaborators (offline receivers), to a centralized database, where data are processed and made visible to all collaborators and the public through visualizations and tables on the Motus website (https://motus.org/). As of June 2025, the Motus community is maintaining more than 2,100 stations across four continents and 34 countries. More than 57,000 individuals representing 358 species have been tracked across more than 964 independent projects. The range of current and potential uses for Motus is vast. There are some limitations with respect to the generally limited geographic precision since birds have to fly through the airspace of the receiver. Nevertheless, with extremely lightweight tags and an expanding array of stations and collaborative partnerships around the globe, Motus is well-poised to provide tracking of individuals throughout many stages of their life cycles.

Archival geolocators, also called light-level geolocators and global location sensors (GLS), are a type of tag that requires subsequent recovery of the bird and the tag for data retrieval. They revolutionized the study of bird migration. Light-level geolocators record light and time of day with an internal clock, and locations are estimated from standard nautical equations. From the first geolocators deployed on seabirds in 1995 (Weimerskirch and Wilson 2000; figure 2.1, table 1), weighing in at 20 g, tag size has decreased substantially. The smallest light-level geolocators now weigh between 0.3 and 0.4 g and can be deployed on birds as small as 7 g, attached with a leg-loop harness. However, spatial accuracy

FIGURE 2.5 *Motus tags with unique coded signals, placed on the backs of small birds (here on a Swainson's Thrush), are detected by fixed location receivers with directional antennas when the animals fly/move within a certain detection radius, depending on topography or vegetative cover.*

remains a major limitation (±200 km, Lisovski et al. 2012), especially when precise knowledge of unknown breeding and nonbreeding areas is needed. Additionally, latitudinal estimates are unreliable for a period of 6–8 weeks surrounding the spring and fall equinoxes (Hill and Braun 2001), and breeding locations at polar latitudes are not possible because geolocators cannot reveal locations in conditions of 24-hour daylight (Bridge et al. 2011). In addition, interpreting stationary endpoints—breeding and nonbreeding locations—derived from multiple, single-day geolocation estimates, provide a decently accurate latitude and longitude estimate with some error. In contrast, points during migration are typically single-point estimates and are not reliable to use for estimating migratory routes. A suite of biases such as these exists in the interpretation of the light data collected with geolocators, so users and interpreters of these data should proceed with caution. That said, archival tag technology has rapidly advanced and continues to provide important insights into migratory connectivity, including more recent tags that collect data on movement (using accelerometers), temperature, and atmospheric pressure, which continues to improve temporal and spatial resolution (Vansteelant et al. 2017; Liechti et al. 2018).

More recently a pressure sensor has been added to archival tags and, in early studies, has provided geolocation data that is three times more accurate than light-level geolocators (Nussbaumer et al. 2023). In both new developments, like light-level geolocators, they require the tagged bird to be recaptured. Archival tags are not useful for those bird species that are not site faithful or that breed in inhospitable or unknown places (the high Arctic), or those species for which targeted recapture of tagged individuals is unlikely.

Tag and Harness Effects

With smaller and smaller tracking technologies and exponential interest in tracking more and more birds, minimizing the impacts to the animal is essential—impacts related not only to device size but also to how the device is attached. The harness type and material, as well as the position of the tag can sometimes, though rarely, negatively impact survival and reproduction (Bodey et al. 2018; Geen et al. 2019). Therefore, in addition to the human user being properly trained and permitted, it is crucial to select the most appropriate tag type and attachment method.

Attachment methods can vary depending on the tag type, habitat occupied by the bird, and the species itself. Tracking devices can be attached to a bird using two general methods: mounted, when the tracking device is directly attached to the bird's body, or harnessed, when a harness holds or fastens the tag on the bird's body (Scarpignato and Stanley et al. 2026). At least six harness types have been used on nearly 250 bird species to study migration and migratory connectivity (table 2).

— INTRINSIC MARKERS

Because *extrinsic markers*, such as the tagging methods described above, are costly or require that the marked individuals be relocated at some point, either through a resighting, recapture or with a receiver and antenna, some researchers have used *intrinsic markers* to determine geographic origin—that is, markers or indicators that come from the animal itself (figure 2.1, table 3). Two of the most commonly used intrinsic markers include genetics and stable isotopes. In addition, looking at organisms within individual birds, such as parasites like malaria, can also be used to assess migratory connectivity. Below we review the strengths and weaknesses of each intrinsic method and how they can be used separately or together to inform our understanding of migratory connectivity.

Genoscape Approaches

Molecular genetic approaches are useful because although only some birds have leg bands (or other extrinsic tags), they all have DNA. Genetic approaches to understanding migratory connectivity rely on first assessing how genetic variation is distributed across the breeding ground of a species and then using DNA from a bird captured anywhere outside of its breeding range to trace it back to its breeding population of origin (Rundel et al. 2013; Ruegg et al. 2017). DNA-based approaches have evolved as DNA sequencing technologies have improved, and with this evolution has come an increase in the precision of the resulting genetic markers for tracking populations. Early methods that were focused on single genes were only able to distinguish deep genetic splits within species and therefore were of limited utility for connectivity studies (figure 2.1, table 3; Kimura et al. 2002). Later methods using multigene approaches were only marginally better at identifying

populations at finer spatial scales (Clegg et al. 2003). More recently, the advent of Next Generation sequencing has led to a rapid increase in the amount of sequence data that can be generated and to a corresponding decrease in the cost. This explosion in genetic data has improved our ability to assign birds to breeding regions as well as to understand the genetic basis of migratory behavior (Ruegg et al. 2014). An example of the power of genome-wide genetic data for identifying patterns of migratory connectivity comes from a recent study of the Common Loon (*Gavia immer*). Using 300 samples from across the breeding range, Brenda Larison and colleagues (2020) developed a panel of 158 single nucleotide polymorphisms (SNPs) loci, to identify genetically distinct populations across the breeding grounds at finer spatial scales than had been demonstrated previously and then subsequently track these populations across the annual cycle (figure 2.6). The resulting map of how population genetic variation is distributed across geographic space using genome-wide data is called a genoscape, and coordinated efforts to create genoscapes

TABLE 2 EXAMPLES OF TRACKING TAG HARNESSES (TRACKING DEVICES CAN ALSO BE MOUNTED DIRECTLY TO THE BIRD'S BODY; ADAPTED FROM SCARPIGNATO AND STANLEY ET AL. 2026)

HARNESS	DESCRIPTION AND REFERENCE	ILLUSTRATION	GROUP
Device placed on the back on the synsacrum			
Leg-loop	Two-loop harness in the shape of a figure-eight. The device sits on the synsacrum with two loops placed around the top of the femur, resting in the hip crease (Rappole and Tipton 1991; Mallory and Gilbert 2008; Streby et al. 2015)		Passeriformes, Charadriiformes, Accipitriformes, Caprimulgiformes, Cathartiformes, Coraciiformes, Cuculiformes, Falconiformes, Galliformes, Gruiformes, Bucerotiformes, Pelecaniformes, Piciformes, Procellariiformes, Strigiformes, Suliformes
Device placed on the back between the shoulders			
Full-body	Sometimes called a neck loop harness, consists of a smaller loop around the neck and a larger body loop connected by a knot or strap. The harness knot typically sits at the top of the sternum (Åkesson et al. 2012; Chan et al. 2016)	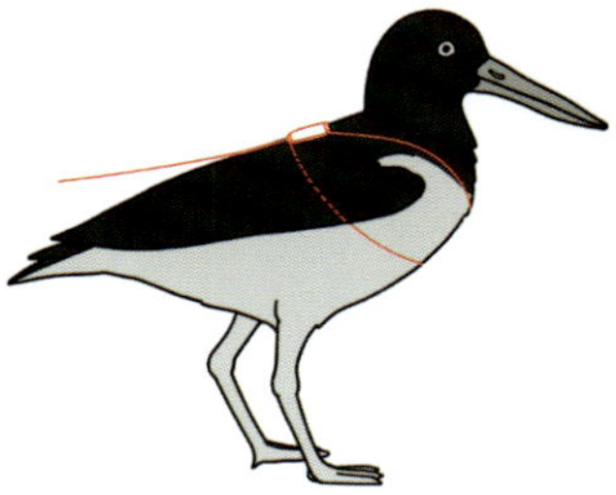	Charadriiformes, Accipitriformes, Falconiformes, Anseriformes, Apodiformes, Caprimulgiformes, Cuculiformes, Gruiformes, Passeriformes, Pelecaniformes
Breast-and-body loop (with or without keel strap)	Two loops of equal or different size, positioned around the body posterior to the wings and anteriorly around the breast. It can be connected ventrally by a short segment of ribbon along the keel (Dunstan 1972)		Anseriformes, Charadriiformes, Passeriformes, Pelecaniformes, Procellariiformes, Strigiformes, Suliformes
Backpack	A neck loop and body loop of equal size joined mid-sternum. The neck loop begins at the top of the device and is connected ventrally at the sternum to the body loop that passes under the wings from the back of the device (D.A. Buehler et al. 1995)		Accipitriformes, Falconiformes, Strigiformes, Anseriformes, Caprimulgiformes, Cathartiformes, Charadriiformes, Ciconiiformes, Galliformes, Gruiformes, Otidiformes, Passeriformes, Pelecaniformes, Phoenicopteriformes, Piciformes, Podicipediformes, Psittaciformes, Suliformes
Wing-loop	Two loops that extend under the wings on either side without a central breast strap or loop around the body (Cooper and Cooper 2013)		Galliformes, Columbiformes, Passeriformes, Pelecaniformes, Strigiformes
Device placed on the chest			
Necklace	A loop around the base of the neck positioned over the bird's crop (Shields and Mueller 1982; Riley and Fistler 1992)		Galliformes, Anseriformes, Charadriiformes, Pelecaniformes, Podicipediformes

ILLUSTRATIONS BY **NICK GARNHART**

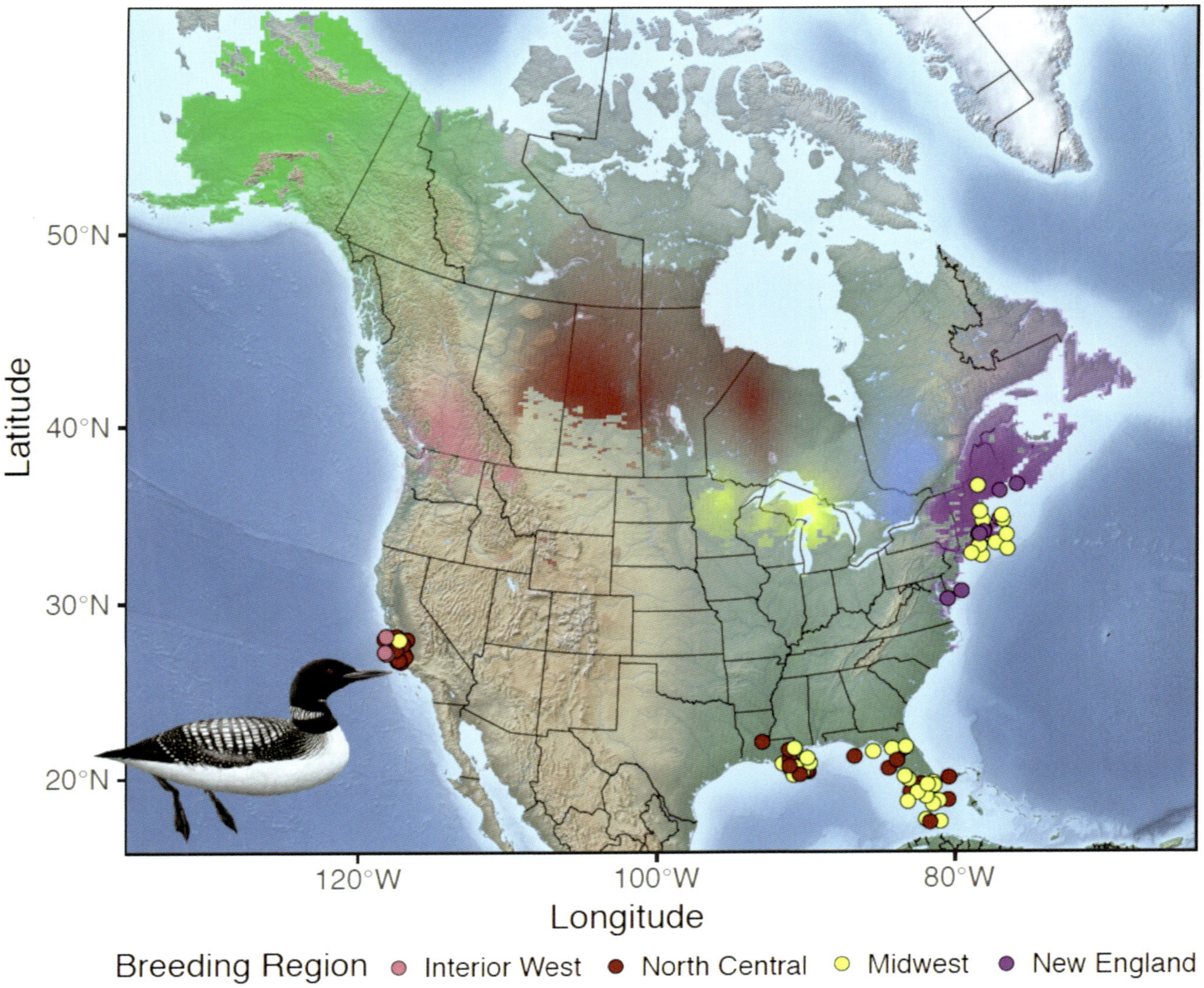

FIGURE 2.6 *Genoscape map for the Common Loon* (Gavia immer) *with six breeding regions (www.birdgenoscape.org). The breeding regions of migrating and nonbreeding birds were identified using SNP-based genetic markers.*

across many species of migratory birds have fallen under the umbrella of the Bird Genoscape Project (www.birdgenoscape.org).

Although migratory connectivity information from genoscape approaches lack the precision of extrinsic tracking technologies, they provide a number of advantages that are challenging or impossible to attain using other tracking methods. The first is that, similar to stable isotope analysis (see below), individuals can be screened using biological information contained within a single feather without the need to recapture the bird at a later date. The second major advantage is the ability to identify locally adapted populations across the breeding range and then track how these populations are connected across the annual cycle (Bay et al. 2018, 2021; Ruegg et al. 2021). When range-wide genomic data are combined with environmental data, it is possible to identify genetic signatures of local adaptation. For example, Bay et al. (2021) demonstrated that the genetic profile of Yellow Warblers that breed in dry areas is distinctly different from that of populations that breed in wetter areas. Further, individuals that breed in drier locations also spend their nonbreeding season in drier locations. Thus, the ability to use whole-genome genetic data to identify genetic signatures of local adaptation in migratory birds makes it possible to begin investigating the potential for natural selection across the full annual cycle.

An important distinction of genetic approaches over extrinsic or other intrinsic tracking methods is that management of biological diversity often requires specifying conservation units (CUs), or intraspecific groupings that can guide conservation and management decisions (Allendorf et al. 2012). As genetic distinctiveness is a primary guideline for the delineation of CUs, using genetic data to track populations of migratory birds across the annual cycle provides a framework for linking two key aspects of full life cycle conservation—namely, connectivity information and delineation of CUs. An example of the importance of conservation unit delineation is best illustrated by work on the Willow Flycatcher (*Empidonax traillii*). While Willow Flycatchers are a fairly common, cross-continentally distributed species, their range is divided into four main subspecies: the Pacific Northwestern form, *E. t. brewsteri*; the Western Central form, *E. t. adastus*; the Eastern form, *E. t. traillii*; and the federally endangered Southwestern form *E. t. extimus*. The ability to connect populations of the endangered subspecies across the full annual cycle is important for conservation, as described in the previous chapter. Kristen Ruegg and colleagues (2021), using whole genome sequencing, were able to show strong population structure between the four main subspecies and to show that Southwestern individuals spend the nonbreeding season in concentrated areas in Nicaragua and Costa Rica. Overall, this work helps demonstrate the value of a genoscape approach for prioritizing full life cycle conservation efforts toward the Conservation Units and geographic areas with the greatest need.

A NEW AGE OF ANIMAL TRACKING

Stable Isotope Approaches

Another technique that relies on intrinsic markers in biological tissues to trace the origin of migratory animals is stable isotope analysis (table 3). Stable isotopes are nonradioactive forms of elements that have similar chemical properties but vary in their atomic mass due to differences in the number of neutrons. Although approximately two-thirds of the elements have more than one stable isotope, currently only the stable isotopes of carbon (^{12}C, ^{13}C), nitrogen (^{14}N, ^{15}N), sulphur (^{32}S, ^{33}S, ^{34}S, ^{36}S), and hydrogen (^{1}H, ^{2}H) have applications in migratory connectivity research. Analyses of all these stable isotopes have yielded important biological insights, but studies with ^{2}H have been especially important. This is largely because the abundance of this stable isotope varies strikingly with latitude, making it an ideal intrinsic marker for measuring the origin of individuals that make north-south migrations. In addition, global geospatial maps of values of the stable isotope of hydrogen in precipitation are freely available online (G.J. Bowen et al. 2005; http://waterisotopes.utah.edu/waterisotopes/) and have led to the explosion of this isotope in migratory connectivity research.

Keratin-based tissues, such as mammal hair, insect wings, and bird feathers, have high natural abundances of $\delta^{2}H$ values, making them ideal for stable isotope investigations of migratory connectivity. It is critical to understand the natural history of a species when sampling $\delta^{2}H$ values because isotopic signatures of foods are incorporated into tissues at the time of growth and remain mostly inert once stored. In mammals, hair replacement is continuous, although some species entirely shed their coat once a year. Insect's wings grown after emergence are kept throughout life, whereas most species of migratory birds undergo a complete molt once each year between July and September on or near breeding areas. $\delta^{2}H$ values in animal tissues generally track the isotopic composition of precipitation but may deviate due to ecological factors such as diet, foraging substrate, and migration strategy (Hobson et al. 2012). Where possible, fractionation rates at the site where the tissue is grown should be measured empirically for each study species or else taken from previously published research. This alignment, if predictable, can be used to sample feathers on nonbreeding areas and then assign individuals back to areas where they molted the feathers the previous breeding season. Migratory connectivity research with $\delta^{2}H$ values has yielded numerous insights, including identifying the breeding origins of hibernating bats (A.R. Sullivan et al. 2012), discovering the natal origins of migrating monarch butterflies (Flockhart et al. 2013), tracing range-wide links between breeding and nonbreeding areas in long-distance migratory birds (e.g., Rubenstein et al. 2002; Studds et al. 2021), and differentiating spatial

TABLE 3 DETAILED HISTORY OF INTRINSIC APPROACHES FOR TRACKING BIRDS

1993	Stable Isotopes	Stable isotopes are used to determine Lesser Snow Goose nonbreeding distribution; the intrinsic method proves useful in the tracking of small and non-game animals and does not require the recovery of marked individuals.	Hobson 1990; Alisauskas and Hobson 1993; Chamberlain et al. 1997; Marra et al. 1998
1997	RAPD	Genetic markers are used to identify shorebird populations, including Long-billed and Short-billed Dowitchers, throughout the annual cycle. The markers—randomly amplified polymorphic DNA (RAPD)—clearly differentiate all shorebird species; thus RAPDs are useful in identifying geographical and difficult-to-differentiate populations of migratory species.	Haig et al. 1997
2002	Single Locus Genetic Markers	Mitochondrial DNA (mtDNA) is used to link populations of the Wilson's Warbler.	Kimura et al. 2002
2003	Stable Isotopes and Multi-Locus Genetic Markers	Clegg uses analysis of microsatellites (nuclear DNA) in combination with hydrogen isotopes to refine migratory connectivity in Wilson's Warblers.	Clegg et al. 2003
2006	Parasites	Over 1,000 Black-throated Blue Warblers tested across species range for malarial parasites and found parasite lineages are too widespread with no site-specific information for connectivity. In contrast, a study on American Redstarts using malarial parasites found some limited structure on breeding areas, suggesting this could be a tool for assessing migratory connectivity.	Fallon et al. 2006; Durrant et al. 2008
	Trace Elements	Used two methods to assess trace elemental signatures of natal feathers in 7 species across 27 sites across the eastern United States. Overall, natal feather locations were not identified with high accuracy using feather elements.	Donovan et al. 2006
2014	Genome Sequencing	>100,000 genetic markers are used to identify fine-scale population structure and population connectivity in the Wilson's Warbler. The method does not require feathers, as prior genetic and isotopic marker approaches do, and thus avoids the low resolution and/or associated technical issues. Thus, genome sequencing proves to be more reliable, precise, and amenable to high throughput screening than previously described intrinsic marking techniques, making it broadly applicable to large-scale monitoring and conservation of migratory organisms.	Ruegg et al. 2014

and temporal migration routes (Cohen et al. 2019) for long-distance migratory birds. That said, despite the advances that stable isotopes have provided, the spatial resolution of these signatures of origin are extremely coarse, to the degree of hundreds of kilometers, and only apply to latitudinal connections,

Other intrinsic techniques have been attempted with mixed success. For example, populations of blood parasites within migratory species, such as malaria and bacteria, have been used in studies of migratory connectivity, but these approaches have had mixed success (Fallon et al. 2006; Durrant et al. 2008; Von Rönn et al. 2015; figure 2.1, table 3). Trace elements have also been explored for their utility in examining migratory connectivity, but their utility has thus far been quite limited (Donovan et al. 2006; figure 2.1). The growing number of studies showing differences in trace element concentrations among spatially discrete bird populations underscores the future potential of this technique. However, unlike the most often used stable isotopes, trace element signals are not known to change in continuous fashion across physical or environmental gradients. Thus, leveraging trace element data to advance our understanding of migratory connectivity will likely first require detailed mapping of these elements across the breeding and/or nonbreeding ranges of migratory species.

Morphology

Many species exhibit natural geographic variation in plumage or other morphological traits (i.e., wing length in Western Sandpipers; O'Hara et al. 2006a), and this variation may be continuous (i.e., clinal) or relatively discrete, where two or more subspecies often have been recognized and named to represent geographically distinct populations (American Ornithologists Union 1957; Winker and Haig 2010). When migratory species exhibit morphological variation, and specimens can be diagnosed in both breeding and nonbreeding areas, important insights into patterns of migratory connectivity may be apparent. These patterns based on morphology can serve as hypotheses to be tested with genoscape analyses or by tracking individuals from known populations. For example, the migratory divide and allopatric (geographically separate) nonbreeding areas between Pacific- and inland-breeding Swainson's Thrush (*Catharus ustulatus*) populations was well documented from studies of museum specimens (Phillips 1991) before genetic and tracking studies provided greater detail on population structure and movements of individuals (Delmore et al. 2012). In other examples, breeding and nonbreeding grounds for populations of high conservation concern were linked based on subspecific identification of specimens for species ranging from the Southwestern Willow Flycatcher (E.H. Paxton et al. 2011) to the Aleutian Cackling Goose (Mini et al. 2011).

For the many species that lack tracking data of any kind, and especially for the vast majority of small passerines (songbirds), morphological variation and subspecific diagnosis of museum specimens remains our only means of assessing possible patterns of migratory connectivity. In this Atlas, we synthesize and interpret published patterns of morphological variation wherever possible. In highly variable species such as the Common Yellowthroat (*Geothlypis trichas*), Fox Sparrow (*Passerella iliaca*), and Hermit Thrush (*Catharus guttatus*), morphological studies reveal strong migratory connectivity across identifiable subspecies. Even in cases where the validity of subspecies has been questioned, underlying variation may provide clues for assessing migratory connectivity. For example, in the Yellow-throated Warbler (*Setophaga dominica*), morphological and genetic analysis suggested that subspecific designation may not be warranted (McKay 2008, 2009), yet careful examination of white- versus yellow-lored individuals from museum specimens (and >10,000 photographs archived in eBird) show clearly differentiated migration and nonbreeding areas for eastern and western breeding populations (Dunn and Garrett 1997; this Atlas).

Even where subspecies are not recognized, clinal variation in a species' morphology may provide insights into population-level migratory connectivity. Clark Rushing and colleagues (2014) used predictable variation in wing length in the Wood Thrush (*Hylocichla mustilena*) across latitudes to investigate how combining stable isotope data with wing length improved migratory connectivity assignments. They found that combining markers significantly increased the precision of geographic assignment compared to using a single marker. Overall, subspecific and other morphological variation remains a seriously underutilized tool for studies of migratory connectivity. As the recognition and use of subspecies has become more controversial (Haig and Winker 2010) and morphological studies of museum specimens have become less fashionable, a century or more of ornithological study is increasingly buried in older literature. This treasure trove of knowledge about intraspecific variation can serve as the basis for new studies using cutting-edge genetic analysis and tracking technologies, greatly expanding our understanding of population origin and connectivity across species.

INTEGRATING MULTIPLE DATA TYPES

The Rushing et al. (2014) effort wasn't the first or the last to combine various markers to estimate migratory connectivity. Integrating multiple markers may allow us to differentiate populations and thereby also examine individual and population origin, and migratory connectivity. This Atlas integrates data types whenever possible when modeling overall migratory connectivity for several species. Some studies have also done this by combining banding and tracking data (Huysman et al. 2022), combining stable isotopes and morphology (Rushing et al. 2014), and combining stable isotopes and abundance data (Rushing et al. 2017). Other studies have combined intrinsic measures (stable isotopes and genetics) with habitat suitability measures (Ruegg et al. 2017) and explored the potential of combining tracking data with banding datasets (Korner-Nievergelt et al. 2017; Hostetler et al. 2025). For example, when birds breed in the same place that they molt their feathers, the two sources of information can be combined to yield higher quality information than either method used in isolation (Clegg et al. 2003; Kelly and Hutto 2005; Ruegg et al. 2017). Alternatively, when birds molt outside of the breeding area, isotopes can be used to identify migratory stopover or nonbreeding areas, and genetics can be used to identify breeding sites. Further, habitat information can add additional resolution to resulting assignments (Ruegg et al. 2017). The power of such a combined genetic, isotopic, and habitat approach for assigning individuals to breeding locations was demonstrated in the case of the Wilson's Warbler; the authors showed that the best accuracy was found when all three datasets were integrated (Ruegg et al. 2017).

New methods have been developed to integrate extrinsic tracking data (GPS, geolocator) and intrinsic data (genetics, stable isotope)—methods that incorporate the often uneven and multidirectional distribution of data types, as well as the error and bias associated with each data type, into estimates of migratory connectivity pattern and strength (Hostetler et al. 2025). A powerful recent demonstration of such combined approaches is an investigation of the evolution of migration around a major mountain barrier, the Karakoram range (Turbek et al. 2022). In this study on Barn Swallows, stable isotope approaches and geolocators revealed a continental migration divide of Chinese-breeding birds, whereby one population spends the nonbreeding season in East Africa and the other in south India. DNA sequencing indicated strong divergence between these populations and suggested behavioral mechanisms as the primary driver.

QUANTIFYING THE STRENGTH OF MIGRATORY CONNECTIVITY

As described in chapter 1, the strength of migratory connectivity is defined as the extent to which individuals and their populations remain together between seasons of the annual cycle. When migratory connectivity is strong, animals remain geographically close and temporally synchronized over the annual cycle. In contrast, when migratory connectivity is weak, animals that are geographically or temporally synchronized in one season are not during other seasons. For example, breeding populations that use the same migration routes, stopovers, or staging areas, and together use them at the same time, are considered to have strong migratory connectivity during this period. Conversely, breeding populations that use different migration routes, or pass through them at different times, have weak connectivity during migration (Bauer et al. 2016; Cohen et al. 2019). Data is now mounting to better understand the range of natural variability in the strength of migratory connectivity and how it is influenced by long-distance dispersal, annual and within-season environmental variation, and interactions between these periods, as well as individual heterogeneity itself. Therefore, methods to quantify the strength of migratory connectivity should be comparable among taxonomic groups, range sizes and shapes, and seasons of the annual cycle, as well as all of the methodological approaches available to measure migratory movements.

That said, estimating the strength of migratory connectivity is also dependent on the scale of tag deployment relative to the range and distribution of a species. Often populations or geographic locations are chosen arbitrarily, or largely for logistical reasons (e.g., convenience, permit availability, study sites) rather than on biology. Ideally, tag deployment would align with a predetermined and biological justification based on true population structure or at least on a hypothesized population structure. For example, studies could use morphological differentiation (e.g., subspecies; see the "Morphology" section above), demographic attributes (natural populations based on fluctuations in abundance; Rushing et al. 2016), or actual genetic differentiation from genoscapes (see the "Genoscape Approaches" section above) as a first layer to establish ideal locations for tag deployment. Such a design would not only provide clarity on the necessary scale, it would eventually provide insights into the mechanisms underlying ecological and evolutionary processes driving migratory connectivity itself.

For now, we cannot let the desire for perfection be the enemy of the good, especially when we are in a race against time to save so many declining species. Much research has taken place, and there exist tremendous amounts of data on migratory connectivity that need to be compiled. As in many cases, the measurement of anything, including the seasonal distributions of populations, most often involves incomplete or biased sampling and use of technologies that vary in accuracy and precision. The MigConnectivity package in the R statistical programming language was created to estimate the strength of migratory connectivity across studies and taxa with data derived from many data types and with uneven sampling (Hostetler et al. 2018; Cohen et al. 2018). The strength of the migratory connectivity metric, MC, is negative when populations are farther apart between seasons, positive when populations remain together between seasons, and zero when populations have no patterns in distribution between seasons and can be estimated with uncertainty (Cohen et al. 2018). MC incorporates the likelihood that an individual from a specific breeding population moves to a specific nonbreeding region (or migratory flyway), given that sampling may not be proportional to the population size (i.e., the number of individuals banded or tagged may not be proportional to the abundance in that population; Cohen et al. 2018). While range-wide measures of MC are ideal, this requires more complete sampling of populations than has been possible from most single data types. Therefore, recent updates to the package integrate the diverse measures of migratory movements from multiple data types with methodology to account for differences in sampling design, directionality, effort, precision, and bias inherent to each data type (MigConnectivity package version 0.5.0).

Knowledge of the pattern and strength of migratory connectivity is urgently needed to fully understand and conserve migratory species. For example, adaptive management approaches for North America's harvested waterfowl species will be most effective when they account for the extent to which flyway boundaries match the observed migratory movement between flyways and try to manage them in a way that explicitly recognizes the strength of migratory connectivity (Roberts et al. 2023). This book, in part, pulls together in one place all we know about species-specific patterns and strength of migratory connectivity for the birds of North America, including quantitative estimates when possible. This allows for clear comparisons of the strengths of migratory connectivity across species and brings us closer to understanding not only more of the basic biology of these remarkable species but also how to conduct more effective and impactful conservation. Similarly, it identifies the gaps in our knowledge—the species where nothing or very little is understood—to help guide future research and conservation.

CHAPTER 3

ATLAS METHODS

Compiling this Atlas involved organizing a massive amount of data. In total, we assessed the current state of knowledge and compiled data for 564 species of migratory birds. We did this by building 1,343 maps, using 16,410 individual birds tracked with electronic tags by 1,047 contributors; we plotted a total of 232,428 recoveries (based on banding and Motus Wildlife Tracking System data); and we cited nearly 2,500 peer reviewed papers. The amount of data, tools, and unsolved and solved tracking mysteries that have emerged through both the development of the science of migratory connectivity and specifically this Atlas has been nothing short of impressive. Although this effort was exhaustive, we have probably missed data, and new data is emerging rapidly. Regardless, compiling the data we do have sufficiently characterizes migratory connectivity for many of the species, and, equally importantly, we clarify where the data is insufficient. This chapter provides detailed information to understand all the rich material and data presented and illustrated in this Atlas.

Compiling information from many contributors and sources (published and unpublished) from a mounting body of literature and integrating data collected by many types of tools over many years takes an enormous amount of effort and time. For example, the banding and reencounter data from the Bird Banding Laboratory alone, a mostly untapped resource for well over a hundred years, was in itself a tremendous task. Species in the Atlas have been tracked using nine types of different tools and approaches (figure 3.1). Using unpublished tracking data required gathering associated permissions from all data contributors and sometimes conducting additional analyses. Our distilling of this complex array of information into maps and species accounts, so that readers can understand patterns of migratory connectivity, was conducted as objectively as possible to ensure repeatability. To interpret the maps, we provide a detailed explanation of our methodologies, our techniques for displaying data, how data were screened, what data were included, what each type of map is meant to illustrate, and what certain symbols denote.

SPECIES SELECTION AND ASSOCIATED TRAITS

As defined in chapter 1, we consider a species "migratory" if it exhibits repeated and predictable movement between breeding and stationary nonbreeding latitudes or altitudes (that is, migration up and down a mountainside or stream). We further consider a species migratory if a subset of populations (or individuals within a given population) migrate while others remain sedentary (called partial migration). We identified 564 species (78% of the 721 birds regularly occurring in North America; Chesser et al. 2024) that fit these criteria.

We defined the periods of the annual cycle as breeding, stationary nonbreeding (hereafter, nonbreeding), pre-breeding migration, and post-breeding migration (also called spring migration and fall migration, respectively) (Albert and Siegel 2024); for each species we created an annual cycle graph with beginning and ending dates for each period. All information is from the Birds of the World (2022) species accounts and eBird Status and Trends data products (Fink et al. 2023).

For each species, we compiled a "traits" table composed of the following: a characterization of primary breeding and nonbreeding habitats; global population size (Panjabi et al. 2024); population change since 1970 (Rosenberg et al. 2019); and conservation designation defined by five different organizations—the International Union of Conservation of Nature (IUCN) Red List of threatened species (2022), the Road to Recovery (R2R) Tipping Point species list (https://r2rbirds.org/tipping-point-species/), the US Threatened & Endangered Species list (https://ecos.fws.gov/ecp/), the Canadian Species at Risk public registry (Species at Risk Act 2002), and the Partners in Flight Continental Importance Score (formerly Continental Concern Score; PIF CCS; Panjabi et al. 2024). The PIF CCS maximum combined score can range from 4 (for a widespread, numerous, and increasing species that is expected to face even more favorable conditions in the future) to 20 (for a species of the very highest conservation concern).

We classified each species as "landbird" or "seabird" for tracking data and migratory connectivity analyses. There is no single taxonomic grouping that defines a seabird; for the purpose of this Atlas we defined seabirds as species that primarily use marine nonbreeding habitats, especially those regularly using offshore (versus strictly coastal) habitats extending beyond territorial waters (12 nautical miles from shore). We treated as seabirds all Procellariiformes, some Gaviiformes, and some Charadriiformes (we treated as seabirds all Alcidae and Stercorariidae, and some Laridae). The remaining species we called landbirds.

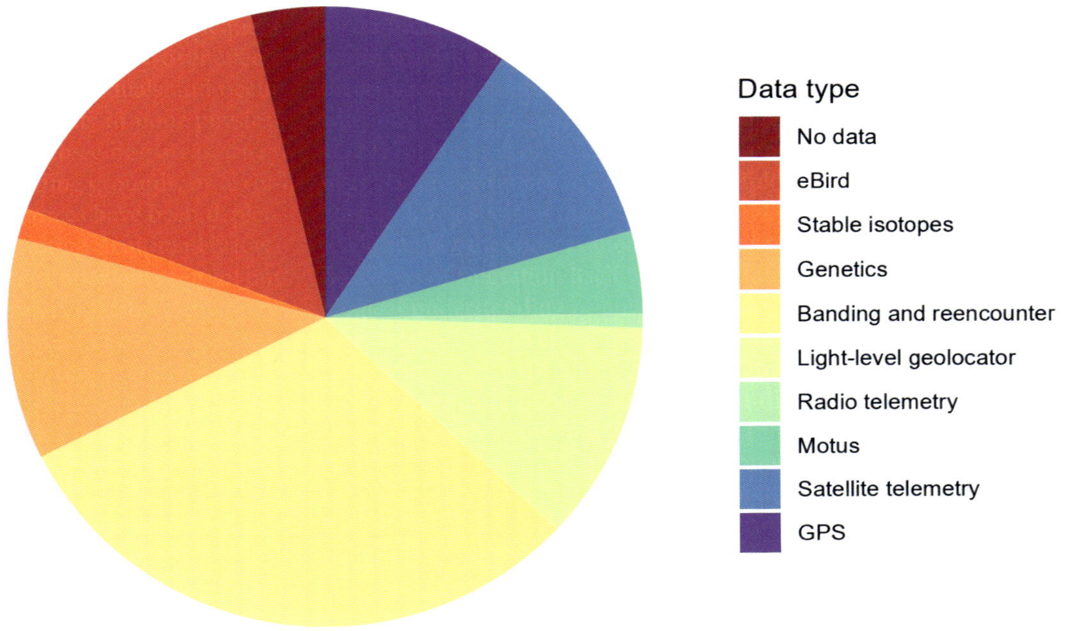

FIGURE 3.1 *The proportion of species (n = 564) within each data type category used to quantify migratory connectivity. Species with no data are also represented in this Atlas.*

SPECIES DATA REVIEW, INCLUSION, AND COLLECTION

We compiled information on migratory connectivity for this Atlas from the literature, data repositories, and direct contributions through a data outreach process. Our first step was to compile all available literature for each species. We used a standardized set of Boolean search terms, "Argos, band encounters, biologging, connectivity, genetic marker, geolocator, GPS, GSM, mark-recapture, migration, migratory connectivity, movement, nano, radar, telemetry, stable isotope, track, transmitter, winter (winter OR nonbreeding) AND (summer OR breeding)," combined with each migratory species' common and scientific name to query the Web of Science Core Collection online database, Google Scholar, and Google for relevant peer-reviewed literature. We screened the resulting articles at three levels (title, abstract, and full article) to decide if they fit our inclusion criteria (Scarpignato et al. 2023).

Second, we used the common name and scientific name to search tracking data repositories such as Movebank (Kays et al. 2022), Movebank Data Repository, Seabird Tracking Database (Carneiro et al. 2024; https://www.seabirdtracking.org/), US Geological Survey's Alaska Science Center wildlife tracking data collection (USGS 2019), and the SEATURTLE.ORG online repository. We included electronic tracking data that estimated the migratory pathways between two or more stages of a bird's annual cycle using locations collected by bird-borne tags, including GPS satellite tags (archival and transmitting); Argos satellite tags, also known as platform transmitting terminal (PTT) tags; and light-level geolocator (LLG) tags (for a review of these technologies, see chapter 2). We also assessed and compiled information on migratory connectivity from two additional data repositories: (1) the banding and reencounter database from the US Geological Survey, Bird Banding Laboratory, North American Bird Banding Program (Nakash et al. 2023), and (2) tracking data from the Motus Wildlife Tracking System (https://motus.org/; Taylor et al. 2017).

In addition to literature and data repository searches, we conducted a data outreach process to obtain data directly from individuals. We corresponded with holders of tracking data and species experts to confirm data found during our online search; this led to the discovery of additional data. Data contributors submitted a data sharing agreement, which outlined the uses of their contributed data and listed three options to contribute tracking data: the full tracking dataset, a subset of the tracking data, or only data on breeding-to-nonbreeding locations. Data contributors were also given the option to share their raw data only without allowing us to map the data. In this case, we only included the data in migratory connectivity analyses and the generalized connectivity map. Tracking data that met our criteria (described above) were included and encompassed the years 1970–2024, although the majority of the data come from the years 2000–2021. All data contributors were acknowledged in three places: on the species account page, in the "Dataset Citations" section, and in the "Data Contributors" section.

For studies where we could not acquire the tracking data itself, we summarized the study in the text or estimated the breeding and nonbreeding locations using, if available, maps, figures, or text from published papers, reports, or websites. Similarly, we summarized in the text stable isotope analysis and genetic studies of migratory connectivity found through our literature review. In addition, we compiled information on migratory connectivity using alternative approaches, including morphology, subspecies classifications, trace element analysis, and at-sea surveys. We also included reviews of studies reporting genetic population differentiation, although not migratory connectivity per se, as it may be a powerful tool for future studies of migratory connectivity.

Geographically, our scope includes deployments within the United States and Canada (including offshore islands within their Exclusive Economic Zones, EEZs; see figures 4.1 and 4.2). We also included deployments outside the United States and Canada if the bird migrated to either country, including their EEZs.

DATA SCREENING AND PROCESSING

We developed data screening and processing protocols for each data source to retain the relevant high-quality data on migratory movements and data that met our criteria for migratory connectivity analyses between breeding and nonbreeding locations.

— BANDING DATA

The Bird Banding Laboratory database contains over 77 million banding records and more than 5 million reencounters (www.pwrc.usgs.gov/bbl). The banding data are digitized from 1955 forward; prior to 1955, only reencounter data are digitized. We searched reencounter data for the years 1914–2015 for all Atlas species. We removed records that could not confirm the real reencounter location, such as a band found, purchased, or traded; then we assigned a season code (breeding, nonbreeding, pre-breeding/spring migration, or post-breeding/fall migration), when possible, to each banding and reencounter location based on the species-specific annual cycle; and finally we calculated a distance between all banding and reencounter locations. Because we were interested in migratory connectivity between breeding and nonbreeding, we only retained records with at least one point in the breeding season and one point in the nonbreeding season that were >200 km between banding and reencounter (Korner-Nievergelt et al. 2012). We excluded reencounters during migration periods; this encompasses the vast majority of recoveries. Consecutive reencounters within one migration period are very limited, and thus these points were often difficult to interpret as part of a full annual cycle. Some species may breed multiple times a year or throughout the year, or have variable breeding phenologies across their range (for example, many tropical seabird species). For these species, it was not always possible to confirm breeding and nonbreeding records, and these connections were shown but omitted from calculations of migratory connectivity.

— MOTUS DATA

We developed a set of selection criteria to query the Motus database for inclusion in both migratory connectivity and migration analyses; a detection of the same tag in both the breeding season and nonbreeding season (as defined by a species' annual cycle dates) >200 km apart (migratory connectivity) or at least three consecutive detections of the same tag between either breeding to post-breeding migration or nonbreeding to pre-breeding migration >200 km apart (migration). We included tracking data that met the above criteria for the years 2014–2021.

In addition, we surveyed the Motus database for detections not meeting the above criteria but that could supplement interpretations of migratory connectivity for individual species. We note this in the text and highlight specific project names where appropriate.

— TRACKING DATA

We processed tracking data differently based on the data type and species classification of landbird or seabird; however, for all data types, we removed data with (1) missing coordinates, (2) missing individual identification, (3) obvious outliers such as data points far away from the main track or species ranges, and (4) locations indicated as erroneous by data contributors or species experts during the review process.

We received light-level geolocator data in multiple formats, each of which was processed differently. Data received as locations, with no noticeably erroneous locations, received no further processing. For data received as estimated locations with erroneous locations, such as large vacillations in latitudinal direction or outliers far outside of species range, we removed data collected surrounding the equinoxes (March 9–April 1; September 11–October 4), which is the time of year with the most unreliable location estimates. In rare cases, we removed whole equinox months (spring: March and April; fall: September and October). Additionally, we removed locations collected during the breeding season for burrow nesters and high Arctic breeders. Similar to the equinoxes, light-level geolocator data collected under these situations are unreliable since entering and exiting a burrow simulates sunrise and sunset, and 24-hour daylight for high-Arctic breeders prevents sunrise/sunset determinations.

We then extracted the distribution of breeding and nonbreeding points, using the annual cycle dates for each species, and determined a breeding and nonbreeding endpoint for each individual. The methods used to determine breeding and nonbreeding endpoints varied depending on the number of available data points and the spatial distribution of locations. If a deployment location was provided and occurred within either the breeding or nonbreeding season, we used that location for the corresponding seasonal endpoint. For the opposite season (nondeployment season) or in cases where no deployment location was provided, we adopted additional methods. For individuals with ≥5 locations in a given season, we constructed a 50% kernel home range (adehabitatHR::kernelUD; Calenge and Fortmann-Roe 2023; Worton 1989, 1995). If the resulting home range was not a single contiguous area, we selected the centroid of the northernmost polygon for the breeding location or the centroid of the southernmost polygon to represent the nonbreeding location. For individuals with <5 locations in the nondeployment season, we used the northernmost point as the breeding endpoint and the southernmost point as the nonbreeding endpoint. When location data included an estimate of uncertainty, we mapped the point using the associated ellipsoid of uncertainty. In the absence of such information, a 200 km buffer was mapped to represent spatial uncertainty. We used both the breeding and nonbreeding locations in the connections map and migratory connectivity analyses (see below), but due to large errors and multiple data formats often with limited metadata, we did not map pathways for light-level geolocator data.

For satellite data, found through our review process or directly contributed by a data holder, we assigned each location to a data type, either PTT or GPS. For PTT data, a location class was also assigned. We included Argos location classes "0," "1," "2," "3" and removed "A," "B," "Z." We fit continuous-time state-space models to filter Argos or GPS data (using function aniMotum::fit_ssm; Jonsen et al. 2020, 2022, 2023) using a taxonomic group-specific speed filter (flight speed references from Alerstam et al. 2007 and Pennycuick et al. 2013), which predicted locations at a 24 hour time-step, one position per day. We included pathways with three or more locations >200 km apart.

For species classified as "seabird," we processed data as above; however, different life histories and conditions present opportunities for additional processing. For example, geolocator location error is often corrected with sea surface temperature (SST) based latitude (e.g., Shaffer et al. 2005, Nielsen et al. 2006). If there still appeared to be systematic latitudinal bias and no SST was provided, we incorporated a 3 degree error in latitude and 1 degree error in longitude into continuous-time state-space models (Jonsen et al. 2023; converted to km within the model based on position on Earth) to predict a most probable path. This error was chosen based on known error reporting in publications across multiple seabird taxa (Phillips et al. 2004; Frederikson et al. 2016; Winship et al. 2012; Harrison et al. 2022) and primarily affected path estimation during the migratory periods, which were not used in migratory connectivity calculations. For all tracking data, we segmented each track into portions of the annual cycle using net squared displacement (NSD; Bunnefeld et al. 2011; for worked example, see Harrison et al. 2022) from colony (or nonbreeding location if the bird was tagged at sea) linked with a change point analysis (PELT algorithm; Killick and Eckley 2014; Killick et al. 2022). Briefly, NSD is near zero when the bird is close to where it was tagged, usually at the colony. The slope of NSD (with time) steeply increases as the bird migrates and quickly distances itself from the colony. The slope then generally plateaus again as the bird enters a nonbreeding "stationary" area displaced far from the colony. A change point analysis is used to detect the inflection points between transitions in the slope of NSD. According to the time of year, we manually reviewed and classified each algorithm-identified segment into breeding, migration, or nonbreeding. We separated the tracking data by colony, derived from metadata, directly from the data contributor, or associated publications. From the first year of tracking data for both juveniles and adults together, we calculated a KDE for each seasonal distribution with 50% and 90% contour (Lascelles et al. 2016) and centroid for each individual and an average KDE for each colony (continuous-time movement modeling; Fleming and Calabrese 2023). When available, we estimated KDEs for both breeding and nonbreeding habitats; however, for Arctic and burrow nesters, breeding distributions were often unavailable for reasons described above.

MIGRATORY CONNECTIVITY

Migratory connectivity—how populations within a species are connected in space and time throughout the year between different stages of the annual cycle—can be quantified in two ways: (1) the *pattern* of migratory connectivity, the movement or "transition" probabilities from the breeding to nonbreeding regions; and (2) the *strength* of migratory connectivity, the extent of co-occurrence between seasons (breeding and nonbreeding; Roberts et al. 2023). The strength of migratory connectivity (MC) ranges from −1 to 1, with 1 indicating that individuals from each population stay together between seasons, 0 indicating complete mixing of populations between seasons, and −1 indicating that individuals close to each other in one season are farther apart during the other season (Ambrosini et al. 2009; Cohen et al. 2019).

We used different methodologies to measure the pattern of migratory connectivity depending on the amount of data and, more specifically, the number of breeding to nonbreeding connections. We did not calculate transition probabilities with <10 connections. When there were 10–30 connections, we calculated transition probabilities as simple percentages (i.e., transition percents) and if there were >30 connections, we calculated transition percents and also transition probabilities which we used to quantify the strength of migratory connectivity (MC). Transition percents, presented in the connectivity table (figure 4.3, found in the "Guide to Species Accounts") and depicted on the generalized connectivity map (figure 4.3) within each species account, are simply the number of total connections, regardless of data type (but see *Seabirds* below), observed from one breeding region to one nonbreeding region divided by the total number of connections from the breeding region. Transition probabilities are a population-level metric of MC patterns that describe the geographic patterns of population movement within and between regions but do not account for variation in marking effort or population size, i.e., the likelihood that an individual from a specific breeding region transitions to a specific nonbreeding region (Roberts et al. 2023). Methods to estimate transition probabilities (MigConnectivity::estTransition) used in MC calculations are included in the online supplement. We combined all data types and did not incorporate error in these estimates, as location error was smaller than the size of our migratory connectivity regions. We defined migratory connectivity regions separately for landbird and seabird groups.

Landbirds. For the purpose of this work, we delimited broad migratory connectivity regions and used the same regions across species to allow for consistency and generality for the visualizations and quantification of migratory connectivity across species. We used four breeding regions (including three North American flyways commonly used to align management priorities): Pacific Flyway, Midcontinent Flyway (combining Central and Mississippi Flyways), Atlantic Flyway, and Greenland (figure 4.1, found in the "Guide to Species Accounts"). We defined 17 nonbreeding regions (figure 4.1). Within the Americas, we included the four breeding regions, as well as Mexico, Central America, and the Caribbean. We subdivided South America into four regions: Northern South America, Andean West, Southern Cone, and Brazil. Each breeding and nonbreeding endpoint was assigned to a migratory connectivity region. This Atlas shows all connections on maps; however, in analyses of migratory connectivity we included only individuals for which we identified both endpoints (n>30) *and* which were breeding within one of our four focal breeding regions (Pacific Flyway, Midcontinent Flyway, Atlantic Flyway, and Greenland) in the connectivity analysis. If relative abundance estimates were available (eBird; Fink et al. 2023), we used the MigConnectivity package in R (Hostetler and Hallworth 2024) to estimate migratory connectivity strength via the function MigConnectivity::calcMC (n<50) or MigConnectivity::estMC (n≥50).

For each individual within a species, we identified one location within the breeding and one location within the nonbreeding season as migratory connectivity endpoints. We defined the endpoints using deployment location or other metadata provided by the data contributor to indicate breeding versus nonbreeding locations. When such data were not available, for Argos and GPS data, we took the northernmost location as the breeding endpoint and the southernmost location as the nonbreeding endpoint within the range and appropriate season using eBird (Fink et al. 2023) or Birds of the World (2026) annual cycle dates. For light-level geolocator data, we took the centroid of the uncertainty breeding and nonbreeding polygons (50% kernel home range or 200 km buffer). When the endpoints or their uncertainty polygons (50% UD or ellipsoids or 200 km) fell outside of land, we shifted the endpoint to the closest land location and increased the uncertainty polygon to be a buffer of radius equal to the distance shifted.

Seabirds. We assigned marine migratory connectivity regions based on Longhurst Provinces (figure 4.2, found in the "Guide to Species Accounts"), biogeographic regions of the upper ocean nested within four broad ocean biomes and partitioned based on unique sets of environmental conditions (Longhurst 2006; Flanders Marine Institute 2009). Longhurst Provinces include such environments as upwelling zones, large current systems, seas, and shelves and are equivalent to ecoregions on land. They span habitats within countries' Exclusive Economic Zones (200 nautical miles from the coast) and within the high seas (areas beyond national jurisdiction, 64% of the ocean's surface area) and are thus applicable to all seabird species whether they use continental shelf habitats or the open ocean. In a broad sense, such habitat regions should be related to seabird distribution, highlighting connectivity differences across species and oceans that are grounded in seabird ecology. We estimated transition percents from breeding colony to nonbreeding Longhurst Province. Because relative abundance data is not readily available for seabirds, we used the Mantel correlation (Mantel; Ambrosini et al. 2009; Cohen et al. 2019 to estimate the strength of migratory connectivity. We included all individuals for which we were able to identify both migratory endpoints in the migratory connectivity analysis, excluding birds hatched during the same calendar year that they were banded or birds recovered by entanglement in fishing gear.

To apply consistency throughout the Atlas and align with the migratory connectivity estimation approach used for landbirds, we identified one location to represent breeding and one location to represent the nonbreeding season as migratory connectivity endpoints for each individual. For banding and reencounter data, we assigned the breeding location to a terrestrial breeding colony and the nonbreeding location to a marine migratory connectivity region (see below). Much of the contributed tracking data include a colony name in their metadata; however, when not assigned, we georeferenced tag deployment locations and assigned colony locations. Thus, for tracking data, we also used the breeding colony as the breeding location. Although some seabird species may range widely during the nonbreeding season, to align with landbird analyses, we used the centroid of the individual's nonbreeding KDE core area (50%) as the nonbreeding location.

We determined the median MC estimate (for landbirds) and the median Mantel estimate (for seabirds) across all species. We called the 0–33rd percentile *weak* migratory connectivity; the 33rd–66th percentile *moderate* migratory connectivity; and above the 66th percentile *strong* migratory connectivity. (For landbirds, the numerical ranges were −0.11 to 0.18 for weak; 0.19 to 0.38 for moderate; and 0.39 to 1 for strong. For seabirds, the numerical ranges were −0.08 to 0.05 for weak; 0.05 to 0.40 for moderate; and 0.41 to 0.77 for strong.) On our MC scales, we illustrate the weak, moderate, and strong ranges using different colors; see examples in the "Guide to Species Accounts" section below. The length of the scale bars are based on the minimum and maximum estimated MC values: −0.11 to +1 for landbirds, and −0.08 to +0.77 for seabirds.

MAPS

For every species, we created a *generalized connectivity map*, an overall summary of all integrated data types, including band reencounter, radio telemetry, light-level geolocator, Argos, and GPS data. If we were able to identify 10 or more breeding to nonbreeding endpoint connections for a species, then this data was used in the map and shown as colored lines and arrows (this is the only map with arrows).

In addition, depending on the amount and type of data available (i.e., published or contributed with permission),, we created two more types of maps: a *connections map* and a *pathways map*. The connections map shows breeding to stationary nonbreeding locations (shown as labeled endpoints and colored lines). This includes band reencounter, radio telemetry, light-level geolocator, and Argos or GPS data contributed as breeding to nonbreeding locations only (excludes migration tracks). The pathways map shows tracks of migration (shown as black lines) as quantified with satellite (e.g., Argos and GPS) and radio (e.g., Motus) telemetry data. We did not include light-level geolocator data in the pathways map, nor did we connect consecutive locations more than 14 days apart. For some species, we split maps into smaller geographical areas to allow for better data visualization.

We used eBird Status and Trends estimates of relative abundance and range to depict the seasonal geography of species populations

(Fink et al. 2023). Specifically, we used the eBird relative abundance map as our base map for the generalized connectivity map; and we used the eBird range map as our base map for both the connections map and the pathways map.

However, when there were <10 breeding to nonbreeding records, we did not create a connectivity map or table, connections map, or pathways map; instead we mapped our connections and pathways data onto the eBird relative abundance map. When no banding or tracking data was found or not contributed for a species, we simply used the eBird relative abundance map as our generalized connectivity map.

If an eBird range map was not available to use as the base map for our connections map and our pathways map, then we used the BirdLife International distribution maps (BirdLife International and *Handbook of the Birds of the World* 2024). In some instances, we used a combination of eBird and BirdLife breeding and nonbreeding maps. Throughout the Atlas, data from eBird Status and Trends were provided by the Cornell Lab of Ornithology. For species with no data or for which we were unable to locate any data, including an eBird relative abundance map, the species account includes only a range map and traits table.

REVIEW PROCESS

We conducted a species expert and data contributor review process for all species accounts. Data contributors reviewed and provided input regarding the text, citations, maps, transition percent, and MC estimates, as well as approval to use their data in this Atlas.

CHAPTER 4

GUIDE TO SPECIES ACCOUNTS

Each species account contains a **species traits table**, a **summary map**, a **watercolor illustration**, and **descriptive text**. Depending on available data, the species account may also include a **migratory connectivity scale**, an **annual cycle diagram**, a **connectivity table**, and **additional maps**. (For species with no data or for which we were unable to locate any data, including an eBird relative abundance map, the species account includes only a traits table and a range map.) The descriptive text includes up to four sections: **Distribution and Abundance**, **Migratory Connectivity**, **Interesting Note**, and **Data Contributors**.

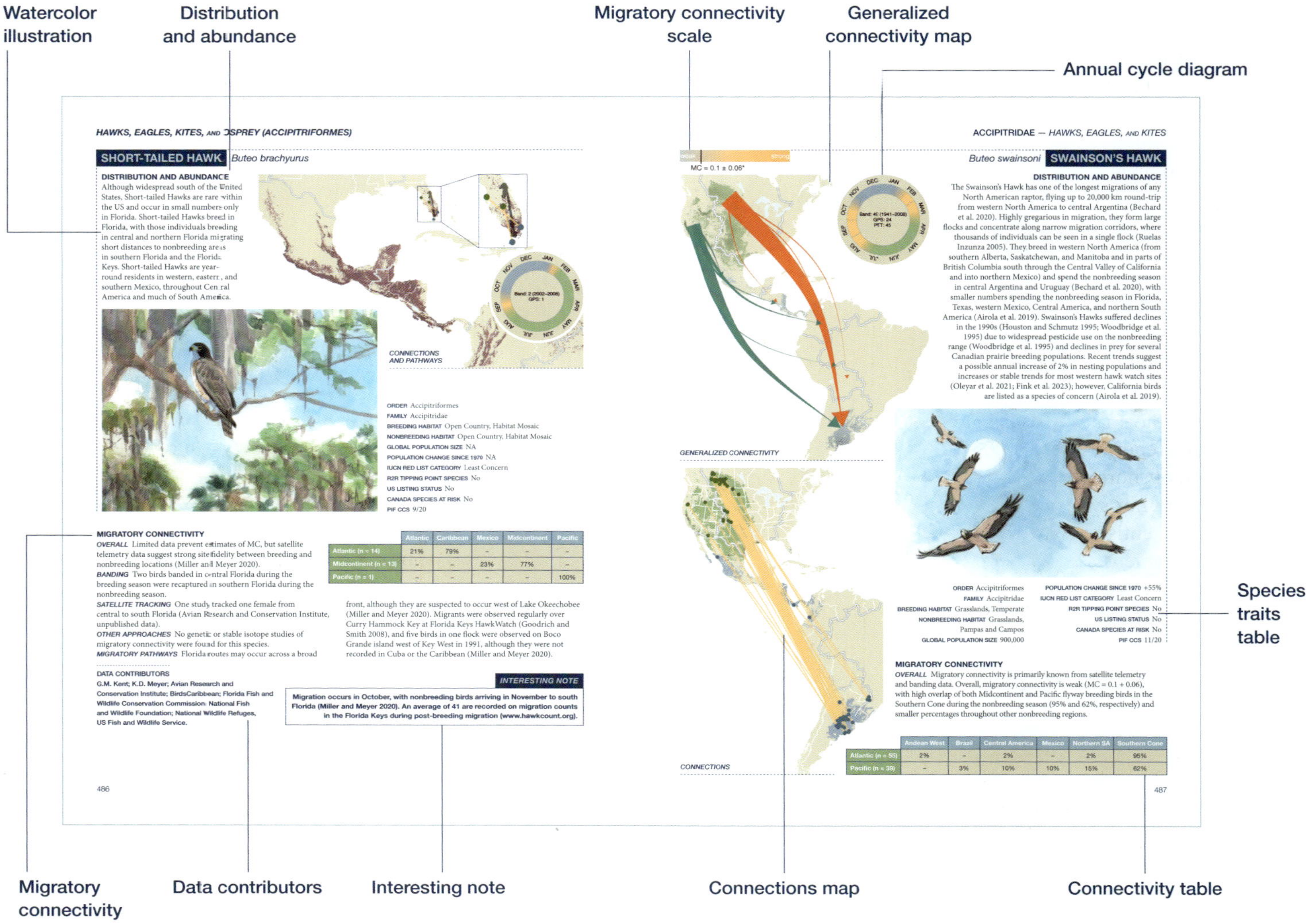

SPECIES TRAITS TABLE

Describes select species characteristics, including taxonomy, habitat associations, population size, population trend, and conservation status (see the "Species Selection and Associated Traits" section above for more details).

ORDER Passeriformes
FAMILY Icteridae
BREEDING HABITAT Grasslands, Temperate
NONBREEDING HABITAT Grasslands, Temperate
GLOBAL POPULATION SIZE 34,000,000
POPULATION CHANGE SINCE 1970 −79%
IUCN RED LIST CATEGORY Near Threatened
R2R TIPPING POINT SPECIES No
US LISTING STATUS No
CANADA SPECIES AT RISK Threatened
PIF CCS 11/20

MIGRATORY CONNECTIVITY SCALE

The estimated strength of migratory connectivity (−1 to 1) is shown as a number on a weak-to-strong continuum and labeled "MC" for land-birds (orange) and "Mantel" for seabirds (blue). Weak, moderate, and strong connectivity ranges for all land birds (or seabirds) are denoted by different colors. The asterisk (*) means that MC is estimated using the MigConnectivity package MCMC model. No asterisk means MC was calculated with CalcMC using the MigConnectivity Package. See the "Migratory Connectivity" section above for more details.

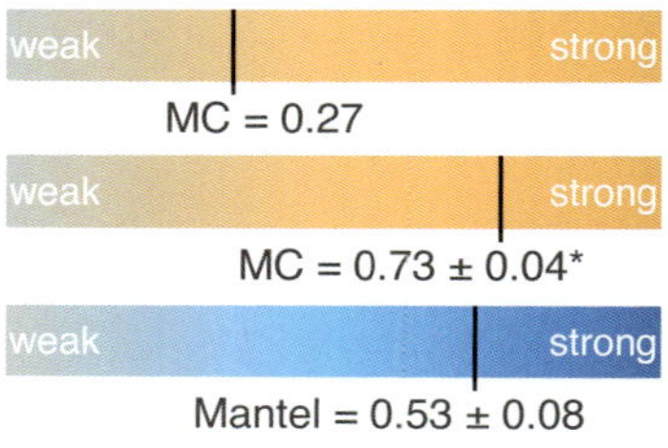

ANNUAL CYCLE DIAGRAM

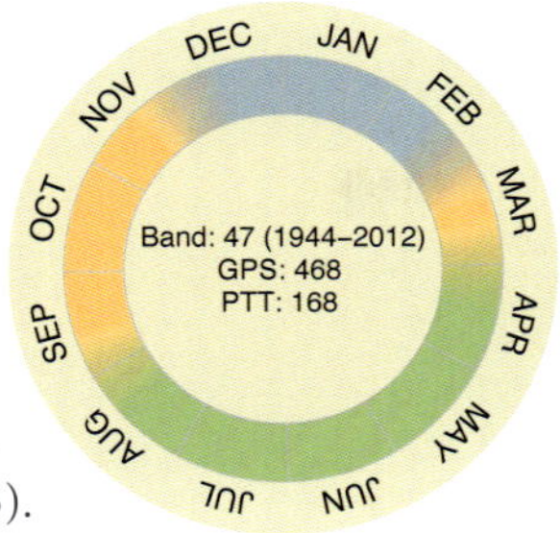

For each species, we defined, with dates, the periods of the annual cycle as breeding (shown in green), nonbreeding (blue), and pre- or post-breeding migration (orange) using the Birds of the World (2026) species accounts and eBird Status and Trends data products (Fink et al. 2023). Sample sizes of all data used (published or contributed with permission only) on the connections map and the pathways map are shown in the center of the annual cycle diagram, as is the source of the data: USGS Bird Banding Laboratory (Band) or the Motus Wildlife Tracking System (Motus); and/or the method by which the tracking data was collected: bands; radio telemetry (VHF); light-level geolocator (LLG) tags; Global Positioning System (GPS) satellite tags (archival and transmitting); or Argos satellite tags, also known as platform transmitting terminal (PTT) tags. For a review of these technologies, see chapter 2.

DISTRIBUTION AND ABUNDANCE

This section describes the distribution and abundance of the species using eBird Status and Trends data products (Fink et al. 2023) when available, as well as information from Birds of the World (2026), the published literature, reports, or other species-specific sources. Many seabird species nest on offshore islands, mainland cliffs, or in the Arctic region and generally are not systematically monitored or regularly accessible to community scientists across their breeding range. In the absence of range-wide assessments for many species, we provided additional detail on seabird distribution and abundance when available, typically from Birds of the World (2026) or recent species status assessments.

MIGRATORY CONNECTIVITY

This section consists of up to nine subsections including: Overall, Banding, Radio tracking, Geolocator tracking, Satellite tracking, Stable isotopes, Genetics, Other approaches, and Migratory pathways.

OVERALL Summarizes the estimates of pattern (transition percents from the connectivity table) and strength (MC/Mantel estimate shown in the traits table) of migratory connectivity using the methods in this Atlas.
BANDING Summarizes banding and reencounter data from the USGS Bird Banding Lab. All data are from either breeding or stationary nonbreeding periods. Reencounters of bands during migration were not used due to the challenges of interpreting these.
RADIO TRACKING Summarizes studies that use radio tracking, including all Motus studies. Studies covered in the text include data contributed to and mapped in the Atlas, as well as data not available to be used in the maps.
GEOLOCATOR TRACKING Summarizes studies that use light-level and newer pressure archival tags. Data from these sources are not used to infer migration routes due to uncertainty of inference from these types of data. Studies covered in the text include data contributed to and mapped in the Atlas, as well as data not available to be used in the maps.
SATELLITE TRACKING Summarizes studies that use archival and transmitting satellite tags such as Argos, GPS, and GSM. Data from these sources are not used to infer migration routes due to uncertainty of inference from these types of data. Studies covered in the text include data contributed to and mapped in the Atlas, as well as data not available to be used in the maps.
STABLE ISOTOPES Summarizes studies that use stable isotope analysis.
GENETICS Summarizes studies that use whole genome sequencing.
OTHER APPROACHES Summarizes studies of migratory connectivity using other known tools such as morphology, subspecies designations, trace elements, or at-sea surveys.
MIGRATORY PATHWAYS Describes the overall migration routes, including those shown on the pathways map, as well as data not available to be used in the maps.

INTERESTING NOTE

A fun fact or remark about the species. This section may also identify gaps and areas of further study for the species.

DATA CONTRIBUTORS

This section lists the name of data contributors that provided and approved tracking data and reviewed species accounts. For more information on data contributors or the contributed data, see chapter 3, and the list of Dataset Citations and Data Contributors at the back of the book.

MAPS

— BASE MAPS

The base maps presented in this Atlas illustrate the seasonal distribution of bird species. For species with sufficient migration data, connections and pathways illustrating seasonal movements are overlaid on these base maps. We used eBird Status and Trends estimates of relative abundance and range, when available, to depict the seasonal geography of species populations (Fink et al. 2023). Specifically, we used the eBird relative abundance map as our base map for our generalized connectivity map; and we used the eBird range maps as our base map for both our connections map and our pathways map. However, when there were less than 10 breeding to nonbreeding records, we did not create a connectivity map or table, connections map, or pathways map; instead, we mapped our connections and pathways data onto the eBird relative abundance map.

When no banding or tracking data was found for a species, we simply used the eBird relative abundance map as our generalized connectivity map. If an eBird range map was not available to use as the base map for our connections map and our pathways map, then we used the BirdLife International distribution maps (BirdLife International and *Handbook of the Birds of the World* 2024). In some instances, we used a combination of eBird and BirdLife breeding and nonbreeding maps. Breeding range is shown in green, nonbreeding range in blue, and year-round range in purple. Throughout the Atlas, data from eBird Status and Trends was provided by the Cornell Lab of Ornithology.

— GENERALIZED CONNECTIVITY MAP AND CONNECTIVITY TABLE

For every species, we created a *generalized connectivity map*, an overall summary of all integrated data types; including band reencounter, radio telemetry, light-level geolocator, Argos, and GPS data. If we were able to identify 10 or more breeding to nonbreeding endpoint connections for a species, then this data was used in the map (shown as colored lines and arrows; this is the only map with arrows) and in the connectivity table (figures 4.3 and 4.4). The sample size shown in the connectivity table is the number of individuals included in the migratory connectivity analysis; this may differ from the total sample size

FIGURE 4.1 *Landbird migratory connectivity breeding (green) and nonbreeding (blue) regions used across all species.*

BELOW: **FIGURE 4.2** *Seabird nonbreeding migratory connectivity regions, Longhurst Provinces, biogeographic regions of the upper ocean nested within four broad ocean biomes (Polar, Coastal, Trades, and Westerlies) and partitioned based on unique sets of environmental conditions (Longhurst 2006; Flanders Marine Institute 2009).*

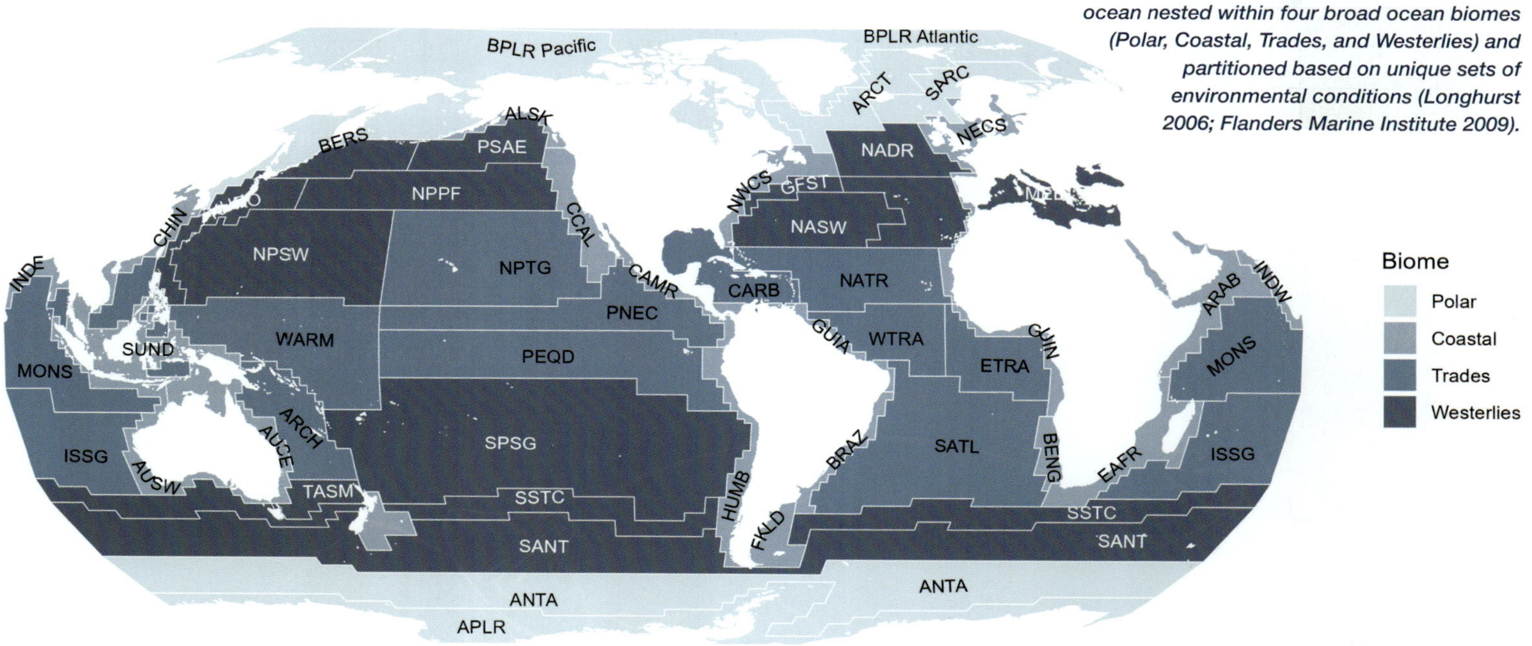

All names in parentheses are acceptable alternate names found in text.

ALSK: Alaska Downwelling Coastal Province
ANTA: Antarctic Province
APLR: Austral Polar Province
ARAB: NW Arabian Upwelling Province
ARCH: Archipelagic Deep Basins Province
ARCT: Atlantic Arctic Province
AUSE: East Australian Coastal Province
AUSW: Australia-Indonesia Coastal Province
BENG: Benguela Current Coastal Province (also known as Benguela Current)
BERS: N. Pacific Epicontinental Province
BPLR Atlantic: Boreal Polar Province
BPLR Pacific: Polar - Boreal Polar Province
BRAZ: Brazil Current Coastal Province (also known as Brazil Coast Current)
CAMR: Central American Coastal Province
CARB: Caribbean Province (also known as Caribbean Sea)
CCAL: California Upwelling Coastal Province (also known as California Current)
CHIN: China Sea Coastal Province
CNRY: Canary Coastal Province (also known as Canary Current)
EAFR: E. Africa Coastal Province
ETRA: Eastern Tropical Atlantic Province
FKLD: SW Atlantic Shelves Province
GFST: Gulf Stream Province (also known as Gulf Stream)
GUIA: Guianas Coastal Province
GUIN: Guinea Current Coastal Province
HUMB: Chile-Peru Current Coastal Province (also known as Humboldt Current)
INDE: E. India Coastal Province
INDW: W. India Coastal Province
ISSG: Indian S. Subtropical Gyre Province
KURO: Kuroshio Current Province
MEDI: Mediterranean Sea, Black Sea Province
MONS: Indian Monsoon Gyres Province
NADR: North Atlantic Drift Province (also known as North Atlantic Drift Current)
NASE: N. Atlantic Subtropical Gyral Province (East)
NASW: N. Atlantic Subtropical Gyral Province (West)
NATR: N. Atlantic Tropical Gyral Province
NECS: NE Atlantic Shelves Province
NEWZ: New Zealand Coastal Province
NPPF: N. Pacific Polar Front Province
NPSW: N. Pacific Subtropical Gyre Province (West)
NPTG: N. Pacific Tropical Gyre Province
NWCS: NW Atlantic Shelves Province
PEQD: Pacific Equatorial Divergence Province
PNEC: N. Pacific Equatorial Countercurrent Province
PSAE: Pacific Subarctic Gyres Province (East)
PSAW: Pacific Subarctic Gyres Province (West)
REDS: Red Sea, Persian Gulf Province
SANT: Subantarctic Province
SARC: Atlantic Subarctic Province
SATL: South Atlantic Gyral Province
SPSG: S. Pacific Subtropical Gyre Province
SSTC: S. Subtropical Convergence Province
SUND: Sunda-Arafura Shelves Province
TASM: Tasman Sea Province
WARM: W. Pacific Warm Pool Province
WTRA: Western Tropical Atlantic Province

shown in the annual cycle diagram. The width of the arrow on the generalized connectivity map corresponds to the strength of the transition probability estimate in the table. For landbirds, the arrow starts within one of four breeding regions (Atlantic Flyway = blue, Midcontinent Flyway = orange, Pacific Flyway = teal, Greenland = Purple) and ends at the centroid of the combined data within the region. For seabirds, the arrow begins at each breeding colony, represented by a different colored arrow, and ends at the centroid of the colony's 50% nonbreeding KDE (kernel density estimate). In cases where individuals migrated outside of the 50% contour, the arrow points to the centroid of the 90% nonbreeding KDE. If a breeding colony only contains banding reencounter data, the arrow ends at the centroid of the nonbreeding locations within each Longhurst Province. The breeding 90% and 50% KDEs, when available, are shown as overlapping transparent green polygons, and the nonbreeding 90% and 50% KDEs are shown as overlapping transparent blue polygons. KDEs are outlined with the colors of the colony (dashed for breeding, solid for nonbreeding). Arrows representing <10 individuals are shown with transparency.

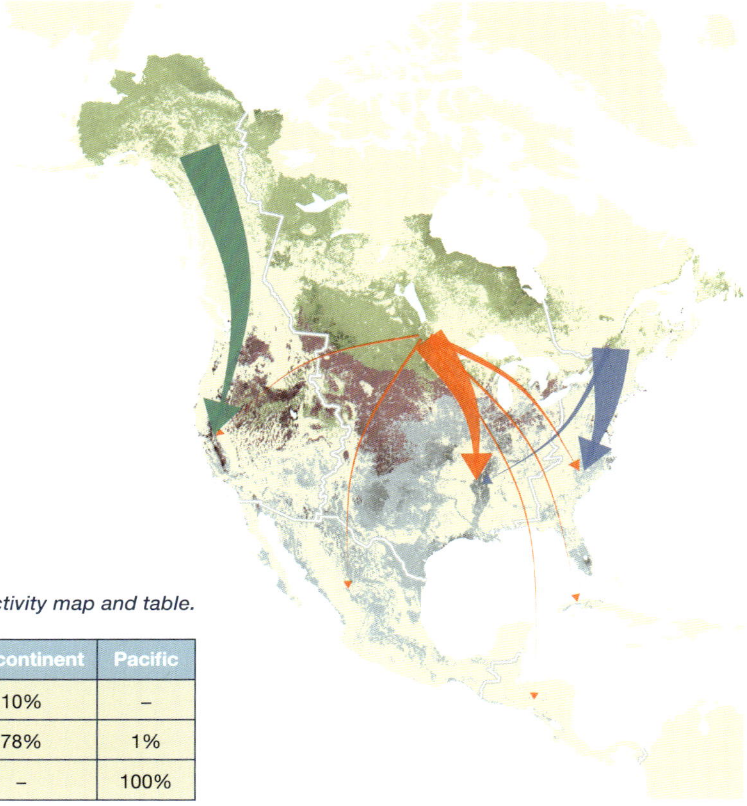

FIGURE 4.3 *Landbird generalized connectivity map and table.*

	Atlantic	Caribbean	Central America	Mexico	Midcontinent	Pacific
Atlantic (n = 10)	90%	–	–	–	10%	–
Midcontinent (n = 77)	14%	3%	1%	3%	78%	1%
Pacific (n = 50)	–	–	–	–	–	100%

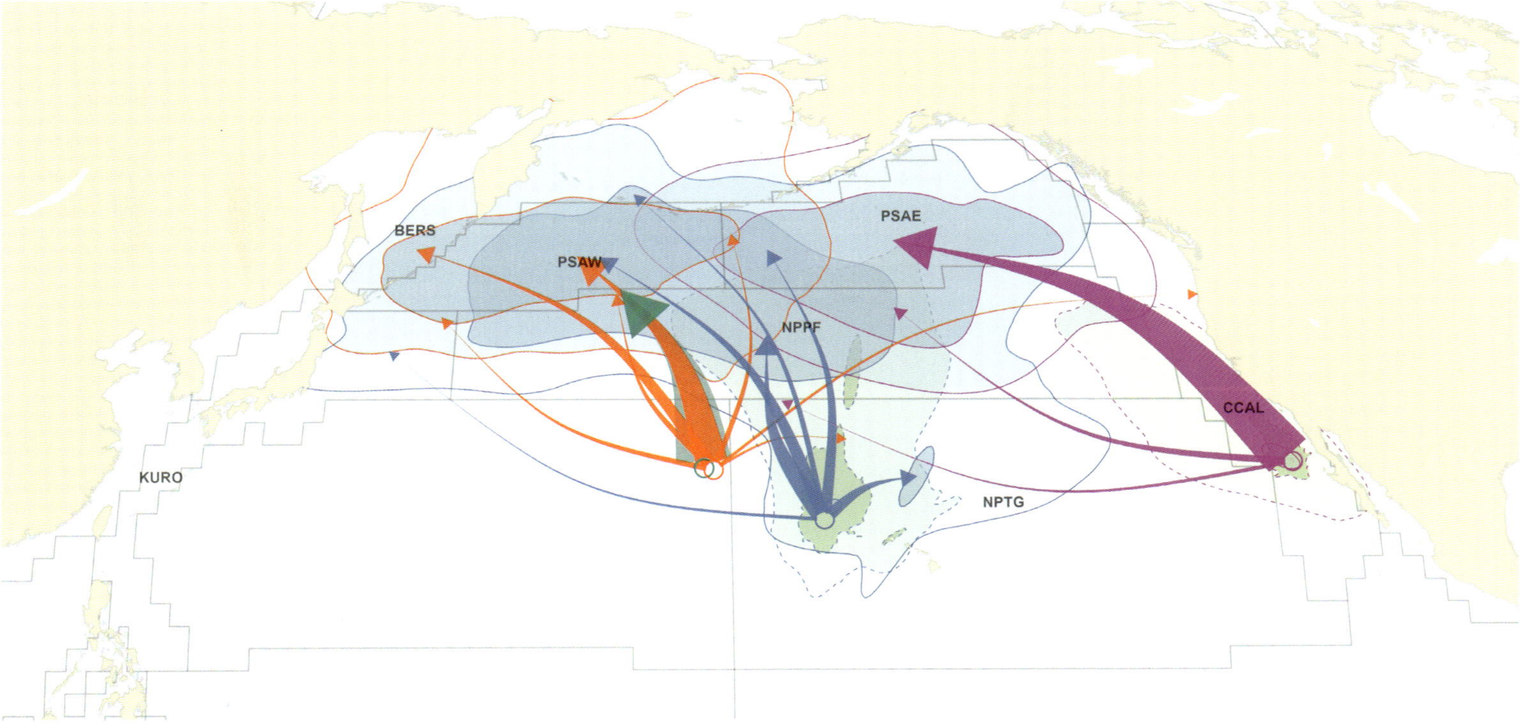

Colony

- Kure Atoll, HI, USA
- Midway Atoll, HI, USA
- French Frigate Shoals, HI, USA
- Isla Guadalupe, Mexico

FIGURE 4.4 *Seabird generalized connectivity map and table.*

	KURO	PSAW	BERS	PSAE	NPPF	CCAL	NPTG
Kure Atoll, HI, USA (n = 2)	–	–	–	–	100%	–	–
Midway Atoll, HI, USA (n = 70)	6%	56%	21%	6%	9%	1%	1%
French Frigate Shoals, HI, USA (n = 42)	2%	21%	7%	17%	33%	–	19%
Isla Guadalupe, Mexico (n = 33)	–	–	–	88%	9%	–	3%

FIGURE 4.5 ***Connections map.*** *The connections map shows breeding to stationary nonbreeding locations (shown as labeled endpoints and colored lines). This includes band reencounter, radio telemetry, light-level geolocator, and PTT or GPS data contributed as breeding to nonbreeding locations only.*

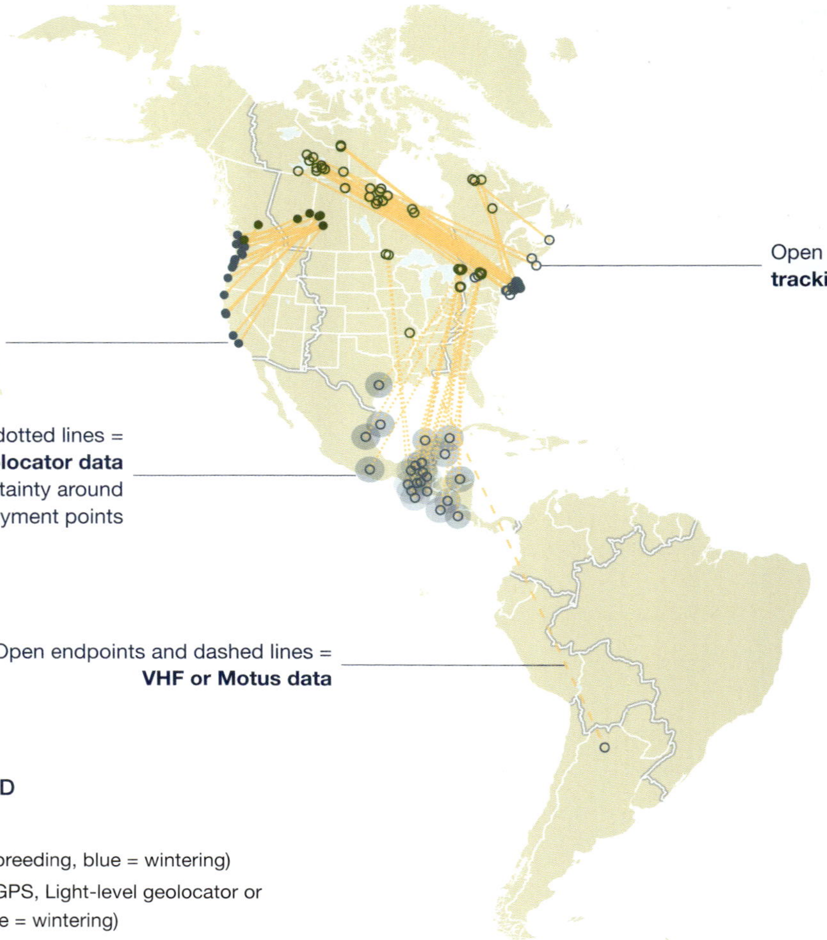

CONNECTIONS MAP LEGEND

ENDPOINTS

Closed circle = Banding (green = breeding, blue = wintering)

Open circle = Tracking from PTT, GPS, Light-level geolocator or Motus/VHF (green = breeding, blue = wintering)

ERRORS

Buffer (transparent around point) = uncertainty area around light-level geolocator enpoints, 200-km uncertainty buffer, or 50%KDE of seasonal points if enough to calculate, or ellipsoid of uncertainty based on latitudinal and longitudinal standard deviations when available

LINES

Solid orange line = Connection between breeding and non-breeding banding (closed circles) or high-resolution tracking (PTT or GPS) points (open circles)

Dotted orange line = Connection between breeding and non-breeding light-level geolocator points

Dashed orange line = Connection between breeding and non-breeding Motus or VHF points

FIGURE 4.6 ***Pathways map.*** *The pathways map shows tracks of migration as quantified with satellite (e.g., PTT and GPS) and radio (e.g., VHF and Motus) telemetry data.*

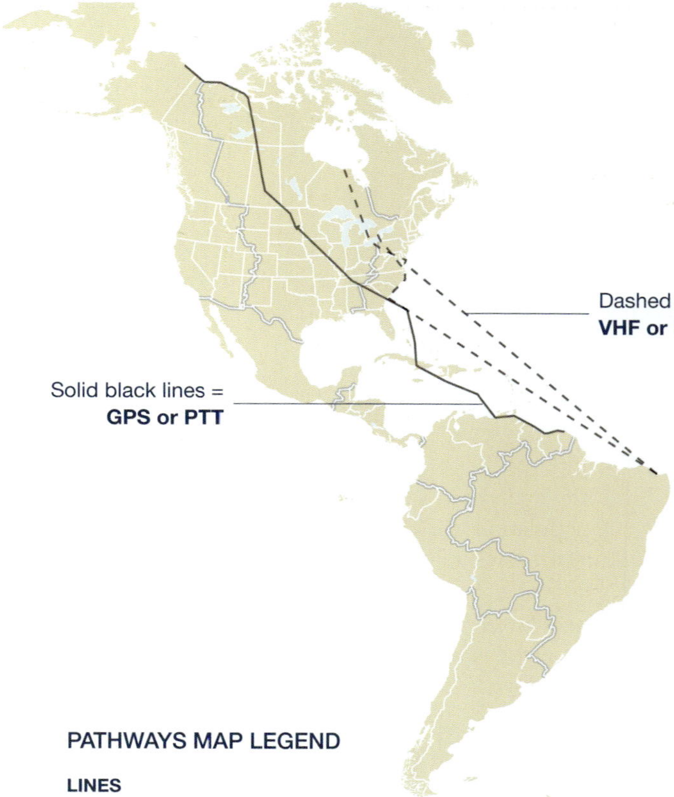

PATHWAYS MAP LEGEND

LINES

Solid = Tracking from PTT or GPS

Dashed = Motus or VHF pathway (usually not full breeding to nonbreeding track)

— CONNECTIONS MAP

Map of breeding to stationary nonbreeding locations are shown with a connecting line (figure 4.5). This includes band reencounter, radio telemetry, light-level geolocator, and satellite telemetry data contributed as breeding to nonbreeding locations only (rather than the full migratory route). Light-level geolocator data are mapped with a 200-km uncertainty buffer (or 50% KDE of seasonal points if enough to calculate, or ellipsoid of uncertainty based on latitudinal and longitudinal standard deviations when available) on the nondeployment location.

— PATHWAYS MAP

Map of satellite and radio telemetry data (figure 4.6). Satellite data are shown as solid lines; radio telemetry data are shown as dashed lines. We did not include light-level geolocator data in the pathways map, nor did we connect consecutive locations more than 14 days apart.

DUCKS, GEESE, AND WATERFOWL (ANSERIFORMES)

WRITING LED BY **ANTHONY J. ROBERTS**

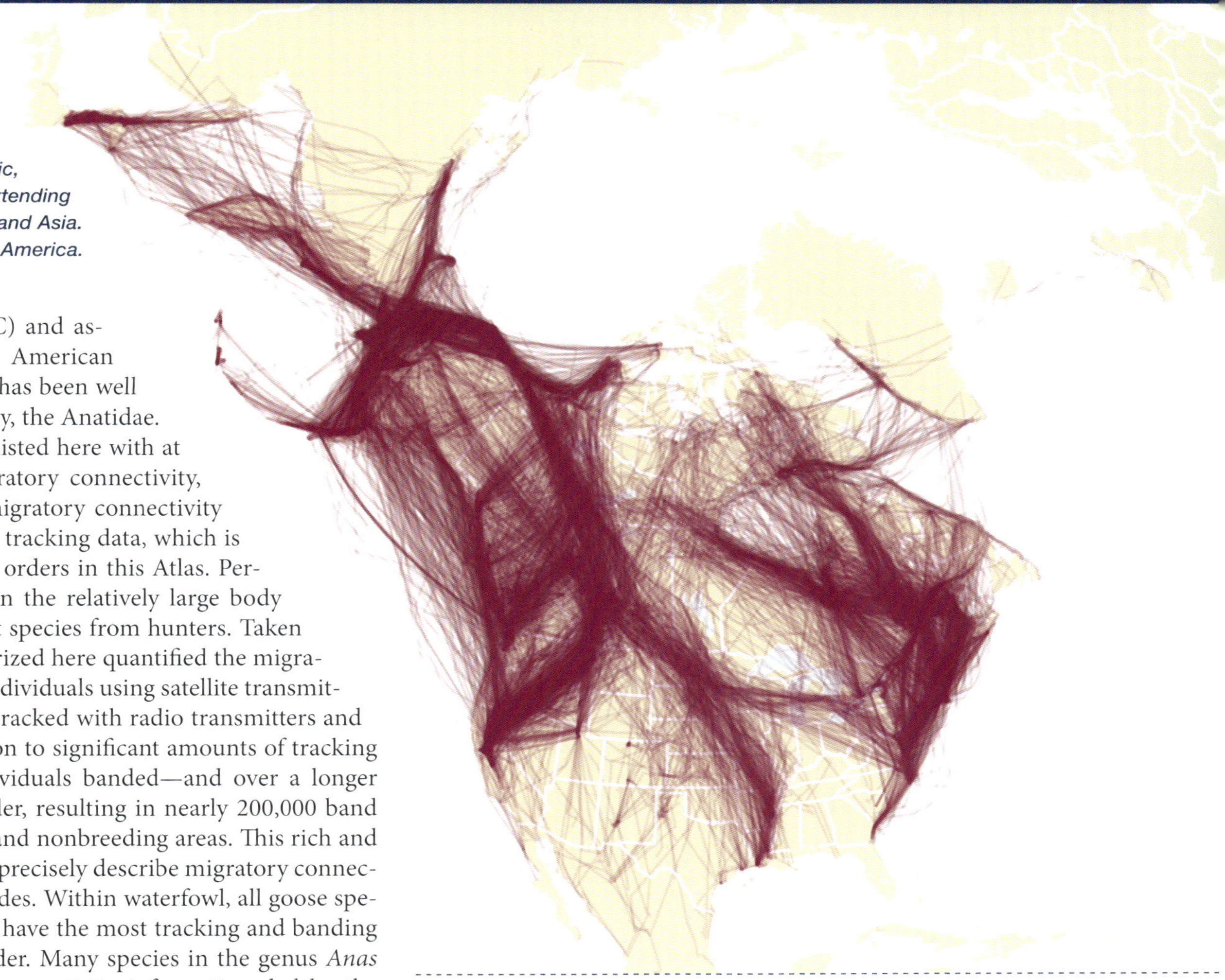

The most heavily tracked groups of birds with satellite and GPS transmitters. Breeding waterfowl have migratory pathways across the continent within the Pacific, Midcontinent, and Atlantic flyways extending to Europe, Central America, Russia, and Asia. Surprisingly, no tracks lead to South America.

The migratory connectivity (MC) and associated ecology of the North American waterfowl (order Anseriformes) has been well studied. It contains a single family, the Anatidae. All species within the order are listed here with at least some information on migratory connectivity, and over 60% of species have migratory connectivity information that is derived from tracking data, which is the highest percentage of all the orders in this Atlas. Perhaps this is not surprising, given the relatively large body sizes and interest in the different species from hunters. Taken together, various studies summarized here quantified the migratory connectivity of over 2,800 individuals using satellite transmitters. Hundreds more have been tracked with radio transmitters and light-level geolocators. In addition to significant amounts of tracking data, waterfowl have more individuals banded—and over a longer period—than any other bird order, resulting in nearly 200,000 band reencounters between breeding and nonbreeding areas. This rich and unparalleled dataset allows us to precisely describe migratory connectivity for many species over decades. Within waterfowl, all goose species have been tracked, and they have the most tracking and banding data of all species within the order. Many species in the genus *Anas* also have abundant migratory connectivity information, led by the most-tracked species, the Mallard. Migratory connectivity is less well-known for species such as Mergansers and Whistling Ducks.

There are tracking data on many species of concern to help managers with conservation efforts. Of the 42 waterfowl species considered here, 7 (17%) are "Tipping Point" species. Tipping Point species are primarily sea ducks (Steller's Eider, Spectacled Eider, King Eider, Long-tailed Duck, and Black Scoter) but also include Northern Pintail. Tracking data from Black Scoters greatly expanded the known breeding range. Prior to 2000, known breeding areas in eastern North America were nearly all east of Hudson Bay; but after tracking more than 100 individuals over 15 years, this proved to be incorrect. Black Scoters were found breeding extensively from the south shore of Hudson Bay north and west through the Northwest Territories. This information allows managers to more efficiently survey core breeding areas to better track population trends.

Waterfowl species have a wide range of strengths of migratory connectivity depending on life history. Species with the strongest MC, above about 0.45, include the swans, Canada Geese, and many species in the genus *Anas*. The strong migratory connectivity was seen in initial banding efforts in the 1960s and resulted in using biological pathways demonstrated by tracking to form the North American administrative flyways in place today. Weak migratory connectivity, below about 0.30, occurs in all species of the genus *Aythya*. Other species with weak migratory connectivity are those that primarily breed in the central Arctic, such as Snow Geese, Greater White-fronted Geese, King Eider, and Long-tailed Duck. These species can have strong connections between restricted breeding and nonbreeding areas, but those areas as described in our work are separate geographical units resulting in a weak MC. Not surprisingly, given their dependence on water bodies, multiple waterfowl species use similar migration routes depending on breeding areas. Birds breeding in central North America often follow the Mississippi River drainage to nonbreeding areas farther south or fly directly to the coasts. Boreal forest breeding birds use stopover sites at the Great Salt Lake or Great Lakes, among others, before moving to coastal areas. Finally, Arctic breeding waterfowl generally follow coastlines or fly overland directly to coasts. The exception are Arctic breeding geese that fly to staging areas in southern Canada prairies before continuing south.

Across all waterfowl species, migratory connectivity is moderate to strong. Most movements are north to south within the flyways, especially in coastal flyways. Atlantic Flyway breeding birds generally stay in the Atlantic Flyway, and a similar pattern occurs in the Pacific Flyway. Waterfowl breeding in the Midcontinent Flyway will migrate to nonbreeding areas in the same flyway, but just as often move to the Atlantic or Pacific Flyways. For ducks this illustrates the importance of the central North American prairies and associated wetlands for waterfowl across the continent. For geese, movements from central to coastal areas are derived from central Arctic breeding geese migrating to nonbreeding areas along the coasts.

BLACK-BELLIED WHISTLING DUCK *Dendrocygna autumnalis*

MC = 0.28 ± 0.1*

DISTRIBUTION AND ABUNDANCE

The Black-bellied Whistling Duck breeds in the southern United States south to South America with concentrations around the Gulf Coast. It appears to be expanding its range in the United States, and in recent years individual vagrants are appearing farther and farther north along the Atlantic coast and the Mississippi River (James and Thompson 2020).

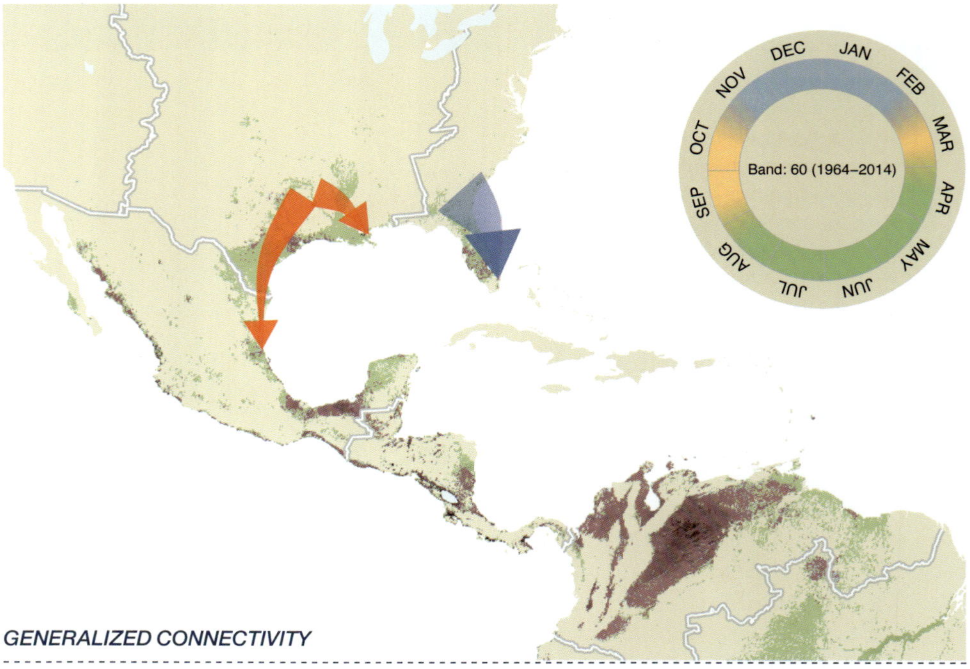

GENERALIZED CONNECTIVITY

ORDER Anseriformes
FAMILY Anatidae
BREEDING HABITAT Wetlands, Generalist
NONBREEDING HABITAT Wetlands, Generalist
GLOBAL POPULATION SIZE 1,000,000
POPULATION CHANGE SINCE 1970 NA
IUCN RED LIST CATEGORY Least Concern
R2R TIPPING POINT SPECIES No
US LISTING STATUS No
CANADA SPECIES AT RISK No
PIF CCS 8/20

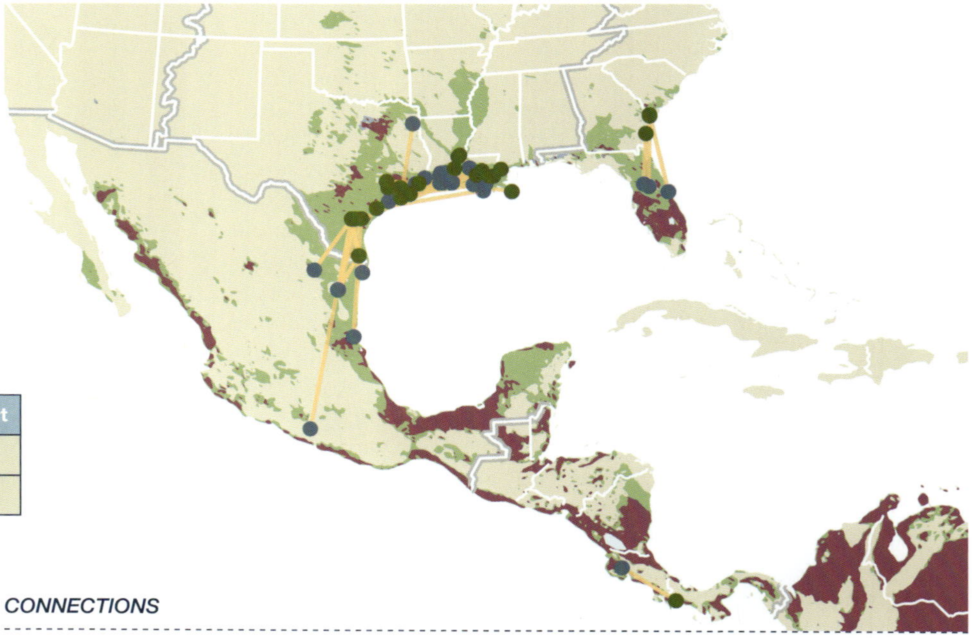

CONNECTIONS

	Atlantic	Mexico	Midcontinent
Atlantic (n = 3)	100%	–	–
Midcontinent (n = 56)	–	50%	50%

MIGRATORY CONNECTIVITY

OVERALL There is no data on migratory connectivity other than several band returns from individuals in the United States and Mexico. The Black-bellied Whistling Duck only appears to be migratory in northernmost populations and exhibits moderate migratory connectivity (MC = 0.28 ± 0.1). Birds breeding in the Midcontinent Flyway migrate to nonbreeding areas within the Midcontinent Flyway (50%) or Mexico (50%); birds breeding in the Atlantic Flyway remain within the Atlantic Flyway for the nonbreeding season (100%). Many birds are residents year-round (Baldassarre 2014).

BANDING Band returns suggest migration of individuals banded in the northern portion of range. Birds breeding in Texas were recovered in Mexico, and those breeding in South Carolina and Georgia were recovered in Florida (Baldassarre 2014). Black-bellied Whistling Ducks banded in Louisiana from 1980 to 2001 were primarily recovered in the same state, with only 2 recoveries outside Louisiana.

SATELLITE TRACKING Twenty adult male birds tracked from Louisiana and Mississippi during pre-breeding migration primarily made regional shifts into the Atchafalaya Basin during the breeding season (Louisiana Department of Wildlife and Fisheries, unpublished data). One bird migrated to breed near Waco, Texas, and another made a brief trip to the Gulf side of Florida before returning to Louisiana (Louisiana Department of Wildlife and Fisheries, unpublished data).

OTHER APPROACHES No genetic or stable isotope studies of migratory connectivity were found for this species.

MIGRATORY PATHWAYS No clear pathways: and in general, movements appear to be nomadic.

INTERESTING NOTE

The population along the southeastern coast of the United States likely originated from releases and escapees from Florida, and has now spread into North Carolina with annual vagrancies farther north into Maryland.

Dendrocygna bicolor

FULVOUS WHISTLING-DUCK

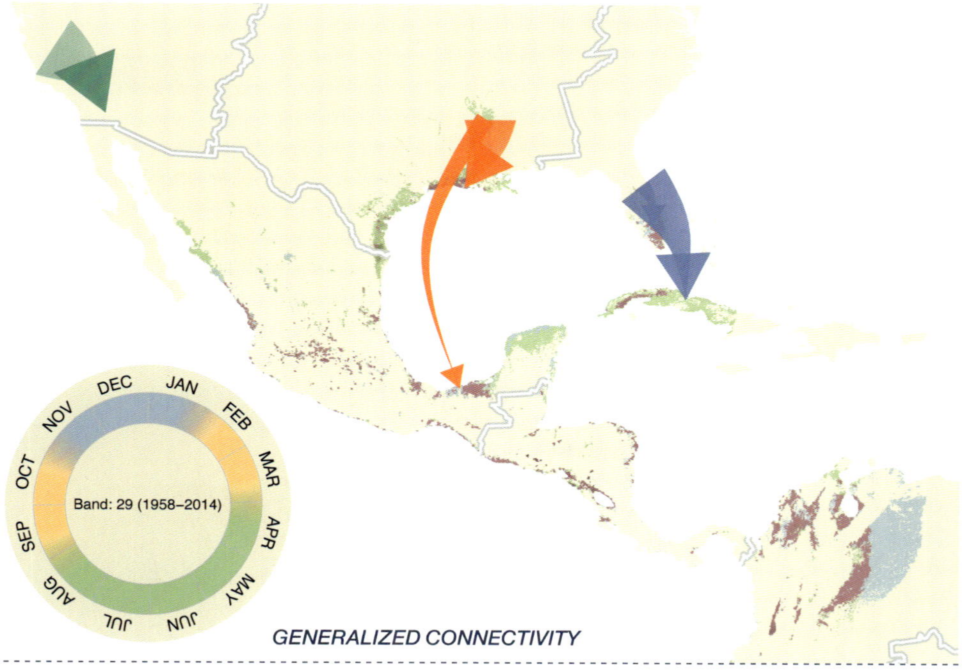

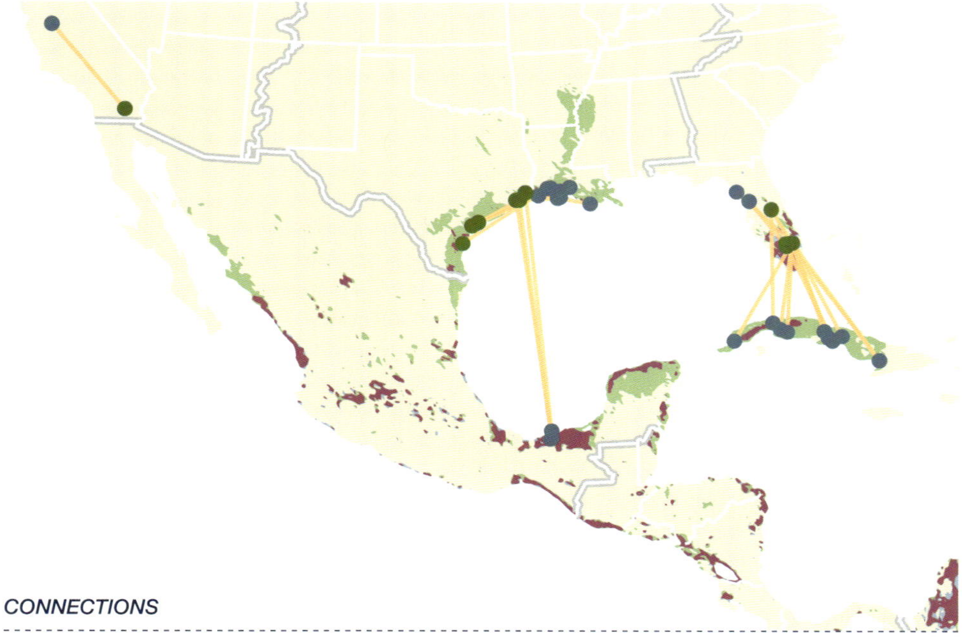

DISTRIBUTION AND ABUNDANCE

Fulvous Whistling-Ducks are found on four continents. In North America they occur along the Gulf Coast, up the Mississippi Alluvial Valley, and in southern Florida. They are short-distance migrants in the northern parts of their range (Hohman and Lee 2020).

ORDER Anseriformes

FAMILY Anatidae

BREEDING HABITAT
Wetlands, Freshwater Marsh

NONBREEDING HABITAT
Wetlands, Freshwater Marsh

GLOBAL POPULATION SIZE 1,400,000

POPULATION CHANGE SINCE 1970 +443%

IUCN RED LIST CATEGORY Least Concern

R2R TIPPING POINT SPECIES No

US LISTING STATUS No

CANADA SPECIES AT RISK No

PIF CCS 9/20

INTERESTING NOTE

Researchers hand-reared 165 Fulvous Whistling-Ducks in southern Texas before subsequently color-marking and releasing them (Flickinger et al. 1973). These birds joined wild flocks and similarly migrated to neighboring Mexican states.

	Atlantic	Caribbean	Mexico	Midcontinent	Pacific
Atlantic (n = 14)	21%	79%	–	–	–
Midcontinent (n = 13)	–	–	23%	77%	–
Pacific (n = 1)	–	–	–	–	100%

MIGRATORY CONNECTIVITY

OVERALL Data from banding and radio transmitters are limited. Movement of breeding birds in the Atlantic Flyway were to nonbreeding areas in the Caribbean (79%) and Atlantic Flyway (21%). Most breeding birds from the Midcontinent Flyway stayed in that flyway during the nonbreeding season (77%), though some migrated to Mexico (23%). MC could not be calculated due to limited data but appears strong.
BANDING Birds banded in Texas and Louisiana were recovered within those states, as well as Mexico, while all recoveries of breeding birds banded in Florida occurred in Cuba during the nonbreeding season. One breeding to nonbreeding recovery in California.
RADIO TRACKING Fulvous Whistling-Ducks (n = 16) captured on nests in Louisiana migrated to Texas or northeast Mexico by September (Hohman and Richard 1994).
OTHER No satellite tracking, genetic, or stable isotope studies of migratory connectivity were found for this species.
MIGRATORY PATHWAYS Short-distance migration along the Gulf Coast, generally following the coastline.

EMPEROR GOOSE *Anser canagicus*

DISTRIBUTION AND ABUNDANCE

Emperor Geese are found year-round almost exclusively in proximity to the Bering Sea in either North America or Russia (Baldassarre 2014). In North America they are restricted to coastal areas in western Alaska. Small numbers sometimes winter in British Columbia, Washington, Oregon, and rarely Northern California (Baldassarre 2014).

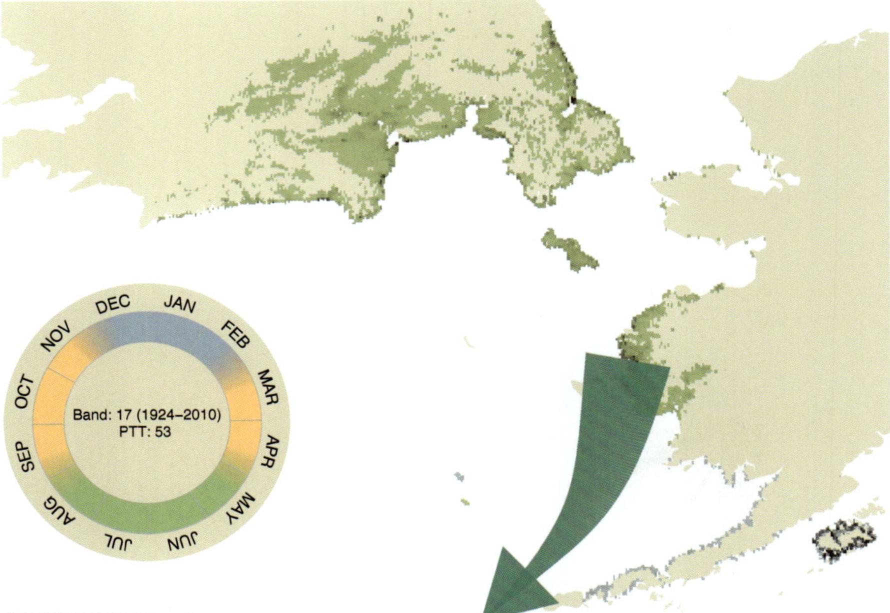

GENERALIZED CONNECTIVITY

ORDER Anseriformes
FAMILY Anatidae
BREEDING HABITAT Coasts, Salt Marsh
NONBREEDING HABITAT Coasts, Marine Waters
GLOBAL POPULATION SIZE 98,000
POPULATION CHANGE SINCE 1970 +13%
IUCN RED LIST CATEGORY Near Threatened
R2R TIPPING POINT SPECIES No
US LISTING STATUS No
CANADA SPECIES AT RISK No
PIF CCS 15/20

	Pacific
Pacific (n = 41)	100%

CONNECTIONS

INTERESTING NOTE

In 1940 an Emperor Goose banded in Russia was recovered in California (Neff 1948).

PATHWAYS

MIGRATORY CONNECTIVITY

OVERALL The migratory connectivity of Emperor Geese is well-known from banding data and satellite tracking, with all birds remaining within the Pacific Flyway during the breeding and nonbreeding season. Because of the single flyway, MC cannot be calculated.
BANDING Juveniles banded in Russia have been recovered during fall migration on the Alaska Peninsula (Schmutz and Kondratyev 1995).
RADIO TRACKING There have been no detections from tags deployed with the Motus Wildlife Tracking System.
SATELLITE TRACKING A study of migration chronology (n = 53) found that some females migrated short distances (650–1,010 km) from the Yukon- Kuskokwim

Delta to the Alaskan Peninsula, while others migrated longer distances (1,600–2,640 km) to the Aleutian Islands (Hupp et al. 2008). Female Emperor Geese who failed to reproduce primarily migrated to Russia to molt (n = 15), while those that were successful molted in the Yukon-Kuskokwim Delta (n = 15; Hupp et al. 2007).

DATA CONTRIBUTORS
D.C. Douglas; B. Uher-Koch.

OTHER APPROACHES No genetic or stable isotope studies of migratory connectivity were found for this species.

MIGRATORY PATHWAYS Relatively short-distance migrants compared to other North American goose species, primarily moving among coastal breeding and nonbreeding areas in Alaska and western Russia (Eisenhauer and Kirkpatrick 1977).

SNOW GOOSE
Anser caerulescens

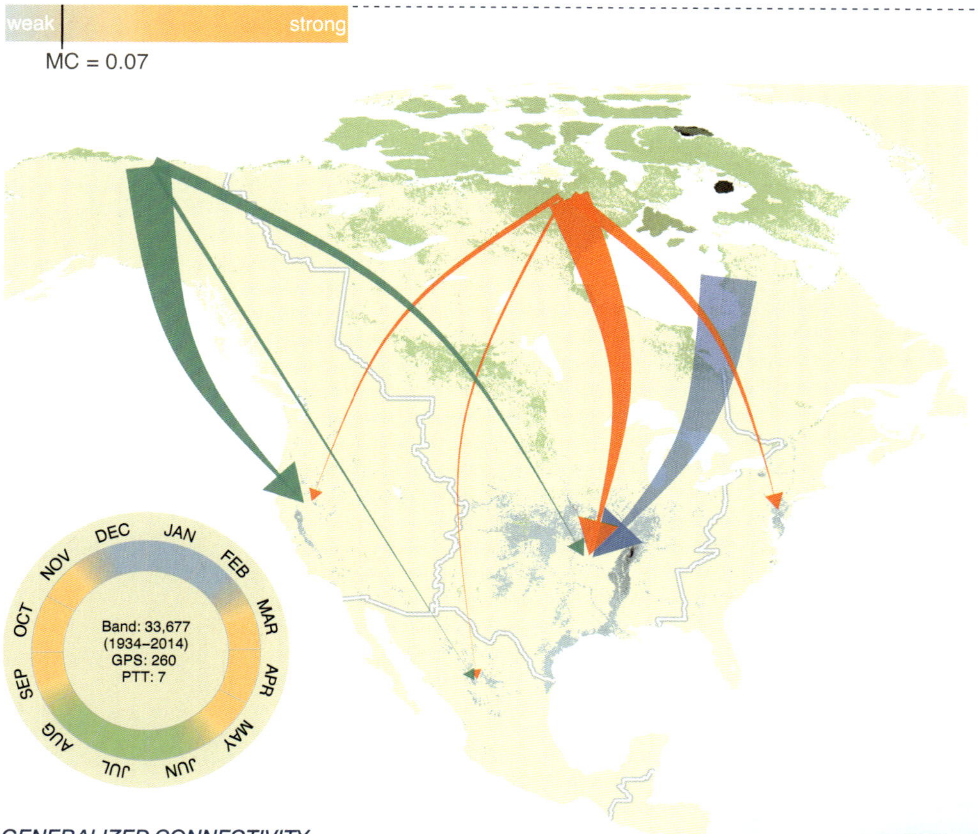

GENERALIZED CONNECTIVITY

	Atlantic	Mexico	Midcontinent	Pacific
Atlantic (n = 4)	–	–	100%	–
Midcontinent (n = 29329)	19%	1%	73%	7%
Pacific (n = 1285)	–	4%	19%	78%

DISTRIBUTION AND ABUNDANCE

Snow Geese breed in dense colonies in the central Arctic and subarctic. Primary nonbreeding areas are the Central Valley of California, the Gulf Coast, the Mississippi River Valley, and the mid-Atlantic coast (Mowbray et al. 2020a). There are two subspecies, the Greater Snow Goose (*C. c. atlantica*) and the Lesser Snow Goose (*C. c. caerulescens*). Both subspecies breed in Arctic Canada, and nonbreeding distributions are distinct. Lesser Snow Goose populations occur at breeding areas on Wrangle Island in far eastern Russia, the western high Arctic, and the central Arctic east of 110 degrees west longitude. The Greater Snow Goose breeding and nonbreeding range is farther east than the Lesser Snow Goose, but both subspecies may overlap within the Midcontinent Flyway.

MIGRATORY CONNECTIVITY

OVERALL Migratory connectivity data from banding and telemetry studies show that across both subspecies, all Atlantic Flyway breeding birds migrate to the Midcontinent Flyway. Midcontinent Flyway breeding populations primarily migrate to nonbreeding areas within the Midcontinent Flyway (73%), and smaller numbers to the Atlantic Flyway (19%). Pacific Flyway breeding populations also primarily migrate to nonbreeding areas within the Pacific Flyway (78%) and also to the Midcontinent Flyway (19%) and Mexico (4%). Although the subspecies' nonbreeding distributions may be distinct, across the species, mixing of breeding populations in nonbreeding areas results in weak migratory connectivity (MC = 0.07).

GENETICS There is no evidence of mitochondrial DNA (mtDNA) distinctions across the species range (Avise et al. 1992). Mitochondrial DNA sampled from Lesser Snow Goose populations in the typical nonbreeding area for this subspecies, Louisiana, and the typical nonbreeding area of Greater Snow Geese, Maryland, did not show genetic differences, though there were significant population differences between the two subspecies (Humphries et al. 2009).

DATA CONTRIBUTORS
J. Bêty; M. Casazza; J. Evenson; J. Lefebvre; P. Legagneux; C. Nicholai; C. Overton; V. Patil; M. Petrie; E. Reed; K. Spragens; P. Thomas; C. Williams; M. Wilson; Canadian Wildlife Service; US Geological Survey.

— GREATER SNOW GOOSE *Anser caerulescens atlantica*

- ORDER Anseriformes
- FAMILY Anatidae
- BREEDING HABITAT Wetlands, Tundra
- NONBREEDING HABITAT Wetlands, Generalist
- GLOBAL POPULATION SIZE 16,000,000
- POPULATION CHANGE SINCE 1970 +887%
- IUCN RED LIST CATEGORY Least Concern
- R2R TIPPING POINT SPECIES No
- US LISTING STATUS No
- CANADA SPECIES AT RISK No
- PIF CCS 10/20

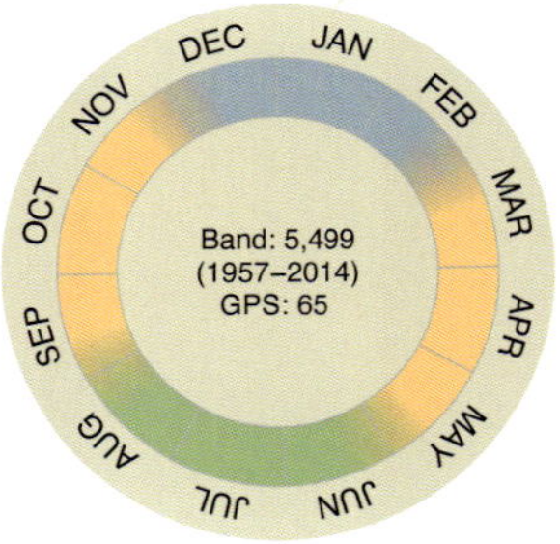

GENERALIZED CONNECTIVITY

	Atlantic	Midcontinent
Midcontinent (n = 5524)	98%	2%

OVERALL The migratory connectivity of Greater Snow Geese is well studied through banding, telemetry, and genetics. Birds breeding in the Midcontinent Flyway spend the nonbreeding season along the mid-Atlantic coast of the Atlantic Flyway (98%). Because of the single breeding flyway, MC cannot be calculated.
BANDING Greater Snow Geese breeding on two islands in the Arctic, Bylot Island and Ellesmere Island, mix during staging in Quebec and spend the nonbreeding season along the mid-Atlantic coast of the United States (Souchay et al. 2015).
RADIO TRACKING A study of 121 radio-collared females showed that molt migration is common among nonbreeders and failed nesters, while only 2% of successful nesters undergo a molt migration (Reed et al. 2003).
SATELLITE TRACKING Individuals tracked from their breeding grounds at Bylot Island migrated to a staging area along the St. Lawrence River, Quebec, then continued to nonbreeding areas along the mid-Atlantic coast (Blouin et al. 1999; Bêty et al. 2004; Legagneux et al., unpublished data).
GENETICS Though many studies suggest high dispersal and gene flow across the range of the subspecies, there is evidence of genetic clustering of familial groups within breeding colonies (Lecomte et al. 2009). There is no evidence that these clusters persist geographically throughout the annual cycle (Desnoyers et al. 2012)

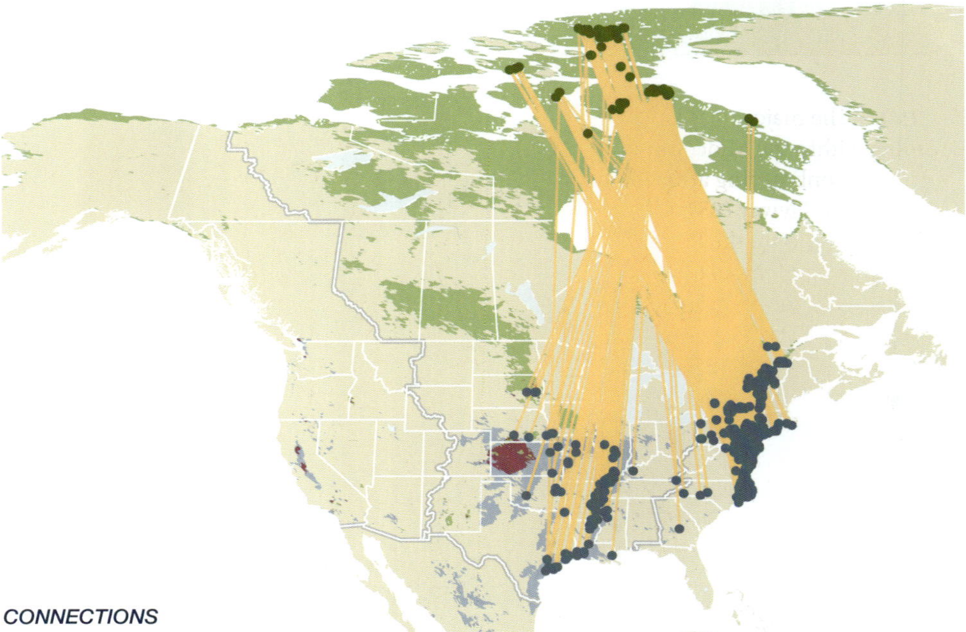

CONNECTIONS

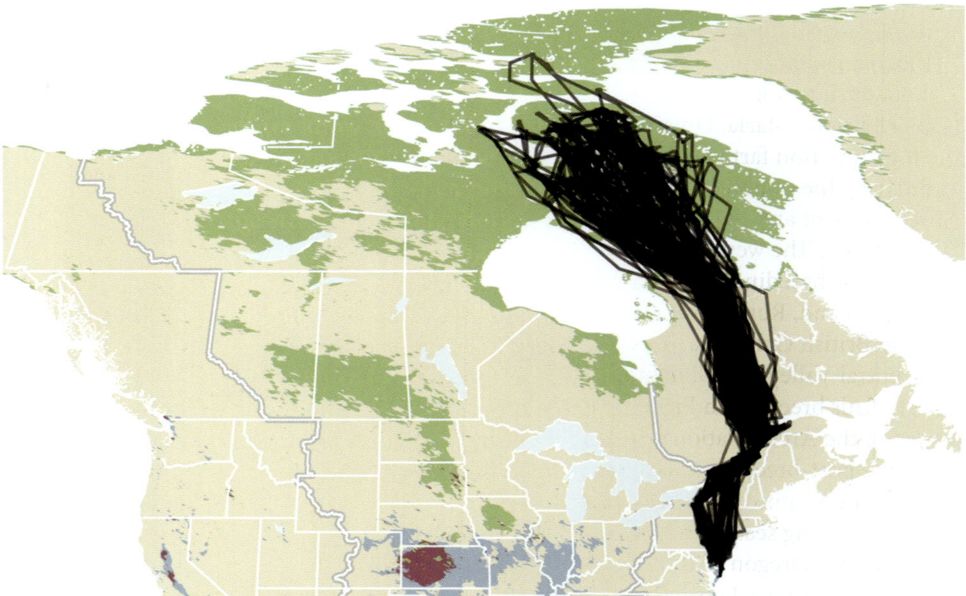

PATHWAYS

— LESSER SNOW GOOSE *Anser caerulescens caerulescens*

MC = 0.14

ORDER Anseriformes
FAMILY Anatidae
BREEDING HABITAT Wetlands, Tundra
NONBREEDING HABITAT Wetlands, Generalist
GLOBAL POPULATION SIZE 16,000,000
POPULATION CHANGE SINCE 1970 +887%
IUCN RED LIST CATEGORY Least Concern
R2R TIPPING POINT SPECIES No
US LISTING STATUS No
CANADA SPECIES AT RISK No
PIF CCS 10/20

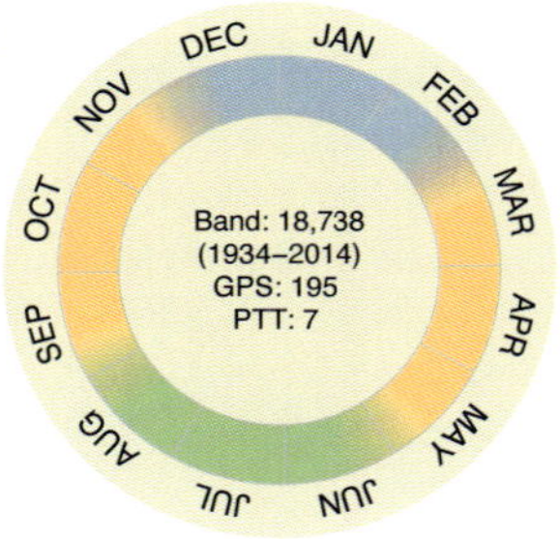

	Atlantic	Mexico	Midcontinent	Pacific
Atlantic (n = 1)	–	–	100%	–
Midcontinent (n = 159659)	<1%	2%	85%	13%
Pacific (n = 1263)	–	4%	17%	79%

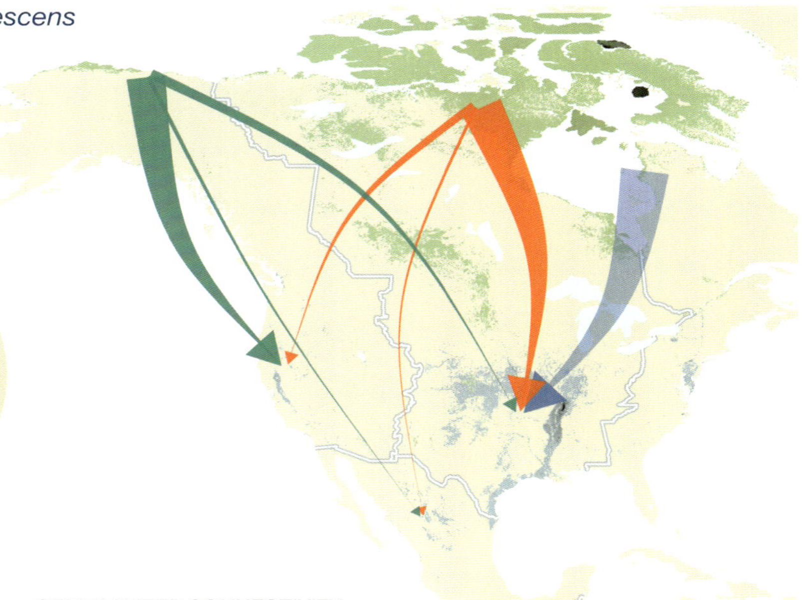

GENERALIZED CONNECTIVITY

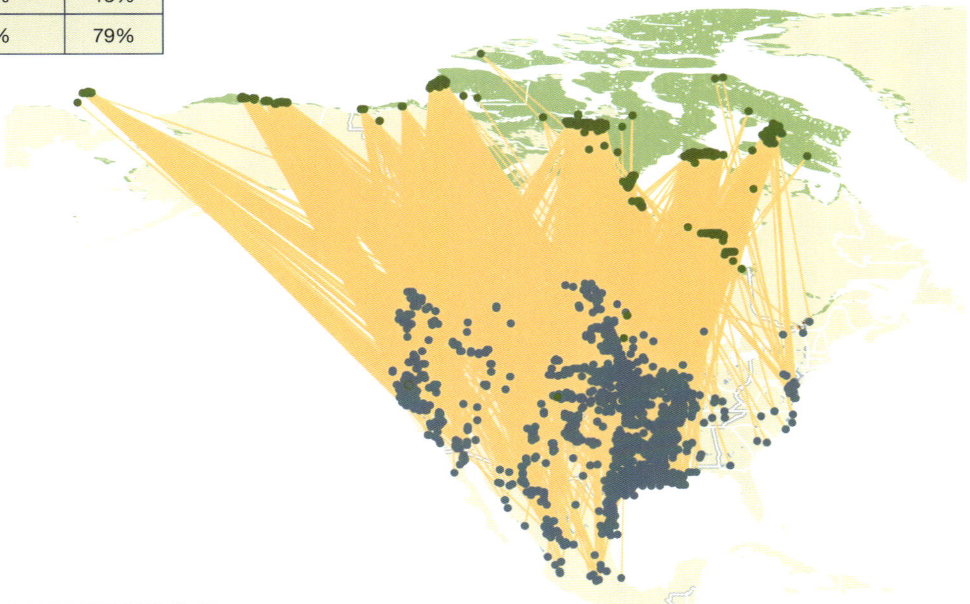

CONNECTIONS (A–E)

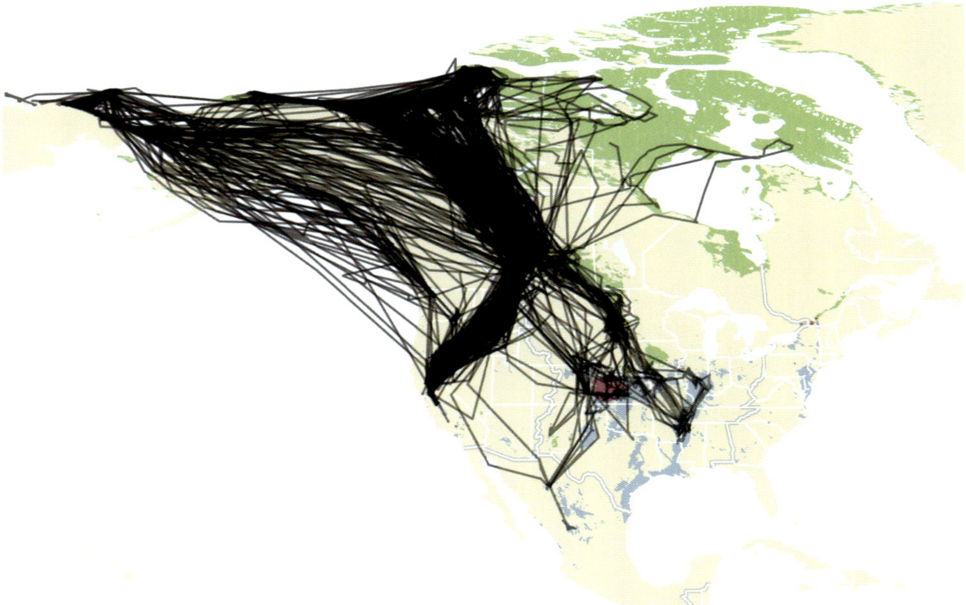

PATHWAYS

OVERALL The majority of Lesser Snow Geese breeding within the Midcontinent Flyway migrate to nonbreeding areas within the Midcontinent Flyway (85%), especially along the US Gulf Coast, or to the Pacific Flyway (13%). Western breeding populations in the Pacific Flyway also mainly migrate to nonbreeding areas within the flyway (79%), especially the Central Valley of California, or to the Midcontinent Flyway (17%) and Mexico (4%). Although within flyway movement is high, overlap of breeding populations across all nonbreeding areas results in weak connectivity within the subspecies (MC = 0.14).

BANDING Bands recovered from Lesser Snow Geese harvested in the midcontinent of the United States (n = 10,190) originated from banding efforts across the central Arctic. Lesser Snow Geese banded on breeding areas in La Pérouse Bay, Manitoba, were recovered in large numbers in southern Manitoba and North Dakota during migration, while geese banded at Cape Henrietta Maria, Ontario, were recovered during migration farther east, around James Bay, and during the nonbreeding season along the Missouri River and the Gulf Coast (Francis and Cooke 1992). The western populations of Lesser Snow Geese breeding on Banks Island and Wrangel Island, Russia, migrate to nonbreeding areas in British Columbia, Washington, and the Central Valley of California (Williams et al. 2008). Birds breeding in Prudhoe Bay, Alaska, staged in the Arctic National Wildlife Refuge, spent the post-breeding season in southeastern Alberta and southwestern Saskatchewan, and the nonbreeding season in Northern California and southern Oregon (S.R. Johnson 1996). Banding data shows Lesser Snow Geese that spend the nonbreeding season along the Atlantic

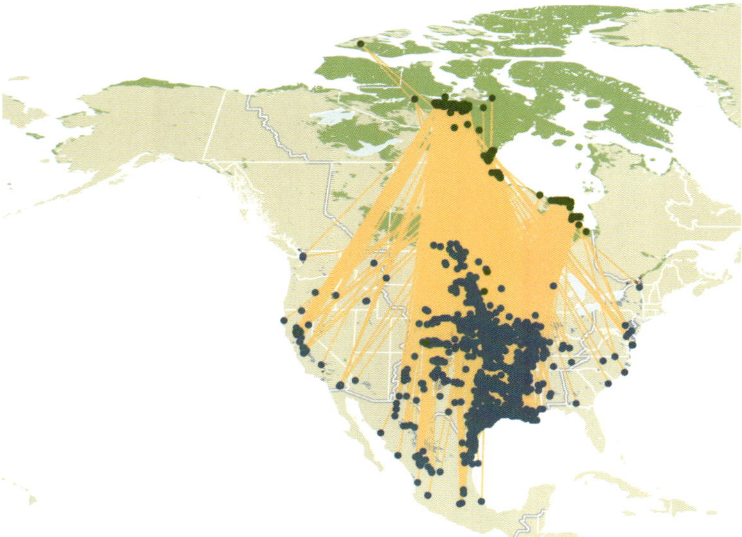

CONNECTIONS A

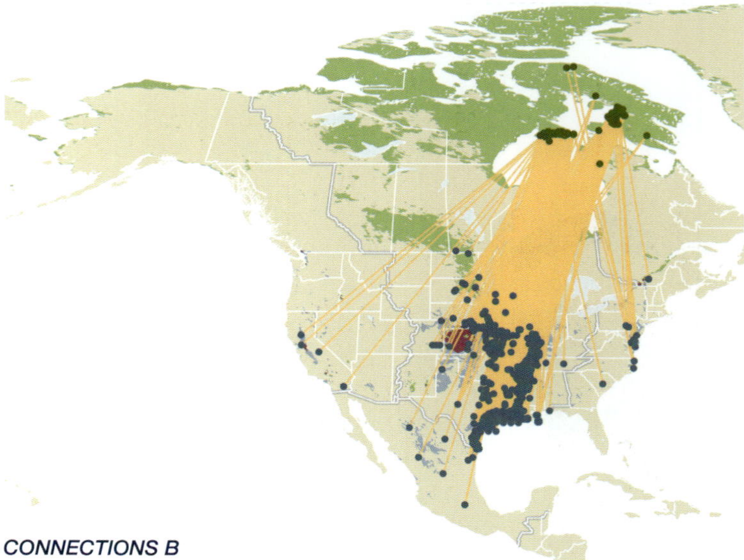

CONNECTIONS B

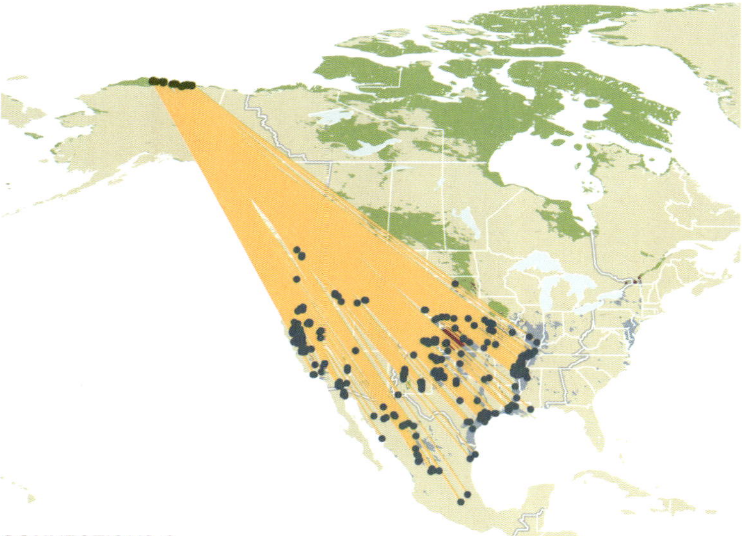

CONNECTIONS C

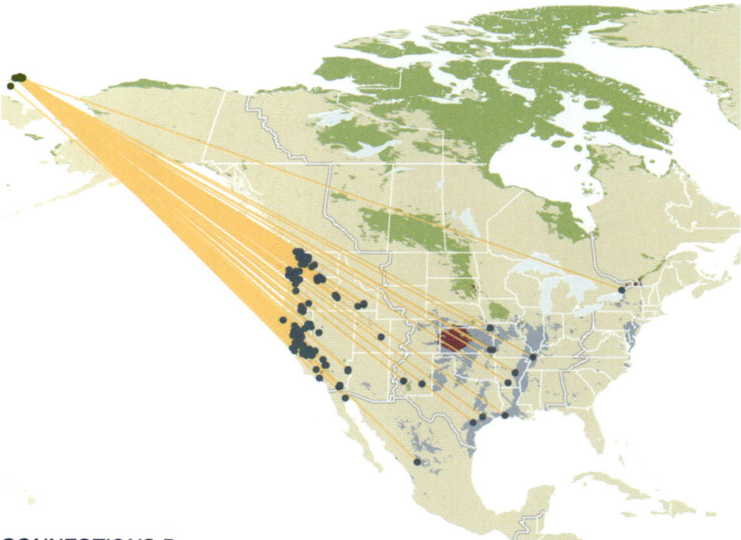

CONNECTIONS D

coast have the same breeding distribution as birds spending the nonbreeding season along the Gulf Coast (Humphries et al. 2009).

SATELLITE TRACKING Birds marked during the nonbreeding season in Washington state (Evenson et al., unpublished data) and the Central Valley of California (US Geological Survey and Canadian Wildlife Service, unpublished data) migrated to areas across the western breeding range. This included the western Canadian Arctic, Alaska, and Russia.

STABLE ISOTOPES Stable isotopes of hydrogen (deuterium) measured in feathers varied strongly with increasing latitude, confirming that deuterium is a useful tool for determining breeding population origin and migratory connectivity (Fowler et al. 2018). Individuals sampled during pre-breeding migration in the Rainwater Basin of Nebraska were assigned, using carbon and nitrogen isotopes from muscle tissue, to nonbreeding grounds in Louisiana and the Texas Gulf Coast (Hénaux et al. 2012).

GENETICS Genetic sampling of Lesser Snow Geese breeding at Wrangel Island in Russia and Banks Island in Canada revealed significant differences in mtDNA and homogeneity in nuclear DNA. Gene flow among these breeding populations is likely due to pair formation on nonbreeding grounds (Shorey et al. 2011). There was some genetic differentiation in the nonbreeding populations from Wrangel Island, though the origin of these differences was not clear (Kuznetsov et al. 1998).

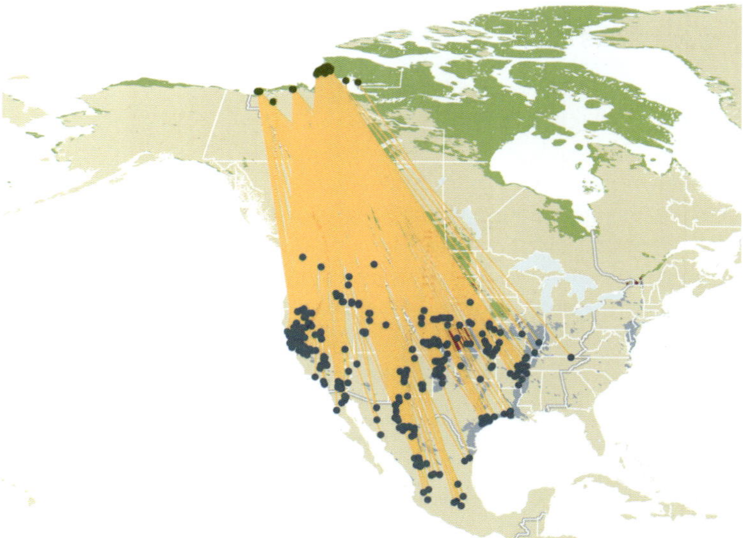

CONNECTIONS E

MIGRATORY PATHWAYS During fall migration, birds stage along southern Hudson Bay and the northern Great Plains before migrating south to nonbreeding areas.

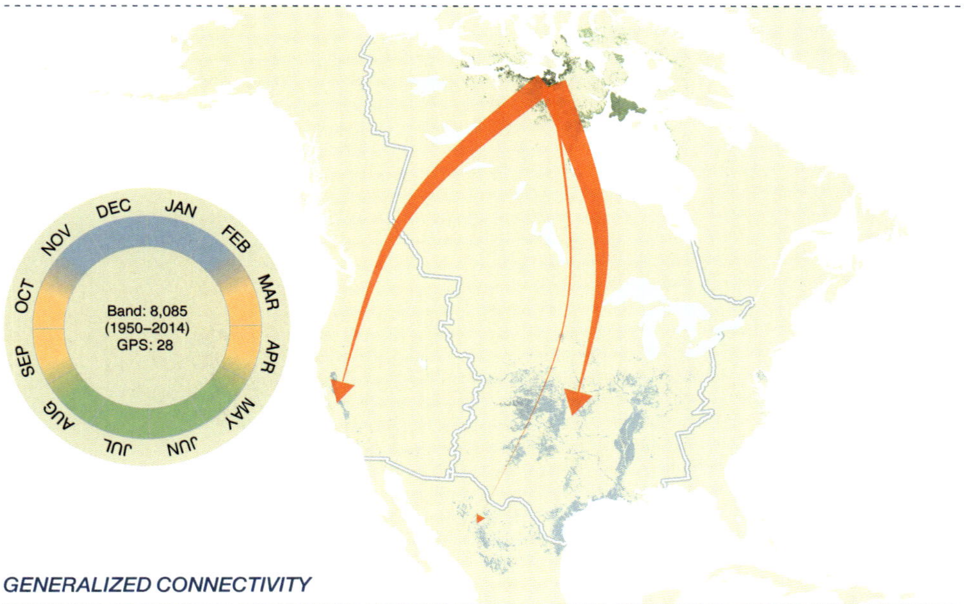

GENERALIZED CONNECTIVITY

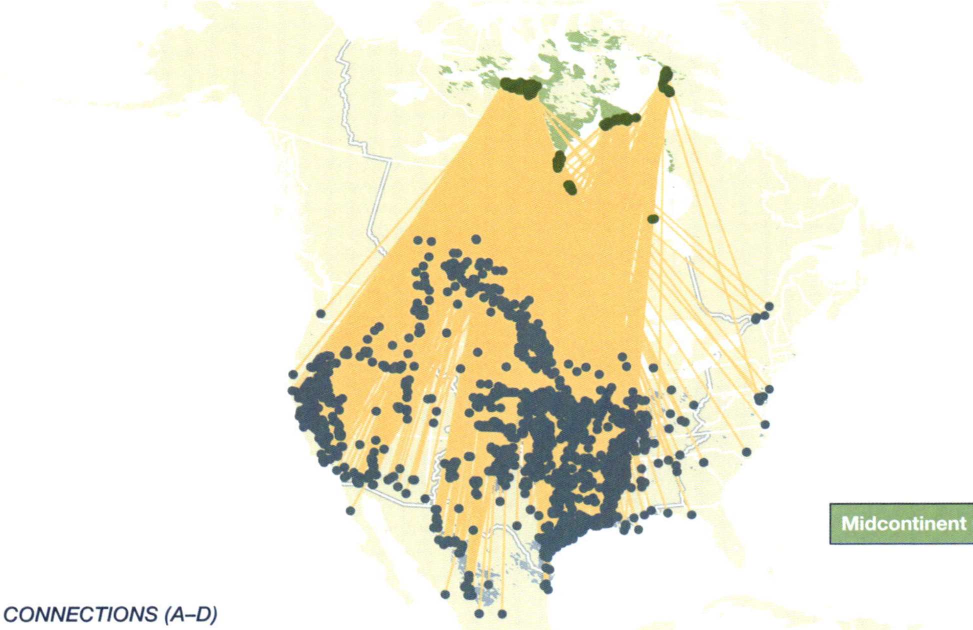

CONNECTIONS (A–D)

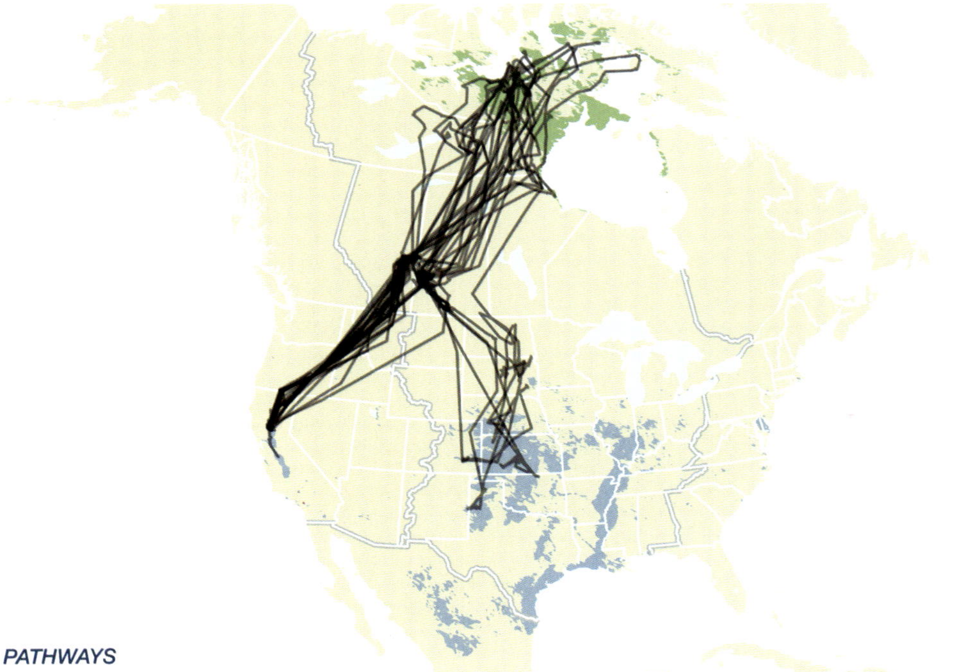

PATHWAYS

Anser rossii ROSS'S GOOSE

DISTRIBUTION AND ABUNDANCE

Ross's Geese nest in colonies, primarily in the central Canadian Arctic around Queen Maud Gulf. The main nonbreeding areas are in California's Central Valley, the southern Great Plains, and the Mississippi Valley, but they can be found in other parts of the United States and Mexico (Baldassarre 2014).

ORDER Anseriformes
FAMILY Anatidae
BREEDING HABITAT Wetlands, Tundra
NONBREEDING HABITAT Wetlands, Generalist
GLOBAL POPULATION SIZE 2,100,000
POPULATION CHANGE SINCE 1970 +7354%
IUCN RED LIST CATEGORY Least Concern
R2R TIPPING POINT SPECIES No
US LISTING STATUS No
CANADA SPECIES AT RISK No
PIF CCS 12/20

	Atlantic	Mexico	Midcontinent	Pacific
Midcontinent (n = 8085)	<1%	1%	55%	44%

MIGRATORY CONNECTIVITY

OVERALL Ross's Geese are a relatively well-studied species, and migratory connectivity data is based on banding and satellite tracking data. All birds breed in the Midcontinent Flyway and migrate to nonbreeding areas within the Midcontinent Flyway (55%) or to the Pacific Flyway (44%), and a small percentage to Mexico (1%). Because of the single breeding flyway, MC cannot be calculated.

BANDING Individuals (n = 2,251) banded in the central Arctic were recovered on nonbreeding areas in the United States (68%), Canada (30%), and Mexico (2%). Between 1989 and 2001, banding recoveries shifted eastward from the Pacific Flyway to the midcontinent United States (Alisauskas et al. 2006). Birds banded around Queen Maud Gulf, Canada, were recovered in Canada, California, and the midcontinent United States, while individuals banded in west Hudson Bay were recovered farther east (Alisauskas et al. 2012).

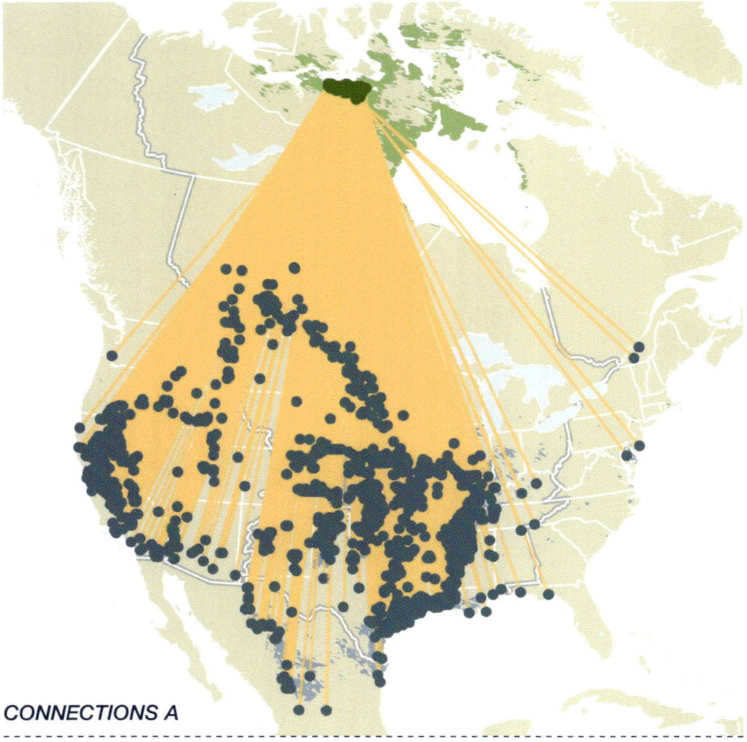

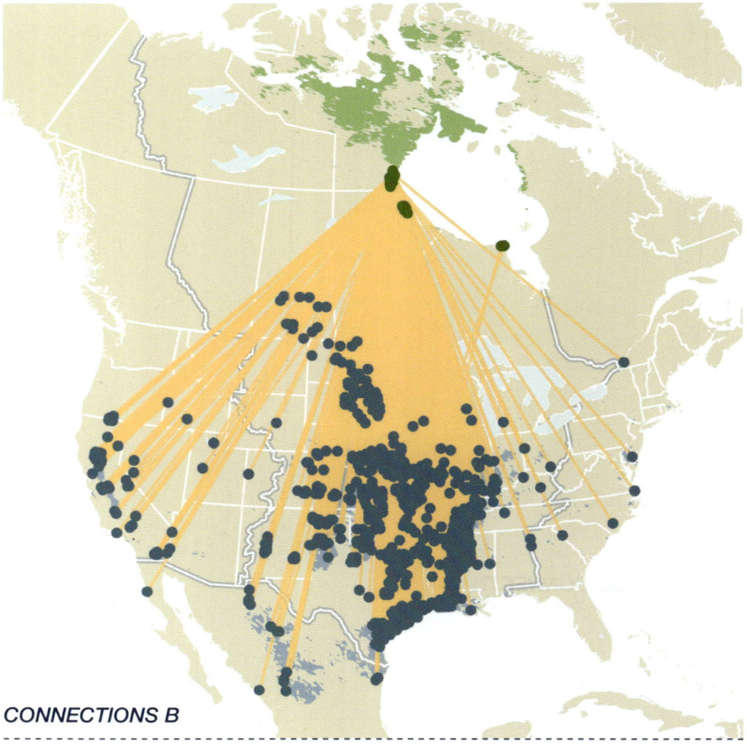

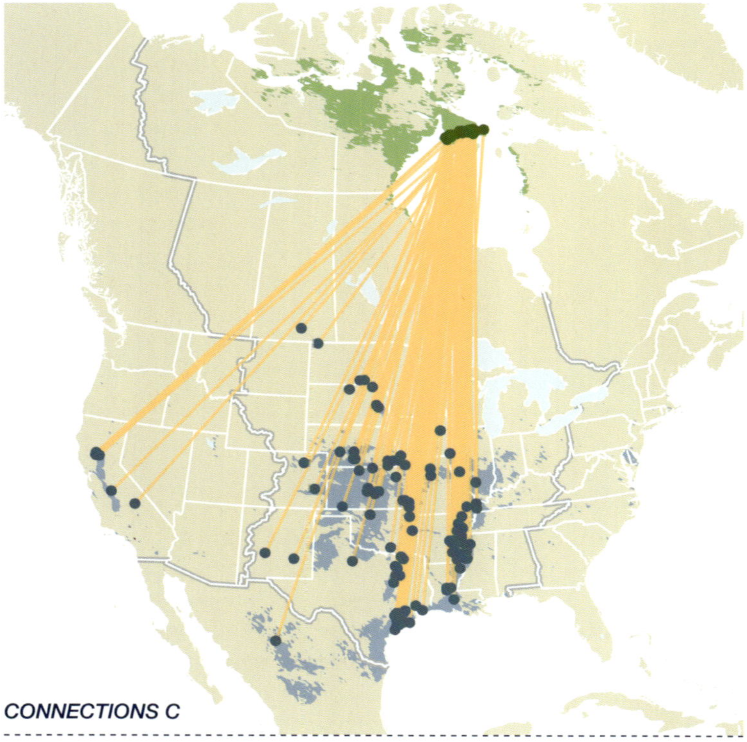

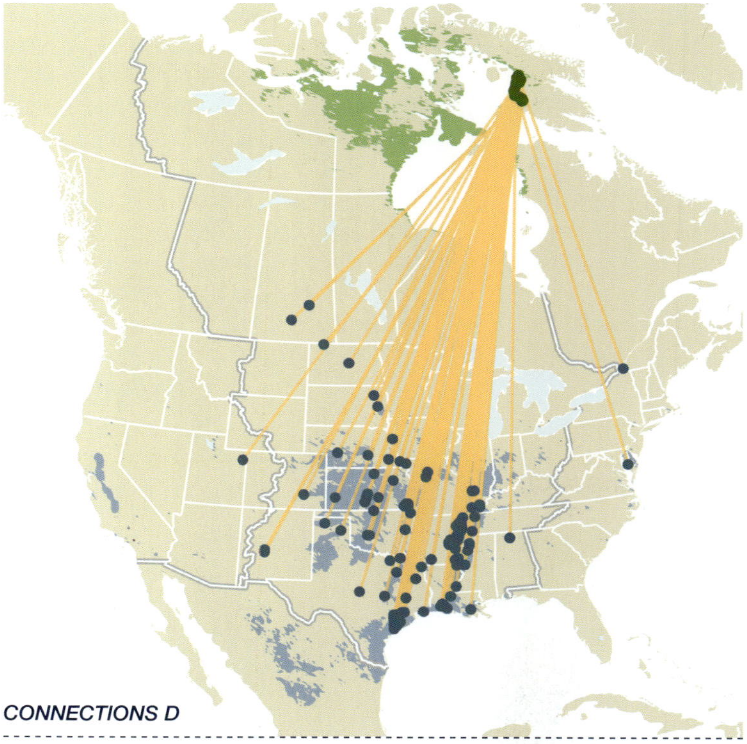

SATELLITE TRACKING Ross's Geese marked in California migrated to the area around Queen Maud Gulf, while those marked in the southern Great Plains migrated to breed on the west coast of Hudson Bay (US Geological Survey, unpublished data). All birds staged in southern Alberta and Saskatchewan.

OTHER APPROACHES No genetic or stable isotope studies of migratory connectivity were found for this species.

MIGRATORY PATHWAYS There are two primary migration paths from central Arctic breeding areas. The westerly pathway goes through Saskatchewan, then west over the Rocky Mountains to nonbreeding areas in California's Central Valley. The more easterly route follows the west coast of Hudson Bay south through Manitoba, then down the Mississippi River Valley to midcontinent nonbreeding areas.

DATA CONTRIBUTORS

M. Casazza; C. Nicholai; C. Overton; V. Patil; M. Petrie; E. Reed.

weak strong

MC = 0.13 ± 0.01*

Anser albifrons

GREATER WHITE-FRONTED GOOSE

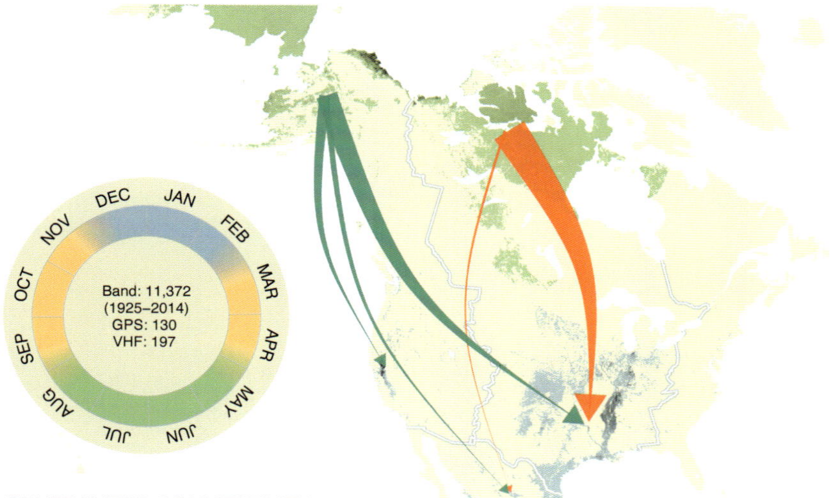

GENERALIZED CONNECTIVITY

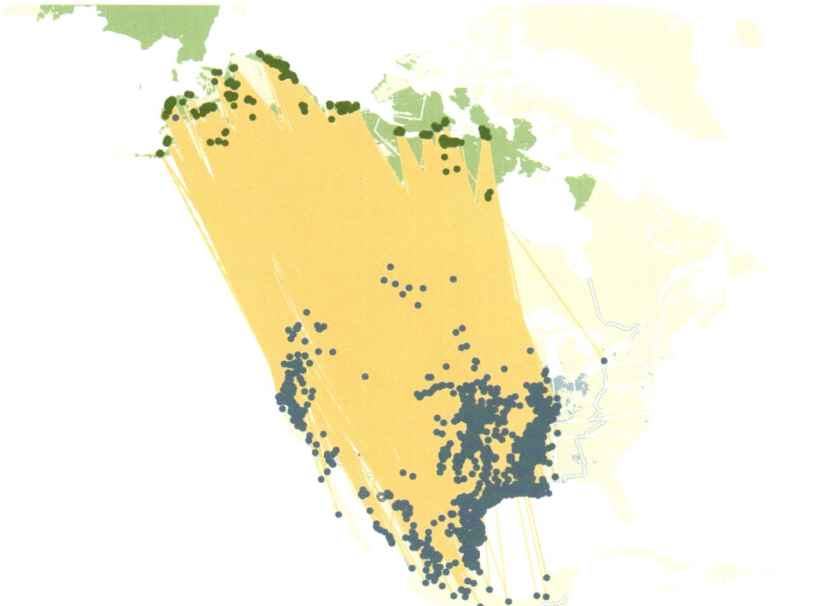

CONNECTIONS (A–C)

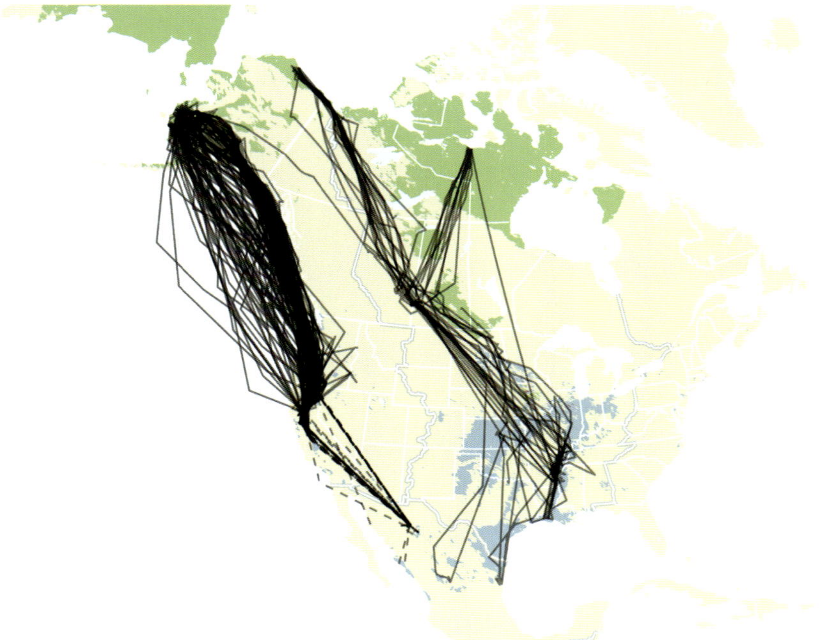

PATHWAYS

DISTRIBUTION AND ABUNDANCE

Greater White-fronted Geese have a circumpolar Arctic breeding distribution that consists of four subspecies. Two of the subspecies occur in North America: the Pacific Greater White-fronted Goose and the Tule Greater White-fronted Goose. Nonbreeding concentrations occur in California's Central Valley, the Mississippi Alluvial Valley, and the southern Great Plains of the United States and Mexico (Fink et al. 2023).

ORDER Anseriformes
FAMILY Anatidae
BREEDING HABITAT Wetlands, Tundra
NONBREEDING HABITAT Wetlands, Generalist
GLOBAL POPULATION SIZE 5,500,000
POPULATION CHANGE SINCE 1970 +544%
IUCN RED LIST CATEGORY Least Concern
R2R TIPPING POINT SPECIES No
US LISTING STATUS No
CANADA SPECIES AT RISK No
PIF CCS 9/20

	Mexico	Midcontinent	Pacific
Midcontinent (n = 6964)	3%	97%	<1%
Pacific (n = 4659)	14%	58%	27%

MIGRATORY CONNECTIVITY

OVERALL Banding records, telemetry, and genetics studies all inform migratory connectivity of this species. Greater White-fronted Geese that breed in the Arctic of the Midcontinent Flyway nearly all transition to nonbreeding areas in the same flyway (97%). Birds breeding in the Pacific Flyway transition to the Midcontinent Flyway (58%), Pacific Flyway (27%), and Mexico (14%) during the nonbreeding season, resulting in a weak migratory connectivity (MC = 0.13 ± 0.01).

STABLE ISOTOPES No stable isotope studies of migratory connectivity were found for this species.

GENETICS There was no support for subspecies designations across the global range except for the Tule Greater White-fronted Goose, according to two analyses of mtDNA (Eda et al. 2013; R.E. Wilson et al. 2018). Differences in mtDNA do suggest sparse inter-flyway movement (Eda et al. 2013). At a local scale, mtDNA analysis of birds breeding along the Alaska coast found genetic separation between

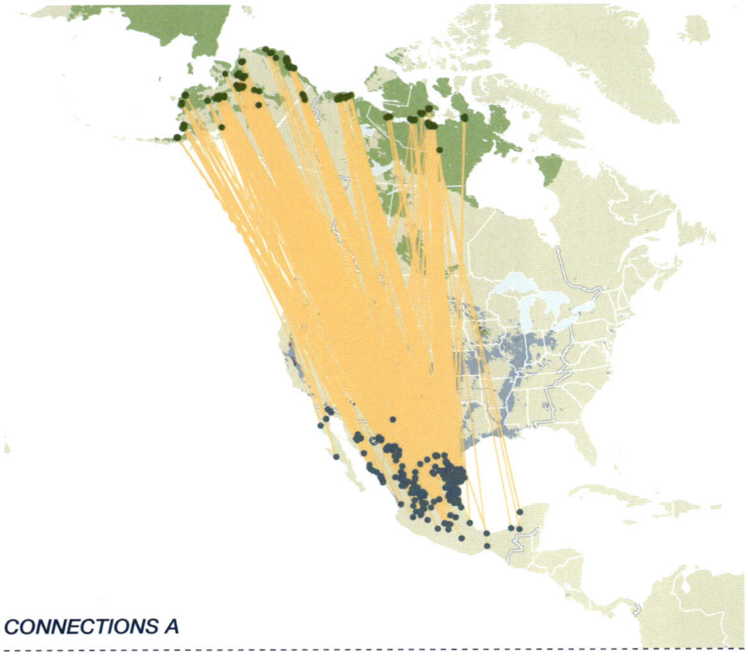

CONNECTIONS A

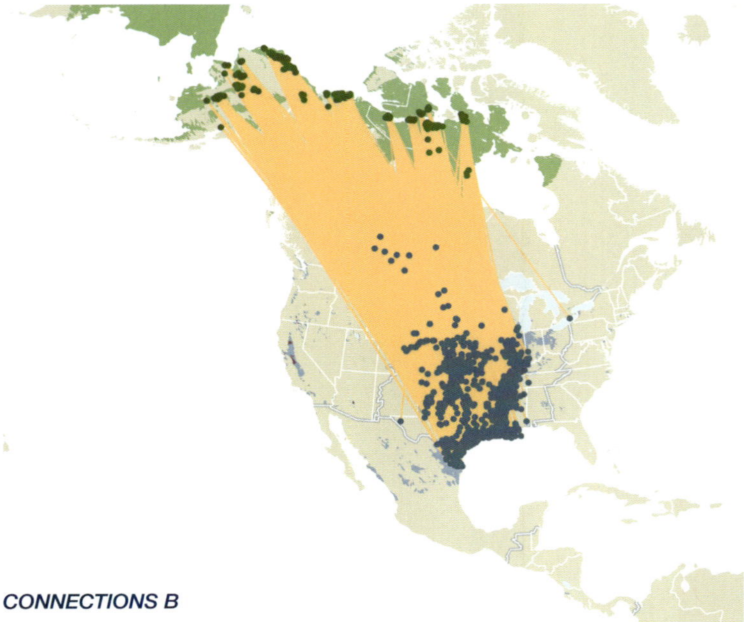

CONNECTIONS B

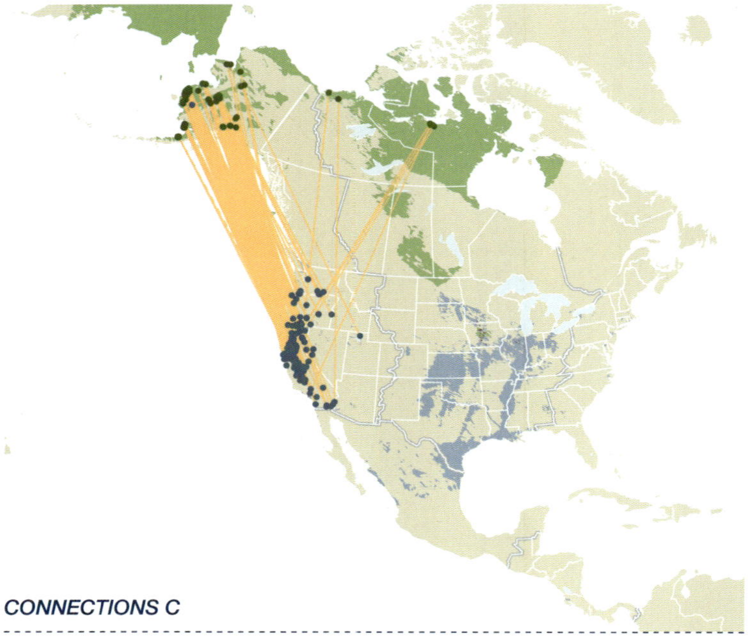

CONNECTIONS C

the breeding areas of the Pacific Greater White-fronted Goose versus the Tule Greater White-fronted Goose (Ely et al. 2017). There was evidence of some gene flow between the subspecies.

— PACIFIC GREATER WHITE-FRONTED GOOSE

Anser albifrons frontalis

OVERALL This subspecies is the most widespread in North America. Birds breeding in western Alaska migrate to the Central Valley of California and Mexico, while birds breeding farther east migrate to the midcontinent of the US, along the Gulf Coast and the Mississippi Alluvial Valley.

BANDING Breeding birds banded across the midcontinent were recovered primarily along the Gulf Coast and the Mississippi Alluvial Valley (Moore et al. 2023). There is no geographic variation in recoveries of birds banded in interior Alaska, the Alaska coastal plain, and Arctic Canada (Central Flyway Council et al. 2015). Band resightings of birds marked across Alaska suggest two populations. Coastal Alaska breeding birds migrated to California and Mexico, with birds that bred farther inland remaining farther inland during the nonbreeding season (Ely et al. 2013).

RADIO TRACKING Radio-marked Greater White-fronted Geese ($n = 244$) that bred on the Yukon-Kuskokwim Delta and Bristol Bay Lowlands were both found to use the Klamath Basin during post-breeding migration, then migrated to the Central Valley of California and Mexico for the nonbreeding season (Ely and Takekawa 1996; Ely et al. 2021).

SATELLITE TRACKING Birds marked in the Central Valley of California migrated to breeding areas in western Alaska (US Geological Survey and Canadian Wildlife Service, unpublished data). Greater White-fronted Geese marked in western Arctic Canada and northern Alaska ($n = 20$) migrated to staging areas in southern Alberta and Saskatchewan, then to nonbreeding areas in the lower Mississippi Alluvial Valley and along the Gulf Coast (Askren et al. 2022). The same pattern was seen in five birds marked in eastern Arctic Canada (US Geological Survey and Canadian Wildlife Service, unpublished data).

— TULE GREATER WHITE-FRONTED GOOSE

Anser albifrons gambelli

OVERALL Tule Greater White-fronted Geese move between breeding areas around Cook Inlet, Alaska, and nonbreeding areas in the Central Valley of California.

BANDING Tule Greater White-fronted Geese banded on nonbreeding areas in California and breeding areas in Alaska were recovered in the same areas, confirming strong migratory connectivity of this subspecies (Pacific Flyway Council 1991). A small portion (6%) of banded birds were recovered in Texas.

RADIO TRACKING Breeding movements of Tule White-fronted Geese ($n = 90$) found that 50% of birds went through a molt migration to wetlands 400–600 km west of Cook Inlet, but returned before post-breeding migration (Ely et al. 2006).

DATA CONTRIBUTORS

C. Brady; M. Casazza; D.C. Douglas; C. Ely; B. Meixell; A. Mott; D. Osborne; C. Overton; D. Skalos; M. Weaver; Canadian Wildlife Service; US Geological Survey.

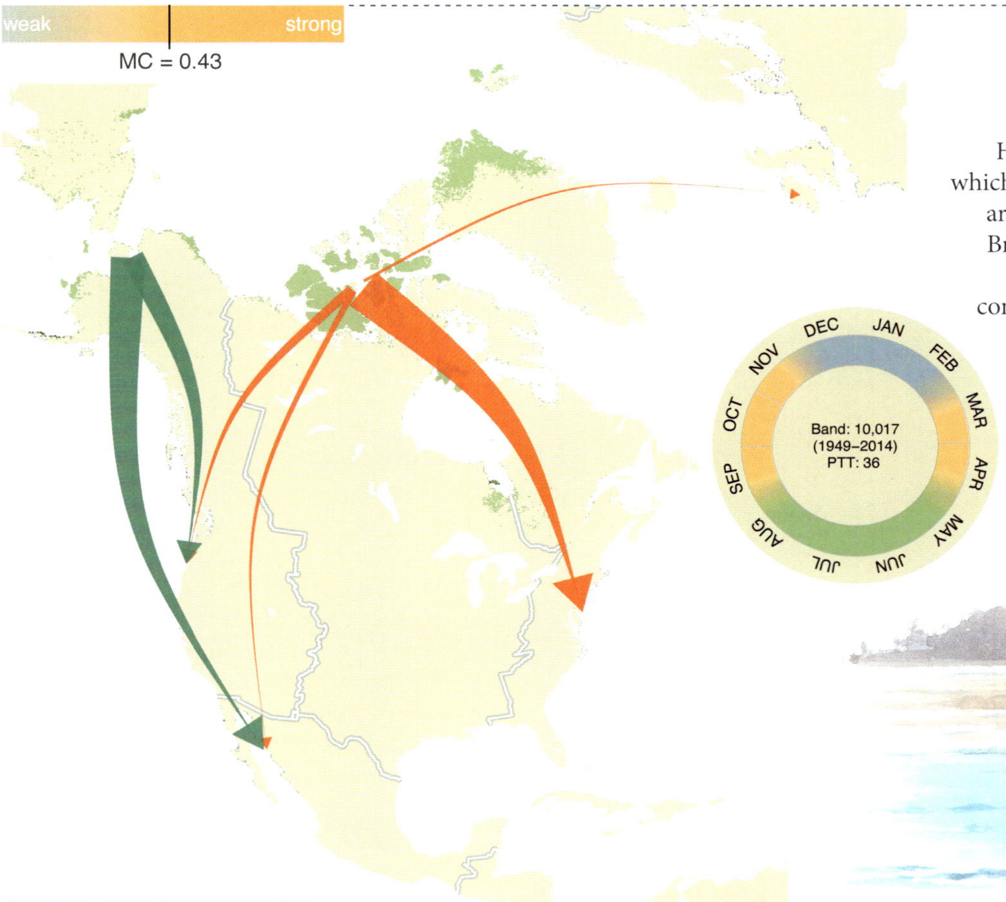

GENERALIZED CONNECTIVITY

Branta bernicla **BRANT**

DISTRIBUTION AND ABUNDANCE

Brant are found throughout the Northern Hemisphere and include three subspecies, two of which breed in North America. These two subspecies are further divided into four populations. Atlantic Brant are composed of Atlantic Brant and Eastern high Arctic Brant populations. Black Brant are composed of Black Brant and Western High Arctic Brant populations, though the latter is often considered an integrate between the two subspecies. All populations breed in the North American Arctic and migrate to coastal nonbreeding areas as far south as Baja California on the west coast of North America and the mid-Atlantic on the east coast (Lewis et al. 2020).

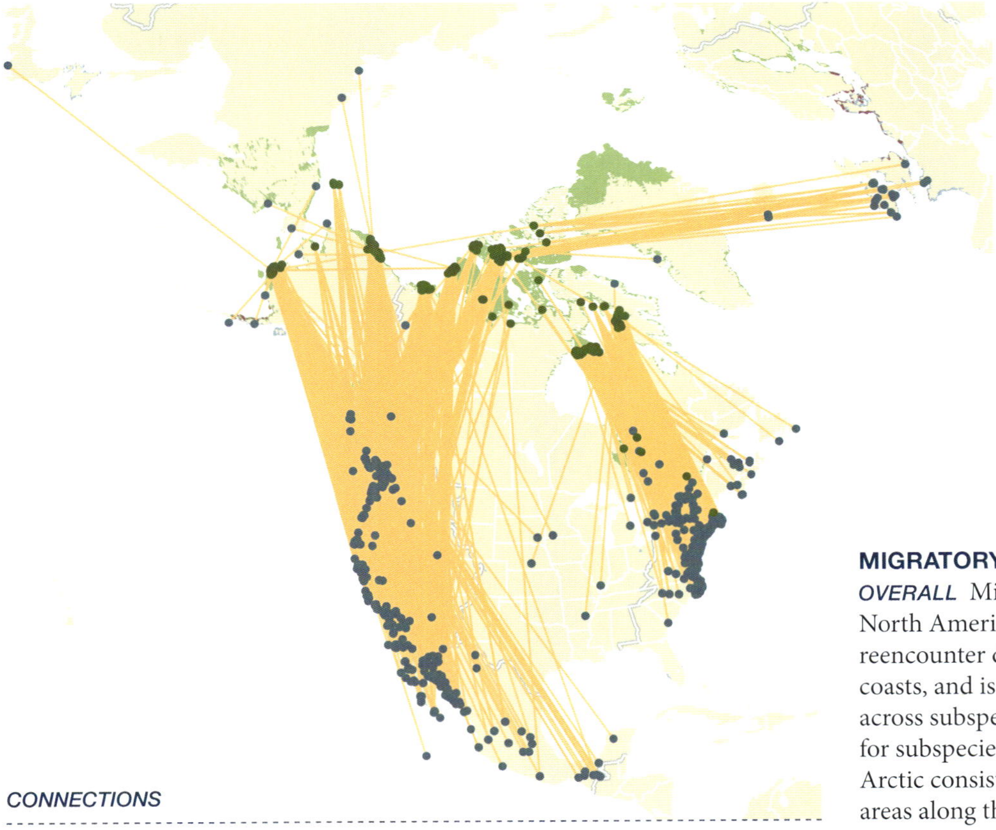

CONNECTIONS

ORDER Anseriformes
FAMILY Anatidae
BREEDING HABITAT Coasts, Salt Marsh
NONBREEDING HABITAT Coasts, Marine Waters
GLOBAL POPULATION SIZE 490,000
POPULATION CHANGE SINCE 1970 +10%
IUCN RED LIST CATEGORY Least Concern
R2R TIPPING POINT SPECIES No
US LISTING STATUS No
CANADA SPECIES AT RISK No
PIF CCS 13/20

MIGRATORY CONNECTIVITY

OVERALL Migratory connectivity for the entire species in North America is based on satellite transmitter and band reencounter data of birds from the central Arctic to both coasts, and is strong (MC = 0.43). Migratory connectivity across subspecies also appears strong, but MC is not calculated for subspecies. Brant breeding in Alaska through the central Arctic consist of two populations that migrate to nonbreeding areas along the Pacific coast, from southern British Columbia

	Asia	Atlantic	Europe	Greenland	Mexico	Midcontinent	Pacific	Russia
Midcontinent (n = 5654)	–	77%	1%	<1%	7%	<1%	15%	–
Pacific (n = 3662)	<1%	–	<1%	–	61%	<1%	38	<1%

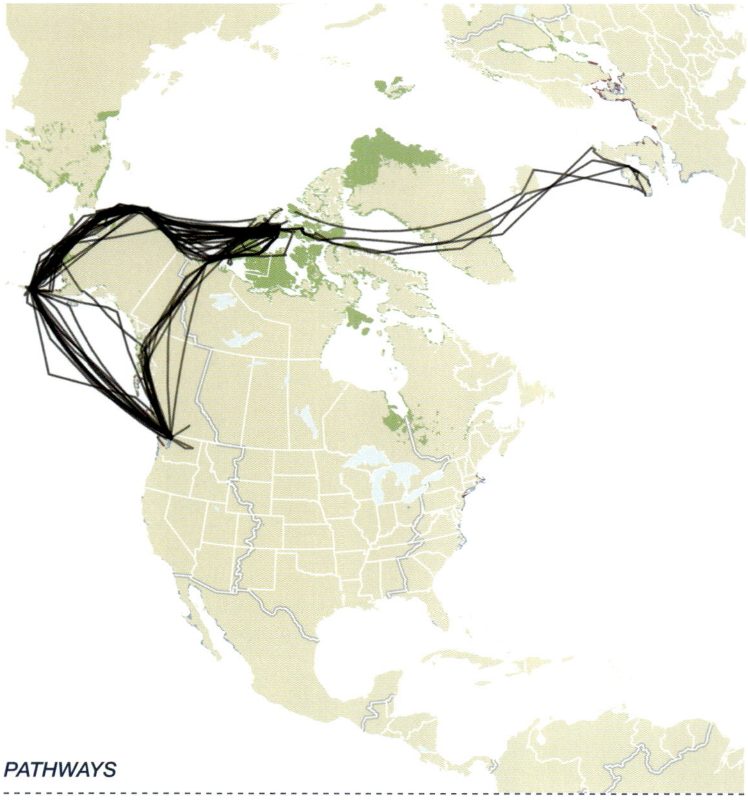

PATHWAYS

to Baja California. Movement of breeding birds in the Pacific Flyway is either to Mexico (61%) or within the flyway (38%). Both Atlantic Brant populations breed in the eastern Arctic but have entirely disjunct nonbreeding distributions; the Atlantic Brant population spends the nonbreeding season along the Atlantic coast of North America, while the Eastern High Arctic Brant population migrates to the United Kingdom. Arctic breeding birds in the Midcontinent Flyway primarily spend the nonbreeding season in the Atlantic Flyway (77%), as well as within the Pacific Flyway (15%) and Mexico (7%).
OTHER APPROACHES No genetic or stable isotope studies of migratory connectivity were found for this species.

— ATLANTIC BRANT *Branta bernicla hrota*

ATLANTIC BRANT POPULATION

OVERALL Limited tracking data are available to assess migratory connectivity of nonbreeding Atlantic Brant in North America. Banding data are the primary source of information for this species. Atlantic Brant banded on breeding areas have strong connectivity to the area from Long Island Sound south to the mid–New Jersey coast. Survey data suggest a contraction of the nonbreeding range from much of the mid-Atlantic to areas along the Long Island and New Jersey coasts (Kirby and Obrecht 1982).
BANDING Atlantic Brant have been banded extensively on their breeding areas around the Foxe Basin, Nunavut, Canada, and on nonbreeding areas in the United States. Band recoveries suggest movement between those areas, with no migration to nonbreeding areas of other populations.
RADIO TRACKING Atlantic Brant were marked with radio transmitters (n = 59) and located using flights over nonbreeding and breeding areas. Individuals used the same areas as those marked with satellite transmitters (Castelli et al. 2010).
SATELLITE TRACKING Atlantic Brant (n = 22) were tracked from nonbreeding areas in the mid-Atlantic states to breeding areas in the Foxe Basin, Nunavut, Canada (Castelli et al. 2010). Migrating birds used western Long Island, New York, and James Bay, as primary staging areas during both pre- and post-breeding migration.
MIGRATORY PATHWAYS Early reports (Lewis 1937) of two migratory pathways between nonbreeding and breeding areas, pieced together from observer accounts, have been confirmed by subsequent satellite tracking data. Atlantic Brant travel from nonbreeding areas either along the coast in a direct route to James Bay, then up the Hudson Bay coast to breeding areas, or along the coast to the mouth of the St. Lawrence River, then west toward Hudson Bay.

EASTERN HIGH ARCTIC BRANT POPULATION

OVERALL This population has been assessed primarily with banding data, with resightings occurring on the primary nonbreeding areas in Europe.
BANDING Band returns show Eastern High Arctic Brant breeders occasionally spending the nonbreeding season with Atlantic Brant in the mid-Atlantic United States.
SATELLITE TRACKING Most tracking studies are of Eastern High Arctic Brant breeding in high Arctic and spending the nonbreeding season in Europe (Robinson et al. 2006). Eastern High Arctic Brant spending the nonbreeding season in Denmark (n = 33) were tracked to breeding areas in Greenland, the eastern edge of this population's range (Clausen et al. 2003).
MIGRATORY PATHWAYS North American birds stop over in both Greenland and Iceland en route to nonbreeding areas. From nonbreeding areas, birds stage in Iceland for about one month before returning to breeding grounds (Lewis et al. 2020).

— BLACK BRANT *Branta bernicla nigricans*

BLACK BRANT POPULATION

OVERALL Black Brant migratory connectivity has been well-studied through band reencounter data between breeding, staging, and nonbreeding areas.
BANDING Black Brant banded in southwest Alaska spent the nonbreeding season at sites from British Columbia to Baja California. Sex ratios were similar at all sites, though most juveniles spent the nonbreeding season at more northern locations (Schamber et al. 2007). Brant that were marked at breeding colonies in Alaska, Russia, and Canada were observed at four nonbreeding sites in Baja California and California and showed high fidelity to nonbreeding sites (Lindberg et al. 2007). Alternatively, nonbreeding Black Brant may have weak site fidelity overall, but older age classes have higher site fidelity (Reed et al. 1998). Marked females from breeding grounds in Alaska were resighted at three nonbreeding locations in Baja California, and nonbreeding location had carry-over effects to breeding probability between years (Sedinger et al. 2011).
RADIO TRACKING A project through the Motus Wildlife Tracking System showed no individual breeding to nonbreeding detections but did show one detection from Alaska and one from Baja California, Mexico (Motus project, Black Brant Migration, unpublished data).
MIGRATORY PATHWAYS A particularly important staging area is Izembek Lagoon on the Alaska Peninsula (Purcell and Brodin 2007). Banding reencounter data shows that Humboldt Bay, California, is also an important stopover location for Black Brant migrating from Alaska to nonbreeding areas farther south (Lee et al. 2007).

WESTERN HIGH ARCTIC BRANT POPULATION

OVERALL This population breeds farther east than the Black Brant population. There is weak migratory connectivity between the two populations.
BANDING A combination of band recoveries and analysis of plumage differences between the two populations suggested that most individuals spending the nonbreeding season in Padilla Bay, Washington, were from the Western High Arctic population (Reed et al. 1989a).
SATELLITE TRACKING Birds breeding on Melville Island, Canada,

were tracked with satellite transmitters along a coastal route around Alaska, to nonbreeding sites in Padilla Bay and Samish Bay (of the Salish Sea) near Puget Sound, Washington (Boyd et al. 2013).

MIGRATORY PATHWAYS Western High Arctic Brant travel from breeding areas in the high Arctic, primarily along the Alaska coast, south to nonbreeding areas in the Salish Sea. Whether they spend the nonbreeding season at Izembek or continue on to Baja California may depend on weather conditions and food availability during staging (Purcell and Brodin 2007). Izembek Lagoon on the Alaska Peninsula is an important post-breeding staging area (Reed et al. 1989b; Boyd et al. 2013). Radio telemetry data showed 80% of 33 radio-marked adult Western High Arctic Brant from four separate breeding areas of the Canadian Arctic and one area in Alaska used this stopover location (Reed et al. 1989a).

DATA CONTRIBUTORS

W.S. Boyd; J. Evenson; K. Spragens; M. Wilson.

INTERESTING NOTE

Color-banded individuals from Alaska and western Canada have been resighted at staging and nonbreeding sites throughout Japan (Derksen et al. 1996).

MC = 0.18 ± 0.11*

Branta hutchinsii CACKLING GOOSE

DISTRIBUTION AND ABUNDANCE

Cackling Geese breed across the North American coastal Arctic and subarctic tundra. They migrate to nonbreeding areas primarily in the western United States and southern Great Plains states (Mowbray et al. 2020b). Four subspecies of Cackling Goose are recognized (listed by primary breeding area, from west to east); Aleutian, Minima, Taverner's, and Richardson's.

ORDER Anseriformes
FAMILY Anatidae
BREEDING HABITAT Wetlands, Tundra
NONBREEDING HABITAT Wetlands, Generalist
GLOBAL POPULATION SIZE 4,500,000
POPULATION CHANGE SINCE 1970: +946%
IUCN RED LIST CATEGORY Least Concern
R2R TIPPING POINT SPECIES No
US LISTING STATUS No
CANADA SPECIES AT RISK No
PIF CCS 10/20

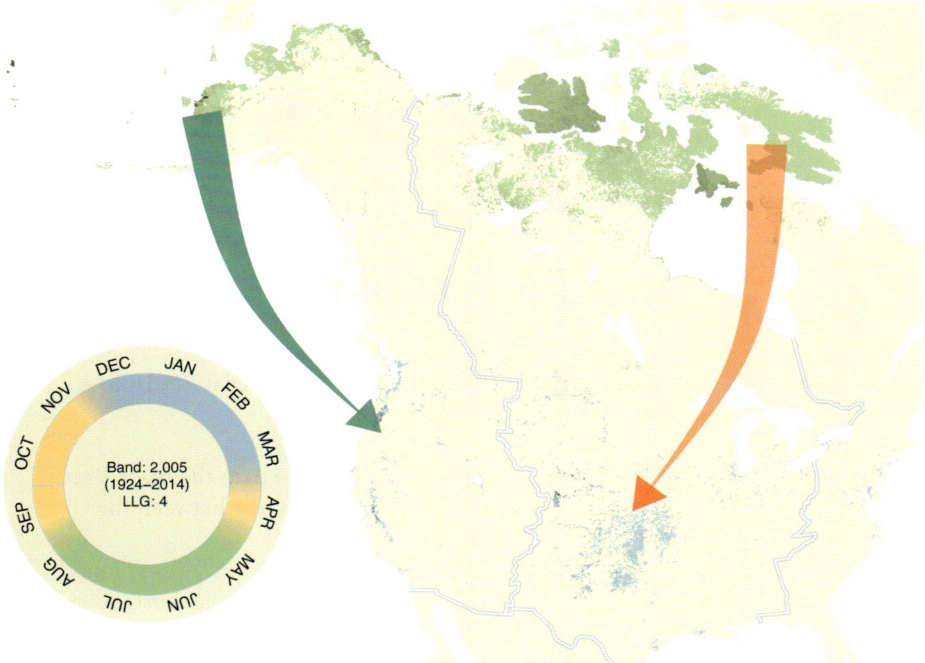

GENERALIZED CONNECTIVITY

	Mexico	Midcontinent	Pacific
Midcontinent (n = 4)	–	100%	–
Pacific (n = 1638)	<1%	<1%	100%

MIGRATORY CONNECTIVITY

OVERALL In 2004, the American Ornithologists Union (AOU) recognized Cackling Goose as a separate species from Canada Goose, and despite the vast research on the latter species, migratory connectivity data is relatively limited for the four subspecies of Cackling Goose but suggests limited cross-flyway movement of subspecies and weak migratory connectivity across the species (MC = 0.18 ± 0.11).

BANDING Banding and neck-collar resight efforts are primarily conducted to inform population boundaries and management actions (Mississippi Flyway Council 2013; Pacific Flyway Council 2016); however, data also show a latitudinal correspondence between breeding and nonbreeding areas.

— ALEUTIAN CACKLING GOOSE

Branta hutchinsii leucopareia

OVERALL The Aleutian Cackling Goose population has recovered from fewer than 400 birds in the 1960s. They breed on many Aleutian Islands and the Alaska Peninsula and spend the nonbreeding season primarily in central California (Baldassarre 2014). During migration, they use sites along coastal Oregon and California, including Humboldt Bay (Mini et al. 2011).
GENETICS Canada and Cackling Geese have been sampled across their breeding range in Alaska, and Aleutian Cackling Geese nesting among various islands were more closely related to each other than to those nesting in mainland Alaska, based on molecular genetics (Pierson et al. 2000).

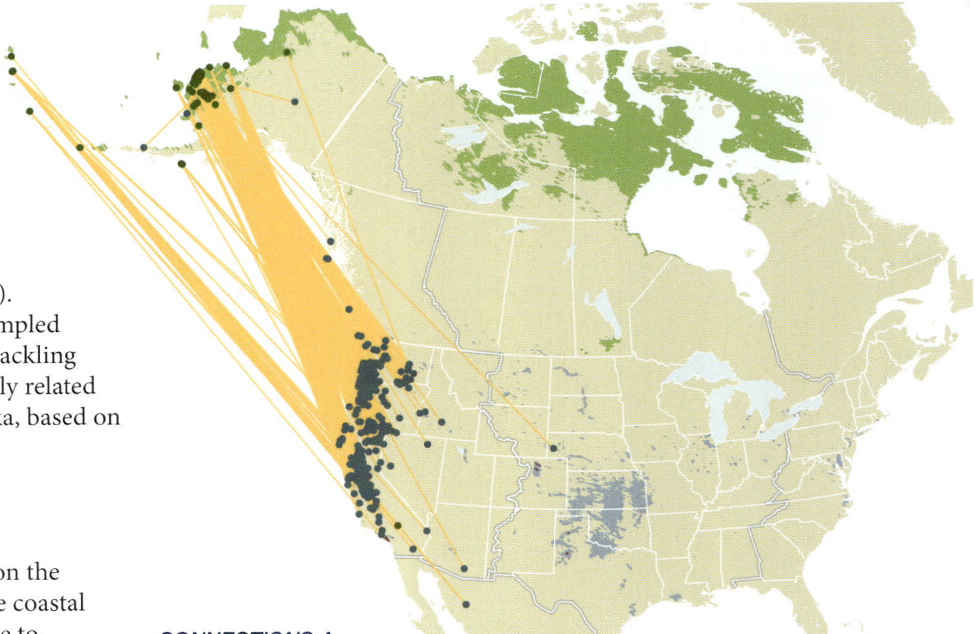

CONNECTIONS A

— MINIMA CACKLING GOOSE

Branta hutchinsii minima

OVERALL Minima Cackling Geese breed exclusively on the Yukon-Kuskokwim Delta in Alaska, generally in more coastal areas than do Taverner's Cackling Geese. They migrate to nonbreeding areas in the lower Columbia River Valley and the Willamette Valley of Oregon (Pacific Flyway Council 2016).
BANDING Minima Cackling Geese banded and marked with neck collars on staging areas had high fidelity to staging areas in the northern Alaska Peninsula and nonbreeding areas in Washington and Oregon (Pacific Flyway Council 2016).

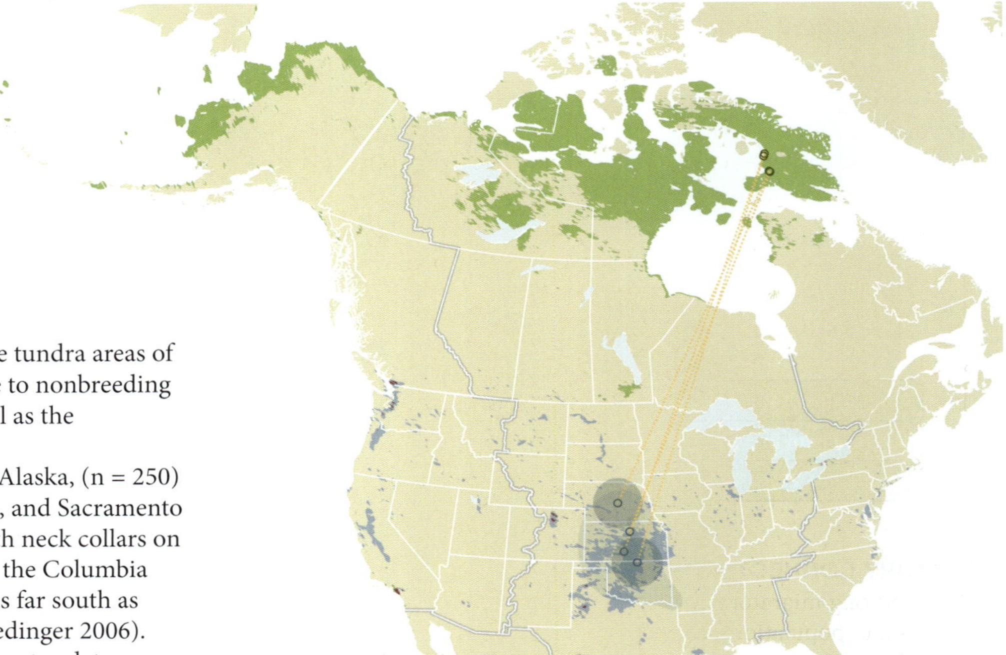

CONNECTIONS B

— TAVERNER'S CACKLING GOOSE

Branta hutchinsii taverneri

OVERALL Taverner's Cackling Geese breed in the tundra areas of the North Slope and western Alaska and migrate to nonbreeding areas in western Oregon and Washington, as well as the midcontinent of the United States.
BANDING Banding in the 1920s near Tule Lake, Alaska, (n = 250) resulted in 39 recoveries in Washington, Oregon, and Sacramento Valley, California (Lincoln 1926). Later work with neck collars on staging areas in Alaska resulted in resightings in the Columbia River Valley, though some birds were observed as far south as the Central Valley of California (Eichholz and Sedinger 2006).
SATELLITE TRACKING Unpublished satellite telemetry data from birds nesting on the North Slope in Alaska show migration to the midcontinent US (C. Latty, US Fish and Wildlife Service, unpublished data).

— RICHARDSON'S CACKLING GOOSE

Branta hutchinsii hutchinsii

OVERALL Richardson's Cackling Geese breed in the high Arctic from the Mackenzie River delta east to Baffin and Southampton islands (Baldassarre 2014) and spend the nonbreeding season in the interior United States, particularly in Kansas, Oklahoma, and western Texas.
BANDING Banding data show movements from breeding areas across the high Arctic to nonbreeding areas in the shortgrass prairie of the central United States, particularly the panhandle of Texas (Hines et al. 2000).
SATELLITE TRACKING Recent tracking of Richardson's Cackling Geese breeding on Southampton and Baffin islands show migration to nonbreeding areas in the United States (F. Baldwin, Canadian Wildlife Service, unpublished data).

DATA CONTRIBUTORS

F. Baldwin; K.C. Fraser; L. Neufeld.

Branta canadensis **CANADA GOOSE**

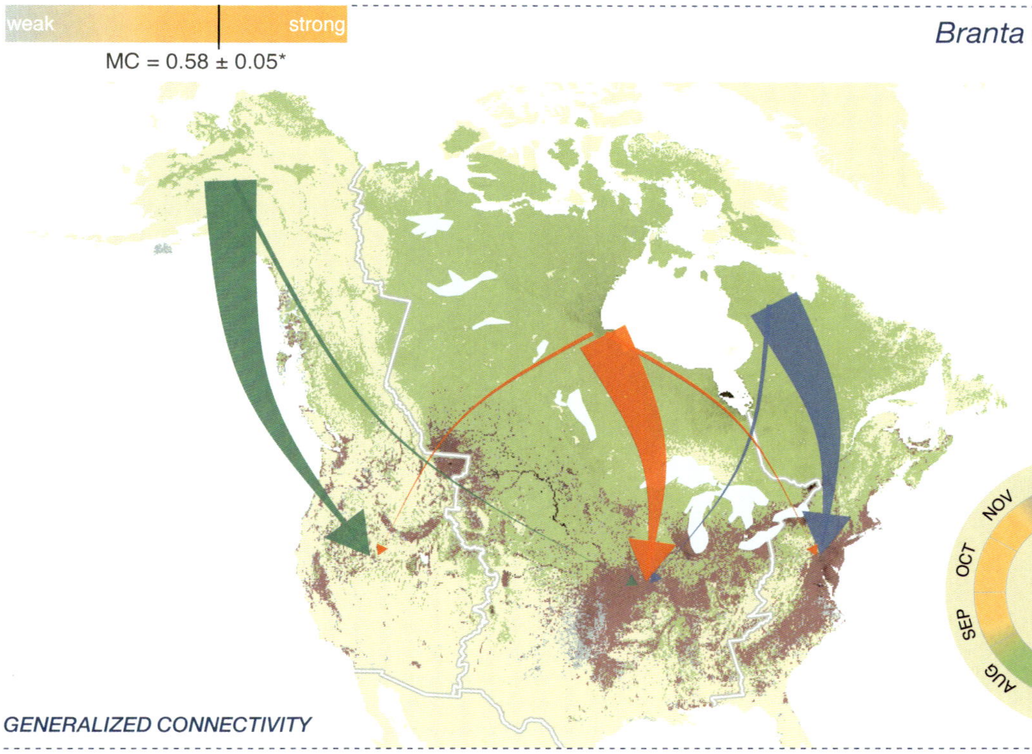

GENERALIZED CONNECTIVITY

DISTRIBUTION AND ABUNDANCE

The Canada Goose is widespread in North America, with the breeding range spanning Canada and the continental United States, and the nonbreeding range spanning the continental United States where open water is available (Mowbray et al. 2020c). After large declines in population size in the early 1900s and subsequent introductions of populations, Canada Geese are now abundant and breed in areas not occupied by pre-European settlements. There are currently seven subspecies recognized (listed by primary breeding area, from west to east): Dusky, Vancouver, Western, Lesser, Interior, Giant, and Atlantic.

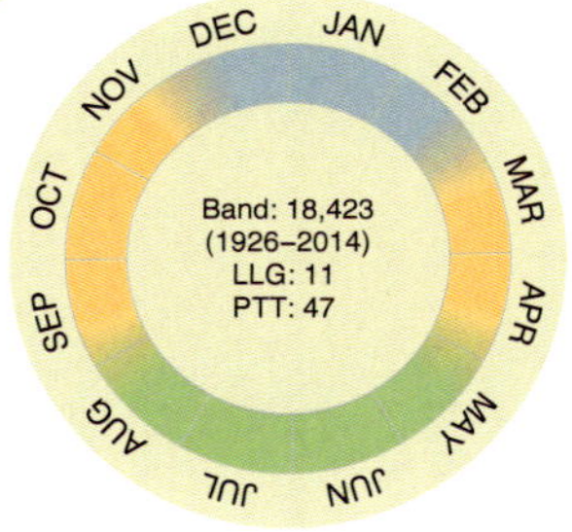

ORDER Anseriformes
FAMILY Anatidae
BREEDING HABITAT Wetlands, Generalist
NONBREEDING HABITAT Wetlands, Generalist
GLOBAL POPULATION SIZE 7,600,000

POPULATION CHANGE SINCE 1970 +100%
IUCN RED LIST CATEGORY Least Concern
R2R TIPPING POINT SPECIES No
US LISTING STATUS No
CANADA SPECIES AT RISK No
PIF CCS 6/20

	Atlantic	Mexico	Midcontinent	Pacific
Atlantic (n = 1457)	95%	–	5%	<1%
Midcontinent (n = 14185)	4%	<1%	93%	3%
Pacific (n = 2073)	–	<1%	5%	94%

MIGRATORY CONNECTIVITY

OVERALL Strong migratory connectivity within subspecies and populations, as well as within flyways. There is high transition probability within the same breeding and nonbreeding region, Atlantic (95%), Midcontinent (93%), and Pacific (94%) flyways, which coincides with population delineation, and thus strong migratory connectivity (MC = 0.58 ± 0.05).

BANDING Canada Geese have been banded extensively across North America; some individuals are resident and some are migratory. An analysis of band and recovery data across North America showed multiple populations with narrow north-south migration routes, except in the southeastern United States, where communities were spatially isolated and generally nonmigratory (Buhnerkempe et al. 2016).

RADIO TRACKING Over 200 tags were deployed using the Motus Wildlife Tracking System. Detections occur throughout the range; yet there have been no individual breeding to nonbreeding detections.

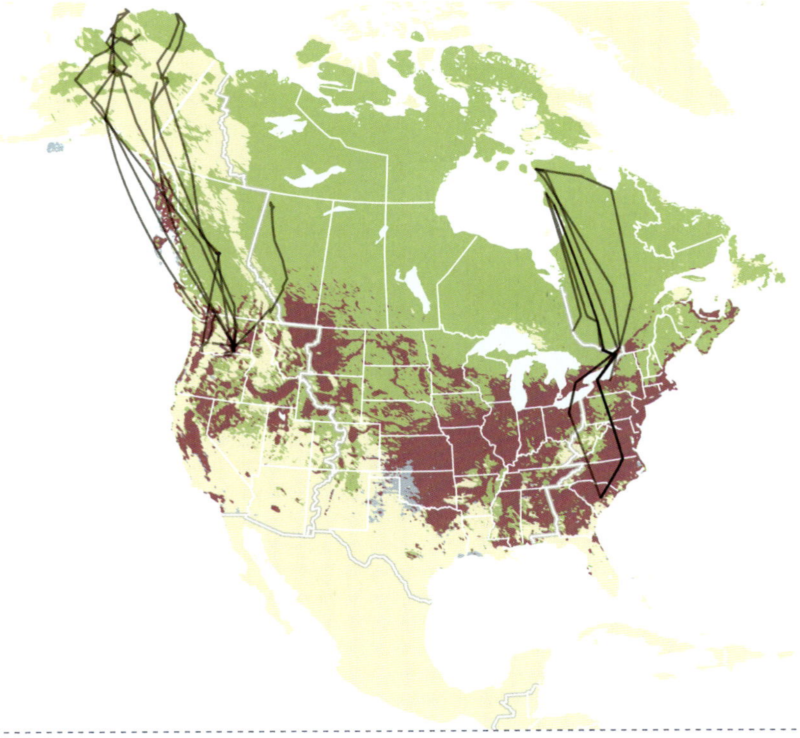

PATHWAYS

DUCKS, GEESE AND WATERFOWL (ANSERIFORMES)

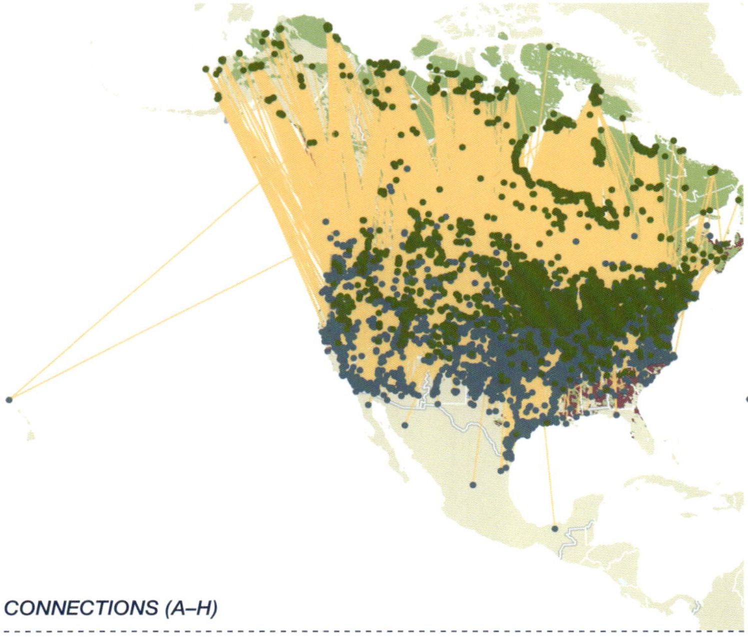

CONNECTIONS (A–H)

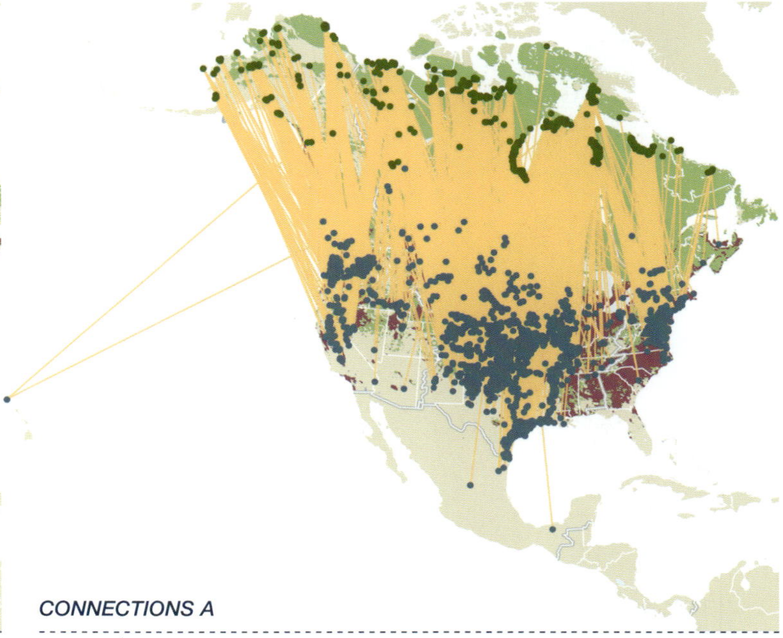

CONNECTIONS A

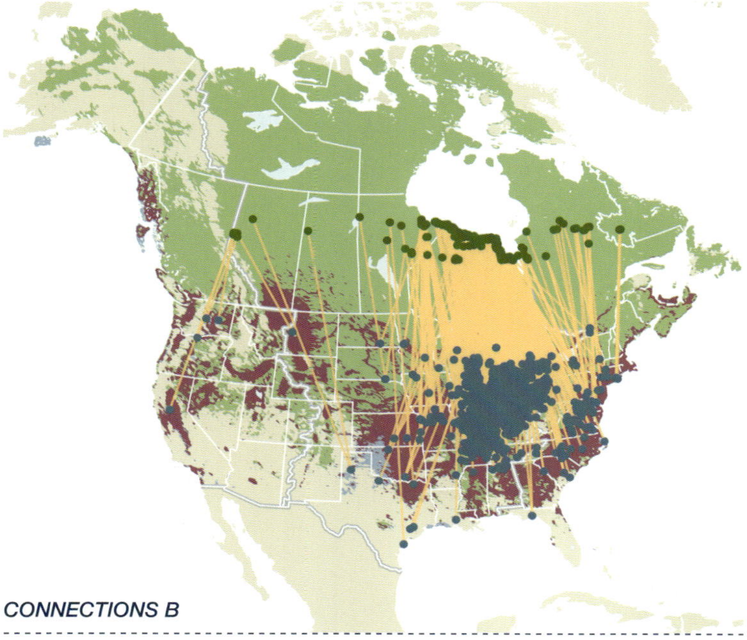

CONNECTIONS B

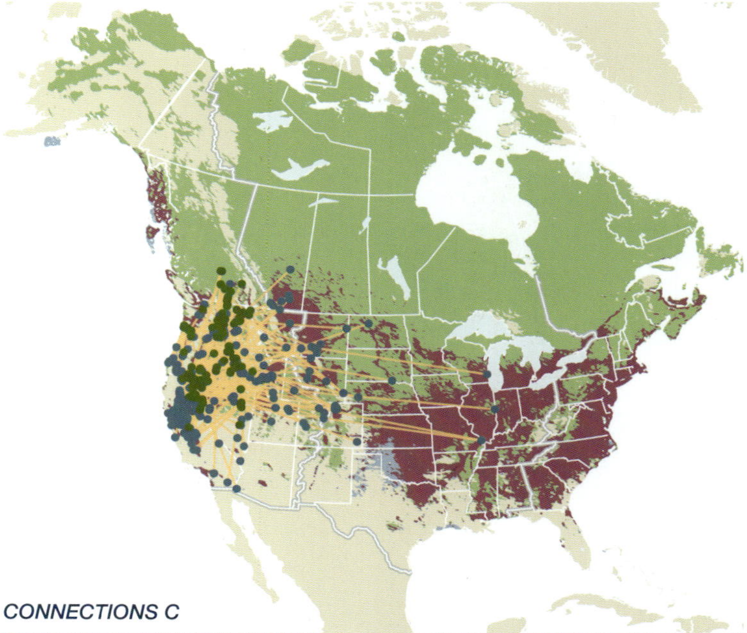

CONNECTIONS C

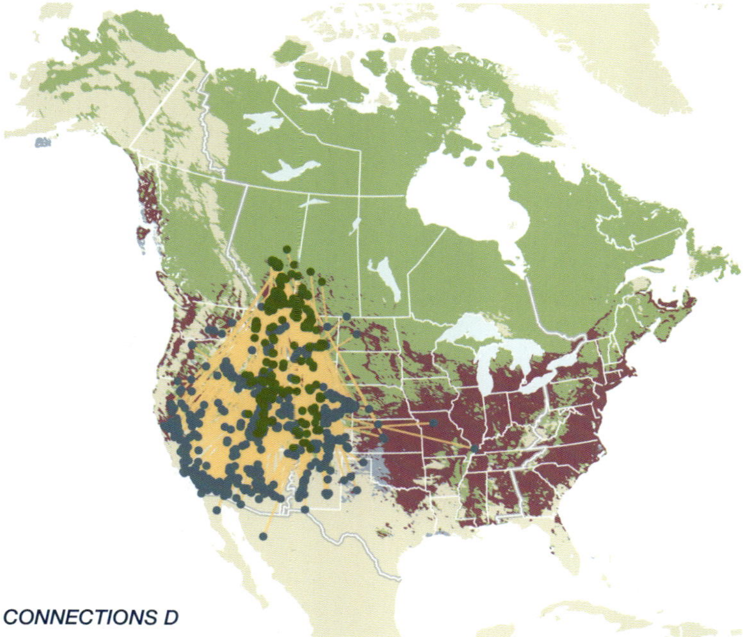

CONNECTIONS D

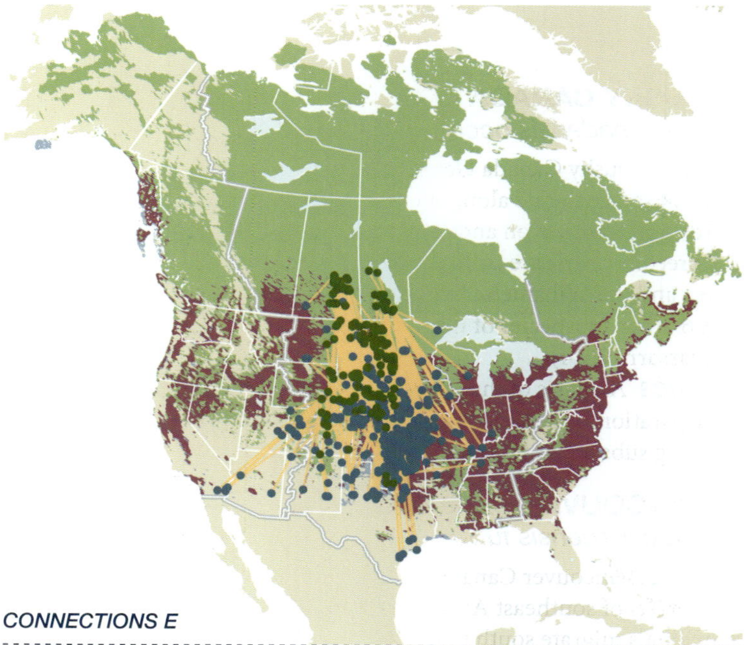

CONNECTIONS E

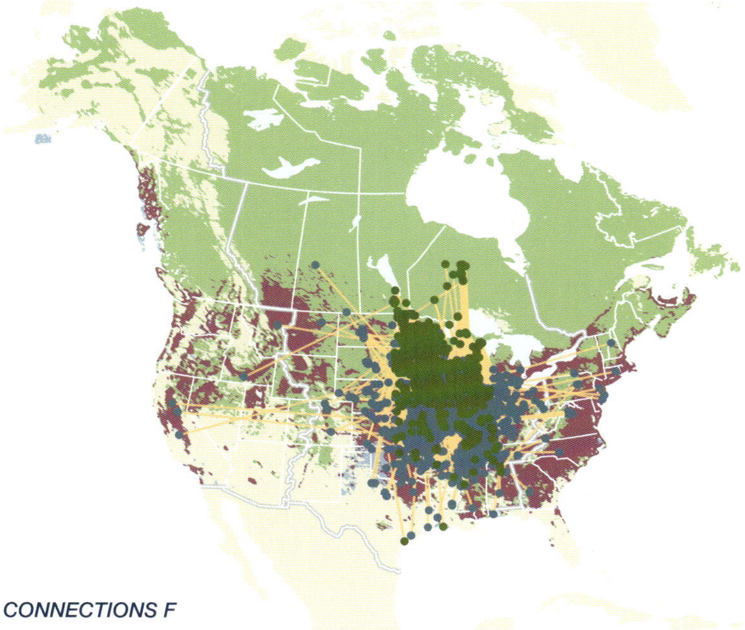

CONNECTIONS F

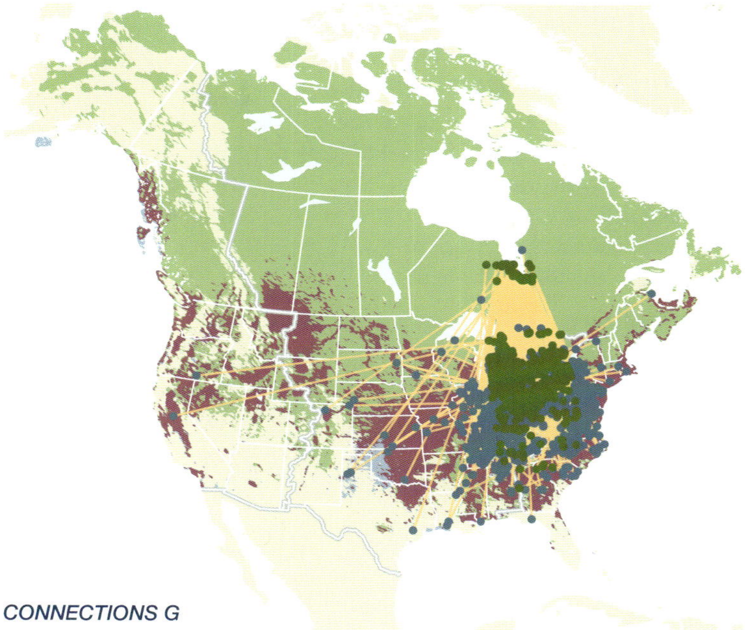

CONNECTIONS G

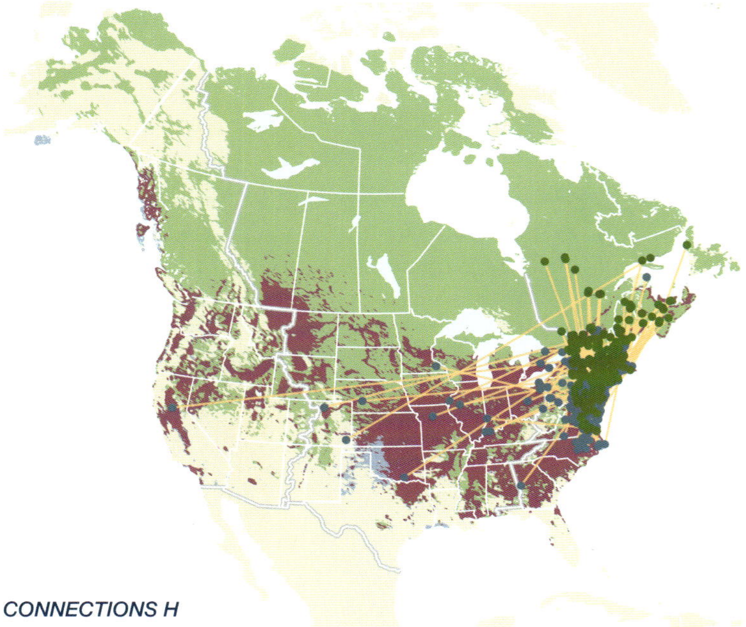

CONNECTIONS H

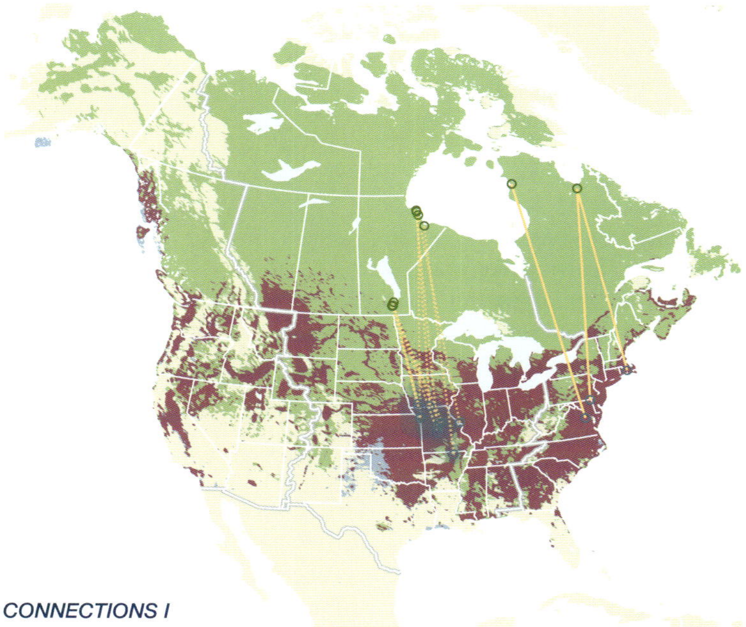

CONNECTIONS I

— DUSKY CANADA GOOSE

Branta canadensis occidentalis

OVERALL Dusky Canada Geese breed on the Copper River Delta in Alaska and migrate along the Pacific coast to nonbreeding areas in western Washington and Oregon (Pacific Flyway Council 2015). Nonbreeding birds are concentrated in the lower Columbia River Valley and the Willamette Valley, where Dusky Canada Geese occur with 6 other subspecies of Canada Goose and Cackling Goose (Baldassarre 2014).

GENETICS Analysis of three classes of genetic markers questions the separation of Dusky Canada Geese from other coastal Alaskan breeding subspecies (Talbot et al. 2003).

— VANCOUVER CANADA GOOSE

Branta canadensis fulva

OVERALL Vancouver Canada Geese occur year-round in temperate rain forests of southeast Alaska and British Columbia. Very few individuals migrate south to coastal Oregon.

BANDING Birds banded in Vancouver, British Columbia, were nearly all (92%) recovered in the same area (Baldassarre 2014). About 2% of Vancouver Canada Geese banded in the Nimpkish Valley, British Columbia, were subsequently recovered in the Willamette Valley, Oregon.

RADIO TRACKING Breeding and nonbreeding Vancouver Canada Geese (n = 166) were implanted with transmitters during breeding and post-breeding (Hupp et al. 2010). Most (70%) birds moved <30 km, and none migrated to coastal British Columbia or western Oregon or Washington.

— WESTERN CANADA GOOSE

Branta canadensis moffitti

OVERALL Western Canada Geese breed throughout the Great Basin of the western United States and in southwestern Canada (Baldassarre 2014). They are widespread and only partial migrants. Birds may be year-round residents or migrate south seeking open water.

— LESSER CANADA GOOSE

Branta canadensis parvipes

OVERALL Lesser Canada Geese breed from interior Alaska east across tundra and taiga vegetation communities to the western shore of Hudson Bay. They migrate to nonbreeding areas across Washington and Oregon, as well as the central Great Plains (Baldassarre 2014).
BANDING Lesser Canada Geese banded and marked with neck collars (n = 884) on staging areas in interior Alaska migrated to Washington and Oregon (Eichholz and Sedinger 2006).

— INTERIOR CANADA GOOSE

Branta canadensis interior

OVERALL Interior Canada Geese spend the breeding and nonbreeding season throughout central and eastern Canada and the United States, with some migratory populations breeding in subarctic areas, principally along Hudson Bay and the Ungava Peninsula. This subspecies is managed as a number of population units based on breeding concentrations (Mississippi Flyway Council 2017). Many populations are relatively short-distance migrants, only going south to reach open water.
BANDING Interior Canada Geese marked with neck collars at Hudson and James bays, Canada, were observed primarily in Wisconsin and Illinois during the nonbreeding season (Craven and Rusch 1983). Individuals banded along the western shore of Hudson Bay were recovered in the Mississippi Valley and eastern Great Plains of the United States (Vaught and Arthur 1965). Birds banded during the breeding season in the Midcontinent Flyway (84%; Mississippi Flyway—80%, Central Flyway—4%), the Atlantic Flyway (3%), and Canada (13%) were recaptured in coastal areas of James Bay and Hudson Bay after a molt migration (Abraham et al. 1999).
RADIO TRACKING Interior Canada Geese marked during breeding in Ontario migrated along the Mississippi River floodplain to nonbreeding areas in the Mississippi Alluvial Valley (Tacha et al. 1991).
SATELLITE TRACKING Female Interior Canada Geese, representing the Atlantic population, fitted with transmitters (n = 34) on breeding areas on the Ungava Peninsula, Quebec, were tracked to nonbreeding areas on the Atlantic coast around Chesapeake and Delaware bays (Malecki et al. 2001). Females marked in New York and Pennsylvania with unsuccessful nests performed a northward molt migration to areas around Hudson Bay (Sheaffer et al. 2007). Interior Canada Geese spending the nonbreeding season in eastern South Carolina (n = 8) were tracked with satellite transmitters north to breeding areas on the Ungava Peninsula (Giles et al. 2013), and birds marked with satellite transmitters in Greenland migrated through the Ungava Peninsula (Scribner et al. 2003).
GENETICS Populations of two subspecies of Canada Geese (*B. c. interior* and *B. c. maxima*) sampled across Ontario and Quebec had little genetic differentiation, except for high relatedness at fine-scale spatial distances (Finnegan et al. 2013). Nuclear DNA markers and mtDNA from Canada Geese sampled in the Ungava Peninsula, Labrador, and Greenland suggest Greenland breeding birds may be more closely related to *B. c. interior* than *B. c. canadensis* (Scribner et al. 2003).

— GIANT CANADA GOOSE

Branta canadensis maxima

OVERALL Giant Canada Geese are the largest geese in the world and are the common breeding subspecies in the eastern half of the United States. Most birds in the eastern United States are a result of intentional introductions of this subspecies and are only partial migrants. Molt migrations often occur to Canada from breeding areas farther south (Mississippi Flyway Council 2017).
BANDING Giant Canada Geese originating from nonbreeding areas in Missouri, Arkansas, Louisiana, Illinois, and Ohio underwent a molt migration to the west coast of Hudson Bay (Davis et al. 1985). Individuals captured in the molting period on the breeding grounds in eastern South Dakota underwent molt migrations within and north of South Dakota (Dieter and Anderson 2009). The majority (80%) of band recoveries of individuals banded in Ohio between 1990 and 2015 were from Ohio, with fewer recoveries in the Mississippi and Atlantic flyways (Shirkey et al. 2018).
RADIO TRACKING The population of geese that nest in lakes in Manitoba, Canada, and migrate to Rochester, Minnesota, were studied with neck collars and radio transmitters and generally flew directly between main breeding and nonbreeding areas (Wege and Raveling 1983).
SATELLITE TRACKING Females (n = 227) tracked from their nests in southern Michigan migrated >900 km to molting locations in the western Ungava Peninsula, Cape Henrietta Maria, James Bay, and Hudson Bay. After molt migration, females returned to their nesting areas in the fall (Luukkonen et al. 2008).
STABLE ISOTOPES Populations of Canada Geese near New Jersey and Ungava Bay, Canada, were sampled for stable isotopes to determine whether they were migratory (*B. c. interior*) or resident (*B. c. maxima*). Stable isotopes were able to distinguish populations in Canada as migratory and in New Jersey as residents (Caccamise et al. 2000).

— ATLANTIC CANADA GOOSE

Branta canadensis canadensis

OVERALL Atlantic Canada Geese breed in eastern Quebec, the Maritime provinces, and Greenland. They primarily migrate to nonbreeding areas in southern New England and Long Island, New York, though can be found from Nova Scotia south to North Carolina (Baldassarre 2014).
BANDING Atlantic Canada Geese marked with bands and neck collars on breeding areas were observed mostly in New England (77%) and the remainder in the mid-Atlantic states. Analyses of neck-collar and band resightings confirmed migration from Atlantic Canada to New England states, and the proportion of birds moving to more southern nonbreeding areas has diminished (Bateman and Daury 1994). Recoveries of Atlantic Canada Geese in Delaware, Maryland, Virginia, and North Carolina declined from 30% of recoveries in 1920–1986 to only 6% in 1987–1995 (Hestbeck and Bateman 2000). Birds marked in western Greenland (n = 135) were recovered during post-breeding migration in Labrador and New Brunswick and during the nonbreeding season in Connecticut, New York, and Pennsylvania (Kristiansen et al. 1999). Newfoundland breeding birds may be more likely to remain within Canada during the nonbreeding season than those breeding in Labrador (Atlantic Flyway Council 2008).
RADIO TRACKING Four Atlantic Canada Geese radio-marked in Greenland all migrated to western Long Island before moving to the Chesapeake Bay area in late January (Atlantic Flyway Council 2008).
SATELLITE TRACKING Birds marked with satellite transmitters in Greenland migrated through the Ungava Peninsula, and these movements, in association with genetic evidence, suggest Greenland breeding birds may be more closely related to *B. c. interior* (Scribner et al. 2003). Atlantic Canada Geese captured in coastal North Carolina and equipped with satellite transmitters migrated to breeding areas in southern Labrador (Atlantic Flyway Council 2008).

DATA CONTRIBUTORS

F. Baldwin; J. Evenson; K.C. Fraser; M. Giles; L. Neufeld; K. Spragens; M. Wilson; Clemson University; South Carolina Cooperative Fish and Wildlife Research Unit; South Carolina Department of Natural Resources; US Fish and Wildlife Service.

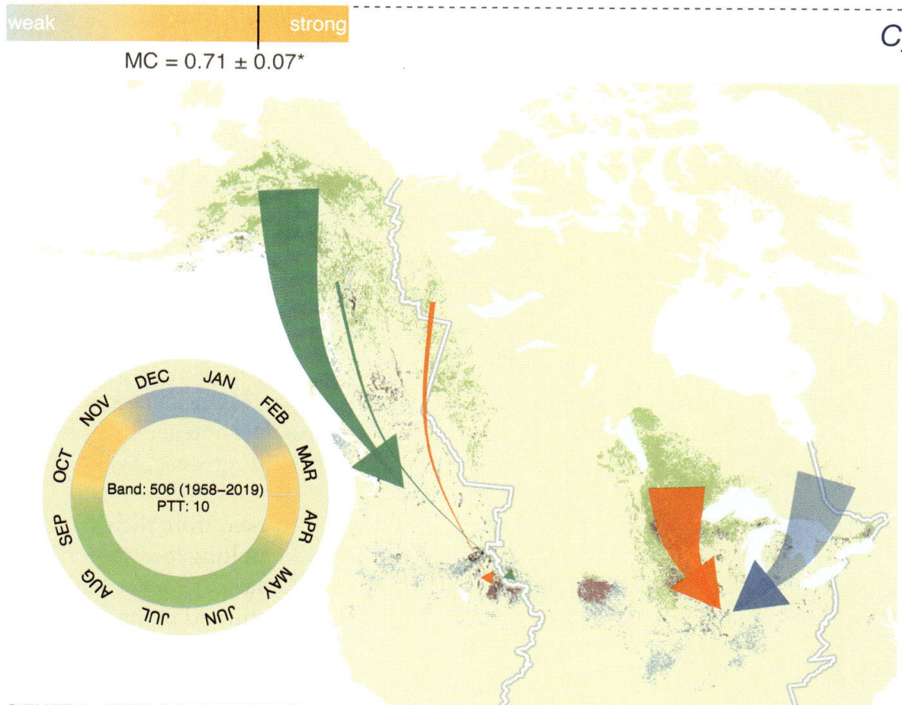

GENERALIZED CONNECTIVITY

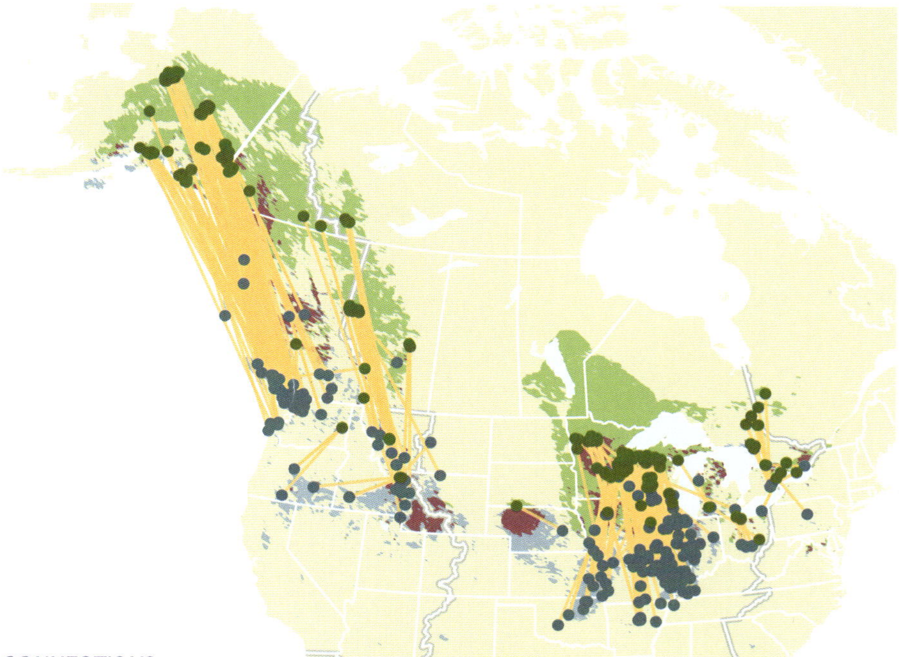

CONNECTIONS

Cygnus buccinator **TRUMPETER SWAN**

DISTRIBUTION AND ABUNDANCE

There are three geographically distinct populations of Trumpeter Swans. Birds breeding in Alaska spend the nonbreeding season near the Pacific coast with concentrations in southeast Alaska. Birds breeding in the Rocky Mountains of western Canada and north-central US migrate to nonbreeding areas in the Greater Yellowstone Ecosystem, where ice-free conditions exist, or in Rocky Mountain states farther south (Gale et al. 1987). Individuals that breed in central North America have a large breeding and nonbreeding range in the eastern United States.

ORDER Anseriformes
FAMILY Anatidae
BREEDING HABITAT Wetlands, Lakes, and Rivers
NONBREEDING HABITAT Wetlands, Generalist
GLOBAL POPULATION SIZE 63,000
POPULATION CHANGE SINCE 1970 +1663%
IUCN RED LIST CATEGORY Least Concern
R2R TIPPING POINT SPECIES No
US LISTING STATUS No
CANADA SPECIES AT RISK No
PIF CCS 11/20

	Atlantic	Midcontinent	Pacific
Atlantic (n = 2)	–	100%	–
Midcontinent (n = 374)	<1%	95%	4%
Pacific (n = 137)	–	1%	99%

MIGRATORY CONNECTIVITY

OVERALL Banding and satellite tracking data, along with reintroduction efforts, confirm strong migratory connectivity (Mitchell and Eichholz 2020). Transition rates within the same breeding and nonbreeding flyway are high (95%–100%) across most breeding populations, with little overlap. This geographic separation results in strong migratory connectivity (MC = 0.71 ± 0.07) and the strongest MC of all species within the Anseriformes.

BANDING Trumpeter Swans banded on breeding areas on the Copper River Delta, Alaska, were resighted most frequently on nonbreeding grounds on Vancouver Island, British Columbia, though some also migrated to Idaho and Montana (Bailey et al. 1990; Sladen and Whissel 2007). Birds banded at Cook Inlet, Alaska, spent the nonbreeding season in the Skagit Valley, Washington (Sladen and Whissel 2007). Trumpeter Swans that breed in interior British Columbia, Alberta, and Saskatchewan migrate to the Greater Yellowstone Ecosystem or the Great Plains for the nonbreeding season (Mackay 1957).

SATELLITE TRACKING Individuals (n = 8) tagged in Alaska all migrated south along the Rocky Mountains to nonbreeding areas in the Pacific Northwest (Dallin 2004; Lance and Mallek 2004). Six breeding males marked at Red Rock Lakes National Wildlife Refuge, Montana, stayed locally or within the Greater Yellowstone Ecosystem during the nonbreeding season (Olson et al. 2009). Trumpeter Swans (n = 16) marked in Wisconsin during the breeding season migrated within the same state or to neighboring states, with a maximum migration distance of birds breeding west of the Great Lakes of <1,000 km (Diebold et al. 2008).

GENETICS Early genetic evidence (nuclear DNA) suggested no differences between populations at the three breeding areas (Barrett and Vyse 1982). However, more recent genetic analysis of Trumpeter

Swans from the Pacific Flyway and interior Rocky Mountain populations shows strong genetic differentiation between those populations (Oyler-McCance et al. 2007). Genetic data also confirm that each population experienced a genetic bottleneck during periods of low population abundance and prior to restoration efforts. The interior population is derived from individuals of both western populations, so it is genetically similar to each (Baldassarre 2014). *MIGRATORY PATHWAYS* Trumpeter Swans in the eastern two populations are short-distance migrants, if they migrate at all. Birds marked with satellite telemetry appear to migrate in a fairly direct path between breeding and nonbreeding areas, where long-distance migration occurs.

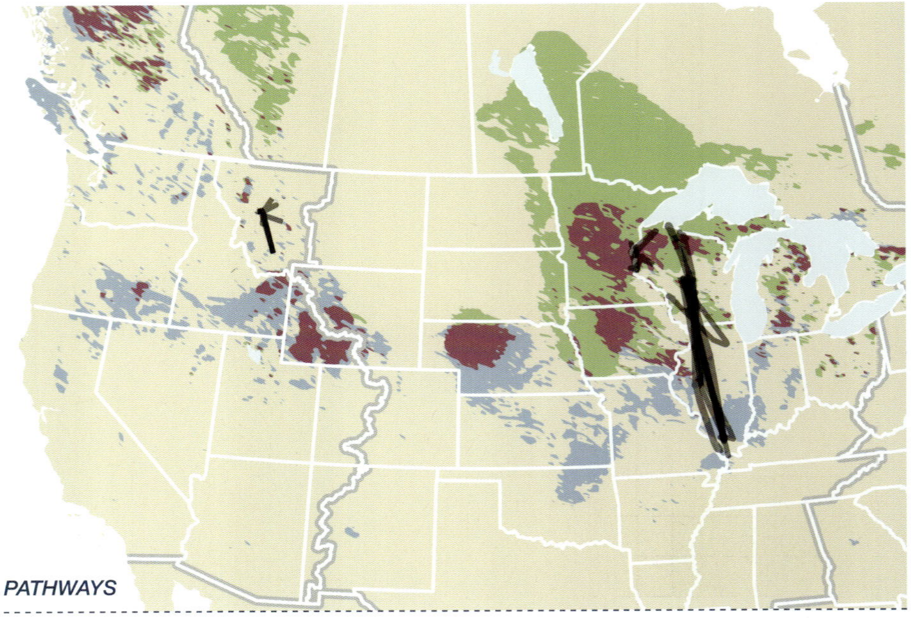

PATHWAYS

DATA CONTRIBUTORS

E. Caton; E. Diebold; C. Koontz; F. Koontz; P. Manthey; S. Matteson; G. Neudecker.

INTERESTING NOTE

Reintroduced Trumpeter Swans initially are often nonmigratory. As the population becomes more established, they migrate longer distances, suggesting migration is a learned behavior in Trumpeter Swans (Mitchell and Eichholz 2020).

TUNDRA SWAN *Cygnus columbianus*

DISTRIBUTION AND ABUNDANCE

Tundra Swans breed in the Arctic, with concentrations in coastal Alaska and the central Arctic. Western breeders occur exclusively in Alaska and migrate to nonbreeding areas primarily in the Sacramento Valley of California and the marshes around the Great Salt Lake, Utah. Birds breeding north and east of the Brooks Range in Alaska and across the Canadian Arctic spend the nonbreeding season along the mid-Atlantic coast of the US.

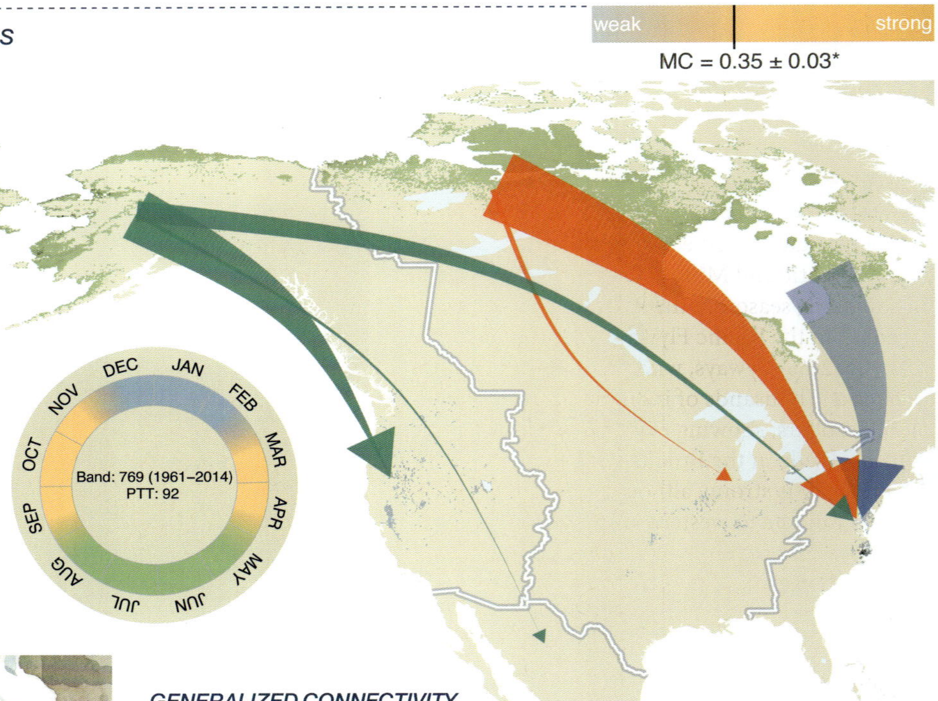

GENERALIZED CONNECTIVITY

ORDER Anseriformes
FAMILY Anatidae
BREEDING HABITAT Wetlands, Tundra
NONBREEDING HABITAT Wetlands, Generalist
GLOBAL POPULATION SIZE 280,000
POPULATION CHANGE SINCE 1970 +6%
IUCN RED LIST CATEGORY Least Concern
R2R TIPPING POINT SPECIES No
US LISTING STATUS No
CANADA SPECIES AT RISK No
PIF CCS 11/20

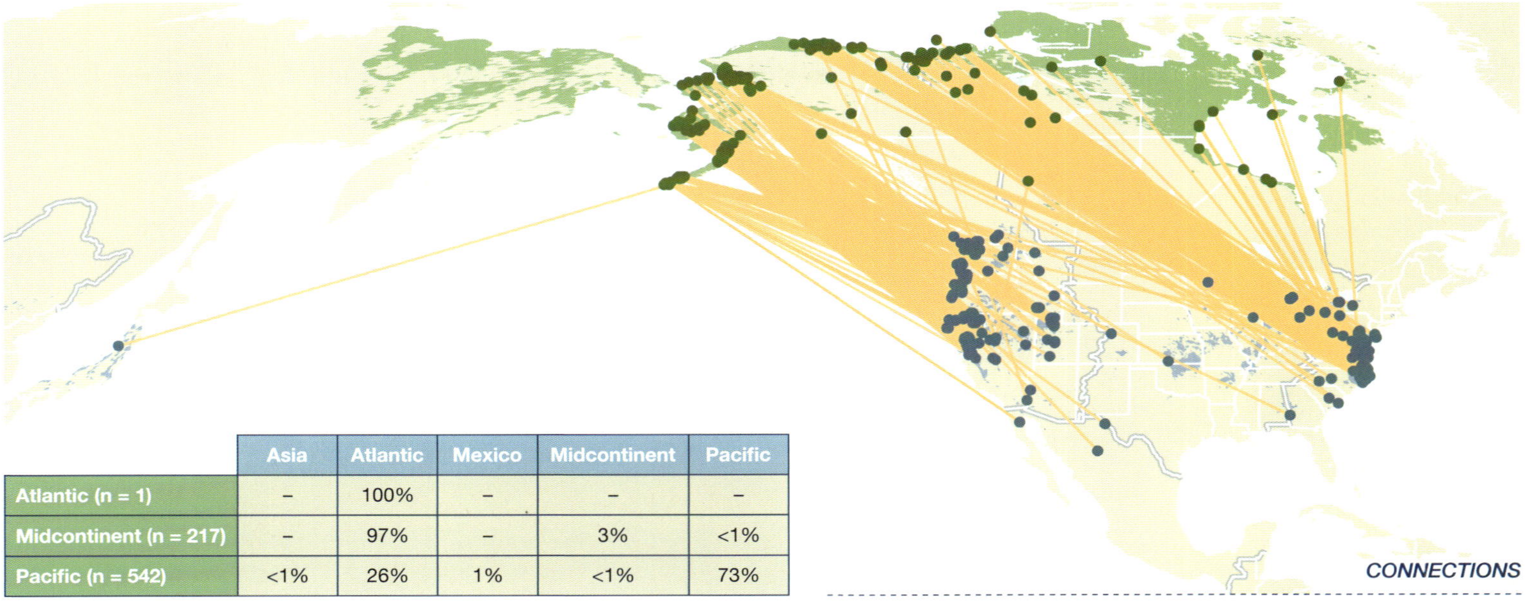

	Asia	Atlantic	Mexico	Midcontinent	Pacific
Atlantic (n = 1)	–	100%	–	–	–
Midcontinent (n = 217)	–	97%	–	3%	<1%
Pacific (n = 542)	<1%	26%	1%	<1%	73%

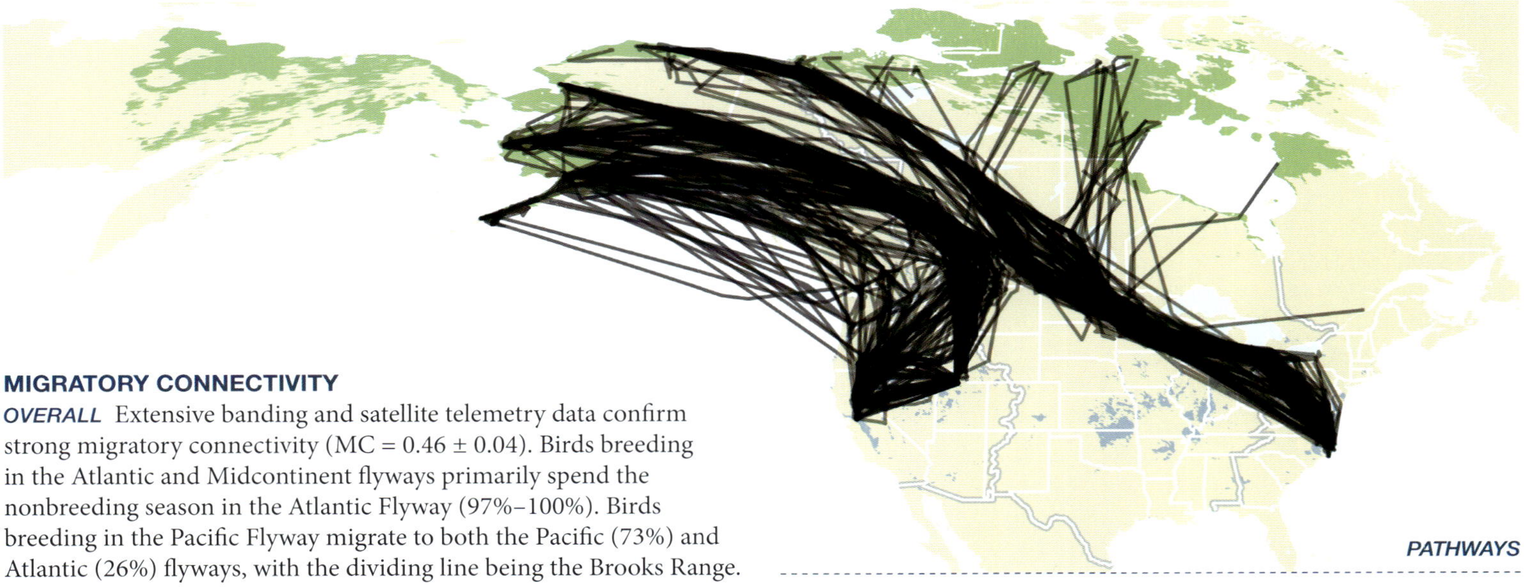

MIGRATORY CONNECTIVITY

OVERALL Extensive banding and satellite telemetry data confirm strong migratory connectivity (MC = 0.46 ± 0.04). Birds breeding in the Atlantic and Midcontinent flyways primarily spend the nonbreeding season in the Atlantic Flyway (97%–100%). Birds breeding in the Pacific Flyway migrate to both the Pacific (73%) and Atlantic (26%) flyways, with the dividing line being the Brooks Range.

BANDING Thousands of individuals have been fitted with bands and/or neck collars, showing evidence for longitudinal migration with a major divide in the Brooks Range in northern Alaska. Migratory connectivity is strong, although it is not unusual for eastern breeding birds to migrate to western states and vice versa (Ely et al. 1997, 2014; Dau and Sarvis 2002). Tundra Swans (n = 190) trapped on nonbreeding grounds in Oregon were banded and dyed yellow for visual resighting. Resightings demonstrated migration to breeding grounds in western Alaska (Paullin and Kridler 1988). Another 119 birds were fitted with neck collars on breeding areas in northwestern Alaska and tracked to nonbreeding areas almost entirely in Northern California (Moermond and Spindler 1997). Tundra Swans marked during breeding on the lower Alaska Peninsula sometimes remained local throughout the nonbreeding season (Dau and Sarvis 2002).

SATELLITE TRACKING Tundra Swans (n = 63) that were satellite tracked from their breeding grounds northeast of the Brooks Range in Alaska southeast to Ontario all migrated to the mid-Atlantic during the nonbreeding season (Weaver 2013). Tundra Swans breeding in western Alaska (n = 50) were tracked to nonbreeding sites in the Central Valley of California and eastern Oregon (Ely and Meixell 2016). Individuals from the western population migrated inland and spent months on staging areas in interior Yukon, Montana, and Idaho, before finishing migration to nonbreeding areas in Northern California (Ely et al. 1997). One study satellite tracked 43 birds from their nonbreeding grounds in the mid-Atlantic, where they migrated northwest to breeding grounds from northeastern Alaska to eastern Canada. Satellite-tagged birds (n = 12) captured mid-migration in Ontario showed similar patterns, migrating between the mid-Atlantic in the nonbreeding season and the central part of their breeding range (Petrie and Wilcox 2003).

OTHER APPROACHES No genetic or stable isotope studies of migratory connectivity were found for this species.

MIGRATORY PATHWAYS Western birds migrated south along the Rockies and the Cascades to nonbreeding areas in valleys in the western United States, while eastern birds congregated on the northern Great Plains from spatially disperse breeding grounds, then migrated past the southern Great Lakes to major nonbreeding areas in the mid-Atlantic. Eastern Tundra Swans follow similar post-breeding and pre-breeding migration routes in an *L* shape, whereby birds migrate west to the Great Lakes, then north toward various breeding locations (Wilkins et al. 2010).

DATA CONTRIBUTORS

D.C. Douglas; C. Ely; J. Terenzi; T.L. Tibbitts.

WOOD DUCK *Aix sponsa*

MC = 0.51 ± 0.01*

DISTRIBUTION AND ABUNDANCE

Wood Ducks are endemic to North America and occur east of the Rocky Mountains with a disjunct population in the west from Washington south to California. The eastern population is large and disperse, while the western population is small but locally abundant. Western birds migrate from breeding areas in Oregon, Washington, Nevada, and California to ice-free wetlands generally in California (Baldassarre 2014). They have migratory and nonmigratory populations (Bellrose and Holm 1994; Hepp and Bellrose 2020).

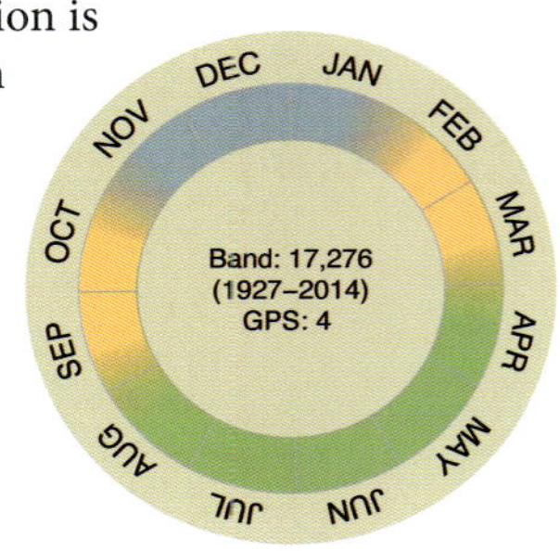

ORDER Anseriformes
FAMILY Anatidae
BREEDING HABITAT Wetlands, Forested
NONBREEDING HABITAT Wetlands, Forested
GLOBAL POPULATION SIZE 4,600,000
POPULATION CHANGE SINCE 1970 +133%
IUCN RED LIST CATEGORY Least Concern
R2R TIPPING POINT SPECIES No
US LISTING STATUS No
CANADA SPECIES AT RISK No
PIF CCS 7/20

	Atlantic	Midcontinent	Pacific
Atlantic (n = 4060)	80%	20%	–
Midcontinent (n = 12551)	17%	83%	–
Pacific (n = 1)	–	–	100%

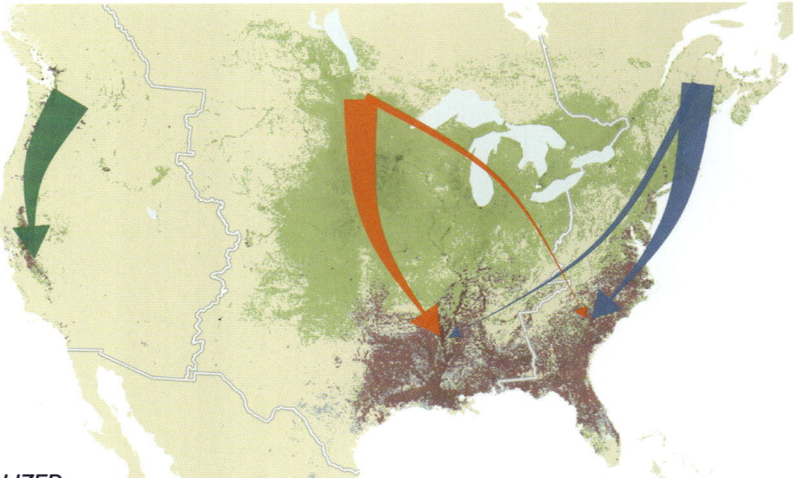

GENERALIZED CONNECTIVITY

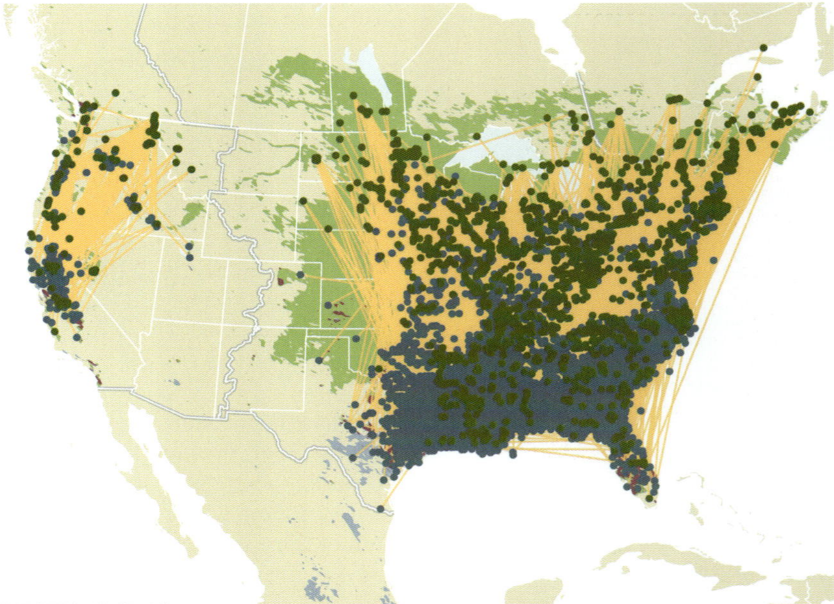

CONNECTIONS (A–H)

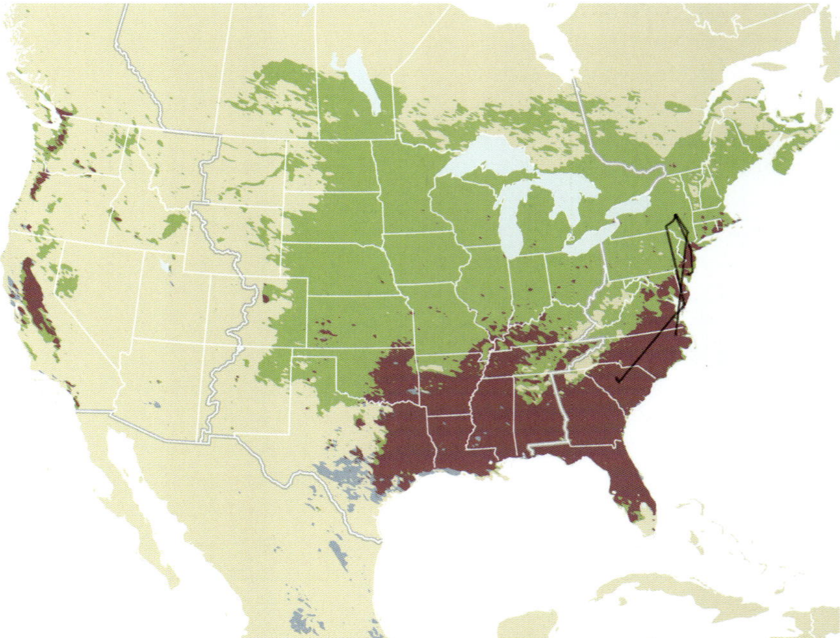

PATHWAYS

MIGRATORY CONNECTIVITY

OVERALL Wood Ducks are one of the most studied waterfowl species, and although some information about migratory connectivity is available from banding and genetics data, overall, we know very little. There is a separation of eastern and western populations, as well as high transition rates within breeding to nonbreeding flyways, resulting in strong migratory connectivity (MC = 0.51 ± 0.01). There is some overlap in the eastern populations, with Atlantic Flyway breeding birds migrating to the Midcontinent Flyway (20%) and Midcontinent Flyway birds migrating to the Atlantic Flyway (17%).

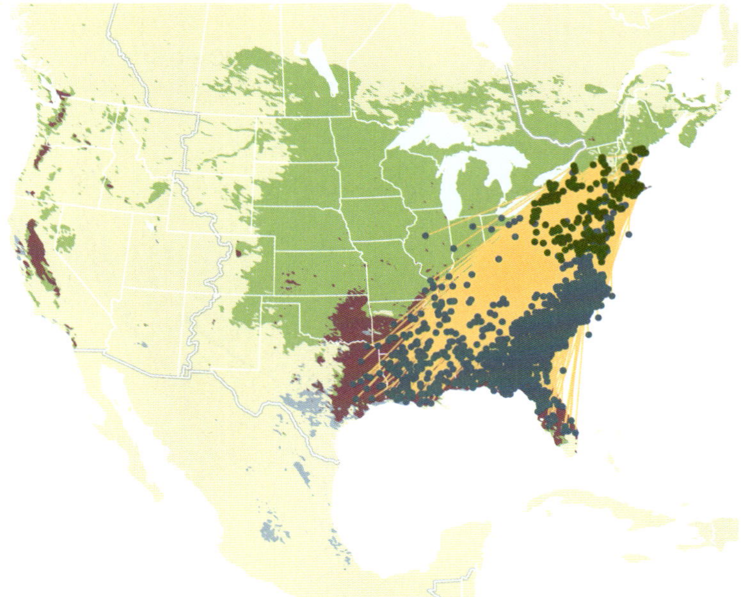

CONNECTIONS A

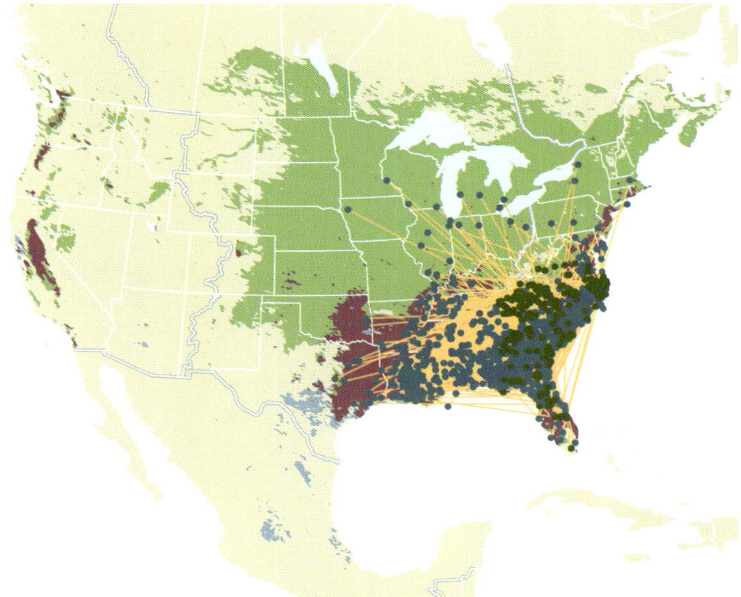

CONNECTIONS B

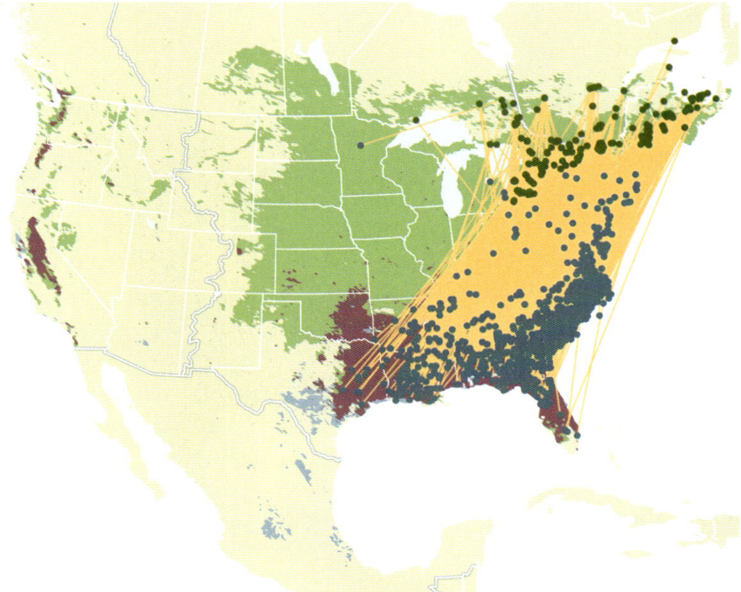

CONNECTIONS C

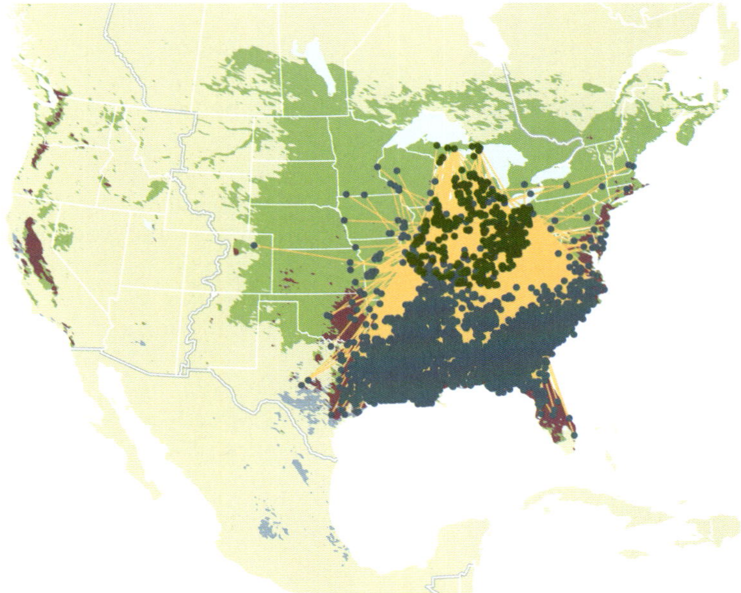

CONNECTIONS D

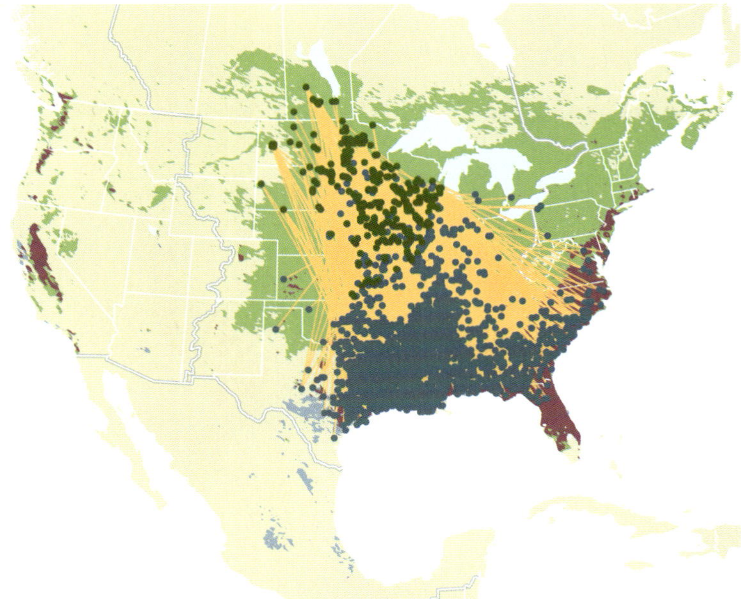

CONNECTIONS E

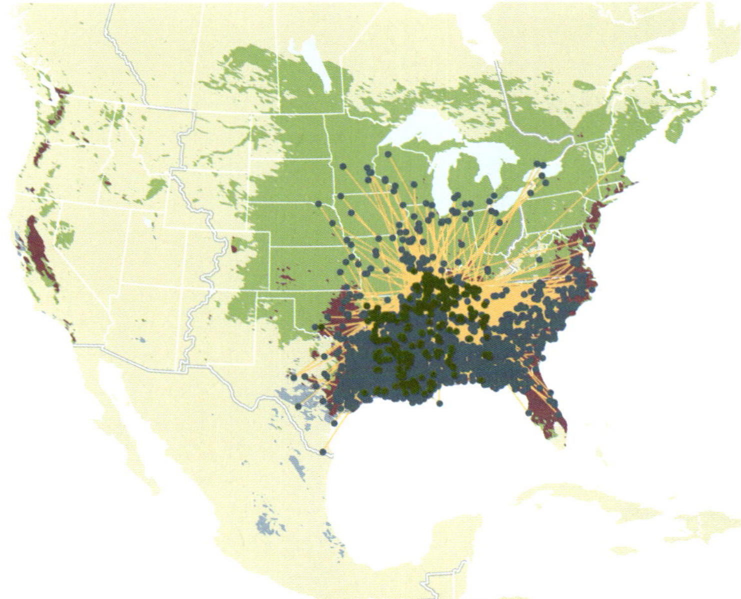

CONNECTIONS F

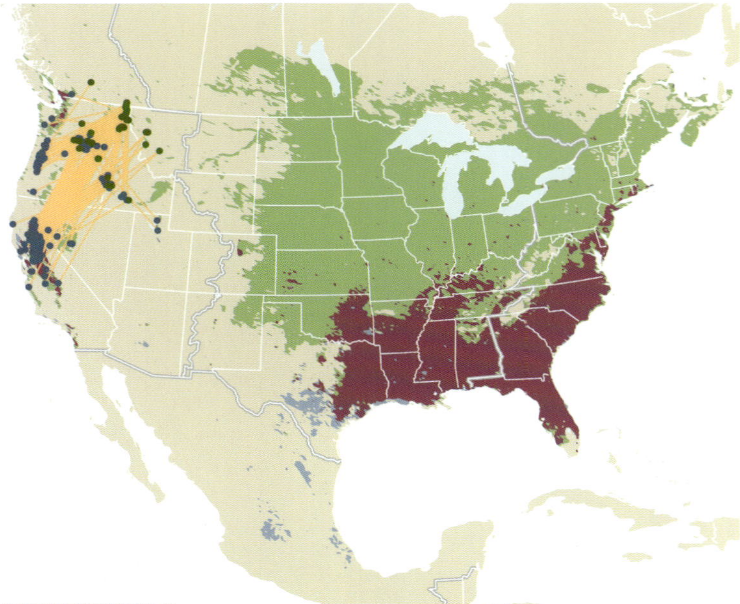

CONNECTIONS G

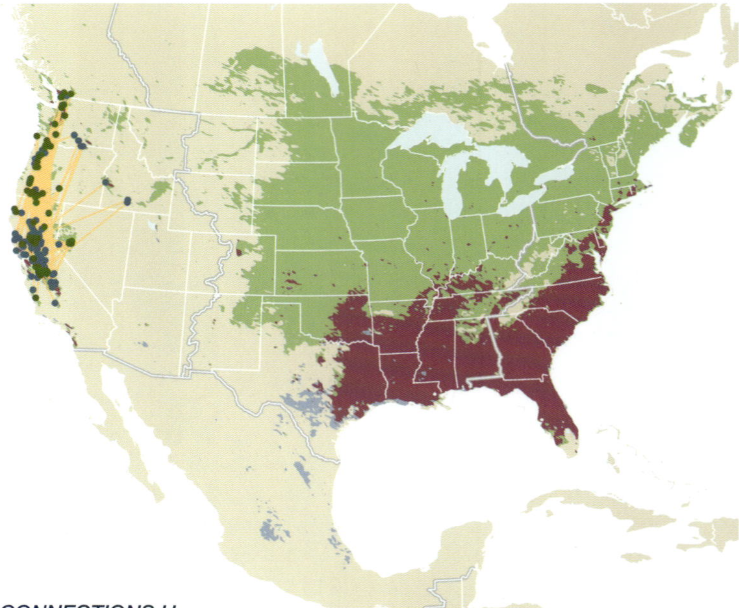

CONNECTIONS H

BANDING Banding data suggest about a third of eastern Wood Ducks are nonmigratory (Bellrose and Holm 1994).
RADIO TRACKING A few tags were deployed through the Motus Wildlife Tracking System; there were no breeding to nonbreeding detections.
SATELLITE TRACKING Limited data exists from satellite tracking. Wood Ducks tracked from New York migrated to nonbreeding areas in New Jersey, North Carolina, and South Carolina (M. Losito et al., unpublished data).
STABLE ISOTOPES No stable isotope studies of migratory connectivity were found for this species.
GENETICS Genetic evidence from mtDNA shows strong differentiation between eastern and western populations (Peters et al. 2005). The eastern population has high genetic diversity and has experienced population expansion, while the western population is more genetically homogeneous and has not seen the same growth. Differentiation likely occurred during the last North American glaciation (Peters et al. 2005).
OTHER APPROACHES Differences existed between parasitic helminth diversity of Wood Ducks collected from the northeastern portion of their breeding range versus the southeastern portion, suggesting some population differentiation (Thul et al. 1985).
MIGRATORY PATHWAYS Wood Ducks breeding in northeastern states and eastern Canadian provinces generally migrate directly south to nonbreeding areas, based on a pattern of band recoveries.

DATA CONTRIBUTORS
M. Losito; R. Masse; C. Slezak; E. Struening.

INTERESTING NOTE

The recovery of Wood Ducks from near extinction to current abundance is one of the more remarkable wildlife recovery success stories, and their populations have continued to increase. The process of captive-rearing and releasing Wood Ducks was key to the revival of the population, and released birds in the northern portion of their range readily learned to migrate and return to release locations (Lee and Nelson 1965).

BLUE-WINGED TEAL *Spatula discors*

weak strong

MC = 0.08 ± 0.02*

DISTRIBUTION AND ABUNDANCE

The Blue-winged Teal is a common breeder in the Prairie Pothole Region of North America and is present in low abundance in boreal forest breeding areas. Nonbreeding season concentrations occur along the Gulf Coast and in Mexico and Central America (Rohwer et al. 2020).

ORDER Anseriformes
FAMILY Anatidae
BREEDING HABITAT Wetlands, Seasonally Wet Grasslands
NONBREEDING HABITAT Coasts, Salt Marsh
GLOBAL POPULATION SIZE 7,800,000

POPULATION CHANGE SINCE 1970 +98%
IUCN RED LIST CATEGORY Least Concern
R2R TIPPING POINT SPECIES No
US LISTING STATUS No
CANADA SPECIES AT RISK No
PIF CCS 8/20

MIGRATORY CONNECTIVITY

OVERALL Migratory connectivity data are available from satellite transmitters, band recoveries, and stable isotopes. Blue-winged Teal have been considered a major carrier of avian influenza, resulting in significant research on their migration routes and distribution. Birds transition from their breeding grounds across their nonbreeding range. Atlantic Flyway breeding birds migrate to nonbreeding areas across the range including the Caribbean (40%), Atlantic Flyway (24%), and Northern South America (24%). Breeding birds from the Midcontinent Flyway also migrate to many nonbreeding regions, with the highest percentage in the Midcontinent Flyway (38%). Pacific Flyway breeding birds primarily migrate to Mexico (44%), the Midcontinent Flyway (20%), or nonbreeding areas within the Pacific Flyway (22%). This diffuse migration results in weak migratory connectivity (MC = 0.08 ± 0.02).

BANDING Large numbers of individuals banded in Alberta and Saskatchewan have been recovered across their nonbreeding range in Mexico, Cuba, Colombia, Venezuela, and Nicaragua (Ramey et al. 2016).

SATELLITE TRACKING Blue-winged Teal tracked by satellite transmitters (n = 42) from their breeding grounds in Canada migrated to the southeastern United States, Mexico, and the Caribbean (Ramey et al. 2019). A sample of 12 individuals migrated to both the Neotropics and the coast of Texas and Louisiana (Ramey et al. 2016). Birds tracked from nonbreeding areas in California's Central Valley migrated to breeding areas in western Canada and the western United States (US Geological Survey, unpublished data).

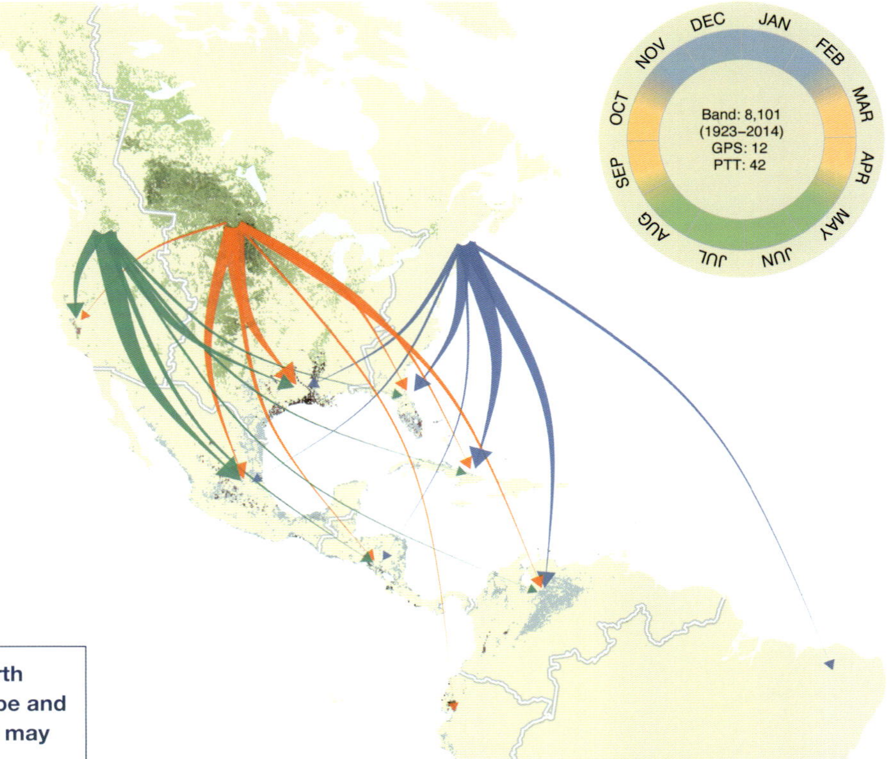

GENERALIZED CONNECTIVITY

INTERESTING NOTE

A small but regular number of Blue-winged Teal from North and South America are found across the Atlantic in Europe and Africa. These were previously thought to be vagrants but may actually be regular migrants (De Juana and Garcia 2010).

	Andean West	Atlantic	Brazil	Caribbean	Central America	Mexico	Midcontinent	Northern SA	Pacific
Atlantic (n = 302)	–	24%	4%	40%	1%	1%	7%	24%	–
Midcontinent (n = 7681)	1%	11%	<1%	10%	7%	22%	38%	11%	1%
Pacific (n = 91)	–	3%	–	3%	7%	44%	20%	1%	22%

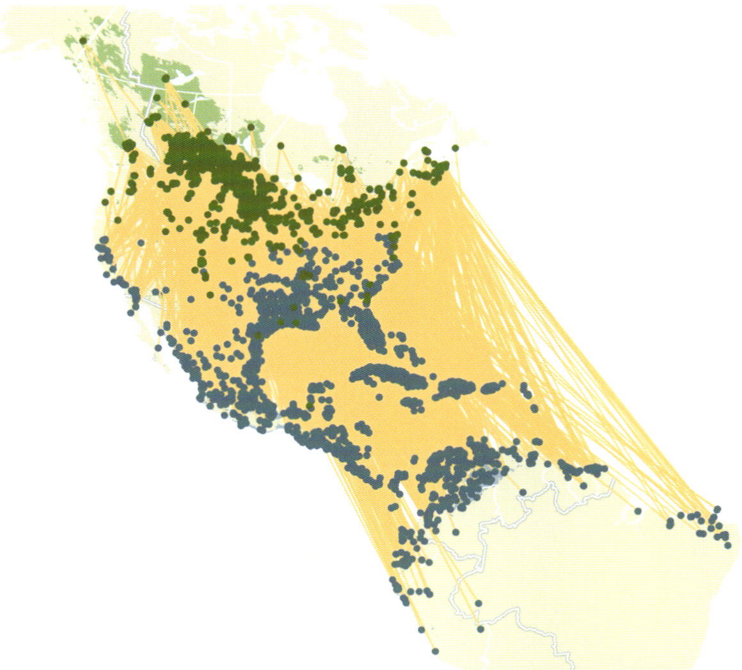

CONNECTIONS (A–I)

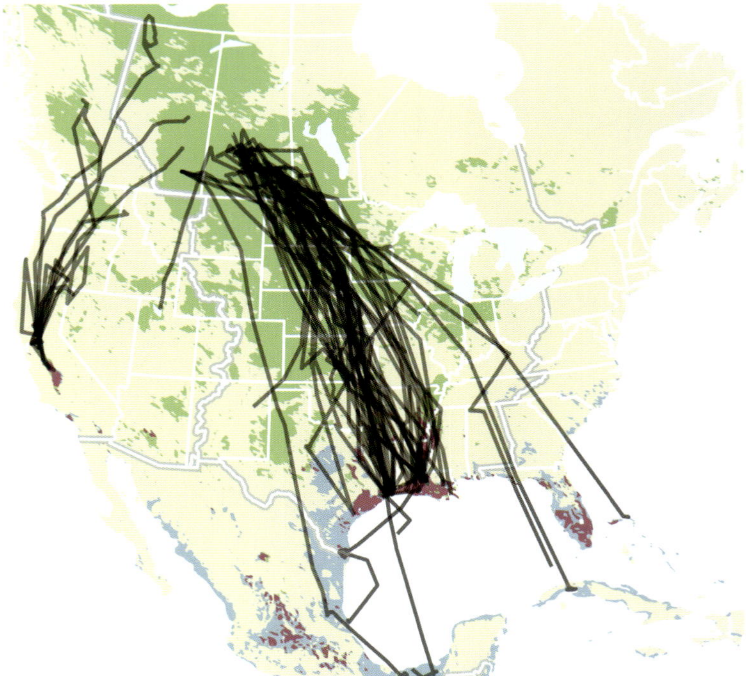

PATHWAYS

DUCKS, GEESE AND WATERFOWL (ANSERIFORMES)

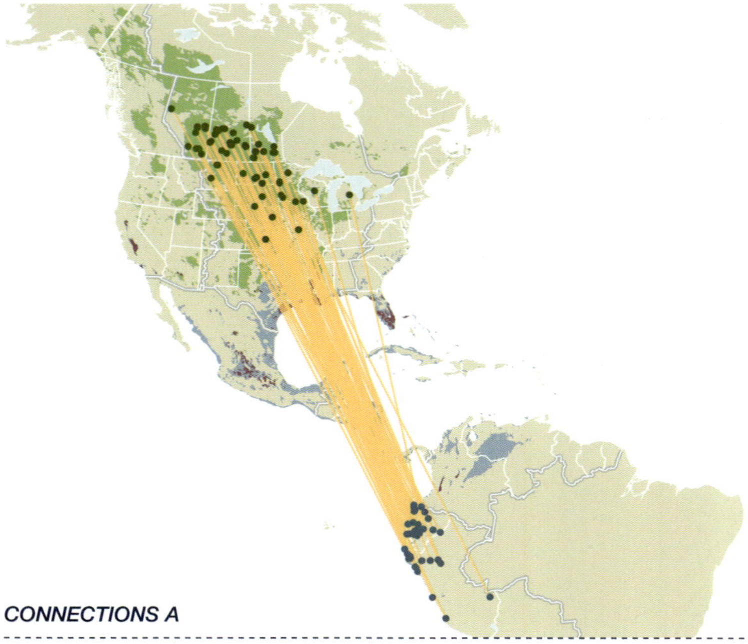

CONNECTIONS A

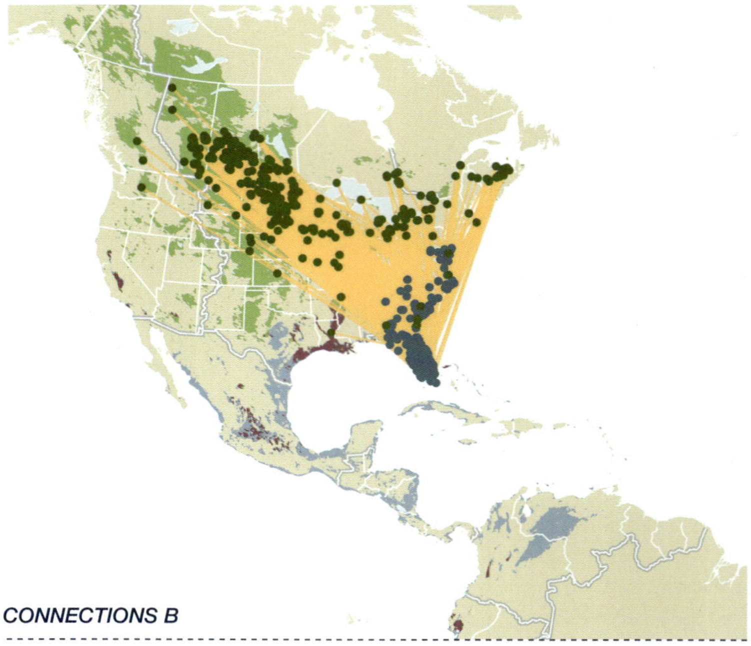

CONNECTIONS B

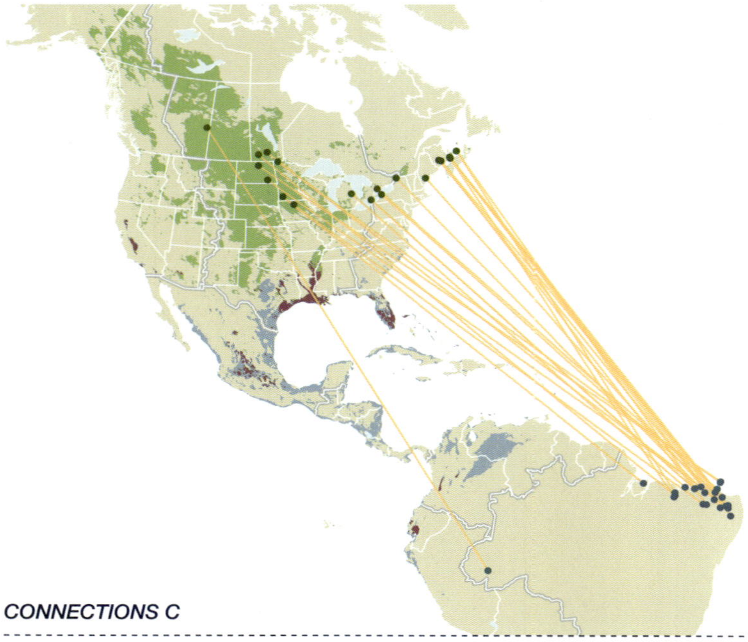

CONNECTIONS C

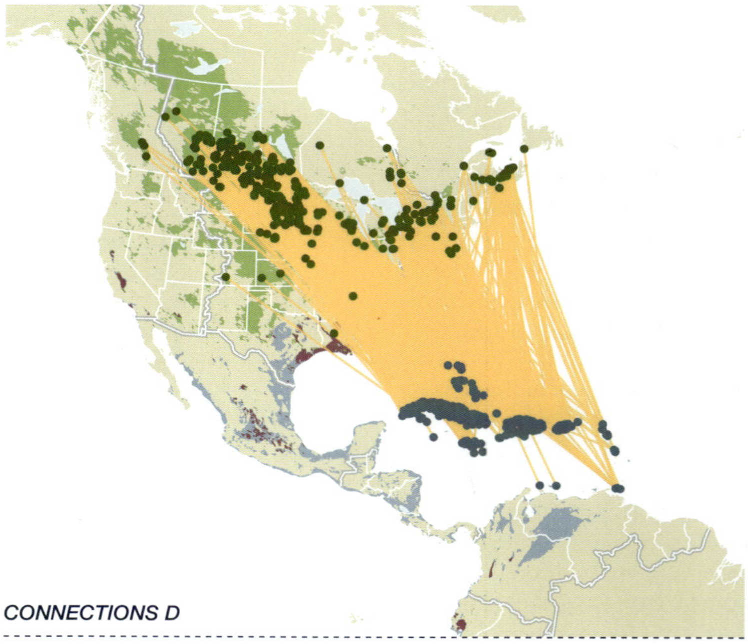

CONNECTIONS D

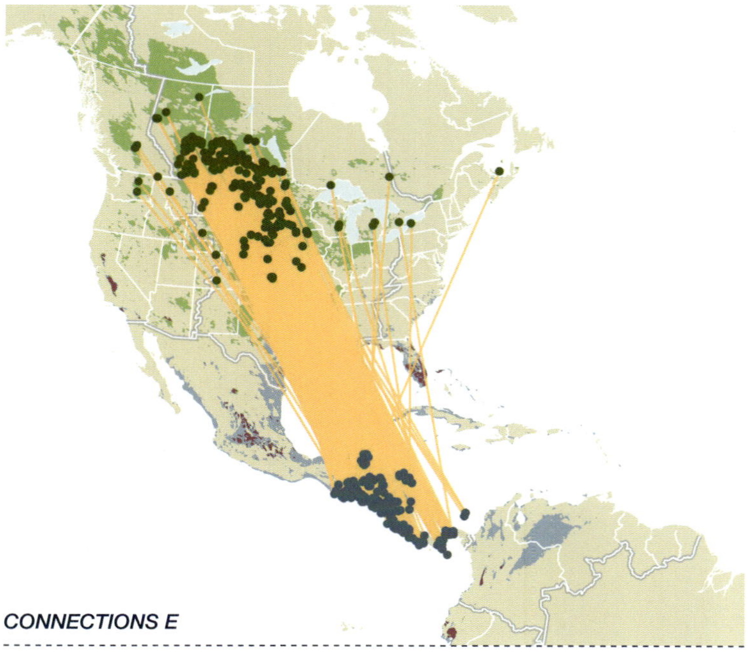

CONNECTIONS E

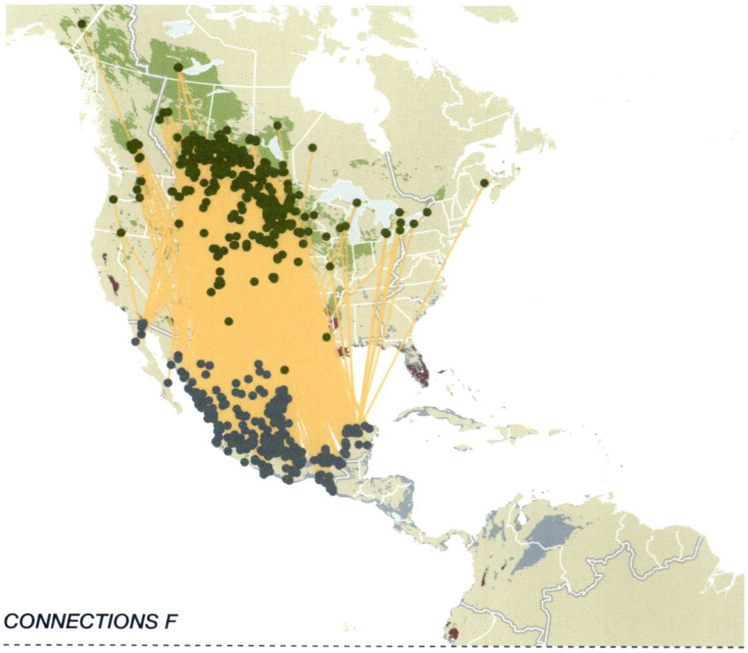

CONNECTIONS F

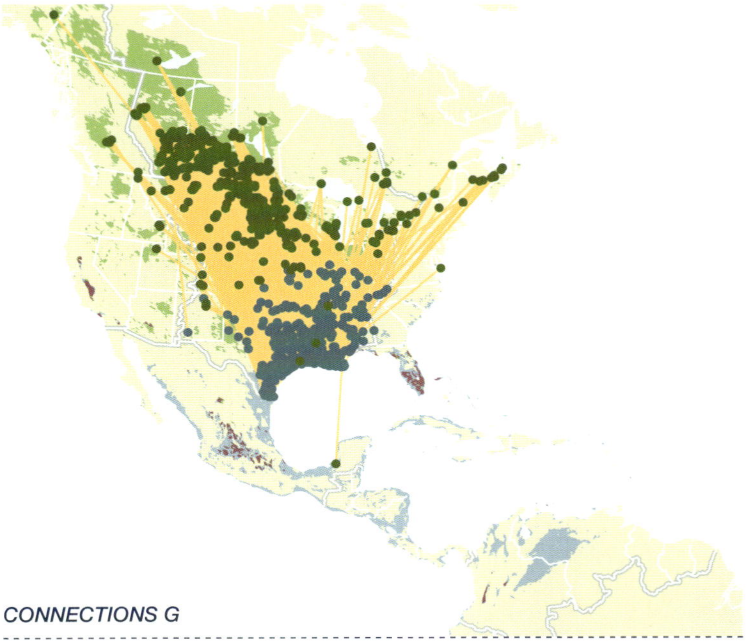

CONNECTIONS G

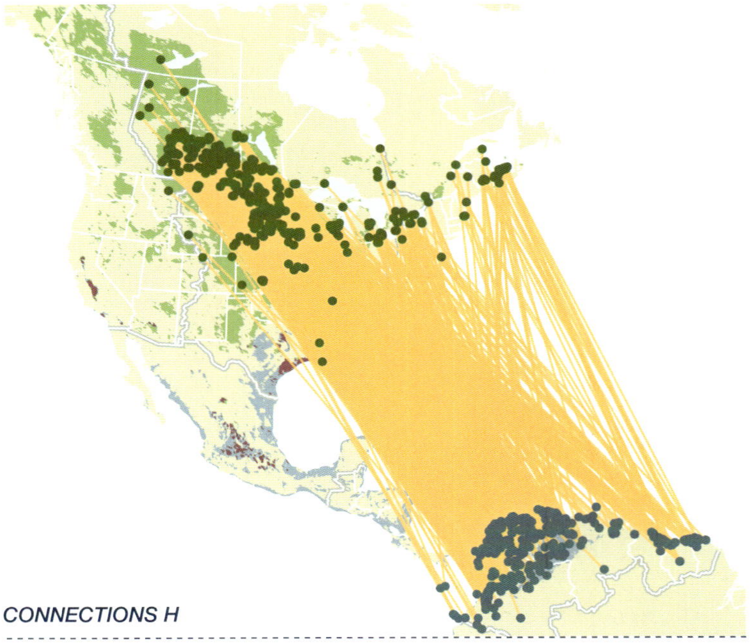

CONNECTIONS H

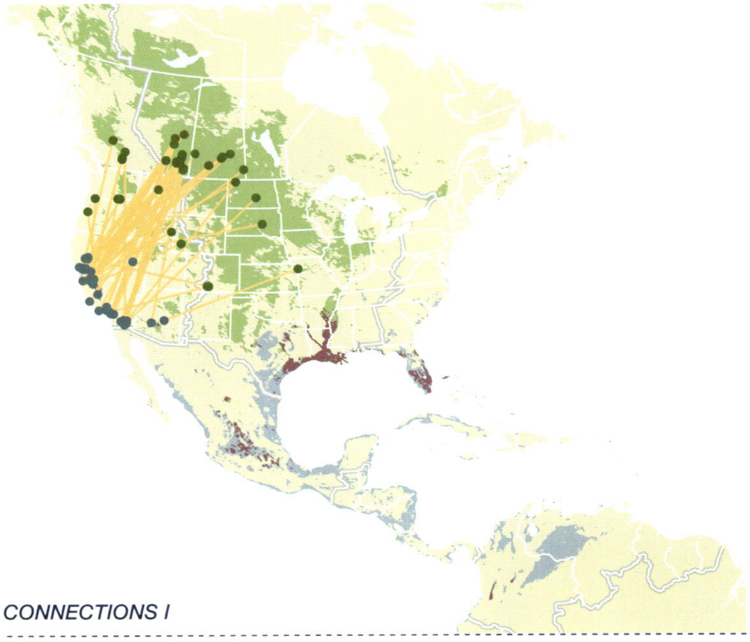

CONNECTIONS I

GENETICS No genetic studies of migratory connectivity were found for this species.

STABLE ISOTOPES Birds collected in the Saskatchewan River Delta, Canada, (n = 52) were assigned to breeding grounds north of Saskatchewan and to the south in the Prairie Pothole Region using stable isotope analysis (Asante et al. 2017). Stable isotope analysis suggests Blue-winged Teal harvested in Ontario come from north and west of the recovery range, in contrast to banding data showing breeding birds originating farther east. This discrepancy is likely a result of available banding sites (Palumbo et al. 2020).

OTHER APPROACHES Blue-winged Teal collected in the northern breeding areas and along the Gulf Coast had parasites known from South America (Ramey et al. 2016). These birds had numerous parasites that were genetically similar to parasites from both North and South America, indicating a clear link between continents. Helminth communities of host Blue-winged Teal differed among birds using a north-south migratory pathway versus those using an eastern pathway through Florida (Garvon et al. 2011).

MIGRATORY PATHWAYS Blue-winged Teal utilize two primary pathways during migration. The first starts in the northern Great Plains and continues south through Texas and Louisiana. The second is a more easterly route, starting in Manitoba and the Great Lakes states and then heading southeast through Florida and over the Atlantic Ocean to the Caribbean and South America (Sharp 1972; Baldassarre 2014). The Saskatchewan River Delta appears to be an important stopover site for individuals using the Central Flyway (Asante et al. 2017). Recent tracking of western populations will help discover migratory pathways in the western US (US Geological Survey, unpublished data).

DATA CONTRIBUTORS

M. Casazza; D.C. Douglas; C. Feldheim; C. Overton; A. Ramey.

CINNAMON TEAL *Spatula cyanoptera*

DISTRIBUTION AND ABUNDANCE

Cinnamon Teal consist of five subspecies, one of which inhabits North America (*S. c. septentrionalium*). This subspecies breeds in the western United States and Canada, particularly the Great Basin. The nonbreeding range encompasses the Central Valley of California, the Pacific coast of Mexico, and central Mexico, with large concentrations in Sonora, Sinaloa, and Nayarit, Mexico (Gammonley 2020).

ORDER Anseriformes
FAMILY Anatidae
BREEDING HABITAT Wetlands, Freshwater Marsh
NONBREEDING HABITAT Wetlands, Freshwater Marsh
GLOBAL POPULATION SIZE 380,000
POPULATION CHANGE SINCE 1970 −63%
IUCN RED LIST CATEGORY Least Concern
R2R TIPPING POINT SPECIES No
US LISTING STATUS No
CANADA SPECIES AT RISK No
PIF CCS 12/20

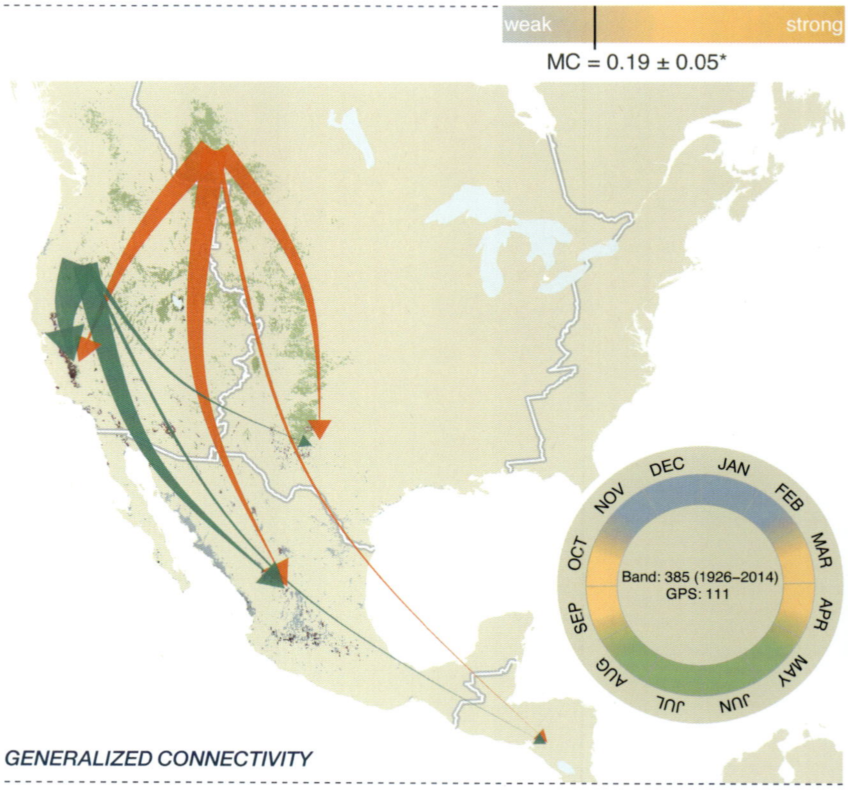

GENERALIZED CONNECTIVITY

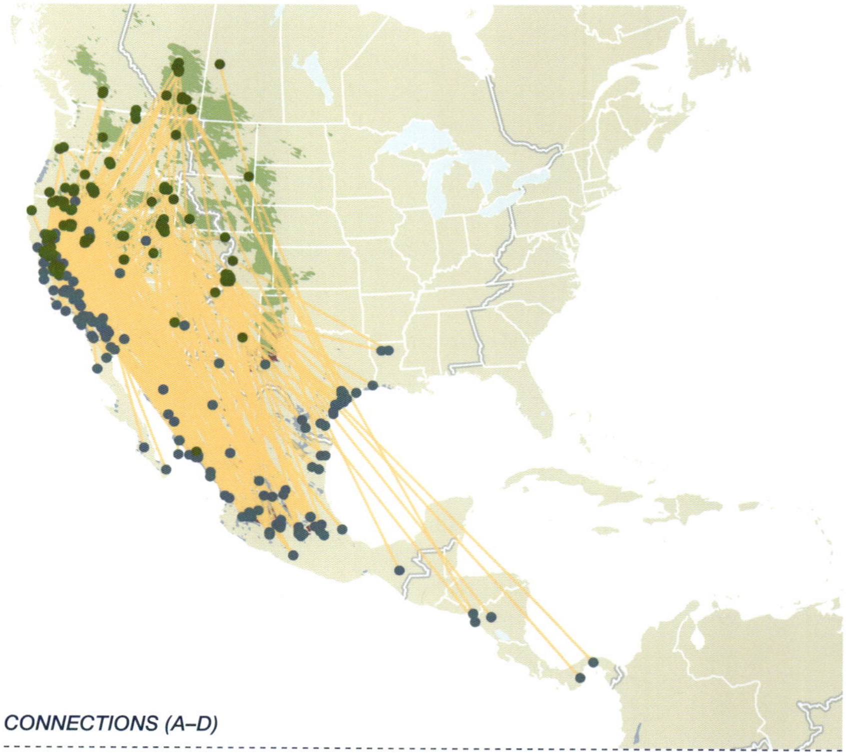

CONNECTIONS (A–D)

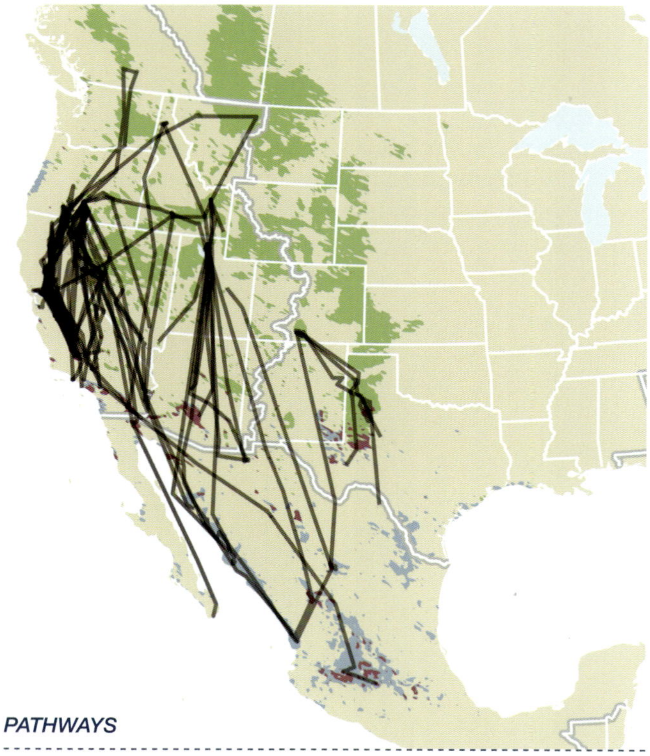

PATHWAYS

	Central America	Mexico	Midcontinent	Pacific
Midcontinent (n = 58)	3%	36%	24%	36%
Pacific (n = 436)	1%	36%	3%	60%

MIGRATORY CONNECTIVITY

OVERALL Birds breeding in the Midcontinent Flyway migrate to nonbreeding areas in the Pacific Flyway (36%), Mexico (36%), and the Midcontinent Flyway (24%). Pacific Flyway breeding birds were more likely to migrate to nonbreeding areas within the Pacific Flyway (60%) or Mexico (36%). Mixing of some breeding populations across the nonbreeding range results in moderate migratory connectivity (MC = 0.19 ± 0.05).

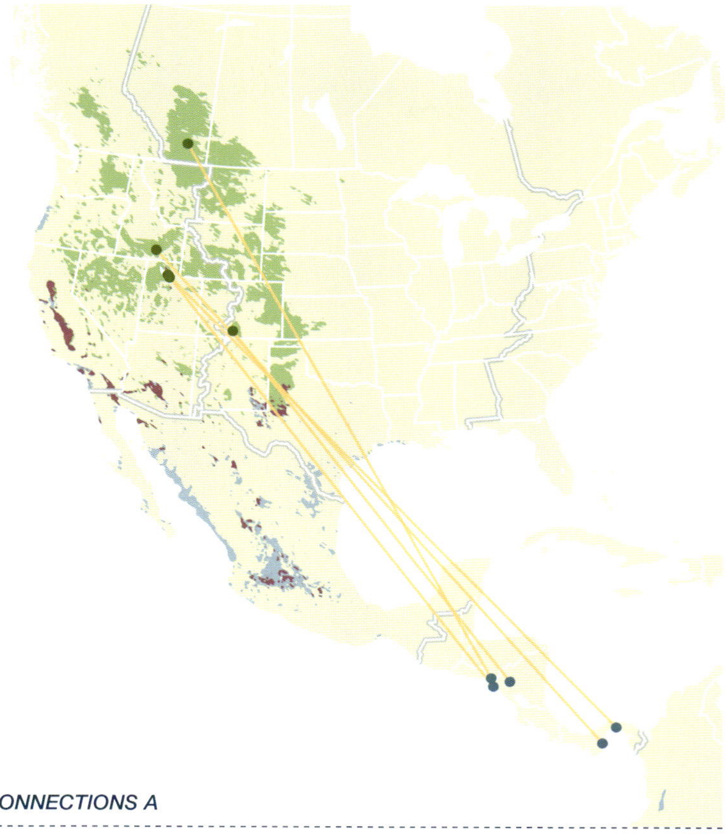

CONNECTIONS A

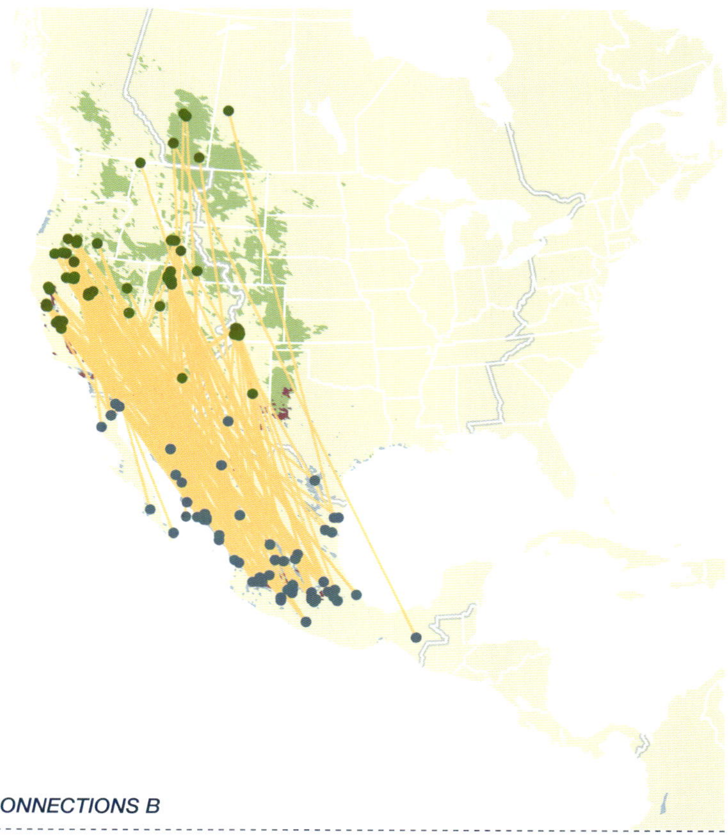

CONNECTIONS B

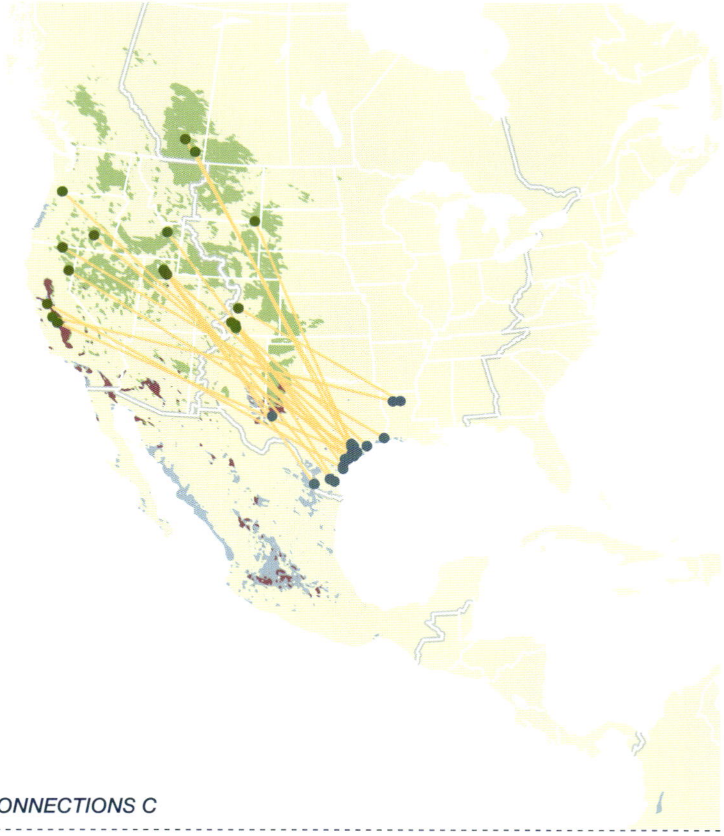

CONNECTIONS C

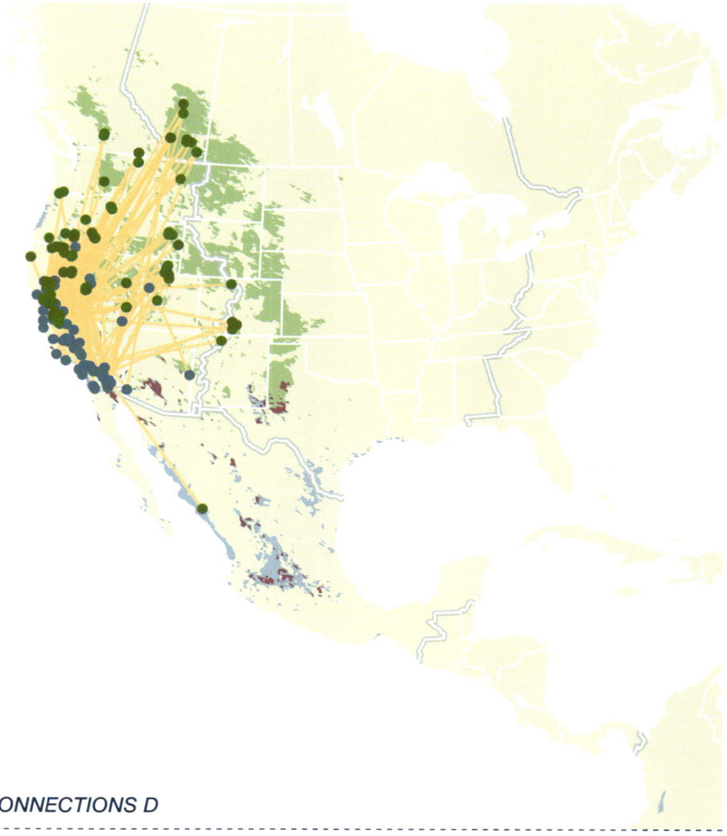

CONNECTIONS D

RADIO TRACKING A few tags were deployed through the Motus Wildlife Tracking System; there were no breeding to nonbreeding detections.
SATELLITE TRACKING Breeding Cinnamon Teal (n = 223) were tracked from breeding areas across the western United States (Mackell et al. 2021). Some birds were resident in California's Central Valley (n = 43), while migratory movements of other individuals were primarily to nonbreeding areas in northern and central Mexico (US Geological Survey, unpublished data).
OTHER APPROACHES No stable isotope or genetic studies of migratory connectivity were found for this species.

DATA CONTRIBUTORS
M. Casazza; D. Mackell; D. Olson; C. Overton.

NORTHERN SHOVELER *Spatula clypeata*

weak strong

MC = 0.18 ± 0.04*

DISTRIBUTION AND ABUNDANCE

Northern Shovelers have a circumpolar distribution, but in North America they breed primarily in the Prairie Pothole Region. Nonbreeding concentrations occur in the Central Valley of California, the Great Salt Lake, the Gulf Coast, and interior Mexico (Dubowy et al. 2020).

ORDER Anseriformes
FAMILY Anatidae
BREEDING HABITAT Wetlands, Seasonally Wet Grasslands
NONBREEDING HABITAT Wetlands, Generalist
GLOBAL POPULATION SIZE 5,900,000
POPULATION CHANGE SINCE 1970 +188%
IUCN RED LIST CATEGORY Least Concern
R2R TIPPING POINT SPECIES No
US LISTING STATUS No
CANADA SPECIES AT RISK No
PIF CCS 7/20

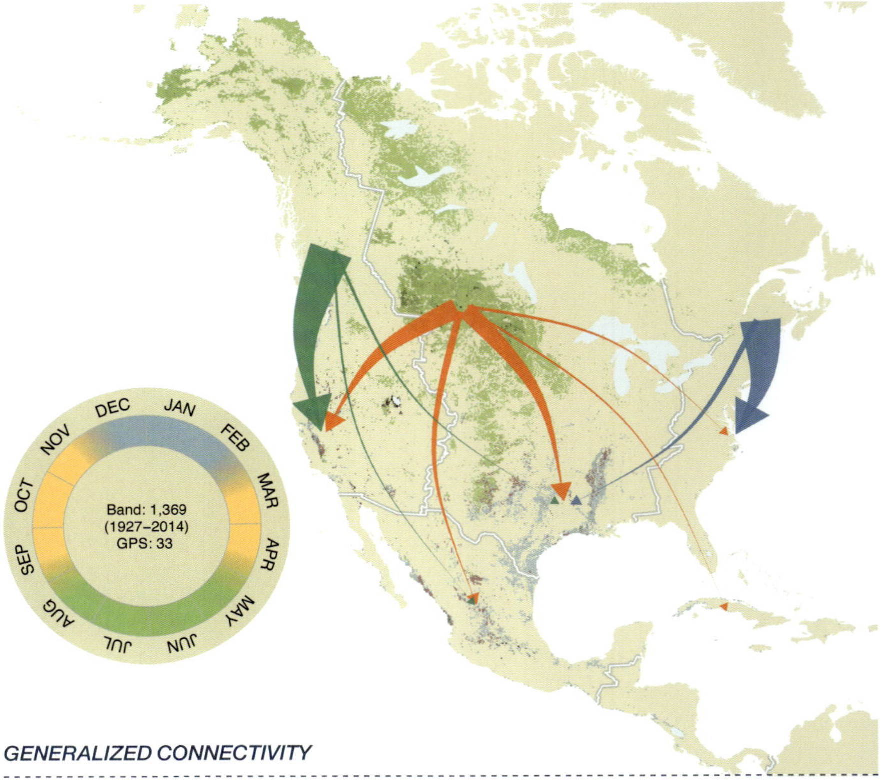

GENERALIZED CONNECTIVITY

	Atlantic	Caribbean	Central America	Mexico	Midcontinent	Pacific
Atlantic (n = 57)	91%	–	–	–	9%	–
Midcontinent (n = 816)	3%	1%	<1%	13%	38%	44%
Pacific (n = 527)	–	–	–	1%	2%	97%

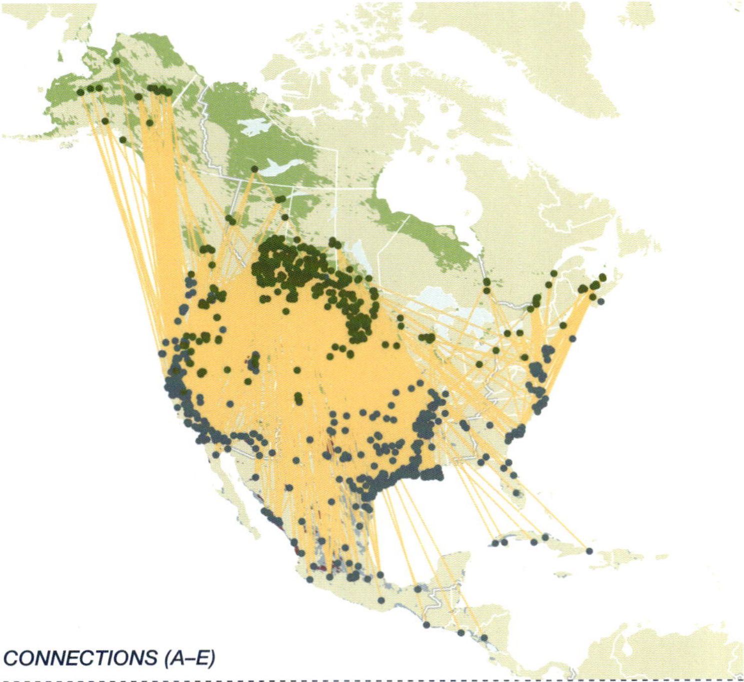

CONNECTIONS (A–E)

PATHWAYS

MIGRATORY CONNECTIVITY

OVERALL Migratory connectivity is known primarily from banding data and limited tracking studies. Most birds breeding in the Atlantic and Pacific flyways migrate within the flyway. However, Midcontinent Flyway breeding birds spread out across nonbreeding areas within the flyway (38%), as well as to the Pacific Flyway (44%) or Mexico (13%), resulting in weak migratory connectivity (MC = 0.18 ± 0.04).

BANDING Breeding Northern Shovelers were recovered at primary staging and nonbreeding areas, including the Great Salt Lake, the Gulf of Mexico, and the Playa Lakes region (Dubowy et al. 2020).

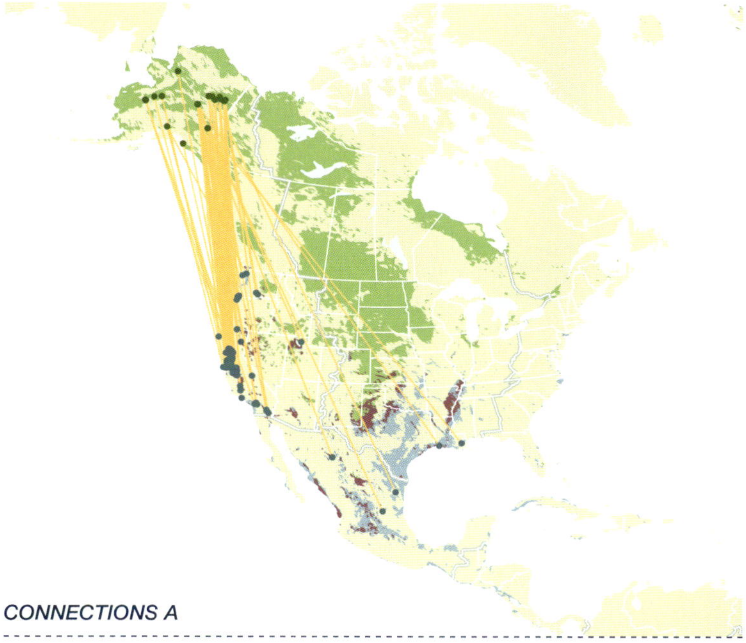

CONNECTIONS A

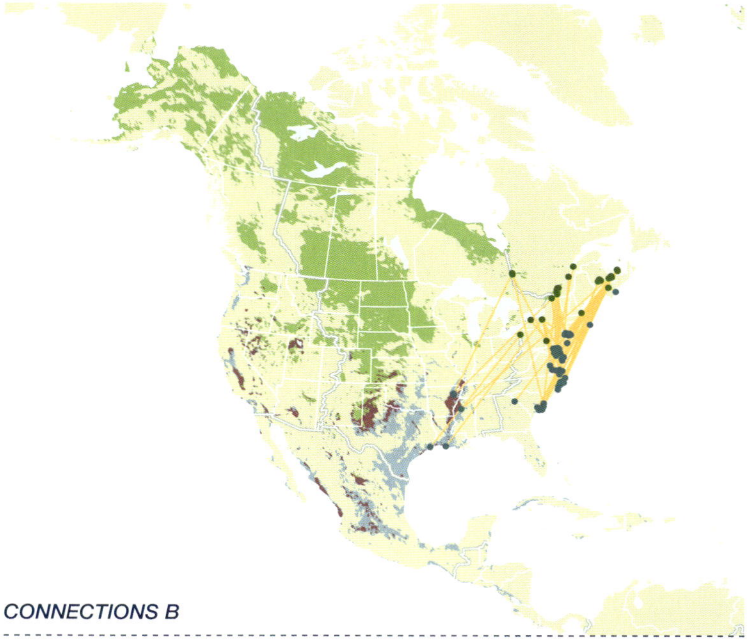

CONNECTIONS B

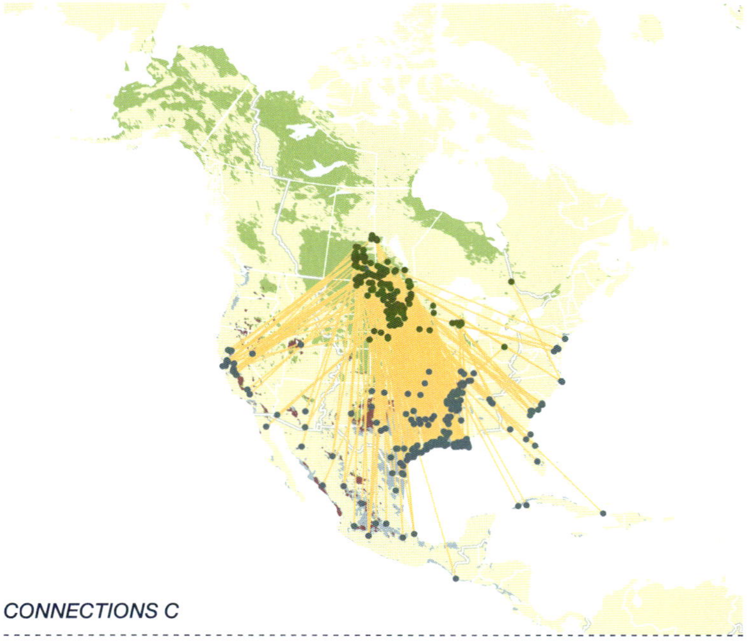

CONNECTIONS C

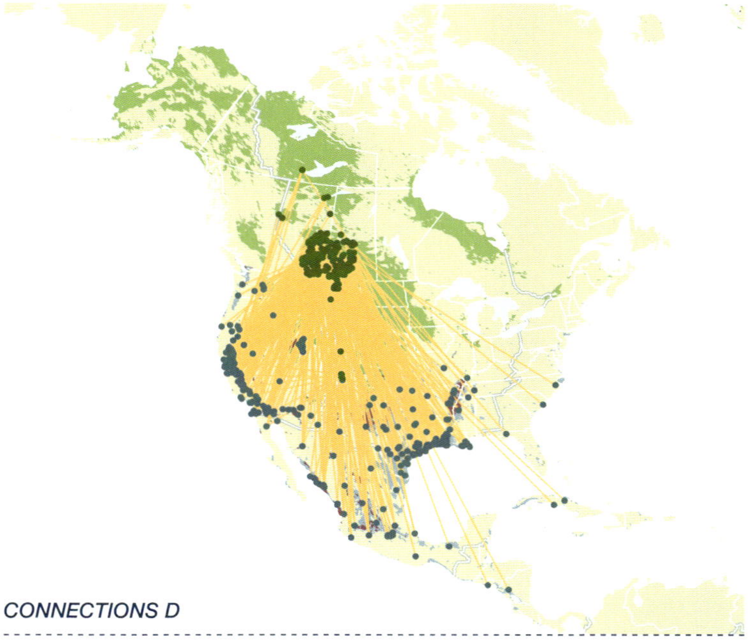

CONNECTIONS D

RADIO TRACKING A few tags were deployed through the Motus Wildlife Tracking System; there were no breeding to nonbreeding detections.

SATELLITE TRACKING Initial results from Northern Shovelers marked during the nonbreeding season in California's Central Valley show migration to breeding areas in the Prairie Pothole Region, Yukon Territories, and Alaska (US Geological Survey, unpublished data).

STABLE ISOTOPES Nonbreeding Northern Shovelers on the Great Salt Lake, Utah, were assigned to breeding locations in the western United States and Canada using stable hydrogen isotopes (Roberts and Conover 2015).

GENETICS No genetic studies of migratory connectivity were found for this species

MIGRATORY PATHWAYS Migration occurs through corridors along the Mississippi and Missouri rivers and through the Great Plains (Baldassarre 2014).

DATA CONTRIBUTORS

M. Casazza; C. Feldheim; C. Overton.

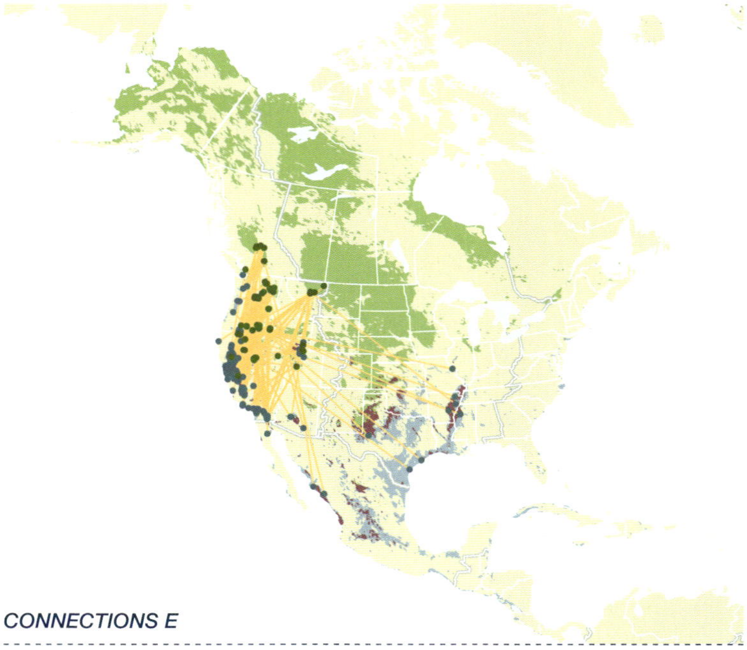

CONNECTIONS E

GADWALL *Mareca strepera*

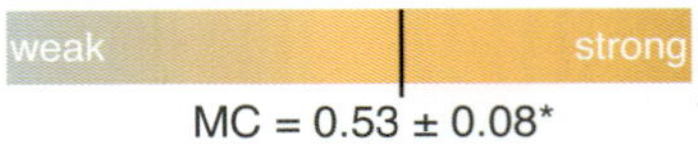

DISTRIBUTION AND ABUNDANCE

In North America, breeding is concentrated in the Prairie Pothole Region of central Canada to the north-central United States, though also found in western states. The nonbreeding range spans the southern United States, particularly the Mississippi Alluvial Valley, the southern Great Plains, and along the Gulf of Mexico (Leschack et al. 2020).

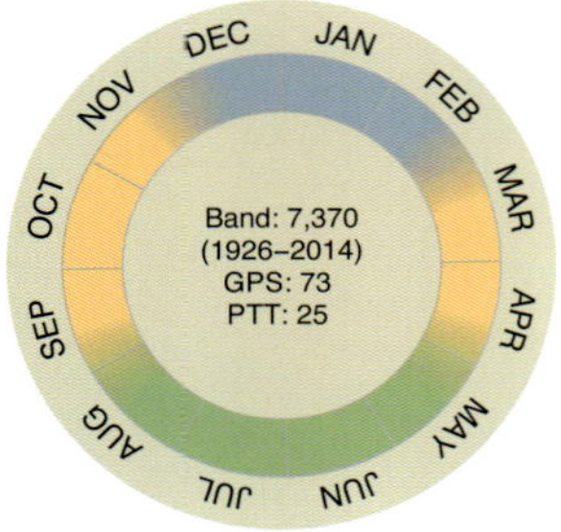

ORDER Anseriformes
FAMILY Anatidae
BREEDING HABITAT Wetlands, Seasonally Wet Grasslands
NONBREEDING HABITAT Wetlands, Generalist
GLOBAL POPULATION SIZE 4,400,000
POPULATION CHANGE SINCE 1970 +204%
IUCN RED LIST CATEGORY Least Concern
R2R TIPPING POINT SPECIES No
US LISTING STATUS No
CANADA SPECIES AT RISK No
PIF CCS 8/20

	Atlantic	Caribbean	Mexico	Midcontinent	Pacific
Atlantic (n = 194)	91%	79%	–	9%	–
Midcontinent (n = 2999)	4%	<1%	5%	87%	3%
Pacific (n = 4263)	<1%	–	2%	3%	95%

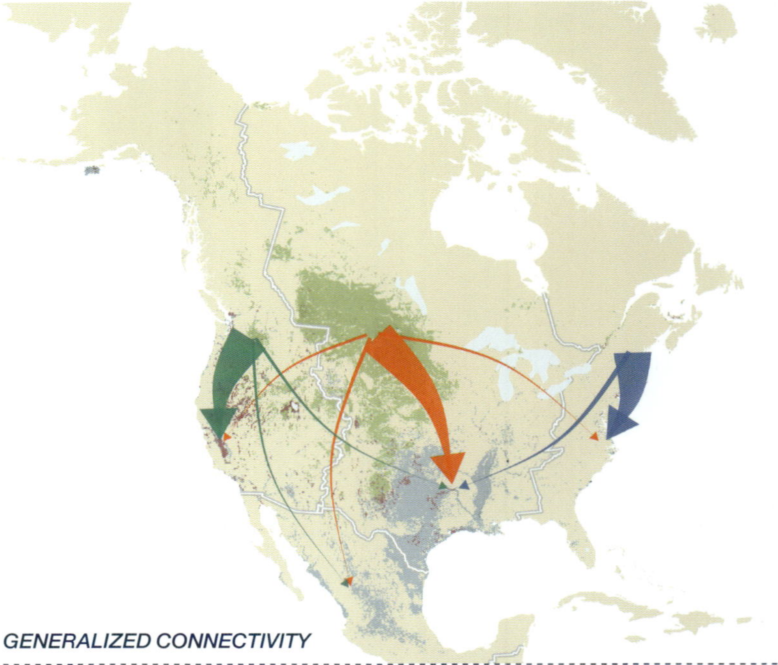

GENERALIZED CONNECTIVITY

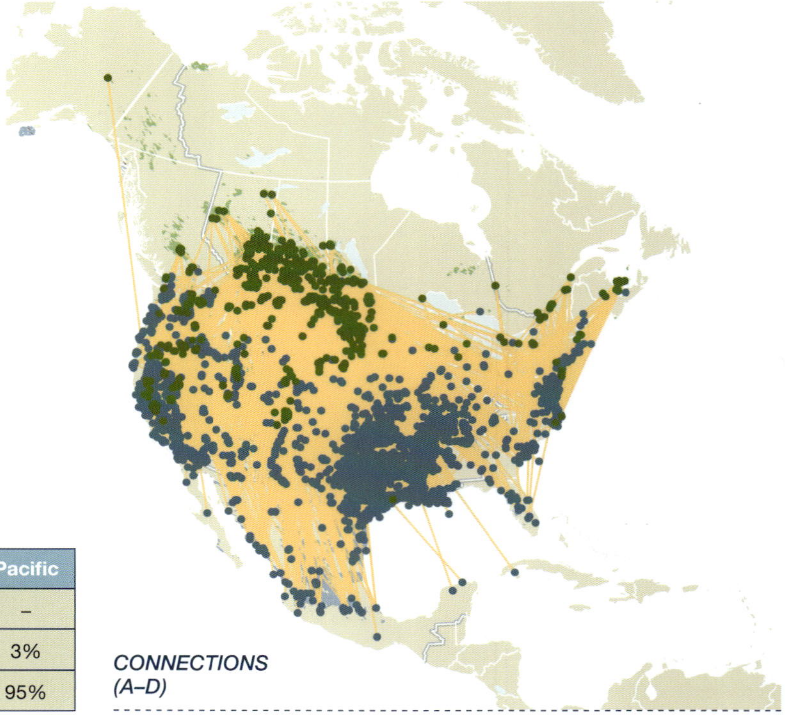

CONNECTIONS (A–D)

MIGRATORY CONNECTIVITY

OVERALL Migratory connectivity is inferred from banding data and limited satellite transmitter and genetics information, and appears strong within flyways. Transition probability within the Atlantic, Midcontinent, and Pacific flyways is near 90%. Transition probabilities from the Midcontinent, where nearly all Gadwall breed, to the Atlantic Flyway, Pacific Flyway, and Mexico are each about 5%. Resulting MC is one of the strongest of all waterfowl (MC = 0.53 ± 0.08).

BANDING Banding recoveries are extensive, and Gadwall banded on breeding areas in central and eastern North America have been recovered across the species' nonbreeding range (Baldassarre 2014). Males marked during breeding in Manitoba had a high return rate in subsequent years (Blohm 1978).

RADIO TRACKING Few tags were deployed through the Motus Wildlife Tracking System with local detections only.

SATELLITE TRACKING Female Gadwall (n = 25) were tracked with satellite transmitters from nonbreeding sites along the Gulf Coast of Louisiana and Texas to breeding areas throughout the Prairie Pothole Region (Gray 2010). A few individuals have been tracked by GPS-GSM transmitters in California's Central Valley and are shown to be residents (US Geological Survey, unpublished data), though molt migration does occur to the Klamath and Great basins (Kohl et al. 2022). A single Gadwall captured in southern Colorado migrated to the playa lakes of Texas (Setash et al., unpublished data).

STABLE ISOTOPES No stable isotope studies of migratory connectivity were found for this species.

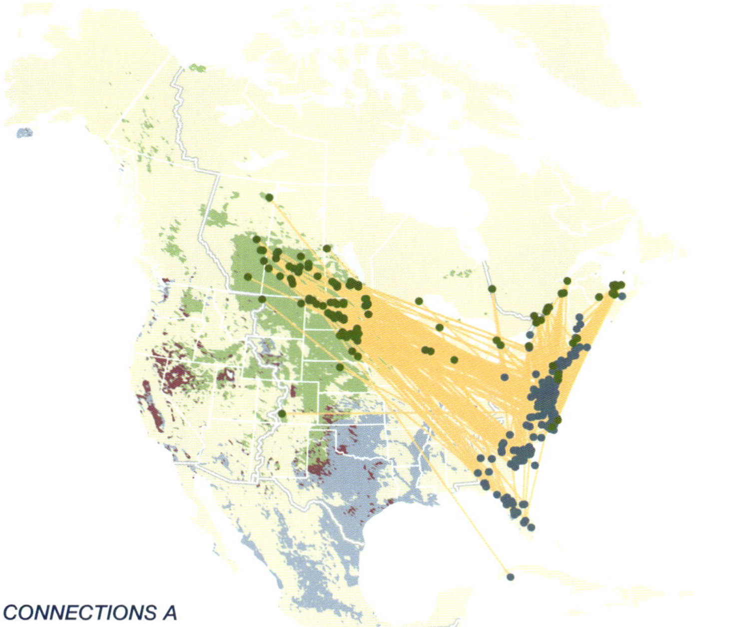

CONNECTIONS A

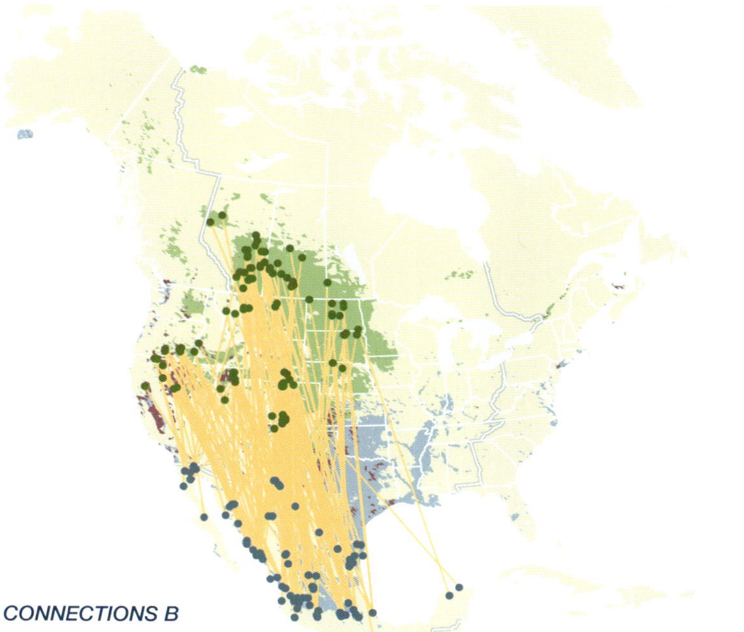

CONNECTIONS B

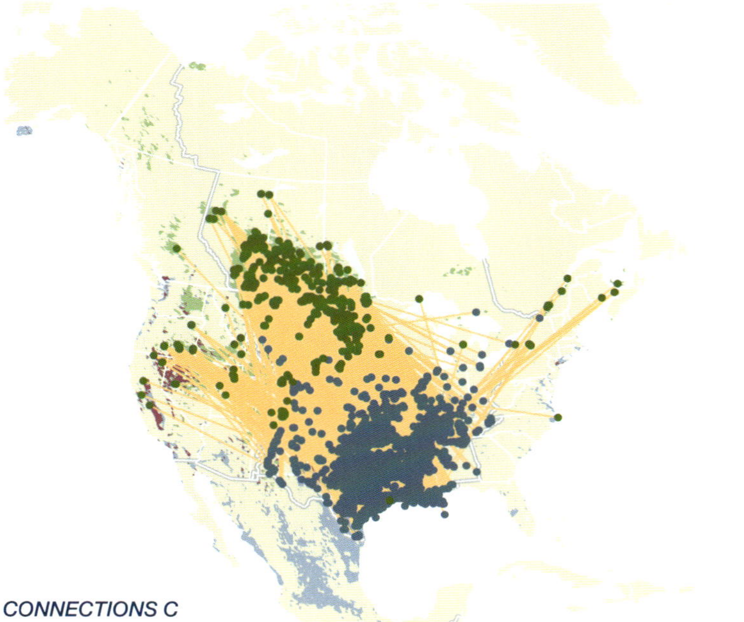

CONNECTIONS C

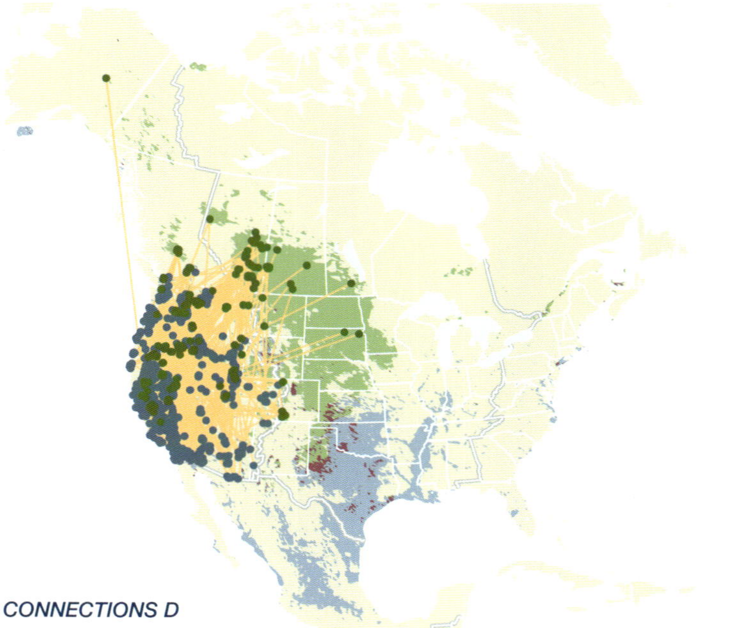

CONNECTIONS D

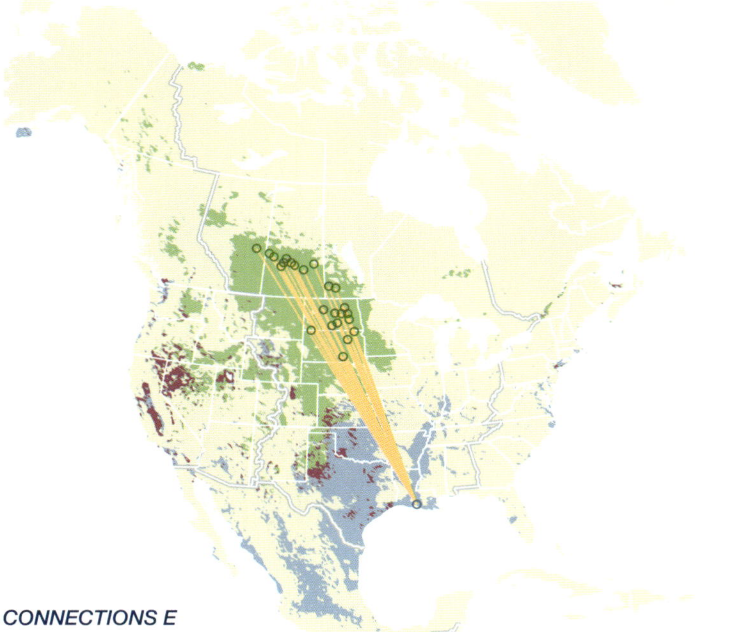

CONNECTIONS E

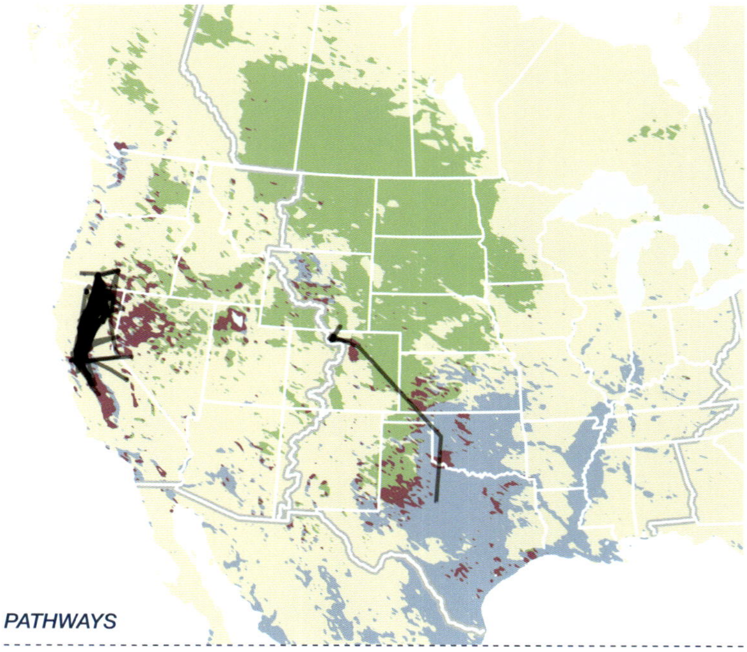

PATHWAYS

GENETICS Peters and Omland (2007), using mtDNa, found no population structure in North America, though there was some evidence of differentiation between Gadwall sampled on breeding areas in Alaska and Washington versus Gadwall on the remainder of the breeding areas.

DATA CONTRIBUTORS

A. Behney; M. Casazza; C. Feldheim; J.H. Gammonley; J. Gray; D.N. Koons; C. Overton; M. Reddy; C. Setash.

MIGRATION PATHWAYS Birds followed major river systems or moved between areas with high densities of wetlands, which is similar to other dabbling ducks.

INTERESTING NOTE

Eastern expansion of breeding Gadwall in North America during the early 1900s may have been related to increased availability of wetland breeding habitats (Henny and Holgerson 1974).

AMERICAN WIGEON *Mareca americana*

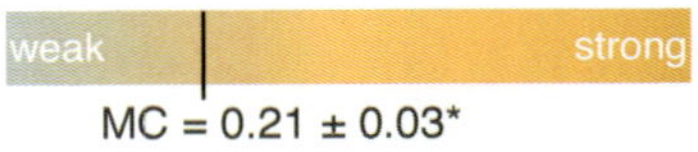

DISTRIBUTION AND ABUNDANCE

American Wigeon breed throughout the Prairie Pothole Region north through the boreal and into Alaska. Birds migrate as far south as Panama in the nonbreeding season, but concentrations occur on lakes in Texas and New Mexico and along the Pacific coast from Washington to Mexico (Fink et al. 2023).

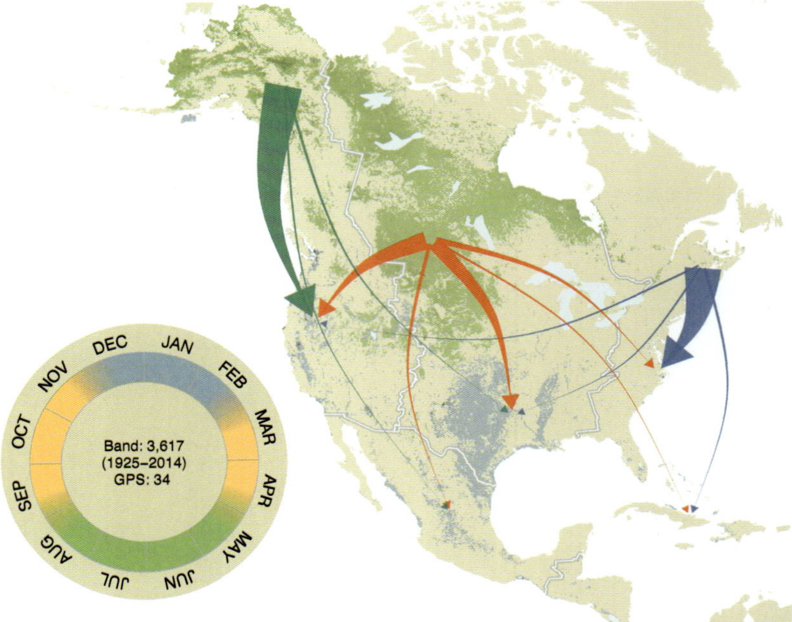

GENERALIZED CONNECTIVITY

ORDER Anseriformes
FAMILY Anatidae
BREEDING HABITAT Wetlands, Seasonally Wet Grasslands
NONBREEDING HABITAT Wetlands, Generalist
GLOBAL POPULATION SIZE 2,700,000
POPULATION CHANGE SINCE 1970 −10%
IUCN RED LIST CATEGORY Least Concern
R2R TIPPING POINT SPECIES No
US LISTING STATUS No
CANADA SPECIES AT RISK No
PIF CCS 10/20

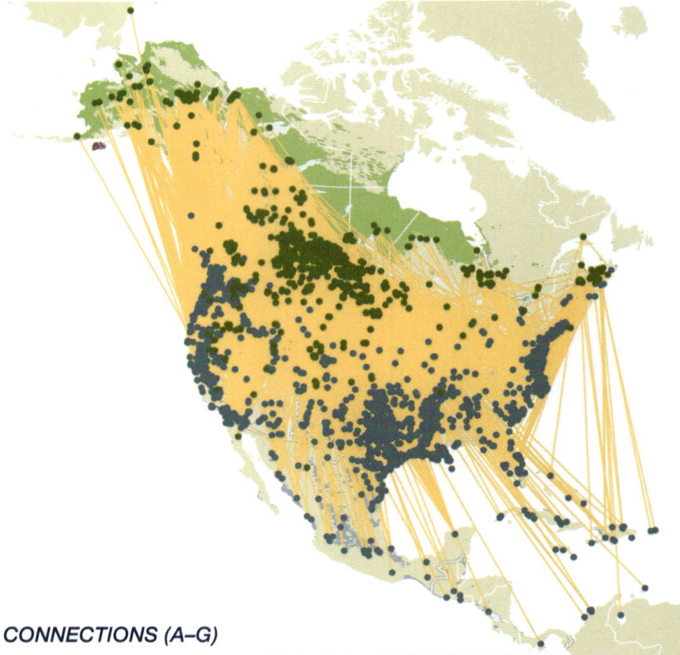

CONNECTIONS (A–G)

MIGRATORY CONNECTIVITY

OVERALL Migratory connectivity is known from banding and tracking data; however, tracking data are limited. American Wigeon that were breeding in the two coastal flyways mostly stayed within the same flyway, Atlantic (95%) and Pacific (96%), during the nonbreeding season. Birds breeding in the Midcontinent Flyway moved within the same flyway (37%), but a large percentage migrated to the Pacific Flyway (45%). Despite high transition probabilities within the coastal flyways, movement from breeding areas within the Midcontinent Flyway to all nonbreeding regions leads to moderate migratory connectivity (MC = 0.21 ± 0.03).

BANDING Band returns suggest birds use all four North American flyways during migration (Mini et al. 2020). Of 800 American Wigeon banded during breeding in western Canada, the most common recovery location was California (36%). The remainder were recovered across the species' nonbreeding range in the United States (Baldassarre 2014).

	Atlantic	Caribbean	Central America	Mexico	Midcontinent	Northern SA	Pacific
Atlantic (n = 335)	95%	3%	–	–	1%	<1%	1%
Midcontinent (n = 2253)	11%	1%	<1%	5%	37%	<1%	45%
Pacific (n = 1007)	<1%	–	–	1%	3%	<1%	96%

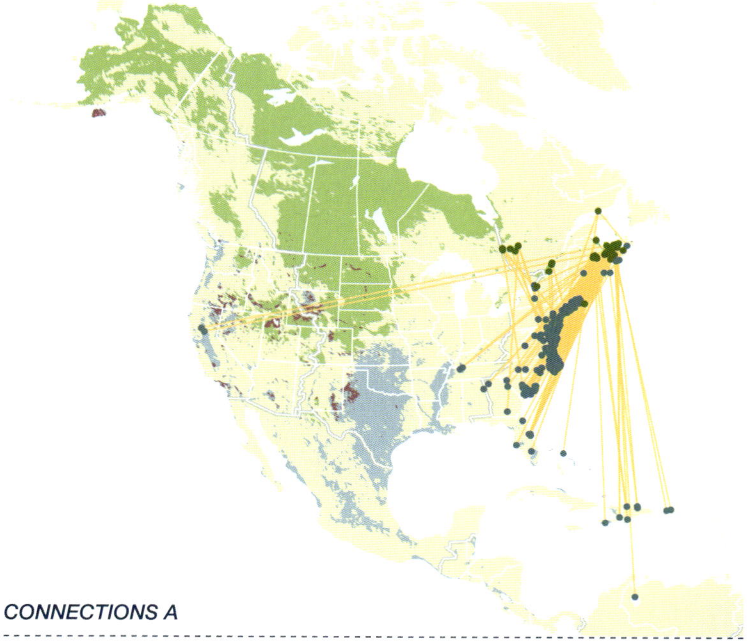

CONNECTIONS A

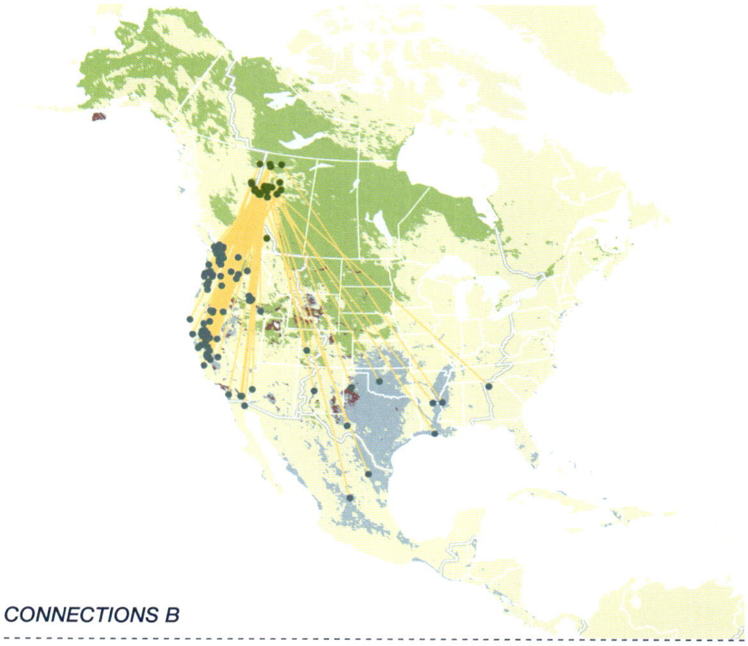

CONNECTIONS B

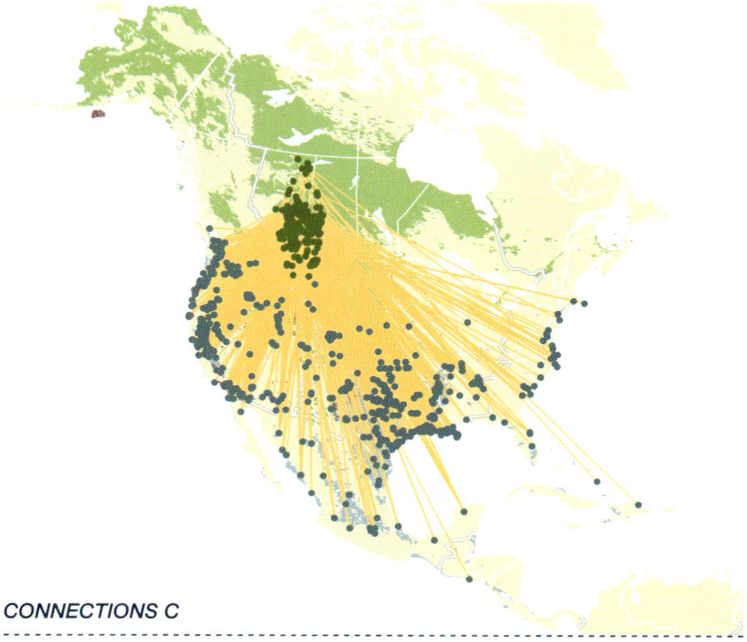

CONNECTIONS C

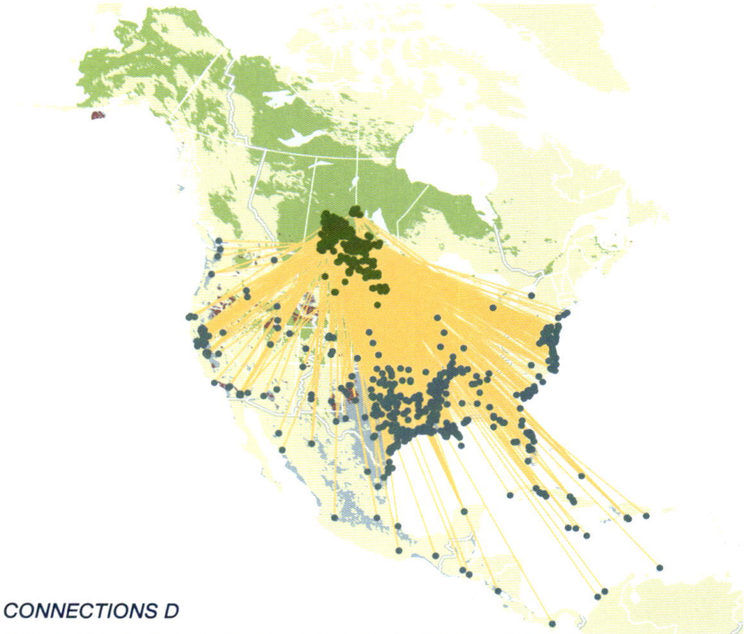

CONNECTIONS D

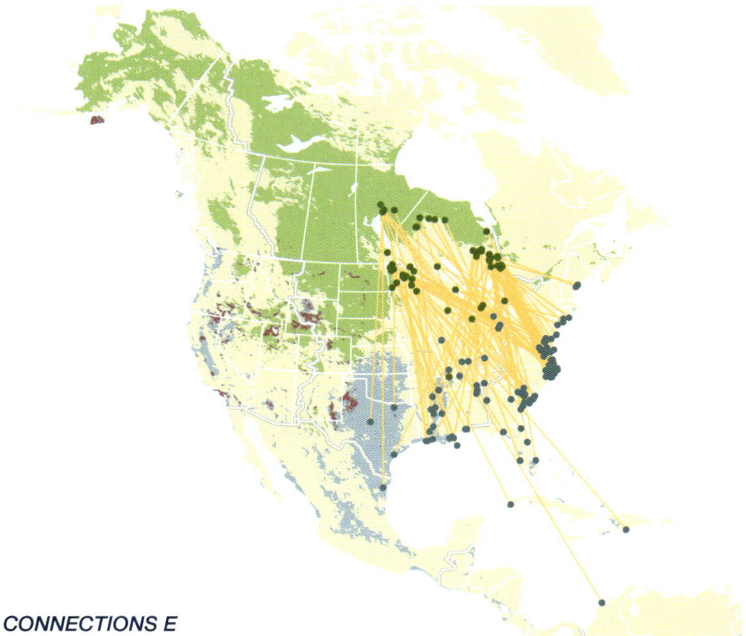

CONNECTIONS E

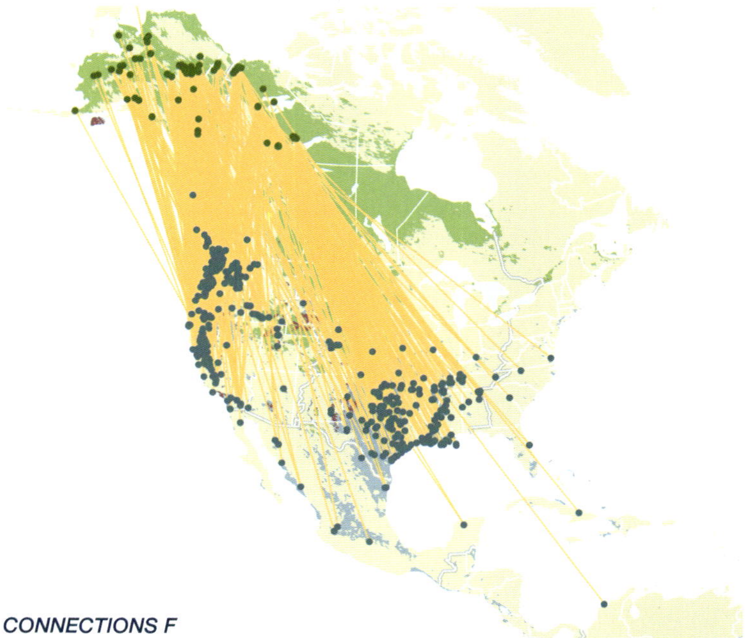

CONNECTIONS F

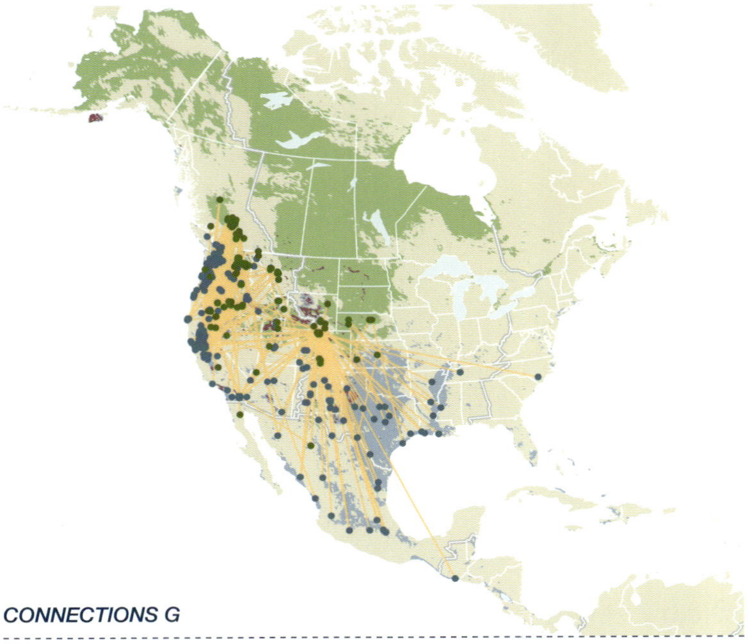

CONNECTIONS G

PATHWAYS

SATELLITE TRACKING Birds marked during the nonbreeding season in the Central Valley of California migrated to breeding areas in Alaska, the western boreal forest, and the Northwest Territories (US Geological Survey, unpublished data). Some birds subsequently spent the nonbreeding season in the Central Flyway.
STABLE ISOTOPES Most (85%) American Wigeon sampled in central Saskatchewan during post-breeding migration originated in northern tundra breeding areas (Asante et al. 2017).
GENETICS Genetic variation of American Wigeon collected in the Texas panhandle (n = 629) showed temporal changes through the nonbreeding season (Rhodes et al. 1993). This suggests there is some genetic differentiation among breeding populations due to differential arrival to nonbreeding areas.
MIGRATORY PATHWAYS American Wigeon are most abundant in western North America during breeding and nonbreeding periods, making a general north-south migration most likely. They are one of the more northerly breeding dabbling ducks; birds breeding across western North America migrate to the Pacific coast through interior wetlands, while more central breeding birds migrate south to nonbreeding areas in the southern Great Plains and Mississippi Alluvial Valley, jumping between major wetlands or following river systems (Baldasarre 2014).

DATA CONTRIBUTORS
M. Casazza; C. Overton; C. Feldheim.

MALLARD *Anas platyrhynchos*

MC = 0.42 ± 0.01*

DISTRIBUTION AND ABUNDANCE

Mallards are widespread and found year-round throughout the continental United States, extending into Mexico in the nonbreeding season and breeding throughout Canada and Alaska (Drilling et al. 2020). They are a partial to medium-distance migrant. Highest breeding concentrations are found in the Prairie Pothole Region, and nonbreeding concentrations in the southeast and central United States.

ORDER Anseriformes
FAMILY Anatidae
BREEDING HABITAT Wetlands, Generalist
NONBREEDING HABITAT Wetlands, Generalist
GLOBAL POPULATION SIZE 19,000,000
POPULATION CHANGE SINCE 1970 −37%
IUCN RED LIST CATEGORY Least Concern
R2R TIPPING POINT SPECIES No
US LISTING STATUS No
CANADA SPECIES AT RISK No
PIF CCS 7/20

MIGRATORY CONNECTIVITY

OVERALL More bands have been put on Mallards than any other bird species in North America. In addition, satellite transmitter, stable isotope, and genetic data are available throughout its range. Mixing occurs across the range, but transition rates are high within the three flyways, Atlantic (75%), Midcontinent (75%), and Pacific (98%). Movement from the Midcontinent to the Pacific flyway (20%) and from the Atlantic to the Midcontinent flyway (25%) is also common. Migratory connectivity is relatively strong for Mallards (MC = 0.42 ± 0.01).

INTERESTING NOTE

Analysis of banding data in the mid-1900s identified four separate flyways used by Mallards. These became the basis of the administrative flyways used today to manage migratory birds (Lincoln 1935).

	Atlantic	Mexico	Midcontinent	Pacific
Atlantic (n = 745)	75%	–	25%	<1%
Midcontinent (n = 9287)	5%	<1%	75%	20%
Pacific (n = 1818)	–	<1%	2%	98%

BANDING Banding data suggest three distinct breeding populations in the Pacific, Midcontinent, and Atlantic flyways of North America (Buhnerkempe et al. 2016).
RADIO TRACKING Few tags were deployed through the Motus Wildlife Tracking System with local detections only.
SATELLITE TRACKING Satellite transmitters (n = 143) tracked pre-breeding migration from Arkansas to breeding sites in the Prairie Pothole Region and areas north and east (Krementz et al. 2011). Mallards from the same study tracked through the following post-breeding migration (n = 55) spent the nonbreeding season in Missouri, Arkansas, and Canada (Krementz et al. 2012). Mallards tracked from nonbreeding areas along the Gulf Coast in Louisiana and Arkansas migrated to breeding sites in the Prairie Pothole Region (Masto et al. 2022; P. Link, unpublished data; D. Osborne, unpublished data), while west coast nonbreeding birds in the Central Valley of California migrated to breeding areas in the prairies of Alberta or were residents (US Geological Survey, unpublished data).
STABLE ISOTOPES Stable isotope analyses were able to assign natal origin of nonbreeding birds to breeding areas largely within the same nonbreeding flyway, supporting strong migratory connectivity within the Pacific and Midcontinent flyways (Hebert and Wassenaar 2005a; Coulton et al. 2010). Mallards using the Saskatchewan River Delta as a stopover site were assigned to breeding areas in eastern Yukon and Nunavut (n = 48) and the Prairie Pothole Region (n = 19; Asante et al. 2017). Natal origins of Mallards harvested in the Atlantic Flyway were primarily southern Canada and the northeast United States (Kucia et al. 2023).
GENETICS Two mitochondrial DNA clades exist, differentiating populations in Eurasia and North America, with little population differentiation among flyways within the respective continents (Kraus et al. 2011). Mallards harvested in the eastern United States in the last century seem to be descended from Eurasian birds, likely from captive releases, in contrast to museum specimens collected in the prior century showing that eastern United States Mallards were genetically similar to others in North American (Lavretsky et al. 2020).
MIGRATORY PATHWAYS Some populations are resident, and migratory populations generally follow distinct north-south routes. The largest abundance of Mallards use a migration path from breeding areas in the south-central Canadian Prairies, southeast, then south along the Mississippi River Valley.

DATA CONTRIBUTORS

A. Behney; M. Casazza; C. Feldheim; J. Gammonley; D. Koons; P. Link; D. Osborne; C. Overton; M. Reddy; C. Setash.

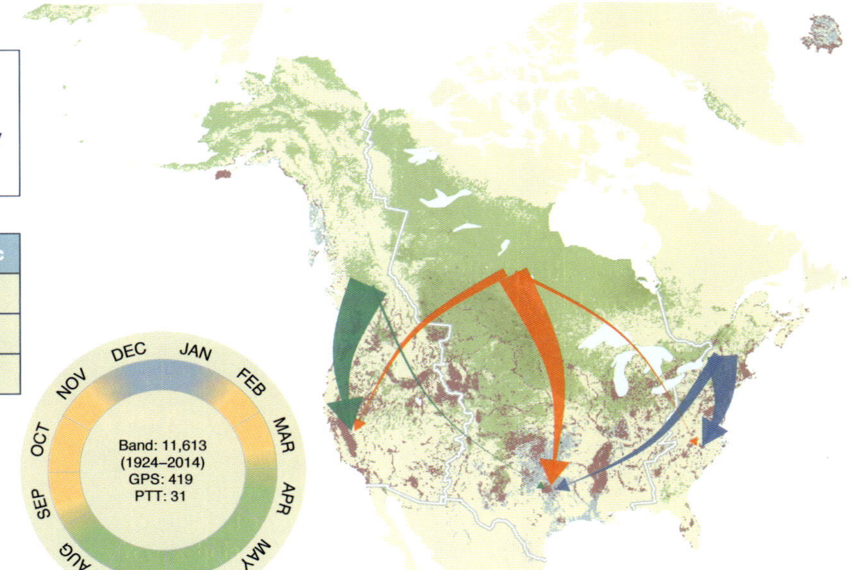

GENERALIZED CONNECTIVITY

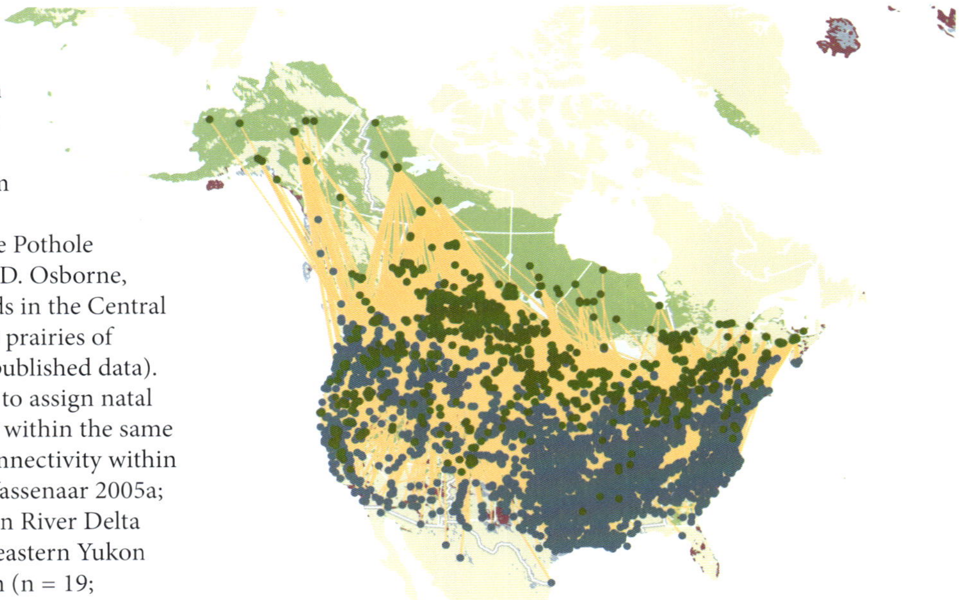

CONNECTIONS (A–I)

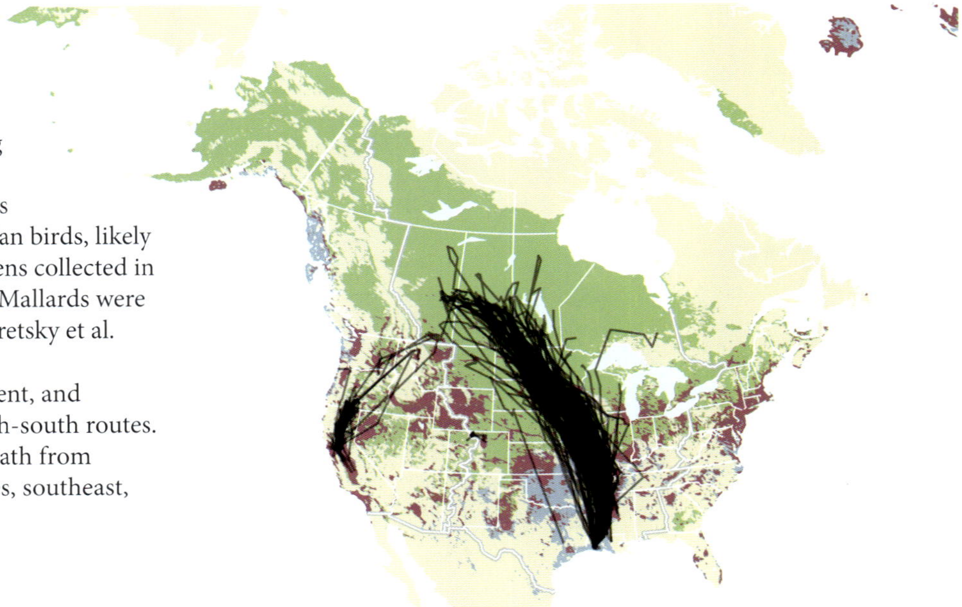

PATHWAYS

DUCKS, GEESE AND WATERFOWL (ANSERIFORMES)

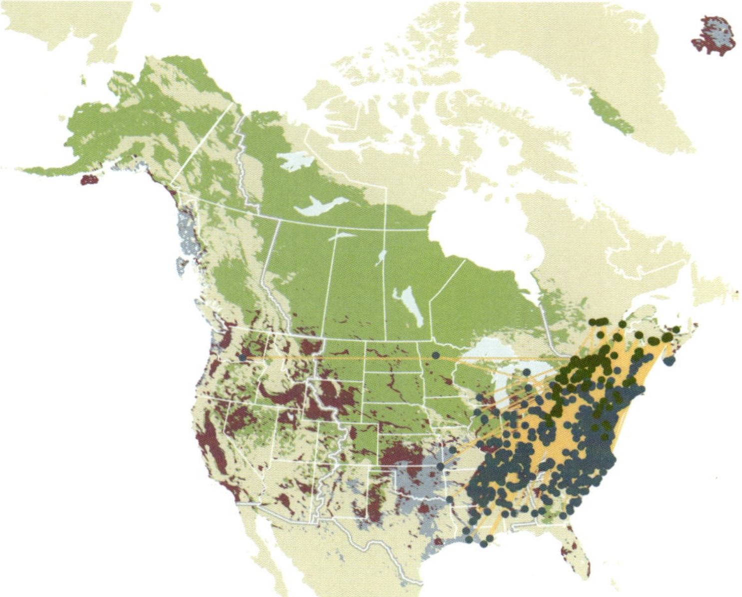

CONNECTIONS A

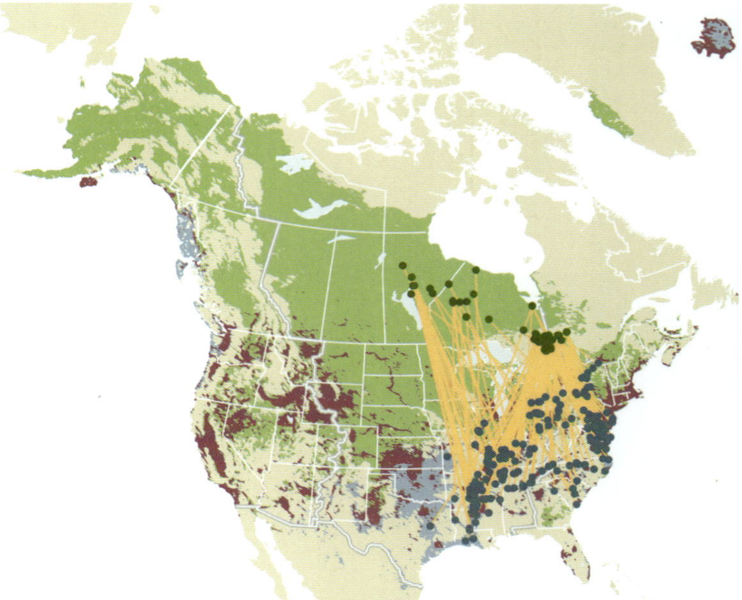

CONNECTIONS B

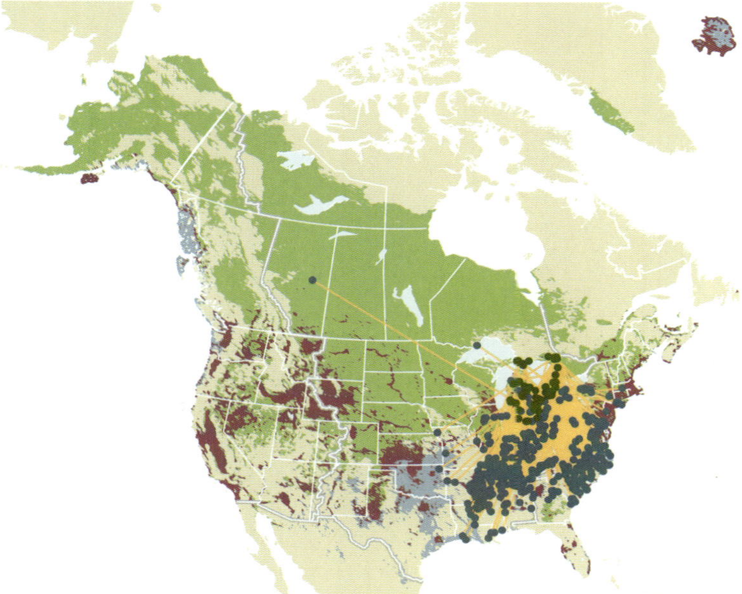

CONNECTIONS C

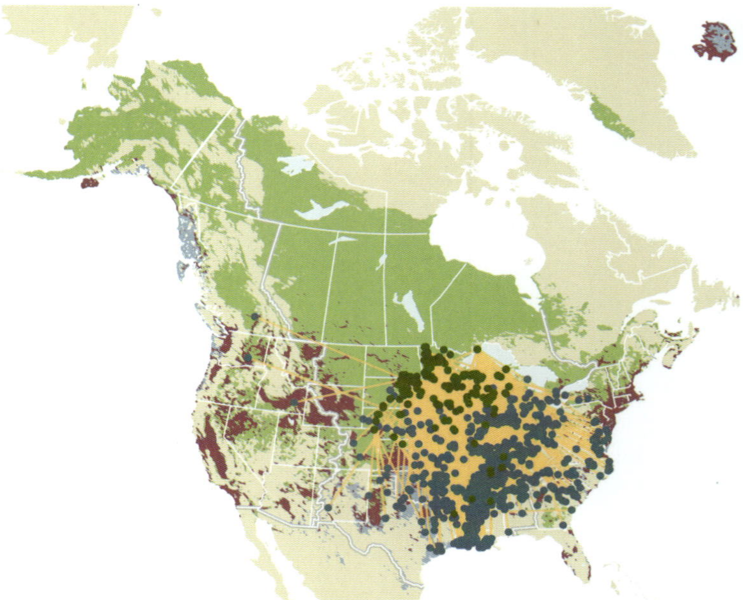

CONNECTIONS D

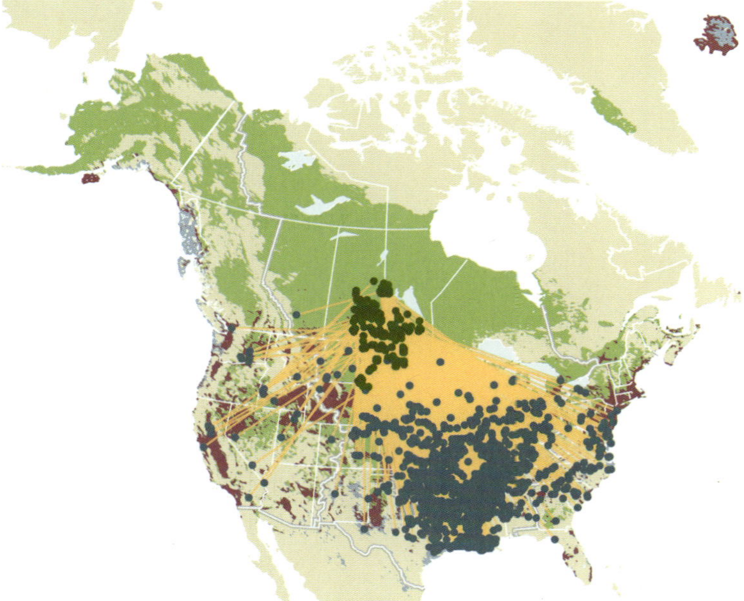

CONNECTIONS E

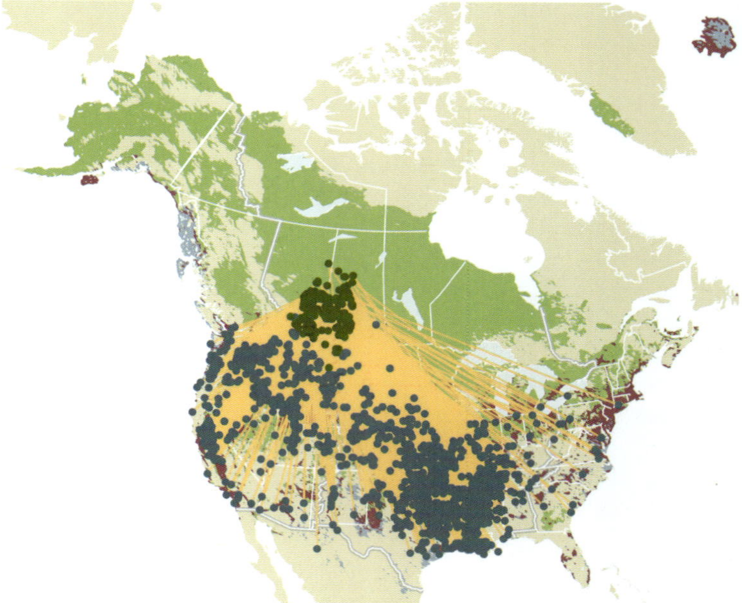

CONNECTIONS F

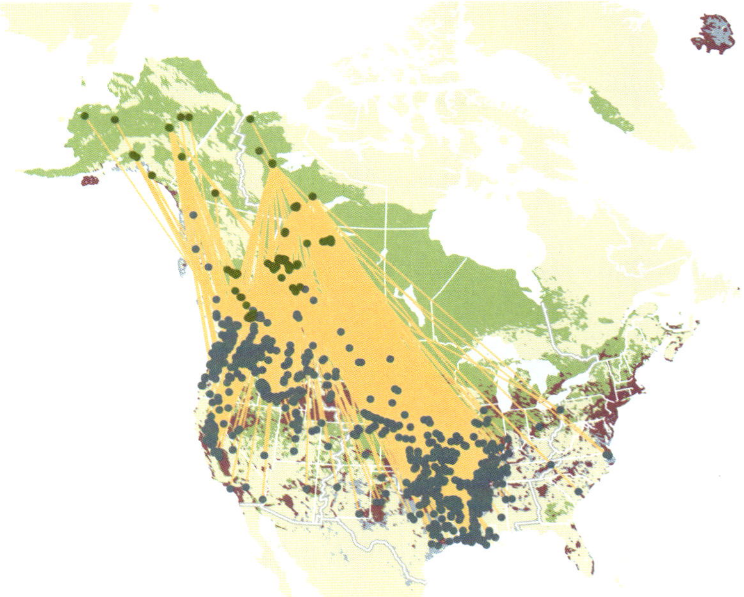

CONNECTIONS G

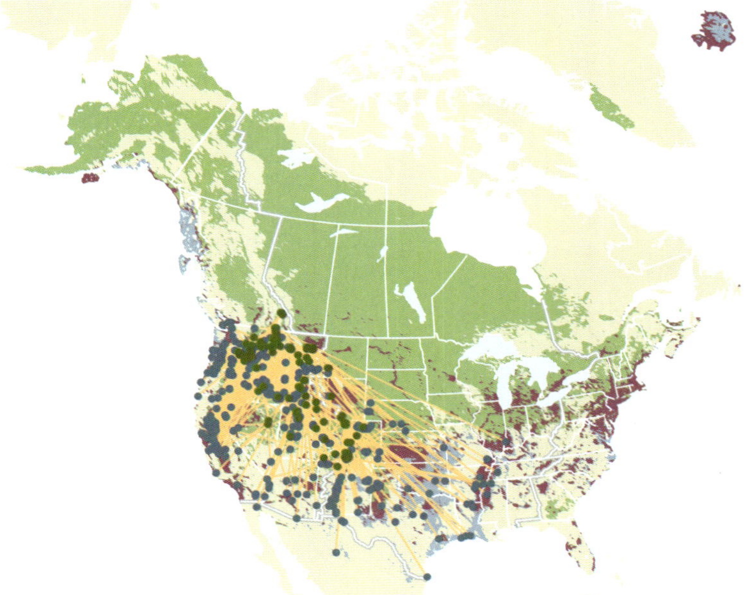

CONNECTIONS H

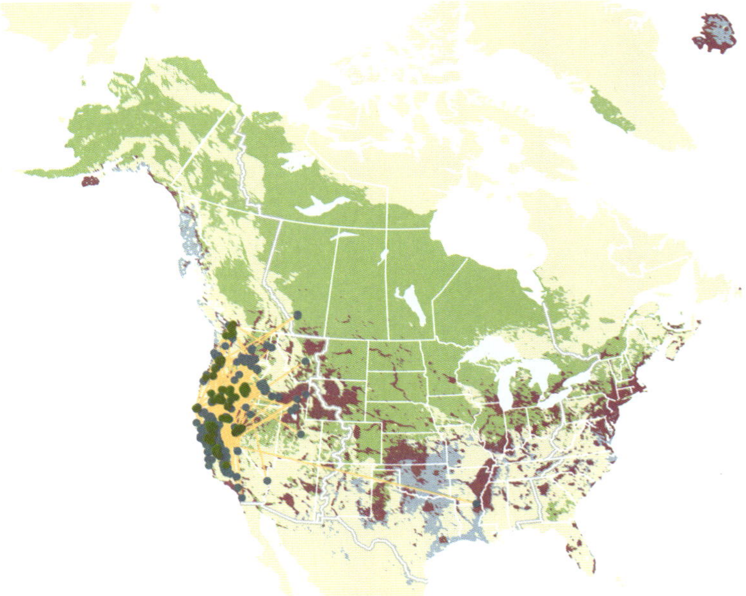

CONNECTIONS I

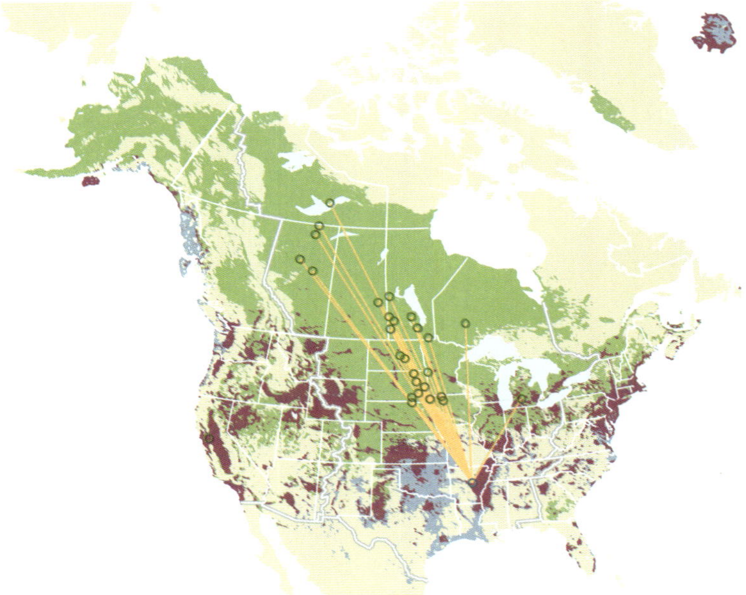

CONNECTIONS J

MC = 0.45 ± 0.02*

Anas rubripes **AMERICAN BLACK DUCK**

DISTRIBUTION AND ABUNDANCE

The American Black Duck breeds primarily in northeastern North America in the boreal forest, with breeding occurring south and east along the Atlantic coast of the United States. It spends the nonbreeding season along the Atlantic coast from Atlantic Canada south to North Carolina.

American Black Ducks experienced significant declines during the 1960s to the 1990s and have since stabilized (Longcore et al. 2020). Over several decades, populations have been increasing during the stationary nonbreeding season in Nova Scotia, consistent with the sense that there are more birds remaining farther north than in the past (Robertson et al. 2017).

ORDER Anseriformes
FAMILY Anatidae
BREEDING HABITAT Wetlands, Boreal
NONBREEDING HABITAT Wetlands, Lakes, and Rivers
GLOBAL POPULATION SIZE 700,000

POPULATION CHANGE SINCE 1970 −37%
IUCN RED LIST CATEGORY Least Concern
R2R TIPPING POINT SPECIES No
US LISTING STATUS No
CANADA SPECIES AT RISK No
PIF CCS 12/20

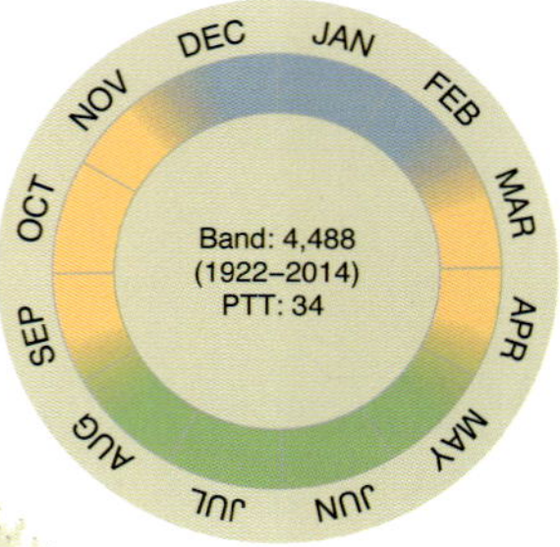

	Atlantic	Midcontinent
Atlantic (n = 3300)	75%	25%
Midcontinent (n = 1143)	5%	75%

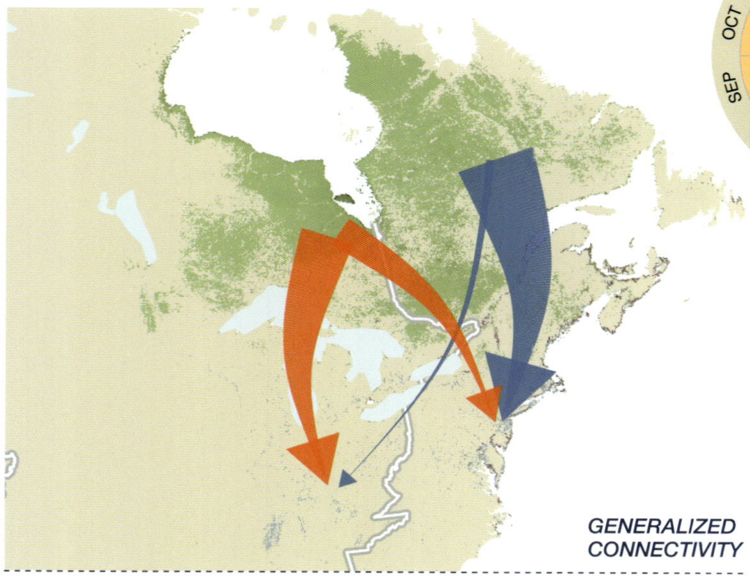
GENERALIZED CONNECTIVITY

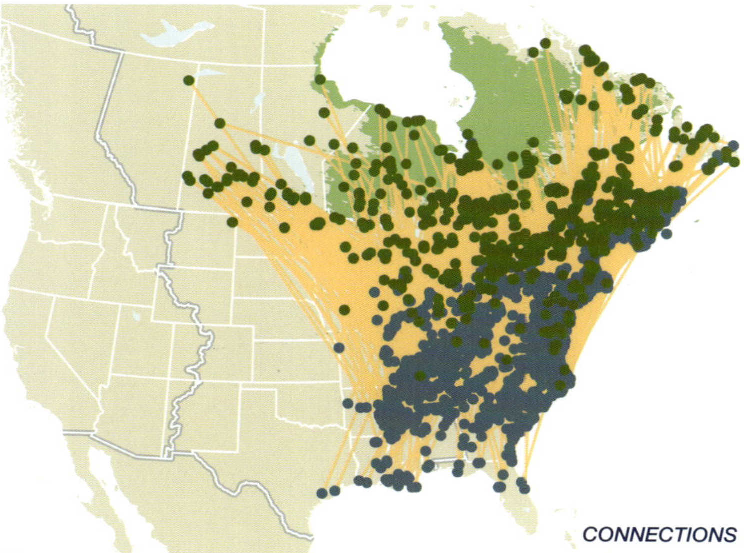
CONNECTIONS

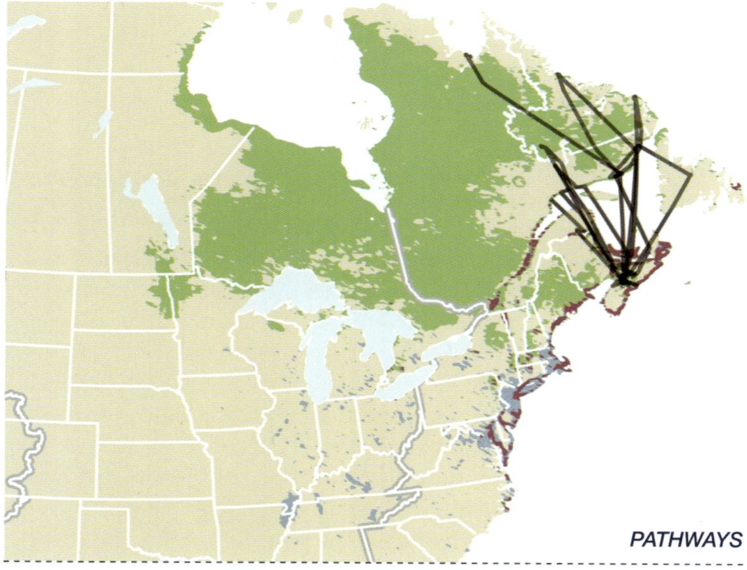
PATHWAYS

INTERESTING NOTE

In 1977, an American Black Duck wearing a US Geological Survey band was recovered at Kwang Ju Air Force Base in Korea (Banks 1985).

MIGRATORY CONNECTIVITY

OVERALL Migratory connectivity information is primarily from band recoveries and telemetry studies. Almost all birds breeding in the Atlantic Flyway remain in the Atlantic Flyway during the nonbreeding season (95%). Breeding birds in the Midcontinent either stay in that flyway (64%) or migrate to the Atlantic Flyway (36%) for the nonbreeding season. Some birds move only short distances, and other populations remain sedentary throughout the year (Longcore et al. 2020). Although there is abundant movement from the Midcontinent east, within-flyway connectivity is strong, resulting in strong overall migratory connectivity (MC 0.45 ± 0.02).

BANDING Banding data suggests flyway fidelity is strong (Robinson et al. 2016a). Recoveries show that American Black Ducks remain within a flyway during pre- and post-breeding migration, and a significant portion of the breeding population moves southeast to nonbreeding areas along the Atlantic coast (Robinson et al. 2016b). American Black Ducks breeding in the western portion of their range generally spend the nonbreeding season in the Midcontinent and southern Atlantic flyways. This is in comparison to birds breeding in the central and eastern portion of the range migrating to nonbreeding areas in the Atlantic Flyway (Zimpfer and Conroy 2006; Lavretsky et al. 2014). American Black Ducks do not appear to be facultative in their migratory behavior (Diefenbach et al. 1988). Nonbreeding birds in southern Lake Erie showed strong site fidelity between years (Robb 2002).

SATELLITE TRACKING Eastern nonbreeding American Black Ducks tracked with satellite transmitters migrated throughout the known breeding range, while birds caught farther west migrated to breeding areas farther west (n = 31; Anderson 2013; Coluccy et al. 2020). Nonbreeding birds in Atlantic Canada may not migrate across international boundaries, as seen in Nova Scotia (Peck et al. 2022).

STABLE ISOTOPES Isotopic analysis of 200 juvenile American Black Ducks showed birds harvested farther east were assigned to breeding areas farthest east (Ashley et al. 2010). Juveniles harvested in southern Quebec primarily originated from nonagricultural breeding areas, in contrast to the agricultural area where they were recovered (Ashley et al. 2010; Cousineau et al. 2014).

GENETICS No genetic studies of migratory connectivity were found for this species.

OTHER APPROACHES Marked females (20%–30%) returned to natal or former home ranges the following breeding season in Nova Scotia (Seymour 1991), New England (Coulter and Miller 1968), and Quebec (Reed 1970).

MIGRATORY PATHWAYS Weakly defined migration paths of American Black Ducks north to south funnel to separate nonbreeding areas (Bellrose and Crompton 1970). Generally, birds migrate southeast to the primary nonbreeding destination of the mid-Atlantic coast of the United States. Many migrant populations stop over in the Gulf of St. Lawrence, Georgian Bay, Long Island Sound, Saginaw Bay, coastal New Jersey, and Chesapeake Bay (Anderson 2013; Longcore et al. 2020).

DATA CONTRIBUTORS
K. Anderson; J. Coluccy; S. Craik; M. English; M.L. Mallory; L. Peck; G.J. Robertson.

NORTHERN PINTAIL

Anas acuta

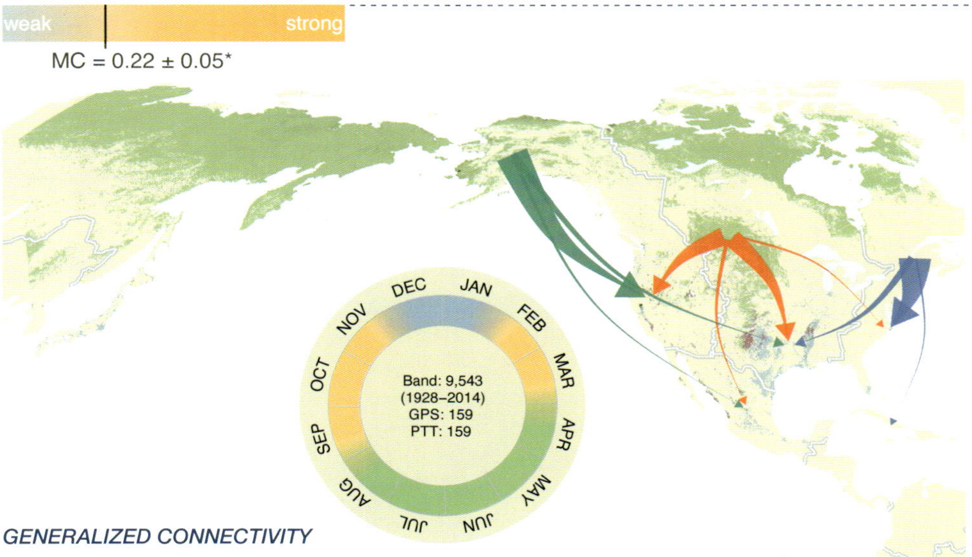

GENERALIZED CONNECTIVITY

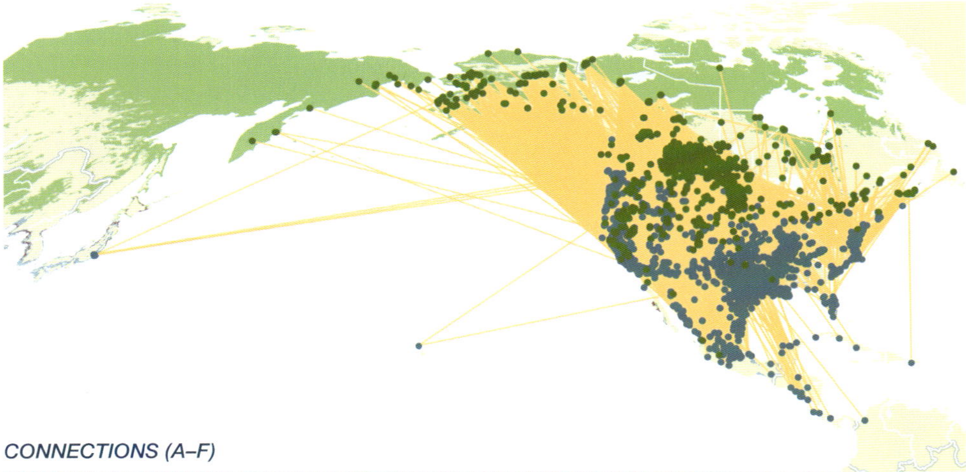

CONNECTIONS (A–F)

DISTRIBUTION AND ABUNDANCE

Northern Pintails have a circumpolar range with a breeding distribution concentrated in Alaska and the Prairie Pothole Region (Baldassarre 2014). Nonbreeding concentrations occur in California's Central Valley and along the Gulf Coast.

ORDER Anseriformes
FAMILY Anatidae
BREEDING HABITAT Wetlands, Seasonally Wet Grasslands
NONBREEDING HABITAT Wetlands, Generalist
GLOBAL POPULATION SIZE 5,100,000
POPULATION CHANGE SINCE 1970 −44%
IUCN RED LIST CATEGORY Least Concern
R2R TIPPING POINT SPECIES Yellow-Alert
US LISTING STATUS No
CANADA SPECIES AT RISK No
PIF CCS 13/20

PATHWAYS

MIGRATORY CONNECTIVITY

OVERALL Abundant banding data, in addition to information from satellite transmitters, stable isotopes, and genetics, have been used to describe Northern Pintail migratory connectivity. Birds are widely dispersed from breeding to nonbreeding areas, especially in the Midcontinent Flyway. Birds breeding in the Atlantic Flyway migrate within the flyway (84%) or transition primarily to the Midcontinent Flyway (15%). Midcontinent Flyway breeding birds mainly migrate within the flyway (46%) or to the Pacific Flyway (46%). Pacific Flyway breeding birds stay within the flyway (82%), or migrate to the Midcontinent Flyway (13%). Within coastal flyways, transitions are high, but mixing of Midcontinent Flyway breeding populations among nonbreeding areas results in moderate migratory connectivity (MC = 0.22 ± 0.05).

BANDING Band recoveries across the range of Northern Pintails showed three main north-south transitions, within the Pacific Flyway, within the Atlantic Flyway, and within the Midcontinent Flyway (Buhnerkempe et al. 2016). Individuals banded on the Yukon-Kuskokwim Delta in Alaska between 1990 and 2001 (n = 13,645) were recovered primarily in the Pacific Flyway, with most recoveries in California's Central Valley (Nicolai et al. 2005).

	Asia	Atlantic	Caribbean	Central America	Mexico	Midcontinent	Northern SA	Pacific
Atlantic (n = 74)	–	84%	1%	–	–	15%	–	–
Midcontinent (n = 6710)	<1%	2%	<1%	<1%	6%	46%	<1%	46%
Pacific (n = 2803)	<1%	<1%	–	–	5%	13%	–	82%

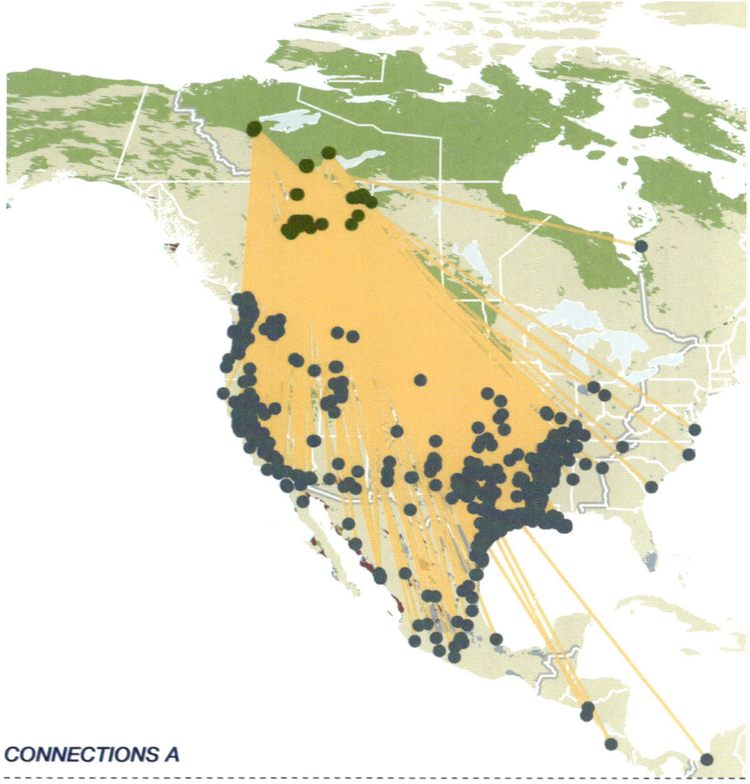

CONNECTIONS A

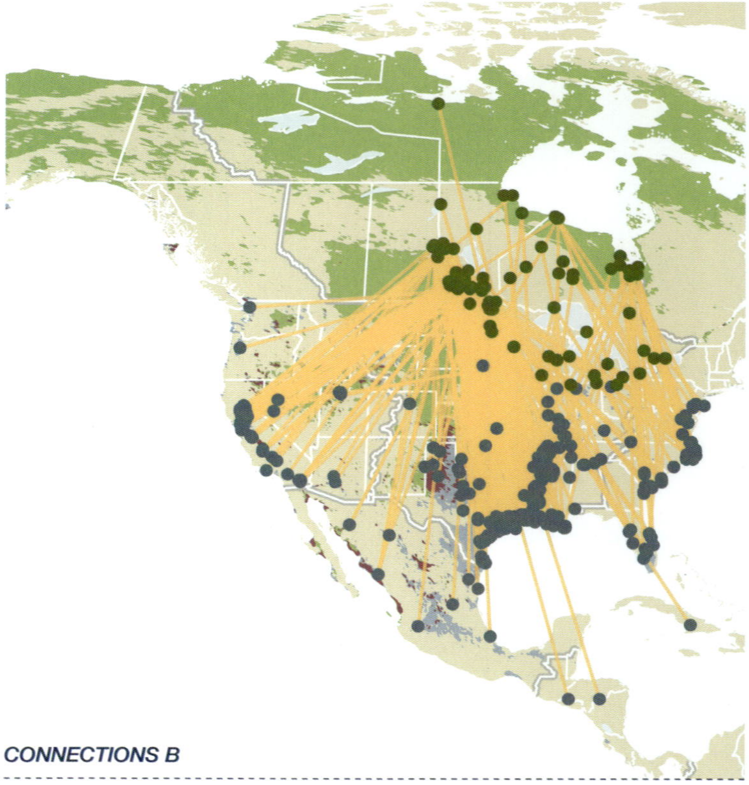

CONNECTIONS B

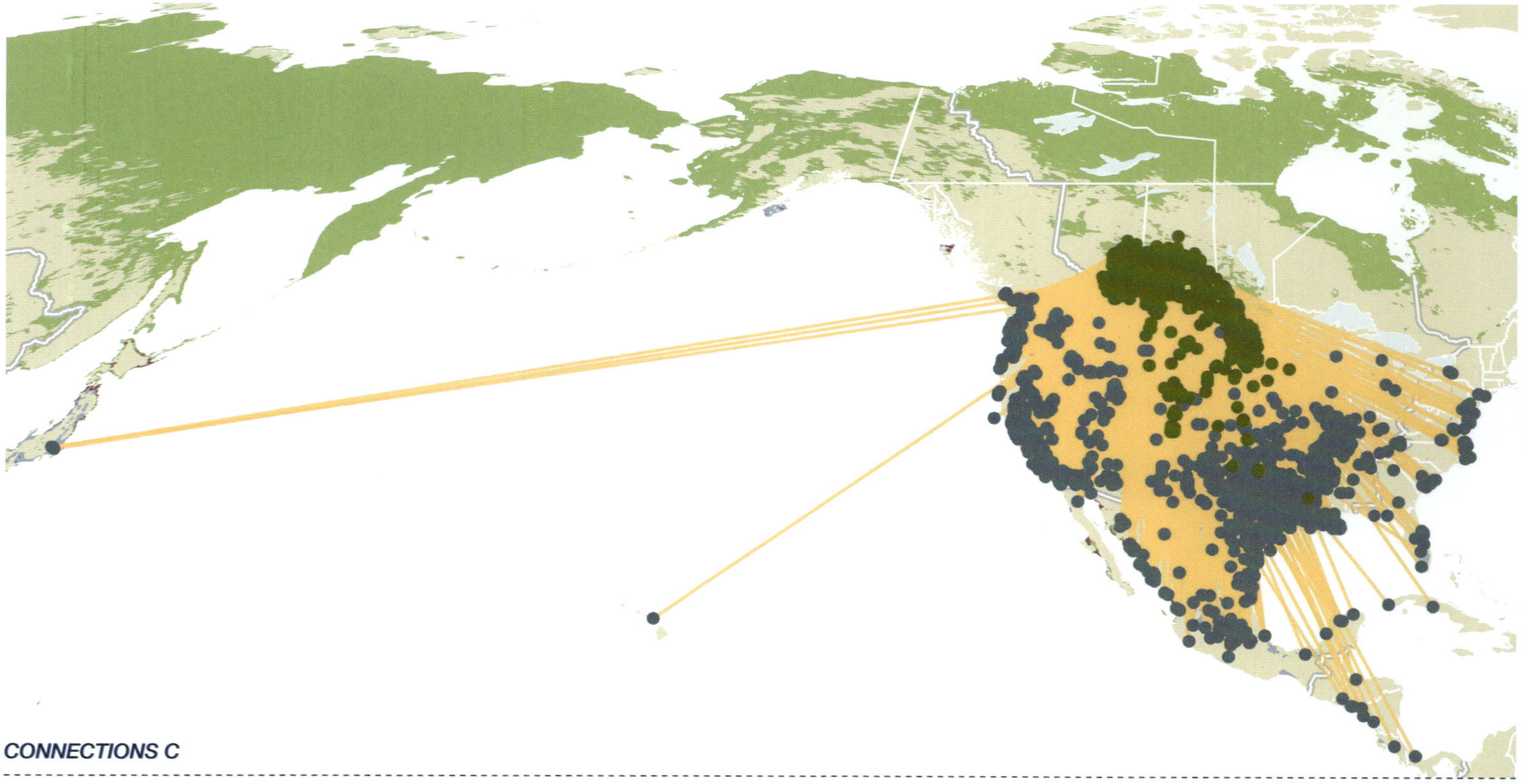

CONNECTIONS C

RADIO TRACKING Few tags were deployed through the Motus Wildlife Tracking System; there were no breeding to nonbreeding detections.

SATELLITE TRACKING Satellite transmitters (n = 31) deployed during the nonbreeding season in Texas and New Mexico showed movement to breeding grounds in North and South Dakota and the Canadian Prairies (Haukos et al. 2006). Female Northern Pintails tracked from California's Central Valley migrated to northern Canada, the Canadian Prairies, and Alaska (Miller et al. 2005; US Geological Survey, unpublished data). Females marked along the Atlantic coast during the nonbreeding season (n = 68) migrated to breeding areas around the Great Lakes. Females migrating farther west generally originated from nonbreeding locations south of North Carolina, while those migrating farther east originated from Delaware Bay and north (Malecki et al. 2006).

STABLE ISOTOPES Northern Pintails collected in the Saskatchewan River Delta were assigned by stable isotopes to origins in the northern boreal forest (69%), and the rest were equally divided between origins in southern prairies and locally (Asante et al. 2017). Hatch-year birds from Alberta and Saskatchewan originated from the Canadian Prairies less than expected by Breeding Bird Surveys, suggesting low productivity in the prairie region (Hebert and Wassenaar 2005b).

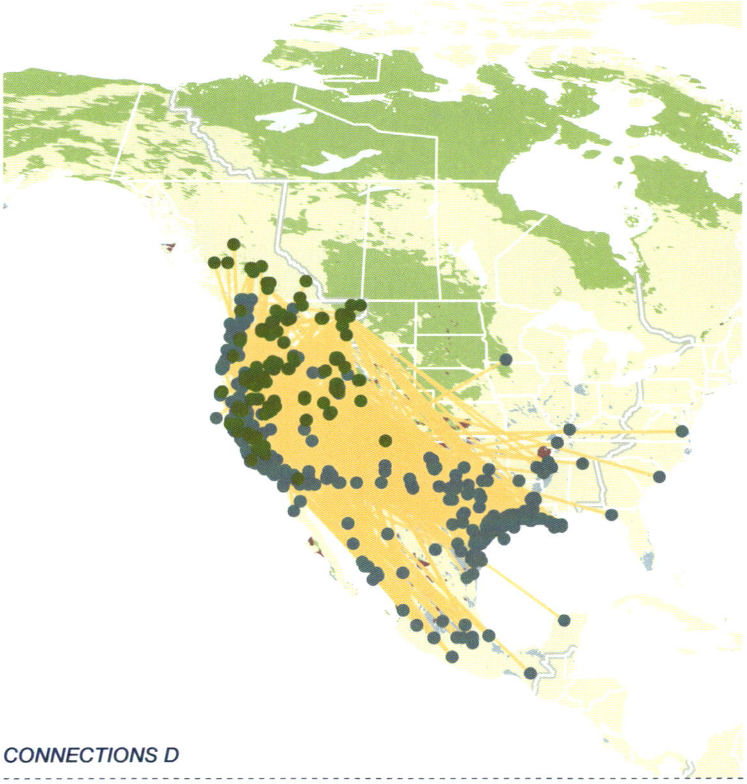

CONNECTIONS D

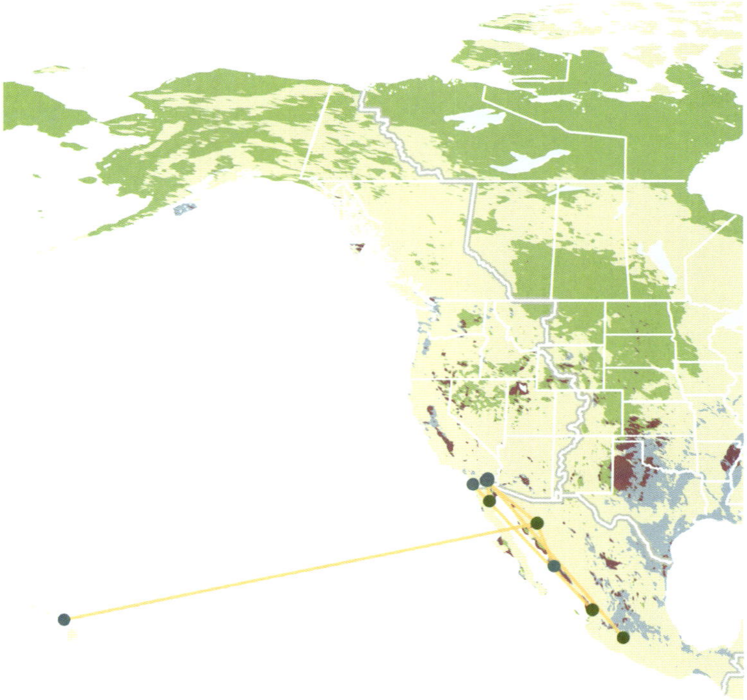

CONNECTIONS E

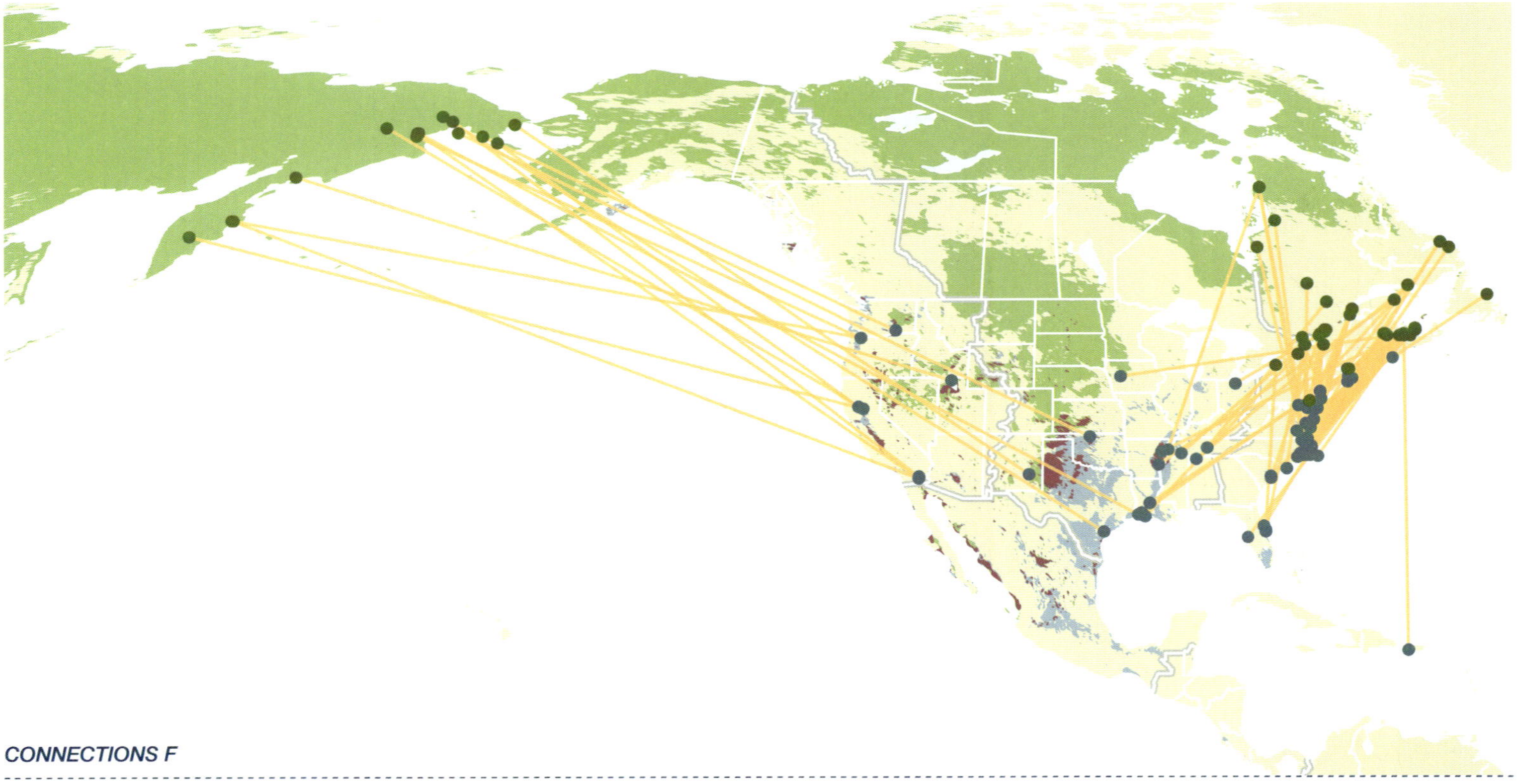

CONNECTIONS F

Stable isotopes from multiple tissues of females breeding in Alaska showed connections to nonbreeding areas across their range in the United States and Mexico (Yerkes et al. 2008).

GENETICS Female Northern Pintails sampled in three nonbreeding regions of California's Central Valley had little genetic structuring based on mtDNA (Fleskes et al. 2010).

MIGRATORY PATHWAYS Similar to other prairie-nesting ducks, Northern Pintails often follow major river systems or the coasts, or jump between major wetlands during pre- and post-breeding migrations. One bird tagged in Japan migrated through Alaska on its return to the nonbreeding grounds (Hupp et al. 2019). Birds tagged in eastern Washington during migration continued north to breeding areas in Alberta and Alaska (S.G. Hayes and M. Wilson, unpublished data).

DATA CONTRIBUTORS

C. Brady; M. Casazza; D.C. Douglas; D. Haukos; S. Hayes; J.W. Hupp; C. Overton; T.L. Tibbitts; M. Wilson.

GREEN-WINGED TEAL *Anas crecca*

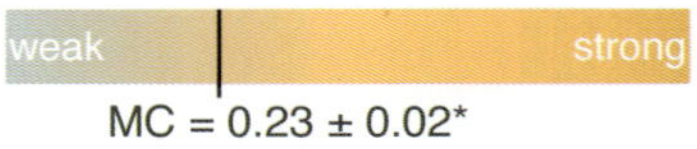

DISTRIBUTION AND ABUNDANCE

The Green-winged Teal has one of the largest ranges of all dabbling ducks. It is a widespread breeder in northern North America, with high abundance in central North America grasslands and the boreal forest. During the nonbreeding season, birds can be found from southern Canada south to Central America. It is particularly abundant in the southern United States and central Mexico (Fink et al. 2023).

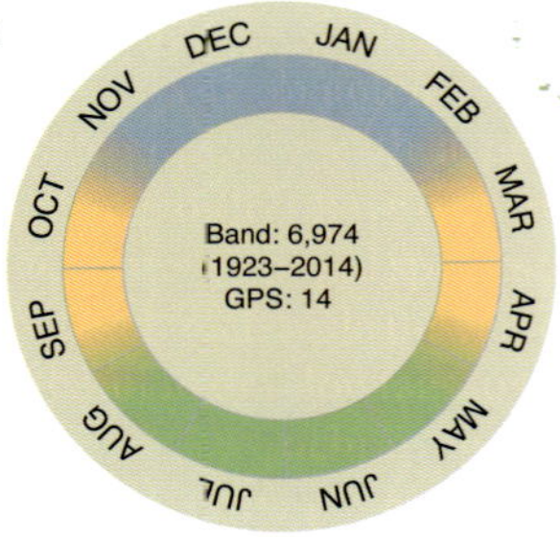

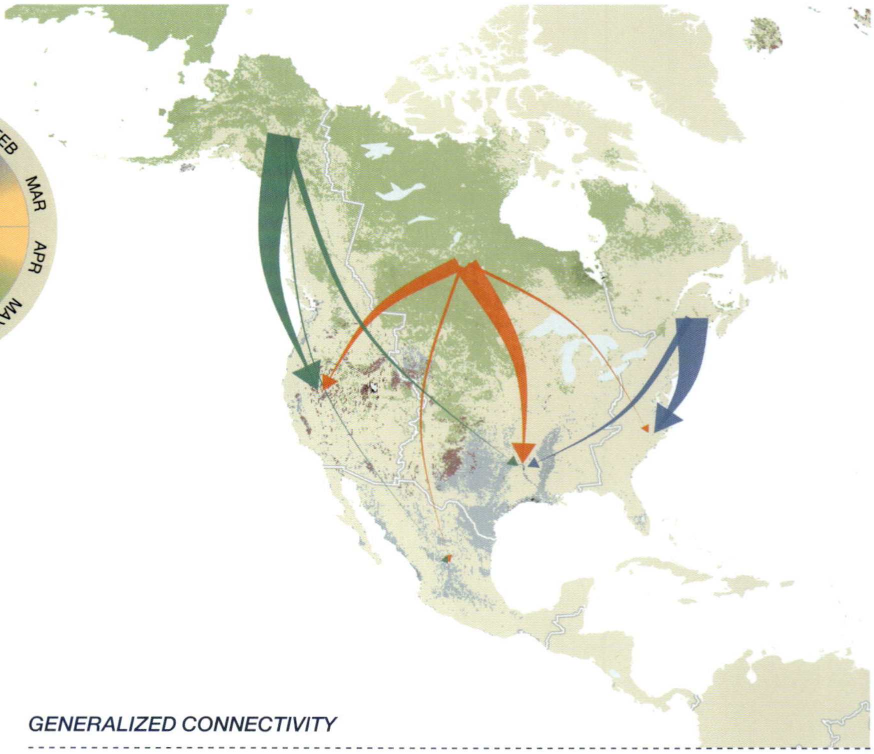

GENERALIZED CONNECTIVITY

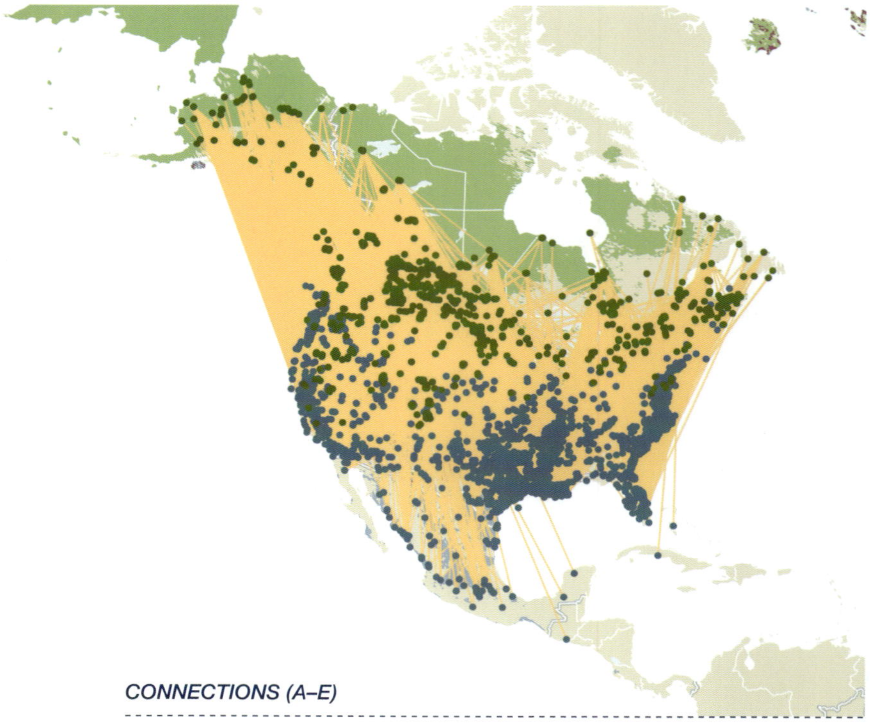

CONNECTIONS (A–E)

ORDER Anseriformes
FAMILY Anatidae
BREEDING HABITAT
Wetlands, Generalist
NONBREEDING HABITAT
Wetlands, Generalist
GLOBAL POPULATION
SIZE 6,700,000
POPULATION CHANGE SINCE 1970 +119%
IUCN RED LIST CATEGORY Least Concern
R2R TIPPING POINT SPECIES No
US LISTING STATUS No
CANADA SPECIES AT RISK No
PIF CCS 9/20

MIGRATORY CONNECTIVITY

OVERALL Not much research has been done on migratory connectivity of Green-winged Teal in the United States, and nearly all data is from banding records. Birds breeding in the two coastal flyways, the Atlantic and the Pacific, tend to remain within the flyway during the nonbreeding season (84%). Birds breeding in the Midcontinent Flyway are also likely to stay in the same flyway (58%), though many migrate to coastal flyways (Atlantic 4%, Pacific 35%) and to Mexico (4%). Temperate breeding populations may be sedentary (K. Johnson et al. 2020). Primary migration within a flyway leads to moderate migratory connectivity (MC = 0.23 ± 0.02).

BANDING Using banding data, Buhnerkempe et al. (2016) found four weakly defined populations within North American Green-winged

	Atlantic	Caribbean	Central America	Mexico	Midcontinent	Northern SA	Pacific
Atlantic (n = 74)	84%	1%	–	–	15%	–	–
Midcontinent (n = 6710)	2%	<1%	<1%	6%	46%	<1%	46%
Pacific (n = 2803)	<1%	–	–	5%	13%	–	82%

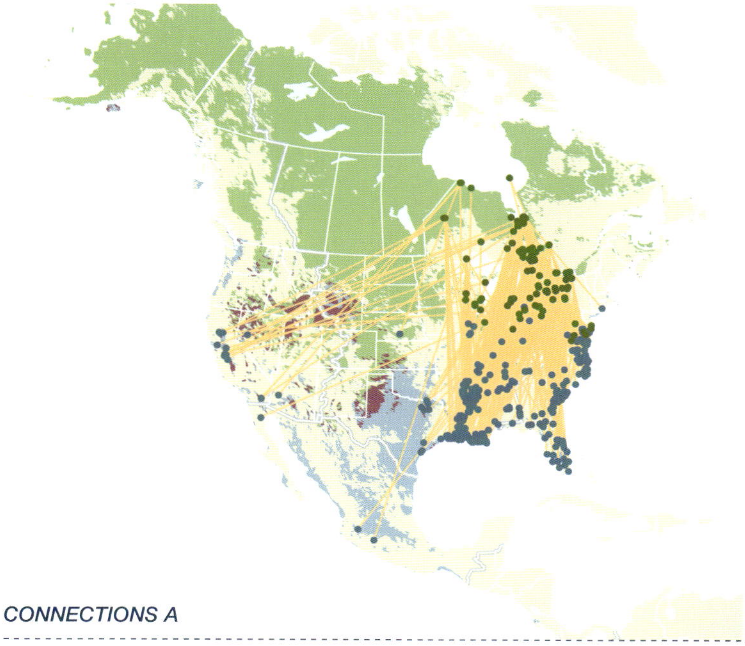

CONNECTIONS A

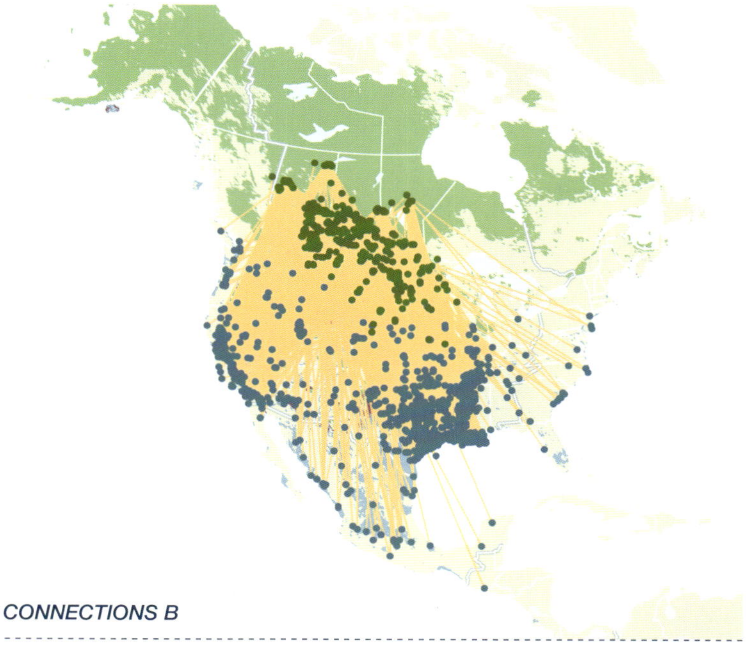

CONNECTIONS B

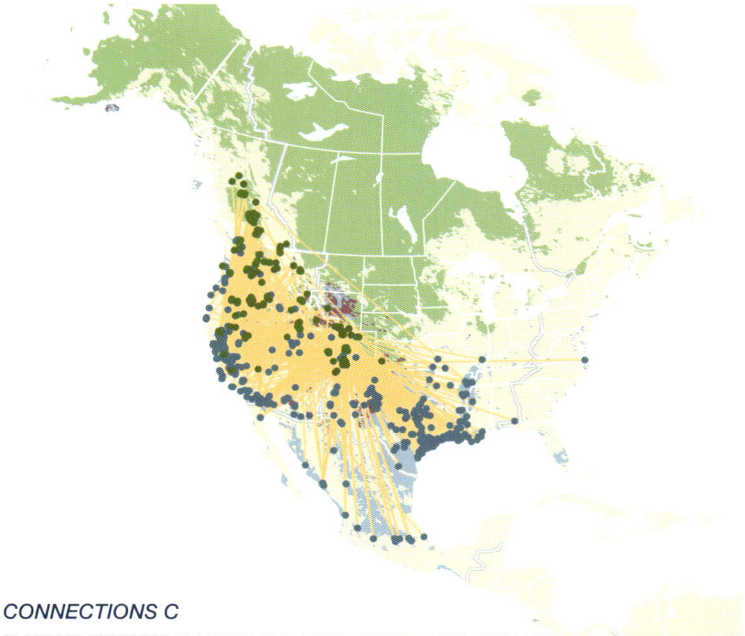

CONNECTIONS C

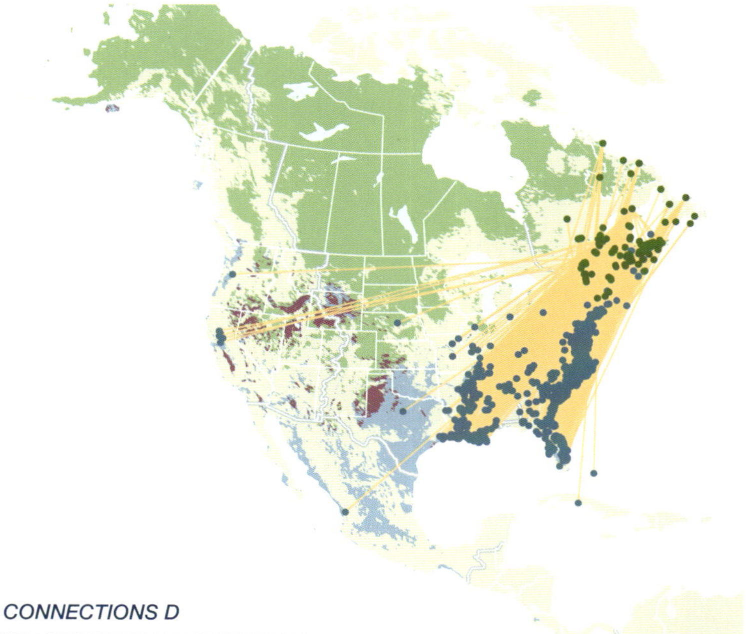

CONNECTIONS D

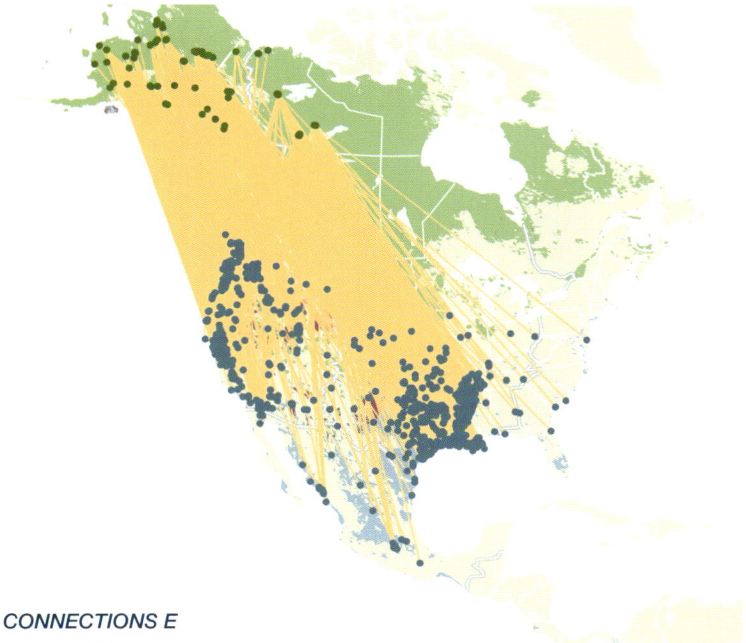

CONNECTIONS E

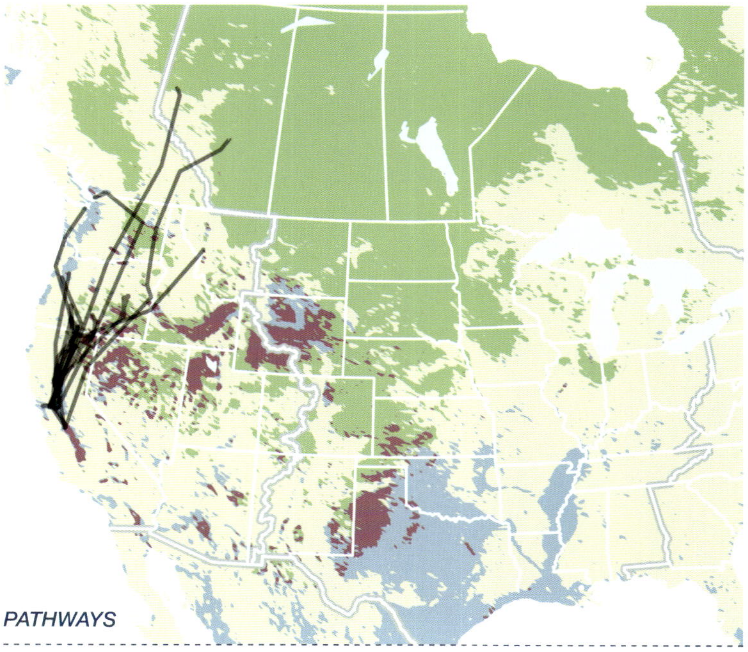

PATHWAYS

Teal, consisting of west coast, central, Great Lakes, and east coast regions.

RADIO TRACKING Few tags were deployed through the Motus Wildlife Tracking System with local detections only.

SATELLITE TRACKING Individuals tracked from California's Central Valley during the nonbreeding season migrated to breeding areas in western states and provinces (US Geological Survey, unpublished data). Work is ongoing in the Atlantic Flyway.

INTERESTING NOTE

Banding data suggest there is an annual migration from North America to Europe (De Juana and Garcia 2010).

STABLE ISOTOPES Birds sampled during post-breeding migration in central Saskatchewan, Canada, originated nearly equally from the surrounding area and from northern boreal forest breeding areas (Asante et al. 2017).

GENETICS No genetic studies of migratory connectivity were found for this species.

MIGRATORY PATHWAYS Typically latitudinal migration, with birds breeding farther west also spending the nonbreeding season farther west, with no clear pathway (Baldassarre 2014).

DATA CONTRIBUTORS

M. Casazza; C. Feldheim; C. Overton.

CANVASBACK *Aythya valisineria*

MC = 0.18 ± 0.03*

DISTRIBUTION AND ABUNDANCE

Canvasbacks breed in the highest densities across the Prairie Pothole Region, and in lower densities through the boreal forest and into Alaska. Nonbreeding concentrations occur on large lakes and rivers in the southern Great Plains and Mississippi Alluvial Valley, as well as coastal areas of the United States and Mexico (Mowbray 2020a). Despite the historic importance of Chesapeake Bay and other Atlantic coast nonbreeding sites, recent surveys have suggested these areas no longer support large nonbreeding concentrations.

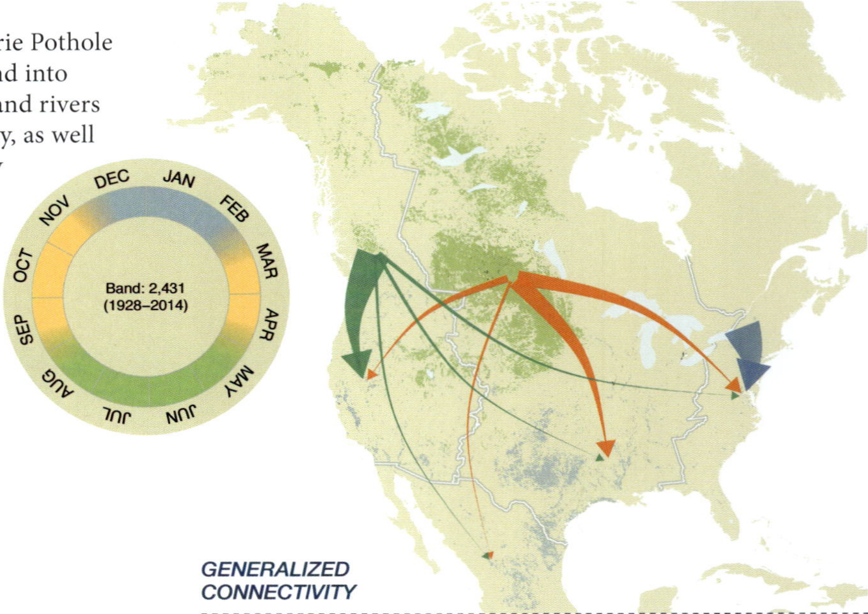

GENERALIZED CONNECTIVITY

ORDER Anseriformes

FAMILY Anatidae

BREEDING HABITAT Wetlands, Seasonally Wet Grasslands

NONBREEDING HABITAT Wetlands, Lakes, and Rivers

GLOBAL POPULATION SIZE 690,000

POPULATION CHANGE SINCE 1970 +43%

IUCN RED LIST CATEGORY Least Concern

R2R TIPPING POINT SPECIES No

US LISTING STATUS No

CANADA SPECIES AT RISK No

PIF CCS 10/20

MIGRATORY CONNECTIVITY

OVERALL Though Canvasbacks have been extensively studied in the past, there is very little recent work on migratory connectivity. Migratory connectivity information is primarily from banding data. Birds breeding in the Atlantic and Pacific flyways mostly migrated to nonbreeding areas within the flyway (Atlantic 100%, Pacific 87%). The largest percentage of Midcontinent Flyway breeding birds also migrated to nonbreeding areas within the Midcontinent Flyway (45%), while others migrated to the Atlantic (36%) or Pacific (16%) Flyway. Mixing on the nonbreeding grounds results in weak migratory connectivity (MC = 0.18 ± 0.03).

BANDING Birds banded in the southern breeding range in Nevada were mostly (92%) recovered in the Pacific Flyway. However the distribution of band recoveries changed through time, suggesting the breeding site hosted individuals from across the range (Kruse et al. 2003).

GEOLOCATOR TRACKING Eight Canvasbacks fitted with geolocators during pre-breeding migration in Nevada subsequently traveled to breeding areas in the Prairie Pothole Region of Canada (n = 6) and Alaska (n = 2; Cook et al. 2021).

	Atlantic	Caribbean	Mexico	Midcontinent	Pacific
Atlantic (n = 8)	100%	–	–	–	–
Midcontinent (n = 1652)	36%	<1%	3%	45%	16%
Pacific (n = 719)	6%	–	3%	5%	87%

INTERESTING NOTE

On Delta Marsh in Manitoba, Canada, in the 1930s, Canvasback were hatched in captivity, hand-reared, and released. These birds were recovered on traditional nonbreeding areas, suggesting they migrated with wild birds (Brakhage 1953).

OTHER APPROACHES Serie et al. (1983) colored feathers of males during post-breeding staging in Wisconsin and Iowa. Resightings of these birds occurred in similar areas to band recoveries, with concentrations in traditional nonbreeding areas of the lower Mississippi River Valley and the Chesapeake Bay. No stable isotope or genetic studies of migratory connectivity were found for this species.

MIGRATORY PATHWAYS Serie et al. (1983) used banding data to surmise two migration paths from staging areas in the upper Midwest to nonbreeding areas. Some Canvasbacks migrate directly south from breeding areas, following the Mississippi River to the Mississippi Alluvial Valley and along the Gulf Coast. Others move east to the southern Great Lakes before ending up along the Atlantic coast (Serie et al. 1983).

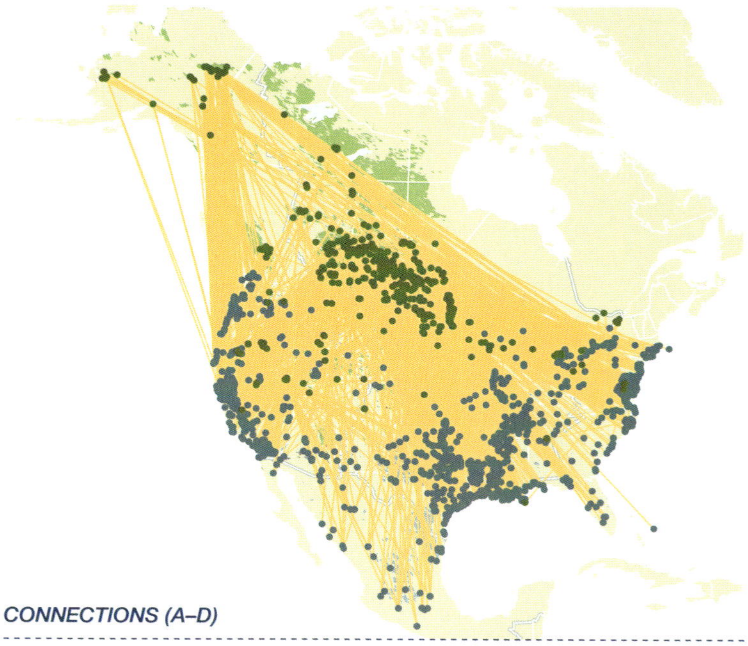

CONNECTIONS (A–D)

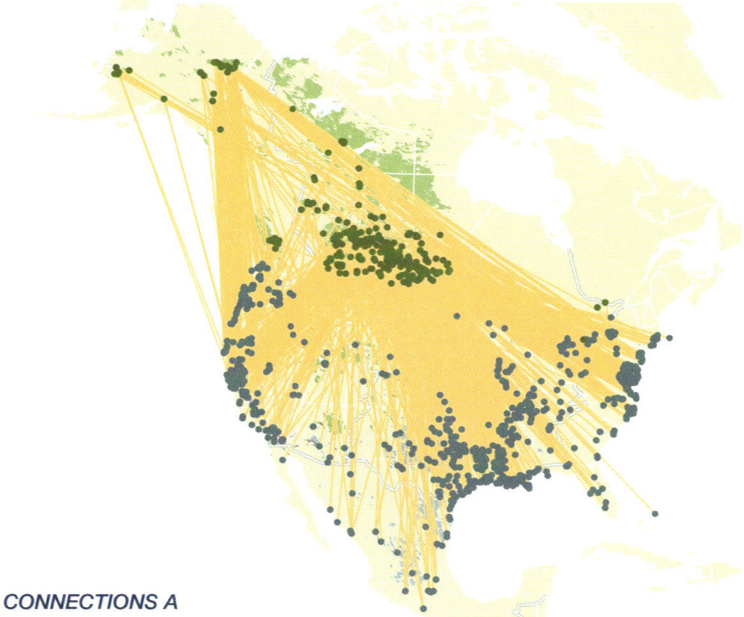

CONNECTIONS A

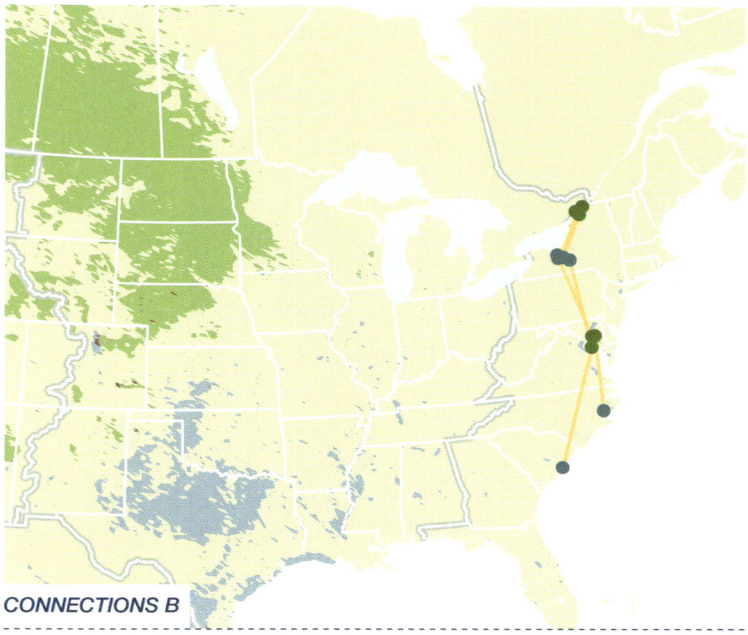

CONNECTIONS B

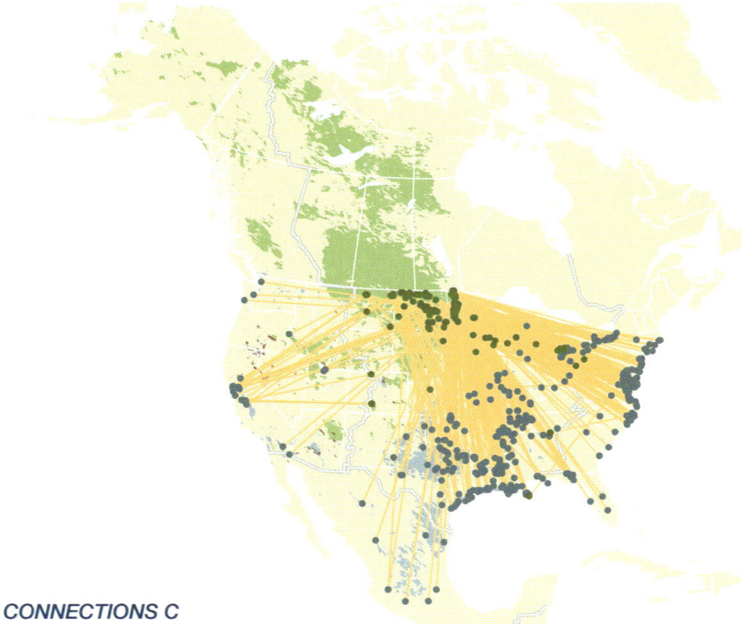

CONNECTIONS C

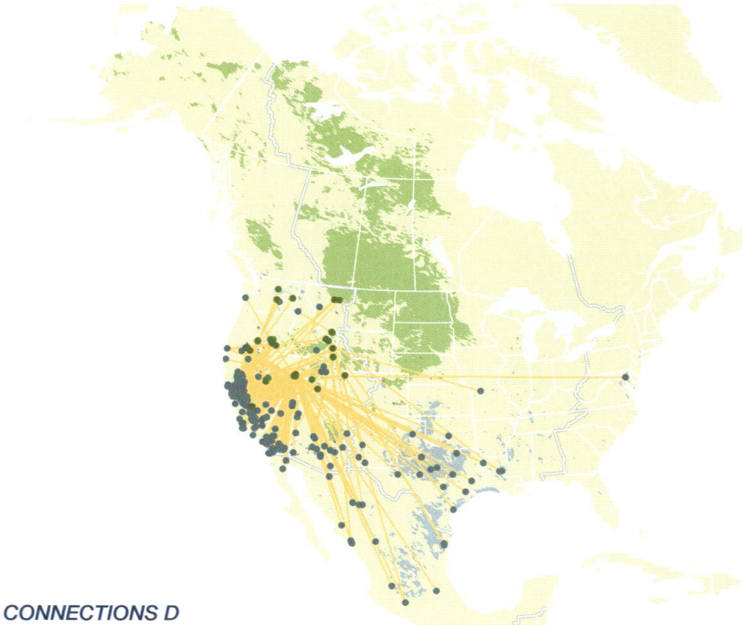

CONNECTIONS D

REDHEAD *Aythya americana*

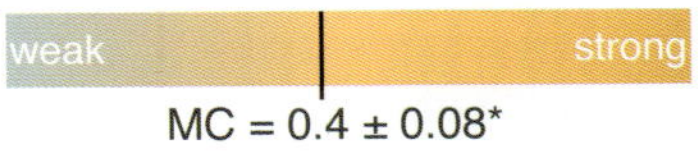

DISTRIBUTION AND ABUNDANCE

Endemic to North America, Redheads breed throughout the Prairie Pothole Region north to Alaska. Nonbreeding concentrations occur in the Gulf of Mexico from northern Mexico to western Florida, though they can be found along both coasts (Woodin and Michot 2020).

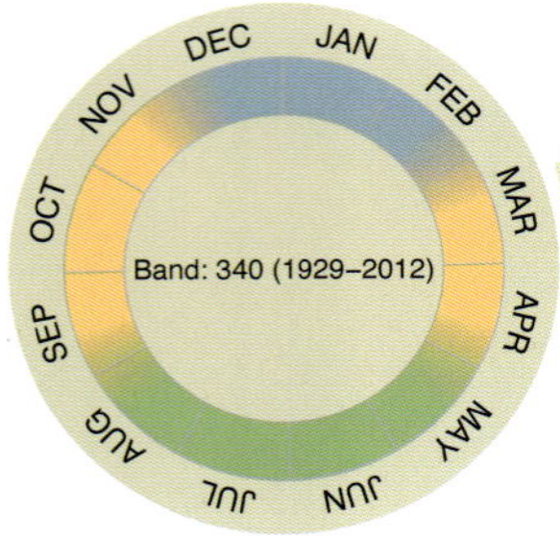

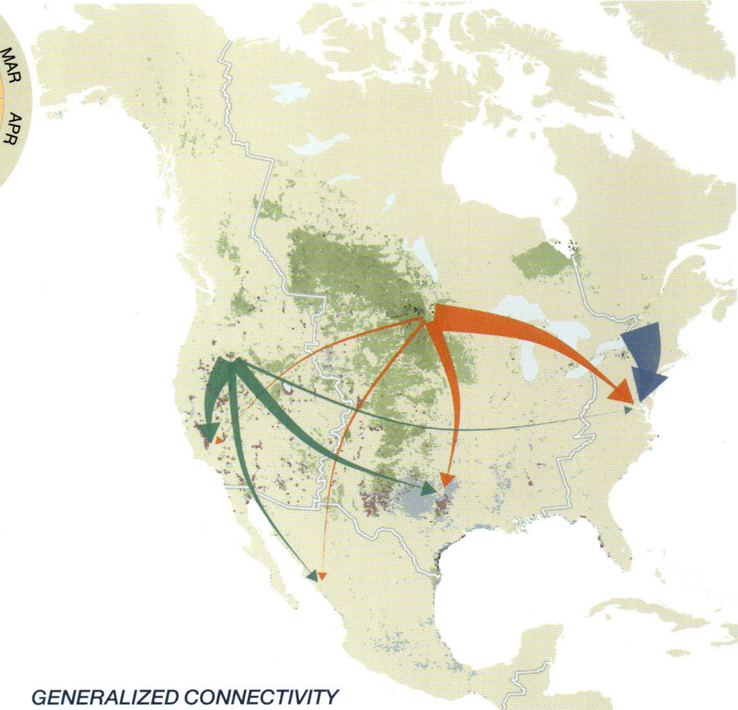

GENERALIZED CONNECTIVITY

ORDER Anseriformes
FAMILY Anatidae
BREEDING HABITAT Wetlands, Seasonally Wet Grasslands
NONBREEDING HABITAT Coasts, Marine Waters
GLOBAL POPULATION SIZE 1,200,000
POPULATION CHANGE SINCE 1970 +109%
IUCN RED LIST CATEGORY Least Concern
R2R TIPPING POINT SPECIES No
US LISTING STATUS No
CANADA SPECIES AT RISK No
PIF CCS 8/20

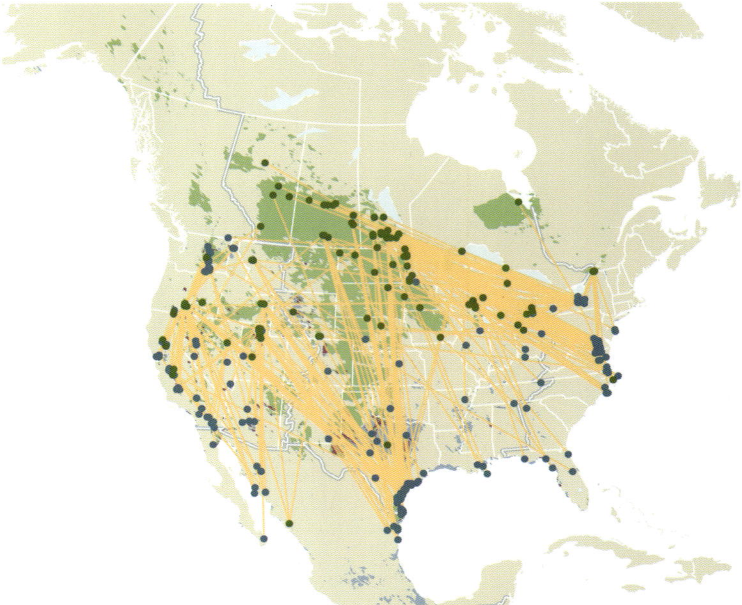

CONNECTIONS

	Atlantic	Mexico	Midcontinent	Pacific
Atlantic (n = 11)	100%	–	–	–
Midcontinent (n = 184)	65%	4%	29%	3%
Pacific (n = 118)	3%	17%	30%	51%

MIGRATORY CONNECTIVITY

OVERALL Migratory connectivity is known from banding data. Redheads breeding in the Atlantic Flyway spend the nonbreeding season within the same flyway. Midcontinent Flyway breeding birds remain within the flyway (29%) but mostly migrate to nonbreeding areas in the Atlantic Flyway (65%). More mixing occurs from Pacific Flyway breeding birds, which transition to nonbreeding areas in the Pacific Flyway (51%), Midcontinent Flyway (30%), Mexico (17%), and the Atlantic Flyway (3%). However, migratory connectivity remains strong (MC = 0.4 ± 0.08).

BANDING Banding recoveries are plentiful and encompass the species' range in North America (Weller 1964). Redheads banded from 1950 to 2010 in Saskatchewan (n = 2,870) were recovered in Texas (33%) or on the migration route to the Gulf Coast. Other recoveries occurred farther east in Minnesota (8%) and New York (3%; Baldassarre 2014). During the same period, Redheads banded in North and South Dakota (n = 584) were recovered in Texas or en route (30%), Michigan (5%), New York (5%), and Florida (2%). Of 1,414 Redheads banded in Alberta from 1950 to 2010, recoveries were primarily from Texas (26%) and Mexico (10%; Baldassarre 2014).

RADIO TRACKING Few tags were deployed through the Motus Wildlife Tracking System with local detections only.

OTHER APPROACHES No satellite tracking, genetic, or stable isotope studies of migratory connectivity were found for this species.

MIGRATORY PATHWAYS Redheads typically migrate due south from breeding areas in the Prairie Pothole Region to nonbreeding areas on the Gulf Coast. Northern and eastern breeding birds are more likely to migrate to nonbreeding areas along the east coast of North America (Weller 1964).

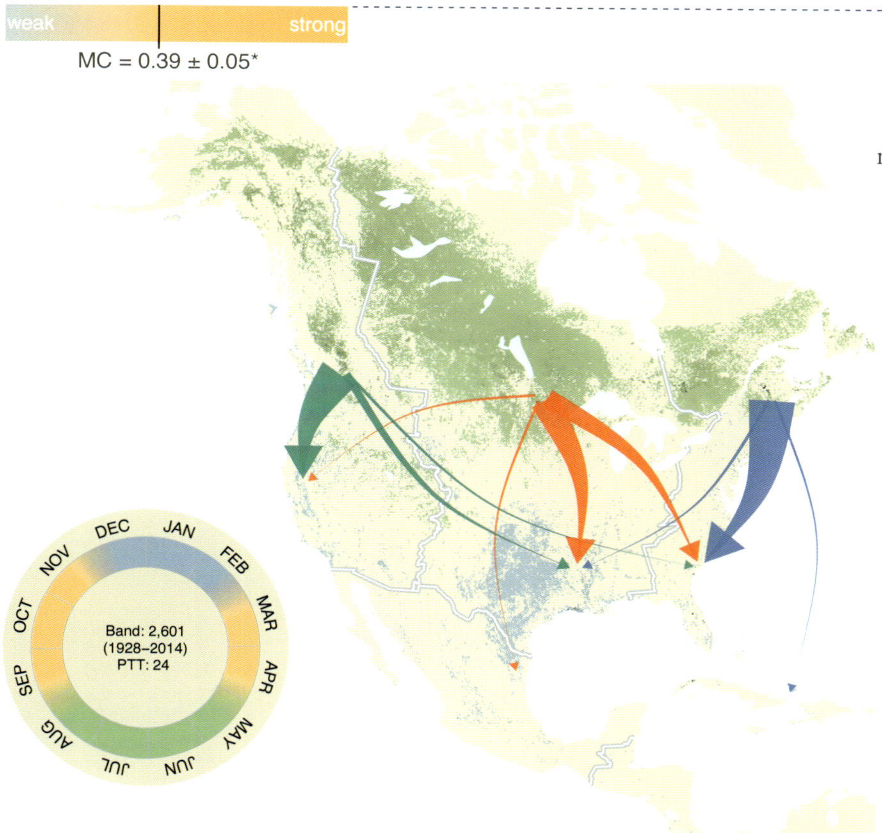

GENERALIZED CONNECTIVITY

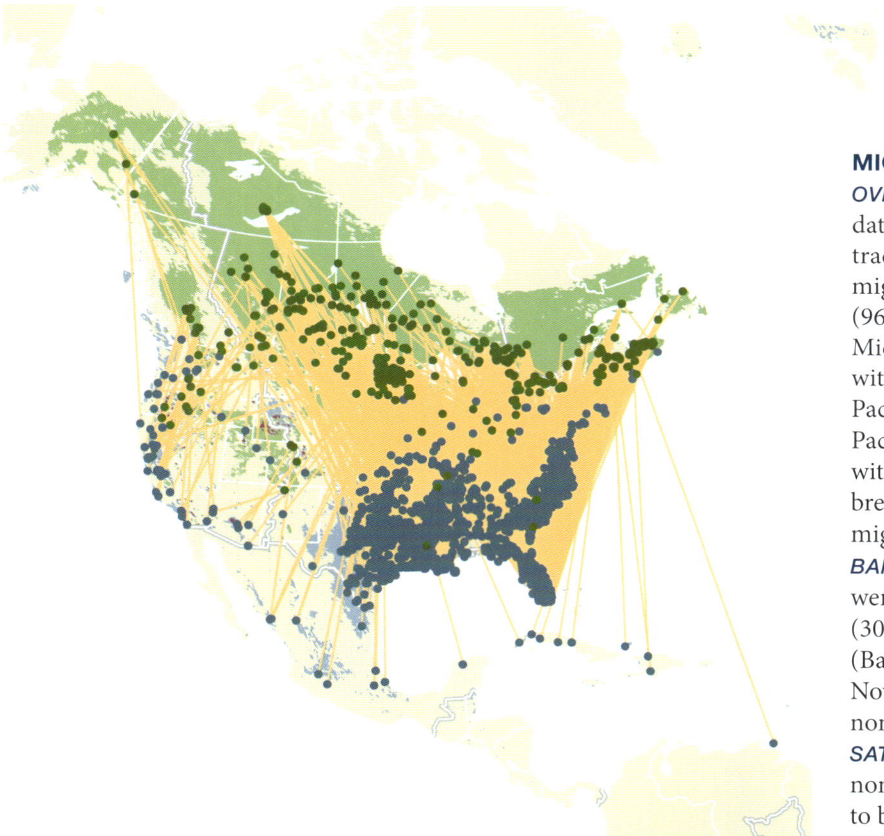

CONNECTIONS (A–E)

	Atlantic	Caribbean	Mexico	Midcontinent	Pacific
Atlantic (n = 547)	96%	1%	–	3%	–
Midcontinent (n = 2022)	38%	<1%	1%	60%	1%
Pacific (n = 47)	2%	–	–	13%	85%

Aythya collaris

RING-NECKED DUCK

DISTRIBUTION AND ABUNDANCE

Ring-necked Ducks breed throughout Canada and the northern United States, concentrated in central and eastern Canada, and winter on inland marshes near both coasts and in the southern Great Plains (Roy et al. 2020)

ORDER Anseriformes
FAMILY Anatidae
BREEDING HABITAT Wetlands, Boreal
NONBREEDING HABITAT Wetlands, Lakes, and Rivers
GLOBAL POPULATION SIZE 2,000,000
POPULATION CHANGE SINCE 1970 +318%
IUCN RED LIST CATEGORY Least Concern
R2R TIPPING POINT SPECIES No
US LISTING STATUS No
CANADA SPECIES AT RISK No
PIF CCS 8/20

MIGRATORY CONNECTIVITY

OVERALL Migratory connectivity information is from banding data available from most of the range and from limited satellite tracking data. Most birds breeding in the Atlantic Flyway migrate to nonbreeding areas within the Atlantic Flyway (96%). Midcontinent Flyway breeding birds migrate within the Midcontinent Flyway (60%) and to the Atlantic Flyway (38%), with smaller percentages across other nonbreeding regions. Pacific Flyway breeding birds primarily remain within the Pacific Flyway (85%) or migrate to the Midcontinent (13%), with some transitions to the Atlantic Flyway (2%). Mixing of breeding populations on nonbreeding areas results in strong migratory connectivity (MC = 0.39 ± 0.05).

BANDING Of 770 Ring-necked Ducks banded in Ontario, most were recovered along the Atlantic coast, especially in Florida (30%), though 6% were recovered in Minnesota and Michigan (Baldassarre 2014). Ring-necked Ducks banded in Quebec, Nova Scotia, New Brunswick, and Maine were recovered in nonbreeding areas in the southeastern United States.

SATELLITE TRACKING Ring-necked Ducks (n = 25) at a single nonbreeding site in the southern Atlantic Flyway migrated to breeding areas in the upper Midwest of the United States and central Canada (Mezebish et al. 2022). Ring-necked Ducks marked farther north in the same study bred in eastern Canada (Delta Waterfowl, unpublished data).

OTHER APPROACHES No genetic or stable isotope studies of migratory connectivity were found for this species.

MIGRATORY PATHWAYS Ring-necked Ducks often use the Great Lakes and surrounding marshes during both pre- and post-breeding migration.

DUCKS, GEESE AND WATERFOWL (ANSERIFORMES)

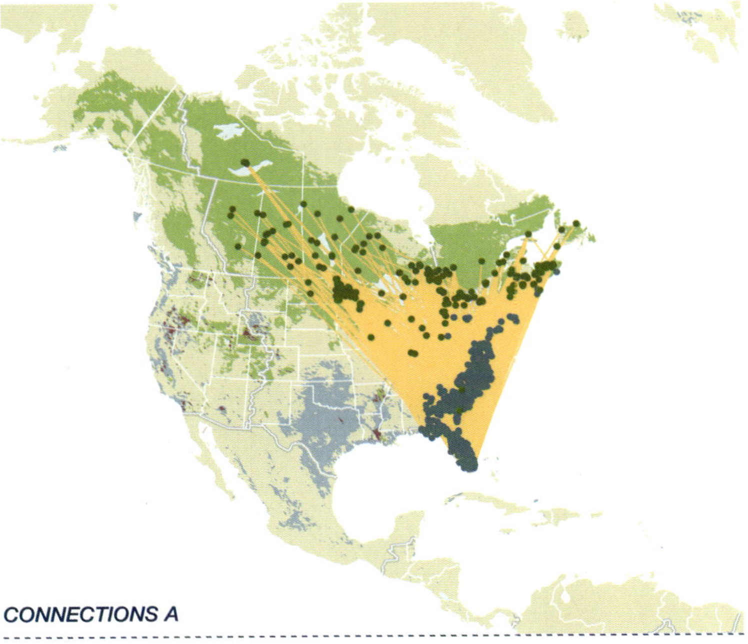

CONNECTIONS A

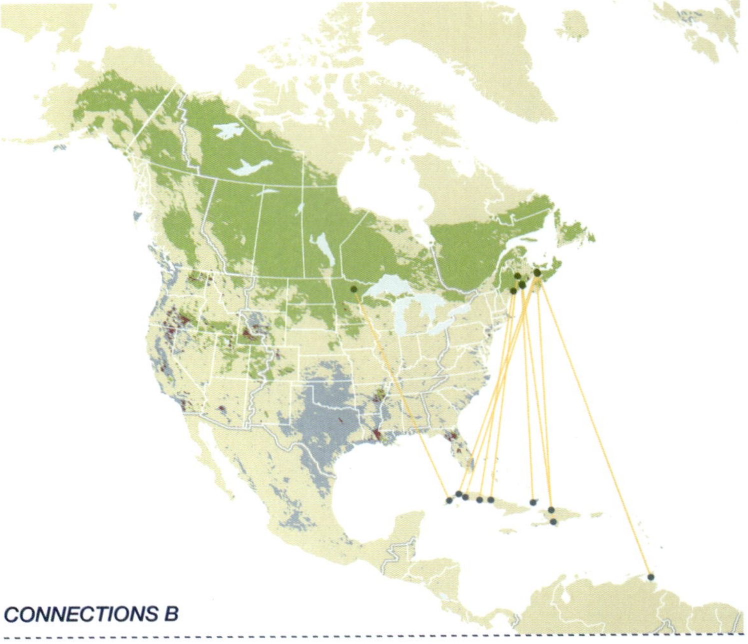

CONNECTIONS B

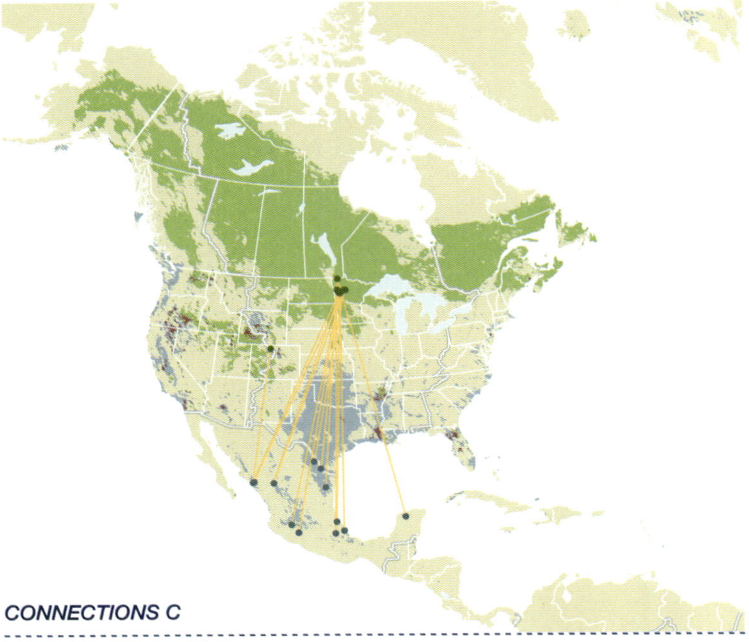

CONNECTIONS C

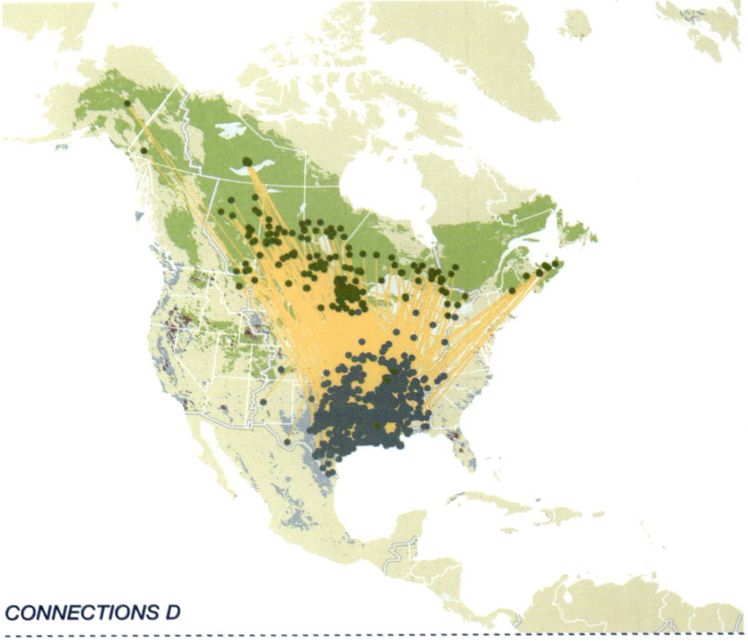

CONNECTIONS D

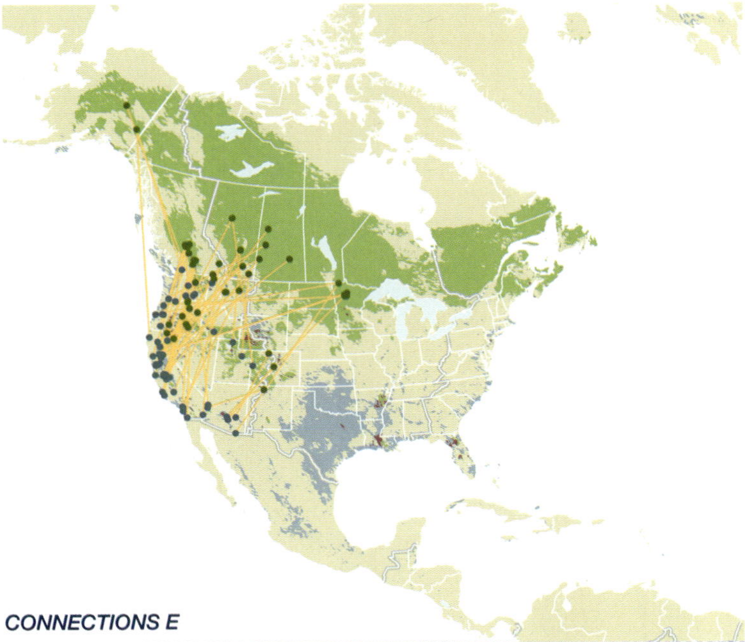

CONNECTIONS E

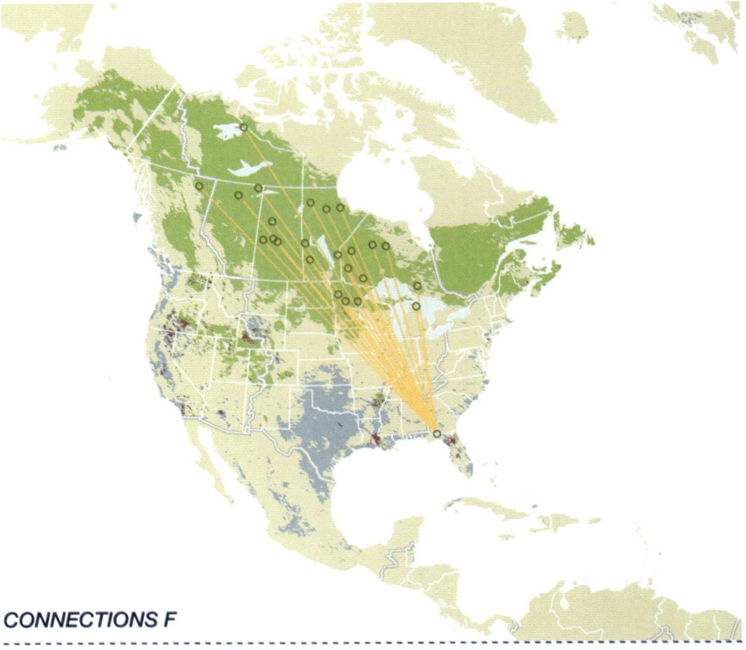

CONNECTIONS F

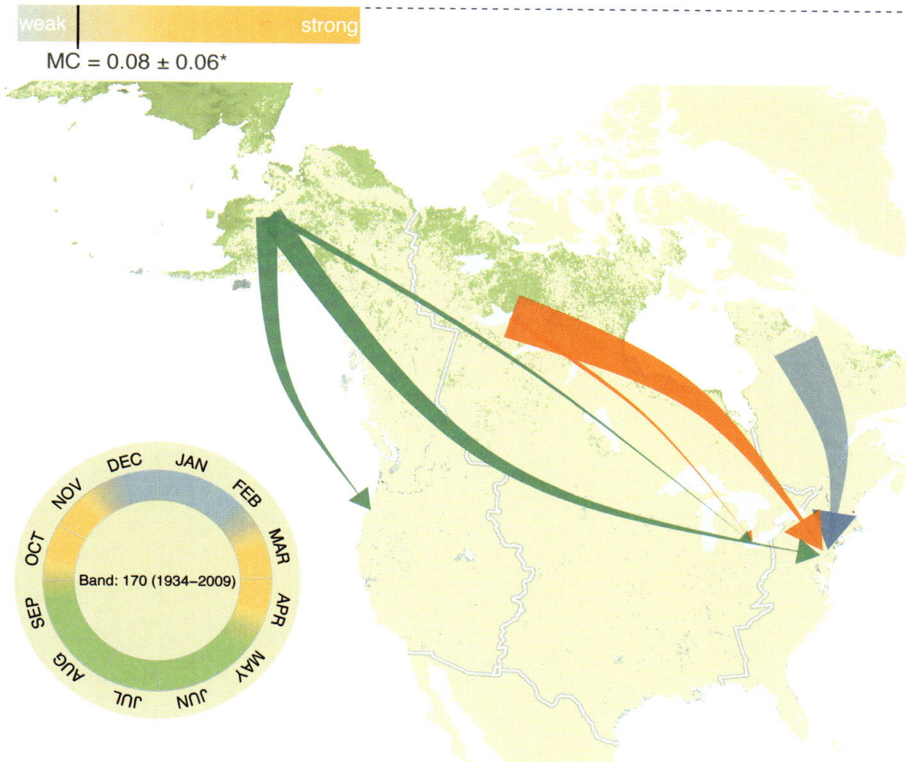

GENERALIZED CONNECTIVITY

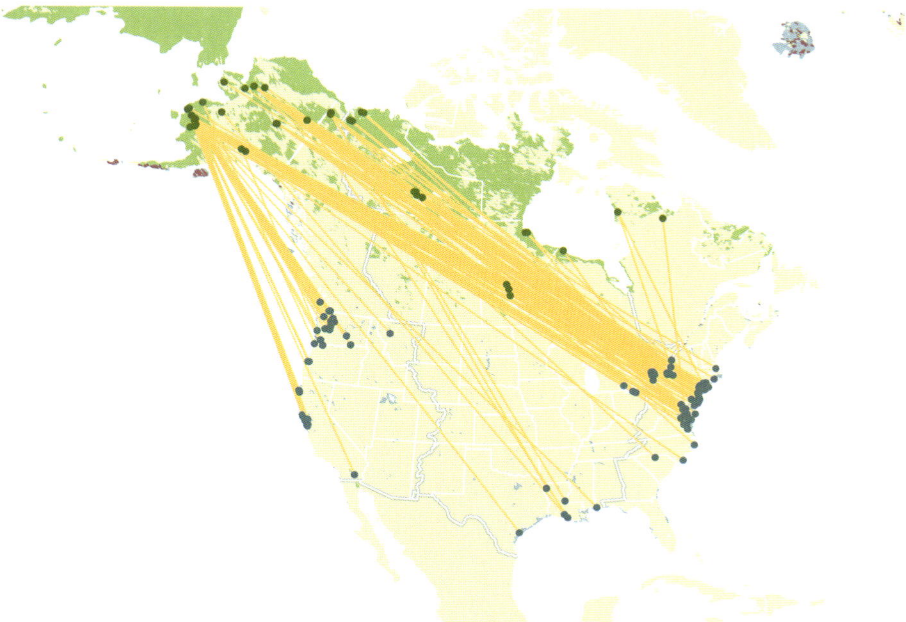

CONNECTIONS

Aythya marila **GREATER SCAUP**

DISTRIBUTION AND ABUNDANCE

Greater Scaup are a circumpolar species, with the North American population breeding in the northern boreal forest, with concentrations along the Alaska coast. The nonbreeding range includes both Pacific and Atlantic coasts and in the Great Lakes, with highest concentrations in the Pacific in southern Alaska and in the Atlantic around Long Island, New York. The breeding and nonbreeding range is farther north than the closely related Lesser Scaup.

ORDER Anseriformes
FAMILY Anatidae
BREEDING HABITAT Wetlands, Tundra
NONBREEDING HABITAT Coasts, Marine Waters
GLOBAL POPULATION SIZE 1,200,000
POPULATION CHANGE SINCE 1970 −39%
IUCN RED LIST CATEGORY Least Concern
R2R TIPPING POINT SPECIES No
US LISTING STATUS No
CANADA SPECIES AT RISK No
PIF CCS 10/20

	Atlantic	Midcontinent	Pacific
Atlantic (n = 3)	100%	–	–
Midcontinent (n = 16)	94%	6%	–
Pacific (n = 141)	48%	10%	42%

MIGRATORY CONNECTIVITY

OVERALL Movement data is primarily from banding records. Birds breeding in the Atlantic and Midcontinent flyways nearly all migrate to nonbreeding areas in the Atlantic Flyway. Breeding birds in the Pacific Flyway transitioned nearly equally to the Atlantic (48%) and Pacific (42%) flyways. Migrations from many of the breeding populations to the Atlantic coast result in weak migratory connectivity (MC = 0.08 ± 0.06).

BANDING Greater Scaup banded on the Yukon-Kuskokwim Delta, Alaska, from 1948 to 2009 (n = 188) were recovered during the nonbreeding season on both the Atlantic and Pacific coasts (Baldassarre 2014).

OTHER APPROACHES No satellite tracking, genetic, or stable isotope studies of migratory connectivity were found for this species.

MIGRATORY PATHWAYS The Great Lakes are an important stopover site for migrating Greater Scaup as they make their way to the Atlantic coast (Kessel et al. 2020). Greater Scaup breeding in interior Alaska likely follow major rivers and large lakes through Canada to the Great Lakes and eventually the Atlantic coast.

LESSER SCAUP *Aythya affinis*

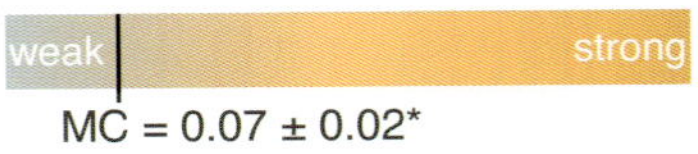

DISTRIBUTION AND ABUNDANCE

Lesser Scaup are one of the most geographically widespread diving ducks in North America. They breed throughout interior Alaska, Canada, and south and east through the Prairie Pothole Region (Baldassarre 2014). Nonbreeding concentrations occur in the southern United States, along the Atlantic coast, and in coastal Mexico.

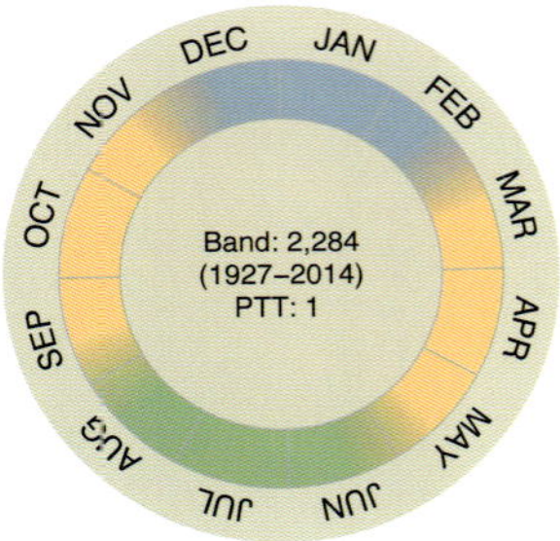

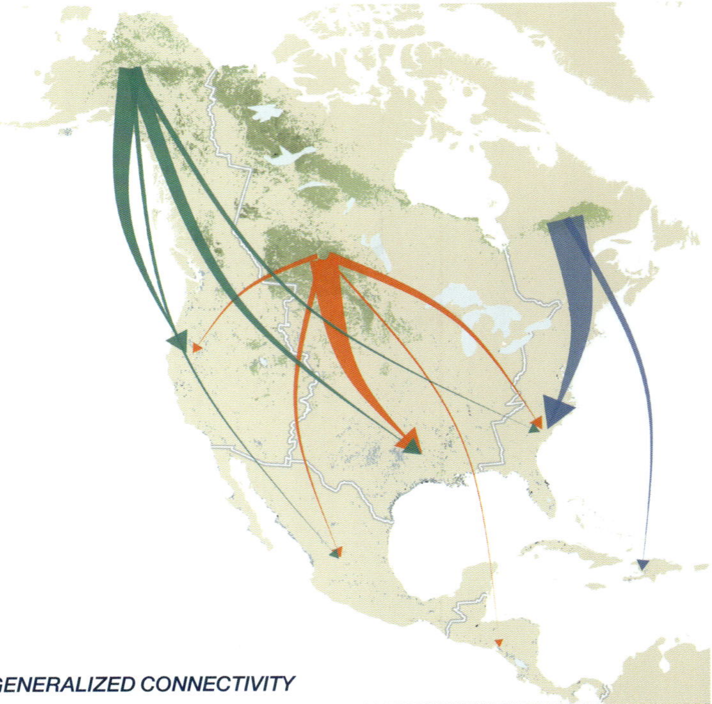

GENERALIZED CONNECTIVITY

ORDER Anseriformes
FAMILY Anatidae
BREEDING HABITAT Wetlands, Lakes, and Rivers
NONBREEDING HABITAT Coasts, Marine Waters
GLOBAL POPULATION SIZE 3,700,000
POPULATION CHANGE SINCE 1970 −41%
IUCN RED LIST CATEGORY Least Concern
R2R TIPPING POINT SPECIES No
US LISTING STATUS No
CANADA SPECIES AT RISK No
PIF CCS 9/20

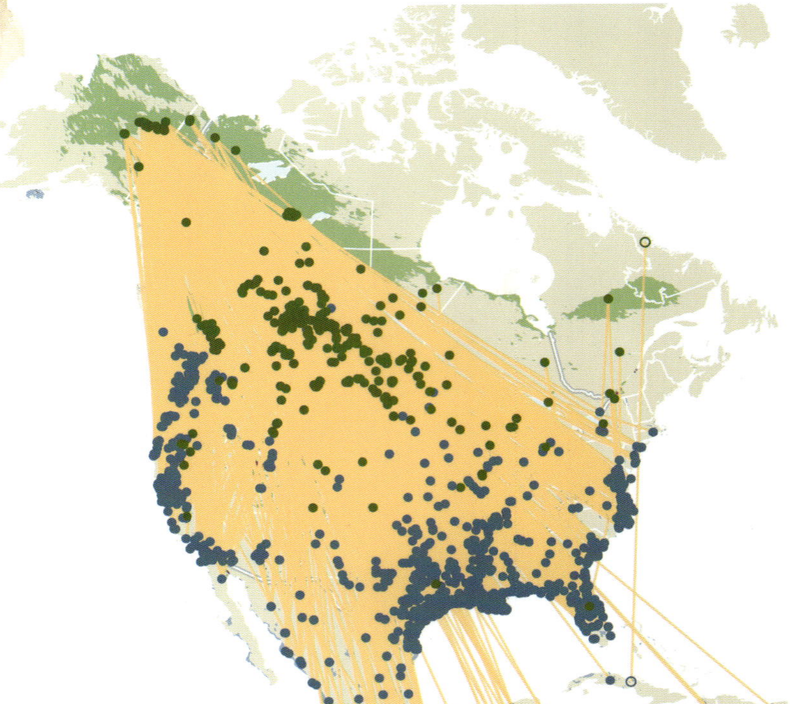

CONNECTIONS (A–D)

	Atlantic	Caribbean	Central America	Mexico	Midcontinent	Pacific
Atlantic (n = 7)	86%	14%	–	–	–	–
Midcontinent (n = 968)	18%	<1%	1%	10%	63%	8%
Pacific (n = 1214)	10%	<1%	<1%	7%	33%	50%

MIGRATORY CONNECTIVITY

OVERALL Migratory connectivity data is available from banding, satellite transmitters, and stable isotopes. The population intermingles extensively throughout its range. Birds breeding in the Atlantic Flyway migrate within the Atlantic Flyway (86%) or to the Caribbean (14%). Midcontinent Flyway breeding birds transition within the flyway (63%), to the Atlantic Flyway (18%), or to Mexico (10%), while those breeding in the Pacific Flyway migrate within the Pacific Flyway (50%) or to the Midcontinent Flyway (33%), and a small portion to the Atlantic (10%). Widespread movements, especially of breeding birds in the Midcontinent and Pacific flyways, result in weak migratory connectivity (MC = 0.07 ± 0.02).

BANDING Banding recoveries show extensive mixing, with no separation of breeding and nonbreeding populations (Baldassarre 2014). Lesser Scaup banded on breeding areas in Montana were recovered across the nonbreeding range, with the largest number of

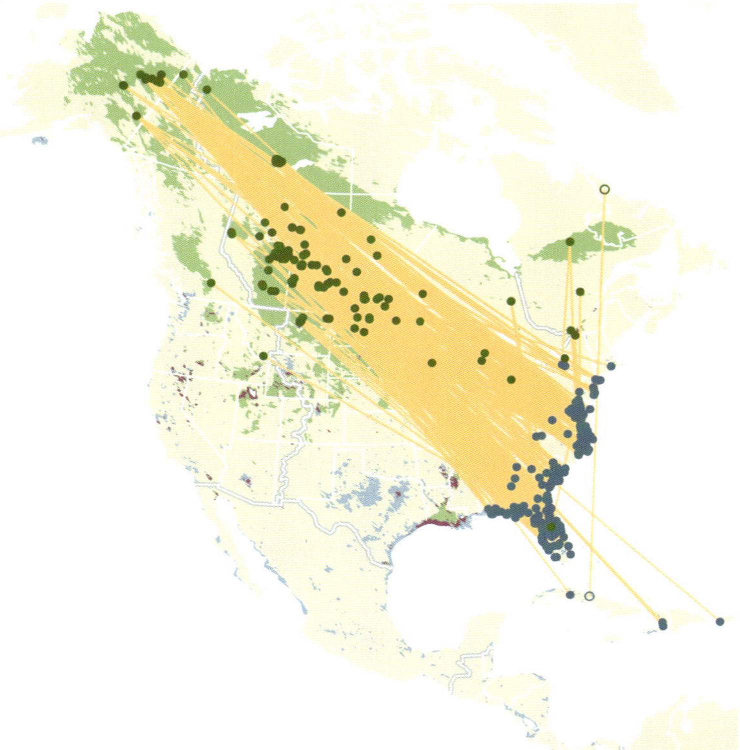

CONNECTIONS A

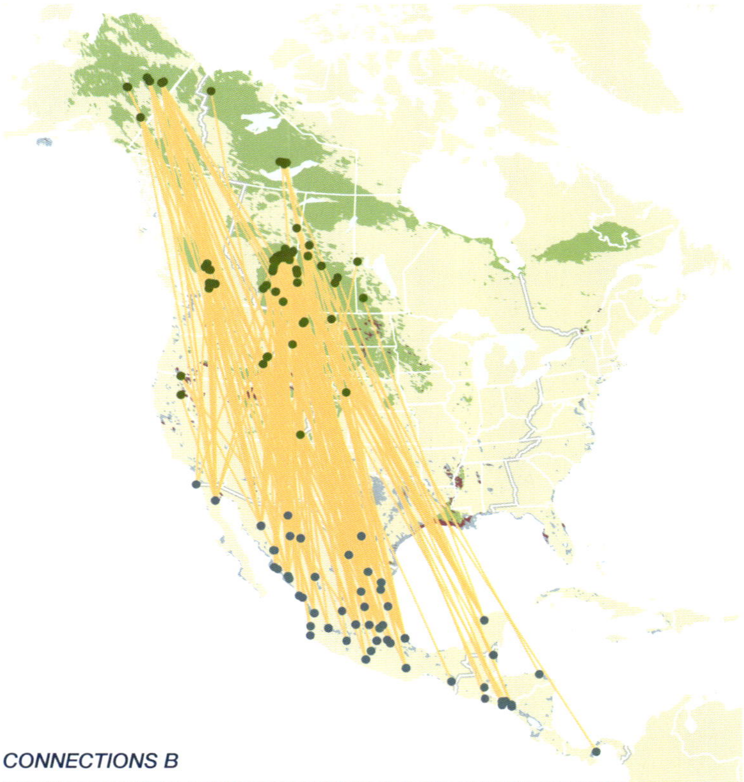

CONNECTIONS B

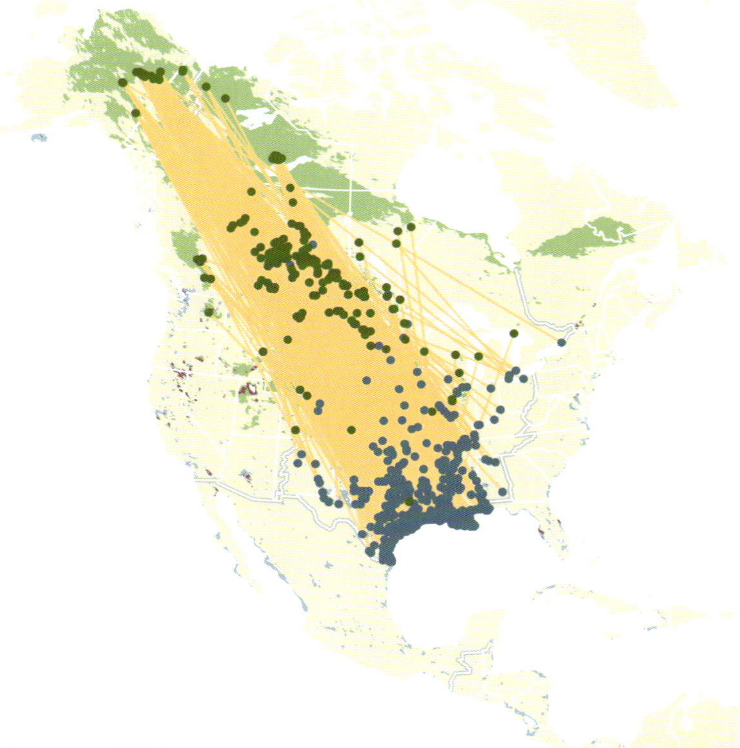

CONNECTIONS C

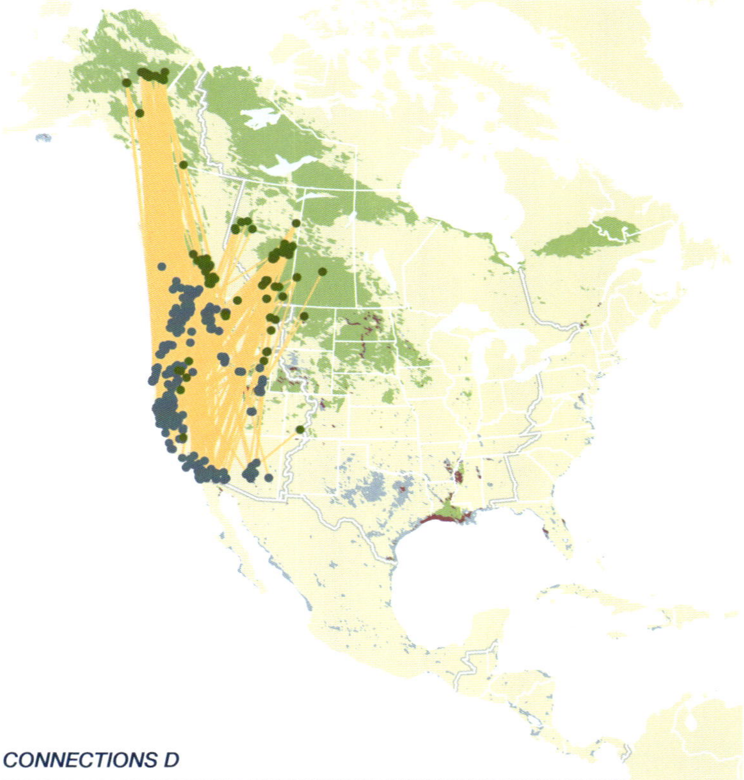

CONNECTIONS D

recoveries in the Central Valley of California, the Gulf Coast, and interior Mexican states (Hopkins et al. 2013).

SATELLITE TRACKING Lesser Scaup (n = 78; pathways not shown) were tracked with satellite transmitters from two pre-breeding staging areas: the Mississippi River in Iowa and Illinois and northern Lake Erie marshes (Finger et al. 2016; Schummer et al. 2018). These birds subsequently migrated to the Gulf Coast, Mexico, and the Caribbean during the nonbreeding season. Lesser Scaup trapped on Lake Erie bred in eastern and central Canada, and those trapped along the Mississippi River bred across the species range from Hudson Bay to Alaska (Finger et al. 2016).

STABLE ISOTOPES Hatch-year Lesser Scaup harvested by hunters from all North American flyways (n = 867) originated throughout their breeding range (Hobson et al. 2011).

GENETICS No genetic studies of migratory connectivity were found.

MIGRATORY PATHWAYS No clear primary migration route is used.

STELLER'S EIDER *Polysticta stelleri*

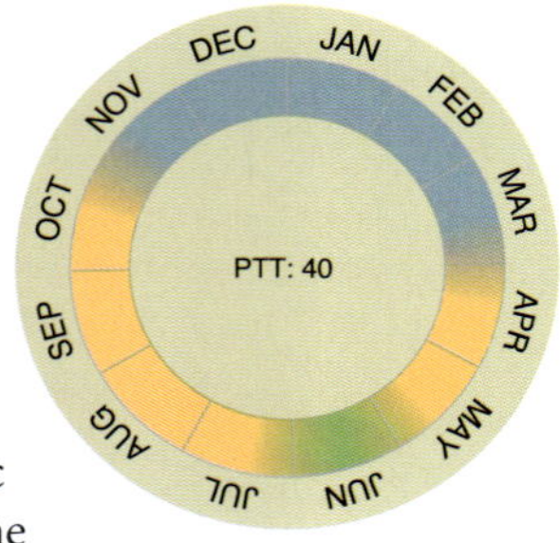

DISTRIBUTION AND ABUNDANCE

Steller's Eider have a circumpolar distribution and are comprised of three breeding populations: the Atlantic, the Pacific Russian, and the Pacific Alaskan. The Atlantic population breeding and nonbreeding range includes Arctic Scandinavia and Russia, and the Pacific Russian population breeds in western Arctic Russia and spends the nonbreeding season along the Aleutian Islands, Alaska (Baldassarre 2014). The Pacific Alaskan population is small and breeds mostly near Utqiaġvik, Alaska, with a small number in the Yukon-Kuskokwim Delta. These birds also spend the nonbreeding season at the edge of the sea ice along the Aleutian Islands.

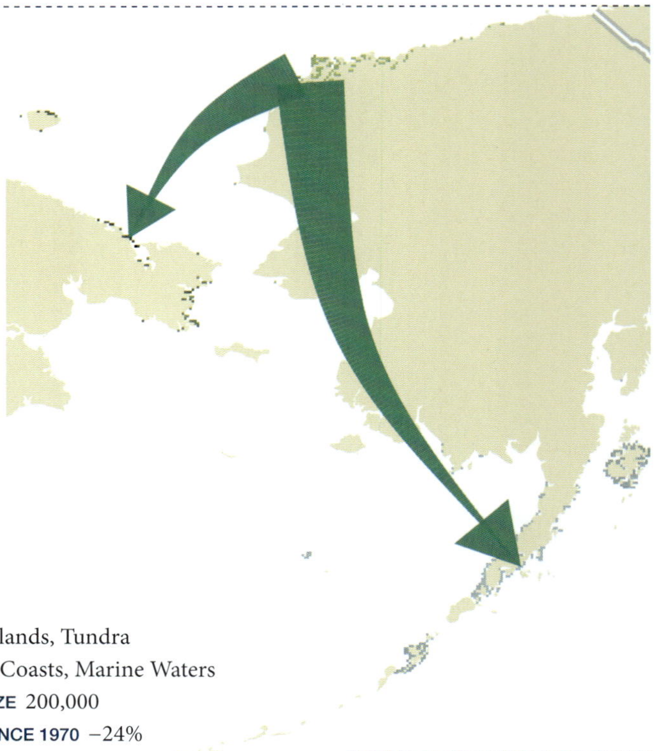

GENERALIZED CONNECTIVITY

ORDER Anseriformes
FAMILY Anatidae
BREEDING HABITAT Wetlands, Tundra
NONBREEDING HABITAT Coasts, Marine Waters
GLOBAL POPULATION SIZE 200,000
POPULATION CHANGE SINCE 1970 −24%
IUCN RED LIST CATEGORY Vulnerable
R2R TIPPING POINT SPECIES Red-Alert
US LISTING STATUS Threatened
CANADA SPECIES AT RISK No
PIF CCS 19/20

	Pacific	Russia
Pacific (n = 22)	59%	41%

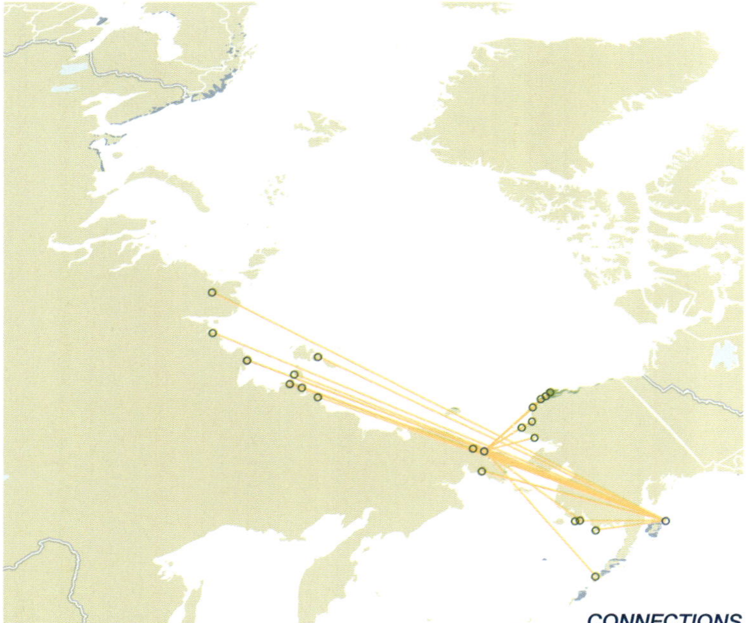

CONNECTIONS

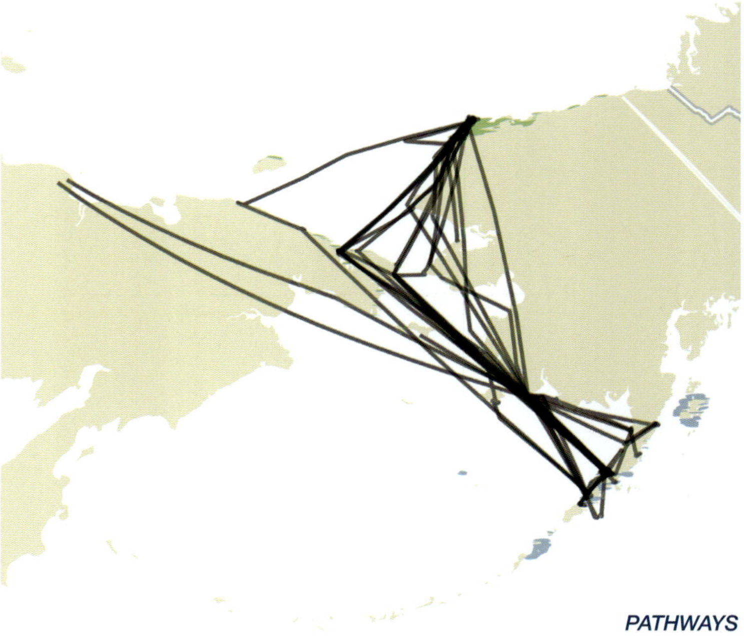

PATHWAYS

MIGRATORY CONNECTIVITY

OVERALL Steller's Eider migration has been studied using satellite transmitters and genetics in the Alaskan and Russian breeding populations. All birds breeding in the Pacific Flyway migrated to nonbreeding areas within the Pacific Flyway (59%) or to Russia (41%). The single breeding region prevents MC from being estimated.

BANDING In the 1960s, over 800 Steller's Eider were banded in Izembek Bay along the Alaska Peninsula during post-breeding molt. Of these, 17 band recoveries occurred during the subsequent breeding season, including one from Point Barrow, Utqiaġvik, Alaska, and the other 16 in Arctic Russia (Jones 1965). A total of 347 of 52,985 Steller's Eiders banded along the lower Alaska Peninsula between 1961 and 1998 were recovered. These recoveries suggested that individuals molting along the Alaska Peninsula were equally from Russian and Alaskan breeding sites (Dau et al. 2000).

SATELLITE TRACKING Steller's Eiders from the Atlantic population in Norway have been tracked (Petersen et al. 2006), but we only report the North American Steller's Eiders (n = 14) tracked from their breeding grounds near Utqiaġvik, Alaska, to nonbreeding areas along the lower Alaska Peninsula (Martin et al. 2015).

STABLE ISOTOPES No stable isotope studies of migratory connectivity were found for this species.
GENETICS Analysis of mtDNA from Russian and Alaskan Steller's Eiders showed little population differentiation between breeding populations (Pearce et al. 2005b).

MIGRATORY PATHWAYS Individuals (n = 24) tracked from their nonbreeding grounds on Kodiak Island, Alaska, all used the same migration corridor to either breeding areas along the coast of Russia or nearshore waters of Russia and Alaska, though routes and timing of migration varied by year and by individual (Rosenberg et al. 2014).

DATA CONTRIBUTORS
D.C. Douglas; P. Martin.

Somateria fischeri SPECTACLED EIDER

DISTRIBUTION AND ABUNDANCE

Three distinct breeding populations of Spectacled Eiders in coastal northern and western Alaska and Russia show distinct migratory patterns, though all populations use the Bering Sea, near St. Lawrence Island, during the nonbreeding season (Baldassarre 2014; Petersen et al. 2020). They molt in the Arctic Ocean, particularly the Bering and Chukchi seas (Petersen et al. 2020).

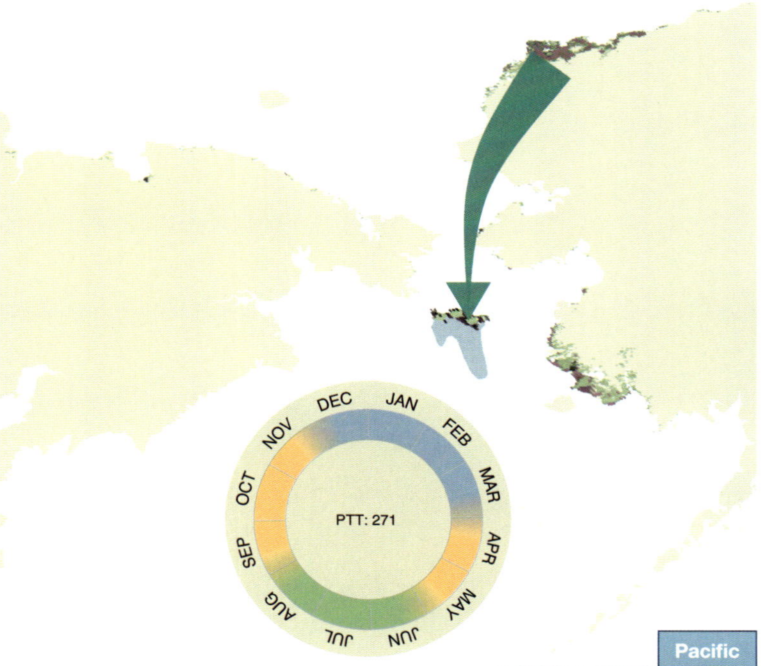

	Pacific
Pacific (n = 60)	100%

GENERALIZED CONNECTIVITY

PATHWAYS

ORDER Anseriformes
FAMILY Anatidae
BREEDING HABITAT Wetlands, Tundra
NONBREEDING HABITAT Oceans, Arctic Polynyas
GLOBAL POPULATION SIZE 250,000
POPULATION CHANGE SINCE 1970 NA
IUCN RED LIST CATEGORY Near Threatened
R2R TIPPING POINT SPECIES Red-Alert
US LISTING STATUS Threatened
CANADA SPECIES AT RISK No
PIF CCS 19/20

MIGRATORY CONNECTIVITY

OVERALL The majority of data for migratory connectivity comes from satellite transmitters deployed at breeding sites in Alaska. All birds breeding in the Pacific Flyway migrate to nonbreeding areas within the Pacific Flyway. The single breeding and nonbreeding area prevent MC from being estimated.
SATELLITE TRACKING Adults (n = 82) breeding at three sites, the Alaska North Slope, the Yukon-Kuskokwim Delta, and Arctic Russia, were tracked to molting and nonbreeding grounds in the Bering, Chukchi, and Beaufort seas. Molting and staging sites differed based on sex and breeding location. Females from the Yukon-Kuskokwim Delta molted on eastern Norton Sound, and those from the North Slope molted on Ledyard and Mechigmensky bays. Males from all breeding grounds distributed equally across molting sites (Petersen et al. 1999). Individuals tracked from the Yukon-Kuskokwim Delta, Alaska, breeding area migrated to the Bering Sea, including St. Lawrence Island (Petersen et al. 1995).

GENETICS Despite the entire population being in a single area during the nonbreeding season, mtDNA evidence suggests the three primary breeding areas are genetically distinct due to high philopatry of females (Scribner et al. 2001).

STABLE ISOTOPES No stable isotope studies of migratory connectivity were found for this species.

MIGRATORY PATHWAYS A study of 129 individuals tracked from northern and western Alaska found the post-breeding migration route in the northern Bering Sea varied by sex and breeding location, with strong migratory connectivity among populations. During migration, Norton Sound was used exclusively by breeders from the Yukon-Kuskokwim Delta, and the eastern Chukchi Sea was used exclusively by breeders in northern Alaska and Russia (Sexon et al. 2014).

DATA CONTRIBUTORS

D.C. Douglas; M. Petersen; M. Sexson.

KING EIDER *Somateria spectabilis*

weak strong

MC = 0.03 ± 0.04*

DISTRIBUTION AND ABUNDANCE

One of the farthest northern breeding birds, King Eiders nest in the high Arctic and spend the nonbreeding season in northern oceans, following the southern extent of ice (Powell and Sudam 2020). There are thought to be two North American populations of King Eiders, a Pacific population and an Atlantic population. The dividing line between populations is often demarcated by the Taymyr Peninsula in Russia and Victoria Island in central Canada (Bustnes et al. 2010; D.L. Dickson 2012a). Breeding concentrations occur on the Alaska North Slope and the central Canadian Arctic, and nonbreeding concentrations occur in southwest Alaska; southeast Chukotka, along the Kamchatka Peninsula and the northern Atlantic; and in southwest Greenland (Merkel et al. 2019).

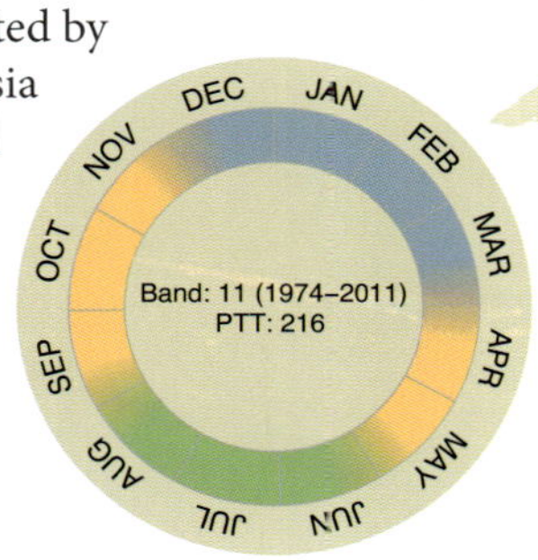

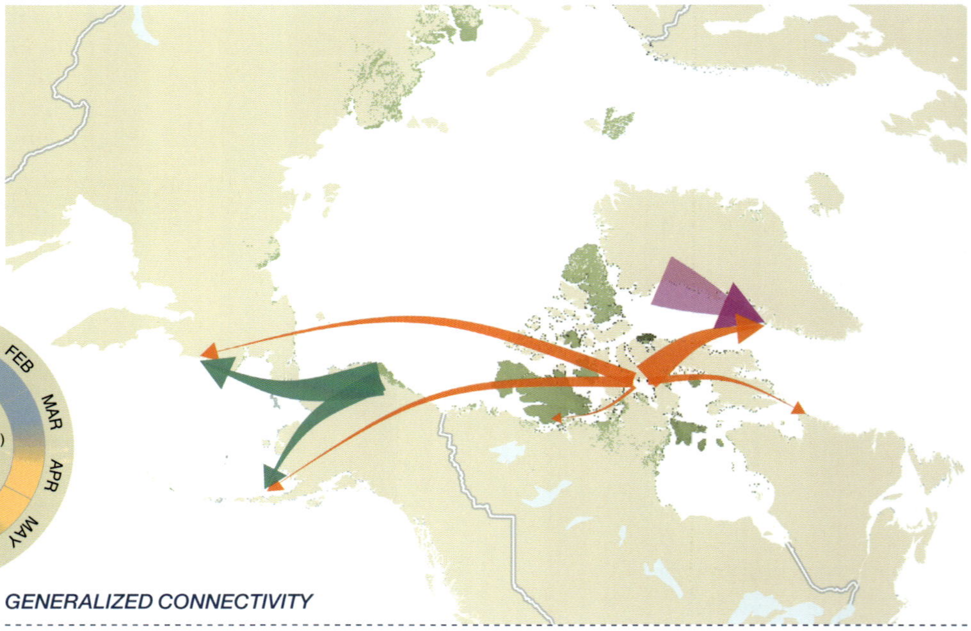

GENERALIZED CONNECTIVITY

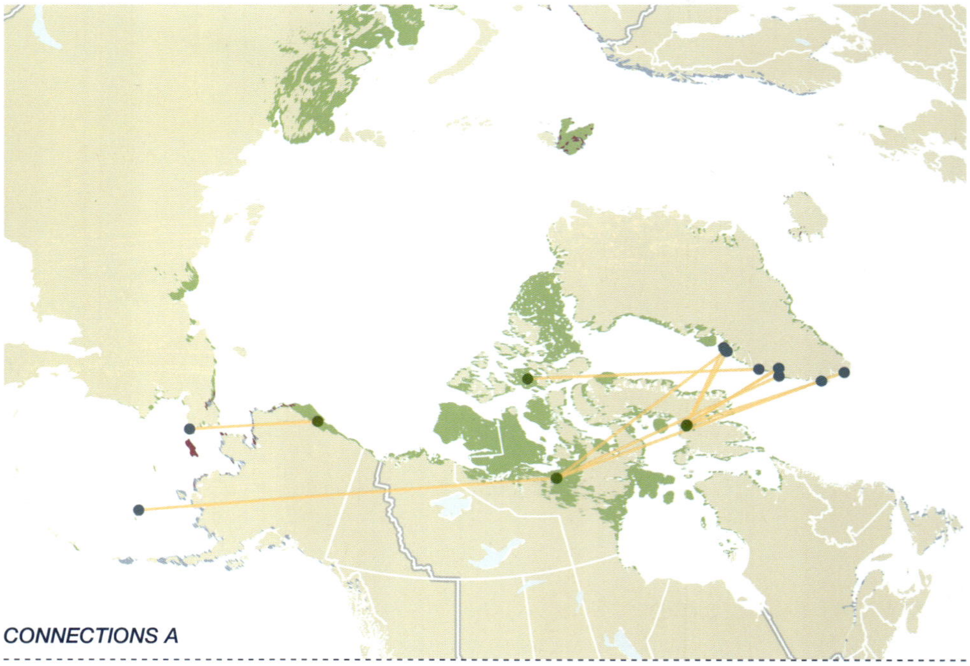

CONNECTIONS A

ORDER Anseriformes

FAMILY Anatidae

BREEDING HABITAT Wetlands, Tundra

NONBREEDING HABITAT Coasts, Marine Waters

GLOBAL POPULATION SIZE 830,000

POPULATION CHANGE SINCE 1970 −67%

IUCN RED LIST CATEGORY Least Concern

R2R TIPPING POINT SPECIES Orange-Alert

US LISTING STATUS No

CANADA SPECIES AT RISK No

PIF CCS 15/20

	Atlantic	Greenland	Midcontinent	Pacific	Russia
Greenland (n = 3)	–	100%	–	–	–
Midcontinent (n = 51)	10%	51%	2%	16%	22%
Pacific (n = 92)	–	–	–	38%	62%

CONNECTIONS B

PATHWAYS

MIGRATORY CONNECTIVITY

OVERALL King Eiders are well-studied using banding data, satellite transmitters, and stable isotopes. Greenland breeding birds migrate to nonbreeding areas within Greenland (100%). Midcontinent Flyway Arctic breeding birds migrate either east to Greenland (51%) and the Atlantic Flyway (10%) or west to the Pacific Flyway (16%) and Russia (22%). Pacific Flyway birds migrate within the flyway (38%) and to Russia (62%). Movements of a large percentage of the breeding populations to multiple nonbreeding areas result in a weak MC (0.03 ± 0.04).

BANDING A combination of banding records and stable isotope analysis from King Eiders breeding in Nunavut, Canada, suggested these birds spent the nonbreeding season in the Pacific (Mehl et al. 2004). Recoveries of 41 banded birds from the Queen Maud Gulf occurred in both Pacific and Atlantic ocean nonbreeding areas, as well as in Russia (Pearce et al. 2019). Generally, birds banded in Alaska were recovered farther west, and birds banded in the eastern Canadian Arctic were recovered farther east.

SATELLITE TRACKING Satellite transmitters have revealed that the species has diffuse migratory connectivity, with breeding colonies dispersing across nonbreeding areas (Oppel et al. 2008). Western Canadian Arctic and Alaska coastal plain breeding King Eiders (n = 96) migrated to similar molting and nonbreeding areas; but females returned to natal breeding areas, while males dispersed throughout the breeding range (Oppel et al. 2009; Bustnes et al. 2010; D.L. Dickson 2012b; Bentzen and Powell 2015). Birds breeding in western Canada and northern Alaska were tracked to molting and nonbreeding sites in the Bering, Chukchi, and Beaufort seas (Phillips et al. 2007; Oppel et al. 2009; D.L. Dickson 2012b). Juvenile King Eiders (n = 63) dispersed more broadly during their first nonbreeding season than their 17 attendant adult females, contributing to gene dispersal (Bentzen and Powell 2015). Molting sites and nonbreeding sites in western Greenland are connected to breeding sites in Arctic Canada, though migration routes and staging areas vary (Mosbech and Boertmann 1999; Mosbech et al. 2006a; Merkel et al. 2017). King Eiders marked during the nonbreeding season in Norway (n = 6) moved throughout the Arctic, including to breeding areas in Arctic Canada (Bustnes et al. 2010).

STABLE ISOTOPES Stable isotopes were used to distinguish between western and eastern North American nonbreeding populations, with western populations associated with the Bering Sea and eastern populations associated with Greenland (Mehl et al. 2005). Individuals captured on their breeding grounds in northern Alaska were assigned to nonbreeding areas offshore of Russia and in southwest Alaska (Knoche et al. 2007; Oppel and Powell 2008).

GENETICS Nuclear DNA and mtDNA sequence data suggested no evidence of population genetic structure in King Eiders (Pearce et al. 2004). It is likely that high levels of gene flow seen in satellite telemetry and stable isotope data homogenized gene frequencies across the range, leading to no genetic differentiation among breeding and nonbreeding areas.

MIGRATORY PATHWAYS Migration occurs along Arctic coasts to open ocean nonbreeding areas. Birds stay near the ice-edge during the nonbreeding season. Heavier hunting in eastern nonbreeding areas (i.e., Greenland) likely biased previous migration pathways toward use of eastern over western nonbreeding sites (Mehl et al. 2004).

DATA CONTRIBUTORS

R. Bentzen; D.L. Dickson; H.G. Gilchrist; F. Merkel; A. Mosbech; S. Oppel; L.M. Phillips; A. Powell; C. Sonne.

COMMON EIDER *Somateria mollissima*

MC = 0.11 ± 0.04*

DISTRIBUTION AND ABUNDANCE

This species is a common colony breeder in the North American Arctic and along the northern Atlantic coast. Breeding concentrations occur in coastal Alaska, the central Arctic, Atlantic Canada, and the northern Atlantic coast of the United States. Nonbreeding areas vary by population but are in northern coastal areas. There are four subspecies of Common Eider in North America: Pacific (*S. m. v-nigrum*), Northern (*S. m. borealis*), Hudson Bay (*S. m. sedentaria*), and American (*S. m. dresseri*; Goudie et al. 2020). Populations of the Pacific, Northern, and Hudson Bay have declined (Goudie et al. 2020).

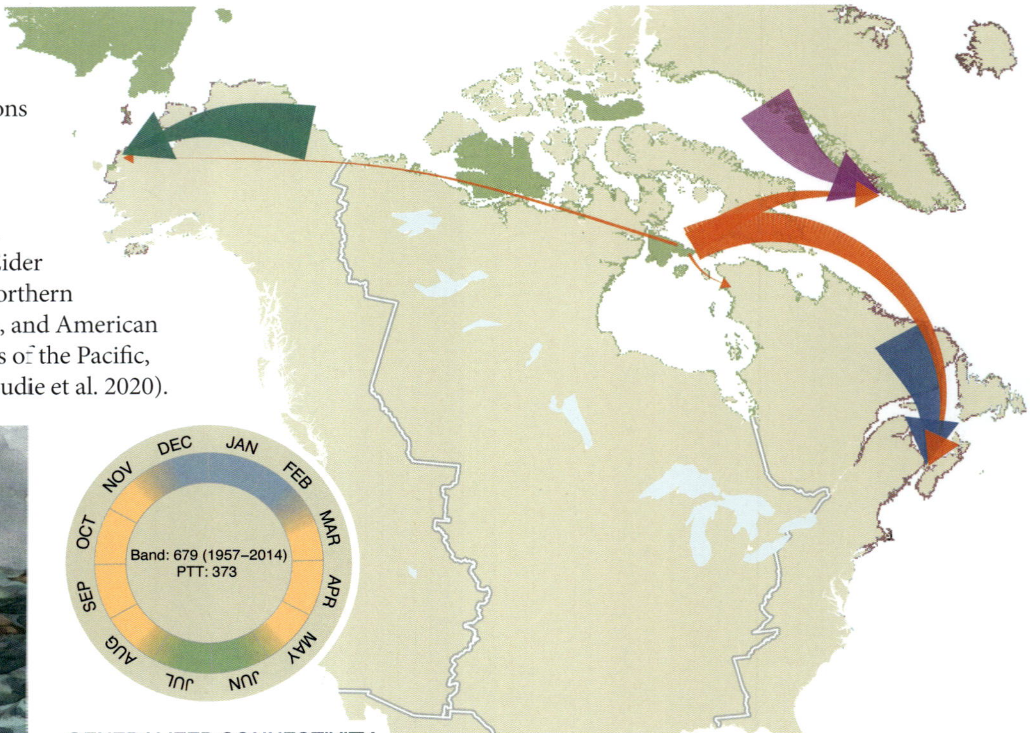

GENERALIZED CONNECTIVITY

ORDER Anseriformes
FAMILY Anatidae
BREEDING HABITAT
Coasts, Cliffs, and Islands
NONBREEDING HABITAT
Coasts, Marine Waters
GLOBAL POPULATION SIZE 2,300,000
POPULATION CHANGE SINCE 1970 −20%
IUCN RED LIST CATEGORY Near Threatened
R2R TIPPING POINT SPECIES No
US LISTING STATUS No
CANADA SPECIES AT RISK No
PIF CCS 12/20

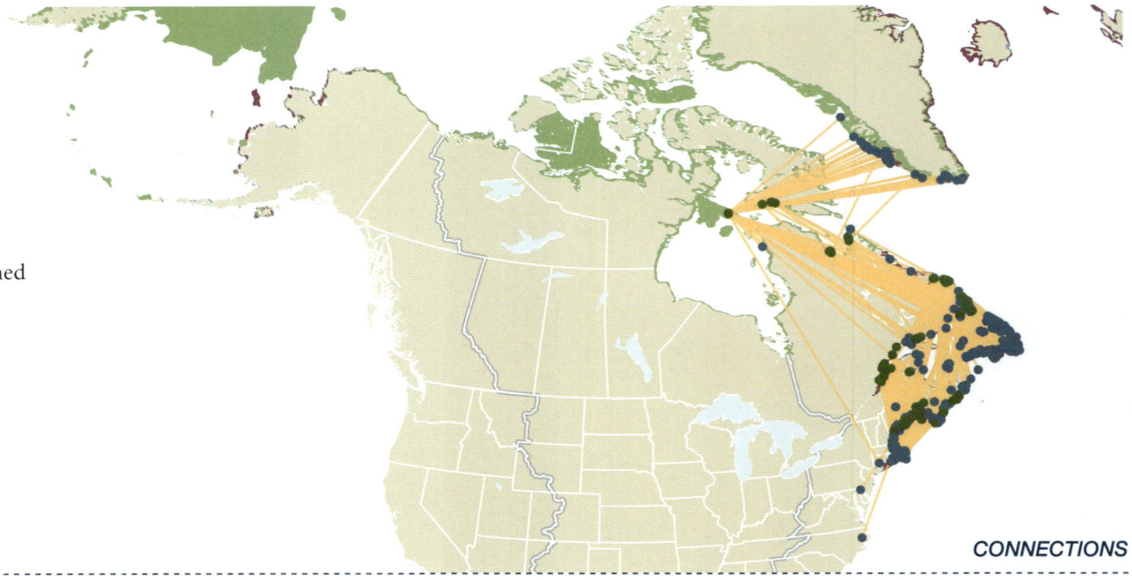

CONNECTIONS

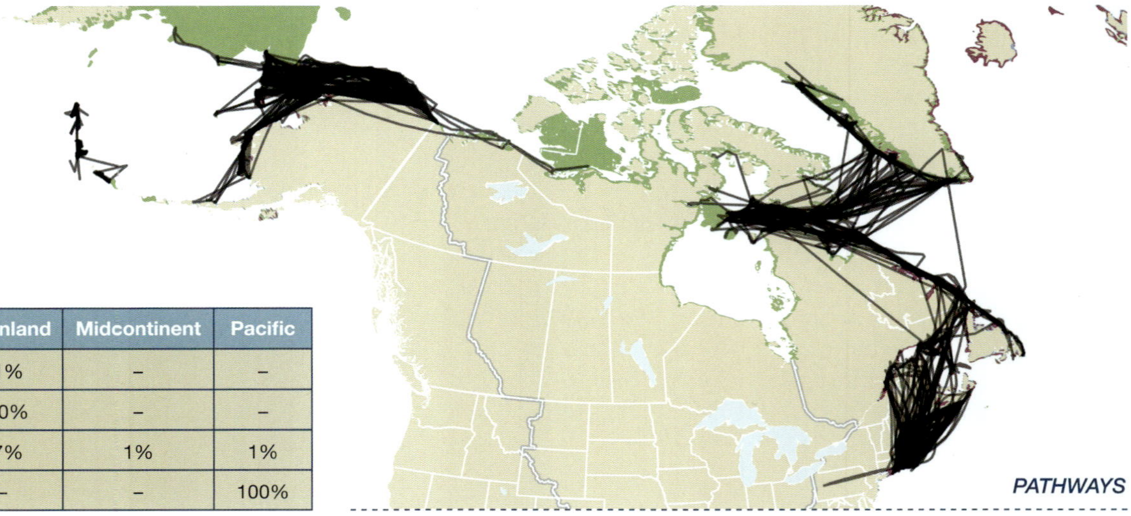

PATHWAYS

	Atlantic	Greenland	Midcontinent	Pacific
Atlantic (n = 634)	100%	<1%	–	–
Greenland (n = 6)	–	100%	–	–
Midcontinent (n = 163)	61%	37%	1%	1%
Pacific (n = 216)	–	–	–	100%

MIGRATORY CONNECTIVITY

OVERALL Migratory connectivity is derived from banding, genetic, and satellite transmitter data. Atlantic Flyway breeding birds spend the nonbreeding season within the Atlantic Flyway (100%). Birds breeding in Greenland remain within Greenland during the nonbreeding season (100%). Birds breeding in Arctic areas of the Midcontinent Flyway migrate to the Atlantic Flyway (61%) or Greenland (37%), while birds breeding in the Pacific Flyway stay in that geography (100%) during the nonbreeding season. Overlap of breeding populations, especially in the Atlantic Flyway and Greenland, during the nonbreeding season results in weak migratory connectivity (MC = 0.11 ± 0.04).

— PACIFIC COMMON EIDER

Somateria mollissima v-nigrum

OVERALL Pacific Common Eider breed in Alaska and the western high Arctic and spend the nonbreeding season in the Bering Sea and other northern coastal areas, as well as Russia, and do not overlap with other subspecies of Common Eider.

BANDING Pearce et al. (2019) showed banded Pacific Common Eider marked in northern Alaska and Canada (n = 11) were recovered in coastal areas of northern Russia and Alaska.

SATELLITE TRACKING Breeding populations can be delineated based on the movement of females marked on the breeding areas in the Yukon-Kuskokwim Delta and the North Slope of Alaska and their movement to differing nonbreeding areas in the Bering Sea (Petersen and Flint 2002). Pacific Common Eiders marked (n = 70) on central Canadian Arctic breeding grounds were found to migrate to ice leads southeast of Chukotsk Peninsula, Russia, and near St. Lawrence Island, Canada. These birds exhibited fidelity to nonbreeding areas; females returned to natal breeding grounds, while males dispersed (D.L. Dickson 2012c). Pacific Common Eiders (n = 39 females) on the Yukon-Kuskokwim Delta appear largely nonmigratory, remaining nearby in Bristol Bay, Alaska, during the nonbreeding season (Petersen and Flint 2002).

GENETICS Using nuclear DNA and mtDNA, Sonsthagen et al. (2011) found evidence that Pacific Common Eiders had genetic structuring among breeding locations in the Arctic.

— NORTHERN COMMON EIDER

Somateria mollissima borealis

OVERALL Northern Common Eider breed in northern Atlantic Canada and migrate to nonbreeding areas locally and in coastal waters around Greenland.

BANDING Banding records suggest some Northern Common Eider overlap with American Common Eider in eastern Canada during the nonbreeding season (Baldassarre 2014).

SATELLITE TRACKING Northern Common Eiders marked (n = 26) at a breeding colony on Southampton Island migrated to two primary nonbreeding areas: the coast of Atlantic Canada (40%) and southwestern Greenland (60%; Mosbech et al. 2006b). Northern Common Eiders breeding in the Gyrfalcon archipelago of Ungava Bay were tracked (n = 15) to nonbreeding areas in Greenland (67%) and Canada (33%; Savard et al. 2011).

STABLE ISOTOPES Carbon isotopes collected from samples of Northern Common Eiders on breeding colonies support satellite telemetry data showing two separate nonbreeding areas, southwest Greenland and Maritime Canada (Steenweg et al. 2017).

GENETICS Sonsthagen et al. (2011) found, based on nuclear and mtDNA, that the three eastern Canada Common Eider subspecies (Northern, Hudson Bay, American) are not genetically differentiated.

— HUDSON BAY COMMON EIDER

Somateria mollissima sedentaria

OVERALL Hudson Bay Common Eider are primarily nonmigratory and remain within Hudson Bay during the breeding and nonbreeding seasons. They breed along the Hudson Bay coast and move farther west in ice-free areas post-breeding.

OTHER Morphology data is the primary way to differentiate the Hudson Bay Common Eider from other subspecies in North America (Goudie et al. 2020).

— AMERICAN COMMON EIDER

Somateria mollissima dresseri

OVERALL American Common Eider breed in eastern Canada and the northeastern United States. They migrate to nonbreeding areas primarily along the Atlantic coast but may stay in ice-free areas of the St. Lawrence River Estuary. Numbers of nonbreeding eiders around Nova Scotia increased for years, but now are declining (Robertson et al. 2021), consistent with overall patterns of declines in numbers of birds in the center of the range (Noel et al. 2021).

BANDING Pearce et al. (2019) mapped 3,219 banding records. Birds banded within the recognized geographic range of the American Common Eider were recovered with the expected geographical distribution of Maritime Canada and the Atlantic coast of the United States from Maine south to Long Island, New York.

SATELLITE TRACKING Nonbreeding females (n = 24) off the coast of southern New England were found to breed in Maine, the Gulf of St. Lawrence, the St. Lawrence Estuary, and Nova Scotia, and returned to the same breeding areas two consecutive years (Beuth 2013). American Common Eiders breeding in Boston Harbor are currently being monitored with satellite telemetry and have been found to molt from Boston Harbor north to the Gulf of St. Lawrence, then return to the Boston Harbor and Nantucket Sound regions during the nonbreeding season (Meattey 2017).

MIGRATORY PATHWAYS Eiders breeding in Nova Scotia migrated along the Gulf of Maine to nonbreeding areas principally around waters off of Cape Cod (Mallory et al. 2020b), a common pattern among sea ducks in northeastern North America (Lamb et al. 2019).

DATA CONTRIBUTORS

B. Allen; D.C. Douglas; C. Dwyer; H.G. Gilchrist; H. Heusmann; J.S. Lamb; M.L. Mallory; D. McAuley; A. Mosbech; D. Meattey; F. Merkel; G. Olsen; M. Petersen; L. Savoy; C. Sonne; K.M. Sullivan.

HARLEQUIN DUCK *Histrionicus histrionicus*

MC = 0.72 ± 0*

DISTRIBUTION AND ABUNDANCE

The North American population of Harlequin Ducks has a disjunct distribution. A western population has breeding concentrations in Alaska and the interior Rocky Mountains. The nonbreeding range extends along the Pacific coast from Alaska south to Oregon. An eastern population breeds in Quebec and Labrador, with nonbreeding concentrations along the Atlantic coast from Labrador south to New York, as well as in Greenland.

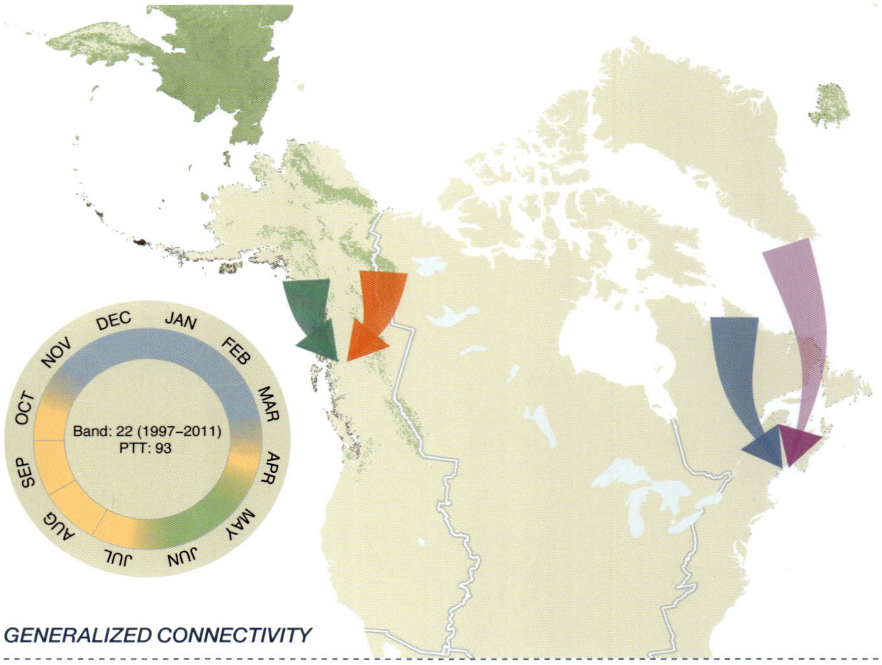

GENERALIZED CONNECTIVITY

ORDER Anseriformes
FAMILY Anatidae
BREEDING HABITAT Wetlands, Lakes, and Rivers
NONBREEDING HABITAT Coasts, Marine Waters
GLOBAL POPULATION SIZE 230,000
POPULATION CHANGE SINCE 1970 −20%
IUCN RED LIST CATEGORY Least Concern
R2R TIPPING POINT SPECIES No
US LISTING STATUS No
CANADA SPECIES AT RISK No
PIF CCS 11/20

	Atlantic	Pacific
Atlantic (n = 7)	100%	–
Midcontinent (n = 1)	100%	–
Midcontinent (n = 25)	–	100%
Pacific (n = 61)	–	100%

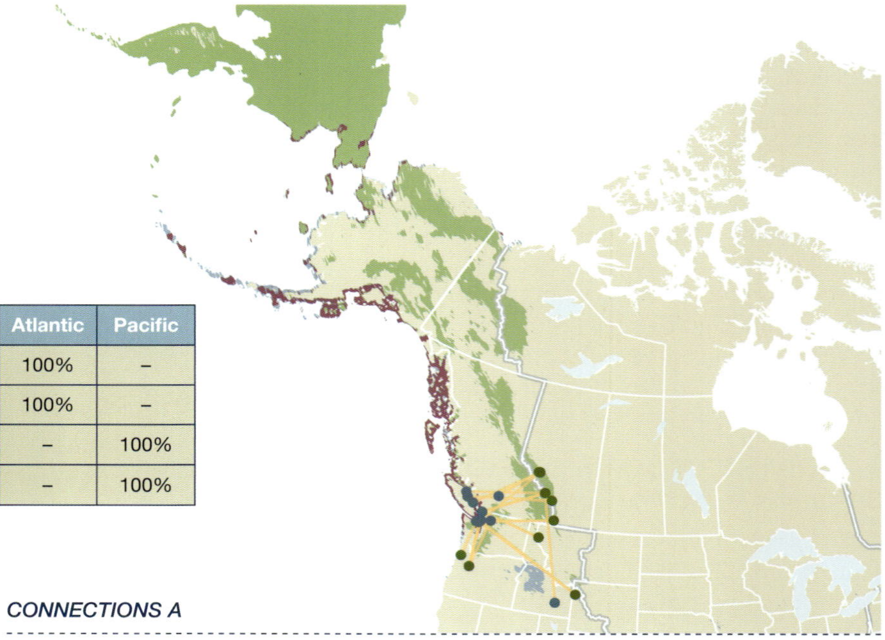

CONNECTIONS A

MIGRATORY CONNECTIVITY

OVERALL Banding, satellite transmitter, and genetic data inform migratory connectivity. There is a distinct separation between eastern and western populations; however, there is significant mixing of breeders within each population. In eastern North America, breeding birds migrate to two nonbreeding areas with substantial overlap (Baldassarre 2014). Greenland and Atlantic Flyway breeding birds spend the nonbreeding season within the Atlantic Flyway (100%), migrating to the coastal United States, especially Maine, but appearing regularly as far south as Maryland. In western North America, breeding birds from the Pacific Flyway and western Midcontinent Flyway overlap during the nonbreeding season along the coast of the Pacific Flyway. Migratory connectivity is strong within this species (MC = 0.72 ± 0).

BANDING Band recovery data from eastern North America show no movement to western populations. Banding data do not support separate populations in eastern North American and Greenland nonbreeding areas (Thomas et al. 2008; Pearce et al. 2019).

SATELLITE TRACKING Male Harlequin Ducks (n = 8) captured during pre-breeding migration along the shores of Forillon National Park in Quebec traveled through Labrador and spent the nonbreeding season in Maine; whereas males (n = 17) captured in northern Quebec and Labrador during breeding migrated to Greenland for molting and the nonbreeding season (Brodeur et al. 2002). Male Harlequin Ducks (n = 8) tagged on the nonbreeding grounds in Maine were found to molt in Greenland and along the Labrador coast (n = 2), before returning to Maine for the next nonbreeding season (Robert et al. 2008). Harlequin Ducks breeding in central Labrador likely represent birds along both the northern Atlantic coast and Greenland nonbreeding areas (Chubbs et al. 2008), though this is an open question. Harlequin Ducks tracked from Montana migrated to nonbreeding areas along the coast of southeastern Alaska and south through British Columbia to Washington state (L. Bate and C. Hammond, unpublished data).

GENETICS Across North America there is strong evidence for genetic differentiation between western and eastern populations. Within the western population, no spatial genetic structuring was found between populations of Harlequin Ducks in nonbreeding areas within Prince William Sound, the Alaska Peninsula, or Kodiak Archipelago

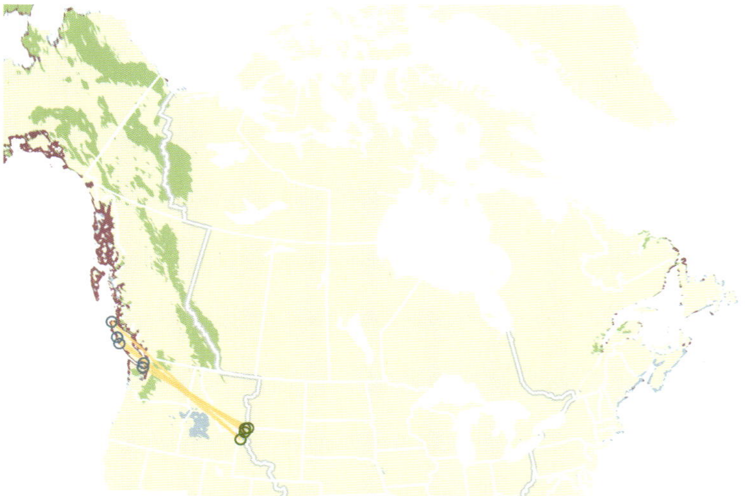

CONNECTIONS B

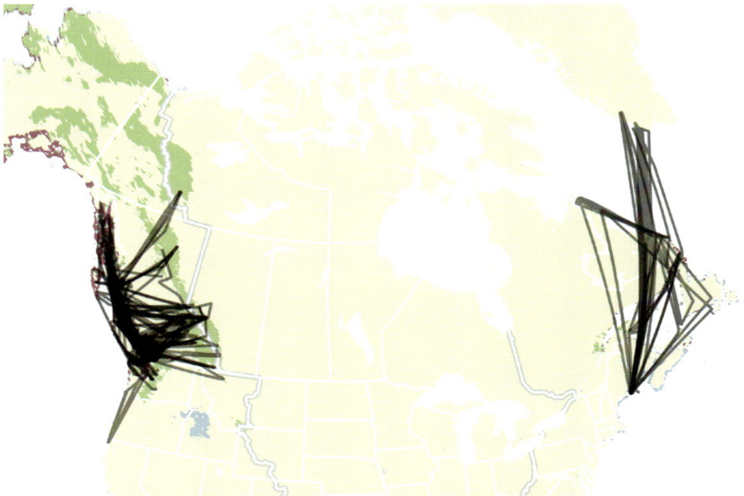

PATHWAYS

(Lanctot et al. 1999). There is some evidence of genetic structure in Harlequin Ducks breeding in eastern North America, with separation between birds in Greenland versus those in Maritime Canada and the northeast United States during the nonbreeding season (Robertson and Goudie 2020).

STABLE ISOTOPES No stable isotope studies of migratory connectivity were found for this species.

OTHER APPROACHES Adult Harlequin Ducks marked in the Strait of Georgia, British Columbia, had strong site fidelity, and annual movement distances between nonbreeding sites decreased with age (Regehr 2011). Nonbreeding site fidelity of Harlequin Ducks in Prince William Sound, Alaska, was high, with a return rate >0.92 (Iverson et al. 2004).

MIGRATORY PATHWAYS There is evidence for general longitudinal migration from interior breeding areas to coastal nonbreeding areas.

DATA CONTRIBUTORS

W.S. Boyd; J. Evenson; G. Fitzgérald; B. Jobin; P. Lamothe; B. MacCallum; M. McAdie; D. Meattey; G.H. Mittehauser; A. Orabona; S. Patla; M. Robert; L. Savoy; D. Smith; K. Spragens; J. Stephenson; M. Wilson.

INTERESTING NOTE

Return rates of radio-tagged juvenile Harlequin Ducks were the same between male and females (Regehr 2003).

Melanitta perspicillata SURF SCOTER

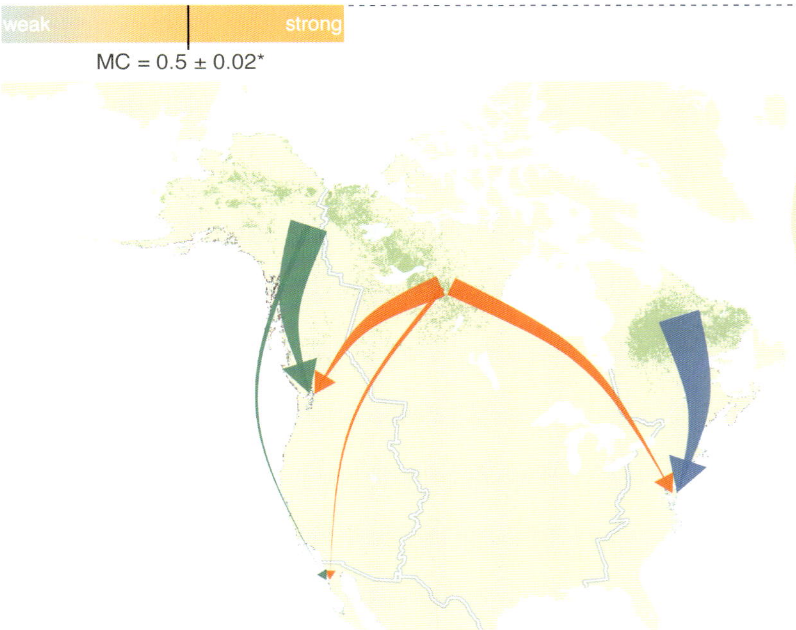

GENERALIZED CONNECTIVITY

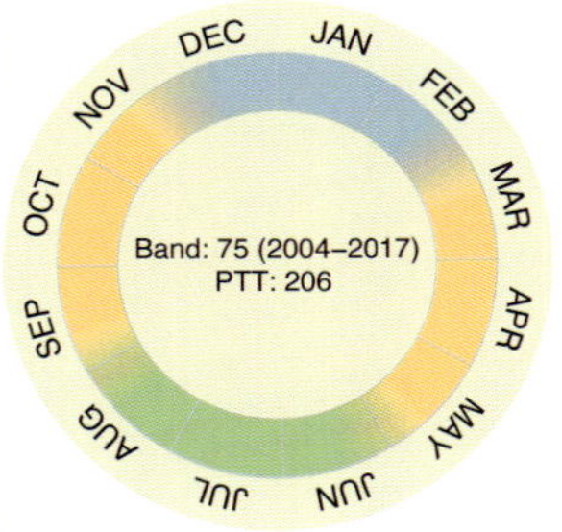

DISTRIBUTION AND ABUNDANCE

Surf Scoters are endemic to North America. They breed in the boreal forest of central Canada, in Alaska, and north of the Gulf of St. Lawrence and spend the nonbreeding season along the Atlantic coast south to the Carolinas, and along the Pacific coast south to Mexico (E.M. Anderson et al. 2020).

ORDER Anseriformes
FAMILY Anatidae
BREEDING HABITAT Wetlands, Boreal
NONBREEDING HABITAT Coasts, Marine Waters
GLOBAL POPULATION SIZE 470,000

POPULATION CHANGE SINCE 1970 +10%
IUCN RED LIST CATEGORY Least Concern
R2R TIPPING POINT SPECIES No
US LISTING STATUS No
CANADA SPECIES AT RISK No
PIF CCS 11/20

	Atlantic	Mexico	Pacific
Atlantic (n = 132)	100%	–	–
Midcontinent (n = 74)	42%	7%	51%
Pacific (n = 11)	–	9%	91%

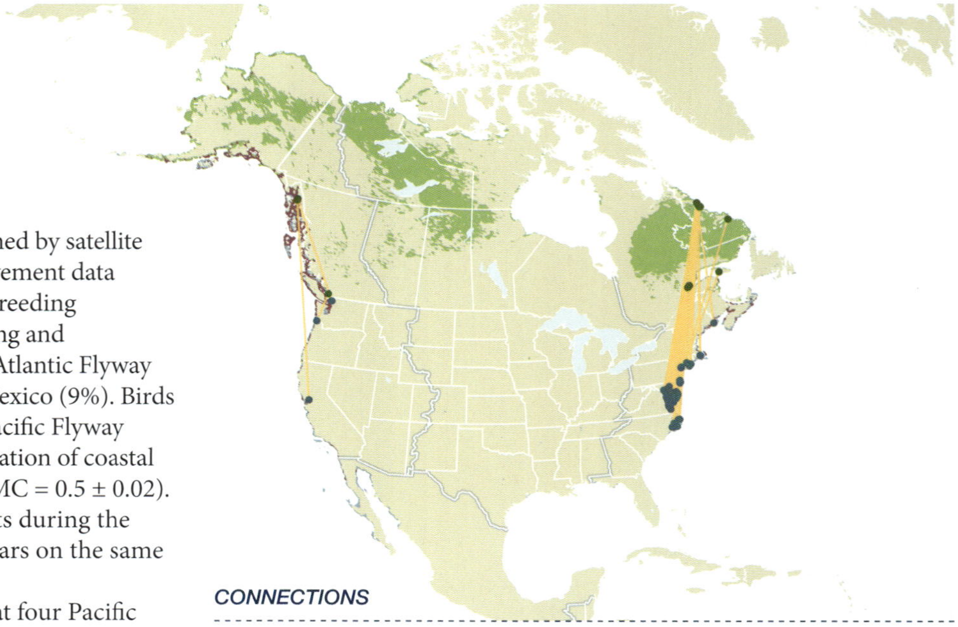

CONNECTIONS

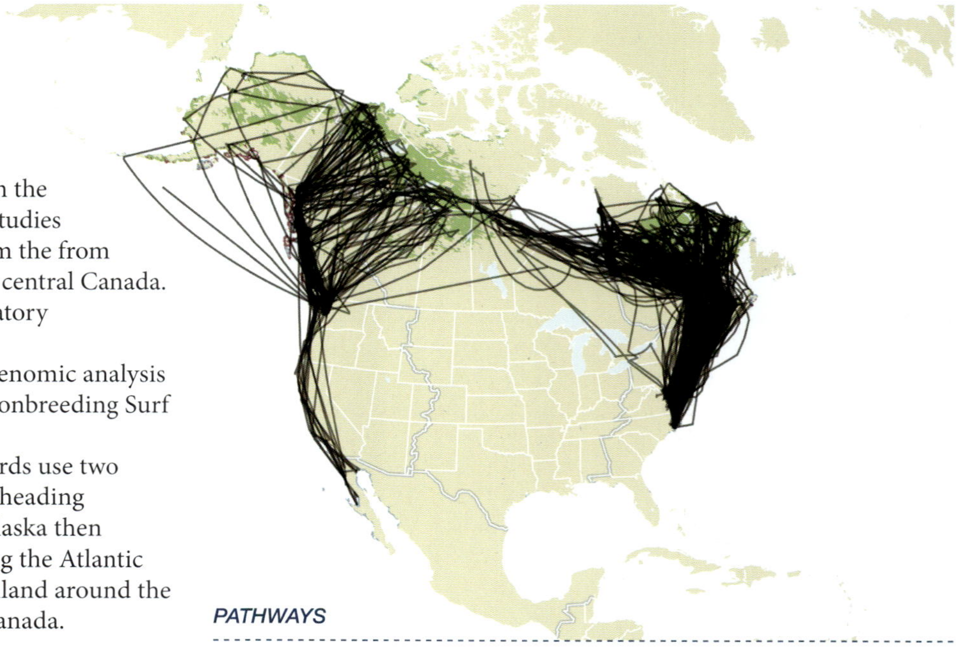

PATHWAYS

MIGRATORY CONNECTIVITY

OVERALL Surf Scoter migratory connectivity is informed by satellite telemetry, genetic analysis, and minimal banding. Movement data suggest the separation of Pacific and Atlantic flyways breeding populations, with high transition rates between breeding and nonbreeding within the Pacific Flyway (91%) and the Atlantic Flyway (100%). Pacific Flyway breeding birds also winter in Mexico (9%). Birds breeding in the Midcontinent Flyway migrate to the Pacific Flyway (51%), Atlantic Flyway (42%), and Mexico (7%). Separation of coastal populations results in strong migratory connectivity (MC = 0.5 ± 0.02).
BANDING Surf Scoters (n = 98) banded on both coasts during the nonbreeding season were recovered in subsequent years on the same coast where they were banded (Pearce et al. 2019).
SATELLITE TRACKING Surf Scoters marked (n = 35) at four Pacific coast nonbreeding sites in Baja California, San Francisco Bay, Puget Sound, and Strait of Georgia showed strong connectivity to unique breeding sites (De La Cruz et al. 2009), demonstrating potential populations structure in the Pacific population. Individuals tracked (n = 202) during pre-breeding migration from the Atlantic coast migrated to breed in the eastern boreal forest (Sea Duck Joint Venture 2015). Studies showed that during the nonbreeding season birds from the from Pacific and Atlantic coasts overlap during breeding in central Canada.
STABLE ISOTOPES No stable isotope studies of migratory connectivity were found for this species.
GENETICS Despite high geographic differentiation, genomic analysis suggests high gene flow among Pacific and Atlantic nonbreeding Surf Scoters (Sonsthagen et al. 2019).
MIGRATORY PATHWAYS Pacific coast nonbreeding birds use two main pre-breeding migration routes, an inland route heading northeast from Puget Sound and a coastal route to Alaska then inland. Birds that spend the nonbreeding season along the Atlantic coast migrate north along the coast before heading inland around the mouth of the St. Lawrence River to breed in boreal Canada.

DATA CONTRIBUTORS

W.S. Boyd; S. De La Cruz; D.C. Douglas; J. Eadie; D. Esler; J. Evenson; M. Kirk; J.S. Lamb; E. Lok; D. Nyeswander; E. Palm; J. Pearce; K. Spragens; J. Takekawa; D. Ward; M. Wilson; J. Yee.

WHITE-WINGED SCOTER *Melanitta deglandi*

weak strong

MC = 0.35 ± 0.05*

DISTRIBUTION AND ABUNDANCE

White-winged Scoters have a wide breeding distribution, occurring throughout central Canada and Alaska. Breeding birds also occur in northeast Canada. Breeding density is generally low. The nonbreeding range extends along the northern Pacific and Atlantic coasts of North America, with some birds occurring on the Great Lakes during the nonbreeding season (Baldassarre 2014).

ORDER Anseriformes
FAMILY Anatidae
BREEDING HABITAT Wetlands, Lakes, and Rivers
NONBREEDING HABITAT Coasts, Marine Waters
GLOBAL POPULATION SIZE 400,000
POPULATION CHANGE SINCE 1970 −19%
IUCN RED LIST CATEGORY Least Concern
R2R TIPPING POINT SPECIES No
US LISTING STATUS No
CANADA SPECIES AT RISK No
PIF CCS 12/20

	Atlantic	Midcontinent	Pacific
Atlantic (n = 49)	92%	8%	–
Midcontinent (n = 101)	62%	6%	32%
Pacific (n = 7)	–	–	100%

MIGRATORY CONNECTIVITY

OVERALL Data of several forms characterize the migratory connectivity of White-winged Scoters, including banding, satellite tracking, stable isotope, and genetic data. They have two primary nonbreeding areas, the Pacific and Atlantic coasts, but there is evidence of mixing on breeding areas due to cross-continent migrations of individuals (Pearce et al. 2019), especially from the midcontinent. White-winged Scoters breeding in the Atlantic and Pacific flyways nearly all migrate to nonbreeding areas within the same flyway (92%–100%), while birds breeding in the Midcontinent Flyway migrate to nonbreeding areas in the Atlantic (62%) and Pacific (32%) flyways, as well as the Midcontinent Flyway (6%). High coastal flyway transitions with overlap from the midcontinent breeding population result in moderate migratory connectivity (MC = 0.35 ± 0.05).

BANDING White-winged Scoters banded on Redberry Lake, Saskatchewan, were recovered on both Pacific and Atlantic coasts (Houston and Brown 1983; Swoboda 2007). Recoveries of birds banded farther north in Canada were subsequently recovered along the Atlantic coast, but also in Alaska, while nearly all birds banded in Alaska were recovered in Alaska (Pearce et al. 2019).

SATELLITE TRACKING White-winged Scoters were tracked from nonbreeding and staging areas in eastern North America (New York to James Bay) to breed along the southeastern coast of Hudson Bay across central Canada to the Yukon Territories (Sea Duck Joint Venture 2015; Meattey et al. 2018; Lamb et al. 2019). White-winged Scoters tracked from nonbreeding areas in Prince William Sound, Alaska, migrated to breeding areas in interior Alaska and the Yukon and Northwest territories of Canada (R.D. Dickson et al. 2012). Fifteen of the 22 birds returned to nonbreeding locations the next year, with over half of those returning to locations where they were marked. All other birds spent the nonbreeding season in other parts of coastal Alaska, particularly the Alaska Peninsula and northern Gulf of Alaska.

STABLE ISOTOPES White-winged Scoter females breeding on Redberry Lake, Saskatchewan, were assigned to nonbreeding areas on both the Atlantic and Pacific coasts using carbon and nitrogen signatures in head feathers (Gurney et al. 2014). Females breeding farther north and west, at Cardinal Lake, Northwest Territories, mostly came from nonbreeding areas on the Pacific coast, though some Atlantic coast nonbreeding birds bred there as well.

GENETICS Analysis of genetic markers from White-winged Scoter and from Velvet Scoters from Europe show a clear differentiation between the recently split species. Within North America, there was no population differentiation of White-winged Scoters, according to genomic analysis, likely due to dispersal of males across the breeding area and both coasts, and to females occasionally switching to a different coast during the nonbreeding season (Sonsthagen et al. 2019).

MIGRATORY PATHWAYS Migratory pathways utilize stopover sites at major water bodies including the Great Lakes, Hudson Bay, and St. Lawrence River (Baldassarre 2014).

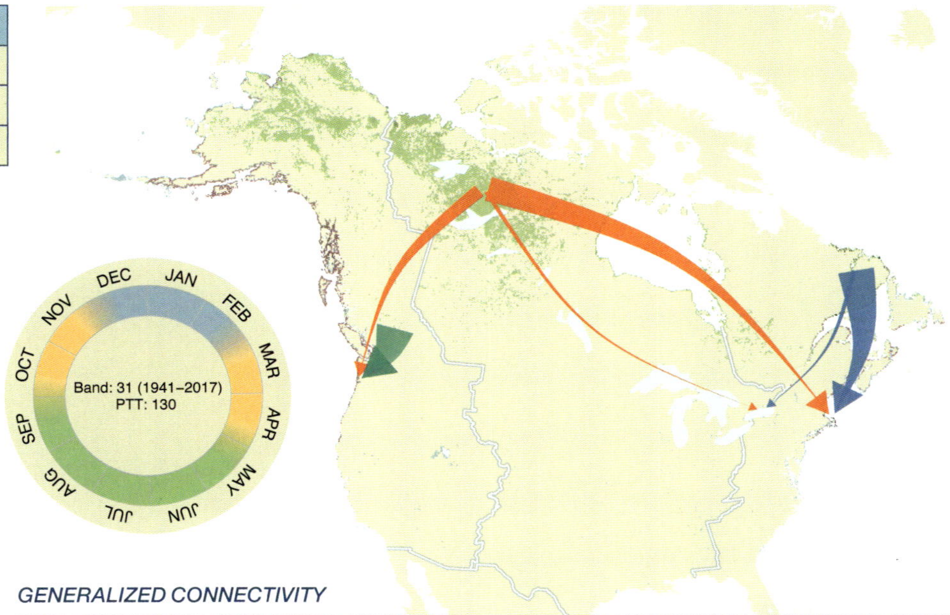

GENERALIZED CONNECTIVITY

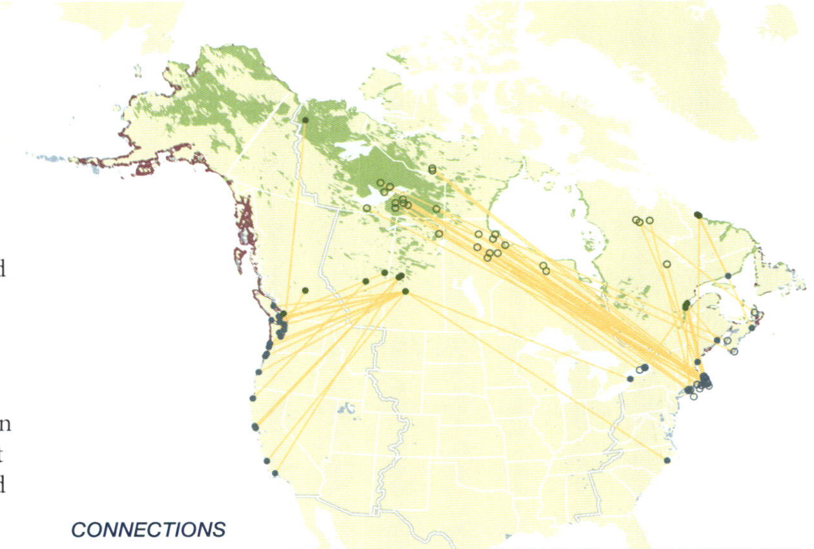

CONNECTIONS

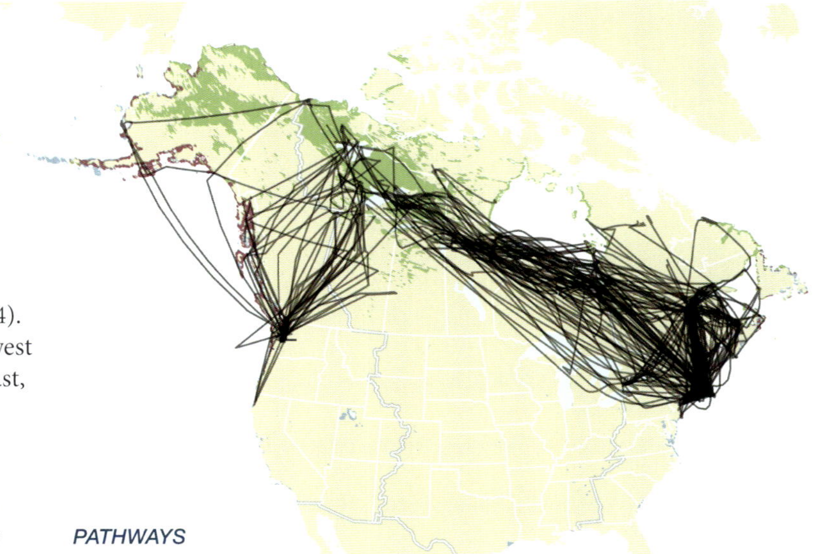

PATHWAYS

DATA CONTRIBUTORS

J. Evenson; S. Gilliland; J.S. Lamb; C. Lepage; D. Meattey; S. McWilliams; G. Olsen; J. Osenkowski; P. Paton; L. Savoy; K. Spragens; M. Wilson.

BLACK SCOTER *Melanitta americana*

DISTRIBUTION AND ABUNDANCE

Black Scoters breed primarily in Alaska and in boreal forests from the Northwest Territories north of Great Slave Lake east along the southern coast of Hudson Bay into northern Quebec and Labrador. Black Scoters spend their nonbreeding season over a large latitudinal gradient on both coasts (Bordage and Savard 2020).

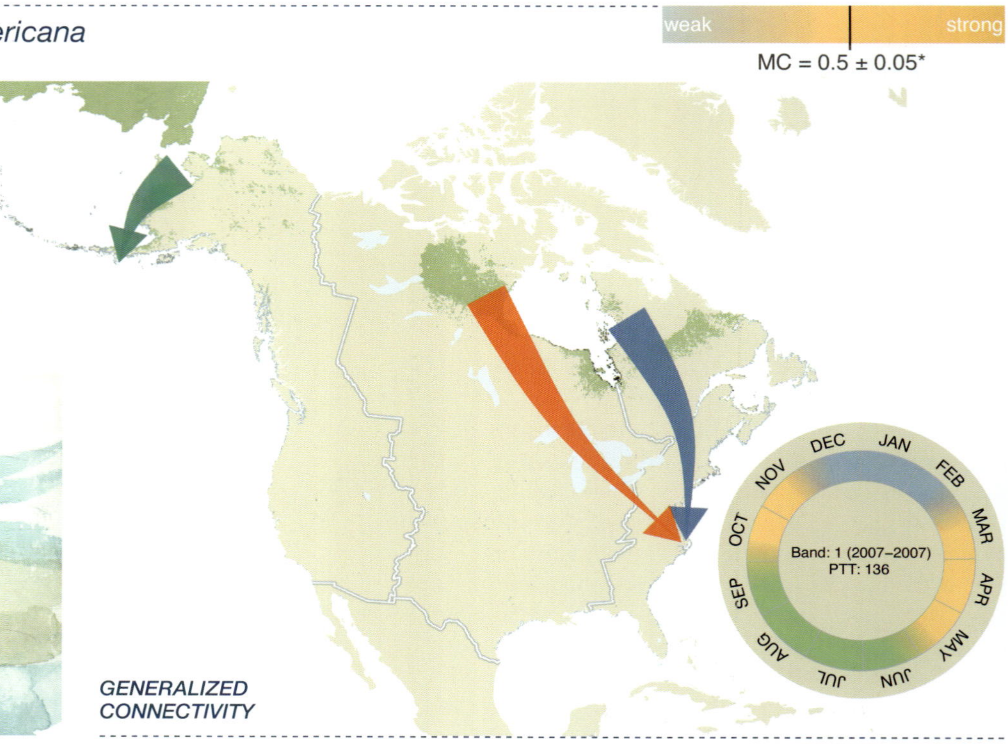

GENERALIZED CONNECTIVITY

ORDER Anseriformes
FAMILY Anatidae
BREEDING HABITAT Wetlands, Tundra
NONBREEDING HABITAT Coasts, Marine Waters
GLOBAL POPULATION SIZE 900,000
POPULATION CHANGE SINCE 1970 −22%

IUCN RED LIST CATEGORY Near Threatened
R2R TIPPING POINT SPECIES Yellow-Alert
US LISTING STATUS No
CANADA SPECIES AT RISK No
PIF CCS 13/20

	Atlantic	Pacific
Atlantic (n = 24)	100%	–
Midcontinent (n = 46)	100%	–
Pacific (n = 42)	–	100%

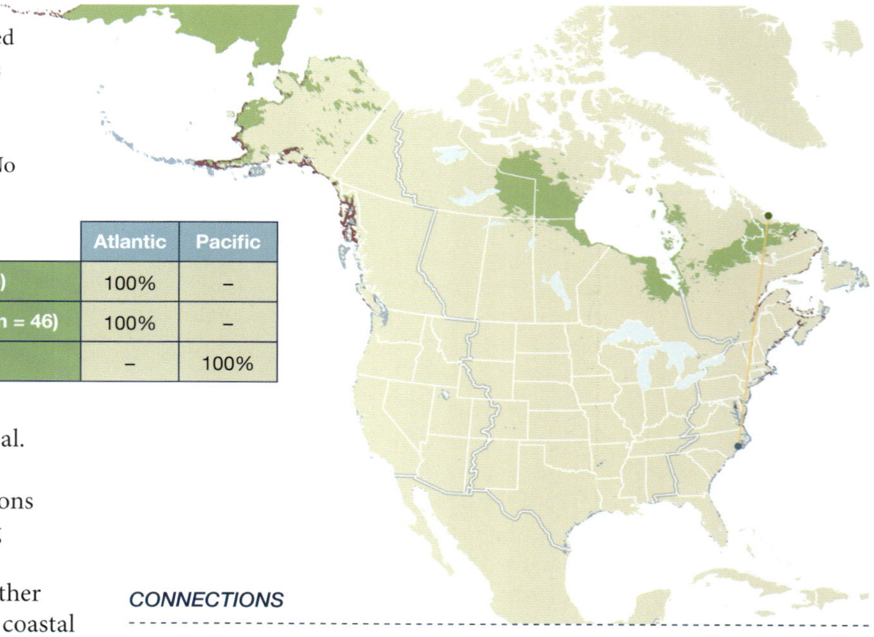

CONNECTIONS

MIGRATORY CONNECTIVITY

OVERALL Migratory connectivity is informed primarily by satellite telemetry and genetics (Bowman et al. 2007; Sea Duck Joint Venture 2014; Bordage and Savard 2020; Lamb et al. 2021). There are two disjunct populations in North America, and no regular interchange has been found between populations (Bowman et al. 2007; Sea Duck Joint Venture 2014), resulting in strong migratory connectivity (MC = 0.50 ± 0.5). The two populations are Alaska breeding birds and those breeding farther east. Birds breeding in the Pacific Flyway (Alaska) migrate to coastal areas within the Pacific Flyway (100%), while those breeding in northern Canada in the Midcontinent and Atlantic flyways spend the nonbreeding season along the Atlantic Flyway coast (100% for both breeding populations).

BANDING Records of 47 recovered Black Scoters showed no movement from nonbreeding areas in eastern North America to western North America (Pearce et al. 2019).

SATELLITE TRACKING Over 100 Black Scoters tracked from nonbreeding and molting areas in eastern North America moved to breeding areas that were previously unknown in interior Canada. These birds extended the known breeding range from the south shore of Hudson Bay north and west through the Northwest Territories (Sea Duck Joint Venture 2014). There was significant intermixing of nonbreeding areas within the eastern population (Lamb et al. 2021). Black Scoters tracked with satellite telemetry from nonbreeding locations in British Columbia and Alaska migrated to breeding areas in the Yukon-Kuskokwim Delta region of Alaska. Tracking data from these Alaska breeding grounds showed females migrated to

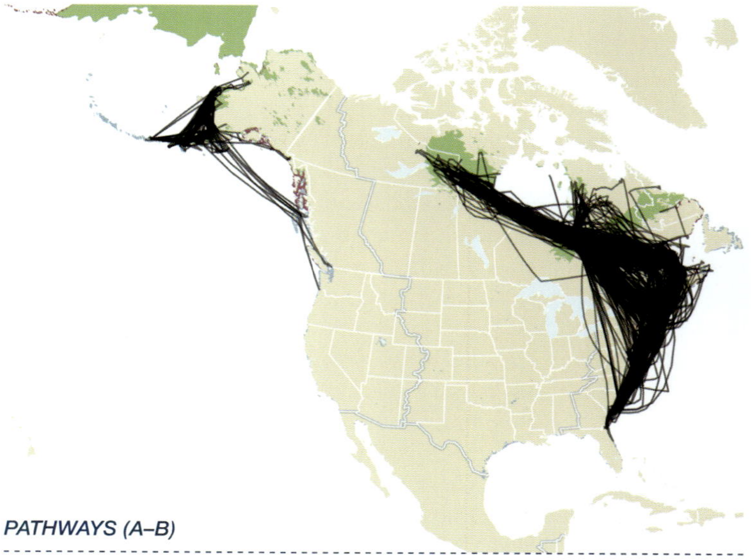

PATHWAYS (A–B)

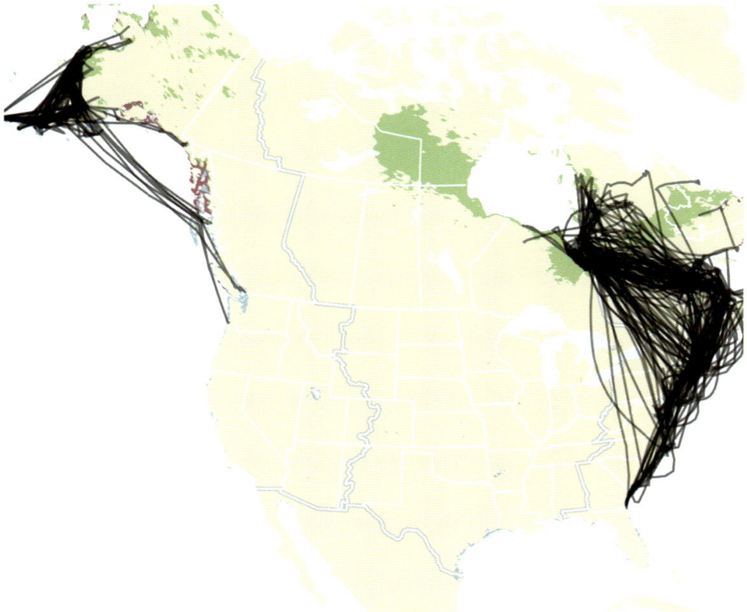

PATHWAYS A

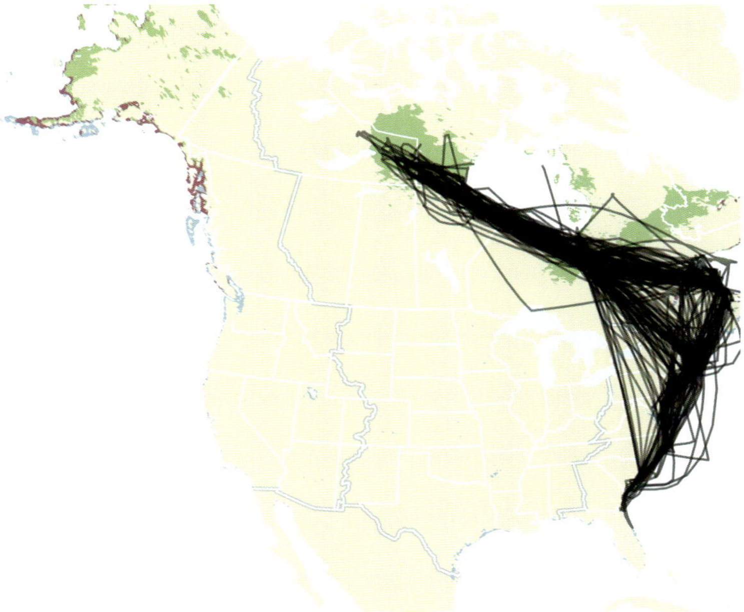

PATHWAYS B

nonbreeding areas along the Pacific coast (Bowman et al. 2007).
GENETICS Sonsthagen et al. (2019), using nuclear DNA, found a clear genetic differentiation between Black Scoters breeding in Alaska and those breeding in central and eastern Canada.
STABLE ISOTOPES No stable isotope studies of migratory connectivity were found for this species.

DATA CONTRIBUTORS
D.C. Douglas; P. Flint; J.S. Lamb; J. Schamber.

MIGRATORY PATHWAYS Western Black Scoters travel directly to the coast from interior breeding areas and follow coastal routes to nonbreeding areas. Eastern Black Scoters follow a southeasterly route from primary breeding areas through Hudson and James bays and the Great Lakes to nonbreeding areas along the Atlantic coast. Some eastern birds travel more directly east to the coast before migrating south.

Clangula hyemalis LONG-TAILED DUCK

weak strong
MC = 0.02 ± 0.02*

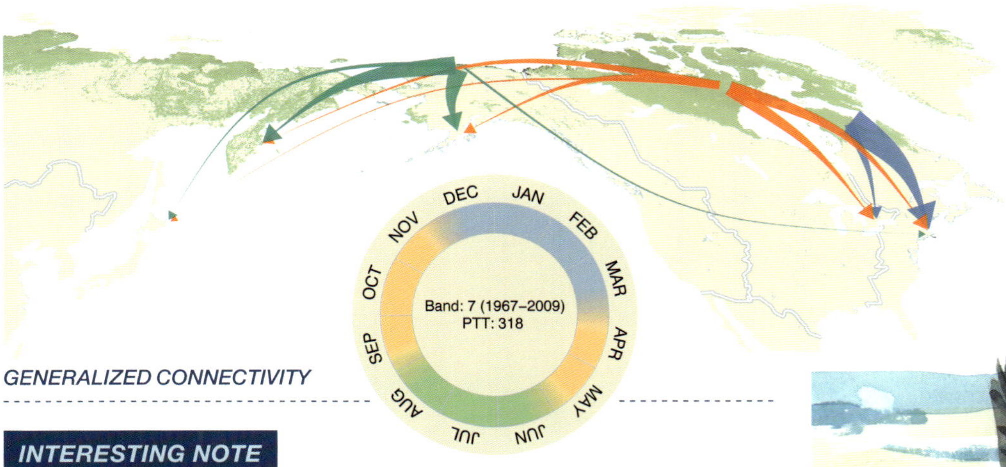

GENERALIZED CONNECTIVITY

INTERESTING NOTE

Habitat needs of Long-tailed Ducks throughout the annual cycle were similar to multiple sea ducks, suggesting they could be an effective umbrella species (Lamb et al. 2019).

ORDER Anseriformes
FAMILY Anatidae
BREEDING HABITAT Wetlands, Tundra
NONBREEDING HABITAT Coasts, Marine Waters
GLOBAL POPULATION SIZE 3,200,000
POPULATION CHANGE SINCE 1970 −51%

IUCN RED LIST CATEGORY Vulnerable
R2R TIPPING POINT SPECIES Orange-Alert
US LISTING STATUS No
CANADA SPECIES AT RISK No
PIF CCS 13/20

DISTRIBUTION AND ABUNDANCE
This species has a circumpolar distribution, and within North America, Long-tailed Ducks breed throughout the Arctic, with high densities along the Alaska coastal plain and around Queen Maud Gulf (Baldassarre 2014). Major nonbreeding areas are the southwest Alaska coast, New England, and the Great Lakes.

MIGRATORY CONNECTIVITY

	Asia	Atlantic	Midcontinent	Pacific	Russia
Atlantic (n = 9)	–	78%	22%	–	–
Midcontinent (n = 88)	1%	32%	33%	19%	15%
Pacific (n = 110)	8%	1%	–	46%	45%

OVERALL Banding, transmitter, and genetic evidence exist to describe migratory connectivity. Banding and genetic data suggest a panmictic population, while transmitter data suggest a possible separation between western and eastern populations (Pearce et al. 2019). Atlantic Flyway breeding birds primarily migrate within the flyway (78%) or transition to the Midcontinent Flyway (22%). On the other coast, Pacific Flyway breeding birds migrate within the Pacific Flyway (46%) or to Russia (45%), and smaller numbers migrate to Asia (8%). Midcontinent Flyway breeding birds migrate to multiple nonbreeding regions including the Midcontinent Flyway (33%), Atlantic Flyway (32%), Pacific Flyway (19%), or Russia (15%). Although separation between western and eastern populations is high, mixing of the Midcontinent Flyway breeding population with coastal flyway breeding populations across nonbreeding areas results in weak migratory connectivity (MC = 0.02 ± 0.02).

BANDING Recoveries of Long-tailed Ducks (n = 72) banded in Alaska in 1950–2011 occurred in Alaska, Russia, Nunavut, Utah, and Massachusetts (Baldassarre 2014). Lawson (2006) concluded that hunter-killed, nonbreeding Long-tailed Ducks (n = 61) on the western and eastern coasts of North America and the Great Lakes region shared similar breeding areas.

SATELLITE TRACKING Long-tailed Ducks (n = 57) tracked from the western Canadian Arctic moved through the Beaufort Sea after molting to the North Pacific south to British Columbia and west to Japan (Bartzen et al. 2016). Satellite transmitters on individuals breeding in the Yukon-Kuskokwim Delta of Alaska (n = 54) molted throughout Alaska and Russia and spent the nonbreeding season in the coastal waters of the North Pacific (Petersen et al. 2003). The Sea Duck Joint Venture (2015) marked Long-tailed Ducks in multiple areas of eastern North America and in the Great Lakes during the nonbreeding season and tracked birds to breeding areas in Nunavut and northern Quebec, particularly around Hudson Bay and the Foxe Basin. Three individuals tracked from a nonbreeding site on Lake Ontario bred around Hudson Bay (Mallory et al. 2006).

STABLE ISOTOPES No stable isotope studies of migratory connectivity were found for this species.

GENETICS Allison and colleagues (2013) found no evidence, based on museum specimens, for genetic differentiation among western and eastern populations. Two clades were suggested from an analysis of mtDNA haplotypes; one was comprised of all sampling locations (eastern Russia, Alaska, and northern Canada), and the other was comprised exclusively of samples from western Alaska (R.E. Wilson et al. 2016).

MIGRATORY PATHWAYS Movement of birds breeding farthest east or west tend toward nonbreeding areas in the Atlantic and Pacific, respectively. They utilize coastal areas and the Great Lakes as travel routes.

DATA CONTRIBUTORS

B. Bartzen; T. Bowman; D.L. Dickson; D.C. Douglas; P. Flint; J.S. Lamb; M.L. Mallory; M. Petersen.

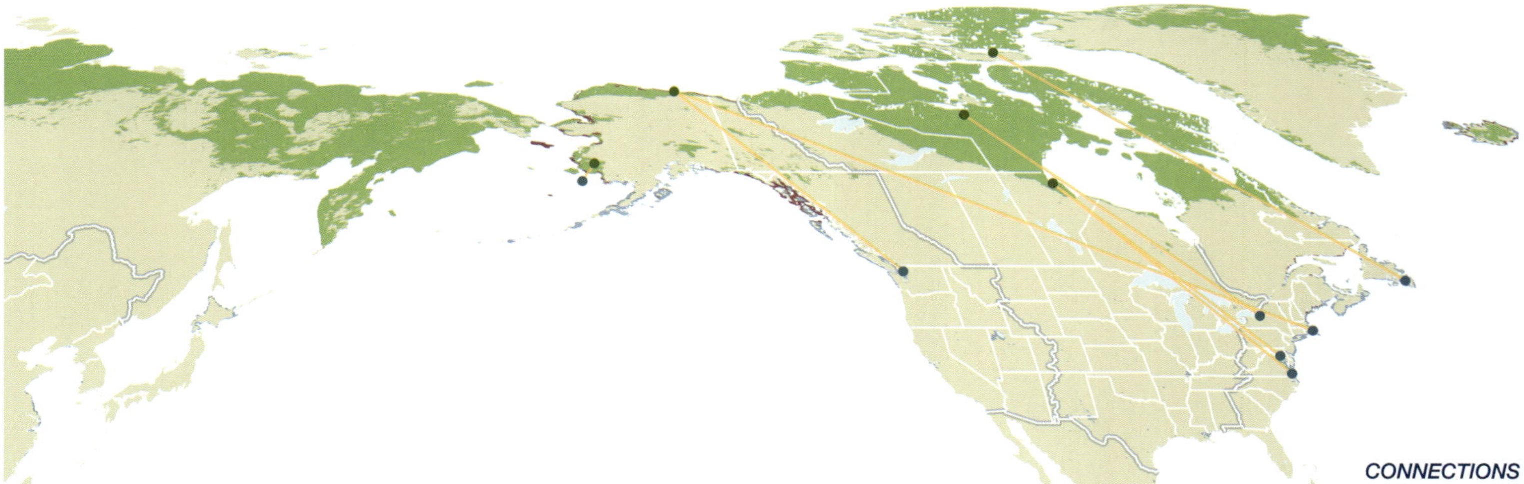

CONNECTIONS

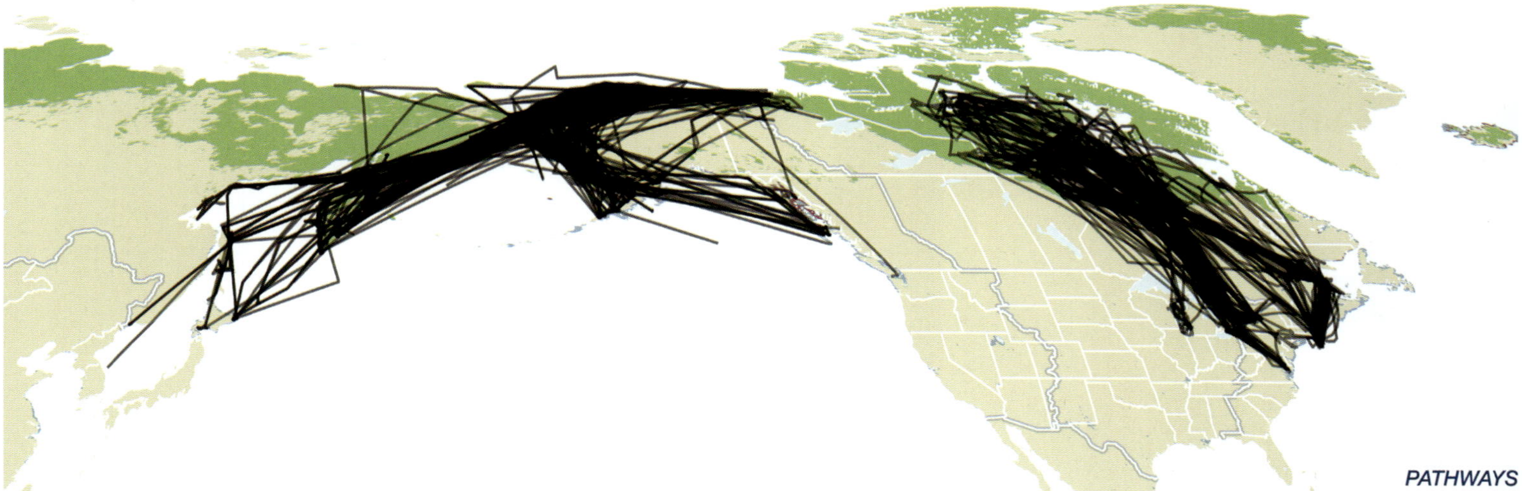

PATHWAYS

Bucephala albeola BUFFLEHEAD

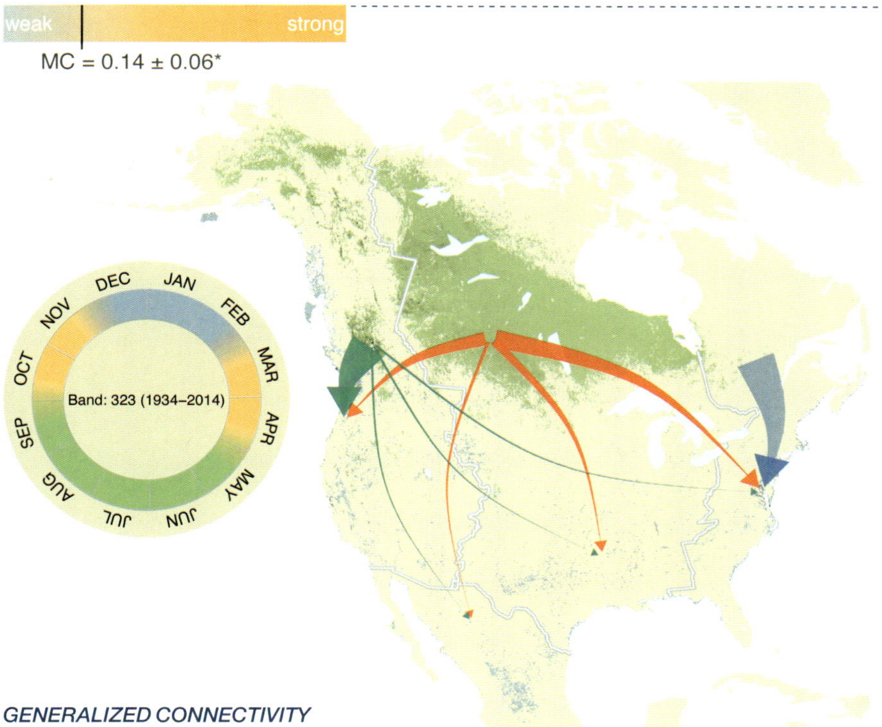

GENERALIZED CONNECTIVITY

DISTRIBUTION AND ABUNDANCE

Bufflehead have a wide breeding range extending from the northern United States throughout Canada and Alaska. The nonbreeding range extends over a large area of North America, with concentrations along both coasts and on large rivers and lakes as far north as open water exists (Fink et al. 2023).

ORDER Anseriformes
FAMILY Anatidae
BREEDING HABITAT Wetlands, Wet Forested Areas
NONBREEDING HABITAT Coasts, Marine Waters
GLOBAL POPULATION SIZE 1,300,000
POPULATION CHANGE SINCE 1970 +357%
IUCN RED LIST CATEGORY Least Concern
R2R TIPPING POINT SPECIES No
US LISTING STATUS No
CANADA SPECIES AT RISK No
PIF CCS 8/20

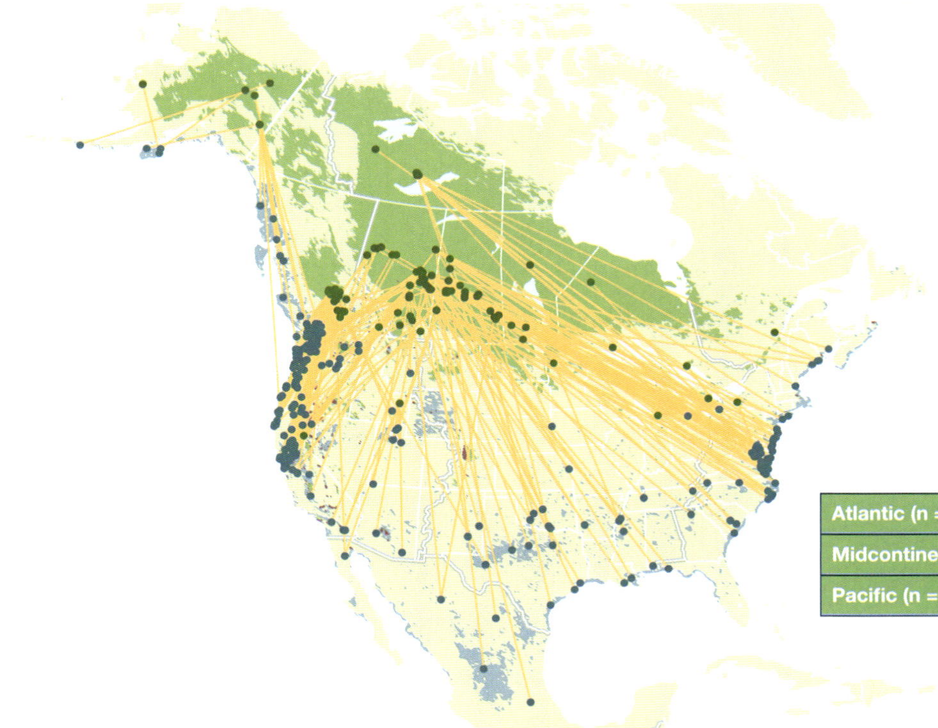

CONNECTIONS

	Atlantic	Mexico	Midcontinent	Pacific
Atlantic (n = 2)	100%	–	–	–
Midcontinent (n = 141)	45%	4%	18%	33%
Pacific (n = 177)	1%	1	1	98%

MIGRATORY CONNECTIVITY

OVERALL Migratory connectivity is described by band recoveries and genetic analysis from both breeding and nonbreeding locations. Both datasets show a monomorphic distribution, with no population structure (Pearce et al. 2019). Most birds that breed in the Atlantic and Pacific flyways migrate to nonbreeding areas within the same flyway as they bred (Atlantic 100%, Pacific 98%). Migration from breeding areas in the Midcontinent Flyway is more diffuse, which results in overlap of breeding populations across nonbreeding areas, i.e., weak migratory connectivity (MC = 0.14 ± 0.06).

BANDING Most birds banded in Alaska during the breeding season spent the nonbreeding season along the Pacific coast of the United States and Canada. Birds banded during the breeding season in Alberta and Saskatchewan migrated to nonbreeding areas throughout the United States and Mexico and showed no clear distinction of separate populations (Pearce et al. 2019).

GENETICS Genetic information from mtDNA agreed with band recovery data for populations sampled during the nonbreeding season; there was no evidence of genetic differentiation among west and east coast individuals (Pearce et al. 2014).

OTHER APPROACHES No satellite tracking or stable isotope studies of migratory connectivity were found for this species.

MIGRATORY PATHWAYS General southward migration to nonbreeding areas along rivers and lakes, but many birds migrate toward the coasts from across the breeding range.

COMMON GOLDENEYE *Bucephala clangula*

MC = 0.37 ± 0.04*

DISTRIBUTION AND ABUNDANCE

Common Goldeneye have a circumpolar breeding distribution, and in North America they breed from Alaska east across the continent to Maritime Canada. Breeding concentrations occur around James Bay and through the southern boreal forest of Canada. Nonbreeding concentrations occur along the Atlantic coast, in the Great Lakes, and along the Pacific coast from Alaska to southern British Columbia (Fink et al. 2023).

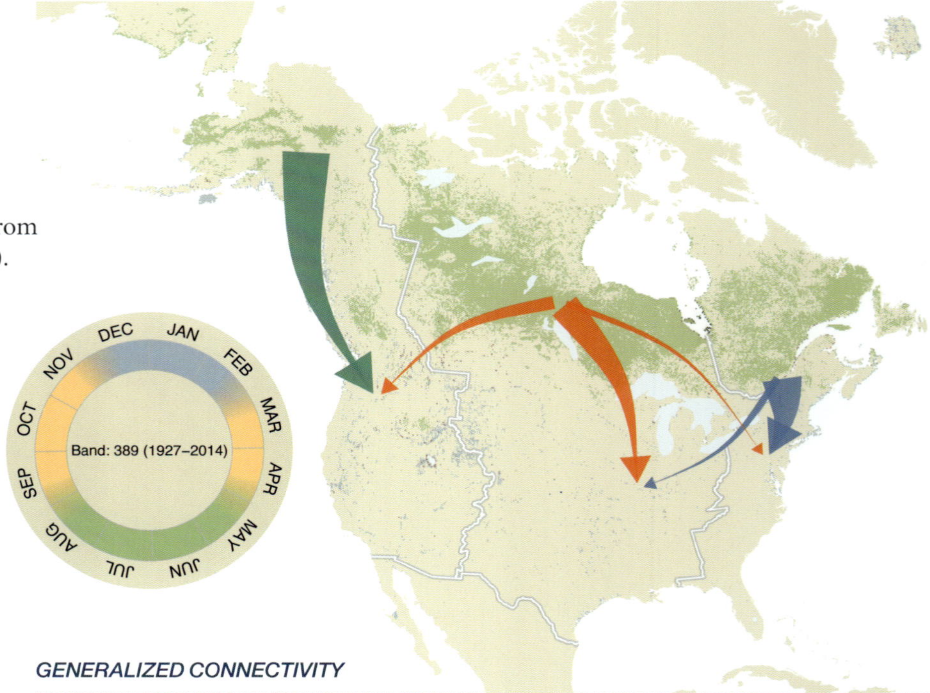

GENERALIZED CONNECTIVITY

ORDER Anseriformes
FAMILY Anatidae
BREEDING HABITAT Wetlands, Wet Forested Areas
NONBREEDING HABITAT Coasts, Marine Waters
GLOBAL POPULATION SIZE 2,500,000
POPULATION CHANGE SINCE 1970 +40%
IUCN RED LIST CATEGORY Least Concern
R2R TIPPING POINT SPECIES No
US LISTING STATUS No
CANADA SPECIES AT RISK No
PIF CCS 9/20

	Atlantic	Midcontinent	Pacific
Atlantic (n = 23)	87%	13%	–
Midcontinent (n = 349)	10%	64%	26%
Pacific (n = 14)	–	–	100%

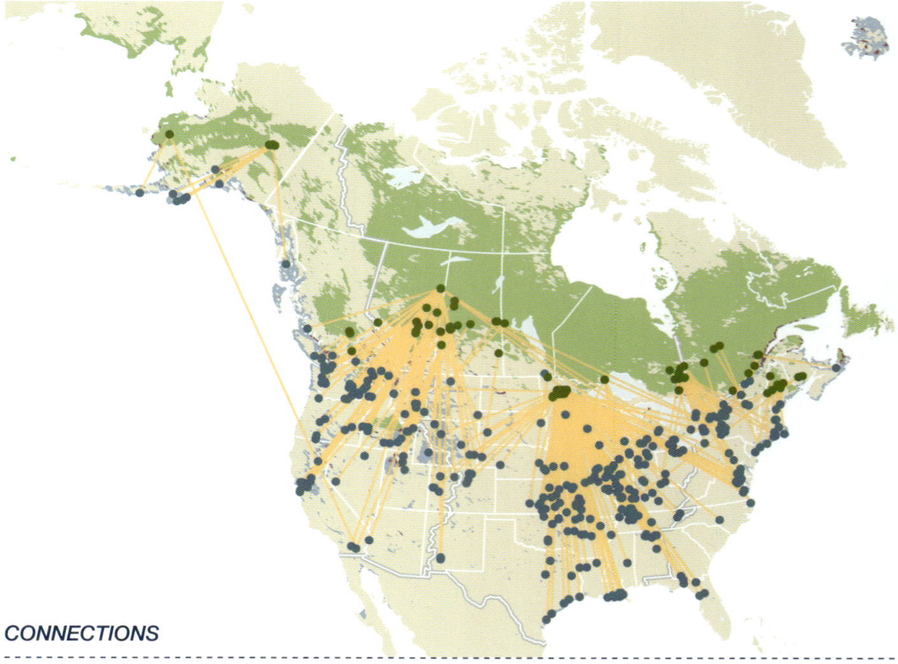

CONNECTIONS

MIGRATORY CONNECTIVITY

OVERALL Migratory connectivity information is derived from band recovery data and genetics. Some separation between eastern and western populations is suggested. Birds breeding in the Atlantic Flyway mostly stay in that flyway (87%) during the nonbreeding season, with some movement to the Midcontinent Flyway (13%). Midcontinent Flyway breeders also primarily stay within the same region (64%) during the nonbreeding season, but do migrate to the Atlantic (10%) and Pacific (26%) coasts. Breeding birds in the Pacific Flyway mostly stay in the Pacific Flyway (100%). Migration to the coastal flyways of Midcontinent Flyway breeding birds moderates migratory connectivity (MC = 0.37 ± 0.04).

BANDING Banding data suggests very little movement of Common Goldeneye breeding in Alaska or eastern North America to areas outside those respective ranges (Pearce et al. 2019). Common Goldeneye banded in central North America were recovered along both coasts, the Gulf of Mexico, and other large water bodies.

STABLE ISOTOPES No stable isotope studies of migratory connectivity were found for this species.

GENETICS Analysis of mtDNA combined with banding data shows differentiation between Old and New World populations, as well as regional differences among groups in Alaska, the Pacific Northwest, and eastern North America (Pearce et al. 2014). More recent work using nuclear DNA suggests no differentiation among Common Goldeneye in North America (J.I. Brown et al. 2020).

MIGRATORY PATHWAYS Common Goldeneye breeding in the interior of the continent move to both coasts through areas containing large lakes and rivers, such as the upper Mississippi River and the Great Salt Lake (Baldassarre 2014). Some individuals make a molt migration to northern lakes, rivers, and bays (Eadie et al. 2020).

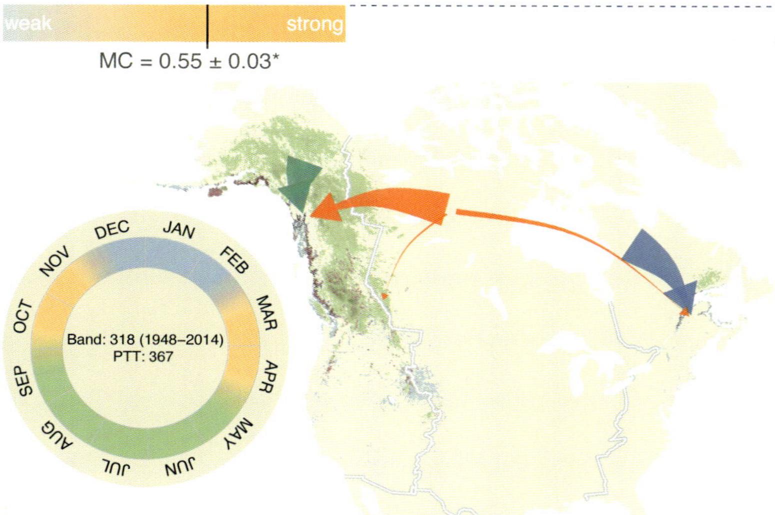

GENERALIZED CONNECTIVITY

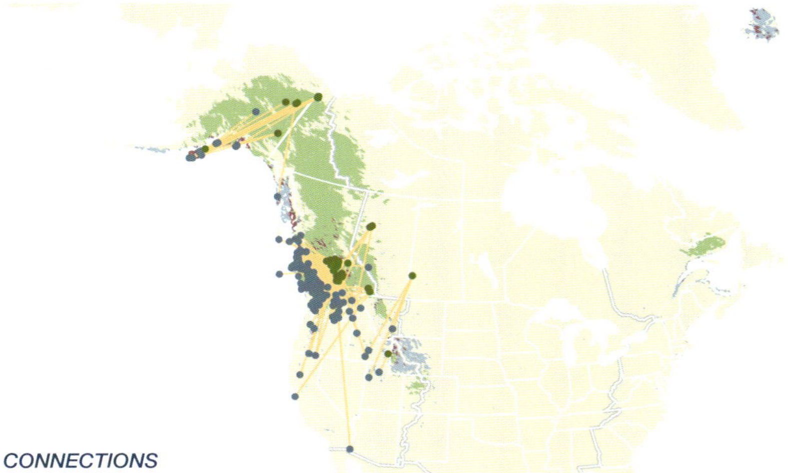

CONNECTIONS

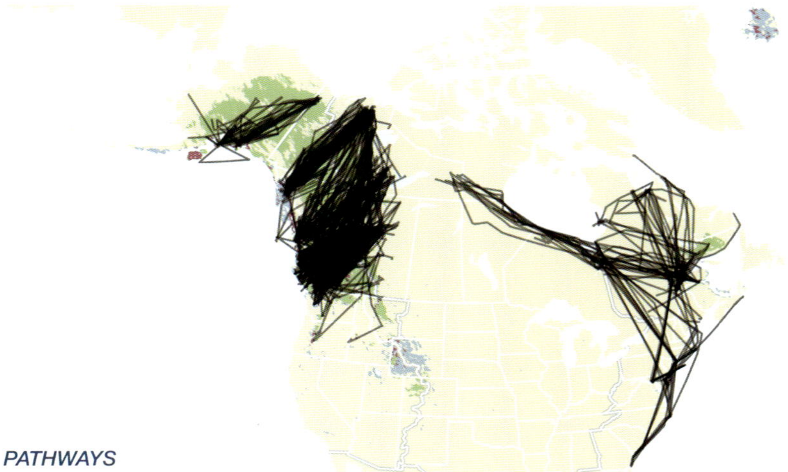

PATHWAYS

	Atlantic	Midcontinent	Pacific
Atlantic (n = 19)	100%	–	–
Midcontinent (n = 120)	11%	1%	88%
Pacific (n = 388)	–	<1%	100%

INTERESTING NOTE

Differences in genetic markers examined suggest male dispersal is the primary method of gene flow in Barrow's Goldeneye (J.I. Brown et al. 2020).

Bucephala islandica

BARROW'S GOLDENEYE

DISTRIBUTION AND ABUNDANCE

There are three distinct populations of Barrow's Goldeneye recognized in North America with highest abundances west of the Rocky Mountains. In the west, one population breeds in interior Alaska and spends the nonbreeding season in southwest Alaska, and one breeds in interior British Columbia and spends the nonbreeding season from coastal southeastern Alaska through British Columbia to Washington state. A small population breeds in eastern North America, in the boreal forest north of the estuary and Gulf of St. Lawrence (see Robert 2019 for exact information on breeding distribution), and spends the nonbreeding season in the St. Lawrence Marine Corridor and along the Atlantic coast.

ORDER Anseriformes
FAMILY Anatidae
BREEDING HABITAT Wetlands, Wet Forested Areas
NONBREEDING HABITAT Coasts, Marine Waters
GLOBAL POPULATION SIZE 180,000
POPULATION CHANGE SINCE 1970 −45%
IUCN RED LIST CATEGORY Least Concern
R2R TIPPING POINT SPECIES No
US LISTING STATUS No
CANADA SPECIES AT RISK No
PIF CCS 13/20

MIGRATORY CONNECTIVITY

OVERALL Migratory connectivity data is from band recoveries and satellite tracking. Coastal breeding birds nearly all migrate to nonbreeding areas along the same coast, with Pacific Flyway breeding birds remaining in the Pacific Flyway (100%). Atlantic Flyway birds stay in the east (100%), and birds breeding in the Midcontinent Flyway nearly all migrate to nonbreeding areas in the Pacific Flyway (88%). Strong migratory connectivity (MC = 0.55 ± 0.03) reflects the limited movement of populations across breeding and nonbreeding regions.

BANDING Pearce et al. (2019) found no overlap of recoveries among birds banded in Alaska and British Columbia. Most Alaska band reencounters were concentrated around Kodiak Island and the northern portion of the Gulf of Alaska, though this could be confounded by higher hunting pressure in these areas compared to other parts of Alaska (Pearce et al. 2014).

SATELLITE TRACKING Boyd and Esler (2012) and Forstner et al. (2022) tracked Barrow's Goldeneye from the western population and found longitudinal movements from breeding to nonbreeding areas with no mixing of populations in Alaska and the Pacific Northwest, suggesting strong migratory connectivity within the western population. Savard and Robert (2013) tracked 5 adult females from the eastern population and found that females from a single breeding site molt-migrated across a wide area, indicating population mixing in the eastern population. Males from the eastern population (n = 18) also showed low migratory connectivity but did undertake a molt migration to locations north of the primary breeding area (Robert et al. 2002).

STABLE ISOTOPES No stable isotope studies of migratory connectivity were found for this species.

GENETICS Using mtDNA sequencing, Pearce et al. (2014) found genetic differentiation between the eastern and western populations, and within the western population, they found genetic differentiation between birds inhabiting Alaska versus those inhabiting British Columbia and Washington. In contrast, using nuclear DNA sequencing, Brown et al. (2020) did not find genetic differentiation within the western population and suggest that dispersing males maintain genetic connectivity.

MIGRATORY PATHWAYS The western population migrates longitudinally from breeding areas in the interior of the continent to nonbreeding areas on the Pacific coast (Hogan et al. 2011; Forstner et al. 2022). The eastern population migrates primarily from south to north and back (thus, latitudinal rather than longitudinal).

DATA CONTRIBUTORS

R. Benoit; T. Bowman; W.S. Boyd; D. Esler; D. Hogan; M. Robert; J.P. Savard; J. Schamber; J. Thompson.

HOODED MERGANSER *Lophodytes cucullatus*

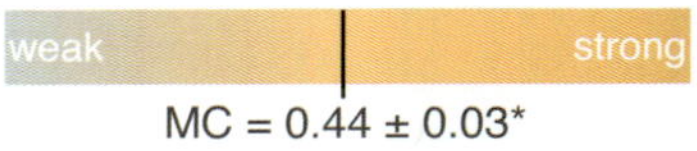

DISTRIBUTION AND ABUNDANCE

The Hooded Merganser is a widespread breeder in eastern North America where tree cavities are available. The species is present year-round in the southern portion of their breeding range. Hooded Mergansers are thought to have eastern and western population segments, though there is likely significant dispersal between the two (Dugger et al. 2020). Nonbreeding concentrations occur in coastal marshes and in the southern Mississippi River Basin.

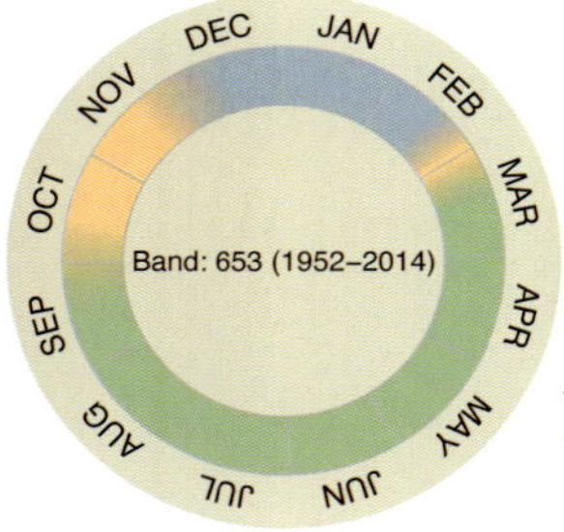

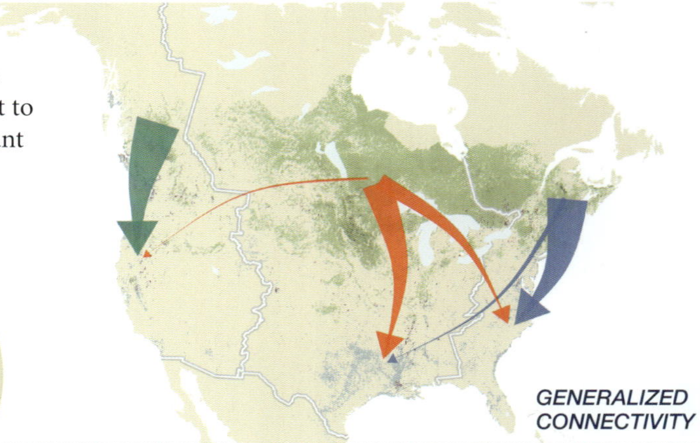

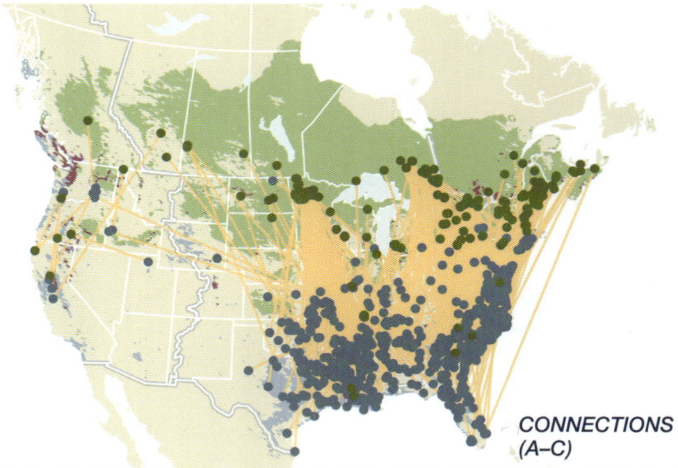

ORDER Anseriformes
FAMILY Anatidae
BREEDING HABITAT Wetlands, Wet Forested Areas
NONBREEDING HABITAT Wetlands, Lakes, and Rivers
GLOBAL POPULATION SIZE 1,100,000
POPULATION CHANGE SINCE 1970 +703%
IUCN RED LIST CATEGORY Least Concern
R2R TIPPING POINT SPECIES No
US LISTING STATUS No
CANADA SPECIES AT RISK No
PIF CCS 8/20

INTERESTING NOTE

Hooded Mergansers often utilize human-made nest boxes installed for Wood Ducks, where the two species overlap.

	Atlantic	Midcontinent	Pacific
Atlantic (n = 128)	92%	8%	–
Midcontinent (n = 516)	35%	64%	1%
Pacific (n = 8)	–	–	100%

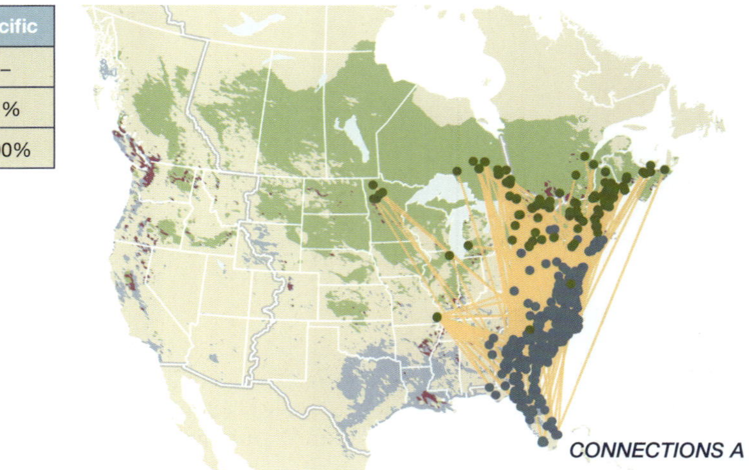

MIGRATORY CONNECTIVITY

OVERALL Migratory connectivity of Hooded Mergansers is inferred from banding and genetic data. Hooded Mergansers are often short-distance migrants and remain within the same flyway during both the breeding and nonbreeding season, resulting in strong migratory connectivity (MC = 0.44 ± 0.03).

BANDING Hooded Merganser banding data from across the eastern portion of the range shows migration within the east (Pearce et al. 2019). Of those mergansers banded in New York and the New

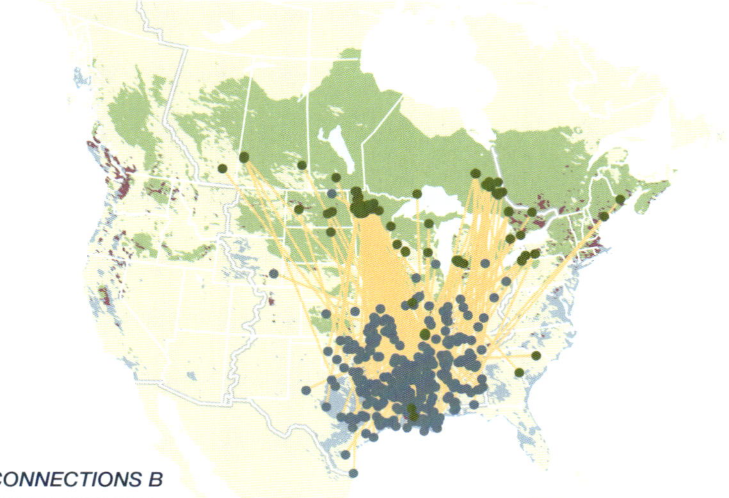

CONNECTIONS B

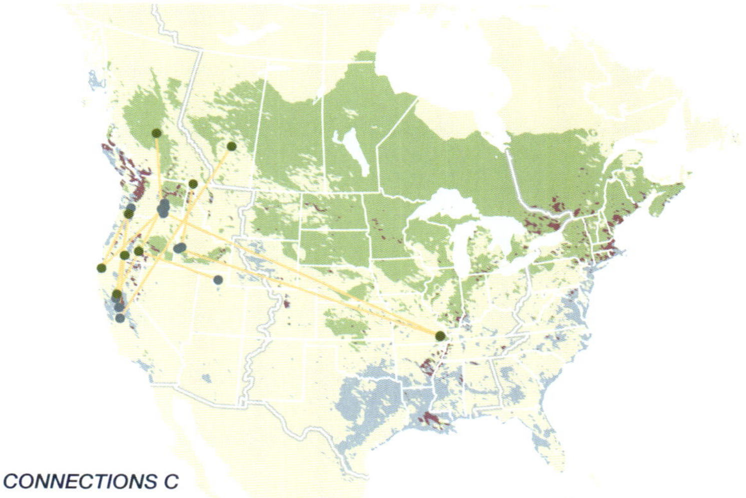

CONNECTIONS C

England states from 1941 to 2009, 80% were recovered in the Atlantic Flyway of the United States, whereas only 17% were recovered in the provinces of Ontario and Quebec (Baldassarre 2014). Birds banded in Missouri were recovered across the species range, with recoveries in Wyoming and Washington, indicating dispersal between eastern and western populations (Pearce et al. 2008).

STABLE ISOTOPES No stable isotope studies of migratory connectivity were found for this species.

GENETICS A combination of nuclear and mitochondrial DNA showed gene flow throughout the range of Hooded Mergansers, facilitated by juvenile dispersal. Females had high breeding site fidelity, a potential mechanism for separation of western and eastern populations (Pearce et al. 2008).

MIGRATORY PATHWAYS Hooded Mergansers generally follow coasts or river systems between breeding and nonbreeding areas.

Mergus merganser

COMMON MERGANSER

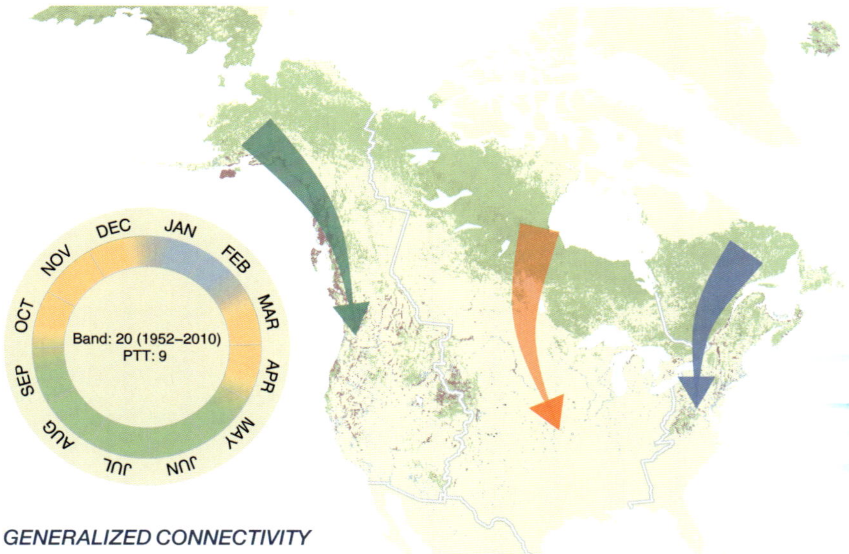

GENERALIZED CONNECTIVITY

DISTRIBUTION AND ABUNDANCE

Common Mergansers have a broad breeding distribution in the boreal forest of North America (Baldassarre 2014), with the highest densities near the Hudson Bay and along the Pacific coast. Their nonbreeding distribution is also broad, but concentrations occur on the Great Lakes, along large rivers, and along the northern Pacific coast (Pearce et al. 2020).

MIGRATORY CONNECTIVITY

OVERALL Migratory connectivity information is from banding, tracking, and genetics data. In the Pacific, Midcontinent, and Atlantic flyways, all observed transitions occur within the same flyway. In addition, populations breeding in Alaska are differentiated from those breeding in areas farther east. Limited data prevents the calculation of MC, but migratory connectivity appears strong.

	Atlantic	Midcontinent	Pacific
Atlantic (n = 12)	100%	–	–
Midcontinent (n = 4)	–	100%	–
Pacific (n = 8)	–	–	100%

BANDING Breeding birds banded in central and eastern North America were nearly all recovered in the same regions in the nonbreeding season (Pearce et al. 2019). Band-recovery data (n = 87) found that migratory distance varied by age and sex cohort, with juvenile males moving the farthest (Pearce et al. 2005a).

ORDER Anseriformes
FAMILY Anatidae
BREEDING HABITAT Wetlands, Wet Forested Areas
NONBREEDING HABITAT Wetlands, Lakes, and Rivers
GLOBAL POPULATION SIZE 1,900,000
POPULATION CHANGE SINCE 1970 −18%
IUCN RED LIST CATEGORY Least Concern
R2R TIPPING POINT SPECIES No
US LISTING STATUS No
CANADA SPECIES AT RISK No
PIF CCS 10/20

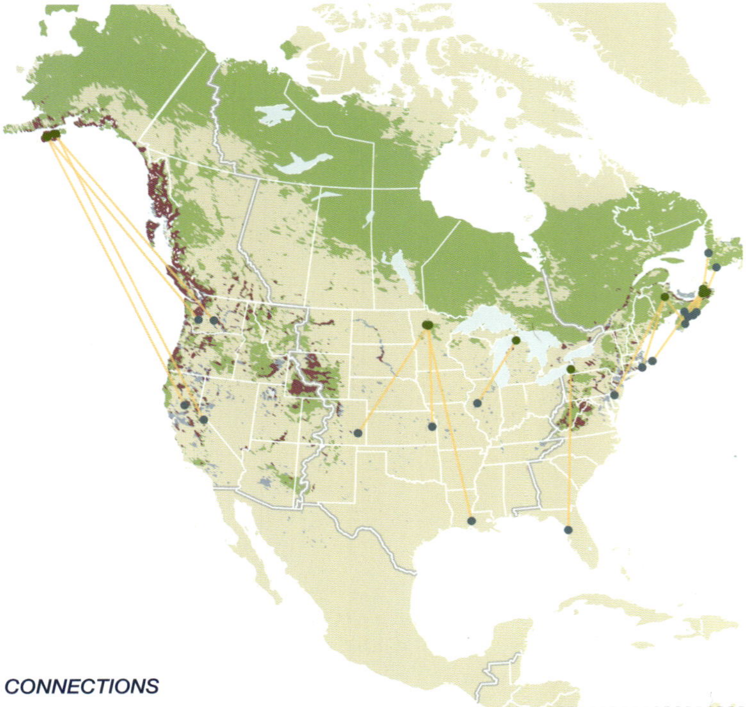

CONNECTIONS

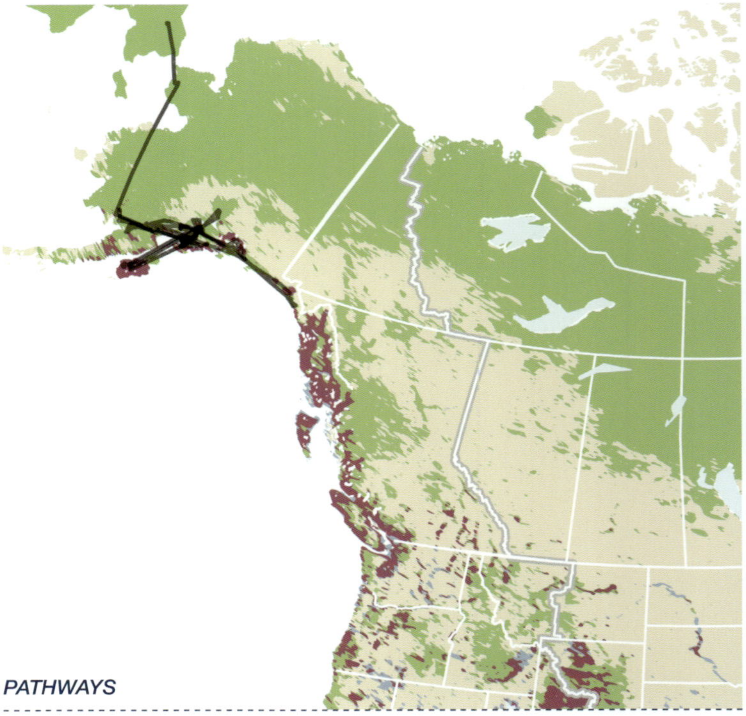

PATHWAYS

SATELLITE TRACKING Satellite telemetry of juvenile Common Mergansers (n = 8) from a natal river in south-central Alaska showed movements up to 750 km within Alaska (Pearce and Petersen 2009).
STABLE ISOTOPES No stable isotope studies of migratory connectivity were found for this species.
GENETICS Genetic analysis of nonbreeding Common Mergansers suggests that individuals move across phylogeographic boundaries along the Pacific coast (Pearce et al. 2009a). Band recovery data, in combination with mtDNA data, suggest that males molting at Kodiak Island, Alaska, are from various natal and breeding grounds (Pearce et al. 2009b). Common Merganser populations were found to be structured at regional and continental scales within North America (Pearce et al. 2009a, 2009b), likely a result of female philopatry (Peters et al. 2012). Genetic evidence suggests three populations: western Alaska, the Pacific Northwest, and eastern North America (Pearce et al. 2009a, 2009b, Peters et al. 2012).
MIGRATORY PATHWAYS Migratory movements follow large rivers or occur directly to the closest large, open water when ice conditions exist.

DATA CONTRIBUTORS
D.C. Douglas; J. Pearce; M. Petersen.

RED-BREASTED MERGANSER *Mergus serrator*

DISTRIBUTION AND ABUNDANCE

This species has broad breeding and nonbreeding distributions in North America. They breed in boreal regions of Canada and Alaska, primarily in coastal areas. Concentrations occur during the nonbreeding season along both coasts and the Great Lakes (Craik et al. 2020). Red-breasted Mergansers have a more northerly distribution than other mergansers as they do not require tree cavities for nesting.

ORDER Anseriformes
FAMILY Anatidae
BREEDING HABITAT Wetlands, Tundra
NONBREEDING HABITAT Coasts, Marine Waters
GLOBAL POPULATION SIZE 600,000
POPULATION CHANGE SINCE 1970 −71%
IUCN RED LIST CATEGORY Least Concern
R2R TIPPING POINT SPECIES No
US LISTING STATUS No
CANADA SPECIES AT RISK No
PIF CCS 9/20

MIGRATORY CONNECTIVITY

OVERALL There are few banding records for Red-breasted Mergansers and little additional data to inform migratory connectivity. The coastal concentrations suggest there may be separate populations, and research on migratory connectivity is sorely needed.

BANDING Thirty-four Red-breasted Mergansers banded in central and eastern portions of the breeding range were recovered along the Gulf Coast and the Atlantic coast during the nonbreeding season (Pearce et al. 2019).

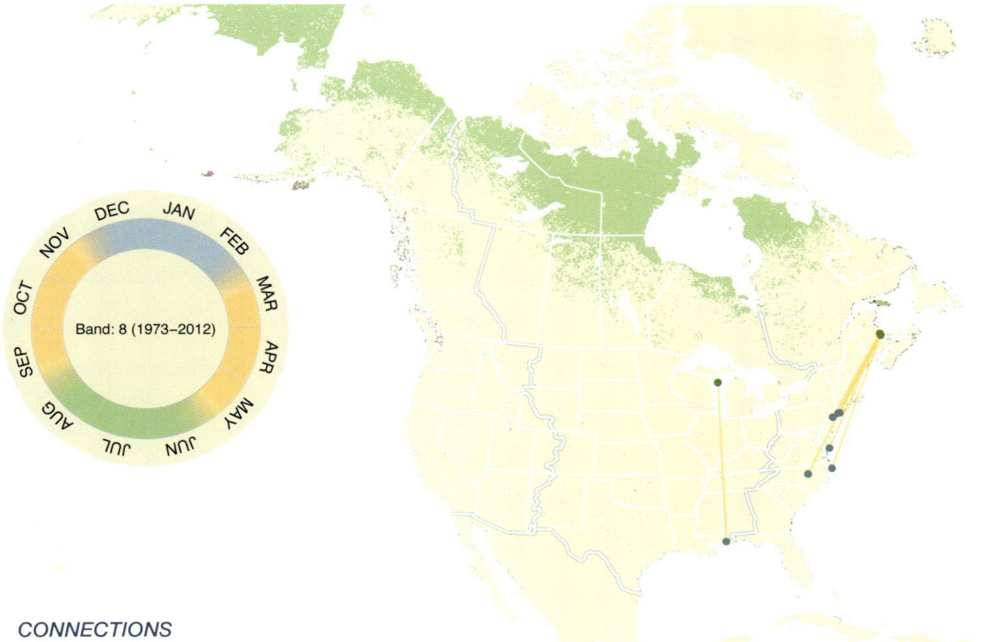

CONNECTIONS

GENETICS mtDNA analysis on samples collected from breeding areas across their range in North America showed little population differentiation (Pearce et al. 2009a). Samples from hunter-collected Red-breasted Mergansers during the nonbreeding season suggest some separation of populations on the Pacific and Atlantic coasts (Pearce et al. 2009a).

OTHER No satellite tracking or stable isotope studies of migratory connectivity were found for this species.

MIGRATORY PATHWAYS No information available.

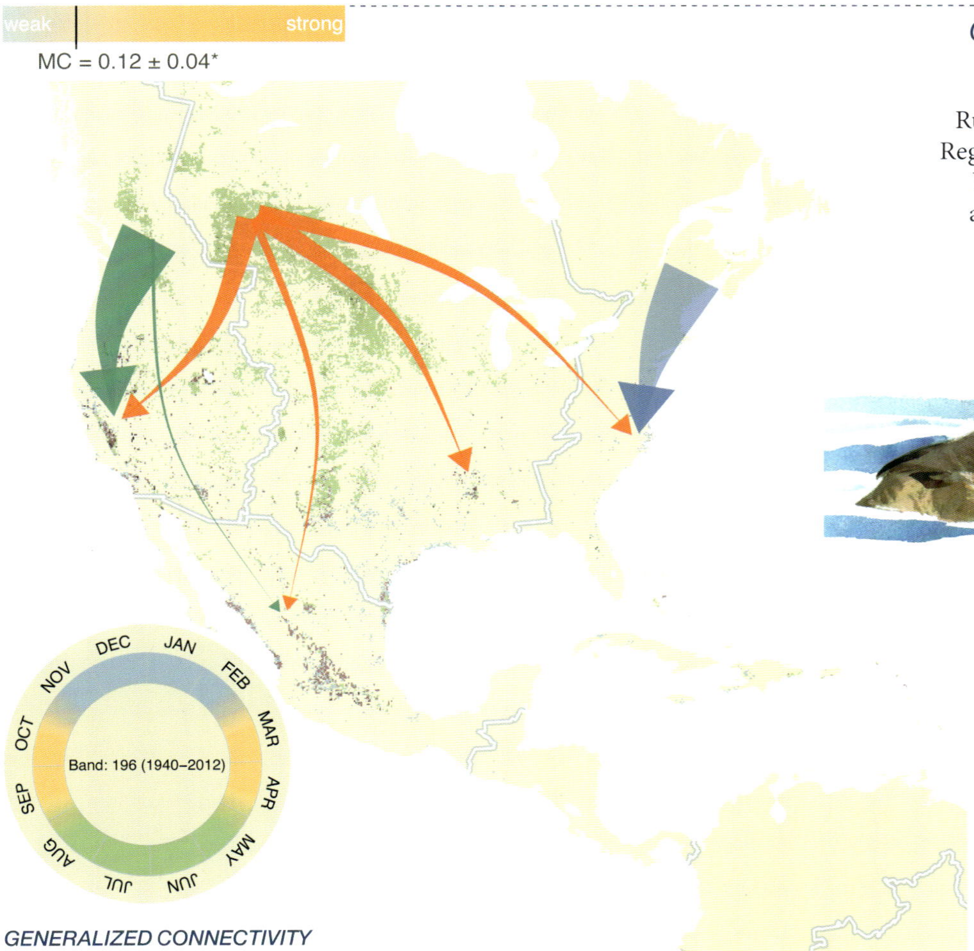

GENERALIZED CONNECTIVITY

Oxyura jamaicensis RUDDY DUCK

DISTRIBUTION AND ABUNDANCE

Ruddy Ducks breed primarily in the Prairie Pothole Region. Nonbreeding concentrations are found along both coasts, generally south of 40 degrees latitude, as well as in the southern Great Plains and interior Mexico (Baldassarre 2014).

ORDER Anseriformes
FAMILY Anatidae
BREEDING HABITAT Wetlands, Seasonally Wet Grasslands
NONBREEDING HABITAT Wetlands, Lakes, and Rivers
GLOBAL POPULATION SIZE 1,700,000
POPULATION CHANGE SINCE 1970 +120%
IUCN RED LIST CATEGORY Least Concern
R2R TIPPING POINT SPECIES No
US LISTING STATUS No
CANADA SPECIES AT RISK No
PIF CCS 8/20

MIGRATORY CONNECTIVITY

OVERALL There are no published studies of Ruddy Duck migratory connectivity. Band recoveries are available but are sparse relative to other waterfowl. Transitions within the coastal flyways are high between breeding and nonbreeding seasons, while birds breeding in the Midcontinent Flyway migrated to all nonbreeding regions throughout the range. Migratory connectivity appears weak for this species (MC = 0.12 ±

	Atlantic	Mexico	Midcontinent	Pacific
Atlantic (n = 3)	100%	–	–	–
Midcontinent (n = 117)	23%	9%	32%	36%
Pacific (n = 72)	–	1	–	99%

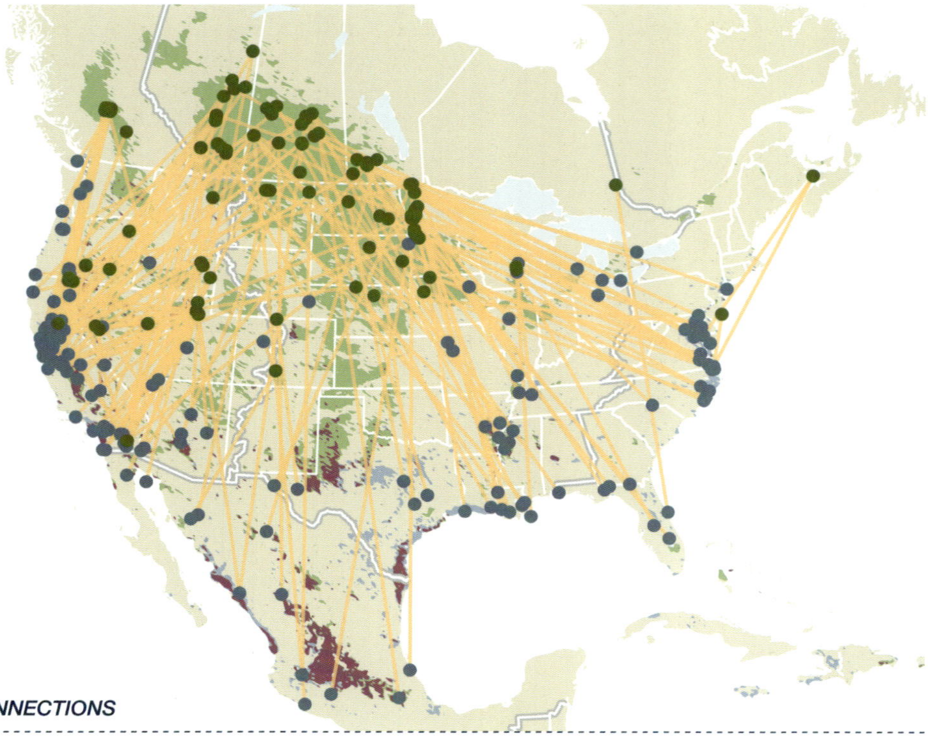

CONNECTIONS

0.04), though this may be inconclusive due to low sample sizes in some of the breeding regions.

RADIO TRACKING Few tags were deployed through the Motus Wildlife Tracking System, with local detections only.

OTHER APPROACHES No tracking, genetic, or stable isotope studies of migratory connectivity were found for this species.

MIGRATORY PATHWAYS Ruddy Ducks breeding the farthest west travel from prairie breeding sites through Idaho and Utah to the Pacific coast. Ruddy Ducks in the western and central portion of their breeding range may take a more southwestern route through the Great Basin to nonbreeding areas along the southern Pacific coast. Finally, a third corridor extends from eastern breeding areas through the Great Lakes to nonbreeding sites along the Gulf Coast and into Mexico (Brua 2020).

PIGEONS AND DOVES (COLUMBIFORMES)

WRITING LED BY **MELANIE A. SMITH, TOM WILL, TIMOTHY GUIDA**

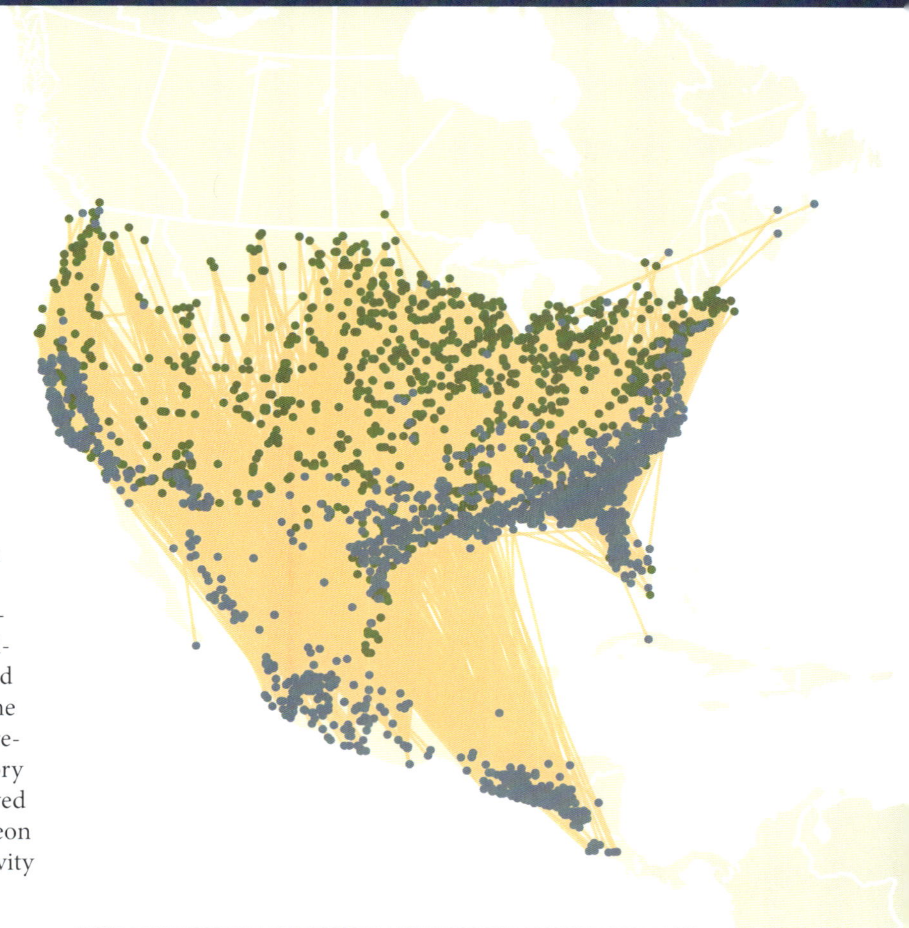

Given that this group is hunted, band reencounter data are abundant throughout the range (n = 3,647) and highlight interior breeding and coastal nonbreeding areas.

The migratory connectivity of North American doves and pigeons, order Columbiformes, all species in the family Columbidae, has been well studied. Forty percent of species have migratory connectivity information derived from tracking data, and 60% of species have connectivity information from band reencounter data, for a total 80% of species with migratory connectivity information (4 out of 5 migratory species in the order). The one species (Red-billed Pigeon) we did not find data for is presented with an eBird abundance map only (Fink et al. 2023). Band reencounter data are well distributed across the ranges of these species, and are supplemented by limited tracking data that add richness to select areas. Although most dove and pigeon populations have fared well, the Band-tailed Pigeon (−53%) has seen concerning declines in the last 50 years.

Within Columbiformes, the Band-tailed Pigeon and the White-crowned Pigeon have limited tracking data. The Band-tailed Pigeon, the Mourning Dove, and the White-winged Dove have band reencounter data available, while the Red-billed Pigeon and the White-crowned Pigeon do not. By far, the majority of banding records are for Mourning Doves. The species with strong migratory connectivity include the White-winged Dove (MC = 0.47), followed by the Mourning Dove (MC = 0.33), while the Band-tailed Pigeon has weak migratory connectivity (MC = 0.06). Regional connectivity varied across species with no clear order-level patterns identified.

COLUMBIDAE | *PIGEONS AND DOVES*

Patagioenas leucocephala **WHITE-CROWNED PIGEON**

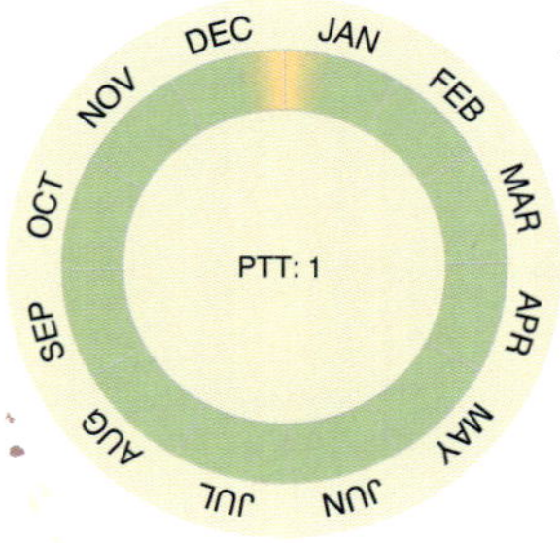

DISTRIBUTION AND ABUNDANCE

White-crowned Pigeon live year-round in coastal areas of the Caribbean Sea including southern Florida, the Bahamas, Cuba, Hispaniola, Jamaica, and Puerto Rico, as well as along the coasts of Panama and Colombia, and sparsely from Mexico's Yucatán Peninsula to Honduras. Generally considered nonmigratory or nomadic; limited data does indicate short-distance migration of some individuals.

ORDER Columbiformes
FAMILY Columbidae
BREEDING HABITAT Coasts, Mangroves
NONBREEDING HABITAT Coasts, Mangroves
GLOBAL POPULATION SIZE 550,000
POPULATION CHANGE SINCE 1970 NA
IUCN RED LIST CATEGORY Near Threatened
R2R TIPPING POINT SPECIES No
US LISTING STATUS No
CANADA SPECIES AT RISK No
PIF CCS 15/20

MIGRATORY CONNECTIVITY

OVERALL Limited data indicate connectivity between the Atlantic Flyway and the Caribbean region.
BANDING Banding records for recoveries of Florida birds showed migration between Florida and the Bahamas (Kent et al. 2016).
SATELLITE TRACKING Birds tracked from Florida did not migrate to the Bahamas as expected, based on older banding records, but instead migrated to Cuba (Kent et al. 2016). Birds tracked from various Caribbean locations were found to be resident throughout the year (Cambrone 2020). White-crowned Pigeons are seasonally absent from the Lower Antilles. It has been thought that they migrate to the Greater Antilles, but their whereabouts remain unknown (Cambrone 2020).
OTHER APPROACHES No geolocator, radio telemetry, stable isotope, or genetic studies of migratory connectivity were found for this species.
MIGRATORY PATHWAYS The only migration pathway data show flights crossing the Straits of Florida between the Florida Keys and Cuba (Kent et al. 2016).

DATA CONTRIBUTORS
G.M. Kent; K.D. Meyer; Avian Research and Conservation Institute; BirdsCaribbean; Florida Fish and Wildlife Conservation Commission; National Fish and Wildlife Foundation; National Wildlife Refuges, US Fish and Wildlife Service.

RED-BILLED PIGEON *Patagioenas flavirostris*

DISTRIBUTION AND ABUNDANCE

Breeds along both coasts of Mexico and throughout Central America through Costa Rica, most densely along the Mexican Gulf Coast and throughout the Yucatán Peninsula; in the US, found only in the lower Rio Grande Valley (Fink et al. 2023). It is resident year-round throughout its range, with the exception of populations in the US and northeast Mexico that leave during the nonbreeding season, presumably migrating south (Lowther et al. 2020a).

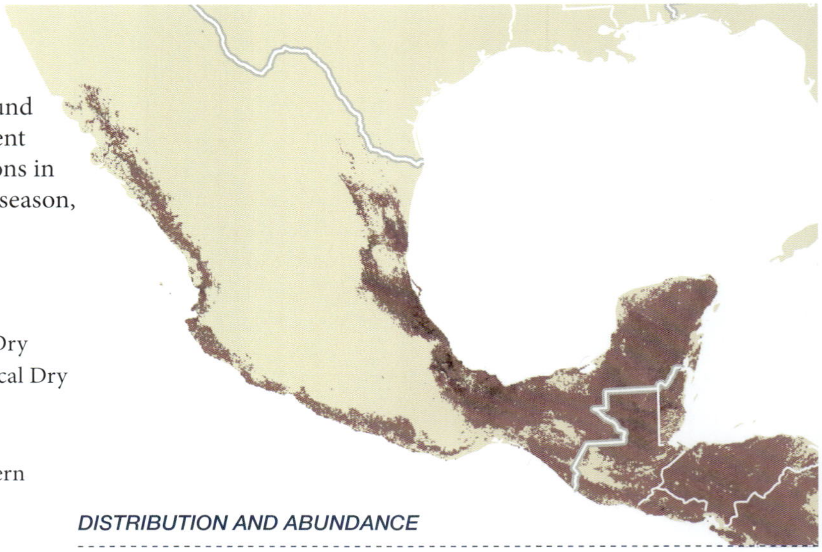

ORDER Columbiformes
FAMILY Columbidae
BREEDING HABITAT Forests, Tropical Dry
NONBREEDING HABITAT Forests, Tropical Dry
GLOBAL POPULATION SIZE 2,000,000
POPULATION CHANGE SINCE 1970 NA
IUCN RED LIST CATEGORY Least Concern
R2R TIPPING POINT SPECIES No
US LISTING STATUS No
CANADA SPECIES AT RISK No
PIF CCS 12/20

MIGRATORY CONNECTIVITY

OVERALL Migratory connectivity is unknown. No studies of migratory connectivity were found for this species.
MIGRATORY PATHWAYS There is limited information, but this species was observed during migration along both the east and west coasts of Mexico and inland through Central America (Fink et al. 2023).

BAND-TAILED PIGEON *Patagioenas fasciata*

MC = 0.06 ± 0.03*

DISTRIBUTION AND ABUNDANCE

Band-tailed Pigeon breed in higher elevation forested regions extending from southwestern British Columbia in Canada, through the western United States and Mexico, and continuing along the Andes Mountains through Central America to northern Argentina. In the US, this area includes the Cascade and Sierra Nevada ranges, as well as parts of the US desert Southwest. They are present year-round across most of Mexico, Guatemala, Honduras, and Costa Rica and northern Panama. In South America they are present year-round throughout the Andes from Venezuela to Peru. Their range in Bolivia and northern Argentina is primarily a nonbreeding range.

ORDER Columbiformes
FAMILY Columbidae
BREEDING HABITAT Forests, Temperate Western
NONBREEDING HABITAT Forests, Mesoamerican Pine-Oak
GLOBAL POPULATION SIZE 6,100,000
POPULATION CHANGE SINCE 1970 −52%
IUCN RED LIST CATEGORY Least Concern
R2R TIPPING POINT SPECIES No
US LISTING STATUS No
CANADA SPECIES AT RISK No
PIF CCS 11/20

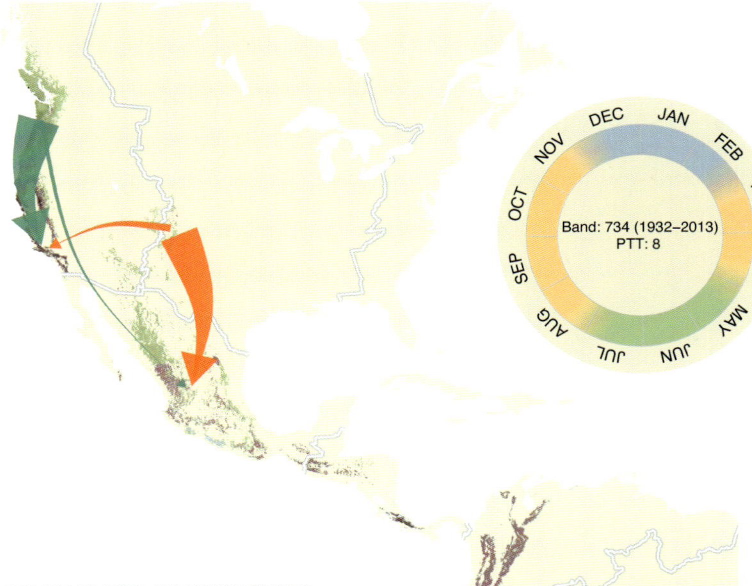

GENERALIZED CONNECTIVITY

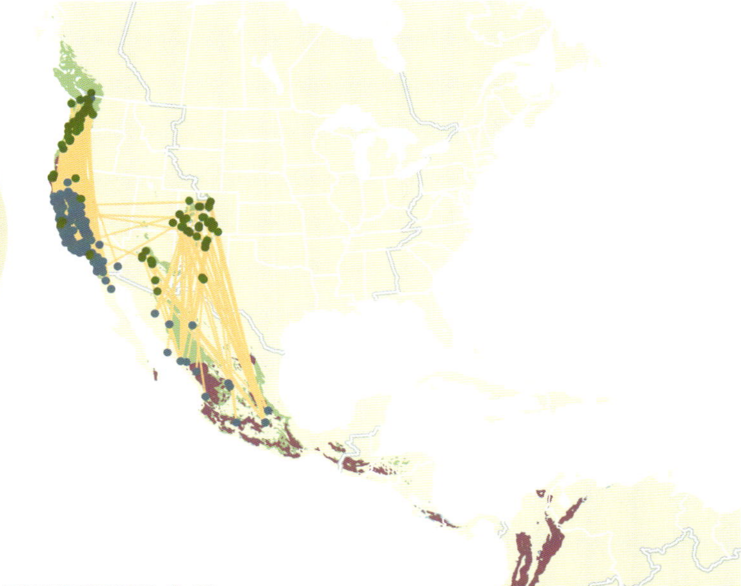

CONNECTIONS (A–B)

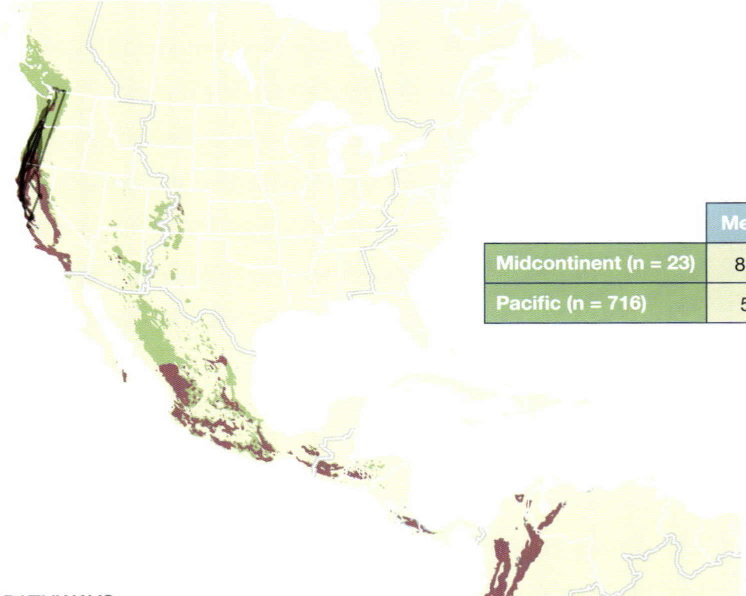

PATHWAYS

MIGRATORY CONNECTIVITY

OVERALL Migratory connectivity is well-known, mainly from band reencounter data and one tracking study (M.L. Casazza and C.T. Overton, unpublished data). Birds breeding in the Midcontinent Flyway migrate out of the region to Mexico (87%) and the Pacific Flyway (13%). The majority of birds breeding in the Pacific Flyway migrate to nonbreeding areas within the Pacific Flyway (95%), with 5% migrating to Mexico. Overlap of Band-tailed Pigeon breeding populations on nonbreeding areas results in weak migratory connectivity (MC = 0.06 ± 0.03).

	Mexico	Pacific
Midcontinent (n = 23)	87%	13%
Pacific (n = 716)	5%	95%

BANDING Band reencounters show moderate connectivity between the western and central breeding populations. Most individuals remain within the flyway during both breeding and nonbreeding, but there is some mixing across the populations.

RADIO TRACKING A few tags were deployed through the Motus Wildlife Tracking System; there were no breeding to nonbreeding detections.

SATELLITE TRACKING Band-tailed Pigeon tracked with solar-powered PTT tags (n = 21) in Washington, Oregon, California, Nevada, and

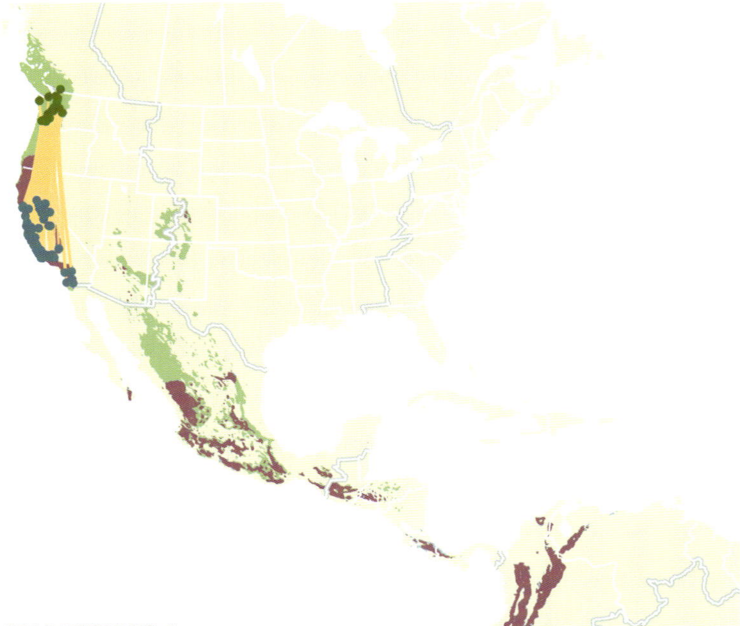

CONNECTIONS A

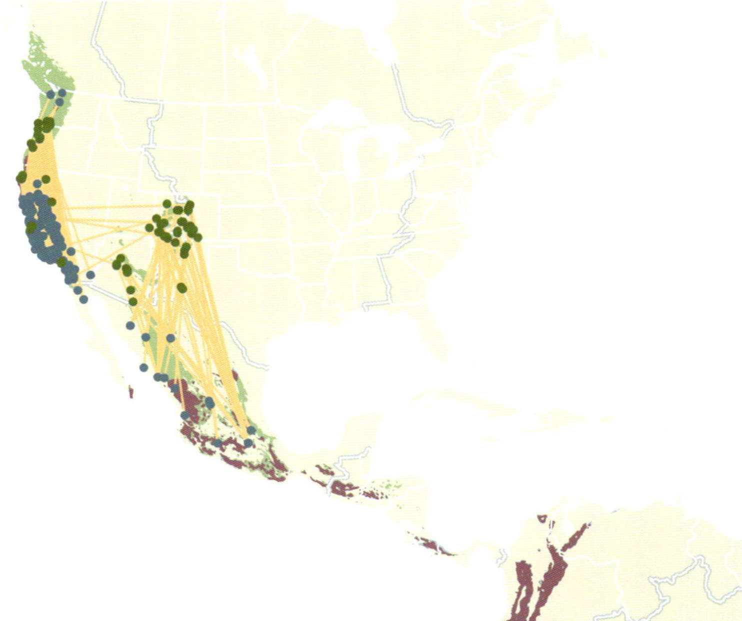

CONNECTIONS B

British Columbia, Canada, remained within the Pacific region, ranging from British Columbia to California during the nonbreeding season (Casazza and Overton, unpublished data).

OTHER APPROACHES No geolocator, isotope, or genetic studies of migratory connectivity were found for this species.

MIGRATORY PATHWAYS Western birds migrate along the Cascade and Sierra Nevada ranges, splitting into two pathways—one coastal and one inland—that avoid the Central Valley of California. Birds in the midcontinent migrate in and out of Mexico and Guatemala via the Sierra Madre mountains with greater abundance in the western Sierra Madre. Southern populations tend to be present year-round, with some birds migrating as far south as northern Argentina during the nonbreeding season (Fink et al. 2023).

DATA CONTRIBUTORS
M. Casazza; C. Overton.

WHITE-WINGED DOVE *Zenaida asiatica*

weak strong
MC = 0.47 ± 0.07*

DISTRIBUTION AND ABUNDANCE

Historically ranging from the southwestern US to Mexico, the Caribbean, and Central America, White-winged Doves have expanded their range in recent decades, likely due to human landscape changes. This species is currently found year-round on many Caribbean islands, from central Panama north to Las Vegas, Nevada, and Santa Fe, New Mexico, and east throughout the Gulf Coast and Florida. Migratory behavior varies by population, with many breeding areas inhabited year-round. During the breeding season, birds expand their range to additional islands in the Caribbean, to a wider swath of the American Southwest, and northward as far as Nebraska. This species is particularly abundant in Puerto Rico, the American Southwest, northern Mexico, the northern Yucatán, and the Pacific Slope of Central America from Guatemala to Nicaragua (Schwertner et al. 2020; Fink et al. 2023)

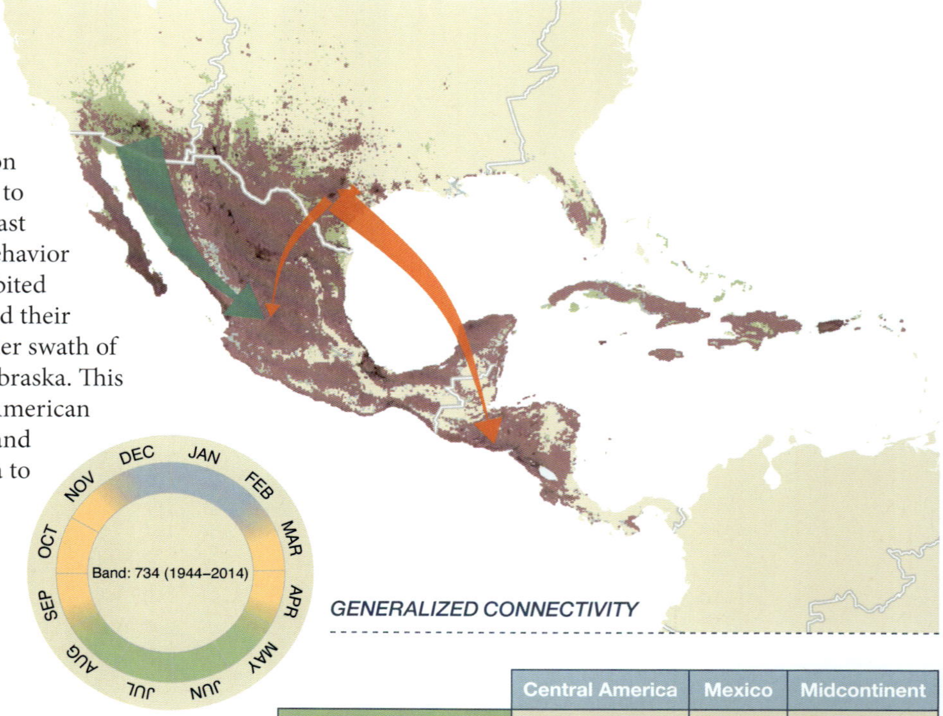

GENERALIZED CONNECTIVITY

	Central America	Mexico	Midcontinent
Midcontinent (n = 156)	62%	21%	17%
Pacific (n = 286)	<1%	100%	–

ORDER Columbiformes
FAMILY Columbidae
BREEDING HABITAT Arid Lands, Desert Scrub
NONBREEDING HABITAT Arid Lands, Desert Scrub
GLOBAL POPULATION SIZE 14,000,000
POPULATION CHANGE SINCE 1970 +67%
IUCN RED LIST CATEGORY Least Concern
R2R TIPPING POINT SPECIES No
US LISTING STATUS No
CANADA SPECIES AT RISK No
PIF CCS 7/20

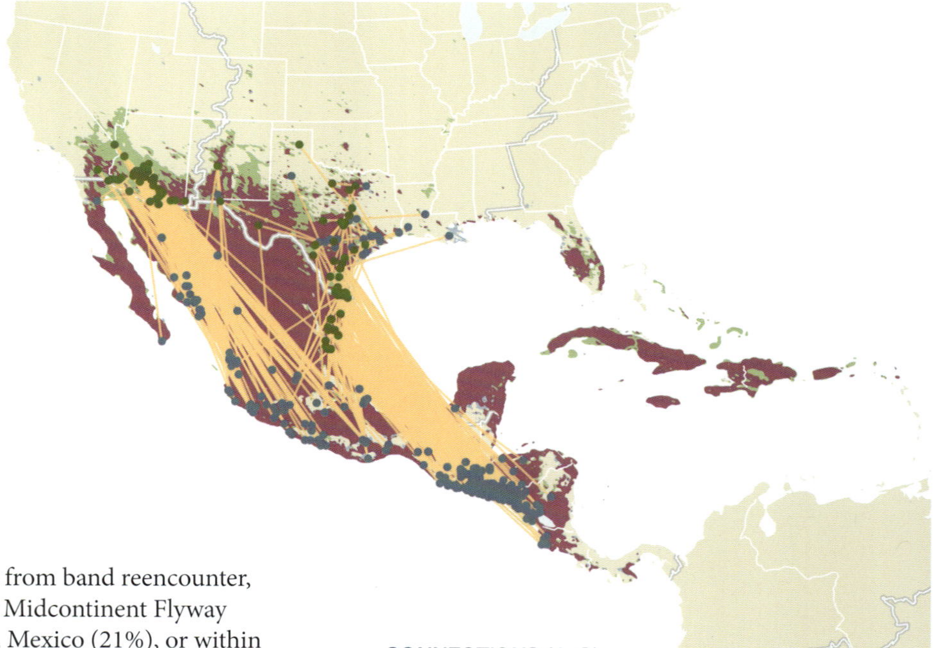

CONNECTIONS (A–C)

MIGRATORY CONNECTIVITY

OVERALL Migratory connectivity information is known from band reencounter, genetic, and stable isotope data. Breeding birds from the Midcontinent Flyway migrate to nonbreeding areas in Central America (62%), Mexico (21%), or within the Midcontinent Flyway (17%). Birds breeding in the Pacific Flyway nearly always

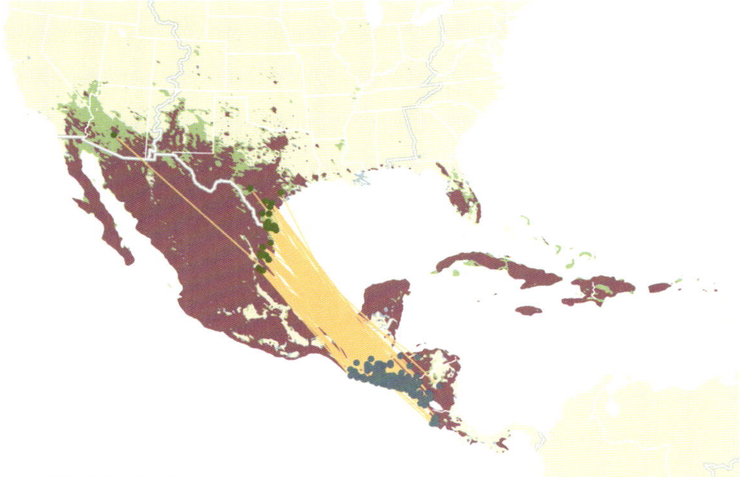

CONNECTIONS A

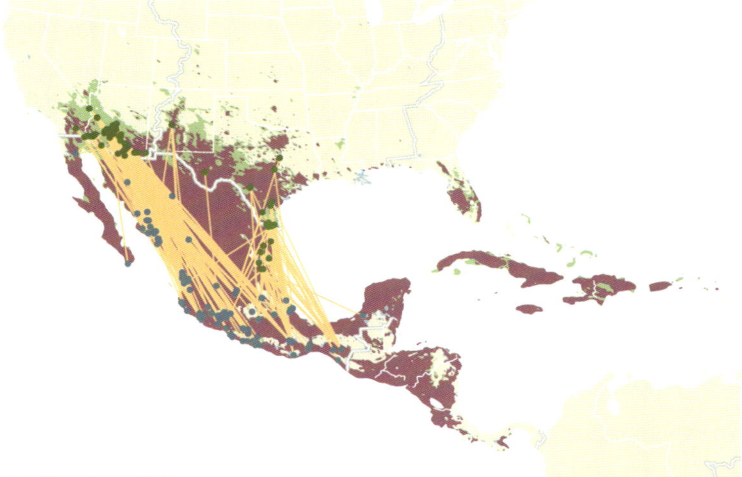

CONNECTIONS B

migrate to Mexico. Separation of breeding populations on nonbreeding areas results in strong migratory connectivity (MC = 0.47 ± 0.07)

BANDING Band reencounters indicate a distinctive split of western and midcontinent breeding birds during the nonbreeding period. Some midcontinent birds have been found as far east as peninsular Florida (Carleton and Martinez del Rio 2010; Carleton et al. 2015).

RADIO TRACKING A few tags were deployed through the Motus Wildlife Tracking System; there are no breeding to nonbreeding detections.

STABLE ISOTOPES In a study across the southwestern US range, deuterium (δ^2H) and δ^{13}C were shown to be useful in differentiating regional populations. Deuterium levels were also able to differentiate between doves foraging in agricultural versus desert habitats. Despite the longitudinal gradient, much of the variability is likely explained by the different isotopic signatures between agricultural and desert food items (Carleton and Martinez del Rio 2010; Carleton et al. 2015).

GENETICS A study of the genetics of this species across the southwestern US showed a minor level of differentiation between White-winged Dove populations, indicating genetic mixing. Additionally, there was not a clear pattern between distance and genetic separation for these populations (Carleton and Martinez del Rio 2010).

OTHER APPROACHES No satellite telemetry studies of migratory connectivity were found for this species.

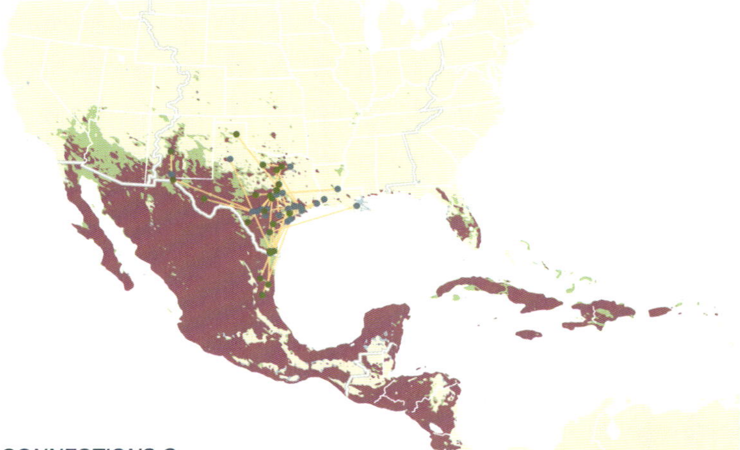

CONNECTIONS C

MIGRATORY PATHWAYS Many individuals of this species undergo short- to medium-range north-south migrations from Mexico and Central America, spending the nonbreeding period in tropical and subtropical areas, mirroring species breeding habitats. Some individuals, especially from the midcontinent population, undertake less predictable movements, variable in direction and distance. Landscape change in recent decades is likely affecting the movement ecology of this species (Schwertner et al. 2020; Fink et al. 2023).

Zenaida macroura MOURNING DOVE

MC = 0.33 ± 0.03*

ORDER Columbiformes
FAMILY Columbidae
BREEDING HABITAT Open Country, Habitat Mosaic
NONBREEDING HABITAT Open Country, Developed/Disturbed
GLOBAL POPULATION SIZE 150,000,000
POPULATION CHANGE SINCE 1970 −15%
IUCN RED LIST CATEGORY Least Concern
R2R TIPPING POINT SPECIES No
US LISTING STATUS No
CANADA SPECIES AT RISK No
PIF CCS 7/20

DISTRIBUTION AND ABUNDANCE

Mourning Doves are widespread and well-distributed breeders from southern Canada across the US, Mexico, and the Caribbean. They are year-round residents across much of this area, including the southwestern, midwestern, and eastern US and north and central Mexico. Regions that tend to have highest abundance are the arid western portions of North America, particularly the Great Plains and Great Basin, and southwestern deserts in the US and northern Mexico (Aldrich and Duvall 1958; Fink et al. 2023). Nonbreeding range extends from southern Mexico to Panama.

There are five distinguishable races of Mourning Dove across this area: the long-winged dark population of the eastern US and Canada; the long-winged pale population of the western US, Canada, and temperate Mexico; the short-winged pale population of Panama and Central America; the short-winged dark to medium, buff-bellied population of the Florida Keys and Caribbean islands; and the large-footed, large-billed dark brownish population off the Pacific coast of Mexico (Aldrich and Duvall 1958).

MIGRATORY CONNECTIVITY

OVERALL Migratory connectivity is known from band reencounter data. Mourning Doves breeding in the Atlantic Flyway primarily migrate to nonbreeding areas within the flyway (93%). Midcontinent Flyway breeding birds disperse across all nonbreeding regions but largely migrate to Mexico (36%), within the Midcontinent Flyway (32%), and within the Atlantic Flyway (23%). Pacific Flyway breeding birds largely migrate to Mexico (76%) or to nonbreeding areas within the Pacific Flyway (20%). The within flyway transition is high for the Atlantic Flyway (93%) but low to moderate for the Pacific Flyway (20%), resulting in moderate migratory connectivity (MC = 0.33 ± 0.03).

BANDING Mourning Dove connectivity is well-informed by almost 2,200 banding records connecting breeding to nonbreeding grounds.

RADIO TRACKING A few tags were deployed through the Motus Wildlife Tracking System; there are no breeding to nonbreeding detections.

OTHER APPROACHES No satellite transmitter, geolocator, stable isotope, or genetic studies of migratory connectivity were found for this species.

MIGRATORY PATHWAYS Migration is generally north-south in both seasons (Fink et al. 2023), with Atlantic birds staying within region and the majority of Midcontinent and Pacific birds funneling into Mexico (Schroeder and Ely 1972).

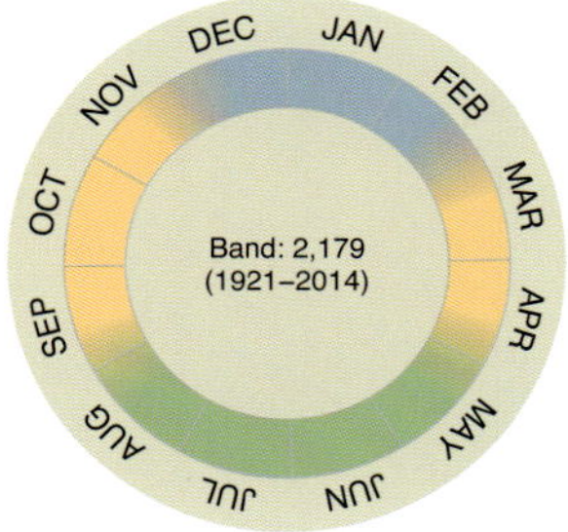

	Atlantic	Caribbean	Central America	Mexico	Midcontinent	Pacific
Atlantic (n = 471)	93%	–	–	<1%	7%	–
Midcontinent (n = 1446)	23%	<1%	8%	36%	32%	<1%
Pacific (n = 238)	–	–	1%	76%	2%	20%

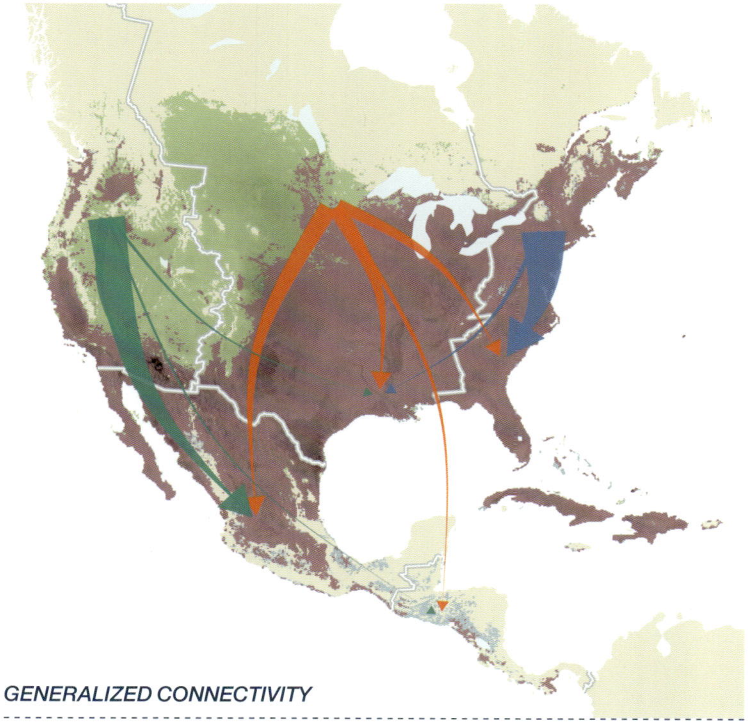

GENERALIZED CONNECTIVITY

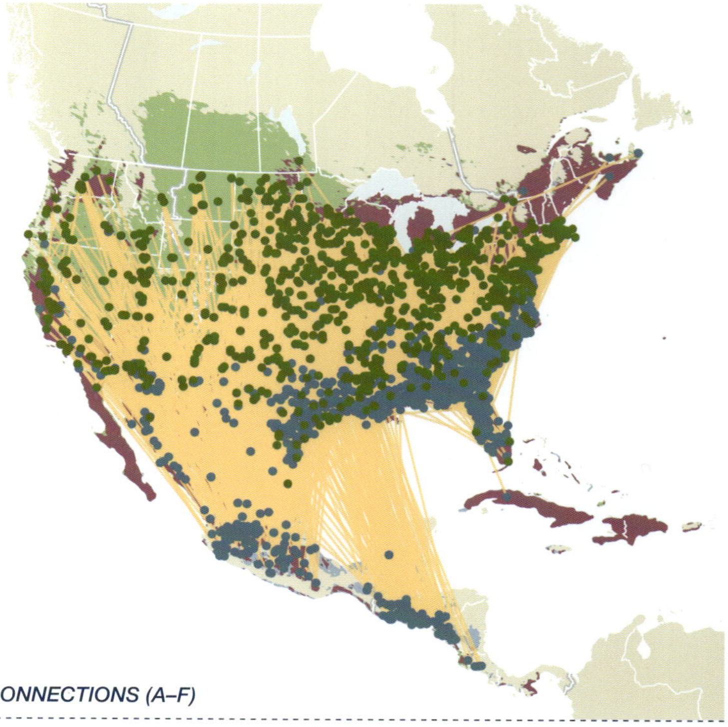

CONNECTIONS (A–F)

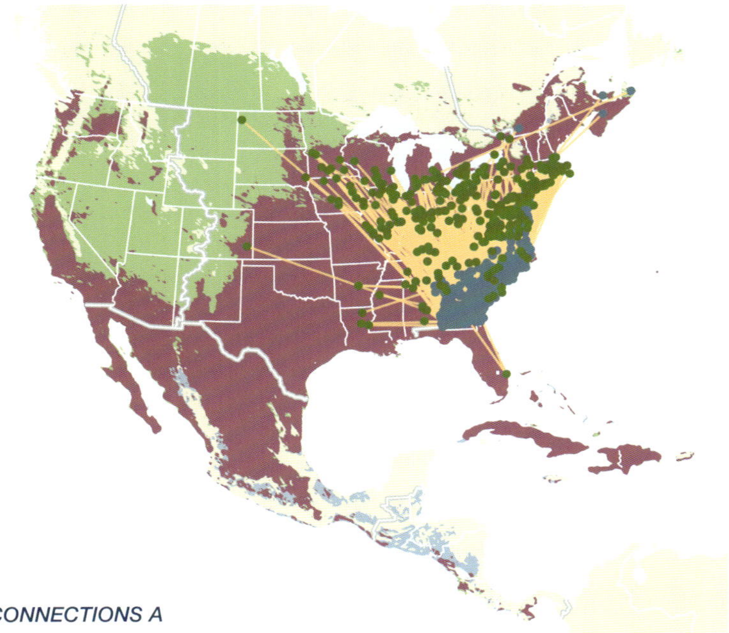

CONNECTIONS A

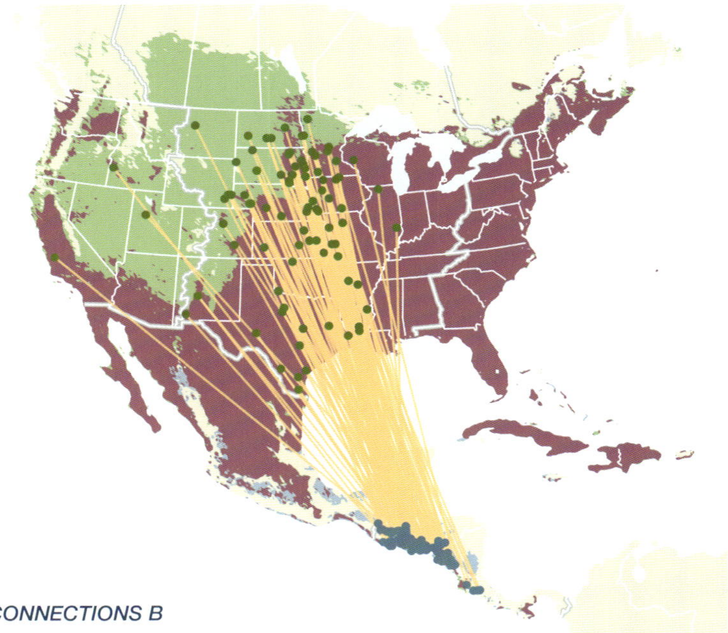

CONNECTIONS B

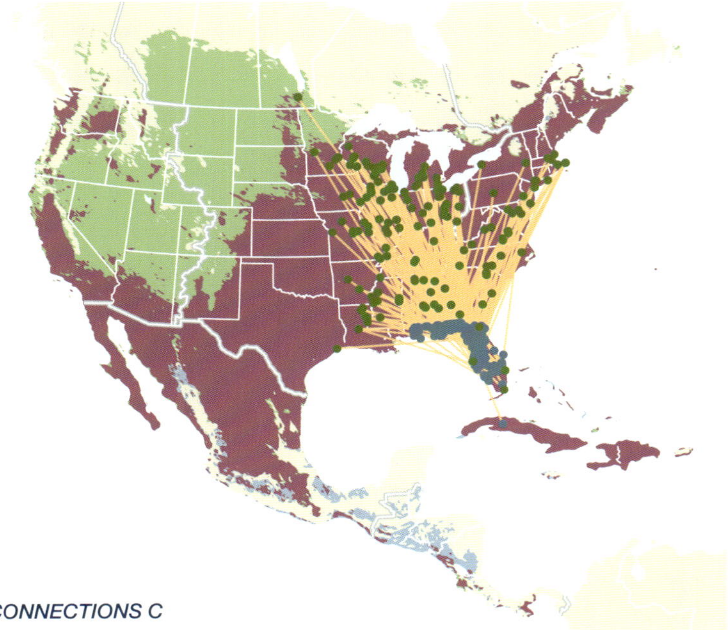

CONNECTIONS C

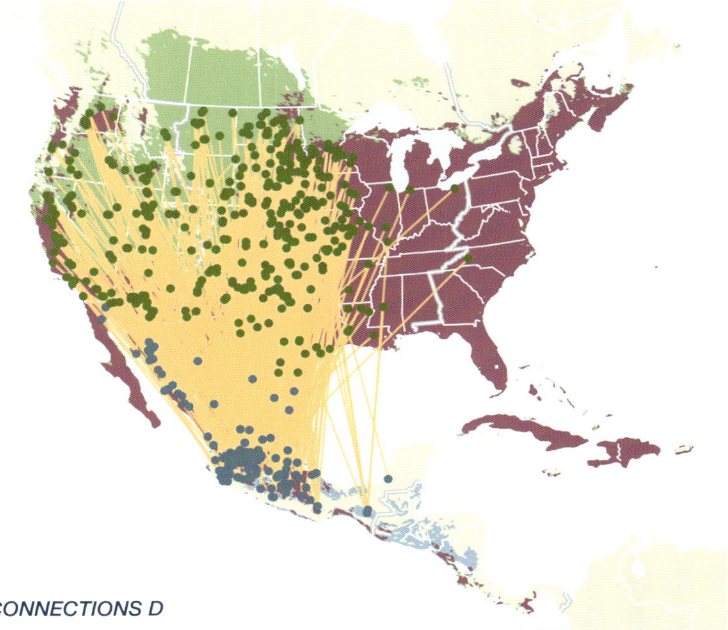

CONNECTIONS D

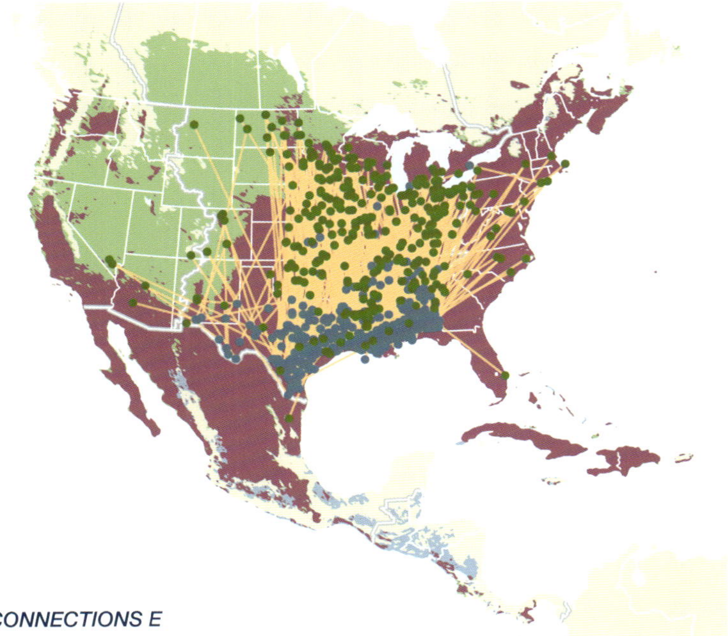

CONNECTIONS E

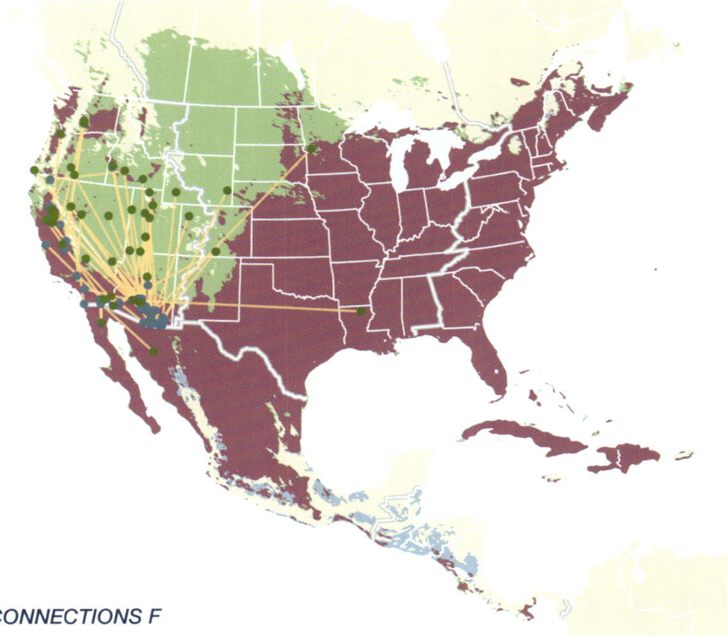

CONNECTIONS F

CUCKOOS (CUCULIFORMES)

WRITING LED BY **MELANIE A. SMITH, TOM WILL, AND PETER P. MARRA**

The migratory connectivity and associated ecology of the order Cuculiformes, which in this Altas includes the Groove-billed Ani, Yellow-billed Cuckoo, and Black-billed Cuckoo, remains poorly understood. Among these, only the Yellow-billed Cuckoo has received significant recent attention through satellite tracking, while migratory connectivity data for the others remain extremely limited due to their secretive behavior and low detectability throughout the annual cycle.

Of the North American Cuculiformes, the Yellow-billed Cuckoo is of high conservation concern, particularly its western distinct population segment (DPS), which has declined dramatically over the past century (Hughes 2020b). Historically breeding as far north as southern British Columbia, the western DPS now persists in only a few remnant populations in the southwestern United States and is federally listed as threatened (https://ecos.fws.gov/ecp/). Eastern populations are more widespread but also show broad-scale declines. Additional tracking, demographic monitoring, habitat assessments, and conservation coordination across international boundaries are urgently needed to fill these knowledge gaps and inform conservation efforts for declining cuckoo populations across North America.

CUCULIDAE | *CUCKOOS*

GROOVE-BILLED ANI *Crotophaga sulcirostris*

DISTRIBUTION AND ABUNDANCE

This species breeds in open savanna and pastures in south Texas, along both slopes of Mexico, throughout Central America, in northern South America in Colombia and Venezuela, and along the Pacific coast of Ecuador and Peru. The species is most abundant along the southern coast of the Gulf of Mexico, in northern Guatemala, and in central Honduras and Nicaragua. It is a year-round resident throughout most of its range, but most individuals breeding in southern Texas and northeast and northwest Mexico migrate farther south during the nonbreeding season (Fink et al. 2023).

DISTRIBUTION AND ABUNDANCE

ORDER Cuculiformes
FAMILY Cuculidae
BREEDING HABITAT Open Country, Developed/Disturbed
NONBREEDING HABITAT Open Country, Developed/Disturbed
GLOBAL POPULATION SIZE 2,000,000
POPULATION CHANGE SINCE 1970 −38%
IUCN RED LIST CATEGORY Least Concern
R2R TIPPING POINT SPECIES No
US LISTING STATUS No
CANADA SPECIES AT RISK No
PIF CCS 8/20

MIGRATORY CONNECTIVITY

OVERALL Migratory connectivity is unknown. No studies of migratory connectivity were found for this species.

MIGRATORY PATHWAYS There is no information on migratory pathways, but this species was observed along the northern coasts of Mexico during pre- and post-breeding migration (Fink et al. 2023).

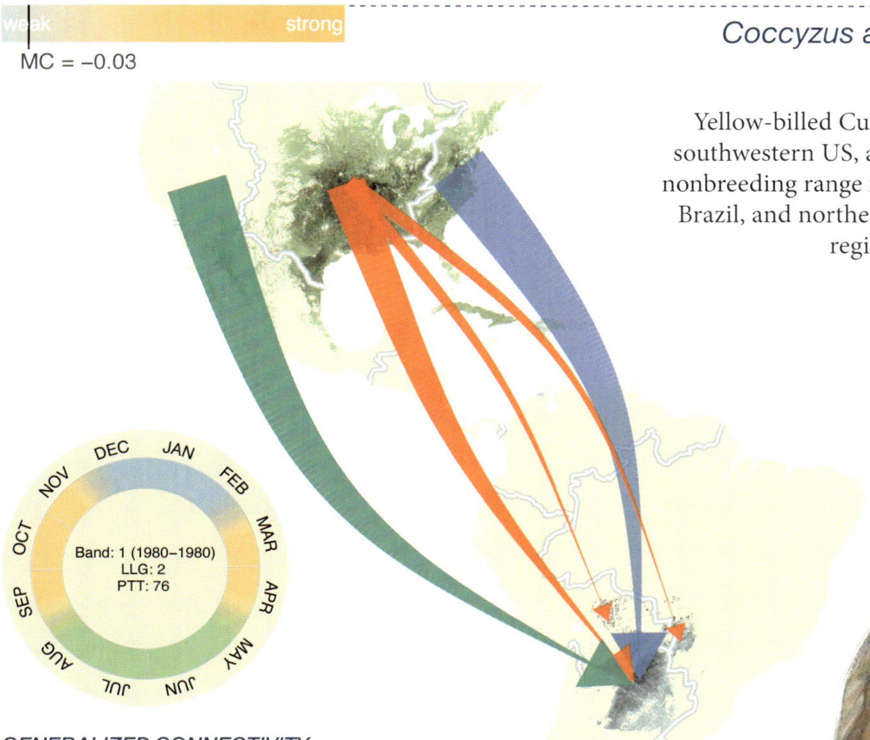

GENERALIZED CONNECTIVITY

Coccyzus americanus YELLOW-BILLED CUCKOO

DISTRIBUTION AND ABUNDANCE

Yellow-billed Cuckoos breed extensively across the eastern, midwestern, and southwestern US, as well as much of Mexico and islands of the Caribbean. The nonbreeding range includes parts of Colombia, Bolivia, Paraguay, southwestern Brazil, and northern Argentina. Highest abundance occurs in the Gran Chaco region, which includes Argentina, Paraguay, Bolivia, and Brazil.

ORDER Cuculiformes
FAMILY Cuculidae
BREEDING HABITAT Forests, Temperate Eastern
NONBREEDING HABITAT Forests, Tropical Dry
GLOBAL POPULATION SIZE 9,600,000
POPULATION CHANGE SINCE 1970 −53%
IUCN RED LIST CATEGORY Least Concern
R2R TIPPING POINT SPECIES No
US LISTING STATUS Threatened western US
CANADA SPECIES AT RISK No
PIF CCS 11/20

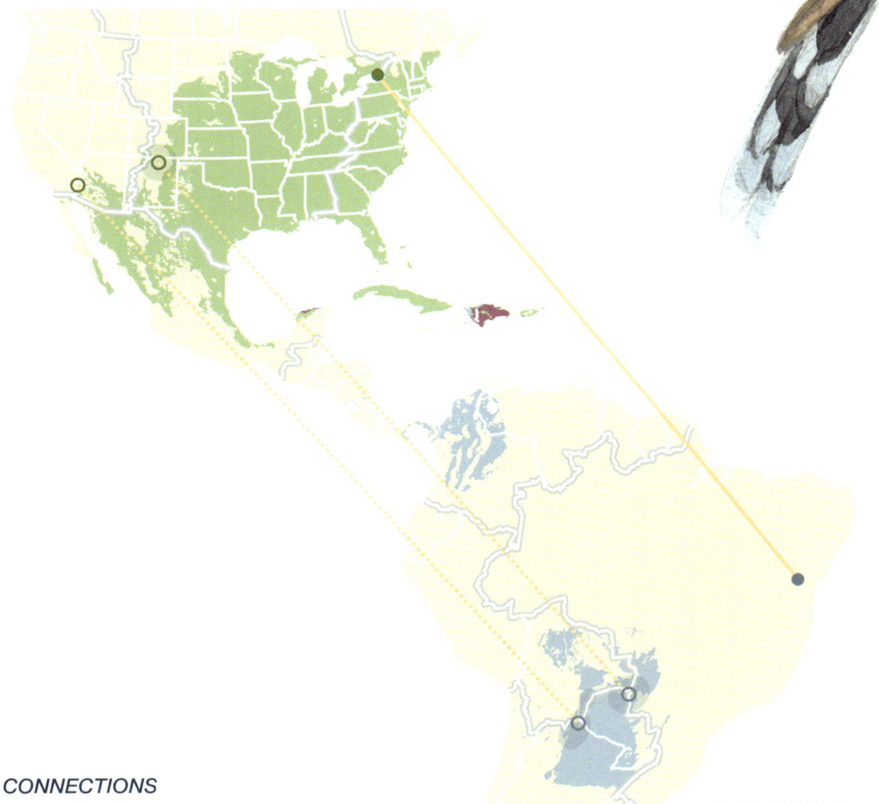

CONNECTIONS

	Andean West	Brazil	Southern Cone
Atlantic (n = 7)	–	–	100%
Midcontinent (n = 27)	19%	19%	63%
Pacific (n = 11)	–	–	100%

INTERESTING NOTE

Yellow-billed Cuckoos never seem to stay in one place for too long. On the breeding grounds, this species does multiple within-season, large-scale itinerant movements in what could be multiple breeding attempts at geographically distinct locations. Similarly, on the nonbreeding grounds, cuckoos shift almost continuously from November to April, possibly in response to local resource peaks.

MIGRATORY CONNECTIVITY

OVERALL Migratory connectivity is known from geolocator and satellite transmitter studies (McNeil et al. 2015; C. Stanley, unpublished data). The majority of all breeding populations migrate to the Southern Cone region of South America. Midcontinent Flyway birds may also be found in the Andean West region (19%) and Brazil (19%) during the nonbreeding season. High mixing on the nonbreeding grounds of breeding populations across the range results in weak migratory connectivity (MC = −0.03). The negative MC estimate indicates Yellow-billed Cuckoo nonbreeding populations are less likely to remain together during the breeding season, i.e., nonbreeding populations include individuals from breeding populations across the range.

BANDING A single breeding to nonbreeding banding record connected Ontario to Brazil.

RADIO TRACKING Data from several tracking projects through the Motus Wildlife Tracking System suggest different migration routes for eastern and western populations. There were no individual breeding to nonbreeding detections.

PATHWAYS (A–C)

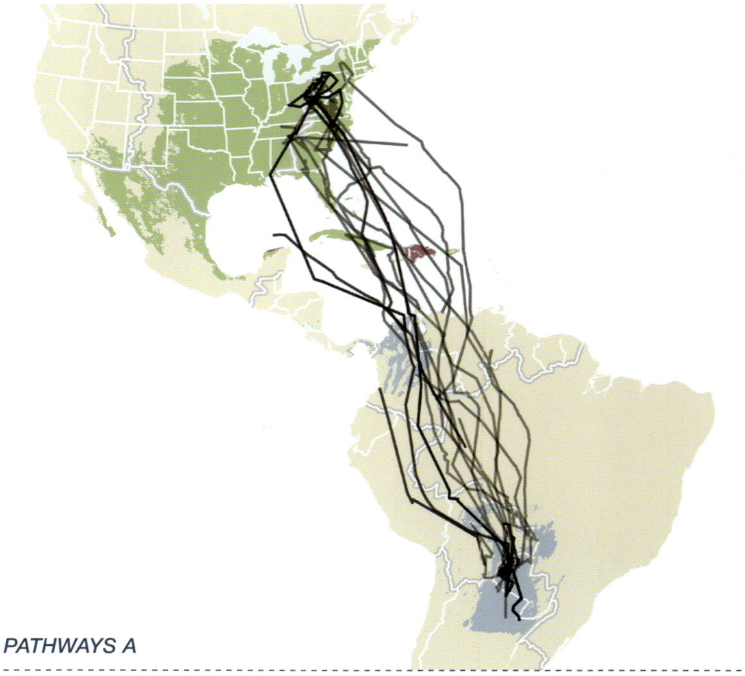

PATHWAYS A

PATHWAYS B

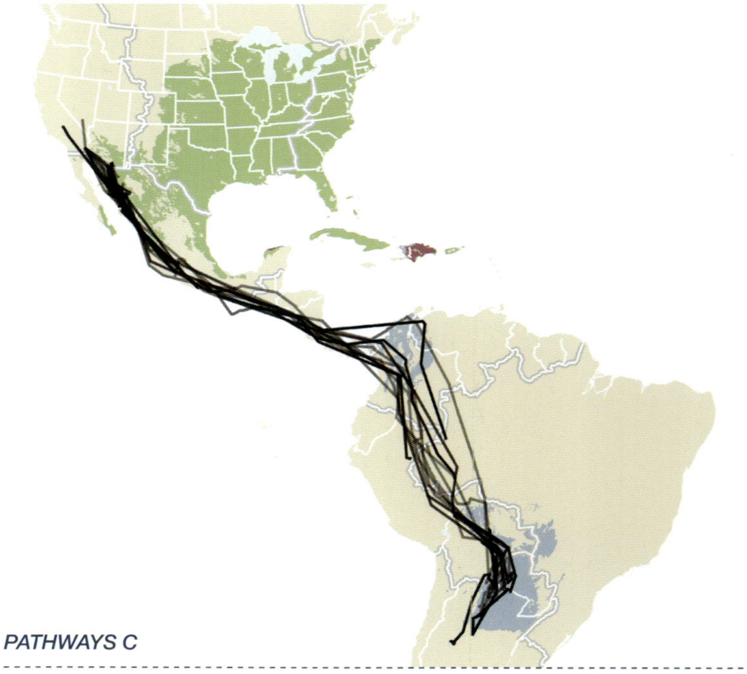

PATHWAYS C

GEOLOCATOR TRACKING McNeil et al. (2015) tracked one Yellow-billed Cuckoo using light-level geolocation from California to the Gran Chaco area of South America. Sechrist et al. (2012) documented a similar migration pathway connecting New Mexico to the Gran Chaco region.

SATELLITE TRACKING Individuals tracked with satellite telemetry (n = 56) from 8 locations across the breeding range (Arizona, California, Illinois, Oklahoma, Pennsylvania, Texas, Tennessee), confirmed geolocator results as all birds migrated to nonbreeding areas in the Gran Chaco and Cerrado regions of South America (C. Stanley and P.P. Marra, unpublished data).

OTHER APPROACHES No stable isotope or genetic studies of migratory connectivity were found for this species.

DATA CONTRIBUTORS

P.P. Marra; S.E. McNeil; C.Q. Stanley; D. Tracey.

MIGRATORY PATHWAYS Western Yellow-billed Cuckoos make a loop migration (McNeil et al. 2015; C. Stanley and P.P. Marra, unpublished data). During southward migration, western populations migrate south along the Pacific Slope of Mexico and Central America into South America. To return north, birds follow a similar route to the east of the southward migration. Eastern populations appear to follow an out-and-back pattern of migration, with midcontinent birds migrating via eastern Mexico and Central America and eastern breeders migrating across the Caribbean to the Yucatán (Fink et al. 2023; C. Stanley and P.P. Marra, unpublished data). Loop migration has been documented for some eastern breeders (C. Stanley and P.P. Marra, unpublished data). Southbound individuals from all three regions converge in Colombia/Venezuela before crossing the upper Amazon Basin to the Gran Chaco.

Coccyzus erythropthalmus **BLACK-BILLED CUCKOO**

PATHWAYS

DISTRIBUTION AND ABUNDANCE

Black-billed Cuckoos breed in Canada and the US, from southern Saskatchewan to Nova Scotia and across the northern Great Plains from the Dakotas south into Appalachia and north to Maine. The nonbreeding distribution is poorly known but thought to occur in western Colombia, Ecuador, and in Peru and Bolivia east of the Andes Range (Hughes 2020a).

ORDER Cuculiformes
FAMILY Cuculidae
BREEDING HABITAT Forests, Temperate Eastern
NONBREEDING HABITAT Forests, Tropical Dry
GLOBAL POPULATION SIZE 880,000
POPULATION CHANGE SINCE 1970 −69%
IUCN RED LIST CATEGORY Least Concern
R2R TIPPING POINT SPECIES No
US LISTING STATUS No
CANADA SPECIES AT RISK No
PIF CCS 12/20

MIGRATORY CONNECTIVITY

OVERALL There is very little information on migratory connectivity.
RADIO TRACKING A single record from the Motus Wildlife Tracking System links Ontario, Canada, with central Colombia. Other tracking projects have few detections close to areas of deployment.
MIGRATORY PATHWAYS Not well documented. Black-billed Cuckoos likely depart the US Gulf Coast before crossing the Gulf of Mexico and arriving in Central America from the Yucatán Peninsula to Panama before heading south to their nonbreeding grounds. During pre-breeding this pattern is reversed, although some individuals may transit through Mexico rather than crossing the Gulf (Fink et al. 2023).
OTHER APPROACHES No satellite transmitter, geolocator, banding, stable isotope, or genetic studies of migratory connectivity were found for this species.

DATA CONTRIBUTORS
N. Bayly; C. Gómez.

NIGHTJARS AND ALLIES (CAPRIMULGIFORMES)

WRITING LED BY **TOM WILL, MELANIE A. SMITH, CAMILLA GÓMEZ, TIMOTHY GUIDA,** *AND* **AMY L. SCARPIGNATO**

GPS tracks of Common Nighthawks and Eastern Whip-poor-wills. Common Nighthawks from across the breeding range converge in the Amazon and Cerrado regions of Brazil during the nonbreeding season. Eastern Whip-poor-wills tracked from northeastern breeding populations spent the nonbreeding season throughout Central America and Southern Mexico.

The order Caprimulgiformes contains the nighthawks, nightjars, and poorwills, all contained in the family Caprimulgidae. Although most nightjars are large enough to carry tracking devices, large gaps exist in our knowledge about the migratory connectivity of Caprimulgiformes given, in part, their nocturnal and secretive behavior, which make them a difficult group to capture and study. Only two out of eight migratory species in this order (25%) have migratory connectivity information or have been tracked: Common Nighthawk and Eastern Whip-poor-will. However, unlike other groups, dozens of individuals for each species, have high-resolution tracking data, making our knowledge about their migratory connectivity slightly more complete. Within the Caprimulgiformes, the Eastern Whip-poor-will is considered a Tipping Point species for having lost >50% of its populations from 1970 to 2019 and continuing a negative trajectory. Common Nighthawks have also declined by almost 50% over that same time.

Evidence from tracked individuals of the two species Common Nighthawk and Eastern Whip-poor-will suggest weak migratory connectivity between breeding and nonbreeding populations. However, further research is necessary to understand the migratory connectivity of unstudied populations of these two species, as well as the other species in the group about which we currently have no information.

CAPRIMULGIDAE | *NIGHTJARS AND ALLIES*

LESSER NIGHTHAWK *Chordeiles acutipennis*

DISTRIBUTION AND ABUNDANCE

The Lesser Nighthawk breeds in arid and semiarid habitats in the southwestern US from south Texas to the Sierras and coastal mountains of California. It is most abundant in southwestern Arizona and northern and western Mexico, and south locally along both coasts of Mexico and Central America to northern South America (Venezuela, Guyana, and northern Brazil) and the Pacific coast of Ecuador and Peru. During the nonbreeding season, northern populations migrate to stationary nonbreeding grounds from central Mexico south to at least Panama (Latta and Baltz 2020), with concentrations along the Pacific coast of Central America (Fink et al. 2023).

ORDER Caprimulgiformes
FAMILY Caprimulgidae
BREEDING HABITAT Arid Lands Aerial, Desert Scrub
NONBREEDING HABITAT Open Country Aerial, Habitat Mosaic
GLOBAL POPULATION SIZE 14,000,000
POPULATION CHANGE SINCE 1970 +1%
IUCN RED LIST CATEGORY Least Concern
R2R TIPPING POINT SPECIES No
US LISTING STATUS No
CANADA SPECIES AT RISK No
PIF CCS 9/20

MIGRATORY CONNECTIVITY

OVERALL Migratory connectivity is unknown. No studies of migratory connectivity were found for this species. ***MIGRATORY PATHWAYS*** eBird animations (Fink et al. 2023) suggest that south Texas and possibly eastern New Mexico breeders migrate south along the Gulf Coast, while Arizona, California, and northwest Mexican breeders migrate south along the Pacific coast.

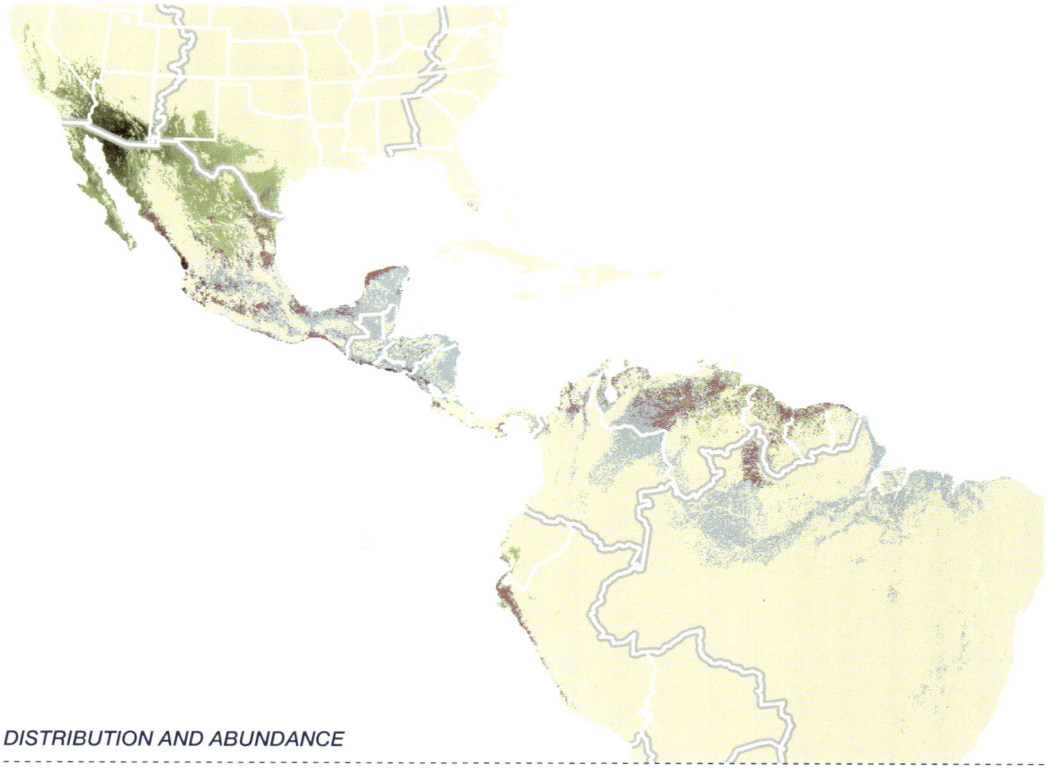

DISTRIBUTION AND ABUNDANCE

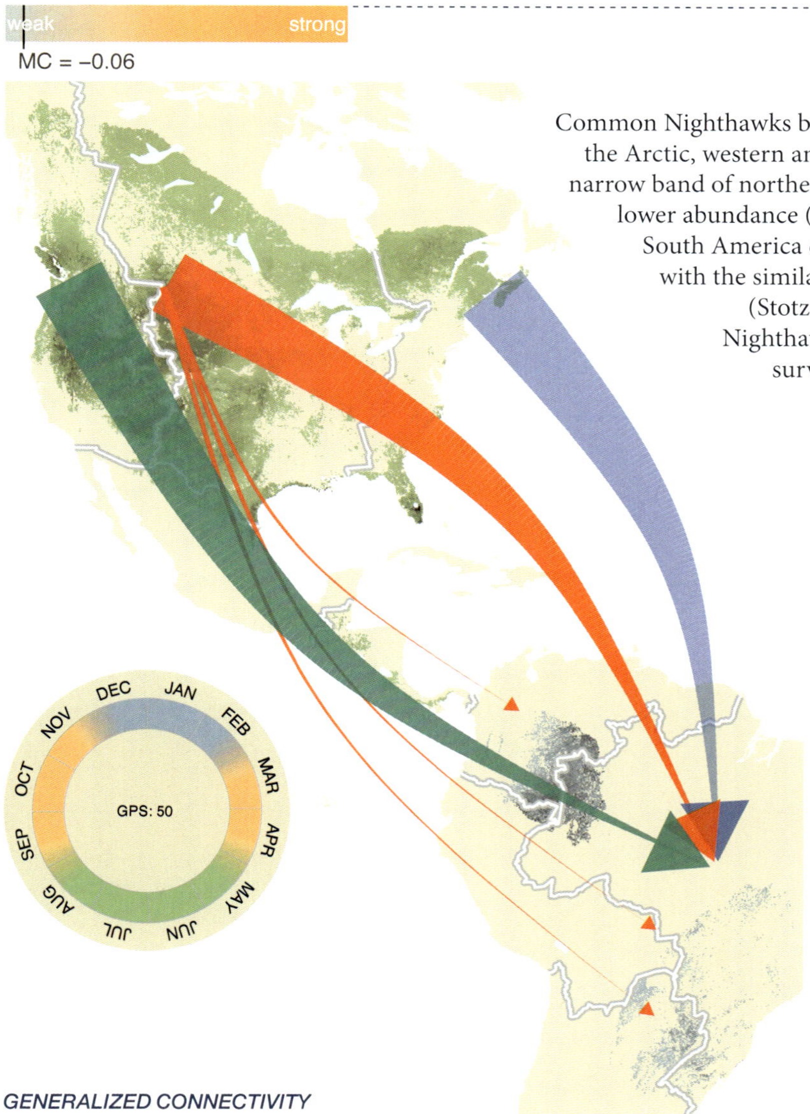

GENERALIZED CONNECTIVITY

Chordeiles minor COMMON NIGHTHAWK

DISTRIBUTION AND ABUNDANCE

Common Nighthawks breed throughout Canada and the United States (excluding Alaska, the Arctic, western and Southern California, and southwestern Arizona), as well as in a narrow band of northern Mexico, and throughout much of Central America in relatively lower abundance (Fink et al. 2023). Range maps indicate they are widespread across South America during the nonbreeding season (observations may be confounded with the similar Lesser Nighthawk), but the actual range may be more restricted (Stotz et al. 1992). The specific distribution and abundance of Common Nighthawks is not well known due to low probability of detection and low survey coverage in areas of peak abundance on both the nonbreeding and breeding grounds (Knight et al. 2021a, 2021b).

ORDER Caprimulgiformes
FAMILY Caprimulgidae
BREEDING HABITAT Open Country Aerial, Habitat Mosaic
NONBREEDING HABITAT Open Country Aerial, Habitat Mosaic
GLOBAL POPULATION SIZE 23,000,000

POPULATION CHANGE SINCE 1970 −57%
IUCN RED LIST CATEGORY Least Concern
R2R TIPPING POINT SPECIES No
US LISTING STATUS No
CANADA SPECIES AT RISK No
PIF CCS 11/20

MIGRATORY CONNECTIVITY

OVERALL Data on the nonbreeding range, migration, and migratory connectivity for Common Nighthawks are known from three satellite tracking studies (Ng et al. 2018; Knight et al. 2021a; Cockle et al. 2023). All individuals tracked from the Atlantic and Pacific flyways migrated to nonbreeding areas in Brazil. The majority of Midcontinent Flyway breeding birds (90%) also migrated to Brazil. Breeding populations may overlap on the nonbreeding grounds but are farther apart from each other in the breeding season; therefore migratory connectivity is not only weak but also negative (MC = −0.06).

RADIO TRACKING One bird was detected through the Motus Wildlife Tracking System during the breeding season in North Dakota and at Reserva Natural Rincón de Santa María in Argentina during the nonbreeding season (Motus project, Connecting Conservation for Declining Pacific Migrants, unpublished data). Several other individuals were detected during migration.

SATELLITE TRACKING One study tracked seven male Common Nighthawks from the boreal forest of eastern Alberta, Canada, to central Brazil (Amazon and Cerrado regions; Ng et al. 2018). A later study expanded the effort in Alberta to a range-wide study, tracking 52 Common Nighthawks from 12 breeding populations (Knight et al. 2021b). This study found low spatial connectivity and mixing of populations on the nonbreeding grounds, primarily in the Amazon and Cerrado biomes of Brazil. Several birds (n = 7) used secondary nonbreeding areas later in the season. Three of nine female Common Nighthawks tracked from nonbreeding locations in Argentina spent the breeding season in Florida (Cockle et al. 2023). They returned to South America the following nonbreeding season, spending about three months in the northern Cerrado of Brazil, before returning to

	Andean West	Brazil	Northern SA	Southern Cone
Atlantic (n = 2)	–	100%	–	–
Midcontinent (n = 29)	3%	90%	3%	3%
Pacific (n = 8)	–	100%	–	–

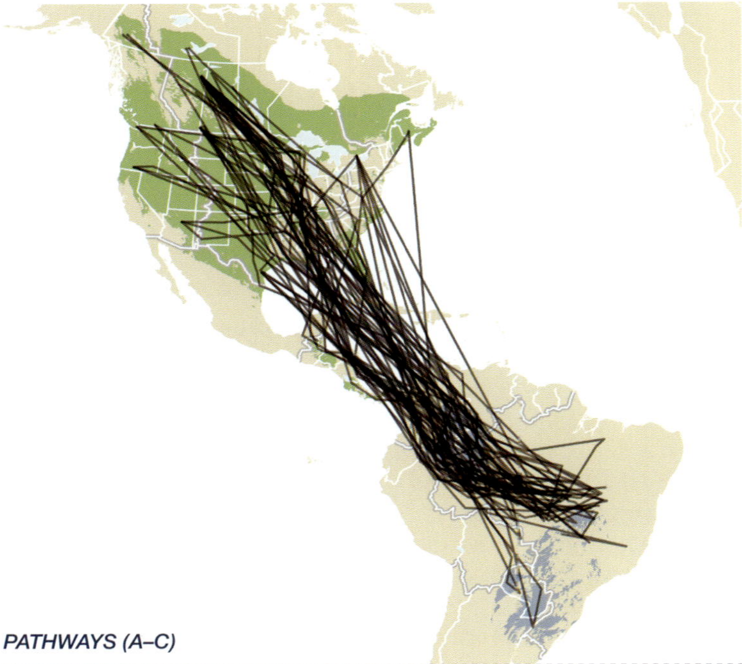

PATHWAYS (A–C)

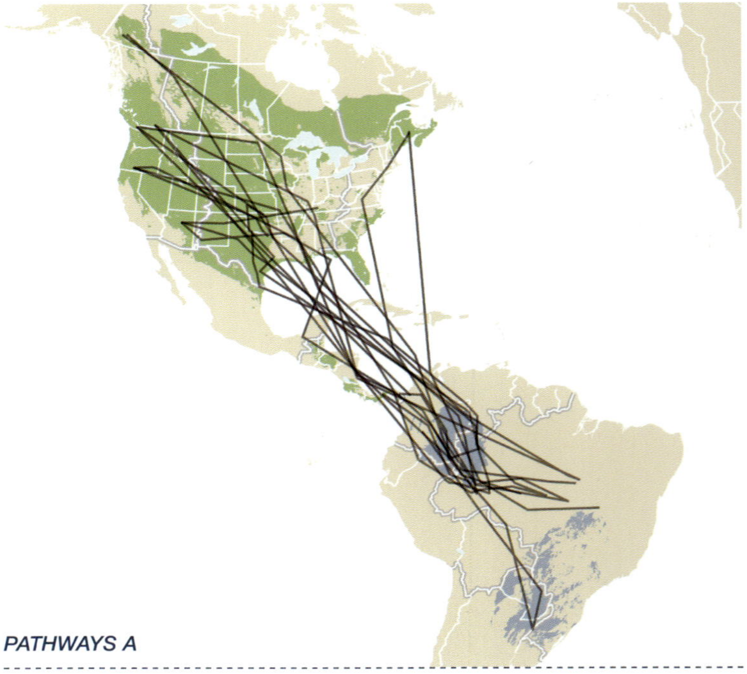

PATHWAYS A

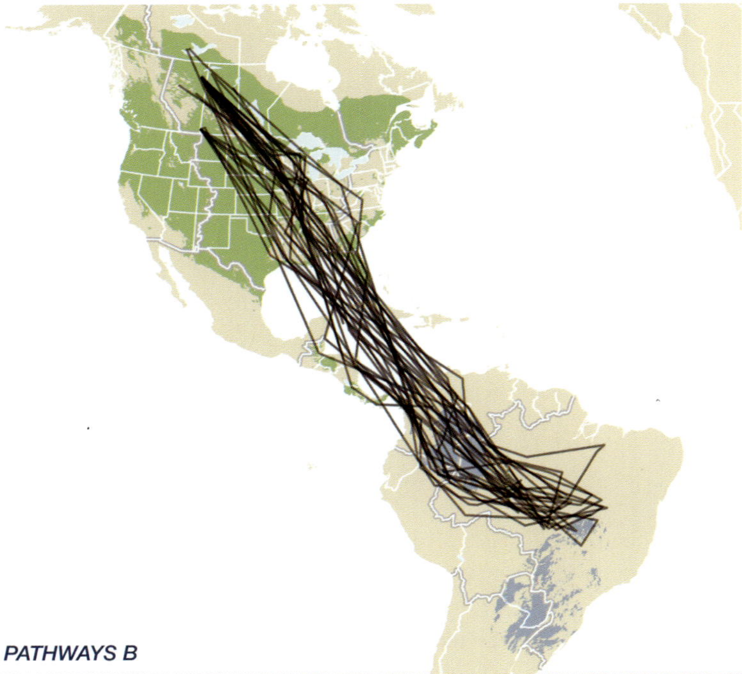

PATHWAYS B

PATHWAYS C

Reserva Natural Rincón de Santa María, where they were tagged the previous year (Cockle et al. 2023).

OTHER APPROACHES No banding, stable isotope, or genetic studies of migratory connectivity were found for this species.

MIGRATORY PATHWAYS Common Nighthawks from across the breeding range use a shared route for much of their north and southward migrations. Post-breeding individuals migrate toward the Midcontinent Flyway and then south across the Gulf of Mexico. All individuals then funnel along a narrow route in the Colombian Andes and continue across the Amazon basin toward nonbreeding areas. Pre-breeding individuals reverse this pattern heading, north and departing for the Gulf of Mexico from northern Colombia (Knight et al. 2021b). Common Nighthawks tracked from Alberta's boreal forest exhibited a loop migration pattern: in general, birds flew from Alberta to southeast US, across the Caribbean or Gulf of Mexico, and along the Andes to the Brazilian Amazon and Cerrado regions; spring birds had more direct flights over the Amazon, across the Caribbean to the Mississippi River Valley, and back to Alberta (Ng et al. 2018). Four out of five birds stopped over in Cuba during either pre- or post-breeding migration. Data from eBird also suggest a loop migration for some individuals (Fink et al. 2023). When traveling northward from Argentina, tracked birds took a fairly direct route through the lowlands of Bolivia, western Amazon, and central Colombia (Cockle et al. 2023). At least two (and possibly all three) individuals stopped for 2–3 weeks in the Orinoco region of Colombia, before crossing the Caribbean from northern Colombia, northern Venezuela, or off the east coast of Central America. Birds departed breeding areas in Florida and traveled south through Colombia and the western Amazon of Brazil and then eastward across the southern Amazon / northern Cerrado of Brazil (Cockle et al. 2023).

DATA CONTRIBUTORS

E.M. Bayne; A.-L. Harrison; E.C. Knight; P.P. Marra; J.W. Ng; A.L. Scarpignato; S.L. Van Wilgenburg; Migratory Connectivity Project.

ANTILLEAN NIGHTHAWK

Chordeiles gundlachii

DISTRIBUTION AND ABUNDANCE

DISTRIBUTION AND ABUNDANCE

This species breeds in arid lowland scrub, nesting on the ground and foraging in open areas, in the westernmost Florida Keys, the Greater Antilles, the Bahamas, and the Virgin Islands. Nonbreeding grounds were unknown until recently but have now been confirmed to be located in South America (Perlut and Levesque 2020; Mlodinow et al. 2024).

ORDER Caprimulgiformes
FAMILY Caprimulgidae
BREEDING HABITAT Open Country Aerial, Habitat Mosaic
NONBREEDING HABITAT Unknown
GLOBAL POPULATION SIZE 200,000
POPULATION CHANGE SINCE 1970 NA
IUCN RED LIST CATEGORY Least Concern
R2R TIPPING POINT SPECIES No
US LISTING STATUS No
CANADA SPECIES AT RISK No
PIF CCS 13/20

MIGRATORY CONNECTIVITY

OVERALL Migratory connectivity is unknown. No information on migratory connectivity within the US was found for this species.

MIGRATORY PATHWAYS There is little information. Migrants observed moving southbound through the southern Caribbean and northbound in northern Venezuela, as well as Curaco, Grenada, and Barbados (Mlodinow et al. 2024).

INTERESTING NOTE

A female bird was tracked from its breeding grounds on Guadeloupe, in the Caribbean, to nonbreeding grounds in a remote area of Amazonas, Brazil (Perlut and Levesque 2020). The single tracked bird followed a north-south migration route from the Caribbean to the Amazon. Post-breeding migration included an extended stopover on an island off the Venezuelan coast, while the pre-breeding migration followed a similar route but did not include an extended stopover (Perlut and Levesque 2020).

COMMON POORWILL *Phalaenoptilus nuttallii*

DISTRIBUTION AND ABUNDANCE

This species breeds in arid or semiarid plains and brushlands from southern British Columbia, throughout the western US, and south to Central Mexico, most abundantly in southern Arizona, northern Baja California, and Sonora, where it is present year-round (Fink et al. 2023). Because the species is inconspicuous outside the breeding season, the nonbreeding distribution is less well known, with northern birds apparently migrating to the southern portion of the species range (Woods et al. 2020).

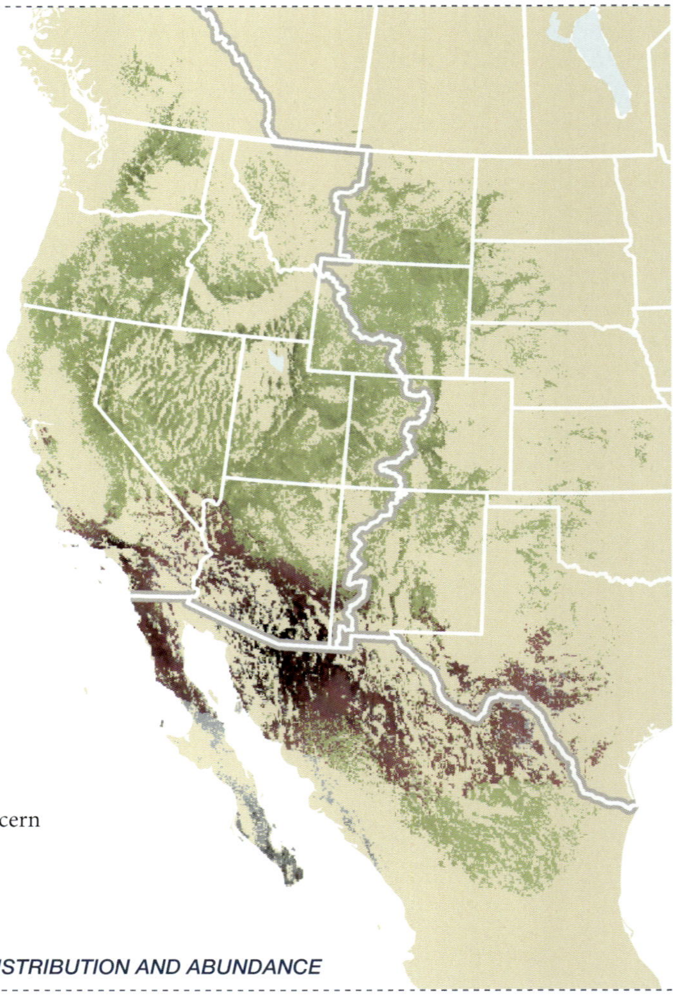

DISTRIBUTION AND ABUNDANCE

ORDER Caprimulgiformes
FAMILY Caprimulgidae
BREEDING HABITAT Arid Lands, Desert Scrub
NONBREEDING HABITAT Arid Lands, Desert Scrub
GLOBAL POPULATION SIZE 1,700,000
POPULATION CHANGE SINCE 1970 +19%
IUCN RED LIST CATEGORY Least Concern
R2R TIPPING POINT SPECIES No
US LISTING STATUS No
CANADA SPECIES AT RISK No
PIF CCS 10/20

MIGRATORY CONNECTIVITY

OVERALL Migratory connectivity is unknown.
RADIO TRACKING Detections from the Motus Wildlife Tracking System connect the Intermountain West to southern coastal California (Motus project, Intermountain West Birds, unpublished data).
OTHER APPROACHES No banding, radio telemetry, geolocator, satellite telemetry, stable isotope, or genetic studies of migratory connectivity were found for this species.
MIGRATORY PATHWAYS Northern birds apparently undergo north-south migrations from much of the American West to the far Southwest and northern Mexico in the nonbreeding season (Fink et al. 2023), but details of pathways are unknown.

CHUCK-WILL'S-WIDOW *Antrostomus carolinensis*

DISTRIBUTION AND ABUNDANCE

Breeds in forested regions throughout the southeastern US except in the Appalachians and the Mississippi Alluvial Plain. During the nonbreeding season, it is known to occur along the Gulf Coast of Mexico and south along the Caribbean slope throughout Mexico and Central America (except the Yucatán Peninsula) to Colombia east of the Andes, also in southern Florida, the Bahamas, the Greater Antilles, and the Virgin Islands (Straight and Cooper 2020). Nonbreeding eBird detections, on the other hand, begin along the coast of Belize and extend along both coasts from Honduras through Panama (Fink et al. 2023).

ORDER Caprimulgiformes
FAMILY Caprimulgidae
BREEDING HABITAT Forests, Temperate Eastern
NONBREEDING HABITAT Open Country, Habitat Mosaic
GLOBAL POPULATION SIZE 5,700,000
POPULATION CHANGE SINCE 1970 −66%
IUCN RED LIST CATEGORY Near Threatened
R2R TIPPING POINT SPECIES No
US LISTING STATUS No
CANADA SPECIES AT RISK No
PIF CCS 12/20

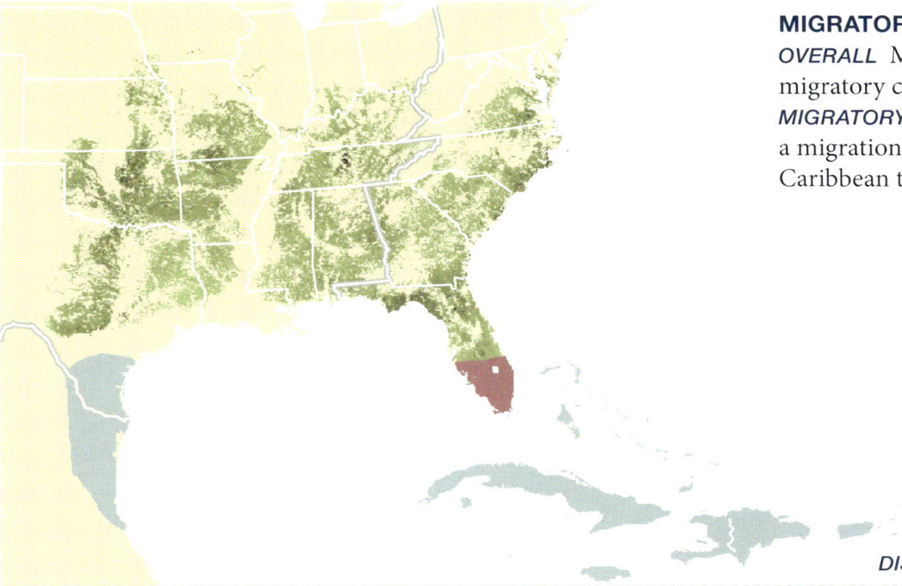

DISTRIBUTION AND ABUNDANCE

MIGRATORY CONNECTIVITY

OVERALL Migratory connectivity is unknown. No studies of migratory connectivity were found for this species.

MIGRATORY PATHWAYS eBird animations (Fink et al. 2023) suggest a migration pathway through Florida, over Cuba, then across the Caribbean to the coast of Belize and Honduras.

Antrostomus ridgwayi BUFF-COLLARED NIGHTJAR

DISTRIBUTION AND ABUNDANCE

A primarily Mexican and Central American species, found only in a few locations in the American Southwest. Populations are year-round throughout most of the species' range but appear migratory in the most northern and southern populations. Movements of individual birds are unknown, and the secretive nature of this species makes it difficult to study. Sparse in the US and northernmost Mexico; the highest densities of this species are found in western Mexico from central Sonora south to Oaxaca (Bowers and Dunning 2020; Fink et al. 2023).

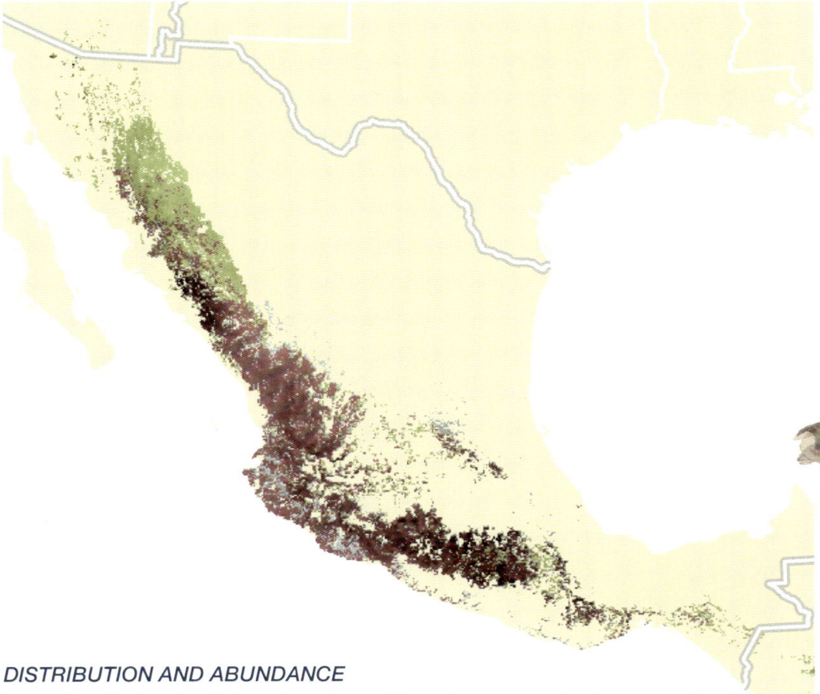

DISTRIBUTION AND ABUNDANCE

MIGRATORY CONNECTIVITY

OVERALL Migratory connectivity is unknown. No studies of migratory connectivity were found for this species.

MIGRATORY PATHWAYS Birds breeding at the northernmost and southernmost extent of the range are probably migratory, moving latitudinally to the year-round range in the nonbreeding period. There is also evidence of latitudinal migrations in some Mexican populations (Bowers et al. 2020; Fink et al. 2023).

ORDER Caprimulgiformes
FAMILY Caprimulgidae
BREEDING HABITAT Forests, Tropical Dry
NONBREEDING HABITAT Forests, Tropical Dry
GLOBAL POPULATION SIZE 2,000,000
POPULATION CHANGE SINCE 1970 NA
IUCN RED LIST CATEGORY Least Concern
R2R TIPPING POINT SPECIES No
US LISTING STATUS No
CANADA SPECIES AT RISK No
PIF CCS 12/20

EASTERN WHIP-POOR-WILL *Antrostomus vociferus*

weak strong
MC = 0.09 ± 0.08*

DISTRIBUTION AND ABUNDANCE

Eastern Whip-poor-wills breed in eastern North America, along the southern edge of Canada east of central Saskatchewan, and in the eastern US from the eastern Great Plains across much of the Midwest to the Atlantic Ocean. Nonbreeding areas include portions of the southeastern US (coastal South Carolina and Georgia; Florida; and the Gulf of Mexico coastline), northeastern and southern Mexico excluding the Yucatán Peninsula, and Central America from southern Belize to Costa Rica (Fink et al. 2023).

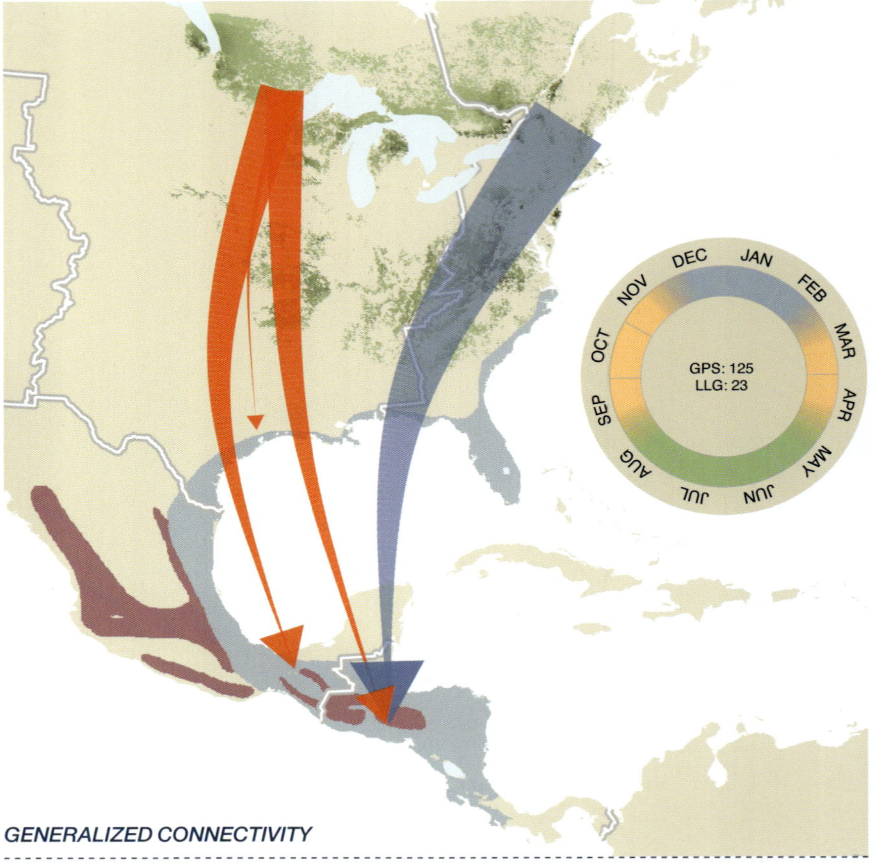

GENERALIZED CONNECTIVITY

ORDER Caprimulgiformes
FAMILY Caprimulgidae
BREEDING HABITAT Forests, Temperate Eastern
NONBREEDING HABITAT Forests, Tropical Lowland Evergreen
GLOBAL POPULATION SIZE 1,800,000
POPULATION CHANGE SINCE 1970 −72%
IUCN RED LIST CATEGORY Near Threatened
R2R TIPPING POINT SPECIES Yellow-Alert
US LISTING STATUS No
CANADA SPECIES AT RISK Threatened
PIF CCS 14/20

	Central America	Mexico	Midcontinent
Atlantic (n = 4)	100%	–	–
Midcontinent (n = 87)	44%	53%	3%

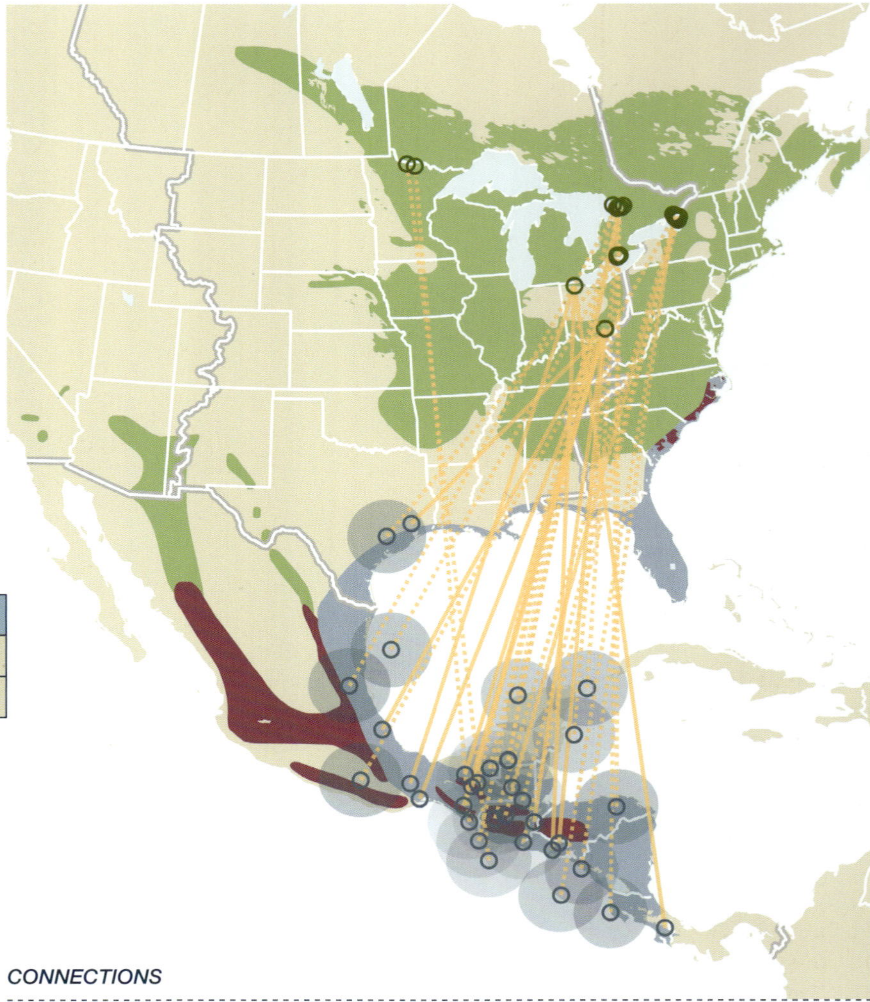

CONNECTIONS

MIGRATORY CONNECTIVITY

OVERALL Several recent tracking studies have greatly increased our knowledge of Eastern Whip-poor-will migratory connectivity (English et al. 2017; Korpach et al. 2019, 2022; Tonra et al. 2019a; Skinner et al. 2022; M.H. Bakermans and A.V. Vitz, unpublished data). Birds breeding within the Atlantic Flyway migrated to nonbreeding areas in Central America (100%).

INTERESTING NOTE

The high spatial and temporal overlap of breeding populations throughout the range at two regions, the southeastern United States during migration and nonbreeding locations in southern Mexico and northern Central America, could be considered potentially important as both positive and negative conditions experienced at these locations affect individuals from across the breeding range (English et al. 2017; Korpach et al. 2022; Skinner et al. 2022).

Midcontinent Flyway breeding populations also migrated to nonbreeding areas in Central America (44%); however, a larger percentage migrated to Mexico (53%). Although each breeding population slightly favored different nonbreeding areas, high regional overlap of breeding populations from the Atlantic and Midcontinent flyways on nonbreeding areas results in weak connectivity (0.09 ± 0.08).

RADIO TRACKING Ongoing studies of Eastern Whip-poor-will breeding populations continue in southern Canada and the northern US to examine migratory routes, stopovers, and habitat use.

GEOLOCATOR TRACKING Similar to GPS data, light-level geolocator data revealed that Eastern Whip-poor-wills breeding within the Great Lakes region of Ontario, Canada (n = 22), spent the nonbreeding season from central Mexico south through Costa Rica (English et al. 2017). Three of four males appeared to migrate to the same nonbreeding area in two consecutive years. Tracked Eastern Whip-poor-wills showed some evidence of leapfrog migration as more northerly breeding individuals and females migrated to nonbreeding areas farther south as compared to more southerly breeding individuals and the majority of males (English et al. 2017).

SATELLITE TRACKING Using GPS technology, studies from throughout the Great Lakes region (Tonra et al. 2019a; Korpach et al. 2022; Skinner et al. 2022) and New England (Massachusetts; M.H. Bakermans and A.V. Vitz, unpublished data) tracked breeding populations to nonbreeding areas from southern Mexico (Veracruz, Oaxaca, Chiapas) to Central America. Breeding to nonbreeding migratory connectivity was low, with extensive overlap on nonbreeding areas. Southern Mexico and northern Central America, particularly Guatemala and El Salvador, were identified as important nonbreeding areas, although individuals spanned from Texas to Costa Rica (Tonra et al. 2019a, 2019b; Korpach et al. 2022; Skinner et al. 2022; M.H. Bakermans and A.V. Vitz, unpublished data). Although connectivity was low regionally, fine-scale differences between breeding populations were documented; more northern and western breeding birds in Manitoba spent the nonbreeding period at sites farther south than eastern-breeding birds from southern Ontario, consistent with a leapfrog pattern of migration (Korpach et al. 2022).

OTHER APPROACHES No banding, genetic, or stable isotope studies of migratory connectivity were found for this species.

MIGRATORY PATHWAYS Spatial connectivity weakens successively during migration as Eastern Whip-poor-wills follow a south to southwesterly route to converge in Texas and circumnavigate the Gulf of Mexico before continuing south through Mexico (Korpach et al. 2022; Skinner et al. 2022; Bakermans and Vitz, unpublished data). Post-breeding patterns appear to be a reversal of pre-breeding pathways; however, the US Gulf Coast is a frequent stopover region during southward migration but not while traveling north (English et al. 2017).

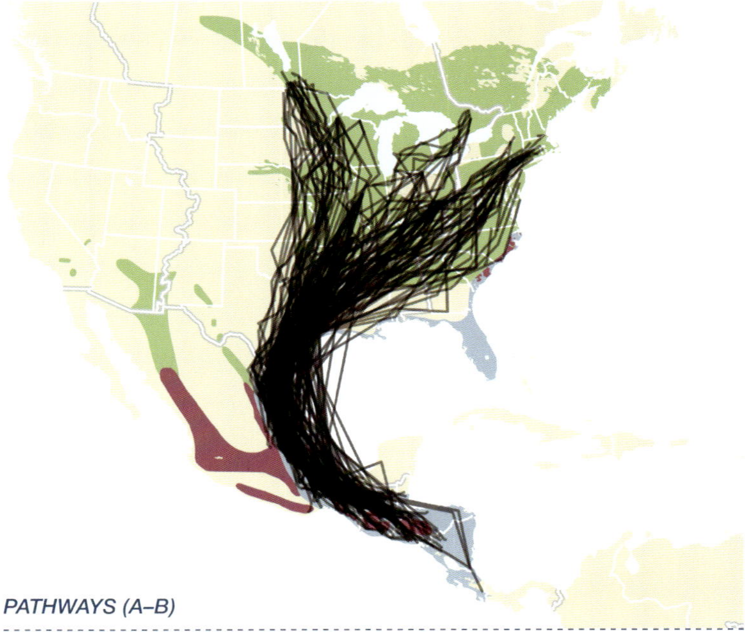

PATHWAYS (A–B)

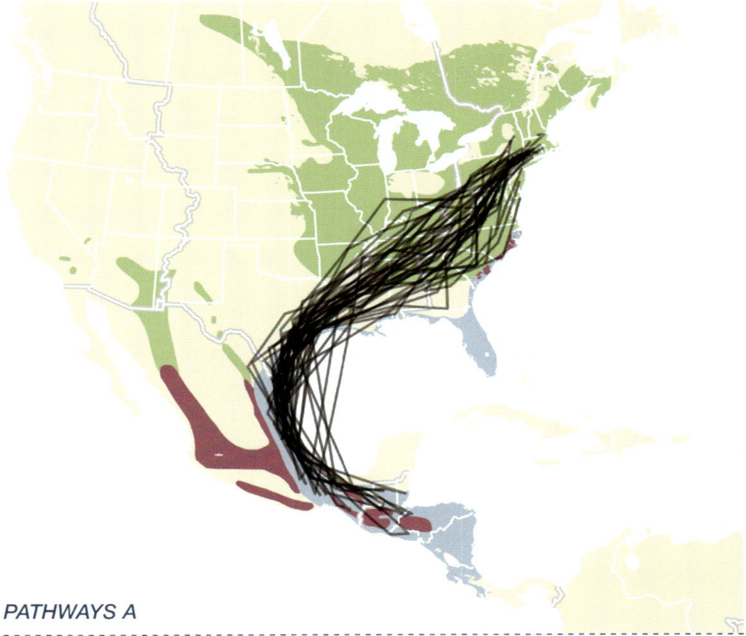

PATHWAYS A

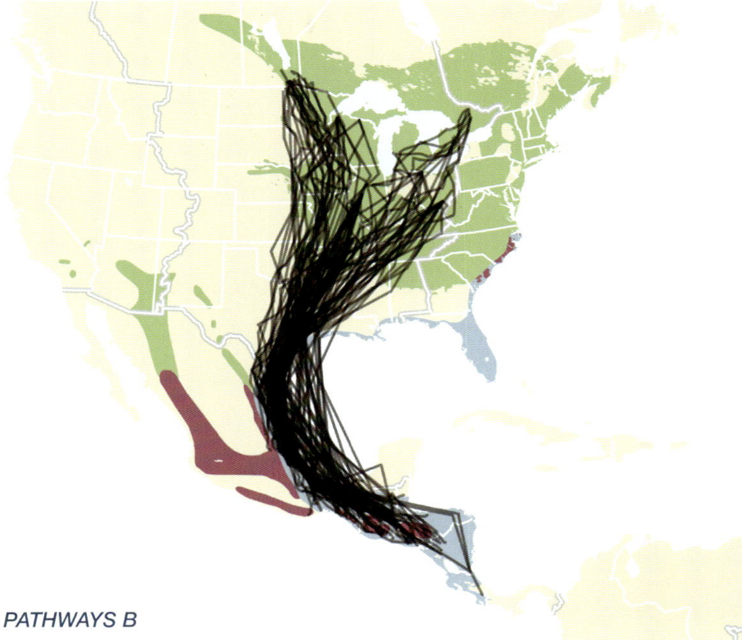

PATHWAYS B

DATA CONTRIBUTORS

M. Bakermans; M.D. Cadman; C. Davy; P. English; K.C. Fraser; D.J. Green; A.E. Heagy; A. Korpach; A.M. Mills; J.J. Nocera; G.J. Rand; A.A. Skinner; C.M. Tonra; A. Vitz.

MEXICAN WHIP-POOR-WILL *Antrostomus arizonae*

DISTRIBUTION AND ABUNDANCE

This species breeds in humid to semiarid forest from central Arizona and New Mexico south via the Mexican *cordilleras* (mountain ranges) to southwest Honduras. During the nonbreeding season, populations in the US and northern Mexico migrate to areas south of at least Durango, Mexico, where they share geographies with the presumably resident populations of southern Mexico and Central America (Fink et al. 2023).

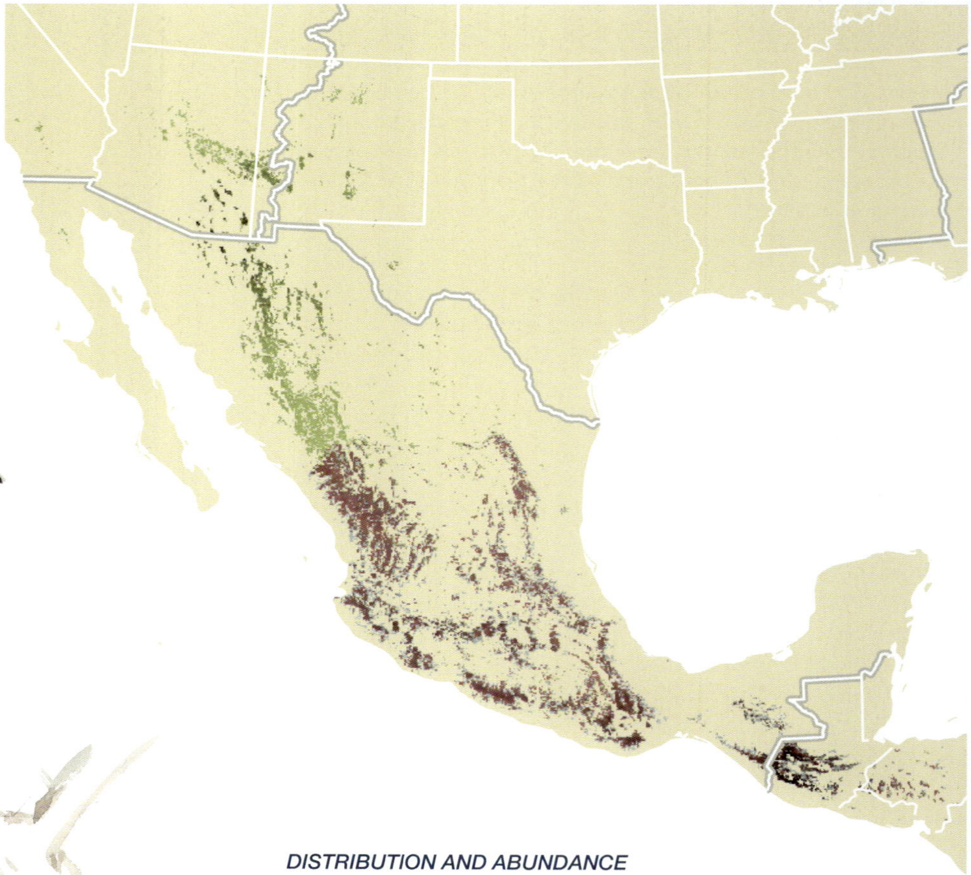

DISTRIBUTION AND ABUNDANCE

ORDER Caprimulgiformes
FAMILY Caprimulgidae
BREEDING HABITAT Forests, Mesoamerican Pine-Oak
NONBREEDING HABITAT Forests, Mesoamerican Pine-Oak
GLOBAL POPULATION SIZE 190,000
POPULATION CHANGE SINCE 1970 NA
IUCN RED LIST CATEGORY Least Concern
R2R TIPPING POINT SPECIES No
US LISTING STATUS No
CANADA SPECIES AT RISK No
PIF CCS 14/20

MIGRATORY CONNECTIVITY

OVERALL Migratory connectivity is unknown. Migratory connectivity is unknown. No information on migratory connectivity or migratory pathways were found for this species.

SWIFTS AND HUMMINGBIRDS (APODIFORMES)

WRITING LED BY **MELANIE A. SMITH, TOM WILL,** *AND* **CAMILLA GÓMEZ**

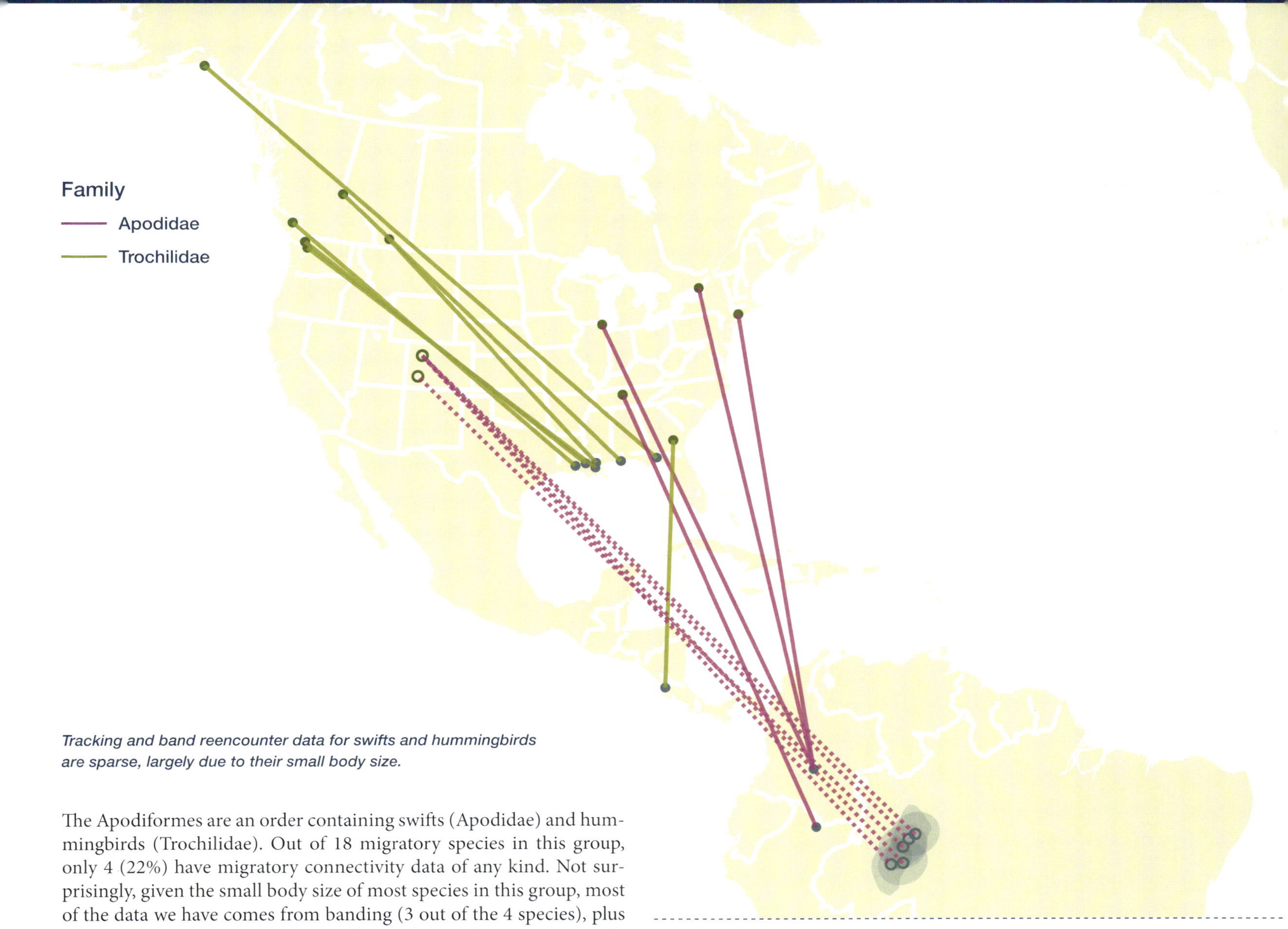

Tracking and band reencounter data for swifts and hummingbirds are sparse, largely due to their small body size.

The Apodiformes are an order containing swifts (Apodidae) and hummingbirds (Trochilidae). Out of 18 migratory species in this group, only 4 (22%) have migratory connectivity data of any kind. Not surprisingly, given the small body size of most species in this group, most of the data we have comes from banding (3 out of the 4 species), plus one species with Motus (radio tracking) data and one species with light-level geolocator data. Three species in this group, the Chimney Swift, Rufous Hummingbird, and Allen's Hummingbird, are considered Tipping Point species, and an additional one, the Black Swift, is "on alert" for its persistent population declines. Of these species of conservation concern, only the Chimney Swift and the Black Swift have any tracking data, and it is very limited. Chimney Swifts have mostly been tracked on their northeastern North American breeding grounds, and current knowledge about the migratory connectivity of Black Swifts relies on the geolocator tracks of only three individuals. It is not surprising that many species in this group remain to be tracked. Their unique ecology makes them either especially hard to capture and study (swifts) or too small to track for long distances with our current technology (hummingbirds).

Patterns of migratory connectivity in this poorly studied group suggest moderate migratory connectivity, with birds either converging in western Brazil (Black Swift), northern South America (Chimney Swift), or the southern US, Mexico, and Central America in the nonbreeding season (hummingbirds). Much more work is necessary for any conclusions to be made regarding the strength of migratory connectivity. As tracking devices continue to decrease in size and more researchers expand their geographic networks of collaborations, we expect more tracking and connectivity data for this group to emerge in the near future. Similarly, given the difficulty in capturing and the small size of many of these species, it would be beneficial to expand the use of intrinsic tracking markers, such as stable isotopes and genetic tools, to improve our knowledge about the connectivity of species in this group.

BLACK SWIFT *Cypseloides niger*

DISTRIBUTION AND ABUNDANCE

Black Swift breed on cliffs, waterfalls, and in caves primarily in northwestern North America, but also in several western states and the Caribbean. The nonbreeding range was relatively unknown until a recent tracking study identified areas in South America, but it still remains unmapped (Lowther et al. 2020c).

ORDER Apodiformes
FAMILY Apodidae
BREEDING HABITAT Forest Aerial, Temperate Western
NONBREEDING HABITAT Forest Aerial, Tropical Generalist
GLOBAL POPULATION SIZE 170,000
POPULATION CHANGE SINCE 1970 −89%
IUCN RED LIST CATEGORY Vulnerable
R2R TIPPING POINT SPECIES Orange-Alert
US LISTING STATUS No
CANADA SPECIES AT RISK Endangered
PIF CCS 15/20

CONNECTIONS

INTERESTING NOTE

Beason et al. (2012) estimated that the Black Swifts traveled on average 340 km/day during pre-breeding migration.

MIGRATORY CONNECTIVITY

OVERALL Migratory connectivity is poorly understood and limited to one geolocator study (Beason et al. 2012).

GEOLOCATOR TRACKING Three light-level geolocators were recovered (four were deployed) from Black Swifts breeding in Colorado. All three individuals migrated to nonbreeding areas in western Brazil (Beason et al. 2012), approximately 7,000 km away.

OTHER APPROACHES No satellite transmitter, banding, genetic, or stable isotope studies of migratory connectivity were found for this species.

MIGRATORY PATHWAYS Based on limited geolocator data, Black Swifts migrate north from the upper Amazon Basin to the western United States via Central America, and some individuals may fly more directly over the Pacific or Gulf of Mexico to shorten their route (Beason et al. 2012).

DATA CONTRIBUTORS

R. Sparks; Bird Conservancy of the Rockies.

Chaetura pelagica **CHIMNEY SWIFT**

DISTRIBUTION AND ABUNDANCE

The Chimney Swift is common in the eastern United States, especially from Illinois east to the Atlantic states from Massachusetts south to Florida. It is well known for using chimneys during migration and as artificial nesting cavities. This species spends the nonbreeding season in the Amazon Basin of South America in countries including Peru, Ecuador, Colombia, Venezuela, and Brazil (Steeves et al. 2020).

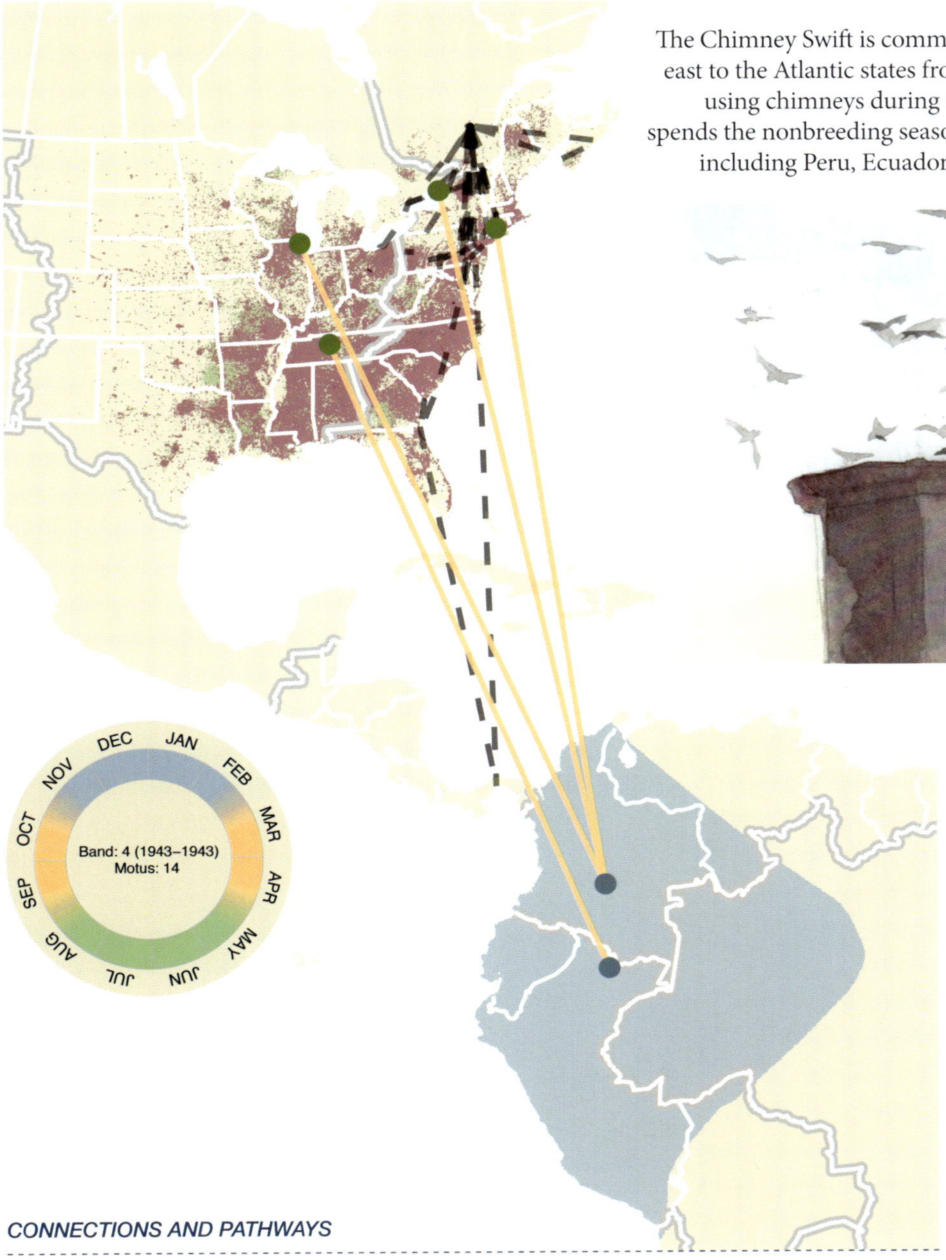

CONNECTIONS AND PATHWAYS

ORDER Apodiformes
FAMILY Apodidae
BREEDING HABITAT Open Country Aerial, Developed/Disturbed Areas
NONBREEDING HABITAT Forest Aerial, Tropical Lowland Evergreen
GLOBAL POPULATION SIZE 8,800,000
POPULATION CHANGE SINCE 1970 −71%
IUCN RED LIST CATEGORY Vulnerable
R2R TIPPING POINT SPECIES Orange-Alert
US LISTING STATUS No
CANADA SPECIES AT RISK Threatened
PIF CCS 13/20

MIGRATORY CONNECTIVITY

OVERALL There is no tracking data connecting breeding and nonbreeding locations. Limited band reencounters show migratory connectivity between eastern US breeding populations (Atlantic and Midcontinent flyways) and the nonbreeding range in northern South America and the Andean West, but there are too few to determine the strength of migratory connectivity among different populations.

BANDING Band reencounters connect breeding sites in the northeastern United States to nonbreeding sites in the upper Amazon Basin of South America.

RADIO TRACKING Motus detections are limited to the migratory period, showing migration between the northeastern United States and Panama, just short of reaching the nonbreeding grounds.

OTHER APPROACHES No satellite transmitter, genetic, or stable isotope studies of migratory connectivity were found for this species.

MIGRATORY PATHWAYS Chimney Swifts migrate south through the Atlantic and Midcontinent flyways during post-breeding, and the majority of individuals cross the Gulf of Mexico before reaching Central America and continuing into northern South America and beyond. Some individuals fly overland along the Gulf Coast through Mexico and Central America. During pre-breeding, this pattern is reversed with a greater number of individuals flying overland (Steeves et al. 2020; Fink et al. 2023).

DATA CONTRIBUTORS
V. Lamarre; J. Tixier; J. Tremblay.

VAUX'S SWIFT *Chaetura vauxi*

INTERESTING NOTE

This species uses chimneys and hollow trees to nest and roost, and spends most of its life on the wing (Schwitter et al. 2020).

DISTRIBUTION AND ABUNDANCE

Vaux's Swift breeds from British Columbia into the northwestern United States south to California, with some year-round populations in Mexico and Central America (Schwitters et al. 2020). Nonbreeding distribution extends from southern Mexico into Central America to northern Panama.

ORDER Apodiformes
FAMILY Apodidae
BREEDING HABITAT Forest Aerial, Temperate Western
NONBREEDING HABITAT Forest Aerial, Tropical Generalist
GLOBAL POPULATION SIZE 870,000
POPULATION CHANGE SINCE 1970 −40%
IUCN RED LIST CATEGORY Least Concern
R2R TIPPING POINT SPECIES No
US LISTING STATUS No
CANADA SPECIES AT RISK No
PIF CCS 14/20

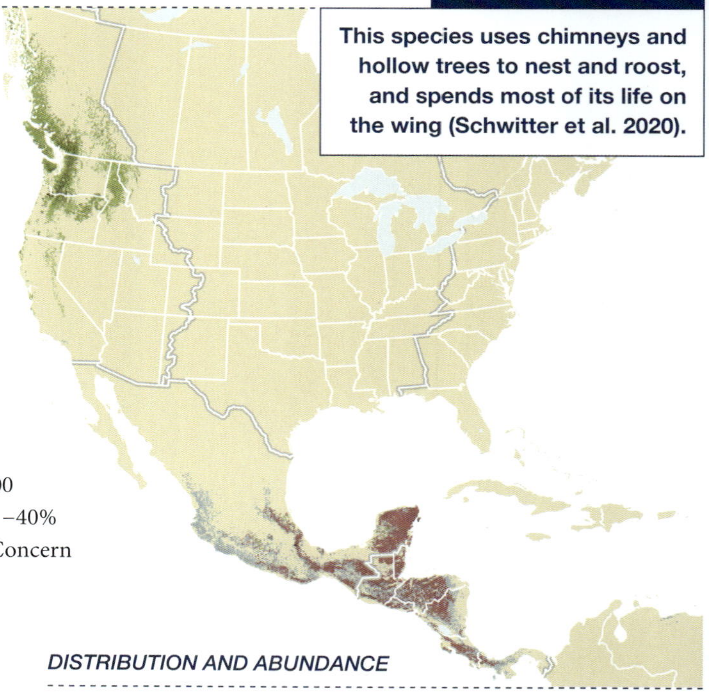

DISTRIBUTION AND ABUNDANCE

MIGRATORY CONNECTIVITY

OVERALL Migratory connectivity information is extremely limited. Stable isotopes suggest that migrants in British Columbia spend the nonbreeding season in two or three distinct habitats or locations (Reudink et al. 2015), though more studies are needed to validate and build upon these results.

STABLE ISOTOPES A mass mortality event at a pre-breeding migration roost in British Columbia allowed for the analysis of stable isotopes of 98 individuals to determine if these individuals had spent the nonbreeding season in similar habitats. Though it was not possible to determine exact nonbreeding locations, stable isotopes indicated that these individuals were in two or three distinct clusters on nonbreeding grounds (Reudink et al. 2015).

OTHER APPROACHES No banding, tracking, or genetic studies of migratory connectivity were found for this species.

MIGRATORY PATHWAYS Vaux's Swifts migrate through the Pacific Flyway during post-breeding overland to nonbreeding grounds in southern Mexico and Central America. Pre-breeding migration patterns appear to mirror post-breeding migration (Fink et al. 2023).

WHITE-THROATED SWIFT *Aeronautes saxatalis*

DISTRIBUTION AND ABUNDANCE

Breeds in mountainous regions of western North America (excluding the coastal ranges and Sierras north of San Francisco) from southern British Columbia to Guatemala, most densely in the southern Rockies and in the coastal ranges of Southern California, northern Baja California, and the Sierra Madre Occidental of Mexico. The species is resident in the latter three areas and is presumably augmented by the remainder of the western US population during the nonbreeding season (Fink et al. 2023).

ORDER Apodiformes
FAMILY Apodidae
BREEDING HABITAT Forest Aerial, Temperate Western
NONBREEDING HABITAT Forest Aerial, Tropical Generalist
GLOBAL POPULATION SIZE 870,000
POPULATION CHANGE SINCE 1970 −40%
IUCN RED LIST CATEGORY Least Concern
R2R TIPPING POINT SPECIES No
US LISTING STATUS No
CANADA SPECIES AT RISK No
PIF CCS 14/20

DISTRIBUTION AND ABUNDANCE

MIGRATORY CONNECTIVITY

OVERALL Migratory connectivity is unknown. No studies of migratory connectivity were found for this species.

MIGRATORY PATHWAYS Migratory northern populations undergo complete north-south migrations between breeding and nonbreeding areas, but the routes of these movements are unknown (Ryan and Collins 2020).

Eugenes fulgens RIVOLI'S HUMMINGBIRD

DISTRIBUTION AND ABUNDANCE

DISTRIBUTION AND ABUNDANCE

Breeds primarily in dry pine–oak forests from southeast Arizona (locally in southwest New Mexico) south through most of Mexico to north-central Nicaragua. The species is resident throughout most of its range, but birds in the US, northern Mexico, and parts of central Mexico apparently migrate to other parts of the range during the nonbreeding season (D.R. Powers et al. 2020; Fink et al. 2023).

ORDER Apodiformes
FAMILY Trochilidae
BREEDING HABITAT Forests, Mesoamerican Pine-Oak
NONBREEDING HABITAT Forests, Mesoamerican Pine-Oak
GLOBAL POPULATION SIZE 2,000,000
POPULATION CHANGE SINCE 1970 NA
IUCN RED LIST CATEGORY Least Concern
R2R TIPPING POINT SPECIES No
US LISTING STATUS No
CANADA SPECIES AT RISK No
PIF CCS 12/20

INTERESTING NOTE

Rivoli's is one of the largest hummingbirds found in the US, second only to Blue-throated Mountain-gem.

MIGRATORY CONNECTIVITY

OVERALL Migratory connectivity is unknown. No studies of migratory connectivity were found for this species.
MIGRATORY PATHWAYS Migratory individuals of this species remain close to their montane forest habitats, undertaking north-south movements along the Sierra Madre Occidental and Sierra Madre Oriental ranges. While many will move south during the nonbreeding period, some individuals may remain in the northern Sierra Madre Occidental as far as Arizona. Populations in the northern Sierra Madre Oriental range undergo a more complete departure during the nonbreeding season, retreating at least as far as central Tamaulipas, Mexico (Fink et al. 2023).

Lampornis clemenciae BLUE-THROATED MOUNTAIN-GEM

DISTRIBUTION AND ABUNDANCE

DISTRIBUTION AND ABUNDANCE

The Blue-throated Mountain-gem has an extremely limited range, primarily in the Sierra Madre mountains of Mexico. It is nonmigratory throughout most of its range. However, some individuals appear to be altitudinal migrants, moving into lower elevations during the breeding season. The species also extends northward, with small breeding populations in southeastern Arizona and southwestern Texas (Fink et al. 2023).

ORDER Apodiformes
FAMILY Trochilidae
BREEDING HABITAT Forests, Mesoamerican Pine-Oak
NONBREEDING HABITAT Forests, Mesoamerican Pine-Oak
GLOBAL POPULATION SIZE 2,000,000
POPULATION CHANGE SINCE 1970 NA
IUCN RED LIST CATEGORY Least Concern
R2R TIPPING POINT SPECIES No
US LISTING STATUS No
CANADA SPECIES AT RISK No
PIF CCS 12/20

INTERESTING NOTE

A single vagrant band reencounter connected Southern California to southeastern Arizona.

MIGRATORY CONNECTIVITY

OVERALL Migratory connectivity is unknown. No information on migratory connectivity or migratory pathways was found for this species.

LUCIFER HUMMINGBIRD *Calothorax lucifer*

DISTRIBUTION AND ABUNDANCE

Breeds from southeast Arizona and West Texas (Big Bend) south through the Central Mexican Plateau to the Trans-Mexican Volcanic Belt. During the nonbreeding season, the population concentrates in the eastern portion of the Trans-Mexican Volcanic Belt and disperses toward the Pacific coast in the west Mexican states from Nayarit to Oaxaca (Fink et al. 2023).

ORDER Apodiformes
FAMILY Trochilidae
BREEDING HABITAT
Arid Lands, Desert Scrub
NONBREEDING HABITAT
Forests, Tropical Dry
GLOBAL POPULATION SIZE 200,000
POPULATION CHANGE SINCE 1970 NA
IUCN RED LIST CATEGORY Least Concern
R2R TIPPING POINT SPECIES No
US LISTING STATUS No
CANADA SPECIES AT RISK No
PIF CCS 14/20

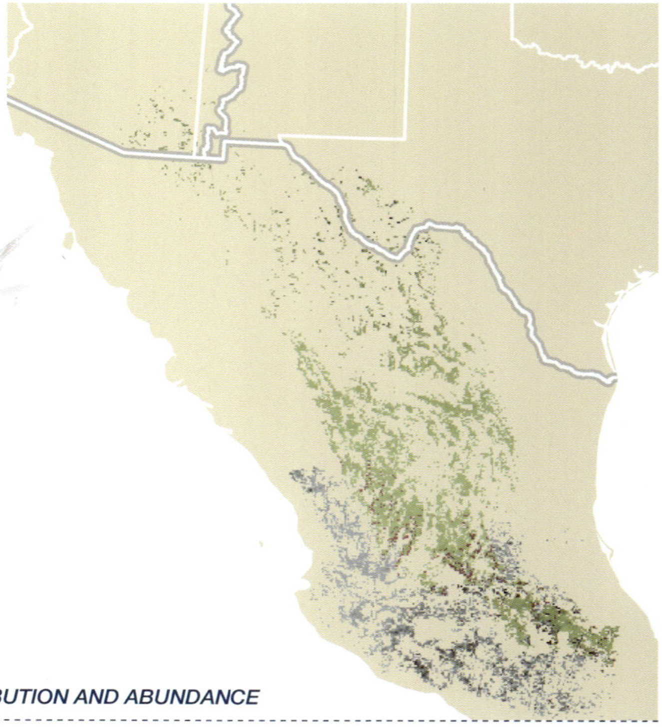

DISTRIBUTION AND ABUNDANCE

MIGRATORY CONNECTIVITY

OVERALL Migratory connectivity is unknown. No studies of migratory connectivity were found for this species.
MIGRATORY PATHWAYS Undergoes north-south migrations away from southwestern US and northern Mexico breeding areas. In central Mexico, there is evidence of altitudinal as well as latitudinal migration, with limited overlap of breeding and nonbreeding ranges (Fink et al. 2023).

RUBY-THROATED HUMMINGBIRD *Archilochus colubris*

DISTRIBUTION AND ABUNDANCE

The Ruby-throated Hummingbird is the only Trochilidae to breed in the eastern United States and Canada. It breeds widely from Florida north to southern Quebec, Canada, and west to the Great Plains from eastern Texas, to Alberta, Canada. Spends the nonbreeding season from central Mexico to northern Panama as well as a small portion of southern Florida.

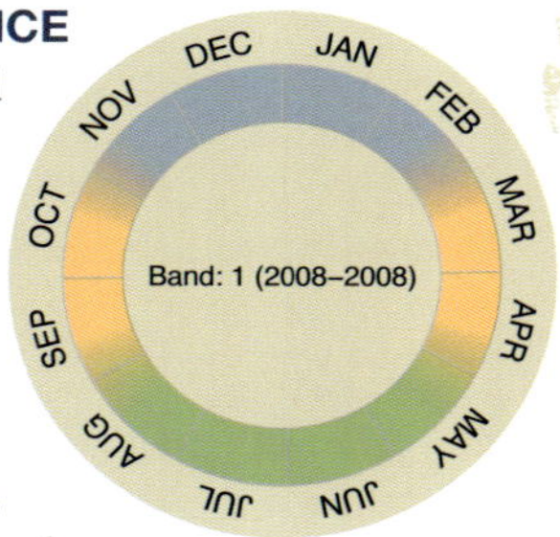

ORDER Apodiformes
FAMILY Trochilidae
BREEDING HABITAT Forests, Temperate Eastern
NONBREEDING HABITAT Forests, Tropical Dry
GLOBAL POPULATION SIZE 36,000,000
POPULATION CHANGE SINCE 1970 +91%
IUCN RED LIST CATEGORY Least Concern
R2R TIPPING POINT SPECIES No
US LISTING STATUS No
CANADA SPECIES AT RISK No
PIF CCS 8/20

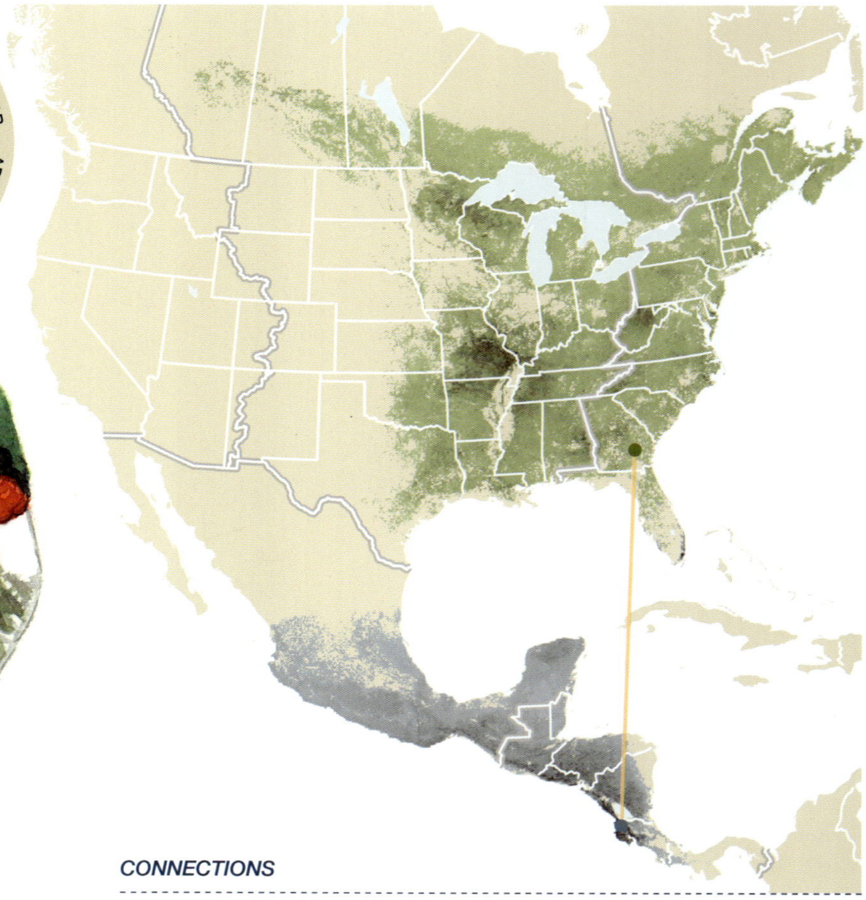

CONNECTIONS

MIGRATORY CONNECTIVITY

OVERALL Migratory connectivity of Ruby-Throated Hummingbirds is not well understood, as data are limited. Further research of individual movements should be a research priority.

BANDING A single female Ruby-throated Hummingbird banded on January 26, 2008, in Costa Rica was found dead on June 19, 2008, in southern Georgia.

STABLE ISOTOPES Stable hydrogen isotope values of feathers from hatch-year birds collected in El Salvador (n = 7) and Costa Rica (n = 2) suggest that the natal origins of these two nonbreeding populations differ significantly, with Costa Rican hummingbirds originating farther south. This pattern is consistent with chain migration and some level of migratory connectivity (Hutcheson et al. 2010). Stable hydrogen isotope ratios of individuals caught at a stopover site in Fort Morgan, Alabama, are consistent with a chain migration pattern (Zenzal et al. 2018).

OTHER No radio telemetry, geolocator, satellite telemetry, or genetic studies of migratory connectivity found for this species.

MIGRATORY PATHWAYS Birds migrate north during the prebreeding season via Mexico's Yucatán Peninsula and Caribbean coast before crossing the Gulf of Mexico and arriving first in the Midcontinent Flyway and soon after in the Atlantic Flyway. Post-breeding migration is primarily overland, with birds moving into the Midcontinent Flyway and south into eastern Mexico. Late arrivals to the Yucatán Peninsula may take a more direct route across the Gulf (Fink et al. 2023).

BLACK-CHINNED HUMMINGBIRD

Archilochus alexandri

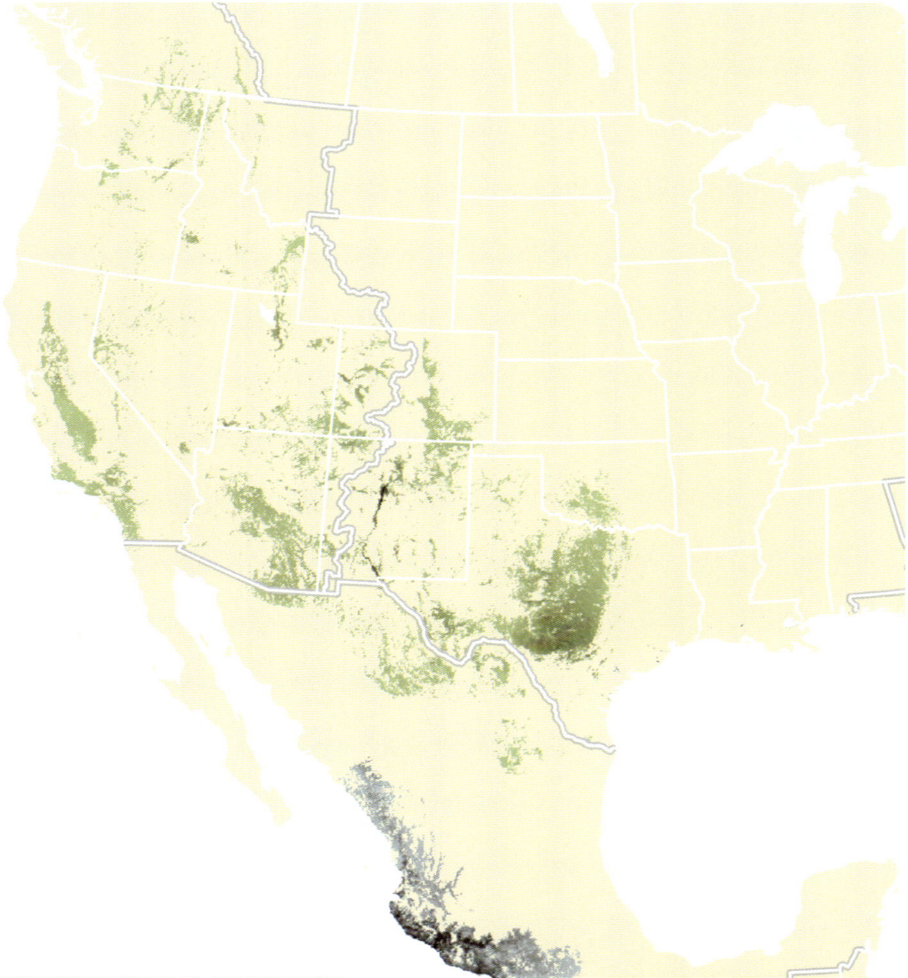

DISTRIBUTION AND ABUNDANCE

DISTRIBUTION AND ABUNDANCE

The Black-chinned Hummingbird breeds in the low country of the western United States from Washington southeast to Texas and west to California, as well as in a small region of southern British Columbia, Canada, and a wide band across northern Mexico. It spends the nonbreeding season throughout central Mexico with highest abundance near the Pacific coast, as well as a few small pockets across the southern United States.

ORDER Apodiformes
FAMILY Trochilidae
BREEDING HABITAT Forests, Temperate Western
NONBREEDING HABITAT Forests, Tropical Dry
GLOBAL POPULATION SIZE 8,800,000
POPULATION CHANGE SINCE 1970 +55%
IUCN RED LIST CATEGORY Least Concern
R2R TIPPING POINT SPECIES No
US LISTING STATUS No
CANADA SPECIES AT RISK No
PIF CCS 10/20

MIGRATORY CONNECTIVITY

OVERALL Migratory connectivity is unknown. No studies of migratory connectivity were found for this species.

MIGRATORY PATHWAYS Birds migrate north from Mexico via the Midcontinent Flyway and, to a lesser extent, the Pacific Flyway. During post-breeding, earlier migrants move through the more eastern Sierra Madre Oriental, and later migrants choose the more western Sierra Madre Occidental (Fink et al. 2023).

ANNA'S HUMMINGBIRD *Calypte anna*

DISTRIBUTION AND ABUNDANCE

DISTRIBUTION AND ABUNDANCE

Breeds on the Pacific coast from Vancouver Island south to northwest Baja California, locally in some interior valleys and foothills, and across Southern California and the southern third of Arizona. During the nonbreeding season, birds may remain on or migrate within the breeding range, with some individuals reaching areas in southern Texas and northern Mexico (Clark and Russell 2020; Fink et al. 2023).

ORDER Apodiformes
FAMILY Trochilidae
BREEDING HABITAT Arid Lands, Chaparral
NONBREEDING HABITAT Arid Lands, Chaparral
GLOBAL POPULATION SIZE 9,600,000
POPULATION CHANGE SINCE 1970 +192%
IUCN RED LIST CATEGORY Least Concern
R2R TIPPING POINT SPECIES No
US LISTING STATUS No
CANADA SPECIES AT RISK No
PIF CCS 8/20

MIGRATORY CONNECTIVITY

OVERALL Migratory connectivity is unknown. No studies of migratory connectivity were found for this species.

OTHER APPROACHES The Bird Genoscape Project has constructed a genoscape for the species, with a focus on understanding range expansion and population boundaries (https://www.birdgenoscape.org/annas-hummingbird/).

MIGRATORY PATHWAYS Movements of the species are complex and not well understood. While individuals may be present year-round throughout the breeding range, nonbreeding birds in any given locality may not be the same individuals which bred there, and local breeders may migrate to other localities within the known breeding range (Clark and Russell 2020).

COSTA'S HUMMINGBIRD *Calypte costae*

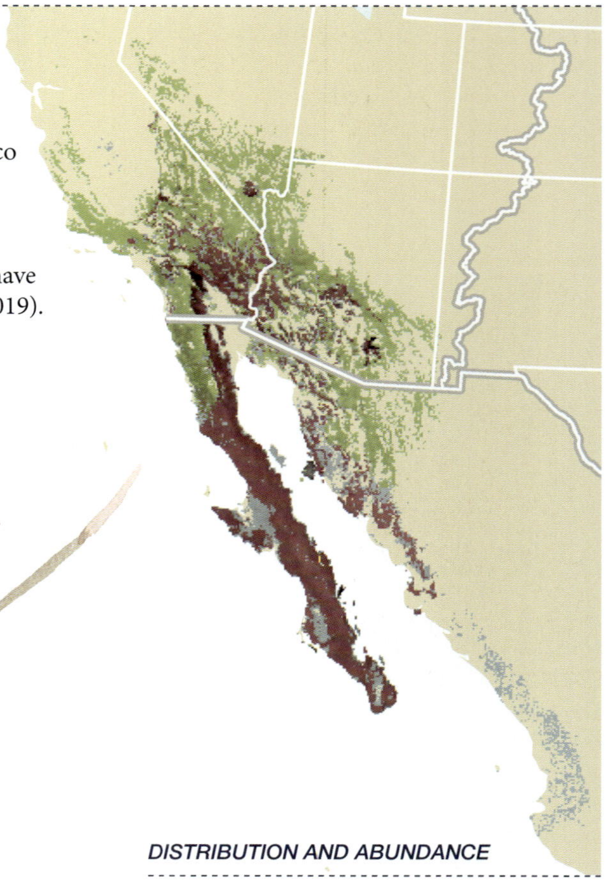

DISTRIBUTION AND ABUNDANCE

DISTRIBUTION AND ABUNDANCE

Breeds in arid desert and scrublands west of the Continental Divide and south of the Great Basin from Reno south through the Baja California peninsula and western Mexico to Sinaloa, with the greatest abundance in the Sonoran Desert of California, southwest Arizona, and Baja California. During the nonbreeding season, birds retreat to the Sonoran Desert regions or move farther south into Baja California Sur, Mexico, and along the Pacific coast of Mexico to Nayarit and Jalisco (Fink et al. 2023). Populations have declined predominantly in the northeastern portion of the species range (Sauer et al. 2019).

ORDER Apodiformes
FAMILY Trochilidae
BREEDING HABITAT Arid Lands, Desert Scrub
NONBREEDING HABITAT Arid Lands, Desert Scrub
GLOBAL POPULATION SIZE 3,400,000
POPULATION CHANGE SINCE 1970 −4%
IUCN RED LIST CATEGORY Least Concern
R2R TIPPING POINT SPECIES No
US LISTING STATUS No
CANADA SPECIES AT RISK No
PIF CCS 11/20

MIGRATORY CONNECTIVITY

OVERALL Migratory connectivity is unknown. No studies of migratory connectivity were found for this species.

MIGRATORY PATHWAYS Costa's Hummingbird migrates through Baja California, Mexico, in both pre- and post-breeding migration. It is commonly observed in cities in the southwestern US during its post-breeding migration (Fink et al. 2023).

Selasphorus calliope

CALLIOPE HUMMINGBIRD

DISTRIBUTION AND ABUNDANCE

DISTRIBUTION AND ABUNDANCE

Breeds in mountains from central British Columbia and southwest Alberta south to southwest Wyoming, and in the Sierras to Southern California. Migrates to mountains of western Mexico during the nonbreeding season, from Sinaloa to Guerrero. Its population has declined 16% since 1970 (Rosenberg et al. 2019), with the steepest declines in Southern California (Sauer et al. 2019).

ORDER Apodiformes
FAMILY Trochilidae
BREEDING HABITAT Forests, Temperate Western
NONBREEDING HABITAT Forests, Mesoamerican Pine-Oak
GLOBAL POPULATION SIZE 4,500,000
POPULATION CHANGE SINCE 1970 −16%
IUCN RED LIST CATEGORY Least Concern
R2R TIPPING POINT SPECIES No
US LISTING STATUS No
CANADA SPECIES AT RISK No
PIF CCS 14/20

INTERESTING NOTE

The Calliope Hummingbird is the smallest breeding bird in North America and the smallest long-distance avian migrant on the planet, potentially traveling 9,000 km annually (Calder and Calder 2020).

MIGRATORY CONNECTIVITY

OVERALL Limited available information suggests weak migratory connectivity between the nonbreeding range in southwestern Mexico and the breeding range throughout western North America.

STABLE ISOTOPES Feathers from adults from breeding sites in Washington, California, Alabama, and Alberta, British Columbia, did not have any differences in their deuterium values, suggesting weak migratory connectivity (Brown 2012).

OTHER APPROACHES No banding, radio telemetry, geolocator, satellite telemetry, or genetic studies of migratory connectivity were found for this species.

MIGRATORY PATHWAYS eBird animations (Fink et al. 2023) reveal pre-breeding migratory movements in the southwestern US and extensive post-breeding movement throughout the central and southern Rockies, north-central Mexico, and southern Mexico to western Veracruz and Oaxaca.

Selasphorus rufus

RUFOUS HUMMINGBIRD

DISTRIBUTION AND ABUNDANCE

The northernmost-breeding hummingbird, this species breeds from Alaska to the northwestern continental United States, with highest abundances in the Pacific Northwest. Rufous Hummingbirds spend the nonbreeding season primarily in central Mexico, with extremely small, scattered pockets of low-abundance in suitable habitats throughout the eastern United States (Fink et al. 2023).

ORDER Apodiformes
FAMILY Trochilidae
BREEDING HABITAT Forests, Temperate Western
NONBREEDING HABITAT Forests, Mesoamerican Pine-Oak
GLOBAL POPULATION SIZE 22,000,000
POPULATION CHANGE SINCE 1970 −62%
IUCN RED LIST CATEGORY Near Threatened
R2R TIPPING POINT SPECIES Orange-Alert
US LISTING STATUS No
CANADA SPECIES AT RISK No
PIF CCS 14/20

INTERESTING NOTE

A comparison of historical migration patterns and recent eBird data suggests that Rufous Hummingbirds are shifting their migration to arrive earlier on northern breeding grounds and avoiding coastal areas to take advantage of more consistent inland migratory routes (Couter 2017).

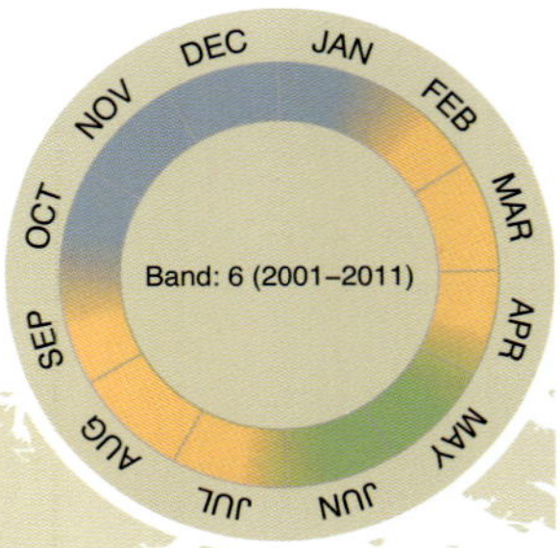

MIGRATORY CONNECTIVITY

OVERALL Migratory connectivity is relatively well-studied in Rufous Hummingbirds compared to other hummingbird species. A combination of stable isotope, banding, and genetic studies show weak structure among distinct breeding populations, suggesting some intermixing along migratory routes or between breeding sites. Stable isotope data shows birds from the Pacific Flyway spend the nonbreeding season within Mexico, while banding data show Pacific Flyway breeding birds spend the nonbreeding season in the Midcontinent and Atlantic flyways.

BANDING Several band recoveries connect the Pacific Northwest with the Gulf of Mexico coast in the southeastern US.

STABLE ISOTOPES Feathers collected from 229 adults at breeding sites in British Columbia, Alberta, and Alaska were used to determine sex and spatial differences in migratory connectivity. Males had higher deuterium (hydrogen isotopes) values than females at more western breeding sites, which indicates allopatric altitudinal separation on nonbreeding grounds, though this pattern was not present at all sampling sites. Females at the western breeding sites had deuterium values that correspond with nonbreeding sites at a higher altitude in Mexico compared to eastern females. Males had less variation among breeding sites, though differences suggested that western breeding males share nonbreeding sites and eastern males can be found at lower altitudes (Moran et al. 2013).

GENETICS Genomic analysis showed weak genetic structure; populations in British Columbia could be distinguished from those in Vancouver Island and Alberta (Bailey et al. 2013).

OTHER APPROACHES No radio telemetry, geolocator, or satellite telemetry studies of migratory connectivity were found for this species.

MIGRATORY PATHWAYS Migration is elliptical, flowing from the Mexican nonbreeding grounds northward via coastal lowlands in the Pacific Flyway during pre-breeding migration, and southward via highlands in the interior western US during post-breeding migration, tracking the availability of abundant flowers (Healy and Calder 2020).

CONNECTIONS

ALLEN'S HUMMINGBIRD *Selasphorus sasin*

DISTRIBUTION AND ABUNDANCE

There are two distinct subspecies. The migratory population (*S. s. sasin*) breeds in a narrow band along the Pacific coast from southwest Oregon to Southern California. It migrates to a limited wintering range in central Mexico, Morelos, and Puebla. The nonmigratory population (*S. s. sedentarius*) occurs in the Channel Islands and the greater Los Angeles area south to San Diego (Clark and Mitchell 2020). The two subspecies differ markedly in their population trajectories. Nonmigratory *S. s. sedentarius* has been steadily increasing and rapidly expanding its breeding range in urbanized Southern California (C.J. Clark 2017), with growth rates of up to 14.9% per year based on urban-focused Christmas Bird Count analyses (K. Rosenberg, personal communication). In contrast, the migratory *S. s. sasin* has been declining range-wide by approximately 3.12% per year, as measured by the nonurban-focused Breeding Bird Survey (Sauer et al. 2019).

ORDER Apodiformes
FAMILY Trochilidae
BREEDING HABITAT Arid Lands, Chaparral
NONBREEDING HABITAT Forests, Mesoamerican Pine-Oak
GLOBAL POPULATION SIZE 1,500,000
POPULATION CHANGE SINCE 1970 −78%
IUCN RED LIST CATEGORY Least Concern
R2R TIPPING POINT SPECIES Red-Alert
US LISTING STATUS No
CANADA SPECIES AT RISK No
PIF CCS 17/20

MIGRATORY CONNECTIVITY

OVERALL Migratory connectivity is unknown. No studies of migratory connectivity were found for this species.

MIGRATORY PATHWAYS An exceptionally early migrant, populations move northward toward breeding areas starting in January, confining their movements to the coast where blooming flowers are available. Post-breeding, they begin their southward movements as early as May, moving along a slightly broader front, which includes the coast and nearby mountains (Clark and Mitchell 2020; Fink et al. 2023).

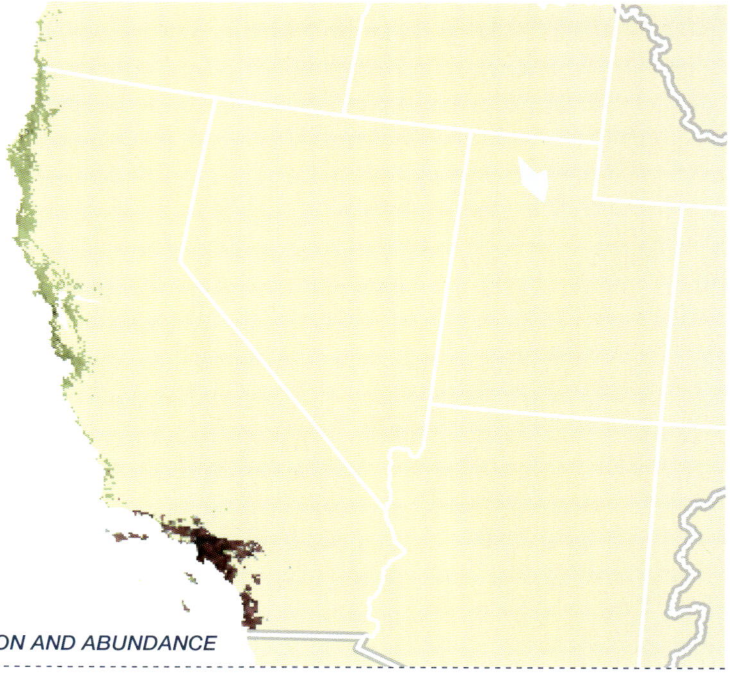

DISTRIBUTION AND ABUNDANCE

Selasphorus platycercus BROAD-TAILED HUMMINGBIRD

DISTRIBUTION AND ABUNDANCE

The Broad-tailed Hummingbird breeds in the Rocky Mountains of the United States, west to the southern Sierra Nevada, and south to northern Mexico. This species spends the nonbreeding season in central and southern Mexico and Guatemala, with an additional nonbreeding population in southeastern Arizona.

DISTRIBUTION AND ABUNDANCE

MIGRATORY CONNECTIVITY

OVERALL Migratory connectivity is unknown. No studies of migratory connectivity were found for this species.

MIGRATORY PATHWAYS Little is known about migration pathways. During pre-breeding migration, birds migrate north from Mexico via the Midcontinent Flyway and to a lesser extent the Pacific Flyway, and this pattern is reversed in southward post-breeding migration (Fink et al. 2023).

ORDER Apodiformes
FAMILY Trochilidae
BREEDING HABITAT Forests, Temperate Western
NONBREEDING HABITAT Forests, Mesoamerican Pine-Oak
GLOBAL POPULATION SIZE 9,800,000
POPULATION CHANGE SINCE 1970 −54%
IUCN RED LIST CATEGORY Least Concern
R2R TIPPING POINT SPECIES No
US LISTING STATUS No
CANADA SPECIES AT RISK No
PIF CCS 13/20

BROAD-BILLED HUMMINGBIRD *Lampornis clemenciae*

DISTRIBUTION AND ABUNDANCE

The Broad-billed Hummingbird occurs year-round throughout much of central and northern Mexico. It breeds in northern Mexico, southern Arizona, and southeastern Texas. It spends the nonbreeding season in eastern Mexico, in a portion of the Sierra Madre Occidental, and in limited pockets in the southeastern US (Fink et al. 2023).

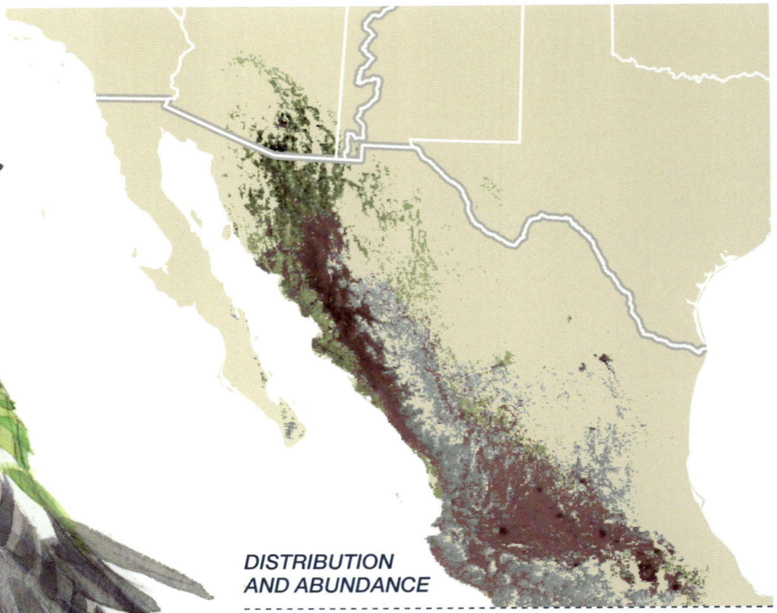

ORDER Apodiformes
FAMILY Trochilidae
BREEDING HABITAT Forests, Tropical Dry
NONBREEDING HABITAT Forests, Tropical Dry
GLOBAL POPULATION SIZE 2,200,000
POPULATION CHANGE SINCE 1970 NA
IUCN RED LIST CATEGORY Least Concern
R2R TIPPING POINT SPECIES No
US LISTING STATUS No
CANADA SPECIES AT RISK No
PIF CCS 10/20

MIGRATORY CONNECTIVITY

OVERALL Migratory connectivity is unknown. No studies of migratory connectivity were found for this species.
MIGRATORY PATHWAYS Migrates throughout the breeding range to nonbreeding areas. Northward migration is more concentrated in the eastern portion of the range (Fink et al. 2023).

INTERESTING NOTE

A single vagrant band reencounter connected Colorado in summer to Louisiana in winter.

VIOLET-CROWNED HUMMINGBIRD *Leucolia violiceps*

DISTRIBUTION AND ABUNDANCE

It breeds in arid to semiarid scrub forests and oak woodlands from southeastern Arizona and southwestern New Mexico south through Mexico and then along the Pacific Slope to Oaxaca, with greatest abundance in the Madrean pine–oak woodlands of Mexico (Wethington 2021; Fink et al. 2023). Presumed to be mostly resident throughout most of its range, the species withdraws from the US and northern Mexico during the nonbreeding season (Wethington 2021), but eBird animations (Fink et al. 2023) suggest substantial population concentrations from August to November in the Central Trans-Mexican Volcanic Belt (Jalisco to Hidalgo and Morelos to Puebla).

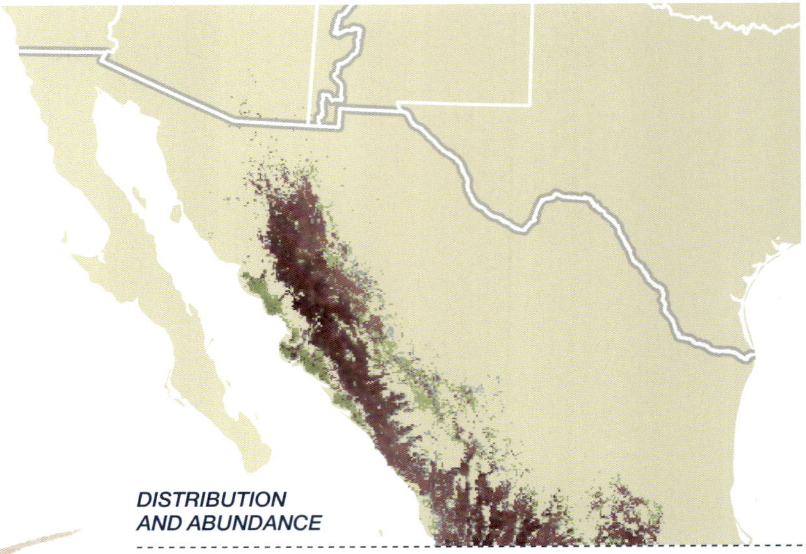

ORDER Apodiformes
FAMILY Trochilidae
BREEDING HABITAT Forests, Tropical Dry
NONBREEDING HABITAT Forests, Tropical Dry
GLOBAL POPULATION SIZE 2,000,000
POPULATION CHANGE SINCE 1970 NA
IUCN RED LIST CATEGORY Least Concern
R2R TIPPING POINT SPECIES No
US LISTING STATUS No
CANADA SPECIES AT RISK No
PIF CCS 10/20

MIGRATORY CONNECTIVITY

OVERALL Migratory connectivity is unknown. No information on migratory connectivity or migratory pathways were found for this species.

INTERESTING NOTE

In its restricted breeding range in the US, this species nests almost exclusively in Arizona Sycamore trees (*Platanus wrightii*) limited to the riparian zones of the arid Southwest (Wethington 2021).

weak strong

MC = 0.28 ± 0.1*

BUFF-BELLIED HUMMINGBIRD

Amazilia yucatanensis

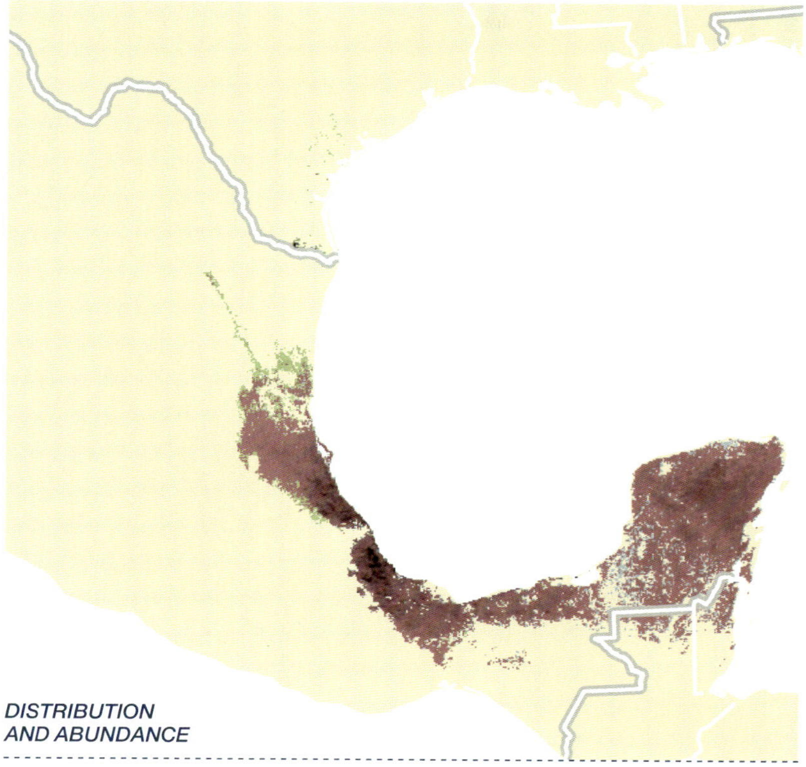

DISTRIBUTION AND ABUNDANCE

DISTRIBUTION AND ABUNDANCE

It breeds along the Gulf Coast from extreme south Texas to northern Guatemala and Belize, most abundantly along the southern Mexican Gulf Coast and in the northern Yucatán Peninsula (Fink et al. 2023). During the nonbreeding season, most individuals remain resident in the breeding range, but some migrate north and northeast to scattered localities in the Rio Grande Valley, in Florida, and along the US Gulf Coast (Chavez-Ramirez and Moreno-Valdez 2020).

ORDER Apodiformes
FAMILY Trochilidae
BREEDING HABITAT Forests, Tropical Dry
NONBREEDING HABITAT Forests, Tropical Dry
GLOBAL POPULATION SIZE 610,000
POPULATION CHANGE SINCE 1970 NA
IUCN RED LIST CATEGORY Least Concern
R2R TIPPING POINT SPECIES No
US LISTING STATUS No
CANADA SPECIES AT RISK No
PIF CCS 12/20

MIGRATORY CONNECTIVITY

OVERALL Migratory connectivity is unknown. No information on migratory connectivity or migratory pathways were found for this species.

RAILS, COOTS, AND CRANES (GRUIFORMES)

WRITING LED BY **M. CLAY GREEN, TIMOTHY GUIDA, ALLISON E. HUYSMAN,** *AND* **AMY L. SCARPIGNATO**

Breeding birds in family Gruidae tracked with GPS and satellite transmitters (n = 121; 90 between breeding and nonbreeding) show strong use of Pacific and Midcontinent flyways. Birds in the Great Lakes region move southward to peninsular Florida. Rallidae breeding birds, tracked with satellite and Motus transmitters (n = 56; 13 between breeding and nonbreeding), show strong use of the Midcontinent and Atlantic flyways.

The migratory connectivity and ecology of the order Gruiformes, which includes rails and cranes, has been moderately well studied, with over 70% of species tracked in some capacity. Within the order, both species of crane and 67% of rails have migratory connectivity information, although the connectivity information for many species of rails is primarily known through band reencounter data. Sandhill Cranes, across all six breeding populations, and the migratory population of Whooping Cranes have been extensively tracked and studied. However within the family Rallidae, little is known about the movement and migratory connectivity of Virginia Rail, Common Gallinule, Purple Gallinule, Yellow Rail, and Black Rail. Within the order Gruiformes, four species are Tipping Point species: Whooping Crane, King Rail, Yellow Rail, and Black Rail. Other species within the order also warrant conservation attention based on recent assessments of the continental populations, including Clapper Rail, Common Gallinule, and Purple Gallinule.

Across the order, Gruiformes exhibit overall weak to moderate migratory connectivity, although many species have not been tracked sufficiently to provide estimates of migratory connectivity. Within the family Gruidae, both species of crane exhibit strong north-south movements. Across Rallidae, most species generally exhibit north-south movements between breeding and nonbreeding regions. Rails within the Midcontinent Flyway move north-south to spend the nonbreeding season along the Gulf Coast, Mexico, and the Caribbean. Some rails within the Midcontinent and Atlantic flyways move between flyways during their annual movements. In some species like King Rail, there may be leapfrog migration as northern-breeding birds spend the nonbreeding season farther south in Mexico and the Caribbean.

Rallus elegans **KING RAIL**

DISTRIBUTION AND ABUNDANCE

King Rail breed in low densities across disjunct populations throughout the central and eastern United States. Northern populations are migratory, while southern coastal populations along the Atlantic coast through the Gulf of Mexico and the Caribbean are present year-round, most densely in Texas, Louisiana, and Florida (Pickens and Meanley 2020).

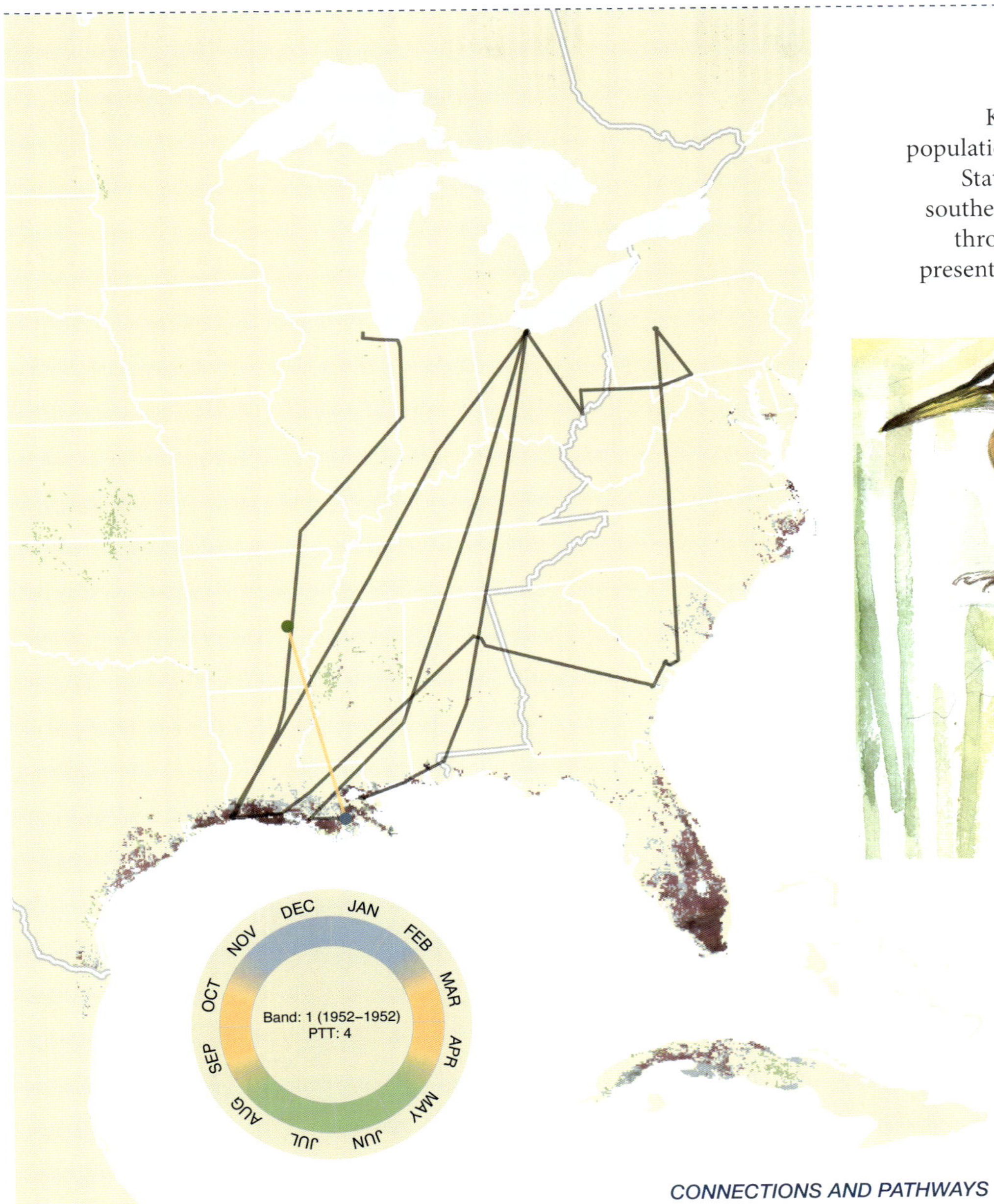

CONNECTIONS AND PATHWAYS

ORDER Gruiformes
FAMILY Rallidae
BREEDING HABITAT Wetlands, Freshwater Marsh
NONBREEDING HABITAT Wetlands, Freshwater Marsh
GLOBAL POPULATION SIZE 69,000
POPULATION CHANGE SINCE 1970 −86%
IUCN RED LIST CATEGORY Near Threatened
R2R TIPPING POINT SPECIES Orange-Alert
US LISTING STATUS No
CANADA SPECIES AT RISK Endangered
PIF CCS 15/20

MIGRATORY CONNECTIVITY

OVERALL As secretive marsh birds, King Rails are difficult to study, but several studies have used transmitters and stable isotopes to learn about the species' migratory connectivity. Data suggest that populations tend to be a mixture of residents and migratory individuals, with northern and interior populations showing evidence of migration. Midcontinent Flyway breeding birds were tracked to nonbreeding areas within the Midcontinent and Atlantic flyways. MC is not calculated due to limited sample size.

BANDING Band reencounters are limited, but one shows connection between an inland breeding location and a nonbreeding location on the Gulf Coast.

RADIO TRACKING A portion of individuals breeding in North Carolina along the mid-Atlantic coast were resident, while others appeared to be migratory (Kolts and McRae 2017). A few tags were deployed through the Motus Wildlife Tracking System, but there are no breeding to nonbreeding detections.

SATELLITE TRACKING Nine King Rails breeding in Ohio were tracked to the Gulf Coast and the southern Atlantic coast. For post-breeding migration, birds (n = 4) tracked completed migration within 5 days; whereas for pre-breeding migration, birds (n = 2) took longer and had more stopovers (Kane et al. 2019).

STABLE ISOTOPES Feathers collected on the coast of Louisiana and Texas suggested that the sample of nonbreeding King Rails were residents to that area (Perkins 2007).

OTHER APPROACHES No genetic studies of migratory connectivity were found for this species.

MIGRATORY PATHWAYS Midcontinent individuals migrated mostly along north-south directions including the Mississippi River Valley from the midwestern US to the Gulf Coast, primarily Louisiana. Some individuals moved more longitudinally between Atlantic and Gulf coasts. Limited evidence suggests possible leapfrog migration of more northern birds spending the nonbreeding season south of the US in Mexico and the Caribbean (Pickens and Meanley 2020).

DATA CONTRIBUTORS
B. Shirkey.

CLAPPER RAIL *Rallus crepitans*

DISTRIBUTION AND ABUNDANCE

Clapper Rail is distributed on the Atlantic and Gulf coasts of the United States and Mexico, as well as the Caribbean. It is resident throughout much of its range, though individuals in the northernmost part of the range are migratory (Rush et al. 2020).

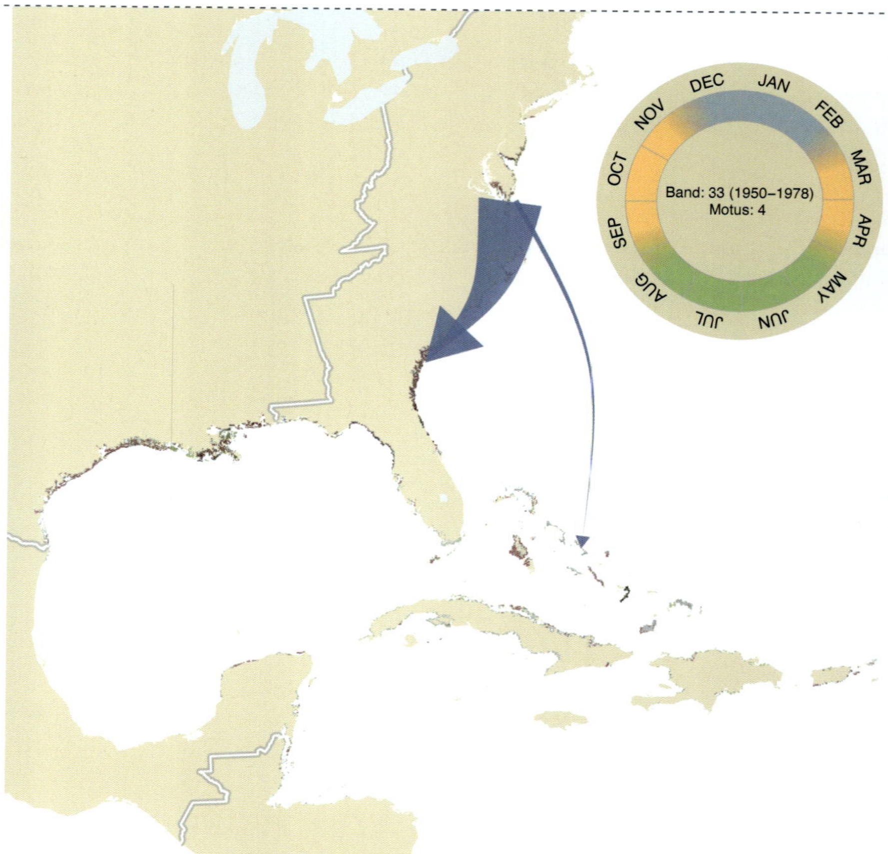

GENERALIZED CONNECTIVITY

ORDER Gruiformes
FAMILY Rallidae
BREEDING HABITAT Coasts, Salt Marsh
NONBREEDING HABITAT Coasts, Salt Marsh
GLOBAL POPULATION SIZE 210,000
POPULATION CHANGE SINCE 1970 −32%

IUCN RED LIST CATEGORY Least Concern
R2R TIPPING POINT SPECIES No
US LISTING STATUS No
CANADA SPECIES AT RISK No
PIF CCS 13/20

	Atlantic	Caribbean
Atlantic (n = 34)	97%	3%

CONNECTIONS

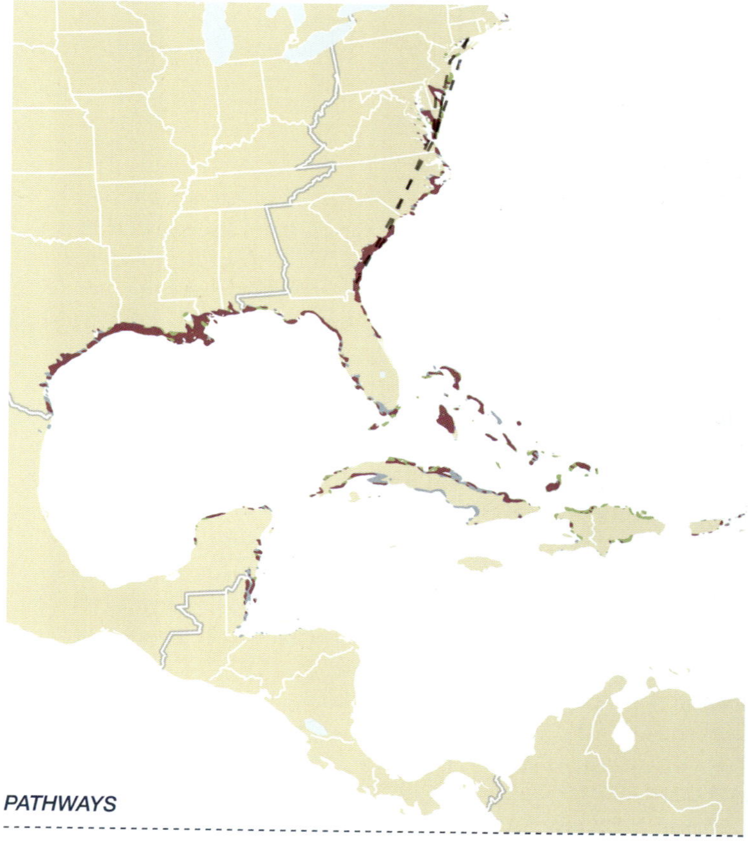

PATHWAYS

MIGRATORY CONNECTIVITY

OVERALL Migratory connectivity data on Clapper Rails are limited to banding recoveries within the Atlantic Flyway; therefore MC is not calculated. Rails breeding within the Atlantic Flyway migrate to nonbreeding areas within the Atlantic Flyway (97%) and the Caribbean (3%).

BANDING Band reencounter data show connections between breeding sites in the northeastern coast of the United States and nonbreeding locations along the southeastern coast. Individuals banded in eastern Virginia have been recovered in North Carolina, South Carolina, Georgia, and Florida (Stewart 1954). Some individuals breeding in the southeastern US may be short-distant migrants (Stewart 1954).

RADIO TRACKING Tags deployed through the Motus Wildlife Tracking System include both individual breeding to nonbreeding detections (M. Huang, unpublished data) and detections during migration in the eastern portion of the range (M. Huang, unpublished data; E. Elizondo, unpublished data).

OTHER APPROACHES No geolocator, satellite telemetry, genetic, or stable isotope studies of migratory connectivity were found for this species.

MIGRATORY PATHWAYS While migration routes are unknown for Clapper Rails, given species distribution and connecting with coastal areas, routes for migratory populations within the Atlantic region are likely north-south movements following the coastline, although some individuals have been documented (< 330 km) inland during post-breeding migration.

DATA CONTRIBUTORS
E. Elizondo; M. Huang.

VIRGINIA RAIL

Rallus limicola

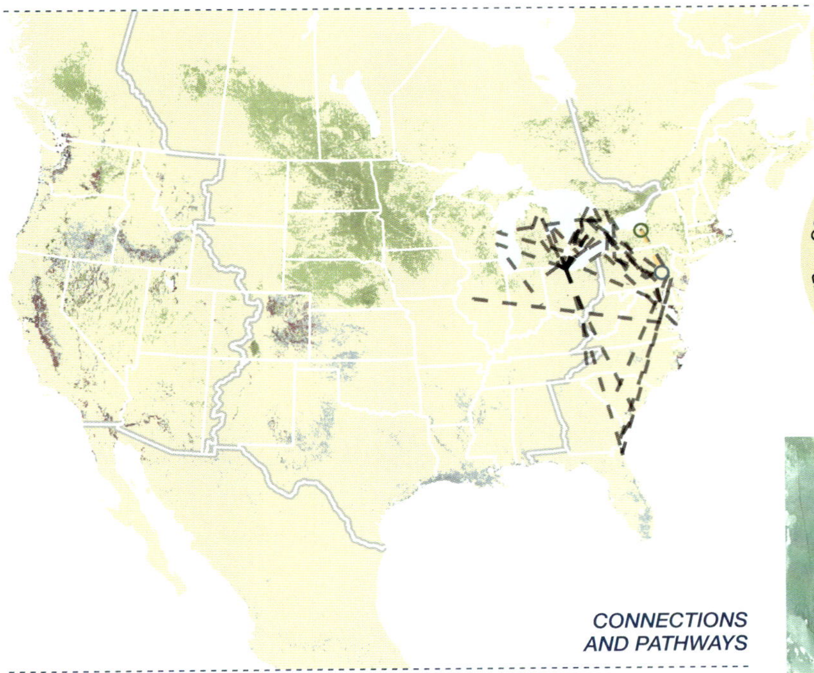

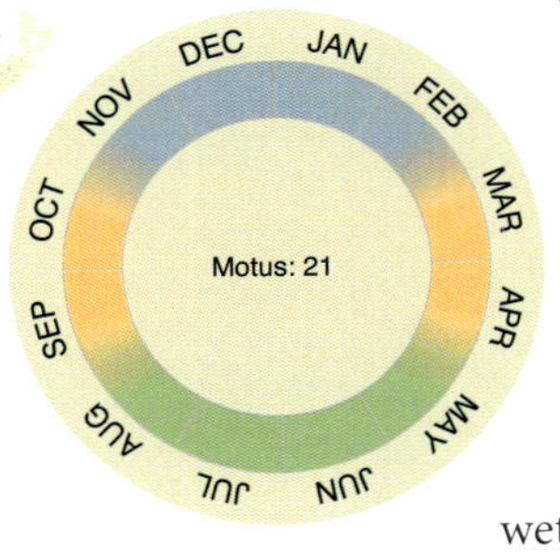

DISTRIBUTION AND ABUNDANCE

The Virginia Rail breeds densely in the Prairie Pothole Region and spends the nonbreeding season along the Pacific, Atlantic, and Gulf coasts of the US and Mexico (Conway 2020), most densely along the coasts of Louisiana and Texas (Fink et al. 2023). There are some resident populations on Atlantic and Pacific coasts and interior wetlands in southwestern US and Mexico.

ORDER Gruiformes
FAMILY Rallidae
BREEDING HABITAT Wetlands, Freshwater Marsh
NONBREEDING HABITAT Wetlands, Freshwater Marsh
GLOBAL POPULATION SIZE 240,000
POPULATION CHANGE SINCE 1970 +98%
IUCN RED LIST CATEGORY Least Concern
R2R TIPPING POINT SPECIES No
US LISTING STATUS No
CANADA SPECIES AT RISK No
PIF CCS 10/20

MIGRATORY CONNECTIVITY

OVERALL As a secretive marsh bird, migratory connectivity information on Virginia Rails is limited. Stable isotope estimates have been refined by combining them with citizen science observations.

STABLE ISOTOPES Stable isotope data collected throughout the species' range, combined with citizen scientist observations, connected Virginia Rails that spend the nonbreeding season on the Gulf Coast to breeding locations in the Great Lakes, Chesapeake Bay, and the Prairie Pothole Region of North Dakota. Migrating individuals from the Great Lakes were assigned to nonbreeding locations on the Gulf Coast of Texas and Louisiana (Fournier et al. 2017a), while migrating individuals in Missouri, Minnesota, Michigan, South Carolina, Ohio, and Arkansas and nonbreeding individuals in Louisiana and Mississippi were connected to breeding locations in Colorado, New Mexico, California, Saskatchewan, and Manitoba (Fournier et al. 2017b.).

OTHER APPROACHES No banding, tracking, or genetic studies of migratory connectivity were found for this species.

MIGRATORY PATHWAYS Limited Motus data shows breeding birds in Ohio moved mostly in north-south directions during migration along the Atlantic coast and Florida. Little else is known about migration routes and stopover sites.

DATA CONTRIBUTORS
D. Brinker; A. Fournier; G. Kearns; A. Patterson; C. Tonra.

SORA *Porzana carolina*

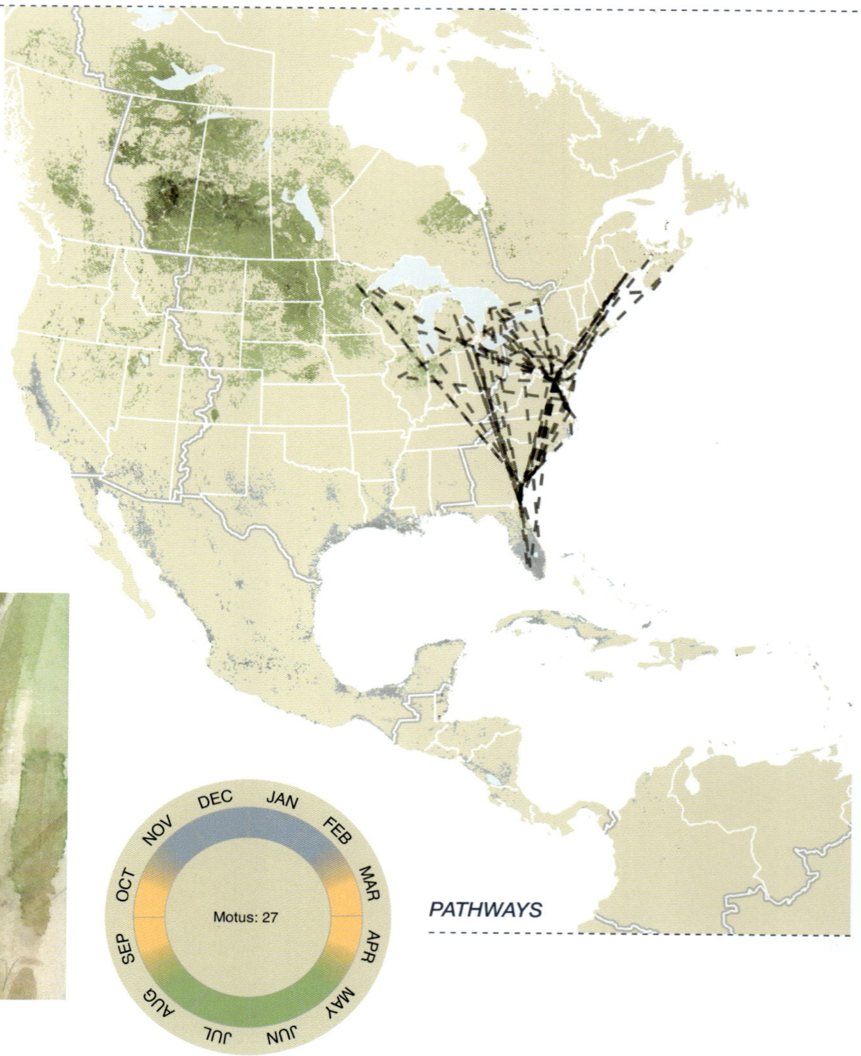

DISTRIBUTION AND ABUNDANCE

The Sora has a wide distribution throughout its annual cycle, breeding in much of Canada and the northern United States, and migrating to the southern US, Mexico, the Caribbean, Central America, and northern South America during the nonbreeding period. This species' highest breeding abundance by far is within prairie wetlands extending from South Dakota to the Northwest Territories, with smaller breeding areas spread from Nova Scotia to Illinois in the east, and from the Yukon to Arizona in the west. During the nonbreeding period, Sora are found in wetlands from Oregon to Ecuador but are particularly abundant in southern Florida, the central Gulf Coast, California's Central Valley, the Salton Sea, the lower Colorado River, and parts of central Mexico and the Yucatán (Fink et al. 2023).

ORDER Gruiformes
FAMILY Rallidae
BREEDING HABITAT Wetlands, Freshwater Marsh
NONBREEDING HABITAT Wetlands, Freshwater Marsh
GLOBAL POPULATION SIZE 4,400,000
POPULATION CHANGE SINCE 1970 +27%
IUCN RED LIST CATEGORY Least Concern
R2R TIPPING POINT SPECIES No
US LISTING STATUS No
CANADA SPECIES AT RISK No
PIF CCS 9/20

INTERESTING NOTE

Few Soras marked on the Patuxent River in Maryland have been recovered, suggesting that the species is no longer popular among hunters and more effort may be needed to recover or resight banded Soras along the Atlantic Flyway (Haramis and Kearns 2007). Vagrancy occurrences of Soras during the breeding season (June–July) include regions (e.g., Norway, Sweden) outside the species distribution, as well as regions (e.g., Caribbean, South America) typically encompassing the nonbreeding range (Fournier 2021).

MIGRATORY CONNECTIVITY

OVERALL No data are available on the migratory connectivity of Soras.
RADIO TRACKING Soras marked with radio during migration on the Patuxent River in Maryland were tracked to North Carolina, South Carolina, Georgia, Florida, and one to the Bahamas (G. Kearns, unpublished data). After departure, birds were simultaneously tracked by ground teams and by airplane. The birds were detected on a southeast trajectory toward the Outer Banks, North Carolina, and were eventually pushed out over open ocean (G. Kearns, personal communication). Motus data across many individuals show connections between breeding sites on the eastern Great Lakes and along the Atlantic coast of the United States and nonbreeding locations along the southeastern coast and Florida.
STABLE ISOTOPES Feathers collected from Soras during post-breeding migration in Missouri, Minnesota, Michigan, South Carolina, Ohio, and Arkansas and during the nonbreeding season in Louisiana and Mississippi were assigned to breeding areas in much of the western United States and Canada (Fournier et al. 2017a, 2017b).
OTHER APPROACHES No satellite telemetry, banding, or genetic studies of migratory connectivity were found for this species.
MIGRATORY PATHWAYS Soras moved through Maryland during post-breeding migration along a north-south direction, including flights over open ocean, to nonbreeding sites along the southeastern US coast and the Caribbean (Haramis and Kearns 2007). Interior populations likely move in a north-south direction along the Midcontinent Flyway between breeding regions in Canada and the northern US and nonbreeding sites along the Gulf Coast and in Mexico and the Caribbean.

DATA CONTRIBUTORS
D. Brinker; A. Fournier; G. Kearns; M.A. Lewis; A.G. Pellegrini.

COMMON GALLINULE

Gallinula galeata

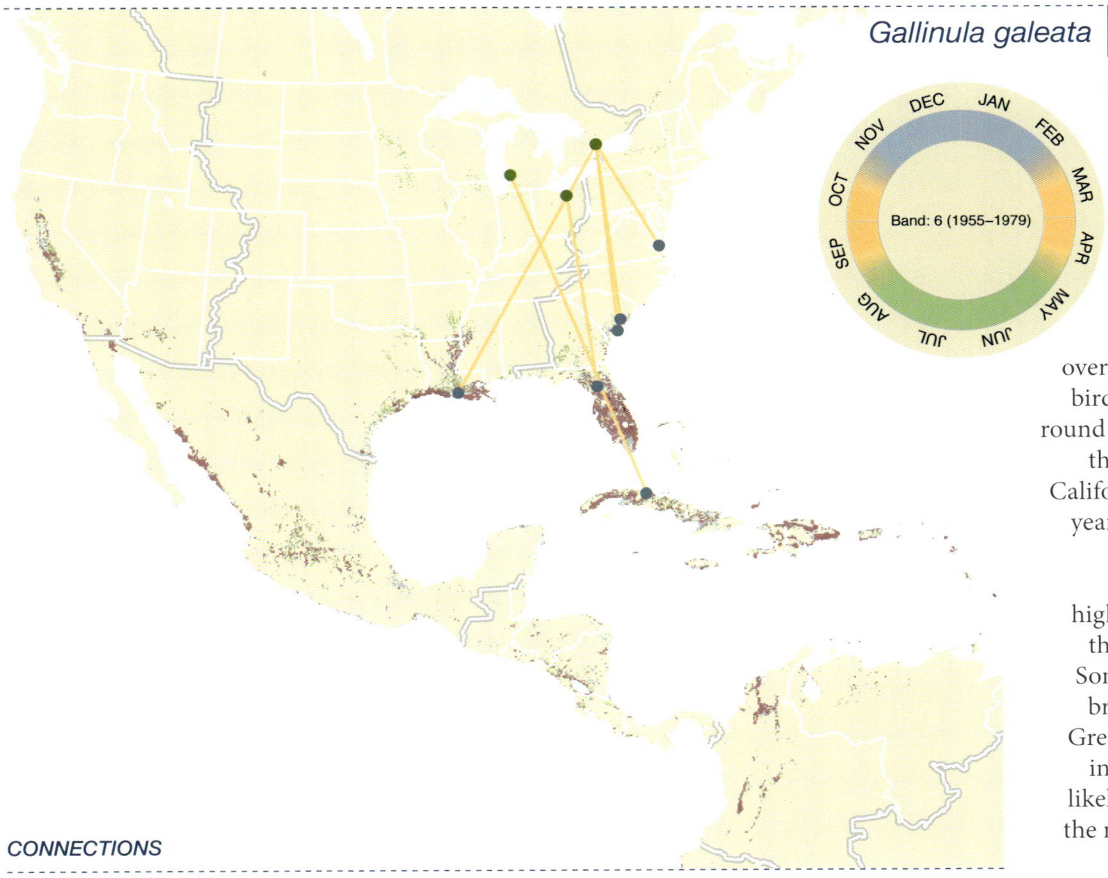

CONNECTIONS

DISTRIBUTION AND ABUNDANCE

Common Gallinules are widespread in warm wetlands of the Western Hemisphere, occurring from southeastern Canada to central Argentina. This species is a partial short- to medium-distant migrant. Understanding its migration and movement is complicated by the overlap of nonbreeding birds and resident birds. In the US, this species occurs year-round in the deep southeastern states, along the Gulf Coast, in central and Southern California, and in southern Arizona. Other year-round populations occur in Mexico, the Caribbean, and Central America. Year-round populations reach their highest abundance in peninsular Florida, the Louisiana coast, and the Caribbean. Some individuals move north during the breeding season, occupying areas of the Great Lakes region and southeast Canada in low numbers. Migratory populations likely join year-round populations during the nonbreeding period (Fink et al. 2023).

ORDER Gruiformes
FAMILY Rallidae
BREEDING HABITAT Wetlands, Freshwater Marsh
NONBREEDING HABITAT Wetlands, Freshwater Marsh
GLOBAL POPULATION SIZE 5,100,000
POPULATION CHANGE SINCE 1970 −40%
IUCN RED LIST CATEGORY Least Concern
R2R TIPPING POINT SPECIES No
US LISTING STATUS No
CANADA SPECIES AT RISK No
PIF CCS 8/20

MIGRATORY CONNECTIVITY

OVERALL No published studies on migratory connectivity were found. The Atlantic migratory populations likely concentrate during the nonbreeding season in Florida, and the midcontinent populations occur along the Gulf Coast with concentrations in Louisiana (Bannor and Kiviat 2020). Movement between Atlantic and Midcontinent flyway populations are known based on limited banding data, but the extent of movement is unknown.

BANDING Banding reencounter data are available from eastern North America, where breeding locations in the northeastern United States are connected to nonbreeding locations along the southeastern Atlantic and Gulf coasts and the Caribbean.

RADIO TRACKING A few tags were deployed through the Motus Wildlife Tracking System; there are no breeding to nonbreeding detections.

MIGRATORY PATHWAYS Little is known about migration routes. Transoceanic migration occurs as Common Gallinules have been documented during post-breeding migration in Bermuda, and individuals have been documented in trans-Gulf flights between the Yucatán Peninsula and Louisiana (Bannor and Kiviat 2020).

AMERICAN COOT *Fulica americana*

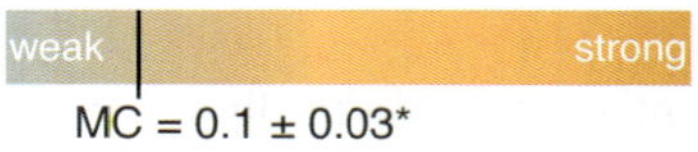

DISTRIBUTION AND ABUNDANCE

The American Coot is distributed widely throughout North America, breeding in interior North America, with migratory populations occurring most densely in the Prairie Pothole Region of the US and throughout western and parts of eastern Canada. There are some year-round populations in the western and southern portions of its range, including California, Nevada, Louisiana, Texas, and Mexico (Brisbin and Mowbray 2020; Fink et al. 2023).

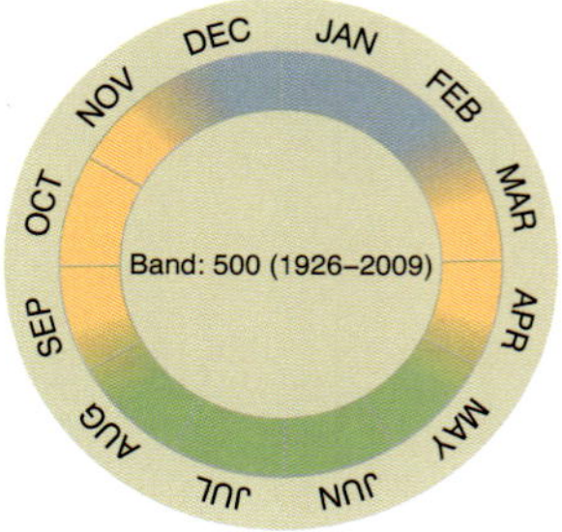

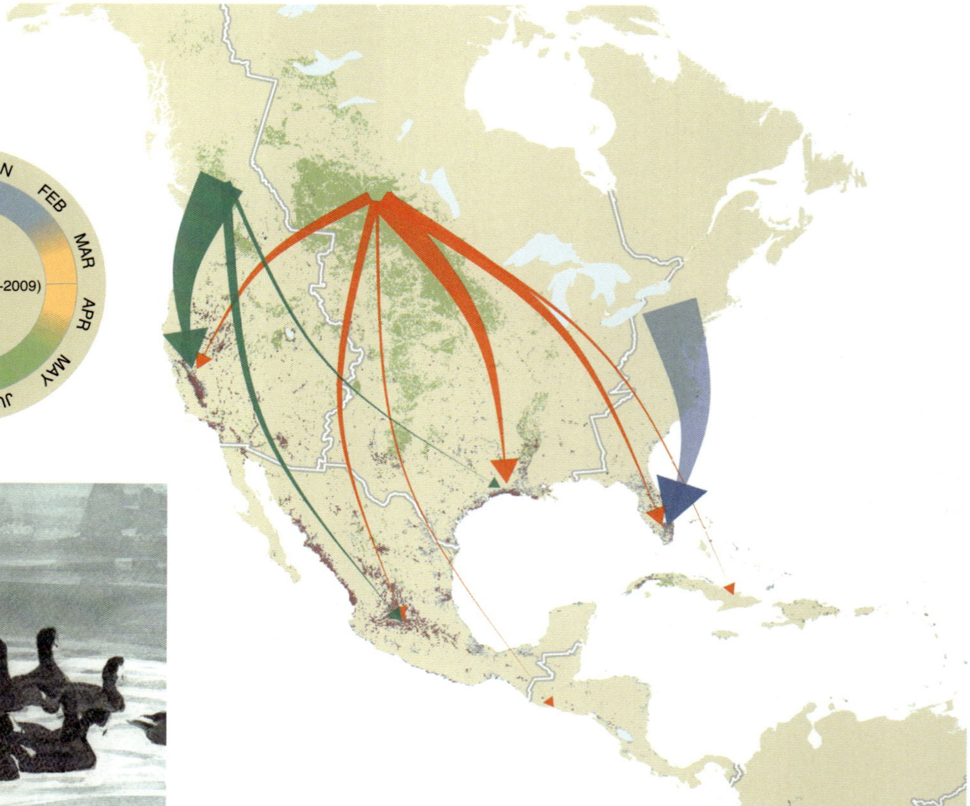

GENERALIZED CONNECTIVITY

ORDER Gruiformes
FAMILY Rallidae
BREEDING HABITAT Wetlands, Freshwater Marsh
NONBREEDING HABITAT Wetlands, Lakes and Rivers
GLOBAL POPULATION SIZE 7,100,000
POPULATION CHANGE SINCE 1970 +8%
IUCN RED LIST CATEGORY Least Concern
R2R TIPPING POINT SPECIES No
US LISTING STATUS No
CANADA SPECIES AT RISK No
PIF CCS 10/20

CONNECTIONS (A–E)

	Atlantic	Caribbean	Central America	Mexico	Midcontinent	Pacific
Atlantic (n = 1)	100%	–	–	–	–	–
Midcontinent (n = 392)	24%	4%	1%	12%	42%	18%
Pacific (n = 91)	–	–	–	14%	3%	82%

MIGRATORY CONNECTIVITY

OVERALL No published studies on migratory connectivity were found, but band reencounters are numerous from across the central and western range of the species. Data from the eastern portions of the range is lacking. Birds breeding in the Midcontinent Flyway vary greatly in nonbreeding locations, with some remaining in the flyway (42%) and the remainder distributed across the Atlantic Flyway (24%), Pacific Flyway (18%), Mexico (12%), Caribbean (4%), and

Central America (1%). Within the Pacific Flyway, breeding birds migrate within the flyway (82%), with the remainder migrating to Mexico (14%) and the Midcontinent Flyway (3%). Movement of the species is complex, with overlap of breeding populations across many nonbreeding regions, resulting in weak migratory connectivity (MC = 0.1 ± 0.03).

BANDING Band reencounters show general north-south movement during migration as well as large movements across flyways that can span most of North America.

OTHER APPROACHES No tracking, genetic, or stable isotope studies of migratory connectivity were found for this species.

MIGRATORY PATHWAYS Birds use all major flyways in their migratory movements, with movement across flyways evident from band reencounters. Based on eBird data, breeding populations in interior North America likely follow the Prairie Pothole Region southward down the Mississippi River Valley to nonbreeding sites. The Midcontinent Flyway has seen an increase in use by American Coots, with subsequent decline in use of the Pacific Flyway (Brisbin and Mowbray 2020).

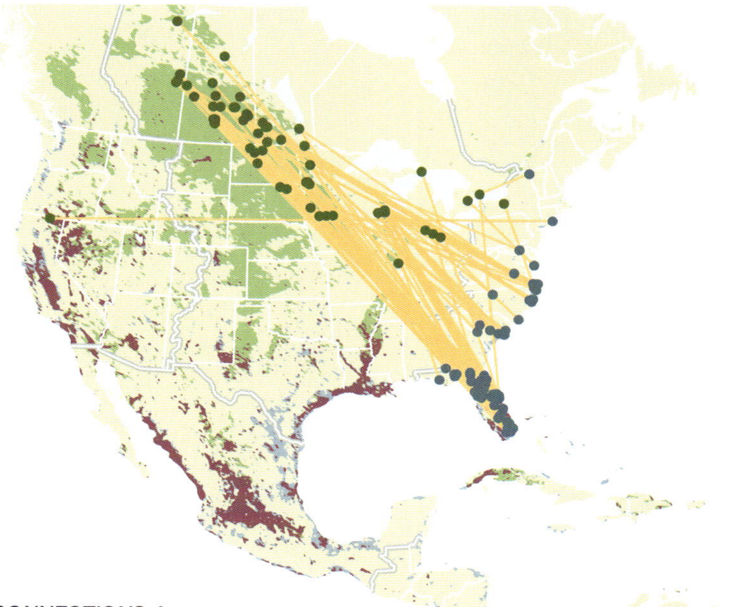

CONNECTIONS A

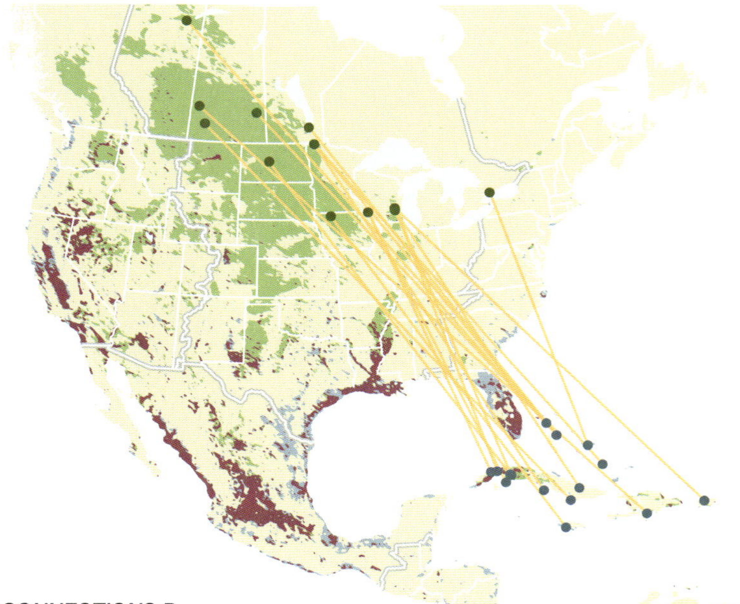

CONNECTIONS B

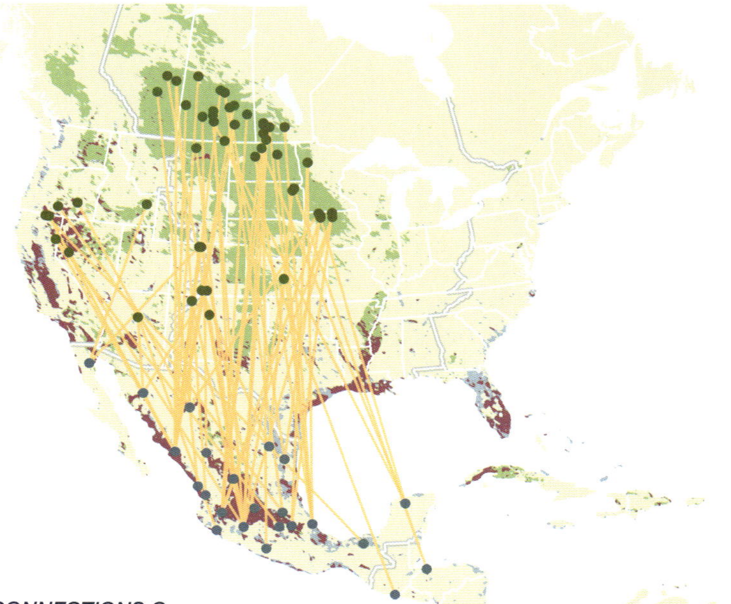

CONNECTIONS C

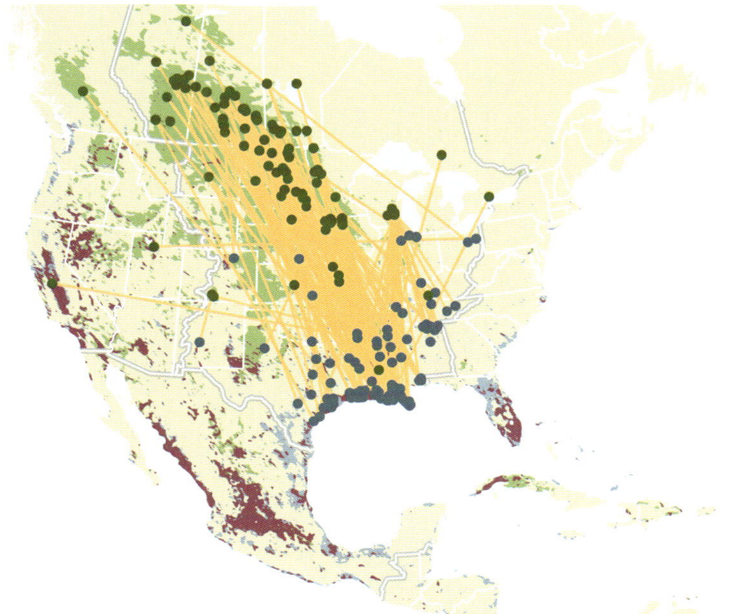

CONNECTIONS D

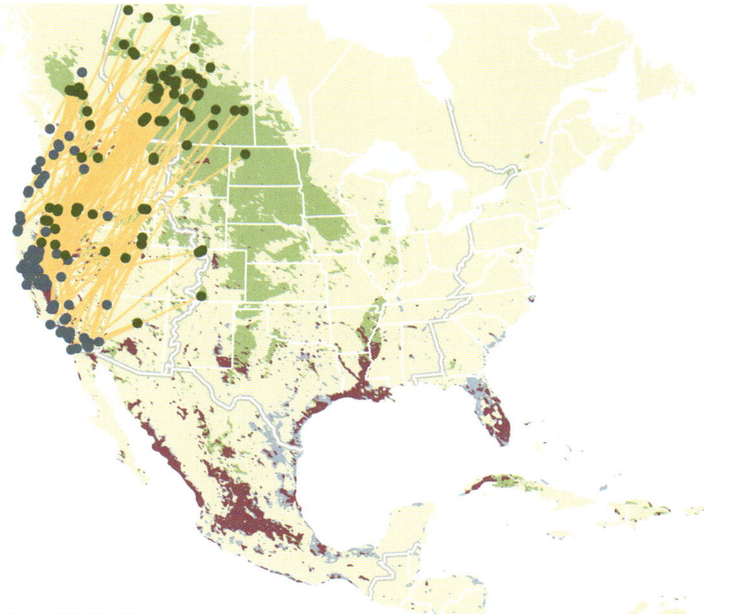

CONNECTIONS E

PURPLE GALLINULE *Porphyrio martinica*

DISTRIBUTION AND ABUNDANCE

Purple Gallinules breed along the coastal plains of the US Gulf and Atlantic coasts, ranging north to South Carolina and west to the Mexican border. Breeding populations also occur inland as far as central Arkansas, especially along the Mississippi River and other water bodies with well-vegetated edges. During the breeding period, US populations are densest from western Louisiana to east Texas, and in southern Florida, where they are present year-round. During the nonbreeding period, this species moves to Florida, Mexico, the Caribbean, and Central and South America, where migrant birds likely overlap with widespread resident populations (Fink et al. 2023).

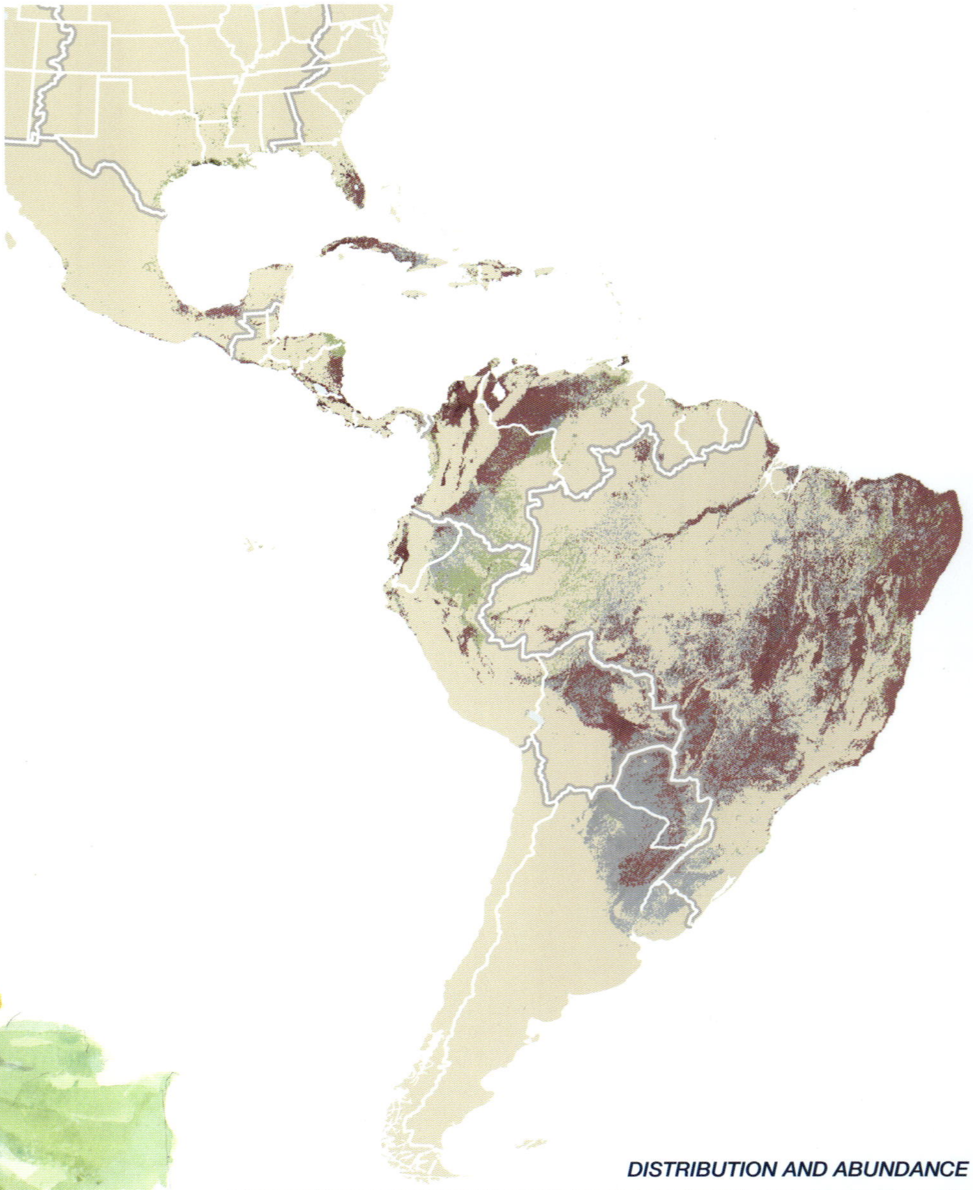

DISTRIBUTION AND ABUNDANCE

ORDER Gruiformes
FAMILY Rallidae
BREEDING HABITAT Wetlands, Freshwater Marsh
NONBREEDING HABITAT Wetlands, Freshwater Marsh
GLOBAL POPULATION SIZE 390,000
POPULATION CHANGE SINCE 1970 −50%
IUCN RED LIST CATEGORY Least Concern
R2R TIPPING POINT SPECIES No
US LISTING STATUS No
CANADA SPECIES AT RISK No
PIF CCS 11/20

MIGRATORY CONNECTIVITY

OVERALL Migratory connectivity is unknown. No studies of migratory connectivity were found for this species.

MIGRATORY PATHWAYS Suspected short-distance, partial migrant species (West and Hess 2020). Little is known about migration routes, but trans-Gulf flights have been documented between the Caribbean and Central America to the US Gulf Coast (West and Hess 2020).

Coturnicops noveboracensis YELLOW RAIL

DISTRIBUTION AND ABUNDANCE

DISTRIBUTION AND ABUNDANCE

The Yellow Rail's breeding range largely occurs in two disjunct populations, including one in central Canada and parts of the northern United States, and a second in Ontario along Hudson and James bays and in Manitoba, Saskatchewan, and Alberta. This species spends the nonbreeding season along the Atlantic and Gulf coasts of the US (Fink et al. 2023).

ORDER Gruiformes
FAMILY Rallidae
BREEDING HABITAT Wetlands, Boreal
NONBREEDING HABITAT Coasts, Salt Marsh
GLOBAL POPULATION SIZE 12,000
POPULATION CHANGE SINCE 1970 NA
IUCN RED LIST CATEGORY Least Concern
R2R TIPPING POINT SPECIES Yellow-Alert
US LISTING STATUS No
CANADA SPECIES AT RISK No
PIF CCS 15/20

MIGRATORY CONNECTIVITY

OVERALL A secretive and difficult species to study. Yellow Rail migratory connectivity information is currently limited to stable isotopes. Feathers have been collected throughout the migratory and nonbreeding ranges to assign individuals to breeding grounds (Butler et al. 2016; Fournier et al. 2017a, 2017b).

STABLE ISOTOPES Stable isotope analysis of nonbreeding individuals from both Oklahoma and Texas were assigned to breeding regions in interior Canadian marshes, from southeastern Alberta to James Bay in Ontario and Quebec, Canada (Butler et al. 2016). Data from Yellow Rail feathers collected from migrating individuals in Missouri, Minnesota, Michigan, South Carolina, Ohio, and Arkansas, and from nonbreeding locations in Louisiana and Mississippi, were combined with citizen science observations to assign Yellow Rails to origins along Hudson and James bays and other parts of Quebec, Minnesota, Wisconsin, and Michigan (Fournier et al. 2017a, 2017b).

OTHER APPROACHES No banding, tracking, or genetic studies of migratory connectivity were found for this species. A few tags were deployed through the Motus Wildlife Tracking System, with local detections only.

MIGRATORY PATHWAYS Little is known about migratory routes and stopover sites. Breeding birds in Canada and the northern US likely move generally north-south through the Midcontinent and Mississippi flyways toward the US Gulf and southeast Atlantic coasts. Data from eBird suggest movements from midcontinent breeding regions across to the Atlantic coast (Fink et al. 2023).

BLACK RAIL *Laterallus jamaicensis*

DISTRIBUTION AND ABUNDANCE

The Black Rail has several disjunct breeding populations distributed throughout the Americas. The California subspecies appears to be largely resident, while the subspecies in the eastern United States is a partial migrant, with breeding and nonbreeding populations overlapping on the Atlantic and Gulf coasts and the Caribbean (Eddleman et al. 2020). The midcontinent subspecies includes small inland breeding populations in Colorado, Kansas, and Oklahoma. The midcontinent populations are likely migratory, as the species is not recorded inland during boreal winter. These populations may spend the nonbreeding season on the Gulf Coast in the Caribbean (Eddleman et al. 2020).

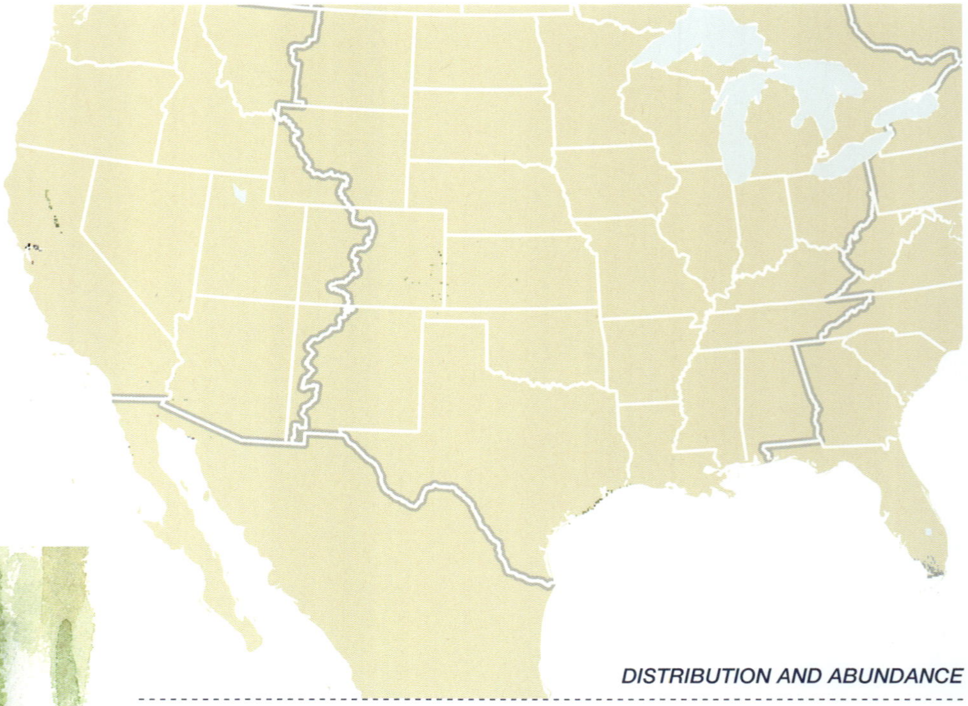

DISTRIBUTION AND ABUNDANCE

ORDER Gruiformes
FAMILY Rallidae
BREEDING HABITAT Coasts, Salt Marsh
NONBREEDING HABITAT Coasts, Salt Marsh
GLOBAL POPULATION SIZE 52,000
POPULATION CHANGE SINCE 1970 NA

IUCN RED LIST CATEGORY Endangered
R2R TIPPING POINT SPECIES Red-Alert
US LISTING STATUS Threatened eastern subspecies
CANADA SPECIES AT RISK No
PIF CCS 17/20

INTERESTING NOTE

Occupancy modeling of the California Black Rail subspecies (*L. j. coturniculus*) showed that populations are largely isolated and disperse over small spatial scales (Hall et al. 2018).

MIGRATORY CONNECTIVITY

OVERALL Black Rails are known for being extremely secretive; thus data on the species are extremely limited. Migratory connectivity studies using genetic markers have focused on the California subspecies, which is generally considered resident.

STABLE ISOTOPES Black Rails in two populations of California were mostly determined to be residents based on stable isotope analyses, though two individuals were migrants that moved more than 100 km (Hall and Beissinger 2017).

GENETICS Nuclear DNA and mitochondrial DNA analyses showed divergence between California populations in the northern and southern parts of the state, even more strongly than divergence from east coast individuals (Girard et al. 2010).

OTHER APPROACHES No banding reencounter data or tracking studies of migratory connectivity were found for this species.

MIGRATORY PATHWAYS Little is known about migration routes of populations. The Atlantic population is partially migratory, following the coastline in north-south movements from the northern extent of the range to nonbreeding areas in the southeastern US and the Caribbean that overlap with the resident population range (Eddleman et al. 2020; Fink et al. 2023).

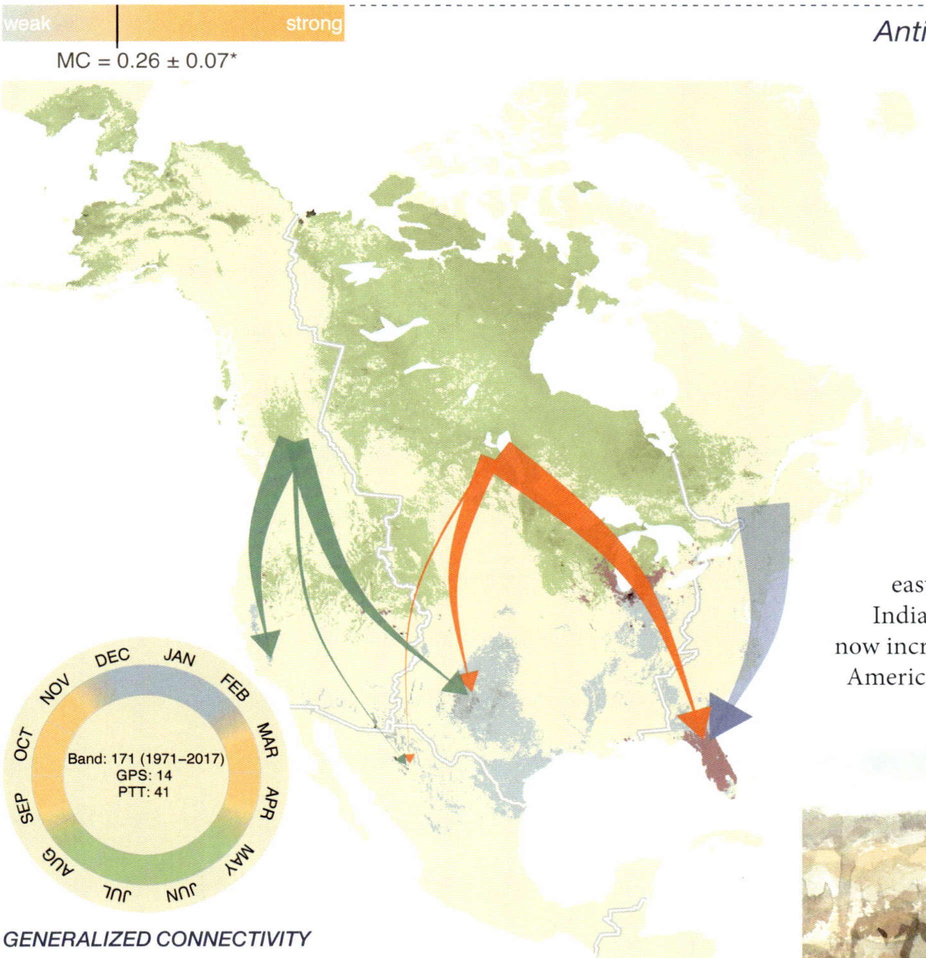

GENERALIZED CONNECTIVITY

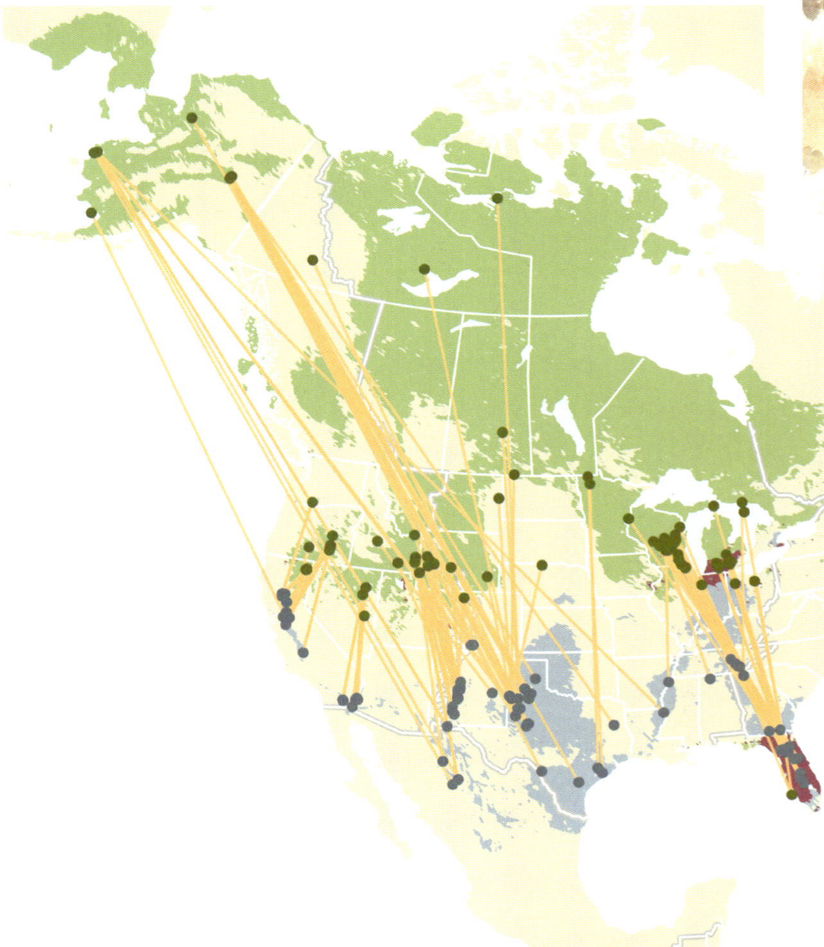

CONNECTIONS A

Antigone canadensis **SANDHILL CRANE**

DISTRIBUTION AND ABUNDANCE

Sandhill Cranes are divided into six subspecies and nine populations, both resident and migratory. The three resident populations reside in Florida, Mississippi, and Cuba. The six migratory populations are geographically delineated by breeding areas and include the Eastern Flyway population, Midcontinent population, Rocky Mountain population, Lower Colorado River Valley population, Central Valley population, and Pacific Flyway population (Gerber et al. 2020). The collective breeding range extends from the Great Lakes northwest through Canada to Alaska, and in the Intermountain West of the US (Fink et al. 2023). The nonbreeding range includes California's Central Valley, Texas, Oklahoma, northern Mexico, and the eastern United States from the Great Lakes south through Indiana and Kentucky to Florida. Although populations are now increasing, the breeding and nonbreeding ranges in North America were historically more extensive (Gerber et al. 2020).

ORDER Gruiformes
FAMILY Gruidae
BREEDING HABITAT Wetlands, Generalist
NONBREEDING HABITAT Wetlands, Generalist
GLOBAL POPULATION SIZE 560,000
POPULATION CHANGE SINCE 1970 +909%
IUCN RED LIST CATEGORY Least Concern
R2R TIPPING POINT SPECIES No
US LISTING STATUS Endangered Mississippi subspecies
CANADA SPECIES AT RISK No
PIF CCS 8/20

	Atlantic	Mexico	Midcontinent	Pacific
Atlantic (n = 1)	100%	–	–	–
Midcontinent (n = 81)	65%	1%	33%	–
Pacific (n = 119)	–	5%	43%	52%

INTERESTING NOTE

Greater (*A. c. tabida*) and Lesser (*A. c. canadensis*) Sandhill Cranes tagged with radios and tracked throughout the nonbreeding period in the Sacramento–San Joaquin Delta region of California indicated Greater Sandhill Cranes showed strong fidelity to nonbreeding sites and mainly occupied one nonbreeding area throughout the season. Lesser Sandhill Cranes were less likely to return to the previous year's nonbreeding location and occupied more than one nonbreeding area within a season (M. Casazza, unpublished data).

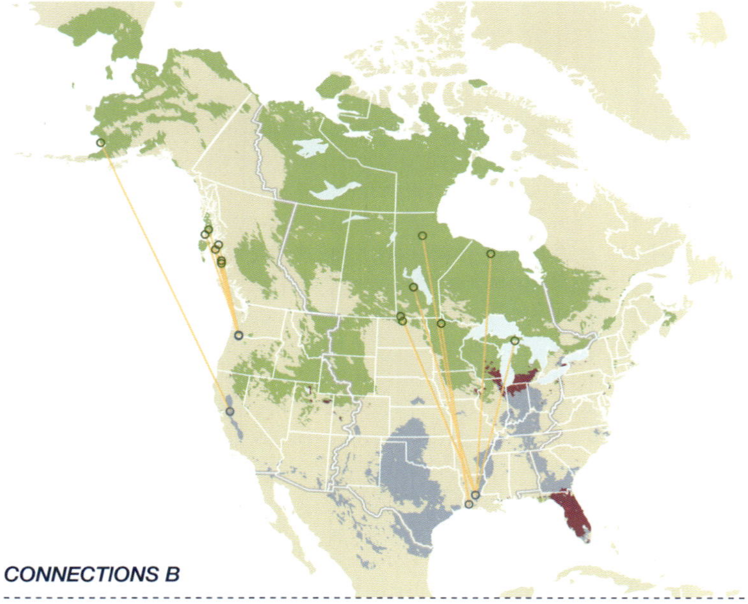

CONNECTIONS B

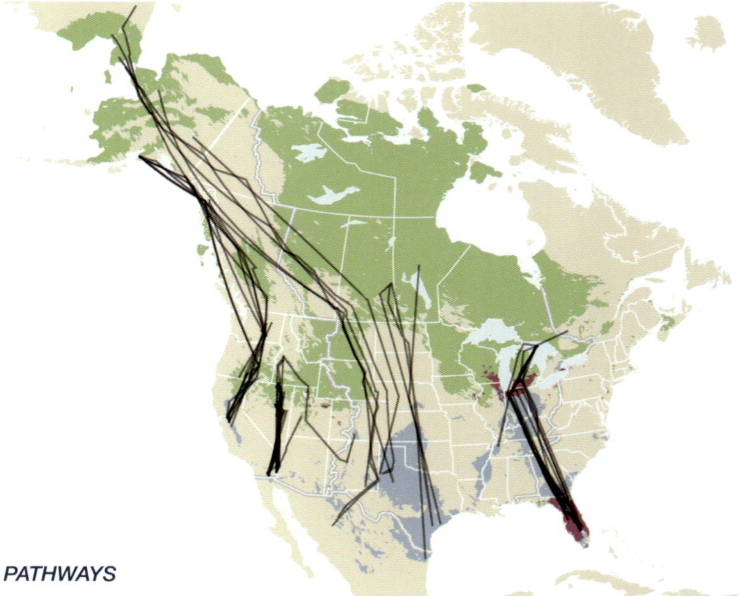

PATHWAYS

MIGRATORY CONNECTIVITY

OVERALL Sandhill Cranes have been researched extensively, and migratory connectivity information is available from tracking, genetic, stable isotope, and banding data. Atlantic Flyway individuals migrate within the flyway. Individuals that breed in the Midcontinent Flyway primarily spend the nonbreeding season in the Atlantic Flyway (65%) and Midcontinent Flyway (33%), while Pacific Flyway breeding individuals are more widely distributed across nonbreeding areas in the Pacific Flyway (52%) and the Midcontinent Flyway (43%), with some migrating to Mexico (5%). Although there are high transitions within flyways, breeding populations overlap on nonbreeding areas, especially within the Atlantic and Midcontinent flyways, resulting in moderate migratory connectivity (MC = 0.26 ± 0.07).

SATELLITE TRACKING Sandhill Cranes have been tracked extensively using satellite transmitters throughout their range. Most studies focus on one of the six migratory populations. The populations also reflect the migratory patterns of the species throughout its range.

Eastern Flyway Population

GPS satellite transmitters deployed on cranes in the eastern population linked breeding sites in Minnesota, Wisconsin, Michigan, and Ontario with nonbreeding sites in Indiana, Kentucky, Tennessee, Georgia, and Florida (Fronczak et al. 2017; E. Hanna, unpublished data). Individuals follow a direct migratory pathway southeast to nonbreeding areas. GPS-GSM transmitters deployed on midcontinent and eastern populations in Minnesota were tracked primarily to nonbreeding grounds in Texas and Florida, with some individuals using both Atlantic and Midcontinent flyways during migration (Wolfson et al. 2017).

Midcontinent Population

Four individuals tracked from east-central Saskatchewan migrated to the Gulf Coast of Texas for the nonbreeding season (Hjertaas et al. 2001). Sandhill Cranes on nonbreeding areas in Louisiana migrated north along the Central Flyway (Midcontinent Flyway) to breeding areas in Ontario, Michigan, Minnesota, and Manitoba (King et al. 2010). Cranes marked at the Central Platte River Valley, Nebraska, during pre-breeding migration belonged to several different breeding subpopulations of the Midcontinent population but had some overlap on nonbreeding areas. The east-central Canada-Minnesota breeding subpopulation spent the nonbreeding season in Texas. The west-central Canada-Alaska breeding subpopulation spent the nonbreeding season in Texas and northern Mexico. The western Alaska-Siberia breeding subpopulation migrated to nonbreeding areas in Arizona, New Mexico, Texas, and Mexico. The northern Canada-Nunavut breeding subpopulation also spent the nonbreeding season in Texas (Krapu et al. 2014).

Rocky Mountain Population

Sandhill Cranes tracked from breeding areas in Idaho migrated to nonbreeding areas in Southern California and Arizona (Collins et al., unpublished data). Other breeding populations in the Rocky Mountain population are known to migrate to New Mexico, Arizona, and Mexico, with stopovers in the San Luis Valley, Colorado, during pre- and post-breeding migration (Gerber et al. 2020).

Lower Colorado River Valley Population

Cranes were tracked between breeding areas in Nevada and Idaho and nonbreeding locations in California, Arizona, and New Mexico (Collins et al. 2016). Individuals from this population used direct northbound and southbound migratory routes between breeding and nonbreeding locations (Conring et al. 2019).

Central Valley and Pacific Flyway Populations

Sandhill Cranes tracked from stopover sites in Washington state bred in Alaska and British Columbia and spent the nonbreeding season in

Oregon and Northern California (Ivey et al. 2005). Satellite transmitters at breeding areas at Cook Inlet and Bristol Bay, Alaska, tracked individuals to nonbreeding areas in California's Central Valley (Petrula and Rothe 2005). Similarly, populations breeding near Homer, Alaska, spend the nonbreeding season in the Sacramento–San Joaquin River Delta region of California (M. Casazza, unpublished data). These individuals migrated closer to the Pacific coast through British Columbia and south to nonbreeding grounds in the Central Valley of California. Other satellite transmitters at breeding sites in eastern Russia tracked migrations through Alaska and central Canada inland to Texas and Mexico and through western Canada and the central United States to nonbreeding sites in northern Texas and Mexico (Pearse et al. 2017).

BANDING Band reencounters and resightings of color-marked individuals show similar connections between breeding and nonbreeding locations, as does tracking data. Sandhill Cranes marked on nonbreeding areas in California's Central Valley were resighted at breeding areas in Alaska, British Columbia, Oregon, and California (Pogson et al. 1988). Resightings from the breeding grounds in Alaska have also been documented on nonbreeding areas in California (Pogson 1987). Two color-marked individuals were documented migrating between breeding areas in Alaska and Manitoba and nonbreeding areas in western Texas and the Texas Gulf Coast, respectively. Both individuals made a pre-breeding stopover in Nebraska (Tacha et al. 1984). Color-marked cranes spending the nonbreeding period in Florida were resighted in the Great Lakes region of the US (Nesbitt and Williams 1979).

RADIO TRACKING Data from radio tracking studies at breeding sites in Yellowstone National Park and Jackson Hole, Wyoming, showed that cranes (n = 9) used staging areas in the Teton Basin in eastern Idaho and spent the nonbreeding season in the Middle Rio Grande Valley, New Mexico (Drewien et al. 1999). Radio tracking has also shown that the eastern and midcontinent populations overlap in breeding distribution in Minnesota and have some overlap in migration routes and nonbreeding areas (D. Wolfson, personal communication).

STABLE ISOTOPES Feathers sampled from Sandhill Cranes during the hunting season in Saskatchewan, North Dakota, Kansas, Oklahoma, and Texas showed that hatch-year individuals originated from Ontario, Manitoba, and Alaska, while after-hatch-year individuals originated from throughout the breeding range. Adults sampled in Saskatchewan originated farther north than individuals sampled in Texas (Hobson et al. 2006).

GENETICS Genomic analysis shows that there is significant gene flow between subspecies of Sandhill Cranes (Jones et al. 2005).

MIGRATORY PATHWAYS See "Satellite tracking" of each population for migratory pathways information.

DATA CONTRIBUTORS

M. Casazza; D.P. Collins; E.E. Hanna; A.T. Pearse; S.A. Petrie; Birds Canada; Canadian Wildlife Service; Idaho Department of Game and Fish; Intermountain West Joint Venture; Long Point Waterfowl; New Mexico Department of Game and Fish; New Mexico State University; The Ontario Federation of Anglers and Hunters; Texas Tech University; University of Western Ontario; WEST, INC; Wildlife Habitat Canada.

Grus americana WHOOPING CRANE

DISTRIBUTION AND ABUNDANCE

Whooping Cranes are found in disjunct populations in the United States and Canada. Intensive management actions saved the species from extinction in the twentieth century, and the only remaining wild population breeds in the Northwest Territories and spends the nonbreeding season on the Texas coast. Reintroduced populations include a nonmigratory population in Louisiana and an eastern migratory population (Urbanek and Lewis 2020).

ORDER Gruiformes
FAMILY Gruidae
BREEDING HABITAT Wetlands, Boreal
NONBREEDING HABITAT Coasts, Salt Marsh
GLOBAL POPULATION SIZE 280

POPULATION CHANGE SINCE 1970 NA
IUCN RED LIST CATEGORY Endangered
R2R TIPPING POINT SPECIES Orange-Alert
US LISTING STATUS Threatened
CANADA SPECIES AT RISK Endangered
PIF CCS 18/20

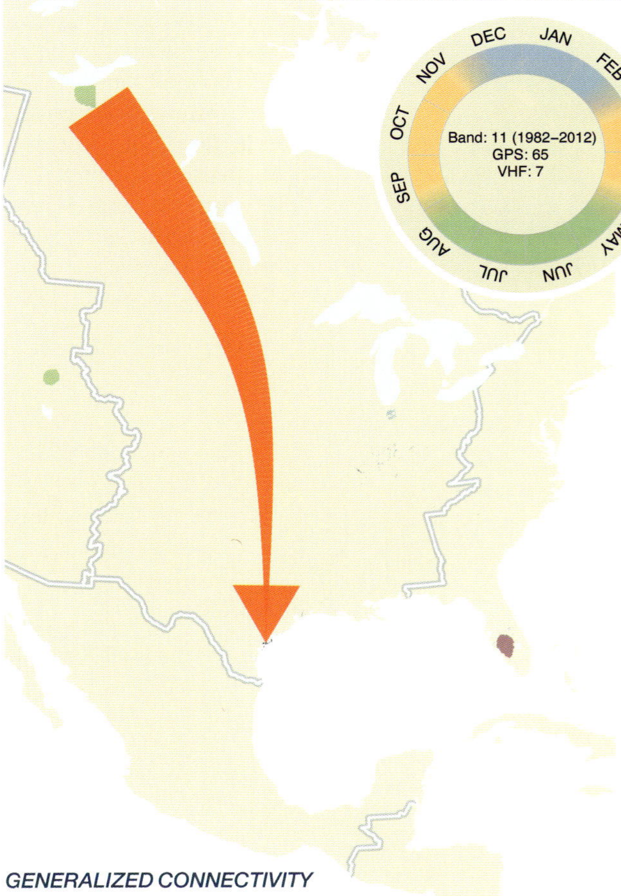

GENERALIZED CONNECTIVITY

	Midcontinent
Midcontinent (n = 77)	100%

INTERESTING NOTE

The unique management history of Whooping Cranes, which has led to the introduction of captive-bred individuals into the eastern migratory population, has been used to assess the importance of social learning in migration. In this population, both social learning and innate programs appear to be important for the establishment of migratory routes, and older individuals in the population contribute experience and improve migratory performance (Mueller et al. 2013; Teitelbaum et al. 2016).

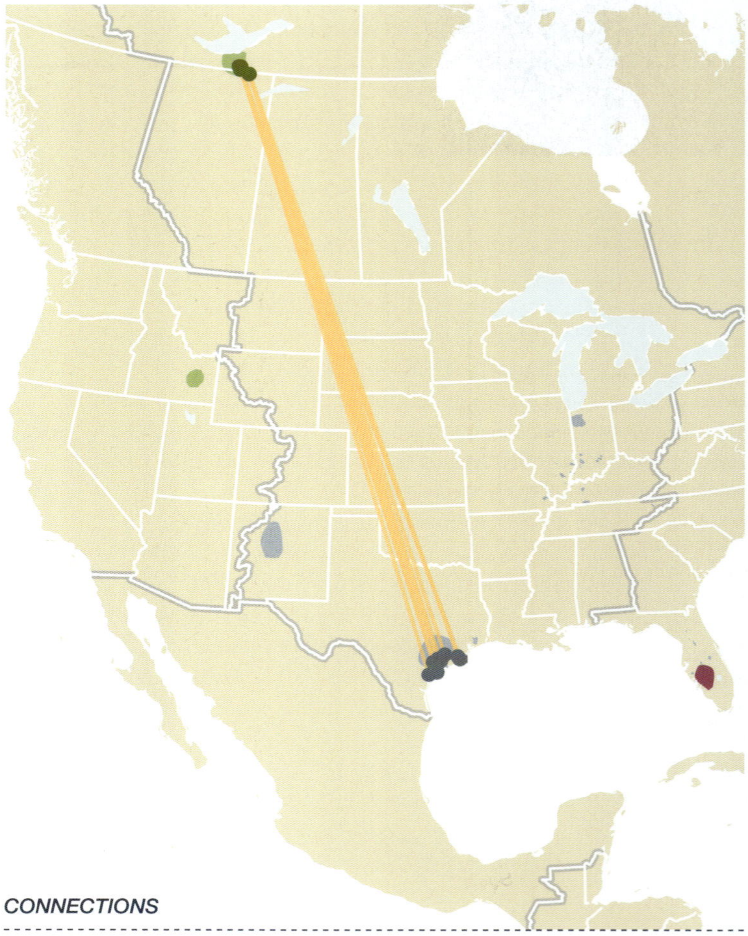

CONNECTIONS

PATHWAYS

MIGRATORY CONNECTIVITY

OVERALL The wild population of this species has been studied extensively using color bands, radio, and satellite telemetry. Results of these studies confirm that Midcontinent Flyway breeders all migrate to nonbreeding areas within the Midcontinent Flyway (100%). Because the population breeds within only one of our migratory connectivity regions, MC could not be calculated.

BANDING Band reencounters showed connections between breeding grounds at Wood Buffalo National Park in Canada and nonbreeding grounds at Aransas National Wildlife Refuge in Texas, but prior to color banding of individuals in the 1980s, knowledge of the movements of individuals was limited (Stehn 1992). Additional band reencounters connect breeding areas in Michigan to nonbreeding areas in Florida.

RADIO TRACKING Radio telemetry was used to track 15 juvenile Whooping Cranes from their breeding grounds at Wood Buffalo National Park in Canada to their nonbreeding grounds at Aransas National Wildlife Refuge in Texas between 1981 and 1983 (Kuyt 1992).

SATELLITE TRACKING Years after the first radio tracking study, the Whooping Crane Tracking Partnership was formed to use satellite telemetry and GPS technology to track individuals from their breeding grounds in Wood Buffalo National Park to their nonbreeding grounds at Aransas National Wildlife Refuge (Pearse et al. 2020). In addition, a separate organization, the Whooping Crane Eastern Partnership, has tracked adult and juvenile Whooping Cranes in the eastern migratory population with satellite, GPS, and GSM transmitters since 2001 (not included on the tracking map). Most Whooping Cranes spent the summer in Wisconsin, with a few in the surrounding states and provinces. Nonbreeding areas were scattered throughout the Midwest and southeastern US, with sites in Wisconsin, Illinois, Indiana, Kentucky, Tennessee, Alabama, Arkansas, North Carolina, Louisiana, Georgia, and Florida (H. Thompson, unpublished data).

OTHER APPROACHES No geolocator, stable isotope, or genetic studies of migratory connectivity were found for this species.

MIGRATORY PATHWAYS Intensive management of Whooping Crane populations has contributed greatly to knowledge of the species' migratory patterns, and many stopover areas are well documented. Several decades of sightings of color-banded Whooping Cranes have been used to collect locations along the Aransas–Wood Buffalo population's migratory corridor. Most of these sightings occurred in Montana, North Dakota, South Dakota, Nebraska, Kansas, Oklahoma, and Texas (Tacha et al. 2010) Satellite transmitters confirmed migratory stopover locations throughout the Great Plains, with intensive use in regions along the border of Texas and Oklahoma, north-central Nebraska, northwestern North Dakota, the border of North Dakota and Saskatchewan, and south-central Saskatchewan (Pearse et al. 2015). A comparison of sightings from 1942 to 2016 with telemetry data collected from 58 individuals between 2010 and 2016 showed an easterly shift in the population's migratory corridor (Pearse et al. 2018).

DATA CONTRIBUTORS

A.T. Pearse.

SHOREBIRDS, AUKS, GULLS, AND TERNS (CHARADRIIFORMES)

WRITING LED BY RICHARD B. LANCTOT, JASON P. LOGHRY, AUTUMN-LYNN HARRISON, AMY L. SCARPIGNATO, TIMOTHY GUIDA, TOM WILL, *AND* CAMILLA GÓMEZ

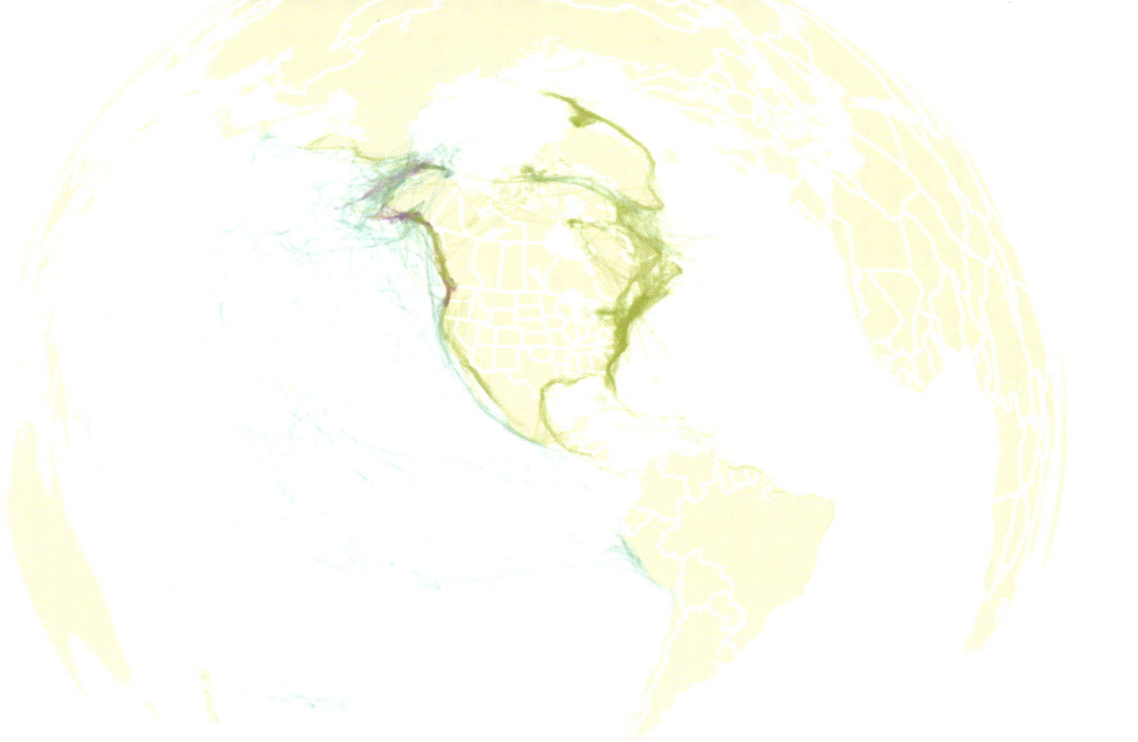

Breeding seabirds tracked with Argos and GPS within the Alcidae (n = 124; 41 breeding to nonbreeding connections), Laridae (n = 322; 188 connections), and Stercorariidae (n = 50; 43 connections) families showed strong use of Pacific and Atlantic flyways and Greenland (Alcidae and Laridae), and the open ocean (Stercoraridae, some Larids), with some additional data within the midcontinent, including along the coast of the Gulf of Mexico. Generally, Alcids and Larids remained within their flyway, but some species breeding in the Pacific Flyway migrated to nonbreeding sites in the western Pacific, and some species breeding in the Atlantic Flyway migrated to nonbreeding sites in the eastern Atlantic. Jaegers undertook north-south and figure-eight migrations to nonbreeding sites in the Southern Hemisphere.

The migratory connectivity and associated ecology of most species within the order Charadriiformes are quite well studied. This includes 96 of the 113 species belonging to seven families, including the Recurvirostridae (stilts and avocets), Haematopodidae (oystercatchers), Charadriidae (plovers), Scolopacidae (sandpipers), Stercorariidae (skuas and jaegers), Alcidae (auks, murres, and puffins), and Laridae (gulls, terns, and skimmers). There is migratory connectivity information for 85% of species across the order, and tracking data for 100% of the species from four of the seven families. The exceptions include Scolopacidae (80%), Alcidae (65%), and Laridae (92%). Across the order, most migratory connectivity information is based on band reencounters (90%), followed by geolocator tracking (7%), satellite tracking (Argos, 1%, and GPS tracking, 2%), and radio tracking (<1%). A comparison of the seven families indicates larids are the most data-rich, with more than 30,000 breeding to nonbreeding connections across 38 species. Compared to the gulls (the larids), the alcids and the sandpipers (the family Scolopacidae) have less than one-tenth the number of breeding to nonbreeding connections (2,647 and 1,406, respectively) but also fewer species to track overall (20 and 36 species, respectively). The remaining four families have 7–255 breeding to nonbreeding connections but also have only 2–10 species each.

Three families within this order are generally thought of as being "seabirds," spending most of their time in marine areas, including the auks, the gulls and terns, and the skuas. The alcids have been moderately studied (13 of 20 species) with both bands (n = 1,749) and either geolocator (1,329 light-level geolocator connections) or satellite tracking devices (120 Argos connections). Six of the 20 species are considered Tipping Point species, including the Dovekie and five murrelets (Marbled, Kittlitz's, Scripps's, Guadalupe, Craveri's). The gulls and terns have been very well studied (35 of 38 species have migratory connectivity data), with bands being the dominant method of monitoring (29,146 breeding to nonbreeding connections), followed by light-level geolocators (1,187) and much smaller numbers of Argos (107) and GPS (81) connections. Generally, temperate species are better studied than Arctic breeders. Six of the 38 species are considered Tipping Point species, including five gulls (Ivory, Short-billed, Western, Glaucous, Great Black-backed) and the Least Tern. All five species of skuas and jaegers have some tracking information, although only 75 individuals across all species have information linking breeding and nonbreeding connections. These connections are band reencounters (17), light-level geolocators (15), and Argos or GPS tracks (43). The largest data gap is for the Pomarine Jaeger—no connectivity data for this species exists in the Atlantic Ocean portion of its range, including the Gulf of Mexico, where sightings are common. Data gaps remain for Short-billed Gull, Scripps's Murrelet, Craveri's Murrelet, and Guadalupe Murrelet—all have limited banding and tracking data.

The three seabird families each exhibit a variety of overall migratory connectivity. Alcids generally exhibit regionally weak migratory connectivity, which becomes stronger when comparing eastern and western populations across an ocean, or populations across an ocean basin. They typically undertake short-distance migrations either north-south or east-west, but some cross-ocean migrations occur (Ancient Murrelet). Gulls and terns (the Larids) exhibit a variety of migration patterns between breeding and nonbreeding sites that include coastal and continental migrations, nomadic range shifts, and the longest migration in the world. Their migratory connectivity also varies. For example, some gulls remain generally within their flyway (Great Black-backed Gull and Herring Gull), while other species show high population mixing on nonbreeding sites (Laughing Gull). Some terns show strong connectivity at the continental scale (Common Tern), while other terns may mix extensively across populations at the oceanic scale (Arctic Tern). The skuas and jaegers generally move in traditional north-south and figure-eight migrations across the open ocean from Arctic to Southern Hemisphere nonbreeding sites and demonstrate weak to

Breeding shorebirds tracked with Argos and GPS within the Charadriidae (n = 110; 58 breeding to nonbreeding locations), Haematopodidae (n = 18; 15 connections), and Scolopacidae (n = 1,622; 670 connections) families showed strong use of the Pacific, Midcontinent, and Atlantic flyways, with a small number of species using the Central Pacific, East Asian–Australasian, and East Atlantic flyways. Most species used continental or coastal routes, although a few used the open ocean for either both north and south migrations, or for their southward migrations, with either a coastal or a continental route for their northward migrations. Midcontinental migrants had the most variation in migration patterns both within and among species. Arctic-breeding shorebirds flew the farthest—to southern South America—while temperate breeders spent the nonbreeding season farther north within the US or Central America.

moderate connectivity that generally stays confined to specific ocean basins but with some movements between high Arctic Canada and the Arctic, Pacific, Atlantic, and Indian oceans.

The remaining four families are shorebirds, as they spend most of their time on land or coastal environments, including the stilts and avocets, oystercatchers, plovers, and sandpipers. All ten plover species have some tracking information, including 255 individual breeding to nonbreeding connections. Most breeding to nonbreeding information is from band resightings (140), followed by radio tracks (4), light-level geolocator connections (53), and GPS and/or Argos connections (58). Six of the ten plover species are considered Tipping Point species, including the Black-bellied, American Golden-, Piping, Mountain, Snowy, and Wilson's Plover.

Both the American Oystercatcher and the Black Oystercatcher have been tracked using bands (38 individual breeding to nonbreeding connections) and/or Argos satellite transmitters (15). Only the American Oystercatcher is considered a Tipping Point species. Among the least studied shorebirds are the stilts and avocets, with only two of the seven species tracked. For both species, only bands have been used to connect breeding and nonbreeding locations. Most (29 of 36) of the sandpipers have been tracked, including 1,406 breeding to nonbreeding connections. Unlike many of the other families, sandpipers have been tracked in many different ways, and migratory connectivity information is known from using bands (487 breeding to nonbreeding connections), radio transmitters (54), light-level geolocators (195), and satellite transmitters (Argos, 222, and GPS, 448). The sandpipers without tracking data are generally too small, are too difficult to equip with tags or recapture for archival tags, or breed and/or spend the nonbreeding season in remote areas. An astounding 17 of the 36 sandpipers are Tipping Point species, including five sandpipers (Stilt, Rock, Buff-breasted, Pectoral, Semipalmated), Bristle-thighed Curlew, Hudsonian Whimbrel, Bar-tailed and Hudsonian Godwit, Ruddy Turnstone, Red Knot, Sanderling, Short-billed and Long-billed Dowitcher, Wandering Tattler, and Lesser and Greater Yellowlegs. Shorebirds are declining across North America, and recent studies suggest that those migrating over the midcontinent are more likely to be declining. However, species-specific traits, such as migrating to nonbreeding areas on remote islands in the South Pacific that are prone to sea level rise and rat introductions, can override these general patterns.

The few species within the four shorebird families with sufficient data to generate migratory connectivity scores generally have weak to moderate connectivity scores. Only one species, the Willet, exhibits very strong connectivity. Weak migratory connectivity (<0.30) occurs in the Dunlin, Lesser Yellowlegs, and Long-billed Dowitcher. Species with moderate connectivity (0.3 to 0.7) include Black-bellied Plover, Piping Plover, American Woodcock, Marbled Godwit, and Semipalmated Sandpiper. The shorebird families exhibit a variety of migration patterns. Plovers breed in the Arctic and temperate regions, migrate short and long distances, and use all five flyways from the Central Pacific to the eastern Atlantic. Some have discrete migrations back and forth (e.g., Pacific Golden-Plover), others have broad migrations (e.g., Killdeer), and still others exhibit elliptical migrations (e.g., American Golden-Plover). The two oystercatchers are limited to north-south migrations along the Pacific and Atlantic coasts of North and Central America. Little information is available on migrations of stilts and avocets, but generally they have short migrations within interior regions of North America. The sheer number of sandpiper species equals a plethora of migration patterns, including broad-front, shorter migrations within North America (e.g., American Woodcock, Long-billed Curlew), long-distance migrations generally through the Midcontinent Flyway to southern South America (e.g., Buff-breasted Sandpiper), long-distance migrations over the central Pacific (e.g., Bristle-thighed Curlew) or Atlantic oceans (e.g., American Golden-Plover), coastal migrations (e.g., Black Turnstone, Red Knot, Red-necked Phalarope), and migrations to Asia (e.g., Dunlin) or Iceland/Europe (e.g., Purple Sandpiper). Many species have multiple populations or subspecies that breed around the globe and travel in parallel north-south migrations (e.g., Ruddy Turnstone, Dunlin, Red Knot). Perhaps most notable is the migration of the Bar-tailed Godwit, which flies nonstop from Alaska to New Zealand in eight or more days, and that of the Pectoral Sandpiper, which travels as far within the Arctic as it travels south to southern South America.

Himantopus mexicanus **BLACK-NECKED STILT**

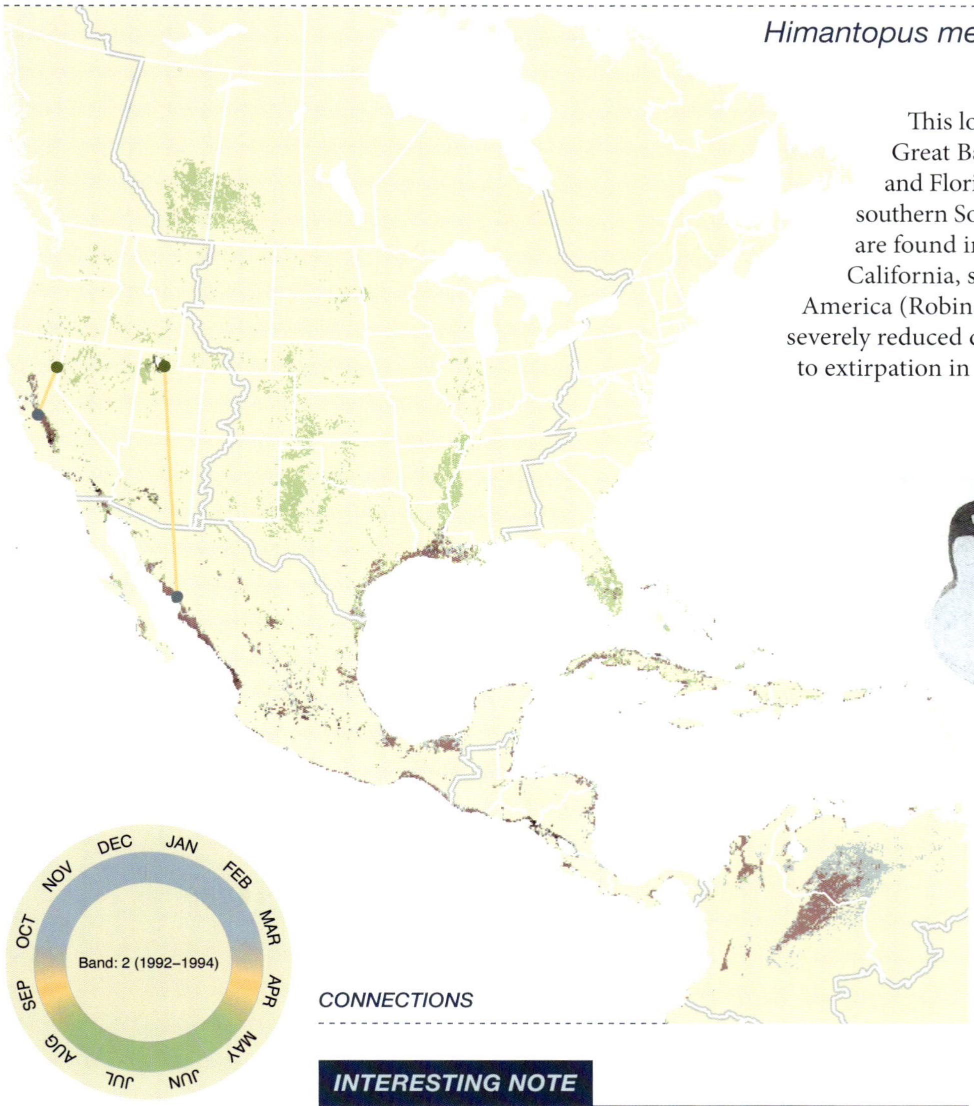

DISTRIBUTION AND ABUNDANCE

This long-legged shorebird breeds in distinct areas of the Great Basin, California, northwestern Texas, New Mexico, and Florida. It occupies much of Mexico, the Caribbean, and southern South America year-round. Nonbreeding populations are found in the Central Valley and Southern California, Baja California, small parts of Texas, and coastal regions of Central America (Robinson et al. 2020). The range size for this species was severely reduced due to loss of wetland habitat and hunting, leading to extirpation in parts of its historical range (Robinson et al. 2020).

INTERESTING NOTE

This species wanders widely and is sensitive to handling; therefore, there is much still unknown about population demography and movements. Natal philopatry has been observed but is not strong (Robinson et al. 2020).

ORDER Charadriiformes
FAMILY Recurvirostridae
BREEDING HABITAT Wetlands, Freshwater Marsh
NONBREEDING HABITAT Wetlands, Generalist
GLOBAL POPULATION SIZE 900,000
POPULATION CHANGE SINCE 1970 +189%
IUCN RED LIST CATEGORY No
R2R TIPPING POINT SPECIES No
US LISTING STATUS No
CANADA SPECIES AT RISK No
PIF CCS 8/20

MIGRATORY CONNECTIVITY

OVERVIEW Little is known, but it probably has weak migratory connectivity. Breeding site fidelity is not strong, and the species moves widely to follow habitat availability (Robinson et al. 2020). It is resident or undergoes short seasonal movements in Hawaii and tropical parts of the species range (Robinson et al. 2020).

BANDING Extensive banding of breeding adults and chicks in the Honey Lake Valley (northeastern California) and Great Salt Lake Basin indicated a hatch-year bird traveled to Navojoa, Sonora, to spend the nonbreeding season, and a two-year-old stilt traveled to the Tulare Basin, California.

RADIO TRACKING A single tag was deployed through the Motus Wildlife Tracking System with local detections only.

OTHER No genetic, stable isotope, or tracking studies of migratory connectivity were found for this species.

MIGRATORY PATHWAYS No information, but habitat use is restricted to inland sites, especially on wetlands of central California as stopover sites on their way to nonbreeding sites in western Mexico (Robinson and Oring 1996).

AMERICAN AVOCET *Recurvirostra americana*

DISTRIBUTION AND ABUNDANCE

This large, distinctive shorebird breeds over a wide area of the western United States and spends the nonbreeding period along the Pacific (California to Baja California) and Atlantic coasts, primarily; and to Mexico, the US Gulf states, and Florida, with lower densities along the Atlantic coast states north to approximately Delaware (Ackerman et al. 2020). A year-round population resides in central Mexico.

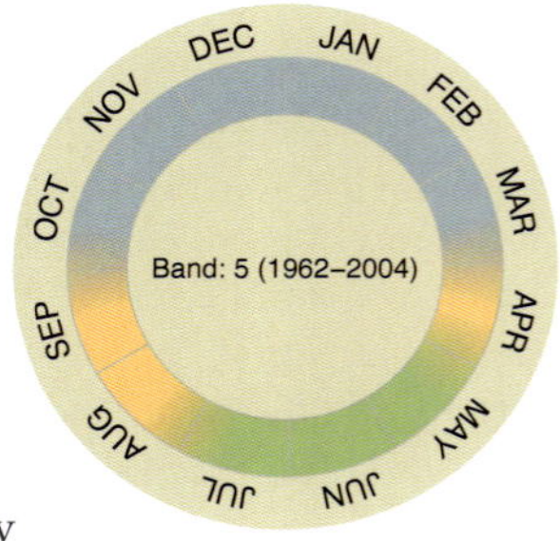

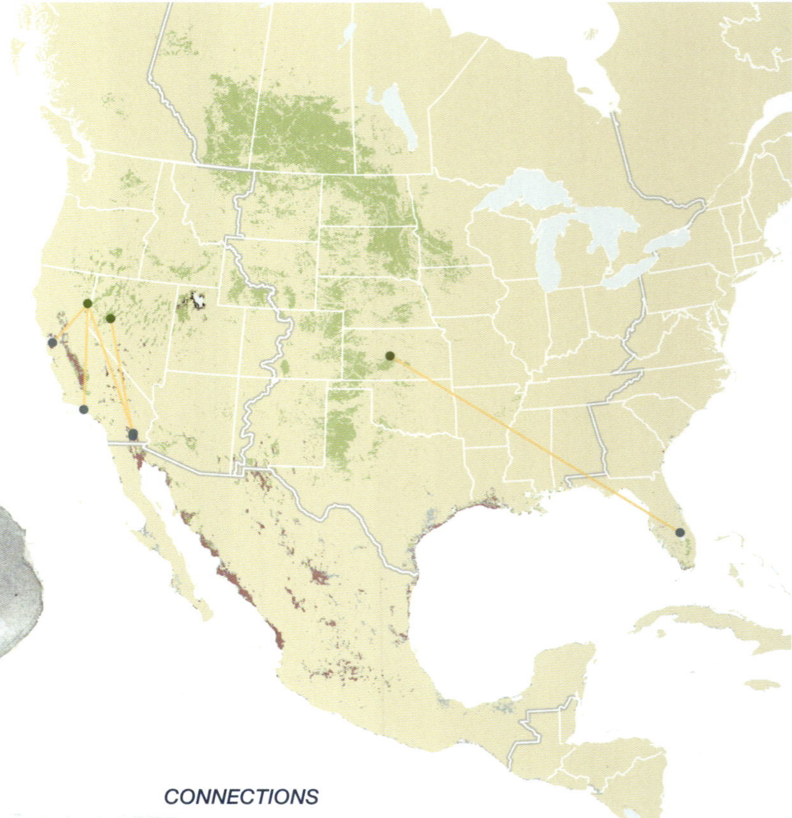

ORDER Charadriiformes
FAMILY Recurvirostridae
BREEDING HABITAT Wetlands, Freshwater Marsh
NONBREEDING HABITAT Coasts, Beach and Estuary

GLOBAL POPULATION SIZE 460,000
POPULATION CHANGE SINCE 1970 +29%
IUCN RED LIST CATEGORY Least Concern
R2R TIPPING POINT SPECIES No

US LISTING STATUS No
CANADA SPECIES AT RISK No
PIF CCS 11/20

MIGRATORY CONNECTIVITY

OVERALL Locations of banded and radio-tagged birds suggests separate eastern and western populations might exist. American Avocets breeding in the western United States tend to spend the nonbreeding season along the Pacific coast of California and Mexico, and birds breeding in more interior regions of United States are in central Mexico and the Atlantic and Gulf coasts for the nonbreeding season. Available information suggests strong intermixing among breeding birds in the western population. Confirmatory information on nonbreeding individuals in Mexico and along the Atlantic and Gulf coasts is needed. Due to limited sample size, estimates of MC are not available.

BANDING Banding of breeding adults and chicks in Honey Lake Valley, California, indicate they spend the nonbreeding season along the California coast from Arcata, California, southward along the west coast of Sinaloa, Mexico. Breeding Avocets banded in the Great Salt Lake, Utah, also spent the nonbreeding season in Mexico, indicating the potential of some degree of population mixing (Robinson and Oring 1996). One banded bird captured in Kansas was resighted in Florida.

RADIO TRACKING Breeding adults (n = 185) were equipped with conventional radio transmitters at five breeding areas in Oregon, California, and Nevada in 1996 and 1997 (Plissner et al. 2000). Post-breeding movements at 8 weeks after capture indicated 56% were still in the region. A few tags were deployed through the Motus Wildlife Tracking System with local detections only.

SATELLITE TRACKING GPS satellite transmitters were deployed on breeding birds near the Great Salt Lake and on nonbreeding birds on the Gulf Coast of Texas and Louisiana (Clements 2022; J. Cavitt and J. Loghry, personal communication). Great Salt Lake birds migrated south to the Gulf of California on the Pacific coast (https://departments.weber.edu/avianecologylab/AMAV/about.html). Fourteen of 27 birds with devices deployed in Texas and Louisiana completed northward migration to breeding areas in North and South Dakota, Manitoba, and Saskatchewan; others exhibited partial migration or remained on nonbreeding areas in coastal Texas and Mexico.

OTHER APPROACHES No stable isotope or genetic studies of migratory connectivity were found for this species.

MIGRATORY PATHWAYS Banding data indicate birds captured in the Great Basin stop at Mono Lake and the Tulare Basin before traveling to nonbreeding areas along the Pacific coast of California and Mexico. In both cases, little information on routes is available. However, first-year birds were more likely along the coast of California, and adults were more likely inland (Robinson and Oring 1996). Satellite-tagged birds in Texas and Louisiana migrated through the North American midcontinent over an approximately 20-day period, traveling medium distances between stops (Clements 2022; S. Clements and J. Loghry, personal communication). Migration patterns of birds that are found at stopover sites along the Atlantic coast and that spend the nonbreeding season in central Mexico are needed (Ackerman et al. 2020).

Haematopus palliatus AMERICAN OYSTERCATCHER

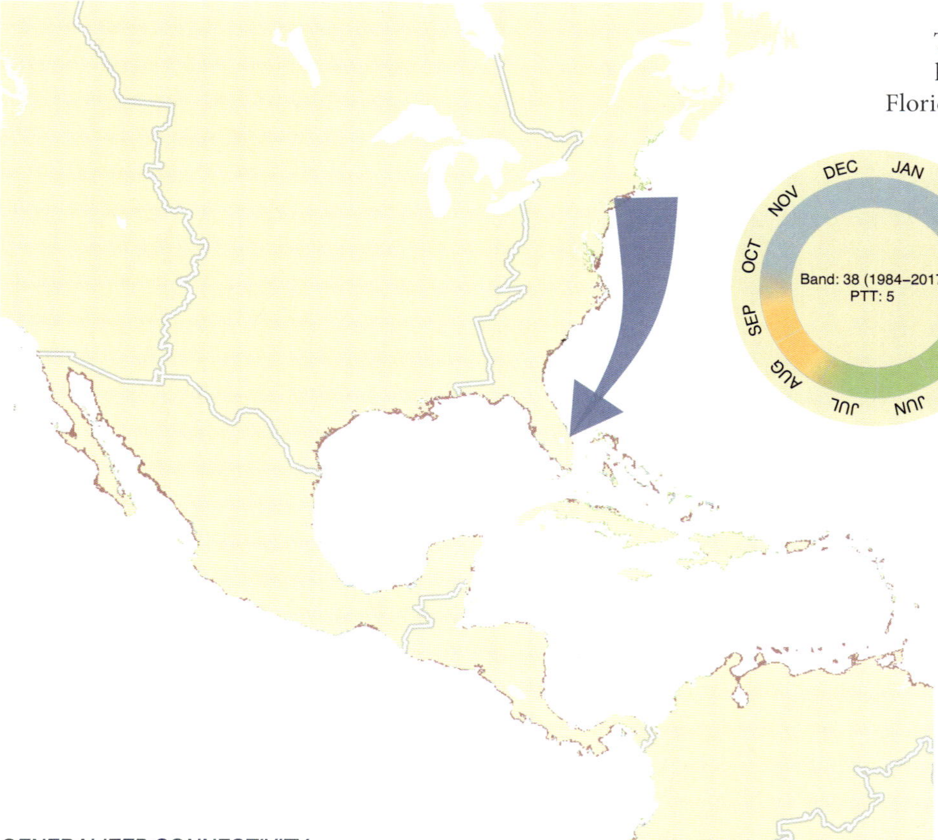

GENERALIZED CONNECTIVITY

DISTRIBUTION AND ABUNDANCE

The American Oystercatcher occupies coastal marine habitats along the Atlantic coast south from Maine to Florida and across the Gulf of Mexico, as well as along the Pacific coast from Baja California to Ecuador. The Atlantic and Pacific populations are thought to represent two races, with the Atlantic coast much more studied (Working Group et al. 2020). Nonbreeding areas include the mid-Atlantic states and south across the Gulf of Mexico, where resident individuals likely mix with migratory birds.

INTERESTING NOTE

Breeding and nonbreeding site fidelity is high but with flexibility due to a very dynamic ecosystem (S.P. Murphy et al. 2017; Working Group et al. 2020). Natal philopatry was observed in several populations, with birds returning to the area about 50–60 km from where they hatched (Working Group et al. 2020).

	Atlantic
Atlantic (n = 40)	100%

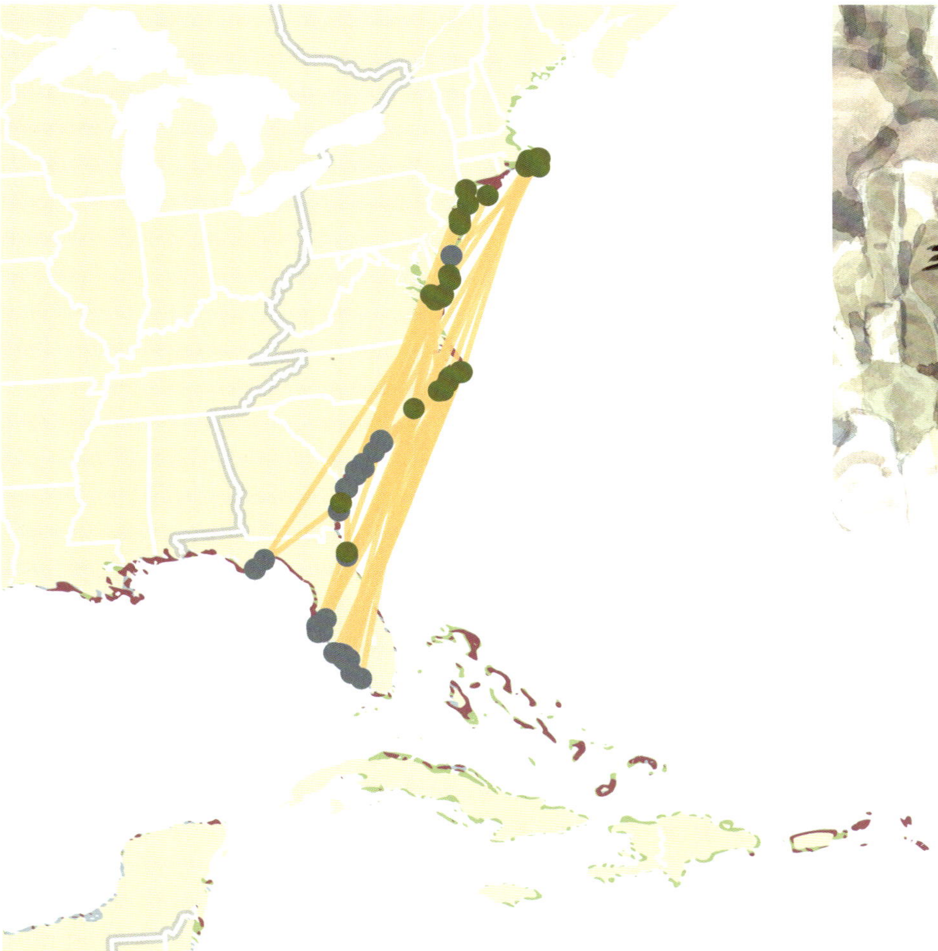

CONNECTIONS

ORDER Charadriiformes
FAMILY Haematopodidae
BREEDING HABITAT Coasts, Beach and Estuary
NONBREEDING HABITAT Coasts, Beach and Estuary
GLOBAL POPULATION SIZE 43,000
POPULATION CHANGE SINCE 1970 +57%
IUCN RED LIST CATEGORY Least Concern
R2R TIPPING POINT SPECIES Yellow-Alert
US LISTING STATUS No
CANADA SPECIES AT RISK No
PIF CCS 14/20

MIGRATORY CONNECTIVITY

OVERVIEW Migratory connectivity is mainly based on resightings of marked birds along the Atlantic coast. All birds breeding within the Atlantic Flyway migrate to nonbreeding areas within the Atlantic Flyway. Little information is available for birds in other parts of the range. Due to a single breeding and nonbreeding region, MC could not be estimated.

BANDING Large-scale banding of birds on the Atlantic and Gulf coasts indicates the degree of migration varies with latitude. Birds breeding in Massachusetts are fully migratory, spending the nonbreeding period in every state of the overall range but primarily in Florida (Addison et al. 2017; Working Group et al. 2020). Northern breeding birds use "leapfrog" migration, often bypassing Atlantic coastal sites to spend the nonbreeding season on Florida's west coast. In contrast, about half the birds breeding from North Carolina to New Jersey do not migrate, and the remainder spend the nonbreeding season in areas from South Carolina to Florida (Addison et al. 2017; Working Group et al. 2020). Birds breeding from South Carolina to Florida generally are nonmigratory but may disperse from territories in the nonbreeding season (Addison et al. 2017; Working Group et al. 2020).

RADIO TRACKING A few tags were deployed through the Motus Wildlife Tracking System with local detections only.

SATELLITE TRACKING Birds tagged with satellite transmitters were used to estimate length of stay and home ranges of breeding birds near Nantucket. One bird retained the transmitter long enough to be tracked to nonbreeding grounds in the southeastern United States (Loring et al. 2017).

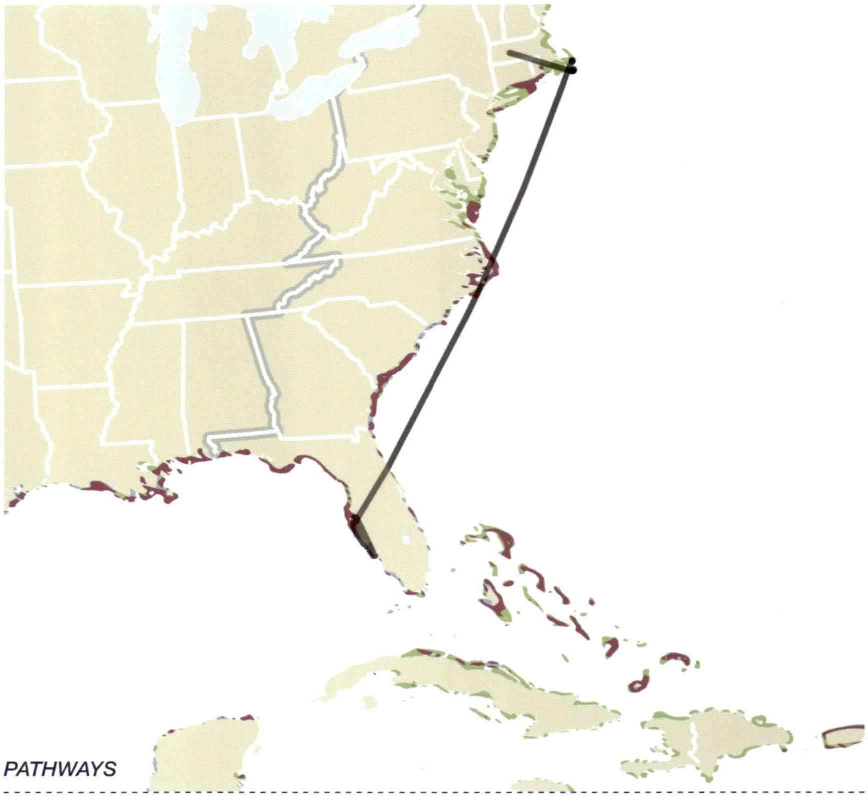

PATHWAYS

MIGRATORY PATHWAYS Migratory populations move north and south along the Atlantic coast, with the most northern breeders eventually crossing over the Florida mainland to reach the western shore. Nonmigratory birds may move short distances up and down the Atlantic coast. No information available for the Pacific coast.

DATA CONTRIBUTORS
P.H. Loring.

BLACK OYSTERCATCHER *Haematopus bachmani*

DISTRIBUTION AND ABUNDANCE

The Black Oystercatcher is distributed unevenly along coastal areas from the Aleutian Islands south to Baja California on the Pacific coast of North America (Andres and Falxa 2020). A portion of Alaska's breeding population leaves Alaska to join others on nonbreeding areas in British Columbia, while others aggregate at nonbreeding sites on Kodiak Island and south-central and Southeast Alaska (M. Johnson et al. 2010; D. Green and L. Ware, unpublished data).

ORDER Charadriiformes
FAMILY Haematopodidae
BREEDING HABITAT Coasts, Rocky Intertidal
NONBREEDING HABITAT Coasts, Rocky Intertidal
GLOBAL POPULATION SIZE 10,000
POPULATION CHANGE SINCE 1970 −9%
IUCN RED LIST CATEGORY No
R2R TIPPING POINT SPECIES No
US LISTING STATUS No
CANADA SPECIES AT RISK No
PIF CCS 13/20

MIGRATORY CONNECTIVITY

	Pacific
Pacific (n = 13)	100%

OVERALL Black Oystercatchers are comprised of both migratory and nonmigratory individuals; those that migrate appear to use the same breeding and nonbreeding territories between years, consistent with strong migratory connectivity. All birds breeding within the Pacific Flyway spend the nonbreeding season within the same flyway. Due to a single breeding and nonbreeding region, MC could not be estimated.

BANDING Banding data indicate strong fidelity to both breeding and stationary nonbreeding locations, but birds may mix with other populations in the nonbreeding season (Tessler et al. 2014).

SATELLITE TRACKING Black Oystercatchers tracked with satellite telemetry from breeding locations in Prince William Sound, Middleton Island, and Juneau, Alaska, migrated to British Columbia, while those from Kodiak Island and Vancouver Island did not migrate and appear to be year-round residents (M. Johnson et al. 2010). A more recent tracking study indicates only half of tagged birds from south-central Alaska migrated to British Columbia, with heavier birds appearing to be more likely to migrate (D. Green and L. Ware, unpublished data).

STABLE ISOTOPES Analyses are underway to explore how diet may explain why some individuals migrate to British Columbia and others remain in Alaska year-round (D. Green and L. Ware, unpublished data).

GENETICS Genomic analysis of Black Oystercatchers from five locations in Alaska and British Columbia indicated that only the population on Middleton Island in the Gulf of Alaska was genetically differentiated from other populations (Guzzetti 2008).

MIGRATORY PATHWAYS Post-breeding birds migrate short distances and aggregate to form large flocks during the nonbreeding season. A portion of the overall population migrates long distances, always along the coast; this does not include those breeding on Middleton Island, which must cross open ocean to reach mainland areas.

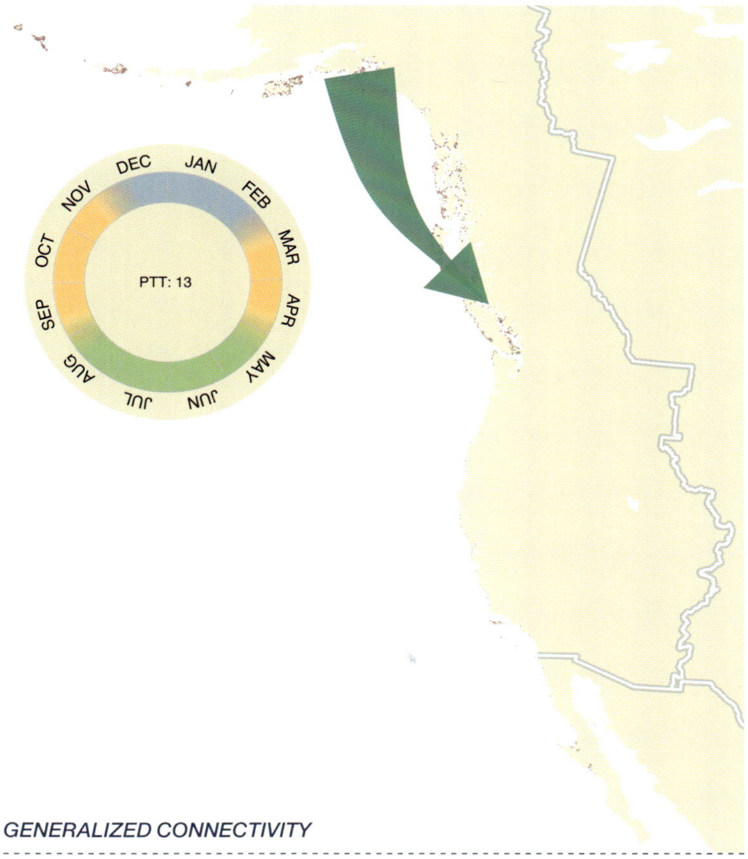

GENERALIZED CONNECTIVITY

INTERESTING NOTE

Additional tracking work is needed, especially for the southern portion of the species range.

CONNECTIONS

PATHWAYS

BLACK-BELLIED PLOVER *Pluvialis squatarola*

DISTRIBUTION AND ABUNDANCE

Black-bellied Plovers breed across a large portion of the world's Arctic region and are also known as Grey or Silver Plovers in other parts of their range. In North America, they breed in a fairly narrow latitudinal band across the Arctic and subarctic regions. Highest breeding abundance appears to be in southwest Alaska and Prince Charles Island, Nunavut, Canada (Fink et al. 2023). The tendency to breed in areas of low human density may mask population declines, compared to shorebird species that are more easily observed and counted. During the nonbreeding season, Black-bellied Plovers can be found along coastlines from southern Canada south to Brazil and Chile and inland in the Central Valley of California. High abundances occur along the coasts of Baja California, the west coast of Mexico and Central America, the Gulf of Mexico, the Caribbean, the southeastern US, northern South America, Brazil, and Peru (Fink et al. 2023).

ORDER Charadriiformes
FAMILY Charadriidae
BREEDING HABITAT Wetlands, Tundra
NONBREEDING HABITAT Coasts, Beach and Estuary
GLOBAL POPULATION SIZE 840,000
POPULATION CHANGE SINCE 1970 −48%

IUCN RED LIST CATEGORY Least Concern
R2R TIPPING POINT SPECIES Orange-Alert
US LISTING STATUS No
CANADA SPECIES AT RISK No
PIF CCS 13/20

INTERESTING NOTE

Migrations range from relatively short distances along the Pacific coast of North America to very long migrations between the North American Arctic and Patagonia. Specimen collections across the nonbreeding range suggest that males make shorter-distance migrations than females (Poole et al. 2020b). During migration, Black-bellied Plovers typically migrate in either small flocks or join in small numbers with mixed flocks containing other species. Individuals appear to be territorial throughout the year (Poole et al. 2020b).

	Andean West	Brazil	Caribbean	Central America	Mexico	Midcontinent	Northern SA	Pacific
Midcontinent (n = 15)	–	20%	53%	7%	–	7%	13%	–
Pacific (n = 19)	5%	–	–	–	53%	11%	–	32%

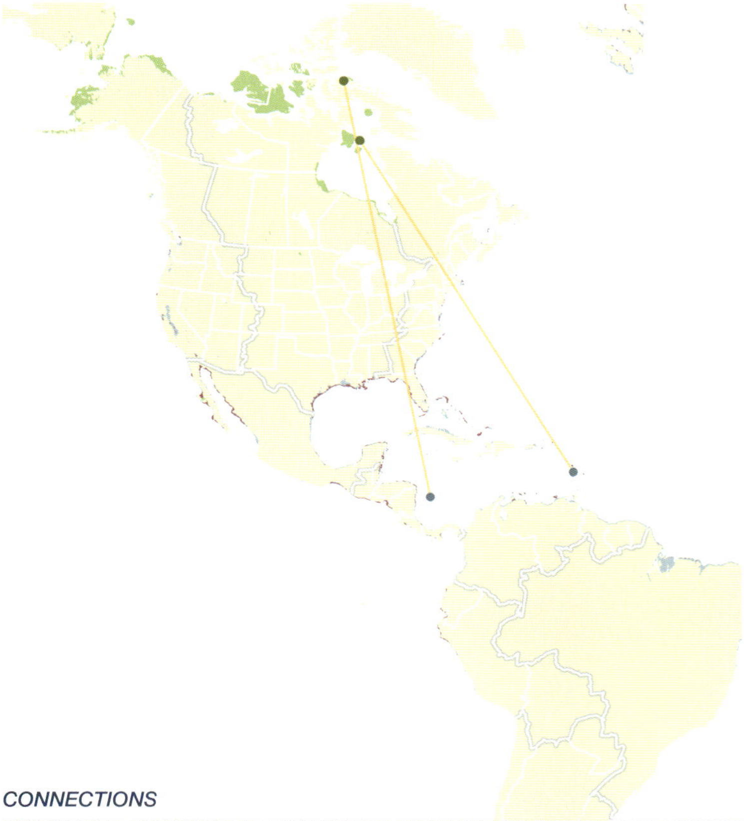

CONNECTIONS

PATHWAYS

MIGRATORY CONNECTIVITY

OVERALL Black-bellied Plovers have strong migratory connectivity and clear migratory divides across their full annual cycle. Birds breeding in the Midcontinent Flyway primarily spend the nonbreeding season either in the Caribbean (53%) or Brazil (20%), with a small proportion also in northern South America (13%), the Midcontinent Flyway (7%), and Central America (7%). Pacific Flyway breeders primarily spend the nonbreeding season within Mexico (53%) and the Pacific Flyway (32%). Separation of breeding populations throughout nonbreeding areas results in strong migratory connectivity (MC = 0.49).

BANDING There are few resightings of banded birds away from their capture sites, and only two that connect breeding sites to stationary nonbreeding areas (from sites in eastern Canada to Caribbean islands).

RADIO TRACKING Detections from the Motus Wildlife Tracking System during migration connect Nunavut, Ontario, Quebec, and Nova Scotia, Canada, breeding populations to east coast migration areas and into South America. There are no breeding to nonbreeding detections.

SATELLITE TRACKING Large-scale deployment of Argos satellite tags since 2014 on birds at five breeding sites (two in Alaska and three in Nunavut, Canada), one migration site (Quebec, Canada), and one nonbreeding site (Texas) in North America yielded 33 movement paths providing migratory connectivity data (A.-L. Harrison, unpublished data; J. Rausch, unpublished data). These data provide the first connectivity information for birds breeding west of Hudson Bay and confirmed a migratory divide across the continent.

MIGRATORY PATHWAYS Tracking data confirmed broad elliptical or loop migrations on both coasts, often over ocean on southward migration, and coastal or inland on northward migration; thus clockwise on the east coast, and counterclockwise on the west coast and a tighter loop migration within the midcontinent. Most birds migrating through the midcontinent stopped within the Prairie Pothole Region before continuing onward to either the western coast of the Gulf of Mexico or the eastern coast of the Gulf of California; whereas birds migrating through the Atlantic and Pacific flyways used a larger variety of stopover and nonbreeding sites. A study that deployed 24 GPS tags at nonbreeding sites in Texas and Louisiana (Clements et al. 2022) match these patterns.

DATA CONTRIBUTORS

P. Bruner; M. Fowler; A.-L. Harrison; S. Koch; P. Loring; A. MacDonald; P.P. Marra; K. O'Brien; N. Pau; J. Rausch; D. Ruthrauff; A.L. Scarpignato; P. Smith; T.L. Tibbitts; P. Woodard.

AMERICAN GOLDEN-PLOVER *Pluvialis dominica*

DISTRIBUTION AND ABUNDANCE

This species breeds throughout much of the Arctic and subarctic regions of Alaska and northwestern Canada and migrates long distances to its nonbreeding grounds in southern South America, primarily in the Pampas (low grasslands) of southern Brazil, Uruguay, and northern Argentina (O.W. Johnson et al. 2021a). American and Pacific Golden-Plovers overlap in distribution in western Alaska but take different migration routes (the latter to the Pacific Ocean).

ORDER Charadriiformes
FAMILY Charadriidae
BREEDING HABITAT Tundra, Arctic
NONBREEDING HABITAT Grasslands, Pampas and Campos
GLOBAL POPULATION SIZE 500,000
POPULATION CHANGE SINCE 1970 −53%
IUCN RED LIST CATEGORY Least Concern
R2R TIPPING POINT SPECIES Orange-Alert
US LISTING STATUS No
CANADA SPECIES AT RISK No
PIF CCS 15/20

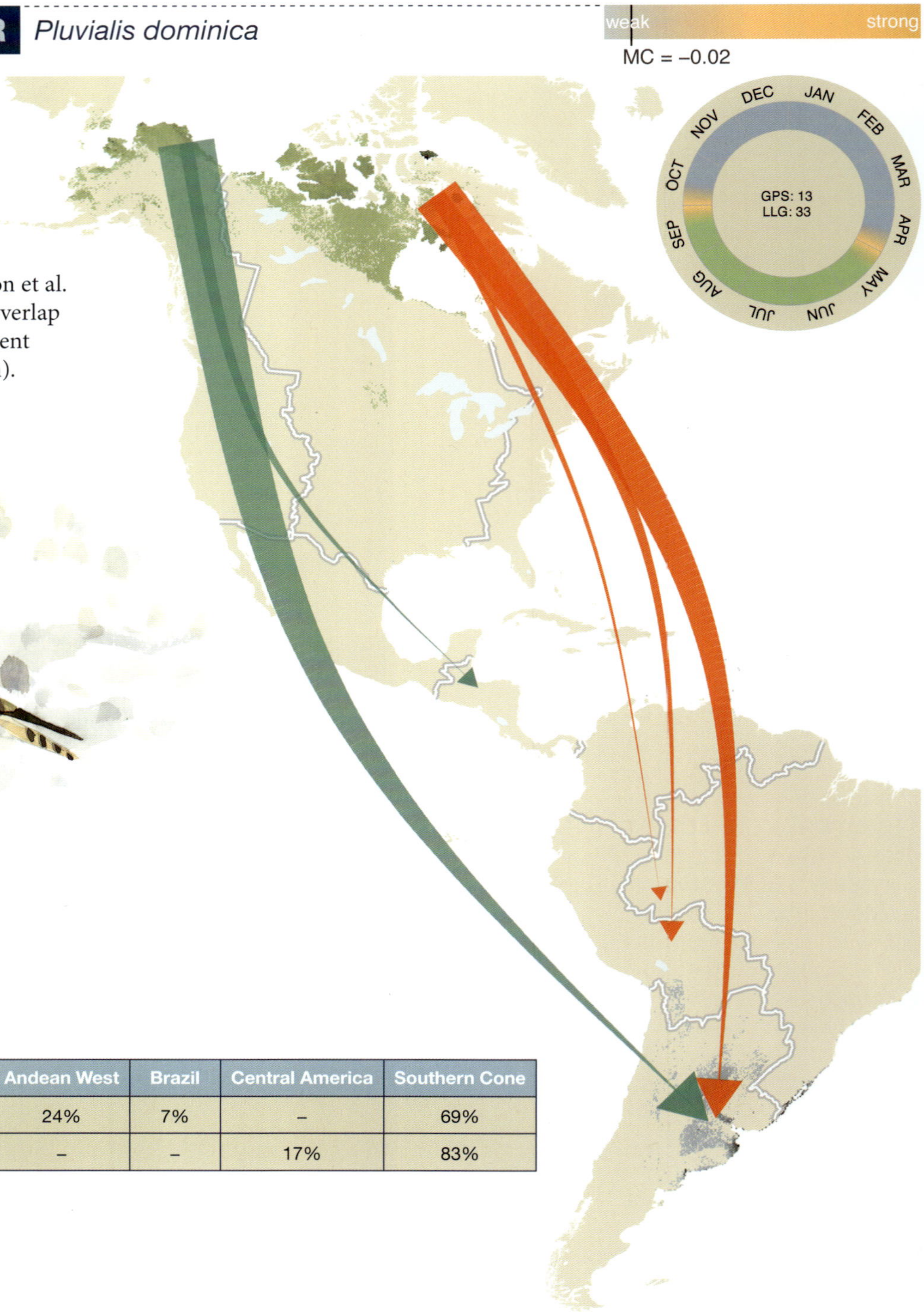

	Andean West	Brazil	Central America	Southern Cone
Midcontinent (n = 29)	24%	7%	–	69%
Pacific (n = 6)	–	–	17%	83%

GENERALIZED CONNECTIVITY

MIGRATORY CONNECTIVITY

OVERALL Available information indicates American Golden-Plovers have weak migratory connectivity (−0.02), with birds breeding throughout the Arctic overlapping in time and space during the south and north migrations and on nonbreeding grounds. Birds breeding in the Midcontinent Flyway primarily spend the nonbreeding season in the Southern Cone (69%) or Andean West (24%) regions of South America. Pacific Flyway breeders also migrate to the Southern Cone region (83%), as well as to Central America (17%).

BANDING There were few resightings of banded birds away from their capture sites, and none from breeding to stationary nonbreeding areas have been documented. However, banding of birds reveals that males are much more site faithful to breeding grounds than females, suggesting females move to new areas of the Arctic each breeding season (O.W. Johnson et al. 2021a).

RADIO TRACKING A few tags were deployed through the Motus Wildlife Tracking System with local detections only.

GEOLOCATOR TRACKING American Golden-Plovers equipped with geolocators and recovered (n = 25) at seven sites across the Alaskan and Canadian breeding grounds used common stationary nonbreeding areas in the Pampas region of southern South America (J-F. Lamarre, unpublished data).

SATELLITE TRACKING Large-scale deployment (n = 75) of GPS tags on breeding birds at four sites in Alaska and five sites in Canada confirmed all birds, regardless of breeding location in the Arctic, spent the nonbreeding season on coastal portions of southeast Brazil, Uruguay, and Argentina (R. Lanctot, unpublished data).

MIGRATORY PATHWAYS Geolocator and satellite transmitter data revealed an elliptical migration with post-breeding migrants moving east across the Arctic and then south across the Atlantic Ocean to

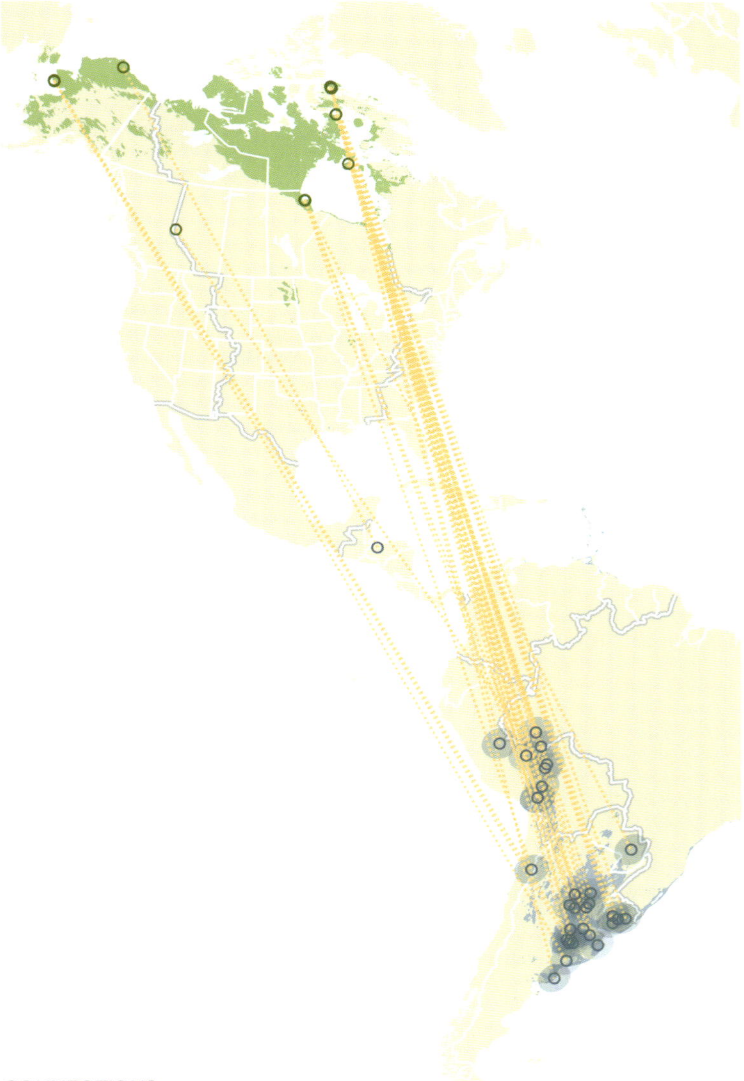

CONNECTIONS

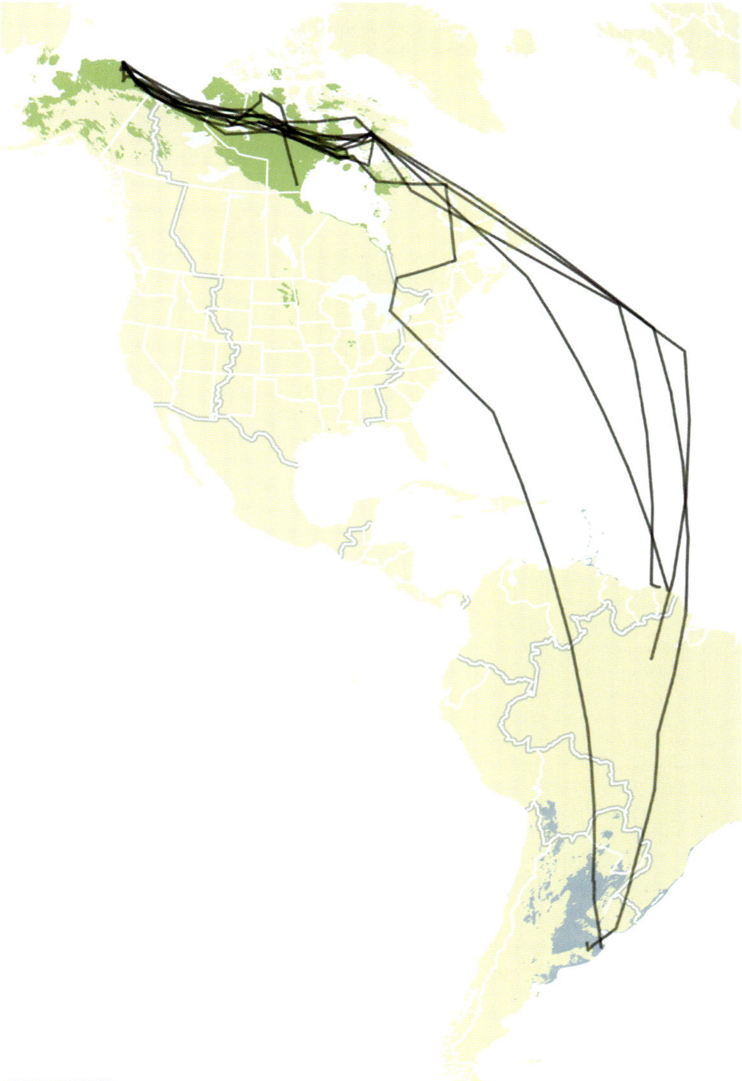

PATHWAYS

northeastern Brazil before continuing to the Pampas in southern South America. Post-breeding migration consisted of northward (for birds captured in subarctic regions) and then eastward movement across the Arctic before going south across the Atlantic Ocean to reach Guyana, Suriname, French Guiana, and northeastern Brazil. Some birds diverted inland to the middle of North America, apparently in response to tropical storms, and then migrated south across the Gulf of Mexico. Some birds also stopped briefly along rivers in the Amazon basin. Birds take a different route to return to the Arctic, traveling either across the Amazon or the Pacific Ocean (near Central America) to reach a major stopover in the central portion of the United States, and then migrating north through the midcontinent to reach their breeding grounds again. Birds used similar routes during southward migration, but the species migration north was more varied; all stopped in the central United States (South Dakota, Nebraska, and Kansas) before continuing to their Arctic breeding territories (Lamarre et al. 2021).

STABLE ISOTOPES Juvenile birds shot in Barbados were determined by stable isotope analysis of deuterium to be from either the eastern Canada or Alaskan breeding populations (Reed et al. 2018). Stable isotope analysis of breeding and nonbreeding grown feathers showed differentiation but did a poor job assigning individuals to areas of geographic origin (Rocque et al. 2006).

DATA CONTRIBUTORS

S. Bauer; J. Bêty; S. Brown; K.H. Elliott; T. Emmenegger; W. English; S.A. Flemming; G. Gauthier; M.-A. Giroux; A.-L. Harrison; J. Liebezeit; O.W. Johnson; L. Koloski; J.-F. Lamarre; R.B. Lanctot; N. Lecomte; P.P. Marra; R. McGuire; L. McKinnon; E. Nol; O.P. Love; S.T. Saalfeld; F. Sanders; S. Schulte; P.A. Smith; E. Reed; M. Russell.

INTERESTING NOTE

Hunted heavily in the eighteenth and nineteenth centuries, the species is now largely protected but still hunted, with unknown impact in the Caribbean (Reed et al. 2018) and the Amazonia region of Brazil during migration (R. Lanctot and J.F. Lamarre, unpublished data).

PACIFIC GOLDEN-PLOVER *Pluvialis fulva*

DISTRIBUTION AND ABUNDANCE
This species breeds in tundra regions of Chukotka, Russia, and western Alaska (Nome to the Alaska Peninsula) and spends the nonbreeding season from coastal California and islands and atolls of the Pacific Ocean to Australia and areas west (O.W. Johnson et al. 2021b).

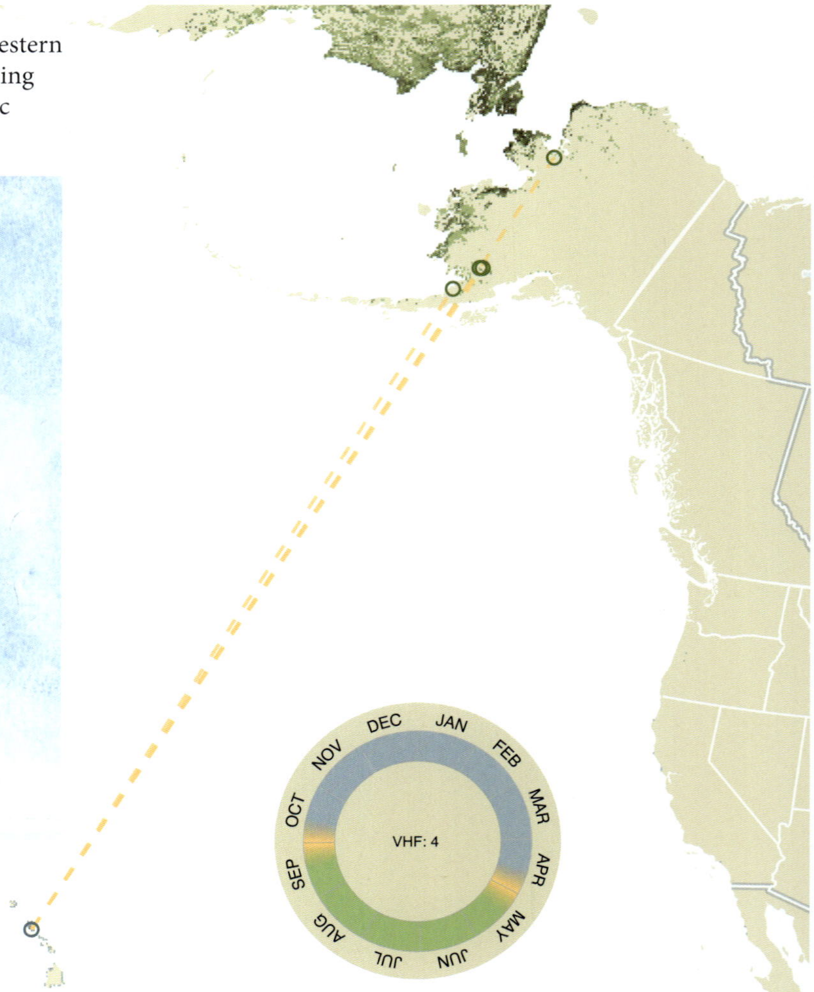

CONNECTIONS

ORDER Charadriiformes
FAMILY Charadriidae
BREEDING HABITAT Wetlands, Tundra
NONBREEDING HABITAT Coasts, Beach and Estuary
GLOBAL POPULATION SIZE 220,000
POPULATION CHANGE SINCE 1970 −29%
IUCN RED LIST CATEGORY Least Concern
R2R TIPPING POINT SPECIES No
US LISTING STATUS No
CANADA SPECIES AT RISK No
PIF CCS 12/20

INTERESTING NOTE

Morphometrics, molt strategy, and migratory patterns suggest that Alaskan breeding birds are a distinct population from the Siberian breeders (Jukema et al. 2015).

MIGRATORY CONNECTIVITY

OVERALL Limited data may suggest strong migratory connectivity between southwestern Alaska and Hawaii, and between northwestern Alaska and areas farther south of Hawaii on islands of the Pacific Ocean.

BANDING Birds banded on nonbreeding areas on Oahu, Hawaii, were later resighted on breeding grounds in south-central and western Alaska (O.W. Johnson and P. Bruner, unpublished data). Four Pacific Golden-Plovers banded during post-breeding migration on St. George Island (Pribilof Islands), Alaska, were later observed at Kure Atoll (northwestern Hawaiian Islands), Hokkaido island, the Solomon Islands, and New South Wales, Australia (O.W. Johnson et al. 2021b).

RADIO TRACKING Birds equipped with radio tags on nonbreeding sites on Oahu, Hawaii, were detected on breeding areas in Alaska (O.W. Johnson et al. 1997; O.W. Johnson, unpublished data). A few tags were deployed through the Motus Wildlife Tracking System with local detections only.

GEOLOCATOR TRACKING Birds tracked with light-level geolocators from nonbreeding sites in Oahu, Hawaii, traveled directly to breeding grounds in the Alaska Peninsula and the Yukon-Kuskokwim Delta; whereas breeding birds tagged in Nome, Alaska, were tracked to a variety of Pacific islands and Australia (O.W. Johnson et al. 2011).

SATELLITE TRACKING Birds tracked with GPS-satellite tags from south Pacific nonbreeding grounds were observed to breed farther north in Alaska than those on nonbreeding areas on the Hawaiian Islands (O.W. Johnson et al. 2012, 2020).

STABLE ISOTOPES Analysis of breeding and nonbreeding grown feathers showed differentiation but did a poor job assigning individuals to geographic areas of origin (Rocque et al. 2006).

MIGRATORY PATHWAYS Southernmost breeding birds in western Alaska migrate in direct and nonstop flights back and forth to the Hawaiian Islands, while more northern breeding birds in western Alaska migrate farther south to islands in the Pacific Ocean and return via Japan to Alaska. A plover captured nesting on the Seward Peninsula, Alaska, in 2008 was resighted in Japan; this was the first indication of an Alaska link with the East Asian–Australasian Flyway (O.W. Johnson, unpublished data) This three-stage circular route, from nonbreeding areas in the Pacific Ocean, first flying to Japan and then migrating to Alaska, was also observed in birds tagged with geolocators (O.W. Johnson et al. 2012) and GPS devices (O.W. Johnson et al. 2020).

Charadrius vociferus **KILLDEER**

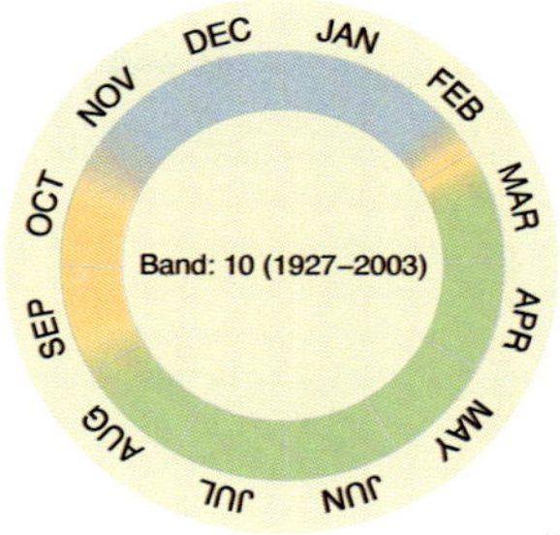

DISTRIBUTION AND ABUNDANCE

The familiar and noisy Killdeer is the most widespread plover in North America; the nominate subspecies, *C. v. vociferus*, breeds from coast to coast and from southern portions of the Northwest Territories to central Mexico. It reaches highest abundance from Louisiana north through the Midwest and east to the Great Lakes and west to Calgary. Nonbreeding areas include the southern US, central California south through Mexico and Central America and the Caribbean, especially Cuba and the western Bahamas (Jackson and Jackson 2020). Two resident subspecies also exist: *C. v. ternominatus*, which resides in the West Indies, and *C. v. peruvianus*, which resides in Ecuador.

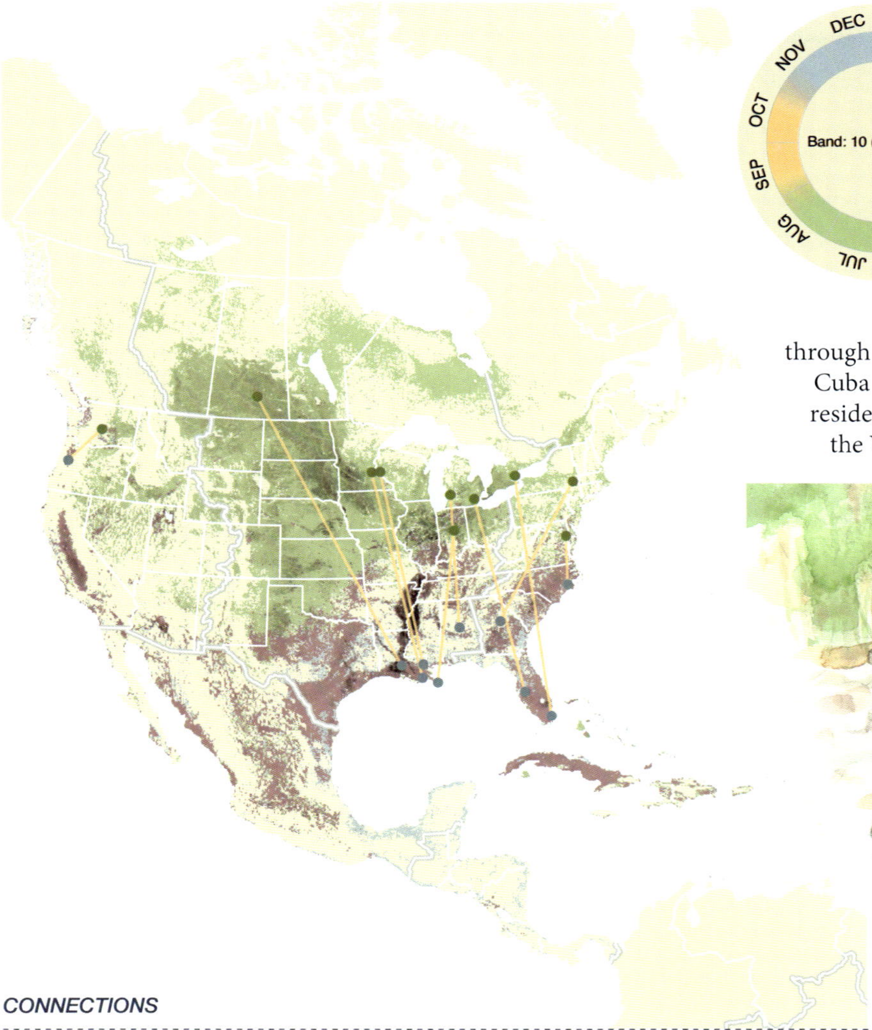

CONNECTIONS

ORDER Charadriiformes
FAMILY Charadriidae
BREEDING HABITAT Open Country, Habitat Mosaic
NONBREEDING HABITAT Open Country, Habitat Mosaic
GLOBAL POPULATION SIZE 2,300,000
POPULATION CHANGE SINCE 1970 −37%
IUCN RED LIST CATEGORY Least Concern
R2R TIPPING POINT SPECIES No
US LISTING STATUS No
CANADA SPECIES AT RISK No
PIF CCS 10/20

INTERESTING NOTE

Site fidelity of breeding birds was observed in Minnesota and Mississippi, but there is no information on nonbreeding birds. A few young from an area are seen to return, but natal philopatry does not appear strong (Jackson and Jackson 2020).

MIGRATORY CONNECTIVITY

OVERALL Despite how common this species is across North America, migratory behaviors are complicated and poorly understood. Connectivity of the *C. v. vociferus* subspecies is poorly known, but banding data suggest possible longitudinal separation in north-south movements. Due to limited sample size, MC could not be calculated.

BANDING A handful of resightings, mainly in the eastern and central part of the range, connecting breeding and nonbreeding sites of birds suggest possible longitudinal separation.

RADIO TRACKING A few tags were deployed through the Motus Wildlife Tracking System with local detections only.

OTHER APPROACHES No genetic, stable isotope, geolocator, or tracking studies of migratory connectivity were found for this species.

MIGRATORY PATHWAYS There are no obvious pathways followed by the migratory subspecies *C. v. vociferus*. Skagen et al. (1999) describe the species migration in the midcontinent of the US as short and widespread. The other two subspecies are nonmigratory.

COMMON RINGED PLOVER *Charadrius hiaticula*

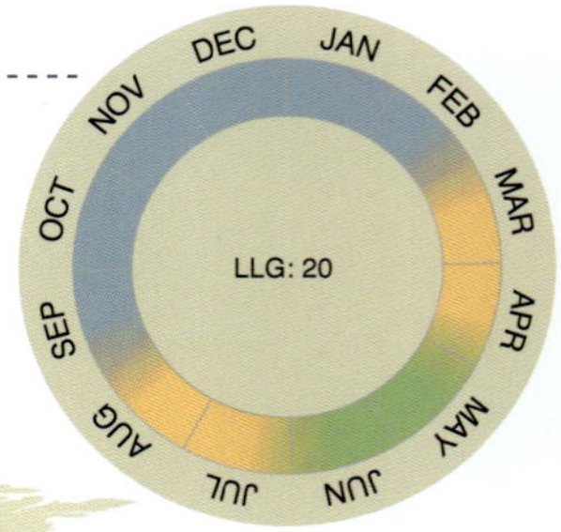

	Africa
Midcontinent (n = 20)	100%

DISTRIBUTION AND ABUNDANCE

The *C. h. hiaticula* subspecies (one of three found across the Arctic) breeds in northeastern Arctic Canada, Greenland, Iceland, Spitsbergen (Norway), and southern Scandinavia to France (Wiersma et al. 2020). Limited information suggests all birds spend the nonbreeding season in western Africa (Lislevand et al. 2016; Léandri-Breton et al. 2019).

GENERALIZED CONNECTIVITY

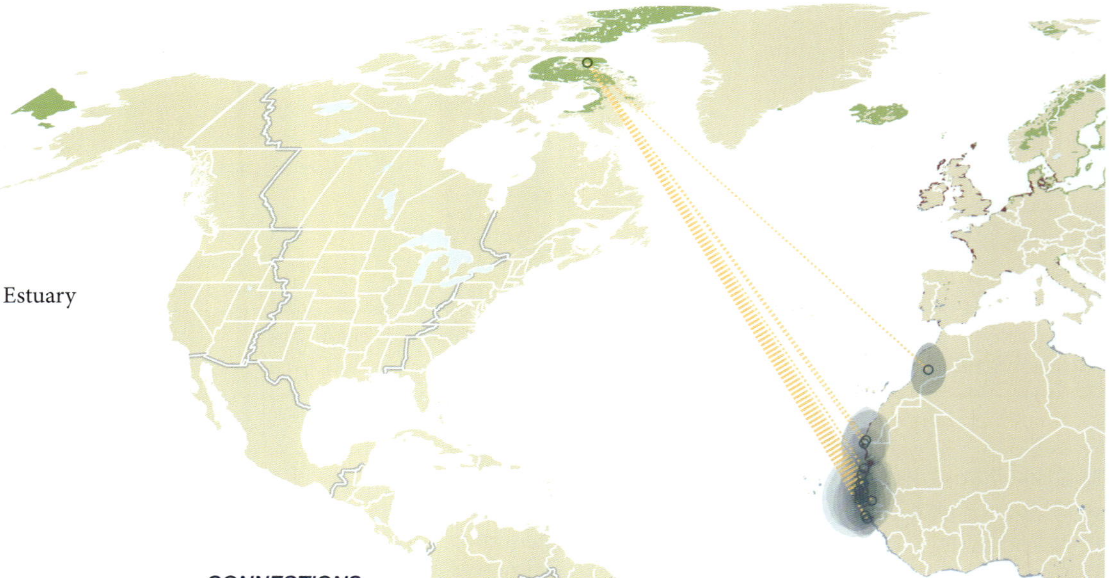

CONNECTIONS

ORDER Charadriiformes
FAMILY Charadriidae
BREEDING HABITAT Wetlands, Tundra
NONBREEDING HABITAT Coasts, Beach and Estuary
GLOBAL POPULATION SIZE 910,000
POPULATION CHANGE SINCE 1970 NA
IUCN RED LIST CATEGORY Least Concern
R2R TIPPING POINT SPECIES No
US LISTING STATUS No
CANADA SPECIES AT RISK No
PIF CCS 10/20

MIGRATORY CONNECTIVITY

OVERALL Geolocator data show breeding birds from the Midcontinent Flyway all migrate to nonbreeding areas in Africa, suggesting low migratory connectivity. MC not calculated due to limited sample size.

GEOLOCATOR TRACKING Birds were tracked from breeding sites at Bylot Island, northern Nunavut, to nonbreeding areas in western Africa (Léandri-Breton et al. 2019).

OTHER No banding, satellite tracking, genetic, or stable isotope studies of migratory connectivity were found for this species.

MIGRATORY PATHWAYS Geolocator tracks indicate northern Canada breeding birds use different routes during pre- and post-breeding to travel to and from nonbreeding areas in western Africa. During pre-breeding migration, birds traveled north along the coast of Europe and Britain before detouring to Iceland and then crossing the Greenland ice cap to reach breeding grounds. During post-breeding migration, plovers first stopped at southern portions of Greenland before crossing the ocean in one direct flight to reach western Europe, and then moving on to western Africa (Léandri-Breton et al. 2019).

DATA CONTRIBUTORS
J. Bêty; J.-F. Lamarre; D.-J. Léandri-Breton.

SEMIPALMATED PLOVER

Charadrius semipalmatus

DISTRIBUTION AND ABUNDANCE

The Semipalmated Plover breeds across the North American Arctic, subarctic, and to a lesser extent, the Atlantic region of Canada and the temperate northwest. During the nonbreeding season, they are distributed along Pacific and Atlantic coastal regions extending along marine coastal areas from southern North America along both coasts of South America almost to Uruguay and Argentina.

ORDER Charadriiformes
FAMILY Charadriidae
BREEDING HABITAT Coasts, Beach and Estuary
NONBREEDING HABITAT Coasts, Beach and Estuary
GLOBAL POPULATION SIZE 200,000

POPULATION CHANGE SINCE 1970 −17%
IUCN RED LIST CATEGORY Least Concern
R2R TIPPING POINT SPECIES No
US LISTING STATUS No
CANADA SPECIES AT RISK No
PIF CCS 13/20

INTERESTING NOTE

From 2005 to 2017, Semipalmated Plovers (n = 15) have been recorded as Nearctic vagrants on islands of the Azores (Alfrey et al. 2018).

MIGRATORY CONNECTIVITY

OVERALL There is limited information about migratory connectivity. A weak to moderate level of migratory connectivity has been described to exist between breeding and nonbreeding areas, based on relatively disjunct Atlantic and Pacific coast migration routes (although mixing occurs), analysis of trace elements (Storm-Suke 2012; Nol and Blanken 2020), and band recoveries (Nol and Blanken 2020).

BANDING There was one resighting of a Semipalmated Plover banded on breeding grounds in Alaska during the nonbreeding season in coastal Southern California.

RADIO TRACKING Nearly a thousand tags were deployed through the Motus Wildlife Tracking System. At this time, there are no individual breeding to nonbreeding connections. However, numerous detections, especially from Nunavut and Hudson Bay to the northeastern US, help inform migration and stopover ecology (G. Brown et al., unpublished data; E. Nol et al., unpublished data; P. Smith et al., unpublished data).

OTHER APPROACHES The majority of birds that originate in the MacKenzie Delta in Canada have trace element signatures similar to those in South Carolina. Birds from Egg Island, Alaska, have trace element signatures that match a wider distribution, including South Carolina and French Guiana. Birds from Akimiski Island and Churchill, Manitoba, are similar to those found in French Guiana. Finally, birds from the eastern breeding areas, such as Southampton Island, Iqaluit (Nunavut), and Newfoundland, match all three nonbreeding areas that were tested (Storm-Suke 2012). No satellite transmitter, stable isotope, or genetic studies of migratory connectivity were found for this species.

MIGRATORY PATHWAYS Migration occurs over a broad front throughout the Americas, but concentrations are found primarily at marine coastal sites (Nol and Blanken 2020). In North America, birds migrate along the Atlantic and Pacific coasts but are also widely distributed across inland sites throughout (Paulson 1993; Stevenson and Anderson 1994). Records suggest that many birds travel from northeast North America to northeast South America via the western Atlantic during post-breeding migration, while returning by a more westerly route (Veit and Petersen 1993; Raffaele et al. 1998; Amos 1991). Considerable migration takes place within the interior of North America during the northward and southward migrations (Skagen et al. 1999), with the St. Lawrence Estuary serving as an important staging area for juveniles (Turcotte et al. 2013). This

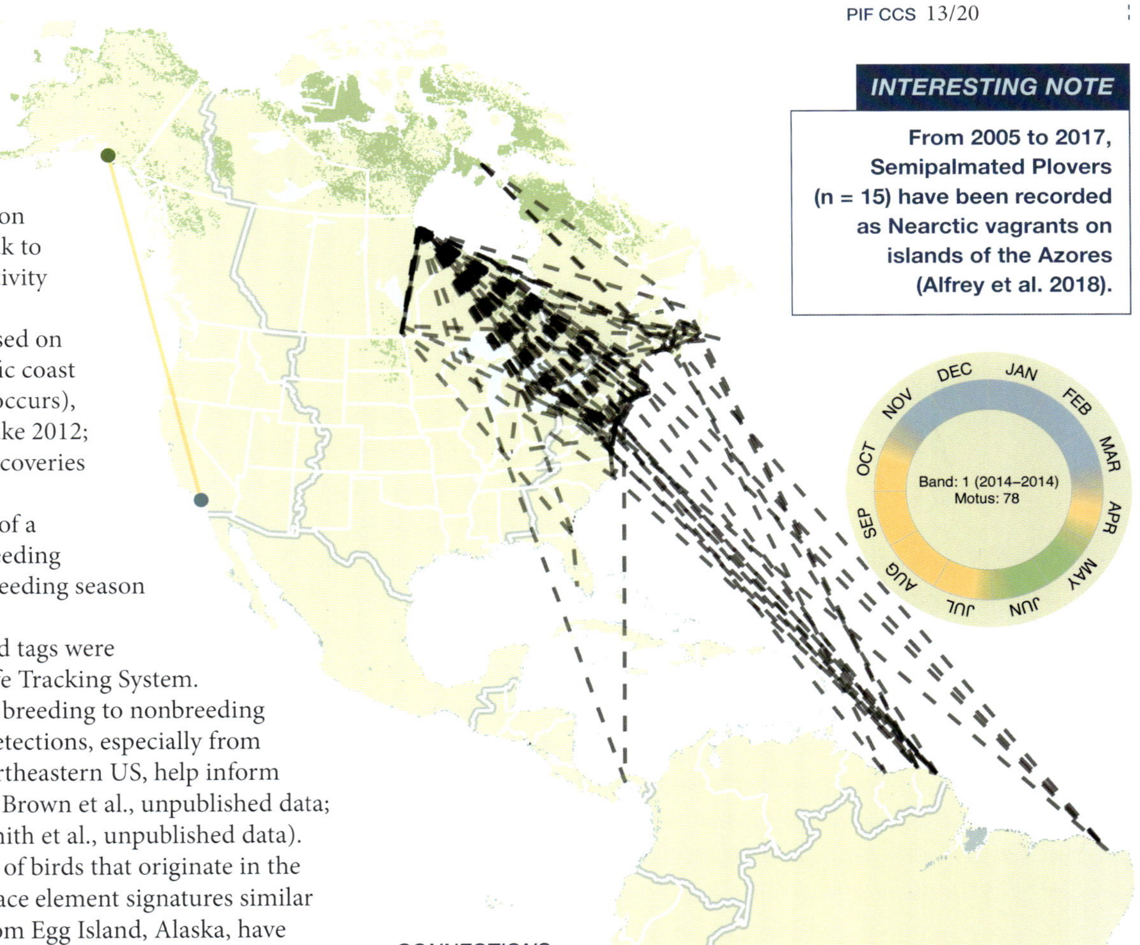

CONNECTIONS AND PATHWAYS

species uses an energy-minimizing strategy; its short-distance migration was matched with shorter stopover lengths, departure with lower relative fuel loads, and slower ground and air speeds (A.M. Anderson et al. 2019).

DATA CONTRIBUTORS

K. Bennett; G. Brown; A. MacDonald; A. McKellar; L. McKinnon; E. Nol; P. Smith.

PIPING PLOVER *Charadrius melodus*

DISTRIBUTION AND ABUNDANCE

This species has two distinct breeding distributions. Interior populations nest along prairie rivers and in alkali wetlands in the northern Great Plains, while Atlantic populations breed along beaches from North Carolina to New Brunswick. All populations spend their stationary nonbreeding season along the Atlantic coast south of North Carolina and to the Bahamas, and across the Gulf Coast, with the majority of birds in Texas (Elliot-Smith and Haig 2020).

ORDER Charadriiformes
FAMILY Charadriidae
BREEDING HABITAT Coasts, Beach and Estuary
NONBREEDING HABITAT Coasts, Beach and Estuary
GLOBAL POPULATION SIZE 8,400
POPULATION CHANGE SINCE 1970 NA
IUCN RED LIST CATEGORY Near Threatened
R2R TIPPING POINT SPECIES Orange-Alert
US LISTING STATUS Endangered: Great Lakes watershed population;
Threatened: all other populations
CANADA SPECIES AT RISK Endangered:
C. m. circumcinctus subspecies
and *C. m. melodus* subspecies
PIF CCS 16/20

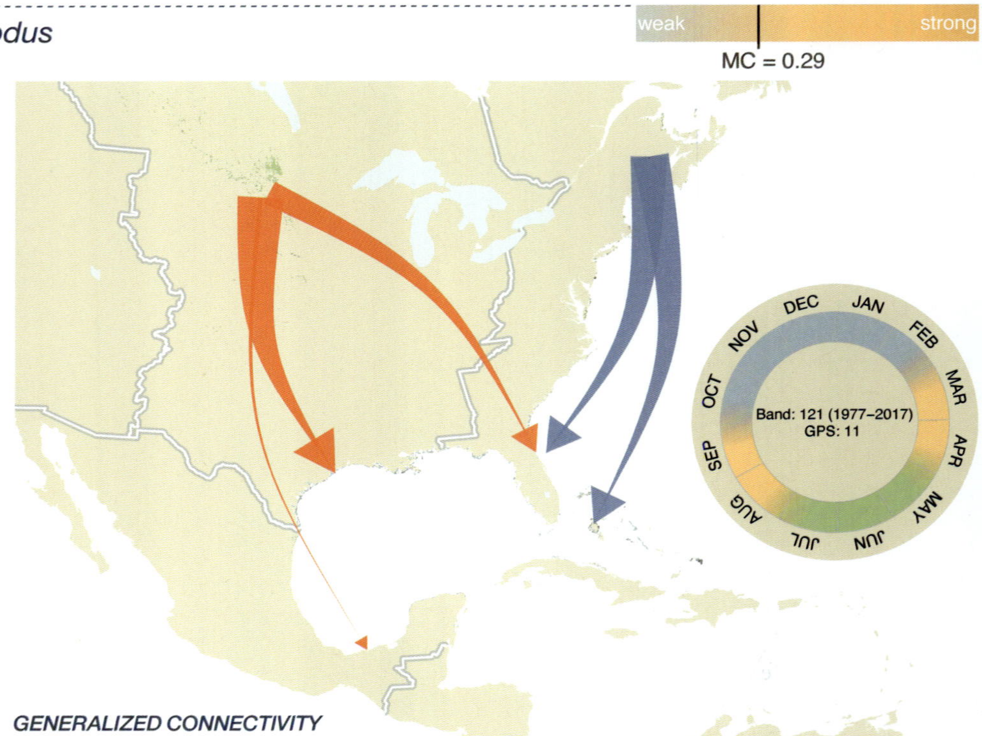

GENERALIZED CONNECTIVITY

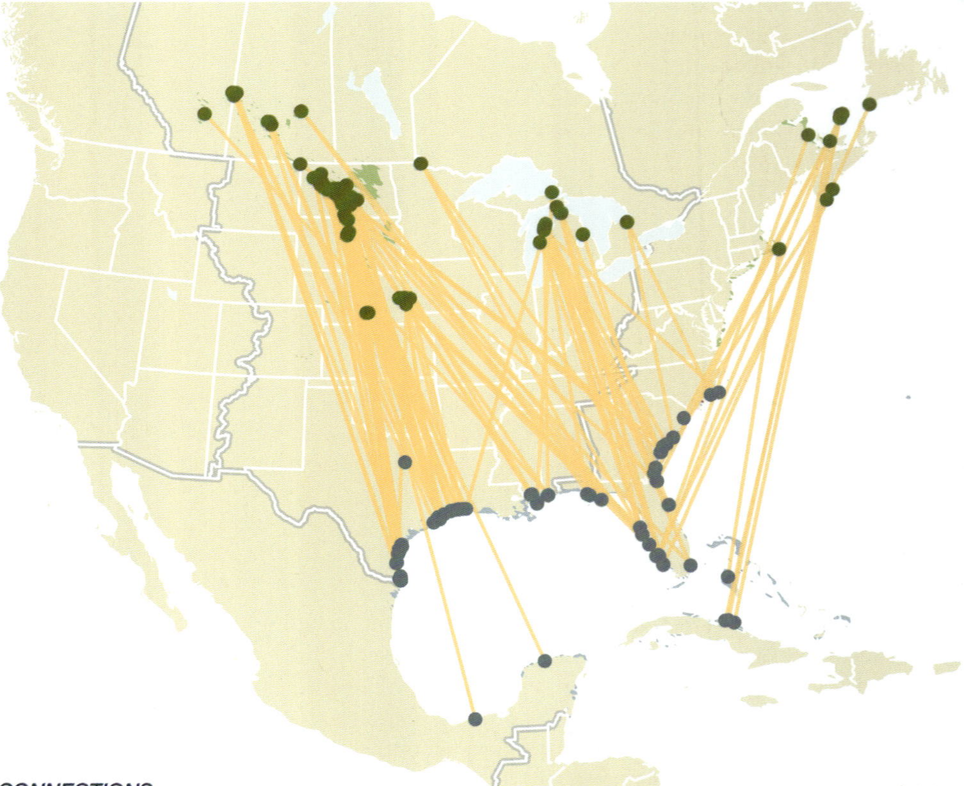

CONNECTIONS

	Atlantic	Caribbean	Mexico	Midcontinent
Midcontinent (n = 12)	50%	50%	–	–
Pacific (n = 110)	32%	–	2%	66%

MIGRATORY CONNECTIVITY

OVERALL Banding data, satellite tracking, and genetic analyses suggest moderate migratory connectivity (MC = 0.29), with little mixing of breeding populations, genetic differentiation into two subspecies, and high breeding and nonbreeding season site fidelity among adults. Atlantic Flyway breeding populations spend the nonbreeding season within the Atlantic Flyway (50%) or the Caribbean (50%). Midcontinent Flyway breeding populations spend the nonbreeding season primarily within the Midcontinent Flyway (66%) with other individuals migrating to the Atlantic Flyway (32%).
BANDING Band resightings indicate little overlap of populations during the breeding season and relatively high adult site fidelity (approximately 50%), but during the nonbreeding season these populations have large apparent overlap in distribution (Haig and Oring 1988; Gratto-Trevor et al. 2012a). However, despite this considerable overlap, distinct patterns in nonbreeding distributions have been observed, along with high nonbreeding site fidelity (Gratto-Trevor et al. 2012a), suggestive of at least moderate migratory connectivity.
RADIO TRACKING Several projects tracked Piping Plover with the Motus Wildlife Tracking System across the range. There are no individual breeding to nonbreeding locations, but there appear to be many detections during migration (Motus project, Piping Plovers—VA, unpublished data; Motus project, NJ Chick Survival, unpublished data; Motus project, Piping Plover Migratory Behavior, unpublished data; Motus project, New England Shorebirds, unpublished data; Motus project, Fire Island PIPL Fledgling Survival and Migration, unpublished data; C.L. Gratto-Trevor, unpublished data).
SATELLITE TRACKING Three birds tracked from New Jersey with GPS tags migrated to different islands within the Bahamas—Andros, New Providence, and Great Abaco (Stantial and Cohen 2020).
GENETICS Recent genetic analyses support the genetic subclassification of Piping Plover into two subspecies—the Atlantic (*C. m. melodus*) and the Interior (*C. m. circumcinctus*; M.P. Miller et al. 2010).

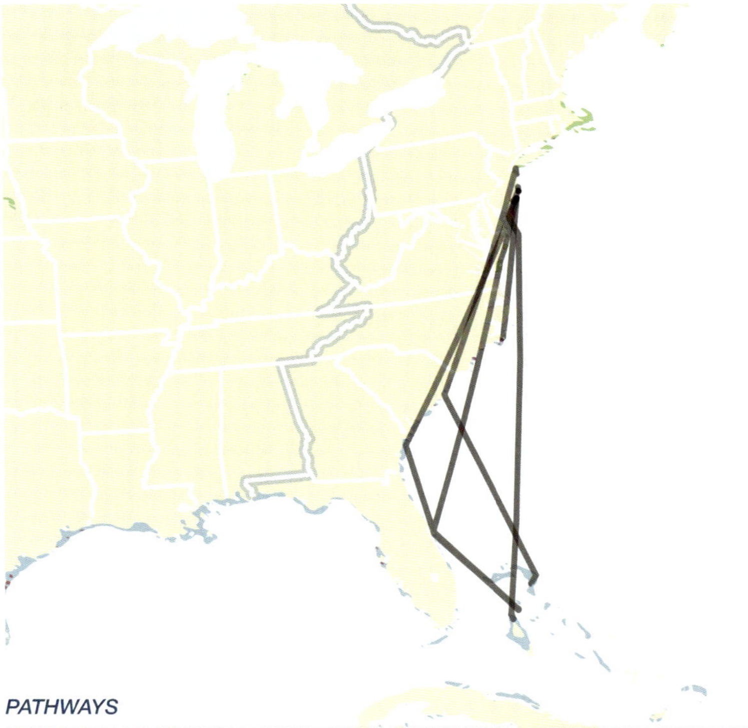

PATHWAYS

MIGRATORY PATHWAYS Atlantic breeding birds migrate southward offshore along the Atlantic coast, with at least some stopping to refuel at sites along the way (Loring et al. 2020; Stantial and Cohen 2020), while interior breeders likely travel nonstop to the Gulf of Mexico or Atlantic coast. Northward migration is likely the reverse course.

DATA CONTRIBUTORS
J.B. Cohen; M. Stantial.

Anarhynchus montanus MOUNTAIN PLOVER

DISTRIBUTION AND ABUNDANCE

The Mountain Plover breeds in the high plains of western North America, from southeastern Alberta and Saskatchewan south to northeastern New Mexico and some of the Texas panhandle, with small breeding populations in the Davis Mountains of West Texas and in La Noria de la Soledad, Nuevo Leon, Mexico (Fink et al. 2023). During the nonbreeding season, Mountain Plovers can be found from Southern California throughout northern and central Mexico, with scattered populations in Arizona and Texas (Knopf and Wunder 2020).

ORDER Charadriiformes
FAMILY Charadriidae
BREEDING HABITAT Grasslands, Temperate
NONBREEDING HABITAT Grasslands, Chihuahuan
GLOBAL POPULATION SIZE 20,000
POPULATION CHANGE SINCE 1970 −75%
IUCN RED LIST CATEGORY Near Threatened
R2R TIPPING POINT SPECIES Red-Alert
US LISTING STATUS No
CANADA SPECIES AT RISK Endangered
PIF CCS 17/20

MIGRATORY CONNECTIVITY

	Mexico	Midcontinent	Pacific
Midcontinent (n = 22)	32%	36%	32%

OVERALL Limited data are available to evaluate migratory connectivity in the Mountain Plover. However, Midcontinent Flyway breeding birds migrate to varied nonbreeding areas, including sites within the Midcontinent Flyway (36%), Pacific Flyway (32%), and Mexico (32%) suggesting weak connectivity.

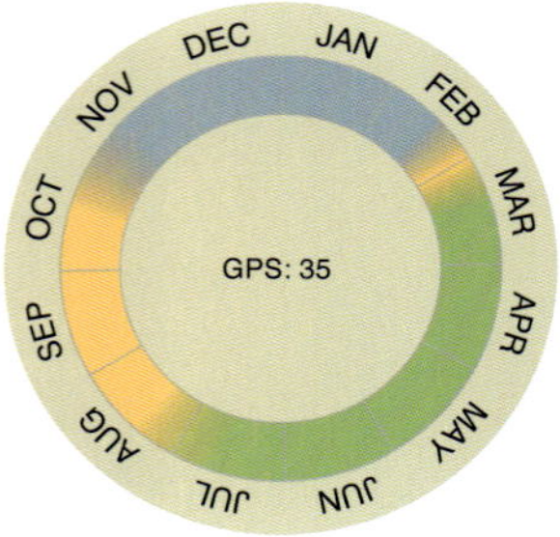

BANDING Adult and hatch-year birds at various breeding locations in Montana, Wyoming, Colorado, and New Mexico were banded with color combinations. Color-marked birds from every breeding location were observed and/or recaptured during the nonbreeding season in Imperial Valley, California, sometimes in the same flocks. Birds were also color-marked on the nonbreeding grounds in Imperial Valley and observed and/or recaptured at nine different breeding locations across Montana, Wyoming, Colorado, and New Mexico (Wunder 2007).

GEOLOCATOR TRACKING Geolocators were deployed at breeding areas in Phillips County, Montana. Four of the five recovered were on plovers that spent the nonbreeding season in Texas; the fifth was recovered from a plover that spent the nonbreeding season in Arizona (Pierce et al. 2017).

SATELLITE TRACKING GPS tags were deployed in Phillips County, Montana, and a track of a full annual cycle movement revealed migration from breeding areas through Wyoming, Colorado, the Texas panhandle, New Mexico, and Arizona to stationary nonbreeding areas in Southern California. GPS archival and satellite tags deployed in Park County, Colorado, and in Thunder Basin National Grassland, Wyoming, revealed birds departed the breeding areas for a stopover in southeast Colorado and southwest Kansas before migrating to nonbreeding areas in Texas, the Central Valley of Southern California, southern Arizona, and across northern Mexico (A. Pierce, unpublished data).

STABLE ISOTOPES Stable isotope results provided support for weak migratory connectivity at the breeding population level (Wunder et al. 2005; Wunder 2007).

GENETICS Analysis of mitochondrial DNA (Oyler-Mccance et al. 2005) and of nuclear DNA (Oyler-McCance et al. 2008) from among four separate breeding populations suggested minimal genetic differentiation.

MIGRATORY PATHWAYS Mountain Plovers depart their stationary nonbreeding areas and migrate northward or eastward and then northward to breeding areas in Colorado, Montana, and Wyoming. They may either fly nonstop over the Sierra Nevada, Great Basin, and Rocky Mountains, or move eastward across Arizona and New Mexico before flying northward. After breeding, they migrate southward along the western edge of the North American grasslands into New Mexico, Texas, and Mexico, with some plovers moving west and then north into Arizona and California (Knopf and Wunder 2020). Tracking data indicate relatively linear migrations (Pierce et al. 2017; R. Siller, unpublished data). Few data exist for estimating pathway fidelity. There are only two known cases of archival tags recovered in two sequential years. In both cases, the plover migrated between the same breeding location in Colorado and the same nonbreeding locations (one in California, the other in Mexico) in each year. The plover that spent the nonbreeding season in California in both years used a similar post-breeding migration route in both years but a different pre-breeding migration route each year: once returning east across Arizona and New Mexico, then north to the breeding area; the other route was northeast across the Great Basin to northwestern Colorado, then back south to the breeding location in central Colorado. The plover that spent the nonbreeding season in Mexico both years used the same spring migration route in each year (A. Pierce, unpublished data).

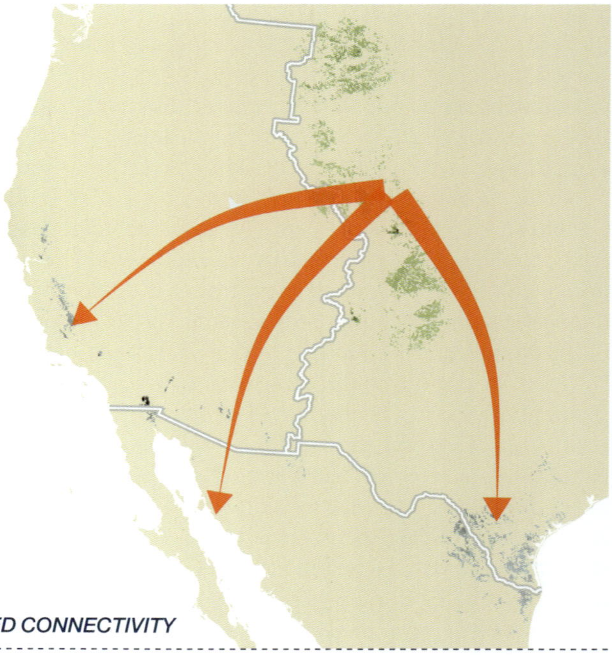

GENERALIZED CONNECTIVITY

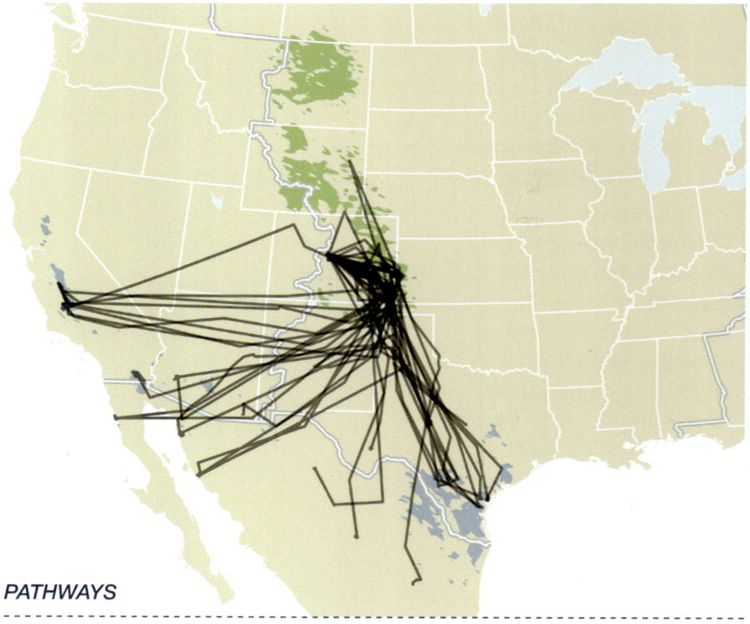

PATHWAYS

DATA CONTRIBUTORS
A.K. Pierce; M.B. Wunder.

INTERESTING NOTE

A color-banded Mountain Plover with the color combination of the Mexican flag was the first US-banded plover resighted in Mexico, in 2017, in San Luis Potosi, south of La Noria de la Soledad. The bird was also equipped with a GPS data logger, which was unfortunately not recovered, but additional plovers have since been tracked migrating to locations across northern Mexico (A. Pierce, unpublished data).

Anarhynchus nivosus **SNOWY PLOVER**

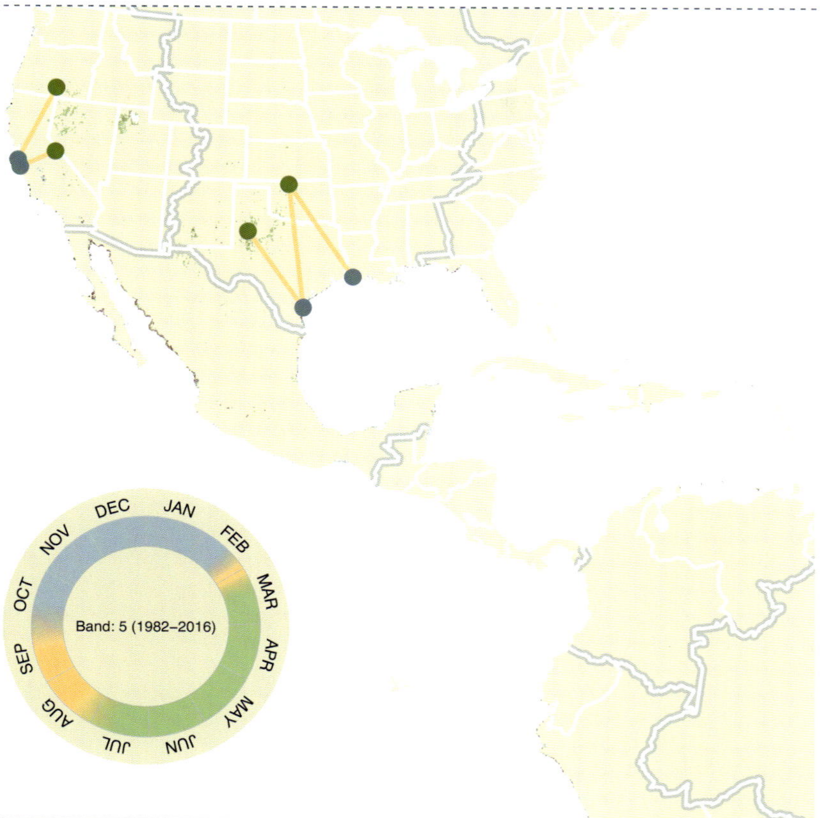

DISTRIBUTION AND ABUNDANCE

The Snowy Plover is a partial migrant endemic to the Americas and is one of our least abundant shorebird species. Populations of these plovers are found locally in the American West, along the Pacific and Gulf coasts of the US and Mexico, on select Caribbean islands, and along much of the Pacific coast of South America (Fink et al. 2023). In North America, coastal populations consist of both migrant and resident individuals, while inland breeding populations migrate to coastal areas during the nonbreeding period. Exceptions occur at some inland sites, such as the Salton Sea, that hold a year-round population. Though uncommon and local throughout their range, Gulf Coast abundance of this species is highest from central Tamaulipas, Mexico, to Corpus Christi, and from coastal Mississippi to the Florida panhandle. In the Pacific, the species is most abundant along the coasts of Southern California, Baja California, and the Gulf of California. Inland breeding populations are widely scattered but are most abundant in areas with large sand or salt flats, such as the Great Salt Lake, the Salton Sea, and Salt Plains National Wildlife Refuge (Fink et al. 2023).

INTERESTING NOTE

A high degree of breeding site fidelity was observed between years, but also significant movement was noted within and between years to form new mating pairs. Snowy Plovers have also demonstrated strong site fidelity to nonbreeding areas (Page et al. 1995). Natal dispersal and philopatry are complex and vary by population, with some birds returning and others dispersing widely (Page et al. 2020).

CONNECTIONS

ORDER Charadriiformes
FAMILY Charadriidae
BREEDING HABITAT Coasts, Beach and Estuary
NONBREEDING HABITAT Coasts, Beach and Estuary
GLOBAL POPULATION SIZE 31,000
POPULATION CHANGE SINCE 1970 +18%

IUCN RED LIST CATEGORY Near Threatened
R2R TIPPING POINT SPECIES Yellow-Alert
US LISTING STATUS Threatened western subspecies
CANADA SPECIES AT RISK No
PIF CCS 14/20

MIGRATORY CONNECTIVITY

OVERALL Migratory connectivity has yet to be fully characterized, but evidence of site fidelity, gene flow, inbreeding, and mixing of populations during the stationary nonbreeding period may suggest moderate connectivity across populations.

BANDING There are few band reencounters between breeding and stationary nonbreeding areas. Two birds from different western interior breeding populations both migrated to nonbreeding areas along Monterey Bay, California. One bird migrated from its breeding area in interior Texas to the Texas Gulf Coast.

RADIO TRACKING There is little mixing between breeding populations located in Oklahoma and populations in Texas and New Mexico on nonbreeding areas, but there is some mixing between Texas and New Mexico breeding populations on nonbreeding areas (Heath-Acre et al. 2021). Detections from the Motus Wildlife Tracking System during migration connect the United States to Mexico, the Bahamas, French Guiana, and Chile (Motus project, Southern Great Plains Snowy Plover, Great Salt Lake Shorebirds, unpublished data).

MIGRATORY PATHWAYS Inland breeding populations tend to migrate short distances toward coastal habitats for the nonbreeding season. Plovers from the western Great Basin region migrate to the California and Baja California coasts during nonbreeding, where they mix with Snowy Plovers that breed along the Pacific coast. Meanwhile, Snowy Plovers breeding in the Great Plains region migrate to the Gulf of Mexico coast (Page et al. 2020). Snowy Plovers in Oklahoma were found to make migratory movements to Texas and the Louisiana Gulf Coast (Heath-Acre 2021).

WILSON'S PLOVER *Charadrius wilsonia*

DISTRIBUTION AND ABUNDANCE

This species has a wide distribution and can be found in coastal and island habitats along the Atlantic and Pacific regions from North America to the coasts of Brazil and Peru, with occasional sightings in Chile. In the US, the northernmost limits of its breeding range are contracting along the Atlantic coast (Zdravkovic et al. 2020). The majority of breeding individuals in the United States leave North America, although a small number spend the nonbreeding season in the southern part of the United States, particularly in Florida. Studies suggest that there may be migratory individuals among the Caribbean breeding populations (Raffaele et al. 1998). Some observations indicate that breeding birds from northern areas migrate post-breeding to stationary nonbreeding grounds as far south as Brazil.

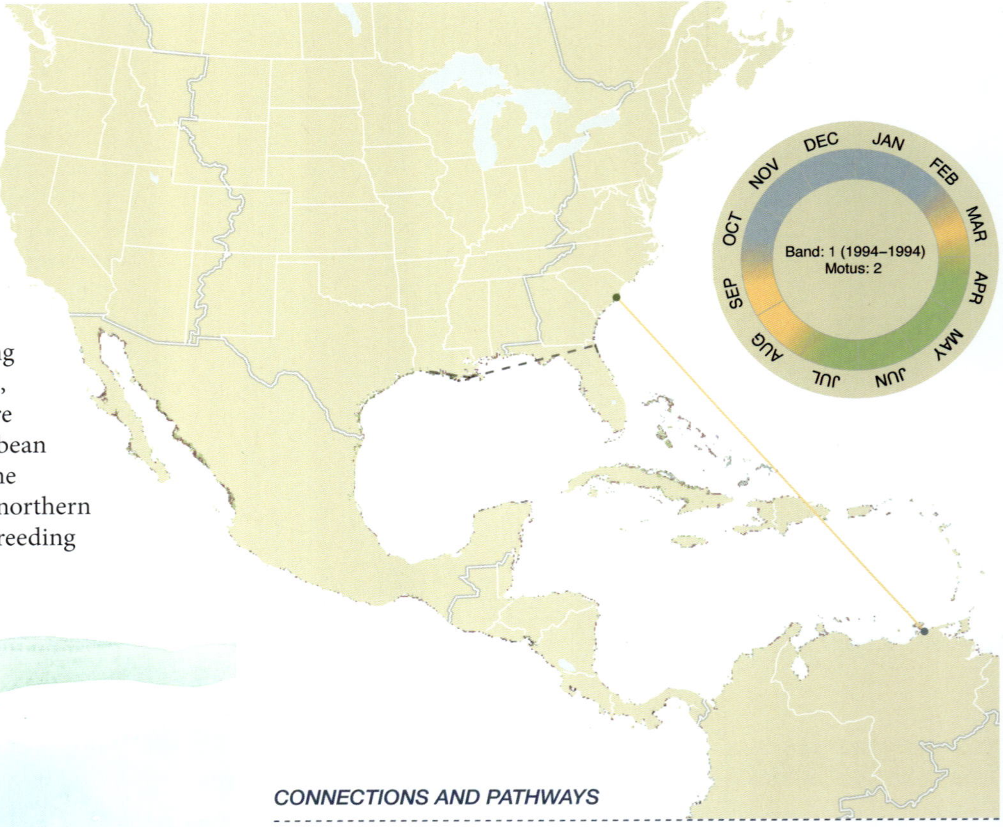

CONNECTIONS AND PATHWAYS

ORDER Charadriiformes
FAMILY Charadriidae
BREEDING HABITAT Coasts, Beach and Estuary
NONBREEDING HABITAT Coasts, Beach and Estuary
GLOBAL POPULATION SIZE 22,000
POPULATION CHANGE SINCE 1970 −37%
IUCN RED LIST CATEGORY Least Concern
R2R TIPPING POINT SPECIES Red-Alert
US LISTING STATUS No
CANADA SPECIES AT RISK No
PIF CCS 17/20

MIGRATORY CONNECTIVITY

OVERALL There is little information available regarding the migratory connectivity of this species.

BANDING One individual banded during the breeding season on the coast of Georgia was reencountered during the stationary nonbreeding season on the coast of Venezuela.

RADIO TRACKING Detections from the Motus Wildlife Tracking System during migration connect Florida to Louisiana (K. Williams, unpublished data). There were no breeding to nonbreeding detections.

GEOLOCATOR TRACKING An unpublished study by Newstead et al. suggests relatively high breeding site fidelity by adults in the northern Gulf of Mexico, and it is likely that a significant proportion of these birds spends the nonbreeding season on the Pacific coast of Central America, thus overlapping in the nonbreeding range with Pacific breeding birds (D. Newstead et al., unpublished data).

OTHER APPROACHES No satellite tracking, stable isotope, or genetic studies of migratory connectivity were found for this species.

MIGRATORY PATHWAYS Major migratory pathways remain unknown for the species. One bird was tracked during migration from the US Gulf Coast to the east coast of Florida.

DATA CONTRIBUTORS
K. Williams.

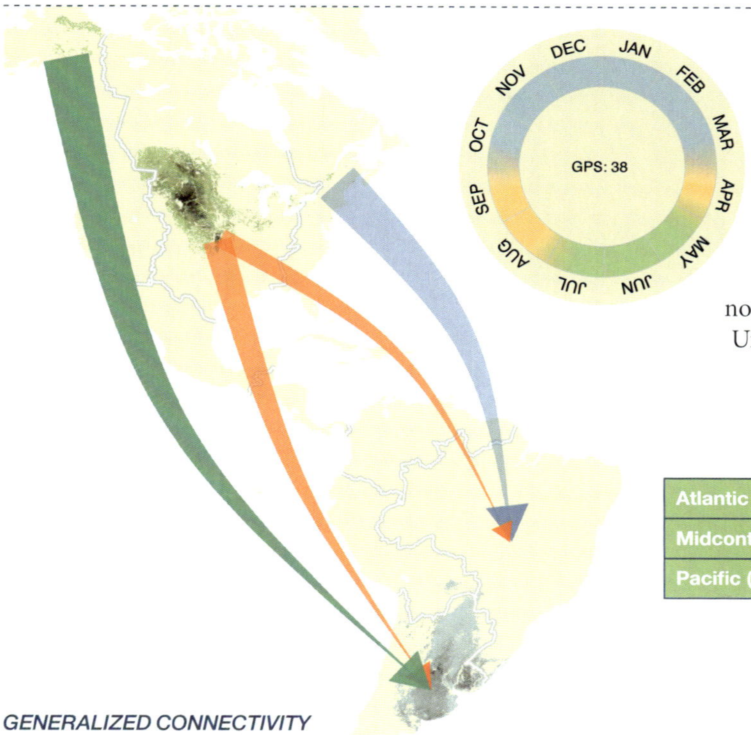

GENERALIZED CONNECTIVITY

Bartramia longicauda

UPLAND SANDPIPER

DISTRIBUTION AND ABUNDANCE

Upland Sandpipers are long-distance migrants traveling from grassland breeding areas in the United States from the upper Midwest from Kansas north to North Dakota and Montana and east to Illinois and patchily distributed into northeastern North America. Also in Canada, from southern Alberta, Saskatchewan, and Manitoba and eastward, but patchy, in southern Quebec. It spends the nonbreeding season in the Pampas ecoregions in northeastern Argentina, Uruguay, southern Brazil, Paraguay, and eastern Bolivia, and in northern Brazil (Blanco and López-Lanús 2008; Houston et al. 2020).

	Brazil	Southern Cone
Atlantic (n = 1)	100%	–
Midcontinent (n = 5)	40%	60%
Pacific (n = 14)	–	100%

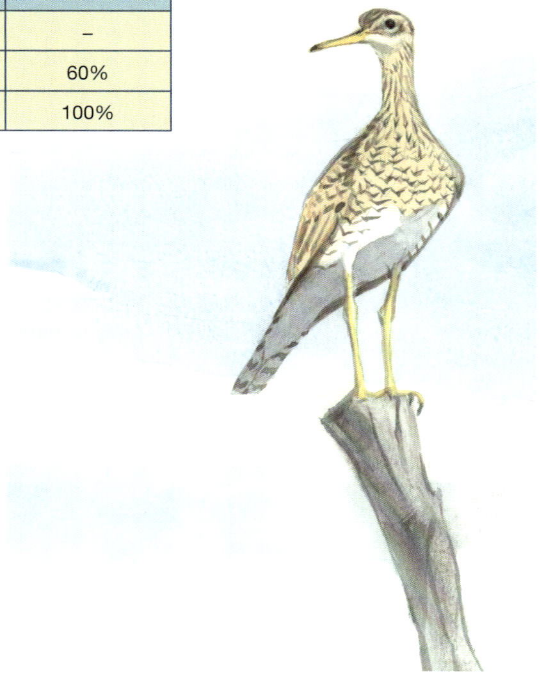

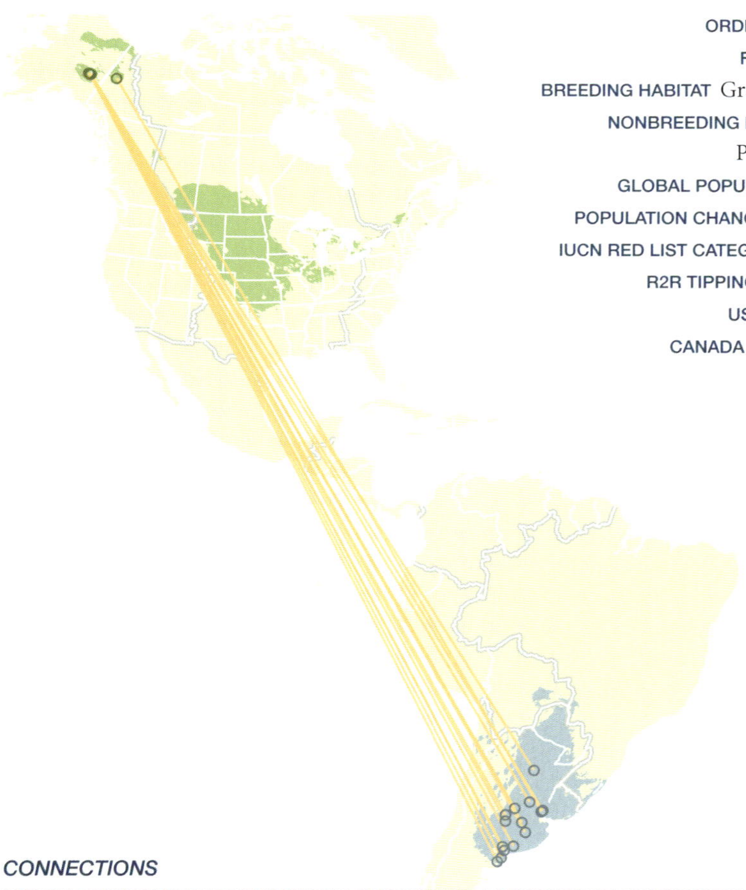

CONNECTIONS

ORDER Charadriiformes
FAMILY Scolopacidae
BREEDING HABITAT Grasslands, Temperate
NONBREEDING HABITAT Grasslands, Pampas and Campos
GLOBAL POPULATION SIZE 750,000
POPULATION CHANGE SINCE 1970 +25%
IUCN RED LIST CATEGORY Least Concern
R2R TIPPING POINT SPECIES No
US LISTING STATUS No
CANADA SPECIES AT RISK No
PIF CCS 11/20

INTERESTING NOTE

Mong and Sandercock (2007) reported an annual return rate of 38.1% (n = 189) of color-banded adults, indicating some fidelity to their Kansas breeding sites. Similarly, Hill et al. (2019) found that tagged birds returned to breeding locations near where they were captured the previous year.

MIGRATORY CONNECTIVITY

OVERALL Migratory connectivity has not been well understood due to limited banding efforts and infrequent band recoveries (Garber et al. 1997; Houston et al. 1999). However, several recent satellite telemetry studies across the range have improved our understanding of migratory connectivity. Birds breeding in the Atlantic Flyway spend the nonbreeding season in Brazil (100%). A portion of birds breeding in the Midcontinent Flyway also migrate to Brazil (40%), while the rest spend the nonbreeding season in the Southern Cone region of South America (60%). Birds breeding in the Pacific Flyway spend the nonbreeding season only in the Southern Cone (100%). Due to low breeding to nonbreeding sample sizes, MC cannot be calculated.

SATELLITE TRACKING GPS satellite tracking of birds breeding in Kansas and New England showed birds from the eastern breeding range spent the nonbreeding season solely in Brazil, while birds breeding in the midcontinent migrated to Argentina and Uruguay as well as Brazil (Hill et al. 2019). In contrast, breeding adults tracked from central Alaska migrated to stationary nonbreeding areas in Argentina (J. Johnson, unpublished data).

OTHER APPROACHES No banding, stable isotope, or genetic studies of migratory connectivity were found for this species.

MIGRATORY PATHWAYS Birds breeding in central Alaska and Kansas migrate in the Midcontinent Flyway along a narrow inland corridor in the Great Plains to reach the Texas coast and then cross the Gulf of Mexico and Central America to reach stopover sites in the Andes of Colombia and Ecuador. In contrast, birds breeding in the northeastern US migrate within the Atlantic Flyway and make shorter regional movements along the Atlantic coast before making long flights over the Caribbean Sea to inland sites in northeast Venezuela. Birds migrated over a broad front with no shared stopover sites (Hill et al. 2019). After arrival in South America, birds follow one of two migration routes, either crossing the Amazon Basin to reach nonbreeding sites in Brazil or continuing down the eastern slopes of the Andes with stopovers in Bolivia, Brazil, and Paraguay before reaching nonbreeding areas in Uruguay and northern Argentina (Hill et al. 2019). All six birds have elliptical migration routes in South America, with northbound paths mostly west of their southbound routes (Hill et al. 2019). Records also suggest the western slope of the Andes as part of a regular migration route (Medrano et al. 2018). An Upland Sandpiper tracked from Mexico crossed the Andes in northern Ecuador and continued south along the eastern side of the Peruvian Andes until the signal was lost (Grosselet et al. 2019). Birds tracked from Alaska migrated south along a narrow corridor east of the Rocky Mountains to the Texas coast. Many of the tracked birds crossed the western Gulf of Mexico and the Isthmus of Tehuantepec and continued offshore of Central America until they made landfall in northwestern South America. Birds continued to migrate south across the Amazon Basin until reaching their nonbreeding sites in central Argentina. Similar to populations breeding in Kansas, birds tracked from Alaska displayed an elliptical migration, with northbound routes mostly west of their southbound routes (J. Johnson, unpublished data).

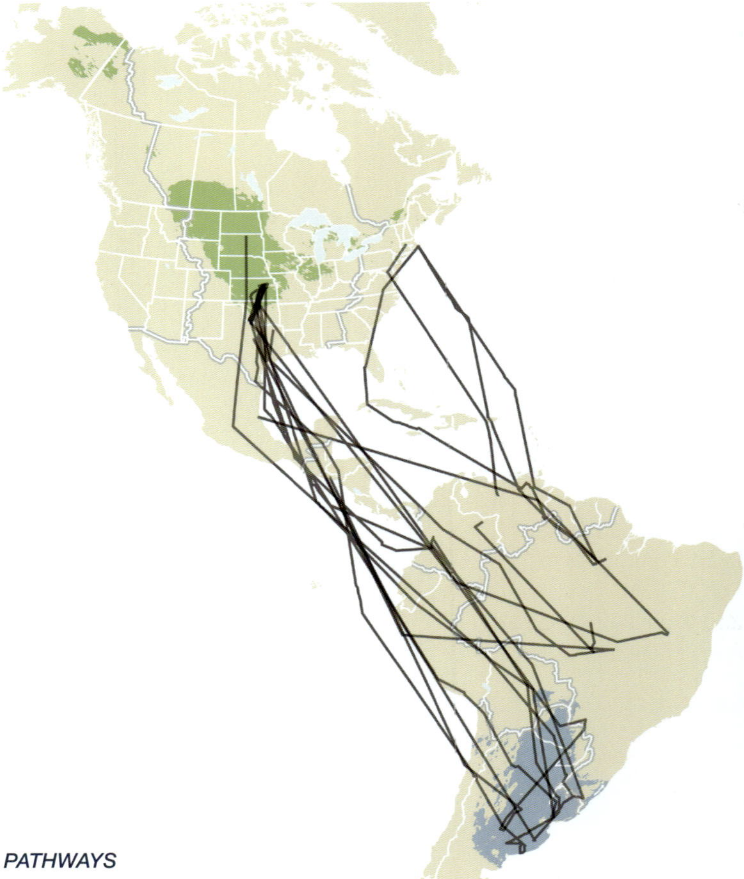

PATHWAYS

DATA CONTRIBUTORS

P. Caspari; C.F. Gesmundo; J.M. Hill; J.A. Johnson; Z.M. Pohlen; R.B. Renfrew; J. Tope.

BRISTLE-THIGHED CURLEW *Numenius tahitiensis*

DISTRIBUTION AND ABUNDANCE

The Bristle-thighed Curlew breeds in two disjunct areas of western Alaska (Nulato Hills and Seward Peninsula) and spends the nonbreeding season in the central Pacific or Oceania (Marks et al. 2020b).

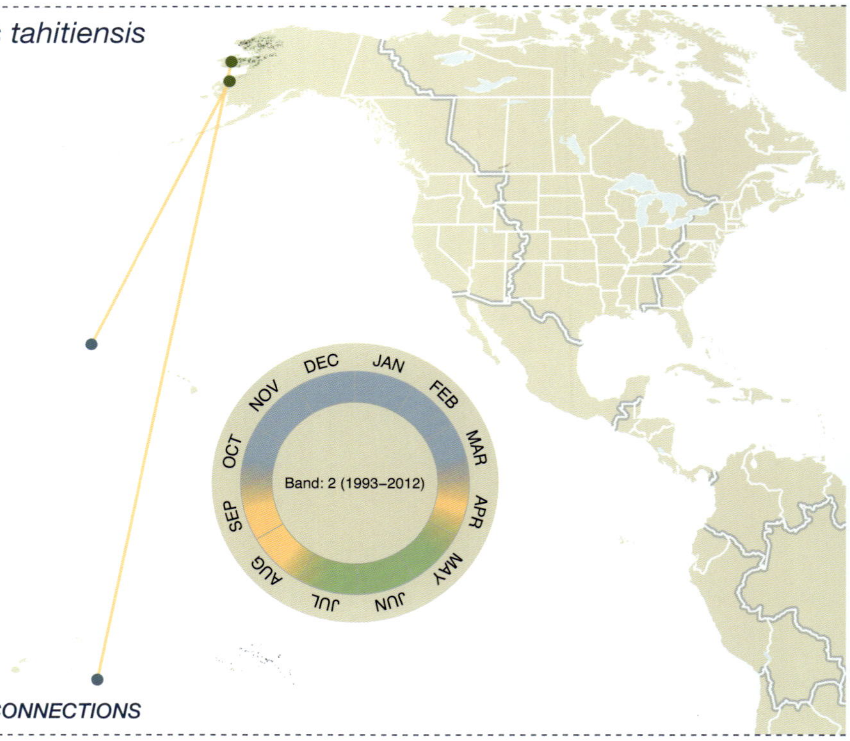

CONNECTIONS

ORDER Charadriiformes

FAMILY Scolopacidae

BREEDING HABITAT Tundra, Arctic

NONBREEDING HABITAT
Coasts, Beach and Estuary

GLOBAL POPULATION SIZE 10,000

POPULATION CHANGE SINCE 1970 NA

IUCN RED LIST CATEGORY Near Threatened

R2R TIPPING POINT SPECIES Red-Alert

US LISTING STATUS No

CANADA SPECIES AT RISK No

PIF CCS 17/20

MIGRATORY CONNECTIVITY

OVERALL Migratory connectivity appears strong based on limited banding and satellite transmitter data, with northern and southern breeding populations separating on the nonbreeding grounds. However, lack of genetic structure between populations on the nonbreeding range suggests some gene flow is occurring or has occurred in the recent past. Limited data prohibit MC estimates.

BANDING Band recoveries of breeding birds come only from the Hawaiian Islands (the northern extent of the nonbreeding range). Band resightings indicate the northwestern Hawaii Islands are a migratory terminus for some portion of the population, while curlews that spend the nonbreeding season farther south overfly the islands (Marks et al. 2020b).

SATELLITE TRACKING Birds tagged with PTT transmitters on breeding grounds indicated two breeding populations use different portions of the central Pacific. Seward Peninsula breeders migrated to islands in the southeastern Pacific Basin, the Marquesas Islands, the Tuamotu Archipelago of French Polynesia, and Vostok Island in Kiribati. Birds tagged in the Nulato Hills flew to the west-central Pacific Basin, Lisianski Island, the Marshall Islands, Gilbert Islands, and Nauru. Deployment of PTT transmitters on nonbreeding grounds in Oahu, Hawaii, revealed birds made direct flights from breeding sites in Nulato Hills, Alaska (R. Gill and L. Tibbitts, unpublished data, not presented in the Atlas).

GENETICS Analyses found variation in allelic and haplotypic frequencies between the two breeding areas in Alaska but not throughout its nonbreeding range in the central (Sonsthagen et al. 2014).

MIGRATORY PATHWAYS The species undergoes a long-distance, transoceanic flight to reach its stationary nonbreeding areas, where it undergoes a flightless wing molt. Four adult birds were radio-tagged on breeding grounds in 1989 staged on the Yukon-Kuskokwim Delta in eastern Alaska for 2–3 weeks (T.L. Tibbitts, R.E. Gill, B.J. McCaffery, unpublished data).

INTERESTING NOTE

Resightings of banded birds indicate strong site fidelity to breeding (Jung 2014; T.L. Tibbitts and R. Gill, unpublished data) and nonbreeding territories (Marks and Redmond 1996). No natal fidelity has been recorded. Attachment of external satellite transmitters to birds indicated there was a potential negative effect on the survival of curlews, dependent on age and sex (Ruthrauff et al. 2019b).

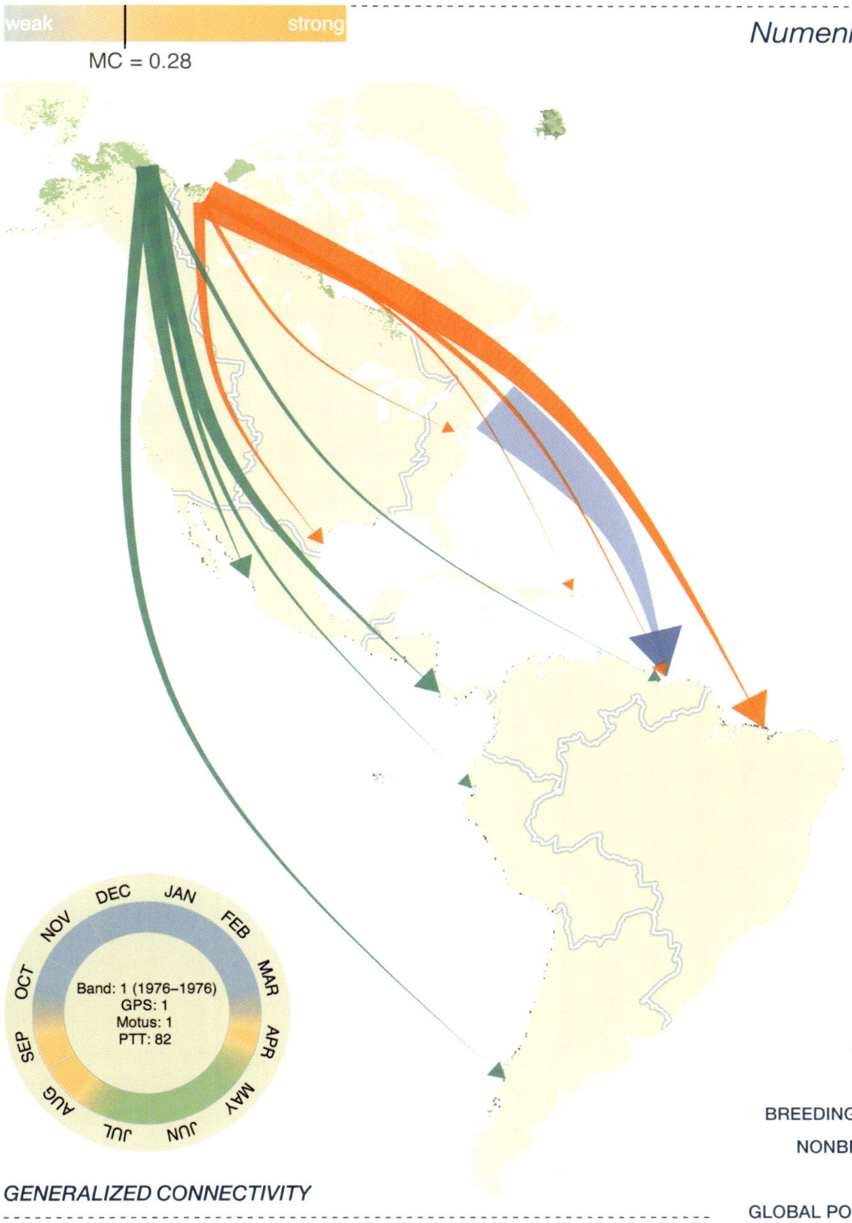

GENERALIZED CONNECTIVITY

Numenius hudsonicus **HUDSONIAN WHIMBREL**

DISTRIBUTION AND ABUNDANCE

The Hudsonian Whimbrel has three disjunct breeding populations, which spend the nonbreeding season primarily in Mexico and Central and South America, with particularly large populations in Chile (Andres et al. 2009). The two western populations breed in Alaska and throughout the Mackenzie River Delta in western Canada, while the eastern population breeds in the lowlands of western Hudson Bay (Engelmoer and Roselaar 1998; Skeel and Mallory 2020). In 2025, the Whimbrel was divided into two species: the Hudsonian Whimbrel, found across the Americas, and the Eurasian Whimbrel (*Numenius phaeopus*), distributed throughout Europe, Asia, Africa, and Australia.

ORDER Charadriiformes
FAMILY Scolopacidae
BREEDING HABITAT Tundra, Arctic
NONBREEDING HABITAT Coasts, Beach and Estuary
GLOBAL POPULATION SIZE 1,800,000

POPULATION CHANGE SINCE 1970 −58%
IUCN RED LIST CATEGORY Least Concern
R2R TIPPING POINT SPECIES Orange-Alert
US LISTING STATUS No
CANADA SPECIES AT RISK No
PIF CCS 13/20

CONNECTIONS

PATHWAYS

	Andean West	Atlantic	Brazil	Caribbean	Central America	Mexico	Midcontinent	Northern SA	Southern Cone
Atlantic (n = 1)	–	–	–	–	–	–	–	100%	–
Midcontinent (n = 26)	–	4%	65%	4%	–	–	15%	12%	–
Pacific (n = 22)	9%	–	–	–	36%	32%	–	5%	18%

MIGRATORY CONNECTIVITY

OVERALL Collectively, band reencounter, Motus, geolocator, and satellite transmitter data from all three breeding populations inform our knowledge of migratory connectivity for the Hudsonian Whimbrel. Birds breeding in the Atlantic Flyway primarily migrate to nonbreeding areas in northern South America (100%), while birds breeding in the Midcontinent Flyway primarily migrate to Brazil (65%), within the Midcontinent Flyway (15%), and northern South America (12%). Birds breeding in the Pacific Flyway spend the nonbreeding season in Central America (36%), Mexico (32%), and the Southern Cone region (18%). Overall migratory connectivity for the species is moderate (MC = 0.28).

BANDING There was one breeding to nonbreeding recovery of a Hudsonian Whimbrel banded in Manitoba and found shot in Suriname. There were several resightings of both Alaska breeding birds and Hudsonian Whimbrels banded during the nonbreeding season in Chile along the Pacific Flyway (D. Ruthrauff, personal communication).

RADIO TRACKING Many Motus tags were deployed through the Motus Tracking Wildlife System in the eastern part of the range. There were limited detections of individual birds between breeding and nonbreeding areas.

GEOLOCATOR TRACKING Nine complete migration tracks indicate that breeding birds from Churchill, Manitoba, displayed strong connectivity with all birds migrating to nonbreeding areas along the northern coasts of Brazil and Suriname, and then the birds exclusively used two mid-Atlantic staging sites during migration: the Chesapeake Bay / Delmarva Peninsula and the coast of Georgia and South Carolina (A.S. Johnson et al. 2016).

SATELLITE TRACKING Breeding Hudsonian Whimbrels from the Alaska population were tracked to locations along the Pacific coast from Canada to South America. Of 28 birds tracked from Alaska to nonbreeding locations, 9 (32%) went to Mexico, 9 (32%) to Central America (Costa Rica, Honduras, Panama), 5 (18%) to northern South America (Colombia, Ecuador), and 5 (18%) to Chile (Ruthrauff et al. 2021).

OTHER APPROACHES No genetic or stable isotope studies of migratory connectivity were found for this species.

MIGRATORY PATHWAYS Hudsonian Whimbrels use two contrasting strategies for long, open-water flights over the Atlantic Ocean: a long, nonstop flight around the core of storm activity with a low likelihood of encountering storms but no safety net of islands to land on, and a shorter flight through the core of storm activity with a high likelihood of encountering storms but with a safety network of islands to use in the event of an encounter. Mackenzie River Delta Whimbrels departed North America from Atlantic Canada and made long. nonstop flights out to sea that took several days to complete, while Hudson Bay Whimbrels departed North America from the south Atlantic coast and made shorter, nonstop flights across the Caribbean that took less time (Watts et al. 2021b). Satellite tracking data of Whimbrels breeding on the Mackenzie River Delta showed that birds staged near the Beaufort Sea, then flew east across the continent to reach terminal staging areas in the North Atlantic and at Hudson and James bays, before making nonstop, transoceanic flights to their nonbreeding territories along the east coast of northern South America. On northward pre-breeding migration, Whimbrels of the Mackenzie River Delta population used two major routes, one with staging along the south Atlantic coast and

INTERESTING NOTE

The Center for Conservation Biology (a research group within the College of William & Mary in Williamsburg, Virginia) attached a satellite transmitter on a Hudsonian Whimbrel, which they named "Hope," in May 2009, on the eastern shore of Virginia. Hope would eventually help tell the story of shorebird migration throughout the annual cycle and was eventually tracked for over 50,000 miles (80,000 km) back and forth four times between her breeding site on the Mackenzie River in far western Canada and her nonbreeding site on Great Pond, St. Croix, US Virgin Islands. In September 2012, the transmitter was removed (due to antenna loss), but the coded leg flag was left on. Hope was last resighted on September 11, 2017, just prior to Hurricane Maria hitting St. Croix (https://ccbbirds.org/2019/04/03/farewell-to-hope/).

another with staging along the western Gulf Coast. Birds of the Atlantic coast route flew toward the Great Lakes and then moved west to the Mackenzie River Delta. Birds of the Gulf Coast route flew directly over the Great Plains to the Mackenzie River Delta, moving along the eastern edge of the Rocky Mountains. Of birds tracked from breeding areas in the Arctic National Wildlife Refuge, some migrate like Mackenzie River birds (Atlantic Flyway) and some like other Alaska-breeding birds (Pacific Flyway) (S. Schulte, unpublished data).

Whimbrels of the Hudson Bay population staged at breeding areas around Hudson and James bays, and then flew southeast to stage along the south Atlantic coast, before flying across the Caribbean Sea to the northern coast of South America. There, they slowly moved east along the coast to nonbreeding territories. On northward pre-breeding migration, Hudson Bay Whimbrels retraced their southward route north to stage along the Atlantic coast before flying over the Great Lakes and on to the Hudson Bay (Watts et al. 2021a, 2021b). Whimbrels tracked for multiple years had high fidelity to staging areas (Watts et al. 2021a).

One Hudsonian Whimbrel was tracked leaving a staging site along the Atlantic coast (Delmarva Peninsula) and migrating overland to the Mackenzie River Delta, where it staged before moving to its probable breeding area on the North Slope of Alaska, suggesting a portion of the western population may undergo an elliptical migration and overlap with the eastern population along the south Atlantic coast (Watts et al. 2008). Also, unlike other tracked birds from the Mackenzie River Delta population, one individual flew across northern Mexico to stage on the Gulf of California for two consecutive years and then flew north along the Pacific Flyway to the Mackenzie River Delta (Watts et al. 2021a).

After breeding, Hudsonian Whimbrels from the Alaska population moved to coastal sites in western and northwestern Alaska. From there they generally followed the Pacific coast, with some movements up to 1600 km offshore, such as crossings of the Gulf of Alaska and movements between Chile and North America. On northward pre-breeding migration, birds largely retrace their southbound movements. Once in Alaska, they fly overland, traversing the Alaska and Brooks mountain ranges rather than approaching via the nearest coastline (Ruthrauff et al. 2021). Alaska-breeding Whimbrels generally exhibited fidelity to breeding and nonbreeding areas but typically exhibited fidelity only to staging areas used prior to long, nonstop migratory flights (Ruthrauff et al. 2021).

DATA CONTRIBUTORS

K. Bennett; G. Brown; A.S. Johnson; A. Kneidel; A. McKellar; E. Nol; J. Paquet; J. Rausch; D.R. Ruthrauff; N.R. Senner; F. Smith; P. Smith; T.L. Tibbitts; B. Watts; B. Winn.

LONG-BILLED CURLEW

Numenius americanus

MC = 0.1 ± 0.04*

ORDER Charadriiformes
FAMILY Scolopacidae
BREEDING HABITAT Grasslands, Temperate
NONBREEDING HABITAT Coasts, Beach and Estuary
GLOBAL POPULATION SIZE 140,000

POPULATION CHANGE SINCE 1970 +2%
IUCN RED LIST CATEGORY Least Concern
R2R TIPPING POINT SPECIES No
US LISTING STATUS No
CANADA SPECIES AT RISK No
PIF CCS 12/2

DISTRIBUTION AND ABUNDANCE

Long-billed Curlews are known to breed in three large regions, from west to east, of the United States and Canada: Great Basin, Intermountain West, and Great Plains. These regions encompass large grasslands in Montana, Wyoming, Nebraska, South Dakota, and Alberta and Saskatchewan, and west to Nevada, Oregon, Idaho, and south into New Mexico. During the nonbreeding period, this species can be found on the Pacific coast, from California all the way to Colima, Mexico, and, more rarely, as far south as El Salvador. Along the Gulf Coast, it can be found from Florida to northern Mexico, with the greatest concentrations along coastal Texas, and very small and declining numbers also residing along the Atlantic coast of Georgia, Florida, and occasionally South Carolina.

INTERESTING NOTE

Multiple published and unpublished tracking studies based on individuals caught across the species' breeding and nonbreeding ranges indicate strong site fidelity to both breeding and nonbreeding sites, as well as to migratory stopover areas (Page et al. 2014; Olalla-Kerstupp et al. 2015; Sesser 2013; Coates 2018; A.J. Boyce et al., unpublished data; Migratory Connectivity Project, unpublished data).

	Atlantic	Mexico	Midcontinent	Pacific
Midcontinent (n = 61)	2%	67%	31%	–
Pacific (n = 65)	–	49%	–	51%

MIGRATORY CONNECTIVITY

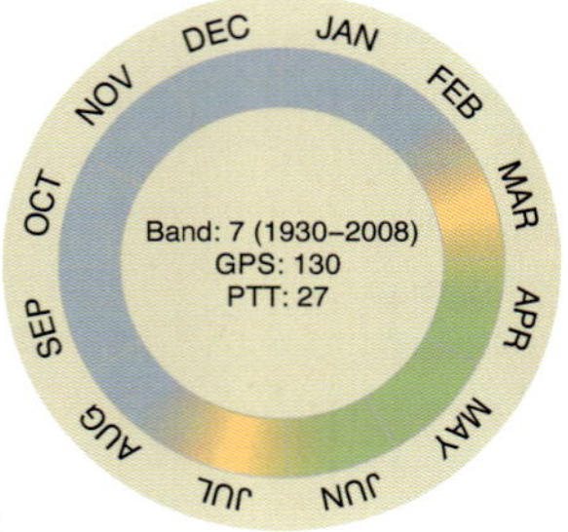

OVERALL This is a well-tracked species; most information about migratory connectivity comes from satellite transmitter data. Curlews breeding within the Midcontinent Flyway migrate to Mexico (67%) or nonbreeding areas within the Midcontinent Flyway (31%). Curlews breeding within the Pacific Flyway also spend the nonbreeding season in Mexico (49%) or migrate to nonbreeding areas within the Pacific Flyway (51%). Within the Atlas breeding and nonbreeding regions, migratory connectivity is weak (MC = 0.1 ± 0.04) due to mixing of Midcontinent and Pacific flyways breeding populations on the nonbreeding grounds, especially in Mexico (but see Knight et al. 2025).

BANDING There are limited banding reencounters, but they show a parallel migration pattern similar to the tracking data.

SATELLITE TRACKING Satellite transmitter data suggest two population divides that delineate three breeding populations (from west to east): Great Basin, Intermountain West, and Great Plains. These breeding populations migrate to different nonbreeding areas, with minimal mixing; Knight et al. (2025) estimate strong migratory connectivity (MC = 0.68) within these regions. Curlews breeding in Oregon and Washington spend the nonbreeding season at inland sites in central California, Idaho, western Wyoming, Utah, and Nevada. Populations also migrate to inland areas of California, as well as coastal areas in Baja, California, and northwestern Mexico. About 75% of southwestern Idaho breeding birds migrate to the Central Valley of California, and 25% migrate to the Imperial Valley, while eastern Idaho and western Wyoming birds migrate to mainly Mexico and the Imperial Valley / Mexicali area (Page et al. 2014; Coates 2018; J. Carlisle et al., unpublished data; Migratory Connectivity Project, unpublished data). Curlews breeding east of the Rockies, including Montana, the Dakotas, southern Alberta, and Saskatchewan, spend the nonbreeding season inland from the Texas panhandle south to the Central Mexican Plateau and Gulf of Mexico (Page et al. 2014; Olalla-Kerstupp et al. 2015; A.J. Boyce et al., unpublished data; Migratory Connectivity Project, unpublished data; J. Carlisle and K. Ellison, unpublished data). One male curlew tracked from Georgia nonbreeding grounds flew directly west to Oklahoma, and then straight north to breeding grounds in Saskatchewan but appeared to be a nonbreeding male (Migratory Connectivity Project, unpublished data).

MIGRATORY PATHWAYS Long-billed Curlews exhibit a parallel migration strategy and use three primary pathways, with the exception of the individual that migrated between Saskatchewan and Georgia (Knight et al. 2025). Western breeding birds migrate north and south through California and along the California-Nevada border. Intermountain West breeding birds span a range of directions from a southwest route over central Nevada to the Central Valley of California to south-southwest tracks crossing eastern Nevada and southeast California (Page et al. 2014; Coates 2018; Migratory Connectivity Project, unpublished data). Northbound migratory

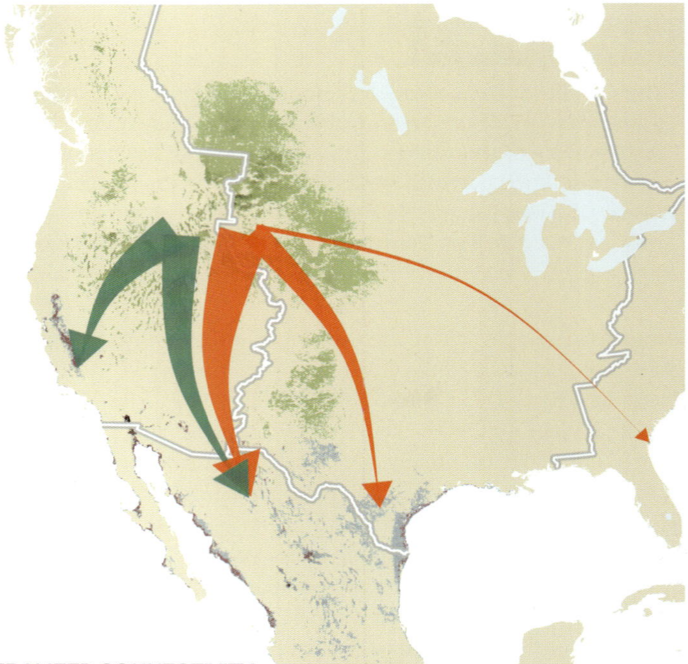

GENERALIZED CONNECTIVITY

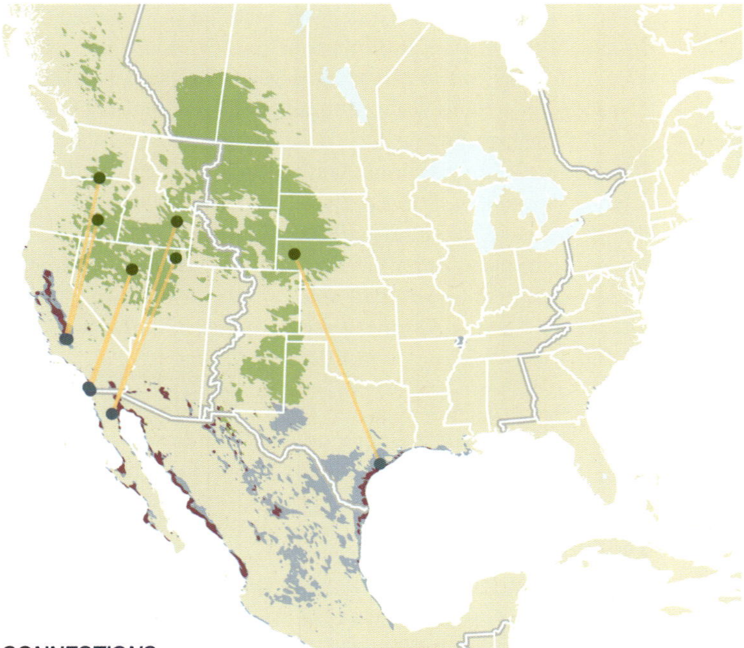

CONNECTIONS

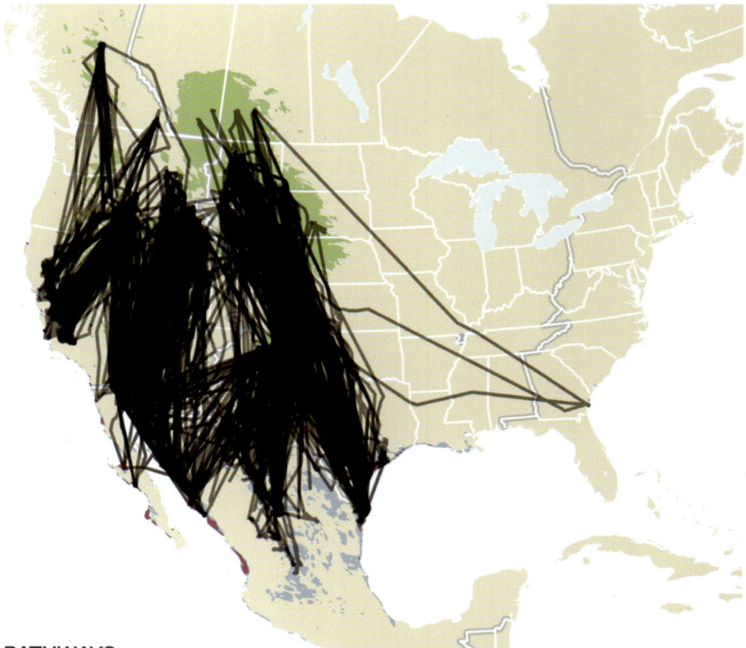

PATHWAYS

routes are similar to southbound routes. Birds breeding in the Canadian Prairies migrate through the midcontinent—Montana, Wyoming, and Colorado—to the Texas panhandle or northeast New Mexico, with north- and southbound tracks largely overlapping (Page et al. 2014; Olalla-Kerstupp et al. 2015; Boyce et al., unpublished data; Migratory Connectivity Project, unpublished data).

DATA CONTRIBUTORS

A.J. Boyce; D. Bradley; J. Carlisle; P. Cimprich; S. Coates; K. Goodenough; J.I. González-Rojas; A. Guzmán-Velaso; A.-L. Harrison; J. Kelly; T. Keyes; P.P. Marra; D.J. Newstead; A. Olalla-Kerstupp; G. Ruiz-Aymá; A.L. Scarpignato; N. Warnock; W. Wehtje; Migratory Connectivity Project.

Limosa lapponica baueri BAR-TAILED GODWIT

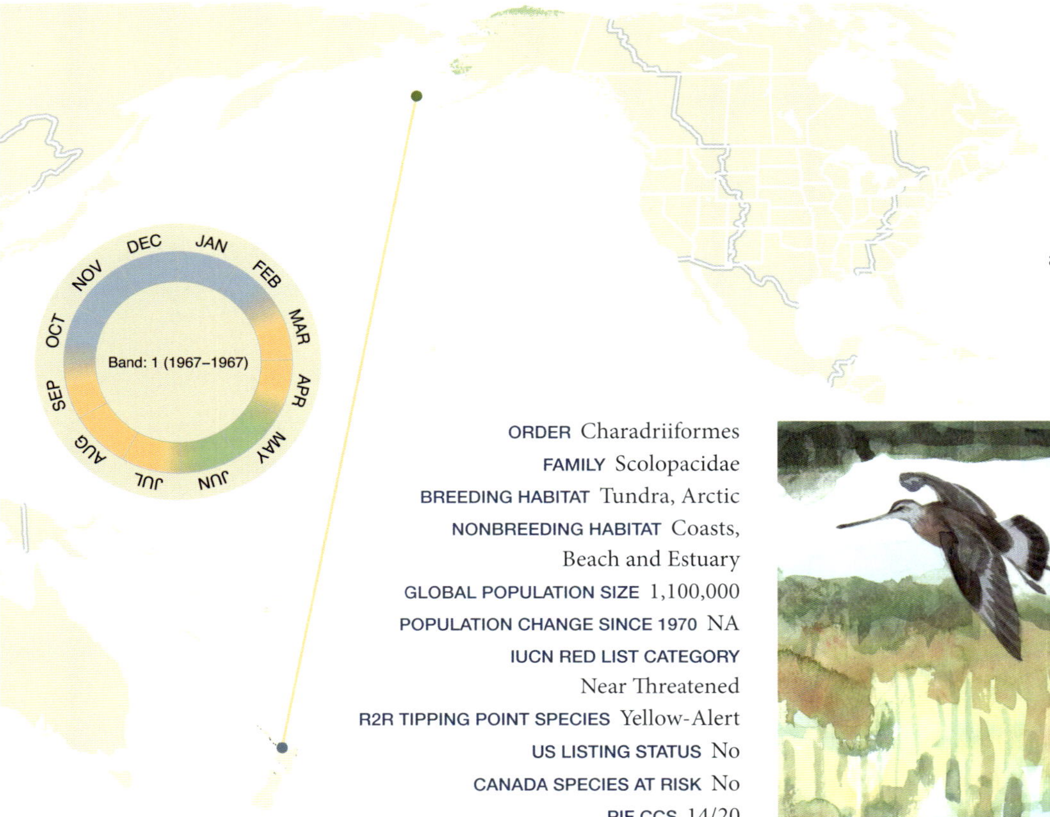

CONNECTIONS

ORDER Charadriiformes
FAMILY Scolopacidae
BREEDING HABITAT Tundra, Arctic
NONBREEDING HABITAT Coasts, Beach and Estuary
GLOBAL POPULATION SIZE 1,100,000
POPULATION CHANGE SINCE 1970 NA
IUCN RED LIST CATEGORY Near Threatened
R2R TIPPING POINT SPECIES Yellow-Alert
US LISTING STATUS No
CANADA SPECIES AT RISK No
PIF CCS 14/20

DISTRIBUTION AND ABUNDANCE

The *L. l. baueri* subspecies is an endemic breeder to Alaska, residing from the Yukon-Kuskokwim Delta to the Seward Peninsula and east across the North Slope of Alaska. The species spends the nonbreeding season in New Zealand and eastern Australia. Species declines are attributed to subsistence harvest in Alaska; degradation of habitat, especially in the Yellow Sea area; and collisions with power lines and other stational objects (Studds et al. 2017; McCaffery and Gill 2020).

MIGRATORY CONNECTIVITY

OVERALL Despite apparent population structure on breeding grounds based on plumage and morphology, migratory connectivity appears weak based on similar migration routes and mixing on the nonbreeding area for individuals breeding throughout Alaska.

BANDING There was one recovery of a bird in New Zealand banded on breeding grounds in Alaska.

RADIO TRACKING Detections from the Motus Wildlife Tracking System during migration connect Sweden and Germany to the Netherlands, United Kingdom, and Mauritania (Motus project, Waders, unpublished data).

SATELLITE TRACKING Birds equipped with PTT transmitters on breeding grounds travel to nonbreeding areas in New Zealand and Australia (R.E. Gill et al. 2009, 2014). Similarly, satellite-tagged birds in New Zealand traveled to the Yellow Sea and then Alaska (Battley et al. 2012).

OTHER APPROACHES No genetic or stable isotope studies of migratory connectivity were found for this species. Latitudinal variation in body size and body plumage coloration suggest population structuring in Alaska, but separation is not maintained on nonbreeding grounds in New Zealand (Conklin et al. 2011).

MIGRATORY PATHWAYS Banding and satellite transmitter data indicate elliptical migration with a long, nonstop flight from Alaska across the Pacific to New Zealand, and two flights northward, stopping first at the Yellow Sea area of China and then to breeding grounds in Alaska. Band resightings indicate staging occurs on the Yukon-Kuskokwim Delta in western Alaska and the subspecies uses the East Asian–Australasian Flyway only during north migration (R.E. Gill et al. 2005).

INTERESTING NOTE

There is extremely strong site fidelity to nonbreeding (Conklin et al. 2016a) and breeding areas, but data are sparse (R. Gill, B.J. McCaffery, unpublished data). The Bar-tailed Godwit's eleven-day, nonstop flight, at a distance of 8,425 miles (13,560 km) from Alaska to Tasmania, Australia, is considered the world record for the longest nonstop bird flight.

HUDSONIAN GODWIT *Limosa haemastica*

DISTRIBUTION AND ABUNDANCE

Hudsonian Godwits breed in boreal habitats in three disjunct areas: south-central and western Alaska, northern Northwest Territories, and the western Hudson Bay. Individuals migrate during pre-breeding through the central US and during post-breeding mainly along both coasts. All appear to spend the nonbreeding season in marine coastal areas along the South American Atlantic and Pacific coasts (Walker et al. 2020).

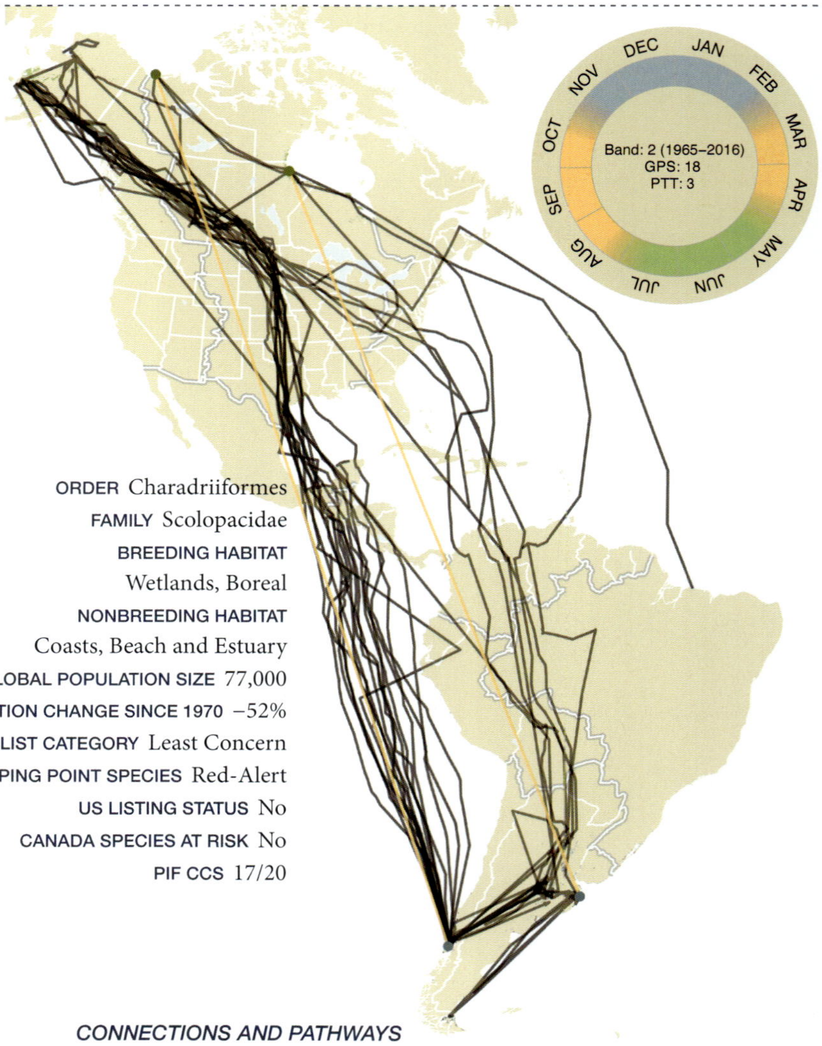

ORDER Charadriiformes
FAMILY Scolopacidae
BREEDING HABITAT
Wetlands, Boreal
NONBREEDING HABITAT
Coasts, Beach and Estuary
GLOBAL POPULATION SIZE 77,000
POPULATION CHANGE SINCE 1970 −52%
IUCN RED LIST CATEGORY Least Concern
R2R TIPPING POINT SPECIES Red-Alert
US LISTING STATUS No
CANADA SPECIES AT RISK No
PIF CCS 17/20

CONNECTIONS AND PATHWAYS

INTERESTING NOTE

Adult fidelity to breeding areas is high, and natal fidelity has been documented (Walker et al. 2020).

MIGRATORY CONNECTIVITY

OVERALL Migratory connectivity is known from genetic, banding, geolocator, and satellite tracking data. Birds breeding within the Midcontinent and Pacific flyways all spend the nonbreeding season within the Southern Cone region of South America. MC is not calculated due to limited breeding to nonbreeding sample size.

BANDING Godwits banded in Alaska were observed in Tierra del Fuego (J. Johnson and N. Senner in Andres et al. 2012), and one Hudson Bay godwit was observed in the Pampas region of Argentina.

RADIO TRACKING A few tags were deployed through the Motus Wildlife Tracking System; there are no individual breeding to nonbreeding detections.

GEOLOCATOR TRACKING Twenty-nine godwits tagged at Beluga Lake, Alaska, spent the nonbreeding period at Chiloé Island, Chile (Senner et al. 2014).

SATELLITE TRACKING Available evidence suggests Alaska and Hudson Bay populations separate on breeding and nonbreeding grounds, although species share some stopovers during north and south migration. Almost all birds from eastern populations migrate to three main nonbreeding areas of the South American Atlantic coast, whereas Alaskan breeders migrate to nonbreeding areas along the Chilean coast, in just 16 bays (N. Senner, unpublished data, in Walker et al. 2020).

GENETICS Genetic markers indicate divergence between birds breeding at Churchill, Manitoba, and Mackenzie, Northwest Territories, although birds from southern and western Alaska do not differ from one another and are closer genetically to the Mackenzie population (Haig et al. 1997).

MIGRATORY PATHWAYS This species has an elliptical migration with pre-breeding migration along the Pacific and Midcontinent flyways, and post-breeding migration along the Atlantic Flyway to reach northern South America, before traveling through interior portions of South America to reach nonbreeding areas in coastal southern South America (Walker et al. 2020). Geolocator data from Alaska breeders showed an elliptical migration route, migrating east across North America before crossing the Atlantic, through interior South America to Chile, then north across the Pacific and through the midcontinent of North America and back to Alaska (Senner et al. 2014). Satellite tracking data indicate travel within South America is through the interior during southbound migration and more coastal during northbound migration. Trackers have also shown that additional stopovers for Alaskan breeding birds occur in the interior of northern South America (N. Senner, unpublished data, in Walker et al. 2020). During pre-breeding, major stopovers occur at Cheyenne Bottoms, Kansas; Eastern Rainwater Basin, Nebraska; and Lake Thompson, South Dakota. During post-breeding, birds stop at Aropuk Lake, Alaska (western Alaska breeders only); the Quill Lakes and other saline lakes of Saskatchewan; James Bay, Ontario; the Amazon Basin, Colombia; and Buenos Aires Province, Argentina (Senner 2012; Senner et al. 2014).

DATA CONTRIBUTORS

B.M. Ballard; J.A. Linscott; J.G. Navedo; J. Rausch; J. Ruiz; N.R. Senner; F. Smith; B. Watts; M.D. Weegman.

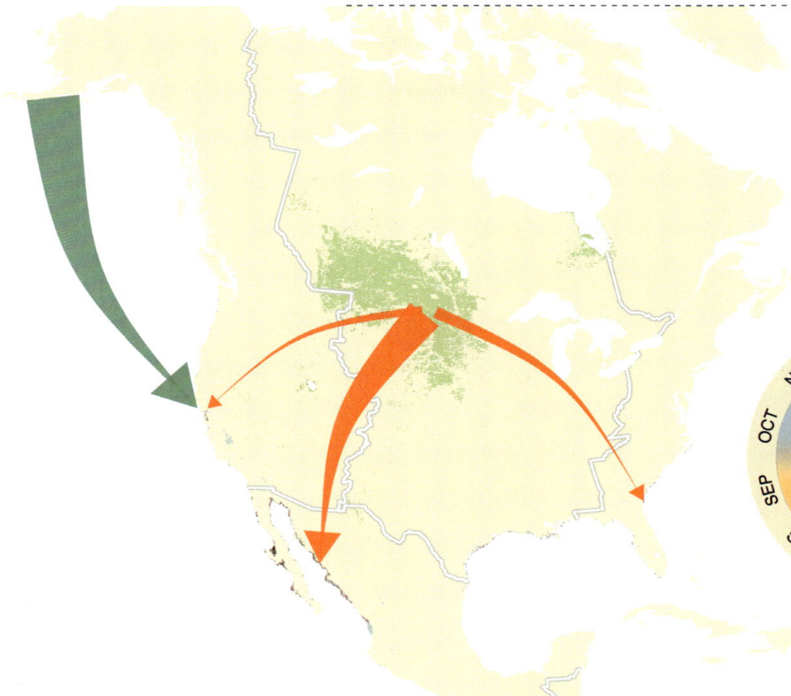

GENERALIZED CONNECTIVITY

Limosa fedoa MARBLED GODWIT

DISTRIBUTION AND ABUNDANCE

The majority of Marbled Godwits breed in the grasslands of the northern Great Plains, but there are also two isolated breeding populations, one on the Bristol Bay coast of Alaska and the other on the southwest coast of James Bay, Ontario. The Pacific population spends the nonbreeding season along the Pacific coast from Washington to El Salvador, as well as interior valleys in California, while the Atlantic population spends the nonbreeding season along the Atlantic coast, from Massachusetts south to Florida. Interior populations can spend the nonbreeding season on either the Pacific or Atlantic coasts, as well as along the entire coast of the Gulf of Mexico (Gratto-Trevor 2020).

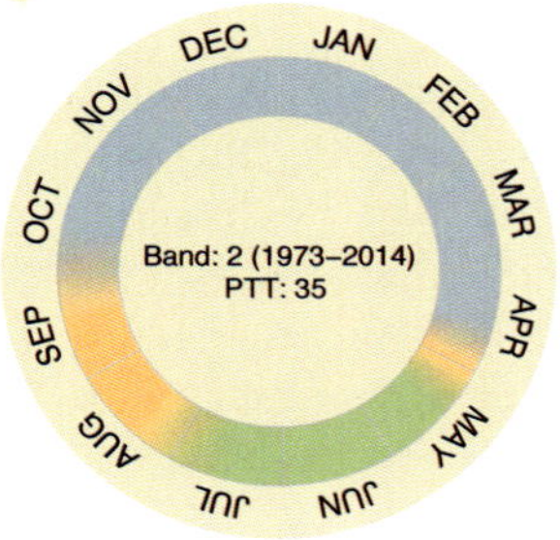

	Atlantic	Mexico	Pacific
Midcontinent (n = 16)	19%	69%	12%
Pacific (n = 9)	–	–	100%

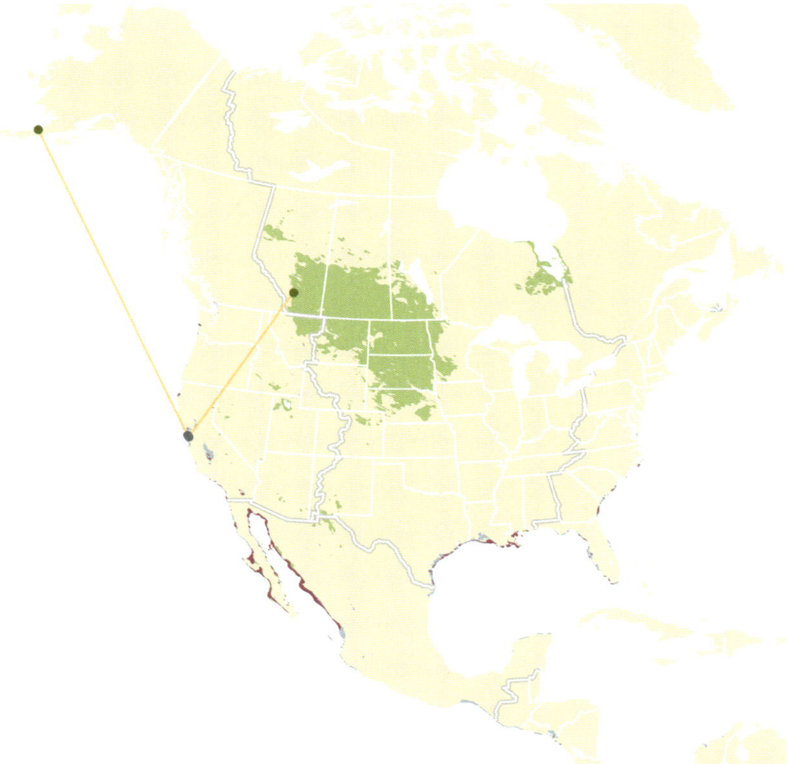

CONNECTIONS

ORDER Charadriiformes
FAMILY Scolopacidae
BREEDING HABITAT Wetlands, Seasonally Wet Grasslands
NONBREEDING HABITAT Coasts, Beach and Estuary
GLOBAL POPULATION SIZE 170,000
POPULATION CHANGE SINCE 1970 −28%
IUCN RED LIST CATEGORY Least Concern
R2R TIPPING POINT SPECIES No
US LISTING STATUS No
CANADA SPECIES AT RISK No
PIF CCS 15/20

MIGRATORY CONNECTIVITY

OVERALL Tracking and banding data suggest moderate migratory connectivity for Marbled Godwits, with little overlap of breeding populations at nonbreeding areas in the Pacific Flyway. Midcontinent Flyway breeders migrate to widely distributed nonbreeding areas along the coasts of Mexico (69%) and the Atlantic (19%) and Pacific (12%) flyways. Pacific Flyway breeders migrate to nonbreeding areas within the Pacific Flyway. MC is not calculated due to limited sample size.

BANDING Two birds, one breeding in Alaska and the other in Alberta, both spent the nonbreeding period along the coast of California, near San Francisco Bay.

RADIO TRACKING A few tags were deployed through the Motus Wildlife Tracking System; there were no breeding to nonbreeding detections.

SATELLITE TRACKING A tracking study on 8 Alaska breeding birds indicated all birds spent the nonbreeding season on the Pacific coasts of Washington, Oregon, and California. Repeat tracking on 5 of these 8 individuals demonstrated their high rate of fidelity to breeding, migratory, and nonbreeding sites (Ruthrauff et al. 2019a). Another tracking study of 23 individuals from different breeding and nonbreeding areas showed central grassland breeders in California and northern Mexico and birds breeding in eastern Canada migrated to the Gulf of California, while birds spending the nonbreeding season along coastal Georgia bred in North and South Dakota (Olson et al. 2014). Breeding Marbled Godwits (4 males and 4 females) satellite tagged in southern Alberta in 2013 and 2014 spent the nonbreeding season in California and Mexico (C.L. Gratto-Trevor, unpublished data). Nine satellite transmitters deployed on Marbled Godwits on nonbreeding areas along the Gulf Coast of Texas failed to transmit due to tag failure; however, 2 birds were resighted the following year near the original banding location (Scarpignato et al. 2016).

MIGRATORY PATHWAYS Alaska breeding birds cross the Gulf of Alaska on southbound migration to nonbreeding sites in Washington, Oregon, and California and follow the coast of the Gulf of Alaska on northbound migration to their breeding grounds on the Alaska Peninsula (Ruthrauff et al. 2019a). The majority of interior birds migrate down the Central Flyway, although a tracking study revealed an interesting crisscross pattern for some populations, where birds breeding in North and South Dakota traveled southeast to coastal Georgia, while birds breeding in eastern Alberta flew southwest to stationary nonbreeding sites in California and Mexico (Olson et al. 2014).

DATA CONTRIBUTORS
A. Farmer; C. Gratto-Trevor; B. Olson; D.R. Ruthrauff; T.L. Tibbitts; K. Sullivan.

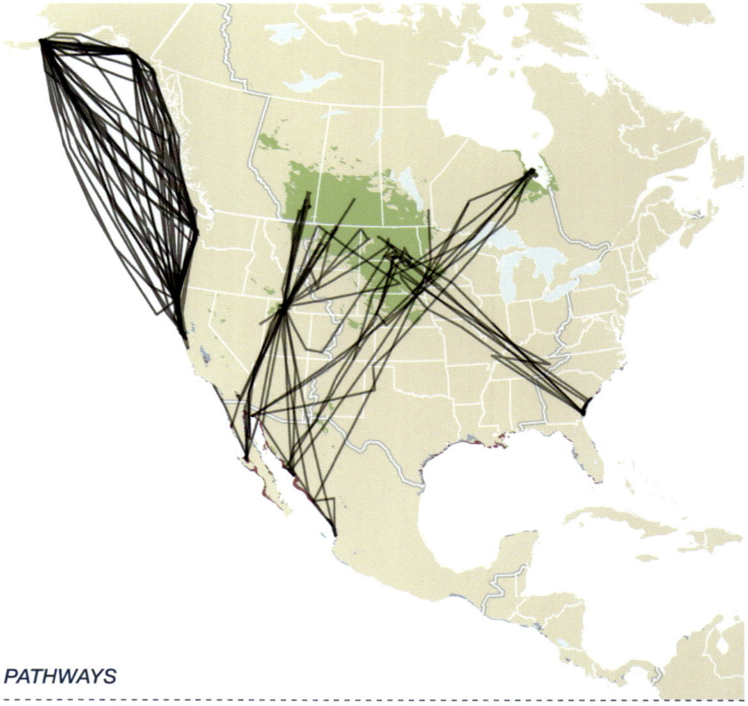

PATHWAYS

RUDDY TURNSTONE *Arenaria interpres*

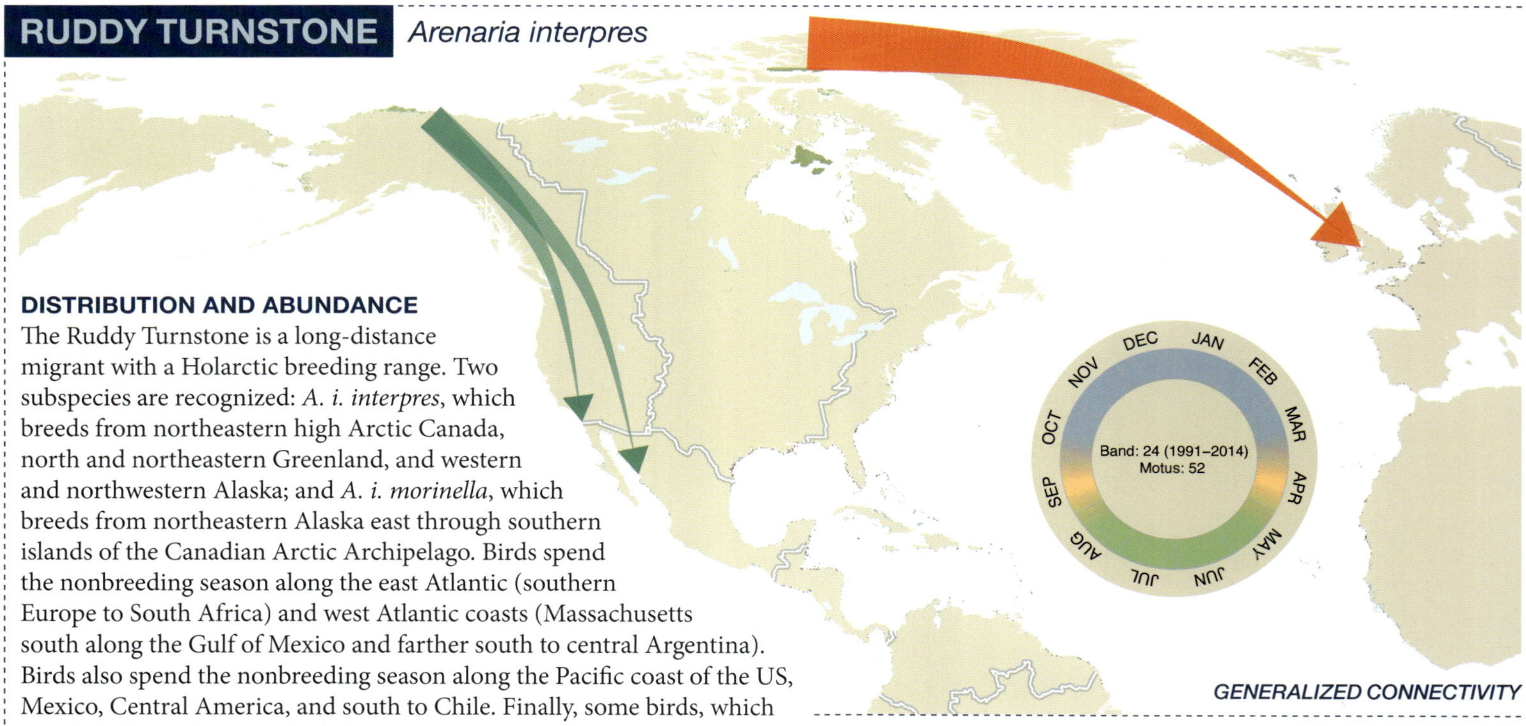

GENERALIZED CONNECTIVITY

DISTRIBUTION AND ABUNDANCE

The Ruddy Turnstone is a long-distance migrant with a Holarctic breeding range. Two subspecies are recognized: *A. i. interpres*, which breeds from northeastern high Arctic Canada, north and northeastern Greenland, and western and northwestern Alaska; and *A. i. morinella*, which breeds from northeastern Alaska east through southern islands of the Canadian Arctic Archipelago. Birds spend the nonbreeding season along the east Atlantic (southern Europe to South Africa) and west Atlantic coasts (Massachusetts south along the Gulf of Mexico and farther south to central Argentina). Birds also spend the nonbreeding season along the Pacific coast of the US, Mexico, Central America, and south to Chile. Finally, some birds, which may breed or only stage in Alaska, are known to spend the nonbreeding season in the central Pacific and as far south as New Zealand and Australia (Nettleship 2020).

	Europe	Mexico	Pacific
Midcontinent (n = 22)	100%	–	–
Pacific (n = 2)	–	50%	50%

ORDER Charadriiformes
FAMILY Scolopacidae
BREEDING HABITAT Coasts, Tundra
NONBREEDING HABITAT Coasts, Beach and Estuary
GLOBAL POPULATION SIZE 600,000
POPULATION CHANGE SINCE 1970 −80%

IUCN RED LIST CATEGORY Least Concern
R2R TIPPING POINT SPECIES Orange-Alert
US LISTING STATUS No
CANADA SPECIES AT RISK No
PIF CCS 14/20

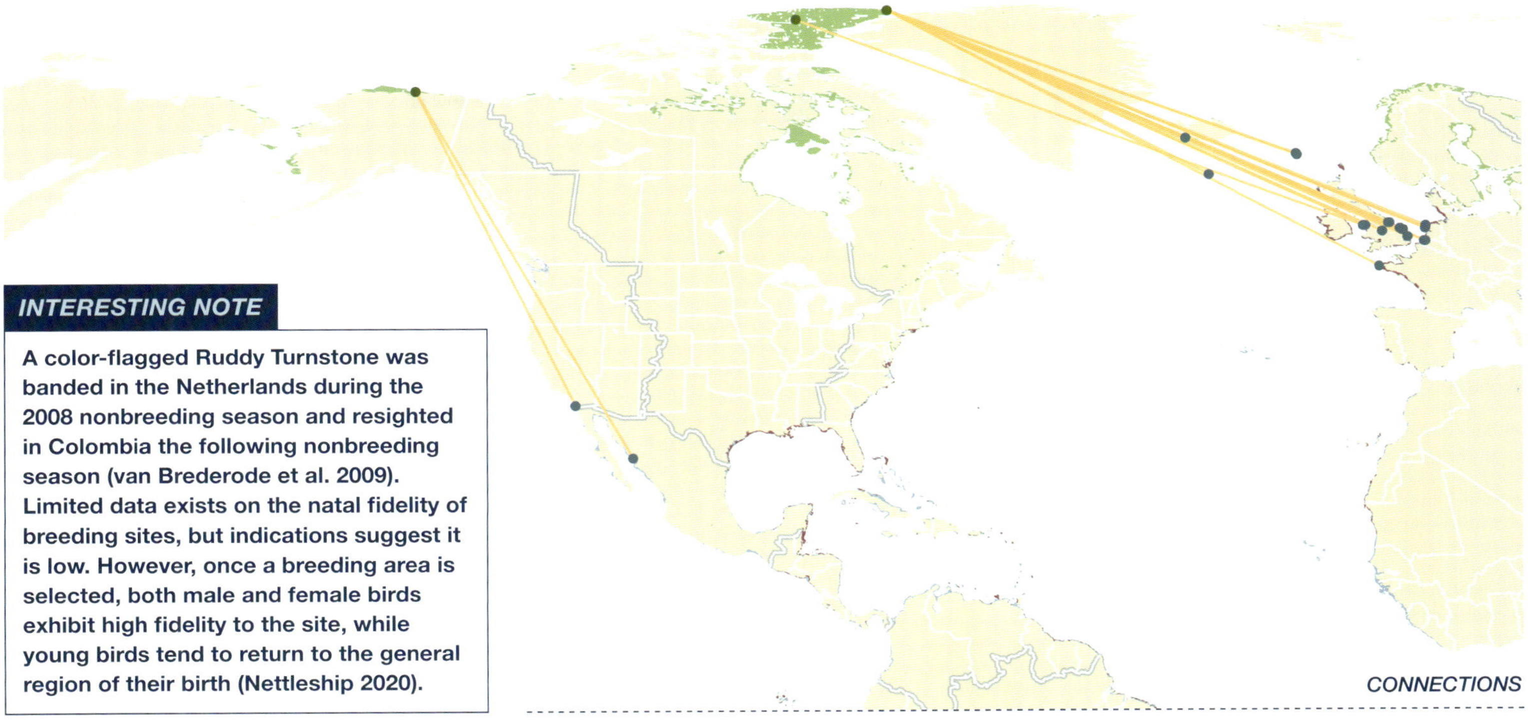

CONNECTIONS

INTERESTING NOTE

A color-flagged Ruddy Turnstone was banded in the Netherlands during the 2008 nonbreeding season and resighted in Colombia the following nonbreeding season (van Brederode et al. 2009). Limited data exists on the natal fidelity of breeding sites, but indications suggest it is low. However, once a breeding area is selected, both male and female birds exhibit high fidelity to the site, while young birds tend to return to the general region of their birth (Nettleship 2020).

MIGRATORY CONNECTIVITY

OVERALL Geolocator and banding data suggest moderate connectivity, but MC was not calculated due to low sample sizes. Within the Midcontinent Flyway, all breeding populations of high Arctic Canada and Greenland migrated to coastal areas in Europe. In the Pacific Flyway, birds breeding in Alaska spend the nonbreeding season along coastal areas within the Pacific Flyway (50%) or Mexico (50%).

BANDING Birds banded on breeding areas within the high Arctic of Canada and Greenland were resighted along coastal areas in Britain, Ireland, and France. Two birds banded on the North Slope of Alaska were resighted during the nonbreeding season: one on the west coast of California and the other on the west coast of Mexico. Band reencounters from birds captured on the Pribilof Islands, Alaska, confirm main nonbreeding areas in the southwest Pacific islands (Davidson and Gill 2008).

RADIO TRACKING Several hundred tags were deployed through the Motus Wildlife Tracking System, leading to many detections during migrations but no breeding to nonbreeding connections.

GEOLOCATOR TRACKING Data from geolocators placed on breeding birds on Southampton Island, Nunavut, Canada, indicated nonbreeding areas in northern and northeastern South America. Birds equipped with geolocators in northern Alaska migrated both south and north along the Pacific coast of North America, staging in Prince William Sound, Alaska, and spending the nonbreeding season in Southern California or northwestern Mexico (D. Ward, unpublished data). One bird tagged on its nonbreeding grounds in southern Australia was documented two years in a row traveling north along the Asian Pacific Rim to breed in Siberia, after which it crossed east to Alaska and returned to Australia across the Pacific, making a stop in Kiribati the first year, and a stop in the Marshall Islands the second year (Minton et al. 2011).

OTHER APPROACHES No genetic, stable isotope, or satellite tracking studies of migratory connectivity were found for this species.

MIGRATORY PATHWAYS Breeding birds on Southampton Island exhibited two distinct migration paths. One population went south through Maine, crossed the Atlantic, and stopped in northeastern South America, where the birds spent the nonbreeding season; these birds then returned north, stopping at Delaware Bay, and then flying directly to breeding areas. The second population of birds exhibited a larger, elliptical migration across the Atlantic to nonbreeding areas in northern South America before migrating north through the midcontinent (Porter and Smith 2013). The high Arctic Canada and Greenland population migrates to and from nonbreeding areas either directly by transatlantic crossing or with a stopover in Iceland or

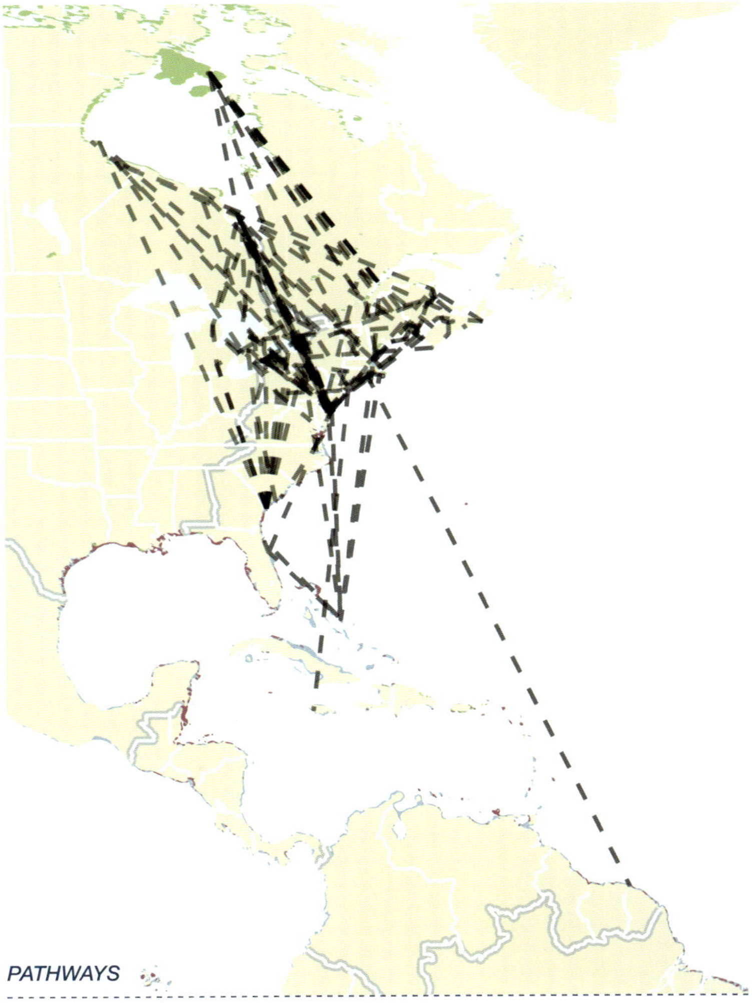

PATHWAYS

Norway (Gudmundsson et al. 1991). Western Alaskan birds display a circular migratory movement, traveling from the Pribolof and Aleutian islands to central Pacific islands, Australia, and New Zealand, and then returning via the Asian Pacific Rim, including staging in Taiwan, China, Japan, and Kamchatka (Thompson 1973; Davidson and Gill 2008; Minton et al. 2010, 2011). Data from geolocators and radio telemetry (Motus) suggest the breeding population from low Arctic Canada migrates to the New England states and then travels over the Atlantic to northeastern South America before returning along the US east coast or midcontinent (Clark et al. 1993; Porter and Smith 2013). Birds breeding in northern Alaska travel to the Pacific coast of North America (D. Ward, unpublished data).

DATA CONTRIBUTORS

S. Flemming; E. Khazan; S. Koch; P. Loring; A. MacDonald; D. Marain; L.J. Niles; F. Sanders; A. Smith; P. Smith.

BLACK TURNSTONE *Arenaria melanocephala*

DISTRIBUTION AND ABUNDANCE

The Black Turnstone is an endemic breeder of western and northwestern Alaska, and spends the nonbreeding period along intertidal areas of the Pacific coast of North America from Cook Inlet and Prince William Sound, Alaska, south to Baja California, Mexico (Handel and Gill 1992, 2020). The nonbreeding range may be moving north as climate warms (Taylor et al. 2022).

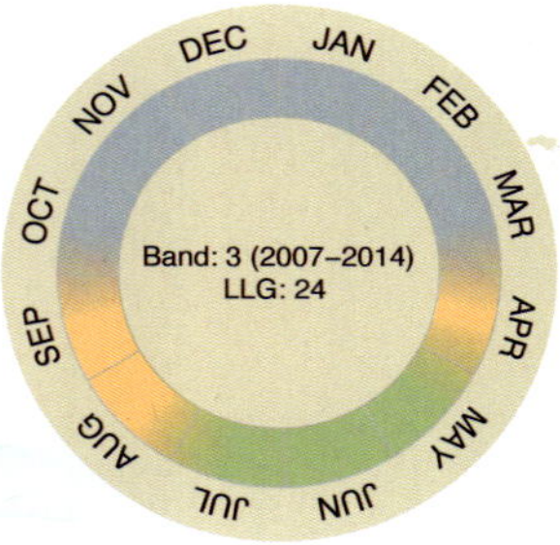

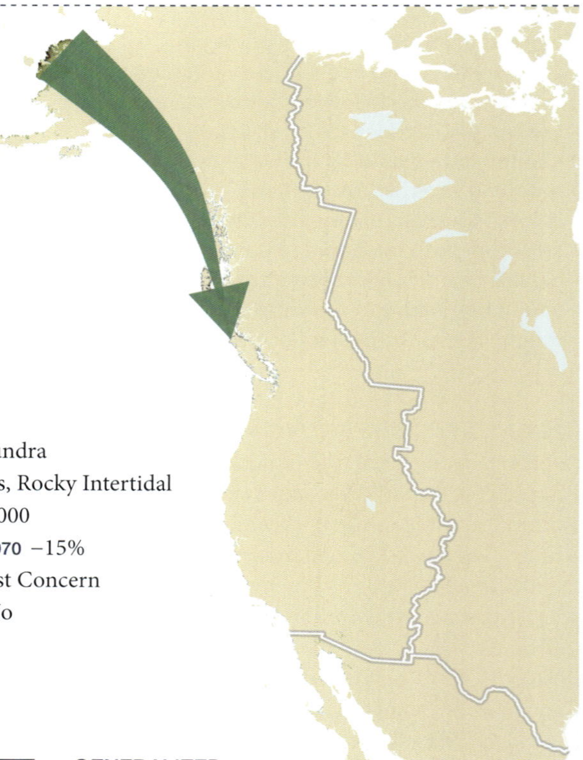

ORDER Charadriiformes
FAMILY Scolopacidae
BREEDING HABITAT Coasts, Tundra
NONBREEDING HABITAT Coasts, Rocky Intertidal
GLOBAL POPULATION SIZE 95,000
POPULATION CHANGE SINCE 1970 −15%
IUCN RED LIST CATEGORY Least Concern
R2R TIPPING POINT SPECIES No
US LISTING STATUS No
CANADA SPECIES AT RISK No
PIF CCS 14/20

MIGRATORY CONNECTIVITY

OVERVIEW Tracking data indicate potential for some population structure with limited evidence of leapfrog migration; i.e., northern breeding birds spend the nonbreeding season farther south. More information from individuals outside the core breeding range are needed to verify this pattern. Birds breeding within the Pacific Flyway also spend the nonbreeding season within the Pacific Flyway (100%), and the majority of the population appears to intermix during all stages of the annual cycle (Handel and Gill 2020; Taylor et al. 2022). MC was not calculated due to limited sample size.

BANDING Three birds banded on the Yukon-Kuskokwim Delta in western Alaska were later resighted in California (Handel and Gill 2020). Natal site fidelity appears weak; 2 of 150 banded nestlings returned to the natal area, but breeding was not confirmed (Handel and Gill 2000).

GEOLOCATOR TRACKING Twenty-three adults equipped with geolocators in western and northwest Alaska were tracked to Prince William Sound, British Columbia, and Oregon. Two adults tagged in Oak Harbor, Washington, migrated to western Alaska (Taylor et al. 2022).

	Pacific
Pacific (n = 27)	100%

GENERALIZED CONNECTIVITY

CONNECTIONS

INTERESTING NOTE

Interannual return rates to breeding grounds by banded adults average 84% (Handel and Gill 2000), and some level of nonbreeding site fidelity is also suggested from a geolocator study (Taylor et al. 2022). Historically, the species has been hunted and still today a small number of birds and eggs are taken by Alaskan natives (Naves et al. 2019).

MIGRATORY PATHWAYS Two high-use regions were identified during migration: one is the western Cook Inlet and Shelikof Strait in southern Alaska, along with the Alexander Archipelago in southeast Alaska; and the other high-use region is Haida Gwaii archipelago in British Columbia. Geolocator data indicate routes similar for northbound and southbound migration. Described as short-distance, short-hop migrants, post-breeding birds travel south along inland routes across the Aleutian Range, where they stopped occasionally before transiting along the outer coast and the Inside Passage of Alaska and British Columbia. A few post-breeding birds made nonstop flights to and from nonbreeding areas within Alaska (Taylor et al. 2022).

DATA CONTRIBUTORS
M.A. Bishop; A.R. Taylor.

Calidris canutus RED KNOT

DISTRIBUTION AND ABUNDANCE

The Red Knot is a Holarctic breeder, with three of six subspecies (*C. c. islandica*, *rufa*, and *roselaari*) distributed in North America from Greenland to northern Alaska. Each subspecies displays only slight variations in physical characteristics, but migration strategies are notably distinct (Piersma and Davidson 1992; Buehler and Piersma 2008). The *C. c. rufa* subspecies spend the nonbreeding season in Tierra del Fuego, coastal Patagonia, Brazil, Florida, South Carolina, Georgia, the Caribbean, and Texas (Baker et al. 2020). *C. c. roselaari* nonbreeding areas are mainly on the Pacific coast of Baja California, the Pacific Northwest coast of Mexico in the Gulf of California, and California. Smaller numbers have been reported along the Texas coast and along the Pacific coast south of Mexico, to southern Chile, suggesting the possibility of an area of overlap with *C. c. rufa* (Carmona et al. 2013; Conklin et al. 2022). During the nonbreeding season, *C. c. islandica* can be found on the European seaboard, including the British Isles, and countries surrounding the Wadden Sea (Baker et al. 2020).

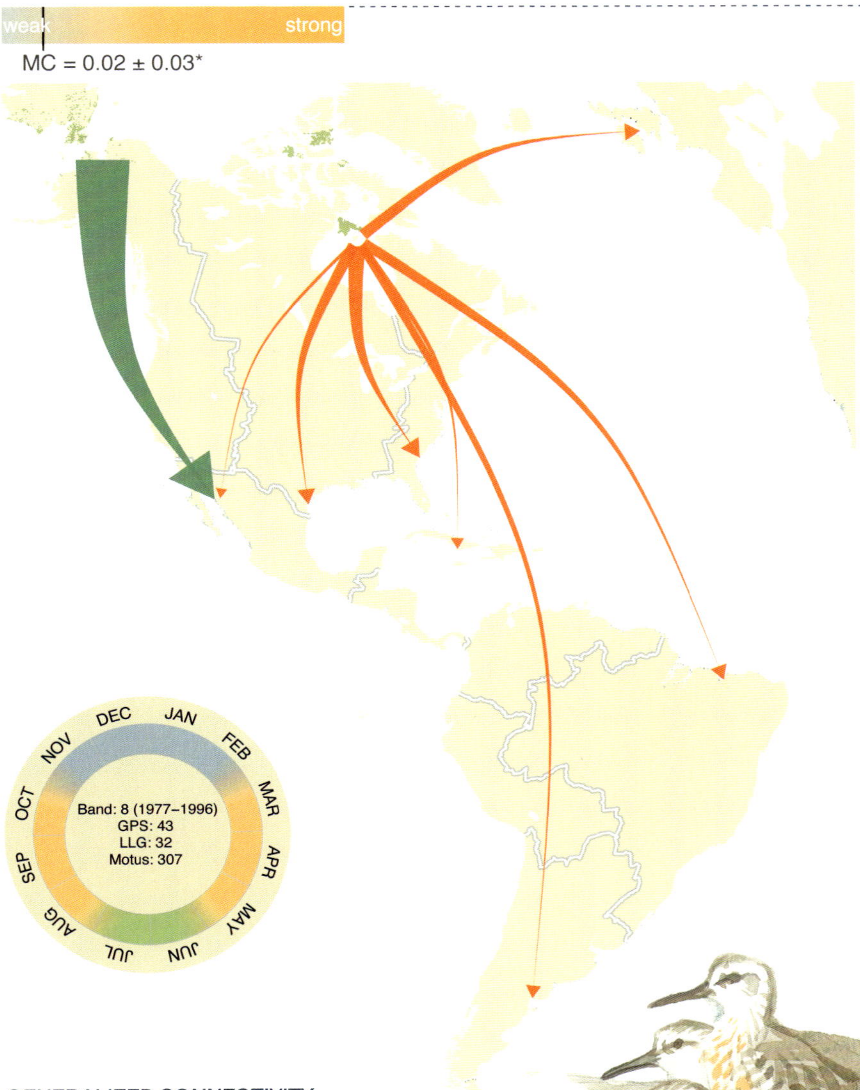

GENERALIZED CONNECTIVITY

ORDER Charadriiformes
FAMILY Scolopacidae
BREEDING HABITAT Tundra, Arctic
NONBREEDING HABITAT
Coasts, Beach and Estuary
GLOBAL POPULATION SIZE 1,000,000
POPULATION CHANGE SINCE 1970 −82%
IUCN RED LIST CATEGORY Near Threatened
R2R TIPPING POINT SPECIES Orange-Alert
US LISTING STATUS Threatened *C. c. rufa* subspecies
CANADA SPECIES AT RISK Endangered *C. c. rufa* subspecies.
Threatened *C. c. roselaari* subspecies
PIF CCS 13/20

	Atlantic	Brazil	Caribbean	Europe	Mexico	Midcontinent	Southern Cone
Midcontinent (n = 44)	27%	16%	7%	18%	2%	18%	11%
Pacific (n = 15)	–	–	–	–	100%	–	–

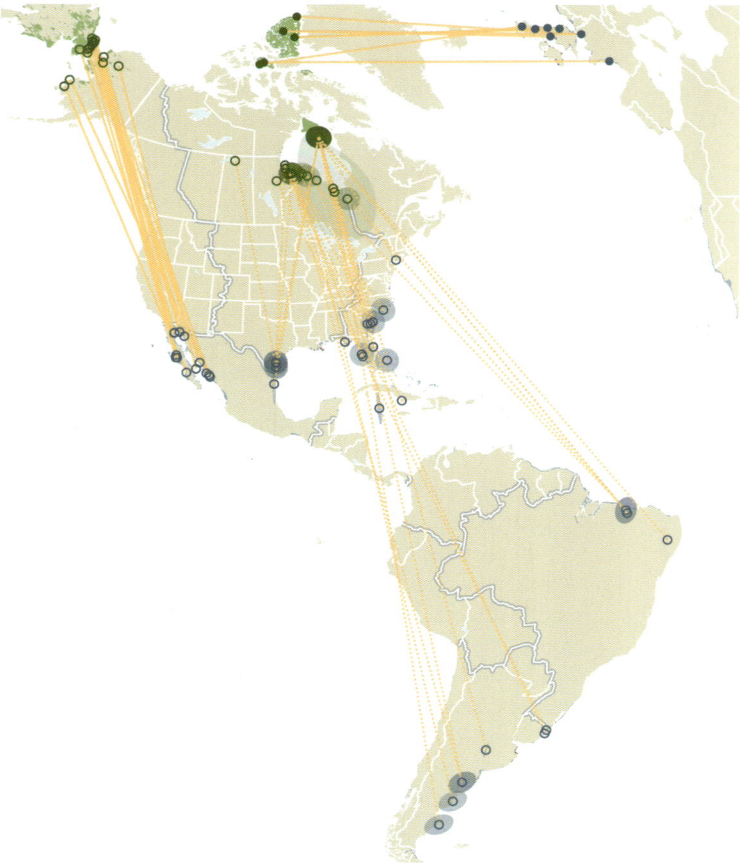

CONNECTIONS

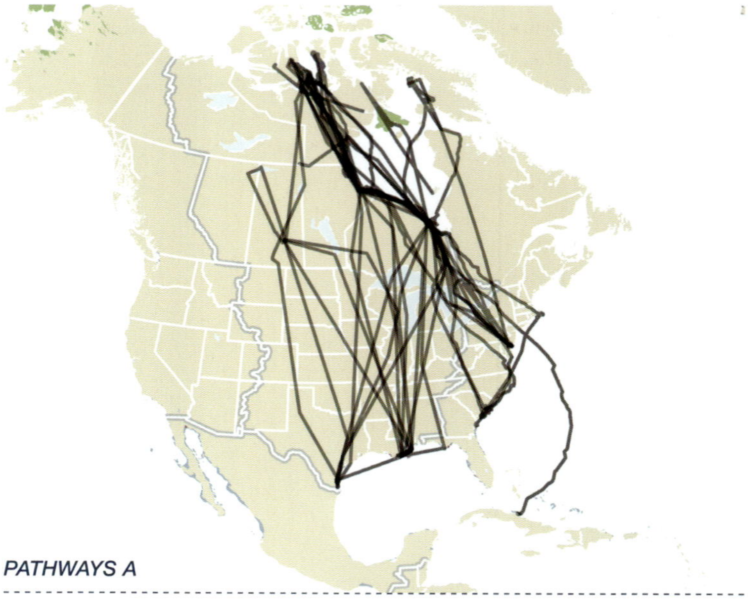

PATHWAYS A

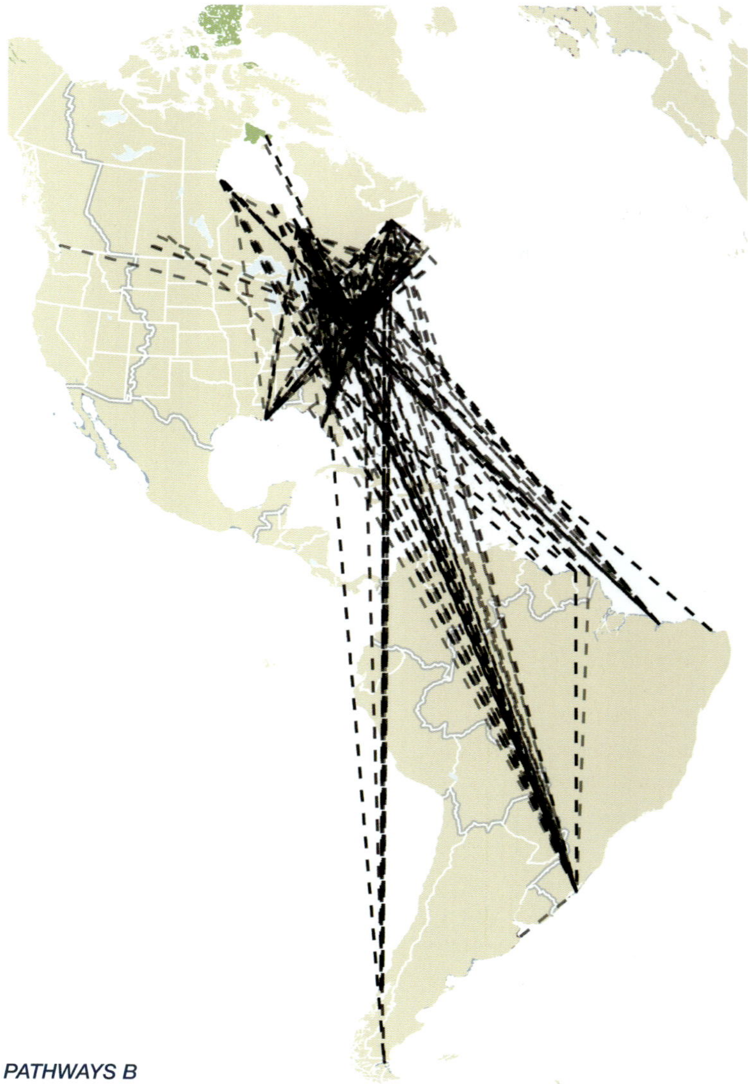

PATHWAYS B

MIGRATORY CONNECTIVITY

OVERALL Existing banding, geolocator, satellite, and GPS tracking data suggest weak migratory connectivity (MC = 0.02 ± 0.03), as well as observations of overlap during migration and at nonbreeding areas. Birds breeding in the Arctic regions of the Midcontinent Flyway inhabit several regions during the nonbreeding season, including the Atlantic Flyway (27%), Midcontinent Flyway (18%), Brazil (16%), and Southern Cone region of South America (11%). Birds breeding in high Arctic Canadian regions of the Midcontinent Flyway were later found in Europe (18%). Birds breeding in Alaska of the Pacific Flyway spend the nonbreeding season in Mexico (100%).

BANDING Resightings of *C. c. roselaari* identified strong connectivity of birds using sites in northwest Mexico to sites in Washington and revealed an area of possible overlap in nonbreeding range between *C. c. rufa* and *C. c. roselaari* in southern Mexico (Carmona et al. 2013). Resightings of birds stopping over in the Mingan Archipelago were used to estimate a significant passage population of *C. c. rufa*, thus identifying this area as an important post-breeding stopover for the species (Lyons et al. 2018b). Resighting of *C. c. islandica* birds between the high Arctic of Canada and Europe confirmed this east-west migration pattern.

RADIO TRACKING Motus tracking of *C. c. rufa* birds captured near Hudson Bay and northeastern Canada indicated movements to the southeastern US, northeastern and southeastern Brazil, and Tierra del Fuego.

GEOLOCATOR TRACKING Due to 24-hour daylight at Arctic breeding areas, geolocator data are unable to determine the northernmost part of Red Knots' migratory journeys and breeding locations. However, geolocator deployments at Delaware Bay gathered three complete migration tracks, with all birds (*C. c. rufa*) going to the Arctic, and then the southern US, Caribbean, and South America for the nonbreeding season (Burger et al. 2012, 2024). The longest-migrating bird migrated to southern South America and took a terrestrial route over the Pampas to its nonbreeding area (Niles et al. 2010). Geolocator deployments at Monomoy Island, Massachusetts, garnered eight tracks, with four birds (*C. c. rufa*) shown to spend the nonbreeding period between Virginia and Florida, and the remaining four in the Caribbean (Burger et al. 2012). Data from eight geolocators deployed on the Texas Gulf Coast showed the tagged birds (*C. c. rufa*) spent a large majority of the nonbreeding season in this area (Newstead et al. 2013). Geolocators deployed on more than 20 adult *C. c. roselaari* from Nome showed that the birds spent the nonbreeding season from Southern California to northwest Mexico (J. Johnson, unpublished data).

SATELLITE TRACKING Tracking of *C. c. rufa* with GPS transmitters along the Texas and New Jersey coasts led to the identification of a potentially new breeding location in the Arctic, on Victoria Island, and confirmation of a breeding location on Prince Charles Island (Scarpignato et al. 2016). Similarly, tracking of *C. c. rufa* birds from both the Atlantic and Gulf coasts show connections to the central Arctic (B. Marsh, unpublished data; D. Newstead, unpublished data; South Carolina Department of Natural Resources, unpublished data). Data from GPS transmitters showed strong connections of *C. c. roselaari* to Alaska and Wrangel Island, Russia, and northwest Mexico, including the Baja California peninsula (J. Johnson, unpublished data).

STABLE ISOTOPES Stable isotopes and resighting data suggest that the Altamaha River area of Georgia is an important post-breeding stopover location, and that 82%–96% of birds stopping there spend the nonbreeding season in the southeast US, the Caribbean, and northern Brazil, with the remainder in Patagonia (Lyons et al. 2017). Stable isotope data also estimated that the birds stopping over in Delaware Bay originated from nonbreeding areas in Bahía Lomas, Chile (58%), Florida (30%), and Rio Grande, Argentina (6%), though some remain unclassified (7%; Atkinson et al. 2005).

GENETICS Three nonbreeding populations for *C. c. rufa* were identified in 2011 using band resightings and genetics, from Tierra del Fuego, northern Brazil, the Caribbean, and Florida (Baker et al. 2020). However, these populations have some overlap at staging sites in North America and the Caribbean. Recent genetic analysis of *C. c. roselaari* found that Alaskan and Russian breeding populations are genetically distinct, despite sharing a migratory flyway (Conklin et al. 2022). On the other hand, very weak genetic differentiation was found between *C. c. rufa* and *C. c. islandica*, which have almost no contact throughout the annual cycle (Conklin et al. 2022).

MIGRATORY PATHWAYS The general pathway for northward and southward migration of *C. c. islandica* involves long-distance flights between breeding and nonbreeding grounds, stopovers at staging sites in the Wadden Sea coasts of the Netherlands and Germany, and transoceanic flights to Iceland and northern Norway, followed by a return to breeding grounds in Greenland and northeastern Canada. *C. c. rufa* breeds in the central Canadian Arctic and has well-known stopover areas during northward migration along the Atlantic coast of the Americas, including Río Gallegos, Península Valdés, San Antonio Oeste, Bahía Blanca, Punta Rasa, and Bahía de Samborombón; Lagoa do Peixe; Maranhão; Florida through the Carolinas; the Virginia Barrier Islands; and Delaware Bay and Hudson Bay (Burger et al. 2024). Data from 119 tracks indicated the three major migratory pathways of *C. c. rufa* knots, and that birds from all three pathways stopped over at several places during migration on both the southward and northward migration (Burger et al. 2024). *C. c. rufa* from the Tierra del Fuego nonbreeding population stopped at San Antonio Oeste, Argentina, and then Delaware Bay; most *C. c. rufa* that spent the nonbreeding season in either northern or southern South America also stopped in Delaware Bay on their northbound migration, whereas those that were migrating from the US or the Caribbean were less likely to stop over. Almost all of the tagged *C. c. rufa* in the western Atlantic Flyway stopped in southern Hudson Bay and/or James Bay in Canada before going on to their Arctic breeding grounds (Burger et al. 2024). Eight birds equipped with geolocators in Massachusetts all stopped over in the Nelson River estuary during northward migration (Niles et al. 2012). During southward migration, staging areas include the Hudson Bay, James Bay, the Mingan Archipelago, the North Atlantic coast, the mouth of the Altamaha River, the Caribbean, and the northern coast of South America from Brazil to Guyana. Red Knots from nonbreeding areas in Texas migrated through the Midcontinent Flyway on both north- and southward migrations with stopover sites in the northern Great Plains of the US and Canada and the Nelson River Delta and Hudson Bay area (Newstead et al. 2013). Geolocator tracks and band returns indicate that *C. c. rufa* stopping over along the Gulf Coast may spend the nonbreeding season along the Pacific coast, and use the Isthmus of Tehuantepec as a migratory corridor (Lamb et al. 2017). *C. c. roselaari* are distributed during northward migration across a few key stopover sites: Guerrero Negro, Ojo de Liebre, and Golfo de Santa Clara, Mexico; San Francisco Bay, California; Grays Harbor and Willapa Bay, Washington; and Copper River and Yukon-Kuskokwim River deltas, Alaska (Carmona et al. 2013). Other than Guerrero Negro, Mexico, less is known about stopovers during southward migration due to few resightings north of Mexico (Carmona et al. 2013).

DATA CONTRIBUTORS

A. Anderson; Y. Aubry; K. Bennett; J. Burger; A. Desrochers; A.D. Dey; S. Feigin; C. Friis; S. Heath; J.A. Johnson; E. Khazan; S. Koch; K. Lefevre; P. Loring; A. MacDonald; S. Mackenzie; D. Marain; B. Marsh; A. McKellar; C. Morrissey; D.J. Newstead; L.J. Niles; K. O'Brien; N. Pau; G. Perkins; R.R. Porter; F. Sanders; A. Smith; P. Smith; A.L. Scarpignato; J. Tremblay; R. Wood; Delaware Shorebird Project; Migratory Connectivity Project; South Carolina Department of Natural Resources; US Fish and Wildlife Service.

SURFBIRD *Calidris virgata*

DISTRIBUTION AND ABUNDANCE

The Surfbird has the longest and narrowest nonbreeding range of any North American shorebird, extending more than 17,500 km along the Pacific coastline (Senner and McCaffery 2020). From Kodiak Island in Alaska to the Strait of Magellan in Chile, they inhabit rocky coastal shores during the nonbreeding season; their range typically only reaches a few meters inland from the tide line. A significant proportion of their population have been estimated to spend the nonbreeding season along the rocky beaches of the northern Chilean shoreline (Garcia-Walther 2016). During the breeding season, they are found above the tree line on the mountains of Alaska and western and southwestern Yukon.

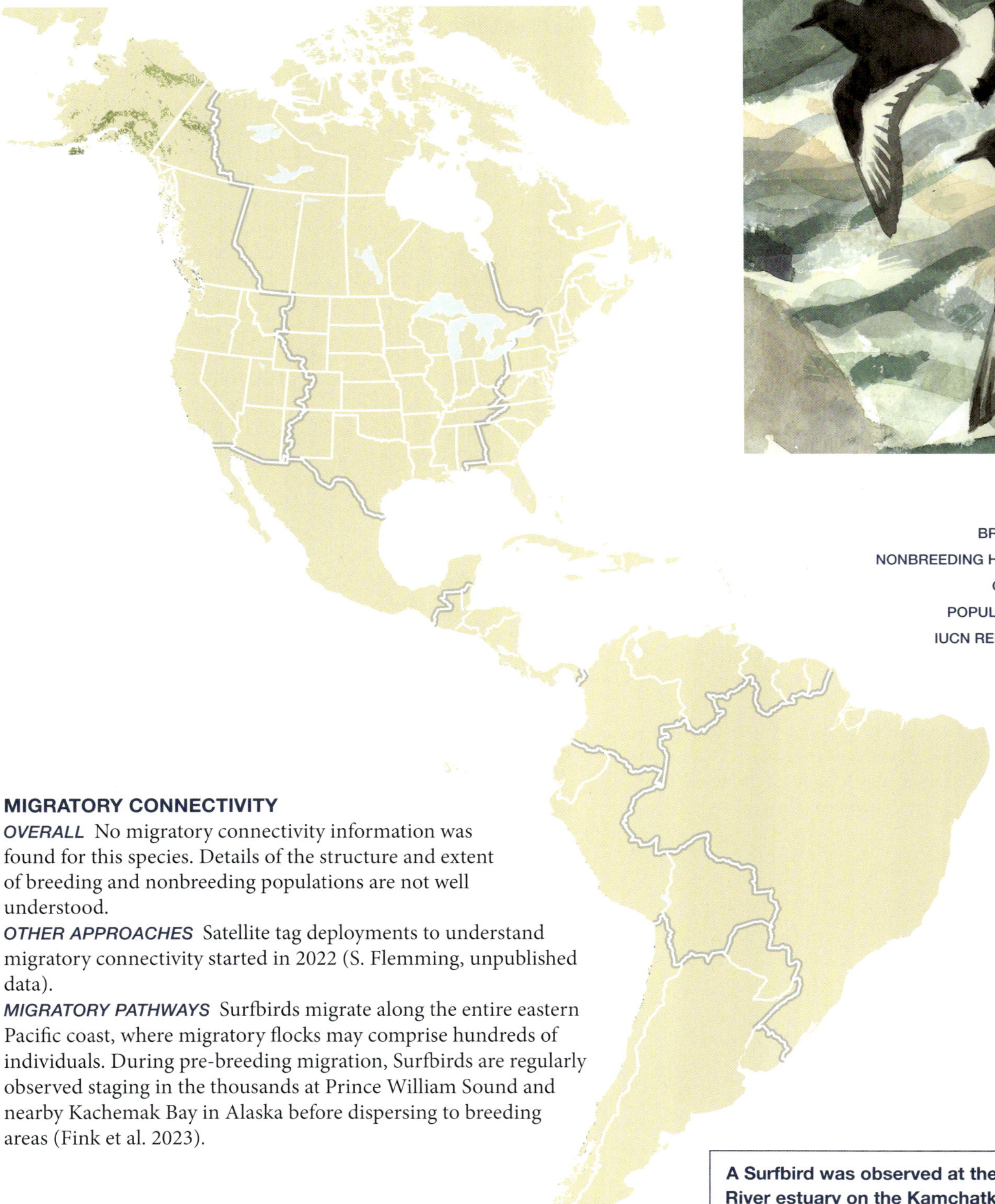

DISTRIBUTION AND ABUNDANCE

ORDER Charadriiformes
FAMILY Scolopacidae
BREEDING HABITAT Tundra, Alpine
NONBREEDING HABITAT Coasts, Rocky Intertidal
GLOBAL POPULATION SIZE 70,000
POPULATION CHANGE SINCE 1970 −53%
IUCN RED LIST CATEGORY Least Concern
R2R TIPPING POINT SPECIES No
US LISTING STATUS No
CANADA SPECIES AT RISK No
PIF CCS 15/20

MIGRATORY CONNECTIVITY

OVERALL No migratory connectivity information was found for this species. Details of the structure and extent of breeding and nonbreeding populations are not well understood.

OTHER APPROACHES Satellite tag deployments to understand migratory connectivity started in 2022 (S. Flemming, unpublished data).

MIGRATORY PATHWAYS Surfbirds migrate along the entire eastern Pacific coast, where migratory flocks may comprise hundreds of individuals. During pre-breeding migration, Surfbirds are regularly observed staging in the thousands at Prince William Sound and nearby Kachemak Bay in Alaska before dispersing to breeding areas (Fink et al. 2023).

INTERESTING NOTE

A Surfbird was observed at the Khairusova-Belogolovaya River estuary on the Kamchatka Peninsula in the Russian Far East in July 2019, which marks the first record of the Surfbird in Eurasia (Stubbings et al. 2020).

Calidris himantopus **STILT SANDPIPER**

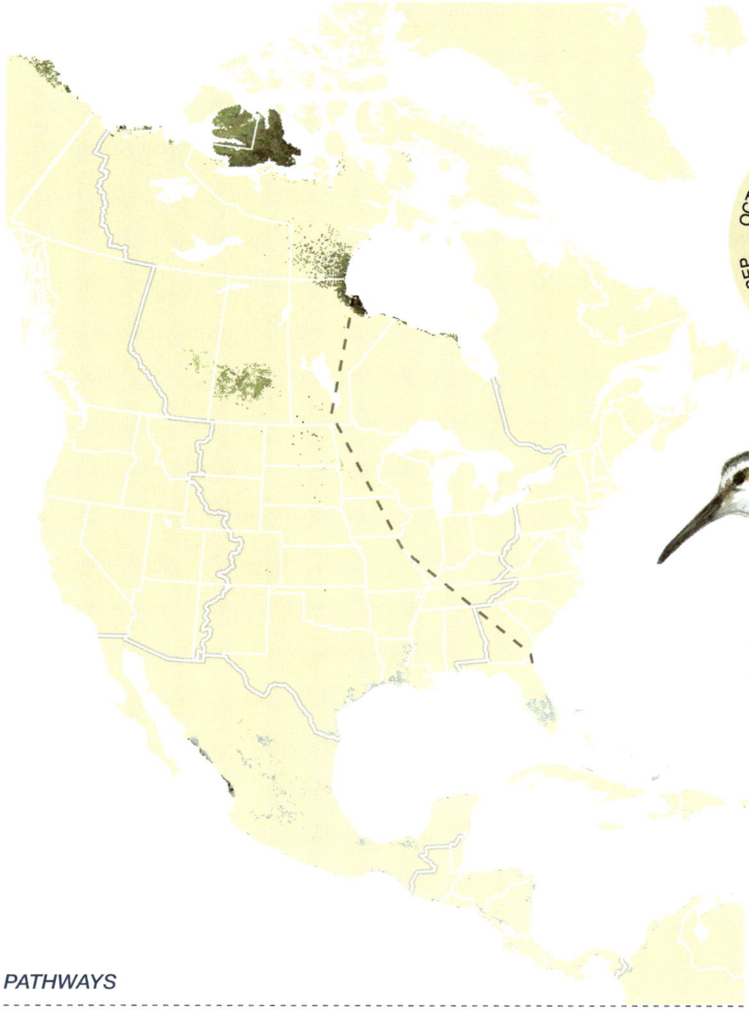

PATHWAYS

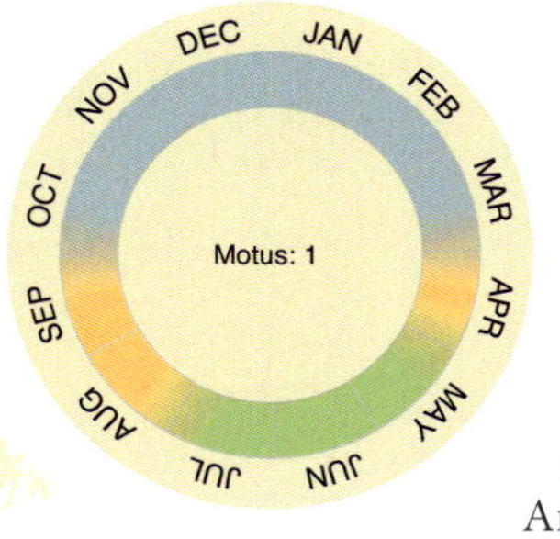

DISTRIBUTION AND ABUNDANCE

The Stilt Sandpiper breeds exclusively in the low Arctic and subarctic tundra regions of North America, from James Bay to northeast Alaska. During the nonbreeding season, the Stilt Sandpiper can be found along the Gulf Coast of Texas and Louisiana; in small numbers on the shores of the Salton Sea, California; locally in Florida; in coastal areas of Mexico; and in Central America, the Caribbean, Columbia, Venezuela, Suriname, and Peru. Its primary nonbreeding area is located in the Pampas of Argentina and Uruguay (Klima and Jehl 2020).

ORDER Charadriiformes
FAMILY Scolopacidae
BREEDING HABITAT Wetlands, Tundra
NONBREEDING HABITAT Wetlands, Lakes and Rivers
GLOBAL POPULATION SIZE 1,200,000
POPULATION CHANGE SINCE 1970 −33%
IUCN RED LIST CATEGORY Least Concern
R2R TIPPING POINT SPECIES Orange-Alert
US LISTING STATUS No
CANADA SPECIES AT RISK No
PIF CCS 15/20

INTERESTING NOTE

In the salt production impoundments of the Bay of La Unión, a group of seven Stilt Sandpipers were sighted on August 6, 1994, marking the first documented occurrence in El Salvador (Komar 2003).

MIGRATORY CONNECTIVITY

OVERALL Migratory connectivity is relatively unknown; currently few data are available for Stilt Sandpiper.

RADIO TRACKING In 2021 and 2022, 10 Stilt Sandpipers were equipped with nanotags near Churchill, Manitoba. Five tags were detected by the Motus Wildlife Tracking System during southbound migration (O. Maillet, unpublished data). Since then, several other projects are tracking Stilt Sandpiper with Motus; there are no individual breeding to nonbreeding detections, but there are several detections throughout migration.

STABLE ISOTOPES Stable isotope data from birds hunted in Barbados suggested that 13 juvenile Stilt Sandpipers likely came from the North Slope of Alaska and/or Nunavut and Manitoba (Reed et al. 2018).

OTHER APPROACHES No satellite telemetry, banding, or genetic studies of migratory connectivity were found for this species.

MIGRATORY PATHWAYS This species migrates northward from northern South America with nonstop flights to the Gulf Coast, then through the central states to North Dakota, Saskatchewan, and Alberta before continuing nonstop to central and west Arctic. From breeding areas on the Atlantic coast, it is thought to turn northwestward toward Hudson Bay on pre-breeding migration. Important stopover areas include Cheyenne Bottoms, Kansas; the Quill Lakes, Saskatchewan; and the Prairie Pothole Region. Post-breeding, most individuals take a southerly migration route along a broad front through the Great Plains and Midwest, though some birds use the Atlantic coast for stopovers, and fewer still are found in western states. Similarly, birds equipped with nanotags were detected first in Manitoba, then in Missouri along the Mississippi River. Following the Missouri stopover, birds were detected in the southeastern US and Costa Rica (O. Maillet, unpublished data).

DATA CONTRIBUTORS
L. McKinnon; E. Nol; P. Smith.

SANDERLING *Calidris alba*

DISTRIBUTION AND ABUNDANCE

Sanderling are circumpolar breeders in the high Arctic tundra. Within North America, they can be found in the Canadian Arctic Archipelago. During the nonbreeding period, they are among the most widespread of shorebird species, associated strictly with marine coastal sandy beaches, extending from British Columbia on the Pacific coast and Massachusetts on the Atlantic coast south throughout virtually all of temperate US and Central and South America (Macwhirter et al. 2020).

ORDER Charadriiformes
FAMILY Scolopacidae
BREEDING HABITAT Tundra, Arctic
NONBREEDING HABITAT Coasts, Beach and Estuary
GLOBAL POPULATION SIZE 650,000
POPULATION CHANGE SINCE 1970 −74%
IUCN RED LIST CATEGORY Least Concern
R2R TIPPING POINT SPECIES Orange-Alert
US LISTING STATUS No
CANADA SPECIES AT RISK No
PIF CCS 13/20

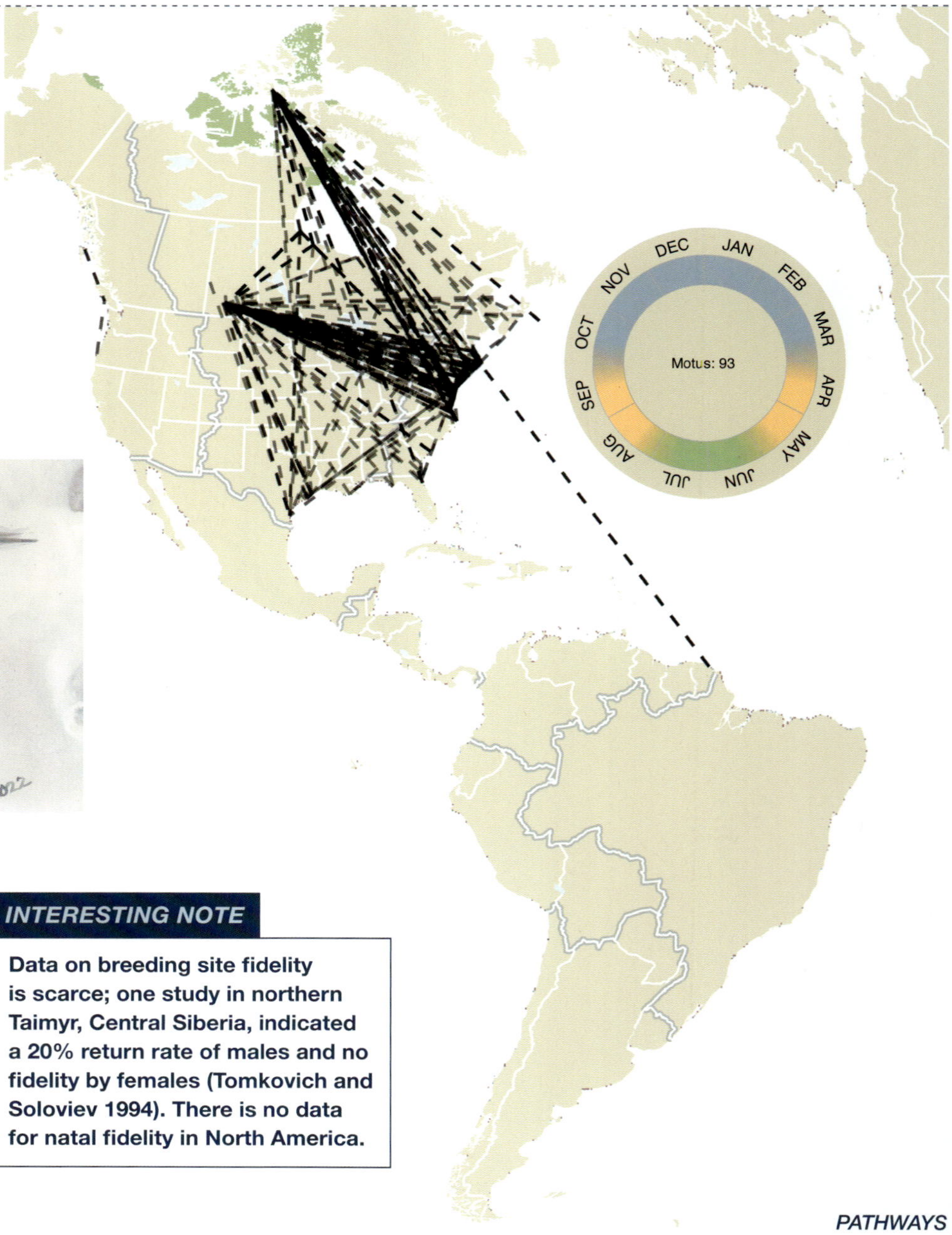

INTERESTING NOTE

Data on breeding site fidelity is scarce; one study in northern Taimyr, Central Siberia, indicated a 20% return rate of males and no fidelity by females (Tomkovich and Soloviev 1994). There is no data for natal fidelity in North America.

PATHWAYS

MIGRATORY CONNECTIVITY

OVERALL Current stable isotope and genetic data suggest that Sanderling exhibit a low degree of migratory connectivity. However, tracking or other types of data are lacking during the nonbreeding season and from across the range, including from birds breeding in the high Arctic.

RADIO TRACKING Several hundred tags were deployed through the Motus Wildlife Tracking System, leading to many detections during migration but no breeding to nonbreeding connections.

STABLE ISOTOPES Stable isotope analyses on feathers collected from Sanderling staging at Chaplin Lake, Saskatchewan, suggested possible nonbreeding origins in southern South America, northwestern South America, and the Gulf of Mexico, indicating a high degree of overlap between these nonbreeding populations during migratory stopover (Labarrère 2016).

GENETICS Studies to date have not found any clear genetic differentiation between Sanderling breeding populations (Conklin et al. 2016b; Sallaberry 1990).

MIGRATORY PATHWAYS Sanderling spending the nonbreeding season on the Gulf of Mexico and the Pacific coast of South America most often use the Central Flyway during northward migration and the Atlantic route during southward migrations. Sanderling spending the nonbreeding season on the Atlantic coast have been observed using Atlantic, Central, and Mississippi flyways during their migration (Myers et al. 1990; Sallaberry and Mann 2007). Band resightings of 35 Sanderling that were captured at Chaplin Lake, Saskatchewan—a key stopover site during their northward migration—along with detections from Motus tracking towers, show that these birds used both the Atlantic and Gulf coasts of the US during migration. Additionally, they were detected along the Atlantic and Pacific coasts of South America, suggesting that Sanderlings from different nonbreeding populations mix at the Chaplin Lake staging site (Sallaberry and Mann 2007). Sanderling equipped with radio nanotags at nonbreeding sites across the Gulf Coast of Texas and Louisiana were also detected by automated Motus towers at Chaplin Lake (Bianchini et al. 2020). Studies from across this species distribution suggest that such mixing is typical (Myers et al. 1990; Lisovski et al. 2016; Reneerkens et al. 2009).

DATA CONTRIBUTORS

M. Drever; W. Easton; S. Flemming; S. Koch; P. Loring; V. Loverti; A. McKellar; C. Morrissey; L.J. Niles; P. Smith; P. Woodard; Y. Zharikov.

Calidris alpina **DUNLIN**

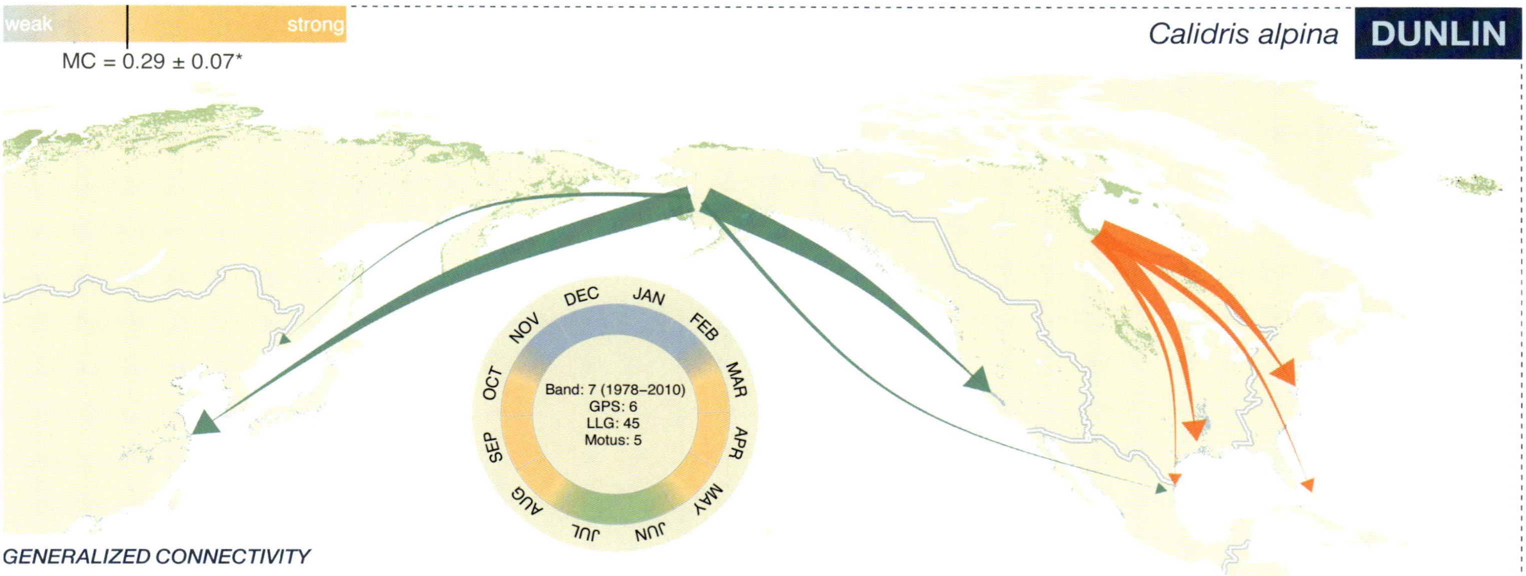

GENERALIZED CONNECTIVITY

	Asia	Atlantic	Caribbean	Mexico	Midcontinent	Pacific	Russia
Midcontinent (n = 26)	–	46%	12%	8%	35%	–	–
Pacific (n = 28)	46%	–	–	4%	–	46%	4%

DISTRIBUTION AND ABUNDANCE

Within North America there are three subspecies of Dunlin: *C. a. pacifica*, which breeds in western Alaska and spends the nonbreeding period along the Pacific coast of the US and Mexico; *C. a. arcticola*, which breeds in northern Alaska and northwestern Canada and spends the nonbreeding period along the East Asian–Australasian Flyway; and *C. a. hudsonia*, which breeds in central Canada and spends the nonbreeding period along the Atlantic coast and the coast of the Gulf of Mexico (Warnock and Gill 2020).

ORDER Charadriiformes
FAMILY Scolopacidae
BREEDING HABITAT Wetlands, Tundra
NONBREEDING HABITAT Coasts, Beach and Estuary
GLOBAL POPULATION SIZE 5,500,000
POPULATION CHANGE SINCE 1970 −58%
IUCN RED LIST CATEGORY Least Concern
R2R TIPPING POINT SPECIES No
US LISTING STATUS No
CANADA SPECIES AT RISK No
PIF CCS 12/20

MIGRATORY CONNECTIVITY

OVERVIEW Band reencounter and geolocator data indicate there is strong migratory connectivity among the three subspecies; however, there is mixing of populations on the nonbreeding grounds within the *C. a. arcticola* and *C. a. hudsonia* subspecies (Gill et al. 2013; Lagassé et al. 2020, 2022; Wright et al. 2022), and a chain migration pattern of connectivity within the *C. a. pacifica* subspecies (R. Lanctot and S. Yezerinac, unpublished data) resulting in moderate migratory connectivity (MC = 0.29 ± 0.07). Dunlin breeding in the Midcontinent Flyway migrate to nonbreeding areas in the Atlantic (46%) and Midcontinent (35%) flyways, while Pacific Flyway breeding birds migrate to nonbreeding areas within the Pacific Flyway (46%) and Asia (46%).

BANDING Band resightings confirmed mixing of *C. a. arcticola* and *C. a. pacifica* during post-breeding in western Alaska, and chain migration of *pacifica*, with birds marked on the Yukon-Kuskokwim Delta resighted mostly in southern British Columbia and Washington, and birds marked on the Alaska Peninsula resighted primarily in Northern California (Gill et al. 2013). Resightings of *arcticola* Dunlin were primarily in Japan, with smaller numbers in the Yellow and China seas (Lagassé et al. 2020).

RADIO TRACKING Nearly a thousand tags were deployed through the Motus Wildlife Tracking System; there were limited breeding to nonbreeding detections (E. Nol et al., unpublished data) and numerous detections during migration (Nol et al., unpublished data; Drever et al. unpublished data).

GEOLOCATOR TRACKING Dunlin were tracked with both light-level and GPS geolocators. *Hudsonia* Dunlin tracked from Churchill, Manitoba, spread out evenly across the entire known nonbreeding range covering the Gulf of Mexico and Atlantic coast (Wright et al.

SHOREBIRDS, AUKS, GULLS, AND TERNS (CHARADRIIFORMES)

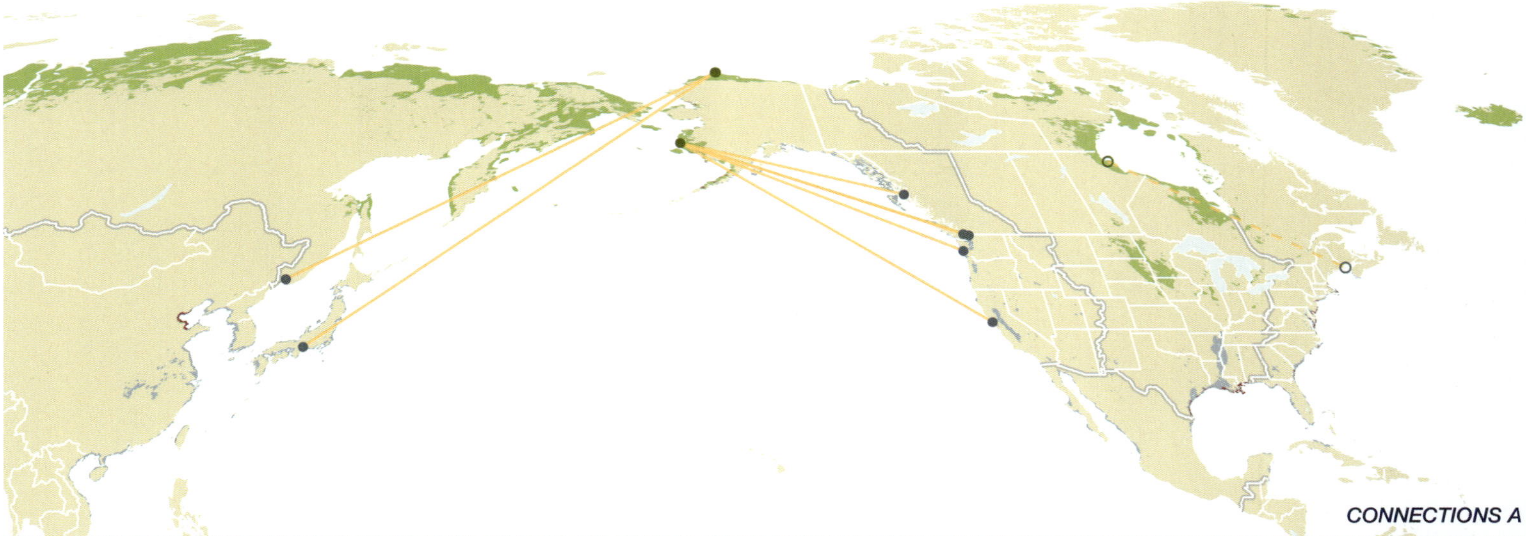

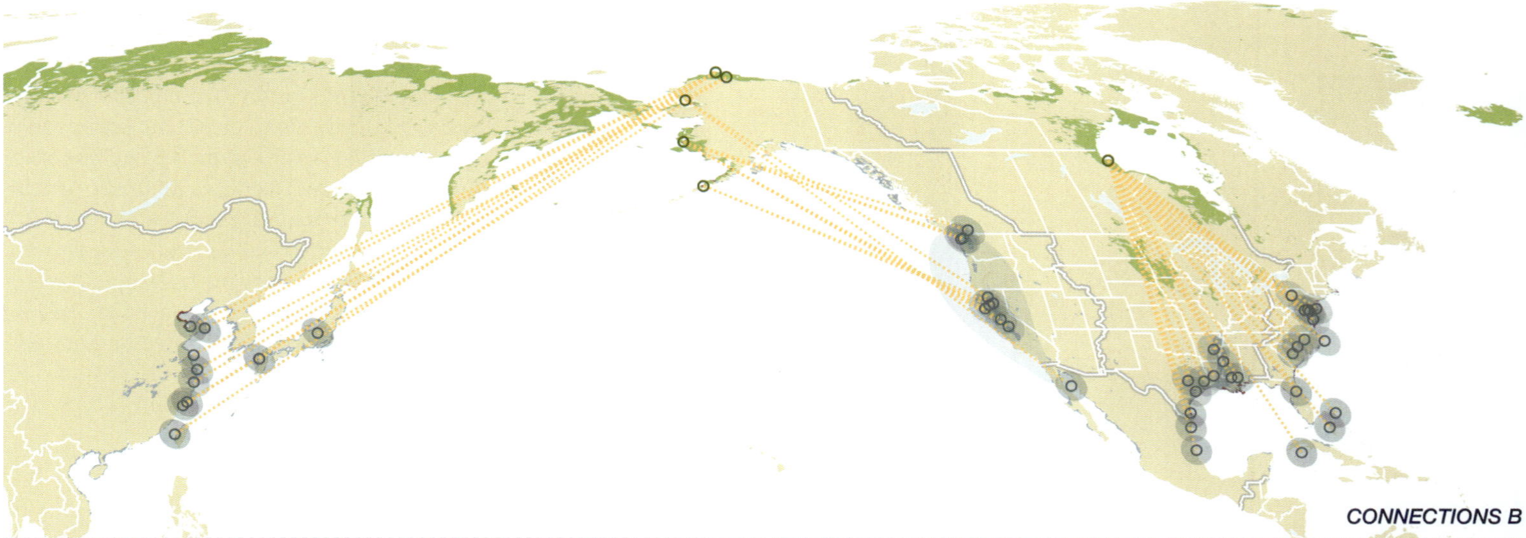

INTERESTING NOTE

There is strong annual fidelity to breeding sites (Saalfeld and Lanctot 2015; Warnock and Gill 2020) and occasionally to nonbreeding sites (Lagassé et al. 2020; Warnock and Gill 2020). The species may also show fidelity to natal areas (Warnock and Gill 2020). There appears to be a negative effect on the start of migration for birds equipped with glue-on radio transmitters (Warnock et al. 2004) and a lower return rate for birds equipped with geolocators attached to leg flags (Weiser et al. 2016) and GPS archival tags attached with leg-loop harnesses (R. Lanctot, unpublished data).

2022). A geolocator study involving *C. a. arcticola* and three other subspecies of Russia-breeding Dunlin indicated that *C. a. arcticola* exhibit weak connectivity between breeding and nonbreeding locations (Lagassé et al. 2022).
GENETICS mtDNA analyses indicated genetic differentiation between *C. a. hudsonia* and the other two subspecies, but indicated some gene flow between *C. a. arcticola* and *C. a. pacifica*, perhaps by *C. a. arcticola* stopping on the Yukon-Kuskokwim Delta to breed on their way to breeding grounds on the North Slope of Alaska (M.P. Miller et al. 2015). A genome-wide analysis from 80 individuals indicated no genetic structure within or among nonbreeding *C. a. pacifica* Dunlin at sites in British Columbia, California, and Baja California, Mexico. However, there was limited evidence to support chain migration (Popovic et al. 2019).

MIGRATORY PATHWAYS Deployment of radio transmitters on adults at San Francisco Bay, California, and Gray's Harbor, Washington, revealed key stopover sites along the Pacific coast from Humboldt Bay, California, to the Yukon-Kuskokwim Delta, Alaska (Warnock et al. 2004). Resightings of banded birds from all three subspecies indicate *C. a. arcticola* migrate along the East Asian–Australasian Flyway from Japan to coastal central China, and *C. a. pacifica* migrate along the Pacific Flyway from southeast Alaska to California (Gill et al. 2013; Lagassé et al. 2020). Geolocator data also indicated *C. a. arcticola* tended to use the eastern edge of the East Asian–Australasian Flyway compared to other subspecies (Lagassé et al. 2022). *C. a. hudsonia* migrate through the midcontinent or along the Atlantic coast, with major staging sites at Delaware and Hudson bays.

DATA CONTRIBUTORS

M. Barter; S. Brown; C.-Y. Chiang; C-.Y. Choi; M. Drever; W. Easton; S. Flemming; H.R. Gates; S. Kendall; J. Liebezeit; B. Lagassé; R.B. Lanctot; V. Loverti; L. McKinnon; E. Nol; S.T. Saalfeld; P. Smith; L.L. Wright; S. Yezerinac; Y. Zharikov.

Calidris ptilocnemis ROCK SANDPIPER

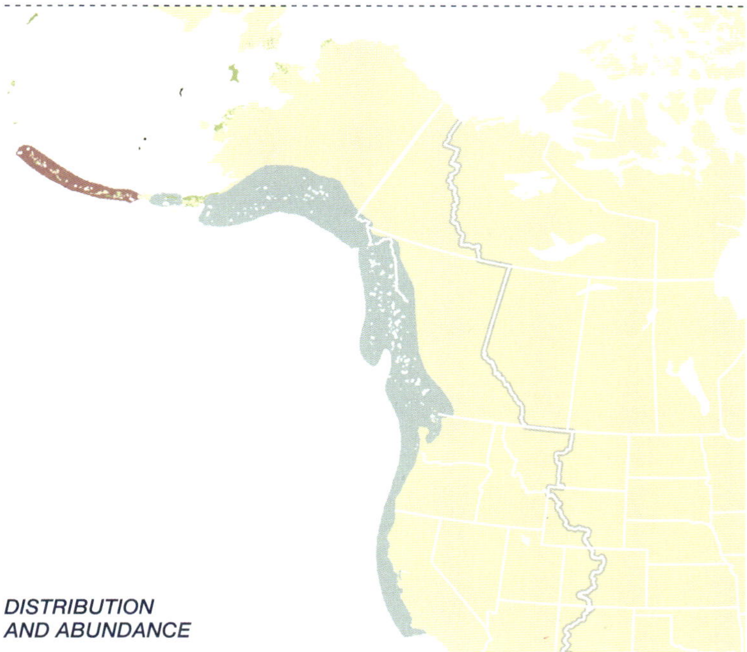

DISTRIBUTION AND ABUNDANCE

DISTRIBUTION AND ABUNDANCE

The Rock Sandpiper is endemic to Beringia, where it breeds in upland tundra in coastal regions encircling the Bering Sea. It has the most northerly nonbreeding distribution of any North American shorebird, spending the nonbreeding season from the coasts of the Aleutian Islands to Northern California. It is known for its variability in migration strategies, with some populations being largely sedentary, while others exhibit migratory behavior (Pruett and Winker 2005).

ORDER Charadriiformes
FAMILY Scolopacidae
BREEDING HABITAT Tundra, Arctic
NONBREEDING HABITAT Coasts, Rocky Intertidal
GLOBAL POPULATION SIZE 140,000
POPULATION CHANGE SINCE 1970 −43%
IUCN RED LIST CATEGORY Least Concern
R2R TIPPING POINT SPECIES Yellow-Alert
US LISTING STATUS No
CANADA SPECIES AT RISK No
PIF CCS 15/20

MIGRATORY CONNECTIVITY

OVERALL No studies of migratory connectivity were found for this species. Four subspecies have been identified based on morphology and genetics (Conover 1944a; Pruett and Winker 2005). Although there may be some overlap, Gill et al. (2020) note that geographic separation is apparent among all subspecies throughout the breeding season. This suggests there may be moderate to strong connectivity, but the level of connectivity is uncertain due to the complexity of subspecies migration and unknown intermixing.
GENETICS DNA analyses reveals that two subspecies, *C. p. quarta* and *C. p. ptilocnemis*, exhibit significant genetic and morphological differentiation, possibly as a result of geographical isolation, while *C. p. couesi* and *C. p. tschuktschorum* display distinctive morphological traits but limited genetic divergence, which could be due to gene flow or insufficient time for genetic differentiation (Pruett and Winker 2005). A putative fifth subspecies, *C. p. kurilensis* (Lobkov 1986, 2001), may exist in birds breeding in Kamchatka and spending the nonbreeding season in Japan (Ornithological Society of Japan 2000), but insufficient specimens and distribution data hinder accurate subspecies assignment (Gill et al. 2020).
MIGRATORY PATHWAYS Two subspecies, *C. p. couesi* and *C. p. quarta*, are considered primarily sedentary as they move locally between upland breeding locations and coastal nonbreeding areas. *C. p. ptilocnemis* is a third subspecies that moves relatively short distances, covering <1,500 km between breeding and nonbreeding sites. A fourth subspecies, *C. p. thscuktschorum*, is the most northern subspecies (Gill et al. 2020; Ruthrauff 2014), and it migrates farthest south with distances of about 4,000 km, exhibiting an intraspecific leapfrog migration (Boland 1990).

PURPLE SANDPIPER *Calidris maritima*

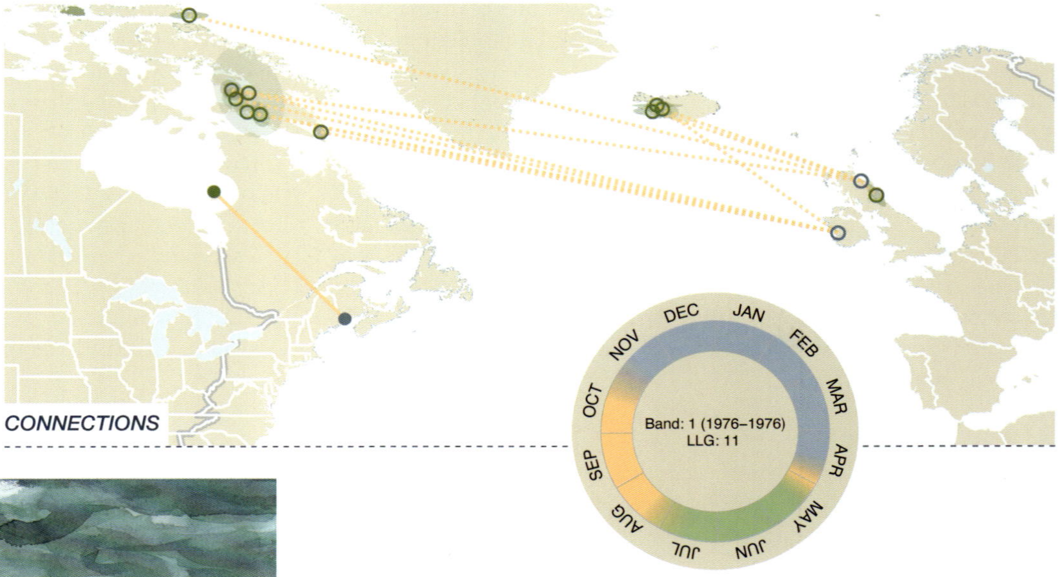

DISTRIBUTION AND ABUNDANCE

The breeding range of Purple Sandpiper extends from northeast Canada to Greenland, Iceland, Svalbard, and Scandinavia, eastward to the Severnaya Zemlya islands north of Arctic Siberia (Payne and Pierce 2020). Its nonbreeding range includes the rocky coastlines on both sides of the North Atlantic Ocean, including above the Arctic Circle in Norway and Russia, making it the northernmost nonbreeding distribution of all shorebird species.

ORDER Charadriiformes
FAMILY Scolopacidae
BREEDING HABITAT Wetlands, Tundra
NONBREEDING HABITAT Coasts, Rocky Intertidal
GLOBAL POPULATION SIZE 250,000
POPULATION CHANGE SINCE 1970 −43%
IUCN RED LIST CATEGORY Least Concern
R2R TIPPING POINT SPECIES No
US LISTING STATUS No
CANADA SPECIES AT RISK No
PIF CCS 12/20

MIGRATORY CONNECTIVITY

OVERALL Despite the potential for quantifying the migratory connectivity between breeding and nonbreeding populations, overall connections still remain unclear (Summers et al. 2014; LeBlanc et al. 2017). MC is not calculated due to limited data.

BANDING A bird banded on the nonbreeding grounds in New Brunswick was recovered during the breeding season in Nunavut (Belcher Islands). Birds breeding in Hudson Bay and the southern part of the Canadian Arctic spend the nonbreeding season in eastern Canada.

GEOLOCATOR TRACKING Geolocators deployed in Scotland and Ireland tracked birds to breeding areas on Baffin and Devon islands in Canada (Summers et al. 2014), but breeding grounds farther west are possible because some geolocators did not provide data in continuous daylight during June and early July.

GENETICS Barisas et al. (2015) examined genetic and morphometric data to assess the validity of the three proposed subspecies. They concluded that, although there were notable differences in morphology and DNA sequence data, differences were not large enough to warrant distinct subspecific status. Further evidence from an assessment of population structure suggests that populations from Iceland and Svalbard are distinct, but overall, there is low genetic diversity and shallow phylogeny, which may indicate a recent divergence of populations (LeBlanc et al. 2017).

OTHER APPROACHES Geographic variation in morphological features has warranted recognition of three subspecies: *C. m. belcheri*, which breed in northeast Canada; *C. m. maritima*, which breed from the coast of Canada to Siberia; and *C. m. littoralis* of Iceland. But the high degree of overlap in morphology among the three groups questions the validity of the designations of subspecies (Barisas et al. 2015). Individuals from different breeding populations have been observed to overlap at nonbreeding areas (Corse and Summers 1999; Nicoll et al. 1988; Summers et al. 2014). No satellite transmitter or stable isotope studies of migratory connectivity have been carried out for this species.

MIGRATORY PATHWAYS North American Purple Sandpipers that breed in the northern part of the Canadian Arctic migrate between breeding and stationary nonbreeding areas along a transatlantic route, from the Canadian Arctic in summer to Scotland and Ireland in the nonbreeding period (Summers et al. 2014). The birds stopped over in Iceland and southwest Greenland during pre-breeding migration, while on post-breeding migration, during November and December, they generally took nonstop transatlantic flights.

DATA CONTRIBUTORS
R. Summers.

INTERESTING NOTE

Geolocators used for Purple Sandpipers were attached with a leg-loop harness made of Silastic (silicone rubber) tubing and elastic thread to allow for stretching if birds gain weight. The geolocator was glued to a Zotefoam cushion, and the final weight was 1.4 g with the harness. Of the 50 tagged birds, 12 were recovered (24%), which did not differ significantly from the 26% recovery rate for birds that were only banded. Some feather loss on the thighs was observed, and there was a small (0.9 g) but significant mass decline for tagged birds versus nontagged birds (−1.2% versus +1.5%; Summers et al. 2014).

Calidris bairdii

BAIRD'S SANDPIPER

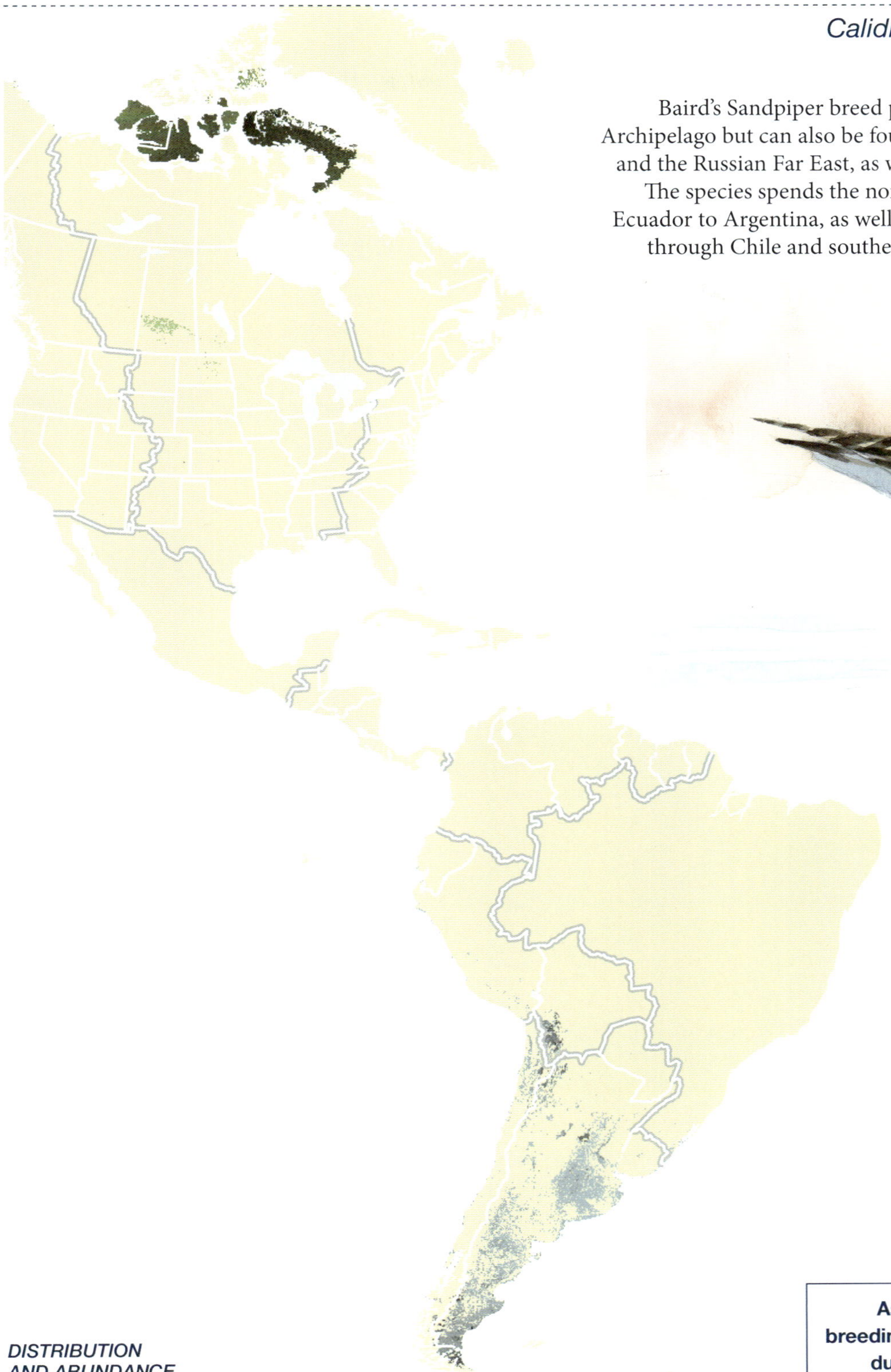

DISTRIBUTION AND ABUNDANCE

DISTRIBUTION AND ABUNDANCE

Baird's Sandpiper breed primarily across the high Arctic in the Canadian Archipelago but can also be found in the low Arctic regions of Alaska, Canada, and the Russian Far East, as well as alpine areas in other portions of Beringia. The species spends the nonbreeding season in the Andes Mountains, from Ecuador to Argentina, as well as the lowlands of Paraguay and Uruguay south through Chile and southern Argentina (Moskoff and Montgomerie 2020).

ORDER Charadriiformes
FAMILY Scolopacidae
BREEDING HABITAT Tundra, Arctic
NONBREEDING HABITAT Grasslands, Pampas and Campos
GLOBAL POPULATION SIZE 300,000
POPULATION CHANGE SINCE 1970 +57%
IUCN RED LIST CATEGORY Least Concern
R2R TIPPING POINT SPECIES No
US LISTING STATUS No
CANADA SPECIES AT RISK No
PIF CCS 13/20

INTERESTING NOTE

Assumed to be a panmictic population, given low breeding site fidelity and common migratory pathways during both northward and southward migrations.

MIGRATORY CONNECTIVITY

OVERALL Migratory connectivity is unknown. No studies of migratory connectivity were found for this species.
BANDING Few birds banded overall, with no band reencounters. Norton (1973) documented that only 3 of 251 birds banded at Utqiaġvik (formerly Barrow) returned to nest in consecutive years.
RADIO TRACKING Detections from the Motus Wildlife Tracking System during migration connect Nebraska to Texas (Cape Horn International Center Ornithological Studies, unpublished data).
STABLE ISOTOPES Analyses of feathers did not work well to establish linkages between nonbreeding grounds in Argentina and breeding grounds in the Arctic, or to identify molting origins at a small spatial scale in northern Argentina (Farmer et al. 2004; Torres-Dowdall et al. 2009).
MIGRATORY PATHWAYS eBird and other observational data suggest that patterns of migration vary with age during southward movements, as adults are primarily in the Central Flyway and juveniles travel over a broader front that goes from coast to coast. Northward migration back to breeding areas occurs over a narrower corridor; nonbreeding birds fly over the Amazon through Venezuela and Colombia, then north through the Mississippi River Valley, and into Canada west of Hudson Bay (Skagen et al. 1999; Moskoff and Montgomerie 2020).

LEAST SANDPIPER *Calidris minutilla*

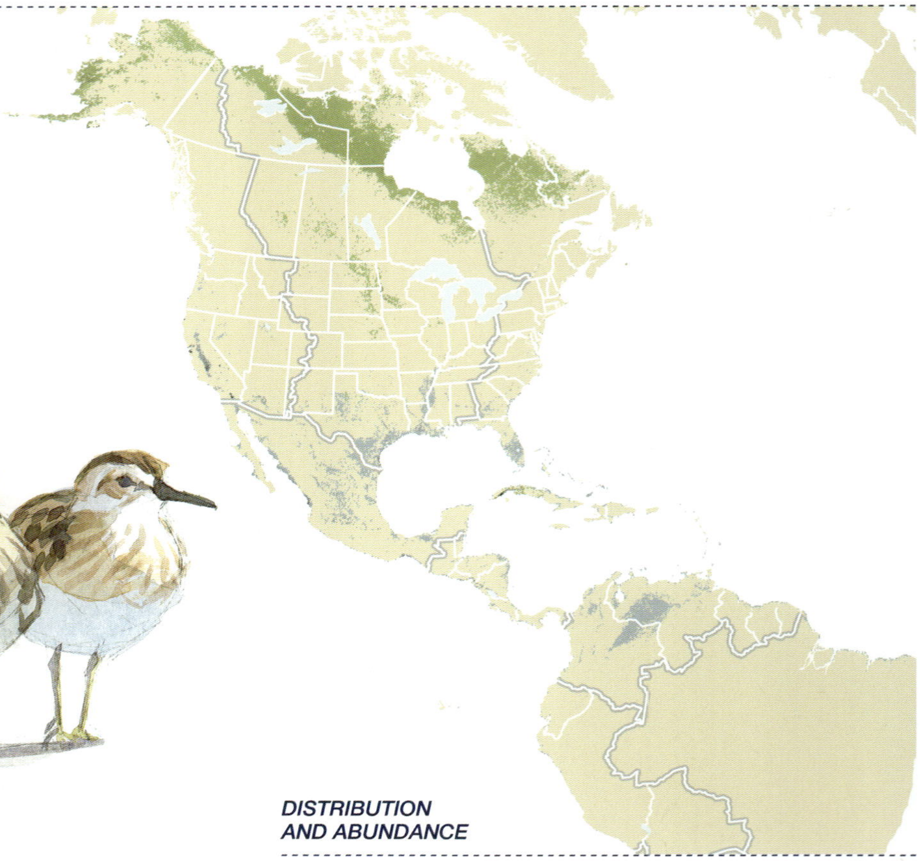

DISTRIBUTION AND ABUNDANCE

This species is the world's smallest shorebird, breeding in subarctic tundra and far northern boreal forest throughout Alaska and in mid-Arctic regions across Canada. Least Sandpipers spend the nonbreeding season from the southern US and Greater Antilles to middle portions of South America (Nebel and Cooper 2020).

ORDER Charadriiformes
FAMILY Scolopacidae
BREEDING HABITAT Wetlands, Tundra
NONBREEDING HABITAT Coasts, Beach and Estuary
GLOBAL POPULATION SIZE 700,000
POPULATION CHANGE SINCE 1970 −1%
IUCN RED LIST CATEGORY Least Concern
R2R TIPPING POINT SPECIES No
US LISTING STATUS No
CANADA SPECIES AT RISK No
PIF CCS 11/20

INTERESTING NOTE

There is some natal site fidelity, but yearling individuals do not nest as close to areas from the previous breeding season as returning adults. There is evidence of adult fidelity to breeding and nonbreeding locations (Nebel and Cooper 2020). Females spend the nonbreeding season farther south than males, as shown by sex ratio data from California, Suriname, Panama, and Ecuador; no difference was found in age ratios among those sites (Nebel 2006). Bill and wing length of both sexes increased with latitude between California, Panama, and Ecuador, suggesting birds partition resources (Nebel 2006).

MIGRATORY CONNECTIVITY

OVERALL No data for breeding to nonbreeding migratory connectivity was found for this species. Limited tracking data suggest Atlantic, midcontinent, and Pacific populations separate. However, the species is known to migrate over a broad front in the midcontinent, suggesting population mixing may occur.

RADIO TRACKING Detections from the Motus Wildlife Tracking System during migration connect Nunavut (Alaska), Ontario (Canada), and Atlantic Canada to the Caribbean and South America (Motus project, New England Shorebirds, unpublished data; Motus project, Atlantic Canada Shorebirds, unpublished data; Motus project, James Bay Shorebirds, unpublished data; Motus project, Arctic Shorebirds—CWS Yellowknife, unpublished data; Motus project, California Shorebird Migration and Drought Study, unpublished data).

MIGRATORY PATHWAYS Least Sandpipers in the eastern portion of their range undertake an elliptical pattern of migration, with post-breeding migrants taking transoceanic flights to South America from Atlantic coastal sites in both Canada and the US and more coastal routes north during pre-breeding migration. Migration also occurs across North America, with the species widely distributed across the landscape and therefore less reliant than other *Calidris* species on key staging sites. Resight data indicate birds banded at James Bay migrate to staging areas in southern Ontario, the Maritime provinces, and New England states, before continuing onto northeast South America (Morrison 1978). Post-breeding migrants and breeders banded in Nova Scotia and Quebec also traveled along the Atlantic coast before flying over the Atlantic Ocean to the Lesser Antilles and northeast South America (McNeil and Burton 1977). Resightings from Manitoba and Saskatchewan indicated birds migrate down the Mississippi or Central flyways and spend the nonbreeding season primarily in Texas and Mexico. About 25% of post-breeding migrants banded at Bolinas Lagoon, California, spend the nonbreeding season there (Page 1974). Forty-one of 62 juveniles equipped with nanotags at a stopover site in James Bay, Ontario, provided information on movements. Few of these birds traveled to the Canadian Maritimes, and several stopped in the Great Lakes region (while other species in the genus *Calidris* bypassed this area); most stopped along the Atlantic coast from Massachusetts to New Jersey, but none were detected south of North Carolina (A.M. Anderson et al. 2019; C. Friis and A. Anderson, unpublished data). More tracking data are needed to identify major stopover and nonbreeding locations.

Calidris fuscicollis WHITE-RUMPED SANDPIPER

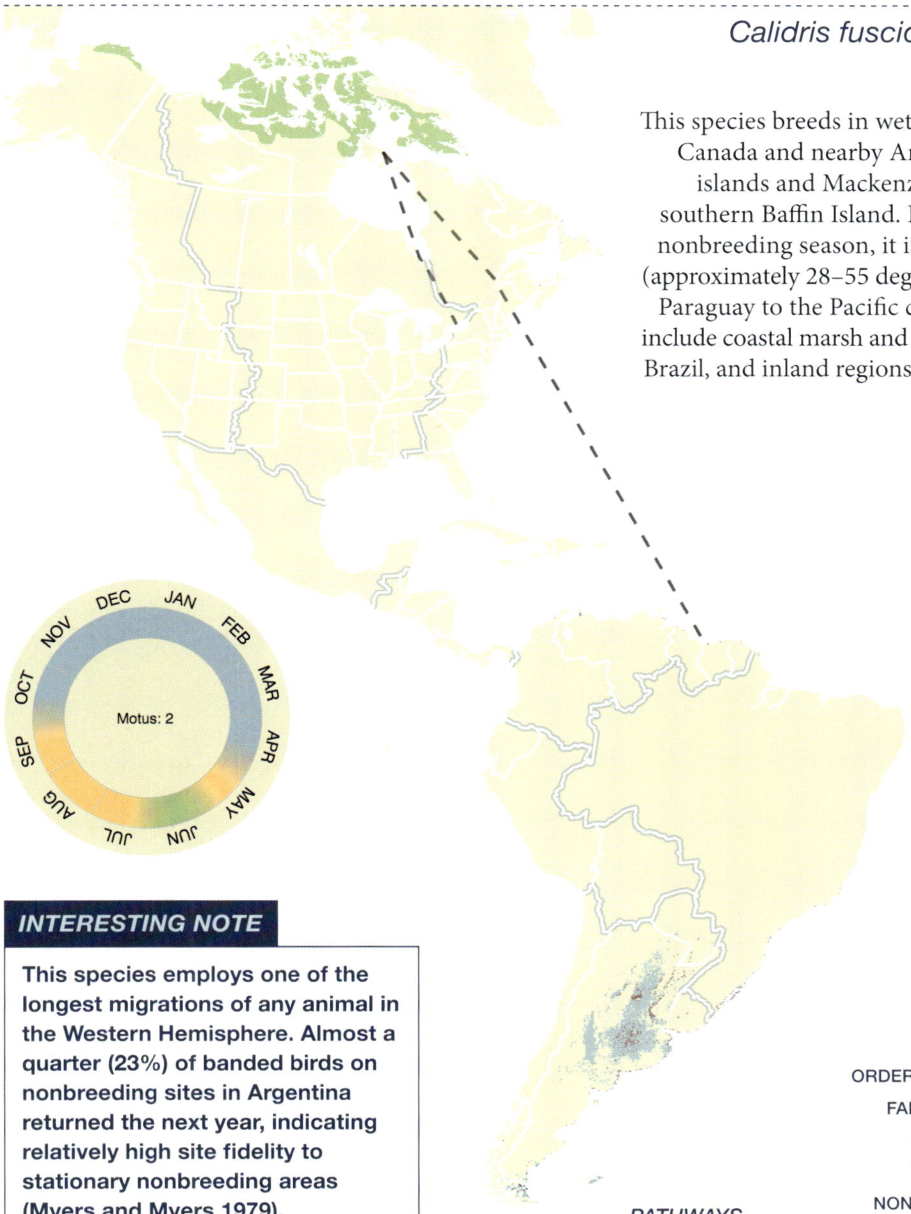

DISTRIBUTION AND ABUNDANCE

This species breeds in wet, low-lying meadows across the high Arctic in mainland Canada and nearby Arctic islands. Its range extends from northwestern Arctic islands and Mackenzie south to parts of mainland Canada, Hudson Bay, and southern Baffin Island. It sometimes also breeds in northern Alaska. During the nonbreeding season, it is typically found in a relatively narrow range of latitudes (approximately 28–55 degrees south) to the east of the Andes, including areas from Paraguay to the Pacific coasts of Tierra del Fuego. Additional nonbreeding areas include coastal marsh and lagoons in Lagoa do Peixe in Rio Grande do Sul, southern Brazil, and inland regions of Buenos Aires Province in Argentina (Parmelee 2020).

INTERESTING NOTE

This species employs one of the longest migrations of any animal in the Western Hemisphere. Almost a quarter (23%) of banded birds on nonbreeding sites in Argentina returned the next year, indicating relatively high site fidelity to stationary nonbreeding areas (Myers and Myers 1979).

ORDER Charadriiformes
FAMILY Scolopacidae
BREEDING HABITAT Wetlands, Tundra
NONBREEDING HABITAT Coasts, Beach and Estuary

GLOBAL POPULATION SIZE 1,700,000
POPULATION CHANGE SINCE 1970 +16%
IUCN RED LIST CATEGORY Least Concern
R2R TIPPING POINT SPECIES No
US LISTING STATUS No
CANADA SPECIES AT RISK No
PIF CCS 14/20

MIGRATORY CONNECTIVITY

OVERALL Migratory connectivity is relatively unknown. During migration, birds follow relatively well-defined pathways to more diffuse breeding and nonbreeding areas; however, no studies to date have shown any signs of fine-scale population structure, indicating weak migratory connectivity.

RADIO TRACKING Radio tags that were paired with automated radio telemetry through the Motus Wildlife Tracking System provided further evidence that some birds attempt nonstop transoceanic flights from the subarctic to the Caribbean or South America (A.M. Anderson et al. 2019). For instance, a single White-rumped Sandpiper completed a nonstop flight (approximately 4,800 km) from James Bay to Barbados, via the Bay of Fundy, in five days. Several other tracking projects have deployed hundreds of tags through the Motus Wildlife Tracking System. There are no individual breeding to nonbreeding locations, but there appear to be many detections during migration.

OTHER APPROACHES No satellite transmitter, banding, genetics, or stable isotope studies of migratory connectivity were found for this species.

MIGRATORY PATHWAYS When migrating northward, White-rumped Sandpipers undertake extended nonstop flights from northern South America, across the Caribbean and Gulf of Mexico, before crossing the midcontinent region through a narrow band to reach breeding areas in the Arctic (Skagen et al. 1999). While most northward migration staging sites are unknown, a crucial stopover area has been identified—Cheyenne Bottoms, Kansas—where birds are thought to embark on a last leg of their journey across the central plains and boreal forest (Harrington et al. 1991). In contrast, southbound migrants make transoceanic flights over the Atlantic Ocean to South America (McNeil and Burton 1973), departing from eastern Canada and the western and northeastern Atlantic region, while others may follow the Atlantic coastline. After arriving in South America, they move southeast along the coast, then turn inland to reach areas east of the Amazon River mouth. During this journey, they make short flights and frequent stops on river bars and banks before reaching nonbreeding areas (Antas 1983).

DATA CONTRIBUTORS
A. MacDonald; P. Smith.

BUFF-BREASTED SANDPIPER *Calidris subruficollis*

DISTRIBUTION AND ABUNDANCE

Buff-breasted Sandpipers have a highly restricted breeding range, limited primarily to the northern coast of Alaska and the northern Canadian Arctic. They migrate south and north through the midcontinent of North and South America to spend the nonbreeding period on the Pampas of Argentina, Brazil, and Uruguay (McCarty et al. 2020). Efforts to understand population size are underway in the Texas coastal plain (J. Lyons, K. Stone, L. Wolfenbarger, unpublished data) and Bolivia (T. Boorsma, unpublished data). Repeated annual surveys on the nonbreeding grounds indicate species population size may be stable over recent years, but additional analysis is needed.

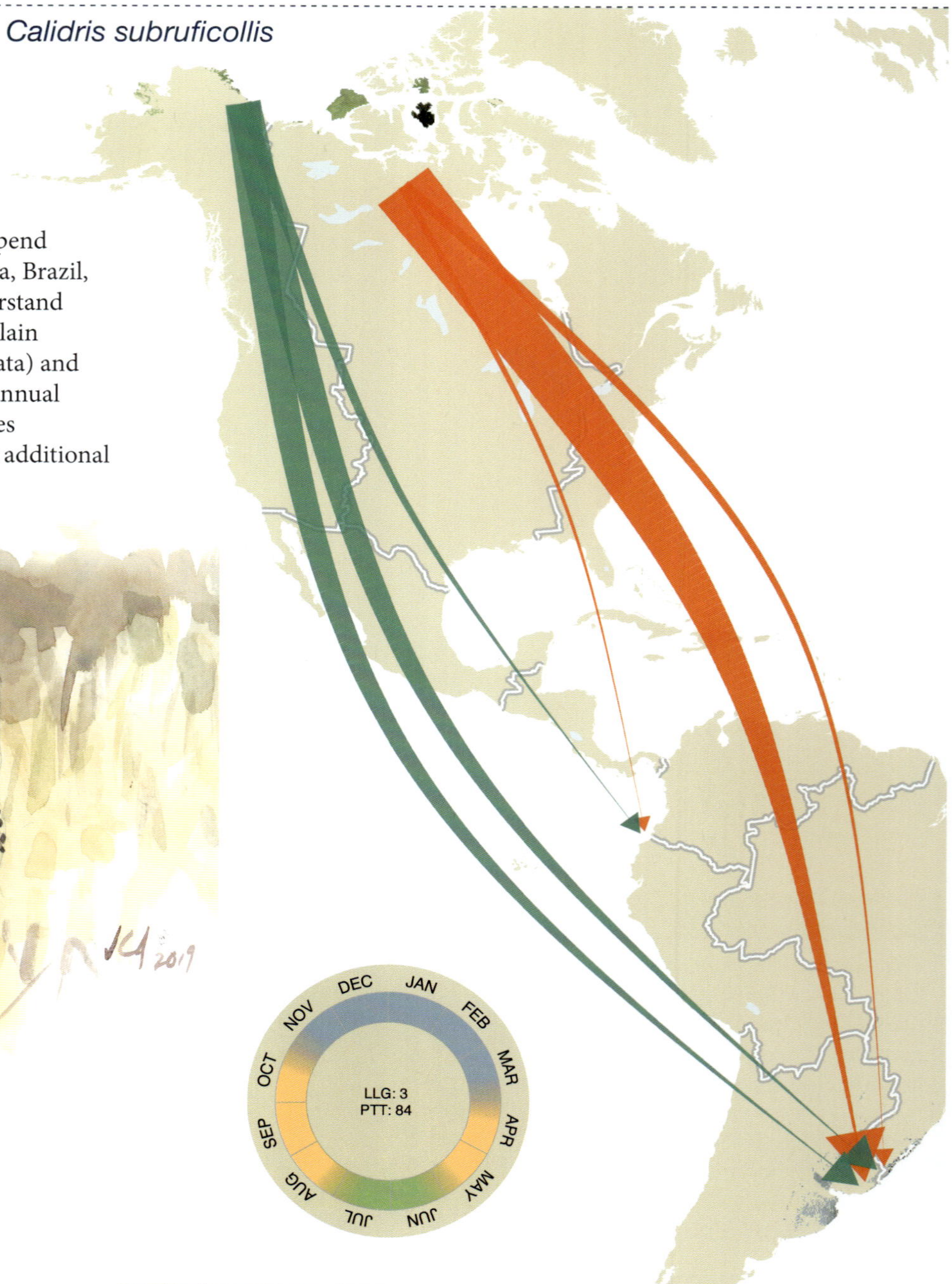

ORDER Charadriiformes
FAMILY Scolopacidae
BREEDING HABITAT Tundra, Arctic
NONBREEDING HABITAT Grasslands, Pampas and Campos
GLOBAL POPULATION SIZE 56,000
POPULATION CHANGE SINCE 1970 +61%
IUCN RED LIST CATEGORY Near Threatened
R2R TIPPING POINT SPECIES Orange-Alert
US LISTING STATUS No
CANADA SPECIES AT RISK No
PIF CCS 15/20

	Andean West	Brazil	Southern Cone
Midcontinent (n = 19)	5%	11%	84%
Pacific (n = 7)	14%	43%	43%

GENERALIZED CONNECTIVITY

MIGRATORY CONNECTIVITY

OVERALL Available evidence of migratory connectivity from geolocator and satellite transmitter data are limited, but birds breeding within Arctic areas of the Midcontinent Flyway primarily spend the nonbreeding season in the Southern Cone region (84%), while birds breeding in the Pacific Flyway are divided between the Southern Cone region (43%) and Brazil (43%). MC could not be calculated due to low sample size, but data indicates little to no migratory connectivity, with birds from throughout the breeding range using the same migration routes and nonbreeding area.

BANDING About 400 birds have been banded at a few sites from the species breeding and nonbreeding range, but few resightings have been recorded: one from Deadhorse, Alaska, to Lagoa do Peixe, Brazil; two from Asunción Bay, Paraguay, to Lagoa do Peixe, Brazil (Almeida 2009; J. Almeida and A. Lesterhuis, unpublished data).

GEOLOCATOR TRACKING Three adults recovered with geolocators indicated birds spending the nonbreeding season in Brazil and Argentina bred in northern Canada and Alaska (Lanctot et al. 2016).

SATELLITE TRACKING Deployment of 2 g PTT microwave technology tags and 3.7 g GPS Lotek PinPoint tags supported geolocator data on migration routes and major migration stopover sites but also found a stopover on Victoria Island, Canada, during post-breeding migration and greater dependence on the Llanos region of Colombia and Venezuela (Tibbitts et al. 2023). Additional GPS tags have been applied to better understand habitat use in Victoria Island, Texas, and Colombia (T. Rodkey, unpublished data).

GENETICS A range-wide genetic analysis revealed little population structure (Lounsberry et al. 2013) and no loss of genetic diversity due to historic population decline (Lounsberry et al. 2014).

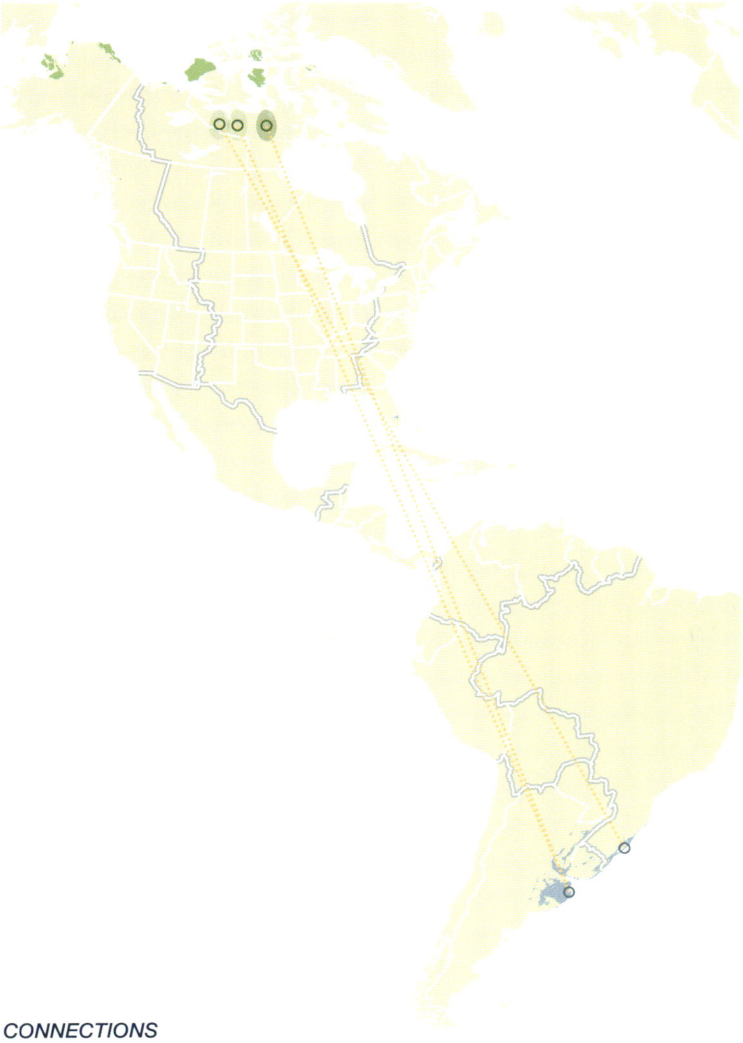

CONNECTIONS

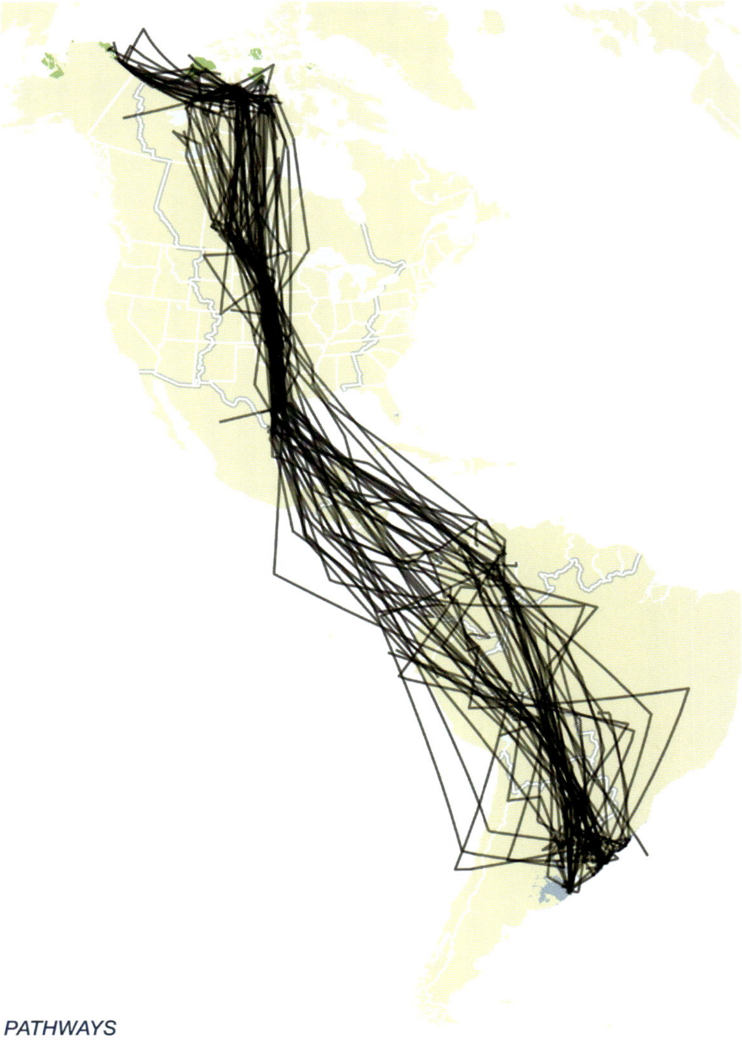

PATHWAYS

OTHER APPROACHES No stable isotope studies of migratory connectivity were found for this species.

MIGRATORY PATHWAYS Geolocator and satellite transmitter tracks indicate birds from throughout the breeding range use the same migration routes, stopping in Bolivia and Paraguay and at waterways in the Amazon, but less so in Colombia and Venezuela during southbound travels to South America (Lanctot et al. 2016; Tibbitts et al. 2023). From the nonbreeding grounds, birds undergo a nonstop flight to the Llanos region of Colombia, then to coastal Texas, then to a few sites in the midcontinent of North America, and then a direct flight to breeding areas over the boreal region of Canada. More refined tracks from GPS transmitters suggest birds make a figure-eight migration pattern with a narrow midcontinental migration path through North America (Tibbits et al. 2023).

DATA CONTRIBUTORS

J. Aldabe; J. Bosi de Almeida; G. Castellanos; G. Castresana; R.B. Lanctot; R. McGuire; T.L. Tibbitts.

INTERESTING NOTE

The species is the only North American shorebird that displays male lekking behavior (Lanctot and Weatherhead 1997). Extensive market hunting in the late 1800s and early 1900s in the midcontinent of North America and habitat modification throughout its migration and nonbreeding range led to a substantial population decline.

PECTORAL SANDPIPER *Calidris melanotos*

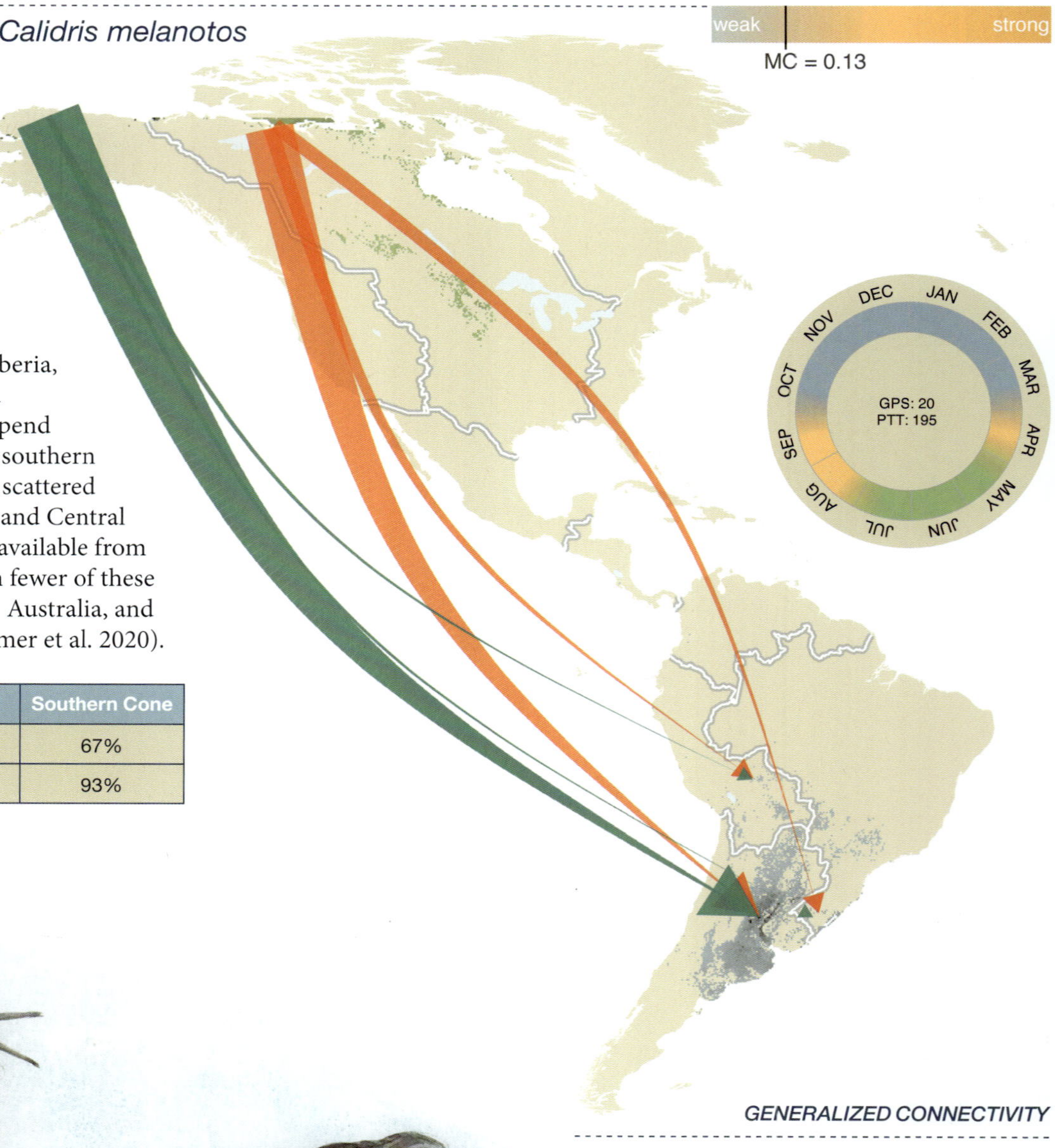

DISTRIBUTION AND ABUNDANCE

This species breeds in Arctic regions of Siberia, Russia, western and northern Alaska, and western and central Canada. Most birds spend the nonbreeding season in the Pampas of southern South America, although a few are found scattered through the southern US, the Caribbean, and Central and South America. Little information is available from these small but isolated populations. Even fewer of these sandpipers are found in the South Pacific, Australia, and western Europe, and rarely in Africa (Farmer et al. 2020).

	Andean West	Brazil	Southern Cone
Midcontinent (n = 6)	17%	17%	67%
Pacific (n = 27)	4%	4%	93%

ORDER Charadriiformes
FAMILY Scolopacidae
BREEDING HABITAT Wetlands, Tundra
NONBREEDING HABITAT Grasslands, Pampas and Campos
GLOBAL POPULATION SIZE 1,600,000
POPULATION CHANGE SINCE 1970 −53%

IUCN RED LIST CATEGORY Least Concern
R2R TIPPING POINT SPECIES Orange-Alert
US LISTING STATUS No
CANADA SPECIES AT RISK No
PIF CCS 14/20

INTERESTING NOTE

This species is considered an opportunistic breeder, exhibiting large fluctuations in breeding densities and low annual breeding site fidelity (Moiteret et al. 1996; Lesku et al. 2012; Saalfeld and Lanctot 2015). Males tagged at Utqiaġvik exhibited nomadic behavior throughout the Arctic breeding range, moving at any time and in any direction between successive breeding sites, dependent on wind conditions (Krietsch et al. 2020) and on local mating opportunities (Kempenaers and Valcu 2017). In contrast, females tagged at nest sites remained there until nests failed, young died, or young fledged (R.B. Lanctot et al., unpublished data).

MIGRATORY CONNECTIVITY

OVERALL Migratory connectivity is known from satellite telemetry and band reencounter data. Both Midcontinent and Pacific flyway breeding birds primarily migrate to the Southern Cone region of South America (67% and 93%, respectively). Smaller percentages from both breeding populations overlap in several other nonbreeding regions farther north, suggesting that this species has weak migratory connectivity (MC = 0.13).

BANDING Two adults banded in Saskatchewan and Kansas were recovered in Siberia, and one chick banded in Siberia was recovered in South Dakota (Farmer et al. 2020).

RADIO TRACKING Many Motus tags were deployed in the eastern part of the range. There were limited detections of individual birds between breeding and nonbreeding locations.

SATELLITE TRACKING Male Pectoral Sandpipers have been tracked with Argos satellite telemetry from breeding locations in Nome and Utqiaġvik, Alaska (120 birds; Kempenaers and Valcu 2017),

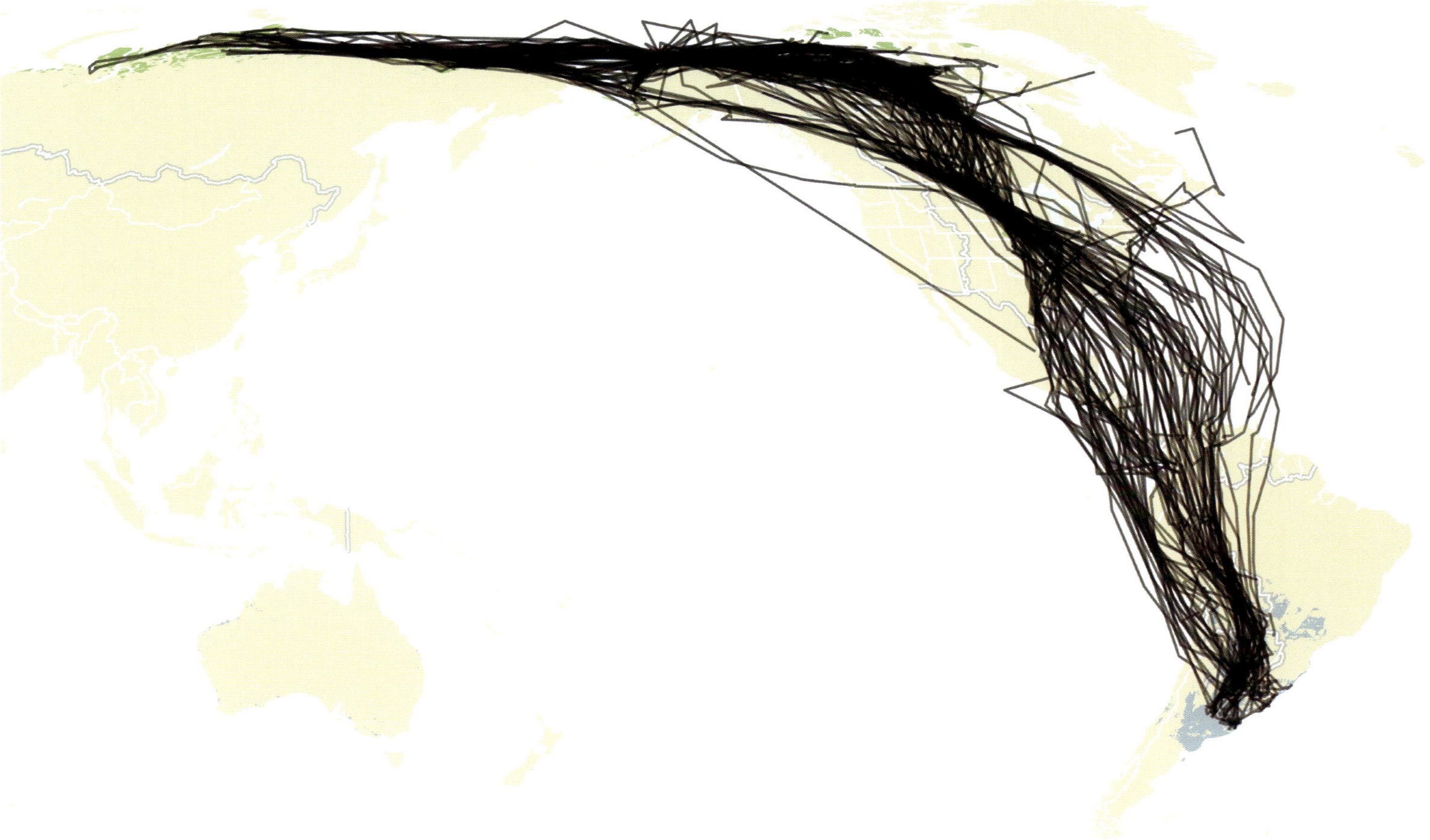

PATHWAYS

whereas females have been tracked with GPS satellite telemetry from four sites in Alaska and Cambridge Bay, Nunavut (37 females; R.B. Lanctot et al., unpublished data). Both males and females migrated to nonbreeding areas in the Pampas of South America.

OTHER APPROACHES No genetic or stable isotope studies of migratory connectivity were found for this species. Analysis of trace elements from Pectoral Sandpipers in Argentina failed to find a large-scale gradient and suggested that trace-element profiles from feathers changed as feathers age. These facts limited the application of trace elements to study migratory connectivity (Torres-Dowdall et al. 2010).

MIGRATORY PATHWAYS Males traveled northward over the Andes and coastal Pacific before returning through the Central Flyway to reach breeding areas (Kempenaers and Valcu 2017). Birds migrate east from Russia and Alaskan breeding areas to the central Canadian Arctic before transiting south through a broad front east of the Rockies and along the Central Flyway of North and South America to reach the Pampas region of Argentina, Brazil, Paraguay, and Uruguay. Some birds migrate farther east across the Arctic before using a transatlantic route to reach Brazil, whereas others (35% of tagged birds) stop migration in the southern US, the Caribbean, and northern South America. La Sorte and Fink (2017) suggest global warming is unlikely to negatively affect individuals using the transatlantic route.

DATA CONTRIBUTORS

S. Brown; B. Kempenaers; J.-F. Lamarre; R.B. Lanctot; C. Latty; R. McGuire; D.R. Ruthrauff; S.T. Saalfeld; S. Schulte; M. Valcu.

SEMIPALMATED SANDPIPER *Calidris pusilla*

DISTRIBUTION AND ABUNDANCE

Semipalmated Sandpiper is a Nearctic-Neotropical migratory bird that breeds across the northern Arctic from Alaska to eastern Canada and spends the nonbreeding season primarily along the eastern and western coasts of South America, both Central American coasts, and the Caribbean (Hicklin and Gratto-Trevor 2020). Significant population declines have been documented at nonbreeding sites in northeastern South America and at breeding areas in the eastern North American Arctic, while populations appear stable or increasing in the central and western Arctic (S. Brown et al. 2017).

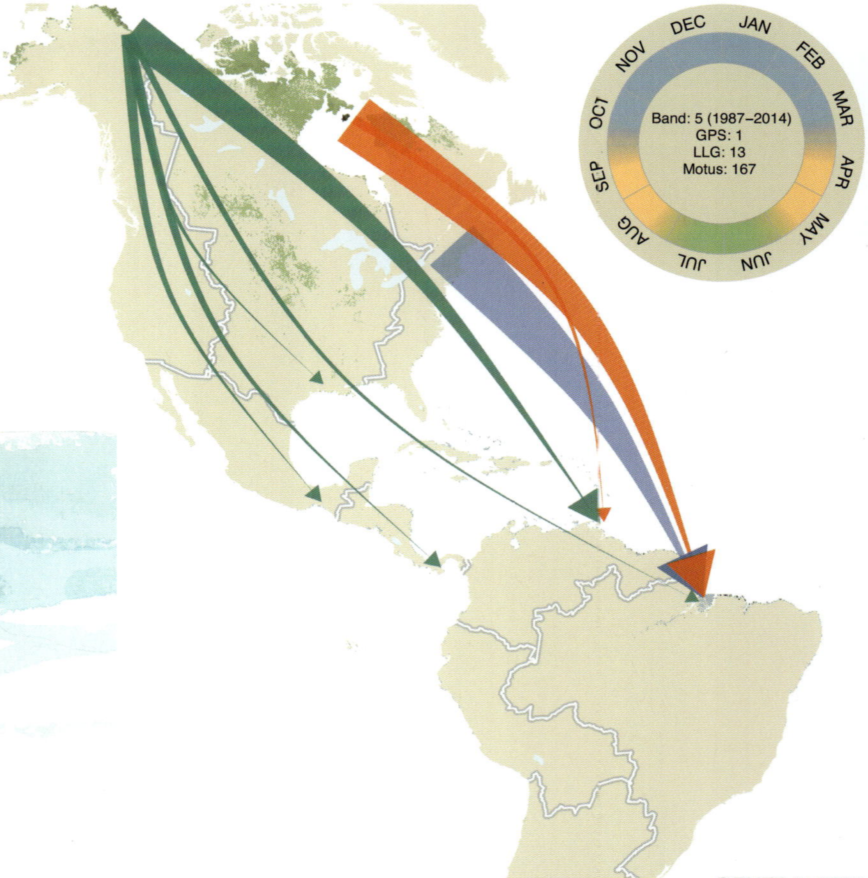

GENERALIZED CONNECTIVITY

ORDER Charadriiformes
FAMILY Scolopacidae
BREEDING HABITAT Wetlands, Tundra
NONBREEDING HABITAT Coasts, Beach and Estuary
GLOBAL POPULATION SIZE 2,300,000
POPULATION CHANGE SINCE 1970 −64%
IUCN RED LIST CATEGORY Near Threatened
R2R TIPPING POINT SPECIES Orange-Alert
US LISTING STATUS No
CANADA SPECIES AT RISK No
PIF CCS 14/20

MIGRATORY CONNECTIVITY

OVERALL Eastern and western breeding populations have relatively disjunct nonbreeding ranges, as well as migration routes, suggested by tracking and banding data. Breeding populations within the Atlantic Flyway spend the nonbreeding season in Brazil. Midcontinent Flyway breeding populations also primarily spend the nonbreeding season in Brazil (89%), with additional birds migrating to northern South America (11%). Pacific Flyway breeding individuals migrate to northern South America (57%), Central America (14%), Mexico (14%), Brazil (7%), and the Midcontinent Flyway (7%). MC is not calculated due to limited sample size, but migratory connectivity appears moderate.

	Brazil	Central America	Mexico	Midcontinent	Northern SA
Atlantic (n = 3)	100%	–	–	–	–
Midcontinent (n = 9)	89%	–	–	–	11%
Pacific (n = 14)	7%	14%	14%	7%	57%

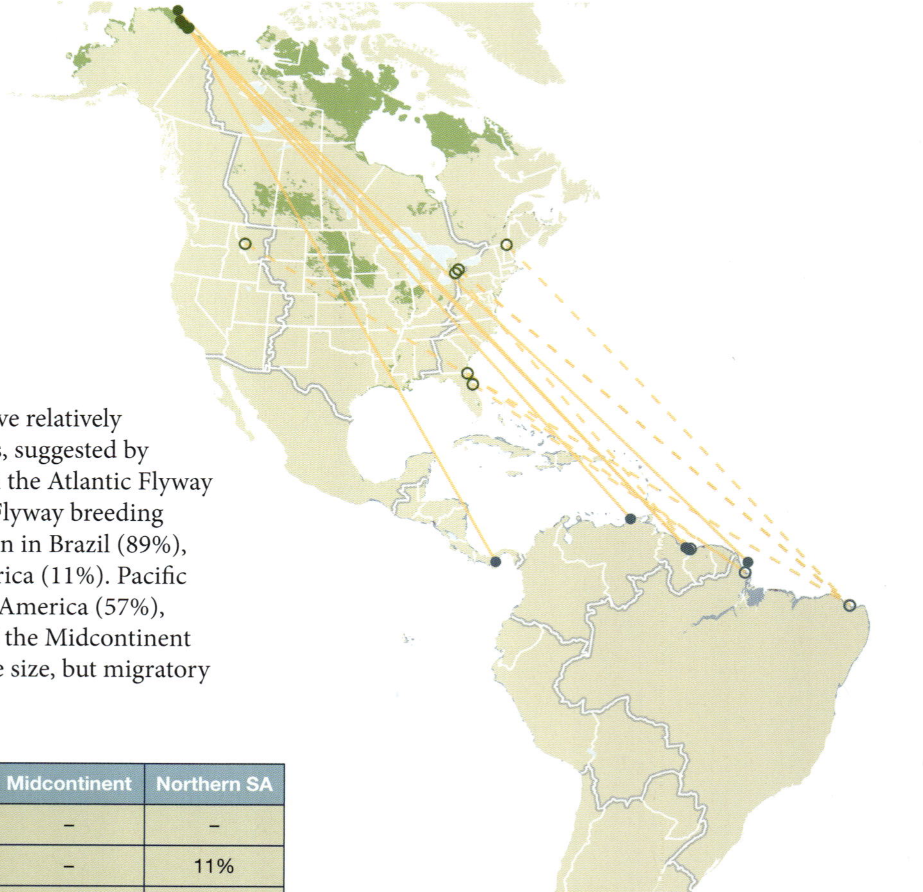

CONNECTIONS A

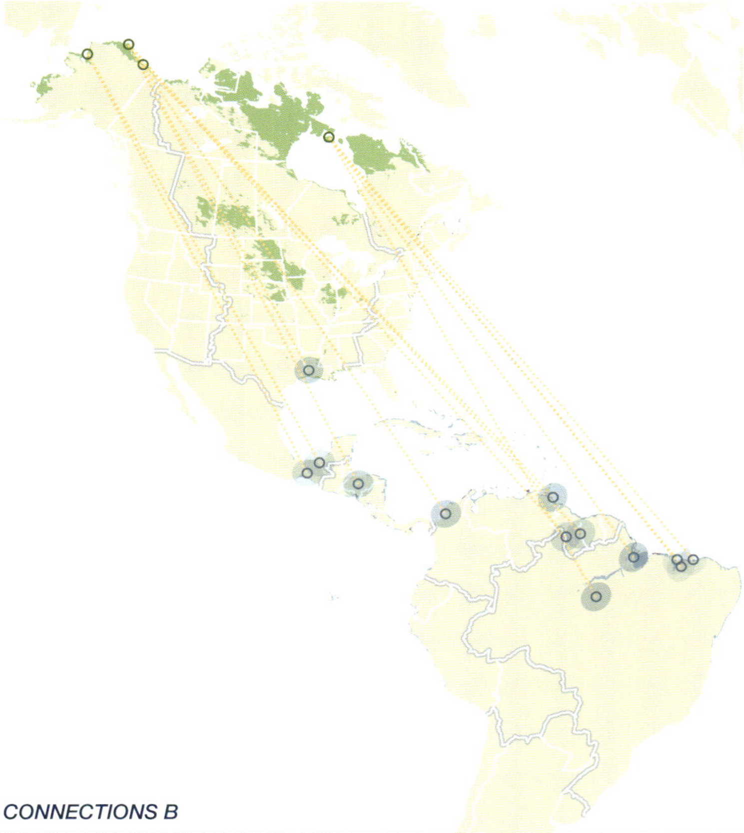

CONNECTIONS B

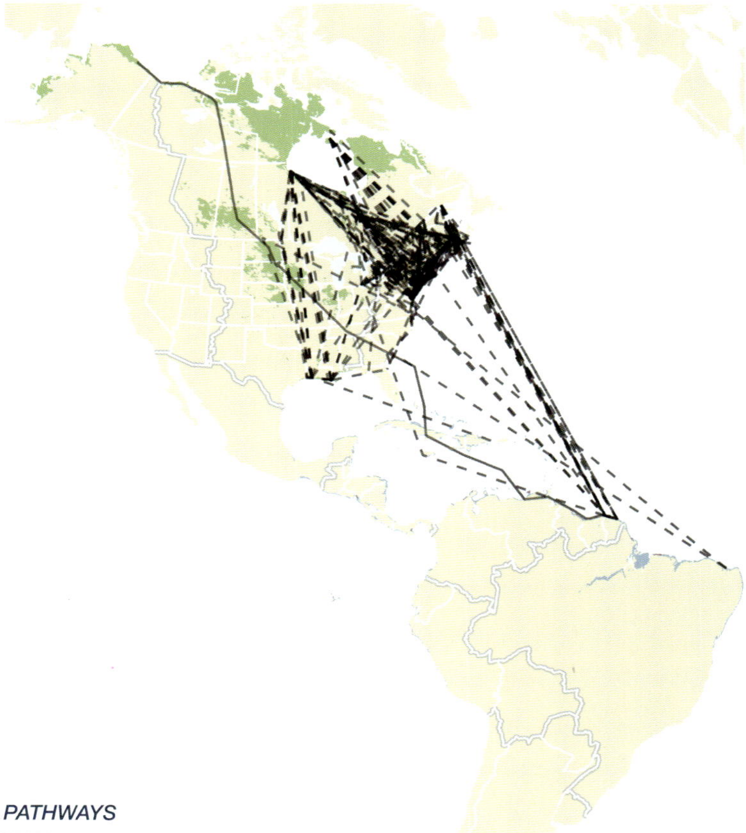

PATHWAYS

INTERESTING NOTE

Geolocators attached to leg bands with flags reduced annual return rates by 63% and reduced nest success by 42% for Semipalmated Sandpipers (Weiser et al. 2016).

BANDING Birds banded in northern Alaska migrated to coastal nonbreeding areas in central and northern South America.
RADIO TRACKING Over a thousand tags were deployed through the Motus Wildlife Tracking System, leading to many detections during migration and several breeding to nonbreeding connections. All Motus nonbreeding detections are on the coast of northern South America and Brazil.
GEOLOCATOR TRACKING A range-wide migratory connectivity study revealed that the western Arctic breeding population is connected to nonbreeding areas on the west coast of South America and in Central America, with some individuals ranging from Peru to French Guiana. The central Arctic breeding population spends the nonbreeding season primarily on the northeastern coast of South America, as well as in western South America and the Caribbean. The eastern Arctic breeding population also spends the nonbreeding season in northeastern South America, specifically Suriname, French Guiana, and Brazil (S. Brown et al. 2017).
SATELLITE TRACKING One bird, equipped with an archival, 1.5 g GPS tag at Prudhoe Bay, Alaska, and recovered in 2018, migrated by first traveling east along the Beaufort Sea coast until reaching Nunavut, where it turned south to move through Saskatchewan and then the south Atlantic coast, crossing the Caribbean to Venezuela and eventually arriving at French Guiana (R. Lanctot, unpublished data).
GEOLOCATOR TRACKING A range-wide migratory connectivity study revealed that the western Arctic breeding population is connected to nonbreeding areas on the west coast of South America and in Central America, with some individuals ranging from Peru to French Guiana. The central Arctic breeding population spends the nonbreeding season primarily on the northeastern coast of South America, as well as in western South America and the Caribbean. The eastern Arctic breeding population also spend the nonbreeding season in northeastern South America, specifically Suriname, French Guiana, and Brazil (S. Brown et al. 2017).
GENETICS DNA analyses demonstrated low levels of genetic differentiation, with no evidence to support the existence of subspecies, and thus were not useful to measure migratory connectivity between breeding and nonbreeding areas (M.P. Miller et al. 2013).
OTHER APPROACHES Gratto-Trevor et al. (2012b) examined migratory connectivity in breeding populations by combining banding data with morphological differences. They found that bill length varies across the species' Arctic breeding range, with smaller bills in western populations and larger bills in the eastern populations. Average bill length data indicate there are three primary breeding populations, which are also separated geographically: western, central, and eastern.
MIGRATORY PATHWAYS Brown et al. (2017) noted differences in migration routes used by birds breeding in the eastern and western portions of the Arctic. Breeders from sites west of Canning River, Alaska, followed similar routes during both northward and southward migration, with stopovers throughout the Central Flyway prairies. Although it was suggested that birds using the Gulf of Mexico use the Central Flyway to connect to the western breeding range, Herbert et al. (2022) found more eastern flight paths using the Mississippi and Atlantic flyways to connect to the western breeding range. Birds breeding at Canning River and the Mackenzie River Delta also used Central Flyway prairies but migrated farther east during southbound migration. Birds that bred in the central and western Arctic made stopovers on the coasts of Texas and Louisiana on both northward and southward migration. In contrast, eastern breeders have a migration route distinct from western breeders. From Coats Island they tended to follow an elliptical migration. During southward migration, they staged at James Bay before flying across the Atlantic to South America, while returning along the Atlantic

coast. Using a combination of morphometrics, the Motus network, and stable isotopes, Herbert et al. (2022) found that during prebreeding migration, birds using the northern Gulf of Mexico coast represented all breeding populations and potentially a large portion of the nonbreeding range. While breeding population estimates were skewed toward the western region, one-fifth of the birds were from the eastern breeding population. Stable isotope analysis could not definitively identify where individuals spent the nonbreeding season due to a lack of known deuterium variation across South America, but it is possible that they were along the Guianas coast.

DATA CONTRIBUTORS

A. Anderson; Y. Aubry; K. Bennett; R. Bentzen; M. Boldenow; S. Brown; R. Clay; A. Desrochers; S. Elbin; W. English; S. Flemming; S. Freeman; C. Friis; M.-A. Giroux; C. Gratto-Trevor; D. Hamilton; J. Herbert; S. Koch; J. Kolodzinski; D. Kriensky; E. Kwon; R.B. Lanctot; D.B. Lank; N. Lecomte; J. Liebezeit; R. Linhart; P. Loring; V. Loverti; A. MacDonald; H. Mann; D. Mizrahi; J. Mobley; K. O'Brien; J. Paquet; K. Parkins; D. Partridge; N. Pau; R.R. Porter; J. Rausch; S.T. Saalfeld; B.K. Sandercock; S. Schulte; P. Smith; A.R. Taylor; E. Tobon; J. Tremblay; E.L. Weiser; B. Winn; R. Wood; S. Yezerinac.

WESTERN SANDPIPER *Calidris mauri*

DISTRIBUTION AND ABUNDANCE

The Western Sandpiper breeds in Alaska, with most birds in the Yukon-Kuskokwim Delta and a smaller number breeding in northwestern and western Alaska, and in eastern Siberia on the Chukotka Peninsula. During the nonbreeding season, the species occupies areas on the Fraser River Delta, and along the Pacific coast from California to Peru, with large numbers in several estuaries in northwest Mexico. In addition, significant nonbreeding populations are found in the upper Panama Bay and the Laguna Madre on the Gulf of Mexico. Along the Atlantic coast, the Gulf Coast, and inland Central Valley and Salton Sea, California, the species is locally common during the nonbreeding season, and it is also present along the coast of Central America, Colombia, Venezuela, and Suriname, and in the Caribbean (Franks et al. 2020).

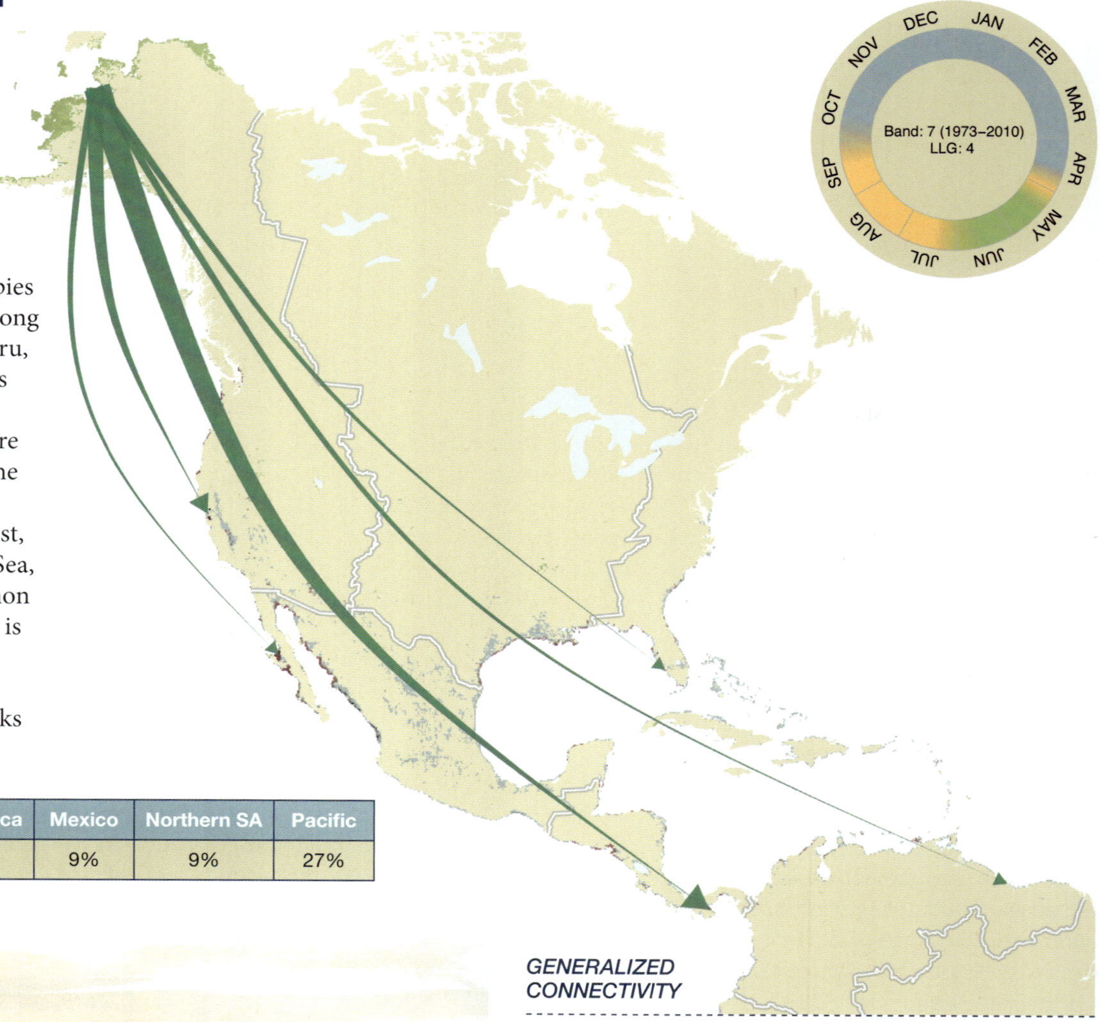

	Atlantic	Central America	Mexico	Northern SA	Pacific
Pacific (n = 11)	9%	45%	9%	9%	27%

ORDER Charadriiformes
FAMILY Scolopacidae
BREEDING HABITAT Wetlands, Tundra
NONBREEDING HABITAT Coasts, Beach and Estuary
GLOBAL POPULATION SIZE 3,500,000
POPULATION CHANGE SINCE 1970 −6%
IUCN RED LIST CATEGORY Least Concern
R2R TIPPING POINT SPECIES No
US LISTING STATUS No
CANADA SPECIES AT RISK No
PIF CCS 14/20

MIGRATORY CONNECTIVITY

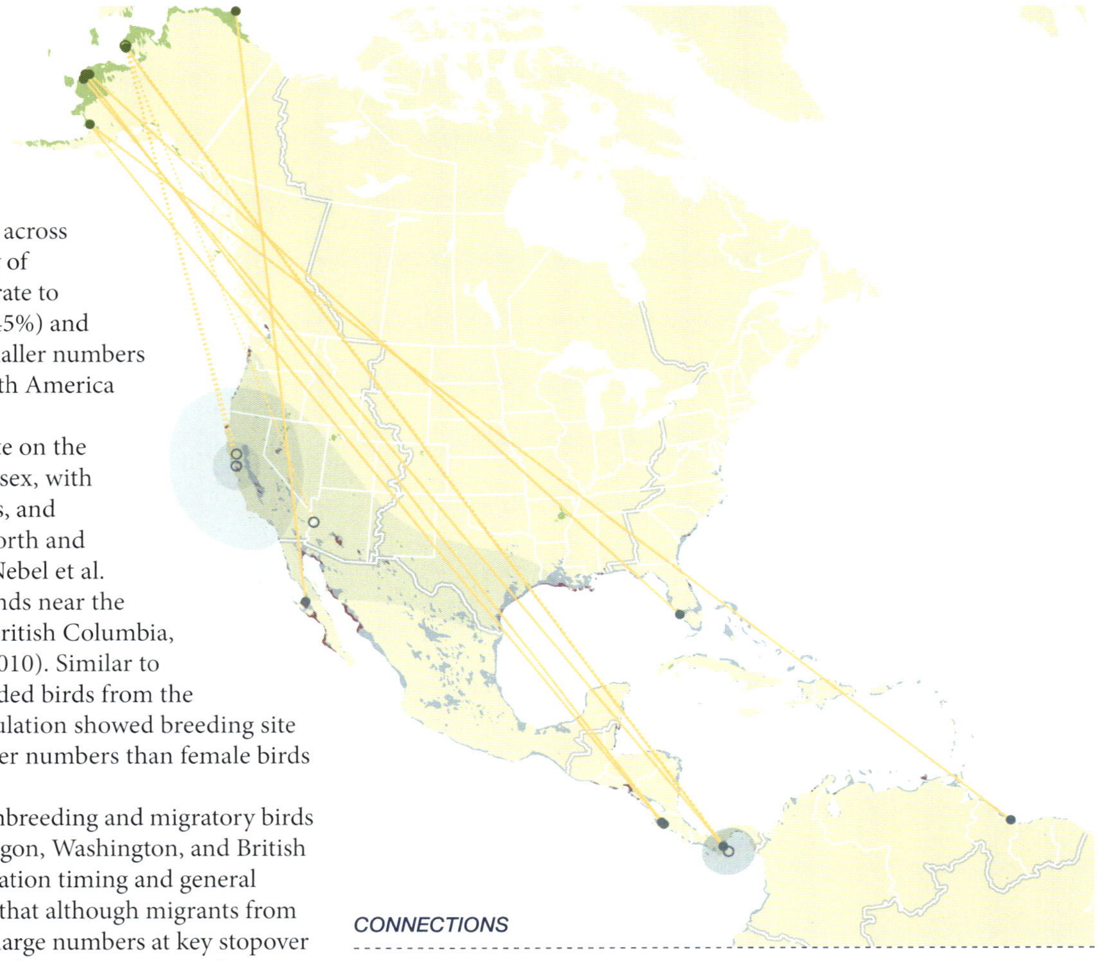

CONNECTIONS

OVERALL Although data are from geolocators deployed only in the Pacific Flyway, and thus MC cannot be calculated, evidence indicates overall weak migratory connectivity given that breeding populations disperse across many nonbreeding regions. The majority of birds breeding in the Pacific Flyway migrate to nonbreeding areas in Central America (45%) and within the Pacific Flyway (27%), with smaller numbers migrating to Mexico (9%), northern South America (9%), and the Atlantic Flyway (9%).

BANDING The species appears to separate on the nonbreeding grounds based on age and sex, with females staying farther south than males, and juveniles staying in areas farther both north and south than adults (Naranjo et al. 1994; Nebel et al. 2002). Birds banded at freshwater wetlands near the central Texas coast were recaptured in British Columbia, Alaska, and Washington state (Ortega 2010). Similar to previous studies, around half of the banded birds from the Yukon-Kuskokwim Delta breeding population showed breeding site fidelity, and male birds returned in higher numbers than female birds (M. Johnson and Walters 2008).

RADIO TRACKING Radio tracking of nonbreeding and migratory birds along the Pacific coast of California, Oregon, Washington, and British Columbia revealed information on migration timing and general stopover patterns. These studies suggest that although migrants from several nonbreeding locations gather in large numbers at key stopover sites, the species also uses short flights with minor stopovers during pre-breeding migration (Iverson et al. 1996; Bishop and Warnock 1998; Warnock and Bishop 1998). Currently there are several tracking projects through the Motus Wildlife Tracking System; hundreds of tags were deployed to understand migration along the Pacific Flyway.

GEOLOCATOR TRACKING Western Sandpipers equipped with geolocators breeding on the Seward Peninsula in northwest Alaska flew along the Pacific coast of western Alaska before either traveling across the Gulf of Alaska to reach California and continuing south to Panama, or continuing along the Pacific coast of Alaska, Canada, the US, and Mexico. One bird traveled east from Baja California to spend the nonbreeding season along the coast in the Gulf of Mexico (E. Kwon, unpublished data). A female departed the Seward Peninsula, staged in southwest Alaska, and then flew nonstop to the Fraser River Delta, where it spent the nonbreeding season. On northward migration, the bird made three coastal stops and reunited with its mate at the same breeding area (Alaska Shorebird Group, unpublished data).

STABLE ISOTOPES Although Yukon-Kuskokwim Delta's breeding population and western Mexico's nonbreeding population show moderate connectivity, stable isotope data support overall weak connectivity due to significant mixing between breeding and nonbreeding populations from both sexes (Franks et al. 2012).

GENETICS Genetic data collected at a stopover site in Mexico found weak demographic differentiation but with two lineages separated temporally over the season (Enríquez-Paredes et al. 2012).

OTHER APPROACHES No satellite transmitter studies of migratory connectivity were found for this species.

MIGRATORY PATHWAYS During northward migration, Western Sandpipers spending the nonbreeding season on the Pacific follow a coastal route, stopping at major sites like San Francisco Bay, Grays Harbor, and the Copper River Delta. More than 90% of Pacific Flyway migrants stage at the latter site. For those spending the nonbreeding season on Atlantic, Gulf, and Caribbean coasts, the migration route is unclear, but they may travel through the midcontinent to meet Pacific Flyway migrants in Washington and British Columbia (Butler et al. 1996). During southward migration, the Yukon-Kuskokwim Delta is a main staging area before birds make direct flights to coastal British Columbia, with the Salish Sea being a first stopover site for fall migrants. They also use stopover sites in California and the Great Basin. Nearly a million stop over at the Gulf Coast, with many spending the stationary nonbreeding season there, while others continue south to coasts of the Caribbean and Central and South America.

DATA CONTRIBUTORS
E. Kwon; B.K. Sandercock.

INTERESTING NOTE

Early studies using radio tracking, with tags glued to the birds' backs after clipping feathers, found that birds without tags departed earlier than tagged birds, suggesting a tagging effect (Iverson et al. 1996; Bishop and Warnock 1998; Warnock and Bishop 1998).

SHORT-BILLED DOWITCHER *Limnodromus griseus*

DISTRIBUTION AND ABUNDANCE

Short-billed Dowitcher is separated into three subspecies with mostly distinguishable distributions. *L. g. caurinus* breeds from southern Alaska to southern Yukon Territory and northwest British Columbia, and spends the nonbreeding season on the Pacific coast from the central US to southern Peru and the Galapagos Islands. *L. g. griseus* breeds in northeast Canada and spends the nonbreeding season on the Atlantic coast from southern US to Brazil. *L. g. hendersoni* breeds in central Canada and spends the nonbreeding season on the mid-Atlantic and Gulf state coasts south to Panama. The breeding range of *L. g. caurinus* is separate from the other subspecies, whereas the ranges of *L. g. griseus* and *L. g. hendersoni* likely intergrade along the western shores of James and Hudson bays (Pyle 2008; Jehl et al. 2020).

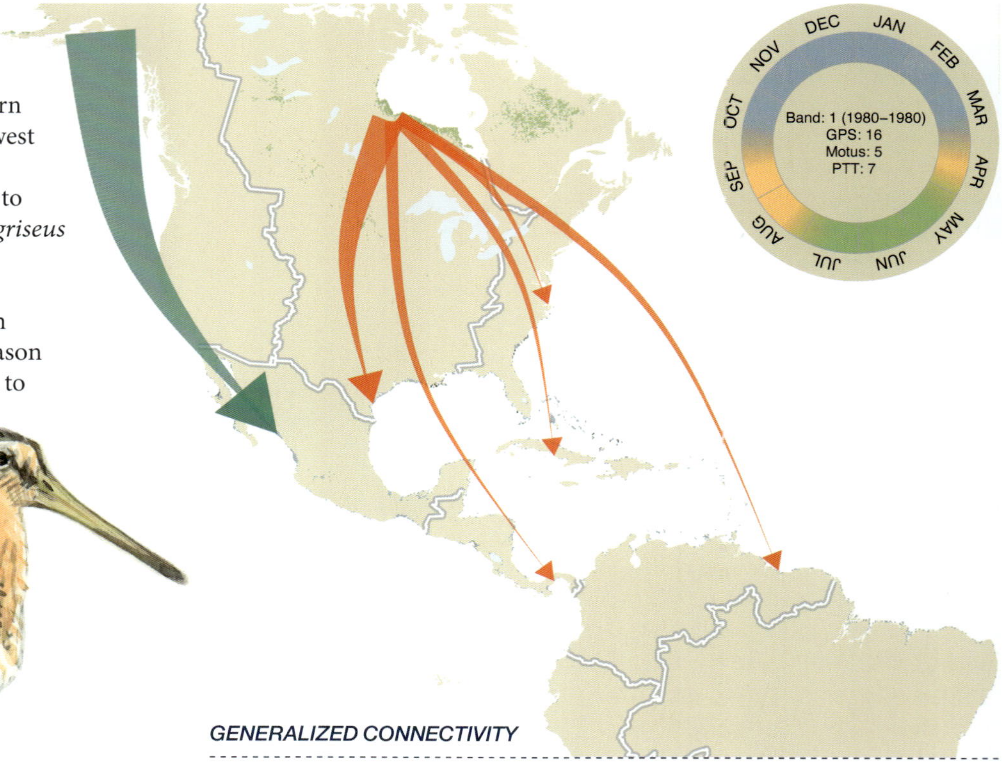

GENERALIZED CONNECTIVITY

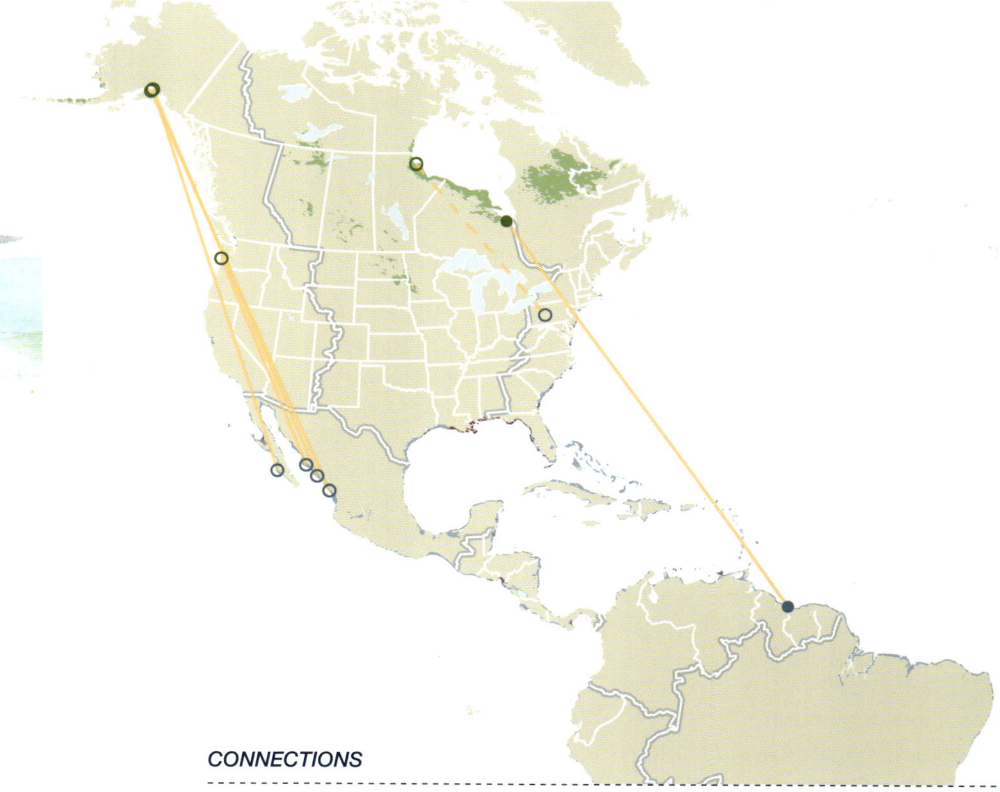

CONNECTIONS

ORDER Charadriiformes
FAMILY Scolopacidae
BREEDING HABITAT Wetlands, Boreal
NONBREEDING HABITAT Coasts, Beach and Estuary
GLOBAL POPULATION SIZE 150,000
POPULATION CHANGE SINCE 1970 −55%
IUCN RED LIST CATEGORY Least Concern
R2R TIPPING POINT SPECIES Orange-Alert
US LISTING STATUS No
CANADA SPECIES AT RISK No
PIF CCS 15/20

	Atlantic	Caribbean	Central America	Mexico	Midcontinent	Northern SA
Midcontinent (n = 7)	14%	14%	14%	–	43%	14%
Pacific (n = 4)	–	–	–	100%	–	–

MIGRATORY CONNECTIVITY

OVERALL Several sources of data suggest disjunct breeding and nonbreeding areas by *L. g. caurinus* and *L. g. hendersoni*. Remaining data suggests some overlap among *L. g. hendersoni* and *L. g. griseus* on the nonbreeding grounds. Thus, there appears to be strong connectivity between the westernmost subspecies and the other two subspecies but low connectivity between the central and eastern subspecies. Overall, birds breeding within the Midcontinent Flyway migrate to nonbreeding areas within the Midcontinent Flyway (43%), as well as several other regions throughout the nonbreeding range. Birds breeding within the Pacific Flyway spend the nonbreeding season in Mexico (100%). MC is not calculated due to limited sample size.

BANDING One bird banded on breeding grounds in Ontario was recovered during the stationary nonbreeding season in Guyana.

RADIO TRACKING Many Motus tags were deployed through the Motus Wildlife Tracking System. There were limited detections of individual birds between breeding and nonbreeding locations.

SATELLITE TRACKING GPS transmitters were deployed on Short-billed Dowitchers at three locations in southern Alaska in 2021. All birds moved south along the Pacific coast before arriving at their nonbreeding locations in Baja California Sur and Sinaloa, Mexico (J. Johnson, unpublished data). Short-billed Dowitchers tracked from the Gulf Coast of Texas during northward migration spent the breeding season in different locations (Migratory Connectivity Project, unpublished data): Northwest Territories near Great Bear Lake and along the southwest shore of Hudson Bay in Manitoba and Ontario, Canada. The majority of birds spent the nonbreeding season on the Texas Gulf Coast. One bird spent the breeding season in Manitoba, the nonbreeding season in Panama, and the following breeding season just east along the shoreline in Ontario (Migratory Connectivity Project, unpublished data).

STABLE ISOTOPES Samples collected from birds hunted in Barbados indicated that the majority of the birds were from the eastern population, with natal origins likely in southeastern Manitoba, northeastern Ontario, or central Quebec. However, there were also birds whose breeding origin may have been in the far western portion of the species range—i.e., Kodiak Island, Alaska (E.T. Reed et al. 2018).

GENETICS Birds sampled in Texas and Alaska indicated at least two subspecies, causing Baker and Strauch (1988) to characterize the Short-billed Dowitcher as a panmictic species.

MIGRATORY PATHWAYS During northward migration, *L. g. hendersoni* migration occurs along the Atlantic coast north to New Jersey and then through the interior, while also occurring along the western Gulf Coast through the midcontinent (Jaramillo et al. 1991). Subspecies *L. g. griseus* stages along coasts of the Guianas, moves through the Caribbean and along the Atlantic coast before making nonstop flights to breeding areas in northeastern Canada (Jehl et al. 2020). During southward migration, there is overlap between *L. g. griseus* and *L. g. hendersoni*, mostly along the Atlantic coast (Jaramillo et al. 1991), with some birds recorded on Barbados (Amos 1991). Finally, the migratory pathway of subspecies *L. g. caurinus* is almost exclusively restricted to the Pacific coast region. Birds tagged on the Texas Gulf Coast during northward migration used a similar route north until the Dakotas and then diverged. Three birds continued north and west up to the Northwest Territories, while the other two took a northeasterly route toward Hudson Bay. The Northwest Territories birds took a similar southbound route back to Texas, but the Hudson Bay birds migrated south through Florida to Cuba and Central America (Migratory Connectivity Project, unpublished data). Radio nanotags were attached to birds on breeding areas near Churchill, Manitoba, beginning in 2021. Motus detections indicated all birds breeding near Churchill passed through the Great Lakes region (O. Maillet and E. Nol, unpublished data), with later detections either in the Midwest US, the interior northeast US, or along the coast of North Carolina or Florida (O. Maillet and E. Nol, unpublished data).

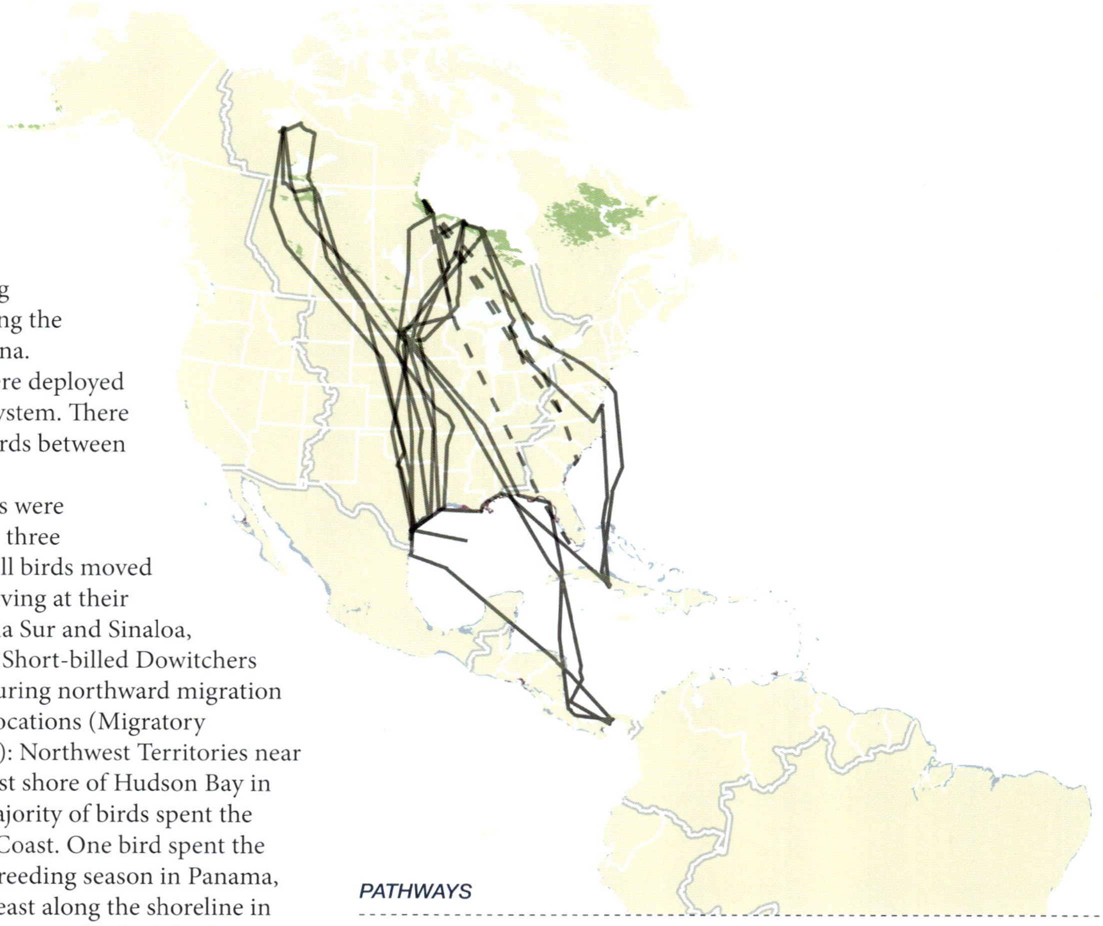

PATHWAYS

DATA CONTRIBUTORS

R. Bathrick; A.-L. Harrison; J.A. Johnson; O. Maillet; L.A. McDuffie; L. McKinnon; D.J. Newstead; E. Nol; D.R. Ruthrauff; N.R. Senner; P. Smith; Migratory Connectivity Project.

INTERESTING NOTE

Annual fidelity to breeding areas has been suggested by a return sighting of a territorial male (Jehl et al. 2020). There also appears to be a moderate level of site fidelity to breeding areas in Alaska (J. Johnson, personal communication). A bird captured and marked in Costa Rica was found again in the same location two years later, suggesting some fidelity to migratory stopover or nonbreeding sites (Smith and Stiles 1979). Also, a second-year migrant bird banded in Gaspé Peninsula, Quebec, was recaptured eleven years later.

LONG-BILLED DOWITCHER *Limnodromus scolopaceus*

DISTRIBUTION AND ABUNDANCE

Long-billed Dowitchers breed in northern Northwest and Yukon territories, northern and western Alaska, and northern Russia. They spend the nonbreeding season in southern portions of the US from the Atlantic coasts, along the Gulf of Mexico, and throughout most of Mexico (Takekawa and Warnock 2020).

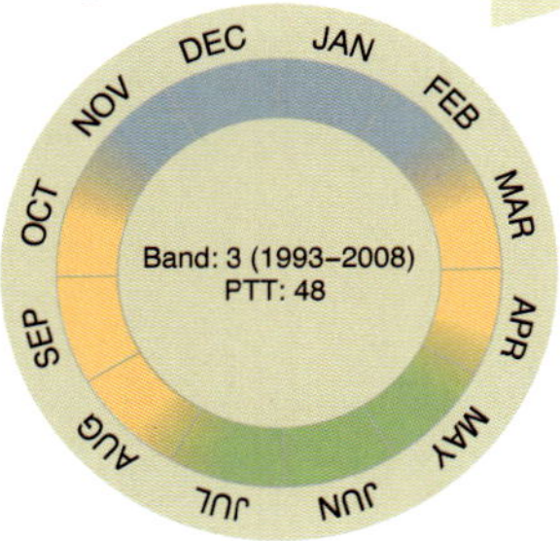

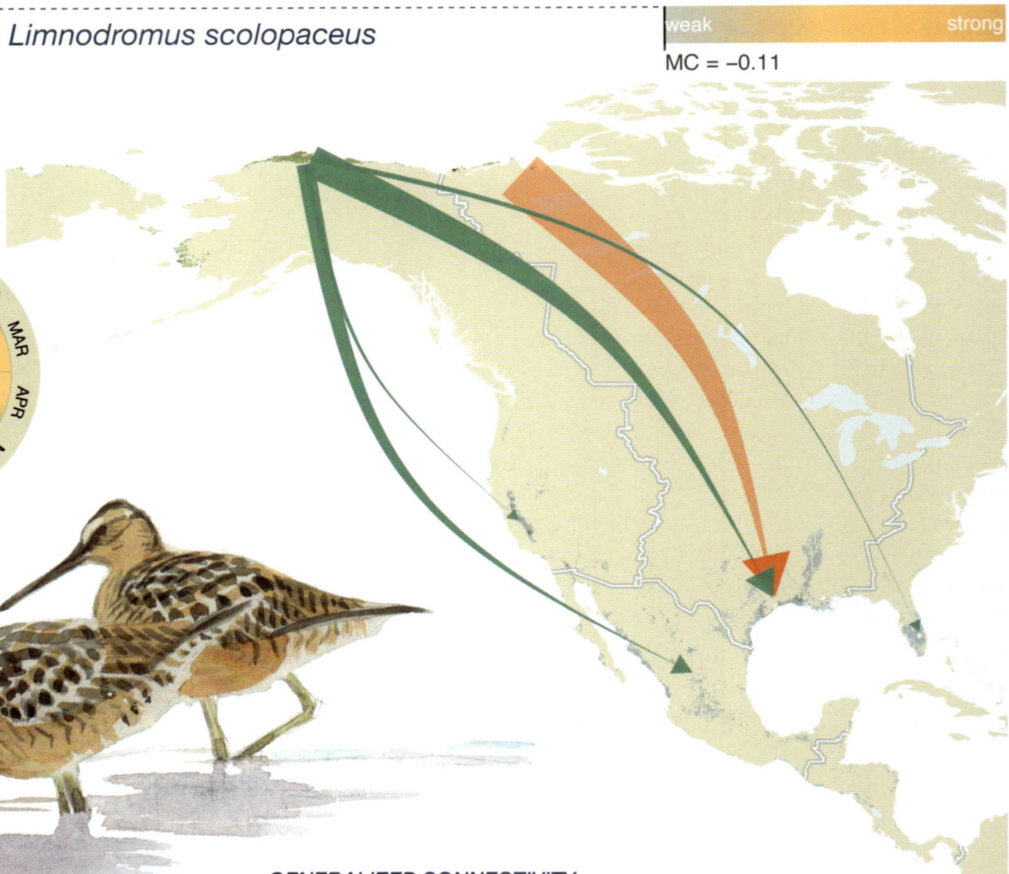

GENERALIZED CONNECTIVITY

ORDER Charadriiformes
FAMILY Scolopacidae
BREEDING HABITAT Wetlands, Tundra
NONBREEDING HABITAT Wetlands, Generalist
GLOBAL POPULATION SIZE 650,000
POPULATION CHANGE SINCE 1970 +2%
IUCN RED LIST CATEGORY Least Concern
R2R TIPPING POINT SPECIES Orange-Alert
US LISTING STATUS No
CANADA SPECIES AT RISK No
PIF CCS 14/20

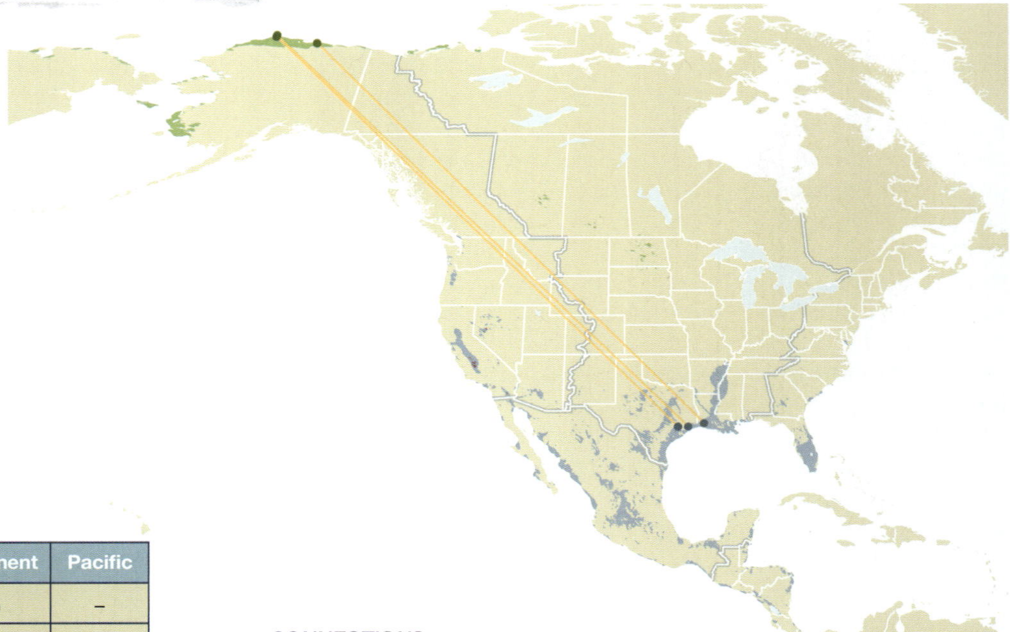

CONNECTIONS

	Atlantic	Mexico	Midcontinent	Pacific
Midcontinent (n = 1)	–	–	100%	–
Pacific (n = 33)	6%	30%	58%	6%

MIGRATORY CONNECTIVITY

OVERALL Migratory connectivity information, primarily from satellite transmitter data and limited band reencounters, suggest weak migratory connectivity, with birds that breed together less likely to remain together during the nonbreeding season (MC = −0.11); however, data are limited from the Midcontinent Flyway. Birds breeding in the Midcontinent Flyway migrate to nonbreeding areas within the Midcontinent Flyway. Pacific Flyway breeding birds also migrate to the Midcontinent Flyway (58%) and Mexico (30%), as well as the Pacific (6%) and Atlantic flyways (6%). Little is known about the migratory connectivity or migration of Northwest and Yukon Territory breeders.

BANDING Several birds banded on breeding areas in northern Alaska were recovered at coastal Texas and Louisiana during the stationary nonbreeding season.

RADIO TRACKING Several Motus tags were deployed in the western part of the range. There are limited detections of individual birds between breeding and nonbreeding areas.

SATELLITE TRACKING A PTT tagging study in Alaska indicated that birds spend the nonbreeding season in coastal regions of Texas, Mexico, and Florida (Kwon and Kempenaers 2023). Birds tracked from Texas bred in Alaska (n = 2) and Russia (n = 2; D. Newstead, unpublished data).

MIGRATORY PATHWAYS
During post-breeding migration, the majority of Alaska breeding birds travel along the east side of the Mackenzie Mountains and south through the midcontinent US. Alternatively, Kwon and Kempenaers (2023) found that some Alaska breeders migrate south through the Pacific Flyway to California and Mexico. Less is known about birds observed on the Pacific coast. Forty-six of 81 birds tracked with nanotags during a stopover in Klamath Basin on the Washington/Oregon border were later detected in California's Central Valley (Barbaree et al. 2016).

DATA CONTRIBUTORS
B. Kempenaers; E. Kwon.

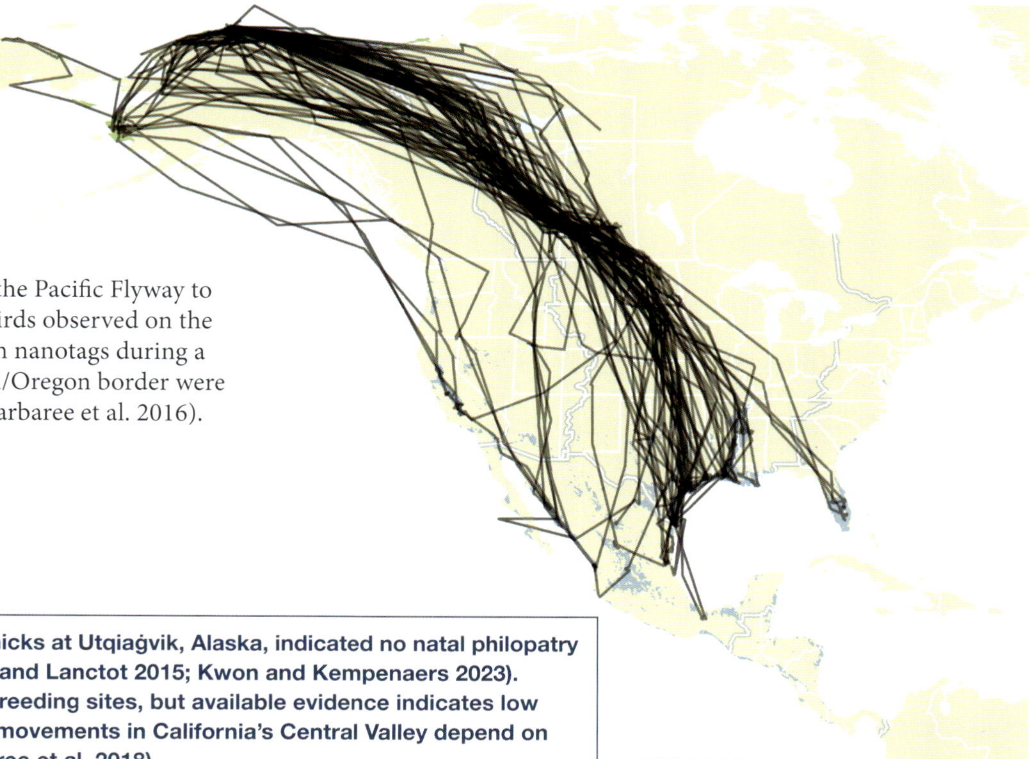

PATHWAYS

INTERESTING NOTE

Extensive banding of breeding adults and chicks at Utqiaġvik, Alaska, indicated no natal philopatry and very low rates of adult fidelity (Saalfeld and Lanctot 2015; Kwon and Kempenaers 2023). There is little information on fidelity to nonbreeding sites, but available evidence indicates low rates (Takekawa and Warnock 2020). Local movements in California's Central Valley depend on the predictability of water resources (Barbaree et al. 2018).

weak strong
MC = 0.27 ± 0.04*

Scolopax minor AMERICAN WOODCOCK

ORDER Charadriiformes
FAMILY Scolopacidae
BREEDING HABITAT
Forests, Temperate Eastern
NONBREEDING HABITAT
Forests, Temperate Eastern
GLOBAL POPULATION SIZE 3,500,000
POPULATION CHANGE SINCE 1970 −31%
IUCN RED LIST CATEGORY Least Concern
R2R TIPPING POINT SPECIES No
US LISTING STATUS No
CANADA SPECIES AT RISK No
PIF CCS 12/20

DISTRIBUTION AND ABUNDANCE

Famous for its dramatic early pre-breeding display flights, this unique shorebird breeds in early successional habitats with open areas throughout much of the central and eastern United States from Virginia north into the Atlantic provinces and west through southern Canada into Manitoba and south through Minnesota, and south through Wisconsin into Kentucky (McAuley et al. 2020).

Nonbreeding areas include most of the southeastern states from Virginia south into Florida and west to east Texas. A popular game bird, this species was once widely market hunted and continues to be sport hunted today (McAuley et al. 2020). As such, the species is managed as having a central population and an eastern population. Ongoing declines are suspected to be due to habitat loss as early successional forests decrease. Additional survey coverage is needed to better estimate range-wide trends; research into habitat effects on population dynamics is also needed (McAuley et al. 2020).

MIGRATORY CONNECTIVITY

OVERALL Overall, this species has been fairly heavily tracked in recent years, although not explicitly for connectivity. A variety of data sources indicate moderate range-wide migratory connectivity (MC = 0.27 ± 0.04). Within-flyway transitions are high for both Atlantic (70%) and Midcontinent (94%) flyway breeding populations. However, a portion of Atlantic Flyway breeding birds migrate to nonbreeding areas within the Midcontinent Flyway (30%).

BANDING Birds banded on breeding grounds in Michigan, Minnesota, and Wisconsin were recovered in Louisiana (Myatt and Krementz 2007a). Analysis of band reencounters by Moore and Krementz (2017) indicated birds remain in the central region during the nonbreeding season, but birds in the eastern region travel throughout the nonbreeding range, mixing widely with central region birds.

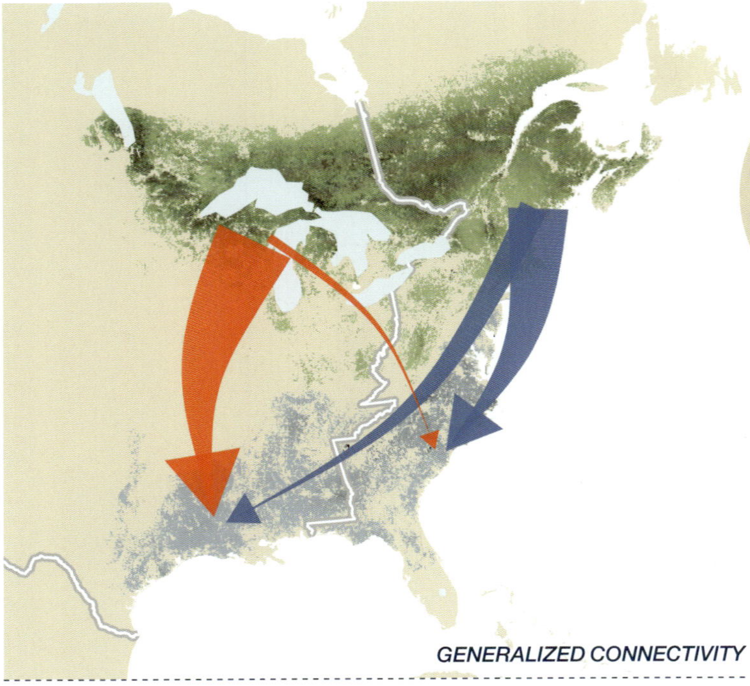

GENERALIZED CONNECTIVITY

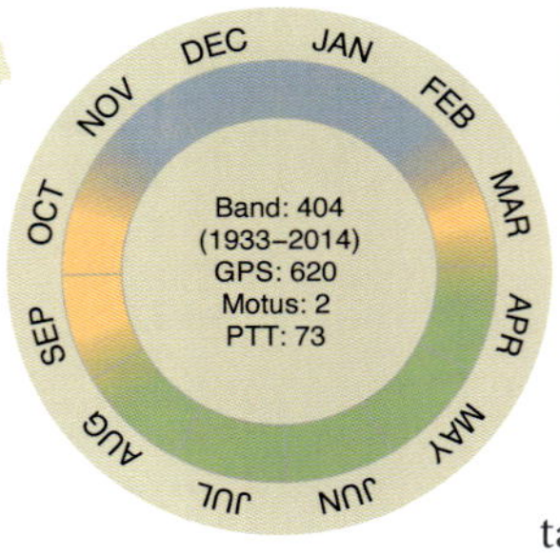

	Atlantic	Midcontinent
Atlantic (n = 341)	70%	30%
Midcontinent (n = 257)	6%	94%

RADIO TRACKING Woodcock tracked from breeding areas in Minnesota and Wisconsin migrated to Gulf Coast states (Myatt and Krementz 2007b). A few hundred tags were deployed through the Motus Wildlife Tracking System; there are limited breeding to nonbreeding detections (D. McAuley, unpublished data) and numerous detections during migration.

SATELLITE TRACKING Moore et al. (2021) successfully tracked 60 Woodcocks using PTT devices in the northern and southern portions of the Midcontinent Flyway of the United States. Graham et al. (2022), using GPS PinPoint tags, demonstrated that the southern New England breeding population (Rhode Island) of Woodcocks were facultative partial migrants, with some birds remaining resident and others migrating short and long distances. The same study demonstrated breeding birds in Rhode Island spent the nonbreeding season in Georgia, South Carolina, Virginia, and Washington, DC. In a recent study using over 200 GPS satellite tags, 36 state, federal, and nongovernmental organizations tagged female Woodcock in eastern North America from Florida northward to Atlantic Canada (Slezak et al. 2024). Over 80% of female Woodcock bred more than once and some up to six times, as females moved northward between subsequent nesting attempts. Example tracks from Slezak et al. (2024) confirm a parallel migration strategy, with individuals in the west staying west and eastern birds staying more east.

STABLE ISOTOPES Using band recovery data and stable isotopic composition of feathers, Sullins et al. (2016) estimated population origins and migratory connectivity among natal, early post-breeding, and nonbreeding ranges of hunter-harvested juveniles. Overall, the majority of all Woodcock origins were assigned to the northernmost portion of both the central and eastern regions, suggesting some evidence of leapfrog migration. Also, assignments were more uniformly distributed along the Atlantic coast in the eastern region, whereas most Woodcock were assigned to origins within and north

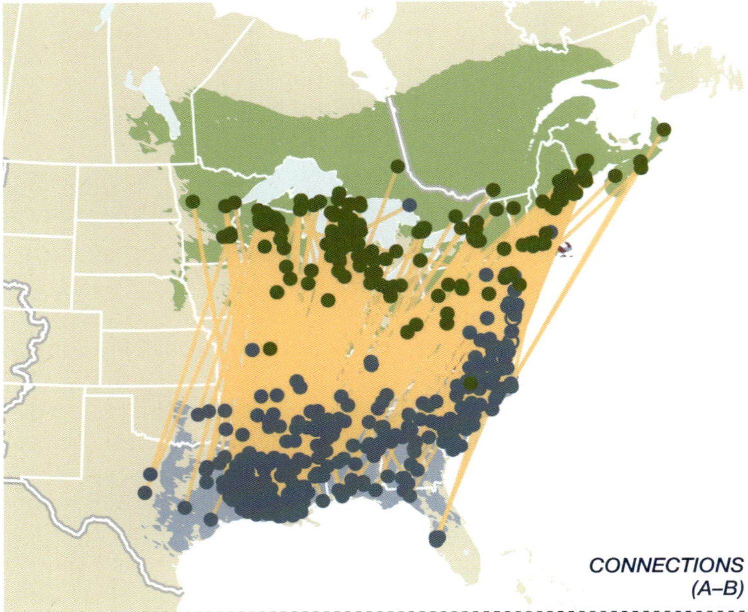

CONNECTIONS (A–B)

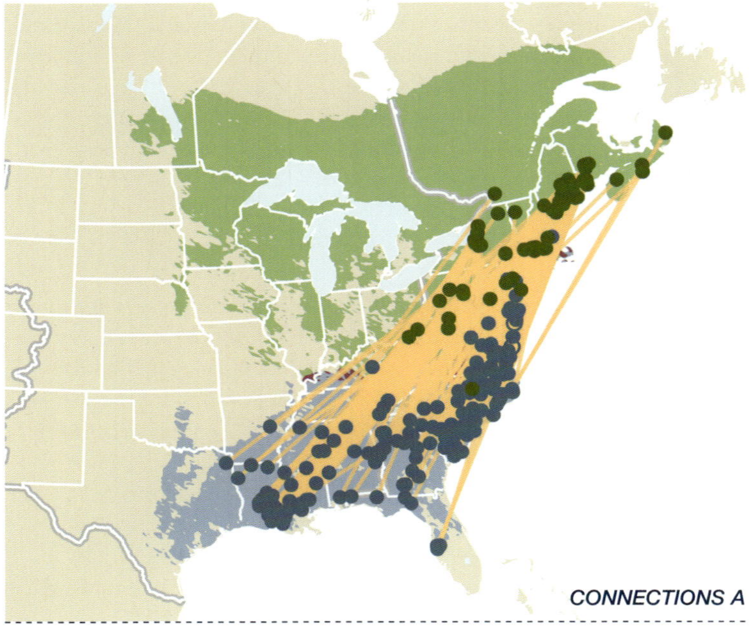

CONNECTIONS A

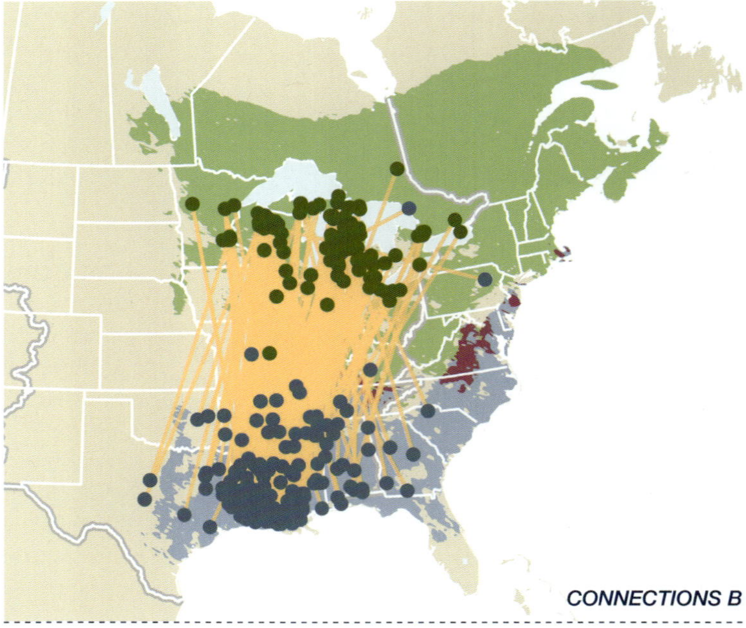

CONNECTIONS B

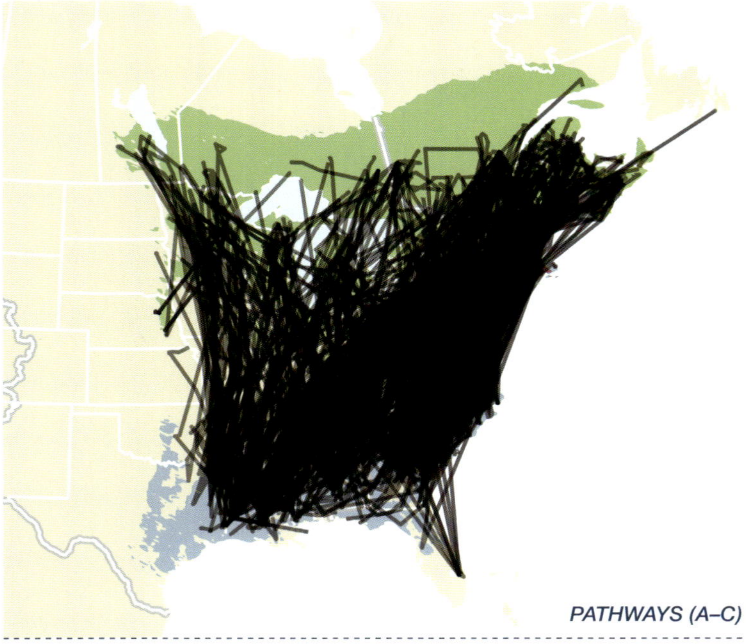

PATHWAYS (A–C)

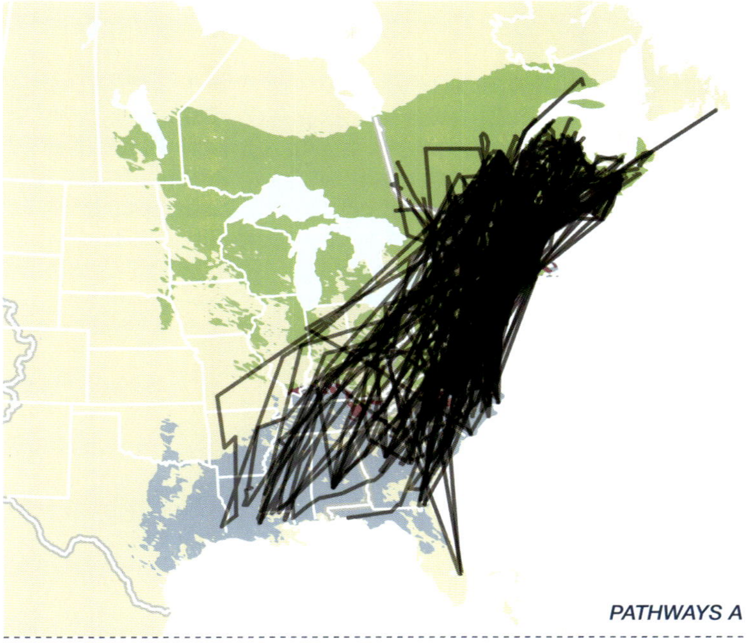

PATHWAYS A

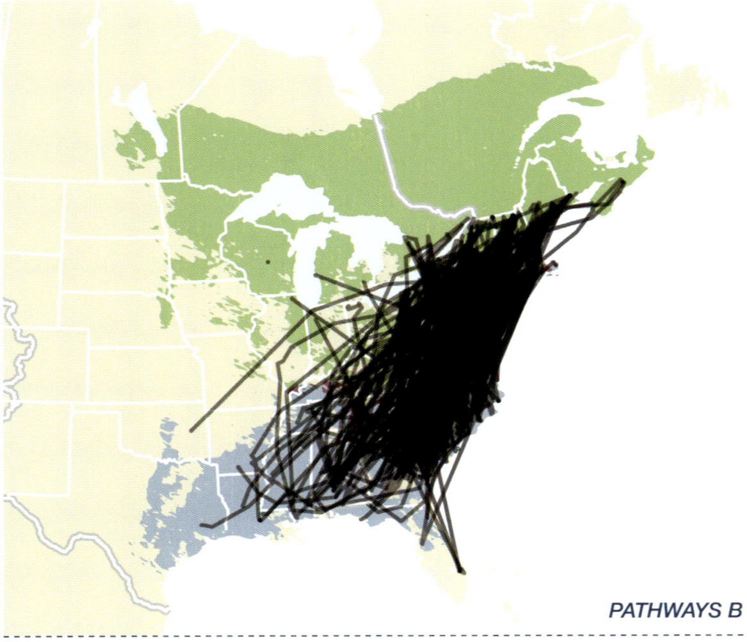

PATHWAYS B

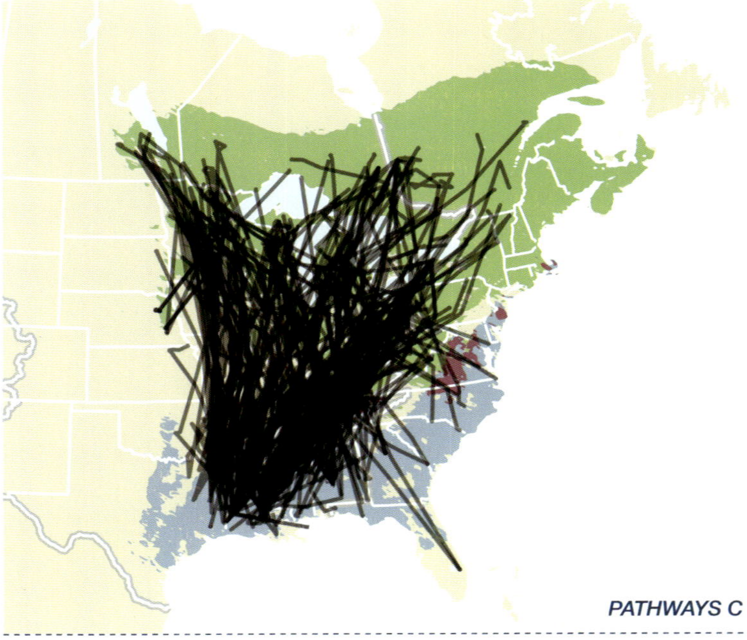

PATHWAYS C

of the Great Lakes region, suggesting an eastern and central region divide.

GENETICS Based on mitochondrial DNA (mtDNA) sequencing, there was no phylogeographic structure among populations across the species range. Results suggest rapid, post-glacial population expansion from southern breeding; or asymmetric movement of birds between migration flyways (Rhymer et al. 2005).

MIGRATORY PATHWAYS Migration routes generally follow a north to south or northeast to southwest post-breeding pattern with a reverse pattern in pre-breeding migration (Moore et al. 2021; Slezak et al. 2024). In the midcontinent, Woodcock generally used few stopovers (average 2.4) for short durations (average 5.1 days; Moore et al. 2021). Migration is more synchronous in pre-breeding than in post-breeding, with post-breeding migration duration shorter than pre-breeding migration (Moore et al. 2021).

DATA CONTRIBUTORS

L.A. Berigan; E.J. Blomberg; A.M. Bracey; A.C. Fish; A. Grinde; J.G. Jorgensen; D. Krementz; D. McAuley; J. Moore; A.M. Roth.

INTERESTING NOTE

There is evidence of natal site fidelity, with birds banded as chicks recaptured within 5 miles of their natal site (McAuley et al. 2020). Graham et al. (2022) found strong evidence for carry-over effects from winter/post-breeding conditions on the timing of post-breeding migration and arrival on nonbreeding grounds.

WILSON'S SNIPE *Gallinago delicata*

DISTRIBUTION AND ABUNDANCE

Its widespread breeding range extends from the Aleutian Islands and northern Alaska east through the taiga and boreal forest of Canada to Newfoundland and south to central California, extending as far south as North Dakota, Michigan, and New York. It spends the nonbreeding season locally from southern Alaska and the coast of British Columbia in the north, down through the central United States and Virginia, south through Central America, Bermuda, and the Caribbean, Colombia, Venezuela, and the Guianas. Resident populations found locally in the northwestern US and in some areas in the Great Lakes region (Mueller 2020).

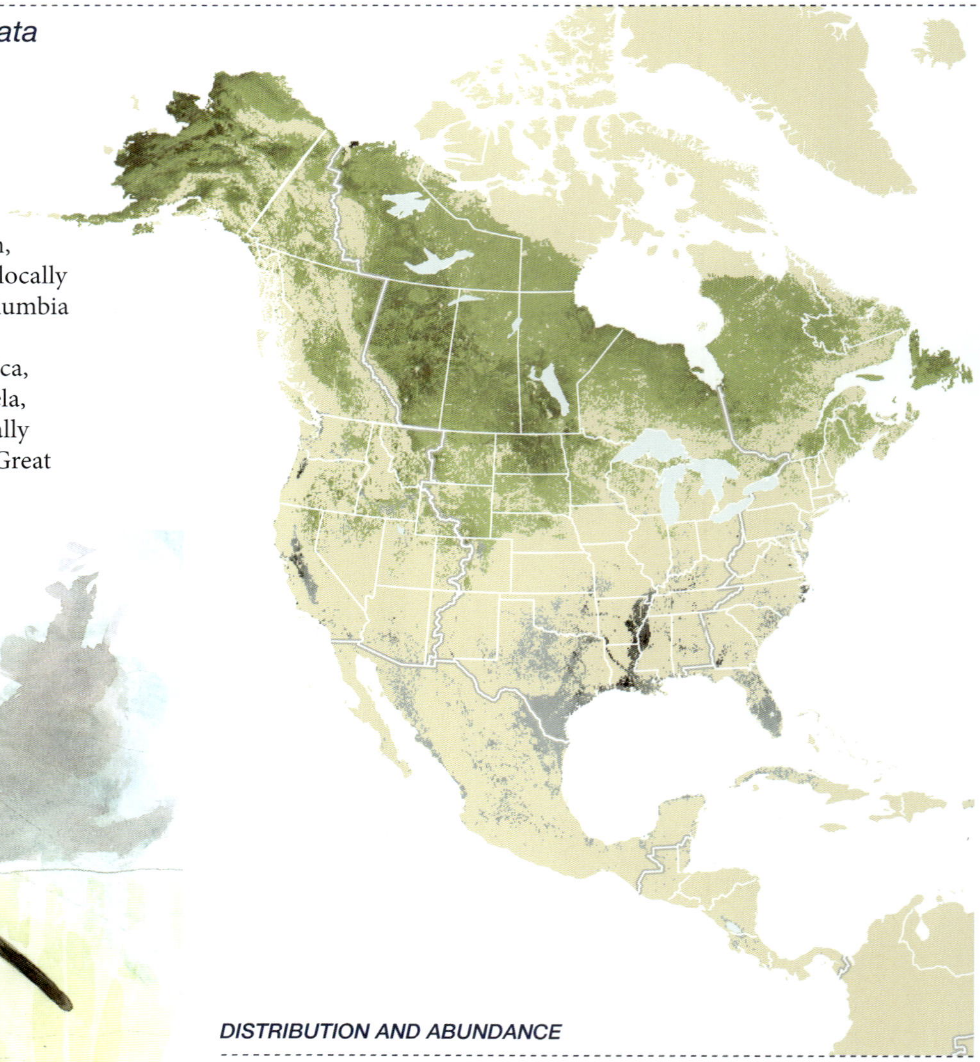

DISTRIBUTION AND ABUNDANCE

ORDER Charadriiformes
FAMILY Scolopacidae
BREEDING HABITAT
Wetlands, Generalist
NONBREEDING HABITAT
Wetlands, Generalist
GLOBAL POPULATION SIZE 2,000,000

POPULATION CHANGE SINCE 1970 +17%
IUCN RED LIST CATEGORY
Least Concern
R2R TIPPING POINT SPECIES No
US LISTING STATUS No
CANADA SPECIES AT RISK No
PIF CCS 9/20

INTERESTING NOTE

A Wilson's Snipe was observed and recorded for the first time in June 2007 near the Anadyr airport in Chukotka, Russia (Arkhipov 2009).

MIGRATORY CONNECTIVITY

OVERALL The species' migratory connectivity across North America is poorly known as tracking data are limited. Available information from a single telemetry study suggests weak migratory connectivity.

RADIO TRACKING A study in the Willamette Valley, Oregon (n = 37), revealed a complex regional snipe population structure: 74% nonbreeding residents, 14% migrants, 9% breeding residents, and 3% year-round resident breeders. Results suggested weak migratory connectivity, partial migration, and between-season fidelity (Cline and Haig 2011). Few tags were deployed through the Motus Wildlife Tracking System, with local detections only.

OTHER APPROACHES No satellite tracking, geolocator, banding, genetic, or stable isotope studies of migratory connectivity were found for this species.

MIGRATORY PATHWAYS Wilson's Snipe undertake a broad-front migration during pre- and post-breeding, and pathways vary depending on their breeding location. Some birds fly across the Atlantic, while others move overland or along coastal routes (Fink et al. 2023).

Actitis macularius **SPOTTED SANDPIPER**

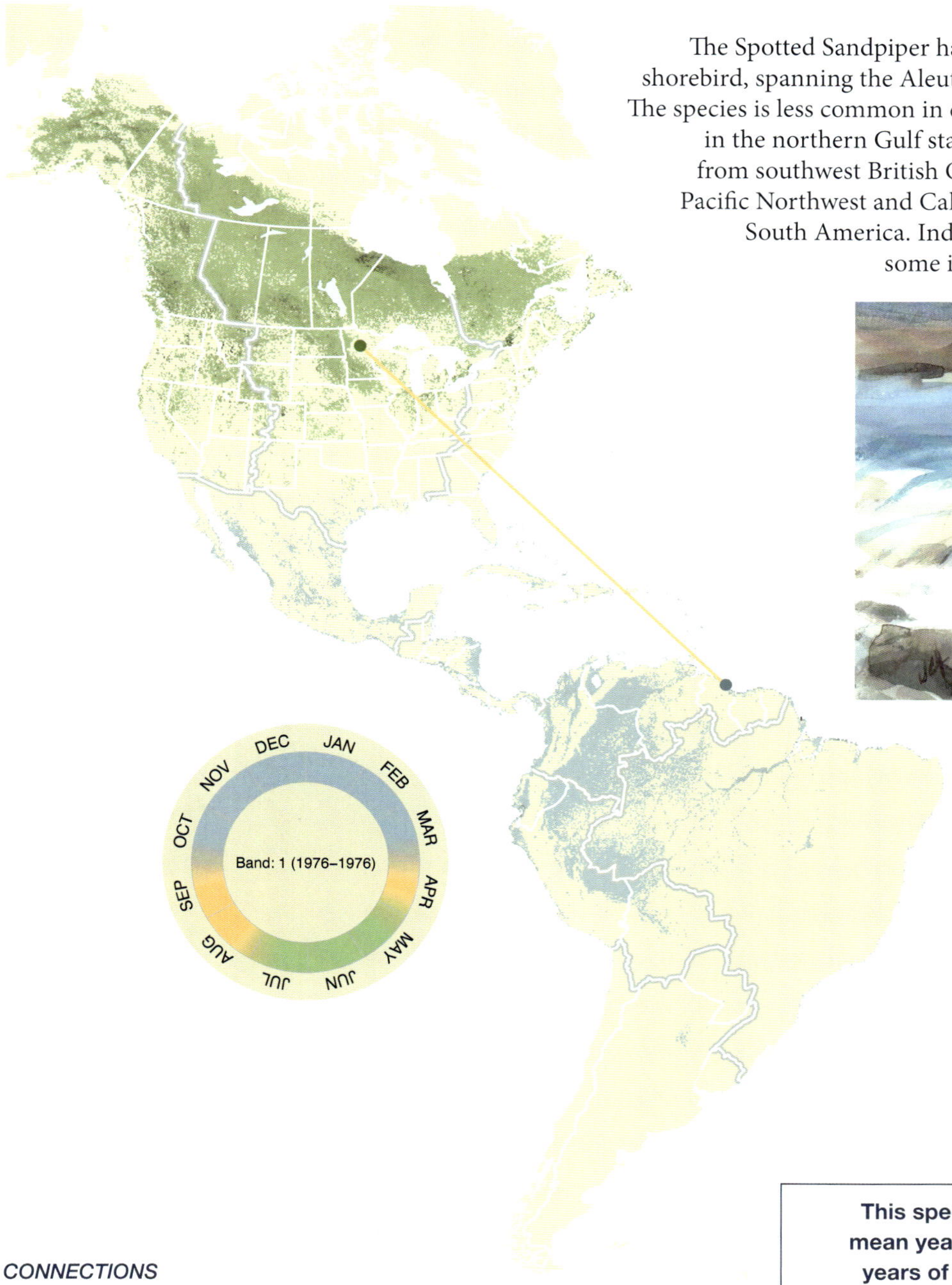

CONNECTIONS

DISTRIBUTION AND ABUNDANCE

The Spotted Sandpiper has the largest breeding range of any North American shorebird, spanning the Aleutian Islands south to California, and east to Virginia. The species is less common in eastern North America and only occasionally breeds in the northern Gulf states. During the nonbreeding season, it can be found from southwest British Columbia to the coastal and near-coastal areas of the Pacific Northwest and California, and south throughout Central America into South America. Individuals of all North American populations migrate, some in small groups, others singly (J.M. Reed et al. 2020).

ORDER Charadriiformes
FAMILY Scolopacidae
BREEDING HABITAT Wetlands, Generalist
NONBREEDING HABITAT Wetlands, Lakes and Rivers
GLOBAL POPULATION SIZE 660,000
POPULATION CHANGE SINCE 1970 −49%
IUCN RED LIST CATEGORY Least Concern
R2R TIPPING POINT SPECIES No
US LISTING STATUS No
CANADA SPECIES AT RISK No
PIF CCS 12/20

INTERESTING NOTE

This species has shown some degree of natal fidelity with a mean yearling return rate of approximately 17% across eight years of a long-term study (Alberico et al. 1992). Successful breeders have higher annual site fidelity, although females who lay multiple clutches may move breeding sites within a season (J.M. Reed et al. 2020). In Venezuela, two birds were recaptured in the same lagoon as the previous winter, suggesting some level of nonbreeding site fidelity (McNeil 1982).

MIGRATORY CONNECTIVITY

OVERALL Although little information about migratory connectivity is available, the Spotted Sandpiper displays some level of fidelity to breeding and nonbreeding areas, while also exhibiting individual variation in migratory pathways and behavior. This suggests a low to moderate level of migratory connectivity, but information is poor.

BANDING One bird banded during the breeding season in Minnesota was recovered shot during the nonbreeding season in Guyana. Birds banded while breeding or migrating through the Great Lakes region and eastern states have been resighted in the southern US and the Caribbean, and along the northern coastline of South America.

RADIO TRACKING Few tags were deployed through the Motus Wildlife Tracking System, with local detections only.

GENETICS DNA analyses from two populations of Spotted Sandpiper at the center of their geographic distribution indicate a lack of genetic differentiation between the populations and a high mutation rate—a degree of differentiation consistent with regular movement between sampled populations. However, an absence of marked birds observed moving between populations despite regular surveys may suggest some population-specific migration behavior (J.M. Reed et al. 1996).

OTHER APPROACHES No stable isotope studies of migratory connectivity were found for this species.

MIGRATORY PATHWAYS This species exhibits a wide-ranging migration pattern throughout the Americas, preferring inland freshwater habitats, although it can also be seen in small numbers in coastal areas. During southward migration, eastern populations are thought to migrate southeast along the Atlantic states or via a transoceanic route from New England to the eastern Neotropics (Cramp and Simmons 1983).

SOLITARY SANDPIPER *Tringa solitaria*

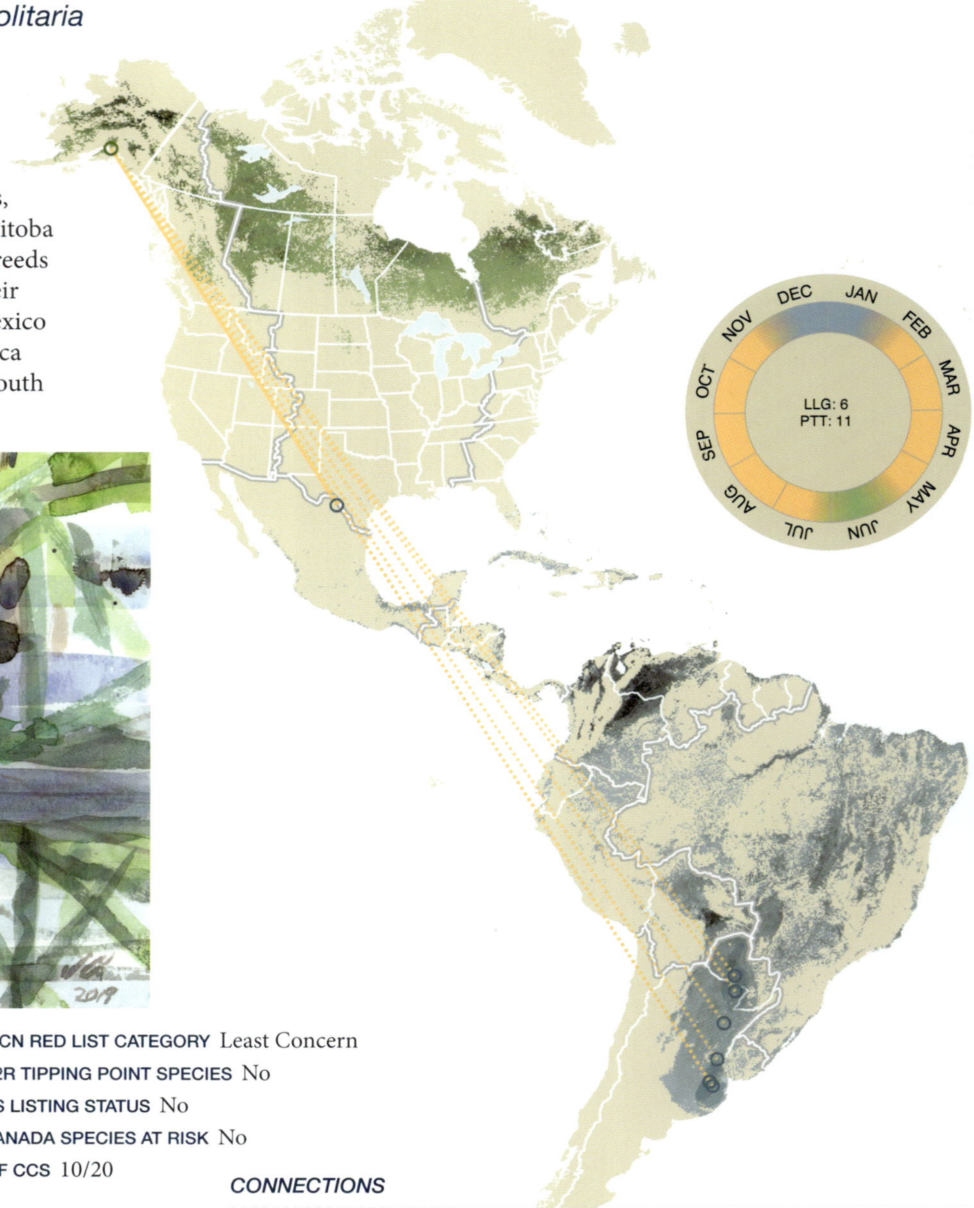

DISTRIBUTION AND ABUNDANCE

The breeding range of the Solitary Sandpiper spans the boreal forest from Alaska and northern Canada to central and southern Labrador. Two subspecies are recognized. The western subspecies, *T. s. cinnamomea*, breeds farther north, from Manitoba to Alaska. The eastern subspecies, *T. s. solitaria*, breeds from northeast British Columbia to Labrador. Their nonbreeding range extends south mostly from Mexico to Central America, and throughout South America east of the Andes to Paraguay and Uruguay, and south to north-central Argentina (Moskoff 2020).

ORDER Charadriiformes
FAMILY Scolopacidae
BREEDING HABITAT Wetlands, Boreal
NONBREEDING HABITAT Wetlands, Lakes and Rivers
GLOBAL POPULATION SIZE 190,000
POPULATION CHANGE SINCE 1970 +27%

IUCN RED LIST CATEGORY Least Concern
R2R TIPPING POINT SPECIES No
US LISTING STATUS No
CANADA SPECIES AT RISK No
PIF CCS 10/20

MIGRATORY CONNECTIVITY

OVERALL This species appear to exhibit moderate migratory connectivity, although due to the low number of individuals tracked between breeding and stationary nonbreeding areas, no formal estimates are available.

RADIO TRACKING Few tags were deployed through the Motus Wildlife Tracking System, with no breeding to nonbreeding detections.

GEOLOCATOR TRACKING Between 2016 and 2018, light-level geolocators were deployed on the western subspecies of Solitary Sandpipers in Anchorage, Alaska. Data from seven birds indicated they traveled as far as the Pampas, Argentina, with the majority of birds migrating to northern Argentina and Paraguay (J. Johnson, unpublished data).

SATELLITE TRACKING In 2021, GPS archival tags (2.5 g; Lotek PinPoint 50) were deployed on the western subspecies of the Solitary Sandpiper in interior Alaska (n = 15) and Anchorage, Alaska (n = 5). One bird migrated to Paraguay (J. Johnson, unpublished data).

MIGRATORY PATHWAYS The western subspecies, *T. s. cinnamomea*, migrates largely west of the Mississippi River into Central America, Bolivia, Paraguay, and Argentina to nonbreeding areas generally south of nonbreeding areas of *T. s. solitaria* (Conover 1944b; Moskoff 2020). Birds tagged in Anchorage, Alaska, used the Central Flyway during southward migration, whereas northward migration occurred through the Mississippi Flyway, staying east of their southbound tracks until reaching Alaska (J. Johnson, unpublished data). Another bird migrated south along the eastern edge of the Rocky Mountains through the Great Plains, south over Central America and east of the Andes before stopping in South America (J. Johnson, unpublished data). The eastern subspecies, *T. s. solitaria*, migrates east of the Mississippi River, with some likely flying southeast over New England and the Atlantic, and others likely following the eastern coastline, through the Caribbean and Central America, to northeastern Argentina (Drury and Keith 1962; Moskoff 2020). The Solitary Sandpiper was included as a transatlantic migrant based on prior radar studies in coastal New England and evidence of their regular presence on Bermuda during pre- and post-breeding migration (Drury and Keith 1962; Amos 1991).

DATA CONTRIBUTORS
L.H. DeCicco; C.F. Gesmundo; J.A. Johnson; L.A. McDuffie.

Tringa incana **WANDERING TATTLER**

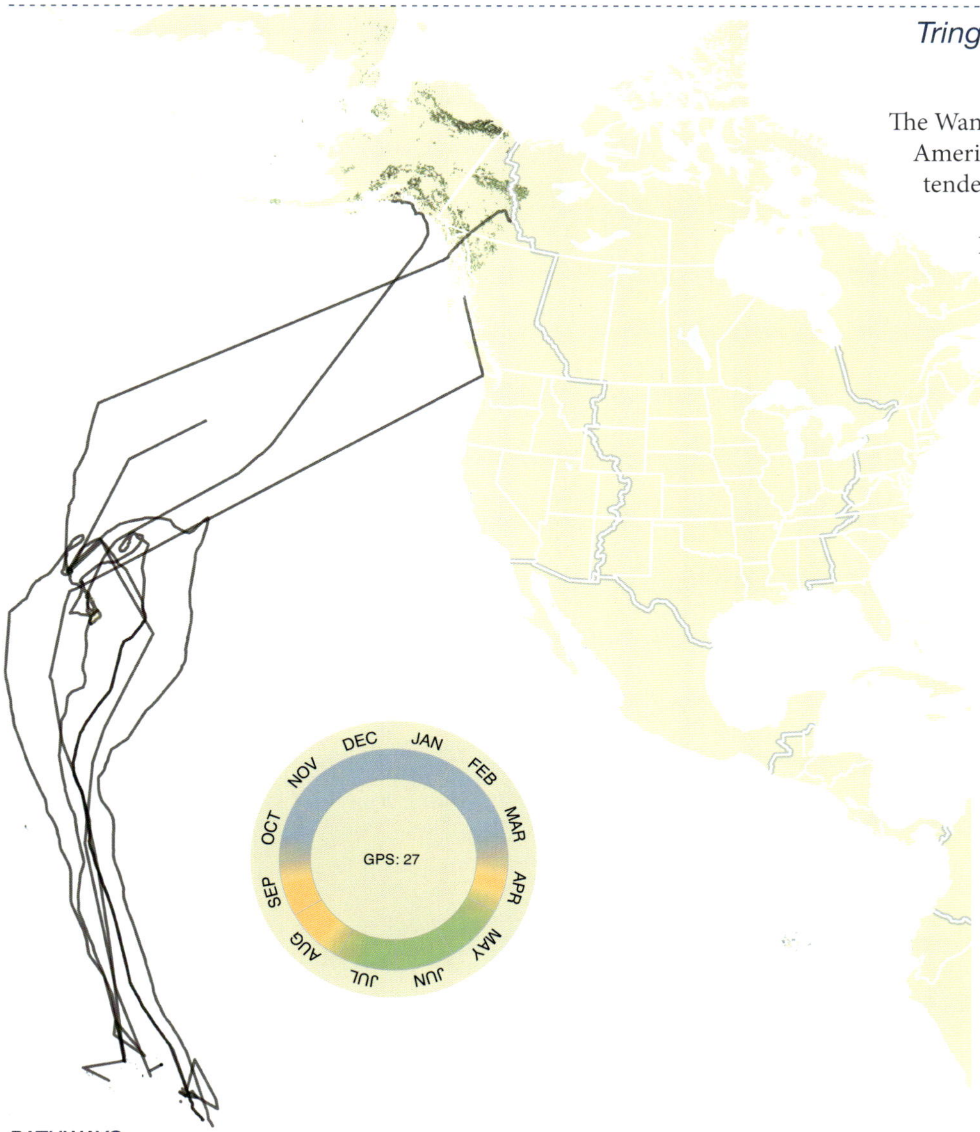

PATHWAYS

DISTRIBUTION AND ABUNDANCE

The Wandering Tattler is one of the least understood North American shorebirds due to its small population size and tendency to occur at low density throughout its lifecycle (R.E. Gill et al. 2020a). It is known to breed in the Koryak Highlands of far eastern Russia, the mostly montane regions of Alaska and northwestern North America, and possibly Asia. The species spends the nonbreeding season on Pacific islands, where it is in sharp decline (Meehan et al. 2018).

ORDER Charadriiformes
FAMILY Scolopacidae
BREEDING HABITAT Tundra, Alpine
NONBREEDING HABITAT Coasts, Beach and Estuary
GLOBAL POPULATION SIZE 18,000
POPULATION CHANGE SINCE 1970 −63%
IUCN RED LIST CATEGORY Least Concern
R2R TIPPING POINT SPECIES Yellow-Alert
US LISTING STATUS No
CANADA SPECIES AT RISK No
PIF CCS 14/20

INTERESTING NOTE

There is little data available, but it seems site fidelity may be high throughout the annual cycle. At Turquoise Lake, Alaska, 89% of color-banded males (n = 9) and 56% of color-banded females (n = 9) returned in consecutive years to the breeding grounds (R.E. Gill et al. 2010). On their nonbreeding grounds, marked individuals returned to the same sites in Hawaii for three consecutive years, and a bird that was banded on Lisianski Island was recaptured in the same location the following year (Clapp and Wirtz 1975).

MIGRATORY CONNECTIVITY

OVERALL Only one study of migratory connectivity was found for this species. Birds captured at nonbreeding areas on the islands of French Polynesia migrated to breeding areas in Alaska and Yukon, Canada (F. Jiguet, unpublished data).

SATELLITE TRACKING The Kivi Kuaka scientific project is the first to use ICARUS (International Cooperation for Animal Research Using Space) GPS transmitters on the species (22 deployed in 2021 on French Polynesia), in order to study possible behavioral responses to cyclones, earthquakes, and tsunamis (F. Jiguet, unpublished data).

MIGRATORY PATHWAYS Wandering Tattlers have a complex migratory pattern that includes continental and oceanic routes to and from their primary nonbreeding areas along the Pacific coast of North America and south-central Pacific Ocean. GPS tracking data revealed a stopover in Hawaii for birds migrating between French Polynesia and the northwestern US and Canada. They also migrate along the northeast coast of Asia. Further research is needed to understand the timing and specific migratory pathways of Wandering Tattlers throughout Oceania (R.E. Gill et al. 2020a).

DATA CONTRIBUTORS
F. Jiguet.

LESSER YELLOWLEGS *Tringa flavipes*

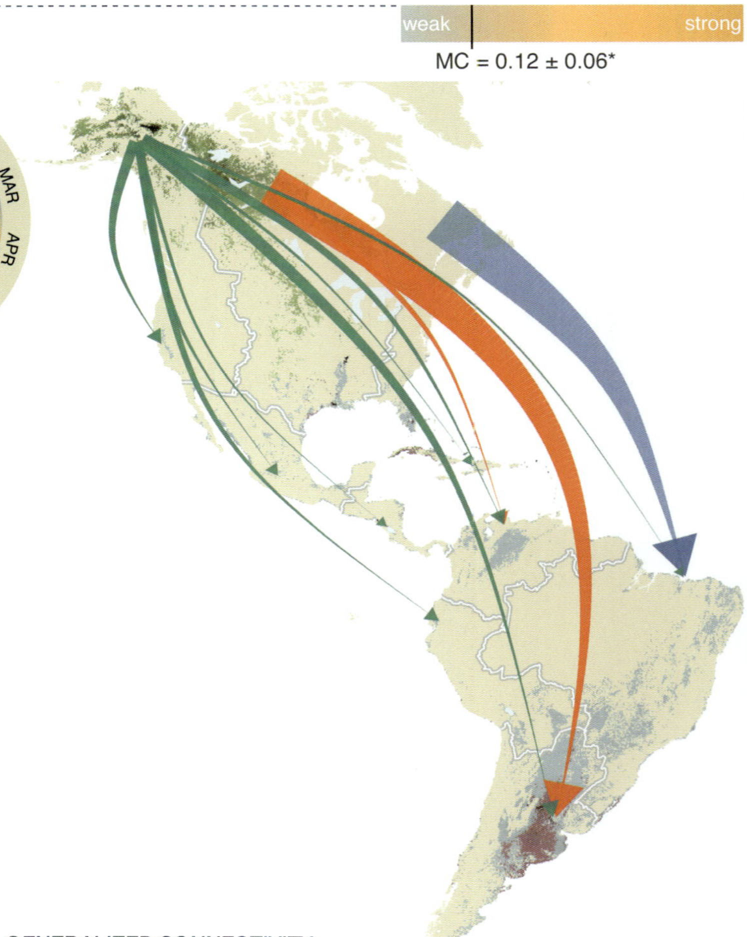

GENERALIZED CONNECTIVITY

DISTRIBUTION AND ABUNDANCE

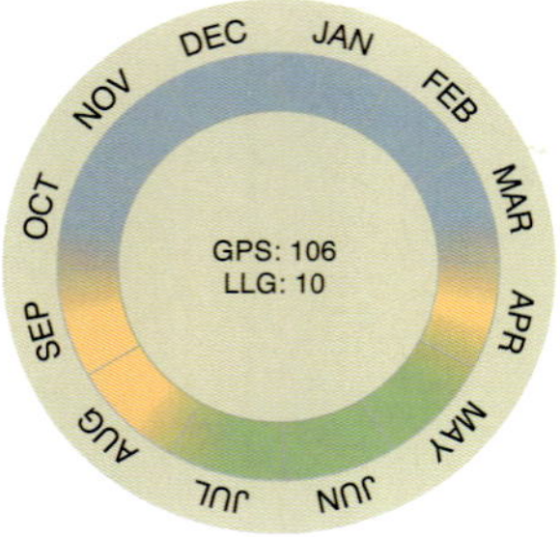

The Lesser Yellowlegs breeds throughout the boreal forest east of Hudson Bay to Alaska. They spend the nonbreeding season in the Southern Cone of South America in Argentina, Bolivia, Paraguay, and southern Brazil, and are patchily distributed elsewhere in wetlands of the Llanos region of Venezuela and Colombia, and along the coasts of South and Central America, Mexico, and the Caribbean (Fink et al. 2023). Only small numbers spend the nonbreeding season in the US, mostly in southern Florida and the western Gulf Coast (the rice-growing region), along the Pacific coast to Southern California, and inland Southern California (also rice fields; Tibbitts and Moskoff 2020; Fink et al. 2023).

ORDER Charadriiformes
FAMILY Scolopacidae
BREEDING HABITAT Wetlands, Boreal
NONBREEDING HABITAT Wetlands, Generalist
GLOBAL POPULATION SIZE 660,000
POPULATION CHANGE SINCE 1970 −59%
IUCN RED LIST CATEGORY Least Concern
R2R TIPPING POINT SPECIES Orange-Alert
US LISTING STATUS No
CANADA SPECIES AT RISK No
PIF CCS 13/20

MIGRATORY CONNECTIVITY

OVERALL Studies from geolocators and satellite telemetry indicate that Lesser Yellowlegs from the same geographic regions follow similar migratory pathways but mix during the nonbreeding season, which suggests weak migratory connectivity (MC = 0.12 ± 0.06). Birds breeding within the Atlantic Flyway migrate to Brazil, while Midcontinent Flyway breeding birds primarily spend the

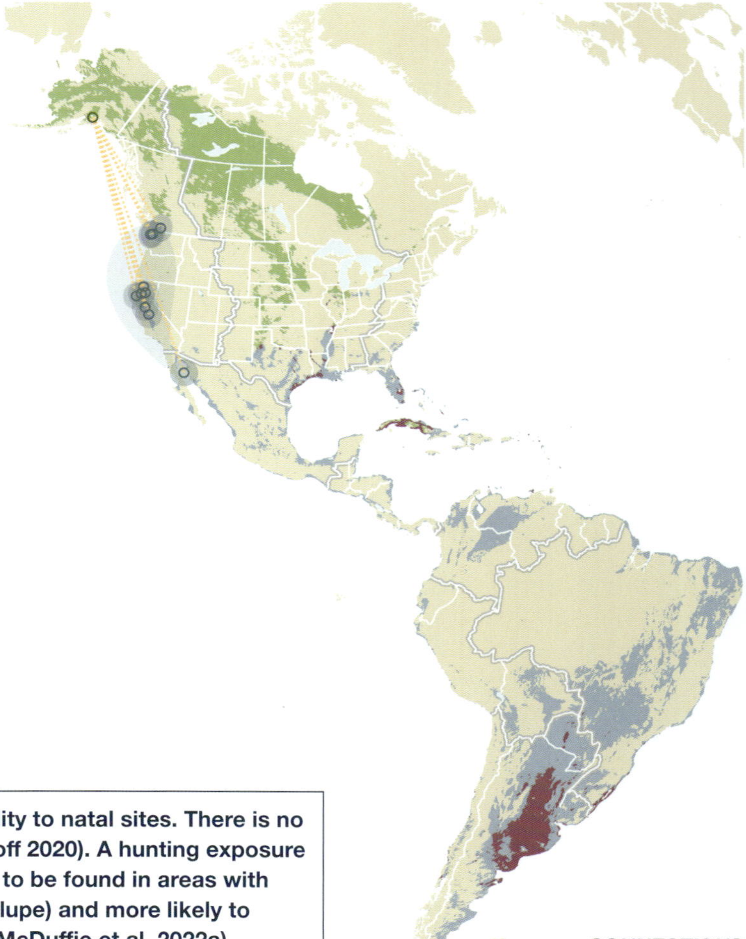

CONNECTIONS

INTERESTING NOTE

Adults show strong breeding site fidelity, while young show some fidelity to natal sites. There is no information about nonbreeding season site fidelity (Tibbitts and Moskoff 2020). A hunting exposure analysis found that birds breeding in eastern Canada were more likely to be found in areas with high shorebird hunting (some islands in the Caribbean, such as Gaudalupe) and more likely to stay there longer than those breeding in western Canada and Alaska (McDuffie et al. 2022a).

	Andean West	Brazil	Caribbean	Central America	Mexico	Northern SA	Pacific	Southern Cone
Atlantic (n = 4)	–	100%	–	–	–	–	–	–
Midcontinent (n = 15)	–	–	–	–	–	13%	–	87%
Pacific (n = 53)	15%	4%	4%	4%	11%	19%	19%	25%

PATHWAYS (A–C)

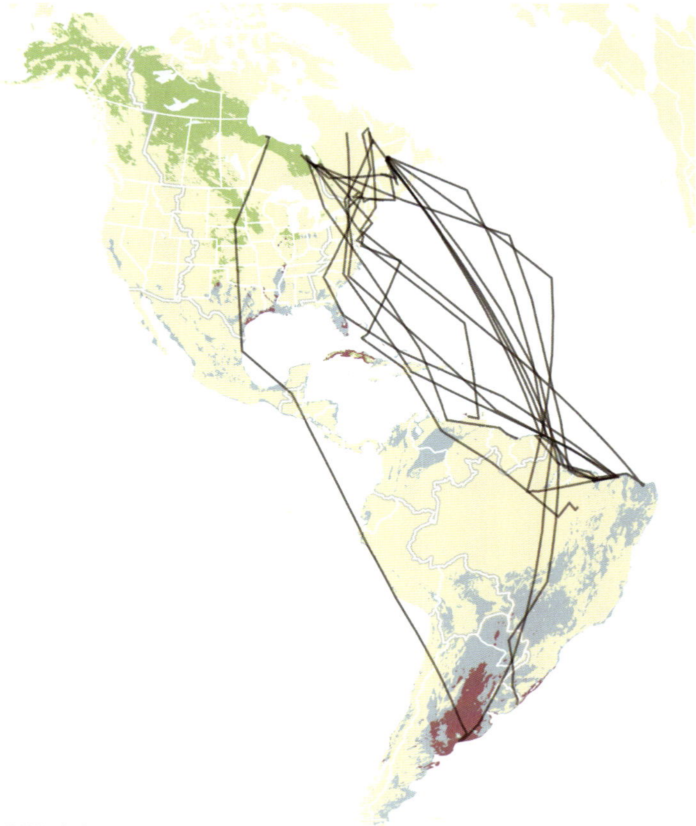

PATHWAYS A

PATHWAYS B

PATHWAYS C

nonbreeding season within the Southern Cone region (87%) and northern South America (13%). Pacific Flyway breeding birds migrate across all nonbreeding regions, including within the Southern Cone region (25%), northern South America (19%), and the Pacific Flyway (19%), as well as the Andean West region (15%) and Mexico (11%).
RADIO TRACKING Detections from the Motus Wildlife Tracking System during migration connect Manitoba, Ontario, and Atlantic Canada to the United States, the Caribbean, and South America (Motus project, Nol—Churchhill Shorebirds, unpublished data; Motus project, James Bay Shorebirds, unpublished data; Motus project, Atlantic Canada Shorebirds, unpublished data; Motus project, National Audubon Society, unpublished data).
GEOLOCATOR TRACKING Birds tagged with geolocators recovered between 2010 and 2016 showed southward movements east along south-central Alaska to the interior of Canada and then either south through the midcontinent or more southeasterly to the Atlantic coast of the US. From here birds traveled primarily to northern South America, although a few traveled as far south as Patagonia, Argentina (L. Tibbitts, R. Lanctot, and L. McDuffie, unpublished data).
SATELLITE TRACKING From 2018 to 2022, a study tracked birds that originated from five breeding locations and two stopovers located from Alaska to eastern Canada. Nonbreeding season locations varied in latitude from Southern California to southern Argentina. Upon reaching nonbreeding areas, birds of all populations overlapped, suggesting weak connectivity (McDuffie et al. 2022b.
STABLE ISOTOPES Feathers collected from harvested juvenile birds in Barbados revealed Lesser Yellowlegs to be mainly from the eastern parts of their breeding range in eastern Canada, with potential origins in the Churchill (Manitoba) and Yellowknife regions (E.T. Reed et al. 2018).
GENETICS Genetic data uncovered significant interconnectedness across the species' range and indicated that large-scale juvenile dispersal maintains connectivity across the range (Christie et al. 2023)
OTHER APPROACHES No banding studies of migratory connectivity were found for this species.
MIGRATORY PATHWAYS Migratory pathways for Lesser Yellowlegs are widespread throughout North America, with primary routes through the Midcontinent and Atlantic flyways during northward migration and along the Atlantic coast and midcontinent during southward migration, with eastern breeding birds more likely to use the Atlantic Flyway compared to central and western breeding birds (McDuffie et al. 2022a, 2022b). Birds from eastern Canada used the Atlantic Flyway, staging in a few key areas before crossing the Atlantic to the Caribbean and South America, with some completing multiday, transoceanic flights across the Atlantic between North and South America. Birds from Alaska and central Canada followed a similar southward migration using the midcontinent, stopping in the Prairie Pothole Region, where high concentration of this species stage in both northward and southward migration, as well as the lower Mississippi Valley and western Gulf Coast. The Pampas and Chaco regions have also been identified as important stopover areas. In the American West, they are almost completely restricted to a few key inland wetlands (e.g., Great Salt Lake, Salton Sea) and, although more abundant during southward migration compared to northward migration (Tibbitts and Moskoff 2020), are rare along the Pacific coast in any season.

DATA CONTRIBUTORS

K. Christie; C. Friis; C.F. Gesmundo; C.M. Harwood; J.A. Johnson; R.B. Lanctot; L.A. McDuffie; E. Nol; J. Rausch; A.R. Taylor; T.L. Tibbitts; J.R. Wright; Alaska Science Center; Alaska Department of Fish and Game; James Bay Shorebird Project; Migratory Connectivity Project; US Geological Survey.

WILLET *Tringa semipalmata*

weak — strong

MC = 0.62 ± 0.07*

DISTRIBUTION AND ABUNDANCE

Eastern and western breeding populations of migratory Willets are recognized as separate subspecies. Eastern Willet (*T. s. semipalmata*) breed in marshes along the Atlantic and Gulf coasts of the US; whereas Western Willet (*T. s. inornata*) breed at interior wetlands and prairies of western North America (Oswald et al. 2016). During the breeding season, observations of the eastern subspecies were limited to the Atlantic and Gulf coasts of the US, while they left the country during the nonbreeding season. The Pacific coasts of Mexico, Central America, and South America are also occupied year-round, mostly by Western Willets based on eBird and connectivity data (Huysman et al. 2022).

MIGRATORY CONNECTIVITY

OVERALL Migratory connectivity is known from banding, radio transmitter, and light-level geolocator data. Willets exhibit strong migratory connectivity throughout their range (MC = 0.62 ± 0.07). Birds breeding along the Atlantic coast primarily spend the nonbreeding season in Brazil (91%), whereas birds breeding in the Midcontinent Flyway migrate to coastal Central America (50%) and the Andean West region of South America (50%). Western Willets

ORDER Charadriiformes
FAMILY Scolopacidae
BREEDING HABITAT Wetlands, Seasonally Wet Grasslands
NONBREEDING HABITAT Coasts, Beach and Estuary
GLOBAL POPULATION SIZE 250,000
POPULATION CHANGE SINCE 1970 −23%
IUCN RED LIST CATEGORY Least Concern
R2R TIPPING POINT SPECIES No
US LISTING STATUS No
CANADA SPECIES AT RISK No
PIF CCS 14/20

INTERESTING NOTE

Due to differences between subspecies in behavior, habitat use, and ranges during both breeding and nonbreeding seasons, conservation planning tailored to individual populations is recommended, particularly regarding hunting (Huysman et al. 2022).

	Andean West	Brazil	Central America	Northern SA	Pacific
Atlantic (n = 35)	–	91%	–	9%	–
Midcontinent (n = 10)	50%	–	50%	–	–
Pacific (n = 45)	–	–	–	–	100%

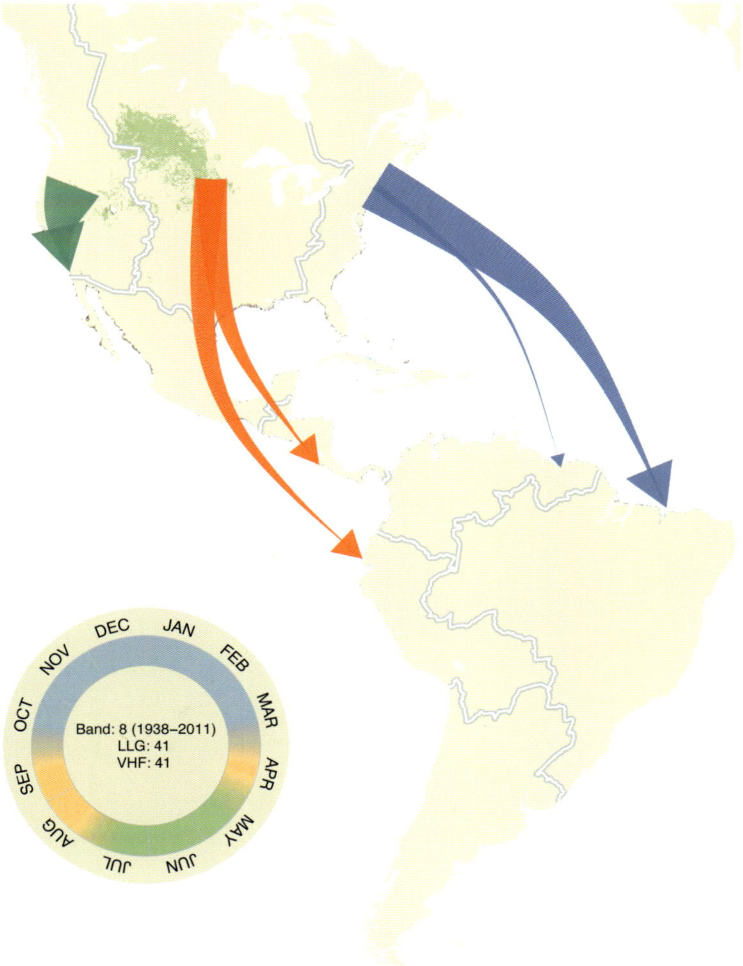

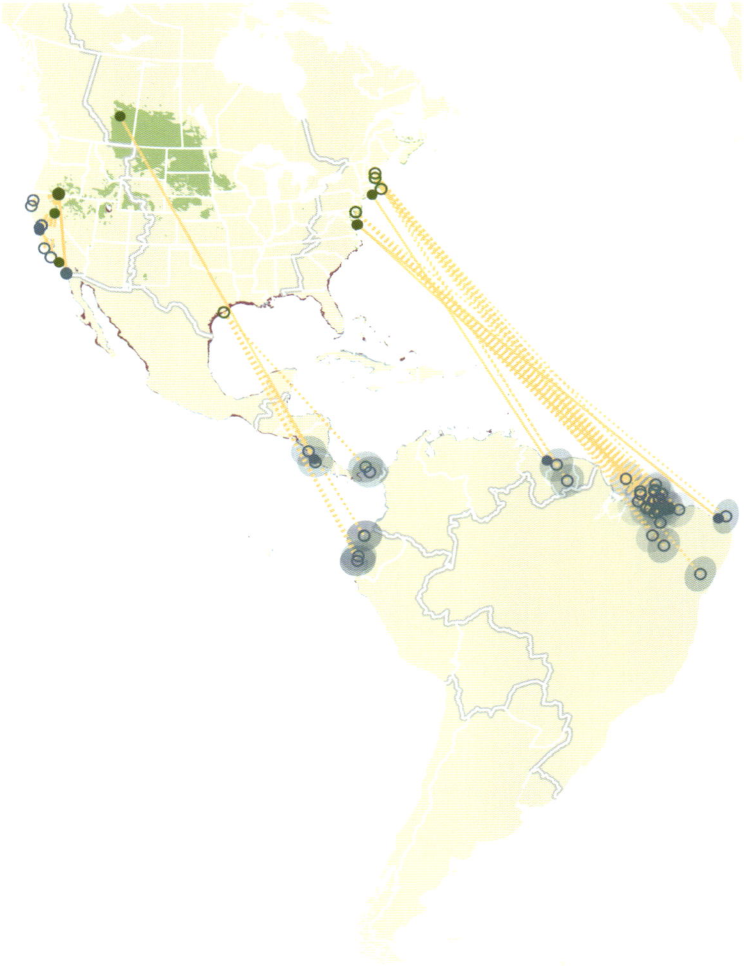

GENERALIZED CONNECTIVITY

CONNECTIONS

breeding in the Pacific Flyway migrate to nonbreeding areas along the coast within the Pacific Flyway (100%).

BANDING Limited band reencounters from across the range show that different breeding regions have minimal overlap on the nonbreeding grounds.

RADIO TRACKING Willets breeding in the Great Basin region of the United States migrated primarily to the San Francisco Bay Area in California and appeared to be site faithful in subsequent winters (Haig et al. 2002). Detections from the Motus Wildlife Tracking System during migration connect Saskatchewan and Manitoba, Canada, to Sinaloa, Mexico (Motus project, Coastal Motus Network in northwest Mexico, unpublished data). There are no individual breeding to nonbreeding detections.

GEOLOCATOR TRACKING Data from light-level geolocators found that breeding populations along the northeastern coast of the United States spend the nonbreeding season exclusively on the Atlantic coast of northern South America (J.A. Smith et al. 2020). Conversely, Gulf Coast breeding populations migrate to either the Pacific coast of Central or northern South America (Heath et al. 2022).

OTHER APPROACHES No satellite transmitter, genetic, or stable isotope studies of migratory connectivity were found for this species. eBird observations of Willets found that ranges of the western and eastern subspecies overlapped spatially but not temporally, except for some overlap in Central America (Fink et al. 2023).

MIGRATORY PATHWAYS Data from resightings of marked birds and radio transmitters show Willets that breed in the western US and Canada migrate through the Pacific Flyway region southward along the Pacific coast to the western US, Mexico, and Central and South America. Birds that migrate through the midcontinent region migrate southward through the central US to stationary nonbreeding areas in the southern US, Mexico, and Central and South America. Birds that use the Atlantic Flyway region migrate southward along the Atlantic coast to nonbreeding areas in the southeastern United States, the Caribbean, and Central and South America (Lowther et al. 2020e). Geolocator data also indicate that many Atlantic coast Willets make direct flights to South America over the Atlantic, either proceeding directly from the breeding grounds to their nonbreeding site or with an initial stopover on the northern coast of South America, primarily in Guyana, Suriname, or French Guiana (J.A.M. Smith et al. 2020).

DATA CONTRIBUTORS

S. Heath; L. Johnson; L. Olson; P.P. Marra; K. Regan; J.A.M. Smith.

GREATER YELLOWLEGS *Tringa melanoleuca*

DISTRIBUTION AND ABUNDANCE

Greater Yellowlegs are a widespread breeder across the entire boreal region of North America, from Newfoundland to western Alaska. The species spends the nonbreeding season throughout the southern United States, especially in Florida and along the Gulf Coast in Louisiana and Texas, with lower numbers traveling north to Massachusetts. The species also spends the nonbreeding season throughout most of Central America, the Caribbean (concentrated in the Greater Antilles), and all of coastal and south-central South America.

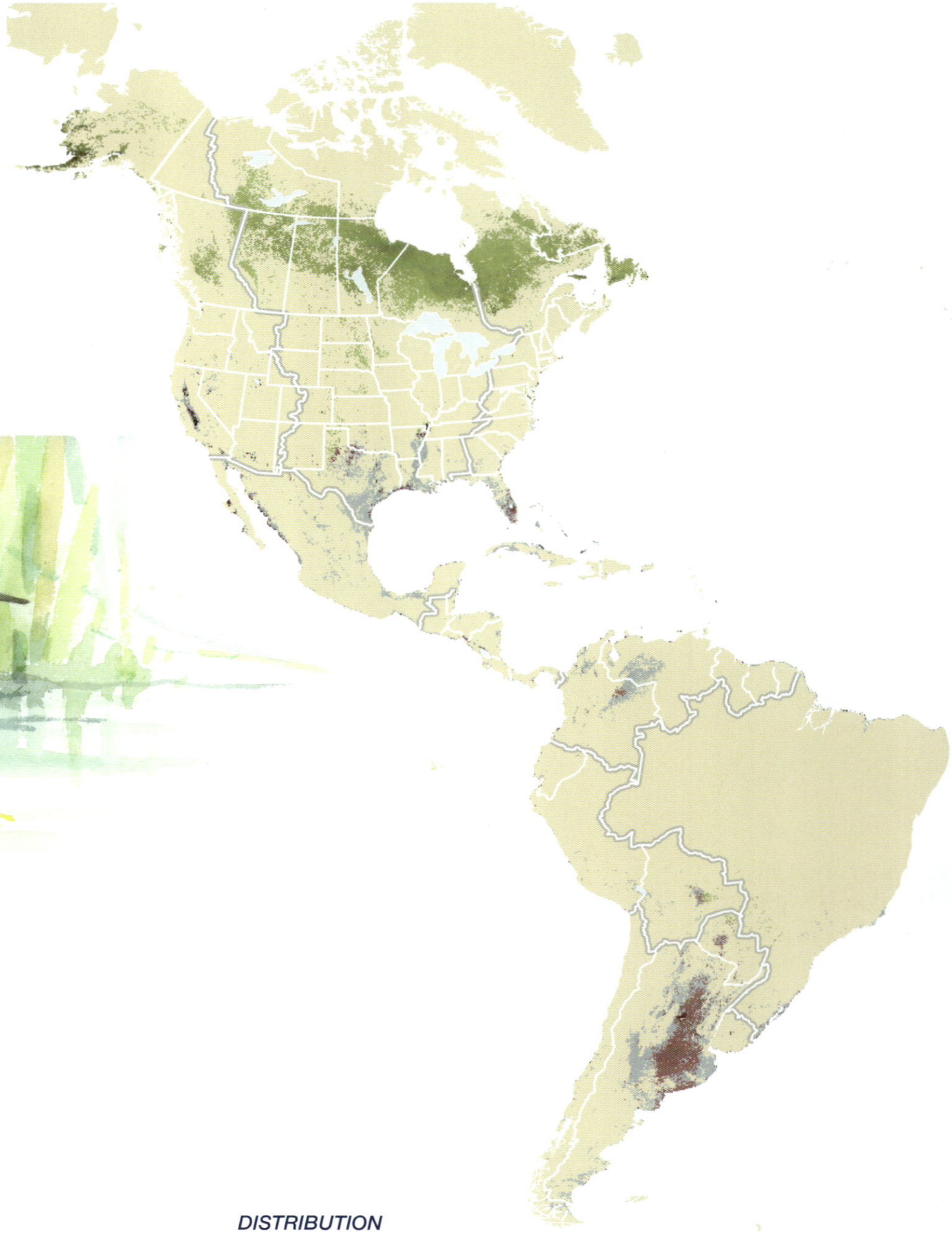

DISTRIBUTION AND ABUNDANCE

ORDER Charadriiformes
FAMILY Scolopacidae
BREEDING HABITAT Wetlands, Boreal
NONBREEDING HABITAT Wetlands, Generalist
GLOBAL POPULATION SIZE 140,000
POPULATION CHANGE SINCE 1970 +28%
IUCN RED LIST CATEGORY Least Concern
R2R TIPPING POINT SPECIES Orange-Alert
US LISTING STATUS No
CANADA SPECIES AT RISK No
PIF CCS 14/20

MIGRATORY CONNECTIVITY

OVERALL Migratory connectivity is unknown. No information on migratory connectivity was found for this species.

RADIO TRACKING A few tags were deployed through the Motus Wildlife Tracking System, with no breeding to nonbreeding detections.

MIGRATORY PATHWAYS Information on routes of migration are not available from banding or tracking data. Northward migration occurs across North America (Bent 1927), and flight range estimates suggest direct flights across the Caribbean from South America to the southeastern United States (McNeil 1970). Southward migration has reduced numbers of birds in interior North America and increased birds in eastern Canada, Atlantic North America, and Caribbean islands (Elphick and Tibbitts 2020). eBird animations show heavy migrations of Gulf Coast birds north in the pre-breeding season through the Midwest beginning in March and April. Other pre-breeding migrations occur along the Pacific and Atlantic coasts of the United States (Fink et al. 2023).

INTERESTING NOTE

This species migrates earlier than most other shorebirds. It is common for nonbreeders to remain throughout the nonbreeding range during the breeding season, with juveniles presumed to skip the first breeding season (Elphick and Tibbitts 2020). Three birds were resighted within a few kilometers of breeding areas the following year (Elphick and Tibbitts 2020).

Phalaropus tricolor WILSON'S PHALAROPE

DISTRIBUTION AND ABUNDANCE

Wilson's Phalarope is a temperate zone breeder found only in the Americas, breeding primarily in central Canada and the United States. During the nonbreeding season, they can be found at saline lakes of the Andes and across the lowlands of the Southern Cone as far south as Tierra del Fuego.

ORDER Charadriiformes
FAMILY Scolopacidae
BREEDING HABITAT Wetlands, Seasonally Wet Grasslands
NONBREEDING HABITAT Wetlands, Lakes and Rivers
GLOBAL POPULATION SIZE 1,500,000
POPULATION CHANGE SINCE 1970 −9%
IUCN RED LIST CATEGORY Least Concern
R2R TIPPING POINT SPECIES No
US LISTING STATUS No
CANADA SPECIES AT RISK No
PIF CCS 11/20

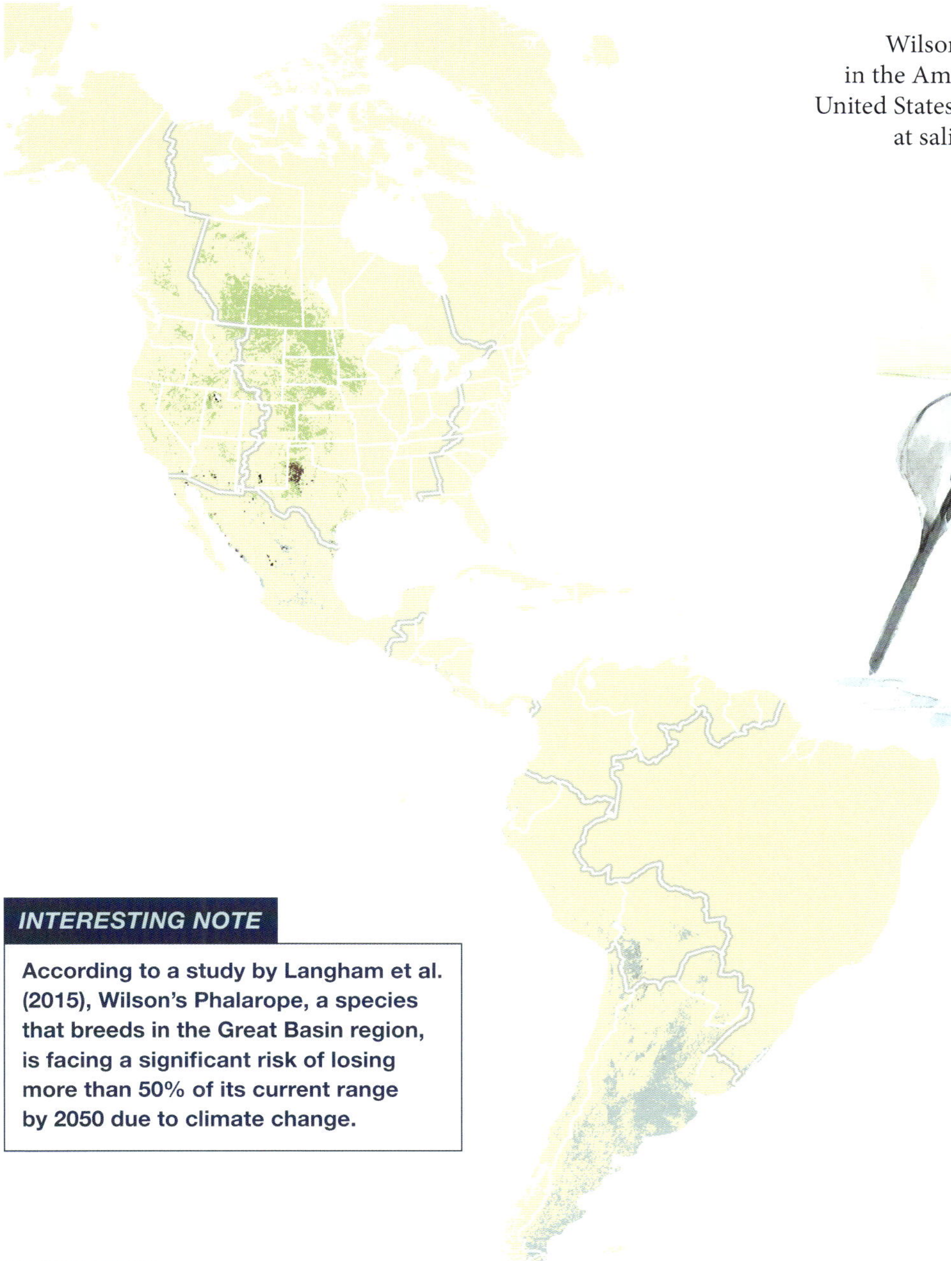

DISTRIBUTION AND ABUNDANCE

INTERESTING NOTE

According to a study by Langham et al. (2015), Wilson's Phalarope, a species that breeds in the Great Basin region, is facing a significant risk of losing more than 50% of its current range by 2050 due to climate change.

MIGRATORY CONNECTIVITY

OVERALL Migratory connectivity is unknown, but the species exhibits weak site fidelity and diverse migratory pathways.

RADIO TRACKING There are several projects tracking Wilson's Phalarope with the Motus Wildlife Tracking System across the range. There are no individual breeding to nonbreeding detections, but there are many detections during migration.

OTHER APPROACHES No banding, tracking, genetic, or stable isotope studies of migratory connectivity were found for this species. Wilson's Phalarope was included in a study investigating the use of trace elements for determining migratory connectivity in shorebirds that spend the nonbreeding period in Argentina. Results suggest that trace elements in tissues can be misleading because chemical profiles of feathers may change with time (Torres-Dowdall et al. 2010).

MIGRATORY PATHWAYS During pre-breeding migration, this species travels largely overland, northward through the highlands of South America, then through Central America or across the Gulf of Mexico, and finally through the Intermountain West and southern plains, where in some years, up to 90% of the population passes through Cheyenne Bottoms, Kansas (Oring and Davis 1966; Morrison and Meyers 1987; Skagen and Knopf 1993). Post-breeding adults undergo molt migration and gather in large flocks at hypersaline or alkaline lakes, largely in the western United States, before migrating south. The southward migration of adults is characterized by a direct and rapid, nonstop trans-equatorial flight from staging areas in the US to the western coast of South America using a great circle route across the Pacific, while juvenile birds take a slower overland route through the western and central US, northern Mexico, and perhaps Central America to reach northern South America.

RED-NECKED PHALAROPE *Phalaropus lobatus*

DISTRIBUTION AND ABUNDANCE

Red-necked Phalaropes are circumpolar low Arctic or subarctic breeders, primarily in Alaska and the western Arctic but also east to Baffin Bay and Greenland. They spend their nonbreeding season at sea, living in the open ocean off the northwestern coast of South America. The largest congregations of this species have been observed from the Humboldt Current, off the coast of Peru, but can occur as far south as coastal Chile and, more irregularly, as far north as the Pacific coast of Mexico (Rubega et al. 2020).

ORDER Charadriiformes
FAMILY Scolopacidae
BREEDING HABITAT Wetlands, Tundra
NONBREEDING HABITAT Oceans, Pelagic
GLOBAL POPULATION SIZE 4,100,000
POPULATION CHANGE SINCE 1970 −62%
IUCN RED LIST CATEGORY Least Concern
R2R TIPPING POINT SPECIES No
US LISTING STATUS No
CANADA SPECIES AT RISK No
PIF CCS 11/20

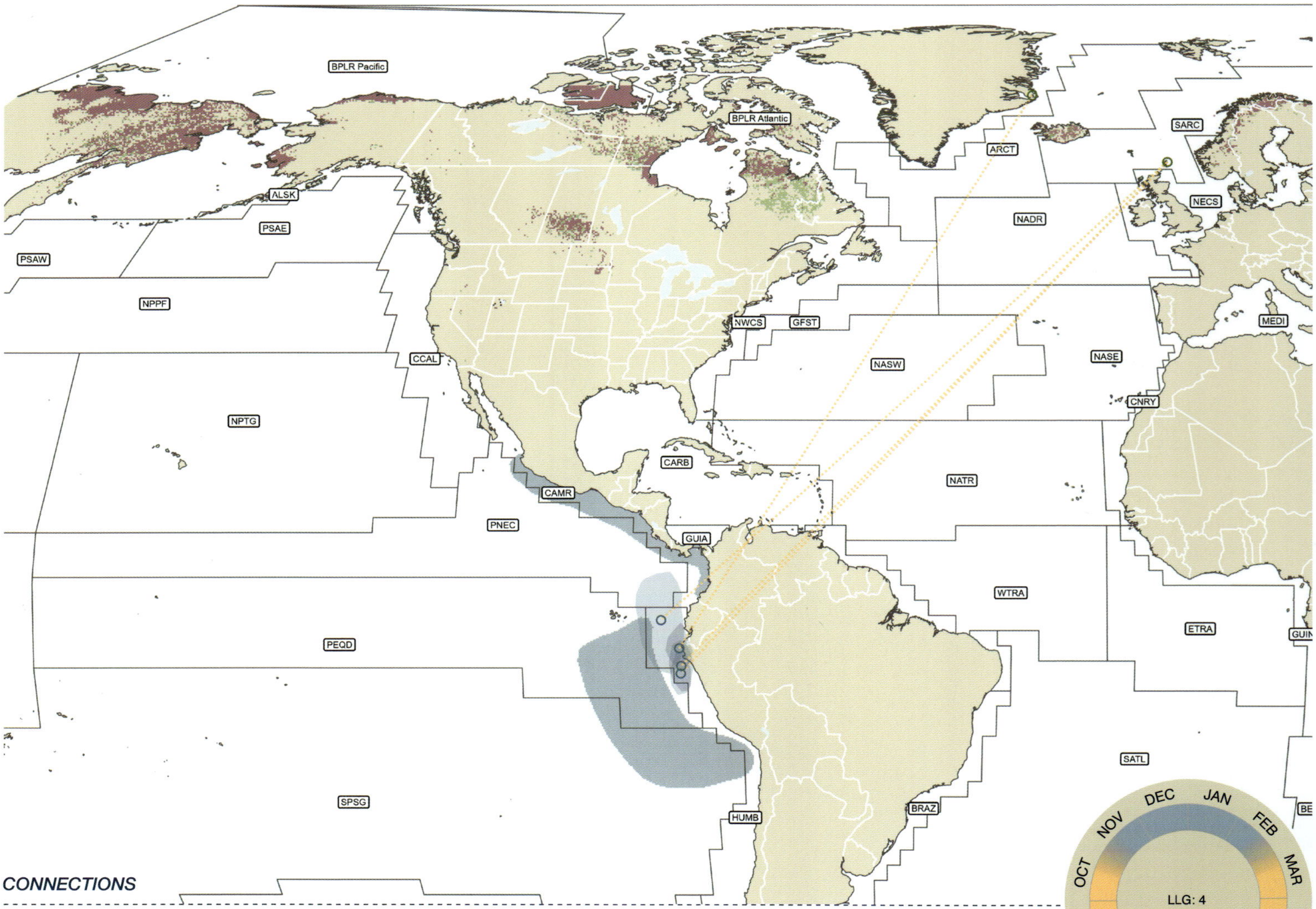

MIGRATORY CONNECTIVITY

OVERALL More studies are required to better quantify the degree of migratory connectivity, determine the extent of mixing during the nonbreeding season, and determine the migratory divide between birds spending the nonbreeding season in either the southwest Pacific or the eastern Pacific, as well as the divide between those of the southwest Pacific and the Arabian Sea (van Bemmelen et al. 2019).

RADIO TRACKING Data from several tracking projects through the Motus Wildlife Tracking System suggest different migration routes for eastern and western populations. There are no individual breeding to nonbreeding detections as of printing.

GEOLOCATOR TRACKING Geolocator studies provided evidence of a transatlantic migration route for this species, with a journey from a northern Scotland breeding site to the Pacific coast of South America and back again (Smith et al. 2014, 2018). A study of Red-necked Phalaropes breeding in Europe and Russia revealed that all tracked individuals captured on breeding sites in North Atlantic islands spent the nonbreeding season off the Pacific coast of South America (van Bemmelen et al. 2019).

OTHER APPROACHES No satellite tracking, genetic, stable isotope, or banding studies of migratory connectivity were found for this species.

MIGRATORY PATHWAYS There are four to five major pathways. Alaskan breeders likely migrate north along the Pacific coast of Central and North America, while central Canadian breeders begin flying northeast overland through Oregon, Washington, and British Columbia (Isleib 1979). Many others begin to fly overland at Southern California or Arizona (Howell and Webb 1995), taking a northeasterly flight through the Great Basin and Prairie provinces of Canada (Behle et al. 1985; Ryser 1985; Paton et al. 1992). Large numbers have also been observed traveling northward off the Atlantic coast of the Carolinas, New England, and Canada, presumedly to North Atlantic breeding areas (Veit and Petersen 1993). Finally, some make a transoceanic migration, veering off their Pacific coast path at the Isthmus of Tehuantepec to cross the Atlantic toward their breeding areas in the Atlantic islands on a spectacular (approximately 22,000 km) journey (van Bemmelen et al. 2019; M. Smith et al. 2014). During post-breeding migration there are major stopovers for the species near Newfoundland, Great Basin, and Mono Lake, California.

DATA CONTRIBUTORS

M. Bolton; O. Gilg; P. Harris; J.D. Okill; G. Petrie; R. Riddington; M. Smith; J. Wilson; French Polar Institute; Groupe de Recherche en Ecologie Arctique; Université de Bourgogne Franche Comté.

INTERESTING NOTE

Genetic studies indicate a low level of genetic differentiation between populations and no presence of subpopulation structure (Haig et al. 1997).

Phalaropus fulicarius RED PHALAROPE

ORDER Charadriiformes
FAMILY Scolopacidae
BREEDING HABITAT Wetlands, Tundra
NONBREEDING HABITAT Oceans, Pelagic
GLOBAL POPULATION SIZE 2,200,000
POPULATION CHANGE SINCE 1970 NA
IUCN RED LIST CATEGORY Least Concern
R2R TIPPING POINT SPECIES No
US LISTING STATUS No
CANADA SPECIES AT RISK No
PIF CCS 13/20

DISTRIBUTION AND ABUNDANCE

Red Phalaropes are circumpolar breeders across coastal tundra terrestrial areas of the Arctic Ocean. As the most pelagic of the phalarope species and in contrast to other migrating shorebirds, they spend almost eleven months of the year at sea and rarely migrate overland. At sea, they largely stay along ocean fronts: in the Pacific, most birds stay along the Humboldt Current, and few stay farther north, along the California Current; in the Atlantic, birds spend the nonbreeding season along the Gulf Stream.

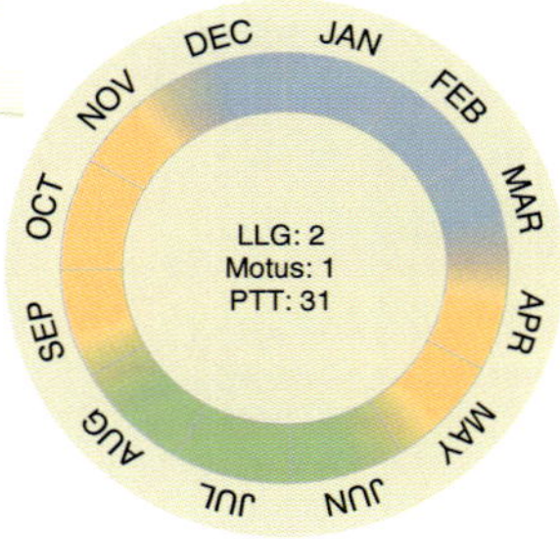

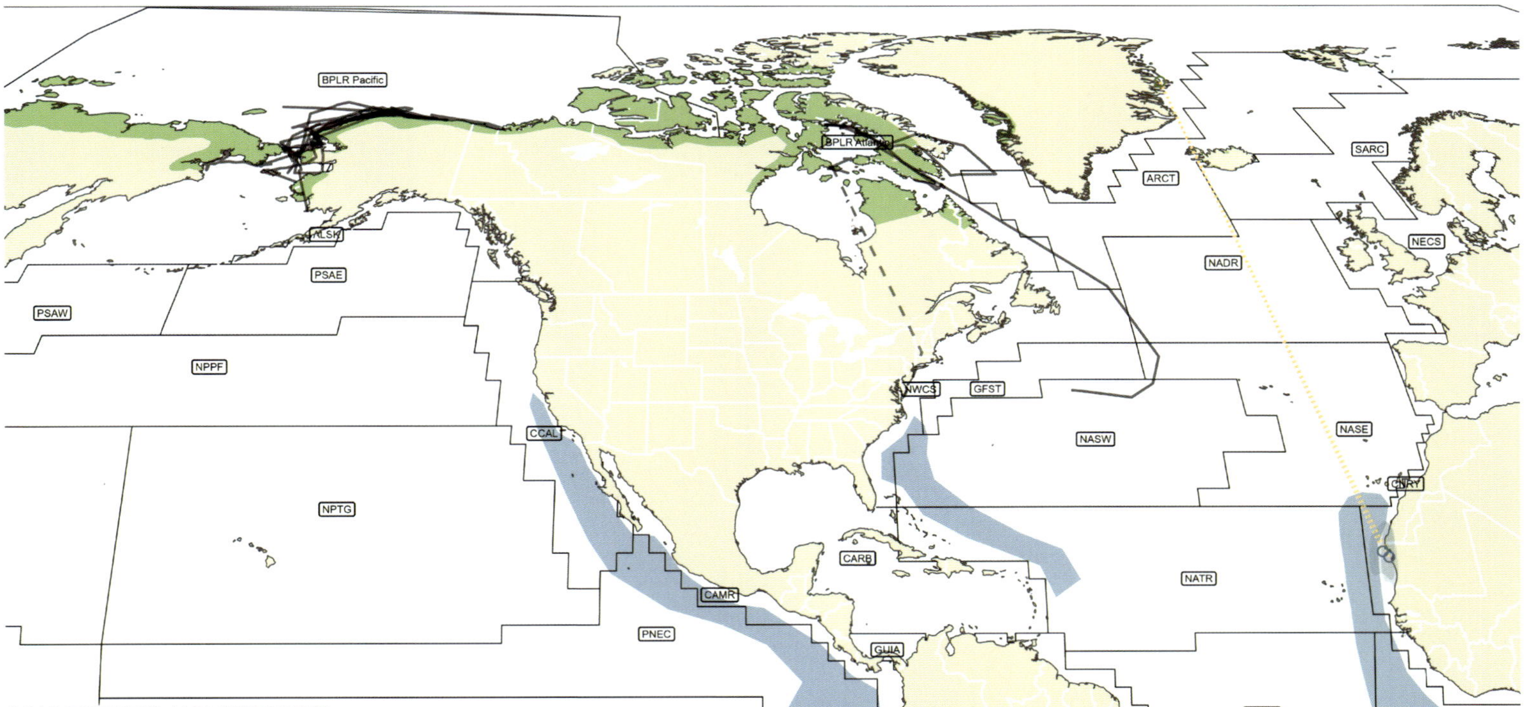

CONNECTIONS AND PATHWAYS

MIGRATORY CONNECTIVITY

OVERALL Since Red Phalaropes spend most of their life at sea, studies on the migration and nonbreeding ecology of this species are sparse, and MC is not calculated due to limited data between breeding and nonbreeding regions.. Those studies conducted during the breeding period indicate Red Phalarope have the lowest site fidelity of any of the phalarope species, which may correlate with low migratory connectivity; however, observational and tracking data indicate that Atlantic and Pacific coast migrants have different nonbreeding and breeding sites, suggestive of stronger migratory connectivity. More research on movement, migration patterns, and migratory connectivity is needed.

RADIO TRACKING There are several projects tracking the Red Phalarope through the Motus Wildlife Tracking System. There are no individual breeding to nonbreeding locations, but there appear to be many detections during migration.

GEOLOCATOR TRACKING Birds breeding in northeast Greenland and Iceland were shown to migrate to three nonbreeding areas across a wide latitudinal range in the Atlantic, including the North Atlantic, as well as off the western and southern coasts of Africa (van Bemmelen 2019). One-third of individuals with more than one year of data switched between nonbreeding areas in subsequent years (van Bemmelen 2019).

SATELLITE TRACKING A tracking study based in Alaska and central Canada indicated highly variable migration patterns, which allowed birds to adjust to annual changes in resource availability. Central Canadian birds exhibited a similar pattern, but data were limited. This study also found that females frequently stopped within the marginal ice zone of the Arctic Ocean, whereas males tended to stay more on land or in open water during their southward nonbreeding period (Saalfeld et al. 2024).

OTHER APPROACHES No genetic, stable isotope, or banding studies of migratory connectivity were found for this species.

MIGRATORY PATHWAYS Red Phalaropes migrate primarily over water; inland observations were usually ascribed to vagrancy caused by extreme weather or confusion. Satellite tracking of birds from Alaska suggest they fly west along the Chukchi Sea and then south through the Bering Sea and eventually southward off the coast of North America and down to the Galapagos (Saalfeld et al. 2024). Northward migration is thought to also be along the Pacific coast of South America, Central America, and North America (Gould et al. 1982; Campbell et al. 1990), through the Bering Strait and toward the Chukchi and Beaufort seas, where they congregate near sea ice as they wait for snow to melt off of their tundra nesting habitat (Bailey 1925; Divoky 1979). Those migrating from nonbreeding areas in Africa cross the Atlantic and travel up the coast to the Scotian Shelf and eastern Canadian Arctic (R.G.B. Brown 1986). Depending on nonbreeding location, birds breeding in Greenland and Iceland were associated with different pre- and post-breeding migration routes and stopover sites (van Bemmelen 2019). Detections from Motus tags showed birds breeding in Nunavut migrating along the Atlantic coast (R. Ronconi et al., unpublished data).

DATA CONTRIBUTORS

L. Bollache; S. Brown; K.H. Elliott; W. English; O. Gilg; M.-A. Giroux; A.-L. Harrison; J.-F. Lamarre; R.B. Lanctot; J. Lang; C. Latty; N. Lecomte; A. MacDonald; R. McGuire; R. Ronconi; S.T. Saalfeld; A.L. Scarpignato; S. Schulte; B. Sittler; P. Smith; R. van Bemmelen.

INTERESTING NOTE

Banding studies in Alaska have demonstrated very low natal and adult site fidelity (Schamel and Tracy 1977, 1991; Colwell et al. 1988; Saalfeld and Lanctot 2015); however, a study in Iceland indicated all identifiable males (n = 4) returned to the same breeding location the following year (Whitfield 1995).

Stercorarius skua **GREAT SKUA**

DISTRIBUTION AND ABUNDANCE

A seabird of the North Atlantic Ocean, this species breeds primarily on northern European islands, with highest abundance in Scotland and Iceland, and much lower densities on the Faroe Islands (Furness et al. 2020a). There has been a recent range expansion to Ireland, Svalbard, Norway, and Jan Mayen, Russia (Furness et al. 2020a). Population trends are variable across this range, with increasing populations in the northern British Isles, stable populations in Iceland, and declining populations in the Faroes (Furness et al. 2020a). It spends the nonbreeding season over the open ocean offshore of Europe, northwest Africa, and eastern Canada, but it may also be found near the equator (Furness et al. 2020a).

ORDER Charadriiformes
FAMILY Stercorariidae
BREEDING HABITAT Coasts, Cliffs and Islands
NONBREEDING HABITAT Oceans, Pelagic
GLOBAL POPULATION SIZE 32,000

POPULATION CHANGE SINCE 1970 NA
IUCN RED LIST CATEGORY Least Concern
R2R TIPPING POINT SPECIES No
US LISTING STATUS No
CANADA SPECIES AT RISK No
PIF CCS 15/20

MIGRATORY CONNECTIVITY

OVERALL Most of the individuals tracked with geolocators from Bjørnøya, Svalbard, Norway (SEATRACK, unpublished data) used nonbreeding areas in the northwest Atlantic Ocean (Arctic, Gulf Stream, North Atlantic Drift, and Northwest Atlantic Shelves provinces; 86% total), suggesting relatively

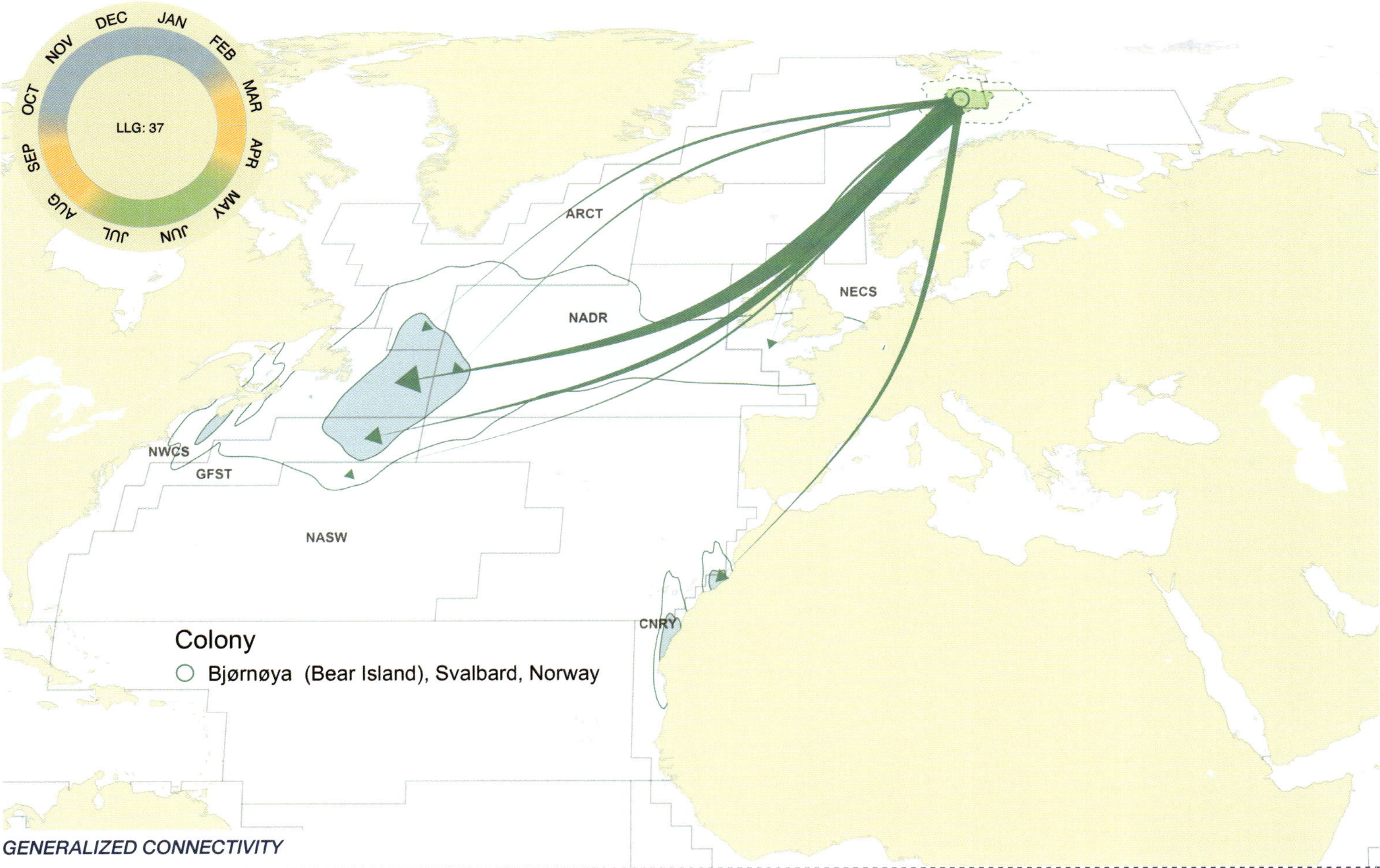

GENERALIZED CONNECTIVITY

	NASW	GFST	NWCS	ARCT	NECS	NADR	CNRY
Bjørnøya (Bear Island), Svalbard, Norway (n = 37)	3%	24%	49%	5%	3%	5%	11%

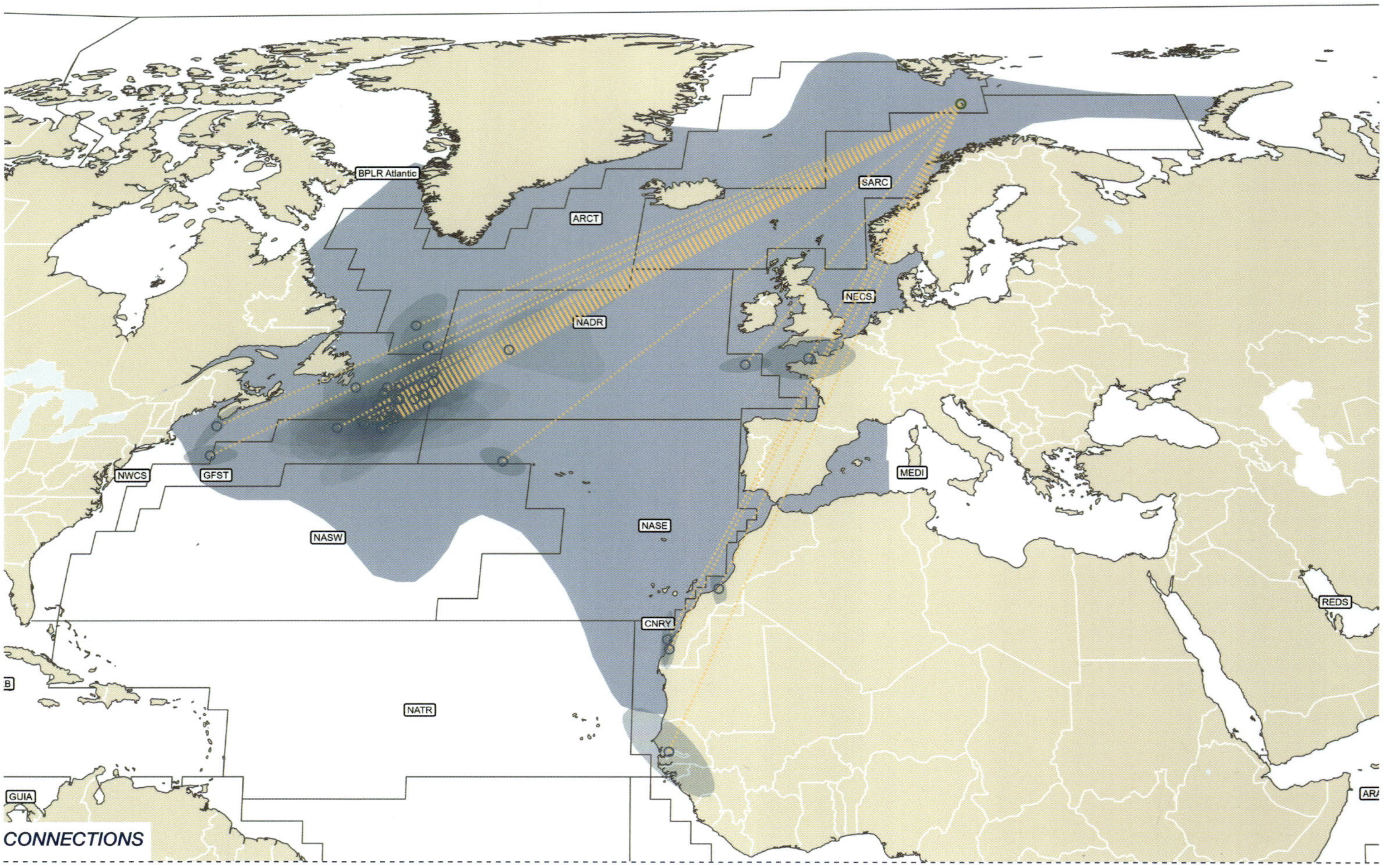

CONNECTIONS

weak connectivity, while a small sample remained off Europe (3%) or spent the nonbreeding season off western Africa (Canary Coastal Province, 14%). MC could not be calculated due to contributed data from only one breeding population. However, weak migratory connectivity was suggested from additional tracking studies across three breeding populations.

GEOLOCATOR TRACKING Within the northwest Atlantic, birds ranged widely during the nonbreeding period. There is a limited sample size but some evidence to suggest nonbreeding ranges are smaller within the Canary Coastal Province and European nonbreeding areas than in the northwest Atlantic (Magnusdottir et al. 2012).

GENETICS Although there is substantial interest in refining the phylogeny of skua species (for a recent example, see Mikkelsen and Weir 2023), no publications using genetics to infer migratory connectivity specifically were found for this species.

OTHER APPROACHES No stable isotope studies of migratory connectivity were found, although in a breeding study, Käkelä et al. (2006) discussed the potential for identifying the nonbreeding diet (and potentially the nonbreeding area) from stable isotopes for this species. For 67 Great Skuas tracked with geolocators in 2015–2017 from Bjørnøya, Svalbard, Albert et al. (2022) showed that mercury concentration in feathers was highest in birds on nonbreeding grounds off Europe, followed by birds in the northwest Atlantic and off Africa. This work suggests trace elements could be used in the absence of tracking data to measure migratory connectivity.

MIGRATORY PATHWAYS Most birds took a due southwest pathway from Svalbard to the northwest Atlantic before either remaining in that region or continuing onward to West Africa or Europe.

DATA CONTRIBUTORS
Hallvard Strøm; SEATRACK.

INTERESTING NOTE

Magnusdottir et al. (2012) noted that their sample was strongly biased toward females because males do little of the incubation; therefore, they could not infer connectivity for males. However, data from Norway included 11 males and 19 females, with no obvious differences in connectivity between the sexes (SEATRACK, unpublished data).

Stercorarius maccormicki **SOUTH POLAR SKUA**

DISTRIBUTION AND ABUNDANCE

This predatory seabird breeds in coastal Antarctic and subantarctic islands and undertakes long migrations to the Indian Ocean and the North Pacific and North Atlantic oceans. It is highly pelagic but may also target seabird colonies for kleptoparasitic activities and eggs, especially penguin colonies during the breeding season. The global population is relatively small (6,000–15,000 mature individuals) and can experience large interannual fluctuations at colonies. The highest breeding abundance is in the Ross Sea area (2,000–6,000 pairs) with other significant colonies in Wilkes Land and on the Antarctic Peninsula. The species' relative abundance across nonbreeding areas is unknown.

MIGRATORY CONNECTIVITY

OVERALL Geolocator studies (Kopp et al. 2011; Weimerskirch et al. 2015) suggest South Polar Skuas exhibit moderately strong connectivity across three Antarctic populations spanning about 200 degrees of longitude. Only a small amount of nonbreeding season overlap occurred across populations. Four adult band recoveries to North American waters first indicated that birds from the Antarctic Peninsula (southeast of South America) spend the nonbreeding season in the central North Pacific, northwest Atlantic, and Caribbean Oceans.

BANDING The North American bird banding dataset includes many individuals banded across Antarctic sites and shows some

ORDER Charadriiformes
FAMILY Stercorariidae
BREEDING HABITAT Coasts, Cliffs and Islands
NONBREEDING HABITAT Oceans, Pelagic
GLOBAL POPULATION SIZE 11,000

POPULATION CHANGE SINCE 1970 NA
IUCN RED LIST CATEGORY Least Concern
R2R TIPPING POINT SPECIES No
US LISTING STATUS No
CANADA SPECIES AT RISK No
PIF CCS 12/20

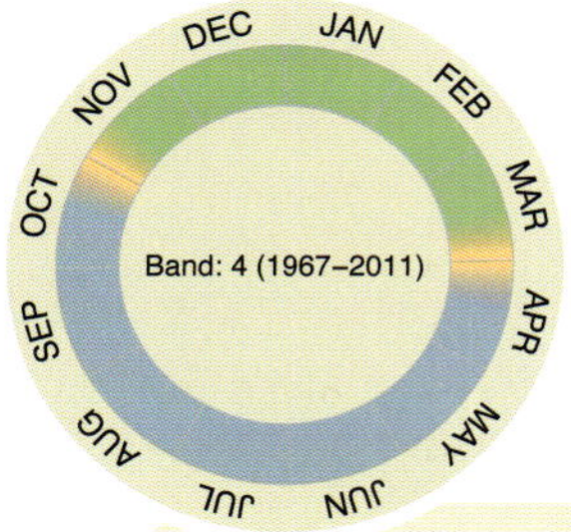

INTERESTING NOTE

In contrast to the trans-equatorial migrations taken by most tracked South Polar Skuas, two individuals tracked by geolocator (8% of study population) took short migrations within the Southern Hemisphere to nonbreeding destinations in Peru and west of Gough Island (Kopp et al. 2011). A hypothesis that these individuals could be hybrids with Brown Skua (*Catharacta antarctica lonnbergi*), which spend the nonbreeding season in the Southern Hemisphere, was not supported by genetic analysis (Ritz 2009).

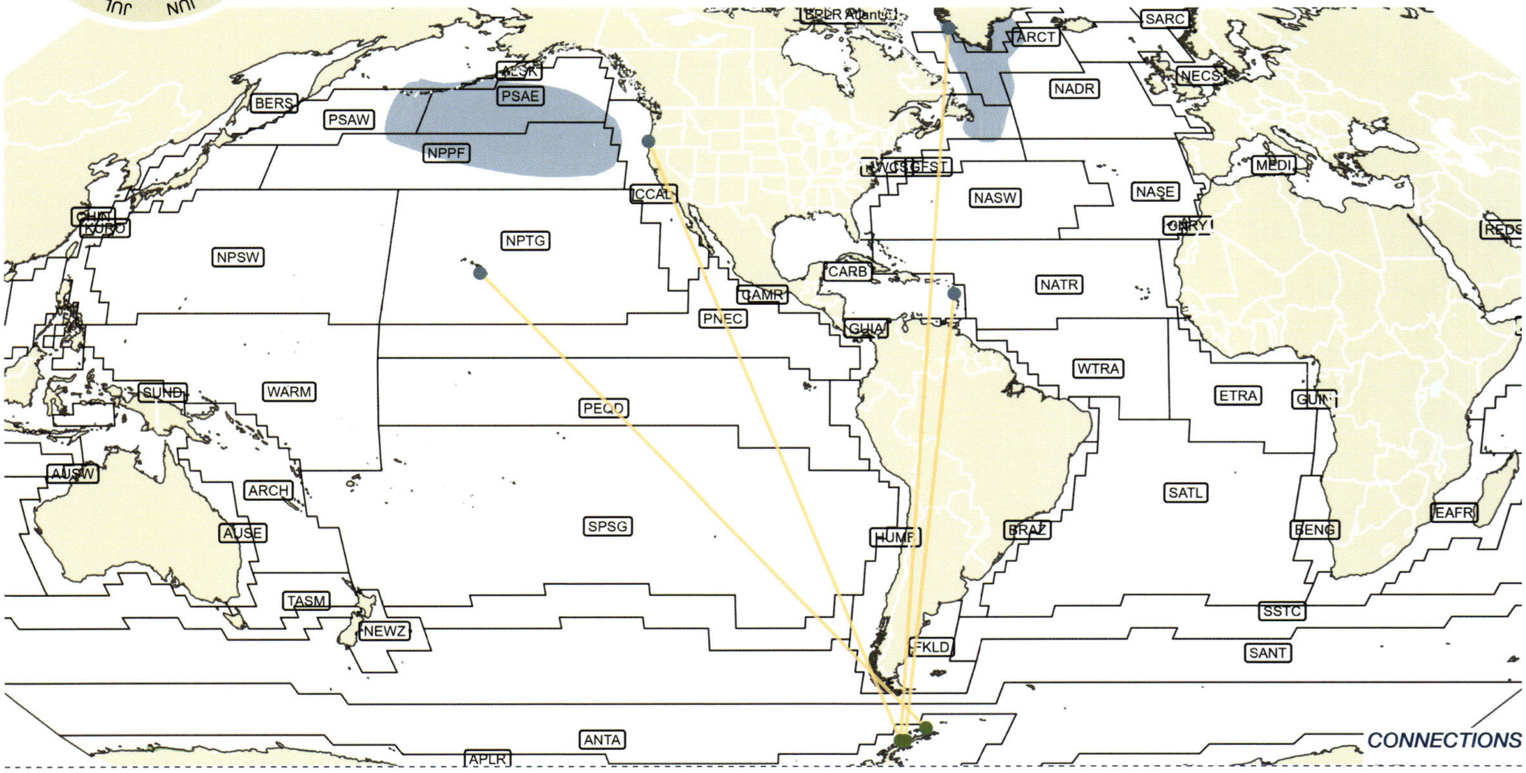

similar patterns to tracking data (Kopp et al. 2011; Weimerskirch et al. 2015). Three birds banded from the Antarctic Peninsula (southeast of South America) were recovered coastally in Hawaii, Oregon, and Greenland, showing similar connectivity to both the Pacific and Atlantic oceans as birds tracked in Kopp et al. (2011). Banding data are coastally biased.

GEOLOCATOR TRACKING Of 52 birds tracked from King George Island, Antarctica, southeast of South America (Kopp et al. 2011), 72% migrated to the Atlantic Ocean and 28% to the Pacific Ocean. Key nonbreeding areas included Newfoundland Basin and south of Georges Bank in the North Atlantic, and the Hawaiian ridge, Gulf of Alaska, and adjacent high seas areas in the North Pacific. A small proportion also used nonbreeding areas off Mauritania, Peru, and Gough Island.

STABLE ISOTOPES Weimerskirch et al. (2015) used feather isotopic values to assign individuals tracked with geolocators to a given molting area, but they were unable to reliably assign signatures to nonbreeding areas. Signatures of fully grown feathers from individuals sampled on King George Island (unique from individuals equipped with geolocators; Kopp et al. 2011) indicated molting habitats near 45 degrees north when compared to ocean isoscapes. About half of those sampled were also actively molting during the breeding season (unusual for seabirds), and these feathers reflected maritime Antarctic foraging during the breeding season (Grilli et al. 2017).

OTHER APPROACHES No genetic studies of migratory connectivity were found for this species, but little structure exists; there is gene flow between South Polar Skua and other southern skua species through hybridization (Ritz et al. 2008).

MIGRATORY PATHWAYS Individuals tracked by Weimerskirch et al. (2015) to the Atlantic and east Pacific oceans took figure-eight pathways, while those tracked to the Indian Ocean and west Pacific seemed to use loop migrations. All individuals tracked by geolocator from King George Island, off the Antarctic Peninsula, took a figure-eight pattern to their nonbreeding destinations, regardless of whether these were in the Pacific or Atlantic oceans (Kopp et al. 2011).

POMARINE JAEGER *Stercorarius pomarinus*

DISTRIBUTION AND ABUNDANCE

Pomarine Jaegers are an irruptive breeder throughout much of the North American Arctic (except Greenland), making abundance comparisons difficult within and among sites. They nest in high numbers in years of high prey density, primarily cyclic lemmings. During the nonbreeding period, they are found throughout the Pacific, Atlantic, and Indian oceans, but due to the difficulty in identifying jaeger species at sea, their relative preference for coastal versus high seas habitats is unknown (Wiley and Lee 2020c). In North America, they are most frequently reported during nonbreeding along the Atlantic and Pacific coasts during migration, or in the Gulf of Mexico during the nonbreeding period (Wiley and Lee 2020c).

ORDER Charadriiformes
FAMILY Stercorariidae
BREEDING HABITAT Wetlands, Tundra
NONBREEDING HABITAT Oceans, Pelagic
GLOBAL POPULATION SIZE 1,100,000
POPULATION CHANGE SINCE 1970 NA
IUCN RED LIST CATEGORY Least Concern
R2R TIPPING POINT SPECIES No
US LISTING STATUS No
CANADA SPECIES AT RISK No
PIF CCS 10/20

MIGRATORY CONNECTIVITY

OVERALL The migratory connectivity of the Pomarine Jaeger is little understood due to limited data. Most migratory connectivity information available for the Pomarine Jaeger comes from Argos. The few North American breeding birds studied spent the nonbreeding period in the Pacific Ocean and were broadly distributed, suggesting weak connectivity. Birds tracked from an Alaskan site spent the nonbreeding period across seven provinces spanning the central and south Pacific Ocean. No birds have been tracked in the Atlantic Ocean. The breeding origins of birds migrating along the Atlantic coast of North America, or of those spending the nonbreeding period in the Gulf of Mexico, are completely unknown and could conceivably include birds from Canada or Russia.

SATELLITE TRACKING An Alaskan individual that spent the nonbreeding period off the coast of southeastern Australia (Troy 2007) and a single individual tracked from the central Canadian high Arctic exhibited nomadic behavior during the nonbreeding period. This latter individual made a 19,000 km loop over the western Pacific and used multiple coastal and pelagic habitats throughout Micronesia and the high seas (Harrison et al. 2022). An ongoing study from the Migratory Connectivity Project found that Alaskan birds also spend the nonbreeding period off Mexico, in Peru's Humboldt Current, and in the high seas (A. Harrison, unpublished data). An individual

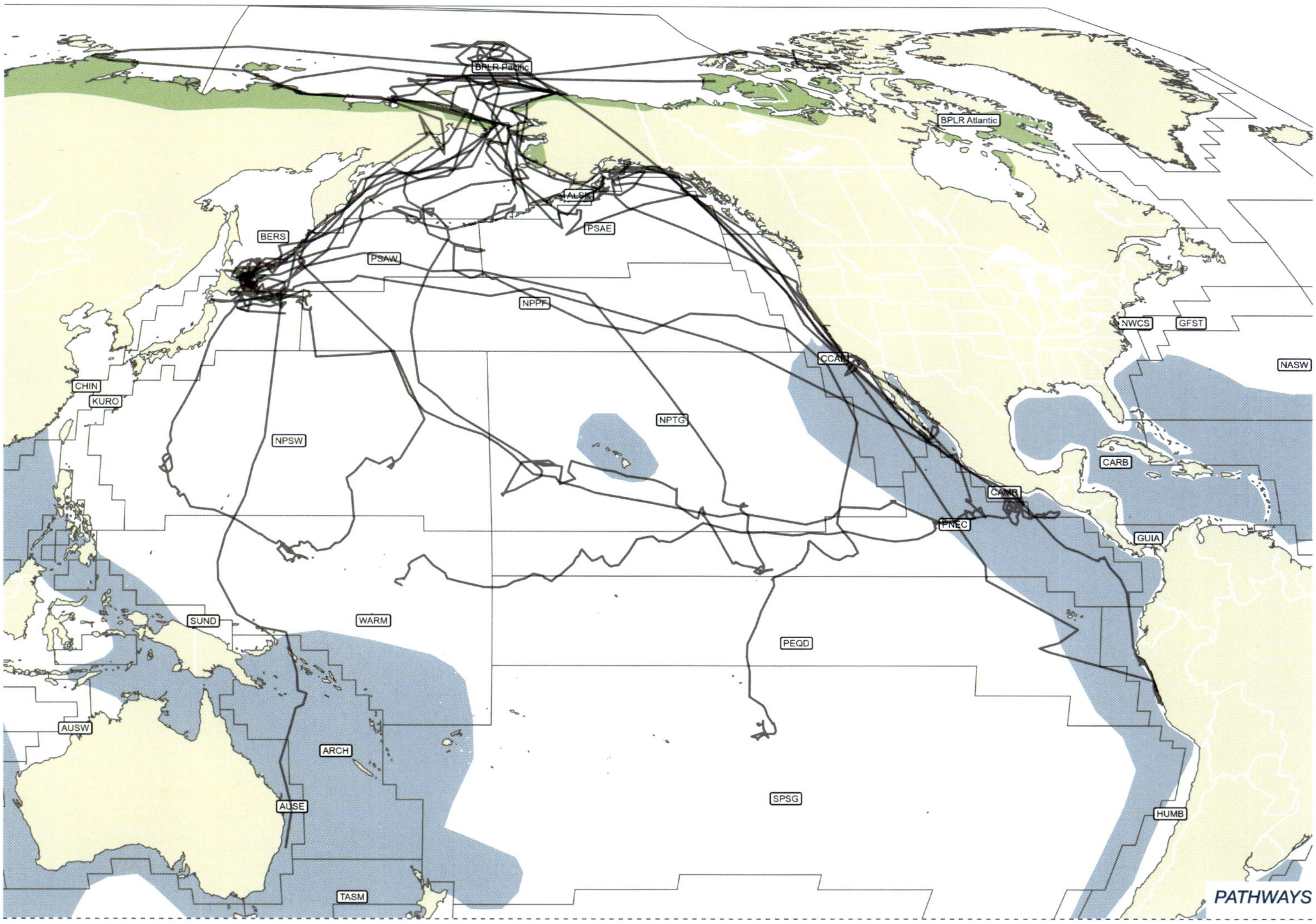

tracked with a high-frequency remote-download GPS from Utqiaġvik, Alaska, spent the nonbreeding period off the Pacific Coast of Mexico (Myles and Lanctot, unpublished data).

OTHER APPROACHES No banding, genetic, or isotope studies of migratory connectivity were found for this species.

MIGRATORY PATHWAYS Migration paths vary. Routes from Alaska include corridors and stopover areas along the western (Russia, Japan) and eastern coasts (Alaska, California, Mexico) of the Pacific Ocean. One bird migrated south through the central Pacific. Individuals tracked by Argos and GPS spent most of their post-breeding period in the eastern Pacific before crossing over to the western Pacific Ocean during their pre-breeding period. The single individual tracked from central Canada first migrated west across the high seas of the Arctic Ocean before turning south along the coasts of Russia and Japan (Harrison et al. 2022).

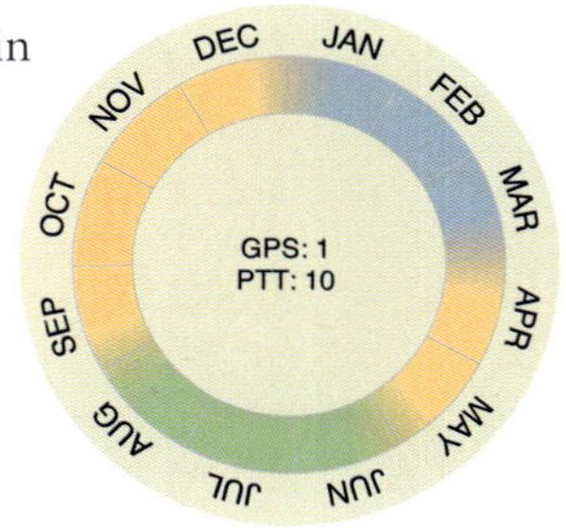

DATA CONTRIBUTORS

A.-L. Harrison; R.B. Lanctot; J. Myles; J. Rausch; P. Woodard; Migratory Connectivity Project.

INTERESTING NOTE

The first full-annual-cycle track of a Pomarine Jaeger (Harrison et al. 2022) provided the first direct evidence for breeding nomadism; this was reinforced by the single bird tracked by GPS (J. Myles and R.B. Lanctot, unpublished data). Both birds made overland migrations during the pre-breeding period, visiting multiple Siberian Russian terrestrial sites separated by a linear distance of over 2,000 km before making overland migrations to Alaska and Canada in their June breeding period. Connectivity for this species should be viewed from the context of large nomadic movements during the breeding as well as nonbreeding periods.

PARASITIC JAEGER *Stercorarius parasiticus*

DISTRIBUTION AND ABUNDANCE

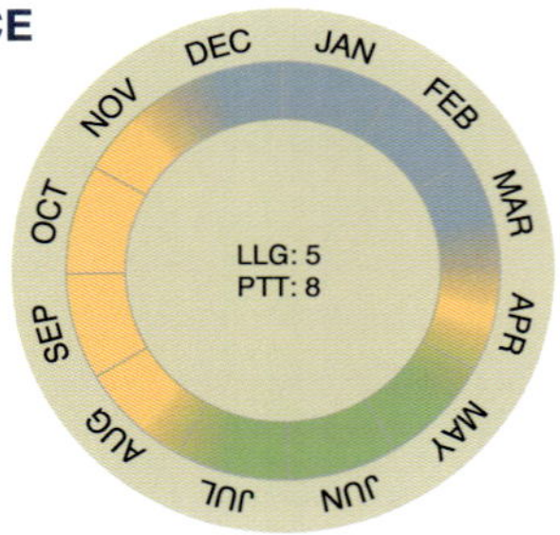

Parasitic Jaegers (also known as Arctic Skuas) have a Holarctic breeding distribution and are notable for their predatory, kleptoparasitic behaviors and high degrees of polymorphism (Wiley and Lee 2020b). In Alaska, this species breeds along the western and northern coasts, in inland but not montane areas, and in the Aleutian Islands; it is scarcer along the southern coast (Wiley and Lee 2020b). It is widely scattered in northern Canada across the mainland and islands as well as some deltas, lakes, and bays (Wiley and Lee 2020b). Its global population is not known with certainty (400,000–560,000 estimated), but has high abundances in Europe (80,000–112,000 mature individuals), especially Iceland and Scotland, and in the Nearctic (Wiley and Lee 2020b). There are no abundance estimates for Alaska as a whole (49,000 individuals estimated for northern Alaska; Bart et al. 2012) and 100,000–200,000 estimated for Canada; Environment and Climate Change Canada 2019. Its nonbreeding distribution is poorly understood because of similarity across jaegers in basic plumage, but the distribution is generally subtropical and temperate and some tropical seas.

ORDER Charadriiformes
FAMILY Stercorariidae
BREEDING HABITAT Wetlands, Tundra
NONBREEDING HABITAT Oceans, Pelagic
GLOBAL POPULATION SIZE 3,500,000
POPULATION CHANGE SINCE 1970 NA

IUCN RED LIST CATEGORY Least Concern
R2R TIPPING POINT SPECIES No
US LISTING STATUS No
CANADA SPECIES AT RISK No
PIF CCS 10/20

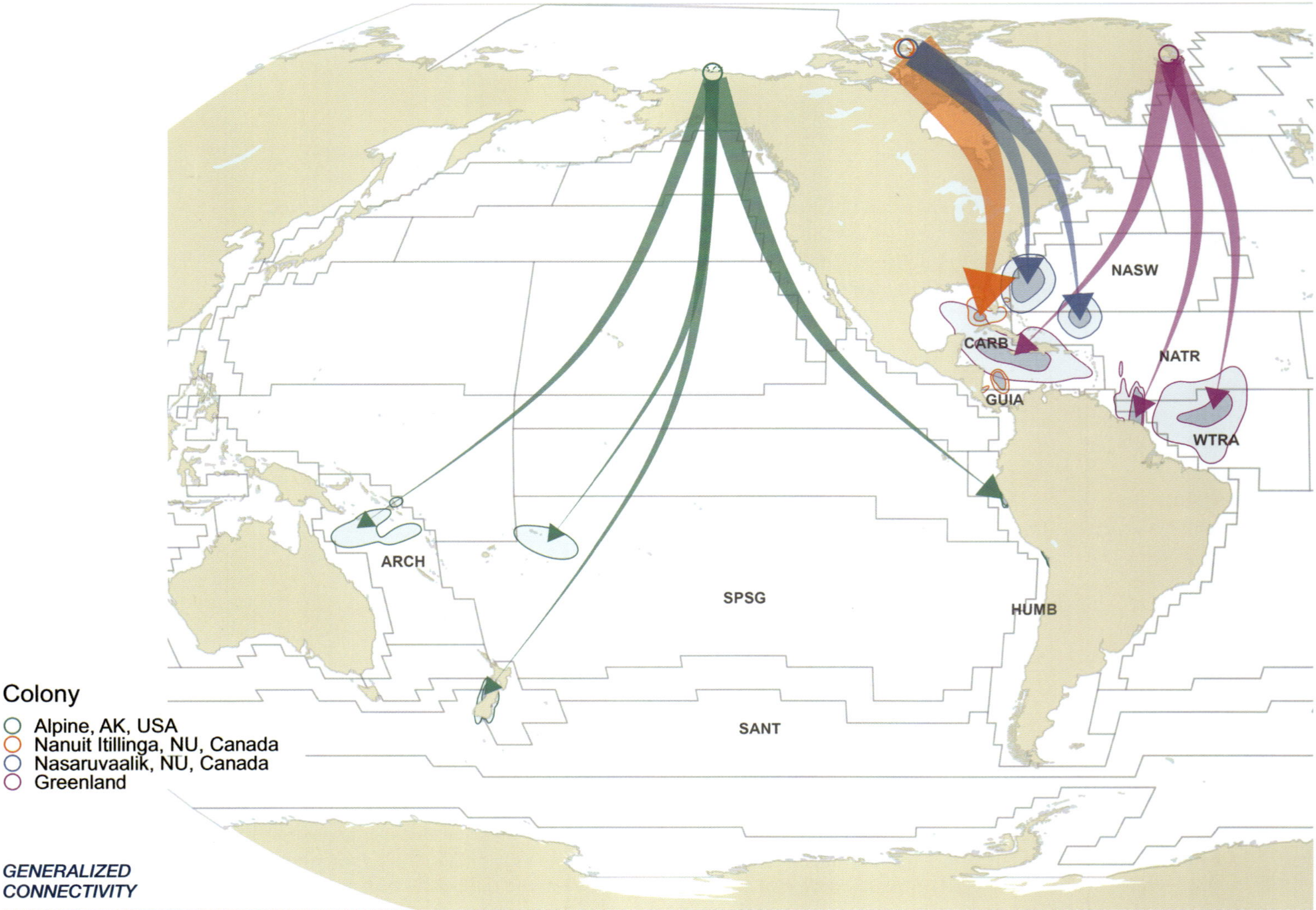

GENERALIZED CONNECTIVITY

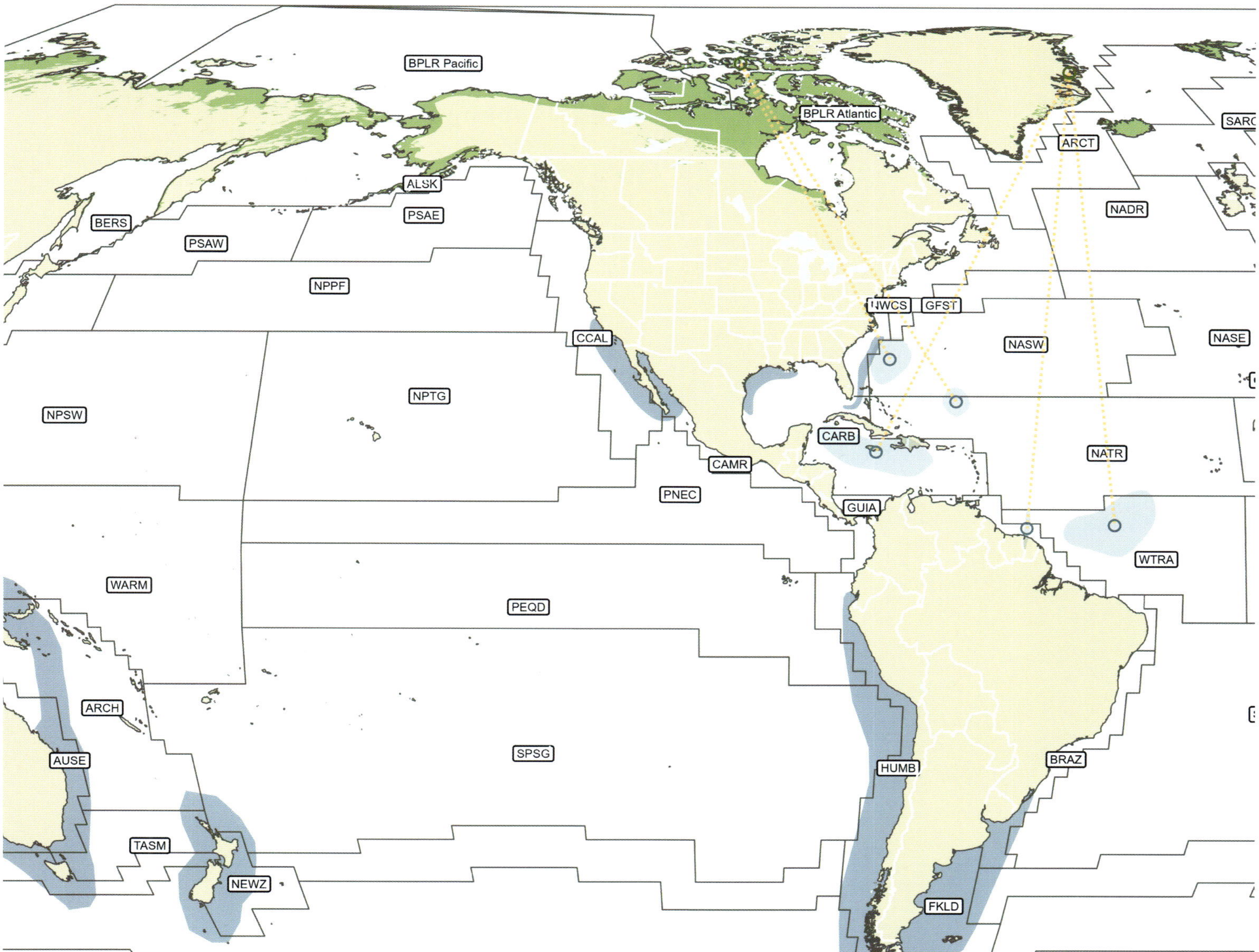

CONNECTIONS

	ARCH	HUMB	SPSG	SANT	CARB	NATR	NASW	GUIA	WTRA
Alpine, AK, USA (n = 5)	20%	40%	20%	20%	–	–	–	–	–
Nanuit Itillinga, NU, Canada (n = 2)	–	–	–	–	100%	–	–	–	–
Nasaruvaalik, NU, Canada (n = 2)	–	–	–	–	–	50%	50%	–	–
Greenland (n = 3)	–	–	–	–	33%	–	–	33%	33%

MIGRATORY CONNECTIVITY

OVERALL Migratory connectivity estimates could not be calculated due to limited sample size, but it is thought to be strong across oceans, with a divide between Norway and western Russian breeding populations distributing either to the Atlantic or Indian oceans, and a Nearctic divide between Alaska and central high Arctic Canada distributing to either the Pacific or Atlantic oceans. There is likely weak connectivity within the Pacific Ocean based on tracking data of one population to both eastern and western Pacific Ocean provinces (40% Humboldt Current, 60% across three western Pacific provinces). In the Atlantic Ocean, weak connectivity of Canadian and Greenland birds to western Atlantic provinces is suggested by limited tracking data.

GEOLOCATOR TRACKING Birds tracked with geolocators from Greenland (O. Gilg, unpublished data) used nonbreeding areas in the Caribbean Sea and off northern South America, while two birds tracked from Canada visited the Caribbean Sea, including the Gulf of Mexico (Harrison et al. 2022). A comprehensive picture of weak migratory connectivity is available for 155 individuals representing Greenland, European, and Russian populations (van Bemmelen et al. 2024). Although no tracked North American birds used eastern Atlantic provinces (O. Gilg, unpublished data; Harrison et al. 2022), a proportion of European birds in this study spent the nonbreeding period in western Atlantic provinces. European birds demonstrated weak connectivity to staging sites (Hanlon et al. 2024), and Atlantic Ocean birds distributed across multiple nonbreeding areas, including the Mediterranean Sea, Caribbean Sea, Patagonia, Canary Coastal Province, Gulf of Guinea, and the Benguela Current (van Bemmelen et al. 2024).

SATELLITE TRACKING Birds from northern Alaska (n = 5) migrated to both the southwest and southeast Pacific Ocean and spent the nonbreeding period largely within Exclusive Economic Zones of

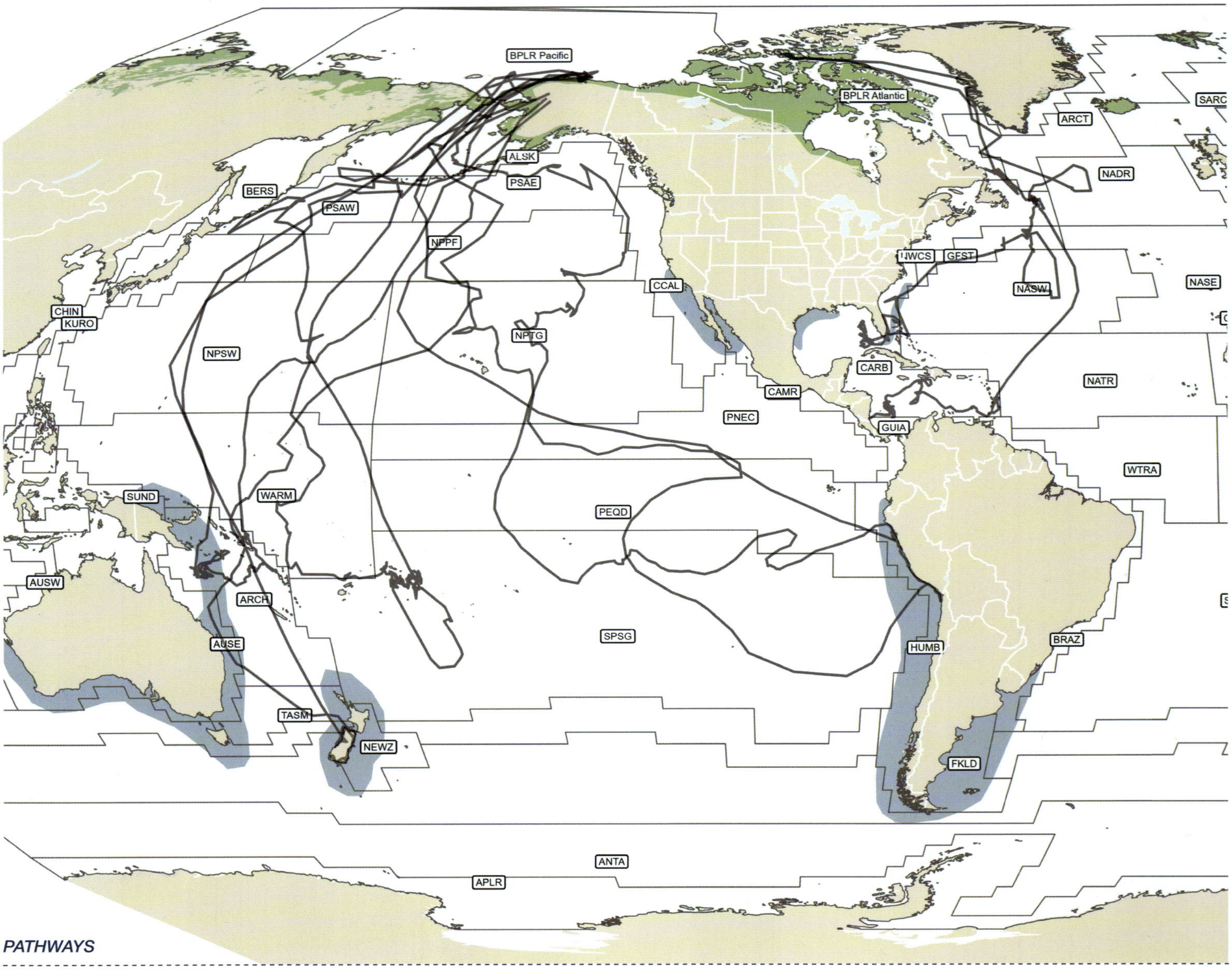

PATHWAYS

Peru, New Zealand, and the Pacific islands (Harrison et al. 2022). Two individuals from high Arctic Canada took relatively short migrations to the Caribbean Sea, including the Florida coast and into the Gulf of Mexico, and the Guianas coast (Harrison et al. 2022).

STABLE ISOTOPES Stable isotopic niches differed according to the nonbreeding areas used by 58 tracked birds from three European colonies, including birds that migrated to the Caribbean Sea (Rypdal 2018). This suggests stable isotopic studies could be used to assess migratory connectivity in the absence of tracking data. This study also demonstrated high individual specialization—niches varied across individuals within nonbreeding areas—and high individual repeatability across years.

OTHER APPROACHES No genetic studies of migratory connectivity were found for this species.

MIGRATORY PATHWAYS Given the wide variety of nonbreeding areas visited by individuals within any single population, migration pathways also show extensive variation. Generally, birds from Greenland (O. Gilg, unpublished data) and Canada (Harrison et al. 2022) took open-ocean southward migrations (one bird migrated along the US southeastern coast), and birds in the Pacific Ocean made highly pelagic migrations, in either the western or central Pacific before turning due east toward Peru (Harrison et al. 2022).

DATA CONTRIBUTORS

L. Bollache; O. Gilg; A.-L. Harrison; J. Lang; M. Mallory; J. Rausch; B. Sittler; R. van Bemmelen; P. Woodard; Migratory Connectivity Project.

INTERESTING NOTE

Improved assessment of population status is needed across the Nearctic, and increased tracking in the Pacific Ocean would help increase knowledge of migratory connectivity and nonbreeding biology for this species.

Mantel = 0.49 ± 0.12

Stercorarius longicaudus **LONG-TAILED JAEGER**

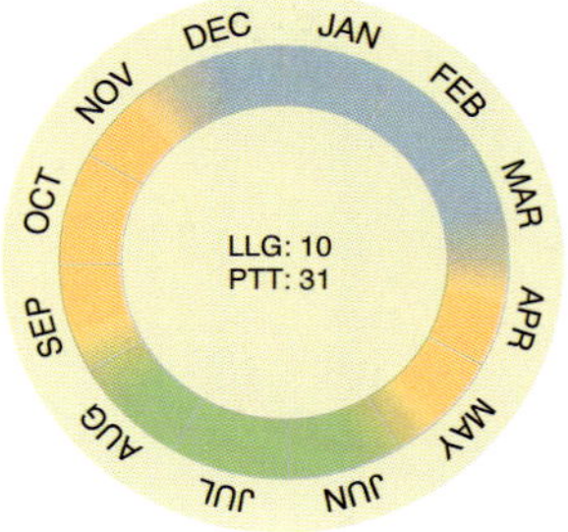

DISTRIBUTION AND ABUNDANCE

Long-tailed Jaegers, a Holarctic species and the smallest of the skuas, feed primarily on small mammals during breeding before switching to a marine lifestyle and undertaking trans-equatorial migrations to pelagic temperate and subantarctic seas (Wiley and Lee 2020a). Widely distributed across dry coastal and montane tundra in North America, they extend farther north than other jaegers but have centers of distribution within their range associated with voles and lemmings in Alaska and from Hudson Bay and Southampton Island north to Ellesmere Island in Canada (Wiley and Lee 2020a). There are no detailed estimates of total numbers in North America, but it is the most abundant jaeger in North America. The European population is estimated at 20,000–53,000 pairs (Wiley and Lee 2020a).

ORDER Charadriiformes
FAMILY Stercorariidae
BREEDING HABITAT Tundra, Arctic
NONBREEDING HABITAT Oceans, Pelagic
GLOBAL POPULATION SIZE 1,700,000
POPULATION CHANGE SINCE 1970 NA
IUCN RED LIST CATEGORY Least Concern
R2R TIPPING POINT SPECIES No
US LISTING STATUS No
CANADA SPECIES AT RISK No
PIF CCS 10/20

INTERESTING NOTE

Seyer et al. (2021) linked breeding propensity to migration metrics derived from light-level geolocators for a Canadian population. Increased time spent flying during nonbreeding and early arrival at the breeding site reduced breeding propensity.

Colony

- Nome, AK, USA
- Denali National Park, AK, USA
- Alpine, AK, USA
- Nanuit Itillinga, NU, Canada
- Hochstetter/Karupelv, Greenland

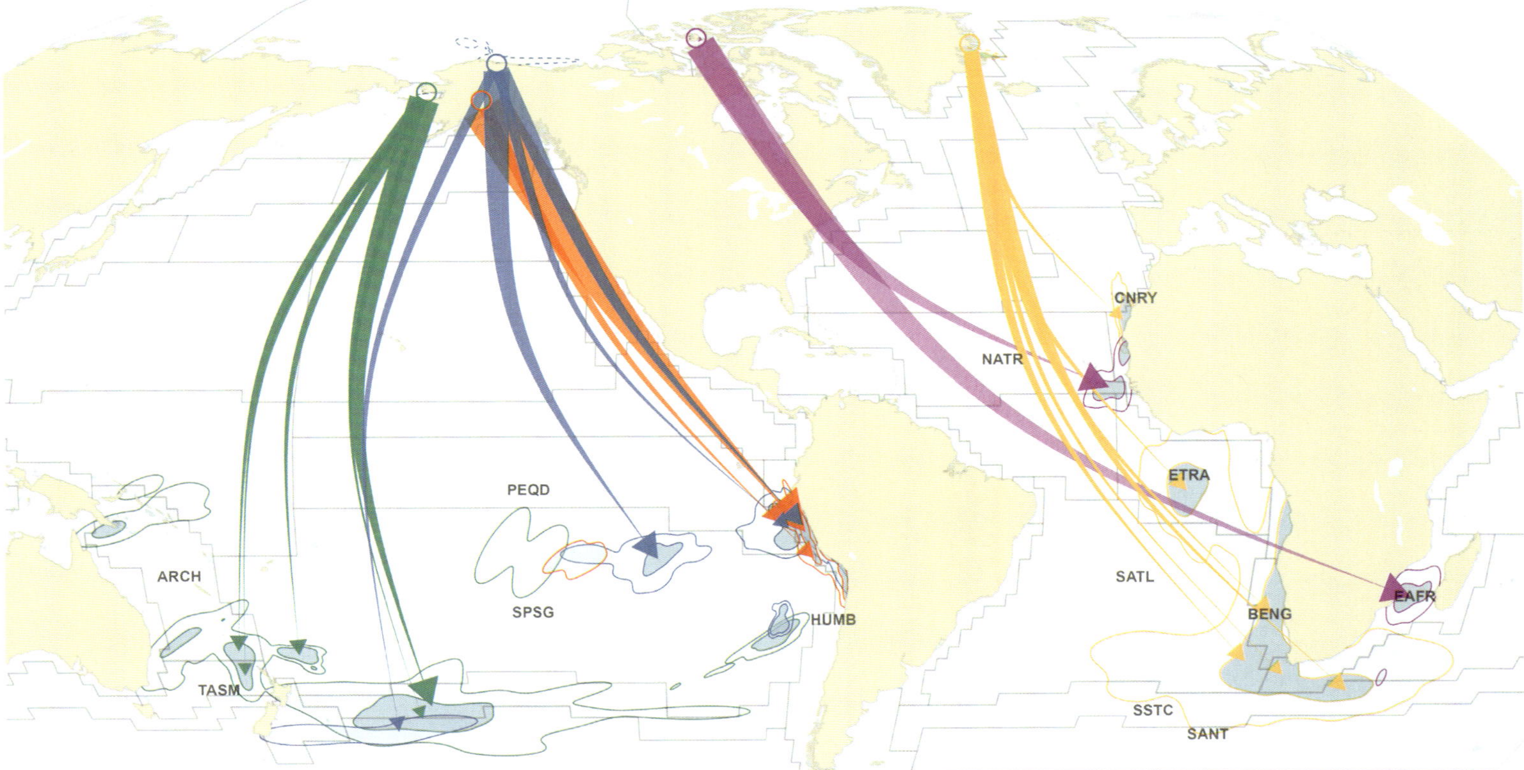

GENERALIZED CONNECTIVITY

SHOREBIRDS, AUKS, GULLS, AND TERNS (CHARADRIIFORMES)

	ARCH	HUMB	PEQD	SPSG	TASM	SSTC	SANT	NATR	CNRY	ETRA	BENG	SATL	EAFR
Nome, AK, USA (n = 12)	17%	–	–	17%	8%	50%	8%	–	–	–	–	–	–
Denali National Park, AK, USA (n = 6)	–	83%	–	17%	–	–	–	–	–	–	–	–	–
Alpine, AK, USA (n = 7)	–	29%	14%	43%	–	–	14%	–	–	–	–	–	–
Nanuit Itillinga, NU, Canada (n = 2)	–	–	–	–	–	–	–	50%	–	–	–	–	50%
Hochstetter/Karupelv, Greenland (n = 10)	–	–	–	–	–	20%	–	–	10%	20%	20%	20%	10%

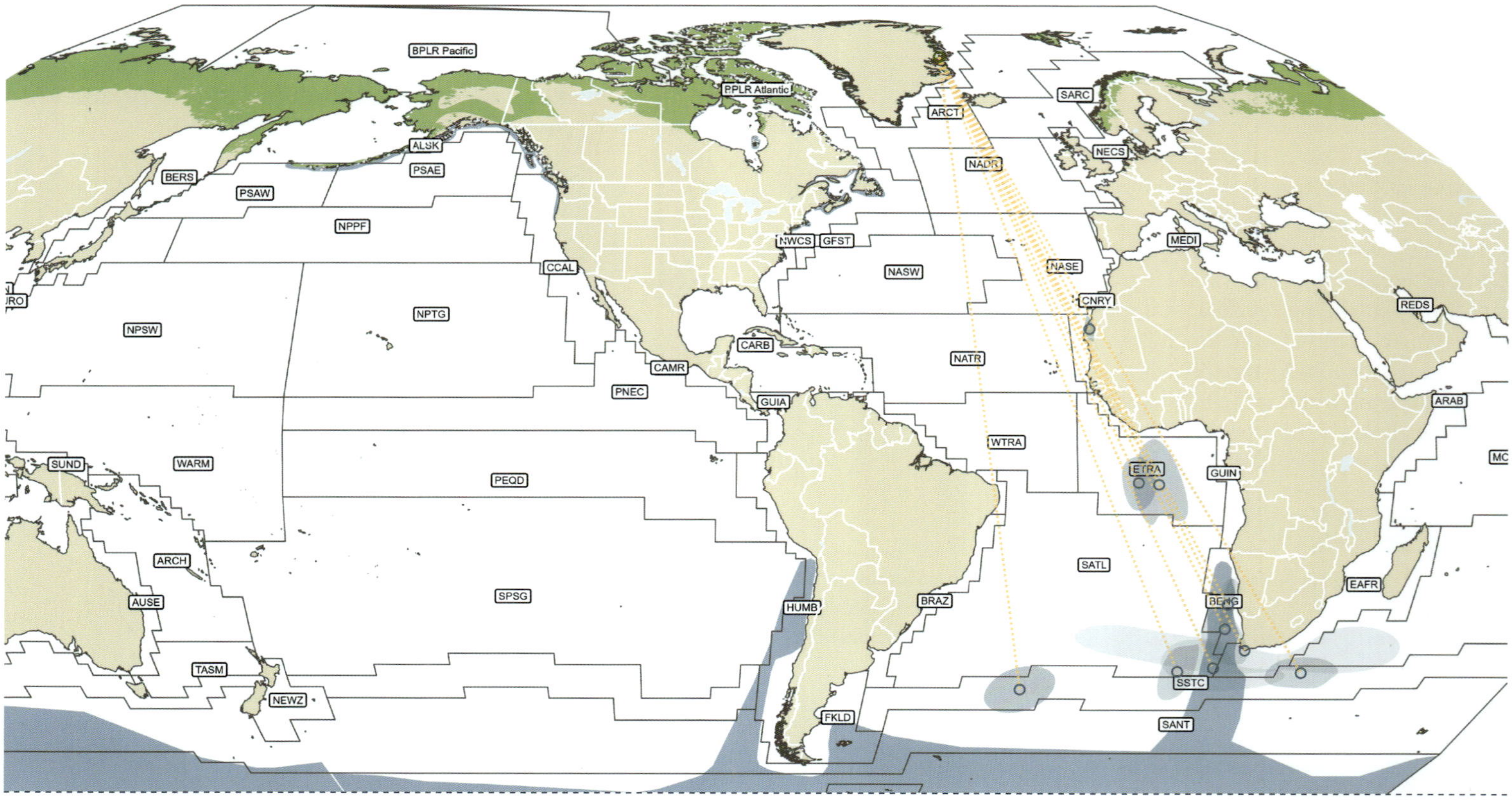

CONNECTIONS

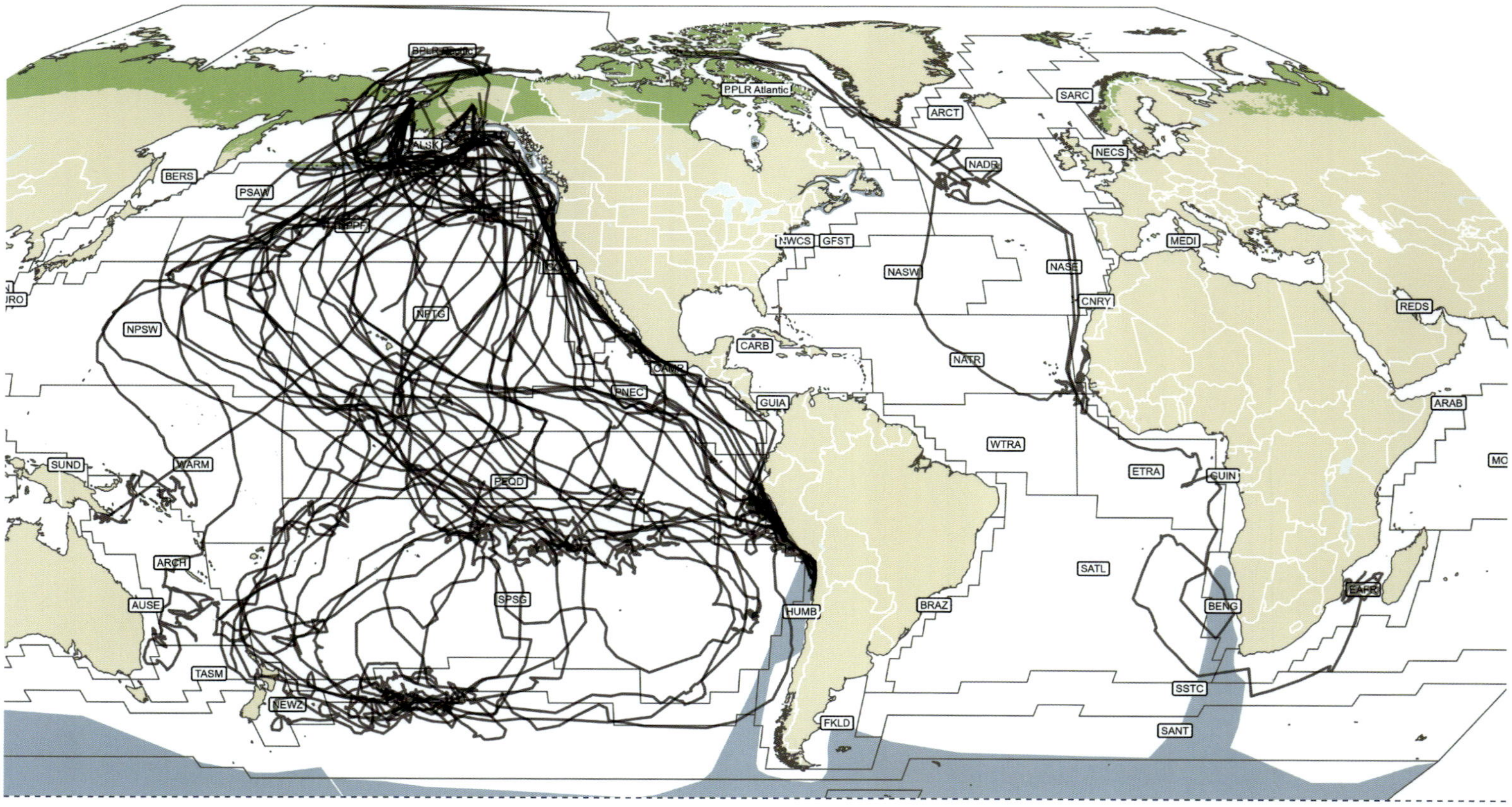

PATHWAYS

MIGRATORY CONNECTIVITY

OVERALL There is generally strong migratory connectivity across ocean basins (Mantel = 0.49 ± 0.12); however, connectivity is weak within the Atlantic Ocean. Birds tracked from Greenland used multiple provinces in the South Atlantic Ocean. Published literature indicate extensive population mixing, possibly with greater use of the southwest Atlantic Ocean by Canadian populations. Tracking data in the Pacific Ocean hint at moderate connectivity, but larger sample sizes are needed. Eighty-three percent of birds tracked from an inland breeding site in Alaska (Denali National Park) used nonbreeding sites in the Humboldt Current province, while no birds from Nome spent the nonbreeding season in this province (50% from Nome spent the nonbreeding season in the South Subtropical Convergence Zone). Birds tracked from northern Alaska (Alpine) distributed across multiple nonbreeding areas, including the Humboldt Current (29%) and the South Pacific Gyre (43%), but excluded the South Subtropical Convergence Province.

GEOLOCATOR TRACKING The first tracking of Long-tailed Jaegers was from northeast Greenland and Svalbard, Norway, starting in 2010 (Gilg et al. 2013). Birds spent the nonbreeding season off the west coast of Africa, in the Indian Ocean, and farther south in a subantarctic high seas area (Gilg et al. 2013). A Canadian study (Seyer et al. 2021) collected 43 tracks from 34 individuals from Bylot and Igloolik islands (2014–2018). These birds also used habitats off Africa but, notably, some individuals spent the nonbreeding season in the western South Atlantic off South America. Birds tracked from Sweden used similar staging areas in the North Atlantic and similar nonbreeding areas off Namibia and South Africa (van Bemmelen et al. 2024).

SATELLITE TRACKING Sittler et al. (2011) and Seyer et al. (2021) deployed satellite transmitters using full body harnesses in Greenland and Canada, respectively; however, these tags (9.5–10 g) did not yield full migrations, and the attachment/tag combination was thought to have had detrimental effects. Later deployments used 5 g tags with leg-loop harnesses (Harrison et al. 2022; Fowler et al. 2023) to track Long-tailed Jaegers from multiple Alaskan sites (2017–2024). Six inland-nesting birds from Denali National Park used nonbreeding sites in the eastern South Pacific, largely off Peru (Fowler et al. 2023). Long-tailed Jaegers tracked from Nome, Alaska, and the Arctic coastal plain primarily distributed across the South Pacific with extensive high seas use (Harrison et al. 2022). Two birds tracked from high Arctic Canada spent the nonbreeding season off the African continent, including one bird tracked to the Indian Ocean (Harrison et al. 2022).

OTHER APPROACHES No genetic or stable isotope studies of migratory connectivity were found for this species.

MIGRATORY PATHWAYS From the Canadian Arctic, Long-tailed Jaegers migrated through Baffin Bay and Davis Strait, then staged in an area extending from the Grand Banks and into the North Atlantic high seas. They then crossed the Atlantic to the western African coast and generally traveled south through the east Atlantic (Harrison et al. 2022; Seyer et al. 2021). Birds from Greenland, Svalbard, and Sweden also staged in the North Atlantic high seas (arriving there either from the northwest or northeast), before using similar southward migrations along the east Atlantic (Gilg et al. 2013; van Bemmelen et al. 2017). Return migrations generally were more oceanic through the central Atlantic. Long-tailed Jaegers in the Pacific Ocean take broad, figure-eight pathways, and many use staging areas in the Gulf of Alaska on southward migration. Some individuals take an overland migration to the North Slope on return migration.

DATA CONTRIBUTORS

L. Bollache; O. Gilg; A.-L. Harrison; J. Lang; B. Sittler; R. van Bemmelen; Migratory Connectivity Project.

DOVEKIE *Alle alle*

Mantel = 0.52 ± 0.04

DISTRIBUTION AND ABUNDANCE

This species breeds in the high Arctic (its highest abundance is in Greenland, with colonies also in Canada, Norway, and Russia) and is the most abundant alcid in the northwest Atlantic Ocean (Montevecchi and Stenhouse 2020). A small breeding population is also on islands in the Bering Sea, where nonbreeding distribution is largely unknown but seen regularly off Alaska and western Canada (Montevecchi and Stenhouse 2020).

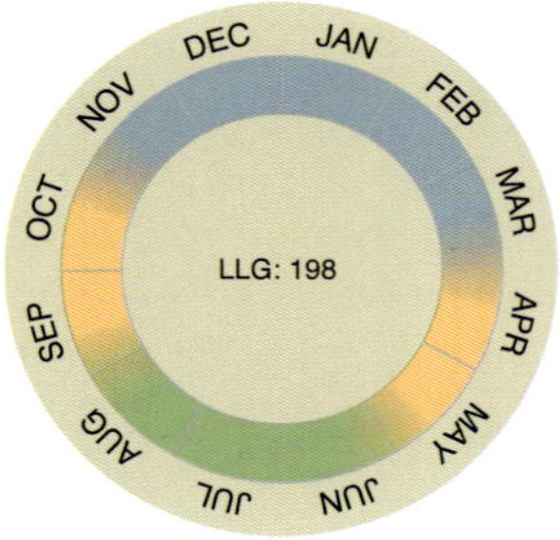

ORDER Charadriiformes
FAMILY Alcidae
BREEDING HABITAT
Coasts, Cliffs and Islands
NONBREEDING HABITAT
Coasts, Marine Waters
GLOBAL POPULATION SIZE 17,000,000
POPULATION CHANGE SINCE 1970 −38%
IUCN RED LIST CATEGORY
Least Concern
R2R TIPPING POINT SPECIES
Orange-Alert
US LISTING STATUS No
CANADA SPECIES AT RISK No
PIF CCS 13/20

MIGRATORY CONNECTIVITY

OVERALL There is some level of mixing of colonies from Greenland; Svalbard Archipelago, Norway; Bjørnøya (Bear Island), Norway; and Franz Joseph Land, Russia, during the nonbreeding season, but there is a higher percent of nonbreeding core areas in the Northwest Atlantic Shelves (45% versus 2%–3%) and the North Atlantic Drift Current (37% versus 3%) for birds tracked from Greenland than from Norway or Russia. Norwegian birds are more likely to use the Boreal Polar region (the Gulf of St. Lawrence, 19%–21%) than are Russian and Greenland birds, although insights from Russian birds are based on only 5 individuals. All birds, except for Greenland birds, overlapped extensively in the Atlantic Arctic Zone (72%–100%). Migratory connectivity across ocean basins is strong (Mantel = 0.52 ± 0.04), but there is some suggestion that the small population of Pacific Ocean breeders largely spend the nonbreeding season in the Bering Sea. Based on variability in breeding phenology by colony (Montevecchi and Stenhouse 2020), strong temporal connectivity is likely.

	NASW	GFST	NWCS	BPLR Atlantic	ARCT	SARC	NADR
Eastern Greenland (n = 51)	2%	8%	45%	2%	6%	–	37%
Svalbard, Norway (n = 106)	–	–	2%	21%	75%	3%	–
Bjørnøya (Bear Island), Svalbard, Norway (n = 36)	–	–	3%	19%	72%	3%	3%
Franz Josef Land, Russia (n = 5)	–	–	–	–	100%	–	–

Colony
- Eastern Greenland
- Svalbard, Norway
- Bjørnøya, (Bear Island), Svalbard, Norway
- Franz Josef Land, Russia

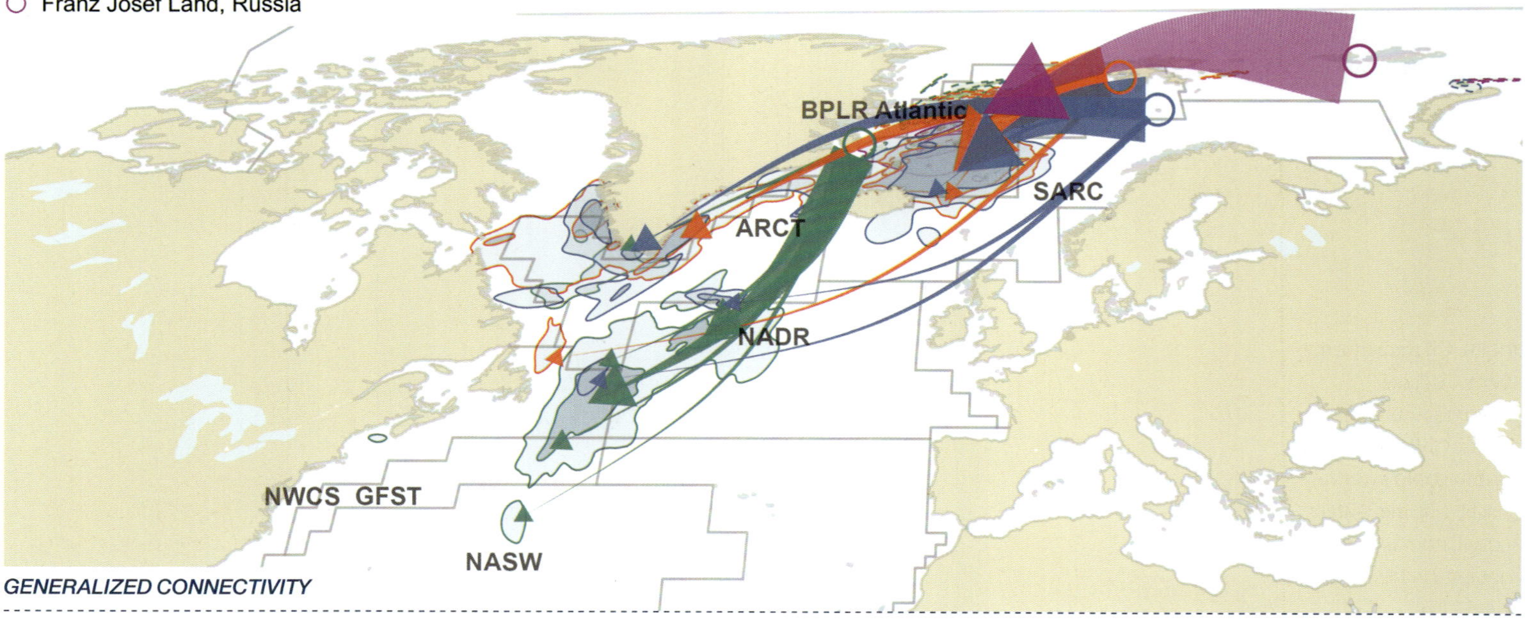

GENERALIZED CONNECTIVITY

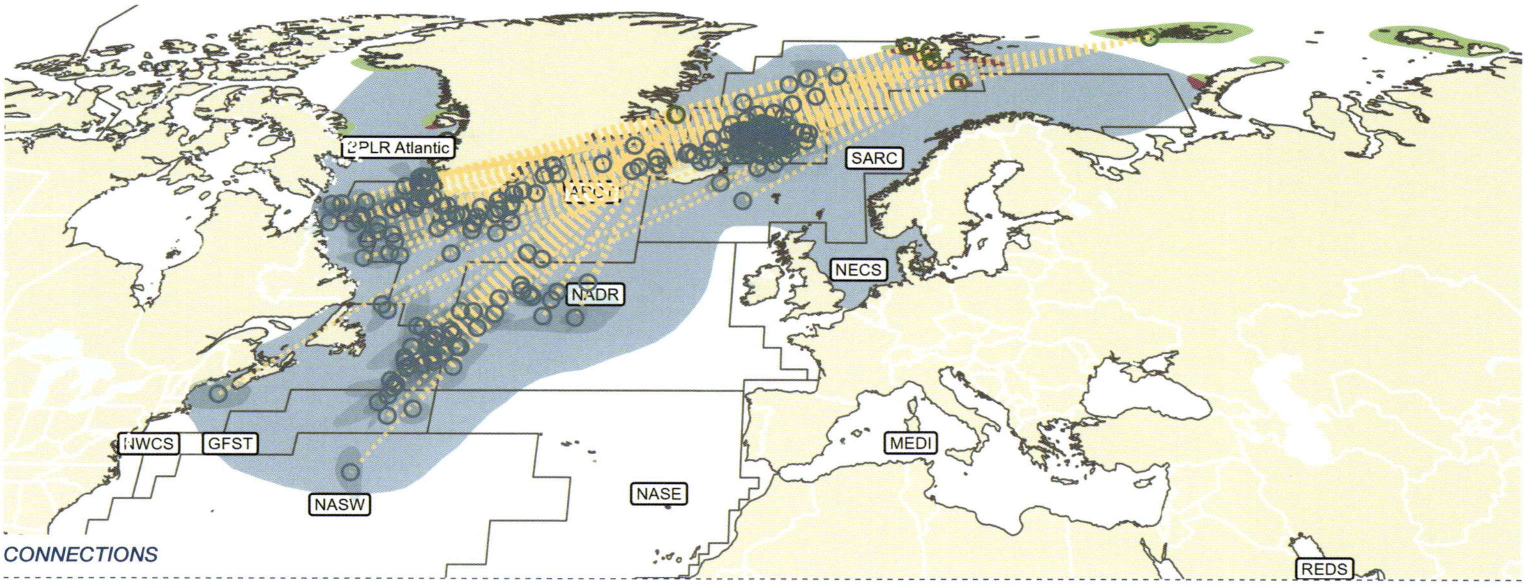

GEOLOCATOR TRACKING A large sample of birds tracked by light-level geolocator from Greenland and Norway shows that birds from Greenland are more likely to use the coastal shelves of North America (Newfoundland and Nova Scotia, Canada, and Maine, USA; D. Grémillet and J. Fort, unpublished data; SEATRACK, unpublished data). On average, individuals use three or four distinct stationary areas during their nonbreeding period (Amélineau et al. 2021). ***OTHER APPROACHES*** No satellite tracking, stable isotope, or genetic studies of migratory connectivity were found for this species. ***MIGRATORY PATHWAYS*** From Greenland and Norway, birds migrate using general southwest pathways to nonbreeding areas in the northwest Atlantic Ocean (D. Grémillet and J. Fort, unpublished data; SEATRACK, unpublished data).

DATA CONTRIBUTORS
S. Descamps; J. Fort; D. Grémillet; M.V. Gavrilo; B. Moe; H. Strøm; K. Wojczulanis-Jakubas.

INTERESTING NOTE

Birds tracked by geolocator from Greenland and Norwegian colonies overlapped in the Atlantic Arctic zone during the late breeding season (SEATRACK, unpublished data). Archival GPS tracking during the chick-rearing period from a single colony on Spitsbergen island across multiple years (Jakubas et al. 2020) showed little difference in foraging distance or duration in warm versus cold years. Birds generally foraged within 100 km of the colony.

Mantel = 0.44 ± 0.1

Uria aalge COMMON MURRE

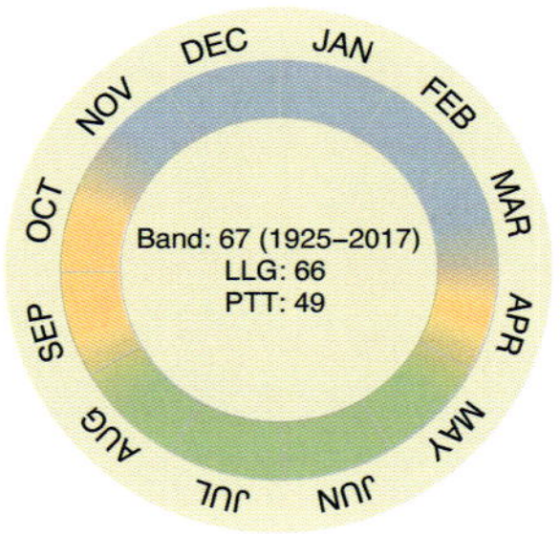

DISTRIBUTION AND ABUNDANCE

The Common Murre is one of the most abundant and well-studied seabirds in the world. In North America, the highest abundance is in the west (7.4 million), with an additional 1.8 million in eastern North America and 1.5 million in Greenland and Iceland (Ainley et al. 2021b).

ORDER Charadriiformes
FAMILY Alcidae
BREEDING HABITAT Coasts, Cliffs and Islands
NONBREEDING HABITAT Coasts, Marine Waters
GLOBAL POPULATION SIZE 12,000,000
POPULATION CHANGE SINCE 1970 −21%
IUCN RED LIST CATEGORY Least Concern
R2R TIPPING POINT SPECIES No
US LISTING STATUS No
CANADA SPECIES AT RISK No
PIF CCS 9/20

MIGRATORY CONNECTIVITY

OVERALL Banding data, tracking efforts, and genetics indicate strong connectivity across ocean basins (Mantel = 0.44 ± 0.1) and moderate to strong connectivity within basins. For example, within the Atlantic Ocean, murres tracked from Canadian colonies remained primarily within the Northwest Atlantic Shelves province and used the Gulf Stream, while only 3% of birds tracked from Norway and no Icelandic birds migrated to the Northwest Atlantic Shelves province,

	BERS	ALSK	PSAE	CCAL	GFST	NWCS	BPLR Atlantic	ARCT	SARC	NECS	NADR
Northwest Alaska, USA (n = n4)	100%	–	–	–	–	–	–	–	–	–	–
St. Lawrence Island, AK, USA (n = 3)	33%	33%	33%	–	–	–	–	–	–	–	–
Southeast Alaska, USA (n = 1)	100%	–	–	–	–	–	–	–	–	–	–
Oregon, USA (n = 10)	–	30%	–	70%	–	–	–	–	–	–	–
Farallon Islands, CA, USA (n = 1)	–	–	–	100%	–	–	–	–	–	–	–
Jan Mayen, Norway (n = 34)	–	–	–	–	–	3%	3%	59%	29%	–	6%
Iceland (n = 18)	–	–	–	–	–	–	–	28%	44%	6%	22%
Gannet Island, NL, Canada (n = 11)	–	–	–	–	–	100%	–	–	–	–	–
Funk Island, NL, Canada (n = 2)	–	–	–	–	50%	50%	–	–	–	–	–
Gulf of St. Lawrence, NB, Canada (n = 0)	–	–	–	–	–	–	–	–	–	–	–

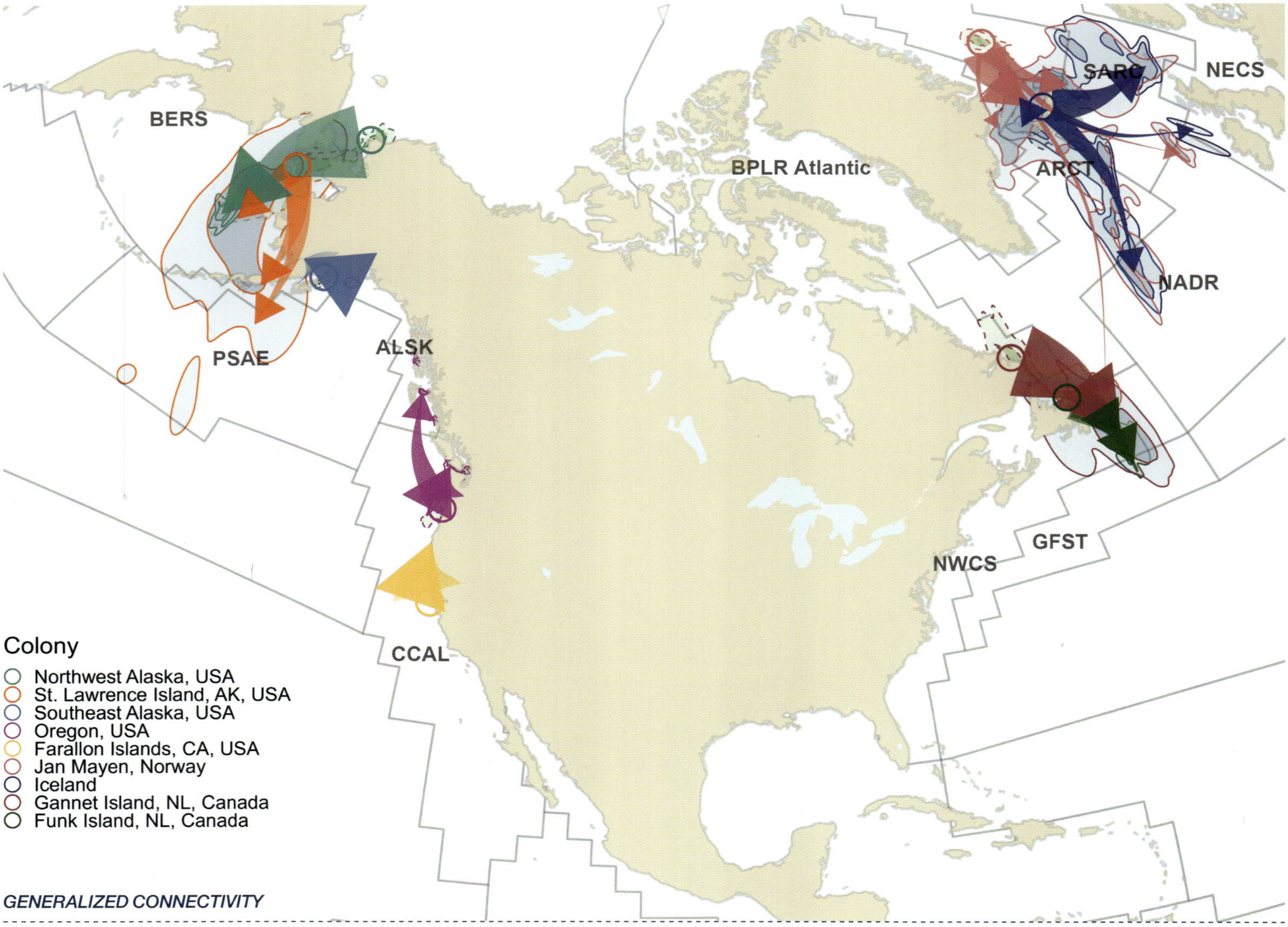

GENERALIZED CONNECTIVITY

instead remaining in the Northeast Atlantic. In the Pacific Ocean, no birds from Alaskan colonies used the California Current province during the nonbreeding period, in contrast to 70%–100% of birds from colonies in Oregon and California. However, 30% of individuals tracked from Oregon migrated north to the Alaska Coastal Downwelling Province.

GEOLOCATOR TRACKING Three birds tracked in 2016–2017 with geolocators deployed on St. Lawrence Island, Alaska, northern Bering Sea, stayed in the eastern Bering Sea and the eastern North Pacific (Takahashi et al. 2021). Murres tracked from Atlantic colonies in Newfoundland and Nunavut in 2007–2011 had high home range overlap among breeding populations (50%–67%), with most nonbreeding core areas concentrated on the Grand Banks and Orphan Basin (Montevecchi et al. 2012; McFarlane et al. 2014). Additional deployments by researchers contributing to the SEATRACK project in Iceland and Norway (SEATRACK, unpublished data) had little overlap with Canadian colonies.

SATELLITE TRACKING In the Pacific Ocean, in 1994–1996, during the months of July–September, Common Murres were tracked with implanted satellite transmitters from four Alaskan colonies (S.A. Hatch et al. 2000). One from Middleton Island migrated to waters near the Barren Islands, where birds were also tracked. All birds

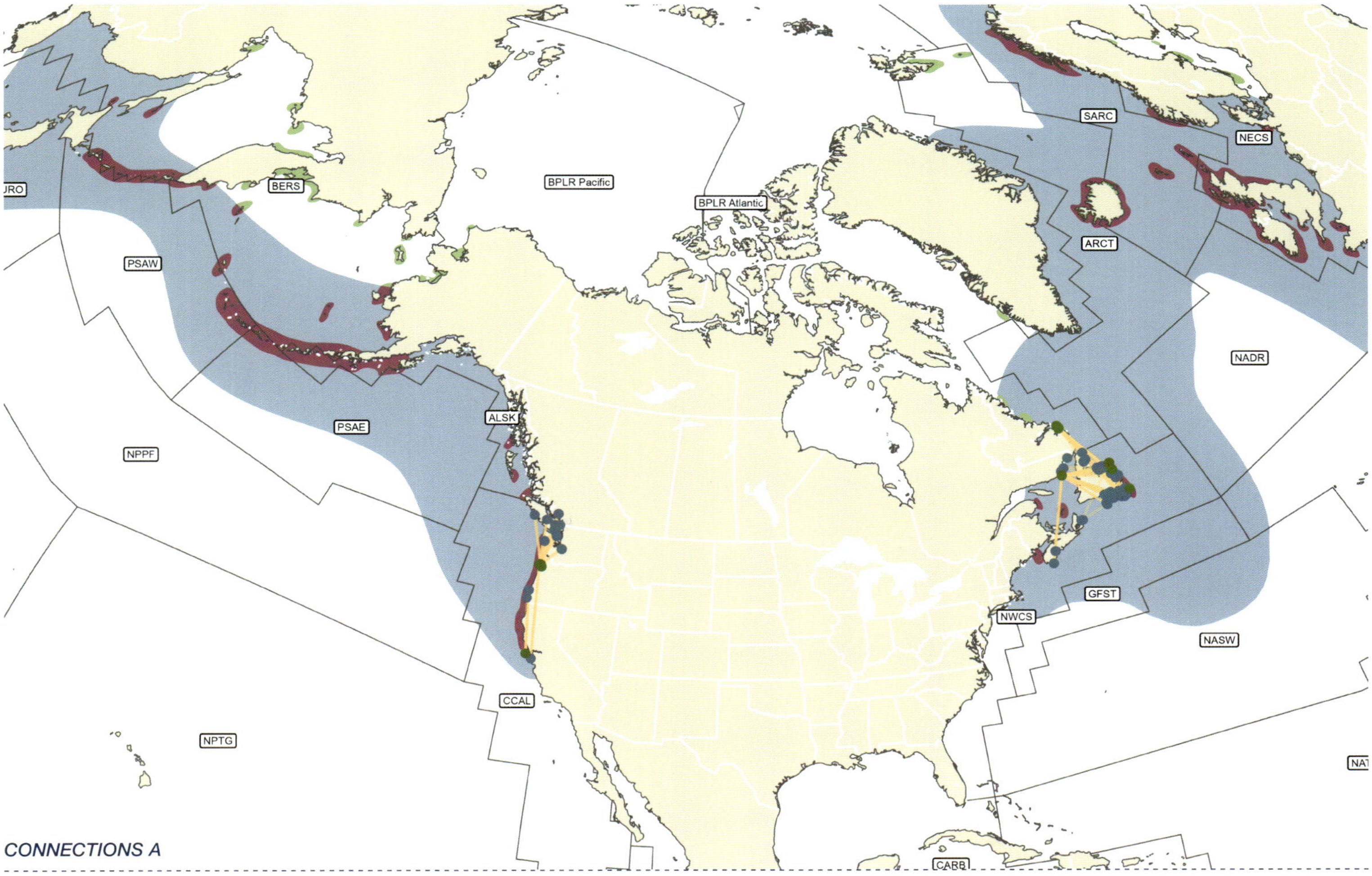

CONNECTIONS A

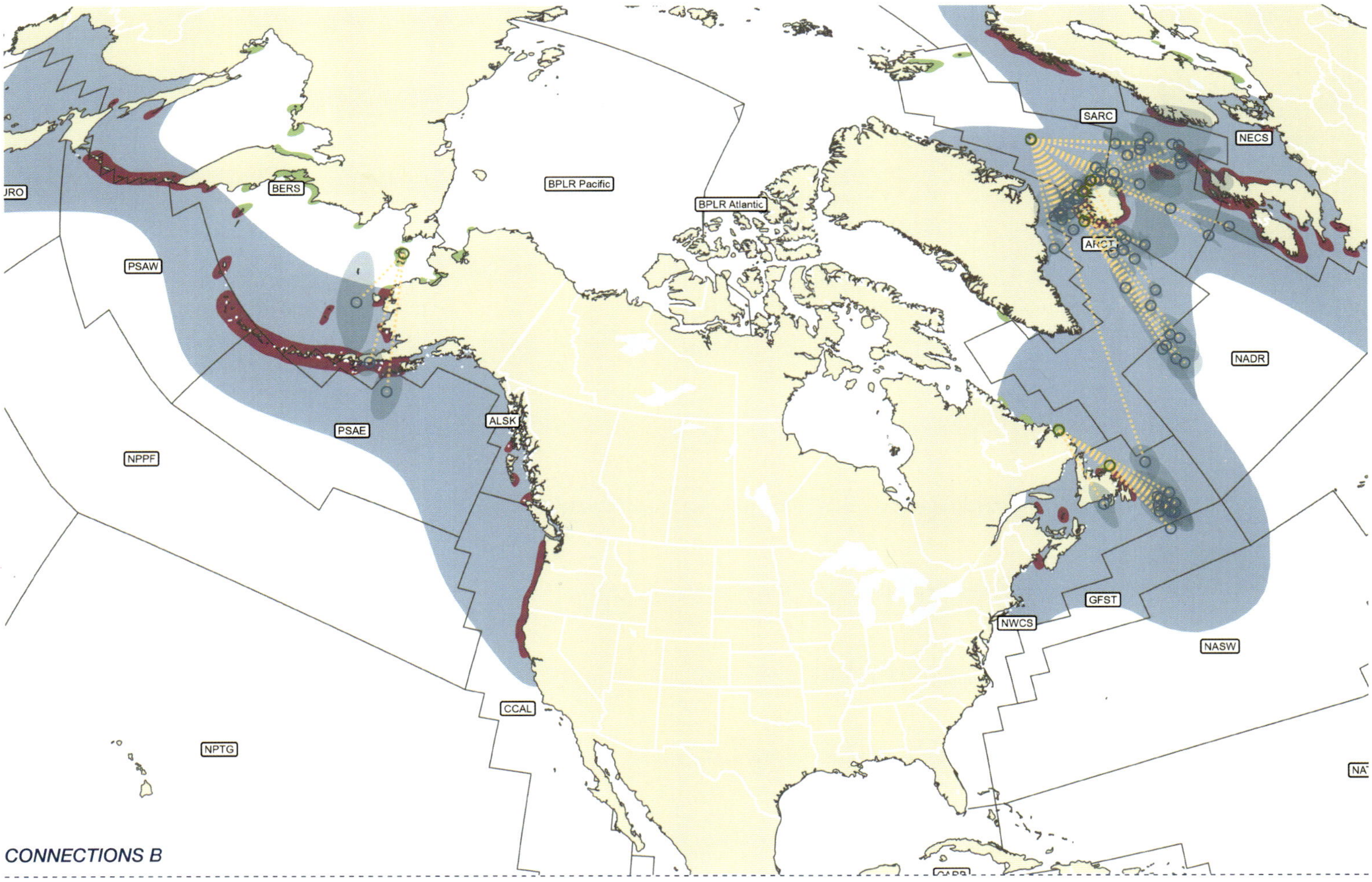

CONNECTIONS B

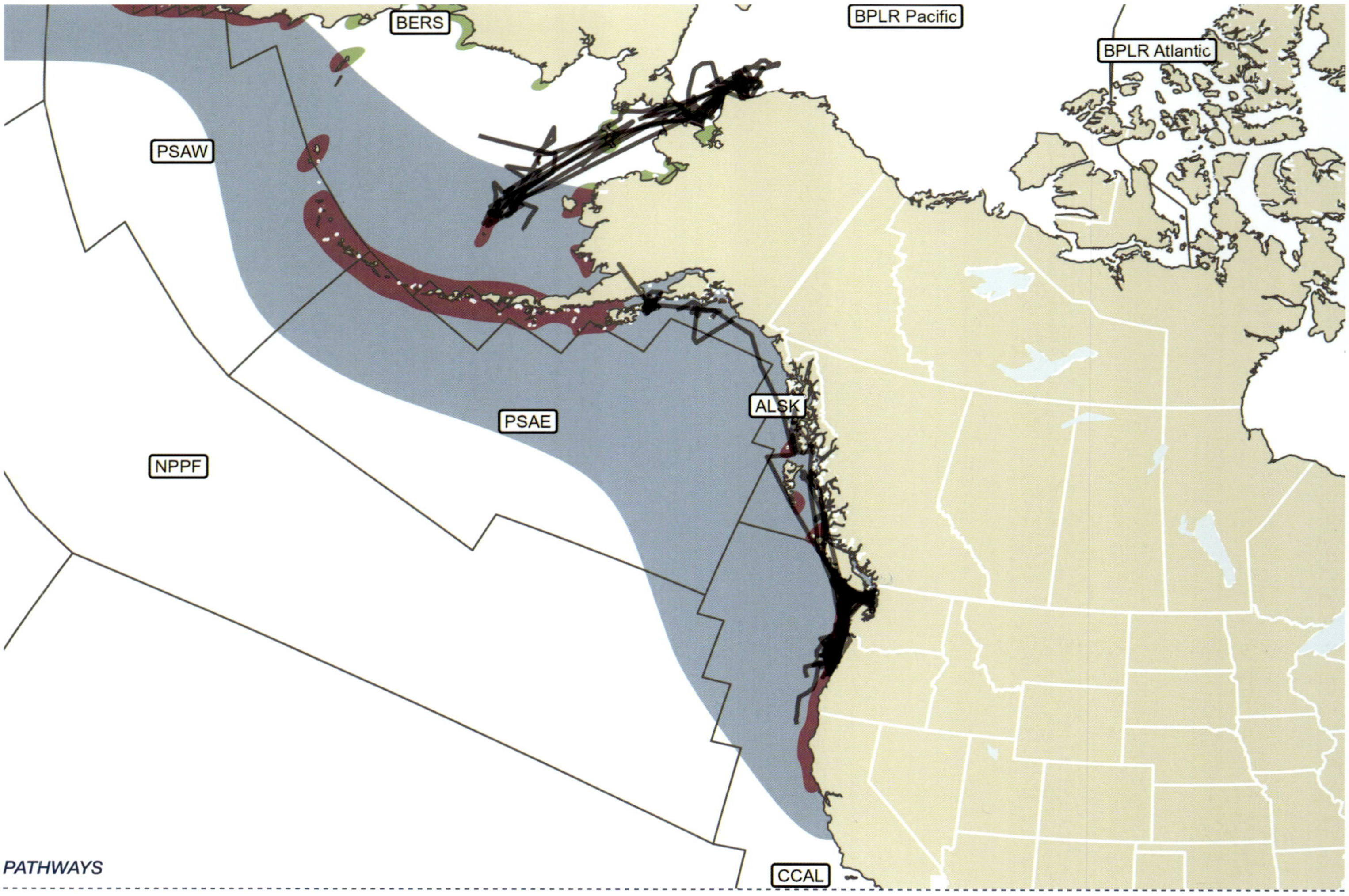

PATHWAYS

remained within a 100 km radius of the island. Birds tracked from more northern colonies (Cape Thompson and Cape Lisburne) made larger movements, from the Chukchi Sea to the Bering Sea near the Pribilof Islands. Birds tracked by satellite from Yaquina Head, Oregon, in 2015 (n = 24) largely migrated north to Washington and British Columbia, but a small proportion used habitats in Alaska (Loredo et al. 2019, 2020).

GENETICS Friesen et al. (1996) conducted mitochondrial DNA sequencing for 160 Common Murres from 10 colonies distributed throughout the Northern Hemisphere and found significant population genetic differentiation between Atlantic and Pacific populations. Within oceans, there is little population genetic structure within the Pacific, but there is significant east-west structuring among Atlantic colonies, as indicated by mitochondrial DNA, nuclear DNA, and intron loci (Morris-Pocock et al. 2008).

STABLE ISOTOPES Moody and Hobson (2007) evaluated interannual variability in foraging trophic level of birds spending the nonbreeding season off Newfoundland and Labrador, suggesting a high level of flexibility in relationship to oceanography and available habitat, which could be reflected in connectivity patterns.

MIGRATORY PATHWAYS Specific pathways are not well resolved from light-level geolocator data; however, satellite-tracked birds from colonies in the Chukchi Sea migrated southwest through the Bering Strait to spend the nonbreeding season near the Pribilof Islands in the Bering Sea (S.A. Hatch et al. 2000). Birds tracked from Oregon took northward coastal migration pathways (Loredo et al. 2019, 2020).

DATA CONTRIBUTORS

S.A. Hatch; A. Kitaysky; Y. Kolbeinsson; W.A. Montevecchi; R.A. Orben; H. Strøm; A. Takahashi; J.-B. Thiebot; T.L. Thorarinsson; A. Will.

INTERESTING NOTE

In North America, research is needed to understand the population-specific demography of Common Murres under different and shifting oceanographic conditions (Ainley et al. 2021b).

Mantel = 0.42 ± 0.03

Uria lomvia **THICK-BILLED MURRE**

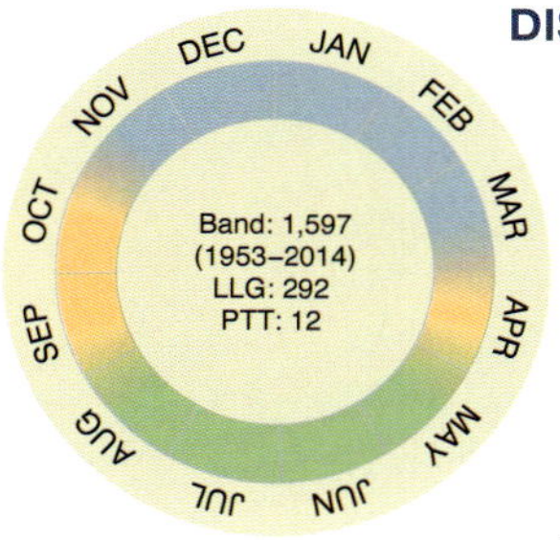

DISTRIBUTION AND ABUNDANCE

This species is one of the most abundant Arctic seabirds in the Northern Hemisphere. It nests in open sites on steep sea cliffs and in very large colonies, sometimes numbering over 1 million birds (Gaston and Hipfner 2020). During the breeding season, it distributes over the continental shelf of the Arctic Ocean and the connected seas on both the Pacific and Atlantic sides of North America, before shifting south for the nonbreeding season. As a generalist feeder, it can be found both over the continental shelf and the slope during nonbreeding, sometimes in dense concentrations. In the Atlantic Ocean, it is often seen within sight of land near Newfoundland, Canada, but in the northwest Pacific Ocean, it is often far from the shelf (Gaston and Hipfner 2020). Of the estimated 15 million global population, approximately 3 million breed in Canada and an uncertain estimate of 5 million breed in Alaska (Gaston and Hipfner 2020)

ORDER Charadriiformes
FAMILY Alcidae
BREEDING HABITAT Coasts, Cliffs and Islands
NONBREEDING HABITAT Coasts, Marine Waters
GLOBAL POPULATION SIZE 15,000,000
POPULATION CHANGE SINCE 1970 −3%

IUCN RED LIST CATEGORY Least Concern
R2R TIPPING POINT SPECIES No
US LISTING STATUS No
CANADA SPECIES AT RISK No
PIF CCS 9/20

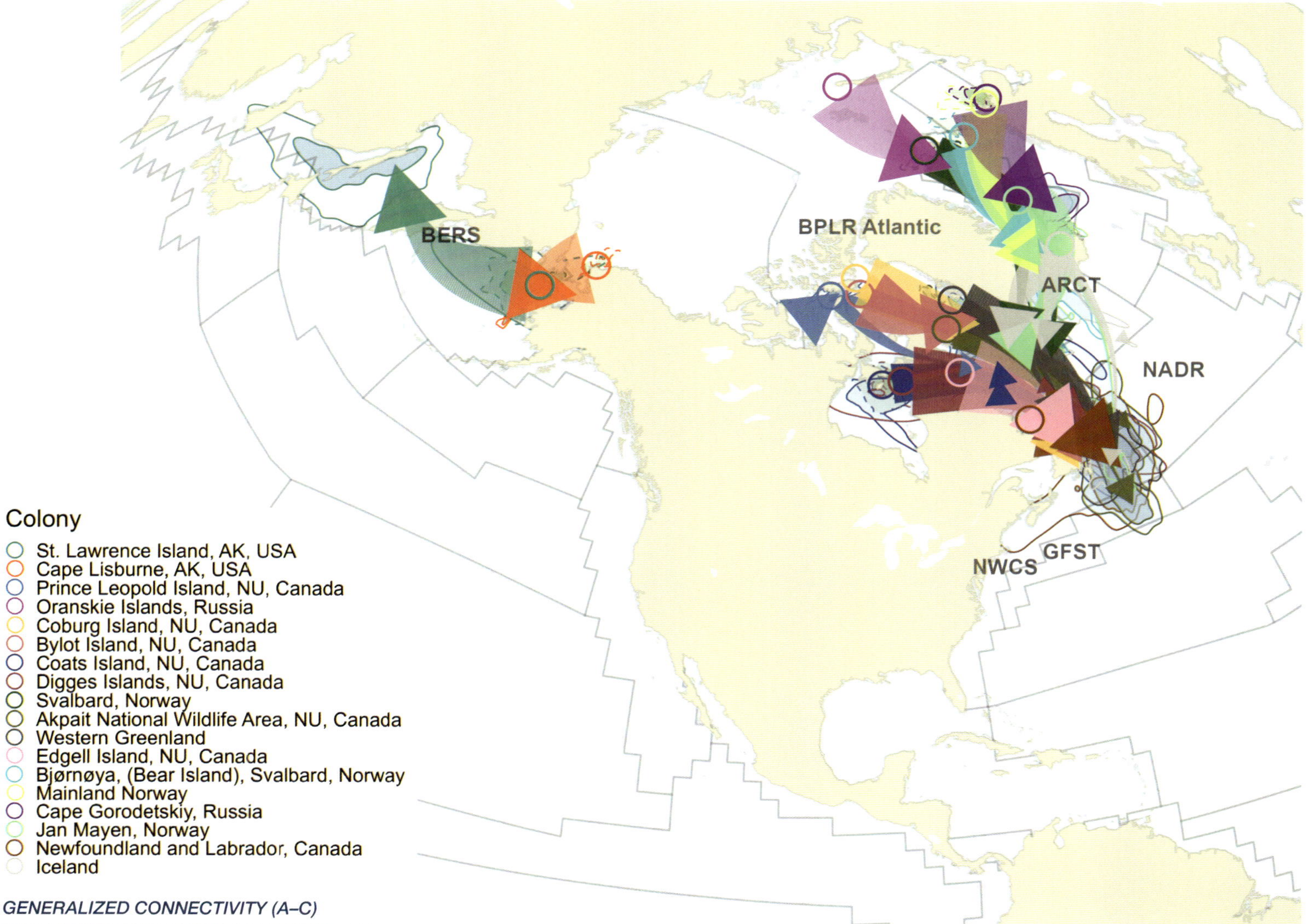

GENERALIZED CONNECTIVITY (A–C)

	BERS	GFST	NWCS	BPLR Atlantic	ARCT	NADR
St. Lawrence Island, AK, USA (n = 3)	100%	–	–	–	–	–
Cape Lisburne, AK, USA (n = 1)	100%	–	–	–	–	–
Prince Leopold Island, NU, Canada (n = 23)	–	–	9%	74%	17%	–
Oranskie Islands, Russia (n = 1)	–	–	–	–	100%	–
Coburg Island, NU, Canada (n = 6)	–	–	67%	33%	–	–
Bylot Island, NU, Canada (n = 94)	–	–	21%	79%	–	–
Coats Island, NU, Canada (n = 99)	–	–	49%	27%	–	–
Digges Islands, NU, Canada (n = 57)	–	–	89%	5%	–	–
Svalbard, Norway (n = 30)	–	–	–	53%	–	–
Akpait National Wildlife Area, NU, Canada (n = 14)	–	29%	43%	–	21%	7%
Western Greenland (n = 15)	–	–	60%	20%	20%	–
Edgell Island, NU, Canada (n = 3)	–	–	100%	–	–	–
Bjørnøya (Bear Island), Svalbard, Norway (n = 24)	–	–	–	33%	67%	–
Mainland Norway (n = 7)	–	–	–	57%	43%	–
Cape Gorodetskiy, Russia (n = 2)	–	–	–	–	100%	–
Jan Mayen, Norway (n = 61)	–	–	5%	66%	30%	–
Newfoundland and Labrador, Canada (n = 14)	–	–	100%	–	–	–
Iceland (n = 43)	–	–	9%	33%	56%	2%

MIGRATORY CONNECTIVITY

OVERALL Strong connectivity by ocean basin (Mantel = 0.42 ± 0.03) and weak to moderate connectivity within the Atlantic Ocean, shown with banding and tracking data. Much more is known about connectivity in the Atlantic Ocean than in the Pacific Ocean. Despite extensive population mixing in the Atlantic, there are some areas of concentration for specific populations; for example, greater proportions of birds from Iceland, Norway, and Russia were estimated to use the Atlantic Arctic Province during nonbreeding (30–100%) than birds from Canada and Greenland (0–23%).

BANDING This is a heavily banded species in Nunavut, Canada, starting in the 1950s (banding in Newfoundland began in the 1920s). Of the records providing breeding to nonbreeding connections, 97% of recoveries were from birds being shot

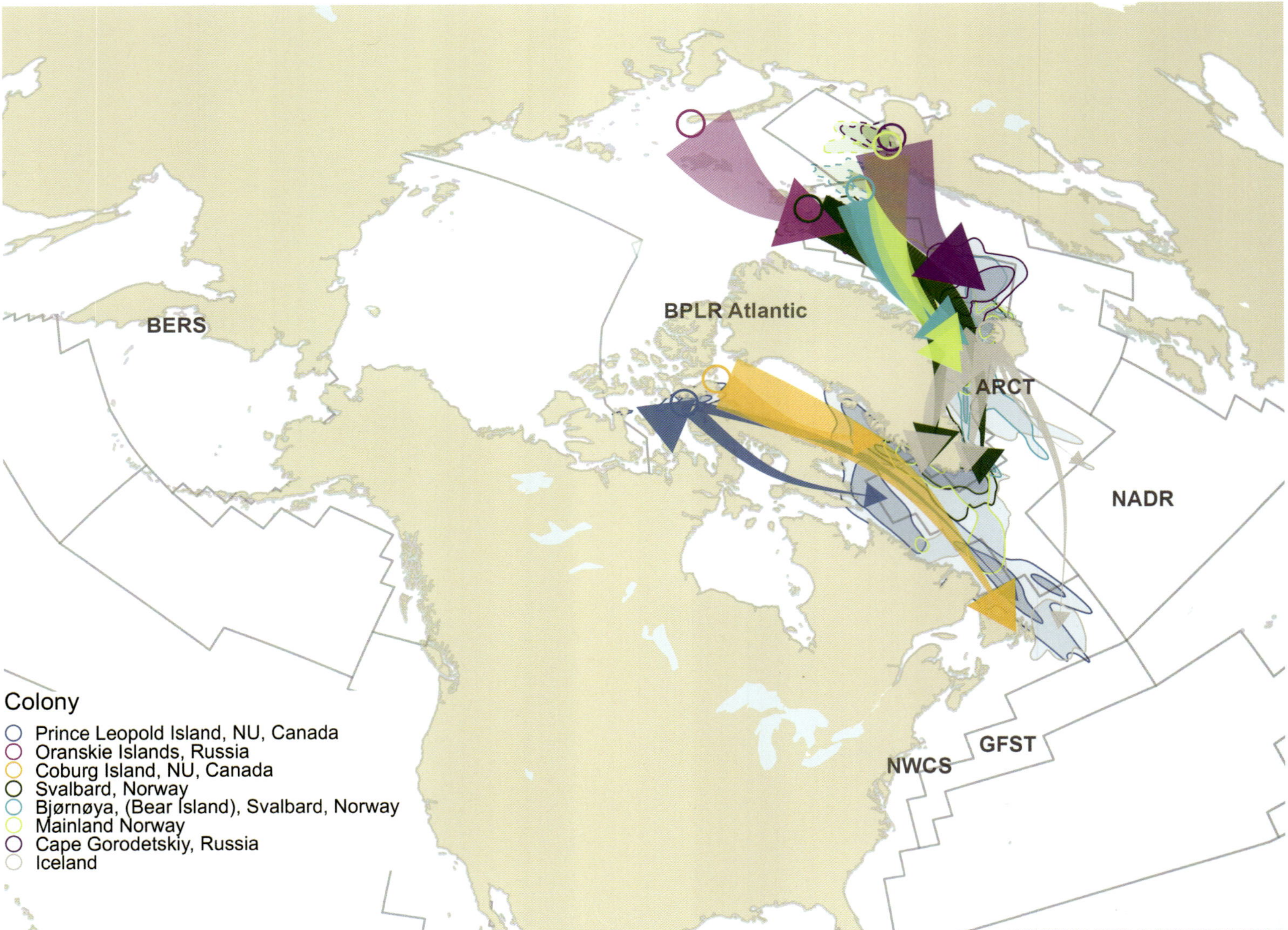

GENERALIZED CONNECTIVITY A

GENERALIZED CONNECTIVITY B

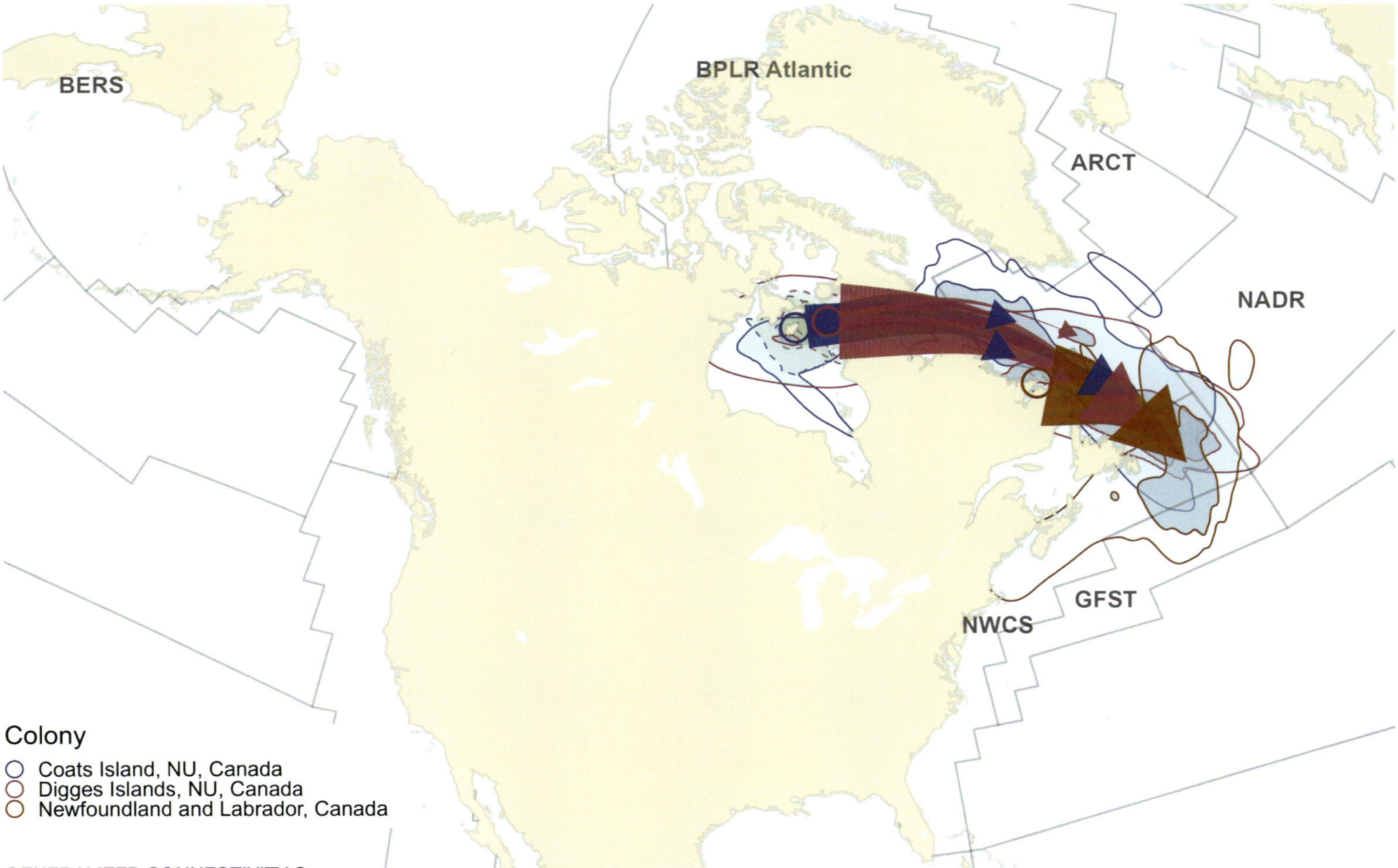

GENERALIZED CONNECTIVITY C

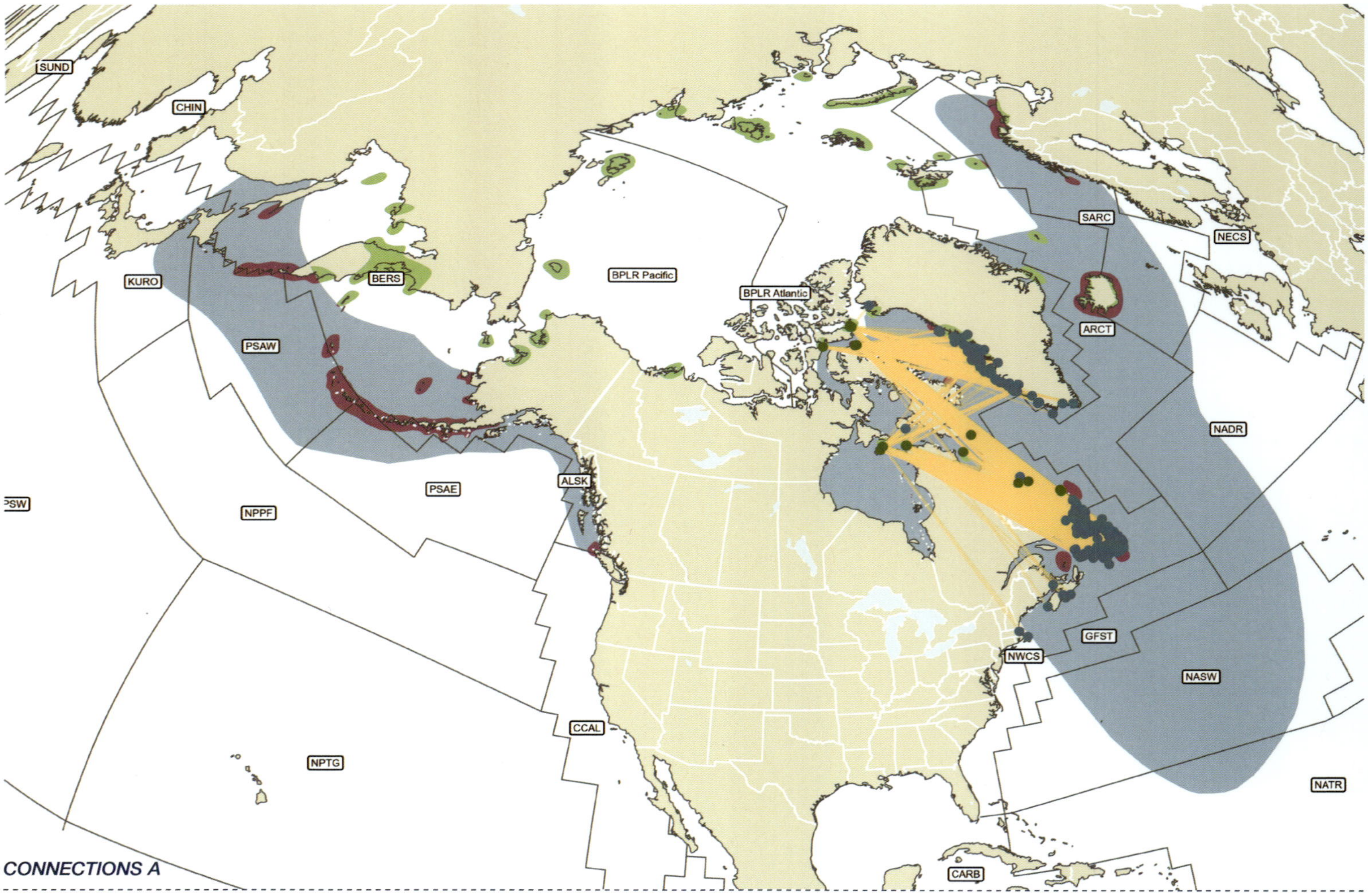

CONNECTIONS A

> **INTERESTING NOTE**
>
> **Time-depth recorders show that Thick-billed Murres are one of the deepest divers of all flighted birds. They can dive to 210 m and stay submerged for 3 minutes (Croll et al. 1992).**

(usually in Newfoundland and Labrador, Canada, or Greenland). Thus, connectivity patterns inferred from banding reflect the distribution of hunting as well as the distribution of birds. All connectivity information reported in this Atlas from Bylot Island, Canada, come from the 94 bands recovered during the post-breeding period. Other colonies with a high proportion of connectivity information from banding include Digges Island (82%) and Coats Island (41%).

GEOLOCATOR TRACKING Three birds tracked in 2016–2017 with geolocators deployed on St. Lawrence Island, Alaska, northern Bering Sea, migrated west to Japan (Takahashi et al. 2021). Orben et al. (2015b) found colony differences between nonbreeding habitats used by tracked birds from three Pribilof Island colonies in the Bering Sea. A large-scale geolocator tracking study spanning the Atlantic Ocean range showed main nonbreeding areas were at the interface of high and low Arctic conditions (Frederiksen et al. 2016). Overlap occurred with several populations congregating at distinct nonbreeding areas. Despite population mixing, there are some areas of concentration for specific populations that were found to be correlated with increasing or decreasing population status (Frederiksen et al. 2016). Tracking of birds from two Canadian colonies showed different strategies by colony, with one closely tracking ice cover and staying in the Hudson Bay for most of the year, and the other moving away from the colony regardless of ice levels (Gaston et al. 2011). Individuals tracked across years had generally consistent nonbreeding movements but also demonstrated flexibility in some years (Montevecchi et al. 2012; Frederiksen et al. 2016).

SATELLITE TRACKING In the Pacific Ocean, one bird was tracked using an implanted satellite transmitter, 1994–1996, from northwest Alaska in the Chukchi Sea to a nonbreeding area in the southeastern Bering Sea (S.A. Hatch et al. 2000).

STABLE ISOTOPES Only one study was found that related specifically to migratory connectivity. Orben et al. (2015b) tracked Thick-billed Murres with light-level geolocators / time-depth recorders from three island colonies in the Bering Sea, arranged north to south and across a gradient from continental shelf to deep-water habitats. Stable isotopes from feathers were used to group tracked individuals into three nonbreeding foraging strategies. The colonies maintained a large degree of spatial structure during the nonbreeding period, and this was associated with differences in body size (affecting diving ability) and diet. Birds from St. Paul Island, Alaska, had larger average body size, took deep daytime dives, and fed on higher trophic-level prey; 31% of tracked birds remained in the Bering Sea during the nonbreeding season. In contrast, smaller murres from St. George and Bogoslof islands left the Bering Sea from November through January and generally took shallower nocturnal dives. This study and others (e.g., Moody and Hobson 2007) also evaluated interannual variability in foraging trophic level during the post-breeding period, suggesting a high level of flexibility in relationship to oceanography and available habitat, which could be reflected in connectivity patterns.

GENETICS Despite some morphometric differences, colonies across the Atlantic population were found to be relatively uniform based on mitochondrial DNA analysis, but clear differentiation was found between Atlantic and Pacific populations (Birt-Friesen et al. 1992).

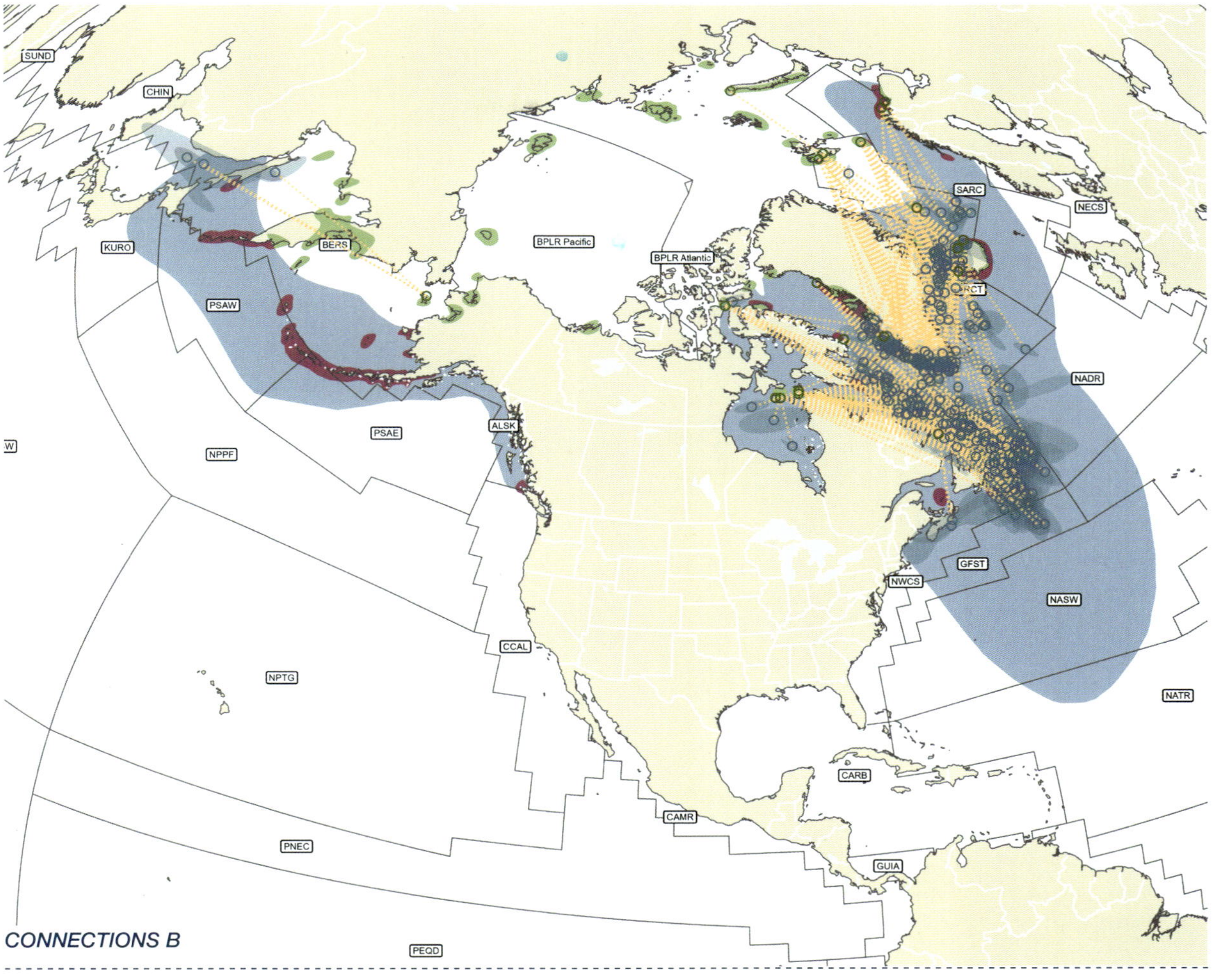

CONNECTIONS B

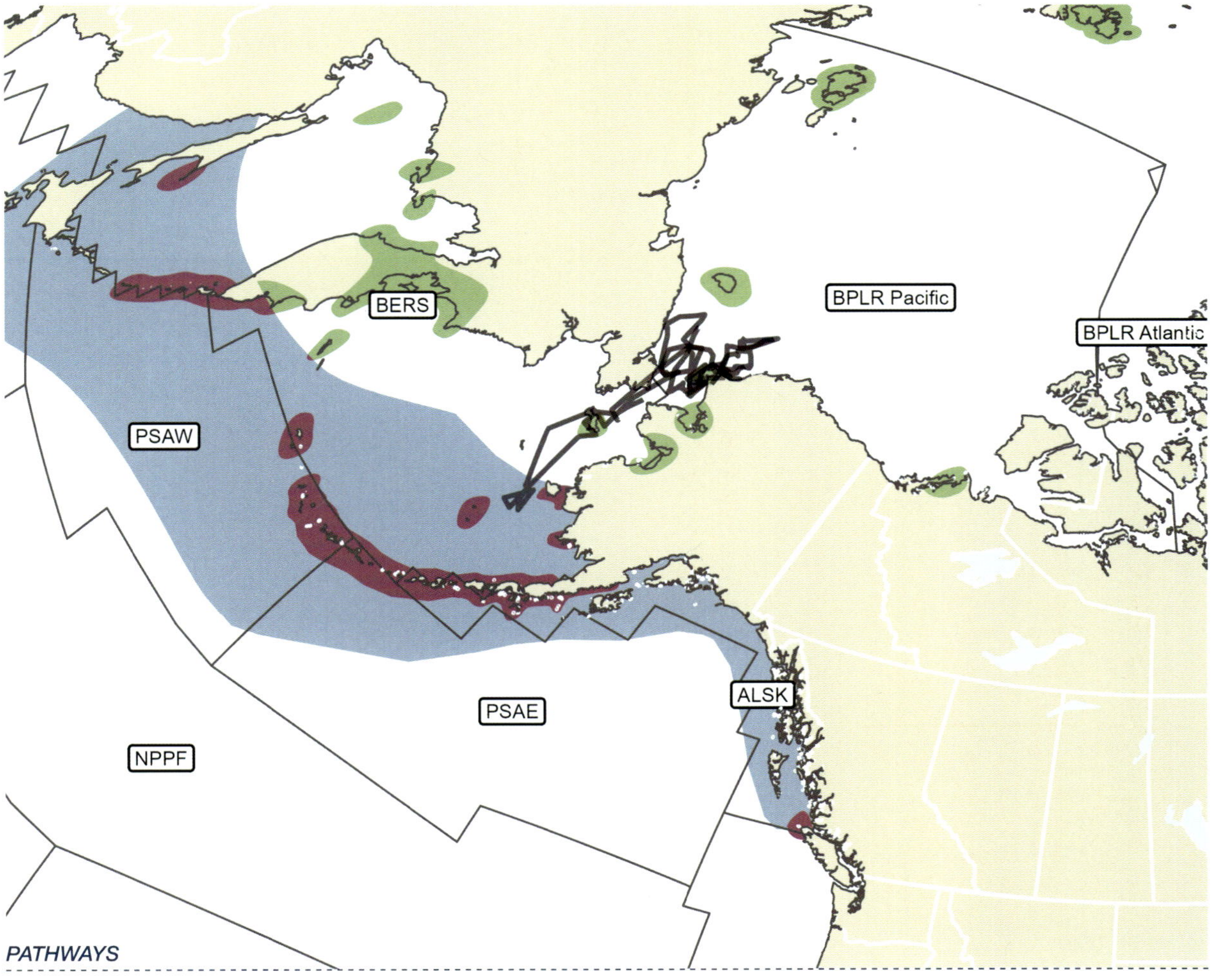

PATHWAYS

Birds from the Hudson Bay that stay there most of the year may also be genetically unique; a genetic analysis grouped them separately from other Canadian colonies, but overall found no genetic population structure (Tigano et al. 2017).

MIGRATORY PATHWAYS Specific pathways are not well resolved from light-level geolocator data at the scale of the relatively short-distance and ranging movements of Thick-billed Murres (i.e., see Montevecchi et al. 2012 in the Atlantic Ocean, and Takahashi et al. 2021 in the Pacific Ocean, where only core areas are shown). The one satellite-tracked individual from Alaska migrated from the Chukchi Sea through the Bering Strait and slightly southwest, staying relatively offshore, to its nonbreeding area.

DATA CONTRIBUTORS

D.C. Douglas; K.H. Elliott; K.E. Erikstad; M. Frederiksen; H.G. Gilchrist; S.A. Hatch; A. Kitaysky; Y. Kolbeinsson; A. Labansen; M.L. Mallory; F. Merkel; W.A. Montevecchi; T. Reiertsen; H. Strøm; A. Takahashi; J.-B. Thiebot; T.L. Thorarinsson; A. Will.

RAZORBILL *Alca torda*

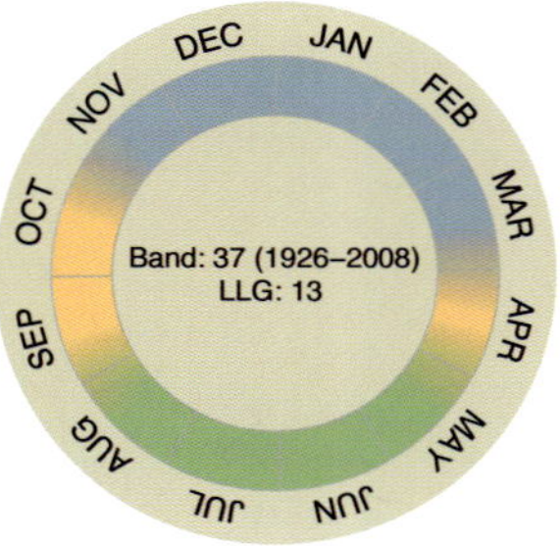

DISTRIBUTION AND ABUNDANCE

This species is a seabird of the North Atlantic Ocean, where it nests in the boreal and Arctic zones, from the United States to western Russia; 60%–70% of the global breeding population is on Iceland (Lavers et al. 2020). In North America, the largest colony is in Labrador, Canada, with colonies in Quebec, Greenland, and south along the North American coast to Maine, US (Lavers et al. 2020). It spends the nonbreeding season in coastal areas of the North Atlantic, with some overlap with breeding distribution but primarily in ice-free areas as far south as the subtropics (Lavers et al. 2020).

	GFST	NWCS
Kitsissut Avalliit, Greenland (n = 13)	8%	92%
Gannet Island, NL, Canada (n = 1)	–	100%
Gulf of St. Lawrence, NB, Canada (n = 6)	–	100%
Grand Manan, NB, Canada (n = 1)	–	100%

ORDER Charadriiformes
FAMILY Alcidae
BREEDING HABITAT Coasts, Cliffs and Islands
NONBREEDING HABITAT Coasts, Marine Waters
GLOBAL POPULATION SIZE 1,200,000
POPULATION CHANGE SINCE 1970 +248%

IUCN RED LIST CATEGORY Least Concern
R2R TIPPING POINT SPECIES No
US LISTING STATUS No
CANADA SPECIES AT RISK No
PIF CCS 10/20

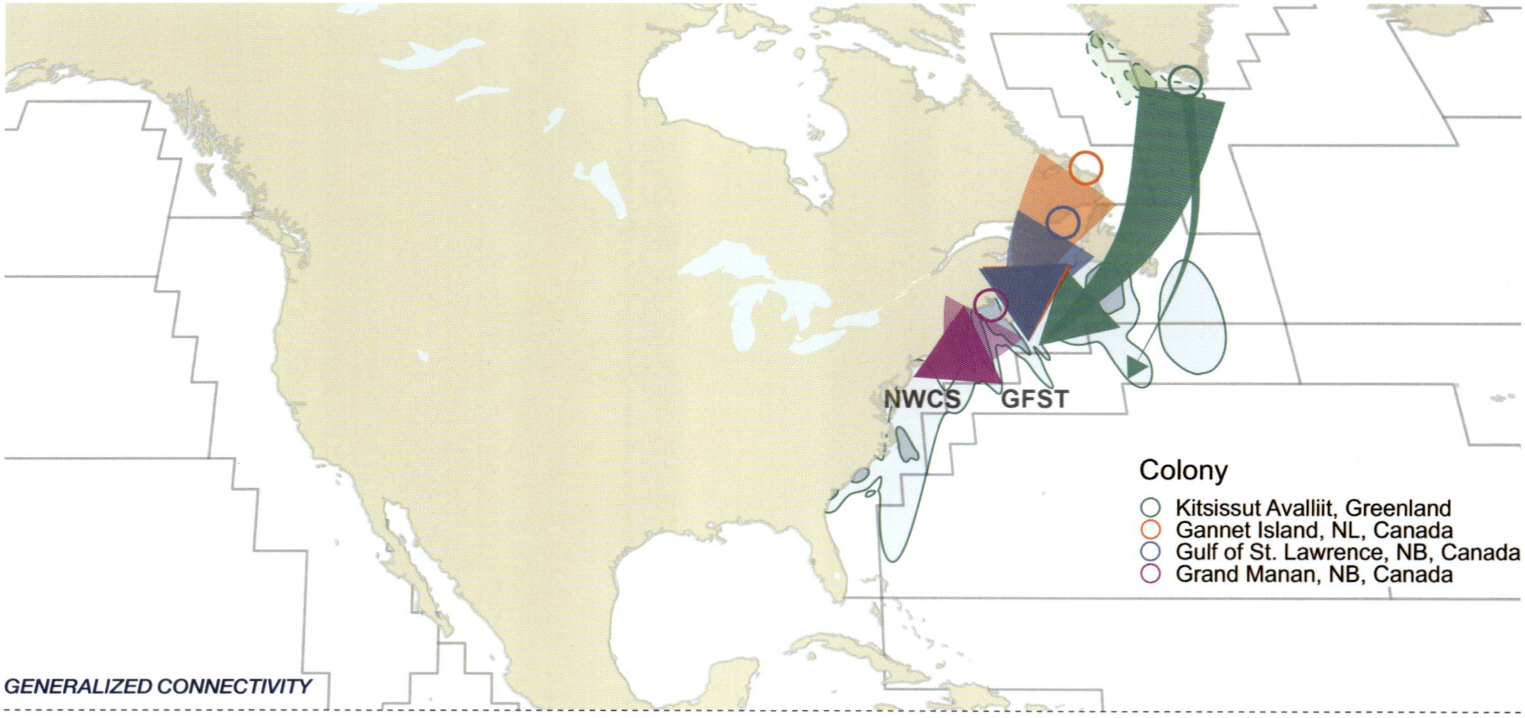

GENERALIZED CONNECTIVITY

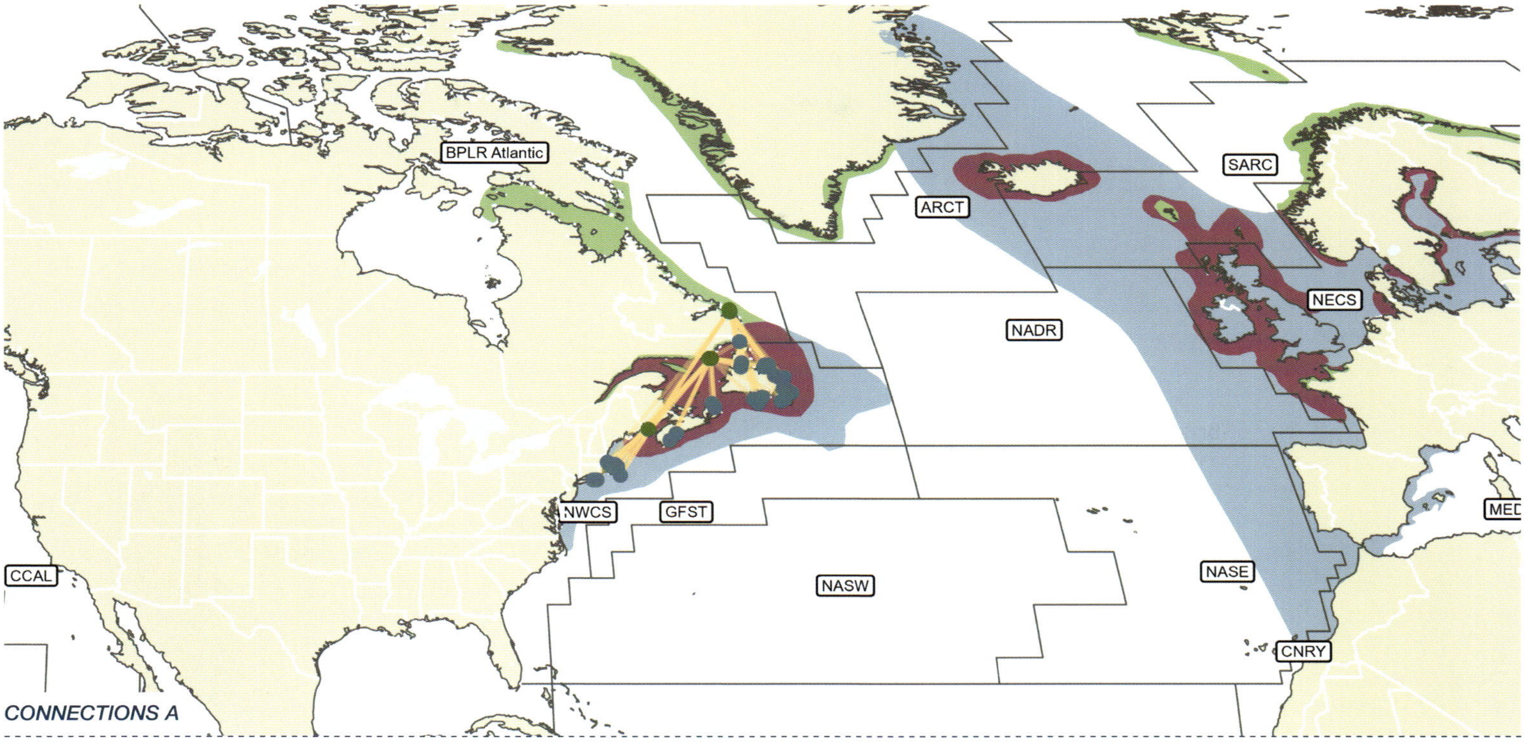

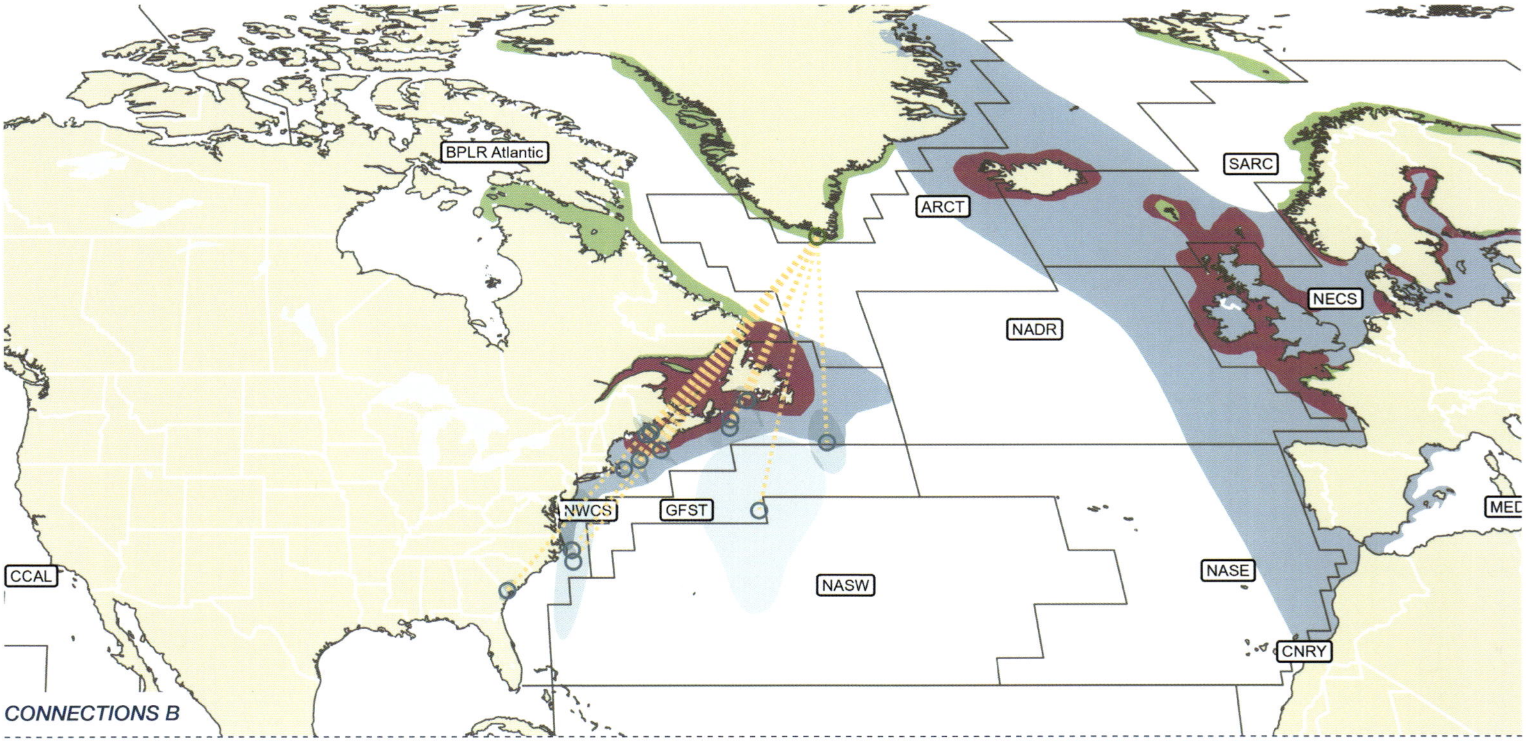

MIGRATORY CONNECTIVITY

OVERALL Although much is known about the migratory connectivity of eastern and central Atlantic breeding populations, until recently, little was known about North American populations. Limited sample size prevents the estimation of migratory connectivity, but it is likely weak to moderate due to lack of genetic differentiation among colonies and mixing on nonbreeding grounds. In total, 92%–100% of tracked core areas and band returns from three North American colonies (Gulf of St. Lawrence and Gannet Islands Ecological Reserve, Canada, and Kitsissut Avalliit, Greenland) were in the Northwest Atlantic Shelves province, but there is some differentiation of colony post-breeding areas suggested by a stable isotope and tracking study with data too recent to be included in this Atlas.

BANDING About half of band recoveries included here are from birds that were shot or entangled in fisheries. Thus, compared to tracking data, band returns are likely biased toward the coast and are biased toward locations where shooting and fishing results in Razorbills being encountered. Nonbreeding recoveries include locations primarily in the Canadian Maritime and New England, US, with one location as far south as waters off of Florida, US, in the Caribbean.

RADIO TRACKING Birds from three separate Canadian colonies were fitted with radio transmitters. Subsequent nonbreeding surveys found some members of each colony in the Bay of Fundy, with 83% of birds tagged at a nearby colony present, and 50% and 14% of tagged birds from farther colonies also detected (Clarke et al. 2010).

GEOLOCATOR TRACKING A study from Greenland tracked 13 birds with geolocators in 2009–2011 (Linnebjerg et al. 2013; J.F. Linnebjerg and M. Frederiksen, unpublished data). Nonbreeding areas spanned the northwest Atlantic, particularly the Gulf Stream and the Northwest Atlantic Shelves (Maftei et al. 2015). A recent master's thesis (Dodds 2024) also used geolocators to track birds in 2011–2019 from five colonies in Quebec and Atlantic Canada. While birds from multiple colonies migrated south to spend the nonbreeding season along the US coast, including New England south to the Carolinas (and similar to birds tracked from Greenland), nine individuals breeding at Machias Seal Island (on the border of the US and Canada between the Gulf of Maine and Bay of Fundy) remained resident to the Gulf of Maine area during the entire nonbreeding period. Additionally, two birds undertook a post-breeding migration north to the Gulf of St. Lawrence, Canada, before returning again to the Gulf of Maine for the remainder of the post-breeding periods, suggesting some colony differences that could lead to more moderate connectivity. A tracking study from Iceland showed that most birds also spent their nonbreeding season locally off the Icelandic coast (Linnebjerg et al. 2018), suggesting some unique differences between birds nesting on Greenland (Linnebjerg et al. 2013; J.F. Linnebjerg and M. Frederiksen, unpublished data) and Iceland, despite their adjacent waters.
SATELLITE TRACKING A study using abdominal implant Argos transmitters deployed on eight birds breeding at Machias Seal Island and tracked through the following nonbreeding season into the early post-breeding period showed that all birds remained within the Bay of Fundy, Canada (Clarke 2009).
STABLE ISOTOPES Most of the many stable isotope studies assess dietary niche at single colonies during single periods of the annual cycle or in comparison to sympatric species (for example, see Pratte et al. 2017 for a breeding season study in Labrador, Canada, of four auk species, including razorbill). However, a recent master's thesis (Dodds 2024) provides insights on North American migratory connectivity. Feathers from individuals tagged with global location sensors (GLS) at five colonies provided stable isotope values for nonbreeding areas identified by tracking. This information in turn allowed migratory connectivity predictions to be made for a larger sample from seven colonies of untagged individuals with feather samples only. The study suggested weak to moderate spatial connectivity, but there may be stronger temporal connectivity.
GENETICS Genetic evidence does not support either strong connectivity or a proposed classification of three subspecies (west Atlantic, Icelandic, Baltic) based on observed morphological and ecological differences (Salomonsen 1944). An analysis of mitochondrial DNA from 123 individuals representing colonies across their breeding range revealed only moderate genetic structure consistent with either relatively high gene flow among colonies and/or association on historical/evolutionary timescales (Moum and Árnason 2001).
MIGRATORY PATHWAYS After breeding, most individuals moved north along the West Greenland coast and then moved south to waters off Atlantic Canada and US (Linnebjerg et al. 2013; J.F. Linnebjerg and M. Frederiksen, unpublished data). Pathways were generally southwest from Greenland offshore of the Canadian and US southeast coast.

INTERESTING NOTE

It is suggested that Razorbills are moving farther north into the Canadian high Arctic as the ice season becomes reduced due to climate change (Gaston and Woo 2008). Short-term GPS tracking and habitat modeling generally suggest local foraging distribution during the breeding season across multiple colonies (Symons and Diamond 2018; Delord et al. 2020; Lavoie et al. 2022; Ronconi et al. 2022) and thus likely strong connectivity during this time.

DATA CONTRIBUTORS
M. Frederiksen; J.F. Linnebjerg.

BLACK GUILLEMOT *Cepphus grylle*

DISTRIBUTION AND ABUNDANCE

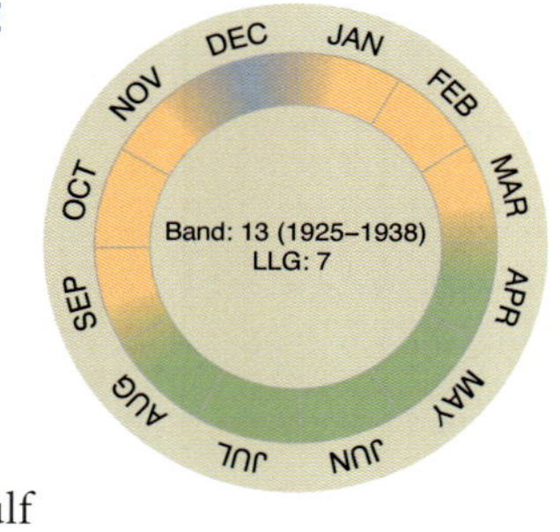

A medium-sized alcid with a nearly circumpolar distribution from the high Arctic south to temperate areas (in North America, south to Maine, US). Breeding colonies are scattered, but larger colonies are generally found farther north, with the majority breeding pairs in North America to be north of the Gulf of St. Lawrence in Canada and along the west coast of Greenland (Butler et al. 2020). Associated within patches of open water and at ice edge, Black Guillemots forage near-shore year-round, often staying close to breeding colonies (Butler et al. 2020).

MIGRATORY CONNECTIVITY

OVERALL Migratory connectivity appears to be weak to moderate within ocean basins, and strong across basins, based on limited banding, geolocator data, and genetic inference. The Mantel correlation coefficient could not be calculated due to limited sample size. Data contributed to this Atlas encompass only one region (Atlantic Canada) and from a single presumed subspecies (*C. g. arcticus*). Birds from these colonies (Gulf of St. Lawrence,

ORDER Charadriiformes
FAMILY Alcidae
BREEDING HABITAT Coasts, Cliffs and Islands
NONBREEDING HABITAT Coasts, Marine Waters
GLOBAL POPULATION SIZE 750,000
POPULATION CHANGE SINCE 1970 +14%
IUCN RED LIST CATEGORY Least Concern
R2R TIPPING POINT SPECIES No
US LISTING STATUS No
CANADA SPECIES AT RISK No
PIF CCS 9/20

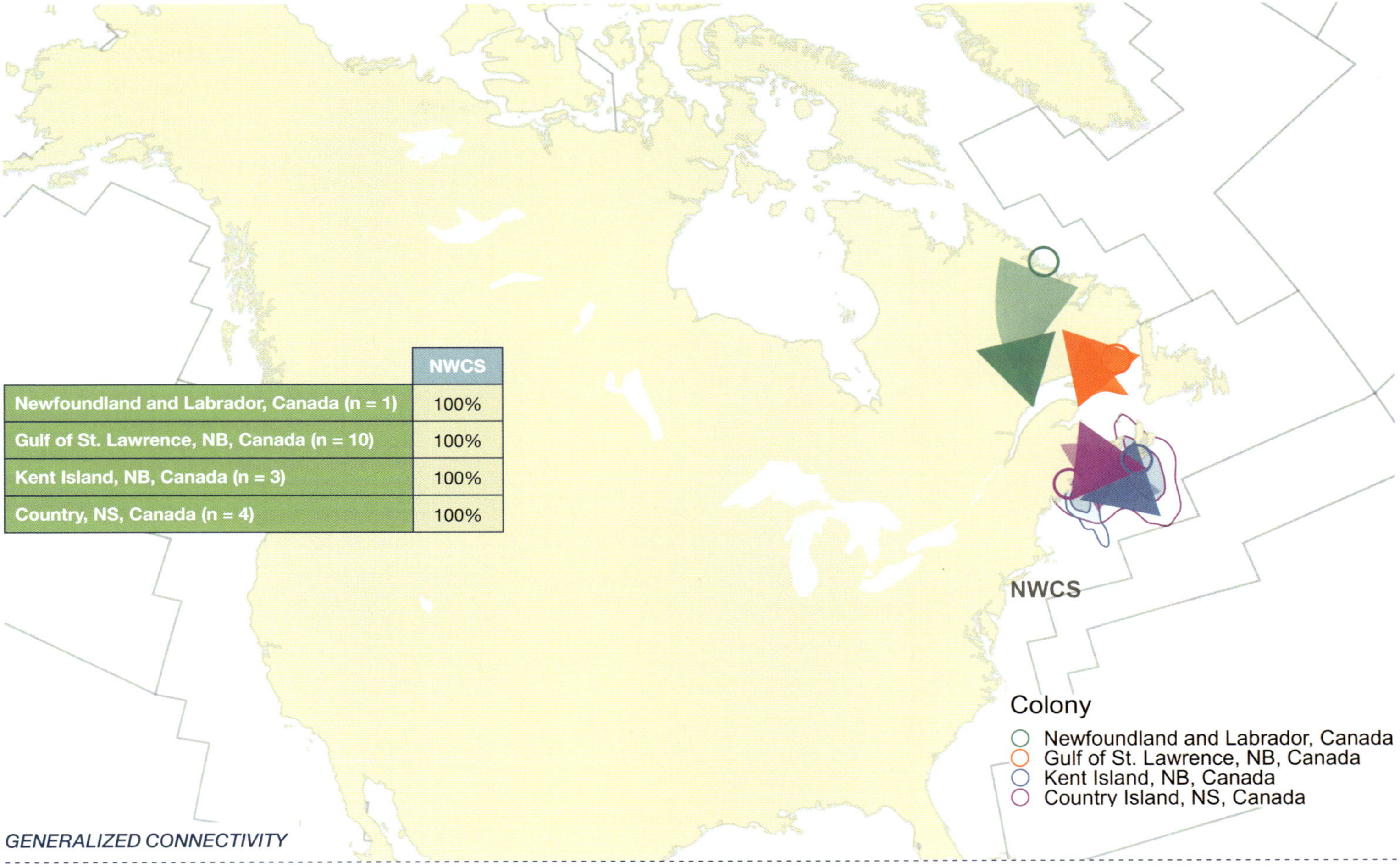

	NWCS
Newfoundland and Labrador, Canada (n = 1)	100%
Gulf of St. Lawrence, NB, Canada (n = 10)	100%
Kent Island, NB, Canada (n = 3)	100%
Country, NS, Canada (n = 4)	100%

GENERALIZED CONNECTIVITY

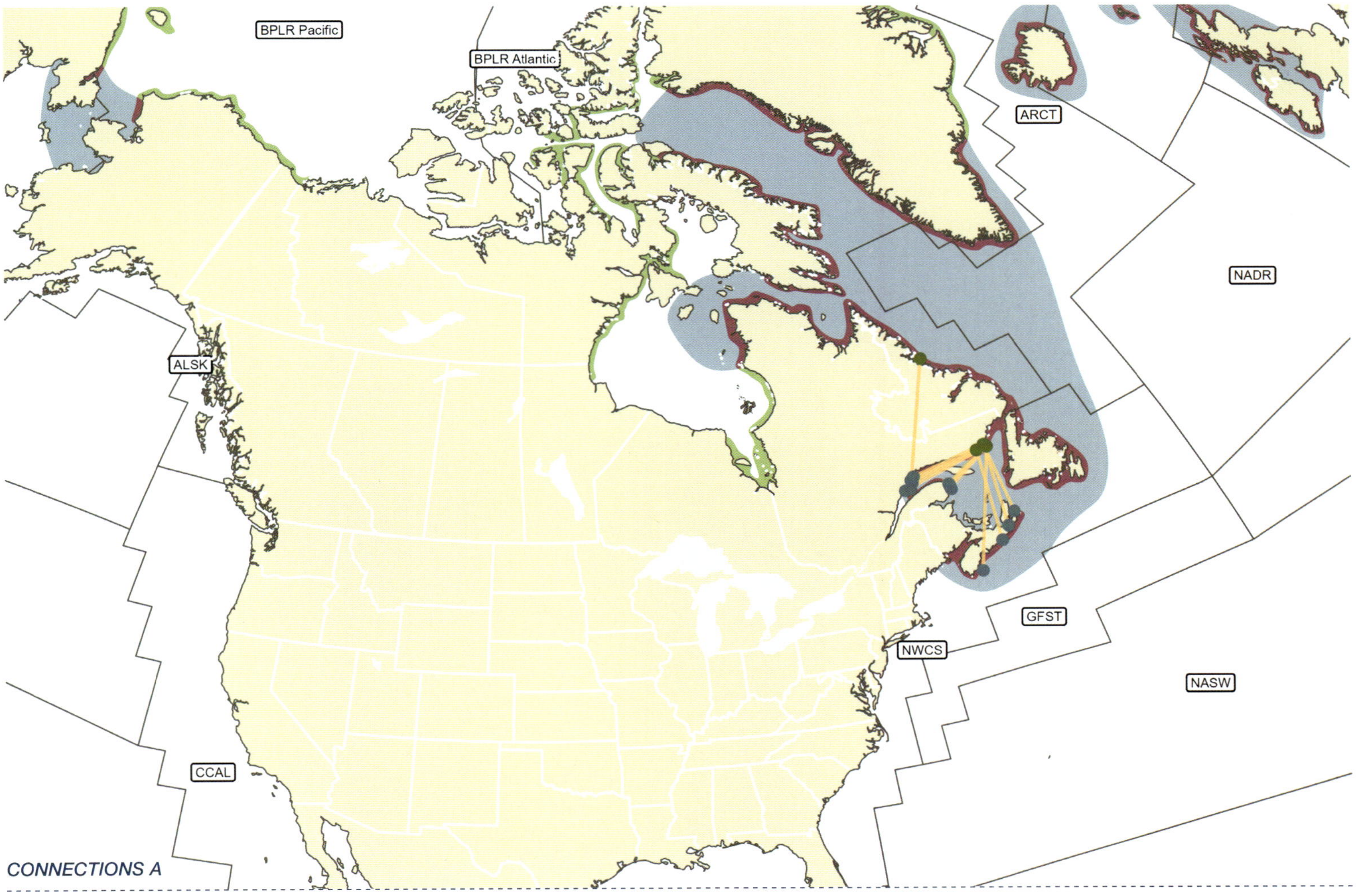

CONNECTIONS A

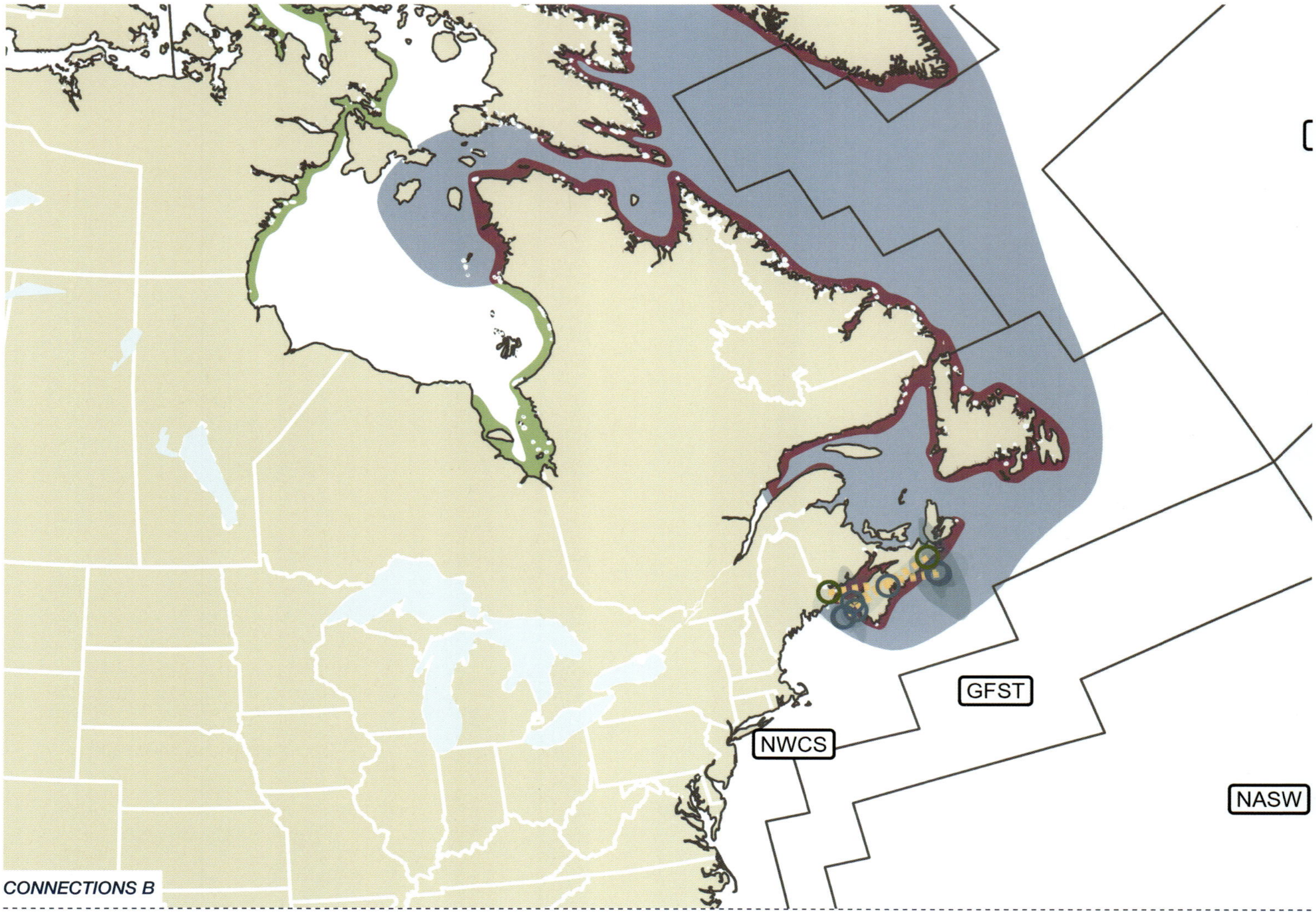

CONNECTIONS B

Newfoundland and Labrador, Country Island, and Kent Island) all shared similar post-breeding habitats within the Northwest Atlantic Shelves province. There may be latitudinal differentiation, from temperate to high Arctic populations, due to birds remaining close to breeding areas during the nonbreeding season when ice conditions permit. Genetics and published tracking data from Alaska, US, also indicate separation from Canadian birds.

BANDING About half of band recoveries during the post-breeding period were due to birds being shot. Recoveries support a pattern of colony mixing within the Atlantic Canada and US New England region.

GEOLOCATOR TRACKING Geolocator tracking of seven birds in 2017–2018 from Atlantic Canada (New Brunswick in the Bay of Fundy and Nova Scotia on the Atlantic Ocean coast) showed mixing of the two colonies during the nonbreeding season (Baak et al. 2021a). In Alaska, 45 individuals were also tracked with geolocators in 2011–2015. This Arctic population first moved north into the Beaufort Sea, followed by a retreat into the Chukchi Sea and south to the Bering Sea (Divoky et al. 2016).

GENETICS Kidd and Friesen (1998) conducted the first genetic assessment of Black Guillemots globally, including three North American colonies. Glacial isolation of populations on evolutionary timescales resulted in morphological differences leading to six subspecies. From a mtDNA analysis, the authors concluded that guillemots in Maine, US, and Finland remained genetically isolated, but contemporary exchange was thought to occur between populations in the Chukchi Sea of Alaska, Nunavut, and Iceland and Norway. This study had limitations, and a more recent study using multiple markers with samples from nine North American colonies found regional population genetic structure that was correlated with geographic separation (Harkness 2017). Populations that could be considered management units include Alaska (*C. g. mandtii*), Nunavut (*C. g. ultimus*), and eastern colonies in Canada and the US (*C. g. arcticus*). Within the latter subpopulation (*C. g. arcticus)*, no genetic structure was found in birds from New Brunswick, Canada, and birds from Maine, US. Therefore, genetics could be used to determine breeding origin from nonbreeding birds and thus migratory connectivity.

MIGRATORY PATHWAYS Birds from Atlantic Canada underwent short-distance migrations that rarely extended beyond the continental shelf (Baak et al. 2021a). In Alaska, birds closely followed areas of 30%–60% sea ice cover year-round rather than direct pathways between breeding and nonbreeding areas (Divoky et al. 2016).

DATA CONTRIBUTORS

J.E. Baak; R.A. Ronconi.

INTERESTING NOTE

Baak et al. (2021a) also used stable isotope methods on Black Guillemots, and although the two Atlantic Canada colonies overlapped during the nonbreeding season, birds from Nova Scotia foraged at higher trophic levels than birds from New Brunswick. This indicates some level of niche partitioning, providing interesting avenues for future studies.

Cepphus columba **PIGEON GUILLEMOT**

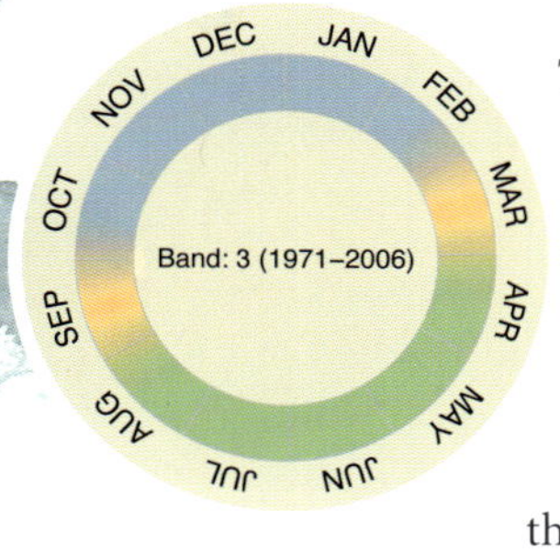

DISTRIBUTION AND ABUNDANCE

This medium-sized alcid of the North Pacific inhabits areas closer to shore compared to most members of its family, feeding on bottom-dwelling organisms along rocky coasts (Ewins 2020). In North America, most breeding locations are south of the Bering Strait to Southern California (Ewins 2020), with the largest colonies on the Farallon Islands, California, US, and in the Chukotka Peninsula in the northwestern Pacific (Ewins 2020). Unlike many seabirds, its nonbreeding range is smaller than its breeding range (Ewins 2020).

ORDER Charadriiformes
FAMILY Alcidae
BREEDING HABITAT Coasts, Cliffs and Islands
NONBREEDING HABITAT Coasts, Marine Waters
GLOBAL POPULATION SIZE 360,000
POPULATION CHANGE SINCE 1970 +43%
IUCN RED LIST CATEGORY Least Concern
R2R TIPPING POINT SPECIES No
US LISTING STATUS No
CANADA SPECIES AT RISK No
PIF CCS 10/20

MIGRATORY CONNECTIVITY

OVERALL There is little information, but migratory connectivity is probably weak.
BANDING Only three band recoveries (1971–2006)—all from the Farallon Islands off California to Oregon—connect adult breeding to nonbreeding locations. Recoveries of immature birds also show dispersal as far north as British Columbia in their first winter (Ainley and Boekelheide 1990, as cited in Ewins 2020).
GEOLOCATOR TRACKING A recent three-year study (Johns and Warzybok 2022) tracked 58 Pigeon Guillemots from Southeast Farallon Island, California, to nonbreeding areas in coastal British Columbia and southeast Alaska.

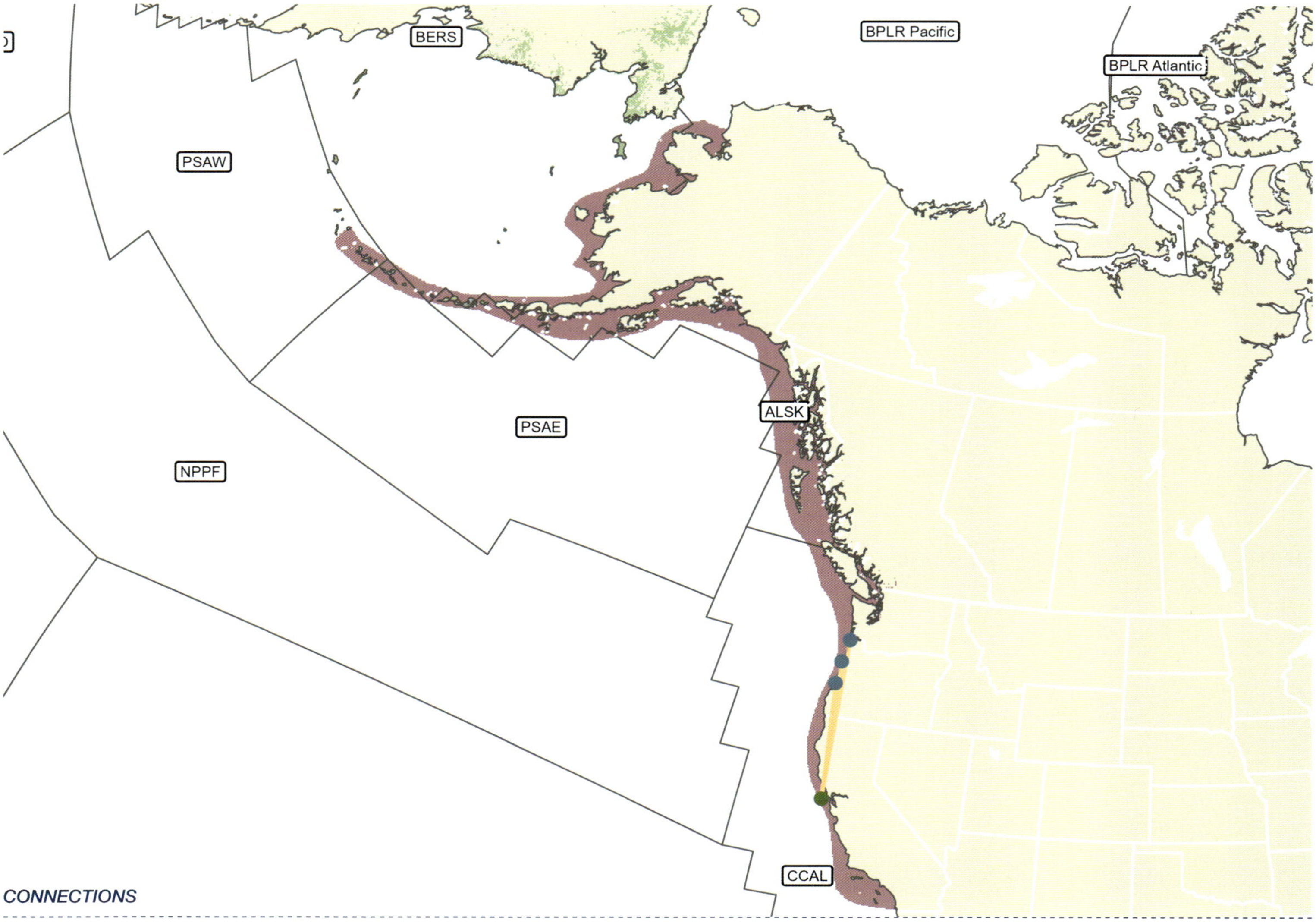

CONNECTIONS

GENETICS Harkness et al. 2024 describe the relatively strong population differentiation in Pigeon Guillemots inferred from mitochondrial and nuclear DNA compared to other high-latitude North American seabirds. Overlapping nonbreeding distributions are thought to be consistent with weak genetic structure in seabirds—Pigeon Guillemots populations likely do have moderate to high nonbreeding overlap.

OTHER APPROACHES No stable isotope studies of migratory connectivity were found for this species. Along the west coast of North America, geographic patterns exist in the length of an individual's wing and culmen (upper ridge of the beak), leading to four subspecies of Pigeon Guillemot recognized in North America (Storer 1950). Wing and tarsal (lower leg) lengths decrease from California to Washington and British Columbia, and increase from there to northern Alaska. Culmen length decreases south to north.

MIGRATORY PATHWAYS At-sea surveys provided the first information on migration. For example, surveys from Washington detected a large increase in numbers of this species during the nonbreeding season, when birds are absent from California (Wahl and Tweit 2000). In the first tracking study to confirm the migratory patterns of California Pigeon Guillemots inferred from at-sea surveys, birds tracked from California breeding colonies made a northward coastal migration in the fall, engaging in stopovers near Haida Gwaii, Canada, likely associated with molt, before continuing on to nonbreeding sites. The coastal southward migration occurred without stops (Johns and Warzybok 2022).

INTERESTING NOTE

Incomplete survey methods and fluctuations due to high mortality events in El Niño years make population numbers difficult to assess. Additional tracking would be helpful at the northern limits of the range in Alaska, as well as in British Columbia and Southern California, to obtain a complete picture of migratory connectivity for this species.

MARBLED MURRELET *Brachyramphus marmoratus*

DISTRIBUTION AND ABUNDANCE

This species is a secretive alcid that nests in trees in older forests (and also on the ground in rocky areas) from the Aleutian Islands south to central California. Marbled Murrelets may undertake both small-scale and longer-scale movements post-breeding, but also may remain near nesting areas or undertake seasonal distribution shifts (Nelson 2020). Population size is uncertain, but Alaska is thought to have 67%–90% of the breeding population, British Columbia about 6%–20% of the population, and Washington, Oregon, and California, each <10,000, although some estimates for Oregon are 20,000 (Nelson 2020). Highest post-breeding concentrations are in waters of Alaska, British Columbia, and Washington (Nelson 2020).

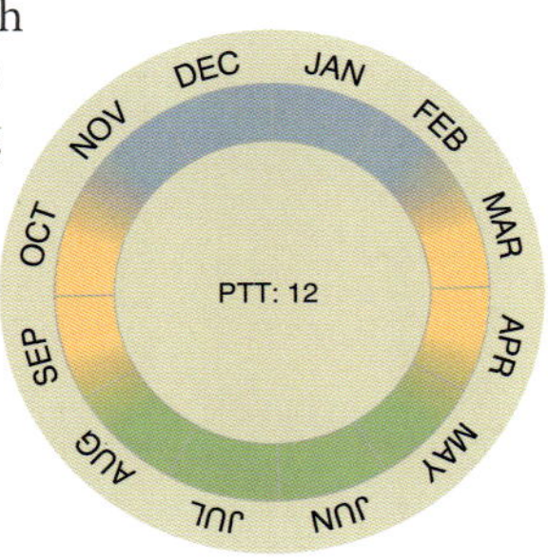

ORDER Charadriiformes
FAMILY Alcidae
BREEDING HABITAT Forests, Temperate Western
NONBREEDING HABITAT Coasts, Marine Waters
GLOBAL POPULATION SIZE 360,000
POPULATION CHANGE SINCE 1970 −15%
IUCN RED LIST CATEGORY Endangered
R2R TIPPING POINT SPECIES Yellow-Alert
US LISTING STATUS Threatened Pacific Northwest
CANADA SPECIES AT RISK Threatened
PIF CCS 15/20

MIGRATORY CONNECTIVITY

OVERALL Little is known about the migratory connectivity of Marbled Murrelets due to limited tracking across the range, limited band resightings, and the variety of migratory behaviors murrelets employ. Limited satellite, banding, and genetic evidence support connectivity between the eastern Aleutian Islands to British Columbia and Washington (Alaska Coastal Downwelling Province and northern California Current), while genetic and radio-tracking evidence suggest that murrelets from central California (a disjunct breeding area) may be genetically distinct and have limited nonbreeding overlap in the central California Current with other populations.

BANDING A breeding Marbled Murrelet banded in 1995 in British Columbia, Canada, was recaptured in fall 1996 in basic plumage in the San Juan Islands, Washington, US, providing the first evidence for long-distance movements in the species (Beauchamp et al. 1999). Seven other color-marked individuals from the same population were

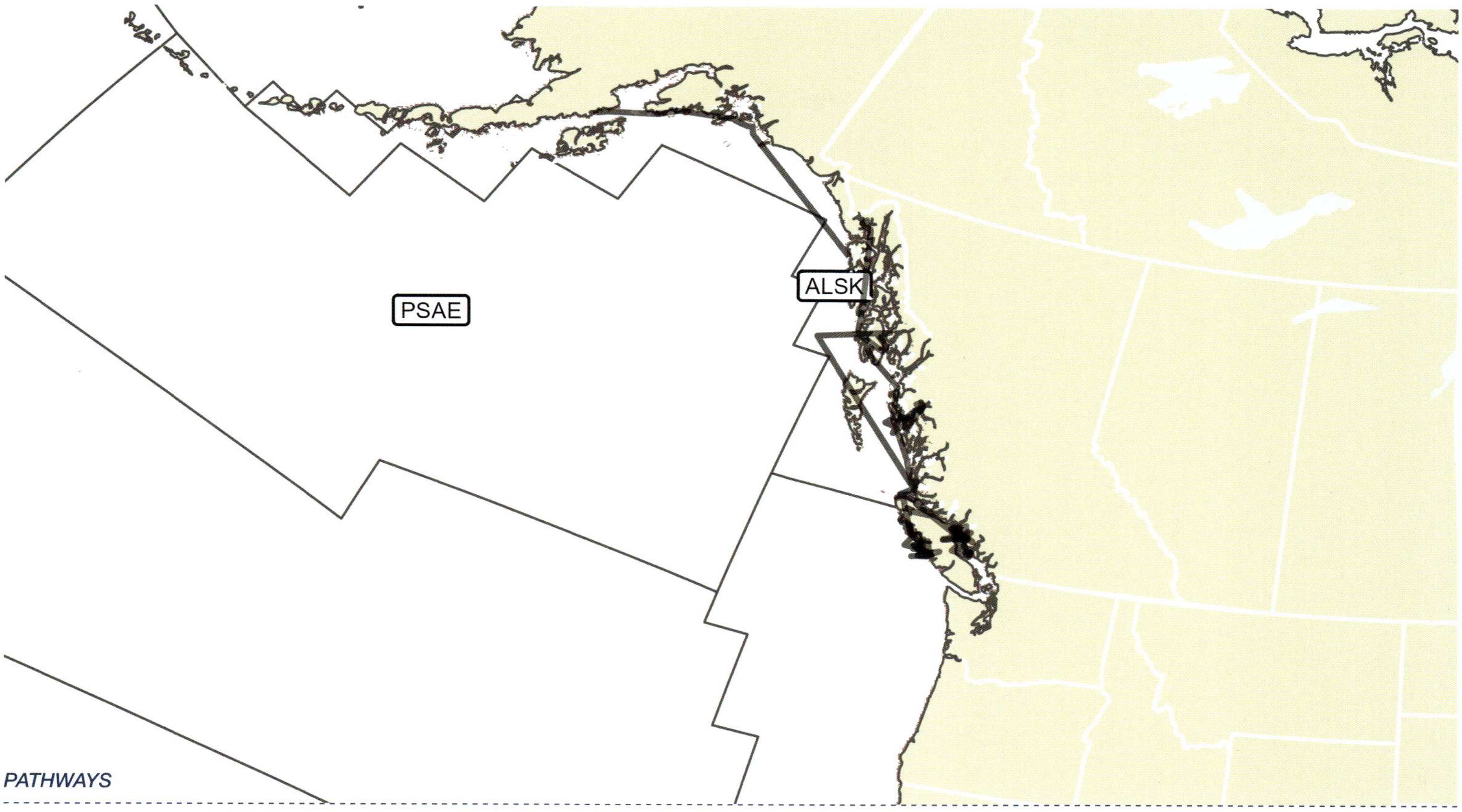

PATHWAYS

instead seen near their breeding sites during the nonbreeding period, suggesting partial migration (Beauchamp et al. 1999).

RADIO TRACKING Similar to the satellite-tracked birds in the same study, one bird tracked with a geolocator from a colony in the Canadian high Arctic spent the nonbreeding season within Canada near the northern Labrador Sea (Maftei et al. 2015).

SATELLITE TRACKING Satellite tracking data shows connectivity between British Columbia and Alaska Marbled Murrelet populations (Bertram et al. 2016, 2023). Five tagged birds moved from British Columbia to Alaska during breeding or post-breeding and originated from three different Marbled Murrelet conservation regions of British Columbia (Bertram et al. 2016, 2023).

GENETICS An analysis of mitochondrial DNA revealed little genetic divergence or morphometric distinction between ground-nesting and tree-nesting populations of Marbled Murrelets in Alaska (Pitocchelli et al. 1995). The most definitive study of genetic structure (Piatt et al. 2007) used multiple lines of evidence, including DNA samples from 282 individuals collected in 1989–2005 from the entire breeding range. The study suggested that murrelets comprise three genetic units with very little contemporary gene flow among the peripheral western and southern populations: western and central Aleutian Islands; eastern Aluetian Islands to Northern California; and central California.

OTHER APPROACHES Many stable isotope studies of both juvenile and adult Marbled Murrelets exist; however, no studies were found comparing stable isotopes across the breeding range to understand migratory connectivity. Radio-tracking studies of Marbled Murrelets focused primarily on breeding movements and nest-site identification (e.g., Whitworth et al. 2000; Lorenz et al. 2017).

MIGRATORY PATHWAYS The species exhibits both short and long migratory movements. Short-distance movements consist of shifts from outer coastal to protected waters, while long-distance movements are generally north-south along coastal areas (Nelson 2020).

INTERESTING NOTE

An Oregon study used small satellite tags to track murrelets during the breeding season. Of interest, however, was one individual that traveled over 300 km in a few days, relatively longer-distance movements during the breeding season than previously thought (Northrup et al. 2018).

DATA CONTRIBUTORS

D.F. Bertam; Environment and Climate Change Canada, Science and Technology Branch, Wildlife Research Division.

KITTLITZ'S MURRELET *Brachyramphus brevirostris*

DISTRIBUTION AND ABUNDANCE

This species is an uncommon and cryptic Beringian alcid associated with glaciated areas. Most of the North American breeding population is dispersed and clustered in Alaska, US, with additional breeding in Yukon Territory and possibly in British Columbia, Canada (Day et al. 2020a). It also breeds in the Russian Far East. It is unusual among seabirds in that it is a ground-nester on or near mountaintops (Day et al. 2020a). It is coastally distributed during nonbreeding, undertaking relatively short migrations.

ORDER Charadriiformes
FAMILY Alcidae
BREEDING HABITAT Coasts, Cliffs and Islands
NONBREEDING HABITAT Coasts, Marine Waters
GLOBAL POPULATION SIZE 44,000
POPULATION CHANGE SINCE 1970 NA
IUCN RED LIST CATEGORY Near Threatened
R2R TIPPING POINT SPECIES Red-Alert
US LISTING STATUS No
CANADA SPECIES AT RISK No
PIF CCS 17/20

MIGRATORY CONNECTIVITY

OVERALL Due to small sample size, migratory connectivity is poorly known but is thought to be weak among some Alaskan colonies. Birds tracked from Icy Bay and Prince William Sound, Alaska, generally overlapped on nonbreeding grounds in the Alaska Coastal Downwelling Province (50%–75%) and Bering Sea (25%) provinces. However, two birds (25%) from Prince William Sound also used habitats in the Boreal Polar province. A single bird from Atka, Aleutian Islands, spent the nonbreeding season in the Bering

	BERS	BPLR Pacific	ALSK
Atka, AK, USA (n = 1)	100%	–	–
Prince William Sound, AK, USA (n = 8)	25%	25%	50%
Icy Bay, AK, USA (n = 8)	25%	–	75%

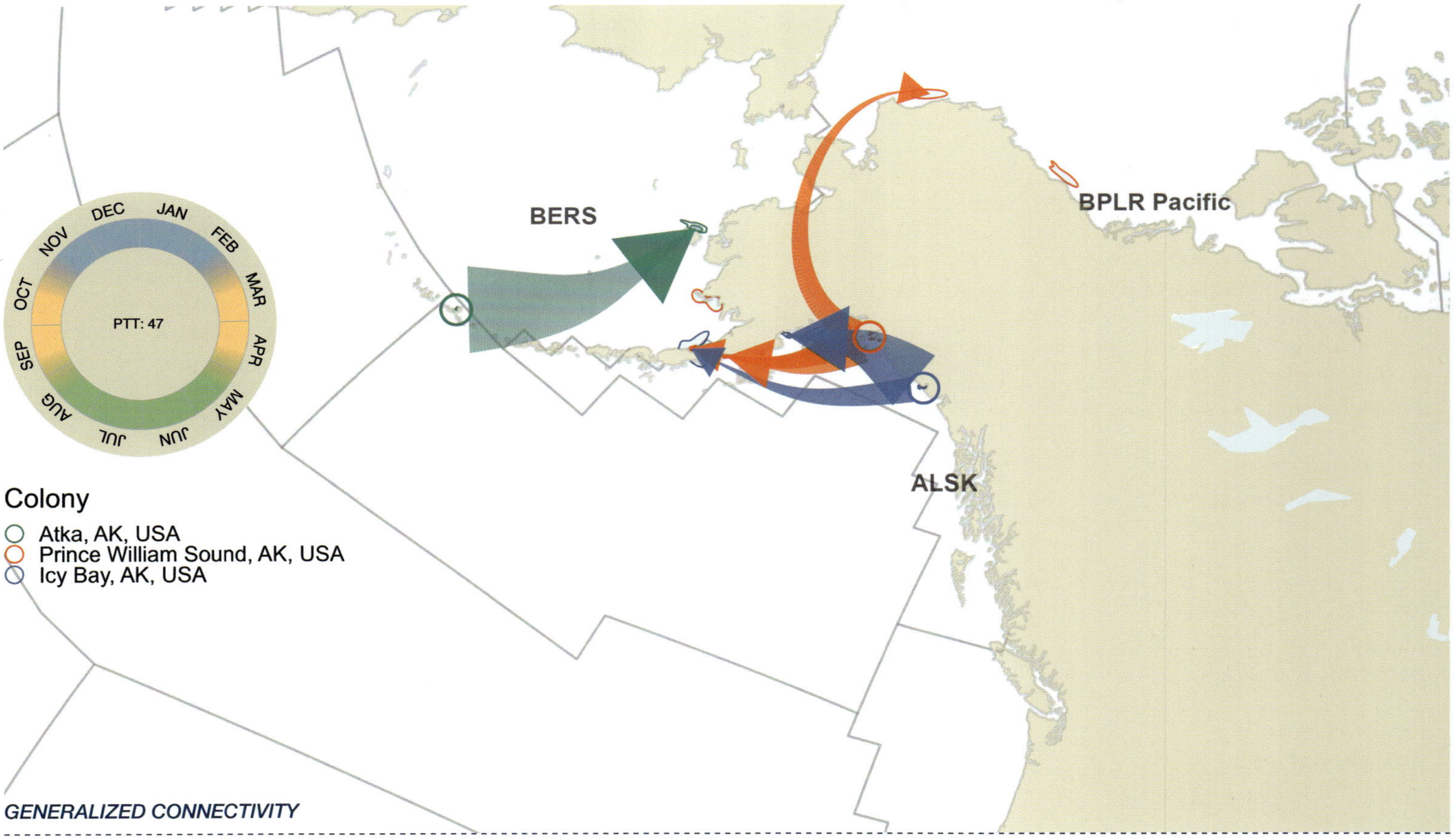

GENERALIZED CONNECTIVITY

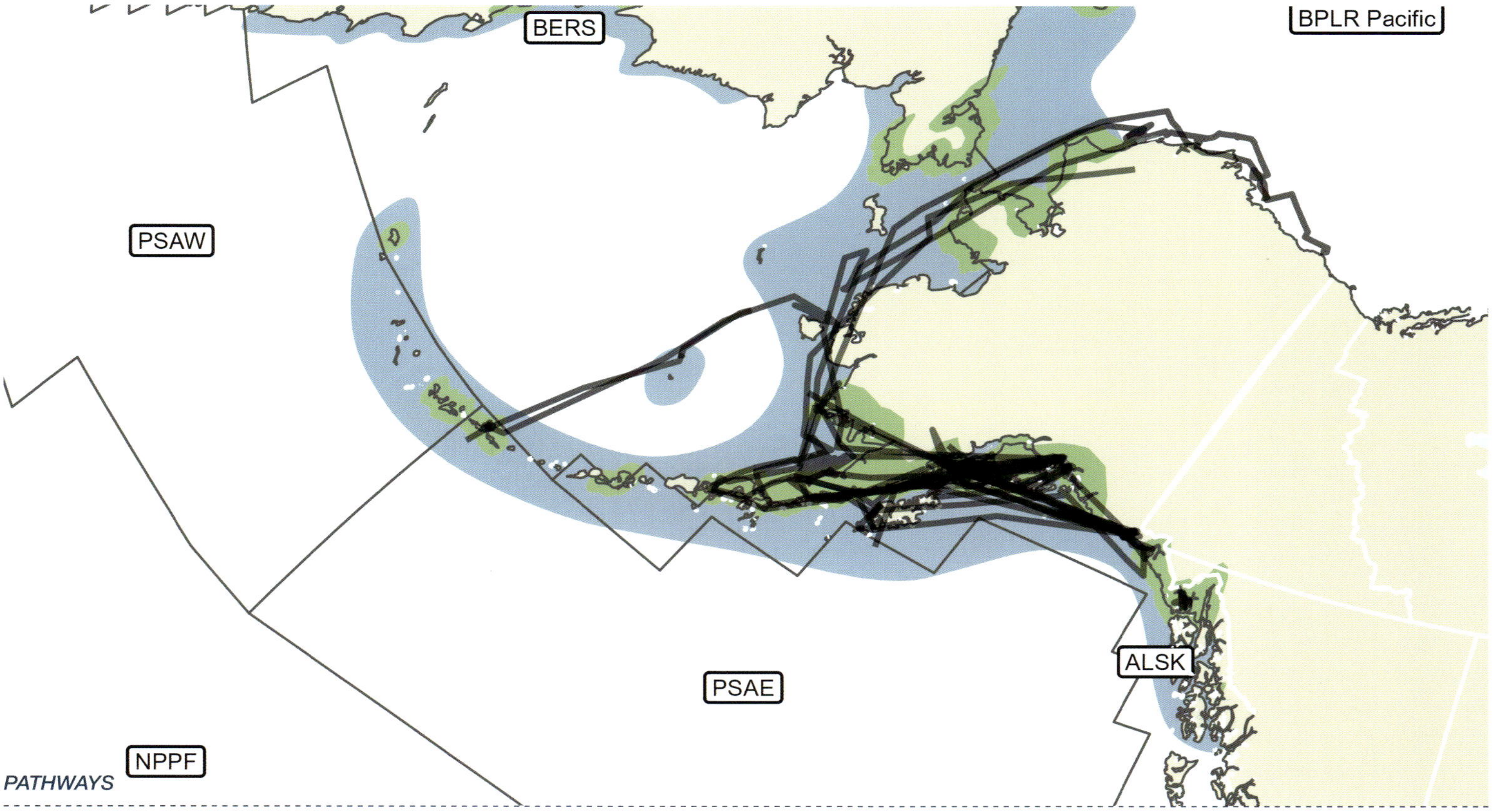

PATHWAYS

INTERESTING NOTE

Additional research is urgently needed for Kittlitz's Murrelet to determine conservation management units, including the collection of genetic data from the remainder of the breeding range and additional tracking, especially from the Aleutian Island population. Given low breeding propensity, and the relationship between Kittlitz's Murrelets and glaciated areas, there is concern for the species under climate change.

Sea province, but its core areas did not overlap with those of other populations. Genetic evidence suggests differentiation of Aleutian Island breeders from colonies in the Gulf of Alaska.

SATELLITE TRACKING Piatt et al. (2021) used 5 g satellite tags to track birds from three breeding populations to nonbreeding sites that spanned the entire coastal shelf regions of Alaska, as well as habitats in the Bering, Chukchi, and Beaufort seas. Tags were deployed in June and transmitted as long as October. Thus, insights on nonbreeding habitat use could be incomplete, as some sites interpreted as final nonbreeding destinations could be staging sites. The US Forest Service Pacific Northwest Research Station has also tracked four Kittlitz's Murrelets from Columbia Bay within the Chugach National Forest, Alaska (A. Duarte, unpublished data) to nonbreeding sites (tags transmitted through October) in Cook Inlet, Bristol Bay, near the Seward Peninsula, and as far north as the Beaufort and Chukchi seas of the North Slope of Alaska, providing further evidence for population mixing of breeding birds from the Gulf of Alaska.

GENETICS A mitochondrial DNA and nuclear DNA study of 53 Alaskan Kittlitz's Murrelets sampled from Attu Island (Aleutian Islands) and Glacier Bay and Kachemak Bay (Gulf of Alaska) suggests regional differentiation (Birt et al. 2011). There was significant genetic differentiation between birds from the Aleutian Islands and birds from the Gulf of Alaska, and little evidence of gene flow between the regions (Birt et al. 2011).

OTHER APPROACHES No stable isotope studies of migratory connectivity were found for this species; however, N. Hatch (2011) evaluated changes in stable carbon and nitrogen isotopic signatures throughout the annual cycle, indicating that Kittlitz's Murrelets rely on distinctly different prey resources during different periods. The first tracking studies were radio-tracking surveys generally within local areas and did not provide connectivity information. For example, studies included pre-breeding aerial tracking for 10 days throughout Glacier Bay, Alaska (Romano et al. 2007), and radio-tagging of a large sample across years in Icy Bay, Alaska, to determine breeding propensity (Kissling et al. 2015).

MIGRATORY PATHWAYS Migration pathways of satellite-tracked birds were generally coastal, first moving west from southeastern breeding colonies around the Gulf of Alaska to Cook Inlet or the Alaskan peninsula into the southern Bering Sea (Piatt et al. 2021). Birds may either remain there or move northward over the mid-eastern Bering Sea shelf, into the northern Bering Sea, through the Bering Strait into the Chukchi Sea, and then into the Beaufort Sea (J. Piatt, personal communication). This journey takes place over weeks to months, and Kittlitz's Murrelets travel through 4 large, distinct marine ecosystems (Gulf of Alaska, Bering, Chukchi, Beaufort). Some appear to stop en route to molt for a month or so, and all seem to stage briefly at foraging hotspots along the way (Piatt et al. 2021). Tags stopped transmitting prior to pre-breeding migration (Piatt et al. 2021).

DATA CONTRIBUTORS
D.C. Douglas; J.F. Piatt.

SCRIPPS'S MURRELET

Synthliboramphus scrippsi

ORDER Charadriiformes
FAMILY Alcidae
BREEDING HABITAT Coasts, Cliffs and Islands
NONBREEDING HABITAT Coasts, Marine Waters
GLOBAL POPULATION SIZE 15,000
POPULATION CHANGE SINCE 1970 NA
IUCN RED LIST CATEGORY Vulnerable
R2R TIPPING POINT SPECIES Red-Alert
US LISTING STATUS No
CANADA SPECIES AT RISK No
PIF CCS 18/20

GUADALUPE MURRELET

Synthliboramphus hypoleucus

ORDER Charadriiformes
FAMILY Alcidae
BREEDING HABITAT Coasts, Cliffs and Islands
NONBREEDING HABITAT Coasts, Marine Waters
GLOBAL POPULATION SIZE 5,000
POPULATION CHANGE SINCE 1970 NA
IUCN RED LIST CATEGORY Endangered
R2R TIPPING POINT SPECIES Red-Alert
US LISTING STATUS No
CANADA SPECIES AT RISK No
PIF CCS 19/20

DISTRIBUTION AND ABUNDANCE

Both species are endemic breeders to Baja California, Mexico, where they nest on Guadalupe and San Benito islands; Nettleship and Kirwan 2020). Guadalupe Murrelet and Scripps's Murrelet were formerly classified together as Xantus's Murrelet (Birt et al. 2012) and are sympatric breeders on San Benito Island (Nettleship and Kirwan 2020). After breeding, Guadalupe Murrelets move north throughout the California Current, especially offshore in central California (Nettleship and Kirwan 2020).

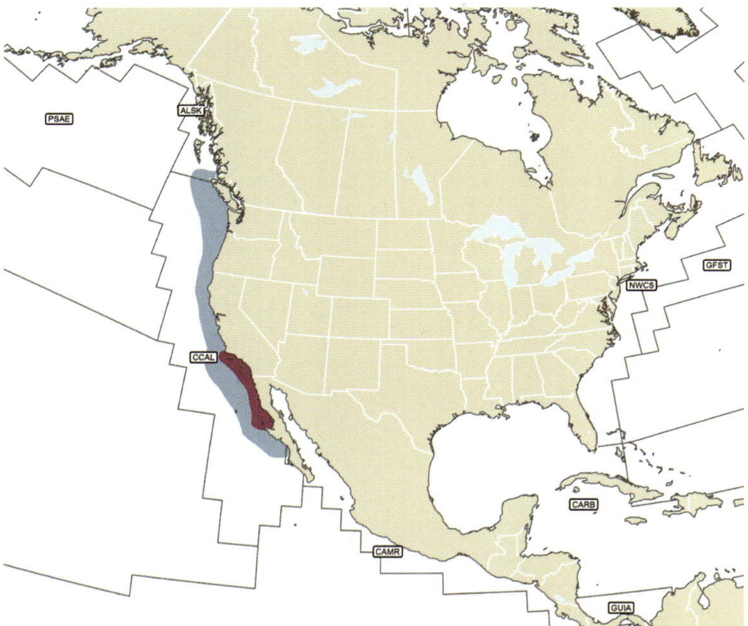

DISTRIBUTION

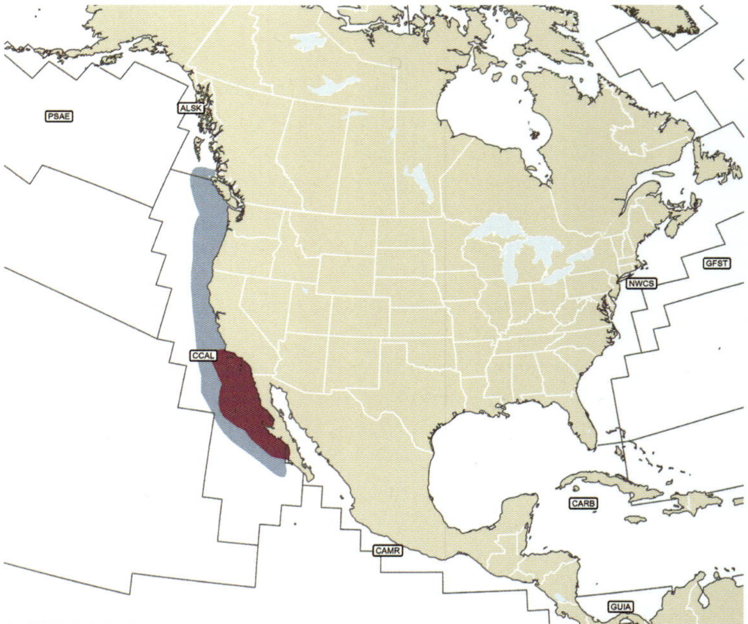

DISTRIBUTION

MIGRATORY CONNECTIVITY

OVERALL Migratory connectivity is unknown. Connectivity between colonies and specific core habitats within the California Current are not well known due in part to difficulty separating Guadalupe Murrelet from Scripps's Murrelet at sea; there is likely overlap between species, and there is no tracking information.

GENETICS Birt et al. (2012) assessed range-wide population structure of the former Xantus's Murrelet using mitochondrial DNA and nuclear DNA analysis on samples from 13 breeding populations. This study resulted in taxonomic subdivision on evidence of strong genetic divergence between the two subspecies recognized at the time (Scripps's Murrelet and Guadalupe Murrelet), and there is little genetic differentiation across breeding populations within subspecies.

OTHER APPROACHES No eBird banding, tracking, or stable isotope studies of migratory connectivity were found for this species.

Synthliboramphus craveri **CRAVERI'S MURRELET**

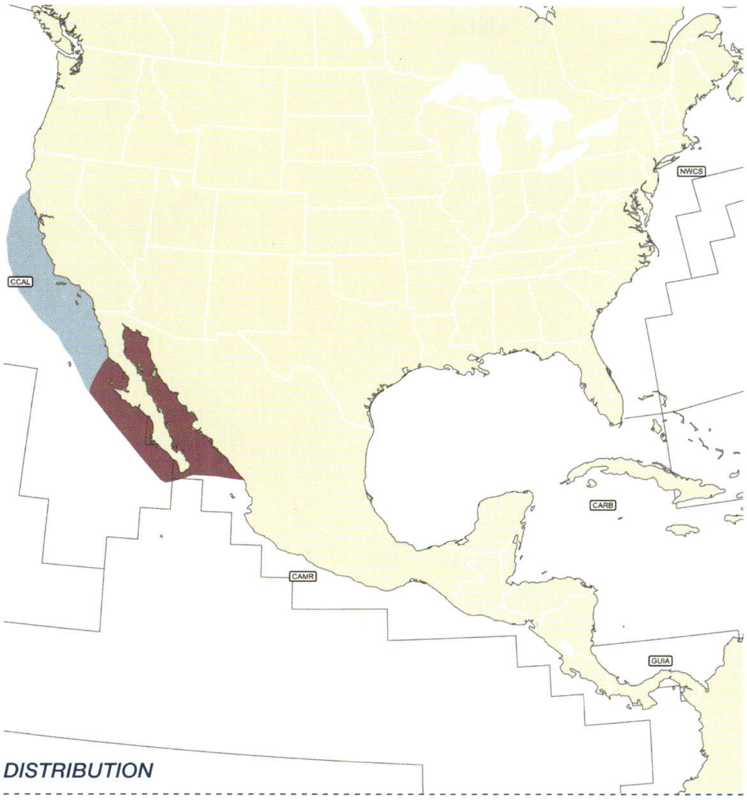

DISTRIBUTION

ORDER Charadriiformes
FAMILY Alcidae
BREEDING HABITAT Coasts, Cliffs and Islands
NONBREEDING HABITAT Coasts, Marine Waters
GLOBAL POPULATION SIZE 8,000
POPULATION CHANGE SINCE 1970 NA
IUCN RED LIST CATEGORY Vulnerable
R2R TIPPING POINT SPECIES Red-Alert
US LISTING STATUS No
CANADA SPECIES AT RISK No
PIF CCS 17/20

MIGRATORY CONNECTIVITY

OVERALL Only a BirdLife distribution range map is available; no eBird data or migratory connectivity information were found for this species.

Mantel = 0.06 ± 0.08

Synthliboramphus antiquus **ANCIENT MURRELET**

DISTRIBUTION AND ABUNDANCE

A small auk of the North Pacific, this species breeds on offshore islands in temperate and subarctic waters from China to Canada (primarily Haida Gwaii) and spends the nonbreeding season offshore of both Asia and North America. Nestlings follow their parents to sea soon after hatching to be the only seabird species reared entirely at sea (Gaston and Shoji 2020). Because of this early imprinting, it is thought that birds from the eastern side of Haida Gwaii may be more likely to use nonbreeding habitats in the California Current than those from the western side of Haida Gwaii.

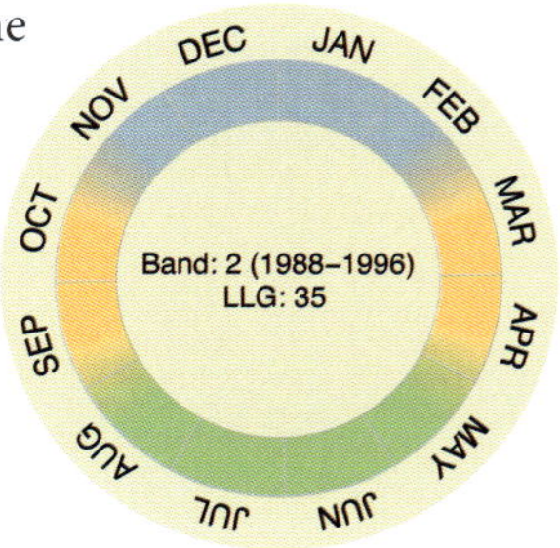

ORDER Charadriiformes
FAMILY Alcidae
BREEDING HABITAT Coasts, Cliffs and Islands
NONBREEDING HABITAT Coasts, Marine Waters
GLOBAL POPULATION SIZE 1,000,000
POPULATION CHANGE SINCE 1970 +139%
IUCN RED LIST CATEGORY Least Concern
R2R TIPPING POINT SPECIES No
US LISTING STATUS No
CANADA SPECIES AT RISK No
PIF CCS 13/20

MIGRATORY CONNECTIVITY

OVERALL There is moderate migratory connectivity (Mantel = 0.06 ± 0.08) of birds tracked and banded from colonies on Haida Gwaii, British Columbia, Canada, to core habitats in marginal seas and coastal currents of both North America (primarily the Bering Sea, 56%–80%) and Asia (especially the East China Sea, 10%–25%). No movement data are available for colonies in Asia or the US. A genetic study suggests that the Japanese and Aleutian populations share some genetic markers. Thus, connectivity may be stronger at the species level when other colonies are considered.

BANDING Two birds nesting on the eastern side of Haida Gwaii were reencountered in Oregon, US during the nonbreeding period.

GEOLOCATOR TRACKING Birds were tracked from colonies on both the eastern and western sides of Haida Gwaii: the western side bordering the open ocean and the eastern side bordering the shallow

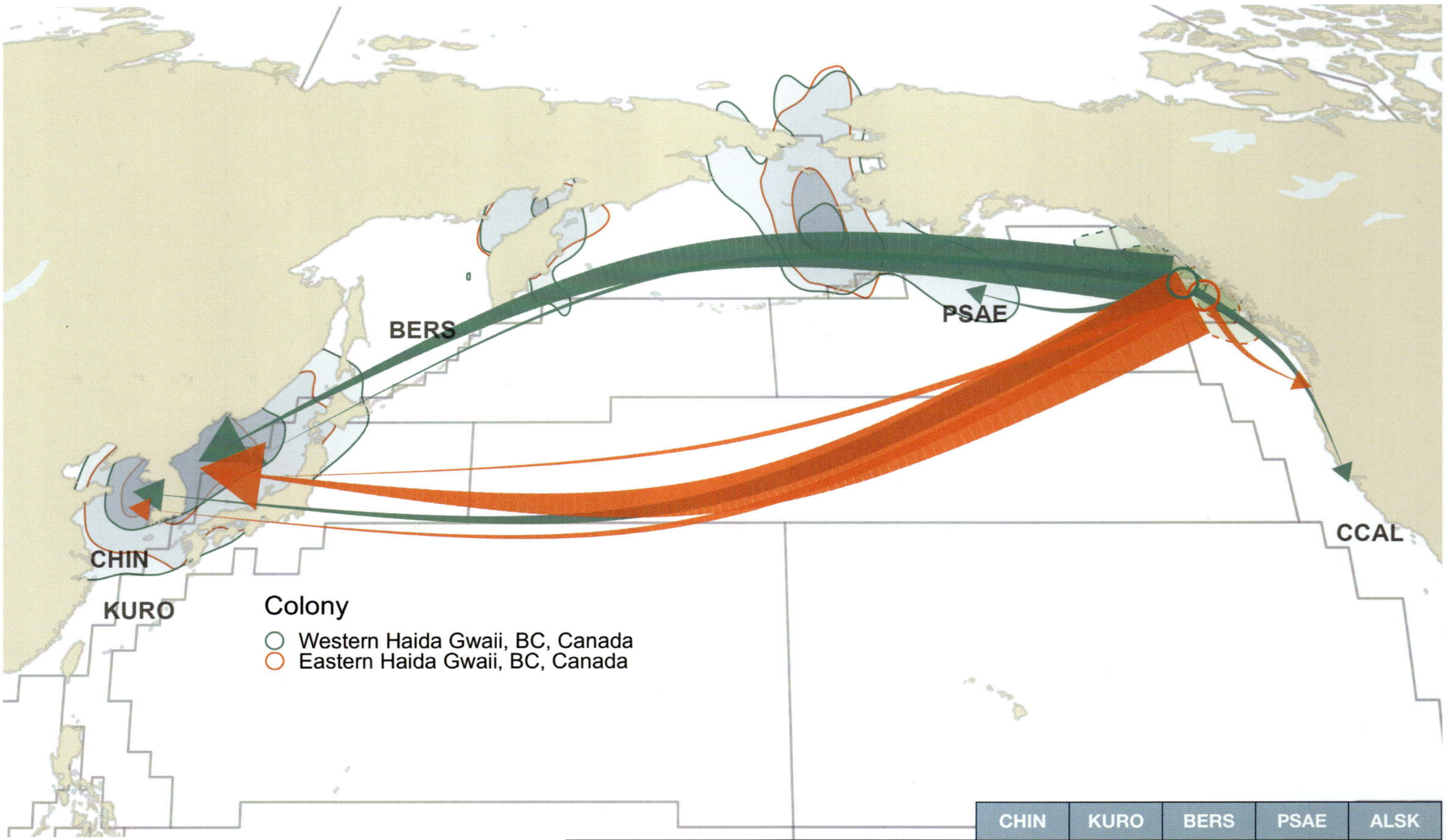

GENERALIZED CONNECTIVITY

	CHIN	KURO	BERS	PSAE	ALSK
Western Haida Gwaii, BC, Canada (n = 16)	25%	6%	56%	6%	6%
Eastern Haida Gwaii, BC, Canada (n = 20)	10%	5%	80%	–	5%

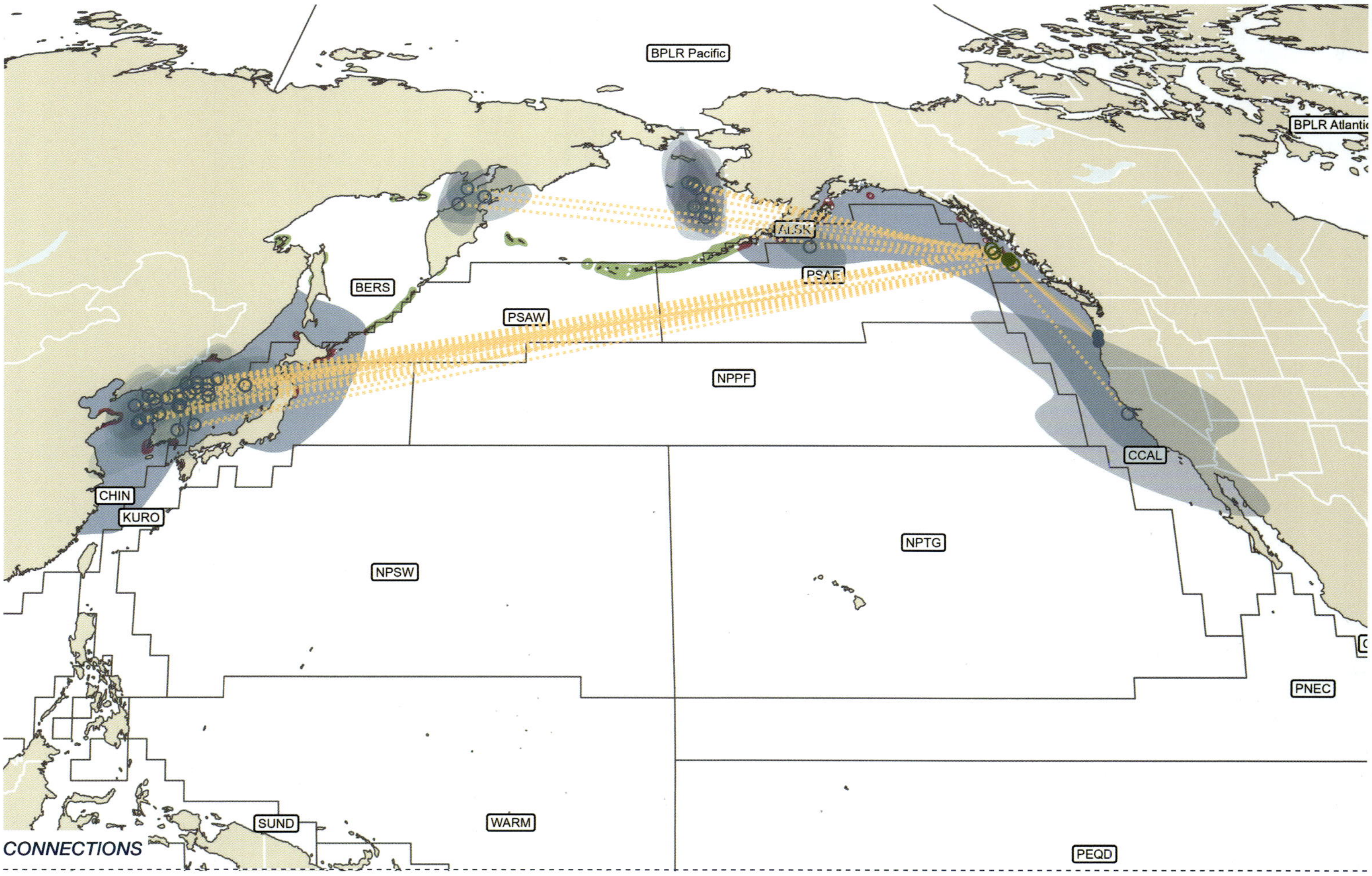

CONNECTIONS

Hecate Strait. But there is no distinction between the two colonies in proportion of birds that migrated either to Asia (primarily off the coast of Japan and South Korea) or remained within North America (coastal waters off the Aleutian Islands, Alaska; Gaston et al. 2015, 2017).

GENETICS An analysis of mitochondrial DNA revealed that Ancient Murrelets from colonies in British Columbia, Canada, showed little genetic differentiation with Asian birds caught at sea during the breeding period off the Kamchatka Peninsula and Kuril Islands (R.L. Pearce et al. 2002). This homogeneity could be due to historical association from a possible range expansion from Asia, but this was not confirmed. A more recent study of nuclear DNA showed high gene flow between a small Japanese population (Teuri Island) and the Aleutian breeding population; together they possessed haplotypes not present in the British Columbian and Kamchatka populations (Sugita et al. 2021).

OTHER APPROACHES No stable isotope studies of migratory connectivity were found for this species.

MIGRATORY PATHWAYS Tracking data showed that birds tracked from Haida Gwaii throughout their full annual cycle first moved according to local current patterns, especially into the Gulf of Alaska, perhaps due to having flightless young and a molt period. By October and November, most birds migrated rapidly to nonbreeding grounds. Birds either migrated west to Asian waters (Gaston et al. 2015) or, more commonly, remained in the Gulf of Alaska or migrated south to coastal North America. During the pre-breeding season, birds returned rapidly to breeding areas (Gaston et al. 2017).

DATA CONTRIBUTORS
A.J. Gaston; Y. Hashimoto; L. Wilson.

INTERESTING NOTE

A bird banded as a chick on May 24, 2003, was brought up drowned in a fishing net in Japan on February 10, 2011 (Gaston et al. 2015).

Mantel = –0.03 ± 0.09

Ptychoramphus aleuticus

CASSIN'S AUKLET

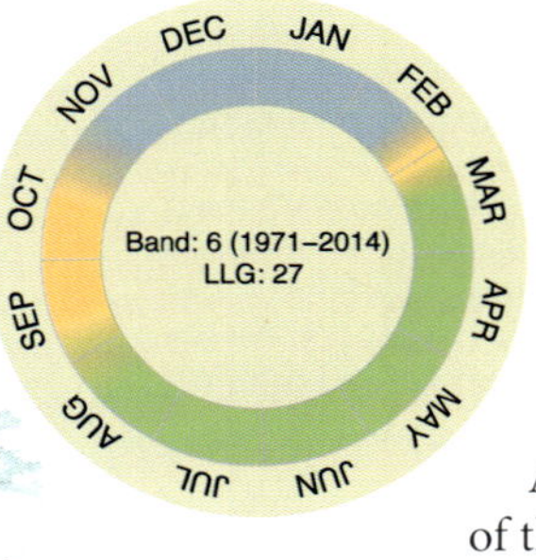

DISTRIBUTION AND ABUNDANCE

This species is a small Pacific Ocean alcid that nests primarily on Beringian islands, but its North American breeding range spans south to Baja California, Mexico, and to Kuril Islands in the western Pacific (Ainley et al. 2020). British Columbia (76%) and Alaska (17%) have the largest proportion of the North American breeding population (Ainley et al. 2020). Other large populations include the Sea of Okhotsk, Kuril Islands, and the US and Mexican Pacific coasts. During the nonbreeding season, it is widely distributed in pelagic waters across the North Pacific.

ORDER Charadriiformes
FAMILY Alcidae
BREEDING HABITAT Coasts, Cliffs and Islands
NONBREEDING HABITAT Coasts, Marine Waters
GLOBAL POPULATION SIZE 3,600,000
POPULATION CHANGE SINCE 1970 +1%
IUCN RED LIST CATEGORY Near Threatened
R2R TIPPING POINT SPECIES No
US LISTING STATUS No
CANADA SPECIES AT RISK No
PIF CCS 14/20

	ALSK	CCAL
Haida Gwaii, BC, Canada (n = 4)	–	100%
Triangle Island, BC, Canada (n = 27)	4%	96%
Morro Bay, CA, USA (n = 1)	–	100%

INTERESTING NOTE

Data on natal site fidelity is lacking for most of the species' range, but it is strong at Farallon Islands. Birds return at two or three years of age (Ainley et al. 2020). There were no breeding to nonbreeding connections reported either through banding or tracking from colonies at the edges of the species range (Alaska, US, or Mexico).

MIGRATORY CONNECTIVITY

OVERALL Weak connectivity of US and Canadian colonies is suggested from tracking and limited banding data (Mantel = –0.03 ± 0.09). The majority of all colonies (Haida Gwaii and Triangle Islands, British Columbia, and Morro Bay, California) migrated to and generally mixed during the nonbreeding period in the California Current. However, there is also clear genetic structure that aligns with northern and southern subspecies designations.

BANDING Breeding birds from British Columbia, Canada, were found in British Columbia and Oregon and Washington, US, during nonbreeding. An individual banded during the nonbreeding season in the Farallon Islands, California, was found at a Morro Bay breeding colony, while an individual banded at its breeding colony in the Farallon Islands was found off Ragged Point, California, during nonbreeding. None of the birds were banded or reencountered south of Point Conception, the genetic divide for subspecies.

RADIO TRACKING Birds from Channel Islands, California (n = 56) were tracked to nonbreeding areas with radio telemetry by aircraft (Adams et al. 2004; data are not included in this Atlas). These birds

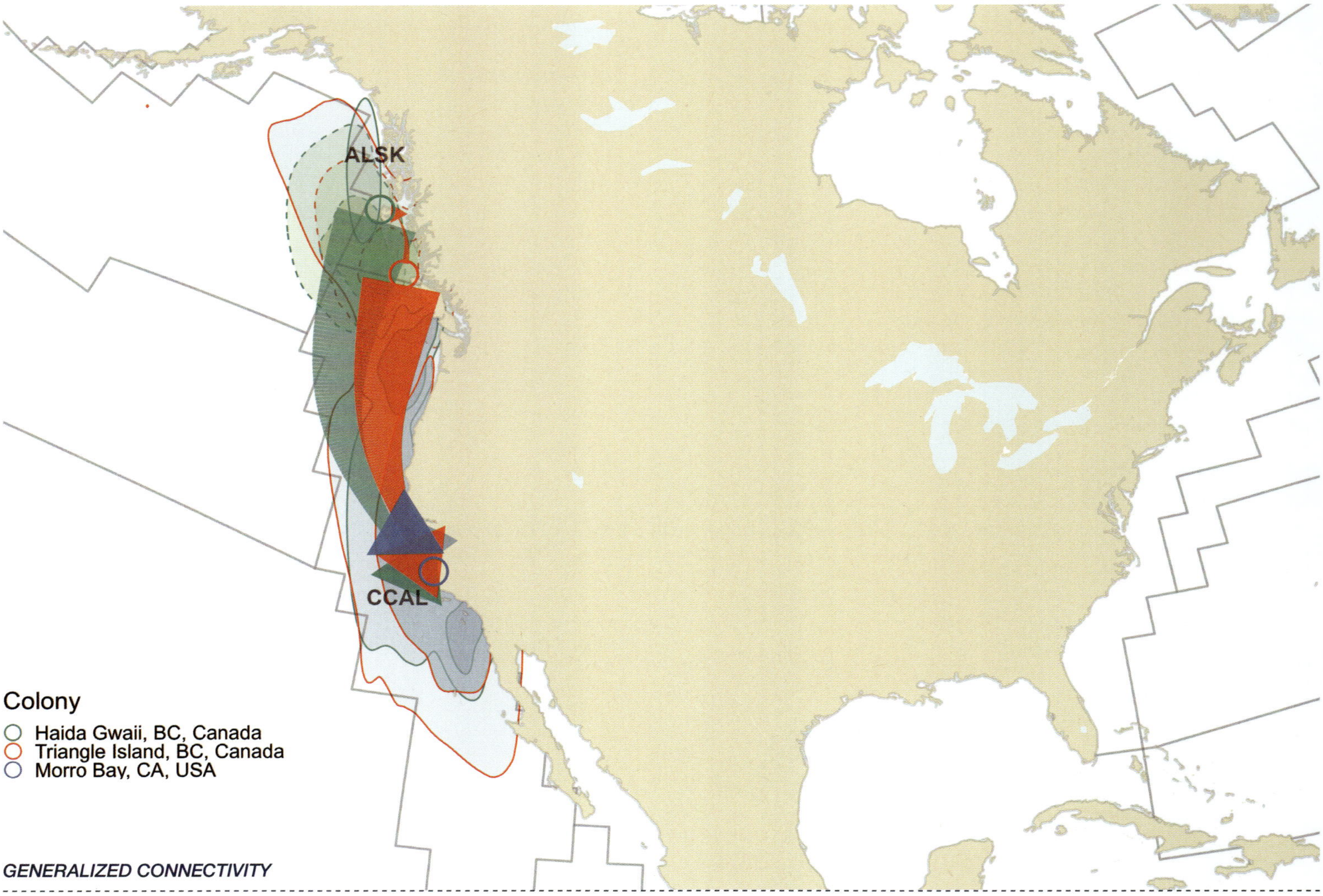

GENERALIZED CONNECTIVITY

moved north along the coast as far as the Farallon Islands when upwelling ceased in the Southern California Current during the nonbreeding period (Adams et al. 2004). Thus, northern and southern colonies are thought to converge during the nonbreeding period, but movements of southern individuals may depend on interannual variation in oceanographic conditions (Adams et al. 2004).

GEOLOCATOR TRACKING Cassin's Auklets (27 birds with breeding to nonbreeding connections) were tracked with light-level geolocators from two colonies in British Columbia in 2014–2017 (Studholme et al. 2019). Three additional birds were tracked from Frederick Island off the northwest coast of Haida Gwaii, Canada. Nonbreeding areas spanned the Aleutian Islands in the western Gulf of Alaska, Alaska Current (2,757 km from colony) to Baja California, Mexico (3,110 km from colony).

GENETICS A mitochondrial DNA and nuclear DNA study of 287 auklets breeding from the Aleutian Islands, Alaska, to Baja California, Mexico, demonstrated genetic differentiation across a barrier at Point Conception, California (Wallace et al. 2015). This structure mostly matches subspecies designations, except that birds breeding in the Channel Islands, California, which are typically grouped with the northern subspecies (*P. a. aleuticus*), are more genetically similar to the southern subspecies (*P. a. australe*). In contrast to the strong connectivity that would be predicted by such structure (Friesen et al. 2007), breeding populations from the northern and southern populations mix during the nonbreeding season. Such mixing should provide opportunity for population admixture, reducing genetic differentiation. Wallace et al. (2015) indicated that some gene flow has occurred from northern to southern populations but may not have occurred in the opposite direction. This could be explained by southern populations initiating breeding much earlier than northern populations do, which could prevent gene flow (a form of temporal connectivity). Additionally, southern breeders may not always move north to converge with northern populations that move south; rather, they may do so in years when local upwelling ceases (Adams et al. 2004).

OTHER APPROACHES No stable isotope studies of migratory connectivity were found for this species.

MIGRATORY PATHWAYS Colonies in different parts of the species range employ different migratory strategies. Individuals tracked with light-level geolocators from northern colonies generally took pathways south along the coast and continental shelf; however, a small number of birds undertook a post-breeding migration northwest across the Gulf of Alaska to the Aleutian Islands, Alaska, before returning south to the California Current for an ultimate nonbreeding destination (Studholme et al. 2019). The Farallon Islands, California, population is thought to be nonmigratory (Ainley et al. 2020). Southern birds tracked by radio telemetry migrated north along the coast as far as the Farallon Islands when upwelling ceased in the Southern California Current (Adams et al. 2004).

DATA CONTRIBUTORS

J.M. Hipfner; K.R. Studholme.

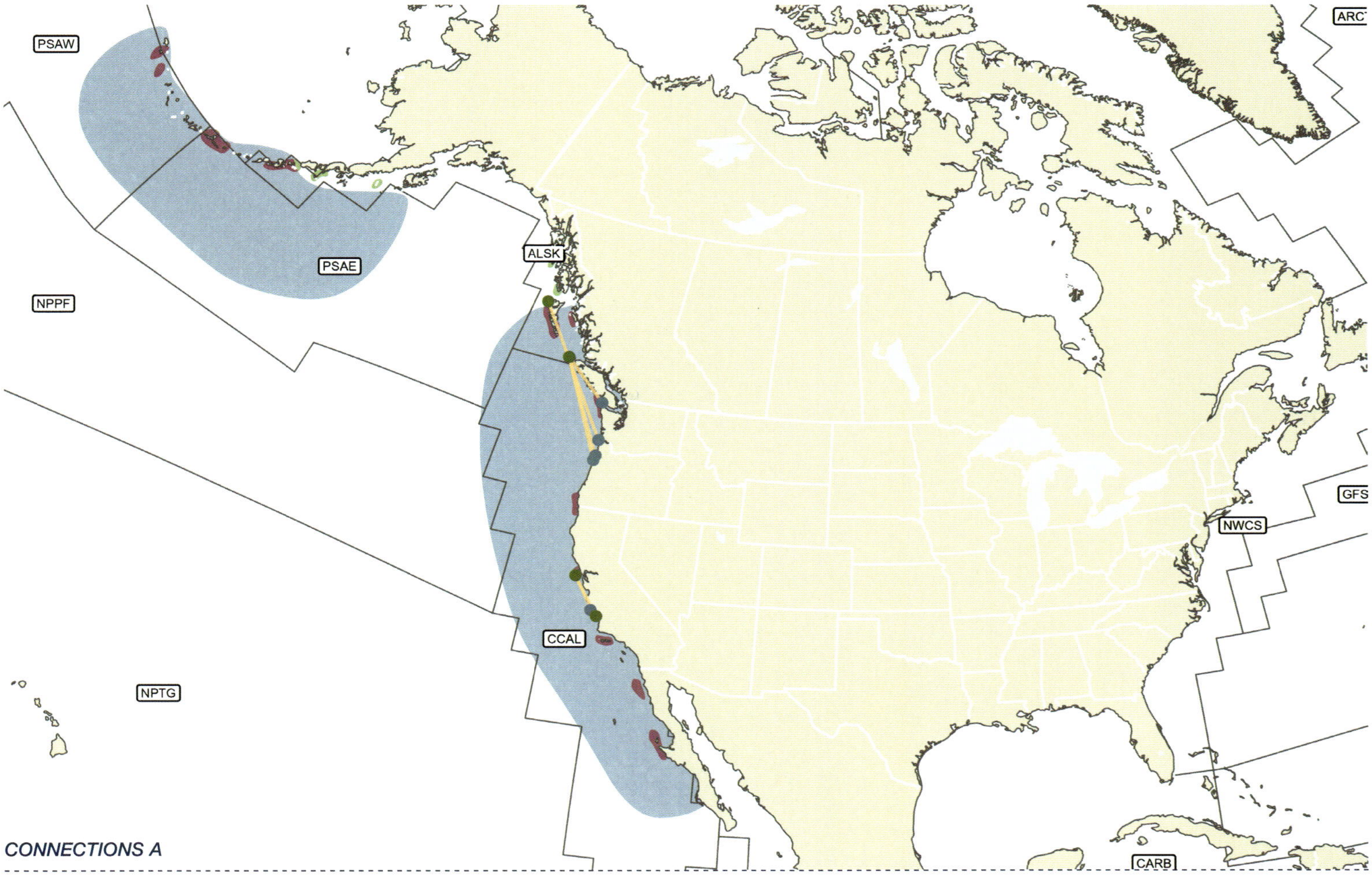

CONNECTIONS A

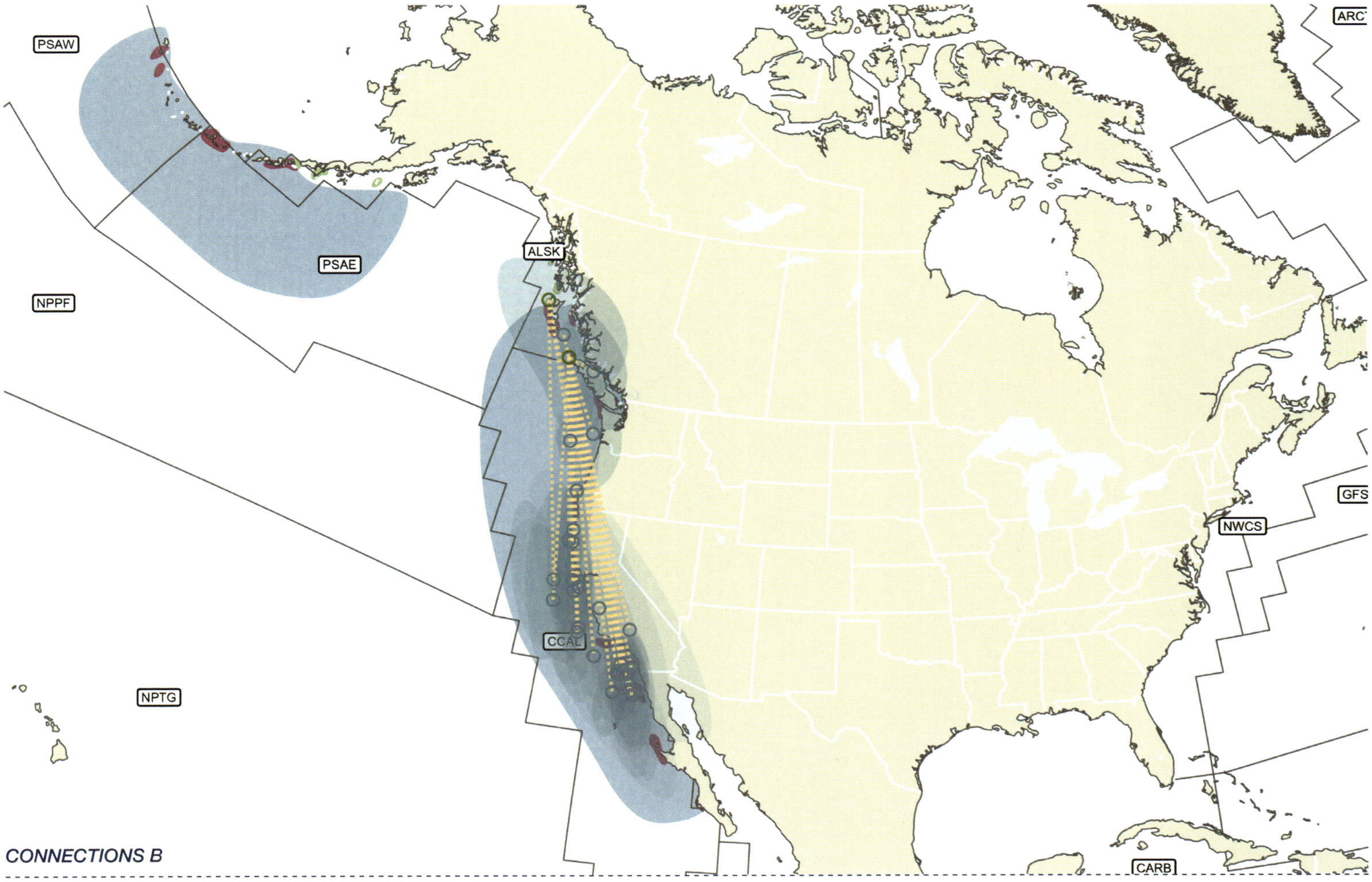

CONNECTIONS B

PARAKEET AUKLET *Aethia psittacula*

DISTRIBUTION AND ABUNDANCE

This highly recognizable, planktivorous seabird of the North Pacific Ocean nests at hundreds of small colonies throughout the Aleutian Islands, Bering Sea, and islands and coasts of the Asian North Pacific. It spends most of the year at sea. It engages in diffusive nonbreeding movements offshore in the north-central and northwestern Pacific rather than long-distance migrations, likely tracking seasonality of prey distribution (Jones et al. 2020; Schacter 2017). It is also regular along the Pacific coast of Canada and as far south as California in the east and Japan in the west.

ORDER Charadriiformes
FAMILY Alcidae
BREEDING HABITAT Coasts, Cliffs and Islands
NONBREEDING HABITAT Coasts, Marine Waters
GLOBAL POPULATION SIZE 1,400,000
POPULATION CHANGE SINCE 1970 NA
IUCN RED LIST CATEGORY Least Concern
R2R TIPPING POINT SPECIES No
US LISTING STATUS No
CANADA SPECIES AT RISK No
PIF CCS 12/20

INTERESTING NOTE

Multiple reports of vagrancy include auklets found in the northwestern Hawaiian Islands and Sweden (Jones et al. 2020). A study of the effects of geolocators on this species found them to be tolerant of 1–2 g tags; there was some reduction of reproductive success but no negative effects on return rate or chick growth rate (Schacter and Jones 2017).

MIGRATORY CONNECTIVITY

OVERALL Likely weak connectivity to nonbreeding areas based on tracking data from two Aleutian Islands colonies that overlapped regions adjacent to other Asian and Alaskan colonies, including the Bering Sea and the North Pacific Subtropical Gyre. However, the strength of migratory connectivity could not be estimated due to limited data contributed to this Atlas. Genetic and morphological evidence suggests Aleutian colonies may be slightly differentiated from others across the range. There is also evidence to suggest that eastern Aleutian colonies of Parakeet Auklets have different nonbreeding areas than those from the western Aleutian colonies (Schacter 2017). Eastern populations may have a more coastal

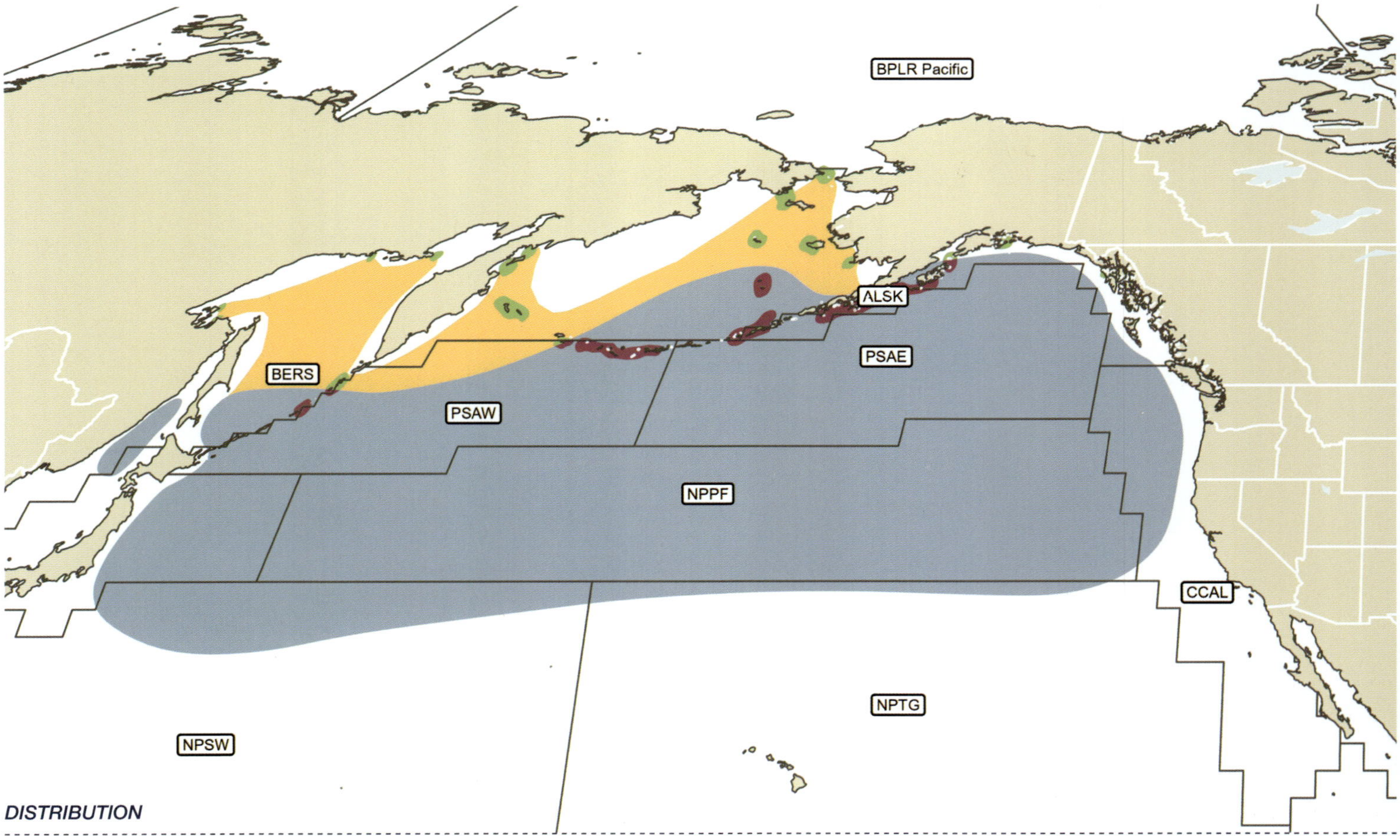

DISTRIBUTION

southward migration, and thus are frequently observed offshore from Washington and California during the nonbreeding season (Jones et al. 2020).

GEOLOCATOR TRACKING Geolocators deployed at two colonies in the Aleutian Islands, Alaska (Buldir, n = 46; Gareloi, n = 18), illustrated an intermediate migration and shifting range throughout the nonbreeding period (Schacter 2017). Parakeet Auklets spent most of the nonbreeding period in the deep Aleutian Basin, with similar overall distributions of birds from the two colonies, but there were some differences, reflecting a slight east-west shift (Schacter 2017). Data are not presented in the Atlas.

GENETICS Pshenichnikova et al. (2021) found slight but significant differentiation between Parakeet Auklets from the Aleutian Islands and the remaining colonies based on mitochondrial DNA and nuclear DNA analysis and morphometrics (the birds from the Aleutian Islands had significantly smaller measurements and better body condition). The authors compared the results to analyses of congeneric species to suggest that foraging and migration patterns determine population structure in small planktivorous alcids; i.e., highly mobile species have little population differentiation, resident species have stronger differentiation, and dispersive species (like the Parakeet Auklet) have an intermediate level of differentiations.

OTHER No banding, satellite tracking, or isotope studies of migratory connectivity were found for this species.

MIGRATORY PATHWAYS Geolocator data showed that, after leaving the breeding colony, Parakeet Auklets made a gradual migration, with most birds making an initial flight northeast, while others moved east along the Aleutian Islands or north to the Chukchi Sea (Schacter 2017).

Aethia pusilla LEAST AUKLET

MIGRATORY CONNECTIVITY

OVERALL Only a BirdLife distribution range map is available; no eBird data or migratory connectivity information were found for this species.

ORDER Charadriiformes
FAMILY Alcidae
BREEDING HABITAT Coasts, Cliffs and Islands
NONBREEDING HABITAT Coasts, Marine Waters
GLOBAL POPULATION SIZE 20,000,000
POPULATION CHANGE SINCE 1970 NA
IUCN RED LIST CATEGORY Least Concern
R2R TIPPING POINT SPECIES No
US LISTING STATUS No
CANADA SPECIES AT RISK No
PIF CCS 13/20

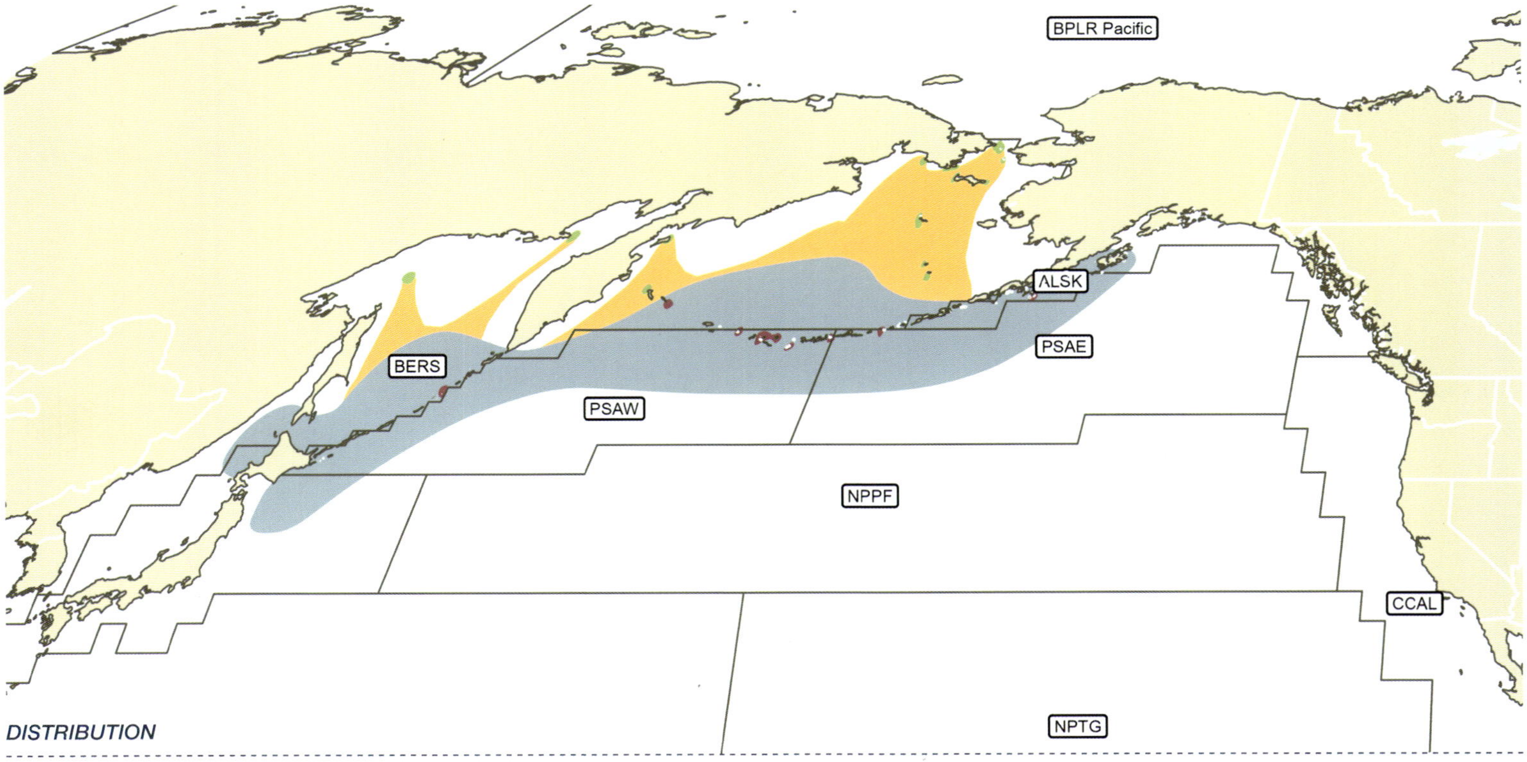

DISTRIBUTION

CRESTED AUKLET *Aethia cristatella*

DISTRIBUTION AND ABUNDANCE

This species is a common pelagic auklet of the North Pacific that breeds on islands and coastlines of the Bering Sea, Aleutian Islands, and Kuril Islands (Jones 2020). It nests in highest abundance (>300,000 individuals) in the northern Sea of Okhotsk, Russia; St. Lawrence Island, Alaska, in the Bering Sea; and Kiska Island in the Aleutians, Alaska (Jones 2020). It is known for huge flocks over colonies and at sea. It ranges as far south as Japan, and there are sighting records in British Columbia, Canada, and California, US, as well as in Arctic waters (Jones 2020). It stays near some colonies year-round, but there is also evidence of migration (Jones 2020).

ORDER Charadriiformes
FAMILY Alcidae
BREEDING HABITAT Coasts, Cliffs and Islands
NONBREEDING HABITAT Coasts, Marine Waters
GLOBAL POPULATION SIZE 5,500,000
POPULATION CHANGE SINCE 1970 NA
IUCN RED LIST CATEGORY Least Concern
R2R TIPPING POINT SPECIES No
US LISTING STATUS No
CANADA SPECIES AT RISK No
PIF CCS 11/20

INTERESTING NOTE

There are rare vagrant records confirmed for Iceland and Greenland in the Atlantic Ocean (one unconfirmed record in Massachusetts, US; Sealy and Carter 2012). This species is notable for smelling like tangerines (Hagelin et al. 2003).

MIGRATORY CONNECTIVITY

OVERALL Migratory connectivity is unknown, as well as the level of partial migration, but it is assumed that colonies with ice-free periods include residents due to year-round observations and that colonies on northern islands surrounded by ice in the nonbreeding season are long-distance migrants (Jones 2020). Genetic and at-sea survey evidence suggest weak connectivity during the nonbreeding period, but connections from specific breeding colonies to specific oceanographic provinces are unknown.

GENETICS No differentiation was found across nine colonies spanning the breeding range based on mitochondrial DNA and nuclear DNA analysis, as well as barks and trumpet calls, and morphometric and ornamentation measurements (Pshenichnikova et al. 2015). The authors suggest this lack of separation is driven by the nonbreeding distribution, where young birds mix at sea with other colonies and may not return to their natal colony: however this is unconfirmed (Pshenichnikova et al. 2015).

OTHER APPROACHES No banding, tracking, or stable isotope studies of migratory connectivity were found for this species, although there is reporting that Crested Auklets were equipped with time-depth recorders and sampled for stable isotope studies at St. Paul Island, Alaska (Harten 2009). Maftei and Russ (2014) conducted a census of a tight passage flock (400 m wide) early in the post-breeding period of >8.5–10 million individuals moving northwest in the Chukchi Sea at the rate of 1,000 individuals per second over four hours. The passage route was previously known, but this huge estimated flock size indicates that the congregation must have been comprised of birds from multiple colonies as it is orders of magnitude larger than any single colony.

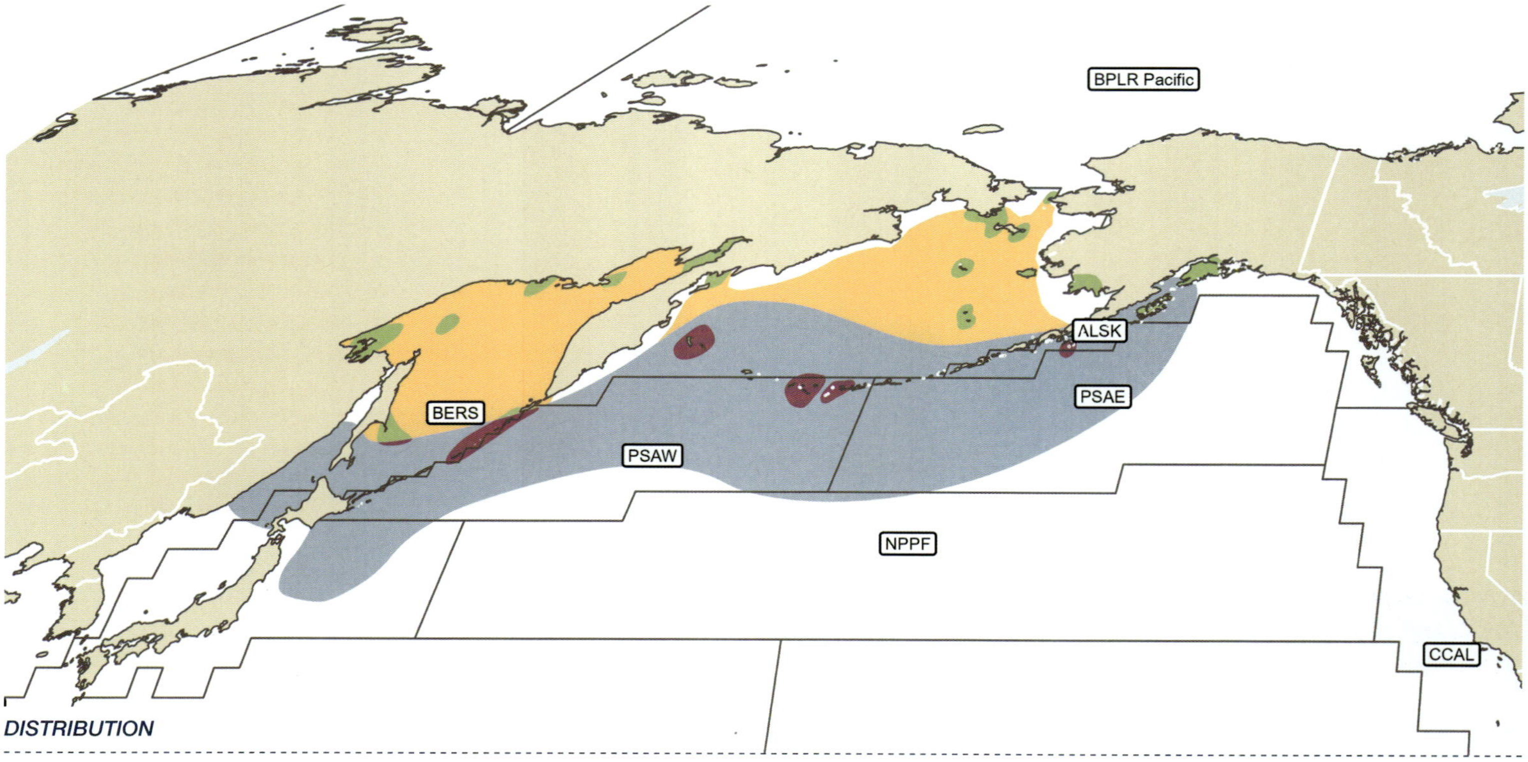

DISTRIBUTION

weak strong

Mantel = –0.07 ± 0.03

Cerorhinca monocerata

RHINOCEROS AUKLET

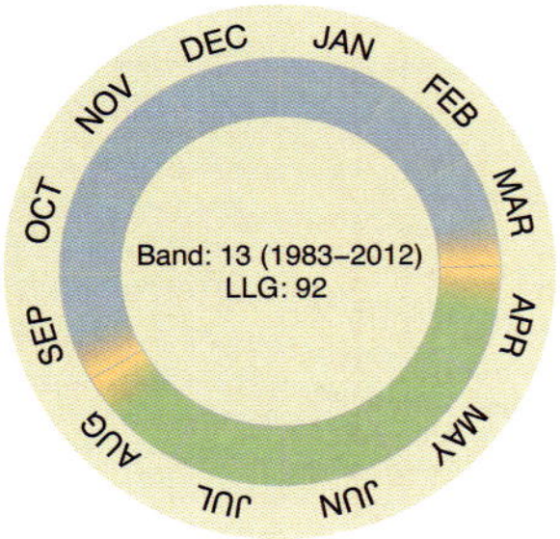

DISTRIBUTION AND ABUNDANCE

This species is a medium-sized and stocky alcid of the temperate North Pacific with a prominent horn-like extension of the bill during breeding. Most of the North American population (approximately 1 million breeding pairs) is in British Columbia, Canada (75%), with other colonies estimated at tens of thousands in Washington (13%) and Southeast Alaska (12%; Gaston and Dechesne 2020). As a nocturnal burrow nester on remote islands, many population estimates are uncertain and outdated (Gaston and Dechesne 2020). It feeds close to shore during the breeding season and spends the nonbreeding season both offshore and inshore. Not all colonies are fully migratory.

ORDER Charadriiformes
FAMILY Alcidae
BREEDING HABITAT Coasts, Cliffs and Islands
NONBREEDING HABITAT Coasts, Marine Waters
GLOBAL POPULATION SIZE 1,500,000
POPULATION CHANGE SINCE 1970 +178%
IUCN RED LIST CATEGORY Least Concern
R2R TIPPING POINT SPECIES No
US LISTING STATUS No
CANADA SPECIES AT RISK No
PIF CCS 13/20

	ALSK	PSAE	NPPF	CCAL
Chowiet Island, AK, USA (n = 4)	25%	50%	–	25%
Middleton Island, AK, USA (n = 17)	24%	35%	29%	12%
Saint Lazaria Islands, AK, USA (n = 5)	–	20%	20%	60%
British Columbia, Canada and Washington, USA (n = 68)	15%	4%	3%	78%
Central California, USA (n = 6)	–	–	–	100%

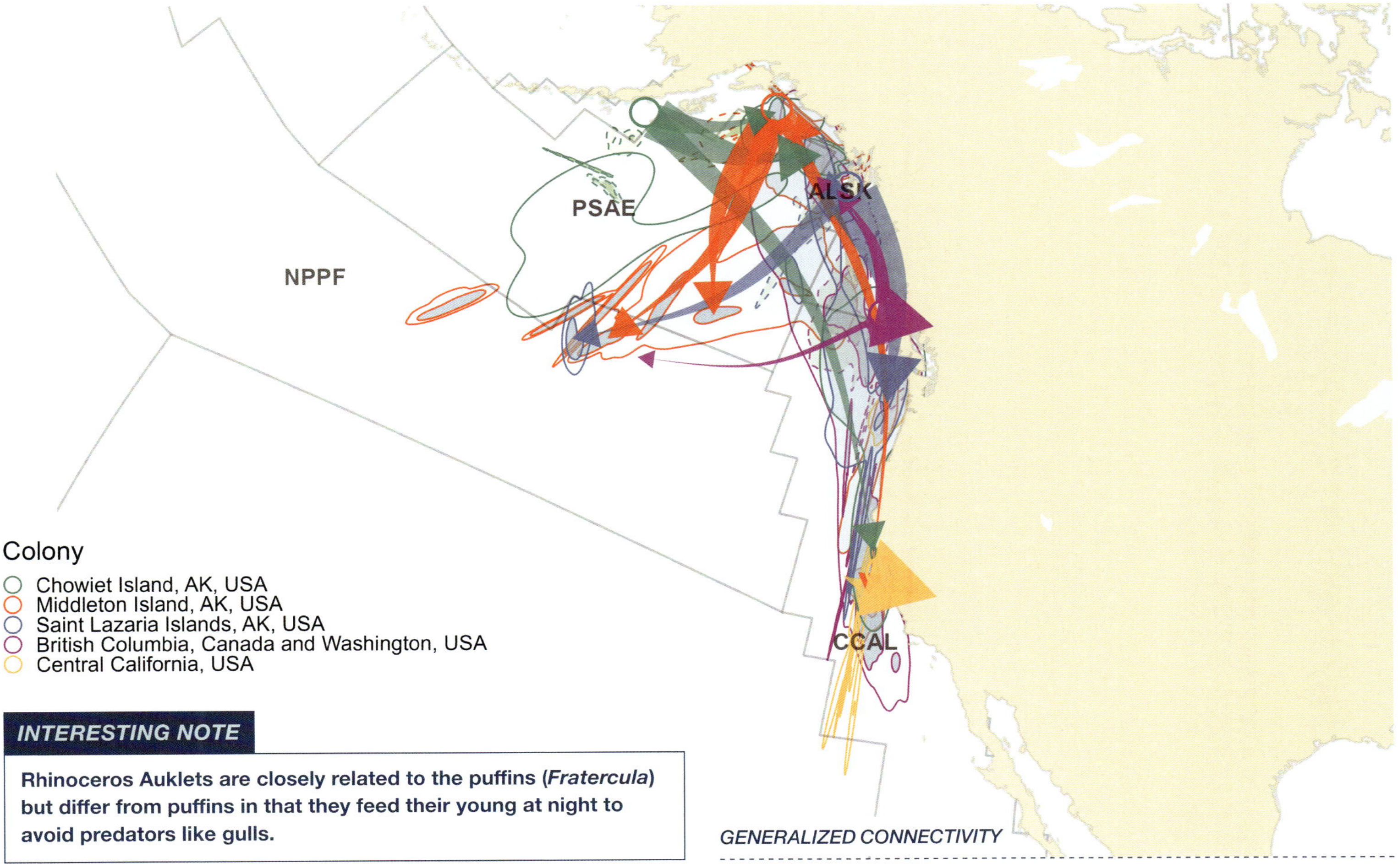

INTERESTING NOTE

Rhinoceros Auklets are closely related to the puffins (*Fratercula*) but differ from puffins in that they feed their young at night to avoid predators like gulls.

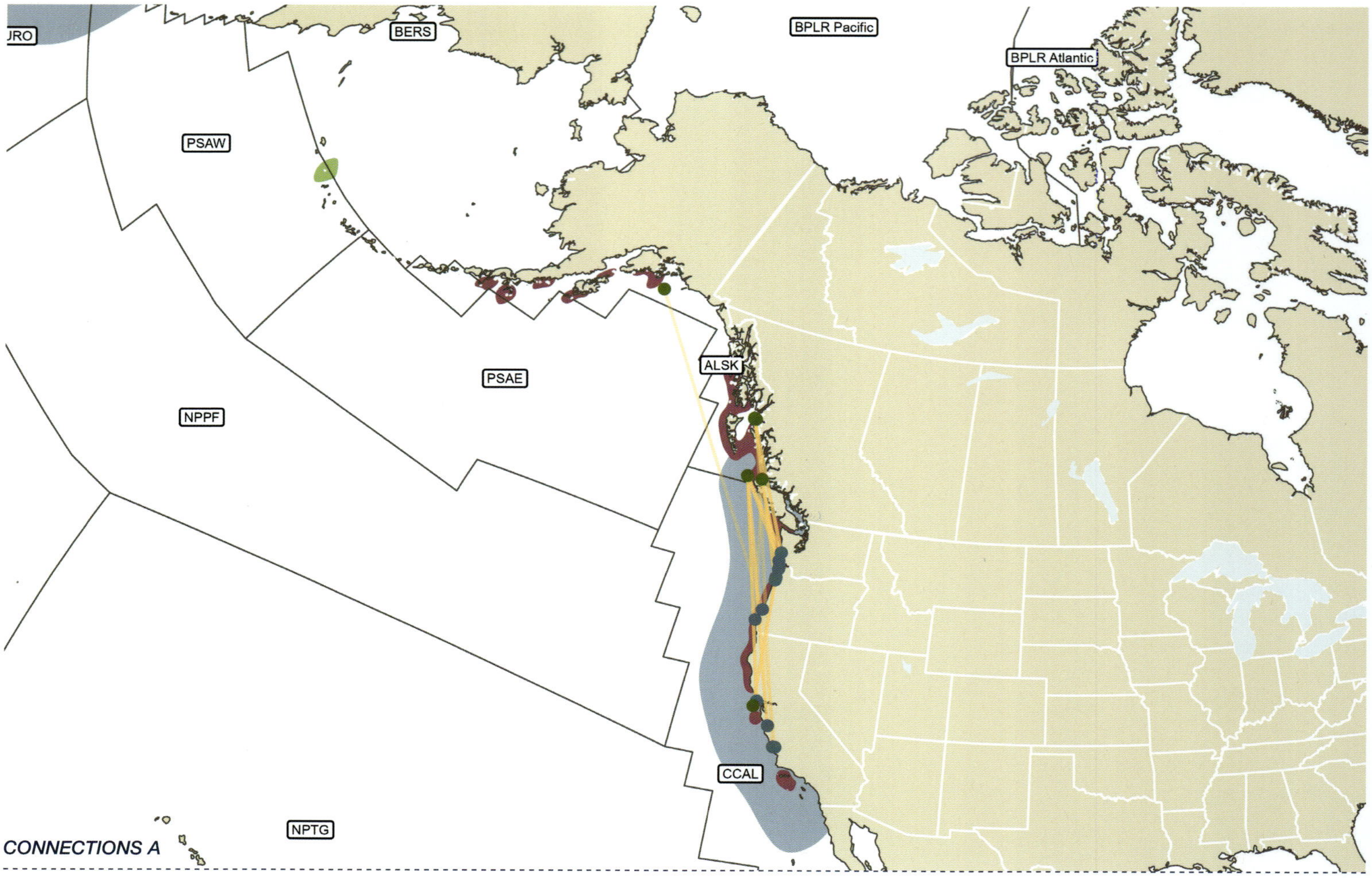

CONNECTIONS A

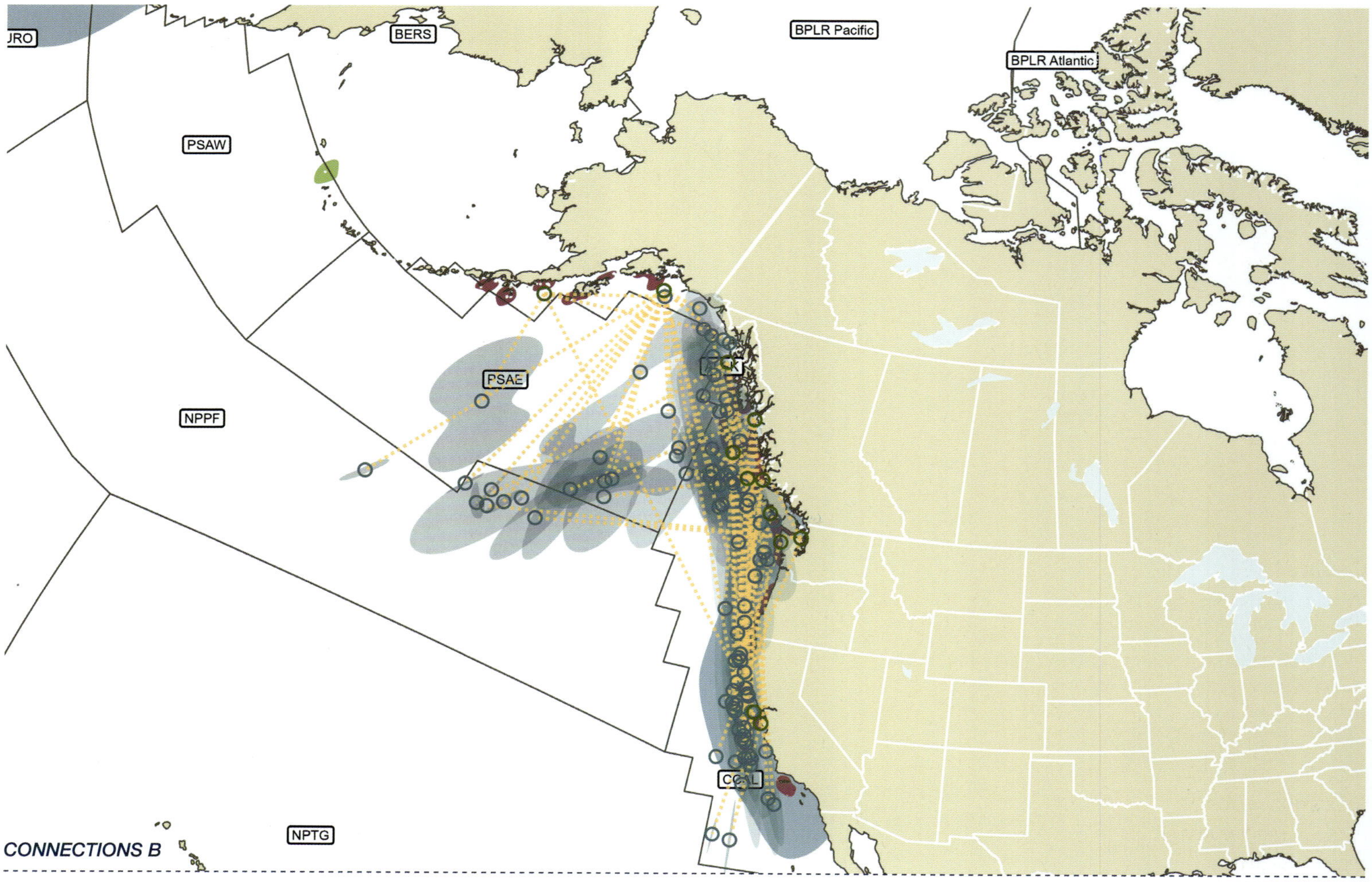

CONNECTIONS B

MIGRATORY CONNECTIVITY

OVERALL Weak migratory connectivity (Mantel = −0.07 ± 0.03) within North American colonies is indicated from tracking data and genetics information. Colonies from across the North American range mix within the Alaska Coastal Downwelling Province, California Current, North Pacific Polar Front, and Pacific Subarctic Gyres during the nonbreeding period. However, there is some regional structure. For example, birds from Alaska used California Current habitats (12%–60%) less than birds from British Columbia (78%).

BANDING Eleven of 13 band recoveries to nonbreeding areas are from British Columbia south to Washington, Oregon, and California, US. However, one breeding bird from the Farallon Islands, California, in 1988 was recovered north in Oregon in 1990. Most birds were found dead on the beach and imply a much more coastal nonbreeding distribution than is recorded from tracking data, indicating the coastal bias of band recoveries for this species.

GEOLOCATOR TRACKING Hipfner et al. (2020) deployed light-level geolocators on 12 colonies in the eastern Pacific. These birds generally used continental shelf habitats during the nonbreeding season. Using a home range overlap approach to assess connectivity, the authors found moderate overlap overall and high overlap within regions—California colonies shared core areas; Alaskan colonies also had high overlap. Some structure in overlap also existed within British Columbia colonies, and birds from two colonies in Washington had little overlap with each other. Patterns for Alaskan birds tracked with geolocators were shown also in Shoji et al. (2021). Birds tracked during the nonbreeding period at a Japanese colony (Takahashi et al. 2015) remained within the western Pacific in the Sea of Okhotsk and the Sea of Japan.

STABLE ISOTOPES Stable isotopes in feathers grown during the nonbreeding season have been used to identify nonbreeding latitude (Elliott et al. 2021), to evaluate foraging trophic level in nonbreeding (Takahashi et al. 2015) and breeding populations (Carle et al. 2015), as well as to determine the influence of pre-breeding diet composition on the size of the bill extension (Sorensen et al. 2010).

GENETICS Nuclear DNA analysis across 18 breeding colonies (n = 424 individuals) indicated strong genetic differentiation between populations in the eastern versus western Pacific Ocean (Hipfner et al. 2020). This supported previous findings comparing British Columbia, Alaska, and Japan colonies (Abbott et al. 2014). Hipfner et al. (2020) also found some within-region structure in populations on both sides of the Pacific, which could be attributed to moderate structure in nonbreeding distributions found by tracking.

OTHER APPROACHES The nonbreeding latitude has been associated with mercury and persistent organic pollutants in eggs. Elliott et al. (2021) found that the eggs of birds spending the nonbreeding season at higher latitudes had increased levels of mercury and mirex but decreased levels of DDE and PCBs. Given some regional structure among colonies on a latitudinal gradient (Hipfner et al. 2020), colonies appear to vary in their exposure to pollutants.

MIGRATORY PATHWAYS Migration pathways are typically either coastal and southward or southwest to offshore pelagic habitats with variability in the extent of migration (Hipfner et al. 2020).

DATA CONTRIBUTORS

K.H. Elliott; S.A. Hatch; J.M. Hipfner; A. Shoji; S. Whelan.

ATLANTIC PUFFIN

Fratercula arctica

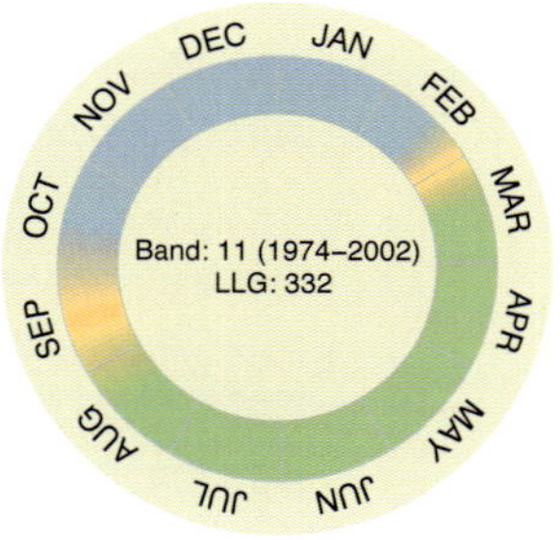

DISTRIBUTION AND ABUNDANCE

This species is a medium-sized auk with a famously colorful bill that breeds widely in the North Atlantic (Lowther et al. 2020d). Atlantic Puffins are rarely encountered during the nonbreeding period; they usually spend time far from shore throughout the high seas, but also south to the middle latitudes of North America and the Mediterranean (Fayet et al. 2017).

ORDER Charadriiformes
FAMILY Alcidae
BREEDING HABITAT Coasts, Cliffs and Islands
NONBREEDING HABITAT Coasts, Marine Waters
GLOBAL POPULATION SIZE 12,000,000
POPULATION CHANGE SINCE 1970 +31%
IUCN RED LIST CATEGORY Vulnerable
R2R TIPPING POINT SPECIES No
US LISTING STATUS No
CANADA SPECIES AT RISK No
PIF CCS 9/20

MIGRATORY CONNECTIVITY

OVERALL Geolocator and genetic data suggest a moderate to strong degree of connectivity (Mantel = 0.66 ± 0.03) across Atlantic Puffin populations, but they overlap regionally. For example, almost all puffins from Newfoundland and Labrador, and New Brunswick, Canada, are estimated to use core habitats in the western provinces (Northwest Atlantic Shelves and Gulf Stream); the majority of puffins from Ireland and the Faroe Islands are estimated to use Northeast Atlantic Shelves and the North Atlantic Drift provinces; and over 50% of individuals from mainland Norway and other places in Norway used the Atlantic Arctic Current province.

	NASW	GFST	NWCS	BPLR Atlantic	ARCT	SARC	NECS	NADR	NASE	CNRY	MEDI
Machias Seal Island, NB, Canada (n = 47)	4%	26%	70%	–	–	–	–	–	–	–	–
Witless Bay, NL, Canada (n = 37)	–	14%	84%	–	3%	–	–	–	–	–	–
Dalrymple Rock, Greenland (n = 8)	–	12%	–	38%	38%	–	–	12%	–	–	–
Iceland (n = 84)	–	–	7%	–	49%	–	–	43%	1%	–	–
Faroe Islands (n = 9)	–	–	–	–	–	–	11%	67%	22%	–	–
Ireland (n = 27)	–	–	–	–	4%	–	26%	37%	19%	4%	11%
Bjørnøya and Svalbard, Norway (n = 3)	–	–	–	33%	67%	–	–	–	–	–	–
United Kingdom (n = 1)	–	–	–	–	–	100%	–	–	–	–	–
Runde and Ålesund, Norway (n = 17)	–	–	–	6%	65%	12%	6%	12%	–	–	–
Mainland Norway (n = 102)	–	–	–	27%	52%	20%	–	1%	–	–	–

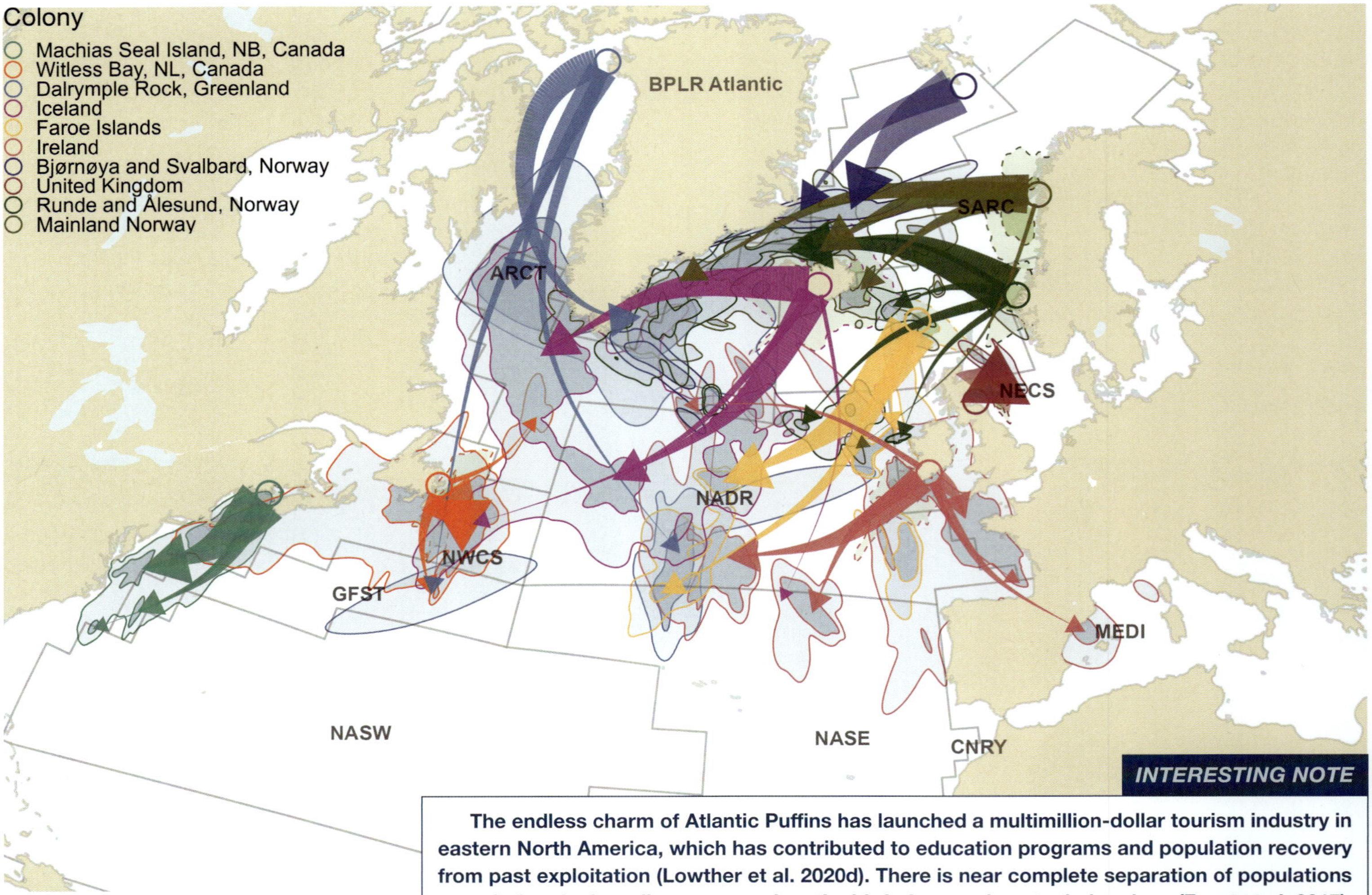

INTERESTING NOTE

The endless charm of Atlantic Puffins has launched a multimillion-dollar tourism industry in eastern North America, which has contributed to education programs and population recovery from past exploitation (Lowther et al. 2020d). There is near complete separation of populations during the breeding season when the birds forage close to their colony (Fayet et al. 2017).

GENERALIZED CONNECTIVITY

BANDING Compared to co-occurring alcids in North America, there are relatively few band reencounters to connect breeding colonies to their nonbreeding ranges (n = 14). This likely reflects their offshore distribution. Birds banded in 1972–2002 in Maine, US, and New Brunswick and Newfoundland and Labrador, Canada, were predominantly reencountered in Massachusetts, US, or in Newfoundland and Labrador, with one in Quebec, Canada.

GEOLOCATOR TRACKING Fayet et al. (2017) synthesized light-level geolocator tracks from 270 puffins across most of the species' breeding and nonbreeding ranges. Colonies had distinct nonbreeding ranges but always overlapped with one to three other colonies. Puffins from larger colonies, and where local nonbreeding conditions were poorer, migrated farther to more distant nonbreeding areas. Key colonies from the southern and northern breeding range in North America were not represented in the above study. Birds from three colonies in the Gulf of Maine (Baran et al. 2022) mixed on nonbreeding sites along the Atlantic coast; the northern part of their nonbreeding region overlapped with birds tracked from Newfoundland and Labrador, Canada (Fayet et al. 2017). In general, birds from Atlantic Canada made much shorter migrations than did high-latitude colonies. Burnham et al. (2021) presented the first data from northwest Greenland on the large-bodied high Arctic subspecies (*F. a. naumanni*). Two tagged females spent the nonbreeding season in the mid-Atlantic Ocean and toward the southern limit, while six males remained at the northern limit, using the sea-ice edge at colder nonbreeding areas than previously studied

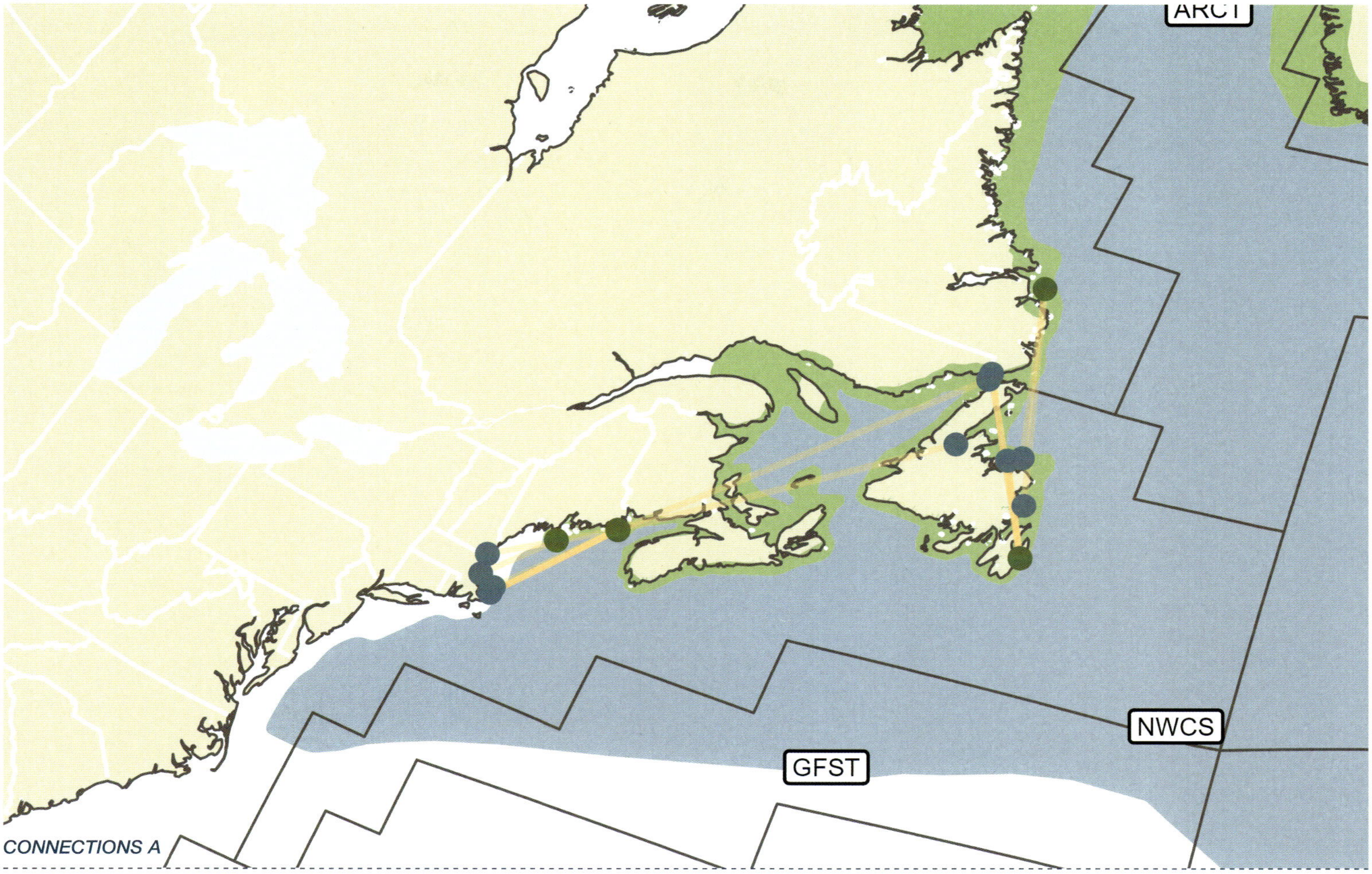

CONNECTIONS A

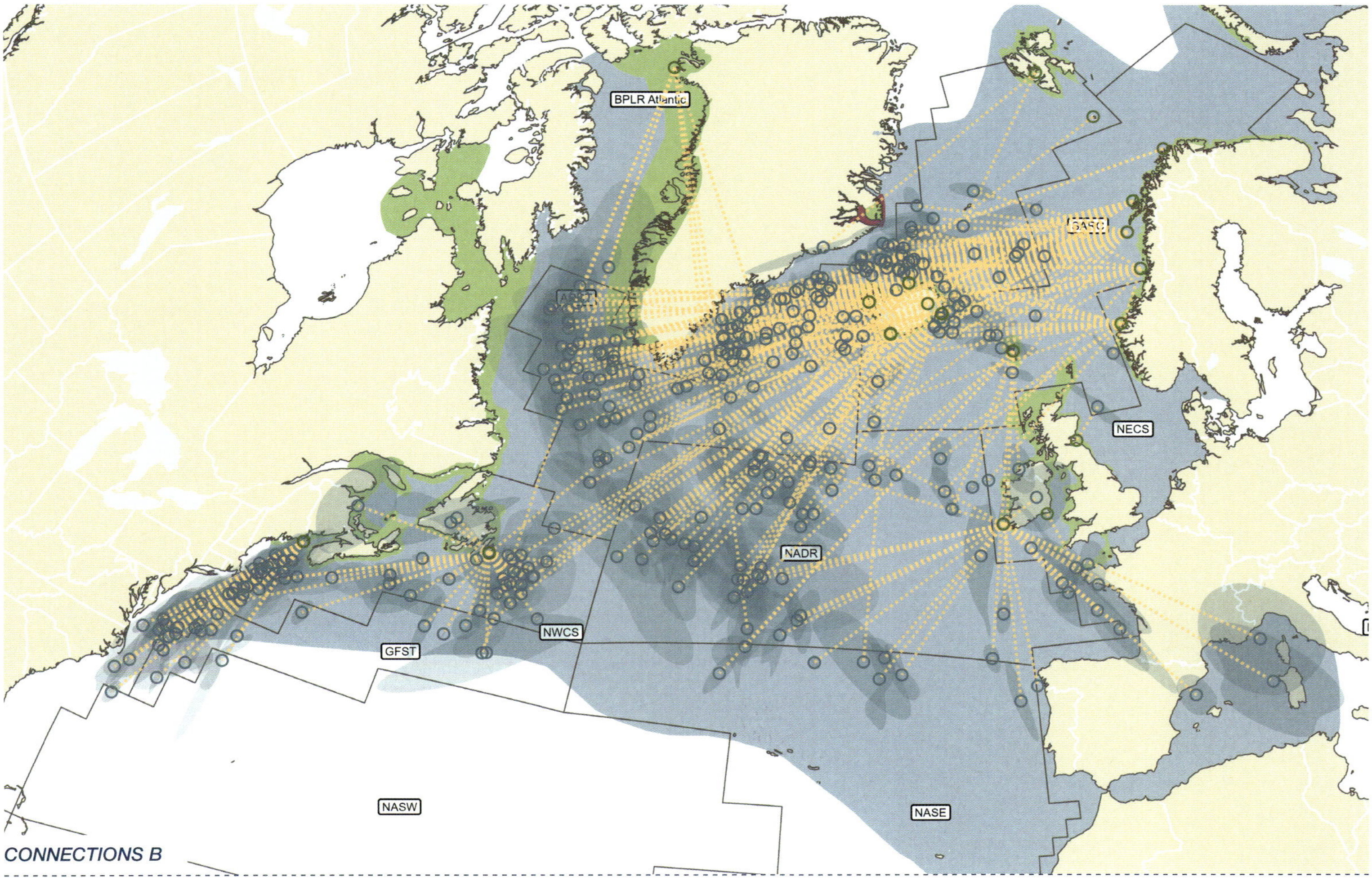

CONNECTIONS B

populations. This study showed that some individuals from this same colony spent nonbreeding seasons thousands of kilometers apart.

STABLE ISOTOPES Baran et al. (2022) found that stable isotopes of claw tissue were not feasible for identifying nonbreeding areas of birds from three colonies in the Gulf of Maine (also tracked with geolocators) due to the similarity of nonbreeding areas used across colonies.

GENETICS A whole-genome study of 72 puffins across 12 colonies concluded that Atlantic Puffins form four genetic groups: Spitsbergen (high Arctic), Canada, Scotland (Isle of May), and one group consisting of colonies across Iceland, the Faroe Islands, and Norway (Kersten et al. 2021). Carvey et. al (2024) suggest that Machias Seal Island in the Gulf of Maine represents a fifth genetic cluster and should be managed separately from other Atlantic Canada and Quebec colonies. Hybrids also exist (Kersten et al. 2023). Kersten et al. (2021) suggested that philopatry alone cannot explain the structure and that nonbreeding distribution plays an important role in this separation. This study did not include US, northern Canada, or Greenland populations. A unique situation occurs in northwestern Greenland, where puffins from two size phenotypes and three distinct genetic clusters breed sympatrically (Leigh et al. 2023).

MIGRATORY PATHWAYS Precise migration routes are difficult to resolve from light-level geolocator data. Birds tagged in the Gulf of Maine (Baran et al. 2022) generally stayed in the Gulf of Maine from August to January, then began a movement over the continental shelf as far south as Cape Hatteras, North Carolina. However, two individuals departed immediately after breeding. One first migrated to the coast of Newfoundland before continuing to the Gulf of St. Lawrence through December. The other likely took an overland migration between the Gulf of Maine and the Gulf of St. Lawrence, corroborated by the tag immersion data (Baran et al. 2022). Males tracked from northwest Greenland (Burnham et al. 2021) followed a coastal pathway to southeast Greenland, while the pathways that females took varied widely, including a route to the mid-Atlantic south of the Azores, one to Iceland where the tag ceased recording, and another to northeast of the Azores, but the following year, the bird remained in the western North Atlantic. For 100 individuals with migration data tracked from the United Kingdom, Fayet et al. (2017) found that routes were variable between individuals from the same colony but that there was inter-individual consistency between years (up to six years of tracking for some individuals). Some birds stayed local, some migrated southeast to the Mediterranean, and some migrated across the North Atlantic to North America.

DATA CONTRIBUTORS

M.A. Baran; K.K. Burnham; J. Danielsen; A.W. Diamond; D.A. Fifield; A. Hedd; M. Jessopp; H.L. Major; B. Olsen; A. Petersen.

HORNED PUFFIN *Fratercula corniculate*

ORDER Charadriiformes
FAMILY Alcidae
BREEDING HABITAT Coasts, Cliffs and Islands
NONBREEDING HABITAT Coasts, Marine Waters
GLOBAL POPULATION SIZE 800,000
POPULATION CHANGE SINCE 1970 NA
IUCN RED LIST CATEGORY Least Concern
R2R TIPPING POINT SPECIES No
US LISTING STATUS No
CANADA SPECIES AT RISK No
PIF CCS 11/20

MIGRATORY CONNECTIVITY

OVERALL Only a BirdLife distribution range map is available; no eBird data or migratory connectivity information were found for this species.

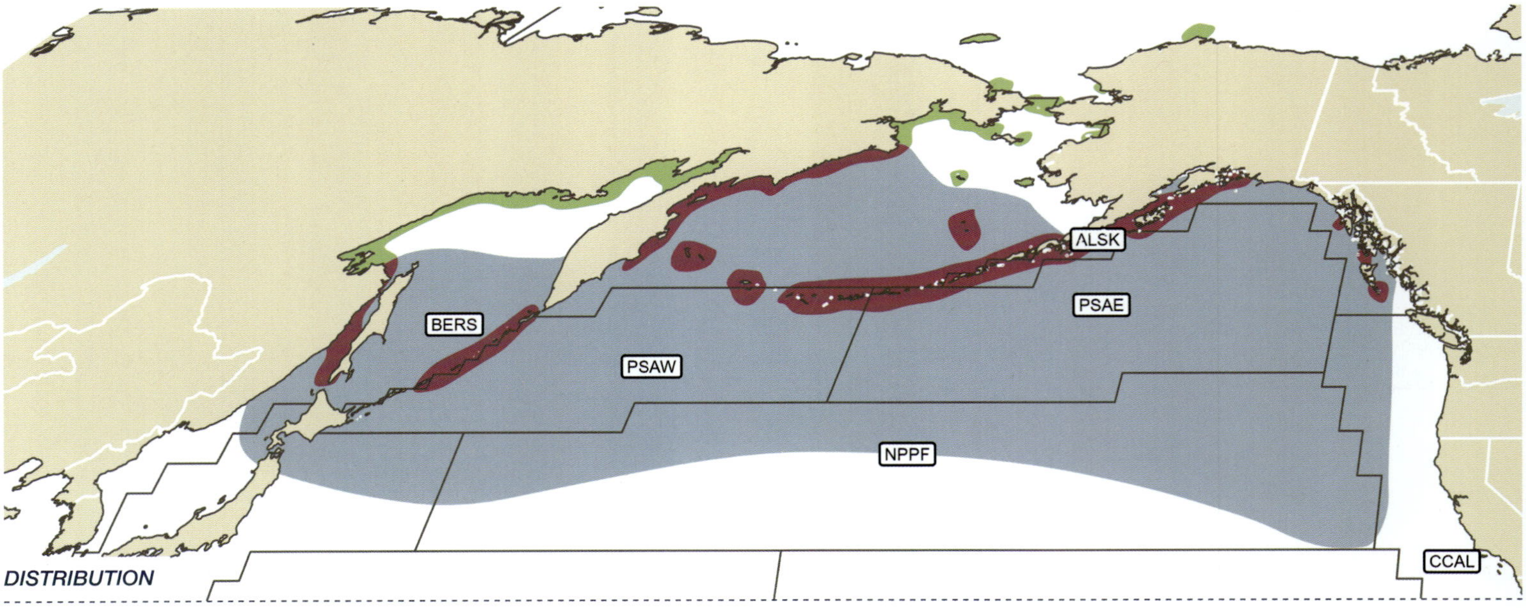

DISTRIBUTION

TUFFTED PUFFIN

Mantel = 0.53 ± 0.08

Fratercula cirrhata **TUFTED PUFFIN**

ORDER Charadriiformes
FAMILY Alcidae
BREEDING HABITAT Coasts, Cliffs and Islands
NONBREEDING HABITAT Coasts, Marine Waters
GLOBAL POPULATION SIZE 2,300,000
POPULATION CHANGE SINCE 1970 NA
IUCN RED LIST CATEGORY Least Concern
R2R TIPPING POINT SPECIES No
US LISTING STATUS No
CANADA SPECIES AT RISK No
PIF CCS 11/20

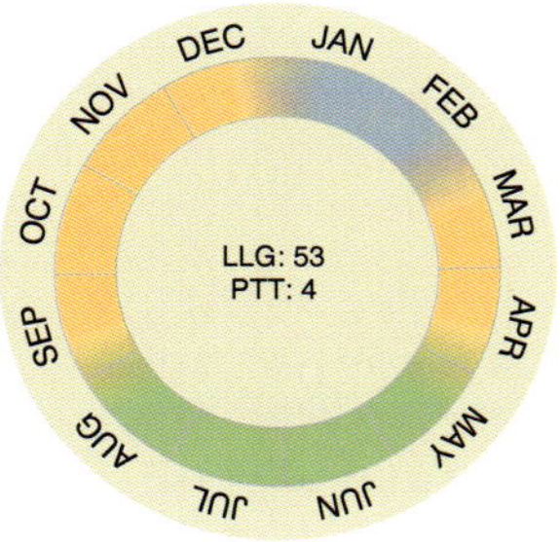

DISTRIBUTION AND ABUNDANCE

This species nests in burrows primarily in Beringia, but colonies located as far south as California, US, in the east and Hokkaido, Japan, in the west. The Alaskan Peninsula and the Aleutian Islands contain 67% of the global population (approximately 2.3 million breeding birds), with other significant populations in the Sea of Okhotsk (500,000), the Kuril Islands (200,000), and approximately 100,000 birds along the Pacific coast of the US and Canada (Piatt and Kitaysky 2020). Considered the most pelagic of the alcids, they spend the nonbreeding season across the North Pacific (Piatt and Kitaysky 2020).

INTERESTING NOTE

Long-term declines have been confirmed in the California Current and the Gulf of Alaska, regions that constitute three-quarters of the species' North American range, and for the Gulf of Alaska, a significant portion of the North American breeding population (Pearson et al. 2023). A project to track puffins as they leave breeding grounds in Washington state, as a step in investigating population declines, is ongoing (J. Gaydos, S. Pearson, and P. Hodum, personal communication).

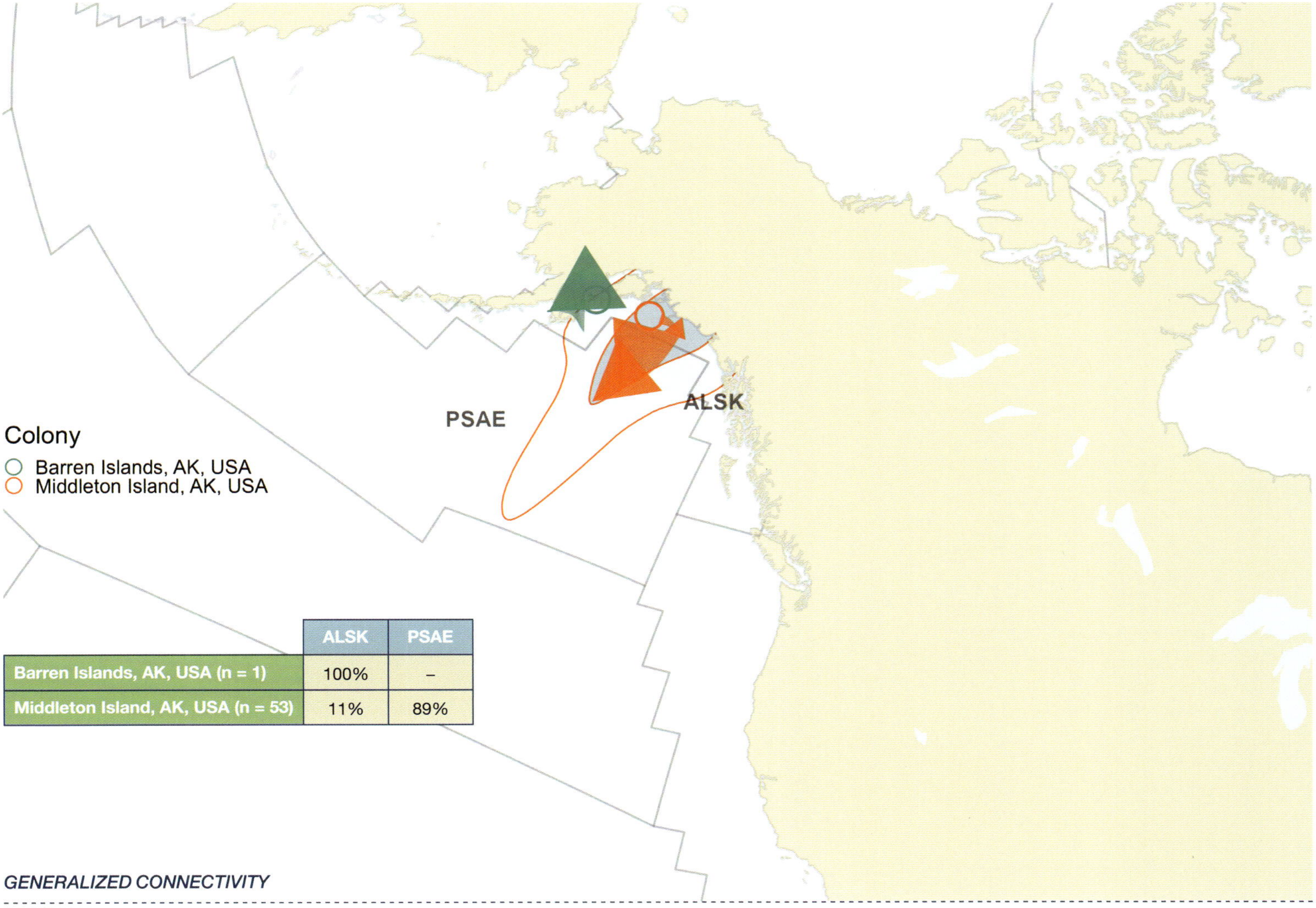

	ALSK	PSAE
Barren Islands, AK, USA (n = 1)	100%	–
Middleton Island, AK, USA (n = 53)	11%	89%

GENERALIZED CONNECTIVITY

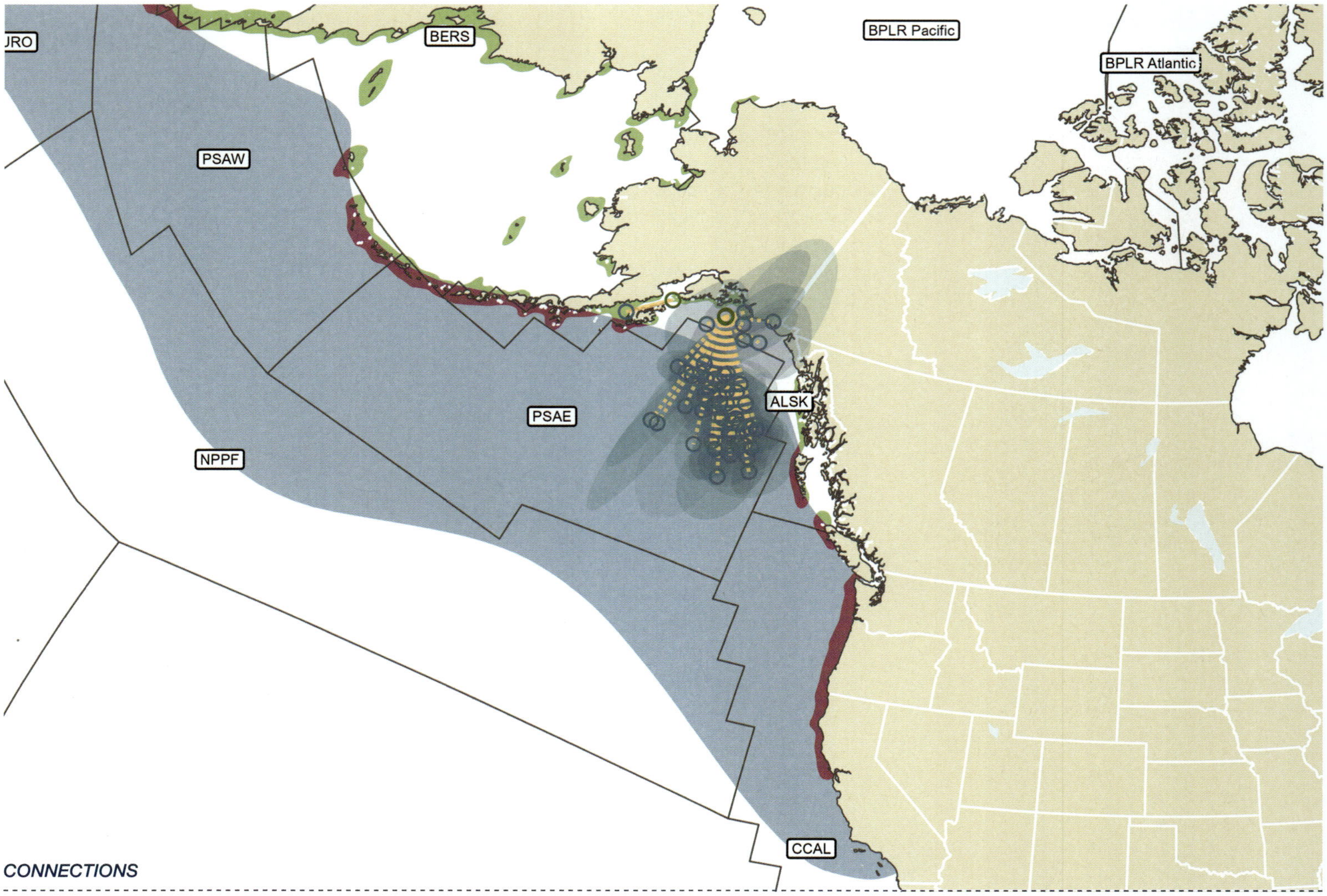

CONNECTIONS

MIGRATORY CONNECTIVITY

OVERALL Genetic evidence indicates low differentiation among North American colonies, as well as high dispersal among Russian and Alaskan colonies, although nonbreeding mixing of colonies across the oceanic range is unknown. Tracking data are available only from two relatively close Alaskan colonies. Migratory connectivity is strong (Mantel = 0.53 ± 0.08), but sample size is highly skewed toward Middleton Island, Alaska (n = 53), with only a single individual tracked from the Barren Islands, Alaska. Thus, inference across colonies is limited. Individuals from Middleton Island migrated to nonbreeding areas within the Eastern Pacific Subarctic Gyres (89%) and Alaska Coastal Downwelling Province (11%).

GEOLOCATOR TRACKING Light-level geolocator data from Middleton Island in 2018–2020 (n = 45; Schaefer et al. 2022) and in 2017–2020 (n = 8; Shoji et al. 2023) showed birds spending the nonbreeding season throughout the Gulf of Alaska.

SATELLITE TRACKING One of five birds tracked with surgically implanted tags in 1995 from Barren Islands, Alaska, transmitted from mid-July through late September (Hatch et al. 2000, 2020). The bird left the waters nearby its colony on August 20, after which it moved to a pelagic area about 150 km southeast of its colony, where it remained for about a month before it ceased transmission. This Atlas considered this a nonbreeding area based on migration duration and timing.

STABLE ISOTOPES Stable carbon (spatial signature) and nitrogen (trophic level) isotopic signatures differed significantly for two colonies of Tufted Puffins during the breeding season in British Columbia, Canada, separated by only 1.5 km (Hipfner et al. 2007). These differences were linked with annual breeding success. This result suggested that even nearby colonies can show spatial and diet separation within the breeding season.

GENETICS A mitochondrial and nuclear DNA study of captive birds originally from far eastern Russia, southeastern Alaska, and Kodiak Island, Alaska, suggested no genetic differentiation (Sakurayama et al. 2021). Contemporary at-sea mixing of these populations is not known. A study of seven North American breeding colonies spanning Oregon to the western Aleutian Islands, Alaska, also found low genetic differentiation, with mitochondrial DNA showing extensive historical gene flow (Graham et al. 2023). The authors interpret their findings to suggest dispersal among colonies is high, "likely facilitated by an overlap in nonbreeding ranges among colonies" (Graham et al. 2023).

OTHER APPROACHES No banding studies of migratory connectivity were found for this species.

MIGRATORY PATHWAYS Puffins generally migrated from their colonies south to southeast to the eastern Gulf of Alaska, departing their breeding ground from August to September, and their nonbreeding grounds in late March (Schaefer et al. 2022). Like other alcids, both males and females tracked throughout the nonbreeding period shifted their distributions south during the nonbreeding season to track oceanographic patterns (Schaefer et al. 2022).

DATA CONTRIBUTORS

M.A. Bishop; D.C. Douglas; K.H. Elliott; K.B. Gorman; S.A. Hatch; D.M. Mulcahy; P.M. Myers; A.L. Schaefer; A. Shoji; S. Whelan.

Mantel = 0.75 ± 0.02

Rissa tridactyla **BLACK-LEGGED KITTIWAKE**

DISTRIBUTION AND ABUNDANCE

This species is a small, circumpolar gull that nests in colonies on cliffs of the northern oceans. Two subspecies are recognized, the smaller *R. t. tridactyla* of the North Atlantic, and the larger *R. t. pollicaris* of the North Pacific. The Pacific population is estimated to be 7 million individuals (S.A. Hatch et al. 2020b). Estimates for Atlantic populations include around 350,000 pairs in eastern North America (S.A. Hatch et al. 2020b). Highly pelagic over northern oceans during the nonbreeding period where they feed on small fish and macrozooplankton at the surface (S.A. Hatch et al. 2020).

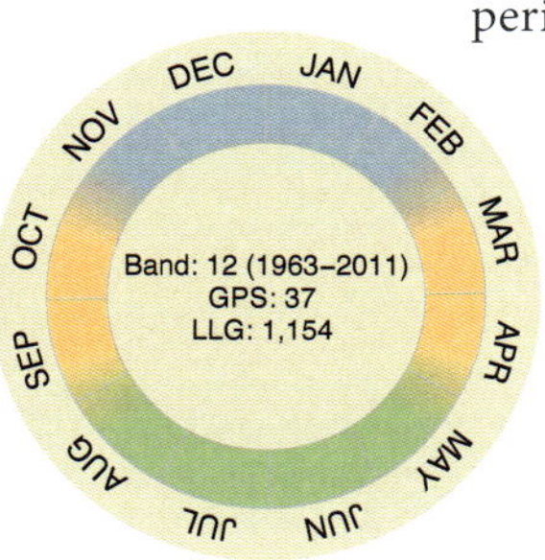

ORDER Charadriiformes
FAMILY Laridae
BREEDING HABITAT Coasts, Cliffs and Islands
NONBREEDING HABITAT Coasts, Marine Waters
GLOBAL POPULATION SIZE 10,000,000
POPULATION CHANGE SINCE 1970 +3%
IUCN RED LIST CATEGORY Vulnerable
R2R TIPPING POINT SPECIES No
US LISTING STATUS No
CANADA SPECIES AT RISK No
PIF CCS 12/20

MIGRATORY CONNECTIVITY

OVERALL Migratory connectivity across ocean basins is strong (Mantel = 0.75 ± 0.02), but may be weak (Atlantic Ocean) or moderate (Pacific Ocean) within ocean basins. In the Atlantic, there is large overlap of birds from different colonies, especially in the western Atlantic, including the Atlantic Arctic, North Atlantic Drift, and the Northwest Atlantic Shelves provinces. Pacific Ocean colonies had stronger separation of westerly and easterly colonies. Birds from southeast Alaska primarily used nonbreeding habitats in the Alaska Coastal Downwelling Province (61%) and California Current (21%).

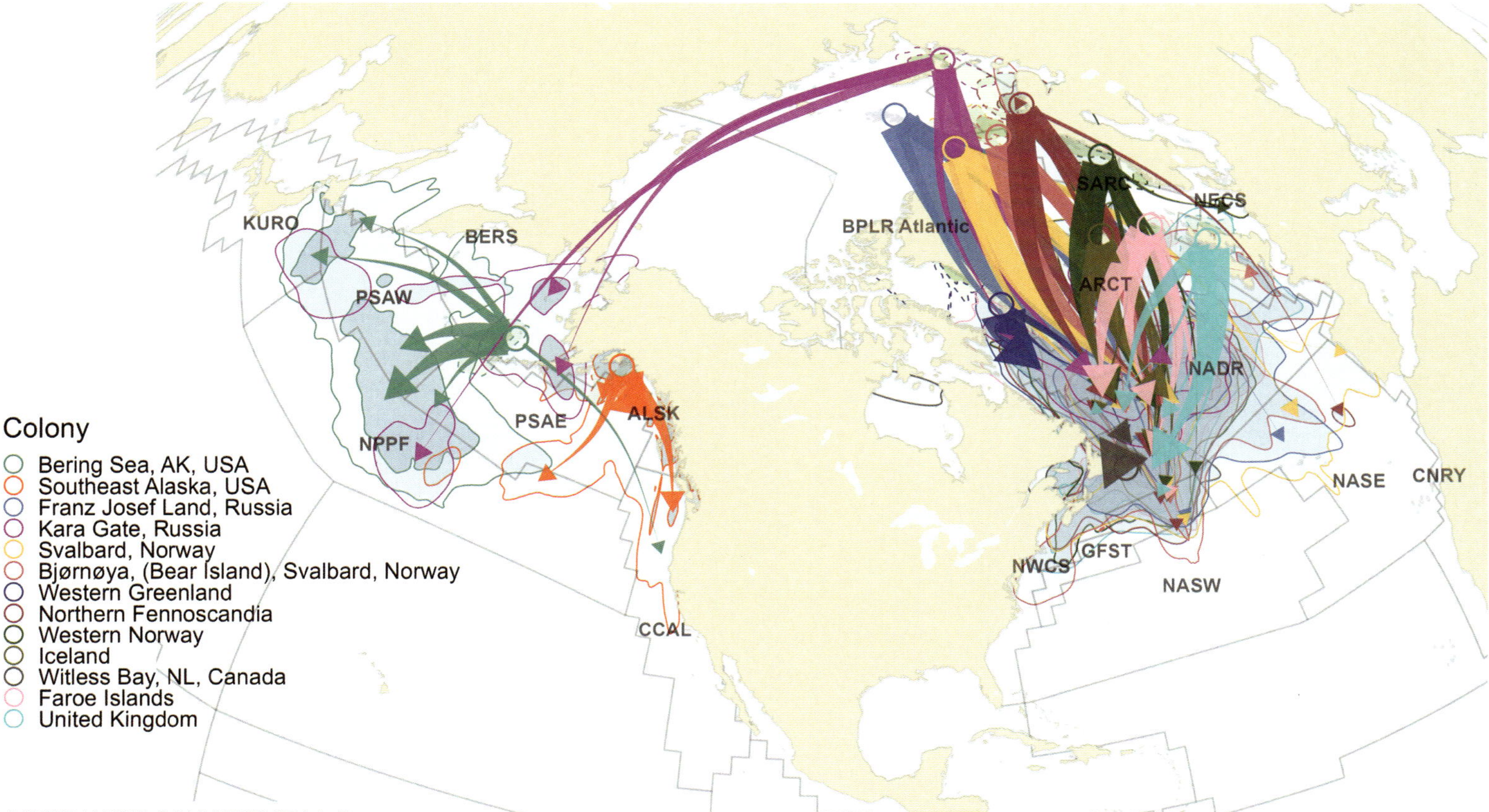

GENERALIZED CONNECTIVITY A–B

In contrast, the primary nonbreeding provinces of birds from Bering Sea colonies included the North Pacific Polar Front (42%) and the Pacific Subarctic Gyres (30%). Limited connectivity has also been observed to both the Atlantic and Pacific oceans by birds from a western Russian colony.

BANDING Two-thirds of the 12 breeding (1957–2005) to nonbreeding band recoveries (1963–2011) are from Alaskan breeding birds, with three from colonies in Newfoundland and Labrador, and one from Quebec. Nonbreeding recoveries are largely coastal (only one was recovered at sea), reflecting observation bias given this species' highly pelagic post-breeding distribution.

GEOLOCATOR TRACKING The Black-legged Kittiwake is perhaps one of the most tracked seabird species in terms of absolute number of individuals and number of colonies studied. Nonbreeding strategy is complex, and varies with location, breeding success, and sex (Bogdanova et al. 2011; McKnight et al. 2011; González-Solís 2011; Frederiksen et al. 2012; Orben et al. 2015c; Whelen et al. 2020; Ezhov et al. 2021). Carry-over effects have been found in both the Atlantic (Bogdonova et al. 2011) and the Pacific oceans (Whelen et al. 2020). Birds from a Russian colony in the southeastern Barents Sea have been tracked to both the Atlantic (78%) and Pacific (22%) oceans (Ezhov et al. 2021).

STABLE ISOTOPES Stable isotopes have successfully predicted nonbreeding location, and these have been confirmed by geolocators (González-Solís et al. 2011).

GENETICS Analysis of nuclear DNA (McCoy et al. 2005) and

	KURO	PSAW	BERS	ALSK	PSAE	NPPF	CCAL	NASW	GFST	NWCS
Bering Sea, AK, USA (n = 120)	11%	30%	8%	–	9%	42%	1%	–	–	–
Southeast Alaska, USA (n = 82)	–	–	–	61%	9%	10%	21%	–	–	–
Franz Josef Land, Russia (n = 45)	–	–	–	–	–	–	–	2%	–	27%
Kara Gate, Russia (n = 16)	–	–	12%	12%	–	12%	–	–	–	–
Svalbard, Norway (n = 199)	–	–	–	–	–	–	–	1%	3%	24%
Bjørnøya, (Bear Island), Svalbard, Norway (n = 24)	–	–	–	–	–	–	–	–	12%	25%
Western Greenland (n = 73)	–	–	–	–	–	–	–	–	–	7%
Northern Fennoscandia (n = 162)	–	–	–	–	–	–	–	1%	1%	28%
Western Norway (n = 249)	–	–	–	–	–	–	–	–	1%	25%
Iceland (n = 49)	–	–	–	–	–	–	–	–	2%	22%
Witless Bay, NL, Canada (n = 4)	–	–	–	–	–	–	–	–	–	75%
Faroe Islands (n = 44)	–	–	–	–	–	–	–	–	5%	25%
United Kingdom (n = 98)	–	–	–	–	–	–	–	–	5%	58%

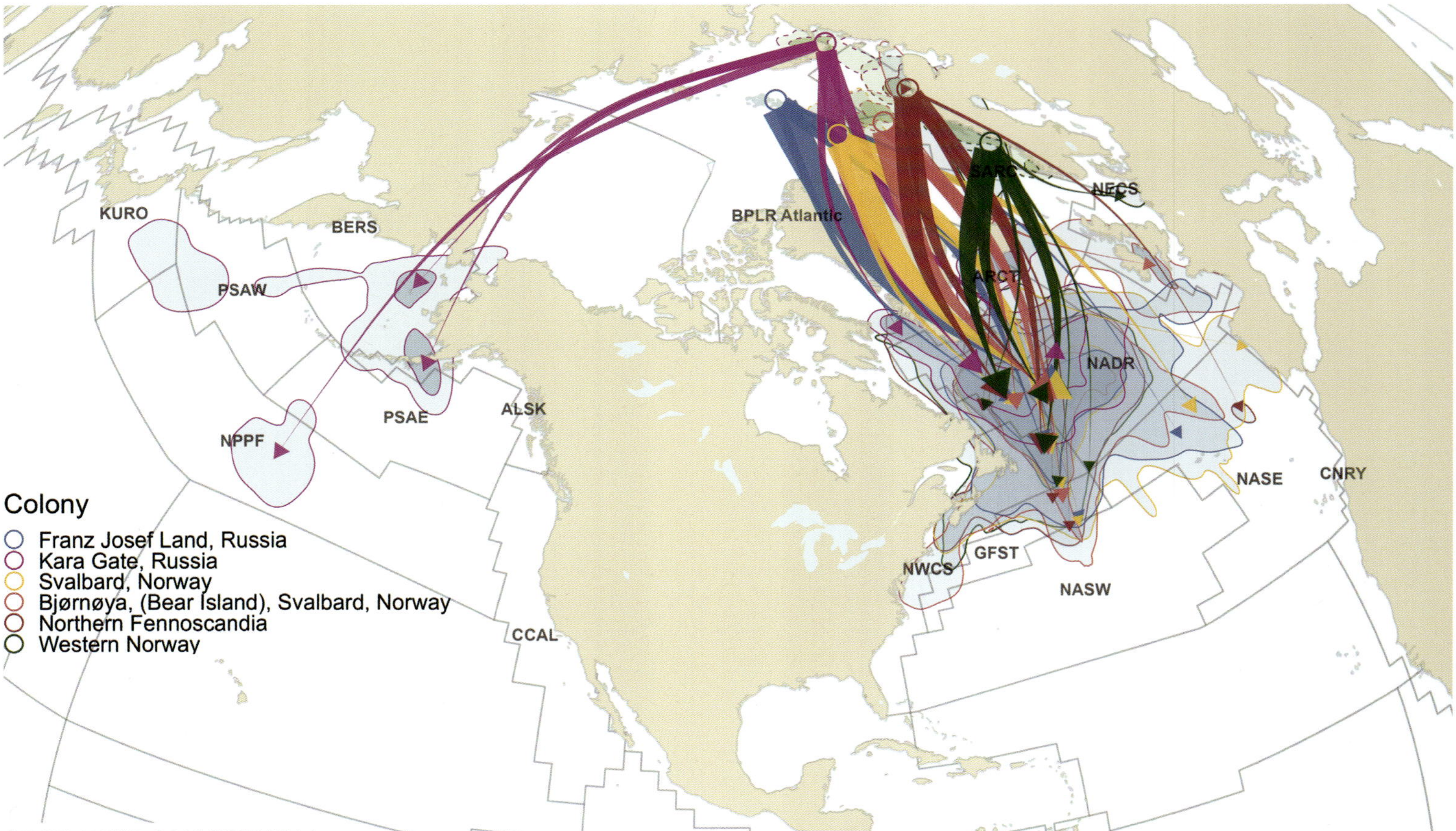

GENERALIZED CONNECTIVITY A

mitochondrial DNA (Sauve et al. 2019) in large multicolony studies confirm genetic differentiation between Atlantic and Pacific subspecies. Sauve et al. (2019) also found strong genetic structure of the mitochondrial DNA of Atlantic colonies, while nuclear markers suggested little genetic structure (McCoy et al. 2005), noting that mitochondrial DNA sorts four times faster than nuclear DNA. Such structure is not reflected in the overlapping nonbreeding distributions of most Atlantic colonies.

MIGRATORY PATHWAYS Birds tracked from a colony in Southeast Alaska showed three migration strategies, with birds remaining local, moving south along the Pacific coast, or moving far out to sea. In contrast, a colony in the Pribilof Islands made westerly or southwesterly migrations to habitats far offshore, and few remained local or migrated east (Orben et al. 2015c). Atlantic Ocean colonies include high-latitude sites, and birds from these colonies undertake longer southward migrations than has been observed in the Pacific Ocean. Birds from a southeastern Barents Sea colony travel either east to the Pacific Ocean or west and south to the Atlantic (Ezhov et al. 2021).

BPLR Atlantic	ARCT	SARC	NECS	NADR	NASE	CNRY
–	–	–	–	–	–	–
–	–	–	–	–	–	–
–	11%	–	–	56%	4%	–
12%	31%	–	–	19%	–	–
–	11%	–	–	49%	12%	–
–	17%	–	4%	42%	–	1%
25%	67%	–	–	1%	–	–
4%	40%	1%	–	24%	1%	–
3%	44%	<1%	1%	24%	1%	<1%
2%	29%	2%	–	43%	–	–
–	25%	–	–	–	–	–
–	45%	–	–	–	–	–
1%	20%	–	2%	–	–	–

DATA CONTRIBUTORS

J.A. Akearok; A.J. Allyn; T. Anker-Nilssen; T. Boulinier; I.S. Bringsvor; O. Chastel; S. Christensen-Dalsgaard; J. Danielsen; F. Daunt; N. Dehnhard; S. Descamps; K.H. Elliott; K.E. Erikstad; A. Ezhov; M. Frederiksen; A.J. Gaston; M.V. Gavrilo; J. González-Solís; D. Grémillet; S.A. Hatch; A. Hedd; D.B. Irons; E. Kern-Lovick; Y. Kolbeinsson; A. Labansen; J.F. Linnebjerg; S.-H. Lorensten; M.L. Mallory; A. McKnight; F.R. Merkel; B. Moe; B. Olsen; R.A. Orben; K.M. Sullivan; R.M. Suryan; H. Strøm; R. Swann; G.H. Systad; T.L. Thorarinsson; S. Whelan.

INTERESTING NOTE

Atlantic populations have experienced large declines, while in the Alaskan portion of the Pacific, population trends have been difficult to accurately determine, with mixed results from different colonies and possible emigration/immigration to different colonies obscuring overall trends (S.A. Hatch et al. 2020). Long-term studies summarized in Hatch et al. (2020) show large differences between life history traits of Pacific and Atlantic kittiwakes, with birds in the Atlantic Ocean generally more productive but with shorter lives.

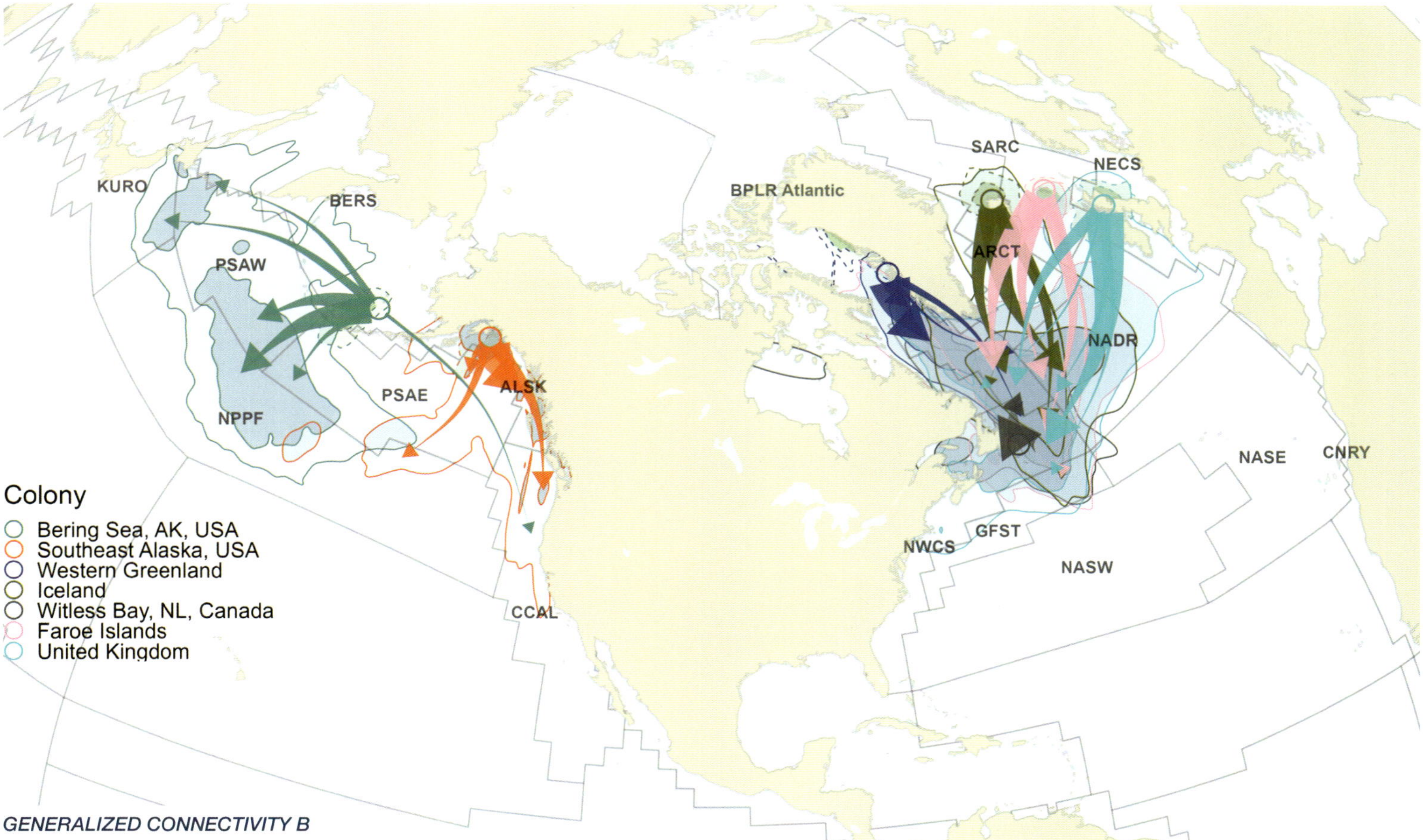

GENERALIZED CONNECTIVITY B

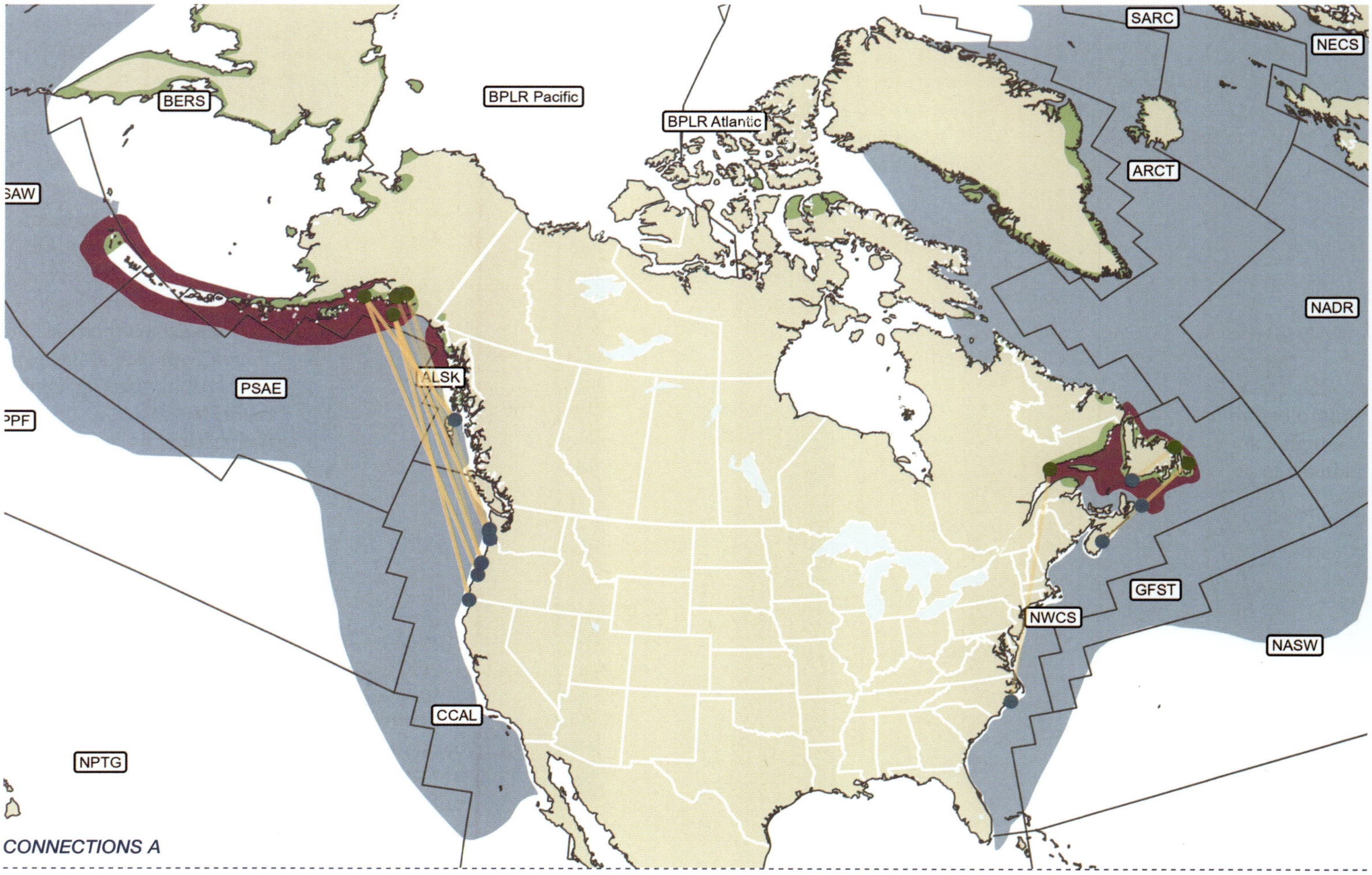

CONNECTIONS A

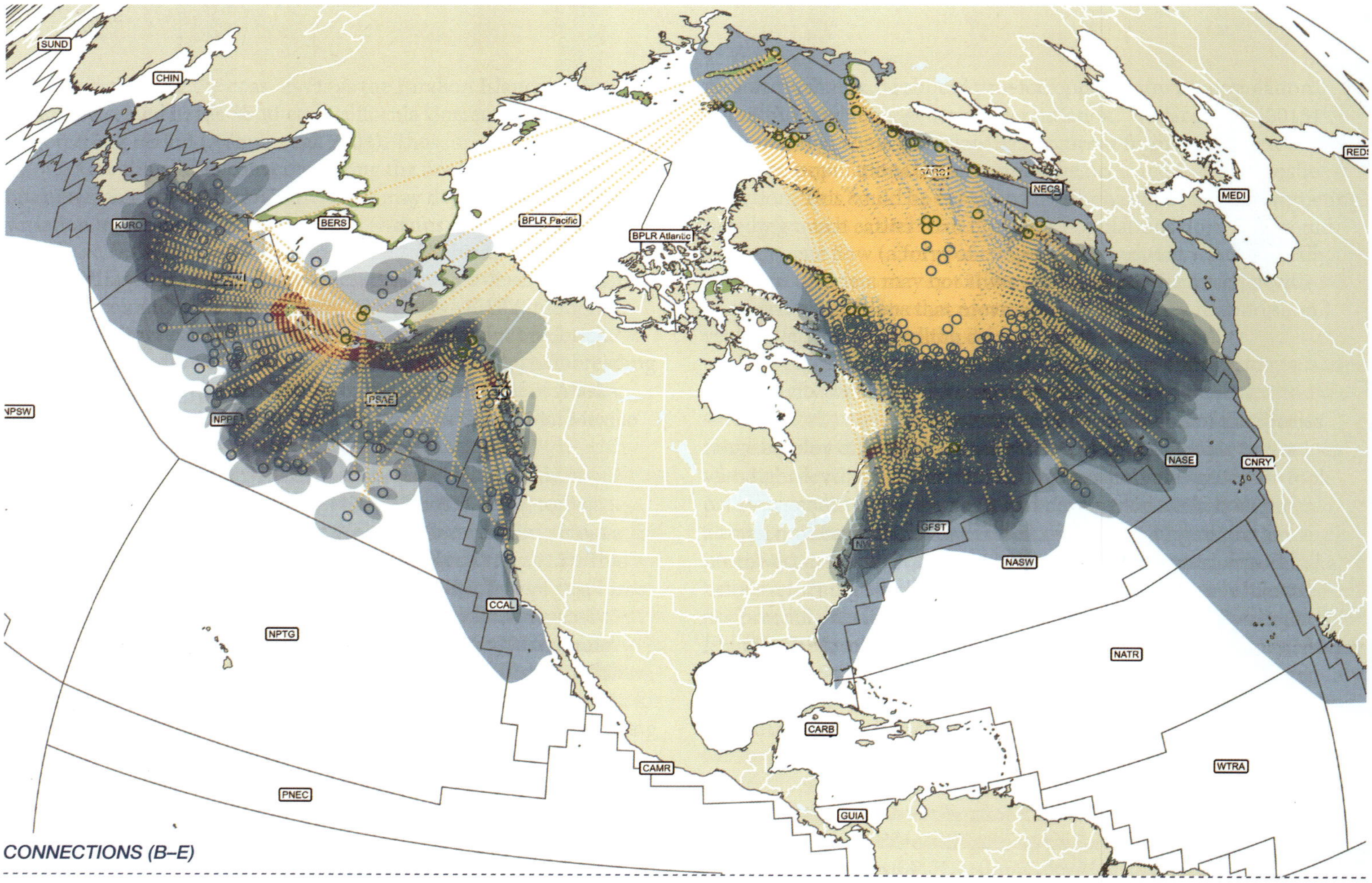

CONNECTIONS (B–E)

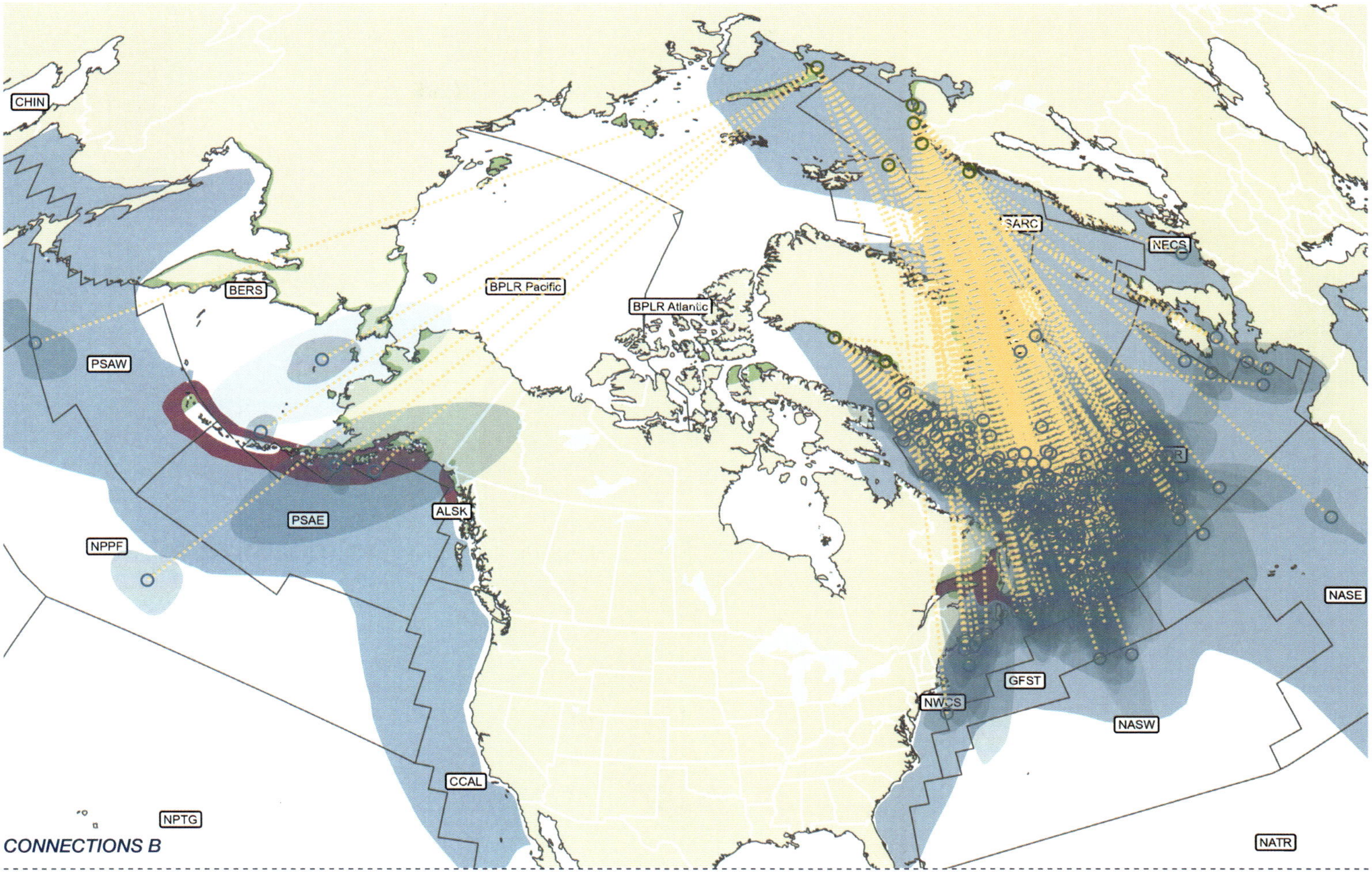

CONNECTIONS B

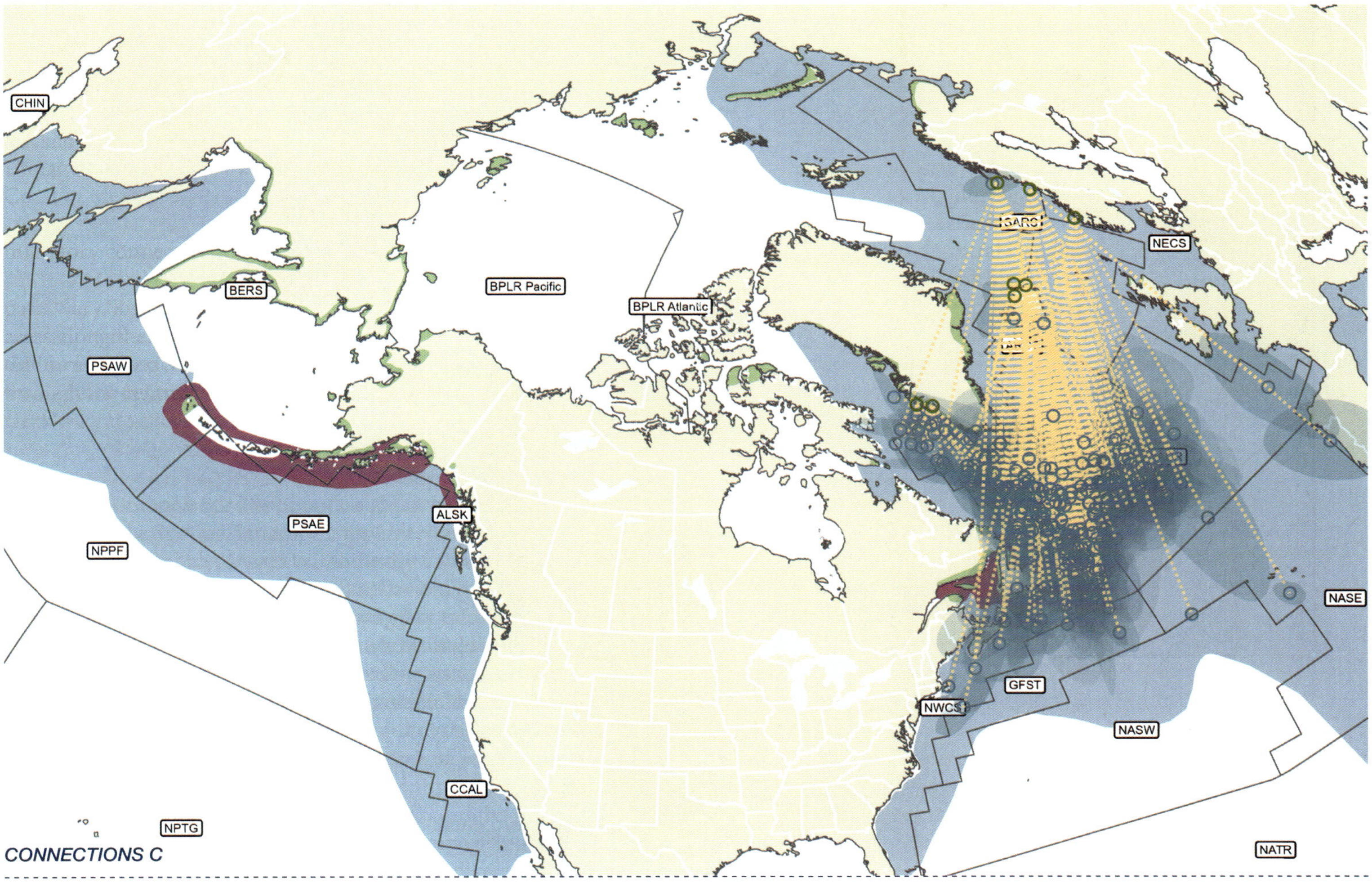

CONNECTIONS C

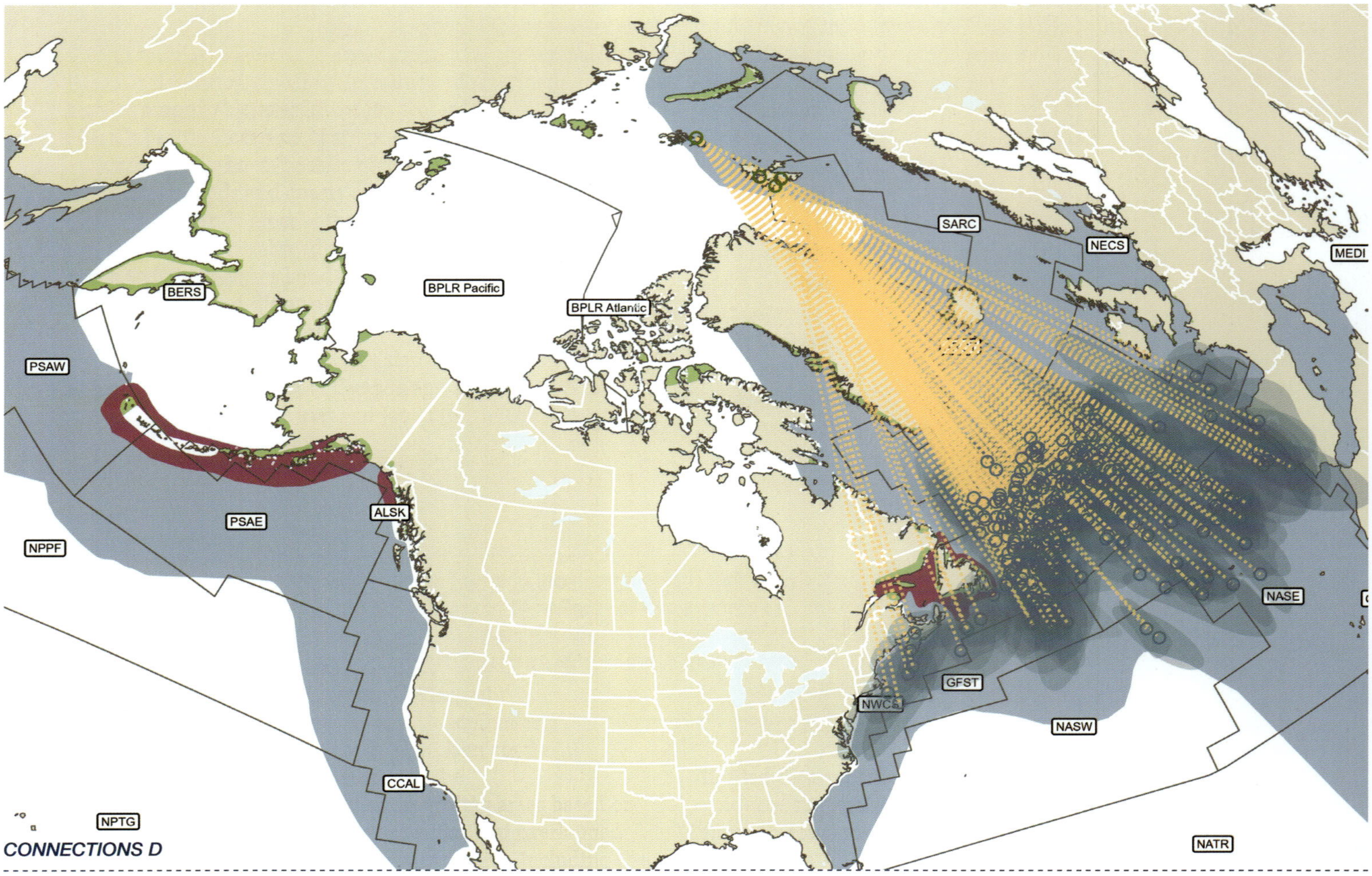

CONNECTIONS D

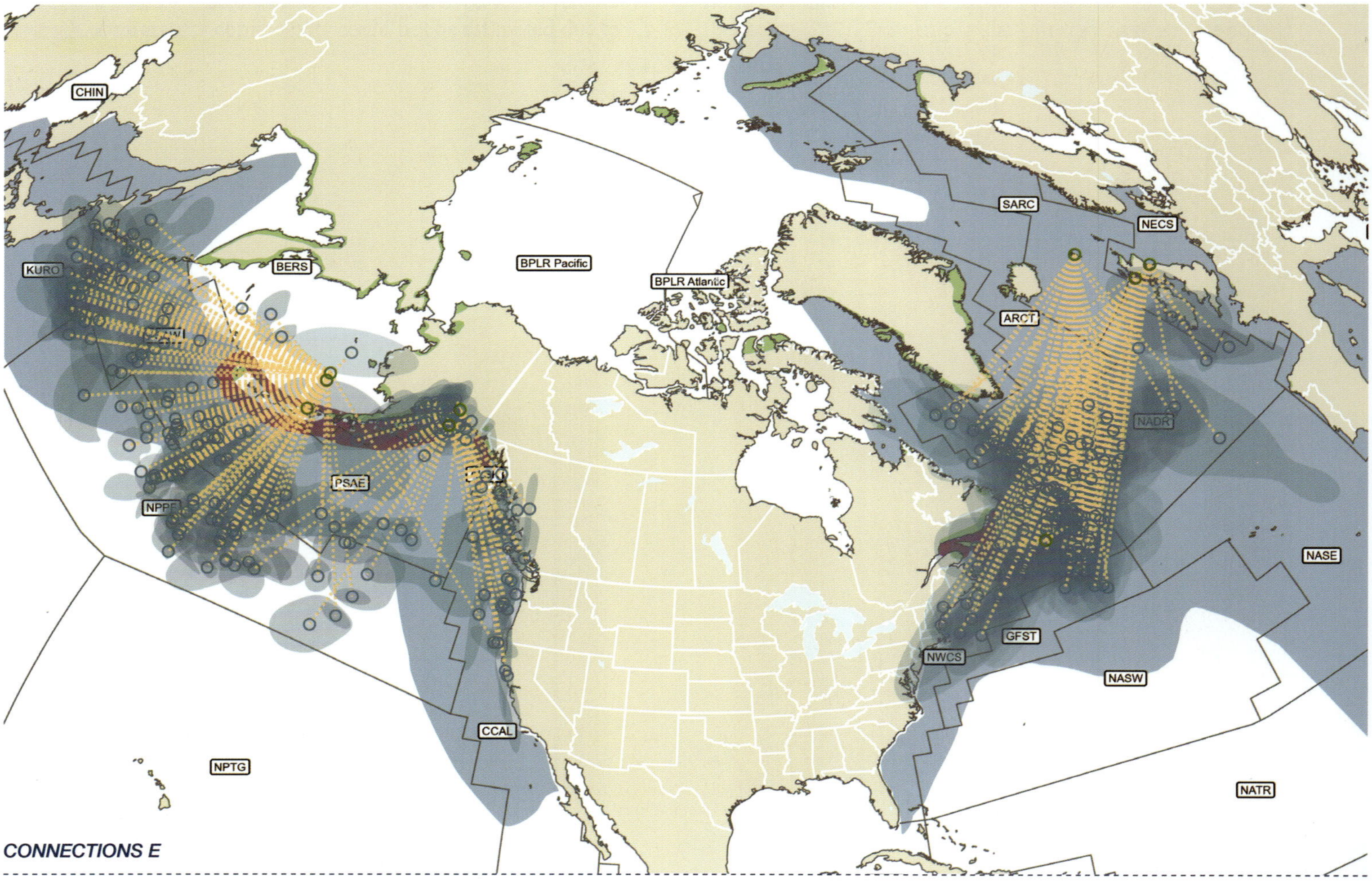

CONNECTIONS E

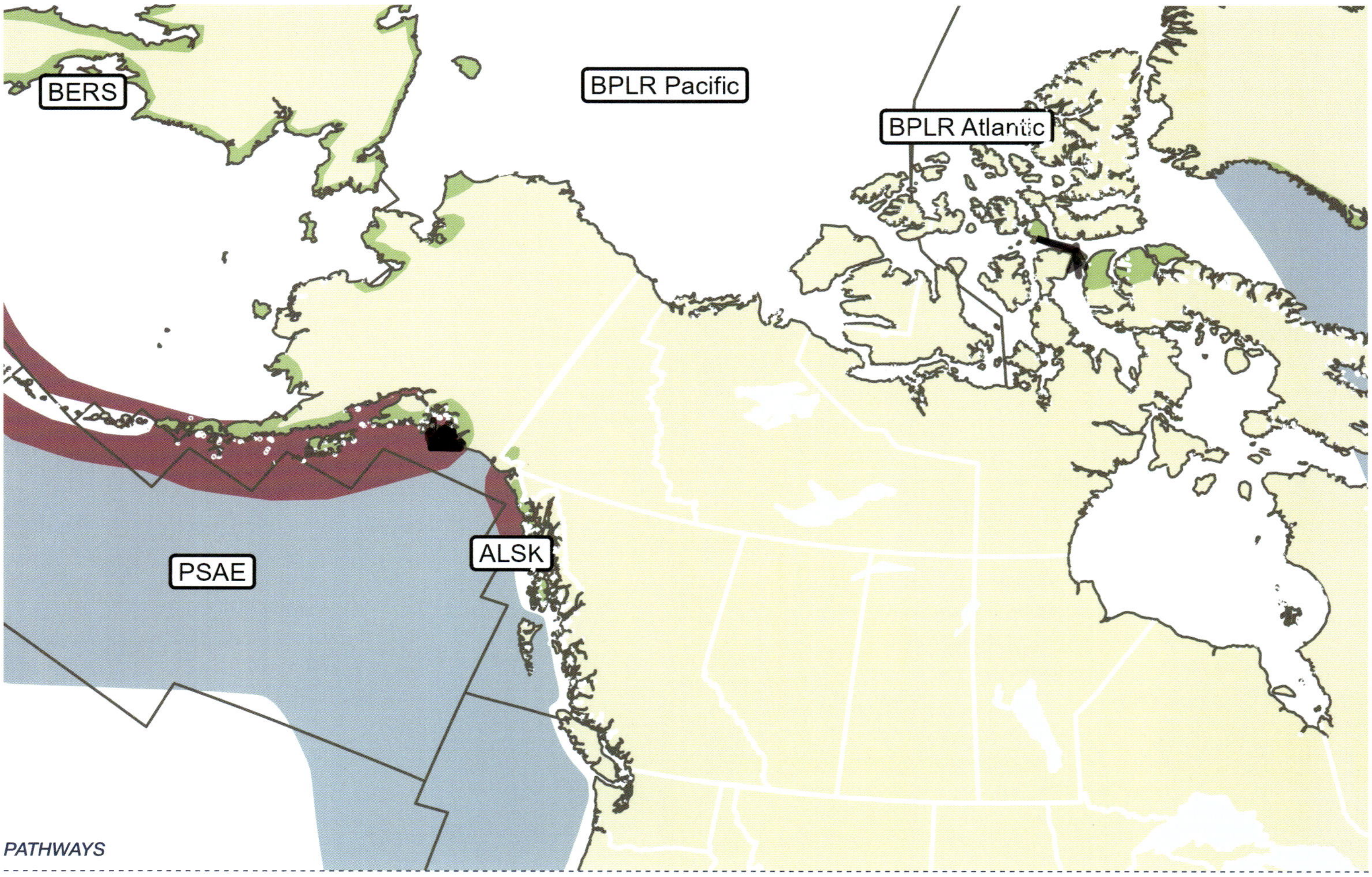

PATHWAYS

weak strong
Mantel = 0.12 ± 0.09

Rissa brevirostris **RED-LEGGED KITTIWAKE**

ORDER Charadriiformes
FAMILY Laridae
BREEDING HABITAT Coasts, Cliffs and Islands
NONBREEDING HABITAT Coasts, Marine Waters
GLOBAL POPULATION SIZE 200,000
POPULATION CHANGE SINCE 1970 NA

IUCN RED LIST CATEGORY Vulnerable
R2R TIPPING POINT SPECIES No
US LISTING STATUS No
CANADA SPECIES AT RISK No
PIF CCS 16/20

DISTRIBUTION AND ABUNDANCE

This small oceanic gull nests on cliffs only at remote island colonies in Alaska and Russia. Populations have declined in the St. George Island colony in the Bering Sea (about 80% of the known global population), but may be increasing in Buldir Island in the western Aleutian Islands (Dragoo et al. 2019). At-sea surveys (Kuletz and Labunski 2017) and tracking studies (Orben et al. 2015a; Drummond et al. 2021) indicate birds, after breeding, move to the ice edge and throughout the northwest Pacific Ocean over deep water.

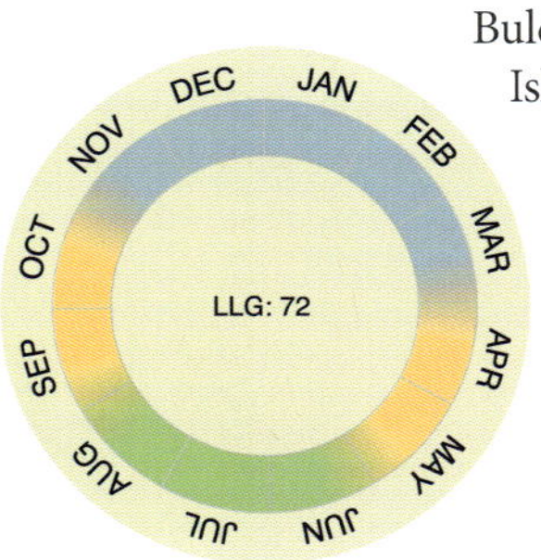

MIGRATORY CONNECTIVITY

OVERALL Some spatial and temporal connectivity inferred from geolocator tracking from two Alaskan island breeding colonies suggests moderate migratory connectivity (Mantel = 0.12 ± 0.09). Breeding birds from Buldir Island in the western Aleutian chain arrived earlier to and were more likely in the western Pacific (92%) during nonbreeding compared to birds from the Bering Sea's St. George Island (59%), while birds from the St. George Island colony were more likely in the Bering Sea (36%) than birds from the Buldir Island colony (8%).

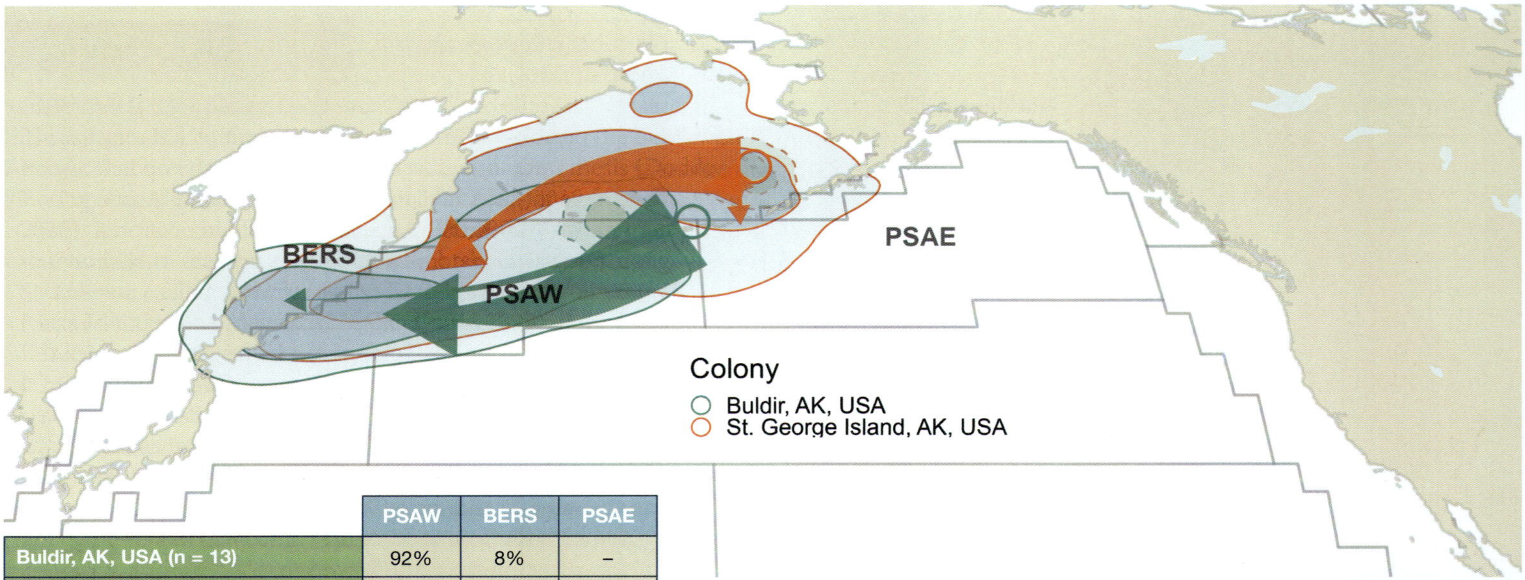

	PSAW	BERS	PSAE
Buldir, AK, USA (n = 13)	92%	8%	–
St. George Island, AK, USA (n = 59)	59%	36%	5%

GENERALIZED CONNECTIVITY

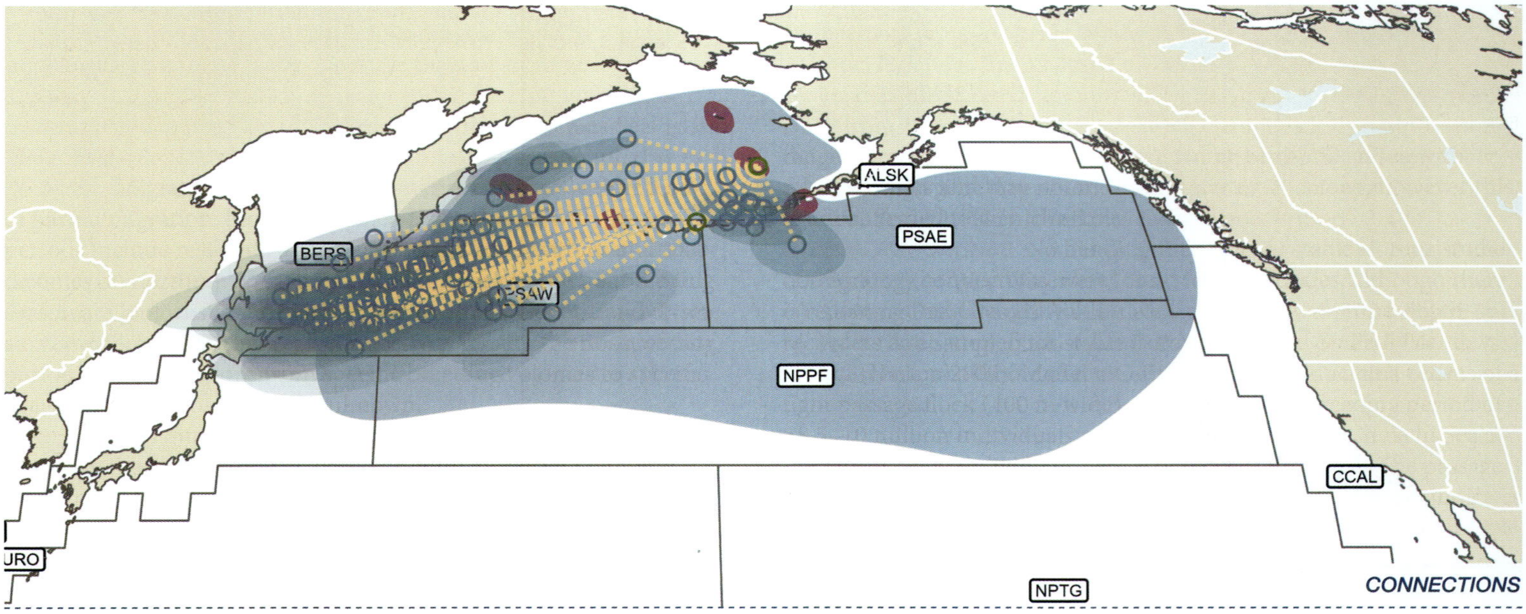

CONNECTIONS

INTERESTING NOTE

Methylmercury concentration within Red-legged Kittiwakes tracked from St. George Island has been linked to their nonbreeding locations. Birds that spent more time in the Western Pacific Subarctic Gyres had elevated methylmercury concentrations compared to birds that remained within the Bering Sea (Fleishman et al. 2019). Fidelity to breeding sites is high for adults (Byrd and Williams 1993).

GEOLOCATOR TRACKING Birds tracked from St. George Island for four years used shelf habitats in the northern and eastern Bering Sea for much of the nonbreeding season, with a short movement to the western Pacific Ocean (Orben et al. 2015a). The general nonbreeding region did not change, but core areas of usage changed annually (Orben et al. 2018). A 2016–2018 study compared birds tracked from Buldir Island and St. George Island simultaneously (Drummond et al. 2021). Birds from Buldir Island went directly to the western Pacific after breeding, while birds from St. George Island spent the first few months of their post-breeding period in the Bering Sea, before shifting into the western Pacific. In the early post-breeding period, there was little spatial overlap of core habitats for birds from the two colonies.

STABLE ISOTOPES Stable isotope signatures of Red-legged Kittiwakes tracked from St. George Island showed separation from their sympatrically breeding congener, the Black-legged Kittiwake (Orben et al. 2015a, 2015c). This separation indicated a greater degree of foraging specialization for the Red-legged Kittiwake.

OTHER APPROACHES No genetic studies of migratory connectivity were found for this species.

MIGRATORY PATHWAYS While birds tracked from both colonies took southwestern migratory pathways to core habitats in the western Pacific (Orben et al. 2015a; Drummond et al. 2021), many individuals could also be considered to simply shift core habitats across the year, with short-distance travel between ranges. Satellite tracking, in addition to geolocator-tracking, could help refine pathways.

DATA CONTRIBUTORS

B.A. Drummond; N. Kokubun; R.A. Orben; T. Yamamoto.

Pagophila eburnea **IVORY GULL**

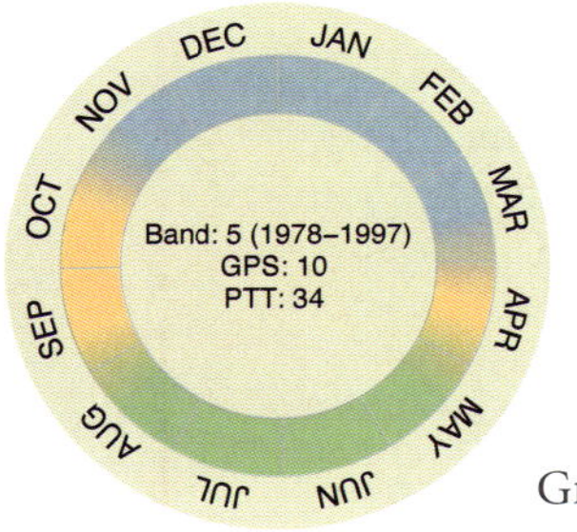

DISTRIBUTION AND ABUNDANCE

This rare Arctic species is usually found over ice or at ice margins year-round. It breeds in the high Arctic of Nunavut, Canada, and northern Greenland in North America; and in Svalbard, Norway, and archipelagos of northwest Russia (Mallory et al. 2020c). Its nonbreeding distribution is nearly circumpolar across the Arctic, Pacific, and Atlantic oceans but with main nonbreeding grounds in Davis Strait and Bering Strait (Mallory et al. 2020c). Ivory Gulls rarely move south of the Bering Sea and Maritime provinces of Canada (Mallory et al. 2020c). Declines have been reported in Canada by Inuit hunters and at-sea surveys (Mallory et al. 2020c). This population may be as few as 250 pairs (Gilchrist and Mallory 2005).

ORDER Charadriiformes
FAMILY Laridae
BREEDING HABITAT Coasts, Cliffs and Islands
NONBREEDING HABITAT Oceans, Arctic Polynyas
GLOBAL POPULATION SIZE 21,000
POPULATION CHANGE SINCE 1970 NA
IUCN RED LIST CATEGORY Near Threatened
R2R TIPPING POINT SPECIES Yellow-Alert
US LISTING STATUS No
CANADA SPECIES AT RISK Endangered
PIF CCS 16/20

MIGRATORY CONNECTIVITY

OVERALL Weak migratory connectivity across global distribution is indicated from band recoveries, satellite tracking, at-sea surveys, and genetic evidence; however, Mantel is not calculated due to limited sample size. Birds tracked/banded from Nunavut, Canada, and birds tracked from northern Greenland mixed within the Boreal Polar (50%) and Atlantic Arctic Currents (38%–42%). Birds from Svalbard, Norway, and Franz Josef Land, Russia, were tracked to the Davis Strait and the Labrador Sea (Gilg et al. 2010), and a summary of at-sea surveys also support this as a core nonbreeding range for colonies outside North America (Spencer et al. 2016). There is little genetic differentiation among colonies, but less is known about connectivity to the Bering Strait, the Pacific Arctic portion of its nonbreeding range. Some genetic evidence suggests that birds spending the nonbreeding season off Alaska show differentiation (Royston and Carr 2016), but four birds have been tracked there from three colonies (Russia, Svalbard, and Greenland; Gilg et al. 2010).
BANDING Five birds banded in Nunavut, Canada, in 1976–1983 were

	BERS	NWCS	BPLR Atlantic	ARCT
Nunavut, Canada (n = 12)	–	8%	50%	42%
Greenland (n = 8)	12%	–	50%	38%

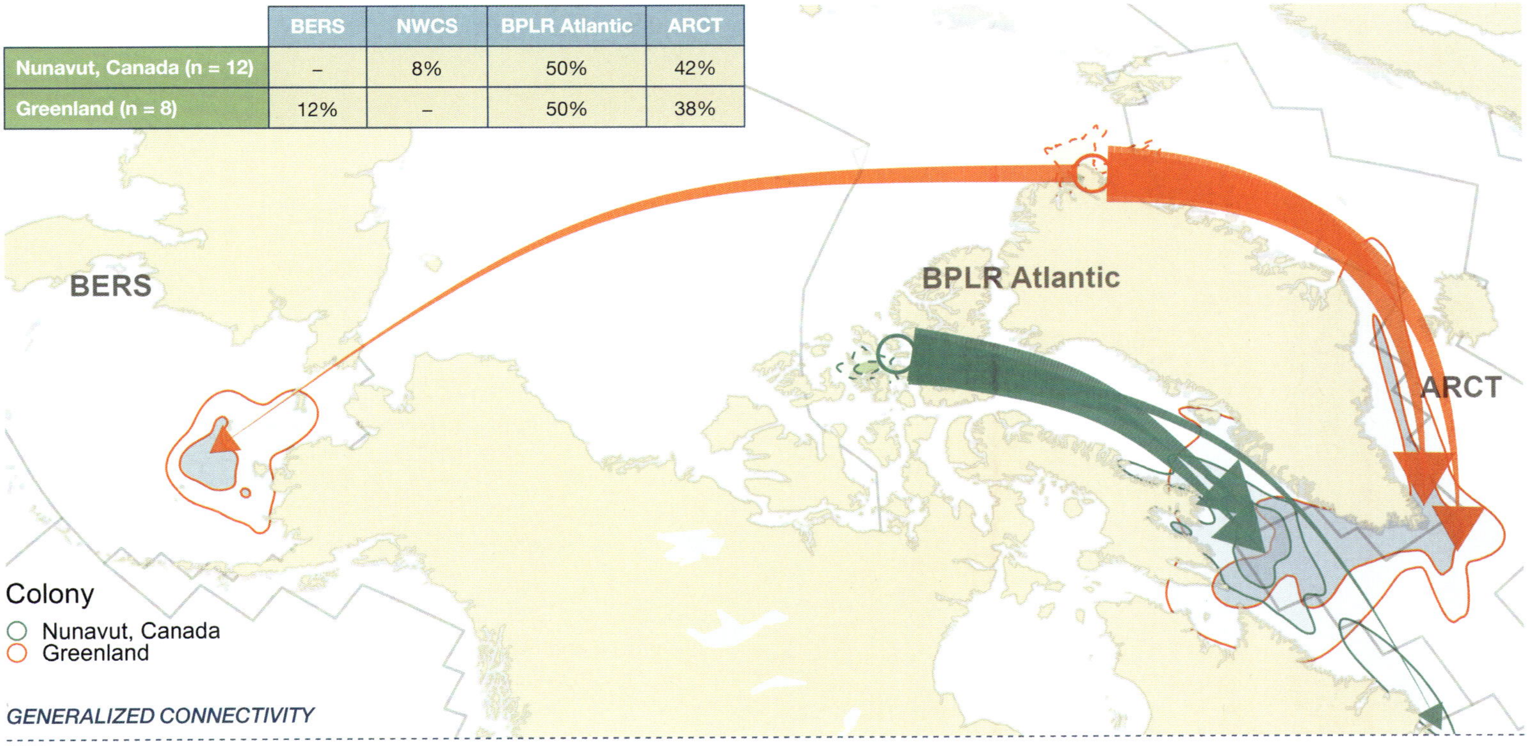

GENERALIZED CONNECTIVITY

recovered during the nonbreeding period in 1978–1997 in Nunavut, Newfoundland and Labrador, and Greenland. Of these, the four birds recovered outside Nunavut were shot (see Stenhouse et al. 2004 for a full review of recoveries and survival rates of banded birds). Banding recoveries of Ivory Gulls from the Alaska nonbreeding ground indicate that they are capable of long-distance movement to and from the Siberian breeding grounds in the Kara Sea (Salomonsen 1967, 1979; Tomkovich 1990), but more study is needed on this population.

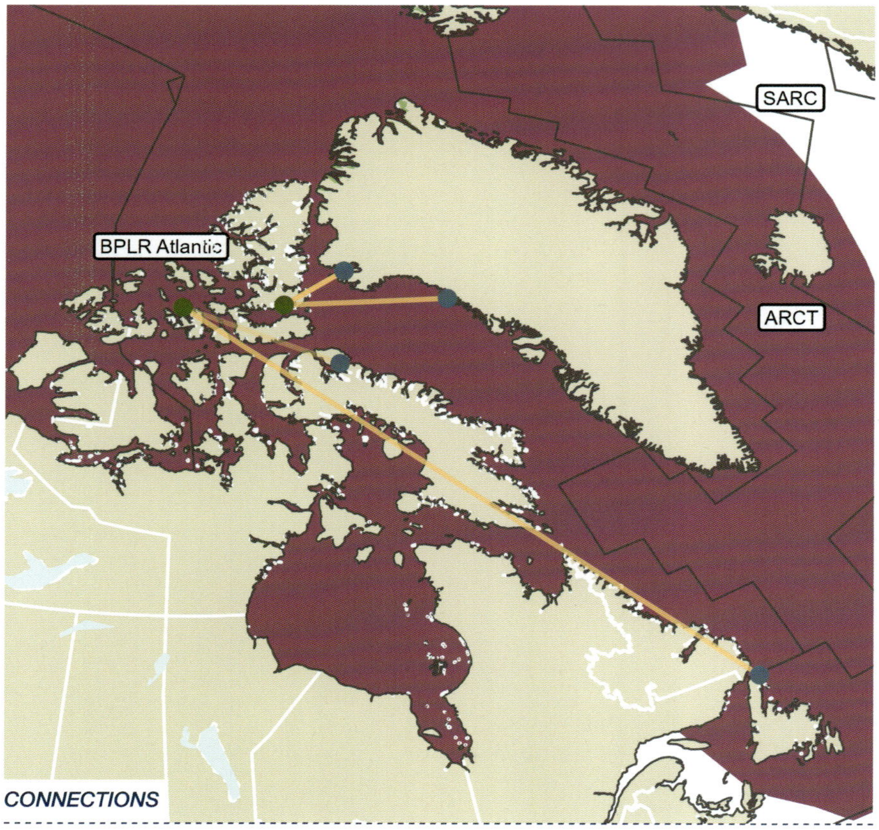

CONNECTIONS

SATELLITE TRACKING Two main research efforts to track birds in 2007–2013 with solar satellite tags from Greenland; Svalbard, Norway; and Franz Josef Land, Russia (17–180 days; Gilg et al. 2010) and from Nunavut, Canada (Spencer et al. 2014) were assembled together (n = 104) in Gilg et al. (2016). Overall, birds undertook circumpolar movements post-breeding, before concentrating in two major nonbreeding areas (Davis Strait and Bering Strait), with substantial overlap between different colonies (Gilg et al. 2010, 2016; Spencer et al. 2014, 2016). Gilg et al. (2010) found that birds from Greenland reached nonbreeding areas earlier than birds from Russia and Svalbard. For birds tracked across years in Spencer et al. (2016), nonbreeding site fidelity was high for individuals across years. Two birds (of 22 tagged) tracked by GPS tags from northeast Greenland spent the nonbreeding season in the Davis Strait (Frederiksen et al. 2021).

MIGRATORY PATHWAYS Birds satellite-tracked from colonies in Greenland; Svalbard, Norway; and Franz Josef Land, Russia (Gilg et al. 2010, 2016), showed a loop migration and two different strategies. Most were observed to move eastward from their colonies post-breeding, before most returned westward and then moved south to nonbreeding areas near Greenland and the Labrador Sea. Four birds (two from Russia, one from Svalbard, and one from Greenland) continued eastward to Russia and Alaska into the Bering Sea (Gilg

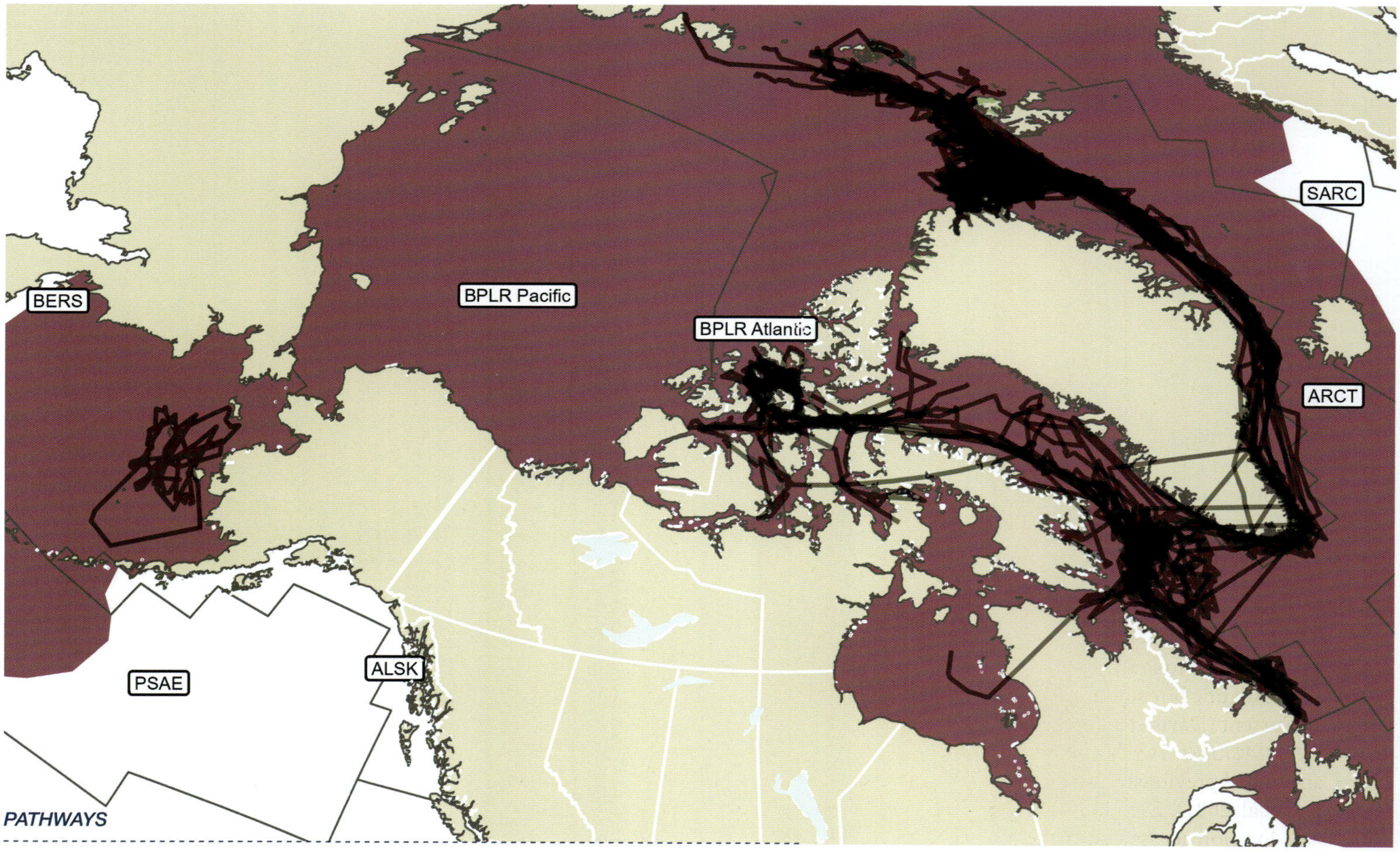

PATHWAYS

et al. 2010). Spencer et al. (2014) described three nonbreeding routes from Nunavut, Canada, that sometimes varied across years within an individual: either south via Davis Strait, south via Foxe Basin Strait, or cutting across Baffin Island between Foxe Basin and Davis Strait. Frederiksen et al. (2021) reported an unusual overland crossing of the Greenland ice cap (1,345 km in 29 hours) of an individual tracked by GPS in 2019 from a nonbreeding area in Davis Strait to its breeding site in northeast Greenland.

STABLE ISOTOPES No stable isotopes studies of migratory connectivity found for this species.

GENETICS Mitochondrial evidence from 127 museum specimens suggested no significant differentiation among birds collected during the breeding season in Canada, Greenland, and Norway (Royston and Carr 2016). However, this study also sampled birds collected from nonbreeding areas and found that birds in the Bering Strait (most collected from Point Barrow, Utqiaġvik, Alaska) were differentiated from other locations and perhaps from a distinct Siberian population.

DATA CONTRIBUTORS

A. Aebischer; M. Frederiksen; M.V. Gavrilo; H.G. Gilchrist; O. Gilg; M.L. Mallory; H. Strøm; G. Yannic.

INTERESTING NOTE

This species has been historically and is currently hunted in several places, especially Greenland, but this practice has declined (Mallory et al. 2020c). As a sea ice specialist (Gilg et al. 2016), this species is extremely vulnerable to the effects of climate change; and as the species feeds high on the food chain (Karnovsky et al. 2009), it has been observed to have potentially detrimental levels of several contaminants (Bond et al. 2015; Lucia et al. 2017). A larger and better coordinated global surveying effort is needed to better determine population trends in this declining species (Mallory et al. 2020c).

SABINE'S GULL *Xema sabini*

Mantel = 0.28 ± 0.17

ORDER	Charadriiformes
FAMILY	Laridae
BREEDING HABITAT	Wetlands, Tundra
NONBREEDING HABITAT	Oceans, Pelagic
GLOBAL POPULATION SIZE	340,000
POPULATION CHANGE SINCE 1970	NA
IUCN RED LIST CATEGORY	Least Concern
R2R TIPPING POINT SPECIES	No
US LISTING STATUS	No
CANADA SPECIES AT RISK	No
PIF CCS	12/20

DISTRIBUTION AND ABUNDANCE

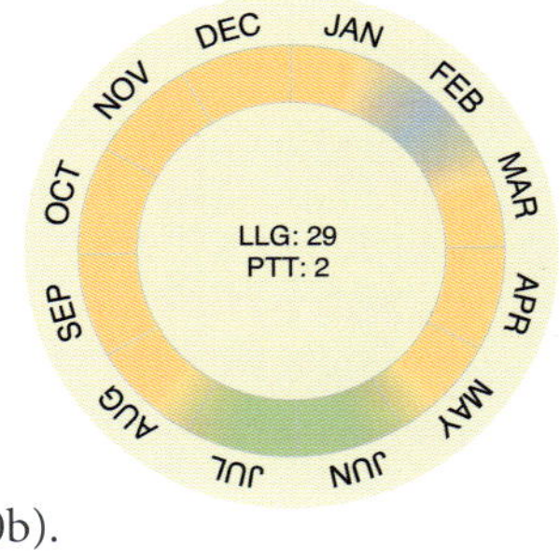

A small, Holarctic breeder that migrates to nonbreeding areas in the eastern boundary currents of South America and Africa. There is limited demographic data from breeding grounds, but there are an estimated tens of thousands of breeding pairs in each of Alaska and Canada, and a lower density of about 1,000–2,000 pairs in Greenland (Day et al. 2020b). There is higher abundance in eastern than western Canada (Day et al. 2020b).

	HUMB	SPSG	BENG	EAFR
Nasaruvaalik Island, NU, Canada (n = 23)	74%	13%	9%	4%
Sand Island, Greenland (n = 8)	–	–	88%	12%

MIGRATORY CONNECTIVITY

OVERALL Sabine's Gull exhibits moderate migratory connectivity by ocean basin (Mantel = 0.28 ± 0.17) on either side of a presumed migratory divide near Nasaruvaalik Island, Nunavut, Canada (96 degrees west), from where 13% of tracked birds migrated to the Atlantic Ocean, while 87% migrated to the Pacific Ocean. All birds tracked from Greenland spent the nonbreeding season in the Atlantic Ocean with birds migrating to nonbreeding grounds in the Benguela Current (88%) and East Atlantic Frontal Zone (12%). No additional connectivity information is available from North American colonies outside of Nasaruvaalik Island to the Pacific Ocean, and thus connectivity is unknown within that basin. However, all birds tracked from Canada to the Pacific Ocean used the same two provinces: the Humboldt Current (74%) and the South Pacific Subtropical Gyre (13%).

GEOLOCATOR TRACKING From Canada, Pacific birds tracked with geolocators migrated to Peru (Davis et al. 2016; Gutowsky et al. 2021), while Atlantic birds migrated to nonbreeding areas in the Benguela Current (Davis et al. 2016).

SATELLITE TRACKING Two birds tracked with satellite telemetry from Canada had previously also been tracked with geolocators (Gutowsky et al. 2021). Each individual spent the nonbreeding season in the same location across years (one in the Atlantic Ocean and one in the Pacific Ocean). Two additional birds were later tracked by satellite from the same colony and, again, one visited the Atlantic Ocean and one the Pacific (M. Maftei, unpublished data).

OTHER APPROACHES No stable isotope or genetic studies of migratory connectivity were found for this species.

MIGRATORY PATHWAYS Birds typically took coastal routes within exclusive economic zones (EEZs), but satellite tracking confirms overland crossing of a small proportion of birds between the Pacific coast from high Arctic Canada (M. Maftei, unpublished data). Pacific

INTERESTING NOTE

A tracked nesting pair from Nunavut, Canada, spent the nonbreeding season in different oceans, and was observed to successfully breed for multiple years (Davis et al. 2016).

SHOREBIRDS, AUKS, GULLS, AND TERNS (CHARADRIIFORMES)

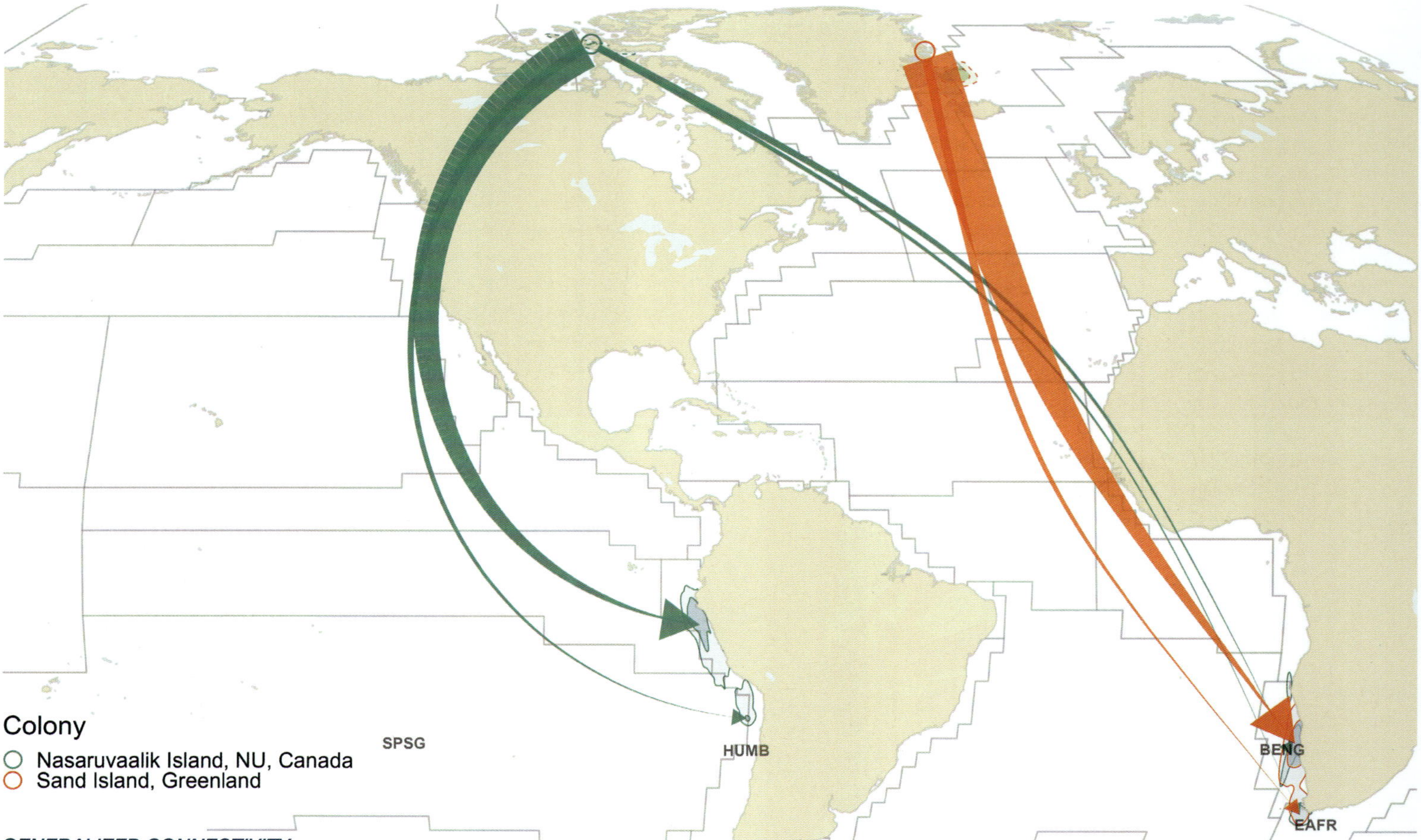

GENERALIZED CONNECTIVITY

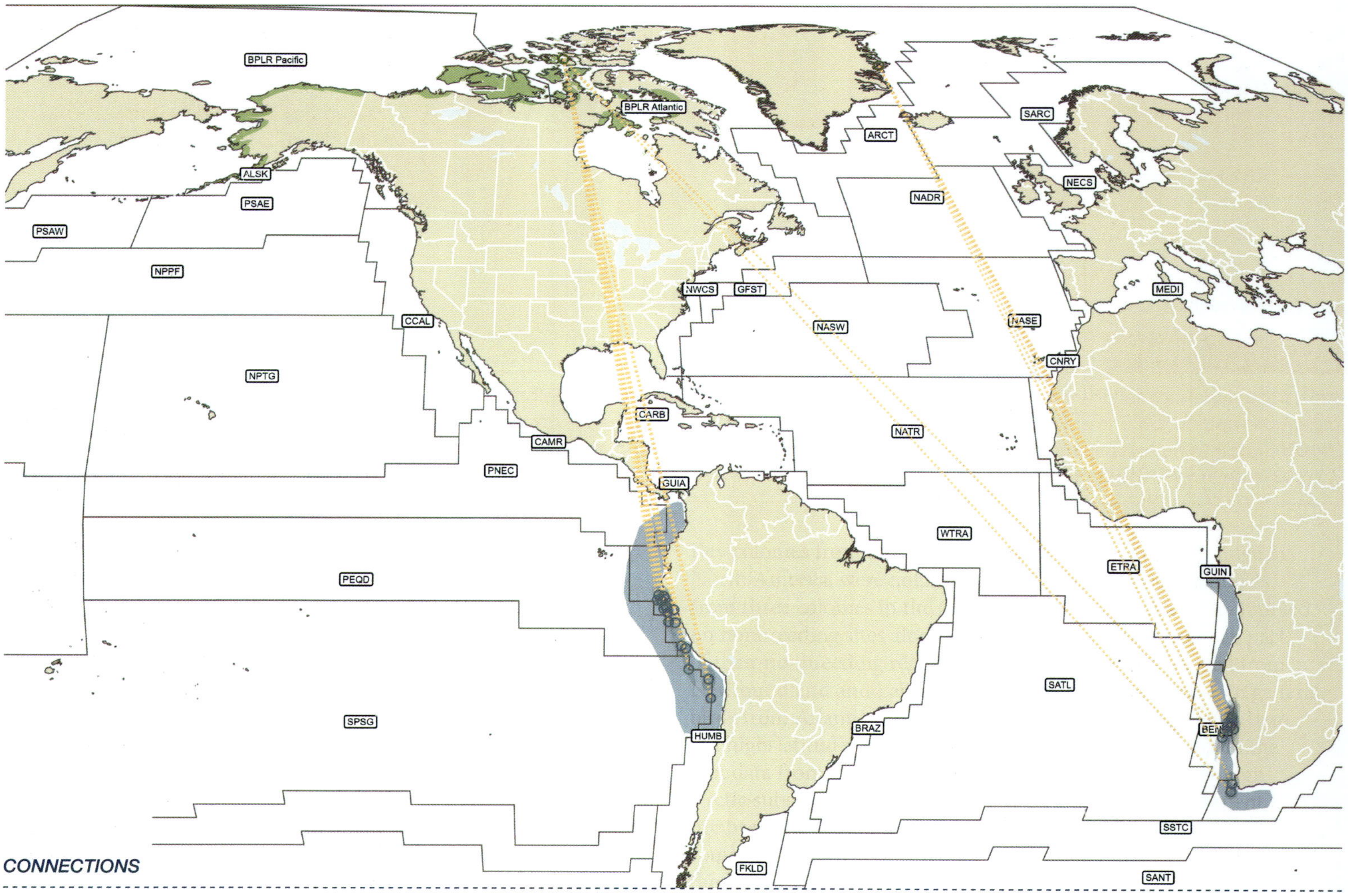

CONNECTIONS

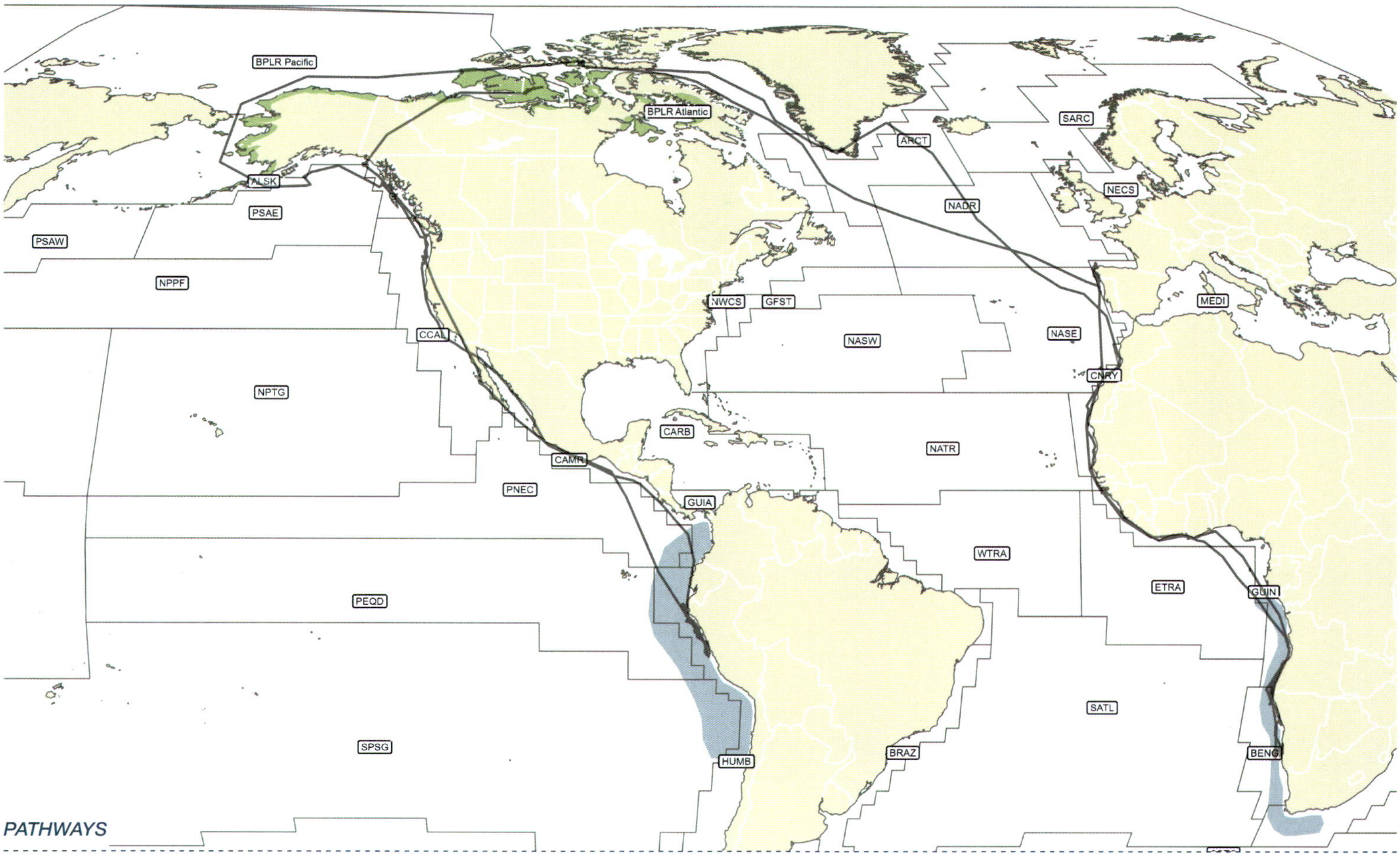

PATHWAYS

birds tracked with geolocators migrated west across the Beaufort Sea before either flying south around or over Alaska or Canada, then crossing the Gulf of Alaska and following the Pacific coast of the Americas south to nonbreeding areas (Davis et al. 2016; Gutowsky et al. 2021). Northbound migrations included an overland route from the Gulf of Alaska back to the breeding colony in Nunavut, Canada, via either Alaska, USA; Yukon, Canada; or central Canada toward James Bay (Gutowsky et al. 2021). Atlantic birds flew southeast down the Gulf of St. Lawrence, crossed the North Atlantic, and then followed the coast south around Africa to the Benguela Current. Flight distance, speed, and timing were similar for all birds tracked from Canada, regardless of which ocean they spent the nonbreeding season in (Davis et al. 2016). Individuals tracked from northeast Greenland made similar migrations that included a prolonged stopover in the Bay of Biscay and returning via stopover near West Africa in the pre-breeding season (Stenhouse et al. 2012).

DATA CONTRIBUTORS

S.E. Davis; C. Egevang; M.S. Maftei; M.L. Mallory; I.J. Stenhouse.

Chroicocephalus philadelphia

BONAPARTE'S GULL

ORDER Charadriiformes

FAMILY Laridae

BREEDING HABITAT Wetlands, Boreal

NONBREEDING HABITAT Coasts, Marine Waters

GLOBAL POPULATION SIZE 790,000

POPULATION CHANGE SINCE 1970 −12%

IUCN RED LIST CATEGORY Least Concern

R2R TIPPING POINT SPECIES No

US LISTING STATUS No

CANADA SPECIES AT RISK No

PIF CCS 10/20

DISTRIBUTION AND ABUNDANCE

It breeds from Alaska, across Canada to the mouth of the St. Lawrence River, and locally around northern shores of the Great Lakes, with the densest concentrations around the shores of Hudson Bay (Fink et al. 2023). It migrates south during the nonbreeding season to the southeastern US and California and Mexican Pacific coast, with the greatest concentrations from January to March along the Atlantic coast from Virginia to northern Florida and along the Pacific shoreline of Baja California and Sonora (Fink et al. 2023).

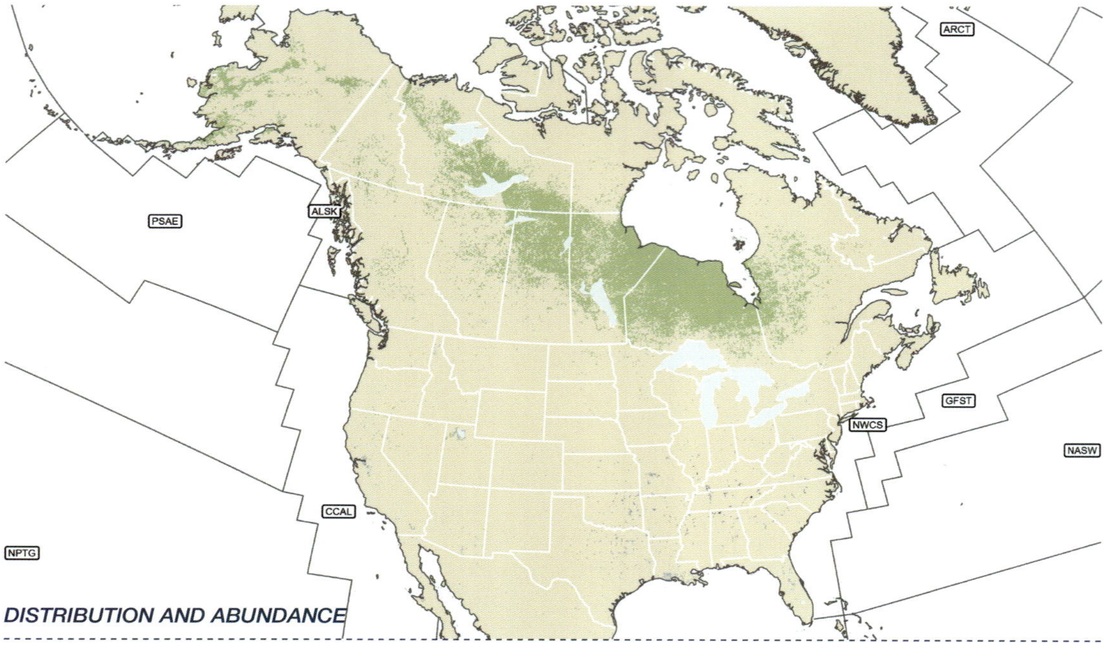

DISTRIBUTION AND ABUNDANCE

MIGRATORY CONNECTIVITY

OVERALL Migratory connectivity is unknown. No studies of migratory connectivity were found for this species.

MIGRATORY PATHWAYS Alaska and Pacific Northwest breeders migrate to nonbreeding areas in California and western Mexico, while most Canadian taiga breeders migrate southeastward to the southern US and Atlantic seaboard (Fink et al. 2023).

LITTLE GULL *Hydrocoloeus minutus*

DISTRIBUTION AND ABUNDANCE

Breeds primarily in northern Europe and central Eurasia and in very low density around the shores of James Bay, Gulf of St. Lawrence, and the northern Lake Erie and Ontario shorelines. During the nonbreeding season in North America, it occurs sporadically and locally in the east, most dependably in the eastern Great Lakes and on open water in southern Indiana/Illinois (Fink et al. 2023).

ORDER Charadriiformes
FAMILY Laridae
BREEDING HABITAT Wetlands, Freshwater Marsh
NONBREEDING HABITAT Coasts, Marine Waters
GLOBAL POPULATION SIZE 120,000
POPULATION CHANGE SINCE 1970 NA
IUCN RED LIST CATEGORY Least Concern
R2R TIPPING POINT SPECIES No
US LISTING STATUS No
CANADA SPECIES AT RISK No
PIF CCS 11/20

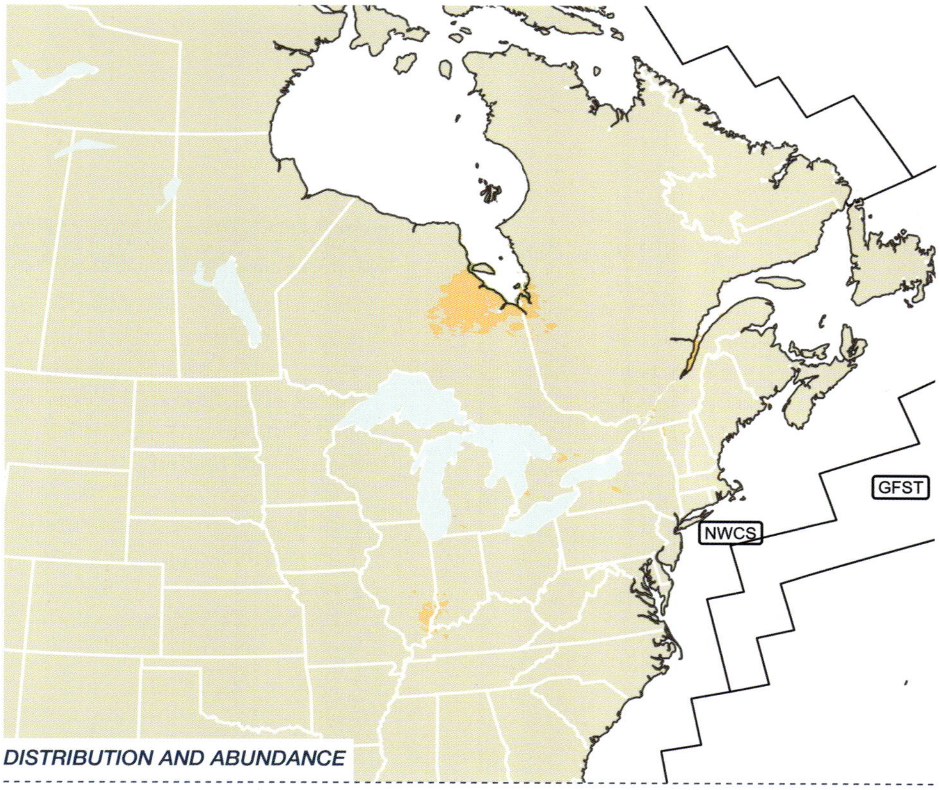

DISTRIBUTION AND ABUNDANCE

MIGRATORY CONNECTIVITY

OVERALL Migratory connectivity is unknown. No information on migratory connectivity or migratory pathways were found for this species.

INTERESTING NOTE

The Little Gull is the world's smallest gull species. Breeding in North America was first documented somewhat recently (1962). The origin and longevity of the Little Gull's presence in North America remains a mystery, or at least a subject of considerable debate.

Rhodostethia rosea **ROSS'S GULL**

ORDER Charadriiformes
FAMILY Laridae
BREEDING HABITAT Wetlands, Tundra
NONBREEDING HABITAT Coasts, Marine Waters
GLOBAL POPULATION SIZE 42,000
POPULATION CHANGE SINCE 1970 NA
IUCN RED LIST CATEGORY Least Concern
R2R TIPPING POINT SPECIES No
US LISTING STATUS No
CANADA SPECIES AT RISK Threatened
PIF CCS 15/20

DISTRIBUTION AND ABUNDANCE

This gull has a circumpolar breeding range with high abundances of geographically distinct populations in Canada and Greenland in the Nearctic and Russia in the Palearctic. It spends the nonbreeding season in the Arctic, but limited data exist about its distribution and abundance.

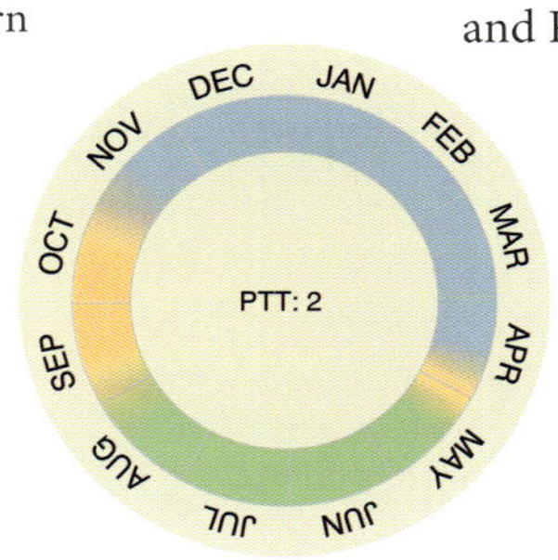

MIGRATORY CONNECTIVITY

OVERALL Very little is known about the migratory connectivity of Ross's Gull due to limited data. A few birds were tracked from each of the Nearctic and Palearctic breeding populations. These birds spent the nonbreeding season primarily within the Boreal Polar, Arctic, and Bering Sea Longhurst provinces.

GEOLOCATOR TRACKING Similar to the satellite-tracked birds in the same study, one bird tracked with a geolocator from a colony in the Canadian high Arctic spent the nonbreeding period within Canada near the northern Labrador Sea (Maftei et al. 2015).

SATELLITE TRACKING Two birds tracked from a colony in the Canadian high Arctic spent the nonbreeding period within Canada near the northern Labrador Sea (Maftei et al. 2015).

GENETICS An analysis of mitochondrial DNA revealed that Ross's Gulls from Canada and Alaska (although passage migrants from Siberia) have strong genetic differentiation (Royston and Carr 2016), which supports the idea that the Nearctic population is discrete and reproductively isolated from the Palearctic population (Maftei et al. 2015). Further tracking of this species is necessary to determine if these populations are truly disjunct and whether birds breeding in Canada spend the nonbreeding season completely within Canada (Maftei et al. 2015).

OTHER APPROACHES No stable isotope studies of migratory connectivity were found for this species.

MIGRATORY PATHWAYS All three tagged birds from the Canadian high Arctic moved east through Lancaster Sound into Davis Strait and then southeast toward the Labrador Sea (Maftei et al. 2015). A female tracked by satellite in multiple years took a similar path back to the breeding grounds, but after deferred breeding in the second year, she wandered widely, traveling west across the Arctic Ocean to Wrangel Island, Russia, before returning east for post-breeding migration to the Labrador Sea that followed a similar route as the prior year (Maftei et al. 2015). A male tracked by satellite tag from the Kolyma Delta, Russia (161 degrees east), spent its post-breeding months in the Laptev Sea before migrating in October through Alaskan waters north of Cape Lisburne (approximately 100 km from land) to the Chukotka Peninsula of Russia, where it stopped transmitting in November (Gilg et al. 2016).

DATA CONTRIBUTORS
S.E. Davis; M.S. Maftei; M.L. Mallory.

INTERESTING NOTE

A satellite-tagged female Ross's Gull flew more than 40,000 km, almost entirely north of 50 degrees north latitude in one year, which is greater than the circumference of the Earth (Maftei et al. 2015).

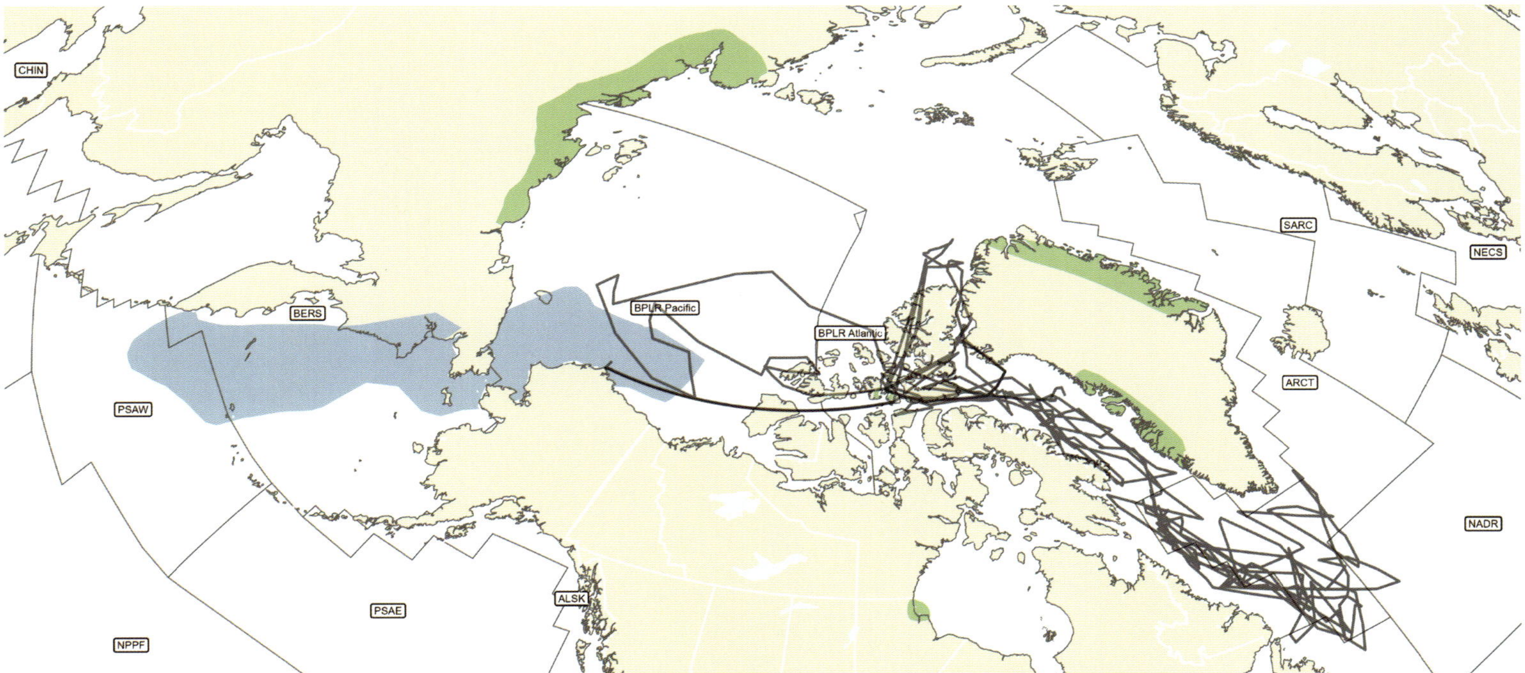

PATHWAYS

LAUGHING GULL *Leucophaeus atricilla*

MC = 0.11 ± 0.03*

DISTRIBUTION AND ABUNDANCE

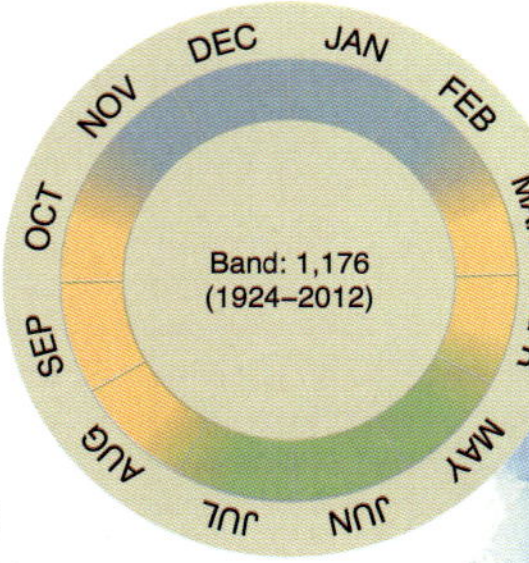

This attractive gull is one of the most abundant and familiar birds along the US Atlantic and Gulf coasts. It breeds colonially from Maine south to islands off the coast of South America, with the largest concentrations along the Atlantic coast of the US and the northern Gulf of Mexico. In the nonbreeding season, this species continues to inhabit much of its breeding range, but departs from the northern areas of the Atlantic coast (Burger 2015; Fink et. al 2023).

ORDER Charadriiformes
FAMILY Laridae
BREEDING HABITAT Coasts, Beach and Estuary
NONBREEDING HABITAT Coasts, Beach and Estuary
GLOBAL POPULATION SIZE 1,600,000

POPULATION CHANGE SINCE 1970 +126%
IUCN RED LIST CATEGORY Least Concern
R2R TIPPING POINT SPECIES No
US LISTING STATUS No
CANADA SPECIES AT RISK No
PIF CCS 11/20

	Andean West	Atlantic	Caribbean	Central America	Mexico	Midcontinent	Northern SA
Atlantic (n = 937)	3%	42%	4%	38%	3%	2%	9%
Midcontinent (n = 109)	1%	3%	–	51%	39%	3%	4%

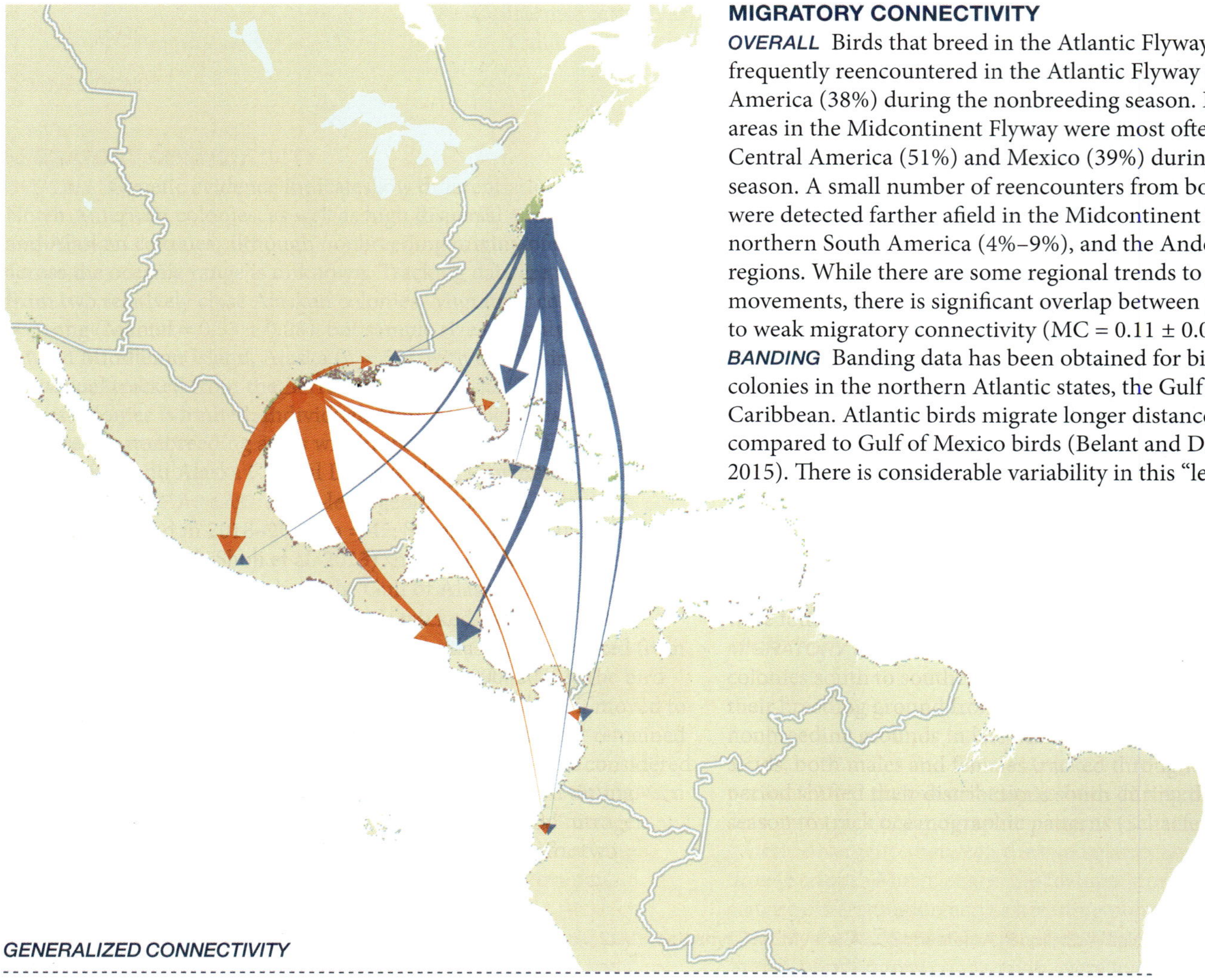

GENERALIZED CONNECTIVITY

MIGRATORY CONNECTIVITY

OVERALL Birds that breed in the Atlantic Flyway were most frequently reencountered in the Atlantic Flyway (42%) and Central America (38%) during the nonbreeding season. Birds from breeding areas in the Midcontinent Flyway were most often encountered in Central America (51%) and Mexico (39%) during the nonbreeding season. A small number of reencounters from both populations were detected farther afield in the Midcontinent Flyway (2%–3%), northern South America (4%–9%), and the Andean West (1%–3%) regions. While there are some regional trends to population movements, there is significant overlap between populations, leading to weak migratory connectivity (MC = 0.11 ± 0.03).

BANDING Banding data has been obtained for birds breeding at colonies in the northern Atlantic states, the Gulf Coast, and the Caribbean. Atlantic birds migrate longer distances on average compared to Gulf of Mexico birds (Belant and Dolbeer 1993; Burger 2015). There is considerable variability in this "leapfrog" pattern,

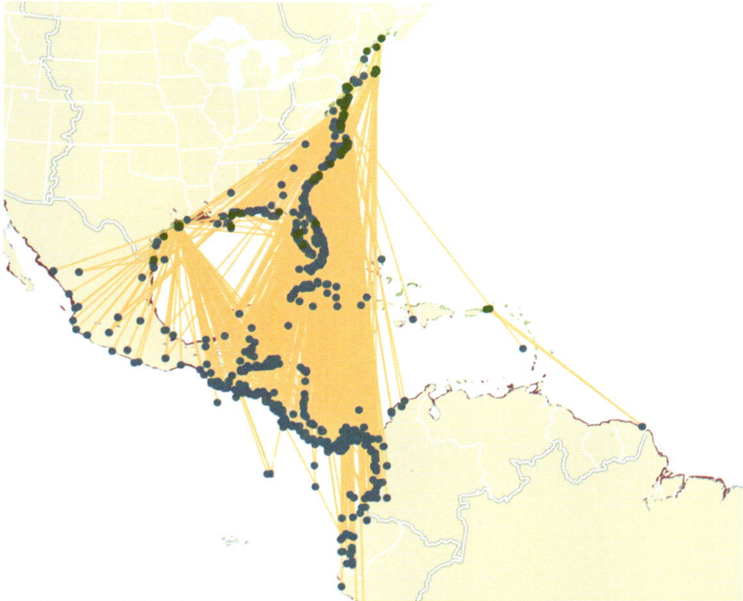

CONNECTIONS (A–G)

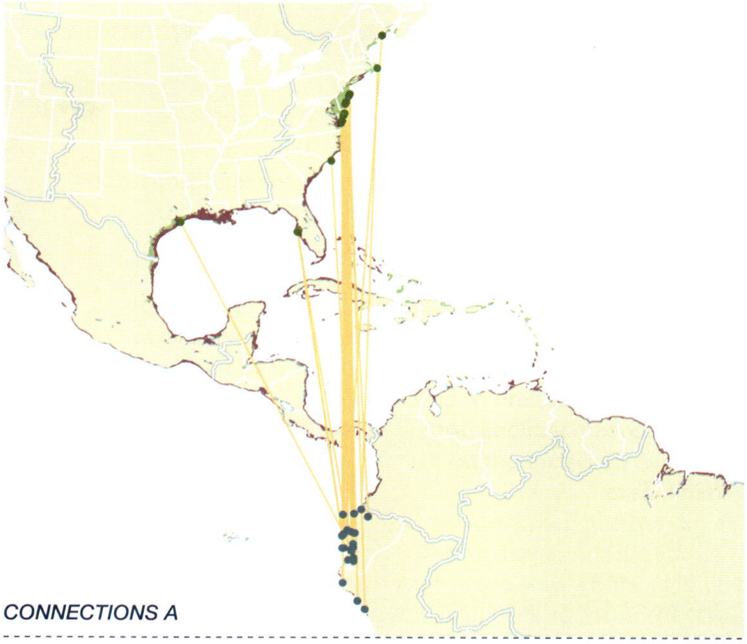

CONNECTIONS A

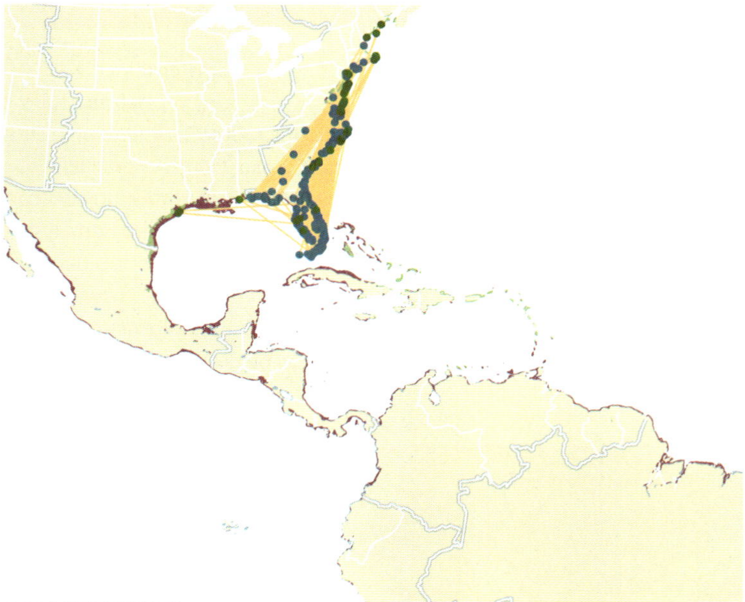

CONNECTIONS B

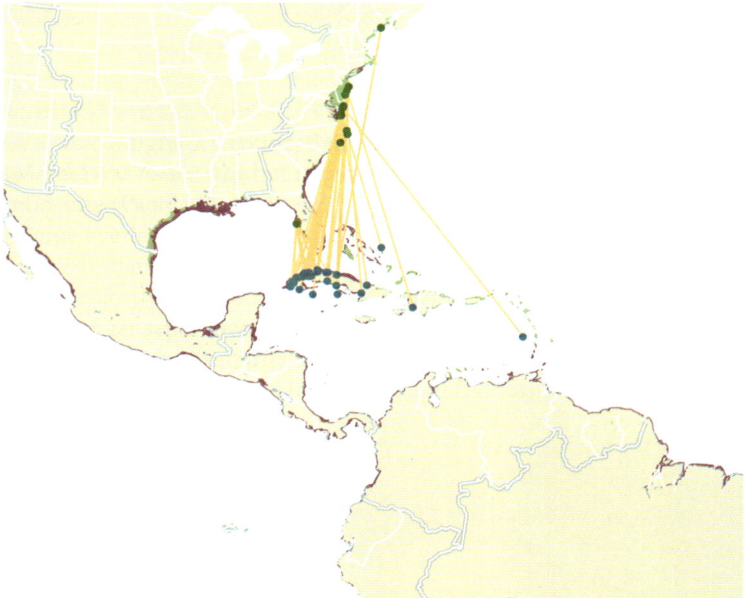

CONNECTIONS C

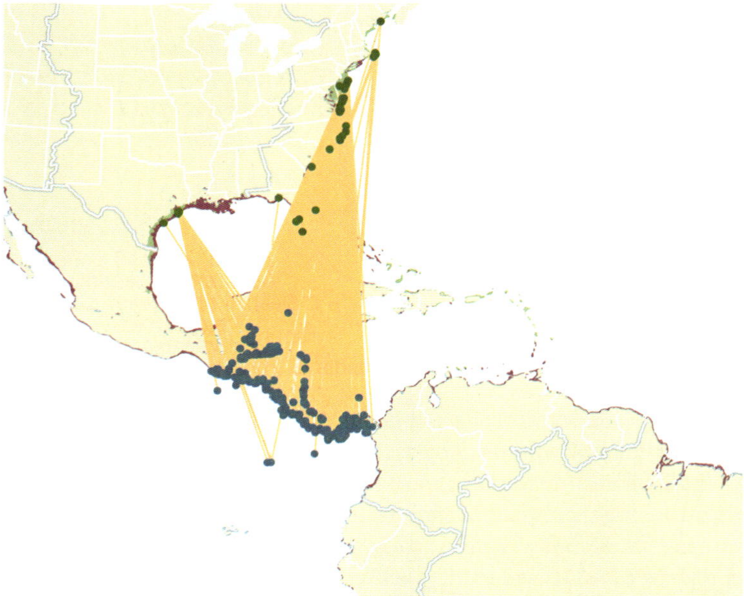

CONNECTIONS D

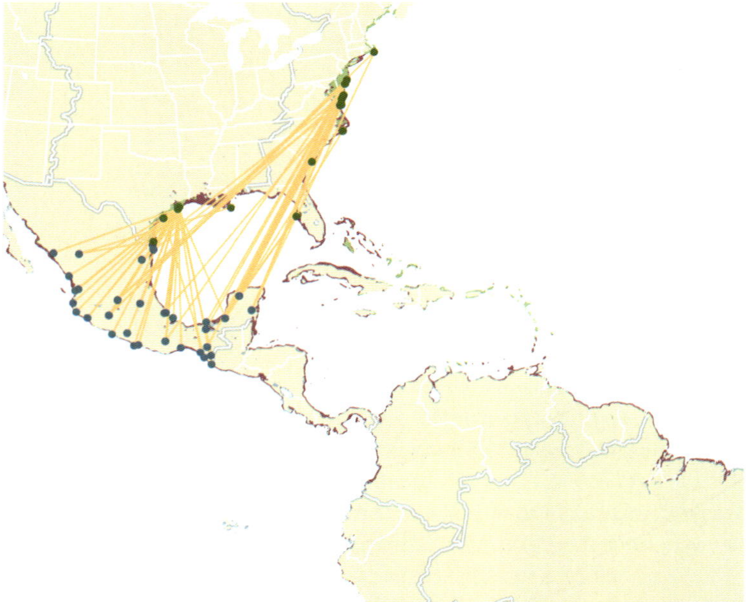

CONNECTIONS E

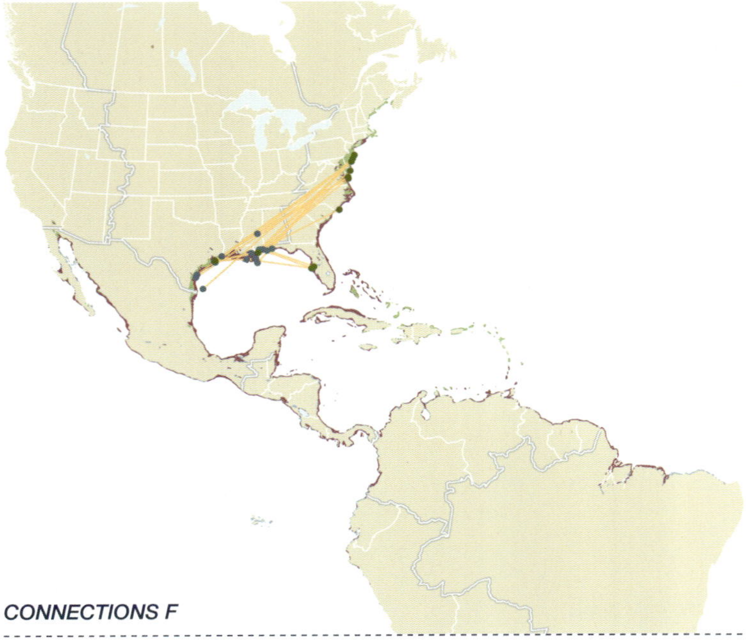

CONNECTIONS F

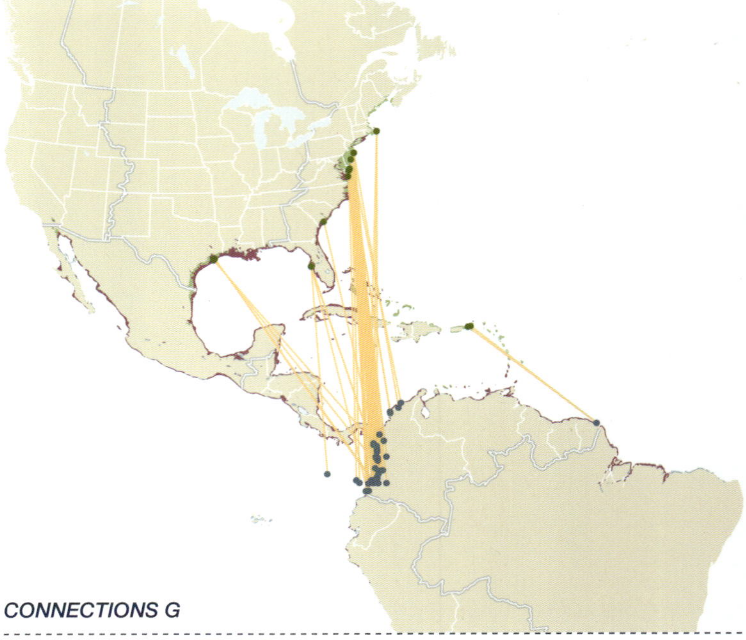

CONNECTIONS G

however, and migratory connectivity is low. This species is also prone to vagrancy, and a fledgling banded in New Jersey was later recovered during the nonbreeding season in Hawaii (Telfer and Shisler 1981). Midcontinent breeders primary spend the nonbreeding season in Mexico and Central America, but may also disperse far from this core nonbreeding range (Belant and Dolbeer 1993; Burger 2015).

OTHER APPROACHES No satellite, genetic, or stable isotope studies of migratory connectivity were found for this species.

MIGRATORY PATHWAYS Birds from Atlantic breeding populations disperse widely in the nonbreeding season, but they tend to concentrate in Florida and Central America.

FRANKLIN'S GULL *Leucophaeus pipixcan*

DISTRIBUTION AND ABUNDANCE

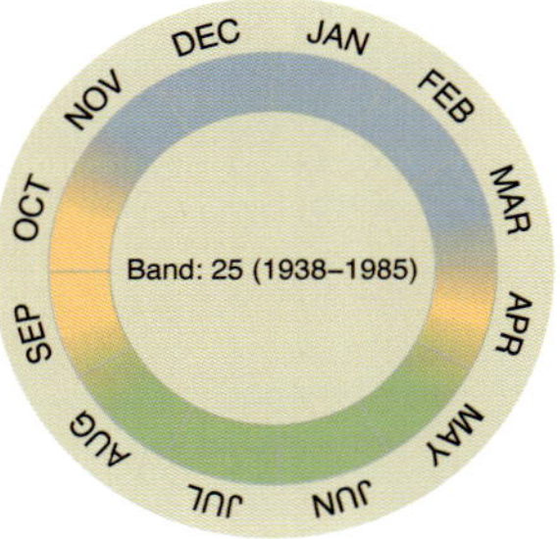

Franklin's Gull is a black-hooded gull that breeds in marshes within the prairies of inland North America. This species shifts colonies frequently in response to water availability (Burger and Gochfeld 2020). Taken from colony counts, the North American population is estimated to be over a million birds, with 278,000–400,000 pairs estimated in the United States in the 1990s (but only about 100,000 pairs verified in 2005–2007) and 500,000 in Canada (Burger and Gochfeld 2020). It disperses widely over the prairies after breeding and before making long-distance migrations to the coasts of Peru and Chile for the nonbreeding season.

ORDER Charadriiformes
FAMILY Laridae
BREEDING HABITAT Wetlands, Freshwater Marsh
NONBREEDING HABITAT Coasts, Beach and Estuary
GLOBAL POPULATION SIZE 2,300,000
POPULATION CHANGE SINCE 1970 −58%
IUCN RED LIST CATEGORY Least Concern
R2R TIPPING POINT SPECIES No
US LISTING STATUS No
CANADA SPECIES AT RISK No
PIF CCS 11/20

MIGRATORY CONNECTIVITY

OVERALL There is sparse historical data on migratory connectivity from band recoveries, but it appears weak to moderate. Due to limited sample size, MC was not calculated for this species. Birds breeding within the Midcontinent Flyway primarily migrated to nonbreeding areas in the Andean West region of South America (93%), while Pacific Flyway breeding birds migrated to either the Andean West (67%) or Southern Cone (33%) regions of South America.

	Andean West	Southern Cone
Midcontinent (n = 14)	93%	7%
Pacific (n = 6)	67%	33%

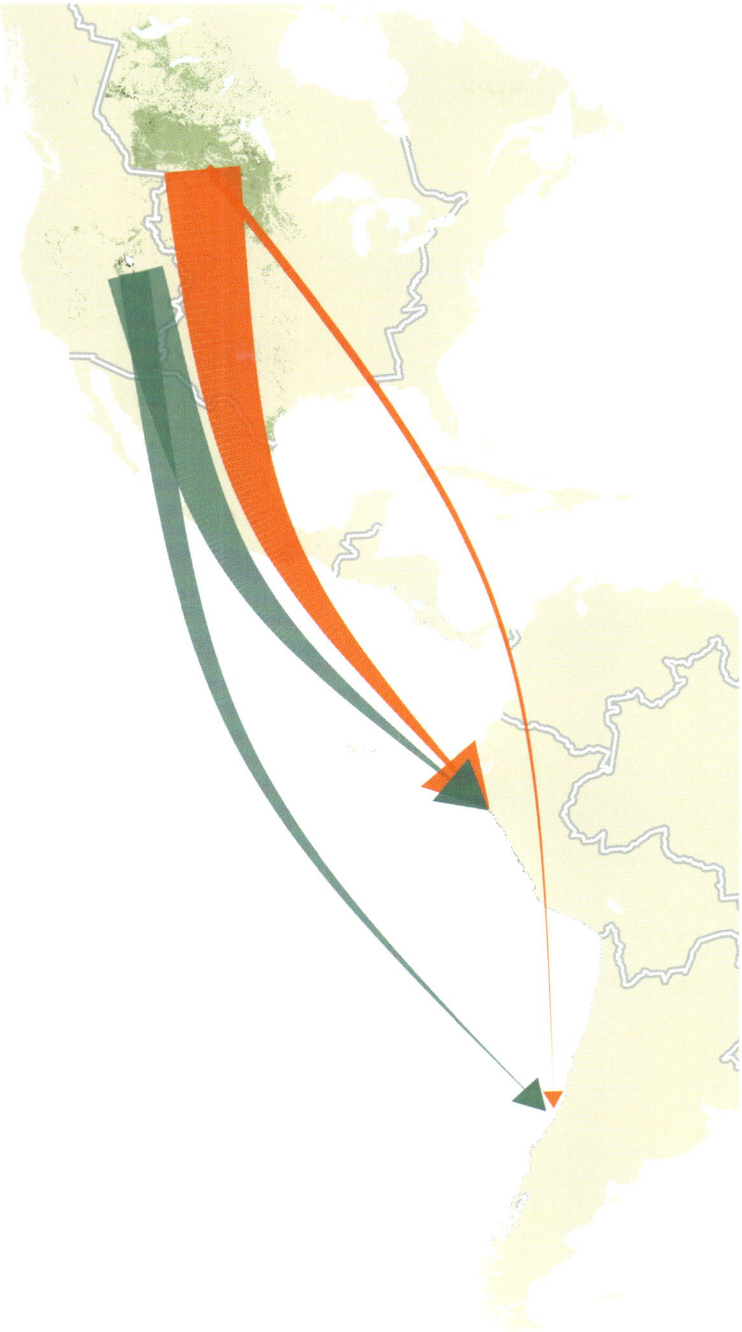

GENERALIZED CONNECTIVITY

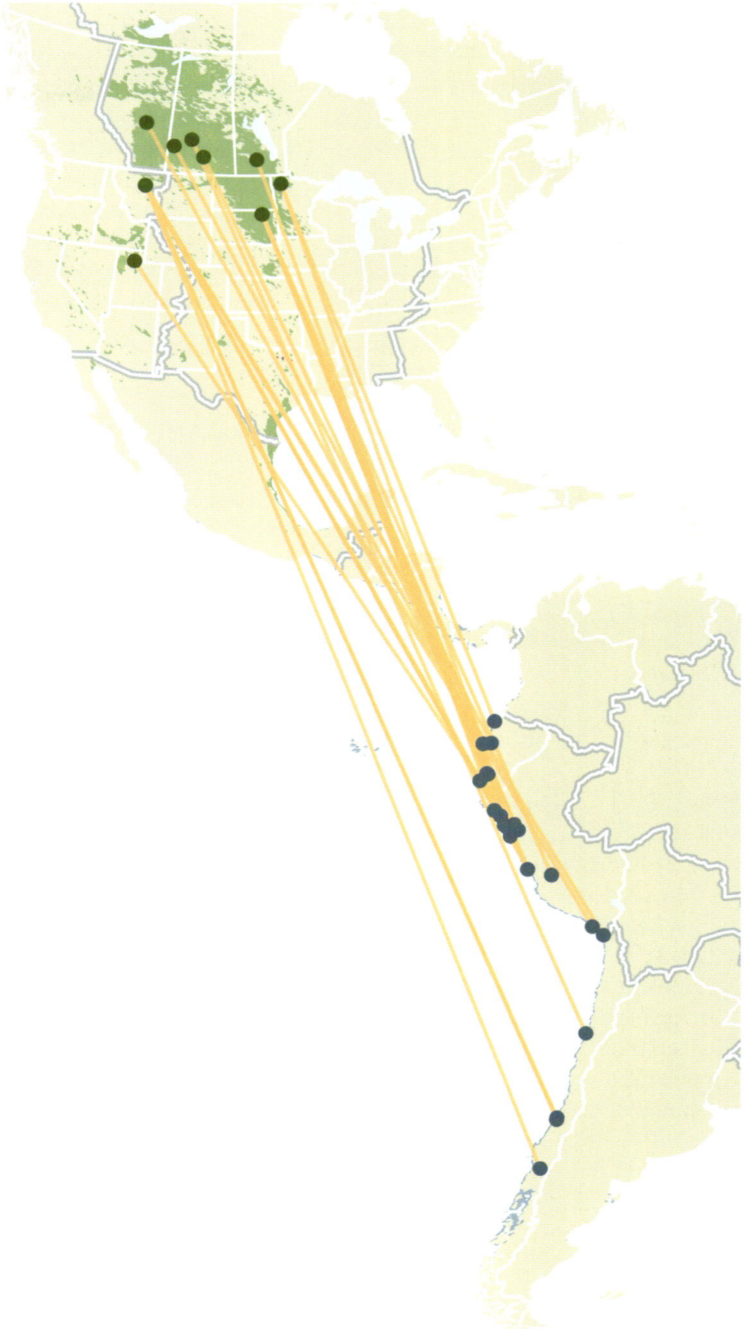

CONNECTIONS

BANDING Of the 20 adult banded birds recovered during the nonbreeding season, December through May, all were banded in 1938–1972 and recovered in 1985 or earlier. There were nonbreeding recoveries in South America (Ecuador, Peru, and Chile) from birds banded in Alberta, Manitoba, Minnesota, Montana, Saskatchewan, South Dakota, and Utah.

GENETICS Franklin's Gulls in the US breeding range have high levels of genetic variation and should be considered a panmictic population (Krmpotich 2012).

OTHER APPROACHES No tracking or stable isotope studies of migratory connectivity were found for this species.

MIGRATORY PATHWAYS Although no tracking data were found for this species, extensive observation data has demonstrated a migratory corridor through Mexico. Most Franklin's Gulls cross between the Atlantic and Pacific coasts of Central America via the Isthmus of Tehuantepec, before continuing their migrations (Burger and Gochfeld 2020).

INTERESTING NOTE

Franklin's Gulls are a priority for additional study, including tracking research, to understand habitat use in South America.

HEERMANN'S GULL *Larus heermanni*

DISTRIBUTION AND ABUNDANCE

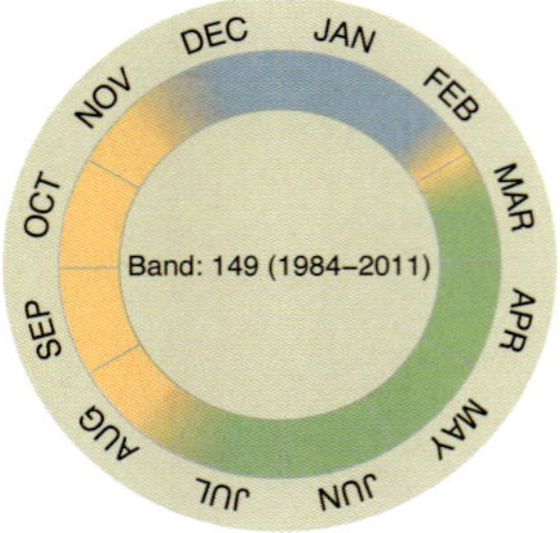

This uniquely patterned gull generally nests on arid offshore islands of the Gulf of California with 90%–95% of the global population (350,000 individuals) on Isla Rasa, Mexico (Islam and Velarde 2020). The Marieta Islands (4,500) and Partida Island (8,000) also hold significant numbers of individuals (Islam and Velarde 2020). This species rarely nests, but is doing so more frequently, in the United States (e.g., 20 pairs in 2018 in Monterey Bay; Islam and Velarde 2020). Overall, it is a partial short-distance migrant, with post-breeding distribution from British Columbia, Canada, to Guatemala, mainly in offshore marine waters but also in a variety of coastal aquatic habitats.

ORDER Charadriiformes
FAMILY Laridae
BREEDING HABITAT Coasts, Beach and Estuary
NONBREEDING HABITAT Coasts, Beach and Estuary
GLOBAL POPULATION SIZE 350,000

POPULATION CHANGE SINCE 1970 NA
IUCN RED LIST CATEGORY Near Threatened
R2R TIPPING POINT SPECIES No
US LISTING STATUS No
CANADA SPECIES AT RISK No
PIF CCS 16/20

INTERESTING NOTE

This species is among the least understood *Larus* gulls of North America in many aspects of their biology.

MIGRATORY CONNECTIVITY

OVERALL There is little migratory connectivity information.

BANDING Most reports of band reencounters during the nonbreeding season are from juveniles (and not included in connectivity analyses) from coastal California and Mexico, with very few offshore or in their northern range. There is likely a coastal (versus offshore) bias in these data, as Heermann's Gulls are often encountered in offshore seabird surveys (especially interacting with Pacific sardine fisheries). For example, seabird surveys conducted in 1996–2004 encountered Heermann's Gulls offshore (within 5 degrees longitude, n = 238) in the Strait of Juan de Fuca between Canada and the US (O'Hara et al. 2006b). The actual proportion of individuals from each colony that use offshore or coastal waters, or that migrate north, south, or remain resident, is unknown.

GENETICS Mitochondrial DNA analysis of 296 gulls from Cardonosa, Rasa, and Isabel islands found higher genetic differentiation by distance (Mancilla-Morales et al. 2022). Isabel Island off the Pacific coast of Nayarit, mainland Mexico, was more strongly differentiated from Rasa and Cardonosa islands within 4 nautical miles of each other in the northern Gulf of Mexico. Nest monitoring in this study also indicated significant male-biased philopatry. Tracking studies across colonies are needed to understand how the observed genetic structure may or may not relate to or be useful to understanding contemporary migratory connectivity.

OTHER APPROACHES No tracking or stable isotope studies of migratory connectivity were found for this species.

MIGRATORY PATHWAYS Individual routes have not been recorded, but pathways are suggested from seasonal distribution changes along the Pacific coast. Those individuals that are migratory take short-distance pathways in May and June north from main breeding areas in the Gulf of California, Mexico, along the Pacific coast to British Columbia, and south to Guatemala. Birds return to nest sites in February and March (Islam and Velarde 2020). The extent of coastal versus offshore pathways is unknown.

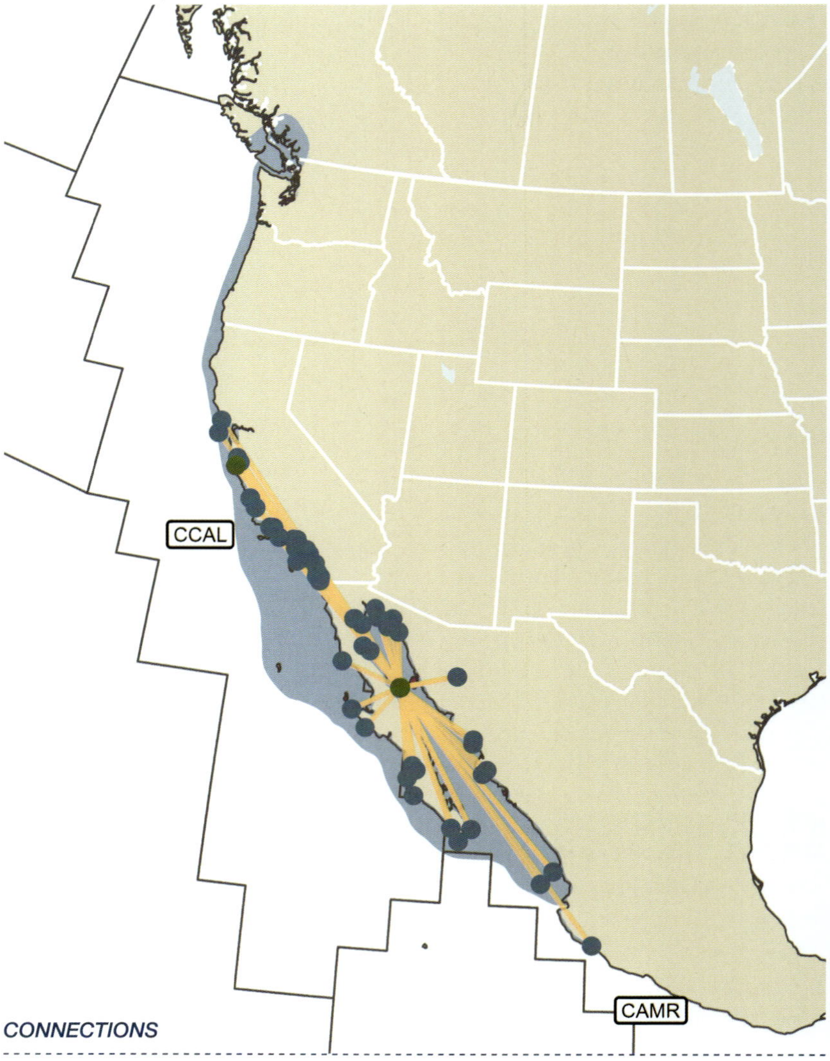

CONNECTIONS

Larus brachyrhynchus SHORT-BILLED GULL

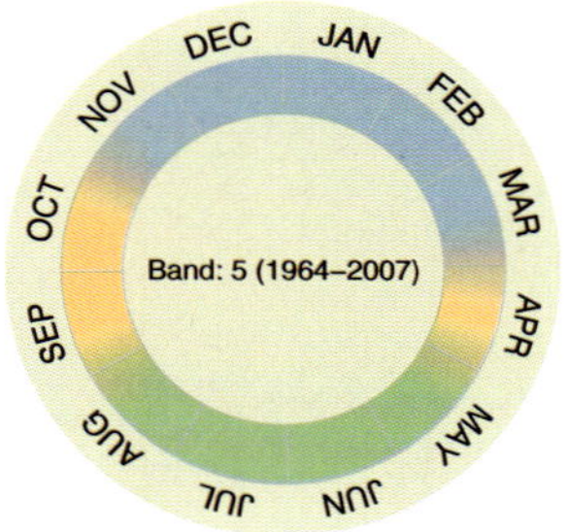

DISTRIBUTION AND ABUNDANCE

The Short-billed Gull, known as the Mew Gull in North America until recently, is a small, white-headed gull common along beaches and estuaries of the Pacific coast in the nonbreeding season. During the breeding season, it can be found in high numbers along the coasts of Alaska and British Columbia (Fink et al. 2023) and in lower numbers throughout interior Alaska, and the Yukon and Northwest Territories (Fink et al. 2023). There are no precise or regional abundance estimates for North America (possibly up to 360,000; Wetlands International 2015) other than the Audubon Christmas Bird Counts (Moskoff and Bevier 2021).

ORDER Charadriiformes
FAMILY Laridae
BREEDING HABITAT Wetlands, Generalist
NONBREEDING HABITAT Coasts, Beach and Estuary
GLOBAL POPULATION SIZE 1,300,000
POPULATION CHANGE SINCE 1970 −63%

IUCN RED LIST CATEGORY Least Concern
R2R TIPPING POINT SPECIES Orange-Alert
US LISTING STATUS No
CANADA SPECIES AT RISK No
PIF CCS 13/20

INTERESTING NOTE

Breeding biology and habitat selection is well-studied (Moskoff and Bevier 2021), but studies of migratory connectivity and population status should be priorities.

MIGRATORY CONNECTIVITY

OVERALL Migratory connectivity is unknown, but it is likely weak within the Pacific Flyway. Limited band recoveries show Alaskan birds from two regions spend the nonbreeding season on the coast of the Pacific Flyway.

BANDING Nonbreeding band recoveries from interior Alaska (n = 2) occurred in British Columbia and Oregon, and those from coastal Gulf of Alaska (n = 3) occurred in British Columbia, Washington, and Oregon.

OTHER APPROACHES No tracking, genetics, or stable isotope studies of migratory connectivity were found for this species.

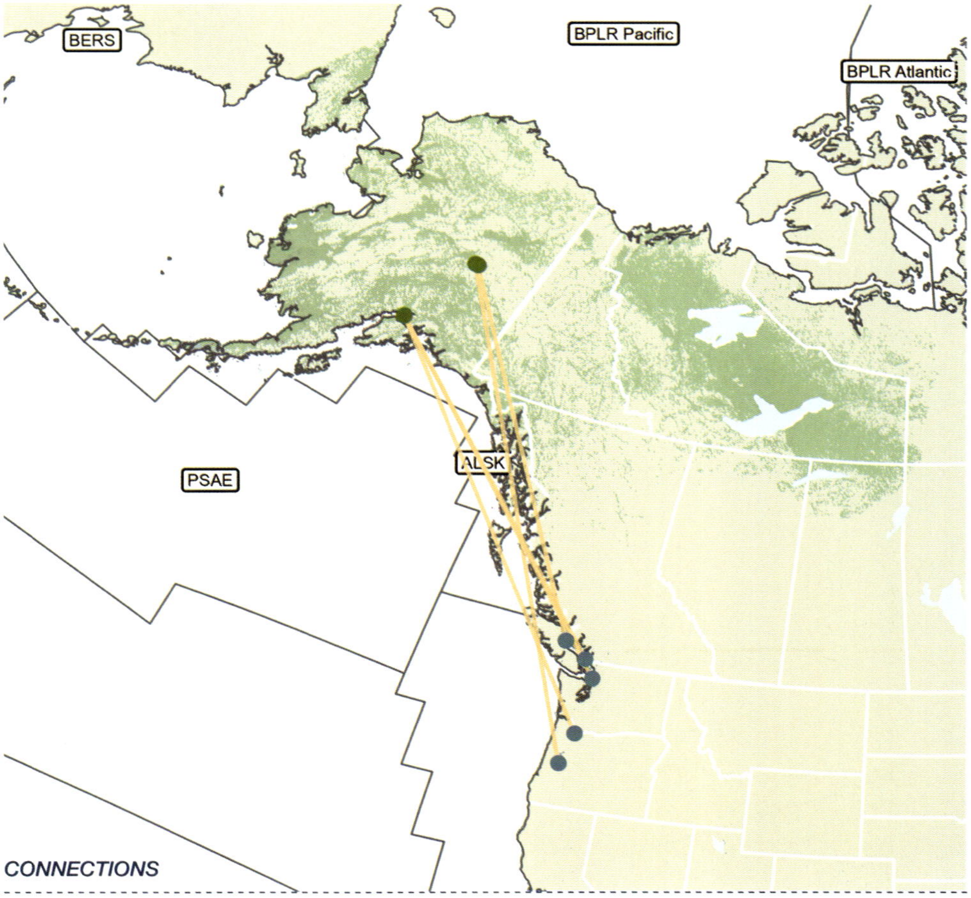

RING-BILLED GULL *Larus delawarensis*

MC = 0.36 ± 0.03*

DISTRIBUTION AND ABUNDANCE

The Ring-billed Gull is common throughout much of North America and is a familiar site at beaches, lakes, rivers, and parking lots. High concentrations of Ring-billed Gulls breed along the shores of the Great Lakes, St. Lawrence River, coastal areas of Quebec and Newfoundland and Labrador, Great Salt Lake, and throughout the Prairie Pothole Region of the United States and Canada, especially south of Lake Winnipeg in Manitoba (Fink et al. 2023). Ring-billed Gulls can be found in high numbers during the nonbreeding season along the Atlantic coast, the Gulf coast, San Francisco Bay and the Central Valley of California, the west coast of Mexico, and along the Mississippi River in Arkansas (Fink et al. 2023).

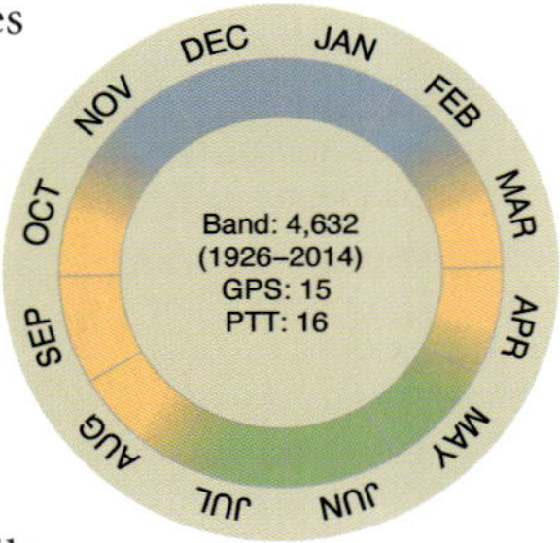

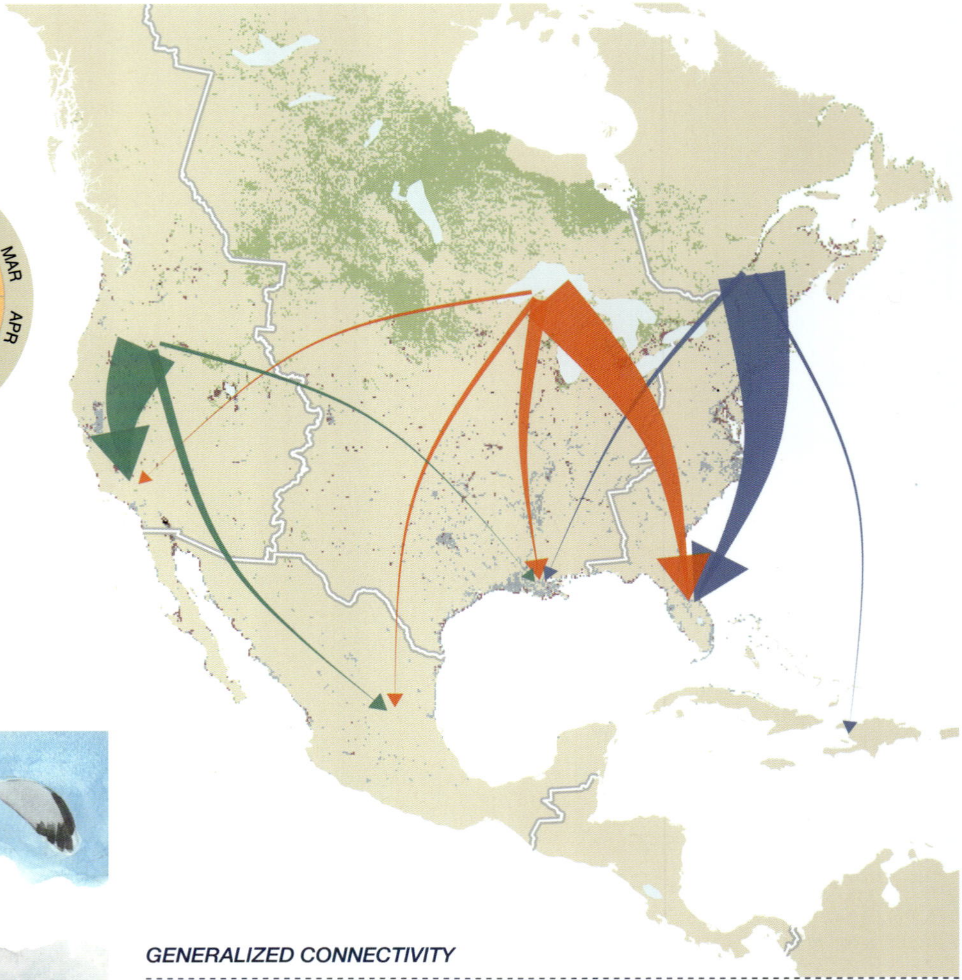

GENERALIZED CONNECTIVITY

ORDER Charadriiformes
FAMILY Laridae
BREEDING HABITAT Wetlands, Generalist
NONBREEDING HABITAT Wetlands, Lakes and Rivers
GLOBAL POPULATION SIZE 3,800,000
POPULATION CHANGE SINCE 1970 +103%
IUCN RED LIST CATEGORY Least Concern
R2R TIPPING POINT SPECIES No
US LISTING STATUS No
CANADA SPECIES AT RISK No
PIF CCS 8/20

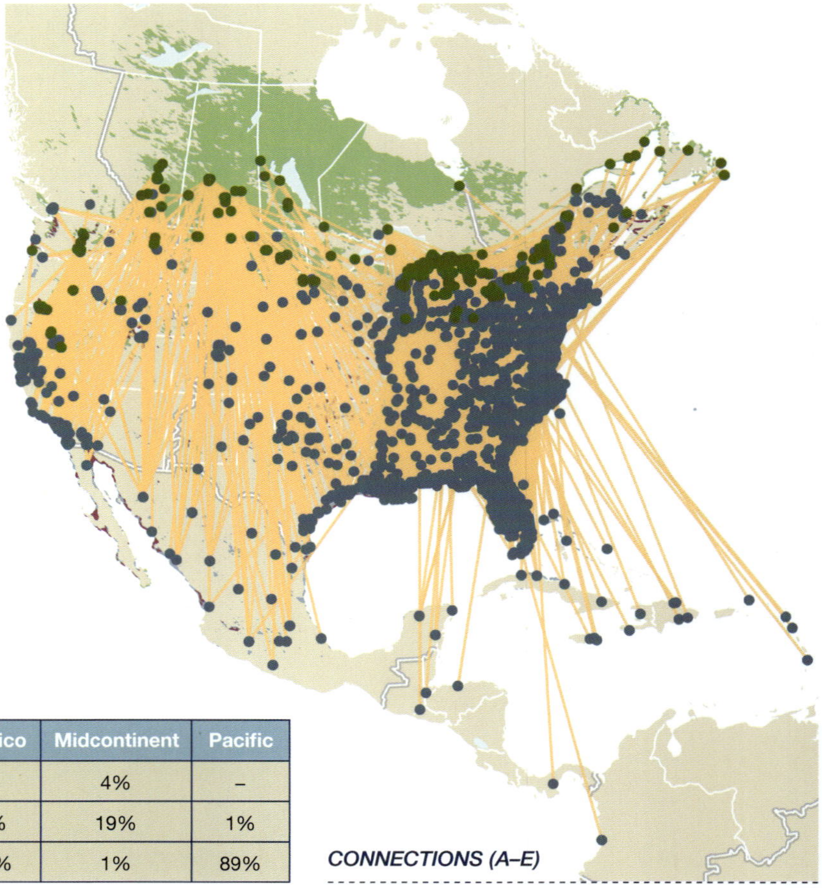

CONNECTIONS (A–E)

	Atlantic	Caribbean	Central America	Mexico	Midcontinent	Pacific
Atlantic (n = 453)	94%	1%	<1%	–	4%	–
Midcontinent (n = 3459)	76	<1%	<1%	4%	19%	1%
Pacific (n = 152)	–	–	–	10%	1%	89%

MIGRATORY CONNECTIVITY

OVERALL Substantial data on migratory connectivity exists for Ring-billed Gulls from banding reencounters and some satellite transmitters. Birds breeding in the Atlantic Flyway mainly migrate to nonbreeding areas within the Atlantic Flyway (94%). Midcontinent Flyway breeding populations migrate to nonbreeding areas along the coast of the Atlantic (76%) and Midcontinent (19%) flyways. Pacific Flyway breeding populations migrate to nonbreeding areas within the Pacific Flyway (89%) or Mexico (10%). The Atlantic and Pacific flyways breeding populations largely remain in the same flyway during the nonbreeding season; however, there is also mixing, especially with the Midcontinent Flyway breeding populations throughout nonbreeding areas, resulting in moderate migratory connectivity (0.36 ± 0.03).

BANDING Data from band returns exist for breeding colonies throughout the range. Eastern and western breeding populations spend the nonbreeding season within eastern and western portions of the nonbreeding range, respectively. There appears to be more mixing of central breeding populations throughout nonbreeding areas. Birds breeding in the Great Lakes primarily migrate to Florida and the

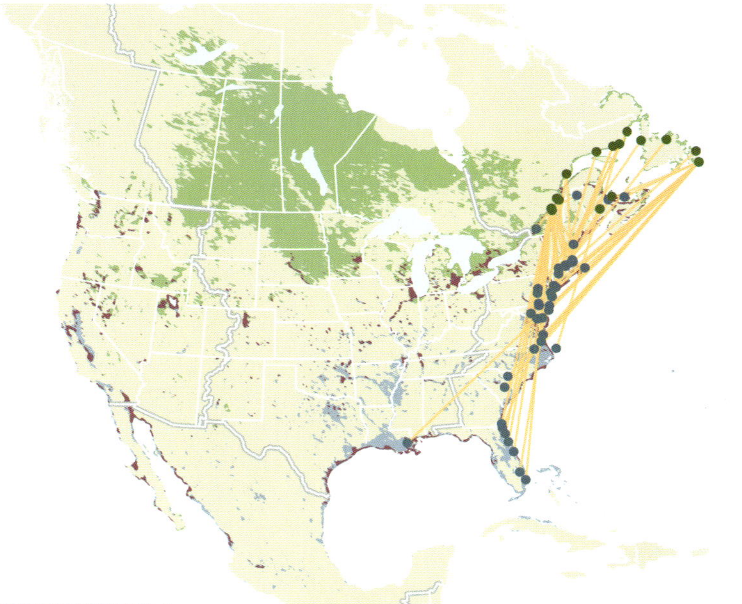

CONNECTIONS A

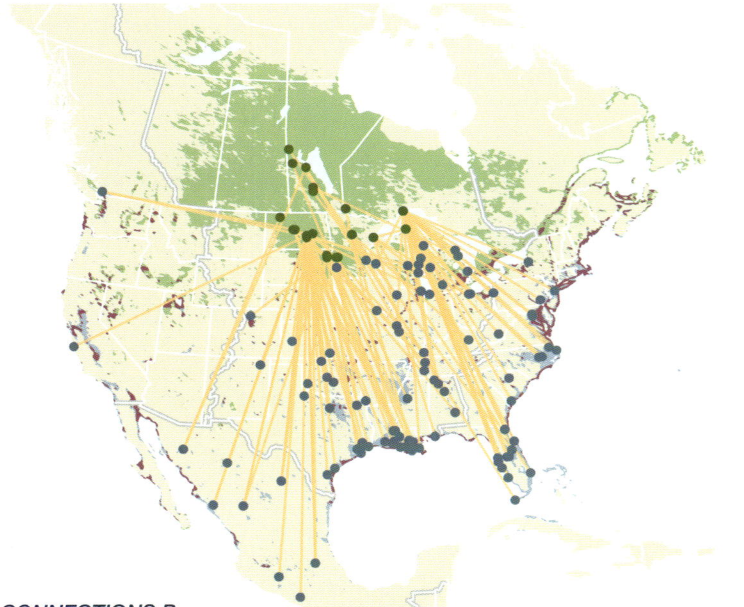

CONNECTIONS B

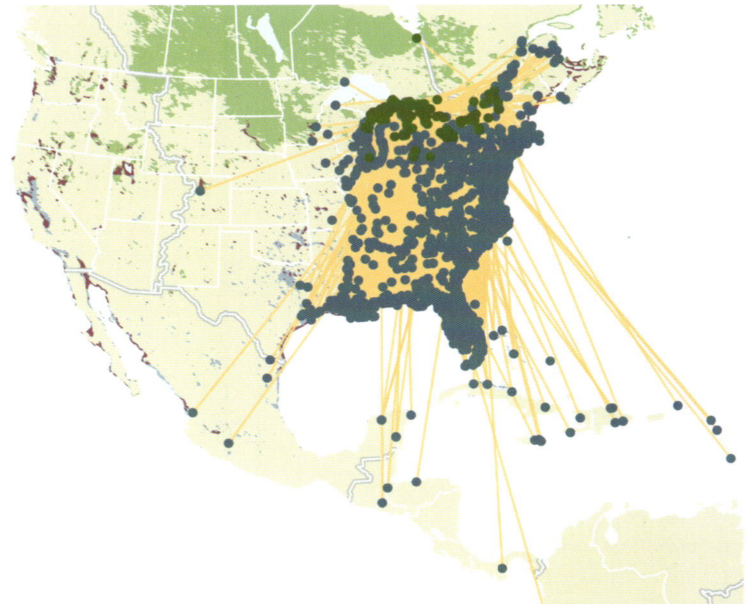

CONNECTIONS C

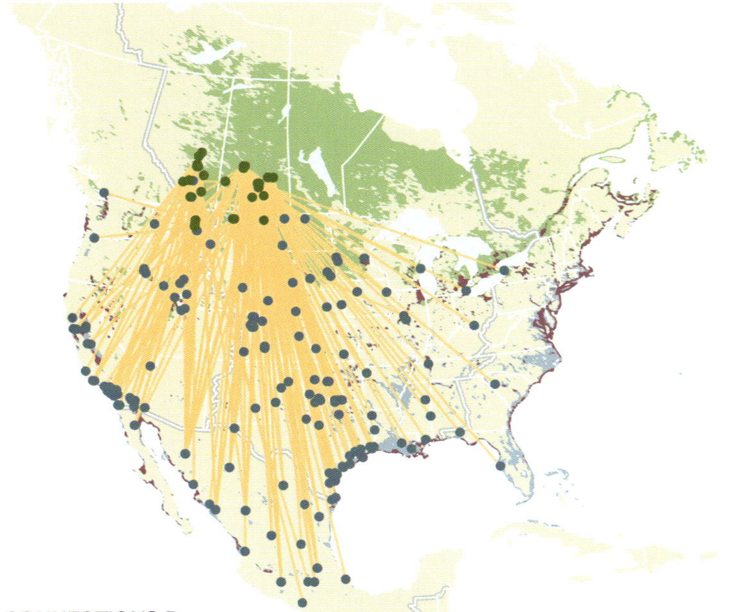

CONNECTIONS D

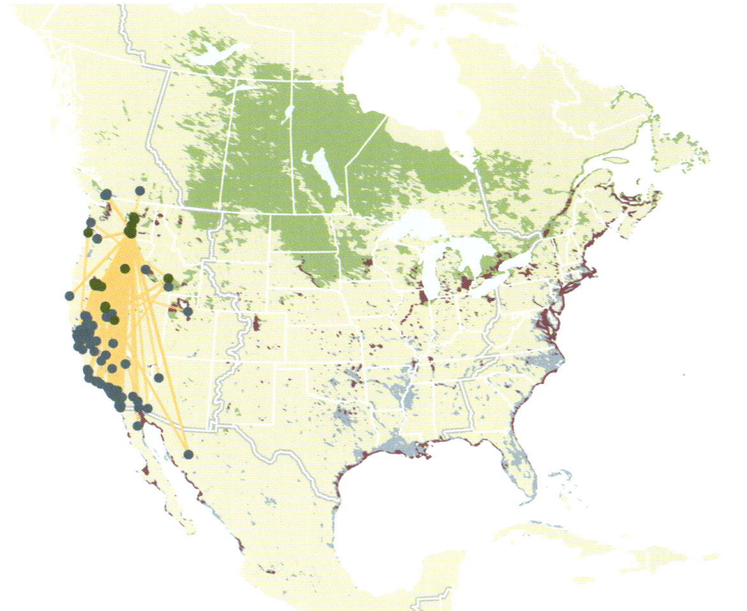

CONNECTIONS E

Gulf region, especially the Atlantic coast of Florida (Southern 1974; Gabrey 1996). Birds banded at breeding colonies in Quebec occurred all along the northern Atlantic coast of North America and as far inland as the Great Lakes during the nonbreeding season (Ineson et al. 2022).

SATELLITE TRACKING Birds tagged on nonbreeding grounds in Massachusetts (n = 17) were found to breed in several northern locations (Quebec, Ontario, New York, and Vermont; Clark et al. 2016). Most returned to Massachusetts the following year, but did not remain the entire season and used several nonbreeding locations throughout the mid-Atlantic, and as far south as Georgia and Florida (Clark et al. 2016). Birds tagged on breeding grounds in Ontario (n = 15) migrated to nonbreeding area across Virginia, North Carolina, and Florida, and one bird to the Bahamas (D. Moore and J. Costa, unpublished data). Most returned to Ontario to breed. Tracking data are lacking for birds in the western portion of the range.

OTHER APPROACHES No genetic or stable isotope studies of migratory connectivity were found for this species.

MIGRATORY PATHWAYS Gulls tracked from Ontario used two main pathways to nonbreeding areas—either inland, flying directly south from breeding areas, or southeast to the coast and then continuing a coastal route south (D. Moore and J. Costa, unpublished data). One bird tracked to the Bahamas from Ontario, flew inland to the Carolinas, then over the open ocean to the Bahamas during post-breeding migration, but traveled a coastal route north during pre-breeding migration, from Florida to South Carolina and then inland to return to Ontario (D. Moore and J. Costa, unpublished data).

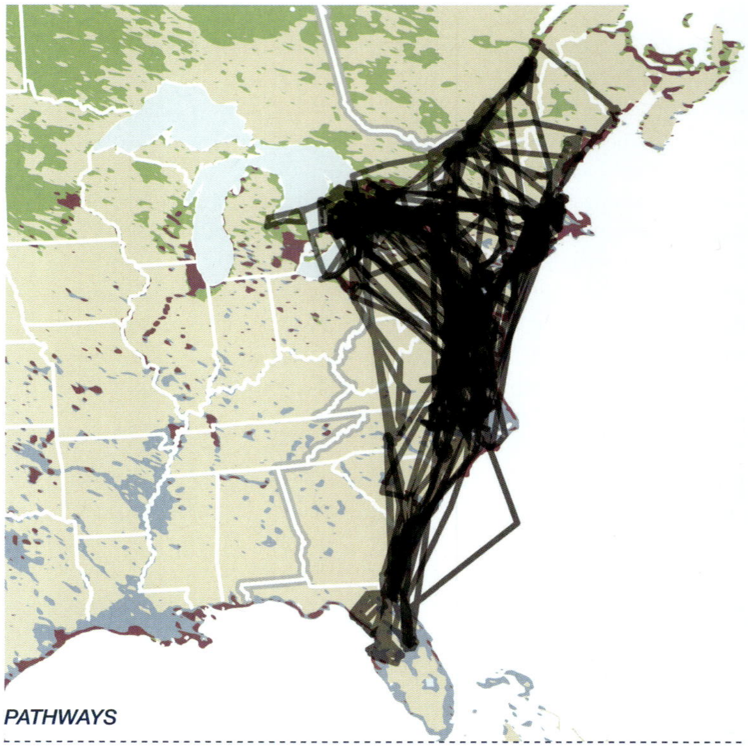

PATHWAYS

DATA CONTRIBUTORS

D.E. Clark; J. Costa; K.K.G. Koenen; K.G. MacKenzie; D. Moore; J.J. Whitney.

INTERESTING NOTE

There is some site fidelity to individual colonies, but it is stronger at the population level, with little immigration/emigration recorded (Gabrey 1996; Pollet et al. 2020).

WESTERN GULL *Larus occidentalis*

DISTRIBUTION AND ABUNDANCE

A large gull of the Pacific North American coast, where it breeds from Washington, US, south to Baja California, Mexico. It's post-breeding distribution is only slightly larger, extending north to Vancouver Island, Canada, and south to the tip of the Baja California peninsula (Pierotti et al. 2024). Western Gulls are rarely seen away from the coast or sheltered bays. Despite major declines at southeast Farallon Islands, California, it remains one of the largest colonies, hosting more than 30% of the global population (Pierotti et al. 2024). More than 16,000 breeding pairs can be found on Pacific islands of Baja California (Pierotti et al. 2024). Western Gulls have declined 55% on Breeding Bird Surveys since 1970 (US Geological Survey 2014), but some populations seem to be increasing (Alcatraz and Año Nuevo Island, California; Pierotti et al. 2024). More research is needed to clarify the population status throughout the species range (Pierotti et al. 2024).

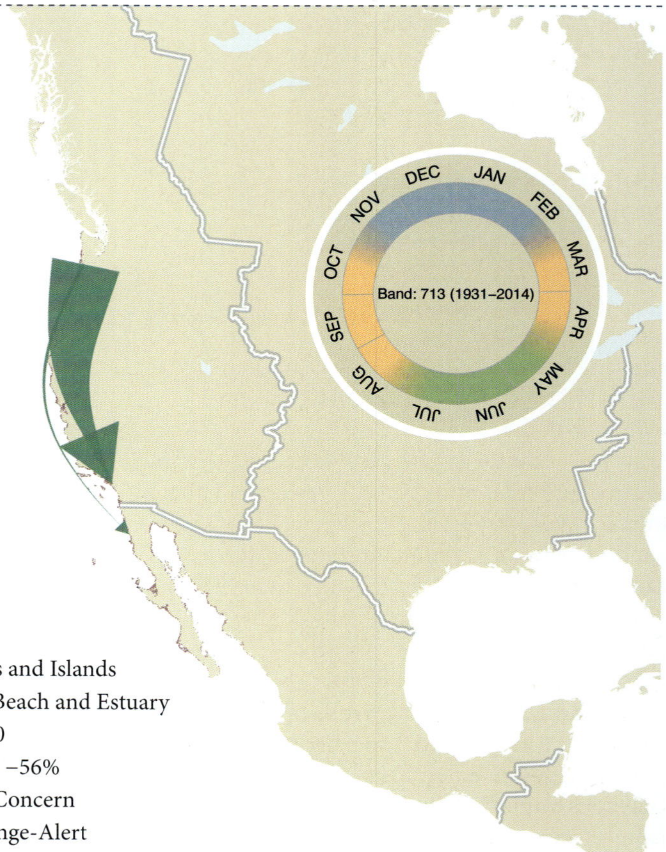

ORDER Charadriiformes
FAMILY Laridae
BREEDING HABITAT Coasts, Cliffs and Islands
NONBREEDING HABITAT Coasts, Beach and Estuary
GLOBAL POPULATION SIZE 86,000
POPULATION CHANGE SINCE 1970 −56%
IUCN RED LIST CATEGORY Least Concern
R2R TIPPING POINT SPECIES Orange-Alert
US LISTING STATUS No
CANADA SPECIES AT RISK No
PIF CCS 15/20

	Mexico	Pacific
Pacific (n = 639)	1%	99%

GENERALIZED CONNECTIVITY

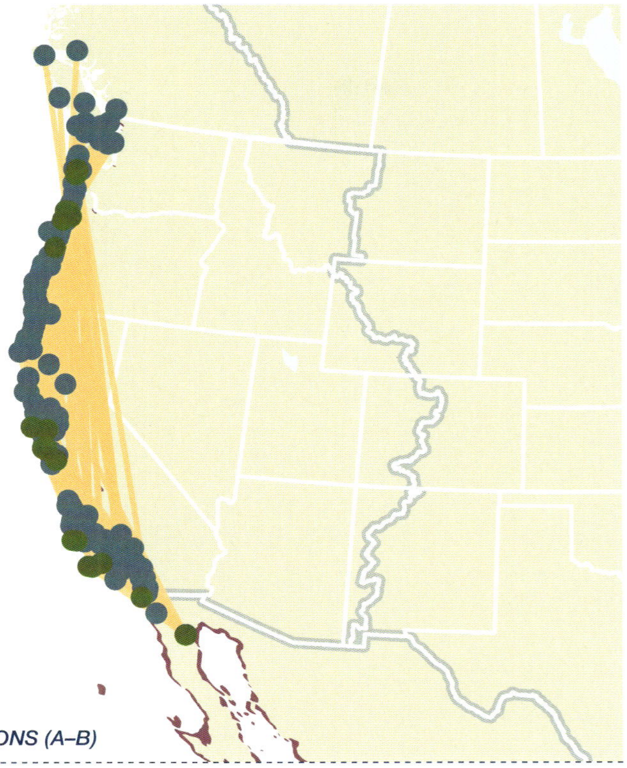

CONNECTIONS (A–B)

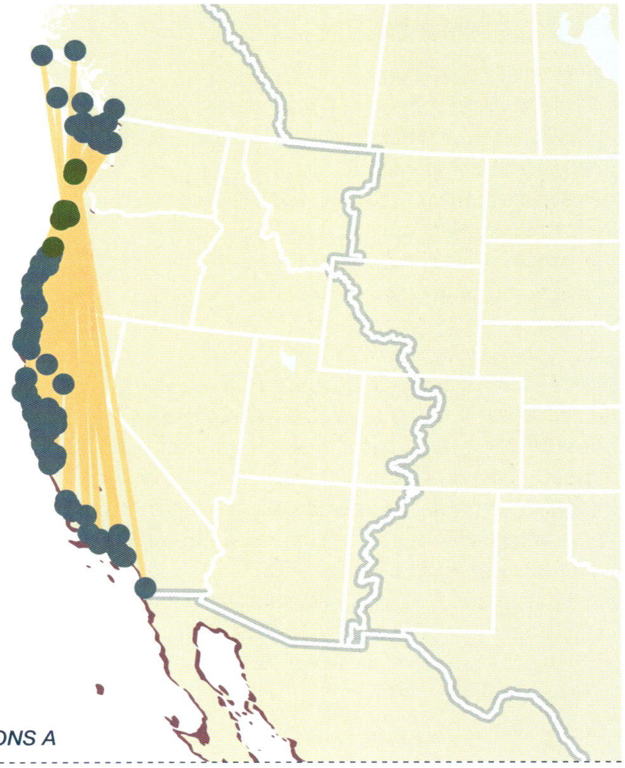

CONNECTIONS A

INTERESTING NOTE

GPS tracking data collected during the incubation period (5–8 days tracking duration) from two California colonies showed marked differences in breeding foraging distributions. Gulls from Southeast Farallon Island (30 km offshore) largely foraged over oceanic waters, while gulls from Año Nuevo Island, California, made 71% of foraging trips to terrestrial sites along the coast, including heavy use of a specific landfill. There was no overlap across colonies during the breeding season.

MIGRATORY CONNECTIVITY

OVERALL This species is well-studied, and its relatively short-distance movements are fairly well understood due to band reencounter data, although these could contain observer bias limiting recoveries in marine habitats away from the coast. Migratory connectivity appears weak; however, MC was not calculated as available data comes from only one breeding population. Birds breeding in the Pacific Flyway mainly migrate to nonbreeding areas within the Pacific Flyway (99%), with a small percentage to Mexico (1%).

BANDING Adult band recoveries range from British Columbia to Mexico for birds banded in Washington, Oregon, California, and Mexico. Western Gull post-breeding movements as inferred from banding have been summarized in multiple papers (Ferris 1940; Coulter 1975).

GENETICS Western Gulls have high genetic variability across their range (Bell 1996), including recognized northern and southern subspecies that vary in morphology (Pierotti et al. 2024). Much genetic work has been done on hybridization with the Glaucous-winged Gull across many colonies (e.g., Bell 1996). Geng et al. (2024) suggest that a persistent hybrid zone of Western Gulls and Glaucous-winged Gulls along the coast in Washington, US, has been maintained since the early twentieth century because each species and their hybrids occupy significantly different ecological niches.

OTHER APPROACHES No tracking or stable isotope studies of migratory connectivity were found for this species.

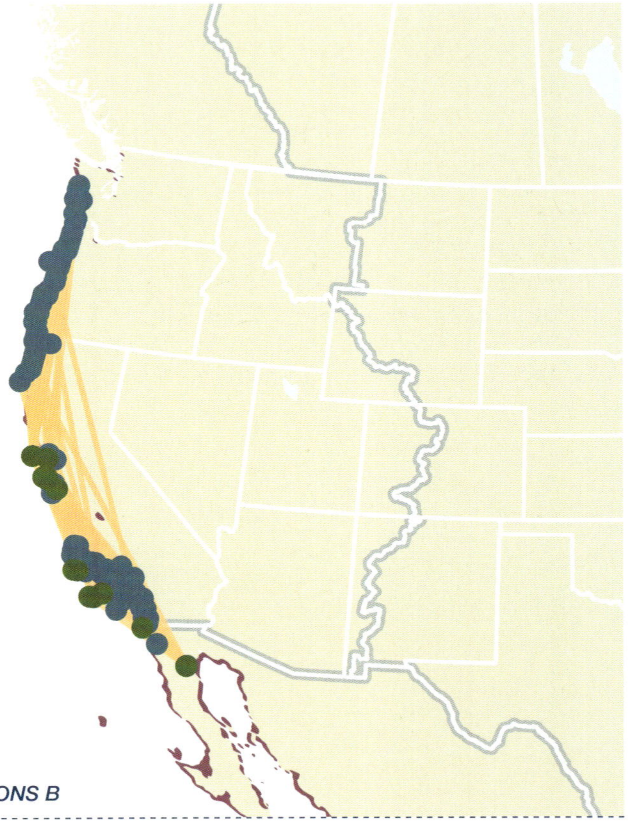

CONNECTIONS B

MIGRATORY PATHWAYS There are no published post-breeding migratory pathways available, but both northward and southward post-breeding coastal routes are inferred from at-sea survey and eBird data showing an influx of birds at the northern and southern parts of the range in October and November and then departure in March and April for pre-breeding migration (Pierotti et al. 2024). However, birds may also remain resident, migrate from the coastal zone to offshore for the post-breeding period without making large latitudinal movements, or migrate to inland habitats (Pierotti et al. 2024).

CALIFORNIA GULL *Larus californicus*

GENERALIZED CONNECTIVITY

DISTRIBUTION AND ABUNDANCE

This gull breeds on arid and prairie inland lakes in central and western Canada and the US. Breeding strongholds are in Saskatchewan and Alberta in Canada, and the Dakotas in the US. It spends the nonbreeding season mostly along the Pacific coast from Vancouver, Canada, to southern Sinaloa, Mexico.

ORDER Charadriiformes
FAMILY Laridae
BREEDING HABITAT Coasts, Beach and Estuary
NONBREEDING HABITAT Coasts, Beach and Estuary
GLOBAL POPULATION SIZE 1,100,000
POPULATION CHANGE SINCE 1970 −48%
IUCN RED LIST CATEGORY Least Concern
R2R TIPPING POINT SPECIES No
US LISTING STATUS No
CANADA SPECIES AT RISK No
PIF CCS 12/20

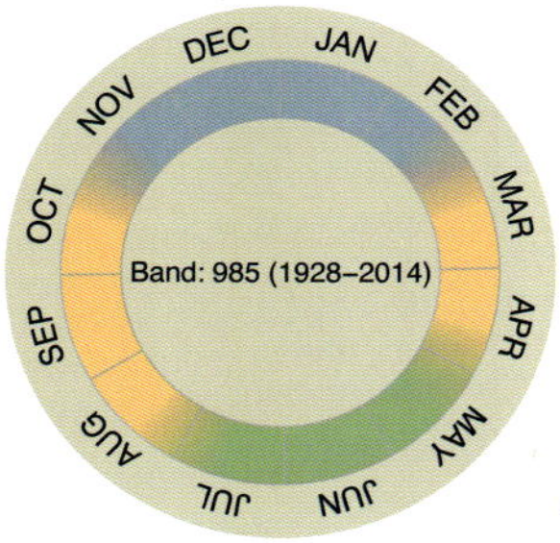

	Mexico	Midcontinent	Pacific
Midcontinent (n = 267)	12%	4%	84%
Pacific (n = 525)	12%	1%	87%

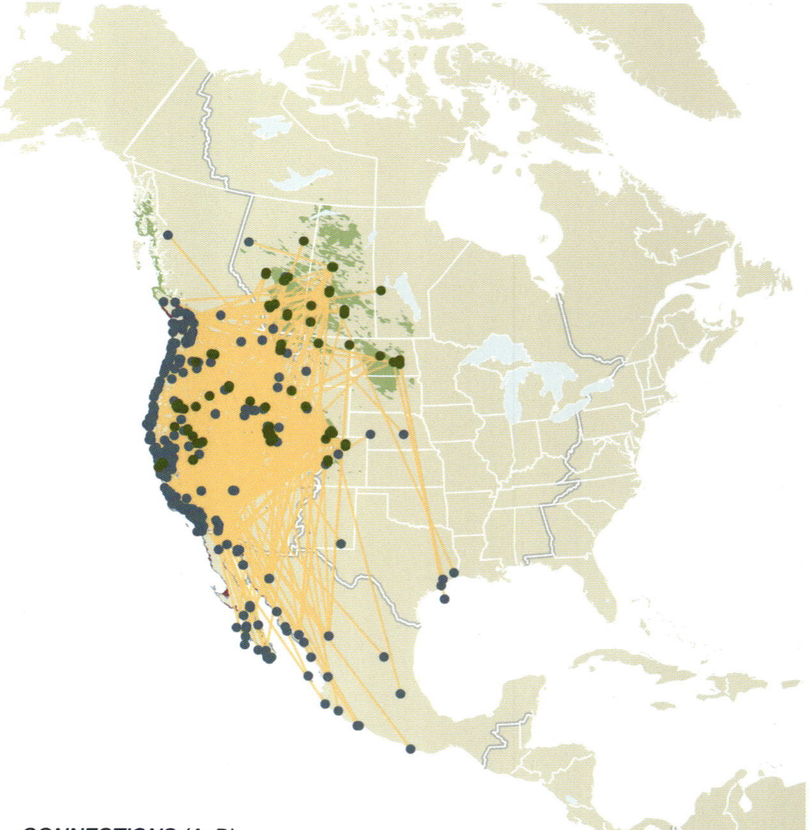

CONNECTIONS (A–D)

MIGRATORY CONNECTIVITY

OVERALL Connectivity information is mostly from banding and patagial wing tag resighting. Both Midcontinent and Pacific flyway breeders spend the nonbreeding season mostly within the Pacific Flyway (84% and 87%, respectively), Mexico (12% and 12%, respectively), and the Midcontinent Flyway (4% and 1%, respectively). Although within-flyway transition probabilities are high, breeding populations overlap on all nonbreeding areas, which results in weak migratory connectivity (MC = 0.03 ± 0.03).

BANDING Results from band and patagial wing tag resighting have shown that birds disperse to the Pacific coast (especially from San Francisco to Vancouver), and then most move south throughout the nonbreeding season. Breeders then return to breeding colonies, while immature birds tend to stay by the coast (Houston 1977; Pugesek et al. 1999).

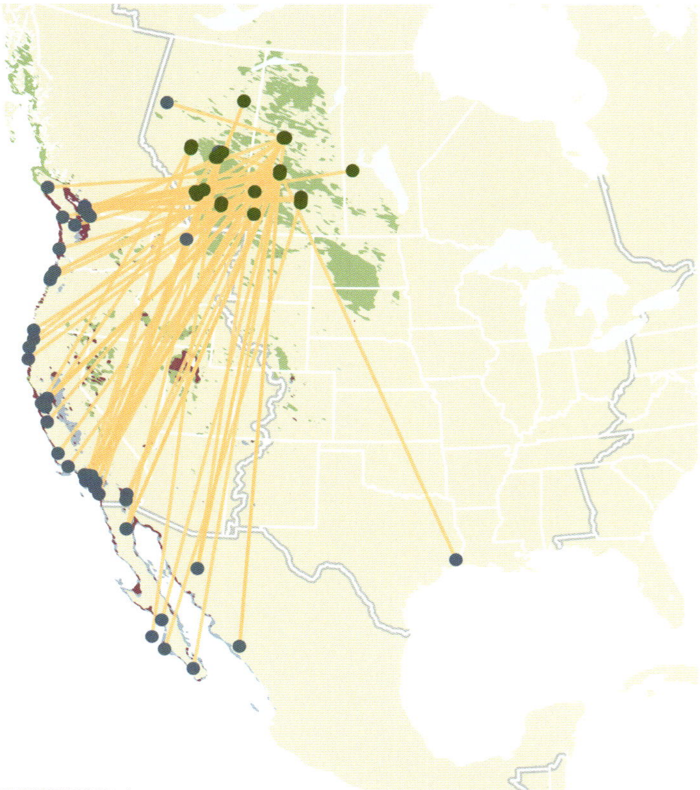

CONNECTIONS A

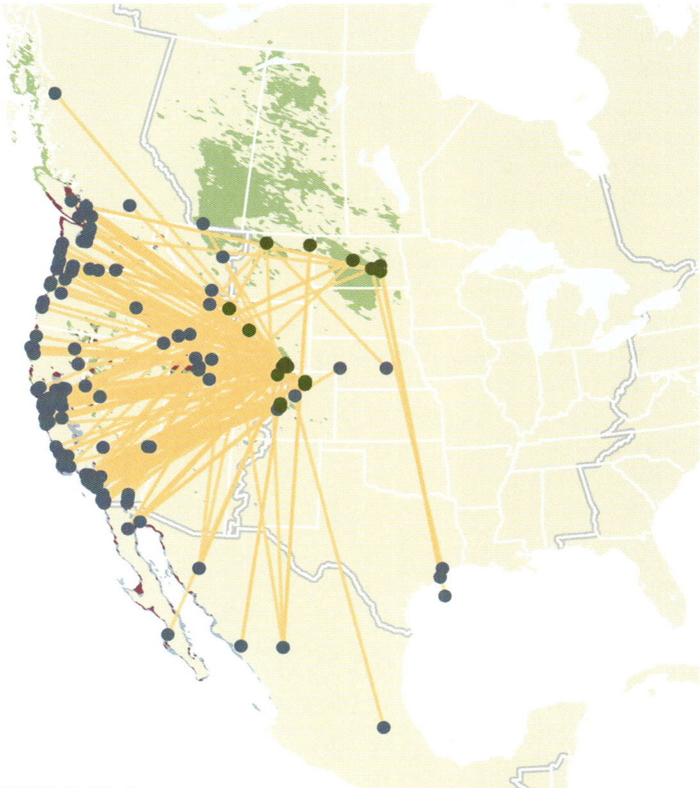

CONNECTIONS B

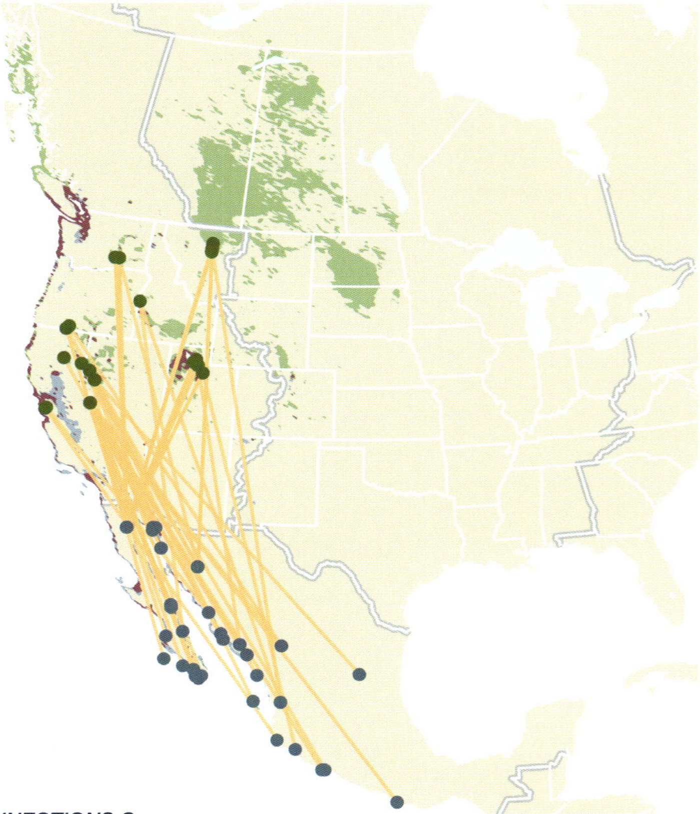

CONNECTIONS C

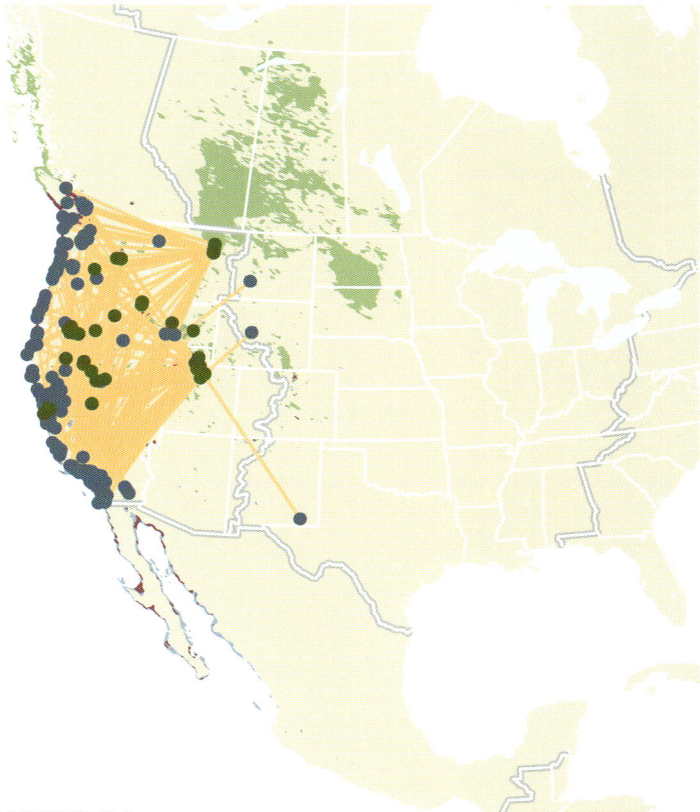

CONNECTIONS D

OTHER APPROACHES No genetic or stable isotope studies of migratory connectivity were found for this species.

MIGRATORY PATHWAYS Birds from the Canadian Prairies cross the Rockies in south British Columbia, and large numbers of California Gulls that follow the Columbia River and its tributaries through a Cascade-Sierra axis to the Pacific coast appear to be derived from colonies throughout the northern Great Basin and intermountain regions (Diem and Condon 1967; Littlefield 1990; Gilligan et al. 1994). Farther south, birds from a large colony at Mono Lake, California, cross the Sierra directly (Winkler 2020). California Gulls carry out slow but sustained migration up and down the Pacific coast in both pre- and post-breeding migrations. Birds of all ages apparently move northward along the coast briefly, perhaps in response to northward movement of front of upwelling events (Shuford et al. 1989).

HERRING GULL *Larus argentatus*

weak strong
MC = 0.4 ± 0.04*

DISTRIBUTION AND ABUNDANCE

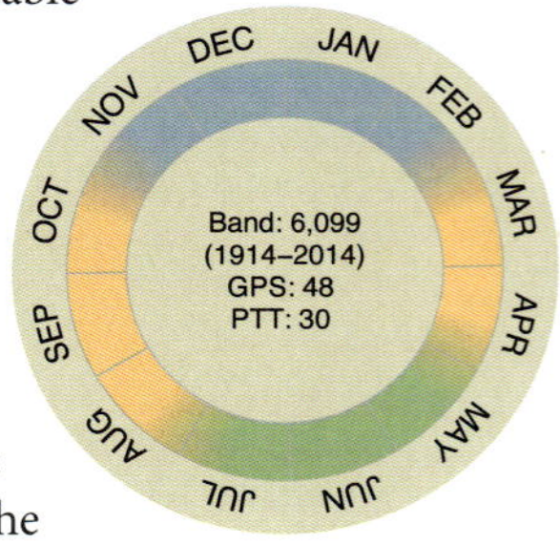

This large gull is extremely adaptable and can be found throughout much of the Northern Hemisphere. In North America, breeding populations are especially concentrated in Arctic Canada and the Maritime provinces, around the Great Lakes, and the Atlantic coast south to North Carolina. The species also breeds in western Canada and Alaska. It can be found throughout much of North America in the nonbreeding season, especially along the Atlantic coast where appropriate resources exist (Nisbet et al. 2017; Fink et al. 2023).

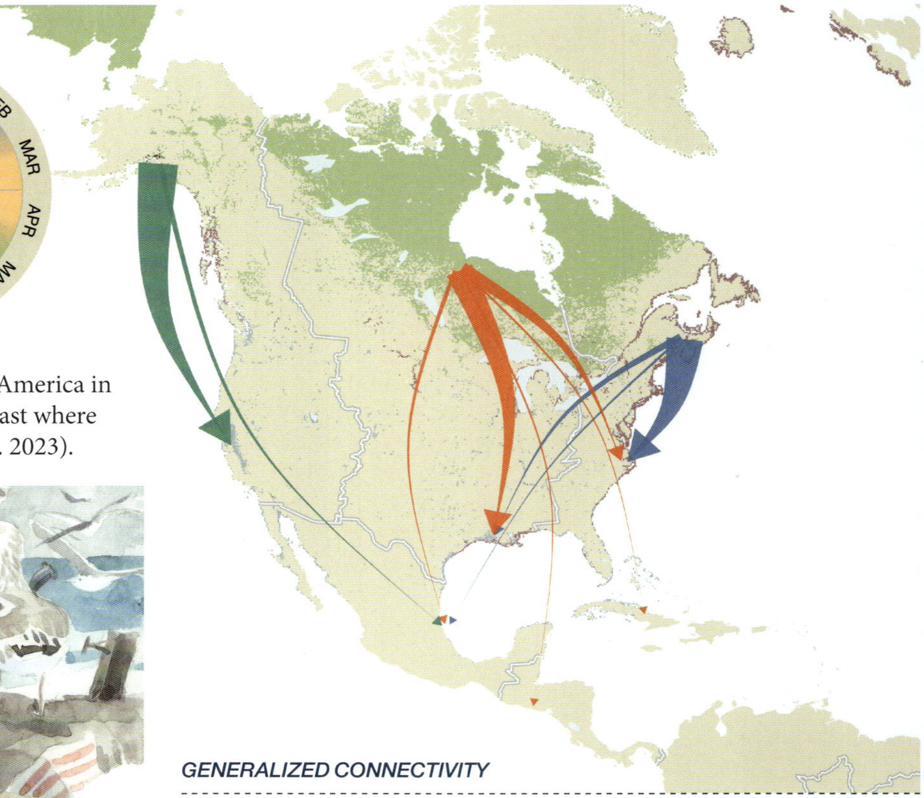

GENERALIZED CONNECTIVITY

ORDER Charadriiformes
FAMILY Laridae
BREEDING HABITAT
Coasts, Beach and Estuary
NONBREEDING HABITAT
Coasts, Beach and Estuary
GLOBAL POPULATION SIZE 3,900,000
POPULATION CHANGE SINCE 1970 −81%
IUCN RED LIST CATEGORY Least Concern
R2R TIPPING POINT SPECIES No
US LISTING STATUS No
CANADA SPECIES AT RISK No
PIF CCS 12/20

INTERESTING NOTE

C. Anderson et al. (2019) found strong habitat connectivity between breeding and nonbreeding areas for tracked Herring Gulls, with Arctic breeding gulls spending the nonbreeding season in marine habitats, and breeders from Newfoundland and Sable Island preferring urban habitats.

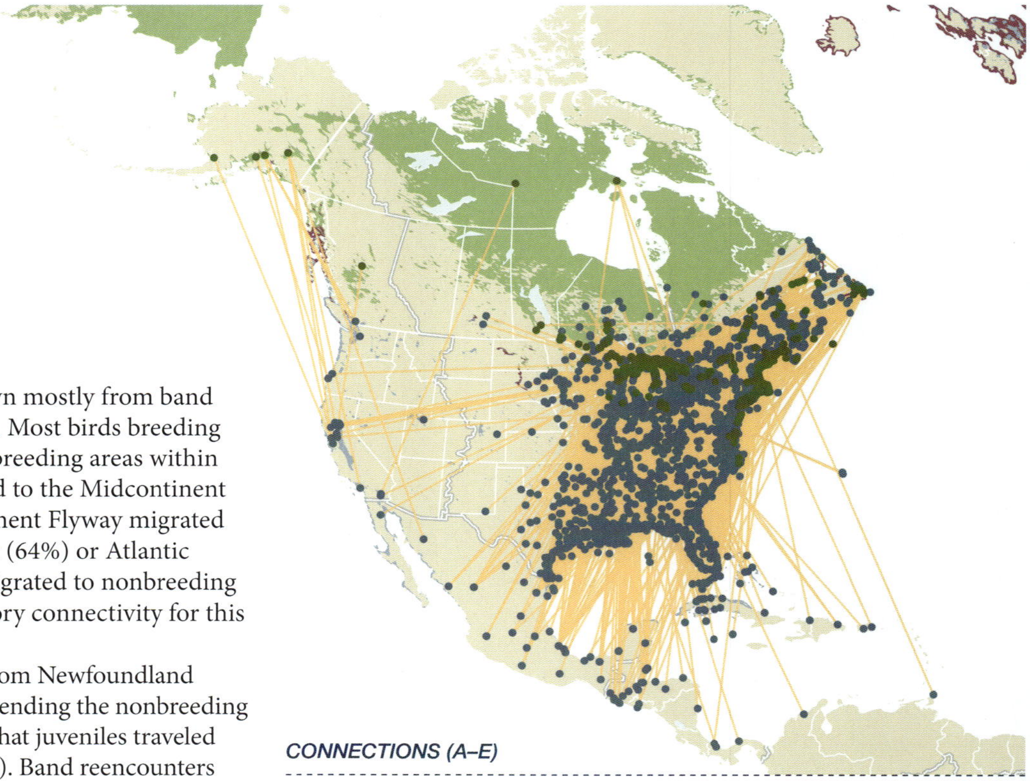

CONNECTIONS (A–E)

MIGRATORY CONNECTIVITY

OVERALL Migratory connectivity is well known mostly from band reencounter data and satellite tracking studies. Most birds breeding in the Atlantic Flyway (86%) migrated to nonbreeding areas within the Atlantic Flyway, while some birds migrated to the Midcontinent Flyway (12%). Birds breeding in the Midcontinent Flyway migrated to nonbreeding areas within the Midcontinent (64%) or Atlantic (29%) flyway. Most Pacific Flyway breeders migrated to nonbreeding areas within the Pacific Flyway (92%). Migratory connectivity for this species is strong (MC = 0.4 ± 0.04).

BANDING Banding records of breeding birds from Newfoundland showed a consistent pattern, with most birds spending the nonbreeding season in New England. This study also noted that juveniles traveled farther than adults on average (Robertson 2016). Band reencounters of birds breeding on the Great Lakes showed most birds spent the nonbreeding period on Lakes Erie, Ontario, and Michigan. Similarly, juveniles from this population migrated farther than adults (Gabrey 1996).

	Atlantic	Caribbean	Central America	Mexico	Midcontinent	Northen SA	Pacific
Atlantic (n = 2814)	86%	<1%	<1%	1%	12%	–	<1%
Midcontinent (n = 2384)	29%	1%	1%	5%	64%	<1%	<1%
Pacific (n = 12)	–	–	–	8%	–	–	92%

SATELLITE TRACKING Tracking studies from the eastern part of the range show strong connectivity at the population level but not at the colony level (Clark et al. 2016; C. Anderson et al. 2019, 2020). Herring Gulls that bred on Southampton Island, Nunavut, spent most of the nonbreeding season ranging from the Great Lakes to the Gulf of Mexico (C. Anderson et al. 2020). Herring Gulls that bred in the Great Lakes spent the majority of the nonbreeding season in the Great Lakes basin, with other individuals migrating to nonbreeding areas in Massachusetts, Ontario, and Quebec. The three populations from Atlantic Canada spent the nonbreeding period in the northeastern US from Cape Cod to Delaware Bay, and two individuals did not migrate (C. Anderson et al. 2020). Similarly, birds tracked from nonbreeding areas in Massachusetts bred in the Canadian Maritime provinces (Clark et al. 2016). One tracking study of breeding birds in Alaska found individuals migrated to nonbreeding areas along the Pacific Northwest coast and San Francisco Bay, California (Ramey et al. 2020).

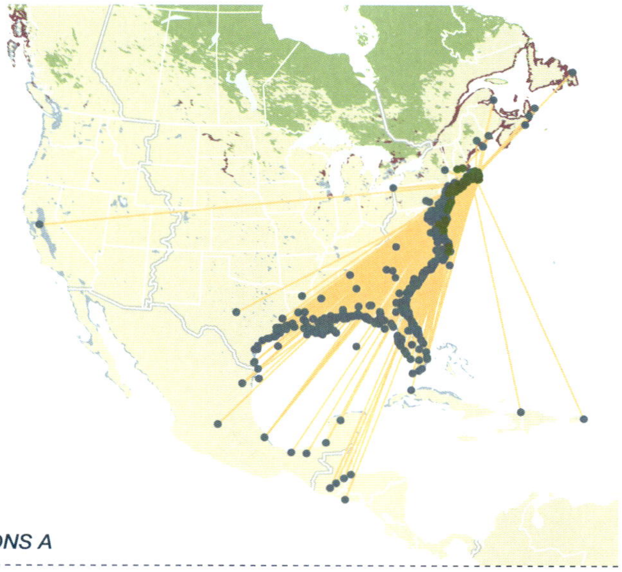

CONNECTIONS A

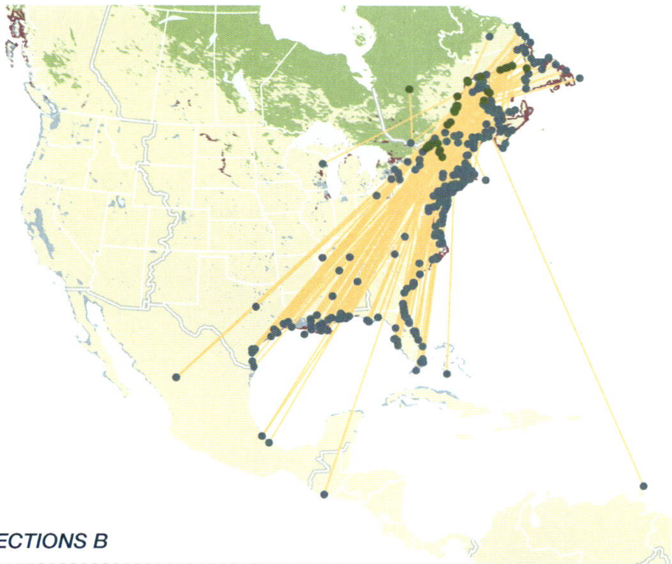

CONNECTIONS B

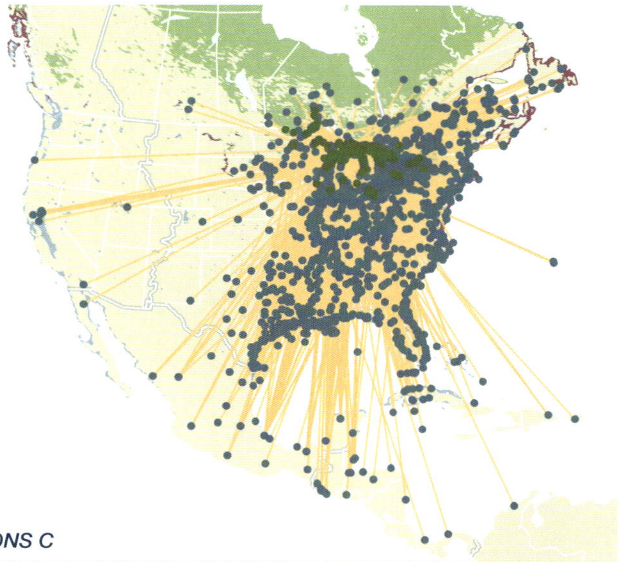

CONNECTIONS C

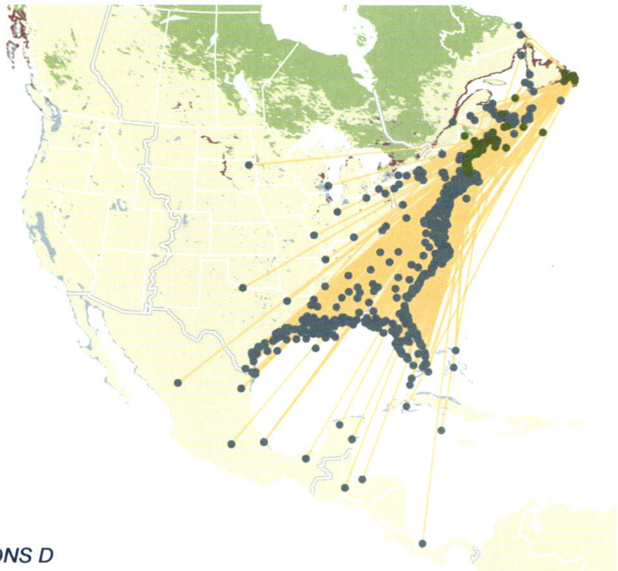

CONNECTIONS D

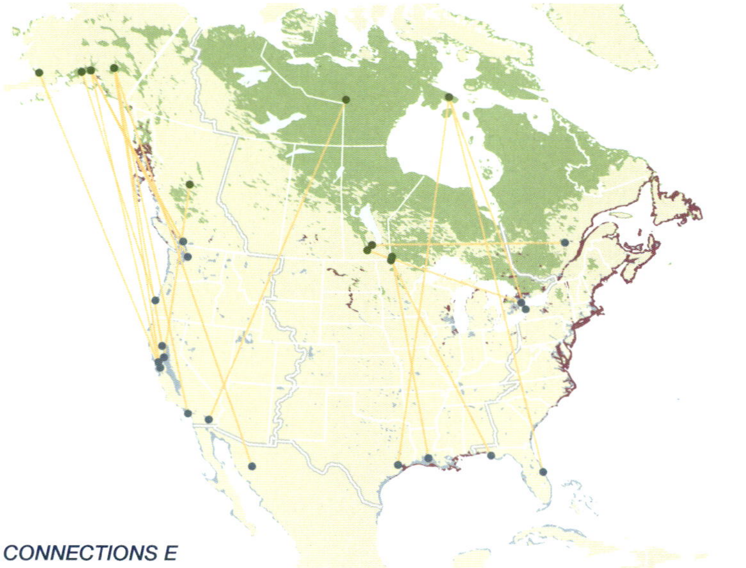

CONNECTIONS E

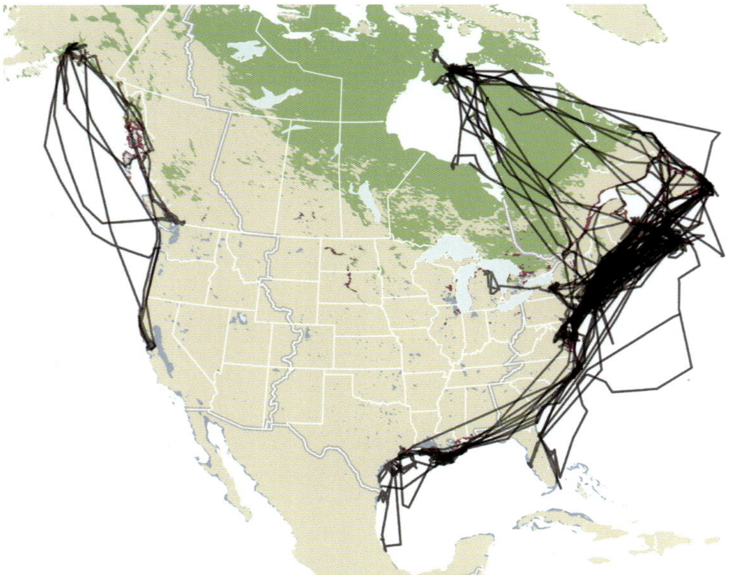

PATHWAYS (A–C)

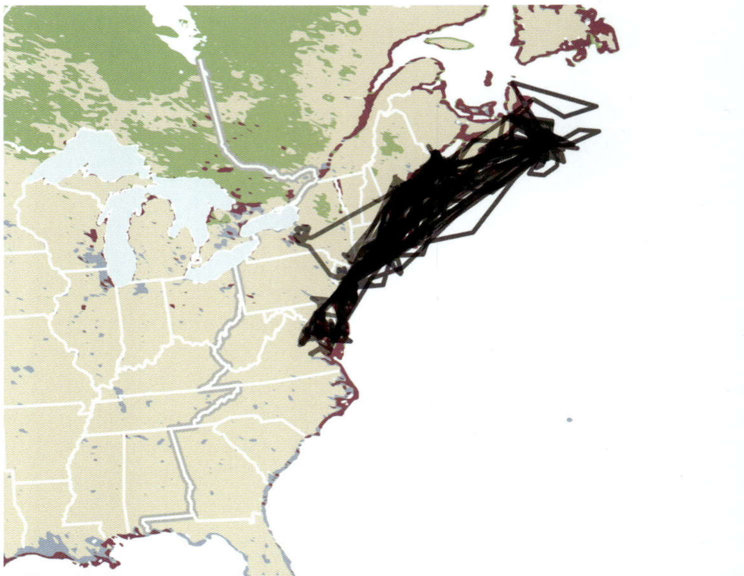

PATHWAYS A

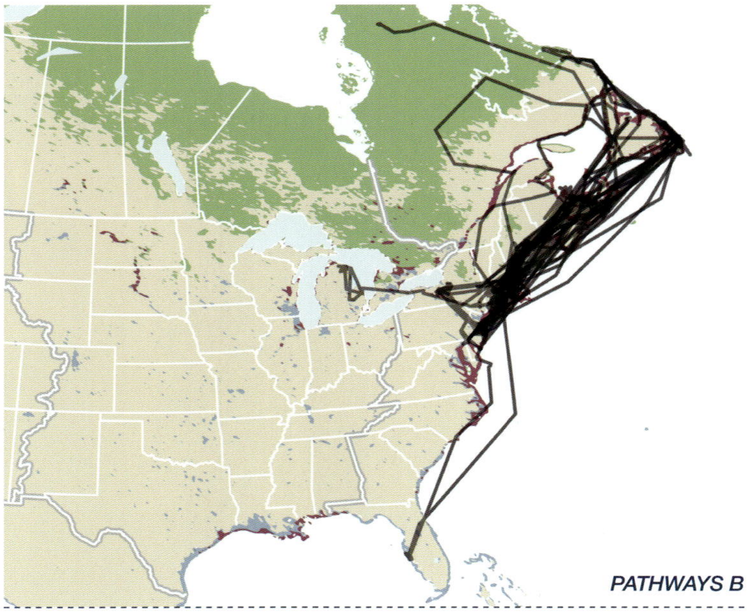

PATHWAYS B

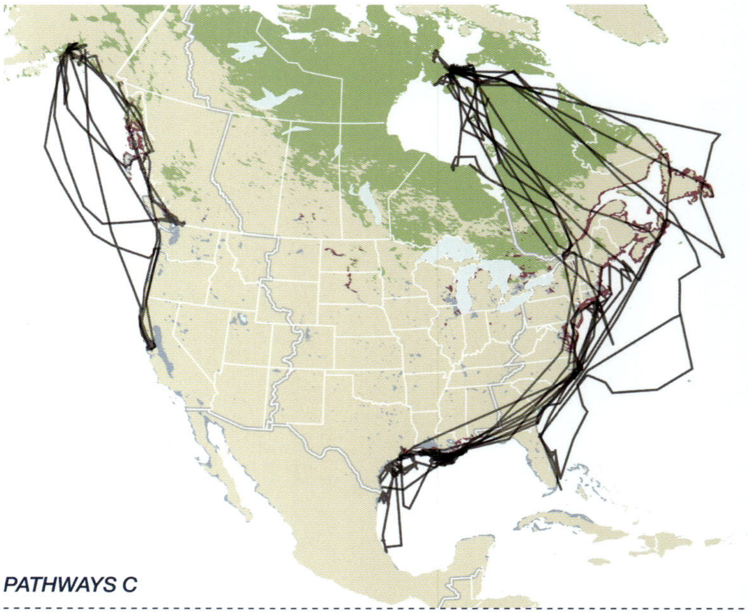

PATHWAYS C

OTHER APPROACHES No genetic or stable isotope studies of migratory connectivity were found for this species.

MIGRATORY PATHWAYS This species takes indirect migratory routes, likely driven by food and habitat availability (C. Anderson et al. 2020). This leads the majority of individuals to travel along the coasts, as well as by large inland water bodies such as the Great Lakes. Many individuals are found away from these major flyways, and this species may turn up wherever resources are available in the continental interior.

DATA CONTRIBUTORS

C.A. Ahlstrom; C.M. Anderson; D.E. Clark; D.C. Douglas; D.A. Fifield; H.G. Gilchrist; K.K.G. Koenen; K.G. MacKenzie; M.L. Mallory; A. Ramey; G.J. Roberston; R.A. Ronconi; K.R. Shlepr; J.J. Whitney.

Larus glaucoides **ICELAND GULL**

DISTRIBUTION AND ABUNDANCE

This species is a high Arctic sea-cliff nester of the Canadian Arctic Archipelago and Greenland in North America. The Iceland Gull is one of the least understood gulls of North America and has a contentious and confusing taxonomy. Currently the species includes three subspecies (*L. g. glaucoides*, *L. g. thayeri*, and *L. g. kumlieni*; Snell et. al. 2020). Thayer's Gull (*L. g. thayeri*) was formerly treated as a separate species (Chesser et al. 2017). Generally, *L. g. glaucoides* breeds in southwestern and eastern Greenland; *L. g. kumlieni* (Kumlien's Gull) breeds in high Arctic Canada from southern and eastern Baffin Island to Coats Island, northern Hudson Bay, and Southampton Island; and *L. g. thayeri* are in scattered colonies in northwestern Greenland and in the Canadian Arctic Archipelago north and west of the distribution of *L.g. kumlieni* (Snell et. al. 2020). Markings help distinguish subspecies. Nonbreeding areas include both North American Arctic habitats and coastal temperate marine habitats along both the Pacific and Atlantic coasts of North America (Snell et. al. 2020). Abundances are highly uncertain due to the difficulty in surveying high Arctic scattered colonies and mismatches between at-sea surveys and breeding surveys. Little is known about trends.

ORDER Charadriiformes
FAMILY Laridae
BREEDING HABITAT Coasts, Cliffs and Islands
NONBREEDING HABITAT Coasts, Beach and Estuary
GLOBAL POPULATION SIZE 220,000
POPULATION CHANGE SINCE 1970 NA
IUCN RED LIST CATEGORY Least Concern
R2R TIPPING POINT SPECIES No
US LISTING STATUS No
CANADA SPECIES AT RISK No
PIF CCS 11/20

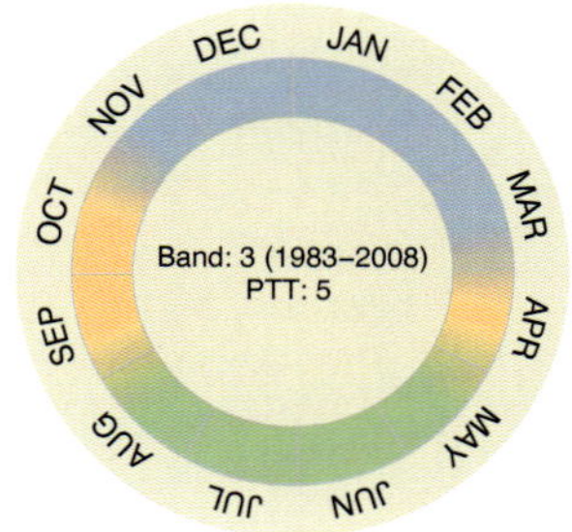

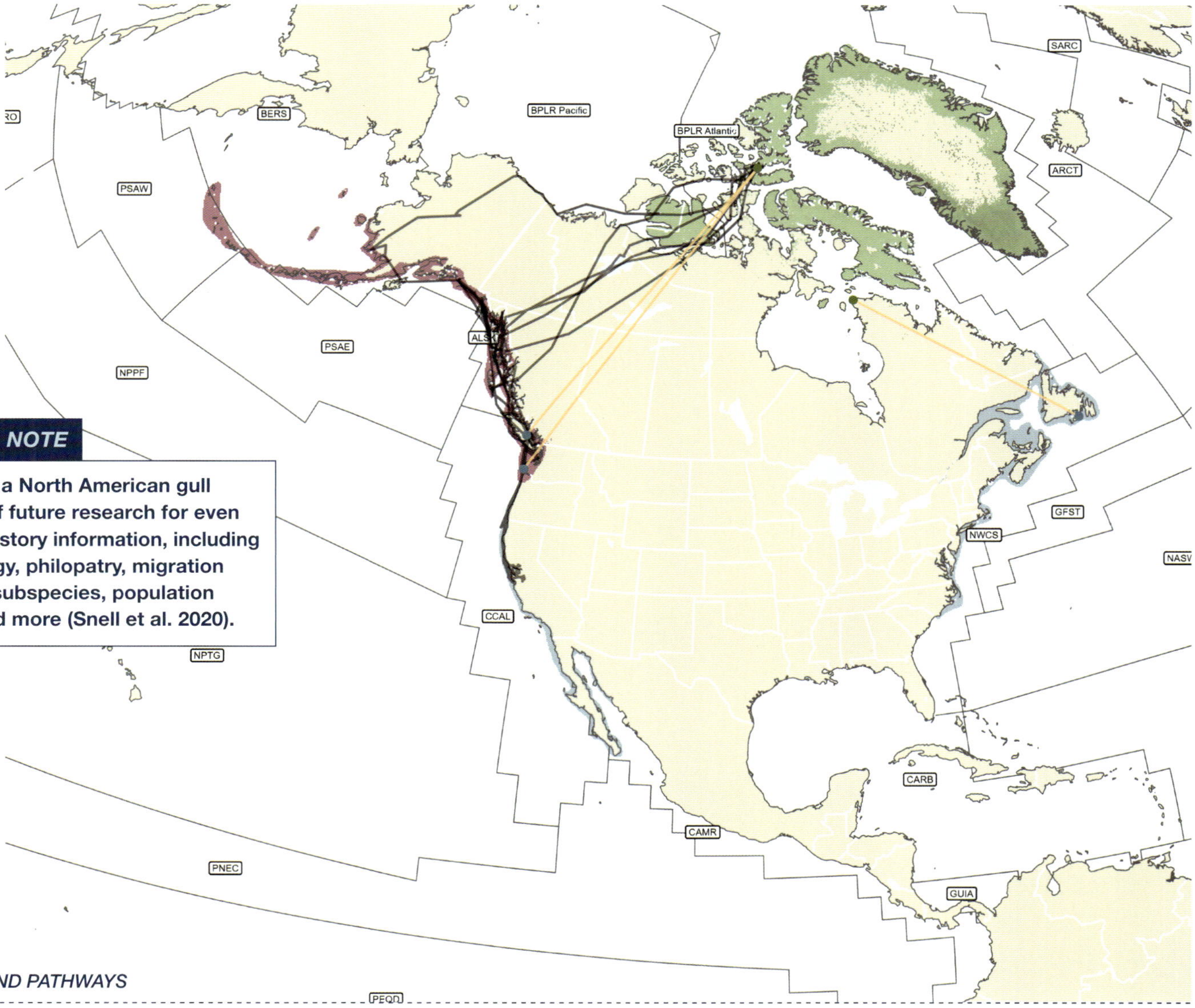

INTERESTING NOTE

This species is a North American gull most in need of future research for even basic natural history information, including foraging ecology, philopatry, migration routes across subspecies, population monitoring, and more (Snell et al. 2020).

CONNECTIONS AND PATHWAYS

MIGRATORY CONNECTIVITY

OVERALL Strong connectivity is presumed by subspecies based on limited tracking and banding data as well as observational data using morphometric differentiation; however, MC was not calculated due to limited sample size. *L.g. thayeri* use habitats along the Pacific coast of North America (Gulf of Alaska and California Current provinces), as evidenced by satellite tracking and banding data. Sightings data (Snell et al. 2020) and one band return of *L.g. kumlieni* suggest that *L.g. glaucoides* primarily remain in the Arctic region near Greenland during the post-breeding season (with limited sightings along the Atlantic coasts south of the Arctic), while *L.g. kumlieni* are more likely to be sighted south of the Arctic along the Atlantic coasts, Great Lakes, and southern Canada. Within subspecies, migratory connectivity is generally unknown, but is likely weak given fairly restricted nonbreeding distributions of each. However, there is likely separation by age class; sighting records of *L.g. glaucoides* and *L.g. kumlieni* in southern nonbreeding distributions (US coast and southern Canada) suggest they are more frequently immature birds than adults (Snell et al. 2020).

BANDING Two birds banded in the 2000s during the breeding season in high Arctic Nunavut, Canada, (*L.g. thayeri*) were found dead in February in British Columbia, Canada, and Washington state. One bird banded in 1982 during breeding in Nunavut (northeast Hudson Bay, presumed *L.g. kumlieni*) was sighted in January 2023 in Newfoundland, Canada.

SATELLITE TRACKING Four *L.g. thayeri* tracked from a colony in the Canadian high Arctic used post-breeding habitats primarily over the continental shelf, ranging from Yakutat Bay, Alaska, to Monterey Bay, California.

OTHER APPROACHES No genetic or stable isotope studies of migratory connectivity were found for this species. Nonbreeding areas for the subspecies generally were determined by differences in wingtip melanism from specimen collection and sightings data as early as the 1800s (reviewed in Weir et al. 2000). This publication also describes potential changes in distribution after 1950, with more *L.g. kumlieni* observed in Europe in recent history.

MIGRATORY PATHWAYS Satellite-tracked birds departed Nunavut, Canada, in September 2016 and migrated southwest overland via the Northwest Territories, Canada, to nonbreeding areas in the Pacific Northwest (n = 4; Gutowksy et al. 2020). Birds arrived to Alaska, US, and British Columbia, Canada, within one month of departure, before some continued south to California. One individual's tag transmitted for 565 days, describing a northward pre-breeding migration similar in its overland route to its southward migration (Gutowsky et al. 2020); however, it spent one month in a different area in Nunavut, before returning to its breeding colony in mid-August. In its second southward migration, it changed routes, migrating west across the Canadian Arctic Archipelago, flying coastally along the Yukon Territory, Canada, and North Slope, Alaska, and moving south to an overland migration across Alaska, before returning to a coastal route around the Gulf of Alaska and south to the Pacific Northwest (Gutowsky et al. 2020).

DATA CONTRIBUTORS
S.E. Davis; J.M. Hipfner.

LESSER BLACK-BACKED GULL *Larus fuscus*

DISTRIBUTION AND ABUNDANCE

The Lesser Black-backed Gull is globally widespread in coastal areas. In North America, the species breeds in high numbers in Greenland, Atlantic Canada, and along the northeast US coast (Fink et al. 2023). During the nonbreeding season, the species can be found commonly in the mid-Atlantic region and along the entire east coast of the United States and in lower abundance along the St. Lawrence River and Gulf Coast (Fink et al. 2023).

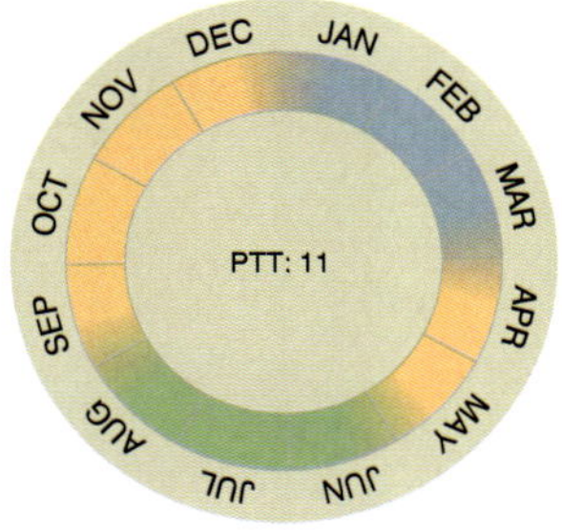

	Atlantic
Greenland (n = 8)	100%
Midcontinent (n = 2)	100%

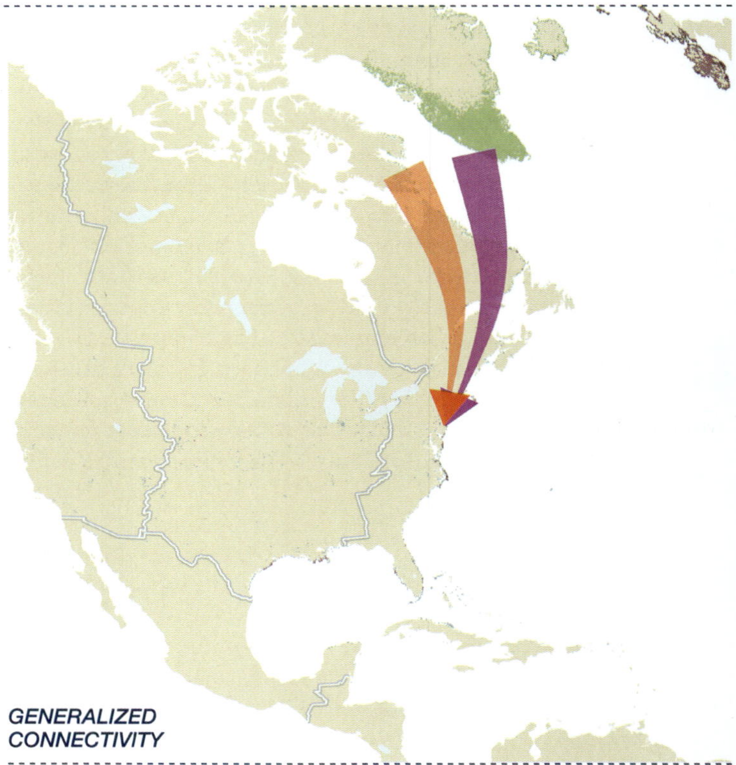

ORDER Charadriiformes
FAMILY Laridae
BREEDING HABITAT Coasts, Beach and Estuary
NONBREEDING HABITAT Coasts, Beach and Estuary
GLOBAL POPULATION SIZE 1,000,000
POPULATION CHANGE SINCE 1970 NA
IUCN RED LIST CATEGORY Least Concern
R2R TIPPING POINT SPECIES No
US LISTING STATUS No
CANADA SPECIES AT RISK No
PIF CCS 7/20

MIGRATORY CONNECTIVITY

OVERALL Little is known about the migratory connectivity of Lesser Black-backed Gulls. MC is not calculated due to limited sample size, but it appears weak as birds breeding in Greenland and the Midcontinent Flyway all spend the nonbreeding season in the Atlantic Flyway.

BANDING A bird banded in Florida during the nonbreeding season was resighted in New York during the breeding season.

SATELLITE TRACKING Lesser Black-backed Gulls tracked with satellite telemetry from nonbreeding locations in Pennsylvania migrated to breeding areas on Baffin Island, Canada (n = 2) and Greenland (n = 8; P. Barber, unpublished data). Another three birds tagged during migration were tracked to breeding areas in Greenland and nonbreeding areas in North Carolina. Gulls tracked in multiple years returned to the same breeding and nonbreeding areas (P. Barber, unpublished data)

OTHER APPROACHES No genetic or stable isotope studies of migratory connectivity were found for this species.

MIGRATORY PATHWAYS There are two main pathways from nonbreeding areas in Pennsylvania and North Carolina to northern breeding areas. One takes a coastal route north to the Gulf of St. Lawrence and then over the Labrador Sea to Baffin Island or Greenland. The interior route follows the St. Lawrence River along the southern coast of Quebec and Newfoundland and Labrador, north to Baffin Island or across the Labrador Sea to Greenland.

DATA CONTRIBUTORS
P. Barber.

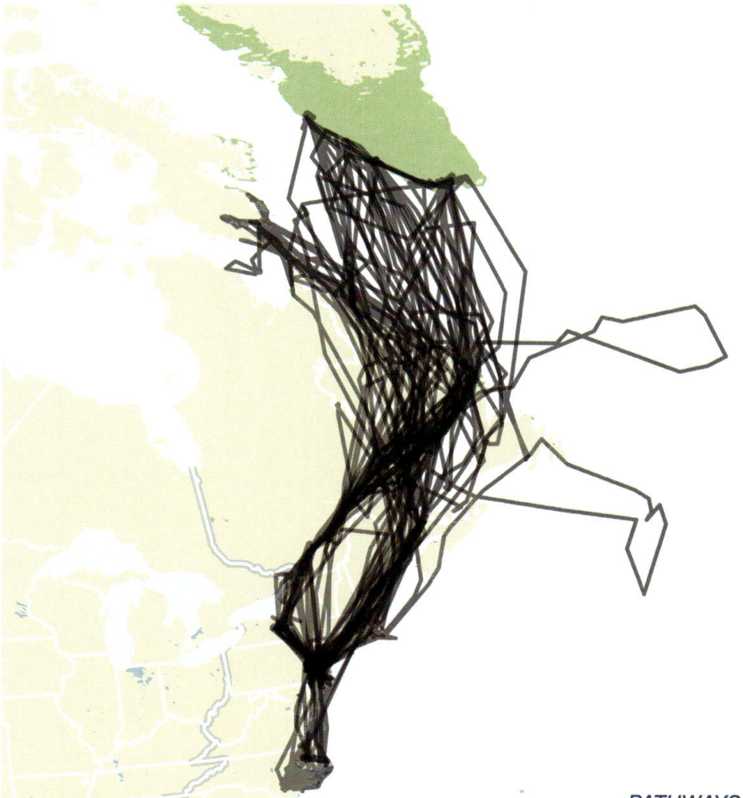

PATHWAYS

Larus glaucescens

GLAUCOUS-WINGED GULL

DISTRIBUTION AND ABUNDANCE

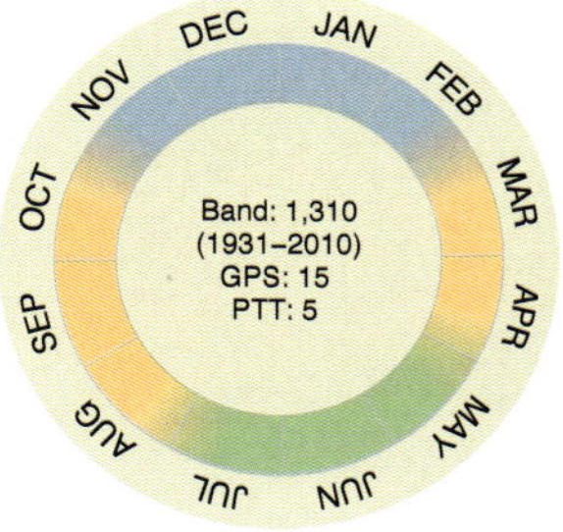

A common and conspicuous omnivore of northwestern North America, Glaucous-winged Gulls breed from the east coast of the Kamchatka Peninsula in Russia, across the Aleutian island chain in the Gulf of Alaska south through coastal British Columbia, Canada, to northwest Oregon (Hayward and Verbeek 2020). Total nesting population in North America is estimated at 200,000 pairs, with about 67% of these in Alaska, 13% in British Columbia, and the remainder in Washington and Oregon. These reported estimates are from the 1980s–1990s, and populations have been dynamic since then, especially in urban areas (Hayward and Verbeek 2020). Some birds remain resident near their breeding colonies during nonbreeding, but others are migratory. Migratory and nonbreeding movements are generally over land or inshore or over waters south to Baja California, Mexico, but birds have been observed to feed offshore.

ORDER Charadriiformes
FAMILY Laridae
BREEDING HABITAT Coasts, Cliffs and Islands
NONBREEDING HABITAT Coasts, Beach and Estuary
GLOBAL POPULATION SIZE 470,000
POPULATION CHANGE SINCE 1970 −22%

IUCN RED LIST CATEGORY Least Concern
R2R TIPPING POINT SPECIES No
US LISTING STATUS No
CANADA SPECIES AT RISK No
PIF CCS 13/20

INTERESTING NOTE

Glaucous-winged Gulls from the Aleutian Islands are significantly larger than those from elsewhere, possibly due to hybridization with larger Glaucous Gulls (*L. hyperboreus*; Bell 1996). Birds from northern and western populations from Alaska to Japan also tend to have slightly darker mantle color (Hayward and Veerbeek 2020).

MIGRATORY CONNECTIVITY

OVERALL Migratory connectivity is weak to moderate based on banding and tracking data. The majority of birds breeding in the Pacific Flyway (99%) migrate to coastal nonbreeding areas in the Pacific Flyway.

BANDING A review of band returns (Hatch et al. 2011) found that Alaskan breeding birds spent the nonbreeding season across a broad range from southeast Alaska to Monterey Bay, California, while southern-nesting birds spanned southern British Columbia to southern California. Northern-breeding birds undertook distinct pre- and post-breeding migrations, whereas southern breeders were

SHOREBIRDS, AUKS, GULLS, AND TERNS (CHARADRIIFORMES)

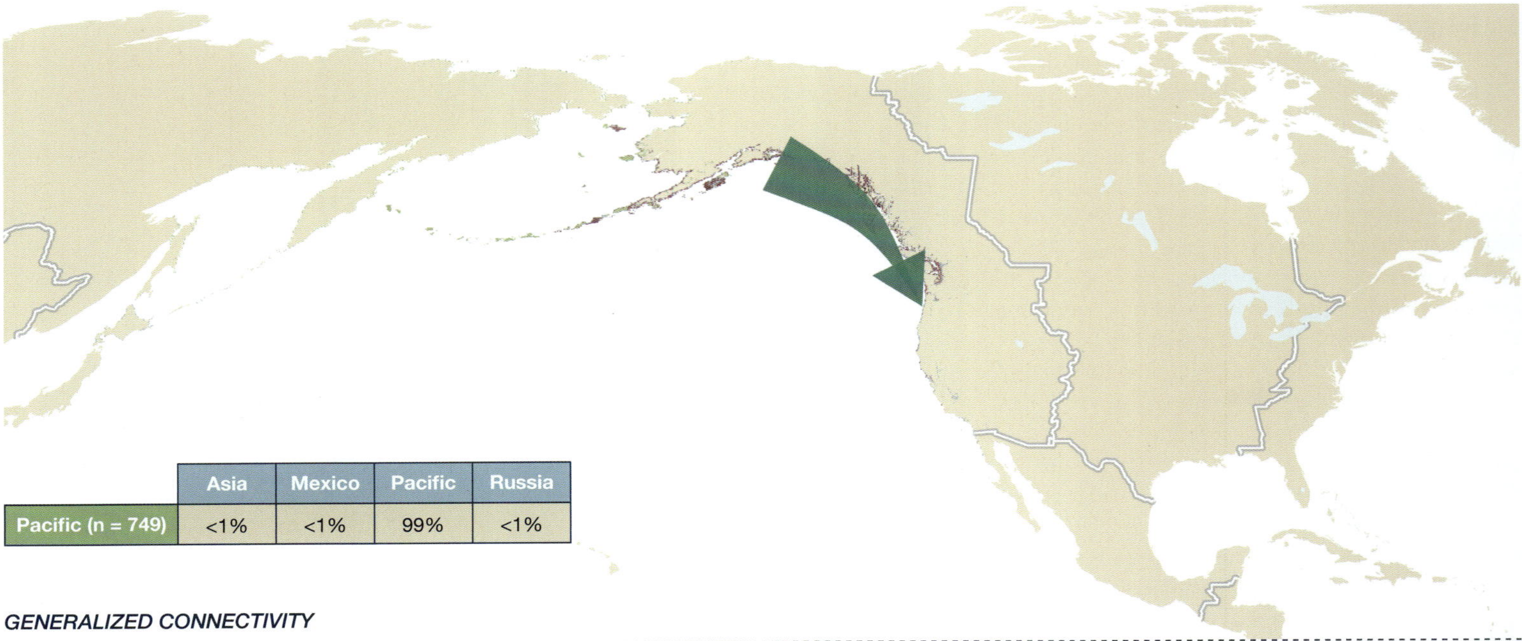

	Asia	Mexico	Pacific	Russia
Pacific (n = 749)	<1%	<1%	99%	<1%

GENERALIZED CONNECTIVITY

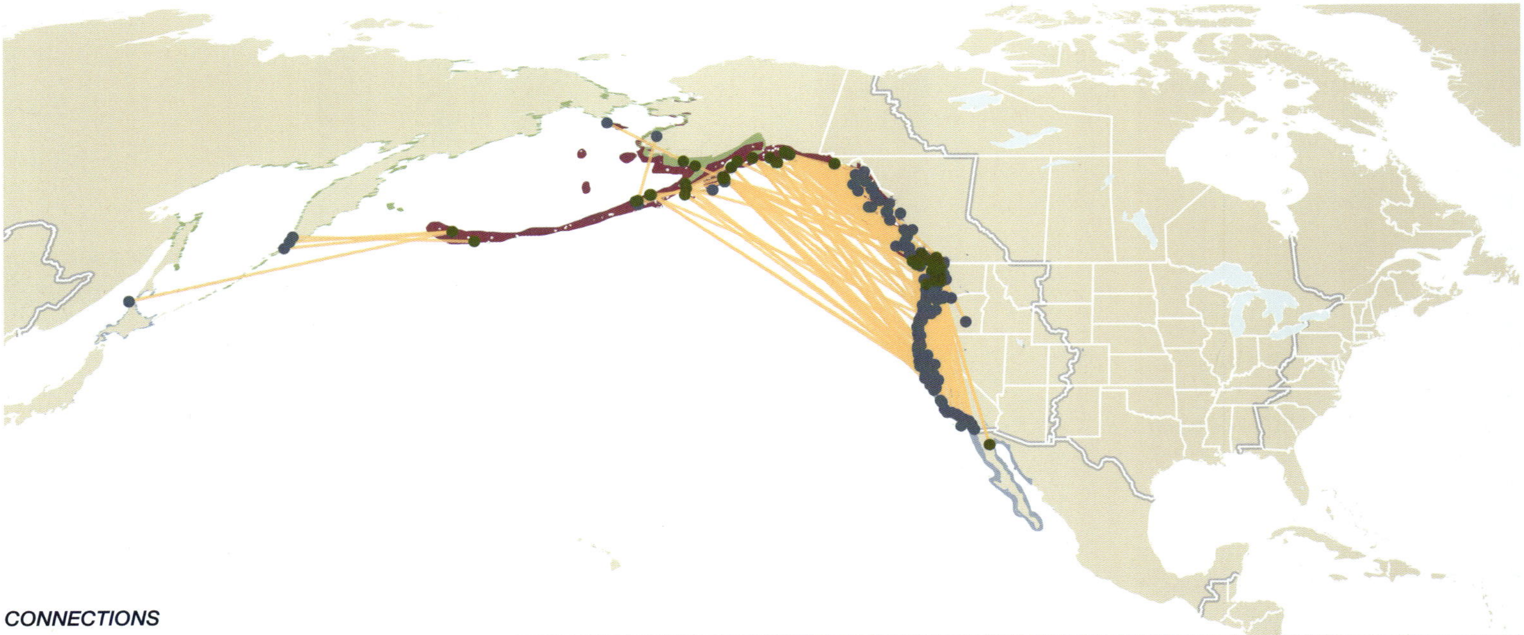

CONNECTIONS

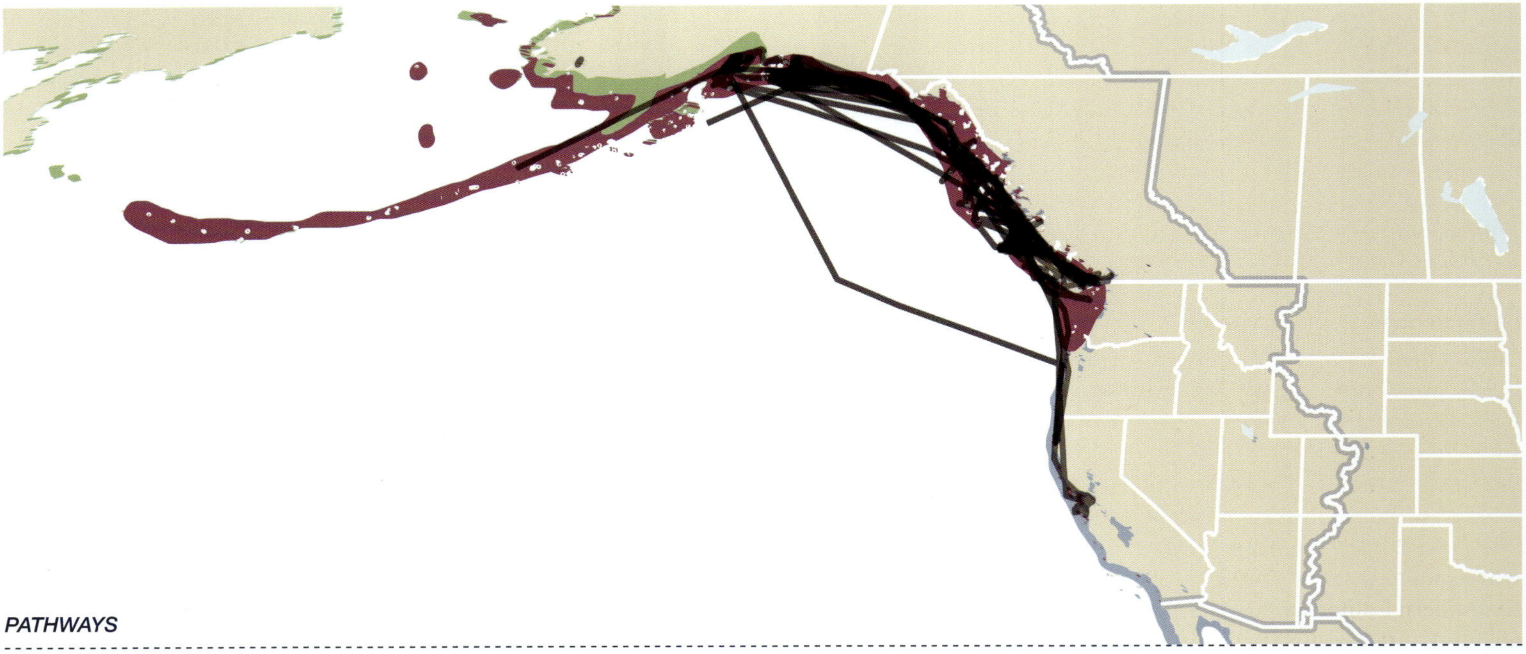

PATHWAYS

dispersive. Three birds banded in the western Aleutian Islands in the early 1960s were shot in Russia, in December through February, 1963–1964. One individual from this same banding effort was found dead off western Hokkaido, Japan, in 1964.

SATELLITE TRACKING Three birds from Middleton Island, Alaska, US, were tracked with implantable tags to nonbreeding areas near Vancouver Island, Canada, with proximity to landfills (Hatch et al. 2011; Ramey et al. 2020). In contrast, four birds tracked from central British Columbia (Domalik et al. 2021) migrated to areas with low levels of anthropogenic activity, mainly nearshore coastal waters (72% of time) or naturally vegetated areas. Three of the birds remained locally throughout the nonbreeding season, while the fourth migrated to Northern California. Residency during the nonbreeding season was also observed within the Salish Sea of British Columbia for seven of eight birds tracked by GPS, while one made a short migration across the Salish Sea to the north coast of Washington, US (Hall 2022). Birds tracked by GPS from a landfill in Soldotna, Alaska (Ramey et al. 2020), migrated to nonbreeding areas from Alaska to Washington, US. Two individuals were tracked between Alaska and British Columbia (K. Elliot and S. Whelan, unpublished data).

GENETICS Much genetic work on hybridization spanning many colonies has found that the Glaucous-winged Gull has high genetic variability (Bell 1996) and has been called a key taxon (Patten 1980) because it potentially links several of the northern Pacific Rim gulls. The species readily hybridizes with Western Gulls, Herring Gulls, Glaucous Gulls, and Slaty-backed Gull (Hayward and Veerbeek 2020). Geng et al. (2024) suggest that a persistent hybrid zone of Western Gulls and Glaucous-winged Gulls along the coast in Washington, US, has been maintained since the early twentieth century because each species and their hybrids occupy significantly different ecological niches.

STABLE ISOTOPES Stable isotope studies of feathers collected from 1860 to 2009 in British Columbia and Washington suggest a shift in habitat connectivity in this oceanographic region. Measures of stable isotopes of nitrogen, carbon, hydrogen, and sulphur indicate the species has shifted from a mid-trophic level marine diet to increased terrestrial and freshwater food sources (Blight et al. 2015; Hobson et al. 2015a), supporting the hypothesis that this is a response to declining forage fish populations. More work is needed to assess nonbreeding diet spanning the breeding range to link with connectivity.

MIGRATORY PATHWAYS The four birds tracked from Middleton Island, Alaska (Hatch et al. 2011), made coastal migrations with multiple stopovers, including an individual tracked to a long stopover in Craig, Alaska, but the tag did not last the full migration to the nonbreeding area. An individual tracked between central British Columbia and Northern California also traveled coastally (Domalik et al. 2021). Birds tracked by GPS from Soldotna, Alaska (Ramey et al. 2020), largely followed the coast, with the exception of one individual that returned north to its colony via a fully marine pathway at the edge of the continental shelf.

DATA CONTRIBUTORS
D.C. Douglas; K.H. Elliott; J.M. Hipfner; A. Ramey; S. Whelan.

Larus hyperboreus **GLAUCOUS GULL**

DISTRIBUTION AND ABUNDANCE

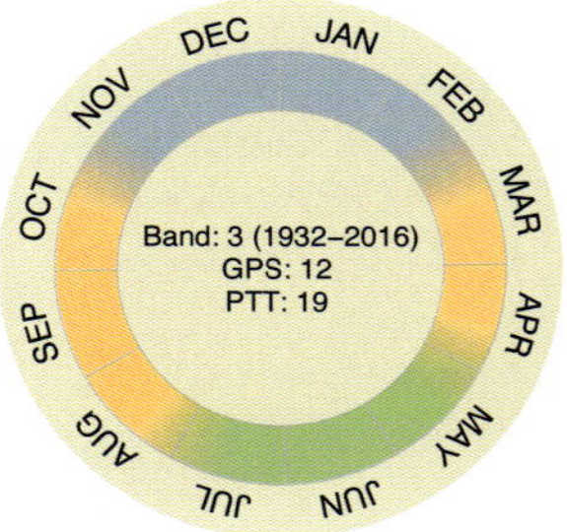

This pale, large, predatory gull has a circumpolar Arctic breeding range. In North America, Glaucous Gulls nest along the north and west coasts of Alaska and the Canadian Arctic and Greenland (Weiser and Gilchrist 2020). North American populations occur along the coastal North Pacific and northwest Atlantic during the nonbreeding period but sometimes can be found along temperate coastal North America and on the Great Lakes (Weiser and Gilchrist 2020). The total Arctic population is estimated at 139,000–219,000 breeding pairs, with 25,000 pairs in Canada; 8,000–100,000 pairs in Alaska; and 20,000–100,000 in Greenland (Petersen et al. 2015).

MIGRATORY CONNECTIVITY

OVERALL There is generally strong migratory connectivity by ocean basin; however, migration between the Atlantic and Pacific oceans has been demonstrated. Limited samples of tracking and banding suggest weak connectivity within oceans. Migratory connectivity was not estimated due to limited sample size. Birds from Alaskan colonies generally used nonbreeding habitats in the Bering Sea (67%–75%), while birds from eastern Canadian colonies visited habitats in the Northwest Atlantic Shelves province (60%–100%).

BANDING Three band recoveries of Glaucous Gulls include one of the oldest records from the Bird Banding Lab (1932) and more recent bands (2004, 2016) and show connections between breeding sites in Nunavut, Canada, and Quebec and Newfoundland, as well as a Labrador breeding site connected to a Newfoundland nonbreeding site.

ORDER Charadriiformes
FAMILY Laridae
BREEDING HABITAT Coasts, Cliffs and Islands
NONBREEDING HABITAT Coasts, Beach and Estuary

GLOBAL POPULATION SIZE 970,000
POPULATION CHANGE SINCE 1970 +26%
IUCN RED LIST CATEGORY Least Concern
R2R TIPPING POINT SPECIES Orange-Alert
US LISTING STATUS No
CANADA SPECIES AT RISK No
PIF CCS 13/20

GEOLOCATOR TRACKING Baak et al. (2021b) provided the first evidence of migration between eastern Canada and Asia. Of five birds tracked with light-level geolocators from Coats Island, Nunavut, Canada, four spent the nonbreeding season in the northwest Atlantic, while one individual migrated to the Sea of Okhotsk, Russia, in the northwestern Pacific Ocean.

Colony
- Utqiagvik, AK, USA
- Bethel, AK, USA
- Nunavut, Canada
- Nunaksaluk Island, NL, Canada

	KURO	BERS	BPLR Pacific	ALSK	NWCS	BPLR Atlantic
Utqiagvik, AK, USA (n = 9)	11%	67%	22%	–	–	–
Bethel, AK, USA (n = 4)	–	75%	–	25%	–	–
Nunavut, Canada (n = 5)	–	20%	–	–	60%	20%
Nunaksaluk Island, NL, Canada (n = 1)	–	–	–	–	100%	–

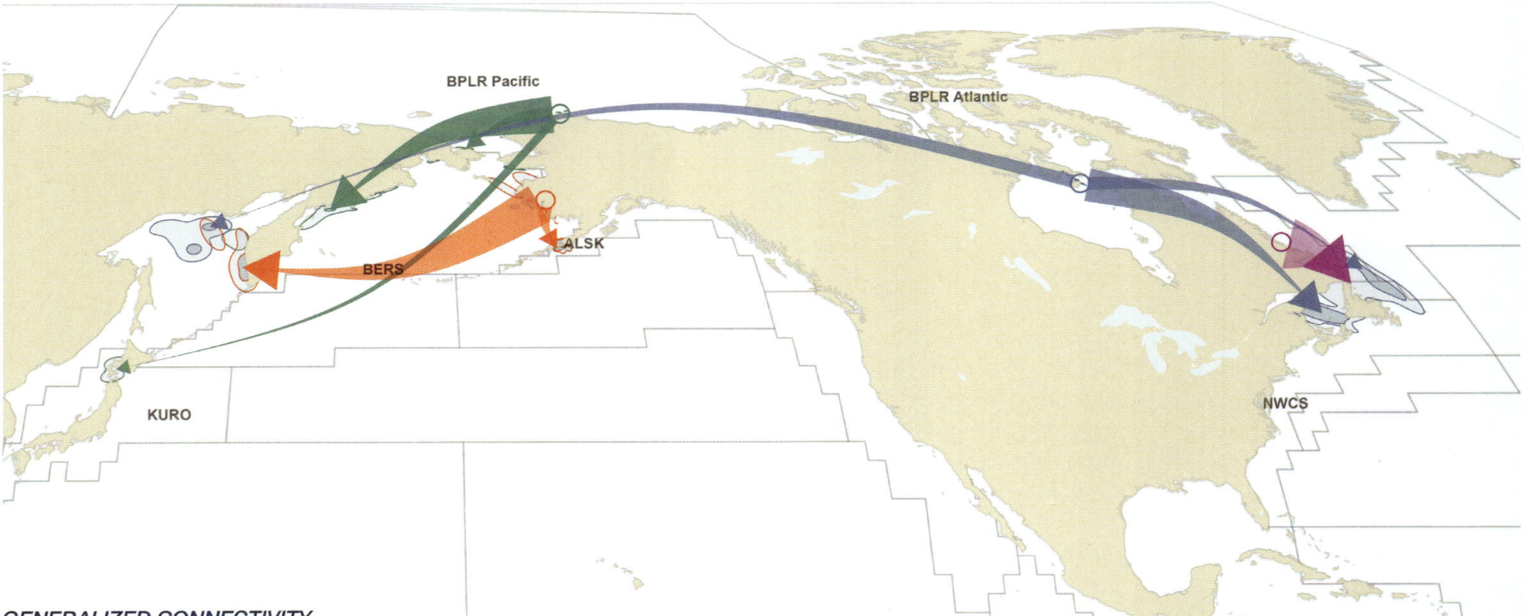

GENERALIZED CONNECTIVITY

CONNECTIONS

SATELLITE TRACKING Adult Glaucous Gulls were tracked by Argos-GPS from Bethel, Alaska (Ramey et al. 2020), to nonbreeding areas in the western Bering Sea (Sea of Okhotsk, Russia) and along the Aleutians in the Gulf of Alaska. Immature birds tracked from Utqiaġvik, Alaska (A.-L. Harrison, unpublished data), were tracked on their first migration and generally spent the nonbreeding season in the western coastal Pacific as far south as Japan. Baak et al. (2021b) also tracked birds with GPS tracking devices (n = 5) from Coats Island, Nunavut, Canada. Data confirmed the light-level geolocator results (see "Geolocator tracking" above); four birds spent the nonbreeding season in the northwest Atlantic, while one individual migrated to the Sea of Okhotsk, Russia, in the northwestern Pacific Ocean.

GENETICS DNA sequencing with an analysis of 621 loci revealed weak population genetic structure between Canadian and European *Larus hyperboreus* spanning the eastern Canadian Arctic and Europe (Linklater et al. 2024). The authors suggested that eastern Canadian Arctic populations are panmictic.

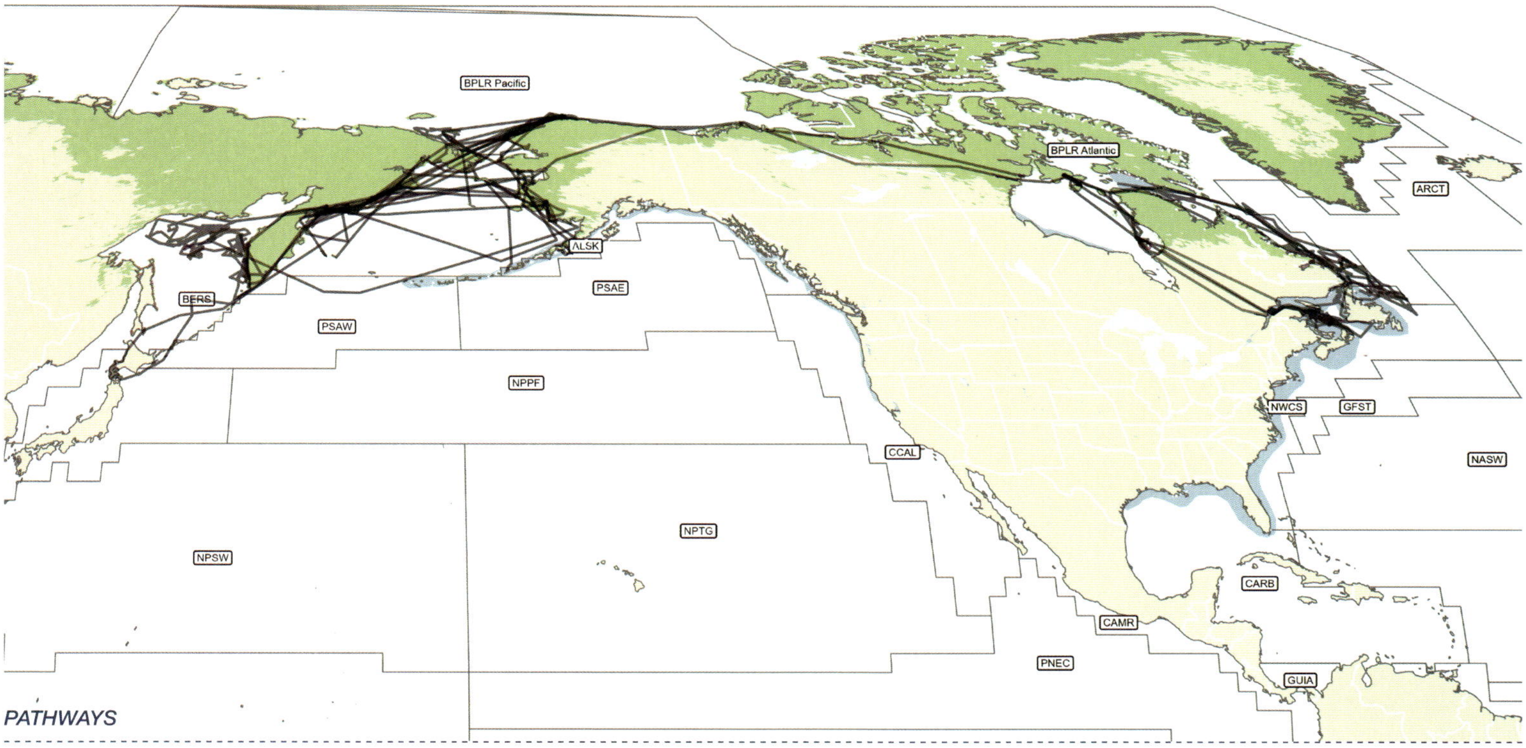

PATHWAYS

OTHER APPROACHES No stable isotope studies of migratory connectivity found were for this species.

MIGRATORY PATHWAYS Alaskan birds generally traveled west or southwest to Asia or southeast to the Aleutian Islands (Ramey et al. 2020; A.-L. Harrison, unpublished data). One immature individual first migrated to Asia, where it stayed for a couple of months, then crossed the North Pacific to Glacier Bay, Southeast Alaska, where the tag stopped transmitting soon after (A.-L. Harrison, unpublished data). Canadian birds used two migration strategies, either a short-distance southward migration to the northwest Atlantic (n = 7), or a westward migration (n = 1) across the Beaufort and Chukchi seas before crossing over land into the Bering Sea and then finally cutting across Kamchatka to the Sea of Okhotsk, Russia (Baak et al. 2021b).

DATA CONTRIBUTORS

K.H. Elliott; H.G. Gilchrist; A.-L. Harrison; Migratory Connectivity Project; A. Patterson; A. Ramey.

Larus marinus GREAT BLACK-BACKED GULL

ORDER Charadriiformes
FAMILY Laridae
BREEDING HABITAT Coasts, Beach and Estuary
NONBREEDING HABITAT Coasts, Beach and Estuary

GLOBAL POPULATION SIZE 370,000
POPULATION CHANGE SINCE 1970 −59%
IUCN RED LIST CATEGORY Least Concern
R2R TIPPING POINT SPECIES Orange-Alert
US LISTING STATUS No
CANADA SPECIES AT RISK No
PIF CCS 15/20

DISTRIBUTION AND ABUNDANCE

Throughout the year, the Great Black-backed Gull is locally common along the Atlantic coast of North America and in the Great Lakes and St. Lawrence riverway regions. Smaller numbers are found in coastal Canadian Arctic and interior populations of the midcontinent US. Breeding populations extend along the coast north from the mid-Atlantic region (Good 2020).

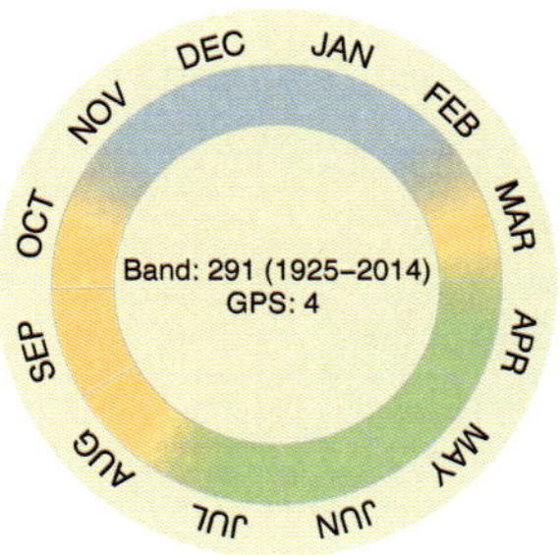

MIGRATORY CONNECTIVITY

OVERALL Banding reencounter and limited tracking data are only available from the Atlantic Flyway; therefore, MC could not be calculated. Birds breeding within the Atlantic Flyway primarily migrate to coastal nonbreeding areas within the Atlantic Flyway (97%), with a small percentage migrating to the Midcontinent Flyway (2%) and Europe (1%).

BANDING Band reencounters throughout the Atlantic coast of North America show northern breeding populations migrating to nonbreeding areas along the mid-Atlantic coast or in the Great Lakes area (Ineson et al. 2022). Populations breeding along the mid-Atlantic coast migrate farther south along the coast for the nonbreeding season.

SHOREBIRDS, AUKS, GULLS, AND TERNS (CHARADRIIFORMES)

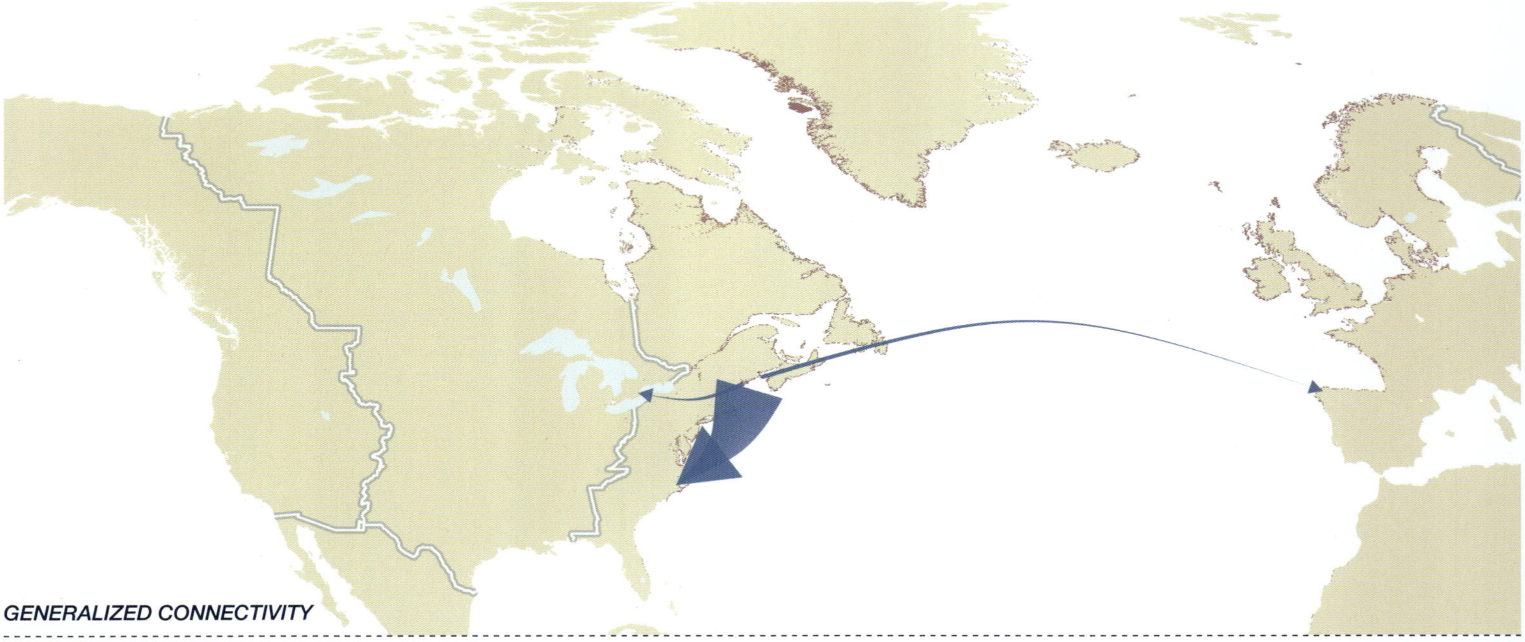

GENERALIZED CONNECTIVITY

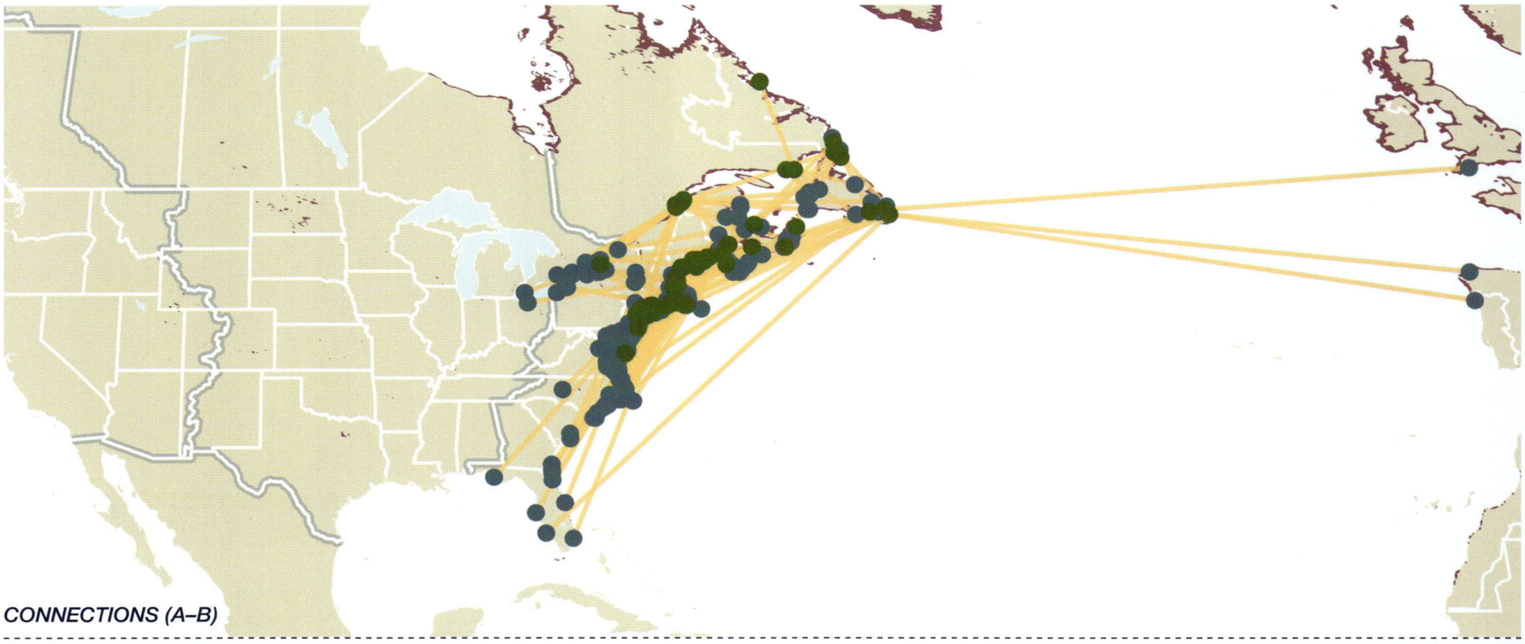

CONNECTIONS (A–B)

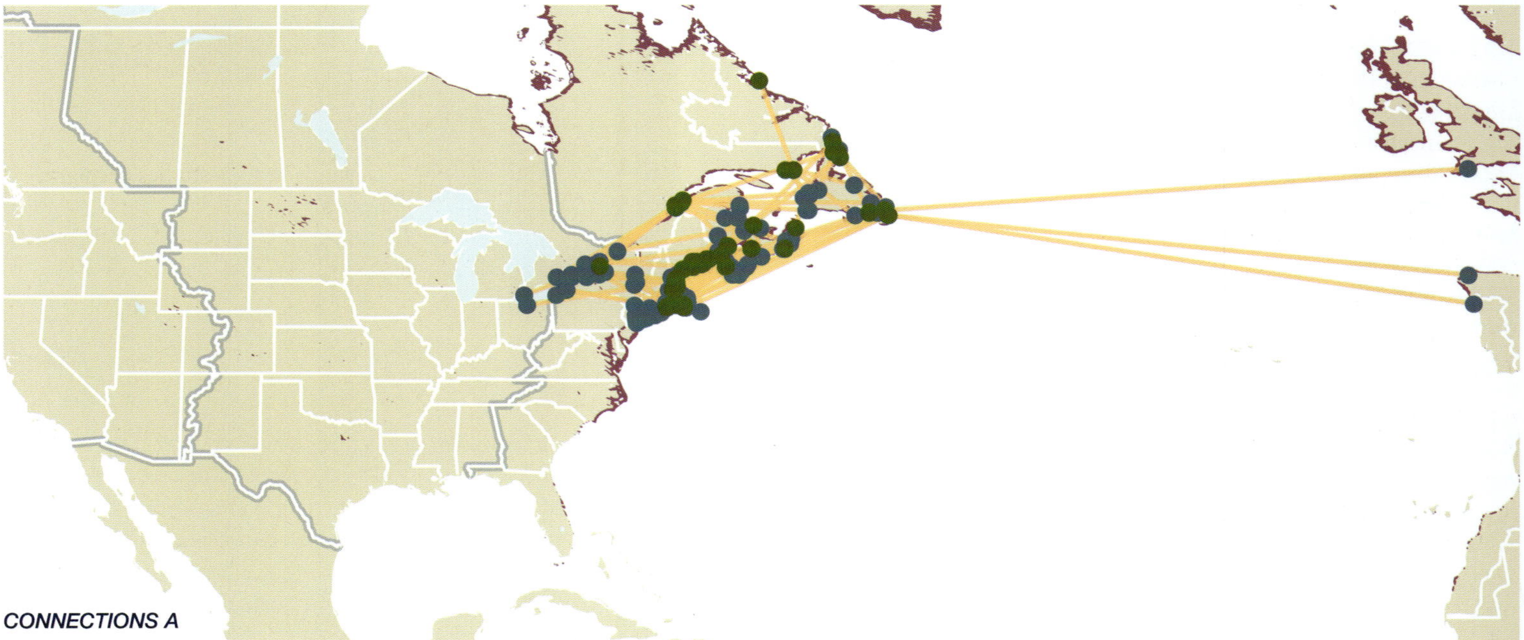

CONNECTIONS A

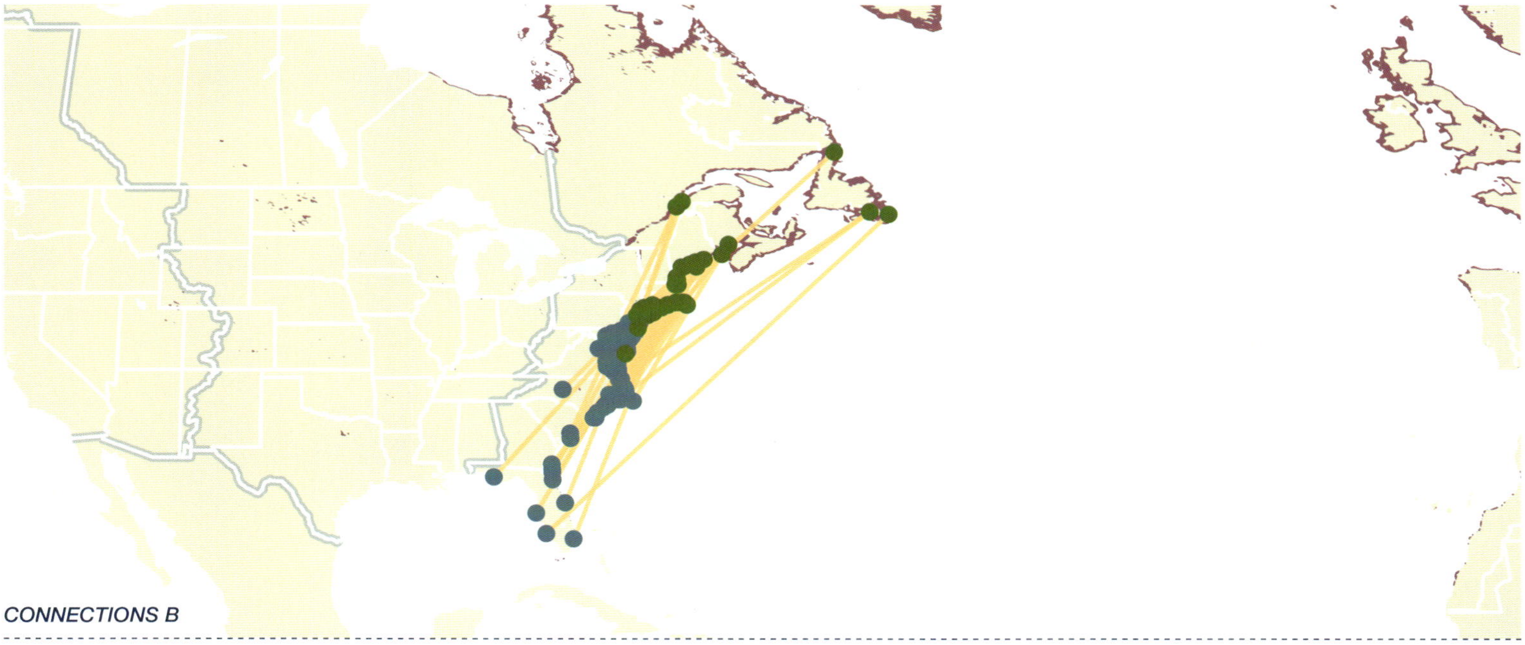

CONNECTIONS B

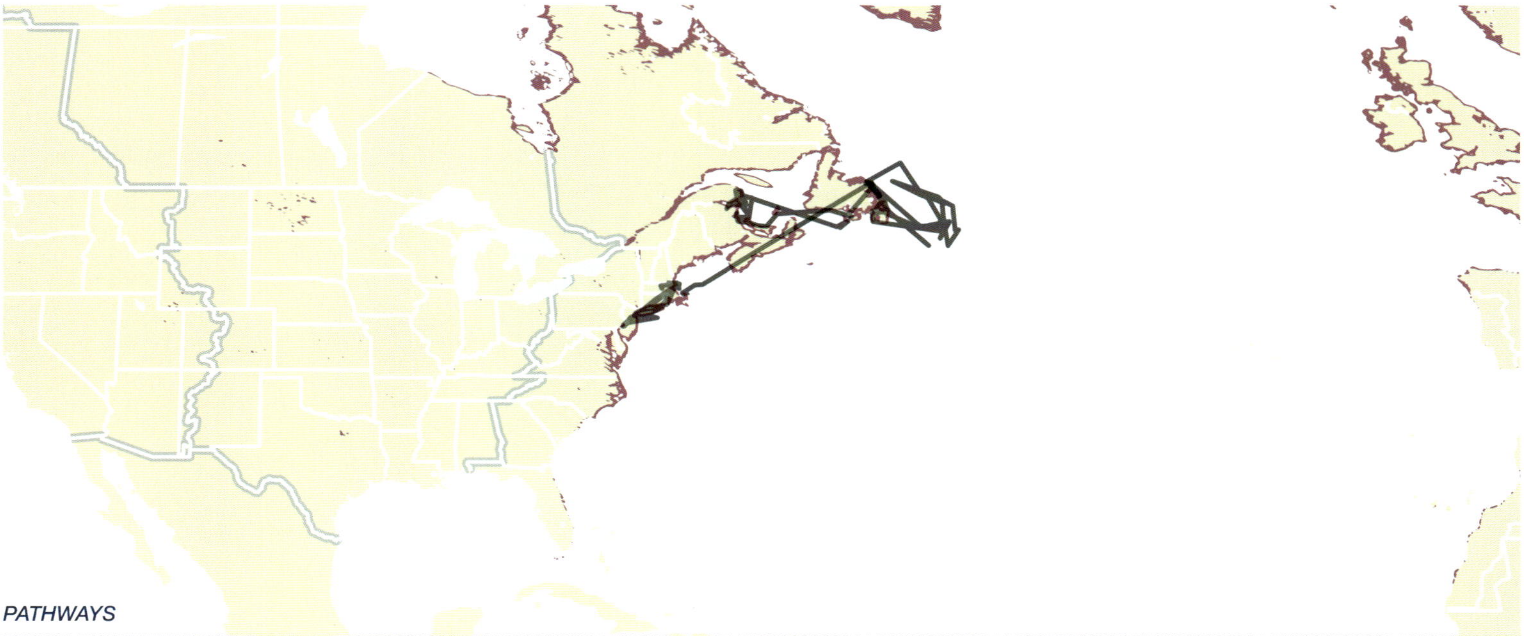

PATHWAYS

SATELLITE TRACKING A Great Black-backed Gull tracked with satellite telemetry from breeding areas in Massachusetts spent the nonbreeding season along the Delaware River between New Jersey and Pennsylvania (D. Clark, unpublished data). Two Great Black-backed Gulls breeding on the coast of Newfoundland migrated to two different nonbreeding areas, and one bird remained local. One bird migrated east to New Brunswick, and the other to Cape Cod, Massachusetts (L.D. Maynard and G.K. Davoren, unpublished data).

	Atlantic	Europe	Midcontinent
Atlantic (n = 273)	97%	1%	2%

OTHER APPROACHES No genetic or stable isotope studies of migratory connectivity were found for this species.

MIGRATORY PATHWAYS There is limited tracking data, so pathways are relatively unknown. Birds most likely migrate along the coast.

DATA CONTRIBUTORS

D.E. Clark; G.K. Davoren; K.K.G. Koenen; K.G. MacKenzie; L.D. Maynard; J.J. Whitney.

INTERESTING NOTE

Great black-backed Gulls are considered partial migrants. Birds from northernmost breeding areas migrate south during the nonbreeding season, while southernmost breeding populations likely remain at or near breeding colonies throughout the year (Good 2020). Other populations show a variety of strategies, with some birds migrating and others year-round residents (Good 2020).

BROWN NODDY *Anous stolidus*

DISTRIBUTION AND ABUNDANCE

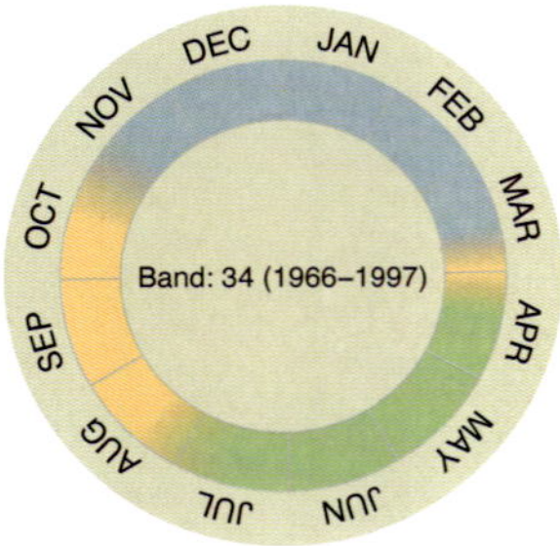

Brown Noddies are the most common and widespread noddy species across islets and islands of tropical seas (Chardine et al. 2020). Away from their breeding colonies, where they nest on the ground, trees, cliffs, and human structures, Brown Noddies can be pelagic wanderers, but some populations exhibit partial migration or only short-distance movements (Chardine et al. 2020). The closest colony to the continental US is in Dry Tortugas, Florida (1,000–2,000 breeding pairs), and thousands breed in Puerto Rico and the US Virgin Islands. Pacific Ocean colonies are larger, including large colonies on Kaula Island (15,000–25,000 pairs) and Nihoa Island (25,000–35,000 pairs, Chardine et al. 2020), Hawaii.

MIGRATORY CONNECTIVITY

OVERALL Banding data and some tracking data suggest strong migratory connectivity across ocean basins and weak to moderate within, but with limited data, MC could not be calculated. More extensive banding and nonbreeding tracking efforts are needed to examine demography and migratory behavior.

BANDING Birds nesting in Dry Tortugas, Florida, distributed across the central and east South American coast, including offshore areas. They spend the nonbreeding season at sea, including over long distances. From the Hawaiian islands and Johnston Atoll, recoveries generally were southwest toward Indonesia; however, one individual was found dead along the California coast. Recoveries would likely be coastally biased.

GEOLOCATOR TRACKING No geolocator data was found for North American populations, but light-level geolocator nonbreeding data from the eastern Indian Ocean (western Australia) indicated

ORDER Charadriiformes
FAMILY Laridae
BREEDING HABITAT Islands, Oceanic
NONBREEDING HABITAT Oceans, Pelagic
GLOBAL POPULATION SIZE 1,600,000
POPULATION CHANGE SINCE 1970 NA
IUCN RED LIST CATEGORY Least Concern
R2R TIPPING POINT SPECIES No
US LISTING STATUS No
CANADA SPECIES AT RISK No
PIF CCS 10/20

	ARCH	WARM	CCAL	ETRA	GUIA
Northwest Hawaiian Islands, USA (n = 4)	–	100 %	–	–	–
Honolulu, HI, USA (n = 4)	25 %	75 %	–	–	–
Johnston Atoll (n = 2)	–	50 %	50 %	–	–
Phoenix Islands, Kiribati (n = 1)	100 %	–	–	–	–
Dry Tortugas, FL, USA (n = 1)	–	–	–	100 %	–
Jamaica (n = 1)	–	–	–	–	100 %

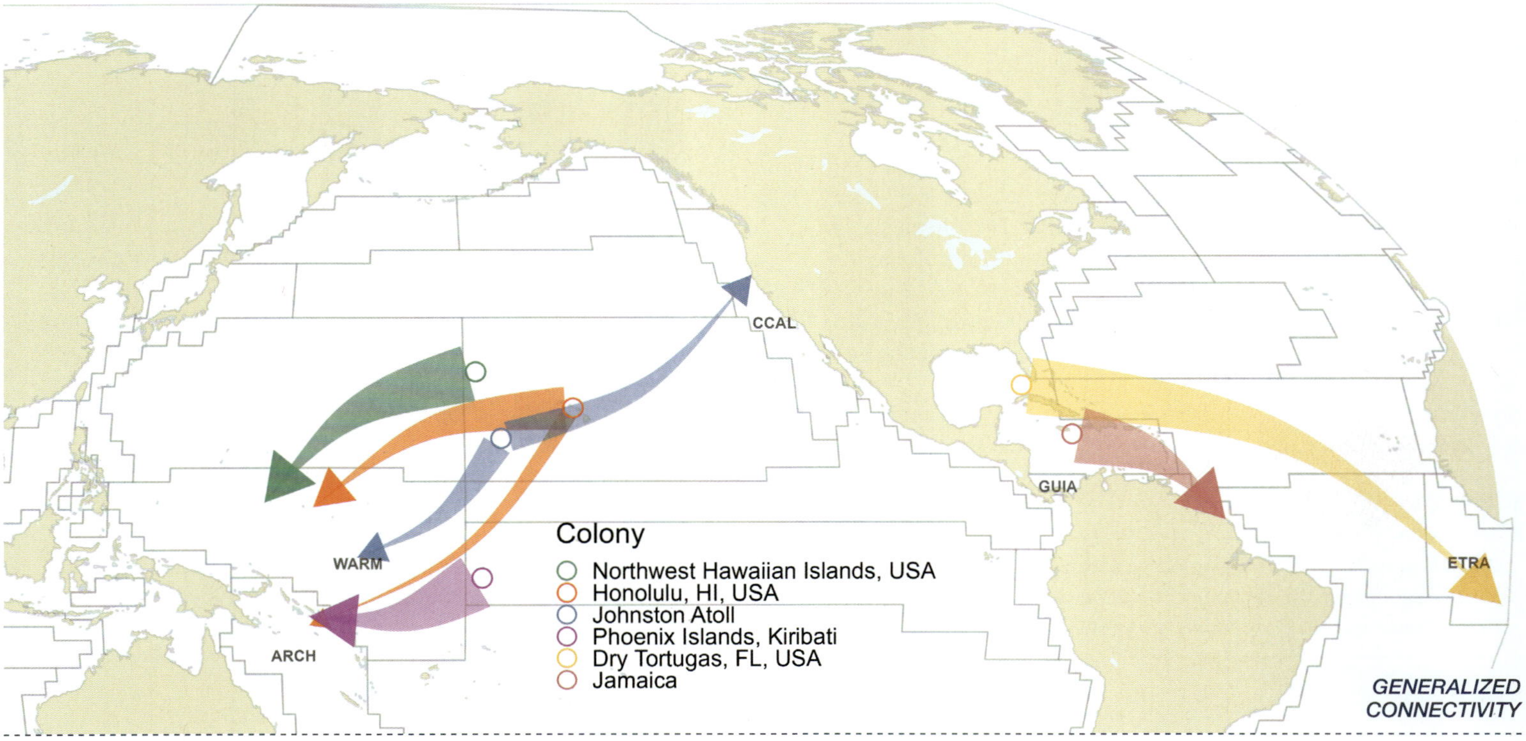

INTERESTING NOTE

Breeding Brown Noddies were tracked by radio transmitter in Hawaii in 1979, perhaps the first pelagic seabird to be tracked by radio telemetry. Of 11 birds tagged, batteries failed in the first week for 7. From bearings taken en route to marine foraging areas beyond the range of receivers for 4 individuals, specific offshore high use areas were hypothesized (Harrison and Stone-Burner 1981). Chick-rearing Brown Noddies were tracked by GPS from the Dry Tortugas National Park, Florida, in 2016 (n = 10). Birds traveled an average of 37.5 km from the colony and primarily southwest over the shelf break and Gulf of Mexico Loop Current. Over 90% of the birds' locations were in Marine Protected Areas, as well as 65% of probable foraging areas.

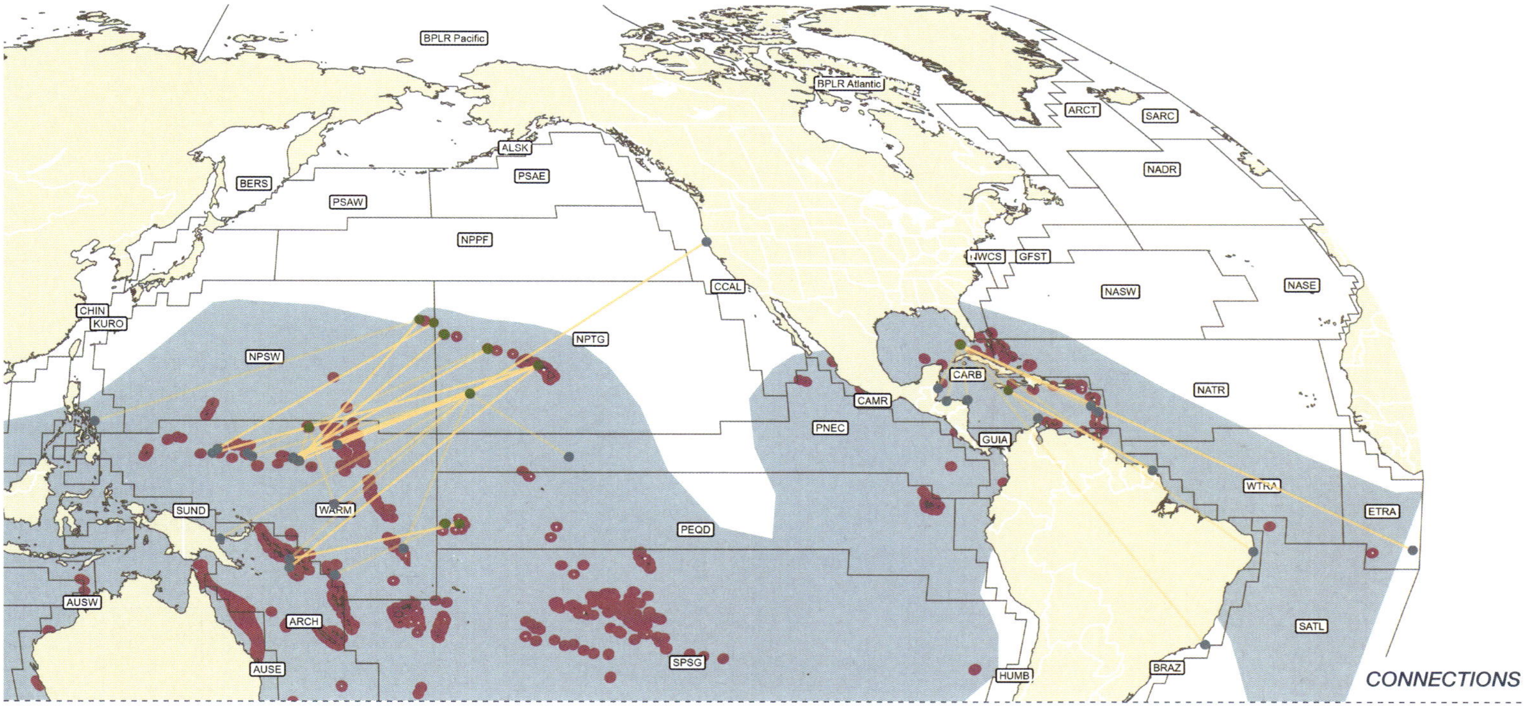

CONNECTIONS

that this population took short migrations and remained within the western Australian exclusive economic zone (EEZ) (Surman et al. 2018). These birds did not cross into the Pacific Ocean, suggesting a lack of mixing with Hawaiian populations.

GENETICS For six colonies in the southwestern Atlantic Ocean that vary in breeding times, morphology, and migratory versus residency behavior across 20 degrees of latitude, a mitochondrial DNA comparison revealed low genetic structure, indicating that all colonies belong to the same genetic population (Mazzochi et al. 2024). However, isotopic niche differed between most population pairs, suggesting that there is foraging separation among populations.

MIGRATORY PATHWAYS Migratory pathways are unknown for North American populations (Mlodinow et al. 2025).

Anous minutus BLACK NODDY

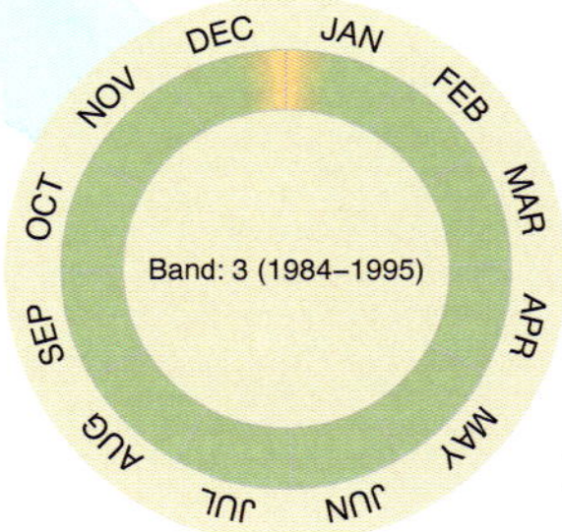

DISTRIBUTION AND ABUNDANCE

Black Noddies are gregarious, tropical seabirds that nest in trees, shrubs, and sometimes cliffs. Nesting colonies are widespread in the central and southwest Pacific Ocean and scattered in the Caribbean Sea and islands of the equatorial Atlantic Ocean (Gauger 2020). They also nest on a small number of islands off Pacific, Caribbean, and Atlantic coasts of central and northern South America (Gauger 2020). The global breeding population is estimated at 1–1.5 million, with the largest population in the Pacific Ocean (Gauger 2020). There are 10,000–20,000 breeding pairs across the Hawaiian archipelago, hundreds in the Caribbean Sea, and about 10,000 off the Brazilian coast (Gauger 2020). Estimates are highly uncertain as little current monitoring occurs.

ORDER Charadriiformes
FAMILY Laridae
BREEDING HABITAT Islands, Oceanic
NONBREEDING HABITAT Oceans, Pelagic
GLOBAL POPULATION SIZE 1,300,000
POPULATION CHANGE SINCE 1970 NA

IUCN RED LIST CATEGORY Least Concern
R2R TIPPING POINT SPECIES No
US LISTING STATUS No
CANADA SPECIES AT RISK No
PIF CCS 10/20

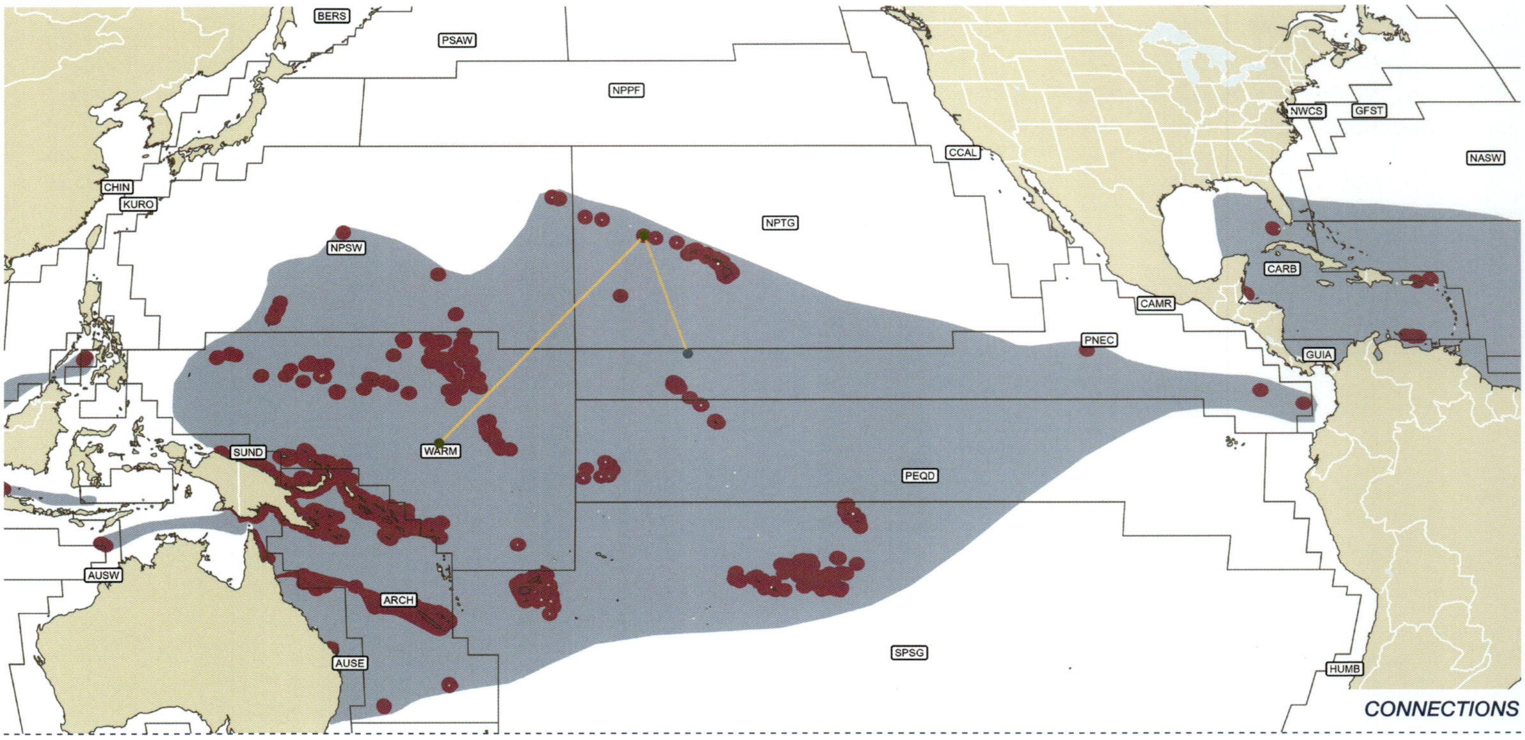

MIGRATORY CONNECTIVITY

OVERALL Migratory connectivity is unknown due to limited banding data. Some populations are thought to be sedentary, while others have a seasonal exodus. Some populations depart nesting islands after breeding and can undergo long-range movements, but it is unclear whether most long-range movements are dispersal or whether birds return to their colony (Gauger 2020).

INTERESTING NOTE

Migratory connectivity studies and population census efforts are clear priorities for this species.

BANDING Three recoveries in 1984–1995 connect breeding at Tern Island, Hawaii, to nonbreeding locations. There were two recoveries off Nauru, and one at sea. There were none in the Caribbean–South American range.

OTHER APPROACHES No tracking or stable isotope studies of migratory connectivity were found for this species, but isotopic signatures have been measured for breeding Black Noddies in Brazil (Mancini and Bugoni 2014) and Palmyra Atoll (Young et al. 2010), mainly to assess trophic niche overlap with other species. GPS tags have successfully tracked Black Noddies during multiple breeding seasons off northeast Australia (James and Cumming 2024).

MIGRATORY PATHWAYS Migratory pathways are unknown for North American populations (Gauger 2020).

SOOTY TERN *Onychoprion fuscatus*

weak strong

Mantel = 0.77 ± 0.04

DISTRIBUTION AND ABUNDANCE

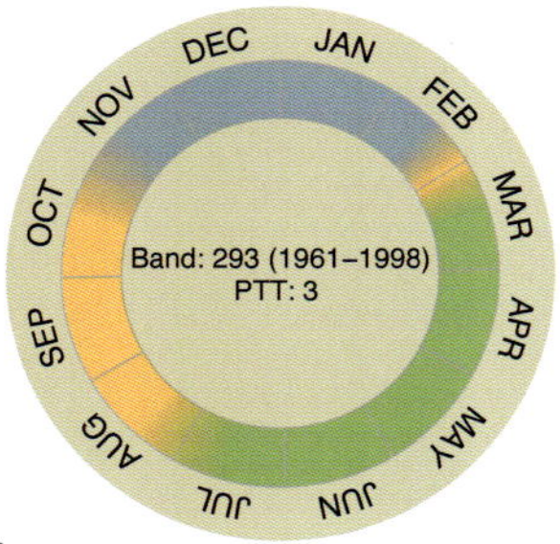

This species is a completely pelagic tern found in warm tropical oceans worldwide. During the nonbreeding season, Sooty Terns do not come to land, but may perch on floating debris or sea turtles, and are famous for spending years over the ocean prior to breeding (Schreiber et al. 2020). One million birds are estimated in the Hawaiian archipelago, but only 230,000–400,000 total in the Caribbean and Gulf of Mexico (Schreiber et al. 2020). There are two recognized subspecies, one from the Atlantic (*O. fuscatus fuscatus*) and the other from the Indo-Pacific (*O. fuscatus nubilosus*).

ORDER Charadriiformes
FAMILY Laridae
BREEDING HABITAT Coasts, Beach and Estuary
NONBREEDING HABITAT Oceans, Pelagic
GLOBAL POPULATION SIZE 47,000,000
POPULATION CHANGE SINCE 1970 NA

IUCN RED LIST CATEGORY Least Concern
R2R TIPPING POINT SPECIES No
US LISTING STATUS No
CANADA SPECIES AT RISK No
PIF CCS 9/20

MIGRATORY CONNECTIVITY

OVERALL Genetics and banding data suggest strong migratory connectivity between ocean basins (Mantel = 0.77 ± 0.04) and likely weak to moderate within, depending on age class. In the Atlantic

	SUND	ARCH	WARM	KURO	NPSW	NPTG	PNEC	PEQD	SPSG	CARB	NASW	NWSC	GUIA	GUIN	WTRA
Northwest Hawaiian Islands, HI, USA (n = 36)	47 %	14 %	–	22 %	6 %	6 %	6 %	–	–	–	–	–	–	–	–
Wake Island (n = 11)	55 %	–	9 %	18 %	9 %	9 %	–	–	–	–	–	–	–	–	–
Johnston Atoll (n = 27)	33 %	–	–	–	41 %	26 %	–	–	–	–	–	–	–	–	–
Palmyra Atoll (n = 1)	–	100 %	–	–	–	–	–	–	–	–	–	–	–	–	–
Howland Island (n = 3)	67 %	33 %	–	–	–	–	–	–	–	–	–	–	–	–	–
Kiribati (n = 7)	–	–	29 %	14 %	29 %	–	–	14 %	14 %	–	–	–	–	–	–
Starbuck Island, Kiribati (n = 1)	–	–	–	–	–	–	–	100 %	–	–	–	–	–	–	–
Dry Tortugas, FL, USA (n = 13)	–	–	–	–	–	–	–	–	–	31 %	8 %	8 %	31 %	15 %	8 %

Colony

- Northwest Hawaiian Islands, HI, USA
- Wake Island
- Johnston Atoll
- Palmyra Atoll
- Howland Island
- Kiribati
- Starbuck Island, Kiribati
- Dry Tortugas, FL, USA

INTERESTING NOTE

Fine-scale GPS tracking of 8 birds from Anguilla in the Caribbean Sea demonstrates international movements during the breeding season (Soanes et al. 2015b). Huang et al. (2017) used geolocators (n = 2) and satellite tags (n = 2) to track breeding season movements from Dry Tortugas National Park, Florida. Individuals spent the most time northwest of Dry Tortugas along the continental shelf, but ranged as far as 415 km away.

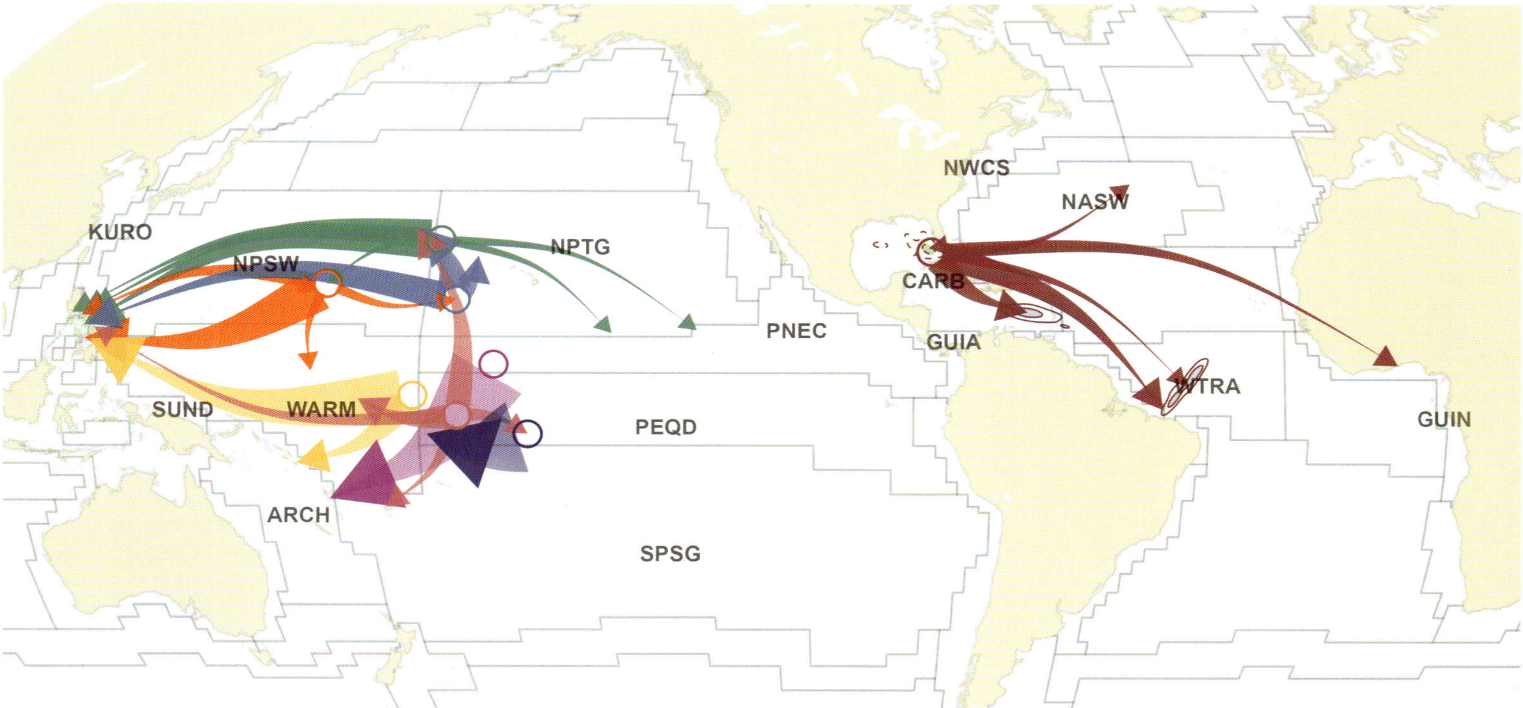

GENERALIZED CONNECTIVITY

Ocean, tracking and banding suggests some moderate connectivity of adults, while banding suggest juveniles tend to overlap during the nonbreeding period with a greater variety of colonies.

BANDING There is extensive banding dating to 1938. Sooty Terns are rarely seen at sea; thus, records are likely coastally biased. Huang et al. (2017) used banding records from Dry Tortugas National Park to infer migratory phenology and ontogeny. They suggested birds use the Gulf of Mexico from April to September, the Caribbean Sea from June to December, and the coast of Africa from October to March. They noted that although 51% of band recoveries were juveniles, 89% of recoveries in West Africa were juveniles. In the central Pacific, a large banding and resighting effort occurred in the 1960s as a part of the Pacific Ocean Biological Survey Program (Gould 1974).

GEOLOCATOR TRACKING Two birds tracked with geolocators from Dry Tortugas National Park spent the nonbreeding season off northeast Brazil (Huang et al. 2017). Sooty Terns tracked from Ascension Island in the Atlantic Ocean did not use North American waters (Reynolds et al. 2021), and a large sample of birds tracked in the Indian Ocean (Jaeger et al. 2017) showed that post-breeding movements remained within that ocean basin. No tracking was found in the Pacific Ocean.

SATELLITE TRACKING One bird tracked from Dry Tortugas National Park, Florida, with a 5 g satellite tag, migrated to deep-water habitats off northeast Brazil (Huang et al. 2017). A second bird's tag ceased transmission before completing its southward migration.

GENETICS An analysis of mitochondrial DNA (from a limited number of colonies) revealed differentiation between Atlantic Ocean and Indo-Pacific rookeries for the Sooty Tern, but little structure within oceans (Avise et al. 2000).

OTHER APPROACHES No stable isotope studies of migratory connectivity were found for this species.

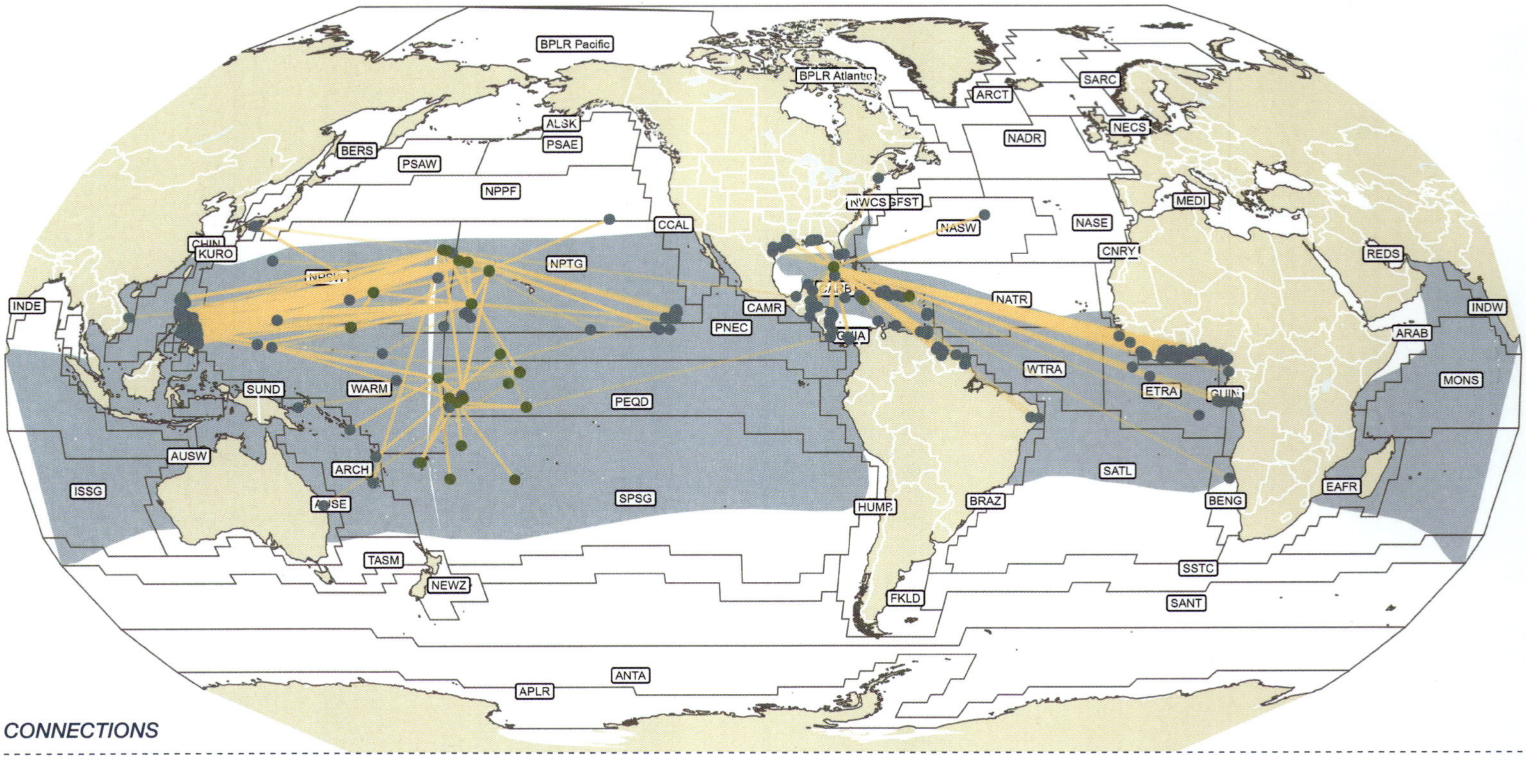

CONNECTIONS

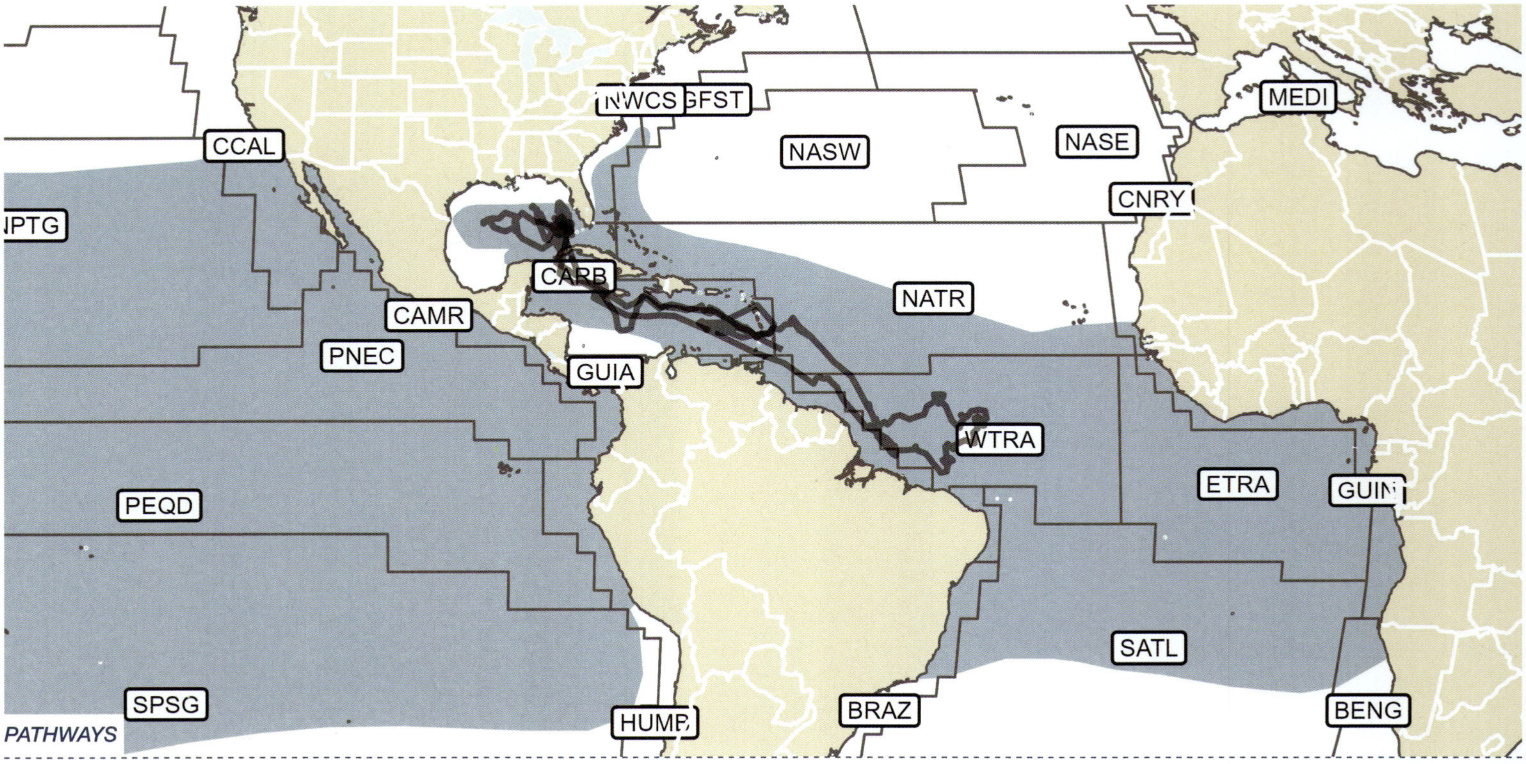

PATHWAYS

MIGRATORY PATHWAYS Transatlantic movements of juveniles were first described from banding records in Robertson (1969), where young travel first southwestward through the Yucatán Channel, then southward and eastward through the Caribbean Sea during August and September, arriving in African waters in October. For three birds tracked through a full migration with geolocators (n = 2) and a satellite tag from Dry Tortugas National Park (Huang et al. 2017), birds departed the Gulf of Mexico in late June or early July through the Yucatán Channel, then southeast through the Caribbean Sea, reaching the Lesser Antilles in early August, and used the deep waters off the northeast coast of Brazil through November. Birds returned to the breeding colony in December and January.

DATA CONTRIBUTORS

R. Huang.

Onychoprion lunatus **GRAY-BACKED TERN**

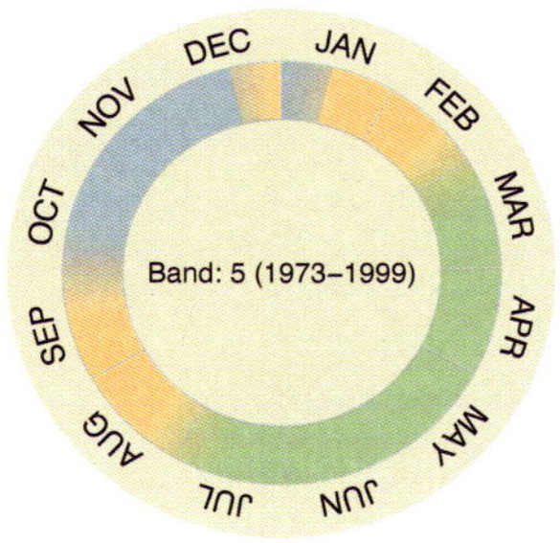

DISTRIBUTION AND ABUNDANCE
This rarely studied tropical tern's breeding range is confined to small islands in the central Pacific Ocean with nonbreeding concentration in the western Pacific Ocean (Mostello et al. 2000).

ORDER Charadriiformes
FAMILY Laridae
BREEDING HABITAT NA
NONBREEDING HABITAT NA
GLOBAL POPULATION SIZE NA
POPULATION CHANGE SINCE 1970 NA

IUCN RED LIST CATEGORY
Least Concern
R2R TIPPING POINT SPECIES No
US LISTING STATUS No
CANADA SPECIES AT RISK No
PIF CCS NA

MIGRATORY CONNECTIVITY
OVERALL This is a rarely studied and remotely located species, but limited banding data shows that some individuals undertake long-distance movements. MC is not able to be generated due to insufficient data.
BANDING Banding data from birds breeding on islands and atolls in the central Pacific Ocean show long-distance migration, with reencounters in the nonbreeding season from the Philippines, New Guinea, and the Solomon Islands.
OTHER APPROACHES No satellite, genetic, or stable isotope studies of migratory connectivity were found for this species. Migration is poorly understood.

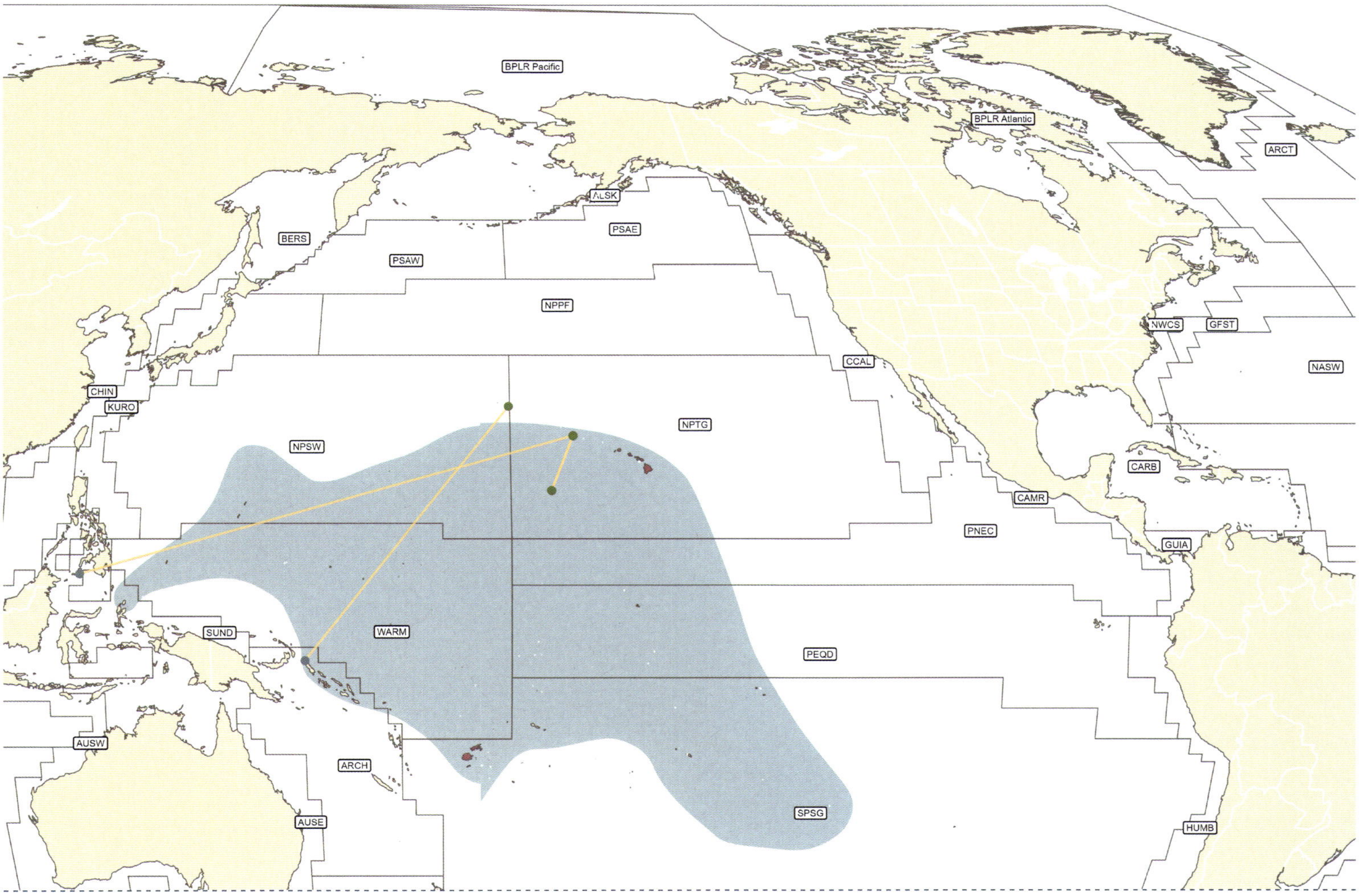

CONNECTIONS

BRIDLED TERN *Onychoprion anaethetus*

ORDER Charadriiformes
FAMILY Laridae
BREEDING HABITAT Islands, Oceanic
NONBREEDING HABITAT Oceans, Pelagic
GLOBAL POPULATION SIZE 700,000

POPULATION CHANGE SINCE 1970 NA
IUCN RED LIST CATEGORY Least Concern
R2R TIPPING POINT SPECIES No
US LISTING STATUS No
CANADA SPECIES AT RISK No
PIF CCS 11/20

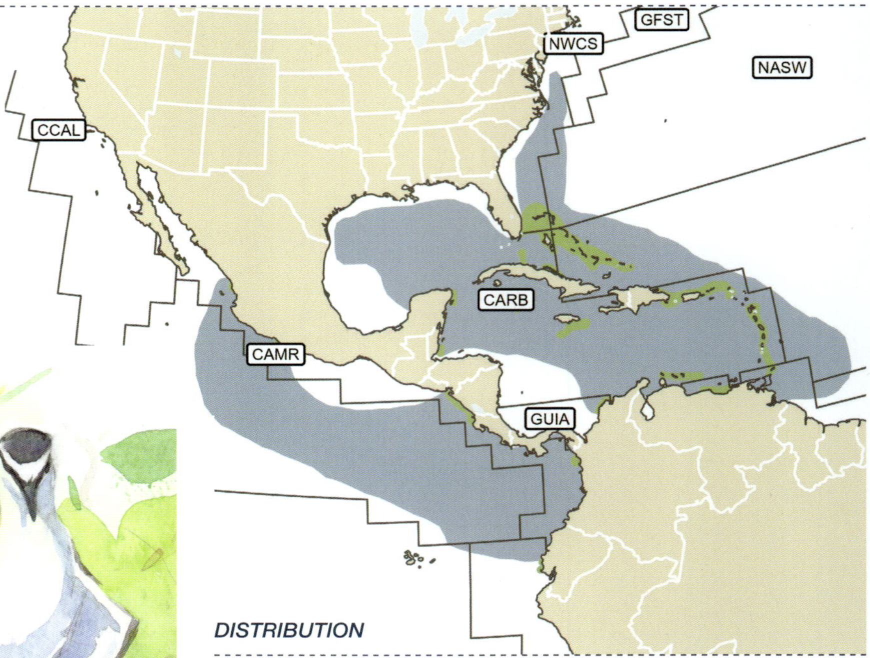

DISTRIBUTION

MIGRATORY CONNECTIVITY

OVERALL No information on migratory connectivity found for this species.

ALEUTIAN TERN *Onychoprion aleuticus*

DISTRIBUTION AND ABUNDANCE

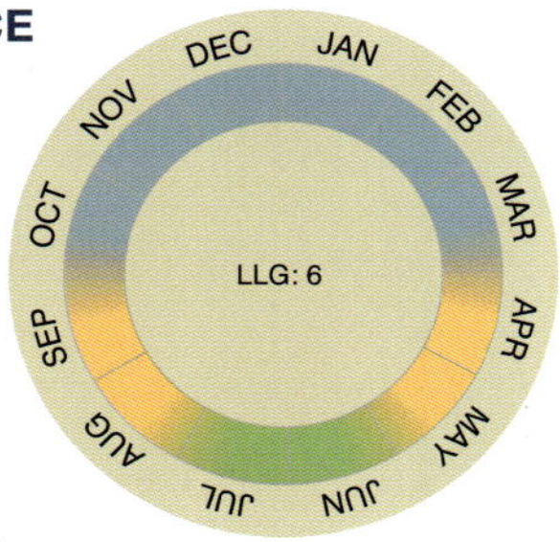

This species is a North Pacific subarctic breeder in Alaska, US, and Russia, ranging along the western Pacific Rim south to Australia (North 2020). In Alaska, it is largely absent from Bering Sea islands like the Pribilofs, but is distributed along the Aleutian chain and coastal Alaska south of the Arctic (North 2020). The 202 known colonies are split between Alaska (n = 110) and Russia (n = 92), but 82% of the total breeding population is estimated to be in Russia (Renner et al. 2015).

ORDER Charadriiformes
FAMILY Laridae
BREEDING HABITAT Coasts, Beach and Estuary
NONBREEDING HABITAT Coasts, Beach and Estuary
GLOBAL POPULATION SIZE 22,000

POPULATION CHANGE SINCE 1970 NA
IUCN RED LIST CATEGORY Vulnerable
R2R TIPPING POINT SPECIES No
US LISTING STATUS No
CANADA SPECIES AT RISK No
PIF CCS 13/20

MIGRATORY CONNECTIVITY

OVERALL There is little data about migratory connectivity, but weak connectivity is expected to nonbreeding regions in southeast Asia and Oceania based on tracking data from two Alaskan colonies.

GEOLOCATOR TRACKING Six Aleutian Terns tracked in 2010–2013 by light-level geolocators (Goldstein et al. 2019) from Yakutat, Alaska, used individually unique nonbreeding ranges spanning a 6,000 km region in Indonesia (n = 3), Malaysia (n = 1), Philippines (n = 1), and Papua New Guinea (n = 1). Nonbreeding areas were generally within 100 km of shore, keeping in mind confidence intervals around geolocator-derived positions.

OTHER APPROACHES No banding, genetics, or stable isotope studies of migratory connectivity were found for this species.

MIGRATORY PATHWAYS Birds tracked by Goldstein et al. (2019) took generally coastal pathways along the East Asian–Australasian Flyway from Yakutat, Alaska. Birds had high synchrony across years in arrival and departure dates and close spatial proximity during southbound migration stopover (Goldstein et al. 2019). Tracks indicate likely coastal stopover areas in the Sea of Japan in close proximity to major Russian colonies on the southbound migration and a pelagic stopover area 1,500 km east of Japan during northbound migration.

DATA CONTRIBUTORS
N. Catterson; D.C. Duffy; J. Frederick; M.I. Goldstein; S. Oehlers; S. Pyare.

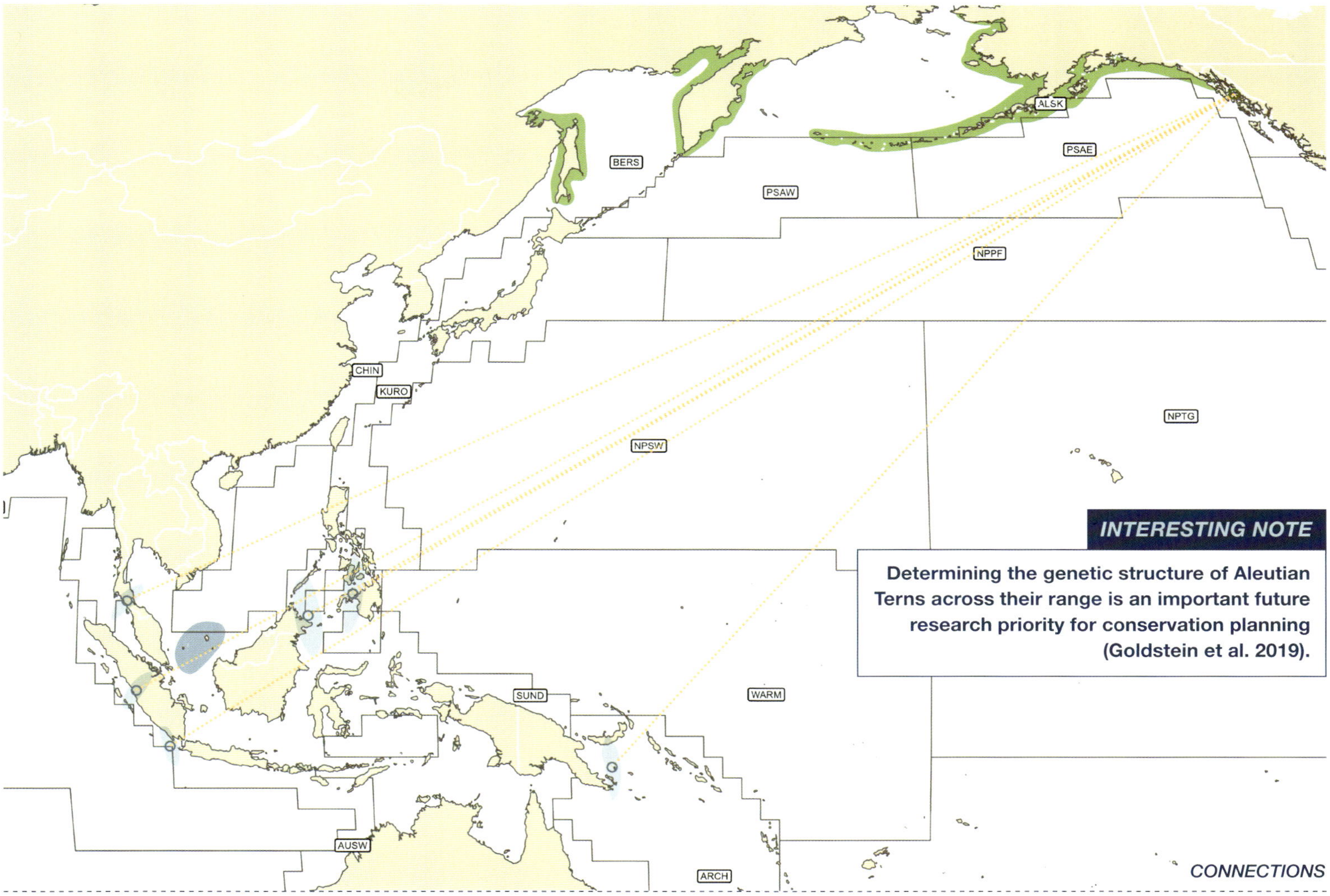

INTERESTING NOTE

Determining the genetic structure of Aleutian Terns across their range is an important future research priority for conservation planning (Goldstein et al. 2019).

LEAST TERN

Sternula antillarum

DISTRIBUTION AND ABUNDANCE

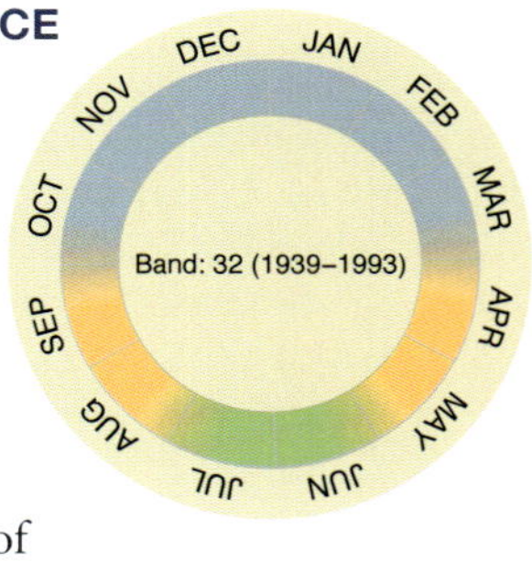

The Least Tern is the smallest North American tern, weighing only about 45 grams. It is widespread along coastal sandy areas in North America, the Caribbean, and northeastern South America. Although the Least Tern is well understood during its breeding period, much of this species' nonbreeding ecology is not well understood (Thompson et al. 2020; Fink et al. 2023). California Least Terns (*Sternula antillarum browni*)—one subspecies of Least Terns in North America—have been listed as endangered by the US federal government since 1970, and the State of California since 1971.

MIGRATORY CONNECTIVITY

OVERALL Migratory connectivity is poorly known. What is known comes from band reencounter and limited geolocator data. Birds breeding in the Atlantic and Midcontinent flyways migrate to northern South America (89% and 100%, respectively) for the nonbreeding season. Atlantic Flyway breeders also migrate to the Caribbean (11%). MC is not calculated due to limited sample size.

ORDER Charadriiformes
FAMILY Laridae
BREEDING HABITAT Coasts, Beach and Estuary
NONBREEDING HABITAT Coasts, Beach and Estuary
GLOBAL POPULATION SIZE 120,000

POPULATION CHANGE SINCE 1970 −79%
IUCN RED LIST CATEGORY Least Concern
R2R TIPPING POINT SPECIES Orange-Alert
US LISTING STATUS Endangered California subspecies
CANADA SPECIES AT RISK No
PIF CCS 15/20

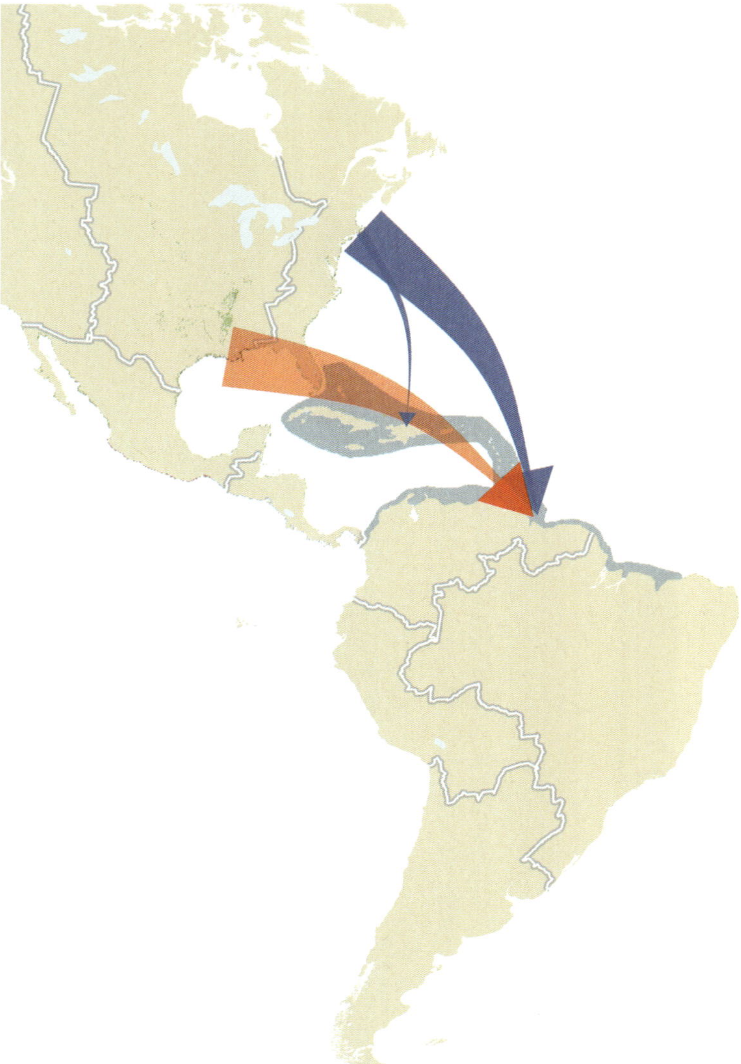

GENERALIZED CONNECTIVITY

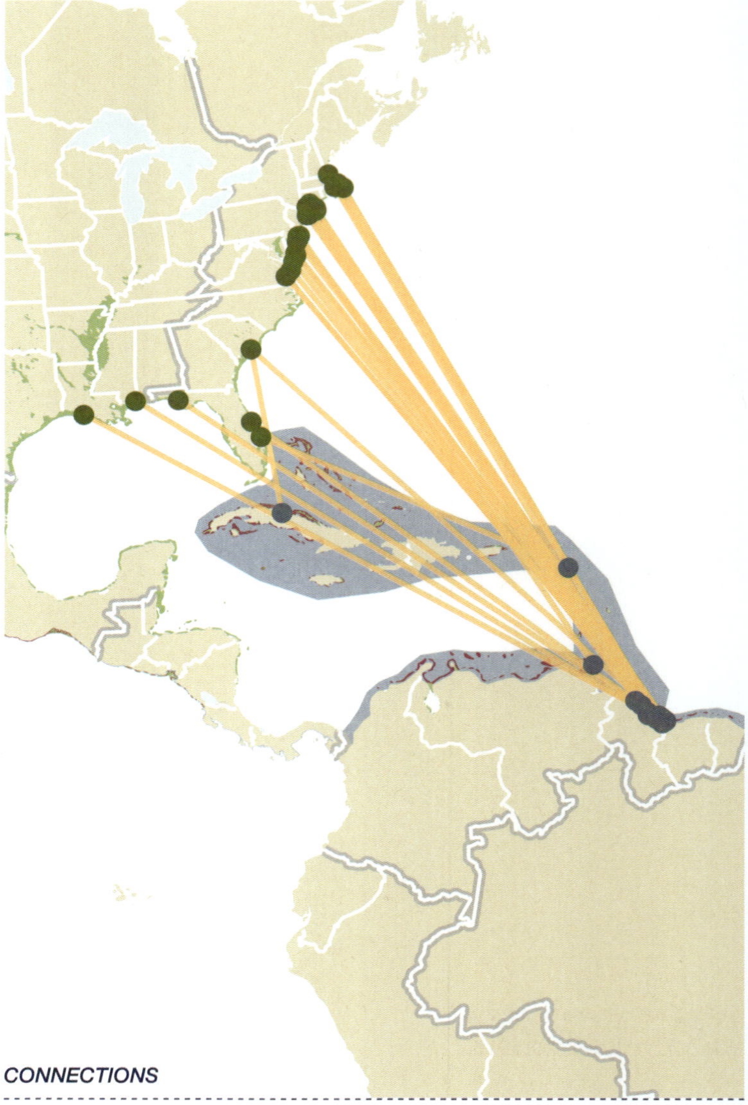

CONNECTIONS

	Caribbean	Northern SA
Atlantic (n = 27)	11%	89%
Midcontinent (n = 2)	–	100%

INTERESTING NOTE

A study deploying GPS loggers and miniaturized radio telemetry tags on endangered California Least Terns recently started in Southern California and will continue over the next several years (N. Vilchis, unpublished data) to identify migratory routes and nonbreeding locations and to understand habitat use during the nonbreeding season and other elements of the nonbreeding ecology to effectively manage its recovery.

BANDING Banding reencounter data for breeding to nonbreeding movements is mostly from birds along the Atlantic and Gulf coasts being recovered along the eastern coast of South America, as well as a few Caribbean recoveries. Banding data shows that birds from breeding areas in the North Atlantic undergo longer migrations than those from the South Atlantic and Gulf of Mexico.

GEOLOCATOR TRACKING A series of deployments of geolocators in California showed that California Least Terns moved south after breeding into the Gulf of California or the Pacific coast of Baja California (Ryan 2016). Tag defects and poor data quality resulted in limited data (n = 5; results not included on maps), but there are future plans to deploy GPS tags.

OTHER APPROACHES No satellite tracking, genetic, or stable isotope studies of migratory connectivity were found for this species.

MIGRATORY PATHWAYS Pathways are relatively unknown, but birds follow major rivers and marine coasts and may migrate across open water. Geolocator data suggests that, by early to mid-August, California Least Terns have reached an area between Guaymas, Sonora, and Los Mochis, Sinaloa, and may then cross to the Baja California peninsula between Bahia de Magdelena and La Paz in Baja California Sur (Ryan 2016). Least Terns breeding along the US Gulf Coast migrate around the Gulf of Mexico; Atlantic coast terns migrate across the ocean through the Caribbean islands, and interior populations follow major river basins south to the Gulf of Mexico, with the remaining routes unknown (Thompson et al. 2020).

Gelochelidon nilotica **GULL-BILLED TERN**

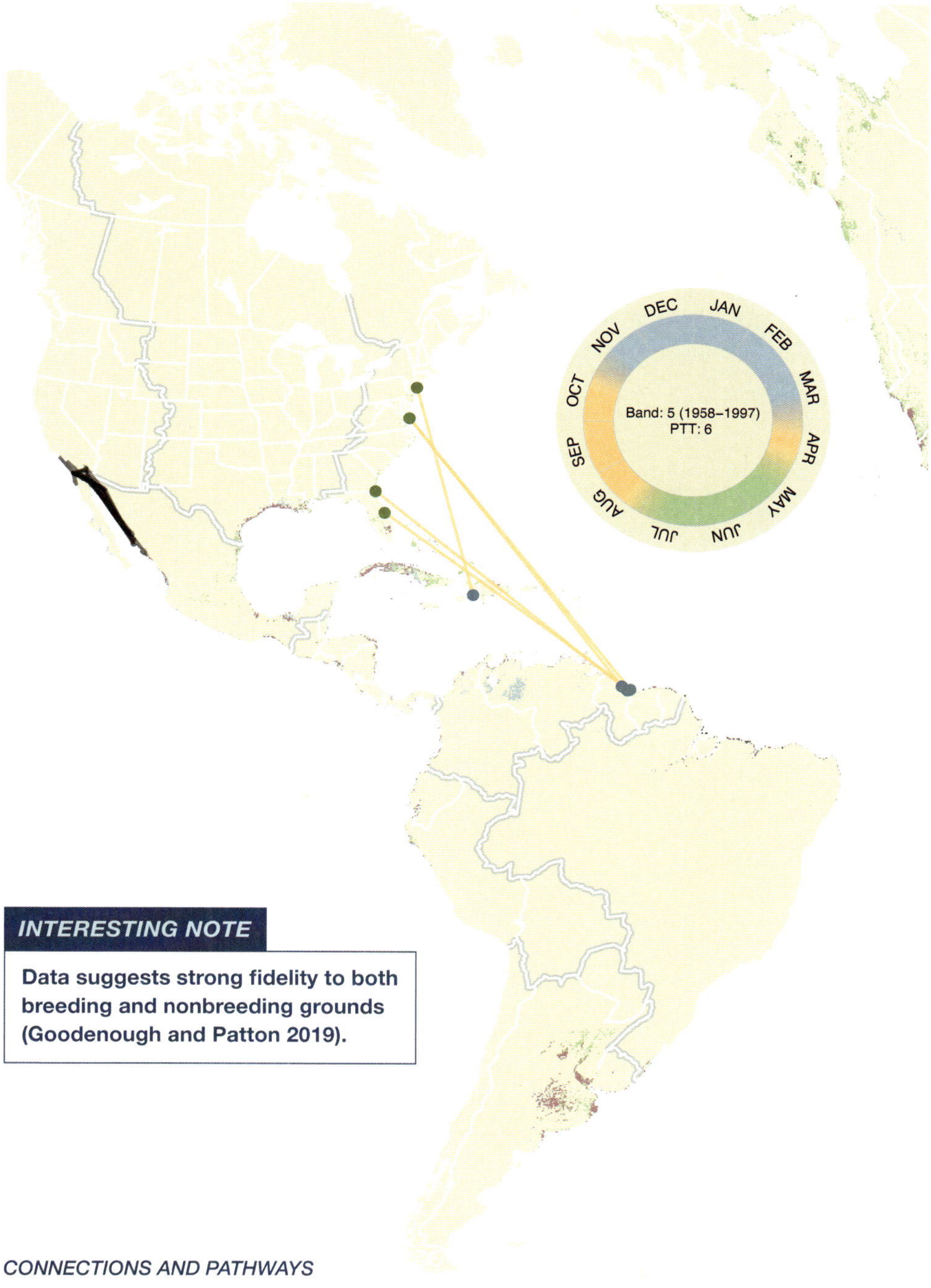

INTERESTING NOTE

Data suggests strong fidelity to both breeding and nonbreeding grounds (Goodenough and Patton 2019).

CONNECTIONS AND PATHWAYS

DISTRIBUTION AND ABUNDANCE

The Gull-billed Tern is globally distributed. In North America, this species breeds along the Atlantic coast at highest densities from Virginia south to Georgia (Fink et al. 2023). It also breeds in high numbers along the Gulf Coast. In California, breeding populations are limited to a single coastal and a single interior site in the southern part of the state. Gull-billed Terns spend the nonbreeding season in high numbers along the Gulf Coast and coastal areas of western Mexico and northeastern South America, and in smaller numbers throughout the Caribbean, Mexico, and Central and South America (Fink et al. 2023).

ORDER Charadriiformes
FAMILY Laridae
BREEDING HABITAT Coasts, Beach and Estuary
NONBREEDING HABITAT Coasts, Beach and Estuary
GLOBAL POPULATION SIZE 190,000
POPULATION CHANGE SINCE 1970 +40%
IUCN RED LIST CATEGORY Least Concern
R2R TIPPING POINT SPECIES No
US LISTING STATUS No
CANADA SPECIES AT RISK No
PIF CCS 13/20

MIGRATORY CONNECTIVITY

OVERALL The migratory connectivity of the Gull-billed Tern is known from one satellite tracking study (Pacific Flyway) and band reencounter data (Atlantic Flyway). No data was found for the Midcontinent Flyway populations. Birds breeding in the Atlantic Flyway migrate to nonbreeding areas in northern South America and the Caribbean, while Pacific Flyway breeding birds migrate to Mexico. MC was not calculated due to limited sample size.

BANDING The only band reencounters were from birds banded on the east coast. Birds breeding in both the mid-Atlantic region and southeast of the US migrated to nonbreeding areas along the coast of Guyana.

SATELLITE TRACKING Western Gull-billed Terns (*G. n. vanrossemi*), considered a California State Species of Special Concern (Shuford and Gardali 2008) and a United States Fish and Wildlife Service Bird of Conservation Concern (US Fish and Wildlife Service 2008), tracked with satellite transmitters from San Diego, California, spent the nonbreeding season along a 250-km stretch of coast in northern Sinaloa, Mexico (Goodenough and Patton 2019).

OTHER APPROACHES No genetic or stable isotope studies of migratory connectivity were found for this species

MIGRATORY PATHWAYS Birds tracked in the Pacific Flyway had similar pre- and post-breeding migration routes, and the Colorado River Delta was an important stopover location for both pre- and post-breeding migrations (Goodenough and Patton 2019).

DATA CONTRIBUTORS
K.S. Goodenough.

CASPIAN TERN *Hydroprogne caspia*

DISTRIBUTION AND ABUNDANCE

This species is the largest tern in the world and can be found on every continent except Antarctica. In North America, Caspian Terns breed primarily along the west coast and Interior West of the United States, central Canada, the Great Lakes, and the Gulf and Atlantic coasts. Breeding populations have recently extended into Alaska. Major nonbreeding areas include coastal areas of Baja, the west coast of Mexico and Central America, the US Gulf coast, Florida, and Cuba.

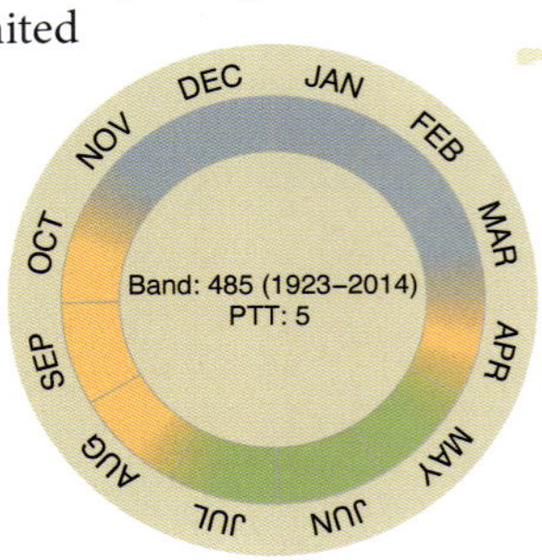

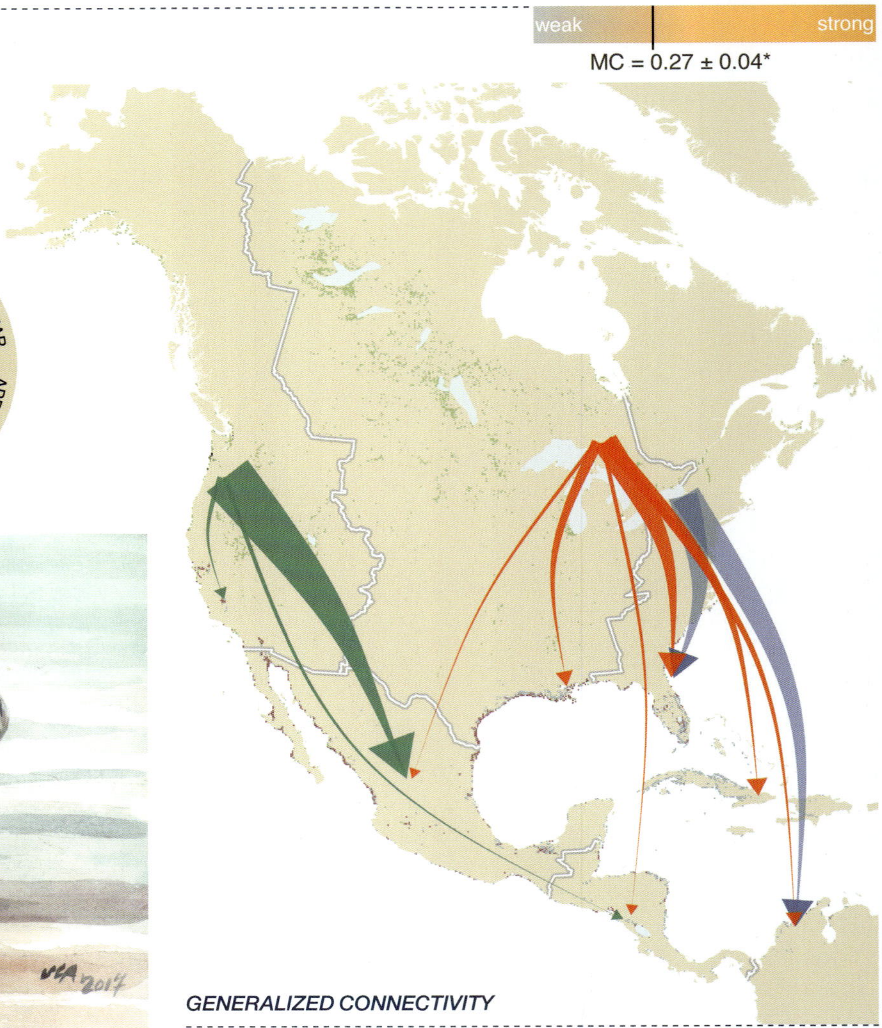

GENERALIZED CONNECTIVITY

ORDER Charadriiformes
FAMILY Laridae
BREEDING HABITAT Wetlands, Lakes and Rivers
NONBREEDING HABITAT Coasts, Beach and Estuary
GLOBAL POPULATION SIZE 530,000
POPULATION CHANGE SINCE 1970 +44%
IUCN RED LIST CATEGORY Least Concern
R2R TIPPING POINT SPECIES No
US LISTING STATUS No
CANADA SPECIES AT RISK No
PIF CCS 9/20

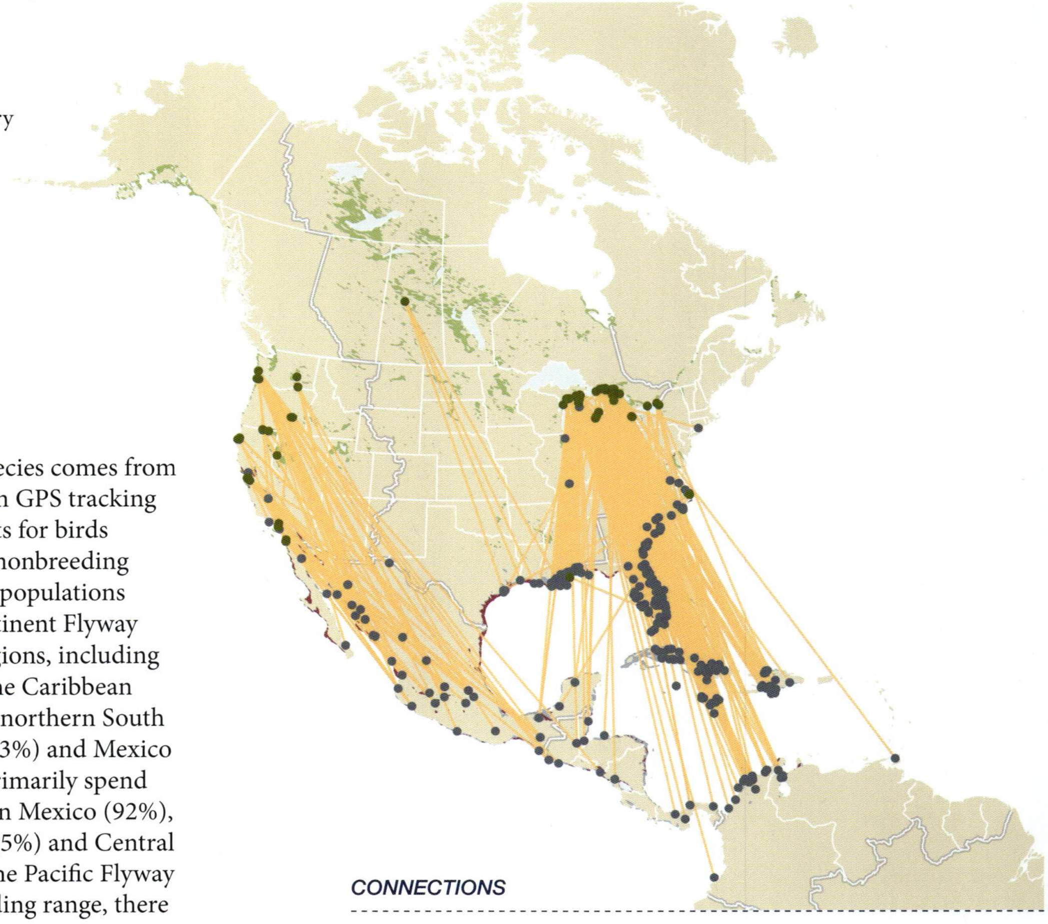

CONNECTIONS

MIGRATORY CONNECTIVITY

OVERALL Migratory information for this species comes from extensive band reencounter records and from GPS tracking of the western population. Limited data exists for birds breeding in the Atlantic Flyway, so primary nonbreeding regions and the degree of mixing with other populations are unknown. Birds breeding in the Midcontinent Flyway spend the nonbreeding season in various regions, including coastal areas of the Atlantic Flyway (43%), the Caribbean (23%), the Midcontinent Flyway (16%), and northern South America (13%), as well as Central America (3%) and Mexico (2%). Birds breeding in the Pacific Flyway primarily spend the nonbreeding season along coastal areas in Mexico (92%), with smaller numbers in the Pacific Flyway (5%) and Central America (3%). Although terns breeding in the Pacific Flyway remain within western areas of the nonbreeding range, there

	Atlantic	Caribbean	Central America	Mexico	Midcontinent	Northern SA	Pacific
Atlantic (n = 2)	50%	–	–	–	–	50%	–
Midcontinent (n = 322)	–	23%	3%	2%	16%	13%	–
Pacific (n = 134)	–	–	3%	92%	–	–	5%

is mixing with other populations in Central America. The high abundance of terns breeding in the Midcontinent Flyway, along with their tendency to distribute to different nonbreeding areas, suggest moderate migratory connectivity (MC = 0.27 ± 0.04).

BANDING Banding reencounter data are predominantly from Great Lakes and coastal Pacific breeding populations, with limited data from Atlantic, interior western, and central Canada breeding populations. Caspian Terns breeding in the Great Lakes spend the nonbreeding period along the Gulf Coast, southeastern United States, the Caribbean, and the Colombian coast. There were few to no recoveries in Mexico or Central America. A study of Caspian Terns breeding in the southern San Francisco Bay identified the west coast of Mexico as an important nonbreeding area, with no indication of Pacific coast birds migrating to the Gulf of Mexico or the Caribbean (Gill and Mewaldt 1979).

SATELLITE TRACKING Satellite tagging of this species in the Pacific Flyway revealed variable migration routes but with very high connectivity to one stopover location, the Salton Sea. In this study, nearly every bird tagged from two Washington colonies used this location for a significant post-breeding (and to a lesser degree, pre-breeding) stopover. Most of these birds spent the nonbreeding season in western Mexico (Lyons et al. 2018a).

GENETICS Boutilier et al. (2014) characterized variation in mitochondrial DNA and found significant range-wide population differentiation. Pacific coast sites differed genetically from sites east of the Rocky Mountains, and sites in central Canada differed from those in the Great Lakes region. Gene flow among the regions appears to be restricted.

OTHER APPROACHES No stable isotope studies of migratory connectivity were found for this species.

MIGRATORY PATHWAYS In North America, the same flyways are generally used between breeding and nonbreeding sites, along the Pacific and Atlantic coasts, the Mississippi Flyway, and other major rivers (Cuthbert and Wires 2020). The Salton Sea is an important stopover area for western birds (Lyons et al. 2018a).

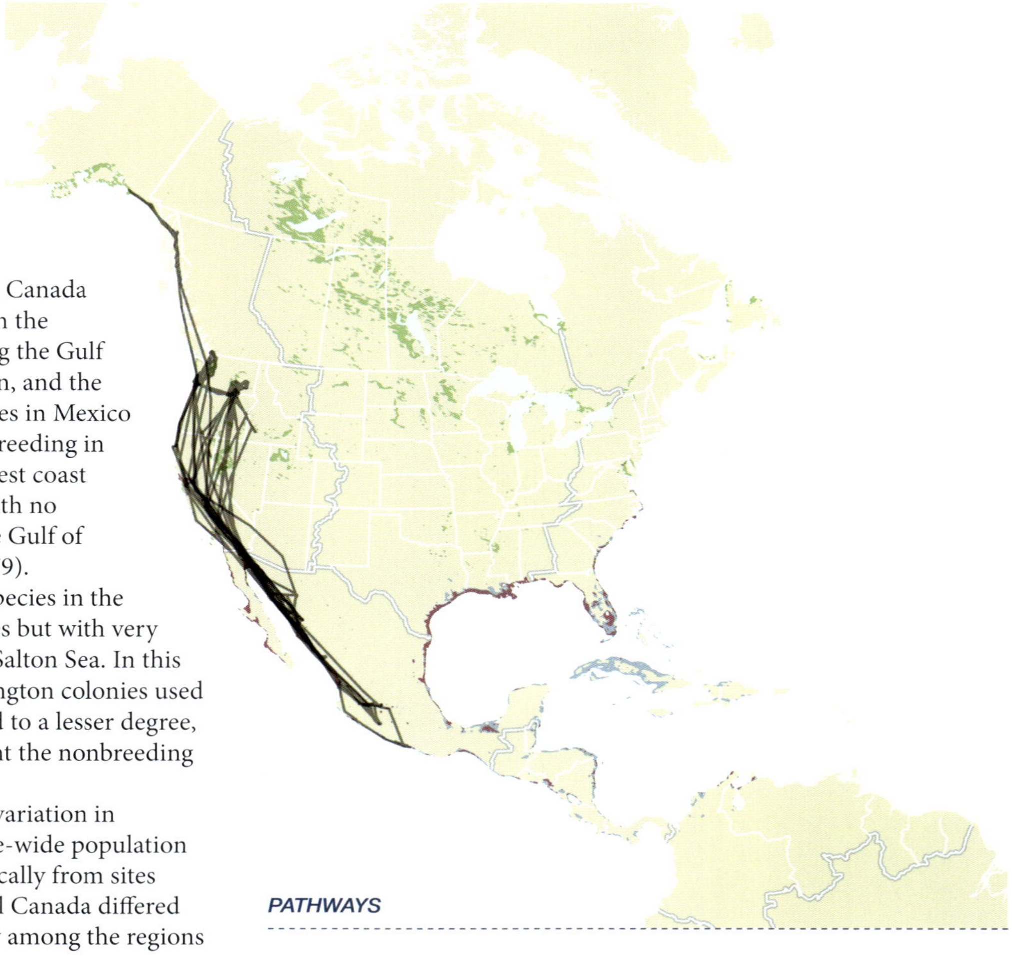

PATHWAYS

INTERESTING NOTE

It is recommended that the three population groups supported by genetic evidence, band recoveries, and tracking data be managed separately (Boutilier et al. 2014; Craig and Larson 2019). More data are needed to examine whether Gulf Coast and Atlantic coastal breeders form distinct populations as well (Craig and Larson 2019). More information about movement and stopover use in the nonbreeding season is a priority for future research (Cuthbert and Wires 2020).

DATA CONTRIBUTORS

T.J. Lawes; D.E. Lyons.

BLACK TERN *Chlidonias niger*

DISTRIBUTION AND ABUNDANCE

This small tern has strikingly different plumage from other terns and is equally unique in its life history. This species breeds throughout much of the interior United States and Canada, including Nebraska, the Dakotas, and Minnesota, and north into Ontario, Manitoba, Saskatchewan, and eastern Alberta. Also, it breeds in Oregon, Nevada and Northern California. This species spends the nonbreeding season primarily in coastal areas of western Mexico and along both coasts of Central America and into northern South America (Heath et al. 2009; Fink et al. 2023).

ORDER Charadriiformes
FAMILY Laridae
BREEDING HABITAT Wetlands, Freshwater Marsh
NONBREEDING HABITAT Oceans, Pelagic
GLOBAL POPULATION SIZE 5,100,000
POPULATION CHANGE SINCE 1970 59%
IUCN RED LIST CATEGORY Least Concern
R2R TIPPING POINT SPECIES No
US LISTING STATUS No
CANADA SPECIES AT RISK No
PIF CCS 12/20

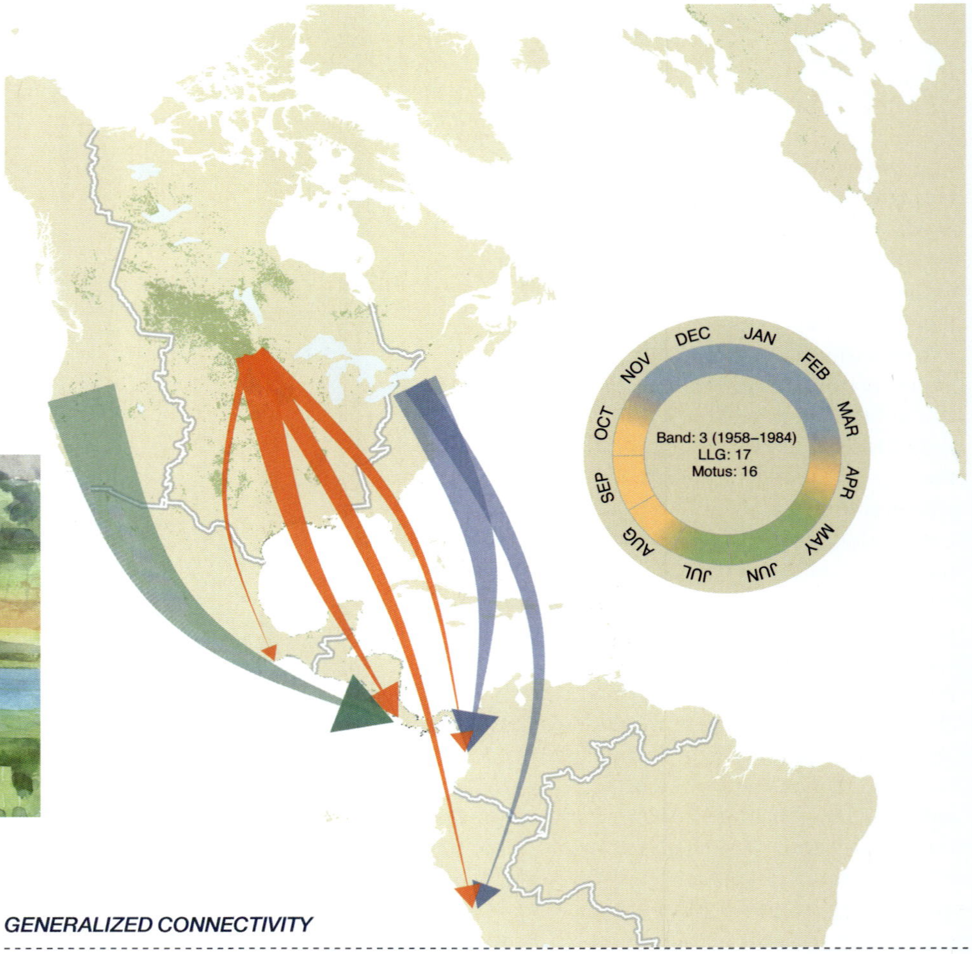

GENERALIZED CONNECTIVITY

MIGRATORY CONNECTIVITY

OVERALL Little is known about migratory connectivity in this species due to insufficient data, but migratory connectivity appears weak due to mixing of breeding populations on nonbreeding areas and weak genetic structure among breeding populations. Atlantic Flyway breeding populations spend the nonbreeding season along coastal areas in northern South America (67%) and the Andean West region of South America (33%). Midcontinent Flyway breeding populations primarily spend the nonbreeding season in Central America (47%), where they may overlap with Pacific Flyway

	Andean West	Central America	Mexico	Northern SA
Atlantic (n = 3)	33%	–	–	67%
Midcontinent (n = 15)	27%	47%	7%	20%
Pacific (n = 1)	–	100%	–	–

CONNECTIONS A

breeding populations. They also migrate to northern South America (20%) and the Andean West (27%) regions, where populations overlap with Atlantic Flyway breeding populations.

BANDING Banding data for this species is limited, with three breeding to nonbreeding banding records. A bird from Oregon and a bird from Minnesota were both located in the area of Guatemala and El Salvador during the nonbreeding season. A third bird from Wisconsin was found in Ecuador during the nonbreeding season.

GEOLOCATOR TRACKING Geolocator data from breeding colonies in Saskatchewan, Minnesota, and Ontario showed shared nonbreeding locations along coastal areas of Mexico, Central America, and South America (n = 19; Shephard et al. 2023). Individuals used multiple nonbreeding locations throughout the season (from Mexico to Peru); however, all tagged individuals spent some of the nonbreeding period along the coast of Panama, suggesting that Panama is a key nonbreeding location for this species (Shephard et al. 2023). Black Terns showed strong connectivity in post-breeding migratory routes; however, migratory connectivity did not hold up on the nonbreeding grounds, with individuals from the two regions overlapping (Shephard et al. 2023).

GENETICS Genetic samples collected from nine breeding colonies across the range (Maine, Ontario, Michigan, Iowa, Wisconsin, Nebraska, Saskatchewan, and Oregon) demonstrated weak genetic structure (Shephard et al. 2023).

OTHER APPROACHES No satellite tracking or stable isotope studies of migratory connectivity were found for this species.

MIGRATORY PATHWAYS Geolocator and Motus data show this species takes a variety of paths southward from breeding areas during the post-breeding season, with a separation of migratory routes between birds breeding in central Canada and birds breeding in the Great Lakes region (Shephard et al. 2023). Eastern birds follow the Atlantic coast, central birds cross overland to the Gulf Coast, and western birds move along the Pacific coast of Mexico and the Gulf of California. Most birds then proceed to both coasts of Central America, especially Nicaragua, Costa Rica, and Panama. Some individuals may proceed to South America for part of the nonbreeding season (Heath et al. 2009; Fink et al. 2023). During northward migration, individuals from the eastern colonies migrated through the Gulf of Mexico, the Mississippi Flyway, and along the Atlantic coast, while individuals from Saskatchewan migrated north through the Central Flyway, with stopovers recorded on the western coast of Mexico and along the Gulf of Mexico (Shephard et al. 2023).

DATA CONTRIBUTORS

J. Costa; A.E. McKellar; D. Moore; N.G. Shephard.

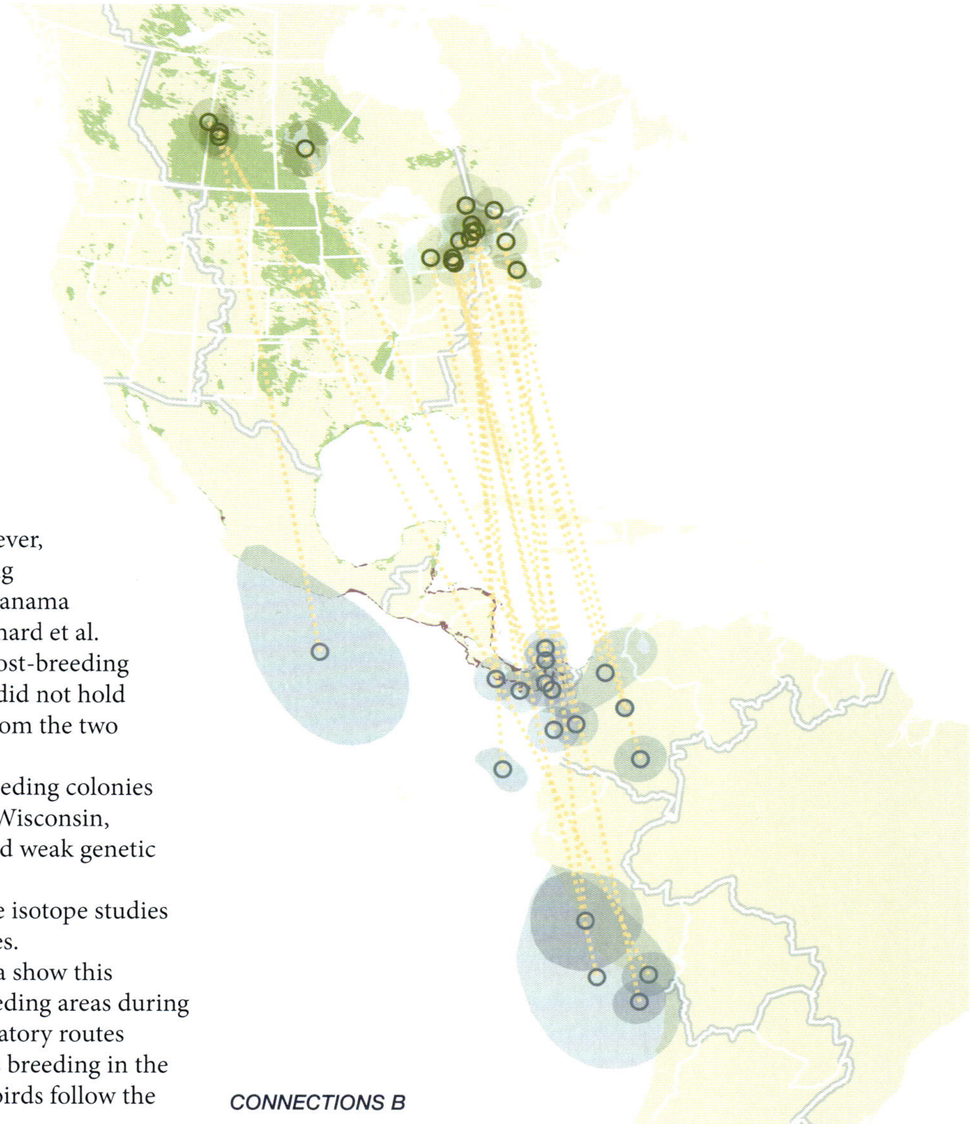

CONNECTIONS B

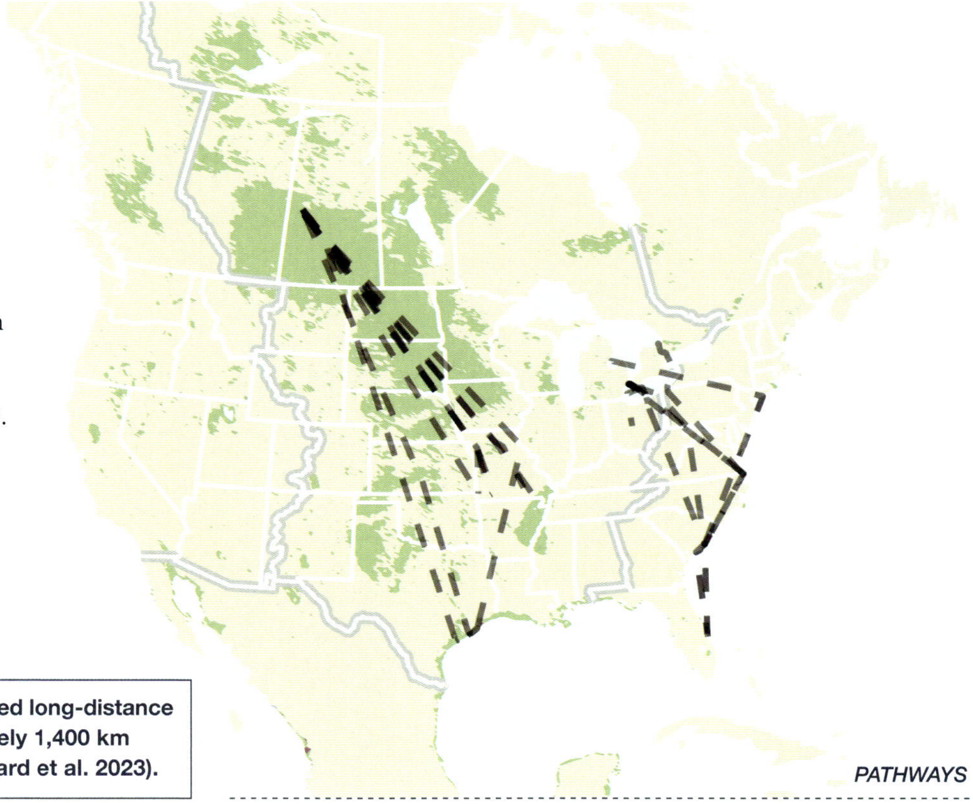

PATHWAYS

INTERESTING NOTE

An individual tracked across multiple years exhibited long-distance breeding dispersal with a distance of approximately 1,400 km between breeding locations across years (Shephard et al. 2023).

ROSEATE TERN *Sterna dougallii*

DISTRIBUTION AND ABUNDANCE

This oceanic tern has a nearly global distribution, but populations are local and disjunct. In North America, this species breeds from New York to Nova Scotia, and on islands near Puerto Rico and through the Caribbean, although data for Caribbean-nesting Roseate Terns is scant (Gochfeld and Burger 2020). Nonbreeding areas include most of the Caribbean and northern South America.

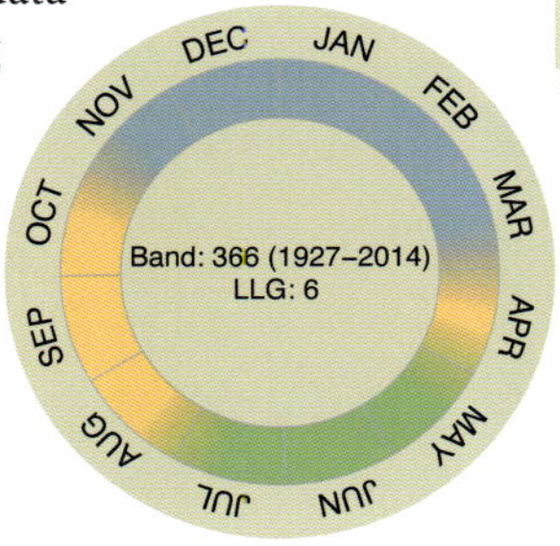

ORDER Charadriiformes
FAMILY Laridae
BREEDING HABITAT Coasts, Beach and Estuary
NONBREEDING HABITAT Oceans, Pelagic
GLOBAL POPULATION SIZE 160,000
POPULATION CHANGE SINCE 1970 NA
IUCN RED LIST CATEGORY Least Concern
R2R TIPPING POINT SPECIES No
US LISTING STATUS Endangered Atlantic coast population Threatened all other populations
CANADA SPECIES AT RISK Endangered
PIF CCS 15/20

	Brazil	Caribbean	Northern SA	Southern Cone
Atlantic (n = 331)	45%	5%	50%	<1%

MIGRATORY CONNECTIVITY

OVERALL Migratory connectivity information is moderate for this species, with most data coming from band reencounters and limited light-level geolocator information. Birds breeding in the Atlantic Flyway spent the nonbreeding season in northern South America (50%), Brazil (45%), or the Caribbean (5%). Due to insufficient data in other breeding regions, an MC value is not available.

BANDING Band reencounters show that northeastern birds spend the nonbreeding season along the Atlantic coast of South America, with strong connectivity to coastal areas of Guyana, Suriname, and Brazil (Cohen et al. 2014).

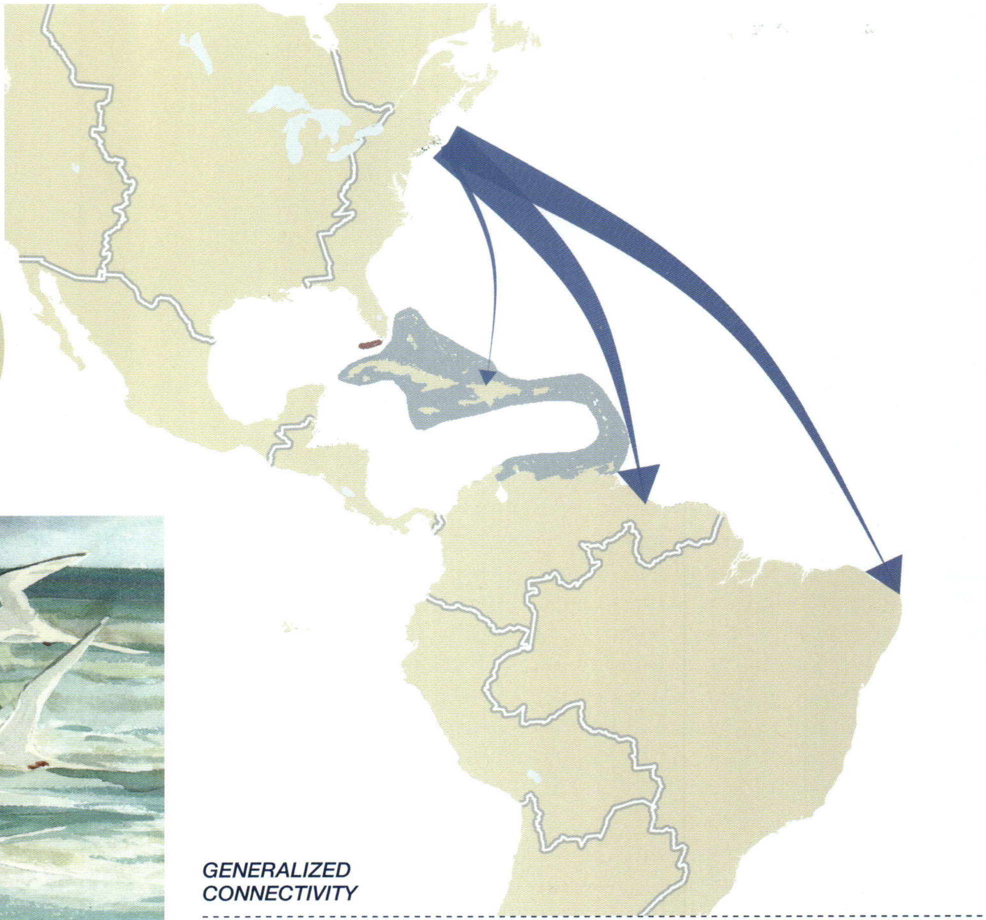

INTERESTING NOTE

At least two studies documented negative effects of tracking devices in this species (Paton et al. 2021; Mostello et al. 2014).

GEOLOCATOR TRACKING Birds from a Massachusetts colony were tracked to several nonbreeding sites along the coast of South America (Mostello et al. 2014). All tracked birds moved one or more times during the nonbreeding season.
SATELLITE TRACKING A study from Great Gull Island, New York, attempted to use satellite transmitters, but none of the transmitters functioned for more than seven days. No migratory connectivity information was obtained (Paton et al. 2021).
GENETICS A genetic study of three populations (Northwest Atlantic, Caribbean, Azores) found significant genetic differentiation between all three populations, indicating limited mixing and dispersal of birds from one population to another (Byerly et al. 2023)
MIGRATORY PATHWAYS Birds from a Massachusetts colony tracked with light-level geolocators showed diverse migratory pathways but with distinct stopover locations. The staging area post-breeding around Cape Cod appears important as birds fly directly over the Atlantic to a second stopover in Puerto Rico or the Dominican Republic before proceeding to nonbreeding areas (Mostello et al. 2014). During pre-breeding migration, birds migrate north through the western Caribbean, eventually returning to breeding islands in April and June (Gochfeld and Burger 2020; Mostello et al. 2014). The migratory routes for Caribbean-breeding birds are unknown. This species often spends significant time at sea during migratory periods.

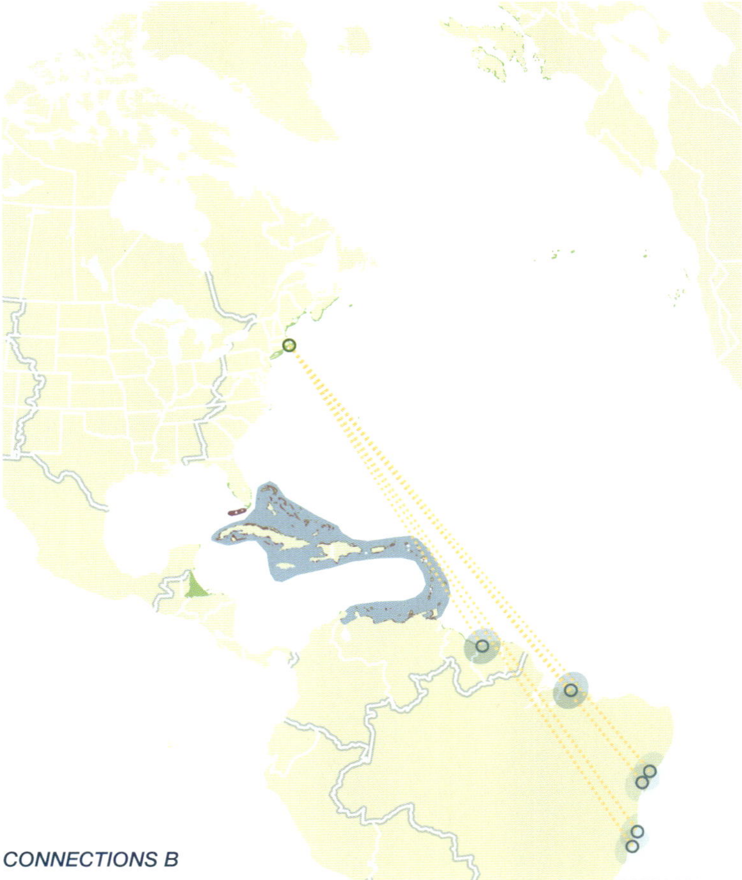

CONNECTIONS B

MC = 0.27

Sterna hirundo COMMON TERN

DISTRIBUTION AND ABUNDANCE

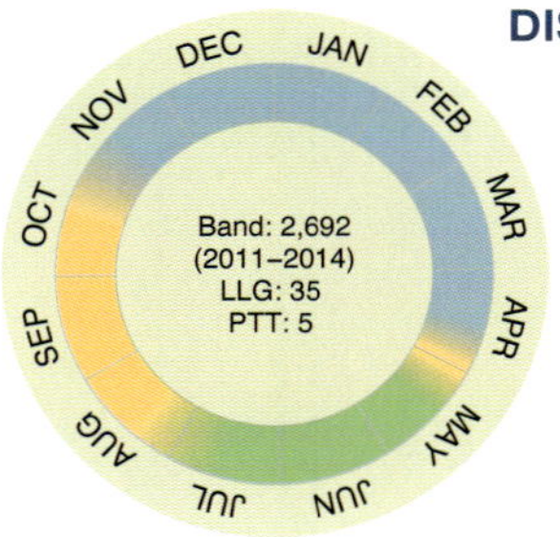

Common Terns breed in large numbers along the Atlantic coast south to North Carolina. Large breeding colonies also exist interiorly, in the Great Lakes region and throughout the Central and Prairie provinces in Canada. Nonbreeding areas span the coasts of Central and South America, with the main nonbreeding areas in Brazil and Peru. Smaller numbers occur on the US Gulf Coast and coastal areas of Florida, the Caribbean, and Mexico.

ORDER Charadriiformes
FAMILY Laridae
BREEDING HABITAT Coasts, Beach and Estuary
NONBREEDING HABITAT Coasts, Beach and Estuary
GLOBAL POPULATION SIZE 2,800,000
POPULATION CHANGE SINCE 1970 −60%
IUCN RED LIST CATEGORY Least Concern
R2R TIPPING POINT SPECIES No
US LISTING STATUS No
CANADA SPECIES AT RISK No
PIF CCS 10/20

MIGRATORY CONNECTIVITY

OVERALL The migratory connectivity of the Common Tern is well known for birds breeding in the northeast Atlantic and Great Lakes regions, mostly from band recovery, satellite transmitter, and geolocator data. Little is known about the migratory connectivity of birds breeding in other regions. In the Atlantic Flyway, Common Terns primarily spend the nonbreeding period along the north and east coast in northern South America (36%), Brazil (27%), and the Southern Cone region (24%). Midcontinent Flyway breeding populations primarily migrate to nonbreeding areas in Central America (30%), Andean West (17%), northern South America (16%), and the Atlantic Flyway (16%) and Mexico (7%). Breeding populations overlap on all nonbreeding areas, but some east/west separation of breeding populations on the nonbreeding areas results in moderate migratory connectivity (MC = 0.27).

SHOREBIRDS, AUKS, GULLS, AND TERNS (CHARADRIIFORMES)

	Andean West	Atlantic	Brazil	Caribbean	Central America	Mexico	Midcontinent	Northern SA	Southern Cone
Atlantic (n = 2243)	1%	1%	27%	11%	1%	<1%	<1%	36%	24%
Midcontinent (n = 244)	17%	16%	<1%	6%	30%	7%	7%	16%	<1%

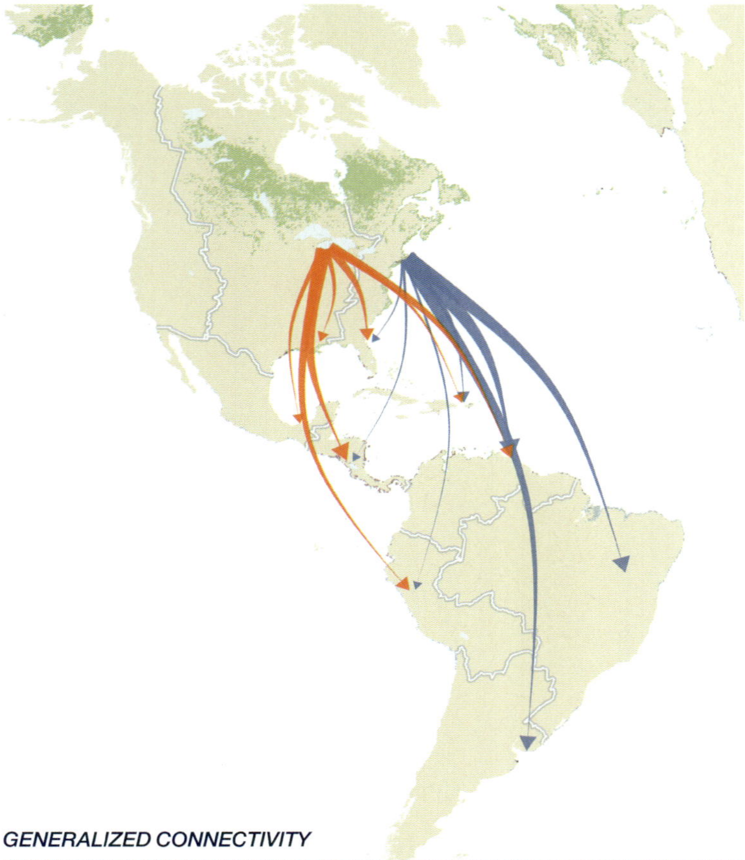

GENERALIZED CONNECTIVITY

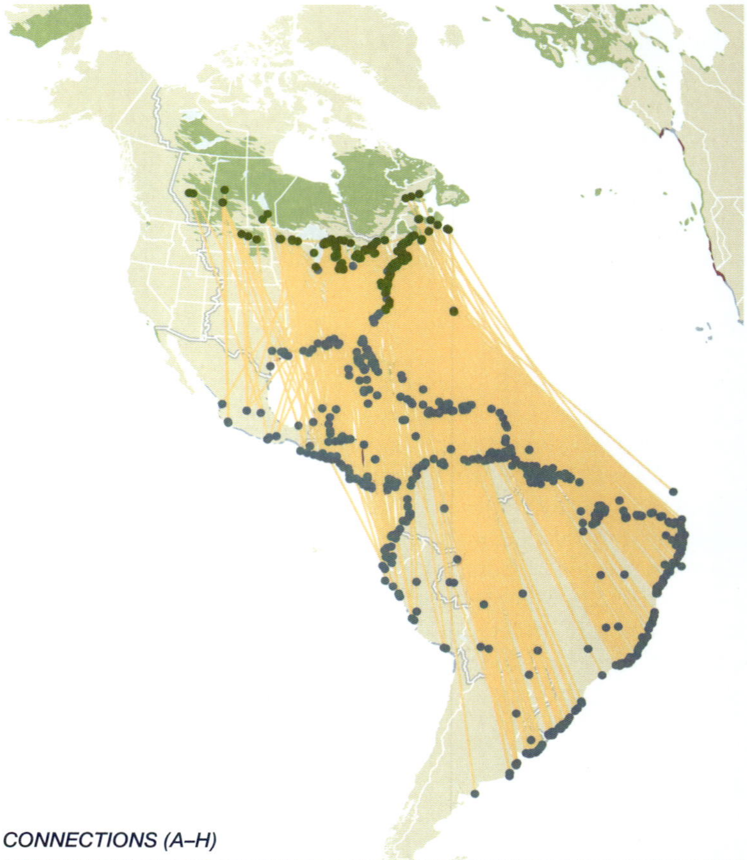

CONNECTIONS (A–H)

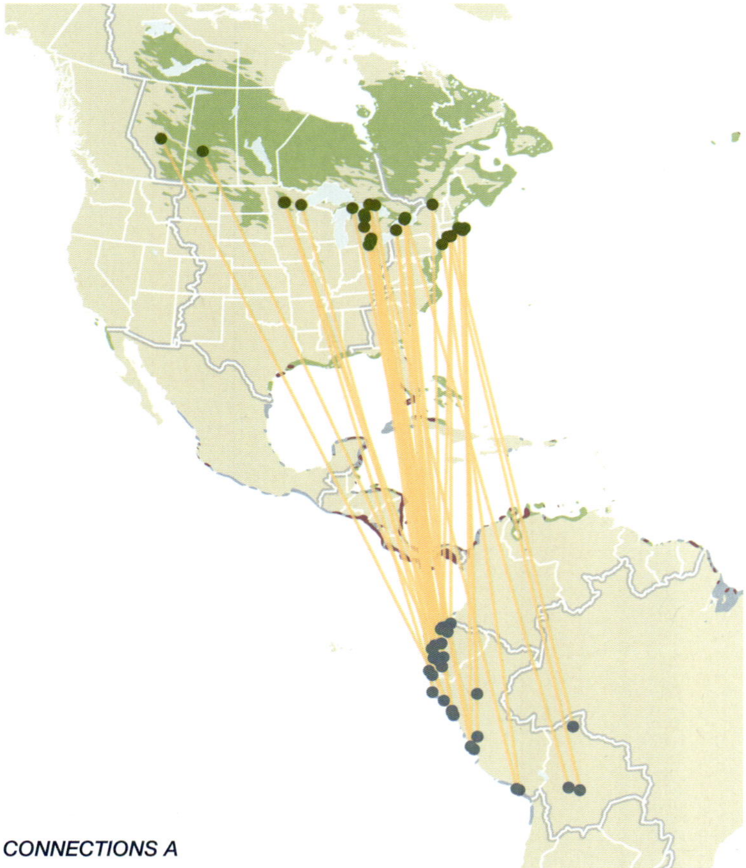

CONNECTIONS A

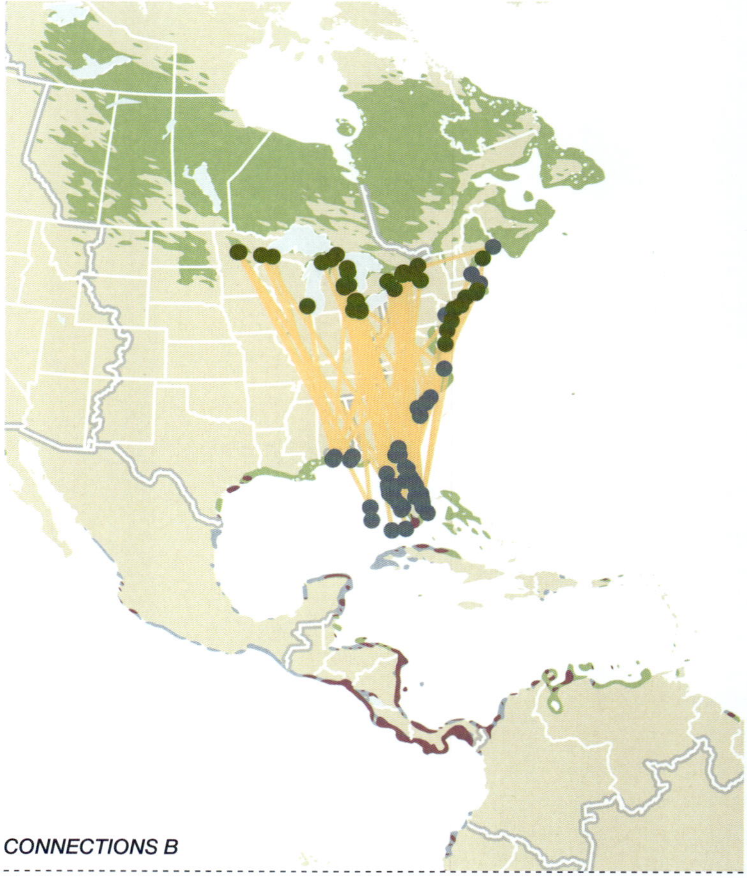

CONNECTIONS B

INTERESTING NOTE

Of the over 1 million Common Terns banded since 1924, most of them as nestlings, 2.5% have been reencountered away from their original banding site in subsequent breeding seasons. One example of a reencounter outside the Americas was a bird banded in July 1986 in New York recovered after entanglement in fishing gear 8,275 km away, off the coast of Ghana/Togo in February 1987.

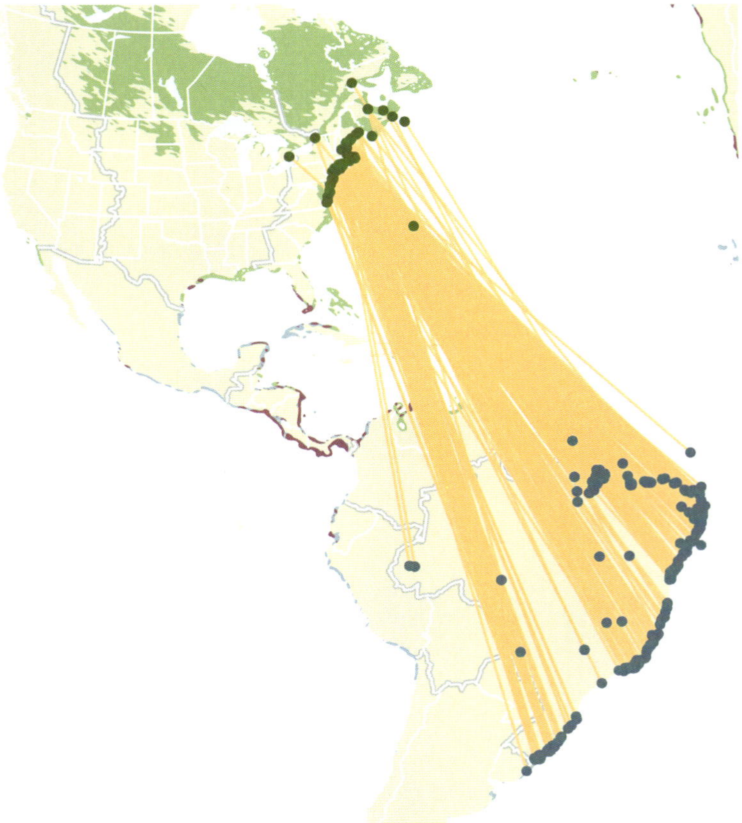

CONNECTIONS C

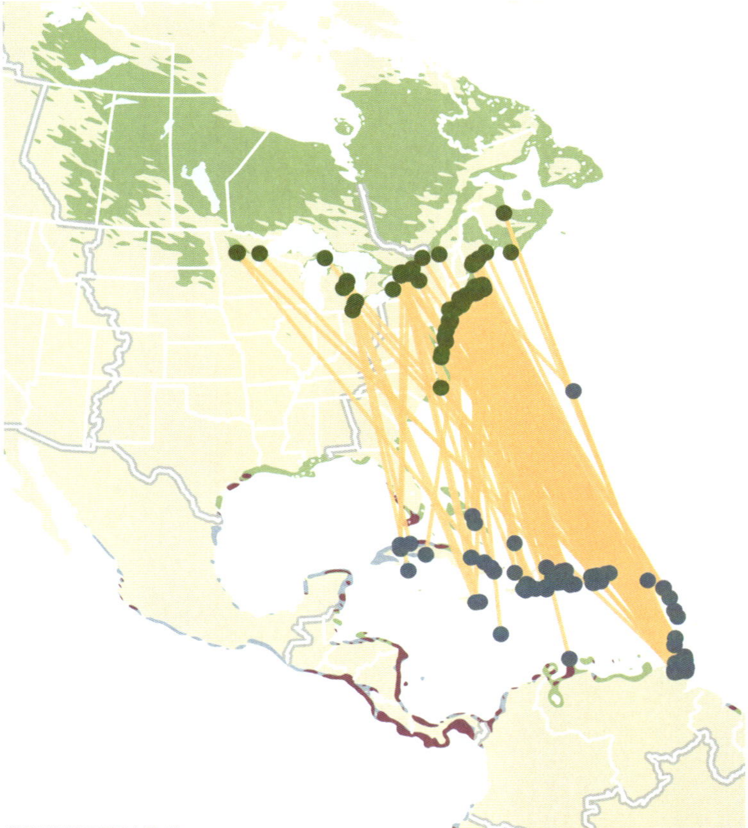

CONNECTIONS D

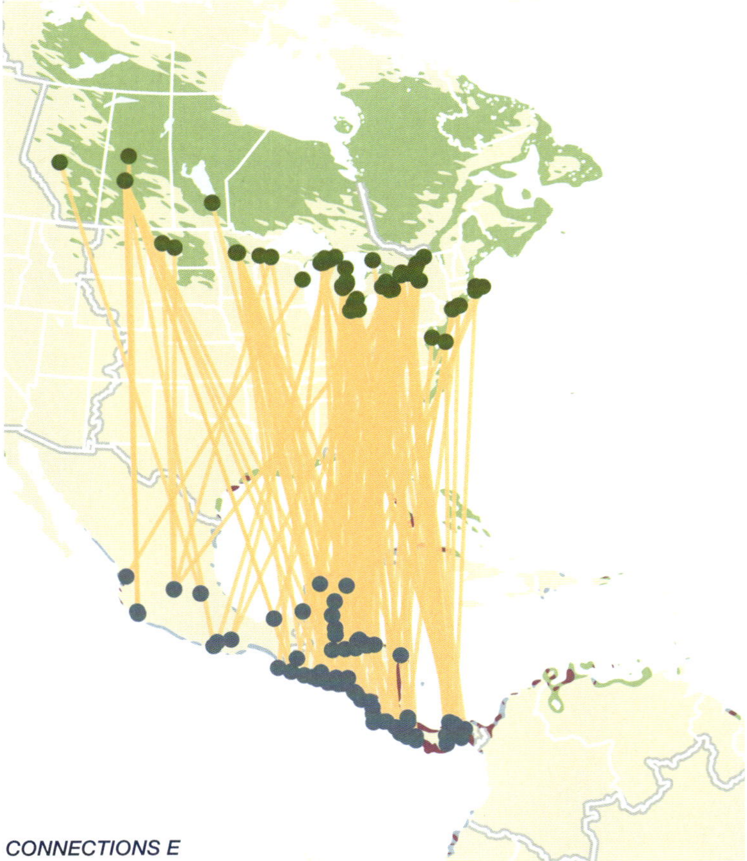

CONNECTIONS E

CONNECTIONS F

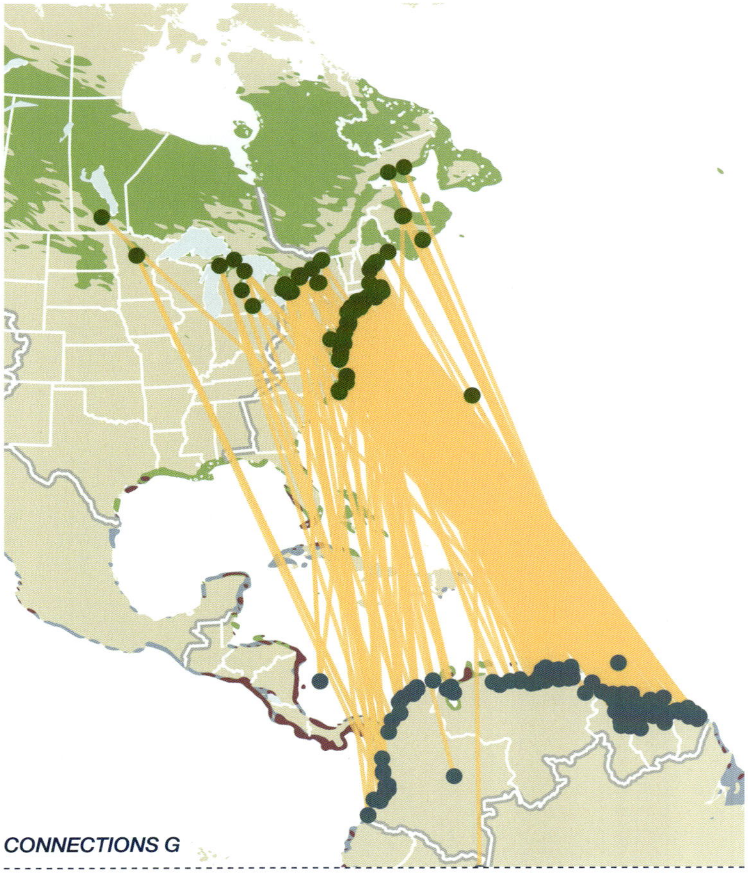

CONNECTIONS G

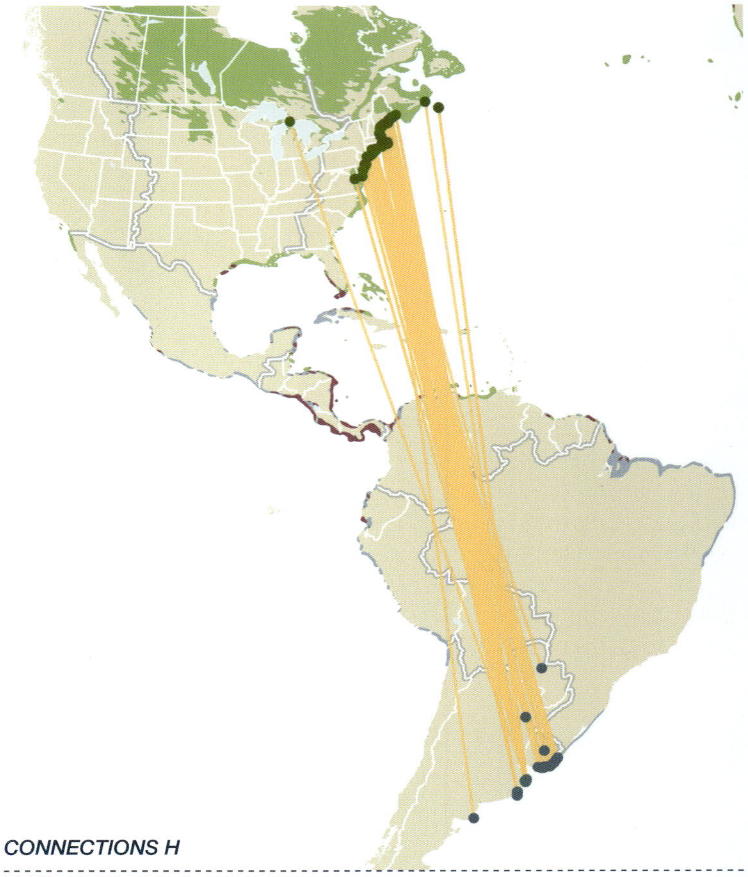

CONNECTIONS H

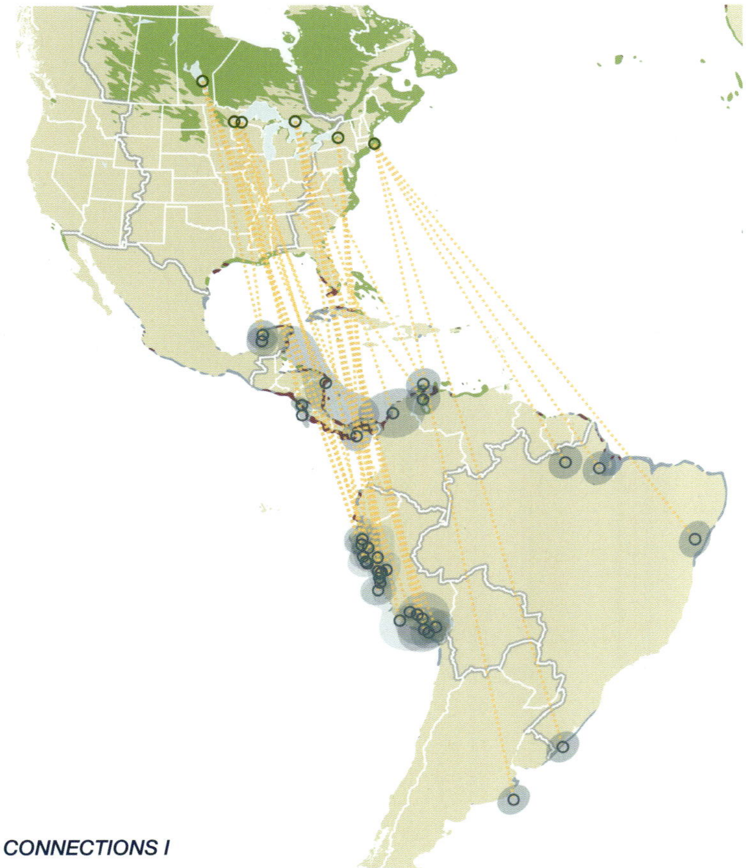

CONNECTIONS I

PATHWAYS

BANDING Banding data primarily comes from Common Terns banded within three main breeding areas: northeast Atlantic coast, Great Lakes region, and the Prairie provinces of Canada. Capture-recapture models (Cohen et al. 2014) from banding recovery data found that northeastern breeding terns were strongly connected to the east coast of South America, and breeding birds in the western part of the range were strongly tied to the western part of their nonbreeding range, with no linkage between western breeding birds and eastern South America. Great Lakes–breeding Common Terns were equally connected to eastern and western portions of the nonbreeding range.

GEOLOCATOR TRACKING Nonbreeding locations of Common Terns tracked from the northeast United States confirms banding data, identifying four nonbreeding areas on the north and east coasts of South America. Three birds remained within discrete areas during the nonbreeding season, whereas two ranged more widely (Nisbet et al. 2011). Common Terns tracked from inland breeding colonies in Manitoba, Canada, and the Great Lakes regions of the United States and Canada primarily migrated to Peru for the nonbreeding season, while the remainder spread throughout the Gulf of Mexico, Central America, and northwestern South America (Bracey et al. 2018).

SATELLITE TRACKING Common Terns tracked from Petit Manan Island, Maine, migrated to Brazil and Argentina (L. Welch, unpublished data).

OTHER APPROACHES No genetic or stable isotope studies of migratory connectivity were found for this species.

MIGRATORY PATHWAYS Birds breeding along the Atlantic coast concentrate at staging areas, especially around Cape Cod, Massachusetts, before flying across the Atlantic Ocean to the West Indies (Arnold et al. 2020). From here, birds may move across the Caribbean or fly directly to nonbreeding areas along the coast of South America (Arnold et al. 2020). Birds tracked from breeding colonies in Maine flew south over the open ocean until reaching Puerto Rico and other islands in the Lesser Antilles, then over to the coast of Venezuela, and followed the coast south to nonbreeding areas (L. Welch, unpublished data). Two strategies for Common Terns breeding in the interior have been documented (Bracey et al. 2018; Arnold et al. 2020). The first is a migration east over land to the Atlantic coast and then south along the coast to nonbreeding areas in Florida and Central and South America. The second is a migration directly south over land from the Great Lakes to the Gulf Coast and then south to nonbreeding areas. Bracey et al. (2018) identified eight staging areas from the Atlantic coast to northeastern Brazil, including major staging areas at the lower Great Lakes, along the Atlantic coast, Florida, and the Gulf of Mexico. Recent Motus studies may help identify other migratory pathways for Common Terns.

DATA CONTRIBUTORS

A.M. Bracey; E. Craig; P. Curtis; A. McKellar; D. Moore; K. Myers; P. Shannon; L. Welch.

Mantel = −0.08 ± 0.04

Sterna paradisaea ARCTIC TERN

ORDER Charadriiformes
FAMILY Laridae
BREEDING HABITAT Wetlands, Tundra
NONBREEDING HABITAT Coasts, Marine Waters
GLOBAL POPULATION SIZE 3,000,000
POPULATION CHANGE SINCE 1970 −40%
IUCN RED LIST CATEGORY Least Concern
R2R TIPPING POINT SPECIES No
US LISTING STATUS No
CANADA SPECIES AT RISK No
PIF CCS 12/20

DISTRIBUTION AND ABUNDANCE

This species is a circumpolar Arctic breeder that generally nests coastally, but also inland, in a variety of habitats and colony types (can be highly diffuse or extremely numerous). Colonies often move from year to year, resulting in limited abundance estimates across its range (J.J. Hatch et al. 2020b); eBird models suggest highest abundance in Canada, Iceland, and Arctic Europe (Fink et al. 2023). The Arctic Tern is a trans-equatorial migrant to Southern Ocean nonbreeding areas, where it is associated with pack ice (J.J. Hatch et al. 2020b).

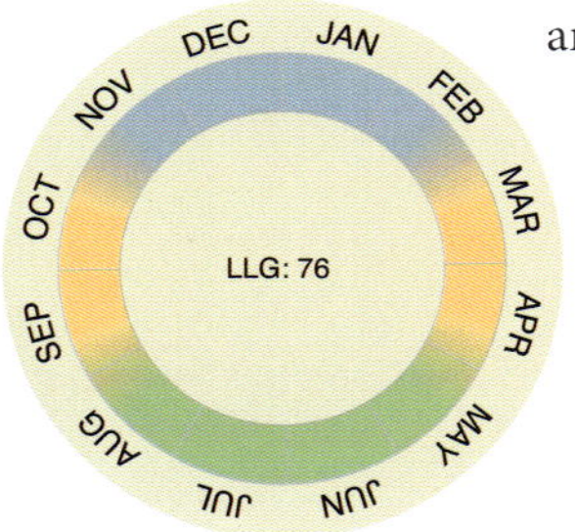

INTERESTING NOTE

Studies from European colonies (Fijn et al. 2013; Alerstam et al. 2019; Redfern et al. 2021) have tracked individuals to core nonbreeding habitats much farther east around Antarctica than those shown here, with presumed weak connectivity for those colonies, but implying stronger connectivity across the Atlantic Ocean colonies. However, gaps in knowledge exist for Russian colonies, and more information about colonies in the Pacific Ocean would be beneficial.

MIGRATORY CONNECTIVITY

OVERALL This is one of the best-studied seabirds from a migratory connectivity perspective due to tracking from multiple colonies, countries, and ocean basins. There is weak connectivity of North American breeding colonies to core nonbreeding regions (Mantel = −0.08 ± 0.04), although there was a relatively small number of individuals tracked from each colony (n = 4–19). All populations used core nonbreeding areas in the Antarctic (20%–100% across colonies), and most used nonbreeding areas in the Austral Polar Zones (32%–77% across colonies). However, birds originating from Atlantic Ocean colonies were predominantly east of the Antarctic Peninsula, while birds from Alpine, Alaska, US, used areas west of the peninsula. Some birds tracked from Harriman Fjord in southeastern Alaska crossed into the Atlantic Ocean and overlapped spatially with core areas of birds tracked from Atlantic Ocean colonies; 20% of their core areas were in the Brazilian Coastal Current.

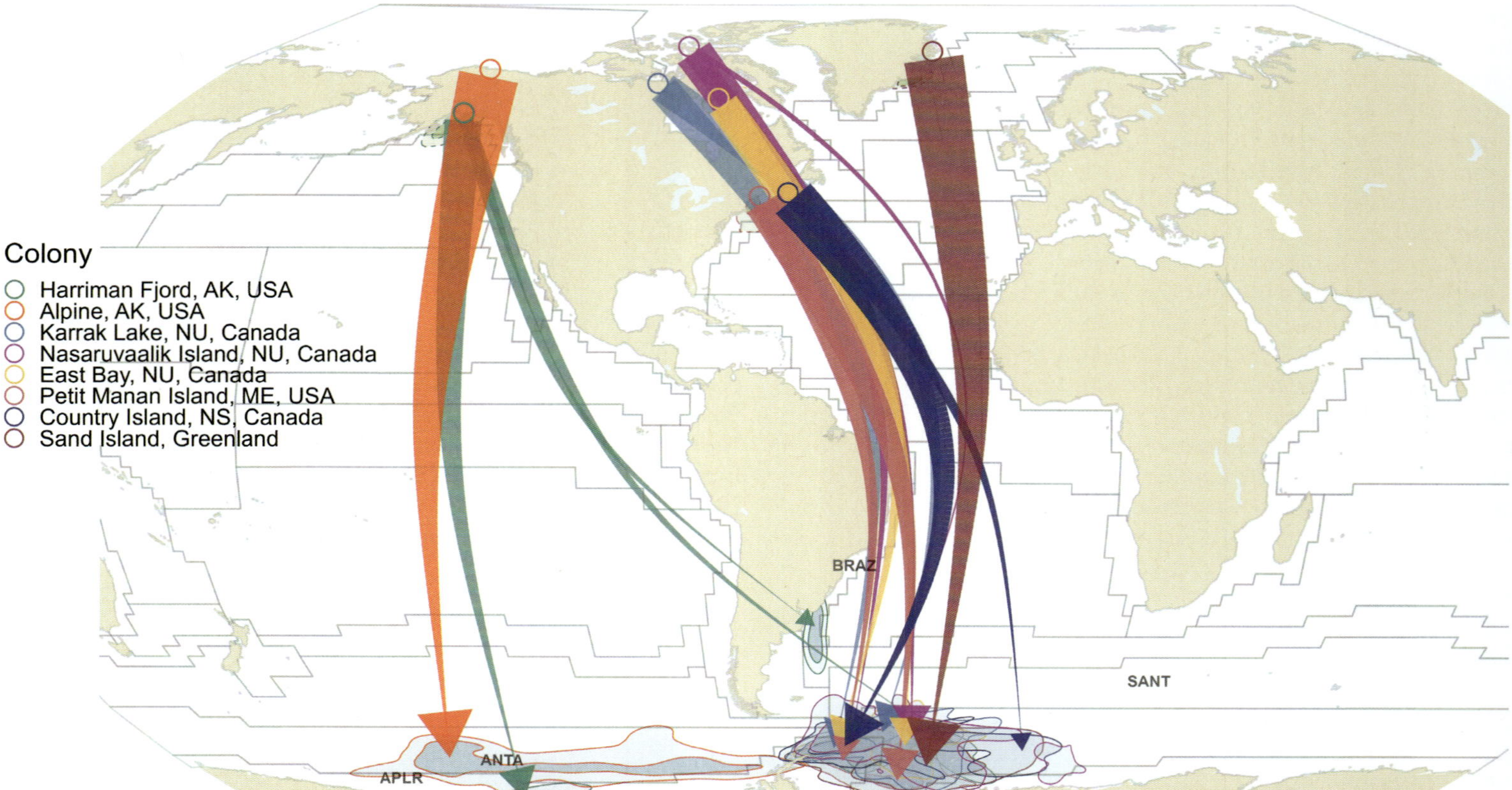

GENERALIZED CONNECTIVITY

GEOLOCATOR TRACKING Geolocator tracking of Arctic Terns from Greenland revealed that this species makes the longest recorded migration of any animal in the world (Egevang et al. 2010).

Additional studies contributed to this Atlas include colonies that span the North American breeding range longitudinally and latitudinally (McNight et al. 2013; Loring et al. 2017b; Wong et al. 2021; Welch

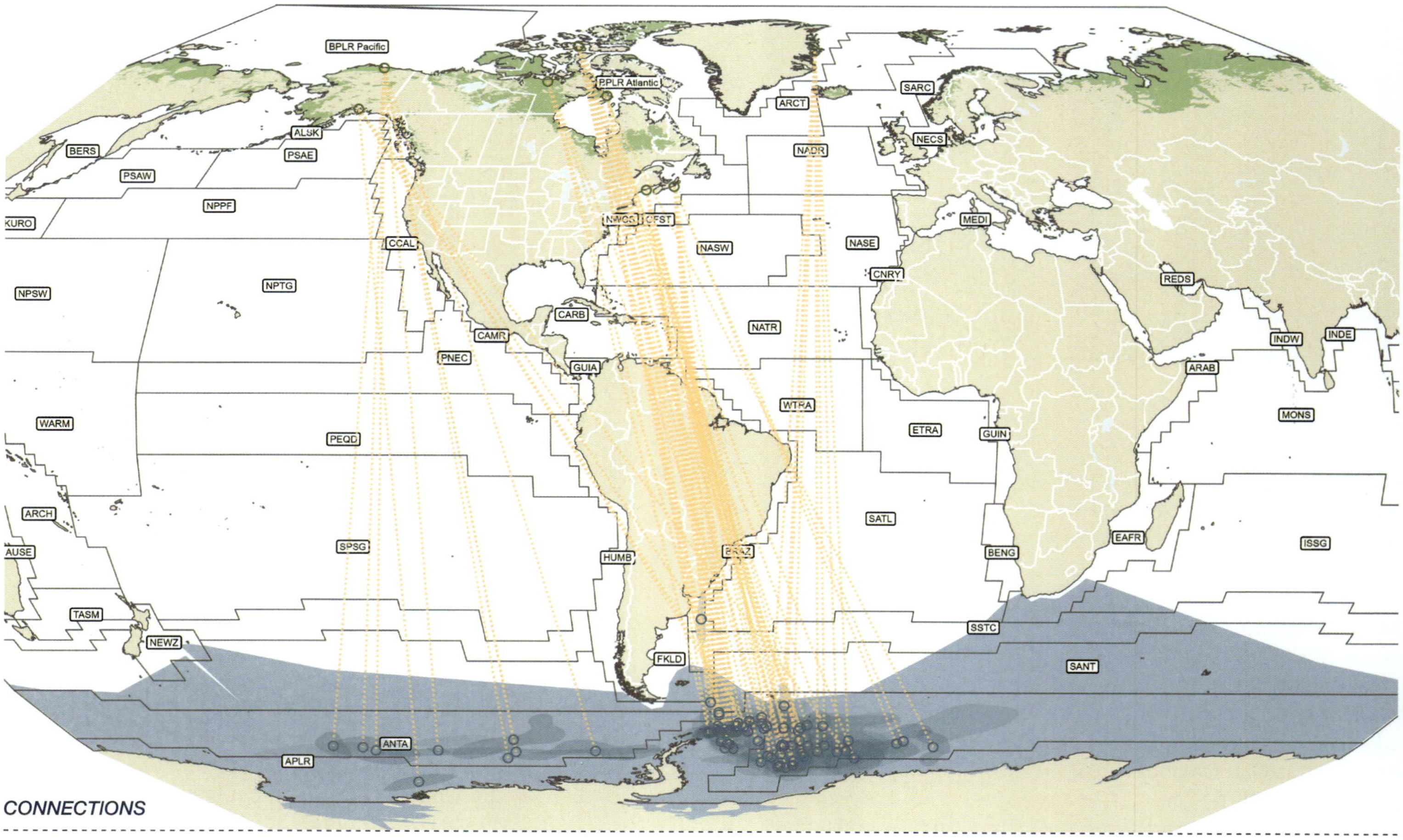

CONNECTIONS

	SANT	ANTA	APLR	BRAZ
Harriman Fjord, AK, USA (n = 5)	–	20%	60%	20%
Alpine, AK, USA (n = 8)	–	100%	–	–
Karrak Lake, NU, Canada (n = 4)	–	50%	50%	–
Nasaruvaalik Island, NU, Canada (n = 19)	5%	63%	32%	–
East Bay, NU, Canada (n = 6)	–	50%	50%	–
Petite Manan Island, ME, USA (n = 13)	–	62%	38%	–
Country Island, NS, Canada (n = 13)	–	23%	77%	–
Sand Island, Greenland (n = 8)	–	100%	–	–

et al., unpublished data). Redfern (2021) demonstrated complete separation of members of breeding pairs during the nonbreeding period.

SATELLITE TRACKING Two Arctic Terns were tracked by GPS from Iceland during the full annual cycle, providing increased accuracy, flight altitude, and further insights into how Arctic Terns use their habitat now and how this may change in the future (Morten et al. 2023).

GENETICS No genetic studies of connectivity were found. The complete mitochondrial genome of the Arctic Tern has been sequenced from a single specimen (Skujina et al. 2019).

OTHER APPROACHES No banding data or stable isotope studies of migratory connectivity were found for this species.

MIGRATORY PATHWAYS Within each ocean basin, Arctic Terns generally share similar migration routes (Wong et al. 2021). Routes split in the central Atlantic to either follow the coast of West Africa or the coast of Brazil. Northward routes are similar across colonies through the central Atlantic. A notable exception to this pattern includes the extensive eastward movements across the southern Indian Ocean that some birds make after following the African coast south (Egevang et al. 2010; Fijn et al. 2013). In the Pacific Ocean, birds fly south along the Americas, but only one study has data for the pre-breeding migration. In the Pacific Ocean, common staging areas are also used in the Gulf of Alaska and throughout the upwelling currents of the eastern Pacific. Arrival to nonbreeding and breeding areas coincides with increases in local feeding conditions (Wong et al. 2022).

DATA CONTRIBUTORS

A.J. Allyn; D.C. Duffy; C. Egevang; A.-L. Harrison; D.B. Irons; P. Loring; M.S. Maftei; M.L. Mallory; A. McKnight; K. Myers; P. Shannon; I.J. Stenhouse; L. Welch.

Sterna forsteri FORSTER'S TERN

ORDER Charadriiformes
FAMILY Laridae
BREEDING HABITAT Wetlands, Freshwater Marsh
NONBREEDING HABITAT Coasts, Beach and Estuary

GLOBAL POPULATION SIZE 150,000
POPULATION CHANGE SINCE 1970 −45%
IUCN RED LIST CATEGORY Least Concern
R2R TIPPING POINT SPECIES No
US LISTING STATUS No
CANADA SPECIES AT RISK No
PIF CCS 13/20

DISTRIBUTION AND ABUNDANCE

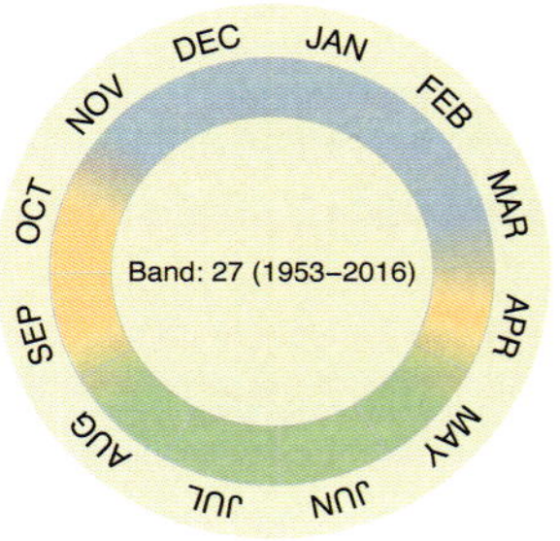

This small tern has breeding concentrations along the Atlantic coasts, the Gulf of Mexico, California, the Gulf of California, and inland sites as far north as central Canada. Inland breeding sites of importance include in the Prairie Pothole Region, the Great Salt Lake, and central Oregon. While it can be found along southern coasts year-round, this species moves out of most inland sites and northern coastal areas during the nonbreeding season and is primarily found along coastal areas of the southeast Gulf and both the Atlantic and Mexican coasts (McNicholl et al. 2001).

MIGRATORY CONNECTIVITY

OVERALL Migratory connectivity is known from band reencounter data. Migratory connectivity appears moderate across flyway breeding regions and strong between eastern and western populations, but MC could not be calculated due to limited sample size. Atlantic Flyway breeding birds mainly migrate to coastal nonbreeding areas within the Atlantic Flyway (85%), as well as the Midcontinent Flyway (15%). Midcontinent Flyway breeding birds also migrate to coastal nonbreeding areas in the Atlantic (89%) and Midcontinent (11%) flyways. Pacific Flyway breeding birds spend the nonbreeding season in Mexico (50%), the Pacific Flyway (25%), and Central America (25%).

BANDING Banding records show a probable east-west divide for this species. Birds breeding in the San Francisco area were found during the nonbreeding season in Southern California and Mexico (Gill and Mewaldt 1979). Breeding birds banded in the vicinity of the Great Lakes and the Atlantic coast were found in the southern Atlantic states and the Gulf Coast during the nonbreeding season.

OTHER APPROACHES No satellite, genetic, or stable isotope studies of migratory connectivity were found for this species.

MIGRATORY PATHWAYS The number of birds breeding in the Atlantic and Gulf of Mexico ebb and flow along these coasts with the seasons, with some birds present year-round in much of the area. Central inland breeders move along a broad front through the central plains, with stopovers in marshy areas throughout much of the central states, and spend the nonbreeding season mainly along the Gulf of Mexico. Birds breeding west of the Rockies concentrate in the nonbreeding season in San Francisco Bay, the Pacific coast from Southern California through Baja California Sur, and the Gulf of California (McNicholl et al. 2001; Fink et al. 2023).

	Atlantic	Central America	Mexico	Midcontinent	Pacific
Atlantic (n = 13)	85%	–	–	15%	–
Midcontinent (n = 9)	89%	–	–	11%	–
Pacific (n = 4)	–	25%	50%	–	25%

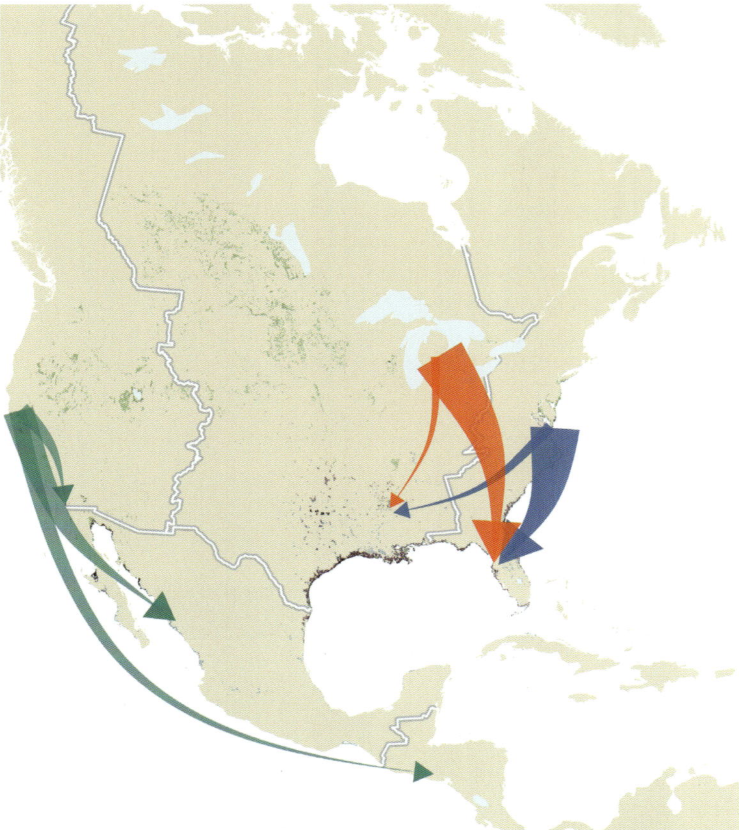

GENERALIZED CONNECTIVITY

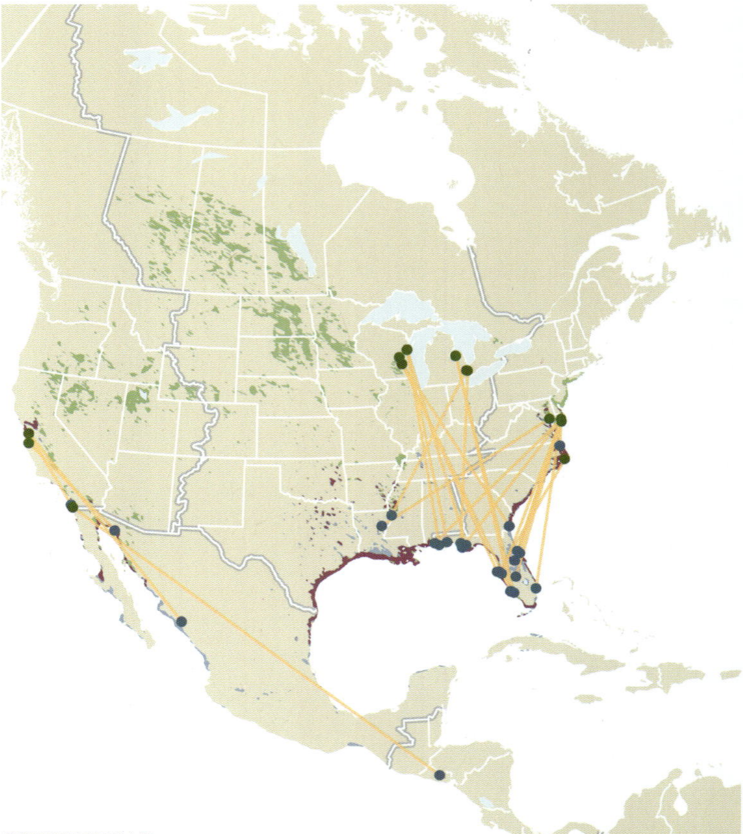

CONNECTIONS

ROYAL TERN *Thalasseus maximus*

MC = 0.02 ± 0.03*

DISTRIBUTION AND ABUNDANCE

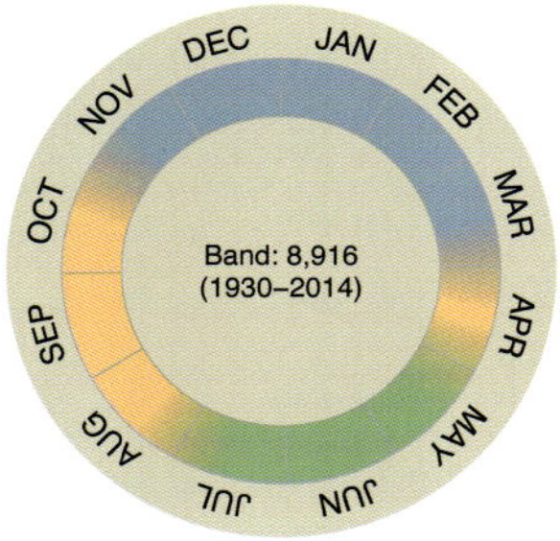

This tern is widespread in the Western Hemisphere, with breeding colonies in both North and South America. In the Northern Hemisphere, this species breeds from Virginia southward in the Atlantic, along the Gulf Coast, in the Caribbean, and along the Pacific coast of Mexico. In the nonbreeding season, this species continues to inhabit much of its southern breeding range but departs from the northern areas of the Atlantic coast (Buckley and Buckley 2002; Fink et. al 2023).

MIGRATORY CONNECTIVITY

OVERALL Migratory connectivity is known from banding reencounter data. Atlantic Flyway breeding birds largely migrate to nonbreeding areas within the Atlantic Flyway (73%), with others migrating to the Caribbean (15%) or South America. Midcontinent Flyway breeding birds do not favor any one nonbreeding area and disperse across all regions. Mixing of breeding populations across all nonbreeding areas results in weak migratory connectivity (MC = 0.02 ± 0.03).

ORDER Charadriiformes
FAMILY Laridae
BREEDING HABITAT Coasts, Beach and Estuary
NONBREEDING HABITAT Coasts, Beach and Estuary
GLOBAL POPULATION SIZE 55,000
POPULATION CHANGE SINCE 1970 +4%
IUCN RED LIST CATEGORY Least Concern
R2R TIPPING POINT SPECIES No
US LISTING STATUS No
CANADA SPECIES AT RISK No
PIF CCS 11/20

	Andean West	Atlantic	Caribbean	Central America	Mexico	Midcontinent	Northern SA
Atlantic (n = 8378)	1%	73%	15%	6%	1%	<1%	4%
Midcontinent (n = 74)	3%	36%	7%	27%	9%	9%	8%

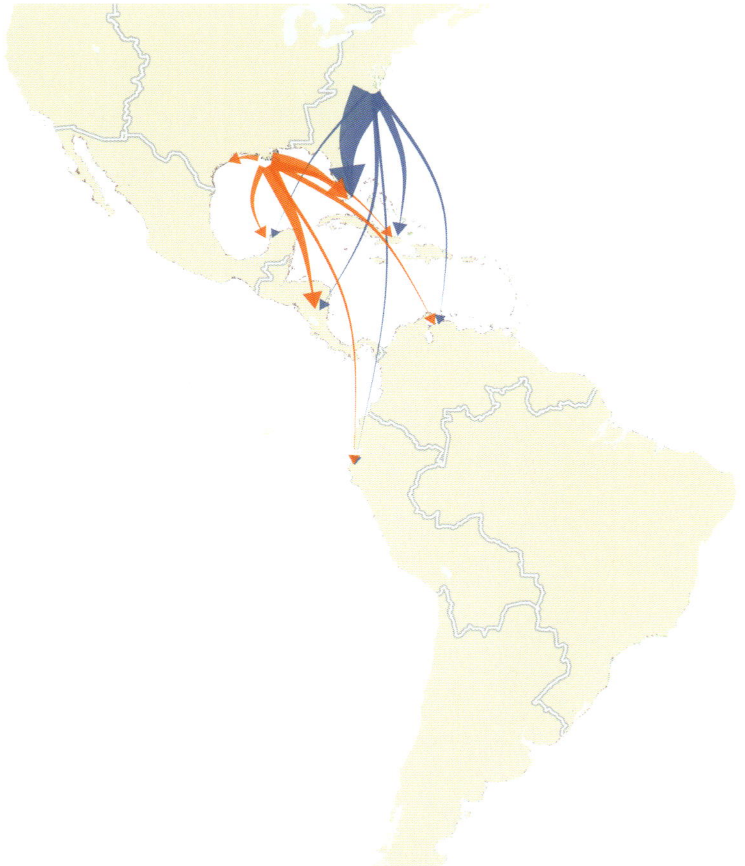

GENERALIZED CONNECTIVITY

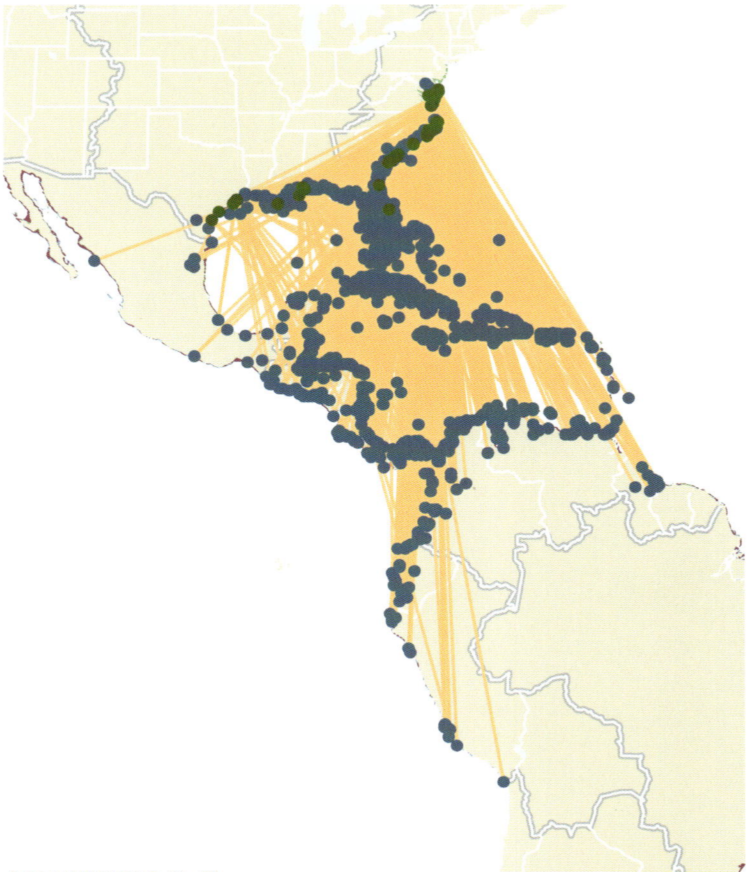

CONNECTIONS (A–F)

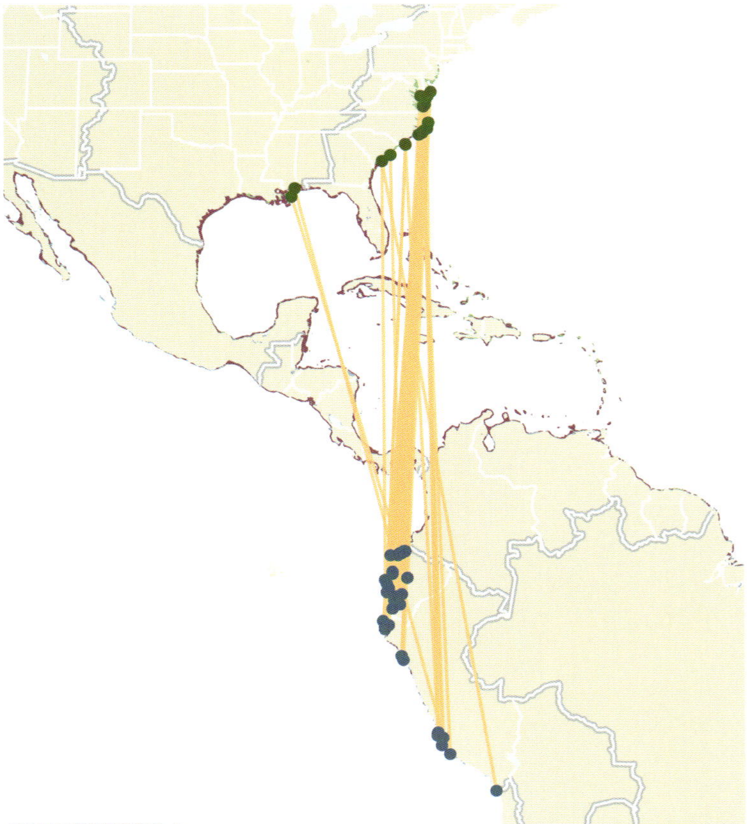

CONNECTIONS A

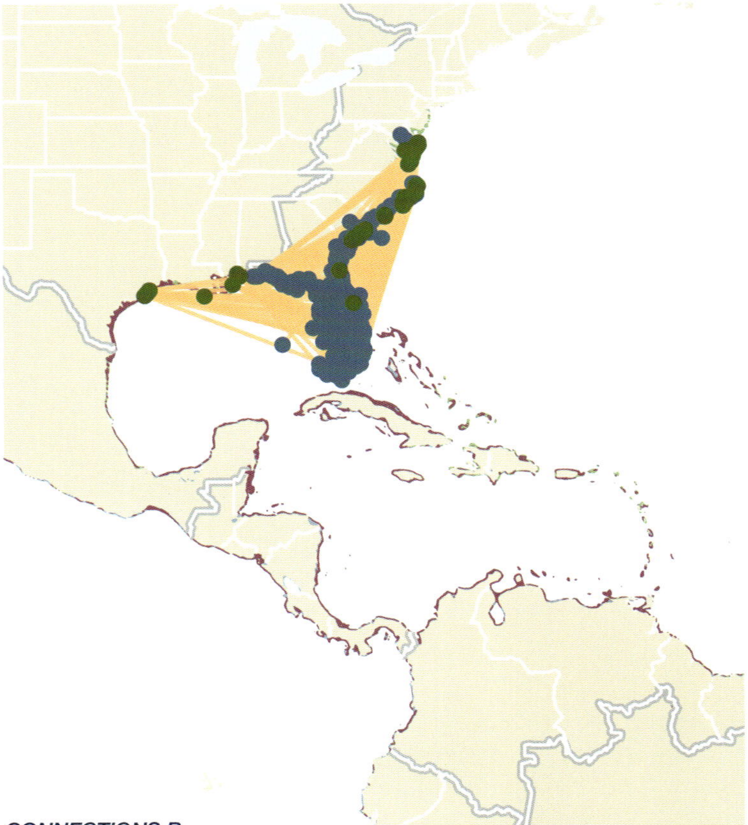

CONNECTIONS B

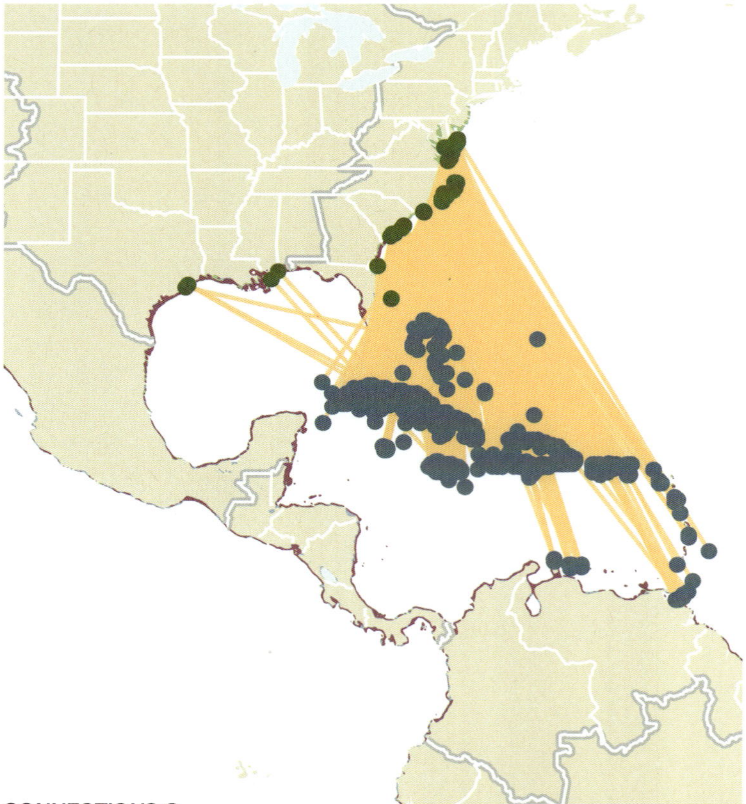

CONNECTIONS C

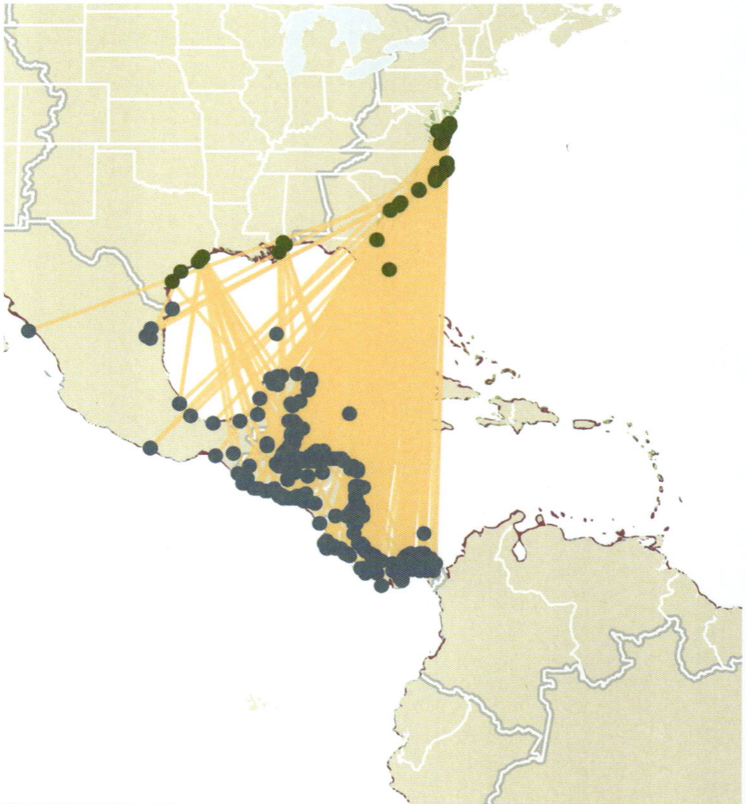

CONNECTIONS D

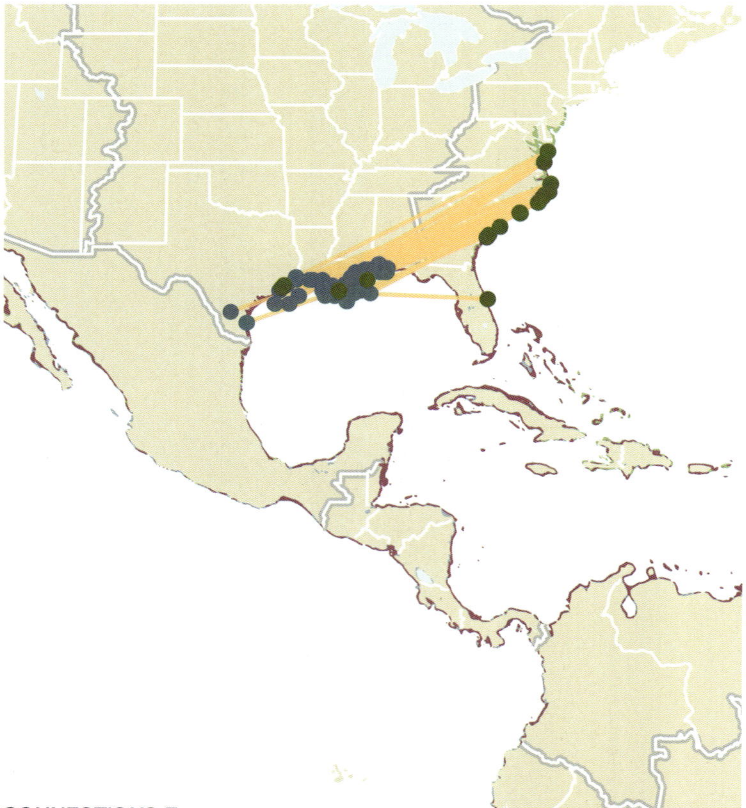

CONNECTIONS E

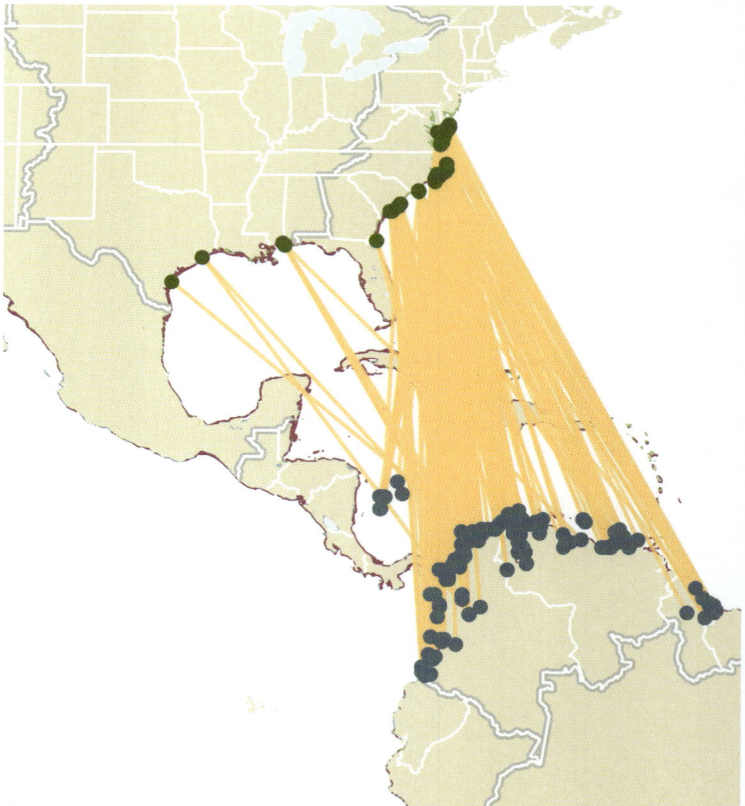

CONNECTIONS F

BANDING Royal Tern have been banded at colonies in the northern part of the Atlantic population and along the Gulf Coast. Birds from the northern Atlantic colonies largely moved south, with Florida and the Caribbean being especially important nonbreeding areas. Many birds from the Gulf of Mexico also moved eastward into the Atlantic during the nonbreeding season. Gulf of Mexico birds were more likely than Atlantic birds to move westward to Mexico, Central America, and northern South America, but both banded populations were represented in these areas.

OTHER APPROACHES No satellite, genetic, or stable isotope studies of migratory connectivity were found for this species.

MIGRATORY PATHWAYS Most data are from banding reencounters, so there is limited information on migratory pathways.

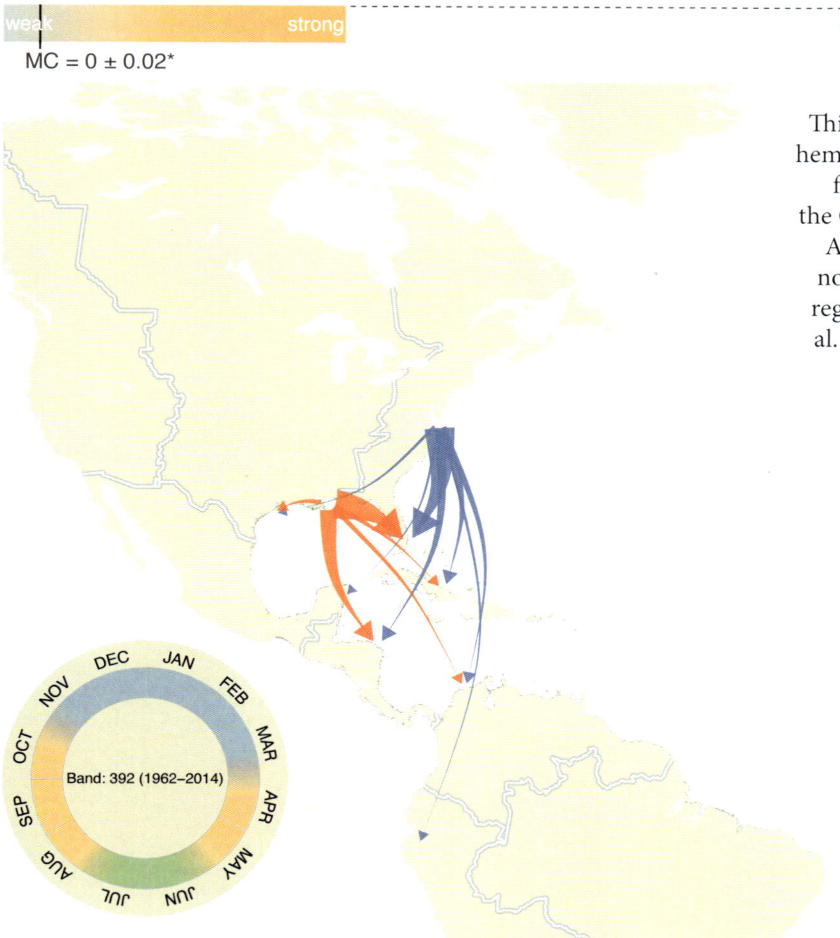

GENERALIZED CONNECTIVITY

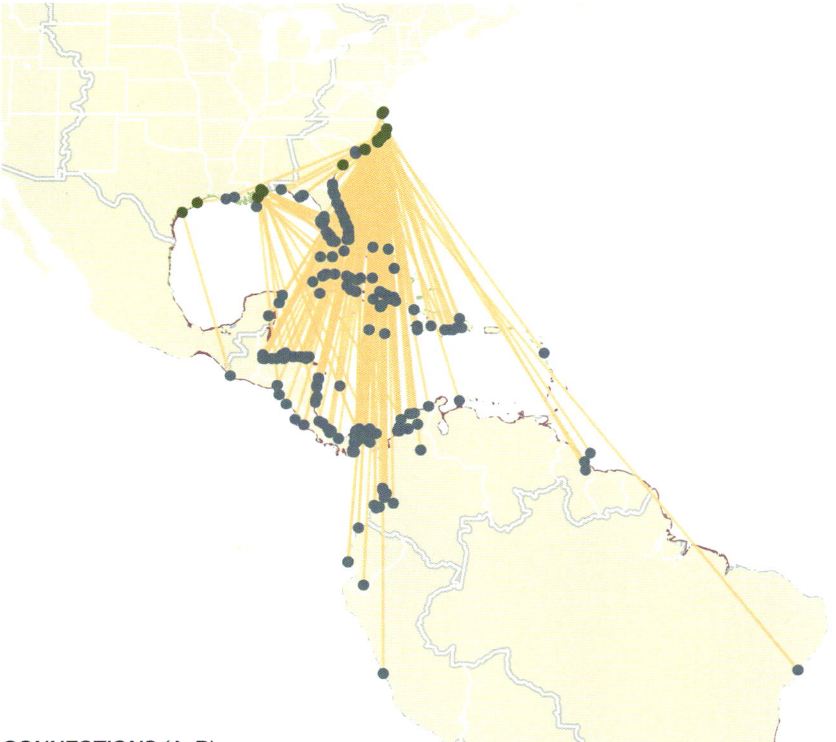

CONNECTIONS (A–B)

Thalasseus sandvicensis **SANDWICH TERN**

DISTRIBUTION AND ABUNDANCE

This very widespread species has a distribution that includes both hemispheres. In the Western Hemisphere, it breeds in coastal areas from Maryland southward along the Atlantic coast, throughout the Gulf of Mexico, Mexico, the Caribbean, and Central and South America. The species also inhabits most of this range during the nonbreeding season, but birds depart from the northern Atlantic region and the central Gulf of Mexico (Shealer et al. 2016; Fink et al. 2023). Florida and the southern Atlantic coast of the US are of importance as a nonbreeding area for this species.

ORDER Charadriiformes
FAMILY Laridae
BREEDING HABITAT Coasts, Beach and Estuary
NONBREEDING HABITAT Coasts, Beach and Estuary
GLOBAL POPULATION SIZE 1,000,000
POPULATION CHANGE SINCE 1970 +81%
IUCN RED LIST CATEGORY Least Concern
R2R TIPPING POINT SPECIES No
US LISTING STATUS No
CANADA SPECIES AT RISK No
PIF CCS 10/20

MIGRATORY CONNECTIVITY

OVERALL Migratory connectivity is known from banding reencounter data. The majority of Atlantic Flyway breeding birds migrate to nonbreeding areas within the Atlantic Flyway (56%); however, the other half disperse across nonbreeding areas, including Central America (16%), the Caribbean (15%), Northern South America (9%), and smaller percentages to all other regions. There is a similar pattern for Midcontinent Flyway breeders, with the majority of breeding birds migrating to the Atlantic Flyway (57%) and Central America (29%), and smaller percentages across other regions. The high degree of mixing across many nonbreeding regions leads to weak migratory connectivity (MC = 0.0 ± 0.02). ***BANDING*** Banding data for this species comes from colonies in the middle and southern Atlantic states and the Gulf of Mexico. Most Atlantic birds moved southward but remained within the Atlantic region. Many Midcontinent breeders moved eastward, overlapping with Atlantic breeders in the same nonbreeding region. Some birds of both populations

	Andean West	Atlantic	Brazil	Caribbean	Central America	Mexico	Midcontinent	Northern SA
Atlantic (n = 349)	2%	56%	<1%	15%	16%	1%	1%	9%
Midcontinent (n = 21)	–	57%	–	5%	29%	–	5%	5%

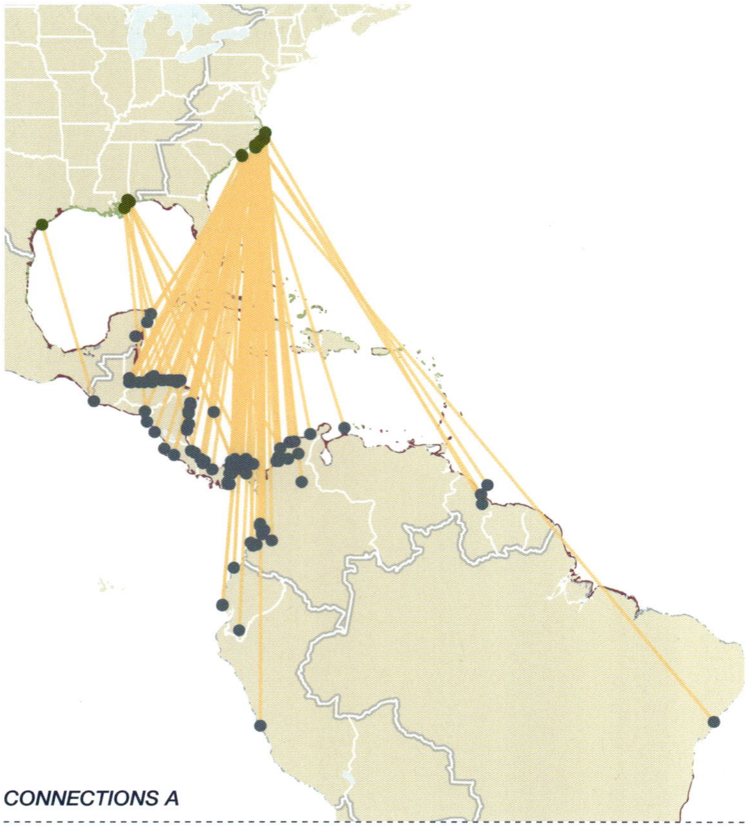

CONNECTIONS A

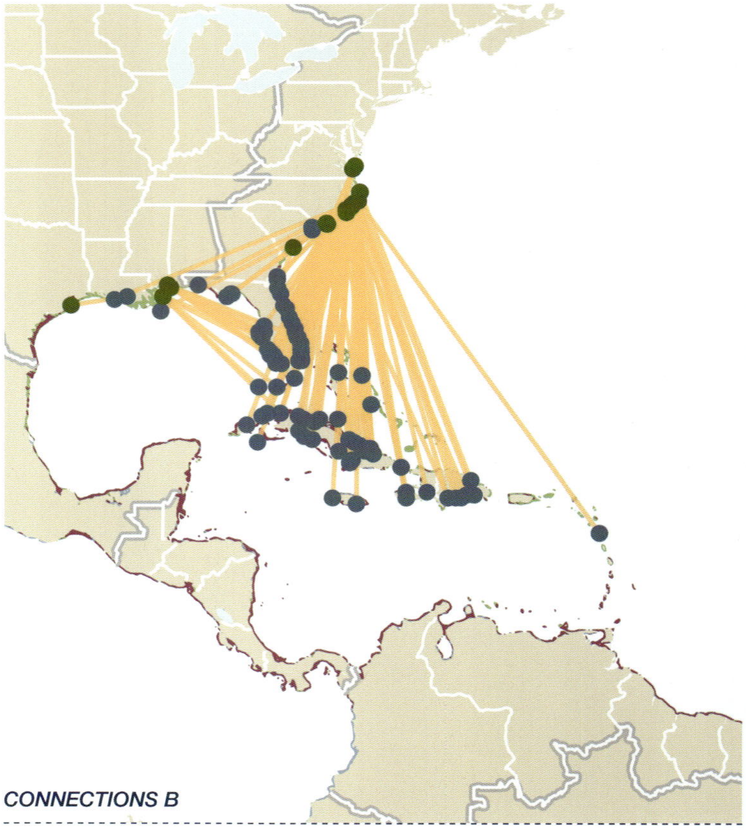

CONNECTIONS B

moved farther south to the Caribbean, Mexico, Central America, and northern South America. A small percentage of individuals from both populations migrate farther still, with records of Midcontinent breeders going to the Andean West, and a record of an Atlantic breeder in Brazil. Midcontinent breeders migrated into the southernmost reaches of the nonbreeding range more often than Atlantic breeders.

OTHER APPROACHES No satellite, genetic, or stable isotope studies of migratory connectivity were found for this species.

MIGRATORY PATHWAYS Atlantic birds often move north post-breeding before returning south. Midcontinent breeders also wander post-breeding, before moving eastward to the Atlantic (overlapping with Atlantic breeders) or southward to the rest of their nonbreeding range. Many birds of both populations also migrate south to the Caribbean, Mexico, Central America, and northern South America, with some individuals crossing over Central America to spend the nonbreeding season on the Pacific coast (Shealer et al. 2016; Fink et al. 2023).

ELEGANT TERN *Thalasseus elegans*

DISTRIBUTION AND ABUNDANCE

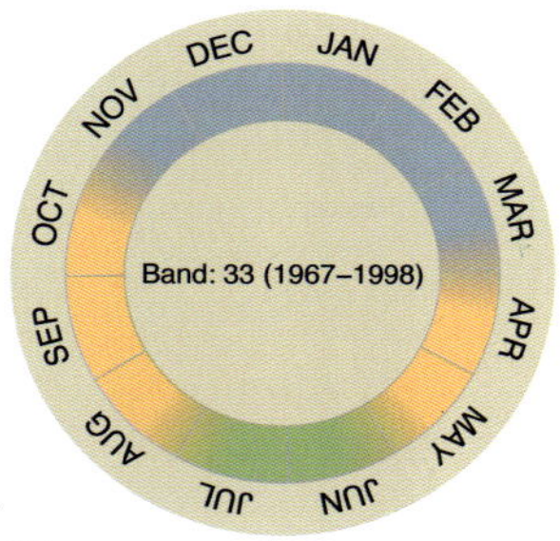

This medium-sized crested tern has a limited breeding range in California and northern Mexico, with one colony, at Isla Rasa, Mexico, containing over 90% of the breeding population. After breeding, this species disperses north and south, with a notable movement up the Pacific coast. Movements are likely tied to prey resources and oceanic factors. During the nonbreeding season, this species occurs widely along the Central American and South American Pacific coasts (Burness et al. 1999).

MIGRATORY CONNECTIVITY

OVERALL Birds breeding at colonies in the Pacific Flyway have a wide distribution of nonbreeding areas, with most birds migrating to Central America (69%) and northern South America (19%). MC is not able to be generated due to insufficient data.

BANDING This species has been banded at breeding colonies in California and Mexico. Nonbreeding band encounters document this

ORDER Charadriiformes
FAMILY Laridae
BREEDING HABITAT Coasts, Beach and Estuary
NONBREEDING HABITAT Coasts, Beach and Estuary
GLOBAL POPULATION SIZE 180,000

POPULATION CHANGE SINCE 1970 NA
IUCN RED LIST CATEGORY Near Threatened
R2R TIPPING POINT SPECIES No
US LISTING STATUS No
CANADA SPECIES AT RISK No
PIF CCS 16/20

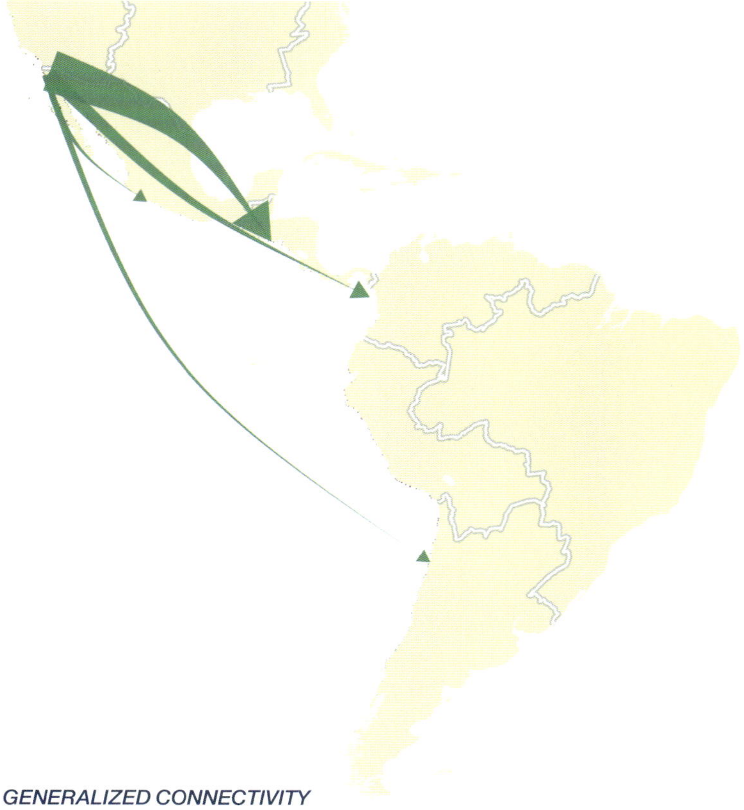

GENERALIZED CONNECTIVITY

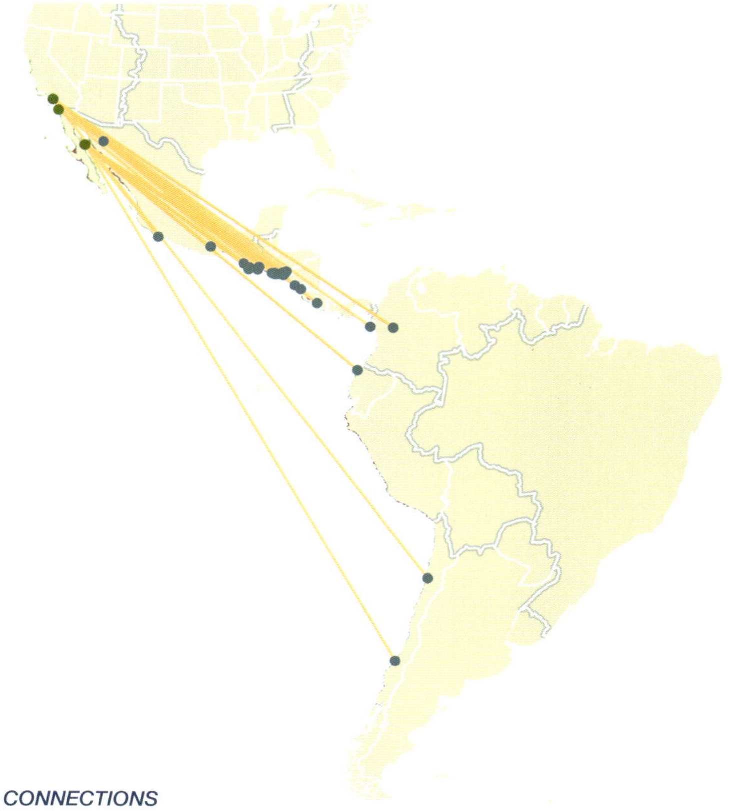

CONNECTIONS

species' southward movement to Mexico, Central America, and the South American coasts.

OTHER APPROACHES No satellite, genetic, or stable isotope studies of migratory connectivity were found for this species.

MIGRATORY PATHWAYS This species' entire annual cycle is closely tied to the Pacific coast. Birds that move northward post-breeding typically range to central California but occasionally reach as far as British Columbia. These northward-traveling birds eventually return south, leaving California by late October (Burness et al. 1999). During the nonbreeding season, the species ranges widely, but with highest concentrations along the coasts of Peru and Chile (Fink et al. 2023).

	Central America	Mexico	Northern SA	Southern Cone
Pacific (n = 16)	69%	6%	19%	6%

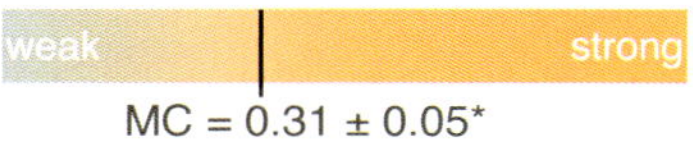

Rynchops niger **BLACK SKIMMER**

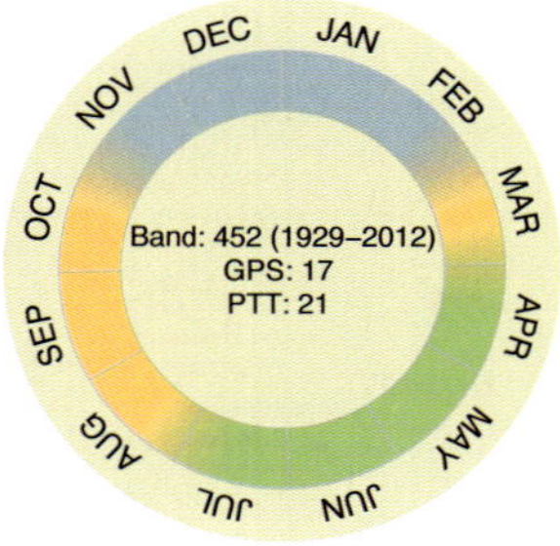

DISTRIBUTION AND ABUNDANCE

In North America, Black Skimmers breed colonially along the Pacific and Atlantic coasts, with the majority of the population along the Atlantic coast, from the Northeast to Florida, and along the Gulf coast. In South America, the Black Skimmer breeds inland along freshwater river systems. During the nonbreeding season, individuals from North American populations may form large flocks, with large numbers occurring along the Gulf of Mexico coast, coastal areas of Florida, the Caribbean, and western Mexico. Individuals from South American populations move from riverine breeding locations to coastal areas in South America.

ORDER Charadriiformes
FAMILY Laridae
BREEDING HABITAT
Coasts, Beach and Estuary
NONBREEDING HABITAT
Coasts, Beach and Estuary
GLOBAL POPULATION SIZE 800,000
POPULATION CHANGE SINCE 1970 −67%
IUCN RED LIST CATEGORY Least Concern
R2R TIPPING POINT SPECIES No
US LISTING STATUS No
CANADA SPECIES AT RISK No
PIF CCS 12/20

MIGRATORY CONNECTIVITY

OVERALL Migratory connectivity information comes from GPS and satellite tracking and band reencounter records. Birds breeding in the Atlantic Flyway primarily remain within the Atlantic Flyway (95%) during the nonbreeding season. Birds breeding in the Midcontinent Flyway migrate to nonbreeding areas in Mexico (46%),

SHOREBIRDS, AUKS, GULLS, AND TERNS (CHARADRIIFORMES)

	Atlantic	Caribbean	Central America	Mexico	Midcontinent	Pacific
Atlantic (n = 395)	95%	5%	<1%	–	<1%	–
Midcontinent (n = 26)	8%	–	15%	46%	31%	–
Pacific (n = 7)	–	–	–	86%	–	14%

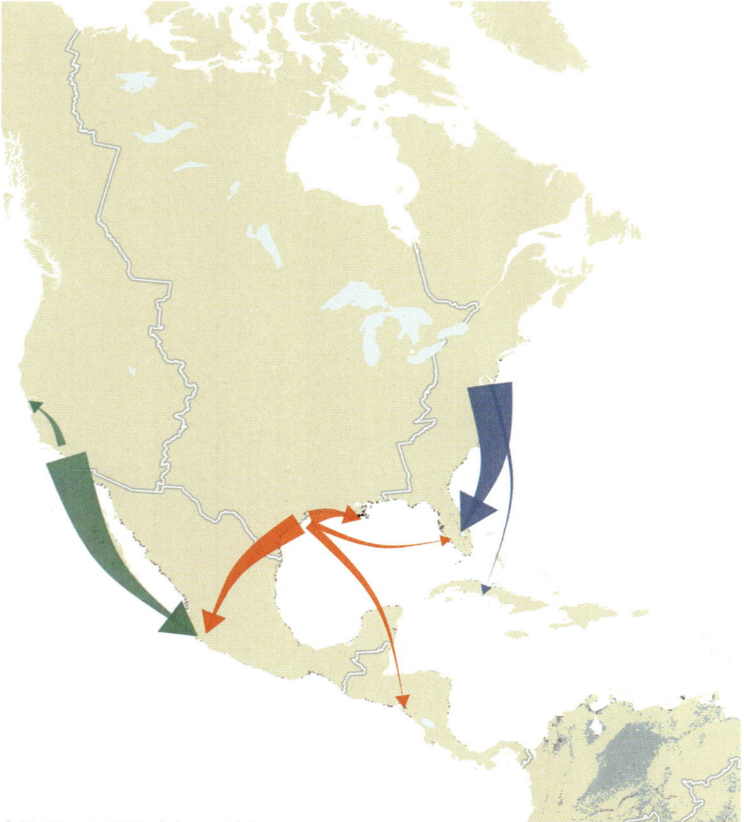

GENERALIZED CONNECTIVITY

CONNECTIONS (A–C)

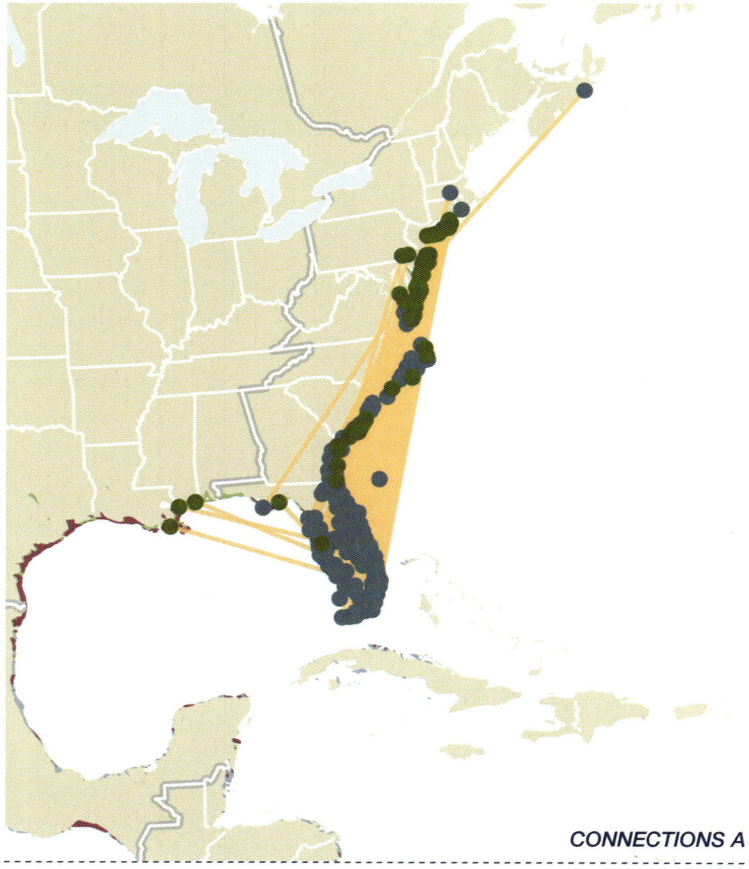

CONNECTIONS A

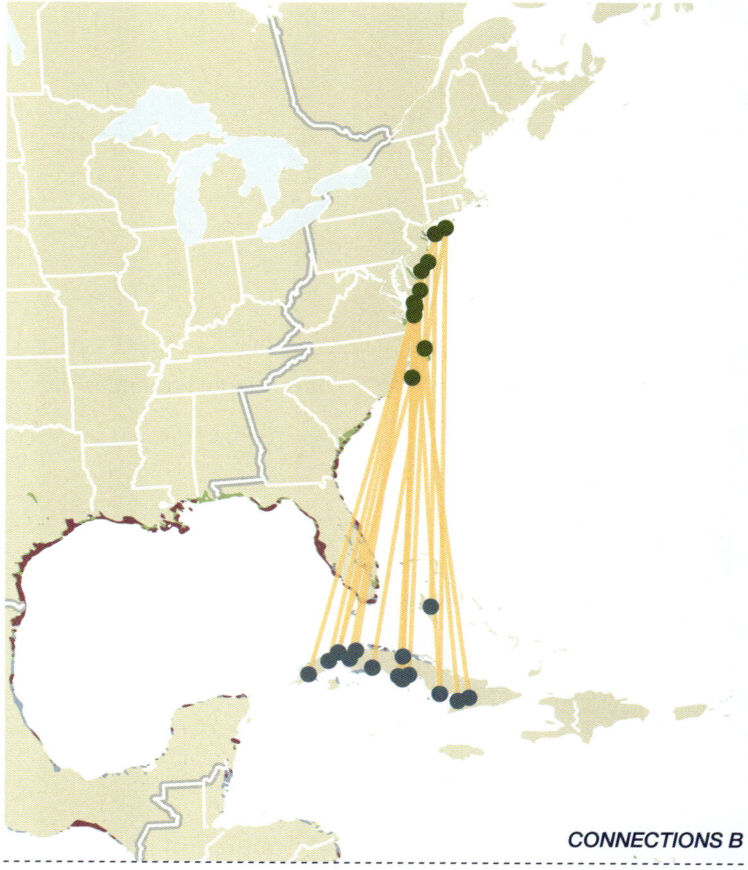

CONNECTIONS B

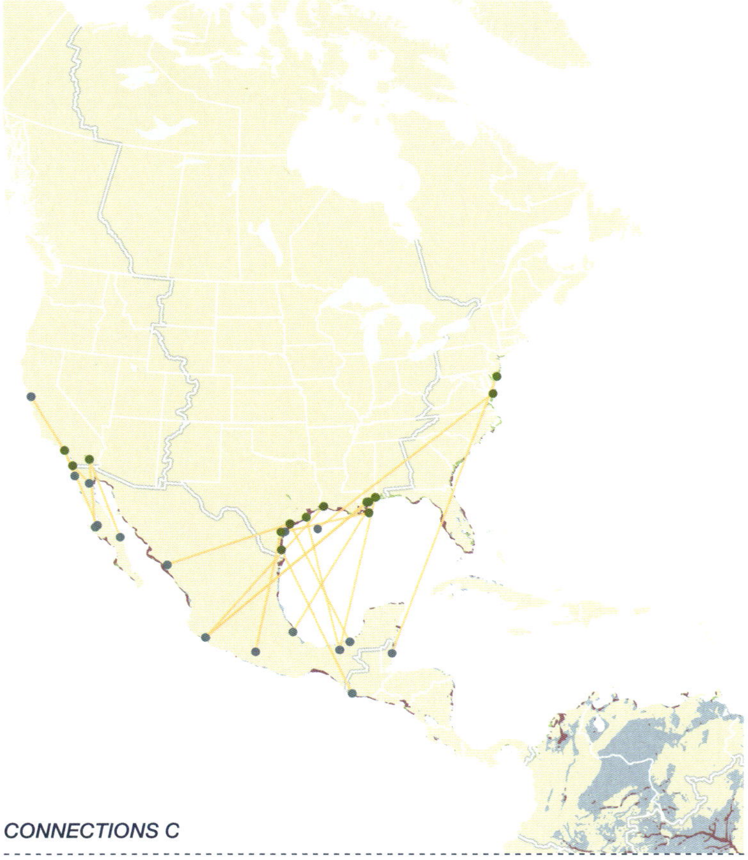

CONNECTIONS C

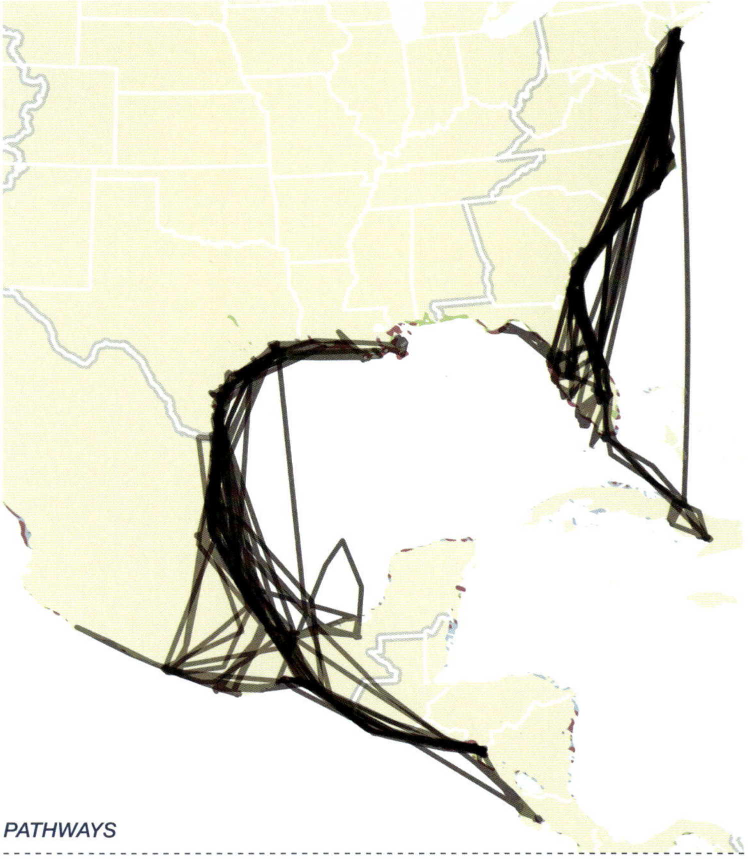

PATHWAYS

the Midcontinent Flyway (31%), Central America (15%), and the Atlantic Flyway (8%). Pacific Flyway–breeding birds migrate to Mexico (86%) or within the Pacific Flyway (14%). Despite a high transition probability for breeding populations within the Atlantic Flyway, the mixing of Midcontinent Flyway–breeding populations with other breeding populations across nonbreeding areas results in moderate migratory connectivity (MC = 0.31 ± 0.05).

BANDING Based on banding records, most eastern birds spend the nonbreeding season along the coast of the Carolinas, Georgia, and Florida, with many records from Cuba as well. Birds from different breeding colonies in Florida were also found to spend the nonbreeding period in Florida; however, flocks often contained birds from more than one breeding colony (Sprandel et al. 1999). The few birds banded in Southern California migrated to Baja California. Southern birds may be sedentary or move south into the Caribbean, Mexico, and Central America.

SATELLITE TRACKING Black Skimmers tracked from breeding colonies in New York migrated to nonbreeding areas in Florida and Cuba (Goodenough 2022). Black Skimmers tracked from the Texas Gulf Coast spent the nonbreeding season in Mexico and along the west coast of Central America (D. Newstead, unpublished data). Satellite telemetry in South America documented near continent-wide dispersal strategies (Davenport et al. 2016).

GENETICS Results of current research suggest there is very little support for connectivity between North and South American populations (K.S. Goodenough et al., unpublished data). However, a weak genetic differentiation between colonies in southern South America was observed, which could be explained by demographic history and/or the effect of an important shared nonbreeding site in southern South America (Mariano-Jelicich and Madrid 2014).

OTHER APPROACHES No stable isotope studies of migratory connectivity were found for this species.

MIGRATORY PATHWAYS Eastern birds followed the Atlantic coast during migration (Goodenough 2022). Texas birds followed the east coast of Mexico, flying around the Gulf of Mexico and crossing over to the west coast after stopovers along the coast of Veracruz, Mexico (D. Newstead, unpublished data).

DATA CONTRIBUTORS
K.S. Goodenough; D.J. Newstead.

INTERESTING NOTE

Fidelity to nesting colonies is strong, but is less so for natal dispersing birds (Sprandel et al. 1999). Two birds of the South American subspecies of Black Skimmer were tracked from breeding areas in the Amazon to the Pacific over the Andes, while a third bird was headed toward the Atlantic when its tag stopped transmitting (Davenport et al. 2016).

GREBES (PODICIPEDIFORMES)

WRITING LED BY **TIMOTHY GUIDA, MELANIE SMITH,** *AND* **TOM WILL**

Most of our insights on movement by grebes as a group are restricted to band reencounter data. The overall trend is for species within this group to breed in freshwater areas in the northern and central parts of the continent, and then migrate to coastal areas during the nonbreeding season.

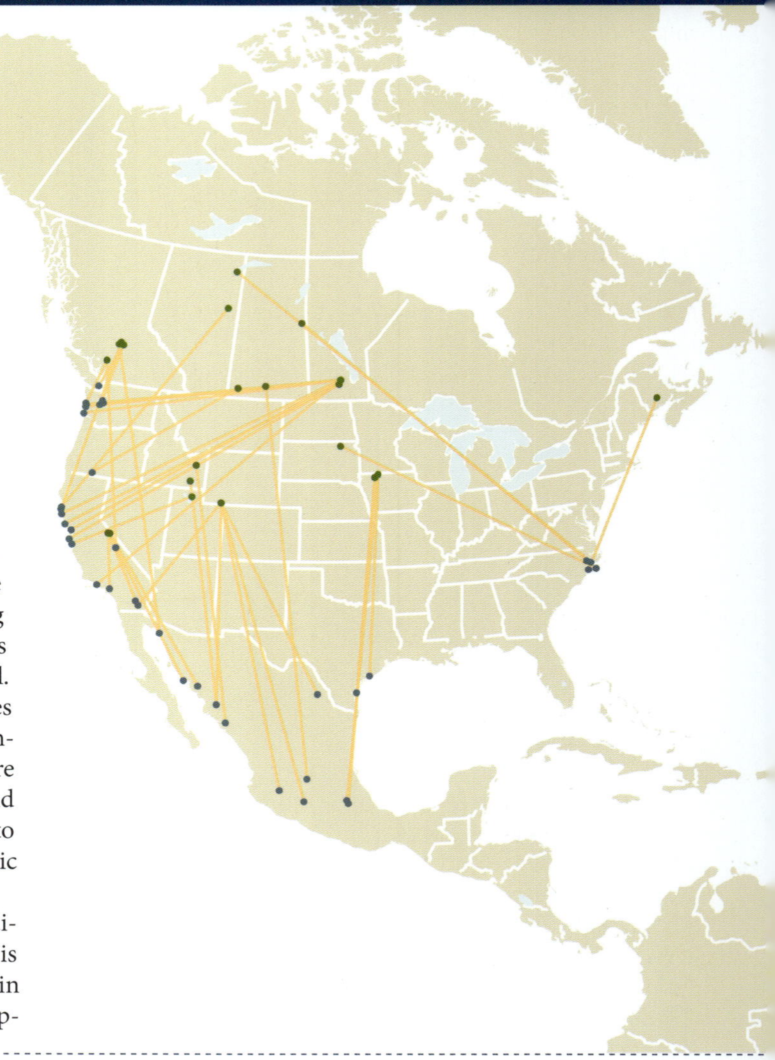

The order Podicipediformes contains a single unique family, the Podicipedidae or grebes. Migratory connectivity information for the Podicipedidae is limited. Despite 83% of species having migratory connectivity information, conclusions are restricted by considerable individual variability and small sample sizes. Data from Argos satellites tracking Western Grebes indicate variable movement strategies between breeding and nonbreeding areas. Data for all other species is limited to sparse band reencounter data (n = 39). Clark's Grebes and Western Grebes were formerly considered conspecific, leading to ambiguity in older banding records. As such, no connectivity information is available for Clark's Grebe, although a pattern similar to that of Western Grebe is probable. Data limitations are at least in part due to the difficulty in capturing grebes, which are aquatic year-round and are capable divers. Additionally, grebes were found to be tracked less frequently than other groups (Scarpignato et al. 2023), possibly due to the negative impacts of harness attachments, with grebes experiencing 100% effect rates, including fatal outcomes (Geen et al. 2019). Within this order some species appear to have increasing populations, while others are strongly declining. The Horned Grebe is considered a Tipping Point species, and both Clark's Grebes and Western Grebes have declined about 68% from 1970 to 2017. Causes of declines have not been determined, but degradation of aquatic habitats and disturbance in breeding areas are likely contributing factors.

Based on the available information, migratory connectivity in Podicipediformes appears moderate to weak. Some species breed colonially, and there is some evidence of breeding site fidelity, but more study is needed. Movements in the nonbreeding period appear to vary considerably between individuals, supporting weak migratory connectivity overall.

PODICIPEDIDAE | *GREBES*

PIED-BILLED GREBE *Podilymbus podiceps*

DISTRIBUTION AND ABUNDANCE

The Pied-billed Grebe has a wide distribution throughout the Americas, with migratory populations in the northern portions of the range throughout much of the United States and Canada, as well as nonmigratory populations in the southern US, Mexico, and Central and South America (Muller and Storer 2020; Fink et al. 2023). The largest concentrations of breeding birds for this species occur in the central plains of North America (Fink et al. 2023). Other important locations for this species include nonbreeding concentrations along the southern Atlantic coast and much of the Pacific coast of the US, as well as a large year-round population in San Francisco Bay and the California Central Valley (Muller and Storer 2020; Fink et al. 2023).

ORDER Podicipediformes
FAMILY Podicipedidae
BREEDING HABITAT Wetlands, Freshwater Marsh
NONBREEDING HABITAT Wetlands, Lakes, and Rivers
GLOBAL POPULATION SIZE 3,100,000
POPULATION CHANGE SINCE 1970 +54%

IUCN RED LIST CATEGORY Least Concern
R2R TIPPING POINT SPECIES No
US LISTING STATUS No
CANADA SPECIES AT RISK No
PIF CCS 8/20

MIGRATORY CONNECTIVITY

OVERALL No studies were available directly assessing migratory connectivity. Limited band reencounter data show migration between northern breeding grounds to the southern United States. MC is not able to be generated due to insufficient data.
BANDING Banding data include two breeding to nonbreeding reencounters. Two breeding season individuals, one from New Brunswick and one from South Dakota, were located in North Carolina during the nonbreeding season.
MIGRATORY PATHWAYS Migrates in a broad front throughout the range (Fink et al. 2023).
OTHER No satellite, genetic, or stable isotope studies of migratory connectivity were found for this species.

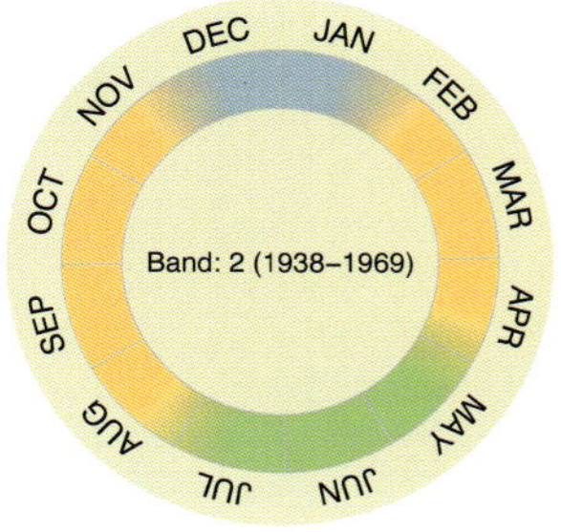

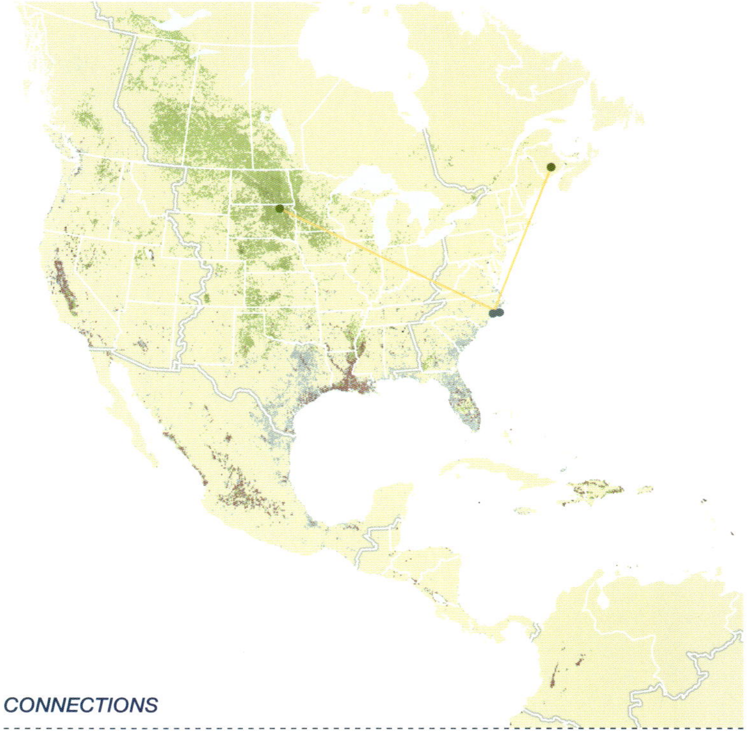

CONNECTIONS

Podiceps auritus **HORNED GREBE**

DISTRIBUTION AND ABUNDANCE

This small grebe breeds widely in the northern latitudes of western North America and Eurasia. In North America, it's breeding range in Canada is concentrated particularly in central Saskatchewan, eastern Alberta, the Northwest Territories, and the Yukon and eastern Alaska south to the Prairie Pothole Region in the United States. During the nonbreeding season this species is found on large inland lakes and reservoirs, as well as both the Atlantic and Pacific coasts (Stedman 2020).

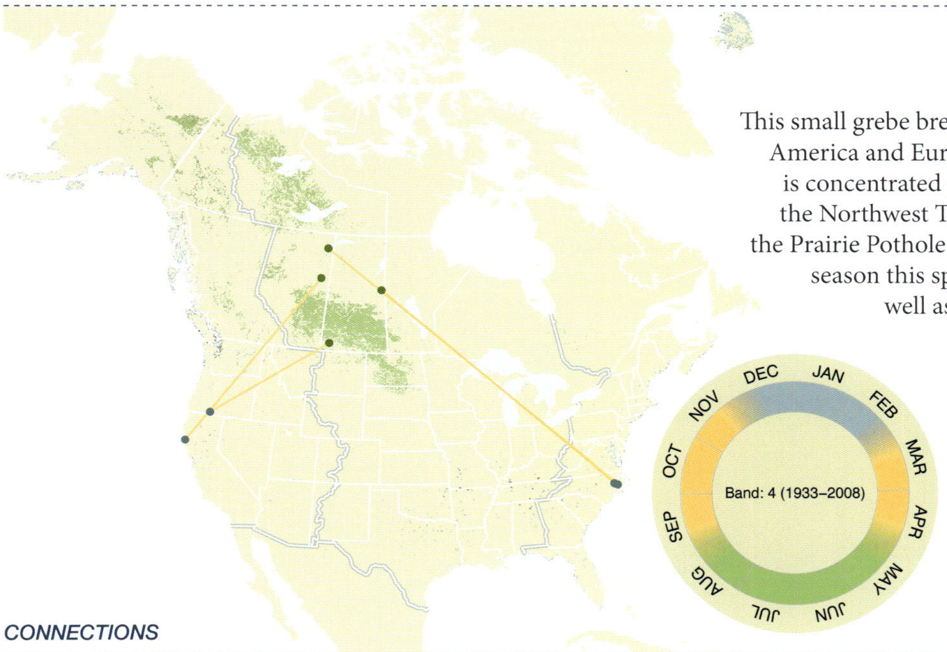

CONNECTIONS

MIGRATORY CONNECTIVITY

OVERALL This species is a medium-distance nocturnal migrant with well-defined migratory patterns, but little is known about migratory connectivity. MC is not able to be generated due to insufficient data.
BANDING Limited band reencounter data shows breeding populations in central Canada move to either the Pacific or Atlantic coasts of the United States during the nonbreeding season. Banding recaptures in Alaska show that individuals are likely faithful to their breeding sites (Stout and Cooke 2003).
OTHER APPROACHES No satellite, genetic, or stable isotope studies of migratory connectivity were found for this species.
MIGRATORY PATHWAYS Birds spending the nonbreeding season in the Pacific move directly to breeding grounds, likely via routes that avoid the high mountain areas. Birds with nonbreeding ranges in the Atlantic, the Gulf of Mexico, and the interior migrate along a broad front northward. The Great Lakes are an important migratory stopover point for eastern birds, but which populations use this stopover is unclear (Stedman 2020; Fink et al. 2023).

ORDER Podicipediformes
FAMILY Podicipedidae
BREEDING HABITAT Wetlands, Freshwater Marsh
NONBREEDING HABITAT Coasts, Marine Waters
GLOBAL POPULATION SIZE 620,000
POPULATION CHANGE SINCE 1970 −48%
IUCN RED LIST CATEGORY Vulnerable
R2R TIPPING POINT SPECIES Yellow-Alert
US LISTING STATUS No
CANADA SPECIES AT RISK Endangered Magdalen Islands population
PIF CCS 13/20

RED-NECKED GREBE *Podiceps grisegena*

DISTRIBUTION AND ABUNDANCE

The Red-necked Grebe is a large grebe that ranges widely in the Northern Hemisphere. In North America, this species breeds on inland water bodies largely in the western Canadian provinces (especially Saskatchewan), Alaska, and the northern United States from Minnesota west to Washington. The nonbreeding range is coastal: the Pacific coast from Alaska to California and the northern Atlantic coast (Stout and Nuechterlein 2020; Fink et al. 2023).

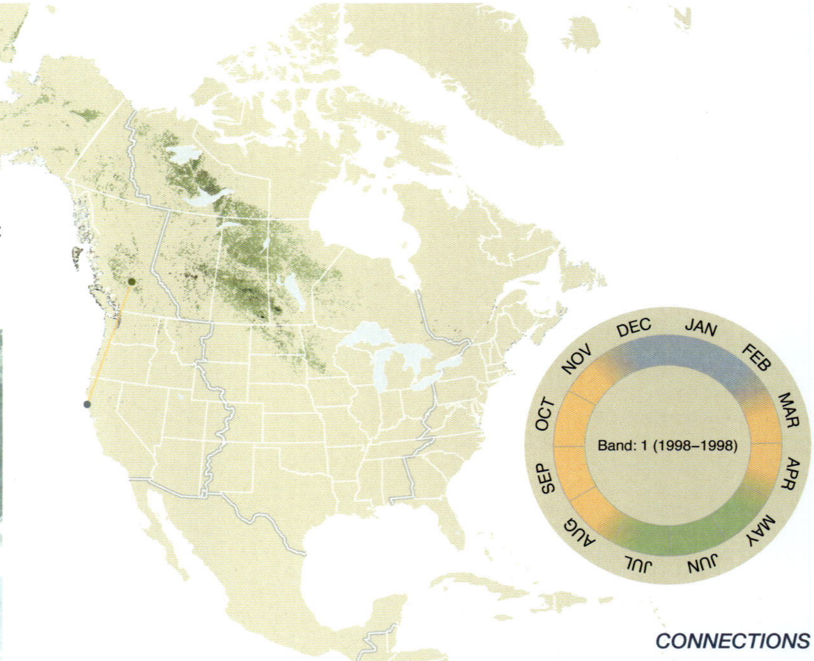

ORDER Podicipediformes
FAMILY Podicipedidae
BREEDING HABITAT Wetlands, Lakes, and Rivers
NONBREEDING HABITAT Coasts, Marine Waters
GLOBAL POPULATION SIZE 2,500,000
POPULATION CHANGE SINCE 1970 +52%
IUCN RED LIST CATEGORY Least Concern
R2R TIPPING POINT SPECIES No
US LISTING STATUS No
CANADA SPECIES AT RISK No
PIF CCS 9/20

MIGRATORY CONNECTIVITY

OVERALL Very little data are available to draw conclusions about migratory connectivity. MC is not able to be generated due to insufficient data.

BANDING Band reencounter data only includes one breeding to stationary nonbreeding recovery, documenting a British Columbia breeding bird encountered in Northern California during the nonbreeding season.

OTHER APPROACHES No satellite, genetic, or stable isotope studies of migratory connectivity were found for this species.

MIGRATORY PATHWAYS In the nonbreeding season, many inland breeding birds fly west to the northern Pacific coast, especially from Washington north to the Aleutian Islands. Fewer birds spend the nonbreeding season along the Atlantic coast, from Massachusetts north to Labrador. The Great Lakes are an important stopover location for this species, especially during post-breeding migration (Stout and Nuechterlein 2020; Fink et al. 2023). Counts of unmarked Red-necked Grebes suggest that the species undergoes a molt migration before moving to nonbreeding sites. These molting sites are usually coastal sites and areas on the Great Lakes (Stout and Cooke 2003).

EARED GREBE *Podiceps nigricollis*

DISTRIBUTION AND ABUNDANCE

This species is widespread in North America as well as in Eurasia and southern Africa. In North America, the Eared Grebe ranges widely west of the Mississippi River, from the northern Yukon to central Mexico. It is particularly common in the Dakotas north into Saskatchewan and eastern Alberta and extending southwest into Baja California and other aquatic areas of Mexico. Nonbreeding populations move southwest throughout California, Baja California, and the Pacific coast of Mexico and inland on bodies of water south to Mexico City.

ORDER Podicipediformes
FAMILY Podicipedidae
BREEDING HABITAT Wetlands, Freshwater Marsh
NONBREEDING HABITAT Coasts, Marine Waters
GLOBAL POPULATION SIZE 3,100,000
POPULATION CHANGE SINCE 1970 +61%
IUCN RED LIST CATEGORY Least Concern
R2R TIPPING POINT SPECIES No
US LISTING STATUS No
CANADA SPECIES AT RISK No
PIF CCS 8/20

MIGRATORY CONNECTIVITY

OVERALL Migratory connectivity data for this species are limited, though the combination of radio transmitters and band reencounter data suggest general migration patterns. Midcontinent Flyway breeding populations migrate to nonbreeding areas in Mexico (60%) or within the Midcontinent Flyway (40%). Pacific Flyway breeding birds were also likely to migrate to nonbreeding areas in Mexico (56%) or within the Pacific Flyway (44%). MC is not calculated due to limited data.

	Mexico	Midcontinent	Pacific
Midcontinent (n = 5)	60%	40%	–
Pacific (n = 16)	56%	–	44%

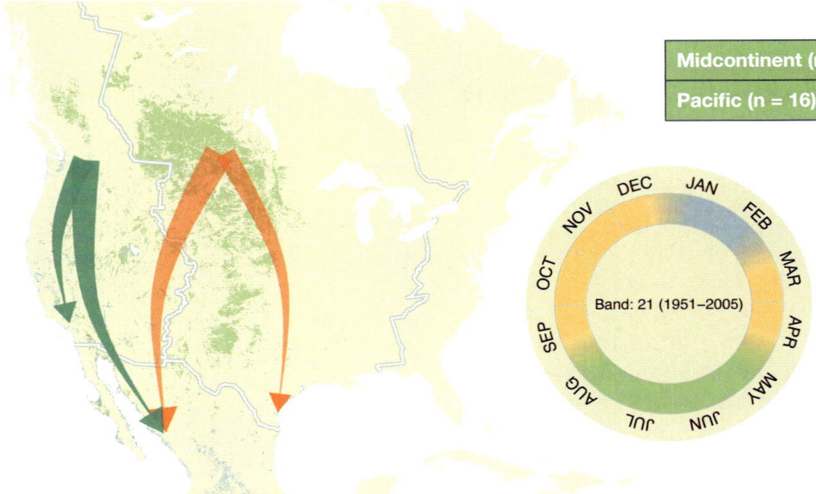

GENERALIZED CONNECTIVITY

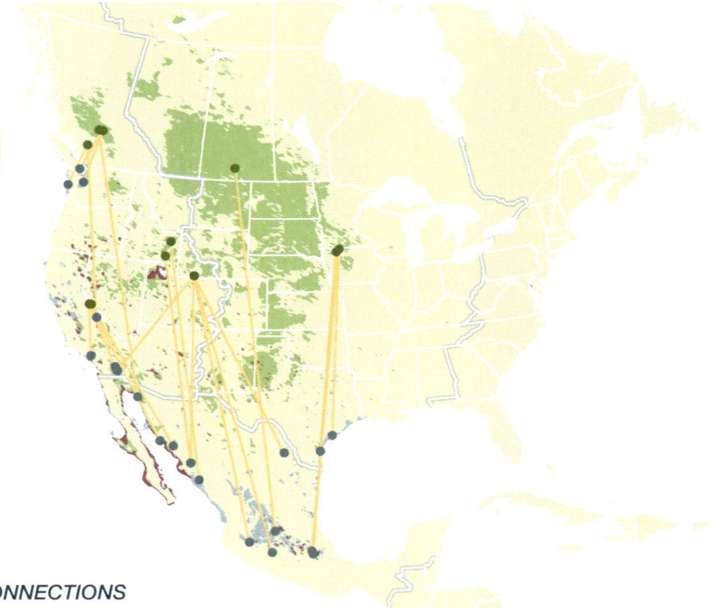

CONNECTIONS

BANDING Banding recoveries suggest that breeders from central North America migrate to nonbreeding areas in Mexico and the Gulf of Mexico, while individuals from western breeding regions in the Great Basin and High Plains spend the nonbreeding season at the Salton Sea or Gulf of California. Many individuals also staged at the Great Salt Lake, Utah, and Mono Lake, California (Jehl and Yochem 1986).
RADIO TRACKING Eared Grebes tracked from their breeding site in British Columbia migrated to Mono Lake, California, and the Great Salt Lake, Utah. There was also some movement between Mono Lake and the Salton Sea in Southern California (Boyd et al. 2000).
OTHER APPROACHES No geolocator, satellite telemetry, stable isotope, or genetic studies of migratory connectivity were found for this species.
MIGRATORY PATHWAYS The species breeds in inland lakes of the western United States and Canada and gathers in high abundance at important stopover locations such as Mono Lake in California and the Great Salt Lake, Utah, before continuing southward to their nonbreeding grounds (Cullen et al. 2020; Fink et al. 2023). Eastern populations appear to be more likely to spend the nonbreeding season near the Gulf of Mexico, and western populations on the Gulf of California (Jehl and Yochem 1986; Boyd et al. 2000; Cullen et al. 20202; Fink et al. 2023).

Aechmophorus occidentalis WESTERN GREBE

DISTRIBUTION AND ABUNDANCE

Western Grebe breed from the Great Plains to Pacific coasts of the US and Canada, with concentrations in the Prairie Pothole Region of the US and Canada. The nonbreeding range spans the Pacific coast from southern Southeast Alaska to the Baja California peninsula and mainland Pacific coast of northern Mexico (LaPorte et al. 2020; Fink et al. 2023).

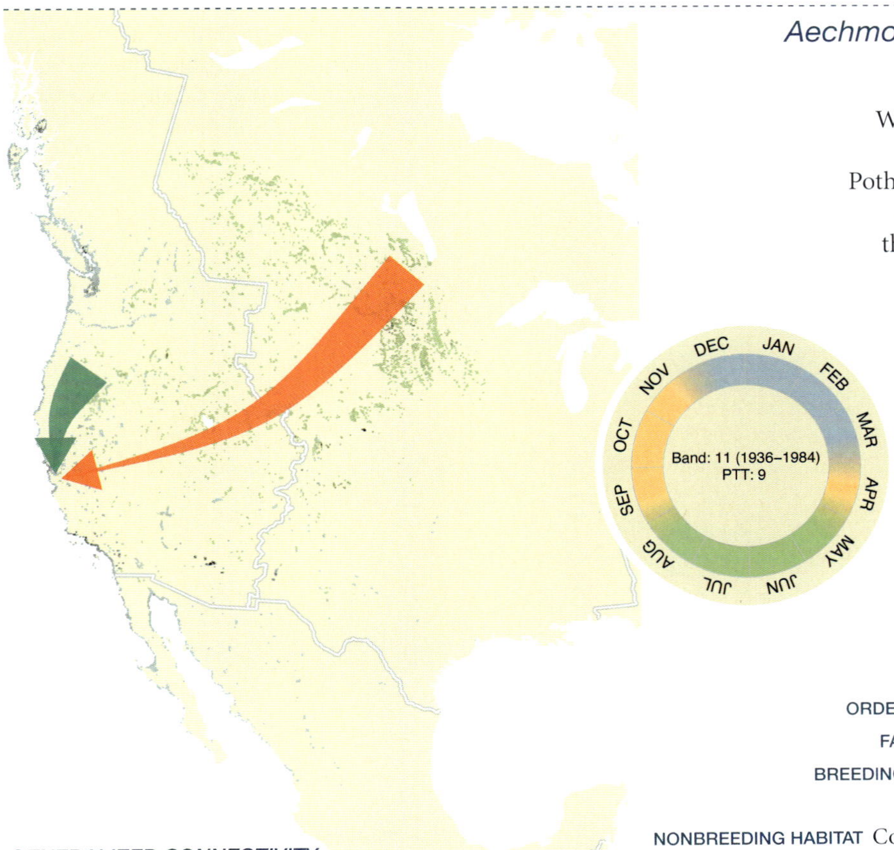

GENERALIZED CONNECTIVITY

ORDER Podicipediformes
FAMILY Podicipedidae
BREEDING HABITAT Wetlands, Freshwater Marsh
NONBREEDING HABITAT Coasts, Marine Waters
GLOBAL POPULATION SIZE 1,000,000

POPULATION CHANGE SINCE 1970 −68%
IUCN RED LIST CATEGORY Least Concern
R2R TIPPING POINT SPECIES No
US LISTING STATUS No
CANADA SPECIES AT RISK No
PIF CCS 12/20

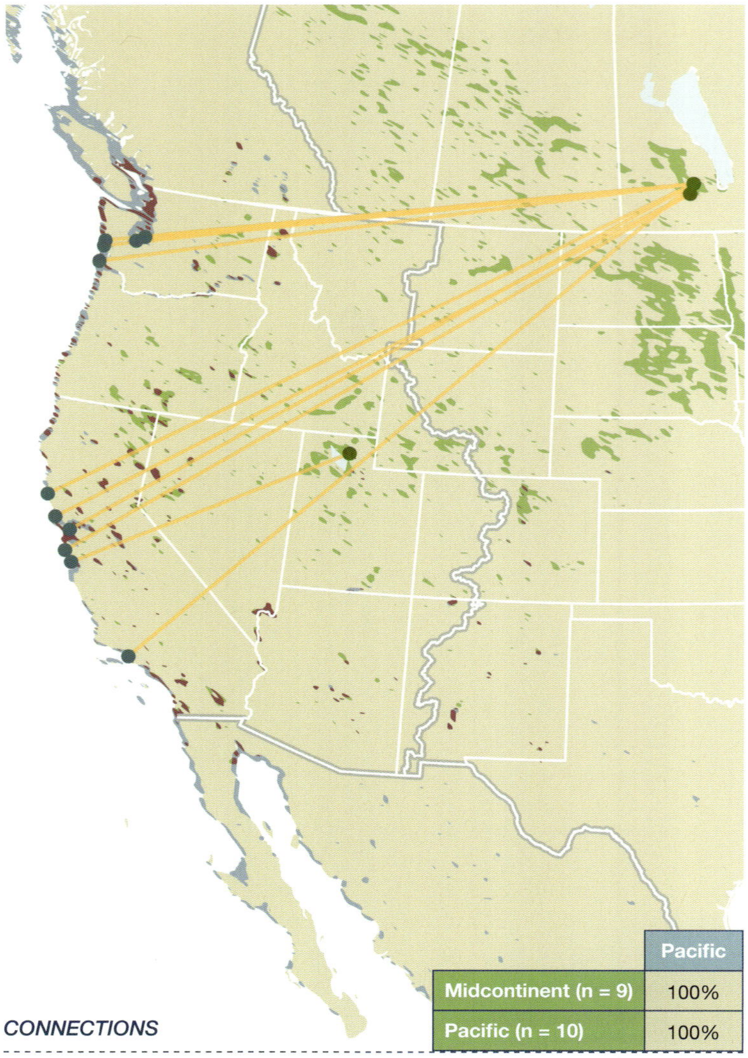

	Pacific
Midcontinent (n = 9)	100%
Pacific (n = 10)	100%

CONNECTIONS

PATHWAYS

INTERESTING NOTE

Western Grebes seem to undergo a molt migration post-breeding in the late summer in which some important sites include Lake Manitoba, Manitoba; Commencement Bay, Washington; Santa Barbara Channel, California; and Boundary Bay, British Columbia (Stout and Cooke 2003).

MIGRATORY CONNECTIVITY

OVERALL Western Grebe migratory connectivity has been studied using several methods, but small sample sizes and individual variation make it difficult to draw general conclusions. Individuals within the Pacific and Midcontinent flyways all migrated to nonbreeding areas in the Pacific Flyway. MC not calculated due to limited sample size.

BANDING An analysis of banding data that did not distinguish between Clark's Grebes and Western Grebes found that many grebes banded in the interior US and Canada during the breeding season were recovered on the Pacific coast in the nonbreeding season (Eichhorst 1992). Studies with data that distinguished Clark's Grebes from Western Grebes found similar patterns, suggesting the strongest migratory connectivity is between the nonbreeding region on the coast along Washington and British Columbia and the breeding region in Manitoba (Humple 2009). Overall, banding encounters suggest that nonbreeding season site fidelity is variable and that migratory connectivity is unlikely to be strong (Humple and Holcomb 2014).

RADIO TRACKING There were no detections from tags deployed with the Motus Wildlife Tracking System.

SATELLITE TRACKING Two individuals tracked with transmitters from San Francisco Bay, California, remained in the area for the nonbreeding season. One individual then migrated to Clear Lake, California, and then Upper Klamath Lake, Oregon, for the breeding season before returning to San Francisco Bay for the nonbreeding season. The other individual remained in San Francisco Bay for a full year before migrating south to San Diego Bay, California, for several days in the nonbreeding season and then returned to San Francisco Bay (Mills et al. 2016).

GENETICS Genomic analysis was not useful to distinguish between individuals from two breeding colonies in California and Alberta and three nonbreeding areas in Washington, Northern California, and Southern California (Stout and Cooke 2003).

OTHER APPROACHES No geolocator or stable isotope studies of migratory connectivity were found for this species.

MIGRATORY PATHWAYS Interior migration pathways are not distinct. In spring, many birds migrate widely across the western US and Canada primarily to the Prairie Pothole Region, while others first migrate north along the Pacific coast into Southeast Alaska and British Columbia coastal waters before heading inland, lagging behind individuals that migrated directly from their nonbreeding area to the interior.

DATA CONTRIBUTORS

S. De La Cruz; C.V. Fiorello; J.K. Gaydos; K.L. Mills; D.M. Mulcahy; L.I. Vilchis; E.R. Whitmer; M.H. Ziccardi.

Aechmophorus clarkii **CLARK'S GREBE**

DISTRIBUTION AND ABUNDANCE

This species breeds locally on freshwater lakes and marshes throughout western North America from the southern Canadian Prairies to central Mexico. During the nonbreeding season, most US breeding populations migrate west along the Pacific coast. Birds breeding in the southwestern US are more sedentary or resident, as are populations found south to the Trans-Mexican Volcanic Belt surrounding Mexico City (Fink et al. 2023).

ORDER Podicipediformes
FAMILY Podicipedidae
BREEDING HABITAT Wetlands, Freshwater Marsh
NONBREEDING HABITAT Coasts, Marine Waters
GLOBAL POPULATION SIZE 81,000
POPULATION CHANGE SINCE 1970 −68%

IUCN RED LIST CATEGORY Least Concern
R2R TIPPING POINT SPECIES No
US LISTING STATUS No
CANADA SPECIES AT RISK No
PIF CCS 11/20

MIGRATORY CONNECTIVITY

OVERALL Migratory connectivity is unknown. No studies of migratory connectivity were found for this species.
RADIO TRACKING Few tags were deployed through the Motus Wildlife Tracking System, with local detections only.

MIGRATORY PATHWAYS Movements are likely similar to those of the closely related Western Grebe, moving along a wide front rather than a distinct pathway, and they migrate between their breeding and nonbreeding areas.

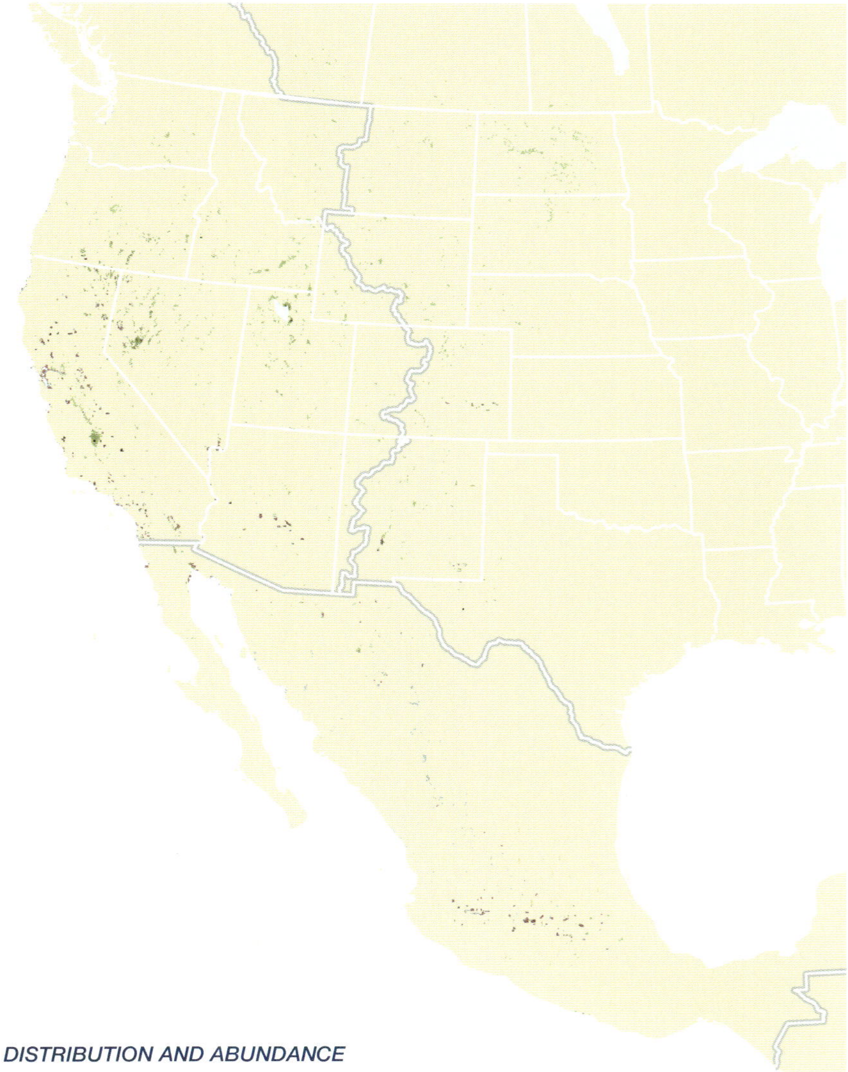

DISTRIBUTION AND ABUNDANCE

INTERESTING NOTE

Band reencounter data show that patterns of movement within and between nonbreeding seasons suggest plasticity in movements possibly due to variability in food resources within marine environments (Humple 2009).

TROPICBIRDS (PHAETHONTIFORMES)

WRITING LED BY **AUTUMN-LYNN HARRISON** *AND* **PETER P. MARRA**

The migratory connectivity and associated ecology of North America's three Phaethontiformes—the White-tailed Tropicbird, the Red-billed Tropicbird, and the Red-tailed Tropicbird—remain poorly understood compared to many other seabird groups. None of these species is currently listed under the US Endangered Species Act or as a Tipping Point species, but comprehensive tracking studies for these species are lacking, leaving major gaps in understanding of migratory connectivity and nonbreeding distributions. Given their reliance on offshore foraging habitats, transboundary movements, and sensitivity to disturbance at nesting sites, improved migratory tracking, population monitoring, and threat assessment are essential for effective conservation planning. Coordinated research across breeding sites, coupled with international marine conservation initiatives, will be critical for maintaining healthy North American tropicbird populations.

PHAETHONTIDAE | *TROPICBIRDS*

WHITE-TAILED TROPICBIRD *Phaethon lepturus*

DISTRIBUTION AND ABUNDANCE

This species is the smallest and most common tropicbird in tropical and subtropical oceans. Breeding population estimates reported in Lee and Walsh-McGee (2020) are decades old, but most well-known colonies reported between 1,000 and 13,000 pairs. For North America, there are about 10,000 pairs in the Caribbean (*P. l. catesbyi*) and 500–3,000 pairs in the Hawaiian Islands (*P. l. dorotheae*; Lee and Walsh-McGee 2020). The species is highly pelagic during the nonbreeding period. In eastern North America, it is most frequently seen in the Gulf Stream (Lee and Walsh-McGee 2020). Off the main Hawaiian Islands, it can be seen coastally all year, even during the post-breeding period (Lee and Walsh-McGee 2020). It is largely absent from the eastern Pacific Ocean, even in tropical and subtropical areas.

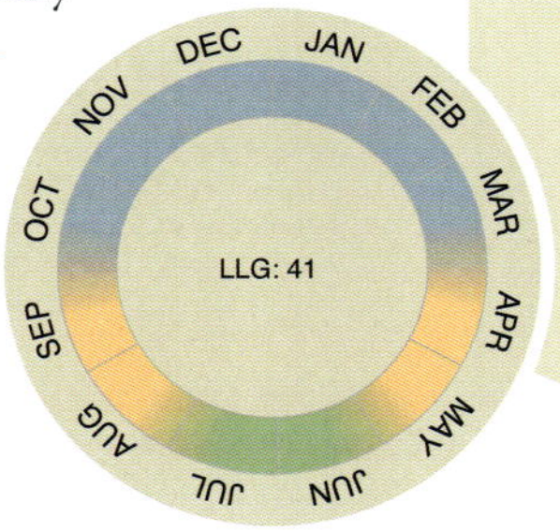

ORDER Phaethontiformes
FAMILY Phaethontidae
BREEDING HABITAT Islands, Oceanic
NONBREEDING HABITAT Oceans, Pelagic
GLOBAL POPULATION SIZE 400,000
POPULATION CHANGE SINCE 1970 NA

IUCN RED LIST CATEGORY Least Concern
R2R TIPPING POINT SPECIES No
US LISTING STATUS No
CANADA SPECIES AT RISK No
PIF CCS 15/20

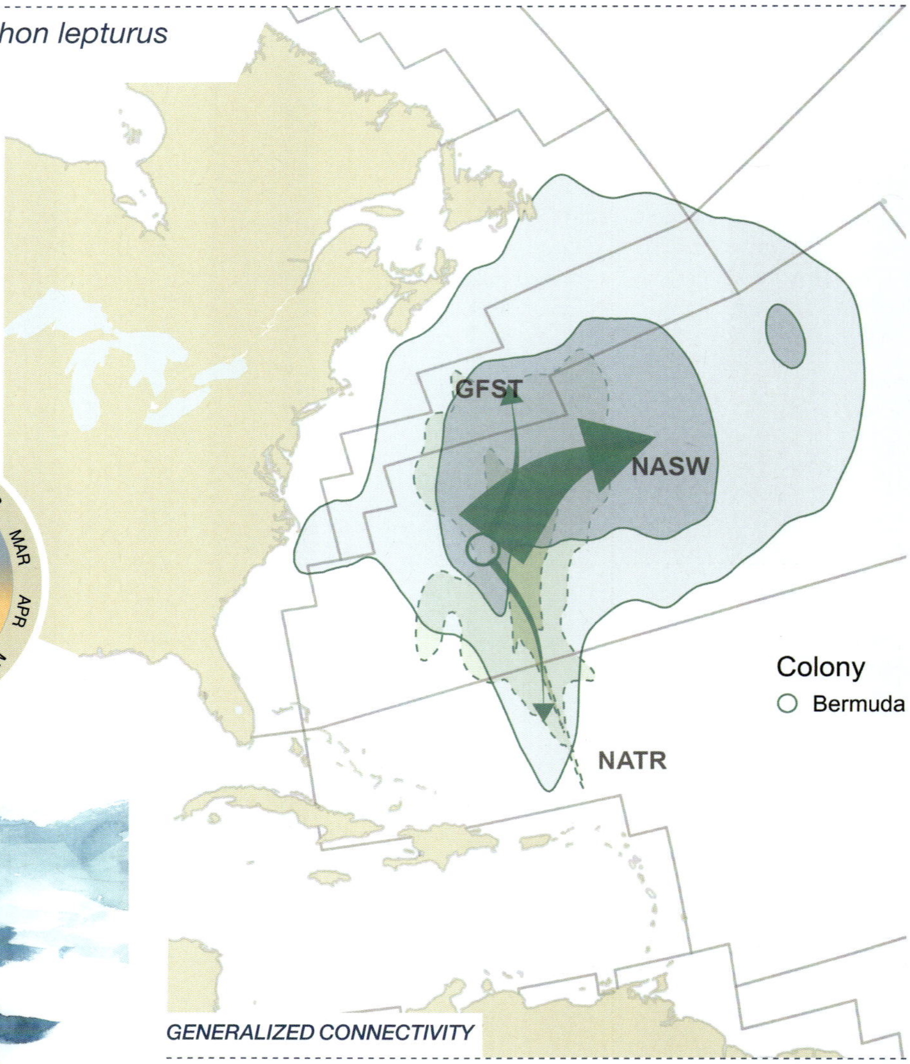

	NATR	NASW	GFST
Bermuda (n = 41)	5%	88%	7%

MIGRATORY CONNECTIVITY

OVERALL There is likely no mixing of North Atlantic Ocean breeding populations (*P. l. catesbyi*) with either Pacific Ocean or Indian Ocean populations given genetic and morphological evidence. Moderate to strong connectivity is also suspected between North and South Atlantic subspecies (*P. l. ascensionis*) given genetic differentiation, although no nonbreeding tracking data are available

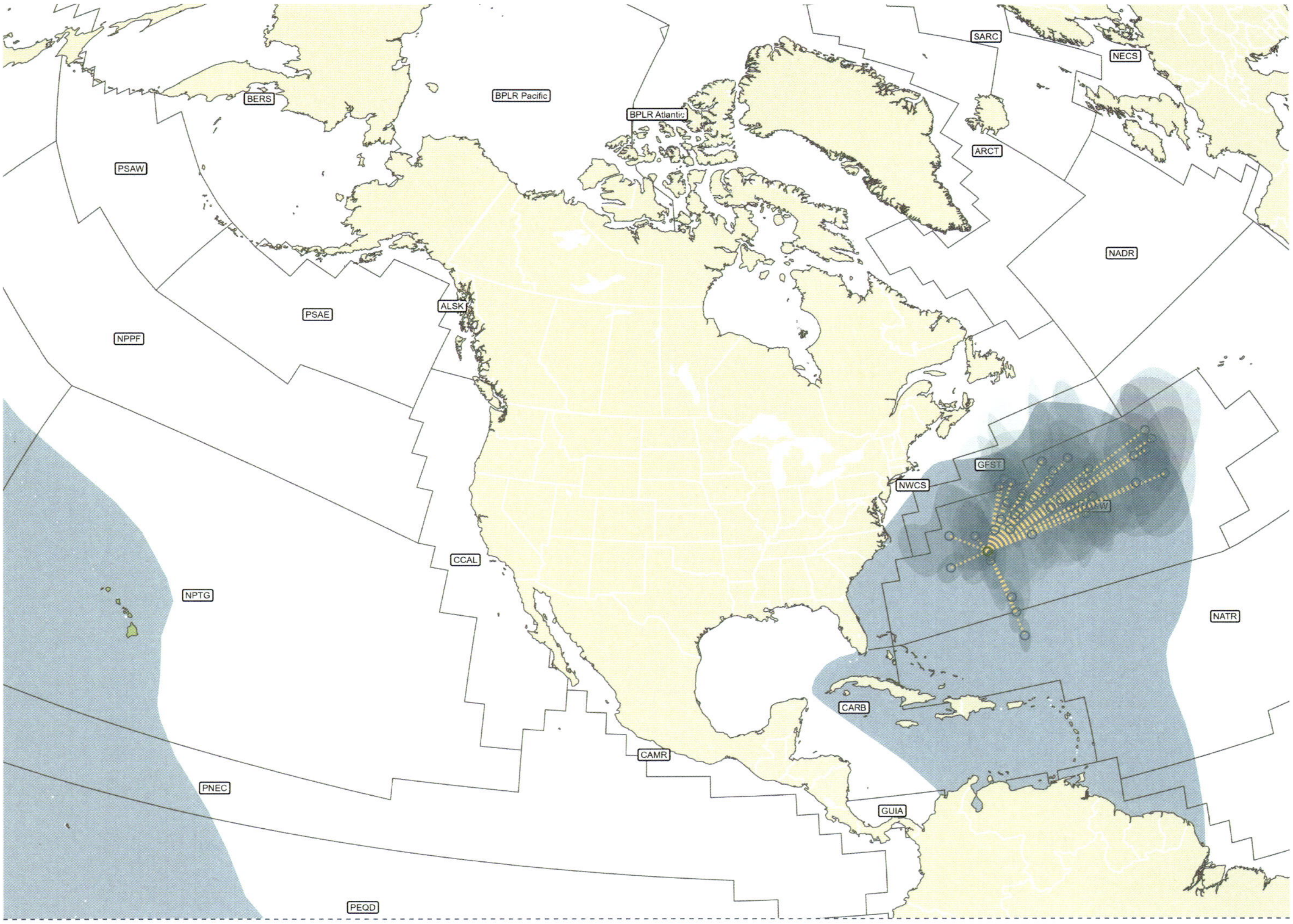

CONNECTIONS

INTERESTING NOTE

Neither genetic nor tracking studies have been published for birds from Hawaii, the most distant colonies from the Indian Ocean, and thus further work in the Pacific Ocean would be helpful to understand contemporary patterns of connectivity among Indo-Pacific populations.

for *P. l. ascensionis* to confirm. Genetic evidence suggests possible weak connectivity of Pacific Ocean and Indian Ocean populations. For North American populations, only birds from Bermuda have been tracked during the nonbreeding period; thus MC is not calculated. Within this sample, most individuals used core areas within the western North Atlantic Subtropical Gyre (88%), with some individual core areas also in the Gulf Stream and the North Atlantic Tropical province.

GEOLOCATOR TRACKING White-tailed Tropicbirds tracked from Bermuda ranged across the northwest Atlantic Ocean, west to the southeastern United States, east to the Mid-Atlantic Ridge, and north offshore Newfoundland, Canada (Mejías et al. 2017). Most birds undertook long-distance ranging migrations that shifted seasonally, although two spent the nonbreeding season close to Bermuda (Mejías et al. 2017). None of the birds crossed the equator into the southern Atlantic Ocean where *P. l. ascensionis* nests.

GENETICS Differences across genetic (mitochondrial DNA and nuclear DNA) and morphological markers support the supposition that birds nesting in Bermuda are isolated from other subspecies (Humeau et al. 2020). In contrast, there was little genetic differentiation of birds sampled from Tahiti, French Polynesia (*P. l. dorotheae*) in the Pacific Ocean, and Indian Ocean birds representing *P. l. lepturus*, suggesting either that gene flow occurs or isolation was recent (Humeau et al. 2020).

OTHER APPROACHES No stable isotope studies of migratory connectivity were found for this species.

DATA CONTRIBUTORS

M.A. Mejías; Y.F. Wiersma.

RED-BILLED TROPICBIRD *Phaethon aethereus*

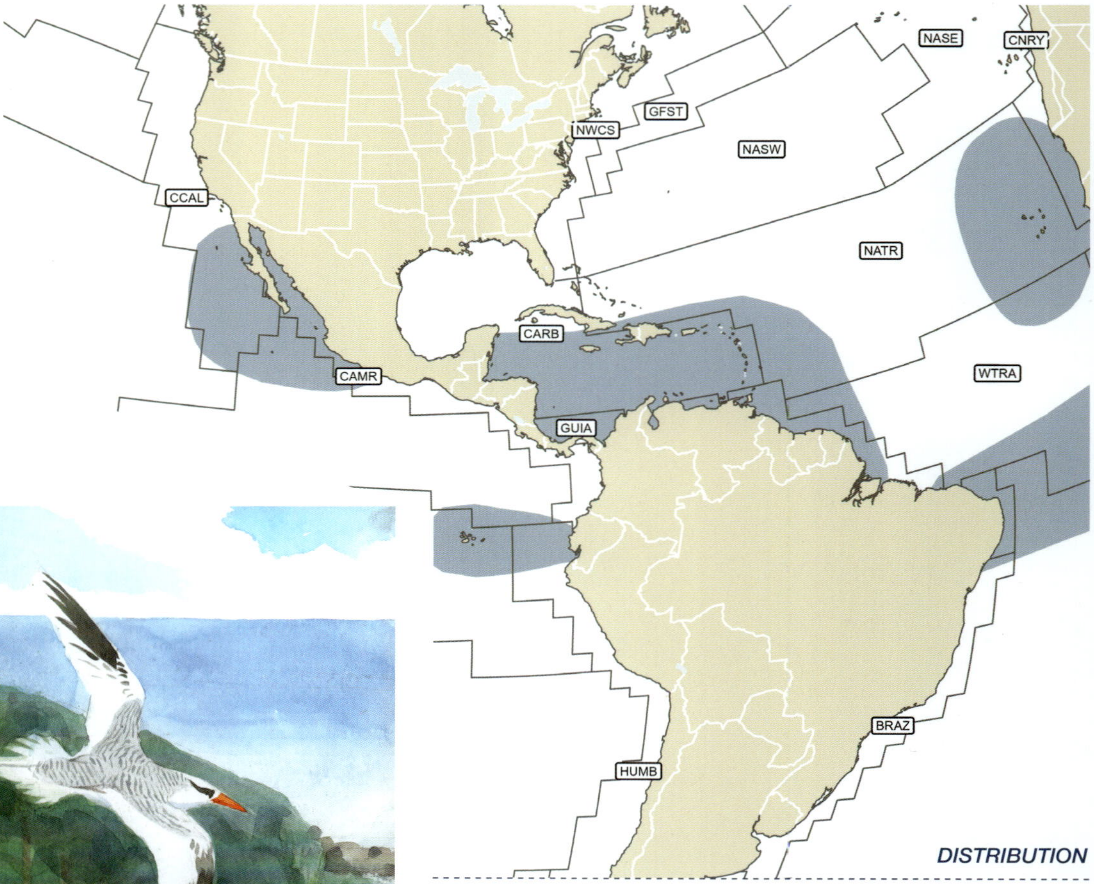

DISTRIBUTION

ORDER Phaethontiformes
FAMILY Phaethontidae
BREEDING HABITAT Islands, Oceanic
NONBREEDING HABITAT Oceans, Pelagic
GLOBAL POPULATION SIZE 8,200
POPULATION CHANGE SINCE 1970 NA
IUCN RED LIST CATEGORY Least Concern
R2R TIPPING POINT SPECIES No
US LISTING STATUS No
CANADA SPECIES AT RISK No
PIF CCS 17/20

MIGRATORY CONNECTIVITY

OVERALL Only a BirdLife distribution map is available; no eBird data or migratory connectivity information were found for this species.

RED-TAILED TROPICBIRD *Phaethon rubricauda*

DISTRIBUTION AND ABUNDANCE

A tropicbird of the Pacific and Indian Oceans, the Red-tailed Tropicbird is the rarest of North America's three species (Schreiber and Schreiber 2020). Large breeding colonies in the Pacific include Kure, Midway, Laysan, and Lisianski of the Hawaiian Islands and Johnston Atoll. There are other small colonies throughout the Hawaiian Island chain. There is limited knowledge of its nonbreeding biology at sea, which ranges widely over the oceans throughout the breeding range but also extends north to Japan and south to New Zealand; it is uncommon in the eastern Pacific off California (Schreiber and Schreiber 2020).

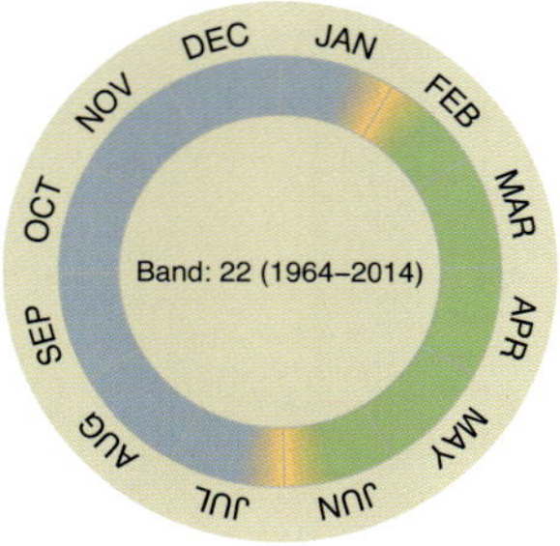

ORDER Phaethontiformes
FAMILY Phaethontidae
BREEDING HABITAT Islands, Oceanic
NONBREEDING HABITAT Oceans, Pelagic
GLOBAL POPULATION SIZE 70,000
POPULATION CHANGE SINCE 1970 NA
IUCN RED LIST CATEGORY Least Concern
R2R TIPPING POINT SPECIES No
US LISTING STATUS No
CANADA SPECIES AT RISK No
PIF CCS 13/20

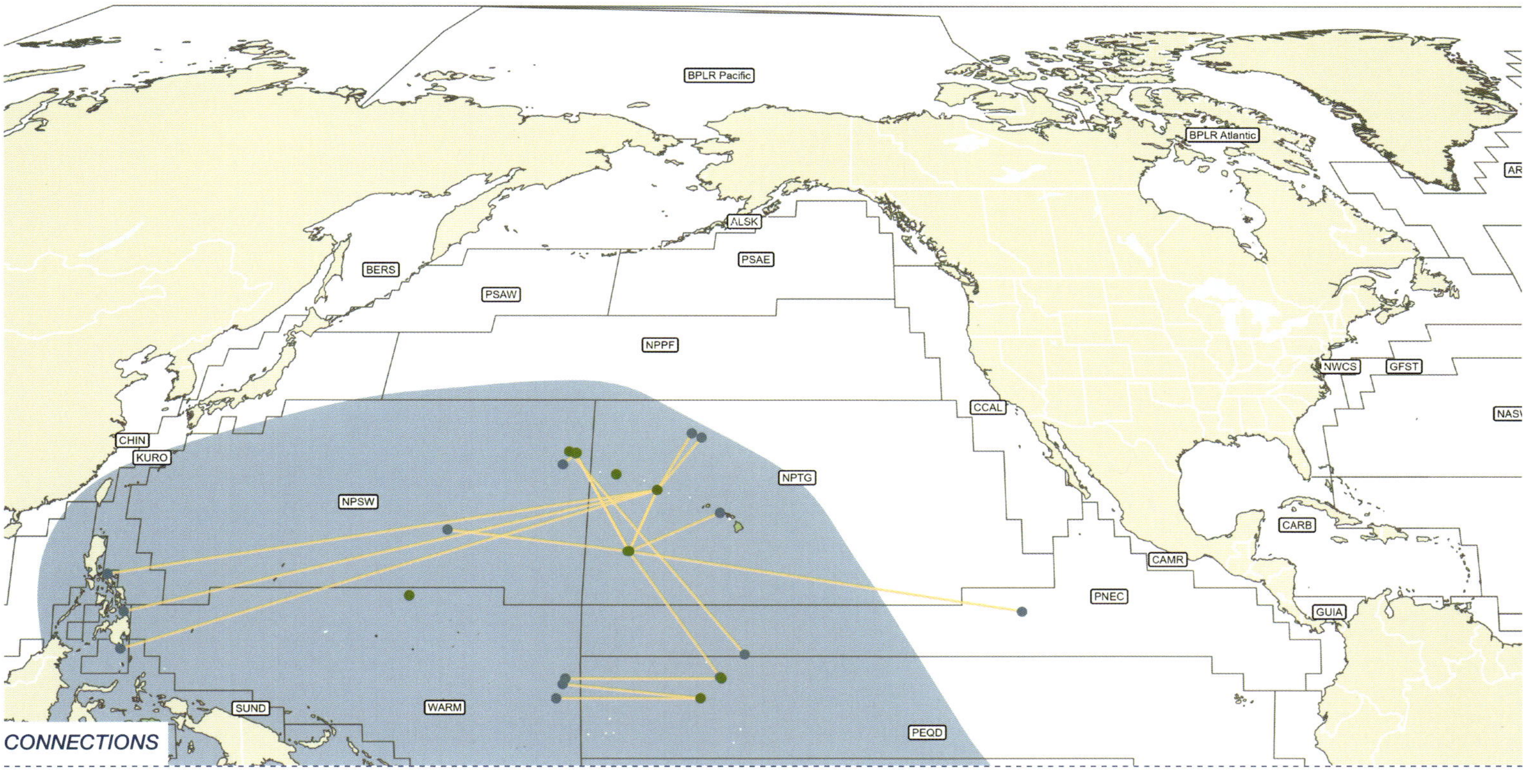

CONNECTIONS

MIGRATORY CONNECTIVITY

OVERALL Very little is known about the migratory connectivity of Red-tailed Tropicbirds, but genetics, at-sea surveys, and banding data suggest regionally strong connectivity within the Pacific Ocean; and banding data suggest no connectivity between Pacific and Indian oceans. Due to limited sample size, MC is not calculated for this species.

BANDING Band reencounter data indicate that birds breeding in northern islands in the Hawaiian chain generally use nonbreeding habitats to the south and southeast, while central Pacific birds, including Johnston Atoll, and Jarvis and Christmas islands, primarily use nonbreeding habitats in a westerly direction, likely due to the prevailing winds (Schreiber and Schreiber 2020).

GENETICS Analyses of mitochondrial DNA (Varela et al. 2021) and single nucleotide polymorphisms (SNPs) for 132 Red-tailed Tropicbirds from six colonies across much of the Pacific Ocean (Varela et al. 2024) suggested significant regional structure across Chile, Australasia, and Hawaii and variable differentiation within regions. For example, there was a lack of significant genetic differentiation between O'ahu and Kaua'i, separated by 200 km, but significant differentiation between two Chilean islands (Varela et al. 2024).

OTHER APPROACHES No tracking or stable isotope studies of migratory connectivity were found for this species in the Pacific Ocean, but multiple tracking studies (geolocator and GPS) are available in the Indian Ocean. Surveys undertaken in 1980 to 1995 in the eastern Pacific found three regions of high densities that were associated with region-specific breeding timing, thus suggesting some population separation (Spear and Ainley 2005). These regions included an area northwest of the leeward Hawaiian Islands, west of 130 degrees west longitude associated with the central Pacific Polynesian islands, and immediately north of the Pitcairn Islands east of 130 degrees west longitude.

MIGRATORY PATHWAYS Highly pelagic away from the colony during post-breeding, but specific migratory routes or even if there are regular migrations are unknown (Schreiber and Schreiber 2020). Direction of post-breeding departures vary depending upon the colony's location within regions of prevailing winds (Schreiber and Schreiber 2020).

INTERESTING NOTE

An improved understanding of at-sea ecology is a major priority for this species (Egerton et al. 2022).

LOONS (GAVIIFORMES)

WRITING LED BY **AUTUMN-LYNN HARRISON** *AND* **TIMOTHY GUIDA**

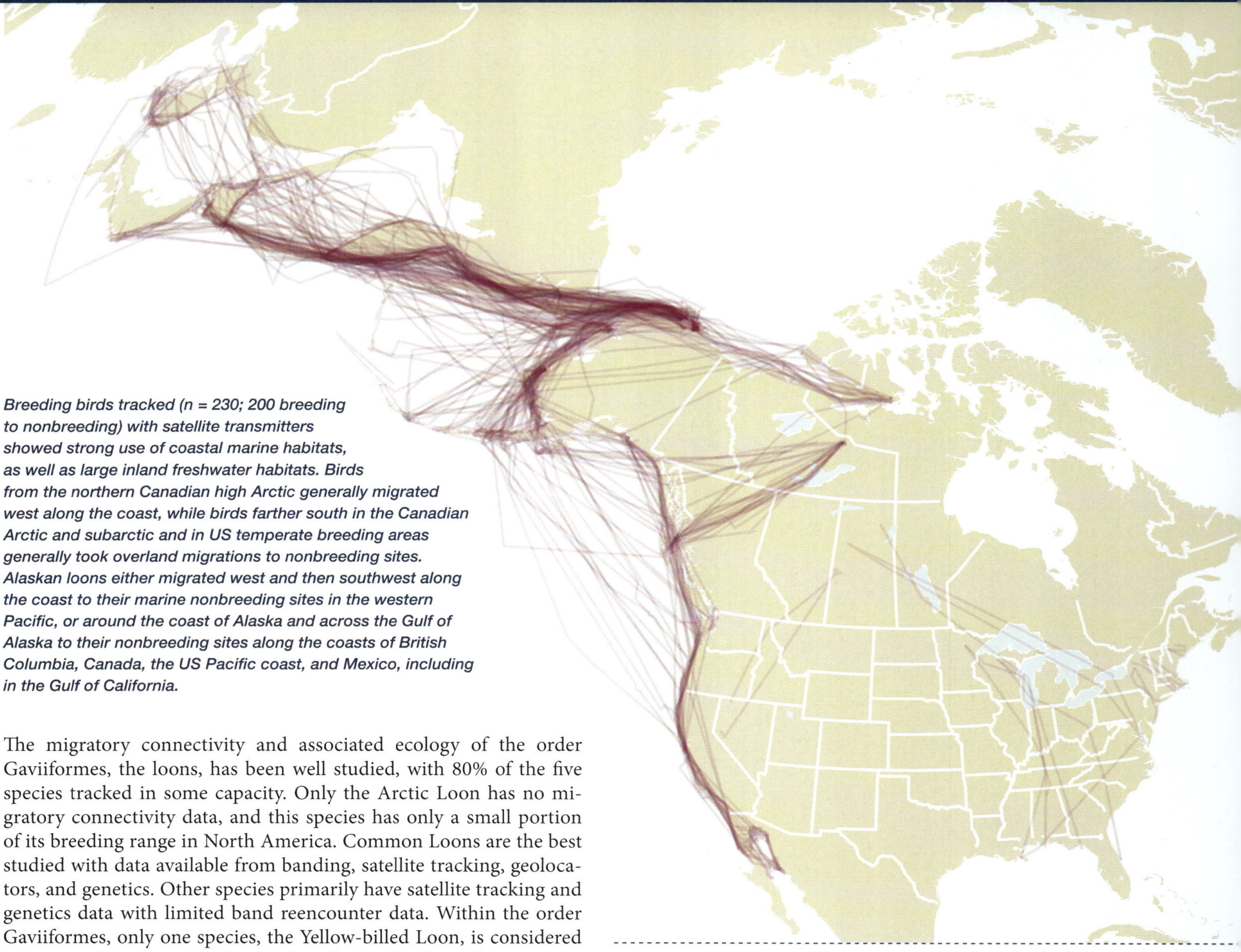

Breeding birds tracked (n = 230; 200 breeding to nonbreeding) with satellite transmitters showed strong use of coastal marine habitats, as well as large inland freshwater habitats. Birds from the northern Canadian high Arctic generally migrated west along the coast, while birds farther south in the Canadian Arctic and subarctic and in US temperate breeding areas generally took overland migrations to nonbreeding sites. Alaskan loons either migrated west and then southwest along the coast to their marine nonbreeding sites in the western Pacific, or around the coast of Alaska and across the Gulf of Alaska to their nonbreeding sites along the coasts of British Columbia, Canada, the US Pacific coast, and Mexico, including in the Gulf of California.

The migratory connectivity and associated ecology of the order Gaviiformes, the loons, has been well studied, with 80% of the five species tracked in some capacity. Only the Arctic Loon has no migratory connectivity data, and this species has only a small portion of its breeding range in North America. Common Loons are the best studied with data available from banding, satellite tracking, geolocators, and genetics. Other species primarily have satellite tracking and genetics data with limited band reencounter data. Within the order Gaviiformes, only one species, the Yellow-billed Loon, is considered a Tipping Point species, classified as "Red-Alert." Satellite tracking studies of Yellow-billed Loons demonstrated population-specific nonbreeding habitats. Breeding birds in western Canada largely migrate to the eastern Pacific, and most birds breeding in Alaska migrate either to the western Pacific or Bering Sea during the nonbreeding season. No migratory connectivity data are available for the eastern Canadian portion of Yellow-billed Loon breeding range, and little reliable breeding and nonbreeding counts of Yellow-billed Loons are available to inform management. Other species within Gaviiformes also warrant conservation attention. Species of the *Gavia* genus are caught in high rates in fisheries of Arctic Canada, so improving understanding of connectivity to these fisheries is a priority. Common Loons are listed in some US states as a species of conservation concern, and declines in their reproductive success have been observed in Ontario, Canada.

This order includes four species treated as seabirds (Red-throated Loon, Arctic Loon, Pacific Loon, and Yellow-billed Loon) and one treated as a landbird (Common Loon). The marine species nest primarily in the Arctic region and migrate from terrestrial freshwater lakes to oceanic nonbreeding habitats. Red-throated Loons and Yellow-billed Loons demonstrate moderate migratory connectivity. Pacific Loons show a complete migratory divide between two Alaskan populations, with northern birds migrating southwest to the western North Pacific and western Alaskan birds migrating southeast to the eastern North Pacific. Red-throated Loons in eastern Canada migrate south to the Atlantic coast, where populations are thought to mix, while Alaskan populations show a similar divide to Pacific Loons. Some population mixing does occur for Yellow-billed Loons breeding in western Canada and Alaska, but there are dominant population-level patterns as described above. Common Loons nest across a large swath of North American "lake country," but are less common in the tundra, and migrate generally in north-south directions both inland to large freshwater lakes and reservoirs and coastally. Common Loons also have moderate migratory connectivity. There is extensive movement from midcontinent breeding areas to Atlantic nonbreeding areas, but relatively little movement away from the Atlantic or Pacific flyways for populations breeding within them.

Mantel = 0.37 ± 0.08

Gavia stellata

RED-THROATED LOON

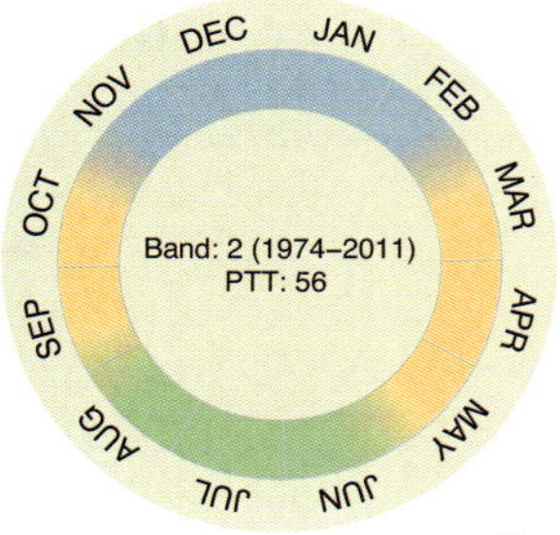

DISTRIBUTION AND ABUNDANCE

The smallest loon (Rizzolo et al. 2020), the Red-throated Loon has a circumpolar distribution and nests along small boreal, subarctic, and Arctic lakes up to high latitudes. In North America, breeding areas span Alaska to Greenland, with nonbreeding distributions along both the Atlantic and Pacific coasts. Recent estimates from decades of breeding and nonbreeding surveys suggest Red-throated Loon populations are stable to negative in Alaska and stable to positive in Canada (Amundson et al. 2019; Pollet et al. 2024).

ORDER Gaviiformes
FAMILY Gaviidae
BREEDING HABITAT Wetlands, Tundra
NONBREEDING HABITAT Coasts, Marine Waters
GLOBAL POPULATION SIZE 1,400,000
POPULATION CHANGE SINCE 1970 +23%
IUCN RED LIST CATEGORY Least Concern
R2R TIPPING POINT SPECIES No
US LISTING STATUS No
CANADA SPECIES AT RISK No
PIF CCS 9/20

MIGRATORY CONNECTIVITY

OVERALL Moderate migratory connectivity exists across ocean basins (Mantel = 0.37 ± 0.08) and within the Pacific Ocean for Alaskan colonies based on satellite tracking data. Birds from northern Alaska

	CHIN	KURO	PSAW	BERS	ALSK	CCAL	CAMR	NWCS
Yukon-Kuskokwim Delta, AK, USA (n = 7)	–	–	–	14%	14%	71%	–	–
Seward Peninsula, AK, USA (n = 4)	–	–	–	–	50%	50%	–	–
Point Lay, AK, USA (n = 7)	–	14%	29%	43%	–	14%	–	–
North Slope, AK, USA (n = 13)	38%	31%	–	23%	–	–	8%	–
Copper River Delta, AK, USA (n = 5)	–	–	–	–	20%	80%	–	–
Nunavut, Canada (n = 7)	–	–	–	–	–	–	–	100%
Greenland (n = 1)	–	–	–	–	–	–	–	100%

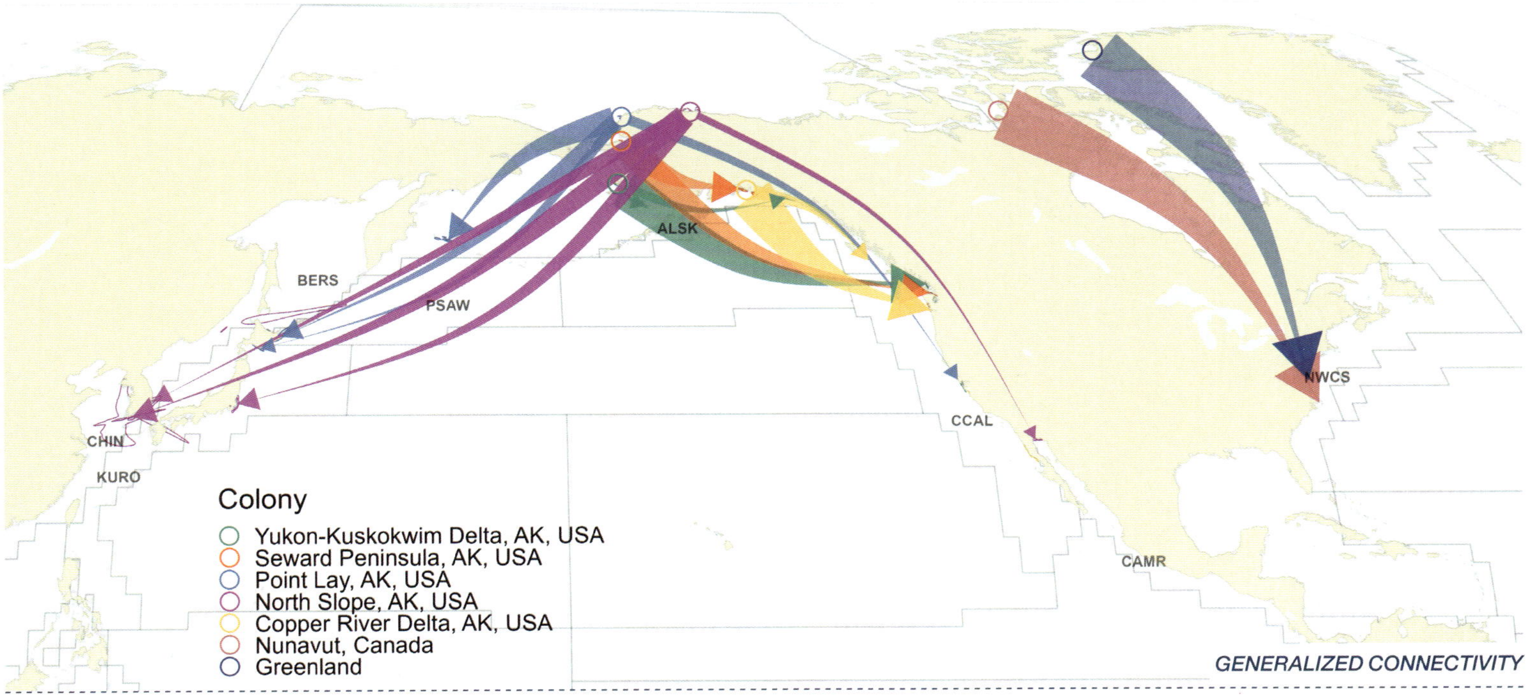

GENERALIZED CONNECTIVITY

(Point Lay and North Slope) distributed across six northwest Pacific provinces (with most within the Bering Sea, Kuroshio Current, and China Sea provinces) during the nonbreeding period, while 50%–80% of birds from southeastern Alaska (Yukon-Kuskokwim Delta, Seward Peninsula, and Copper River Delta) spent the nonbreeding period in the California Current and 14%–50% in the Alaska Coastal Downwelling Province. Notably, however, one individual banded at Point Lay, Alaska (western North Slope), broke the pattern and was reencountered in the California Current. There is limited information on migratory connectivity of Canadian and Greenland populations because most tag deployments have occurred in the US mid-Atlantic (capture sites spanning New Jersey to North Carolina). However, a recent tracking study revealed migratory connectivity of two individuals between nest sites in Greenland and nonbreeding habitats in the North Sea (Kleinschmidt et al. 2022), suggesting more is to be learned about migratory connectivity of Greenland and Canadian breeding populations.

BANDING While banded loons are rarely reencountered on nonbreeding grounds, one individual banded in Nunavut in 1974 was caught entangled in fishing gear in North Carolina, US, in the same year, and one individual banded during breeding in Point Lay, Alaska, in 2008 was sighted in January in California.

SATELLITE TRACKING Like Pacific Loons, Red-throated Loons tracked by satellite from northern Alaska spent the nonbreeding season in East Asia as far west as Korea, while those breeding farther south in Alaska (Seward Peninsula, Yukon-Kuskokwim Delta, and Copper River Delta)

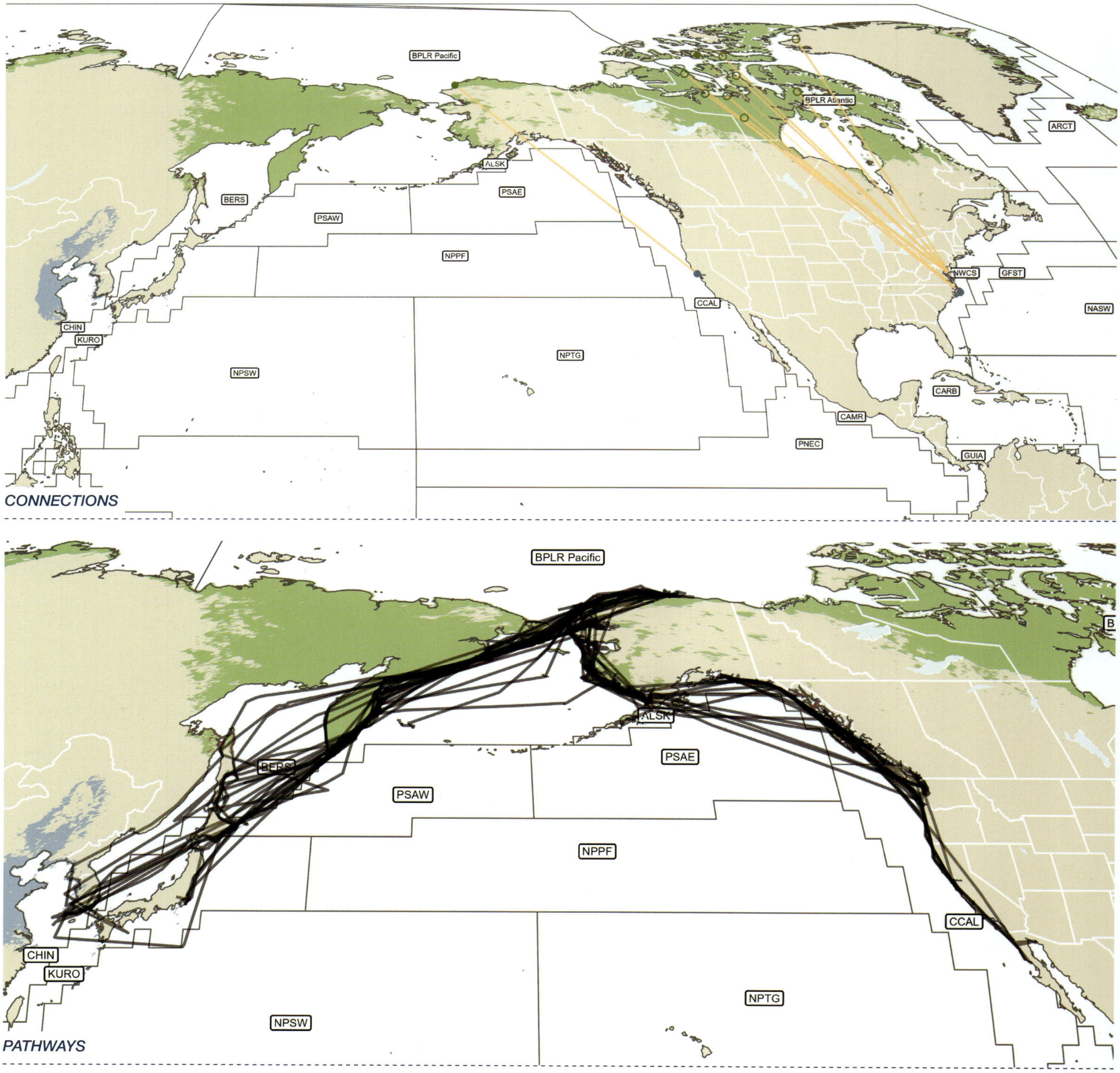

spent the nonbreeding season along the Pacific coast of North America, as far south as Baja California (Uher-Koch et al. 2017). Red-throated Loons have also been tracked from nonbreeding areas on the mid-Atlantic coast of the United States to breeding areas in Nunavut, Canada, and to Greenland (Spiegel et al. 2017).

OTHER APPROACHES No genetics studies of migratory connectivity were found. No North American stable isotope studies were found to indicate post-breeding connectivity.

MIGRATORY PATHWAYS Birds tracked from the North Slope of Alaska (McCloskey et al. 2018) either staged in Alaska first, or continued to follow the coastline west across the Bering Sea to stage in Japan and Russia and then continued southwest to points in Asia. Birds tracked from the Seward Peninsula and Yukon-Kuskokwim Delta generally staged in Alaska before either crossing the Gulf of Alaska or following the coastline south to the coast of Baja California, the Gulf of California, and Mexico. Alaska birds took similar routes in spring. Birds tracked from the mid-Atlantic, US, take four main routes (Spiegel et al. 2017; Gray and Olsen 2023) that include coastal and overland pathways and mixes of the two. Northward migrations include staging areas along the way, including Nantucket Shoals, US, and Prince Edward Island and the St. Lawrence River, Canada, with a small number staging in the Great Lakes. After staging, loons generally took overland migrations to their breeding areas. During their southward migrations, Hudson Bay was a key molting area, before visiting the St. Lawrence River estuary and other sites before taking an overland or coastal migration to the mid-Atlantic coast of the US.

INTERESTING NOTE

Kleinschmidt et al. (2022) demonstrated migratory connectivity between European populations of Red-throated Loons (aka Red-throated Divers) and Greenland. Two of 33 individuals tracked by satellite transmitters from the German Bight (a nonbreeding and staging area) migrated west to Greenland to presumed breeding grounds, while most (25 individuals) migrated east to breed in Russia. One individual crossed the North Sea, migrated over the North Atlantic, crossing Iceland, and then around the southern tip of Greenland to its western side. Another individual migrated north to Norway before crossing the Norwegian and Greenland seas to the eastern coast of Greenland.

DATA CONTRIBUTORS

D.C. Douglas; J.A. Schmutz; B.D. Uher-Koch.

Gavia arctica ARCTIC LOON

DISTRIBUTION AND ABUNDANCE

A widely distributed species across Arctic and subarctic Eurasia that breeds on freshwater lakes and ponds. The breeding distribution extends to western Alaska, US, where they breed only rarely and very locally (the only known breeding is at the tip of the Seward Peninsula and the north side of Kotzebue Sound, possibly near Nome; Russell 2020a). The species spends the nonbreeding season coastally in both the northeastern Atlantic Ocean and in the western Pacific Ocean, and rarely on the Pacific coast of North America (Russell 2020a).

ORDER Gaviiformes
FAMILY Gaviidae
BREEDING HABITAT Wetlands, Lakes and Rivers
NONBREEDING HABITAT Coasts, Marine Waters
GLOBAL POPULATION SIZE 590,000
POPULATION CHANGE SINCE 1970 NA
IUCN RED LIST CATEGORY Least Concern
R2R TIPPING POINT SPECIES No
US LISTING STATUS No
CANADA SPECIES AT RISK No
PIF CCS 11/20

INTERESTING NOTE

This species was treated as conspecific with the Pacific Loon (*Gavia pacifica*) before 1985, and previous studies were published under this name. Almost all aspects of Arctic Loon ecology and biology require further research.

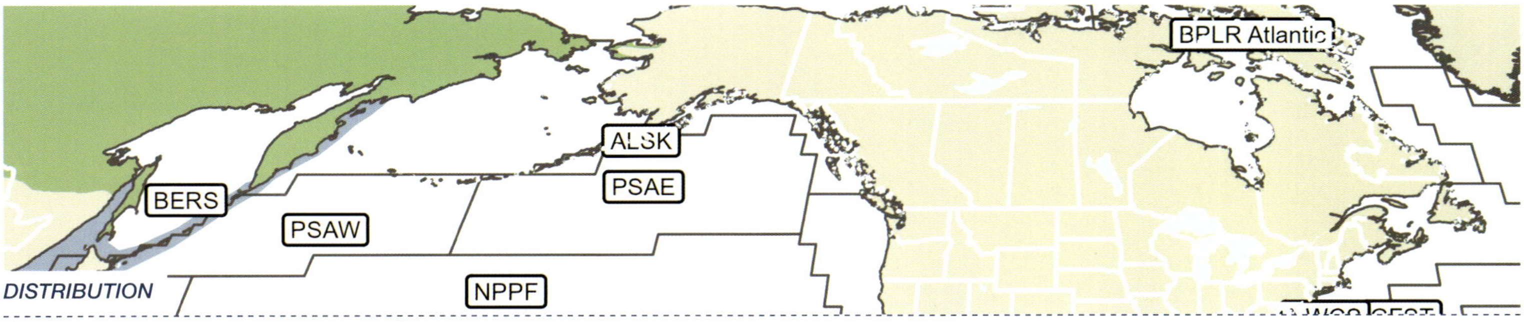

DISTRIBUTION

MIGRATORY CONNECTIVITY

OVERALL Migratory connectivity is unknown as it is unresolved if the small North American breeding population spends the nonbreeding period along the Pacific coast, where some rare sightings have occurred, or along the Asian coast, where the Siberian population spends the nonbreeding period.

OTHER APPROACHES No tracking, banding, genetic, or stable isotope studies of migratory connectivity were found for this species.

MIGRATORY PATHWAYS Migratory pathways are unknown. Alaskan birds were historically thought to migrate west toward the coast of China and Japan to mix with Asian populations (Russell 2020a). However, records of Arctic Loons during the migration and nonbreeding seasons along the US Pacific coast since the 1990s open the question of whether this is a regular migration of Alaskan birds 0or vagrancy (Russell 2020a).

PACIFIC LOON *Gavia pacifica*

DISTRIBUTION AND ABUNDANCE

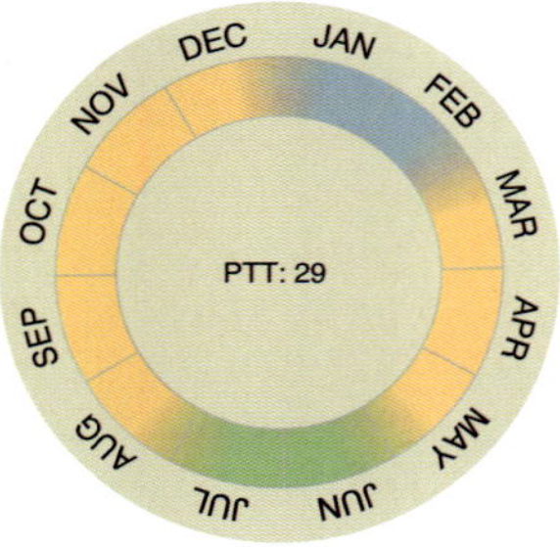

Considered the most abundant loon in North America (Russell 2020b), Pacific Loons breed on freshwater lakes and other wetlands throughout the North American Arctic, subarctic, and taiga regions (Russell 2020b) and spend the nonbreeding season in coastal marine habitats of the northern Pacific Rim. Recent estimates, from decades of breeding and nonbreeding surveys, suggest Pacific Loon populations are stable to increasing across their North American range (Pollet et al. 2024). Breeding population numbers are uncertain due to variability in survey effort and large annual fluctuations but are probably above 162,000 in Alaska. Estimates within Alaska include 30,000 along the Arctic coastal plain (Amundson 2019) and 58,000 in the Yukon-Kuskokwim Delta (Mallek and Groves 2011), with recent suggestions that general surveys underestimate the breeding population (Schmidt et al. 2014). No breeding population estimates for Canada are reported in Russell (2020b).

ORDER Gaviiformes
FAMILY Gaviidae
BREEDING HABITAT Wetlands, Tundra
NONBREEDING HABITAT Coasts, Marine Waters
GLOBAL POPULATION SIZE 840,000
POPULATION CHANGE SINCE 1970 NA
IUCN RED LIST CATEGORY Least Concern
R2R TIPPING POINT SPECIES No
US LISTING STATUS No
CANADA SPECIES AT RISK No
PIF CCS 10/20

	CHIN	KURO	BERS	CCAL	CAMR
Yukon-Kuskokwim Delta, AK, USA (n = 12)	–	–	–	17%	83%
Alpine, AK, USA (n = 8)	38%	25%	38%	–	–

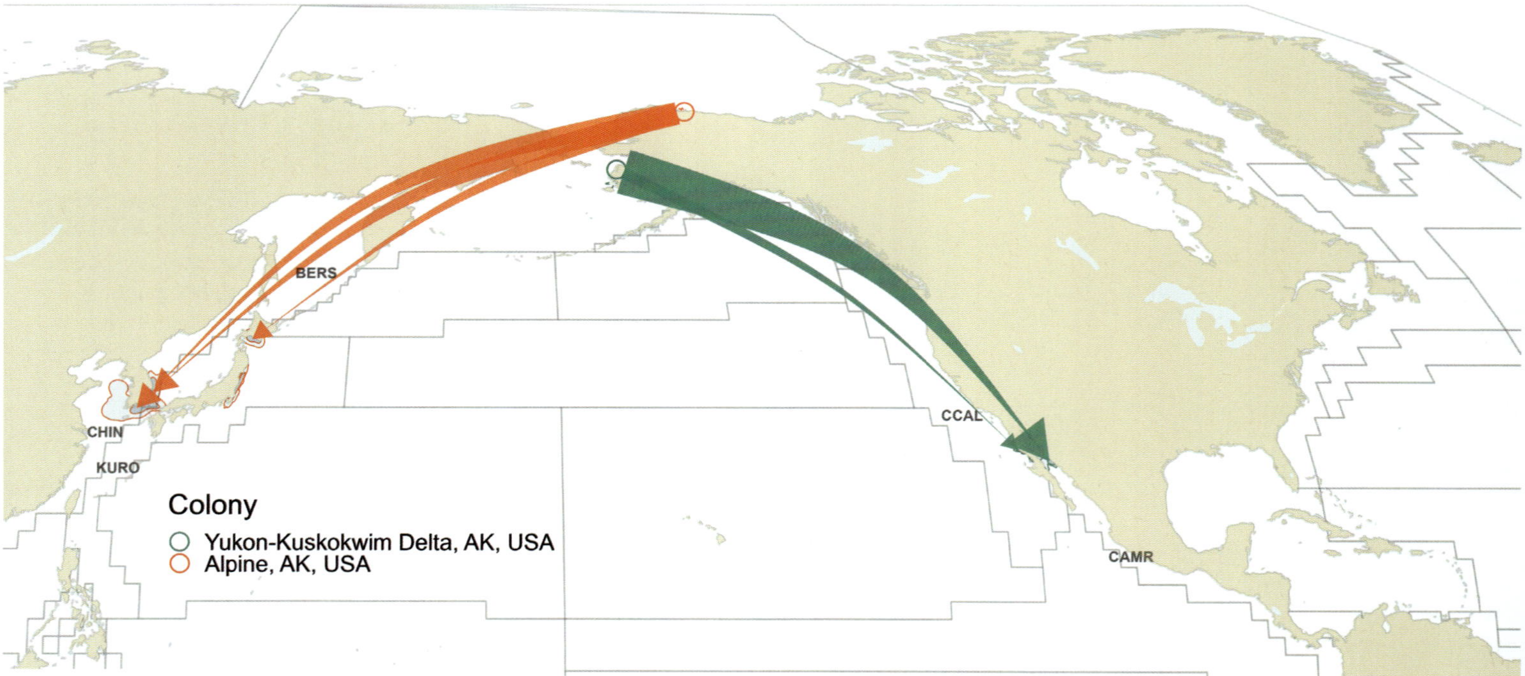

GENERALIZED CONNECTIVITY

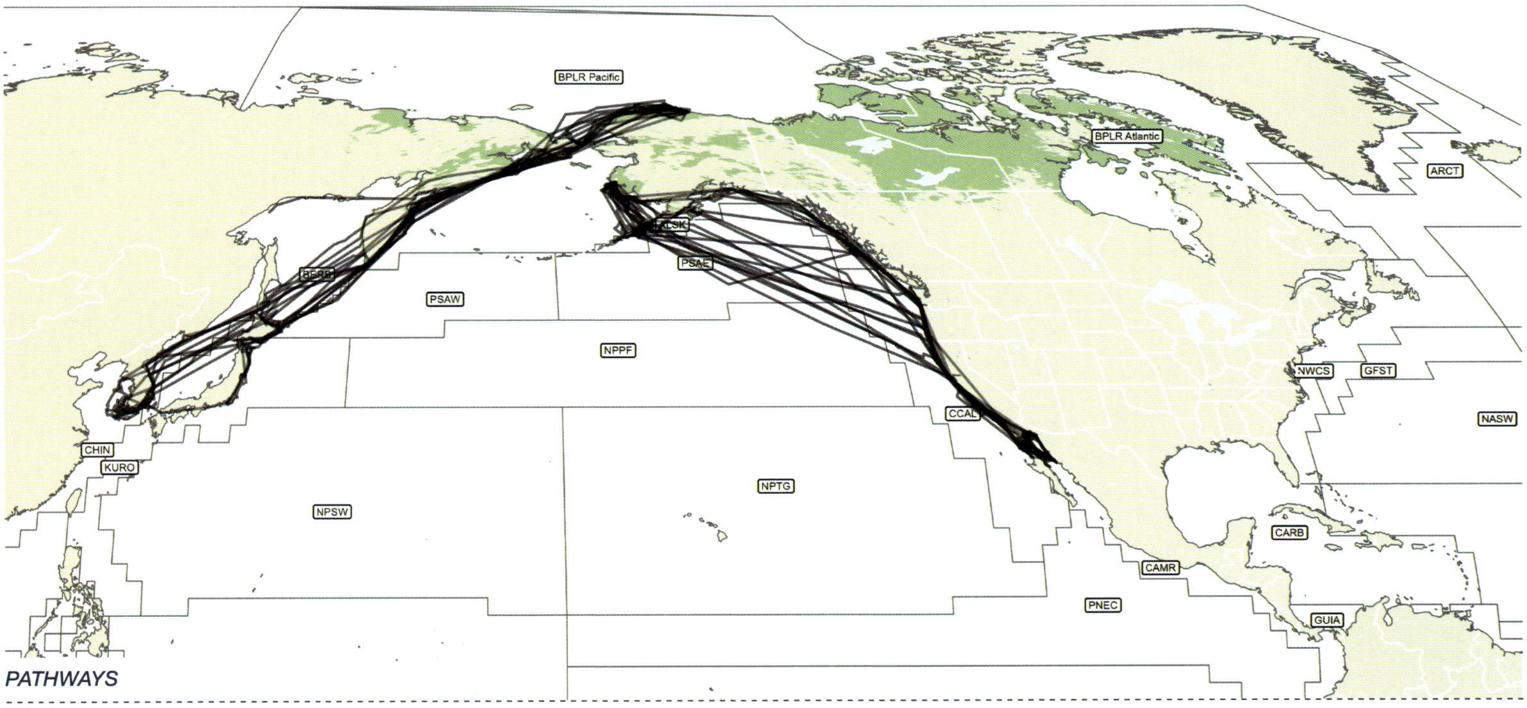

PATHWAYS

MIGRATORY CONNECTIVITY

OVERALL Migratory connectivity of two Alaskan colonies appears strong based on satellite tracking, but limited data prevents estimates of MC. No information is available for Canadian colonies. Birds from northern Alaska (Alpine) distributed across three northwest Pacific provinces (25%–38% of individuals in each) during the nonbreeding period, while all birds from western Alaska (Yukon-Kuskokwim Delta) spent the nonbreeding season in the east Pacific along the Central American Coastal (83%) or the southern California Upwelling Coastal provinces (17%).

SATELLITE TRACKING Birds tracked by satellite by the Migratory Connectivity Project and the US Geological Survey Alaska Science Center (Harrison et al. 2020) from two breeding areas in Alaska exhibited a clear migratory divide. Pacific Loons from northern Alaska spent the nonbreeding season in Asia, while birds from western Alaska spent the nonbreeding season in California and Mexico.

OTHER APPROACHES While the complete genome of the Pacific Loon has been sequenced (confirming separation from the Arctic Loon, *Gavia arctica*; Moon et al. 2019), no genetics studies of migratory connectivity were found. There are no stable isotope studies of migratory connectivity.

MIGRATORY PATHWAYS Birds tracked from the North Slope of Alaska followed the coastline west across the Bering Sea and then southwest to points in Asia. They took similar coastal return routes in pre-breeding migration. Birds tracked from the Yukon-Kuskokwim Delta generally took a first, short flight to the Aleutian Islands, Alaska, for a stopover before crossing the Gulf of Alaska to follow the coastline south to the coast of Baja California, Gulf of California, Mexico. However, birds returned north via a coastal route.

DATA CONTRIBUTORS

D.C. Douglas; A.-L. Harrison; J.A. Schmutz; B.D. Uher-Koch; Migratory Connectivity Project; US Geological Survey, Alaska Science Center.

INTERESTING NOTE

Poessel et al. (2020) evaluated breeding season habitats for Pacific Loons nesting on the Arctic coastal plain (from the satellite tracking data contributed to this study). These loons had very small terrestrial breeding home ranges (0.6 km), but they did use both marine and terrestrial habitats as the season advanced, staging in marine waters offshore northern Alaska during post-breeding before commencing migration.

COMMON LOON *Gavia immer*

DISTRIBUTION AND ABUNDANCE

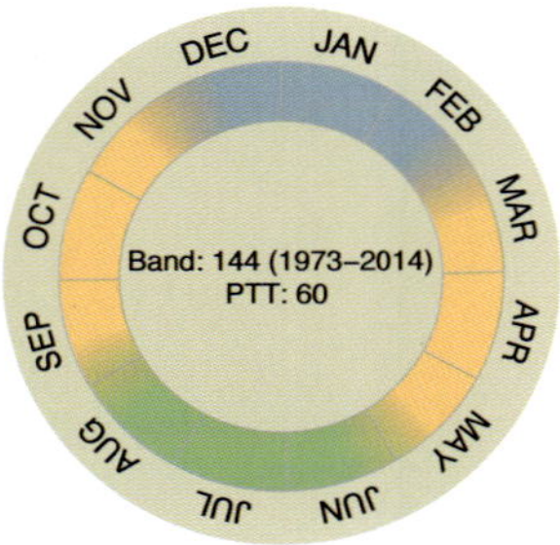

The Common Loon is an iconic bird of northern wilderness, widespread across its breeding range in lakes of Canada and the northern United States. It is also widespread in the nonbreeding season in coastal waters along the Pacific, Atlantic, and Gulf coasts, as well as large, southeastern water bodies (Kenow et al. 2021; Paruk et al. 2021; Fink et al. 2023).

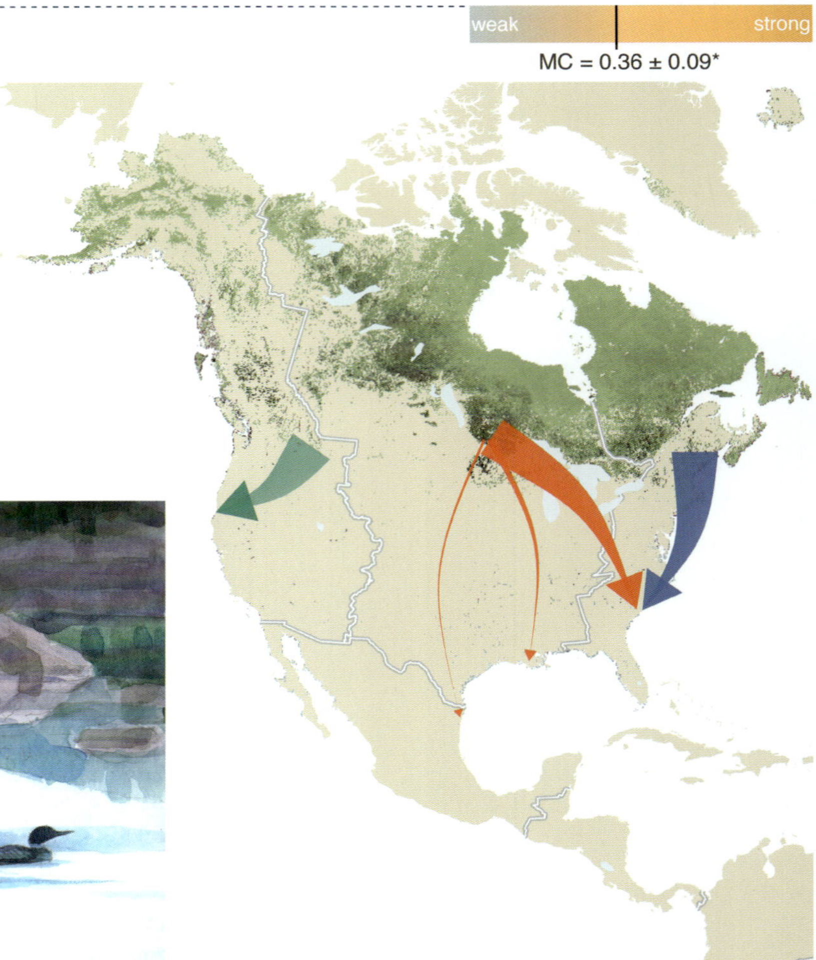

GENERALIZED CONNECTIVITY

	Atlantic	Mexico	Midcontinent	Pacific
Atlantic (n = 35)	100%	–	–	–
Midcontinent (n = 162)	90%	1%	10%	–
Pacific (n = 3)	–	–	–	100%

ORDER Gaviiformes
FAMILY Gaviidae
BREEDING HABITAT Wetlands, Lakes and Rivers
NONBREEDING HABITAT Coasts, Marine Waters
GLOBAL POPULATION SIZE 1,200,000
POPULATION CHANGE SINCE 1970 +43%
IUCN RED LIST CATEGORY Least Concern
R2R TIPPING POINT SPECIES No
US LISTING STATUS No
CANADA SPECIES AT RISK No
PIF CCS 9/20

MIGRATORY CONNECTIVITY

OVERALL Common Loons are a well-studied species, with data available from banding, satellite transmitters, geolocators, and genetics. Atlantic Flyway breeding birds migrate to nonbreeding areas within the Atlantic Flyway. Similarly, Pacific Flyway breeding birds spend the breeding and nonbreeding seasons within the same flyway. The majority of Midcontinent Flyway breeding birds spend the nonbreeding season in the Atlantic flyway (90%). Although within-flyway transitions are high within the coastal flyways, the overlap of Atlantic and Midcontinent flyway breeding populations in the Atlantic Flyway during the nonbreeding season leads to moderate migratory connectivity (MC = 0.36 ± 0.09).

BANDING Banded individuals from the Great Lakes have been recovered along the Gulf Coast in the nonbreeding season (Belant et al. 1991). Other studies have used banding data in combination with satellite transmitter tracking to infer migratory connectivity (Gray et al. 2014; Paruk et al. 2015).

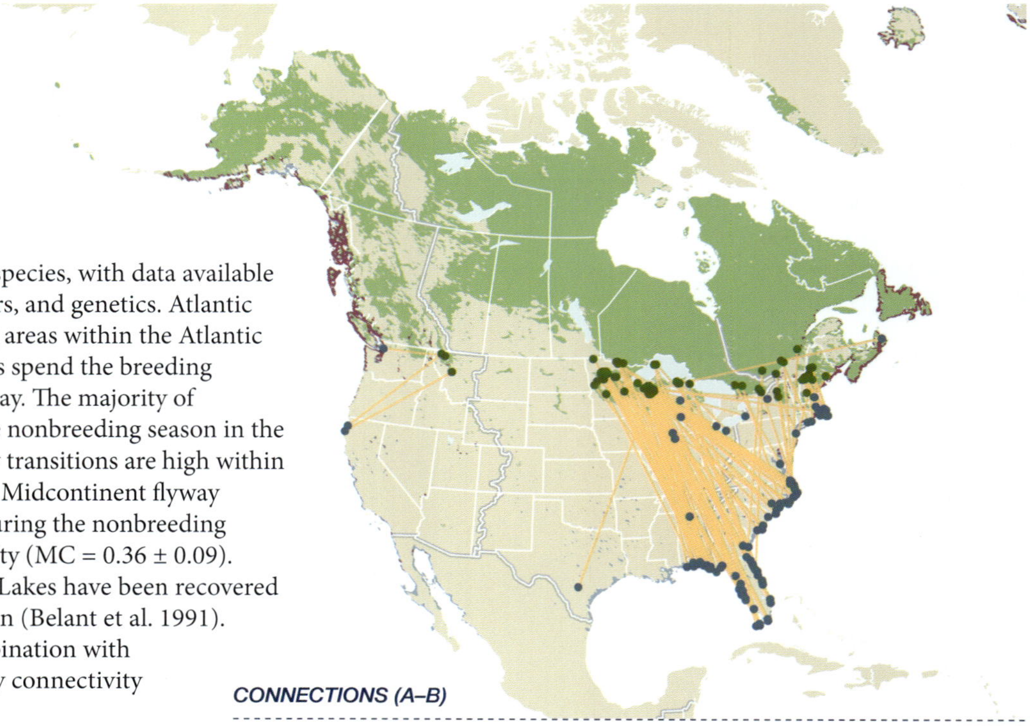

CONNECTIONS (A–B)

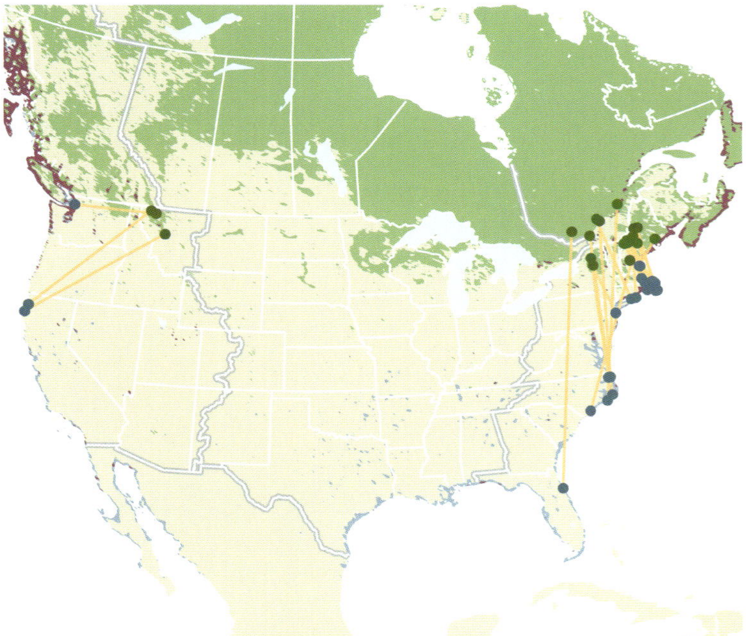

CONNECTIONS A

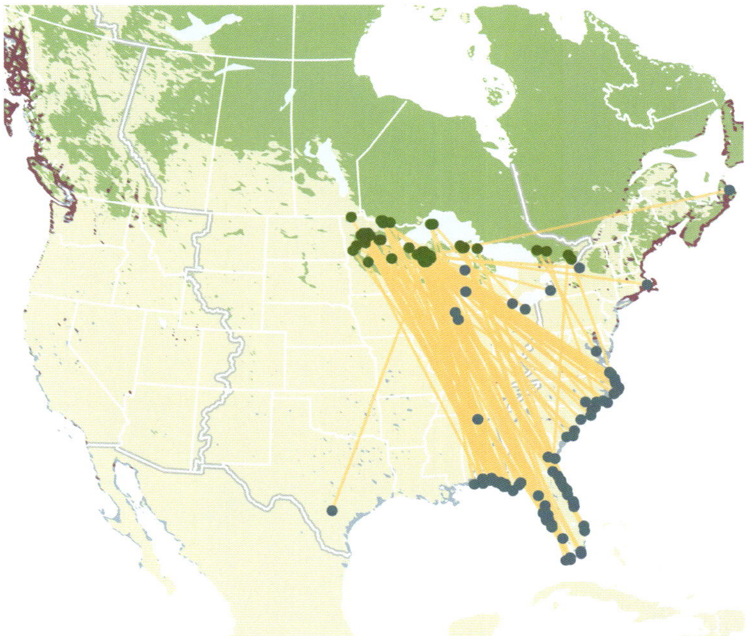

CONNECTIONS B

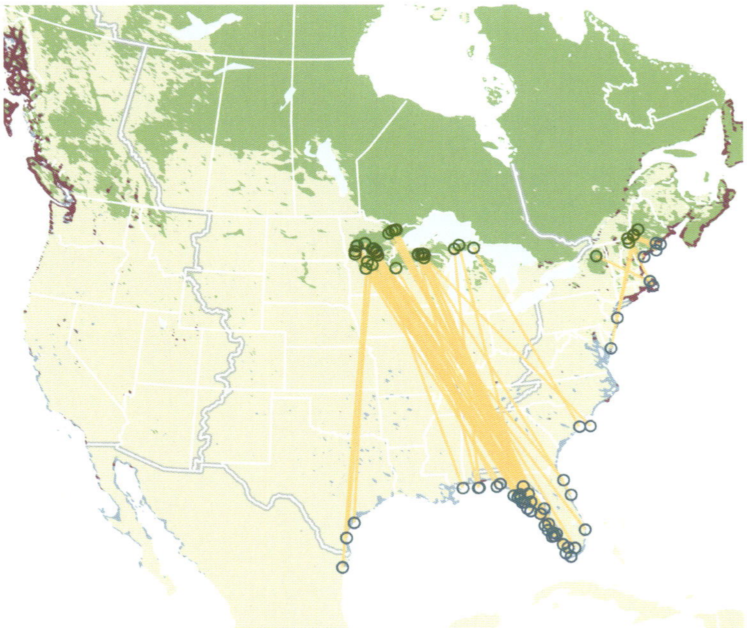

CONNECTIONS C

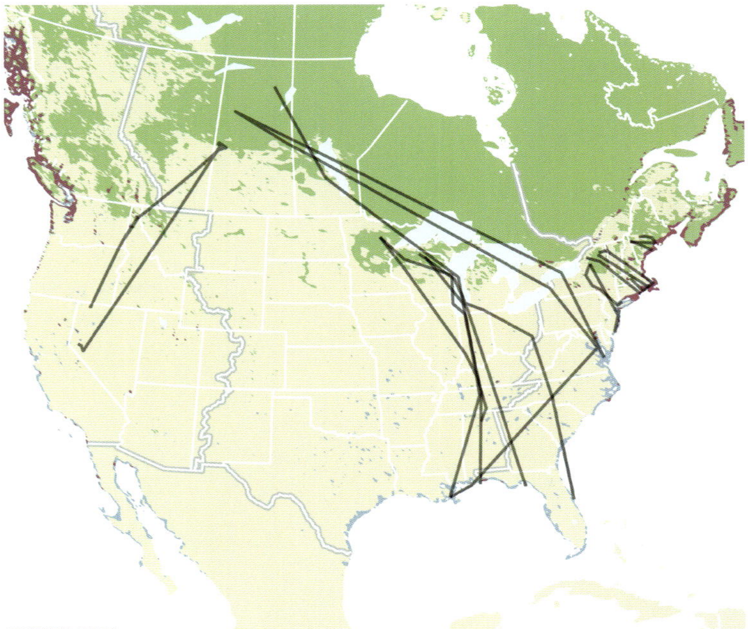

PATHWAYS

INTERESTING NOTE

A summary of banding records combined with satellite tracking found that loons nesting in interior US states and Canadian provinces travel a farther distance to reach their nonbreeding sites than those nesting in coastal states. These different breeding regions are correlated with differences in body mass, with those migrating shorter distances being significantly larger (Gray et al. 2014).

GEOLOCATOR TRACKING Studies have used geolocators in combination with satellite transmitter tracking data (Gray et al. 2014; Paruk et al. 2015; Kenow et al. 2018, 2021).

SATELLITE TRACKING Satellite tracking (combined with resighting of color-marked individuals) showed high site fidelity to nonbreeding locations in California, Washington, Louisiana, Maryland, and Massachusetts (Paruk et al. 2015; Kenow et al. 2021). Two nonbreeding individuals on the coast of Louisiana were tracked to breeding grounds in Saskatchewan, while other loons breeding in Saskatchewan have been tracked to nonbreeding grounds in Nevada, the Gulf of California, and the Pacific Ocean (Yates et al. 1999; Paruk et al. 2014). Breeders from Wisconsin and Minnesota were tracked to nonbreeding grounds in Florida waters of the Gulf of Mexico and in the Atlantic Ocean offshore of southeastern US states (Kenow et al. 2002, 2021). Individuals breeding on lakes in New York, New Hampshire, and Maine were tracked to nonbreeding locations on the Atlantic coast (Kenow et al. 2009). Loons (n = 89) were tracked using satellite transmitters and light-level geolocators from breeding grounds in Minnesota, Wisconsin, and Michigan. Most of these birds staged in Lake Michigan, and 82% went on to spend the nonbreeding season on the Gulf of Mexico, 15% on the Atlantic coast, and 2% in Indiana, Kentucky, Alabama, or Georgia reservoirs (Kenow et al. 2018).

GENETICS A sample of mitochondrial DNA from nonbreeding individuals in Europe was compared to sequences from North American specimens, and there was a lack of genetic differentiation across the breeding range (Bartolomé et al. 2011). However, a selection of loons sampled from across the nonbreeding range in North America detected birds from at least two breeding populations at any given nonbreeding location, with birds from the Midwest and central Canada widely distributed (Larison et al. 2021). The proportion of birds assigned to the central Canada breeding group increased from east to west, and the proportion of birds assigned to the Midwest population decreased from east to west. New England breeders moved only a short distance to the northeast Atlantic, and Pacific Northwest birds were only found in California during the nonbreeding season (Larison et al. 2021).

MIGRATORY PATHWAYS Birds breeding near the northern coasts of the Atlantic and Pacific undertake short migrations and remain along their breeding coasts, while inland breeding birds undertake longer and more variable migrations. The Great Lakes and other large inland water bodies are important stopover sites during longer migrations. Of birds tracked from breeding areas in Minnesota and Wisconsin, juvenile post-breeding migration was initiated later and was more direct compared to adult routes (Kenow et al. 2021).

DATA CONTRIBUTORS
M. Yates.

YELLOW-BILLED LOON *Gavia adamsii*

DISTRIBUTION AND ABUNDANCE

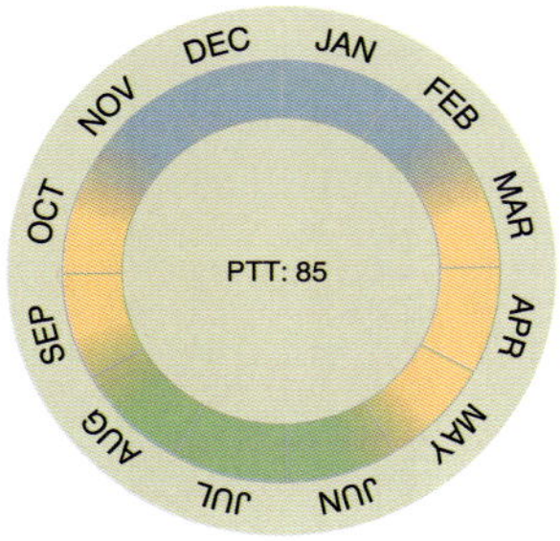

Yellow-billed Loons nest in northern treeless areas, primarily in Arctic tundra regions of Alaska and parts of Canada, as well as in Eurasia. Birds are patchily distributed around or on islands of large lakes and are relatively rare; most areas are not well-surveyed and a global breeding population size is very uncertain (Uher-Koch et al. 2020), as are population trends (Pollet et al. 2024). Despite uncertainties, most surveys suggest Yellow-billed Loon numbers are stable or increasing across the Arctic (Pollet et al. 2024). The species migrates overland in freshwater areas as well as coastally, and far offshore to nonbreeding habitats in protected bays, seas, and nearshore marine habitats (Uher-Koch et al. 2020) of Asia and the Pacific coast of North America.

ORDER Gaviiformes
FAMILY Gaviidae
BREEDING HABITAT Wetlands, Tundra
NONBREEDING HABITAT Coasts, Marine Waters
GLOBAL POPULATION SIZE 16,000

POPULATION CHANGE SINCE 1970 31%
IUCN RED LIST CATEGORY Near Threatened
R2R TIPPING POINT SPECIES Red-Alert
US LISTING STATUS No
CANADA SPECIES AT RISK No
PIF CCS 17/20

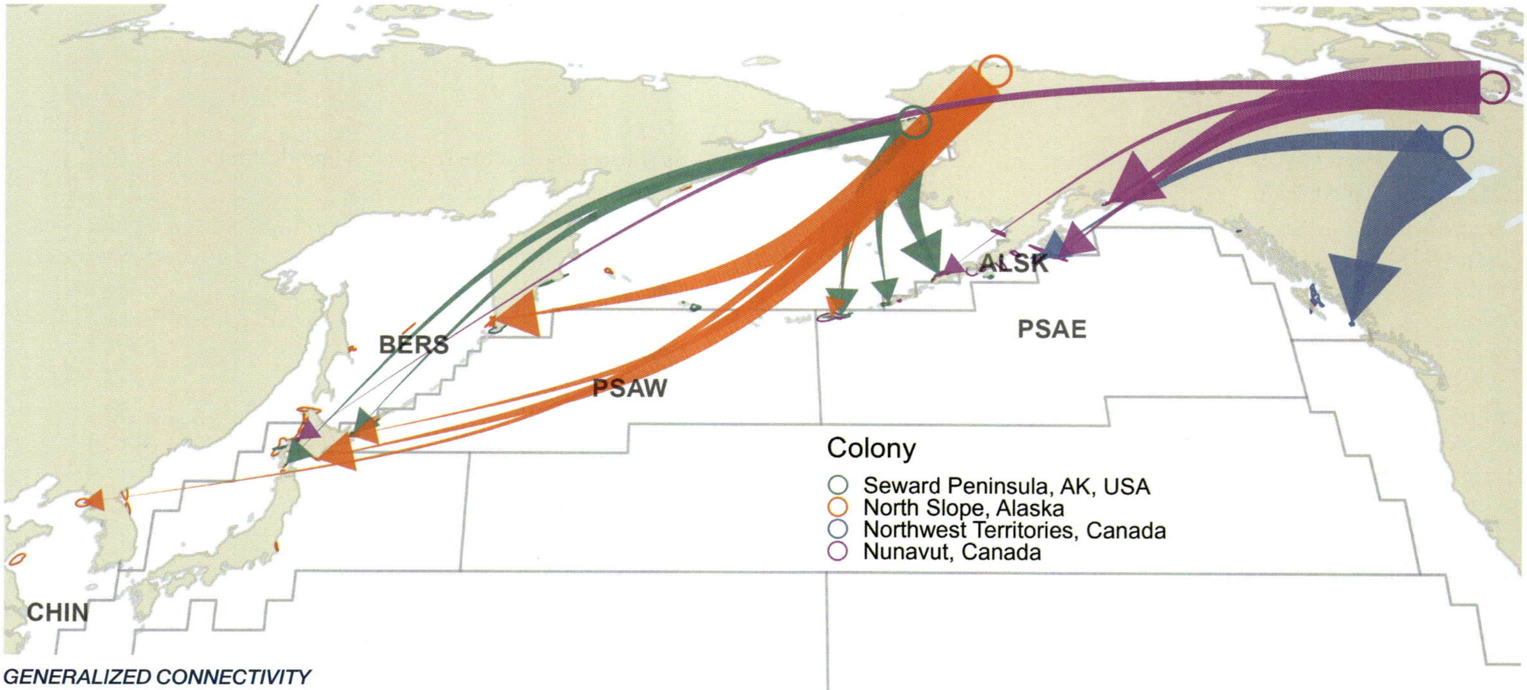

GENERALIZED CONNECTIVITY

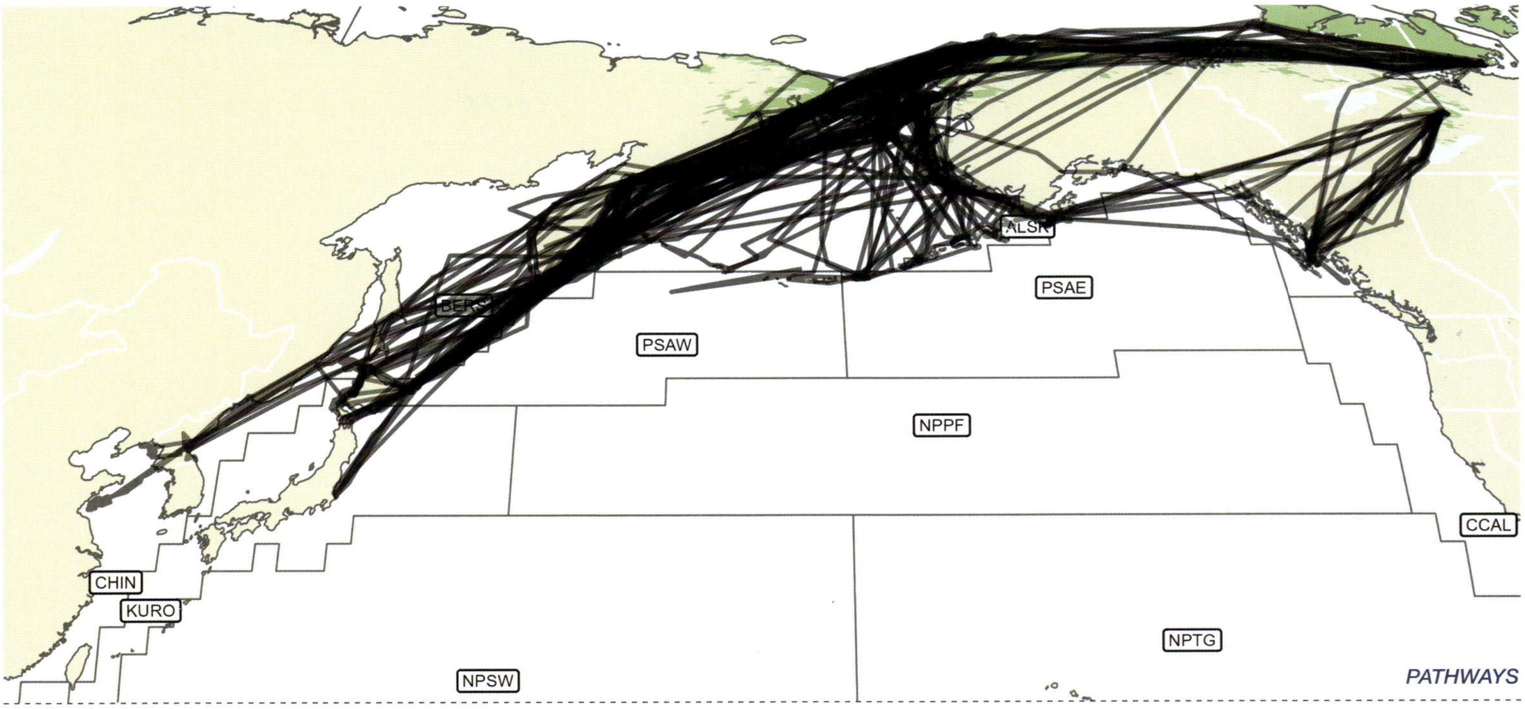

MIGRATORY CONNECTIVITY

	CHIN	KURO	PSAW	BERS	ALSK	PSAE
Seward Peninsula, AK, USA (n = 15)	–	13%	13%	40%	13%	20%
North Slope, Alaska (n = 34)	6%	32%	15%	44%	–	3%
Northwest Territories, Canada (n = 15)	–	–	–	–	80%	20%
Nunavut, Canada (n = 14)	–	7%	–	7%	57%	29%

OVERALL Preliminary genetic evidence provides a hypothesis for population structure (Talbot et al. 2014), and satellite telemetry data suggest some differences between Alaskan and Canadian populations, especially separation in migration routes, with Canadian breeding birds largely migrating to the eastern Pacific and most birds breeding on Alaska's North Slope either in the western Pacific or Bering Sea during the nonbreeding season. However, overlap of tracked breeding populations across nonbreeding areas results in moderate migratory connectivity (Mantel = 0.11 ± 0.06). Alaskan breeders during the nonbreeding season split about 26%–53% in East Asia (Western Pacific Subarctic Gyres, Kuroshio Current, and China Sea provinces), and about 47%–73% in the eastern Pacific Ocean (Bering Sea, Eastern Pacific Subarctic Gyres, and Alaska Coastal Downwelling Current provinces), while Canadian breeders have much higher use of the Alaska Coastal Downwelling Current Province (57%–80%) and other eastern Pacific provinces.

SATELLITE TRACKING Birds tracked by satellite (n = 78) from breeding destinations in Alaska and Canada migrated to nonbreeding areas in either the Alaska Peninsula or Asian waters (Schmutz et al. 2019). Yellow-billed Loons at an inland breeding site in Northwest Territories, Canada, migrated to nonbreeding sites in coastal British Columbia, Canada, or southern Alaska (Schmutz et al. 2019).

GENETICS There is no geographic variation in appearance or morphology, but preliminary results from an ongoing nuclear DNA study of North American loons (Talbot et al. 2014) suggests there may be population structuring between populations in northern Alaska (n = 34) and Canada (Daring Lake, Northwest Territories, n = 14). Given the small sample, more information is needed.

OTHER APPROACHES No stable isotope studies of migratory connectivity were found for this species. Evers et al. (2014) sampled feather mercury concentration from 48 individuals that were also satellite-tracked to nonbreeding areas from three breeding sites (Schmutz et al. 2014; a subset of the data contributed to this Atlas). This study found a significant relationship between feather mercury concentration and the longitude of nonbreeding areas, with higher mercury concentrations in individuals spending the nonbreeding season along the Asian coast.

MIGRATORY PATHWAYS Two migration routes were demonstrated by birds tracked by satellite from breeding destinations in Alaska and Canada (Schmutz et al. 2019). The majority followed coastal migration routes to their nonbreeding destinations. These birds migrated west along the Beaufort Sea coast and south through the Bering Strait. However, most Yellow-billed Loons at an inland breeding site in Northwest Territories, Canada, made short overland migrations to nonbreeding sites.

DATA CONTRIBUTORS

D.C. Douglas; J.A. Schmutz; B.D. Uher-Koch.

INTERESTING NOTE

Since the 1980s, records of Yellow-billed Loons migrating and spending the nonbreeding season inland in North America have increased (Uher-Koch et al. 2020).

ALBATROSSES, PETRELS, AND SHEARWATERS (PROCELLARIIFORMES)

WRITING LED BY **AUTUMN-LYNN HARRISON**

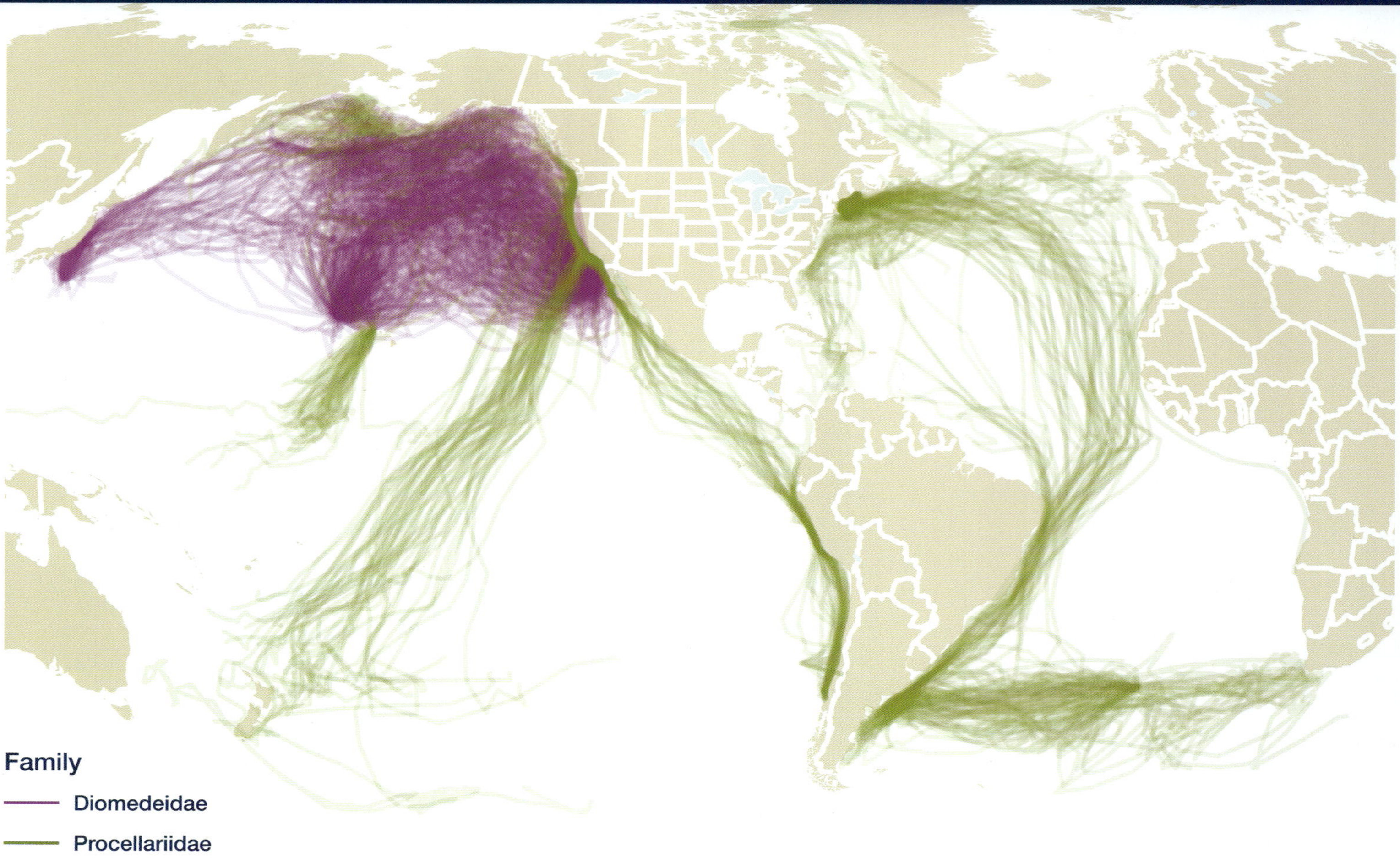

Breeding albatross, shearwaters, petrels, and storm-petrels tracked with satellite and GPS tags (n = 839; n = 178) demonstrate both coastal and pelagic migrations between breeding and nonbreeding areas, as well as long commuting flights during the breeding period that can also span multiple biomes and countries.

The migratory connectivity and associated ecology of the order Procellariiformes, which includes albatrosses, petrels, and shearwaters, and two families of storm-petrels (northern and southern), is well known for some families and little known for others. All albatross (Diomedeidae; n = 3) and 54% of petrels and shearwaters (Procellariidae; n = 28) included in this Atlas have connectivity data of some kind, while only 22% of northern storm-petrels (Hydrobatidae; n = 9) and 0% of southern storm-petrels (Oceanitidae; n = 2) do. Eleven species in this order are considered Tipping Point species, including three storm-petrels (Townsend's, Ashy, Band-rumped), four petrels (Bermuda, Black-capped, Hawaiian, Fea's), three shearwaters (Pink-footed, Manx, Audubon's), and one albatross (Short-tailed Albatross). All of these species are considered poorly monitored species for which urgency is assessed through expert-assigned vulnerability. Two are listed (Hawaiian Petrel and Bermuda Petrel) as US Endangered, and Pink-footed Shearwater is listed as a Canada Species at Risk. All Procellariiformes are considered of conservation concern due to exposure to invasive species while breeding, potential loss of breeding habitat due to sea level rise, at-sea mortality from fisheries bycatch, and power line collisions. Notably, all albatross species have extensive satellite tracking data that have been integral in global conservation efforts to address high at-sea mortality from fisheries bycatch and in monitoring the recovery of translocated populations.

Across the order, Procellariiformes tend to exhibit strong connectivity across ocean basins and weak to moderate within. Origin colonies often overlap to some extent on pelagic nonbreeding ranges that may span multiple countries and oceanographic provinces. A notable exception is Cook's Petrel; birds nesting in southern New Zealand migrate to the southeastern Atlantic Ocean off Peru and Chile, while birds from northern New Zealand migrate to the North Pacific, where their nonbreeding range includes Hawaiian waters and the California Current off the Pacific coast of North America.

We did not find data for 15 species across three families: Band-rumped Storm-Petrel, Black Storm-Petrel, Tristram's Storm-Petrel, and Least Storm-Petrel (Hydrobatidae); Wilson's Storm-Petrel and White-faced Storm-Petrel (Oceanitidae); and Mottled Petrel, Juan Fernandez Petrel, Black-winged Petrel, Bonin Petrel, Buller's Shearwater, and Bulwer's Petrel, Christmas Shearwater, Flesh-footed Shearwater, and Sargasso Shearwater (Procellaridae). These species' range maps are provided by BirdLife International.

weak strong

Mantel = 0.14 ± 0.06

Phoebastria immutabilis **LAYSAN ALBATROSS**

DISTRIBUTION AND ABUNDANCE

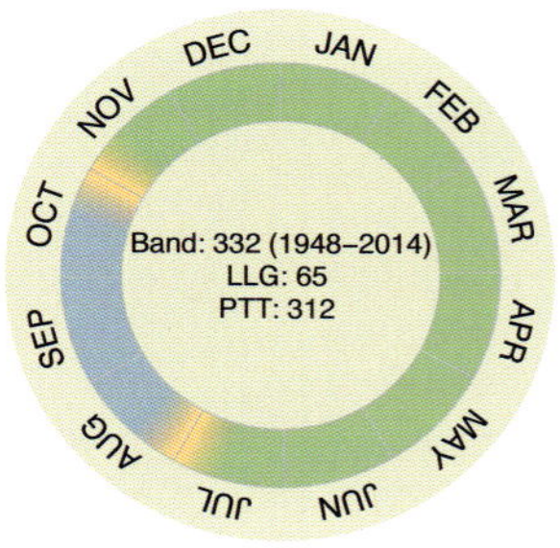

The most populous albatross of the North Pacific, the Laysan Albatross has recovered from 99% population reduction (Arata et al. 2009). Core breeding abundance is in the northwest Hawaiian Islands, but as populations recovered in the late 1970s, Laysan Albatrosses began recolonizing western Pacific sites (Mukokima, Japan, and Wake Islands, US) and the main Hawaiian Islands, where they had been extirpated (Awkerman et al. 2020b). Range expansion has also occurred on Mexican islands of the eastern Pacific where they previously were not documented (Awkerman et al. 2020b). Nonbreeding Laysan Albatrosses range widely across the North Pacific.

ORDER Procellariiformes
FAMILY Diomedeidae
BREEDING HABITAT Islands, Oceanic
NONBREEDING HABITAT Oceans, Pelagic
GLOBAL POPULATION SIZE 1,200,000
POPULATION CHANGE SINCE 1970 NA

IUCN RED LIST CATEGORY Near Threatened
R2R TIPPING POINT SPECIES No
US LISTING STATUS No
CANADA SPECIES AT RISK No
PIF CCS 15/20

MIGRATORY CONNECTIVITY

OVERALL Banding and tracking data from five Hawaiian colonies suggest some regional separation of core habitat use west to east across the Pacific Ocean, resulting in moderate migratory connectivity (Mantel = 0.14 ± 0.06). Midway Atoll albatrosses were more likely encountered in the Western Pacific Subarctic Gyres (56%) and the Bering Sea (21%). French Frigate Shoals albatrosses migrate to the North Pacific Polar Front (33%) and the Western Pacific Subarctic Gyres (21%). Albatrosses in Isla Guadalupe, Mexico, migrate to the Eastern Pacific Subarctic Gyres (88%). Genetics work has found weak population structure range-wide.

BANDING The long recovery record of banded Laysan Albatrosses at sea (1948 onward) is due to their overlap with fishery operations. Banding records are thus fishery-biased. Only one banded bird from Isla Guadalupe, Mexico (a relatively new colony), has been reencountered at sea.

SATELLITE TRACKING Laysan Albatrosses from Tern Island, French Frigate Shoals, have been tracked with satellite tags since 1997 (Fernandez et al. 2001; Kappes et al. 2010; D. Anderson, BirdLife International Seabird Tracking Database). Laysan Albatrosses were also captured at sea off the coasts of the Aleutian Islands in Alaska and tracked during the nonbreeding period to reveal foraging habitat preferences, but these birds were not tracked back to colonies (Fischer et al. 2009; Suryan and Balogh, BirdLife International

	KURO	PSAW	BERS	PSAE	NPPF	CCAL	NPTG
Kure Atoll, HI, USA (n = 2)	–	–	–	–	100%	–	–
Midway Atoll, HI, USA (n = 70)	6%	56%	21%	6%	9%	1%	1%
French Frigate Shoals, HI, USA (n = 42)	2%	21%	7%	17%	33%	–	19%
Isla Guadalupe, Mexico (n = 33)	–	–	–	88%	9%	–	3%

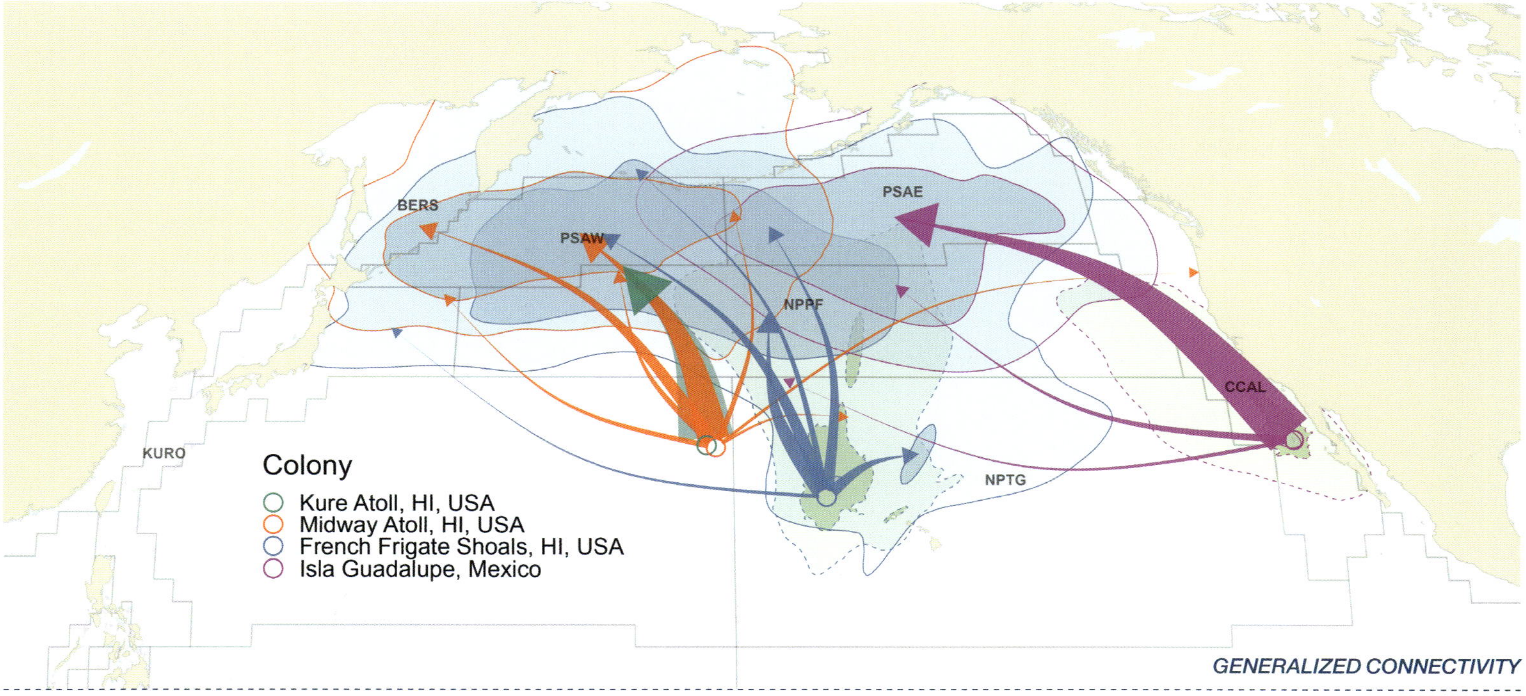

GENERALIZED CONNECTIVITY

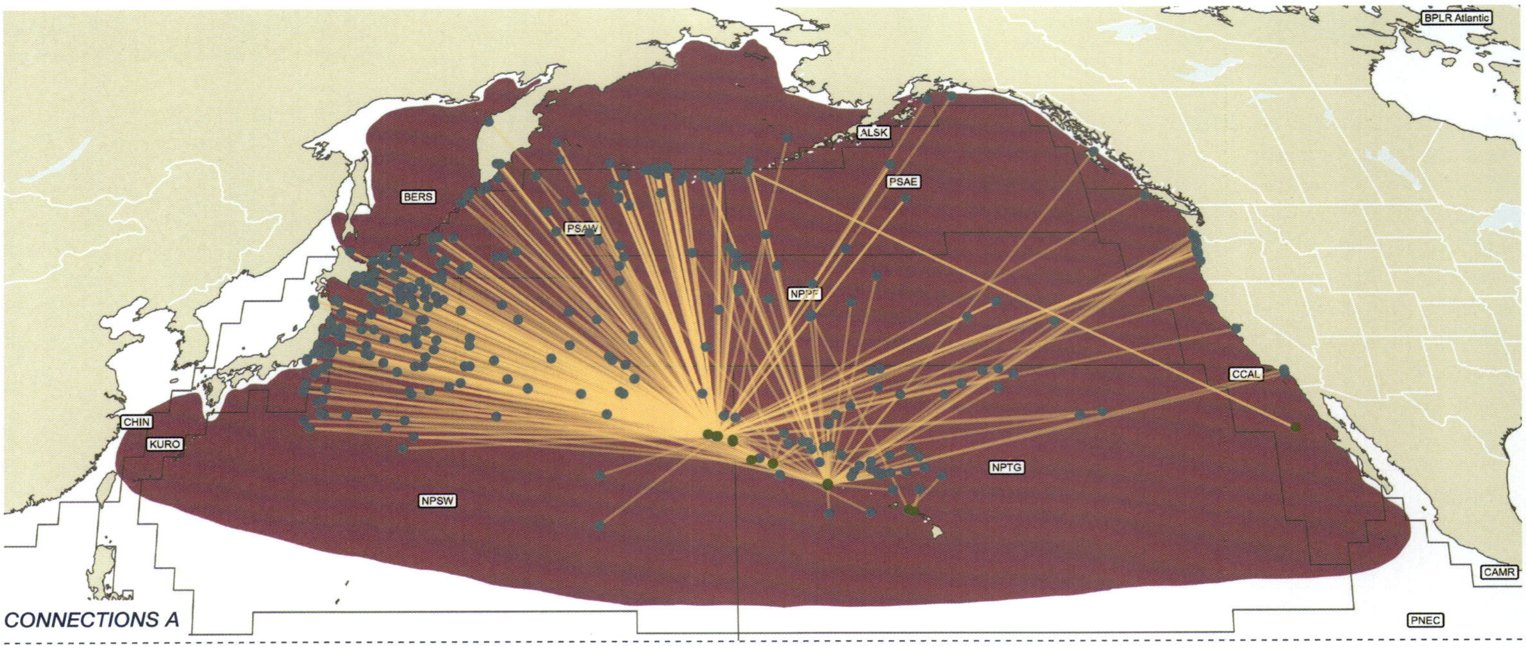

CONNECTIONS A

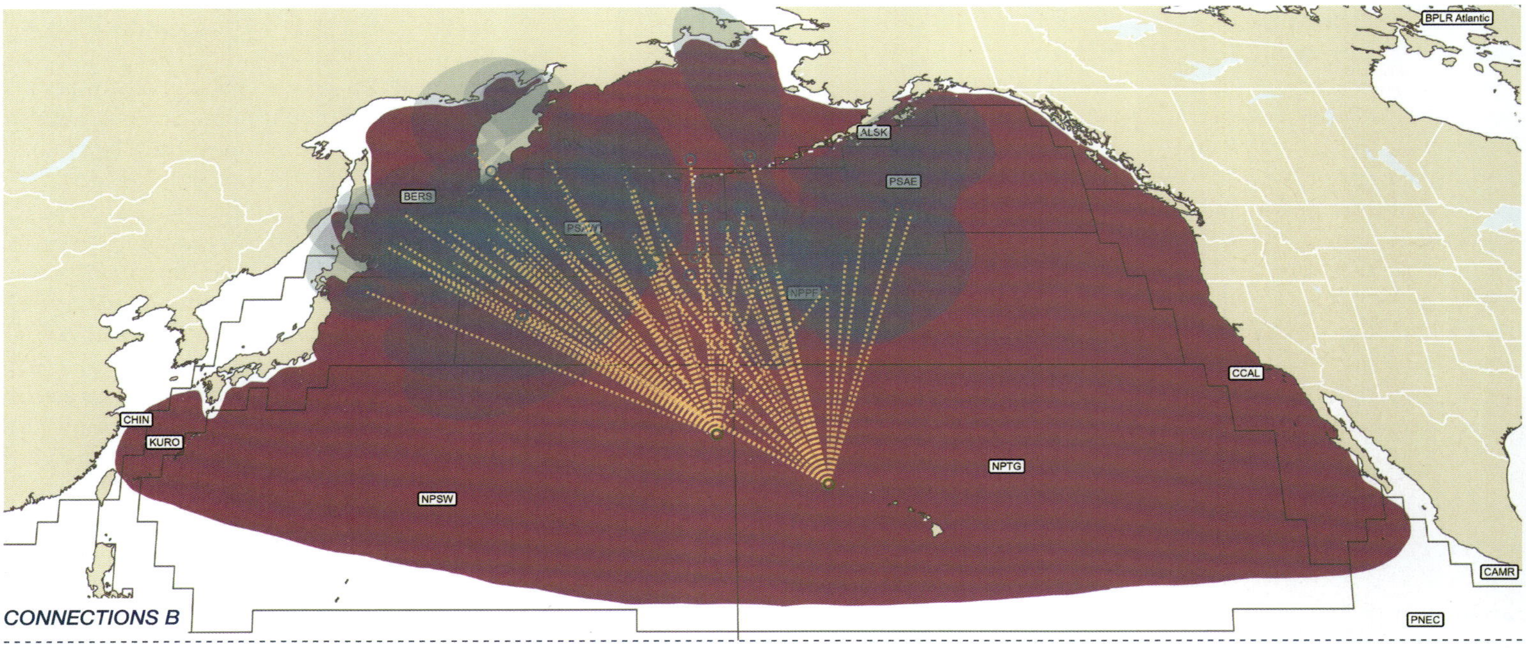

CONNECTIONS B

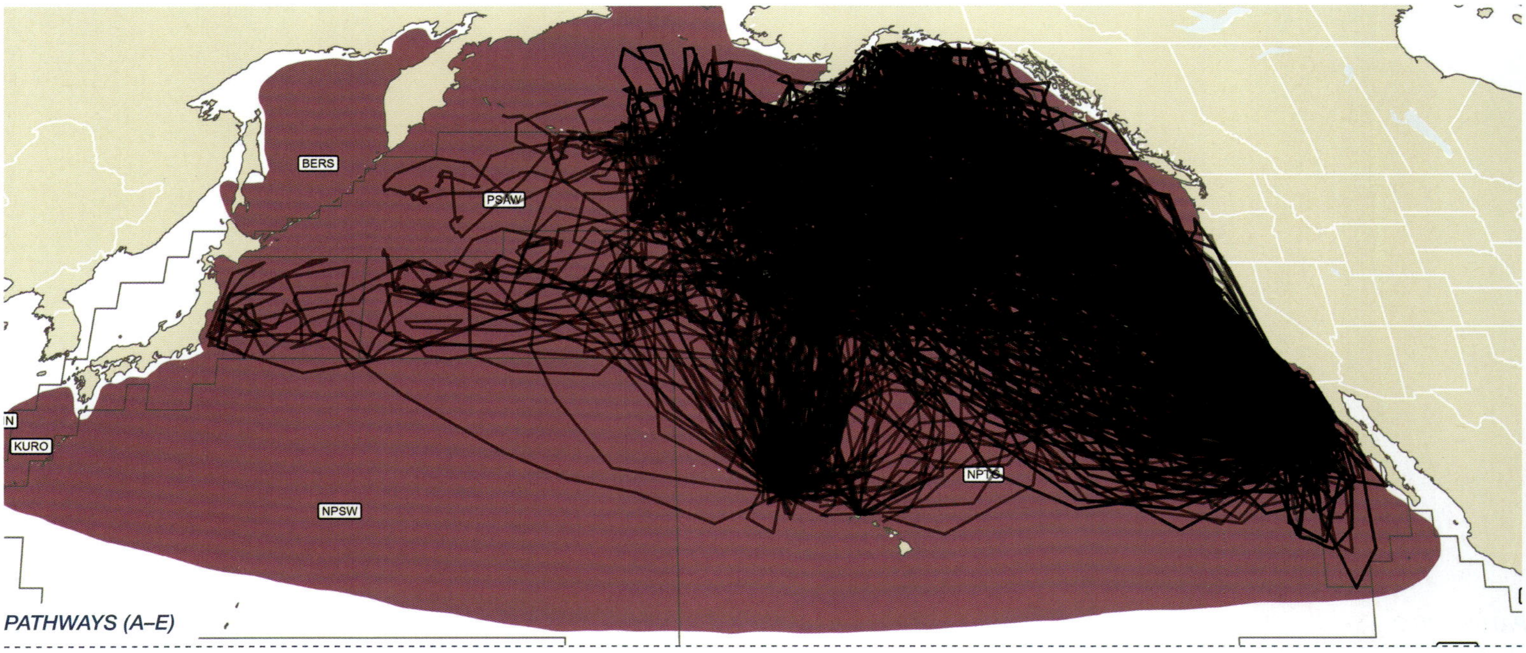

PATHWAYS (A–E)

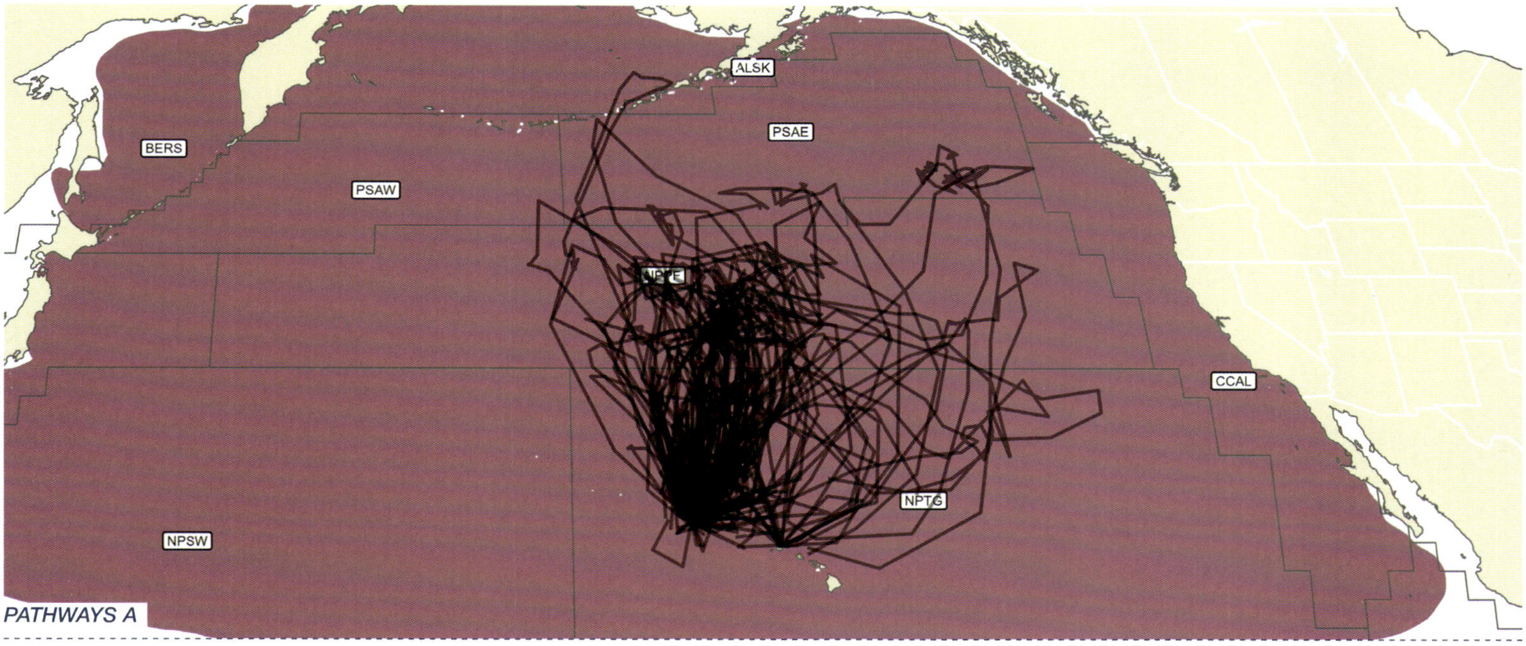
ALSK
BERS
PSAE
PSAW
NPTE
CCAL
NPTG
NPSW
PATHWAYS A

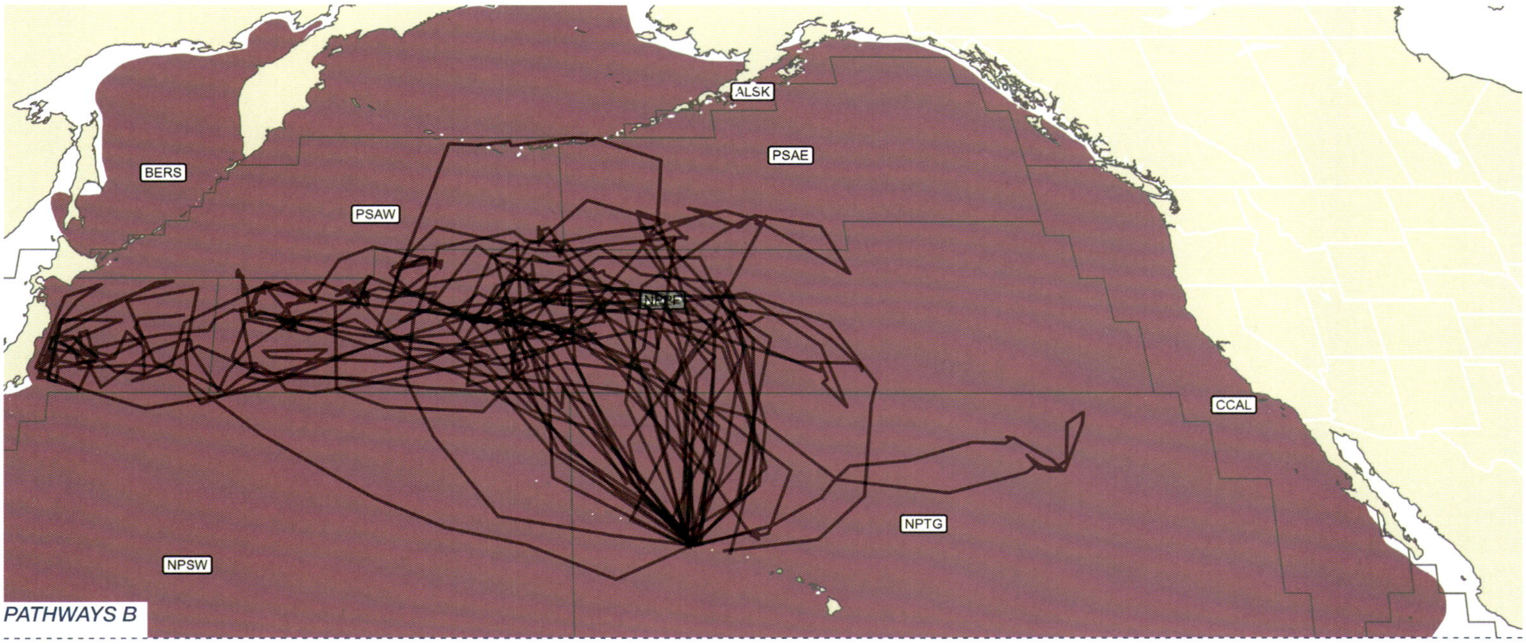
ALSK
BERS
PSAE
PSAW
NPTE
CCAL
NPTG
NPSW
PATHWAYS B

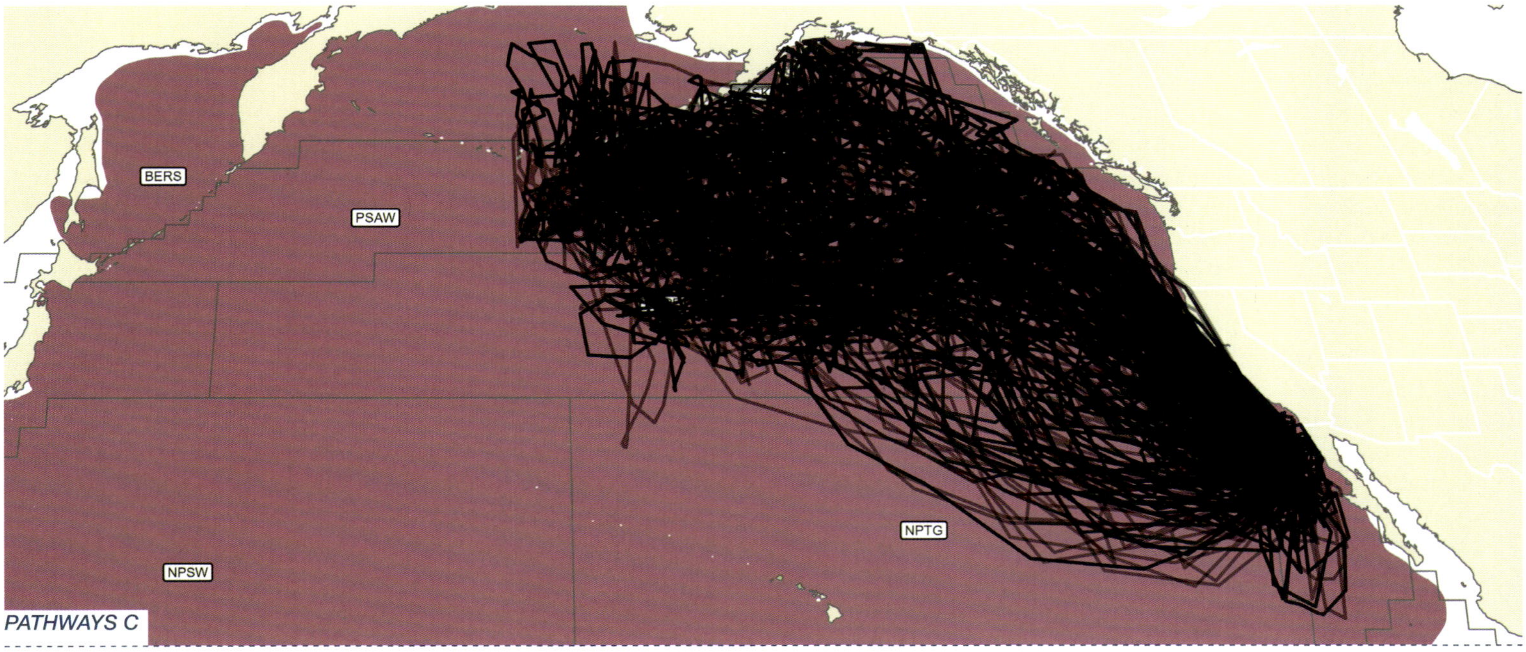
BERS
PSAW
NPTG
NPSW
PATHWAYS C

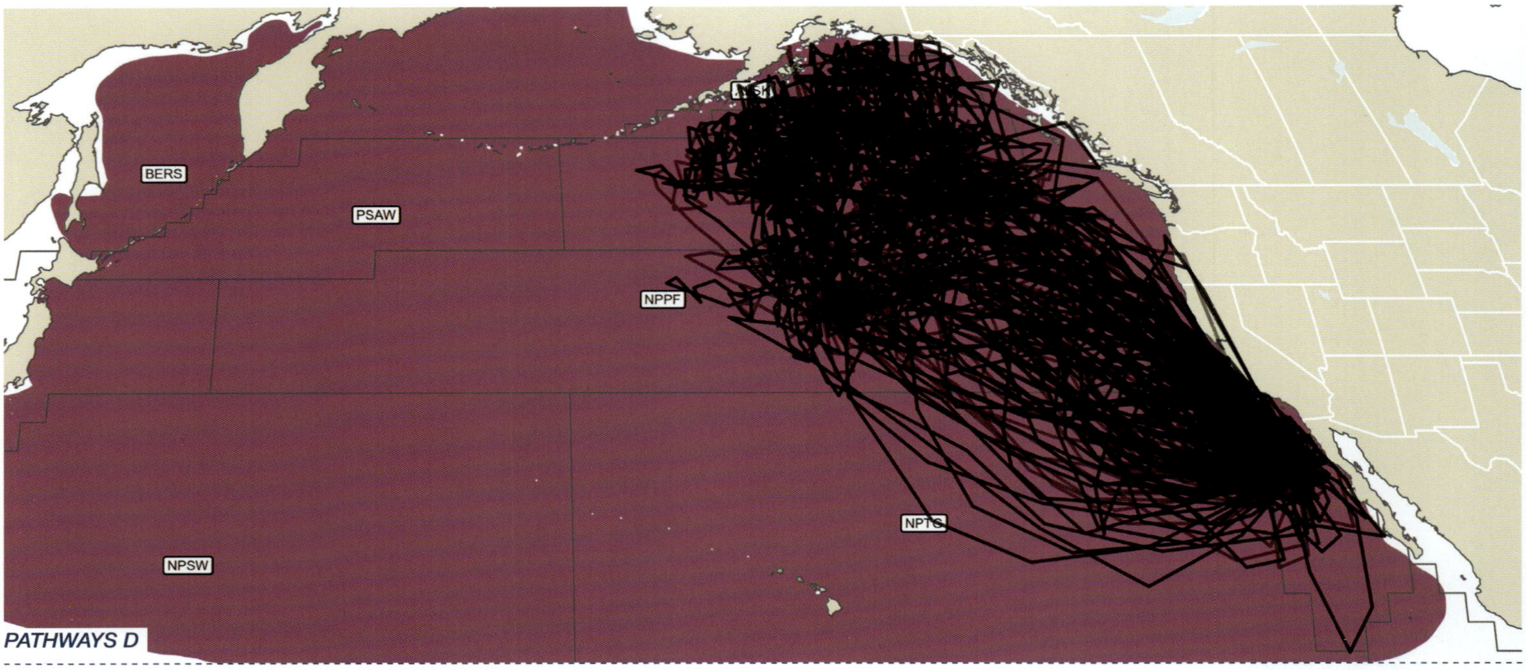

PATHWAYS D

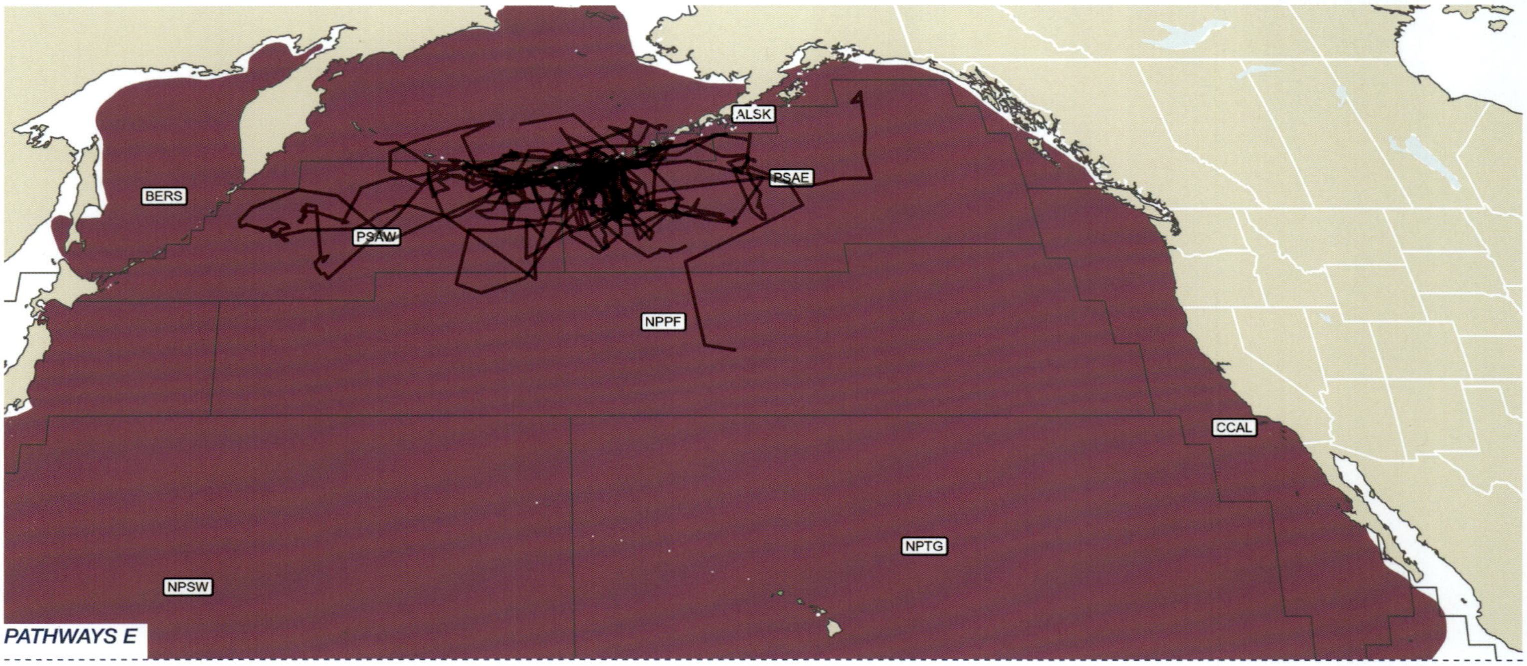

PATHWAYS E

INTERESTING NOTE

Albatrosses are usually highly monogamous. In a relatively new colony on Oahu, Main Hawaiian Islands, 31% of nests monitored 2004–2007 were attended by female-female pairs (Young et al. 2008). This may be the result of a skewed overall sex ratio (59% female) that resulted from female-biased immigration. Future research could address whether a skewed sex ratio at some newly colonized sites is due to differential mortality, connectivity, or dispersal.

Seabird Tracking Database). Multiple studies used both satellite tags and light-level geolocators, sometimes on the same bird to assess geolocator location error and to obtain wet-dry data and sea surface temperature measurements (Shaffer et al. 2005).

GEOLOCATOR TRACKING Light-level geolocators were used to track nonbreeding season movements from Tern Island, French Frigate Shoals (2002–2011) and Midway Atoll (2012–2016; Jordan et al. 2022). Birds from the two colonies had high overlap in the western Pacific, but core areas of birds from Midway were distributed west of those from Tern Island and had a greater proportion of their total distribution in the north Bering Sea. Young et al. (2009) also found high overlap of broad post-breeding distributions across two breeding colonies tracked with geolocators from the main Hawaiian Islands, but <50% overlap of specific core areas used. Birds from Isla Guadalupe in the eastern North Pacific had minimal overlap of core habitat use with Laysan Albatrosses from Tern Island or Midway Island (Henry et al. 2021; Henry and Costa, BirdLife International Seabird Tracking Database).

STABLE ISOTOPES Over 25 years of stable isotope work on post-breeding birds compares Laysan Albatross with its sister species, Black-footed Albatross (*P. immutabilis*). Samples salvaged from fisheries bycatch suggest that Black-footed Albatrosses feed at higher trophic levels (e.g., Gould et al. 1997; Inoue et al. 2023) – the two species – and the two species separate geographically (Finkelstein et al. 2006).

GENETICS Using nuclear DNA and mitochondrial DNA analysis, no significant genetic population structure was detected between Laysan populations on nine Hawaiian Islands, the Bonin Island in Japan, and Guadalupe Island near Mexico (Young 2010). This suggested male-mediated dispersal and high philopatry in females. Multiple source populations contributed to newly colonized sites (Young 2010).

MIGRATORY PATHWAYS After breeding, Laysan Albatrosses move to a region of the North Pacific (to the western, central, or eastern Pacific, depending on the colony; Jordan et al. 2022; Henry et al. 2021), where they are largely nomadic for the post-breeding period. The true migration portion of their pathways from the colony to pelagic habitats is relatively short, and the entire five months the birds wander at sea after breeding is sometimes referred to as their "post-breeding migration."

DATA CONTRIBUTORS

D. Anderson; M. Antolos; G.R. Balogh; M. Conners; D. Costa; R.W. Henry; S.A. Shaffer; R.M. Suryan.

Mantel = –0.02 ± 0.02

Phoebastria nigripes

BLACK-FOOTED ALBATROSS

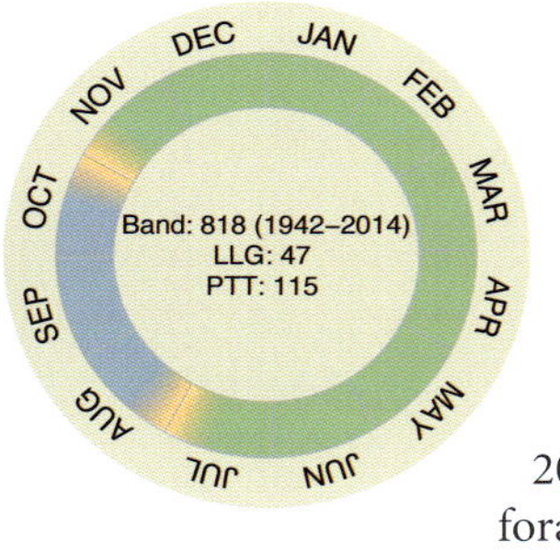

DISTRIBUTION AND ABUNDANCE

This albatross of the North Pacific nests on beaches of the Hawaiian archipelago and on offshore Japanese islands. Hawaiian colonies are 95% of the global breeding population with Midway Atoll and Laysan Island the largest colonies (Awkerman et al. 2020a). This species makes long-distance foraging trips during some breeding stages and travels the entire North Pacific during the nonbreeding period, using both continental margins and deep pelagic waters. It is regularly seen off the North American coast.

ORDER Procellariiformes
FAMILY Diomedeidae
BREEDING HABITAT Islands, Oceanic
NONBREEDING HABITAT Oceans, Pelagic
GLOBAL POPULATION SIZE 130,000
POPULATION CHANGE SINCE 1970 NA

IUCN RED LIST CATEGORY
Near Threatened
R2R TIPPING POINT SPECIES No
US LISTING STATUS No
CANADA SPECIES AT RISK No
PIF CCS 15/20

Colony
- Kure Atoll, HI, USA
- Midway Atoll, HI, USA
- Pearl and Hermes Atoll, HI, USA
- Laysan Island (Kauō), HI, USA
- French Frigate Shoals, HI, USA

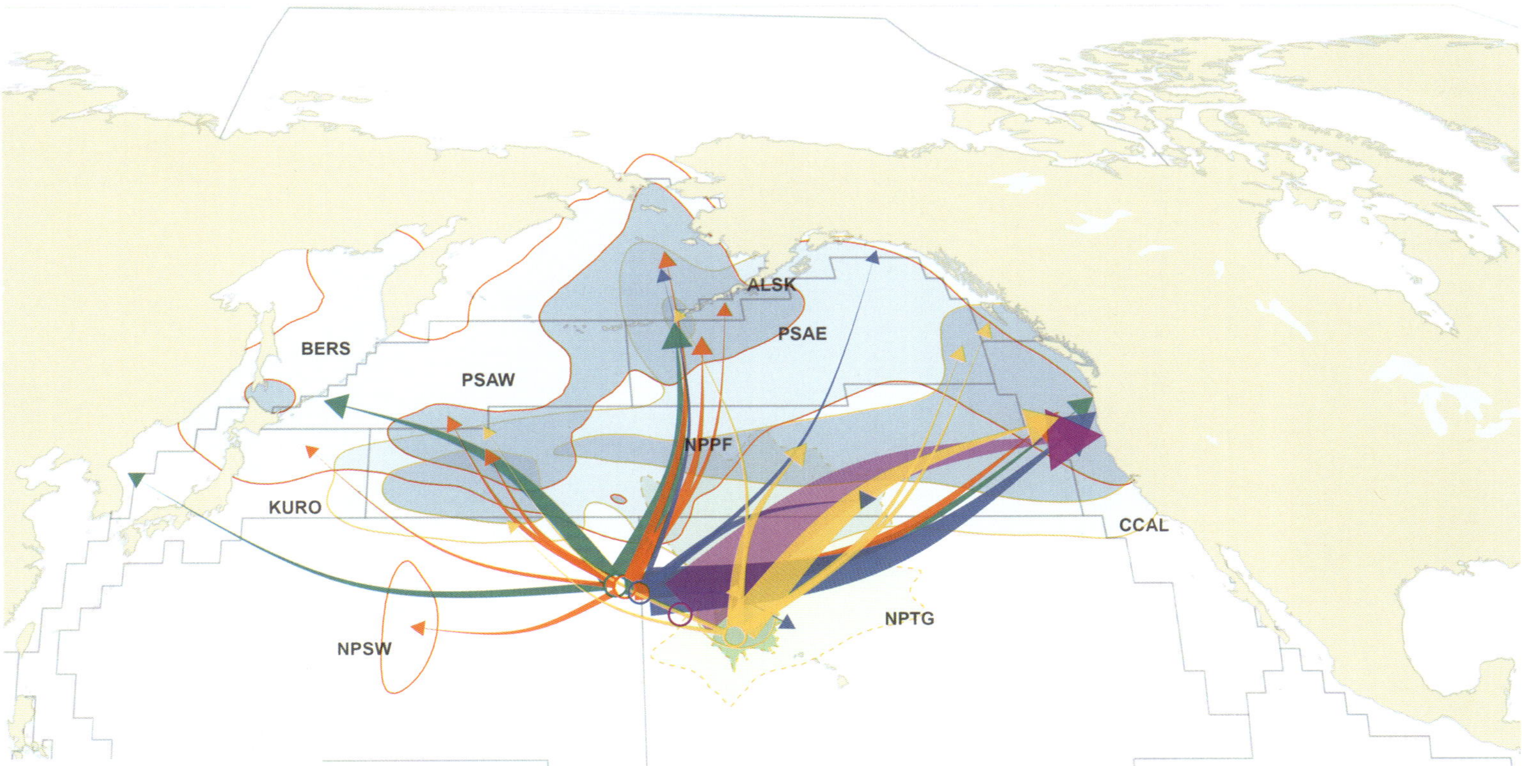

GENERALIZED CONNECTIVITY

MIGRATORY CONNECTIVITY

OVERALL Banding and tracking data from five Hawaiian colonies suggest colonies mix at sea, resulting in weak migratory connectivity (Mantel = −0.02 ± 0.02), but with some apparent regional differences. Birds from western colonies (Kure Atoll and Midway Atoll) were more often observed migrating to northern and western provinces. For example, 63% of Kure Atoll individuals were observed migrating to either the Western or Eastern Pacific Subarctic Gyres provinces, and 9% in the Bering Sea, while 72% of birds from Pearl and Hermes Atoll and 51% of birds from French Frigate Shoals were observed in the California Current provinces in the northeastern Pacific.

	KURO	NPSW	PSAW	BERS	ALSK	PSAE	NPPF	CCAL	NPTG
Kure Atoll, HI, USA (n =11)	–	–	27%	9%	–	36%	–	27%	–
Midway Atoll, HI, USA (n = 124)	2%	6%	8%	11%	5%	17%	10%	40%	2%
Pearl and Hermes Atoll, HI, USA (n = 25)	–	–	–	4%	4%	–	12%	72%	8%
Laysan Island (Kauō), HI, USA (n = 2)	–	–	–	–	–	–	–	100%	–
French Frigate Shoals, HI, USA (n = 53)	–	2%	2%	2%	8%	9%	23%	51%	4%

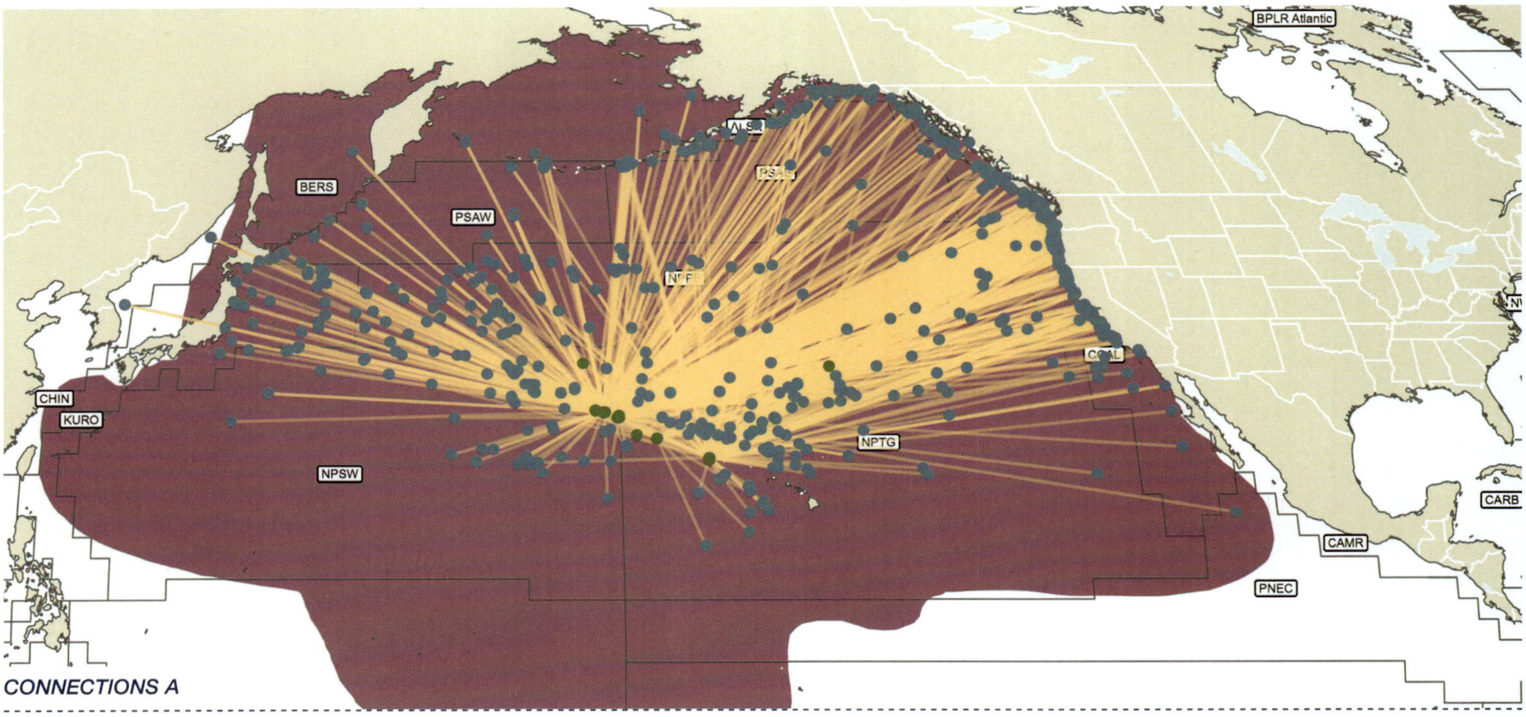

CONNECTIONS A

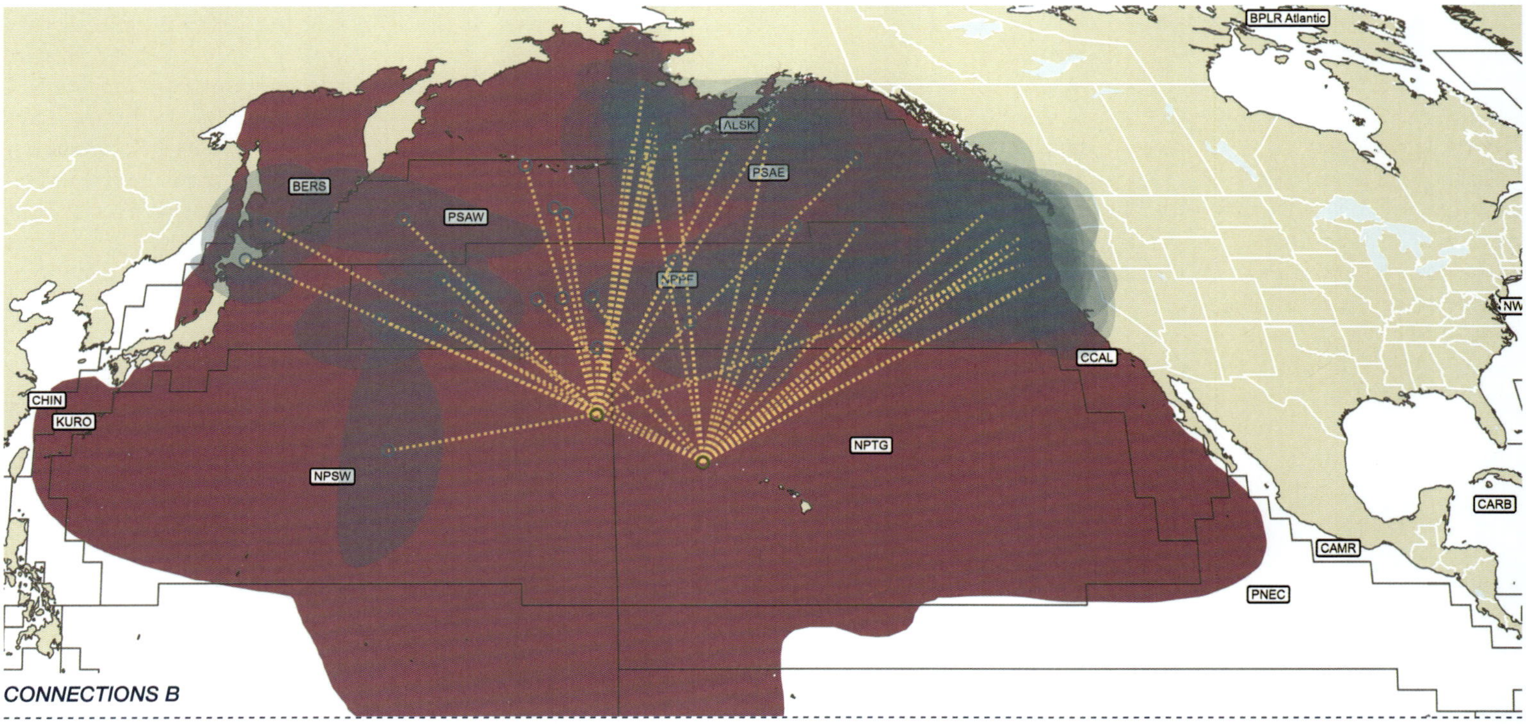

CONNECTIONS B

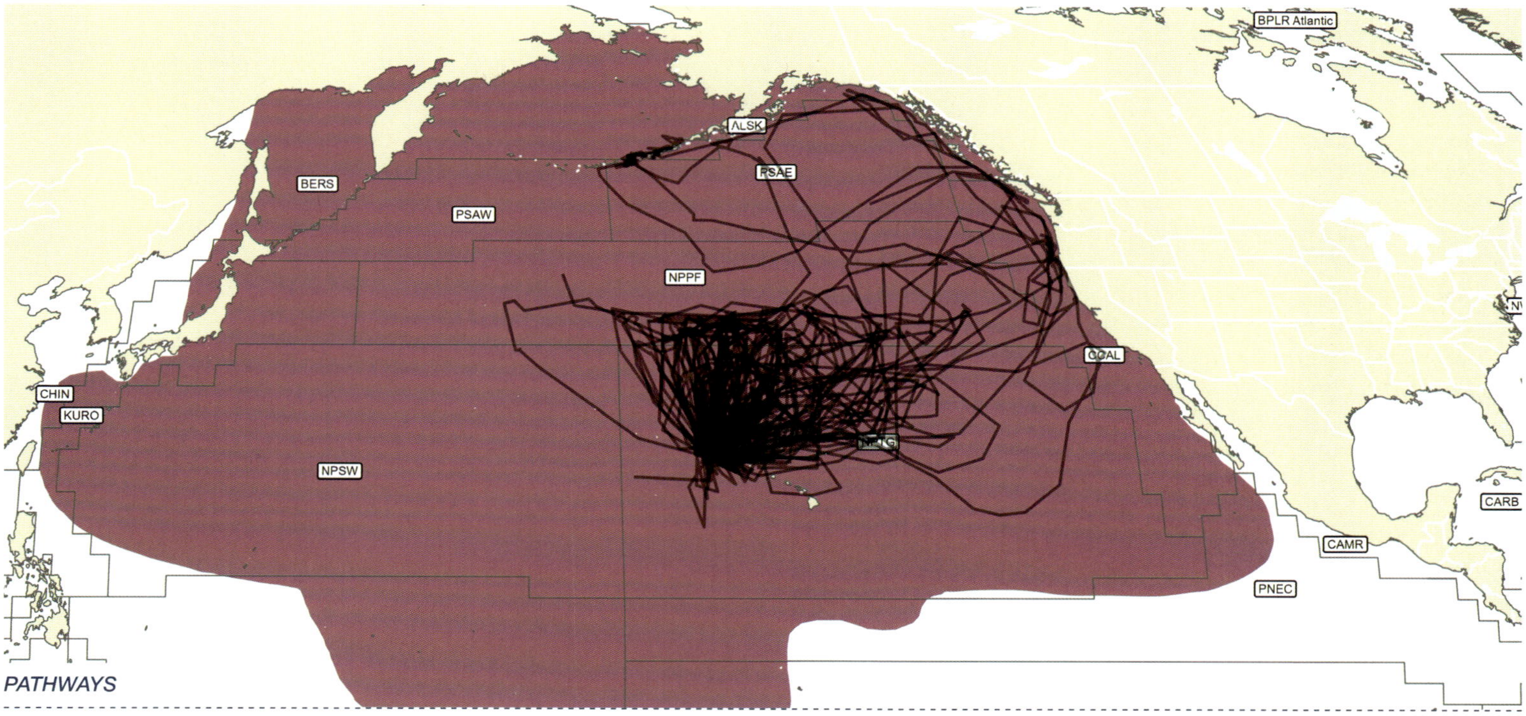

PATHWAYS

BANDING Unlike most seabirds, there is an extensive recovery record of banded Black-footed Albatrosses at sea (from the 1950s onward) due to their overlap with fishery operations and their tendency to be incidentally caught. Thus, banding records are fisheries-biased but still informative.

GEOLOCATOR TRACKING Light-level geolocators were used to track post-breeding movements from Tern Island, French Frigate Shoals (2002–2011), and Midway Atoll (2012–2016; Gutowsky et al. 2014a; Jordan et al. 2022). Birds from Midway used the entire North Pacific, including the coast of Japan, the Bering Sea, and extending into the Chukchi Sea. Birds from Tern Island primarily visited coastal areas of the Gulf of Alaska and the west coast of Canada and the US.

SATELLITE TRACKING Black-footed Albatrosses from Tern Island, French Frigate Shoals, have been tracked with satellite tags since 1997 (Fernandez et al. 2001; Kappes et al. 2010), although only one contributed track was outside the breeding period (D. Anderson, BirdLife International Seabird Tracking Database). Black-footed Albatrosses were also captured at sea off the coasts of the Aleutian Islands in Alaska (Fischer et al. 2009; D. Anderson, BirdLife International Seabird Tracking Database) and off California (Hyrenbach and Dotson 2003) and tracked during the nonbreeding period to reveal foraging habitat preferences, but these birds were not tracked back to colonies. Finally, fledglings tracked with satellite tags dispersed from Midway Atoll to the subtropical gyre (Gutowsky et al. 2014b) rather than the colder and more productive northern and western waters used by post-breeding adults tracked by geolocators.

STABLE ISOTOPES No studies were found using stable isotopes to quantify migratory connectivity.

INTERESTING NOTE

Black-footed Albatrosses were drowned in high numbers in the North Pacific drift net fishery prior to its termination in 1992 (Gould et al. 1997). Bycatch in pelagic longline fleets remains a major source of mortality (Lewison and Crowder 2003), although US regulations for longline fisheries and the implementation of seabird deterrence measures has reduced impacts in North America. Increasing sea levels due to climate change threaten low-elevation nesting beaches (Baker et al. 2006).

GENETICS The mitochondrial DNA of Hawaiian and Japanese colonies have genetic differences that have been used to assign birds caught as fisheries bycatch to their origin population (Walsh and Edwards 2005). Genetics data suggest structure between Japanese and Hawaiian populations, but no nonbreeding movement data were found for Japanese birds to assess migratory connectivity.

OTHER APPROACHES Variability in Hawaiian albatross foraging flight length across three breeding periods is well documented and may indicate that colony connectivity for wanderers could be considered during the breeding period as well (Kappes et al. 2015).

MIGRATORY PATHWAYS Black-footed Albatrosses may typically migrate to a region of the North Pacific (i.e., to the western, northern, or eastern Pacific, depending on the colony), where they will then wander, showing temporal shifts during the nonbreeding season across multiple regions and oceanographic provinces (Gutowsky et al. 2014a; Jordan et al. 2022).

DATA CONTRIBUTORS

D. Anderson; M. Antolos; G.R. Balogh; M. Conners; R.A. Orben; S.A. Shaffer; R.M. Suryan.

SHORT-TAILED ALBATROSS *Phoebastria albatrus*

DISTRIBUTION AND ABUNDANCE

On the brink of extinction in the 1940s and the rarest albatross of the North Pacific Ocean, the Short-tailed Albatross breeds mainly on Torishima Island, Japan (84%), on the Senkaku Islands (also known as the Diaoyu Islands and the Tiaoyutai Islands; the sovereignty is disputed; 16%), and through translocation (Deguchi et al. 2017) on the Ogasawara Islands, Japan (<1% of breeding population; percentages from US Fish and Wildlife Service 2020). This species typically nests on steep, thinly vegetated slopes of volcanic ash (Carboneras et al. 2020c) but occasionally breeds on flat, sandy beaches of Midway Atoll, Hawaii, US, with other limited or attempted breeding on Kure Atoll, US (Carboneras et al. 2020c). It ranges during the nonbreeding season throughout pelagic waters of the northwestern Pacific to Alaska, US, and more rarely along the west coast of the contiguous United States, Canada, or Mexico (Carboneras et al. 2020c).

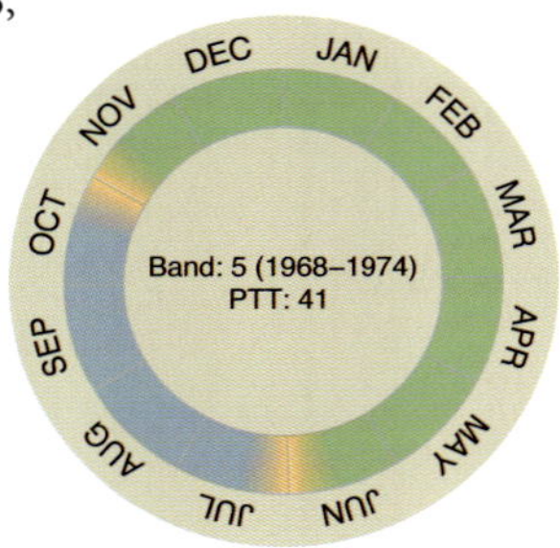

ORDER Procellariiformes
FAMILY Diomedeidae
BREEDING HABITAT Islands, Oceanic
NONBREEDING HABITAT Oceans, Pelagic
GLOBAL POPULATION SIZE 1,600
POPULATION CHANGE SINCE 1970 NA
IUCN RED LIST CATEGORY Vulnerable
R2R TIPPING POINT SPECIES Yellow-Alert
US LISTING STATUS Endangered
CANADA SPECIES AT RISK Threatened
PIF CCS 16/20

MIGRATORY CONNECTIVITY

OVERALL Insights of migratory connectivity are still growing, given the recovery of Short-tailed Albatrosses from a severe bottleneck that occurred in the late nineteenth and early twentieth centuries. Based on genetics, morphology, stable isotope signatures, and tracking, there may be some separation at sea by two genetic types/clades, but only limiting tracking exists for Senkaku-type individuals and not calculated due to limited sample size. Birds from Toroshima Island, and those translocated to Mukojima Islands in the Ogasawara Islands mix at sea. The migratory connectivity table reflects only the tracking and banding data contributed to this Atlas, and further insights from additional tracking studies are described below.

	PSAW	BERS	ALSK	PSAE
Torishima Island, Japan (n = 26)	35%	31%	4%	31%

BANDING Historical data is from birds banded in 1962 to 1964 that were later caught at sea or visually resighted off the coast of Japan, and in the open ocean in the North Pacific Tropical and Subtropical Gyres.
GEOLOCATOR TRACKING Twelve birds representing two distinct genetic populations breeding sympatrically on Toroshima Island were tracked with light-level geolocators (Tomita et al. 2024; tags deployed 2014–2017). Four Toroshima-type birds migrated to the Aleutian Islands and Bering Sea. Of nine Senkaku-type birds tracked, five migrated to the Kuril Islands and Sea of Okhotsk, three birds to the Aleutian Islands and Bering Sea, and three birds used both areas. There was no difference in timing of return to the colony for nesting. Birds tracked for multiple years (n = 8; up to three nonbreeding years) used predominantly the same site each year.
SATELLITE TRACKING Argos tags deployed on Toroshima Island (n = 6) and on nonbreeding grounds in Alaska (n = 5) provided the first insights into nonbreeding connections (Suryan et al. 2007; Suryan and

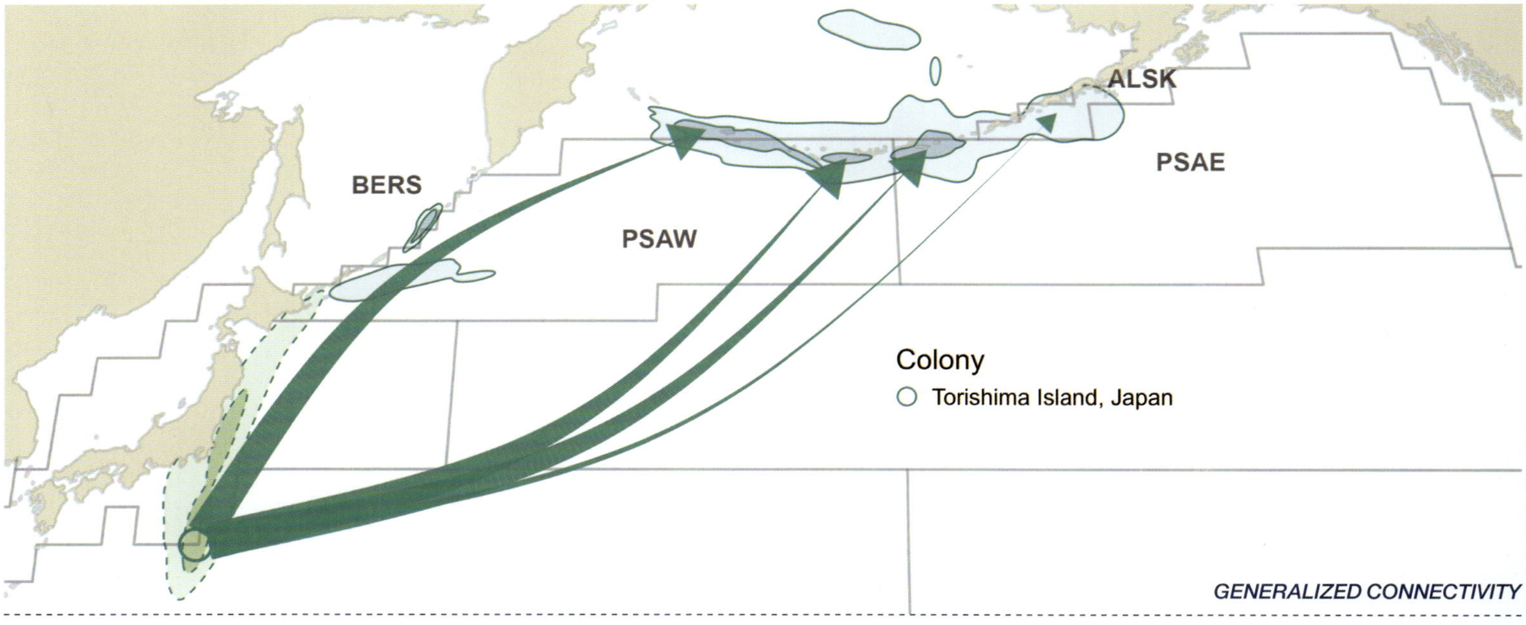

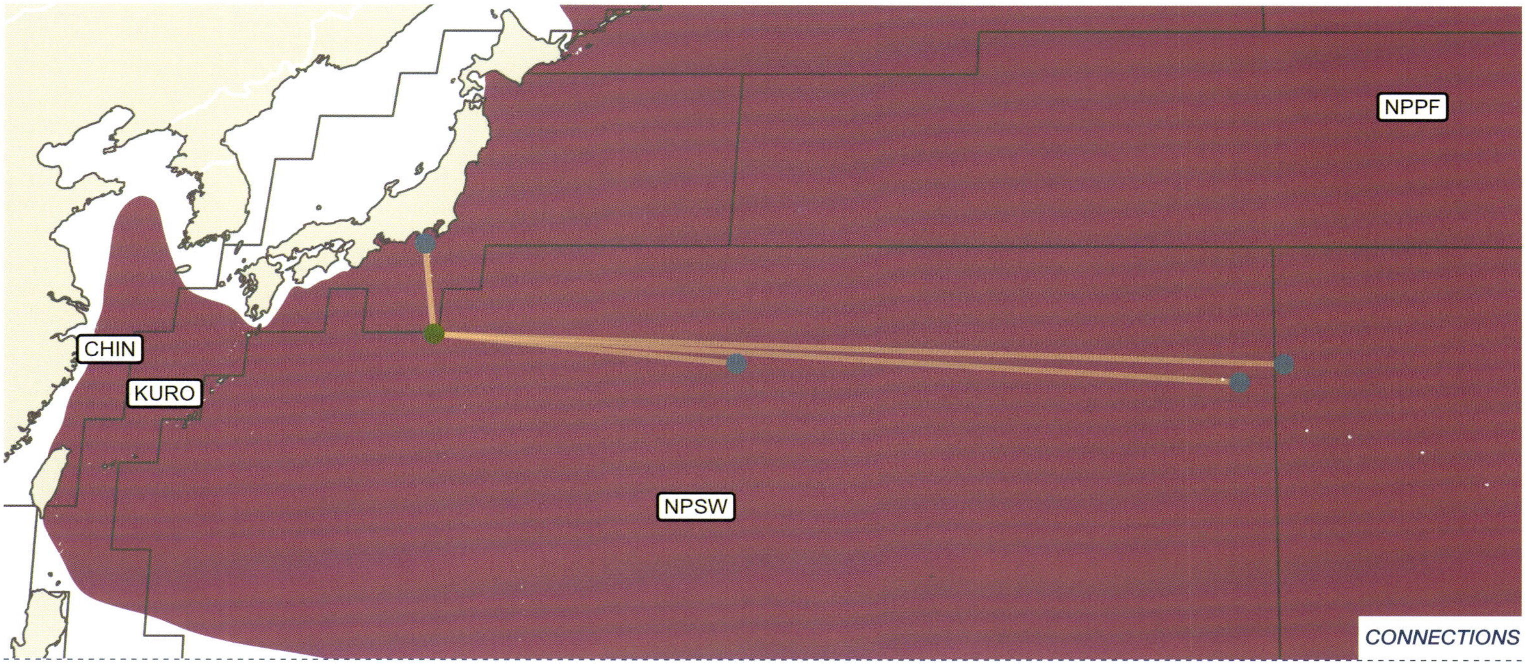

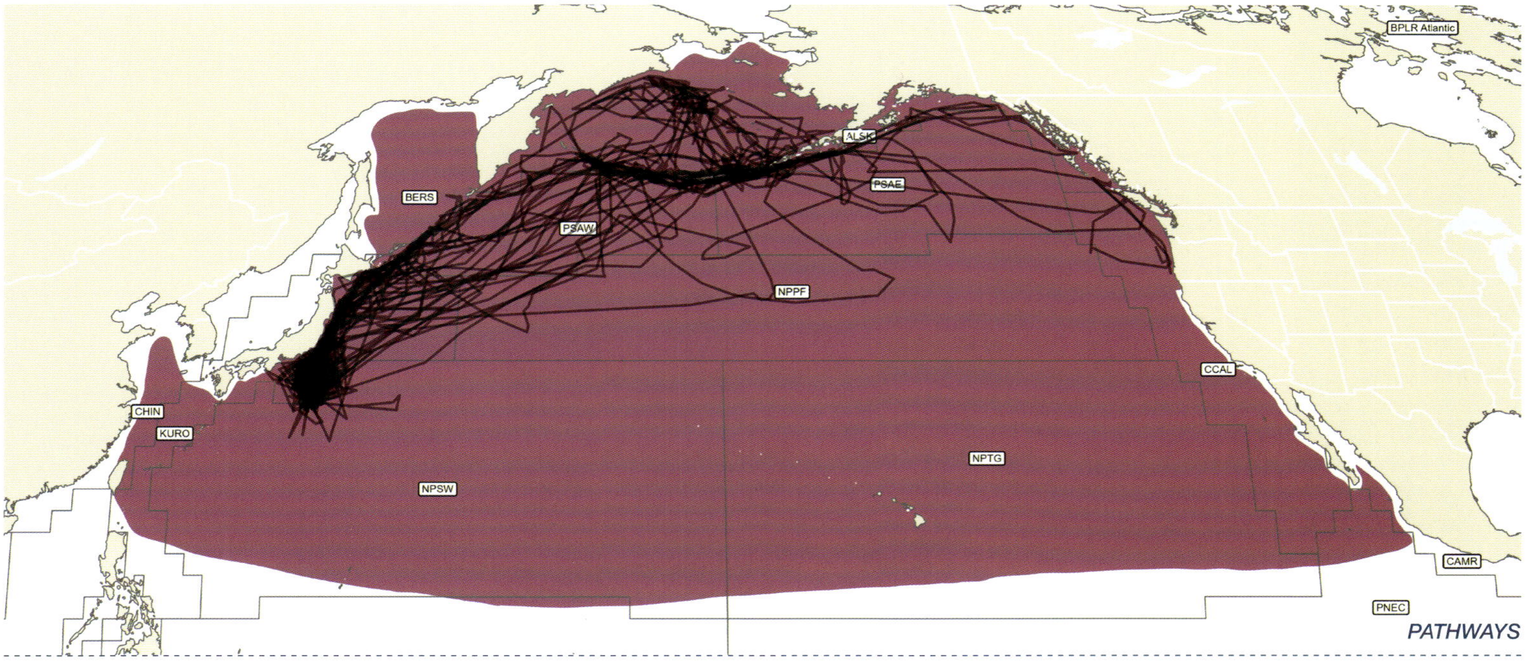

Fischer 2010) to core habitats largely over the continental shelf around the northern Pacific Rim north from the colonies in the western Pacific and as far south as California in the eastern Pacific. A GPS-Argos tracking study of juveniles tagged on Toroshima Island and hand-reared translocated birds from Mukojima Islands (Deguchi et al. 2013; Orben et al. 2018a, not contributed to this Atlas) compared movements to adults previously tracked by Suryan et al. (2007) and Suryan and Fischer (2010). The study suggested that migratory connectivity may shift with age. Juveniles tended to explore a wider range of the North Pacific, including the Sea of Okhotsk, not used by adults. Core areas included most coastal areas north from their colonies and around the Pacific Rim to British Columbia, Canada. Adults spent more time on shelf-break areas with higher productivity (Suryan et al. 2007; Suryan and Fischer 2010; Orben et al. 2018a). No differences were found between sexes or colonies. Juveniles did not migrate south to overlap with the Senkaku Islands.

STABLE ISOTOPES A stable isotope analysis of archaeological specimens from Rebun Island, Japan (Eda et al. 2012), compared carbon and nitrogen stable isotope values to those of modern birds caught as fisheries bycatch. Ancient albatrosses from all four sites foraged east of the Aleutian Islands in the California Current more frequently than modern birds and may have foraged at a lower trophic level. However, samples from archaeological sites off Southern California were isotopically distinct from those found in sites from the western North Pacific. A sample of 95 specimens from an archaeological site in British Columbia, Canada, spanning four millennia of albatross harvest, indicated that birds had a high degree of interannual foraging site fidelity at sea. Birds used the same locations for hundreds of generations (Guiry et al. 2022).

GENETICS Despite being reduced from millions to just a few hundred individuals, the majority of haplotypes were preserved (Kuro-o et al. 2010). A mitochondrial study of individuals from Toroshima Island, Japan, showed higher genetic diversity than most bird species, though it was not very high compared to other albatross species (Kuro-o et al. 2010). Research over the past 20 years from modern and archaeological samples has established two genetic clades (also suggested to be cryptic species; Eda 2020) with distinctive morphologies (Eda et al. 2012) that usually breed on different islands

INTERESTING NOTE

During the Pleistocene (approximately 405,000 years ago), Short-tailed Albatrosses likely formed colonies in the Atlantic Ocean, suggested by remains found near present-day sea level along the southeastern shore of Bermuda (Olson and Hearty 2003). The authors suggested that their extirpation was likely caused by sea level rise reducing available nesting sites. Today, 80% of Short-tailed Albatrosses currently breed on an active volcano (Toroshima Island, Japan). Eruptions in 1902 and 1939 exacerbated extreme losses due to hunting that led the species to near-extinction in the 1940s (Carboneras et al. 2020c) and has caused concern for species recovery. However, stochastic modeling work by Finklestein et al. (2010) suggests that periodic volcanic eruptions are less likely to affect population viability than low-level chronic mortality.

(Clade 1 on Toroshima, and Clade 2 primarily on Senkaku Islands but also observed on Toroshima).

MIGRATORY PATHWAYS Adults tracked from Toroshima Island (Suryan et al. 2007) migrated north, with females spending more time offshore of Japan and the Kuril Islands and Kamchatka Peninsula, Russia, than males, which spent more time within the Aleutian Islands and Bering Sea. Two young birds tagged in this study ranged more widely and traveled a greater distance per day, including one young bird that migrated to the US west coast. These comparisons were later supported by a larger GPS study (Orben et al. 2018a), which showed young birds using both coastal pathways but also pelagic pathways across the North Pacific. Tags deployed on Alaskan post-breeding habitats (Suryan and Fischer 2010) were not of long enough duration to demonstrate full migratory pathways.

DATA CONTRIBUTORS
G.R. Balogh; K. Ozaki; F. Sato; R.M. Suryan.

OCEANITIDAE | *SOUTHERN STORM-PETRELS*

WILSON'S STORM-PETREL *Oceanites oceanicus*

ORDER Procellariiformes
FAMILY Oceanitidae
BREEDING HABITAT Islands, Oceanic
NONBREEDING HABITAT Oceans, Pelagic
GLOBAL POPULATION SIZE 14,000,000
POPULATION CHANGE SINCE 1970 NA

IUCN RED LIST CATEGORY Least Concern
R2R TIPPING POINT SPECIES No
US LISTING STATUS No
CANADA SPECIES AT RISK No
PIF CCS 9/20

MIGRATORY CONNECTIVITY

OVERALL Only a BirdLife distribution map is available; no eBird data or migratory connectivity information were found for this species.

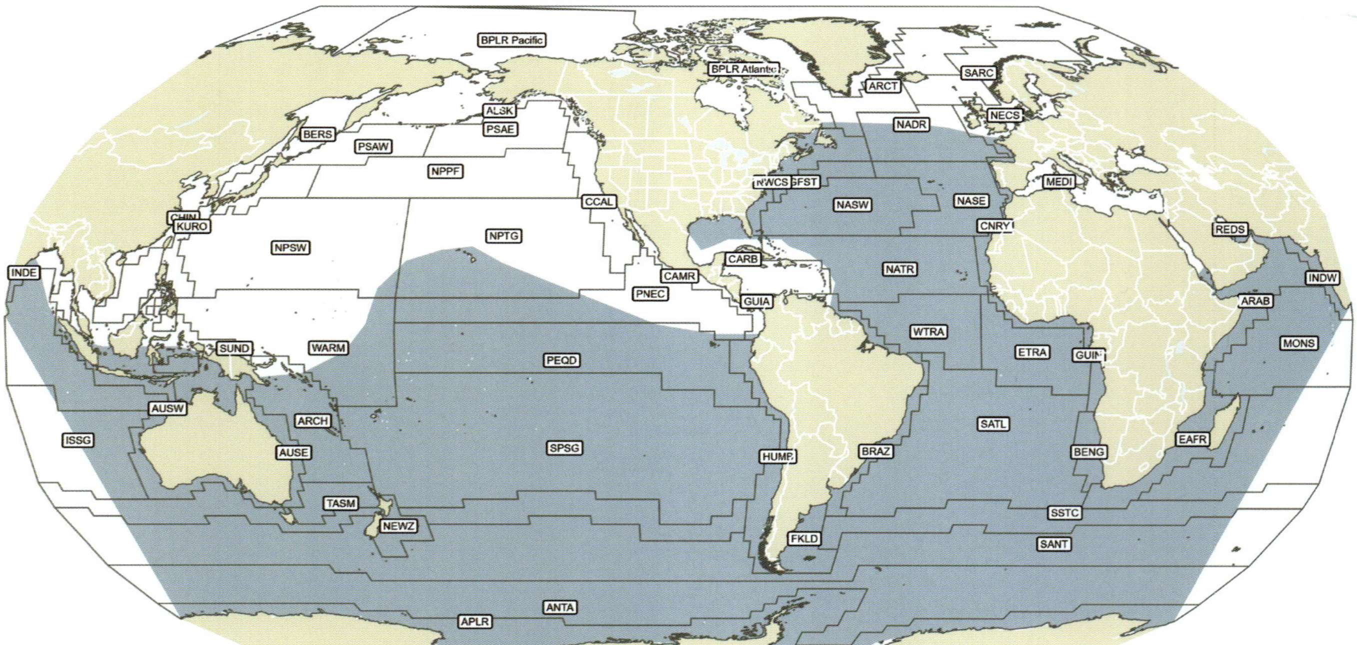

DISTRIBUTION

Pelagodroma marina **WHITE-FACED STORM-PETREL**

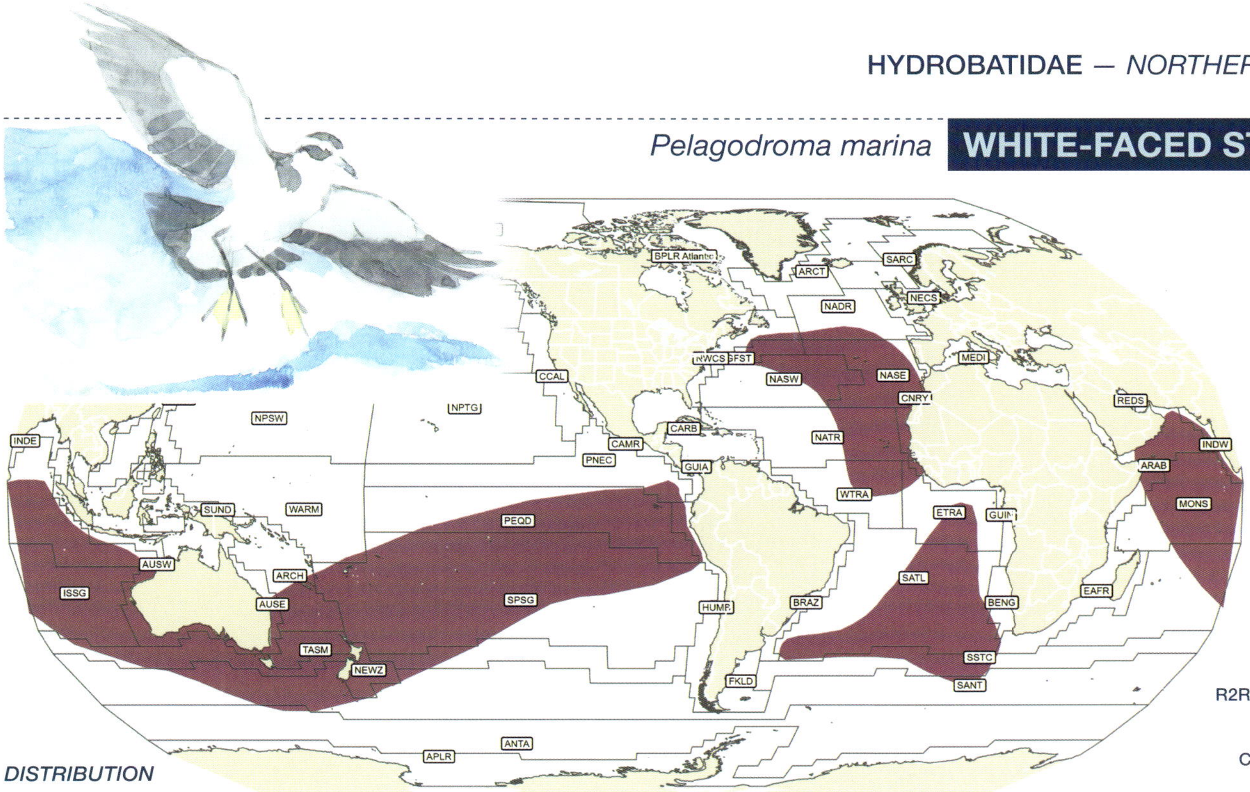

ORDER Procellariiformes
FAMILY Oceanitidae
BREEDING HABITAT Islands, Oceanic
NONBREEDING HABITAT Oceans, Pelagic
GLOBAL POPULATION SIZE 2,700,000
POPULATION CHANGE SINCE 1970 NA
IUCN RED LIST CATEGORY Least Concern
R2R TIPPING POINT SPECIES No
US LISTING STATUS No
CANADA SPECIES AT RISK No
PIF CCS 14/20

MIGRATORY CONNECTIVITY

OVERALL Only a BirdLife distribution map is available; no eBird data or migratory connectivity information were found for this species.

HYDROBATIDAE | *NORTHERN STORM-PETRELS*

Hydrobates furcatus **FORK-TAILED STORM-PETREL**

DISTRIBUTION AND ABUNDANCE

An abundant storm-petrel of the North Pacific Ocean, where it nests on offshore islands from Northern California in the thousands to Alaska where it is most abundant. It also nests along the northeastern Asian coast (Dee Boersma and Silva 2021). It spends the nonbreeding season throughout the North Pacific in pelagic habitats, but surveys suggest it remains predominantly subarctic (Dee Boersma and Silva 2021).

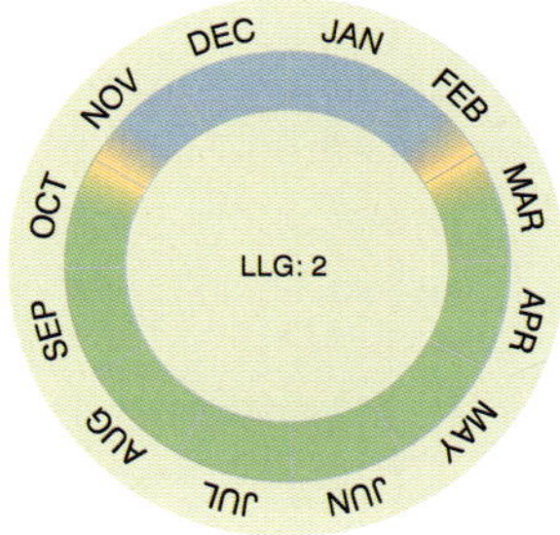

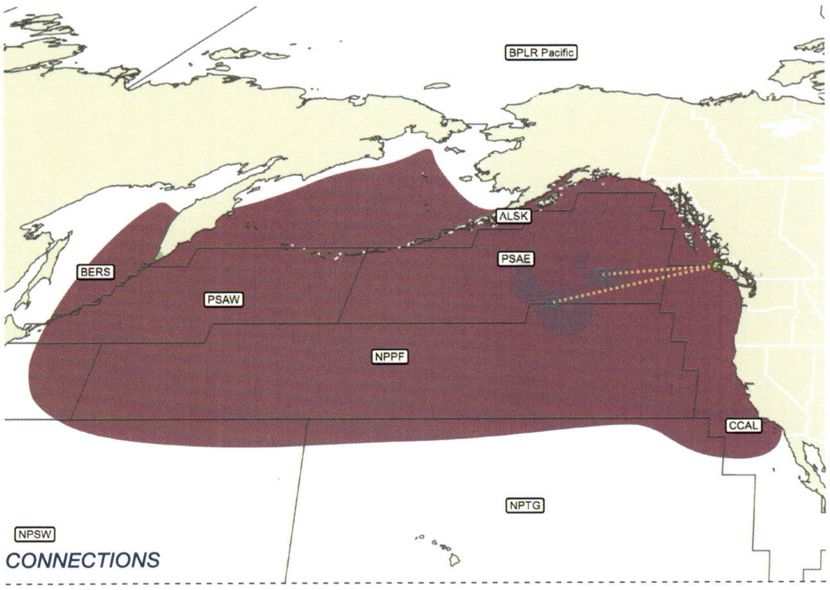

ORDER Procellariiformes
FAMILY Hydrobatidae
BREEDING HABITAT Coasts, Cliffs and Islands
NONBREEDING HABITAT Oceans, Pelagic
GLOBAL POPULATION SIZE 4,000,000
POPULATION CHANGE SINCE 1970 NA
IUCN RED LIST CATEGORY Least Concern
R2R TIPPING POINT SPECIES No
US LISTING STATUS No
CANADA SPECIES AT RISK No
PIF CCS 9/20

INTERESTING NOTE

Harris (1974) suggested that birds nesting in Northern California are resident, remaining near colonies during the winter. In general, Fork-tailed Storm-Petrels are not considered highly migratory and instead are described as dispersing to adjacent pelagic waters (Dee Boersma and Silva 2021). More research is needed to confirm these suggestions and to understand connectivity across populations.

MIGRATORY CONNECTIVITY

OVERALL Very little is known about the migratory connectivity of Fork-tailed Storm-Petrels due to limited data originating from only a single Canadian colony.

BANDING A bird banded at a colony in British Columbia was found dead in Oregon, but it is unknown if this is a migration or nonbreeding location.

GEOLOCATOR TRACKING Halpin et al. (2018) tracked six individuals from British Columbia, Canada, during the breeding season (including one failed breeder) and two individuals post-breeding to the Eastern Pacific Subarctic Gyres and North Pacific Polar Front.

STABLE ISOTOPES Carbon stable isotope signatures of Fork-tailed Storm-Petrel feathers were significantly more depleted than the sympatrically breeding congeneric species, the Leach's Storm-Petrel, reflecting the different marine systems they use during the post-breeding period (Halpin et al. 2018). No stable isotope values are available from other Fork-tailed Storm-Petrel colonies.

OTHER APPROACHES No genetic studies of migratory connectivity were found for this species.

MIGRATORY PATHWAYS Two birds tracked from their breeding colony in British Columbia migrated west to a maximum distance of 3,600 km (Halpin et al. 2018).

DATA CONTRIBUTORS
L.R. Halpin.

LEACH'S STORM-PETREL *Hydrobates leucorhous*

DISTRIBUTION AND ABUNDANCE

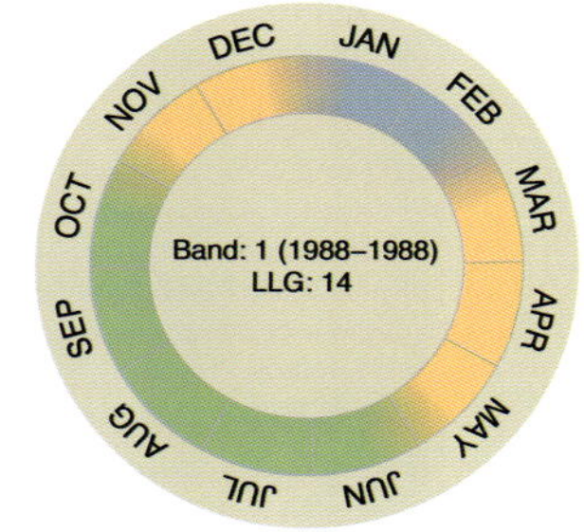

Breeds in northern Atlantic and Pacific oceans with one small, isolated, southern colony off western South Africa. It has the most widespread breeding distribution of any Northern Hemisphere species within the Procellariiformes (Pollet et al. 2021). Disperses throughout the tropical pelagic Atlantic and Pacific oceans during the nonbreeding season. Abundance surveys may be outdated, especially in the Pacific Ocean, but it is estimated that Pacific colonies comprise 50%–60% of breeding pairs. Atlantic Ocean colonies are thought to comprise 40%–50% of the global breeding population and have experienced major declines in recent years (Newson et al. 2008; Pollet and Shutler 2018; Wilhelm et al. 2019; d'Entremont et al. 2020).

ORDER Procellariiformes
FAMILY Hydrobatidae
BREEDING HABITAT Coasts, Cliffs and Islands
NONBREEDING HABITAT Oceans, Pelagic
GLOBAL POPULATION SIZE 13,000,000
POPULATION CHANGE SINCE 1970 NA

IUCN RED LIST CATEGORY Vulnerable
R2R TIPPING POINT SPECIES No
US LISTING STATUS No
CANADA SPECIES AT RISK No
PIF CCS 9/20

MIGRATORY CONNECTIVITY

OVERALL This species has strong connectivity by ocean basins, which is supported by genetic (Birchard 2019) and tracking datasets (Halpin et al. 2018; Pollet et al. 2019; Hansen and Hedd 2021), but migratory connectivity was not calculated due to limited sample size. Within ocean basins, connectivity to nonbreeding habitats is uncertain due to only a small number of colonies linked to nonbreeding habitats. Birds tracked by geolocator from Nova Scotia, Canada (Pollet et al. 2019), used core nonbreeding areas spread across a wide geography and variety of ocean habitats, including the Benguela (17%) and Canary currents (8%), the North (8%) and South Atlantic (17%) Tropical Gyres, the Western Tropical Atlantic (33%), and the Northwest Atlantic Shelves (8%). Two individuals tracked in the Pacific Ocean from British Columbia, Canada (Halpin et al. 2018), migrated to different core nonbreeding areas: the North Pacific Equatorial Countercurrent or the Humboldt Current.

	PNEC	HUMB	NATR	NWCS	NECS	CNRY	BENG	SATL	GUIN	WTRA
Gillam Islands, BC, Canada (n = 2)	50%	50%	–	–	–	–	–	–	–	–
Kent Islands NB, Canada (n = 1)	–	–	–	–	100%	–	–	–	–	–
Nova Scotia, Canada (n = 12)	–	–	8%	8%	–	8%	17%	17%	8%	33%

GEOLOCATOR TRACKING There is recent evidence of weak connectivity within the Atlantic Ocean from geolocator-tracked birds originating from Canada (Pollet et al. 2019) and Iceland (Hansen and Hedd 2021), but differences in nonbreeding connectivity across subspecies within the Pacific Ocean are unknown. Four individuals tracked recently by geolocator from Iceland appear to use similar habitats in the Benguela Current and the North Atlantic Tropical Gyre. Immature birds may remain at sea year-round, sometimes making colony visits (Pollet et al. 2021), but connectivity of immatures is unknown.

STABLE ISOTOPES Stable isotopic signatures of birds tracked by geolocators in the Atlantic Ocean revealed differentiation by nonbreeding location (Pollet et al. 2019). Those migrating to the Southern Hemisphere showed higher nitrogen values than those remaining in the Northern Hemisphere

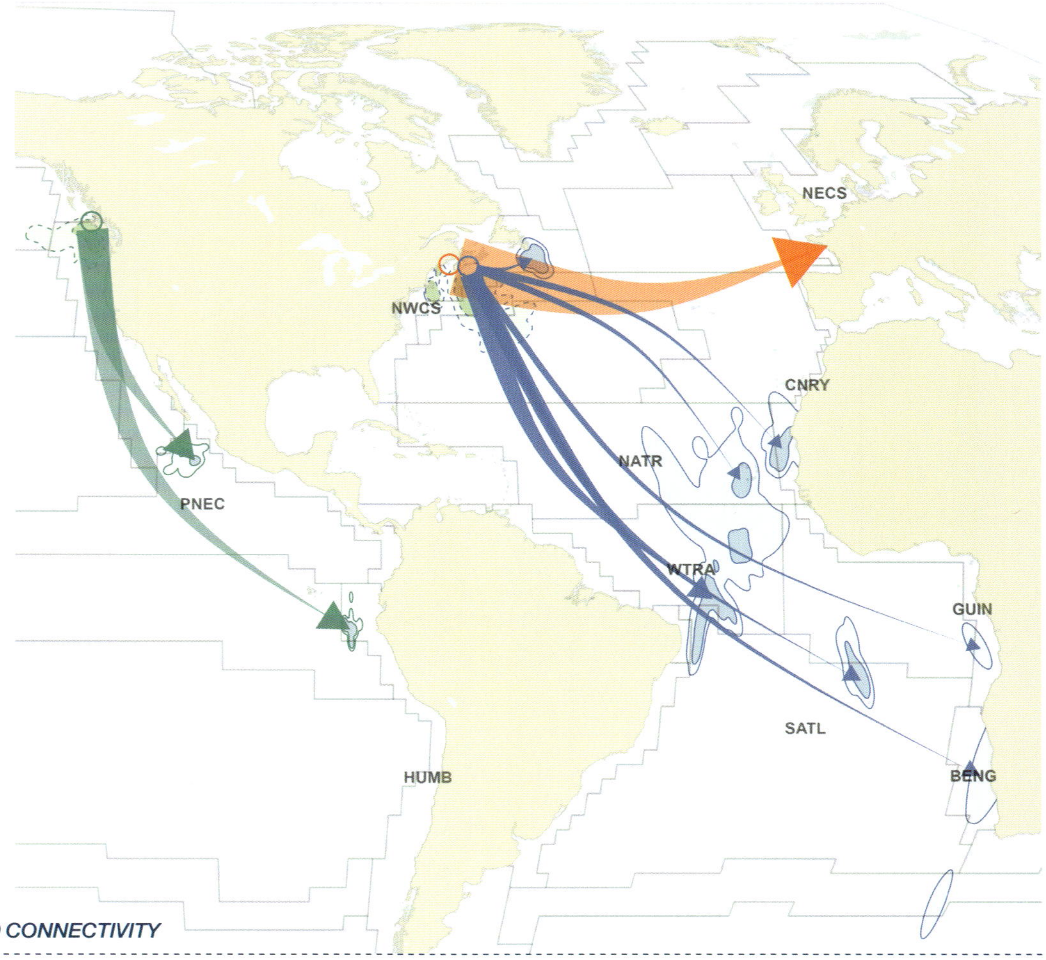

Colony

- Gillam Islands, BC, Canada
- Kent Island, NB, Canada
- Nova Scotia, Canada

GENERALIZED CONNECTIVITY

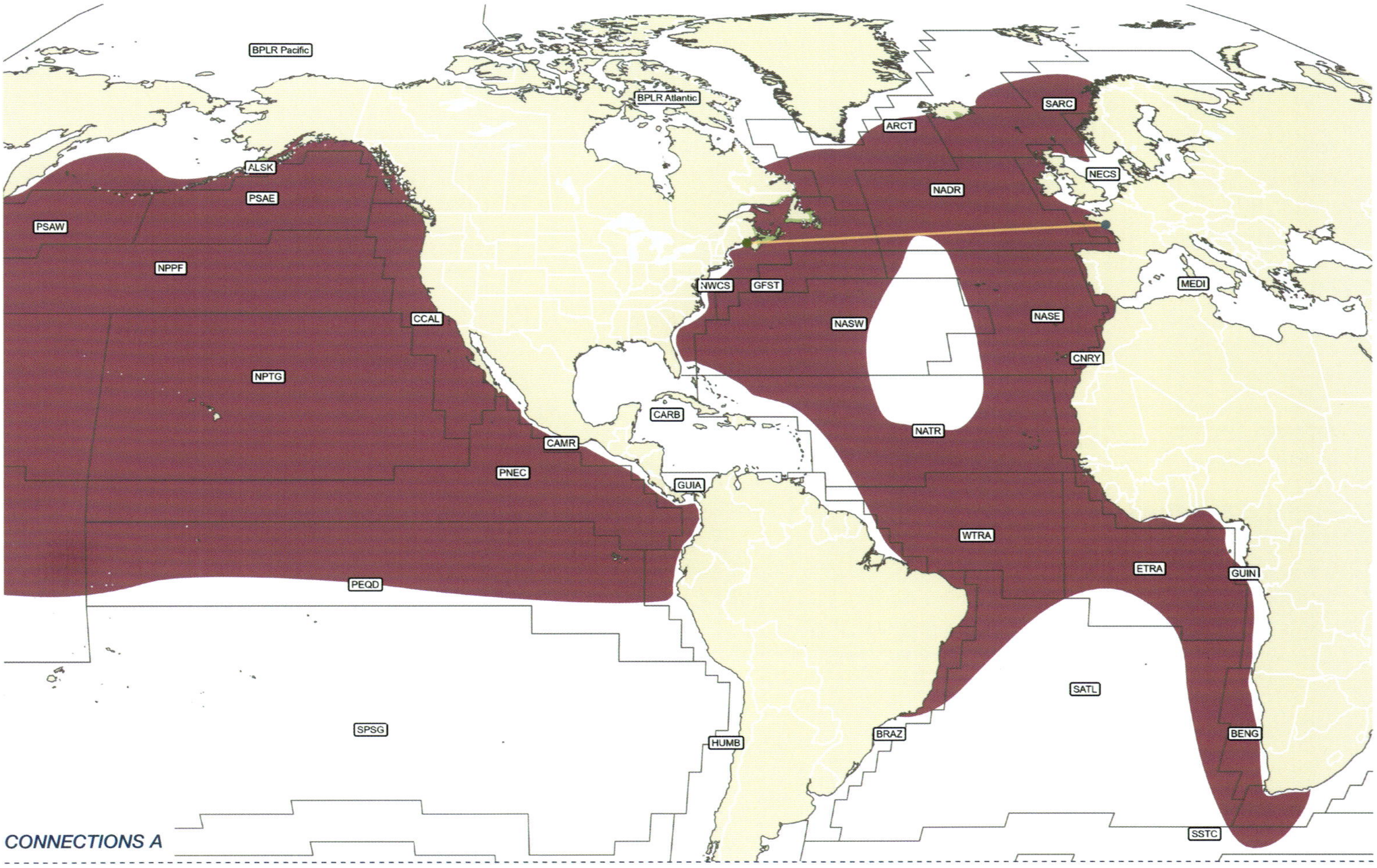

CONNECTIONS A

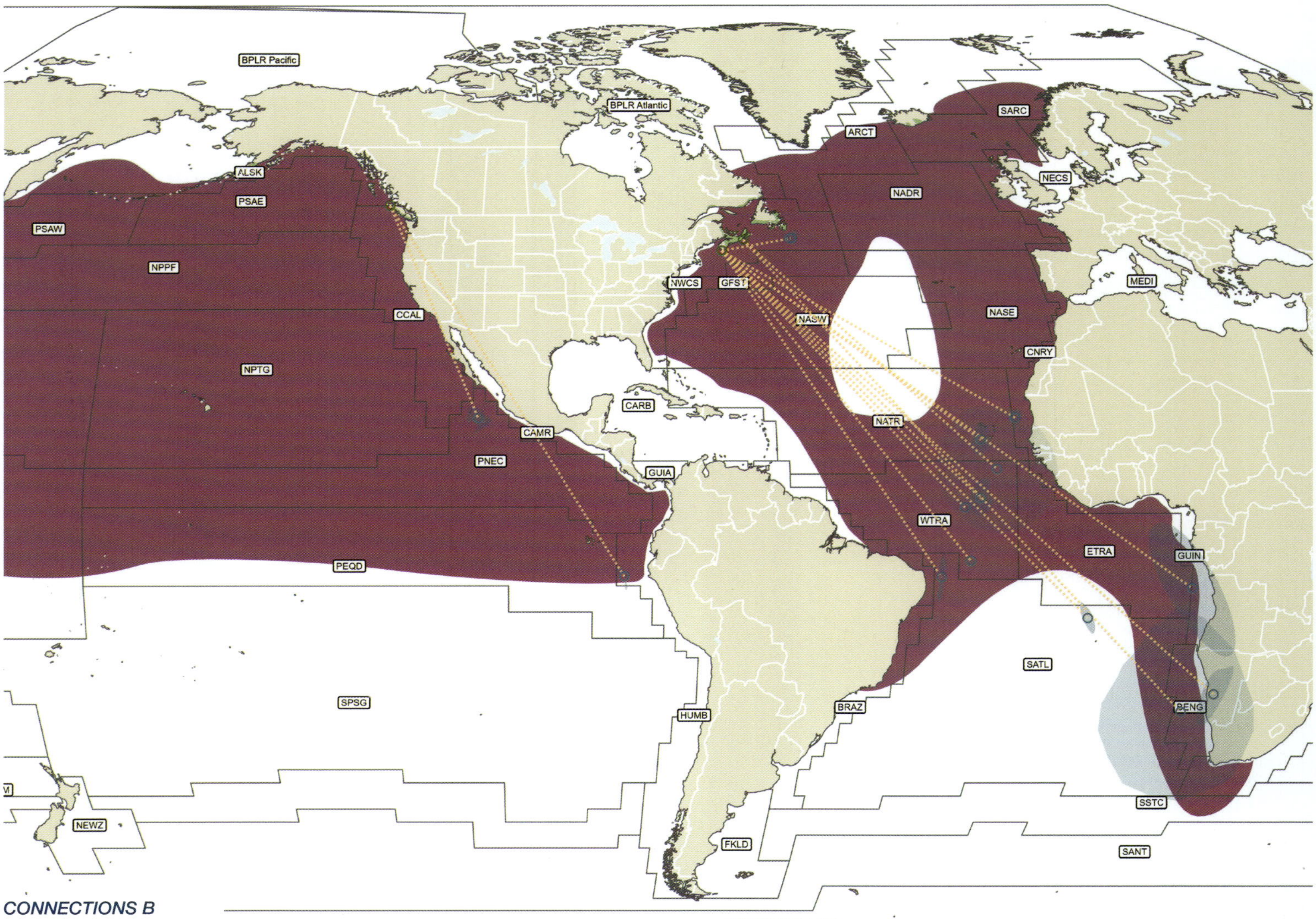

CONNECTIONS B

during nonbreeding. This suggests use of a higher trophic level and potentially a marker that could be used to infer nonbreeding connectivity in lieu of tracking.

GENETICS A genomic analysis of 11 breeding colonies (Birchard 2019) found genetic differentiation between subspecies, and weak but significant separation between Atlantic and Pacific breeding populations of *H. l. leucorhous* supporting a lack of mixing between the two ocean basins.

OTHER APPROACHES Several Motus tags were deployed on Leach's Storm-Petrels in Nova Scotia, with local detections only.

MIGRATORY PATHWAYS In the Pacific Ocean, two tracked individuals migrated south through the California Current and coastal EEZs of the Americas but returned north across areas of the high seas west of their southward route (Halpin et al. 2018). In the Atlantic Ocean, 12 of 13 individuals tracked by geolocator took a southeastern pathway from their breeding colony in Nova Scotia, Canada, across the North Atlantic Ocean toward the Macaronesia archipelagos. One individual instead traveled north toward Greenland before taking a southward route to the central Atlantic. Pre-breeding migration was available for five birds, and all undertook rapid northwest routes (Pollet et al. 2019).

DATA CONTRIBUTORS

L.R. Halpin; I.L. Pollet; P. Taylor.

INTERESTING NOTE

There is evidence of stronger connectivity during the breeding period among seven eastern Canadian colonies during incubation (Pollet et al. 2014; Hedd et al. 2018). For all colonies studied, breeding foraging distribution extended beyond Exclusive Economic Zones (EEZs) into international waters (Pollet et al. 2014; Halpin et al. 2018; Hedd et al. 2018). Future priorities include connectivity studies of *H. l. chapmani* and additional *H. l. leucorhous* colonies, including the interesting South African breeding population.

TOWNSEND'S STORM-PETREL

Hydrobates socorroensis

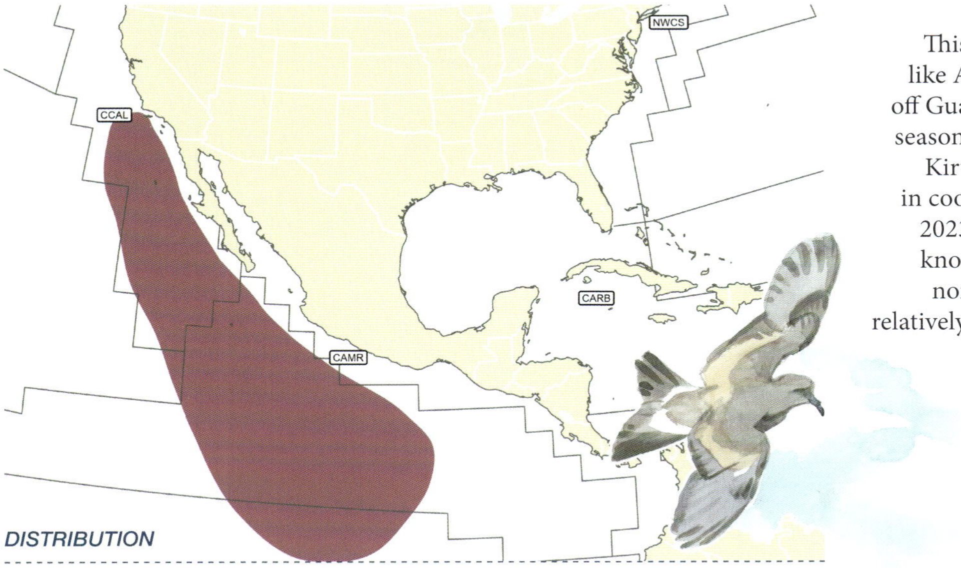

DISTRIBUTION AND ABUNDANCE

This species looks similar to Ainley's Storm-Petrel and, like Ainley's Storm-Petrel, it nests only on volcanic islets off Guadalupe Island, off western Mexico; but its breeding season is in warmer months (May through early October; Kirwan et al. 2023), while Ainley's Storm-Petrel breeds in cooler months (October through April; Medrano et al. 2023). Their at-sea nonbreeding distribution is not well known, but ranges from 35 degrees north to 10 degrees north in the eastern Pacific (Kirwan et al. 2023) and is relatively common in pelagic waters off Southern California during the breeding season.

ORDER Procellariiformes
FAMILY Hydrobatidae
BREEDING HABITAT Islands, Oceanic
NONBREEDING HABITAT Oceans, Pelagic
GLOBAL POPULATION SIZE 7,000
POPULATION CHANGE SINCE 1970 NA

IUCN RED LIST CATEGORY Endangered
R2R TIPPING POINT SPECIES Red-Alert
US LISTING STATUS No
CANADA SPECIES AT RISK No
PIF CCS 17/20

INTERESTING NOTE

At-sea surveys are unreliable for Townsend's Storm-Petrel as the species looks similar to both Ainley's Storm-Petrel and to the more widely distributed Leach's Storm-Petrel.

MIGRATORY CONNECTIVITY

OVERALL Migratory connectivity is unknown, but it is likely weak across breeding islets given the restricted range.

GEOLOCATOR TRACKING Seven Townsend's Storm-Petrels have been tracked by geolocators over their nonbreeding period to the southern Baja California peninsula; they spent most of their time in Mexican waters (Medrano et al. 2024). Most movements matched the distribution suggested in the literature; however, three birds also ventured farther south to the high seas. Data is not presented in the Atlas.

OTHER APPROACHES No banding, satellite tracking, genetic, or stable isotope studies of migratory connectivity were found for this species.

MIGRATORY PATHWAYS Pathways are available from seven birds tracked from Mexico (Medrano et al. 2024). These birds migrated southeast within the Mexican Exclusive Economic Zone to offshore islands, and due south to high-seas sites, where they undertook area-restricted search movements before returning north to breeding sites. Return movements included narrow directed return loops as well as broad loops to the west over the high seas.

AINLEY'S STORM-PETREL

Hydrobates cheimomnestes

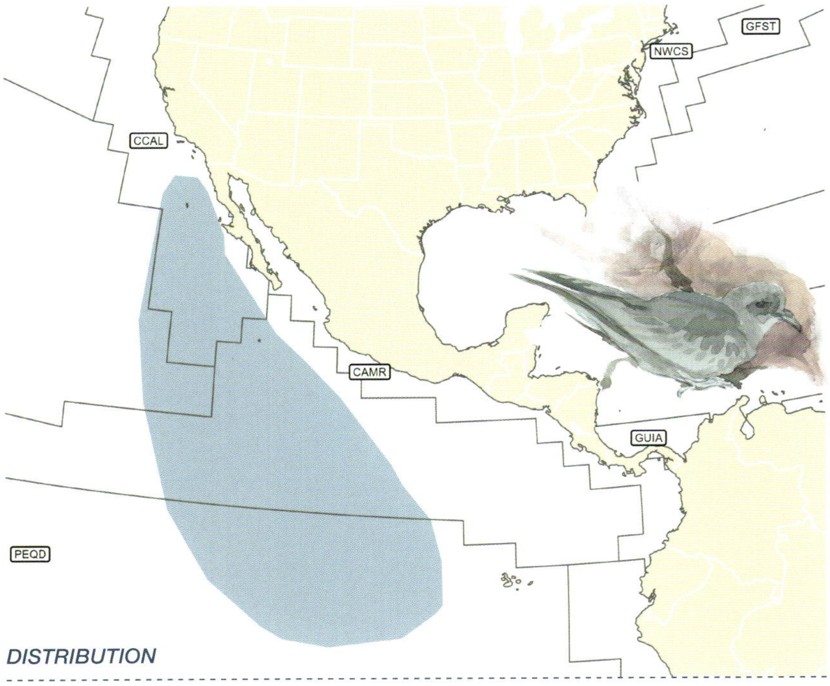

DISTRIBUTION AND ABUNDANCE

Accepted as a species in 2016, Ainley's Storm-Petrel was formerly grouped with Townsend's Storm-Petrel (Medrano et al. 2023). Both species nest in a restricted distribution on volcanic islets off Guadalupe Island, off western Mexico, but Ainley's Storm-Petrel breeds in the cooler months, October through April (Medrano et al. 2023), and Townsend's Storm-Petrel in the warmer months, May through early October (Kirwan et al. 2023), among other distinguishing characteristics. Ainley's Storm-Petrel population is estimated at 2,500–10,000 individuals (Medrano et al. 2023). New natural history information is still being uncovered for Ainley's Storm-Petrel, including their at-sea distribution.

ORDER Procellariiformes
FAMILY Hydrobatidae
BREEDING HABITAT Islands, Oceanic
NONBREEDING HABITAT Oceans, Pelagic
GLOBAL POPULATION SIZE NA
POPULATION CHANGE SINCE 1970 NA

IUCN RED LIST CATEGORY Vulnerable
R2R TIPPING POINT SPECIES No
US LISTING STATUS No
CANADA SPECIES AT RISK No
PIF CCS 17/20

INTERESTING NOTE

Ainley's Storm-Petrels tracked during the breeding season with small GPS tags (n = 37; 54 foraging trips) visited waters of the eastern North Pacific between central Baja California, Mexico, and Southern California, US. This study provided the first known presence of Ainley's Storm-Petrels in the United States. At-sea surveys are unreliable for Ainley's Storm-Petrel as the species looks similar to Townsend's Storm-Petrel and to the more widely distributed Leach's Storm-Petrel. Their cryptic nature is the likely reason Ainley's Storm-Petrels were not thought to be in US waters. Medrano et al. (2023) suggest a future priority is to reestablish a breeding population of Ainley's Storm-Petrels on Guadalupe Island, Mexico.

MIGRATORY CONNECTIVITY

OVERALL Migratory connectivity is unknown but is likely weak across breeding islets given their restricted range.

GEOLOCATOR TRACKING Four Ainley's Storm-Petrels were tracked by geolocators from Guadalupe Island over their nonbreeding period, spending most of the time on the high seas (Medrano et al. 2024). Data is not presented in the Atlas.

OTHER APPROACHES No banding, genetic, or stable isotope studies of migratory connectivity were found for this species.

MIGRATORY PATHWAYS Tracked individuals migrated toward and north of the Hawaiian EEZ (Medrano et al. 2024). It was thought that Ainley's Storm-Petrels predominantly migrate south toward Galapagos as a specimen has been collected there (Medrano et al. 2023), and so even these four movement paths add much to our knowledge of this species.

ASHY STORM-PETREL *Hydrobates homochroa*

DISTRIBUTION AND ABUNDANCE

Endemic to the California Current, it nests mainly on islands from Baja California, Mexico, to Northern California. Highest breeding abundances are likely at South Farallon Islands and Channel Islands in California (Carter et al. 2016). Distribution is at sea, typically over the continental slope (200–2,000 m depth) and pelagic waters (> 2,000 m depth; Spear and Ainley 2007).

ORDER Procellariiformes
FAMILY Hydrobatidae
BREEDING HABITAT Coasts, Cliffs and Islands
NONBREEDING HABITAT Coasts, Marine Waters
GLOBAL POPULATION SIZE 5,100
POPULATION CHANGE SINCE 1970 NA
IUCN RED LIST CATEGORY Endangered
R2R TIPPING POINT SPECIES Red-Alert
US LISTING STATUS No
CANADA SPECIES AT RISK No
PIF CCS 17/20

MIGRATORY CONNECTIVITY

OVERALL The extent of movements or population mixing within the California Current is unknown due to limited data. Weak connectivity is expected based on radio telemetry data of birds from Southern California to Northern California, but it is unknown whether northern nesters also disperse south.

RADIO TRACKING Adams and Takekawa (2008) tracked 57 radio-tagged birds (mostly nonbreeding females) via aerial telemetry surveys. The birds, captured at three colonies in the Channel Islands, California, moved north during post-breeding migration to Monterey Bay and the Gulf of the Farallones (600 km from the colony), likely in association with high productivity in these areas. Adams and Takekawa (2008) noted some colony-specific habitat associations during the breeding season, but birds from all colonies showed persistent use of the Santa Cruz Basin. Data is not presented in the Atlas.

GENETICS No differentiation has been noted in mitochondrial DNA analysis among birds from the Farallon Islands toward the northern part of the range and the Channel Islands in the southern part of the range (Girdeman et al. 1999, as cited in Ainley et al. 2021a).

OTHER APPROACHES No banding, tracking, or stable isotope studies of migratory connectivity have been found for this species.

MIGRATORY PATHWAYS Southern colonies disperse up to 600 km along northward offshore pathways to Monterey Bay and the Gulf of the Farallones in August and September (Adams and Takekawa 2008).

DISTRIBUTION

Hydrobates castro BAND-RUMPED STORM-PETREL

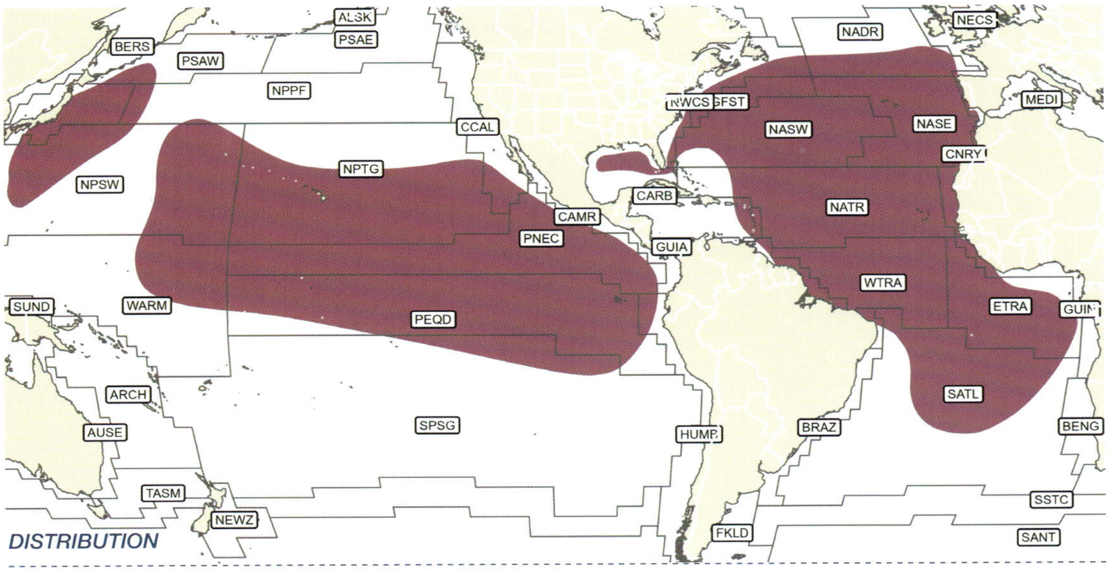

MIGRATORY CONNECTIVITY

OVERALL Only a BirdLife distribution map is available; no eBird data or migratory connectivity information were found for this species.

ORDER Procellariiformes
FAMILY Hydrobatidae
BREEDING HABITAT Islands, Oceanic
NONBREEDING HABITAT Oceans, Pelagic
GLOBAL POPULATION SIZE 110,000
POPULATION CHANGE SINCE 1970 NA
IUCN RED LIST CATEGORY Least Concern
R2R TIPPING POINT SPECIES Red-Alert
US LISTING STATUS Endangered
CANADA SPECIES AT RISK No
PIF CCS 17/20

Hydrobates melania BLACK STORM-PETREL

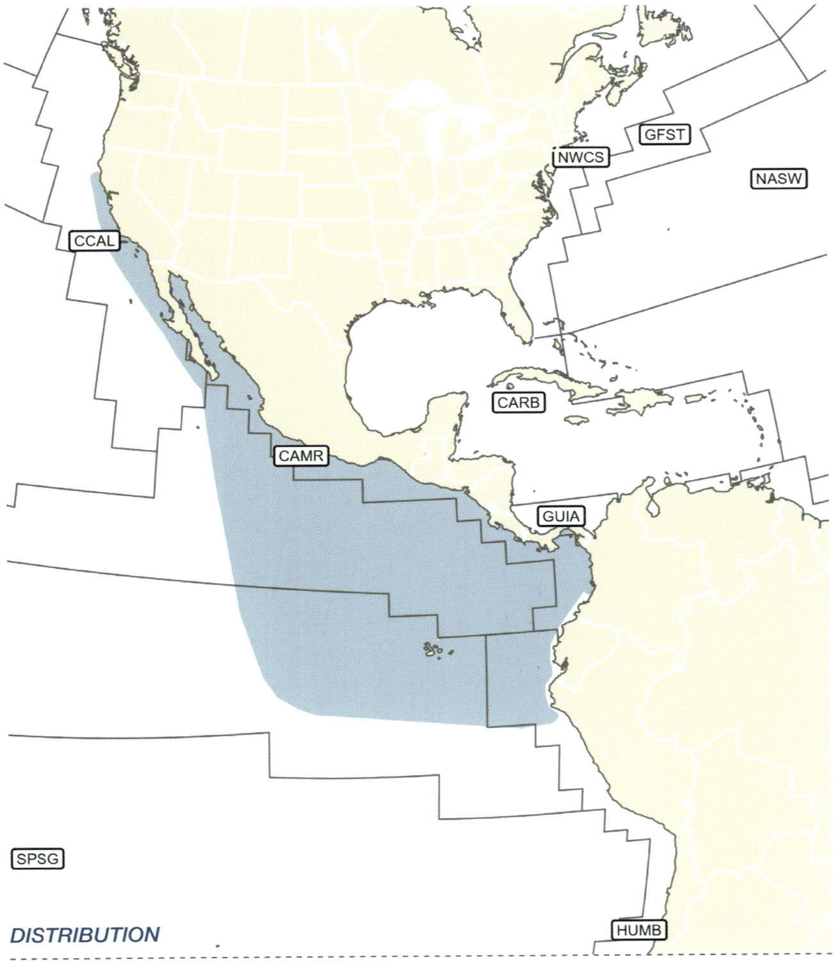

ORDER Procellariiformes
FAMILY Hydrobatidae
BREEDING HABITAT Islands, Oceanic
NONBREEDING HABITAT Oceans, Pelagic
GLOBAL POPULATION SIZE 110,000
POPULATION CHANGE SINCE 1970 NA
IUCN RED LIST CATEGORY Least Concern
R2R TIPPING POINT SPECIES Red-Alert
US LISTING STATUS Endangered
CANADA SPECIES AT RISK No
PIF CCS 17/20

TRISTRAM'S STORM-PETREL *Hydrobates tristrami*

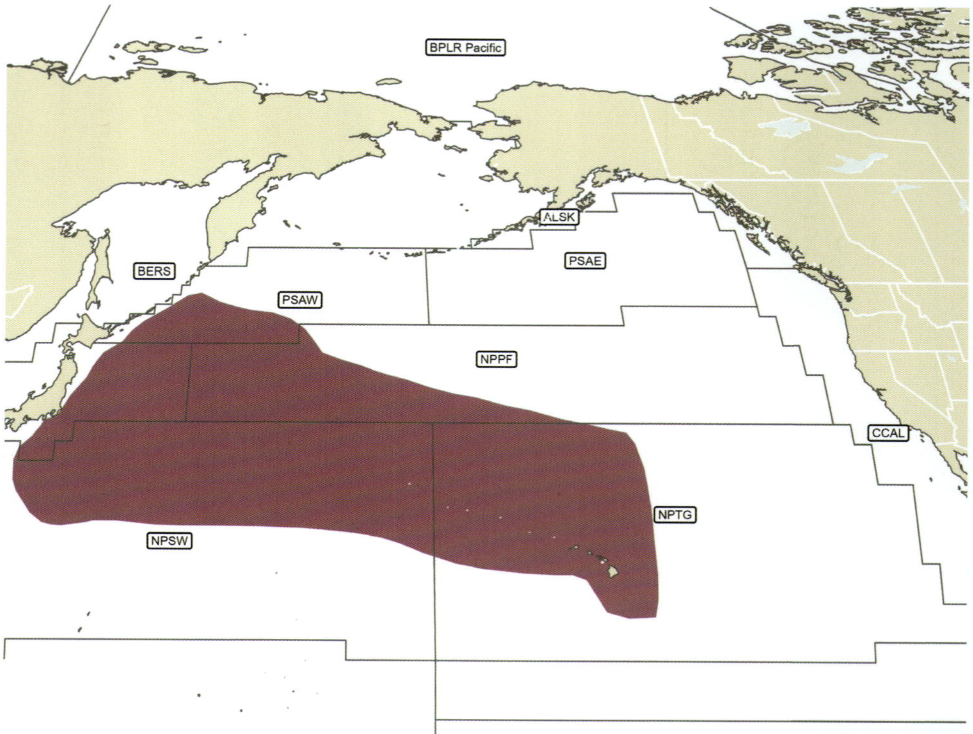

DISTRIBUTION

ORDER Procellariiformes
FAMILY Hydrobatidae
BREEDING HABITAT NA
NONBREEDING HABITAT NA
GLOBAL POPULATION SIZE NA
POPULATION CHANGE SINCE 1970 NA
IUCN RED LIST CATEGORY Least Concern
R2R TIPPING POINT SPECIES No
US LISTING STATUS No
CANADA SPECIES AT RISK No
PIF CCS NA

MIGRATORY CONNECTIVITY

OVERALL Only a BirdLife distribution map is available; no eBird data or migratory connectivity information were found for this species.

LEAST STORM-PETREL *Hydrobates microsoma*

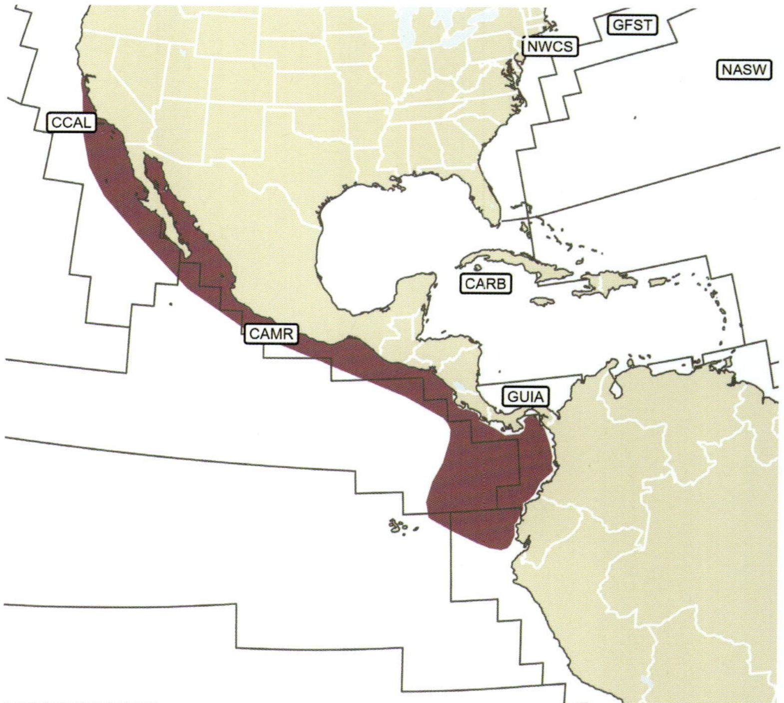

DISTRIBUTION

ORDER Procellariiformes
FAMILY Hydrobatidae
BREEDING HABITAT Coasts, Cliffs and Islands
NONBREEDING HABITAT Oceans, Pelagic
GLOBAL POPULATION SIZE NA
POPULATION CHANGE SINCE 1970 NA
IUCN RED LIST CATEGORY Least Concern
R2R TIPPING POINT SPECIES No
US LISTING STATUS No
CANADA SPECIES AT RISK No
PIF CCS 15/20

MIGRATORY CONNECTIVITY

OVERALL Only a BirdLife distribution map is available; no eBird data or migratory connectivity information were found for this species.

Mantel = −0.02 ± 0.04

Fulmarus glacialis **NORTHERN FULMAR**

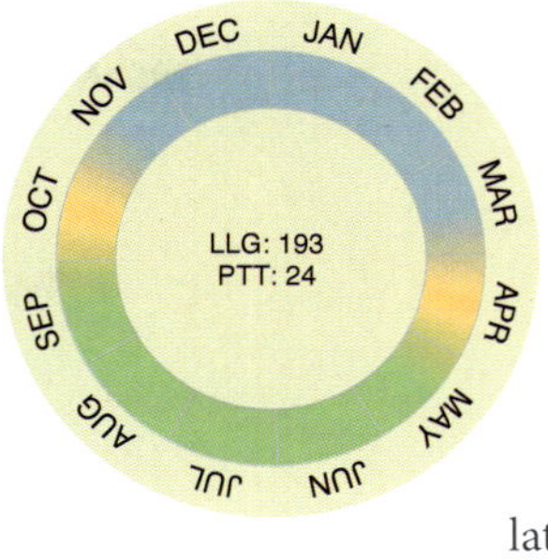

DISTRIBUTION AND ABUNDANCE

Common pelagic seabird of the northern oceans, Northern Fulmars breed from the high Arctic to the boreal zone on island sea cliffs and mainland promontories (Mallory et al. 2020a). In North America, breeding colonies in eastern Canada are mainly north of 65 degrees north latitude, and Alaskan colonies are primarily on islands of the Bering Sea and Gulf of Alaska (Mallory et al. 2020a). About 80% of the total North American population is in Alaska (99% of these breed at four colonies), and 20% is in the Canadian Arctic, with <0.01% in British Columbia and Newfoundland and Labrador (Mallory et al. 2020a). During the nonbreeding period, it ranges over deep cold waters and shelf breaks.

ORDER Procellariiformes
FAMILY Procellariidae
BREEDING HABITAT Coasts, Cliffs and Islands
NONBREEDING HABITAT Oceans, Pelagic
GLOBAL POPULATION SIZE 14,000,000
POPULATION CHANGE SINCE 1970 −8%
IUCN RED LIST CATEGORY Least Concern
R2R TIPPING POINT SPECIES No
US LISTING STATUS No
CANADA SPECIES AT RISK No
PIF CCS 9/20

MIGRATORY CONNECTIVITY

OVERALL Migratory connectivity is strong across ocean basins and weak (Mantel = −0.02 ± 0.04) within, inferred from genetics, banding, and tracking data. Colonies in the Pacific Ocean appear to have stronger connectivity than colonies in the Atlantic Ocean, but more research is still needed, and sample sizes are small for some colonies. All birds from Chowiet Island, Alaska, spent the nonbreeding season in the California

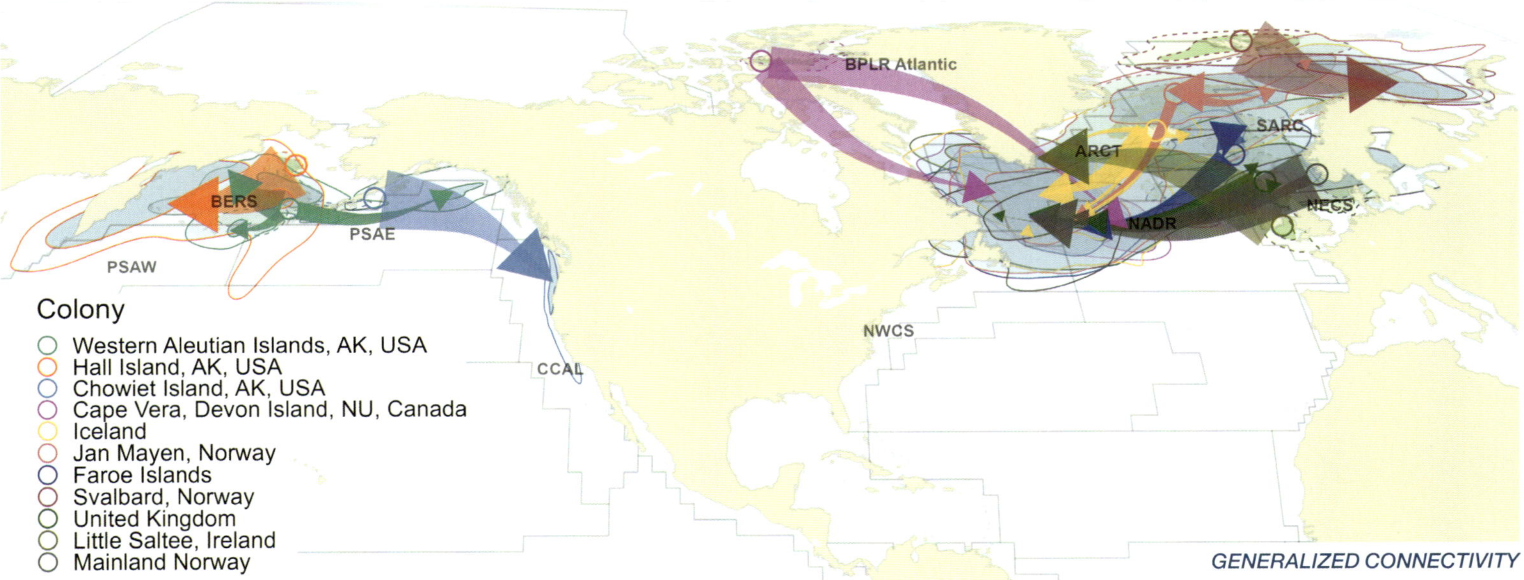

	PSAW	BERS	PSAE	CCAL	NWCS	BPLR Atlantic	ARCT	SARC	NECS	NADR
Western Aleutian Islands, AK, USA (n = 11)	18%	55%	27%	–	–	–	–	–	–	–
Hall Island, AK, USA (n = 4)	–	100%	–	–	–	–	–	–	–	–
Chowiet Island, AK, USA (n = 3)	–	–	–	100%	–	–	–	–	–	–
Cape Vera, Devon Island, NU, Canada (n = 4)	–	–	–	–	–	–	50%	–	–	50%
Iceland (n = 79)	–	–	–	–	5%	5%	59%	8%	–	23%
Jan Mayen, Norway (n = 49)	–	–	–	–	–	2%	65%	16%	–	16%
Faroe Islands (n = 9)	–	–	–	–	–	–	56%	–	–	44%
Svalbard, Norway (n = 2)	–	–	–	–	–	–	–	100%	–	–
United Kingdom (n = 46)	–	–	–	–	–	2%	48%	7%	11%	33%
Little Saltee, Ireland (n = 1)	–	–	–	–	–	100%	–	–	–	–
Mainland Norway (n = 7)	–	–	–	–	–	–	86%	–	–	14%

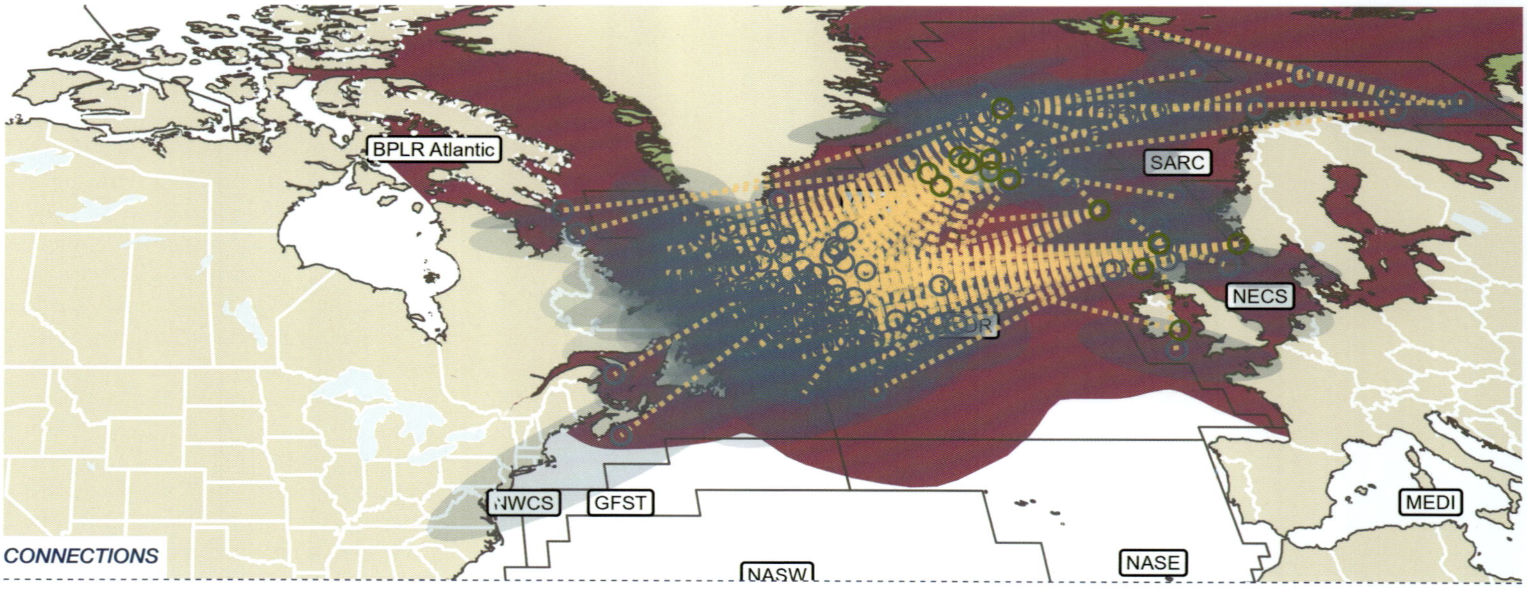

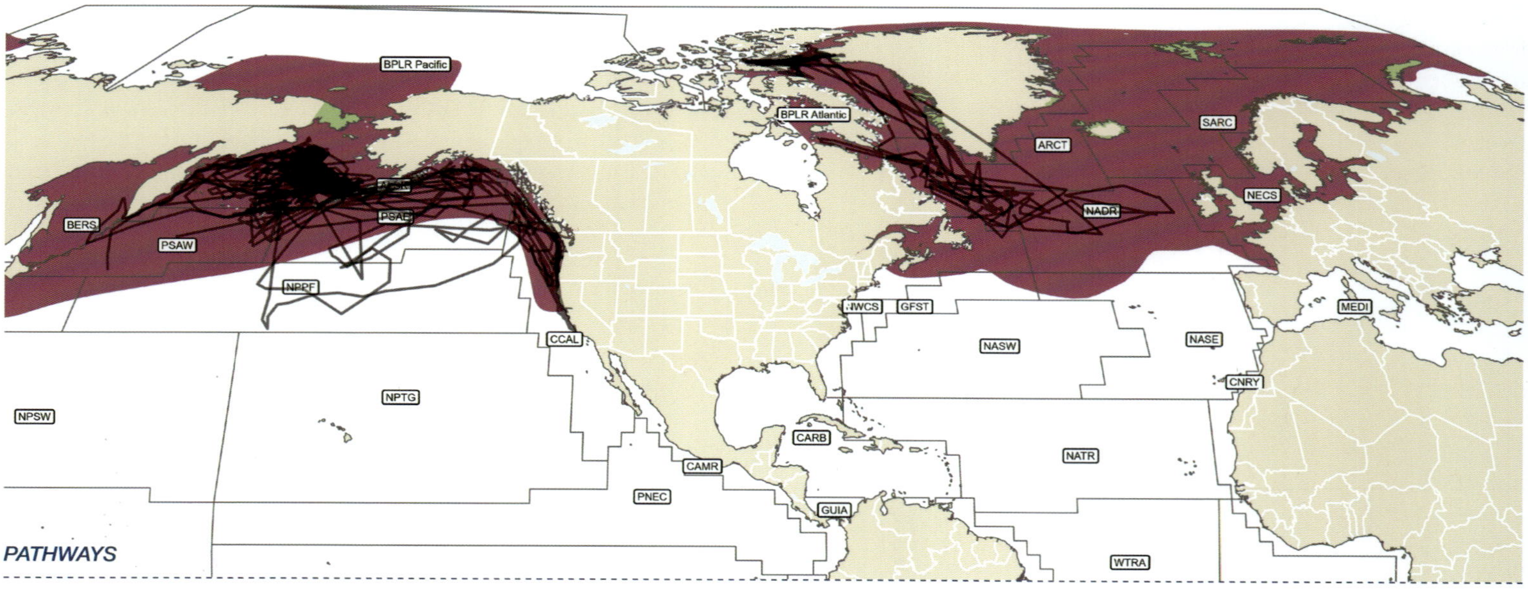

Current, while no birds from any other colony did. In the Atlantic Ocean, birds from many European colonies visited North American and North Atlantic pelagic waters, but some colonies (e.g., from the United Kingdom and from the Faroe Islands) had a higher proportion (33%–44%, respectively) of individuals tracked to the North Atlantic Drift Current. There was high overlap of Canadian and European colonies in the Atlantic province (48%–86% of individuals), but no Canadian birds spent the nonbreeding season in the Atlantic Subartic Province.

BANDING Band reencounter data indicate that, like many seabird species, immature birds may have weaker connectivity (Mallory et al. 2020a), given that they remain at sea for at least three years and range widely during this time. No adult band recoveries were found for this species to include in migratory connectivity estimates.

GEOLOCATOR TRACKING Data from light-level geolocators deployed by researchers participating in the SEATRACK project at 12 colonies within the United Kingdom, Norway, Ireland, Iceland, and the Faroe Islands showed birds tracked from all colonies spread across the North Atlantic during the nonbreeding season (SEATRACK, unpublished data).

SATELLITE TRACKING Four Canadian individuals tracked from Devon Island, Nunavut, Canada, used nonbreeding grounds in the northwest Atlantic Ocean, where they ranged widely (Mallory et al. 2008). Birds tracked from the four main Alaskan colonies (S.A. Hatch et al. 2010) diverged in their nonbreeding habitats. Fulmars from the northern Bering Sea (Hall Island) moved to Russian coastal waters; birds from the Pribilof Islands (southeastern Bering Sea) remained fairly resident; birds from Chagulak Island in the eastern Aleutians used wide areas of the central Bering Sea and North Pacific; and fulmars from the Semidi Islands in the western Gulf of Alaska migrated to the California Current.

STABLE ISOTOPES Quinn et. al. (2016) validated the use of feather isotopes to successfully assign 88% of Scottish breeders tracked with geolocators to two broad nonbreeding regions.

GENETICS Assessment of mitochondrial DNA confirm high genetic divergence between Atlantic and Pacific populations (Kerr and Dove 2013), which is consistent with observations at the species level for Procellariiformes. Variation is associated with geography rather than color morph. Baetscher et al. (2022) used colony-level differentiation in genetic markers to assign fulmars caught as fisheries bycatch (n = 1,507) to their Alaskan breeding colony. Birds from the Pribilof Islands had a proportionally higher incidence of bycatch relative to their proportion of the Alaskan metapopulation, and birds from Chagulak Island had a disproportionately lower incidence. In eastern Canada, populations are distributed across many small colonies. In a similar study, Colston-Nepali et al. (2020) were unable to assign bycaught fulmars to an exact breeding colony due to the lack of differentiation among colonies. Broadly, they inferred that birds caught in Baffin Bay and Davis Strait originated from breeding colonies in Arctic Canada, and birds caught in the Labrador Sea were a mixed group of nonbreeders, including many individuals from European colonies.

OTHER APPROACHES Fulmars have high geographic variation in plumage (light, dark, and intermediate color morphs), bill length, and body size (Mallory et al. 2020a). Movements from European or European Arctic colonies to North American waters of the northwest Atlantic, and the reverse, were historically inferred from morphometrics of collected birds (Mathiasson 1963; Brown 1973).

MIGRATORY PATHWAYS It is generally considered more ranging than purely migratory, but birds breeding on the Semidi Islands, Alaska, migrate east across the Gulf of Alaska to the US Pacific coast (S.A. Hatch et al. 2010), and birds in high Arctic Canada undertake rapid southward migrations from breeding areas through Baffin Bay and Davis Strait in response to a winter freeze-up (Mallory et al. 2008).

DATA CONTRIBUTORS

O. Bjornstad; J. Danielsen; S. Descamps; D.C. Douglas; S.A. Hatch; H.H. Helgason; M. Helberg; Y. Kolbeinsson; M.L. Mallory; H. Strøm; P. Thompson; T.L. Thorarinsson.

Pterodroma arminjoniana TRINDADE PETREL

ORDER Procellariiformes
FAMILY Procellariidae
BREEDING HABITAT Islands, Oceanic
NONBREEDING HABITAT Oceans, Pelagic
GLOBAL POPULATION SIZE 2,250
POPULATION CHANGE SINCE 1970 NA
IUCN RED LIST CATEGORY Vulnerable
R2R TIPPING POINT SPECIES No
US LISTING STATUS No
CANADA SPECIES AT RISK No
PIF CCS 16/20

DISTRIBUTION AND ABUNDANCE

The Trindade Petrel is a seabird of the Atlantic and Indian oceans, where it nests at only two locations: a larger colony at Ilha da Trindade, Brazil and a smaller colony at Round Island, Mauritius. After breeding, it disperses across subtropical and temperate waters, predominantly in the western Atlantic Ocean, including the mid-Atlantic, US (Kirwan et al. 2022). There are few records in the southeast Atlantic. In the Indian Ocean, it is described as having a limited range near its breeding colony, but few data exist (Kirwan et al. 2022).

MIGRATORY CONNECTIVITY

OVERALL It likely has strong migratory connectivity by ocean basins, based on genetic data and distribution of morphs, though the Round Island population is the result of a colonization event from the Atlantic Ocean (R.M. Brown et al. 2010). Within the Atlantic Ocean, tracking data (not presented in this Atlas) suggests strong connection across individuals to a post-breeding core area in the west-central North Atlantic Subtropical and Tropical Gyres. Moderate connectivity to migration routes is suggested from samples of tracked birds based on timing of breeding.

GEOLOCATOR TRACKING There are three geolocator studies, which likely include some overlap of datasets (Ramos et al. 2017; Leal et al. 2017; Leal and Bugoni 2021). Leal and Bugoni (2021) conducted a study that is species-specific and appears to have the largest sample of individuals with complete post-breeding tracks (n = 20) from Ilha de Trindade, Brazil. In this study, 19 of 20 birds used a common nonbreeding area in the high seas, 15–40 degrees north and 40–60 degrees west in the North Atlantic Gyre province. An earlier paper on four individuals from Krüger et al. (2016; likely a separate dataset) from

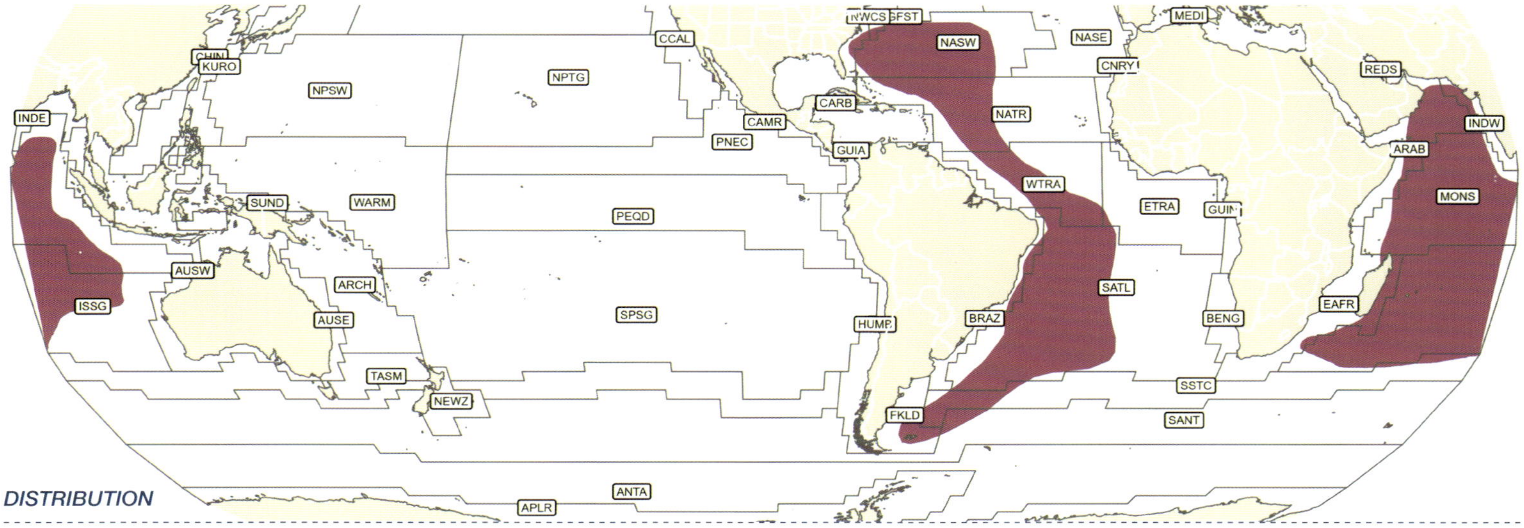

DISTRIBUTION

the same island showed post-breeding core habitat in the same area of the North Atlantic, but also within the Gulf Stream of the US and Canadian Exclusive Economic Zones. Data is not presented in the Atlas.
STABLE ISOTOPES Isotopic signatures of tracked birds indicated separation between migratory routes: signatures showed lower $\delta^{15}N$ values for those birds traveling directly to the nonbreeding area than those that used the stopover habitat. Different stable isotopic niches are also noted between breeding and nonbreeding samples (Krüger et al. 2016; Leal and Bugoni 2021).
GENETICS A comparison of nuclear DNA of Atlantic and Indian populations of Trindade Petrels showed low but significant differentiation and low migration rates between the two populations (R.M. Brown et al. 2010). The authors concluded that the Indian Ocean was colonized from the Atlantic Ocean, and the populations had recently separated and are now isolated (R.M. Brown et al. 2010).
OTHER APPROACHES No breeding to nonbreeding band reencounter data were found for this species. In the Indian Ocean, the intermediate and dark color morphs are the large majority, but in the Atlantic Ocean, the pale morph comprises more than 50% of the population (Kirwan et al. 2022). Observation data is somewhat unreliable in the Atlantic Ocean due to likely breeding of Kermedec Petrel (*Pterodroma neglecta*; predominantly dark-morph birds more commonly found in the Pacific Ocean) on Ilha da Trindade. Thus, in the Atlantic Ocean, some dark morph records could be Kermedec Petrel (Imber 2004).
MIGRATORY PATHWAYS Most interesting was a suggested partial separation of migratory routes by breeding phenology (Leal and Bugoni 2021). All petrels breeding during austral spring-summer (August through early March) traveled directly to the core nonbreeding area over the warm water of the continental ridge (also found in Krüger et al. 2016), while among the austral fall-winter breeders (February through early August), only 16% of birds used this route and instead incorporated a stopover area east of the colony and traveled north through the central Atlantic.

MURPHY'S PETREL *Pterodroma ultima*

DISTRIBUTION AND ABUNDANCE

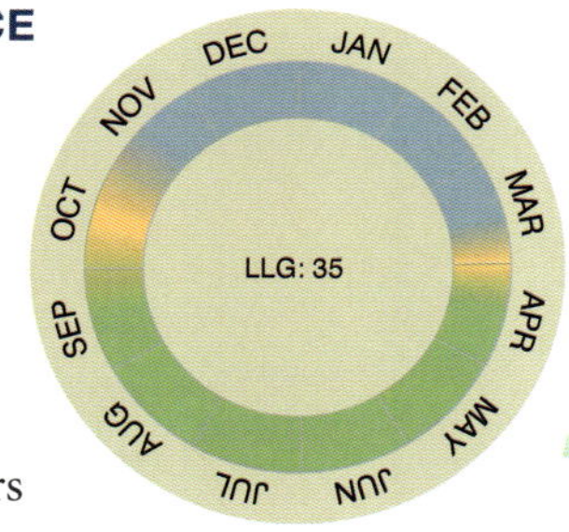

This species breeds on four island groups (Carboneras et al. 2020b) in the tropical central Pacific Ocean (Pitcairn, Austral, Tuamotus, and Easter islands) and is highly pelagic away from the colony. It is most abundant in the Pitcairn Islands (up to 250,000 breeding pairs on Ducie Island), with much smaller (<100 pairs) populations at other colonies (Carboneras et al. 2020b). Distribution is throughout the northeastern, south-central, and southeastern Pacific Ocean, but it is rarely close to land (Carboneras et al. 2020b).

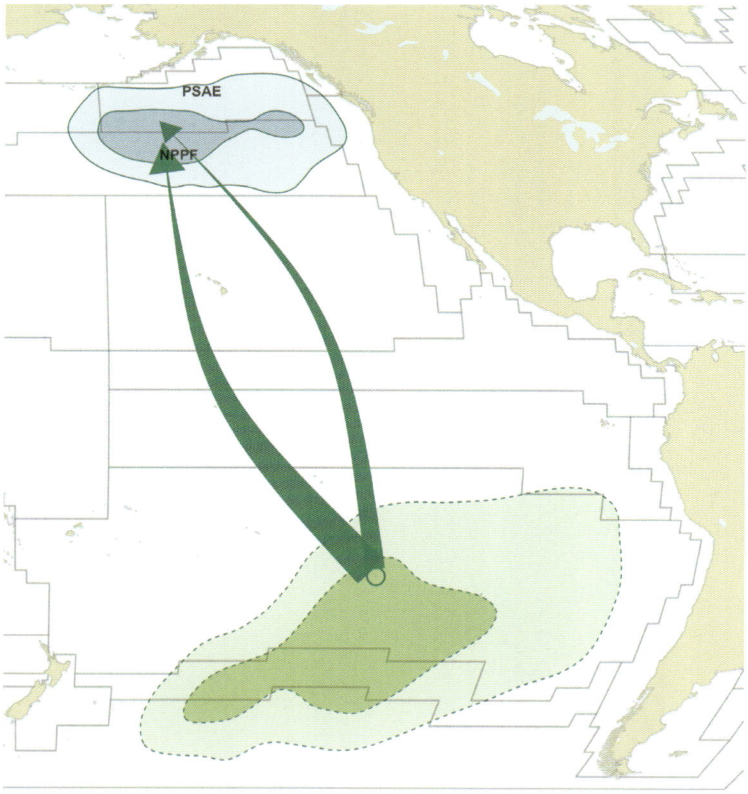

GENERALIZED CONNECTIVITY

ORDER Procellariiformes
FAMILY Procellariidae
BREEDING HABITAT Islands, Oceanic
NONBREEDING HABITAT Oceans, Pelagic
GLOBAL POPULATION SIZE 270,000
POPULATION CHANGE SINCE 1970 NA
IUCN RED LIST CATEGORY Least Concern
R2R TIPPING POINT SPECIES No
US LISTING STATUS No
CANADA SPECIES AT RISK No
PIF CCS 16/20

MIGRATORY CONNECTIVITY

OVERALL There is likely weak migratory connectivity between breeding and nonbreeding core habitats given that the distribution is restricted to the south-central and northeastern Pacific. However, migratory connectivity information is available only from a single colony. Core habitats of individuals tracked from Henderson Island, one of the Pitcairn Islands, were either in the North Pacific Polar Front (66%) or the Eastern Pacific Subarctic Gyres (34%).
GEOLOCATOR TRACKING Geolocator data from Henderson Island showed full separation between South Pacific breeding habitats

Colony

○ Henderson Island, Pitcairn Islands

	PSAE	NPPF
Henderson Island, Pitcairn Islands (n = 35)	34%	66%

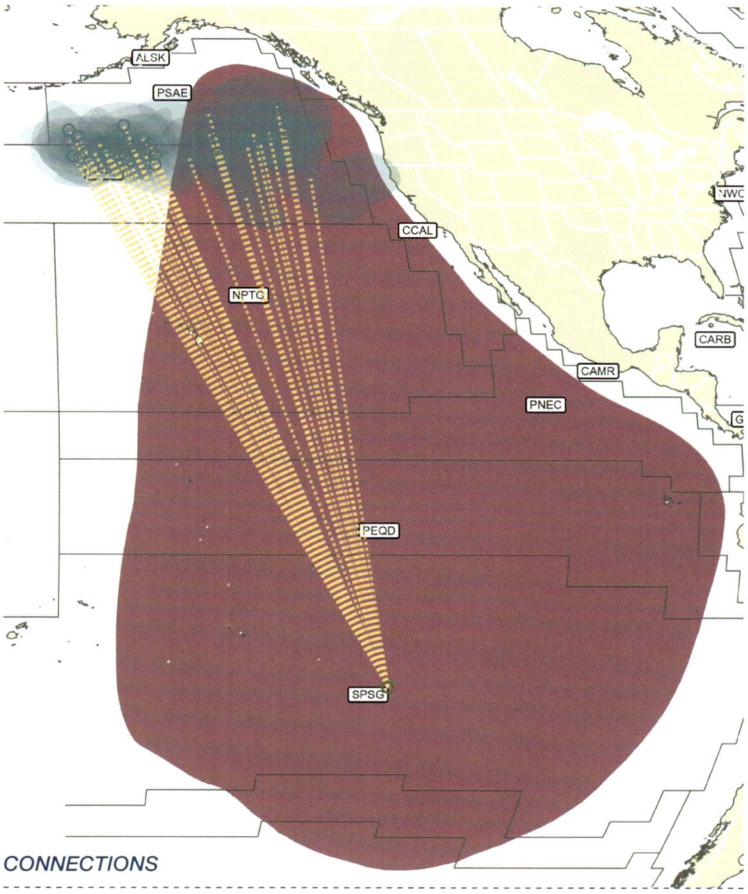

CONNECTIONS

and North Pacific nonbreeding habitats (Clay et al. 2017). No other colonies have been studied.

SATELLITE TRACKING No studies of migratory connectivity however, archival GPS tags were used to track ten nesting birds on Henderson Island during the incubation period (Clay et al. 2019). Two distinct foraging habitats were used by individuals: long (up to 3,700 km away) eastward loops toward the South Pacific Gyre off Peru or loops toward the Subtropical Front southwest of the colony (up to 1,840 km away).

OTHER APPROACHES No banding, stable isotope, or genetic studies of migratory connectivity were found for this species.

MIGRATORY PATHWAYS Like many geolocator studies of seabirds, migration occurred during the equinox periods, and thus exact routes are not well resolved; but a general pathway was from Henderson Island northwest across the Pacific Ocean and a return in a southerly direction (Clay et al. 2017).

DATA CONTRIBUTORS

M. de L. Brooke; T.A. Clay.

INTERESTING NOTE

In March 1998, a Murphy's Petrel was recorded in the South Atlantic Ocean at St. Helena, with observations suggesting that a bird visited in multiple years (Rowlands and Trueman 1999).

Pterodroma inexpectata MOTTLED PETREL

ORDER Procellariiformes
FAMILY Procellariidae
BREEDING HABITAT Islands, Oceanic
NONBREEDING HABITAT Oceans, Pelagic

GLOBAL POPULATION SIZE 780,000
POPULATION CHANGE SINCE 1970 NA
IUCN RED LIST CATEGORY Near Threatened
R2R TIPPING POINT SPECIES No
US LISTING STATUS No
CANADA SPECIES AT RISK No
PIF CCS 16/20

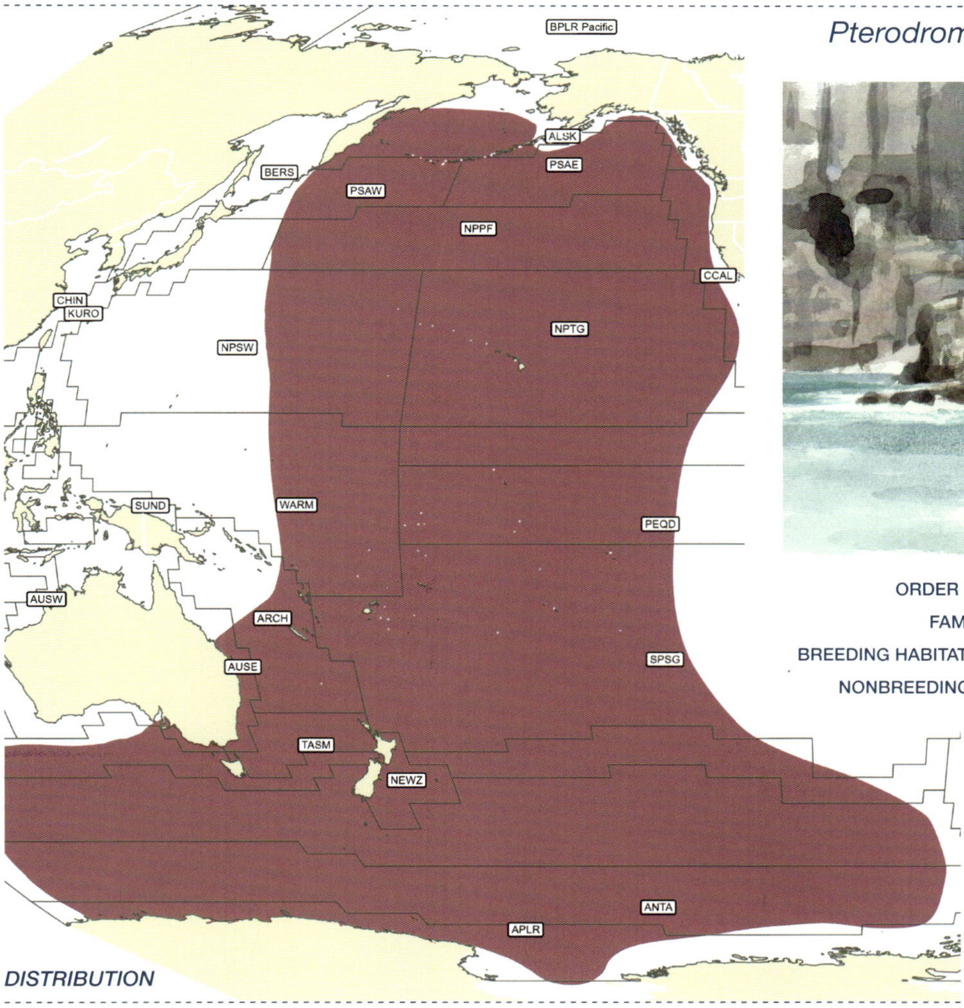

DISTRIBUTION

MIGRATORY CONNECTIVITY

OVERALL Only a BirdLife distribution map is available; no eBird data or migratory connectivity information were found for this species.

BERMUDA PETREL *Pterodroma cahow*

DISTRIBUTION AND ABUNDANCE

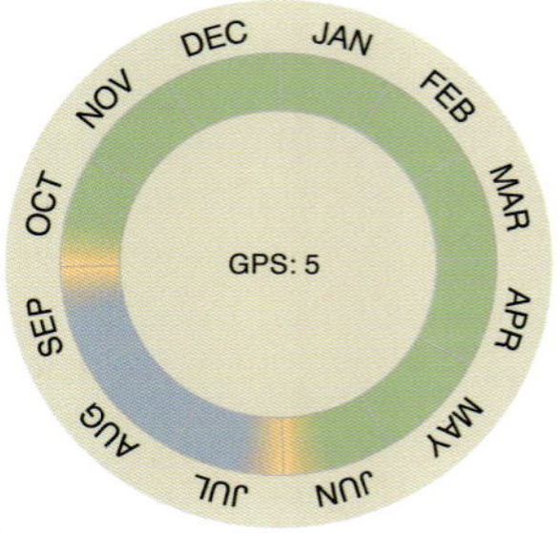

The Bermuda Petrel is a critically endangered pelagic seabird endemic to six islands of Bermuda, including Nonsuch Island, where two translocation efforts resulted in successful breeding (35 breeding pairs; Madeiros 2022) and Horn Rock (47 pairs, 38 of which are in artificial burrows; Raine et al. 2021). It uses pelagic habitats far from the colony during the breeding season. Nonbreeding distribution is still being defined, but includes the coasts of Canada south to North Carolina, US, with range extending over the North Atlantic Ocean to Ireland (Brinkley and Sutherland 2020).

ORDER Procellariiformes
FAMILY Procellariidae
BREEDING HABITAT Islands, Oceanic
NONBREEDING HABITAT Oceans, Pelagic
GLOBAL POPULATION SIZE 140
POPULATION CHANGE SINCE 1970 NA
IUCN RED LIST CATEGORY Endangered
R2R TIPPING POINT SPECIES Red-Alert
US LISTING STATUS Endangered
CANADA SPECIES AT RISK No
PIF CCS 18/20

MIGRATORY CONNECTIVITY

OVERALL Migratory connectivity to nonbreeding habitat was assessed from the Bermuda archipelago (Madeiros et al. 2014). Individuals visited a variety of habitats across the North Atlantic (Gulf Stream, Northwest Atlantic Shelves, North Atlantic Drift Current, North Atlantic Subtropical Gyre), suggesting weak connectivity. Ongoing geolocator projects to track adults and fledglings should provide additional insight (Madeiros 2022).

GEOLOCATOR TRACKING Tracking of 12 individuals from two areas in Bermuda (2009–2012) showed use of pelagic waters around the Azores and northwestern Europe, and habitats west and north of Bermuda that overlap with their marine distribution during breeding (including Canadian and American Exclusive Economic Zones [EEZs]). About one-third of individuals remained within 1,440 km of nesting grounds (Madeiros et al. 2014; Ramos et al. 2017).

OTHER APPROACHES No banding, genetic, or stable isotope studies of migratory connectivity were found for this species.

MIGRATORY PATHWAYS Kernel density maps from geolocator data of nine birds (Madeiros et al. 2014) suggest transoceanic migrants take a slightly northeasterly route to cross toward the Azores, while short-distance migrants migrate north and west toward the US coast. One individual's pathway was mapped in Madeiros et al. (2014). It migrated in March 2010 to the Azores and then northeast toward Ireland in April before looping southwest over the high seas toward the central North Atlantic during April, it returned to the Azores and then the European coast again in May, and returned to Bermuda in June.

DATA CONTRIBUTORS
C. Gjerdrum; J. Madeiros; A.F. Raine.

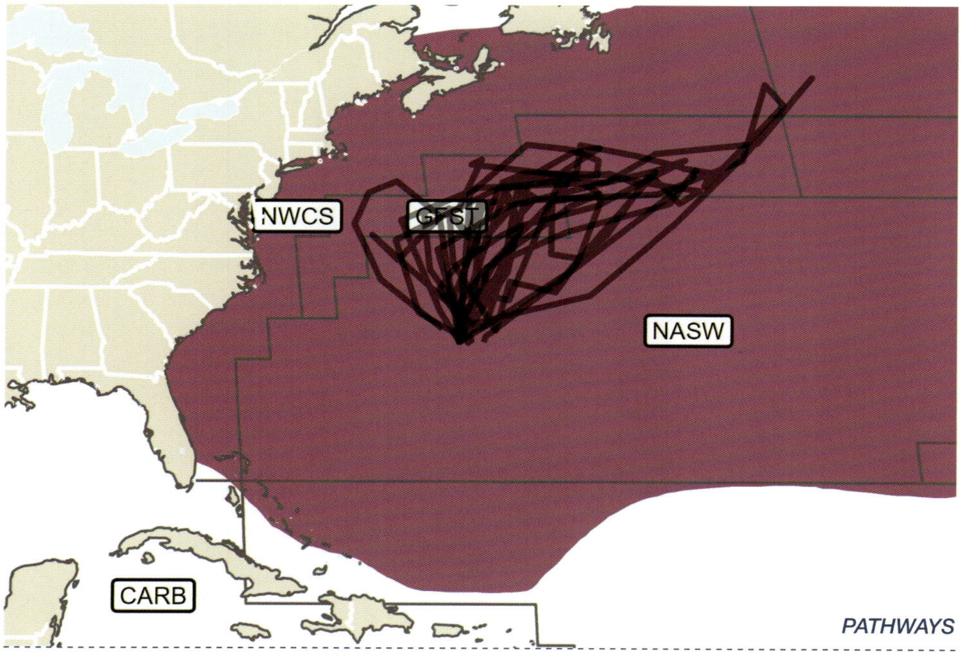

INTERESTING NOTE

From likely hundreds of thousands of nesting pairs in the 1500s, hunting and introduced predators nearly extirpated the Bermuda Petrel. It was thought to be extinct for over 300 years until seven nesting pairs were rediscovered in 1951 (Brinkley and Sutherland 2020). Despite little genetic diversity given their substantial range reduction and apparent bottleneck, the species shows no signs of inbreeding depression (Afonso et al. 2023). Five birds tracked with GPS technology during the chick-rearing period from two colonies (Horn Rock and Nonsuch Island) ranged up to 2,513 km from the colonies over the high seas and into the Canadian and US EEZs (Raine et al. 2021). All tracked individuals used similar areas.

Pterodroma hasitata **BLACK-CAPPED PETREL**

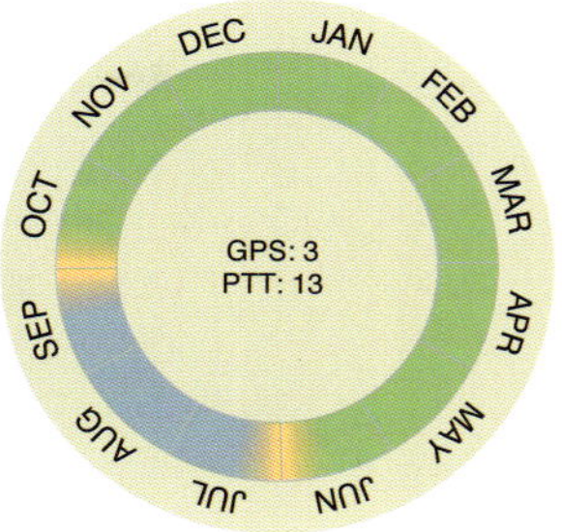

ORDER Procellariiformes
FAMILY Procellariidae
BREEDING HABITAT
Islands, Oceanic
NONBREEDING HABITAT
Oceans, Pelagic
GLOBAL POPULATION SIZE 1,500
POPULATION CHANGE
SINCE 1970 NA

IUCN RED LIST CATEGORY
Endangered
R2R TIPPING POINT SPECIES Red-Alert
US LISTING STATUS Endangered
CANADA SPECIES AT RISK No
PIF CCS 20/20

DISTRIBUTION AND ABUNDANCE

This species is endangered and is the only gadfly petrel (petrels of the genus *Pterodroma*) currently breeding in the Caribbean. Nesting is confirmed only on Hispaniola, but it may also nest on other islands of its former breeding range, including Dominica, Cuba, Guadeloupe, and Jamaica (Satgé et al. 2023a). Based on radar, Haiti likely holds the majority of the population, with the Dominican Republic population estimated to be around 600 individuals (Satgé et al. 2023a). It can be found year-round over near-shore and pelagic waters in the western North Atlantic (usually south of 40 degrees north latitude), the Caribbean Sea, and the Gulf of Mexico, but it is scarce in the Sargasso Sea (Satgé et al. 2023a).

MIGRATORY CONNECTIVITY

OVERALL Satellite tags deployed at colonies (Jodice et al. 2015) and at sea in the Gulf Stream (Satgé et al. 2023b) have yielded insights on the migratory connectivity of the Black-capped Petrel but was not calculated due to limited sample size. Birds from the central Dominican Republic are more likely to spend the nonbreeding season in the Gulf Stream province, while birds from Haiti and western Dominican Republic spend the nonbreeding season within the Northwest Atlantic Shelves and subtropical gyral provinces. No birds have been tracked to or from the Gulf of Mexico, and so the source for this nonbreeding population is unknown. To further elucidate connectivity for this species, future research could include tracking birds from the Gulf of Mexico to identify their colony origin.
SATELLITE TRACKING Three individuals tracked from the Dominican Republic migrated to either international waters, or to the US Exclusive Economic Zone (EEZ), where they used the continental shelf and slope of the southern and mid-Atlantic US (Jodice et al. 2015). They ranged widely during the nonbreeding period. In a study aimed at understanding differences in the distribution of light (n = 5) and dark color morphs (n = 5), petrels were captured off North Carolina, US, during the nonbreeding period (Satgé et al. 2023b). Light morphs had a more northerly and easterly distribution than dark morphs (Satgé et al. 2023b). One bird was tracked back to a presumed Haitian breeding site, providing the first connectivity information to this breeding site (Satgé et al. 2023b). This individual overlapped during the breeding season with birds from the western Dominican Republic. Tracking studies also highlight the significant use of the southern Caribbean Sea by breeding adults (Jodice et al. 2015; Satgé et al. 2019).
OTHER APPROACHES No banding, genetic, or stable isotope studies of migratory connectivity were found for this species.
MIGRATORY PATHWAYS Tracking data show birds depart breeding colonies and fly either overland or around Hispaniola in a northwest direction to nonbreeding areas.

DATA CONTRIBUTORS
P.G.R. Jodice; B. Keitt; Y. Satgé.

INTERESTING NOTE

An analysis of mitochondrial DNA sequences from museum specimens collected off the coast of North Carolina, US, revealed that light and intermediate morphs are significantly genetically differentiated from the dark phenotype, suggesting that the Black-capped Petrel may comprise two genetically distinct and reproductively isolated taxa (Manly et al. 2013). Establishing the genetic identity of birds at specific nesting colonies is a priority for future research (Manly et al. 2013).

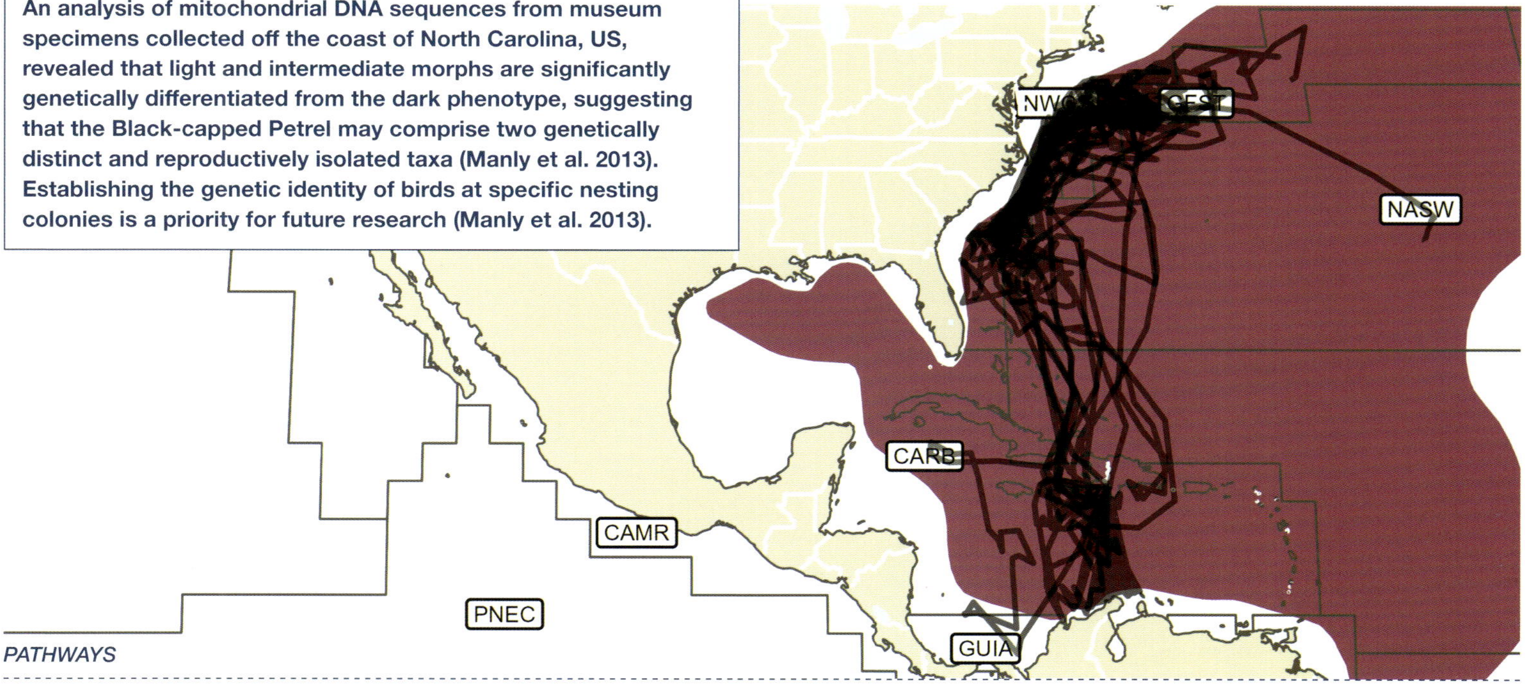

PATHWAYS

JUAN FERNANDEZ PETREL *Pterodroma externa*

ORDER Procellariiformes
FAMILY Procellariidae
BREEDING HABITAT Islands, Oceanic
NONBREEDING HABITAT Oceans, Pelagic
GLOBAL POPULATION SIZE 2,000,000
POPULATION CHANGE SINCE 1970 NA
IUCN RED LIST CATEGORY Vulnerable
R2R TIPPING POINT SPECIES No
US LISTING STATUS No
CANADA SPECIES AT RISK No
PIF CCS 16/20

MIGRATORY CONNECTIVITY

OVERALL Only a BirdLife distribution map is available; no eBird data or migratory connectivity information were found for this species.

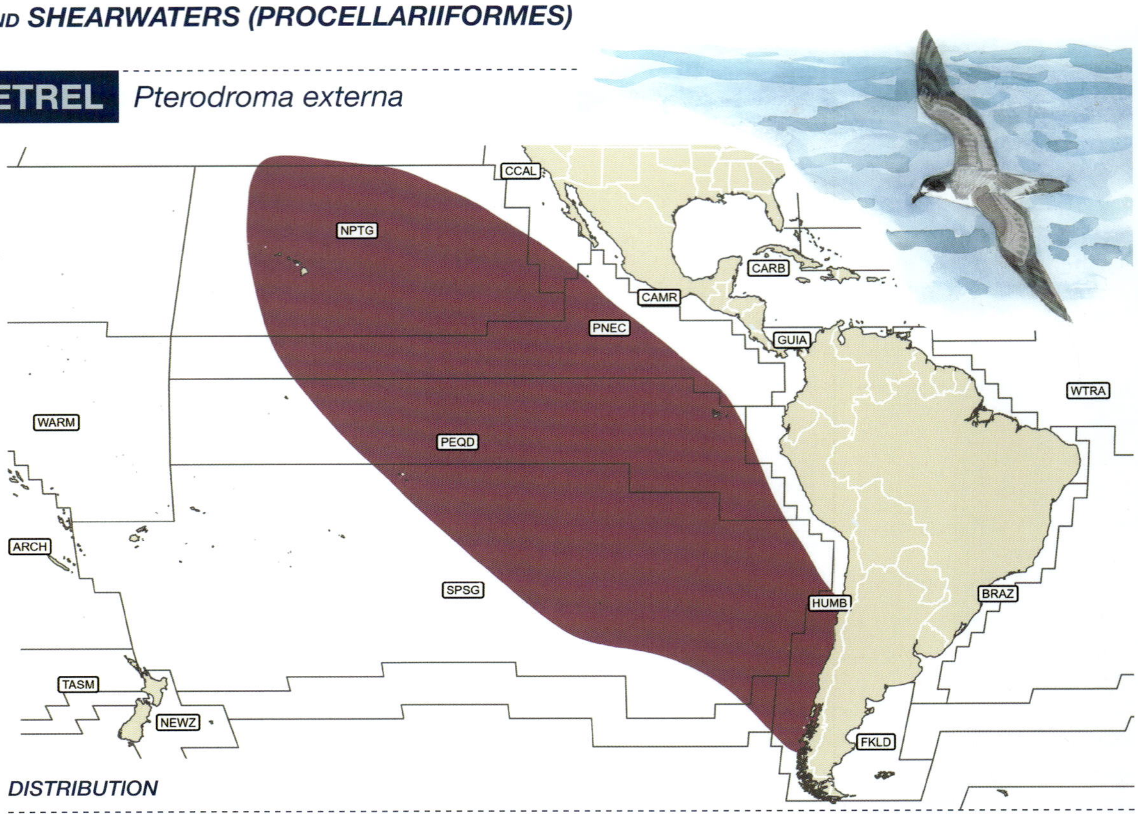

DISTRIBUTION

HAWAIIAN PETREL *Pterodroma sandwichensis*

DISTRIBUTION AND ABUNDANCE

Endemic to Hawaii, this species was formerly considered a subspecies of the Dark-rumped Petrel. The largest nesting population is on Maui, comprising 450–650 breeding pairs (Simons and Bailey 2020), with populations also on Hawai'i, Moloka'i, Lāna'i, and Kaua'i (Simons and Bailey 2020). Some populations are not confirmed and suggested only by calling and grounded fledglings (Simons and Bailey 2020). There is little information on range or distribution at sea, but at-sea surveys and tracking suggest Hawaiian Petrels distribute over pelagic waters of the North Pacific.

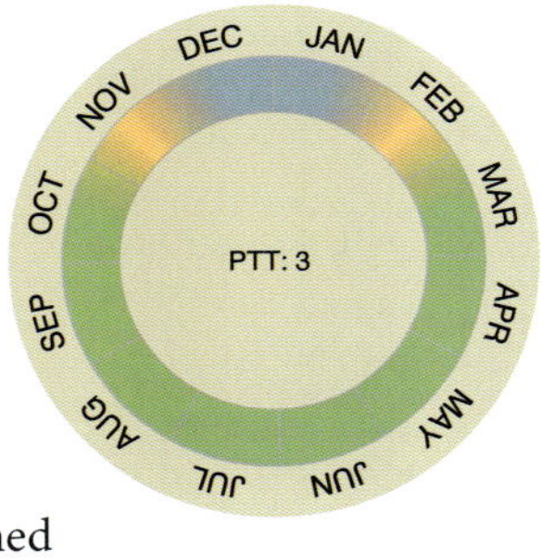

ORDER Procellariiformes
FAMILY Procellariidae
BREEDING HABITAT Islands, Oceanic
NONBREEDING HABITAT Oceans, Pelagic
GLOBAL POPULATION SIZE 8,500
POPULATION CHANGE SINCE 1970 NA
IUCN RED LIST CATEGORY Endangered
R2R TIPPING POINT SPECIES Red-Alert
US LISTING STATUS Endangered
CANADA SPECIES AT RISK No
PIF CCS 18/20

MIGRATORY CONNECTIVITY

OVERALL Stable isotopic studies suggest strong migratory connectivity to nonbreeding areas. Limited tracking data suggest use of the North Pacific Equatorial Countercurrent and western Pacific habitats during the nonbreeding season.

SATELLITE TRACKING Hawaiian Petrels tracked with Argos satellite tags during three breeding seasons (2006–2008) from Lāna'i and Maui (n = 17) made clockwise, circular, long-distance foraging trips (mean 9,716 km; 18 days) in the central North Pacific Ocean as far north as the Gulf of Alaska (J. Adams et al. 2012a). Three individuals tracked that were either not provisioning chicks (n = 2) or not recaptured during the breeding season (n = 1) moved west to the Philippine Sea (J. Adams et al. 2012a). A fledgling was tracked from Kaua'i to the Philippines (A. Raine, unpublished data). Two breeding adults tracked from Kaua'i made long-distance foraging loops to the North Pacific Ocean during the breeding season (A. Raine, unpublished data), but transmission failure prevented the discovery of nonbreeding locations.

STABLE ISOTOPES Petrels sampled from Lāna'i and Maui breeding colonies differed in mean carbon and nitrogen stable isotope values (reflective of the period of molt), total nitrogen isotopic niche width, and extent of individual specialization of nitrogen values, suggesting that birds from the two colonies use different nonbreeding habitats (Wiley et al. 2019). The authors infer strong connectivity to nonbreeding areas between the two colonies, suggesting that the loss of either colony could result in a reduction in ecological diversity (Wiley et al. 2019).

GENETICS An analysis of mitochondrial DNA from subfossil bones, museum specimens, and modern samples revealed high gene flow across colonies and no significant loss of genetic diversity as the species breeding distribution contracted over 3,000 years (Welch

INTERESTING NOTE

The Hawaiian Petrel was nearly extirpated due to hunting and introduced predators (Simons and Bailey 2020), and today they also face other challenges, such as cats. Young et al. (2023) translocated 110 Hawaiian Petrel chicks between 2015 and 2020 to a 2.5-hectare, predator-free enclosure on Kaua'i constructed with artificial burrows. Three breeding pairs have returned as adults, providing the first predator-free colony for the species.

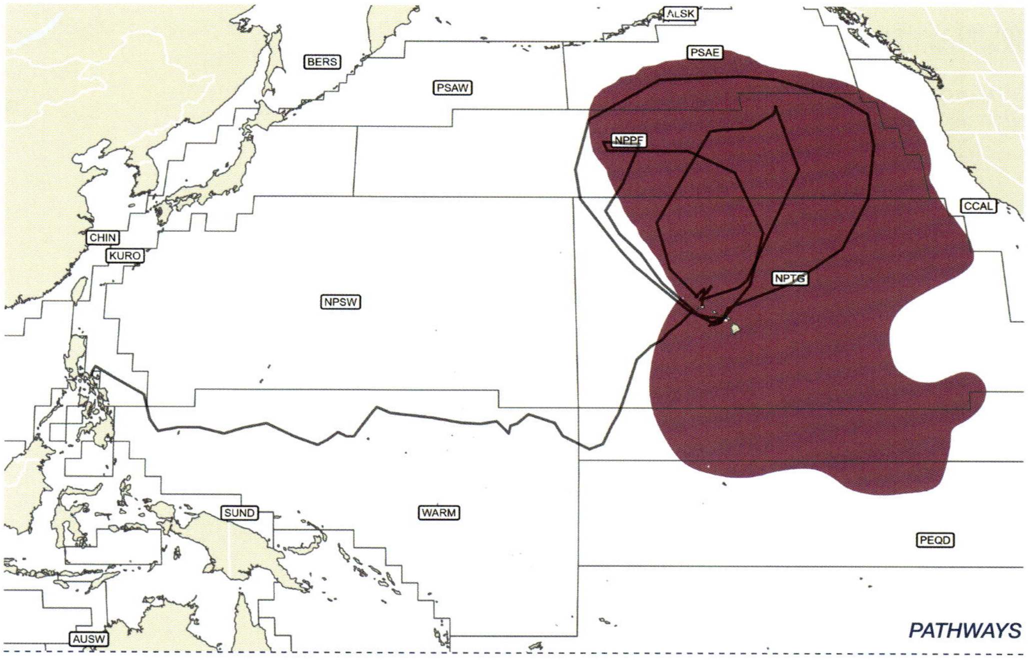

PATHWAYS

et al. 2012). Gene flow was found between extirpated colonies and extant samples from the island of Lāna'i (Welch et al. 2012). Simulations suggested that high dispersal of individuals across colonies, together with the long generation time, avoided a severe genetic bottleneck (Welch et al. 2012).

OTHER APPROACHES No banding studies of migratory connectivity were found for this species.

MIGRATORY PATHWAYS Information is limited. Three tracked individuals migrated southeast of the Hawaiian Islands, then traveled south to the North Pacific Equatorial Countercurrent before moving west toward the Philippine Sea (J. Adams et al. 2012a). Additionally, a fledgling from Kaua'i flew southwest to the North Pacific Equatorial Current and continued due west to the Philippines (A. Raine, unpublished data).

DATA CONTRIBUTORS
A.F. Raine.

Pterodroma hypoleuca BONIN PETREL

DISTRIBUTION AND ABUNDANCE

It nests in burrows on the Northwest Hawaiian Islands and on two small islands of Japan (Volcano and Bonin). The Hawaiian population (230,000 and 358,000 pairs) is estimated to be 99% of global population with colonies ranging from about 32,000 pairs on Midway Atoll (formerly 500,000 before rats were introduced) to the largest colony of 150,000–200,000 pairs on Lisianski Island (Seto and O'Daniel 2020). During the nonbreeding season, it is highly pelagic.

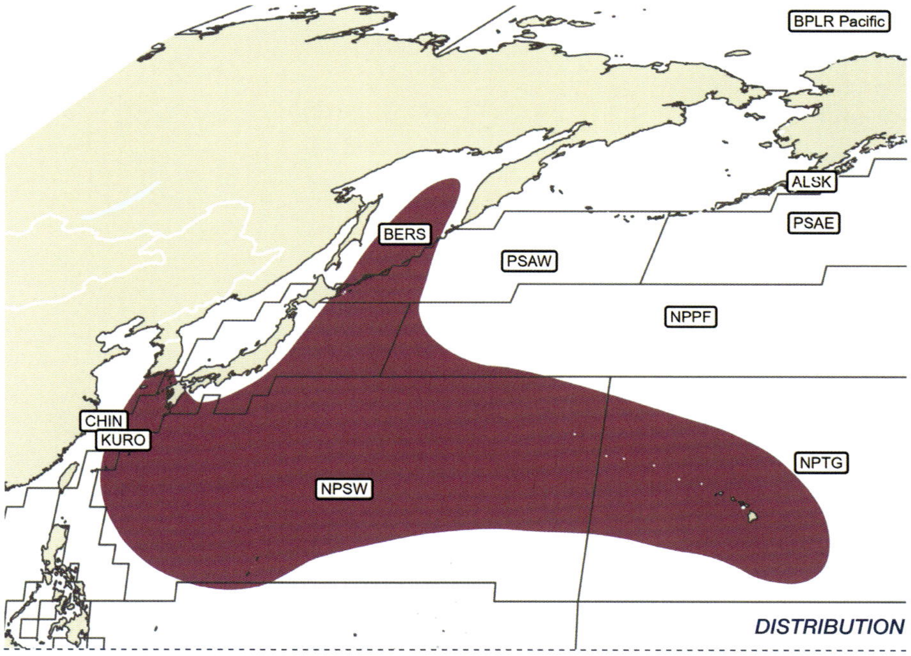

DISTRIBUTION

ORDER Procellariiformes
FAMILY Procellariidae
BREEDING HABITAT NA
NONBREEDING HABITAT NA
GLOBAL POPULATION SIZE NA
POPULATION CHANGE SINCE 1970 NA
IUCN RED LIST CATEGORY Least Concern
R2R TIPPING POINT SPECIES No
US LISTING STATUS No
CANADA SPECIES AT RISK No
PIF CCS NA

MIGRATORY CONNECTIVITY

OVERALL There is currently limited data on migratory connectivity; however, recent tracking studies have the potential to provide valuable insights into their movement patterns.

GEOLOCATOR TRACKING In 2024 and 2025, light-level geolocators (n = 37) were deployed on Bonin Petrels to track post-breeding movements from Midway Atoll National Wildlife Refuge, Hawaii (D. Rapp, R. Orben, S. Shaffer, and A.-L. Harrison, unpublished data). Nineteen of 22 tags deployed in 2024 were recovered in 2025 and are awaiting analysis.

OTHER APPROACHES GPS tags were also deployed on 28 Bonin Petrels in 2024 to track breeding season foraging loops. These movements were generally 1,500 km north and northeast of the colony. Thirty more GPS tags were deployed in 2025 (D. Rapp, R. Orben, S. Shaffer, and A.-L. Harrison, unpublished data).

MIGRATORY PATHWAYS Mostly unknown. Adults and juveniles leave breeding islands and move north to waters off the northeastern coast of Honshu Island, Japan (Seto and O'Daniel 2020). Historic data from Japanese marine cruises suggest Hawaiian birds also migrate west toward Japan to mix with Japanese breeders (Seto and O'Daniel 2020).

BLACK-WINGED PETREL *Pterodroma nigripennis*

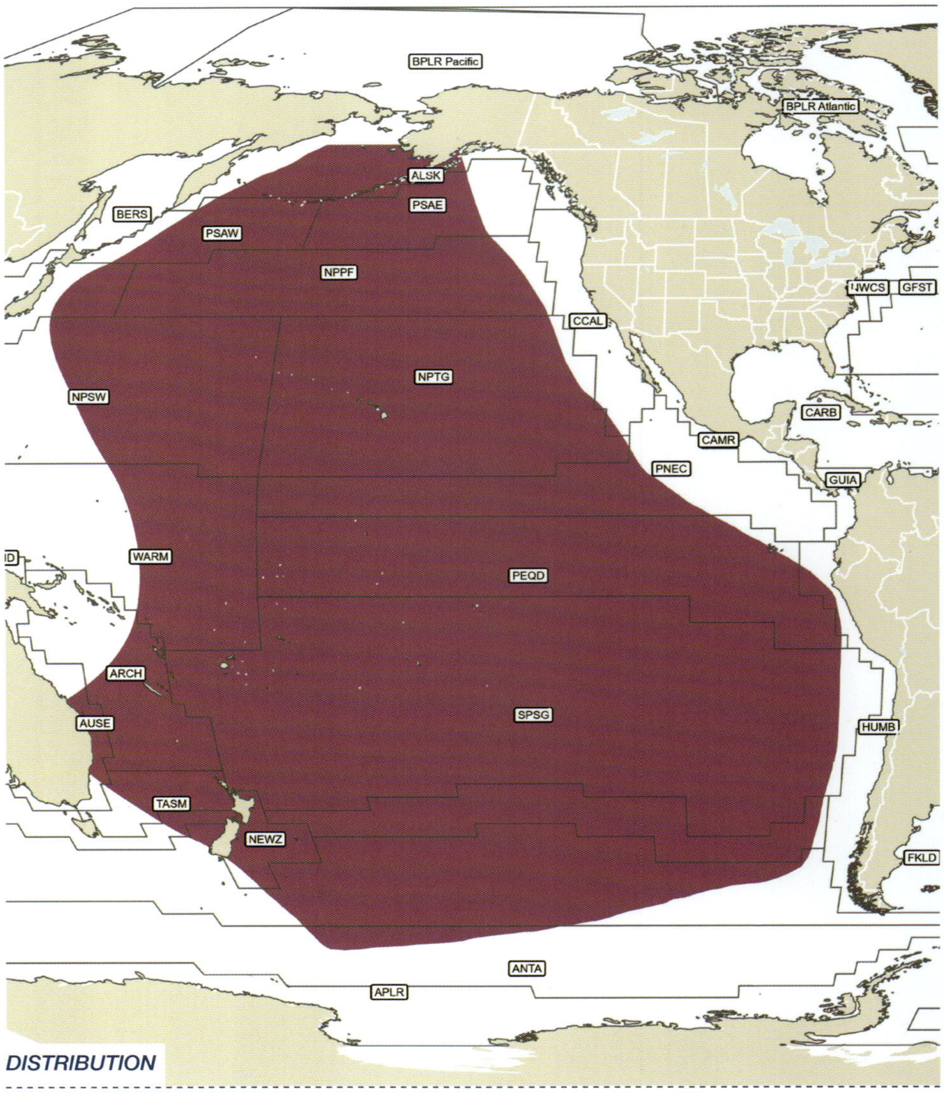

DISTRIBUTION

ORDER Procellariiformes
FAMILY Procellariidae
BREEDING HABITAT NA
NONBREEDING HABITAT NA
GLOBAL POPULATION SIZE NA
POPULATION CHANGE SINCE 1970 NA
IUCN RED LIST CATEGORY Least Concern
R2R TIPPING POINT SPECIES No
US LISTING STATUS No
CANADA SPECIES AT RISK No
PIF CCS NA

MIGRATORY CONNECTIVITY

OVERALL Only a BirdLife distribution map is available; no eBird data or migratory connectivity information were found for this species.

Pterodroma desertas DESERTAS PETREL

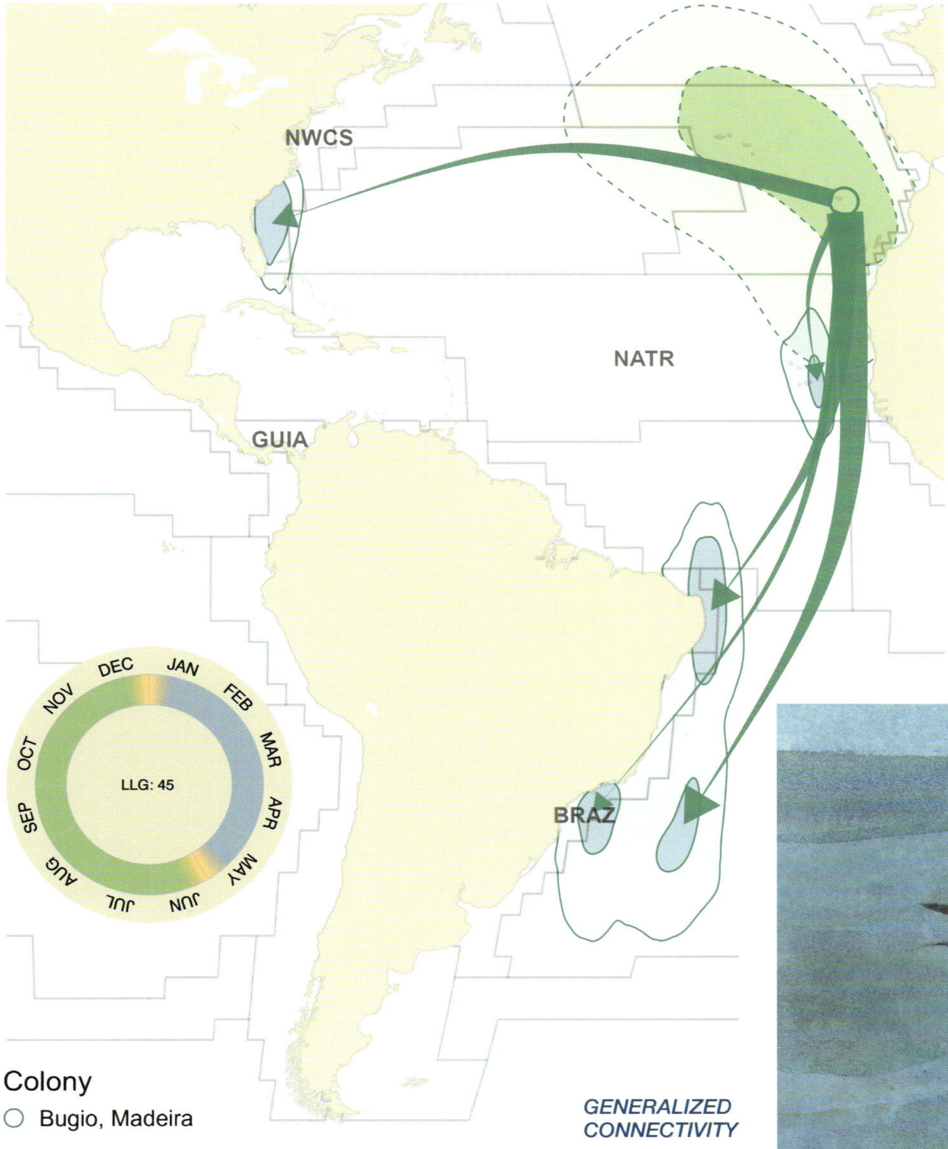

	NATR	NWCS	GUIA	SATL	BRAZ
Bugio, Madeira (n = 45)	7%	18%	27%	40%	9%

DISTRIBUTION AND ABUNDANCE

The Desertas Petrel nests in burrows on islands of Madeira (Fjeldså et al. 2020), and during breeding it forages over a large area of the eastern North Atlantic. The breeding population on Bugio Island off Desertas Island of Madeira is estimated to be 160–180 pairs in 2010. Other colonies possibly exist, for example in the Azores, but more information is needed to ascertain the extent of range restriction. Its nonbreeding distribution is across subtropical waters of the western Atlantic Ocean. The taxonomy of Fea's Petrel and Desertas Petrel is still widely debated, with some sources accepting Desertas Petrel as a separate species on the basis of disjunct breeding timing and molecular data (Fjeldså et al. 2020). The two petrels are likely indistinguishable at sea, but they nest on different island groups. Also, Desertas Petrel and Fea's Petrel are morphologically similar to Zino's Petrel (*P. madeira*), though Desertas Petrel and Fea's Petrel may have a bigger and bulkier body with a much heavier bill (Fjeldså et al. 2020).

ORDER Procellariiformes
FAMILY Procellariidae
BREEDING HABITAT NA
NONBREEDING HABITAT NA
GLOBAL POPULATION SIZE NA

POPULATION CHANGE SINCE 1970 NA
IUCN RED LIST CATEGORY Vulnerable
R2R TIPPING POINT SPECIES No
US LISTING STATUS No
CANADA SPECIES AT RISK No
PIF CCS NA

MIGRATORY CONNECTIVITY

OVERALL Migratory connectivity is unknown and not calculated here as the available information is only from Bugio Island. It is likely weak given the limited breeding distribution. The available information. Individuals tracked from Bugio Island used a variety of core nonbreeding habitats in the western Atlantic Ocean, including the South Atlantic Tropical Gyre (39%), Guianas Coastal Province (27%), the North Atlantic Shelves (18%), and the Brazilian Current (9%). A small proportion of individuals (7%) also spent the nonbreeding season in the eastern North Atlantic Tropical Gyre near the Cape Verde Islands off West Africa.

GEOLOCATOR TRACKING Desertas Petrel tracked from Bugio Island with light-level geolocators either migrated west to core nonbreeding habitats along the continental shelf of subtropical North America (southeastern USA), south to Cape Verde (where Fea's Petrel nests),

INTERESTING NOTE

A study showed that Desertas Petrel and Zino's Petrel leapfrog the nonbreeding habitats of Fea's Petrel, both spending the nonbreeding season primarily in equatorial and subtropical habitats of the western Atlantic Ocean (Ramos et al. 2016).

or southwest to the coastal currents of South America (Ramírez et al. 2013, 2016). Individuals tracked for multiple years had high fidelity to their nonbreeding site. Thus, birds may use a variety of sites, but at the individual level, there is little flexibility. During the breeding season, individuals tracked by both geolocator and GPS from Bugio Island made long foraging trips (up to 12,000 km) with a high degree of spatial overlap primarily over deep water of the Northeast Atlantic Shelves, but also venturing as far as the Northwest Atlantic Shelves off Newfoundland (Ramírez et al. 2013, 2016; Ventura et al. 2020)

STABLE ISOTOPES Individuals from Bugio Island had a high degree of specialization and separation in isotopic niches during both the breeding and nonbreeding seasons (Ramírez et al. 2016).

OTHER APPROACHES No banding or genetic studies of migratory connectivity were found for this species.

MIGRATORY PATHWAYS In the eastern North Atlantic, birds disperse over subtropical waters around Cape Verde (Ramos et al. 2016; Kirwan et al. 2025), while some individuals migrate farther south to equatorial and South Atlantic waters (Ramos et al. 2016).

DATA CONTRIBUTORS
V. Paiva; I. Ramírez.

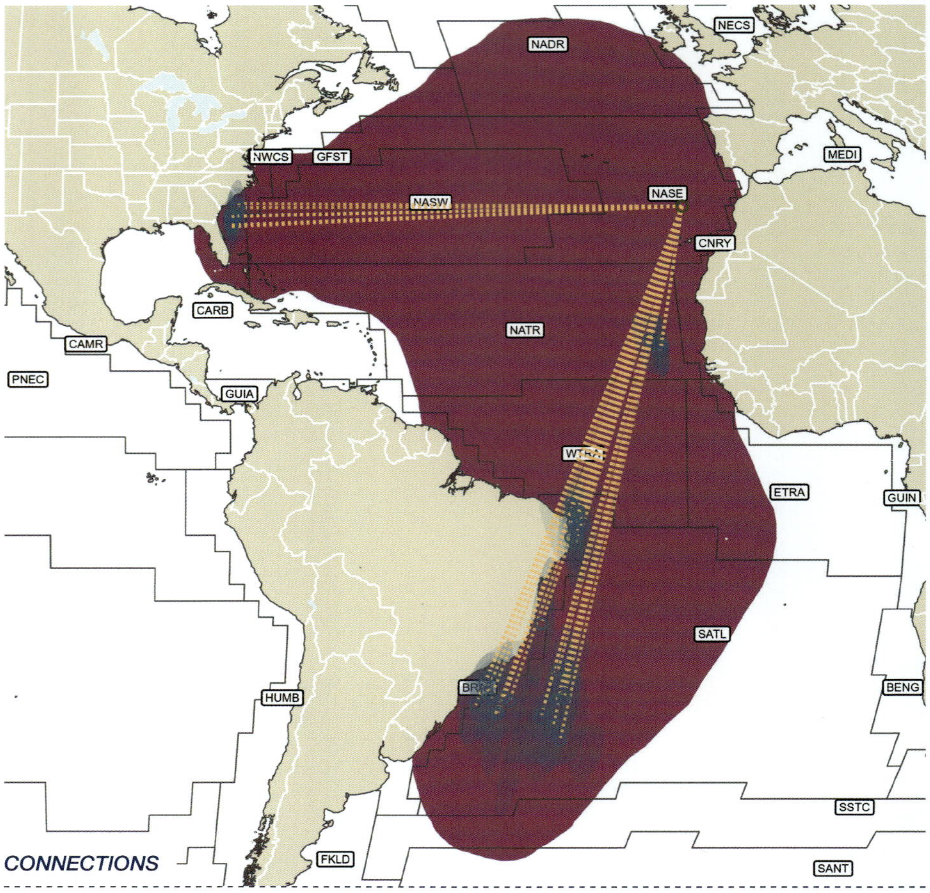

FEA'S PETREL *Pterodroma feae*

DISTRIBUTION AND ABUNDANCE

Fea's Petrel nests on Cape Verde Islands; during the post-breeding season, it is likely in the western Atlantic Ocean (Fjeldså et al. 2020). The taxonomy of Fea's Petrel and Desertas Petrel is still widely debated, with some sources accepting Desertas Petrel as a separate species on the basis of disjunct breeding timing and molecular data (Fjeldså et al. 2020). The two petrels are likely indistinguishable at sea, but they nest on different island groups. Also, Deserta's Petrel and Fea's Petrel are morphologically similar to Zino's Petrel (*P. madeira*), though Deserta's Petrel and Fea's Petrel may have a bigger and bulkier body with a much heavier bill (Fjeldså et al. 2020).

ORDER Procellariiformes
FAMILY Procellariidae
BREEDING HABITAT Islands, Oceanic
NONBREEDING HABITAT Oceans, Pelagic
GLOBAL POPULATION SIZE 1,500
POPULATION CHANGE SINCE 1970 NA

IUCN RED LIST CATEGORY Near Threatened
R2R TIPPING POINT SPECIES Red-Alert
US LISTING STATUS No
CANADA SPECIES AT RISK No
PIF CCS 18/20

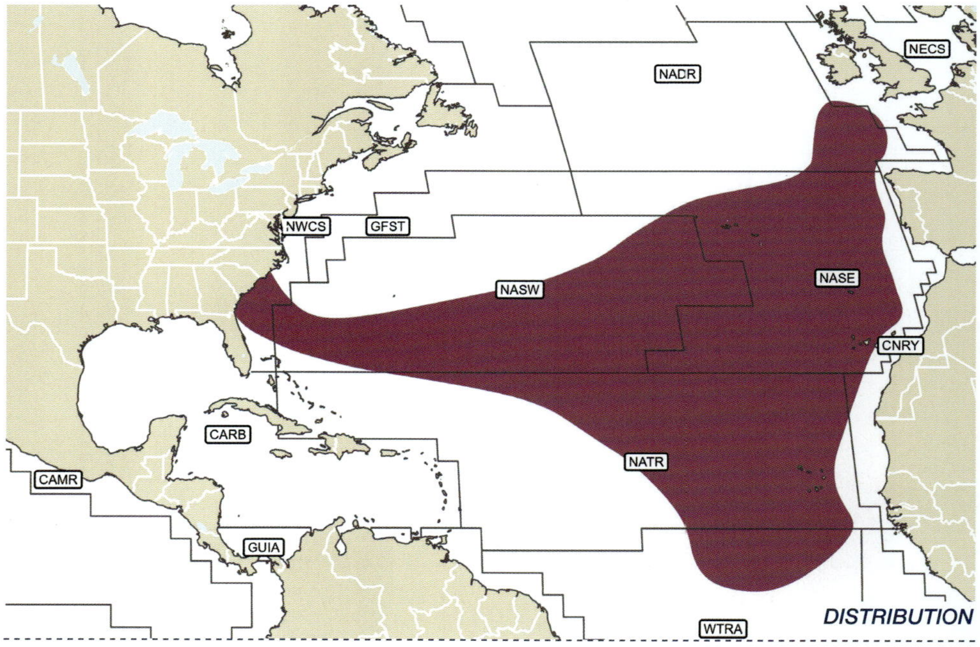

MIGRATORY CONNECTIVITY

OVERALL Only a BirdLife distribution map is available; no eBird data or migratory connectivity information were found for this species.

Pterodroma cookii **COOK'S PETREL**

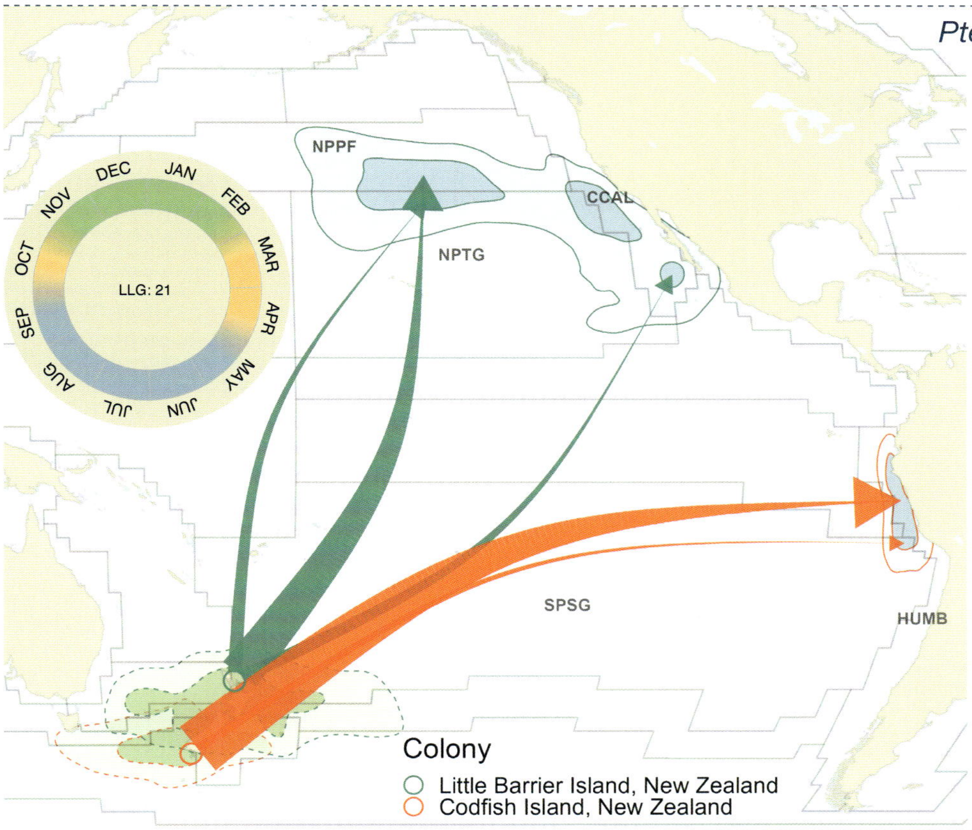

GENERALIZED CONNECTIVITY

	NPPF	CCAL	NPTG	HUMB	SPSG
Little Barrier Island, New Zealand (n = 11)	64%	18%	18%	–	–
Codfish Island, New Zealand (n = 10)	–	–	–	90%	10%

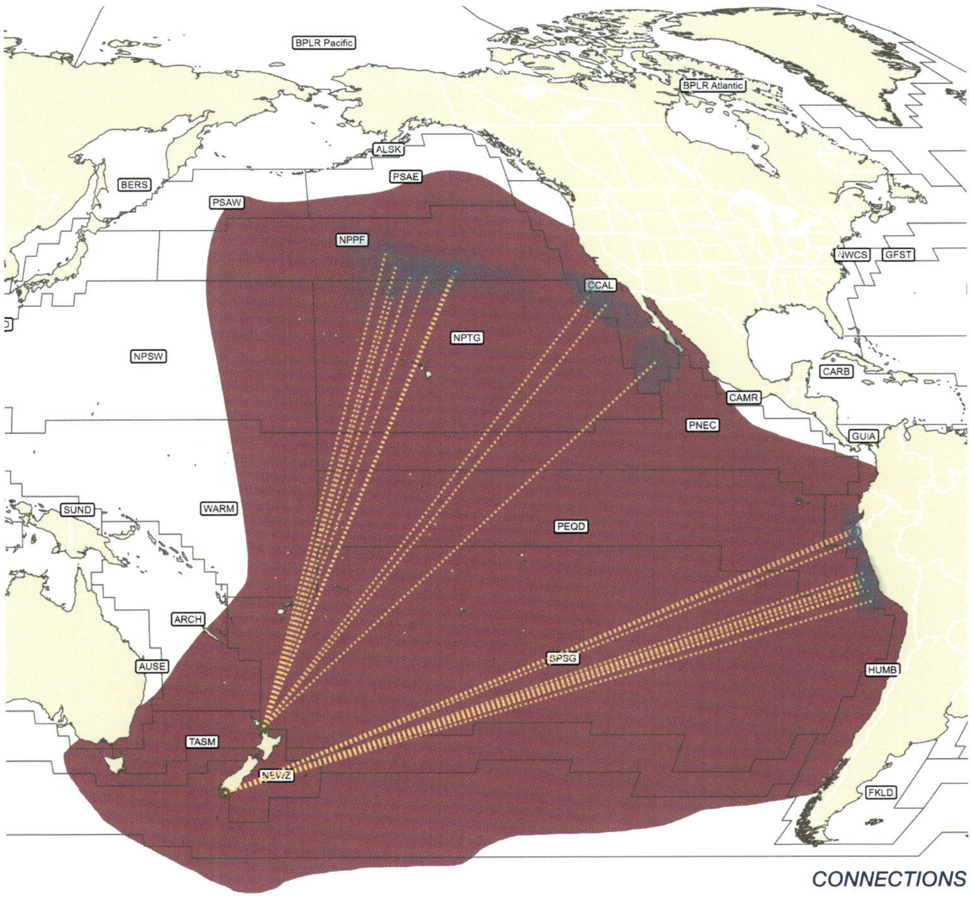

CONNECTIONS

DISTRIBUTION AND ABUNDANCE

It breeds only on three offshore islands of New Zealand on forested ridges; its former mainland populations were extirpated due to introduced predators (Taylor and Rayner 2022). It has much greater abundance on Little Barrier Island off New Zealand's north coast than on Codfish Island off New Zealand's southern coast (Taylor and Rayner 2022). At-sea foraging is pelagic, and breeding occurs over the Tasman Sea and western South Pacific. It spends the nonbreeding season across a wide swath of the Pacific Ocean, rarely within sight of land (Rayner et al. 2011).

ORDER Procellariiformes
FAMILY Procellariidae
BREEDING HABITAT Islands, Oceanic
NONBREEDING HABITAT Oceans, Pelagic
GLOBAL POPULATION SIZE 670,000
POPULATION CHANGE SINCE 1970 NA
IUCN RED LIST CATEGORY Vulnerable
R2R TIPPING POINT SPECIES No
US LISTING STATUS No
CANADA SPECIES AT RISK No
PIF CCS 14/20

MIGRATORY CONNECTIVITY

OVERALL A migratory connectivity metric could not be calculated due to limited sample size; however, tracking, genetic, and isotope data show that this species exhibits very strong connectivity, in contrast to most trans-equatorial migratory seabirds. Birds from northern New Zealand (Little Barrier Island) spend the nonbreeding season in the California Current (18%), North Pacific Tropical Gyre (18%), and North Pacific Polar Front (64%), while birds tracked from the southern population (Codfish Island) spent the nonbreeding season in the Humboldt Current (90%) off Peru and the far-eastern South Pacific Subtropical Gyre (10%). ***GEOLOCATOR TRACKING*** Birds tracked by light-level geolocators between 2007 and 2009 from two New Zealand breeding populations exhibited a post-breeding trans-equatorial separation (Rayner et al. 2011). The northern population (Little Barrier Island) spent the nonbreeding period in core habitats of the North Pacific, while the southern population (Codfish Island) spent the nonbreeding period in core habitats off the Peruvian coast.

STABLE ISOTOPES Isotope signatures from feathers of contemporary tracked and untracked individuals demonstrated petrels were using unique nonbreeding habitats between the northern and southern populations, supporting population separation during the nonbreeding period (Rayner et al. 2011). Signatures of specimens collected in the early 1900s in two primary nonbreeding oceanic habitats (California Current, North Pacific, and Humboldt Current, South Pacific) matched contemporary samples within each population, providing additional evidence for historical as well as contemporary divergence (Rayner et al. 2011).

GENETICS An analysis of mitochondrial DNA haplotypes of historical skins and contemporary samples confirm that genetic divergence of the northern and southern breeding populations has been in place for at least 100 years (Rayner et al. 2011). Despite the trans-equatorial separation of nonbreeding ranges between the genetically distinct populations, habitat specialization during the breeding and nonbreeding periods is thought to promote the evolutionary divergence via local adaptation (Rayner et al. 2011).

MIGRATORY PATHWAYS Post-breeding birds tracked from Little Barrier Island completed a counterclockwise loop migration. They first flew east, then north, crossing the equator to reach nonbreeding core ranges in the North Pacific Ocean, returning prior to breeding in a directly southwest route. In contrast, birds tracked from Codfish Island in the south departed and returned one month after northern birds, staying within the South Pacific Ocean. They flew east and then north to the Humboldt Current, where they spent the nonbreeding season. These birds returned to New Zealand via a southwestern corridor north of their post-breeding migration (Rayner et al. 2011).

DATA CONTRIBUTORS
M.J. Rayner.

BULWER'S PETREL *Bulweria bulwerii*

ORDER Procellariiformes
FAMILY Procellariidae
BREEDING HABITAT Islands, Oceanic
NONBREEDING HABITAT Oceans, Pelagic
GLOBAL POPULATION SIZE 750,000
POPULATION CHANGE SINCE 1970 NA
IUCN RED LIST CATEGORY Least Concern
R2R TIPPING POINT SPECIES No
US LISTING STATUS No
CANADA SPECIES AT RISK No
PIF CCS 14/20

MIGRATORY CONNECTIVITY

OVERALL Only a BirdLife distribution map is available; no eBird data or migratory connectivity information were found for this species.

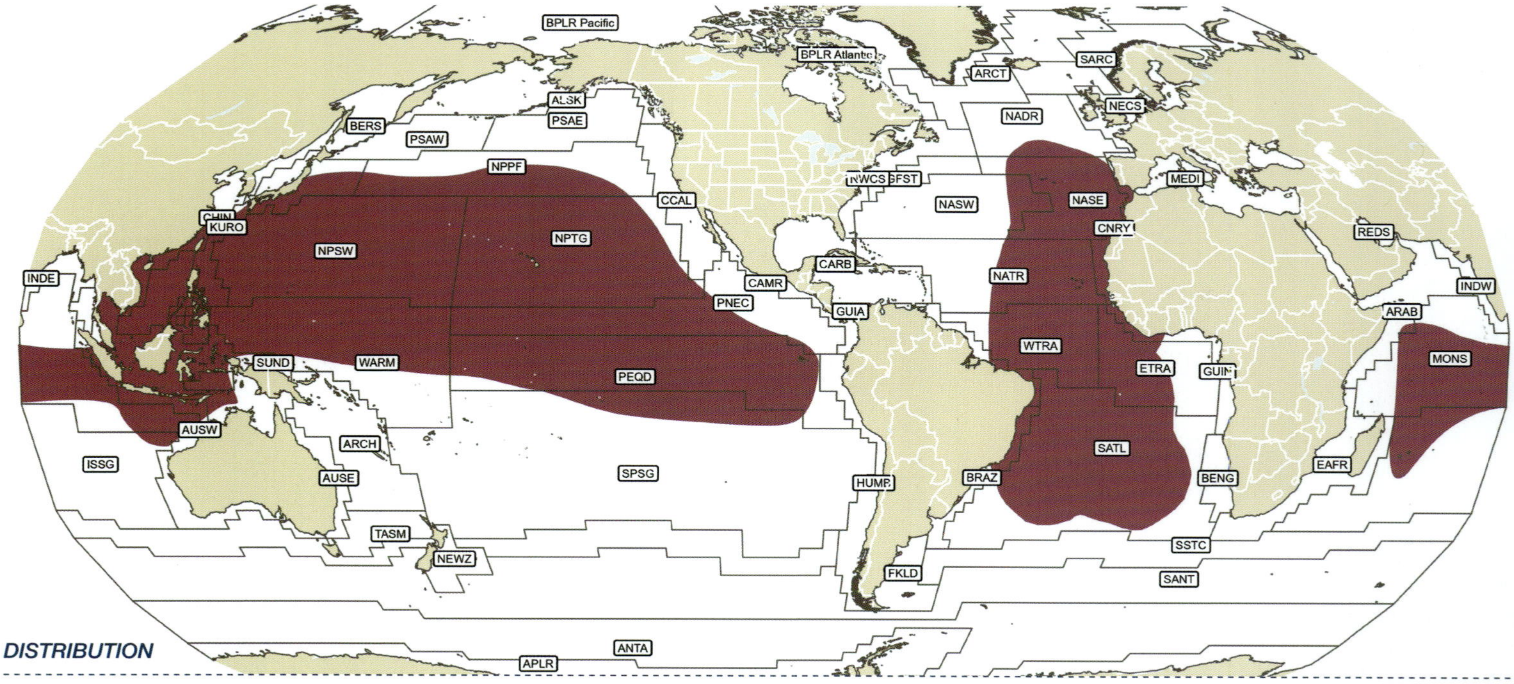

DISTRIBUTION

weak strong

Mantel = 0.03 ± 0.03

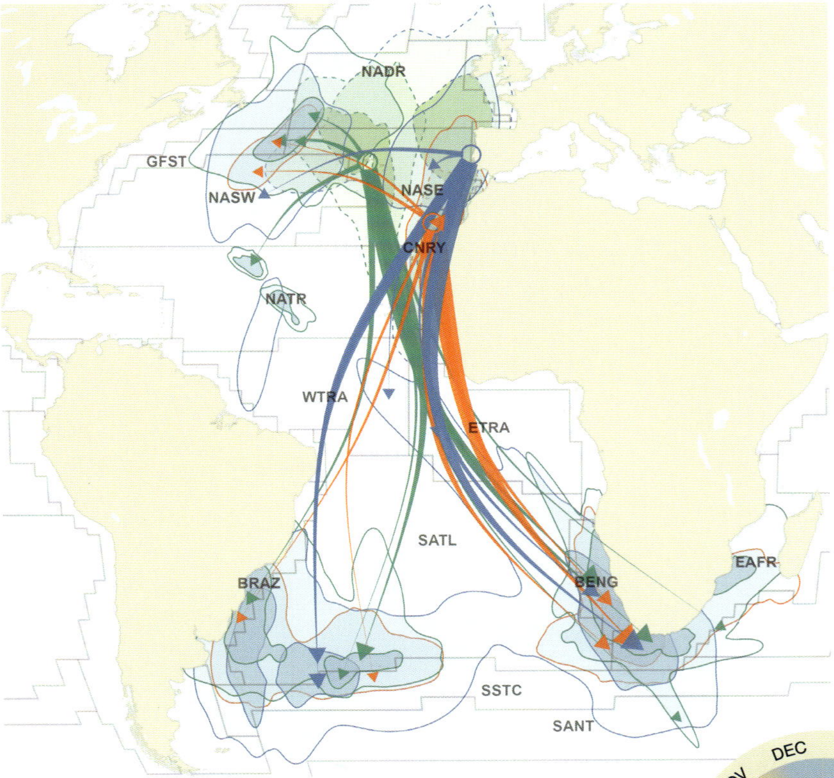

GENERALIZED CONNECTIVITY

Colony

- Azores
- Canary and Savage Islands
- Berlengas, Portugal

Calonectris borealis

CORY'S SHEARWATER

DISTRIBUTION AND ABUNDANCE

This is a large shearwater that nests on islands in the eastern Atlantic Ocean. The Azores hosts the largest breeding population (up to 188,000 pairs; del Hoyo et al. 2020b). During post-breeding, it is found throughout the Atlantic but predominantly in large upwelling areas.

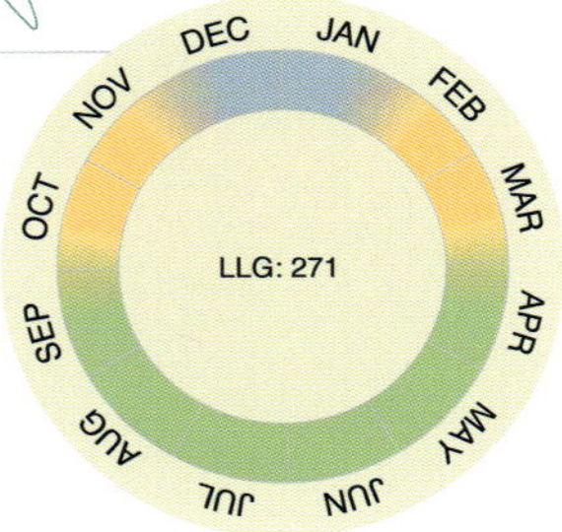

ORDER Procellariiformes
FAMILY Procellariidae
BREEDING HABITAT Islands, Oceanic
NONBREEDING HABITAT Oceans, Pelagic
GLOBAL POPULATION SIZE 510,000
POPULATION CHANGE SINCE 1970 NA
IUCN RED LIST CATEGORY Least Concern
R2R TIPPING POINT SPECIES No
US LISTING STATUS No
CANADA SPECIES AT RISK No
PIF CCS 14/20

MIGRATORY CONNECTIVITY

OVERALL Cory's Shearwater are one of the best-studied seabirds from a migratory connectivity perspective, mainly through geolocator tracking studies (González-Solís et al. 2007; Gómez-Díaz et al. 2009; Ramos et al. 2009; Dias et al. 2011; Morera-Pujol et al. 2023, 2025; P. Catry, unpublished data; E. Wakefield, unpublished data). Weak connectivity (Mantel = 0.03 ± 0.03) has been determined through extensive geolocator tracking across three regions: the Azores, Portugal, and the Canary and Savage Islands off Africa. Individuals from all colonies migrate to core upwelling habitats in the East African Coastal Zone (22%–36%) and the Benguela Current (13%–27%), and to a core area offshore in the South Atlantic Gyre (14%–20%). Birds from Portugal appeared to visit the South Subtropical Convergence zone more than other colonies (17% compared to 1%–3%), but it was also a smaller sample size.

GEOLOCATOR TRACKING González-Solís et al. (2007) first revealed the extensive mixing of Cory's Shearwater populations while migrating, staging, and during the nonbreeding season through the deployment of geolocators at five colonies. Individuals also demonstrate certain flexibility in their nonbreeding site selection across years, including switching from the western to the eastern South Atlantic Ocean, from the Northern to the Southern Hemisphere, and from the Atlantic to the Indian Ocean (Dias et al. 2011).

STABLE ISOTOPES Stable isotope signatures of the secondary feathers grown during the nonbreeding period differed across six main nonbreeding areas (Ramos et al. 2009). Complete geographic

INTERESTING NOTE

Given their long-distance migration, Cory's Shearwaters have an extremely short nonbreeding period: on average 243 days are spent at breeding grounds, and only 80 days at nonbreeding areas (González-Solís 2007).

	SSTC	SANT	NATR	NASW	GFST	NADR	NASE	CNRY	ETRA	BENG	SATL	WTRA	EAFR	BRAZ	ISSG
Azores (n = 60)	3%	2%	5%	–	5%	3%	3%	–	–	27%	20%	–	22%	8%	2%
Canary and Savage Islands (n = 188)	1%	–	–	2%	1%	–	16%	13%	–	13%	14%	–	36%	3%	–
Berlengas, Portugal (n = 23)	17%	–	–	4%	–	–	4%	–	4%	17%	17%	4%	30%	–	–

separation of birds tracked at the Azores, Portugal, and off West Africa was also reflected in unique stable isotope signatures of breeding feathers across colonies (Ramos et al. 2009).

GENETICS From a sample of 14 colonies, no population substructure was found across the geographically distinct regions of the breeding range based on mitochondrial DNA (Gómez-Díaz et al. 2009). Substantial dispersal among colonies was suggested from gene flow estimates.

OTHER APPROACHES No satellite tracking or banding studies of migratory connectivity were found for this species.

MIGRATORY PATHWAYS Individuals follow a classic shearwater figure-eight pattern (González-Solís et al. 2007; P. Catry, unpublished data; E. Wakefield, unpublished data) but with much flexibility in stopover habitats and timing (Dias et al. 2011).

DATA CONTRIBUTORS

P. Catry; M.A. Dias; J. González-Solís, Jacob; J.P. Granadeiro; V. Paiva; J.A. Ramos; R. Ramos; E.D. Wakefield.

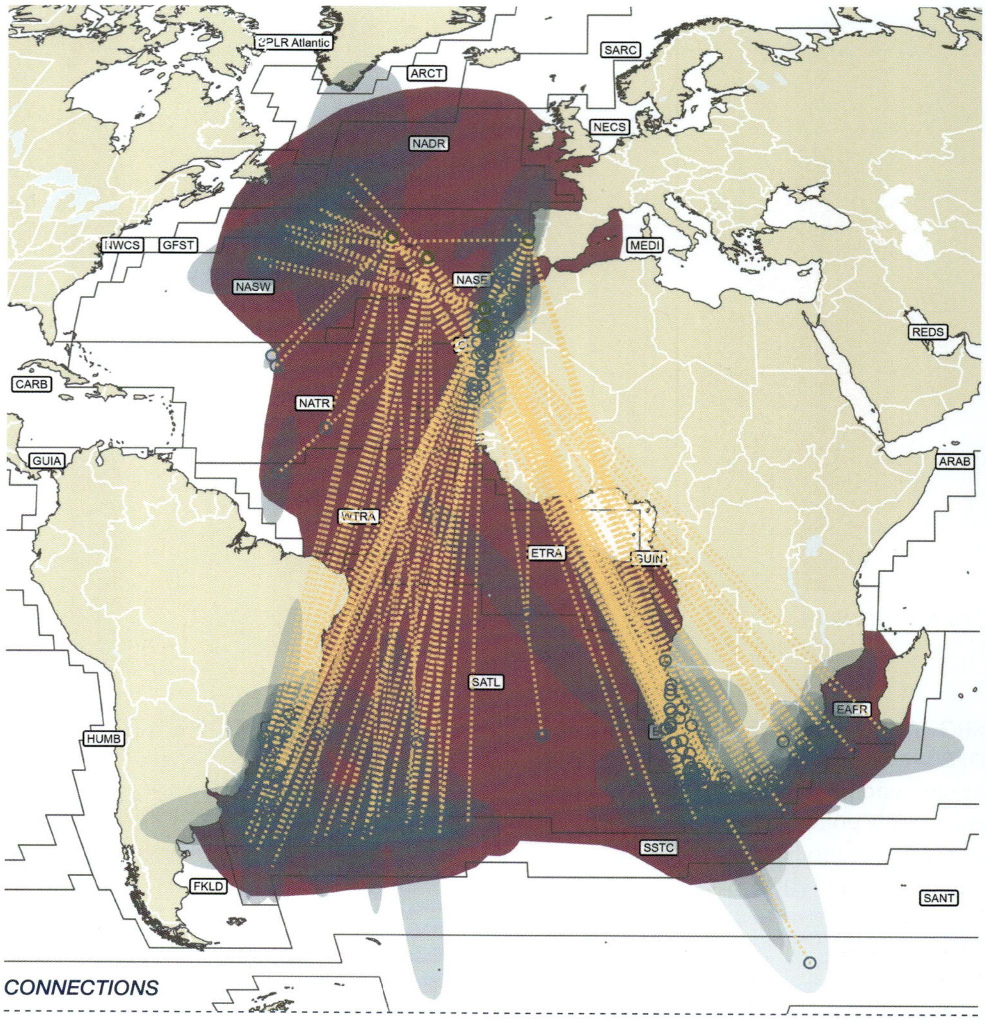

CONNECTIONS

WEDGE-TAILED SHEARWATER *Ardenna pacifica*

DISTRIBUTION AND ABUNDANCE

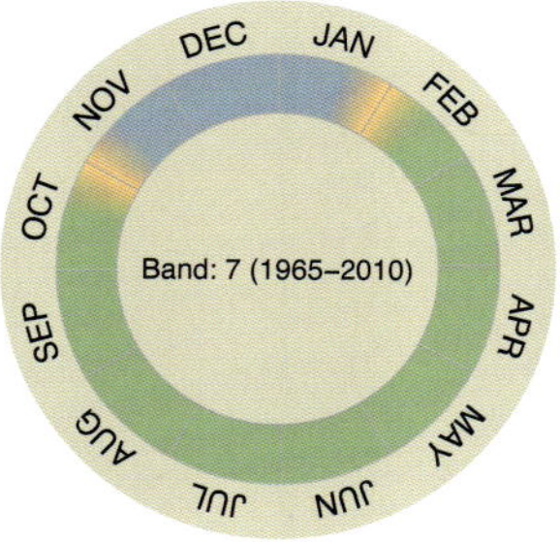

This dusky pelagic seabird (but with much polymorphism) is common in warm tropical waters from Madagascar in the Indian Ocean to Mexico in the Pacific Ocean (Whittow 2020). Wedge-tailed Shearwaters nest on small islands in a burrow. It is one of the most abundant seabirds of the Hawaiian Islands. Post-breeding behavior varies, with some populations largely sedentary, while others at the north and south of the range migrate to warm tropical waters, including trans-equatorial crossings (Whittow 2020). The nonbreeding range is only slightly larger than the breeding range.

ORDER Procellariiformes
FAMILY Procellariidae
BREEDING HABITAT Islands, Oceanic
NONBREEDING HABITAT Oceans, Pelagic
GLOBAL POPULATION SIZE 3,500,000
POPULATION CHANGE SINCE 1970 NA

IUCN RED LIST CATEGORY Least Concern
R2R TIPPING POINT SPECIES No
US LISTING STATUS No
CANADA SPECIES AT RISK No
PIF CCS 12/20

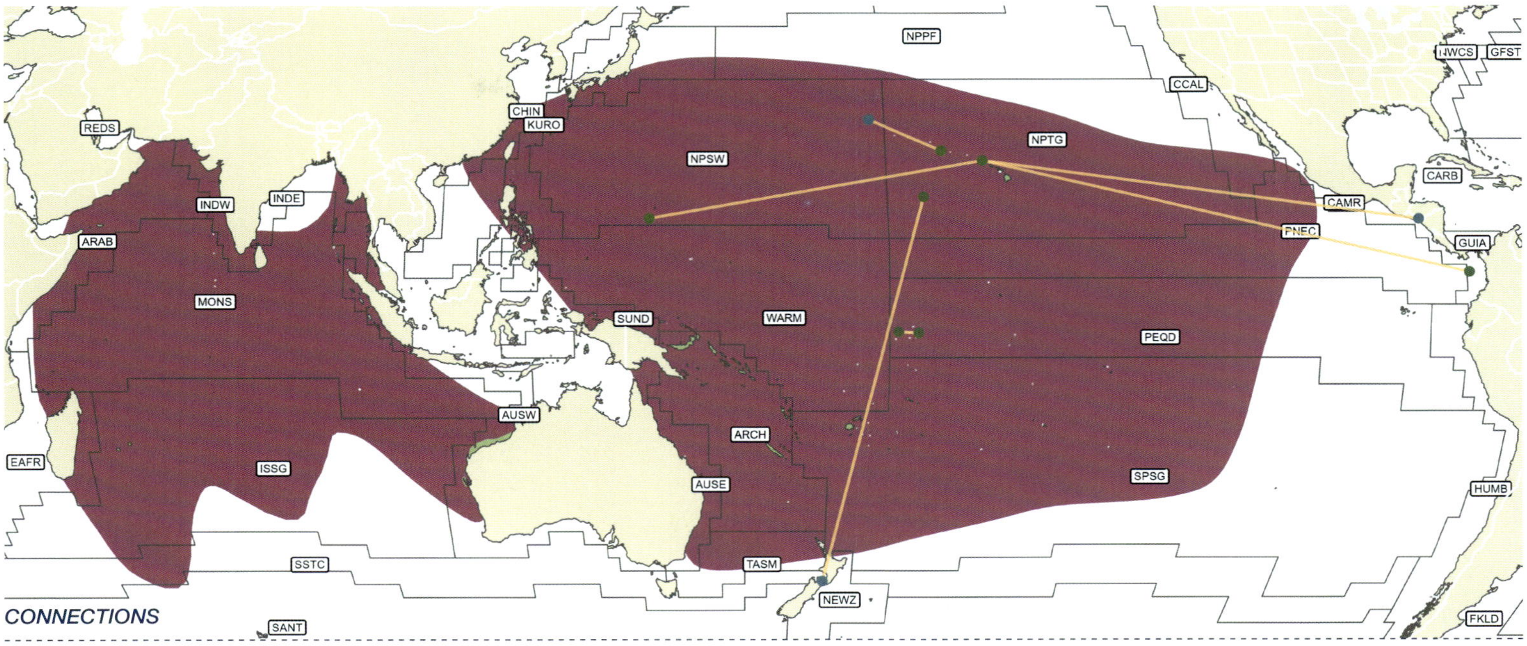

CONNECTIONS

MIGRATORY CONNECTIVITY

OVERALL There is strong connectivity across the Indian and Pacific oceans, and mixed evidence across genetic, tracking, and banding collectively points to regional population groupings within oceans suggestive of moderate to strong connectivity. Birds banded on the Hawaiian Islands have been recovered in the eastern central and western Pacific across multiple Longhurst Provinces. Tracking from Australia and New Caledonia suggests some separation between these colonies during nonbreeding, but there is likely high overlap between New Caledonian and Hawaiian birds. MC could not be calculated due to limited sample size.

BANDING Limited band reencounters from breeding to nonbreeding areas, about half during immature years. These include two short-distance connections among islands of Kiribati; Kaua'i to Tori-shima, Guam, Moloka'i, Honduras, and offshore Columbia; Hawaii (main island) to Oahu and Guam; and Johnston Atoll to Cape Jackson, New Zealand.

GEOLOCATOR TRACKING Birds tracked in 2014–2018 from three New Caledonian colonies (Weimerskirch et al. 2020; n = 35) migrated to nonbreeding areas over oligotrophic waters in the northwestern tropical Pacific in a wide longitudinal band that included Hawaii and other US island territories. There was high overlap among colonies but less overlap with birds tracked from Heron Island, Australia, that spent the nonbreeding season farther west, north of Indonesia (McDuie and Congdon 2016).

GENETICS Using data from sequence capture and high-throughput sequencing of nucleotide polymorphisms sampled from 11 countries across the breeding range, Herman et al. (2022) found marked genetic differentiation among North Pacific, South Pacific, and Indian Ocean individuals. This differentiation allowed for the assignment of nonbreeding or vagrant individuals collected along North America's Pacific coast to the Hawaiian/Guam population (n = 4) or east Australia / Solomon Islands (n = 1 collected on a beach in Washington, US). The conservation units suggested by this genetic structure does not match current subspecies designations.

OTHER APPROACHES No stable isotope studies of migratory connectivity were found for this species. In the Pacific, the differing movements of the more northerly light morphs and the more southerly dark morphs help the identification of distinct populations on nonbreeding areas (Whittow 2020).

MIGRATORY PATHWAYS In the Pacific Ocean, light morphs from Hawaii migrate south to the Equatorial Countercurrent, then east to the eastern Pacific off Central America, and return west via the North Equatorial Current. Dark morphs move northward with warm waters into the central Pacific. Some birds from New Zealand may also migrate eastward toward the eastern Pacific (Whittow 2020). Birds tracked from three New Caledonian colonies undertook fast, northeastern trans-equatorial migrations to nonbreeding areas and a slow westward movement to return to New Caledonia (Weimerskirch et al. 2020).

INTERESTING NOTE

Using GPS tags, M.G. Miller et al. (2018) found separation between the 50% use area of birds chick-rearing on Heron Island versus Lord Howe Island, Australia. Weimerskirch et al. (2020) found similar separation between two pairs located in southern New Caledonia versus two pairs located in northern New Caledonia (separated by about 200 km). A similar study approach across multiple Hawaiian islands would be helpful to evaluate breeding-season connectivity.

BULLER'S SHEARWATER *Ardenna bulleri*

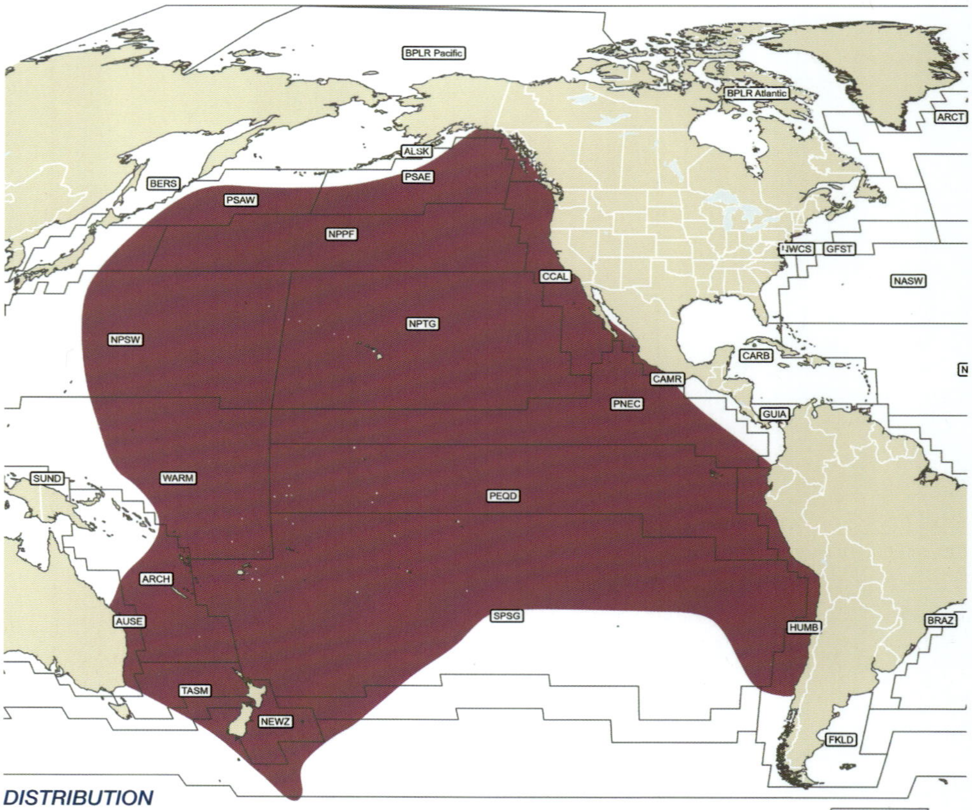

DISTRIBUTION

ORDER Procellariiformes
FAMILY Procellariidae
BREEDING HABITAT Islands, Oceanic
NONBREEDING HABITAT Oceans, Pelagic
GLOBAL POPULATION SIZE 1,700,000
POPULATION CHANGE SINCE 1970 NA
IUCN RED LIST CATEGORY Vulnerable
R2R TIPPING POINT SPECIES No
US LISTING STATUS No
CANADA SPECIES AT RISK No
PIF CCS 15/20

MIGRATORY CONNECTIVITY

OVERALL Only a BirdLife distribution map is available; no eBird data or migratory connectivity information were found for this species.

SHORT-TAILED SHEARWATER *Ardenna tenuirostris*

DISTRIBUTION AND ABUNDANCE

This species is a trans-equatorial migrant that nests on the islands and coastal headlands of southern and eastern Australia and migrates to the Bering Sea during the nonbreeding season (Carboneras et al. 2020d). It is extremely abundant and spread across hundreds of colonies (Carboneras et al. 2020d). Its largest breeding population is in Tasmania (>5 million pairs) and Victoria (Carboneras et al. 2020d).

ORDER Procellariiformes
FAMILY Procellariidae
BREEDING HABITAT Islands, Oceanic
NONBREEDING HABITAT Oceans, Pelagic
GLOBAL POPULATION SIZE 15,000,000
POPULATION CHANGE SINCE 1970 NA
IUCN RED LIST CATEGORY Least Concern
R2R TIPPING POINT SPECIES No
US LISTING STATUS No
CANADA SPECIES AT RISK No
PIF CCS 11/20

INTERESTING NOTE

In the 1970s and 1980s, it is estimated that 40,000 Short-tailed Shearwaters were drowned as bycatch in the North Pacific gillnet fisheries. Since 1991, however, gillnets have been prohibited within international waters by the United Nations Resolution 46/215.

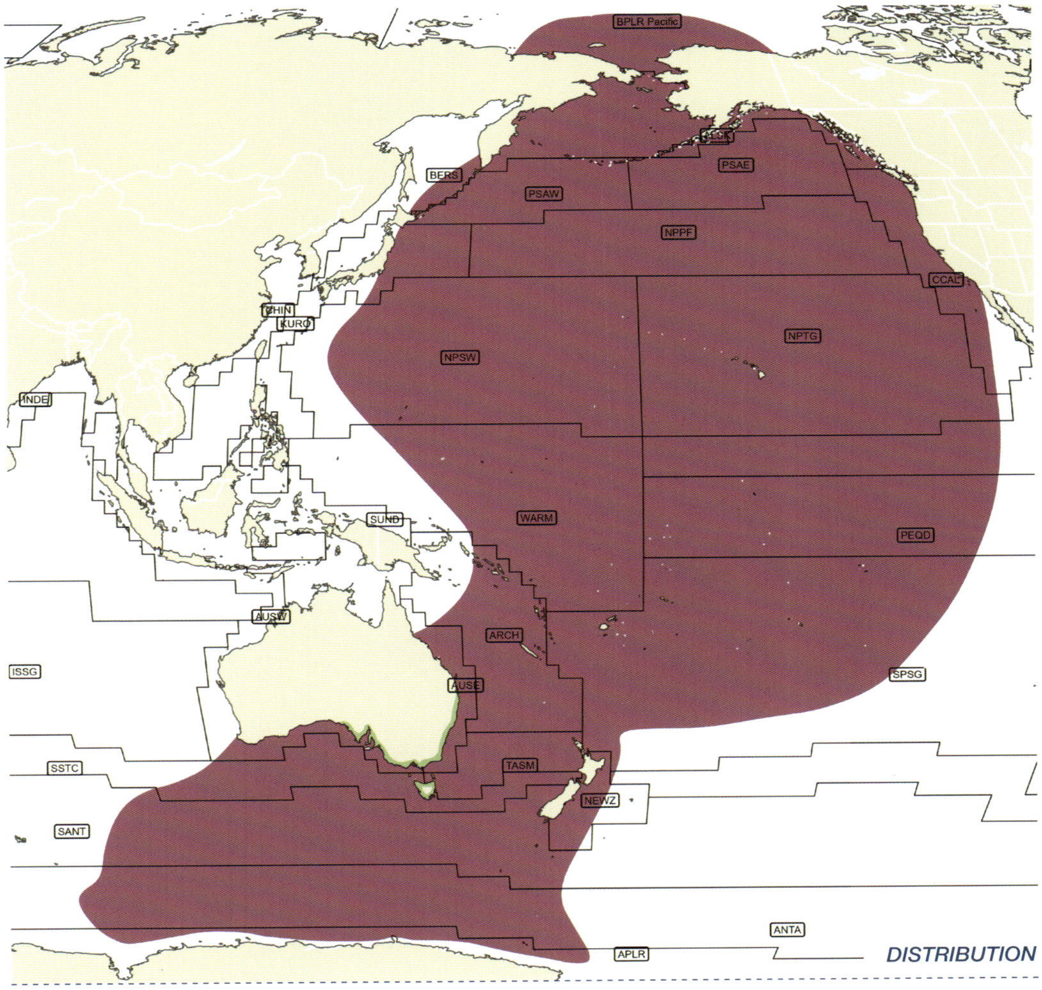

DISTRIBUTION

MIGRATORY CONNECTIVITY

OVERALL Information from the Western Hemisphere is limited, but high levels of mixing in other parts of the range is suggested by tracking and banding data from multiple colonies; and a lack of genetic structure suggests weak migratory connectivity.

BANDING Of 36,127 Short-tailed Shearwaters banded at Phillip Island, Victoria, Australia, in 1956–2010, 14 were recovered during the nonbreeding period (May–August) in the northern Pacific Ocean in 1962–1997 (Boyle et al. 2012, 2018). Ten of these were birds entangled in fishing gear, and the recoveries spanned from Japan to Alaska.

GEOLOCATOR TRACKING One individual from Phillip Island, Victoria, Australia, was tracked for more than three years (D. Boyle and Victorian Ornithological Research Group, unpublished data). This individual showcases the use of the Southern Ocean directly after breeding, as well as the tendency of many individuals that first visit the western Pacific (Kuroshio Current off Japan and the Sea of Okhotsk) to later shift to the eastern Pacific (Aleutian Islands and Bering Sea) as the nonbreeding season advances (Boyle et al. 2012; Carey et al. 2014). Geolocator data from Great Dog Island, Tasmania, Australia (Carey et al. 2014; Watanuki et al. 2015) show similar patterns.

STABLE ISOTOPES Watanuki et al. (2015) sampled feathers for mercury contamination and $\delta^{15}N$ from 28 birds tracked from Great Dog Island, Tasmania, Australia, to the North Pacific. Birds that spent the nonbreeding season in the western Pacific Ocean had a clear signature of higher mercury compared to birds that spent the nonbreeding season in the eastern Pacific or that shifted between the two. Birds spending the nonbreeding season in the eastern Pacific had slightly higher $\delta^{15}N$ values. This study suggests that mercury could be an additional marker of connectivity to nonbreeding habitat for this species.

GENETICS A study of mitochondrial DNA haplotypes found a lack of genetic structure across 11 colonies (n = 335 Shearwater chicks sampled) in southeastern Australia (Austin et al. 1994). Despite high philopatry to nest site, the authors suggest colony founding and recovery occurs via immigration of a large number of individuals (Austin et al. 1994).

MIGRATORY PATHWAYS Tracking data show the characteristic figure-eight migration pathway to and from the North Pacific (northward, northeastern pathway; southward, southwestern pathway; D. Boyle and Victorian Ornithological Research Group, unpublished data). An individual tracked from the breeding grounds in Australia left the Southern Ocean and traveled northeast to the Pacific Ocean, arriving southeast of Japan. After a brief stopover, the bird continued northeast to the waters off the Aleutian Islands and Alaska (Boyle et al. 2012). There are exceptional reports of vagrancy to the western Indian Ocean and to the Atlantic Ocean, and new connections to previously unknown stopover sites and migration routes are frequently discovered.

SOOTY SHEARWATER *Ardenna grisea*

Mantel = −0.08 ± 0.04

DISTRIBUTION AND ABUNDANCE

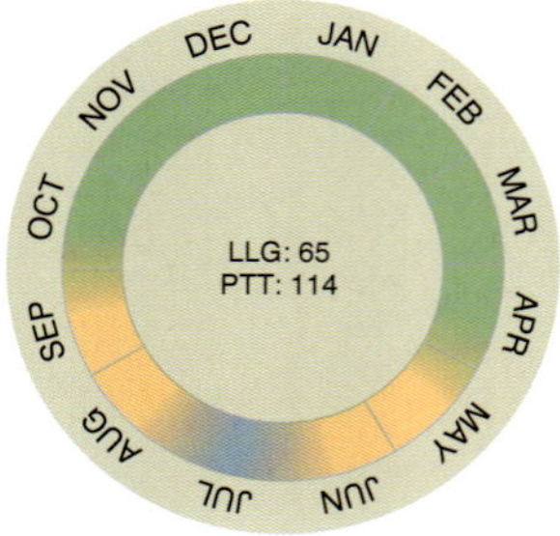

One of the most abundant seabirds in the world, Sooty Shearwaters breed on South Pacific and South Atlantic islands (and some mainland headlands) where some colonies exceed 2.5 million pairs (Carboneras et al. 2020e). It nests on densely vegetated slopes typically close to the ocean but also inland. Most of the global population is in the Pacific Ocean, where New Zealand likely hosts about half of the world's breeding population, and other large colonies are in southern Chile (some colonies are up to 200,000). The largest Atlantic Ocean colonies are in the Falkland Islands and are estimated to be about 170,000 breeding pairs, with about 85% in one colony, Kidney Island (T.J. Clark et al. 2019). Sooty Shearwaters can make large international and high-seas movements during the breeding period (Shaffer et al. 2009; Carboneras et al. 2020e). Nonbreeding, they migrate to their respective northern oceans, where they are generally found in cold offshore and pelagic waters.

Colony

- Mana Island, New Zealand
- South Island, New Zealand
- Kidney Island, Falkland Islands

ORDER Procellariiformes
FAMILY Procellariidae
BREEDING HABITAT Islands, Oceanic
NONBREEDING HABITAT Coasts, Marine Waters
GLOBAL POPULATION SIZE 13,000,000
POPULATION CHANGE SINCE 1970 NA
IUCN RED LIST CATEGORY Near Threatened
R2R TIPPING POINT SPECIES No
US LISTING STATUS No
CANADA SPECIES AT RISK No
PIF CCS 12/20

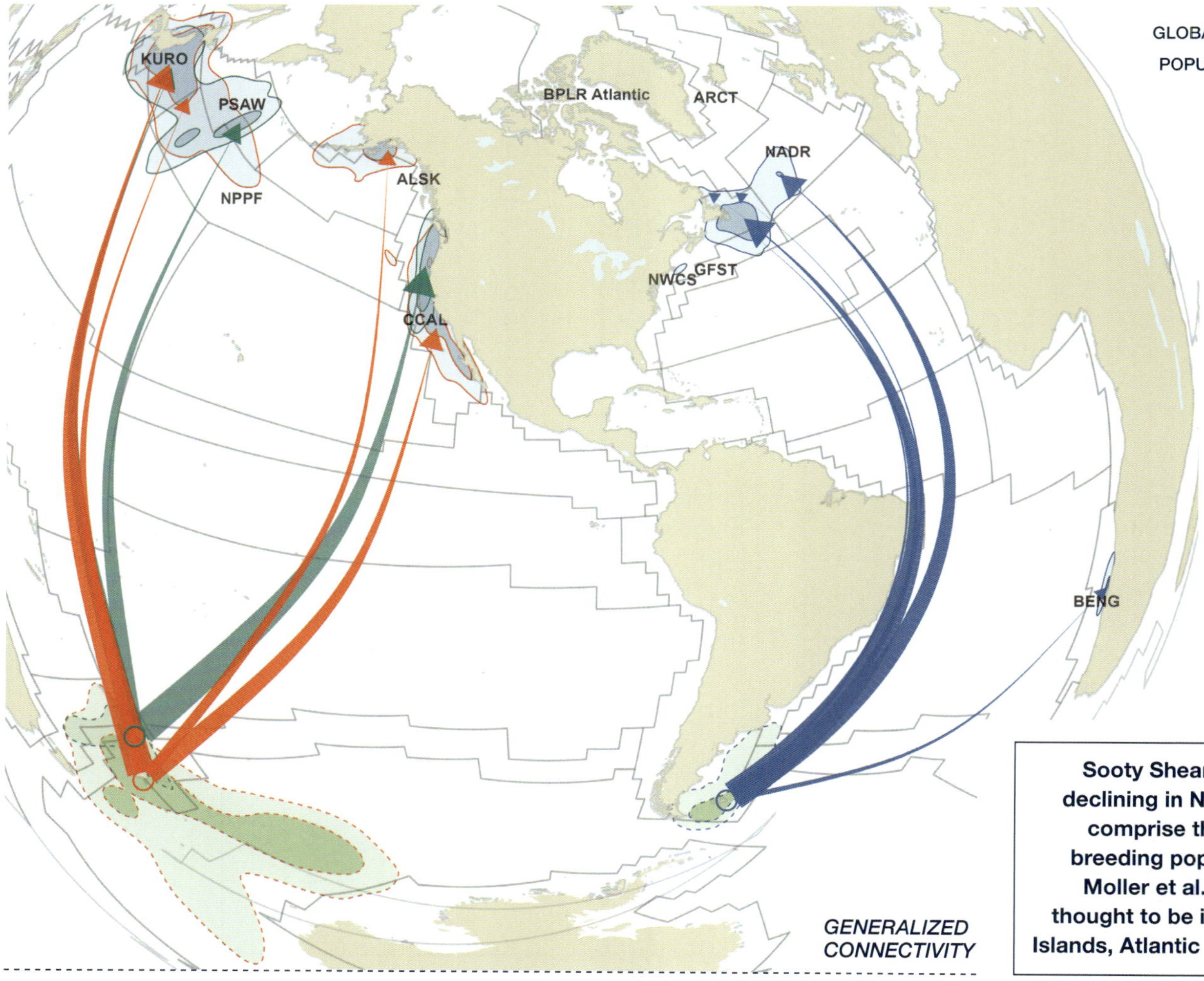

GENERALIZED CONNECTIVITY

INTERESTING NOTE

Sooty Shearwaters are thought to be declining in New Zealand colonies that comprise the majority of their global breeding population (Scott et al. 2008; Moller et al. 2009). However, they are thought to be increasing in the Falkland Islands, Atlantic Ocean (Clark et al. 2019).

	KURO	PSAW	ALSK	NPPF	CCAL	GFST	NWCS	BPLR Atlantic	ARCT	NADR	BENG
Mana Island, New Zealand (n = 4)	25%	25%	–	–	50%	–	–	–	–	–	–
South Island, New Zealand (n = 14)	43%	–	14%	14%	29%	–	–	–	–	–	–
Kidney Island, Falkland Islands (n = 47)	–	–	–	–	–	9%	47%	2%	2%	36%	4%

MIGRATORY CONNECTIVITY

OVERALL Strong connectivity by ocean basin, weak connectivity within (Mantel = −0.08 ± 0.04). There is mixing of birds in the Pacific Ocean during the nonbreeding season. From New Zealand, 25% of birds tracked from Mana Island (off New Zealand's North Island) and 43% of birds tracked from South Island spent the nonbreeding season in the Kuroshio Current. And 50% of birds tracked from Mana Island and 29% of birds tracked from South Island spent the nonbreeding season in the California Current. Note the small sample size from Mana Island (Shaffer et al. 2006; Thompson et al. 2015). Birds tracked from New Zealand are also known to migrate east to Chilean waters, where they presumably mix with Chilean colonies prior to undertaking northward migration. In the Atlantic Ocean, knowledge of connectivity is from only one colony, where 47% of individuals had core nonbreeding areas in the Northwest Atlantic Shelves provinces and 36% in the North Atlantic Drift province (over the mid-Atlantic ridge; Hedd et al. 2012; Wakefield et al. 2021).

GEOLOCATOR TRACKING Sooty Shearwaters tracked from New Zealand visited one of three areas of the North Pacific Ocean during the nonbreeding period: Japan, Alaska, or California (Shaffer et al. 2006; Thompson et al. 2015). Four birds in this study visited the Chilean coast after breeding and before migrating northward, suggesting that shearwaters breeding in Chile could mix with New Zealand birds. Birds tracked from Kidney Island, Falkland Islands, in the South Atlantic (Hedd et al. 2012), predominantly used adjacent nonbreeding habitats off Newfoundland and Labrador or over the Mid-Atlantic Ridge, except for one individual that also visited the northeast Atlantic. Individuals typically spent more time in one nonbreeding area than the other, but most birds visited both.

SATELLITE TRACKING Satellite transmitters were deployed at three nonbreeding areas in the Pacific Ocean basin; Columbia River Plume, Washington; and Monterey Bay Santa Barbara Channel, California (J. Adams et al.

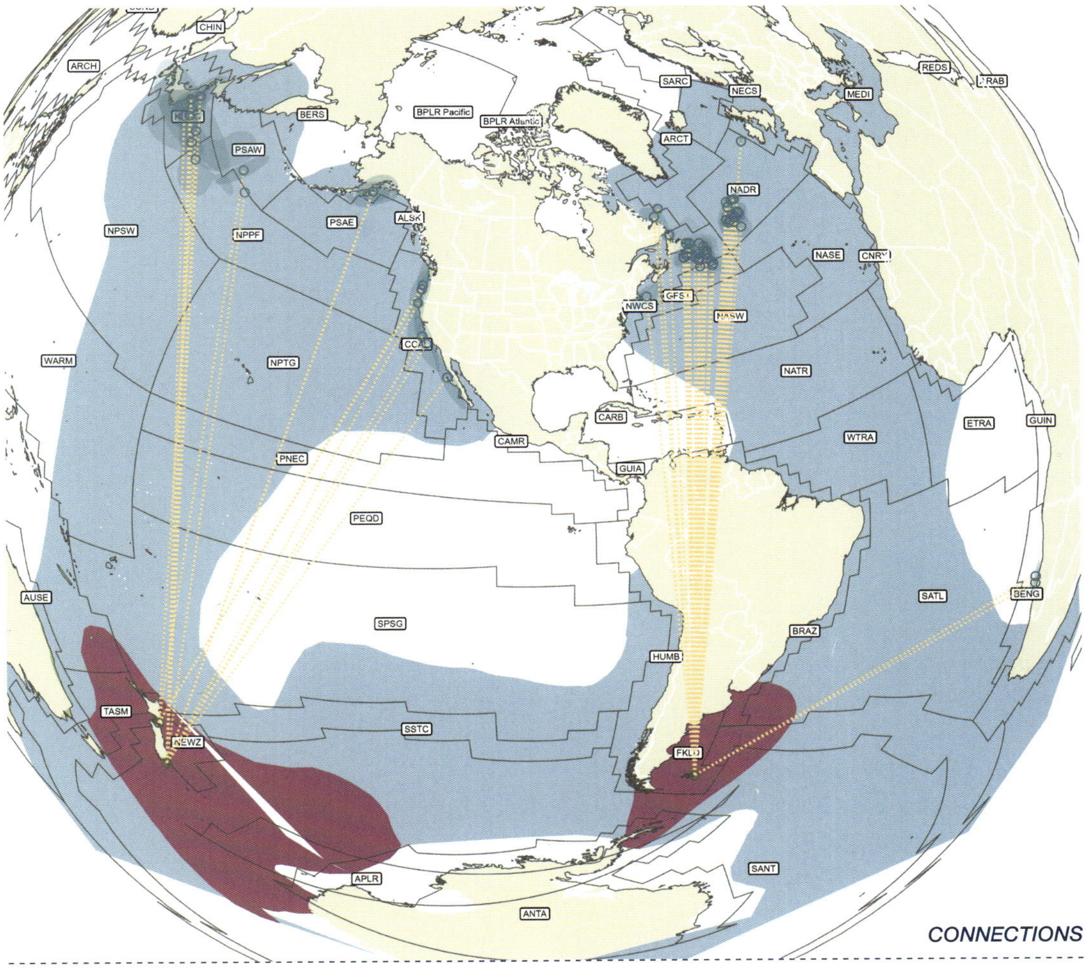

CONNECTIONS

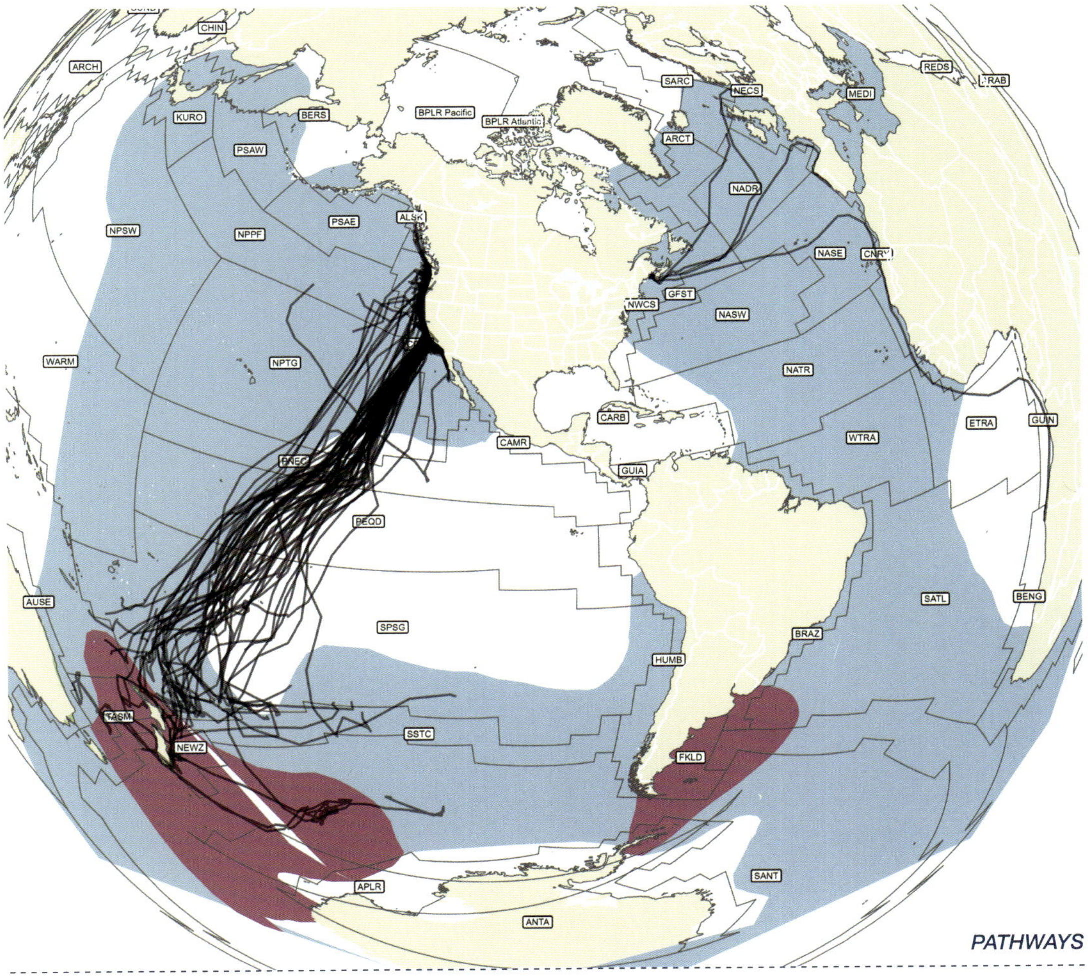

PATHWAYS

2012b). Tracking data identified several important areas, including the Columbia River Plume, Cape Blanco, Monterey Bay, Estero Bay, San Luis Obispo Bay, and the eastern Santa Barbara Channel through the inner Southern California Bight (J. Adams et al. 2012b). In the Atlantic Ocean, satellite transmitters were deployed at a nonbreeding site in the Bay of Fundy near the Grand Manan archipelago, New Brunswick, Canada (R. Ronconi, unpublished data). Birds migrated east to the coast of Europe. One bird traveled southeast to the west coast of Africa and then continued south (R. Ronconi, unpublished data).

STABLE ISOTOPES Stable isotope analysis successfully predicted nonbreeding areas (Thompson et al. 2015) of the same birds tracked by geolocators (Shaffer et al. 2006). Stable isotope analysis separated individuals that used coastal versus offshore habitats off Japan. Similarly, stable isotopes from feathers ($\delta^{15}N$ and $\delta^{13}C$) delineated between birds that predominantly used the Mid-Atlantic Ridge (more negative nitrogen, more positive carbon) versus birds that predominantly spent the nonbreeding season south of the eastern Canadian Grand Banks (Hedd et al. 2012).

GENETICS Friesen et al. (2007) found statistically significant population genetic structure but no phylogeographic structure (suggesting no population-specific genealogical lineages) in mitochondrial DNA of birds from South Pacific colonies. No studies were found assessing genetic structure across Atlantic and Pacific populations.

OTHER APPROACHES No banding migratory connectivity data were found for this species.

MIGRATORY PATHWAYS On equatorial Pacific at-sea surveys, Spear and Ainley (1999) observed differences in Sooty Shearwater seasonal flight direction. They postulated that Sooty Shearwaters make trans-equatorial figure-eight migrations, confirmed by geolocator tracking in Shaffer et al. (2006). Routes are facilitated by prevailing winds; they depart colonies flying east, then migrate generally northwest to nonbreeding areas, shift east, returning southwest to breeding colonies. This pattern is present in both the Pacific and Atlantic oceans; however, a tighter migration corridor has been observed in the Atlantic, likely due to most individuals being tracked during the nonbreeding season in the northwest Atlantic (Hedd et al. 2012) rather than across three zones as in the Pacific.

DATA CONTRIBUTORS

J. Adams; P. Catry; J.J. Felis; A. Hedd; H. Koopman; W.A. Montevecchi; R. Phillips; R.A. Ronconi; S.A. Shaffer; D.R. Thompson; E.D. Wakefield; A.J. Westgate; S.N.P. Wong.

GREAT SHEARWATER *Ardenna gravis*

DISTRIBUTION AND ABUNDANCE

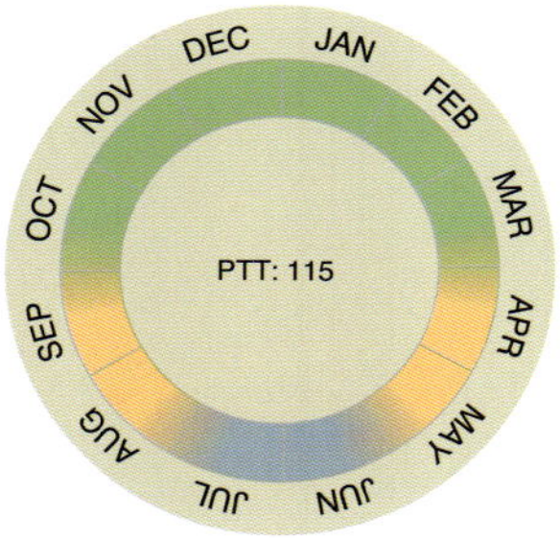

This large and abundant shearwater of the Atlantic Ocean nests in a very restricted breeding range on two islands—Tristan da Cunha (5 million pairs) and Gough Island (600,000–3 million pairs)—and in small numbers (50–100 pairs) on Kidney Island in the Falklands (Carboneras et al. 2020a). Great Shearwaters are abundant offshore Argentina and Brazil during the breeding period but venture as far east as the western Indian Ocean (Carboneras et al. 2020a). Small numbers remain in the South Atlantic all year, but most migrate primarily to offshore, pelagic waters of North America during nonbreeding (Carboneras et al. 2020a).

ORDER Procellariiformes
FAMILY Procellariidae
BREEDING HABITAT Islands, Oceanic
NONBREEDING HABITAT Oceans, Pelagic
GLOBAL POPULATION SIZE 10,000,000
POPULATION CHANGE SINCE 1970 NA
IUCN RED LIST CATEGORY Least Concern
R2R TIPPING POINT SPECIES No
US LISTING STATUS No
CANADA SPECIES AT RISK No
PIF CCS 13/20

MIGRATORY CONNECTIVITY

OVERALL Weak migratory connectivity is inferred from multiple satellite tracking studies, but it is not calculated due to limited data between breeding and nonbreeding areas. The majority of birds tracked from breeding colonies in the Tristan da Cunha island group in the South Atlantic used core nonbreeding habitats in the Northwest Atlantic Shelves province.

SATELLITE TRACKING Great Shearwaters have been tracked by Argos satellites through their migrations from three nonbreeding areas in the northwest Atlantic (New Brunswick, Canada, and Maine and Massachusetts, US) and from breeding colonies at Inaccessible and Gough islands (Powers et al. 2017, 2022; Ronconi et al., 2018).

STABLE ISOTOPES Multiple studies used stable isotope analysis to determine post-breeding molt locations (Carvalho et al. 2022) and dietary niche dynamics (Ronconi et al. 2010; Gulka et al. 2017; Hong et al. 2019).

OTHER No genetic studies of migratory connectivity were found.

MIGRATORY PATHWAYS Birds took coastal northward migrations along Brazil before flying due north over the open ocean to nonbreeding areas in the northwest Atlantic (Powers et al. 2022). Most birds caught on nonbreeding areas were immature, and there seemed to be no differences between colonies. However, older Great Shearwaters tend to migrate earlier than juveniles (Powers et al. 2017, 2022; Ronconi et al. 2018) and encounter different wind conditions that are reflected in two southward migration pathways. Birds returned south to their colonies through the eastern Atlantic Ocean, where they visited Europe and West Africa before either crossing southwest to the coast of South America (later migrants) or migrating farther offshore through a more central south Atlantic route (earlier migrants; Powers et al. 2022).

INTERESTING NOTE

Exceptional vagrancy has been reported for this species, including offshore Australia, French Polynesia, New Zealand, the California Current, and the Gulf of Alaska (Carboneras et al. 2020a). Over 200 Great Shearwaters were collected in 2008–2019 across their annual cycle in the Northwest and South Atlantic to measure plastic ingestion (Robuck et al. 2022). South Atlantic adults and chicks sampled from the breeding colonies had higher plastic burdens but lower occurrence than shearwaters sampled in the northwest Atlantic during the nonbreeding period. Plastic loads in Great Shearwaters in the North Atlantic Ocean increased in 2018 from the 1980s, but the proportion of shearwaters containing plastic decreased (Perold et al. 2024). Tracking of Great Shearwaters during the breeding season via GPS (Schoombie et al. 2018; Gough Island, incubation and chick-rearing) and Argos (Ronconi et al. 2018; Gough and Inaccessible islands, pre-laying exodus and incubation) demonstrated very large breeding distributions, with core areas extending to the Antarctic Polar Front and Subarctic Polar Front, west to Brazil, and east to South Africa. Ronconi et al. (2018) described apparent at-sea separation of core areas between birds breeding on Gough and Inaccessible islands (400 m apart) during incubation, with birds on Gough Island foraging farther east. Adults and immature birds had similar distributions (Ronconi et al. 2018).

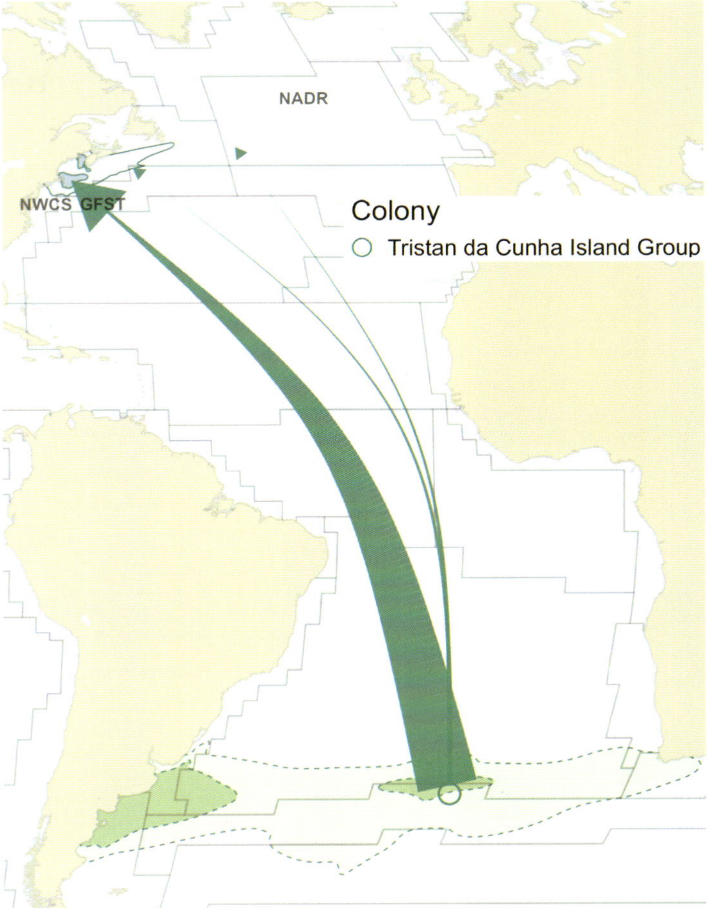

GENERALIZED CONNECTIVITY

	GFST	NWCS	NADR
Tristan da Cunha Island Group (n = 29)	3%	93%	3%

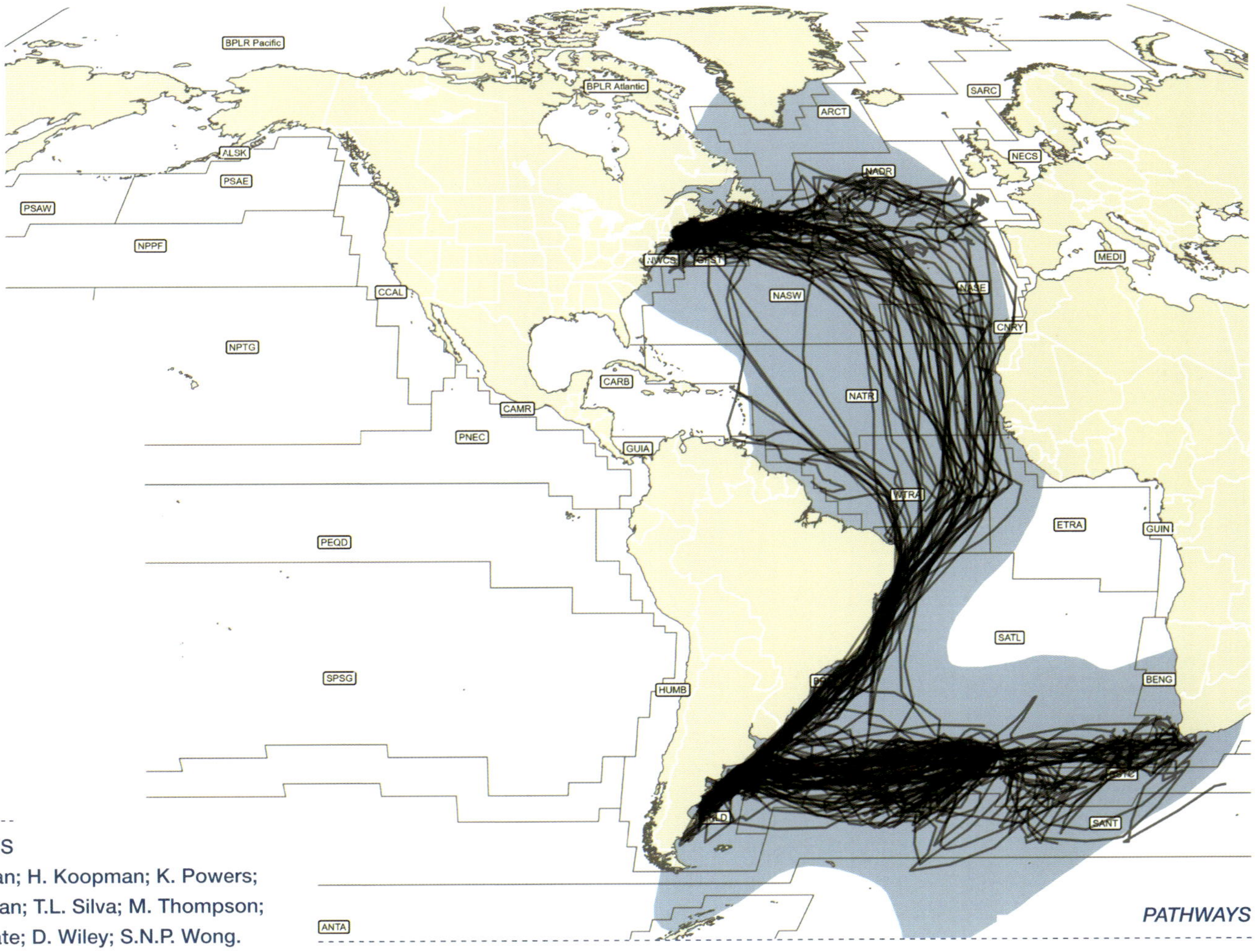

PATHWAYS

DATA CONTRIBUTORS

K.A. Hatch; L. Kaufman; H. Koopman; K. Powers; R.A. Ronconi; P.G. Ryan; T.L. Silva; M. Thompson; L. Welch; J.A. Westgate; D. Wiley; S.N.P. Wong.

PINK-FOOTED SHEARWATER *Ardenna creatopus*

weak strong
Mantel = 0.11 ± 0.13

DISTRIBUTION AND ABUNDANCE

The Pink-footed Shearwater is a pelagic seabird that nests in burrows within a restricted range on three Chilean offshore islands, Isla Mocha, Isla Robinson Crusoe, and Isla Santa Clara (Carle et al. 2022). Population size varies across these three islands (Carle et al. 2022), with the largest colony at Isla Mocha. During the nonbreeding period, the shearwater is a common visitor to productive eastern boundary currents of the Pacific Ocean: the California Current of North America and the Humboldt Current of South America. Here it can be found primarily over the continental shelf and slope (Felis et al. 2019). Pink-footed Shearwaters appear unique among Pacific shearwater species in that they rely exclusively on the two Pacific Ocean eastern boundary current upwelling systems across their entire annual cycle (Felis et al. 2019).

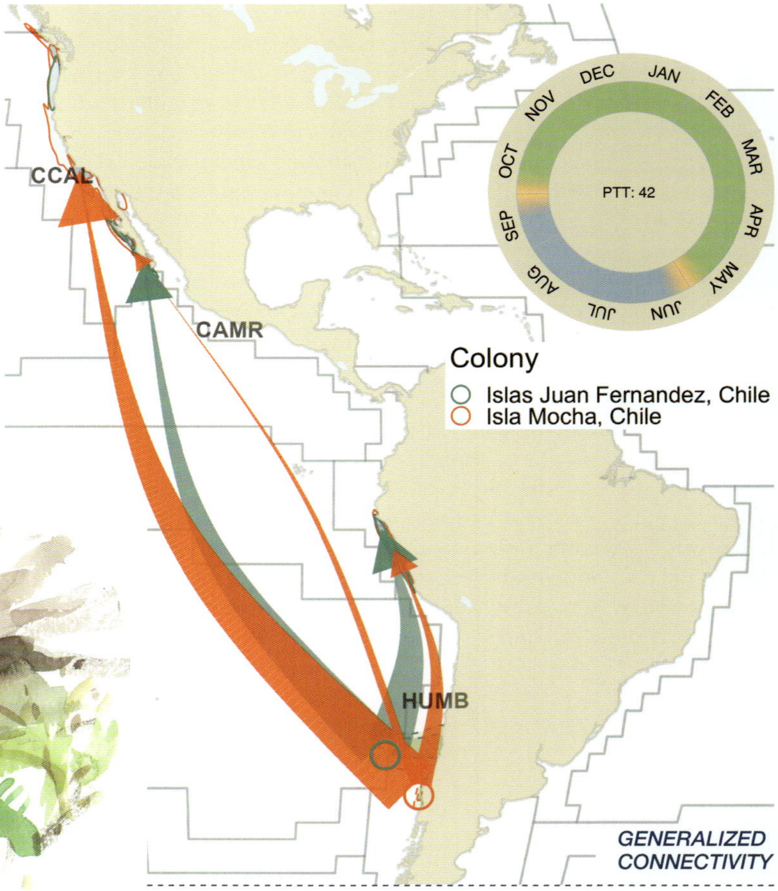

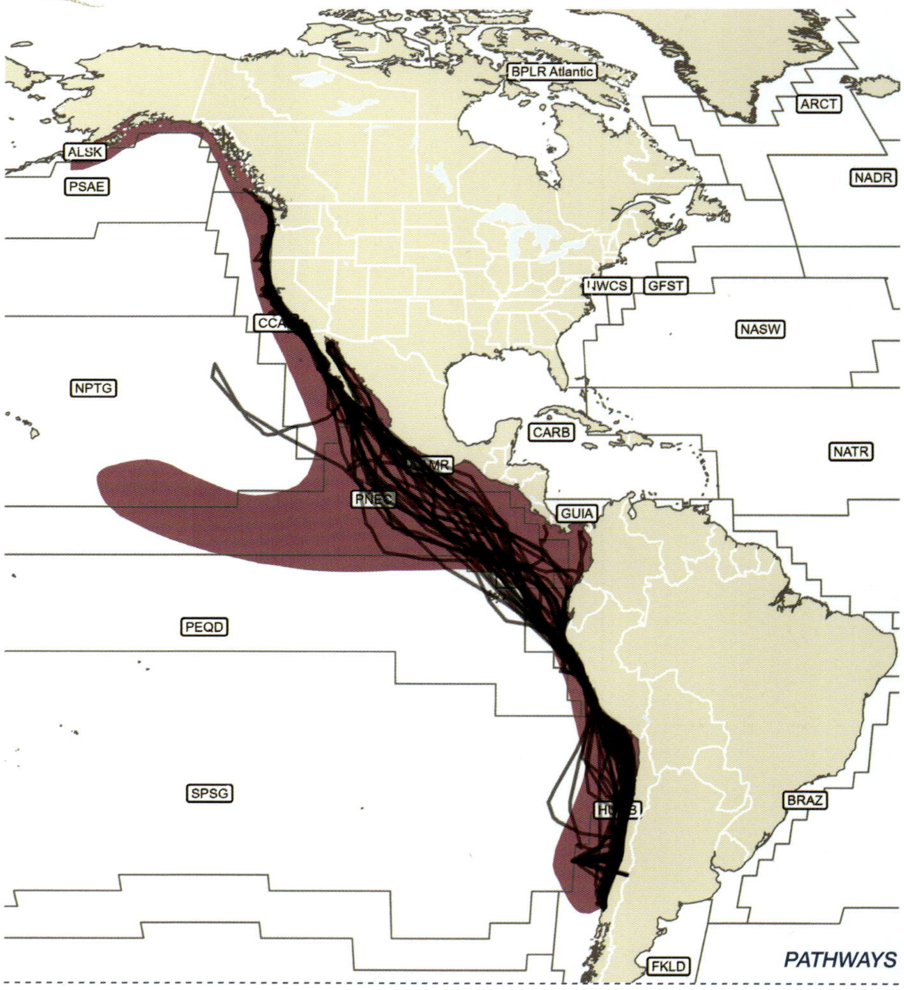

ORDER Procellariiformes
FAMILY Procellariidae
BREEDING HABITAT Islands, Oceanic
NONBREEDING HABITAT Oceans, Pelagic
GLOBAL POPULATION SIZE 40,000
POPULATION CHANGE SINCE 1970 NA
IUCN RED LIST CATEGORY Vulnerable
R2R TIPPING POINT SPECIES Red-Alert
US LISTING STATUS No
CANADA SPECIES AT RISK Endangered
PIF CCS 17/20

	CCAL	CAMR	HUMB
Islas Juan Fernandez (n = 4)	50%	–	50%
Isla Mocha, Chile (n = 33)	73%	6%	21%

MIGRATORY CONNECTIVITY

OVERALL Migratory connectivity of the Pink-footed Shearwater is moderate (Mantel = 0.11 ± 0.13) and known from satellite tracking data (Felis et al. 2019). Birds remain within the eastern Pacific Ocean throughout their annual cycles and mix during the nonbreeding period. The core nonbreeding habitat of 73% of birds breeding at Isla Mocha and 50% at Isla Santa Clara (Islas Juan Fernandez) was in the California Current of North America, while 21% and 50%, respectively, spent the nonbreeding season in Peruvian waters of the Humboldt Current.

SATELLITE TRACKING In 2005–2015, 42 birds were captured at either breeding islands or at sea in their nonbreeding habitats off Southern California and tracked back to their nesting island in Chile (Felis et al. 2019). One individual tagged at sea during the nonbreeding period in Southern California ventured north to the Alaska Downwelling Coastal Province, but its tag ceased transmitting before it could be tracked back to its nesting colony (Felis et al. 2019).

OTHER APPROACHES No banding, genetic, or isotope studies of migratory connectivity were available for this species.

MIGRATORY PATHWAYS Individuals from each island undergo both short- and long-distance migrations (Felis et al. 2019). Birds tracked from breeding islands migrated between central Chile and Peru. Some birds ended their migration in Peru, while others continued north on a trans-equatorial pathway. This northerly route followed the continental shelf and slope until birds reached the Peru-Ecuador border. From here, birds used a high-seas migration corridor but switched back to continental shelf habitats upon arrival to the California Current near Baja California, Mexico.

DATA CONTRIBUTORS

J. Adams; R.D. Carle; V. Colodro; J.J. Felis; P.J. Hodum.

INTERESTING NOTE

During the breeding period, individuals from Isla Santa Clara use oceanic pelagic habitats much more than those from Isla Mocha, likely due to Isla Mocha's much closer proximity to the Chilean coast (Carle et al. 2022).

Ardenna carneipes FLESH-FOOTED SHEARWATER

ORDER Procellariiformes

FAMILY Procellariidae

BREEDING HABITAT Islands, Oceanic

NONBREEDING HABITAT Oceans, Pelagic

GLOBAL POPULATION SIZE 150,000

POPULATION CHANGE SINCE 1970 NA

IUCN RED LIST CATEGORY Near Threatened

R2R TIPPING POINT SPECIES No

US LISTING STATUS No

CANADA SPECIES AT RISK No

PIF CCS 15/20

MIGRATORY CONNECTIVITY

OVERALL Only a BirdLife distribution map is available; no eBird data or migratory connectivity information were found for this species.

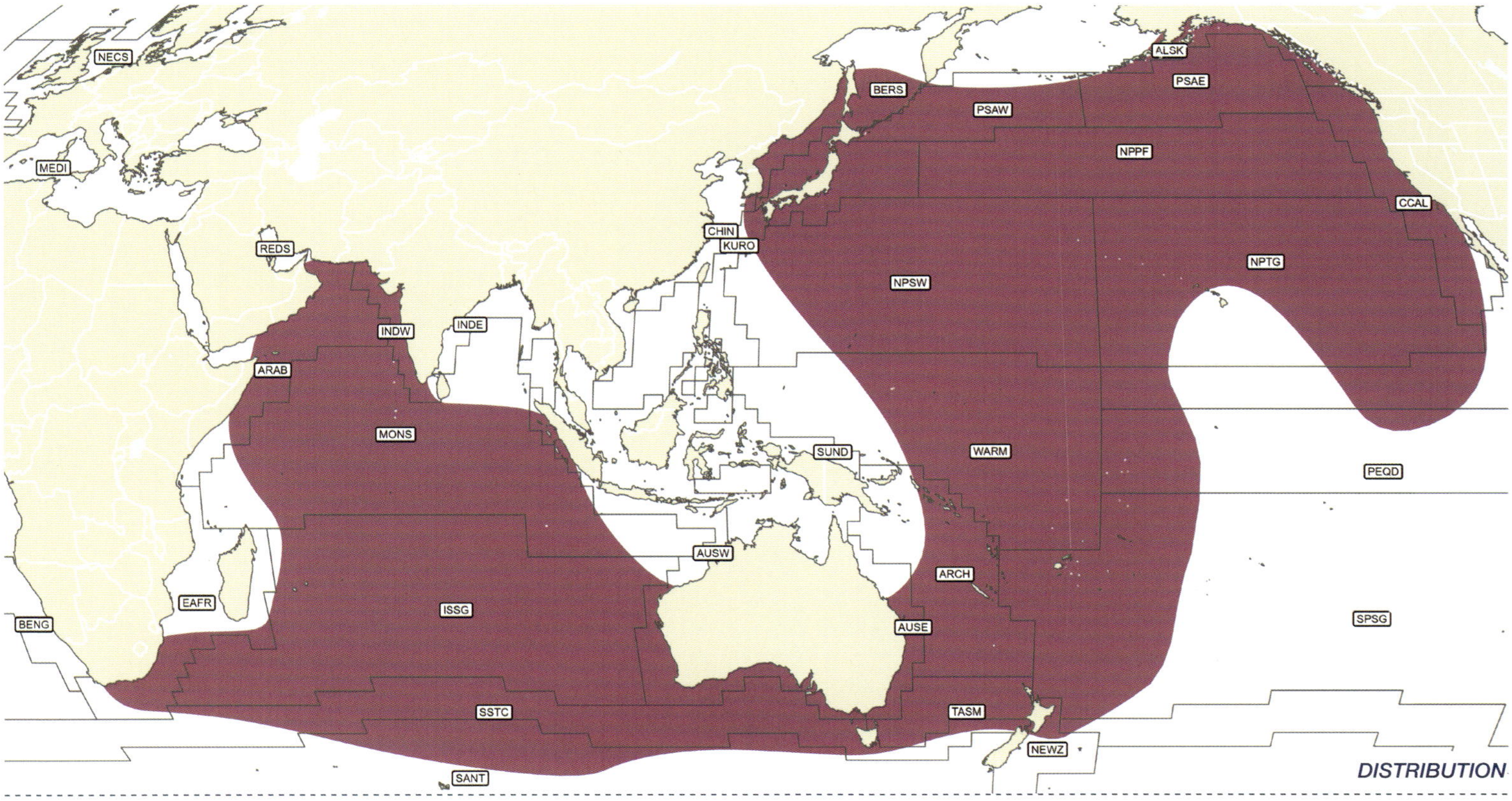

DISTRIBUTION

CHRISTMAS SHEARWATER *Puffinus nativitatis*

ORDER Procellariiformes
FAMILY Procellariidae
BREEDING HABITAT NA
NONBREEDING HABITAT NA
GLOBAL POPULATION SIZE NA
POPULATION CHANGE SINCE 1970 NA
IUCN RED LIST CATEGORY Least Concern
R2R TIPPING POINT SPECIES No
US LISTING STATUS No
CANADA SPECIES AT RISK No
PIF CCS NA

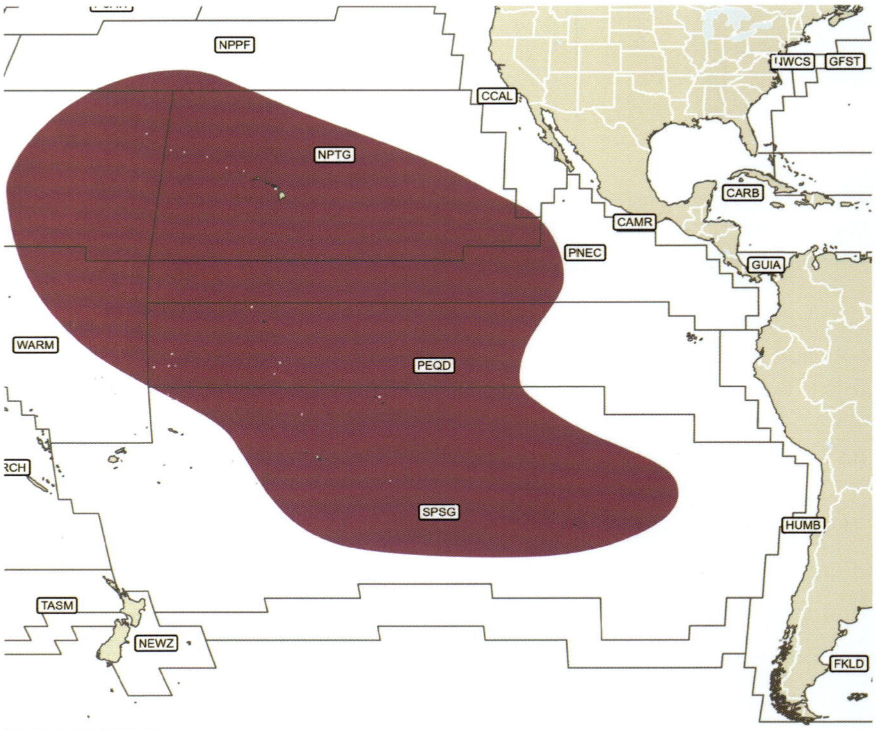

DISTRIBUTION

MIGRATORY CONNECTIVITY

OVERALL Only a BirdLife distribution map is available; no eBird data or migratory connectivity information were found for this species.

MANX SHEARWATER *Puffinus puffinus*

DISTRIBUTION AND ABUNDANCE

An Atlantic Ocean species with the most northern breeding distribution of any shearwater (Lee et al. 2023). The largest colonies are in the United Kingdom (predominantly Scotland and Wales, approximately 600,000 pairs in total; Joint Nature Conservation Committe 2021). Mid-size colonies occur in Iceland, the Faroes, and the Republic of Ireland, and smaller colonies (<1,000) occur on Spanish, French, and Portuguese islands (Lee et al. 2023). A colony established at Middle Lawn Island, Newfoundland, Canada, in the 1970s (Storey and Lien 1985) is estimated to have pairs in the low 100s (Robertson 2002), and a new breeding site on Matinicus Rock, Maine, US, had five pairs as of 2018 (Fayet et al. 2020). Post-breeding distribution is primarily in the southwestern Atlantic Ocean, and during migration this species occurs in much lower densities than other procellarids in North American coastal and slope habitats (Lee et al. 2023).

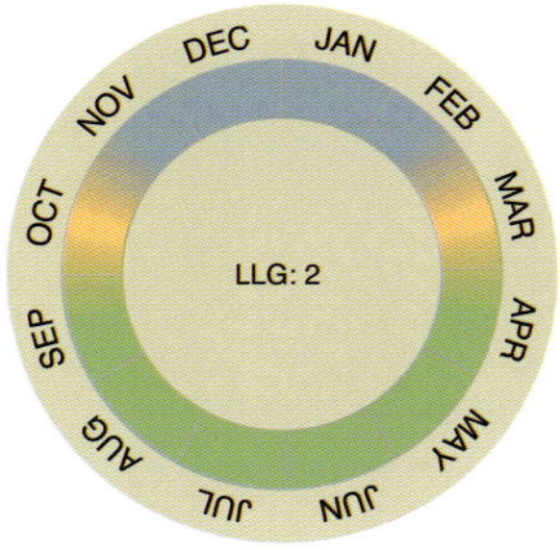

ORDER Procellariiformes
FAMILY Procellariidae
BREEDING HABITAT Islands, Oceanic
NONBREEDING HABITAT Oceans, Pelagic
GLOBAL POPULATION SIZE 740,000
POPULATION CHANGE SINCE 1970 NA
IUCN RED LIST CATEGORY Least Concern
R2R TIPPING POINT SPECIES Yellow-Alert
US LISTING STATUS No
CANADA SPECIES AT RISK No
PIF CCS 15/20

INTERESTING NOTE

Manx Shearwaters are being recorded in the North Pacific (from California northward) with ever increasing frequency since the 1990s (Mlodinow 2004). Breeding has also been suspected on Triangle Island, British Columbia (Mlodinow 2004), and Middleton Island, Alaska, US (Gibson et al. 2008).

MIGRATORY CONNECTIVITY

OVERALL Results from migratory movement studies that were focused on central and northeastern Atlantic Ocean breeding colonies, combined with movement paths of two individuals tracked from Maine, US, suggest weak connectivity to nonbreeding areas; however, this was not calculated due to limited sample size. Birds tracked from breeding colonies spanning the North Atlantic Ocean spent their post-breeding period in the Southwest Atlantic Shelves and Brazilian Coastal Current provinces.

GEOLOCATOR TRACKING Two birds tracked from Maine, US, undertook a western Atlantic coastal migration to post-breeding areas off Argentina and the Falkland Islands (Fayet et al. 2020) within the Southwest Atlantic Shelves and Brazilian Coastal Current provinces.

GENETICS Manx Shearwaters from the Canary Islands (the most southern population) are differentiated from northern breeding colonies based on breeding phenology, morphology, call, plumage color, and genetics, suggesting a long period of isolation (Rodríguez et al. 2020). No migration data are available from the Canary Islands (or other small southern populations), and given that breeding is two to three months earlier than northern populations, tracking is necessary to determine if the southern population has unique spatial or temporal connectivity as compared with the northern populations (Rodríguez et al. 2020).

OTHER APPROACHES No banding, satellite tracking, or stable isotope studies of migratory connectivity were found for this species.

DATA CONTRIBUTORS
A.L. Fayet; D.E. Lyons.

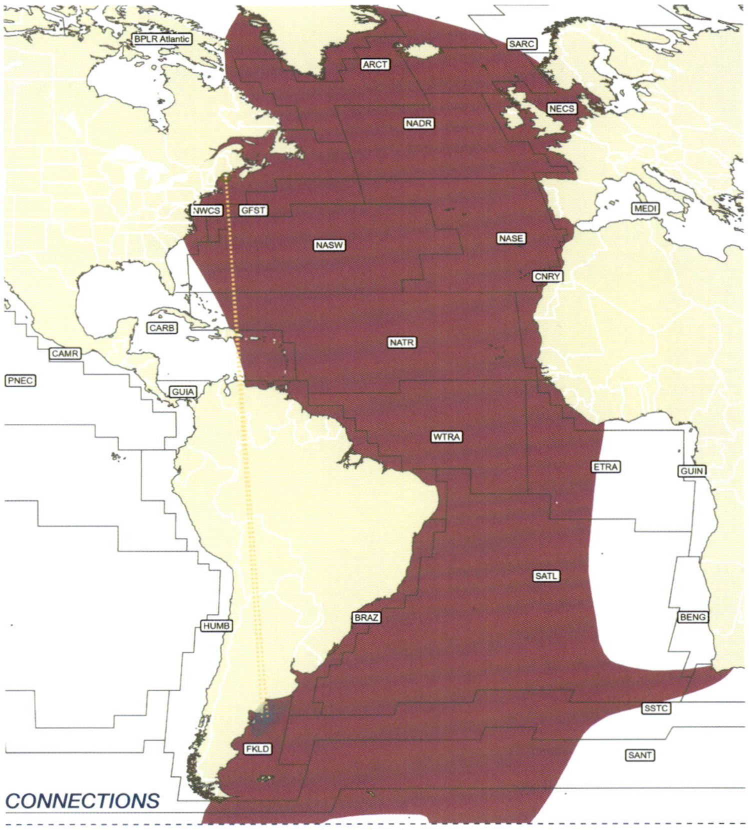

MIGRATORY PATHWAYS The two North American birds tracked from Maine took a different southbound migratory route (farther west and facing headwinds; Fayet et al. 2020) than European birds (clockwise from breeding colonies to Spain and Africa and southwest across the Atlantic (González-Solís et al. 2009; Guilford et al. 2009; Kirk 2017) until reaching Brazil.

Puffinus auricularis

TOWNSEND'S SHEARWATER

DISTRIBUTION AND ABUNDANCE

A small breeding population is limited to Revillagigedo Islands, Mexico, in the Pacific Ocean. Post-breeding, it ranges from Baja California to El Salvador in the eastern tropical Pacific, typically along the continental slope (Ainley et al. 2020b).

ORDER Procellariiformes
FAMILY Procellariidae
BREEDING HABITAT Islands, Oceanic
NONBREEDING HABITAT Coasts, Marine Waters
GLOBAL POPULATION SIZE 620
POPULATION CHANGE SINCE 1970 NA
IUCN RED LIST CATEGORY Critically Endangered
R2R TIPPING POINT SPECIES No
US LISTING STATUS No
CANADA SPECIES AT RISK No
PIF CCS 19/20

MIGRATORY CONNECTIVITY

OVERALL Migratory connectivity is unknown, but when this species is considered as a conspecific with Newell's Shearwater, there is likely strong connectivity of Hawaiian and Mexican subspecies due to a lack of evidence for post-breeding range overlap (Spear et al. 1995). Status as migratory is unknown, and it is generally considered resident to continental slope habitats southeast of its breeding grounds in the Costa Rica Coastal Current—the northern-flowing far eastern current derived from the North Pacific Equatorial Countercurrent. ***GENETICS*** There is little genetic differentiation found in mitochondria sequences between Townsend's Shearwaters (n = 5) and Newell's Shearwaters (n = 6; Martínez-Gómez et al. 2015). ***OTHER APPROACHES*** No banding, tracking, or stable isotope studies of migratory connectivity were found for this species. Spear et. al. (1995) found marine range separation between Townsend's Shearwater and Newell's Shearwater. Townsend's Shearwater were generally around 650 km from colonies in upwelling fronts in the Costa Rica Coastal Current over the continental slope of Mexico (highest density within the Exclusive Economic Zone), while Newell's Shearwater generally used ocean fronts more central to the North Pacific Equatorial Countercurrent in the high seas (see the Newell's Shearwater species account for tracking data from multiple colonies). ***MIGRATORY PATHWAYS*** Migratory status, movement pathways, and phenology are generally unknown (Ainley et al. 2020b).

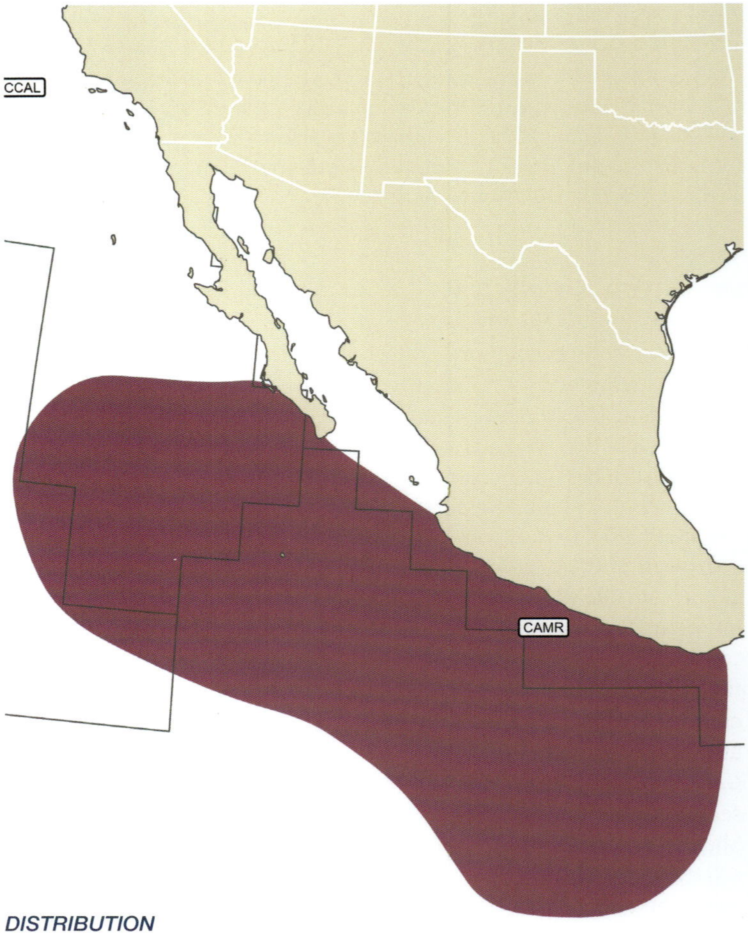

DISTRIBUTION

NEWELL'S SHEARWATER *Puffinus newelli*

DISTRIBUTION AND ABUNDANCE

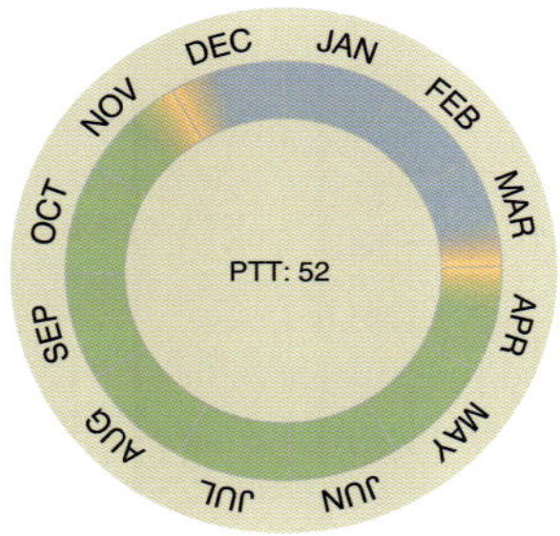

This species breeds on steep mountains and cliffs under tree roots and dense foliage only on Hawaiian Islands. Most of the small population breeds on Kaua'i (90%) with the remainder on Moloka'i, Hawaii, and Maui (Ainley et al. 2020c). At-sea surveys have helped elucidate breeding versus nonbreeding distribution (Ainley et al. 2020c). During the breeding period, it is found within the Hawaiian Exclusive Economic Zone, but it is also highly pelagic, extending to deep water and the high seas; during the nonbreeding period, it migrates south to tropical and subtropical central Pacific. It is often associated with tuna fisheries, and it is rare among tropical seabirds in their plunge-pursuit diving behavior (Ainley et al. 2020c).

MIGRATORY CONNECTIVITY

OVERALL Connections between specific breeding islands and nonbreeding habitat is available only from Kaua'i, mainly for fledglings, and for only a portion of the nonbreeding period (Raine et al. 2020). Most satellite-tracked fledglings used core areas within the North Pacific Equatorial Countercurrent, southwest of Hawaii. A single breeding adult was tracked to an area farther east in the Pacific Equatorial Divergence Zone (A. Raine, unpublished data). See the Townsend's Shearwater account for an assessment of migratory connectivity across subspecies.

ORDER Procellariiformes
FAMILY Procellariidae
BREEDING HABITAT NA
NONBREEDING HABITAT NA
GLOBAL POPULATION SIZE NA
POPULATION CHANGE SINCE 1970 NA

IUCN RED LIST CATEGORY Critically Endangered
R2R TIPPING POINT SPECIES No
US LISTING STATUS Threatened
CANADA SPECIES AT RISK No
PIF CCS NA

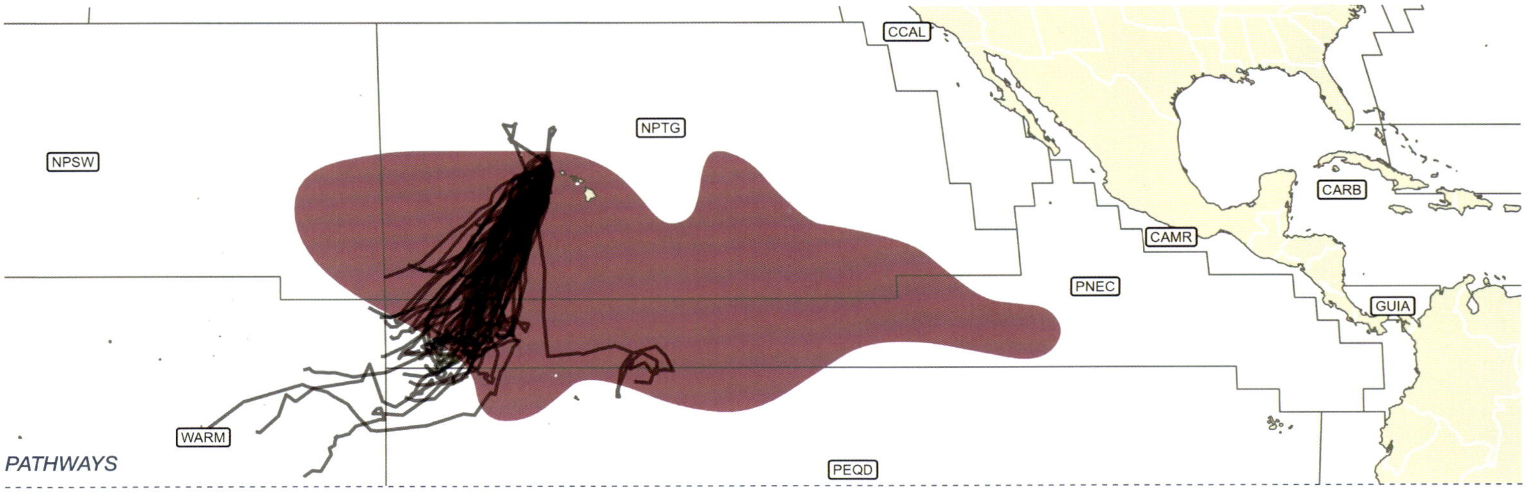

SATELLITE TRACKING Fledglings were tracked to nonbreeding areas southwest of the breeding colony, but full annual cycle information is not available as most of the tags only transmitted on average, between 15 and 34 days. For the four adults tracked in this study, tags transmitted on average, 82 days (one representative track is shown on the map). However, all tracked adults were resighted at the breeding colony the following year without their tags, indicating that tag attachment did not impact survival (Raine et al. 2020).
GENETICS Little genetic differentiation was found in mitochondria sequences between subspecies of Townsend's (n = 5) and Newell's Shearwaters (n = 6; Martínez-Gómez 2015).
OTHER APPROACHES No banding or stable isotope studies of migratory connectivity were found for this species.

MIGRATORY PATHWAYS Fledglings that were either wild caught or released after a rehabilitation program and were tracked with a solar Argos tag (Raine et al. 2020) all migrated southwest of Hawaii. An adult tracked in the same study migrated southeast of Hawaii.

INTERESTING NOTE

A specimen of Newell's Shearwater from coastal California suggests possible migratory movements between the range of Newell's Shearwater and Townsend's Shearwater (Unitt et al. 2009).

DATA CONTRIBUTORS
A.F. Raine.

BLACK-VENTED SHEARWATER

Puffinus opisthomelas

ORDER Procellariiformes
FAMILY Procellariidae
BREEDING HABITAT Coasts, Cliffs and Islands
NONBREEDING HABITAT Coasts, Marine Waters
GLOBAL POPULATION SIZE 160,000
POPULATION CHANGE SINCE 1970 NA
IUCN RED LIST CATEGORY Near Threatened
R2R TIPPING POINT SPECIES No
US LISTING STATUS No
CANADA SPECIES AT RISK No
PIF CCS 16/20

DISTRIBUTION AND ABUNDANCE

This is the only shearwater to nest in the California Current, where it is endemic and nests on three islands off Baja California Sur, Mexico (Keitt et al. 2020). Velarde et al. (2015) provided evidence for an additional colony that exists in the Gulf of California. Natividad Island hosts 95% of the breeding population (Keitt et al. 2020). Although it is rare among shearwaters in being distributed primarily in coastal waters (within 25 km of shore) during the post-breeding period, this species occurs along the Pacific coast of North America, with highest abundance south of central California (Keitt et al. 2020).

MIGRATORY CONNECTIVITY

OVERALL Possible moderate connectivity by sex during some parts of the annual cycle is suggested by tracking and isotopic evidence. But this species is typically not considered to have a classic migration pattern. All evidence from a single breeding colony suggests that individuals mix throughout the restricted nonbreeding range. Many individuals stay near the colony most of the year, but some also shift north and south of the nesting islands during the nonbreeding period.
STABLE ISOTOPES δ^{13}C ratios from feathers of 93 shearwaters (42 males, 39 females, and 12 chicks) sampled from Natividad Island, Mexico, in 2016 suggested that all birds migrated to similar nonbreeding areas (Rosas-Hernández et al. 2022). Sexes demonstrated similarity in δ^{15}N values during the breeding and nonbreeding periods, but differed during

INTERESTING NOTE

Future research priorities include better resolution of southern nonbreeding distribution (where the Black-vented Shearwater is difficult to distinguish from other species) using tracking devices, like geolocators, that will provide full-annual-cycle tracking. During 2016–2019, 56 Black-vented Shearwaters were tracked from Isla Natividad with archival GPS-accelerometers during breeding foraging trips (yearly datasets from 7–22 individuals). The birds used habitats both in the California Current north and south of Natividad Island, traveling a median of 661 km round-trip (range: 144–1759 km; Soldatini et al. 2023). There was sexual separation during chick-rearing in 2016, when there was low productivity at the conclusion of an El Niño; core habitats of males were north of the colony, and core habitats for females were south of the colony (Soldatini et al. 2019). Across all years, generally females foraged over shallower coastal waters, while males foraged over deeper waters (Soldatini et al. 2023).

late chick-rearing and the post-breeding period, when molt begins. This suggests the sexes used different trophic levels for foraging during that period (and possibly some surprising use of pelagic habitats by females). Using isoscapes, the authors suggest that females move farther south than males to the Revillagigedo Islands, Mexico.
OTHER APPROACHES No banding, tracking, or genetic studies of migratory connectivity were found for this species.

DISTRIBUTION

SARGASSO SHEARWATER *Puffinus lherminieri*

ORDER Procellariiformes
FAMILY Procellariidae
BREEDING HABITAT Islands, Oceanic
NONBREEDING HABITAT Oceans, Pelagic
GLOBAL POPULATION SIZE 35,000
POPULATION CHANGE SINCE 1970 NA
IUCN RED LIST CATEGORY Least Concern
R2R TIPPING POINT SPECIES Yellow-Alert
US LISTING STATUS No
CANADA SPECIES AT RISK No
PIF CCS 14/20

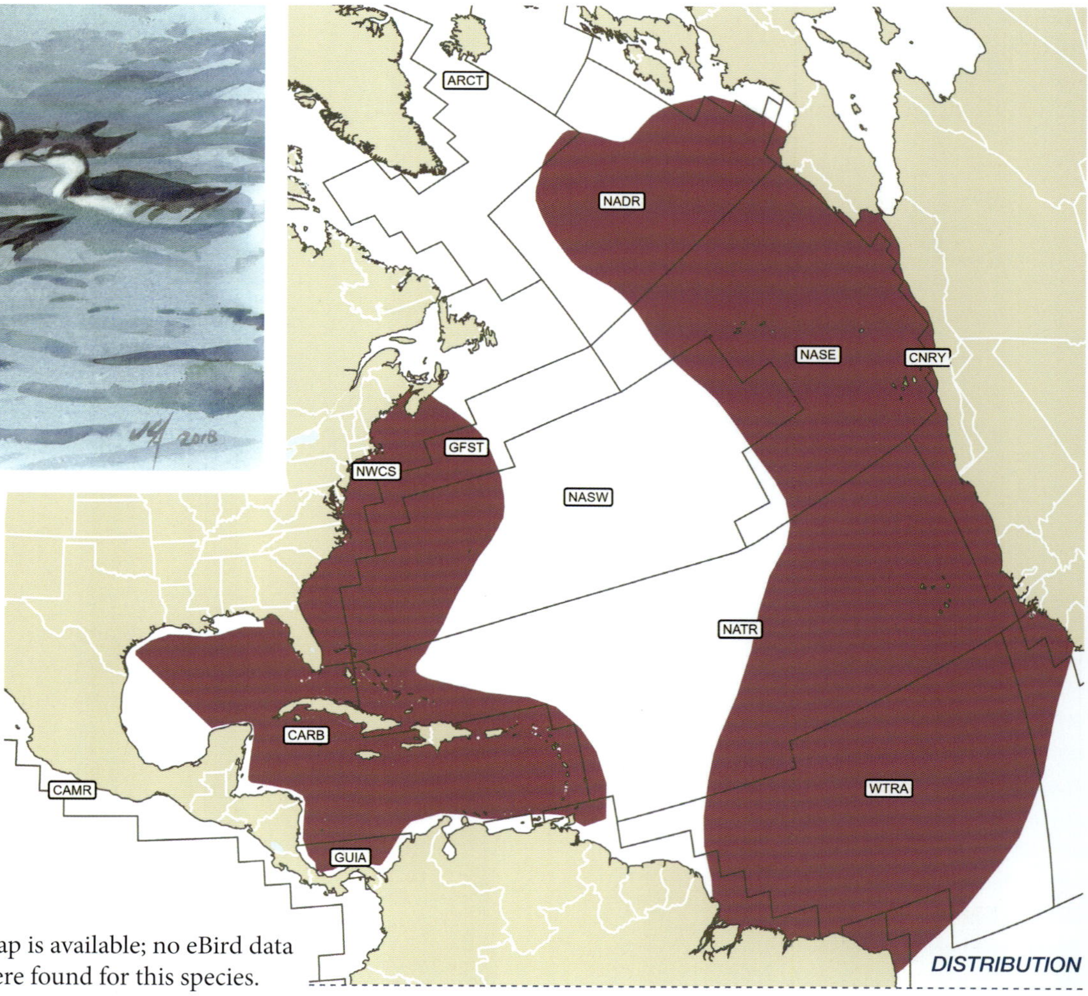

DISTRIBUTION

MIGRATORY CONNECTIVITY

OVERALL Only a BirdLife distribution map is available; no eBird data or migratory connectivity information were found for this species.

STORKS (CICONIIFORMES)

WRITING LED BY **TIMOTHY GUIDA** *AND* **PETER P. MARRA**

The Wood Stork is the only stork species breeding in North America and is an important indicator of wetland ecosystem health. Once federally listed as endangered, the Wood Stork has recovered in parts of its range and is now listed as threatened (https://ecos.fws.gov/ecp/). While its breeding ecology, particularly in Florida, has been well studied, full, annual, cycle data remain incomplete. Migratory connectivity across its range, especially for populations outside Florida, is still poorly resolved, highlighting the importance of international collaboration due to transboundary movements into Mexico and Central America.

CICONIIDAE | *STORKS*

weak strong
MC = 0.03 ± 0.05*

Mycteria Americana **WOOD STORK**

DISTRIBUTION AND ABUNDANCE

The Wood Stork is a large wading bird, widespread in Central and South America, but in North America it is limited to the southeastern United States and Mexico. All breeding colonies in the United States are found in Florida, Georgia, South Carolina, and a newly established colony in North Carolina (Coulter et al. 2020; Fink et al. 2023). Nonbreeding birds gather in high numbers in the Mississippi Delta. This population is largely made up of birds from Mexico, but some breeding individuals from the southeastern US and possibly Central America also spend the nonbreeding season in this region (Basille et al. 2020; Coulter et al. 2020; Picardi et al. 2020).

MIGRATORY CONNECTIVITY

OVERALL Both satellite and radio tracking data provide insights into migratory connectivity for Wood Storks. All Atlantic Flyway breeding birds (100%) remained in that region during the nonbreeding season. Most Midcontinent Flyway breeders (85%) also migrated to the

ORDER Ciconiiformes
FAMILY Ciconiidae
BREEDING HABITAT Wetlands, Forested
NONBREEDING HABITAT Wetlands, Generalist
GLOBAL POPULATION SIZE 450,000
POPULATION CHANGE SINCE 1970 +104%
IUCN RED LIST CATEGORY Least Concern
R2R TIPPING POINT SPECIES No
US LISTING STATUS Threatened
CANADA SPECIES AT RISK No
PIF CCS 10/20

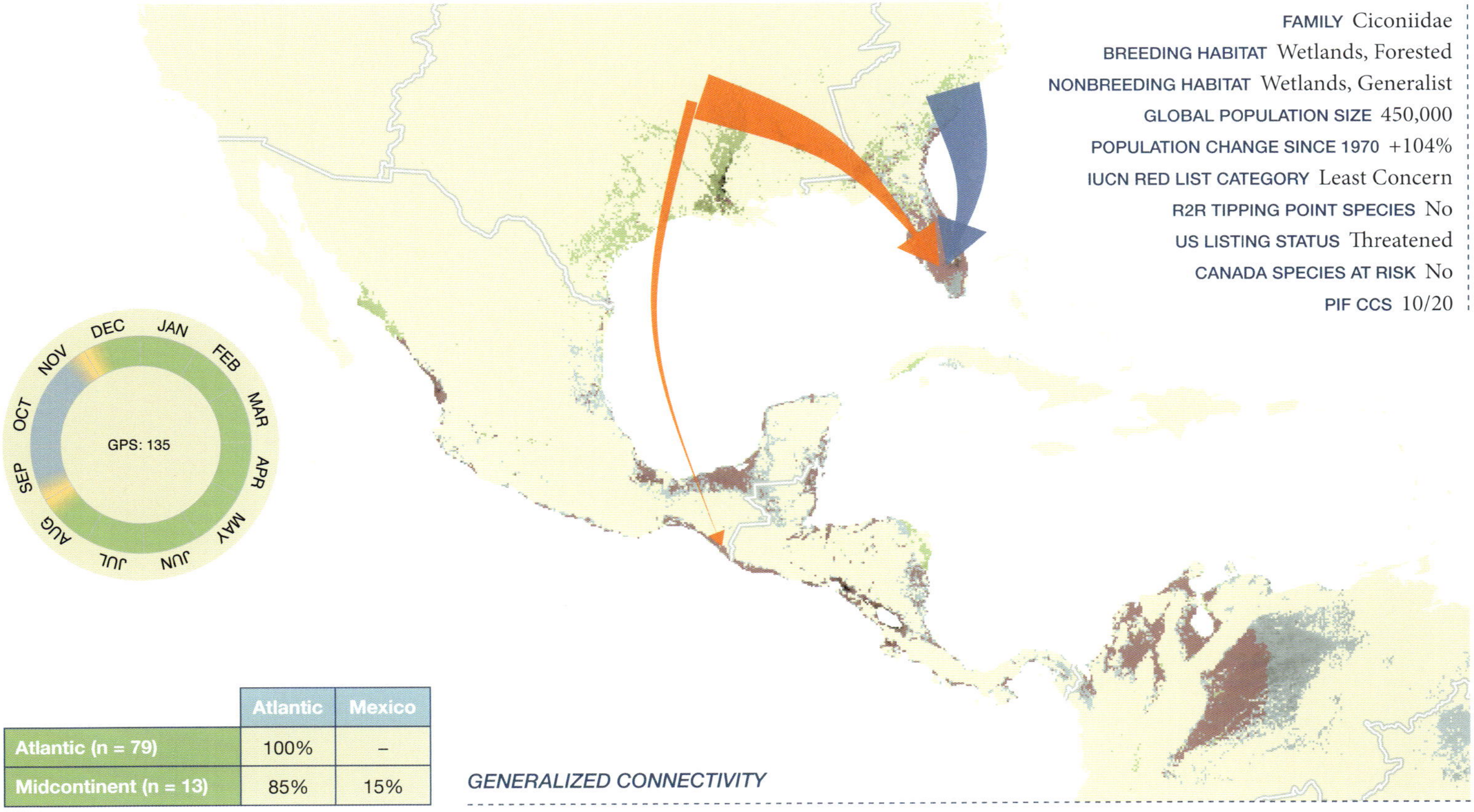

	Atlantic	Mexico
Atlantic (n = 79)	100%	–
Midcontinent (n = 13)	85%	15%

GENERALIZED CONNECTIVITY

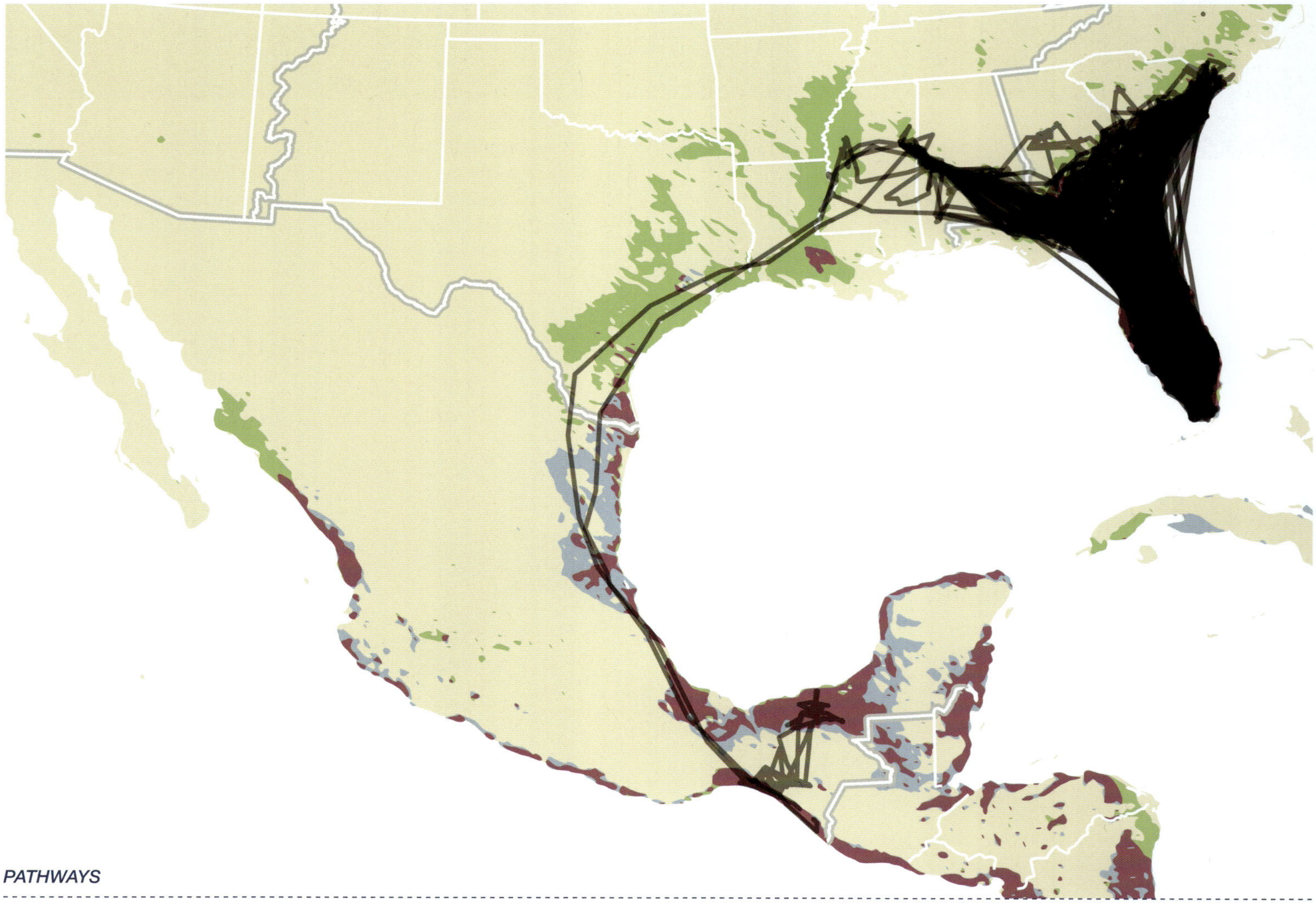

PATHWAYS

Atlantic Flyway for the nonbreeding season, while others migrated to Mexico (15%). Mixing of breeding populations on nonbreeding areas within the Atlantic Flyway leads to weak migratory connectivity (MC= 0.03 ± 0.05).

RADIO TRACKING Wood Storks breeding in a Georgia colony (n = 3) were tracked to southern Georgia and Florida in the nonbreeding season (Comer et al. 1987).

SATELLITE TRACKING One study tracked 133 individuals from 14 breeding colonies in the southeastern US using satellite and GPS technology (Picardi et al. 2018, 2020; Basille et al. 2020). Almost one-third of the population remained in Florida year-round, concentrated in a few hotspots in the southeast and Jacksonville areas. The majority of individuals breeding throughout the southeast concentrated in south Florida during the nonbreeding season (Picardi et al. 2018, 2020; Basille et al. 2020). The exception were two birds that migrated to Mexico from Georgia and Mississippi breeding colonies (Picardi et al. 2018). Bryan and colleagues (2008) also provided evidence of potential mixing of the eastern and western populations rather than the general assumption of spatial separation between the two populations due to the Mississippi River. All birds tagged in east Mississippi migrated to south Florida for the nonbreeding season, birds from western Mississippi migrated to Mexico or Guatemala, and Louisiana birds migrated to Florida or Mexico, indicating population mixing on nonbreeding areas (Bryan et al. 2008). Birds tracked from a breeding colony in North Carolina also migrated to Florida (S. Schweitzer et al., unpublished data).

OTHER APPROACHES No banding, genetic, or stable isotope studies of migratory connectivity were found for this species.

MIGRATORY PATHWAYS Storks followed two general migration routes along the east and west coastline of Florida, with the east one used more during pre-breeding migration and the west during post-breeding migration (Picardi et al. 2020). Individuals tracked from breeding colonies in the US took routes around the Gulf of Mexico (Picardi et al. 2018), south along the Atlantic coast from North Carolina to Florida, and then traveled inland across to southwestern Florida (S. Schweitzer et al., unpublished data).

DATA CONTRIBUTORS

M. Basille; R.R. Borkhataria; A.L. Bryan; J. Brzorad; D.N. Bucklin; P.C. Frederick; R. Kays; S. Picardi; S. Schweitzer.

FRIGATEBIRDS, BOOBIES, GANNETS, ANHINGAS, AND CORMORANTS (SULIFORMES)

WRITING LED BY **TIMOTHY GUIDA** *AND* **AUTUMN-LYNN HARRISON**

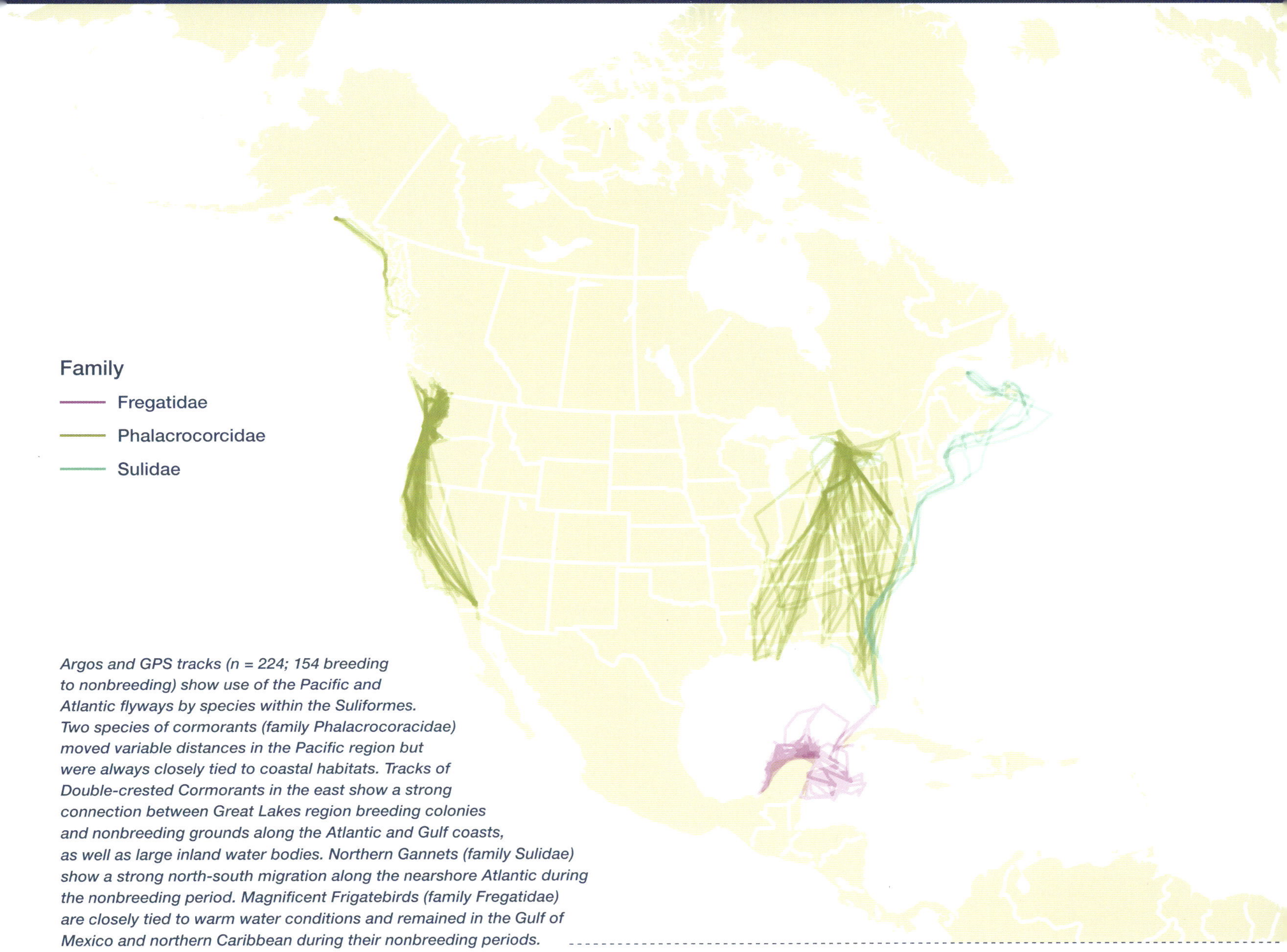

Argos and GPS tracks (n = 224; 154 breeding to nonbreeding) show use of the Pacific and Atlantic flyways by species within the Suliformes. Two species of cormorants (family Phalacrocoracidae) moved variable distances in the Pacific region but were always closely tied to coastal habitats. Tracks of Double-crested Cormorants in the east show a strong connection between Great Lakes region breeding colonies and nonbreeding grounds along the Atlantic and Gulf coasts, as well as large inland water bodies. Northern Gannets (family Sulidae) show a strong north-south migration along the nearshore Atlantic during the nonbreeding period. Magnificent Frigatebirds (family Fregatidae) are closely tied to warm water conditions and remained in the Gulf of Mexico and northern Caribbean during their nonbreeding periods.

The order Suliformes contains four families of water-associated birds: frigatebirds, boobies and gannets, anhingas, and cormorants. This order has been well-studied, with migratory connectivity data available for all species covered in this Atlas. Insights come from multiple methods, including band reencounters (n = 3,897), satellite telemetry (n = 154), and light-level geolocator data (n = 38). Within the order, the movements of the Anhinga and the Magnificent Frigatebird are least understood, and increased tracking of these species would help improve our overall understanding of the order. Within the order Suliformes, the Pelagic Cormorant is considered a Tipping Point species and has experienced a 53% decline from 1970 to 2017. Disturbance at breeding colonies is considered a major threat to this species and is likely a factor affecting the populations of other members of this largely colonial-nesting order overall. Movement data in this order is especially robust for the cormorants, as members of this family are often considered a threat to fisheries.

Available data suggests a moderate level of migratory connectivity within the order, with most species holding ties to colonial breeding areas, but with individual birds undertaking variable nonbreeding movements. Members of the order that live in tropical and subtropical areas, especially the Anhinga, Magnificent Frigatebird, and boobies, undergo limited migrations and may remain near breeding areas year-round, moving only as food availability demands. More northern members of the order, i.e., Northern Gannet and Double-crested Cormorant, undertake more complete north-south annual migrations. Both the oceanic and freshwater members of this order are closely tied to water conditions, and migration distances vary as birds move to avoid icy conditions or other unfavorable foraging situations.

MAGNIFICENT FRIGATEBIRD *Fregata magnificens*

DISTRIBUTION AND ABUNDANCE

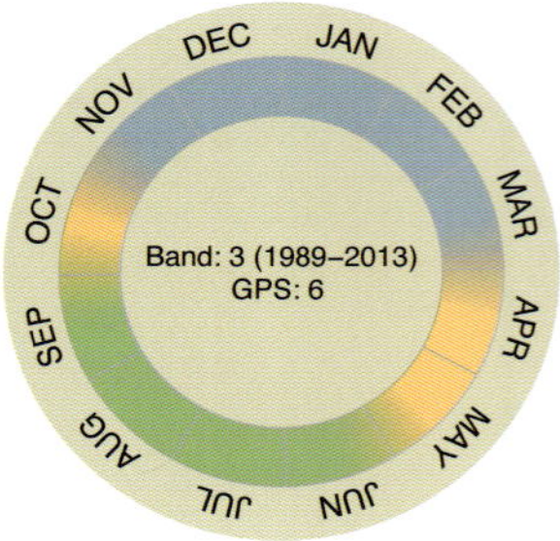

This species nests on islands and coasts of tropical North, Central, and South America. Highest abundance is on the Pacific coast (Diamond and Schreiber 2020), where some colonies comprise 10,000–20,000 pairs (e.g., Isla Santa Margarita off Baja California, Mexico, and Bahia del Pabellon, Mexico). In the Atlantic Ocean, breeding colonies are typically in the hundreds (e.g., Little Cayman Island, 300–500 pairs; and Redonda Island, Brazil, 300 pairs). Some individuals may be permanent residents of their breeding range, while others may undertake longer migrations (Diamond and Schreiber 2020). It has been extirpated from several historic breeding sites in the Caribbean owing to human disturbance and development (Lindsey et al. 2000). During the nonbreeding season, it ranges throughout adjacent coastal oceans, the Caribbean Sea, and the Gulf of Mexico, but has also been observed as far north as Alaska, US, and Newfoundland, Canada (Diamond and Schreiber 2020).

ORDER Suliformes
FAMILY Fregatidae
BREEDING HABITAT Coasts, Mangroves
NONBREEDING HABITAT Coasts, Marine Waters
GLOBAL POPULATION SIZE 130,000
POPULATION CHANGE SINCE 1970 NA
IUCN RED LIST CATEGORY Least Concern
R2R TIPPING POINT SPECIES No
US LISTING STATUS No
CANADA SPECIES AT RISK No
PIF CCS 16/20

MIGRATORY CONNECTIVITY

OVERALL Little is known about the migratory connectivity of the Magnificent Frigatebird in the Americas due to limited tracking and band reencounter data across the entire range. Assessments of migratory connectivity are complicated by the fact that some individuals may remain resident on their breeding range while others migrate. Tracking and band reencounter data from birds marked in the Caribbean Sea suggest weak connectivity at a regional scale. Within populations, spatial separation by sex is thought to be common (Harrington et al. 1972), with females being more coastally restricted, but results from stable isotope studies vary. Genetic studies suggest both high gene flow across ocean basins and isolated populations within basins. It is thus unclear how this genetic structure relates to contemporary migratory connectivity.

SATELLITE TRACKING GPS tracking studies from two breeding populations from the Yucatán Peninsula, Mexico, demonstrated some level of population mixing at the regional level, although sample sizes are small. Birds from Isla Blanca, Mexico (B. Voirin, unpublished data; M. Wikelski, unpublished data), primarily moved into the southern Gulf of Mexico, but also overlapped with birds from Cayo Culebra in the Caribbean Sea (K. Meyer and G. Kent, unpublished data; n = 2). An individual from Cayo Culebra migrated north to Florida, US.

BANDING Three individuals banded on Barbuda, Lesser Antilles, were resighted during the post-breeding period in the Lesser Antilles

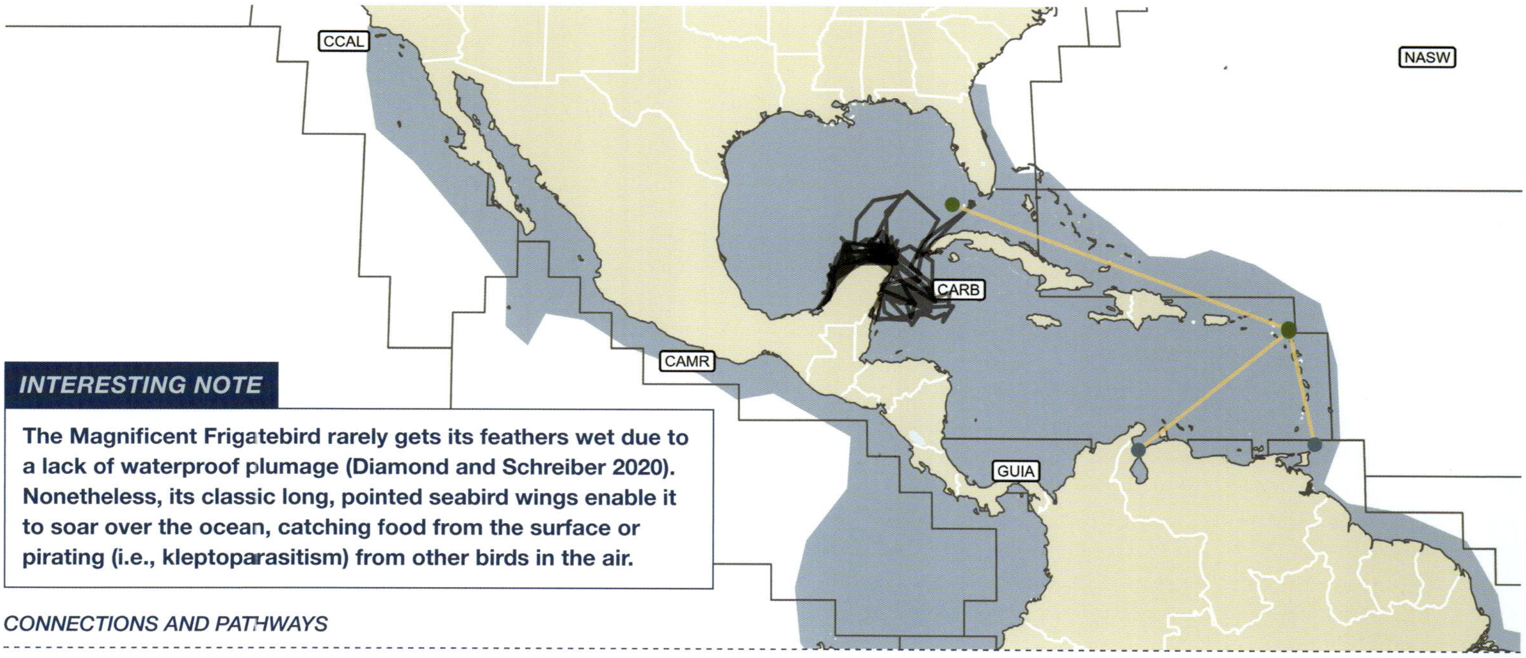

INTERESTING NOTE

The Magnificent Frigatebird rarely gets its feathers wet due to a lack of waterproof plumage (Diamond and Schreiber 2020). Nonetheless, its classic long, pointed seabird wings enable it to soar over the ocean, catching food from the surface or pirating (i.e., kleptoparasitism) from other birds in the air.

CONNECTIONS AND PATHWAYS

(but away from the nesting island), Venezuela, and Florida, US. Chicks banded on Barbuda were also recovered in the Grenadines, St. Martin, Antigua, Martinique, Guadeloupe, Tobago, and Haiti. A chick banded in Brazil was recovered in the Lesser Antilles (Diamond and Schreiber 2020).

STABLE ISOTOPES Males and females from Barbuda, Lesser Antilles, in the Atlantic Ocean differed in nitrogen isotope ratios sampled across multiple years from feathers and claws (reflecting nonbreeding conditions; Trefry and Diamond 2017), suggesting spatial separation of the sexes during the nonbreeding period. For three island colonies in the Galapagos Islands, feathers sampled for carbon, nitrogen, and mercury had similar signatures across two years and across islands within a year (Zarn et al. 2020). This suggests that colonies are likely to mix and may use similar habitats across years, although sample sizes of years and individuals were small. Sexes were not compared in the Galapagos study.

GENETICS Using mitochondrial DNA (mtDNA), nuclear DNA, and intron loci, Hailer et al. (2010) found extensive gene flow among populations in the eastern Pacific and central western Atlantic Ocean, although González-Jaramillo and Rocha-Olivares (2011) found differentiation between eastern Pacific Mexican and Caribbean Mexican populations using mtDNA. Additional exceptions include the Galapagos population, which is isolated from other populations within the eastern Pacific Ocean (Hailer et al. 2010), and the Brazilian population, which appears distinct within the western Atlantic Ocean (Nuss et al. 2016). The population of Grand Connétable, off French Guiana, shares mtDNA haplotypes with both the Caribbean and Brazilian populations, suggesting it as an intermediate between the two regions.

MIGRATORY PATHWAYS Pathways were generally close to breeding colonies and nomadic, with some exceptions. Birds from Isla Blanca, Mexico, primarily moved a short distance west, but also rounded the Yucatán Peninsula (B. Voirin, unpublished data; M. Wikelski, unpublished data). Birds from Cayo Culebra also primarily moved short distances west, while one individual migrated west then north around the western tip of Cuba to the Florida Keys (K. Meyer and G. Kent, unpublished data).

DATA CONTRIBUTORS

G.M. Kent; K.D. Meyer; B. Voirin; M. Wikelski.

SULIDAE | *BOOBIES AND GANNETS*

Sula dactylatra MASKED BOOBY

DISTRIBUTION AND ABUNDANCE

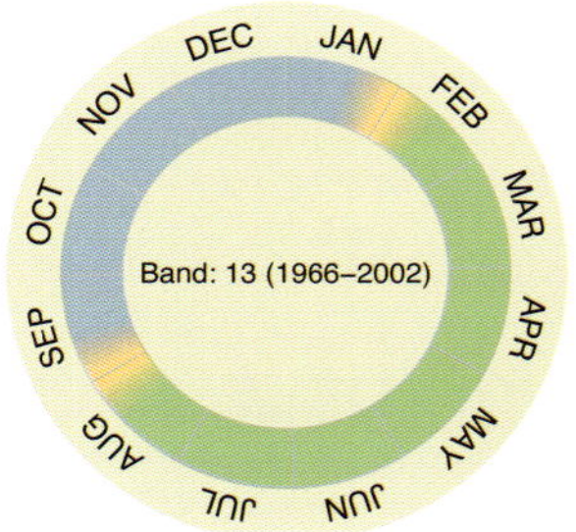

This species is a circumtropical seabird with high abundance near oceanic islands. It breeds on three Hawaiian Islands and many Mexican locations, including San Benedicto Island in the Pacific and the Yucatán Peninsula in the Gulf of Mexico (Grace et al. 2020). It has nested successfully in Dry Tortugas, Florida, in the past (Grace et al. 2020). The largest colony globally is on Clipperton Island in the eastern Pacific, but most colonies are small and remote with unknown populations (Grace et al. 2020). Some populations are resident, but others may migrate thousands of kilometers (Grace et al. 2020).

ORDER Suliformes
FAMILY Sulidae
BREEDING HABITAT Islands, Oceanic
NONBREEDING HABITAT Coasts, Marine Waters
GLOBAL POPULATION SIZE 100,000
POPULATION CHANGE SINCE 1970 NA
IUCN RED LIST CATEGORY Least Concern
R2R TIPPING POINT SPECIES No
US LISTING STATUS No
CANADA SPECIES AT RISK No
PIF CCS 12/20

MIGRATORY CONNECTIVITY

OVERALL Migratory connectivity is relatively unknown due to limited band reencounter data during the nonbreeding period from adult breeding birds. Conclusions about specific colony-level migratory connectivity should not be made; however, strong connectivity across oceans and moderate to strong within is suggested by banding, limited tracking, and genetics. Banding in the Pacific Ocean hints at regional separation; birds from North Pacific islands were reencountered in the North Pacific Tropical Gyre (100%); birds from Howland and Baker Islands, US, across the North Pacific Tropical Gyre (33%), the Pacific equatorial divergence (50%), and South Pacific Gyre (17%); and birds from Kiribati and Jarvis Island across the Western Pacific warm pool (100%) and Pacific equatorial divergence (100%), respectively.

	WARM	NPTG	PEQD	SPSG
Kure Atoll, HI, USA (n = 1)	–	100%	–	–
Laysan Island (Kauō), HI, USA (n = 2)	–	100%	–	–
Howland and Baker Islands (n = 6)	–	33%	50%	17%
Kiribati (n = 3)	100%	–	–	–
Jarvis Island (n = 1)	–	–	100%	–

BANDING Band reencounter data suggest separation between eastern and western Pacific colonies, though data from these regions are limited.

STABLE ISOTOPES Muraro et al. (2024) related isotopic measures to genetic diversity of six southwestern Atlantic colonies (east of Brazil). The authors found low genetic structure across populations but identified differences that related to whether colonies were located inshore or offshore, consistent with morphometric and isotopic niche variations.

GENETICS There is likely no gene flow between the Indo-Pacific and Atlantic populations, indicated by mitochondrial control region sequence variation (Steeves et al. 2003). Within each region, there is also some gene flow restriction likely due to nonphysical barriers, including limited natal dispersal and local adaptations (Steeves et al. 2005).

OTHER APPROACHES No tracking studies of migratory connectivity were found for this species. Many GPS studies using short-duration deployments during the breeding season across the range found that birds cued on distinct bathymetric and oceanographic features and ranged 19–100 km from the colony, suggesting fairly strong connectivity during breeding (Wilkinson et al. 2020).

MIGRATORY PATHWAYS Migratory status varies among populations; and although large movements have been recorded in the Pacific Ocean, no clear routes are known for North American breeding populations (Grace et al. 2020).

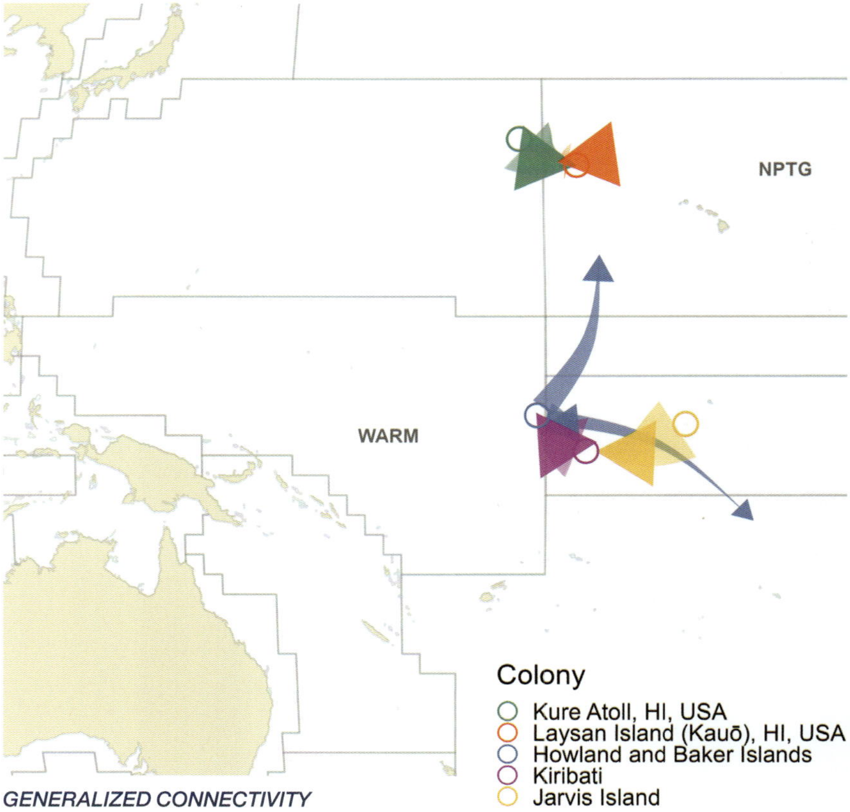

GENERALIZED CONNECTIVITY

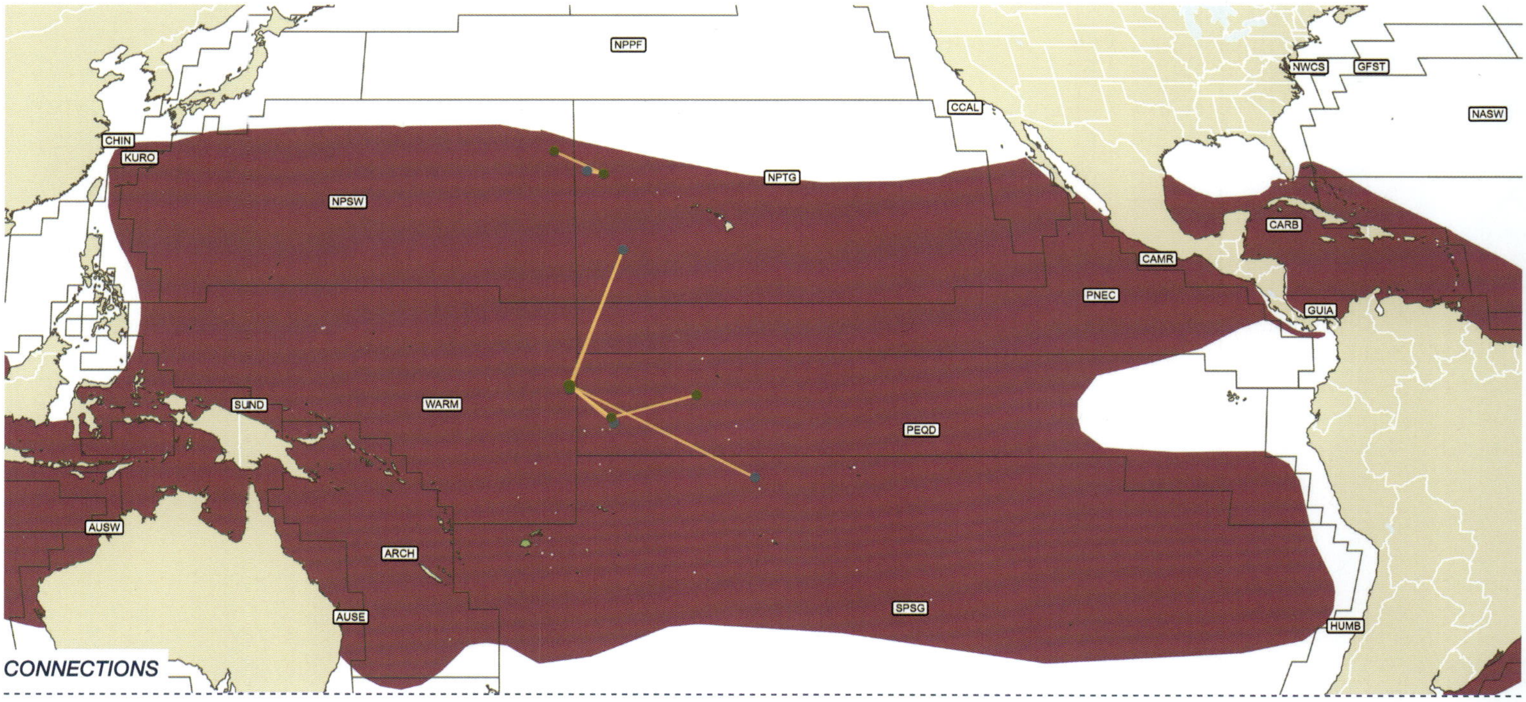

CONNECTIONS

INTERESTING NOTE

Thirty-one birds tracked from Fernando de Noronha archipelago, Brazil, with light-level geolocators indicated this population is resident across their annual cycle and thus do not mix with North American Atlantic Ocean breeders (Roy et al. 2021).

Sula leucogaster **BROWN BOOBY**

DISTRIBUTION AND ABUNDANCE

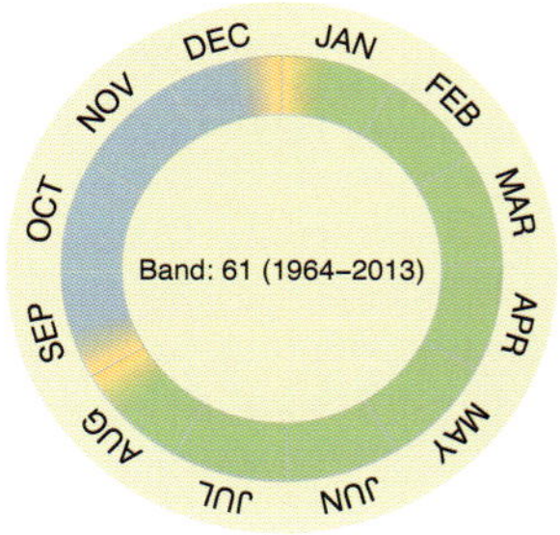

The Brown Booby occurs in pantropical oceans and seas, including the Caribbean Sea, the Mexican coasts, and Pacific Ocean US islands and territories, with highest abundance in Australia and Mexico and increasing numbers on Johnston Atoll (Schreiber and Norton 2020). It is thought to be more coastal than other boobies, but adults typically are not found on breeding islands during the nonbreeding season and can be found thousands of miles away (Schreiber and Norton 2020). It is irregular along the US coasts, including the Gulf of Mexico during the post-breeding season (Schreiber and Norton 2020).

ORDER Suliformes
FAMILY Sulidae
BREEDING HABITAT Coasts, Cliffs and Islands
NONBREEDING HABITAT Coasts, Marine Waters
GLOBAL POPULATION SIZE 500,000
POPULATION CHANGE SINCE 1970 NA
IUCN RED LIST CATEGORY Least Concern
R2R TIPPING POINT SPECIES No
US LISTING STATUS No
CANADA SPECIES AT RISK No
PIF CCS 12/20

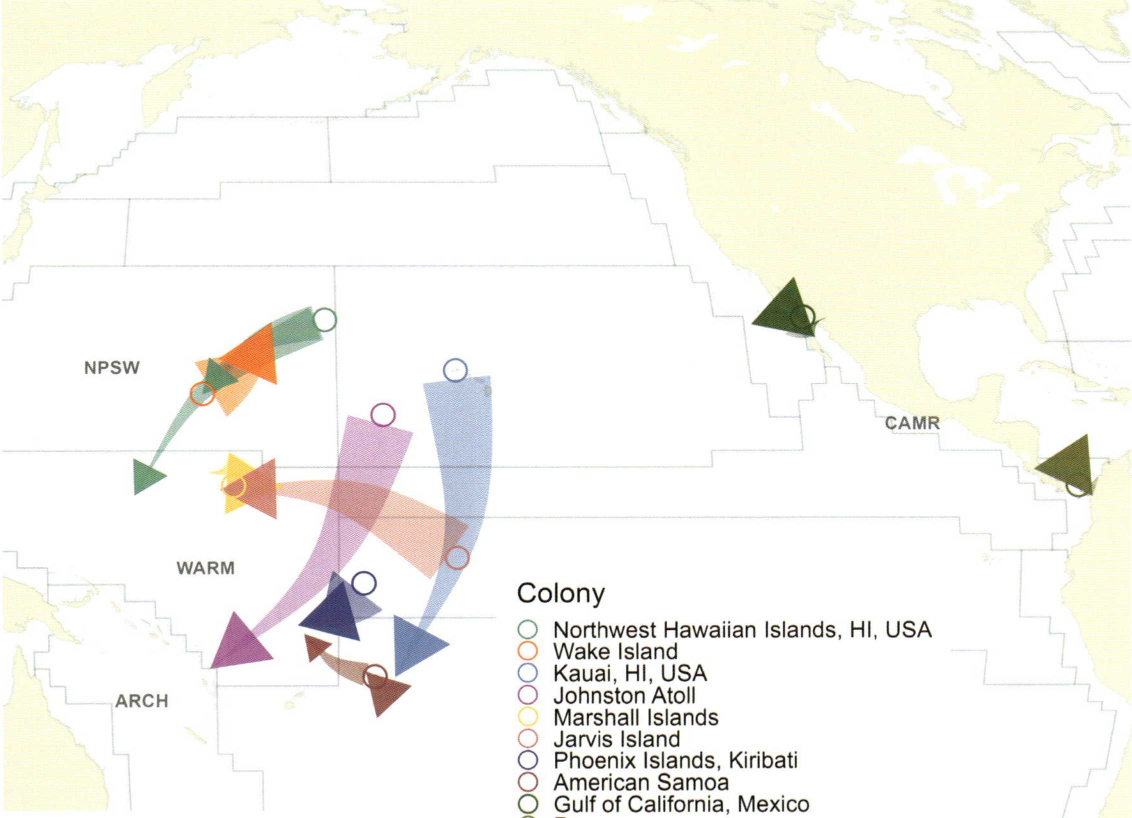

GENERALIZED CONNECTIVITY

MIGRATORY CONNECTIVITY

OVERALL Little is known about the migratory connectivity of Brown Boobies within North America. What we do know comes from band reencounter data; yet sample sizes of adult breeding birds at any one colony are limited, so conclusions about specific colony-level migratory connectivity should not be made and are not calculated here. However, birds banded along mainland North America spent the nonbreeding season in the Central American Coast province, while birds banded in the central Pacific used habitats in the southwest Pacific.

	ARCH	WARM	NPSW	CAMR	SPSG
Northwest Hawaiian Islands, HI, USA (n = 2)	–	50%	50%	–	–
Wake Island (n = 2)	–	–	100%	–	–
Kauai, HI, USA (n = 1)	–	–	–	–	100%
Johnston Atoll (n = 1)	100%	–	–	–	–
Marshall Islands (n = 1)	–	100%	–	–	–
Jarvis Island (n = 1)	–	100%	–	–	–
Phoenix Islands, Kiribati (n = 2)	–	100%	–	–	–
American Samoa (n = 3)	–	33%	–	–	67%
Gulf of California, Mexico (n = 2)	–	–	–	100%	–
Panama (n = 1)	–	–	–	100%	–

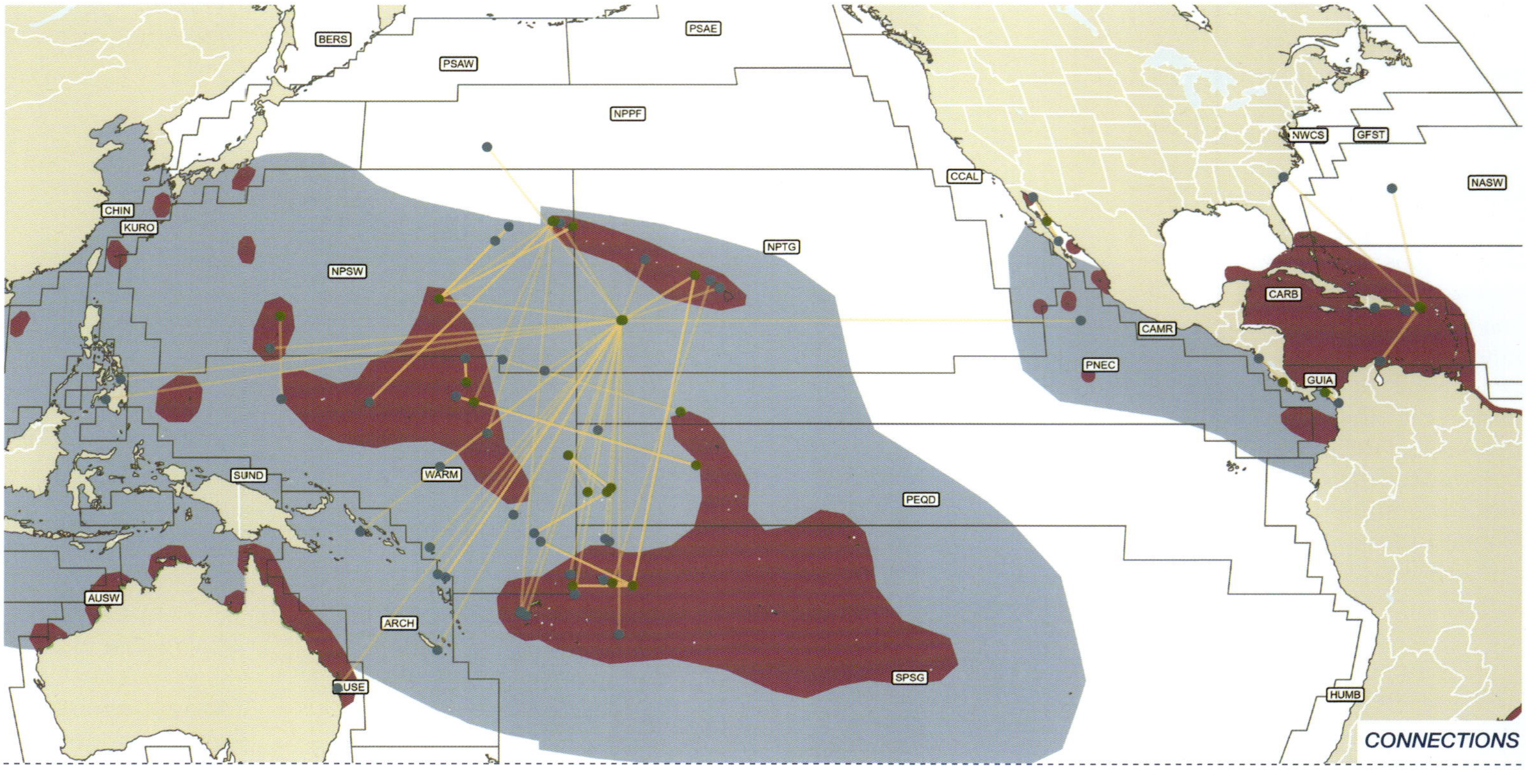

INTERESTING NOTE

A study of Brown Boobies used light-level geolocators to track eight birds in 2009–2015 on Nakanokamishima Island, Japan, at the northern range limit in the western Pacific Ocean (Kohno et al. 2019). All birds remained in the western Pacific, with some migrating large distances, distributing across nonbreeding areas separated by over 6,500 km. None of the tracked birds moved into the central Pacific to overlap with the Hawaiian Exclusive Economic Zone. Storms and El Niño events can cause wide dispersal and vagrancy (Schreiber and Norton 2020). Specimens salvaged in New Mexico and Colorado and genetically typed to have originated in the Caribbean contained multiple freshwater helminth parasites that suggests overland dispersal is not possible for boobies, thus limiting connectivity across oceans.

BANDING Despite substantial banding effort at some locations (e.g., 3,000 on Johnson Atoll; Schreiber and Norton 2020), there are few band reencounters to determine adult nonbreeding connectivity (n = 16), and all of these records are in the Pacific Ocean. Birds from the Hawaiian island chain used habitats in the Western Pacific Warm Pool (Kure Atoll), Northwest Pacific Subtropical province (Midway Atoll), and South Pacific Gyre (Kaua'i). One individual from Johnston Atoll was reencountered farther south in the Archipelagic Deep Basins province of the southwest Pacific Ocean. Some records of immature birds from the central Pacific are being encountered in the northeastern Pacific.

GENETICS From a study of the mitochondrial control region of 242 Brown Boobies across their range, global population genetic structure is extremely high, and female-mediated gene flow among oceans has been restricted by physical barriers (Morris-Pocock et al. 2010). With the exception of one haplotype found in birds at two Mexican colonies, no colonies shared any haplotypes, and many had a single high-frequency haplotype. A later study of the same dataset that also incorporated nuclear DNA also found high differentiation, suggesting colonies could be grouped into four genetic populations: Caribbean Sea, Central Atlantic Ocean, Central Indo-Pacific, and Eastern Pacific (Morris-Pocock et al. 2011).

OTHER APPROACHES No tracking or stable isotope studies of migratory connectivity were found for this species. Birds tracked during the breeding season with archival GPS from Gulf of California, Mexico (Weimerskirsch et al. 2009), and Dog Island, Anguilla (Soanes et al. 2015a), made foraging trips up to 125 km from their colony. In the Caribbean, these included international movements, but no movements from the Pacific or Atlantic crossed oceans to mix with other colonies.

MIGRATORY PATHWAYS Migratory pathways are generally unknown for North American populations (Mlodinow et al. 2024).

Mantel = 0.05 ± 0.08

Morus bassanus **NORTHERN GANNET**

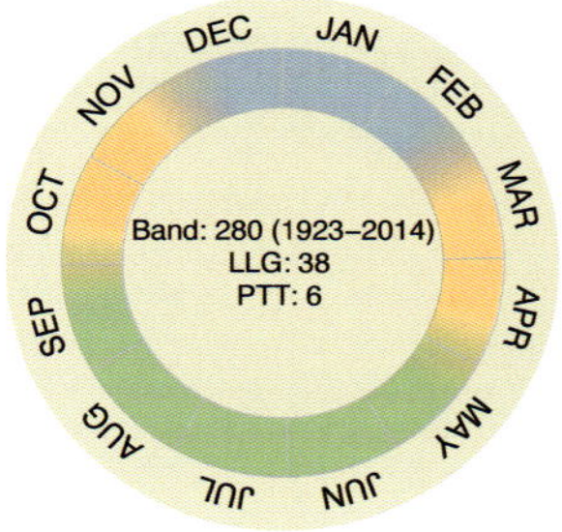

DISTRIBUTION AND ABUNDANCE

A large plunge-diver that breeds in large colonies on both sides of the North Atlantic on sea stacks, islands, and cliffs. Most of the global breeding population is in British waters, with population increases leading to the establishment of new European colonies throughout the 1900s (Mowbray 2020b). Large North American colonies were greatly reduced in the 1800s; however, the population continues to recover with 116,000 pairs estimated in 2009 (Chardine et al. 2013). The Gulf of St. Lawrence is estimated to host 69% of the North American population with the remainder in Atlantic Canada (Mowbray 2020b). During post- and nonbreeding periods Northern Gannets use continental shelf and slope waters.

ORDER Suliformes
FAMILY Sulidae
BREEDING HABITAT Coasts, Cliffs and Islands
NONBREEDING HABITAT Coasts, Marine Waters
GLOBAL POPULATION SIZE 720,000
POPULATION CHANGE SINCE 1970 +210%
IUCN RED LIST CATEGORY Least Concern
R2R TIPPING POINT SPECIES No
US LISTING STATUS No
CANADA SPECIES AT RISK No
PIF CCS 11/20

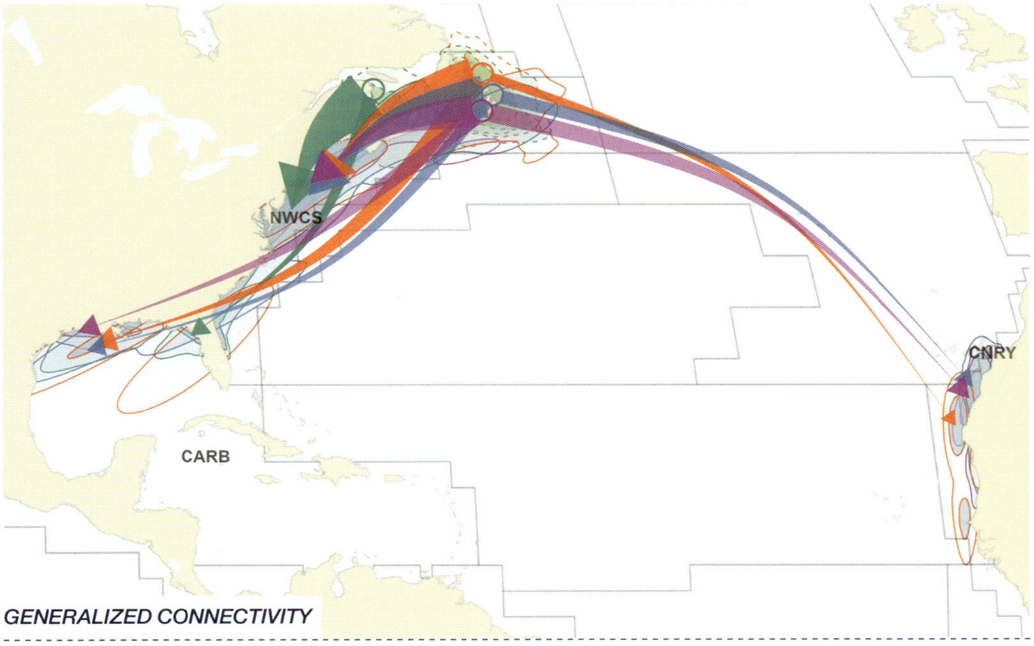

GENERALIZED CONNECTIVITY

Colony
- Bonaventure Island, QC, Canada
- Funk Island, NL, Canada
- Baccalieu Island, NL, Canada
- Cape St. Mary's, NL, Canada

	CARB	NWCS	CNRY
Bonaventure Island, QC, Canada (n = 38)	18%	82%	–
Funk Island, NL, Canada (n = 10)	30%	60%	10%
Baccalieu Island, NL, Canada (n = 6)	17%	67%	17%
Cape St. Mary's, NL, Canada (n = 4)	25%	50%	25%

INTERESTING NOTE

The impact of high pathogenicity avian influenza virus (HPAIV) on breeding connectivity of Northern Gannets has been evaluated in multiple studies with mixed results. Jeglinski et al. (2024) provided evidence that a small sample of gannets tracked from Bass Rock, Scotland, made unprecedented long-distance movements during a disease outbreak beyond known breeding foraging ranges and the first recorded visits of GPS-tagged birds to other colonies. Such connectivity could amplify disease transmission.

MIGRATORY CONNECTIVITY

OVERALL North American and European breeding colonies generally use different nonbreeding areas, suggesting strong connectivity at the species level. However, connectivity estimates from North American band reencounter and tracking data show that weak connectivity occurs within continental colony groupings (Mantel = 0.05 ± 0.08). The majority of individuals migrate to the Northwest Atlantic Shelves (50%–84%), as well as the Caribbean (17%–30%) provinces. However, a small proportion of birds make transatlantic crossings to the Canary Coastal Province (10%–25%), where European colonies spend the nonbreeding season.

BANDING Most band recoveries have been coastal, and about half are from fishing entanglement, extending down the Atlantic coast of North America to the southern tip of Florida and into the Gulf of Mexico as far west as Texas. However, banding recoveries underestimate the importance of the Gulf of Mexico to gannets as indicated by tracking data: 10% of band recoveries came from the Gulf of Mexico, while 28% of birds tracked in Fifield et al. (2014) occurred in the Gulf. Several Northern Gannets banded at Funk Island, Newfoundland, Canada, have been recovered near coastal areas of the eastern Atlantic Ocean from Iceland to Europe and south to Morocco (Gaston et al. 2008; Fifield et al. 2014).

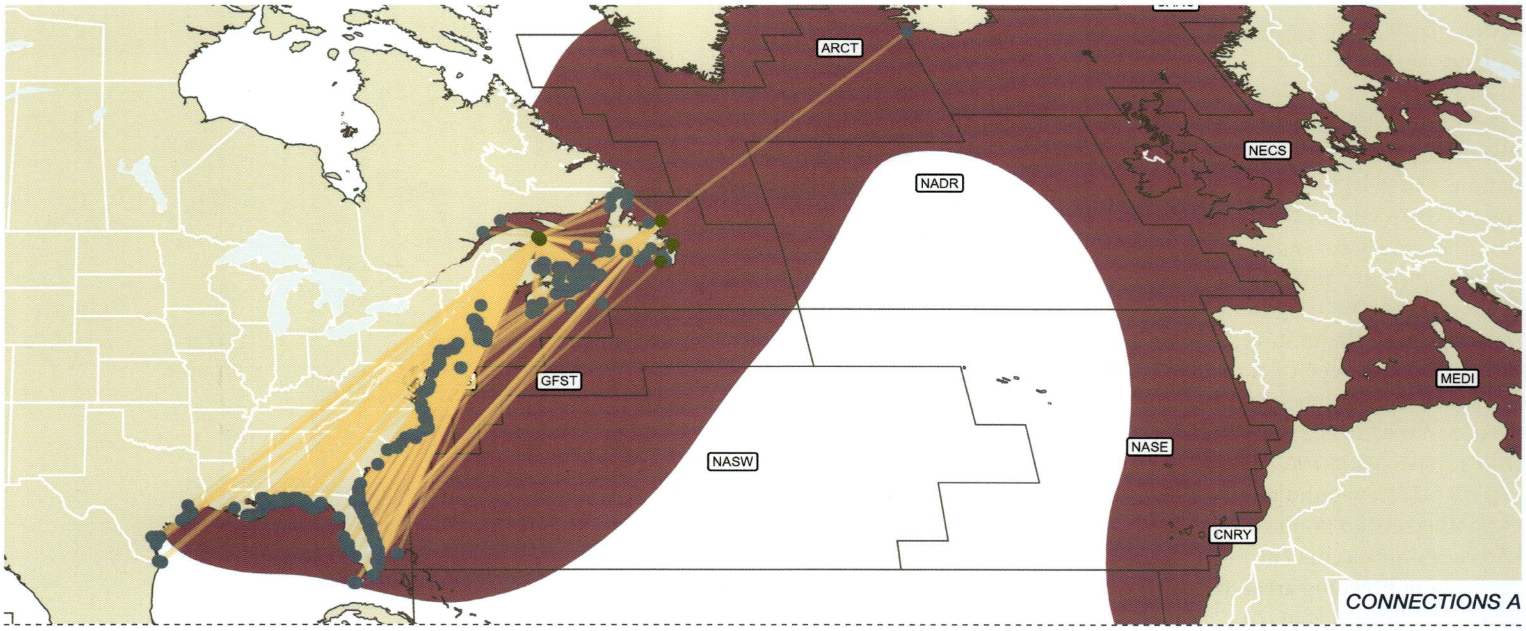
ARCT
NECS
NADR
MEDI
GFST
NASW
NASE
CNRY
CONNECTIONS A

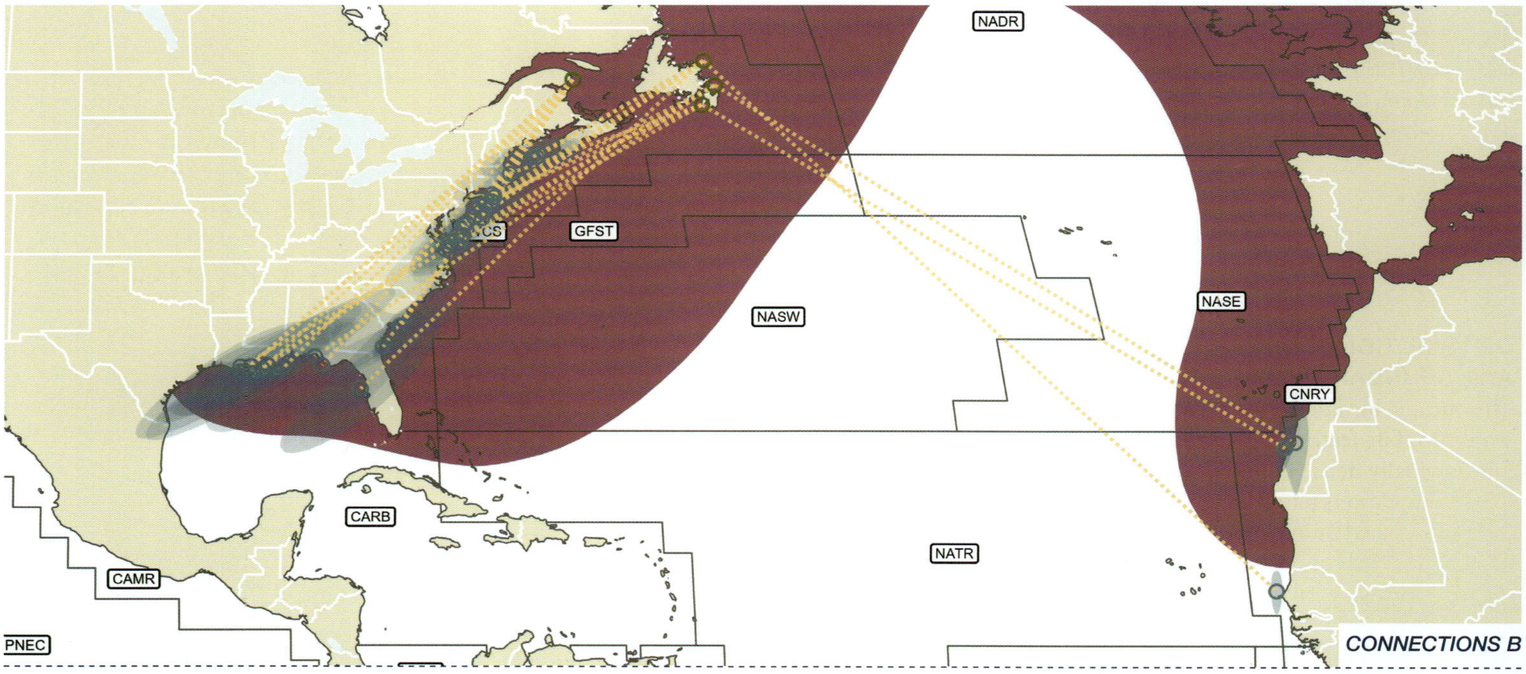
NADR
GFST
NASW
NASE
CNRY
CARB
NATR
CAMR
PNEC
CONNECTIONS B

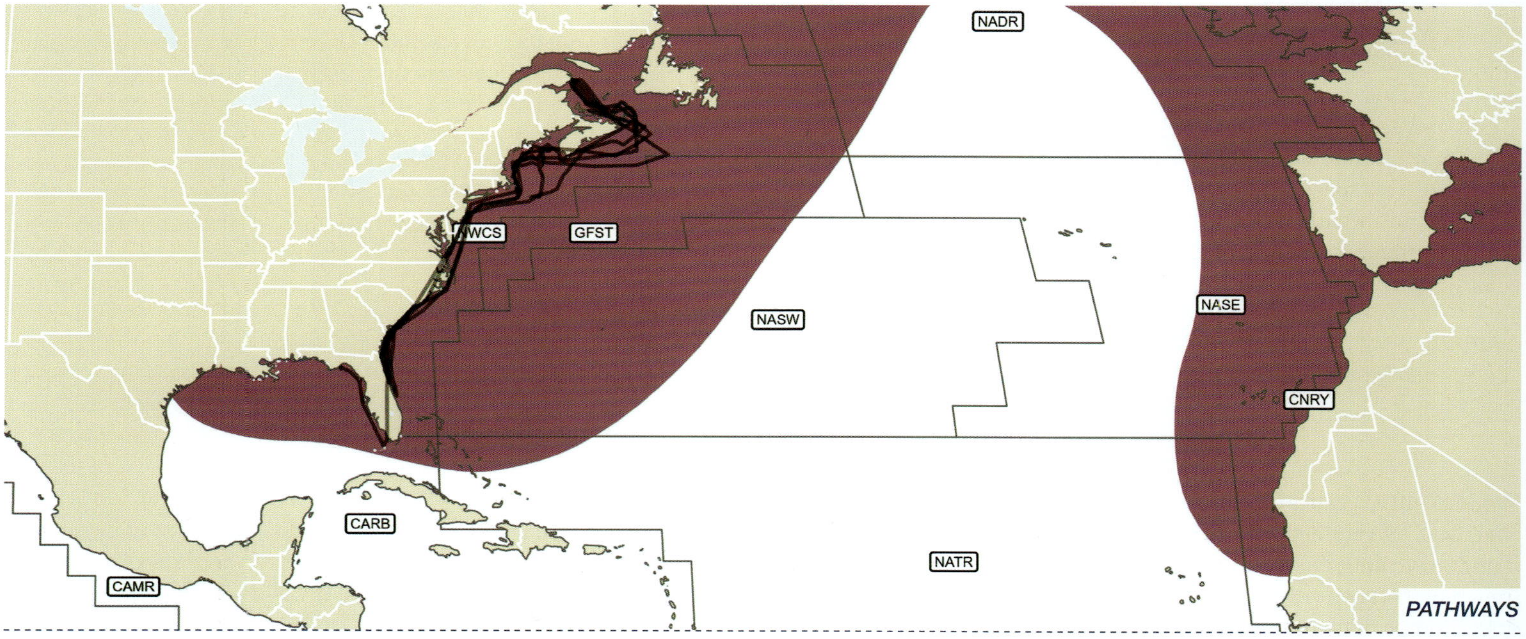
NADR
NWCS
GFST
NASW
NASE
CNRY
CARB
NATR
CAMR
PATHWAYS

GEOLOCATOR TRACKING In a monograph aimed at determining migratory connectivity, Fifield et al. (2014) tracked 46 individuals from four of the six Northern Gannet colonies in North America. The authors found weak migratory connectivity, and those tracked for multiple years had high nonbreeding side fidelity. Most (n = 43) gannets used habitats along the US Atlantic and Gulf of Mexico coasts, especially in the Gulf of Maine and South Atlantic Bight, along the New Jersey coast, the Chesapeake Bay, and the Louisiana coast. Three gannets, one from each colony in Newfoundland, provided the first recorded transatlantic round-trip migrations to Africa, overlapping with nonbreeding areas of European birds. No Atlantic crossings have been documented from any of the several studies from Iceland (Garthe et al. 2016) and Europe (e.g., Fort et al. 2012; Laurenson et al. 2021).
SATELLITE TRACKING Three juvenile Northern Gannets from Cape St. Mary's, Newfoundland, were tracked with satellite tags along the Atlantic coast to nonbreeding areas in the Gulf of Mexico (data not in Atlas; Montevecchi et al. 2012). Satellite tags have also been deployed within nonbreeding areas along the mid-Atlantic coast of North America (Spiegel et al. 2017). Eighty percent of these birds migrated to the Gulf of St. Lawrence, indicating that connectivity could differ between nonbreeding areas to breeding colonies and vice versa.
STABLE ISOTOPES Stable isotope analysis successfully predicted nonbreeding areas (Champoux et al. 2020) of the same birds also tracked by geolocators (Montevecchi et al. 2012; Fifield et al. 2014). Feather corticosterone and plasma thyroid hormone levels were higher for gannets spending the nonbreeding season in the Gulf of Mexico. The authors suggested this could be due to increased energetic demands and/or greater exposure to environmental stress (Champoux et al. 2020).
GENETICS Using genomics methods, N.J. Clark (2017) found significant population genetic structure corresponding to the division between North American and European gannet populations, but no differentiation within continental groupings. Friesen et al. (2021) found little female-mediated gene flow between Canadian and European populations of Northern Gannets.
MIGRATORY PATHWAYS Most gannets tracked follow Atlantic coastal pathways south from colonies, except for three individuals tracked on transatlantic flights (Fifield et al. 2014). These birds crossed the Atlantic in as little as five days. Birds tracked across multiple years had interannual consistency in migration distance, duration, and arrival and departure timing to nonbreeding grounds (Fifield et al. 2014).

DATA CONTRIBUTORS
D.A. Fifield; S. Garthe; U. Kubetzki; W.A. Montevecchi; J.-F. Rail; G.J. Robertson.

ANHINGIDAE | *ANHINGAS*

Anhinga anhinga ANHINGA

DISTRIBUTION AND ABUNDANCE

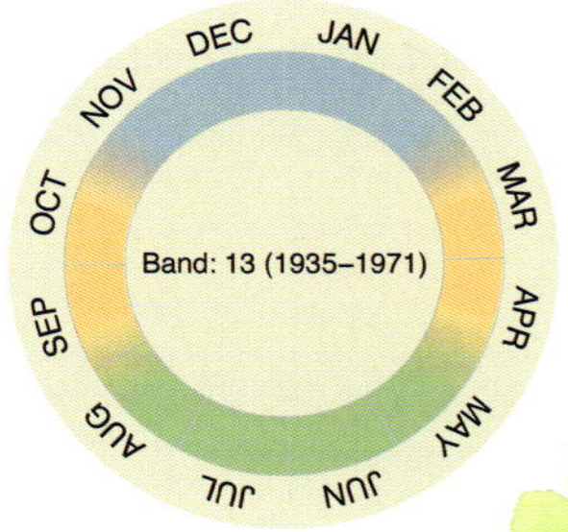

This unique bird does not waterproof its feathers and has unusually dense bones. These adaptations allow them to have reduced buoyancy as they stalk fish underwater. Anhingas are confined to tropical and subtropical areas of the Western Hemisphere, where there is adequate warmth for them to warm their bodies and dry their plumage after hunting (Frederick and Siegel-Causey 2020; Fink et al. 2023). Breeding densities are at their highest in Florida and Louisiana, but its breeding range extends west to Texas and east to other Gulf states and along the southeast Atlantic states north to North Carolina (Fink et al. 2023). This species is known to demonstrate migratory behavior and wanders widely, but these movements are not well understood (Frederick and Siegel-Causey 2020).

	Atlantic	Mexico	Midcontinent
Atlantic (n = 1)	100%	–	–
Midcontinent (n = 11)	–	64%	36%

ORDER Suliformes
FAMILY Anhingidae
BREEDING HABITAT Wetlands, Generalist
NONBREEDING HABITAT Wetlands, Generalist
GLOBAL POPULATION SIZE 2,000,000

POPULATION CHANGE SINCE 1970 +110%
IUCN RED LIST CATEGORY Least Concern
R2R TIPPING POINT SPECIES No
US LISTING STATUS No
CANADA SPECIES AT RISK No
PIF CCS 8/20

MIGRATORY CONNECTIVITY

OVERALL No studies on migratory connectivity were found. Limited band reencounter and observational data indicate the species inhabits much of its range year-round but moves out of the northern part of its US range in the nonbreeding period. There appears to be strong migratory connectivity between individuals breeding within the Atlantic Flyway and those breeding in the Midcontinent Flyway. However, MC was not calculated due to insufficient data.
BANDING Band reencounter data suggest individuals move southward during the nonbreeding period. Several birds from the

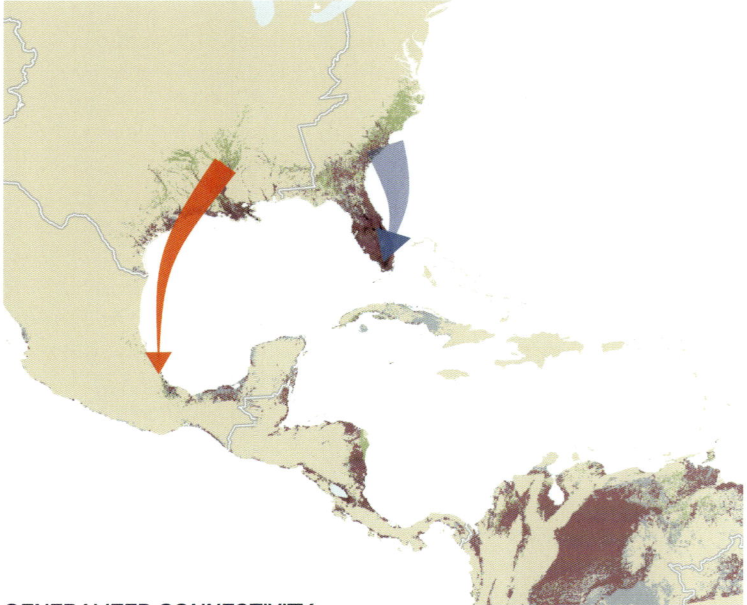

GENERALIZED CONNECTIVITY

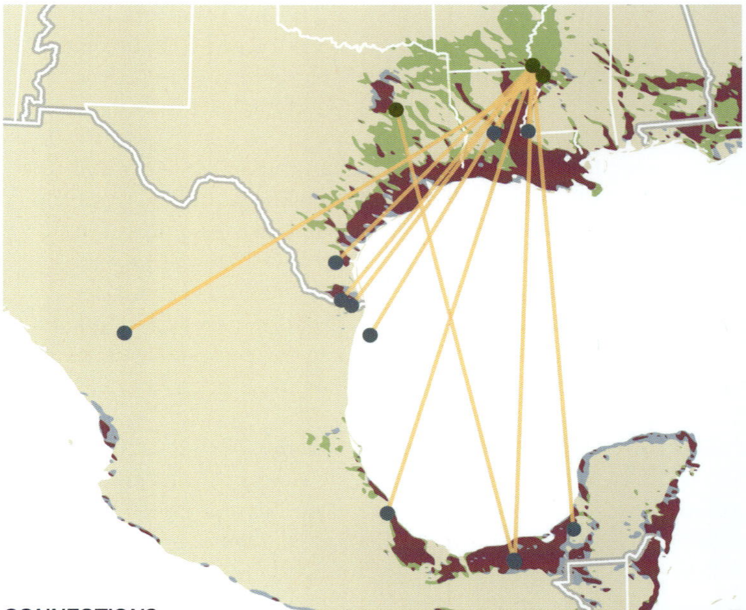

CONNECTIONS

central Gulf states moved south to southern Texas and Mexico. A bird in Florida moved a short distance southward along the Florida peninsula.

OTHER APPROACHES No satellite, genetic, or stable isotope studies of migratory connectivity were found for this species.

MIGRATORY PATHWAYS While likely resident in much of its range, this species does undergo medium-range movements. The connection between central US breeding birds and Mexican nonbreeding grounds is supported by banding data. This species soars easily and likely undergoes most migratory movements by riding thermals along overland routes.

PHALACROCORACIDAE | *CORMORANTS AND SHAGS*

BRANDT'S CORMORANT *Urile penicillatus*

DISTRIBUTION AND ABUNDANCE

Found only along the Pacific coast of North America, this species is closely tied to the upwelling of the California Current system. Breeding colonies are concentrated in California and Oregon, but the species disperses along the Pacific coast from Alaska to Mexico in the nonbreeding season (Wallace and Wallace 2021; Fink et al. 2023).

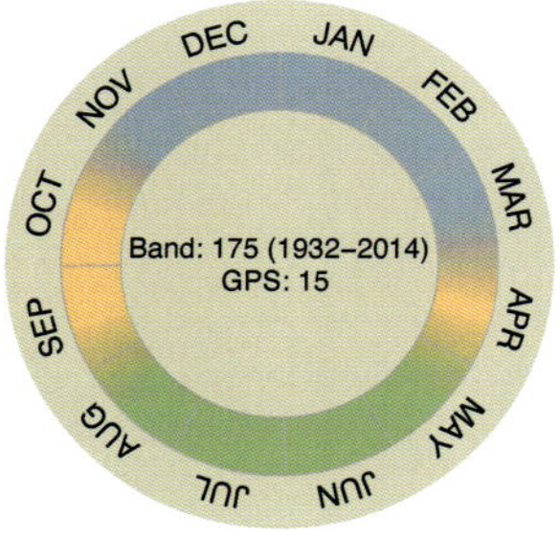

ORDER Suliformes
FAMILY Phalacrocoracidae
BREEDING HABITAT Coasts, Cliffs and Islands
NONBREEDING HABITAT Coasts, Marine Waters

GLOBAL POPULATION SIZE 150,000
POPULATION CHANGE SINCE 1970 −7%
IUCN RED LIST CATEGORY Least Concern
R2R TIPPING POINT SPECIES No
US LISTING STATUS No
CANADA SPECIES AT RISK No
PIF CCS 11/20

	Mexico	Pacific
Pacific (n = 178)	2%	98%

MIGRATORY CONNECTIVITY

OVERALL This species does not follow a classic migration route, but instead undergoes dispersal north and south after breeding. Birds breeding in the Pacific Flyway largely (98%) migrate within the Pacific Flyway, with others (2%) migrating to Mexico during the nonbreeding season. Because the population breeds within only one of our migratory connectivity regions, MC could not be calculated.

BANDING Band reencounter data largely show a pattern of post-breeding dispersal north and south from breeding colonies. Movements of Central California breeders extend as far north as British Columbia and as far south as Mexico. Limited records from breeders farther north show those individuals moving predominantly south in the nonbreeding period.

SATELLITE TRACKING Individuals from Oregon tracked with GPS-GSM transmitters dispersed north into the Salish Sea and the Strait of Georgia. Birds from this region also underwent southward

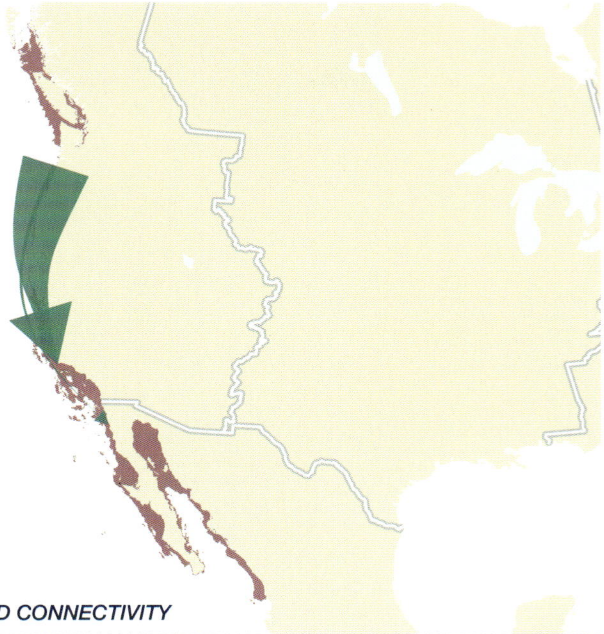

GENERALIZED CONNECTIVITY

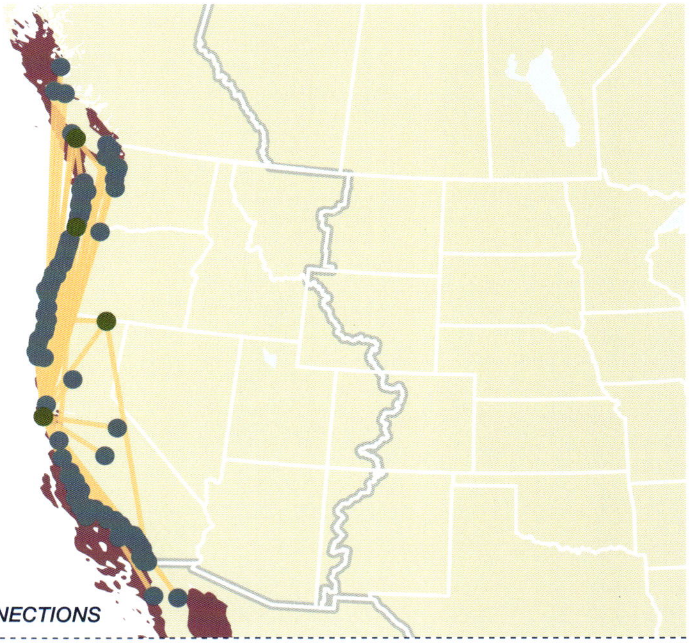

CONNECTIONS

INTERESTING NOTE

Individuals (especially males) show fidelity to colony locations, but entire colonies may break down or relocate based on food availability or disturbance (Wallace and Wallace 2021).

movements extending as far south as Long Beach, California (Peck-Richardson et al. 2024b).

OTHER APPROACHES No genetic or stable isotope studies of migratory connectivity were found for this species.

MIGRATORY PATHWAYS Post-breeding individuals disperse north and south mainly along the Pacific coast in search of food resources, often congregating in sheltered saltwater bays or offshore feeding areas.

DATA CONTRIBUTORS

J. Lerzcak; R.A. Orben; A. Peck-Richardson.

PATHWAYS

Urile pelagicus **PELAGIC CORMORANT**

ORDER Suliformes
FAMILY Phalacrocoracidae
BREEDING HABITAT Coasts, Cliffs and Islands
NONBREEDING HABITAT Coasts, Marine Waters
GLOBAL POPULATION SIZE 400,000
POPULATION CHANGE SINCE 1970 −53%
IUCN RED LIST CATEGORY Least Concern
R2R TIPPING POINT SPECIES Orange-Alert
US LISTING STATUS No
CANADA SPECIES AT RISK No
PIF CCS 14/20

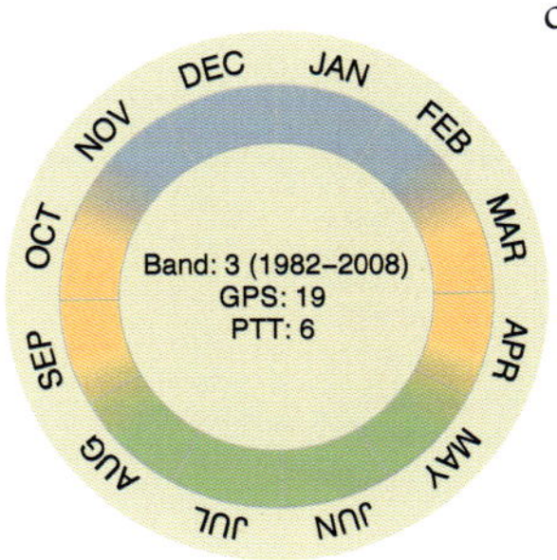

DISTRIBUTION AND ABUNDANCE

Pelagic Cormorants are a widespread species that ranges along both sides of the northern Pacific arc. In North America, it is found from northern Alaska to Baja California, with the highest concentrations in southern Alaska and British Columbia (Hobson 2021; Fink et al. 2023). Despite its name, this cormorant occupies areas close to shore and not pelagic habitats, and is tied to the productive Alaska and California Coastal Current systems.

INTERESTING NOTE

Band recoveries from British Columbia and Washington indicate that birds in those populations disperse locally (S.A. Hatch et al. 2011b)

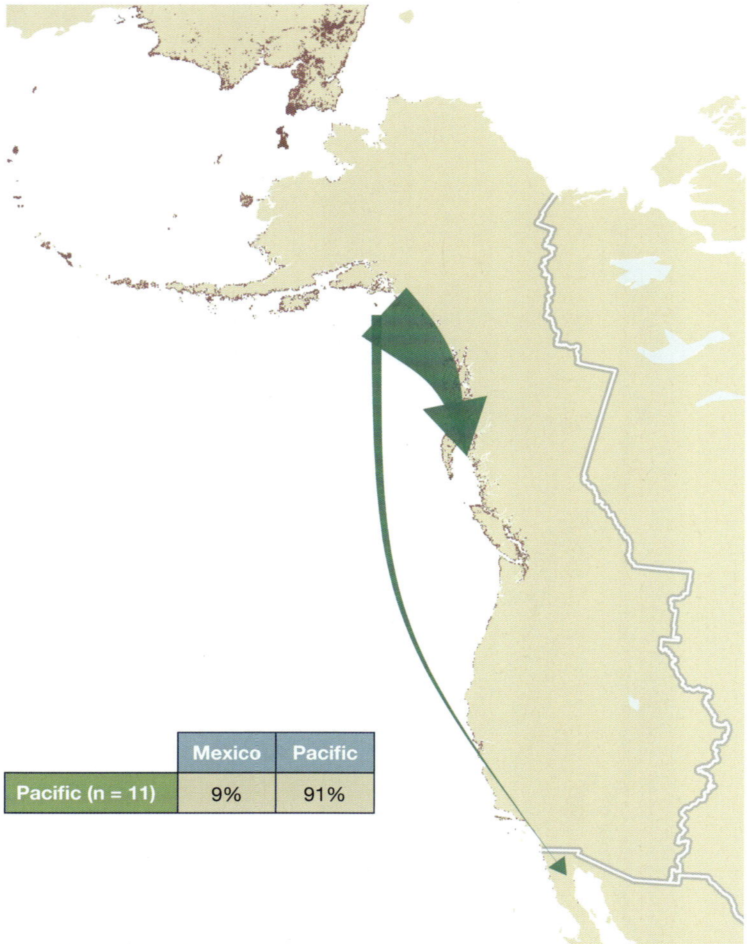

GENERALIZED CONNECTIVITY

CONNECTIONS

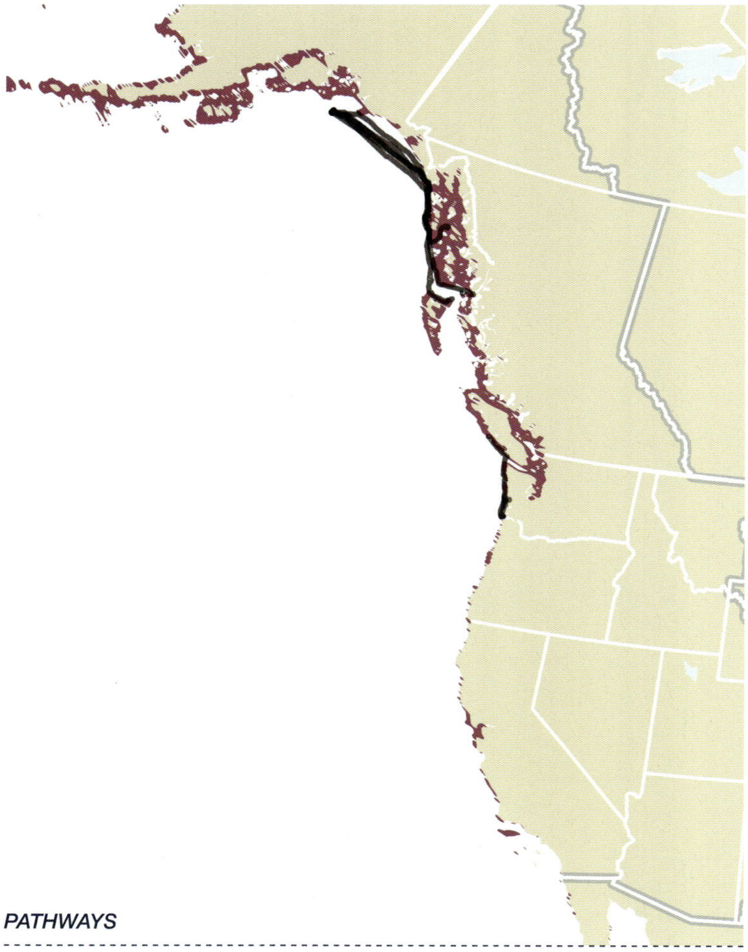

PATHWAYS

MIGRATORY CONNECTIVITY

OVERALL Migratory connectivity is known from both GPS and satellite tracking data and limited band reencounter data. The majority of Pacific Flyway breeders remain within the Pacific Flyway (91%), while others migrate to Mexico (9%) in the nonbreeding period. MC is not able to be generated due to insufficient data.

BANDING Breeding to nonbreeding banding data is limited, with records of one bird moving from a breeding colony in the Gulf of Alaska to British Columbia, and of two birds from central California moving to Southern California and northern Mexico.

SATELLITE TRACKING Birds breeding in the Gulf of Alaska were tracked to nonbreeding sites in Southeast Alaska and northern British Columbia (S.A. Hatch et al. 2011b; Peck-Richardson et al. 2024a, 2024b). Two satellite-tagged juveniles remained at nonbreeding sites for the duration of the study period (S.A. Hatch et al. 2011b). Two birds captured in the Columbia River region remained within the region and moved north to southern British Columbia (Peck-Richardson et al. 2024a, 2024b).

OTHER APPROACHES No genetic or stable isotope studies of migratory connectivity were found for this species.

MIGRATORY PATHWAYS This is an exclusively coastal bird of the North Pacific, and its movements are variable and influenced largely by weather and the availability of prey and nesting sites. When migrating, most individuals retreat from breeding areas in the north to more southerly feeding grounds, traveling along coastal routes during the duration of their movements (Hobson 2021; Fink et al. 2023).

DATA CONTRIBUTORS

D.C. Douglas; S.A. Hatch; J. Lerczak; R.A. Orben; A. Peck-Richardson.

Phalacrocorax carbo **GREAT CORMORANT**

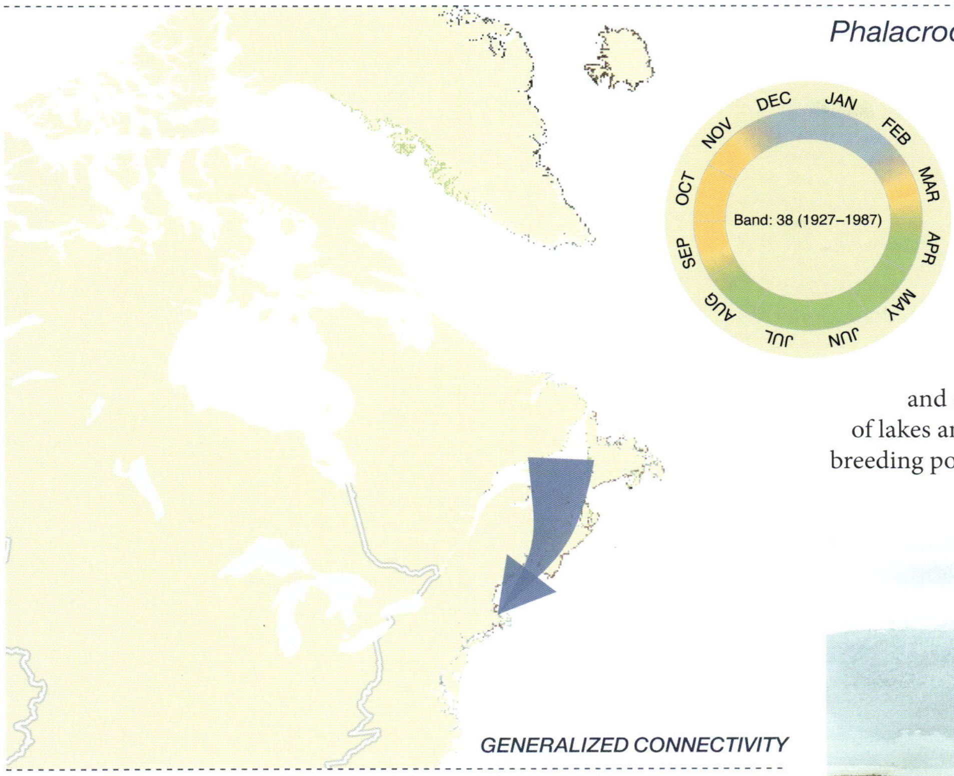

DISTRIBUTION AND ABUNDANCE

The Great Cormorant has the largest global range of any cormorant. In North America, this species is only partially migratory, with some birds present in the breeding range year-round. It spends the breeding season along the coast of Maine and northward into the Canadian Maritime provinces and Greenland. During a typical nonbreeding season, individuals can extend south to Maryland, Virginia, and North Carolina, and even south to Florida. While it is typically a denizen of lakes and rivers across most of its range, the North Atlantic breeding population is found along cold rocky coasts (J.J. Hatch et al. 2020; Fink et al. 2023).

ORDER Suliformes
FAMILY Phalacrocoracidae
BREEDING HABITAT Coasts, Cliffs and Islands
NONBREEDING HABITAT Coasts, Marine Waters
GLOBAL POPULATION SIZE 1,400,000
POPULATION CHANGE SINCE 1970 NA
IUCN RED LIST CATEGORY Least Concern
R2R TIPPING POINT SPECIES No
US LISTING STATUS No
CANADA SPECIES AT RISK No
PIF CCS 10/20

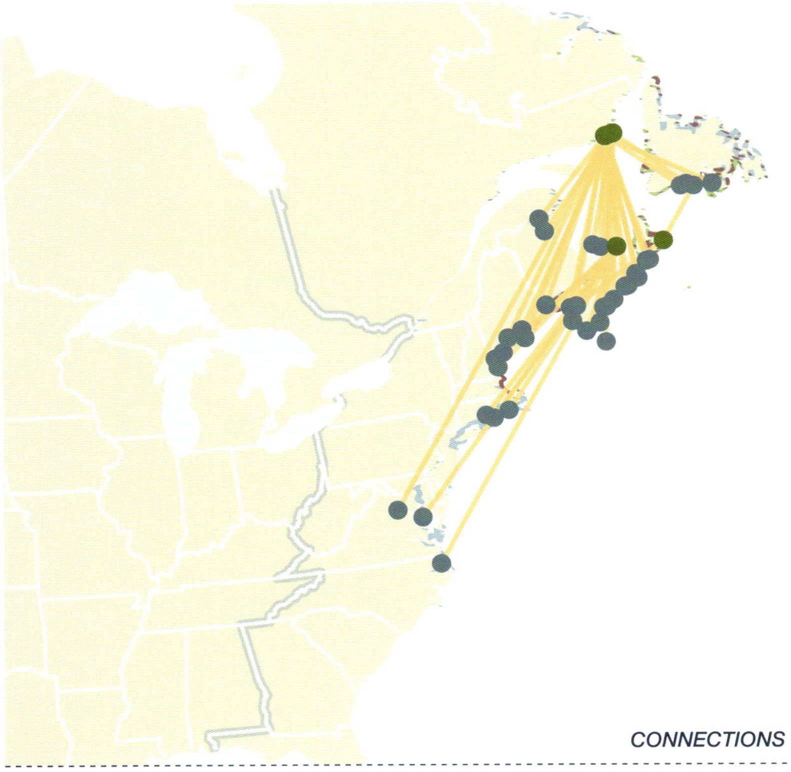

	Atlantic
Atlantic (n = 37)	100%

MIGRATORY CONNECTIVITY

OVERALL No studies on migratory connectivity nor migratory routes were found. The species is closely tied to the Atlantic region, with all (100%) birds breeding in the Atlantic Flyway remaining in the Atlantic Flyway during the nonbreeding period. Data are available within only one of our breeding migratory connectivity regions; thus MC could not be calculated.

BANDING Band reencounter data shows individuals undergoing variable southward migrations in the nonbreeding period, sometimes as far as Florida, but more often short-distance movement southward within the Maritimes, or to New England and the mid-Atlantic states.

OTHER APPROACHES No tracking, genetic, or stable isotope studies of migratory connectivity were found for this species.

MIGRATORY PATHWAYS This species will occasionally frequent freshwater lakes and rivers but is largely confined to the inshore waters of the Atlantic. Some individuals move just a short distance in the nonbreeding period, traveling only to find food. When this species does move over longer distances, it largely travels along the Atlantic coast. Some individuals, mostly juveniles, follow the Atlantic coast on longer migrations, sometimes as far south as Florida (J.J. Hatch et al. 2020).

DOUBLE-CRESTED CORMORANT *Nannopterum auritum*

weak strong
MC = 0.53 ± 0.06*

DISTRIBUTION AND ABUNDANCE

The Double-crested Cormorant is a common aquatic bird found throughout much of North America with highest abundances at inland lakes, including the Great Lakes and Lake Winnipeg and along coastal areas of the northeastern US, the Canadian North Atlantic, the west coast of Florida, Baja California, and the western coast of Mexico (Dorr et al. 2021; Fink et al. 2023). During the nonbreeding period, inland breeding birds most often move to coastal areas. However, areas of high abundance during the nonbreeding season also occur inland within the lower reaches of the Mississippi River and lakes in the southeast (Dorr et al. 2021; Fink et al. 2023). In the western portion of the nonbreeding range, birds are found in high numbers from coastal British Columbia south to Baja California and the west coast of Mexico (Dorr et al. 2021; Fink et al. 2023).

ORDER Suliformes
FAMILY Phalacrocoracidae
BREEDING HABITAT Wetlands, Generalist
NONBREEDING HABITAT Coasts, Marine Waters
GLOBAL POPULATION SIZE 630,000
POPULATION CHANGE SINCE 1970 +482%
IUCN RED LIST CATEGORY Least Concern
R2R TIPPING POINT SPECIES No
US LISTING STATUS No
CANADA SPECIES AT RISK No
PIF CCS 8/20

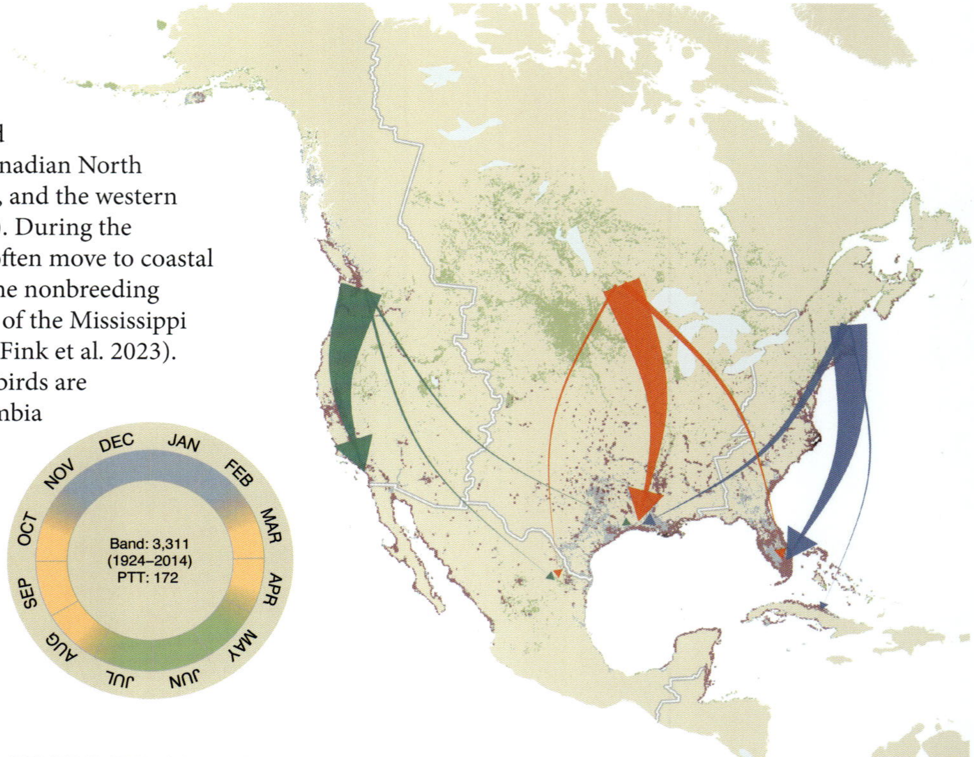

GENERALIZED CONNECTIVITY

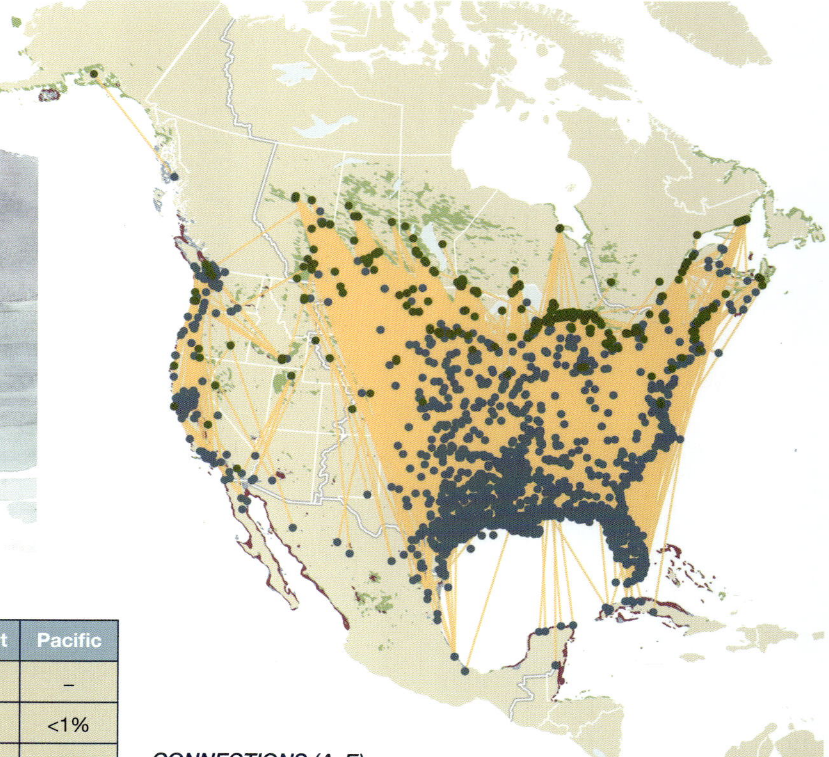

CONNECTIONS (A–E)

	Atlantic	Caribbean	Mexico	Midcontinent	Pacific
Atlantic (n = 523)	82%	1%	<1%	17%	–
Midcontinent (n = 2503)	11%	<1%	2%	87%	<1%
Pacific (n = 253)	–	–	2%	1%	96%

MIGRATORY CONNECTIVITY

OVERALL Migratory connectivity is known from satellite tracking and band reencounter data. Most individuals of this species undertake north-south migrations and remain within the same flyway during the breeding and nonbreeding season, resulting in strong migratory connectivity (MC = 0.53 ± 0.06). Most Atlantic Flyway breeders (82%) migrate within the Atlantic Flyway for the nonbreeding period, while 17% migrate to the Midcontinent Flyway. Midcontinent Flyway breeding birds also primarily migrate within the same flyway during the nonbreeding season (87%), and many fewer migrate to the Atlantic Flyway (11%). Pacific Flyway breeding birds are unlikely to cross the Continental Divide, with the vast majority (96%) remaining in the Pacific Flyway.

BANDING Band reencounter data demonstrate an east-west divide, with few birds crossing the Rocky Mountains. Inland birds often travel longer distances to nonbreeding areas than do

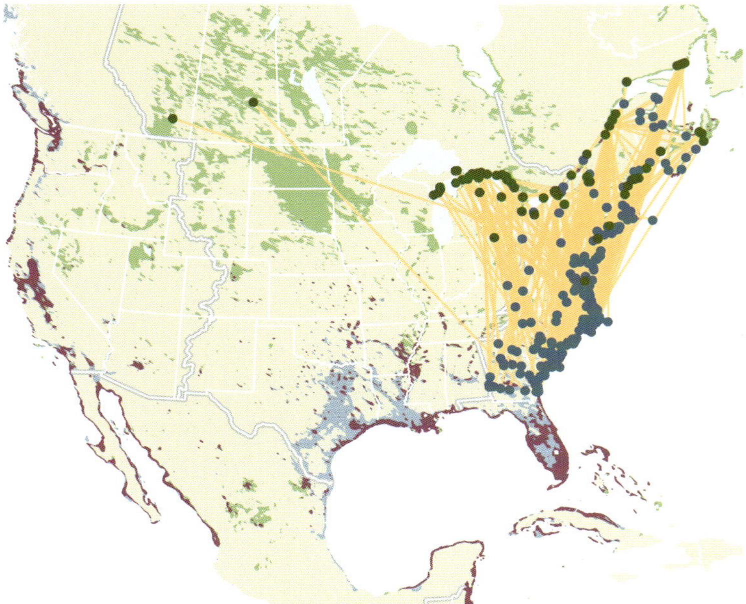

CONNECTIONS A

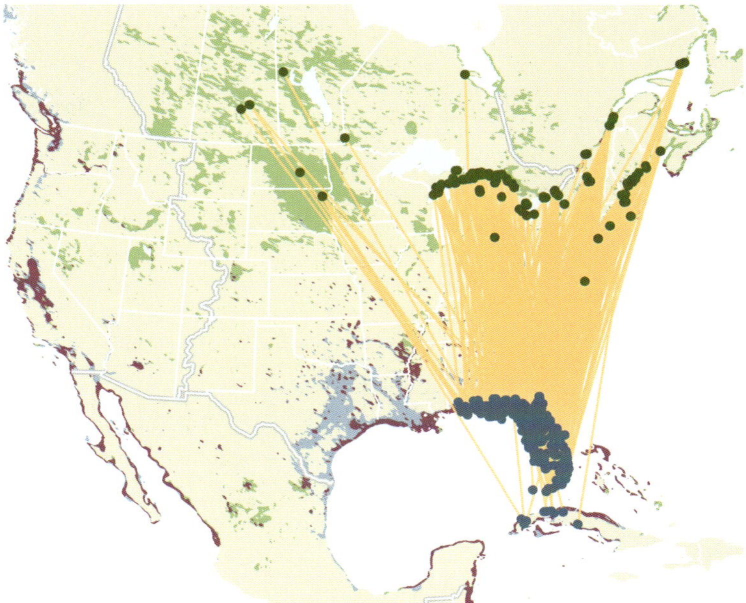

CONNECTIONS B

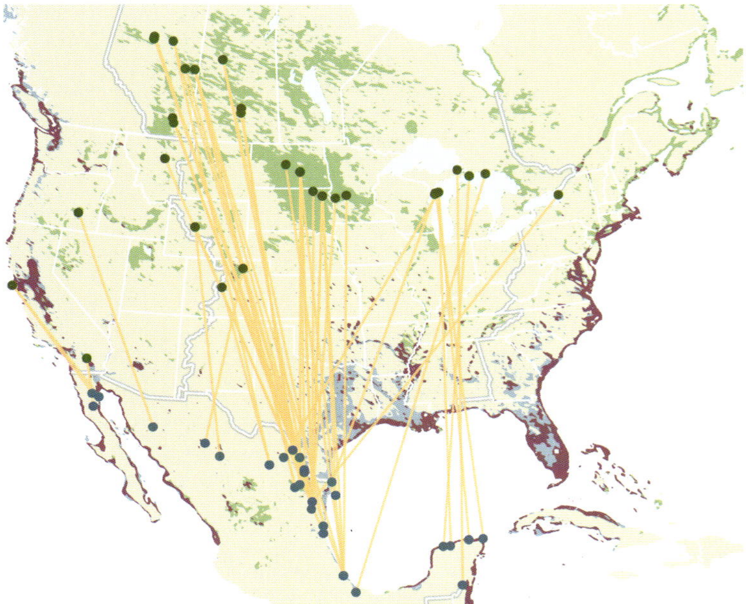

CONNECTIONS C

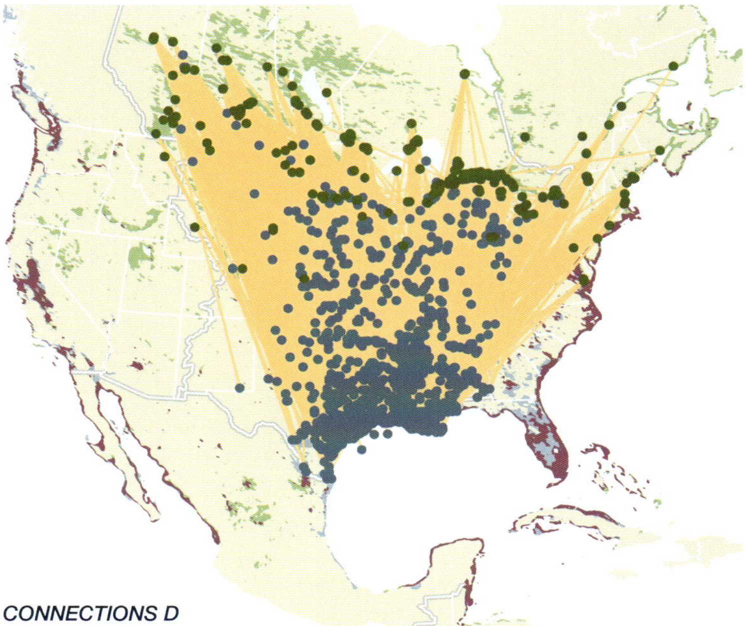

CONNECTIONS D

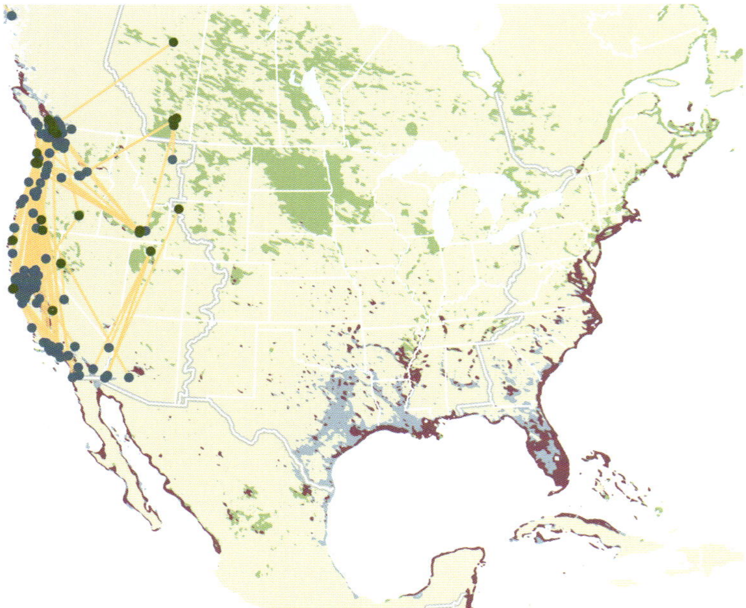

CONNECTIONS E

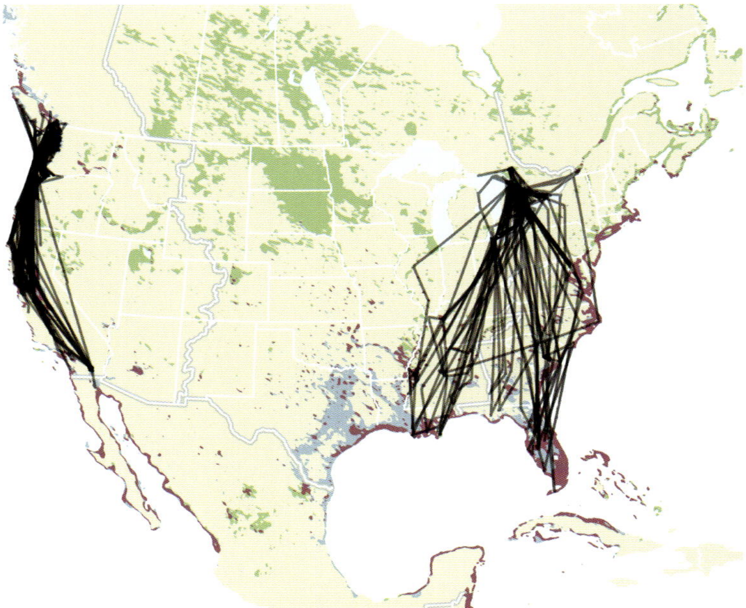

PATHWAYS

individuals breeding near coasts. A banding study of Double-crested Cormorant movements east of the Rocky Mountains found that cormorants nesting from Alberta to the Gulf of St. Lawrence migrated to nonbreeding areas in the southern US between Texas and Florida. Additionally, there was considerable mixing of eastern breeding populations in the nonbreeding range, but little mixing of populations across the western mountains (Dolbeer 1991).

RADIO TRACKING Few tags were deployed through the Motus Wildlife Tracking System, with local detections only.

SATELLITE TRACKING Satellite tracking studies illustrate a divide between eastern and western populations, and the roughly north-south movement of this species' migration. Satellite tracking of 26 birds breeding in Georgian Bay, Ontario, found that most birds spent the nonbreeding period in the southeastern United States, and that nonbreeding habitats were mostly lakes and coasts, with only 26% using aquaculture habitats (Currier et al. 2010). Another satellite tracking study of breeding birds from East Sand Island, Oregon, showed that post-breeding birds mostly stayed west of the Sierra Nevada / Cascade Range and then dispersed widely during the nonbreeding period from British Columbia to northern Mexico. Only two birds traveled east of the mountains (Courtot et al. 2012). Cormorants tracked from the same colony several years later (n = 72) spent the nonbreeding season to the north in Washington or British Columbia (43%), in the Salish Sea region of California (39%), in the vicinity of the Willamette River valley (15%), and in the Columbia River estuary (3%; Peck-Richardson 2017). An additional 52 individuals were satellite tracked from a breeding colony in Lake Ontario, with most (75%) individuals migrating east of the Appalachian Mountains, and with a nonbreeding range extending from southeastern Louisiana, along the coast of the Gulf of Mexico, to the southern portion of coastal Florida (Dorr et al. 2012).

GENETICS An examination of DNA structure of individuals across North America showed strong evidence that Alaskan birds are genetically divergent, and a southwestern lineage had unique haplotypes in Southern California. There was little support for genetic differentiation in the rest of the range, and existing variation was mostly explained by a gradient of distance (Mercer et al. 2013).

MIGRATORY PATHWAYS Inland birds in the northern part of the breeding range primarily travel long distances across the continent to nonbreeding areas along the coasts. Inland travelers tend to avoid mountain ranges, concentrating along the eastern and western seaboards and the central flyway. Tracking of 28 individuals captured in Alabama during the nonbreeding season showed that cormorants migrated east of the Mississippi River and mostly west of the Appalachian Mountains to breeding areas. Cormorants from Arkansas, Louisiana, and Mississippi migrated north along the Mississippi River Valley, the Missouri River Valley, and the Ohio River Valley (King et al. 2013).

DATA CONTRIBUTORS

J.Y. Adkins; J. Bowman; K. Collis; B. Cramer; H. Currier; K. Fischer; D.T. King; R.S. Larsen; D.E. Lyons; A. Peck-Richardson; D.D. Roby; Y. Suzuki; Ontario Ministry of Natural Resources; Oregon State University; Real Time Research; US Geological Survey Oregon Cooperative Fisheries and Wildlife Research Unit.

PELICANS, HERONS, EGRETS, BITTERNS, IBISES, AND SPOONBILLS (PELECANIFORMES)

WRITING LED BY **M. CLAY GREEN, AMY L. SCARPIGNATO,** *AND* **AUTUMN-LYNN HARRISON**

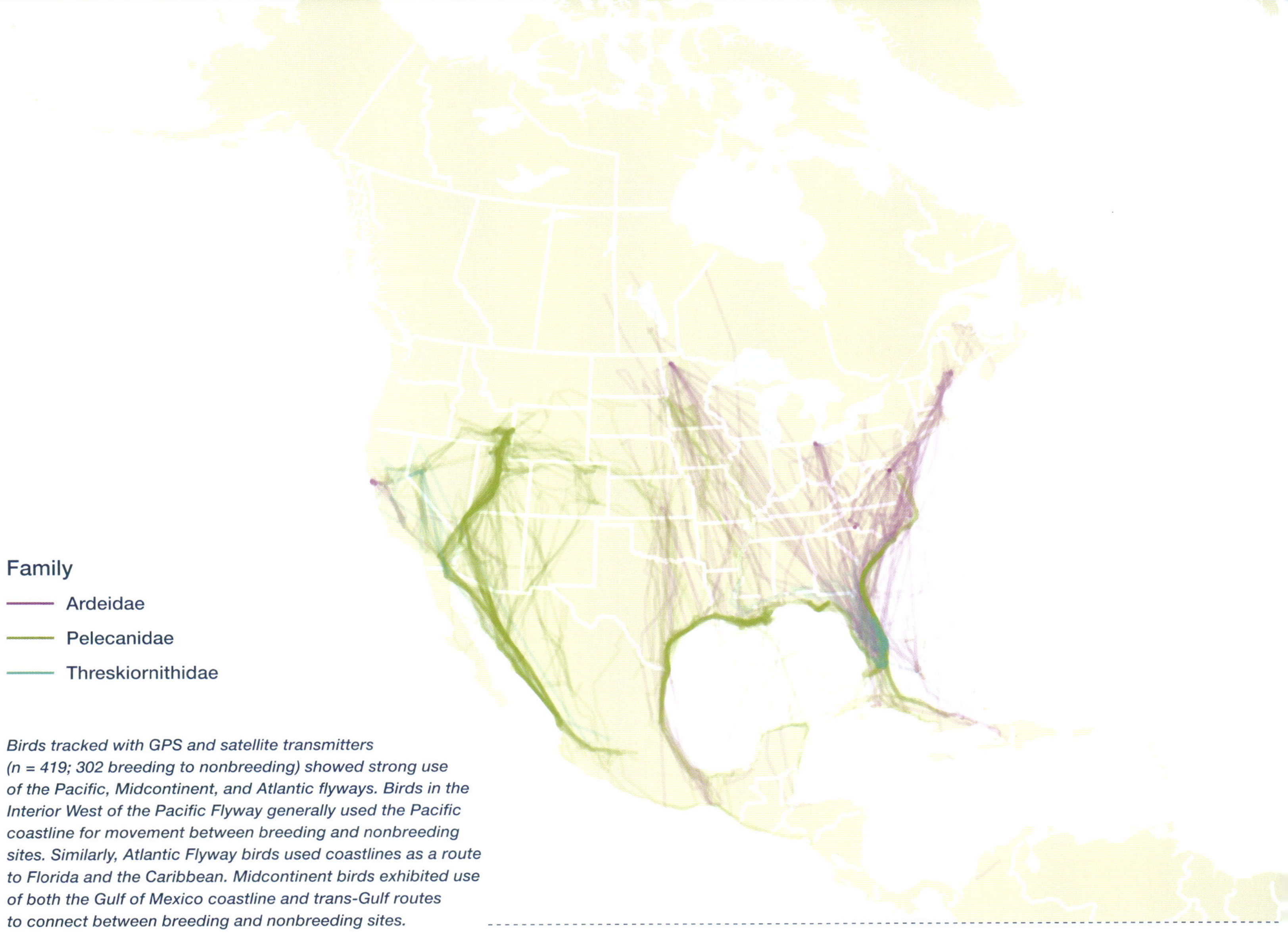

Birds tracked with GPS and satellite transmitters (n = 419; 302 breeding to nonbreeding) showed strong use of the Pacific, Midcontinent, and Atlantic flyways. Birds in the Interior West of the Pacific Flyway generally used the Pacific coastline for movement between breeding and nonbreeding sites. Similarly, Atlantic Flyway birds used coastlines as a route to Florida and the Caribbean. Midcontinent birds exhibited use of both the Gulf of Mexico coastline and trans-Gulf routes to connect between breeding and nonbreeding sites.

The migratory connectivity and associated ecology of the order Pelicaniformes, which includes pelicans, herons and bitterns, and ibises and spoonbills, has been well studied, with migratory connectivity information for 94% of species. However, little is known about some species, such as Green Heron, Least Bittern, and Yellow-crowned Night Heron. Within the order Pelecaniformes, only one species, the Reddish Egret, is a Tipping Point species. Satellite tracking studies of Reddish Egrets demonstrated migratory behavior in western Gulf of Mexico populations, a behavior suspected but not documented until the tracking studies were conducted. Understanding the migratory and nonmigratory populations of Reddish Egrets and their connectivity has been important in conservation actions. Recent assessments of the continental populations of the Little Blue Heron and the American White Pelican, as well as the Caribbean population of Brown Pelicans, show that these species are birds of conservation concern. The Florida subspecies of the Great White Heron also merits attention as a species of concern.

Within the order, pelicans exhibit moderate migratory connectivity with considerably more movement east-west, especially between the Pacific and Midcontinent flyways. The herons, egrets, and bitterns generally move in more traditional north-south directions, with weak to moderate migratory connectivity with some movement between Atlantic and midcontinent populations. Some species (Black-crowned Night Heron and Great Egret) exhibit moderate connectivity, whereas others (like the Tricolored Heron) exhibit strong connectivity. Within the ibises and spoonbills there is a high degree of connectivity variation, with the nomadic White Ibis yielding very weak connectivity, in contrast to Roseate Spoonbills, which exhibit very strong connectivity. Birds that were satellite tracked used coastlines as well as open ocean (e.g., trans-Gulf) routes to move between breeding and nonbreeding sites. Across the nonbreeding range of Pelecaniformes, there is considerable overlap between Pacific and midcontinent populations in Mexico, reflected mostly in the pelicans and to a lesser extent within the herons, egrets, and bitterns.

AMERICAN WHITE PELICAN *Pelecanus erythrorhynchos*

weak — strong

MC = 0.08 ± 0.04*

DISTRIBUTION AND ABUNDANCE

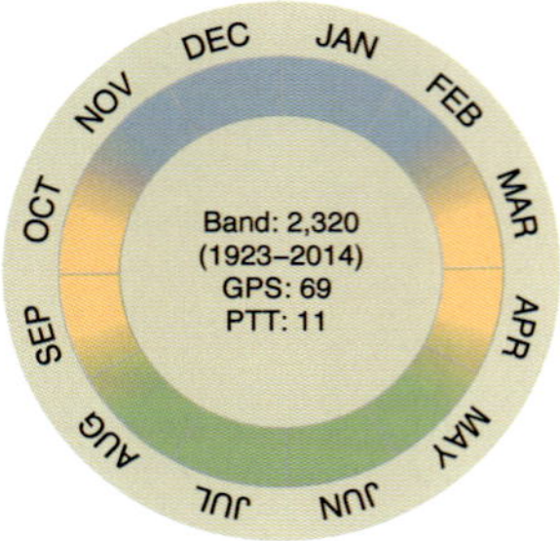

During the breeding season, this wide-ranging species is found as far north as the Northwest Territories, Canada, and as far south as central Mexico (where nonmigratory populations sometimes breed). The highest breeding season abundances are found in in the prairie wetlands extending from the Dakotas north to Saskatchewan. Other notable breeding season populations are found in eastern Wisconsin, the Great Salt Lake, several locations in California, and western Nevada. During the nonbreeding season, this species vacates the northern reaches of its range, but can be found in small numbers lingering as far north as Washington state and central Illinois. Most of the nonbreeding season population concentrates in the southern US, Mexico, and Cuba, with some birds found as far south as Nicaragua. Areas of particular nonbreeding season abundance include the coast of the Gulf of Mexico from the Yucatán to Mobile Bay, peninsular Florida, the Salton Sea, and the lower Colorado River, as well as the Pacific coast and central wetlands of Mexico (Knopf and Evans 2020; Fink et al. 2023).

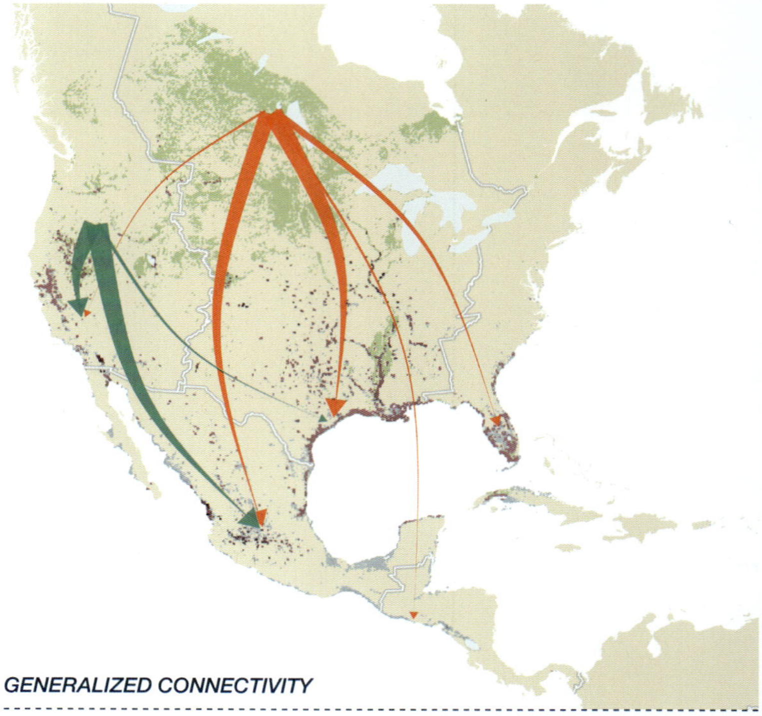

GENERALIZED CONNECTIVITY

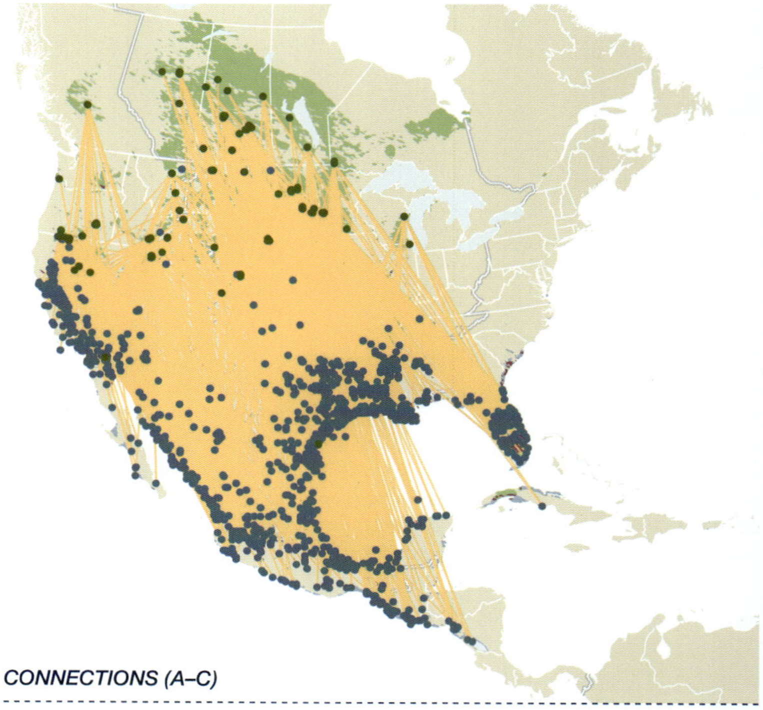

CONNECTIONS (A–C)

ORDER Pelecaniformes
FAMILY Pelecanidae
BREEDING HABITAT Wetlands, Lakes and Rivers
NONBREEDING HABITAT Coasts, Salt Marsh
GLOBAL POPULATION SIZE 450,000
POPULATION CHANGE SINCE 1970 +1315%
IUCN RED LIST CATEGORY Least Concern
R2R TIPPING POINT SPECIES No
US LISTING STATUS No
CANADA SPECIES AT RISK No
PIF CCS 10/20

	Atlantic	Caribbean	Central America	Mexico	Midcontinent	Pacific
Midcontinent (n = 1473)	13%	<1%	3%	37%	45%	1%
Pacific (n = 808)	<1%	–	<1%	50%	4%	45%

MIGRATORY CONNECTIVITY

OVERALL American White Pelican migratory connectivity has been extensively studied with banding, GPS, satellite, stable isotope, and genetic data. Birds from breeding colonies in the Midcontinent Flyway migrate to nonbreeding areas within the Midcontinent Flyway (45%), as well as Mexico (37%) and the Atlantic Flyway (13%), and smaller numbers to other nonbreeding regions. Birds from breeding colonies in the Pacific Flyway spend the nonbreeding season in Mexico (50%) and the Pacific Flyway (45%), with smaller numbers in three other nonbreeding regions. Mixing of individuals throughout nonbreeding regions results in weak migratory connectivity (MC = 0.08 ± 0.04). Stable isotopes and genetics also show mixing of populations through high gene flow among geographically distinct populations (Oomen et al. 2011).

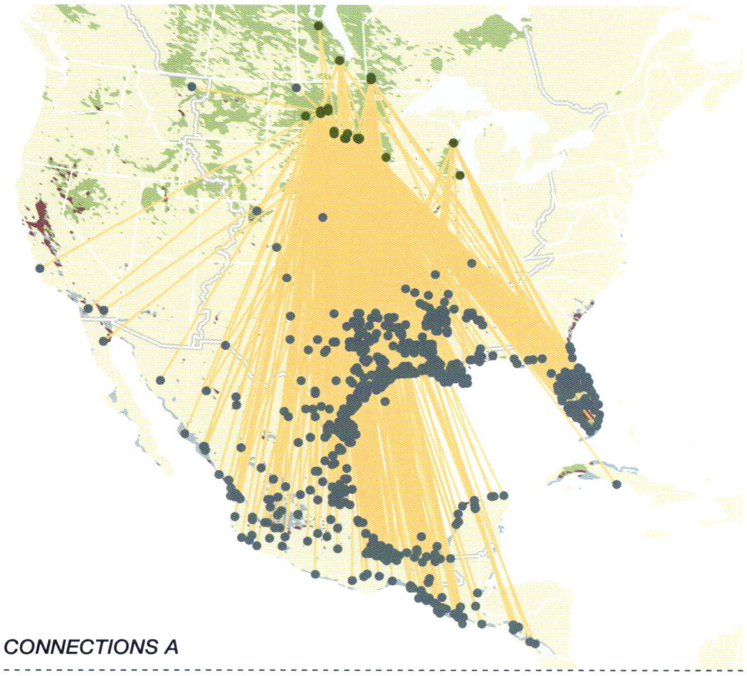

CONNECTIONS A

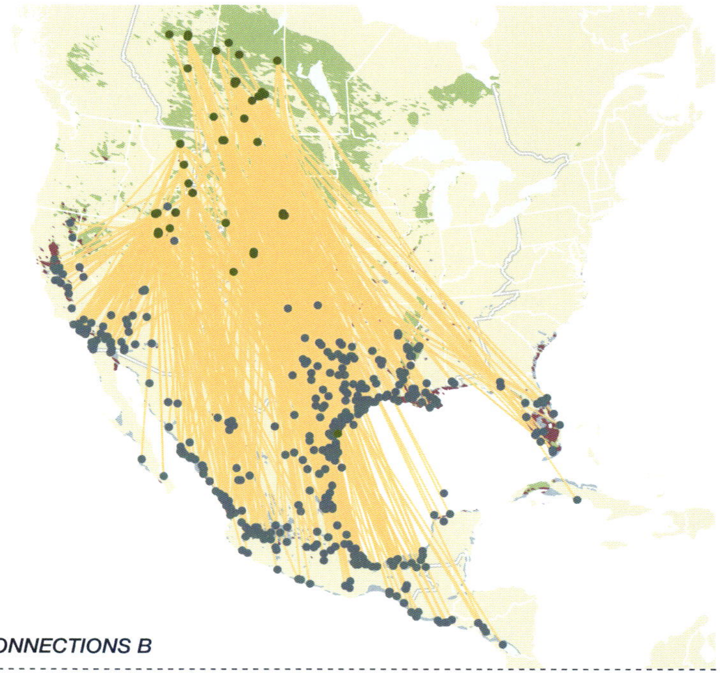

CONNECTIONS B

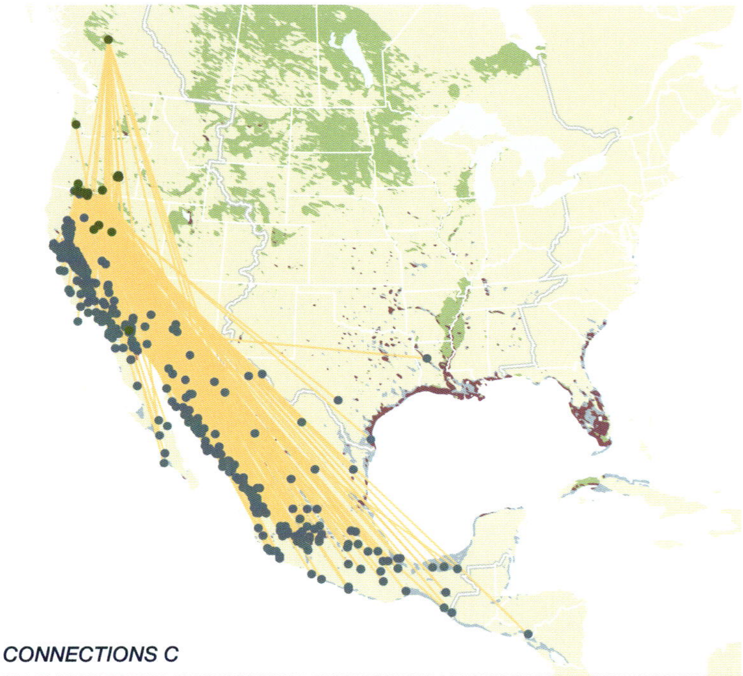

CONNECTIONS C

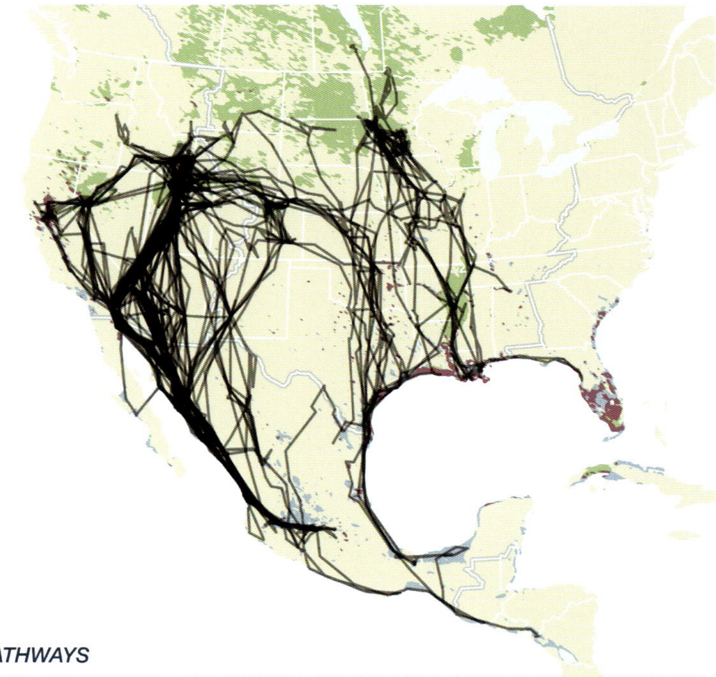

PATHWAYS

BANDING Band reencounters occurred over two main breeding to nonbreeding connections. One connection was between breeding colonies in California, Oregon, and Nevada and nonbreeding locations on the Gulf of California and the Pacific coast of Mexico. Another connection was between breeding colonies in central Canada, the Dakotas, Colorado, Minnesota, and Montana and nonbreeding sites along the Gulf of Mexico, Florida, and Central America (Anderson and Anderson 2005). Recoveries of individuals banded at Marsh Lake in Minnesota between 1972 and 1998 (n = 687) were all in the region of the Mississippi Flyway between the Prairie Pothole Region and Central America (King and Grewe 2001).

SATELLITE TRACKING Pelicans breeding in South Dakota and North Dakota (Sovada et al. 2013) and Minnesota (Audubon Minnesota, unpublished data) were tracked with satellite and GPS transmitters to nonbreeding grounds surrounding the Gulf of Mexico, including Mexico, Texas, Louisiana, Mississippi, and Florida. Satellite transmitters deployed at a breeding colony at Medicine Lake, Montana (n = 5), also spent the nonbreeding season along the Gulf of Mexico (Restani and Madden 2005). Individuals on nonbreeding areas in the Gulf of Mexico and Lower Mississippi Valley were tracked with GPS transmitters (n = 36), 12 of which were resident and 24 of which migrated to breeding grounds in the northern Great Plains (King et al. 2016, 2017). Pelicans breeding in the western US (Nevada, Utah, and Idaho) were tracked to nonbreeding grounds primarily along the west coast of Mexico (Earthspan, Idaho Department of Fish and Game, Utah Division of Wildlife Resources, unpublished data).

STABLE ISOTOPES Stable isotopes sampled from breeding colonies across the North American range of American White Pelicans showed little geographic clustering, suggesting that dispersal has facilitated gene flow of the species across its range (Reudink et al. 2016).

GENETICS Analyses from pelican nestlings (n = 367) using nuclear and mitochondrial DNA reveal low natal philopatry and high levels of gene flow across the North American breeding populations (Oomen et al. 2011; Reudink et al. 2011).

MIGRATORY PATHWAYS Across the breeding regions, individuals exhibit fairly north-south movements along direct corridors between breeding and nonbreeding sites. Coastlines (Pacific and Gulf of Mexico), as well as major river systems, are likely followed by migratory individuals. The Continental Divide does not likely serve as a barrier as breeding individuals from Montana migrate to the Pacific coast and genetic studies support a North American panmictic population.

DATA CONTRIBUTORS
K. Hall; C. Moulton; R. Norvell; C.S. Rushing; A. Van Tatenhove; M. Yates; Audubon Minnesota; Department of Defense Legacy Resource Management Program; Earthspan; Great Salt Lake Institute at Westminster College; Idaho Department of Fish and Game; Minnesota Department of Natural Resources; North Dakota State University; Salt Lake City International Airport; Tracy Aviary Conservation Science; Utah Division of Wildlife Resources.

BROWN PELICAN *Pelecanus occidentalis*

weak strong
MC = 0.29 ± 0.05*

DISTRIBUTION AND ABUNDANCE

The Brown Pelican is a coastal seabird that breeds primarily on islands along the Pacific coast from Southern California to Ecuador, the Atlantic coast from Maryland south to Venezuela, and along the coast of the Gulf of Mexico, where relative abundance is highest. The Brown Pelican nonbreeding distribution overlaps substantially with its nesting distribution due to many populations being partially migratory (King et al. 2013a; Poli 2015; Lamb et al. 2017; Wilkinson and Jodice 2023). After breeding, but before migration, they may range as far north as Canada in both the Atlantic and Pacific flyways. Its Atlantic nesting range expanded north in 1929 to North Carolina and in 1987 to Maryland and Virginia, while its Pacific range has a long history of expansion and contraction in synchrony with ocean temperatures (Shields 2020). It forages in shallow waters and is limited in offshore movements by the need for dry roosting sites along coastal marine and estuarine habitats (Shields 2020).

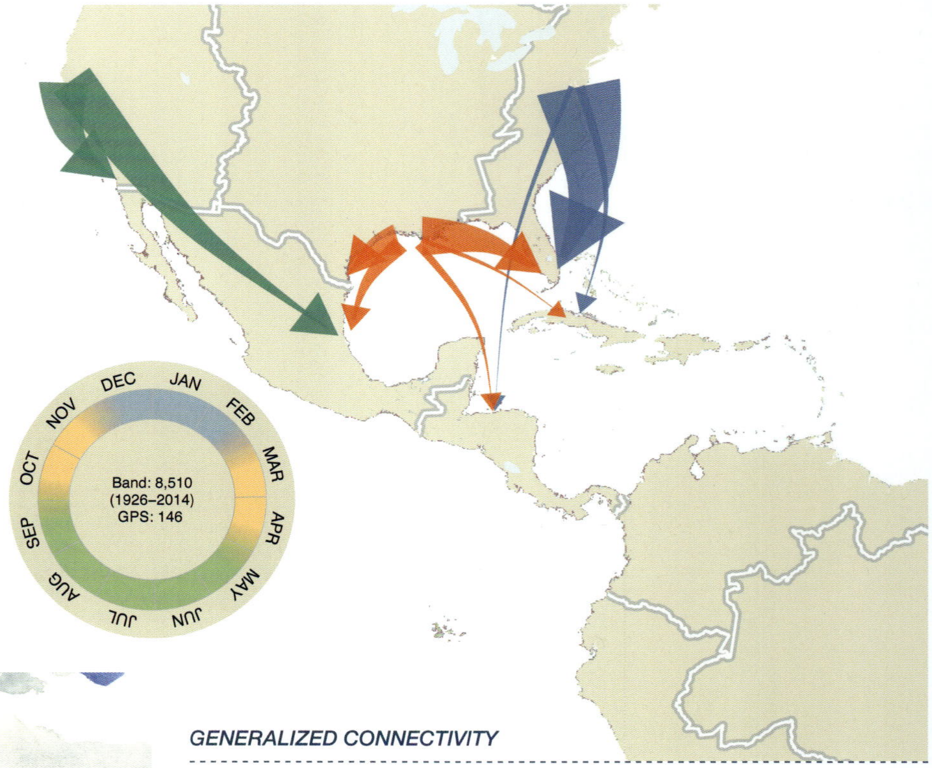

GENERALIZED CONNECTIVITY

ORDER Pelecaniformes
FAMILY Pelecanidae
BREEDING HABITAT Coasts, Beach and Estuary
NONBREEDING HABITAT Coasts, Beach and Estuary
GLOBAL POPULATION SIZE 370,000
POPULATION CHANGE SINCE 1970 +216%
IUCN RED LIST CATEGORY Least Concern
R2R TIPPING POINT SPECIES No
US LISTING STATUS No
CANADA SPECIES AT RISK No
PIF CCS 11/20

MIGRATORY CONNECTIVITY

OVERALL The migratory connectivity of the Brown Pelican is well known from satellite transmitter and banding data. Birds breeding in the Atlantic Flyway primarily migrate to nonbreeding areas within the Atlantic Flyway (93%). Birds breeding in the Midcontinent Flyway (Gulf of Mexico) spend the nonbreeding season within the Atlantic (43%) and Midcontinent (36%) flyways as well as in Mexico (17%), with substantially fewer individuals in other regions. Most birds breeding in the Pacific Flyway migrate to nonbreeding areas within the flyway (55%) or Mexico (45%). Within-flyway movement is high, but the abundant movement to other nonbreeding regions results in a moderate migratory connectivity (MC = 0.29 ± 0.05).

BANDING Banded Brown Pelicans are heavily resighted (Stefan 2008; Schreiber 1976) throughout the range due to their large body size and distributional overlap with humans. Band reencounters show similar connectivity patterns to tracking data.

	Atlantic	Caribbean	Central America	Mexico	Midcontinent	Pacific
Atlantic (n = 7502)	93%	6%	1%	<1%	<1%	–
Midcontinent (n = 453)	43%	1%	4%	17%	36%	–
Pacific (n = 104)	–	–	–	45%	–	55%

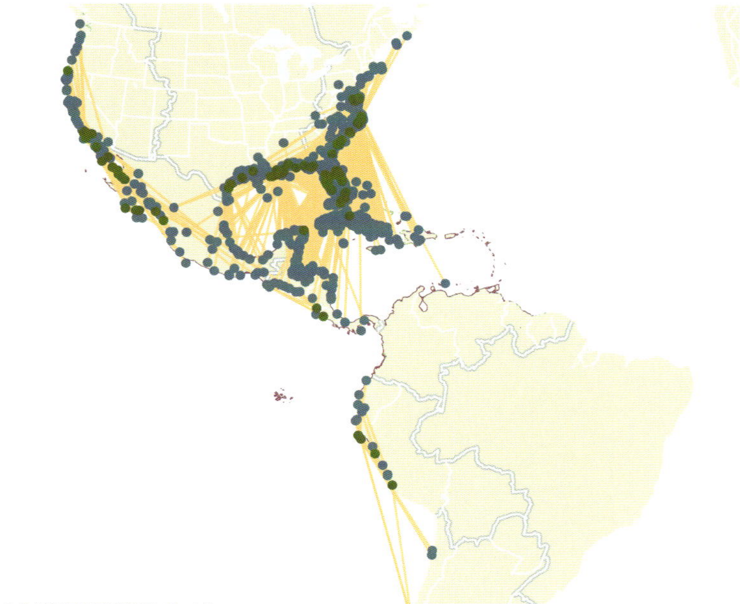

CONNECTIONS (A–H)

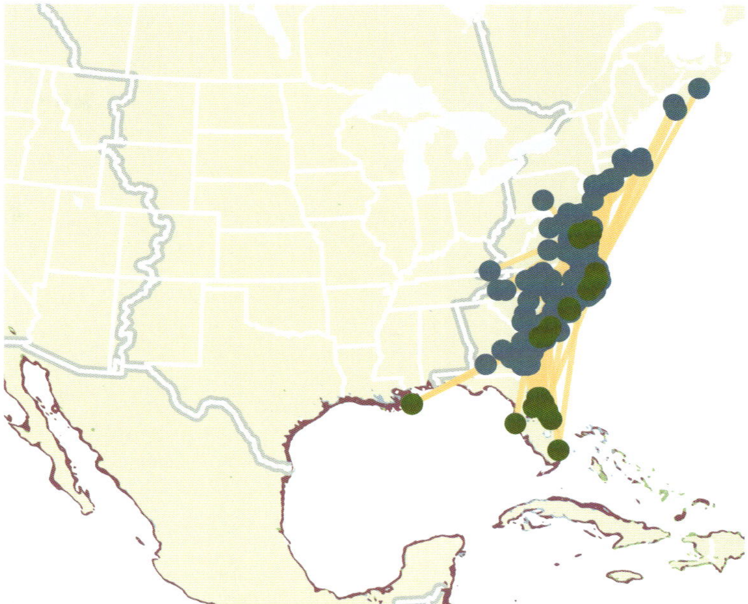

CONNECTIONS A

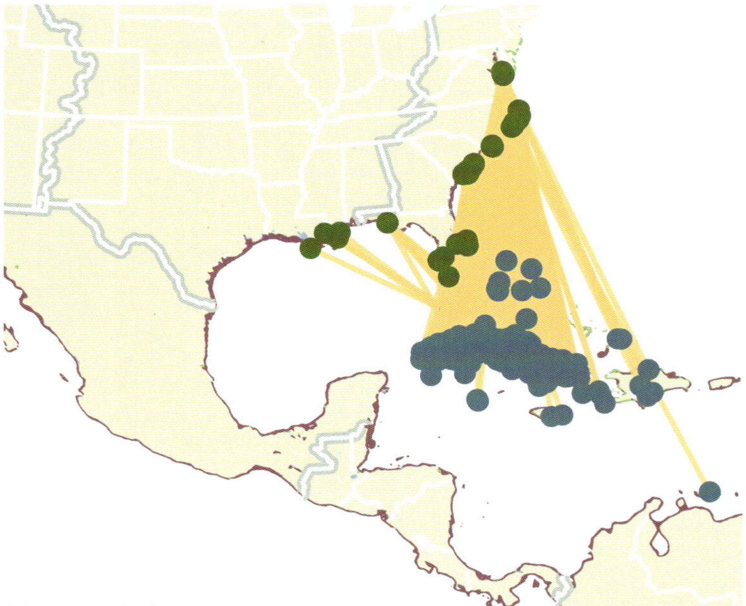

CONNECTIONS B

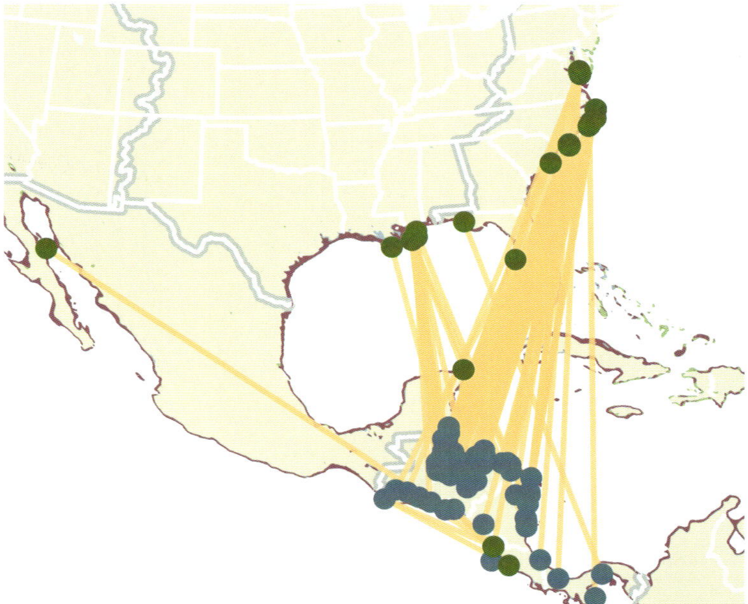

CONNECTIONS C

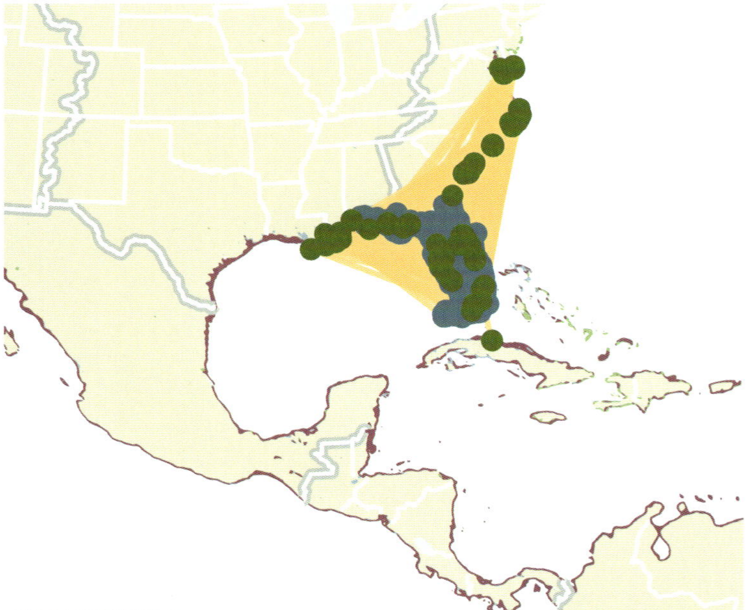

CONNECTIONS D

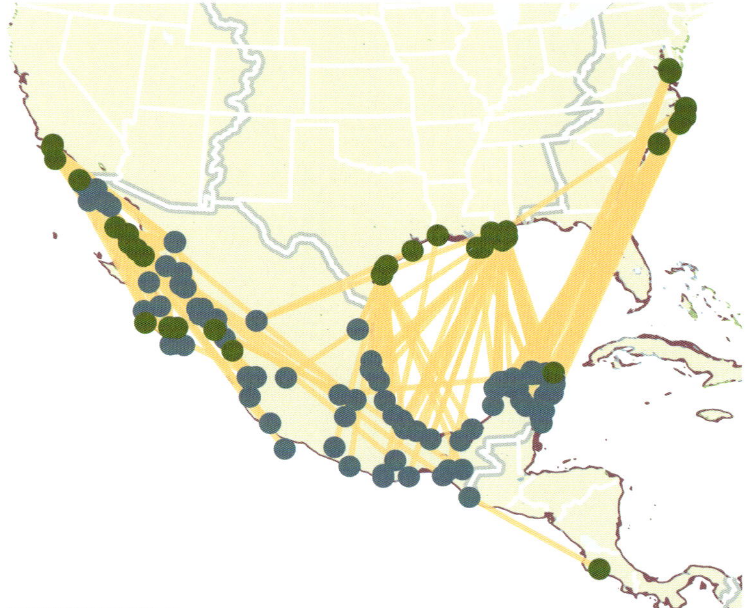

CONNECTIONS E

PELICANS, HERONS, EGRETS, BITTERNS, IBISES, AND SPOONBILLS (PELECANIFORMES)

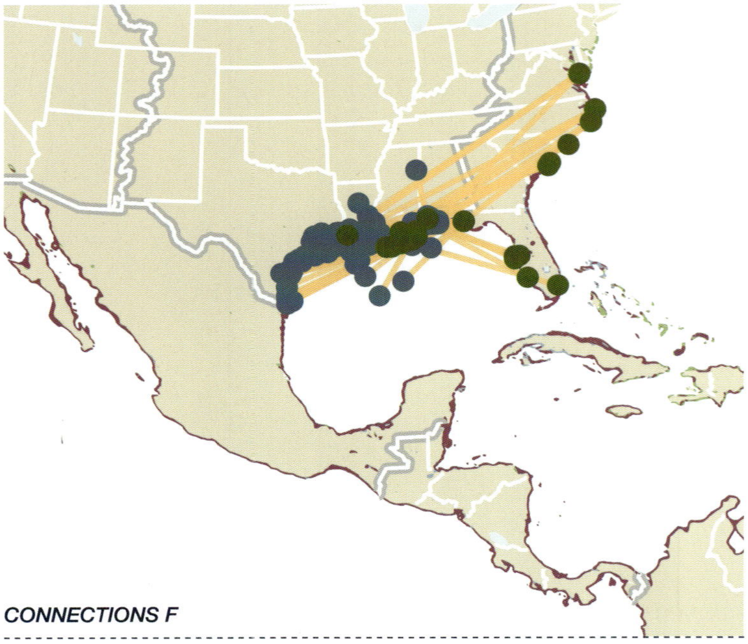

CONNECTIONS F

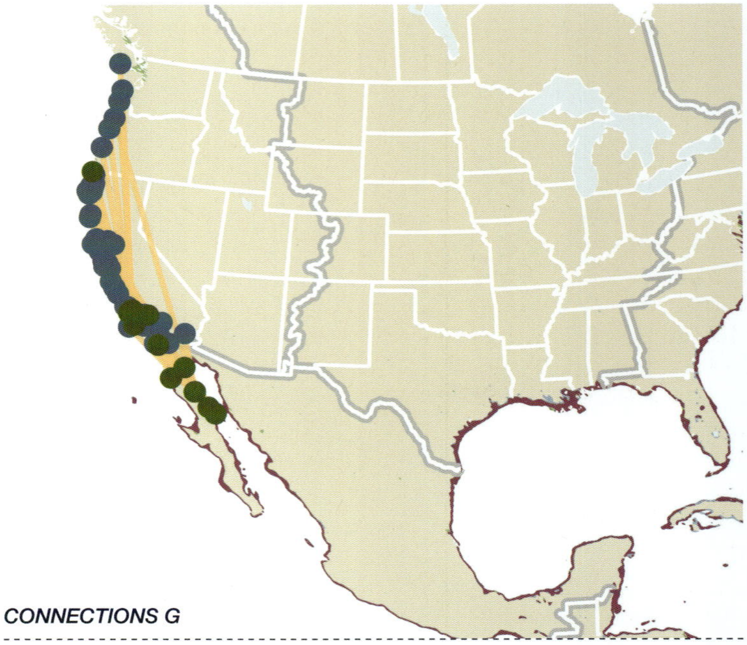

CONNECTIONS G

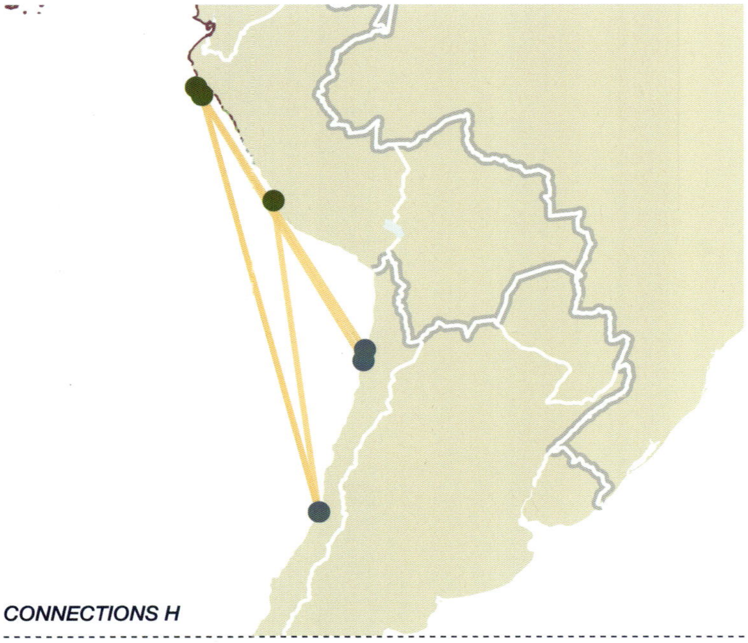

CONNECTIONS H

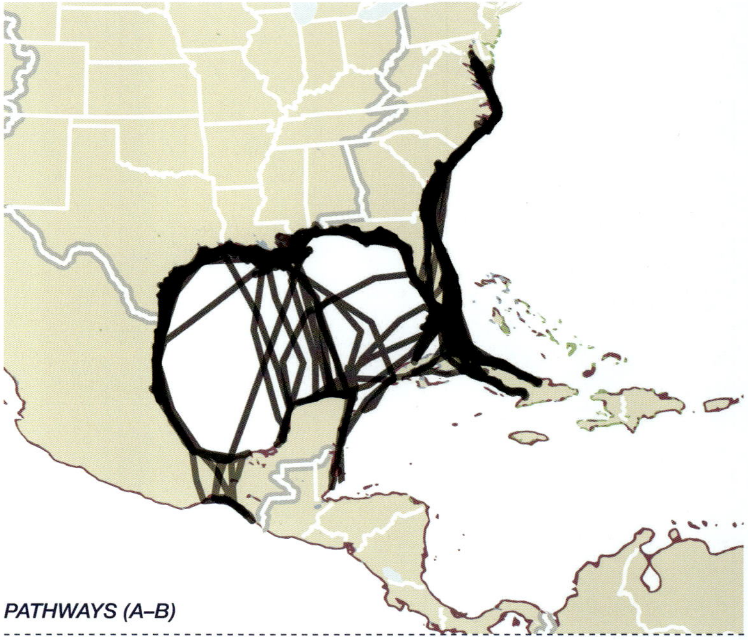

PATHWAYS (A–B)

PATHWAYS A

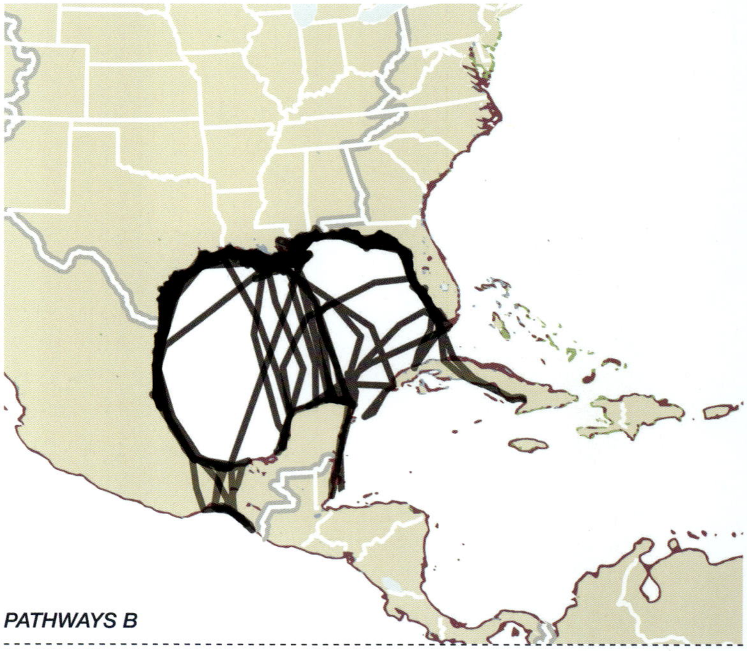

PATHWAYS B

MIGRATORY CONNECTIVITY

OVERALL Migratory connectivity data is mostly available from band reencounter data, which show north-south movement with no overlap between breeding populations, resulting in strong migratory connectivity (MC = 0.64). Atlantic Flyway breeding birds migrate to nonbreeding areas in the Caribbean (72%), within the Atlantic Flyway (17%), and to Central America (10%). Midcontinent Flyway breeding birds migrate to nonbreeding areas in Mexico (67%) or within the Midcontinent Flyway (33%).
BANDING Important nonbreeding areas identified from banding data include the Atlantic coast of the United States, the coast of the Gulf of Mexico, the Caribbean, and Central America (Mikuska et al. 1998).
OTHER APPROACHES No tracking, genetic, or stable isotope studies of migratory connectivity were found for this species.
MIGRATORY PATHWAYS Banding recoveries across the range indicate mostly north-south movement (Melvin et al. 1999). Midcontinent birds likely travel along the Gulf of Mexico coast to and from nonbreeding areas, but the apparent arrival of trans-Gulf migrants has been documented in the pre-breeding season (Frederick 2020). With the Atlantic populations, birds likely follow the coastline to nonbreeding areas, but some individuals may migrate over open ocean, as band recoveries and the regular appearance of migrants occur in Bermuda and there have been offshore sightings between Florida and Cuba (Frederick 2020).

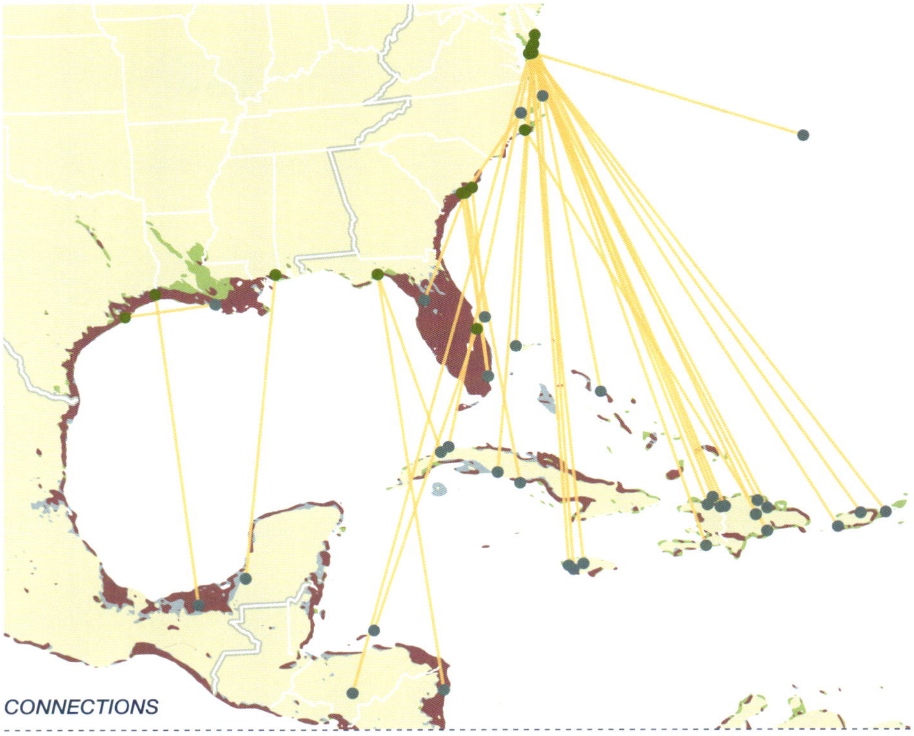

Egretta rufescens **REDDISH EGRET**

DISTRIBUTION AND ABUNDANCE

The Reddish Egret breeding range extends along the Gulf of Mexico, the west coast of Florida, the Pacific coast of western and southern Mexico, and the Caribbean, most densely in Louisiana, Texas, Tamaulipas, coastal Mexico, Chiapas, the Bahamas, and northern Cuba (Fink et al. 2023). The nonbreeding range overlaps with the breeding range and also extends from Florida to Texas and south to Guatemala and El Salvador. The Mexican Gulf Coast was identified as an important nonbreeding area from banding data (Mikuska et al. 1998). The species is partially migratory and only appears to be migratory in the northernmost portion (i.e., Texas and Louisiana) of the breeding range (Koczur et al. 2020).

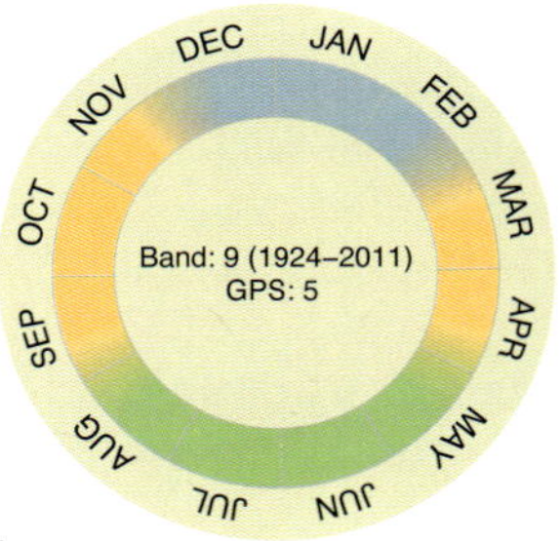

MIGRATORY CONNECTIVITY

OVERALL Migratory connectivity is relatively unknown. While most populations are nonmigratory, some birds breeding in the Midcontinent Flyway (i.e., Texas and Louisiana) migrate to nonbreeding areas in Mexico and Central America. MC was not calculated due to limited sample size.
BANDING Resightings of color-banded birds from Baja California Sur have been documented north to southwestern California and Arizona and south to Chiapas and El Salvador (Koczur et al. 2020). Resightings of color-banded birds from the Yucatán Peninsula have been documented in Texas and Florida.
SATELLITE TRACKING Reddish Egrets breeding in Texas (n = 16) and Florida (n = 5) stayed resident throughout the nonbreeding period; whereas Reddish Egrets breeding in Texas (n = 11) migrated to

ORDER Pelecaniformes
FAMILY Ardeidae
BREEDING HABITAT Coasts, Mangroves
NONBREEDING HABITAT Coasts, Beach and Estuary
GLOBAL POPULATION SIZE 15,000
POPULATION CHANGE SINCE 1970 +78%
IUCN RED LIST CATEGORY Near Threatened
R2R TIPPING POINT SPECIES Yellow-Alert
US LISTING STATUS No
CANADA SPECIES AT RISK No
PIF CCS 15/20

Mexico and El Salvador (Koczur et al. 2018). Individuals breeding in Texas (n = 5) were also found to migrate to nonbreeding areas on the Pacific coast of Mexico (Lamb et al. 2018b).
GENETICS Populations of Reddish Egrets seem to fall into four genetically distinct regions: Baja California, Chiapas/Oaxaca, Texas/Louisiana / Mexico Gulf Coast, and Bahamas/Florida, with limited dispersal between regions. Maternally inherited mitochondrial DNA analyses suggest sex-biased dispersal, with males dispersing greater distances than females (Shahrokhi et al. 2020).
OTHER APPROACHES No stable isotope studies of migratory connectivity were found for this species.
MIGRATORY PATHWAYS Along the northwestern Gulf of Mexico, individuals migrate southward following the coastline to Mexico and Central America, with some individuals making trans-Gulf crossings from Texas to the Yucatán Peninsula and some crossing the Isthmus of Tehuantepec to the Pacific coast of Mexico (Lamb et al. 2018b). The Laguna Madre of Tamaulipas and La Laguna of San Andres are important stopover sites for birds migrating from Texas (Koczur et al. 2020).

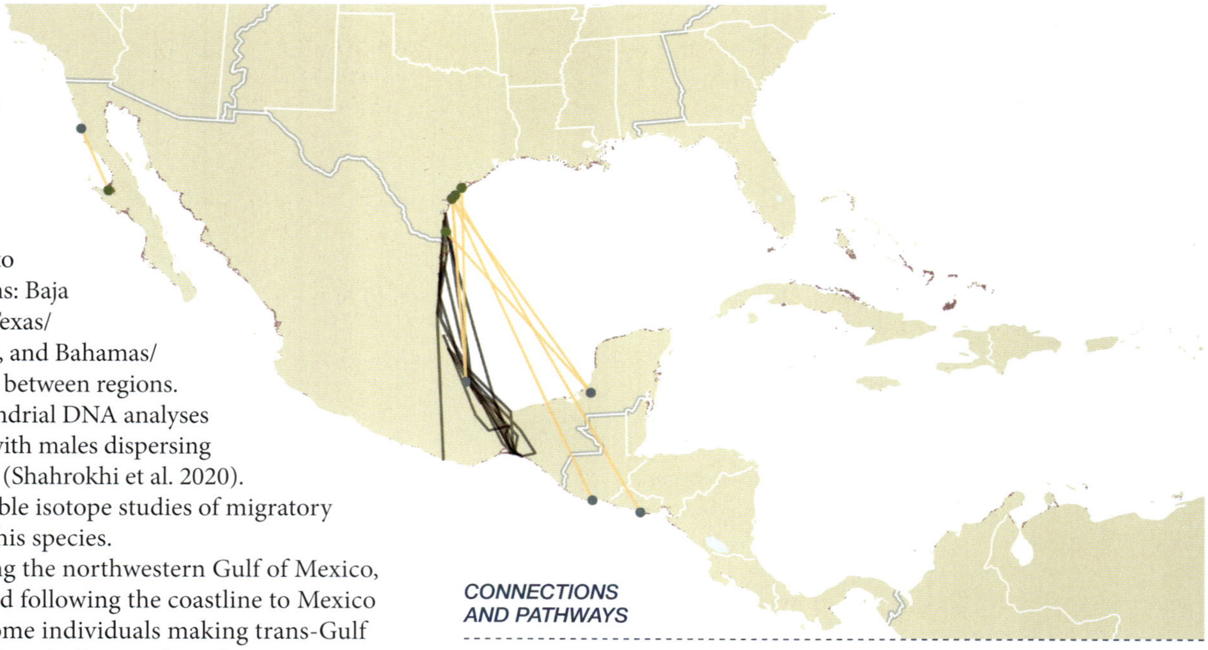

CONNECTIONS AND PATHWAYS

DATA CONTRIBUTORS
J.S. Lamb.

SNOWY EGRET *Egretta thula*

MC = 0.14 ± 0.03*

ORDER Pelecaniformes
FAMILY Ardeidae
BREEDING HABITAT Wetlands, Generalist
NONBREEDING HABITAT Wetlands, Generalist
GLOBAL POPULATION SIZE 2,100,000
POPULATION CHANGE SINCE 1970 +167%
IUCN RED LIST CATEGORY Least Concern
R2R TIPPING POINT SPECIES No
US LISTING STATUS No
CANADA SPECIES AT RISK No
PIF CCS 7/20

DISTRIBUTION AND ABUNDANCE

The Snowy Egret breeds in the southeastern US states, with some scattered breeding to southern New England states; the south, central, and western US; both coasts of Mexico; and the Caribbean. It is a partial migrant, and coastal populations, including much of South America, are present year-round, whereas interior and northeastern populations in the US are completely migratory. The nonbreeding range overlaps much of the breeding range, and important areas include the Atlantic coast, Mexico, the Caribbean, the Gulf Coast, and Central America (Parsons and Master 2020).

MIGRATORY CONNECTIVITY

OVERALL Migratory connectivity data for Snowy Egrets come primarily from band reencounter data. Atlantic Flyway breeding populations primarily spend the nonbreeding season in the Caribbean (75%). Midcontinent Flyway breeding populations largely spend the nonbreeding season in Mexico (69%) and Central America (27%). Pacific Flyway breeding populations migrate to Mexico (78%), within the Pacific Flyway (17%), and to Central America (4%). Breeding to nonbreeding movement occurs mostly in north-south corridors; however, movement is usually to nonbreeding regions outside of the breeding region. This leads to overlap of breeding populations on nonbreeding areas, especially Midcontinent and Pacific flyway populations in Mexico, and thus weak migratory connectivity (MC = 0.14 ± 0.03).
BANDING Important nonbreeding areas identified from banding data include the Pacific coast of California, Mexico and Central America, the Gulf of Mexico, the Caribbean, and the Atlantic coast (Mikuska et al. 1998). Band reencounter data show movement along several north-south corridors across its range in North America, with little movement between Atlantic and Pacific populations (Melvin et al. 1999).
OTHER APPROACHES No tracking, genetic, or isotope studies of migratory connectivity were found for this species.
MIGRATORY PATHWAYS Although specific pathways are unknown, Pacific coast populations likely travel along the Pacific coast into Mexico and Central America. Within western and midcontinent populations, birds exhibit movement along north-south corridors including the Mississippi River Valley into Mexico and Central America, but it is unknown if birds follow coastlines or make trans-Gulf crossings. Atlantic coast populations follow the coastline to peninsular Florida, the Caribbean, and across to Central and South America. Northward migrations are most likely the reverse of southward migrations.

	Atlantic	Caribbean	Central America	Mexico	Midcontinent	Northern SA	Pacific
Atlantic (n = 40)	10%	75%	5%	–	2%	8%	–
Midcontinent (n = 26)	–	4%	27%	69%	–	–	–
Pacific (n = 23)	–	–	4%	78%	–	–	17%

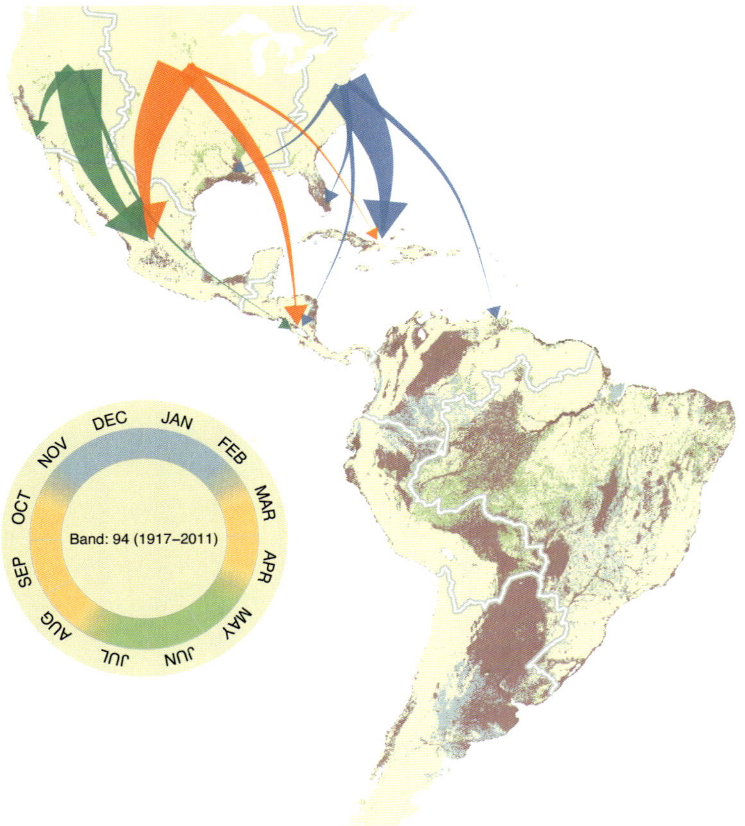

GENERALIZED CONNECTIVITY

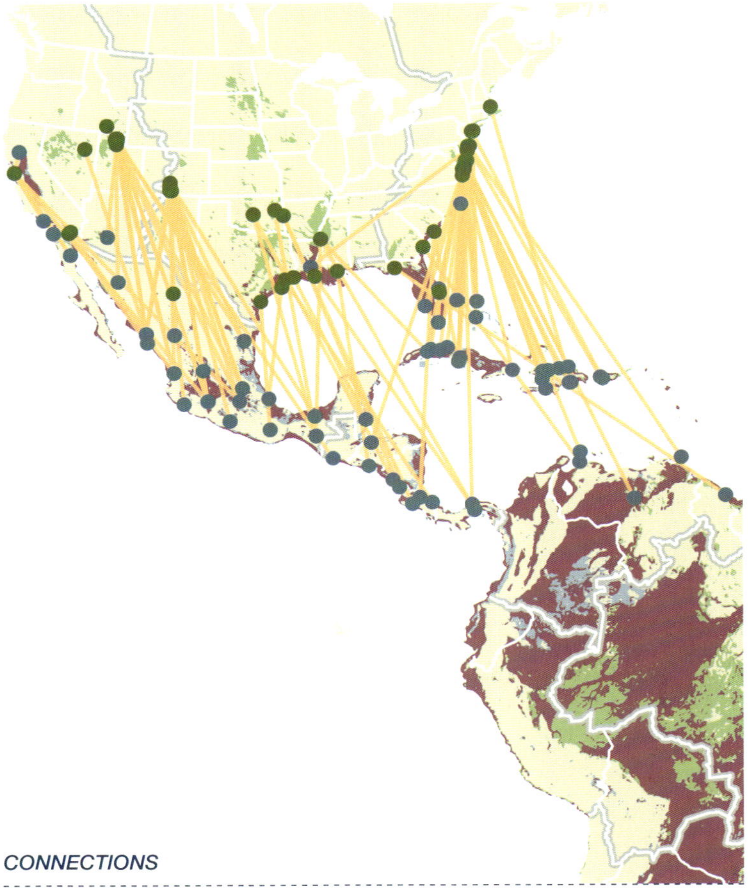

CONNECTIONS

Nyctanassa violacea **YELLOW-CROWNED NIGHT HERON**

DISTRIBUTION AND ABUNDANCE

The breeding range includes the Atlantic and Gulf coasts of the United States; inland to Missouri, Kentucky, and Illinois; and the Pacific coast of Mexico. It is most densely populated in Texas, Louisiana, the Mississippi River Valley, and Florida. It spends the nonbreeding season in Central America, Caribbean, and parts of the US Atlantic coast where crustaceans are available (Watts 2020).

ORDER Pelecaniformes
FAMILY Ardeidae
BREEDING HABITAT Wetlands, Generalist
NONBREEDING HABITAT Coasts, Salt Marsh
GLOBAL POPULATION SIZE 400,000
POPULATION CHANGE SINCE 1970 −17%
IUCN RED LIST CATEGORY Least Concern
R2R TIPPING POINT SPECIES No
US LISTING STATUS No
CANADA SPECIES§ AT RISK No
PIF CCS 12/20

MIGRATORY CONNECTIVITY

OVERALL Very little data is available on the migratory connectivity of the Yellow-crowned Night Heron.

BANDING There is one band recovery between an inland breeding site in the southeastern US and a nonbreeding site on the Gulf Coast.

OTHER APPROACHES No tracking, genetic, or stable isotope studies of migratory connectivity were found for this species.

MIGRATORY PATHWAYS While very little data exists on migratory routes, coastal Atlantic populations likely move in a north-south direction following coastlines, although open-ocean movement occurs as migrant individuals have been documented in Bermuda. Within midcontinent populations, individuals likely follow coastlines along the Gulf of Mexico and Caribbean, with some individuals conducting trans-Gulf crossings to the Caribbean and South America.

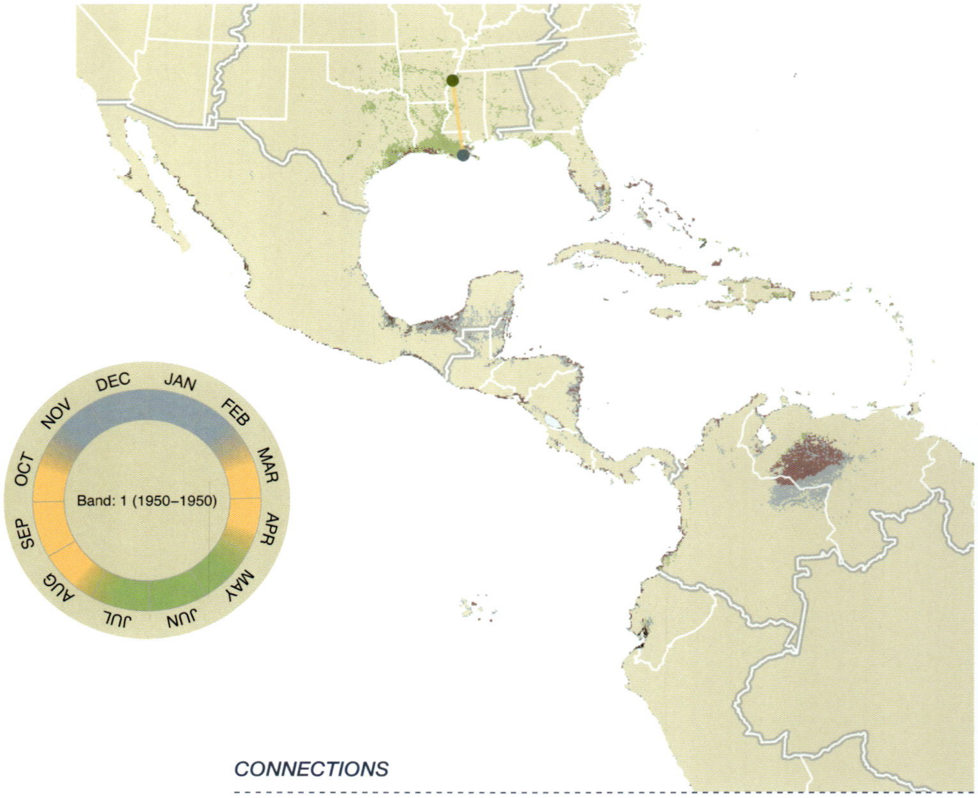

BLACK-CROWNED NIGHT HERON *Nycticorax nycticorax*

DISTRIBUTION AND ABUNDANCE

The Black-crowned Night Heron is a widespread and fairly common heron from the Pacific Northwest east to New Brunswick and southward into Mexico, the Caribbean, and Central and South America. Highest nonbreeding densities are in the southern US, including the Atlantic and Gulf of Mexico coasts, as well as Mexico and the Caribbean (Fink et al. 2023).

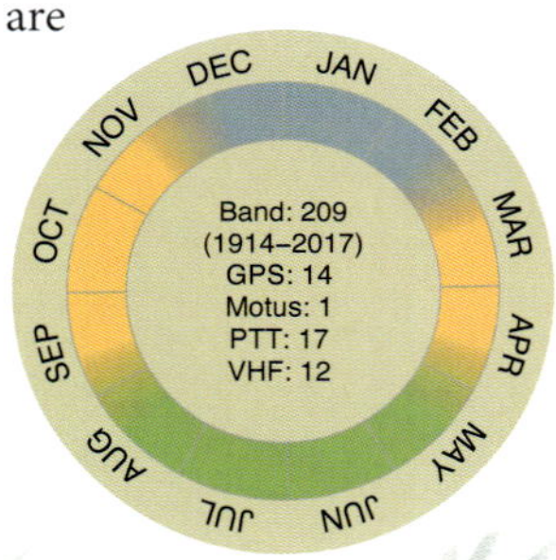

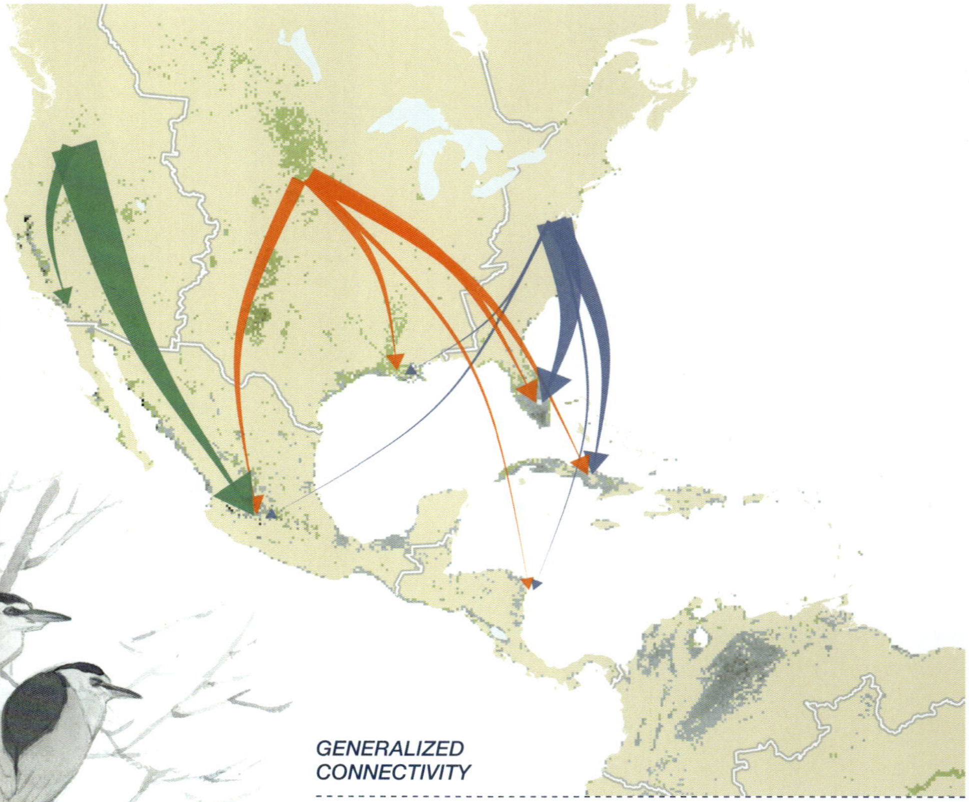

ORDER Pelecaniformes
FAMILY Ardeidae
BREEDING HABITAT Wetlands, Generalist
NONBREEDING HABITAT Wetlands, Generalist
GLOBAL POPULATION SIZE 3,000,000
POPULATION CHANGE SINCE 1970 −17%

IUCN RED LIST CATEGORY Least Concern
R2R TIPPING POINT SPECIES No
US LISTING STATUS No
CANADA SPECIES AT RISK No
PIF CCS 10/20

	Atlantic	Caribbean	Central America	Mexico	Midcontinent	Pacific
Atlantic (n = 86)	60%	34%	2%	1%	2%	–
Midcontinent (n = 118)	31%	24%	6%	22%	17%	–
Pacific (n = 34)	–	–	–	79%	–	21%

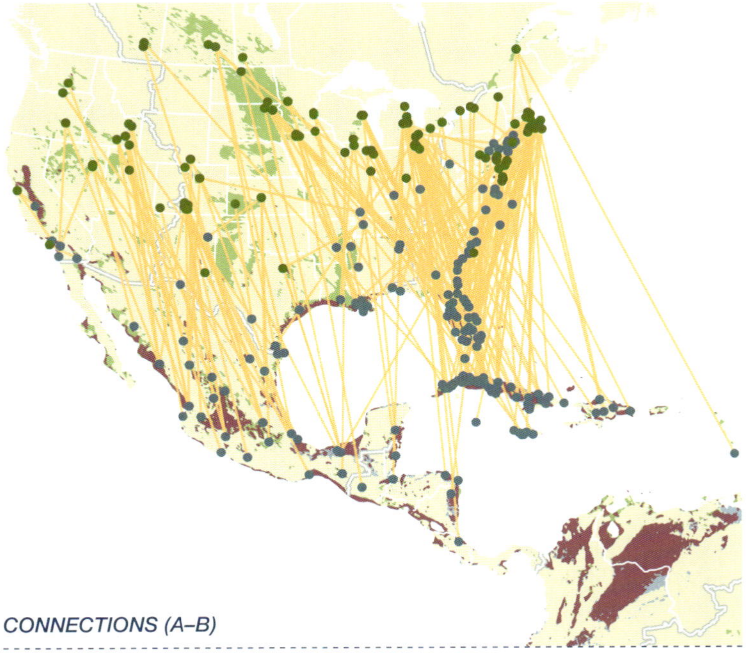

CONNECTIONS (A–B)

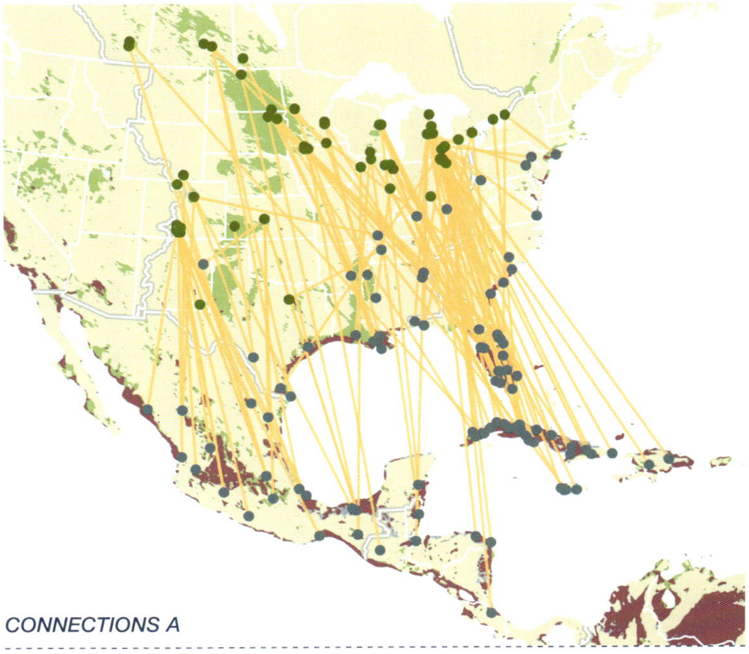

CONNECTIONS A

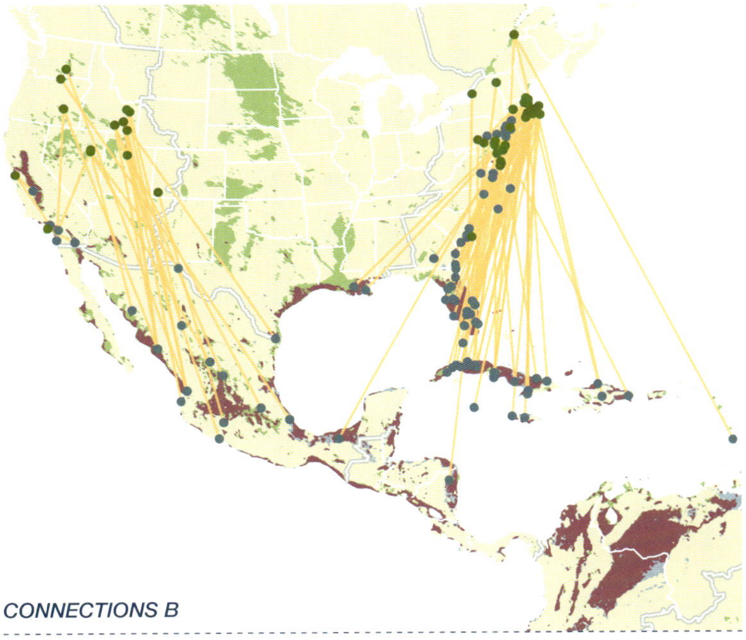

CONNECTIONS B

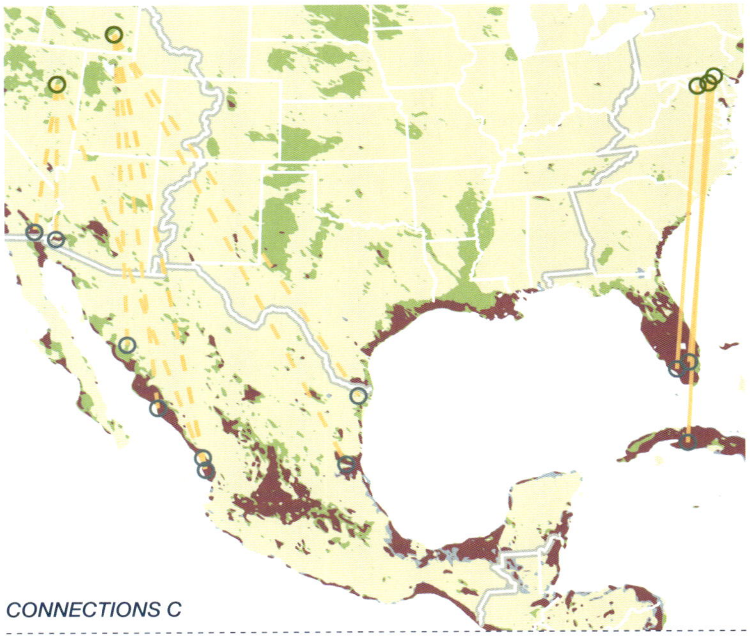

CONNECTIONS C

MIGRATORY CONNECTIVITY

OVERALL Migratory connectivity is considered moderate (MC = 0.31 ± 0.04) and fairly well studied in this species, with information from banding and radio and satellite telemetry data. Birds breeding in the Atlantic Flyway migrate to nonbreeding areas within the Atlantic Flyway (60%) and Caribbean (34%). The Midcontinent Flyway breeding population shows greater variability in nonbreeding locations: the Atlantic Flyway (31%), the Caribbean (24%), Mexico (22%), the Midcontinent Flyway (17%), and Central America (6%). In the Pacific Flyway, breeding birds spend the nonbreeding season in Mexico (79%) and within the Pacific Flyway (21%).

BANDING Band reencounter data demonstrate that eastern and western populations show strong latitudinal and longitudinal migratory connectivity. Eastern breeding populations spent the nonbreeding season along the southeastern US coast (largely Florida) and the Caribbean (Cuba), while breeders in western states migrated to Mexico, and midwestern breeders migrated to nonbreeding areas along the coast of the Gulf of Mexico and into Central America. Capture-recapture data models support latitudinal patterns in connectivity for both adult breeders and individuals banded as hatch years and recovered dead during the nonbreeding season (Scarpignato et al. 2021). Northern birds move to the southern US (54%) or south of the US (31%; Scarpignato et al. 2021). Mid-latitude herons exhibit differential migration; some remain local (41%), whereas others move to the southern US (31%), or south of US (28%; Scarpignato et al. 2021). Herons breeding in the southern US are

INTERESTING NOTE

A banding study in Canada found strong return rate to natal area among individuals banded as nestlings the previous season (L'Arrivee and Blokpoel 1990).

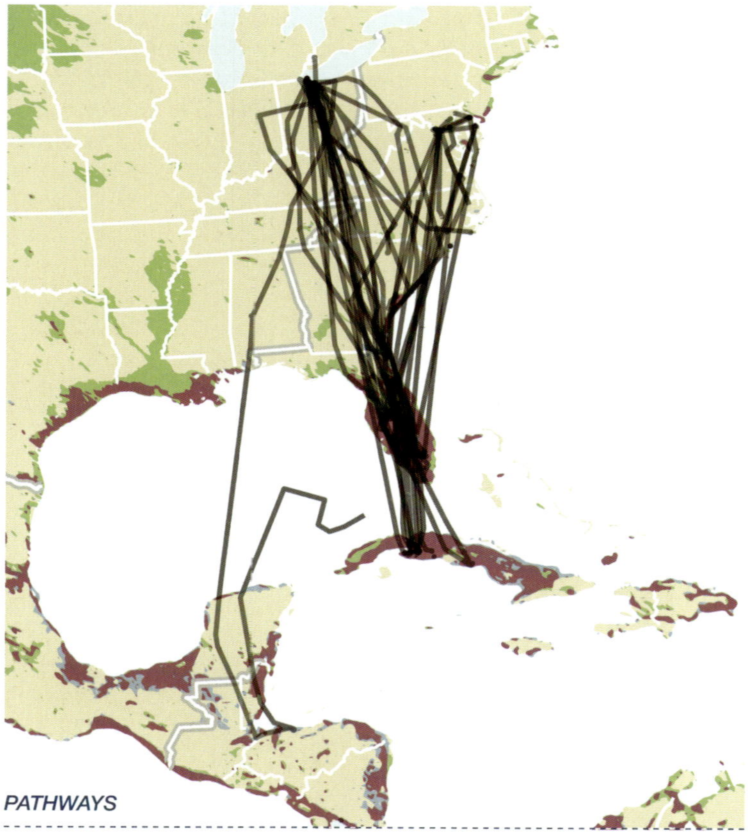

PATHWAYS

primarily resident (96%; Scarpignato et al. 2021). Band recoveries in the San Francisco Bay Area suggest the population breeding there is sedentary or short-distance migrants (Gill and Mewaldt 1979).

RADIO TRACKING Birds tracked by radio breeding in Oregon and Idaho leapfrogged Nevada breeders to spend the nonbreeding season in the coastal zones of central Mexico, while Nevada breeders spent the nonbreeding season in the southwestern US and the interior of northern Mexico (Henny and Blus 1986). There were over 100 Motus tag deployments but limited detections of individual birds between breeding and nonbreeding (C. Tonra, unpublished data). Several detections throughout migration show similar patterns to satellite transmitter data.

SATELLITE TRACKED Black-crowned Night Herons tracked (satellite and GSM tags) from one colony in Washington, DC, exhibited highly variable behavior, with 10% remaining locally in DC (n = 1), and the remaining birds migrating to North Carolina (n = 2), Florida (n = 6), and Cuba (n = 1; Scarpignato et al. 2021). Satellite tagging of birds from an Ohio breeding population indicates these birds migrate to Cuba and the southeastern US (n = 10; Scarpignato et al. 2021). Tracking of individuals breeding in Illinois is ongoing (S. Slayton and M. Ward, unpublished data).

OTHER APPROACHES No genetic or stable isotope studies of migratory connectivity were found for this species.

MIGRATORY PATHWAYS There are three primary pathways for migratory individuals following either Pacific or Atlantic flyways or along the Mississippi River. The Atlantic pathway likely follows the coastline to peninsular Florida and the Caribbean. Similarly, the Pacific pathway likely follows the coastline to nonbreeding locations in California and western Mexico. The midcontinent route follows the Mississippi River Valley southward to the Gulf of Mexico and the Caribbean.

DATA CONTRIBUTORS

L. Kearns; P.P. Marra; A.L. Scarpignato; K. Stein; C. Tonra.

GREEN HERON *Butorides virescens*

DISTRIBUTION AND ABUNDANCE

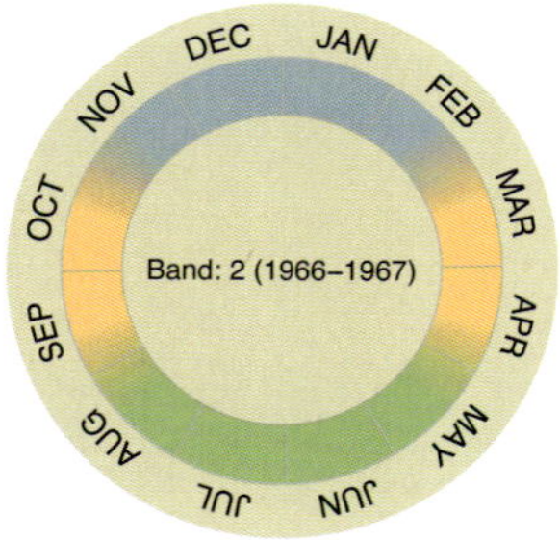

The Green Heron's distribution is comprised of tropical and subtropical areas that the species inhabits year-round, and a vast temperate area that is occupied during the breeding period. Large year-round populations occur in southern Florida, Mexico, the Caribbean, and Central America. A limited number of birds can also be found year-round along the Pacific coast from British Columbia to Mexico, and along the Gulf and southern Atlantic coasts. During the breeding period, this species' range expands widely into the central and eastern United States, extending as far north as Minnesota and Quebec. Within the US breeding range, notable areas of abundance include the Great Lakes region, the Gulf Coast, the lower Mississippi River and its tributaries, and southern Florida. In the nonbreeding season, this species vacates the majority of its eastern and central US range, and the year-round populations of Florida, Mexico, the Caribbean, and Central America are supplemented by more northern breeders. (Davis and Kushlan 2020; Fink et al. 2023).

ORDER Pelecaniformes
FAMILY Ardeidae
BREEDING HABITAT Wetlands, Freshwater Marsh
NONBREEDING HABITAT Wetlands, Generalist
GLOBAL POPULATION SIZE 1,200,000
POPULATION CHANGE SINCE 1970 −58%
IUCN RED LIST CATEGORY No
R2R TIPPING POINT SPECIES No
US LISTING STATUS No
CANADA SPECIES AT RISK No
PIF CCS 12/20

MIGRATORY CONNECTIVITY

OVERALL No studies on migratory connectivity were found for this species, and there are limited band reencounters.

BANDING Two breeding season individuals in the eastern United States were recovered in Central America and Puerto Rico, and a breeding season individual in California was recovered in Mexico in

INTERESTING NOTE

From 2013, when it was first captured as a breeding adult, to 2021, a migratory color-banded Green Heron has been resighted every summer at the same general foraging location in Spring Lake, Hays County, Texas (Haverland and Green 2022).

the nonbreeding season. Other banded individuals from Atlantic populations have been recovered in Florida, the Greater Antilles, and the Cayman Islands; and midcontinent populations have been recovered in Texas, Florida, Mexico, Central America, and the Greater Antilles (Davis and Kushlan 2020).

OTHER APPROACHES No tracking, genetic, or stable isotope studies of migratory connectivity were found for this species.

MIGRATORY PATHWAYS Migration routes are poorly known, but birds likely follow flyways and major river drainages north-south from central and eastern US to nonbreeding sites in Mexico, Caribbean, and Central and South America. Pacific populations likely follow the coastline from British Columbia southward to Mexico (Fink et al. 2023).

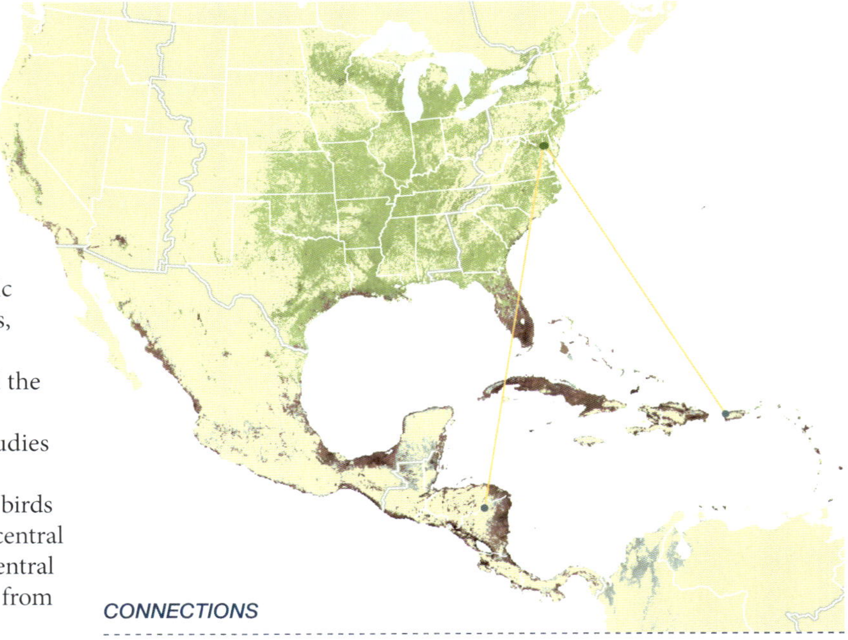

CONNECTIONS

Ardea alba GREAT EGRET

weak — strong

MC = 0.27 ± 0.04*

DISTRIBUTION AND ABUNDANCE

Great Egrets are found year-round throughout much of their range at interior lakes and rivers and along coastlines. Northern breeding areas include eastern and midcontinent portions of the US and Alberta, Saskatchewan, and Manitoba, Canada. Migratory populations in these areas move south to coastal areas in the western and southeastern United States, the Caribbean, Mexico, and Central America for the nonbreeding season (McCrimmon et al. 2020).

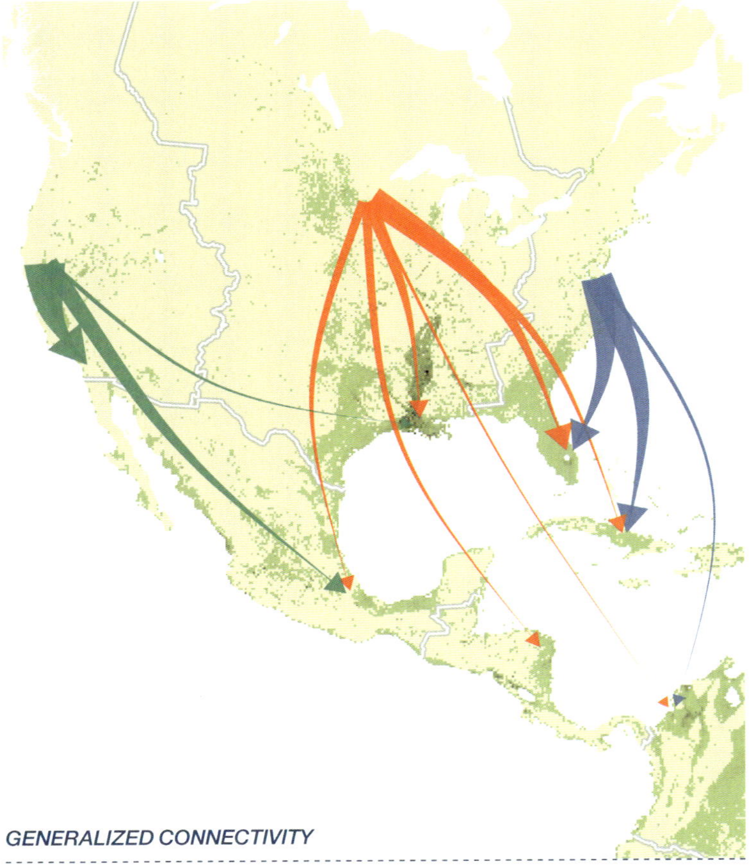

GENERALIZED CONNECTIVITY

ORDER Pelecaniformes
FAMILY Ardeidae
BREEDING HABITAT Wetlands, Generalist
NONBREEDING HABITAT Wetlands, Generalist
GLOBAL POPULATION SIZE 9,500,000
POPULATION CHANGE SINCE 1970 +213%
IUCN RED LIST CATEGORY Least Concern
R2R TIPPING POINT SPECIES No
US LISTING STATUS No
CANADA SPECIES AT RISK No
PIF CCS 6/20

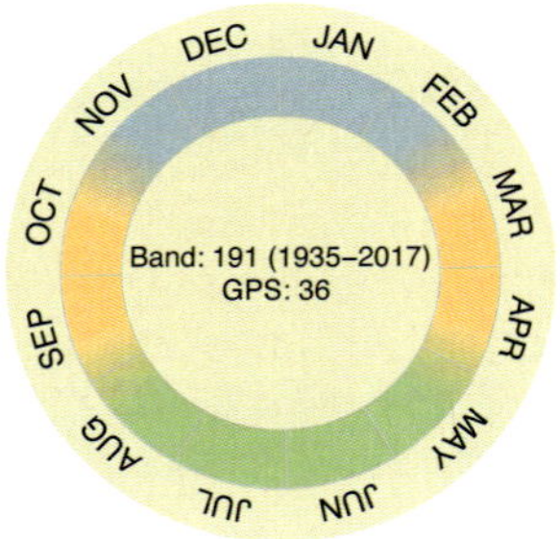

MIGRATORY CONNECTIVITY

OVERALL Band reencounter and tracking data are available for Great Egrets throughout their North American range. These methods show a general north-south movement among populations in the Atlantic and Pacific flyways and variation in movement for the Midcontinent Flyway breeding population, suggesting moderate migratory connectivity (MC = 0.27 ± 0.04). Within the Atlantic

	Atlantic	Caribbean	Central America	Mexico	Midcontinent	Northern SA	Pacific
Atlantic (n = 53)	49%	47%	–	–	–	4%	–
Midcontinent (n = 119)	40%	18%	11%	11%	19%	1%	–
Pacific (n = 33)	–	–	–	27%	3%	–	70%

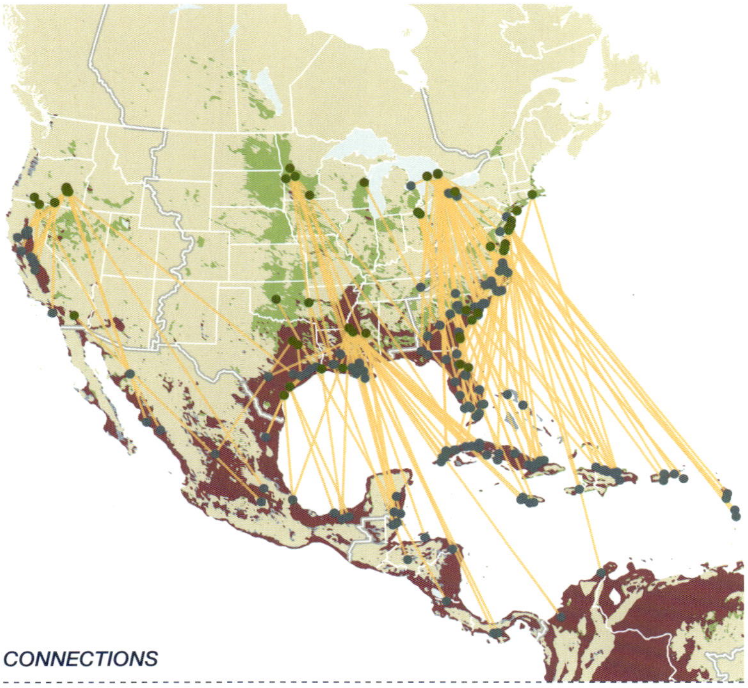

CONNECTIONS

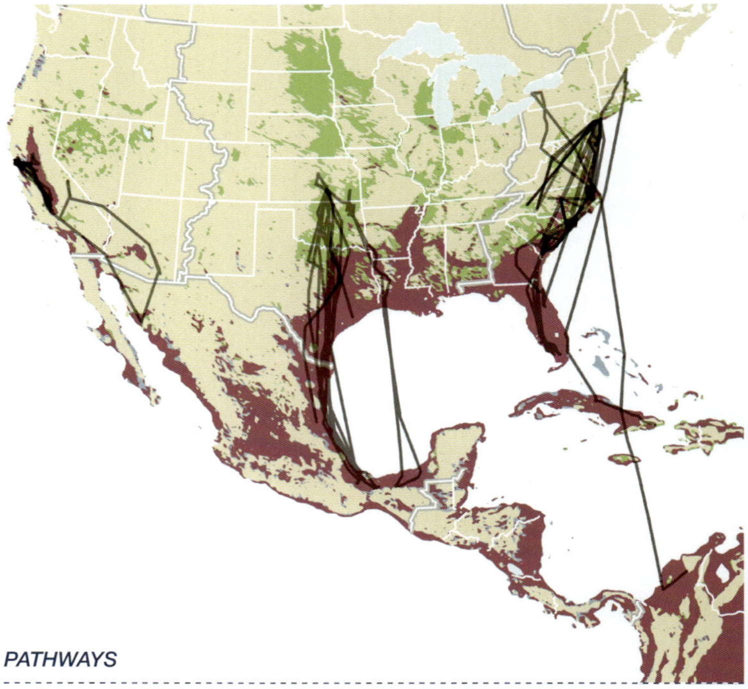

PATHWAYS

INTERESTING NOTE

Birds tracked in multiple years showed strong breeding fidelity but lower fidelity to nonbreeding sites; birds occupied similar regions but not necessarily the same roosts or foraging areas (Brzorad et al. 2021).

Flyway, breeding birds migrate to nonbreeding areas within the flyway (49%) or the Caribbean (47%). Breeding birds within the Midcontinent Flyway primarily migrate to the Atlantic Flyway (40%), with smaller numbers migrating within the flyway (19%) or to the Caribbean (18%), Central America (11%), and Mexico (11%). Pacific Flyway breeding birds migrate to nonbreeding areas within the flyway (70%) or Mexico (27%).

BANDING Important nonbreeding areas identified from banding data include the North American Pacific coast from Oregon to Panama and the Atlantic coast from New Jersey to the Lesser Antilles (Mikuska et al. 1998). A study using breeding-season band recoveries showed primarily north-south movement, though this should be attributed more to dispersal than migratory movements (Melvin et al. 1999). Great Egrets banded and marked with wing tags or leg bands at a long-term banding site in the eastern Great Lakes region were reencountered during the nonbreeding season along the US east coast, from North Carolina to Florida, and south to the Lesser Antilles (Weseloh et al. 2014; D. Weseloh, unpublished data).

SATELLITE TRACKING Data from tracking studies show north-south migratory pathways among populations in California (Lumpkin et al. 2023), the midcontinent (Brzorad 2020 et al.), and on the Atlantic coast (Brzorad 2020 et al.; Kays 2020). Birds in the Midcontinent Flyway spent the nonbreeding season along the northwestern Gulf of Mexico southward to coastal Mexico. Atlantic Flyway birds spent the nonbreeding season in the southeastern United States, as well as Cuba and Hispaniola.

OTHER APPROACHES No genetic or stable isotope studies of migratory connectivity were found for this species.

MIGRATORY PATHWAYS Great Egrets exhibit mostly north-south movement for migration, especially within the Atlantic and Pacific flyways. Within the Atlantic Flyway, migratory birds moved along the Atlantic coastline south to the Florida peninsula, with some individuals crossing over to Cuba, the Bahamas, and Hispaniola. Midcontinent populations followed the flyway south to the coastlines of Texas and Mexico, with some individuals conducting trans-Gulf flights. Migrants in Pacific Flyway populations followed the coastline south to Baja California and Central America.

DATA CONTRIBUTORS

J. Brzorad; S. Jennings; R. Kays; A. Maccarone; C. Weseloh.

Ardea ibis **WESTERN CATTLE-EGRET**

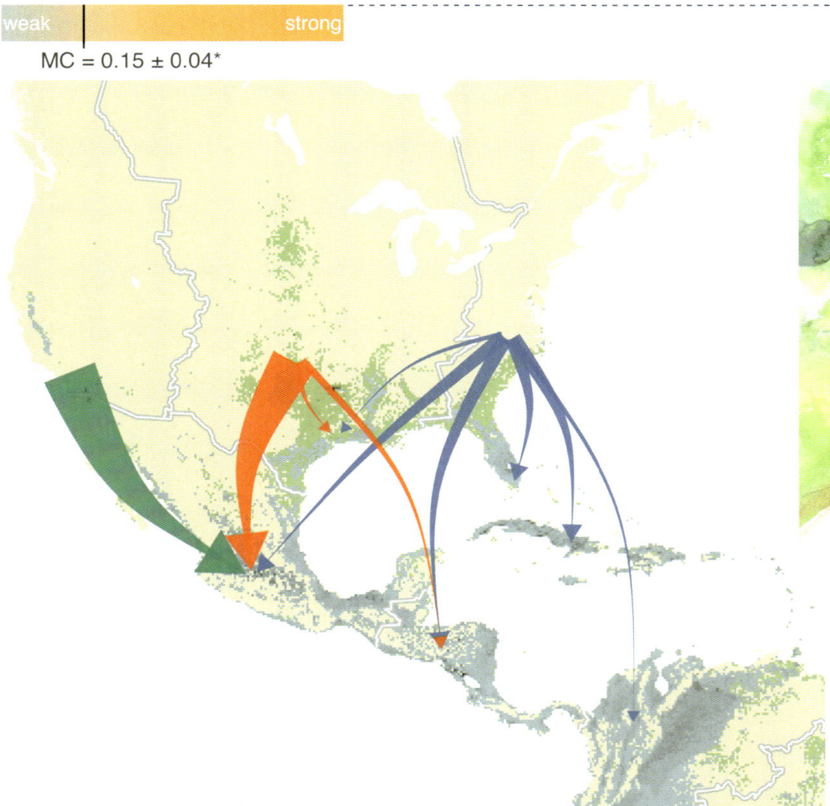

GENERALIZED CONNECTIVITY

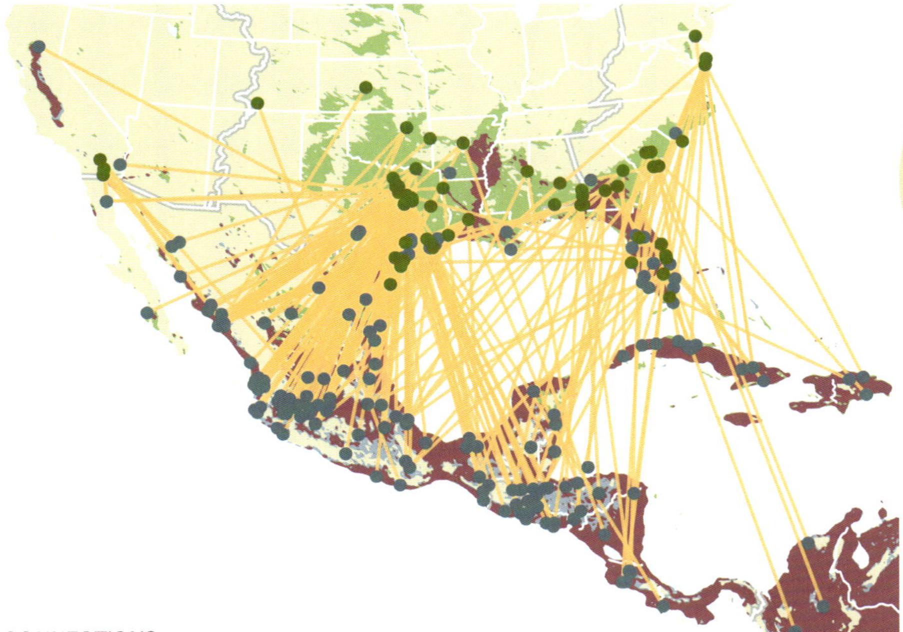

CONNECTIONS

DISTRIBUTION AND ABUNDANCE

Western Cattle-Egrets breed along the east coast of the United States and through the upper Midwest and southern Canada, as well as in Mexico, the Caribbean, and Central and South America. North American breeding populations are both migratory and resident; migratory populations spend the nonbreeding season in Mexico, the Caribbean, and Central America (Telfair 2024).

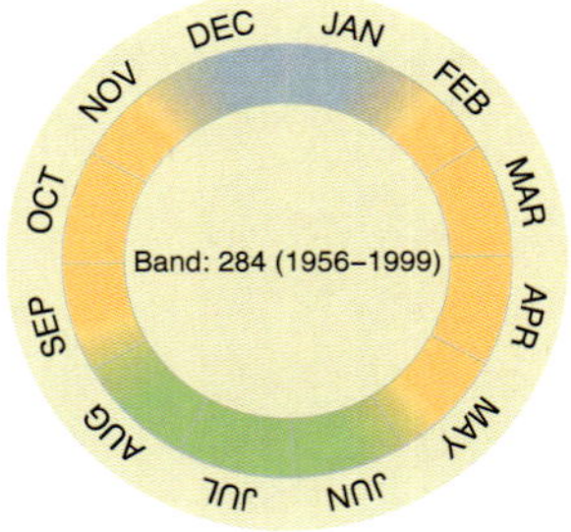

ORDER Pelecaniformes
FAMILY Ardeidae
BREEDING HABITAT Wetlands, Generalist
NONBREEDING HABITAT Open Country, Developed/Disturbed
GLOBAL POPULATION SIZE 48,000,000
POPULATION CHANGE SINCE 1970 −48%
IUCN RED LIST CATEGORY Least Concern
R2R TIPPING POINT SPECIES No
US LISTING STATUS No
CANADA SPECIES AT RISK No
PIF CCS 9/20

MIGRATORY CONNECTIVITY

	Atlantic	Caribbean	Central America	Mexico	Midcontinent	Northern SA	Pacific
Atlantic (n = 55)	22%	24%	24%	22%	4%	5%	–
Midcontinent (n = 204)	<1%	–	10%	82%	7%	–	<1%
Pacific (n = 9)	–	–	–	100%	–	–	–

OVERALL Band reencounters are plentiful, though no studies have directly assessed migratory connectivity. Individuals breeding in the Atlantic Flyway spread out fairly evenly across nonbreeding areas in the Caribbean (24%), Central America (24%), the Atlantic Flyway (22%), and Mexico (22%). Birds breeding within the Midcontinent and Pacific flyways primarily migrate to Mexico (82% and 100%, respectively). High variability and overlap of nonbreeding areas influence weak migratory connectivity (MC = 0.15 ± 0.04).

BANDING While over 66,000 birds have been banded since the 1950s, band recoveries and/or resightings have remained low (0.90%; Telfair 2024). Important nonbreeding areas identified from banding data include Mexico, Central America, the Greater Antilles, Cuba, and South America (Mikuska et al. 1998). Individuals banded in Georgia have been recovered after long-distance movements to

Nova Scotia, Cuba, Mexico, Honduras, and Guatemala (Hopkins 1972).

OTHER APPROACHES No genetic or stable isotope studies of migratory connectivity were found for this species.

MIGRATORY PATHWAYS Along the Atlantic Flyway, individuals primarily migrate southward from Newfoundland along the coast to Florida or across the Gulf to nonbreeding areas in Mexico, Central America, the Caribbean, and northern South America. Within the midcontinent, birds migrate from midwestern states through Texas into Mexico, Central America, and the Caribbean. Western breeding populations migrate southward from Alaska to Mexico and Central America but also move longitudinally to the Midcontinent Flyway as well as the Atlantic Flyway and Caribbean.

GREAT BLUE HERON *Ardea herodias*

weak — strong

MC = 0.21 ± 0.04*

DISTRIBUTION AND ABUNDANCE

Great Blue Herons are present year-round throughout much of its range in North America, with breeding populations extending from Mexico into Canada. It is in high abundance throughout the eastern US, especially along the Mississippi River and coastal areas of the Gulf Coast and Florida. In the west, it is patchily distributed around water sources, with high abundance in the Central Valley of California and the Great Salt Lake of Utah, as well as coastal areas in Washington and Mexico (Fink et al. 2023). The nonbreeding range includes coastal areas of the US, Mexico, the Caribbean, Central America, and northern South America, and interior areas along lakes and rivers (Fink et al. 2023).

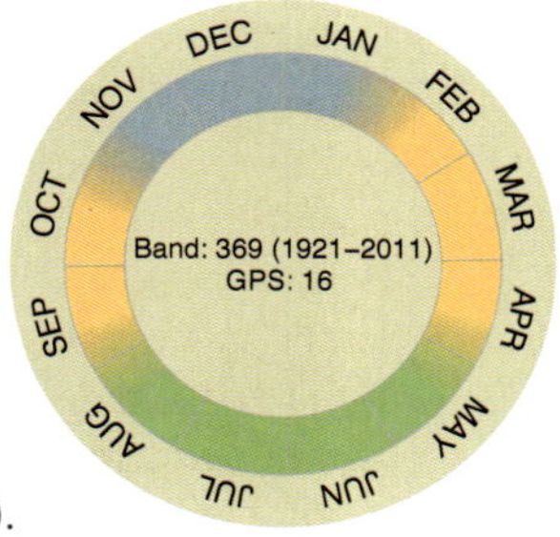

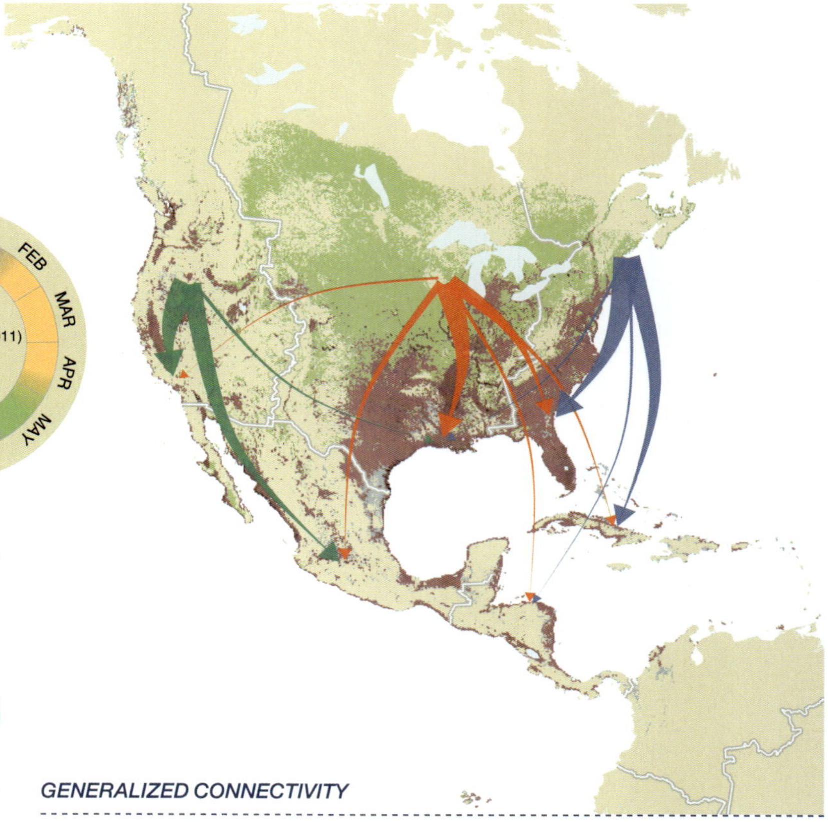

GENERALIZED CONNECTIVITY

ORDER Pelecaniformes
FAMILY Ardeidae
BREEDING HABITAT Wetlands, Generalist
NONBREEDING HABITAT Wetlands, Generalist
GLOBAL POPULATION SIZE 700,000
POPULATION CHANGE SINCE 1970 +37%
IUCN RED LIST CATEGORY Least Concern
R2R TIPPING POINT SPECIES No
US LISTING STATUS No
CANADA SPECIES AT RISK No
PIF CCS 8/20

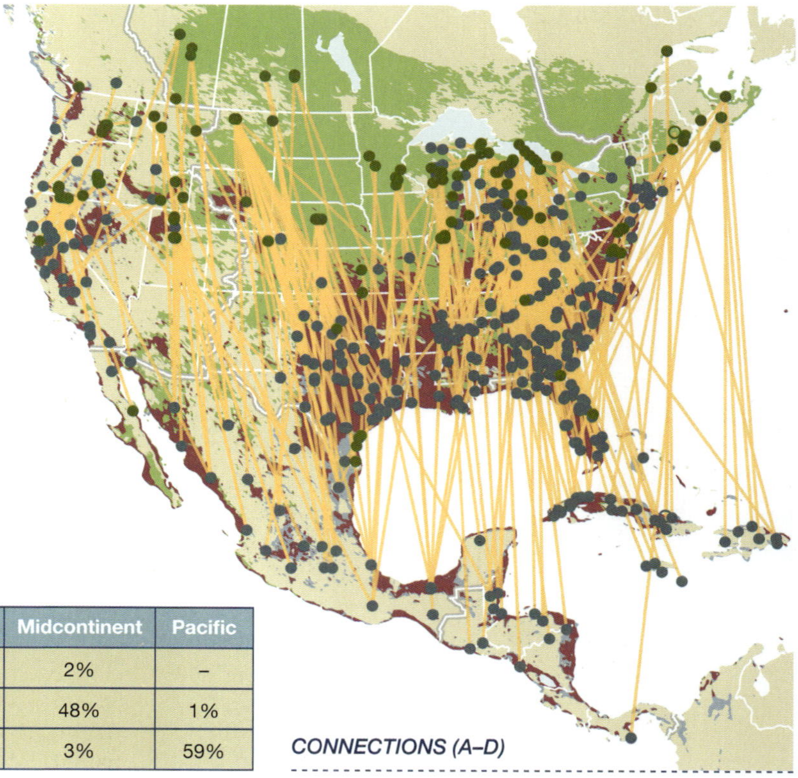

CONNECTIONS (A–D)

	Atlantic	Caribbean	Central America	Mexico	Midcontinent	Pacific
Atlantic (n = 43)	56%	40%	2%	–	2%	–
Midcontinent (n = 251)	28%	7%	4%	12%	48%	1%
Pacific (n = 58)	–	–	–	38%	3%	59%

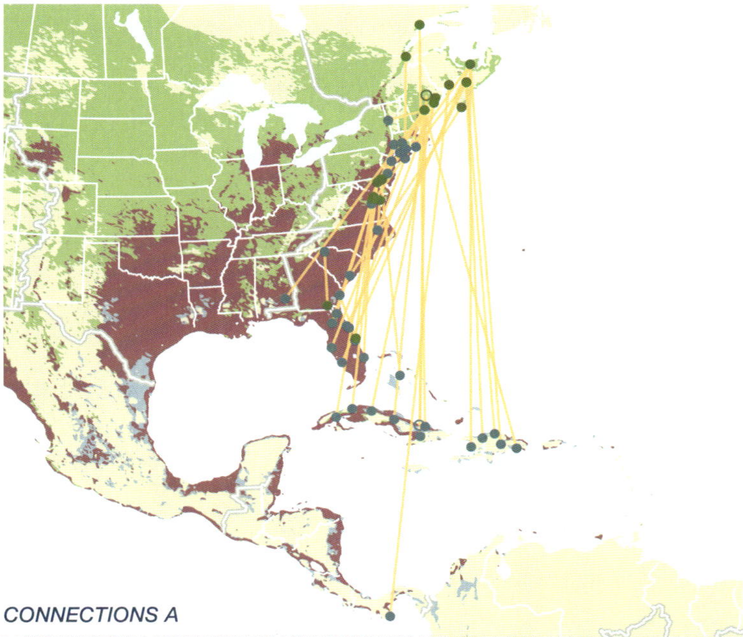

CONNECTIONS A

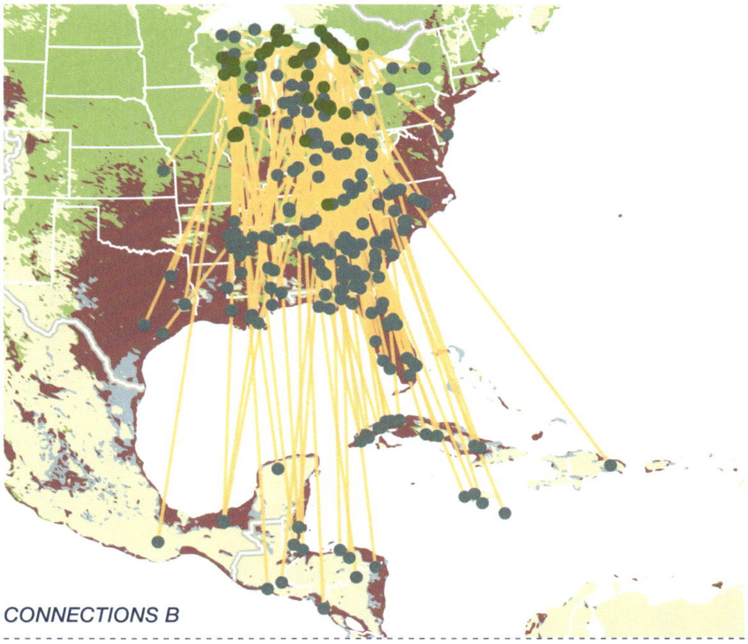

CONNECTIONS B

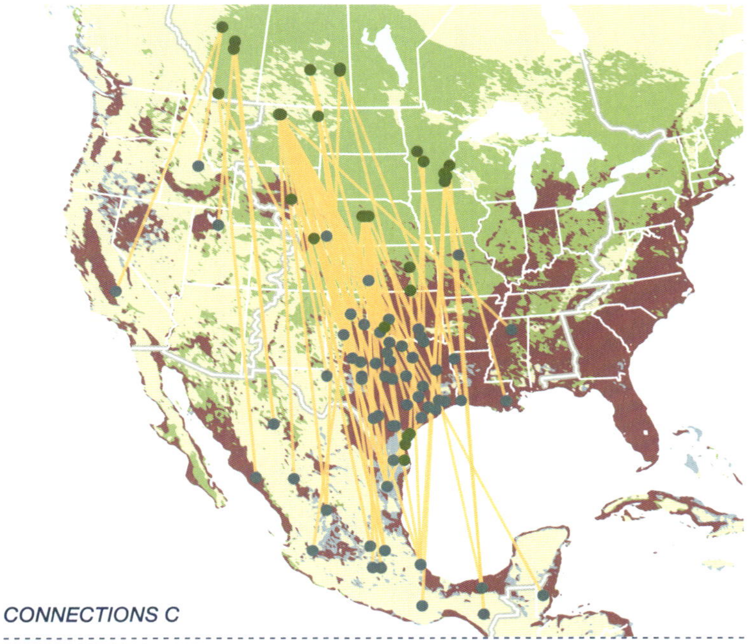

CONNECTIONS C

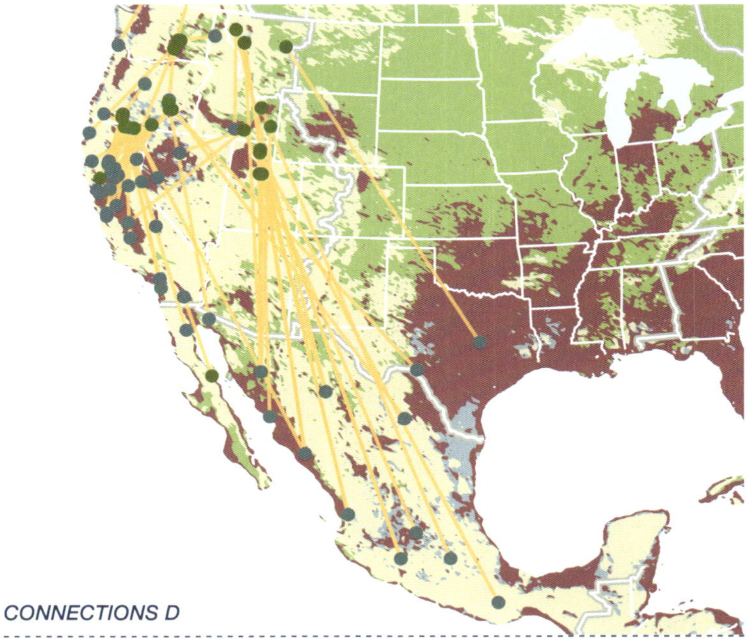

CONNECTIONS D

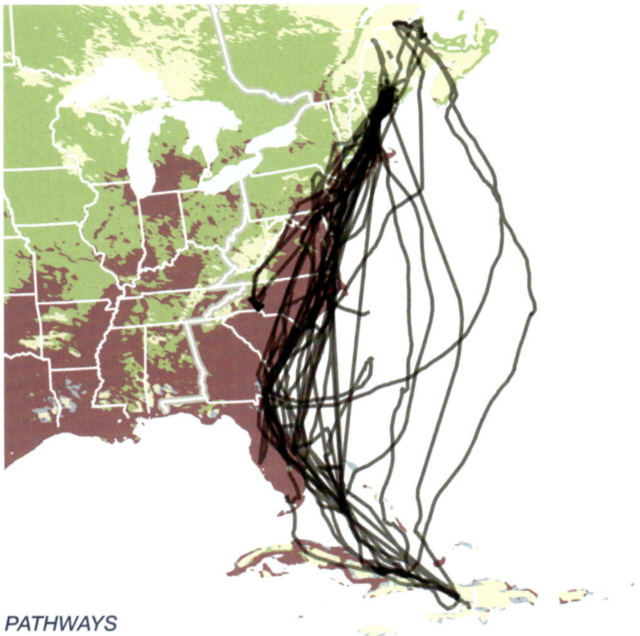

PATHWAYS

MIGRATORY CONNECTIVITY

OVERALL Great Blue Heron migratory connectivity is known from band reencounter and satellite transmitter data from throughout their North American range. Breeding populations generally exhibit north-south migratory movement with some east-west movement resulting in moderate migratory connectivity (MC = 0.21 ± 0.04). Roughly half (56%) of the birds breeding within the Atlantic Flyway migrate to nonbreeding areas within the flyway or migrate farther south to the Caribbean (40%), whereas Midcontinent Flyway breeding populations are more distributed throughout nonbreeding regions, including the Midcontinent (48%) and Atlantic (28%) flyways, Mexico (12%), the Caribbean (7%), and Central America (4%). Pacific Flyway breeding populations generally remain within the Pacific Flyway (59%) or Mexico (38%) during the nonbreeding season.

BANDING Important nonbreeding areas based on banding data include the coasts of Mexico and Hispaniola, though the species' nonbreeding range extends as far north as Canada (Mikuska et al. 1998). A study using breeding season banding recoveries showed primarily north-south movement, though this should be attributed more to dispersal than migratory movements (Melvin et al. 1999).

SATELLITE TRACKING Great Blue Herons breeding on the Atlantic coast of Canada and the United States have been tracked with GPS transmitters migrating along the coast to nonbreeding sites on Hispaniola and Cuba. Several individuals made flights over the Atlantic from Canada southward to Florida and the Caribbean, with one individual exhibiting stopover behavior in Bermuda. Another individual migrated from its breeding area near Philadelphia, Pennsylvania, north to Raritan Bay, New Jersey. Two Great Blue Herons in North Carolina did not leave the vicinity of their respective bodies of water (Lake Norman, Abernathy Pond) for five years. This is part of ongoing research tracking eastern Great Blue Heron migration (Brzorad 2020 et al.).

DATA CONTRIBUTORS
J. Brzorad; D. D'Auria.

OTHER APPROACHES No genetic or stable isotope studies of migratory connectivity were found for the species.
MIGRATORY PATHWAYS Within eastern populations, birds exhibit north-south movement along the Atlantic coast down the Florida peninsula to the Caribbean but also exhibit some movement over open ocean, including potential stopover sites in Bermuda and Bahamas before traveling farther southward. In the west, birds primarily exhibit north-south movement along coasts and interior waterways (Fink et al. 2023).

THRESKIORNITHIDAE | *IBISES AND SPOONBILLS*

WHITE IBIS *Eudocimus albus*

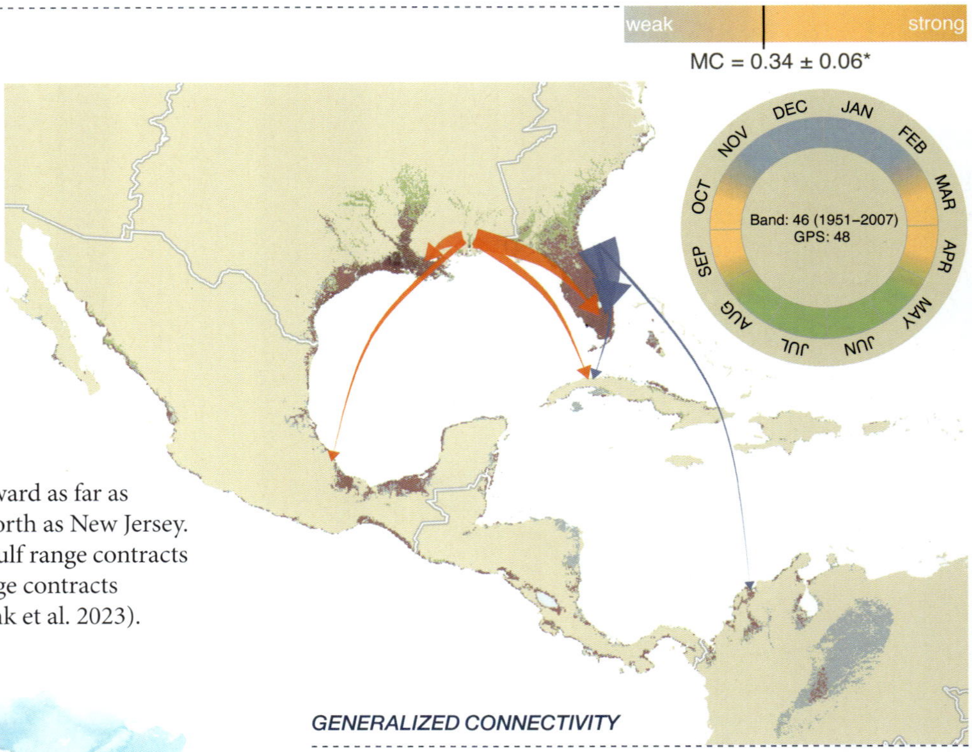

GENERALIZED CONNECTIVITY

DISTRIBUTION AND ABUNDANCE

White Ibis are present year-round throughout most of their range, which includes the US Gulf states and the south Atlantic coast, coastal Mexico and Central America, the large islands of the Caribbean, and parts of northern South America. Areas of especially high abundance for this species include peninsular Florida, and the Louisiana and Texas coasts. This species' largely nomadic movements make it difficult to clearly define their breeding and nonbreeding ranges, but there is some northward range expansion during the breeding period. During the breeding period, birds along the US Gulf coast extend northward as far as Arkansas, while Atlantic coast birds extend as far north as New Jersey. During the subsequent nonbreeding periods, the Gulf range contracts south to northern Louisiana, while the Atlantic range contracts southward to North Carolina (Heath et al. 2020; Fink et al. 2023).

ORDER Pelecaniformes
FAMILY Threskiornithidae
BREEDING HABITAT Wetlands, Generalist
NONBREEDING HABITAT Wetlands, Generalist

GLOBAL POPULATION SIZE 2,400,000
POPULATION CHANGE SINCE 1970 +566%
IUCN RED LIST CATEGORY Least Concern
R2R TIPPING POINT SPECIES No
US LISTING STATUS No
CANADA SPECIES AT RISK No
PIF CCS 9/20

	Atlantic	Caribbean	Mexico	Midcontinent	Northern SA
Atlantic (n = 64)	89%	9%	–	–	2%
Midcontinent (n = 22)	41%	18%	9%	32%	–

MIGRATORY CONNECTIVITY

OVERALL Migratory connectivity information is from satellite tracking and band reencounter data. Within the Atlantic Flyway, breeding birds generally migrate to nonbreeding areas within the flyway (89%), with some individuals migrating to the Caribbean (9%) and northern South America (2%). Midcontinent Flyway breeding birds exhibit more varied directional movement, with some migrating to the Atlantic Flyway (41%), some staying within the region (32%), and some migrating to the Caribbean (18%) or Mexico (9%). Migration to multiple nonbreeding regions along with high within-flyway movement (in the Atlantic Flyway) yields moderate migratory connectivity (MC = 0.34 ± 0.06).
BANDING Banding recoveries in North America include movements

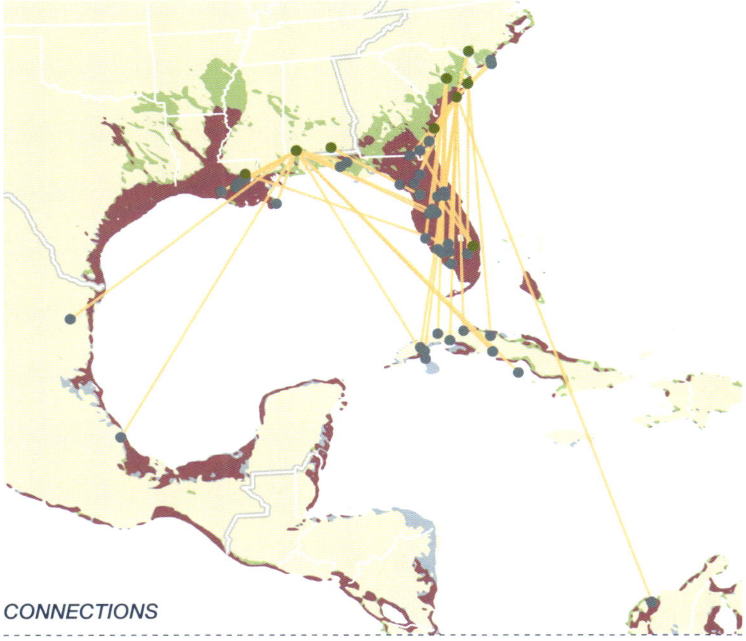

CONNECTIONS

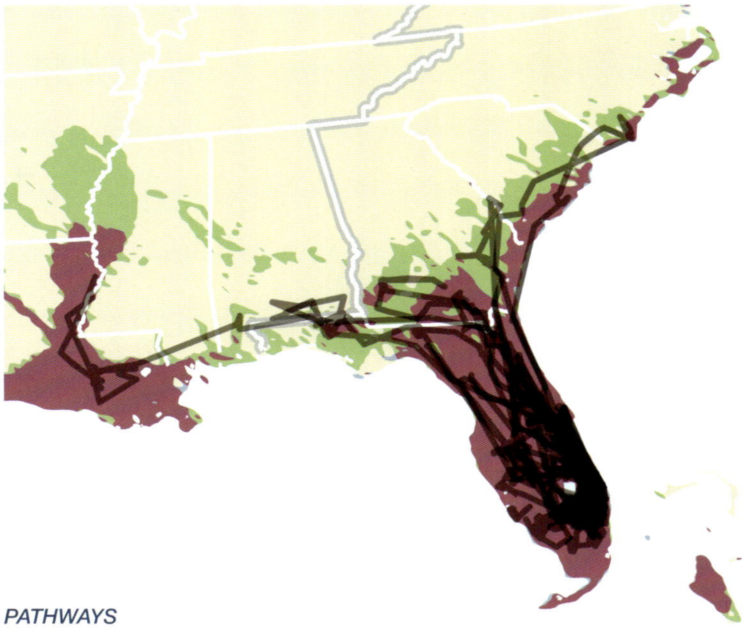

PATHWAYS

between the southeastern United States, especially Florida, and parts of Mexico and Cuba, and one recovery in Colombia (Melvin et al. 1999). Individuals banded in Alabama exhibited movement between Atlantic and Gulf populations, with some individuals recovered in Mexico and others in Florida and Cuba.

SATELLITE TRACKING White Ibis tracked in Florida (n = 41) were found to have their annual cycles affected by the use of urban habitats. The timing of breeding and nonbreeding seasons, as well as movement patterns, were changed to allow the individuals to exploit urban resources. This could result in fewer nomadic behaviors in individuals associated with more urban environments (Kidd-Weaver et al. 2020). US Geological Survey's Florida Cooperative Fish and Wildlife Research Unit is currently tracking adult and juvenile White Ibis using satellite telemetry from natal or breeding colonies in and around Mobile Bay, Alabama (A. Powell, unpublished data).

OTHER No genetic or stable isotope studies of migratory connectivity were found for this species.

MIGRATORY PATHWAYS Birds within US Atlantic and Gulf Coast populations exhibit general north-south movements during migration, although some movement east-west likely occurs based on band recoveries and telemetry data. Trans-Gulf and Caribbean migrations likely occur based on recoveries in the Caribbean and Central and South America, as well as documented flights south of the Florida Keys during post-breeding (Heath et al. 2020). The Florida peninsula is an important corridor for movement of birds within the Atlantic population.

DATA CONTRIBUTORS

J. Hepinstall-Cymerman.

Plegadis falcinellus GLOSSY IBIS

DISTRIBUTION AND ABUNDANCE

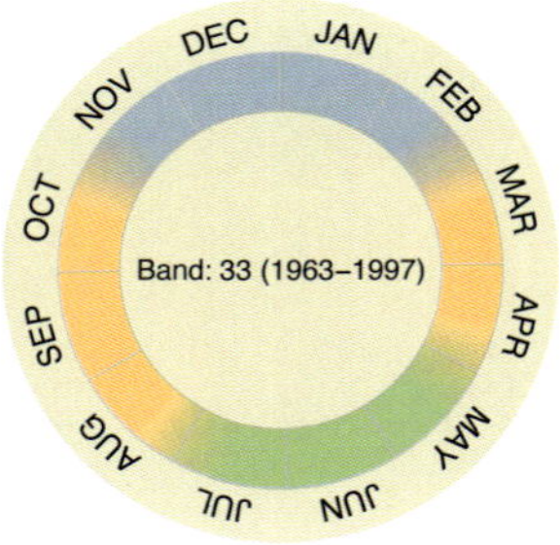

The Glossy Ibis is a globally widespread species but in North America is restricted to the eastern coastal and near-coastal regions of the continent. The species nests colonially in wetlands from New Brunswick to the Caribbean and often wanders after breeding, though the core of the nonbreeding range includes Florida, the Gulf Coast, and the Caribbean (Davis and Kricher 2020).

ORDER Pelecaniformes
FAMILY Threskiornithidae
BREEDING HABITAT Coasts, Salt Marsh
NONBREEDING HABITAT Wetlands, Generalist
GLOBAL POPULATION SIZE 550,000
POPULATION CHANGE SINCE 1970 +680%
IUCN RED LIST CATEGORY Least Concern
R2R TIPPING POINT SPECIES No
US LISTING STATUS No
CANADA SPECIES AT RISK No
PIF CCS 8/20

MIGRATORY CONNECTIVITY

OVERALL Glossy Ibis migratory connectivity information is available from limited band reencounter data; thus MC was not calculated. Birds breeding in the Atlantic Flyway primarily spend the nonbreeding period in the Caribbean (70%), the Atlantic Flyway (21%), and northern South America (9%).

BANDING Band reencounters of Glossy Ibis in eastern North America occurred between the northeastern United States and Florida and the Caribbean islands, mainly Cuba and Hispaniola (Melvin et al. 1999).

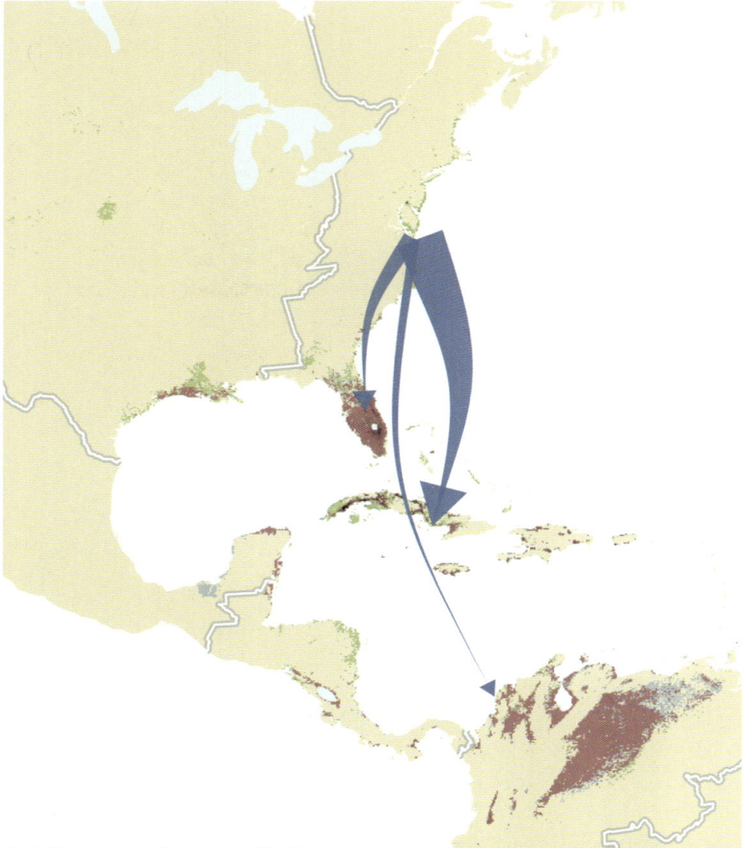

GENERALIZED CONNECTIVITY

	Atlantic	Caribbean	Northern SA
Atlantic (n = 33)	21%	70%	9%

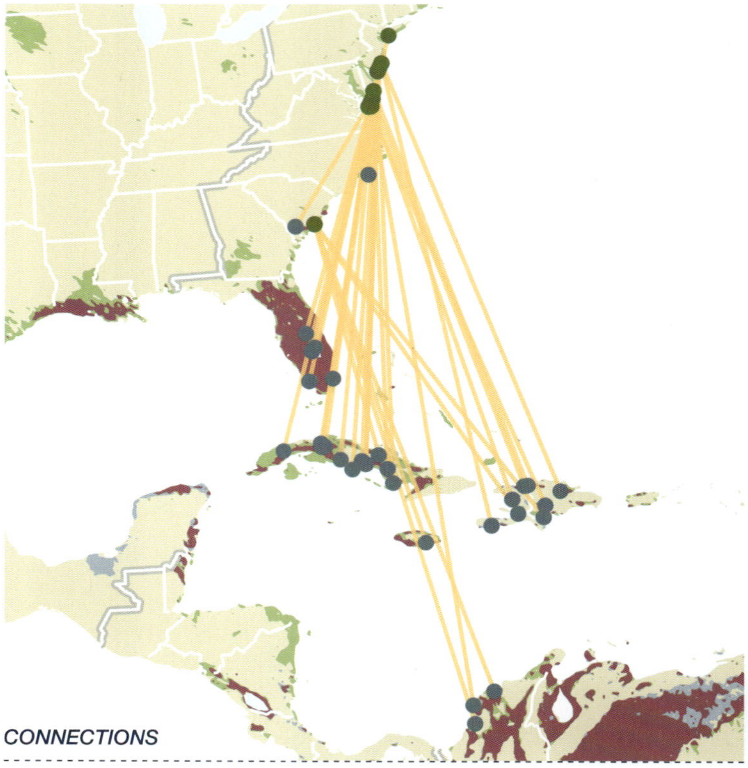

CONNECTIONS

OTHER APPROACHES No tracking, genetic, or stable isotope studies of migratory connectivity exist for this species.
MIGRATORY PATHWAYS Eastern populations exhibit north-south migratory movement likely following coastlines from northern breeding sites to nonbreeding sites in Florida, the Caribbean, and South America. Movement over the Atlantic Ocean likely occurs as well, as banded migrants in Bermuda are a mix of North American and European birds (Melvin et al. 1999).

WHITE-FACED IBIS *Plegadis chihi*

MC = 0.17 ± 0.08*

DISTRIBUTION AND ABUNDANCE

The White-faced Ibis is a colonial breeder in marshes in the western and central United States; it spends the nonbreeding season in Mexico, Texas, and Louisiana, and occurs year-round in coastal regions of Texas and Louisiana, as well as central-south California and in Durango and Jalisco states of Mexico (Ryder and Manry 2020).

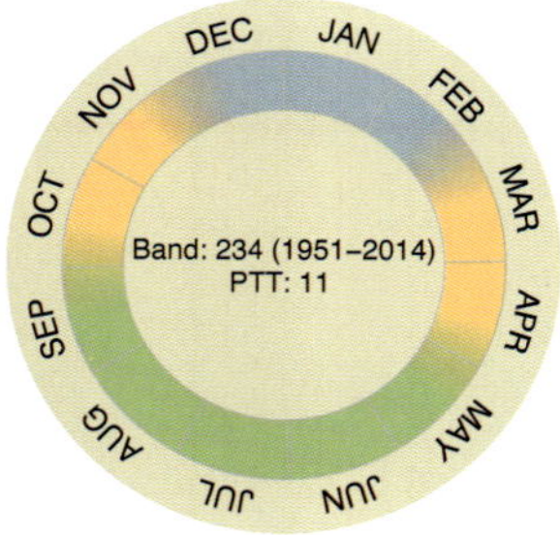

MIGRATORY CONNECTIVITY

OVERALL The migratory connectivity of the White-faced Ibis is known mostly from band reencounter data of breeding colonies in the western US and one satellite transmitter study (Yates et al. 2009). In the Midcontinent Flyway, breeding birds spend the nonbreeding season in Mexico (60%) and within the flyway (40%), whereas the majority of breeding individuals in the Pacific Flyway migrate to Mexico (97%). High overlap of the two breeding populations results in weak migratory connectivity (MC = 0.17 ± 0.08).
BANDING Most White-faced Ibises have been banded in Utah, Nevada, and other breeding colonies in the western United States. Band recoveries of these individuals are usually located in Mexico in the nonbreeding season, with a few individuals spending the nonbreeding season in Texas and Louisiana (Ryder 1967).

ORDER Pelecaniformes
FAMILY Threskiornithidae
BREEDING HABITAT
Wetlands, Freshwater Marsh
NONBREEDING HABITAT
Wetlands, Generalist
GLOBAL POPULATION SIZE
7,200,000

POPULATION CHANGE SINCE 1970 +237%
IUCN RED LIST CATEGORY Least Concern
R2R TIPPING POINT SPECIES No
US LISTING STATUS No
CANADA SPECIES AT RISK No
PIF CCS 7/20

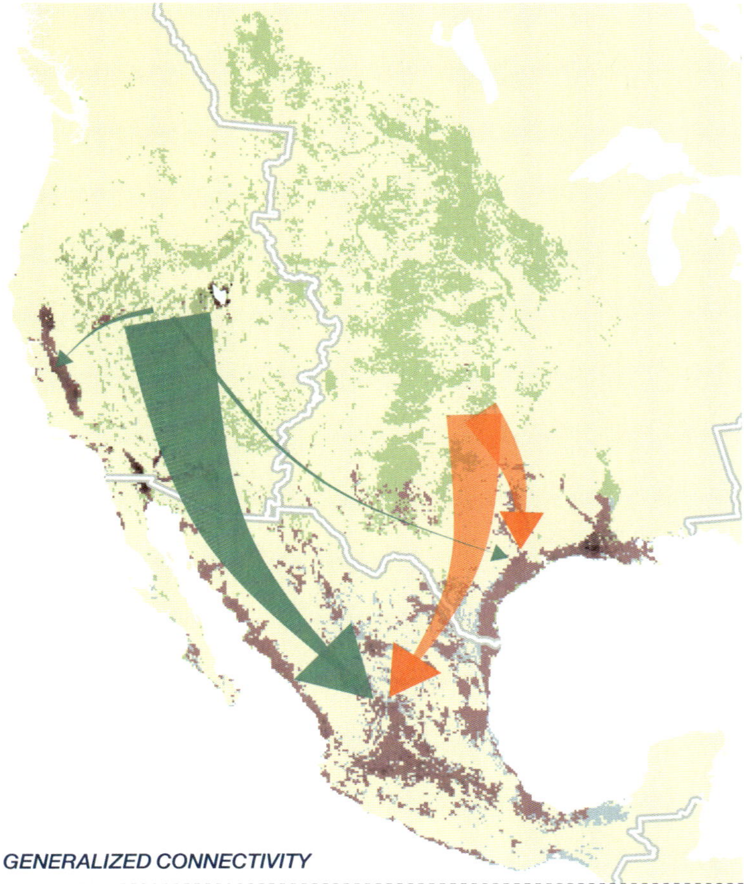

GENERALIZED CONNECTIVITY

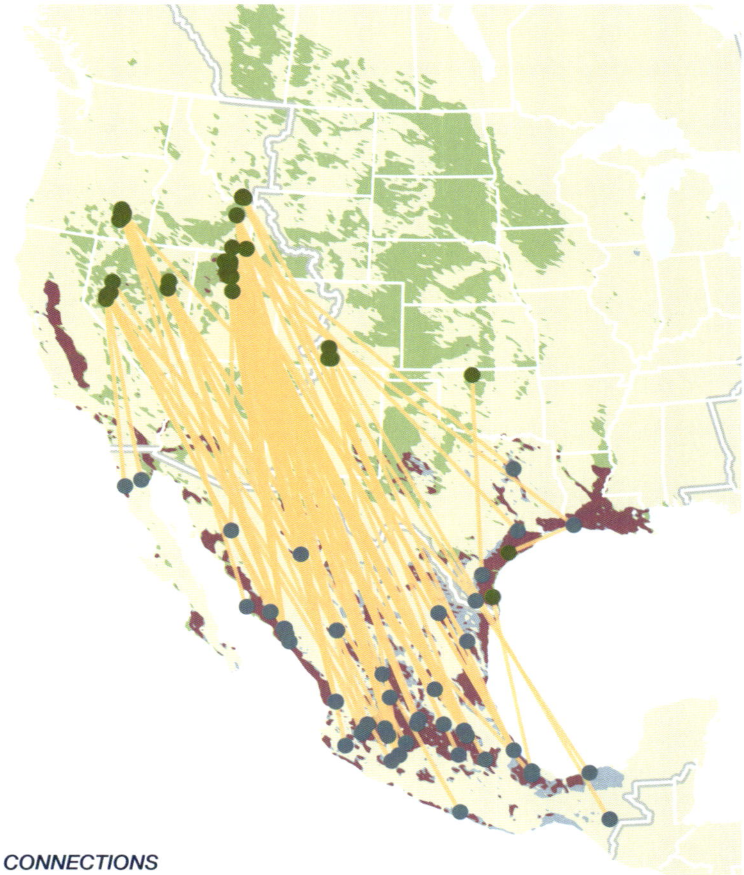

CONNECTIONS

	Mexico	Midcontinent	Pacific
Midcontinent (n = 5)	60%	40%	–
Pacific (n = 233)	97%	1%	2%

SATELLITE TRACKING White-faced Ibis breeding at Carson Lake, Nevada (n = 12), were tracked to nonbreeding grounds in California (n = 6) and Mexico (n = 6), which were likely a source of DDE contamination (Yates et al. 2009).

OTHER APPROACHES No genetic or stable isotope studies of migratory connectivity were found for this species.

MIGRATORY PATHWAYS Birds generally exhibit north-south migratory movements, although some midcontinent individuals show more southeast-northwest movements between western/central US states and nonbreeding locations in Texas and Mexico. Limited telemetry data suggest birds may move along coastal areas southward toward nonbreeding sites (Yates et al. 2009). The Colorado River Valley is likely an important migration route for Pacific breeding individuals (Ryder and Manry 2020).

DATA CONTRIBUTORS
M. Yates.

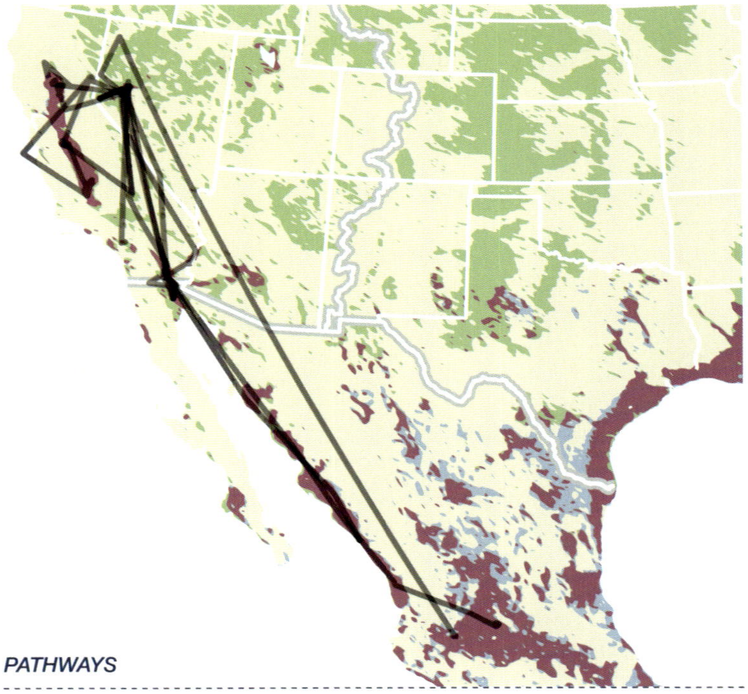

PATHWAYS

ROSEATE SPOONBILL *Platalea ajaja*

DISTRIBUTION AND ABUNDANCE

In North America, the Roseate Spoonbill breeds in Texas, Louisiana, southern Florida, coastal Mexico, and the Caribbean. It spends the nonbreeding period locally along the Gulf Coast, the Pacific coast of Mexico, and Central and South America. Throughout the year, the Roseate Spoonbill uses a variety of freshwater, brackish, and marine habitats mostly along the coast, with some interior nesting and foraging (Dumas 2020).

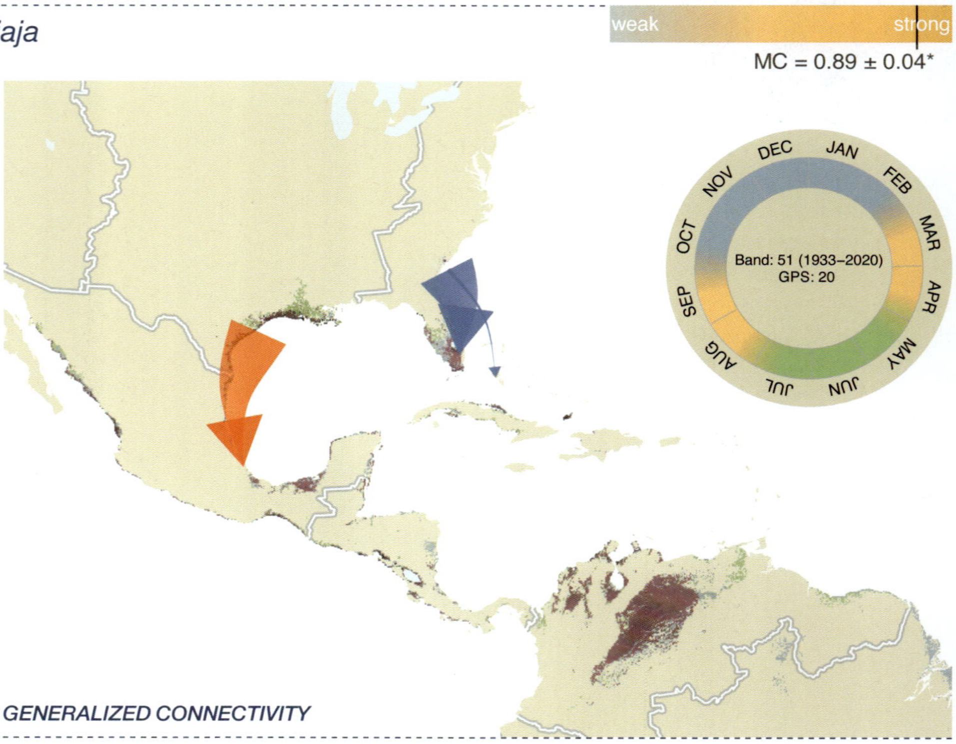

GENERALIZED CONNECTIVITY

ORDER Pelecaniformes
FAMILY Threskiornithidae
BREEDING HABITAT Coasts, Mangroves
NONBREEDING HABITAT Wetlands, Generalist
GLOBAL POPULATION SIZE 170,000
POPULATION CHANGE SINCE 1970 +1865%
IUCN RED LIST CATEGORY Least Concern
R2R TIPPING POINT SPECIES No
US LISTING STATUS No
CANADA SPECIES AT RISK No
PIF CCS 9/20

	Atlantic	Caribbean	Mexico
Atlantic (n = 63)	98%	2%	–
Midcontinent (n = 8)	–	–	100%

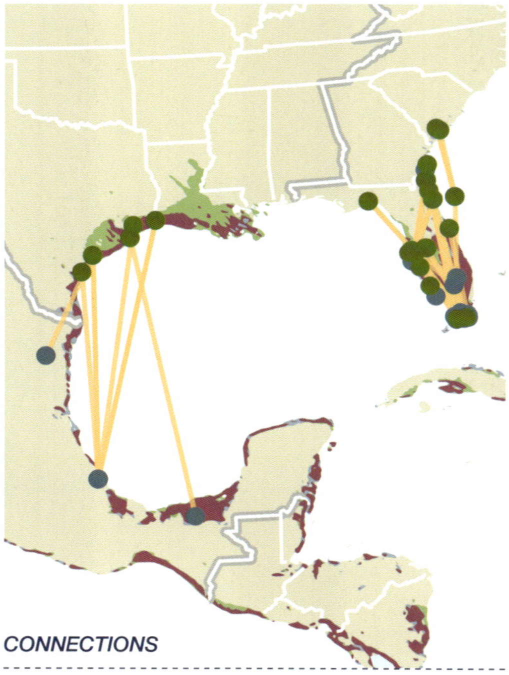

CONNECTIONS

PATHWAYS

MIGRATORY CONNECTIVITY

OVERALL Migration connectivity data for the Roseate Spoonbill include a limited number of band reencounter data and satellite tracking data. Although a small number of birds breeding in the Midcontinent Flyway migrate to Mexico (100%), a larger sample of birds breeding in the Atlantic Flyway remain within the flyway during the nonbreeding season (98%), resulting in strong migratory connectivity (MC = 0.89 ± 0.04).

BANDING Band reencounter data in North America are restricted to the Gulf Coast region: Texas and Florida are connected to the Gulf Coast of Mexico and Florida, respectively.

SATELLITE TRACKING Satellite tracking data indicate that adult birds (n = 20) in Florida exhibited primarily north-south movements within the southeastern US, with one individual traveling east-west between Florida and Bahamas (Andros island; J. Lorenz, unpublished data).

OTHER APPROACHES No genetic or stable isotope studies of migratory connectivity exist for this species.

MIGRATORY PATHWAYS Within the Atlantic and Gulf Coast populations, birds exhibited north-south migratory pathways with little evidence of movement between populations. Post-breeding movements are often northward and inland followed by southward movement toward nonbreeding sites. The occurrence of the species year-round along the Gulf Coast and the southeastern US limits our understanding of movement. Birds have been observed in transit flights between Florida and Cuba, but it is unknown if there are trans-Gulf flights for migratory individuals.

DATA CONTRIBUTORS
J. Lorenz.

NEW WORLD VULTURES (CATHARTIFORMES)

WRITING LED BY **LAURIE J. GOODRICH** *AND* **PETER P. MARRA**

The migratory connectivity and associated ecology of North American vultures, including Turkey Vultures, Black Vultures, and California Condors, remains relatively understudied compared to many other large soaring birds, with detailed migratory connectivity information available for only a subset of populations. Within the Cathartidae family, only the California Condor is listed as critically endangered (https://ecos.fws.gov/ecp/). It is considered largely a resident species and is therefore not included in this Atlas. For Turkey and Black Vultures, the mix of resident and migratory behaviors has important implications for population monitoring and conservation planning, including coordinated continental-scale conservation strategies.

CATHARTIDAE | *NEW WORLD VULTURES*

Coragyps atratus BLACK VULTURE

DISTRIBUTION AND ABUNDANCE

Historically Black Vultures bred from the south-central and eastern US south through Mexico and Central and South America (Buckley et al. 2022). In recent decades, the US breeding range has extended north to western Massachusetts, Vermont, eastern New York, and Pennsylvania, with small numbers in southern Ontario, Canada. High abundances occur in central and southeastern Texas, southeastern Louisiana, Tennessee, and Florida. The species is absent in the Caribbean except in Trinidad (Buckley et al. 2022). Most populations are presumed resident, but some birds show short-distance migration in the northern part of range, and an increasing number of flocks appear to migrate past northern and central US hawk watches and to some sites in Central America.

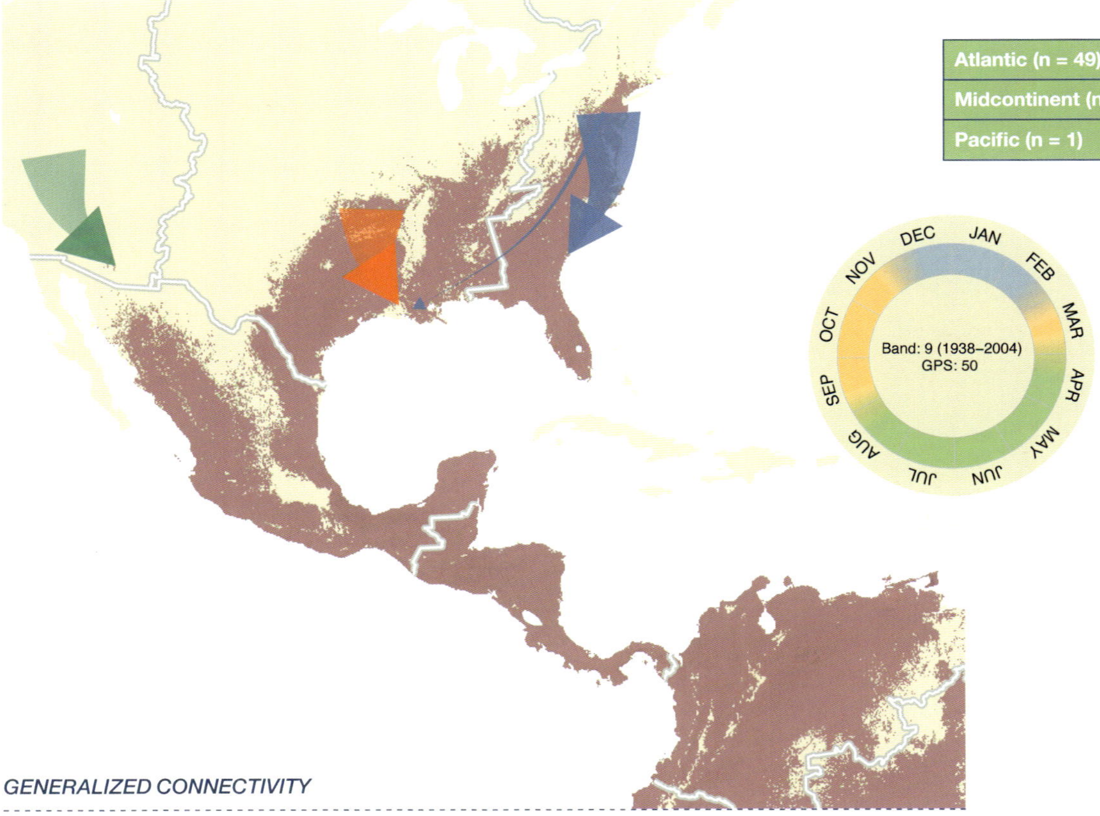

GENERALIZED CONNECTIVITY

	Atlantic	Midcontinent	Pacific
Atlantic (n = 49)	98%	2%	–
Midcontinent (n = 5)	–	100%	–
Pacific (n = 1)	–	–	100%

ORDER Cathartiformes
FAMILY Cathartidae
BREEDING HABITAT
Open Country, Habitat Mosaic
NONBREEDING HABITAT
Open Country, Habitat Mosaic
GLOBAL POPULATION SIZE 190,000,000
POPULATION CHANGE SINCE 1970 +790%
IUCN RED LIST CATEGORY Least Concern
R2R TIPPING POINT SPECIES No
US LISTING STATUS No
CANADA SPECIES AT RISK No
PIF CCS 5/20

INTERESTING NOTE

North American Christmas Bird Counts show Black Vultures increasing at 6% per year throughout their range, and 16% of hawk watches report significant increases in numbers from 2009 to 2019 (Oleyar et al. 2021). Despite these increases, this species appears particularly vulnerable to zoonotic diseases, with high-mortality events reported due to avian influenza and West Nile virus (Vidaña et al. 2020; Erica Miller, University of Pennsylvania, personal communication, 2022). While Black Vulture home ranges were found to be larger in the breeding season months at one site in the southeastern US (Holland et al. 2017), across multiple sites in a similar region, their home range size did not differ between breeding and nonbreeding seasons (Evans et al. 2024).

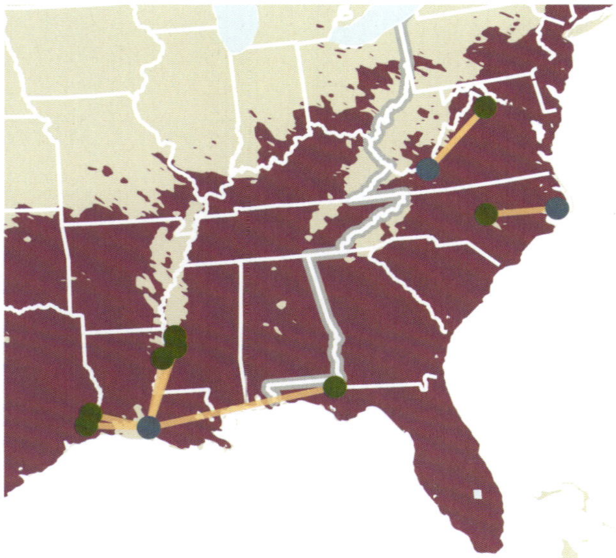

CONNECTIONS

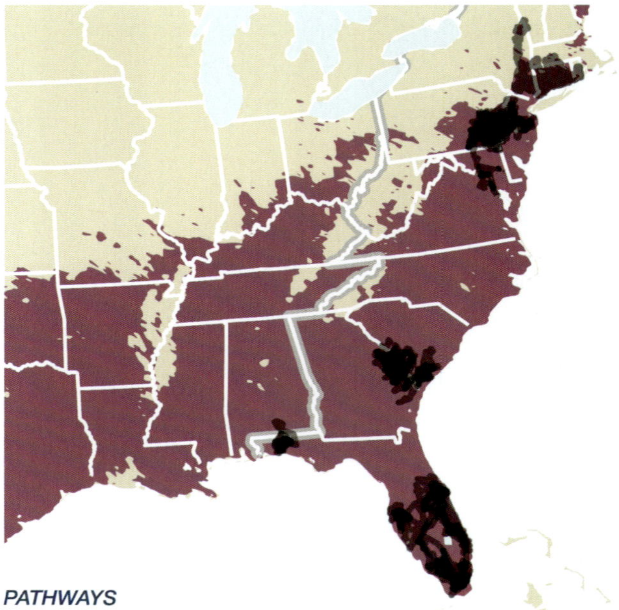

PATHWAYS

MIGRATORY CONNECTIVITY

OVERALL Migratory connectivity is known from banding and tracking data. Breeding populations in all flyways spend the nonbreeding season within the same flyway (98%–100%). Accordingly, migratory connectivity is strong (MC = 1).

BANDING Leg-banding was prohibited due to vultures' urohidrosis behavior (i.e., excreting on their legs). It was shown that the hardening of excreta around the band caused injuries to the tarsi (Houston and Terry 2003); however, wing tagging still occurs. Prior to the ban, banding occurred mostly in the southern United States, where birds are suspected to be resident. Band recoveries in a Louisiana study show local movements of <150 km for most birds (84%), but a few birds moved >900 km east to South Carolina and Florida (Buckley et al. 2022). Data from band recoveries in other southern states from the early part of the 1900s showed that usual movements were <80 km (Buckley et al. 2022).

SATELLITE TRACKING Tagged birds in Pennsylvania and West Virginia showed wider, more dispersed movement patterns in the post-breeding and nonbreeding months than during pre-breeding (Conservation Science Global, Hawk Mountain, unpublished data), but no clear pattern of migration was displayed. Resident birds in southern states are documented moving more than 150 km among roosts on occasion (Buckley et al. 2022). A single juvenile bird tagged near the northern limit of the breeding range in western Massachusetts moved 50 km south into Connecticut in its first nonbreeding season, remaining there for the next year, while on occasion moving east to Rhode Island and west into New York but not exhibiting migration (Conservation Science Global, Hawk Mountain, unpublished data).

OTHER APPROACHES No stable isotope or genetic studies of migratory connectivity were found for this species.

MIGRATORY PATHWAYS Black Vultures rely on thermals to save energy during movements. GPS-tagged birds in Georgia and South Carolina showed a strong relationship between thermal uplift strength and the distance covered and travel rates (Buckley et al. 2022). In eastern hawk watches, highest counts are consistently recorded at eastern Maryland and Virginia, where counts of more than 400 individuals can occur, and in Corpus Christi, Texas, annual fall counts exceed 1,000 vultures.

DATA CONTRIBUTORS

M.L. Avery; D.R. Barber; J. Beasley; K. Bildstein; A.L. Bryan; M.E. Byrne; T.L. DeVault; A.E. Holland; J.S. Humphrey; B.M. Kluever; O.E. Rhodes Jr; J.-F. Therrien; E.A. Tillman; D. Varland.

Cathartes aura TURKEY VULTURE

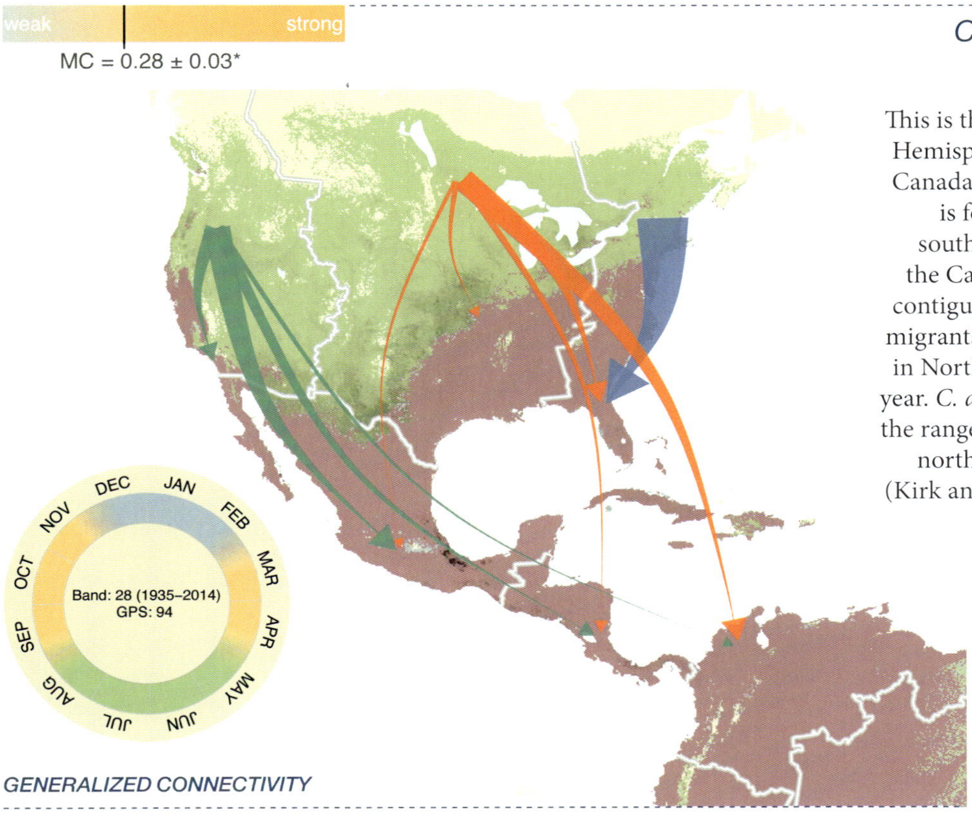

GENERALIZED CONNECTIVITY

DISTRIBUTION AND ABUNDANCE

This is the most widespread vulture species in the Western Hemisphere, with a distribution extending from southern Canada to the southern tip of South America. The species is found year-round from the mid-latitudes of the US south to northern Argentina and Chile and throughout the Caribbean. Northern migrants breed throughout the contiguous US and into southern Canada, while southern migrants reach the tip of South America. Three subspecies in North America occupy different ranges throughout the year. *C. a. septentrionalis* breeds in the northeastern part of the range, west to Minnesota, Kansas, and Oklahoma, with northern populations migrating to the southeastern US (Kirk and Mossman 2020). *C. a. meridionalis* breeds in the western parts of the range, east to Manitoba and south to Texas, migrating to nonbreeding areas in the southwestern US and Central and South America. Birds in the southern parts of the breeding range may remain resident throughout the year (Kirk and Mossman 2020). *C. a. aura* breeds in the southwestern US south to Costa Rica and spends the nonbreeding season from Mexico to Panama (Kirk and Mossman 2020).

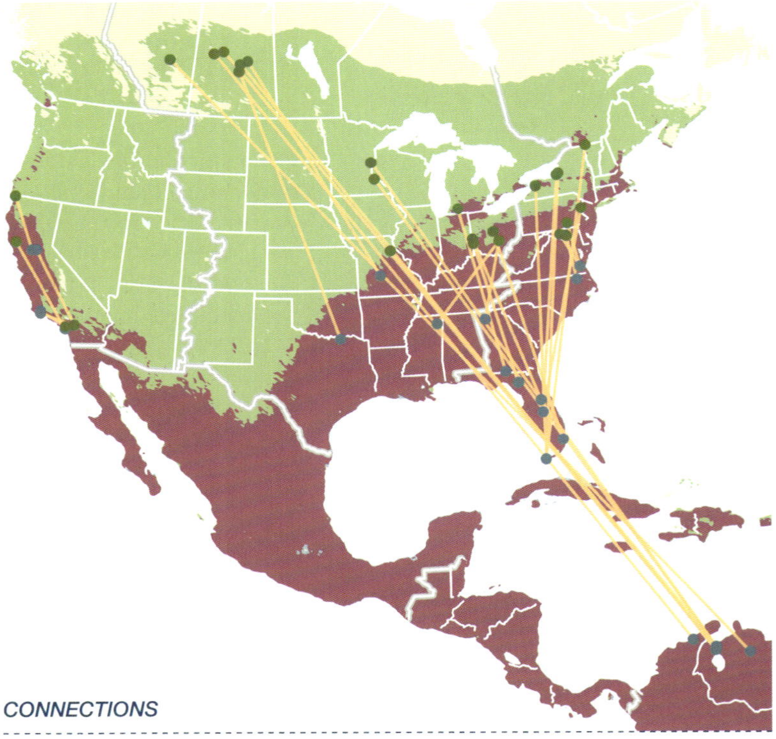

CONNECTIONS

	Atlantic	Central America	Mexico	Midcontinent	Northern SA	Pacific
Atlantic (n = 39)	100%	–	–	–	–	–
Midcontinent (n = 30)	37%	10%	3%	7%	43%	–
Pacific (n = 37)	–	19%	49%	–	5%	27%

ORDER Cathartiformes
FAMILY Cathartidae
BREEDING HABITAT Open Country, Habitat Mosaic
NONBREEDING HABITAT Open Country, Habitat Mosaic
GLOBAL POPULATION SIZE 28,000,000
POPULATION CHANGE SINCE 1970 +217%
IUCN RED LIST CATEGORY Least Concern
R2R TIPPING POINT SPECIES No
US LISTING STATUS No
CANADA SPECIES AT RISK No
PIF CCS 6/20

MIGRATORY CONNECTIVITY

OVERALL The migratory connectivity for Turkey Vultures is well documented from satellite tracking and limited banding data. In North America, all individuals (100%) breeding in the Atlantic Flyway spend the nonbreeding season within the same

INTERESTING NOTE

Turkey Vultures migrate in flocks that can number several thousand birds. Over 1.5 million Turkey Vultures can be observed migrating over Veracruz, Mexico, during post-breeding migration, with one-day counts exceeding 250,000.

flyway. Conversely, individuals that breed in the Midcontinent Flyway spend the nonbreeding season in several regions, including northern South America (43%), the Atlantic Flyway (37%), and Central America (10%). Almost half of the individuals that breed in the Pacific Flyway spend the nonbreeding season in Mexico (49%), while 27% move within the Pacific Flyway, and 19% spend the nonbreeding season in Central America. Overall, migratory connectivity is moderate (MC = 0.28 ± 0.03).

BANDING Thousands of individuals were banded in eastern North America up until 1980, when it was determined that leg bands caused injuries to the tarsi from hardening of excreta around the band, after which leg-banding was prohibited (Houston and Terry 2003). Band reencounters show similar patterns to satellite telemetry data: birds from the northeastern and upper Midwest US migrating to Florida for the nonbreeding season and connections between central Canada and northern South America. In the western US, band reencounters show short-distance movements from Northern to Southern California.

SATELLITE TRACKING Turkey Vulture migration has been well-studied via satellite telemetry (Mandel et al. 2008; Dodge et al. 2014; Grilli et al. 2017; Holland et al. 2017, 2019; Avery et al., unpublished data; Hawk Mountain Sanctuary, unpublished data), and migratory connectivity across the subspecies is strong. In North America, individuals tracked from the Atlantic and upper Midwest region of the US (*C. a. septentrionalis*) primarily spent the nonbreeding season in the southeastern US, particularly Florida (Dodge et al. 2014; Holland et al. 2017, 2019, Avery et al., unpublished data). Individuals from the midcontinent region (*C. a. meridionalis*) migrated to nonbreeding sites in Central America and northern South America (Dodge et al. 2014; Hawk Mountain Sanctuary, unpublished data). Individuals in the west migrated to nonbreeding sites in Mexico and Central America (Dodge et al. 2014; Hawk Mountain Sanctuary, unpublished data). Individuals tagged in the southeastern US were mostly year-round residents, although two individuals migrated to south Florida (Holland et al. 2017, 2019). In South America, individuals migrated between the Southern Cone and Andean West and Brazil regions (Hawk Mountain Sanctuary, unpublished data).

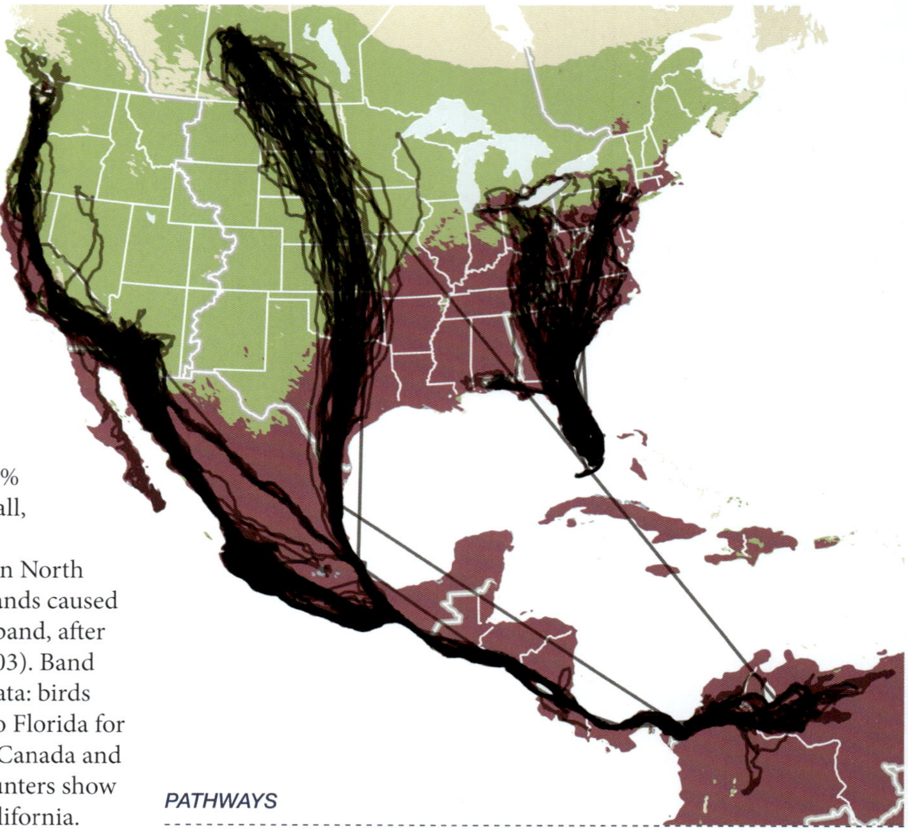

PATHWAYS

MIGRATORY PATHWAYS In North America, three major migratory pathways have been documented via satellite tracking: Atlantic, Midcontinent, and Pacific coast. Individuals tracked from the Atlantic region primarily migrated along the Appalachian Mountains and Atlantic coast. Individuals from the midcontinent region migrated along the Great Plains to the Gulf of Mexico coastal plain and then followed the Central American land corridor to nonbreeding sites. Individuals from the Pacific region migrated along the Pacific coastal plain. Highly reliant on soaring flight, Turkey Vultures avoid large water bodies, over which soaring can be difficult. Thus, migration concentrates along coastlines and over coastal plains or mountain slopes, where thermals are consistent. Turkey Vultures also funnel along major north-south oriented ranges and ridges throughout the continent, using wind updrafts to facilitate their migration. Important mountain corridors include the Sierras, Rockies, and Appalachians. That said, thanks to a wide distribution, Turkey Vultures can be encountered almost anywhere on the continent during migration.

DATA CONTRIBUTORS

M.L. Avery; D.R. Barber; J.C. Beasley; K. Bildstein; A.L. Bryan; M.E. Byrne; T.L. DeVault; A.E. Holland; J.S. Humphrey; B.M. Kluever; O.E. Rhodes Jr; J.-F. Therrien; E.A. Tillman; D. Varland.

HAWKS, EAGLES, KITES, AND OSPREY (ACCIPITRIFORMES)

WRITING LED BY **LAURIE J. GOODRICH, JEAN-FRANÇOIS THERRIEN, AMY L. SCARPIGNATO** *AND* **PETER P. MARRA**

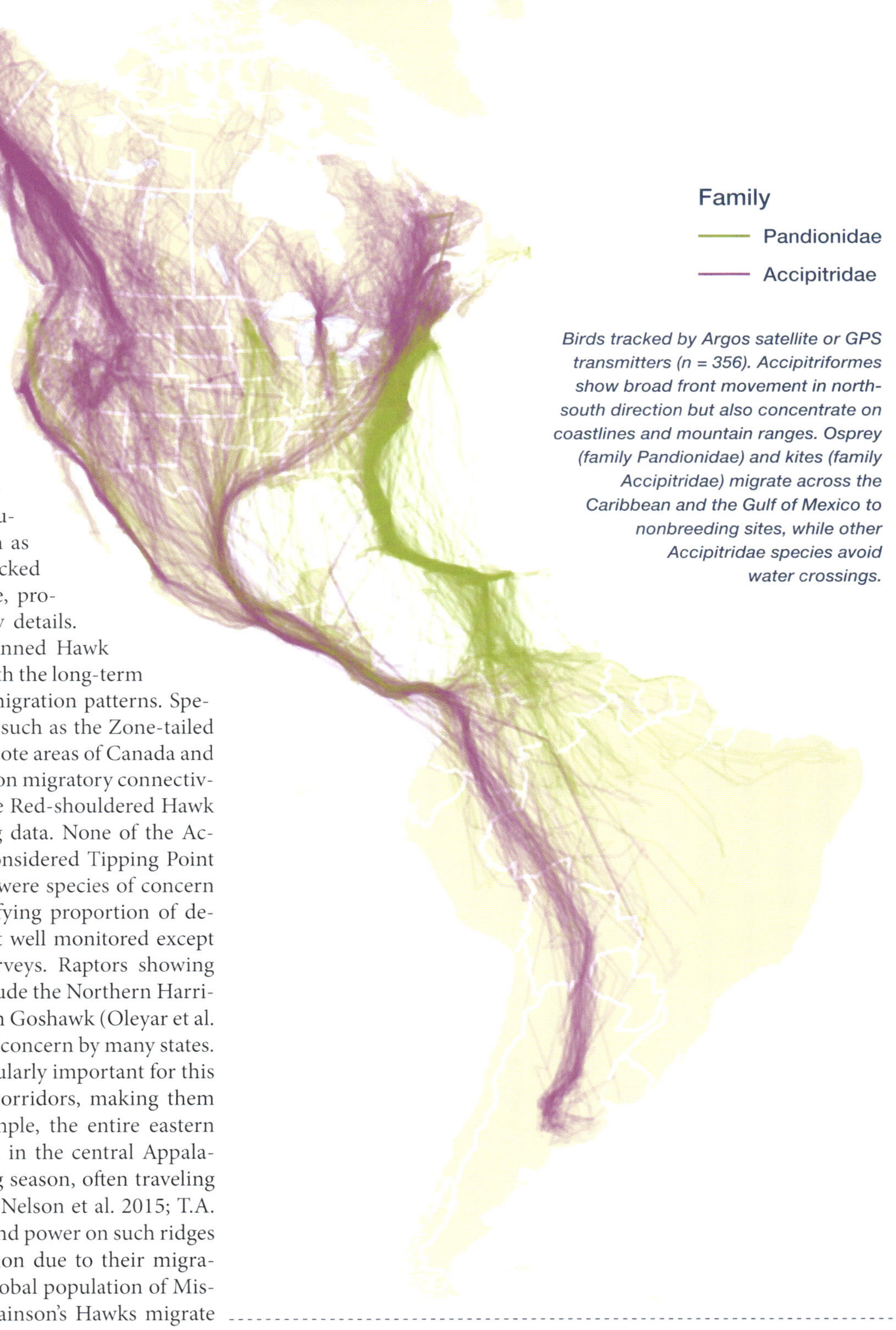

Birds tracked by Argos satellite or GPS transmitters (n = 356). Accipitriformes show broad front movement in north-south direction but also concentrate on coastlines and mountain ranges. Osprey (family Pandionidae) and kites (family Accipitridae) migrate across the Caribbean and the Gulf of Mexico to nonbreeding sites, while other Accipitridae species avoid water crossings.

The migratory connectivity has been well studied in diurnal raptors, with 95% of species having some migratory connectivity data, primarily from band reencounter data. Due to their larger size and ability to carry various transmitters, 58% of the 19 raptor species in this group have been tracked with Argos transmitters, 63% with GPS transmitters, and 5% (one species) with radio transmitters. Some species, particularly those showing declines in the past, such as the Bald Eagle and Osprey, have been well-tracked across their breeding and nonbreeding range, providing comprehensive migratory connectivity details. Easily trapped species such as the Sharp-shinned Hawk and Cooper's Hawk, are frequently banded, with the long-term band recovery data useful in detailing their migration patterns. Species found in the southwestern United States, such as the Zone-tailed Hawk and Gray Hawk, and populations in remote areas of Canada and Alaska show the largest dearth of information on migratory connectivity. Even some widespread species, such as the Red-shouldered Hawk or American Goshawk, have limited tracking data. None of the Accipitriformes species are federally listed or considered Tipping Point species, although the Osprey and Bald Eagle were species of concern through the early 2000s. In addition, quantifying proportion of declines in this group is difficult as they are not well monitored except with migration and nonbreeding season surveys. Raptors showing steady long-term declines continent-wide include the Northern Harrier, the Sharp-shinned Hawk, and the American Goshawk (Oleyar et al. 2021), and these species are listed as species of concern by many states.

Understanding migration patterns is particularly important for this group as they often concentrate in narrow corridors, making them vulnerable to conservation threats. For example, the entire eastern population of the Golden Eagle concentrates in the central Appalachians during migration and the nonbreeding season, often traveling on one or two ridges of the mountain chain (Nelson et al. 2015; T.A. Miller et al. 2016). The growth in industrial wind power on such ridges may pose a particular threat to this population due to their migration behavior. In addition, nearly the entire global population of Mississippi Kites, Broad-winged Hawks, and Swainson's Hawks migrate through a narrow corridor in eastern Mexico almost synchronously during post-breeding migration, making this site, popularly known as the River of Raptors in Veracruz, Mexico, particularly important for their long-term conservation (Ruelas Inzunza 2005).

Accipitriformes display a wide variety of migratory connectivity patterns, from strong to moderate to weak. Most with adequate data show moderate to strong connectivity. The Bald Eagle shows high connectivity within flyways, while the Osprey and Broad-winged Hawk show moderate connectivity patterns. Some species with extensive information, e.g., Cooper's Hawk, show complex patterns among populations, with strong connectivity in Atlantic populations, weak connectivity in midcontinent populations, and moderate connectivity in Pacific populations. For some species, connectivity is unknown due to lack of tracking or extensive banding data. Where tracking data exists within this order, individual birds often return to the same location year after year, showing site fidelity to nonbreeding and breeding sites (e.g., Dodge et al. 2014; McCabe et al. 2020; Hawk Mountain, unpublished data).

OSPREY *Pandion haliaetus*

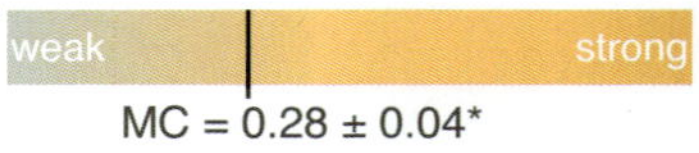

DISTRIBUTION AND ABUNDANCE

Ospreys breed throughout the US and Canada except for the interior US. Highest abundance occurs along the Atlantic, Pacific, and Gulf coasts. Other areas of high abundance include Alaska south of the Brooks Range, northern Canada, and throughout the Great Lakes and northwestern US states (Bierregaard et al. 2020). The breeding range has expanded to include more consistent nesting east and south of the Appalachian Mountains and along waterways in western US states and south to New Mexico (Bierregaard et al. 2020). Highest abundance during the nonbreeding season occurs throughout Florida and the Caribbean, along the US Gulf Coast, coastal areas and waterways throughout Mexico and Central America, and northern South America (Martell et al. 2014). Recent analyses show 36% of 58 eastern hawk watches are experiencing significant declines in migrating Ospreys in northeastern states from 2009 to 2019. Nonbreeding numbers in Gulf and western states have increased during the same period, suggesting possible migration shortstopping (birds shortening their migration distance) in those populations (Oleyar et al. 2021).

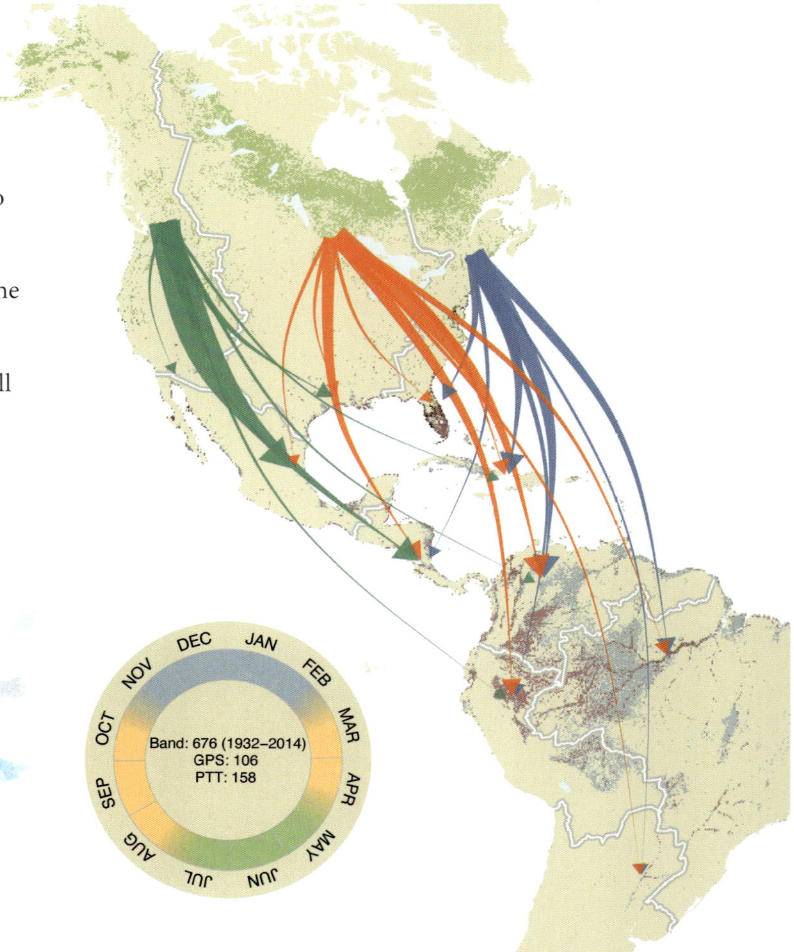

GENERALIZED CONNECTIVITY

ORDER Accipitriformes
FAMILY Pandionidae
BREEDING HABITAT Wetlands, Lakes and Rivers
NONBREEDING HABITAT Wetlands, Generalist
GLOBAL POPULATION SIZE 1,200,000
POPULATION CHANGE SINCE 1970 +304%
IUCN RED LIST CATEGORY Least Concern
R2R TIPPING POINT SPECIES No
US LISTING STATUS No
CANADA SPECIES AT RISK No
PIF CCS 7/20

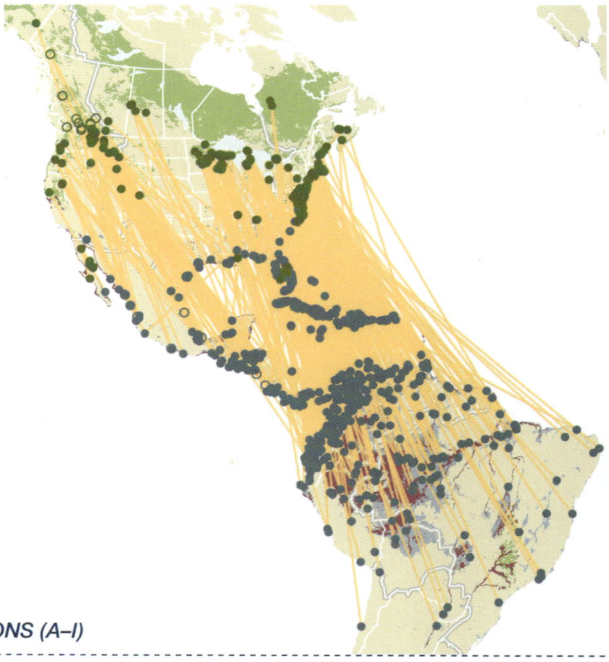

CONNECTIONS (A–I)

MIGRATORY CONNECTIVITY

OVERALL Ospreys breeding in the Atlantic Flyway migrated to nonbreeding areas in northern South America (33%) and the Caribbean (29%), as well as Brazil (15%) and the Atlantic Flyway (12%). Midcontinent breeding birds migrated to areas throughout the nonbreeding range with highest numbers in northern South America (33%), Central America (18%), and the Andean West (17%). Pacific breeding birds primarily spend the nonbreeding season in Mexico (48%) and Central America (34%). Despite mixing of breeding populations on nonbreeding areas, longitudinal patterns persist, resulting in moderate migratory connectivity (MC= 0.28 ± 0.04).
BANDING Banding data shows a similar pattern to satellite telemetry data, with no overlap in nonbreeding band reencounters for birds from

	Andean West	Atlantic	Brazil	Caribbean	Central America	Mexico	Midcontinent	Northern SA	Pacific	Southern Cone
Atlantic (n = 555)	7%	12%	15%	29%	4%	<1%	–	33%	–	1%
Midcontinent (n = 122)	17%	3%	8%	13%	18%	2%	3%	33%	–	2%
Pacific (n = 94)	2%	–	–	1%	34%	48%	48%	2%	1%	–

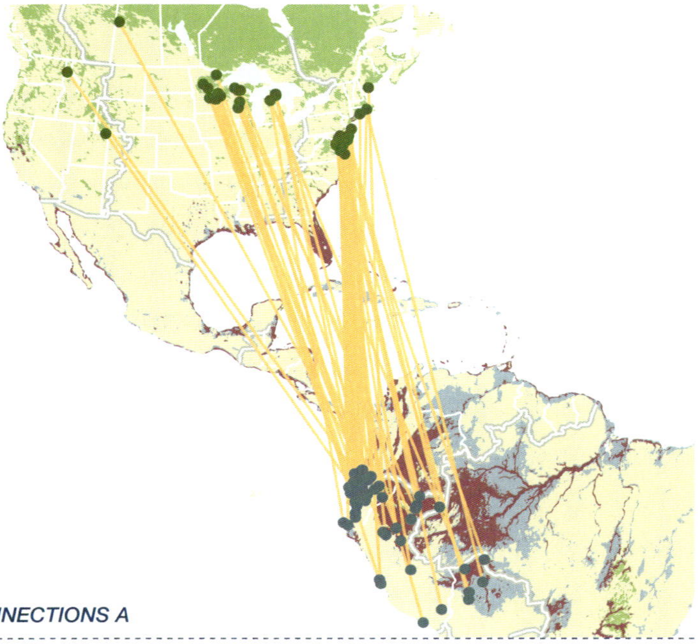

CONNECTIONS A

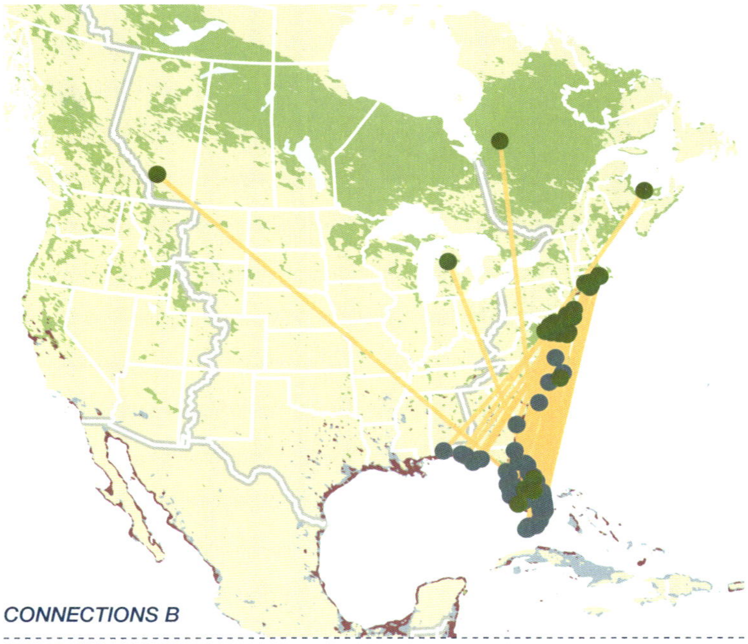

CONNECTIONS B

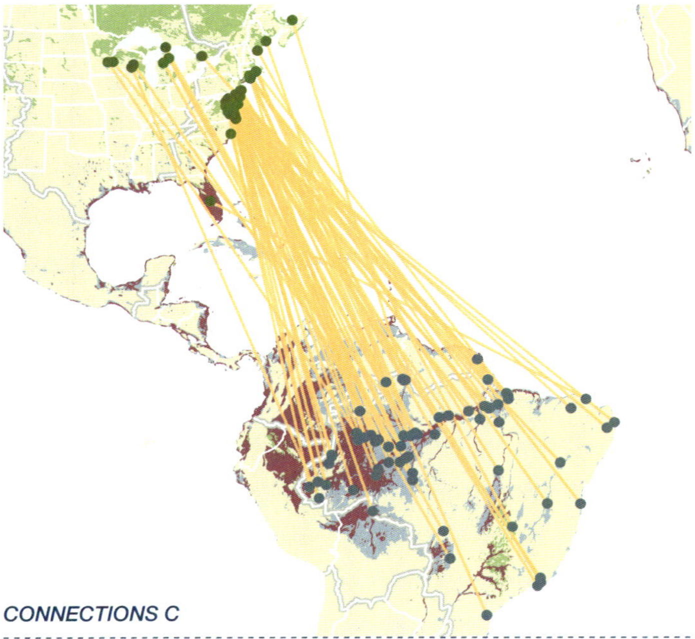

CONNECTIONS C

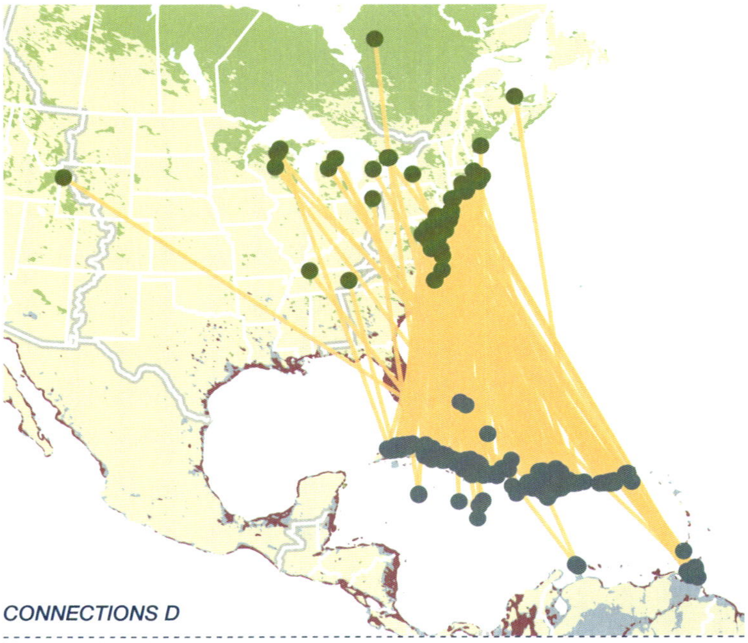

CONNECTIONS D

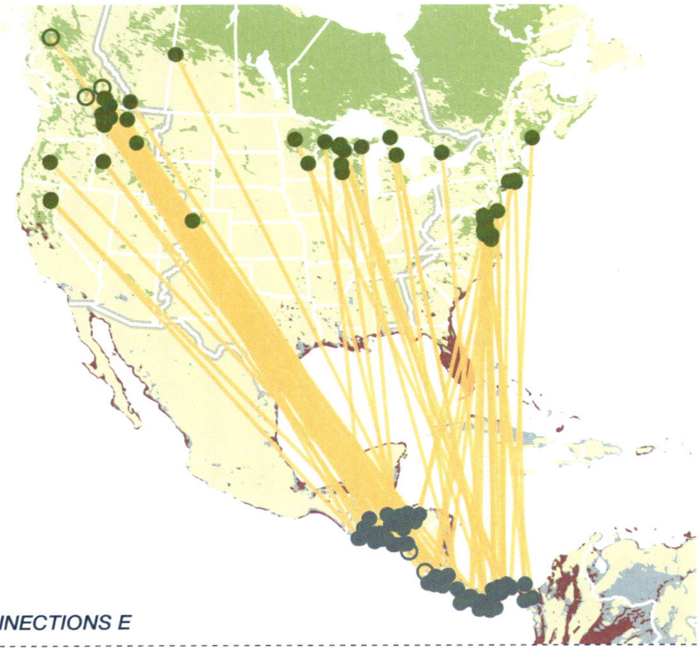

CONNECTIONS E

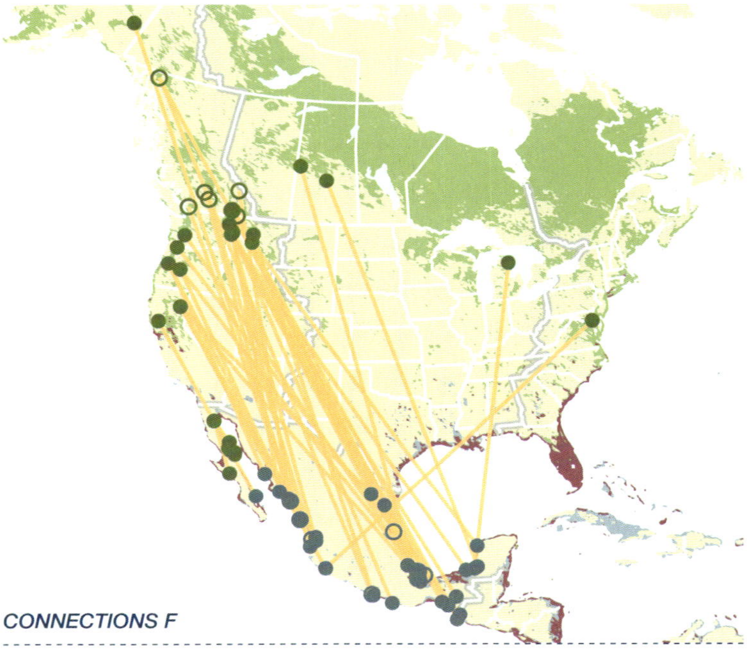

CONNECTIONS F

HAWKS, EAGLES, KITES, AND OSPREY (ACCIPITRIFORMES)

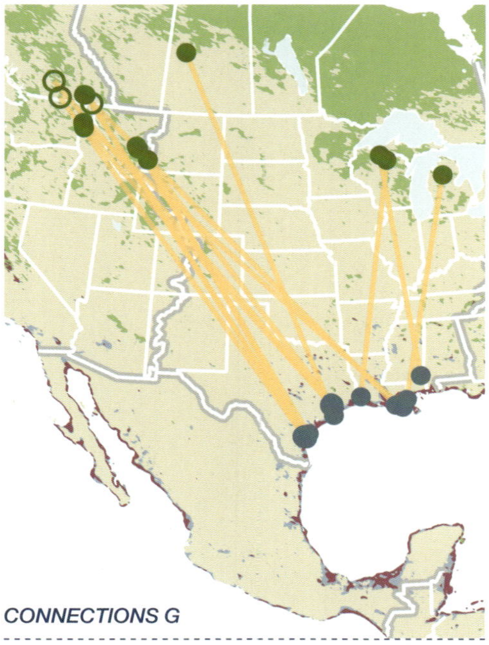

CONNECTIONS G

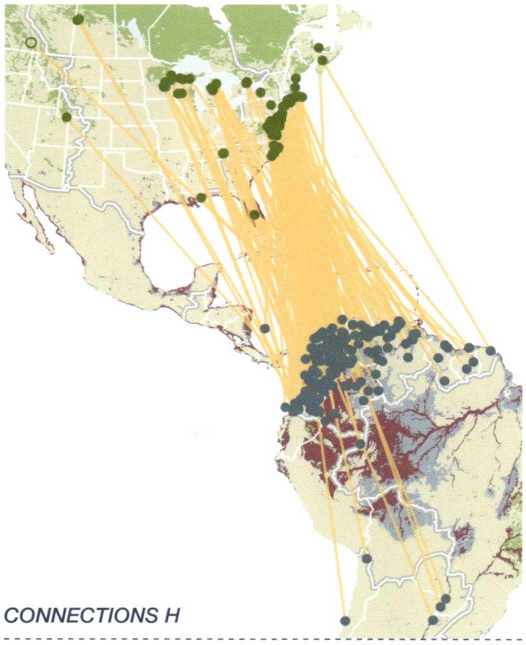

CONNECTIONS H

CONNECTIONS I

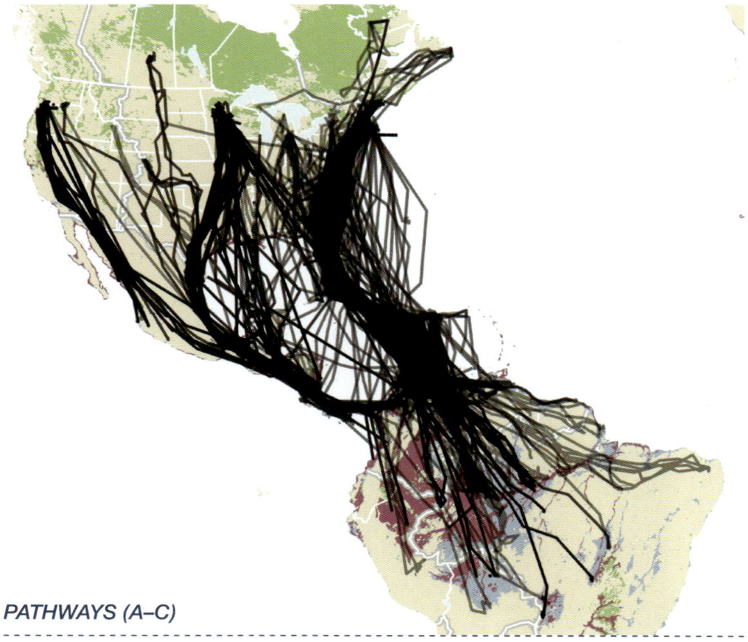

PATHWAYS (A–C)

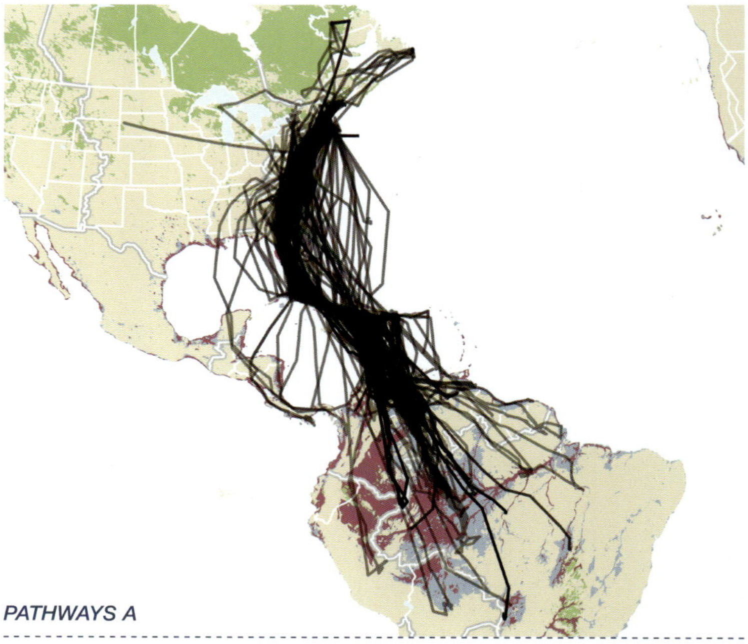

PATHWAYS A

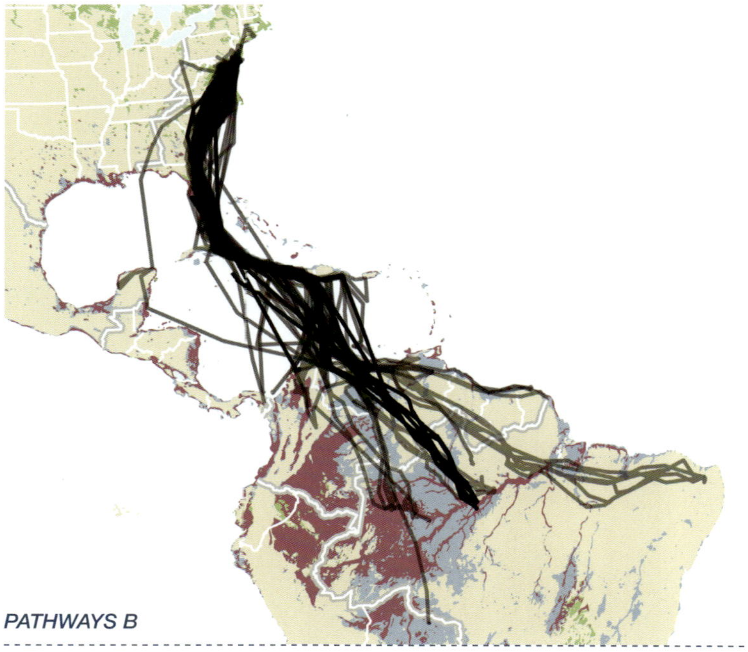

PATHWAYS B

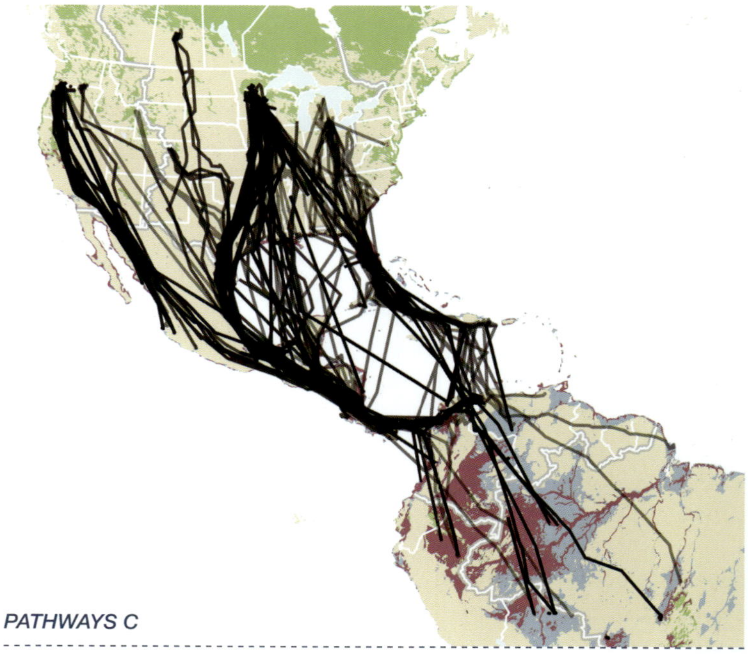

PATHWAYS C

INTERESTING NOTE

Adult Ospreys exhibit strong site fidelity between years (Washburn et al. 2014). During southward migration, females depart earlier than males by nearly a month, but in northward migration, males migrate faster and return earlier than females (Martell et al. 2014). Juvenile North American Ospreys remain on nonbreeding sites until their second year of age, 18 months later, when they return to breed (Bierregaard et al. 2020).

western and eastern populations (Martell et al. 2001). The majority of nonbreeding reencounters from Brazil were along Amazonian riverways, indicating the importance of this habitat to nonbreeding Ospreys (Mestre and Bierregaard 2009).

SATELLITE TRACKING Tracking data are from throughout the contiguous US and North Atlantic Canada (Martell et al. 1998, 2001, 2014, 2019; Houston and Martell 2002; Stout et al. 2009; Washburn 2014; Bedrosian et al. 2015; Henny and Martell 2017; Nye et al. 2018; R. Bierregaard, unpublished data; Bird Conservancy of the Rockies https://birdconservancy.org/a-tale-of-two-osprey/; T.E. Chubbs, unpublished data; Craighead Beringia South, unpublished data; B. Jensen, unpublished data), but data are lacking from populations in Alaska and across the northern Canadian portion of the breeding range. Birds breeding in the west migrate to nonbreeding areas primarily in the Mexico, Central America, and the Gulf Coast region. Females in the west traveled farther than males from the same population, which resulted in females typically spending the nonbreeding season south of males (Martell et al. 2001). Birds breeding in the midcontinent and east migrate primarily to South America, with eastern breeders also migrating to nonbreeding areas in Florida and the Caribbean.

STABLE ISOTOPES No stable isotope studies on migratory connectivity were found.

GENETICS Ospreys from both *P. h. carolinensis* and *ridgwayi* ssp. originate from a single lineage, separate from the other genetic groups throughout the world, which may have implications for conservation strategies (Monti et al. 2015).

MIGRATORY PATHWAYS Ospreys generally migrate along a broad southerly direction during post-breeding migration, with migrants concentrating along leading lines, such as coasts and ridges in North and Central America (Farmer et al. 2010; Bierregaard et al. 2020). Eastern birds migrate south through Florida and cross the Caribbean to Cuba and Hispaniola before flying to Colombia, Venezuela, and northern South America (Martell et al. 2014). Some Atlantic coast juveniles fly due south from nesting sites, compensating for the effects of perpendicular wind drift during open-ocean flights across the Atlantic to the Caribbean and south Florida in a one-way trip exceeding 1,500 km (Horton et al. 2014) before continuing into South America. Midwestern birds migrated south using either the Florida route or along the east coast of Mexico to Central America (Martell et al. 2014; Bedrosian et al. 2015). Western birds traveled southwest into Mexico, either west or east of the Sierra Madre, and then south through Central America. Birds migrate northward using the same pathways taken in autumn (Martell et al. 2014). Compared to post-breeding migrations, all male and all east coast Ospreys spent fewer days on migration and traveled shorter distances overall (on average) each day during pre-breeding migration (Martell et al. 2014).

DATA CONTRIBUTORS

R. Bierregaard; T.E. Chubbs; R. Jennings; B. Jensen; R. Kays; I. MacLeod; M. Martell; P. Nye; Craighead Beringia South; Grand Teton National Park.

ACCIPITRIDAE | *HAWKS, EAGLES, AND KITES*

Elanoides forficatus SWALLOW-TAILED KITE

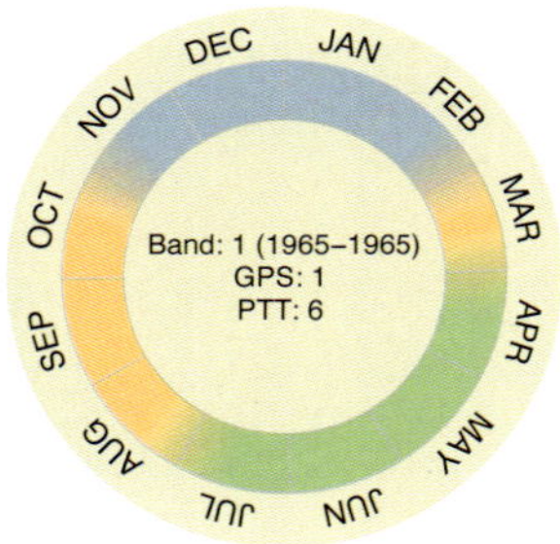

DISTRIBUTION AND ABUNDANCE

The Swallow-tailed Kite has a limited range within North America, but it is widespread across South America. Two-thirds of the North American breeding population (*E. f. forficatus*) nest in Florida (Meyer 2020). The other one-third breed within six states of the southeastern US: South Carolina, Georgia, Alabama, Mississippi, Louisiana, and Texas. Historically, they bred throughout the eastern US (occurring in 20 states), but their numbers and range have greatly contracted, likely due to loss of forested wetland habitat. This subspecies migrates to nonbreeding areas in Brazil, Paraguay, and Bolivia. The other recognized subspecies, *E. f. yetapa*, breeds in southern Mexico, Central America, and the northern half of South America, where it is a year-round resident.

ORDER Accipitriformes
FAMILY Accipitridae
BREEDING HABITAT Forests, Temperate Eastern
NONBREEDING HABITAT Forests, Tropical Lowland Evergreen
GLOBAL POPULATION SIZE 260,000
POPULATION CHANGE SINCE 1970 +1136%
IUCN RED LIST CATEGORY Least Concern
R2R TIPPING POINT SPECIES No
US LISTING STATUS No
CANADA SPECIES AT RISK No
PIF CCS 10/20

MIGRATORY CONNECTIVITY

OVERALL There is some evidence for weak migratory connectivity, with tracked individuals migrating to a small area within Brazil, Paraguay, and Bolivia, but more data is needed.

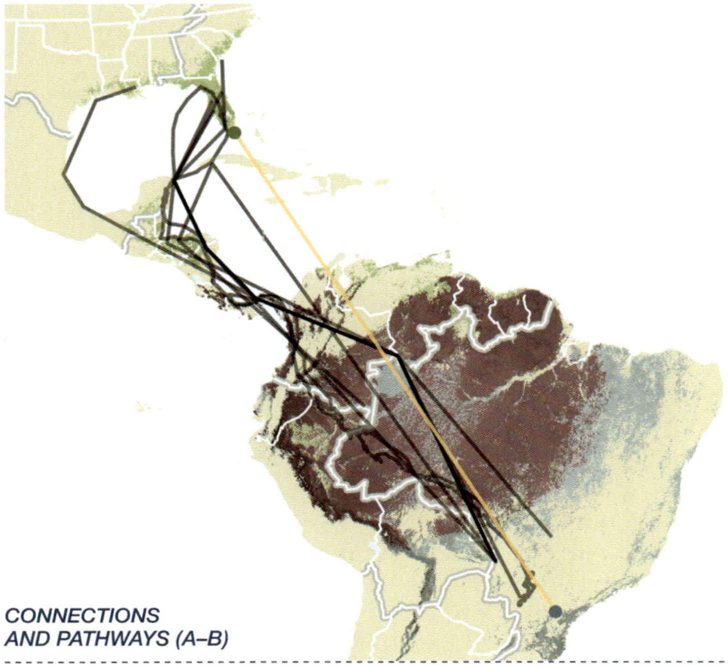

CONNECTIONS AND PATHWAYS (A–B)

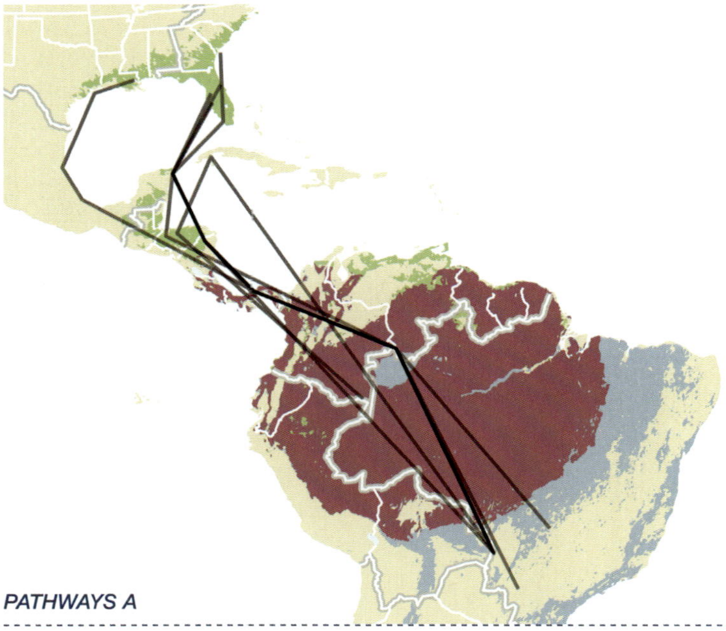

PATHWAYS A

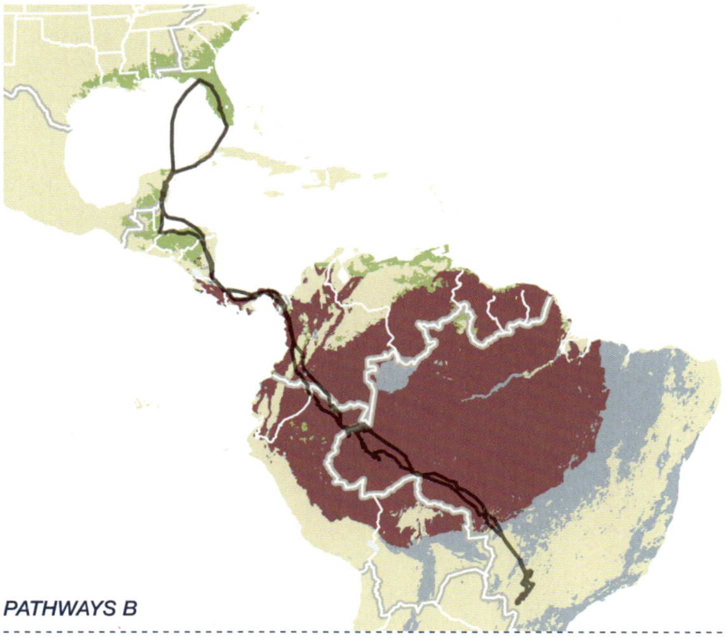

PATHWAYS B

INTERESTING NOTE

Swallow-tailed Kites begin post-breeding migration early, with some leaving mid-July, and most Florida birds showing a median departure date of August 10 and median arrival date of mid to late October to nonbreeding sites in southern Brazil (Zimmerman 2004). Water crossings from Cuba to Yucatán may incur high mortality, with more than 80% of satellite-tagged kites dying during migration (K. Meyer and G. Kent, unpublished data).

BANDING There are few recoveries reported outside of satellite telemetry studies. One juvenile banded in Florida Keys was recovered in Brazil (Meyer 2020).

RADIO TRACKING Individuals from the Florida breeding population migrated to nonbreeding areas in Bolivia, Brazil, and Paraguay (Zimmerman 2004).

SATELLITE TRACKING In 1996, the Avian Research and Conservation Institute began deploying satellite transmitters on breeding Swallow-tailed Kites in Florida, eventually describing the nonbreeding areas in Brazil, Paraguay, and Bolivia (K. Meyer and G. Kent, unpublished data). Following this study, Zimmerman (2004) deployed satellite tags on individuals from the Florida breeding population. The kites also migrated to nonbreeding areas in Bolivia, Brazil, and Paraguay (Zimmerman 2004).

OTHER APPROACHES No genetic or stable isotope studies of migratory connectivity were found for this species.

MIGRATORY PATHWAYS Swallow-tailed Kites follow a narrow, 8,000-km migration corridor in both pre- and post-breeding migrations. The majority of kites pass through peninsular Florida, cross the 200-km Straits of Florida to Cuba, then fly another 200 km to the important stopover area on the Yucatán Peninsula of Mexico and Belize (Bildstein et al. 2002; Zimmerman 2004; K. Meyer and G. Kent, unpublished data). From the Yucatán, birds then follow a narrow migration corridor south through Central America and South America to central South America (Zimmerman 2004; K. Meyer and G. Kent, unpublished data). Other populations of kites nesting farther north and west along the Gulf Coast may migrate west through Texas, around the Gulf of Mexico, and south through Veracruz, Mexico (Goodrich and Smith 2008; K. Meyer and G. Kent, unpublished data). From here, all North American kites continue along the eastern edge of Central America into Columbia, where they cross the Andes and head southeast to nonbreeding sites in southwestern Brazil, eastern Paraguay, and Bolivia (Zimmerman 2004). Swallow-tailed Kites concentrate in largest numbers in the Mesoamerican Biological Corridor south of Belize and Mexico, where Kekoldi, Costa Rica, hawk watches record 1,000–2,500 during each post-breeding migration, often passing in flocks of more than 100 individuals (Goodrich and Smith 2008; hawkcount.org, unpublished data). Smaller numbers are observed along the Andes in sites such as Tolima, Colombia, and Concepción, Bolivia (hawkcount.org, unpublished data). Some hawk counts begin migration counts too late to capture peak flights of this early migrant (e.g., in Veracruz, Mexico).

DATA CONTRIBUTORS

G.M. Kent; K.D. Meyer; Avian Research and Conservation Institute; BirdsCaribbean; Florida Fish and Wildlife Conservation Commission; National Fish and Wildlife Foundation; National Wildlife Refuges, US Fish and Wildlife Service.

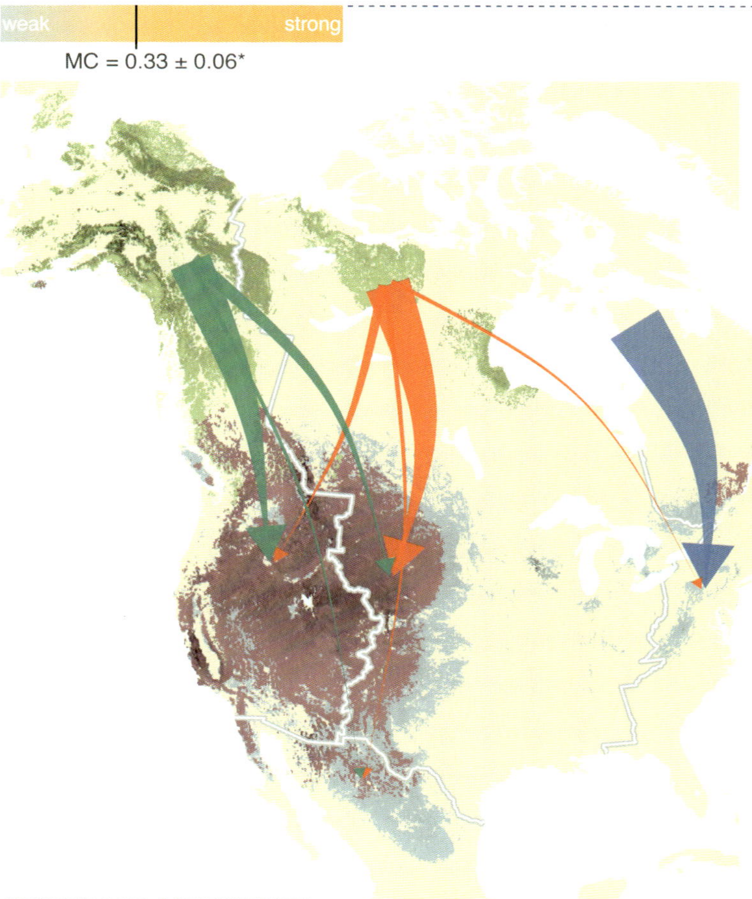

GENERALIZED CONNECTIVITY

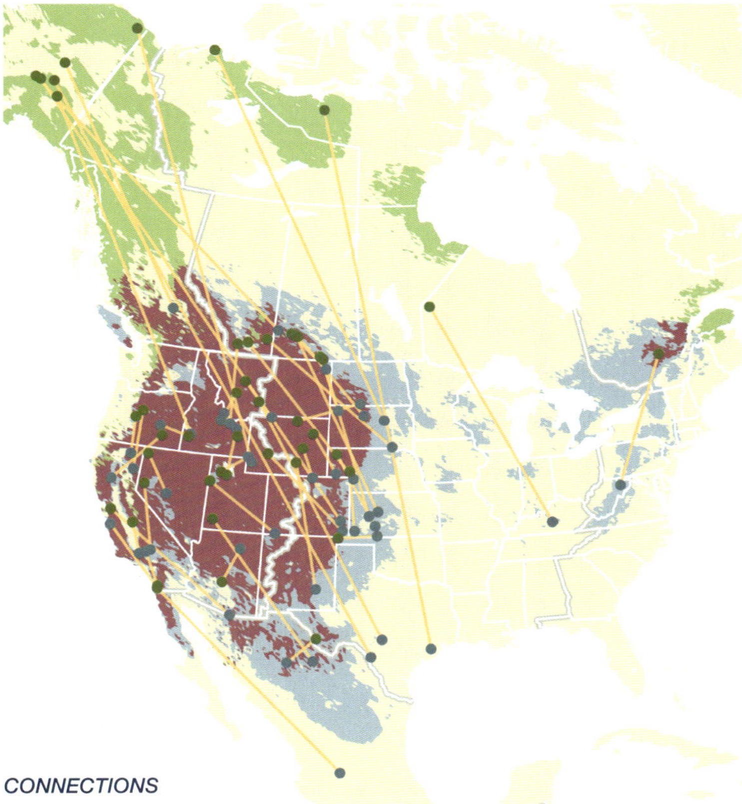

CONNECTIONS

	Atlantic	Mexico	Midcontinent	Pacific
Atlantic (n = 8)	100%	–	–	–
Midcontinent (n = 37)	3%	5%	78%	14%
Pacific (n = 167)	–	4%	23%	73%

Aquila chrysaetos **GOLDEN EAGLE**

DISTRIBUTION AND ABUNDANCE

Golden Eagles breed in much of western North America from Canada and Alaska south to central Mexico. Smaller numbers nest east across the northern region of the Canadian provinces with a small eastern population found breeding in eastern Quebec, Labrador, and New Brunswick. High nonbreeding abundance exists throughout the Intermountain West into Mexico and throughout central California from the coast to the western edge of the Central Valley. Overall, the North American population trends appear predominantly stable (Katzner et al. 2020; Oleyar et al. 2021). Some western United States populations have declined in recent decades (Kochert and Steenhof 2002; Katzner et al. 2020), and Golden Eagles breeding in most eastern provinces and states declined in the 1970s through the 1990s due to DDT and other factors, but may be continuing a slow rebound started in 1980s (Bednarz et al. 1990; Katzner et al. 2020).

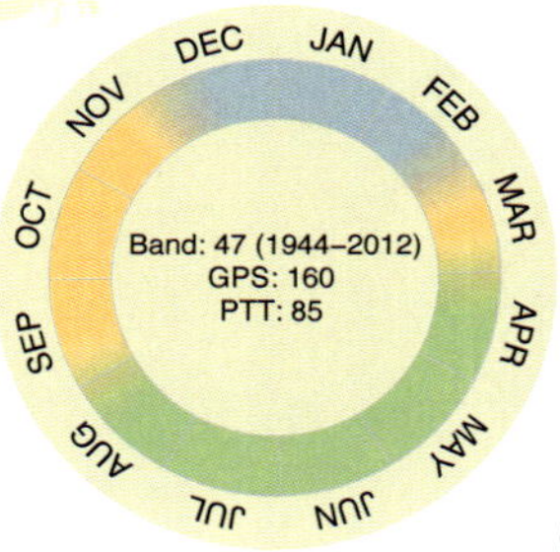

ORDER Accipitriformes
FAMILY Accipitridae
BREEDING HABITAT Open Country, Habitat Mosaic
NONBREEDING HABITAT Open Country, Habitat Mosaic
GLOBAL POPULATION SIZE 130,000
POPULATION CHANGE SINCE 1970 +4%
IUCN RED LIST CATEGORY Least Concern
R2R TIPPING POINT SPECIES No
US LISTING STATUS No
CANADA SPECIES AT RISK No
PIF CCS 10/20

MIGRATORY CONNECTIVITY

OVERALL Golden Eagle migratory connectivity has been studied using several techniques, including satellite telemetry (Brodeur et al. 1996; Goodrich and Smith 2008; McIntyre et al. 2008; J.P. Smith 2010; T.A. Miller et al. 2016; Poessel et al. 2016; J.L. Brown et al. 2017; R.K. Murphy et al. 2017; Bedrosian et al. 2018; Nye et al. 2018; Center for Conservation Biology, unpublished data; J. Barnes, unpublished data; T. Booms, unpublished data; Craighead Beringia South, unpublished data; Earthspan, unpublished data; HawkWatch International, unpublished data; S.B. Lewis and C. McIntyre, unpublished data;

B.W. Smith, unpublished data; Delaware-Otsego Audubon Society, unpublished data; M. Martell and S. Mehus, unpublished data), banding, stable hydrogen isotopes (Domenech et al. 2015; Nelson et al. 2015), and genetics (Craig et al. 2016). Atlantic Flyway breeding birds primarily remain within the Atlantic Flyway (100%). Similarly, Midcontinent Flyway breeding birds mostly remain within Midcontinent Flyway (78%) during the nonbreeding season, while others migrate to the Pacific Flyway (14%), and smaller numbers to

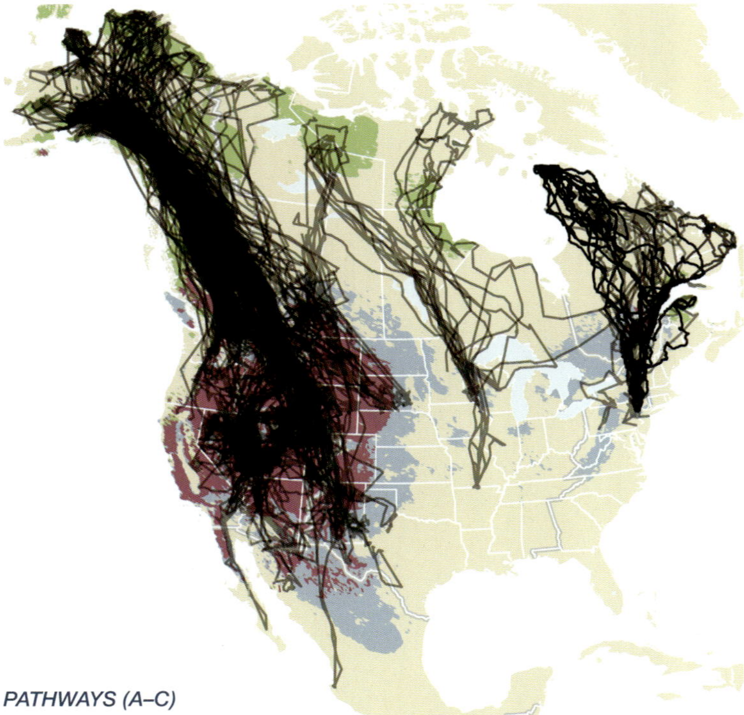

PATHWAYS (A–C)

PATHWAYS A

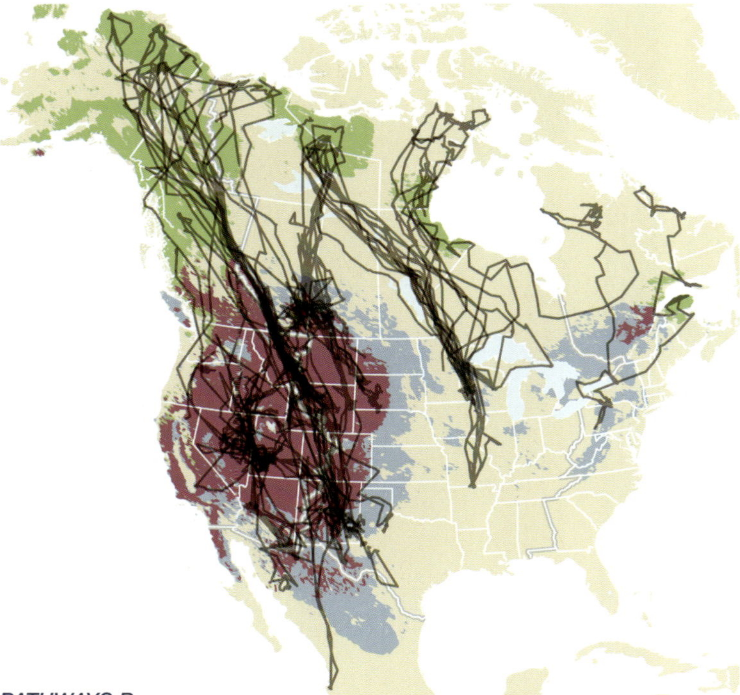

PATHWAYS B

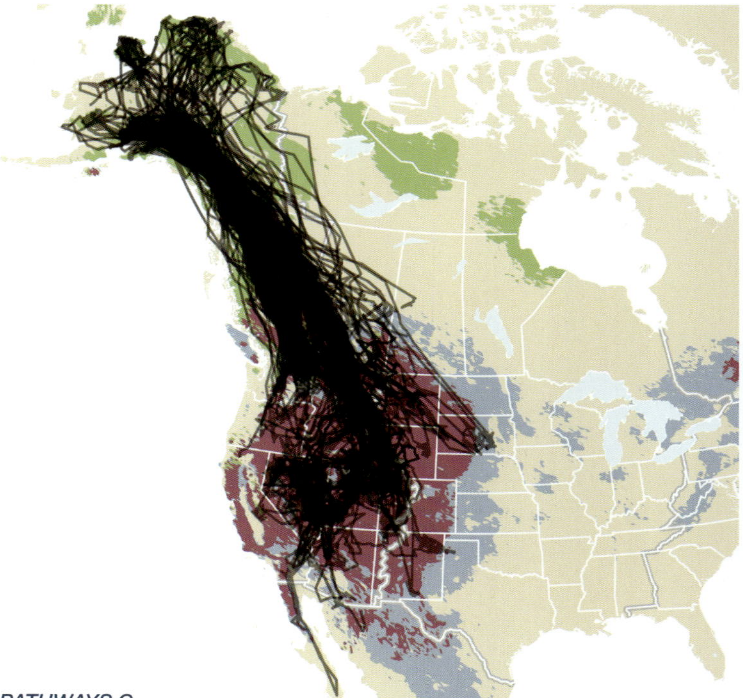

PATHWAYS C

INTERESTING NOTE

The longest migration recorded for Golden Eagles was from the Arctic Slope to central Mexico (about 5,000 km; HawkWatch International, unpublished data), and the highest counts of migrating Golden Eagles are recorded at Mount Lorette, Alberta, where post-breeding counts exceed 3,000 annually (Goodrich and Smith 2008). Golden Eagle adults show slower post-breeding migration, using a time-minimization strategy, whereas in pre-breeding migration, adults move more rapidly north, perhaps more pressed to reclaim nesting sites. Tracked juveniles showed no difference in migration rates between seasons (Miller et al. 2016). Migrating adults tend to return to the same breeding area, but there is less consistency for young birds (Miller et al. 2016; Katzner et al. 2020). Juvenile birds from California showed long-distance movements (up to 1,500 km; Poessel et al. 2016).

Mexico (5%) and the Atlantic Flyway (3%). Pacific Flyway breeding birds primarily spend the nonbreeding season within the Pacific Flyway (73%) as well as the Midcontinent Flyway (23%). Although breeding populations primarily remain within the flyway during the nonbreeding season, the mixing of breeding populations throughout the nonbreeding range moderates the overall strength of migratory connectivity (MC = 0.33 ± 0.06).

BANDING Banding data from Golden Eagles supports satellite telemetry patterns with birds from Alaska, California, and the Intermountain West migrating south-southeast to nonbreeding areas, and birds in the east mostly migrating south to the Appalachian Mountain region. Migration distance varied, with some reencounters in Texas and Mexico from birds that originated from the farthest north breeding areas in Alaska and northern Canada and others from California and the Intermountain West.

SATELLITE TRACKING Satellite tracking suggests moderate longitudinal connectivity (Brodeur et al. 1996; McIntyre et al. 2008; Miller et al. 2016; Poessel et al. 2016; Brown et al. 2017; Murphy et al. 2017; Bedrosian et al. 2018; T. Booms, unpublished data, Center for Conservation Biology, unpublished data; R. Kays, unpublished data; Raptor View Research Institute, unpublished data) and very little to no mixing between eastern (Quebec and east) and western populations (Brown et al. 2017). Tracking data show birds in Alaska migrating to nonbreeding areas in the Pacific Northwest (T. Booms, unpublished data; S.B. Lewis and C. McIntyre, unpublished data), Intermountain West (Goodrich and Smith 2008; McIntyre et al. 2008; Smith 2010; T. Booms, unpublished data; Craighead Beringia South, unpublished data; Earthspan, unpublished data; HawkWatch International, unpublished data; S.B. Lewis and C. McIntyre, unpublished data; B.W. Smith, unpublished data), northern Great Plains (McIntyre et al. 2008), and as far south as Mexico (HawkWatch International, unpublished data). Birds have also been tracked migrating between British Columbia and Oregon (HawkWatch International, unpublished data). In the midcontinent, mixed strategies of migration are evident with both resident behavior and short-distance migrations documented within the Great Basin Desert (J. Barnes, unpublished data) and Intermountain West (HawkWatch International, unpublished data), and south to Texas (HawkWatch International, unpublished data). Long-distance migrations have been documented between the Northwest Territories and Nunavut south to the US midcontinent (M. Martell and S. Mehus, unpublished data) and New Mexico (Smith 2010; HawkWatch International, unpublished data), as well as Alberta to South Dakota, Utah (Goodrich and Smith 2008; Smith 2010; HawkWatch International, unpublished data), and Arizona (B.W. Smith, unpublished data). In the east, populations tracked between breeding areas in northern Quebec and northwestern Newfoundland head to nonbreeding areas in the northeastern US (Nye et al. 2018; Delaware-Otsego Audubon Society, unpublished data; M. Martell and S. Mehus, unpublished data).

GENETICS Genetic studies found poor structure for populations from California, Idaho, and Alaska, indicating the three regions are essentially panmictic, perhaps as a result of natal dispersal, juvenile dispersal, or floating breeding adults (Craig et al. 2016).

STABLE ISOTOPES Stable hydrogen isotope analyses in adult and juvenile birds documented possible leapfrog migration patterns in eastern populations, e.g., northern Quebec and Labrador birds migrating to southern Appalachia, and southern Quebec breeding birds migrating to central Appalachia (Nelson et al. 2015). Stable isotope analyses on feathers from juveniles trapped in migration in Montana suggested these migrants most likely originated from breeding areas in the Yukon, western Northwest Territories, Canada, and eastern Alaska (Domenech et al. 2015).

MIGRATORY PATHWAYS Migrating Golden Eagles concentrate during post-breeding migration along key mountain ranges, while return flights in pre-breeding migrations are more dispersed and often use different routes than during post-breeding (McIntyre et al. 2008; Bedrosian et al. 2018). Based on telemetry data, northwestern breeding Golden Eagles concentrate on migration along the Rocky Mountain Front Range in British Columbia south through Montana, with smaller numbers along the Fraser and Thompson plateaus to the west (McIntyre et al. 2008; Bedrosian et al. 2018; T. Booms, unpublished data; J.P. Smith, unpublished data). South of Montana key routes are not well delineated. Eagles nesting in central Canada migrate south, avoiding the Great Lakes as they head to nonbreeding sites in the central United States. Eastern Golden Eagles largely migrate southwest from eastern Canada along the Appalachian Mountains, with concentrations notable on several Pennsylvania ridges as birds move into West Virginia, Tennessee, and nearby states for the nonbreeding season (Brodeur et al. 1996; Miller et al. 2016; Brown et al. 2017; Nye et al. 2018).

DATA CONTRIBUTORS

C. Barger; J.G. Barnes; T. Booms; J. Hagelin; K. Hall; S.B. Lewis; M. Lockhart; M. Martell; S. Mehus; C. McIntyre; P. Nye; R. Kays; T. Salo; S. Slater; B.W. Smith; J.P. Smith; D. Stahlecker; M. Yates; Alaska Department of Fish and Game; Bureau of Land Management; Craighead Beringia South; Department of Defense Legacy Resource Management Program; Earthspan; Hycroft Gold Mine; Northwest Wildlife Consultants; US Fish and Wildlife Service.

SHARP-SHINNED HAWK *Accipiter striatus*

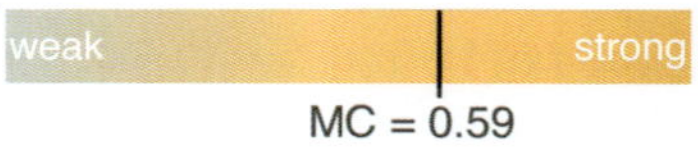

DISTRIBUTION AND ABUNDANCE

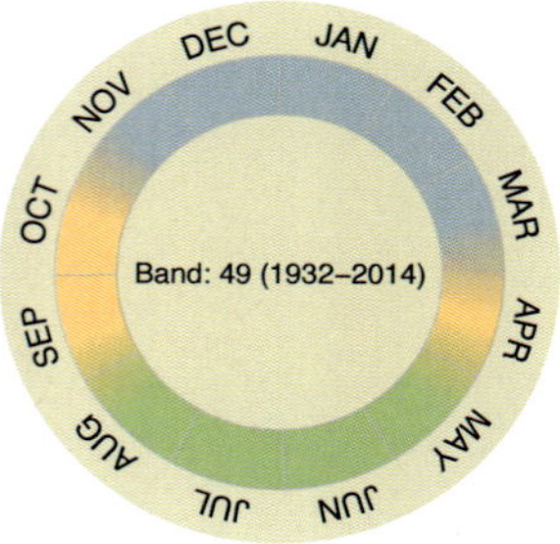

The Sharp-shinned Hawk is a small, secretive forest raptor that specializes on avian prey. Sharp-shinned Hawks breed in forests across Canada and Alaska and south through the western United States to northern Nevada and Arizona and in forests of Great Lakes states, northeast and south through Appalachian states to Tennessee. Sharp-shinned Hawks congregate in large numbers during migration along ridges and coastlines throughout the US and south through Mexico, and spend the nonbreeding season throughout the contiguous US, Mexico, and Central America (Fink et al. 2023). Year-round populations occur in forests of Mexico, Central America, some islands of the Caribbean, and south along the Andes in South America (Fink et al. 2023). Sharp-shinned Hawks underwent historic population declines in the 1960s due to use of the pesticide DDT with rebounds detected in early 1980s (Bednarz et al. 1990). More recent declines have been observed, particularly in the eastern United States populations (Oleyar et al. 2021; Fink et al. 2023); however, central US populations appear stable.

ORDER Accipitriformes
FAMILY Accipitridae
BREEDING HABITAT Forests, Temperate Generalist
NONBREEDING HABITAT Forests, Temperate Generalist
GLOBAL POPULATION SIZE 1,000,000

POPULATION CHANGE SINCE 1970 +75%
IUCN RED LIST CATEGORY Least Concern
R2R TIPPING POINT SPECIES No
US LISTING STATUS Endangered Puerto Rican subspecies
CANADA SPECIES AT RISK No
PIF CCS 9/20

INTERESTING NOTE

A hatch-year female Sharp-shinned Hawk banded at Thunder Cape Bird Observatory on Lake Superior, Ontario, in August 1998, was reencountered 31 days later in coastal Veracruz, Mexico, after traveling 3,250 km (Ruelas Inzunza et al. 2009). Migrant Sharp-shinned Hawks tracked with VHS telemetry showed an energy-minimization strategy on migration, migrating longer distances with tail winds and clear skies and migrating short distances and exhibiting stopover in less favorable weather (Goodrich 2010). During the 1990s, declines at eastern migration hawk watch sites, with increases on some Christmas Bird Counts, suggest Sharp-shinned Hawks were altering their migration patterns by short-stopping, i.e, shortening their migration distance (Viverette et al. 1996).

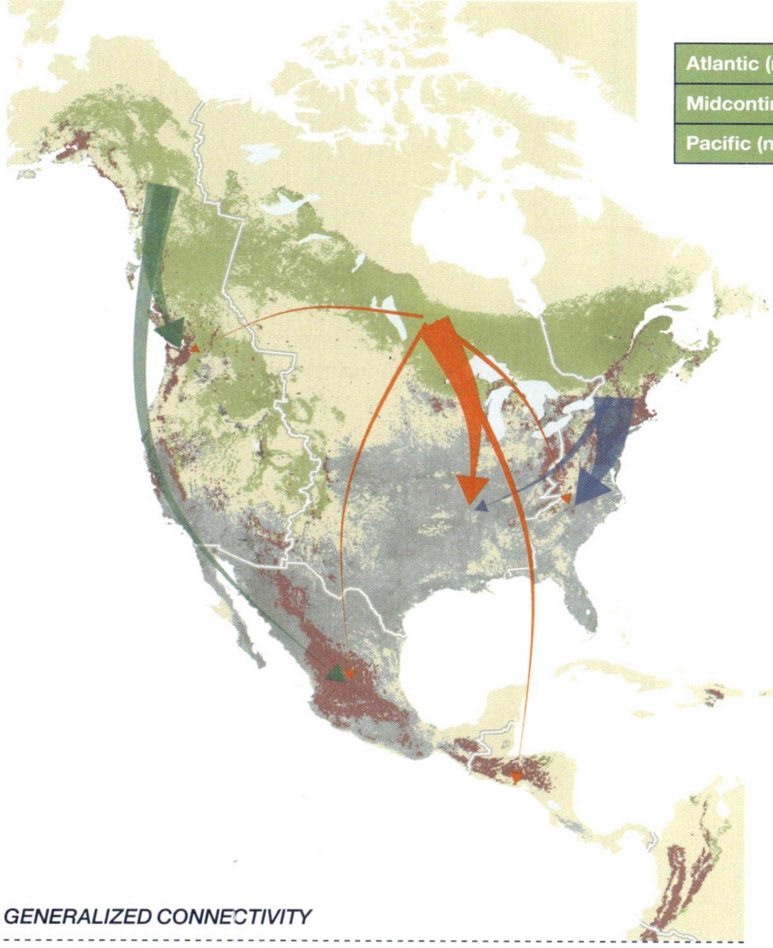

GENERALIZED CONNECTIVITY

	Atlantic	Central America	Mexico	Midcontinent	Pacific
Atlantic (n = 7)	86%	–	–	14%	–
Midcontinent (n = 37)	8%	14%	5%	70%	3%
Pacific (n = 3)	–	–	33%	–	67%

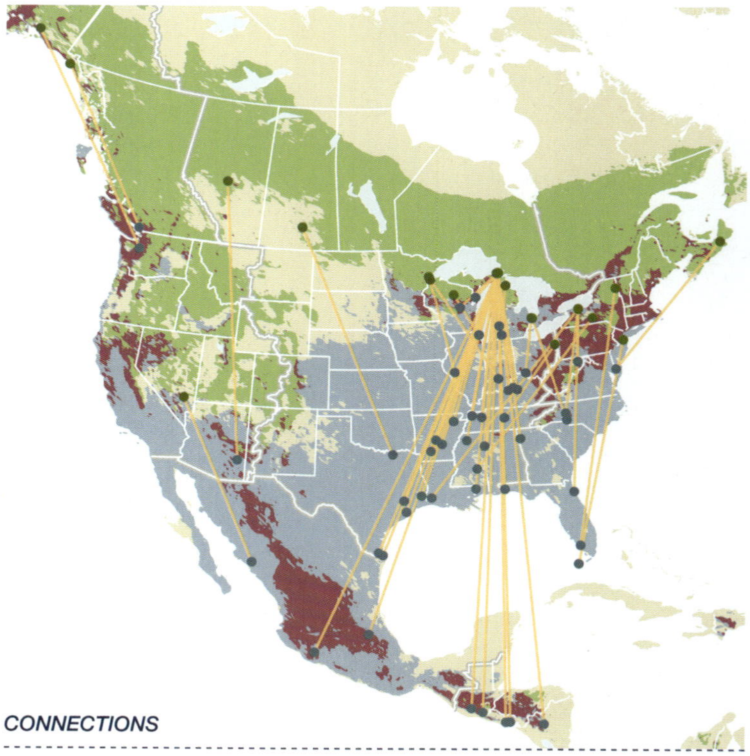

CONNECTIONS

MIGRATORY CONNECTIVITY

OVERALL Band reencounter and stable isotope data suggest Sharp-shinned Hawk migratory connectivity is strong (MC = 0.59). Atlantic Flyway breeding birds remained largely in the Atlantic Flyway during the nonbreeding season (86%). Similarly, the majority of Midcontinent Flyway breeding birds remained within the same flyway during both breeding and nonbreeding seasons (70%). Pacific Flyway breeding birds spent the nonbreeding period primarily in the Pacific Flyway (67%), with the remainder migrating to Mexico (33%).

BANDING Published analyses of banding data indicate moderate to strong migratory connectivity for Sharp-shinned Hawks, with a general separation of east and west populations. Birds breeding in the US Midwest and in southeastern Canada spend the nonbreeding season in the eastern US, Mexico, and Central America (Duncan 1982; Evans and Rosenfield 1985; Devereux et al. 1985; Carpenter et al. 1990; and Viverette et al. 1996). Birds breeding in the western US and Canada spend the nonbreeding season in the western US, Mexico, and Central America (Hoffman et al. 2002; Goodrich and Smith 2008). There is evidence that Sharp-shinned Hawks have a chain migration pattern (R.B. Smith et al. 2003; Goodrich et al. 2012). However, birds using the Midcontinent Flyway migrate farther south into Central America than those using Atlantic and Pacific flyways (Goodrich et al. 2012). Hatch-year birds migrate farther south than older birds (Hoffman et al. 2002; Goodrich et al. 2012), and females migrate farther south than males (Evans and Rosenfield 1985). Birds breeding south of 35 degrees north latitude are mostly year-round residents (Goodrich et al. 2012).

RADIO TRACKING Data from several tracking projects through the Motus Wildlife Tracking System suggest different migration routes for eastern and western populations. There are no individual breeding to nonbreeding detections.

STABLE ISOTOPES Stable hydrogen isotopes show that Sharp-shinned Hawks from south of the US-Canada border migrated earlier during post-breeding compared to birds from north of the border (R.B. Smith et al. 2003); in addition, birds migrating through Nevada bred in the northern Rocky Mountains (i.e., Montana, Idaho, British Columbia) or interior Alaska (Lott and Smith 2006).

OTHER APPROACHES No genetic studies of migratory connectivity were found for this species.

MIGRATORY PATHWAYS Migration occurs across a broad front in a generally southern direction during post-breeding and north during pre-breeding, with concentrations occurring along coastlines and ridges. Birds avoid water crossing in most conditions, leading to diversion and concentrations at the Atlantic and Gulf coasts and along the Great Lakes. The Rockies, Cascades, Appalachians, and other mountain ridge corridors provide additional concentration points (Goodrich and Smith 2008). Juvenile birds in the Atlantic particularly concentrate on the Atlantic coast during post-breeding southbound migration, with numbers exceeding 10,000 in most years. Sharp-shinned Hawk migration shows greater dispersion in Great Lakes migrants or the midcontinent region, with less connectivity than in Pacific and Atlantic corridors (Goodrich and Smith 2008). Spring migrant band recoveries suggest birds retrace their pathways north, using similar routes to return to breeding areas.

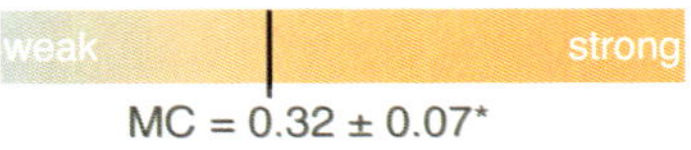

Accipiter cooperii COOPER'S HAWK

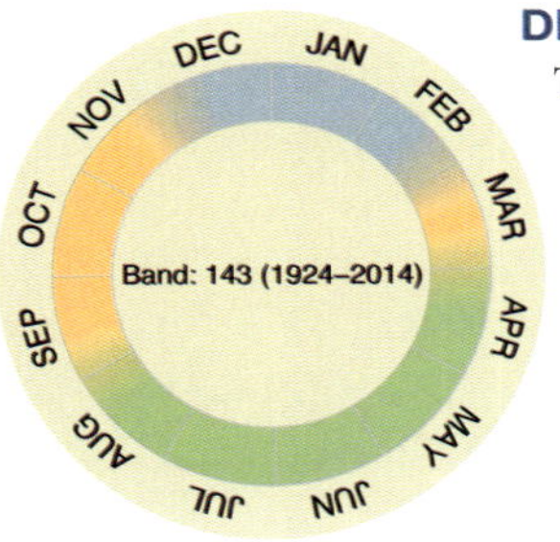

DISTRIBUTION AND ABUNDANCE

The Cooper's Hawk is a widespread, medium-sized forest raptor that nests across North America from southern Canada south through the highlands of the Mexican Sierra Madre Occidental and northern Baja. It is a migrant, and southern breeding birds are often nonmigratory. Migration distance varies, and birds spend the nonbreeding season from the northern United States south into Guatemala. Year-round, Cooper's Hawks can be found in the Mexican Sierra Madre Occidental and across parts of the contiguous US. Historic population declines were due to persecution in the early 1900s, when they were thought to be predators of domestic fowl (Bent 1937), followed by poisoning from DDT (Bednarz et al. 1990; Snyder et al. 1973). North American Cooper's Hawk populations rebounded in 1980s and currently appear to be mostly stable, based on migration and nonbreeding season counts (Oleyar et al. 2021).

ORDER Accipitriformes
FAMILY Accipitridae
BREEDING HABITAT Forests, Temperate Generalist
NONBREEDING HABITAT Open Country, Habitat Mosaic
GLOBAL POPULATION SIZE 1,000,000
POPULATION CHANGE SINCE 1970 +309%
IUCN RED LIST CATEGORY Least Concern
R2R TIPPING POINT SPECIES No
US LISTING STATUS No
CANADA SPECIES AT RISK No
PIF CCS 7/20

MIGRATORY CONNECTIVITY

OVERALL Migratory connectivity is known from banding and limited satellite tracking data. Breeding populations in the Atlantic Flyway primarily spend the nonbreeding season within the Atlantic flyway (67%) as well as the Midcontinent Flyway (33%). Midcontinent Flyway breeding birds primarily spend the nonbreeding season within the Midcontinent Flyway (77%) but also spread out in smaller numbers across other nonbreeding regions. Pacific Flyway breeding birds mostly spend the nonbreeding season within the Pacific Flyway (67%) while others migrate to Mexico (30%). While transition probabilities within the flyways are high, use of diverse nonbreeding areas moderates migratory connectivity (MC = 0.32 ± 0.07).

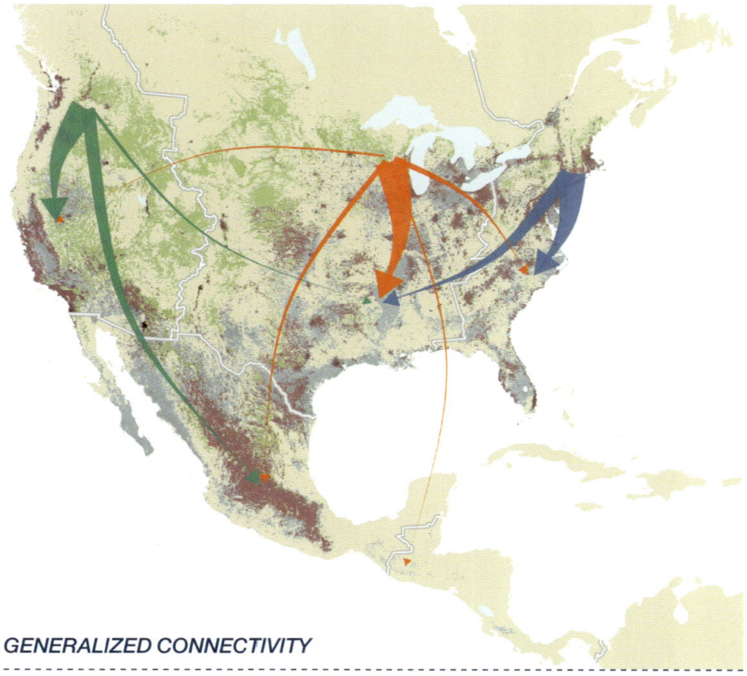

GENERALIZED CONNECTIVITY

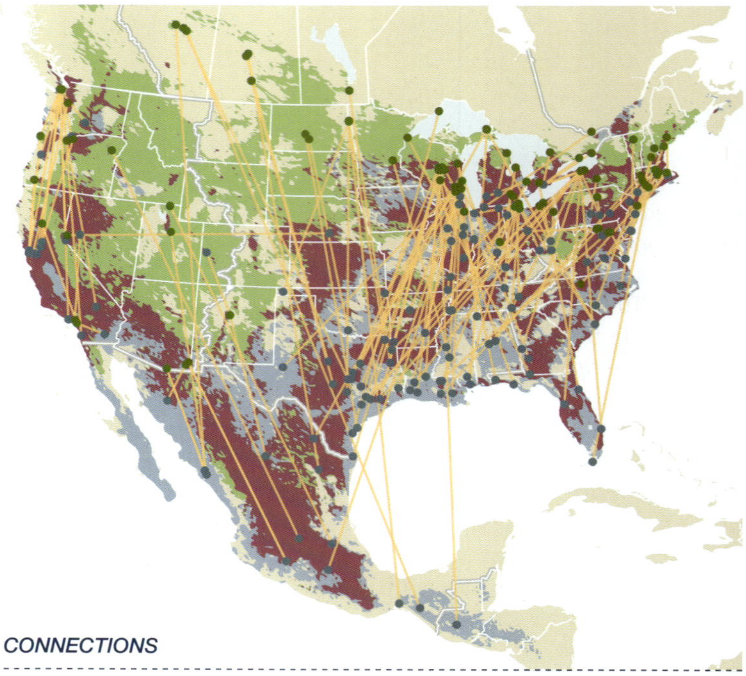

CONNECTIONS

	Atlantic	Central America	Mexico	Midcontinent	Pacific
Atlantic (n = 36)	67%	–	–	33%	–
Midcontinent (n = 79)	11%	1%	9%	77%	1%
Pacific (n = 27)	–	–	30%	4%	67%

INTERESTING NOTE

Genetic differentiation between populations of Cooper's Hawks suggests large-scale restriction in gene flow, possibly due to geographic barriers to natal dispersal (Sonsthagen et al. 2012).

BANDING Northern breeding birds from the Midcontinent and Pacific flyways migrate farther south than Atlantic Cooper's Hawks, with some spending the nonbreeding period in Mexico and Central America (Hoffman et al. 2002; Goodrich et al. 2012). Some proportion of hatch-year Cooper's Hawks from Southern California show possible reverse migration northward exceeding 600 km, in contrast to some southern Arizona birds that move primarily south or move a shorter distance north (Bloom et al. 2017). Some possible northward movement was suggested from band recoveries across the southern breeding range (Goodrich et al. 2012).
RADIO TRACKING There are no detections from tags deployed with the Motus Wildlife Tracking System.
SATELLITE TRACKING Four birds tracked from British Columbia and Alberta spent the nonbreeding season in central to southern Mexico, with three individuals returning to the same breeding location the following year (Raptor View Research Institute, unpublished data).
MIGRATORY PATHWAYS Cooper's Hawks concentrate on migration along the Atlantic coast, Appalachian Mountains, Great Lakes, Rocky Mountains, Pacific coast, and Gulf of Mexico, with highest numbers observed in the Rocky Mountains and Atlantic coast (3,500–4,200 birds each post-breeding migration on average; Goodrich and Smith 2008). Birds from the Intermountain West travel into western Mexico for the nonbreeding season (Hoffman et al. 2002), and band reencounters from birds banded along the Gulf of Mexico in Veracruz often originate from the Great Lakes region of the midcontinent (River of Raptors, unpublished data). Some declining migration counts in the Atlantic region may be due to changes in migration behavior, but further research is needed.

AMERICAN GOSHAWK *Accipiter atricapillus*

weak — strong
MC = 0.98

DISTRIBUTION AND ABUNDANCE

It breeds in boreal and temperate forests across central Alaska, in Canada from the Yukon east to Newfoundland, south through the north-central Appalachians, and the Great Lakes states. It is also resident in western regions of the Intermountain West / Great Basin, Pacific coast, and Sierra Madre Occidental. Some northern birds move south to the prairies of southwestern Canada and the western US, with small numbers in midlatitudes in the Midwest and mid-Atlantic region during the

ORDER Accipitriformes
FAMILY Accipitridae
BREEDING HABITAT Forests, Boreal
NONBREEDING HABITAT Forests, Temperate Generalist
GLOBAL POPULATION SIZE 420,000
POPULATION CHANGE SINCE 1970 +30%
IUCN RED LIST CATEGORY Least Concern
R2R TIPPING POINT SPECIES No
US LISTING STATUS No
CANADA SPECIES AT RISK Threatened *laingi* subspecies
PIF CCS 10/20

	Atlantic	Midcontinent	Pacific
Atlantic (n = 3)	100%	–	–
Midcontinent (n = 1)	–	100%	–
Pacific (n = 28)	–	4%	96%

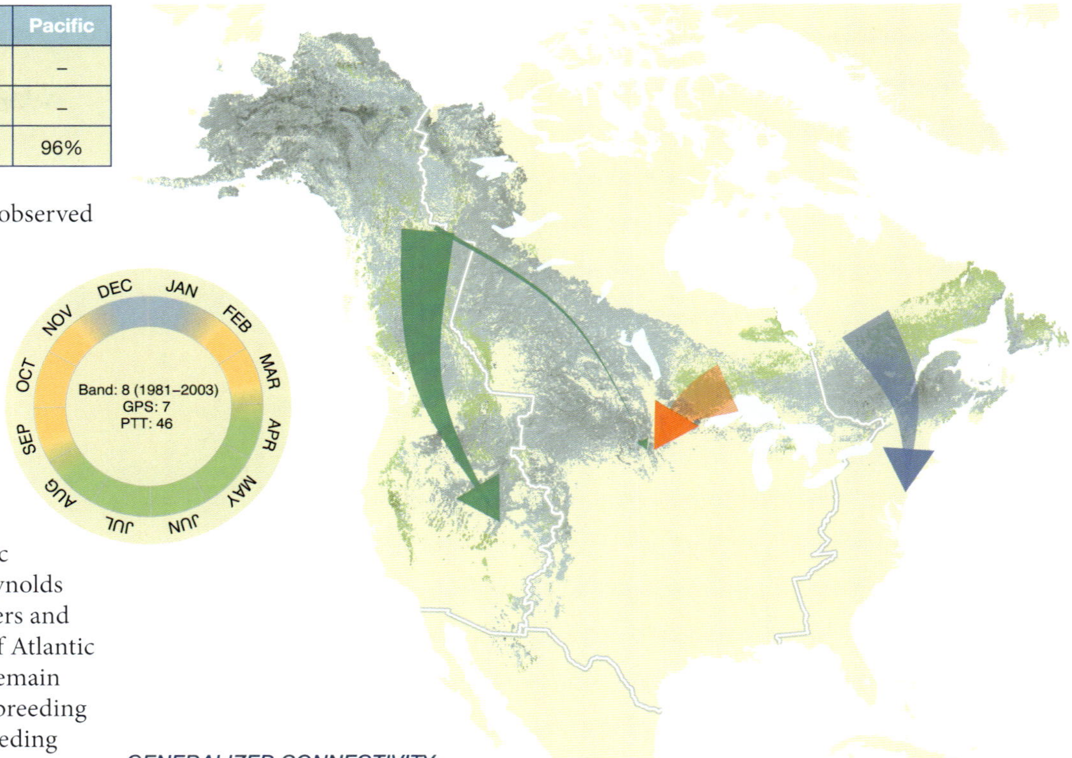

GENERALIZED CONNECTIVITY

nonbreeding season. Greater numbers are observed in the nonbreeding range during periodic irruptions estimated to occur every 10 years. The American Goshawk was recently split from the European Goshawk (*Accipiter gentilis*; Chesser et al. 2023).

MIGRATORY CONNECTIVITY

OVERALL Many populations may be largely sedentary, partially migratory (northern populations), or exhibit periodic irruptions related to prey (Squires and Reynolds 2023). Current data from band reencounters and telemetry, though limited, suggest 100% of Atlantic and Midcontinent flyway breeding birds remain in their respective regions during the nonbreeding season, whereas 96% of Pacific Flyway breeding birds remain within the Pacific Flyway. Migratory connectivity is strong (MC = 0.98).

BANDING Band reencounter data shows mainly short-distance movements within the Pacific and Intermountain West. Longer-distance movements were documented for birds breeding farther north.

SATELLITE TRACKING Satellite telemetry studies of goshawks in Utah found populations to be partially migratory, even within individuals, as one bird tracked for two years migrated in one year and remained resident the other year (Underwood et al. 2006; Sonsthagen et al. 2006). Nonbreeding site fidelity was observed in some birds. Dozens of juvenile birds tracked in other western states show short to moderate distance movement, largely in a north to south direction (Goodrich and Smith 2008; Smith 2010). A study in the Great Basin Desert of Nevada (M. Jeffress and J. Barnes, unpublished data) tracked eight territory-holding adults over a six-year period. Nearly all individuals remained in the broader northeastern Nevada area throughout each tracking year, exhibiting patterns of limited local dispersal outside the breeding season and year-round residency (M. Jeffress and J. Barnes, unpublished data).

INTERESTING NOTE

An adult female banded during post-breeding migration at Hawk Ridge, Duluth, Minnesota, was recovered in Louisiana that same fall, having traveled 1,860 km in 60 days, an average daily flight distance of 31 km (Evans and Sindelar 1974).

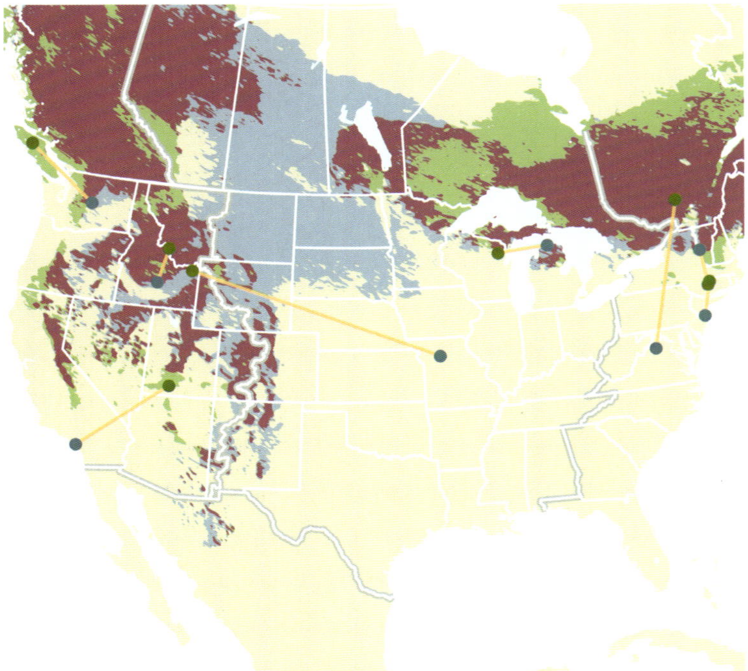

CONNECTIONS A

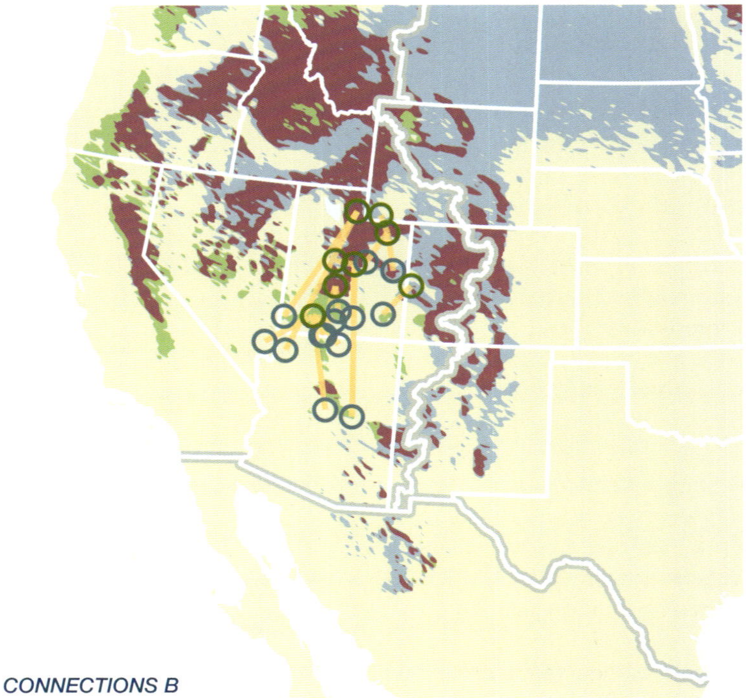

CONNECTIONS B

OTHER APPROACHES No genetic or stable isotope studies of migratory connectivity were found for this species.
MIGRATORY PATHWAYS American Goshawks migrate largely north to south but often follow mountain ranges and coastlines using wind currents. Notable concentrations of migrants (500–1,000) historically occur each post-breeding season along shores of the Great Lakes, e.g., Hawk Ridge in Duluth, Minnesota. Other concentration points include the Goshutes Mountains, Nevada, and Tadoussac, Quebec, in the northern Appalachians (Goodrich and Smith 2008).

DATA CONTRIBUTORS
J. Barnes; M. Jeffress; J.P. Smith; HawkWatch International.

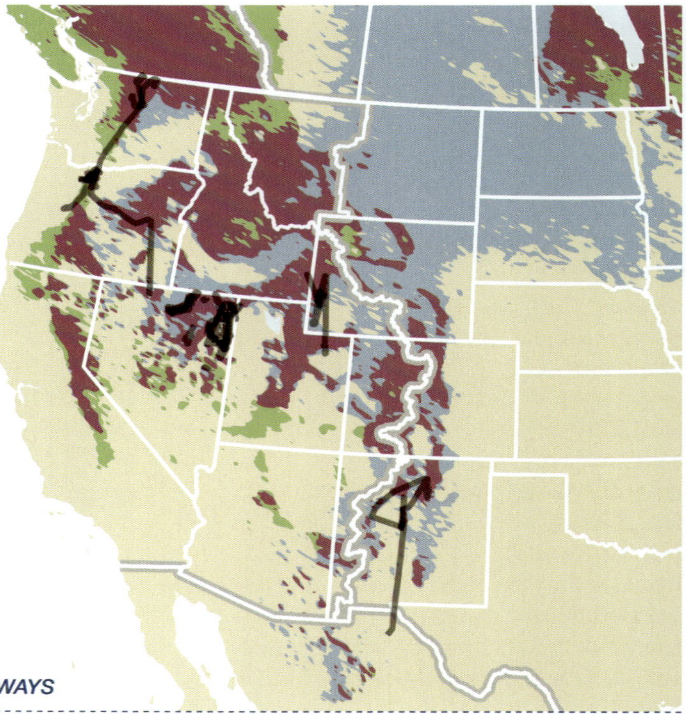

PATHWAYS

NORTHERN HARRIER *Circus hudsonius*

weak strong
MC = 0.53 ± 0.06*

DISTRIBUTION AND ABUNDANCE:
Northern Harriers breed across Canada and Alaska to the Arctic Slope (though not in Nunavut, northern Quebec, or Labrador), south into the upper Midwest, and in New England and the mid-Atlantic states. The nonbreeding range spans the contiguous US and south into Mexico, Central America, and the Caribbean (K.G. Smith et al. 2020b). Northern Harriers are year-round residents across the midlatitudes of the US, including the northern Great Plains, Rocky Mountains, Intermountain West / Great Basin, Pacific coast, and coastal Massachusetts south to Chesapeake Bay (K.G. Smith et al. 2020b). Harriers show a four-year cycle of increased migration counts and nonbreeding numbers in many central and eastern states, which may correlate to cycles of prey abundance in Canadian breeding areas (Schimpf et al. 2020). Recent regional variation in trends showed that 51% of eastern North American migration counts from 2009 to 2019 declined. Correspondingly, Northern Harriers are listed as "threatened" in most northeastern states (e.g., New Jersey, Pennsylvania, and New England states; Oleyar et al. 2021).

ORDER Accipitriformes
FAMILY Accipitridae
BREEDING HABITAT Grasslands, Temperate
NONBREEDING HABITAT Open Country, Habitat Mosaic
GLOBAL POPULATION SIZE 820,000
POPULATION CHANGE SINCE 1970 −30%
IUCN RED LIST CATEGORY Least Concern
R2R TIPPING POINT SPECIES No
US LISTING STATUS No
CANADA SPECIES AT RISK No
PIF CCS 11/20

MIGRATORY CONNECTIVITY

OVERALL Migratory connectivity is known from satellite telemetry and banding data. Atlantic Flyway breeding birds primarily remain within the Atlantic Flyway (90%) during the nonbreeding season. Midcontinent Flyway breeding birds mostly spend the nonbreeding season within the midcontinent (78%), with smaller numbers migrating to other nonbreeding regions. Birds breeding within the Pacific Flyway remain within the Pacific Flyway during the nonbreeding season (100%). Although there is some mixing of Atlantic and Midcontinent flyway breeding populations during the nonbreeding season, most populations remain within the same flyway during the breeding and nonbreeding season, resulting in strong migratory connectivity (MC = 0.53 ± 0.06).
BANDING Banding data support similar patterns to telemetry data and also suggest a leapfrog migration, with northern Canada and Alaska nesting birds migrating farther south than birds from southern Canada and the northern US (Bildstein 2006). Band reencounters from Wisconsin nesting harriers found that most birds spent the nonbreeding season southeast of Wisconsin and as far south as Florida (Beske 1982). More data is needed to understand the breeding origins of harriers that spend the nonbreeding season in Mexico and Central America, but some band reencounters suggest western prairie birds migrate through this corridor.
RADIO TRACKING Seven juvenile Northern Harriers tracked from nesting areas in Wisconsin all moved southeast on migration (Beske 1982).
SATELLITE TRACKING Two birds tracked from the northeast US migrated to the mid-Atlantic (C. DeSorbo, unpublished data; K.

O'Brien, unpublished data). An adult female captured in the Florida Keys, en route to Cuba, migrated to a breeding area in North Dakota and then back to Cuba the following year for the nonbreeding season (G. Kent and K. Meyer, Avian Research and Conservation Institute, unpublished data). Birds tracked from nonbreeding areas in California migrated to breeding areas in Alaska (US Geological Survey, unpublished data).

STABLE ISOTOPES Consistent with patterns represented by banding and telemetry data, stable isotope data linked juvenile harriers sampled on nonbreeding areas in the Florida Keys to natal areas in the northern Atlantic region of the US (Wittenberg and Smith 2009).

MIGRATORY PATHWAYS During migration, harriers concentrate along coastlines (e.g., Cape May, New Jersey; Golden Gate Raptor Observatory, California), shorelines (Great Lakes), and ridges (e.g., Cascade Range and Rocky Mountains; US Geological Survey, unpublished data). Band reencounter and tracking data show migration direction primarily to the south or southeast from the midcontinent (G. Kent and K. Meyer, Avian Research and Conservation Institute, unpublished data), with some Great Lakes birds migrating southwest (K.G. Smith et al. 2020b).

	Atlantic	Caribbean	Central America	Mexico	Midcontinent	Pacific
Atlantic (n = 10)	90%	–	–	–	10%	–
Midcontinent (n = 77)	14%	3%	1%	3%	78%	1%
Pacific (n = 50)	–	–	–	–	–	100%

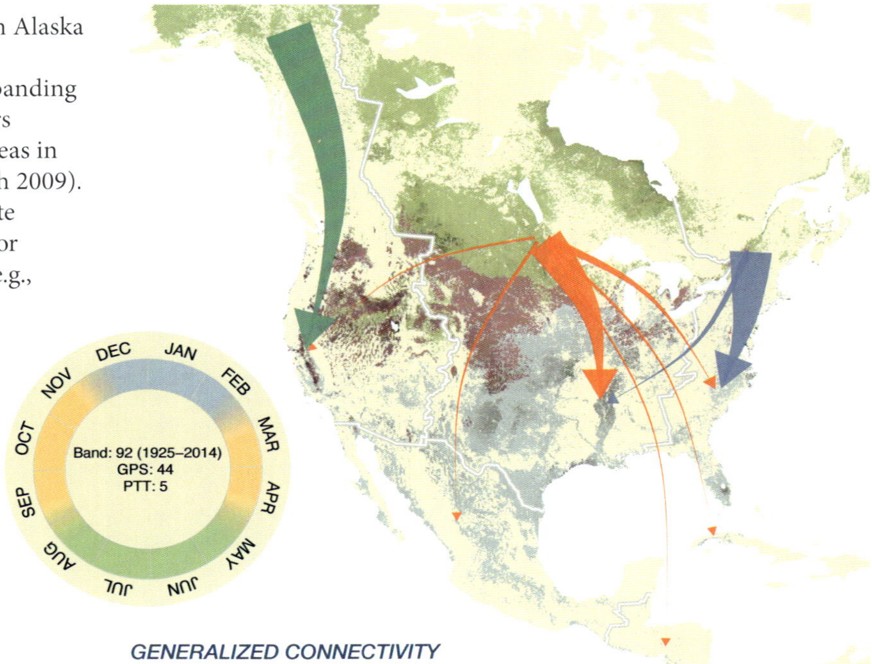

GENERALIZED CONNECTIVITY

DATA CONTRIBUTORS

M. Casazza; C.R. DeSorbo; J. Hull; G.M. Kent; K.D. Meyer; K.M. O'Brien; C. Overton; S. Skalos; Avian Research and Conservation Institute; Biodiversity Research Institute; New York State Department of Environmental Conservation.

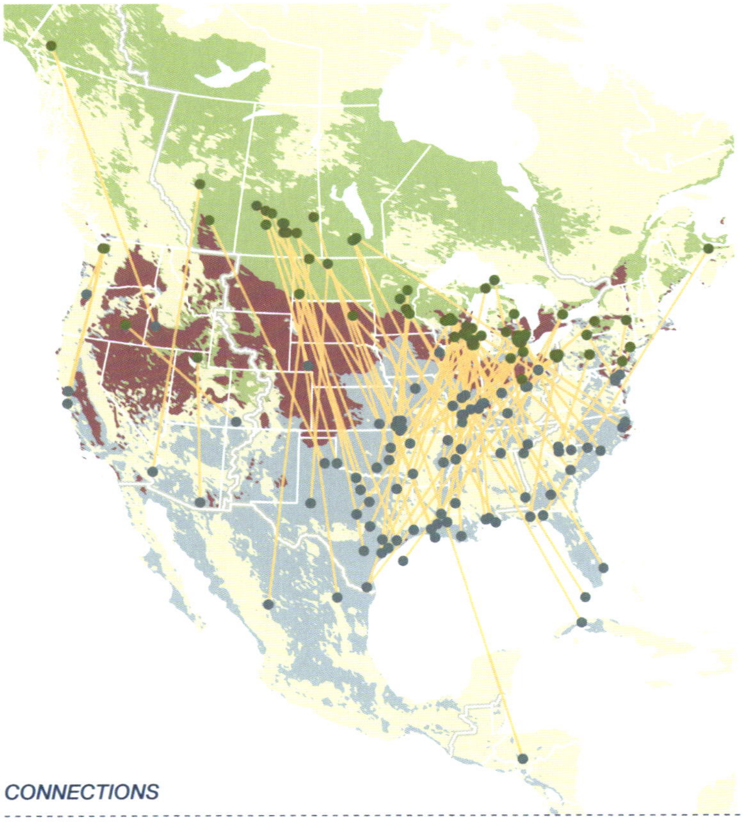

CONNECTIONS

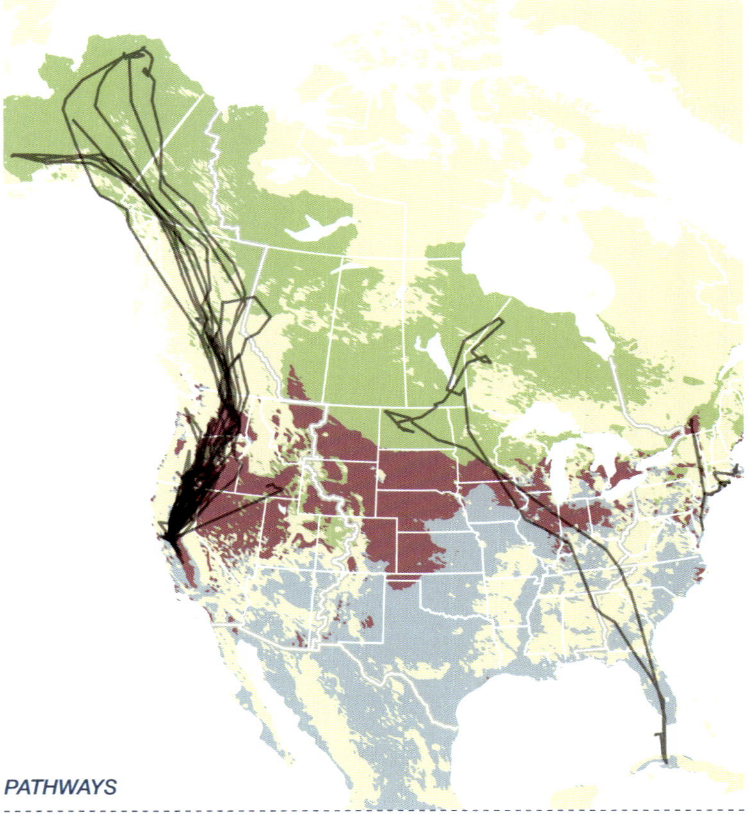

PATHWAYS

INTERESTING NOTE

Juvenile Northern Harriers migrate before adults in post-breeding and pre-breeding. Adult males return earlier than females in pre-breeding migration (Kerlinger 1989). The highest concentrations of migrating harriers in North America occur in Cape May, New Jersey, with a mean of 1,570 harriers per post-breeding season in 1997–2006 (Goodrich and Smith 2008).

BALD EAGLE *Haliaeetus leucocephalus*

weak strong
MC = 0.73 ± 0.04*

DISTRIBUTION AND ABUNDANCE

It breeds in every state in the contiguous United States, with large populations along the Atlantic coast and in the Great Lakes and Pacific Northwest, as well as throughout Alaska (Buehler 2020). In Canada, the breeding range extends from northern Yukon, south to British Columbia and east to the Atlantic provinces, with large populations in coastal British Columbia and southern Quebec. The nonbreeding range spans the contiguous United States, coastal and southern Canada, and southeastern Alaska, with large populations found along the Chesapeake Bay, coastal areas of the southeastern US, Florida, major reservoirs and rivers in the midcontinent and Intermountain West, Pacific Northwest, and Klamath Basin (Buehler 2020). It occurs in small numbers in Mexico during both the breeding and nonbreeding seasons (Buehler 2020).

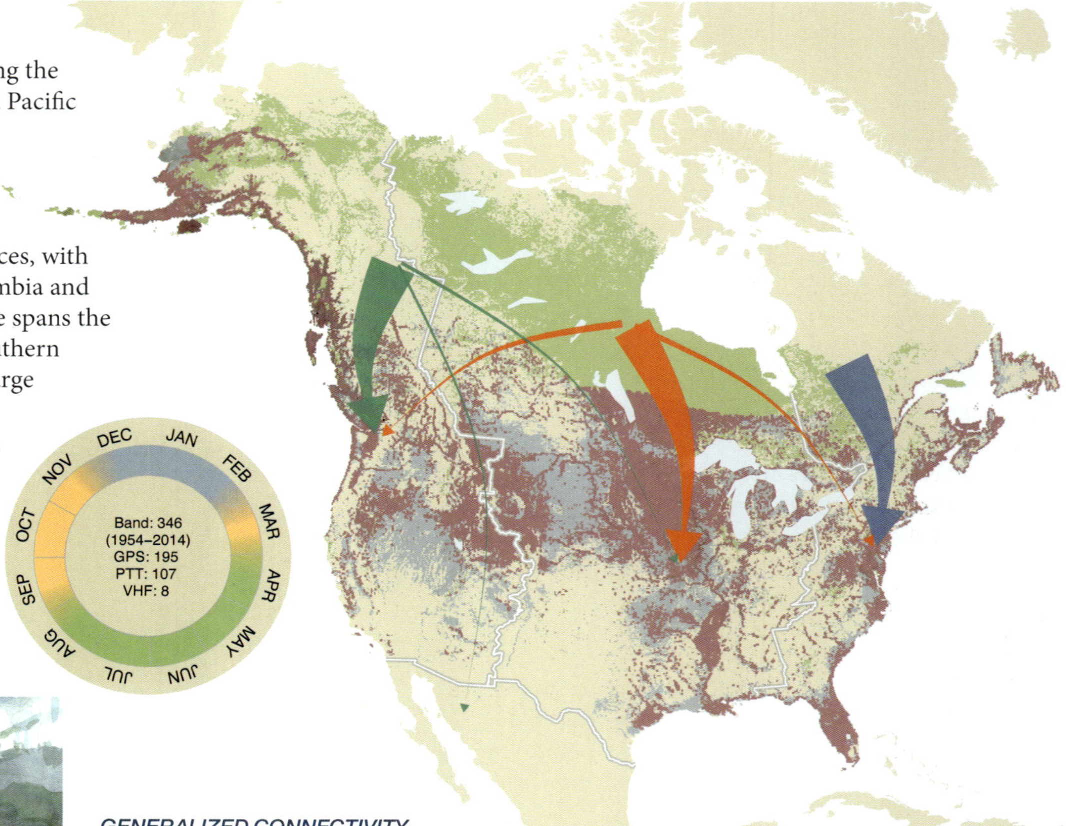

GENERALIZED CONNECTIVITY

ORDER Accipitriformes
FAMILY Accipitridae
BREEDING HABITAT Wetlands, Generalist
NONBREEDING HABITAT Wetlands, Generalist
GLOBAL POPULATION SIZE 200,000
POPULATION CHANGE SINCE 1970 +1105%
IUCN RED LIST CATEGORY Least Concern
R2R TIPPING POINT SPECIES No
US LISTING STATUS No
CANADA SPECIES AT RISK No
PIF CCS 9/20

	Atlantic	Mexico	Midcontinent	Pacific
Atlantic (n = 249)	100%	–	–	–
Midcontinent (n = 274)	8%	–	80%	12%
Pacific (n = 100)	–	1%	3%	96%

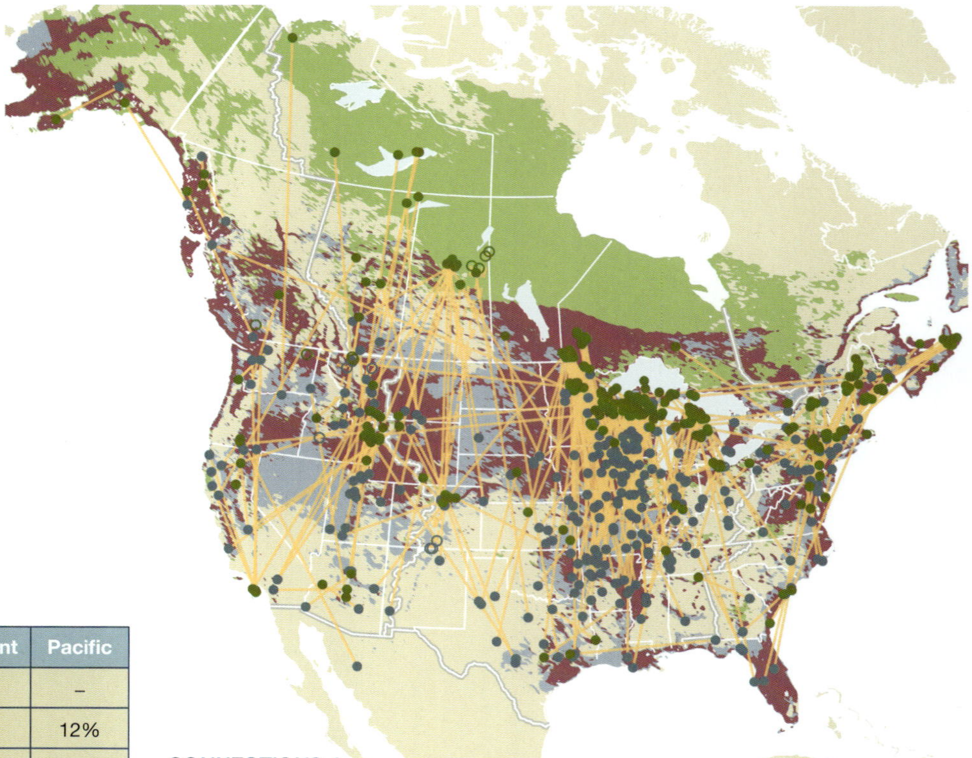

CONNECTIONS A

MIGRATORY CONNECTIVITY

OVERALL The migratory connectivity of the Bald Eagle is well known from a combination of satellite and radio tracking data, as well as band reencounters and resightings. Migratory connectivity is strong (MC = 0.73 ± 0.04), as the majority of individuals remain within the same flyway during the breeding and nonbreeding seasona.

BANDING Several hundred banding encounters suggest similar connectivity patterns as found in radio and satellite transmitter data.

RADIO TRACKING Bald Eagles have been tracked extensively with radio transmitters, with hundreds of individuals tagged and tracked across North America, particularly in the eastern region of the US and Canada. Tracking data show complex and varied migration strategies, depending on population and age. Juveniles tend to have weak or no connectivity. Sub-adults and adults are either migratory with strong connectivity or nonmigratory (Wheat et al. 2017). Eagles from a single location can exhibit both strategies. Bald Eagles in southeastern states, Texas, and Arizona display a reverse migration pattern in which they breed in the typical nonbreeding season (winter) and then migrate north for a summer nonbreeding season, often wandering widely, before returning in Northern Hemisphere fall (Hunt et al. 1992, 2009; Mojica et al. 2008; Smith et al. 2017).

SATELLITE TRACKING Satellite tracking shows movements and connections throughout the Atlantic region, from Quebec to Florida, as well as some movements between the Atlantic and eastern portions of the midcontinent regions (Mojica et al. 2008; Nye et al. 2021; B. Watts et al. 2021, unpublished data; J. Bullis, unpublished data; T.E. Chubbs, unpublished data; K.E. Clark, unpublished data; C.R. DeSorbo, unpublished data; R. Kays and T. Simons, unpublished data; B. Paxton, unpublished data). In the midcontinent, individuals migrated between Nunavut Province of Canada to the southwestern United States and between Manitoba and Illinois (Harmata 1984; McClelland et al. 1996; Madernack et al. 2012; Smith et al. 2017;

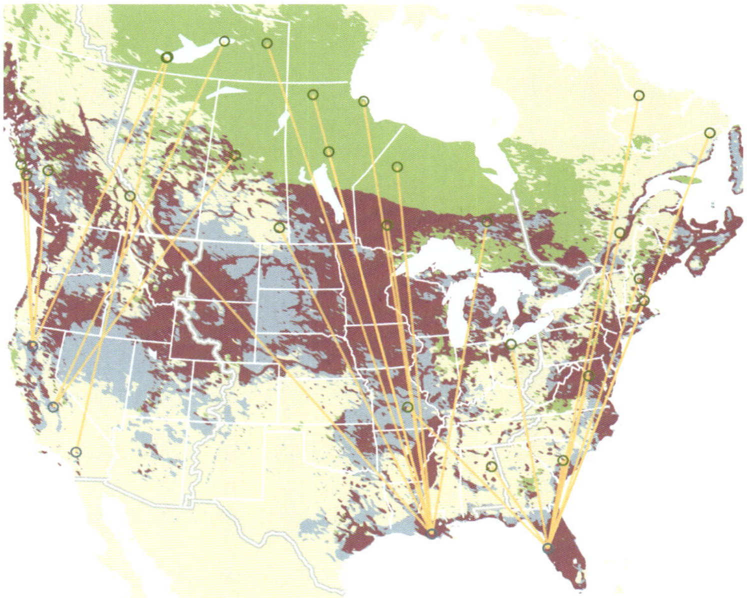

CONNECTIONS B

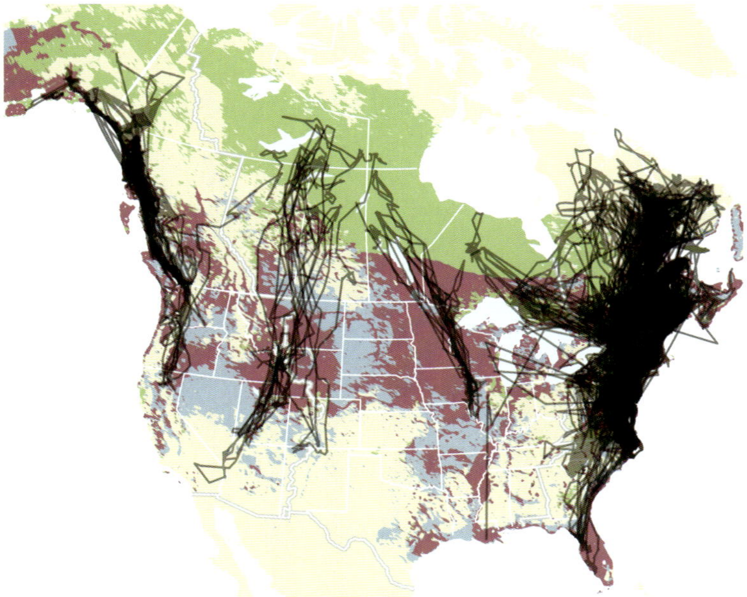

PATHWAYS (A–C)

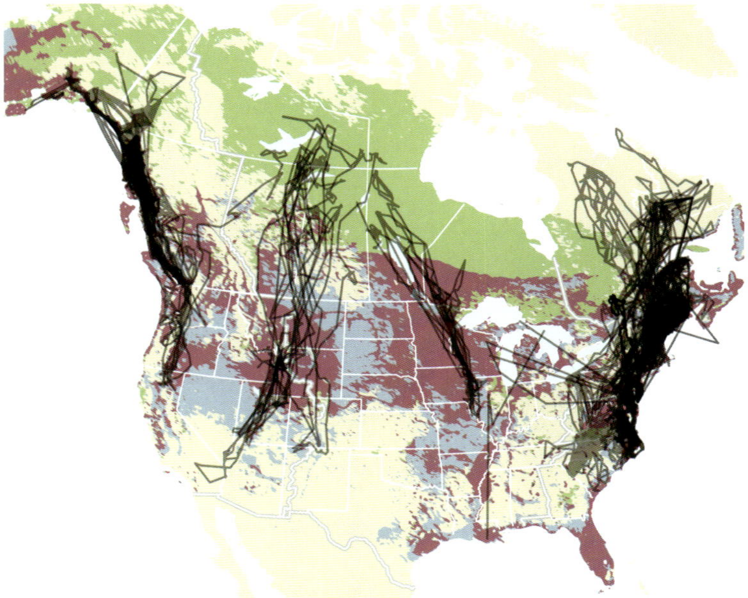

PATHWAYS A

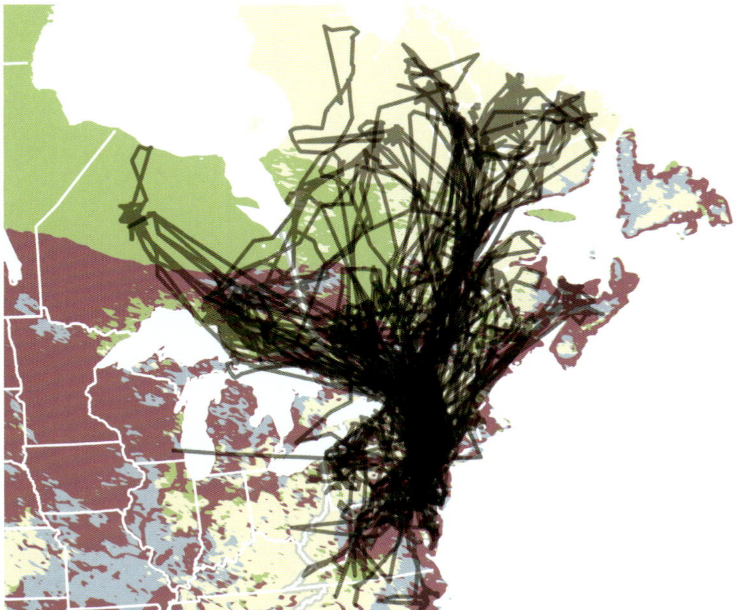

PATHWAYS B

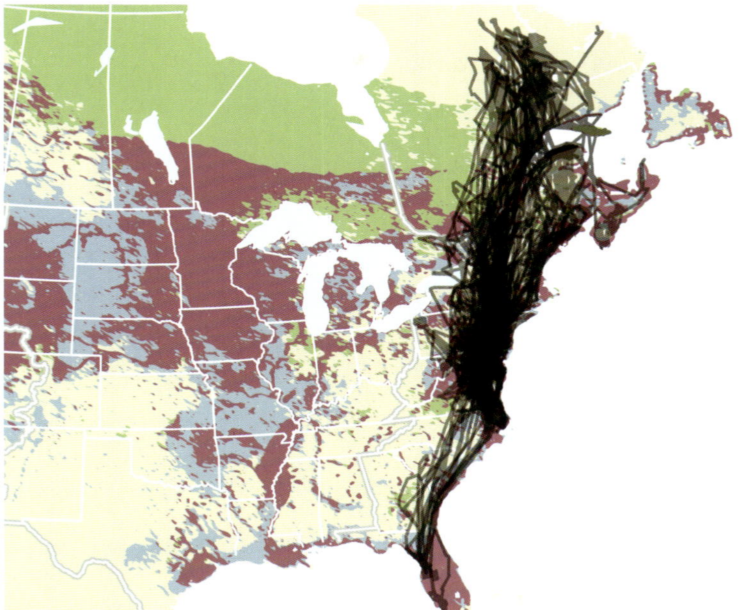

PATHWAYS C

INTERESTING NOTE

The Bald Eagle is an iconic symbol of the United States and among the largest birds in North America. The early studies of its movements have helped to pioneer bird tracking methods, including radio and satellite tracking systems. More than 100 Bald Eagles can be seen in one day on post-breeding migration at Hawk Ridge, Minnesota.

Craighead Beringia South, unpublished data; A. Madernack et al., unpublished data). In the Pacific region, individuals migrated from Northern California to nonbreeding areas along the Pacific coast and British Columbia (Linthicum et al. 2007; Garcia and Associates 2016; M. Lamont and D. Hancock, unpublished data). Alaska Bald Eagles are mostly nonmigratory, but the approximately 25% that do migrate are relatively short-distance migrants that travel down the coast and back up. Birds have been tracked from Sitka, Alaska, to Vancouver Island and Prince Rupert, British Columbia, as well as from Anchorage, Alaska, to coastal British Columbia (Wheat et al. 2017). There can also be wandering into the interior of Alaska and the Yukon (Wheat et al. 2017).

MIGRATORY PATHWAYS Satellite and radio telemetry data indicate that Bald Eagles use several migration corridors, including the Atlantic coast, Appalachian Mountains, Mississippi River / Great Lakes, Rocky Mountains, Intermountain West / Great Basin, Sierra Nevada / Cascade Mountains, and Pacific coast. Important stopover locations during migration include the Great Lakes, Chesapeake Bay, Mississippi River, and many large lakes and rivers in the midcontinent region. Bald Eagles tracked from natal areas in Northern California (n = 5) flew north and likely timed movements with spawning fish runs in British Columbia and Alaska (Hunt et al. 1992). Two juveniles traveled northwest to the Pacific Ocean, then north along the coast, while three eagles migrated north along the Cascade Mountains (Hunt et al. 1992). Similarly, juvenile Bald Eagles tracked from natal areas in Arizona flew north to locations within the Pacific Northwest, Intermountain West, and Great Plains, with reports of spawning fish or other food resources available to inexperienced juveniles (Hunt et al. 2009). About 75% of the tracked juveniles returned to Arizona (Hunt et al. 2009).

DATA CONTRIBUTORS

T.E. Chubbs; K.E. Clark; R. Crandall; C.R. DeSorbo; J.E. Drennan; R.E. Jackman; R. Kays; M. Lamont; S.B. Lewis; B.A. Mandernack; P. Nye; B. Paxton; M. Pitts; R.T. Schmitz; T. Simons; N. Smith; B. Watts; American Eagle Foundation; Canadian Wildlife Service; Center for Conservation Biology, College of William and Mary; Craighead Beringia South; EA Engineering; Exelon Corporation; Florida Fish and Wildlife Conservation Commission; Georgia Department of Natural Resources; Gomez and Sullivan Engineers; Kentucky Department of Fish and Wildlife Research Units; Kleinschmidt Associates; Louisiana Department of Wildlife and Fisheries; Louisiana State University; Merrill Creek Reservoir (J. Bullis); Natural Resources, Ltd.; Norfolk Botanical Gardens; PPL Holtwood; Rockefeller Wildlife Refuge Trust; Simon Fraser University; University of Georgia; US Army Corps of Engineers; US Department of Defense; US Fish and Wildlife Service; US Geological Survey Louisiana Cooperative Fish and Wildlife Research Unit; US National Park Service; Virginia Commonwealth University.

MISSISSIPPI KITE *Ictinia mississippiensis*

DISTRIBUTION AND ABUNDANCE

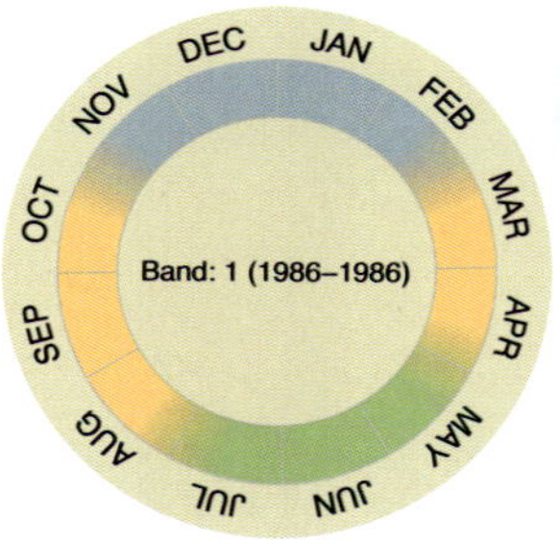

Mississippi Kites breed from South Carolina and Georgia, south through northern Florida and west through Gulf states including Alabama, Mississippi, Louisiana, Arkansas, north through the Mississippi River Valley to Tennessee, Missouri, and southern Illinois, and west to Oklahoma and Texas. Small numbers breed west to New Mexico and Arizona. This species may be expanding its breeding range northward in eastern North America as far as Connecticut, and in the midcontinent region to Indiana, Illinois, Iowa, Missouri, and Nebraska, and as far west as Colorado (Goodrich and Smith 2008). Individuals are regularly sighted as far north as the Great Lakes in migration and dispersal periods, but nesting attempts appear rare to date (Goodrich and Smith 2008). Nonbreeding areas are believed to be largely in northern Argentina and across Paraguay (Fink et al. 2023; Parker 2020), with flocks recorded primarily from 11 degrees south to 32 degrees south latitudes (Juhant and Areta 2013). Juhant and Areta (2013) report birds primarily seen during the nonbreeding season in Gran Chaco forest habitat.

MIGRATORY CONNECTIVITY

OVERALL Migratory connectivity information is relatively unknown. However, the limited nonbreeding range in northern Argentina, Paraguay, and southern Brazil (Juhant and Areta 2013; Fink et al. 2023) suggests connectivity may be weak for this species.

BANDING Few birds have been banded. One recovery from Texas was reported north of the core nonbreeding area. More data is needed.

OTHER APPROACHES No satellite telemetry, genetic, or stable isotope studies of migratory connectivity were found for this species.

ORDER Accipitriformes
FAMILY Accipitridae
BREEDING HABITAT Forests, Temperate Eastern
NONBREEDING HABITAT Forests, Tropical Dry
GLOBAL POPULATION SIZE 700,000

POPULATION CHANGE SINCE 1970 +47%
IUCN RED LIST CATEGORY Least Concern
R2R TIPPING POINT SPECIES No
US LISTING STATUS No
CANADA SPECIES AT RISK No
PIF CCS 9/20

INTERESTING NOTE

Mississippi Kites appear to build up fat prior to migration (Parker 2020) but also display foraging behavior in migration. Migrants are observed to catch insects early and late in the day and often roost in groups at night (Parker 2020). Migration counts in Bolivia are highest one to three days after the passage of a cold front, possibly timed for migrants to take advantage of flying ant swarms that emerge after frontal rains (Olivo 2004).

CONNECTIONS

MIGRATORY PATHWAYS Our understanding of Mississippi Kite migration is largely based on hawk migration counts. Birds begin moving south out of the breeding range early in the Northern Hemisphere fall, with small flocks observed in Texas and along the Gulf Coast in mid-August and peak counts in late August at watch sites such as Corpus Christi, Texas (hawkcount.org; Parker 2020). Loose flocks of more than 10,000 birds are sighted throughout Central and South America in early September (Ruelas Inzunza 2005). Flights appear to concentrate along the eastern coastal plain in Mexico south through Central America during post-breeding migration. Peak flights pass Veracruz, Mexico, in early September, where an average of more than 300,000 are sighted annually. Smaller counts are reported occasionally along the Pacific coast of Mexico (Goodrich and Smith 2008), with flight lines likely converging south of Mexico. In Costa Rica, a hawk watch at the Kekoldi site, located in the Caribbean lowlands of southeast Talamanca, tallies the largest concentration of migrating kites, with a peak count of 688,490 in 2023. The site reported a one-day total of more than 187,000 birds on September 9, 2023. Just south of Kekoldi in northern Panama, large kite flocks are also reported with single-day estimates of more than 200,000 in 2023, at the Bocos del Toro watch sites near the Caribbean coast (hawkcount.org). In South America, post-breeding flights remain mostly unreported until birds move south into central Colombia. In Tolima, Colombia, migrants pass in mid-September east of the Andes, where a watch site in Iguasitos recorded an average of 337,035 kites for 2022 and 2023 and a peak one-day count of 61,446 migrants on September 15, 2023 (Tropicos watch site, hawkcount.org). Farther south in Concepción, Bolivia, a one-season count recorded 118,153 kites during post-breeding migration in 2001, with peak numbers in late September and early October, although some appeared to be moving north and possibly had overshot their nonbreeding destination (Olivo 2004). A few individuals may island-hop across the Caribbean between Florida and South America through the Antilles, but this route is understudied (Goodrich and Smith 2008).

Pre-breeding migration is less well studied, and migrants are suspected to fly higher or use more dispersed routes north. In Veracruz, Mexico, count data from a transect of sites showed kites migrating north through a wider flight path compared to post-breeding, possibly due to greater availability of thermals in higher-elevation areas during pre-breeding (Ruelas Inzunza 2005; Ruelas Inzunza et al. 2010). In contrast to post-breeding counts, Colombia pre-breeding counts have not observed large numbers of Mississippi Kites in Tolima; it is suspected that birds take a different route north than during post-breeding migration. However, through Central America and Mexico, most kites appear to return to breeding sites along the same route used during post-breeding migration. More than 100,000 kites were recorded in spring 2023 in Kekoldi, Costa Rica, and also in Bocos del Toro, Panama (hawkcount.org), with peak flights in early April. In the United States, peak counts of 6,000–8,000 kites are observed migrating north in mid to late April at Bentson State Park, southern Texas. Scattered sightings occur in the eastern United States, with consistent numbers of 8–20 birds sighted along Chesapeake Bay at Fort Smallwood, Maryland, in late April and early May (Goodrich and Smith 2008; hawkcount.org).

Buteogallus anthracinus COMMON BLACK-HAWK

ORDER Accipitriformes
FAMILY Accipitridae
BREEDING HABITAT Wetlands, Forested
NONBREEDING HABITAT Wetlands, Forested
GLOBAL POPULATION SIZE 2,000,000
POPULATION CHANGE SINCE 1970 NA
IUCN RED LIST CATEGORY Least Concern
R2R TIPPING POINT SPECIES No
US LISTING STATUS No
CANADA SPECIES AT RISK No
PIF CCS 11/20

DISTRIBUTION AND ABUNDANCE

The species' distribution extends from northern South America, including Ecuador, Colombia, and Venezuela, throughout Central America and Mexico and into the southwestern United States and south Texas. The species is a year-round resident throughout most of its range but is migratory in North America (L.S. Johnson and Schnell 2024).

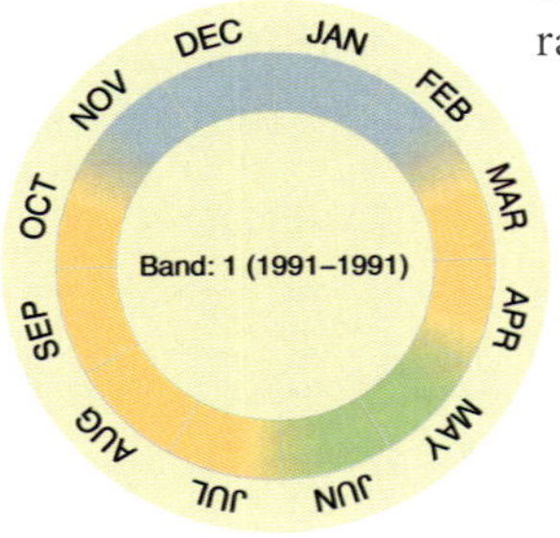

MIGRATORY CONNECTIVITY

OVERALL Very little is known about the migratory connectivity of Common Black-Hawks.

BANDING According to the US Geological Survey's Bird Banding Laboratory, 21 Common Black Hawks have been banded. One recovery of an individual banded on breeding grounds in southern Arizona was made during the nonbreeding season in Durango, Mexico.

OTHER APPROACHES No tracking, stable isotope, or genetic studies of migratory connectivity were found for this species.

MIGRATORY PATHWAYS Migratory pathways have not been studied sufficiently. A raptor migration count site in Tubac, Arizona, counts several hundred migrating Common Black Hawks moving over riparian habitat each March, with a peak one-day count of 134 individuals observed in 2019. Small numbers of migrants are seen along the Gulf of Mexico that possibly originate from Texas or New Mexico breeding birds.

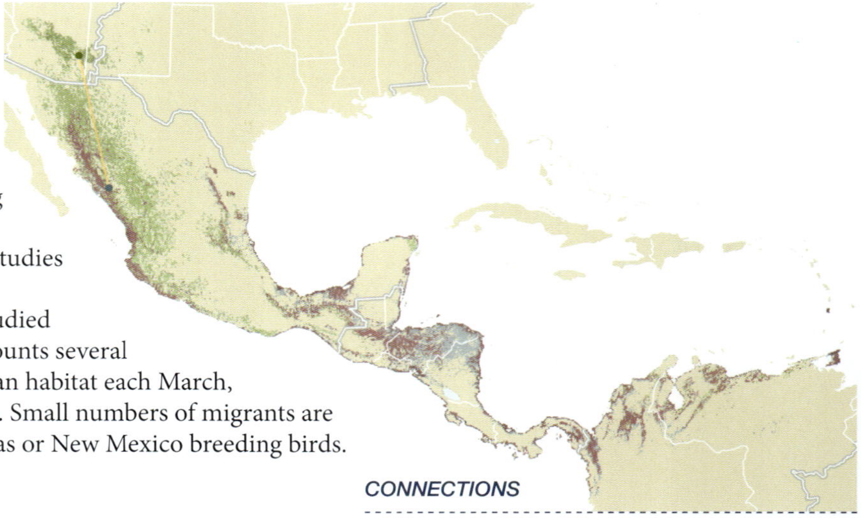

CONNECTIONS

INTERESTING NOTE

Migrants arrive to the southwestern US in March and depart as late as October. Peak flights are seen in mid-March in Arizona.

GRAY HAWK *Buteo plagiatus*

DISTRIBUTION AND ABUNDANCE

Gray Hawks breed from the southwestern United States and the Rio Grande Valley of Texas south through Mexico along both coastal plains and south through Costa Rica (Bibles et al. 2020). They primarily spend the nonbreeding season south of United States, although adults have been observed to remain year-round on territories in the Rio Grande Valley, Texas, while some juveniles migrated south (Stewart et al. 2023).

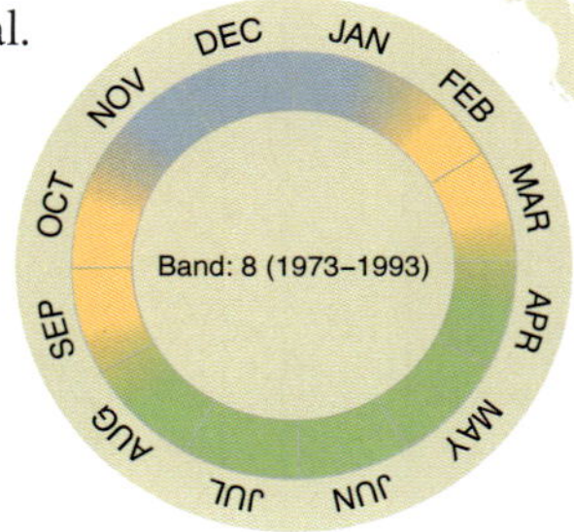

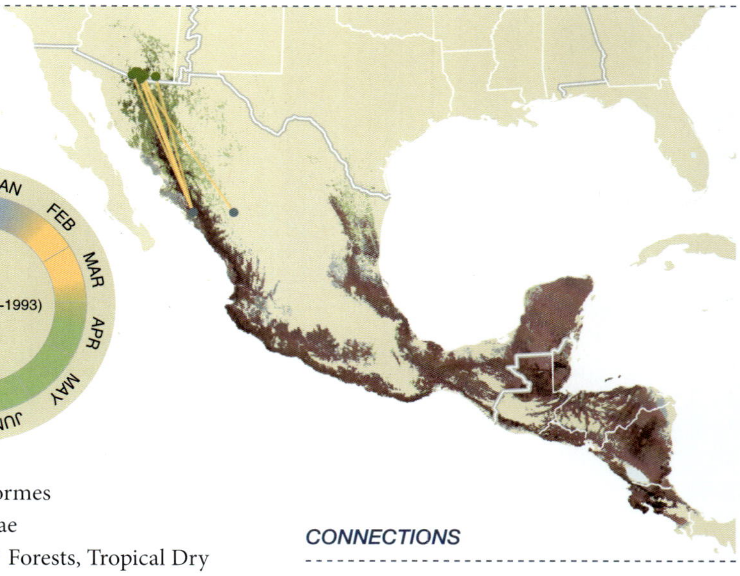

CONNECTIONS

ORDER Accipitriformes
FAMILY Accipitridae
BREEDING HABITAT Forests, Tropical Dry
NONBREEDING HABITAT Forests, Tropical Dry
GLOBAL POPULATION SIZE 2,000,000
POPULATION CHANGE SINCE 1970 NA
IUCN RED LIST CATEGORY Least Concern
R2R TIPPING POINT SPECIES No
US LISTING STATUS No
CANADA SPECIES AT RISK No
PIF CCS 9/20

INTERESTING NOTE

The Veracruz River of Raptors, in Mexico, averages the highest counts, with 524 Gray Hawks on average during each post-breeding migration season. Analyses of counts show a stable trend for 2009–2019, with a record one-day migration count of 1,258 Gray Hawks during post-breeding migration in 2005. Largest daily flights are observed in mid to late October.

MIGRATORY CONNECTIVITY

OVERALL There is little information on migratory connectivity from banding data.

BANDING Limited band reencounters suggest Arizona and northwestern Mexican breeding populations move south into western Mexico.

RADIO TRACKING Few tags were deployed through the Motus Wildlife Tracking System, with local detections only.

SATELLITE TRACKING Only one tracking study to date examined movements of 22 adults and 14 juvenile Gray Hawks in the Rio Grande Valley, Texas. Texas and northeastern Mexico birds appear to remain resident or move south along the Gulf coastal plain into Central America (Stewart et al. 2023). None of the adults migrated. Some of the juveniles dispersed short distances, whereas four of the juveniles tagged in Texas migrated long distances into Mexico and Central America.

MIGRATORY PATHWAYS Gray Hawks concentrate along both coasts from Mexico and throughout Central America during pre- and post-breeding.

Buteo lineatus RED-SHOULDERED HAWK

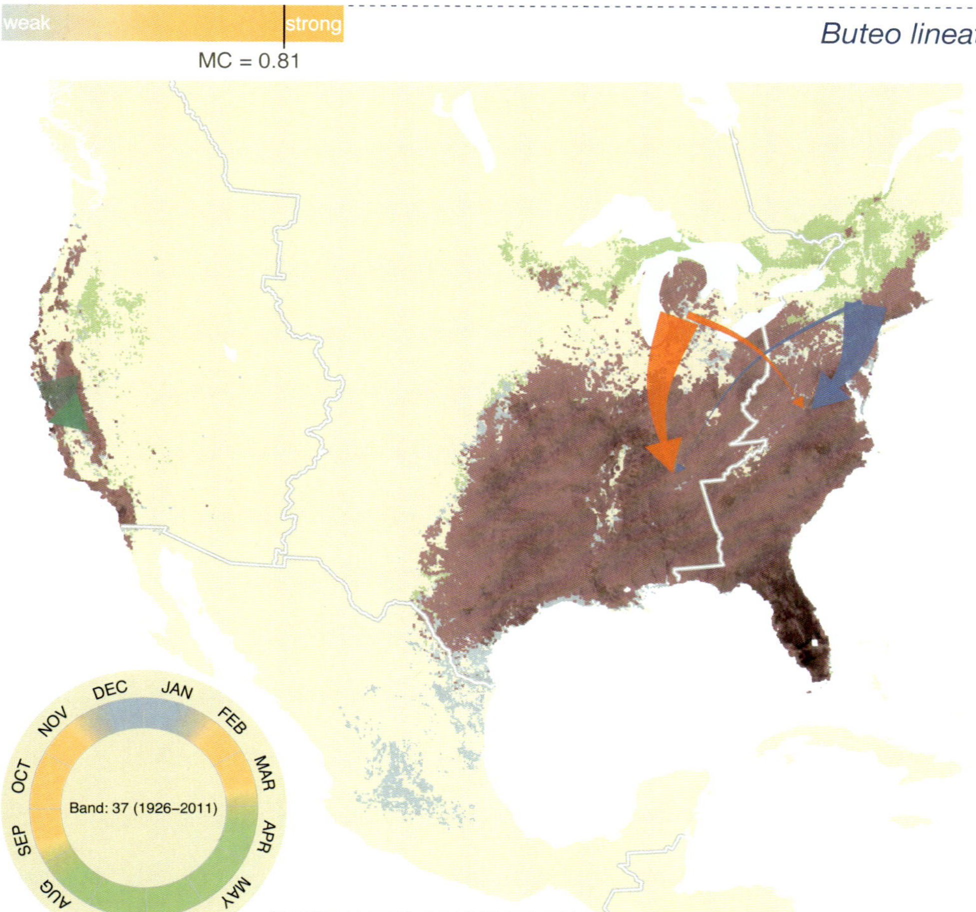

GENERALIZED CONNECTIVITY

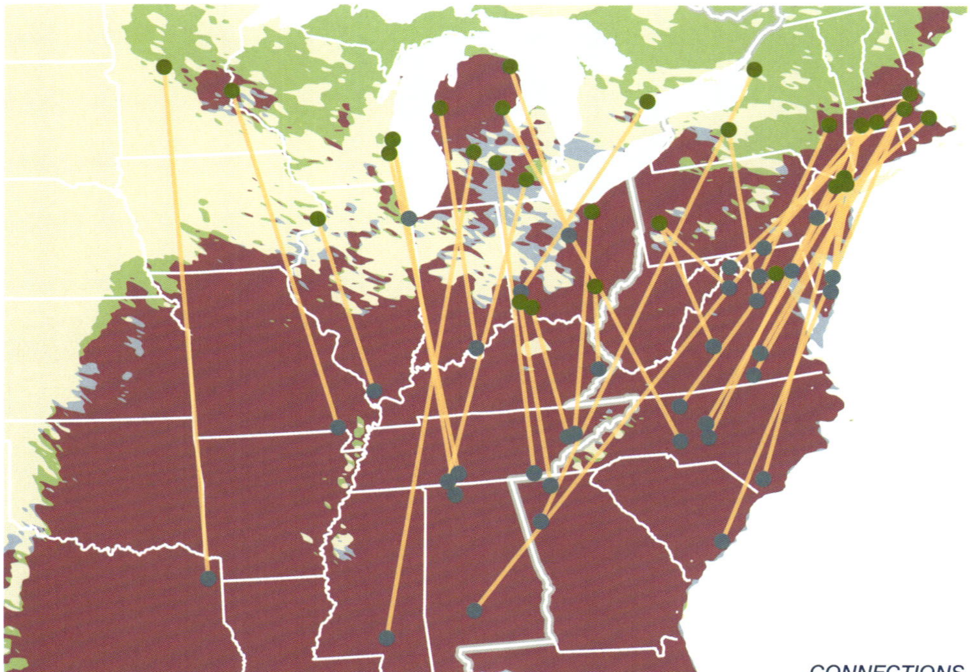

CONNECTIONS

DISTRIBUTION AND ABUNDANCE

Red-shouldered Hawks are a medium- to short-distance partial migrant in the northern part of their range and are primarily resident in the southern and western United States (Goodrich and Smith 2008; Fink et al. 2023). Birds in southern Canada and the northern United States can exhibit long migrations (Hanneman 2021). They breed across eastern North America from central Canada and the Great Lakes east to New Brunswick and south to Florida and west to the Mississippi River (Dykstra et al. 2020). A northward expansion is reported in some Great Lakes states and southern Canada (Dykstra et al. 2020). They spend the nonbreeding season throughout the breeding range but are also found south into northeastern Mexico and Baja California (Dykstra et al. 2020). A western population is found primarily in California and the Pacific Northwest, with small numbers into the Baja region of Mexico.

ORDER Accipitriformes
FAMILY Accipitridae
BREEDING HABITAT Forests, Temperate Eastern
NONBREEDING HABITAT Forests, Temperate Eastern
GLOBAL POPULATION SIZE 1,900,000
POPULATION CHANGE SINCE 1970 +296%
IUCN RED LIST CATEGORY Least Concern
R2R TIPPING POINT SPECIES No
US LISTING STATUS No
CANADA SPECIES AT RISK No
PIF CCS 8/20

	Atlantic	Midcontinent	Pacific
Atlantic (n = 18)	94%	6%	–
Midcontinent (n = 15)	13%	87%	–
Pacific (n = 1)	–	–	100%

MIGRATORY CONNECTIVITY

OVERALL Overall migratory connectivity is strong (MC = 0.81) and primarily known from banding data, although more recent tracking studies do exist. Birds may exhibit chain migration, but this needs more study (Dykstra et al. 2020). Atlantic, Midcontinent, and Pacific flyway breeding birds primarily spend the nonbreeding season within the same flyway as breeding.

BANDING Band reencounters show that Red-shouldered Hawks breeding in northeastern US states move southwest to nonbreeding areas in the mid-Atlantic

INTERESTING NOTE

Bloom et al. (2011) reported that band recoveries of nestling Red-shouldered Hawks dispersing longer distances (>100 km) show lower survival rates, with most dying before breeding age, although one bird survived 8.6 years.

or southeastern US states. Birds banded around the Great Lakes migrate mostly due south to nonbreeding sites in east Texas through the Gulf states and southern Appalachians; however, western Great Lakes nesting birds stay predominantly in the midcontinent. Red-shouldered Hawks in southern Ohio (Dykstra et al. 2019) and California (Bloom et al. 2011) are considered mostly resident, but young birds show dispersal movements both north and south from nesting areas.

SATELLITE TRACKING In 2018–2021, five adult Red-shouldered Hawks were tracked from breeding sites in Wisconsin to nonbreeding sites in Wisconsin, Illinois, Kentucky, Alabama, and Louisiana. Tracked birds moved an average of 919 km, over an average migration period of 28.9 days, with the longest journey being 1,577 km from Wisconsin to Louisiana (Hanneman 2021).

OTHER APPROACHES No genetic or stable isotope studies of migratory connectivity were found for this species.

MIGRATORY PATHWAYS Red-shouldered Hawks concentrate along diversion lines in migration, particularly at coastal sites (e.g., Great Lakes shorelines; Cape May, New Jersey; Golden Gate Raptor Observatory, California). They may preferentially migrate along river corridors in inland migration, although more data is needed (Goodrich and Smith 2008). Highest migration counts are in the eastern United States at Cape May, New Jersey (Goodrich and Smith 2008). California birds disperse widely post-breeding (Bloom et al. 2011). It is not clear if northern birds (Oregon, Washington) exhibit migration, although individuals are regularly recorded at Golden Gate Bird Observatory, San Francisco, during the post-breeding season (hawkcount.org; Goodrich and Smith 2008).

BROAD-WINGED HAWK *Buteo platypterus*

DISTRIBUTION AND ABUNDANCE

The Broad-winged Hawk is a highly gregarious migratory raptor. Breeding grounds are primarily from northeastern British Columbia east across southern Canada through the Atlantic provinces, throughout most of Nova Scotia and south through all northeastern, eastern, and central states to northern Florida and eastern Texas. During the nonbreeding season, individuals can be found from central Mexico south through Panama into Colombia and northeastern South America. Some adults spend the nonbreeding season south through the Andes into Peru, Brazil, and Bolivia. Resident subspecies also occur on some Caribbean islands, including Puerto Rico, Cuba, Hispaniola.

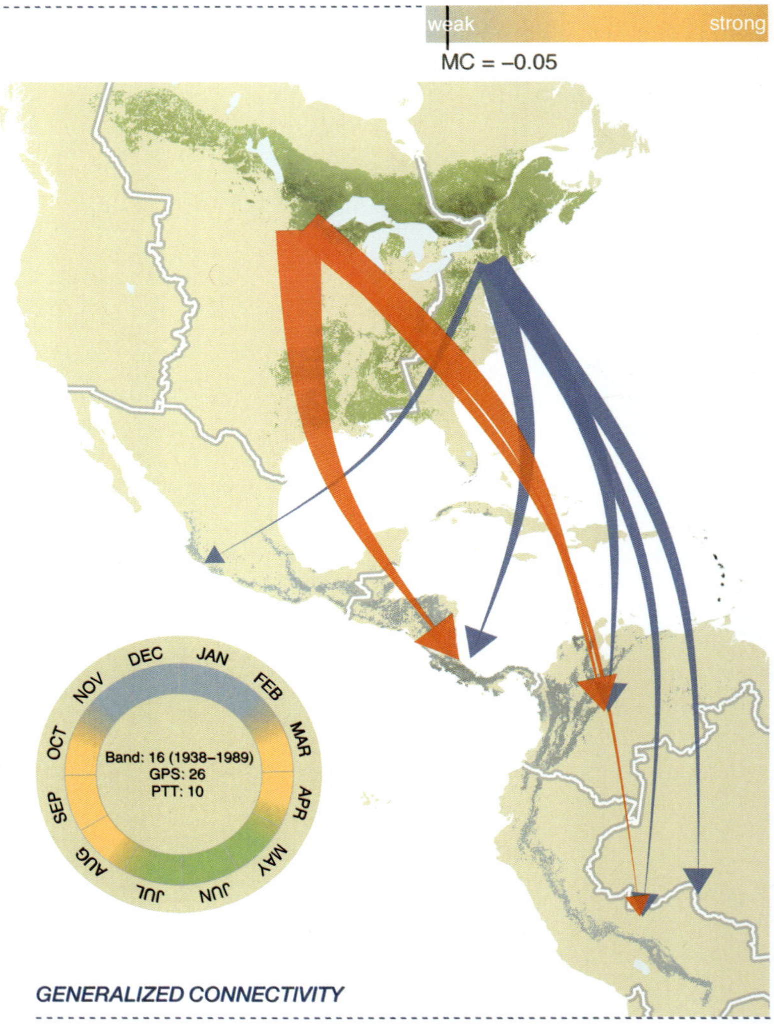

GENERALIZED CONNECTIVITY

ORDER Accipitriformes
FAMILY Accipitridae
BREEDING HABITAT Forests, Boreal
NONBREEDING HABITAT Forests, Tropical Generalist
GLOBAL POPULATION SIZE 1,900,000
POPULATION CHANGE SINCE 1970 +46%
IUCN RED LIST CATEGORY Least Concern
R2R TIPPING POINT SPECIES No
US LISTING STATUS Endangered Puerto Rican subspecies
CANADA SPECIES AT RISK No
PIF CCS 8/20

	Andean West	Brazil	Central America	Mexico	Northern SA
Atlantic (n = 18)	17%	22%	28%	6%	28%
Midcontinent (n = 16)	12%	–	50%	–	38%

MIGRATORY CONNECTIVITY

OVERALL Migratory connectivity is known from band reencounter and satellite tracking data of breeding individuals tagged in Pennsylvania, Minnesota, Maryland, Connecticut, and New Hampshire, as well as Alberta and Ontario, Canada.

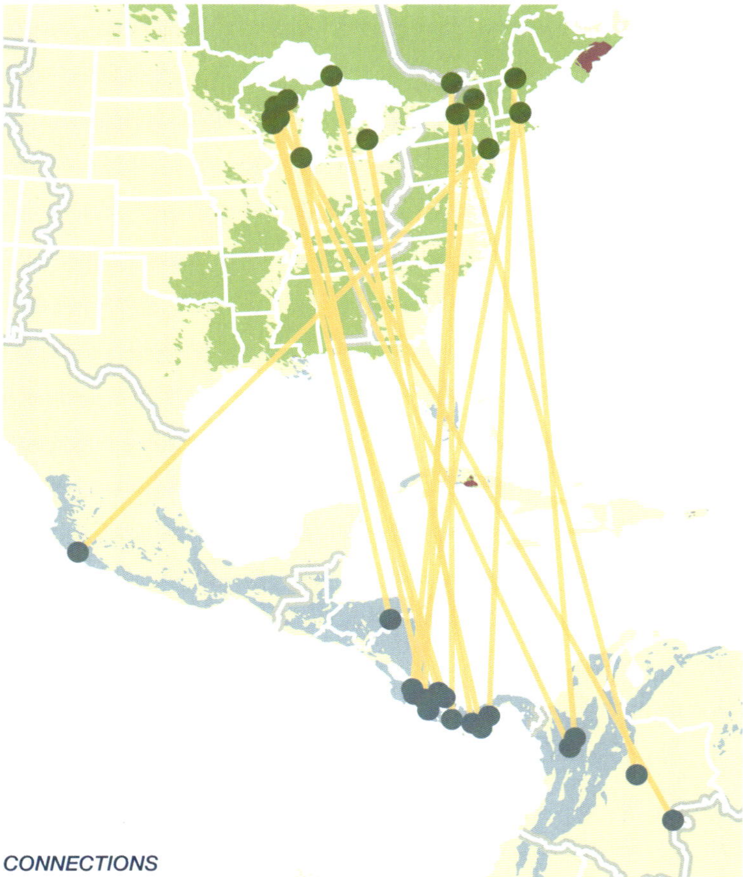

CONNECTIONS

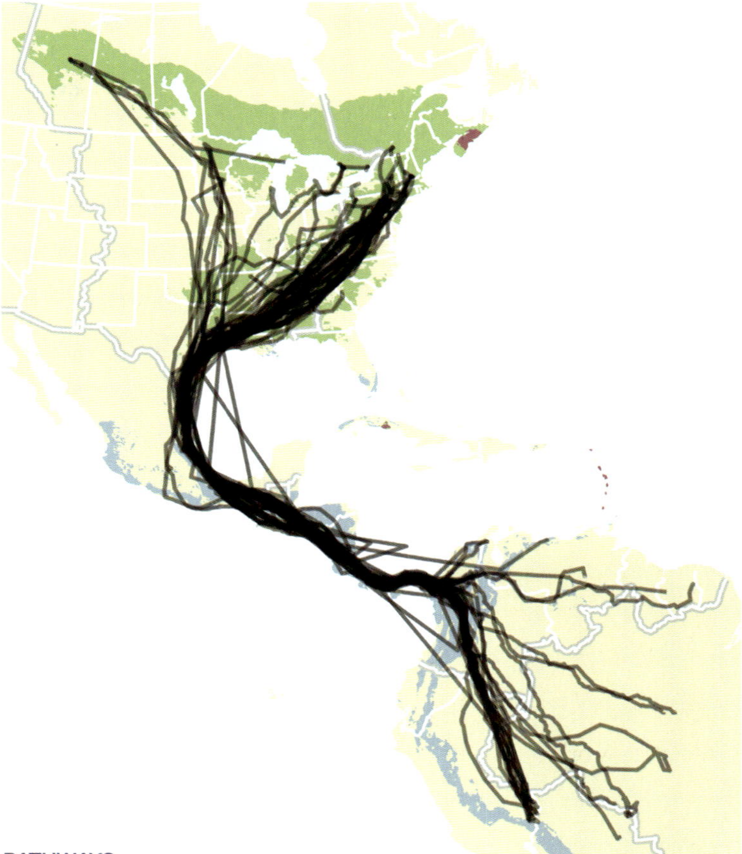

PATHWAYS

Individuals breeding in the Atlantic Flyway spend the nonbreeding season across several regions, including Central America (28%), northern South America (28%), Brazil (22%), and the Andean West (17%). Birds breeding in the Midcontinent Flyway primarily migrate to Central America (50%) and northern South America (38%) for the nonbreeding season. Overall, migratory connectivity is weak for this species (MC = −0.05).

BANDING Band reencounters suggest that birds breeding within a latitudinal band from 40 degrees north to 47 degrees north across the range primarily migrate to nonbreeding areas within Costa Rica, Panama, and Colombia.

SATELLITE TRACKING In the first satellite tracking study of the species, Haines et al. (2003) captured and tagged five adult female Broad-winged Hawks in Minnesota and Maryland. Nonbreeding areas included Panama, Venezuela, Brazil, and Peru. Recent tracks of adults from Pennsylvania, Connecticut, New Hampshire, and Ontario (McCabe et al. 2020; Hawk Mountain Sanctuary, unpublished data) show northeastern adult birds spending the nonbreeding season from Panama south through Brazil, Peru, and Bolivia. Two of three birds tagged in Alberta spent the nonbreeding season in northeastern South America, and the other bird spent it in Brazil. Three immature birds satellite tagged in Pennsylvania (McCabe et al. 2020) spent the nonbreeding season in Central America.

OTHER APPROACHES No genetic or stable isotope studies of migratory connectivity found for this species. Observations from eBird data suggest that immature birds spend the nonbreeding season farther north than adults (Heveran et al. 2022).

MIGRATORY PATHWAYS Migration occurs over a broad front in the United States and South America, with eastern birds concentrating along the Appalachian Mountain corridor in post-breeding movements. Large, conspicuous flocks numbering in hundreds to thousands of individuals can be seen in eastern and central migration paths, with very large congregations observed along the Gulf Coast from south Texas to Mexico, as well as throughout Central America. Smaller concentrations (e.g., up to 100,000 per day), migrate south along the Andes in central Colombia (Tropicos, unpublished data). Individuals from the Atlantic and Midcontinent flyways migrate along the Gulf Coast through Mexico and then fly inland through Central America (Haines et al. 2003; McCabe et al. 2020). Individuals from Alberta, Canada, show similar migration routes along the Mexican Gulf Coast and inland through Central America, with some birds stopping to spend the nonbreeding season in Central America (McCabe et al. 2020; Hawk Mountain Sanctuary, unpublished data).

DATA CONTRIBUTORS

E. Bayne; J.B. Dayton; L. Goodrich; A. Haines; M.B. Henke; P.P. Marra; M. Martell; R. McCabe; M. McGradyt; W. Seegar; J. Watson; Hawk Mountain Sanctuary; Migratory Connectivity Projec.

INTERESTING NOTE

Most of the approximately 2 million Broad-winged Hawks that breed in North America get regularly counted along the Gulf of Mexico in Veracruz, Mexico, within a week's time. This intense spatial bottleneck can sometimes have more than half a million birds passing on a single peak day.

SHORT-TAILED HAWK *Buteo brachyurus*

DISTRIBUTION AND ABUNDANCE

Although widespread south of the United States, Short-tailed Hawks are rare within the US and occur in small numbers only in Florida. Short-tailed Hawks breed in Florida, with those individuals breeding in central and northern Florida migrating short distances to nonbreeding areas in southern Florida and the Florida Keys. Short-tailed Hawks are year-round residents in western, eastern, and southern Mexico, throughout Central America and much of South America.

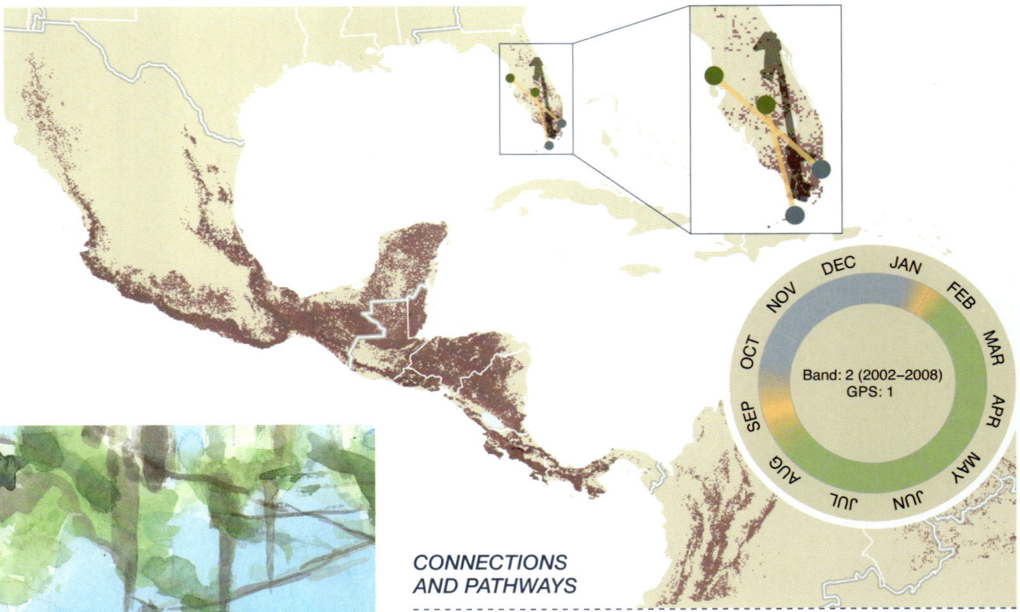

CONNECTIONS AND PATHWAYS

ORDER Accipitriformes
FAMILY Accipitridae
BREEDING HABITAT Open Country, Habitat Mosaic
NONBREEDING HABITAT Open Country, Habitat Mosaic
GLOBAL POPULATION SIZE NA
POPULATION CHANGE SINCE 1970 NA
IUCN RED LIST CATEGORY Least Concern
R2R TIPPING POINT SPECIES No
US LISTING STATUS No
CANADA SPECIES AT RISK No
PIF CCS 9/20

MIGRATORY CONNECTIVITY

OVERALL Limited data prevent estimates of MC, but satellite telemetry data suggest strong site fidelity between breeding and nonbreeding locations (Miller and Meyer 2020).

BANDING Two birds banded in central Florida during the breeding season were recaptured in southern Florida during the nonbreeding season.

SATELLITE TRACKING One study tracked one female from central to south Florida (Avian Research and Conservation Institute, unpublished data).

OTHER APPROACHES No genetic or stable isotope studies of migratory connectivity were found for this species.

MIGRATORY PATHWAYS Florida routes may occur across a broad front, although they are suspected to occur west of Lake Okeechobee (Miller and Meyer 2020). Migrants were observed regularly over Curry Hammock Key at Florida Keys HawkWatch (Goodrich and Smith 2008), and five birds in one flock were observed on Boco Grande island west of Key West in 1991, although they were not recorded in Cuba or the Caribbean (Miller and Meyer 2020).

	Atlantic	Caribbean	Mexico	Midcontinent	Pacific
Atlantic (n = 14)	21%	79%	–	–	–
Midcontinent (n = 13)	–	–	23%	77%	–
Pacific (n = 1)	–	–	–	–	100%

DATA CONTRIBUTORS
G.M. Kent; K.D. Meyer; Avian Research and Conservation Institute; BirdsCaribbean; Florida Fish and Wildlife Conservation Commission; National Fish and Wildlife Foundation; National Wildlife Refuges, US Fish and Wildlife Service.

INTERESTING NOTE

Migration occurs in October, with nonbreeding birds arriving in November to south Florida (Miller and Meyer 2020). An average of 41 are recorded on migration counts in the Florida Keys during post-breeding migration (www.hawkcount.org).

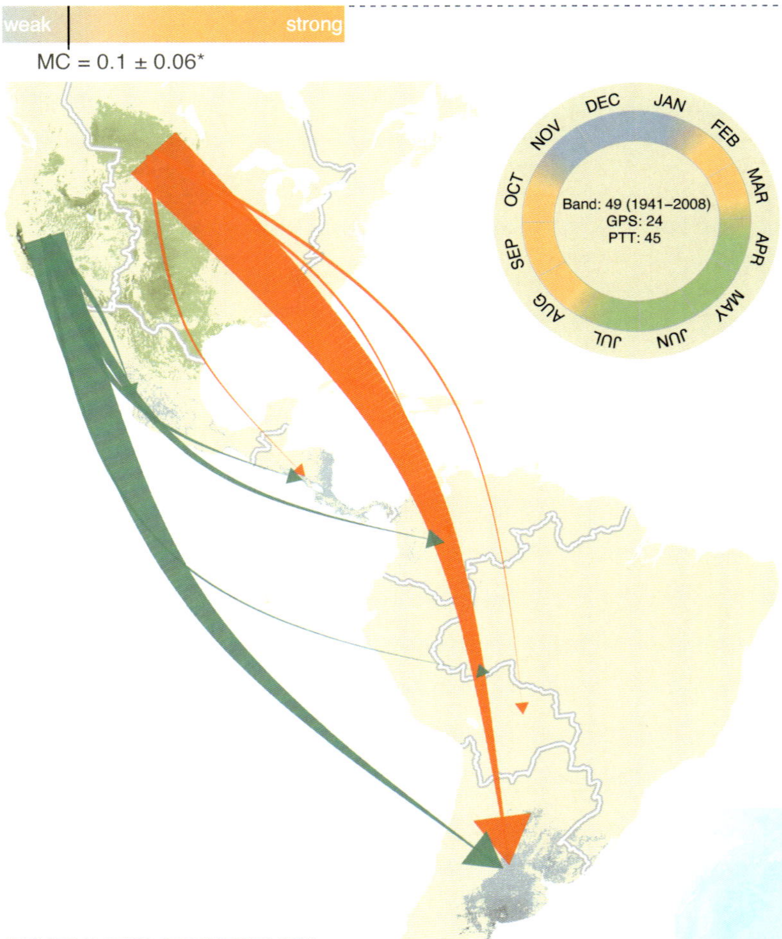

GENERALIZED CONNECTIVITY

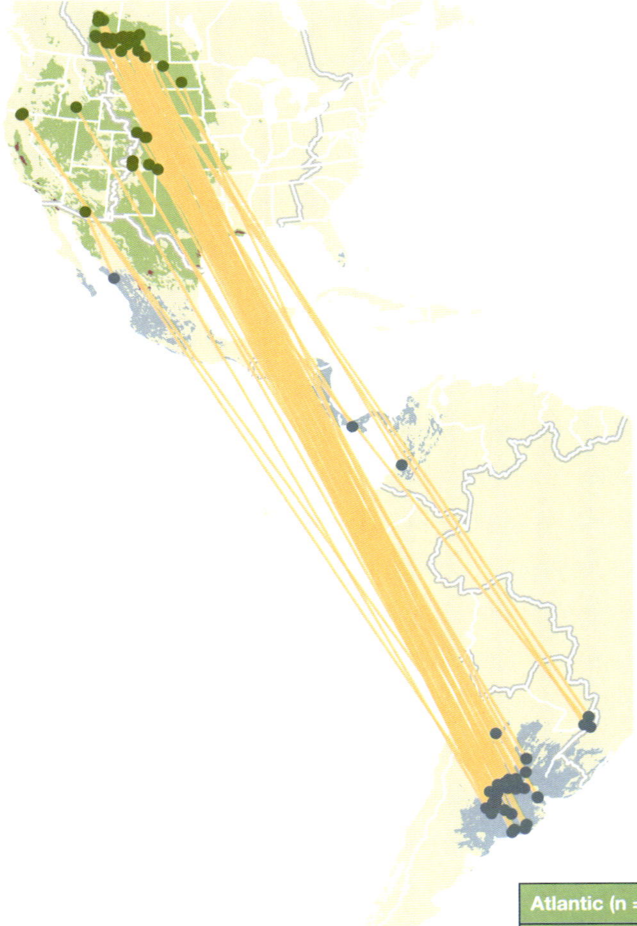

CONNECTIONS

Buteo swainsoni **SWAINSON'S HAWK**

DISTRIBUTION AND ABUNDANCE

The Swainson's Hawk has one of the longest migrations of any North American raptor, flying up to 20,000 km round-trip from western North America to central Argentina (Bechard et al. 2020). Highly gregarious in migration, they form large flocks and concentrate along narrow migration corridors, where thousands of individuals can be seen in a single flock (Ruelas Inzunza 2005). They breed in western North America (from southern Alberta, Saskatchewan, and Manitoba and in parts of British Columbia south through the Central Valley of California and into northern Mexico) and spend the nonbreeding season in central Argentina and Uruguay (Bechard et al. 2020), with smaller numbers spending the nonbreeding season in Florida, Texas, western Mexico, Central America, and northern South America (Airola et al. 2019). Swainson's Hawks suffered declines in the 1990s (Houston and Schmutz 1995; Woodbridge et al. 1995) due to widespread pesticide use on the nonbreeding range (Woodbridge et al. 1995) and declines in prey for several Canadian prairie breeding populations. Recent trends suggest a possible annual increase of 2% in nesting populations and increases or stable trends for most western hawk watch sites (Oleyar et al. 2021; Fink et al. 2023); however, California birds are listed as a species of concern (Airola et al. 2019).

ORDER Accipitriformes
FAMILY Accipitridae
BREEDING HABITAT Grasslands, Temperate
NONBREEDING HABITAT Grasslands, Pampas and Campos
GLOBAL POPULATION SIZE 900,000

POPULATION CHANGE SINCE 1970 +55%
IUCN RED LIST CATEGORY Least Concern
R2R TIPPING POINT SPECIES No
US LISTING STATUS No
CANADA SPECIES AT RISK No
PIF CCS 11/20

MIGRATORY CONNECTIVITY

OVERALL Migratory connectivity is primarily known from satellite telemetry and banding data. Overall, migratory connectivity is weak (MC = 0.1 + 0.06), with high overlap of both Midcontinent and Pacific flyway breeding birds in the Southern Cone during the nonbreeding season (95% and 62%, respectively) and smaller percentages throughout other nonbreeding regions.

	Andean West	Brazil	Central America	Mexico	Northern SA	Southern Cone
Atlantic (n = 55)	2%	–	2%	–	2%	95%
Pacific (n = 39)	–	3%	10%	10%	15%	62%

PATHWAYS

INTERESTING NOTE

Southbound migrating Swainson's Hawks tracked in the 1990s, using satellite telemetry, traveled for up to 98 days and remained on nonbreeding areas for up to 128 days (Kochert et al. 2011). Birds exhibited longer stopovers earlier in their migration, in the southern United States or northern Mexico, lasting nine or more days. Some stopover sites were used in multiple years and may warrant conservation attention (Kochert et al. 2011). Migration behavior and timing was similar for males and females (Kochert et al. 2011). Counts of Swainson's Hawks in Veracruz, Mexico, the largest concentration point for southbound migrants, averaged 812,419 ± 287,822 (standard deviation) for the period 1995–2004 (Ruelas Inzunza 2005).

BANDING Band reencounter data show movement from western North America nesting areas south and east through Mexico and Central America to nonbreeding sites in Central and South America.
RADIO TRACKING Few tags were deployed through the Motus Wildlife Tracking System, with local detections only.
SATELLITE TRACKING Most nesting Swainson's Hawks appear to migrate to nonbreeding areas in Argentina and Uruguay (Woodbridge et al. 1995; Fuller et al. 1998; Kochert et al. 2011; Fleishman et al. 2016; Watson et al. 2018; Salt Lake City Airport Wildlife Department, unpublished data); however, birds nesting in the Central Valley of California also migrate to nonbreeding areas in western Mexico, Central America, and northern South America (Fleishman et al. 2016).
GENETICS No genetic studies of migratory connectivity were found.
STABLE ISOTOPES Stable hydrogen isotope analyses of nonbreeding birds found widespread mixing of breeding populations on the Argentina nonbreeding grounds (Sarasola et al. 2008). No analyses of other nonbreeding areas were found.
MIGRATORY PATHWAYS Swainson's Hawks appear to migrate across a broad front early in their migration, taking advantage of thermal lift. In the southern United States, flocks converge, forming large streams of migrants in southern Texas and along the east and west coastal plain of Mexico (Woodbridge et al. 1995; Fuller et al. 1998; Kochert et al. 2011; Watson et al. 2017; Salt Lake City Airport Wildlife Department, unpublished data), with largest concentrations east of the Sierra Madre range in Veracruz (Kochert et al. 2011; Airola et al. 2019). A second migration pathway, particularly for California birds, is through California and western Mexico (Fleishman et al. 2016; Airola et al. 2019), with both flight lines converging in Central America. From there, birds funnel through Central America, continuing along the Andes Mountains and south through Colombia, Bolivia, and Argentina (Fuller et al. 1998; Kochert et al. 2011; Watson et al. 2018; E. Vallejos, unpublished data; Salt Lake City Airport Wildlife Department, unpublished data). The same route is used on the return journey (Fuller et al. 1998; Kochert et al. 2011). Occasional sightings in eastern North America, Florida, and rarely Cuba are of unknown origin but may derive from central Canada (Goodrich and Smith 2008).

DATA CONTRIBUTORS

C. Deavila; E. Fleishman; M.N. Kochert; Boise State University; Canadian Wildlife Service; US Fish and Wildlife Service; US Geological Survey.

ZONE-TAILED HAWK *Buteo albonotatus*

ORDER Accipitriformes
FAMILY Accipitridae
BREEDING HABITAT Forests, Tropical Generalist
NONBREEDING HABITAT Forests, Tropical Generalist
GLOBAL POPULATION SIZE 2,000,000
POPULATION CHANGE SINCE 1970 NA
IUCN RED LIST CATEGORY Least Concern
R2R TIPPING POINT SPECIES No
US LISTING STATUS No
CANADA SPECIES AT RISK No
PIF CCS 10/20

DISTRIBUTION AND ABUNDANCE

The breeding range includes Arizona north to the Grand Canyon region, southern New Mexico, southwestern Texas, and southwestern California (R.R. Johnson et al. 2020a). This species also breeds in higher elevations of Baja California and northern Mexico (R.R. Johnson et al. 2020a). It is reportedly rare as a nesting bird in the United States, with possibly 125 nesting pairs in Arizona (R.R. Johnson et al. 2020a). However, nesting occurrence in the United States increased by 23% from 2012 to 2022 (Fink et al. 2023). Nonbreeding areas are primarily from northern Mexico south along both Pacific and Gulf of Mexico coastal plains and south into Central America. South American birds are thought to be resident throughout their range (R.R. Johnson et al. 2020a).

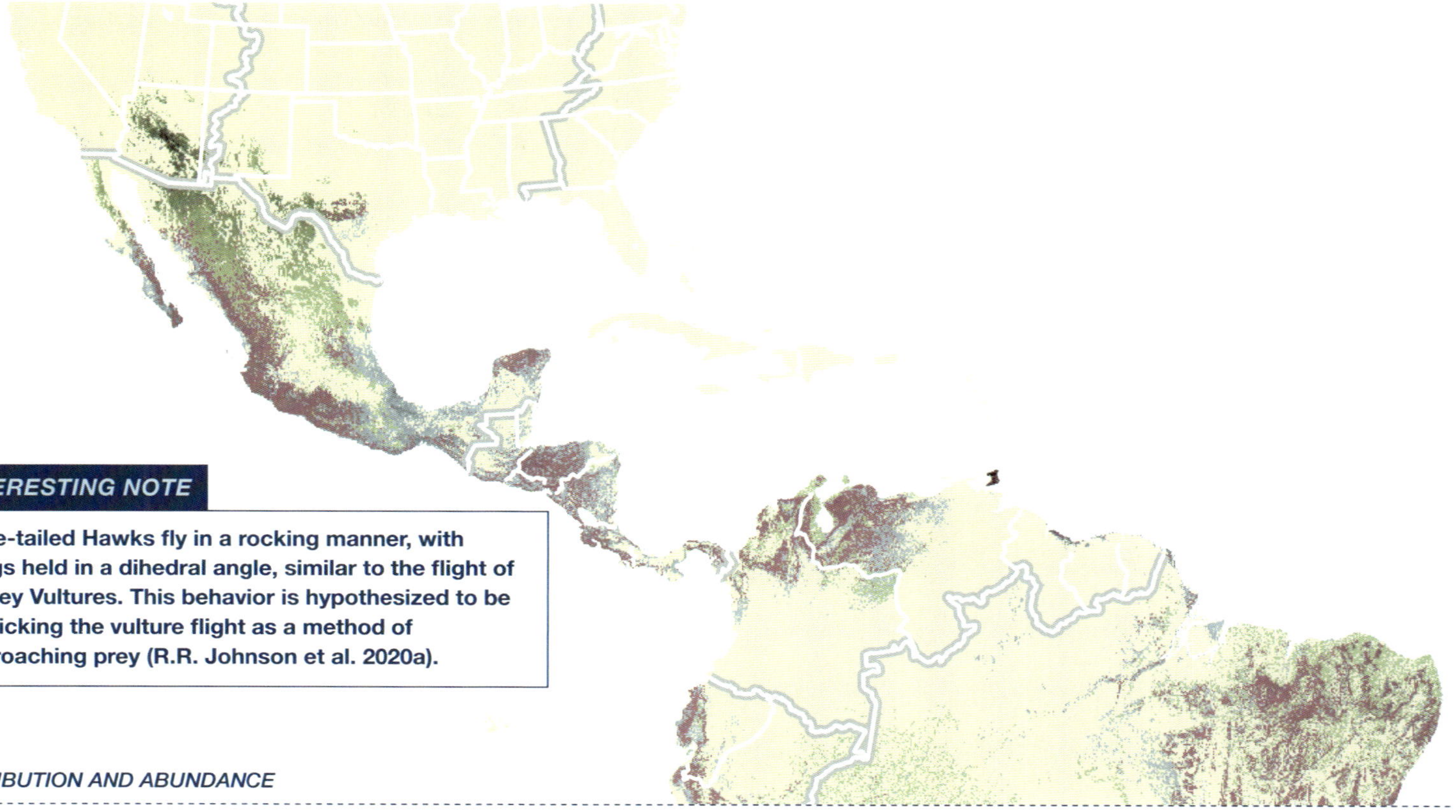

INTERESTING NOTE

Zone-tailed Hawks fly in a rocking manner, with wings held in a dihedral angle, similar to the flight of Turkey Vultures. This behavior is hypothesized to be mimicking the vulture flight as a method of approaching prey (R.R. Johnson et al. 2020a).

DISTRIBUTION AND ABUNDANCE

MIGRATORY CONNECTIVITY

OVERALL Migratory connectivity is unknown due to limited data. A recent pilot study deploying GPS transmitters on Zone-tailed Hawks in the Trans-Pecos region of Texas may shed light on the movements of this species (C. Boal, unpublished data). Data is not presented in the Atlas.

RADIO TRACKING There are no detections from tags deployed with the Motus Wildlife Tracking System.

OTHER APPROACHES No banding, radio tracking, geolocator, stable isotope, or genetic studies of migratory connectivity were found for this species.

MIGRATORY PATHWAYS Zone-tailed Hawks are soaring migrants that may join flocks of Swainson's Hawk in migration (R.R. Johnson et al. 2020a). Largest concentrations of migrating Zone-tailed Hawks in the southwestern United States were reported during pre-breeding at Tubac, Arizona, with an average count of 70 birds, with peak counts in late March. In post-breeding migration, the largest concentrations of southbound migrants occur in Veracruz, Mexico, with a mean count of 189 migrants per year at two sites combined (Goodrich and Smith 2008). Peak one-day counts across these sites recorded a combined total of 75 birds in late September (hawkcount.org). Central American count sites in Costa Rica and Panama report <10 sightings per year, suggesting most migrants may stop in southern Mexico and Guatemala (hawkcount.org).

MC = 0.34 ± 0.04*

Buteo jamaicensis RED-TAILED HAWK

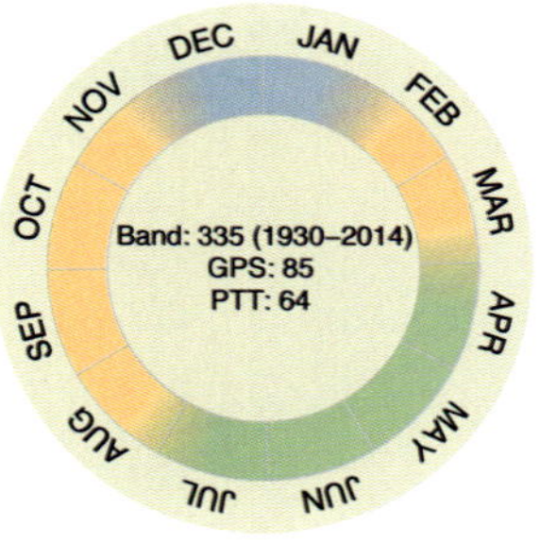

ORDER Accipitriformes
FAMILY Accipitridae
BREEDING HABITAT Open Country, Habitat Mosaic
NONBREEDING HABITAT Open Country, Habitat Mosaic
GLOBAL POPULATION SIZE 3,100,000
POPULATION CHANGE SINCE 1970 +115%

IUCN RED LIST CATEGORY Least Concern
R2R TIPPING POINT SPECIES No
US LISTING STATUS No
CANADA SPECIES AT RISK No
PIF CCS 7/20

DISTRIBUTION AND ABUNDANCE

The Red-tailed Hawk is a widespread raptor nesting across urban to rural environments. The continental subspecies breeds from Canadian and Alaskan boreal forests south through northern and central Mexico, and in parts of Central America. They spend the nonbreeding season across most of the US (excluding northern reaches of the Rocky Mountains, Great Plains, and Midwest), northern and central Mexico, and parts of Central America (Preston and Beane 2020). Populations found in the northern US and throughout Canada may be mostly migratory, whereas populations in southern and central states display partial migration, with adults particularly being year-round residents (Bloom et al. 2015; Preston and Beane 2020). North American Red-tailed Hawk breeding populations increased by an estimated 1.6% annually from 1970 to 2017 (95% confidence interval of 1.5%–1.8%; J.R. Sauer), but

	Atlantic	Central America	Mexico	Midcontinent	Pacific
Atlantic (n = 48)	73%	–	–	27%	–
Midcontinent (n = 233)	11%	–	3%	85%	1%
Pacific (n = 128)	–	2%	38%	9%	52%

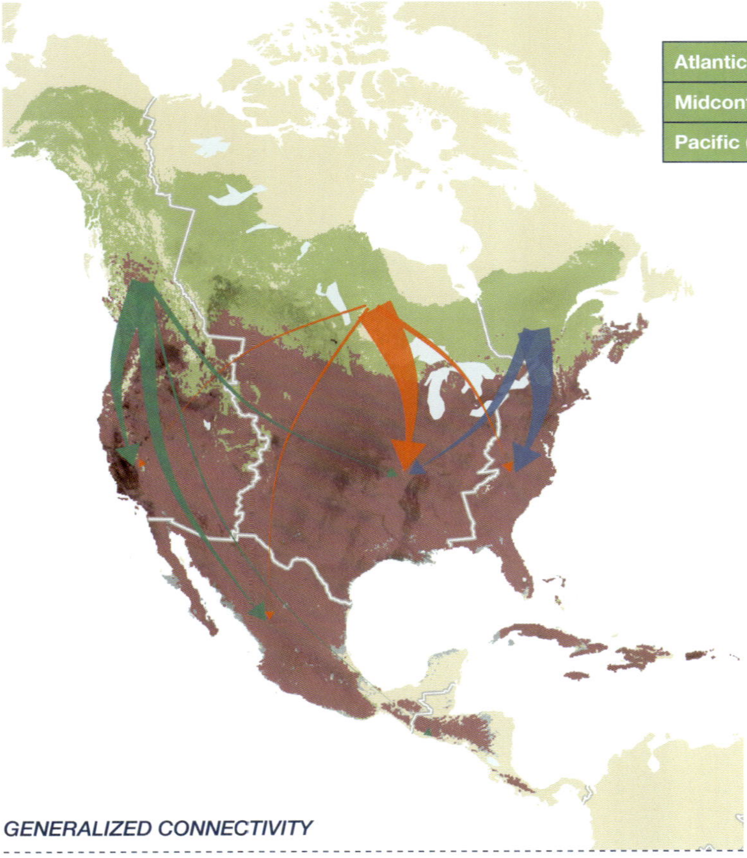

GENERALIZED CONNECTIVITY

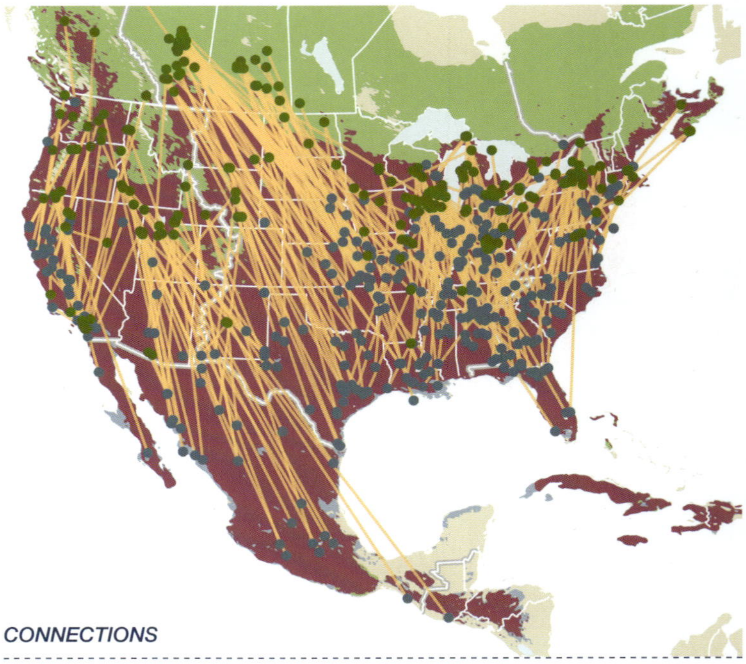

CONNECTIONS

recent eBird data suggest a decline of 6% per year range-wide (Fink et al. 2023). Christmas Bird Count data from 2009 to 2019 also show declines in Ontario (Oleyar et al. 2021) and the southeastern US during the nonbreeding season. Migration counts show declines of 41% at 76 sites across North America. However, increases in the West and northern New England during the nonbreeding season suggest some populations may be migrating shorter distances, which may alter migration count trends (Paprocki et al. 2017; Oleyar et al. 2021). This shift could reflect either reduced migration or declines in migratory populations (Morrison and Baird 2016).

MIGRATORY CONNECTIVITY

OVERALL Migratory connectivity is well known from satellite telemetry and banding data, as well as stable isotope and genetic data. However, most studies are from within the Pacific and Midcontinent flyways. Midcontinent Flyway breeding birds primarily remain within the flyway during the nonbreeding season (85%). Similarly, a large portion of breeding birds in the other flyways remain within the flyway during the nonbreeding season (Atlantic Flyway 73%, Pacific Flyway 52%), resulting in moderate migratory connectivity (MC = 0.34 + 0.04). Outside of breeding flyways, Atlantic Flyway breeding birds also migrate to nonbreeding areas in the Midcontinent Flyway (27%), and Pacific Flyway breeding birds migrate to Mexico (38%). Leapfrog migration patterns have been documented in several studies where the most northern breeding birds migrate farther, passing over shorter-distance migrants at lower latitudes (Craighead et al. 2016; Morrison and Baird 2016; Steenhof et al. 1984).

BANDING Band reencounter data collected in the northwest (Pacific and Rockies) showed movement south or southeast to nonbreeding areas in the United States, Mexico, and Central America. In contrast, juvenile birds from south of 38 degrees north latitude in the Pacific region largely move north or northeast for the nonbreeding season (Bloom et al. 2017; Hoffman et al. 2002). In the midcontinent and Atlantic regions, movements appear south or southwest, with most birds spending the nonbreeding season within their natal region (Morrison and Baird 2016). Band recoveries of birds reencountered at Golden Gate Raptor Observatory showed that later post-breeding migrants may have come from farther north compared to birds trapped earlier in the season (Hull et al. 2009).

RADIO TELEMETRY Few tags were deployed through the Motus Wildlife Tracking System, with local detections only.

SATELLITE TRACKING Satellite tracking from multiple studies in the western US largely support western breeding populations migrating to nonbreeding areas within California and Mexico (Hoffman et al. 2002; Craighead et al. 2016; Bloom et al. 2017; Craighead Beringia South, unpublished data; C. Deavila, unpublished data; Golden Gate Raptor Observatory and C. Briggs, unpublished data; J. Smith unpublished data). Juvenile Red-tailed Hawks (n = 16) tagged in California migrated northward in their first year (Bloom et al. 2017). However, birds breeding in Alaska and British Columbia spend the nonbreeding season within the central US (N. Bickford, unpublished data; B. Robinson, unpublished data). Birds outfitted with transmitters during pre-breeding migration at the Mackinac Straits, Michigan, raptor watch migrated between breeding areas in Manitoba, Ontario, and Quebec, and nonbreeding areas in Indiana, Kentucky, and as far south as Alabama (N. Alioto, unpublished data). Similarly, birds tagged during migration through the Great Lakes migrated due north to spend the breeding season in Ontario and Quebec (B. Robinson, unpublished data).

GENETICS Analyses of nuclear DNA suggest birds observed in central California during post-breeding migration originate from both the central California and Intermountain West regions, with early migrants being solely from California and later migrants a mix of the two regions (Hull et al. 2009).

STABLE ISOTOPES Donahue (2009) found evidence for differential migration by natal latitude for Red-tailed Hawks migrating in the Goshutes Mountains, Nevada, using stable hydrogen isotopes; however, stable isotopic patterns did not clearly match satellite data

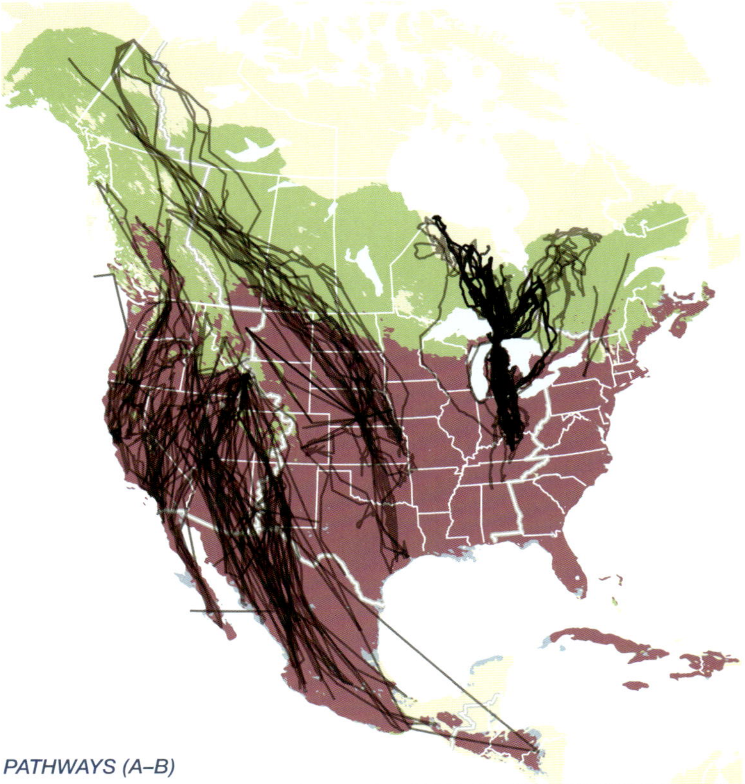

PATHWAYS (A–B)

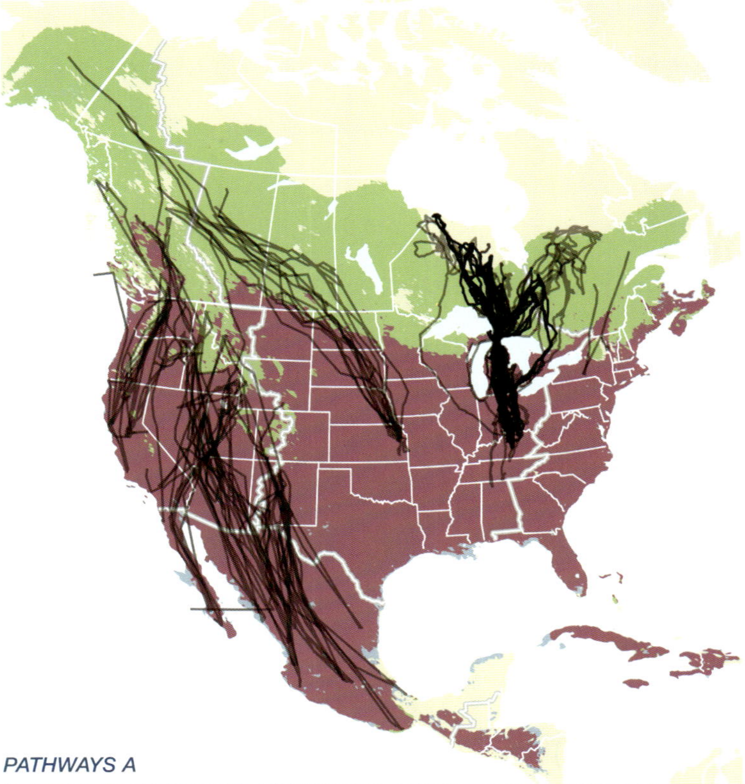

PATHWAYS A

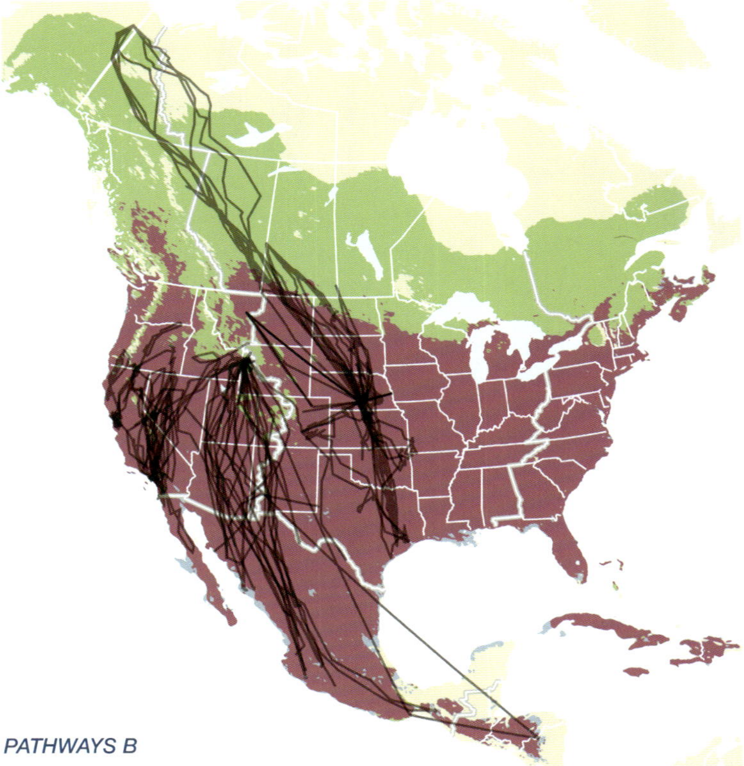

PATHWAYS B

collected for the same region. Pavlovic (2022) also found that more northern birds migrated later through Duluth, Minnesota, in the post-breeding season, using hydrogen isotopes, but longitudinal assignment was less clear. In the same study, a single dark morph was predicted to originate from the Yukon (Pavlovic 2022).

MIGRATORY PATHWAYS Red-tailed Hawks concentrate in migration along ridges and shorelines, often avoiding water crossings (Goodrich and Smith 2008). Broad front movement is evident in banding and satellite tracking data and may be more prevalent in the West (Preston and Beane 2020). Banding data from migrants recaptured at migration watch sites indicate that most individuals originate from natal areas located north of the watch sites within the same broad geographic region (Morrison and Baird 2016; Hull et al. 2009; Bonerbo and Goodrich 2020).

Most migration movement appears to be in a southerly or southwesterly direction in the east and central regions, and south or southeast in the west. One exception is that juveniles and young adults from south of 38 degree north latitude in the Pacific region often display northward migration, while adults are largely resident. Some eastern birds also show northward migration in the first year, but the proportion may be less than in the west (Bloom et al. 2015; Morrison and Baird 2016; for eastern migration, see Braddock Bay Raptor Research, hawkcount.org, unpublished data). Spatial and annual variability in migration counts suggests shifting use of migration routes between seasons (Titus and Fuller 1990; Kirk and Hyslop 1998; Hoffman and Smith 2003).

Red-tailed Hawks that were satellite tracked for more than two years in the western US showed high fidelity to migration pathways between years and seasons (J. Smith, in Preston and Beane 2020). The Red-tailed Hawk Project (redtailedhawkproject.org) has also documented fidelity between post- and pre-breeding migration routes.

DATA CONTRIBUTORS

N. Alioto; N. Bickford; P.H. Bloom; C.W. Briggs; Craighead Beringia South; C. Deavila; L.H. DeCicco; T.E. Ely; A.M. Fish; HawkWatch International; I.J. Lovette; M.D. McCrary; J.M. Papp; J.P. Smith; M.B. Robbins; B.W. Robinson.

INTERESTING NOTE

The highest counts of migrating Red-tailed Hawks in the continental United States occur during post-breeding migration at the Golden Gate Raptor Observatory, San Francisco, California (mean = 9,340) and Hawk Ridge Bird Observatory, Duluth, Minnesota (mean = 8,934; Goodrich and Smith 2008). In 2019, the Mackinac Straits Raptor Watch recorded the largest pre-breeding migration count (Mackinac Straits Raptor Watch, unpublished data).

ROUGH-LEGGED HAWK *Buteo lagopus*

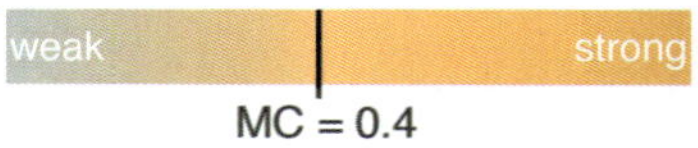

DISTRIBUTION AND ABUNDANCE

In North America, Rough-legged Hawks breed at highest densities in the Arctic from Nunavut west to Victoria Island and the North Slope of Alaska, with lower densities in the subarctic of Alaska and Canada. They spend the nonbreeding season in open landscapes in southern Canada and the contiguous US (excluding the southeast). The breeding and nonbreeding ranges are completely disjunct (Fink et al. 2023).

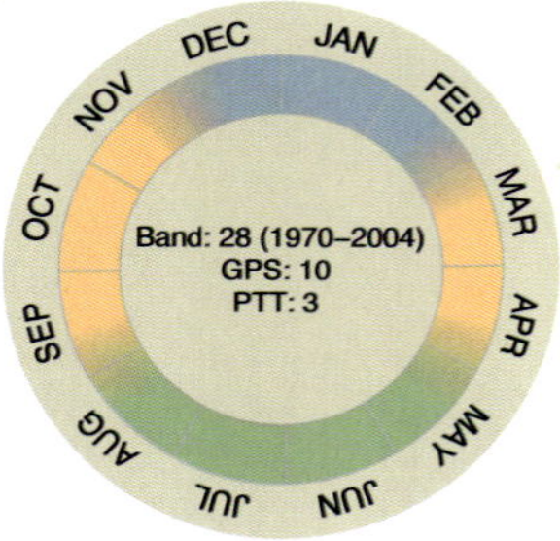

GENERALIZED CONNECTIVITY

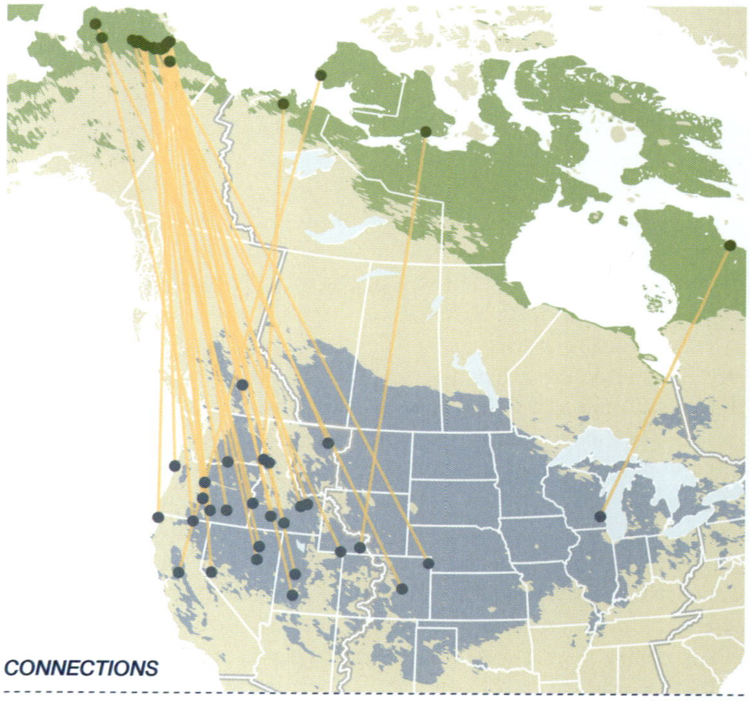

CONNECTIONS

ORDER Accipitriformes
FAMILY Accipitridae
BREEDING HABITAT Tundra, Arctic
NONBREEDING HABITAT Open Country, Habitat Mosaic
GLOBAL POPULATION SIZE 590,000
POPULATION CHANGE SINCE 1970 +3%

IUCN RED LIST CATEGORY Least Concern
R2R TIPPING POINT SPECIES No
US LISTING STATUS No
CANADA SPECIES AT RISK No
PIF CCS 9/20

	Atlantic	Midcontinent	Pacific
Atlantic (n = 2)	50%	50%	–
Midcontinent (n = 8)	–	62%	38%
Pacific (n = 25)	–	8%	92%

MIGRATORY CONNECTIVITY

OVERALL The migratory connectivity of Rough-legged Hawks is not well understood because until recently (see Paprocki et al. 2025) there were few published tracking studies and limited band reencounter data. Birds breeding in the Atlantic Flyway spend the nonbreeding season within the Atlantic (50%) and Midcontinent (50%) flyways. Birds breeding in the Midcontinent Flyway spend the nonbreeding season within the Midcontinent (62%) and Pacific (38%) flyways. Birds breeding in the Pacific Flyway primarily spend the nonbreeding season within the same flyway (92%). Although a portion of Atlantic and Midcontinent flyway breeding populations mix on the nonbreeding grounds, most birds spend the breeding and nonbreeding season within the same flyway, resulting in moderate migratory connectivity (MC = 0.4).

BANDING Band reencounters from breeding areas in Alaska show birds migrated to the westernmost portions of the nonbreeding range, especially Oregon, Idaho, Nevada, and Utah. Additional band reencounters suggest wide dispersal patterns, especially in young birds (Galushin 1974; Dunn et al. 2009).

SATELLITE TRACKING Rough-legged Hawks tracked with satellite telemetry from nonbreeding locations across the western US migrated to breeding areas in the western Canadian Arctic and Alaska (Pokrovsky et al. 2023; Craighead Beringia South, unpublished data; Paprocki et al. 2025). Subsequent tracking from these breeding grounds found additional nonbreeding areas along the Pacific coast (Paprocki et al. 2025). Satellite telemetry from a single study of two individuals over a two-year period suggests some migratory connectivity: both birds spent the nonbreeding season in Wyoming and bred above the Arctic Circle near the Beaufort Sea (Teton Raptor Center, unpublished data). In the eastern part of the range, birds tracked from nonbreeding locations in New York along the St. Lawrence River migrated to breeding areas in Quebec (Pokrovsky et al. 2023).

OTHER APPROACHES No published genetic or isotope studies of migratory connectivity were found for this species.

MIGRATORY PATHWAYS There are few published studies on Rough-legged Hawk migration (but see Paprocki et al. 2025 for extensive telemetry tracking). There are two primary migration paths from central Arctic breeding areas. The westerly pathway goes through Saskatchewan, then west over the Rocky Mountains to nonbreeding areas in California's Central Valley. The more easterly route follows the west coast of Hudson Bay south through Manitoba, then down the Mississippi River Valley to midcontinent nonbreeding areas. Several raptor migration count sites routinely and regularly record Rough-legged Hawks (Tadoussac, Hawk Ridge, Whitefish Point; see hawkcount.org, hawkmigration.org, and https://www.rpi-project.org/). However, they are not typically seen in large numbers at migration and hawk watch stations, suggesting that Rough-legged Hawks generally migrate across a broad front and do not concentrate along ridges or bottlenecks (e.g., Allen et al. 1996; except see Great Lakes migration sites, Titus and Fuller 1990).

DATA CONTRIBUTORS

Craighead Beringia South; Grand Teton National Park; N. Lecomte; I. Pokrovsky; J.-F. Therrien; M. Wikelski.

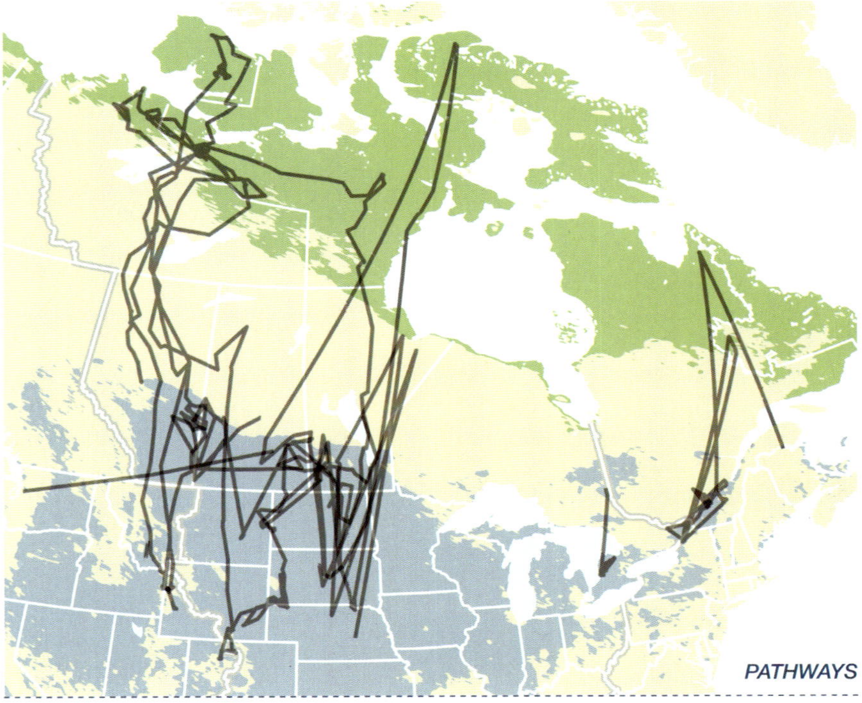

PATHWAYS

INTERESTING NOTE

Rough-legged Hawks seem to follow the progression of snowmelt during their nonbreeding period (Curk et al. 2020; Pokrovsky et al. 2023). Therefore, as the ongoing trend of climate warming leads to a reduction in the extent of snow cover, lemming populations may stabilize, leading more Rough-legged Hawks to return consistently to both breeding and nonbreeding areas and increasing migratory connectivity among populations. Future research needs include, in addition to more general studies of migratory connectivity, understanding how fluctuations in small mammalian prey populations affect where individuals decide to migrate and breed.

weak — strong

MC = 0.21 ± 0.04*

Buteo regalis FERRUGINOUS HAWK

DISTRIBUTION AND ABUNDANCE

The Ferruginous Hawk, North America's largest buteo, breeds west of the 100th meridian west, with two main breeding areas from the southern Canadian provinces south through the Great Plains states, and west of the Rocky Mountains from eastern Washington south through the Great Basin. Highest breeding concentrations range from northwestern Montana north into eastern Alberta and western Saskatchewan. Most hawks spend the nonbreeding season in southern Canada through north-central Mexico, with high densities from southwestern South Dakota into northern Texas, and in California. Year-round populations are found in the center of their range, especially in northern Arizona, northern New Mexico, and eastern Colorado (Ng et al. 2020).

ORDER Accipitriformes
FAMILY Accipitridae
BREEDING HABITAT Grasslands, Temperate
NONBREEDING HABITAT Open Country, Habitat Mosaic
GLOBAL POPULATION SIZE 110,000
POPULATION CHANGE SINCE 1970 +54%
IUCN RED LIST CATEGORY Least Concern
R2R TIPPING POINT SPECIES No
US LISTING STATUS No
CANADA SPECIES AT RISK Threatened
PIF CCS 10/20

HAWKS, EAGLES, KITES, AND OSPREY (ACCIPITRIFORMES)

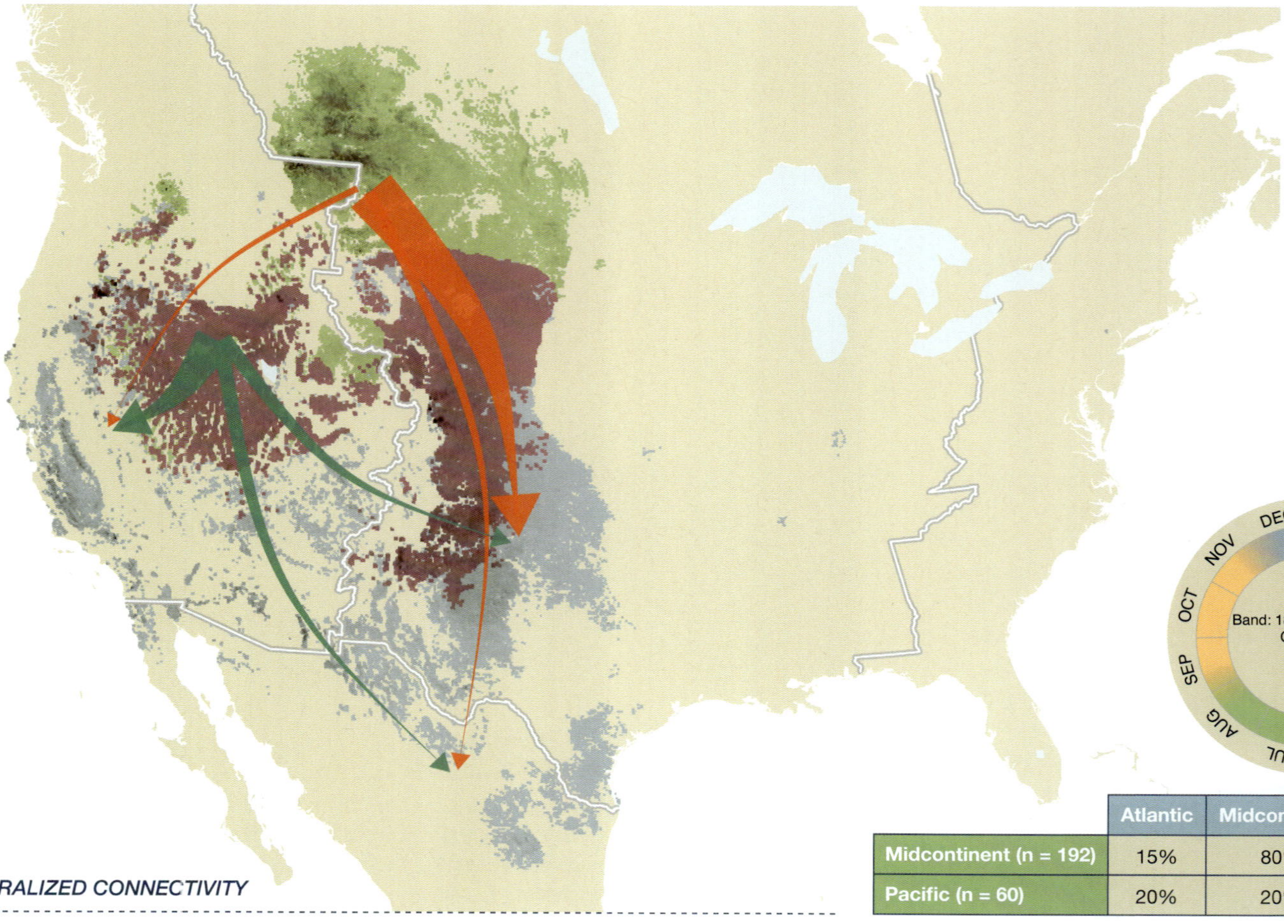

GENERALIZED CONNECTIVITY

	Atlantic	Midcontinent	Pacific
Midcontinent (n = 192)	15%	80%	6%
Pacific (n = 60)	20%	20%	60%

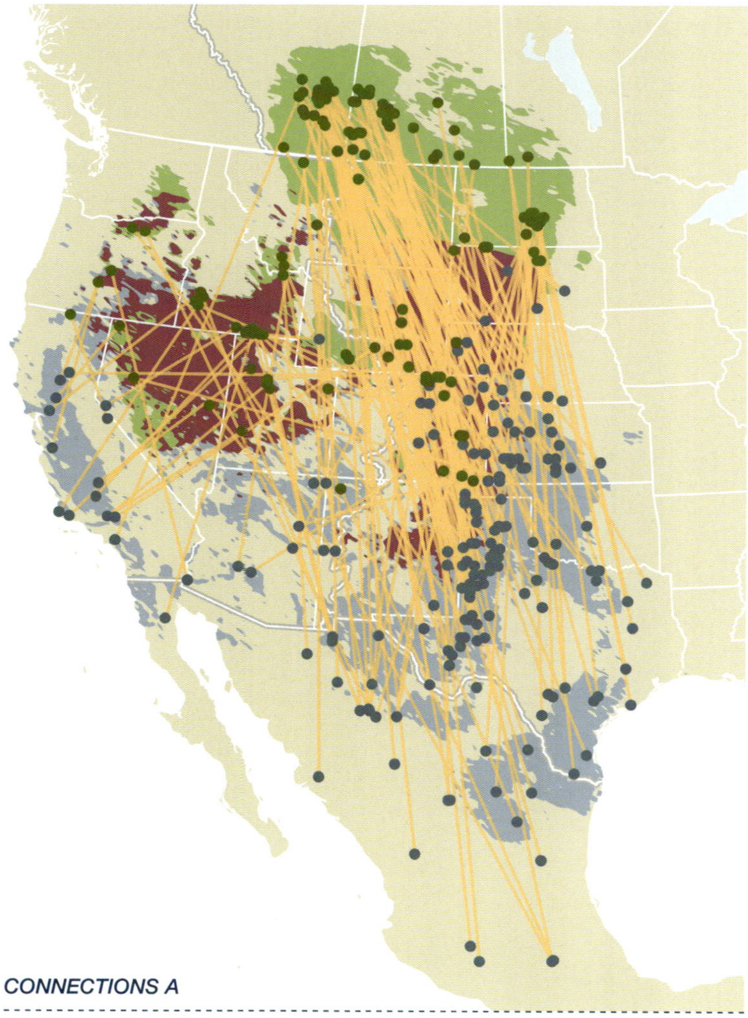

CONNECTIONS A

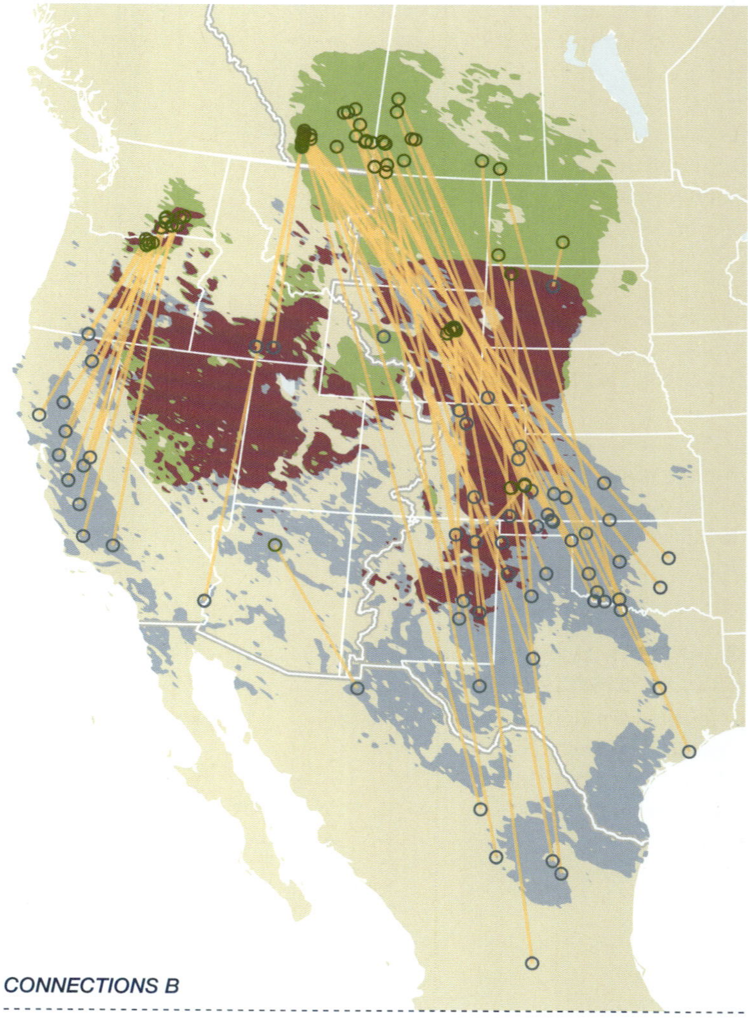

CONNECTIONS B

MIGRATORY CONNECTIVITY

OVERALL Migratory connectivity from across the Ferruginous Hawk's range is fairly well-known, from studies of band reencounters (Harmata 1982; Perkins and Lindsey 1983; Gilmer et al. 1985; Schmutz and Fyfe 1987; Harmata et al. 2001) and studies using satellite telemetry (Watson et al. 2018; Watson 2020; J. Barnes, unpublished data; Earthspan, unpublished data; Salt Lake City Airport Wildlife Department, unpublished data). Breeding populations in both the Midcontinent and Pacific flyways primarily spend the nonbreeding season within the same flyway as their breeding locations. However, mixing of breeding populations on nonbreeding areas in other regions results in moderate migratory connectivity (MC = 0.21 ± 0.04).

BANDING Banding data indicate that birds breeding in regions north of 40 degrees north latitude migrate primarily to the southwest US and Texas, and birds from Canada also migrate to the southwest, with some birds migrating as far south as northern Mexico. Birds breeding in the Pacific Northwest migrate to either California or the southwest US (Harmata 1982; Gilmer et al. 1985; Schmutz and Fyfe 1987; Harmata et al. 2001).

SATELLITE TRACKING Watson et al. (2018) deployed satellite transmitters across the breeding range and found differences in migratory connectivity in populations breeding east versus west of the Continental Divide. Most (94%) Ferruginous Hawks breeding east of the Continental Divide did not cross the divide while migrating, whereas only 12% of birds breeding west of the divide did not cross. Most birds (86%) breeding south of 40 degrees north latitude did not migrate. Other studies have found similar patterns. Individuals tracked from a breeding population in the Great Basin Desert in Nevada (n = 20) first migrated in a northeast direction after the breeding season, before migrating to nonbreeding areas in central and Southern California during the nonbreeding season (J.G. Barnes, unpublished data). One individual from the Great Basin Desert breeding population migrated to Texas (J.G. Barnes, unpublished data). Utah breeding birds (n = 2) also spent the nonbreeding season in Southern California (Salt Lake City Wildlife Department, unpublished data). One bird tracked from Montana migrated to Nevada (Earthspan, unpublished data). Individuals tracked with both satellite and GPS-GSM transmitters from breeding areas in Alberta and Saskatchewan (n = 38) spent the nonbreeding season across the southern Great Plains states and into Mexico (Watson 2020).

MIGRATORY PATHWAYS The Continental Divide appears to divide Ferruginous Hawks into two separate migratory populations: birds east of the Continental Divide migrate during the post-breeding season south through the Dakotas into Wyoming and Colorado into Texas and New Mexico, whereas birds west of the divide spread out into western states and move into California, Nevada, and Utah. However, western birds may cross the divide to occupy post-breeding foraging territories, after which they migrate across the Continental Divide again to travel to nonbreeding locations (Watson et al. 2018). In southern Canada, a substantial population of Ferruginous Hawks gather prior to the start of southward migration.

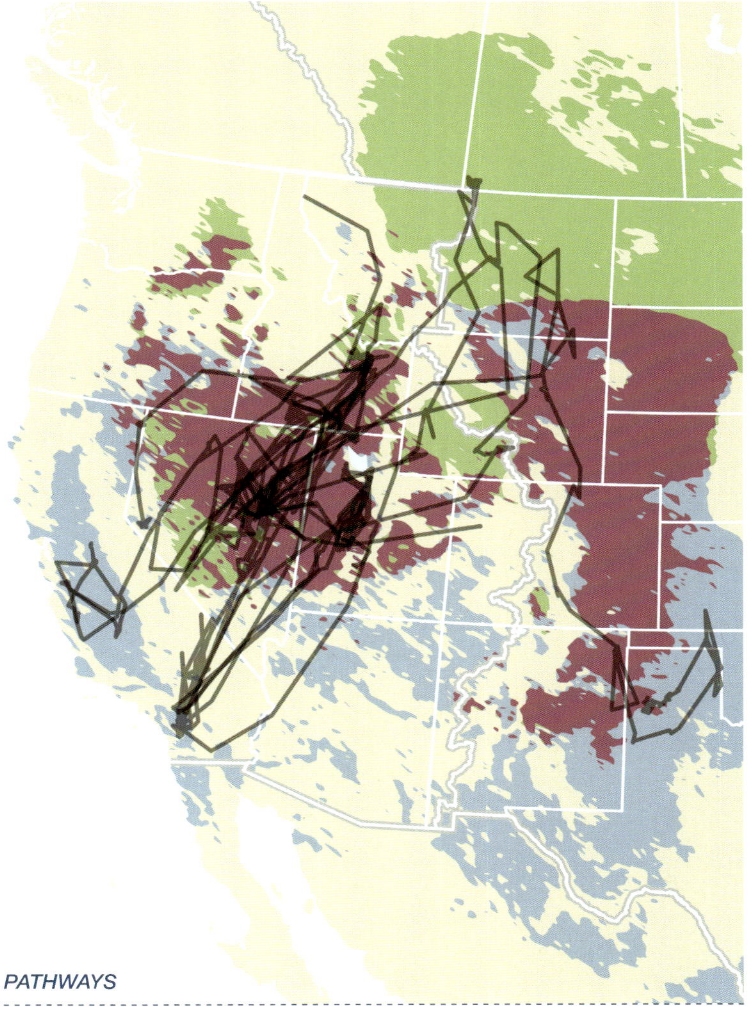

PATHWAYS

DATA CONTRIBUTORS

U. Banasch; J.G. Barnes; E.M. Bayne; T. Byer; M.Á. Cruz; C. Deavila; R. Gerhardt; D. Hanni; A. Lafón; R. McCready; D.N. Svingen; J.W. Watson; J.L. Watson; T.I. Wellicome; M. Yates; Bureau of Land Management; Department of Defense Legacy Resource Management Program; Earthspan; Jerritt Canyon Gold, LLC; US Forest Service.

INTERESTING NOTE

Ferruginous Hawks are detected in small numbers at western hawk migration sites (Goodrich and Smith 2008; Oleyar and Watson 2020), most likely because they are broad-front, prey-driven migrants (Watson et al. 2018) that do not necessarily migrate along ridgelines. Migration of this raptor species has often been described as nomadic; however, Watson and Keren (2019) found adult hawks only exhibited wandering in post-breeding, but not during breeding or nonbreeding, when they consistently returned to the same ranges.

OWLS (STRIGIFORMES)

WRITING LED BY **JEAN-FRANÇOIS THERRIEN, AMY L. SCARPIGNATO,** *AND* **PETER P. MARRA**

Breeding owls tracked (n = 235) with satellite and GPS transmitters showed extended inland use of the entire North American continent, as well as large dispersal movements spanning the continent.

The migratory connectivity and associated ecology of the order Strigiformes, which includes all owls as well as barn owls, has been moderately well studied; 78% of the species have migratory connectivity information, and just over 50% of the species have been tracked in some capacity. A few members of the Strigidae family (e.g., Snowy, Short-eared, and Burrowing owls) have been tracked with GPS and Argos transmitters, while migratory connectivity information for the sole member of the family Tytonidae (American Barn Owl) is known from band reencounter data. Great Horned, Great Gray, and Long-eared owls, as well as the Northern Hawk Owl, are nonmigratory and therefore are not included in this Atlas; rather their movements are highly irruptive and irregular, particularly for the first three species. Satellite tracking studies of Snowy Owls demonstrated large-scale migratory and breeding dispersal behavior across the Northern Hemisphere as well as onto the sea ice for extended periods of time during the nonbreeding season, behaviors that were suspected but not documented in detail until tracking studies were conducted. Understanding the detailed movement ecology and migration connectivity of owls has been essential in making conservation recommendations for these little-studied and potentially declining species. Other species within Strigiformes also warrant conservation attention, including American Barn, Spotted, screech and Pygmy owls, species that could be declining at the moment (the latter two species are considered resident and are not included in the Atlas). The Spotted Owl is categorized as a Yellow-Alert Tipping Point species by the Road to Recovery, and is listed in both the US and Canada as an endangered species.

Across the order, Strigiformes exhibit variable migratory connectivity. Some species, like the Burrowing Owl, exhibit some form of north-south movements, although patterns are far from regular. At the end of the spectrum, Short-eared and Snowy owls display some of the most extreme cases of nondirectional movements, ranging from north-south, east-west, and all other possible geographical orientations and distances, with very limited fidelity to breeding and nonbreeding locations. Telemetry tracked birds used a diverse suite of habitats for traveling between breeding and nonbreeding, from over the sea ice to coastlines to more inland routes.

Tyto furcata **AMERICAN BARN OWL**

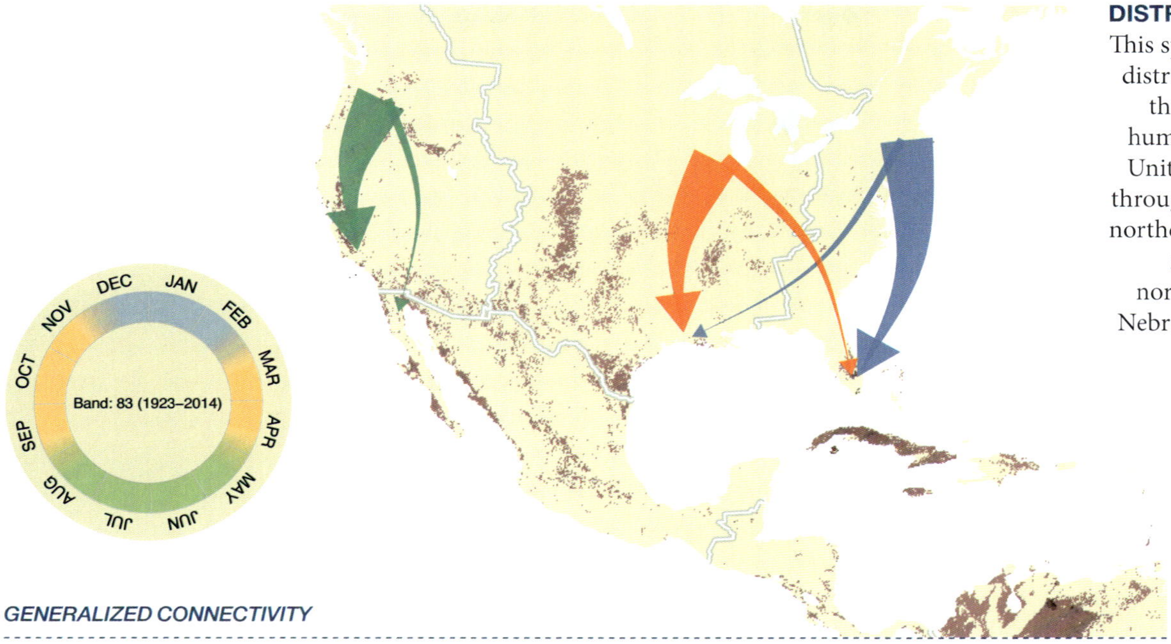

GENERALIZED CONNECTIVITY

DISTRIBUTION AND ABUNDANCE

This species is one of the most widely distributed land bird species around the globe and is highly tolerant of human-modified landscapes. In the United States, it is fairly widespread throughout all states except in the far northern reaches. It is more common in the southeast, the Gulf Coast north into Oklahoma, and western Nebraska and the southwest (Fink et al. 2023). It uses open habitats, particularly grasslands, and nests in cavities including trees, cliffs, barns, and nest boxes (Marti et al. 2020).

The extent of migration in this species remains unresolved, as most North American studies of marked individuals have found no evidence of migration, although migratory movements have been reported in some regions of the northern United States (Marti et al. 2024).

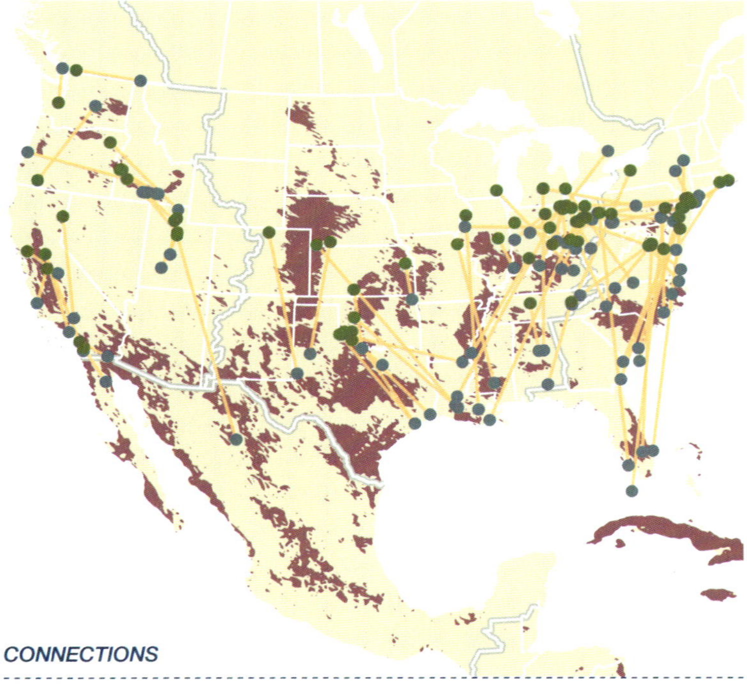

CONNECTIONS

	Atlantic	Mexico	Midcontinent	Pacific
Atlantic (n = 21)	90%	–	10%	–
Midcontinent (n = 37)	19%	–	81%	–
Pacific (n = 19)	–	11%	–	89%

ORDER Strigiformes
FAMILY Tytonidae
BREEDING HABITAT Open Country, Habitat Mosaic
NONBREEDING HABITAT Open Country, Habitat Mosaic
GLOBAL POPULATION SIZE 3,600,000
POPULATION CHANGE SINCE 1970 +203%
IUCN RED LIST CATEGORY Least Concern
R2R TIPPING POINT SPECIES No
US LISTING STATUS No
CANADA SPECIES AT RISK Endangered Eastern population
Threatened Western population
PIF CCS 8/20

MIGRATORY CONNECTIVITY

OVERALL Migratory connectivity is known from band reencounter data and is strong (MC = 0.72), as most breeding individuals migrate to nonbreeding areas within the same breeding region. There is some mixing of breeding populations on nonbreeding areas.

BANDING Band reencounters show some seasonal movement, but there is little information on the extent of migration in the species.

OTHER APPROACHES No tracking, genetic, or stable isotope studies of migratory connectivity were found for this species.

MIGRATORY PATHWAYS No known pathways.

INTERESTING NOTE

Nuclear and mitochondrial DNA sampled from Barn Owls in several locations in western North America and one in the eastern United States showed little to no genetic structure among regions (Huang et al. 2016).

FLAMMULATED OWL *Psiloscops flammeolus*

DISTRIBUTION AND ABUNDANCE

Flammulated Owls occur in mountainous forests from southwestern British Columbia through most western states from Colorado and Wyoming to California south to northern Mexico. Individuals are thought to be migratory in the northern parts of their breeding range (Linkhart and McCallum 2020).

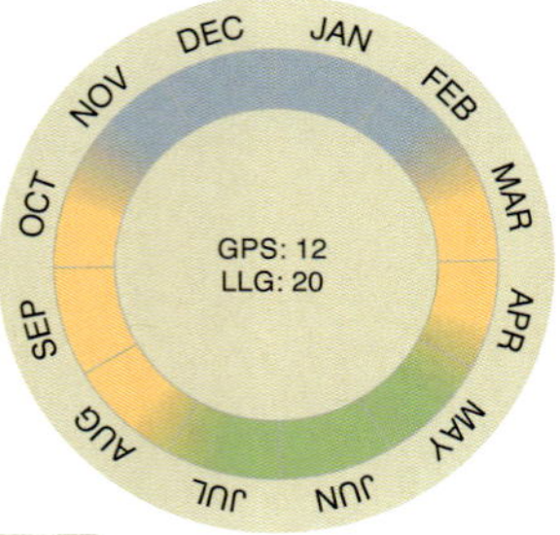

ORDER Strigiformes
FAMILY Strigidae
BREEDING HABITAT Forests, Temperate Western
NONBREEDING HABITAT Forests, Mesoamerican Pine-Oak
GLOBAL POPULATION SIZE 12,000
POPULATION CHANGE SINCE 1970 NA
IUCN RED LIST CATEGORY Least Concern
R2R TIPPING POINT SPECIES No
US LISTING STATUS No
CANADA SPECIES AT RISK No
PIF CCS 15/20

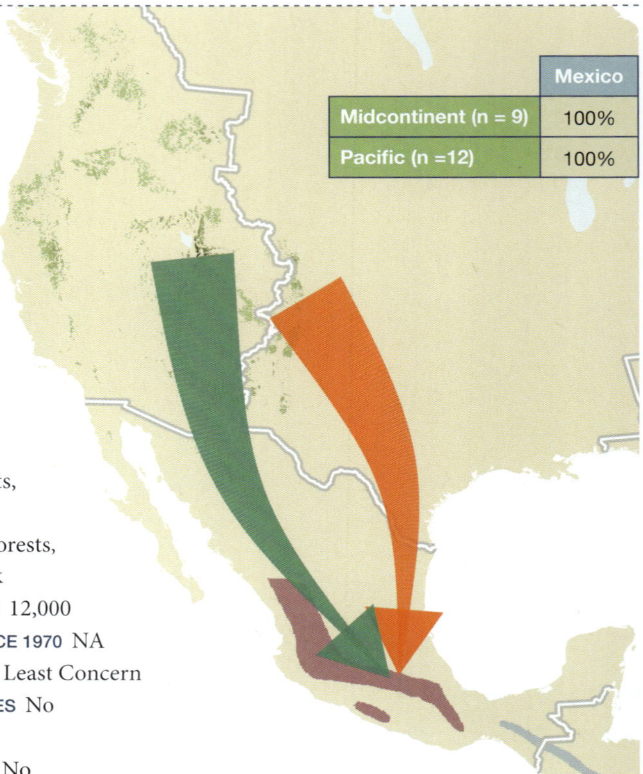

	Mexico
Midcontinent (n = 9)	100%
Pacific (n =12)	100%

GENERALIZED CONNECTIVITY

MIGRATORY CONNECTIVITY

OVERALL Migratory connectivity is known from light-level geolocator and GPS archival tags, mainly for Interior West populations. The migratory connectivity of other breeding populations of this wide-ranging species is unknown. Birds breeding in the Midcontinent and Pacific flyways spend the nonbreeding season in Mexico (100%). MC is not calculated due to limited sample size.

RADIO TRACKING A few tags were deployed through the Motus Wildlife Tracking System; there were no breeding to nonbreeding detections.

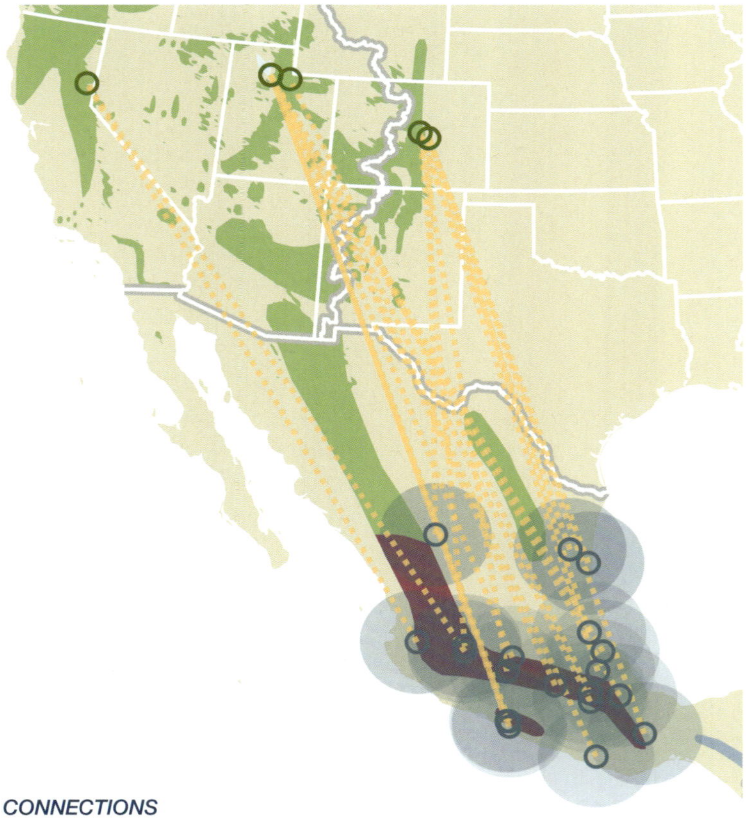

CONNECTIONS

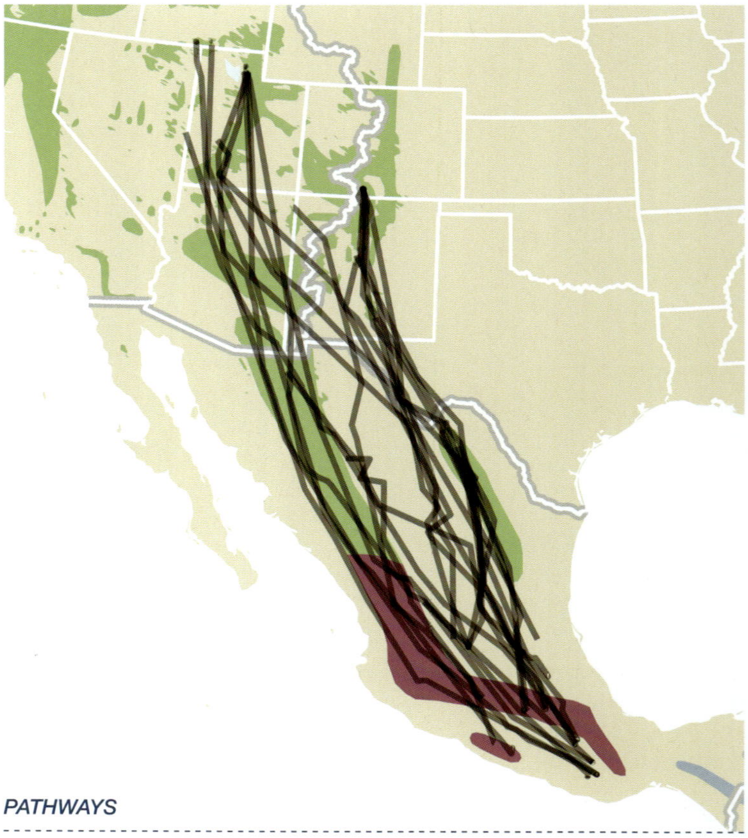

PATHWAYS

GEOLOCATOR TRACKING Light-level geolocator data showed several birds tracked (n = 4) from Colorado all migrated to northern/central Mexico (Linkhart et al. 2016). An additional 16 out of 60 light-level geolocators deployed on adult male and female owls in Washington, Colorado, Utah, and California showed the majority of Utah and Colorado owls migrated to central and southern Mexico but within different geographical areas (Rich 2019; Hawk Watch International, unpublished data). The California birds also migrated to Mexico but to the west of the other breeding populations.

SATELLITE TRACKING GPS archival tags deployed from locations in Colorado, Idaho, Utah, and New Mexico showed owls spent the nonbreeding season in different locations across southern Mexico and northwestern Guatemala (Yanco et al. 2021), revealing an east-west migratory divide.

MIGRATORY PATHWAYS Flammulated Owls migrate in a fairly straight path between breeding and nonbreeding grounds, following mountain ranges from the US into Mexico with a possible stopover site in western Texas for some individuals (Linkhart and McCallum 2020; Yanco et al. 2021). Individuals tracked from Utah, Colorado, and California took a variety of migratory routes but overall exhibited a southwestern route to nonbreeding areas and a northeastern route back to breeding areas (Rich 2018). Individuals tracked with GPS geolocators from two Colorado breeding populations generally followed an eastern route along the Sierra Madre Oriental during both pre- and post-breeding migrations, whereas birds from the Idaho breeding population primarily migrated to the west along the Sierra Madre Occidental. Owls from the Utah breeding population exhibited both strategies (Yanco et al. 2021).

DATA CONTRIBUTORS

B. Linkhart; P.P. Marra; M.B. Wunder; S.W. Yanco; Migratory Connectivity Project.

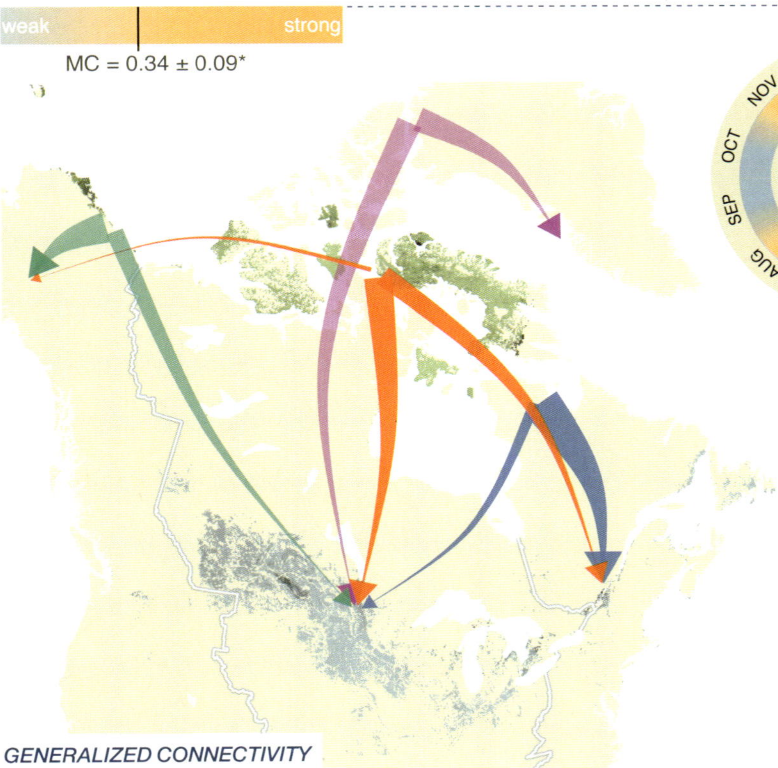

GENERALIZED CONNECTIVITY

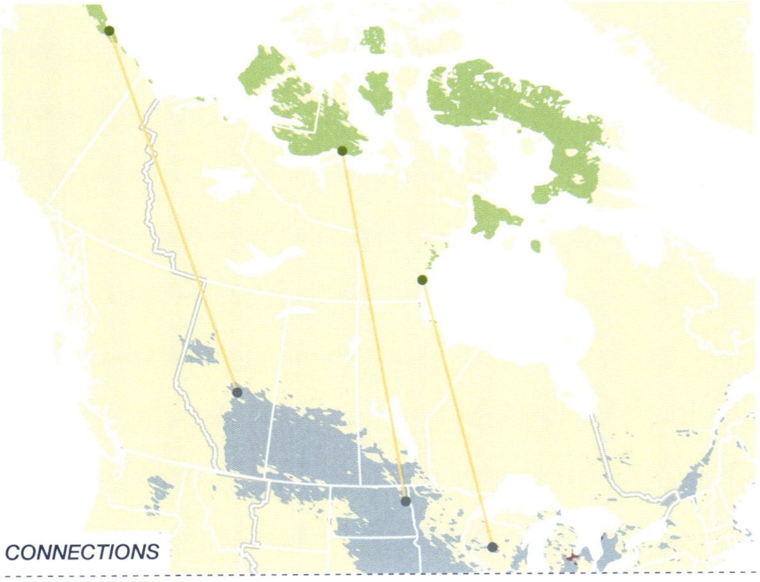

CONNECTIONS

Bubo scandiacus SNOWY OWL

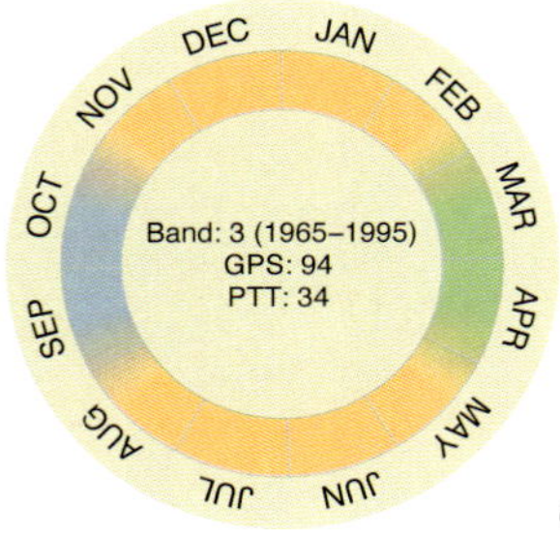

DISTRIBUTION AND ABUNDANCE

The Snowy Owl is the northernmost breeding owl species, nesting in the Arctic tundra to the very northern tip of Ellesmere Island (Nunavut, Canada). Densities are highest along the North Slope of Alaska reaching to the Beaufort Sea. Breeding densities are equally high on Victoria, Bathurst, and Baffin islands in the northernmost regions of the globe (Fink et al. 2023). Snowy Owls are known for irruptive, nomadic migrations that are unpredictable and thought to be in response to prey availability following successful breeding seasons. Because of the irruptive behavior, the nonbreeding range varies annually, from across British Columbia east to the Atlantic (Therrien et al. 2011) and south into the United States, from the Dakotas east into the eastern states (Holt et al. 2020).

ORDER Strigiformes
FAMILY Strigidae
BREEDING HABITAT Tundra, Arctic
NONBREEDING HABITAT Open Country, Habitat Mosaic
GLOBAL POPULATION SIZE 29,000

POPULATION CHANGE SINCE 1970 −57%
IUCN RED LIST CATEGORY Vulnerable
R2R TIPPING POINT SPECIES No
US LISTING STATUS No
CANADA SPECIES AT RISK No
PIF CCS 13/20

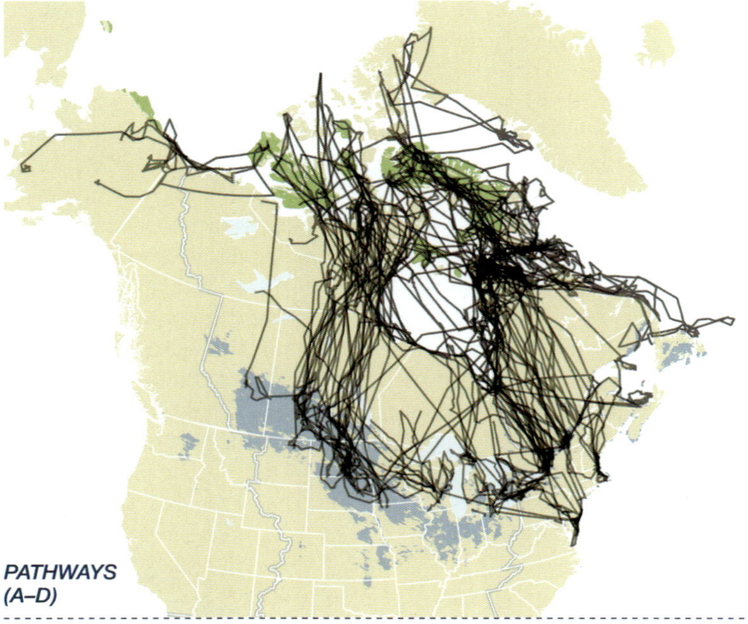

PATHWAYS (A–D)

	Atlantic	Greenland	Midcontinent	Pacific
Atlantic (n = 16)	81%	–	19%	–
Greenland (n = 2)	–	50%	50%	–
Midcontinent (n = 43)	40%	–	56%	5%
Pacific (n = 3)	–	–	33%	67%

MIGRATORY CONNECTIVITY

OVERALL The migratory connectivity of the Snowy Owl is well known, derived mostly from satellite telemetry and banding data. Birds breeding in Greenland migrate to nonbreeding areas within Greenland (50%) and the Midcontinent Flyway (50%). Birds breeding in the Atlantic Flyway spend the nonbreeding season within the Atlantic Flyway (81%) as well as the Midcontinent Flyway (19%). Midcontinent Flyway breeding birds migrate within the flyway (56%) or to the Atlantic Flyway (40%). Birds breeding in the Pacific Flyway largely migrate within the Pacific Flyway (67%) or to the Midcontinent Flyway (33%). High within-flyway transitions, in combination with overlap of breeding populations across

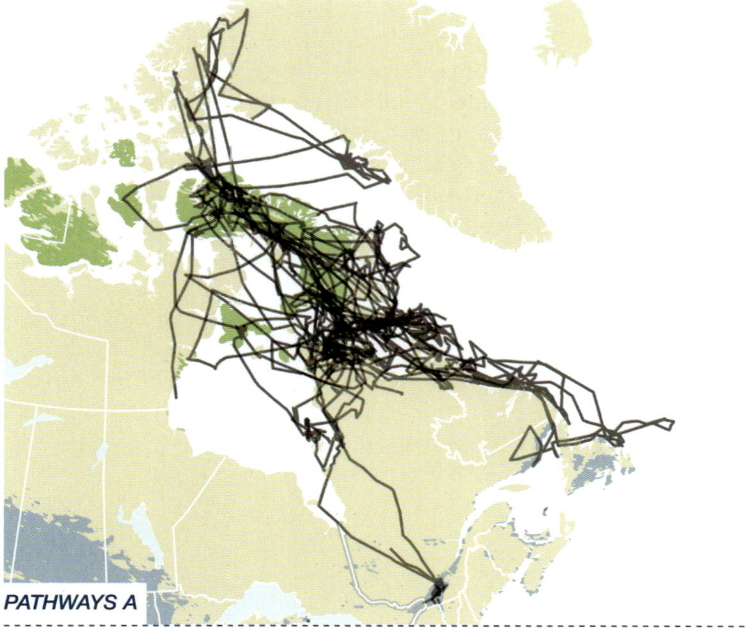

PATHWAYS A

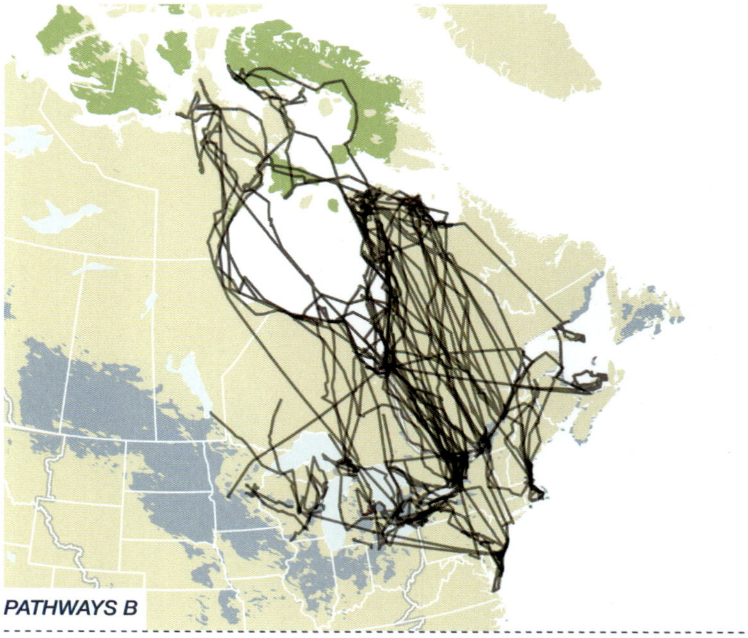

PATHWAYS B

PATHWAYS C

PATHWAYS D

nonbreeding regions, yield moderate migratory connectivity (MC = 0.34 ± 0.09).

BANDING Band reencounters are limited but generally consistent with tracking data.

SATELLITE TRACKING Most tracking studies in the eastern and western Canadian Arctic have shown breeders to stay at high latitudes for the nonbreeding season (Fuller et al. 2003; Doyle et al. 2018) and often near marine or coastal areas (Robillard et al. 2018). Similarly, owls tracked from nonbreeding areas in Saskatchewan and Alberta migrated to areas of the Canadian Arctic throughout Nunavut (G. Holroyd, unpublished data; K. Wiebe, unpublished data). Snowy Owls from a single study site on the breeding grounds in the eastern Canadian Arctic (Bylot Island) were tracked across a wide range of locations during subsequent nonbreeding seasons, including the Canadian Arctic, Greenland and the Greenland coast, the Midwest, Atlantic provinces and states, and around the Great lakes (Therrien et al. 2014). Individual Snowy Owls tracked from their nonbreeding grounds in eastern North America or the Great Plains spread across the Arctic in subsequent breeding periods. Four female owls nesting in the western Canadian Arctic all spent the nonbreeding season in boreal Alaska and the northern Yukon territory (Doyle et al. 2018), while females breeding in the eastern Canadian Arctic spent the nonbreeding season at high latitudes in the Arctic (Therrien et al. 2011). Similarly, females tracked from breeding sites in Alaska all spent the nonbreeding season in the Arctic, in either Russia, Alaska, or Canada (Fuller et al. 2003). Project SNOWstorm, an effort to understand the full annual cycle movements of Snowy Owls, tagged 92 Snowy Owls with GPS-GSM transmitters on their nonbreeding grounds from North Dakota east to Quebec and Maine, and south to Maryland. Snowy Owls migrated to breeding areas primarily in northern Nunavut and the Ungava Peninsula in northern Quebec (Project SNOWstorm, unpublished data).

GENETICS Mitochondrial gene sequencing from three geographic regions within the Snowy Owls' breeding range (Scandinavia, North America, and eastern Siberia) found no phylogeographic structure indicating high levels of gene flow in the recent past and possibly still today (Marthinsen et al. 2009). SNP-based analyses of Snowy Owls sampled across the North American nonbreeding range also found an absence of genetic differentiation among individuals, suggesting high interbreeding and dispersal at the continental scale despite philopatry to nonbreeding sites (Gousy-Leblanc et al. 2023).

MIGRATORY PATHWAYS Because the Snowy Owl is a classic irruptive species, no definite migration pathways can be described.

DATA CONTRIBUTORS

Aurora Research Institute; M. Bertram; A. Blanchford; D. Brinker; J. Burch; S. Creed; C.R. DeSorbo; D. DiFolco; T. Doolittle; F. Doyle; J. Duxbury; G. Gauthier; L. Gilpatrick; M. Hinkes; G.L. Holroyd; S. Huy; G. Jacobs; A. Kenney; M. Lanzone; D. La Puma; N. Mackentley; M. Mance-Coniglio; R. McCabe; T. McDonald; A. McGann; C. McIntyre; T. Miller; D. Mossop; Natural Resources Canada Polar Continental Shelf Program; M. Nelligan; C. Neri; A. Robillard; M. Shieldcastle; N. Smith; M. Solensky; J.-F. Therrien; H. Trefry; E. Wald; D. Weber; S. Weidensaul; K. Wiebe; M. Williams; Yukon's Qikiqtaruk (Herschel Island) Territorial Park.

INTERESTING NOTE

Pre-breeding satellite tracking in the eastern Arctic showed searching behavior likely helps Snowy Owls settle in areas with high lemming density for the breeding season (Therrien et al. 2014). A total of 12 individuals tracked by satellite transmitters for three years resulted in an annual survival rate of 85.2%–92.3% and successful breeding, indicating that transmitters did not affect life history traits of Snowy Owls (Therrien et al. 2012).

Micrathene whitneyi ELF OWL

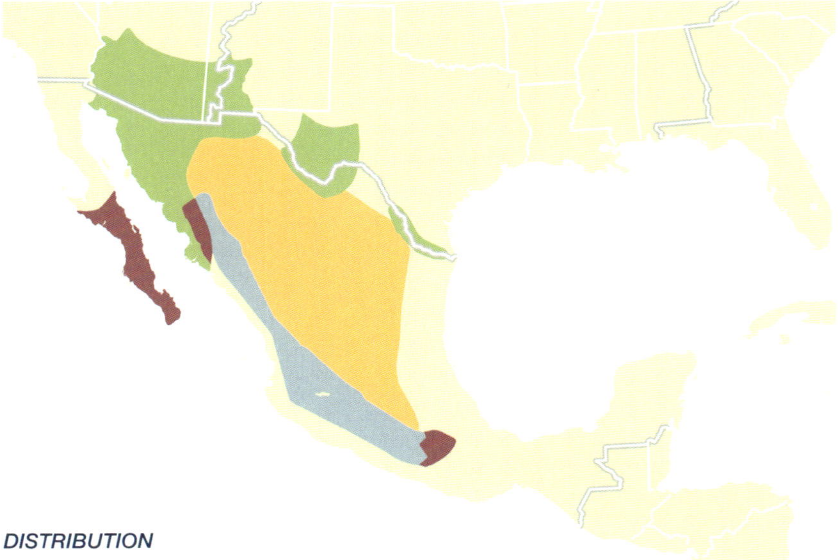

DISTRIBUTION

ORDER Strigiformes
FAMILY Strigidae
BREEDING HABITAT Arid Lands, Desert Scrub
NONBREEDING HABITAT Forests, Tropical Dry
GLOBAL POPULATION SIZE 46,000

POPULATION CHANGE SINCE 1970 NA
IUCN RED LIST CATEGORY Least Concern
R2R TIPPING POINT SPECIES No
US LISTING STATUS No
CANADA SPECIES AT RISK No
PIF CCS 14/20

MIGRATORY CONNECTIVITY

OVERALL Only a BirdLife distribution map is available; no eBird data or migratory connectivity information were found for this species.

BURROWING OWL *Athena cunicularia*

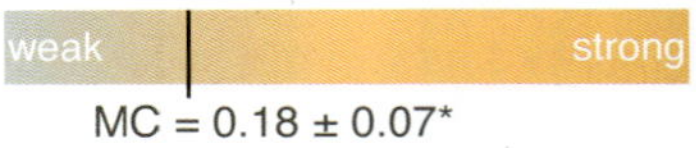

DISTRIBUTION AND ABUNDANCE

Burrowing Owls are widely distributed and found in open habitats throughout the Americas. The breeding range extends throughout the western US, Canada, and into northern Mexico, and the nonbreeding range extends across the southwestern US and Mexico. Burrowing Owls are migratory in the northern part of the breeding range, with varying proportions of resident and migratory individuals in populations throughout the rest of the breeding range (Poulin et al. 2020).

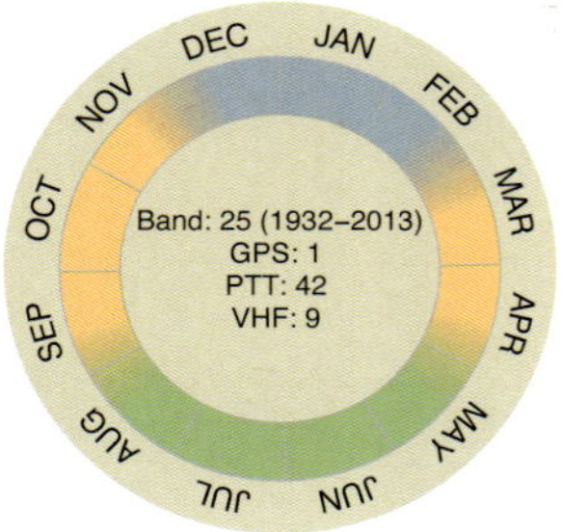

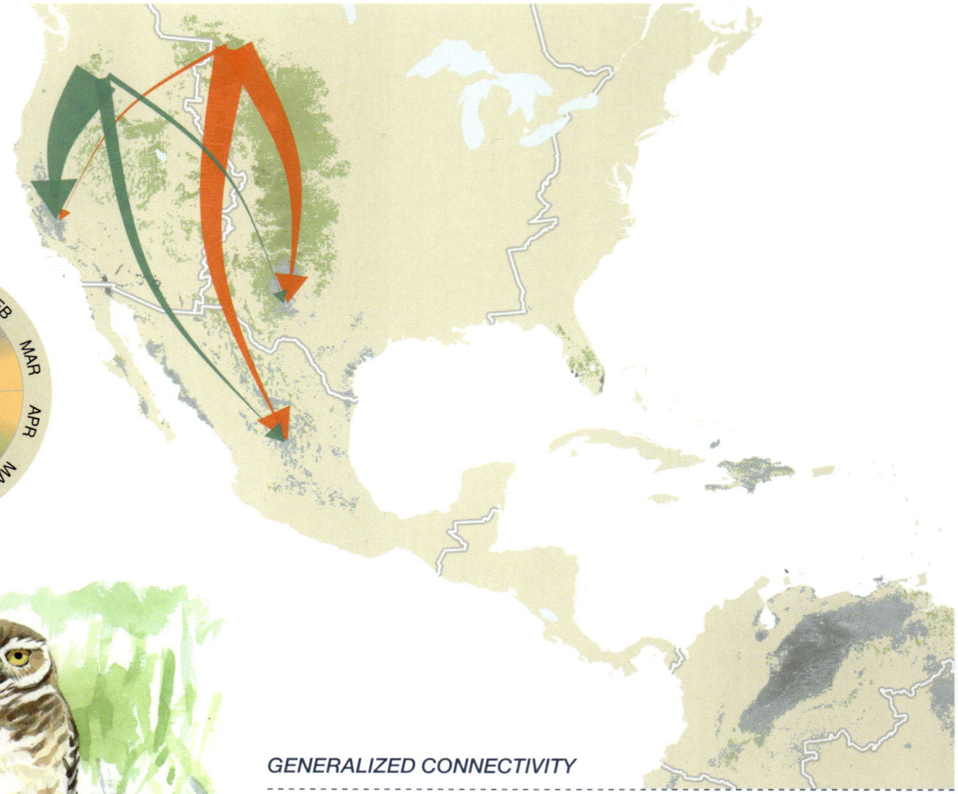

GENERALIZED CONNECTIVITY

ORDER Strigiformes
FAMILY Strigidae
BREEDING HABITAT Grasslands, Temperate
NONBREEDING HABITAT Grasslands, Temperate
GLOBAL POPULATION SIZE 18,000,000
POPULATION CHANGE SINCE 1970 −26%
IUCN RED LIST CATEGORY Least Concern
R2R TIPPING POINT SPECIES No
US LISTING STATUS No
CANADA SPECIES AT RISK Endangered
PIF CCS 11/20

	Mexico	Midcontinent	Pacific
Midcontinent (n = 48)	52%	46%	2%
Pacific (n = 20)	20%	5%	75%

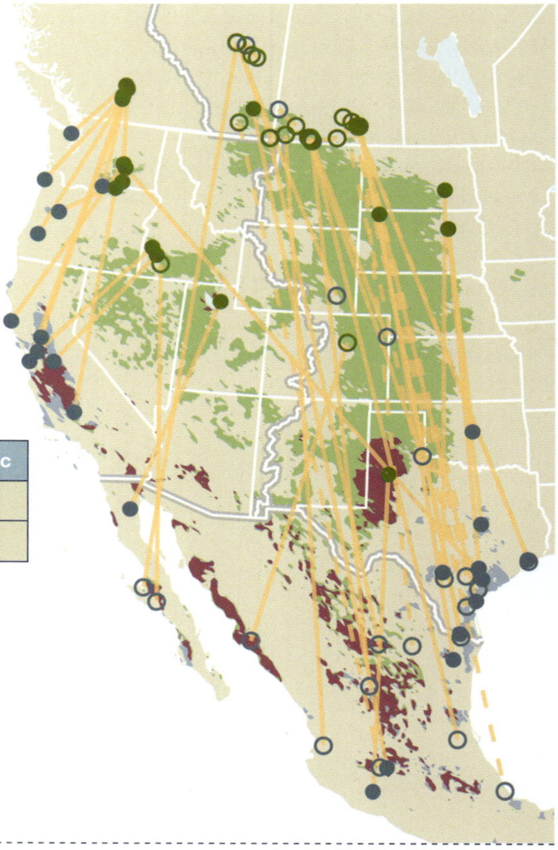

CONNECTIONS

MIGRATORY CONNECTIVITY

OVERALL The migratory connectivity of the Burrowing Owl is understood from satellite and radio telemetry, light-level geolocator, band reencounter, genetic, and stable isotope data. Individuals breeding within the Midcontinent Flyway largely migrate to Mexico (52%) or within the Midcontinent Flyway (46%) for the nonbreeding season. Pacific Flyway breeding individuals migrate to nonbreeding areas within the flyway (75%) or to Mexico (20%). Although a large portion of breeding individuals remain within the same flyway during the nonbreeding season, mixing of breeding populations across all nonbreeding regions results in weak migratory connectivity (MC = 0.18 ± 0.07).

BANDING Band reencounters suggest a diverse pattern of short- to long-distance movements and support similar patterns found with telemetry data.

RADIO TRACKING Data from radio telemetry supported longitudinal connectivity (Holroyd et al. 2010).

GEOLOCATOR TRACKING Data from 15 recovered geolocators showed that breeding birds in Oregon and Washington exhibited a variety of migration strategies (M. Gregg et al., unpublished data; Global Owl Project, unpublished data). The majority of birds

INTERESTING NOTE

A female Burrowing Owl dispersed 1,860 km (from Arizona to Saskatchewan) between two nesting attempts within the same season, the longest recorded for any raptor and possibly any species (Holroyd et al. 2011). Both stable isotope and genetic analyses provided some evidence that dispersal from breeding populations in Canada and the US in the northern part of the range may be leading to declines in these areas (Duxbury 2004; Macías-Duarte et al. 2020)

migrated to nonbreeding areas from central Oregon to southern Nevada and from central to Southern California, while some birds only migrated a short distance within Oregon and Washington. One bird did not migrate. One male tagged in Saskatchewan migrated to south Texas (Global Owl Project, unpublished data). Data from one geolocator showed a bird from southern Alberta followed the high Great Plains south to coastal Texas before turning inland to central Mexico (Holroyd and Trefry 2011).

SATELLITE TRACKING Satellite telemetry data indicate high longitudinal connectivity across the species' range, from Washington breeders migrating to central California, North Dakota breeders migrating to eastern Mexico (D. Johnson, unpublished data), and Alberta and Saskatchewan breeders migrating to central Mexico (G. Holroyd, unpublished data). Occasionally individuals cross over, like one bird breeding in the central part of its range in Alberta migrating to the southwest in central Baja for the nonbreeding season (Holroyd et al. 2011). The Rocky Mountains may create a migratory divide, as Burrowing Owls breeding in the Great Plains spend the nonbreeding season throughout northern and central Mexico and the coastal plains of Texas and Mexico on the Gulf of Mexico, whereas Burrowing Owls breeding west of the Rocky Mountains spend the nonbreeding season in California and the Baja California peninsula (Macías-Duarte et al. 2020; C.J. Conway, unpublished data).

STABLE ISOTOPES Most of the Burrowing Owls sampled during the nonbreeding season in central Mexico appear to be short-distant migrants breeding in areas that include the southern US and possibly northern Mexico. Additional nonbreeding feather samples from both Texas and Mexico provided evidence of birds originating from breeding regions in Canada and the northern US (Duxbury 2004). Stable isotopes of owls in Alberta and Saskatchewan indicate an average annual breeding dispersal of 400 km, with some breeding owls originating in the US and Mexico in the previous breeding seasons (Duxbury 2004).

GENETICS There is some evidence that Burrowing Owl populations in northwestern Mexico and Canada are more genetically differentiated from the rest of the populations in the breeding range, lending some support to the idea that declines near the northern edge of the breeding range may be at least partially caused by migration-driven dispersal to newly created, irrigated agricultural areas in northwestern Mexico (Macías-Duarte et al. 2020). A combination of genomic sequencing and banding data showed a pattern of partial migration in a declining population of western burrowing owls (*A. c. hypugaea*) in Northern California, in which long-distance migrants overlapped with resident birds in the nonbreeding season (Trulio et al. 2023).

MIGRATORY PATHWAYS Migratory routes tend to be relatively straight, north-south routes, without major stopover sites. Two pathways are documented for Canadian breeding birds from banding and satellite data. Owls breeding in the Canadian Great Plains migrate east of the Rocky Mountains through the US Great Plains in a southeastern direction, and British Columbia breeding birds migrate southwest along the west coast of North America (Holroyd et al. 2010).

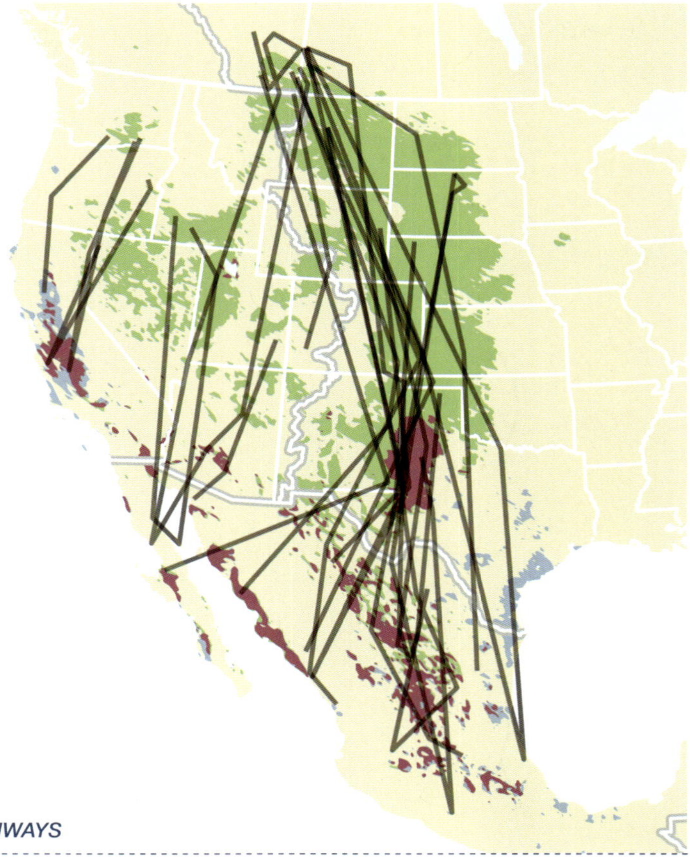

PATHWAYS

DATA CONTRIBUTORS

G.L. Holroyd.

SPOTTED OWL *Strix occidentalis*

DISTRIBUTION AND ABUNDANCE

Spotted Owls are restricted to late successional stage conifer forests of western North America. They range from Washington state south through Oregon and California. Populations also occur in central Arizona, New Mexico, and central Mexico. There are three subspecies, of which the northern spotted owl (*S. o. caurina*) is nonmigratory, the California spotted owl (*S. o. occidentalis*) is typically nonmigratory with occasional altitudinal migration, and the Mexican spotted owl (*S. o. lucida*) is a regular migrant (Gutiérrez et al. 2020).

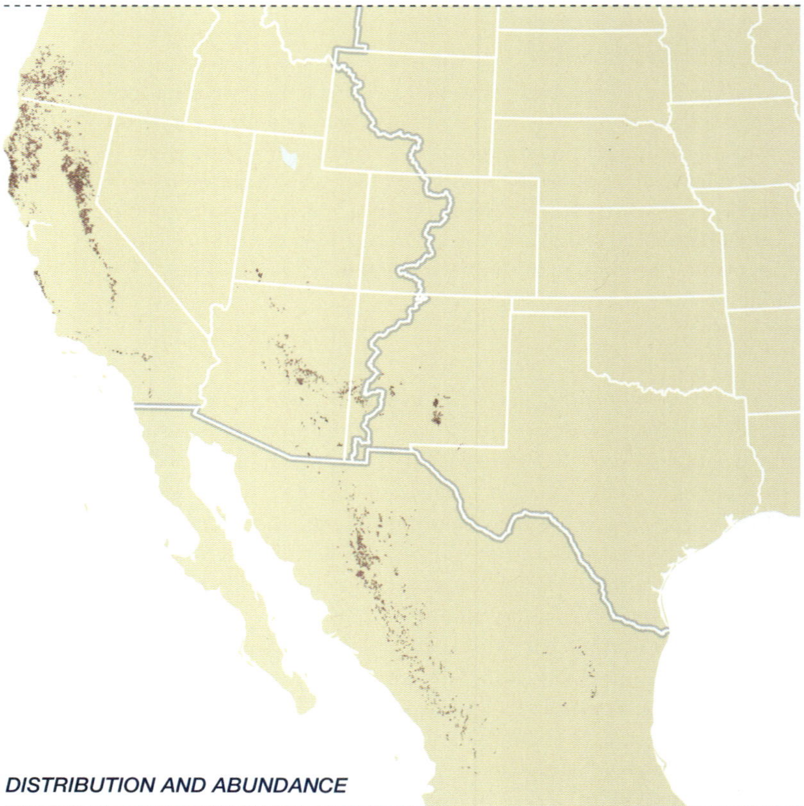

DISTRIBUTION AND ABUNDANCE

ORDER Strigiformes
FAMILY Strigidae
BREEDING HABITAT Forests, Temperate Western
NONBREEDING HABITAT Forests, Temperate Western
GLOBAL POPULATION SIZE 15,000
POPULATION CHANGE SINCE 1970 NA
IUCN RED LIST CATEGORY Near Threatened
R2R TIPPING POINT SPECIES Yellow-Alert
US LISTING STATUS Threatened northern and Mexican subspecies
CANADA SPECIES AT RISK Endangered Northern subspecies
PIF CCS 15/20

MIGRATORY CONNECTIVITY

OVERALL Migratory connectivity data are limited for all three subspecies of Spotted Owls.

BANDING There was one breeding to nonbreeding reencounter within British Columbia, which showed that the individual moved a distance of about 215 km.

RADIO TRACKING Radio transmitter studies in New Mexico, Utah, and Colorado have determined that only a subset of breeding individuals leave the study area for the nonbreeding season, and the number of migratory individuals changes from year to year ranging from 0%–100%. Individuals that left the study area for the nonbreeding season were located 5–50 km away from breeding areas (Ganey and Block 2005). Spotted Owls in Sierra Nevada, California, migrated to lower elevations in post-breeding to spend the nonbreeding season away from areas with heavy snow (Laymon 1989).

GENETICS Genetic analysis of northern spotted owls throughout their Pacific Northwest range showed a high level of dispersal and mixing throughout the region, with a distinct genetic cluster at the southern extent of the region (M.P. Miller et al. 2018).

MIGRATORY PATHWAYS There is limited data on migratory pathways, although they can be variable within and among years (Gutiérrez et al. 2020).

INTERESTING NOTE

Barred Owls (*Strix varia*; not included in this Atlas) are currently spreading throughout the range of the federally threatened northern spotted owl and are considered a major threat to its current persistence (Bierregaard et al. 2024).

Asio otus **LONG-EARED OWL**

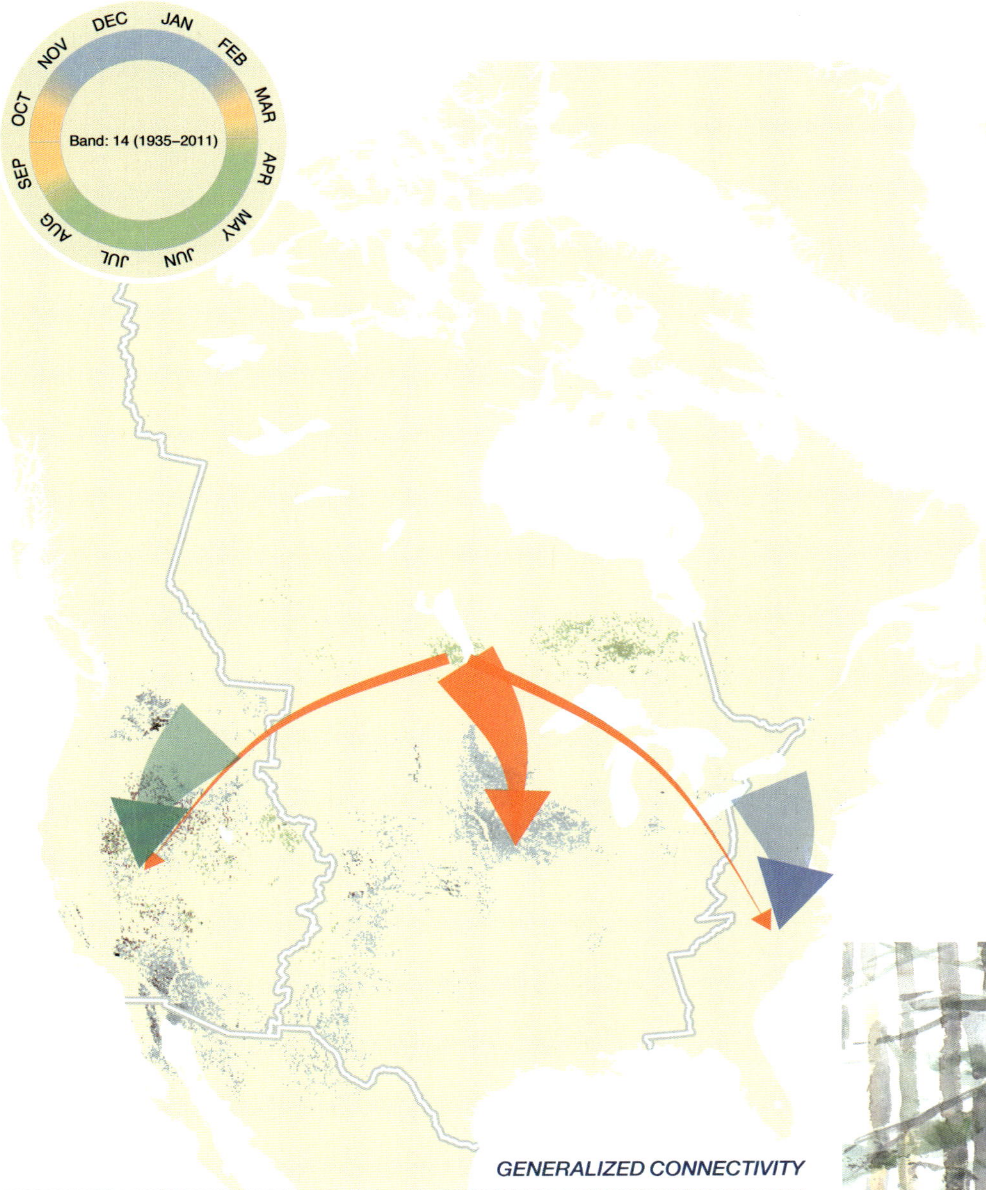

GENERALIZED CONNECTIVITY

DISTRIBUTION AND ABUNDANCE

Long-eared Owls are widely distributed throughout the Northern Hemisphere, in both the Palearctic and Nearctic regions. In North America, during the breeding season they are found most commonly in open habitats and forests in the Canadian provinces of Quebec, Ontario, Manitoba, Saskatchewan, and eastern Alberta. In the US, they are most abundant in northwest Nevada, Utah, eastern California, and southern Arizona; they are also abundant in northern Mexico, primarily Sonora. During the nonbreeding season, some individuals appear to be resident, especially in the southern portion of the range, while others in northern regions migrate to the Rocky Mountain states and upper midwestern states. They (or maybe individuals) can also be irruptive, especially in the northeast, with birds sometimes concentrating along coastal areas in communal roosts (Marks et al. 2020a).

	Atlantic	Midcontinent	Pacific
Atlantic (n = 1)	100%	–	–
Midcontinent (n = 10)	10%	80%	10%
Pacific (n = 1)	–	–	100%

ORDER Strigiformes
FAMILY Strigidae
BREEDING HABITAT Forests, Temperate Generalist
NONBREEDING HABITAT Forests, Temperate Generalist
GLOBAL POPULATION SIZE 520,000
POPULATION CHANGE SINCE 1970 −52%
IUCN RED LIST CATEGORY Least Concern
R2R TIPPING POINT SPECIES No
US LISTING STATUS No
CANADA SPECIES AT RISK No
PIF CCS 10/20

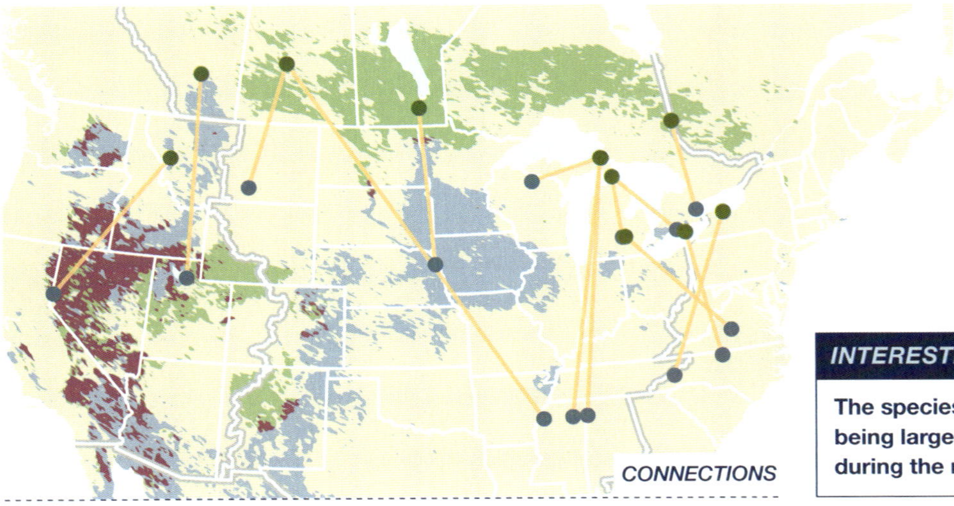

CONNECTIONS

INTERESTING NOTE

The species exhibits reverse sexual dimorphism, with females being larger than males. It is known to roost communally during the nonbreeding season (Marks et al. 2020a).

MIGRATORY CONNECTIVITY

OVERALL The migratory connectivity of the Long-eared Owl is known from band reencounter and limited satellite tracking data. Birds breeding in the Atlantic and Pacific flyways migrate to nonbreeding areas
within the flyway. Birds breeding in the Midcontinent Flyway also migrate to nonbreeding areas within the same flyway (80%); however, they also migrate to the Atlantic (10%) and Pacific (10%) flyways. MC was not calculated due to limited sample size.

BANDING Band reencounter data are limited but show a general movement from a northern breeding range to a southern nonbreeding range, with long-distance recoveries as far south as Mexico. Analysis of band reencounters since 1931 suggests that the species is irruptive and nomadic, depending on food availability. Adults and nestlings banded in the northern portion of the range in Saskatchewan, Ontario, Michigan, Montana, and Minnesota were recovered in Mexico (Houston 2005).

RADIO TRACKING A few tags were deployed through the Motus Wildlife Tracking System; there are no individual breeding to nonbreeding detections.

SATELLITE TRACKING Christenson and Ward (2022) tagged five owls with GPS satellite tags in late February at a communal roost in New Jersey. Three birds subsequently died during the nonbreeding season, and two departed in late March/early April on pre-breeding migration, eventually moving to breeding areas in northern Quebec.

MIGRATORY PATHWAYS Migration itself is not well understood, partially due to their nomadic and irruptive nature, which may be linked to fluctuating prey numbers (Marks et al. 2020a). Little is known about timing and synchronicity of migration and irruptions (Houston 2005).

SHORT-EARED OWL *Asio flammeus*

DISTRIBUTION AND ABUNDANCE

The breeding range of the Short-eared Owl extends from the high Arctic, through Alaska and Yukon, along the Beaufort Sea east around the Hudson Bay to northern latitudes of the United States (Wiggins et al. 2020). A separate disjunct breeding population includes areas in the Intermountain West (Idaho, Utah, Montana, South Dakota) and Prairie Pothole region, especially Manitoba and North Dakota. During the nonbreeding season, highest abundance is throughout the Prairie Potholes and Great Plains, with other high-density areas located in western New York, eastern Washington and Oregon, northwestern Nevada, Sacramento Valley, California, and along the coast of Alaska. The species also extends into Texas and northeastern Mexico (Fink et al. 2023).

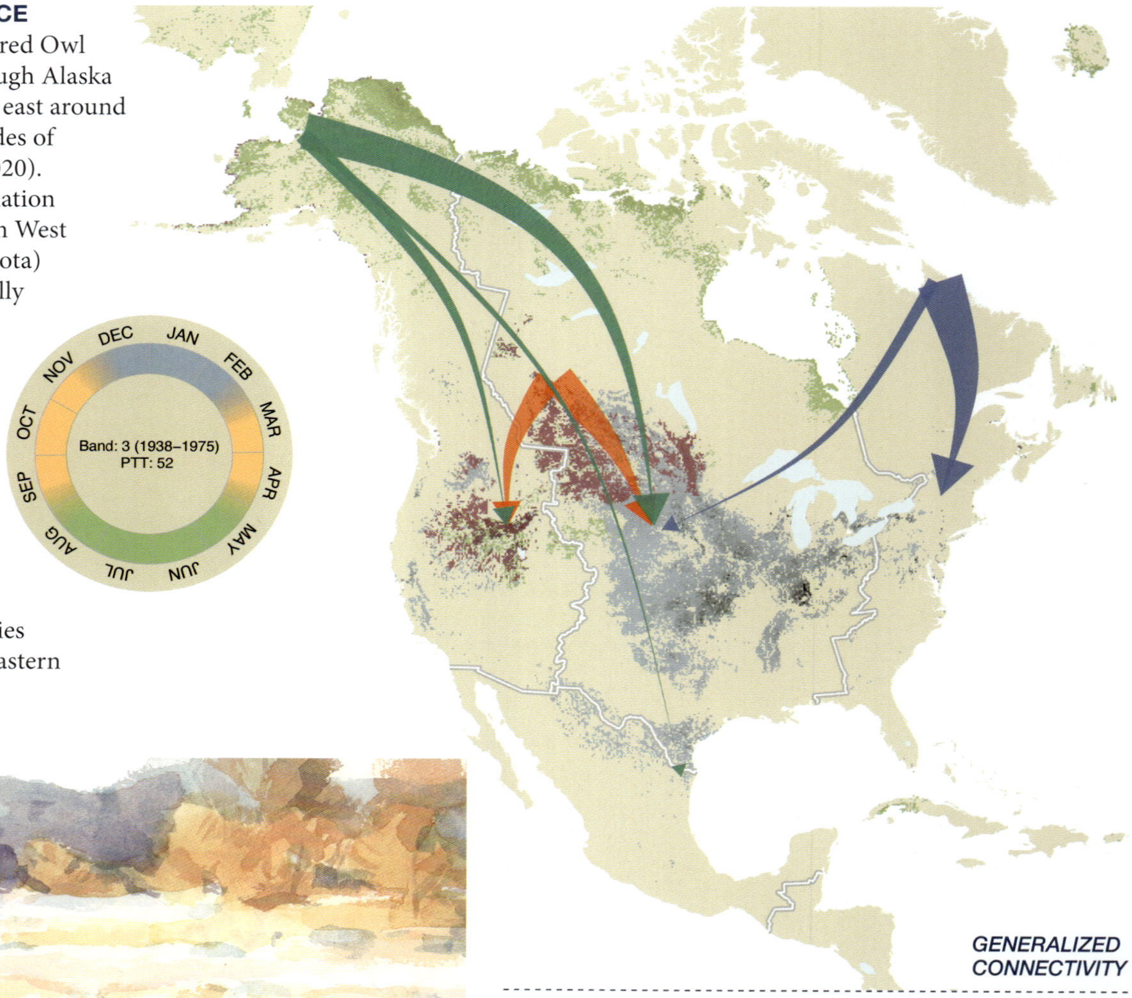

GENERALIZED CONNECTIVITY

ORDER Strigiformes
FAMILY Strigidae
BREEDING HABITAT Tundra, Arctic
NONBREEDING HABITAT Grasslands, Temperate
GLOBAL POPULATION SIZE 2,300,000
POPULATION CHANGE SINCE 1970 −52%
IUCN RED LIST CATEGORY Least Concern
R2R TIPPING POINT SPECIES No
US LISTING STATUS No
CANADA SPECIES AT RISK No
PIF CCS 12/20

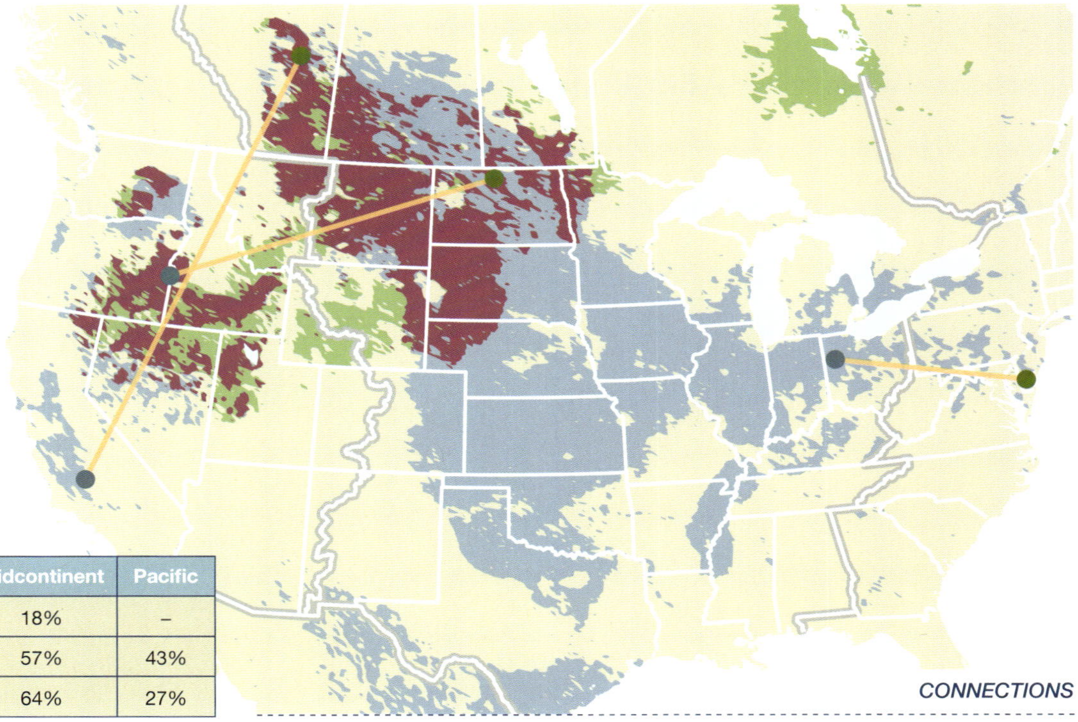

	Atlantic	Mexico	Midcontinent	Pacific
Atlantic (n = 11)	82%	–	18%	–
Midcontinent (n = 7)	–	–	57%	43%
Pacific (n = 11)	–	9%	64%	27%

CONNECTIONS

INTERESTING NOTE

Because this is an irruptive species, its movements are thought to follow the abundance of small mammals throughout their life cycle (Wiggins et al. 2020).

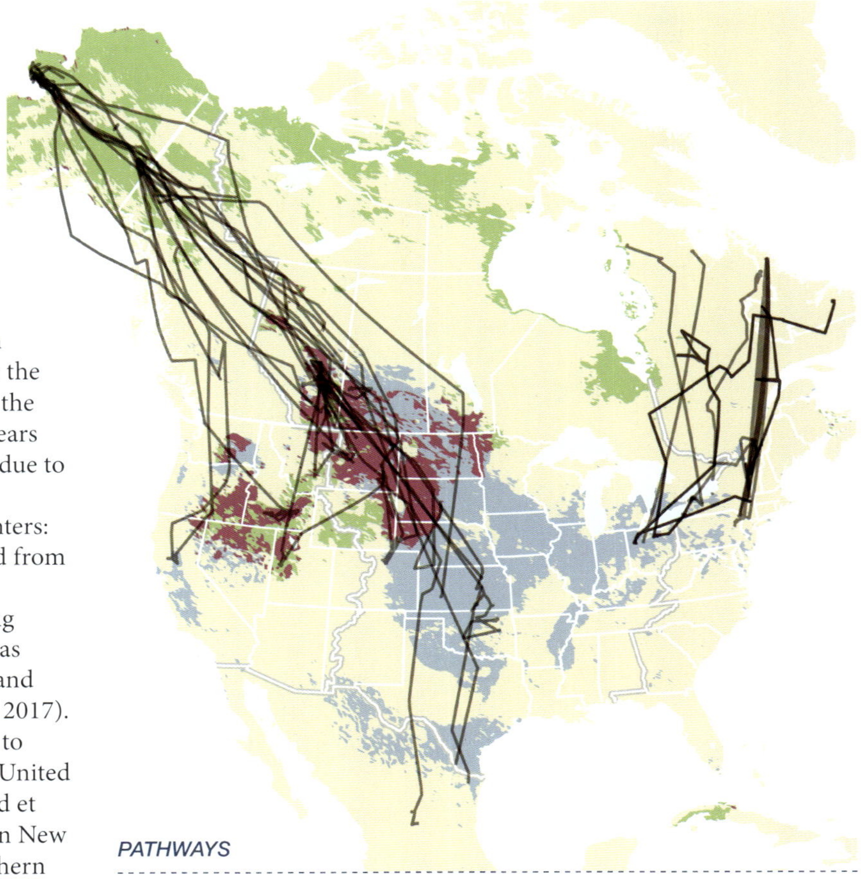

PATHWAYS

MIGRATORY CONNECTIVITY

OVERALL Migratory connectivity is known from satellite tracking and band reencounter data. While Atlantic Flyway breeding birds primarily spend the nonbreeding season within the Atlantic Flyway (82%), there is mixing of birds breeding in the Midcontinent and Pacific flyways at nonbreeding areas within the Midcontinent and Pacific flyways. Migratory connectivity appears moderate across an east-west divide, but MC is not calculated due to small sample size.

BANDING There were three breeding to nonbreeding reencounters: from Alberta to California, from North Dakota to Oregon, and from Delaware to Ohio.

SATELLITE TRACKING Short-eared Owls tracked from breeding sites in Alaska (n = 26) primarily migrated to nonbreeding areas in the Pacific and Midcontinent flyways, including California and Idaho, and from Montana south to Mexico (J.A. Johnson et al. 2017). Birds tracked from breeding areas in Alberta (n = 5) migrated to nonbreeding areas throughout the Intermountain West of the United States (Montana, Kansas, Oklahoma, and the Dakotas; Holroyd et al., unpublished data). Birds tracked from nonbreeding areas in New York migrated to breeding areas in northern Quebec and northern areas of Newfoundland and Labrador (R. Kays and T. Simon, unpublished data). Several tags failed to transmit more than one or two months past deployment (R. Kays and T. Simon, unpublished data). Tag failure was also documented for several tags deployed on Short-eared Owls throughout the western United States.

OTHER APPROACHES No genetic or stable isotope studies of migratory connectivity were found for this species

MIGRATORY PATHWAYS Short-eared Owls tracked from breeding sites in Alaska and Alberta primarily migrated east of the Rocky Mountains to the Prairie provinces and Great Plains states. Several individuals migrated west of the Rocky Mountains along a more coastal route (J.A. Johnson et al. 2017). En route band reencounters suggest similar movements as tracking data (Dunn et al. 2009).

DATA CONTRIBUTORS

T. Booms; L.H. DeCicco; G.L. Holroyd; J.A. Johnson; R. Kays; P. Nye; K.M. O'Brien; New York State Department of Environmental Conservation.

NORTHERN SAW-WHET OWL *Aegolius acadicus*

DISTRIBUTION AND ABUNDANCE

This species breeds across most of Canada, from the Atlantic provinces west to British Columbia. It also breeds patchily in most of the western United States but also in northern Minnesota, Wisconsin, Michigan, Vermont, New Hampshire, and Maine (Fink et al. 2023). Some Saw-whets stay year-round in areas of the United States and Canada, while other populations are highly migratory (Rasmussen et al. 2020). Nonbreeding areas include the eastern US states south to Maryland, the upper Midwest around the Great Lakes, and British Columbia south into most western states (Fink et al. 2023).

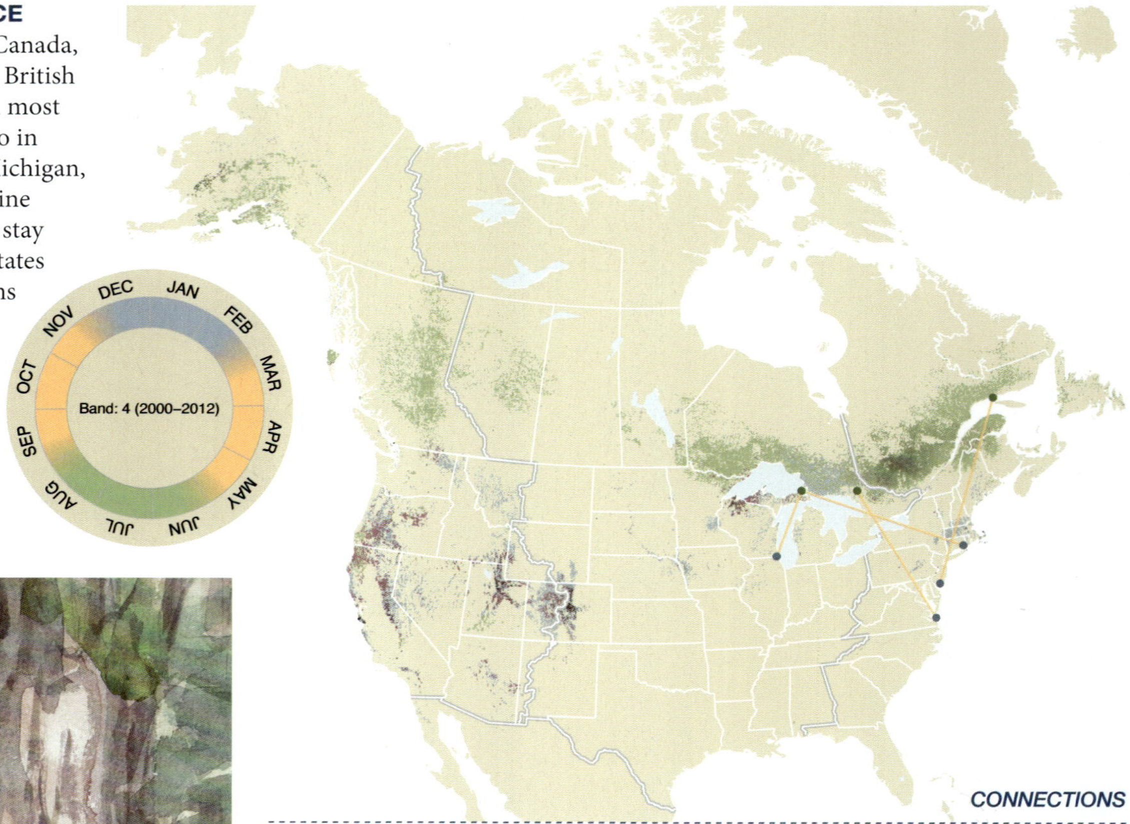

ORDER Strigiformes
FAMILY Strigidae
BREEDING HABITAT Forests, Boreal
NONBREEDING HABITAT Forests, Temperate Generalist
GLOBAL POPULATION SIZE 2,000,000
POPULATION CHANGE SINCE 1970 +162%
IUCN RED LIST CATEGORY Least Concern
R2R TIPPING POINT SPECIES No
US LISTING STATUS No
CANADA SPECIES AT RISK Threatened *brooksi* subspecies
PIF CCS 8/20

MIGRATORY CONNECTIVITY

OVERALL Band reencounter data suggest the general migratory patterns of the species are from breeding sites in southern Canada to nonbreeding ranges in Appalachia and the northern United States. Migration appears to be irruptive, and some individuals are nomadic, making it difficult to determine overall migratory connectivity for the species. MC is not calculated due to limited data.

BANDING Widespread and long-term banding efforts (projectOwlnet.org) have resulted in recaptures of 1,444 banded birds at sites ranging from Manitoba to the Canadian Maritime provinces. The data show that Northern Saw-whets in central and eastern North America are divided into two populations. Northwestern individuals from the upper Great Lakes spend the nonbreeding season primarily in Minnesota and Wisconsin, while individuals from eastern breeding populations spend the nonbreeding season primarily in Appalachia (Confer et al. 2014). Sixty-one owls banded in Alberta and Saskatchewan were recaptured away from the banding site. About half were east or southeast of the breeding range, some did not migrate, and the farthest movement was 2,315 km away in Pennsylvania (Priestley et al. 2010). One owl banded in Montana was recaptured in Massachusetts (3,322 km east), though two other recaptures in Montana originated in Alberta (Frye 2012). Individuals captured in Arkansas in the nonbreeding season appeared to originate in the Great Lakes region (Pruitt and Smith 2016).

RADIO TRACKING There were several studies tracking Northern Saw-whet Owls using the Motus Wildlife Tracking System. There are no breeding to nonbreeding detections.

STABLE ISOTOPES Feathers from Northern Saw-whets migrating through a banding station in Delta Marsh, Manitoba, had hydrogen isotopes higher than expected, making them ineffective in assigning the owls to their origin (De Ruyck et al. 2013).

MIGRATORY PATHWAYS Regularly banded at monitoring stations across North America during post-breeding migration, the Northern Saw-whet Owl does not exhibit regular use of given stopover areas. Differences in stable hydrogen isotopes and capture timing at banding stations suggest that migration patterns differ among age classes (Beckett and Proudfoot 2011; De Ruyck et al. 2013).

TROGONS (TROGONIFORMES)

WRITING LED BY **TOM WILL** *AND* **PETER P. MARRA**

The Elegant Trogon is the only species of Trogoniformes that breeds regularly in the United States, and its migratory connectivity is poorly understood. Elegant Trogons are thought to migrate to nonbreeding grounds in Mexico, but the extent to which US breeders mix with Mexican resident populations remains unknown (Kunzmann et al. 2020). No tracking studies have been conducted, leaving major gaps in our knowledge of connectivity between breeding and nonbreeding sites. The species is not listed as a Tipping Point species or of conservation concern, although without detailed movement and population data, knowledge about the Elegant Trogon remains limited.

TROGONIDAE | *TROGONS*

Trogon elegans ELEGANT TROGON

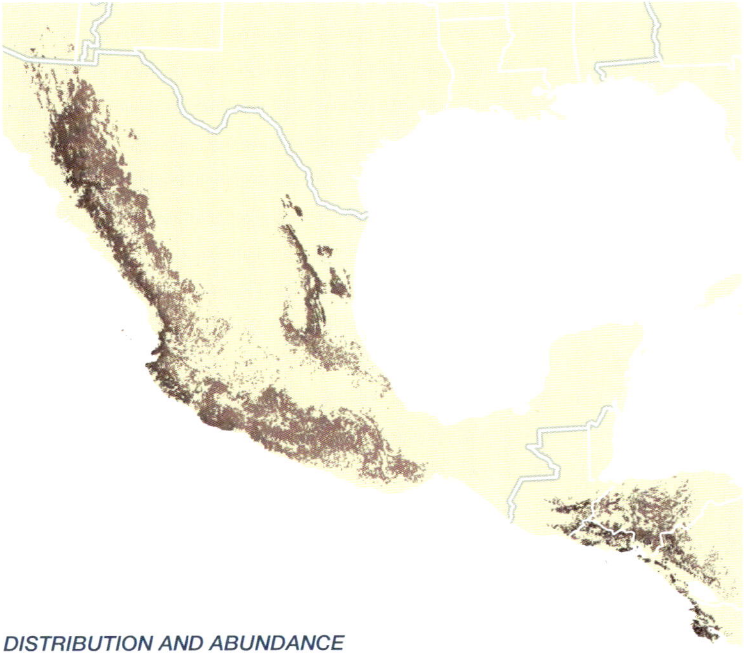

DISTRIBUTION AND ABUNDANCE

ORDER Trogoniformes
FAMILY Trogonidae
BREEDING HABITAT Forests, Mesoamerican Pine-Oak
NONBREEDING HABITAT Forests, Tropical Dry
GLOBAL POPULATION SIZE 200,000
POPULATION CHANGE SINCE 1970 NA
IUCN RED LIST CATEGORY Least Concern
R2R TIPPING POINT SPECIES No
US LISTING STATUS No
CANADA SPECIES AT RISK No
PIF CCS 14/20

DISTRIBUTION AND ABUNDANCE

This species breeds in a wide variety of forest types in the *cordilleras* (mountain ranges) of eastern, western, and southern Mexico; the Pacific Slope of Guatemala, Honduras, Nicaragua, and northwestern Costa Rica; and regularly in four mountain ranges of southern Arizona. The densest breeding concentrations occur in Sonora, Mexico (Fink et al. 2023). Most individuals migrate south from the US and possibly the mountains of northern Mexico during the nonbreeding season; otherwise, the species is thought to be a year-round resident throughout most its range (Kunzmann et al. 2020).

MIGRATORY CONNECTIVITY

OVERALL Migratory connectivity is unknown. No migratory connectivity or migratory pathways information was found for this species.

KINGFISHERS (CORACIIFORMES)

WRITING LED BY **TOM WILL** *AND* **PETER P. MARRA**

The migratory connectivity and associated ecology of the North American kingfishers are relatively poorly documented compared to other riparian or wetland-associated bird groups. In the United States, three species occur regularly. Of these, only the Belted Kingfisher is widespread across the continent and a regular long-distance migrant in northern portions of its range (Kelly et al. 2020). Despite their broad range, relatively few tracking studies have been conducted, and migratory connectivity between breeding and nonbreeding populations remains poorly quantified.

ALCEDINIDAE | *KINGFISHERS*

BELTED KINGFISHER *Megaceryle alcyon*

DISTRIBUTION AND ABUNDANCE

This species breeds throughout most of North America except the Arctic tundra and large portions of the southwestern US deserts. It is sparse in Alaska, interior northwest Canada, southern Arizona, and west Texas; it is most dense in the taiga and boreal forest regions of Canada from Saskatchewan through Quebec (Kelly et al. 2020). During the nonbreeding season, when waters freeze, northern populations migrate south to the unfrozen portions of the eastern US range, the California Central Valley, Florida, the Gulf Coast, Caribbean islands, and both coastal areas of Mexico and Central America (Fink et al. 2023). Some individuals migrate as far south as the northern coast of South America. The species appears stable or is increasing only in the Mississippi Alluvial Valley and Boreal Taiga Plains of Canada (Sauer et al. 2019).

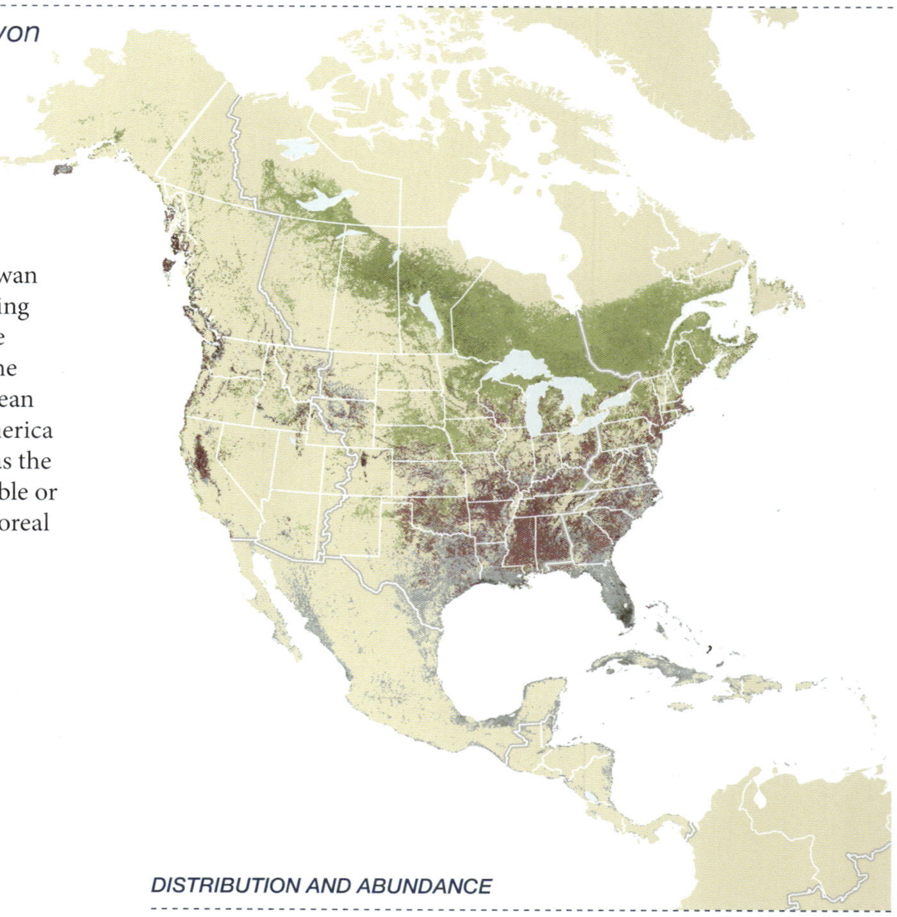

DISTRIBUTION AND ABUNDANCE

ORDER Coraciiformes
FAMILY Alcedinidae
BREEDING HABITAT Wetlands, Generalist
NONBREEDING HABITAT Wetlands, Generalist
GLOBAL POPULATION SIZE 1,800,000
POPULATION CHANGE SINCE 1970 −49%
IUCN RED LIST CATEGORY Least Concern
R2R TIPPING POINT SPECIES No
US LISTING STATUS No
CANADA SPECIES AT RISK No
PIF CCS 10/20

MIGRATORY CONNECTIVITY

OVERALL Migratory connectivity is unknown. No studies of migratory connectivity were found for this species.

MIGRATORY PATHWAYS A partial migrant, this species moves southward or toward the coasts in the nonbreeding season if ice cover limits food availability. Movements are concentrated along the edges of large water bodies. There is a higher proportion of females in the southern extent of the range in the nonbreeding season, indicating that males may undergo shorter migrations or are more likely to remain sedentary (Kelly et al. 2020; Fink et al. 2023).

WOODPECKERS (PICIFORMES)

WRITING LED BY **MELANIE A. SMITH** *AND* **TOM WILL**

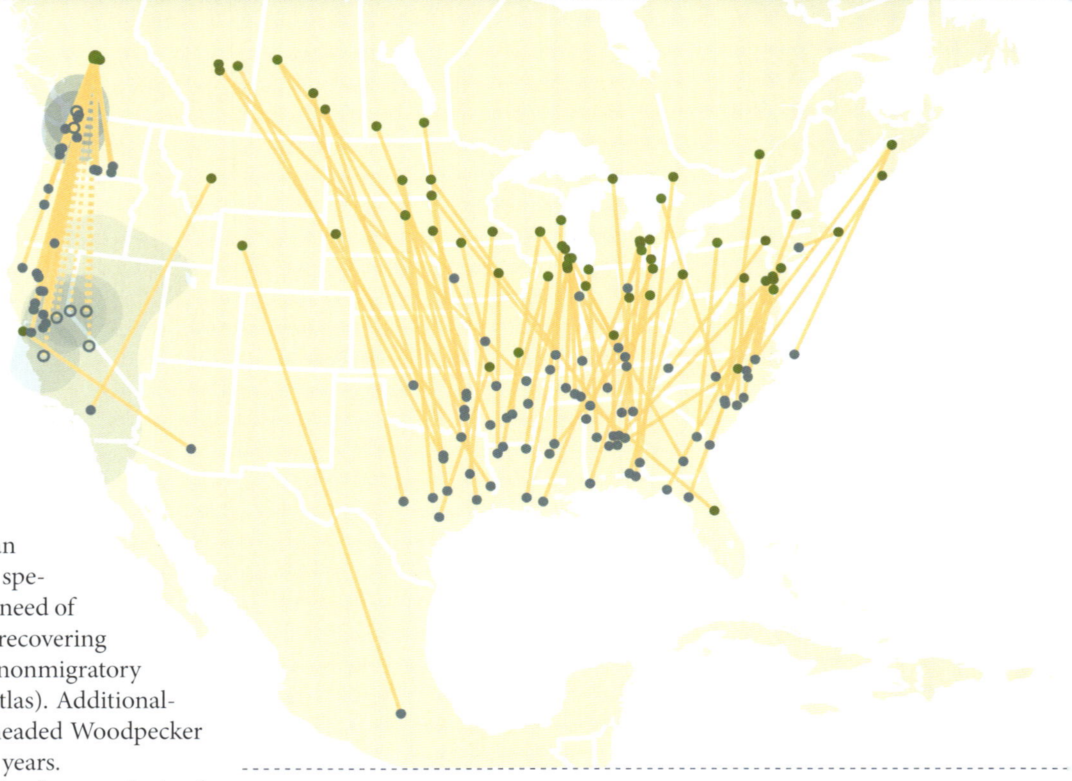

The majority of migratory connectivity data for Piciformes comes from band reencounters (n = 101), with limited data from light-level geolocators (n = 7) in the western US.

The migratory connectivity of North American woodpeckers, order Piciformes, has not been well studied. We have migratory connectivity information for five of the eight species, though it is very limited. Only one of eight migratory species in the order has been tracked and with enough information to derive an MC score. Available studies quantified the movement of seven individuals using light-level geolocators and approximately 100 band reencounter records. Banding records are somewhat well distributed, although this is truer in the eastern than the western parts of the range. Within the order, two species have been identified as Tipping Point species in need of focused and immediate action that is dedicated to recovering their populations: the Lewis's Woodpecker and the nonmigratory Red-cockaded Woodpecker (not included in this Atlas). Additionally, both Lewis's Woodpecker (−59%) and the Red-headed Woodpecker (−65%) have seen concerning declines in the last 50 years.

Within Piciformes, the Northern Flicker is the best-studied of the woodpeckers, with a focus on understanding the ecology of the red-shafted and yellow-shafted populations. Only the Northern Flicker has tracking data, but it is limited. However, current automated telemetry tracking of Lewis's Woodpeckers may add to the understanding of the movement ecology of this order. All species have banding data except for Lewis's Woodpecker, the Red-breasted Sapsucker, and Williamson's Sapsucker. Woodpecker species have just a few banding reencounter records each, and records for the yellow-shafted flicker (a subspecies of the Northern Flicker) make up the majority of connections. There is not enough data to generalize connectivity patterns across species, although data for the Northern Flicker showed strong migratory connectivity (MC = 0.95).

PICIDAE | *WOODPECKERS*

Melanerpes lewis LEWIS'S WOODPECKER

ORDER Piciformes
FAMILY Picidae
BREEDING HABITAT
Forests, Temperate Western
NONBREEDING HABITAT
Forests, Temperate Western
GLOBAL POPULATION SIZE 82,000
POPULATION CHANGE SINCE 1970 −59%
IUCN RED LIST CATEGORY Least Concern
R2R TIPPING POINT SPECIES Yellow-Alert
US LISTING STATUS No
CANADA SPECIES AT RISK Threatened
PIF CCS 13/20

DISTRIBUTION AND ABUNDANCE

This species breeds in the western US mountains in open canopy forests from southern British Columbia to northern New Mexico and central California. During the nonbreeding season, it migrates from the northern portions of the range, except for the Washington/Oregon Cascades, and occurs sparsely and locally as far south as Baja California and northern Mexico (Fink et al. 2023). The species has declined range-wide (Rosenberg et al. 2019), except in the Badlands and Prairie Pothole Region, where it occurs at low density (Sauer et al. 2019).

MIGRATORY CONNECTIVITY

OVERALL Migratory connectivity is unknown. No information on migratory connectivity was found for this species.

RADIO TRACKING Motus tags were deployed across the Bitterroot Valley, Montana, Oregon, and British Columbia (Motus project, Intermountain West

INTERESTING NOTE

Unlike most other woodpeckers, this species catches flying insects in the air for a large portion of its otherwise opportunistic diet (Vierling et al. 2020).

Birds, unpublished data; Motus project, Oregon Department of Fish and Wildlife—Lewis's Woodpecker Migration, unpublished data; Motus project, Lewis's Woodpecker Migratory Connectivity in British Columbia's Southern Interior, unpublished data). Detections during migration connect eastern Montana with southern Oregon, and eastern Montana and Oregon with the Sacramento Valley of California. There are no breeding to nonbreeding detections.

MIGRATORY PATHWAYS Migration in this species is highly variable and changes as resource availability shifts. Birds in the northern reaches of the breeding range likely migrate farther distances to nonbreeding areas than southern birds, which may undertake short flights to the nearest appropriate nonbreeding habitat. Movements are a combination of north-south and altitudinal migration (Vierling et al. 2020).

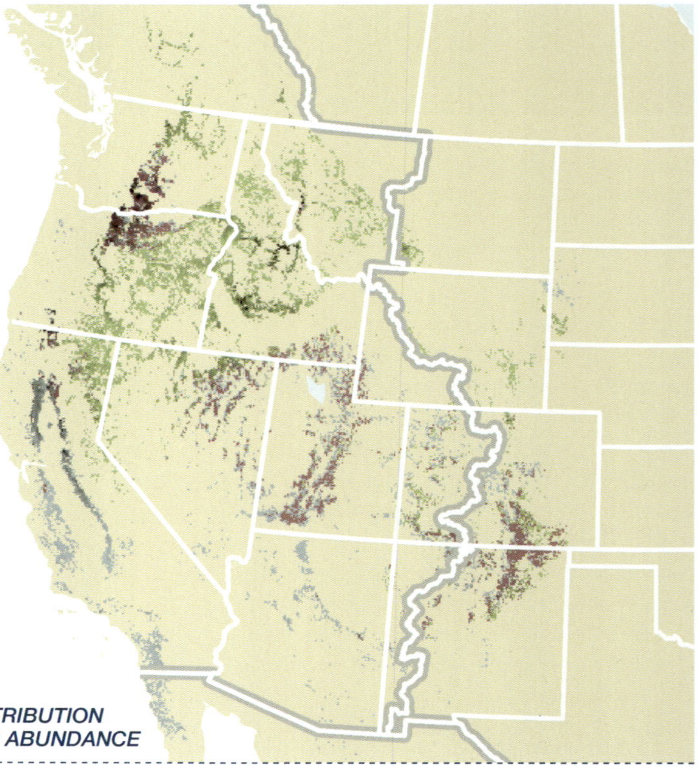

DISTRIBUTION AND ABUNDANCE

RED-HEADED WOODPECKER *Melanerpes erythrocephalus*

DISTRIBUTION AND ABUNDANCE

The Red-headed Woodpecker is distributed widely from the Great Plains side of the Rocky Mountains, with patchy but higher densities in Kansas, Oklahoma, and Iowa, east to the Atlantic coast, especially Virginia south into Florida, and in the Gulf states. Their northern extent includes southern Manitoba and southeastern Ontario, but they are largely absent in the northeastern US. Western and northernmost parts of the range are used during the breeding season, while the southern and eastern portions are used year-round (Frei et al. 2020). Smaller regions of the nonbreeding-only range are found in Texas and Appalachia.

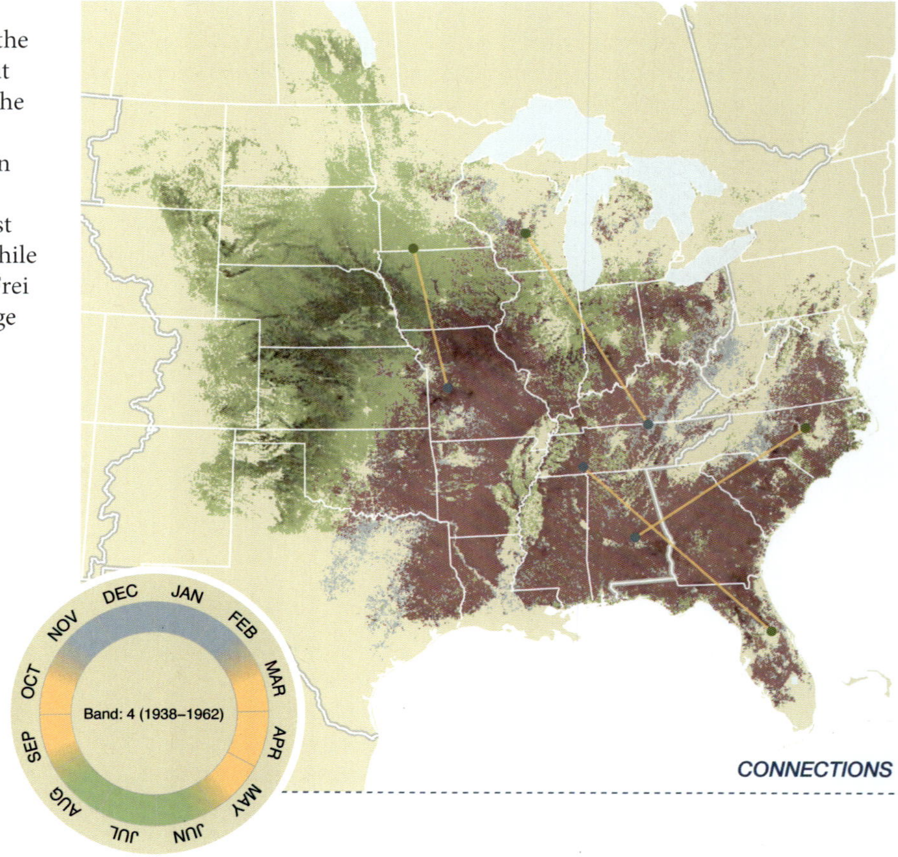

CONNECTIONS

ORDER Piciformes
FAMILY Picidae
BREEDING HABITAT Forests, Temperate Eastern
NONBREEDING HABITAT Forests, Temperate Eastern
GLOBAL POPULATION SIZE 1,800,000

POPULATION CHANGE SINCE 1970 −65%
IUCN RED LIST CATEGORY Least Concern
R2R TIPPING POINT SPECIES No
US LISTING STATUS No
CANADA SPECIES AT RISK Endangered
PIF CCS 12/20

MIGRATORY CONNECTIVITY

OVERALL Migratory connectivity is relatively unknown; however, limited band reencounter, GPS, and radio tracking data exist.
BANDING Limited band recoveries connect breeding locations in the Great Lakes and southeastern US, with nonbreeding locations in the interior midwest.
RADIO TRACKING Radio tags used to track post-breeding migration of individuals that breed in South Carolina showed that rates of migration differed among years of the study. Some birds migrated short distances, while some did not migrate at all, suggesting the population is best described as facultative migrants (Vukovich and Kilgo 2013).
SATELLITE TRACKING GPS tags used to track post-breeding migration of individuals breeding in Minnesota and Ohio found that some Ohio individuals migrated, while Minnesota individuals did not. The northern Ohio birds migrated to central Indiana and southern Ohio (Red-headed Woodpecker Research and Recovery Project, unpublished data).
OTHER APPROACHES No geolocator, stable isotope, or genetic studies of migratory connectivity were found for this species.
MIGRATORY PATHWAYS Western breeding individuals appear to follow an east-west migration pattern as they expand outward in pre-breeding from the nonbreeding grounds to the western breeding range, and contract back east again in the post-breeding season (Fink et al. 2023). Based on banding recoveries, birds in the eastern part of the breeding range likely migrate south during post-breeding and north during pre-breeding.

Melanerpes carolinus

RED-BELLIED WOODPECKER

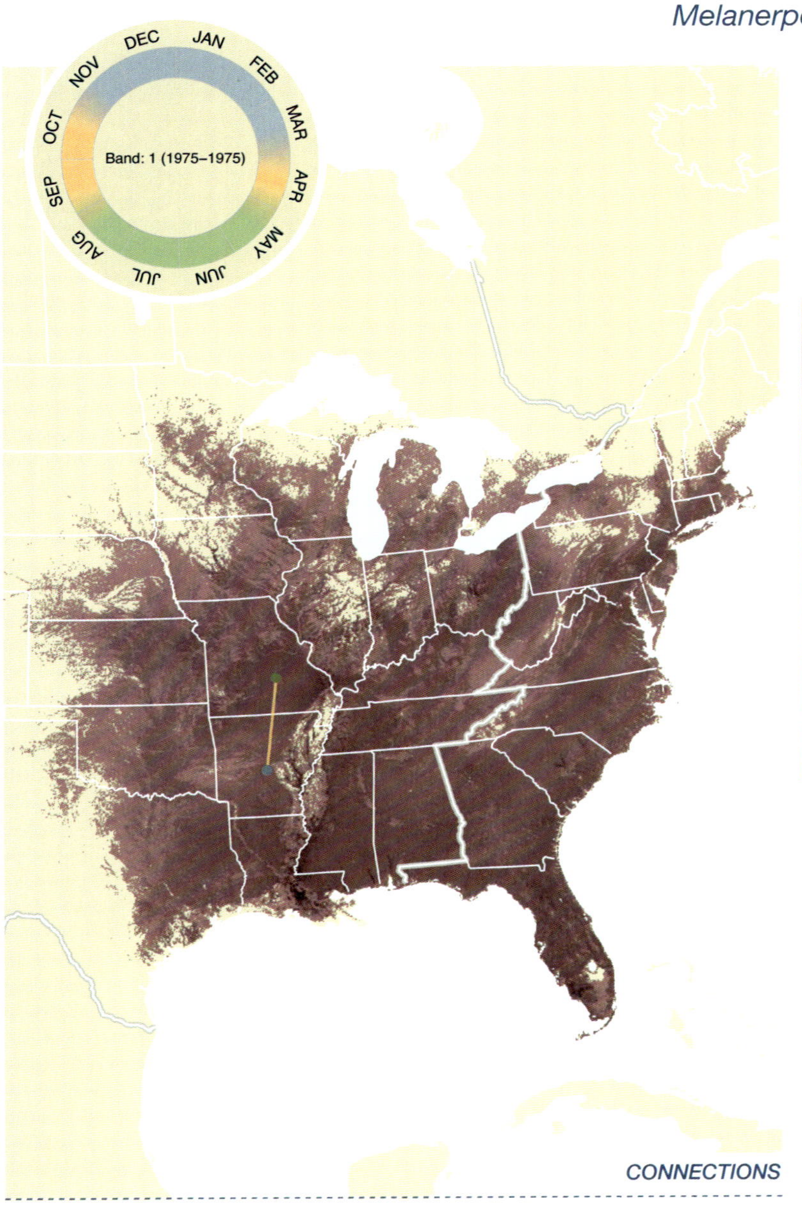

DISTRIBUTION AND ABUNDANCE

The Red-bellied Woodpecker is mostly resident throughout its range in the eastern US ; however, some northern populations may migrate southward (K.E. Miller et al. 2020). They are found widely from the Great Plains states to all along the Atlantic coast, as well as north into southeastern Ontario (K.E. Miller et al. 2020).

ORDER Piciformes
FAMILY Picidae
BREEDING HABITAT Forests, Temperate Eastern
NONBREEDING HABITAT Forests, Temperate Eastern
GLOBAL POPULATION SIZE 16,000,000
POPULATION CHANGE SINCE 1970 +63%
IUCN RED LIST CATEGORY Least Concern
R2R TIPPING POINT SPECIES No
US LISTING STATUS No
CANADA SPECIES AT RISK No
PIF CCS 7/20

MIGRATORY CONNECTIVITY

OVERALL There is very little information on migratory connectivity.
BANDING A single band record links the US states of Missouri and Arkansas.
OTHER APPROACHES No tracking, stable isotope, or genetic studies of migratory connectivity were found for this species.
MIGRATORY PATHWAYS Very little information is available. Limited evidence suggests that small numbers migrate past Cape May, New Jersey, during pre- and post-breeding, as well as along Lake Erie, but these are difficult to distinguish from regional movements (K.E. Miller et al. 2020).

WILLIAMSON'S SAPSUCKER *Sphyrapicus thyroideus*

DISTRIBUTION AND ABUNDANCE

This species breeds in the western mountains in conifer and mixed conifer-deciduous forests (Gyug et al. 2023) from southern British Columbia to central Arizona and Southern California (except for the Intermountain West). During the nonbreeding season, it migrates from the northern portion of its range in the Rockies and northern Cascades but can still be found in Oregon and California, and in Arizona, New Mexico, southern Colorado, and as far south as Durango in northwestern Mexico (Fink et al. 2023).

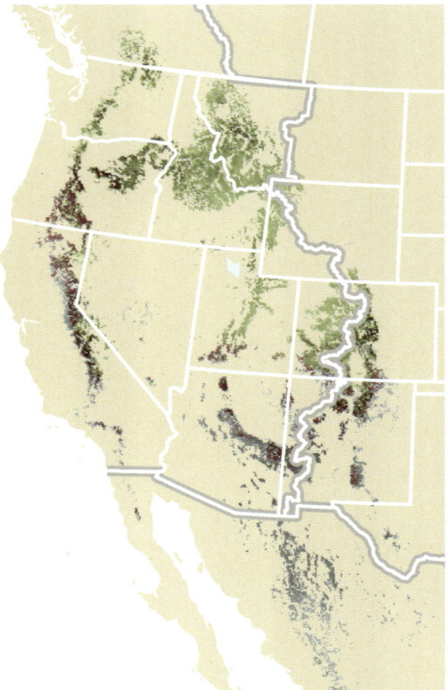

DISTRIBUTION AND ABUNDANCE

MIGRATORY CONNECTIVITY

OVERALL Migratory connectivity is unknown. No studies of migratory connectivity were found for this species.
MIGRATORY PATHWAYS Populations east and west of the Great Basin appear to migrate independently, with the western birds concentrating in central to Southern California and the eastern birds concentrating in Arizona and New Mexico and reaching northern Mexico during the nonbreeding season (Fink et al. 2023).

INTERESTING NOTE

Williamson's Sapsucker is unlike all other woodpeckers in that it exhibits extreme plumage dimorphism between males and females (Gyug et al. 2023).

ORDER Piciformes
FAMILY Picidae
BREEDING HABITAT Forests, Temperate Western
NONBREEDING HABITAT Forests, Temperate Western
GLOBAL POPULATION SIZE 300,000
POPULATION CHANGE SINCE 1970 +13%
IUCN RED LIST CATEGORY Least Concern
R2R TIPPING POINT SPECIES No
US LISTING STATUS No
CANADA SPECIES AT RISK Endangered
PIF CCS 13/20

YELLOW-BELLIED SAPSUCKER *Sphyrapicus varius*

DISTRIBUTION AND ABUNDANCE

The Yellow-bellied Sapsucker occurs widely throughout eastern North America. They breed in woodland habitats, favoring early successional trees, from the Yukon Territory to Nova Scotia, Canada, as well as from eastern North Dakota through the Great Lakes and northeastern US, including higher elevations into the southern Appalachian Mountains. The species spends the nonbreeding season in diverse woodland habitats in the Midwest and from Massachusetts south into the southeastern US, Mexico, Central America, and the Caribbean (Walters et al. 2020b).

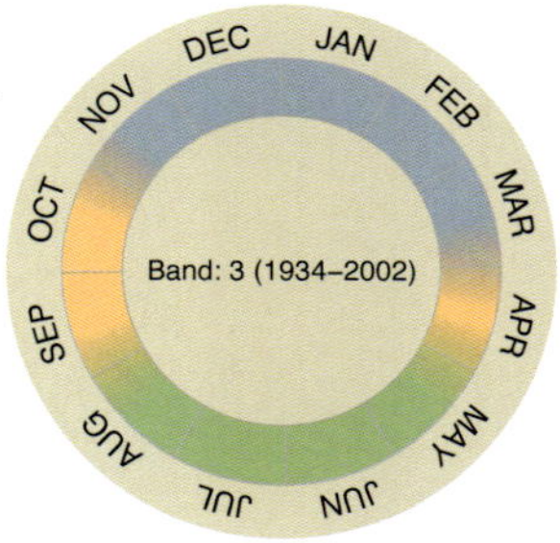

MIGRATORY CONNECTIVITY

OVERALL There is limited information on migratory connectivity and patterns of migration for the species. Band reencounter data show within-region movements in both the Atlantic and Midcontinent flyways.
BANDING Band reencounter data show movements from breeding areas in eastern Canada to nonbreeding locations in the southeastern US, from Ontario to Georgia and Alabama, and from Quebec to South Carolina.
OTHER APPROACHES No tracking, stable isotope, or genetic studies of migratory connectivity were found for this species.

ORDER Piciformes
FAMILY Picidae
BREEDING HABITAT Forests, Boreal
NONBREEDING HABITAT Forests, Temperate Generalist
GLOBAL POPULATION SIZE 14,000,000
POPULATION CHANGE SINCE 1970 +92%
IUCN RED LIST CATEGORY Least Concern
R2R TIPPING POINT SPECIES No
US LISTING STATUS No
CANADA SPECIES AT RISK No
PIF CCS 7/20

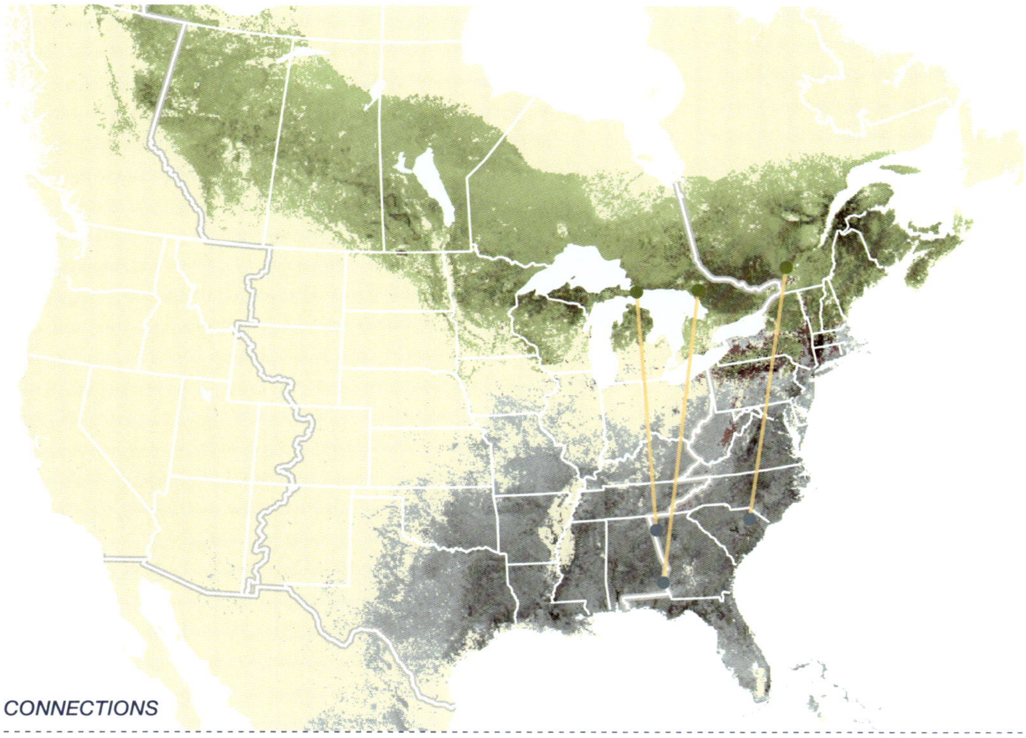

CONNECTIONS

MIGRATORY PATHWAYS Band recoveries from eastern birds indicate typical north-south movement patterns. During pre-breeding migration, birds appear to leave Central America by flying north via Mexico or crossing the Gulf of Mexico, with some stopping on Cuba and nearby islands. Individuals moving into the western breeding area expand northwest from the Great Lakes region. During post-breeding, those individuals migrate southeast back to the Great Lakes and Great Plains, and then south to Texas before moving into Mexico for the nonbreeding season, while eastern birds again cross the Gulf and Caribbean islands to head back to Central America (Fink et al. 2023).

Sphyrapicus nuchalis

RED-NAPED SAPSUCKER

ORDER Piciformes
FAMILY Picidae
BREEDING HABITAT Forests, Temperate Western
NONBREEDING HABITAT Forests, Temperate Western
GLOBAL POPULATION SIZE 2,000,000
POPULATION CHANGE SINCE 1970 +70%
IUCN RED LIST CATEGORY Least Concern
R2R TIPPING POINT SPECIES No
US LISTING STATUS No
CANADA SPECIES AT RISK No
PIF CCS 11/20

CONNECTIONS

DISTRIBUTION AND ABUNDANCE

Red-naped Sapsuckers breed in the Rocky Mountain region of Canada (British Columbia and Alberta) and the US in deciduous and mixed woodlands. They spend the nonbreeding season in diverse woodland habitats in southwest US and Mexico (Walters et al. 2020a).

MIGRATORY CONNECTIVITY

OVERALL There are no published studies on migratory connectivity. Very limited banding data connects the Pacific Flyway to Mexico.

BANDING A single band recovery connects a breeding location in western Colorado, US, to a nonbreeding location near Mexico City, Mexico.

OTHER APPROACHES No tracking, stable isotope, or genetic studies of migratory connectivity were found for this species.

MIGRATORY PATHWAYS Birds appear to follow a typical north-south migration pattern when moving between breeding and nonbreeding grounds. Most birds follow the Rocky Mountains south into mainland Mexico in the nonbreeding season, while others follow the Sierra Nevada Mountains south to the Baja California peninsula (Fink et al. 2023).

RED-BREASTED SAPSUCKER *Sphyrapicus ruber*

DISTRIBUTION AND ABUNDANCE

Red-breasted Sapsuckers breed along the western edge of North America from the Pacific Coast Ranges of southeast Alaska and British Columbia to the Cascades, Sierra Nevada, and Pacific Coast Ranges in the continental US. They are found year-round in the western part of this area, and spend the nonbreeding season from the Central Valley of California to the northern Baja California peninsula in Mexico (Fink et al. 2023).

MIGRATORY CONNECTIVITY

OVERALL Migratory connectivity is unknown. No studies of migratory connectivity were found for this species.

MIGRATORY PATHWAYS Pathways are not well documented. Birds expand their range inland during breeding season and contract coastward during the nonbreeding season. Some move into Southern California and a small portion of northern Mexico in the nonbreeding season, likely following typical north-south movement patterns (Fink et al. 2023).

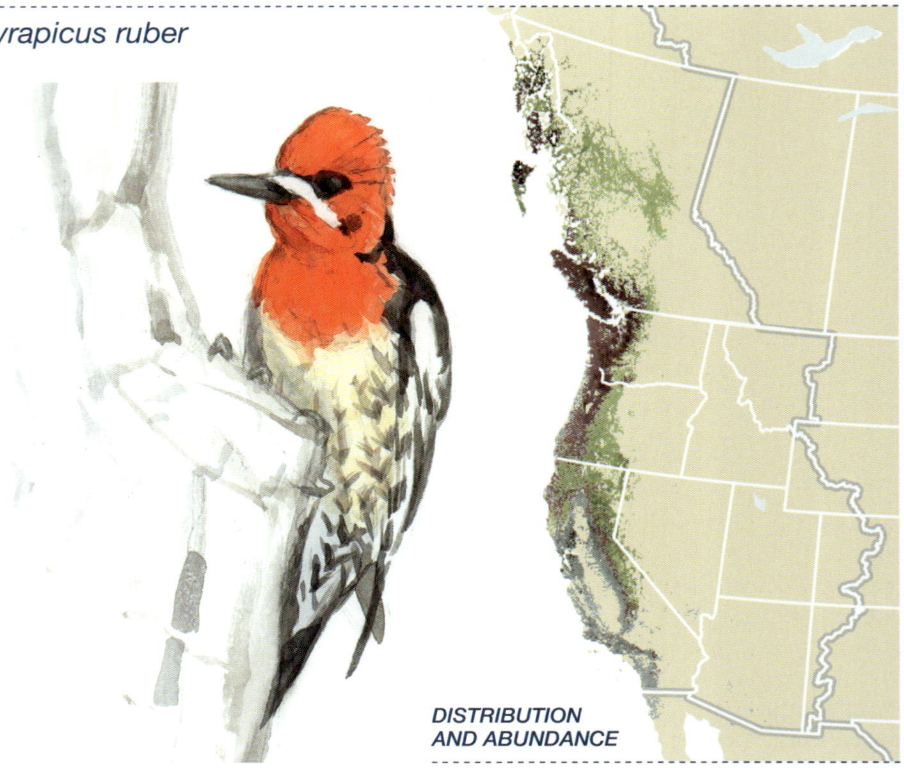

ORDER Piciformes
FAMILY Picidae
BREEDING HABITAT Forests, Temperate Western
NONBREEDING HABITAT Forests, Temperate Western
GLOBAL POPULATION SIZE 2,800,000
POPULATION CHANGE SINCE 1970 +80%
IUCN RED LIST CATEGORY Least Concern
R2R TIPPING POINT SPECIES No
US LISTING STATUS No
CANADA SPECIES AT RISK No
PIF CCS 11/20

NORTHERN FLICKER *Colaptes auratus*

DISTRIBUTION AND ABUNDANCE

The Northern Flicker has a broad geographic range and is found year-round across the United States and Mexico. The species also breeds in Canada and Alaska, roughly following the tree line. Two subspecies are distinguished by differences in their ranges and plumage differences (Aguillon et al. 2018); the yellow-shafted flicker (*C. a. auratus*) inhabits eastern North America, and the red-shafted flicker (*C. a. cafer*) inhabits western North America (Wiebe and Moore 2024).

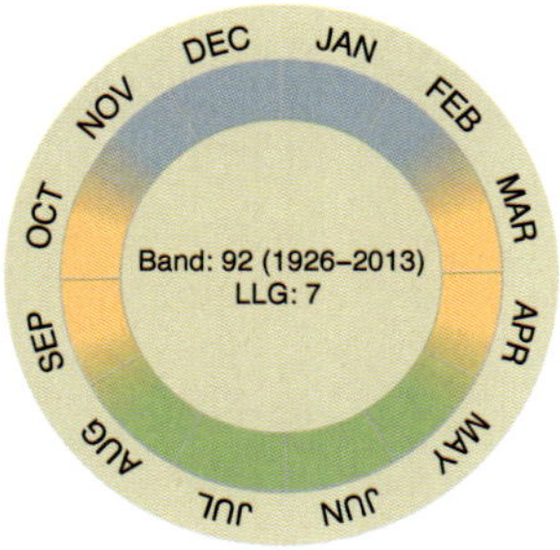

ORDER Piciformes
FAMILY Picidae
BREEDING HABITAT Forests, Generalist
NONBREEDING HABITAT Forests, Temperate Generalist
GLOBAL POPULATION SIZE 12,000,000
POPULATION CHANGE SINCE 1970 −44%
IUCN RED LIST CATEGORY Least Concern
R2R TIPPING POINT SPECIES No
US LISTING STATUS No
CANADA SPECIES AT RISK No
PIF CCS 10/20

INTERESTING NOTE

Though there is little genetic divergence overall, genetic differences exist between red-shafted and yellow-shafted flickers and the other Northern Flicker subspecies (Manthey et al. 2017; Aguillon et al. 2018).

MIGRATORY CONNECTIVITY

OVERALL Northern Flickers have been well-studied throughout their range, both as a species and independently as morphologically different subspecies. Both patterns and strength of connectivity are strong (MC = 0.95). In the Pacific Flyway, 100% of birds stayed within the flyway for both the breeding and nonbreeding seasons. In the Midcontinent Flyway, 94% stayed within the flyway; and in the Atlantic Flyway, 86% stayed within the flyway, while 14% migrated to the Midcontinent Flyway. Each population generally migrates south of its breeding region, creating parallel patterns of migratory connectivity across the range.

BANDING Band reencounters of Northern Flickers follow the same parallel migration patterns as shown with geolocator data: western populations remain west of the Continental Divide in both the breeding and nonbreeding seasons; central breeding populations migrate due south to south-central US for the nonbreeding season; and eastern breeding populations migrate due south to spend the nonbreeding season in the southeastern US (Gow and Wiebe 2014).

GEOLOCATOR TRACKING Data recovered from light-level geolocators deployed in British Columbia showed that individuals that bred farther north traveled farther distances to nonbreeding locations than did individuals breeding farther south and that female nonbreeding locations were farther north than male nonbreeding locations. There was strong migratory connectivity between populations east and west of the Continental Divide (Gow and Wiebe 2014).

OTHER APPROACHES No radio or satellite tracking, stable isotope, or genetic studies of migratory connectivity were found for this species.

MIGRATORY PATHWAYS While many Northern Flickers are residents, those that migrate generally stay within the flyway and follow a north-south pattern (Gow and Wiebe 2014; Wiebe and Moore 2024), except for individuals breeding in the midcontinent of Canada, which appear to follow a northwest-southeast route (Fink et al. 2023).

DATA CONTRIBUTORS
E.A. Gow; K.L. Wiebe.

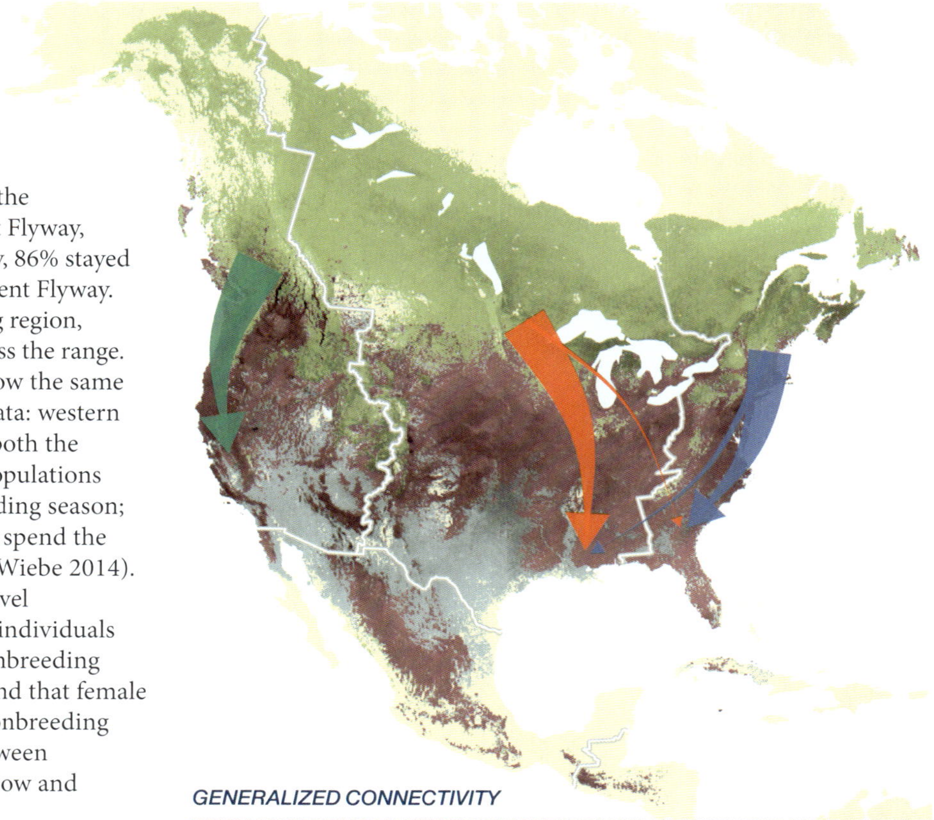

GENERALIZED CONNECTIVITY

	Atlantic	Midcontinent	Pacific
Atlantic (n = 14)	86%	14%	–
Midcontinent (n = 47)	6%	94%	–
Pacific (n = 34)	–	–	100%

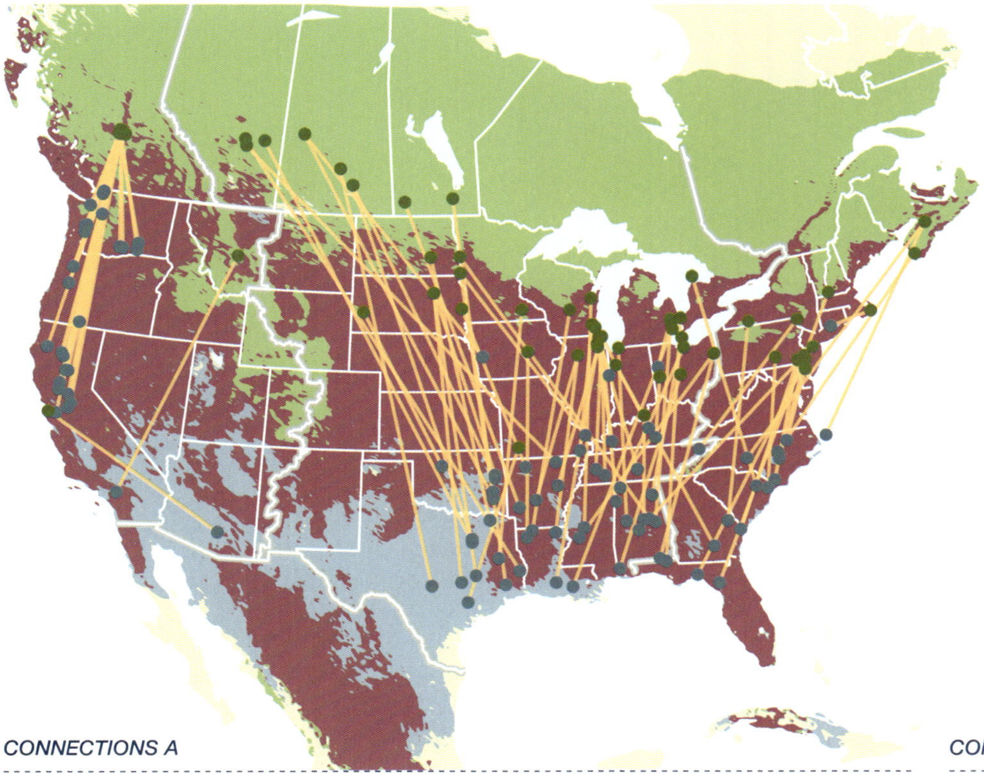

CONNECTIONS A

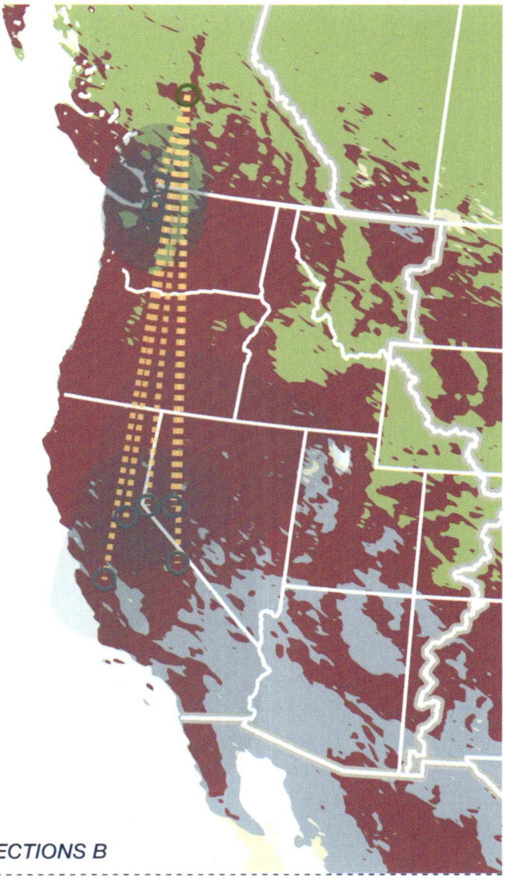

CONNECTIONS B

FALCONS AND CARACARAS (FALCONIFORMES)

WRITING LED BY **LAURIE J. GOODRICH** *AND* **AMY L. SCARPIGNATO**

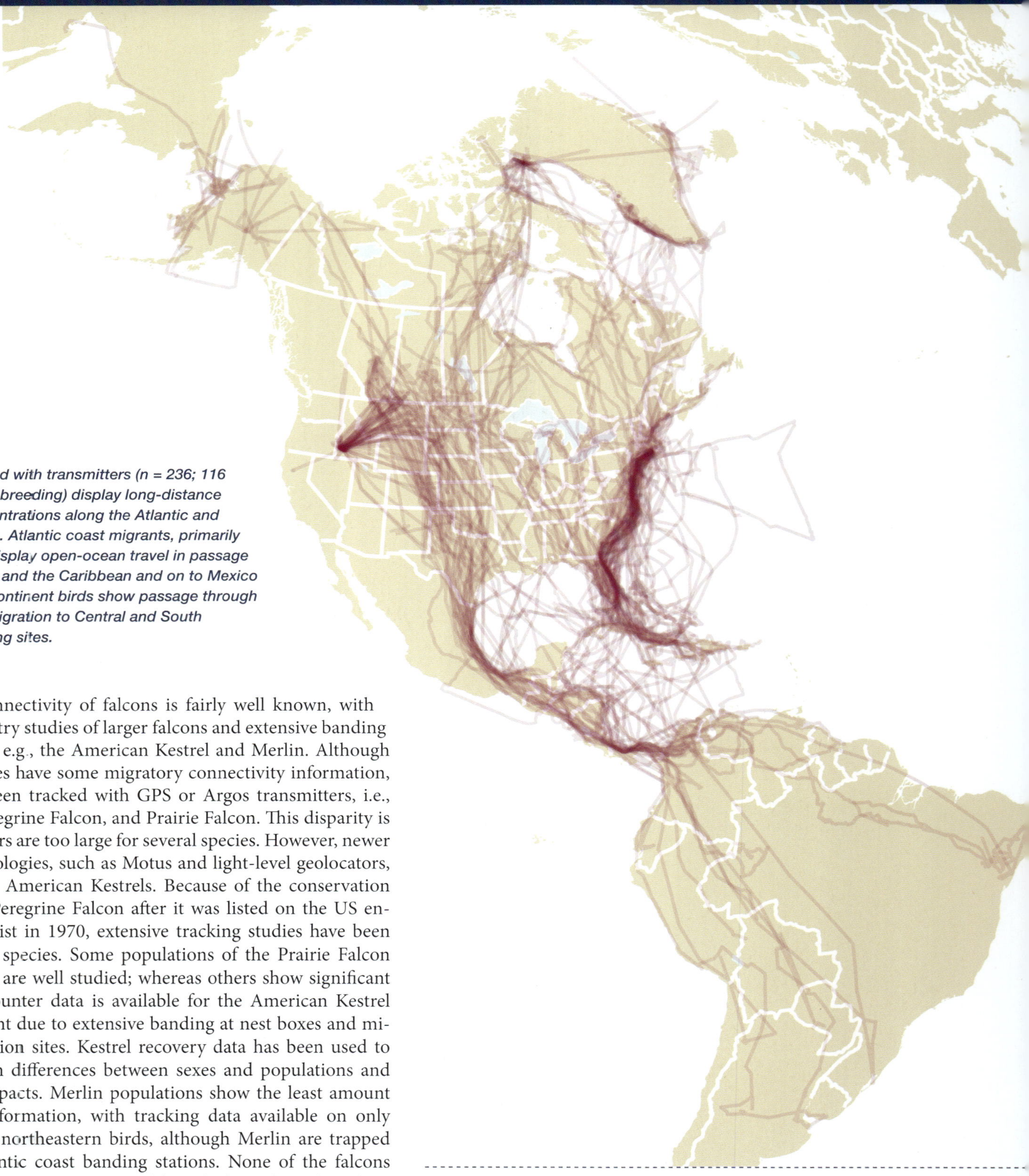

Falconiformes tracked with transmitters (n = 236; 116 from breeding to nonbreeding) display long-distance migration with concentrations along the Atlantic and Midcontinent flyways. Atlantic coast migrants, primarily Peregrine Falcons, display open-ocean travel in passage from Florida to Cuba and the Caribbean and on to Mexico and Venezuela. Midcontinent birds show passage through eastern Mexico on migration to Central and South American nonbreeding sites.

The migratory connectivity of falcons is fairly well known, with widespread telemetry studies of larger falcons and extensive banding of smaller species, e.g., the American Kestrel and Merlin. Although all of the six species have some migratory connectivity information, only three have been tracked with GPS or Argos transmitters, i.e., the Gyrfalcon, Peregrine Falcon, and Prairie Falcon. This disparity is because transmitters are too large for several species. However, newer and smaller technologies, such as Motus and light-level geolocators, are being used on American Kestrels. Because of the conservation concerns for the Peregrine Falcon after it was listed on the US endangered species list in 1970, extensive tracking studies have been conducted on the species. Some populations of the Prairie Falcon and the Gyrfalcon are well studied; whereas others show significant gaps. Band reencounter data is available for the American Kestrel across the continent due to extensive banding at nest boxes and migration concentration sites. Kestrel recovery data has been used to examine migration differences between sexes and populations and climate change impacts. Merlin populations show the least amount of connectivity information, with tracking data available on only small numbers of northeastern birds, although Merlin are trapped frequently in Atlantic coast banding stations. None of the falcons are currently Tipping Point species, although the Aplomado Falcon is listed as threatened by the US. Peregrine Falcon populations are still recovering but no longer listed as of the 2000s, and the Gyrfalcon and American Kestrel are species for which conservation concerns are warranted, with climate and landscape change impacting their numbers. In some states, the American Kestrel is a state-listed species due to widespread, long-term declines in numbers. For the Peregrine Falcon and American Kestrel, a leapfrog migration pattern is shown, with the northernmost nesting birds spending the nonbreeding season the farthest south and therefore having the longest migrations.

Migratory connectivity is weak among most falcons, with overlap of breeding populations across nonbreeding regions. However, American Kestrel migratory connectivity is moderate (0.33 ± 0.05), with most breeding populations migrating to nonbreeding areas within the breeding flyway, reducing overlap.

Falco sparverius **AMERICAN KESTREL**

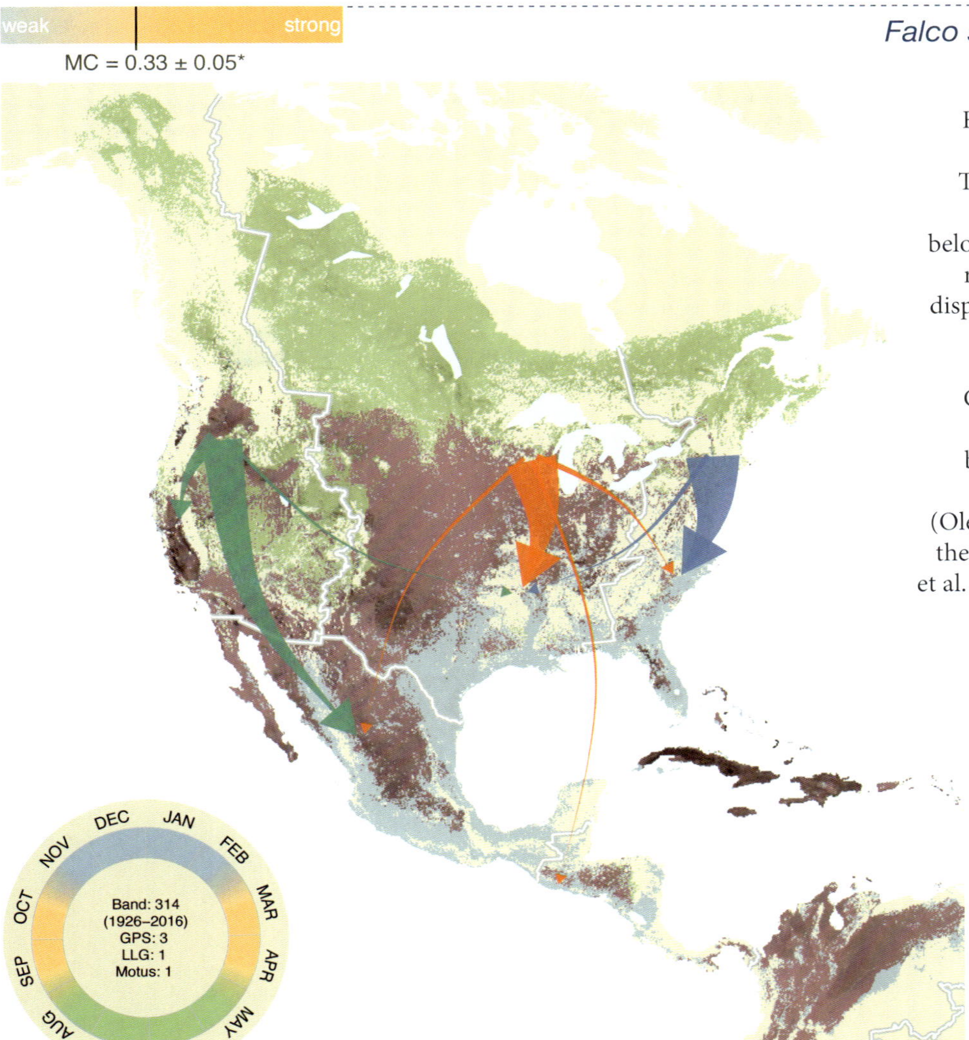

GENERALIZED CONNECTIVITY

DISTRIBUTION AND ABUNDANCE

Found throughout North and South America from Central Alaska east across Canada, and south to Tierra del Fuego, it is absent only from large forests of the Amazon. Partial migrants, kestrels nesting below 35 degrees north latitude are largely year-round residents, whereas the most northern nesting birds display leapfrog migration, migrating the farthest into southern US states, Mexico, and Central America for the nonbreeding season (Hobson et al. 2009; Goodrich et al. 2012). Declining population trends of up to 3% per year over the last thirty years have been documented on post-breeding migration and nonbreeding season counts across North America (Oleyar et al. 2023). Counts of western kestrels during the nonbreeding season have been increasing (Oleyar et al. 2023). In contrast, both eastern and midcontinent populations continue to show declines on recent nonbreeding and post-breeding migration counts (Oleyar et al. 2023).

ORDER Falconiformes
FAMILY Falconidae
BREEDING HABITAT Open Country, Habitat Mosaic
NONBREEDING HABITAT Open Country, Habitat Mosaic
GLOBAL POPULATION SIZE 9,200,000
POPULATION CHANGE SINCE 1970 −42%
IUCN RED LIST CATEGORY Least Concern
R2R TIPPING POINT SPECIES No
US LISTING STATUS No
CANADA SPECIES AT RISK No
PIF CCS 10/20

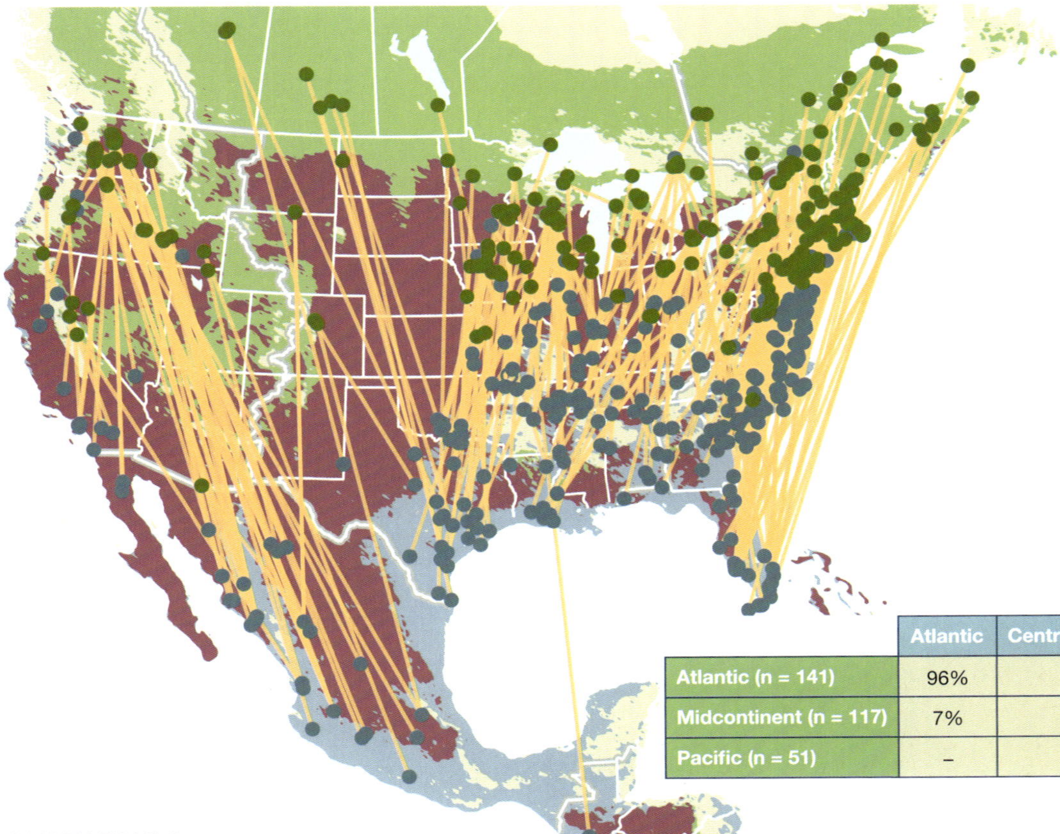

	Atlantic	Central America	Mexico	Midcontinent	Pacific
Atlantic (n = 141)	96%	–	–	4%	–
Midcontinent (n = 117)	7%	2%	3%	89%	–
Pacific (n = 51)	–	–	69%	2%	29%

CONNECTIONS A

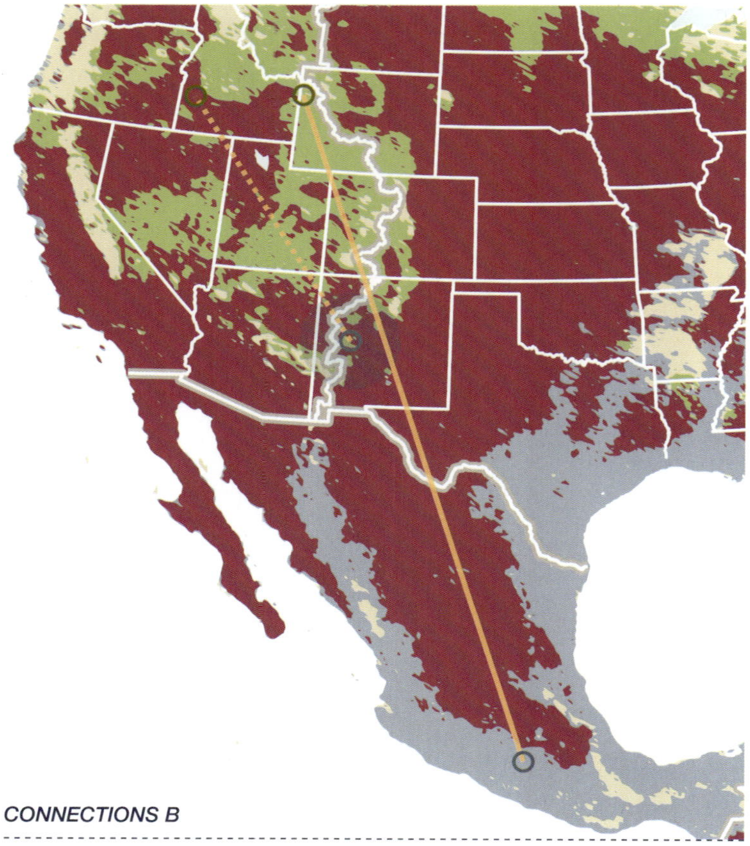

CONNECTIONS B

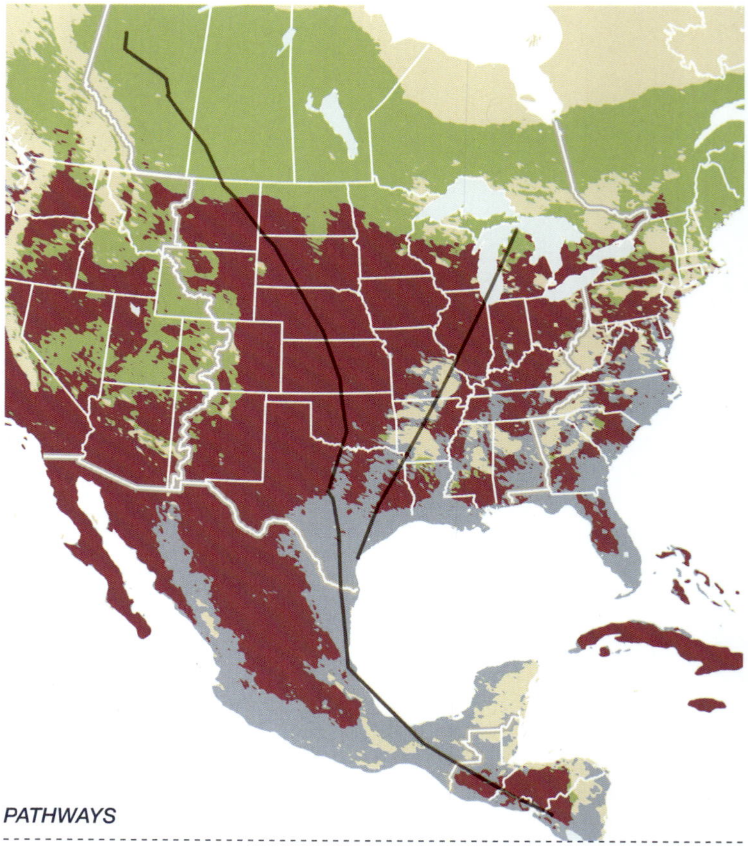

PATHWAYS

MIGRATORY CONNECTIVITY

OVERALL There are limited tracking studies, but numerous band reencounter data suggest moderate migratory connectivity (0.33 ± 0.05) and a leapfrog migration pattern for American Kestrels. Atlantic (96%) and Midcontinent (89%) flyway breeding birds remain within the same flyway during the nonbreeding season, while Pacific Flyway breeding populations migrate to Mexico (69%) or remain within the Pacific flyway (29%) during the nonbreeding season. Smaller numbers are found in other regions.

BANDING Band reencounters show a longitudinal pattern of migration and connectivity, including one study from the western US suggesting 85% of kestrels migrate, with most going to western Mexico (Henny and Brady 1994).

RADIO TRACKING One kestrel breeding in Minnesota near the confluence of the St. Croix and Mississippi rivers was detected again during the nonbreeding season just south of Springfield, Missouri. Several projects tracking American Kestrels with the Motus Wildlife Tracking System suggest different migration routes.

GEOLOCATOR TRACKING Light-level geolocator data found one kestrel traveled from a breeding area in Idaho to a nonbreeding area in western New Mexico, while another five kestrels did not migrate (Hunt et al. 2023).

SATELLITE TRACKING Only 4 of 22 satellite transmitters deployed on female kestrels resulted in data across multiple seasons (Hunt et al. 2023). One kestrel was tracked from its breeding grounds in Alberta to the nonbreeding grounds in Nicaragua, while another kestrel was tracked from its nonbreeding grounds in Texas and traveled at least as far north as Michigan (Hunt et al. 2023). An additional two kestrels were tracked from breeding areas in Saskatchewan and went to South Dakota and Texas, respectively, before both transmitters failed (Hunt et al. 2023). Tracking of kestrels from a breeding population in the Piedmont region of Virginia is ongoing (J. Kolowski, unpublished data). A male kestrel deployed with a GPS tag on the breeding grounds in Teton County, Wyoming, migrated to the state of Guerrero, Mexico, approximately 3,000 km due south from its breeding territory (Crandall and Craighead 2019).

STABLE ISOTOPES The majority of kestrels (>86%) sampled on nonbreeding areas in South Carolina and Tennessee were migrants, but breeding origin could not be identified (Hobson et al. 2009). In a study distinguishing migrant from resident kestrels in a suspected partially migratory population in southwestern Idaho, the proportion of migrants varied between years, with male kestrels almost twice as likely to migrate than females, and in colder than average winters, smaller birds of both sexes were more likely to migrate than larger birds (Ranck et al. 2023). Only 27% of 26 recaptured individuals showed evidence of switching their migration strategies on an annual basis (Ranck et al. 2023).

MIGRATORY PATHWAYS Migrating American Kestrels concentrate along coastlines and ridges, with thousands counted annually along the Atlantic coast (>6,000 in Cape May, New Jersey), the Great Lakes, and Veracruz, Mexico (>7,000; Ruelas Inzunza 2005) during southward migration (Goodrich and Smith 2008). Lower concentrations of kestrels are observed migrating along the Appalachian Mountains, the west coast, and the Rocky Mountains. Consistent post-breeding migration counts in the Florida Keys averaging 800 kestrels per season suggest some may migrate into the Caribbean (Goodrich and Smith 2008).

INTERESTING NOTE

In north Texas, 52% of marked kestrels were resighted the following year after capture (K. Biles and J. Bednarz, unpublished data). In addition, of the kestrels fitted with tracking devices, 50% returned the following nonbreeding season, similar to the return rates for banded kestrels without tracking devices (Biles et al. 2023).

DATA CONTRIBUTORS

K. Hall; J.A. Heath; A.R. Hunt; R.R. Porter; J.L. Watson; J.M. Winiarski.

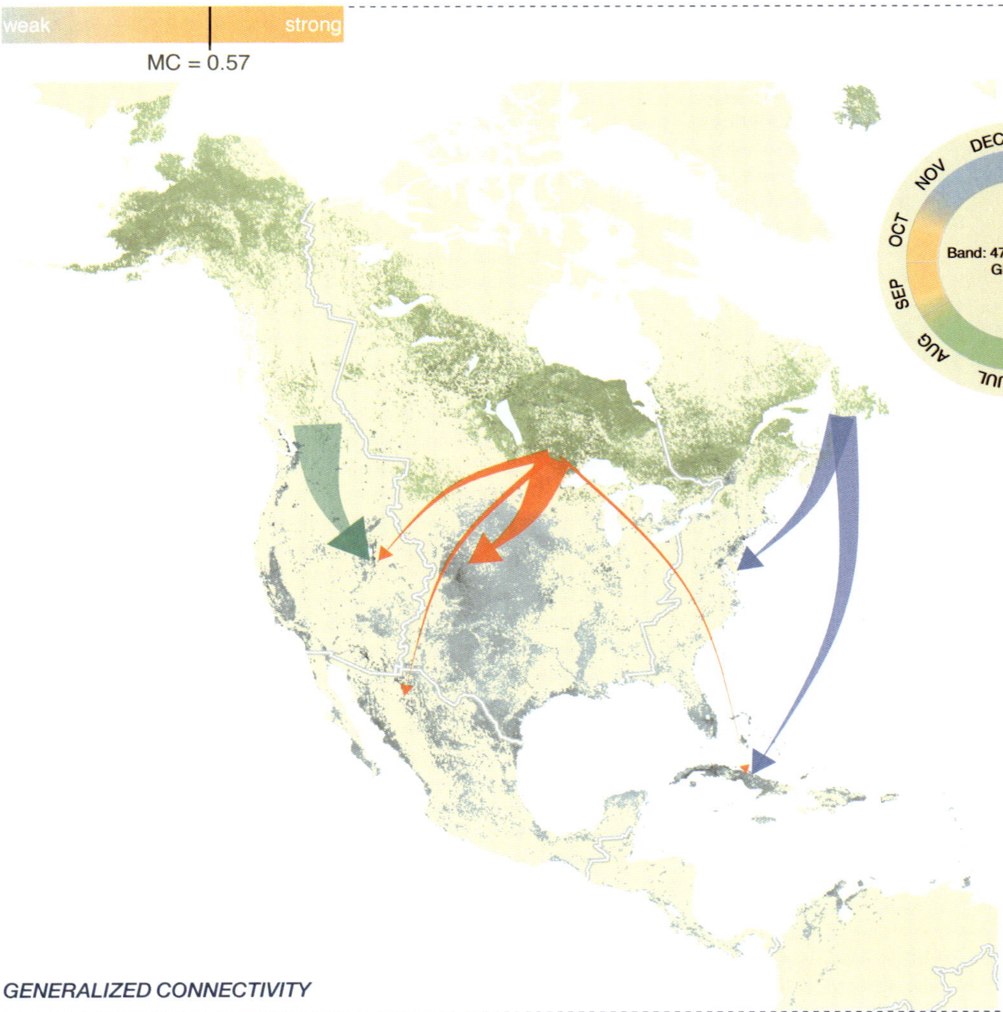

Falco columbarius MERLIN

DISTRIBUTION AND ABUNDANCE

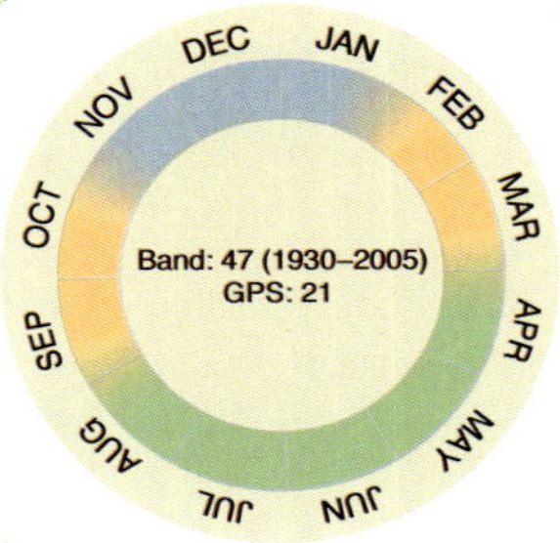

The Merlin is a small falcon that breeds from Alaska east across Canada from the tree line south into the northern US. Recently the breeding range has expanded south along the Great Lakes and east into New England states, as well as into urban and suburban areas of southern New York and throughout Pennsylvania. The migration pattern of the three North American subspecies differs considerably but appears to be a leapfrog style. Black Merlins (*F. c. suckleyi*) of the Pacific Northwest are typically sedentary, with southward movements restricted to more northern populations that generally migrate to southern parts of the subspecies' breeding range. Prairie Merlins (*F. c. richardsonii*) are partial migrants, with nonbreeding areas in the south-central US and northern Mexico. Taiga Merlins (*F. c. columbarius*) are complete migrants spending the nonbreeding season across the western and coastal southeastern US, Mexico, the Antilles, and Central and South America (Warkentin et al. 2020).

GENERALIZED CONNECTIVITY

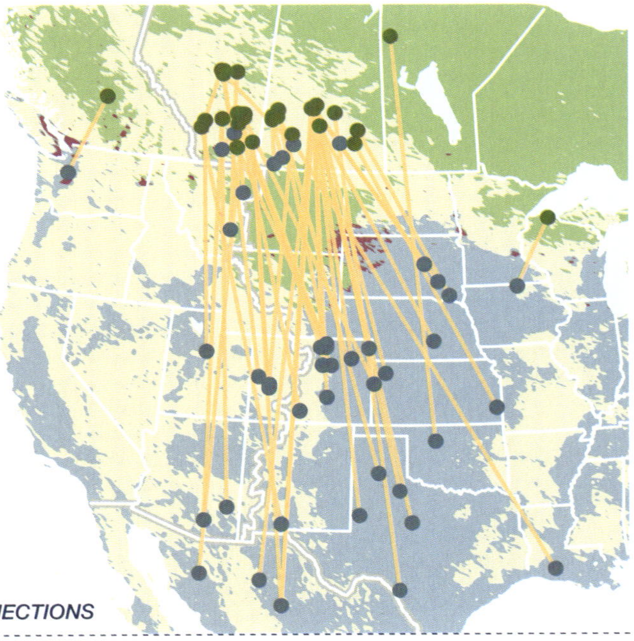

CONNECTIONS

ORDER Falconiformes
FAMILY Falconidae
BREEDING HABITAT Forests, Boreal
NONBREEDING HABITAT Open Country, Habitat Mosaic
GLOBAL POPULATION SIZE 3,200,000

POPULATION CHANGE SINCE 1970 +381%
IUCN RED LIST CATEGORY Least Concern
R2R TIPPING POINT SPECIES No
US LISTING STATUS No
CANADA SPECIES AT RISK No
PIF CCS 8/20

MIGRATORY CONNECTIVITY

OVERALL Merlin migratory connectivity is known from one tracking study in the east and banding data in the west and appears strong (MC = 0.57). Atlantic breeding populations migrate within the Atlantic Flyway (50%) or to the Caribbean (50%) during the nonbreeding season but rarely mix with Midcontinent and Pacific Flyway breeding populations. Midcontinent breeding populations primarily spend the

	Atlantic	Caribbean	Mexico	Midcontinent	Pacific
Atlantic (n = 6)	50%	50%	–	–	–
Midcontinent (n = 48)	–	2%	10%	71%	17%
Pacific (n = 1)	–	–	–	–	100%

nonbreeding period within the Midcontinent Flyway (71%) but may mix with Pacific breeding populations in the Pacific Flyway (17%).

BANDING Two banding studies suggest strong connectivity with birds from northeastern US and eastern Canada primarily migrating to nonbreeding areas in Florida and the Greater Antilles (Clark 1985), birds from Alberta and Saskatchewan migrating to Colorado (Schmutz et al. 1991), and Wisconsin breeders migrating to Colombia, Ecuador, and the Dominican Republic (Clark 1985). Band recoveries continent-wide suggest western Canadian birds spend the nonbreeding season in areas from the northern US south through Mexico and Central America. More northern breeding birds appear to migrate into Central and South America from west and central Canada, while eastern Canada birds migrate to the Caribbean.

RADIO TRACKING Data from Merlin tagged at Block Island, Rhode Island, in 2014–2016 suggest some individuals migrate inland to nonbreeding areas along the Atlantic coast or in other areas where automated telemetry arrays were not present to confirm nonbreeding locations (DeSorbo et al. 2022).

SATELLITE TRACKING Of 17 females (3 adults, 14 hatch-year) tagged during post-breeding migration in Rhode Island, 14 were followed to nonbreeding areas south of the US: Cuba (n = 5) and Puerto Rico (n = 1); and within the US: Connecticut (n = 2), Rhode Island (n = 2), Virginia (n = 1), North Carolina (n = 1), and the Delmarva Peninsula (n = 2; DeSorbo et al. 2022). Birds tracked back to breeding areas returned to eastern Quebec, Newfoundland, and New Brunswick (DeSorbo et al. 2022). Adult Merlins (4 females and 2 males) tracked from the Great Lakes showed that some are year-round residents (n = 2) and some are long-distance migrants (3 females and 1 male; F.G. López and N.W. Cooper, unpublished data). All migrants first made short movements during August and September (south of Michigan, Ohio, and Wisconsin) and then departed during the fall (October 15–November 15) to nonbreeding areas in the south: Alabama (n = 1), Florida (n = 1), and Cuba (n = 2). One female migrated to Cuba in two consecutive years (F.G. López and N.W. Cooper, unpublished data).

MIGRATORY PATHWAYS Eastern Merlin concentrate on the Atlantic coast in large numbers during migration (Clark 1985; Warkentin et al. 2020). Post-breeding migration counts in Cape May, New Jersey, surpass any other hawk watch, averaging 1,805 Merlins annually (Goodrich and Smith 2008). Birds tracked with satellite telemetry from Block Island, Rhode Island, consistent with other limited data from along the Atlantic seaboard, indicated Merlins use offshore habitats during migration. Satellite telemetry also documented an open-water crossing from Maine to Nova Scotia during the northward migration of one individual. Merlins may also use both inland and coastal migration corridors. Merlins tracked from the Great Lakes used the Appalachian corridor (F.G. López and N.W. Cooper, unpublished data). Motus detections confirmed primarily southern movements from Block Island, but both easterly and westerly movements occurred for some individuals after release (DeSorbo et al. 2022). During post-breeding migration, females precede males on migration, while males precede females during pre-breeding migration.

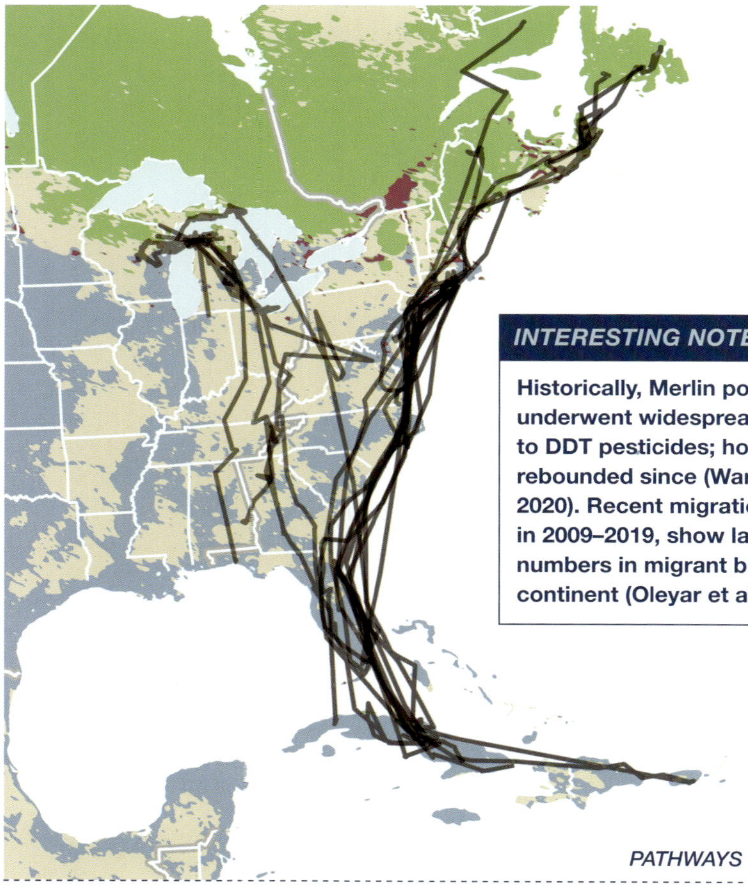

PATHWAYS

INTERESTING NOTE

Historically, Merlin populations underwent widespread declines due to DDT pesticides; however, they have rebounded since (Warkentin et al. 2020). Recent migration count trends, in 2009–2019, show largely stable numbers in migrant birds across the continent (Oleyar et al. 2021).

DATA CONTRIBUTORS
N.W. Cooper; C.R. DeSorbo; F.G. López.

APLOMADO FALCON *Falco femoralis*

DISTRIBUTION AND ABUNDANCE

The Aplomado Falcon is a medium-sized falcon found from southwestern Texas and New Mexico south through Central America and more frequently in northern South America and south of the Amazon basin across South America (Keddy-Hector et al. 2020). Mostly resident throughout its range, the Aplomado Falcon is reported to be a partial or altitudinal migrant, withdrawing from some areas in some regions (Bildstein 2004; Keddy-Hector et al. 2020).

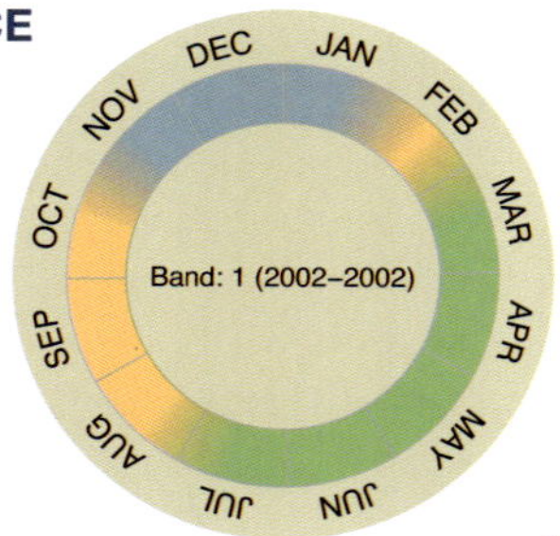

ORDER Falconiformes
FAMILY Falconidae
BREEDING HABITAT Grasslands, Tropical
NONBREEDING HABITAT Grasslands, Tropical
GLOBAL POPULATION SIZE 200,000
POPULATION CHANGE SINCE 1970 NA
IUCN RED LIST CATEGORY Least Concern
R2R TIPPING POINT SPECIES No
US LISTING STATUS Threatened northern subspecies
CANADA SPECIES AT RISK No
PIF CCS 13/20

MIGRATORY CONNECTIVITY

OVERALL Migratory connectivity is unknown. Further research is needed for this species.

BANDING There are few records, with one bird from Texas reencountered in central Mexico.

OTHER APPROACHES No tracking, stable isotope, or genetic studies of migratory connectivity were found for this species.

MIGRATORY PATHWAYS One or two birds are sighted annually at the migration watch site in Corpus Christi, Texas, in September (hawkcount.org).

INTERESTING NOTE

Recovery efforts for this species involved large numbers of captive-reared birds being released in grassland and wetland habitats in Texas and New Mexico in the 1990s and 2000s, with surprisingly few successful pairs being established. Habitat protection efforts and public education continues in the United States and Mexico, although declines in numbers in some regions of Mexico remain concerning (Keddy-Hector et al. 2020).

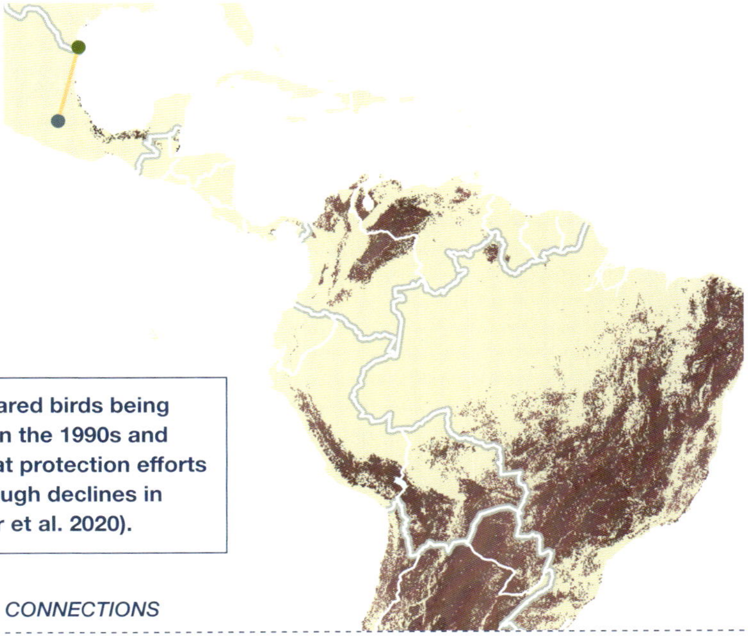

CONNECTIONS

Falco rusticolus GYRFALCON

ORDER Falconiformes

FAMILY Falconidae

BREEDING HABITAT Tundra, Arctic

NONBREEDING HABITAT Open Country, Habitat Mosaic

GLOBAL POPULATION SIZE 83,000

POPULATION CHANGE SINCE 1970 −27%

IUCN RED LIST CATEGORY Least Concern

R2R TIPPING POINT SPECIES No

US LISTING STATUS No

CANADA SPECIES AT RISK No

PIF CCS 12/20v

DISTRIBUTION AND ABUNDANCE

The Gyrfalcon, the largest of the *Falco* genus, breeds across the Arctic and subarctic regions from Alaska east to Greenland. Gyrfalcons consistently spend the nonbreeding season in Alaska and across Canada, with some birds as far south in the northern US as 40 degrees north latitude (Burnham and Newton 2011). High Arctic populations breeding above 70 degrees north latitude are thought to be entirely migratory, while those in the low Arctic are largely resident, with some females and immatures exhibiting partial migration. Data suggest some migrant birds return to nonbreeding sites regularly (Booms et al. 2020). Cyclical patterns in food availability may lead to irruptive movements in some populations as well (Booms et al. 2020).

GENERALIZED CONNECTIVITY

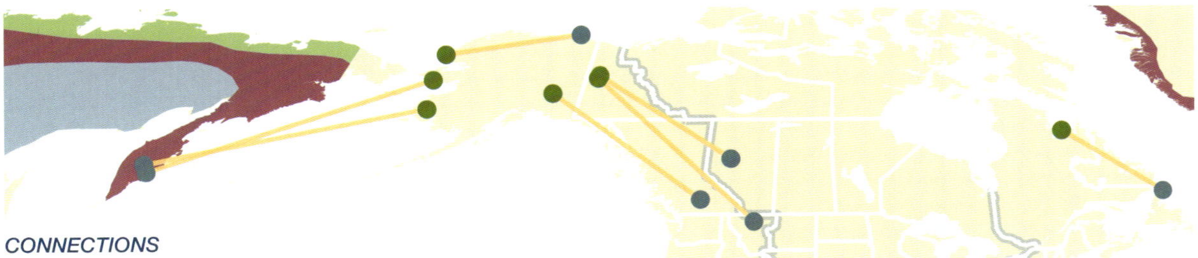

CONNECTIONS

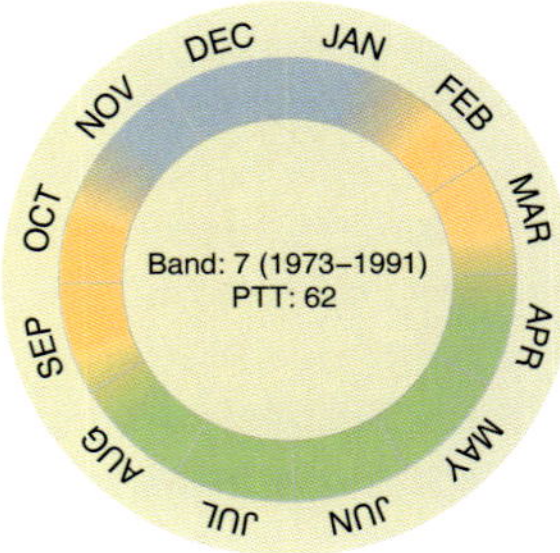

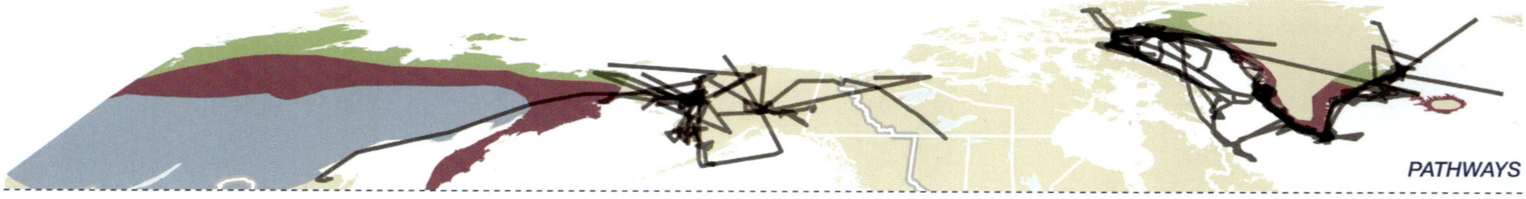

MIGRATORY CONNECTIVITY

	Atlantic	Greenland	Midcontinent	Pacific	Russia
Atlantic (n = 6)	100%	–	–	–	–
Greenland (n = 14)	70%	79%	14%	–	–
Midcontinent (n = 48)	–	50%	–	50%	–
Pacific (n = 14)	–	–	7%	71%	21%

OVERALL Data collected to date suggest moderate to strong migratory connectivity across the North American continent, especially in the eastern part of the range, with 100% of Atlantic Flyway breeding birds and 79% of Greenland breeding birds remaining within the same region during the nonbreeding season. The majority of Pacific Flyway breeding birds remain in the Pacific Flyway (71%), while 21% migrate to Russia for the nonbreeding season. There is variability in the Midcontinent Flyway breeding populations, with birds migrating to either Greenland (50%) or the Pacific Flyway (50%), suggesting that a divide may occur within this flyway. MC is not calculated due to a lack of abundance data; however, appears to exhibit strong migratory connectivity (Mantel 0.8 ± 0.05). ***BANDING*** Banded Gyrfalcons from the Canadian Arctic and subarctic moved 900–2,400 km south into southern Canada for the nonbreeding season (Schmutz et al. 1991). Birds banded in Alaska during the breeding season migrated to a variety of nonbreeding areas, including within Alaska, Russia, and southern British Columbia. Yukon breeding birds migrated to Alberta and Montana. ***SATELLITE TRACKING*** A satellite telemetry study from the east and west coasts of northern Greenland showed Gyrfalcons migrating primarily along the coast to southern Greenland and spending extensive time in the open ocean and on sea ice during the nonbreeding season, presumably preying on seabirds (Burnham and Newton 2011). Juvenile Gyrfalcons tagged on the Seward Peninsula, Alaska, remained in western Alaska or flew to eastern Russia with no movements into interior Alaska. In contrast, Gyrfalcons from Denali National Park remained in interior Alaska or flew to northern and western Alaska, or into northern Alberta (McIntyre et al. 2009). Two birds tagged in southwest Alaska remained nonmigratory (Eisaguirre et al. 2016). Juvenile Gyrfalcons appear more migratory, with recently fledged birds from Alaska traveling between 105 and 4,299 km (McIntyre et al. 2009) and exhibiting highly variable movement patterns, including unidirectional long-distance, multidirectional long- and short-distance, and short-range movements.

INTERESTING NOTE

Gyrfalcon movements appear highly influenced by prey population movements. In South Dakota, researchers report the first Gyrfalcon nonbreeding sightings correlate with increased waterfowl numbers (Sanchez 1993; Booms et al. 2020).

DATA CONTRIBUTORS

K.K. Burnham; J. Duxbury; G.L. Holroyd; H.E. Trefry; C. McIntyre.

PEREGRINE FALCON *Falco peregrinus*

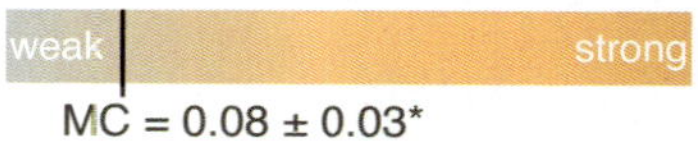

DISTRIBUTION AND ABUNDANCE

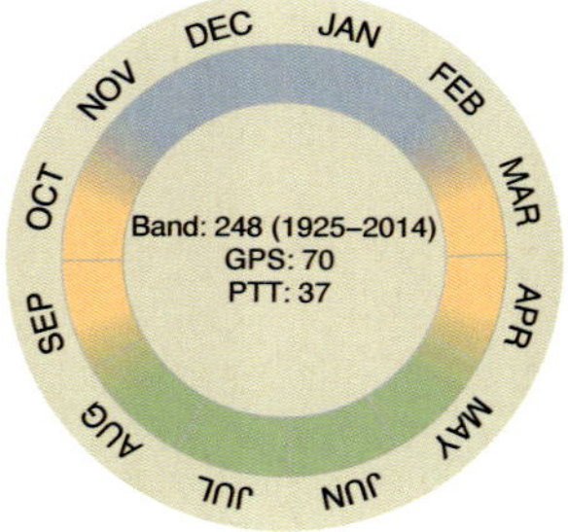

The Peregrine Falcon is the most widely distributed raptor in the world, occurring from the tundra to the tropics and on six continents. The most highly migratory race is the North American tundra peregrine (*F. p. tundrius*), which breeds across the North American Arctic from Alaska to Baffin Island, Greenland. In North America, Peregrine Falcons breed in high abundance in the Arctic from the Aleutian Islands and Alaska east across Canada to Quebec, and to the west coast of Greenland. It is a locally common breeder in the west, especially along the coasts of California and western Mexico, west through the prairies, and south into Mexico, and is patchily distributed in the eastern US and Canada, often using cliffs or city buildings and bridges. Peregrines spend the nonbreeding season primarily along coastal regions of the US (Pacific, Atlantic, and Gulf), Mexico, Cuba, and Central and South America. Year-round populations are found in southern Alaska, including the Aleutian Islands, the western US, the eastern mid-Atlantic region, and western Mexico. Peregrine populations crashed precipitously in 1950–1970, due to contamination from DDT and other organochlorine pesticides (Bednarz et al. 1990; White et al. 2020). By the 1960s, Peregrines were extirpated east of the Rocky Mountains in the US and southern Canada (White et al. 2020) and were listed as

ORDER Falconiformes
FAMILY Falconidae
BREEDING HABITAT Open Country, Habitat Mosaic
NONBREEDING HABITAT Open Country, Habitat Mosaic
GLOBAL POPULATION SIZE 340,000

POPULATION CHANGE SINCE 1970 +973%
IUCN RED LIST CATEGORY Least Concern
R2R TIPPING POINT SPECIES No
US LISTING STATUS No
CANADA SPECIES AT RISK No
PIF CCS 9/20

	Andean West	Atlantic	Brazil	Caribbean	Central America	Mexico	Midcontinent	Northern SA	Pacific	Southern Cone
Atlantic (n = 85)	–	75%	1%	6%	1%	–	14%	1%	–	1%
Greenland (n = 7)	14%	43%	–	–	–	29%	–	–	–	14%
Midcontinent (n = 123)	1%	11%	3%	7%	2%	15%	46%	6%	2%	8%
Pacific (n = 60)	5%	–	13%	–	8%	15%	8%	2%	45%	3%

endangered in 1973 in the US (under the Endangered Species Act) and in 1978 in Canada. Thanks to the effectiveness of that legislation, as well as other recovery strategies, including hacking programs and the banning of DDT, Peregrines were delisted in 1999 in the US and in 2007 in Canada, and are now considered a species of least concern. North American populations are reportedly stabilizing or increasing (5.7% annually, 95% confidence interval of 3.9%–7.3%; from Breeding Bird Survey counts 1970–2017, Sauer et al. 2019; Oleyar et al. 2021); however, in the last few years, significant declines are being reported in several breeding populations (Yukon River, Montana, New Mexico, Nevada), as well as in migration stopover sites (Assateague Island, Maryland; Padre Island, Texas), launching a concerted survey effort (M. Yates, personal communication).

GENERALIZED CONNECTIVITY

MIGRATORY CONNECTIVITY

OVERALL Peregrine Falcons are partial migrants; birds undergo short-distance, long-distance, or no migration. They are considered to have a leapfrog migration, with northern birds experiencing the longest migration from the Arctic to South America, and populations within the midlatitudes experiencing a short-distance migration. The Greenland breeding population migrates to nonbreeding areas in the Atlantic Flyway (43%), Mexico (29%), the Andean West (14%), and the Southern Cone (14%). The majority of Atlantic (75%), Midcontinent (46%), and Pacific (45%) flyway breeding birds were found at nonbreeding areas within the breeding flyway; however, birds are also spread out across numerous (7–10) nonbreeding regions. Mixing of breeding populations throughout nonbreeding areas results in weak migratory connectivity (MC = 0.08 ± 0.03).

BANDING Banding data confirm connectivity patterns shown with satellite tracked birds. Banded birds from northwest Canada (67%; Schmutz et al. 1991) and Alaska migrated to South America, while birds in the northern US migrated to the southern US, Mexico, and the Caribbean, consistent with a leapfrog migration pattern.

SATELLITE TRACKING Starting in 1993, the first satellite transmitters, weighing 27 g, were attached to Tundra Peregrine Falcons in five locations in North America and one location in the western Russian Arctic to identify breeding, migratory, and nonbreeding areas for the conservation of Peregrines (Seegar et al. 1996). Since then, hundreds of satellite transmitters have been deployed across the range, a sample of which are presented in the Atlas. These early studies tracked Tundra Peregrine Falcons captured at nest sites across the Alaskan and Canadian boreal forest and low Arctic or on the migration routes along the mid-Atlantic and Gulf of Mexico coasts to nonbreeding areas as far north as the US mid-Atlantic coast and as far south as central Argentina (Fuller et al. 1998). In the early 2000s, the Falcon Research Group's Southern Cross Peregrine Project satellite tracked some of the longest migrating peregrines in the Western Hemisphere. Tagged on the nonbreeding grounds on the coast of Chile, these Peregrines migrated back and forth from Chile's coast to the tundra cliffs of northern Canada. Earthspan also tracked peregrines breeding in Greenland to nonbreeding areas in Central America, and birds breeding in the tundra of western North America to Tamaulipas, Mexico (Earthspan, unpublished data). Moreover 79% of birds trapped while on nonbreeding areas in Tamaulipas, Mexico, bred in the tundra of western Canada (McGrady et al. 2002). More tracking by Earthspan linked breeding populations in Alaska to Brazil and Argentina, and from Quebec to Colombia. Birds tracked from breeding populations in Alberta also migrated to Colombia, as well as Mexico (G. Holroyd, unpublished data). Peregrines tracked from northern Greenland migrated to Florida, Mexico, Colombia, Paraguay, and Brazil (K. Burnham, unpublished data; C. DeSorbo, unpublished data) and those breeding in southern Greenland migrated to the northeastern US (K. Burnham, unpublished data). Birds tracked from New

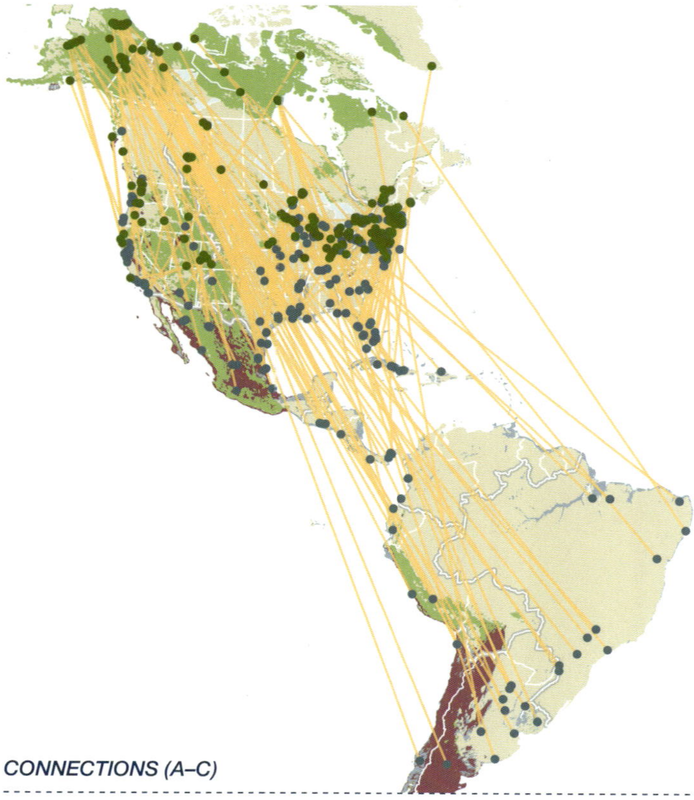

CONNECTIONS (A–C)

Hampshire migrated to Pennsylvania and Connecticut (C. DeSorbo, unpublished data), while birds tracked within the eastern Great Lakes region near Lakes Erie, Huron, and Ontario remained in that area throughout the year (Pokrovsky et al. 2023). Satellite data and capture locations of birds trapped in more than one year suggest fidelity to nonbreeding areas, although perhaps not particular home ranges (McGrady et al. 2002).

Juvenile Peregrine Falcons have also been tracked using satellite transmitters. More than 50% of Peregrine Falcons tracked from the mid-Atlantic region between 2001 and 2011 migrated south of the mid-Atlantic states to the southern United States and Central America. The nonmigratory birds moved north and east of their fledging sites, most spending the nonbreeding season near water in the northern mid-Atlantic Coastal Plain. Individuals did not change their migratory status between years. Fledging location influenced both migratory route and nonbreeding location; birds from coastal areas were the only birds to migrate to the tropics (Falcon Trak, unpublished data). A fledgling bird from New York was tracked to the coast of Massachusetts during its first nonbreeding season. It then spent the next eight years in the Great Lakes region of the US (New York State Department of Environmental Conservation Raptor Tracking, unpublished data).

GENETICS Genetic studies found that tundra (*F. p. tundrius*) and boreal (*F. p. anatum*) breeders are genetically indistinct, suggesting movement and gene flow between breeding regions or perhaps wide-ranging natal dispersal. The nonmigratory coastal population

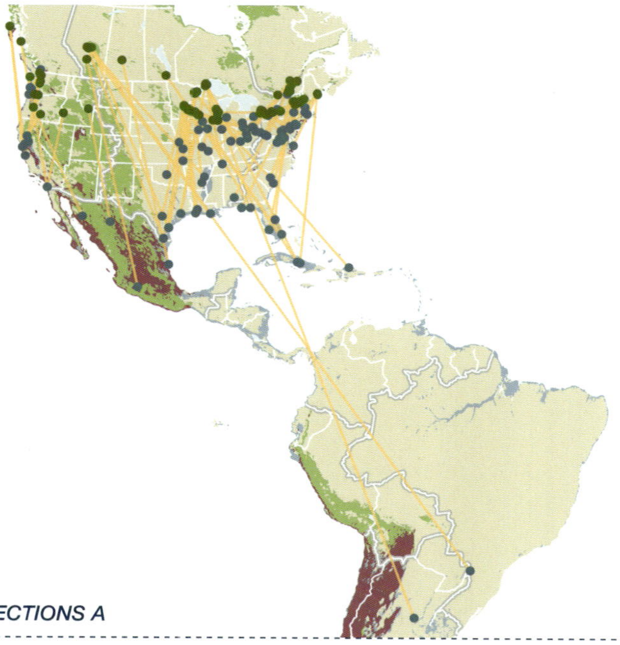

CONNECTIONS A

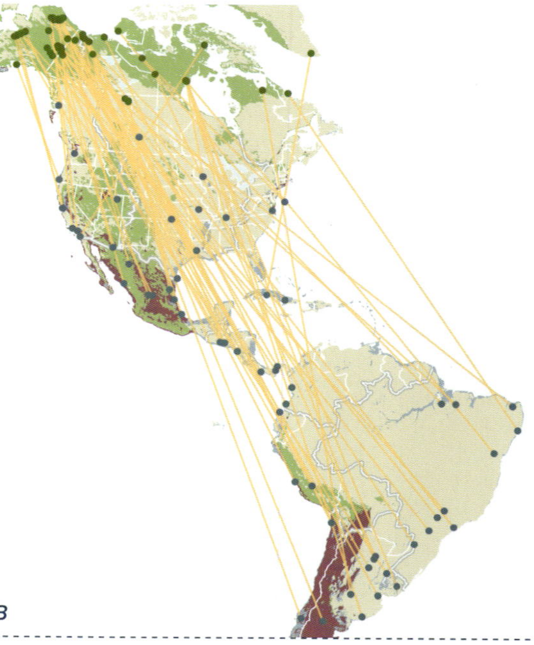

CONNECTIONS B

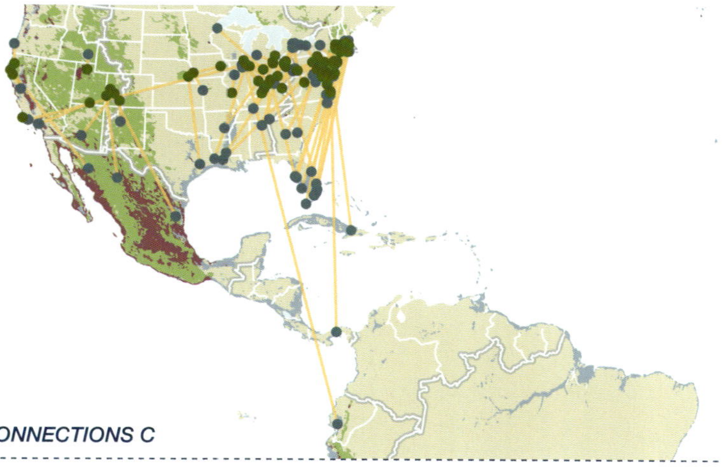

CONNECTIONS C

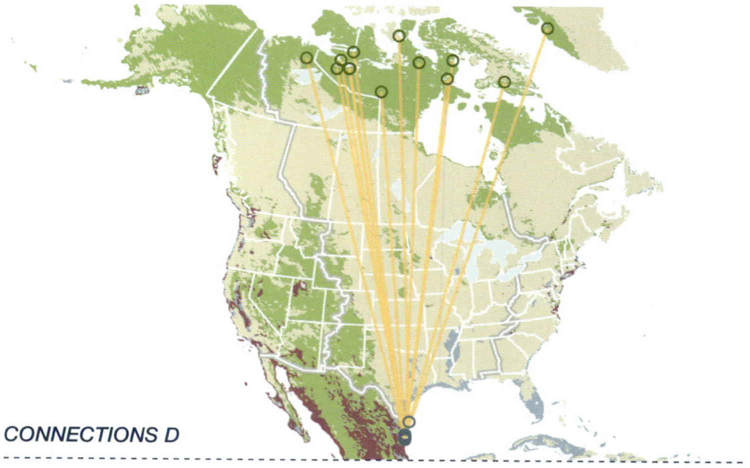

CONNECTIONS D

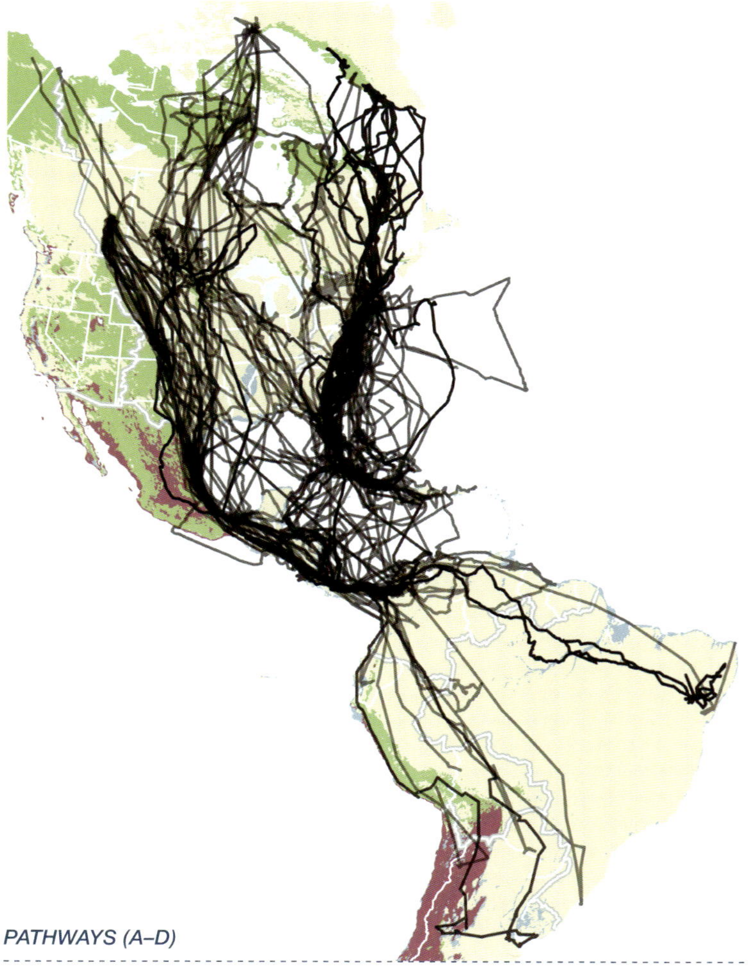

PATHWAYS (A–D)

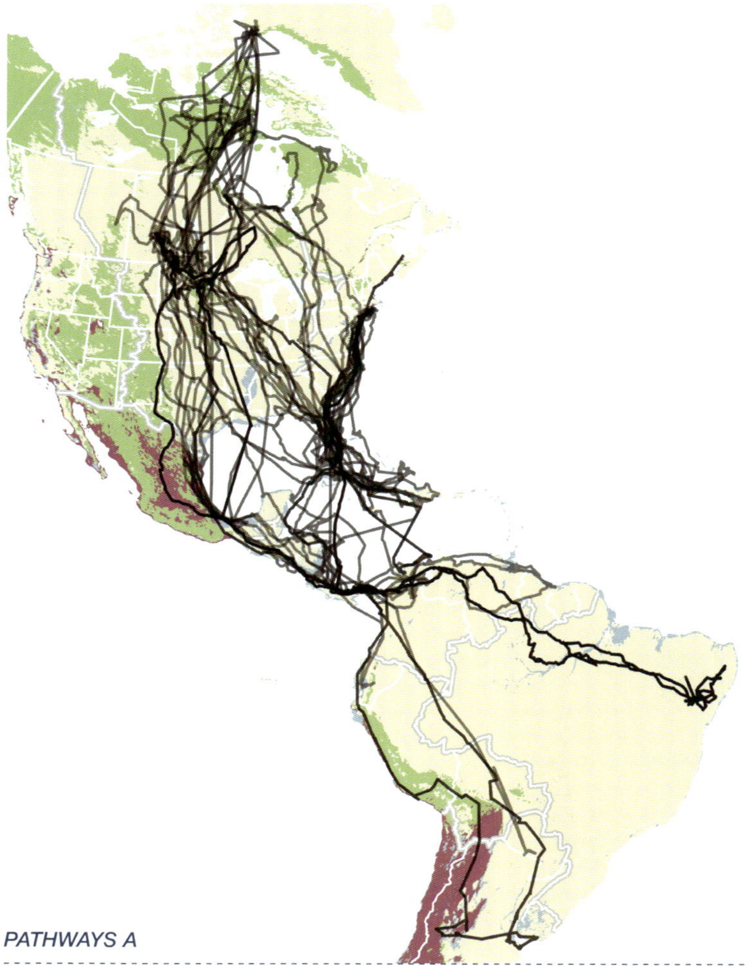

PATHWAYS A

PATHWAYS B

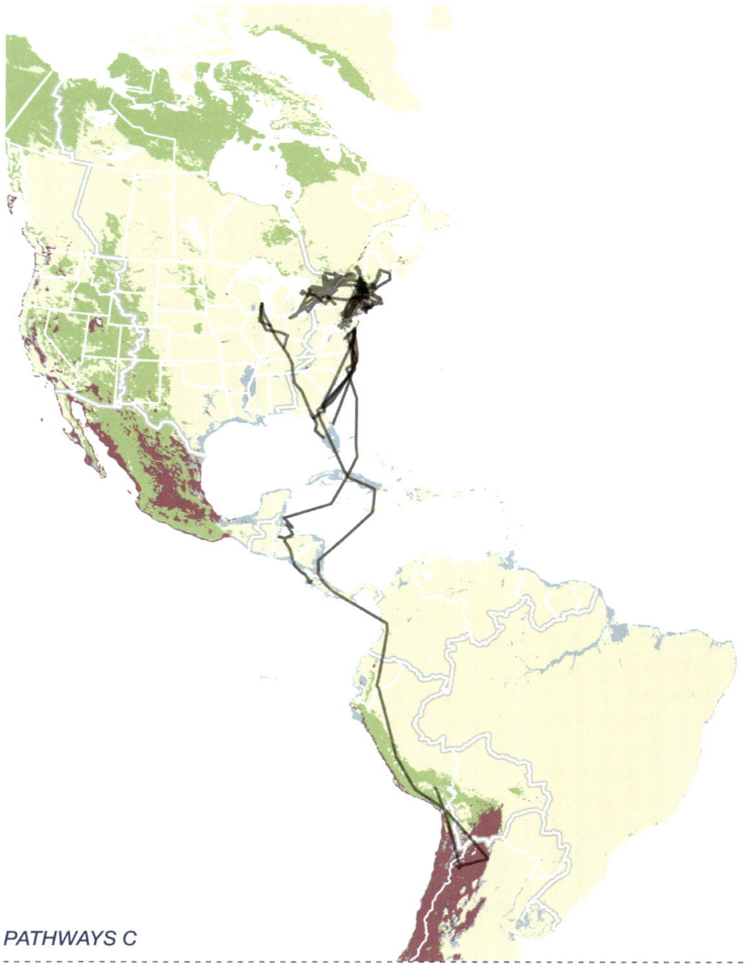

PATHWAYS C

PATHWAYS D

(*F. p. pealei*) is genetically distinct, indicating that these birds are indeed stationary (J.A. Johnson et al. 2010; Talbot et al. 2017). **MIGRATORY PATHWAYS** Tracking data suggests peregrines use three broad routes south from breeding areas: coastal, midcontinental, and water-crossing routes over the Davis Strait and Caribbean Sea (Fuller et al. 1998). Peregrines concentrate in large numbers during southward migration on the Atlantic coast, in the Caribbean, eastern Mexico, and Central America en route to nonbreeding destinations in Central and South America. Hawk watches in the Florida Keys count more than 3,000 migrating falcons each fall (hawkcount.org), presumably crossing the Caribbean on migration. Birds tagged in Alaska and Canada move south through the midcontinent to Mexico and beyond (C. De Sorbo, unpublished data; G. Holroyd, unpublished data; Earthspan, unpublished data). Most peregrines use different routes in pre- and post-breeding. During northward migration, peregrines stopping at Padre Island, Texas, diverge for destinations from central Alaska across the continent to west-central Greenland (Fuller et al. 1998; McGrady et al. 2002).

DATA CONTRIBUTORS

K.K. Burnham; C.R. DeSorbo; J. Duxbury; G.L. Holroyd; R. Kays; N. Lecomte; P. Nye; K.M. O'Brien; I. Pokrovsky; J-F. Therrien; H.E. Trefry; M. Wikelski; M. Yates; New York State Department of Environmental Conservation.

PRAIRIE FALCON *Falco mexicanus*

DISTRIBUTION AND ABUNDANCE

The Prairie Falcon is a medium-sized falcon that breeds from British Columbia, Alberta, and Saskatchewan south into central Mexico and Baja California, with high abundance through the Intermountain West, Badlands, Great Plains (mostly west of 100 degrees west longitude), and interior Southern California. The nonbreeding range overlaps with much of the breeding range, and areas of high abundance include the Intermountain West, southern Great Plains, and interior California to southwestern Arizona (Fink et al. 2023).

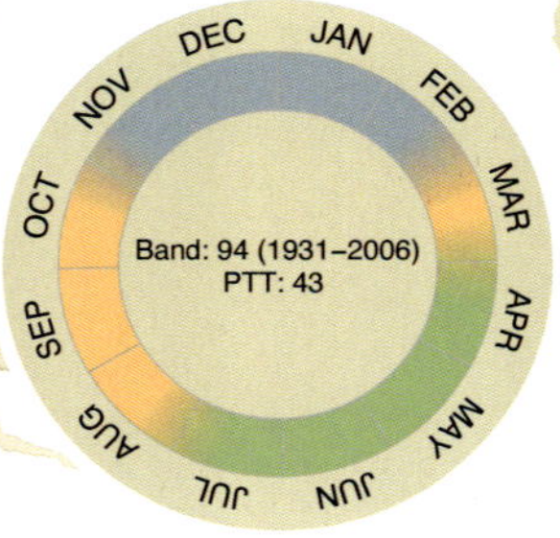

ORDER Falconiformes
FAMILY Falconidae
BREEDING HABITAT Open Country, Habitat Mosaic
NONBREEDING HABITAT Open Country, Habitat Mosaic
GLOBAL POPULATION SIZE 110,000

POPULATION CHANGE SINCE 1970 +80%
IUCN RED LIST CATEGORY Least Concern
R2R TIPPING POINT SPECIES No
US LISTING STATUS No
CANADA SPECIES AT RISK No
PIF CCS 10/20

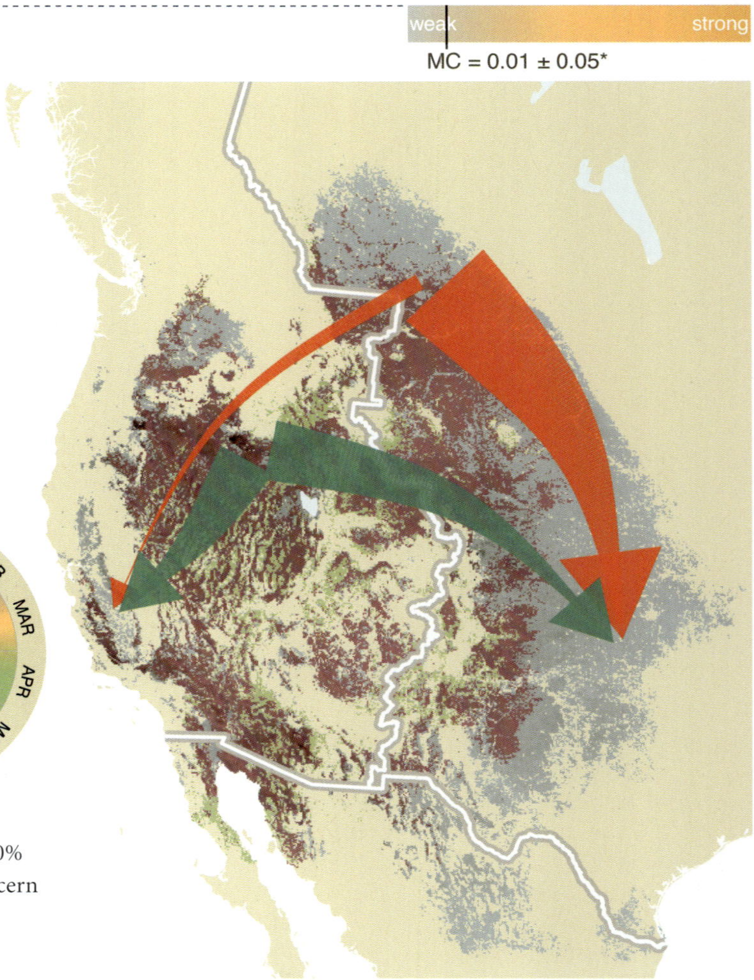

GENERALIZED CONNECTIVITY

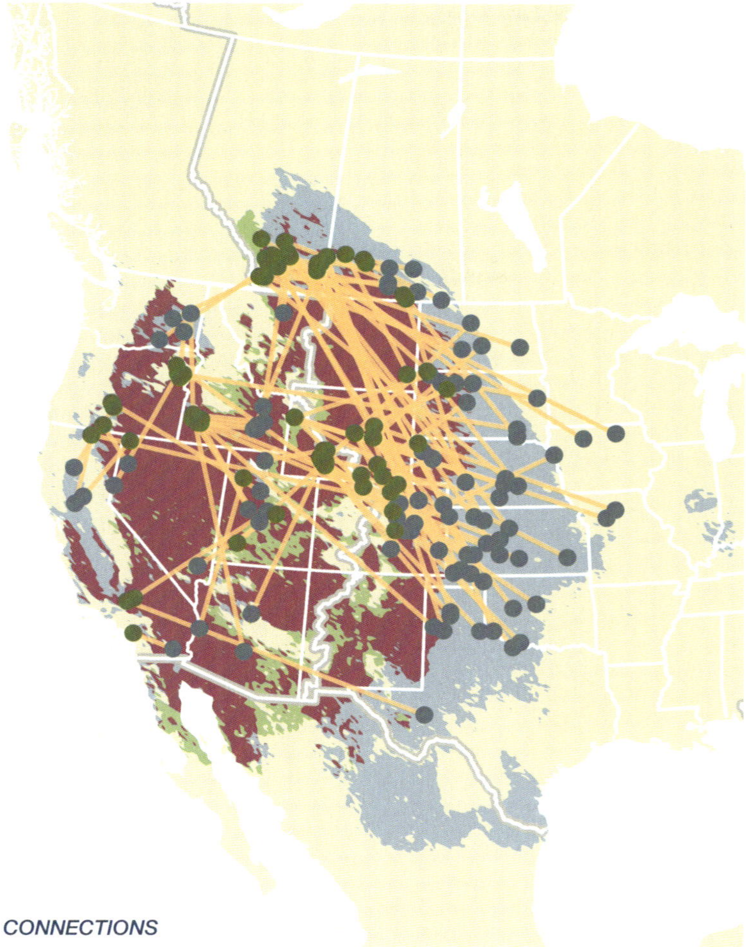

CONNECTIONS

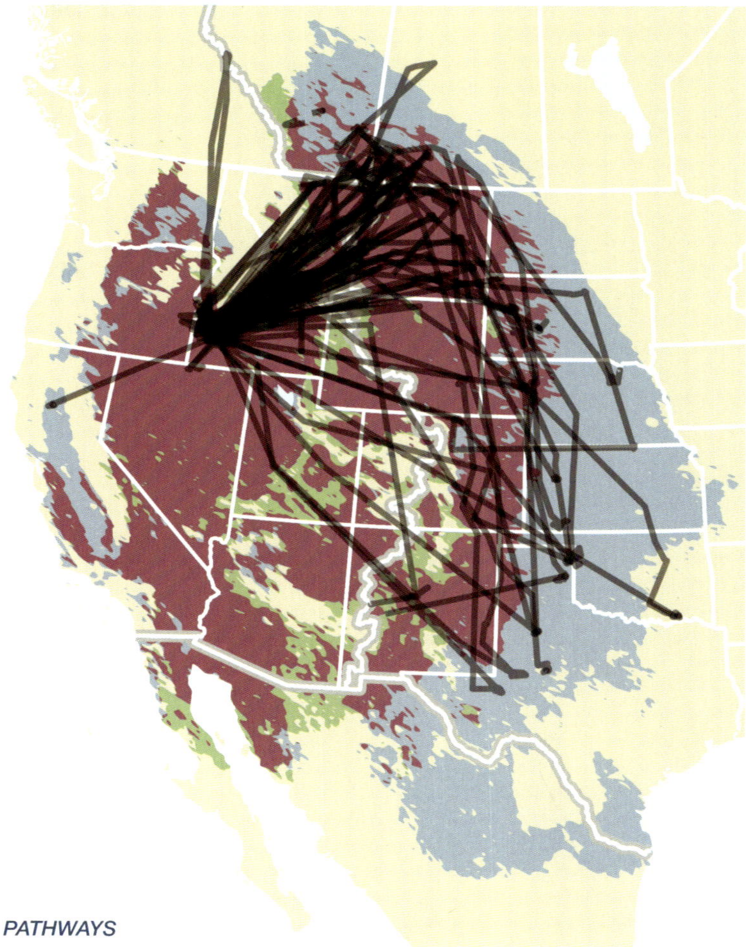

PATHWAYS

	Midcontinent	Pacific
Midcontinent (n = 59)	90%	10%
Pacific (n = 61)	51%	49%

MIGRATORY CONNECTIVITY

OVERALL Prairie Falcon migratory connectivity is known from satellite tracking and banding data. Birds breeding in the Midcontinent Flyway primarily migrate within the Midcontinent Flyway (90%), but may also migrate to the Pacific Flyway (10%) for the nonbreeding season. Pacific Flyway breeding birds migrate within the flyway (49%) or migrate to the Midcontinent Flyway (51%). Mixing of breeding populations on nonbreeding areas results in a weak migratory connectivity (MC = 0.01 ± 0.05)

BANDING Banding data from Canada suggest Prairie Falcons exhibit partial migration, with a handful of nonbreeding recoveries near nest sites; however, most migrated east to Saskatchewan before continuing south through the Great Plains (Schmutz et al. 1991).

SATELLITE TRACKING Prairie Falcons tracked from nesting areas in the Snake River Birds of Prey National Conservation Area in southwest Idaho migrated to nonbreeding areas in the Texas panhandle and southwest Idaho (Steenhof et al. 2005). A few birds moved west from breeding sites into California. Most falcons had three or less use areas during the nonbreeding season, and showed a high degree of fidelity to their use areas during each season, suggesting connectivity to grassland and shrub-steppe habitats (Steenhof et al. 2005). Two of five breeding birds in Alberta were tracked to nonbreeding locations; one migrated to Alabama, while the other remained within Alberta throughout the year (G. Holroyd, unpublished data).

OTHER APPROACHES No stable isotope or genetic studies of migratory connectivity were found for this species. However, individuals sampled in California and Idaho represent a single, panmictic population (Doyle et al. 2018).

MIGRATORY PATHWAYS Tracking studies of Prairie Falcons show a loop pattern of migration and three distinct use areas—breeding, post-breeding, and stationery nonbreeding—that may follow seasonal fluctuations in prey abundance (Enderson 1964; Steenhof et al. 1984, 2005; Schmutz et al. 1991). From nesting areas, falcons moved northeast across the Continental Divide in late summer into the prairies of Montana, Alberta, and the Dakotas for 1–4 months before moving south through the Great Plains to nonbreeding sites (Steenhof et al. 2005; Steenhof 2020). The Great Plains appear to be an important pathway for many southbound Prairie Falcons (Steenhof et al. 2005). Birds nesting in Alberta use two routes south, either the east the slope of the Rocky Mountains or the west side (Schmutz et al. 1991; G. Holroyd, unpublished data).

DATA CONTRIBUTORS

K.K. Bates; J. Duxbury; M.R. Fuller; G.L. Holroyd; M.N. Kochert; K. Steenhof; H.E. Trefry.

INTERESTING NOTE

Prairie Falcons on migration moved an average of 21–299 km per day and at speeds of 13–55 km per hour, very similar to rates observed in tracked Peregrine Falcons (Steenhof et al. 2005). Due to their diet of mostly small mammals, Prairie Falcons did not appear as heavily affected by DDT pesticides as other raptors (Steenhof 2020).

SONGBIRDS (PASSERIFORMES)

WRITING LED BY **KENNETH V. ROSENBERG, TOM WILL, PETER P. MARRA, CAMILLA GÓMEZ, MELANIE A. SMITH, TIMOTHY GUIDA, AMY L. SCARPIGNATO,** *AND* **ALLISON E. HUYSMAN**

Migration pathways for 18 songbird (passerine) species, across 9 families, tracked with GPS transmitters (n = 302) or Argos satellite tags (n = 34). Although representing only 8% of species, collectively these tracks illustrate many of the widespread migration and connections patterns observed within the Passeriformes, including east-west parallel migrations (Painted Bunting, Gray Catbird), eastward post-breeding migration of birds from the western boreal region (Rusty Blackbird), trans-Gulf and trans-Atlantic crossings by birds migrating to the Caribbean (Ovenbird) and Central America (Wood Thrush), movements between the Pacific states and western Mexico (Black-headed Grosbeak, Yellow-breasted Chat), and a unique east-west migration between the Arizona deserts and California coast (Phainopepla). The large-bodied Eastern and Western Meadowlarks (Icteridae) have been studied using both GPS and Argos.

The order Passeriformes is an diverse taxonomic group, constituting 42% of all migratory North American species covered in this Atlas. The 237 passerines, commonly referred to as songbirds, are distributed across 27 families, ranging from several with a single species represented in the Atlas, such as the larks (Alaudidae) and Yellow-breasted Chats (Icteriidae), to families such as the flycatchers (Tyrannidae), sparrows (Passerellidae), and wood warblers (Parulidae). Because the vast majority of passerines are small (<50 g), the tracking data using technologically advanced devices or methods is extremely limited. For example, only 16 species (7%) have been studied using GPS tags, and only 4 species (1 corvid and 3 meadowlark species) have been tracked using Argos. More researchers have taken advantage of archival light-level geolocators to track 53 species (22%), including swallows (5 species), thrushes (8 species), and warblers (18 species). In fact, the desire to study migration patterns of small passerines has helped to spur the development of lighter-weight tracking devices. Not surprisingly, the vast majority of data on migratory connectivity for passerines (84% of all connections) comes from band recoveries. Even so, more than half of the 7,987 breeding-to-nonbreeding band recoveries are from blackbirds (most were trapped to reduce perceived agricultural damage), and sample sizes are small for most species. For example, from the tens of thousands of warblers that have been banded over the past 80 years, we have only 83 recoveries (13 species) that can shed any light on migratory connectivity—an infinitesimally small return on investment. On the other hand, the exponential increase in use of miniaturized tracking devices, including systems such as Motus, even since data-gathering was completed for this Atlas, is encouraging and likely to provide much more data on small passerines in the near future.

For many species that lack tracking and even banding data, geographic variation in morphology or genetics can sometimes provide important insights into migratory connectivity among populations. The ornithological literature has several examples of linked breeding, migratory, and nonbreeding regions from studies of museum specimens that were identified as part of geographically variable subspecies. More recently, genetic variation has been studied directly with whole genomes, with blood or feather samples. Specifically, the Bird Genoscape Project is able to link populations across the annual cycle at finer scales than was possible using morphology alone. The growing list of passerine species with completed year-round genoscapes includes the American Redstart, Canada Warbler, Grasshopper Sparrow, Tricolored Blackbird, Veery, Willow Flycatcher, Yellow-breasted Chat Common Yellowthroat, Wilson's and Yellow Warbler, Hermit and Swainson's Thrush, Willow Flycatcher, and Painted Bunting.

The passerines are an important group of birds to study from a conservation perspective, as they include most of the long-distance, Neotropical migratory species whose declines sparked modern bird-conservation initiatives such as Partners in Flight and the Road to Recovery (R2R; Tipping Point species). Fifty-six passerine species across 13 families are included in Partners in Flight's Watch List for the US and Canada. Of these, 35 are also Tipping Point species, representing nearly a third of species in need of urgent recovery. A high proportion of species with tracking data are those of high conservation concern, as researchers have targeted the need to understand their migratory connectivity. In addition, because of their large population size, common passerine species make up the vast majority of the 2.9 billion birds estimated to be lost from the North American avifauna since 1970. Three diverse families—sparrows, warblers, and blackbirds—constitute nearly 60% of the total population loss.

Because of the sparsity of tracking data on passerines, our knowledge of migratory connectivity across this diverse order is also fragmentary. Among the best studied species, the strength of connectivity ranges from weak to strong, even within the same family or genus. Similarly, among the few well-studied warblers, migratory connectivity is believed to be strong in the American Redstart, Cerulean Warbler, and Golden-winged Warbler but weak in the Canada Warbler, Connecticut Warbler, and Prothonotary Warbler. Other species documented to show relatively strong migratory connectivity from tracking studies include the Barn Swallow, Gray Catbird, and Song Sparrow. Additional species are suggested to show strong connectivity based on subspecies and genetic variation across the range, including the Hermit Thrush, Wilson's Warbler, and Willow Flycatcher. Extensive mixing of individuals from disparate breeding populations is also commonly observed; for example, up to four subspecies of Dark-eyed Juncos can occur together in nonbreeding flocks in Arizona, New Mexico, or Colorado.

The great diversity of passerine species results in a large variety of migration strategies, ranging from short-distance and altitudinal movements to hemispheric migrations involving long water-crossings and multiday stopovers at specific sites, much as in shorebirds. A very commonly observed pattern across passerine families and species is east-west parallel migration, where individuals migrate mostly in a north-south direction between breeding and respective nonbreeding sites. Most long-distance migrants breeding in eastern Canada or the northeast US migrate along the Atlantic Flyway to the southeast US, the Caribbean, or northern South America. Breeders from the Midwest and Great Lakes region typically migrate to the Gulf Coast or to eastern Mexico or Central America. Many prairie-nesting species migrate through the Great Plains to Texas or northern Mexico. In western North America, breeders from the Rocky Mountains and Intermountain regions typically migrate to the southwestern US or western Mexico, whereas breeders in the Pacific region north to Alaska migrate south along the Pacific coast to California or Baja California. In many species that breed across the boreal forest to British Columbia or even Alaska, post-breeding migration is eastward to the Atlantic coast before turning south, whereas pre-breeding migration is northward through the central US—thus exhibiting a clockwise looped migration. An extreme example is the Blackpoll Warbler, for which recent tracking data documented migration from Nome, Alaska, to the Atlantic coast and then a remarkable 90-hour transoceanic flight to northern South America as part of a nearly 20,000 km round-trip journey. Some western species also exhibit a looped migration to and from nonbreeding sites in Mexico, with pre-breeding migration concentrated through the low deserts and Pacific coast and post-breeding migration concentrated farther east through mountainous regions. These migration patterns are often apparent, even in species that lack any tracking or banding data, in the weekly eBird abundance animations showing differential timing and movements of populations between different breeding and nonbreeding regions.

TITYRIDAE | *TITYRAS AND ALLIES*

ROSE-THROATED BECARD

Pachyramphus aglaiae

DISTRIBUTION AND ABUNDANCE

Found largely year-round along both coasts of Mexico, through Central America, and locally to western Panama, with the greatest abundance in the northern Yucatán Peninsula and along the Pacific coast from southern Mexico to Costa Rica (Fink et al. 2023). Irregularly breeding US birds belong to two distinct subspecies, *P. a. albiventris* in southeastern Arizona and western Mexico and *P. a. gravis* in southern Texas and northeastern Mexico (Greenlaw 2020). After breeding, US and northern Mexican birds migrate southward, although individuals from the Mexican northern Gulf Coast subspecies (*P. a. gravis*) may occasionally migrate north to spend the nonbreeding season in south Texas (Greenlaw 2020).

ORDER Passeriformes
FAMILY Tityridae
BREEDING HABITAT Forests, Tropical Dry
NONBREEDING HABITAT Forests, Tropical Dry
GLOBAL POPULATION SIZE 2,000,000
POPULATION CHANGE SINCE 1970 NA
IUCN RED LIST CATEGORY Least Concern
R2R TIPPING POINT SPECIES No
US LISTING STATUS No
CANADA SPECIES AT RISK No
PIF CCS 11/20

MIGRATORY CONNECTIVITY

OVERALL Migratory connectivity is unknown. No information on migratory connectivity was found for this species.
OTHER APPROACHES Plumage of both males and females varies strikingly across eight recognized subspecies.
MIGRATORY PATHWAYS Eastern and western ranges in northern Mexico are separated by the largely unoccupied Central Mexican Plateau; eBird animations (Fink et al. 2023) indicate that movement within those portions of the range are independent of one another.

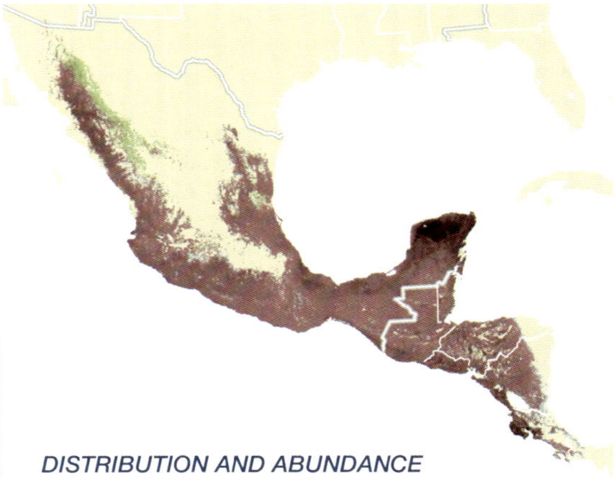

DISTRIBUTION AND ABUNDANCE

INTERESTING NOTE

The Rose-throated Becard is the only species in the tropical family Tityridae that enters the US as a breeder, however irregularly (Greenlaw 2020).

NORTHERN BEARDLESS-TYRANNULET *Camptostoma imberbe*

DISTRIBUTION AND ABUNDANCE

It breeds throughout most of Mexico and Central America to northern Costa Rica, with the densest concentrations along the western Mexican *cordilleras* (mountain ranges) and Pacific coast and Campeche, Mexico, on the Yucatán Peninsula. As a breeding species, it barely reaches into the US in southern Arizona, southwestern New Mexico, and southern Texas. It is a year-round resident throughout most of its range, but at least some of the US and northernmost Mexican breeders appear to move south during the nonbreeding season (Fink et al. 2023).

ORDER Passeriformes
FAMILY Tyrannidae
BREEDING HABITAT Forests, Tropical Dry
NONBREEDING HABITAT Forests, Tropical Dry
GLOBAL POPULATION SIZE 2,000,000
POPULATION CHANGE SINCE 1970 NA
IUCN RED LIST CATEGORY Least Concern
R2R TIPPING POINT SPECIES No
US LISTING STATUS No
CANADA SPECIES AT RISK No
PIF CCS 10/20

DISTRIBUTION AND ABUNDANCE

MIGRATORY CONNECTIVITY

OVERALL Migratory connectivity is unknown. No information on migratory connectivity was found for this species.

MIGRATORY PATHWAYS Although it is not clear how much of the US population migrates south, eBird animations (Fink et al. 2023) support subspecies records (Tenney 2020b) indicating that Arizona birds (*C. i. ridgwayi*) withdraw to western Mexico, while south Texas breeders (*C. i. imberbe*) move south along the Gulf Coast.

DUSKY-CAPPED FLYCATCHER *Myiarchus tuberculifer*

DISTRIBUTION AND ABUNDANCE

It breeds in a wide variety of habitats along eastern and western *cordilleras* (mountain ranges) of Mexico, throughout southern Mexico, Central America, and northern South America south to northern Argentina, reaching into the US only in southern Arizona and locally in western New Mexico (Fink et al. 2023). It is resident year-round throughout most of its range, but the US and northern Mexico population (*M. t. olivascens*) migrates during the nonbreeding season to southern Sonora through eastern Oaxaca (Tweit and Tweit 2020).

ORDER Passeriformes
FAMILY Tyrannidae
BREEDING HABITAT Forests, Tropical Generalist
NONBREEDING HABITAT Forests, Tropical Generalist
GLOBAL POPULATION SIZE 20,000,000
POPULATION CHANGE SINCE 1970 NA
IUCN RED LIST CATEGORY Least Concern
R2R TIPPING POINT SPECIES No
US LISTING STATUS No
CANADA SPECIES AT RISK No
PIF CCS 9/20

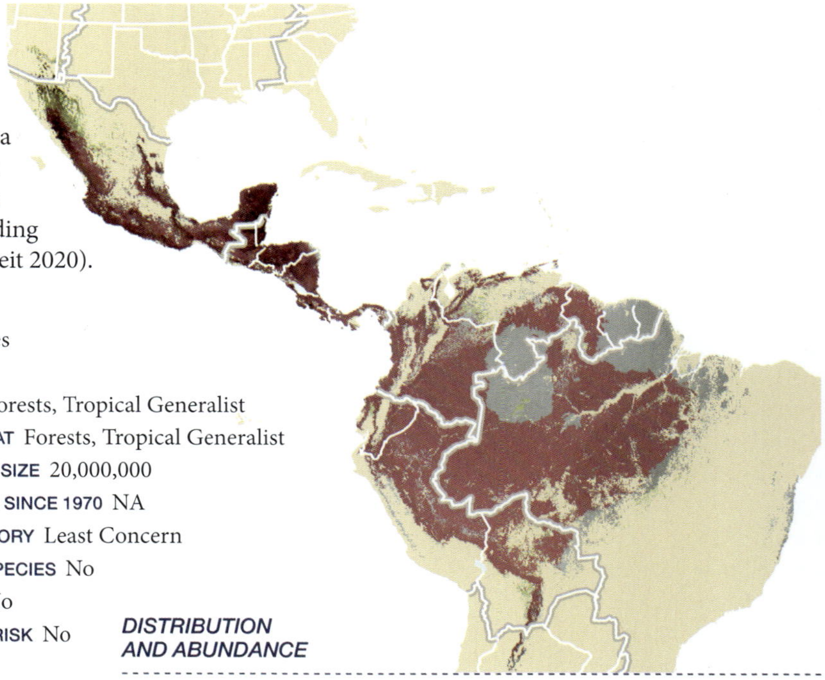

DISTRIBUTION AND ABUNDANCE

MIGRATORY CONNECTIVITY

OVERALL Migratory connectivity is poorly known. Limited information on migratory connectivity was found for this species.

OTHER APPROACHES Of the 13 subspecies that have been recognized, *M. t. olivascens* is differentiated by plumage and vocal characters as the population that breeds in the US and northwestern Mexico south to the northern third of Nayarit, migrating for the nonbreeding season from southern Sonora to Santa Efigenia, Oaxaca (Tweit and Tweit 2020).

MIGRATORY PATHWAYS Migrants from the southwestern United States and northern Mexico remain in forested foothill and montane habitats during migration, avoiding the Pacific coastal lowlands, the Sonoran and Chihuahuan deserts, and the high elevation areas of central Mexico (Tweit and Tweit 2020; Fink et al. 2023).

Myiarchus cinerascens ASH-THROATED FLYCATCHER

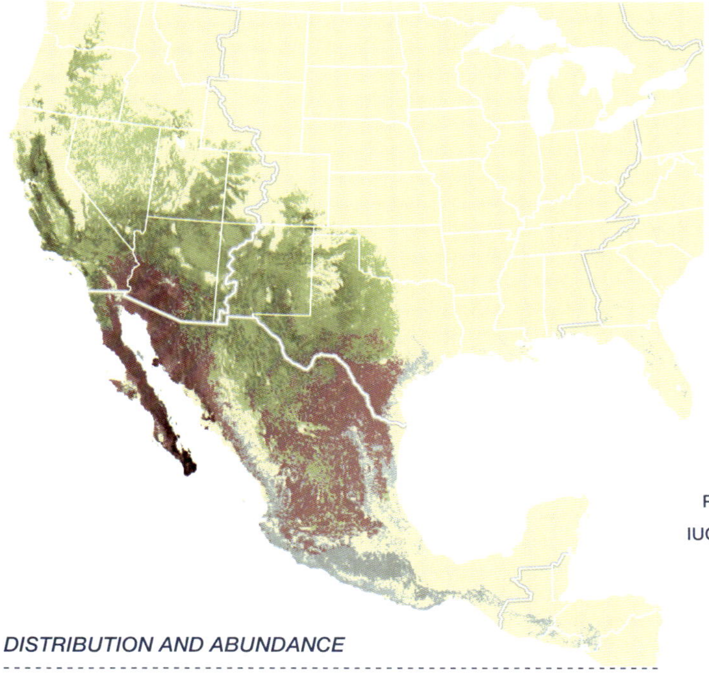

DISTRIBUTION AND ABUNDANCE

DISTRIBUTION AND ABUNDANCE

It breeds in western semiarid scrub and open woodland from central Washington to central Mexico, with greatest abundance throughout California, the Southwest, northwestern Mexico, and the Baja California peninsula. During the nonbreeding season, it retreats from most of the US range except for southeastern California, southern Arizona, and southwestern Texas, and occurs farther south in Mexico and into Central America to northwestern Nicaragua (Fink et al. 2023).

ORDER Passeriformes
FAMILY Tyrannidae
BREEDING HABITAT Arid Lands, Desert Scrub
NONBREEDING HABITAT Forests, Tropical Dry
GLOBAL POPULATION SIZE 10,000,000
POPULATION CHANGE SINCE 1970 +56%
IUCN RED LIST CATEGORY Least Concern
R2R TIPPING POINT SPECIES No
US LISTING STATUS No
CANADA SPECIES AT RISK No
PIF CCS 8/20

MIGRATORY CONNECTIVITY

OVERALL Migratory connectivity is unknown. No information on migratory connectivity was found for this species.

MIGRATORY PATHWAYS It migrates along a wide front through the western US and northern Mexico, with most populations undergoing north-south migrations (Fink et al. 2023). A few populations in the year-round range appear to be resident, with their numbers likely supplemented in the nonbreeding season by post-breeding arrivals and passage migrants (Cardiff and Dittmann 2020a).

Myiarchus crinitus GREAT CRESTED FLYCATCHER

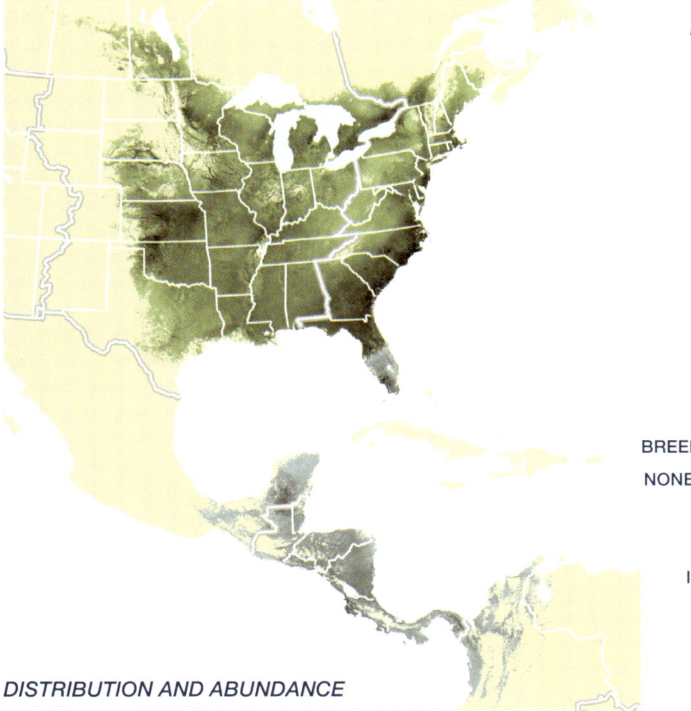

DISTRIBUTION AND ABUNDANCE

DISTRIBUTION AND ABUNDANCE

It breeds in open deciduous and mixed woodlands in southern Canada from Saskatchewan to Quebec and throughout the eastern US; it is less dense in mountainous areas and the lower Mississippi Valley. Spends the nonbreeding season in southern Florida, along the Gulf coastal plain of Mexico in Veracruz and Tabasco, throughout the Yucatán Peninsula and Central America, and at lower elevations in western Colombia. Its highest densities during the nonbreeding season occur in southern Florida, central Yucatán, Nicaragua, and western Costa Rica (Fink et al. 2023).

ORDER Passeriformes
FAMILY Tyrannidae
BREEDING HABITAT Arid Lands, Desert Scrub
NONBREEDING HABITAT Forests, Tropical Dry
GLOBAL POPULATION SIZE 10,000,000
POPULATION CHANGE SINCE 1970 +56%
IUCN RED LIST CATEGORY Least Concern
R2R TIPPING POINT SPECIES No
US LISTING STATUS No
CANADA SPECIES AT RISK No
PIF CCS 8/20

MIGRATORY CONNECTIVITY

OVERALL Migratory connectivity is unknown. No information on migratory connectivity was found for this species.

MIGRATORY PATHWAYS Fink et al. (2023) indicate very few migration records in Cuba, and the extent of migration across the Gulf of Mexico remains unknown (Miller and Lanyon 2020). Most birds appear to migrate from the US along the Mexican Gulf Coast into Central America (Fink et al. 2023).

BROWN-CRESTED FLYCATCHER *Myiarchus tyrannulus*

DISTRIBUTION AND ABUNDANCE

It breeds in a wide variety of relatively open, semiarid and riparian habitats in disjunct ranges from the southwestern US and southern Texas, along both slopes of Mexico, Central America to northwest Costa Rica, and again in northern South America and eastern Brazil, Paraguay, and northwestern Argentina (Cardiff and Dittmann 2020b). In North America, the most abundant populations are in the southwestern US and northwestern Mexico, southern Texas, the Yucatán Peninsula, and southern Nicaragua. US and northern Mexico populations migrate southward during the nonbreeding season to other parts of the species range; the remainder of the population is mostly resident year-round (Fink et al. 2023).

ORDER Passeriformes
FAMILY Tyrannidae
BREEDING HABITAT Forests, Tropical Dry
NONBREEDING HABITAT Forests, Tropical Dry
GLOBAL POPULATION SIZE 14,000,000
POPULATION CHANGE SINCE 1970 +414%
IUCN RED LIST CATEGORY Least Concern
R2R TIPPING POINT SPECIES No
US LISTING STATUS No
CANADA SPECIES AT RISK No
PIF CCS 9/20

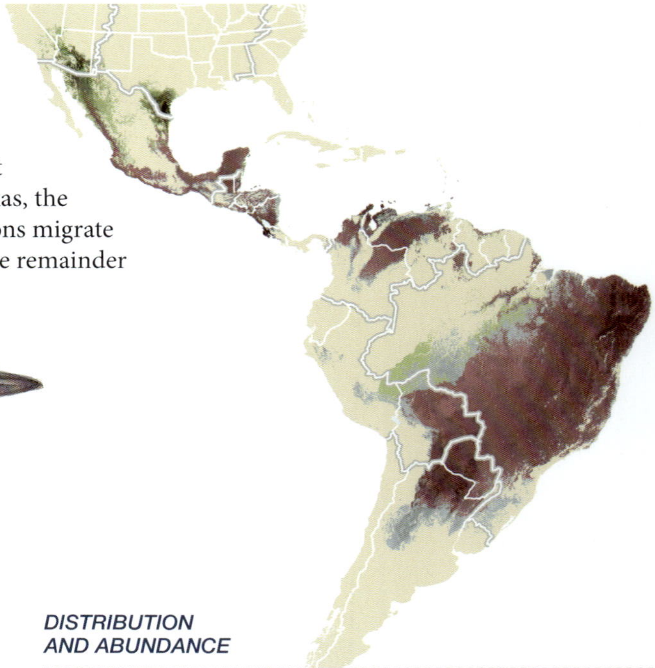

DISTRIBUTION AND ABUNDANCE

MIGRATORY CONNECTIVITY

OVERALL Migratory connectivity is poorly known. Limited information on migratory connectivity was found for this species.
OTHER APPROACHES Although subspecific distinctions are not always clear, southwestern US breeding birds (*M. t. magister*) are believed to spend the nonbreeding season in Guerrero and Oaxaca, Mexico, while southern Texas birds (*M. t. cooperi*) migrate to eastern Oaxaca and Chiapas, Mexico, south to Guatemala, Honduras, and El Salvador (Cardiff and Dittmann 2020b).
MIGRATORY PATHWAYS Populations in the southwestern US migrate along the Pacific coast, while southern Texas populations migrate along the Gulf Coast (Fink et al. 2023).

SULPHUR-BELLIED FLYCATCHER *Myiodynastes luteiventris*

DISTRIBUTION AND ABUNDANCE

It breeds in riparian canyon and tropical deciduous forest edges and clearings (Lowther and Stotz 2020) throughout the eastern and western slopes of Mexico (avoiding the Central Mexican Plateau) and south to Costa Rica, reaching the US only locally in southeastern Arizona. For the nonbreeding season, the entire population migrates to South America, sparsely in eastern Panama, Colombia, eastern Ecuador, and northern Peru, and more densely in the western Amazonia portions of Peru, Brazil, and Bolivia (Lowther and Stotz 2020; Fink et al. 2023).

ORDER Passeriformes
FAMILY Tyrannidae
BREEDING HABITAT Forests, Tropical Dry
NONBREEDING HABITAT Forests, Tropical Lowland Evergreen
GLOBAL POPULATION SIZE 2,000,000
POPULATION CHANGE SINCE 1970 NA
IUCN RED LIST CATEGORY Least Concern
R2R TIPPING POINT SPECIES No
US LISTING STATUS No
CANADA SPECIES AT RISK No
PIF CCS 11/20

DISTRIBUTION AND ABUNDANCE

MIGRATORY CONNECTIVITY

OVERALL Migratory connectivity is unknown. No information on migratory connectivity was found for this species.
MIGRATORY PATHWAYS Arizona breeding birds move south in migration through the western *cordillera* of Mexico (Fink et al. 2023).

INTERESTING NOTE

It is one of the latest tree-nesting birds to commence breeding in Arizona, delaying until the start of the monsoon season (Brandt 1951).

Tyrannus melancholicus TROPICAL KINGBIRD

DISTRIBUTION AND ABUNDANCE

DISTRIBUTION AND ABUNDANCE

Widely distributed throughout the American tropics and subtropics, it breeds mostly along the coasts of Mexico and northern Central America, and more widely from Nicaragua south to central Argentina—along the Pacific coast to southern Ecuador, in the large river valleys of Colombia and Venezuela, and east of the Andes throughout the remainder of South America. In the US, it breeds in southeastern Arizona and the southern Rio Grande Valley of Texas. During the nonbreeding season, it migrates from Arizona, Sonora, and Baja California Sur south to other parts of the range, while populations in southern South America (Argentina and Uruguay) migrate north to northern Amazonia. Texas breeding birds appear to be year-round residents (Stouffer et al. 2020; Fink et al. 2023).

ORDER Passeriformes
FAMILY Tyrannidae
BREEDING HABITAT Open Country, Habitat Mosaic
NONBREEDING HABITAT Open Country, Habitat Mosaic
GLOBAL POPULATION SIZE 200,000,000
POPULATION CHANGE SINCE 1970 NA
IUCN RED LIST CATEGORY Least Concern
R2R TIPPING POINT SPECIES No
US LISTING STATUS No
CANADA SPECIES AT RISK No
PIF CCS 5/20

MIGRATORY CONNECTIVITY

OVERALL Migratory connectivity is unknown. No information on migratory connectivity was found for this species. There were no detections from the few tags deployed through the Motus Wildlife Tracking System.

MIGRATORY PATHWAYS Migratory movements are complex due to the potential presence of permanent residents, as well as boreal and austral migrants in any given region. However, individuals in the far northwestern and far southern edge of the range undergo north-south seasonal migrations through lowland areas and avoid high mountains. The birds in the most northwestern part of the range are found year-round from central Sinaloa southward along the western lowlands of Mexico, but during the breeding season, they migrate along the Pacific lowlands as far north as southern Arizona. Similarly, birds in the Southern Cone of South America have year-round populations as far south as central Paraguay and southern Brazil, but migrate southward during the austral breeding season, moving throughout lowland areas as far south as the Argentinian Rio Negro Province (Stouffer et al. 2020; Fink et al. 2023)

Tyrannus couchii COUCH'S KINGBIRD

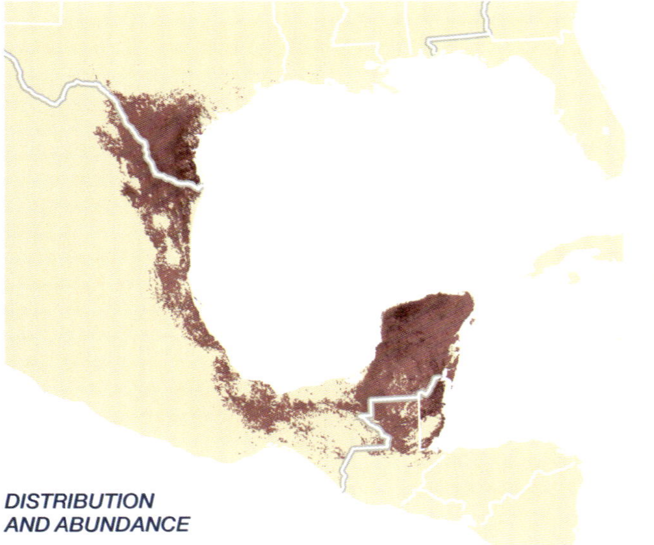

DISTRIBUTION AND ABUNDANCE

DISTRIBUTION AND ABUNDANCE

It breeds in thorn forest and brushy clearings within tropical forests (Brush 2020) from southern Texas south along the Gulf Coast of Mexico to the Yucatán, northern Guatemala, and Belize. It is most abundant in southern Texas, Nuevo León, northern Tamaulipas, and the northern Yucatán Peninsula. It is resident year-round throughout most of its range (Fink et al. 2023).

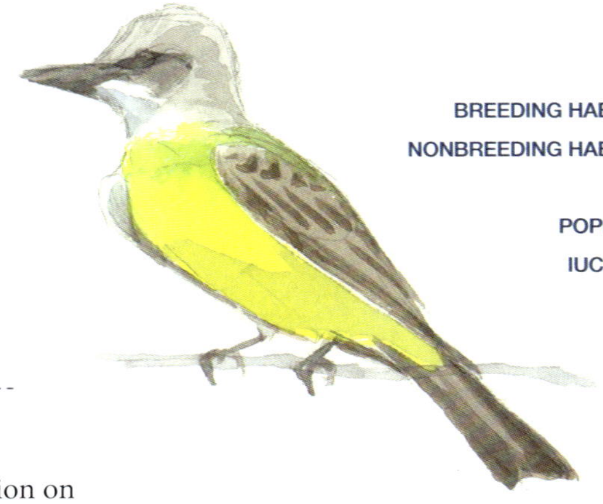

ORDER Passeriformes
FAMILY Tyrannidae
BREEDING HABITAT Open Country, Habitat Mosaic
NONBREEDING HABITAT Open Country, Habitat Mosaic
GLOBAL POPULATION SIZE 1,100,000
POPULATION CHANGE SINCE 1970 +3542%
IUCN RED LIST CATEGORY Least Concern
R2R TIPPING POINT SPECIES No
US LISTING STATUS No
CANADA SPECIES AT RISK No
PIF CCS 11/20

MIGRATORY CONNECTIVITY

OVERALL Migratory connectivity is unknown. No information on migratory connectivity was found for this species.

MIGRATORY PATHWAYS Many birds from southern Texas and northeastern Mexico depart from their breeding areas during the nonbreeding season, presumably undergoing southward migrations to parts of the species' year-round range (Brush 2020; Fink et al. 2023).

INTERESTING NOTE

Couch's Kingbird was long regarded as a subspecies of the Tropical Kingbird (*Tyrannus melancholicus*) until it was recognized as a distinct species in 1979 (Brush 2020).

CASSIN'S KINGBIRD *Tyrannus vociferans*

DISTRIBUTION AND ABUNDANCE

It breeds in a narrow belt from southern Montana to New Mexico, then more broadly west to southern Nevada and south through central and western Mexico to Oaxaca; it also breeds in Southern California. Breeding abundance is greatest in eastern New Mexico and southern Mexico. After breeding, US populations migrate south to the Mexican portion of the breeding range, possibly as far south as central Mexico (eastern Jalisco, Querétaro, Guanajuato, Hidalgo, and Puebla), where the greatest nonbreeding concentrations occur. Some individuals remain in southwestern California and southern Arizona during the nonbreeding season (Fink et al. 2023).

ORDER Passeriformes
FAMILY Tyrannidae
BREEDING HABITAT Open Country, Habitat Mosaic
NONBREEDING HABITAT Open Country, Habitat Mosaic
GLOBAL POPULATION SIZE 8,800,000
POPULATION CHANGE SINCE 1970 +10%
IUCN RED LIST CATEGORY Least Concern
R2R TIPPING POINT SPECIES No
US LISTING STATUS No
CANADA SPECIES AT RISK No
PIF CCS 10/20

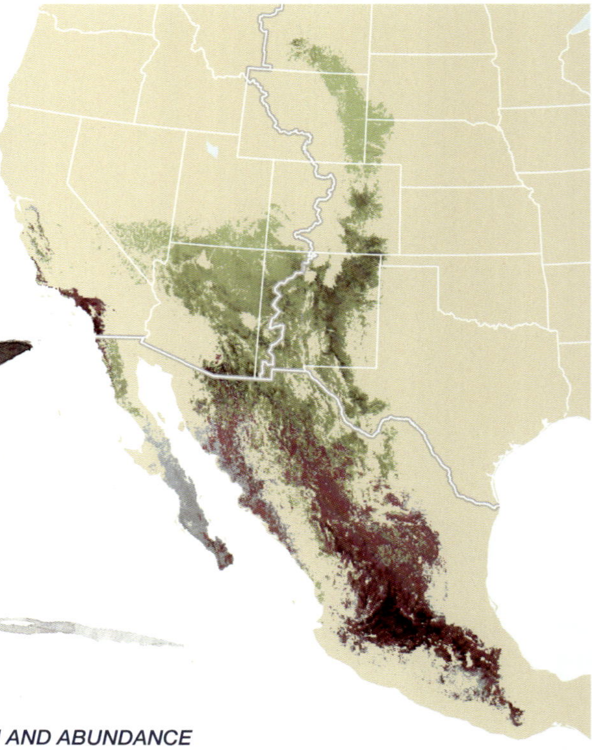

DISTRIBUTION AND ABUNDANCE

MIGRATORY CONNECTIVITY

OVERALL Migratory connectivity is unknown. No information on migratory connectivity was found for this species.
MIGRATORY PATHWAYS California breeding populations remain resident in the Los Angeles area or migrate south to Baja California Sur. The remainder of the population in the US appears to migrate via northern Mexico to the central Mexico nonbreeding range (Fink et al. 2023).

THICK-BILLED KINGBIRD *Tyrannus crassirostris*

DISTRIBUTION AND ABUNDANCE

It breeds from southeastern Arizona through western Mexico (including southern Baja California Sur) to Oaxaca, Mexico. During the nonbreeding season, it migrates to locations in the southern portion of the year-round range; some individuals may move downslope to coastal areas (Lowther et al. 2020h; Fink et al. 2023).

ORDER Passeriformes
FAMILY Tyrannidae
BREEDING HABITAT Forests, Tropical Dry
NONBREEDING HABITAT Forests, Tropical Dry
GLOBAL POPULATION SIZE 2,000,000
POPULATION CHANGE SINCE 1970 NA
IUCN RED LIST CATEGORY Least Concern
R2R TIPPING POINT SPECIES No
US LISTING STATUS No
CANADA SPECIES AT RISK No
PIF CCS 12/20

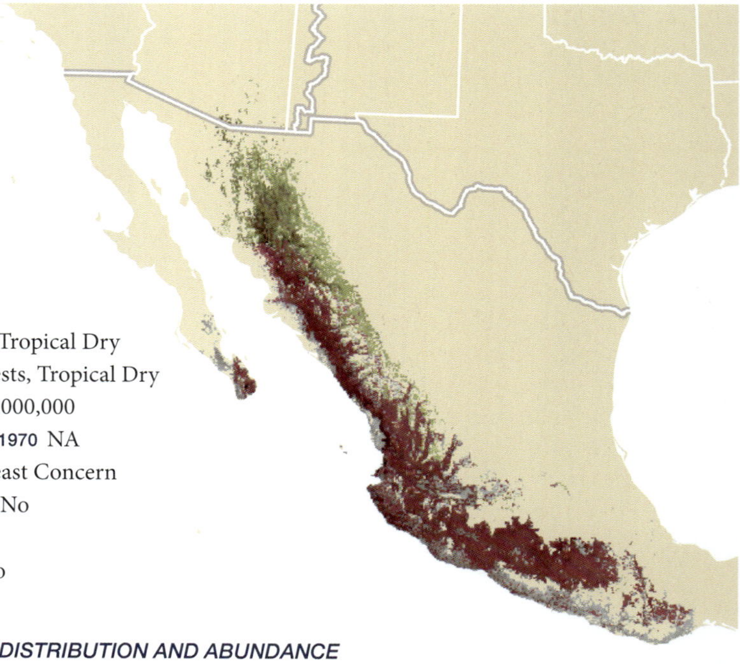

DISTRIBUTION AND ABUNDANCE

MIGRATORY CONNECTIVITY

OVERALL Migratory connectivity is unknown. No information on migratory connectivity was found for this species.
MIGRATORY PATHWAYS During the nonbreeding season, some populations undergo north-south and/or elevational migrations. Birds on the northern edge of the breeding range retreat southward during the nonbreeding season, moving to southern Sonora and beyond. Evidence of elevational movements is also seen along the edges of much of the core range, as birds move higher during the breeding season, and retreat lower during the nonbreeding season (Lowther et al. 2020h; Fink et al. 2023).

Tyrannus verticalis **WESTERN KINGBIRD**

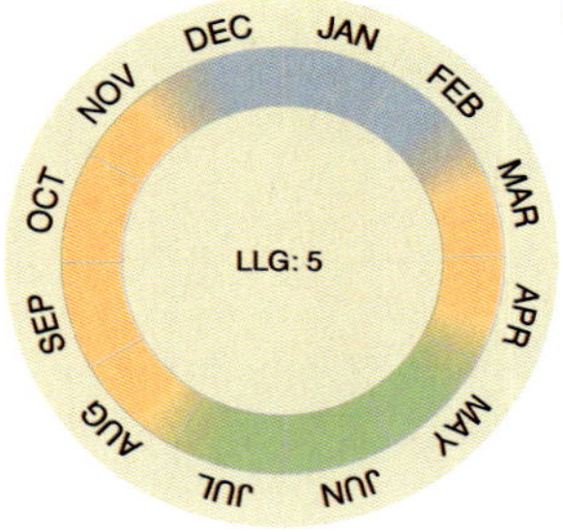

DISTRIBUTION AND ABUNDANCE

Western Kingbirds breed from central Manitoba, Saskatchewan, Alberta, and southern British Columbia, and all through the western US and northern Mexico. The species occupies nonbreeding areas along the Pacific coast from southern Mexico, to Guatemala, El Salvador, Honduras, Nicaragua, and Costa Rica. There have also been records of birds along the Gulf Coast of the US and in Florida during the nonbreeding season (Lowther et al. 2023).

CONNECTIONS

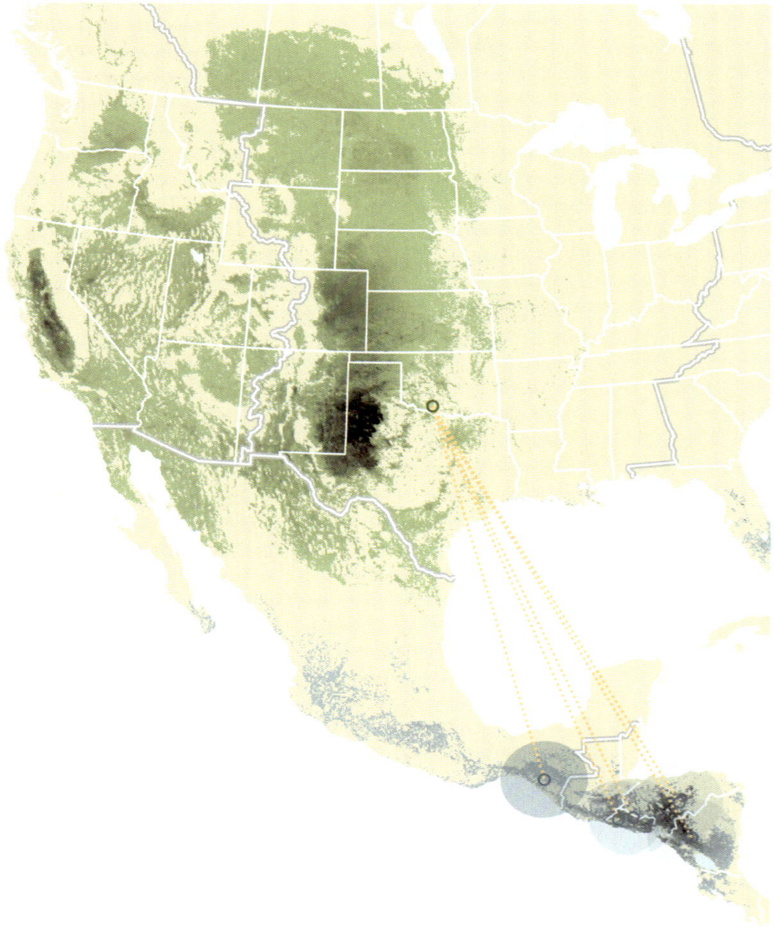

ORDER Passeriformes
FAMILY Tyrannidae
BREEDING HABITAT Grasslands, Temperate
NONBREEDING HABITAT Open Country, Habitat Mosaic
GLOBAL POPULATION SIZE 30,000,000
POPULATION CHANGE SINCE 1970 +7%
IUCN RED LIST CATEGORY Least Concern
R2R TIPPING POINT SPECIES No
US LISTING STATUS No
CANADA SPECIES AT RISK No
PIF CCS 10/20

MIGRATORY CONNECTIVITY

OVERALL The migratory connectivity of Western Kingbirds has not been widely studied and cannot be calculated due to limited sample size. Birds breeding in the Midcontinent Flyway migrated to Mexico and Central America.

GEOLOCATOR TRACKING Data from five geolocators recovered at a breeding site in Oklahoma showed birds migrated to nonbreeding areas in Mexico and Central America (Jahn et al. 2013). The five birds tracked with geolocators occupied several nonbreeding areas: first a molting area in northwestern Mexico and then areas ranging from southern Mexico to northern Central America. One bird flew directly from its molting area in northwestern Mexico to its nonbreeding site in Guatemala, where it remained until northward migration (Jahn et al. 2013).

OTHER APPROACHES No banding, stable isotope, or genetic studies of migratory connectivity were found for this species.

MIGRATORY PATHWAYS Geolocator data showed birds first flew southwest from breeding areas in Oklahoma to northwestern Mexico and then south to nonbreeding areas. Pre-breeding migration for one bird showed an overland route from Mexico through Texas back to Oklahoma (Jahn et al. 2013).

INTERESTING NOTE

Timing of migration appears to be linked to molt as breeding birds in Oklahoma departed breeding areas early to molt in the Sonoran Desert region during the monsoon season (Jahn et al. 2013).

DATA CONTRIBUTORS
M.S. Husak; A.E. Jahn; D.H. Kim; D.V. Landoll; M.T. Murphy.

EASTERN KINGBIRD *Tyrannus tyrannus*

DISTRIBUTION AND ABUNDANCE

Eastern Kingbird has a broad eastern North America breeding distribution, ranging from central and southeastern Alberta, Saskatchewan, and Manitoba, extending through southeastern Canada and throughout the eastern US. West of the Rocky Mountains, there are breeding populations in British Columbia, Washington, Oregon, and Idaho. The nonbreeding range extends from southern Colombia to Bolivia, with higher densities in north and central Bolivia and central western Brazil (Fink et al. 2023).

ORDER Passeriformes
FAMILY Tyrannidae
BREEDING HABITAT Open Country, Habitat Mosaic
NONBREEDING HABITAT Forests, Tropical Lowland Evergreen
GLOBAL POPULATION SIZE 26,000,000
POPULATION CHANGE SINCE 1970 −43%
IUCN RED LIST CATEGORY Least Concern
R2R TIPPING POINT SPECIES No
US LISTING STATUS No
CANADA SPECIES AT RISK No
PIF CCS 11/20

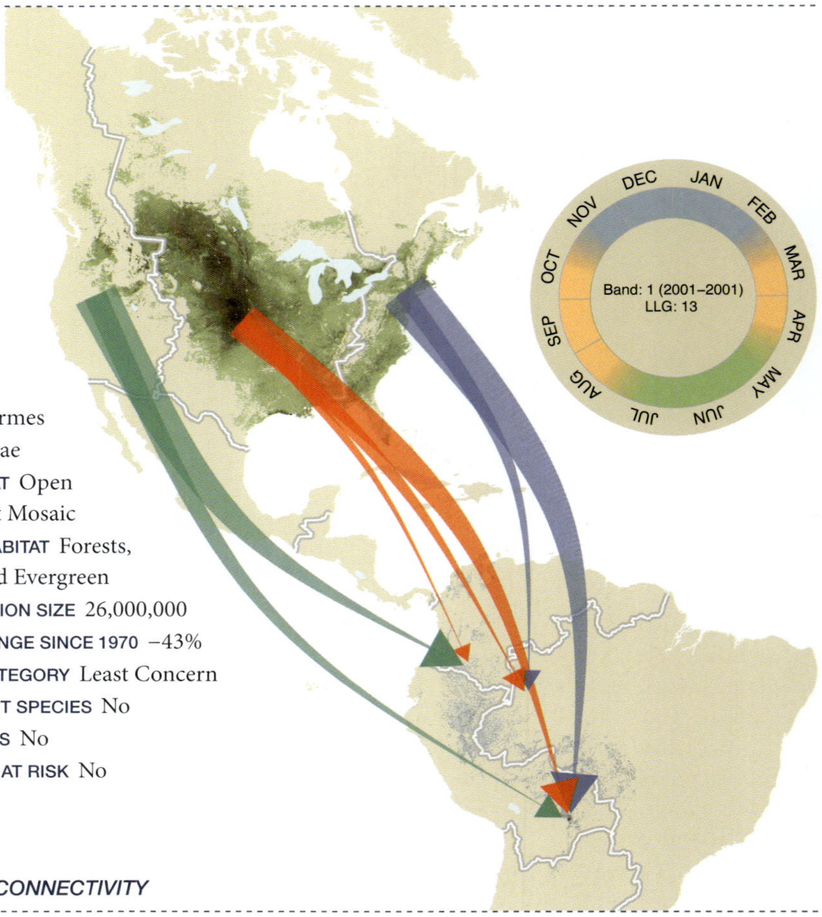

GENERALIZED CONNECTIVITY

	Andean West	Brazil	Northern SA
Atlantic (n = 4)	75%	25%	–
Midcontinent (n = 7)	57%	29%	14%
Pacific (n = 3)	33%	–	67%

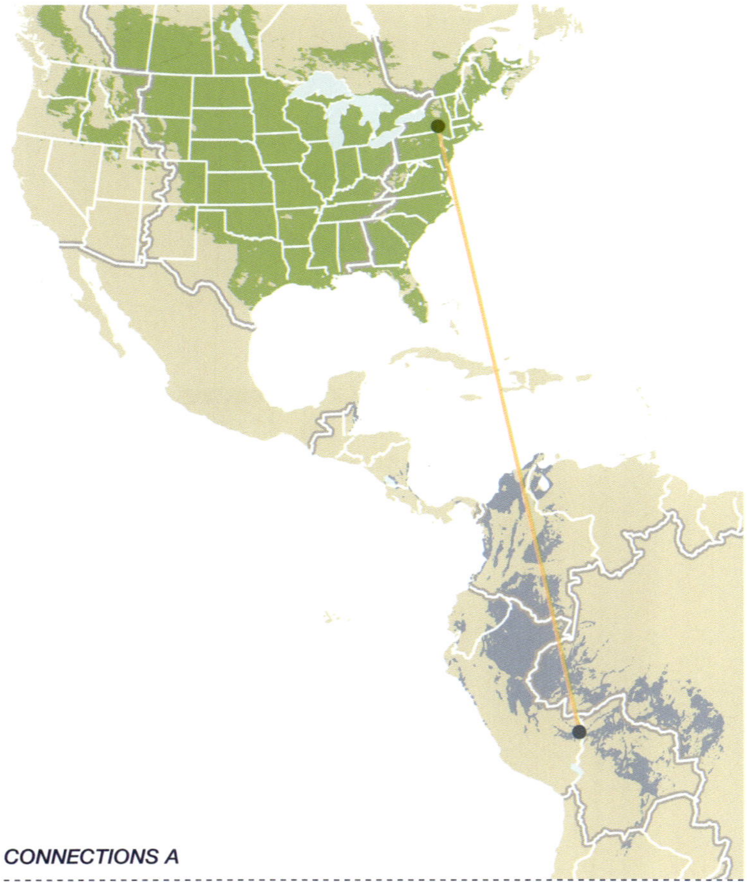

CONNECTIONS A

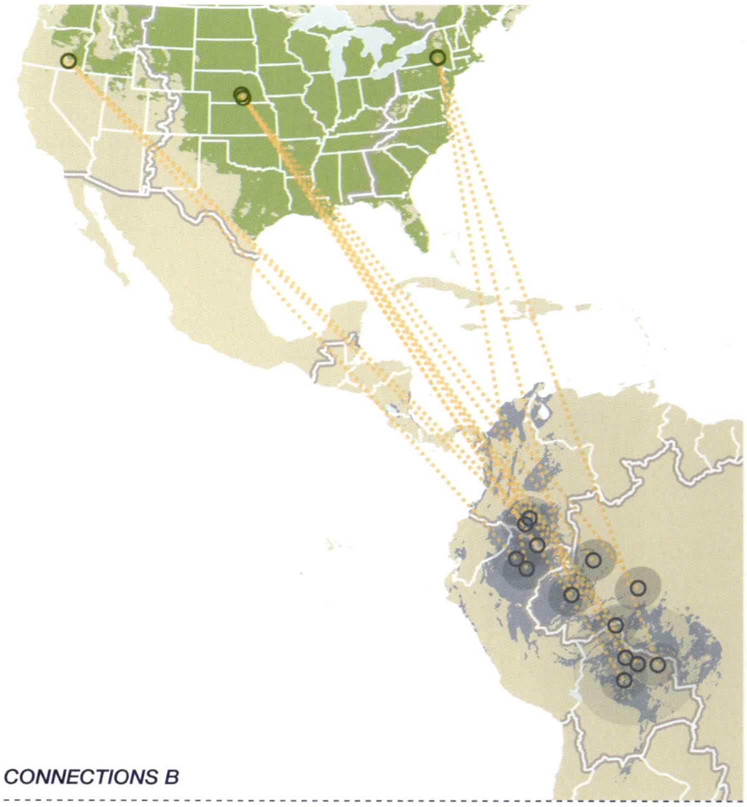

CONNECTIONS B

MIGRATORY CONNECTIVITY

OVERALL The migratory connectivity of Eastern Kingbirds is known from a mixture of band reencounter and geolocator data. Atlantic Flyway breeding populations primarily spend the nonbreeding season in the Andean West (75%) and Brazil (25%). Midcontinent Flyway breeders divide broadly between the Andean West (57%), Brazil (29%), and northern South America (14%). Pacific Flyway breeders primarily migrate to nonbreeding areas in northern South America (67%) and the Andean West (33%). MC could not be calculated due to limited sample size but appears weak.

BANDING There was one breeding to nonbreeding band reencounter from New York to the Peru/Bolivia border.

GEOLOCATOR TRACKING Eastern Kingbirds breeding in New York (n = 3), Nebraska (n = 6), Oklahoma (n = 1), and Oregon (n = 3) exhibited one of two strategies during the nonbreeding season. Birds from Oregon occupied a single nonbreeding area in northwestern Amazonia, while most kingbirds from New York, Nebraska, and Oklahoma were intratropical migrants, occupying two regions during the nonbreeding season. These kingbirds first migrated to nonbreeding areas in Brazil and Bolivia and then moved north to the same area in northwestern Amazonia, where the other kingbirds remained throughout the nonbreeding period (Jahn et al. 2013; Kim et al. 2021). Kim et al. (2021) found migratory connectivity was low, and northwestern Amazonia appears to be a critical area for all Eastern Kingbirds during the nonbreeding season.

OTHER APPROACHES No stable isotope or genetic studies of migratory connectivity were found for this species.

MIGRATORY PATHWAYS Eastern Kingbirds concentrate in the southern Gulf states before migrating around the Gulf of Mexico or attempting Gulf and Caribbean crossings to Central and South America (Jahn et al. 2013). Large numbers of Eastern Kingbirds enter South America through the Darién and the Caribbean (Bayly et al. 2014), and head south to their nonbreeding grounds in Bolivia and Brazil. Birds tracked from New York migrated along the Atlantic coast to Florida and flew either nonstop across the Gulf of Mexico or stopped once along the way in Cuba or the Cayman Islands to land in Yucatán or Central America (Kim et al. 2021). Birds from Nebraska and Oregon most likely took a land route around the Gulf of Mexico to Central America. All New York birds flew nonstop across the Gulf of Mexico on northward migration and, once in North America, took a more inland route than in post-breeding migration (Kim et al. 2021). Trans-Gulf flights were also more common among Nebraska and Oklahoma birds in pre-breeding migration (Jahn et al. 2013; Kim et al. 2021)

INTERESTING NOTE

The duration of post-breeding migration was twice as long as pre-breeding migration (Jahn et al. 2013).

DATA CONTRIBUTORS

M.S. Husak; A.E. Jahn; D.H. Kim; D.V. Landoll; M.T. Murphy.

Tyrannus dominicensis **GRAY KINGBIRD**

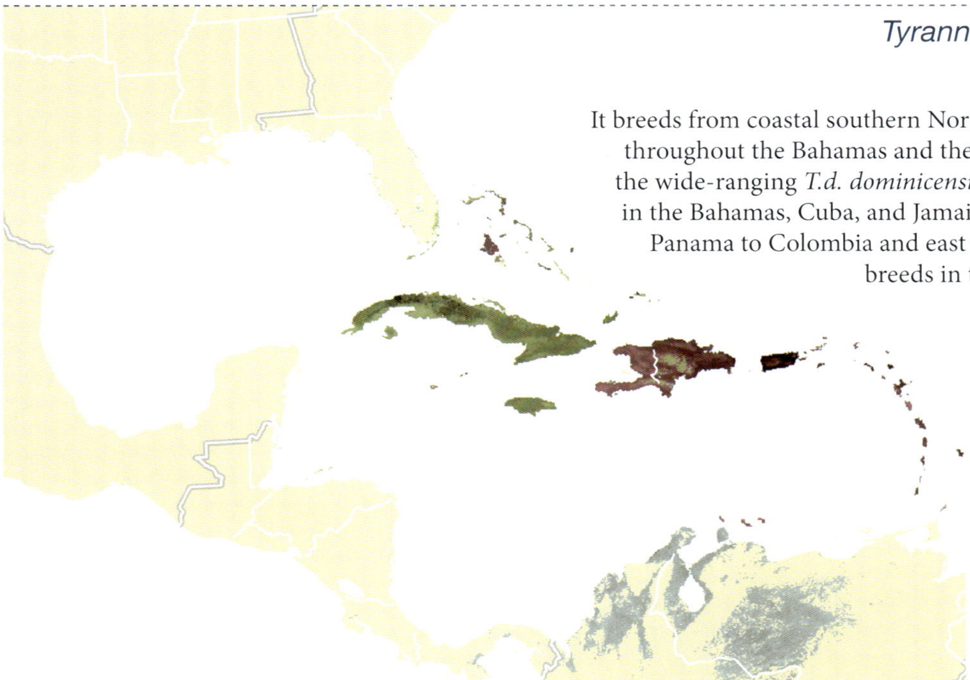

DISTRIBUTION AND ABUNDANCE

DISTRIBUTION AND ABUNDANCE

It breeds from coastal southern North Carolina to Florida and west to Mississippi, and throughout the Bahamas and the Greater and Lesser Antilles. Northern breeders of the wide-ranging *T.d. dominicensis*, including all US breeding birds and populations in the Bahamas, Cuba, and Jamaica, migrate to spend the nonbreeding season from Panama to Colombia and east to French Guiana. A second subspecies, *T.d vorax*, breeds in the Lesser Antilles (G.A. Smith and Jackson 2020).

MIGRATORY CONNECTIVITY

OVERALL Migratory connectivity is unknown. No information on migratory connectivity was found for this species.

MIGRATORY PATHWAYS During the breeding season, this species expands northwestward from its year-round range on Hispaniola and islands to the east. These migrants may island-hop or undergo direct overwater flights, but this is still unknown (G.A. Smith and Jackson 2020).

ORDER Passeriformes
FAMILY Tyrannidae
BREEDING HABITAT Coasts, Mangroves
NONBREEDING HABITAT Forevsts, Tropical Dry
GLOBAL POPULATION SIZE 670,000

POPULATION CHANGE SINCE 1970 +2%
IUCN RED LIST CATEGORY Least Concern
R2R TIPPING POINT SPECIES No
US LISTING STATUS No
CANADA SPECIES AT RISK No
PIF CCS 13/20

SCISSOR-TAILED FLYCATCHER *Tyrannus forficatus*

DISTRIBUTION AND ABUNDANCE

The Scissor-tailed Flycatcher breeds in a relatively small region of the south-central US (primarily Texas, Oklahoma, Kansas) and northeastern Mexico (eastern Coahuila, Nuevo León, and Tamaulipas; Regosin 2020; Fink et al. 2023). The nonbreeding range extends from southern Mexico to northwestern Costa Rica, with strongholds in southern Guatemala, El Salvador, and Nicaragua (Jahn et al. 2013). A few individuals can be found in southern Florida during the nonbreeding season (Fink et al. 2023).

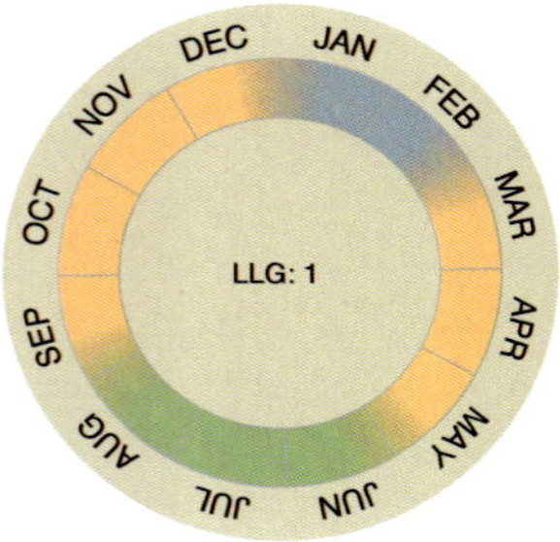

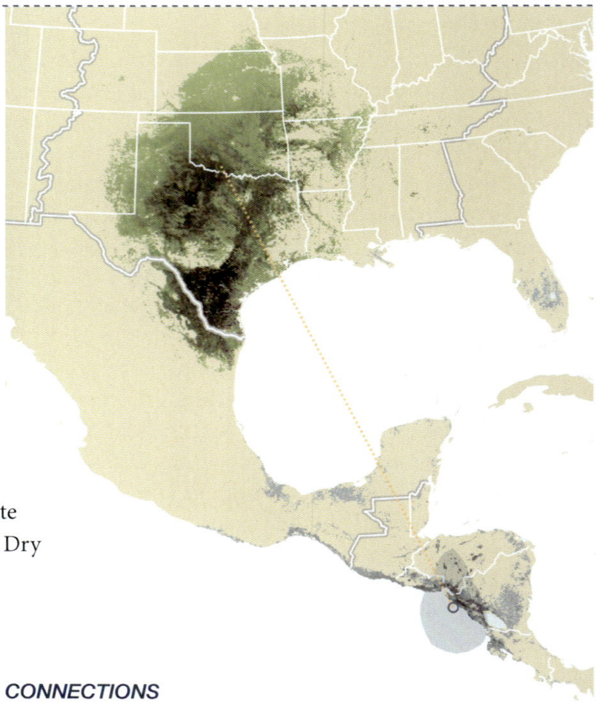

CONNECTIONS

ORDER Passeriformes
FAMILY Tyrannidae
BREEDING HABITAT Grasslands, Temperate
NONBREEDING HABITAT Forests, Tropical Dry
GLOBAL POPULATION SIZE 9,100,000
POPULATION CHANGE SINCE 1970 −22%
IUCN RED LIST CATEGORY Least Concern
R2R TIPPING POINT SPECIES No
US LISTING STATUS No
CANADA SPECIES AT RISK No
PIF CCS 11/20

INTERESTING NOTE

Timing of migration appears to be linked to molt. Scissor-tailed Flycatchers in Oklahoma undergo a complete molt on the breeding grounds when insect abundance is high and thus depart breeding areas later than Western or Eastern Kingbirds (Jahn et al. 2013).

MIGRATORY CONNECTIVITY

OVERALL The migratory connectivity of Scissor-tailed Flycatchers is unknown. Limited data exist from one geolocator study and show that birds breeding in the Midcontinent Flyway migrate to Central America (Jahn et al. 2013).

GEOLOCATOR TRACKING Data recovered from geolocators showed Scissor-tailed Flycatchers migrated from breeding locations in Oklahoma to nonbreeding areas in El Salvador, Guatemala, Honduras, and Nicaragua (Jahn et al. 2013).

OTHER APPROACHES No banding, automated telemetry, stable isotope, or genetic studies of migratory connectivity were found for this species.

MIGRATORY PATHWAYS Scissor-tailed Flycatchers flew over the Gulf of Mexico in post-breeding migration fall, but pre-breeding migration routes were not recorded (Jahn et al. 2013).

DATA CONTRIBUTORS
M.S. Husak; A.E. Jahn; D.H. Kim; D.V. Landoll; M.T. Murphy.

OLIVE-SIDED FLYCATCHER *Contopus cooperi*

DISTRIBUTION AND ABUNDANCE

This large flycatcher is widespread in boreal Canada and the western United States during the breeding season. A notably long-distance migrant, this species spends the nonbreeding season primarily in the Andes, as well as other mountainous areas of Central and South America (Altman and Sallabanks 2020; Fink et al. 2023). The Olive-sided Flycatcher apparently reaches its highest nonbreeding abundance in the Colombian Andes (Hagelin et al. 2021).

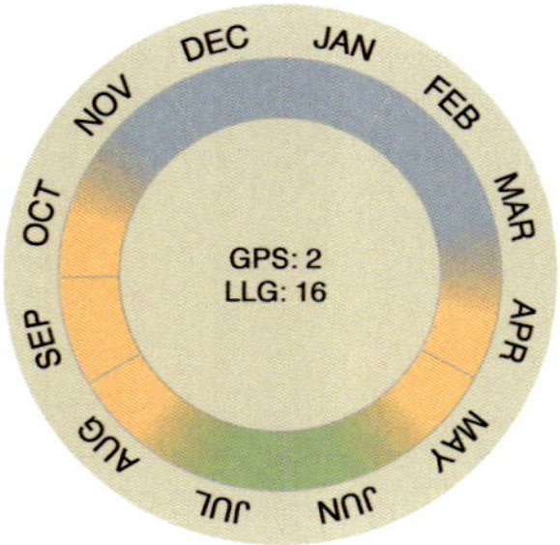

ORDER Passeriformes
FAMILY Tyrannidae
BREEDING HABITAT Forests, Boreal
NONBREEDING HABITAT Forests, Tropical Montane Evergreen
GLOBAL POPULATION SIZE 1,900,000
POPULATION CHANGE SINCE 1970 −76%
IUCN RED LIST CATEGORY Near Threatened
R2R TIPPING POINT SPECIES Yellow-Alert
US LISTING STATUS No
CANADA SPECIES AT RISK No
PIF CCS 13/20

	Andean West	Brazil	Northern SA
Pacific (n = 16)	19%	44%	38%

MIGRATORY CONNECTIVITY

OVERALL Migratory connectivity data only exists for the Pacific Flyway breeding population, specifically birds breeding in boreal Alaska. For this population, birds spend the nonbreeding season in the western edge of Brazil (44%), northern South America (38%), and the Andean West region of South America (19%). MC was not calculated due to lack of data from other breeding regions, but see Hagelin et al. (2021).

GEOLOCATOR TRACKING Light-level geolocators (n = 66) were deployed at seven sites across boreal Alaska. Data from geolocators recovered (n = 14) showed the tagged birds migrated to areas within the western portion of the species' nonbreeding range in South America (Hagelin et al. 2021)—specifically, Colombia and Ecuador, central and southern Peru, and western Brazil (Hagelin et al. 2021). Migratory connectivity among birds in this study was weak, with birds from three deployment areas in Alaska overlapping during the nonbreeding period (Hagelin et al. 2021).

SATELLITE TRACKING In addition to light-level geolocators, archival GPS tags (n = 29) were deployed at seven sites across boreal Alaska (Hagelin et al. 2021). Two of the five recovered GPS tags (three tags recovered without data) spent the nonbreeding season in the Peruvian Andes (Hagelin et al. 2021).

OTHER APPROACHES No banding, genetic, or stable isotope studies of migratory connectivity were found for this species.

MIGRATORY PATHWAYS Moving south from their coniferous breeding grounds, this species tends to stay close to mountains and foothills along its migratory journey. Western breeding birds closely follow mountains into Mexico and Central and South America. Eastern birds cut westward across the center of the continent before following a similar route as western breeders (Fink et al. 2023). This pattern leads to this species being notably uncommon in the southeastern US and Caribbean (Altman and Sallabanks 2020; Fink et al. 2023). Birds tagged in Alaska followed a clockwise, looped migration pattern, with 13 important stopovers identified along the migratory route (Hagelin et al. 2021). The most probable migration route indicated most birds (12 of 14) circumnavigated the Gulf of Mexico rather than crossing it; only 2 (1 in post-breeding, 1 in pre-breeding) exhibited a partial, overwater crossing in the southwestern Gulf (Hagelin et al. 2021). More stopovers were identified near the equator and fewer at higher latitudes; however, during pre-breeding migration, two stopovers, which included montane, forested regions of northeastern California and the Willamette River Valley in Oregon, were identified for their importance to the tagged Alaska birds (Hagelin et al. 2021). In post-breeding, Guatemala/Mexico was identified as an important stopover area and northern Central America an important stopover location during both pre- and post-breeding migration (Hagelin et al. 2021).

DATA CONTRIBUTORS

K.A. DuBour; J.C. Hagelin; J.A. Johnson.

INTERESTING NOTE

Interannual site fidelity to nonbreeding areas in South America was high for three birds tagged in Alaska that carried geolocators for two consecutive years (Hagelin et al. 2021). Individuals also often show high breeding site fidelity (Altman and Sallabanks 2020). The wide range of migratory routes, migration distances, and nonbreeding locations from a single breeding study area illustrates why migratory connectivity may be weak or difficult to measure in the Olive-sided Flycatcher (Hagelin et al. 2021).

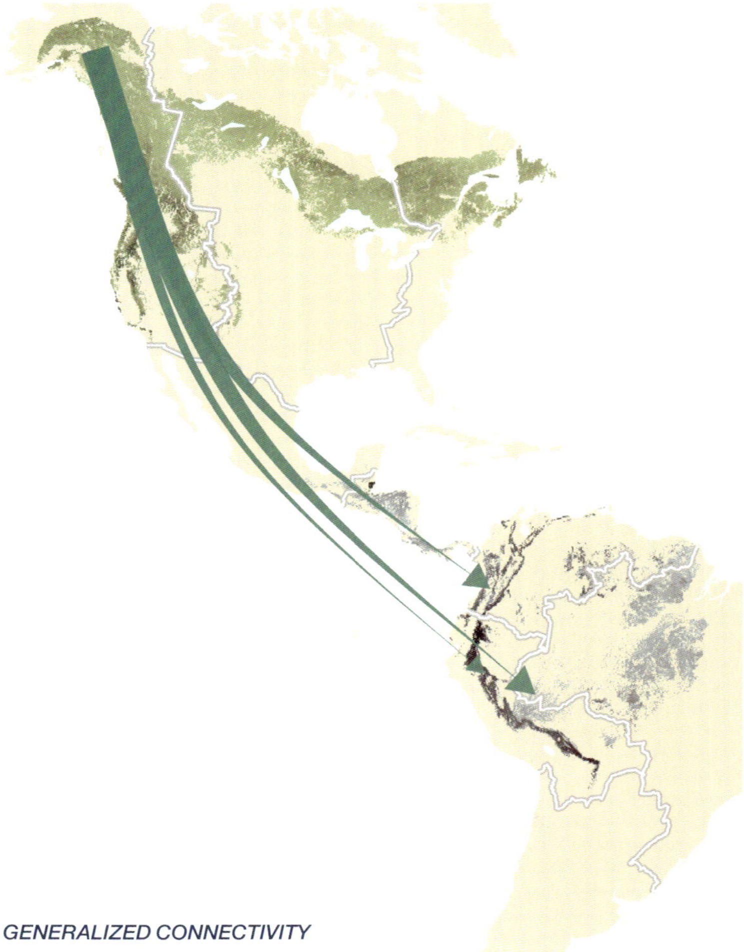

GENERALIZED CONNECTIVITY

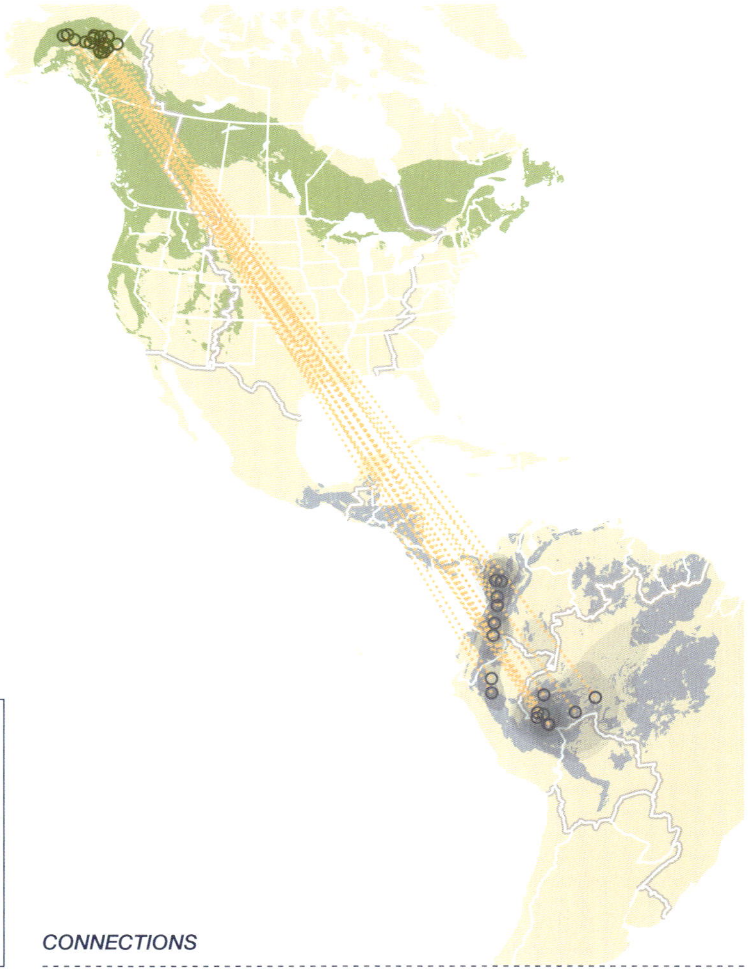

CONNECTIONS

GREATER PEWEE *Contopus pertinax*

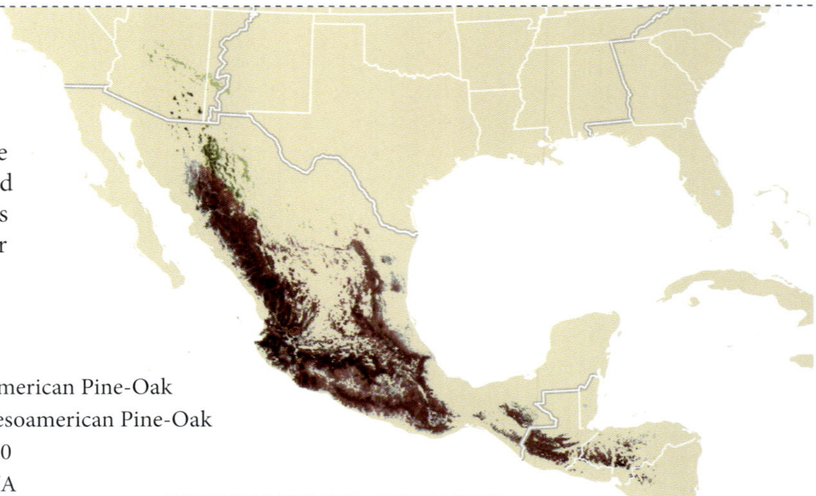

DISTRIBUTION AND ABUNDANCE

DISTRIBUTION AND ABUNDANCE

It breeds in the mountain *cordilleras* of Mexico, Guatemala, Honduras, and northwestern Nicaragua, and more sparsely in the Central Mexican Plateau and in central and southern Arizona and southwestern New Mexico. During the nonbreeding season, birds migrate to nonbreeding areas from northwestern Mexico to other portions of the year-round range (Fink et al. 2023).

ORDER Passeriformes
FAMILY Tyrannidae
BREEDING HABITAT Forests, Mesoamerican Pine-Oak
NONBREEDING HABITAT Forests, Mesoamerican Pine-Oak
GLOBAL POPULATION SIZE 2,000,000
POPULATION CHANGE SINCE 1970 NA
IUCN RED LIST CATEGORY Least Concern
R2R TIPPING POINT SPECIES No
US LISTING STATUS No
CANADA SPECIES AT RISK No
PIF CCS 12/20

MIGRATORY CONNECTIVITY

OVERALL Migratory connectivity is unknown. No information on migratory connectivity was found for this species.
MIGRATORY PATHWAYS In the southern parts of the species range, many individuals remain in breeding areas year-round. Northern breeding and migratory southern populations undergo altitudinal and regional movements to nonbreeding areas in pine–oak and riparian habitats (Chace and Tweit 2020; Fink et al. 2023).

WESTERN WOOD-PEWEE *Contopus sordidulus*

DISTRIBUTION AND ABUNDANCE

DISTRIBUTION AND ABUNDANCE

It breeds in open forest and edge habitats from central Alaska, the western Canadian provinces, the western US (except for the California Central Valley and southwestern deserts), and central Mexico south to Honduras and possibly north-central Nicaragua and Costa Rica (Bemis and Rising 2020). It migrates to spend the nonbreeding season in mature tropical forest, sparsely from central Nicaragua to Colombia and western Ecuador, and more densely along the eastern slopes of the Andes from southern Colombia to Bolivia (Fink et al. 2023). Western Wood-Pewee populations have declined by more than 50% in the US and Canada since 1970 (Rosenberg et al. 2019), with Alaskan, Canadian, Great Basin, and West Coast populations showing the steepest declines (Sauer et al. 2019).

ORDER Passeriformes
FAMILY Tyrannidae
BREEDING HABITAT Forests, Temperate Western
NONBREEDING HABITAT Forests, Tropical Montane Evergreen
GLOBAL POPULATION SIZE 9,600,000
POPULATION CHANGE SINCE 1970 −54%
IUCN RED LIST CATEGORY Least Concern
R2R TIPPING POINT SPECIES No
US LISTING STATUS No
CANADA SPECIES AT RISK No
PIF CCS 11/20

MIGRATORY CONNECTIVITY

OVERALL Migratory connectivity is unknown. No information on migratory connectivity was found for this species.
OTHER APPROACHES Migratory connectivity based on subspecific identification is challenging due to plumage and measurement similarities with the Eastern Wood-Pewee, especially nonbreeding birds. Nonetheless, *C. s. saturatus* of Alaska to western Oregon probably spends the nonbreeding season in Colombia and Venezuela; *C. s. veliei* makes up most of the remainder of the US population and probably spends the nonbreeding season in western South America (Bemis and Rising 2020)
MIGRATORY PATHWAYS This species migrates along a broad front, departing their breeding grounds and migrating throughout the US west of the Great Plains. They continue southward through much of Mexico and Central America, excluding the east Mexican lowlands and the Yucatán. Upon entering South America, the species concentrates their southward movements along the Andean slopes to reach their nonbreeding grounds (Fink et al. 2023).

EASTERN WOOD-PEWEE *Contopus virens*

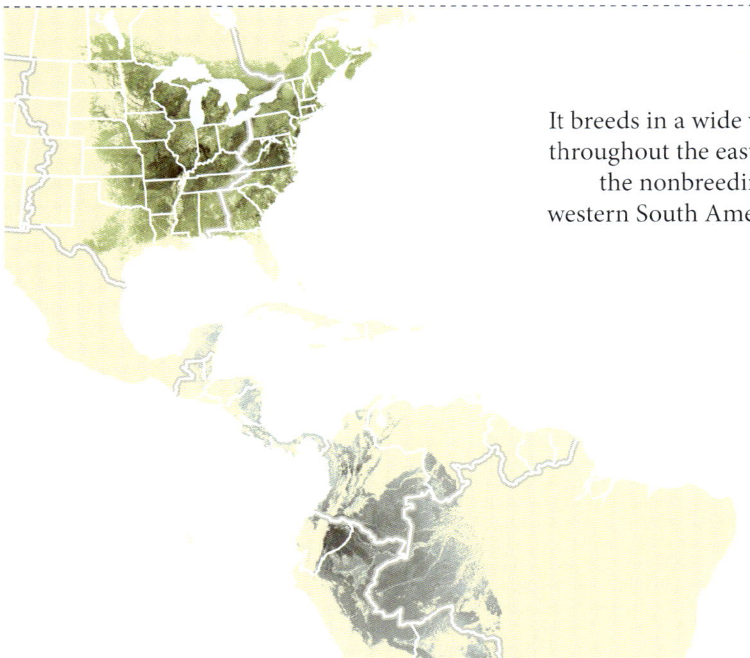

DISTRIBUTION AND ABUNDANCE

DISTRIBUTION AND ABUNDANCE

It breeds in a wide variety of wooded habitats from southern Manitoba to Nova Scotia and throughout the eastern US south to central Texas to northern Florida. It migrates to spend the nonbreeding season sparsely in the Yucatán and Central America and throughout western South America to Bolivia, with the densest concentrations in eastern Ecuador and northeastern Peru (Fink et al. 2023).

ORDER Passeriformes
FAMILY Tyrannidae
BREEDING HABITAT Forests, Temperate Eastern
NONBREEDING HABITAT Forests, Tropical Lowland Evergreen
GLOBAL POPULATION SIZE 6,500,000
POPULATION CHANGE SINCE 1970 −46%
IUCN RED LIST CATEGORY Least Concern
R2R TIPPING POINT SPECIES No
US LISTING STATUS No
CANADA SPECIES AT RISK No
PIF CCS 9/20

MIGRATORY CONNECTIVITY

OVERALL Migratory connectivity is unknown. No information on migratory connectivity was found for this species. One Motus tag was deployed, but there are no breeding to nonbreeding connections.

MIGRATORY PATHWAYS Birds may migrate both along the eastern Gulf Coast of Mexico as well as across the Gulf of Mexico to northern Central America and possibly the northern coast of South America (Fink et al. 2023).

YELLOW-BELLIED FLYCATCHER *Empidonax flaviventris*

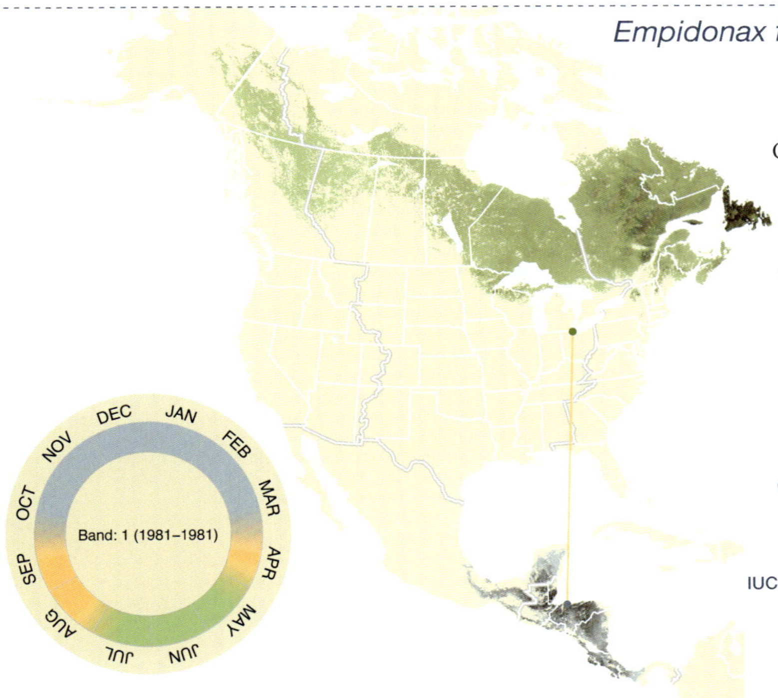

CONNECTIONS

DISTRIBUTION AND ABUNDANCE

Yellow-bellied Flycatchers breed in boreal forests and peatlands of Canada, especially in the provinces of Manitoba, Ontario, Quebec, and Newfoundland and Labrador, and the far northeastern US. Highest breeding abundances are in eastern Canada (Fink et al. 2023). The species spends the nonbreeding season in the humid forests and vegetation of Mexico and Central America (Gross and Lowther 2020).

ORDER Passeriformes
FAMILY Tyrannidae
BREEDING HABITAT Forests, Boreal
NONBREEDING HABITAT Forests, Tropical Lowland Evergreen
GLOBAL POPULATION SIZE 13,000,000
POPULATION CHANGE SINCE 1970 +196%
IUCN RED LIST CATEGORY Least Concern
R2R TIPPING POINT SPECIES No
US LISTING STATUS No
CANADA SPECIES AT RISK No
PIF CCS 10/20

MIGRATORY CONNECTIVITY

OVERALL No published studies on migratory connectivity were found. A single band record connected the Atlantic Flyway with Central America.

BANDING A single band recovery connected a breeding site in Ohio with a nonbreeding site in Honduras.

OTHER APPROACHES No tracking, stable isotope, or genetic studies of migratory connectivity were found for this species.

MIGRATORY PATHWAYS In pre-breeding, Yellow-bellied Flycatchers migrate north through eastern Mexico across the Great Plains and midwestern US to breeding grounds in Canada (Gross and Lowther 2020). In post-breeding migration, many follow this same pattern south, although they also expand east to the Atlantic. Upon reaching the Gulf Coast, some birds fly overland through eastern Mexico, while later birds appear to cross the Gulf of Mexico before making landfall in Central America (Fink et al. 2023).

ACADIAN FLYCATCHER *Empidonax virescens*

DISTRIBUTION AND ABUNDANCE

It breeds in mature deciduous and coniferous forests in the eastern US, except for northern New England, the northern Great Lakes region, and peninsular Florida. In the nonbreeding season, it migrates to Central America (Nicaragua, Costa Rica, and Panama) and the mountains of Colombia and Ecuador, most abundantly in eastern Panama and Colombia (Fink et al. 2023). Populations have declined mostly in the Appalachians and the west Gulf Coastal Plain (Sauer et al. 2019).

ORDER Passeriformes
FAMILY Tyrannidae
BREEDING HABITAT Forests, Temperate Eastern
NONBREEDING HABITAT Forests, Tropical Lowland Evergreen
GLOBAL POPULATION SIZE 5,200,000
POPULATION CHANGE SINCE 1970 −8%
IUCN RED LIST CATEGORY Least Concern
R2R TIPPING POINT SPECIES No
US LISTING STATUS No
CANADA SPECIES AT RISK Endangered
PIF CCS 10/20

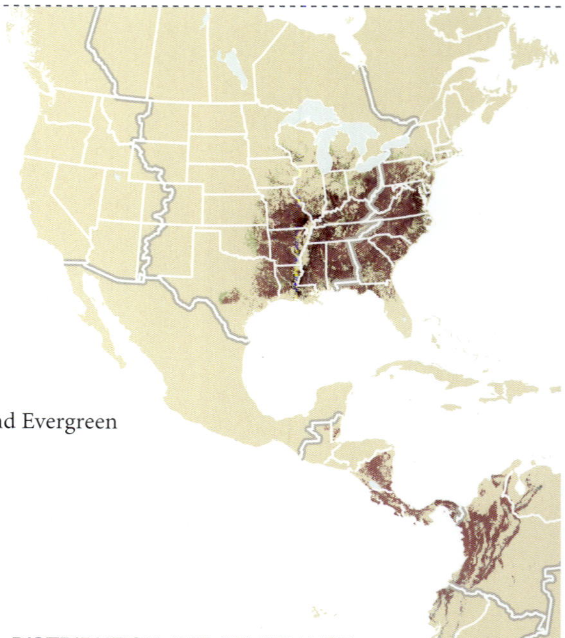

DISTRIBUTION AND ABUNDANCE

MIGRATORY CONNECTIVITY

OVERALL Migratory connectivity is unknown. No information on migratory connectivity was found for this species.

RADIO TRACKING Acadian Flycatchers are currently being tracked through the Motus Wildlife Tracking System from coffee landscapes in Colombia. There are no detections between breeding and nonbreeding areas (R. Bennett, unpublished data).

MIGRATORY PATHWAYS eBird animations (Fink et al. 2023) suggest a trans-Gulf migration to Belize and northern Honduras.

ALDER FLYCATCHER *Empidonax alnorum*

DISTRIBUTION AND ABUNDANCE

It breeds in the Canadian taiga and boreal forest from Alaska to Nova Scotia and in the US in the upper Midwest and northern New England. It is believed to spend the nonbreeding season in western South America, from Colombia south through Bolivia and east to western Brazil, with concentrations along the eastern base of the Andes in Peru and Bolivia (Lowther 2020a; Fink et al. 2023). Until 1973, Alder and Willow Flycatchers were considered a single species, Traill's Flycatcher (Lowther 2020a).

ORDER Passeriformes
FAMILY Tyrannidae
BREEDING HABITAT Wetlands, Boreal
NONBREEDING HABITAT Forests, Tropical Generalist
GLOBAL POPULATION SIZE 120,000,000
POPULATION CHANGE SINCE 1970 −28%
IUCN RED LIST CATEGORY Least Concern
R2R TIPPING POINT SPECIES No
US LISTING STATUS No
CANADA SPECIES AT RISK No
PIF CCS 8/20

INTERESTING NOTE

Because the Alder Flycatcher rarely vocalizes in the nonbreeding season and is difficult to visually separate from the Willow Flycatcher, verifying nonbreeding ranges depends on close examination of museum specimens, including both coloration and morphological measurements (Lowther 2020a).

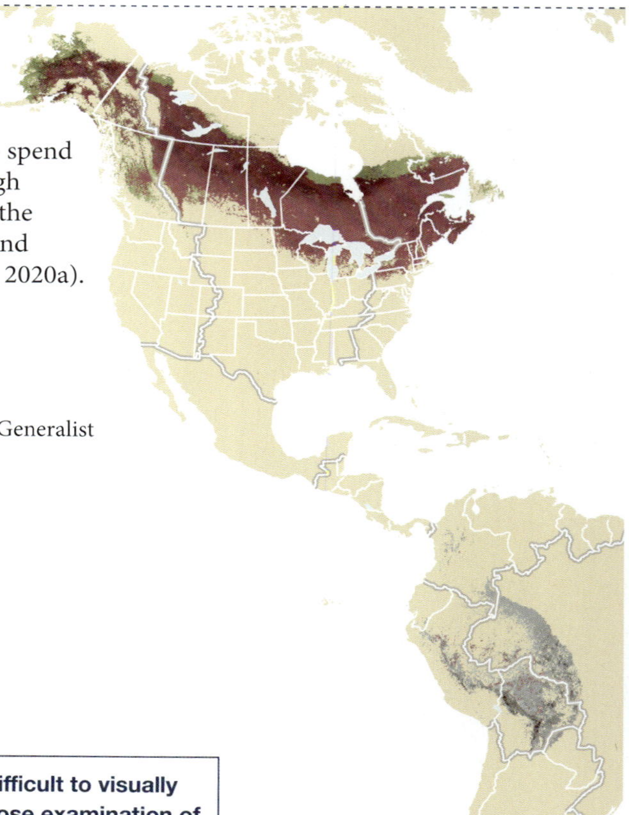

DISTRIBUTION AND ABUNDANCE

MIGRATORY CONNECTIVITY

OVERALL Migratory connectivity is unknown. No information on migratory connectivity was found for this species. A few tags were deployed through the Motus Wildlife Tracking System; there are no individual breeding to nonbreeding detections.

MIGRATORY PATHWAYS Both northbound and southbound migration is mainly through the central US (east Texas and Louisiana), southeastern Mexico, and Central America (Fink et al. 2023), but due to identification issues with nonbreeding *Empidonax*, these inferences should be regarded as hypothetical at best.

Empidonax traillii WILLOW FLYCATCHER

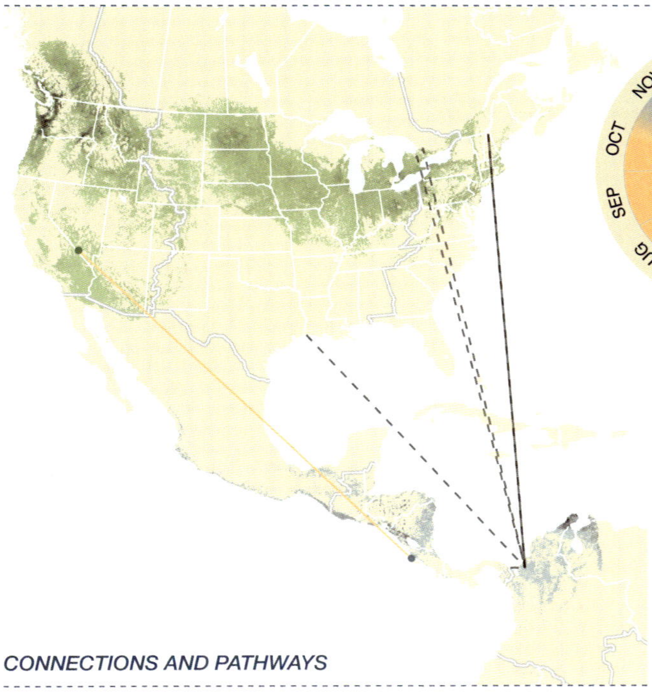

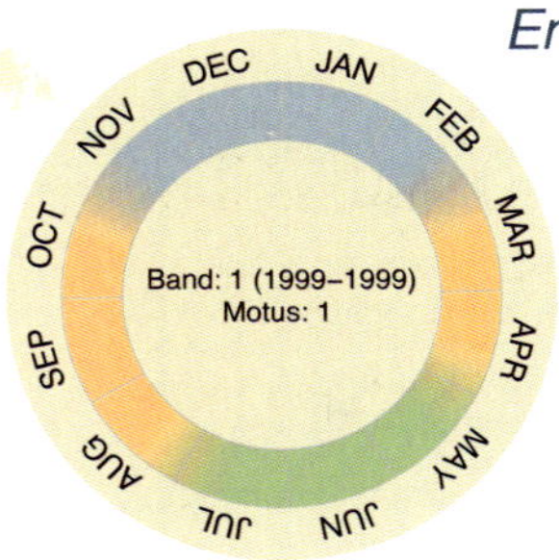

DISTRIBUTION AND ABUNDANCE

Willow Flycatchers breed across the northern US, Great Plains, Midwest, and Northeast, as well as southern British Columbia and Alberta, and the extreme southern portions of Saskatchewan, Manitoba, and Quebec. Willow Flycatchers spend the nonbreeding season in southern Mexico, Central America, and northern South America, specifically Colombia and Venezuela (Sedgwick 2020b).

ORDER Passeriformes
FAMILY Tyrannidae
BREEDING HABITAT Open Country, Habitat Mosaic
NONBREEDING HABITAT Open Country, Habitat Mosaic
GLOBAL POPULATION SIZE 8,100,000
POPULATION CHANGE SINCE 1970 −28%
IUCN RED LIST CATEGORY Least Concern
R2R TIPPING POINT SPECIES No
US LISTING STATUS Endangered southwestern subspecies
CANADA SPECIES AT RISK No
PIF CCS 10/20

MIGRATORY CONNECTIVITY

OVERALL Migratory connectivity for this species is known primarily from genetic data and limited banding or tracking data and appears moderate, as subspecies of Willow Flycatcher occupy separate breeding grounds and distinct but overlapping nonbreeding grounds (Paxton et al. 2011; Ruegg et al. 2021).
BANDING There is one band reencounter from a breeding site on the California-Nevada border to a nonbreeding site in Costa Rica.
RADIO TRACKING Detections from the Motus Wildlife Tracking System during migration connect Vermont and the Gulf Coast, US, and Ontario, Canada, to Colombia (Bayly and Gómez, unpublished data). There are no breeding to nonbreeding detections.
GENETICS Seven genetically distinct populations of Willow Flycatcher have been identified, including Interior West, Pacific Northwest, Eastern, Southwest, Kern (California), Southern California, and White Mountain (Arizona). The Pacific Northwest breeding population spend the nonbreeding period from Mexico to Central America. The Interior West breeding birds spend the nonbreeding period from Central America to Ecuador. The Eastern genetic group also spends the nonbreeding period in Central America to Ecuador. The endangered Southwest Willow Flycatcher nonbreeding grounds are localized to Nicaragua and Costa Rica (Ruegg et al. 2021).
OTHER APPROACHES No satellite transmitter, geolocator, or stable isotope studies of migratory connectivity were found for this species.
MIGRATORY PATHWAYS Willow Flycatchers appear to fly overland in pre-breeding migration from South and Central America through Mexico to the southwestern US before dispersing to northern and eastern breeding grounds; this pattern is reversed in post-breeding migration (Sedgwick 2020b; Fink et al. 2023). The Pacific Northwest breeding population migrates along the Pacific Flyway. Interior West breeding birds use the Pacific and Central flyways. The Eastern genetic group uses the Central, Mississippi, and Atlantic flyways.

DATA CONTRIBUTORS
N. Bayly; C. Gómez.

Empidonax minimus LEAST FLYCATCHER

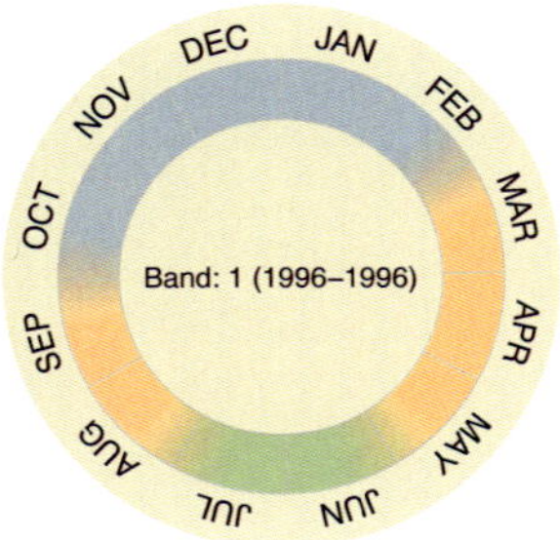

DISTRIBUTION AND ABUNDANCE

The Least Flycatcher is a common flycatcher across the northern United States from Maine west to Montana and across all of Canada, but at lower abundance in British Columbia. During the nonbreeding season, Least Flycatchers are most abundant in Central America from southern Mexico, including the Yucatán, into Guatemala, Belize, El Salvador, and Honduras, and at lower densities into Nicaragua and Costa Rica (Fink et al. 2023)

ORDER Passeriformes
FAMILY Tyrannidae
BREEDING HABITAT Forests, Temperate Eastern
NONBREEDING HABITAT Forests, Tropical Generalist
GLOBAL POPULATION SIZE 27,000,000
POPULATION CHANGE SINCE 1970 −57%
IUCN RED LIST CATEGORY Least Concern
R2R TIPPING POINT SPECIES No
US LISTING STATUS No
CANADA SPECIES AT RISK No
PIF CCS 11/20

MIGRATORY CONNECTIVITY

OVERALL Migratory connectivity is unknown. Based on observations and limited band reencounter data, some degree of migratory connectivity may exist, but further research is necessary (Tarof and Briskie 2020; Fink et al. 2023). MC is not able to be generated due to insufficient data.

INTERESTING NOTE

An analysis of Least Flycatcher tail feathers from the nonbreeding grounds found that stable isotope signatures for hatch-year birds were significantly different than those of adults captured at the same site, which is consistent with the hypothesis that adults grow new flight feathers after migration to Central America, while hatch-year birds grow feathers on the breeding grounds (Hobson and Wassenaar 1996).

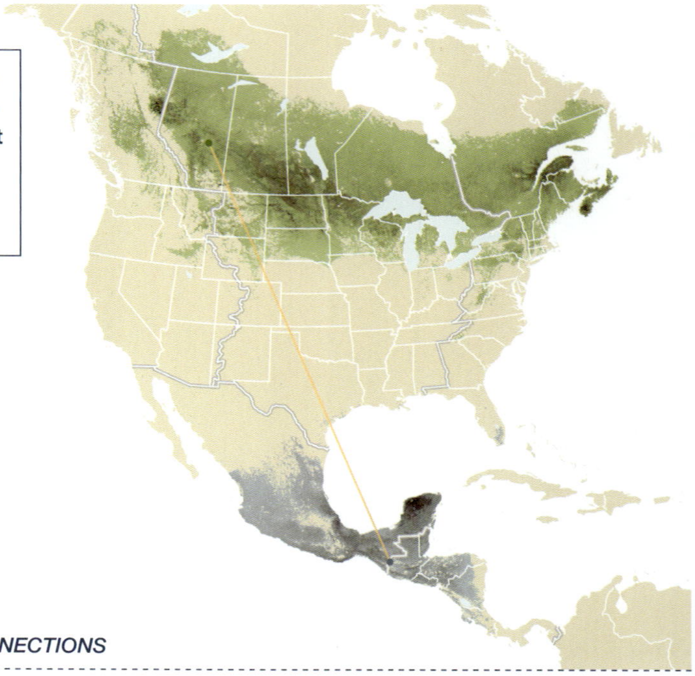

CONNECTIONS

BANDING There is one band reencounter from a breeding site in Alberta to a nonbreeding site in Guatemala.
RADIO TRACKING There were limited tag deployments within the Motus Wildlife Tracking System. There are no breeding to nonbreeding detections.
OTHER APPROACHES No tracking, stable isotope, or genetic studies of migratory connectivity were found for this species.
MIGRATORY PATHWAYS The breeding range of this species is nearly contiguous across northern North America. Animations using eBird during migration, however, suggest that Least Flycatchers concentrate along the Rockies in the Central Flyway as they move toward nonbreeding grounds in Mexico and Central America (Fink et al. 2023). Individuals from different parts of this range therefore travel different routes to this central location, mostly avoiding the far southeastern US and areas west of the Rockies. Many travel overland through eastern Mexico, while at least some proportion of the population undergoes trans-Gulf migrations (Russell 2005; Fink et al. 2023).

HAMMOND'S FLYCATCHER *Empidonax hammondii*

DISTRIBUTION AND ABUNDANCE

It breeds from central Alaska south through the mountains of British Columbia, in the Pacific Coast Ranges south to Northern California, the Cascades, and the Sierra Nevada to Southern California, and from the Rockies to northern New Mexico. It migrates to Southern California, southern Arizona, western New Mexico, most of Mexico, Guatemala, Honduras, and northwestern Nicaragua for the nonbreeding season, with the densest concentrations in the mountain *cordilleras* of Mexico (Fink et al. 2023).

ORDER Passeriformes
FAMILY Tyrannidae
BREEDING HABITAT Forests, Temperate Western
NONBREEDING HABITAT Forests, Mesoamerican Pine-Oak
GLOBAL POPULATION SIZE 20,000,000
POPULATION CHANGE SINCE 1970 +31%
IUCN RED LIST CATEGORY Least Concern
R2R TIPPING POINT SPECIES No
US LISTING STATUS No
CANADA SPECIES AT RISK No
PIF CCS 10/20

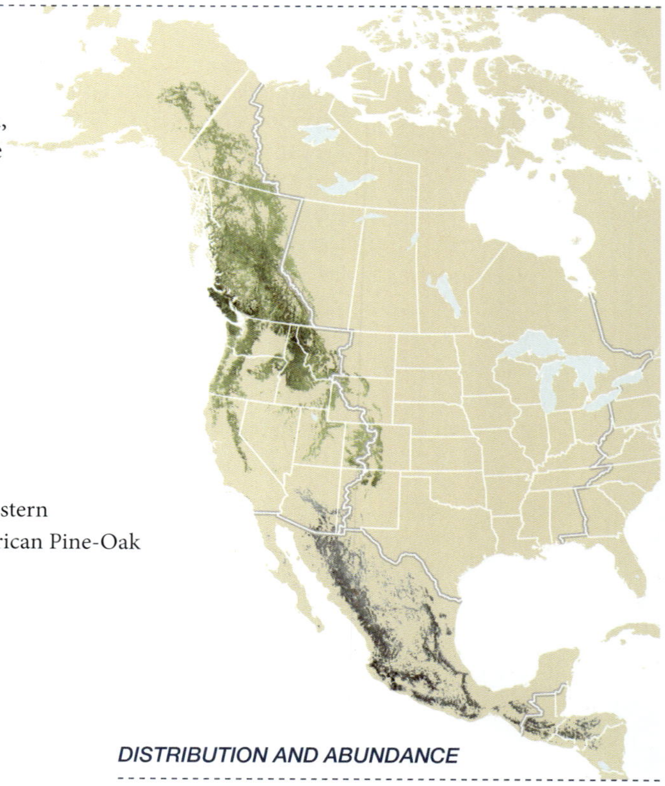

DISTRIBUTION AND ABUNDANCE

MIGRATORY CONNECTIVITY

OVERALL Migratory connectivity is relatively unknown.
RADIO TRACKING A few tags deployed through the Motus Wildlife Tracking System; there are no individual breeding to nonbreeding detections.
STABLE ISOTOPES Stable hydrogen isotope analysis, incorporating relative abundance information from Breeding Bird Surveys, assigned feather samples collected from nonbreeding Hammond's Flycatchers in southern Tamaulipus, Mexico, to the southern portions of the breeding distribution (Hobson et al. 2014).
MIGRATORY PATHWAYS This species is a complete migrant. Upon departure from their breeding grounds, this species moves broadly through a variety of habitats in the American West, with the highest migratory densities in montane areas. For individuals that continue into Mexico, populations are concentrated along the Sierra Madre Occidental and the Sierra Madre Oriental. Further movements bring this species to the Central Mexican Plateau and southward to montane areas of northern Central America (Sedgwick 2020a; Fink et al. 2023).

Empidonax wrightii GRAY FLYCATCHER

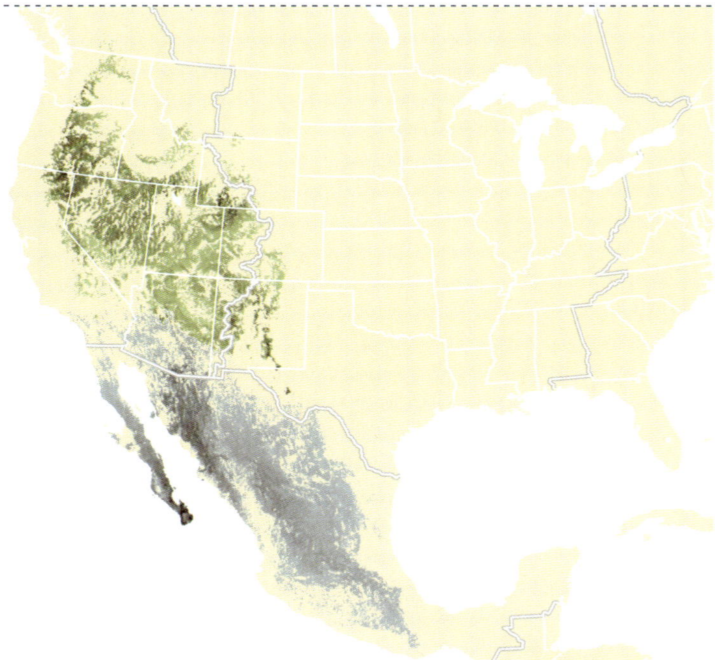

DISTRIBUTION AND ABUNDANCE

DISTRIBUTION AND ABUNDANCE

It breeds from southern British Columbia to southwestern California and from southern Montana to western Texas. It spends the nonbreeding season in low density in the southwestern US and northeastern and southwestern Mexico, and in higher density in Baja California Sur, Sonora, and central Mexico (Fink et al. 2023).

ORDER Passeriformes
FAMILY Tyrannidae
BREEDING HABITAT Forests, Temperate Western
NONBREEDING HABITAT Arid Lands, Desert Scrub
GLOBAL POPULATION SIZE 2,900,000
POPULATION CHANGE SINCE 1970 +166%
IUCN RED LIST CATEGORY Least Concern
R2R TIPPING POINT SPECIES No
US LISTING STATUS No
CANADA SPECIES AT RISK No
PIF CCS 10/2

INTERESTING NOTE

In 1939, Allan Phillips discovered that the type specimen for the Dusky Flycatcher (*Empidonax oberholseri*) was actually a Gray Flycatcher. The two had long been confused, but again it was Phillips who discovered a discriminating field mark: the Gray Flycatcher wags its tail gently downward, whereas all other *Empidonax* flycatchers flick their tails up and then down (Schlossberg and Sterling 2020).

MIGRATORY CONNECTIVITY

OVERALL Migratory connectivity is unknown. No information on migratory connectivity was found for this species.

MIGRATORY PATHWAYS It migrates primarily through the Great Basin, southeastern California, Arizona, New Mexico, and western Texas (Schlossberg and Sterling 2020).

Empidonax oberholseri DUSKY FLYCATCHER

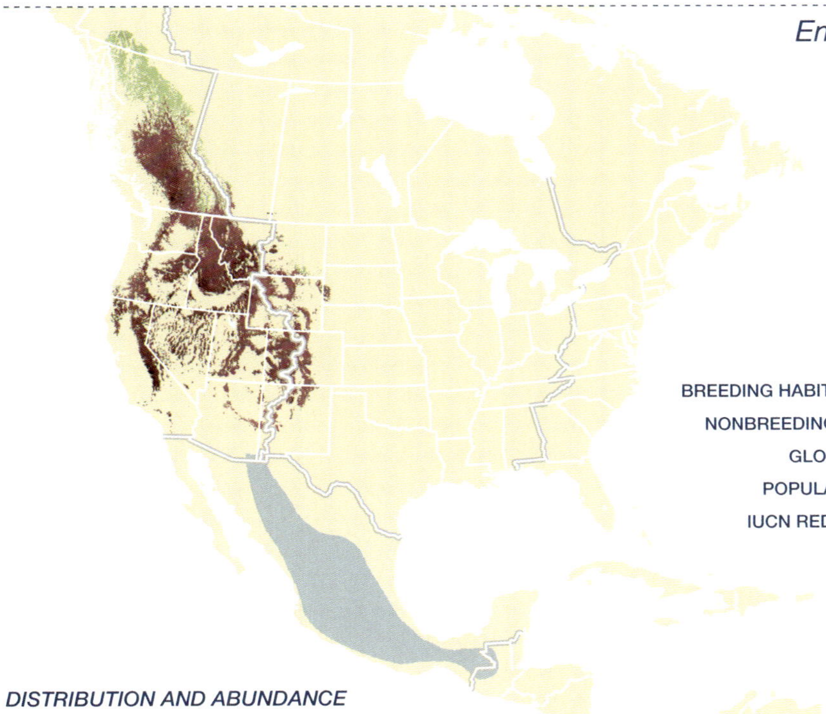

DISTRIBUTION AND ABUNDANCE

DISTRIBUTION AND ABUNDANCE

It breeds in brushy and open areas within mountain forests from the southern Yukon territory in Canada to southwestern California and northern New Mexico. It migrates to nonbreeding grounds from extreme southern Arizona, the Baja California peninsula, and mountainous areas of interior and western Mexico south to Oaxaca (Fink et al. 2023).

ORDER Passeriformes
FAMILY Tyrannidae
BREEDING HABITAT Forests, Temperate Western
NONBREEDING HABITAT Forests, Tropical Dry
GLOBAL POPULATION SIZE 8,800,000
POPULATION CHANGE SINCE 1970 −27%
IUCN RED LIST CATEGORY Least Concern
R2R TIPPING POINT SPECIES No
US LISTING STATUS No
CANADA SPECIES AT RISK No
PIF CCS 10/20

MIGRATORY CONNECTIVITY

OVERALL Migratory connectivity is unknown. No information on migratory connectivity was found for this species. One tag was deployed through the Motus Wildlife Tracking System, but there have been only local detections.

MIGRATORY PATHWAYS There is no information on specific routes. The species undergoes north-south migrations along a broad front in the American West and northern Mexico (Fink et al. 2023).

WESTERN FLYCATCHER *Empidonax difficilis*

Cordilleran Flycatcher (*E. d. occidentalis*) and Pacific-slope Flycatcher (*E. d. difficilis*)—long considered conspecific as Western Flycatcher—have now once again been recognized as races of a single species, Western Flycatcher (Chesser et al. 2023; Lowther et al. 2023). Here we summarize migratory connectivity information for both Cordilleran and Pacific-slope Flycatcher.

— CORDILLERAN FLYCATCHER *Empidonax difficilis occidentalis*

DISTRIBUTION AND ABUNDANCE

It breeds in association with water courses in mountain pine, fir, and spruce forests (Lowther et al. 2023) from extreme southern British Columbia and Alberta, throughout the US Rockies, and along the *cordilleras* (mountain ranges) of Mexico south to Oaxaca. During the nonbreeding season, US and northern Mexican populations migrate to the remainder of the Mexican range (where the species is a year-round resident); some individuals (from unknown breeding origin) move farther south into the mountains of Chiapas, Mexico (Fink et al. 2023). Populations have primarily declined in the western portion of its range (Sauer et al. 2019).

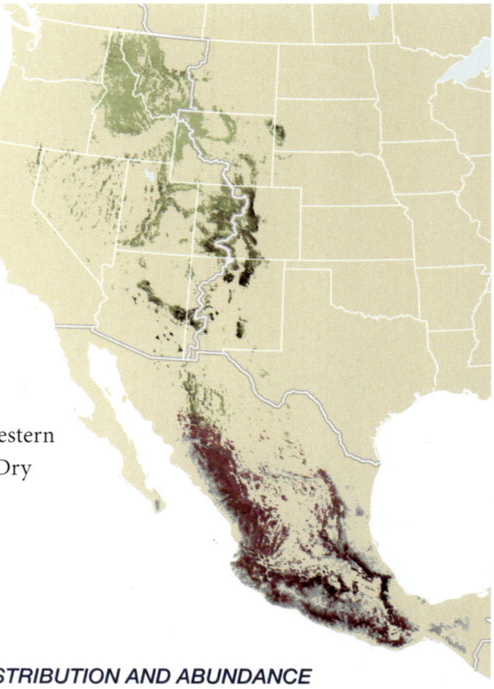

DISTRIBUTION AND ABUNDANCE

ORDER Passeriformes
FAMILY Tyrannidae
BREEDING HABITAT Forests, Temperate Western
NONBREEDING HABITAT Forests, Tropical Dry
GLOBAL POPULATION SIZE 9,500,000
POPULATION CHANGE SINCE 1970 −13%
IUCN RED LIST CATEGORY Least Concern
R2R TIPPING POINT SPECIES No
US LISTING STATUS No
CANADA SPECIES AT RISK No
PIF CCS 11/20

MIGRATORY CONNECTIVITY

OVERALL Migratory connectivity is relatively unknown. Limited information on migratory connectivity was found for this species.

GEOLOCATOR TRACKING Geolocator data shows that birds breeding in the San Juan Mountains of southwest Colorado spent the nonbreeding season in southern Mexico on the Pacific Slope of Guerrero, Oaxaca, and Chiapas (A. Boyce and C. Van Riper, unpublished data). Birds breeding on Mt. Lemmon in southeastern Arizona have been tagged with geolocators (C. Van Riper, unpublished data). Data are not presented in the Atlas.

MIGRATORY PATHWAYS US populations move south through the interior mountains of the US to at least central Mexico (Fink et al. 2023)

— PACIFIC-SLOPE FLYCATCHER *Empidonax difficilis difficilis*

DISTRIBUTION AND ABUNDANCE

It breeds in humid forests in the Pacific Coast Ranges, including the Cascade and Sierra Nevada mountains, from southern Alaska (Juneau area) to northern Baja California, and sparsely in eastern British Columbia, southwestern Alberta, and western Washington. During the nonbreeding season, the species migrates to southern Arizona, southern Baja California, and western Mexico to southern Oaxaca (Fink et al. 2023). Populations have declined primarily in the northern portion of the range (Sauer et al. 2019).

ORDER Passeriformes
FAMILY Tyrannidae
BREEDING HABITAT Forests, Temperate Western
NONBREEDING HABITAT Forests, Tropical Dry
GLOBAL POPULATION SIZE 9,500,000
POPULATION CHANGE SINCE 1970 −13%
IUCN RED LIST CATEGORY Least Concern
R2R TIPPING POINT SPECIES No
US LISTING STATUS No
CANADA SPECIES AT RISK No
PIF CCS 11/20

DISTRIBUTION AND ABUNDANCE

MIGRATORY CONNECTIVITY

OVERALL Migratory connectivity is unknown. No information on migratory connectivity was found for this species.

MIGRATORY PATHWAYS US populations migrate south along the Pacific and Gulf of California coasts (Fink et al. 2023).

Empidonax fulvifrons BUFF-BREASTED FLYCATCHER

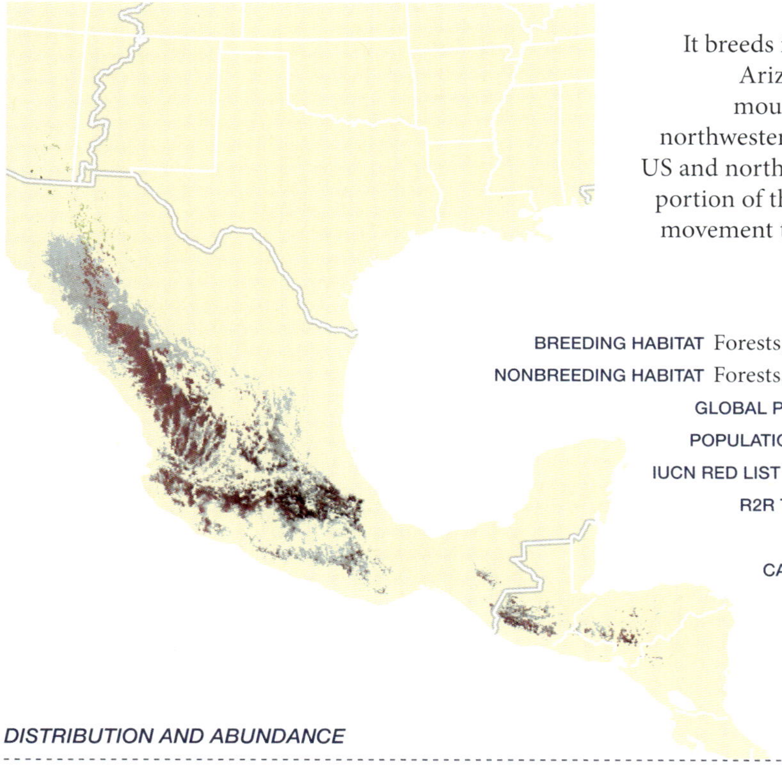

DISTRIBUTION AND ABUNDANCE

DISTRIBUTION AND ABUNDANCE

It breeds in open-canopy pine and pine–oak in the mountains of southeastern Arizona, the Sierra Madre Occidental, Trans-Mexican Volcanic Belt, and mountains of Oaxaca and Chiapas, Mexico, Guatemala, and Honduras to northwestern Nicaragua. During the nonbreeding season, it withdraws from the US and northern Sonora and Chihuahua, presumably to locations in the southern portion of the range, where the species resides year-round. eBird records suggest movement to lower elevations during the nonbreeding season (Fink et al. 2023).

ORDER Passeriformes
FAMILY Tyrannidae
BREEDING HABITAT Forests, Mesoamerican Pine-Oak
NONBREEDING HABITAT Forests, Mesoamerican Pine-Oak
GLOBAL POPULATION SIZE 2,000,000
POPULATION CHANGE SINCE 1970 NA
IUCN RED LIST CATEGORY Least Concern
R2R TIPPING POINT SPECIES No
US LISTING STATUS No
CANADA SPECIES AT RISK No
PIF CCS 12/20

MIGRATORY CONNECTIVITY

OVERALL Migratory connectivity is unknown. No information on migratory connectivity or pathways was found for this species.

Sayornis nigricans BLACK PHOEBE

DISTRIBUTION AND ABUNDANCE

DISTRIBUTION AND ABUNDANCE

It breeds from the US/Canadian border along the west coast, southwestern US, throughout Mexico (except the southern Gulf Coast and Yucatán Peninsula) and Central America, and in western South America to northern Argentina, mainly along the eastern slopes of the Andes. The greatest breeding abundance occurs in California, the Baja California peninsula, and western and central Mexico. The nonbreeding range is similar, except that most individuals migrate south from Nevada, Utah, Colorado, northern Arizona, and New Mexico (Fink et al. 2023). The US population has more than doubled in size since 1970 (Rosenberg et al. 2019).

ORDER Passeriformes
FAMILY Tyrannidae
BREEDING HABITAT Wetlands, Generalist
NONBREEDING HABITAT Wetlands, Generalist
GLOBAL POPULATION SIZE 5,000,000
POPULATION CHANGE SINCE 1970 +215%
IUCN RED LIST CATEGORY Least Concern
R2R TIPPING POINT SPECIES No
US LISTING STATUS No
CANADA SPECIES AT RISK No
PIF CCS 8/20

MIGRATORY CONNECTIVITY

OVERALL Migratory connectivity is unknown. No information on migratory connectivity was found for this species.

MIGRATORY PATHWAYS Patterns of occupancy indicate that this species moves geographically and altitudinally during the nonbreeding season, with increasing numbers of birds found in coastal areas and southern parts of the species' range (Wolf 2020).

EASTERN PHOEBE *Sayornis phoebe*

DISTRIBUTION AND ABUNDANCE

Eastern Phoebes breed throughout the eastern US and Great Plains as well as southern Canada from Nova Scotia to Alberta, in woody vegetation near water, often using bridges, buildings, and other structures for nesting. They spend the nonbreeding period primarily in woodland edges in the southeastern US, where the species is present year-round. Its nonbreeding range extends along the Gulf Coast, up the lower Mississippi River, and south into Mexico (Weeks 2020; Fink et al. 2023).

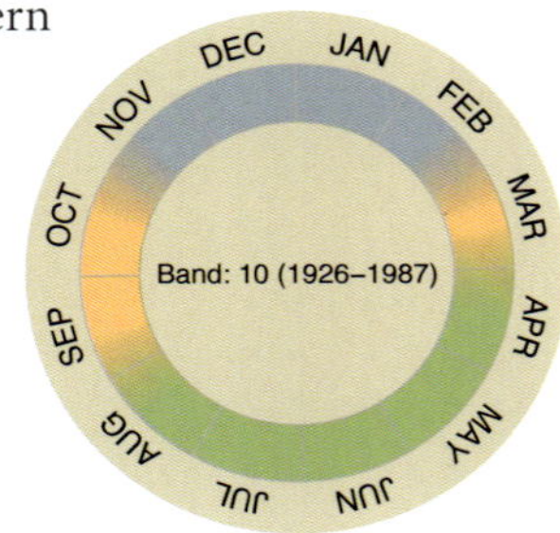

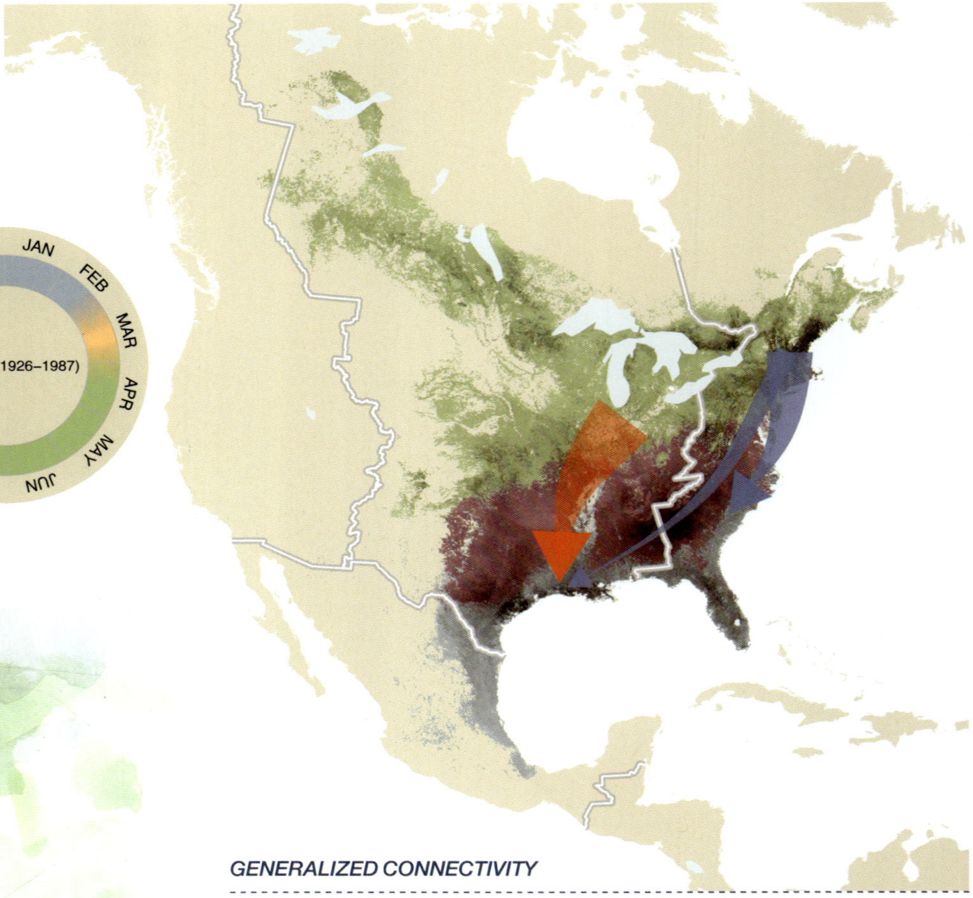

GENERALIZED CONNECTIVITY

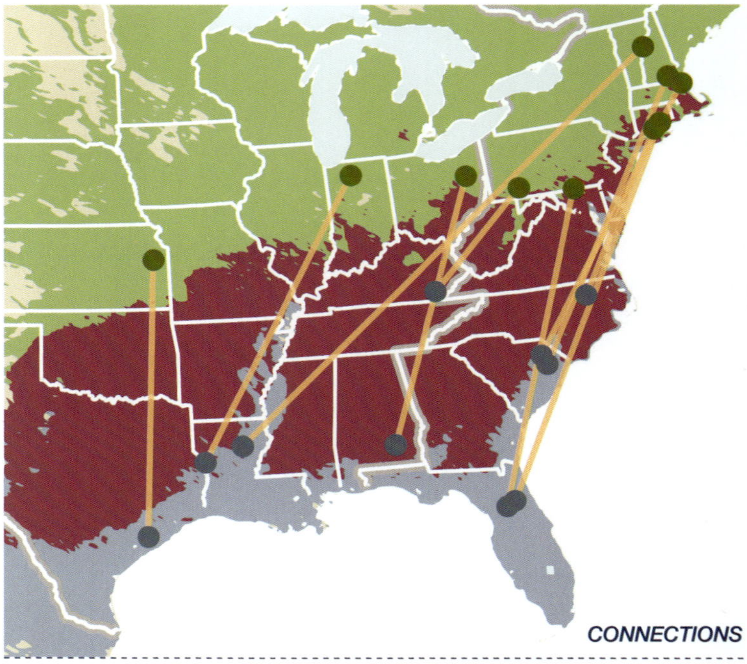

CONNECTIONS

ORDER Passeriformes
FAMILY Tyrannidae
BREEDING HABITAT Open Country, Habitat Mosaic
NONBREEDING HABITAT Open Country, Habitat Mosaic
GLOBAL POPULATION SIZE 35,000,000
POPULATION CHANGE SINCE 1970 +32%
IUCN RED LIST CATEGORY Least Concern
R2R TIPPING POINT SPECIES No
US LISTING STATUS No
CANADA SPECIES AT RISK No
PIF CCS 9/20v

	Atlantic	Midcontinent
Atlantic (n = 7)	71%	29%
Midcontinent (n = 3)	–	100%

MIGRATORY CONNECTIVITY

OVERALL Based on limited band reencounter data records, Eastern Phoebes breeding within the Midcontinent Flyway migrated to nonbreeding areas within the flyway (100%). Atlantic Flyway breeding birds also largely spent the nonbreeding season within the flyway (71%) as well as the Midcontinent Flyway (29%). MC was not calculated due to limited sample size.

BANDING Band reencounters between breeding and nonbreeding periods show connections between breeding sites in the Midwest and nonbreeding sites along or near the Gulf Coast and from the northeastern to southeastern US.

OTHER APPROACHES No tracking, stable isotope, or genetic studies of migratory connectivity were found for this species.

MIGRATORY PATHWAYS Individuals appear to migrate in a straightforward north-south pattern such that Midcontinent Flyway individuals disperse south and along the western Gulf Coast, while Atlantic Flyway individuals disperse south and along the eastern Gulf Coast (Fink et al. 2023).

Sayornis saya SAY'S PHOEBE

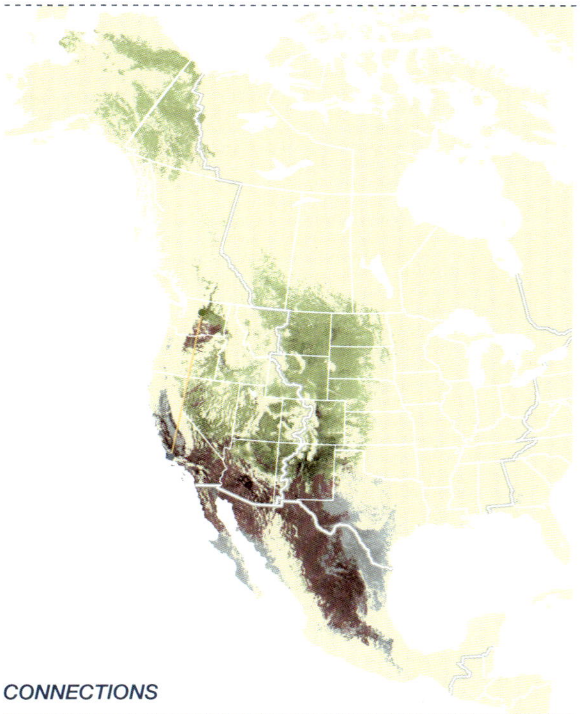

CONNECTIONS

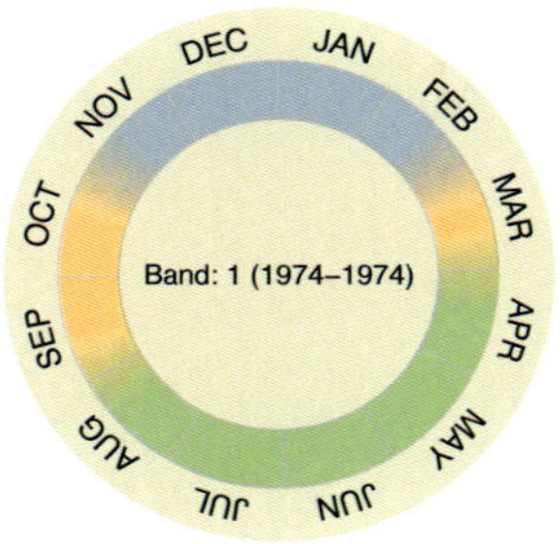

DISTRIBUTION AND ABUNDANCE

Say's Phoebes breed in open habitats of western North America in what appears to be two disjunct ranges, one in Alaska and the Yukon territory and another in most of the western states south to Mexico. They spend the nonbreeding season in the southwestern US from California to Texas and south through Mexico, including Baja California (Schukman and Wolf 2020).

ORDER Passeriformes
FAMILY Tyrannidae
BREEDING HABITAT Arid Lands, Desert Scrub
NONBREEDING HABITAT Open Country, Habitat Mosaic
GLOBAL POPULATION SIZE 5,900,000
POPULATION CHANGE SINCE 1970 +63%
IUCN RED LIST CATEGORY Least Concern
R2R TIPPING POINT SPECIES No
US LISTING STATUS No
CANADA SPECIES AT RISK No
PIF CCS 7/20

MIGRATORY CONNECTIVITY

OVERALL Migratory connectivity unknown. One band reencounter was located within the Pacific Flyway for both the breeding and nonbreeding season.

BANDING A single band reencounter connects a breeding site in central Washington to a nonbreeding site in coastal Southern California, US.

OTHER APPROACHES No tracking, stable isotope, or genetic studies of migratory connectivity were found for this species.

MIGRATORY PATHWAYS Individuals appear to migrate in a straightforward north-south pattern (Fink et al. 2023).

Pyrocephalus rubinus VERMILION FLYCATCHER

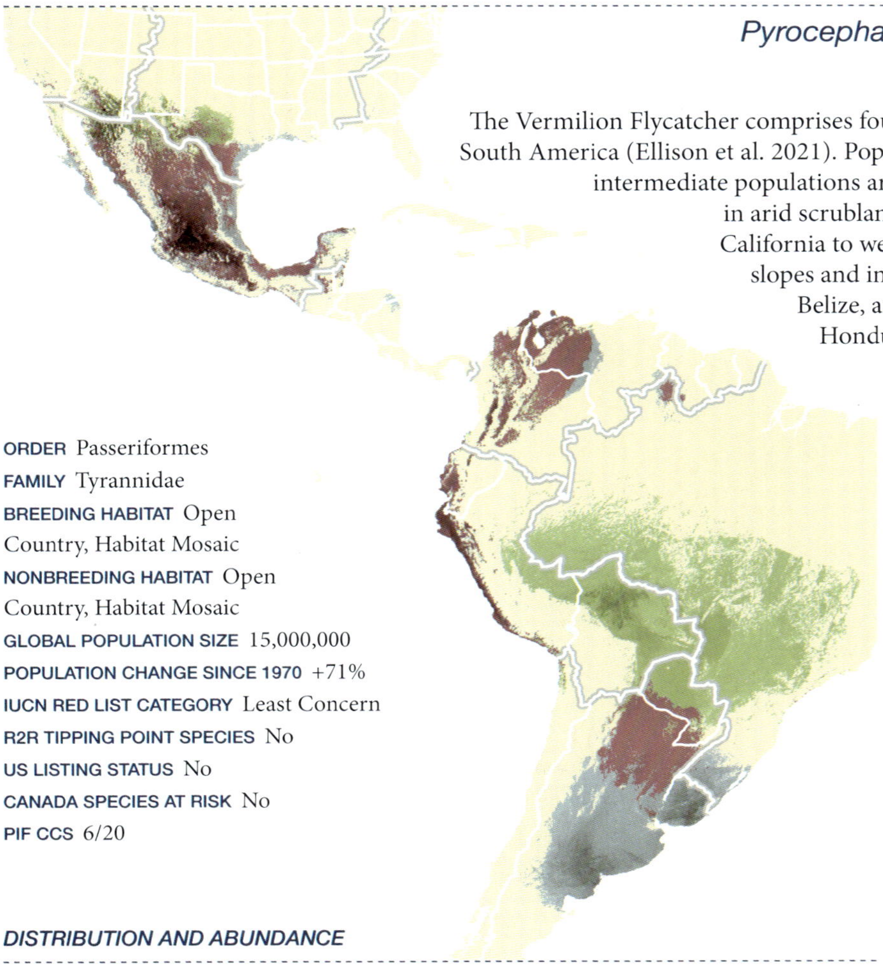

DISTRIBUTION AND ABUNDANCE

DISTRIBUTION AND ABUNDANCE

The Vermilion Flycatcher comprises four distinct populations, one in North America and three in South America (Ellison et al. 2021). Populations at northern and southern extremes are migratory; intermediate populations are resident (Ellison et al. 2021). In North America, it breeds in arid scrublands and other open areas, usually near water, from Southern California to western Texas, then most densely along both northern coastal slopes and in interior central Mexico, and in northern Tabasco, Yucatán, Belize, and Guatemala, and sparsely in the pine savannas of eastern Honduras and northeastern Nicaragua. After breeding, northern Mexican and US birds withdraw irregularly to spend the nonbreeding season farther south, or they disperse (perhaps joining other populations) to coastal Southern California, locally in southwestern deserts, along the US Gulf Coast, throughout mainland Florida, and in additional areas outside the breeding range in the coastal plains of Texas and Mexico (Ellison et al. 2021; Fink et al. 2023).

ORDER Passeriformes
FAMILY Tyrannidae
BREEDING HABITAT Open Country, Habitat Mosaic
NONBREEDING HABITAT Open Country, Habitat Mosaic
GLOBAL POPULATION SIZE 15,000,000
POPULATION CHANGE SINCE 1970 +71%
IUCN RED LIST CATEGORY Least Concern
R2R TIPPING POINT SPECIES No
US LISTING STATUS No
CANADA SPECIES AT RISK No
PIF CCS 6/20

MIGRATORY CONNECTIVITY

OVERALL Migratory connectivity is unknown. No information on migratory connectivity was found for this species. *P. r. flammeus* individuals that breed from southeast California east to western Texas generally migrate south into Mexico for the nonbreeding season, with some lingering in the southern US; *P. r. mexicanus* individuals that breed in south-central Texas may move south during the nonbreeding season but may also spread northward and eastward to southern Louisiana, southern Mississippi, and northwest Florida (Ellison et al. 2021).

MIGRATORY PATHWAYS Overall movements are variable, but most migrants move to the coastal plains of Mexico and the Gulf Coast, or to mild inland locations such as low-elevation cities. Others likely move to parts of the species' core range farther south (Fink et al. 2023).

VIREONIDAE | *VIREOS*

BLACK-CAPPED VIREO *Vireo atricapilla*

DISTRIBUTION AND ABUNDANCE

It breeds in a restricted range from south-central Oklahoma through central Texas to northern Coahuila, Mexico. Migrates to nonbreeding areas in western Mexico from Sinaloa south to Michoacán (Fink et al. 2023). Listed as endangered under the US Endangered Species Act in 1987 (MacGregor-Fors et al. 2012), the species was delisted in 2018 due to successful recovery.

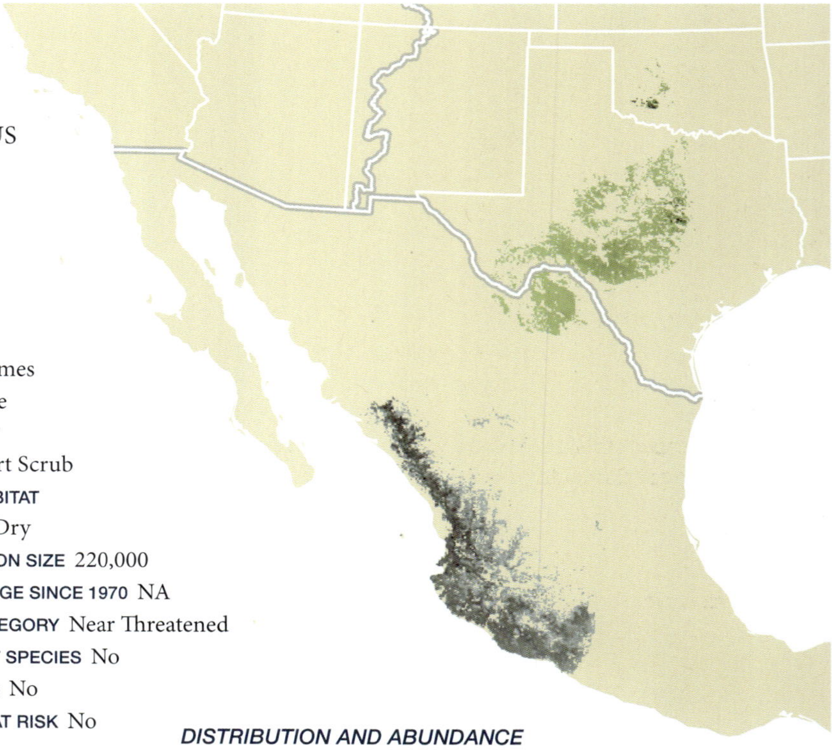

DISTRIBUTION AND ABUNDANCE

ORDER Passeriformes
FAMILY Vireonidae
BREEDING HABITAT Arid Lands, Desert Scrub
NONBREEDING HABITAT Forests, Tropical Dry
GLOBAL POPULATION SIZE 220,000
POPULATION CHANGE SINCE 1970 NA
IUCN RED LIST CATEGORY Near Threatened
R2R TIPPING POINT SPECIES No
US LISTING STATUS No
CANADA SPECIES AT RISK No
PIF CCS 16/20

MIGRATORY CONNECTIVITY

OVERALL Migratory connectivity is unknown. No information on migratory connectivity was found for this species.

MIGRATORY PATHWAYS Based on sight observations, MacGregor-Fors et al. (2012) suggest a migratory pathway along the Sierra Madre Oriental and the Trans-Mexican Volcanic Belt for some individuals.

WHITE-EYED VIREO *Vireo griseus*

DISTRIBUTION AND ABUNDANCE

It breeds in dense scrubby undergrowth usually associated with young forest edge in the southeastern US and northeastern Mexico (Coahuila, Nuevo León, and Tamaulipas), much less densely in the Appalachians and upper Mississippi Valley. Northern breeding birds migrate south for the nonbreeding season to southern portions of the breeding range along the east coast from Virginia to Florida, west through southern Georgia and the southern portions of the Gulf states and also to southeastern Mexico, the Yucatán Peninsula, northern Guatemala and Belize, northern Honduras, the Bahamas, and western Cuba, and Puerto Rico. The densest nonbreeding abundance is in the Yucatán Peninsula and the southern tip of Florida (Fink et al. 2023).

ORDER Passeriformes
FAMILY Vireonidae
BREEDING HABITAT Forests, Temperate Eastern
NONBREEDING HABITAT Forests, Generalist
GLOBAL POPULATION SIZE 24,000,000
POPULATION CHANGE SINCE 1970 +47%
IUCN RED LIST CATEGORY Least Concern
R2R TIPPING POINT SPECIES No
US LISTING STATUS No
CANADA SPECIES AT RISK No
PIF CCS 7/20

MIGRATORY CONNECTIVITY

OVERALL Migratory connectivity is relatively unknown.
BANDING A male White-eyed Vireo banded in Maryland during the breeding season on April 29, 1970 (recaptured at the same location on May 10), was killed with a slingshot on December 24, 1970, at Corozal, Honduras (Woodward 1972).
RADIO TRACKING A few tags were deployed through the Motus Wildlife Tracking System; there are no breeding to nonbreeding detections.
OTHER APPROACHES The two northern subspecies (*V. g. noveboracensis* and *V. g. griseus*) migrate to southeastern Mexico south to Honduras for the nonbreeding season; V. *g. noveboracensis* also occurs in Cuba (Hopp 2022).
MIGRATORY PATHWAYS Northern breeding birds migrate along a broad front southward throughout the southern and central US but avoid the more arid plains and high areas of the Appalachians. As they move southward, the birds concentrate into higher-density corridors as they funnel toward areas of high nonbreeding concentration in southern Texas and northeast Mexico, the Yucatán, and Florida (Fink et al. 2023). This species undergoes both overland and trans-Gulf migrations, as indicated by its presence on oil rigs in the Gulf of Mexico during migratory periods (Russell 2005).

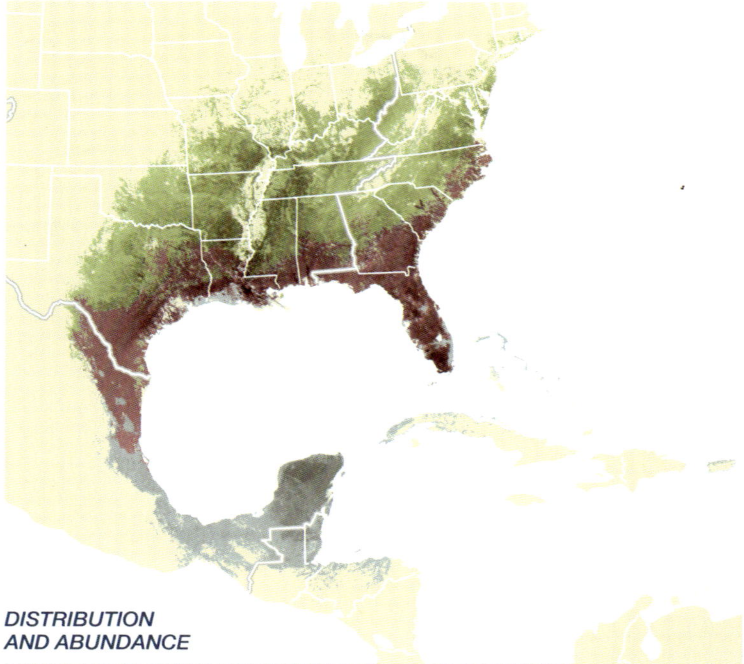

DISTRIBUTION AND ABUNDANCE

Vireo bellii BELL'S VIREO

DISTRIBUTION AND ABUNDANCE

This species has two distinct geographic populations. The eastern population breeds in the central US from South Dakota and southern Minnesota and Wisconsin south to central Texas and into northern Mexico (northeastern Chihuahua and Coahuila). This population of vireos likely spends the nonbreeding season along the central Pacific coast of Mexico and coastal southwestern Mexico south to El Salvador (Kus et al. 2022). The western population breeds from southern Nevada and southwestern Utah south and east to southern Sonora, Chihuahua, and western Texas, densely in southern Arizona and Sonora. This population of vireos probably spends the nonbreeding season in northwestern Mexico (Kus et al. 2022). The population breeding from central California south to northern Baja California, least Bell's vireo (*V. b. pusillus*), spends the nonbreeding season in southern Baja California Sur and is currently listed as endangered.

ORDER Passeriformes
FAMILY Vireonidae
BREEDING HABITAT Arid Lands, Desert Riparian
NONBREEDING HABITAT Forests, Tropical Dry
GLOBAL POPULATION SIZE 5,700,000
POPULATION CHANGE SINCE 1970 +66%
IUCN RED LIST CATEGORY Least Concern
R2R TIPPING POINT SPECIES No
US LISTING STATUS Endangered (least subspecies)
CANADA SPECIES AT RISK No
PIF CCS 10/20

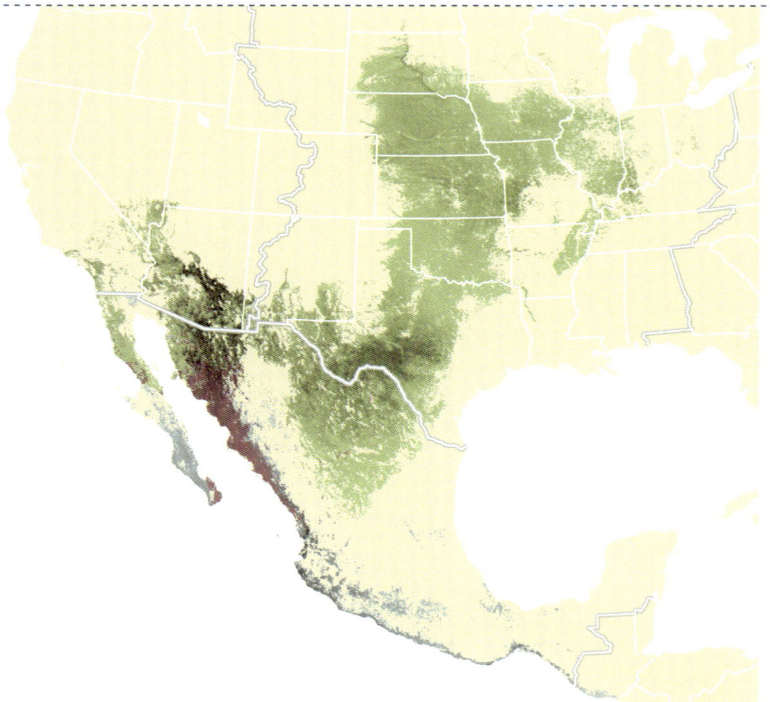

DISTRIBUTION AND ABUNDANCE

MIGRATORY CONNECTIVITY

OVERALL Migratory connectivity is relatively unknown. No information on migratory connectivity was found for this species.
OTHER APPROACHES Two distinct clades have been recognized, an eastern group (*V. b. bellii* and *V. b. medius*) and a western group (*V. b. arizonae* and *V. b. pusillus*), with evidence that *V. b. pusillus* (least Bell's vireo) is genetically distinct (Klicka et al. 2016).
MIGRATORY PATHWAYS eBird animations (Fink et al. 2023) support the independent movement of three populations, especially during the northbound pre-breeding migration: least Bell's vireo in California, the remaining western population, and the eastern population as described above, with an apparent divide in southwestern New Mexico and along the Sonoran/Chihuahua border.

GRAY VIREO *Vireo vicinior*

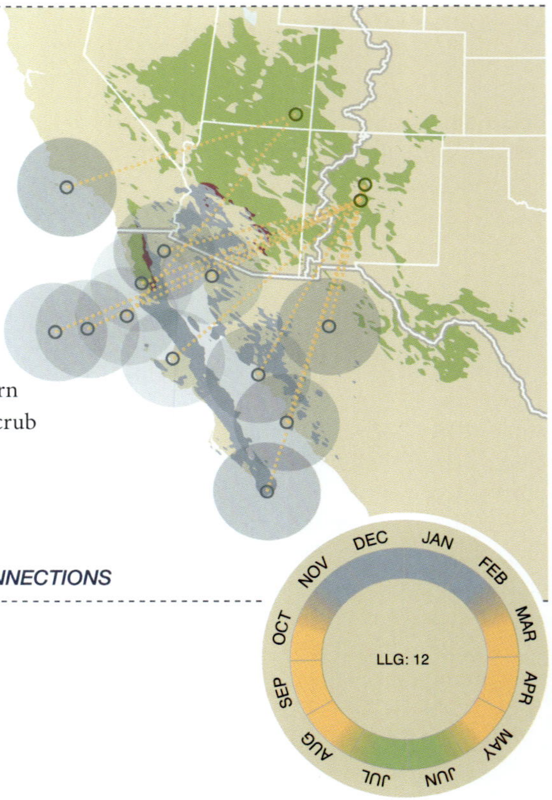

CONNECTIONS

DISTRIBUTION AND ABUNDANCE

It breeds in piñon–juniper / oak scrub and chaparral in hot, arid regions of the American Southwest and northern Mexico from southern Nevada, Utah, and western Colorado southeast through Arizona and New Mexico to southwest Texas and northern Coahuila; it also breeds in Southern California and northern Baja California. For the nonbreeding season, individuals migrate to southwestern Arizona, western Sonora, and the entire Baja California peninsula (Fink et al. 2023).

ORDER Passeriformes
FAMILY Vireonidae
BREEDING HABITAT Forests, Temperate Western
NONBREEDING HABITAT Arid Lands, Desert Scrub
GLOBAL POPULATION SIZE 560,000
POPULATION CHANGE SINCE 1970 +413%
IUCN RED LIST CATEGORY Least Concern
R2R TIPPING POINT SPECIES No
US LISTING STATUS No
CANADA SPECIES AT RISK No
PIF CCS 12/20

MIGRATORY CONNECTIVITY

OVERALL Migratory connectivity information is limited and known from one geolocator tracking study (Fischer et al. 2025), which detected weak to moderate migratory connectivity.

GEOLOCATOR TRACKING Light-level geolocators (n = 23) were deployed at three sites (Utah, and central and north-central New Mexico); 15 were recovered, and 12 had usable data (Fischer 2020). Gray Vireos from Utah and central New Mexico spent the nonbreeding season along the Baja California peninsula, Mexico, or north into Southern California, while Gray Vireos breeding in northern New Mexico migrated to nonbreeding areas in Sonora, Mexico. Tagged females spent the nonbreeding season farther north than males (Fischer et al. 2025).

MIGRATORY PATHWAYS There is no information on specific routes. The species appears to migrate across a broad front, with high abundance during northbound migration, west of the Rocky Mountains. Southbound migration is more widespread, and extends as far east as Nebraska, the Colorado plains, and West Texas (Barlow et al. 2020; Fink et al. 2023).

YELLOW-THROATED VIREO *Vireo flavifrons*

DISTRIBUTION AND ABUNDANCE

It is patchily distributed in the breeding season in southern Manitoba and Ontario and throughout the eastern US south to central Texas and northern Florida. It migrates to spend the nonbreeding season in southern Florida, the Bahamas, the Dominican Republic, and Puerto Rico, and more densely in Cuba, the Yucatán Peninsula, and southern Mexico through Central America, and sparsely again in eastern Panama and Colombia (Fink et al. 2023).

ORDER Passeriformes
FAMILY Vireonidae
BREEDING HABITAT Forests, Temperate Eastern
NONBREEDING HABITAT Forests, Tropical Lowland Evergreen
GLOBAL POPULATION SIZE 4,700,000
POPULATION CHANGE SINCE 1970 +71%
IUCN RED LIST CATEGORY Least Concern
R2R TIPPING POINT SPECIES No
US LISTING STATUS No
CANADA SPECIES AT RISK No
PIF CCS 10/20

DISTRIBUTION AND ABUNDANCE

MIGRATORY CONNECTIVITY

OVERALL Migratory connectivity is unknown. One tag was deployed through the Motus Wildlife Tracking System; there are no breeding to nonbreeding detections.

MIGRATORY PATHWAYS eBird animations (Fink et al. 2023) suggest a trans-Gulf overwater migration, with central Texas birds perhaps first moving east prior to the southbound flight.

Vireo cassinii CASSIN'S VIREO

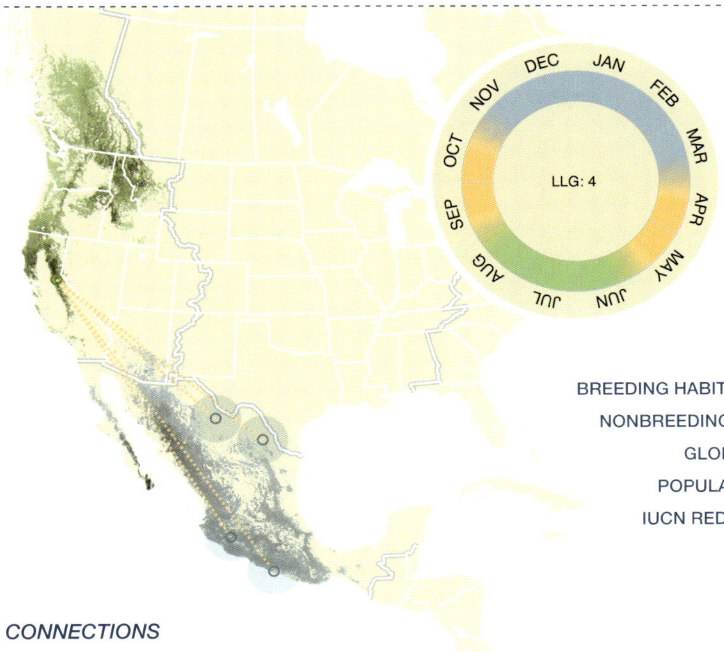

DISTRIBUTION AND ABUNDANCE

Cassin's Vireos are extreme western breeders from central British Columbia and southwestern Alberta to Northern California. There is a year-round resident population in the southern tip of Baja California Sur. The species' nonbreeding range extends through Baja California and western Mexico (Fink et al. 2023)

ORDER Passeriformes
FAMILY Vireonidae
BREEDING HABITAT Forests, Temperate Western
NONBREEDING HABITAT Forests, Tropical Dry
GLOBAL POPULATION SIZE 5,000,000
POPULATION CHANGE SINCE 1970 +66%
IUCN RED LIST CATEGORY Least Concern
R2R TIPPING POINT SPECIES No
US LISTING STATUS No
CANADA SPECIES AT RISK No
PIF CCS 9/20

MIGRATORY CONNECTIVITY

OVERALL Migratory connectivity is not well known. Besides one geolocator tracking study, no information on migratory connectivity was found for this species.
GEOLOCATOR TRACKING One tracking study (Hedley 2019) showed that four individuals with breeding territories in the Sierra Nevada mountains of California spent the nonbreeding season in disparate locations of north-central Mexico (n = 2) or southwestern Mexico (n = 2). Data are not presented.

MIGRATORY PATHWAYS Cassin's Vireos undertake a short-distance migration to molt in higher elevations of the Sierra Nevada before continuing south and southeast through Mexico to nonbreeding areas (Fink et al. 2023).

DATA CONTRIBUTORS
R. Hedley.

Vireo solitarius BLUE-HEADED VIREO

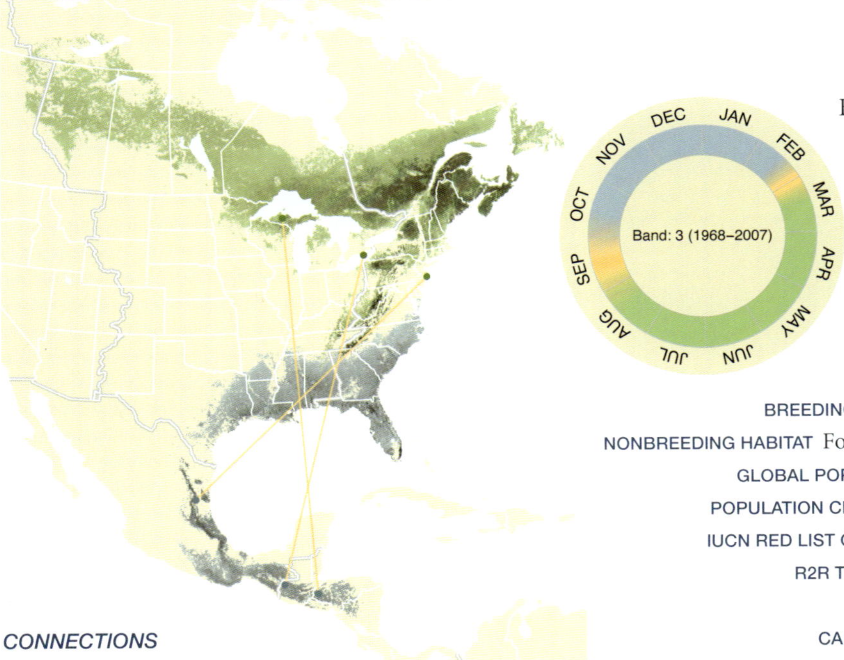

DISTRIBUTION AND ABUNDANCE

Blue-headed Vireos breed across southern Canada from eastern British Columbia to Newfoundland, and in the eastern US, including the Great Lakes, New England, and south through Appalachia. Their nonbreeding range spans the US mid-Atlantic, southeast, and Gulf Coast, as well as eastern Mexico and Central America, mainly Guatemala, Honduras, El Salvador, and Nicaragua (Morton and James 2020).

ORDER Passeriformes
FAMILY Vireonidae
BREEDING HABITAT Forests, Boreal
NONBREEDING HABITAT Forests, Tropical Generalist
GLOBAL POPULATION SIZE 13,000,000
POPULATION CHANGE SINCE 1970 +302%
IUCN RED LIST CATEGORY Least Concern
R2R TIPPING POINT SPECIES No
US LISTING STATUS No
CANADA SPECIES AT RISK No
PIF CCS 8/20

MIGRATORY CONNECTIVITY

OVERALL No published studies on migratory connectivity were found. Limited banding reencounter data connect the Atlantic Flyway to Mexico and the Midcontinent Flyway to Central America.
BANDING Band reencounter data connects New Jersey with northeastern Mexico, southern Ontario with Guatemala, and Michigan with Honduras.

MIGRATORY PATHWAYS Exact routes taken are unknown (Morton and James 2020). In pre-breeding migration, western birds appear to fly overland from Central America through Mexico and across the Great Plains to breeding areas in Canada, while eastern birds leave the southeast for Appalachia and beyond. These patterns appear to be reversed in post-breeding migration (Fink et al. 2023).

PLUMBEOUS VIREO *Vireo plumbeus*

DISTRIBUTION AND ABUNDANCE

It breeds in the US Great Basin and Intermountain ranges from central Montana southwest to extreme western California, south through Arizona and New Mexico, south along the Sierra Madre Occidental to Michoacán, and again in Oaxaca to western Honduras and northwest Nicaragua (Fink et al. 2023). After breeding, the US population migrates south to spend the nonbreeding season from southern Arizona south into the Mesoamerican portions of the breeding range and to the Baja California peninsula, northwestern Sonora, eastern Mexico, and farther east in Honduras and Nicaragua (Fink et al. 2023). Populations have declined range-wide since 1970 (Rosenberg et al. 2019), with the greatest losses in western California and in the northern and eastern portions of the breeding range (Sauer et al. 2019).

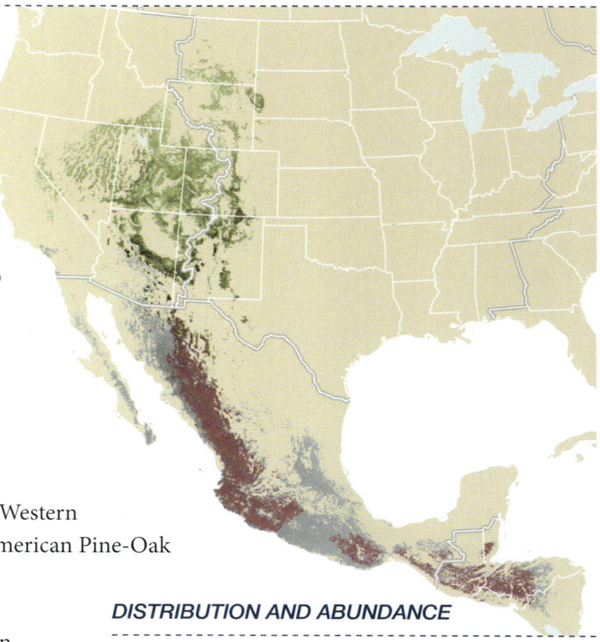

DISTRIBUTION AND ABUNDANCE

ORDER Passeriformes
FAMILY Vireonidae
BREEDING HABITAT Forests, Temperate Western
NONBREEDING HABITAT Forests, Mesoamerican Pine-Oak
GLOBAL POPULATION SIZE 3,500,000
POPULATION CHANGE SINCE 1970 −31%
IUCN RED LIST CATEGORY Least Concern
R2R TIPPING POINT SPECIES No
US LISTING STATUS No
CANADA SPECIES AT RISK No
PIF CCS 11/20

INTERESTING NOTE

The Plumbeous Vireo was first recognized as a species in 1997, when the American Ornithological Society divided The Solitary Vireo into three species: Plumbeous, Blue-headed, and Cassin's (Goguen and Curson 2020).

MIGRATORY CONNECTIVITY

OVERALL Migratory connectivity is unknown. No information on migratory connectivity was found for this species.

MIGRATORY PATHWAYS The movements of northern breeders are largely north-south, and mostly confined within similar longitudes as the breeding range. The species migrates through a variety of vegetated habitats, especially montane forests and riparian corridors (Goguen and Curson 2020).

PHILADELPHIA VIREO *Vireo philadelphicus*

DISTRIBUTION AND ABUNDANCE

It breeds in young, northern hardwood and mixed conifer/deciduous forests from British Columbia to Labrador south to the northern fringe of the US from Minnesota to Maine, most densely populated in Ontario and Quebec (Fink et al. 2023). It migrates through the eastern US to nonbreeding grounds, mainly in Central America, from Guatemala to western Panama, most densely in Honduras, Nicaragua, and Costa Rica, and sparsely in the Yucatán Peninsula and southern Mexico (Fink et al. 2023).

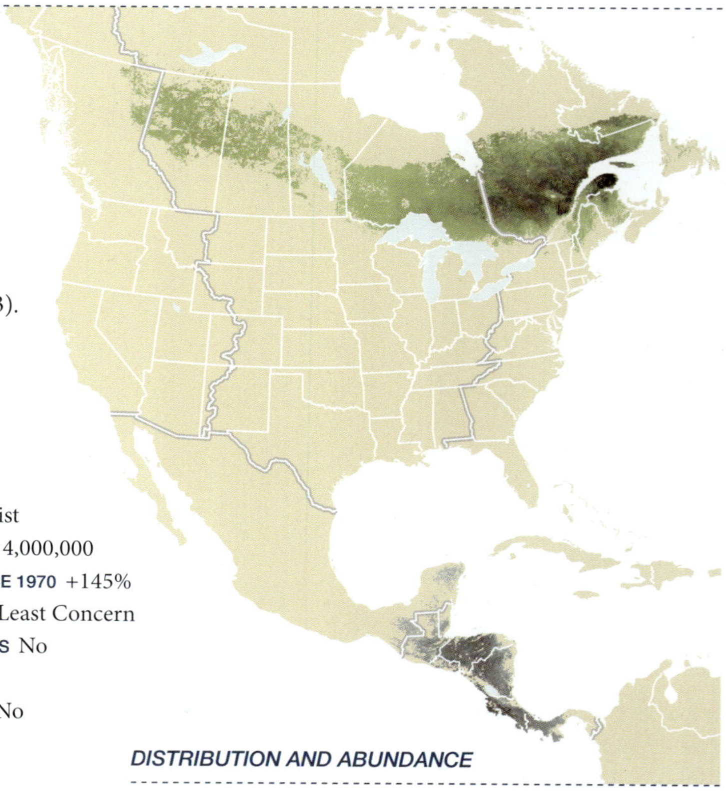

DISTRIBUTION AND ABUNDANCE

ORDER Passeriformes
FAMILY Vireonidae
BREEDING HABITAT
Forests, Boreal
NONBREEDING HABITAT
Forests, Tropical Generalist
GLOBAL POPULATION SIZE 4,000,000
POPULATION CHANGE SINCE 1970 +145%
IUCN RED LIST CATEGORY Least Concern
R2R TIPPING POINT SPECIES No
US LISTING STATUS No
CANADA SPECIES AT RISK No
PIF CCS 9/20

MIGRATORY CONNECTIVITY

OVERALL Migratory connectivity is unknown. No information on migratory connectivity was found for this species.

MIGRATORY PATHWAYS Birds move along the Gulf Coast of Mexico during both seasonal migrations, with some evidence of trans-Gulf flights via Florida during the southbound migration (Fink et al. 2023).

WARBLING VIREO *Vireo gilvus*

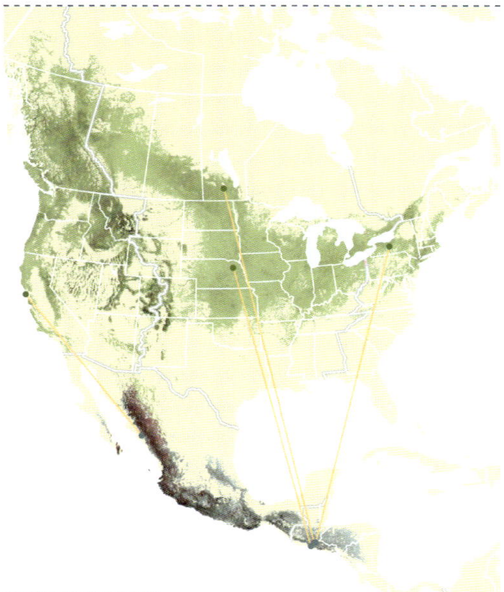

CONNECTIONS

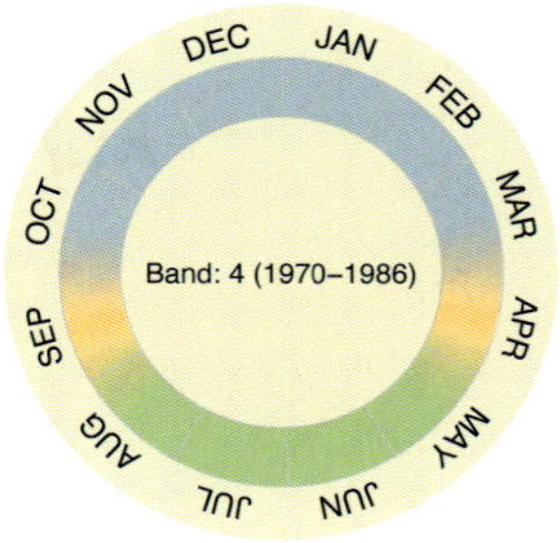

DISTRIBUTION AND ABUNDANCE

The Warbling Vireo breeds broadly throughout deciduous forest habitats in the US from the Atlantic coast across the Midwest and to the Pacific coast and desert southwest, as well as western Canada from Manitoba to the Yukon, and central Mexico in the Sierra Madre Occidental. They spend the nonbreeding season in diverse habitats in western Mexico and Central America to Nicaragua, and are observed year-round in the Sierra Madre Occidental mountains of Mexico (Gardali and Ballard 2020).

ORDER Passeriformes
FAMILY Vireonidae
BREEDING HABITAT Forests, Temperate Eastern
NONBREEDING HABITAT Forests, Tropical Generalist
GLOBAL POPULATION SIZE 53,000,000
POPULATION CHANGE SINCE 1970 +55%
IUCN RED LIST CATEGORY Least Concern
R2R TIPPING POINT SPECIES No
US LISTING STATUS No
CANADA SPECIES AT RISK No
PIF CCS 8/20

MIGRATORY CONNECTIVITY

OVERALL There are no published studies on migratory connectivity. Limited band reencounter data show connections from the Atlantic and Midcontinent flyways to Central America and from the Pacific Flyway to Mexico. This, along with genetic separation within eastern and western groups, may suggest strong migratory connectivity.
BANDING Four band reencounters show one eastern (New York) and two central (Manitoba and Nebraska) breeding individuals spending the nonbreeding season in Central America, and one California breeder in western Mexico.
GENETICS Two genetic groups were found with mitochondrial DNA analysis, splitting eastern and western Warbling Vireos (Carpenter et al. 2022), which may support strong migratory connectivity within the species. An additional isolated genetic group was found in the Black Hills of South Dakota. The Bird Genoscape Project is currently constructing a genoscape for the Warbling Vireo (https://www.birdgenoscape.org/warbling-vireo/) that may reveal more about migratory connectivity for this species.
OTHER APPROACHES No tracking or stable isotope studies of migratory connectivity were found for this species.
MIGRATORY PATHWAYS In pre-breeding migration, nonbreeding birds split along the Sierra Madre Occidental and Sierra Madre Oriental to head to western and eastern breeding grounds, respectively, and this pattern is reversed during post-breeding migration (Fink et al. 2023).

RED-EYED VIREO *Vireo olivaceus*

ORDER Passeriformes
FAMILY Vireonidae
BREEDING HABITAT Forests, Temperate Generalist
NONBREEDING HABITAT Forests, Tropical Lowland Evergreen
GLOBAL POPULATION SIZE 130,000,000
POPULATION CHANGE SINCE 1970 +47%
IUCN RED LIST CATEGORY Least Concern
R2R TIPPING POINT SPECIES No
US LISTING STATUS No
CANADA SPECIES AT RISK No
PIF CCS 5/20

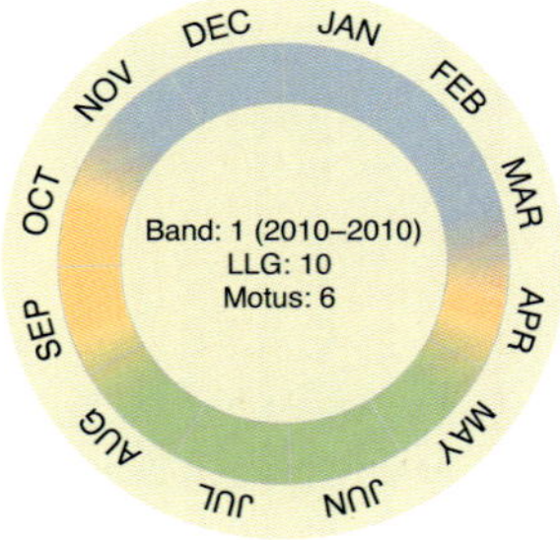

DISTRIBUTION AND ABUNDANCE

Red-eyed Vireos are widespread breeders throughout southern Canada, across all the provinces, mostly south of the boreal, and in the US from the Pacific Northwest to the Great Plains and across to the Atlantic coast (Cimprich et al. 2020). They spend the nonbreeding season in northwestern South America from Bolivia to Venezuela and west to the Pacific coast, with the exception of the high Andes Mountains (Cimprich et al. 2020).

MIGRATORY CONNECTIVITY

OVERALL Migratory connectivity is relatively unknown due to limited light-level geolocator and band reencounter data. Atlantic Flyway breeding individuals migrated to Brazil (90%) and northern South America (10%), and one Midcontinent Flyway breeding individual spent the nonbreeding season in northern South America.
BANDING One bird banded during the breeding season in Minnesota was reencountered in Colombia during the nonbreeding season.
RADIO TRACKING There are several projects tracking Red-eyed Vireos with the Motus Wildlife Tracking System across the range. There are no individual breeding to nonbreeding locations, but there appear to be many detections during migration, including individual tracks from

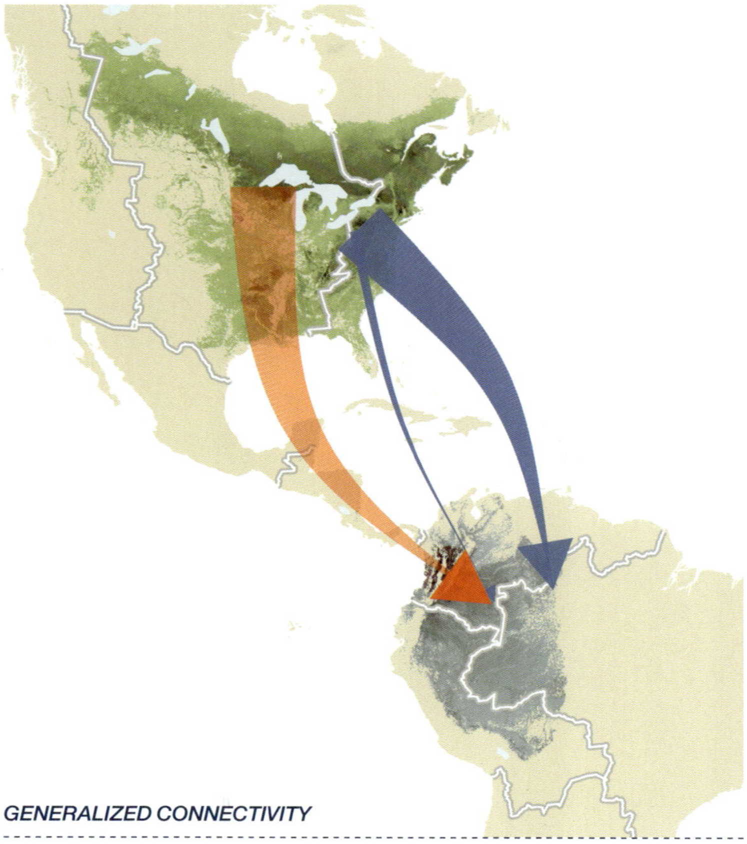

GENERALIZED CONNECTIVITY

	Brazil	Northern SA
Atlantic (n = 10)	90%	10%
Midcontinent (n = 1)	–	100%

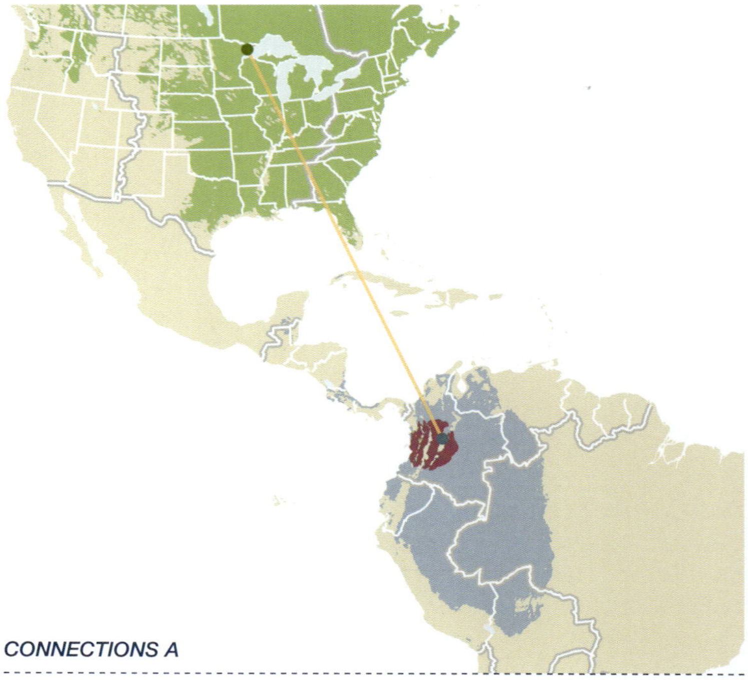

CONNECTIONS A

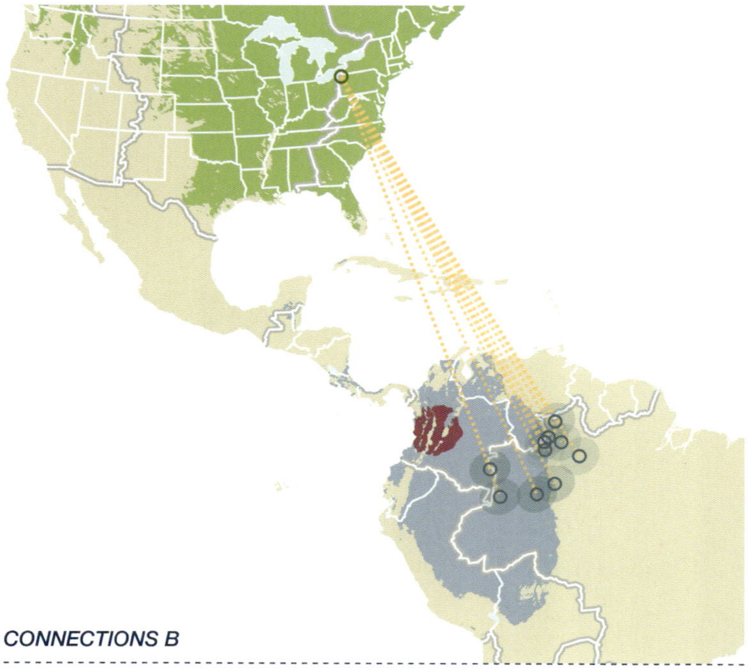

CONNECTIONS B

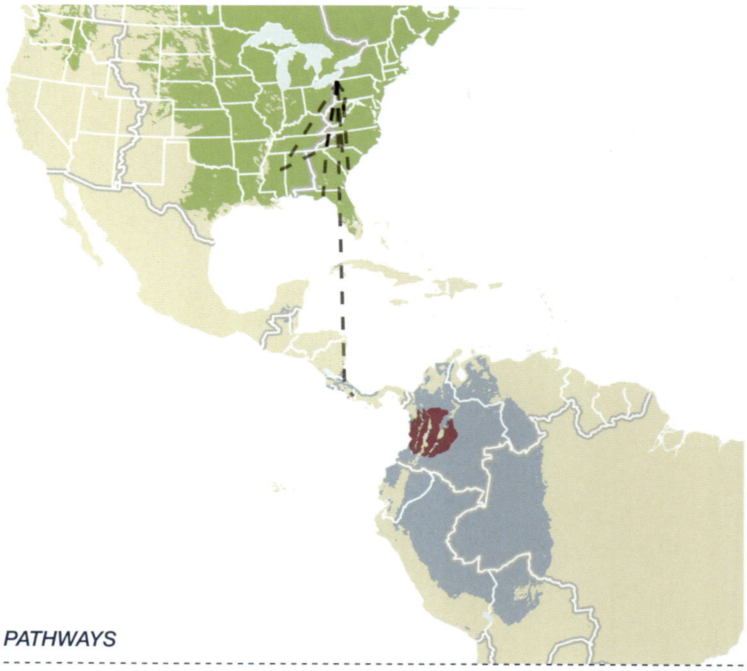

PATHWAYS

the Great Lakes region south to Florida and Central America (B. Stutchbury, unpublished data).

GEOLOCATOR TRACKING Light-level geolocators were used to track Red-eyed Vireos from a breeding site in Pennsylvania to nonbreeding grounds in northwestern South America, including Brazil, Colombia, and Venezuela (Callo et al. 2013).

OTHER APPROACHES No satellite transmitter, stable isotope, or genetic studies of migratory connectivity were found for this species.

MIGRATORY PATHWAYS In one study, Red-eyed Vireos migrated north during pre-breeding by crossing the Gulf of Mexico from the Yucatán Peninsula to Louisiana after a prolonged stopover in Colombia (Callo et al. 2013). More broadly, birds move from South America into Central America north to the Yucatán before departing to cross the Gulf of Mexico, then make landfall from Texas to Florida, with some individuals briefly stopping in Cuba (Fink et al. 2023). This pattern is repeated southward during post-breeding migration, with more individuals utilizing the Caribbean islands than during pre-breeding migration (Fink et al. 2023), which may indicate looped migration that is more westerly in pre-breeding and more easterly in post-breeding (Callo et al. 2013).

DATA CONTRIBUTORS
B.J.M. Stutchbury.

BLACK-WHISKERED VIREO *Vireo altiloquus*

DISTRIBUTION AND ABUNDANCE

It breeds in coastal mangroves and hammocks in southern Florida, and throughout the Caribbean islands in coastal mangroves, tropical dry and limestone forests, and less densely in closed-canopy cloud forests (Chace et al. 2020). It migrates to spend the nonbreeding season in Hispaniola, Puerto Rico, the Lesser Antilles, and northern South America (northeast Colombia, Venezuela, and Guyana) and the northern Amazon basin (Chace et al. 2020).

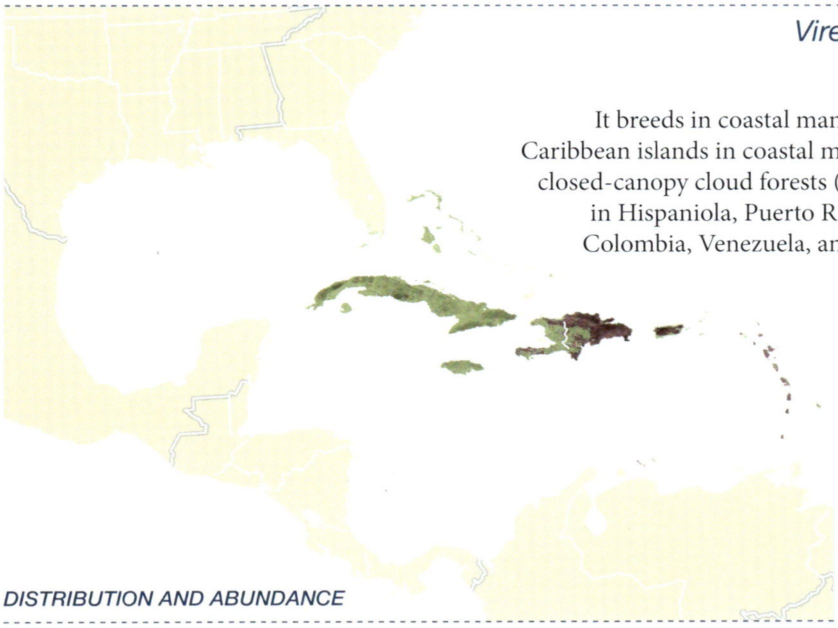

MIGRATORY CONNECTIVITY

OVERALL Migratory connectivity is unknown. No information on migratory connectivity was found for this species.

OTHER APPROACHES The subspecies *V. a. barbatulus*, which breeds in Florida, the Bahamas, and Cuba, presumably migrates to the Amazon basin for the nonbreeding season. The subspecies *V. a. altiloquus*, which breeds in Jamaica, Hispaniola, and Puerto Rico, is believed to spend the nonbreeding season in northern South America (Chace et al. 2020).

MIGRATORY PATHWAYS eBird animations (Fink et al. 2023) suggest that at least some individuals migrate west to the Yucatán Peninsula and then along the Caribbean coast of Central America to South America.

ORDER Passeriformes
FAMILY Vireonidae
BREEDING HABITAT Coasts, Mangroves
NONBREEDING HABITAT Forests, Tropical Lowland Evergreen
GLOBAL POPULATION SIZE 6,200,000
POPULATION CHANGE SINCE 1970 −19%
IUCN RED LIST CATEGORY Least Concern
R2R TIPPING POINT SPECIES No
US LISTING STATUS No
CANADA SPECIES AT RISK No
PIF CCS 12/20

INTERESTING NOTE

Populations on Hispaniola and Puerto Rico may consist of year-round residents, migratory breeders, and nonbreeding-season visitors (Zwartjes 2001).

LANIIDAE | *SHRIKES*

LOGGERHEAD SHRIKE *Lanius ludovicianus*

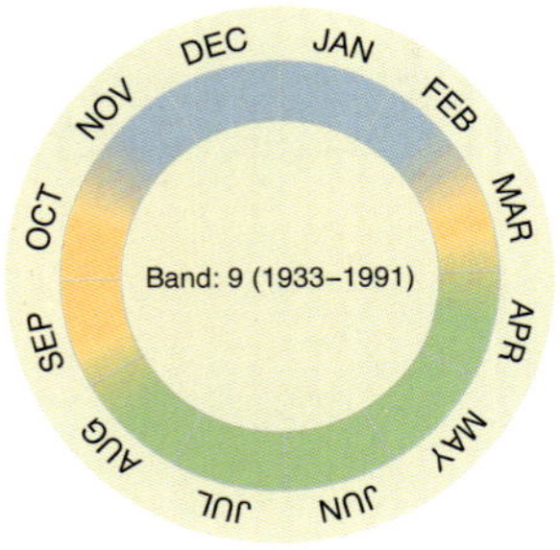

DISTRIBUTION AND ABUNDANCE

Found throughout open habitat of North America, Loggerhead Shrikes can be seen hunting arthropods and small mammals (Yosef 2020). Migratory individuals in the northern part of the breeding range, including the prairies of the central United States and Canada, eastern Washington, Oregon, and isolated portions of Ontario, move to the southern portions of the range for the nonbreeding season. Other populations are present year-round in much of the southern United States, from California to Florida and much of Mexico. Loggerhead Shrikes are in steep decline, possibly due to pesticide use and habitat loss and degradation throughout breeding and nonbreeding grounds (Yosef 2020).

ORDER Passeriformes
FAMILY Laniidae
BREEDING HABITAT Grasslands, Temperate
NONBREEDING HABITAT Grasslands, Temperate
GLOBAL POPULATION SIZE 7,000,000
POPULATION CHANGE SINCE 1970 −72%
IUCN RED LIST CATEGORY Near Threatened
R2R TIPPING POINT SPECIES No
US LISTING STATUS Endangered San Clemente subspecies
CANADA SPECIES AT RISK Endangered Eastern subspecies Threatened Prairie subspecies
PIF CCS 11/20

MIGRATORY CONNECTIVITY

OVERALL Migratory connectivity is known from limited band reencounter data, as well as genetic and stable isotope data.
BANDING Band reencounter data indicate a slight parallel migration pattern, with individuals breeding farther west migrating to nonbreeding areas farther west. Other long-distance connections include Quebec to Virginia, Michigan to Alabama, and Ohio to Florida (Burnside 1987).
RADIO TRACKING Motus Wildlife Tracking System tags deployed on captive-reared juveniles in Ontario showed southwest migratory movements. No nonbreeding season detections have been received, but birds have been detected migrating as far south as Pennsylvania (H. Wheeler, personal communication).
GEOLOCATOR TRACKING Light-level geolocators (n = 190) deployed in Ontario had low recapture rates (n = 8) and tag failure. Two recovered tags that functioned were on individuals that spent the nonbreeding season in Illinois or New York, though the confidence in those assignments is low (H. Wheeler, personal communication).
STABLE ISOTOPES Hydrogen stable isotope values of feather samples from western migratory shrikes spending the nonbreeding season in Mexico indicate the possibility of a leapfrog migration pattern, but further research is required to confirm (Perez and Hobson 2007). Individuals that spend the nonbreeding season in northeastern Mexico have been classified as a combination of both residents and migrants, using stable isotopes (Perez and Hobson 2009).
GENETICS Genetic analyses of San Clemente Island Loggerhead Shrikes suggest that it is a separate subspecies with little gene flow and rare migration to mainland California and Santa Catalina Island (Rutledge et al. 2017). A genoscape is in progress for the Loggerhead Shrike to understand the genetic basis of its migratory behavior (https://www.birdgenoscape.org/loggerhead-shrike/).
OTHER APPROACHES A combination of genetic and stable isotope data from shrikes sampled across the majority of their nonbreeding range identified four main nonbreeding areas: Atlantic coastal United States, east-central United States, west-central United States and eastern Mexico, and western Mexico (Chabot et al. 2018). The data showed that western individuals spend the breeding and nonbreeding season more closely together than do eastern shrikes; there is a pattern of chain migration, mostly in the western nonbreeding range; and there is differential migration based on age (Chabot et al. 2018).
MIGRATORY PATHWAYS Little is known about specific migratory routes. The species is detected across the range during pre- and post-breeding migration (Fink et al. 2023).

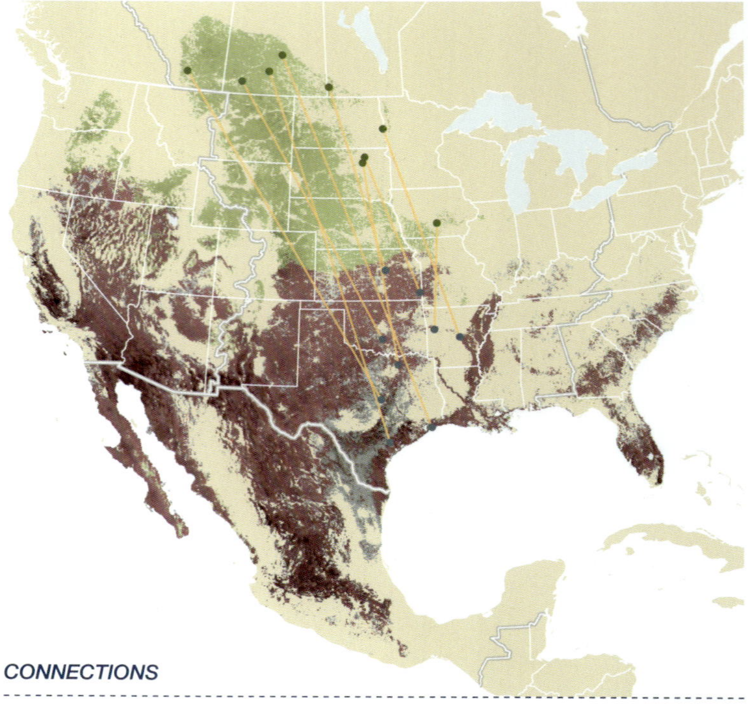

CONNECTIONS

NORTHERN SHRIKE *Lanius borealis*

DISTRIBUTION AND ABUNDANCE

A Holarctic species, it breeds in North America in taiga/tundra willow shrubland across northern Alaska, northwest Canada, the southwest side of Hudson Bay, and more locally in Quebec and Labrador (Paruk et al. 2020). Irruptive (irregular from year to year) in the nonbreeding season, it migrates to the southern portions of British Columbia and the Canadian Prairie provinces and throughout the western US east of the Cascades as far south as northern New Mexico, less densely in the upper Midwest and New England (Fink et al. 2023).

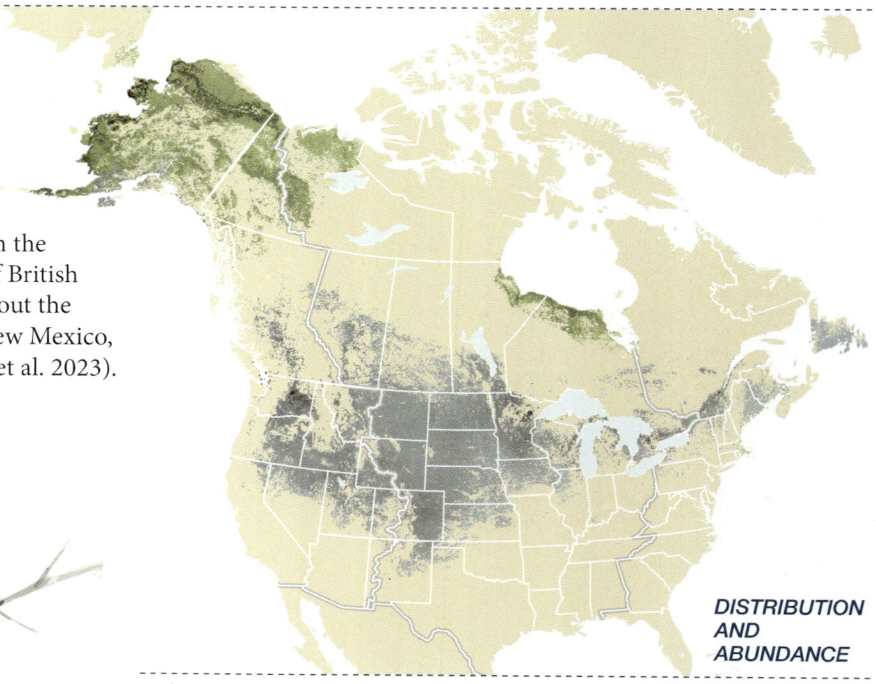

DISTRIBUTION AND ABUNDANCE

ORDER Passeriformes
FAMILY Laniidae
BREEDING HABITAT Forests, Boreal
NONBREEDING HABITAT Open Country, Habitat Mosaic
GLOBAL POPULATION SIZE 180,000
POPULATION CHANGE SINCE 1970 −22%
IUCN RED LIST CATEGORY Least Concern
R2R TIPPING POINT SPECIES No
US LISTING STATUS No
CANADA SPECIES AT RISK No
PIF CCS 13/20

MIGRATORY CONNECTIVITY

OVERALL Migratory connectivity is unknown. No information on migratory connectivity was found for this species.

MIGRATORY PATHWAYS As an irruptive migrant, the migration of this species varies annually. Birds move largely north-south, with concentrations in preferred open-country habitats and avoiding the Pacific coast, the high peaks of the Continental Divide, and much of the eastern US (Fink et al. 2023).

INTERESTING NOTE

Twelve Northern Shrikes were banded during the nonbreeding season and recaptured in subsequent years at or near their original banding locations; this suggests that the species is faithful to nonbreeding season territories and potentially migratory stopover sites (Rimmer and Darmstadt 1996).

CORVIDAE | *CROWS, JAYS, AND MAGPIES*

Cyanocitta cristata BLUE JAY

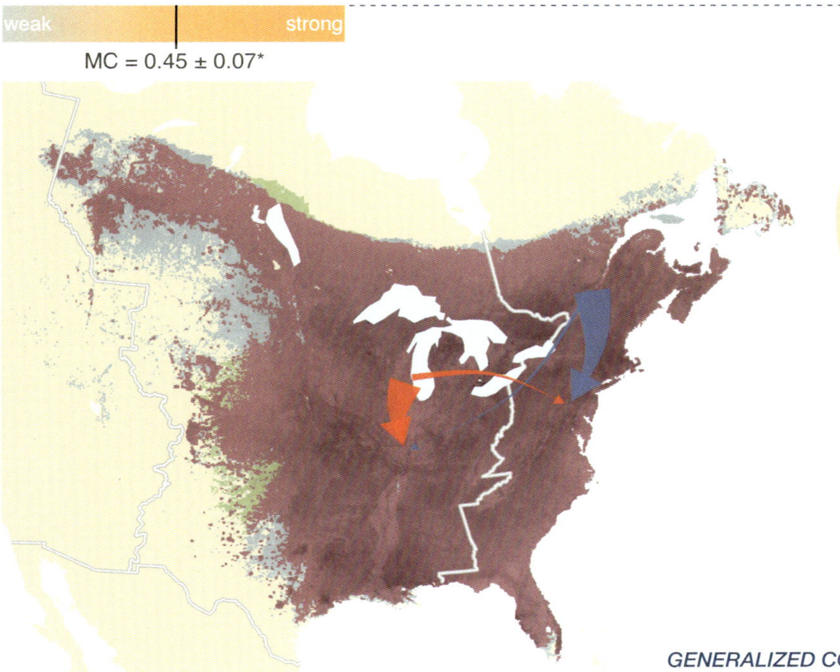

GENERALIZED CONNECTIVITY

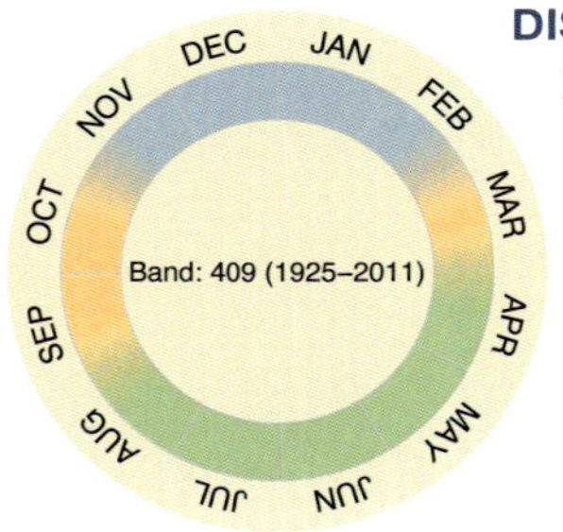

DISTRIBUTION AND ABUNDANCE

Blue Jays breed in southern Canada from Nova Scotia to Alberta and in all US states east of the Rocky Mountains. Blue Jays from all populations remain on their breeding grounds year-round, but individuals from northern populations (above approximately 40 degrees north latitude) undergo short- to medium-distance migrations to nonbreeding areas in the central and southern US (K.G. Smith et al. 2020a).

	Atlantic	Midcontinent
Atlantic (n = 203)	94%	6%
Midcontinent (n = 182)	13%	87%

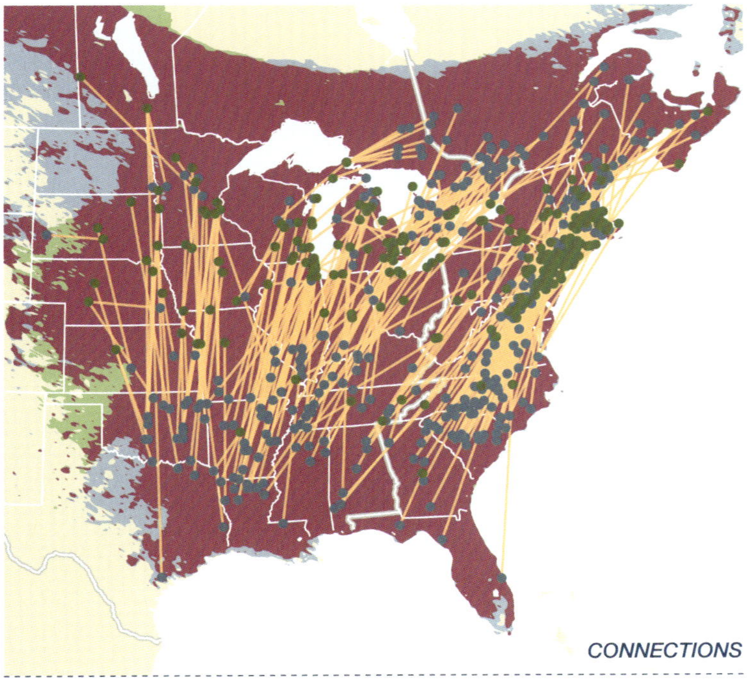

CONNECTIONS

ORDER Passeriformes
FAMILY Corvidae
BREEDING HABITAT Forests, Temperate Eastern
NONBREEDING HABITAT Forests, Temperate Eastern
GLOBAL POPULATION SIZE 17,000,000
POPULATION CHANGE SINCE 1970 −24%
IUCN RED LIST CATEGORY Least Concern
R2R TIPPING POINT SPECIES No
US LISTING STATUS No
CANADA SPECIES AT RISK No
PIF CCS 8/20

INTERESTING NOTE

The Blue Jay is a well-known diurnal migrant. In post-breeding migration, large flocks of Blue Jays can be seen streaming through the air as they move south to their nonbreeding areas.

MIGRATORY CONNECTIVITY

OVERALL The migratory connectivity of Blue Jays is not well understood, and what we know comes from band reencounter data. Birds breeding in the Atlantic and Midcontinent flyways primarily spend the nonbreeding period within same flyway as breeding (94% and 87%, respectively), resulting in strong migratory connectivity (MC = 0.45 ± 0.07).

BANDING Band reencounter data show three clusters across the range: eastern / east of the Appalachian Mountains, central/Midwest, and western / Great Plains.

OTHER APPROACHES No genetic, stable isotope, or tracking studies of migratory connectivity were found for this species.

MIGRATORY PATHWAYS The migration routes of Blue Jays from eastern populations parallel the Atlantic coast and the Appalachian Mountains. Those from central populations move through the Mississippi River Valley in a north-south direction, while the westerly populations seem to orient slightly southeast to reach their nonbreeding grounds (K.G. Smith et al. 2020a).

AMERICAN CROW *Corvus brachyrhynchos*

weak strong

MC = 1

DISTRIBUTION AND ABUNDANCE

American Crows are highly intelligent and adaptable, and are a familiar bird in towns, cities, and open habitats. This species ranges widely throughout the US and Canada, living year-round as far north as southern Alaska and Labrador, and as far south as the Gulf Coast and Baja California. Populations in the inland northwestern US and central Canada are migratory, only inhabiting these areas during the breeding season. This species is especially abundant in the eastern half of the US, in Newfoundland and Nova Scotia, and along the Pacific coast from Baja California to Alaska. Populations are more sparce in the inland west (except for near urban areas such as Salt Lake City), and are largely absent from much of the arid west and the high Arctic (Verbeek et al. 2024; Fink et al. 2023).

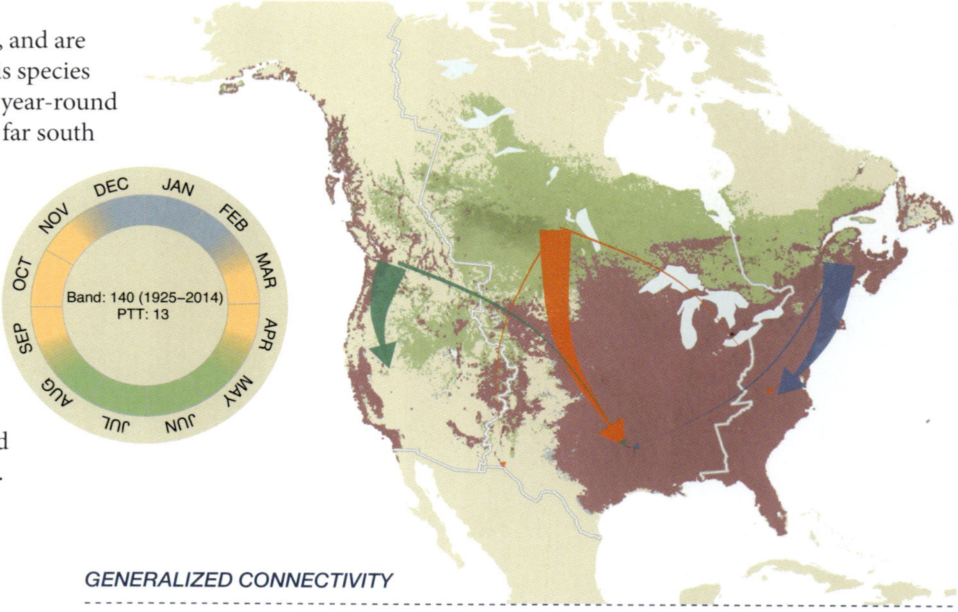

GENERALIZED CONNECTIVITY

ORDER Passeriformes
FAMILY Corvidae
BREEDING HABITAT Open Country, Habitat Mosaic
NONBREEDING HABITAT Open Country, Habitat Mosaic
GLOBAL POPULATION SIZE 29,000,000
POPULATION CHANGE SINCE 1970 +7%
IUCN RED LIST CATEGORY Least Concern
R2R TIPPING POINT SPECIES No
US LISTING STATUS No
CANADA SPECIES AT RISK No
PIF CCS 8/20

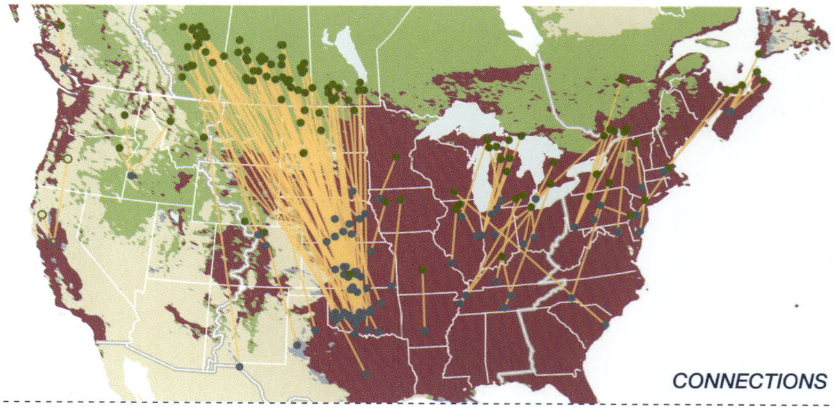

CONNECTIONS

	Atlantic	Mexico	Midcontinent	Pacific
Atlantic (n = 24)	96%	–	4%	–
Midcontinent (n = 116)	2%	1%	97%	–
Pacific (n = 13)	–	–	8%	92%

MIGRATORY CONNECTIVITY

OVERALL Band reencounter and satellite and radio telemetry data show migratory movements are largely oriented north-south and of short to medium distance. American Crows breeding in the Atlantic, Midcontinent, and Pacific flyways primarily migrated to nonbreeding areas within the same flyway as breeding (92%–97%), resulting in strong migratory connectivity (MC = 1) for this species.

BANDING Band reencounter data show disjunct populations. Individuals from the central Canadian provinces migrated south to the central United States. Birds from the west of the Rockies moved south but remained in the western region. Eastern birds divided roughly into Atlantic and Midcontinent flyway populations.

RADIO TRACKING Radio-tagged American Crows (n = 8) were found to be diurnal migrants during pre-breeding migration, and although

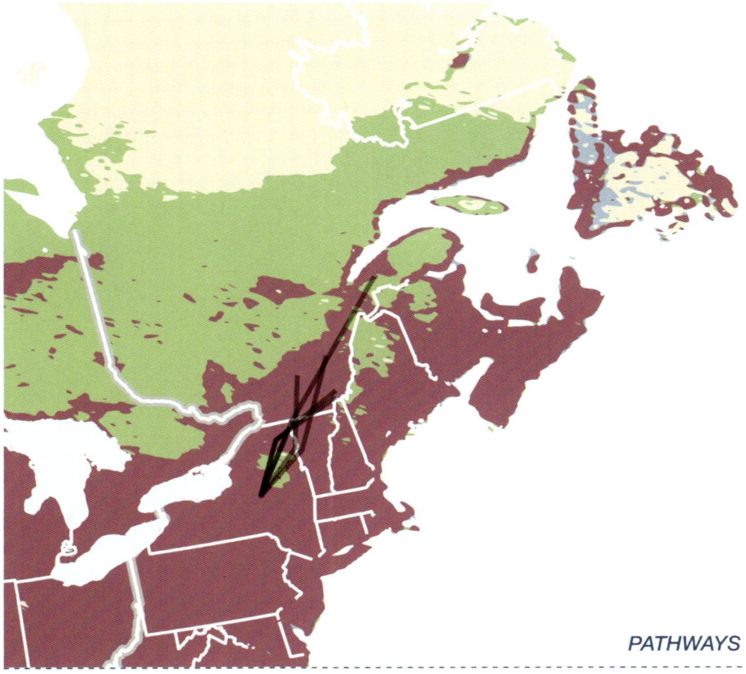

PATHWAYS

they migrated alone, the crows would often stop to forage with conspecifics (Ward and Raim 2011).

SATELLITE TRACKING Of American Crows fitted with satellite transmitters in Davis, California (n = 11), and Utica, New York (n = 7), 14 were migratory, undergoing north-south movements of 177–1,095 km (Townsend et. al 2018).

OTHER APPROACHES Stable isotopic and molecular markers from known-origin birds were usually successful in identifying long-distance migrant populations from resident populations at mixed nonbreeding sites, but were sometimes unsuccessful at identifying short-distance migrants (Townsend et. al 2018). The Bird Genoscape Project is currently working to build a genoscape for the American Crow (https://www.birdgenoscape.org/american-crow/).

MIGRATORY PATHWAYS Migratory American Crows move out of their northern ranges in the nonbreeding season, with a tendency to avoid high mountains. Western birds remained within the Great Basin or along coastal areas, and central Canadian birds shifted southeast, avoiding the Rockies to spend the nonbreeding season in the central plains. Eastern breeders from the Great Lakes region moved south to the Midwest and the South, west of the Appalachians, while birds from the Maritimes and northeastern US moved to southern New England and the mid-Atlantic east of the Appalachians.

Corvus ossifragus FISH CROW

DISTRIBUTION AND ABUNDANCE

This nasal-voiced corvid traditionally ranges along the southeastern coastal plain, but has expanded its range northward in recent decades. It is abundant year-round in most of Florida, as well as coastal areas of the Atlantic, and the Gulf Coast east of the Mississippi River. Year-round populations reach as far north as Massachusetts in the east, and southern Illinois in the central states (McGowan 2020; Fink et. al 2023).

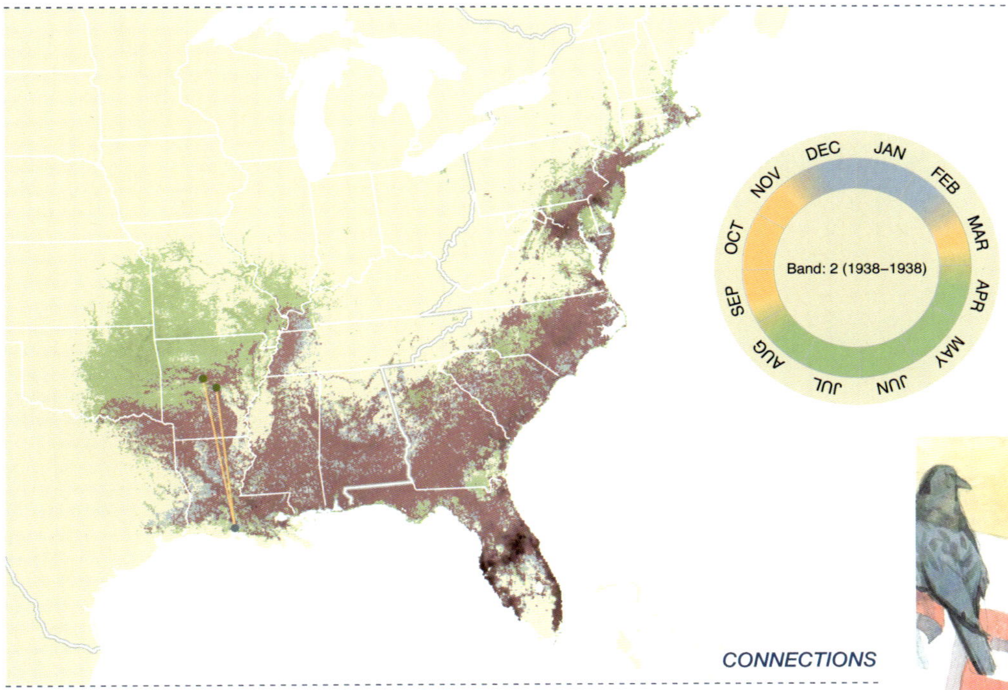

CONNECTIONS

MIGRATORY CONNECTIVITY

OVERALL Migratory connectivity and general movements of this species are poorly understood. Limited band reencounter and observational data suggest short-distance migration may occur, although it is unknown if movement is a true migration (McGowan 2020). We are unable to generate MC due to insufficient data.

BANDING Band reencounter data is from a few birds banded in the nonbreeding season in Louisiana and found shot in Arkansas. All other recoveries were in close proximity to the banding location (McGowan 2020).

OTHER APPROACHES No tracking, genetic, or stable isotope studies of migratory connectivity were found for this species.

MIGRATORY PATHWAYS Little is known about migratory pathways. In pre-breeding migration, movements occur inland as far as Missouri and Kansas, primarily along the Mississippi River and its major tributaries (McGowan 2020; Fink et. al 2023). In post-breeding migration, movement trends back toward coastal areas (Fink et al. 2023). There have been observations of several hundred individuals a day flying along ridges in northwestern New Jersey (Walsh et al. 1999; McGowan 2020).

ORDER Passeriformes
FAMILY Corvidae
BREEDING HABITAT Open Country, Habitat Mosaic
NONBREEDING HABITAT Open Country, Habitat Mosaic
GLOBAL POPULATION SIZE 470,000
POPULATION CHANGE SINCE 1970 +28%
IUCN RED LIST CATEGORY Least Concern
R2R TIPPING POINT SPECIES No
US LISTING STATUS No
CANADA SPECIES AT RISK No
PIF CCS 11/20

HORNED LARK *Eremophila alpestris*

DISTRIBUTION AND ABUNDANCE

Horned Larks are widely distributed throughout North America, Europe, and Asia—primarily in the Northern Hemisphere, although there is a small population at a high elevation in Colombia, South America. The northernmost breeding populations are migratory, while populations that breed south of Canada are residents (Beason 2020). The species is present year-round in much of the United States, with highest abundance in midwestern and western states. The northern limits of the breeding range include Alaska and Arctic regions of Canada. The nonbreeding range extends south of the year-round range into Texas and Mexico (Beason 2020).

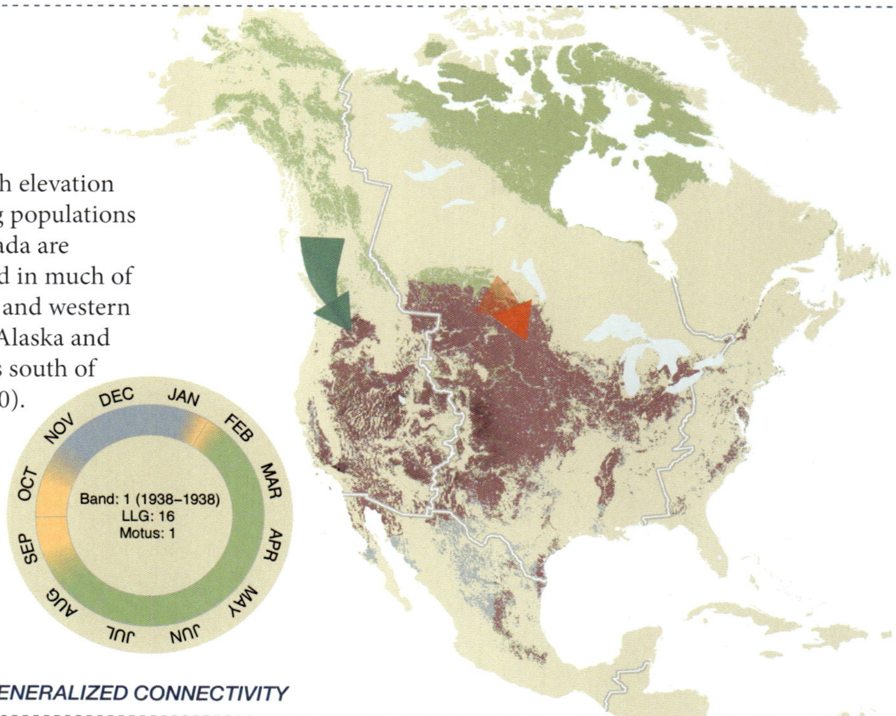

GENERALIZED CONNECTIVITY

ORDER Passeriformes
FAMILY Alaudidae
BREEDING HABITAT Grasslands, Temperate
NONBREEDING HABITAT Open Country, Developed/Disturbed
GLOBAL POPULATION SIZE 140,000,000
POPULATION CHANGE SINCE 1970 −69%
IUCN RED LIST CATEGORY Least Concern
R2R TIPPING POINT SPECIES No
US LISTING STATUS Threatened *E. a. strigata*
CANADA SPECIES AT RISK Endangered *E. a. strigata*
PIF CCS 9/20

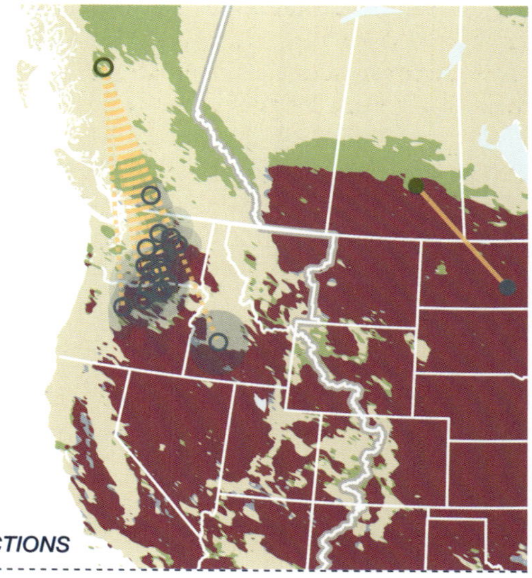

CONNECTIONS

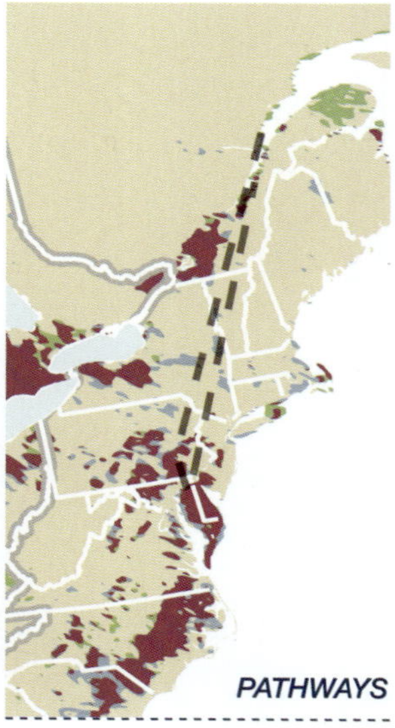

PATHWAYS

	Midcontinent	Pacific
Midcontinent (n = 1)	100%	–
Pacific (n = 16)	–	100%

INTERESTING NOTE

Horned Larks tracked from British Columbia showed sex-specific migration patterns, with males migrating earlier in pre-breeding migration than females (de Zwaan et al. 2019). There are 42 recognized subspecies of Horned Larks worldwide (Gill et al. 2023).

MIGRATORY CONNECTIVITY

OVERALL There is limited data on migratory connectivity, though recent studies have shown connections between breeding and nonbreeding sites within the same flyway.

BANDING There is one breeding to nonbreeding reencounter from Saskatchewan to Montana.

GEOLOCATOR TRACKING Males (n = 8) and females (n = 9) tracked from a single breeding site in British Columbia migrated to nonbreeding areas in the northwestern United States, including Washington, Oregon, and Idaho (de Zwaan et al. 2019).

GENETICS The Bird Genoscape Project is starting to build a genoscape for the Horned Lark (https://www.birdgenoscape.org/horned-lark/).

OTHER APPROACHES No genetic or stable isotope studies of migratory connectivity were found for this species.

MIGRATORY PATHWAYS Limited light-level geolocator data indicate that migration occurs north to south and is short-distance (around 1,000 km), with breeding and nonbreeding locations in the same flyway (de Zwaan et al. 2019). Horned Larks breeding in eastern Canada were detected by the Motus Wildlife Tracking System during migration inland through eastern Canada and the northeastern United States (Walker et al. 2025).

DATA CONTRIBUTORS
D. de Zwaan; E. Gow; K. Martin; J. Tremblay; S. Wilson.

Riparia riparia **BANK SWALLOW**

CONNECTIONS

DISTRIBUTION AND ABUNDANCE

The Bank Swallow has a cosmopolitan distribution across the Holarctic, being one of the most widely distributed swallows of the world. In the Americas, it breeds throughout Alaska, and from southwest Canada and the northern United States all the way to the east coast above approximately 35 degrees north latitude. The nonbreeding range extends throughout a large part of South America, although some individuals spend the nonbreeding period in southern Mexico in the state of Jalisco (Fink et al. 2023).

ORDER Passeriformes
FAMILY Hirundinidae
BREEDING HABITAT Wetlands Aerial, Lakes and Rivers
NONBREEDING HABITAT Open Country Aerial, Habitat Mosaic
GLOBAL POPULATION SIZE 29,000,000
POPULATION CHANGE SINCE 1970 −89%
IUCN RED LIST CATEGORY Least Concern
R2R TIPPING POINT SPECIES No
US LISTING STATUS No
CANADA SPECIES AT RISK Threatened
PIF CCS 11/20

MIGRATORY CONNECTIVITY

OVERALL Migratory connectivity information for Bank Swallows is limited and known from a few banding reencounters, as well as a geolocator and stable isotope study of a northeastern Canadian breeding colony (Imlay et al. 2018).

BANDING There are several breeding to nonbreeding reencounters from the Great Lakes region to Peru/Brazil. There is one reencounter from Saskatchewan to Bolivia.

GEOLOCATOR TRACKING There is one tracking study of breeding individuals in New Brunswick, Canada. One geolocator was recovered. Within one season, the individual used two stationary nonbreeding areas along the southern part of the Colombia-Venezuela border and northwestern Brazil in the northern Amazonian region (Imlay et al. 2018).

RADIO TRACKING There are several projects tracking Bank Swallows registered with the Motus Wildlife Tracking System. There are no breeding to nonbreeding detections.

STABLE ISOTOPES Rectrix and contour feathers from northeastern Canadian breeding Bank Swallows suggest nonbreeding origins are from a broad region in northeastern Argentina, Paraguay, Uruguay, and southernmost Brazil (Imlay et al. 2018).

GENETICS The Bird Genoscape Project is currently working to build a genoscape for the Bank Swallow (https://www.birdgenoscape.org/bank-swallow/).

MIGRATORY PATHWAYS In addition to one individual tracked with a geolocator, birds tracked using the Motus Wildlife Tracking System from northeastern Canada fly southwest following the coastline and possibly turn inland to reach the Gulf Coast (Imlay et al. 2018, 2020). Due to lack of detections, it is unknown if birds flew around or across the Gulf of Mexico (Imlay et al. 2020). After the Gulf of Mexico, birds continued farther south to Central America and entered South America through the Panama-Colombia border in the Darién (Bayly et al. 2014; Imlay et al. 2018), continuing southward to their nonbreeding grounds.

DATA CONTRIBUTORS
K.A. Hobson; T. Imlay; M.L. Leonard; A. Roberto-Charron; S. Saldanha; P. Taylor.

TREE SWALLOW *Tachycineta icolor*

weak | strong

MC = 0.36 ± 0.04*

DISTRIBUTION AND ABUNDANCE

Tree Swallows breed across central and northern North America, with a northern limit coinciding with the tree line. The nonbreeding range concentrates in western Mexico and the entire coastal region surrounding the Gulf of Mexico, including the southern portions of the Gulf states (Texas, Louisiana, Mississippi, Alabama, and Georgia), Florida, Cuba, southeastern Mexico (Tabasco, Veracruz), and the coasts of the Yucatán Peninsula (Fink et al. 2023).

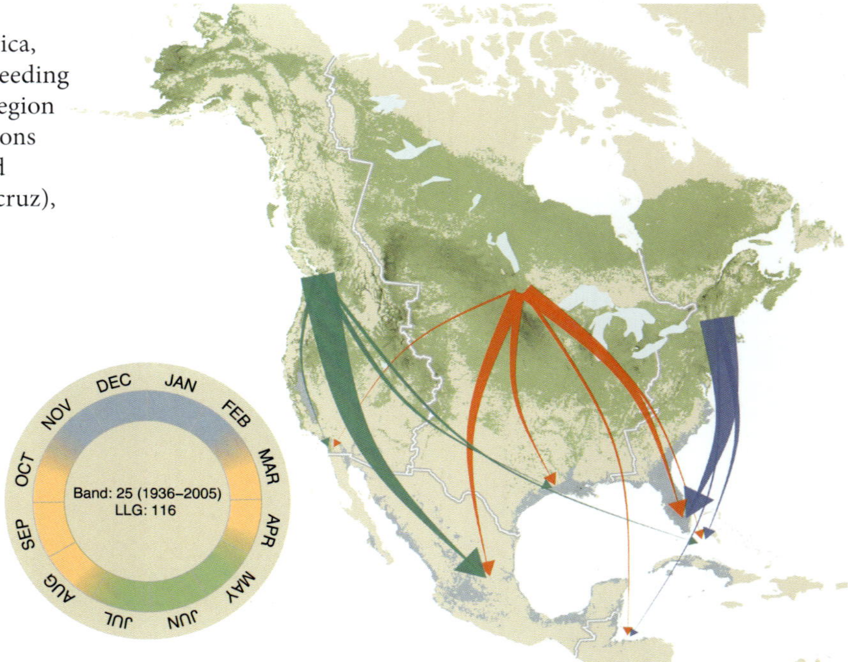

GENERALIZED CONNECTIVITY

ORDER Passeriformes
FAMILY Hirundinidae
BREEDING HABITAT Wetlands Aerial, Generalist
NONBREEDING HABITAT Wetlands Aerial, Generalist
GLOBAL POPULATION SIZE 19,000,000
POPULATION CHANGE SINCE 1970 −42%
IUCN RED LIST CATEGORY Least Concern
R2R TIPPING POINT SPECIES No
US LISTING STATUS No
CANADA SPECIES AT RISK No
PIF CCS 10/20

	Atlantic	Caribbean	Central America	Mexico	Midcontinent	Pacific
Atlantic (n = 43)	81%	16%	−2%	–	–	–
Midcontinent (n = 77)	38%	8%	5%	30%	18%	1%
Pacific (n = 20)	–	5%	–	80%	10%	5%

MIGRATORY CONNECTIVITY

OVERALL The migratory connectivity of Tree Swallows is well known from a mixture of band reencounter data and geolocator tracking studies. Atlantic Flyway breeding populations primarily spend the nonbreeding season within the Atlantic Flyway (81%), while Midcontinent breeding populations span the Atlantic Flyway (38%), Mexico (30%), and the Midcontinent Flyway (18%). Pacific Flyway breeders primarily migrate to nonbreeding areas in Mexico (80%). Within the coastal flyways, transition probabilities are high, yet there is overlap with the Midcontinent Flyway populations on nonbreeding areas, leading to moderate migratory connectivity (MC = 0.36 ± 0.04).

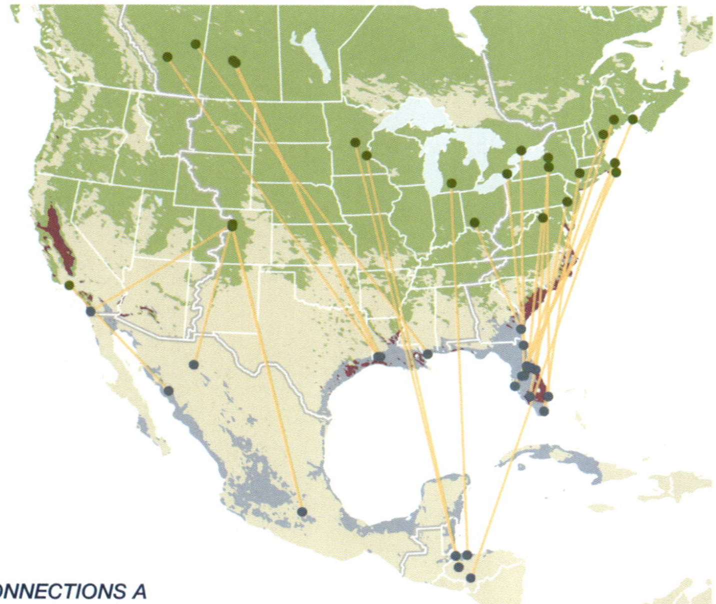

CONNECTIONS A

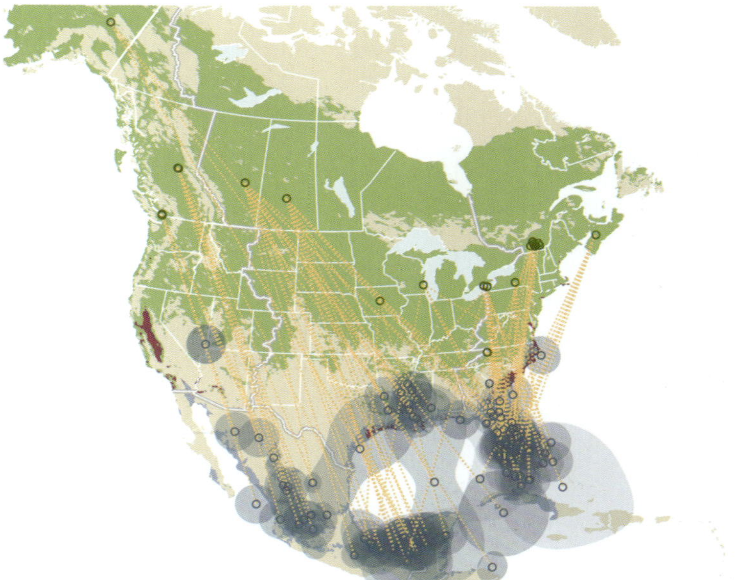

CONNECTIONS B

INTERESTING NOTE

NEXRAD weather radar has been used to map nocturnal multispecies roost sites of swallows (Winkler 2006).

BANDING Band reencounters from across the range show similar patterns of migratory connectivity to tracking studies: western breeders migrate to nonbreeding areas in California and western Mexico; central breeders spend the nonbreeding season along the Gulf Coast of the US and the Yucatán Peninsula; and eastern breeders mostly spend the nonbreeding season in Florida, Cuba, and the Caribbean.

GEOLOCATOR TRACKING Tracking studies have shown that Tree Swallows concentrate in three major flyways (Laughlin et al. 2013; Rakhimberdiev et al. 2015; Knight et al. 2018) with moderately strong migratory connectivity between breeding and nonbreeding sites (Knight et al. 2018). Western breeders migrate to nonbreeding areas in California and western Mexico (Knight et al. 2018). Central breeders spend the nonbreeding season along the Gulf Coast of the US and the Yucatán Peninsula (Laughlin et al. 2013; Knight et al. 2018), and eastern breeders mostly spend the nonbreeding season in Florida, Cuba, and the Caribbean (Laughlin et al. 2013; Rakhimberdiev et al. 2015; Knight et al. 2018).

OTHER APPROACHES No stable isotope or genetic studies of migratory connectivity were found for this species.

MIGRATORY PATHWAYS Although there is some mixing, Tree Swallows use three main flyways and maintain high connectivity between breeding and post-breeding stopover sites (Knight et al. 2018). Western breeders fly south following the Rocky Mountains; central breeders follow the Midcontinent flyway (Laughlin et al. 2013; Knight et al. 2018); and eastern breeders follow the Atlantic coast (Laughlin et al. 2013; Rakhimberdiev et al. 2015; Knight et al. 2018). Important stopover sites were identified within the Midwest, the mid-Atlantic coast, the Dakotas (Knight et al. 2018), and southeastern Louisiana (Laughlin et al. 2013). Swallows spend stopovers in dense roost sites (Winkler 2006).

DATA CONTRIBUTORS
E.A. Gow; S.M. Knight; D.R. Norris.

Tachycineta thalassina VIOLET-GREEN SWALLOW

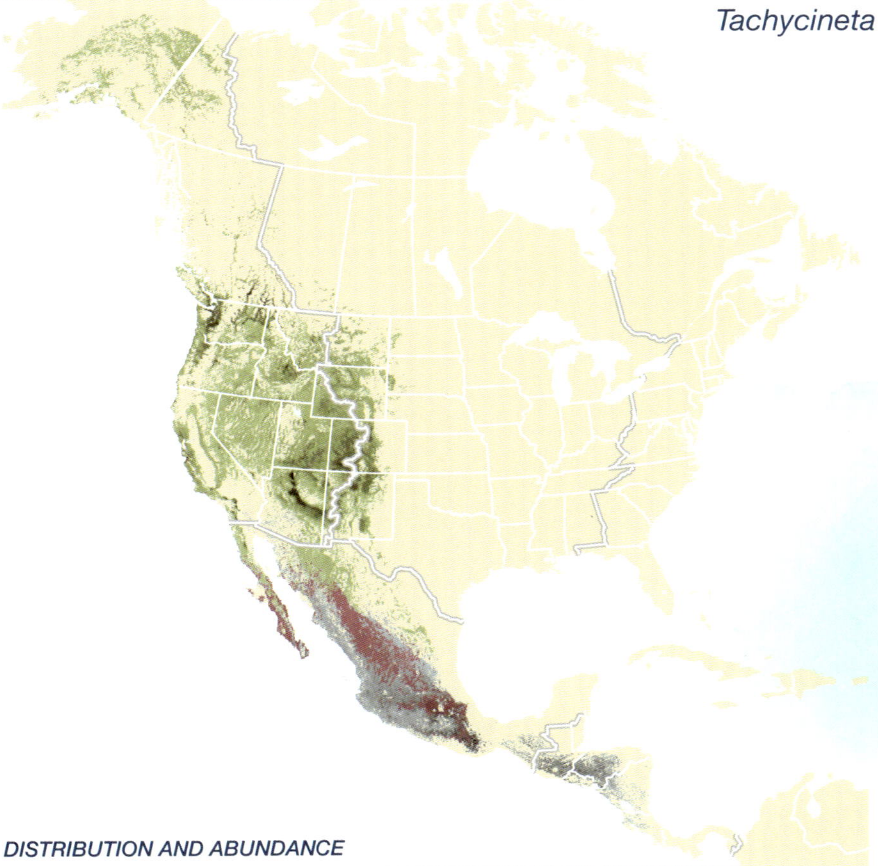

DISTRIBUTION AND ABUNDANCE

DISTRIBUTION AND ABUNDANCE

It breeds primarily in montane coniferous forests throughout the West, patchily in Alaska, and most abundantly in the Cascades, California coastal ranges, central Rockies, and the Trans-Mexican Volcanic Belt and Sierra Madre del Sur of Mexico (Fink et al. 2023). Virtually all birds leave the US/Canada breeding range by December for the nonbreeding range in central Mexico and Guatemala/Honduras, concentrating in the Trans-Mexican Volcanic Belt and Sierra Madre del Sur in December and throughout southwestern Mexico and Baja California Sur in January–February (Fink et al. 2023). US and Canadian populations have decreased since 1970 (Rosenberg et al. 2019), with the greatest losses in the northwestern interior forest (Alaska and northwestern Canada), Northern Pacific rainforest, and Southern Rockies / Colorado Plateau (Sauer et al. 2019).

ORDER Passeriformes
FAMILY Hirundinidae
BREEDING HABITAT Forest Aerial, Temperate Western
NONBREEDING HABITAT Open Country Aerial, Habitat Mosaic
GLOBAL POPULATION SIZE 7,200,000
POPULATION CHANGE SINCE 1970 −29%
IUCN RED LIST CATEGORY Least Concern
R2R TIPPING POINT SPECIES No
US LISTING STATUS No
CANADA SPECIES AT RISK No
PIF CCS 10/20

INTERESTING NOTE

Less is known about the Violet-green Swallow than any other North American swallow (C.R. Brown et al. 2020).

MIGRATORY CONNECTIVITY

OVERALL Migratory connectivity is unknown. No information on migratory connectivity was found for this species.

MIGRATORY PATHWAYS The entire US/Canada population appears to move south via western Mexico (Fink et al. 2023).

NORTHERN ROUGH-WINGED SWALLOW *Stelgidopteryx serripennis*

DISTRIBUTION AND ABUNDANCE

Breeds throughout the US and southern Canada south through Mexico to Costa Rica in a wide range of open habitats with suitable vertical surface nesting cavities or crevices (De Jong 2020). During the nonbreeding season, northern populations migrate to southern Mexico and Central America, where they mix with year-round resident populations (De Jong 2020; Fink et al. 2023).

ORDER Passeriformes
FAMILY Hirundinidae
BREEDING HABITAT Open Country Aerial, Habitat Mosaic
NONBREEDING HABITAT Open Country Aerial, Habitat Mosaic
GLOBAL POPULATION SIZE 27,000,000
POPULATION CHANGE SINCE 1970 −18%
IUCN RED LIST CATEGORY Least Concern
R2R TIPPING POINT SPECIES No
US LISTING STATUS No
CANADA SPECIES AT RISK No
PIF CCS 9/20

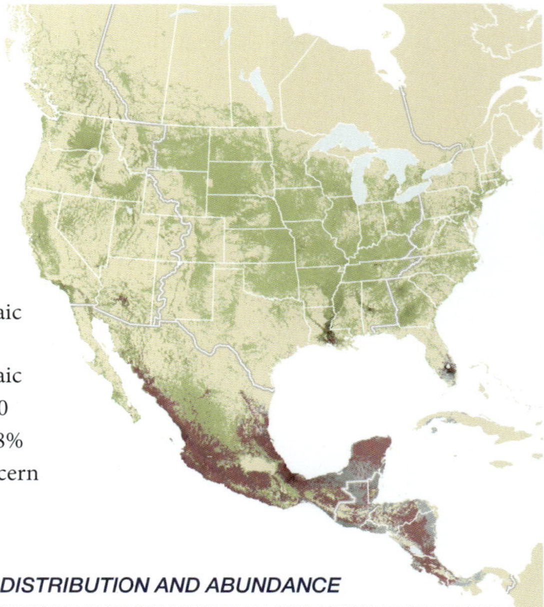

DISTRIBUTION AND ABUNDANCE

MIGRATORY CONNECTIVITY

OVERALL Migratory connectivity is largely unknown. No band reencounter data or tracking studies of migratory connectivity were found for this species.

STABLE ISOTOPES A study of stable isotopes found that Northern Rough-winged Swallows that breed in western North America migrate to the monsoon region of northwest Mexico to complete their primary (post-breeding) wing molt, while individuals that breed in the east replace primaries as they migrate south (Rohwer et al. 2011).

OTHER APPROACHES Substantial geographic co-occurrence of migratory and resident populations in Mexico and Central America is difficult to decipher due to local variations in timing and abundance (De Jong 2020), but at least in the Yucatán, northern Guatemala, and the mountains of Oaxaca and Chiapas, similar migrant and resident subspecies can be distinguished in the field from resident Ridgway's rough-winged swallow (*S. serripennis ridgwayi*; Howell and Webb 1995).

MIGRATORY PATHWAYS eBird animations (Fink et al. 2023) suggest that eastern populations follow the Gulf and Caribbean coastline to nonbreeding grounds, while western populations migrate along the Pacific coast of Mexico.

PURPLE MARTIN *Progne subis*

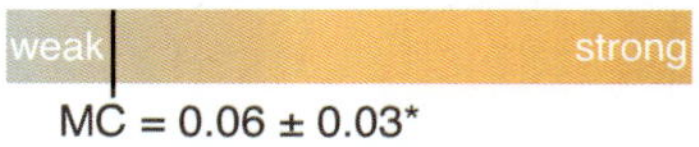

DISTRIBUTION AND ABUNDANCE

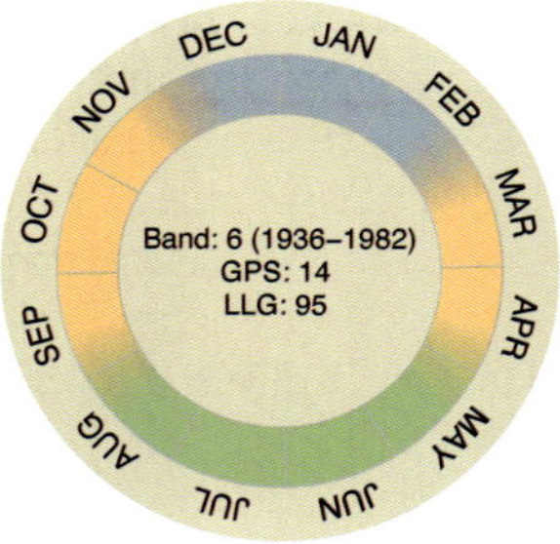

There are three subspecies of Purple Martins, which differ in size, plumage, and geographic range (C.R. Brown et al. 2021). The eastern subspecies, *P. s. subis*, breeds in central and southern Alberta, Saskatchewan, and Manitoba, and throughout all the eastern states of the US, east of the 102nd parallel (102nd meridian west). Breeding populations of the two western subspecies are patchier, and mostly coastal from Vancouver to Baja California (*P. s. arboricolor*), or in the southwestern US into Mexico with breeding concentrations in Arizona, Colorado, New Mexico, and Sonora in northern Mexico (*P. s. hesperia*; Fink et al. 2023; C.R. Brown et al. 2021). The core nonbreeding range is located in northwestern Amazonia in the Brazilian states of Amazonas, Pará, Rondonia, Mato Grosso, and Tocantins, and extends to eastern Bolivia (C.R. Brown et al. 2021).

ORDER Passeriformes
FAMILY Hirundinidae
BREEDING HABITAT Open Country Aerial, Habitat Mosaic
NONBREEDING HABITAT Open Country Aerial, Habitat Mosaic
GLOBAL POPULATION SIZE 9,300,000
POPULATION CHANGE SINCE 1970 −28%
IUCN RED LIST CATEGORY Least Concern
R2R TIPPING POINT SPECIES No
US LISTING STATUS No
CANADA SPECIES AT RISK No
PIF CCS 9/20

MIGRATORY CONNECTIVITY

OVERALL Migratory connectivity information is well known from tracking (light-level geolocator and GPS) and band reencounter data. Breeding populations from all three flyways primarily spend the nonbreeding season in Brazil: Atlantic Flyway (96%), Midcontinent

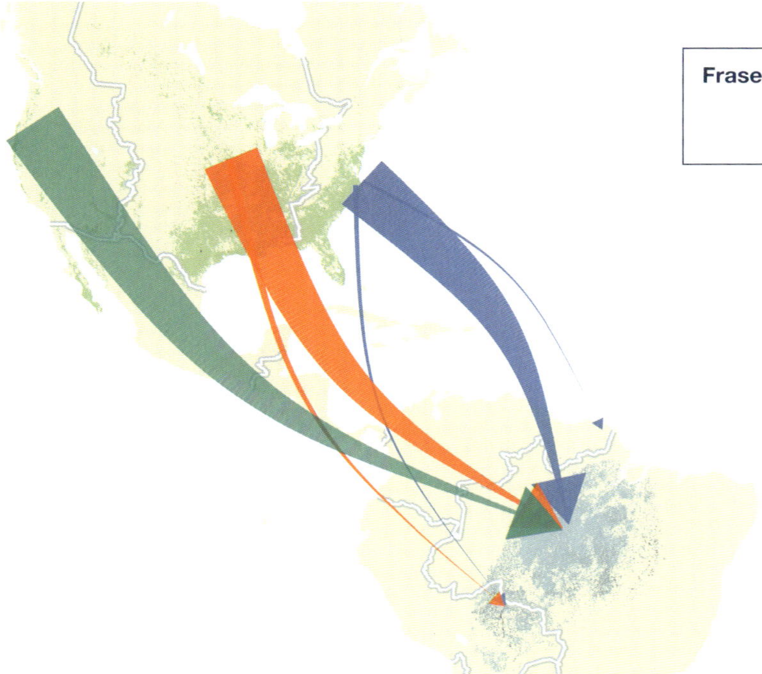

GENERALIZED CONNECTIVITY

INTERESTING NOTE

Fraser et al. (2013) found similar migration rates of Purple Martin breeding populations across the range, and the overall rate was unrelated to breeding origin, timing, route, and distance of the migration.

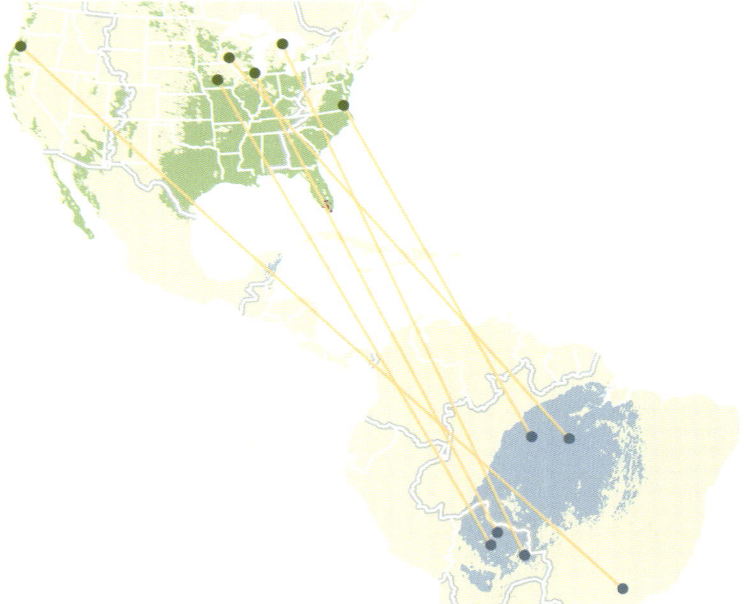

CONNECTIONS A

	Andean West	Brazil	Northern SA
Atlantic (n = 70)	3%	96%	1%
Midcontinent (n = 33)	9%	91%	–
Pacific (n = 6)	–	100%	–

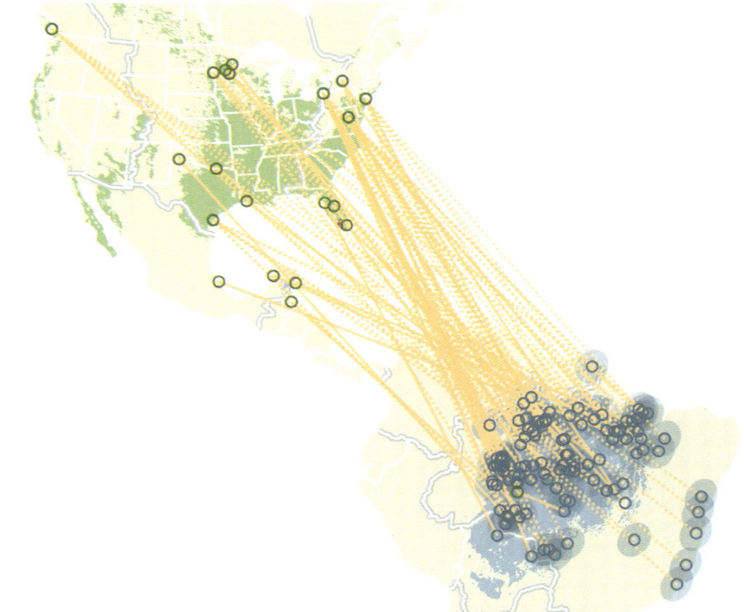

CONNECTIONS B

Flyway (91%), and Pacific Flyway (100%). This results in weak migratory connectivity (MC = 0.06 ± 0.03).

BANDING Band reencounters were primarily from the eastern portion of the range (*P. s. subis*) and show similar patterns of migratory connectivity to geolocator and GPS data.

GEOLOCATOR TRACKING Data from light-level geolocators (Fraser et al. 2012, 2013) showed that Purple Martins from at least seven different breeding populations across the *P. s. subis* range (Alberta, Florida, Minnesota, Ontario, Pennsylvania, and Texas) exhibit weak connectivity. Individuals of the *P. s. arboricola* were tracked from British Columbia to Brazil (Fraser et al. 2012).

SATELLITE TRACKING Archival GPS tags deployed across the range supported weak connectivity, with mixing across all nonbreeding locations, including shared roost sites in the Amazon Basin of Brazil (Fraser et al. 2017). Archival GPS tags were deployed from 2020 to 2022 on adult western purple martins (*P. s. arboricola*) on the central Oregon coast (n = 23). Data from three females revealed nonbreeding sites on the southeastern coast of Brazil, consistent with Fraser et al. (2012). Nonbreeding areas ranged from −16 to −21 degrees latitude, in the states of Bahia and Espírito Santo, Brazil (S. Rockwell, unpublished data). Data not presented in Atlas.

OTHER APPROACHES No genetic or stable isotope studies of migratory connectivity were found for this species.

MIGRATORY PATHWAYS Breeding populations of the *P. s. subis* show a consistent southbound migration pattern where birds reach Central America either by a mostly overland route around the Gulf of Mexico by way of a Florida and Caribbean crossing (*P. s. subis*), by directly crossing the Gulf of Mexico to the Yucatán Peninsula, or by going around the Gulf through Mexico (Fraser et al. 2013). The two other subspecies follow a more westerly route through Mexico and Central America. After crossing the Gulf, birds funnel into South America through the Darién (Bayly et al. 2014) and continue south to their nonbreeding grounds in Amazonia. Birds tracked from coastal Oregon employed extended post-breeding stopovers in northwestern Mexico and in northern Columbia, or inland Bahia or Minas Gerais, Brazil, before reaching the Brazilian coast (S. Rockwell, unpublished data).

DATA CONTRIBUTORS

K. Applegate; J. Barrow; K. Bell; E. Bridge; B.F. Cousens; K.C. Fraser; P.M. Kramer; J.C. Lee; P. Mammenga; N. Mickle; D.M. Morrison; D.J. Newstead; M. Pearman; J.D. Ray; A. Ritchie; A. Savage; T. Shaheen; A. Shave; J. Siegrist; C. Silverio; B.J.M. Stutchbury; J. Tautin.

BARN SWALLOW *Hirundo rustica*

weak strong
MC = 0.34

DISTRIBUTION AND ABUNDANCE

The Barn Swallow has a cosmopolitan distribution. In the Americas, it breeds from Alaska and northern Canada throughout the US and south to central Mexico. Northern Hemisphere breeders spend the nonbreeding season throughout Central and South America to central Chile and northern Argentina (Brown and Brown 2020). In 1980, a population of Barn Swallows were first observed breeding in southern Buenos Aires, Argentina. This established population now carries out an austral migration from their southern breeding grounds to northeastern South America (Venezuela, Guyana, Suriname, French Guiana, and northern Brazil; Winkler et al. 2017).

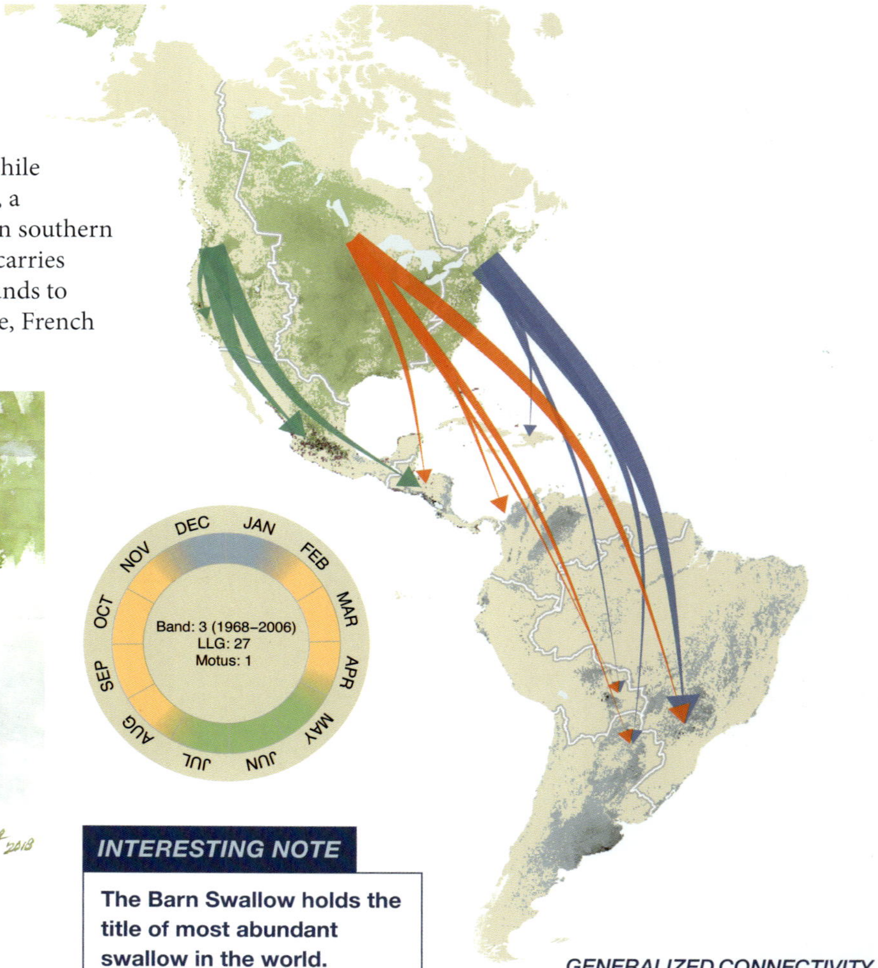

GENERALIZED CONNECTIVITY

ORDER Passeriformes
FAMILY Hirundinidae
BREEDING HABITAT Open Country Aerial, Habitat Mosaic
NONBREEDING HABITAT Open Country Aerial, Habitat Mosaic
GLOBAL POPULATION SIZE 190,000,000
POPULATION CHANGE SINCE 1970 −40%
IUCN RED LIST CATEGORY Least Concern
R2R TIPPING POINT SPECIES No
US LISTING STATUS No
CANADA SPECIES AT RISK Threatened
PIF CCS 9/20

INTERESTING NOTE

The Barn Swallow holds the title of most abundant swallow in the world.

	Andean West	Brazil	Caribbean	Central America	Mexico	Northern SA	Pacific	Southern Cone
Atlantic (n = 11)	9%	64%	9%	–	–	–	–	18%
Midcontinent (n = 12)	8%	33%	–	17%	–	25%	–	17%
Pacific (n = 8)	–	–	–	38%	50%	–	12%	–

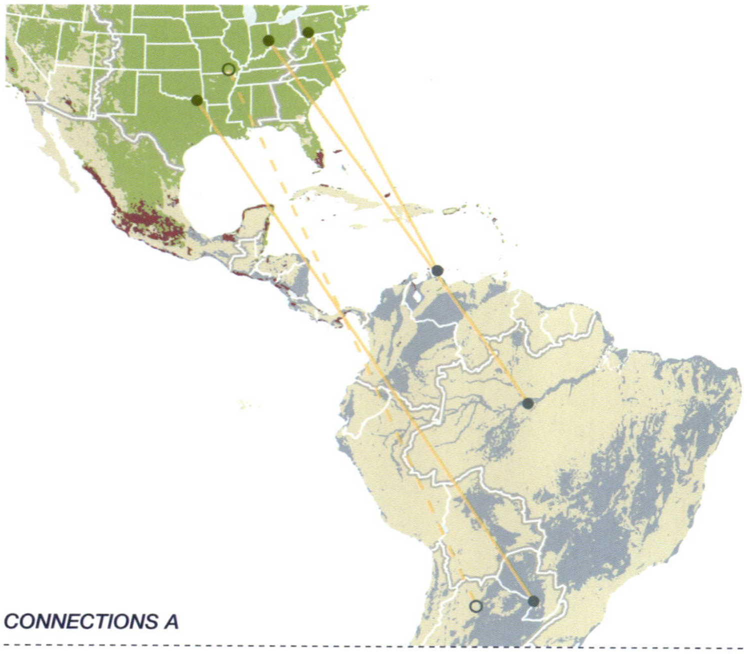

CONNECTIONS A

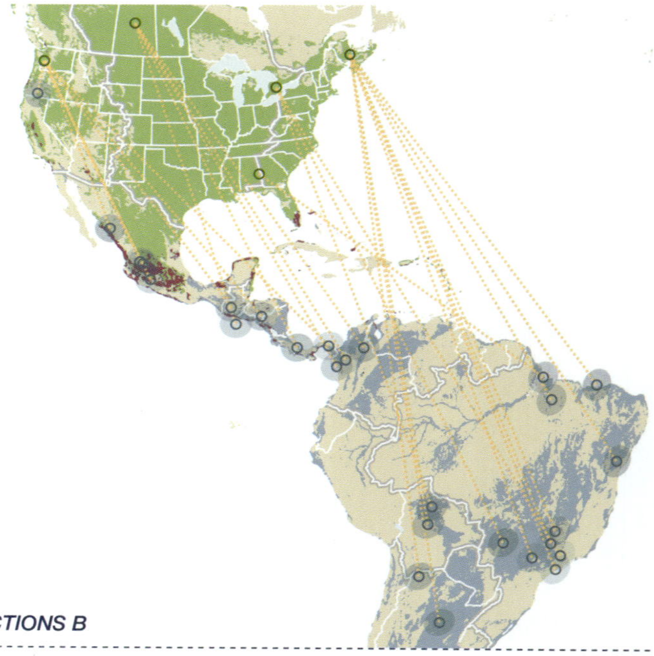

CONNECTIONS B

MIGRATORY CONNECTIVITY

OVERALL The migratory connectivity of Barn Swallows is known from a mixture of band reencounter, geolocator, automated telemetry, and stable isotope data. Atlantic Flyway breeding populations primarily spend the nonbreeding season in Brazil (64%) and the Southern Cone region of South America (18%). Midcontinent Flyway breeders migrate to several nonbreeding regions, including Brazil (33%), northern South America (25%), the Southern Cone region (17%), Central America (17%), and the Andean west (8%). Pacific Flyway breeders primarily migrate to nonbreeding areas in Mexico (50%) and Central America (38%). Breeding populations appear to maintain an east-west divide on nonbreeding areas, yet the Midcontinent Flyway population mixes with both Atlantic and Pacific flyway breeding populations on the nonbreeding grounds, especially in Central and South America. This results in moderate migratory connectivity across the three breeding populations (MC = 0.34).

BANDING Limited band reencounters show a similar east-west divide as geolocator data.

RADIO TRACKING Several studies are tracking Barn Swallows through the Motus wildlife tracking system. There has been one breeding to nonbreeding detection from Missouri to northern Argentina. There have been no other breeding to nonbreeding detections.

GEOLOCATOR TRACKING Tracking studies have shown that Barn Swallows have an east-west migratory divide, in which individuals from western populations migrate to nonbreeding areas farther west and north compared to eastern populations (Hobson et al. 2015b). The austral population of Barn Swallows has been tracked to nonbreeding areas in Venezuela, Guyana, and northern Brazil (Winkler et al. 2017).

STABLE ISOTOPES ^{2}H, ^{13}C, and ^{34}S isotopes, measured from Barn Swallows that bred in eastern Canada, all indicated a potential molting/nonbreeding region in south-central Brazil (Hobson and Kardynal 2016). A more extensive study confirmed this pattern, showing that northern and western breeding populations molt in western and central South America, whereas eastern and southern breeding populations spend the nonbreeding season in northeastern and central-eastern South America (García-Pérez and Hobson 2014). For Barn Swallows breeding in New Brunswick and Nova Scotia, Canada, the most probable nonbreeding areas were eastern and southern Brazil, eastern Bolivia, Paraguay, Uruguay, and northeastern Argentina, although there was evidence of annual variation in nonbreeding areas but use of similar habitats across years (Imlay et al. 2018).

OTHER APPROACHES No genetic studies of migratory connectivity were found for this species.

MIGRATORY PATHWAYS Migration pathways maintain population divides. Birds from Washington state followed the Pacific coast to their nonbreeding grounds; birds from Saskatchewan followed a path through the Great Plains and Central America to reach their nonbreeding grounds; and birds from Ontario flew directly south through the southern states to Central and South America (Hobson et al. 2015b).

DATA CONTRIBUTORS

K. Kardynal.

Petrochelidon pyrrhonota CLIFF SWALLOW

DISTRIBUTION AND ABUNDANCE

The Cliff Swallow has a broad breeding distribution in North America, from Alaska and northwestern Canada to the east coast of the US and south to central Mexico. Breeding strongholds are concentrated in southwest Alaska, in a wide central corridor from eastern Alberta to western Ontario and all the way south to Texas, and from Washington state south through Oregon, Idaho, Nevada, and California (Fink et al. 2023). Nonbreeding grounds extend from Bolivia and south-central Brazil to central Argentina.

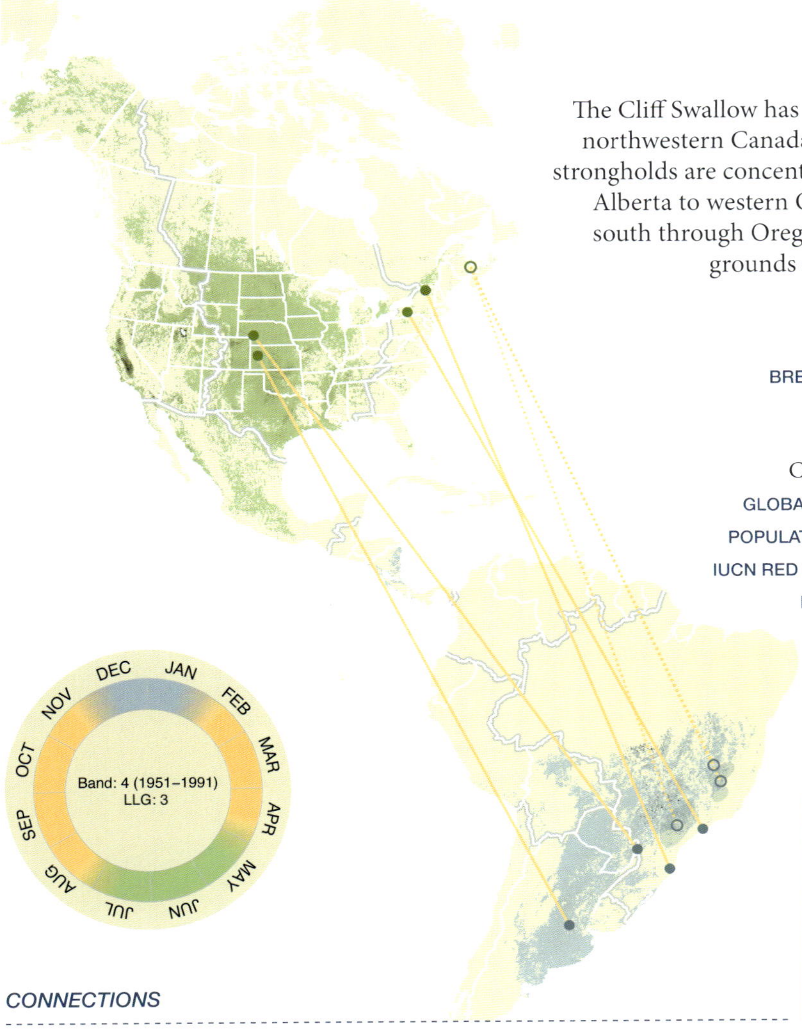

CONNECTIONS

ORDER Passeriformes
FAMILY Hirundinidae
BREEDING HABITAT Open Country Aerial, Habitat Mosaic
NONBREEDING HABITAT Open Country Aerial, Habitat Mosaic
GLOBAL POPULATION SIZE 83,000,000
POPULATION CHANGE SINCE 1970 +49%
IUCN RED LIST CATEGORY Least Concern
R2R TIPPING POINT SPECIES No
US LISTING STATUS No
CANADA SPECIES AT RISK No
PIF CCS 5/20

INTERESTING NOTE

A population of Cliff Swallows started breeding in Argentina in 2015 and were tracked during their austral migrations to three new nonbreeding areas in southern Colombia, Mexico, and the southern US (Areta et al. 2021). Some also spent the boreal summer in the traditional breeding range (Areta et al. 2021).

MIGRATORY CONNECTIVITY

OVERALL The migratory connectivity of the Cliff Swallow is partially known from stable isotope, banding, and geolocator tracking studies. MC was not calculated due to small sample sizes.

BANDING Limited band reencounter data from central and eastern areas of the breeding range show similar patterns of migratory connectivity as geolocator data.

GEOLOCATOR TRACKING Tracking studies have shown that Cliff Swallows from New Brunswick spent the nonbreeding season in southern Brazil (n = 3; Imlay et al. 2018). Two of the three individuals had three stationary periods throughout the nonbreeding season (Imlay et al. 2018).

STABLE ISOTOPES ^{2}H and ^{13}C isotopes measured from Cliff Swallows from New Brunswick indicated potential molting/nonbreeding regions in southern and western Brazil, northern Bolivia, southern Paraguay, and northern Argentina (Imlay et al. 2018).

OTHER APPROACHES No genetic studies of migratory connectivity were found for this species.

MIGRATORY PATHWAYS Pathways are not well known; however, birds from central and possibly eastern populations likely funnel into Central America and enter South America through the Darién (Bayly et al. 2014; Areta et al. 2021), where they carry on in a southern/southeasterly direction to their nonbreeding grounds in the Southern Cone (Fink et al. 2023).

DATA CONTRIBUTORS

K.A. Hobson; T. Imlay; M.L. Leonard; A. Roberto-Charron.

CAVE SWALLOW *Petrochelidon fulva*

DISTRIBUTION AND ABUNDANCE

It is mostly a year-round resident near colony sites in Mexico and the Greater Antilles; however, most individuals breeding in the US are migratory. Breeding areas include Florida, New Mexico, and the northern half of Texas. US breeding populations migrate to areas in the Caribbean, Mexico, Central America, and, more recently, southern Texas (Strickler and West 2020; Fink et al. 2023).

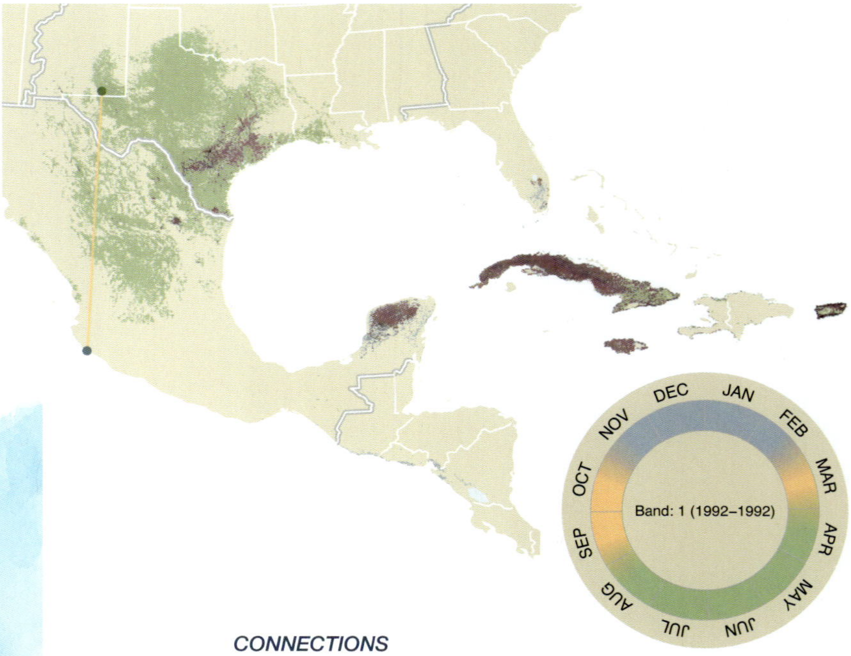

CONNECTIONS

ORDER Passeriformes
FAMILY Hirundinidae
BREEDING HABITAT Arid Lands Aerial, Desert Scrub
NONBREEDING HABITAT Open Country Aerial, Habitat Mosaic
GLOBAL POPULATION SIZE 9,600,000
POPULATION CHANGE SINCE 1970 +110,838%
IUCN RED LIST CATEGORY Least Concern
R2R TIPPING POINT SPECIES No
US LISTING STATUS No
CANADA SPECIES AT RISK No
PIF CCS 9/20

MIGRATORY CONNECTIVITY

OVERALL Movements of this species are not well understood, and patterns are confused by the presence of resident populations throughout much of the nonbreeding range. What little is known is based largely on observational data. MC is not able to be generated due to insufficient data.

BANDING There is only one record of breeding to nonbreeding banding data. A breeding bird from southeastern New Mexico was located in coastal Jalisco, Mexico, during the nonbreeding season.

OTHER APPROACHES No tracking, genetic, or stable isotope studies of migratory connectivity were found for this species.

MIGRATORY PATHWAYS Specific migratory routes are unknown, but birds were observed during pre- and post-breeding migration in Texas, Mexico, the Caribbean, and parts of Florida (Strickler and West 2020; Fink et al. 2023).

PHYLLOSCOPIDAE | *LEAF WARBLERS*

Phylloscopus borealis ARCTIC WARBLER

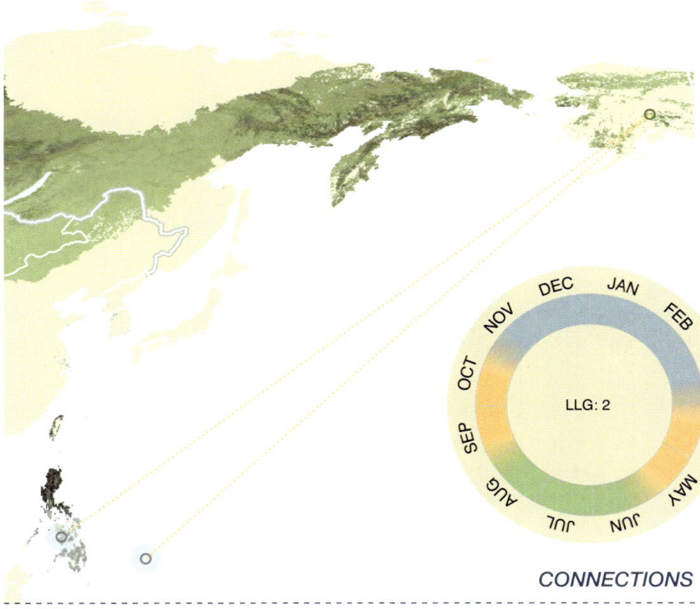

DISTRIBUTION AND ABUNDANCE

It breeds throughout northern Eurasia across the Bering Strait into dense riverine thickets of western (primarily) and central Alaska. It spends the nonbreeding season in southeast Asia, most likely (*P. b. kennicotti* subspecies) in the northern Philippines (Lowther and Sharbaugh 2020).

ORDER Passeriformes
FAMILY Phylloscopidae
BREEDING HABITAT Tundra, Arctic
NONBREEDING HABITAT Forests, Tropical Generalist
GLOBAL POPULATION SIZE 170,000,000
POPULATION CHANGE SINCE 1970 −66%
IUCN RED LIST CATEGORY Least Concern
R2R TIPPING POINT SPECIES No
US LISTING STATUS No
CANADA SPECIES AT RISK No
PIF CCS 9/20

MIGRATORY CONNECTIVITY

OVERALL Migratory connectivity information is from two geolocator recoveries and thus largely unknown.

GEOLOCATOR TRACKING Data from two recovered (of 42) geolocators deployed on male Arctic Warblers breeding in Denali National Park and Preserve, Alaska, showed that both individuals spent the nonbreeding season in the Philippines and on the island of Palau (Adams et al. 2022). Data are not presented in the Atlas.

OTHER APPROACHES The Alaska populations comprise the subspecies *P. b. kennicotti*, a division on which not all taxonomists agree, despite its morphological distinctness (Lowther and Sharbaugh 2020).

MIGRATORY PATHWAYS It migrates west across the Bering Strait, then south to southeast Asia (Lowther and Sharbaugh 2020). Geolocator data showed that two individuals migrated south from Alaska in early September and subsequently staged in southeast Russia and China (Adams et al. 2022).

REGULIDAE | *KINGLETS*

Regulus calendula RUBY-CROWNED KINGLET

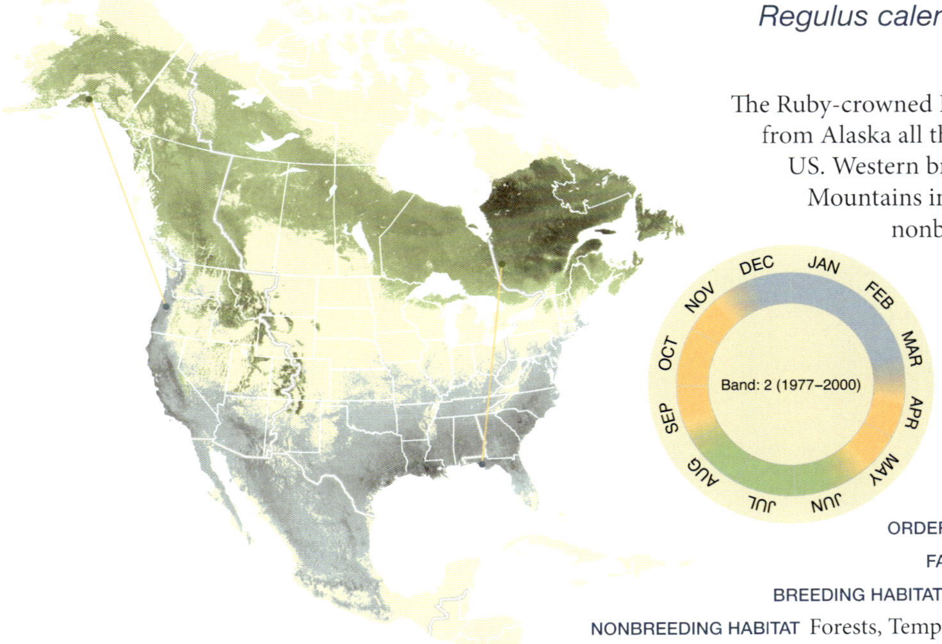

DISTRIBUTION AND ABUNDANCE

The Ruby-crowned Kinglet is a boreal breeding species. Its range extends from Alaska all through the Canadian boreal region and northeastern US. Western breeding populations extend south through the Rocky Mountains into Arizona and New Mexico. The species spends the nonbreeding season in the coastal regions of Washington, Oregon, and California, all through the southern states, north to New York and throughout Mexico south to Mexico City. Populations breeding in the mountainous region of California and Oregon are presumed to be year-round residents and altitudinal migrants (Swanson et al. 2021).

ORDER Passeriformes
FAMILY Regulidae
BREEDING HABITAT Forests, Boreal
NONBREEDING HABITAT Forests, Temperate Generalist
GLOBAL POPULATION SIZE 100,000,000
POPULATION CHANGE SINCE 1970 +24%
IUCN RED LIST CATEGORY Least Concern
R2R TIPPING POINT SPECIES No
US LISTING STATUS No
CANADA SPECIES AT RISK No
PIF CCS 7/20

MIGRATORY CONNECTIVITY

OVERALL The migratory connectivity of the Ruby-crowned Kinglet is unknown. Band reencounter and observational data (Fink et al. 2023) suggest strong migratory connectivity across an east-west divide. More study is needed.

BANDING Two breeding to nonbreeding band reencounters suggest strong migratory connectivity from southern Alaska to northwestern Oregon, and from southwestern Quebec to the panhandle of Florida.
OTHER APPROACHES No tracking, genetic, or stable isotope studies of migratory connectivity were found for this species.

MIGRATORY PATHWAYS Weekly abundance data suggest that western populations keep west of the nonbreeding range, and eastern populations remain east (Fink et al. 2023; Swanson et al. 2021).

GOLDEN-CROWNED KINGLET *Regulus satrapa*

DISTRIBUTION AND ABUNDANCE

It breeds in boreal spruce-fir forest from maritime Alaska across Canada to Nova Scotia, south in the west along the Cascades and Sierras to Southern California and in the Rockies, most abundantly through Idaho; and, in the east, in New England and south in the Appalachians to Georgia. Two disjunct populations are resident in the Trans-Mexican Volcanic Belt and in the mountains of southeast Chiapas, Mexico, and southern Guatemala (Swanson et al. 2020). During the nonbreeding season, the species can be found throughout the US except in the western Great Plains, most abundantly in the southern and western portions of the breeding range and in the southeastern US from Maryland to east Texas (Fink et al. 2023). Despite range extensions in the southeast (Swanson et al. 2020), the species has declined range-wide (Rosenberg et al. 2019), with population loss concentrated in the Northern Rockies, Great Basin, and the northern Pacific rain-forest regions (Sauer et al. 2019).

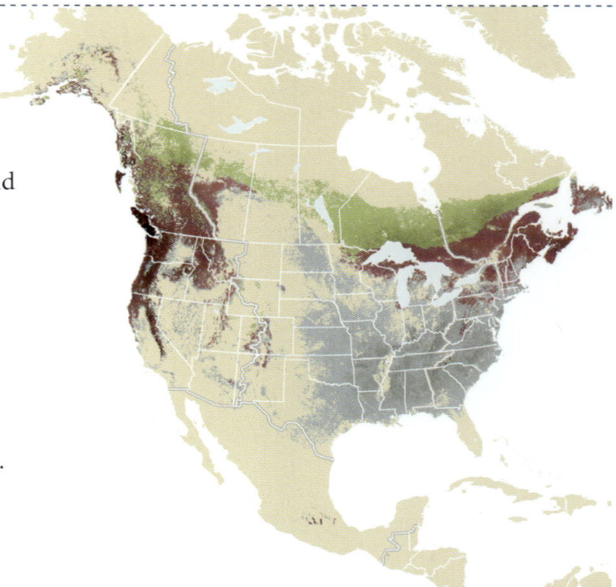

DISTRIBUTION AND ABUNDANCE

ORDER Passeriformes
FAMILY Regulidae
BREEDING HABITAT Forests, Boreal
NONBREEDING HABITAT Forests, Temperate Generalist
GLOBAL POPULATION SIZE 140,000,000
POPULATION CHANGE SINCE 1970 −21%
IUCN RED LIST CATEGORY Least Concern
R2R TIPPING POINT SPECIES No
US LISTING STATUS No
CANADA SPECIES AT RISK No
PIF CCS 10/20

MIGRATORY CONNECTIVITY

OVERALL Migratory connectivity is unknown. No information on migratory connectivity was found for this species.
MIGRATORY PATHWAYS eBird animations (Fink et al. 2023) indicate that eastern and western populations behave independently with respect to migratory movements, separated by the western Great Plains. This difference is corroborated by subspecies breeding/nonbreeding records (Swanson et al. 2020).

BOMBYCILLIDAE | *WAXWINGS*

BOHEMIAN WAXWING *Bombycilla garrulus*

DISTRIBUTION AND ABUNDANCE

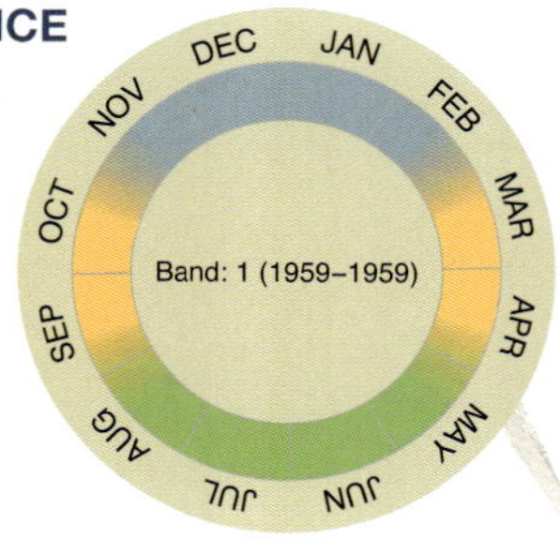

Circumpolar in their distribution, Bohemian Waxwings are named for their nomadic movements, and they move around in the nonbreeding season in large foraging flocks (Witmer 2024). In North America, they breed in coniferous forests during summer in interior Alaska and across northern Canada, migrating to southern Canada, the western United States, and the northeastern United States in the nonbreeding season (Witmer 2024). They can be irruptive in some nonbreeding seasons, extending farther south from their typical nonbreeding range limit. Distributional information is somewhat limited because of low eBird coverage.

ORDER Passeriformes
FAMILY Bombycillidae
BREEDING HABITAT Forests, Boreal
NONBREEDING HABITAT Forests, Boreal
GLOBAL POPULATION SIZE 5,000,000
POPULATION CHANGE SINCE 1970 −57%
IUCN RED LIST CATEGORY Least Concern
R2R TIPPING POINT SPECIES No
US LISTING STATUS No
CANADA SPECIES AT RISK No
PIF CCS 11/20

MIGRATORY CONNECTIVITY

OVERALL There is little information on the migratory connectivity of Bohemian Waxwings, and discovering more about their migration patterns should be a research priority.

BANDING A single band reencounter connects a breeding location in Alberta to a nonbreeding location in Minnesota.

OTHER APPROACHES No tracking, genetic, or stable isotope studies of migratory connectivity exist for this species.

INTERESTING NOTE

Bohemian Waxwings do not defend breeding territories (Witmer 2024).

CONNECTIONS

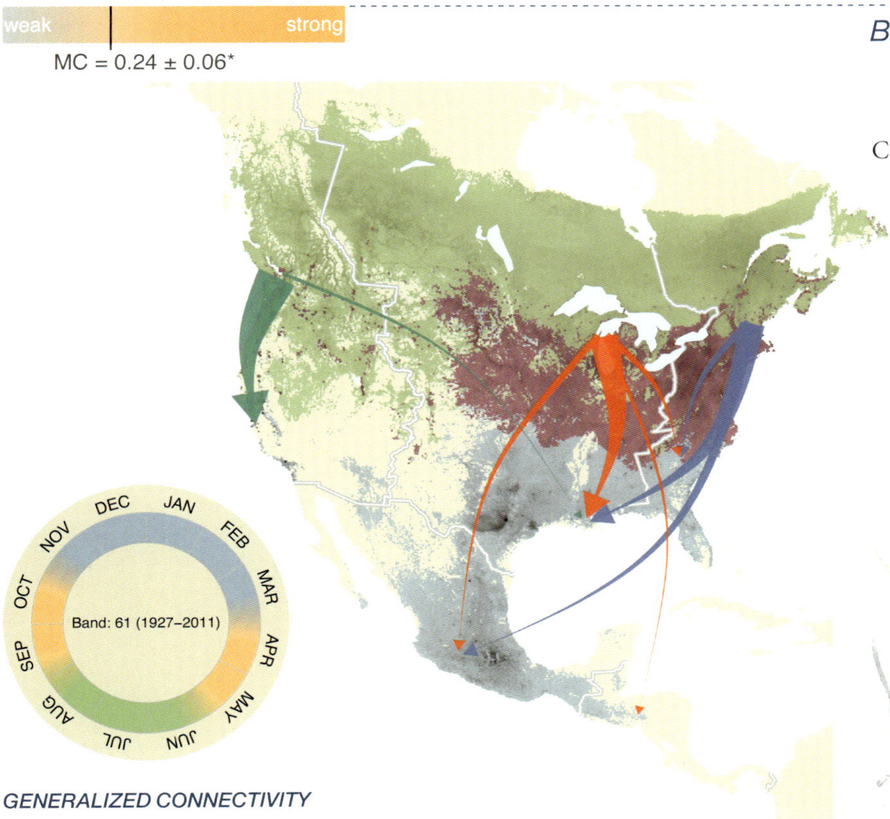

GENERALIZED CONNECTIVITY

Bombycilla cedrorum **CEDAR WAXWING**

DISTRIBUTION AND ABUNDANCE

The Cedar Waxwing breeding range extends across all of Canada and most of the northern United States. During the nonbreeding period, waxwings move south, departing most of Canada. Many remain in the southern United States, while others migrate as far south as Costa Rica and Panama. Waxwings are often found foraging in flocks and are social throughout the year (Witmer 2024).

ORDER Passeriformes
FAMILY Bombycillidae
BREEDING HABITAT Forests, Temperate Generalist
NONBREEDING HABITAT Forests, Temperate Generalist
GLOBAL POPULATION SIZE 64,000,000
POPULATION CHANGE SINCE 1970 +21%
IUCN RED LIST CATEGORY Least Concern
R2R TIPPING POINT SPECIES No
US LISTING STATUS No
CANADA SPECIES AT RISK No
PIF CCS 8/20

MIGRATORY CONNECTIVITY

OVERALL Migratory connectivity information is limited to band reencounter data. Individuals banded in the Pacific coastal portion of their range have been recovered south,

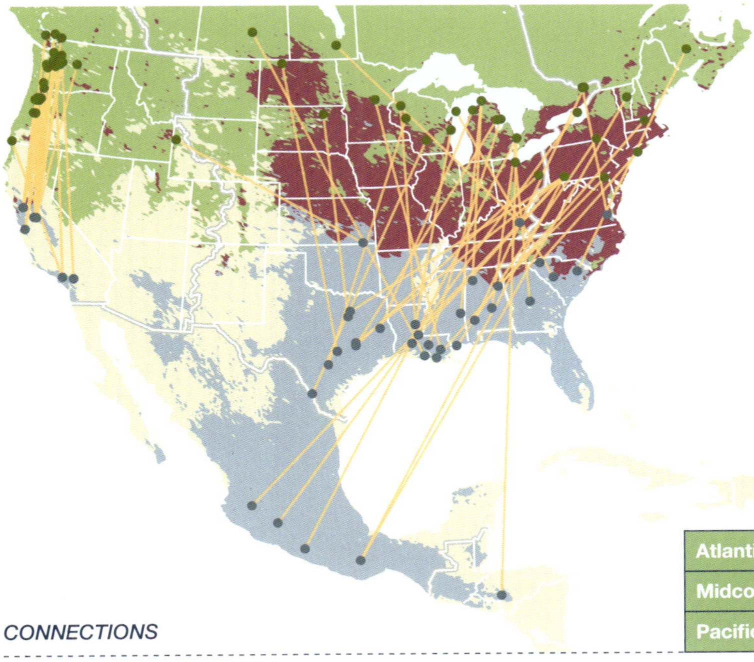

CONNECTIONS

	Atlantic	Central America	Mexico	Midcontinent	Pacific
Atlantic (n = 12)	17%	–	25%	58%	–
Midcontinent (n = 22)	14%	4%	14%	68%	–
Pacific (n = 27)	–	–	–	4%	96%

throughout California (96%). Individuals breeding in the Atlantic Flyway primarily migrate out of the flyway to the Midcontinent Flyway (58%) and Mexico (25%); however, some spend the nonbreeding season within the flyway (17%). Birds breeding in the Midcontinent Flyway remain within the flyway (68%) during the nonbreeding season, as well as migrate to Mexico (14%), the Atlantic Flyway (14%), and Central America (4%). The majority of Pacific Flyway breeding birds migrate to nonbreeding areas within the Pacific Flyway (96%). Although the within-flyway transition is high in the Pacific Flyway, there is more mixing of the Atlantic and Midcontinent flyway breeding populations in high-abundance nonbreeding areas, resulting in moderate migratory connectivity (0.24 ± 0.06).

OTHER APPROACHES No tracking, genetic, or stable isotope studies of migratory connectivity were found for this species.

BANDING Band reencounter data show that individuals do not tend to cross the Rocky Mountains, with individuals breeding in Washington and Oregon migrating to Southern California, while individuals from the eastern range migrate to southeastern states, the Gulf Coast, and Mexico. Cedar Waxwings in Oaxaca, Mexico, were found to have no between-year nonbreeding site fidelity based on captures and recaptures (Monroy-Ojeda et al. 2013).

PTILIOGONATIDAE | *SILKY-FLYCATCHERS*

PHAINOPEPLA *Phainopepla nitens*

DISTRIBUTION AND ABUNDANCE

Phainopeplas exhibit an itinerant breeding strategy, which means breeding at two spatially and temporally distinct areas and migrating between them (Chu and Walsberg 2020). The year-round range spans the southwestern United States and Mexico, including Southern California, the Sonoran Desert of Arizona, southern Nevada, New Mexico, and West Texas (Fink et al. 2023).

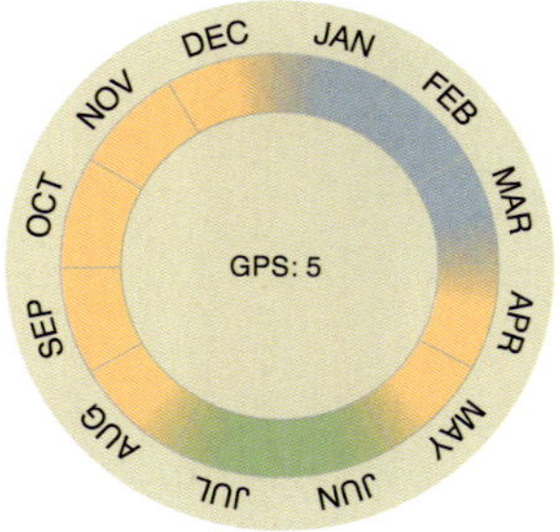

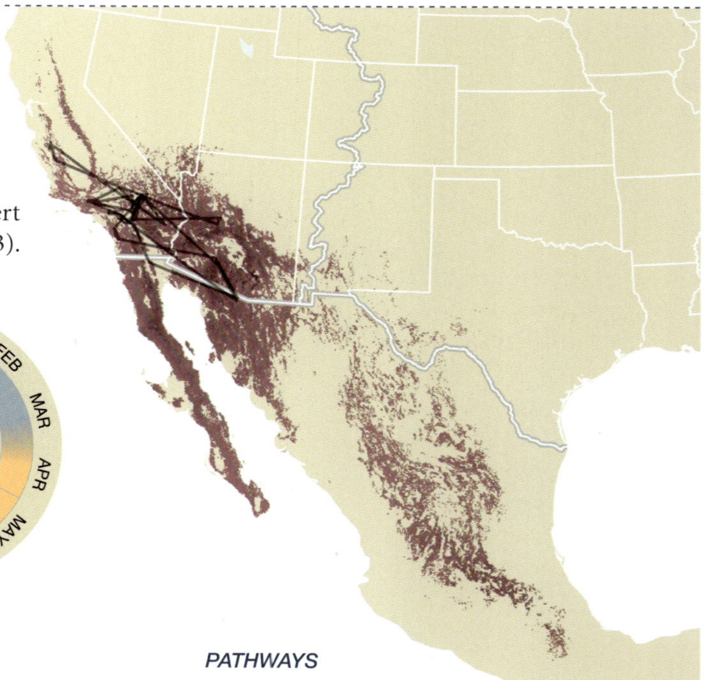

PATHWAYS

ORDER Passeriformes
FAMILY Ptiliogonatidae
BREEDING HABITAT Arid Lands, Desert Scrub
NONBREEDING HABITAT Arid Lands, Desert Scrub
GLOBAL POPULATION SIZE 3,200,000
POPULATION CHANGE SINCE 1970 +23%
IUCN RED LIST CATEGORY Least Concern
R2R TIPPING POINT SPECIES No
US LISTING STATUS No
CANADA SPECIES AT RISK No
PIF CCS 10/20

INTERESTING NOTE

Population genomic analyses showed no genetic differentiation between breeding populations in the Mojave Desert and woodland habitats of California, indicating gene flow between the two breeding regions (Baldassarre et al. 2019).

MIGRATORY CONNECTIVITY

Overall: Migratory connectivity is relatively unknown. There is limited data from one tracking study (Baldassarre et al. 2019).

SATELLITE TRACKING GPS archival tags deployed on individuals breeding in the Mojave Desert of California in the breeding season of March to April were tracked to a second breeding location in woodland habitat on the California coast over 200 km away (Baldassarre et al. 2019). Tags were recovered from the Mojave Desert during the nonbreeding season, showing these individuals returned to their initial breeding locations (Baldassarre et al. 2019).

OTHER APPROACHES No banding, genetic, or stable isotope studies of migratory connectivity were found for this species.

MIGRATORY PATHWAYS Four individuals spent time in a pre-breeding stopover location in the southern Mojave Desert between the desert and woodland breeding habitats, while one individual flew directly (Baldassarre et al. 2019). After potentially breeding in woodland habitats, four individuals spent a considerable amount of time at a post-breeding stopover site before returning, and one individual flew directly to the initial breeding area. One individual's post-breeding stopover was located near its pre-breeding stopover site, but the others traveled southeast to the eastern edge of the Sonoran Desert (Baldassarre et al. 2019).

DATA CONTRIBUTORS
J.W. Atwell; D.T. Baldassarre; L. Campagna; M. Chu; L.H. Crampton; R.C. Fleischer; C. Riehl; H.A. Thomassen.

SITTIDAE | NUTHATCHES

Sitta canadensis RED-BREASTED NUTHATCH

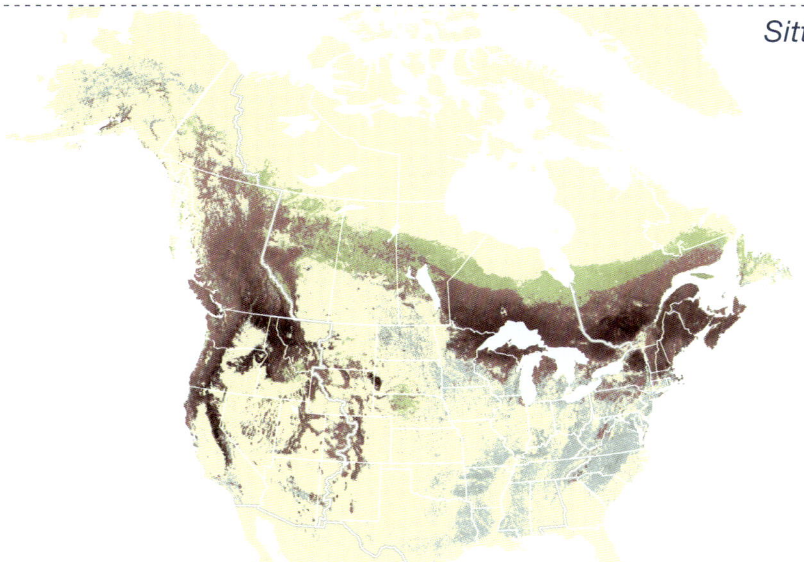

DISTRIBUTION AND ABUNDANCE

DISTRIBUTION AND ABUNDANCE

It breeds in the boreal forest zone across southern Canada and the northern US (except for the northern Great Plains provinces and states), south through the Appalachians and the western mountain ranges of British Columbia and the US. During the nonbreeding season, the species migrates from the northern portion of the boreal forest, with birds concentrating in the remainder of the breeding range and sparsely across the central and eastern US (Fink et al. 2023).

ORDER Passeriformes
FAMILY Sittidae
BREEDING HABITAT Forests, Boreal
NONBREEDING HABITAT Forests, Temperate Generalist
GLOBAL POPULATION SIZE 20,000,000
POPULATION CHANGE SINCE 1970 +85%
IUCN RED LIST CATEGORY Least Concern
R2R TIPPING POINT SPECIES No
US LISTING STATUS No
CANADA SPECIES AT RISK No
PIF CCS 7/20

MIGRATORY CONNECTIVITY

OVERALL Migratory connectivity is unknown. No information on migratory connectivity was found for this species.

MIGRATORY PATHWAYS eBird animations (Fink et al. 2023) suggest that eastern and western populations migrate independently, separated by the divide of the Great Plains.

INTERESTING NOTE

Irruptive migrations occur quite sporadically in this species (Ghalambor and Martin 2020).

CERTHIIDAE | TREECREEPERS

Certhia americana BROWN CREEPER

DISTRIBUTION AND ABUNDANCE

Breeding distribution is quite complex: in the east, it nests in northern forests from central Ontario and Quebec through the upper Midwest states and south in the Appalachians through North Carolina. In the west, it breeds sparsely in Alaska through British Columbia, and then abundantly in Vancouver, through the Cascades, coastal mountain ranges, and Sierras, and in the Northern Rockies of southern British Columbia, southwest Alberta, western Montana, and Idaho. Breeding concentrations also occur in the Colorado Rockies, the mountains of southern Arizona, and the Sierra Madre Occidental of Mexico, with scattered populations in southern Mexico, Guatemala, and Honduras. During the nonbreeding season, populations migrate south from the northern portion of the boreal forest belt in Canada, and eastern populations appear to disperse throughout the entire eastern US, Florida excepted (Fink et al. 2023).

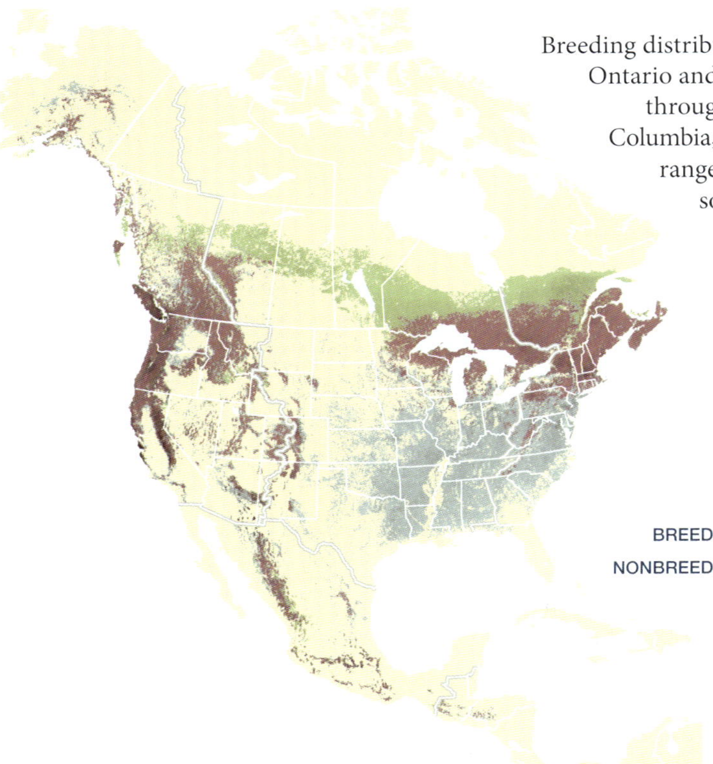

DISTRIBUTION AND ABUNDANCE

ORDER Passeriformes
FAMILY Certhiidae
BREEDING HABITAT Forests, Temperate Generalist
NONBREEDING HABITAT Forests, Temperate Generalist
GLOBAL POPULATION SIZE 11,000,000
POPULATION CHANGE SINCE 1970 +35%
IUCN RED LIST CATEGORY Least Concern
R2R TIPPING POINT SPECIES No
US LISTING STATUS No
CANADA SPECIES AT RISK No
PIF CCS 8/20

MIGRATORY CONNECTIVITY

OVERALL Migratory connectivity is unknown. No information on migratory connectivity was found for this species.
MIGRATORY PATHWAYS eBird animations (Fink et al. 2023) suggest that eastern and western populations migrate independently but that birds breeding in the northern forests of Alberta, Saskatchewan, and Manitoba move eastward to join eastern populations during the nonbreeding season.

POLIOPTILIDAE | *GNATCATCHERS*

BLUE-GRAY GNATCATCHER *Polioptila caerulea*

DISTRIBUTION AND ABUNDANCE

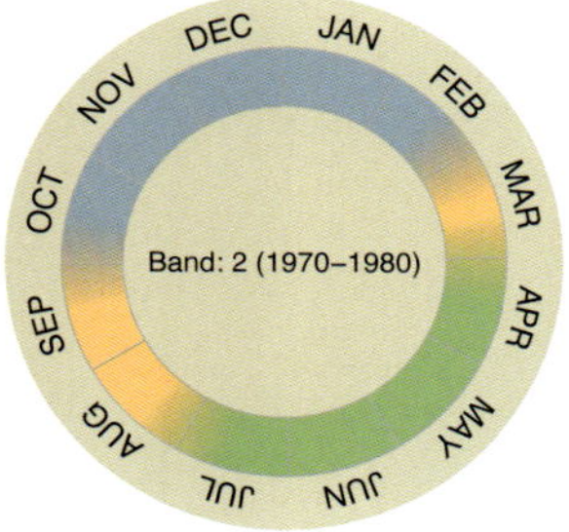

Ranging throughout much of the US and Mexico, this smaller species is the most migratory member of the gnatcatchers. Movements appear complex and population-dependent; populations in the north are long-distance migrants, while southern US breeders migrate shorter distances, and some populations in Central America and the Caribbean are resident. It fully departs from the northern breeding range to the southern United States and into the Bahamas, Cuba, Mexico, and northern Central America for the nonbreeding season (Kershner and Ellison 2020; Fink et al. 2023).

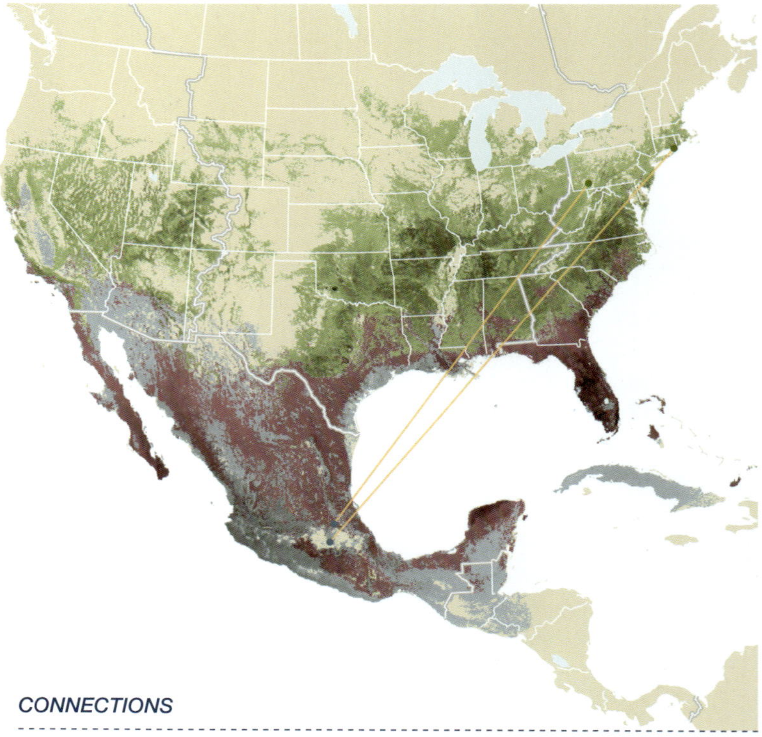

ORDER Passeriformes
FAMILY Polioptilidae
BREEDING HABITAT Forests, Generalist
NONBREEDING HABITAT Forests, Generalist
GLOBAL POPULATION SIZE 260,000,000
POPULATION CHANGE SINCE 1970 +21%
IUCN RED LIST CATEGORY Least Concern
R2R TIPPING POINT SPECIES No
US LISTING STATUS No
CANADA SPECIES AT RISK No
PIF CCS 8/20

MIGRATORY CONNECTIVITY

OVERALL Migratory connectivity and movements are not well understood due to limited data. Two birds breeding in the Atlantic Flyway spent the nonbreeding season in Mexico.
BANDING Band reencounter data is limited. Two individuals from breeding sites in Pennsylvania and Connecticut were reencountered in central Mexico during the nonbreeding season.
OTHER APPROACHES No tracking, genetic, or stable isotope studies of migratory connectivity were found for this species.
MIGRATORY PATHWAYS Movements are variable and poorly understood at the population level. Birds from Florida and the Caribbean move northward along the eastern seaboard to the breeding grounds; nonbreeding birds in the Gulf of Mexico and Mexico move broadly north to breeding areas. Evidence suggests that many individuals of this species migrate north from Mexico and Central America via trans-Gulf migration (Kershner and Ellison 2020; Fink et al. 2023).

Salpinctes obsoletus ROCK WREN

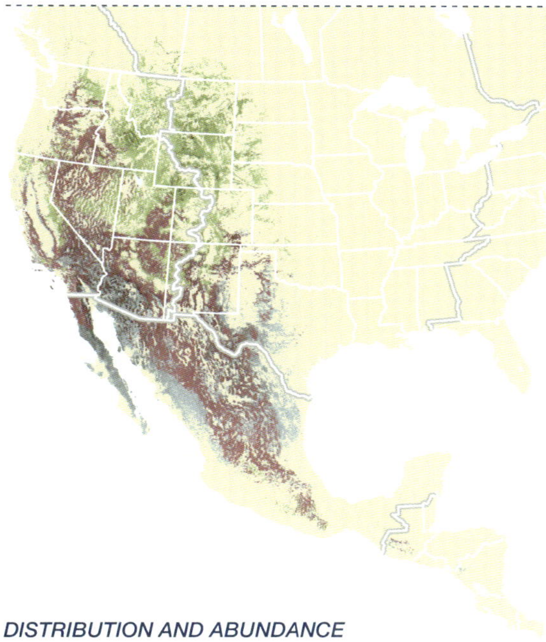

DISTRIBUTION AND ABUNDANCE

DISTRIBUTION AND ABUNDANCE

It breeds throughout the western US, south through the Mexican *cordilleras*, and in the mountains of Chiapas and Guatemala. Most of the northern population appears to migrate south during the nonbreeding season to concentrate in Southern California, Nevada, Arizona, northwestern Mexico, and southwestern Texas (Fink et al. 2023).

ORDER Passeriformes
FAMILY Troglodytidae
BREEDING HABITAT Arid Lands, Desert Scrub
NONBREEDING HABITAT Arid Lands, Desert Scrub
GLOBAL POPULATION SIZE 4,100,000
POPULATION CHANGE SINCE 1970 −45%
IUCN RED LIST CATEGORY Least Concern
R2R TIPPING POINT SPECIES No
US LISTING STATUS No
CANADA SPECIES AT RISK No
PIF CCS 12/20

MIGRATORY CONNECTIVITY

OVERALL Migratory connectivity is unknown. No information on migratory connectivity was found for this species.

MIGRATORY PATHWAYS This species undertakes both north-south and elevational migratory movements, retreating from the northern reaches of its range and higher altitudes during the nonbreeding season. The direction and distance of these movements appear highly variable and are likely influenced by the patchy availability of rocky habitat and weather conditions (Benedict et al. 2021).

Thryomanes bewickii BEWICK'S WREN

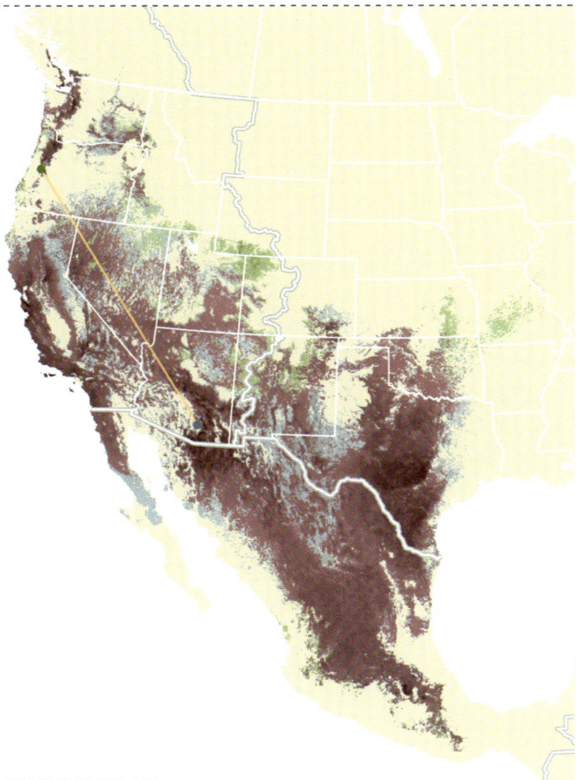

CONNECTIONS

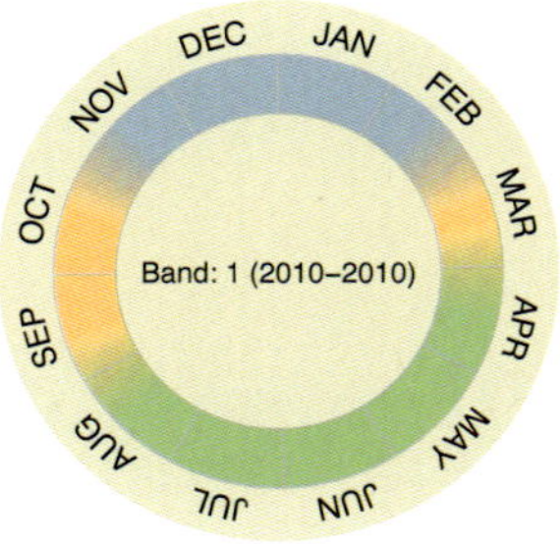

DISTRIBUTION AND ABUNDANCE

This vociferous songbird is common in scrubby habitats of the central and western United States and Mexico. It is a resident or short-distance migrant over most of its range. In the western US, it is a year-round resident with some short-distance migration. In the east, there is some migration from the northern boundary of the breeding range southward to Texas, the Gulf Coast, and Florida, but it is likely resident in the rest of the eastern breeding range (Kennedy and White 2020).

ORDER Passeriformes
FAMILY Troglodytidae
BREEDING HABITAT Arid Lands, Desert Scrub
NONBREEDING HABITAT Arid Lands, Desert Scrub
GLOBAL POPULATION SIZE 7,900,000
POPULATION CHANGE SINCE 1970 −30%
IUCN RED LIST CATEGORY Least Concern
R2R TIPPING POINT SPECIES No
US LISTING STATUS No
CANADA SPECIES AT RISK No
PIF CCS 10/20

MIGRATORY CONNECTIVITY

OVERALL Some migratory movement is demonstrated by limited band reencounter and observational data, as well as television tower kills (Kennedy and White 2020). MC is unknown due to insufficient data.

BANDING Band reencounter data connects a breeding site in western Oregon to a nonbreeding site in southern Arizona.

OTHER APPROACHES No tracking, genetic, or stable isotope studies of migratory connectivity were found for this species.

MIGRATORY PATHWAYS During the nonbreeding season, it recedes from the northern and eastern edges of the range, and expands slightly into additional areas outside the core year-round range (Kennedy and White 2020; Fink et al. 2023).

NORTHERN HOUSE WREN *Troglodytes aedon*

DISTRIBUTION AND ABUNDANCE

One of the most widespread birds in the Western Hemisphere, it has a variety of populations and subspecies throughout its vast range. North American birds are mostly migratory and breed from coast to coast, as far north as northern Alberta and British Columbia, and as far south as the Mexican border (L.S. Johnson 2024).

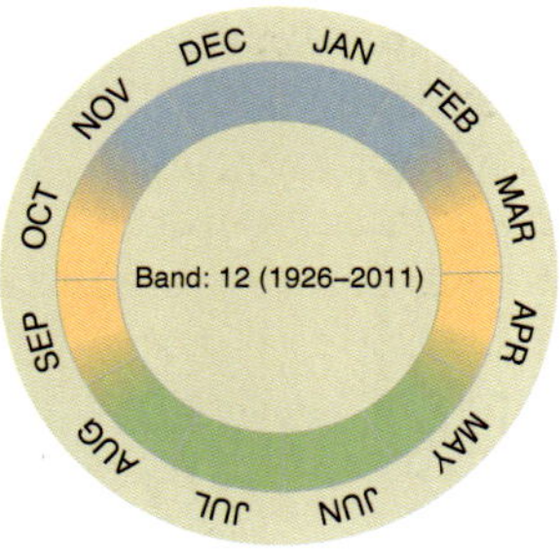

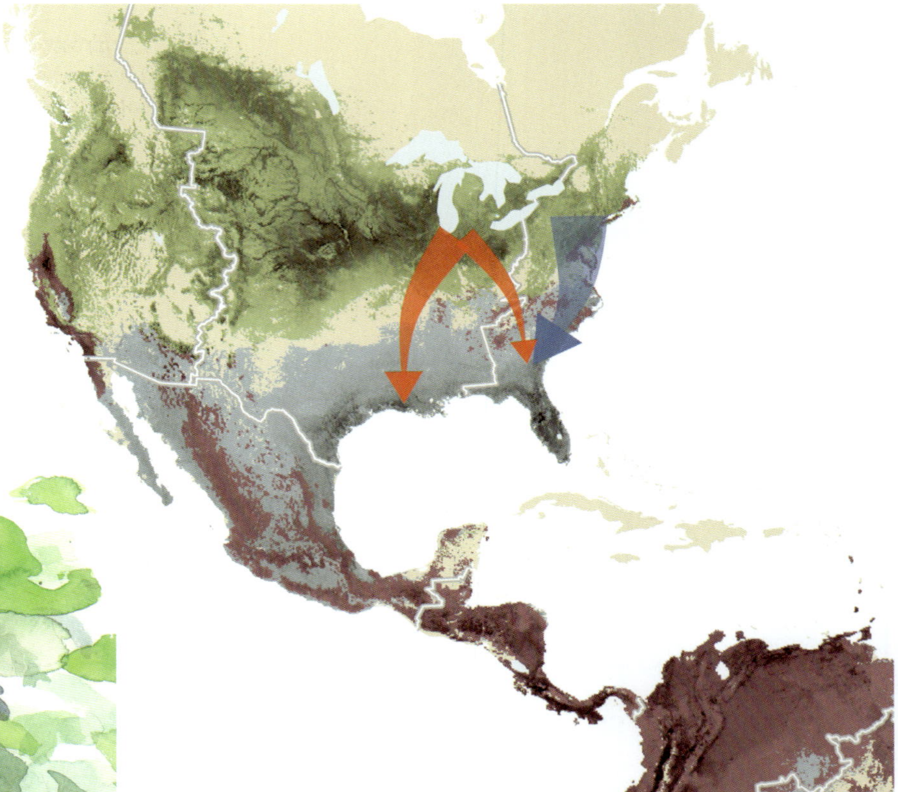

GENERALIZED CONNECTIVITY

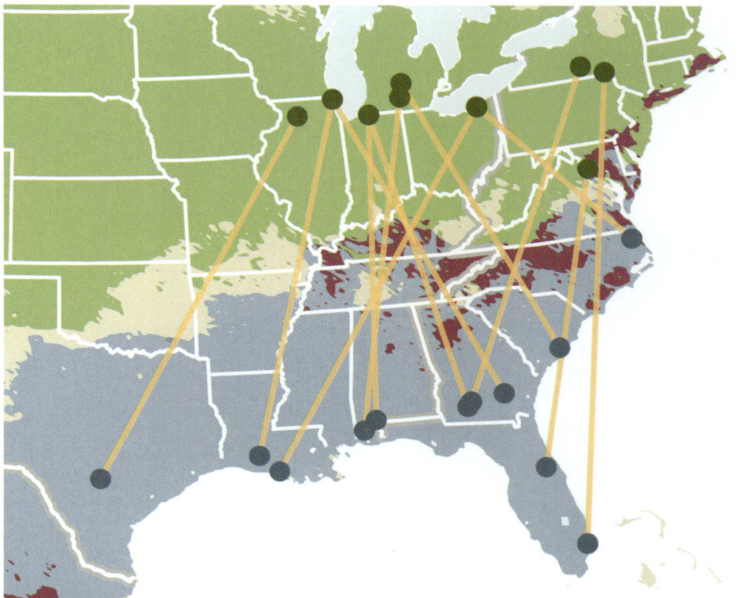

CONNECTIONS

ORDER Passeriformes
FAMILY Troglodytidae
BREEDING HABITAT Forests, Generalist
NONBREEDING HABITAT Open Country, Habitat Mosaic
GLOBAL POPULATION SIZE 190,000,000
POPULATION CHANGE SINCE 1970 +13%
IUCN RED LIST CATEGORY Least Concern
R2R TIPPING POINT SPECIES No
US LISTING STATUS No
CANADA SPECIES AT RISK No
PIF CCS 5/20

	Atlantic	Midcontinent
Atlantic (n = 3)	100%	–
Midcontinent (n = 8)	38%	62%

MIGRATORY CONNECTIVITY

OVERALL Migratory connectivity appears moderate to strong, but further study of individual movements is necessary. Birds breeding in the Atlantic Flyway migrate to nonbreeding areas within the flyway. Midcontinent Flyway breeders migrate to nonbreeding areas within the flyway (62%) and the Atlantic Flyway (38%). MC is not able to be generated due to insufficient data.

BANDING All band reencounter data comes from east of the Rockies, suggesting that House Wrens breeding east of the Appalachian Mountains migrate to nonbreeding areas primarily in Florida, while individuals breeding west of the Appalachian Mountains spend the nonbreeding season from South Carolina to Texas. Some birds breeding on the western Great Plains and eastern slope of the Rocky Mountains appear to migrate southeast rather than directly south (L.S. Johnson and Wise 2000; L.S. Johnson 2024).

RADIO TRACKING There were limited tag deployments through the Motus Wildlife Tracking System. There are no breeding to nonbreeding detections.

OTHER APPROACHES No tracking, genetic, or stable isotope studies of migratory connectivity were found for this species.

MIGRATORY PATHWAYS There is no information on specific routes. Birds appear to migrate over a broad front in both pre- and post-breeding (Fink et al. 2023).

Troglodytes pacificus PACIFIC WREN

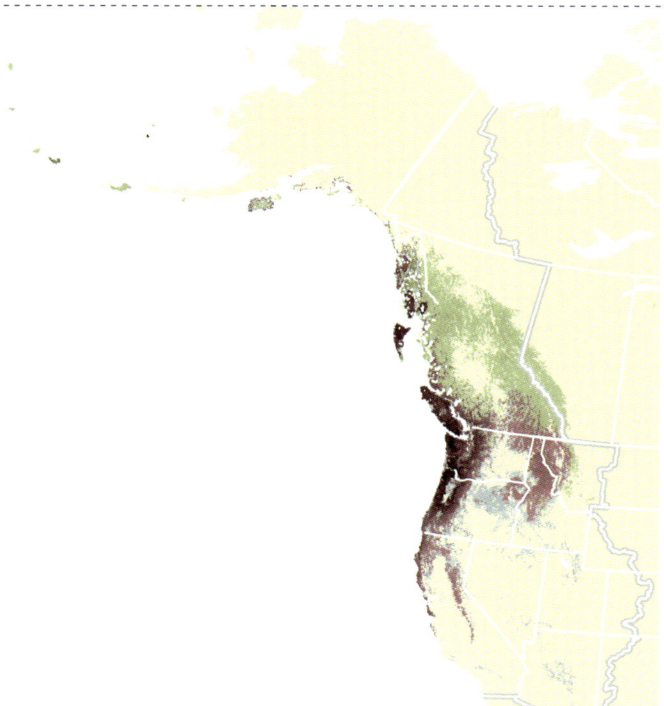

DISTRIBUTION AND ABUNDANCE

DISTRIBUTION AND ABUNDANCE

It breeds in a wide array of terrestrial habitats, primarily in forests—well inland in British Columbia, Washington, and Idaho, and most abundantly in coastal coniferous forests from the Aleutians to central California. During the nonbreeding season, inland populations migrate to the coastal ranges, where the migratory influx augments the resident population (Toews and Irwin 2020; Fink et al. 2023). Across this species' range, some populations can be fully migrant, while others are partially migrant or resident.

ORDER Passeriformes
FAMILY Troglodytidae
BREEDING HABITAT
Forests, Temperate Western
NONBREEDING HABITAT
Forests, Temperate Western
GLOBAL POPULATION SIZE 7,500,000
POPULATION CHANGE SINCE 1970 −23%
IUCN RED LIST CATEGORY Least Concern
R2R TIPPING POINT SPECIES No
US LISTING STATUS No
CANADA SPECIES AT RISK No
PIF CCS 9/20

MIGRATORY CONNECTIVITY

OVERALL Migratory connectivity is unknown. No information on migratory connectivity was found for this species.

MIGRATORY PATHWAYS The details of migratory movements are not well understood but include short- to medium-range regional movement, as well as altitudinal migration away from high peaks during the nonbreeding season (Toews and Irwin 2020; Fink et al. 2023).

Troglodytes hiemalis WINTER WREN

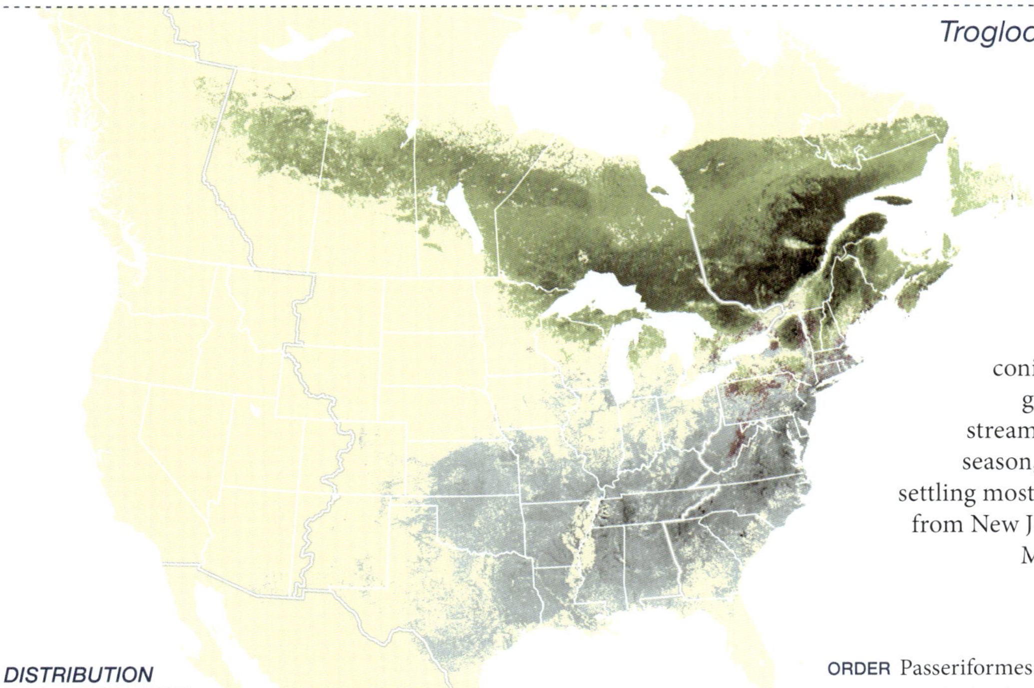

DISTRIBUTION AND ABUNDANCE

DISTRIBUTION AND ABUNDANCE

It breeds in a wide variety of woodland habitats across the northern tier of states in northeastern and north-central US, throughout the boreal forests of Canada from Newfoundland and Labrador to Alberta, and more sparsely along the Appalachian chain. It reaches its highest densities in conifer forests in southern Quebec and Ontario, generally in association with water, especially streams (Hejl et al. 2020). During the nonbreeding season, it retreats southward across the eastern US, settling most densely in a wider array of wooded habitats from New Jersey south to northern Georgia and west to Missouri and eastern Texas (Fink et al. 2023).

ORDER Passeriformes
FAMILY Troglodytidae
BREEDING HABITAT Forests, Boreal
NONBREEDING HABITAT Forests, Temperate Eastern
GLOBAL POPULATION SIZE 11,000,000
POPULATION CHANGE SINCE 1970 +56%
IUCN RED LIST CATEGORY Least Concern
R2R TIPPING POINT SPECIES No
US LISTING STATUS No
CANADA SPECIES AT RISK No
PIF CCS 12/20

MIGRATORY CONNECTIVITY

OVERALL Migratory connectivity is unknown. No information on migratory connectivity was found for this species.

MIGRATORY PATHWAYS eBird animations (Fink et al. 2023) suggest southward migration across a broad front, with heavier concentrations along the Appalachians, Piedmont, and eastern seaboard.

SEDGE WREN *Cistothorus stellaris*

DISTRIBUTION AND ABUNDANCE

The Sedge Wren breeds primarily in the upper midwestern states and the Prairie provinces of Canada, most densely in Minnesota, the Dakotas, and Manitoba, and spends the nonbreeding season in the southern US and along the Atlantic coast, most densely along the Texas and Louisiana Gulf Coast (Fink et al. 2023).

ORDER Passeriformes
FAMILY Troglodytidae
BREEDING HABITAT Grasslands, Temperate
NONBREEDING HABITAT Wetlands, Seasonally Wet Grasslands
GLOBAL POPULATION SIZE 5,000,000
POPULATION CHANGE SINCE 1970 +71%
IUCN RED LIST CATEGORY Least Concern
R2R TIPPING POINT SPECIES No
US LISTING STATUS No
CANADA SPECIES AT RISK No
PIF CCS 9/20

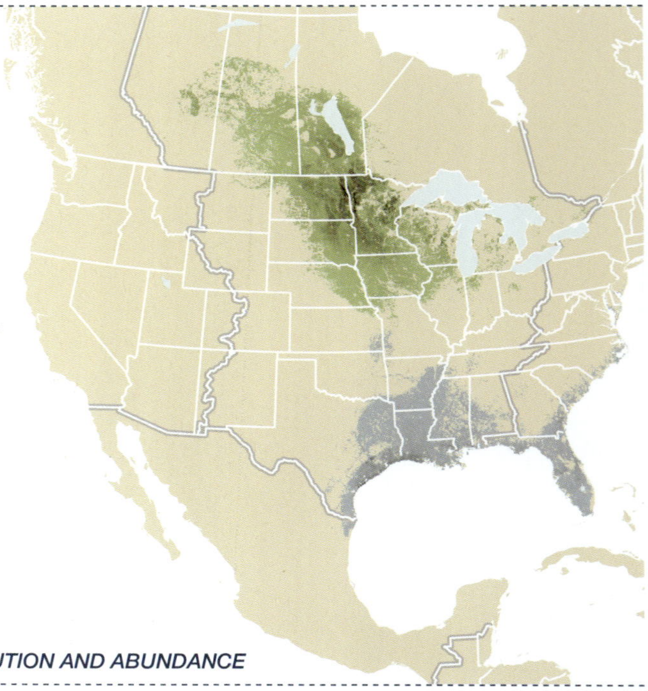

DISTRIBUTION AND ABUNDANCE

MIGRATORY CONNECTIVITY

OVERALL Migratory connectivity is unknown. No information on migratory connectivity was found for this species.

MIGRATORY PATHWAYS There is no information on specific routes, but movement is mostly in a southern and eastern direction during pre-breeding migration (Herkert et al. 2021).

INTERESTING NOTE

Incidental mortality due to communication tower collisions during migration indicate that Sedge Wrens might migrate in flocks with other species of wrens (Herkert et al. 2021).

MARSH WREN *Cistothorus palustris*

DISTRIBUTION AND ABUNDANCE

This widely distributed species breeds in cattail and bullrush marshes across the northern US, the western states, and the prairies of western Canada, with the densest and most extensive populations in the northern US prairies and southern Prairie provinces of Canada. The species spends the nonbreeding season in a wider variety of marshy habitats, including cordgrass and coastal salt marsh, throughout the central US, Florida and the Gulf Coast, and as far south as Veracruz and Michoacán in Mexico (Kroodsma and Verner 2020; Fink et al. 2023). Scattered populations west of the Rockies and along the eastern US and Gulf coasts are resident throughout the year (Fink et al. 2023).

ORDER Passeriformes
FAMILY Troglodytidae
BREEDING HABITAT Wetlands, Freshwater Marsh
NONBREEDING HABITAT Wetlands, Generalist
GLOBAL POPULATION SIZE 11,000,000
POPULATION CHANGE SINCE 1970 +166%
IUCN RED LIST CATEGORY Least Concern
R2R TIPPING POINT SPECIES No
US LISTING STATUS No
CANADA SPECIES AT RISK No
PIF CCS 7/20

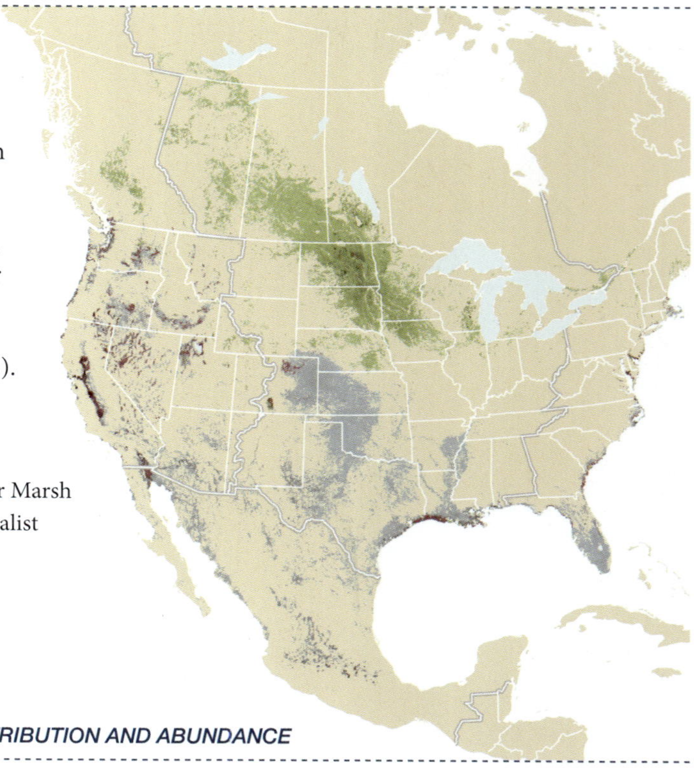

DISTRIBUTION AND ABUNDANCE

INTERESTING NOTE

In a study of eastern Marsh Wren populations, Luttrell and Lohr (2017) found that certain call types were associated with whether individuals belonged to nonmigratory, partially migratory, or migratory populations.

MIGRATORY CONNECTIVITY

OVERALL Migratory connectivity is unknown. No information on migratory connectivity was found for this species.

MIGRATORY PATHWAYS North-central US and Canadian populations migrate through the central US to western Texas and the Gulf Coast (Fink et al. 2023).

MC = 0.26 ± 0.06*

Dumetella carolinensis

GRAY CATBIRD

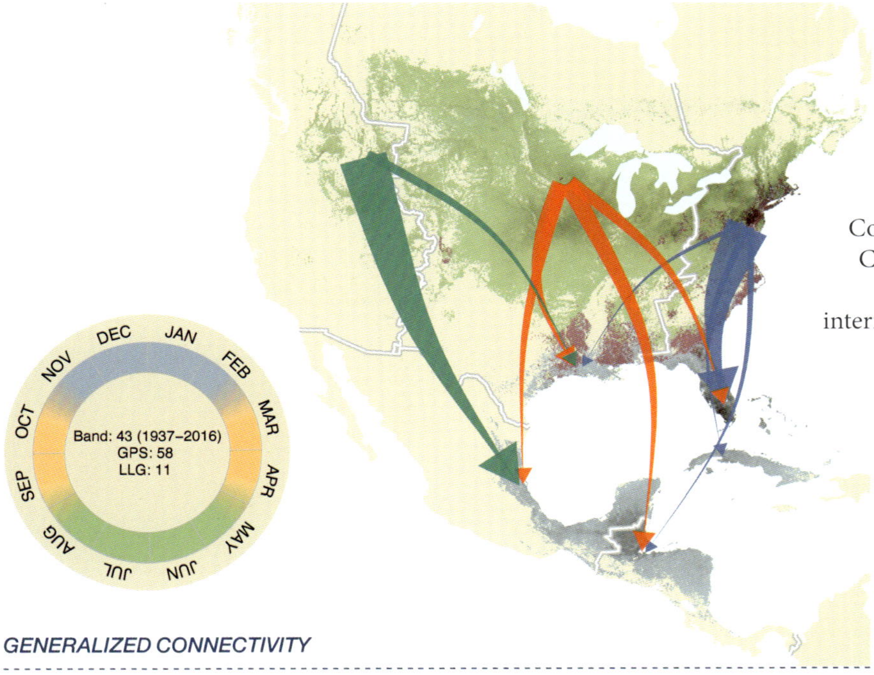

GENERALIZED CONNECTIVITY

DISTRIBUTION AND ABUNDANCE

The Gray Catbird is a widely distributed songbird species. They occur commonly along the eastern seaboard of the United States and west to Washington and Colorado during the breeding season. They are present year-round in the southeastern portion of their range and during the nonbreeding season on the Gulf Coast of the United States, the Caribbean, eastern Mexico, and Central America (R.J. Smith et al. 2020). Breeding abundance is highest in the midwestern, northeastern, and parts of the intermountain western United States. Nonbreeding abundance is highest in Florida, Cuba, the Bahamas, southern Mexico, Belize, and Guatemala (Fink et al. 2023).

	Atlantic	Caribbean	Central America	Mexico	Midcontinent
Atlantic (n = 49)	78%	10%	8%	–	4%
Midcontinent (n = 10)	20%	–	30%	20%	30%
Pacific (n = 22)	–	–	–	86%	14%

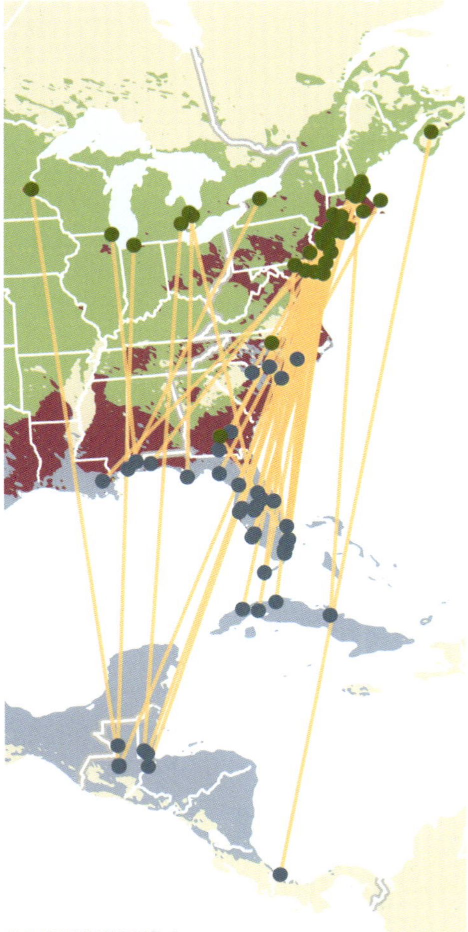

CONNECTIONS A

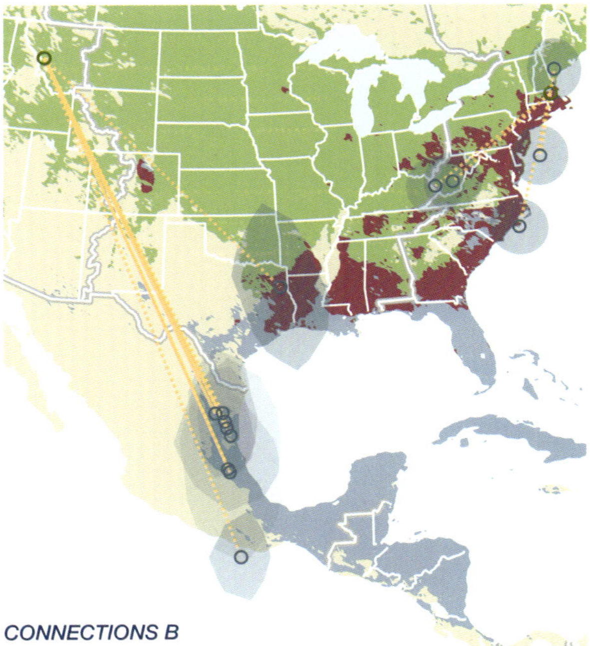

CONNECTIONS B

ORDER Passeriformes
FAMILY Mimidae
BREEDING HABITAT Forests, Temperate Eastern
NONBREEDING HABITAT Forests, Tropical Generalist
GLOBAL POPULATION SIZE 29,000,000
POPULATION CHANGE SINCE 1970 +2%
IUCN RED LIST CATEGORY Least Concern
R2R TIPPING POINT SPECIES No
US LISTING STATUS No
CANADA SPECIES AT RISK No
PIF CCS 8/20

MIGRATORY CONNECTIVITY

OVERALL Migratory connectivity is estimated from band reencounter, archival GPS, and geolocator data. Birds breeding in the Atlantic Flyway largely migrate to nonbreeding areas within the Atlantic Flyway (78%), while birds breeding in the Midcontinent Flyway spread out across multiple nonbreeding regions. Birds breeding in the Pacific Flyway primarily migrate to Mexico (86%). Despite a large number of individuals remaining within the Atlantic and Pacific flyways during the breeding and nonbreeding season, mixing of Midcontinent Flyway breeding populations throughout the nonbreeding range leads to moderate migratory connectivity (MC = 0.26 ± 0.06) across the species' range.

BANDING Banding data confirm light-level geolocator findings and also link midwestern populations to nonbreeding grounds in Central America (Ryder et al. 2011). Band reencounters from the northeast and mid-Atlantic coast of the US show nonbreeding areas in Florida, Cuba, and Central America.
RADIO TRACKING There are many projects tracking Gray Catbirds through the Motus Wildlife Tracking System. There are no individual breeding to nonbreeding connections.
GEOLOCATOR TRACKING Light-level geolocators indicate individuals breeding in mid-Atlantic and northeast regions of the US spend the nonbreeding season in Cuba and southern Florida (Ryder et al. 2011; Adams et al., unpublished data), and individuals breeding in British Columbia, Canada, and western Montana spend the nonbreeding season in Mexico (Mancuso et al. 2021).
SATELLITE TRACKING Further tracking of breeding populations with GPS tags confirms migration of eastern populations to Florida and Cuba (Ryder et al., unpublished data) and western populations to Mexico (Mancuso et al. 2021). Birds from British Columbia, Canada, western Montana, and Colorado also migrated to nonbreeding areas on the Gulf Coast of Texas, with those tracked for multiple years exhibiting some fidelity to nonbreeding locations (Mancuso et al. 2021; Ryder et al., unpublished data).
STABLE ISOTOPES Feathers collected at a post-breeding stopover site on the Gulf Coast of Alabama were assigned to probable breeding origins in the Great Plains or mid-Atlantic regions of the US (Hobson et al. 2007).
OTHER APPROACHES No genetic studies of migratory connectivity were found for this species.

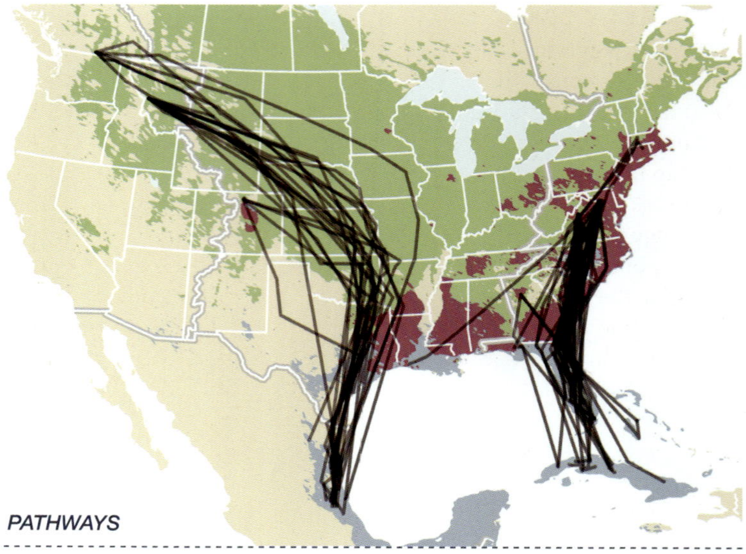

PATHWAYS

MIGRATORY PATHWAYS Birds tagged in the mid-Atlantic and Northeast regions of the US migrated along similar routes through the mid-Atlantic and southeastern US to nonbreeding areas (Ryder et al. 2011; Adams et al., unpublished data). Individuals tracked from Colorado, Montana, and British Columbia followed a similar route through Kansas, Oklahoma, and Texas to nonbreeding grounds on the Gulf Coast of Texas and into Mexico (Mancuso et al. 2021; Ryder et al., unpublished data).

DATA CONTRIBUTORS
E.M. Adams; C. Bishop; B.S. Evans; D. Evers; M. Fylling; K. Hodges; P. Keenan; M.B. Lancaster; S.B. Lerman; K. Mancuso; P.P. Marra; D.L. Moseley; T.B. Ryder; K. Stone; K. Williams.

BROWN THRASHER *Toxostoma rufum*

DISTRIBUTION AND ABUNDANCE

The breeding distribution of Brown Thrashers extends from the Great Plains and prairies of the United States and Canada east to the Atlantic coast. Breeding populations extend north into southern Manitoba, Saskatchewan, and Alberta. Nonbreeding populations contract into the southeastern states east from Texas and Oklahoma to the Atlantic coast (Cavitt and Haas 2020).

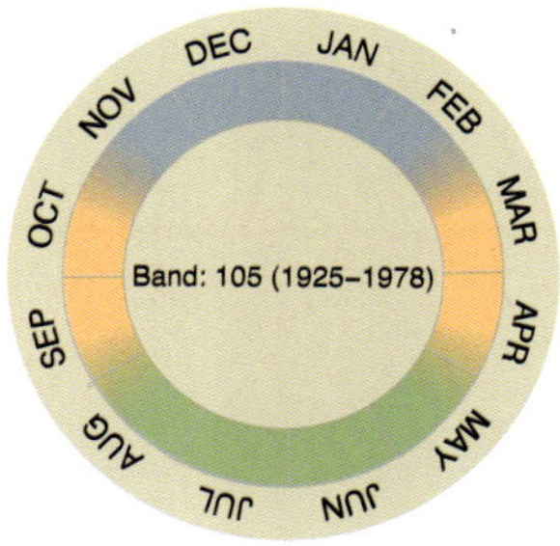

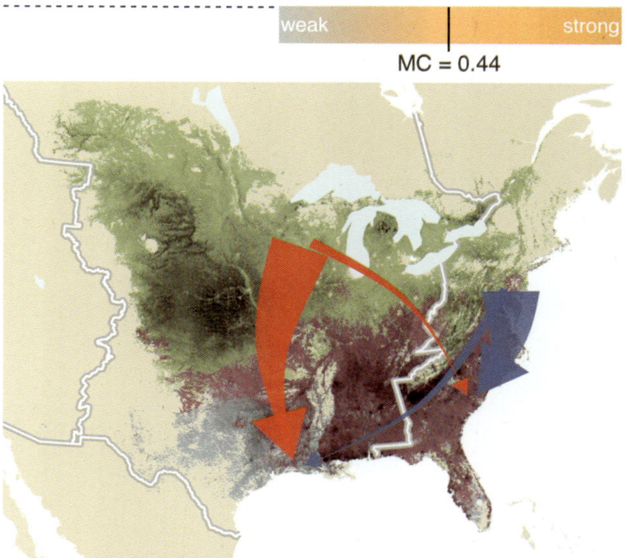

GENERALIZED CONNECTIVITY

ORDER Passeriformes
FAMILY Mimidae
BREEDING HABITAT Forests, Temperate Eastern
NONBREEDING HABITAT Forests, Temperate Eastern
GLOBAL POPULATION SIZE 6,200,000
POPULATION CHANGE SINCE 1970 −39%
IUCN RED LIST CATEGORY Least Concern
R2R TIPPING POINT SPECIES No
US LISTING STATUS No
CANADA SPECIES AT RISK No
PIF CCS 11/20

	Atlantic	Midcontinent
Atlantic (n = 32)	88%	12%
Midcontinent (n = 66)	11%	89%

MIGRATORY CONNECTIVITY

OVERALL Migratory connectivity is known from band reencounter data and is strong (MC = 0.44). Atlantic and Midcontinent flyway breeding populations both remain within the breeding flyway during the nonbreeding season (88% and 89%, respectively) with little overlap in other regions.

BANDING Banding reencounter data show parallel migration across the range with three clusters: eastern, Great Lakes, and central / Great Plains population. Individuals in the eastern group migrate south to Virginia, the Carolinas, and Florida, while individuals in the Great Lakes region migrate south to Louisiana, Mississippi, and Alabama. Those breeding in the Great Plains of the central US and Canada migrate to eastern Texas, Arkansas, and Louisiana.

OTHER APPROACHES No tracking, genetic, or stable isotope studies of migratory connectivity exist for this species. A few tags were deployed through the Motus Wildlife Tracking System, with no breeding to nonbreeding detections.

MIGRATORY PATHWAYS There is no information on specific routes. It appears to migrate over a broad front and was observed in high abundance in the southeastern US (Fink et al. 2023).

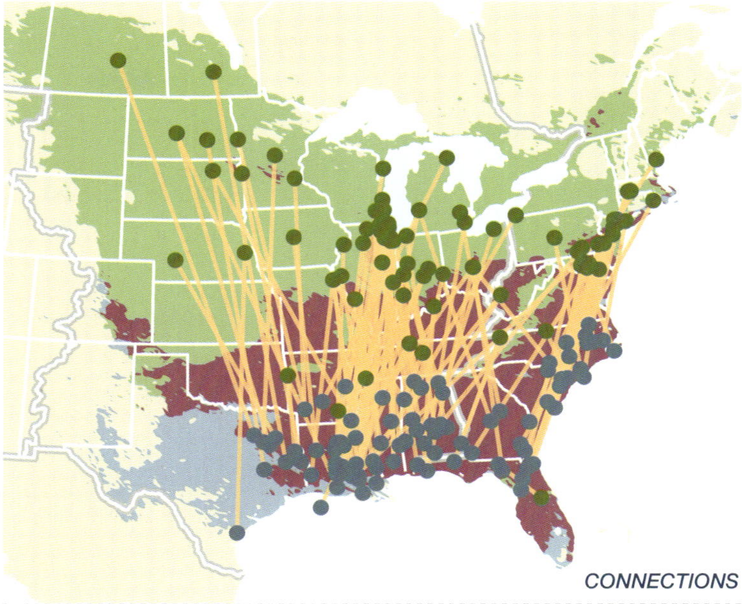

CONNECTIONS

Toxostoma bendirei BENDIRE'S THRASHER

DISTRIBUTION AND ABUNDANCE

It breeds in relatively open desert habitats in the southwestern US and northwestern Mexico, most densely in southwestern Arizona and western Sonora (Fink et al. 2023). In the nonbreeding season, birds from California, Nevada, northern Arizona, and New Mexico are believed to migrate to southern portions of the breeding range where birds are believed to be resident (England and Laudenslayer 2020).

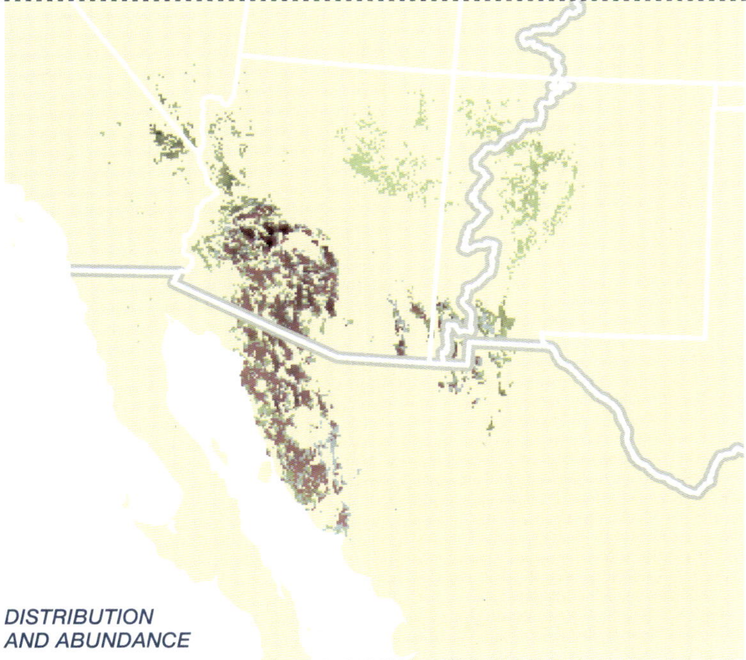

DISTRIBUTION AND ABUNDANCE

MIGRATORY CONNECTIVITY

OVERALL Migratory connectivity is unknown. No information on migratory connectivity or pathways were found for this species. A few tags were deployed through the Motus Wildlife Tracking System, with local detections only.

INTERESTING NOTE

To investigate local Bendire's Thrasher movement, Lotek PinPoint GPS tags were deployed and recovered during the 2019 nonbreeding season in the Tucson and Phoenix areas, revealing that these birds occupied fairly small territories throughout the nonbreeding season (C. Bergman, personal communication).

ORDER Passeriformes
FAMILY Mimidae
BREEDING HABITAT Arid Lands, Desert Scrub
NONBREEDING HABITAT Arid Lands, Desert Scrub
GLOBAL POPULATION SIZE 83,000
POPULATION CHANGE SINCE 1970 −85%
IUCN RED LIST CATEGORY Vulnerable
R2R TIPPING POINT SPECIES Red-Alert
US LISTING STATUS No
CANADA SPECIES AT RISK No
PIF CCS 16/20

SAGE THRASHER *Oreoscoptes montanus*

DISTRIBUTION AND ABUNDANCE

This small thrasher of western North America is a habitat specialist, closely tied to its namesake plant and a few other shrub species. It is a short-distance migrant but with clear separation between breeding and nonbreeding ranges. Breeding is in the central western states from Washington, Oregon, Nevada, east to Montana, Wyoming, Colorado, and south to New Mexico. During the nonbreeding season, individuals move to the southwestern states from California to western Texas and into Mexico, from Baja California to Monterrey, Mexico. Despite being a short-range migrant, this species is prone to wandering and is often detected as a vagrant outside of its normal range in migration (Reynolds et al. 2020).

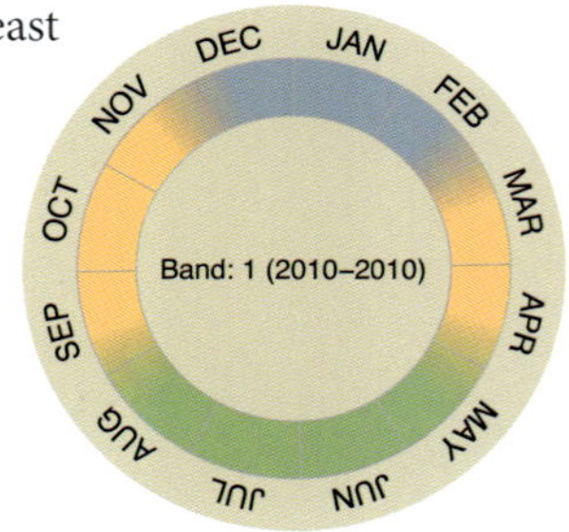

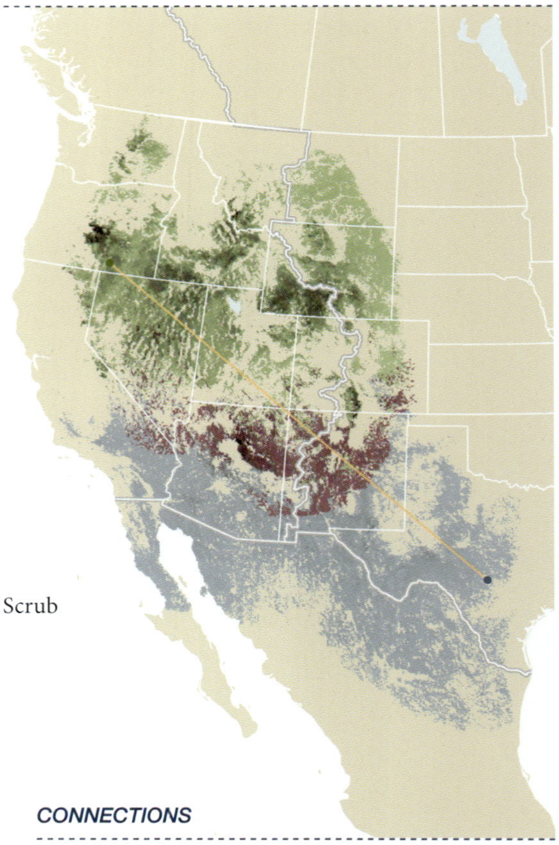

ORDER Passeriformes
FAMILY Mimidae
BREEDING HABITAT Arid Lands, Sagebrush
NONBREEDING HABITAT Arid Lands, Desert Scrub
GLOBAL POPULATION SIZE 6,400,000
POPULATION CHANGE SINCE 1970 −43%
IUCN RED LIST CATEGORY Least Concern
R2R TIPPING POINT SPECIES No
US LISTING STATUS No
CANADA SPECIES AT RISK Endangered
PIF CCS 11/20

MIGRATORY CONNECTIVITY

OVERALL Migratory connectivity is unknown.
BANDING There is one band reencounter between a breeding location in Oregon and a nonbreeding location in Mexico.
OTHER APPROACHES No tracking, genetic, or stable isotope studies of migratory connectivity were found for this species.

MIGRATORY PATHWAYS Most of the breeding range is in the inland western US; during the nonbreeding season, the species undertakes direct migrations to their nonbreeding areas in the US Southwest and northern Mexico (Reynolds et al. 2020; Fink et al. 2023). There was one detection through the Motus Wildlife Tracking System during migration from Wyoming to Nevada.

NORTHERN MOCKINGBIRD *Mimus polyglottos*

DISTRIBUTION AND ABUNDANCE

The Northern Mockingbird is a widely distributed species breeding from coast to coast throughout the lower two-thirds of the United States through most of Mexico and the Greater Antilles. Its nonbreeding range overlaps almost entirely with its breeding range, although individuals in the central US appear to be migratory. A familiar species, it is well adapted to living in human landscapes and is well known for its vocal repertoire and bold disposition (Farnsworth et al. 2020).

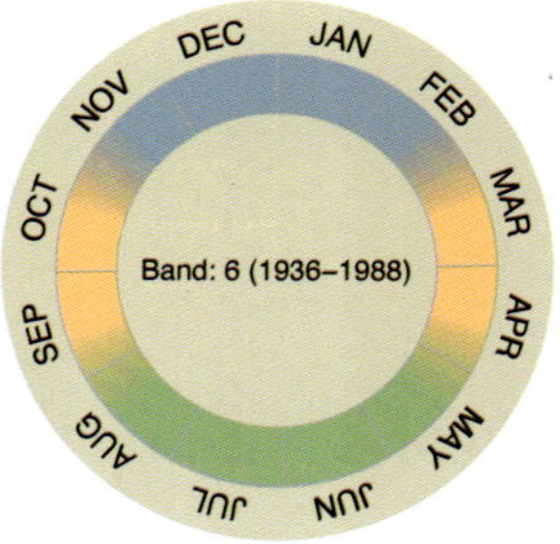

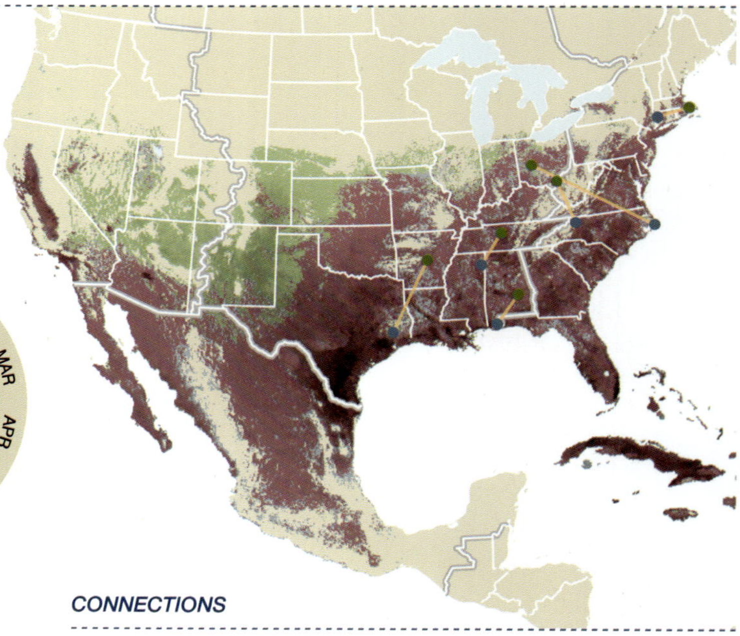

ORDER Passeriformes
FAMILY Mimidae
BREEDING HABITAT Open Country, Habitat Mosaic
NONBREEDING HABITAT Open Country, Habitat Mosaic
GLOBAL POPULATION SIZE 43,000,000
POPULATION CHANGE SINCE 1970 −20%
IUCN RED LIST CATEGORY Least Concern
R2R TIPPING POINT SPECIES No
US LISTING STATUS No
CANADA SPECIES AT RISK No
PIF CCS 9/20

MIGRATORY CONNECTIVITY

OVERALL Migratory connectivity is not well defined for this species. There is limited migratory connectivity data only from band reencounter data records. MC is not calculated due to insufficient data.

BANDING Band reencounter data indicate some individuals may be partly migratory, with individuals making short movements between breeding and nonbreeding season. There are a few records of individuals moving up to 500 miles, but usually not more than 20 miles (Cooke 1946).

OTHER APPROACHES No tracking, genetic, or stable isotope studies of migratory connectivity were found for this species.

MIGRATORY PATHWAYS Some individuals move northward and also into the Southwest from the core range during the breeding season, but the reason for and extent of these movements is poorly understood (Farnsworth et al. 2020; Fink et al. 2023).

TURDIDAE | *THRUSHES*

Sialia sialis EASTERN BLUEBIRD

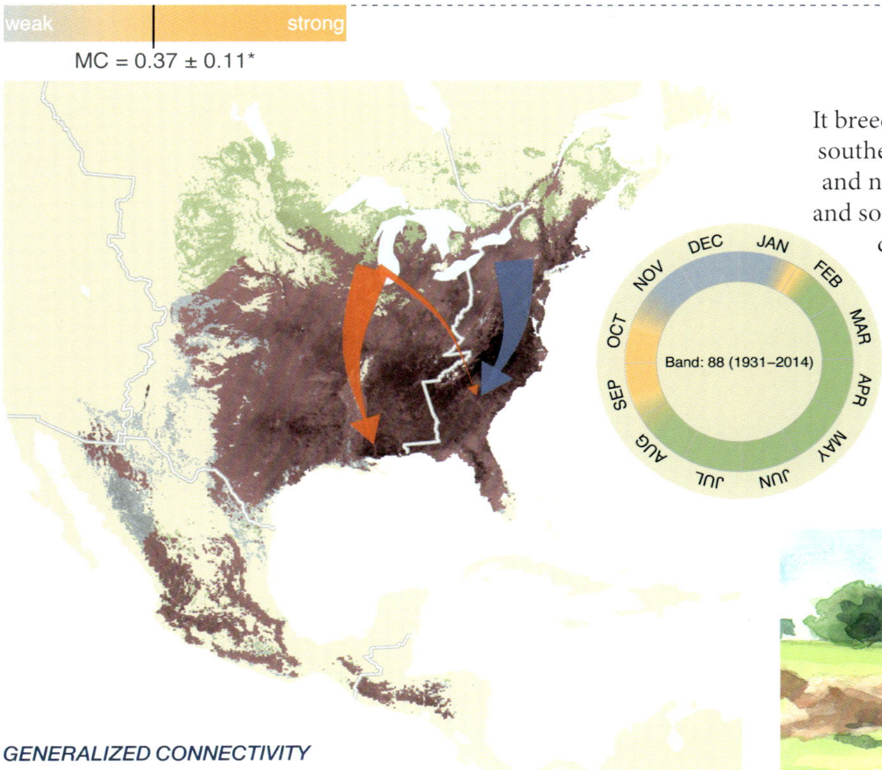

GENERALIZED CONNECTIVITY

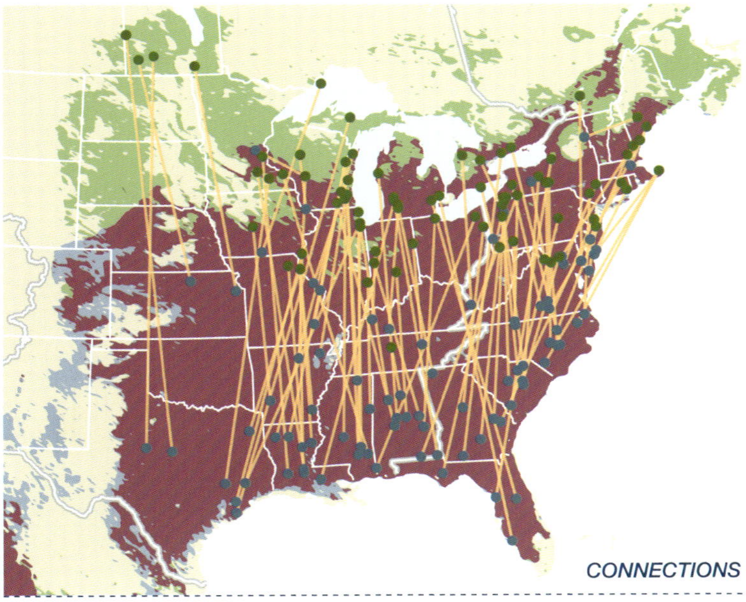

CONNECTIONS

	Atlantic	Midcontinent
Atlantic (n = 32)	100%	–
Midcontinent (n = 55)	15%	85%

DISTRIBUTION AND ABUNDANCE

It breeds throughout eastern North America from Nova Scotia to southeastern Saskatchewan and south to Florida, the Gulf Coast, and northeastern Mexico; highest abundance is in the Piedmont and southeastern and eastern Gulf Coastal Plain. Its nonbreeding distribution encompasses the southern half of the breeding range, with more birds staying farther north in recent decades. A separate population (several subspecies) is resident from southeastern Arizona through the highlands of Mexico to northern Nicaragua; it is also resident (not introduced) on Bermuda. Although long-term trends for this species are positive, recent steep declines have been noted in the southeastern portion of the range, especially in Florida (Sauer et al. 2019).

ORDER Passeriformes
FAMILY Turdidae
BREEDING HABITAT Forests, Temperate Eastern
NONBREEDING HABITAT Forests, Temperate Eastern
GLOBAL POPULATION SIZE 23,000,000
POPULATION CHANGE SINCE 1970 +161%
IUCN RED LIST CATEGORY Least Concern
R2R TIPPING POINT SPECIES No
US LISTING STATUS No
CANADA SPECIES AT RISK No
PIF CCS 7/20

MIGRATORY CONNECTIVITY

OVERALL Migratory connectivity for Eastern Bluebirds is only known from band reencounter data and appears moderate (MC = 0.37 ± 0.11). Atlantic and Midcontinent flyway breeding birds migrate to nonbreeding areas within the same flyway as breeding

(100% and 85%, respectively). About 15% of Midcontinent Flyway breeding birds overlap with Atlantic Flyway breeding populations in the Atlantic Flyway during the nonbreeding season.
BANDING Extensive banding records indicate strong east-west parallel migration, with nonbreeding areas generally directly south of breeding grounds.
RADIO TRACKING A few tags were deployed through the Motus Wildlife Tracking System; there are no breeding to nonbreeding detections.
OTHER APPROACHES No tracking, genetic, or stable isotope studies of migratory connectivity were found for this species.
MIGRATORY PATHWAYS It is considered a partial migrant, with northern populations moving directly south in the nonbreeding season, and mid-Atlantic and more southern populations being year-round residents. Band returns also suggest that northern breeding populations may migrate longer distances and leapfrog resident populations in the southern part of the distribution (Gowaty and Plissner 2020).

WESTERN BLUEBIRD *Sialia mexicana*

DISTRIBUTION AND ABUNDANCE

It breeds in open and burned forests throughout much of western North America but in two largely disjunct regions separated by the Great Basin: from southern British Columbia and southwestern Alberta to northern Baja California, and in mountains from southern Nevada, Utah, and Colorado to central Mexico. Its nonbreeding distribution is within most of the breeding range, but often at lower elevations, as well as to the south and east to southeastern California, southern Arizona, southern New Mexico, western and central Texas, and adjacent regions of northern Mexico (Fink et al. 2023). Although populations have been stable or increasing overall, historic declines and range contractions have been noted in the Pacific Northwest (Guinan et al. 2020).

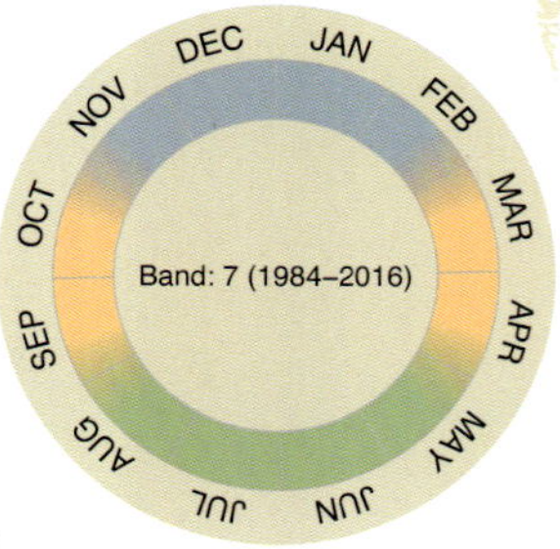

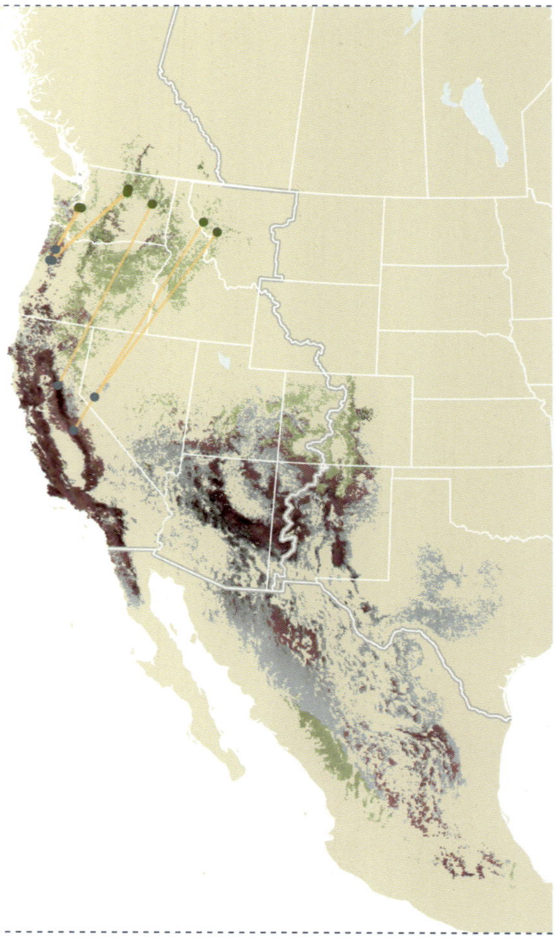

CONNECTIONS

ORDER Passeriformes
FAMILY Turdidae
BREEDING HABITAT Forests, Temperate Western
NONBREEDING HABITAT Forests, Temperate Western
GLOBAL POPULATION SIZE 7,100,000
POPULATION CHANGE SINCE 1970 +45%
IUCN RED LIST CATEGORY Least Concern
R2R TIPPING POINT SPECIES No
US LISTING STATUS No
CANADA SPECIES AT RISK No
PIF CCS 10/20

MIGRATORY CONNECTIVITY

OVERALL Migratory connectivity is based on morphology, limited band reencounter, and eBird data. These sources indicate strong separation, and thus possibly strong migratory connectivity, between disjunct breeding and nonbreeding populations. Northwestern and Pacific breeders remain within the Pacific region during the nonbreeding season, whereas Rocky Mountain breeders migrate eastward, including to central Texas and northern Mexico. MC is not calculated due to limited data.
BANDING A small sample of band reencounters suggests parallel migration within the northwestern population, with connections between breeding sites in western and central Washington to nonbreeding sites in western Oregon, and between breeding sites in eastern Washington and Montana to nonbreeding sites in eastern California and western Nevada.
STABLE ISOTOPES Of 159 Western Bluebirds sampled for hydrogen isotopes (from claw tissue) in British Columbia, 54.4% were migrants and 42.4% were residents; and 5 of 23 birds captured in multiple years switched strategies between migratory and resident. In migratory individuals, isotope signatures were correlated between members of a pair, suggesting that pairs spent the nonbreeding season in the same location (Dale et al. 2019).
OTHER APPROACHES The two disjunct migratory populations are considered separate subspecies (American Ornithologists Union 1957; Phillips 1991); but individual variation is great, and nonbreeding specimens are difficult to assign with certainty. Nevertheless, no mixing of subspecies has been suggested in the separate nonbreeding regions. Resident populations in Mexico comprise several additional subspecies.
MIGRATORY PATHWAYS Migrations are poorly understood, but both latitudinal and altitudinal migration occur within the two disjunct regions. Weekly eBird animations indicate that northwestern breeders migrate through eastern Oregon and western Nevada to California nonbreeding areas, whereas breeders from Utah and Colorado move south into Arizona, New Mexico, western Texas, and northern Mexico (Fink et al. 2023). Stable isotope data from British Columbia suggest that some individuals can switch between migratory and resident strategies (Dale et al. 2019).

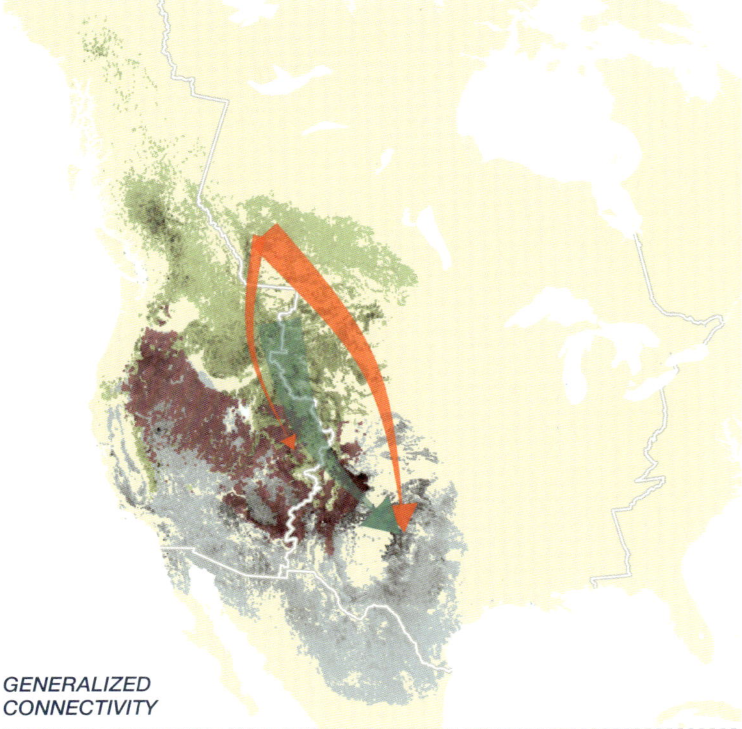

GENERALIZED CONNECTIVITY

MOUNTAIN BLUEBIRD

Sialia currucoides

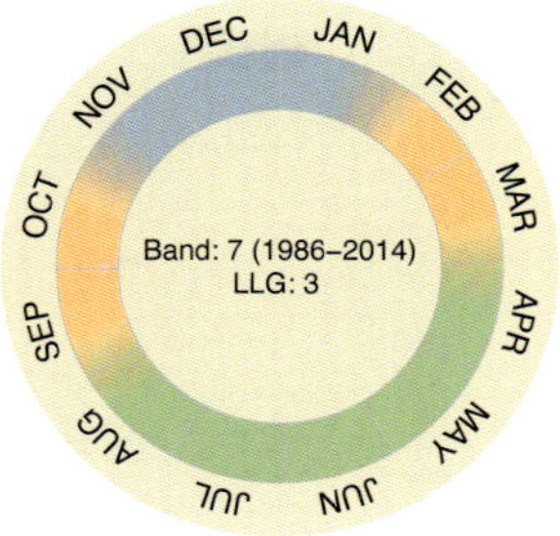

DISTRIBUTION AND ABUNDANCE

It breeds from foothills to the subalpine zone in the mountains of western North America, north sparsely to southwestern Yukon and eastern Alaska, south to central California, northern Arizona, and New Mexico, and eastward in aspen parkland, badlands, and wooded canyons to southwestern Manitoba and the western Dakotas. Nonbreeding distribution is in the southern portions of the breeding range (mostly at lower elevations), as well as southward into southwestern lowlands, the southern Great Plains to central Texas, and to the northern Central Mexican Plateau. Steepest population declines, especially in recent decades, have been in the Canadian Prairie provinces and in the southern Rocky Mountains (Sauer et al. 2019).

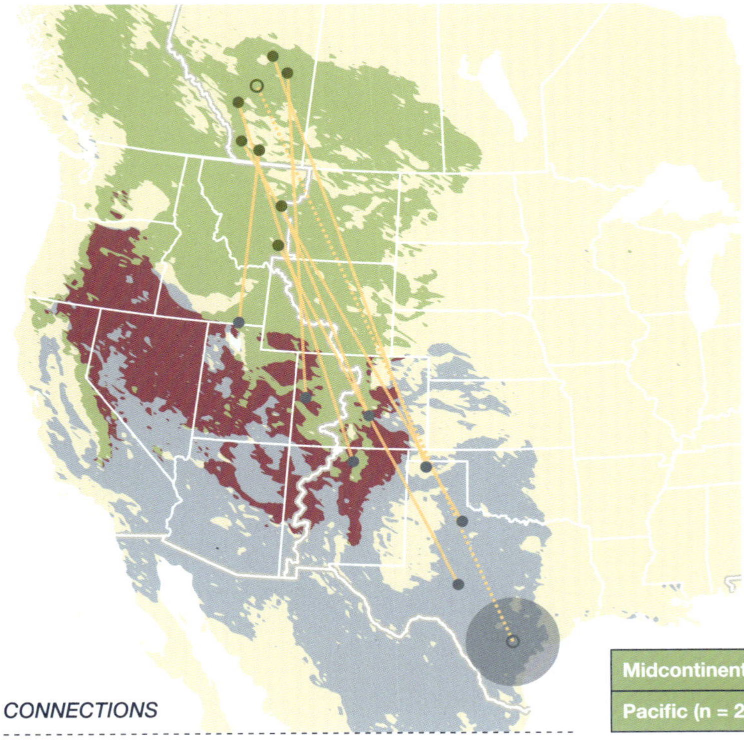

CONNECTIONS

	Midcontinent	Pacific
Midcontinent (n = 8)	75%	25%
Pacific (n = 2)	25%	–

ORDER Passeriformes
FAMILY Turdidae
BREEDING HABITAT Open Country, Habitat Mosaic
NONBREEDING HABITAT Forests, Temperate Western
GLOBAL POPULATION SIZE 5,600,000
POPULATION CHANGE SINCE 1970 −17%
IUCN RED LIST CATEGORY Least Concern
R2R TIPPING POINT SPECIES No
US LISTING STATUS No
CANADA SPECIES AT RISK No
PIF CCS 11/20

MIGRATORY CONNECTIVITY

OVERALL Birds breeding within the Midcontinent Flyway spend the nonbreeding season within the Midcontinent (75%) or Pacific (25%) flyways, while Pacific Flyway breeding birds remain within the Pacific Flyway. MC is not calculated due to limited data.

BANDING Band reencounters show movement between breeding sites in the Rocky Mountains of the western United States and Canada and nonbreeding sites in the Rocky Mountains, Great Basin, and southern Great Plains. Reencounters from breeding sites in the Canadian Prairie provinces were generally due south in the southern Great Plains (mentioned in Stewart 2020).

GEOLOCATOR TRACKING Geolocators have been used to track Mountain Bluebirds from their breeding grounds in Alberta, Canada, but comparisons of analytical methods led to the conclusion that results were highly inconsistent and "biologically implausible" (Roberto-Charron 2018). More recent geolocator data suggest that individuals that breed in south-central Alberta migrate to the southern Great Plains, primarily from eastern Colorado to northern Mexico (Stewart 2020; L.S. Johnson and Dawson 2020).

OTHER APPROACHES No satellite telemetry, genetic, or stable isotope studies of migratory connectivity were found for this species.

MIGRATORY PATHWAYS Most populations appear to be migratory, but migration behavior may be flexible and variable from year to year depending on weather and fruit availability (L.S. Johnson and Dawson 2020). Individuals tracked from central Alberta stayed east of the Rocky Mountains on migration, on routes to and from nonbreeding sites in the southern Great Plains and northern Mexico.

TOWNSEND'S SOLITAIRE *Myadestes townsendi*

DISTRIBUTION AND ABUNDANCE

It breeds in montane coniferous forests from east-central Alaska and Yukon, south through the Rocky Mountain region to New Mexico and northern Arizona and through the Pacific Coast Ranges, Sierra Nevada, and other mountains to Southern California. During the nonbreeding season, it departs from the Alaskan and most Canadian portions of the range, moves to lower elevations in the western US, onto the western Great Plains, and south to northern Mexico. A resident population also occurs in high mountains of northwestern Mexico. Population trends are variable, with declines, especially recently, throughout the Rocky Mountains and increases in Alaska and the Pacific states (Sauer et al. 2019).

ORDER Passeriformes
FAMILY Turdidae
BREEDING HABITAT Forests, Temperate Western
NONBREEDING HABITAT Forests, Temperate Western
GLOBAL POPULATION SIZE 1,100,000
POPULATION CHANGE SINCE 1970 +17%
IUCN RED LIST CATEGORY Least Concern
R2R TIPPING POINT SPECIES No
US LISTING STATUS No
CANADA SPECIES AT RISK No
PIF CCS 11/20

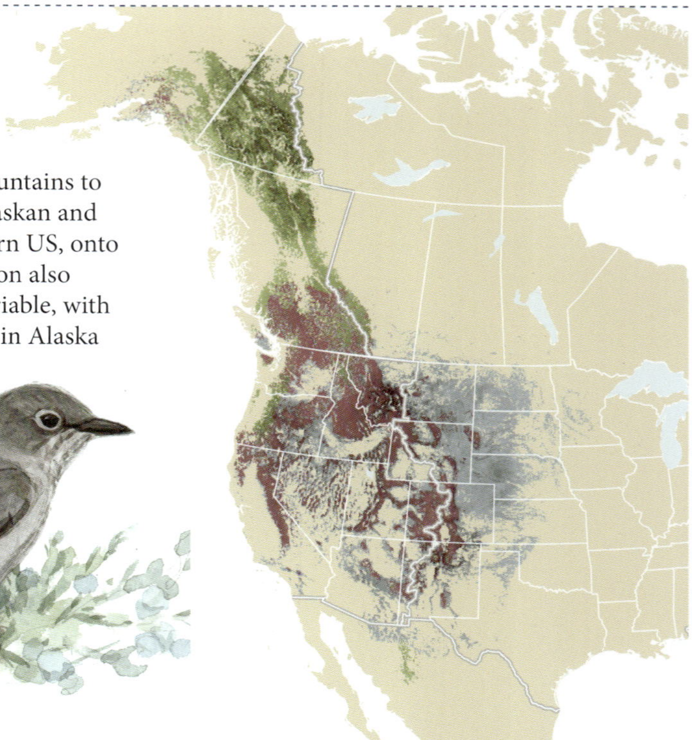

DISTRIBUTION AND ABUNDANCE

MIGRATORY CONNECTIVITY

OVERALL Migratory connectivity is unknown. No information on migratory connectivity was found for this species. Individuals or populations are believed to be either latitudinal migrants, altitudinal migrants, or resident year-round (although the latter is not confirmed with marked individuals; Bowen 2020).

BANDING The only banding study indicated local altitudinal migration among 52 individuals in Lassen County, California (Bowen 2020).

MIGRATORY PATHWAYS Regular post-breeding migration through southern Saskatchewan and widespread occurrence east of the Rocky Mountains in the nonbreeding season suggest that these birds are from migratory populations in the northern part of the range (Fink et al. 2023).

VEERY *Catharus fuscescens*

DISTRIBUTION AND ABUNDANCE

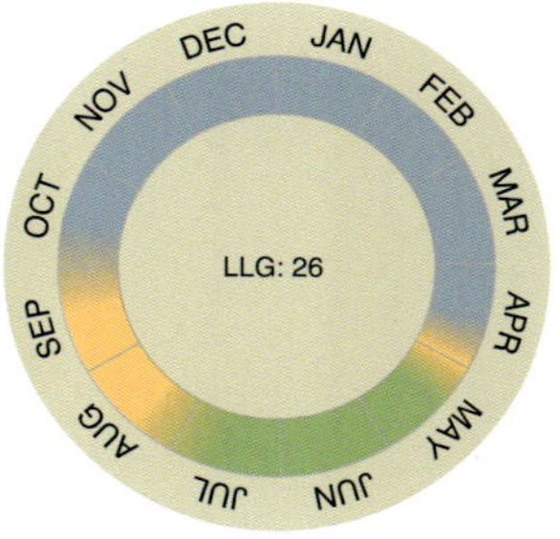

The Veery breeds in two disjunct regions corresponding with eastern and western subspecies groups. The eastern subspecies breeds from the Canadian Maritimes and southwestern Newfoundland south through the northeastern US and in the Appalachian Mountains to northern Georgia, and westward to northern Minnesota and Manitoba and then patchily across the northern Great Plains to eastern Alberta and northeastern Montana. The western subspecies breeds from interior southern British Columbia south through the Rocky Mountains and northern Great Basin to northeastern Oregon, southern Idaho, and Colorado (Fink et al. 2023). Its nonbreeding distribution was only recently identified and involves an unusual midseason, intratropical migration (Heckscher et al. 2020). Most individuals migrate first (December to early February) to areas on the Brazilian Shield or Atlantic forest regions of south-central and southeastern Brazil (there are also records from eastern Brazil); then in late February or March, most individuals move northward to the northern Amazon, the Guiana Shield, and the Orinoco basin region of northern Brazil, Suriname, southern Venezuela, and eastern Colombia (a few westward to Bolivia). Declines have been steepest, especially in recent decades, in Ontario, New York, and New England, and in the northern Rocky Mountains (Sauer et al. 2019).

ORDER Passeriformes
FAMILY Turdidae
BREEDING HABITAT Forests, Temperate Eastern
NONBREEDING HABITAT Forests, Tropical Lowland Evergreen
GLOBAL POPULATION SIZE 11,000,000
POPULATION CHANGE SINCE 1970 −41%
IUCN RED LIST CATEGORY Least Concern
R2R TIPPING POINT SPECIES No
US LISTING STATUS No
CANADA SPECIES AT RISK No
PIF CCS 11/20

MIGRATORY CONNECTIVITY

OVERALL Migratory connectivity appears to be low, based on tracking, stable isotopes, and subspecific identification of museum specimens, with individuals from across the breeding range converging on major stopover sites and across the relatively small nonbreeding distribution in Brazil. MC is not calculated due to limited sample size.

BANDING No breeding to nonbreeding band reencounter data were found for this species.

RADIO TRACKING Data from several tracking projects through the Motus Wildlife Tracking System suggest different migration routes for eastern and western populations. There are no individual breeding to nonbreeding detections.

GEOLOCATOR TRACKING Several studies documented that most Veeries from both eastern (Delaware, Massachusetts, and Maine; Heckscher et al. 2011, 2015; E. Adams, unpublished data; C.M. Heckscher and D. Mendez, unpublished data; D. Shustack, unpublished data) and western (British Columbia; Hobson and Kardynal 2015; Kardynal and Hobson 2017) parts of the breeding distribution spend the early nonbreeding season in south-central Brazil (the Brazilian Shield) and then undergo an intratropical migration to a second nonbreeding area in the northern Amazon Basin or Guiana Shield (Heckscher et al. 2011, 2015; Hobson and Kardynal 2015; Kardynal and Hobson 2017; D. Shustack, unpublished data). Most nonbreeding locations lacked enough precision to identify possible fine-scale migratory connectivity among populations; however, nonbreeding sites in Brazil may be farther to the northwest for western breeders than for eastern breeders. Data from a breeding population in Delaware show that the location of first nonbreeding sites are farther east for females than males and reproductive success influenced the latitude of second nonbreeding sites (Heckscher et al. 2017).

STABLE ISOTOPES Stable isotope ratios from individuals migrating through northern Colombia during post-breeding could not distinguish between eastern and western breeding populations, but suggested that breeders from more southern latitudes arrived earlier at the stopover sites (Gonzalez-Prieto et al. 2011).

GENETICS There are four genetically distinct populations—Eastern, Southern, Interior West, and Pacific Northwest—revealed by whole genome sequencing from samples collected throughout the breeding range (https://www.birdgenoscape.org/veery/). Assignment of general breeding origins of a few nonbreeding site genetic samples demonstrates the potential to assess fine-scale migratory connectivity in this species using genomic data in the future (Kimmitt et al. 2023).

OTHER APPROACHES Based on subtle plumage differences, five subspecies are recognized (Phillips 1991; Heckscher et al. 2020), and study of specimens has documented that both eastern and western subspecies mix on migration and within nonbreeding areas—for example, in Mato Grosso, Brazil (Phillips 1991). The nonbreeding distribution of several subspecies (e.g., *C. f. fuliginosus* from Newfoundland) is still considered unknown.

INTERESTING NOTE

Following Nearctic-Neotropical migration, tagged individuals from both British Columbia and Delaware settled during their first nontransient period in forest communities that are geographically limited and severely threatened between the Brazilian Cerrado and Amazonian lowlands (the Cerrado-Amazonia transition; Heckscher et al. 2021). Veeries were among the most abundant post-breeding migrants in the Sierra Nevada de Santa Marta, Colombia, with most individuals making a 9- to 12-day stopover and acquiring enough fat to fuel the remaining 2,000-km journey to nonbreeding grounds in Brazil (Bayly et al. 2012).

	Andean West	Brazil	Southern Cone
Atlantic (n = 13)	–	100%	–
Pacific (n = 13)	8%	85%	8%

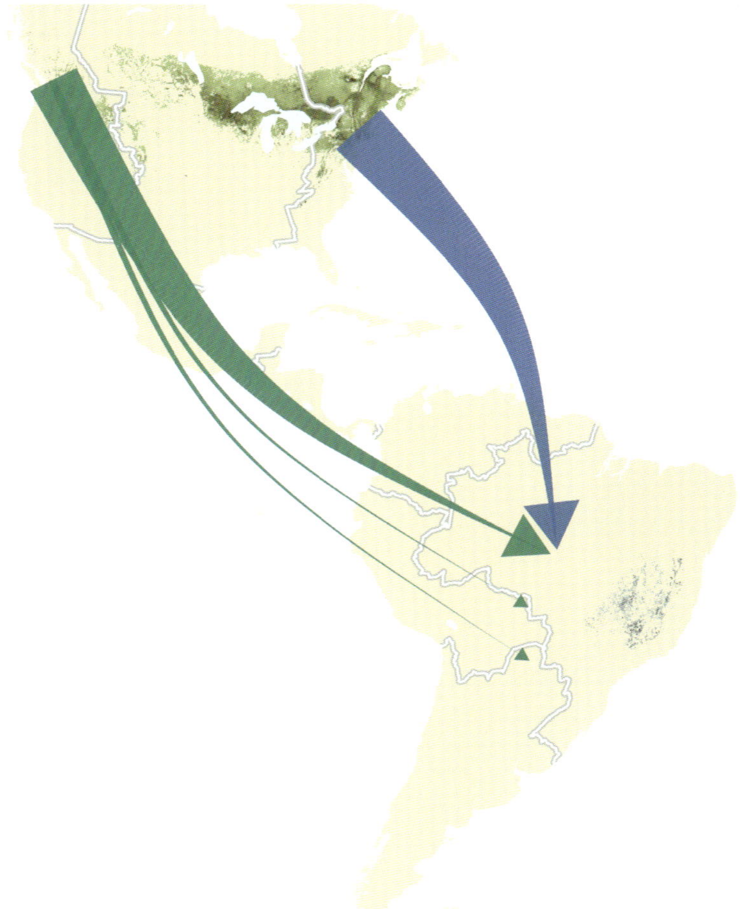

GENERALIZED CONNECTIVITY

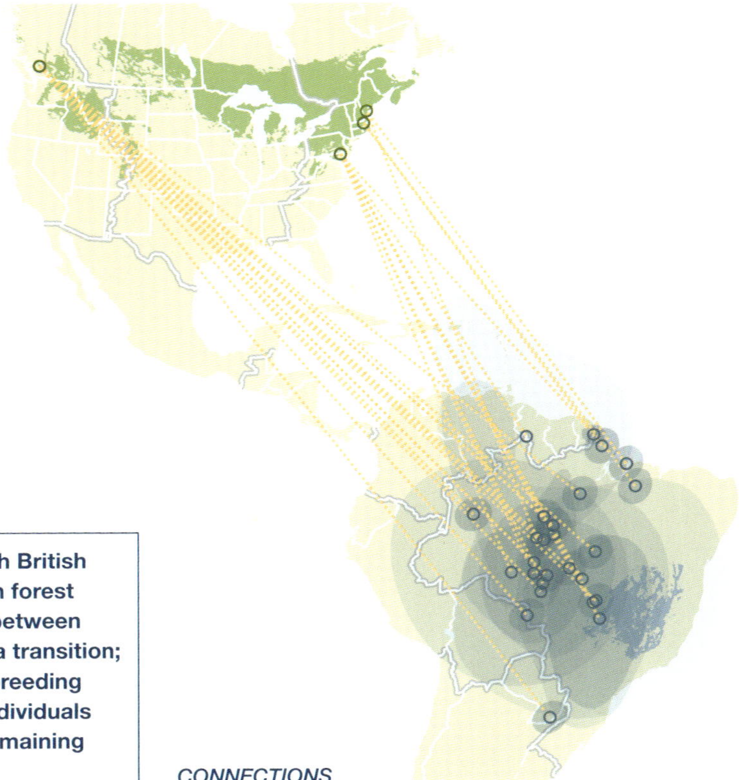

CONNECTIONS

MIGRATORY PATHWAYS It exhibits a clockwise looped migration in eastern North America (La Sorte et al. 2016), with crossings of the Gulf and Caribbean Sea to Central and South America (Heckscher et al. 2017). Within South America, the loop is reversed, with post-breeding migrants arriving in northern Colombia and moving south through the western Amazon Basin, and pre-breeding migrants moving farther east to depart in northern Venezuela (Gómez et al. 2013; Mendez 2018; K. Rosenberg, unpublished data). Based on geolocator tracks, eastern and western Veeries followed subtly different migration routes: individuals from British Columbia migrated east on post-breeding migration and then south through the eastern Gulf states to the Yucatán, Greater Antilles, and northern Colombia (Hobson and Kardynal 2015; Kardynal and Hobson 2017). During pre-breeding, western birds crossed the western Caribbean or moved westward through Central America to the Yucatán, crossing to the Texas Gulf Coast before moving north through the central US. In contrast, Veeries from Delaware and Massachusetts migrated southwesterly during post-breeding to Florida and then to the Yucatán, western Cuba, and the Caribbean coast of Central America before reaching northern Colombia. In pre-breeding migration, Veeries followed a more easterly route across the eastern Caribbean to Florida and the southeastern US (Heckscher et al. 2017; Mendez 2018; C.M. Heckscher and D. Mendez, unpublished data; D. Shustack, unpublished data); much variation was found, however, between individuals, between years, and even for the same individual in multiple years (C.M. Heckscher and D. Mendez, unpublished data).

DATA CONTRIBUTORS

E.M. Adams; D.C. Evers; C.M. Heckscher; K.A. Hobson; K.J. Kardynal; P. Keenan; D.C. Mendez; K.A. Williams.

GRAY-CHEEKED THRUSH *Catharus minimus*

DISTRIBUTION AND ABUNDANCE

It breeds in the Arctic, taiga, and subalpine zones of Alaska, northern Canada, and northeastern Siberia, and south in the northern Rocky Mountains to central British Columbia. Its nonbreeding distribution is in northwestern South America, including lower slopes and valleys of the northern Andes and other mountains of Colombia and adjacent Venezuela, and sparsely through the northern Amazon Basin to eastern Ecuador, northeastern Peru, southern Venezuela, and northwestern Brazil (Fink et al. 2023). It is not well monitored in most of the range, but the Alaskan population has declined by 2% annually since 1993, and the distinct Newfoundland population is steeply declining due to heavy nest predation by introduced red squirrels (Whitaker et al. 2020).

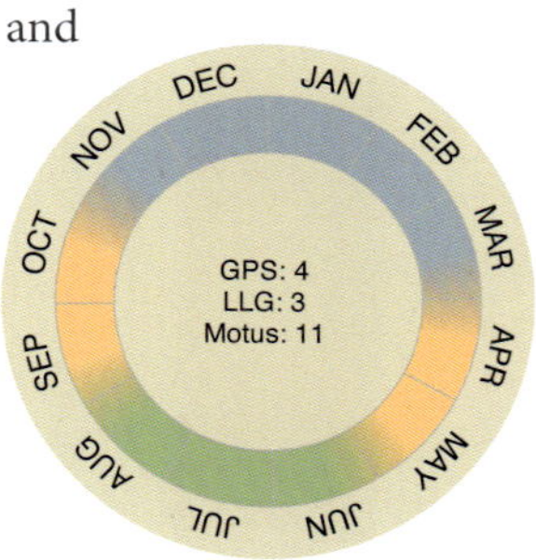

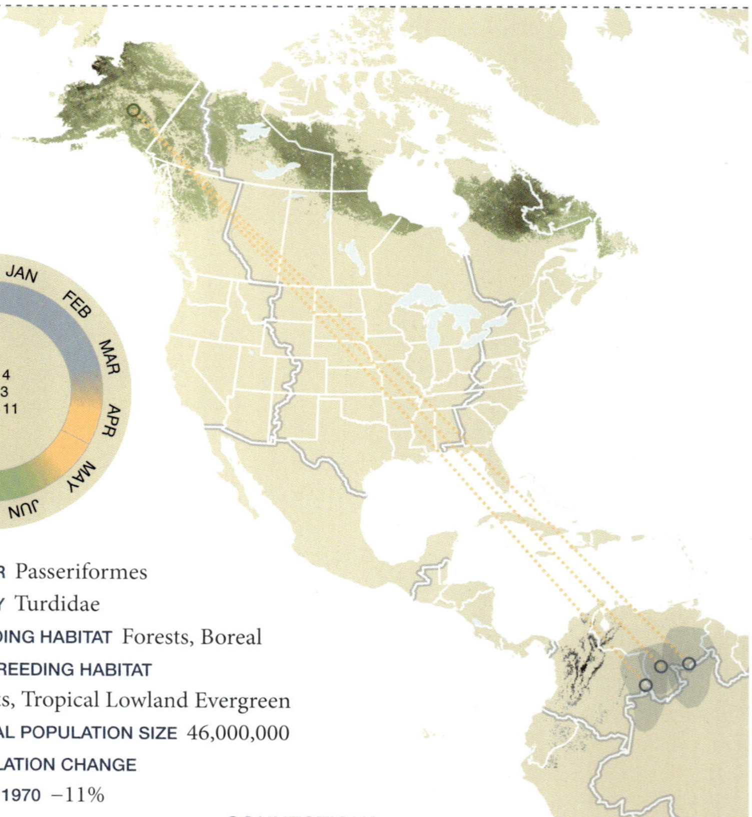

ORDER Passeriformes
FAMILY Turdidae
BREEDING HABITAT Forests, Boreal
NONBREEDING HABITAT Forests, Tropical Lowland Evergreen
GLOBAL POPULATION SIZE 46,000,000
POPULATION CHANGE SINCE 1970 −11%
IUCN RED LIST CATEGORY Least Concern
R2R TIPPING POINT SPECIES No
US LISTING STATUS No
CANADA SPECIES AT RISK No
PIF CCS 9/20

CONNECTIONS

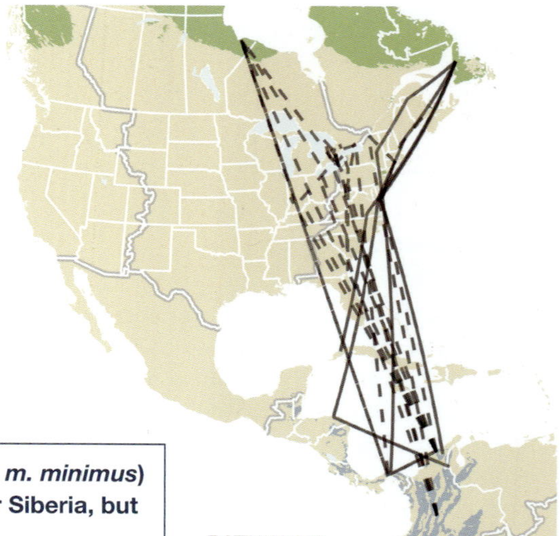

PATHWAYS

MIGRATORY CONNECTIVITY

OVERALL Although the distinctive Newfoundland population maintains largely separate migration routes and nonbreeding areas, migratory connectivity across the remaining breeding and

INTERESTING NOTE

Molecular analysis of nuclear and mitochondrial DNA showed that Newfoundland birds (*C. m. minimus*) shared no haplotypes with individuals from western Labrador, northern Quebec, Alaska, or Siberia, but with a zone of intergradation in south-central Labrador (FitzGerald et al. 2017).

nonbreeding distribution is not well known. MC is not calculated due to limited sample size.

RADIO TRACKING There is one detection from the Motus Wildlife Tracking System between a breeding area in Manitoba and a non-breeding area in Panama (N. Bayly and C. Gomez, unpublished data).

GEOLOCATOR TRACKING Individuals tracked from Denali National Park in Alaska spent the nonbreeding season in Venezuela (National Park Service, Critical Connections Program, unpublished data).

SATELLITE TRACKING Four males breeding in Newfoundland, Canada, were tracked using GPS archival tags to nonbreeding locations in the Sierra Nevada de Santa Marta in northern Colombia and Sierra de Perijá in Venezuela (Whitaker et al. 2018).

STABLE ISOTOPES Individuals that were resident in white sand forests of the Peruvian Amazon during the nonbreeding season were assigned to breeding or natal origins in northwestern Canada or Alaska using stable hydrogen isotope analyses (Ungvari-Martin et al. 2016).

OTHER APPROACHES Based on plumage and soft-part coloration, the Newfoundland breeding population is recognized as a distinct subspecies (*C. m. minimus*), intermediate in characteristics between remaining ("northern") Gray-cheeked Thrushes and Bicknell's Thrush. Wallace (1939) and Phillips (1991) identified specimens of *C. m. minimus* in the nonbreeding season only from northern Colombia (Sierra Nevada de Santa Marta), northern Venezuela (Sierra de Perijá), and Darién, Panama—thus intermediate in distribution between the Caribbean Bicknell's Thrush and the remaining Gray-cheeked Thrush nonbreeding distribution farther south in South America (Whitaker et al. 2020).

MIGRATORY PATHWAYS Gray-cheeked Thrushes follow a clockwise looped migration in North America (La Sorte et al. 2016), but then a counterclockwise elliptical route south of the US (Gómez et al. 2013). In post-breeding, birds move eastward across Canada and northern US, then south to the mid-Atlantic coast, southern Appalachians, and eastern Gulf states, before crossing the Gulf or western Caribbean Sea to the coast of Central America and then entering South America through the Darién Gap. GPS tracks of Newfoundland birds indicated that these birds moved down the Atlantic coast to Maryland and Virginia and then to points on the Caribbean coast of Central America or Colombia; one bird stopped over in western Cuba (Whitaker et al. 2018). In pre-breeding, most of this species concentrates for a prolonged refueling stopover in northern Colombia (or northern Venezuela; National Park Service, Critical Connections Program, unpublished data) before undertaking nonstop flights to the Gulf Coast or midwestern US (Gómez et al. 2019), then north through the northern Great Plains or Great Lakes region to northern breeding areas. A study of radio-tracked individuals using the Motus network demonstrated the critical importance of the Sierra Nevada de Santa Marta as a refueling stopover during pre-breeding migration of Gray-cheeked Thrushes that were later detected following extended nonstop flights by Motus towers on the Gulf Coast, the midwestern United States, and Ontario, Canada (Gómez et al. 2019).

DATA CONTRIBUTORS

E.M. Adams; N. Bayly; R. Boardman; J. Boelsma; A. Gilbert; C. Gomez; A.M. Gonzalez; G. Gress; C. Grigsby; K.A. Hobson; C. McIntyre; G. Mitchell; L.M. Phillips; I.J. Stenhouse; P. Thomas; I. Warkentin; S. Weidensaul; D. Whitaker; E.J. Williams.

Catharus bicknelli

BICKNELL'S THRUSH

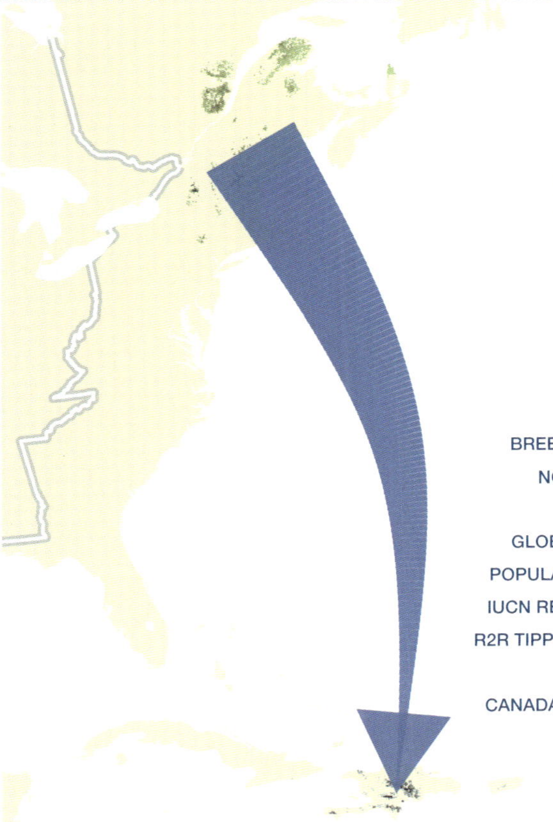

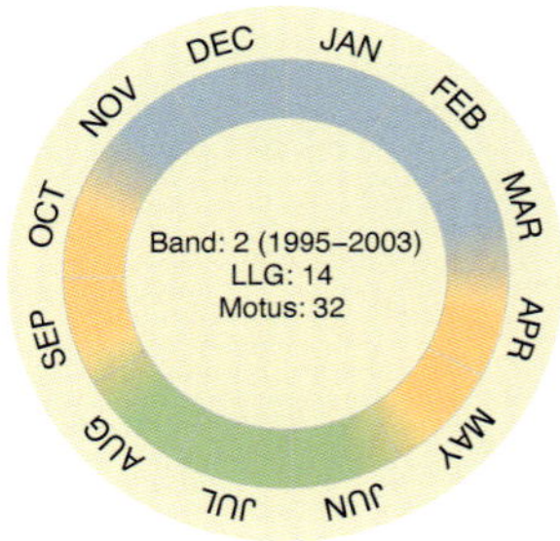

ORDER Passeriformes
FAMILY Turdidae
BREEDING HABITAT Forests, Boreal
NONBREEDING HABITAT Forests, Tropical Montane Evergreen
GLOBAL POPULATION SIZE 110,000
POPULATION CHANGE SINCE 1970 NA
IUCN RED LIST CATEGORY Vulnerable
R2R TIPPING POINT SPECIES Red-Alert
US LISTING STATUS No
CANADA SPECIES AT RISK Threatened
PIF CCS 17/20

GENERALIZED CONNECTIVITY

	Caribbean
Atlantic (n = 16)	100%

DISTRIBUTION AND ABUNDANCE

Bicknell's Thrushes breed in the Canadian Maritimes and on mountaintops in New England and south to the Catskills of New York. The nonbreeding range is restricted to the Greater Antilles, primarily in the Dominican Republic but also in Cuba and Jamaica (Fink et al. 2023).

MIGRATORY CONNECTIVITY

OVERALL Migratory connectivity is known from geolocator, GPS, band reencounter, and stable isotope data. The entire species is restricted to a very small breeding and nonbreeding range, and all individuals in the Atlantic Flyway migrate to the Caribbean. Because the population breeds within only one of our migratory connectivity regions, MC could not be calculated. However, individuals from

CONNECTIONS A

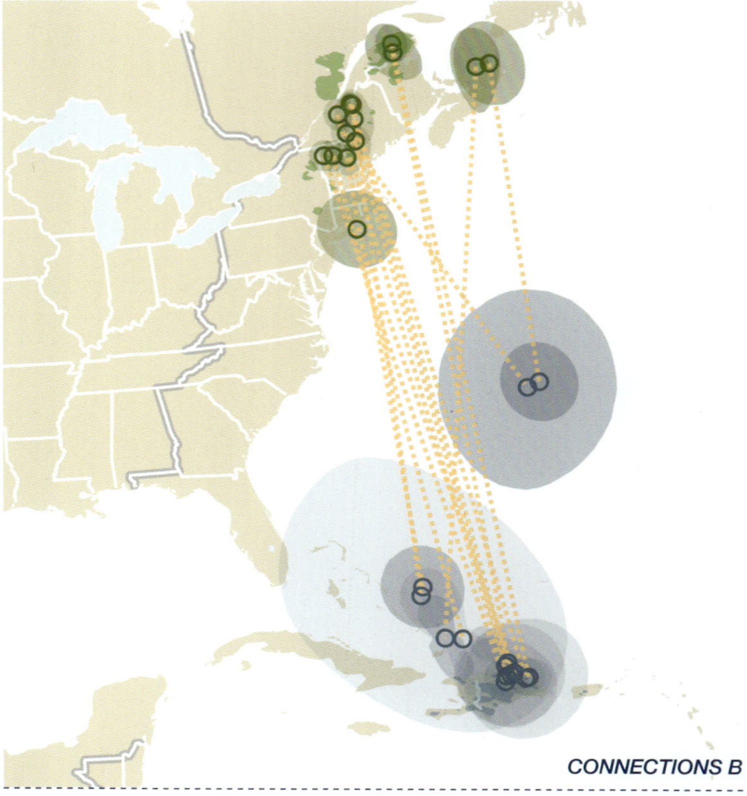

CONNECTIONS B

throughout the breeding range appear to mix in the Dominican Republic; hence migratory connectivity is weak.

BANDING Limited band reencounters confirm that breeders in Vermont spent the nonbreeding season in the Dominican Republic, including one remarkable recapture of an individual six months after being banded on Mt. Mansfield, Vermont (Townsend and Rimmer 2006).

RADIO TRACKING There are several tracking projects through the Motus Wildlife Tracking System. There were numerous detections during migration along the eastern seaboard (A Roth and J. Tremblay, unpublished data); there were no individual breeding to nonbreeding detections.

GEOLOCATOR TRACKING Deployment of light-level geolocators on individuals from New York, Vermont, New Hampshire, and Quebec confirmed that birds from all of these sites spent the nonbreeding season in the Dominican Republic (K. McFarland, unpublished data).

SATELLITE TRACKING Additional deployment of archival GPS tags (n = 13) on birds in Vermont and Quebec revealed more precise details of nonbreeding sites in the Dominican Republic and documented that one individual spent the nonbreeding season in the Sierra Maestra of eastern Cuba (Vermont Center for Ecostudies, unpublished data).

STABLE ISOTOPES Stable isotope analyses of nonbreeding birds in the Dominican Republic indicated limited connectivity of subpopulations from across the breeding range (Hobson et al. 2001). More negative isotope ratios of some individuals captured in the Dominican Republic match those of a previously unknown breeding population on the northern Gaspé Peninsula, Quebec (Hobson et al. 2004).

MIGRATORY PATHWAYS Individuals tracked with geolocators revealed a post-breeding migration route south to the mid-Atlantic coast, then a transoceanic crossing to the Bahamas before moving eastward to the Dominican Republic. Several individuals stopped over in Bermuda, suggesting a longer crossing of the western Atlantic Ocean, and several made stopovers inland (also suggested by banding and specimen records), as far west as Ontario, New York, Kentucky, and Florida (K. McFarland, unpublished data). Based on very limited data, pre-breeding migration appears to be farther west to Florida and the South Atlantic coast and then rapidly to breeding areas without multiday stopovers (Townsend et al. 2020).

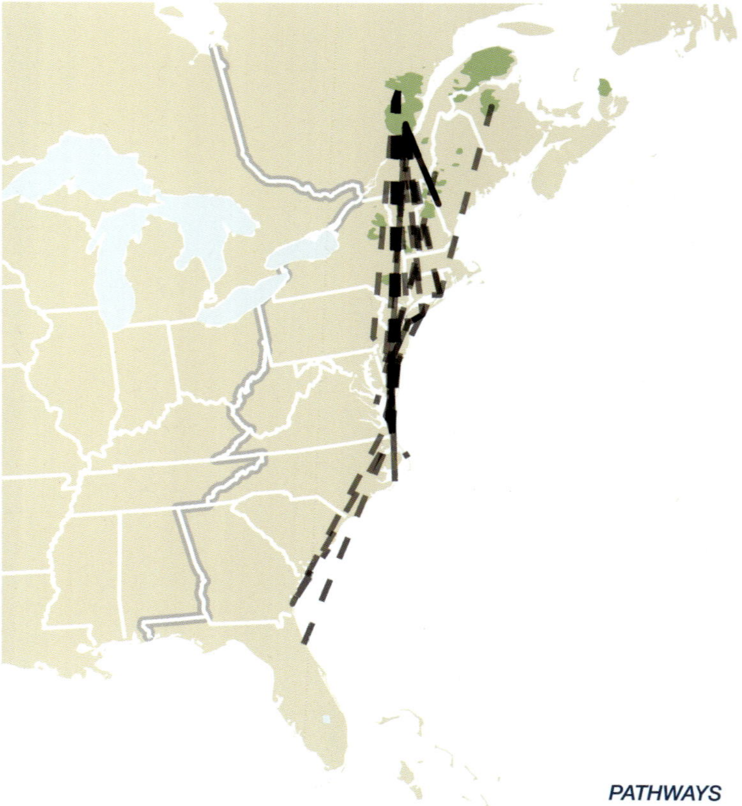

PATHWAYS

DATA CONTRIBUTORS

Y. Aubry; C. Bégin Marchand; D. Bell; D.A. Buehler; G. Campbell; A. Desrochers; J.W. Fox; J.M. Hill; S.A. Mackenzie; P.P. Marra; K.P. McFarland; J.D. Lloyd; S. Paradis; C.C. Rimmer; A. Roth; L.M. Stewart; J.M. Townsend; J. Tremblay; C. Ward; R.M. Whittam.

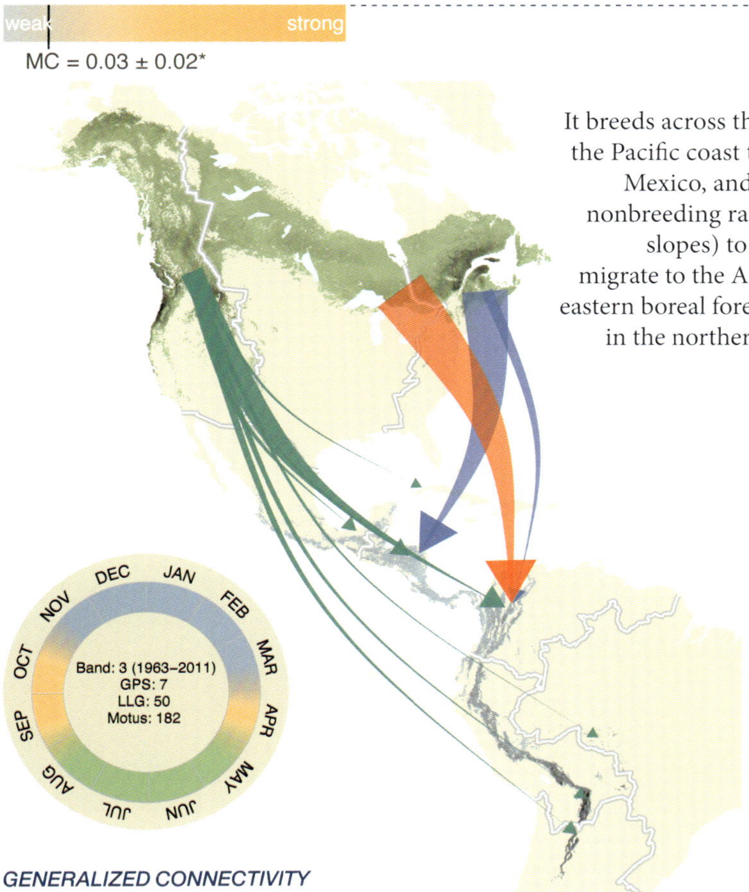

GENERALIZED CONNECTIVITY

Catharus ustulatus **SWAINSON'S THRUSH**

DISTRIBUTION AND ABUNDANCE

It breeds across the boreal region from central Alaska east to Newfoundland, south along the Pacific coast to Southern California, in the Rocky Mountains south to northern New Mexico, and at high elevations in the Appalachian Mountains to West Virginia. The nonbreeding range of Pacific (russet-backed) populations is from central Mexico (both slopes) to western Panama, while interior and eastern (olive-backed) populations migrate to the Andes and other mountains of western South America. Populations in the eastern boreal forest and along the Pacific coast have declined steeply, whereas populations in the northern Rocky Mountains and Alaska have remained stable (Sauer et al. 2019).

	Andean West	Brazil	Caribbean	Central America	Mexico	Northern SA	Southern Cone
Atlantic (n = 6)	–	–	–	83%	–	17%	–
Midcontinent (n = 6)	–	–	–	–	–	100%	–
Pacific (n = 61)	8%	3%	3%	23%	13%	38%	11%

ORDER Passeriformes
FAMILY Turdidae
BREEDING HABITAT
Forests, Boreal
NONBREEDING HABITAT
Forests, Tropical
Montane Evergreen
GLOBAL POPULATION SIZE 120,000,000
POPULATION CHANGE SINCE 1970 −28%
IUCN RED LIST CATEGORY Least Concern
R2R TIPPING POINT SPECIES No
US LISTING STATUS No
CANADA SPECIES AT RISK No
PIF CCS 9/20

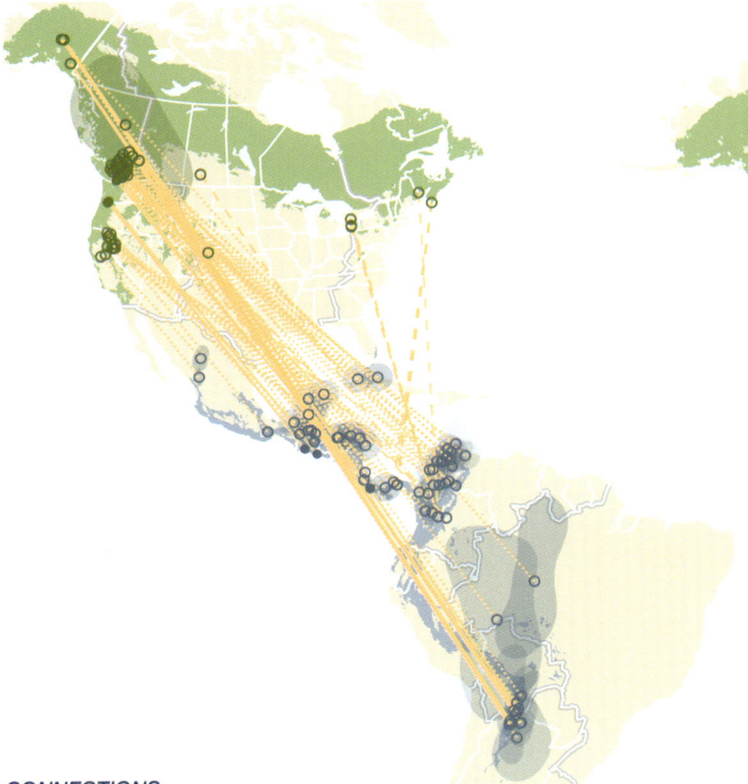

CONNECTIONS

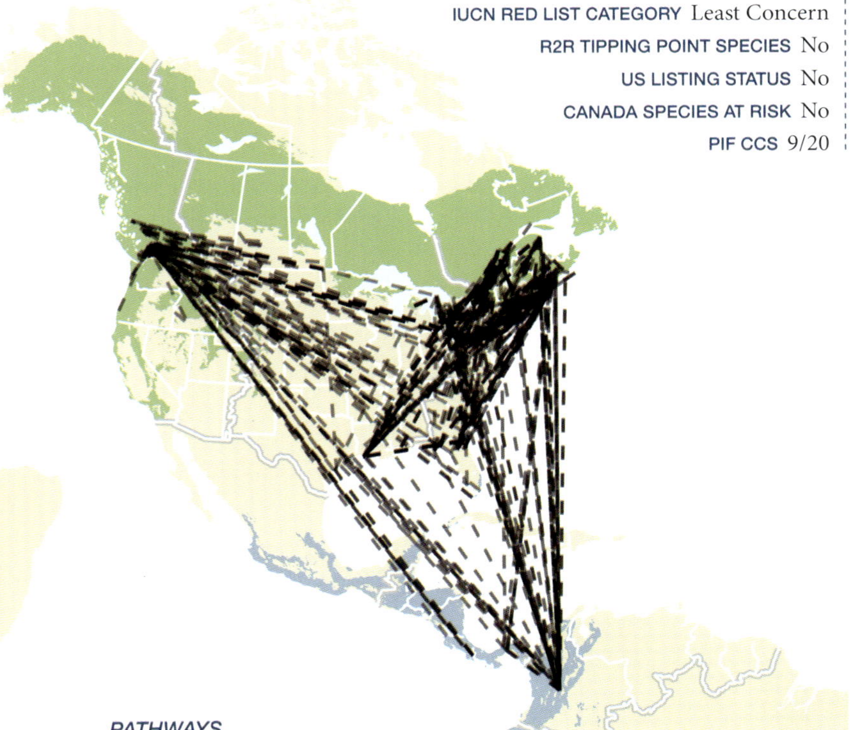

PATHWAYS

MIGRATORY CONNECTIVITY

OVERALL Migratory connectivity is relatively well studied in this species, with morphology, genetics, stable isotopes, banding, telemetry, and geolocator tracking studies, and appears strong between the Pacific "russet-backed" and interior/eastern "Olive-backed" forms, which maintain separate migration systems and nonbreeding regions (russet-backed in Mexico and Central America; olive-backed in South America). However, across the species and flyway regions, mixing of the breeding populations, especially in northern South America and Central America, results in weak migratory connectivity (MC = 0.03 ± 0.02). Atlantic Flyway breeding birds spend the nonbreeding season in Central America (83%) and northern South America (17%), while Midcontinent Flyway breeding birds all spend the nonbreeding season in northern South America. Additionally, 38% of the Pacific Flyway breeding population spend the nonbreeding season in northern South America, as well as Central America (23%), Mexico (13%), and smaller numbers throughout several other nonbreeding regions in South America.

BANDING The few breeding to nonbreeding band reencounters were from (russet-backed) individuals breeding in the northwestern US and migrating to the Pacific Slope of southern Mexico and northern Central America. An analysis of 71 banding recaptures further indicated that breeding individuals in Alaska and northwestern Canada migrate thousands of kilometers to the east before turning south toward South America (Ruegg and Smith 2002).

RADIO TRACKING Thousands of tags were deployed through the Motus Wildlife Tracking System. Data from individual birds detected between breeding and nonbreeding areas (L. Berrigan et al., unpublished data; K. Delmore and W. Easton, unpublished data; A. Gonzalez, unpublished data) supported similar patterns from other studies, including a migratory divide between subspecies.

GEOLOCATOR TRACKING Swainson's Thrushes have been tracked with light-level geolocators, mainly from western areas of the breeding range. Individuals from the same breeding sites showed strong connectivity with nonbreeding areas, as well as migration pathways and stopover sites. A leapfrog pattern of migration is also evident, with inland British Columbia and California birds migrating farther south than coastal British Columbia and California breeders (Delmore et al. 2012; Cormier et al. 2013), and Alaska breeders migrating much farther south in South America (National Park Service, Critical Connections Program, unpublished data) than inland British Columbia, California, or Colorado breeders (Bird Conservancy of the Rockies, unpublished data; K.J. Kardynal and K.A. Hobson, unpublished data). Specifically, birds breeding along the northern coast of California spent the nonbreeding season in Mexico, whereas individuals from the southern Cascades population migrated to Central America (Cormier et al. 2013). Inland British Columbia and California as well as Colorado thrushes spent the nonbreeding season in Colombia and Venezuela (Delmore et al. 2012, 2013; Bird Conservancy of the Rockies, unpublished data). Hybrids were found to use geographically intermediate nonbreeding sites (Delmore and Irwin 2014).

SATELLITE TRACKING Swainson's Thrushes have also been tracked in limited numbers with GPS archival tags from Alaska. Alaska populations spent the nonbreeding season in South America but migrated farther south than other tracked populations to Bolivia, Paraguay, and Argentina (National Park Service, Critical Connections Program, unpublished data).

STABLE ISOTOPES Stable isotope ratios indicate that hatch-year individuals stopping over in Ft. Morgan, Alabama, during post-breeding migration come from west-central regions of the boreal forest (Hobson et al. 2007). Individuals stopping over in southwestern Costa Rica during pre-breeding migration also most likely originated in west-central regions of the boreal forest and northern Rocky Mountains (S. Wilson et al. 2008).

GENETICS Bioclimatic modeling and mtDNA analysis indicate that Swainson's Thrush populations were split during the last glaciation into Pacific coastal and inland groups with completely separate migratory paths and nonbreeding ranges, agreeing with morphological and geolocator data (Ruegg and Smith 2002; Ruegg et al. 2006). A narrow but continuous genetic hybrid zone is maintained in coastal mountains of British Columbia (Ruegg 2008).

OTHER APPROACHES Based on plumage coloration, two subspecies groups were easily distinguished, with "russet-backed" forms breeding along the Pacific Slope from southeastern Alaska to California and migrating to Mexico and the Pacific Slope of Central America, and "olive-backed" forms occupying the remainder of the breeding range and migrating to South America. Three or more subspecies have been recognized within each of these groups (Phillips 1991; Mack and Yong 2020), but analysis of specimens has not clarified to what extent the subspecies within each group separate or mix in the nonbreeding season.

MIGRATORY PATHWAYS Studies of tracked individuals indicate that "russet-backed" Pacific Slope breeders migrate along the Pacific coast of the US and Mexico in both pre- and post-breeding seasons. In contrast, "olive-backed" inland and eastern breeders follow a longer, looped migration, moving south in the post-breeding season through the Great Lakes states and Appalachian Mountains and crossing the eastern Gulf of Mexico to Central America before passing through the Darién Gap to South America (Delmore et al. 2012). Where these two subspecies overlap, hybrids were found to have intermediate or highly variable migration routes (Delmore and Irwin 2014). Motus detections (n = 567) of 241 individuals tagged at stations across eastern Canada during post-bredding migration revealed population-specific migration routes before converging in Florida; only birds from Atlantic Canada migrated along the Atlantic coast (Bégin-Marchand et al. 2021). In the pre-breeding season, migrants circumnavigate or cross the Gulf farther west and move north through the central US before spreading out into boreal breeding grounds. This species makes more frequent short fights and shorter stopovers in northern Colombia than other *Catharus* thrushes (Gómez et al. 2013). Additional Motus detections from several studies (L. Berrigan et al., unpublished data; W. Blake, unpublished data; E. Cohen et al., unpublished data; L. DeGroote et al., unpublished data; K. Delmore and W. Easton, unpublished data; B. Frei et al., unpublished data; A. Gonzalez, unpublished data; C.C. Guglielmo et al., unpublished data; J. Marty et al., unpublished data; G. Mitchell and P. Taylor, unpublished data; S. Sargent, unpublished data; J. Tremblay et al., unpublished data) throughout migration fill gaps in tracking information throughout the range.

DATA CONTRIBUTORS

E.M. Adams; C. Bégin Marchand; L. Berrigan; Bird Conservancy of the Rockies; W. Blake; J. Boelsma; R. Burnett; J. Cepek; E. Cohen; R.L. Cormier; L. DeGroote; K. Delmore; A. Desrochers; K.E. Dybala; W. Easton; K. Elliott; J. Fowler; B. Frei; A. Gilbert; M. Gimpel; A.M. Gonzalez; G. Gress; C. Grigsby; C.C. Guglielmo; T. Guida; K.A. Hobson; D.L. Humple; R. Jacob; K.J. Kardynal; L. Koitsch; T. Krynak; J. Marty; C. McIntyre; G. Mitchell; A. Morales; Y. Morbey; C. Nemes; L.M. Phillips; V. Poirier; J. Rice; T.W. Richardson; S. Sargent; M. Sarver; N.E. Seavy; M. Saymour; I.J. Stenhouse; P. Taylor; J. Tremblay; P. Vasseur; S. Weidensaul; E.J. Williams; D. Yeany.

Catharus guttatus **HERMIT THRUSH**

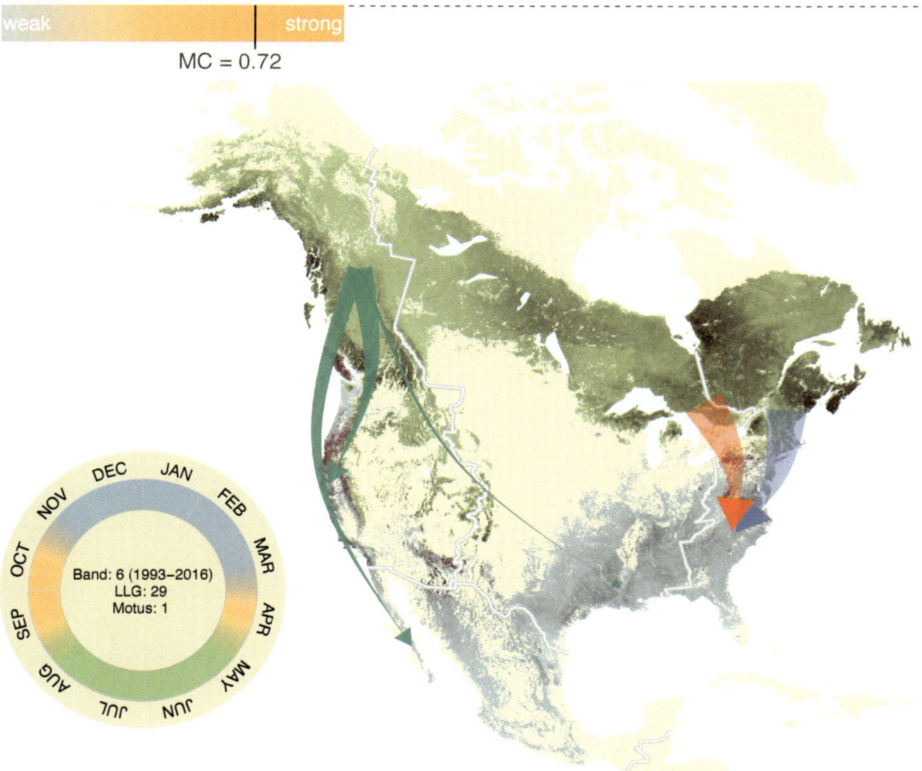

GENERALIZED CONNECTIVITY

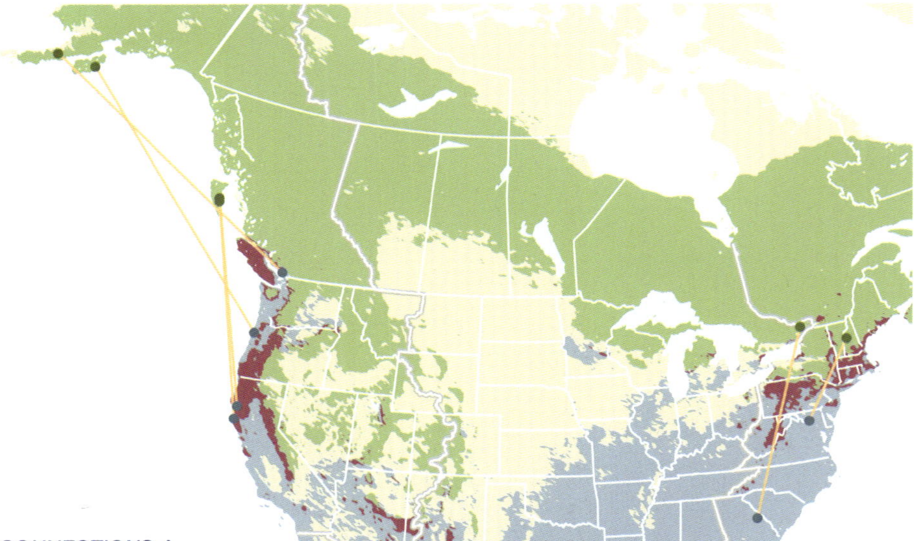

CONNECTIONS A

	Atlantic	Mexico	Midcontinent	Pacific
Atlantic (n = 1)	100%	–	–	–
Midcontinent (n = 1)	100%	–	–	–
Pacific (n = 33)	–	39%	6%	55%

MIGRATORY CONNECTIVITY

OVERALL Migratory connectivity is strong across Hermit Thrush populations (MC = 0.72), corresponding largely with previously described subspecies and documented by genetics, band reencounter, and geolocator tracking data. Atlantic and Midcontinent flyway breeding populations remain within the same flyway during the nonbreeding season. Pacific Flyway breeding populations either remain in the Pacific Flyway (55%) or migrate to nonbreeding areas in Mexico (39%). In addition, this species exhibits a chain migration pattern along the Pacific coast (and perhaps elsewhere), as northern breeders move into nonbreeding areas in central California after local breeders depart for nonbreeding sites farther south (Nelson et al. 2021).

DISTRIBUTION AND ABUNDANCE

It is a widespread breeder from central and southern Alaska east through the boreal region of Canada to Newfoundland, south to the northeastern US and high Appalachian Mountains to North Carolina, and in the west along the Pacific coast to central California and in the mountains south to southern Arizona, New Mexico, western Texas, and northwestern Mexico. Its nonbreeding range is also large, across most of the southern US north along the Pacific coast to Vancouver Island and in the east to southern New England, and south through most of Mexico sparsely to Guatemala and El Salvador. Twelve subspecies fall into three groups: Western Lowland, Western Mountain, and Northern (Dellinger et al. 2020). The Western Lowland group breeds primarily along coastal areas of Alaska, British Columbia, south through the Cascades and into California. The nonbreeding range extends along the Pacific coast from Washington south to Baja California and east to Texas and into Mexico. The Western Mountain group breeds in the mountains of California, east to Utah, and throughout the Rocky Mountain and spends the nonbreeding season in Arizona, Mexico, and south to El Salvador. The Northern group breeds across eastern Canada and US, with a nonbreeding range extending from the southeastern US south to northeastern Mexico (Dellinger et al. 2020). Although long-term trends have been mostly stable or increasing, most regional populations show steep declines over the past two decades, especially so in eastern Canada and the northeastern US (Sauer et al. 2019).

ORDER Passeriformes
FAMILY Turdidae
BREEDING HABITAT Forests, Temperate Generalist
NONBREEDING HABITAT Forests, Temperate Generalist
GLOBAL POPULATION SIZE 72,000,000
POPULATION CHANGE SINCE 1970 +19%
IUCN RED LIST CATEGORY Least Concern
R2R TIPPING POINT SPECIES No
US LISTING STATUS No
CANADA SPECIES AT RISK No
PIF CCS 9/20

BANDING Breeding to nonbreeding band reencounters are consistent with separate migration systems for eastern and western populations of Hermit Thrushes, with western individuals breeding in coastal Alaska and British Columbia migrating to California, whereas individuals in the northeastern US and Canada migrate to the southeastern US.

RADIO TRACKING There are numerous tracking projects through the Motus Wildlife Tracking System. There are no individual breeding to nonbreeding detections. There are abundant detections during migration, especially in the eastern portion of the range, where more information is needed.

GEOLOCATOR TRACKING Nonbreeding Hermit Thrushes in the San Francisco Bay Area (n = 10) were tracked with light-level geolocators to breeding grounds along the Pacific coast from southeastern Alaska to northwestern Washington. Strong migratory connectivity was suggested by separation of breeding sites by individuals from North Bay and South Bay nonbreeding sites, corresponding with separation of previously described subspecies distributions (Nelson et al. 2016). An additional geolocator study of the distinct population breeding in the Santa Cruz Mountains, California (*C. g. slevini*; n = 13), documented their nonbreeding range in a small area of Baja California Sur and adjacent mainland western Mexico (Nelson et al. 2021). Another study used geolocators (n = 5) to confirm separate migration routes and nonbreeding areas across a migratory divide in British Columbia, consistent with genetics and morphology (Alvarado et al. 2014).

GENETICS Analysis of nuclear DNA confirmed genetic differentiation across a migratory divide in British Columbia and broad separation of eastern and western populations throughout the annual cycle (Alvarado et al. 2014).

OTHER APPROACHES Hermit Thrushes show high levels of morphological variation, and 12 geographic subspecies have been recognized (Phillips 1991; Dellinger et al. 2020). Careful study of nonbreeding specimens indicates that, in general, boreal and northeastern breeders migrate to the southeastern US; northwestern breeders (central Alaska to northern Rocky Mountains) migrate to the south-central US; southern Rocky Mountain breeders migrate to southern Mexico and northern Central America; and the remaining western populations exhibit a complex nonbreeding pattern on the Pacific coast, southwestern US, and Mexico. There is evidence for separation by habitat in regions where several subspecies occur in the nonbreeding season (e.g., northwestern Mexico; Unitt and Estrella 1996).

MIGRATORY PATHWAYS The species migrates over several broad fronts corresponding to the separate subspecies groups: throughout the eastern US; through the western Great Plains to Texas and northeastern Mexico; through the Great Basin and Colorado Plateau to the southwestern US and western Mexico; and along the Pacific coast to Baja California (Fink et al. 2023). Limited banding and tracking data suggest a more complex separation of populations during migration, with a major migratory divide between Pacific and interior populations documented in British Columbia (Alvarado et al. 2014). Additional migration-season band recoveries also suggest an east-west parallel migration within the widespread eastern and northern populations. Eastern populations migrate much earlier in pre-breeding and later in post-breeding than most western populations.

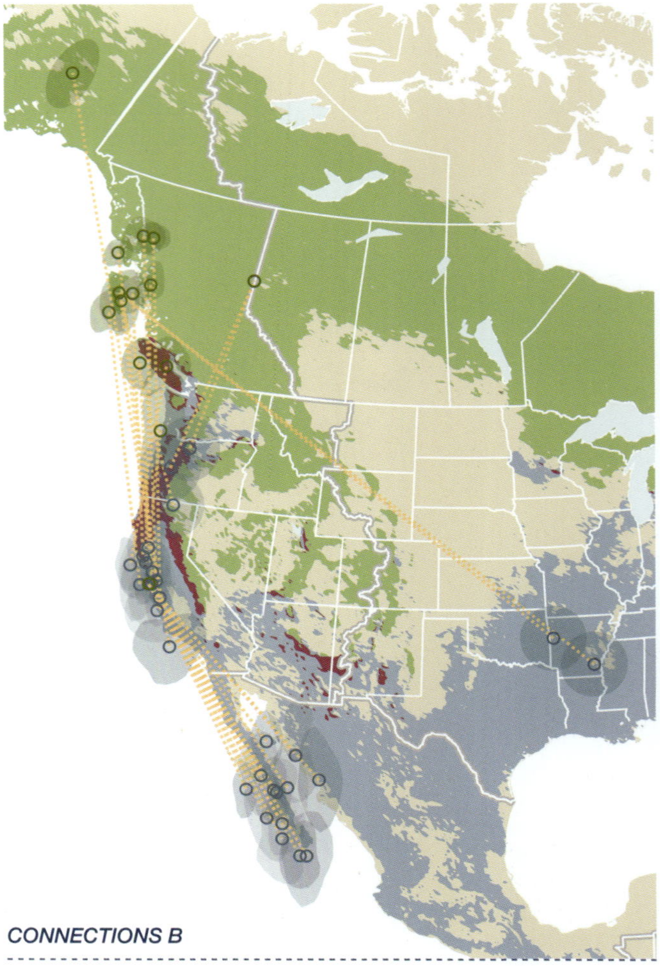

CONNECTIONS B

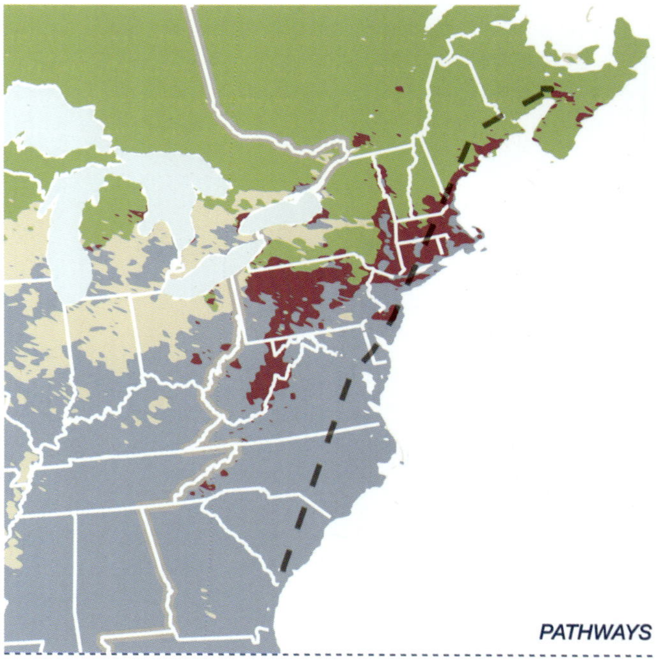

PATHWAYS

DATA CONTRIBUTORS

E.M. Adams; J. Boelsma; A. Gilbert; G. Gress; C. Grigsby; C.C. Guglielmo; C. McIntyre; Y. Morbey; A.R. Nelson; L.M. Phillips; I.J. Stenhouse; S. Weidensaul; E.J. Williams.

INTERESTING NOTE

A recent range-wide genoscape analysis revealed five distinct genetic clusters, broadly corresponding with the major (previously described) subspecies groups (Alvarado et al. 2022). Populations showed no genetic structure across the vast boreal and eastern portion of the range, from central British Columbia to Maine and Pennsylvania. High levels of genetic structure were evident across the four western population clusters; however, temperature was identified as a major predictor of adaptive genomic variation, both at the macrogeographic scale throughout western North America and at the local scale across a local hybrid zone between Pacific Slope and interior populations in British Columbia (Alvarado et al. 2022).

MC = 0.28 ± 0.07*

Hylocichla mustelina **WOOD THRUSH**

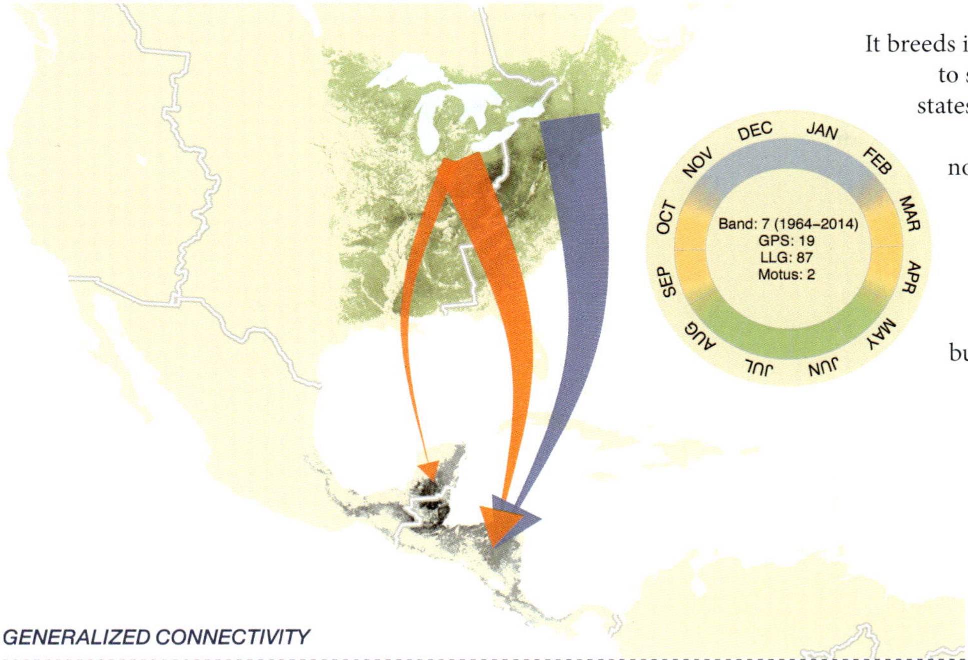

GENERALIZED CONNECTIVITY

DISTRIBUTION AND ABUNDANCE

It breeds in deciduous forests across the eastern US, sparsely to southernmost Canada and west to the Great Plains states; highest abundance is in the central and southern Appalachian Mountains and Gulf Coast states. Its nonbreeding distribution is primarily in wet tropical forests on the Caribbean slope from southeastern Mexico to Costa Rica, sparsely to western Panama and on the Pacific Slope of Central America. Declines have been steepest and more continuous in New England and mid-Atlantic coastal states, but increases have been noted, especially recently, in western parts of the range (Sauer et al. 2019).

MIGRATORY CONNECTIVITY

	Central America	Mexico
Atlantic (n = 84)	100%	–
Midcontinent (n = 24)	75%	25%

OVERALL Wood Thrush migratory connectivity is well known based on tracking and band reencounter data, as well as stable isotope analysis. Birds breeding within the Atlantic Flyway spend the nonbreeding season in Central America (100%). Birds breeding in the midcontinent also spend the nonbreeding season in Central America (75%), with a portion of the population also migrating to nonbreeding areas in Mexico (25%). Wood Thrushes exhibit moderate east-west migratory connectivity (MC = 0.28 ± 0.07) across the range. GPS tracking also revealed fine-scale parallel migratory connectivity across breeding populations (Stanley et al. 2021b).

BANDING Breeding to nonbreeding band reencounters have all been from mid-Atlantic and New England breeding populations and show some degree of east-west parallel migration to nonbreeding sites in Guatemala, Belize, Nicaragua, and Costa Rica.

ORDER Passeriformes
FAMILY Turdidae
BREEDING HABITAT Forests, Temperate Eastern
NONBREEDING HABITAT Forests, Tropical Lowland Evergreen
GLOBAL POPULATION SIZE 12,000,000
POPULATION CHANGE SINCE 1970 –60%
IUCN RED LIST CATEGORY Least Concern
R2R TIPPING POINT SPECIES Yellow-Alert
US LISTING STATUS No
CANADA SPECIES AT RISK Threatened
PIF CCS 14/20

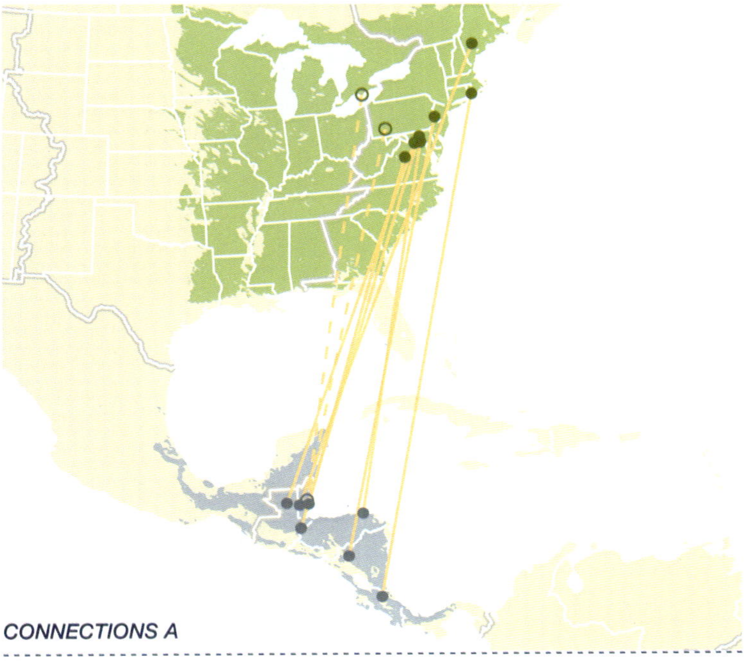

CONNECTIONS A

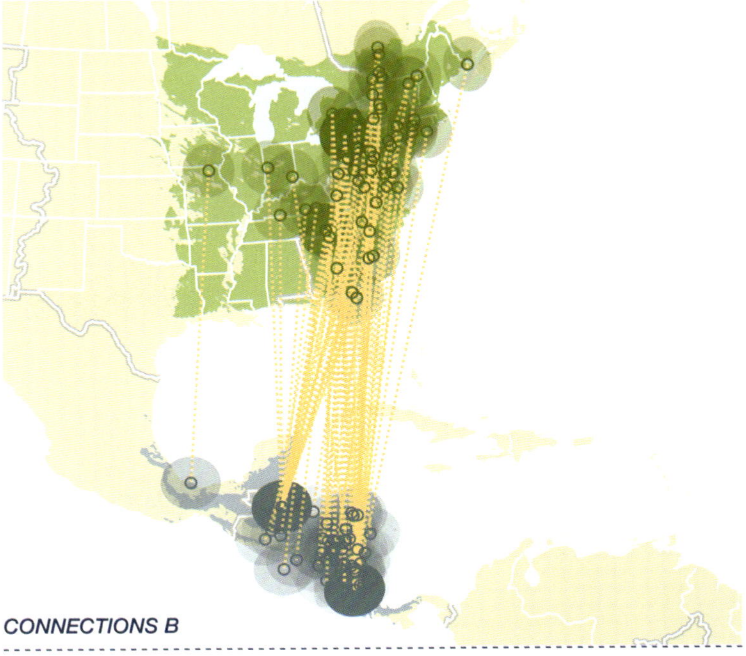

CONNECTIONS B

RADIO TRACKING Over a thousand tags were deployed through the Motus Wildlife Tracking System. At this time, there have been no individual breeding to nonbreeding connections. However, new data can be used to fill gaps in migratory connectivity knowledge throughout the range and help inform migration and stopover ecology. ***GEOLOCATOR TRACKING*** Stanley et al. (2015) deployed (n = 594) and recovered (n = 96) geolocators on both breeding and nonbreeding grounds and showed that Wood Thrushes exhibit fairly strong parallel migration and migratory connectivity, with eastern-breeding birds migrating to nonbreeding sites farther southeast in Central America (Panama / Costa Rica) and midwestern birds migrating to the Belize/Guatemala region or southern Mexico. ***SATELLITE TRACKING*** Additional tracking of Wood Thrushes using archival GPS tags (n = 19) further established that breeders from New York, North Carolina, and Delaware mostly migrated to the Caribbean slope of Central America from Honduras to Costa Rica, whereas most breeders from Indiana and Minnesota migrated to nonbreeding sites farther north and west in Central America and southern Mexico (Stanley et al. 2021b). ***STABLE ISOTOPES*** Rushing et al. (2014) combined morphological and stable isotope data and confirmed western breeding populations are more likely to migrate to western nonbreeding sites in Mexico, and eastern breeding populations are more likely to migrate to southeastern nonbreeding sites in Central America. ***MIGRATORY PATHWAYS*** Wood Thrushes exhibit a clockwise looped migration, based both on observational data (La Sorte et al. 2016) and tracked individuals (Stanley et al. 2012, 2015). A majority of individuals passed through Florida and Cuba or across the eastern Gulf of Mexico during post-breeding migration, whereas during pre-breeding migration, most individuals entered the US by crossing the western Gulf of Mexico via the Yucatán Peninsula (Stanley et al. 2015).

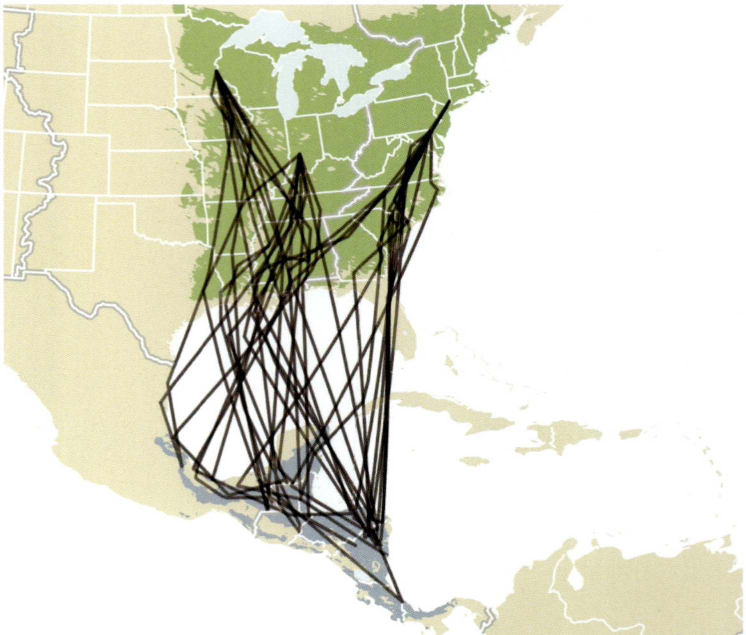

PATHWAYS

DATA CONTRIBUTORS

S. Adalsteinsson; J. Cepek; G. Casbourn; L. DeGroote; N.E. Diggs; M.R. Dudash; J. Fowler; K.C. Fraser; L. Friesen; M. Gimpel; R. Jacob; L. Koitsch; T. Krynak; M.P. MacPherson; P.P. Marra; E.A. McKinnon; J. Rice; T.B. Ryder; S. Sargent; M. Sarver; K. Serno; G.W. Shriver; C.Q. Stanley; C.E. Studds; B.J.M. Stutchbury; D. Yeany.

AMERICAN ROBIN *Turdus migratorius*

weak — strong

MC = 0.47 ± 0.05*

DISTRIBUTION AND ABUNDANCE

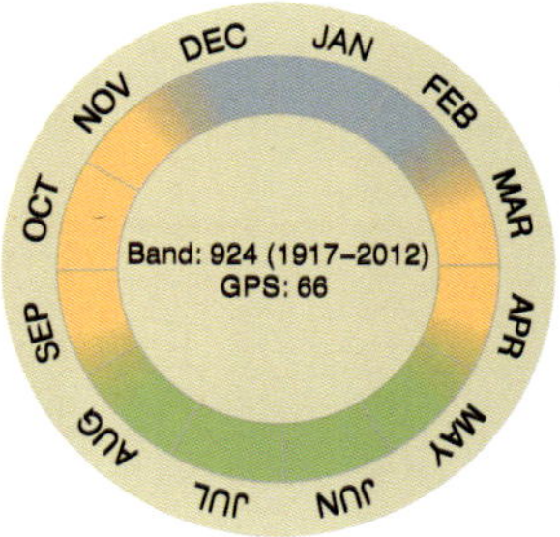

American Robins are widespread and common breeders throughout most of North America, from the Arctic tundra through the northern US and more patchily in the southern US and in mountains south to southern Mexico. The breeding and nonbreeding ranges overlap substantially, belying the fact that many US and Canadian populations are migratory. Breeders across the south-central and southeast US are largely nonmigratory. Nonbreeding abundance fluctuates annually but is generally highest across the southeastern US and in the Pacific states. Long-term declines are apparent in Pacific coast and Rocky Mountain populations and, in recent decades, throughout the northeastern portion of the breeding range (Sauer et al. 2019).

ORDER Passeriformes
FAMILY Turdidae
BREEDING HABITAT Forests, Temperate Generalist
NONBREEDING HABITAT Forests, Temperate Generalist
GLOBAL POPULATION SIZE 370,000,000
POPULATION CHANGE SINCE 1970 +6%
IUCN RED LIST CATEGORY Least Concern
R2R TIPPING POINT SPECIES No
US LISTING STATUS No
CANADA SPECIES AT RISK No
PIF CCS 7/20

MIGRATORY CONNECTIVITY

OVERALL About 80% of individuals breeding in three flyways spend the nonbreeding season within the same flyway as breeding. The other 20% of the Atlantic and Pacific flyway populations migrate to the Midcontinent Flyway. Despite mixing within the Midcontinent Flyway, overall migratory connectivity is strong (MC = 0.47 ± 0.05), with strong east-west

	Atlantic	Mexico	Midcontinent	Pacific
Atlantic (n = 293)	77%	<1%	23%	–
Midcontinent (n = 508)	22%	1%	77%	<1%
Pacific (n = 39)	–	–	21%	79%

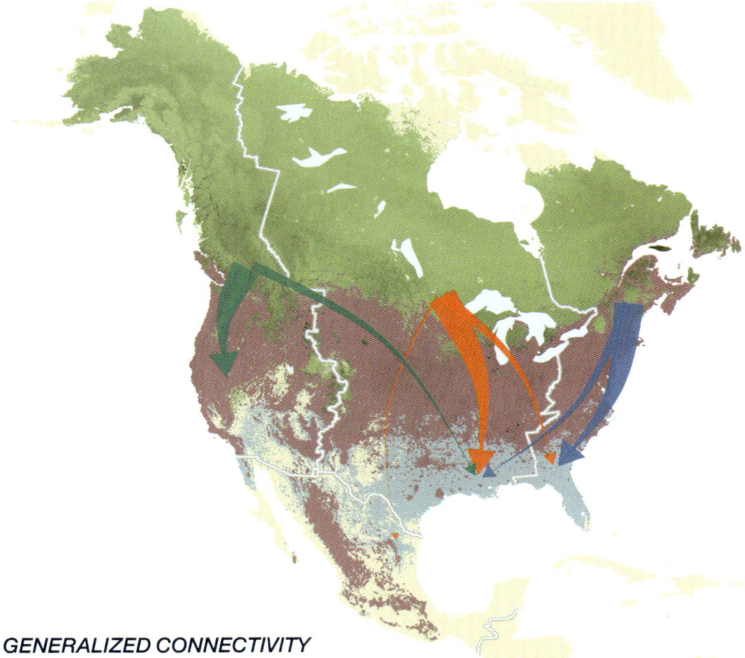

GENERALIZED CONNECTIVITY

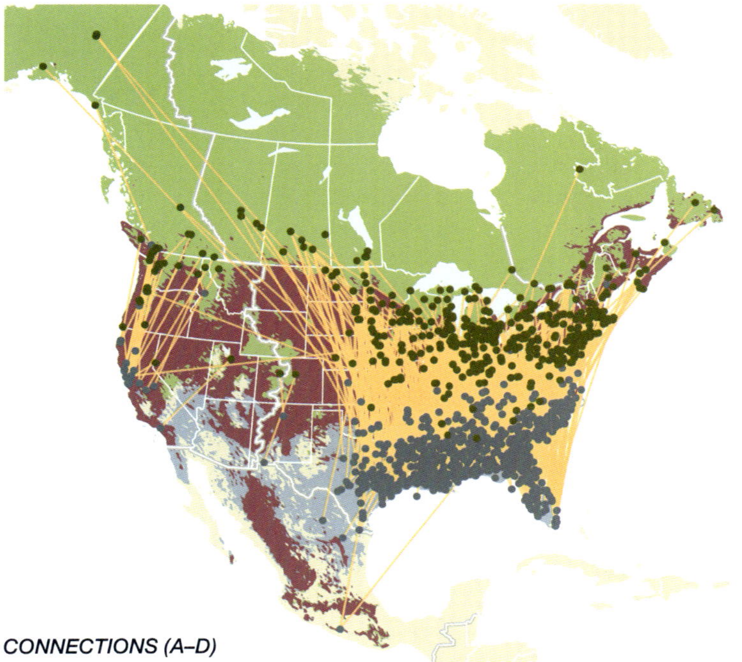

CONNECTIONS (A–D)

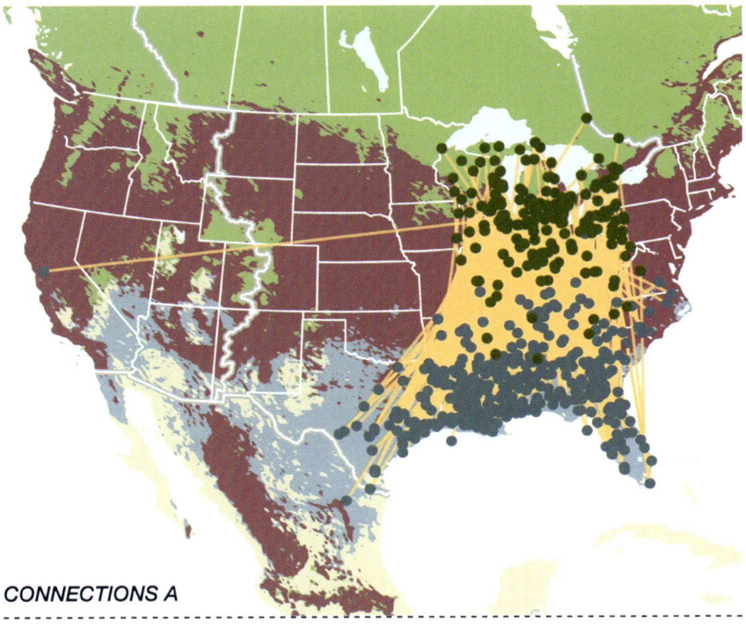

CONNECTIONS A

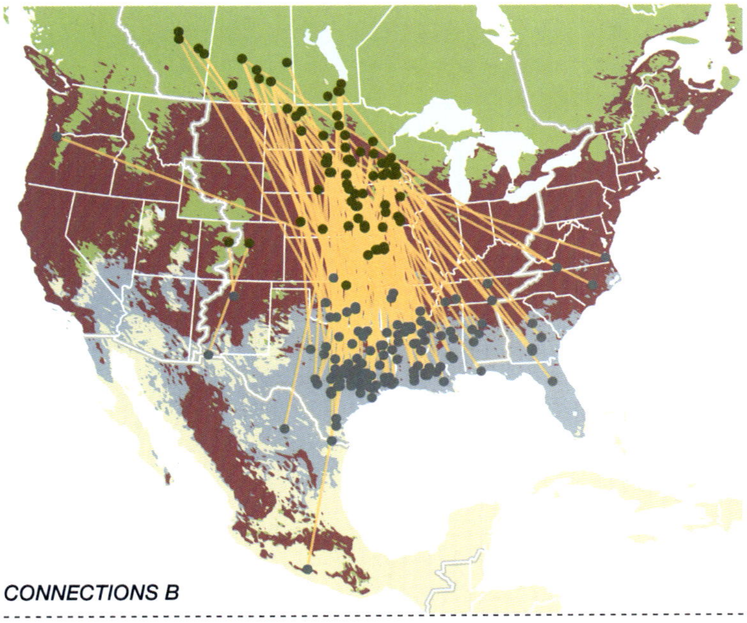

CONNECTIONS B

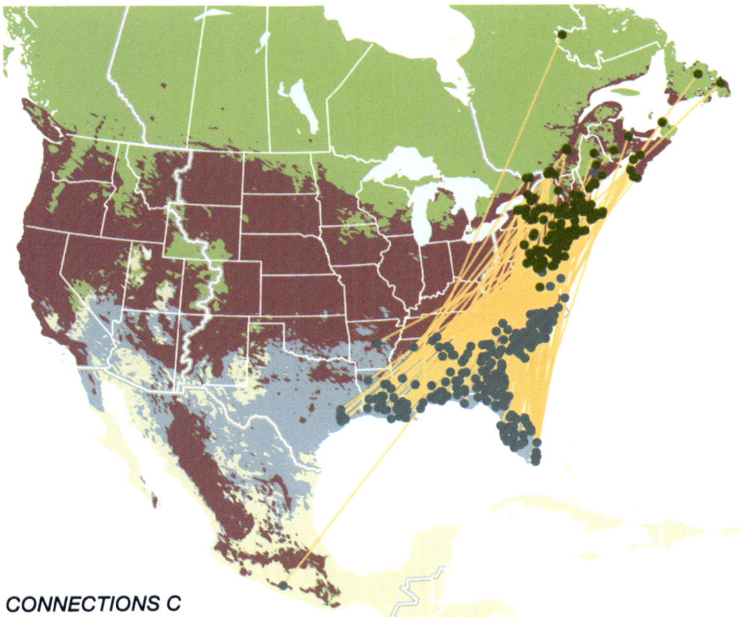

CONNECTIONS C

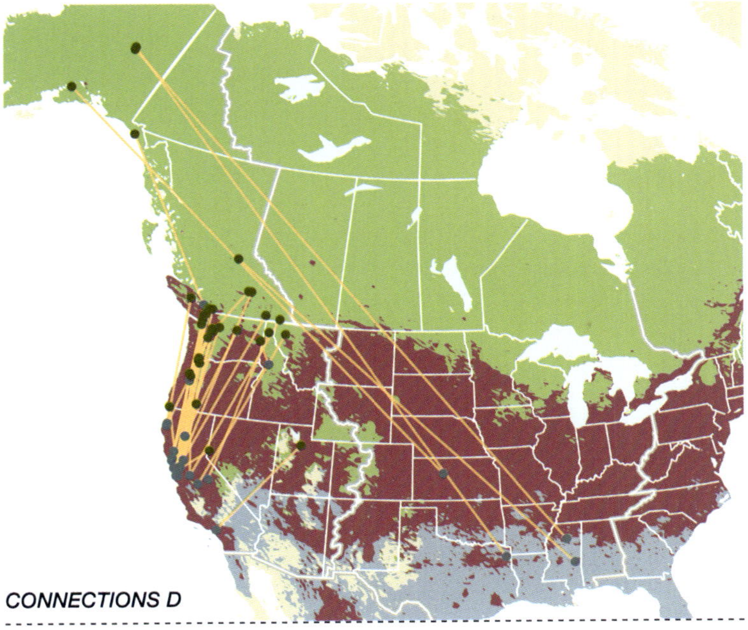

CONNECTIONS D

INTERESTING NOTE

American robins exhibit partial migration, with some proportion of birds south of midlatitudes being regional residents year-round. This line of partial migration may be moving north with warming annual winter temperatures. Some individuals have also been documented to switch between migratory and resident behaviors in different years (E.J. Williams et al., unpublished data).

parallel migration and separation of regional populations throughout the annual cycle, as demonstrated from morphological variation (subspecies), band reencounter, geolocator, and satellite tracking data.

BANDING Banding reencounters indicate a strong separation of eastern and western populations, with few individuals crossing the Rocky Mountains (Brown and Miller 2016). Within the eastern population, there is evidence of moderate connectivity and east-west parallel migration, with most nonbreeding individuals in Florida migrating along the Atlantic coast and most individuals in Texas migrating to the Great Plains to breed. Migratory connectivity within western populations was lower, but based on much smaller samples and fewer locations. About 20% of band recoveries indicated year-round residency, a majority of these being within the breeding distribution of *T. m. archusterus* or on the Pacific coast.

RADIO TRACKING American Robins are tracked through the Motus Wildlife Tracking System in both the eastern and western portions of the range. There are no individual breeding to nonbreeding detections.

SATELLITE TRACKING A study using GPS archival tags (n = 7) also showed that individuals in Alaska migrated to the Great Plains from eastern Montana to northern Texas (n = 4), one bird from Massachusetts migrated to South Carolina, and a male and female from Washington, DC, stayed in the vicinity year-round (Jahn et al. 2019). A study using GPS-satellite tags also found: individuals from Alaska migrated to the Great Plains states; an individual breeding in Massachusetts was tracked to South Carolina; and individuals from the mid-Atlantic region (Maryland, Virginia, Washington, DC) were a mix of year-round residents and short-distance migrants (Jahn et al. 2019; E.J. Williams et al., unpublished data). Another study using GPS-satellite tags found that individuals migrating through central Alberta migrated northwest to breeding sites in western Canada and Alaska (Oliver et al. 2020).

GENETICS The Bird Genoscape Project is currently working to construct a genoscape for the American Robin (https://www.birdgenoscape.org/american-robin-2/).

OTHER APPROACHES Based on study of museum specimens, seven subspecies have been described (Phillips 1991; Vanderhoff et al. 2020), and their year-round distribution reveals strong connectivity among populations. The northeastern subspecies (*T. m. nigrideus*), breeding from northern Quebec to Newfoundland, migrates primarily to the Atlantic states and provinces, south to northern Florida. The most widespread northern subspecies (*T. m. migratorius*), breeding from Alaska to the Canadian Maritimes and northeastern US, has the largest nonbreeding distribution, throughout the Great Plains and eastern US to Florida, Texas, and eastern Mexico. The widespread western subspecies (*T. m. propinquus*) moves south to the southwestern US, southern Great Plains, and northern Mexico. The Pacific Northwest subspecies (*T. m. caurinus*) spends the nonbreeding season along the Pacific coast from southern British Columbia to central California. Breeders across the south-central and southeastern US (*T. m. archusterus*) are thought to be largely resident.

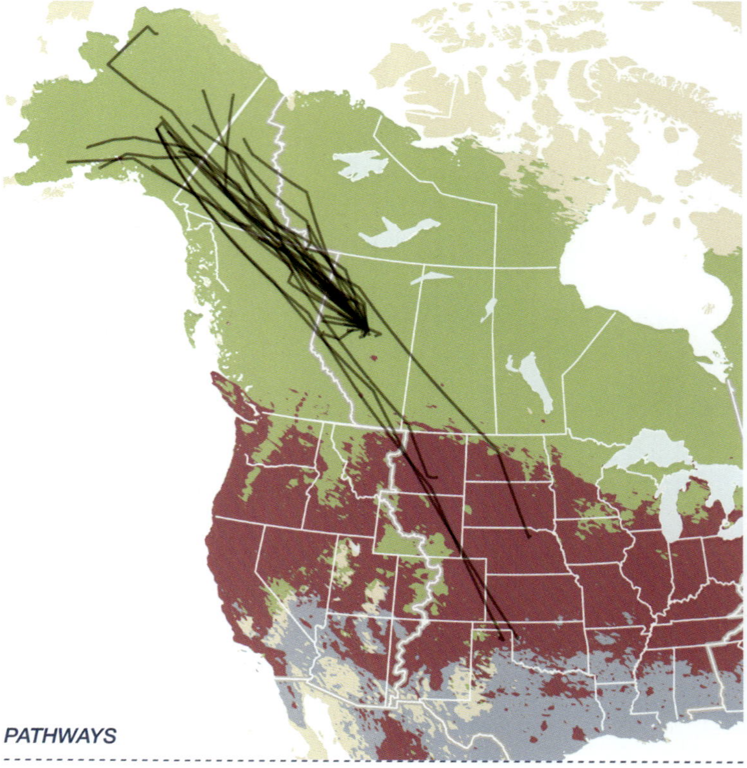

PATHWAYS

MIGRATORY PATHWAYS American robins migrate over a broad front, both diurnally and nocturnally and often in large flocks. Movements of individual populations are complex, however, and often vary with dynamic distributions of fruit availability and snow cover (Oliver et al. 2020). In general, robins migrate to areas south of breeding sites in an east-west parallel migration, except that far-northwestern breeders (Alaska) migrate east during post-breeding through Canada before turning south and return along the same route during pre-breeding (Jahn et al. 2019).

DATA CONTRIBUTORS

N. Boelman; A. Jahn; E. Ketterson; N. Krikun; S. Lerman; P.P. Marra; R.Y. Oliver; L.M. Phillips; T.B. Ryder; B. Weeks; E.J. Williams.

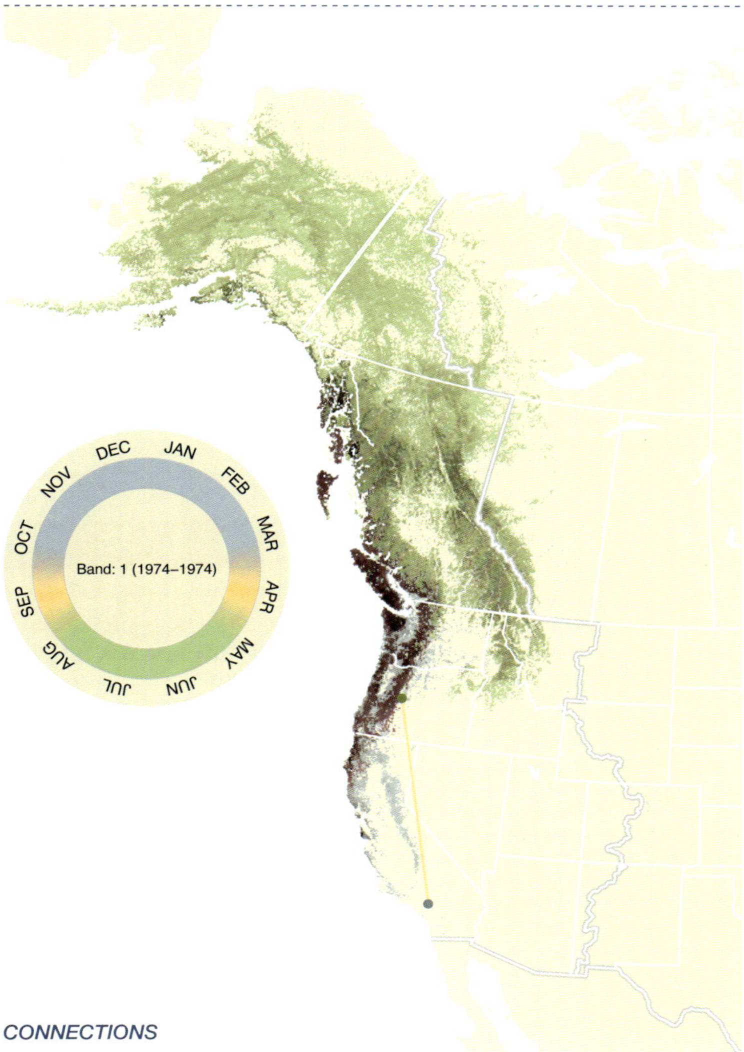

CONNECTIONS

Ixoreus naevius **VARIED THRUSH**

DISTRIBUTION AND ABUNDANCE

Varied Thrush breed in taiga and mature coniferous forests from western and northern Alaska and northern Yukon south along the Pacific coast and coastal mountains to northwestern California, and in the northern Rocky Mountains to northeastern Oregon, northern Idaho, and northwestern Montana. It occurs in the nonbreeding season along the Pacific coast from southeastern Alaska to Southern California and inland to western parts of the Great Basin and Sierra Nevada mountains. Declines have been steepest, especially in recent decades, in northern and inland populations rather than in the Pacific coastal region (Sauer et al. 2019).

ORDER Passeriformes
FAMILY Turdidae
BREEDING HABITAT Forests, Temperate Western
NONBREEDING HABITAT Forests, Temperate Western
GLOBAL POPULATION SIZE 35,000,000
POPULATION CHANGE SINCE 1970 −64%
IUCN RED LIST CATEGORY Least Concern
R2R TIPPING POINT SPECIES No
US LISTING STATUS No
CANADA SPECIES AT RISK No
PIF CCS 10/20

MIGRATORY CONNECTIVITY

OVERALL Migratory connectivity between two recognized subspecies appears strong, with evidence that northern and inland breeding populations leapfrog coastal breeders to spend the nonbreeding season farther south and east (George 2020). No data were found on finer-scale migratory connectivity.

BANDING The single breeding to nonbreeding reencounter is an individual from the eastern Cascades of Oregon that migrated to inland Southern California.

RADIO TRACKING A few tags were deployed through the Motus Wildlife Tracking System, with local detections only.

OTHER APPROACHES No tracking, genetic, or stable isotope studies of migratory connectivity were found for this species. Based on variation in plumage of females, two subspecies are recognized, and these maintain largely separate distributions throughout the annual cycle (Phillips 1991; George 2020). Northern and inland-breeding *I. n. meruloides* migrate longer distances to more southern and inland nonbreeding areas and make up most or all of the irruptive individuals across the eastern US. Coastal-breeding *I. n. naevius* move shorter distances and remain close to the Pacific coast year-round.

MIGRATORY PATHWAYS Migratory behavior is poorly known, but Pacific coastal populations are thought to migrate short distances or altitudinally, whereas northern and inland populations are known to migrate primarily through the western Great Basin to nonbreeding areas in central and Southern California (George 2020; Fink et al. 2023). There are irruptive movements every two to five years to lowland deserts of Southern California and Arizona and eastward across the northern US (Wells et al. 1996).

MUSCICAPIDAE | *OLD WORLD FLYCATCHERS*

BLUETHROAT *Luscinia svecica*

DISTRIBUTION AND ABUNDANCE

The Bluethroat breeds at northern latitudes across Europe and Asia, with North American populations occurring only in northwestern Alaska and the Yukon Territory. The nonbreeding range is primarily in the Middle East, India, and southeast Asia; Alaska breeders are presumed to migrate to southeastern China (Guzy et al. 2020).

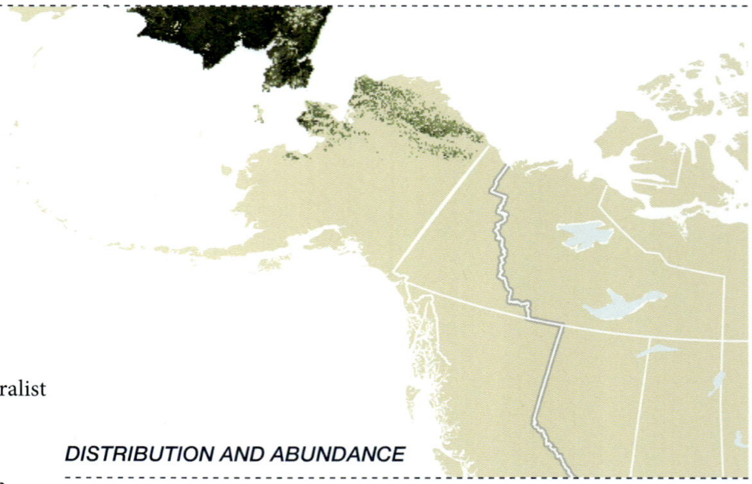

DISTRIBUTION AND ABUNDANCE

ORDER Passeriformes
FAMILY Muscicapidae
BREEDING HABITAT Tundra, Arctic
NONBREEDING HABITAT Wetlands, Generalist
GLOBAL POPULATION SIZE 50,000,000
POPULATION CHANGE SINCE 1970 NA
IUCN RED LIST CATEGORY Least Concern
R2R TIPPING POINT SPECIES No
US LISTING STATUS No
CANADA SPECIES AT RISK No
PIF CCS 8/20

INTERESTING NOTE

Band recoveries and stable isotopes of Bluethroats breeding in central and western Europe are connected to nonbreeding areas in southern Europe and northern Africa (Arizaga and Tamayo 2013; Arizaga 2015, 2016).

MIGRATORY CONNECTIVITY

OVERALL Migratory connectivity is unknown for North American populations but has been well-studied in European populations (e.g., Arizaga and Tamayo 2013; Arizaga 2015, 2016).

GEOLOCATOR TRACKING Twenty light-level geolocators were attached to Bluethroats near Nome, Alaska. One recovery revealed no information because feathers grew over the light sensor (P.P. Marra, unpublished data).

MIGRATORY PATHWAYS Although eBird data are relatively sparse during migration periods, the complete lack of records in southern Alaska and the fact that it is rare even on the northern Alaska coast suggests that migration into and out of Alaska follows an east-west path across the Bering Strait and along the foothills of the Brooks Range. Post-breeding migration is very early, with adult males departing Alaska in July and juveniles lingering only through August; the latest record at Nome is September 5 (Fink et al. 2023).

NORTHERN WHEATEAR *Oenanthe oenanthe*

DISTRIBUTION AND ABUNDANCE

Northern Wheatears have a nearly circumpolar distribution, and in North America breed on rocky Arctic and alpine tundra in two disjunct areas in far northeastern Canada and in Alaska and Yukon. They are one of a few North American passerines to migrate entirely to the Old World, spending the nonbreeding season in sub-Saharan Africa (Dunn et al. 2022).

ORDER Passeriformes
FAMILY Muscicapidae
BREEDING HABITAT Tundra, Arctic
NONBREEDING HABITAT Open Country, Habitat Mosaic
GLOBAL POPULATION SIZE 2,600,000
POPULATION CHANGE SINCE 1970 NA
IUCN RED LIST CATEGORY Least Concern
R2R TIPPING POINT SPECIES No
US LISTING STATUS No
CANADA SPECIES AT RISK No
PIF CCS 9/20

MIGRATORY CONNECTIVITY

OVERALL Migratory connectivity appears strong for this species in North America based on geolocator studies, with complete separation of migration routes and nonbreeding areas for Alaskan and eastern Canadian breeding populations. MC is not calculated due to small sample sizes.

RADIO TRACKING More than 1,500 tags were deployed through the Motus Wildlife Tracking System. There are no individual breeding to nonbreeding detections.

GEOLOCATOR TRACKING Individuals equipped with light-level geolocators in Alaska and northern Canada were confirmed to

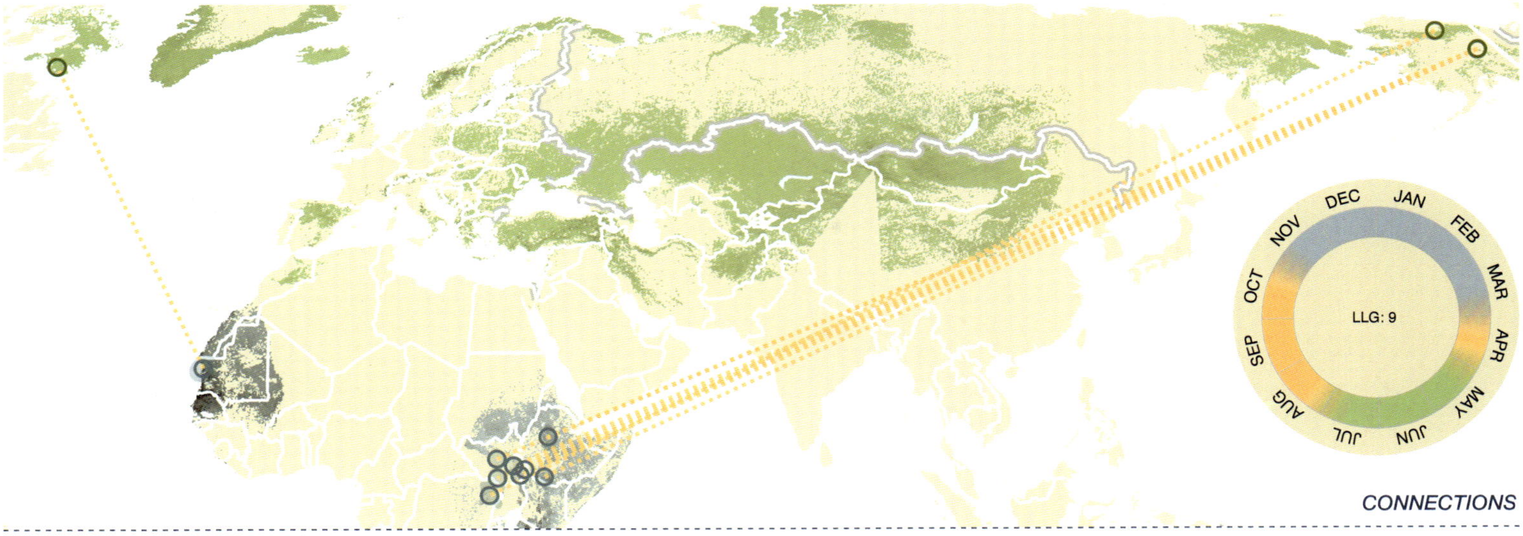

spend the nonbreeding season in sub-Saharan Africa (Bairlein et al. 2012; Schmaljohann et al. 2017). Alaskan breeders (n = 8) migrated west through Asia to reach nonbreeding grounds in eastern Africa (Schmaljohann et al. 2017), whereas one individual from eastern Canada migrated across the Atlantic Ocean to nonbreeding grounds in western Africa (Bairlein et al. 2012).

STABLE ISOTOPES Analysis of stable isotopes in feathers grown during the nonbreeding season failed to accurately identify nonbreeding areas (Rocque et al. 2006) or showed only average differences across regions assigned to Alaskan (East Africa) versus eastern Canadian (West Africa) breeders (Bairlein et al. 2012).

OTHER APPROACHES Alaskan and eastern Canadian populations represent two distinct subspecies; the former belongs to the nominate form linked to populations across Siberia and northern Europe, and the latter belongs to the *O. o. leucorhoa* subspecies aligned with breeders in Greenland and Iceland (Dunn et al. 2022). Although general differences in nonbreeding distributions were known for the two subspecies (eastern and central versus west African), specific nonbreeding areas for North American breeders were not known prior to recent tracking studies.

MIGRATORY PATHWAYS Tracks from eight breeding individuals from Alaska followed a consistent 14,500-km route in both post- and prebreeding migrations through Russia, Kazakhstan, and the Arabian Peninsula to nonbreeding sites in eastern sub-Saharan Africa (Bairlein et al. 2012; Schmaljohann et al. 2012, 2017). The single individual from eastern Canada crossed the North Atlantic Ocean on a 3,400 km, four-day journey to the British Isles, before continuing south through Europe to Mauritania on Africa's west coast (Bairlein et al. 2012). Indirect evidence suggests that some Canadian and Greenland breeders may be able to make a nonstop transoceanic flight directly to West Africa (Thorup et al. 2006).

DATA CONTRIBUTORS
F. Bairlein; H. Schmaljohann.

PEUCEDRAMIDAE | *OLIVE WARBLER*

Peucedramus taeniatus **OLIVE WARBLER**

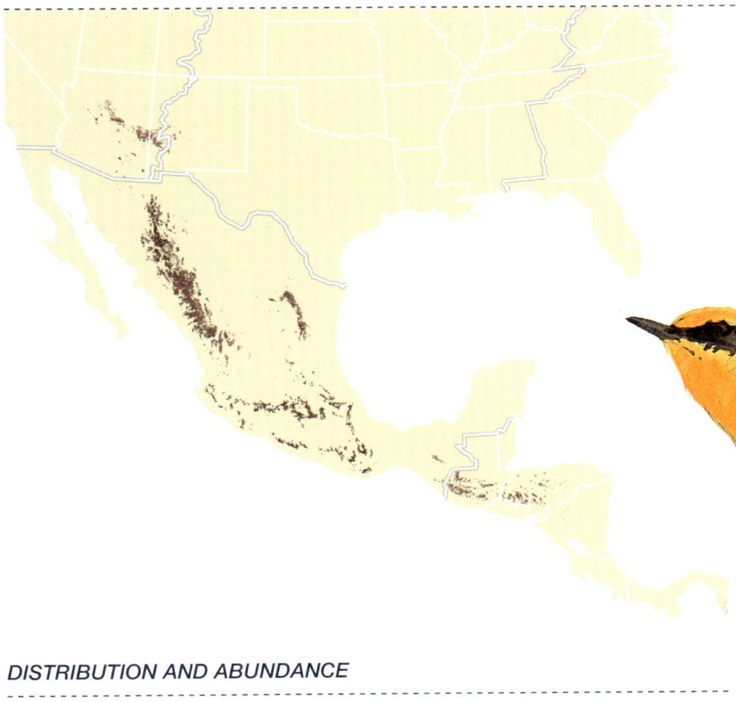

DISTRIBUTION AND ABUNDANCE
It breeds in the US only in pine forests of the Mogollon Rim and Sky Island mountains of Arizona and New Mexico; it is also resident in mountains throughout Mexico and northern Central America to northern Nicaragua (Lowther and Nocedal 2020).

ORDER Passeriformes
FAMILY Peucedramidae
BREEDING HABITAT Forests, Mesoamerican Highland
NONBREEDING HABITAT Forests, Mesoamerican Highland
GLOBAL POPULATION SIZE 1,900,000
POPULATION CHANGE SINCE 1970 NA
IUCN RED LIST CATEGORY Least Concern
R2R TIPPING POINT SPECIES No
US LISTING STATUS No
CANADA SPECIES AT RISK No
PIF CCS 13/20

MIGRATORY CONNECTIVITY

OVERALL Migratory connectivity is unknown. No information on migratory connectivity was found for this species.

OTHER APPROACHES Birds found during the nonbreeding season away from breeding areas in Nayarit, Mexico, have been identified as the northern subspecies, *P. t. arizonae* (Phillips 1991), indicating some migration south of that population's breeding range.

MIGRATORY PATHWAYS It is thought to be at least partly migratory in the US portion of the range (Lowther and Nocedal 2020), although individuals are only rarely detected away from breeding areas at lower elevations or in mountains without suitable breeding habitat (Dunn and Garrett 1997). eBird data show a clear withdrawal from most US breeding areas between October and March, but there are only a few documented observations of presumed migrants from nonbreeding sites in the southwestern US or Mexico (Fink et al. 2023).

MOTACILLIDAE | *WAGTAILS AND PIPITS*

EASTERN YELLOW WAGTAIL *Motacilla tschutschensis*

DISTRIBUTION AND ABUNDANCE

It breeds extensively in northeastern Asia, and in North America in the Arctic tundra in western and northern Alaska and along the Beaufort Sea coast of Canada east to the Mackenzie River Delta (Badyaev et al. 2020a). After breeding, they migrate to nonbreeding areas in Southeast Asia (Fink et al. 2023).

ORDER Passeriformes
FAMILY Motacillidae
BREEDING HABITAT Tundra, Arctic
NONBREEDING HABITAT Open Country, Habitat Mosaic
GLOBAL POPULATION SIZE 20,000,000
POPULATION CHANGE SINCE 1970 NA
IUCN RED LIST CATEGORY Least Concern
R2R TIPPING POINT SPECIES No
US LISTING STATUS No
CANADA SPECIES AT RISK No
PIF CCS 8/20

DISTRIBUTION AND ABUNDANCE

MIGRATORY CONNECTIVITY

OVERALL Migratory connectivity is largely unknown. No banding, tracking, genetic, or stable isotope studies of migratory connectivity were found for this species.

OTHER APPROACHES *M. t. tschutschensis*, the subspecies that breeds in western Alaska, migrates to nonbreeding grounds in southeastern China, the Philippines, and southeastern Asia from Myanmar east to Indochina and south to Indonesia (Badyaev et al. 2020a).

MIGRATORY PATHWAYS North American birds follow their ancestral migration route eastward across the Bering Strait to Russia and then south to nonbreeding grounds (Badyaev et al. 2020a).

WHITE WAGTAIL *Motacilla alba*

DISTRIBUTION AND ABUNDANCE

Breeding throughout an immense range that spans virtually all of Eurasia, the White Wagtail reaches North America only along the shorelines and river basins of western Alaska (Fink et al. 2023). The Alaska breeding subspecies *M. a. ocularis* likely spends the nonbreeding season in south to southeastern Asia (Badyaev et al. 2020b).

ORDER Passeriformes
FAMILY Motacillidae
BREEDING HABITAT Coasts, Cliffs and Islands
NONBREEDING HABITAT Open Country, Habitat Mosaic
GLOBAL POPULATION SIZE 860,000,000
POPULATION CHANGE SINCE 1970 NA
IUCN RED LIST CATEGORY Least Concern
R2R TIPPING POINT SPECIES No
US LISTING STATUS No
CANADA SPECIES AT RISK No
PIF CCS 7/20

DISTRIBUTION AND ABUNDANCE

MIGRATORY CONNECTIVITY

OVERALL Migratory connectivity is unknown. No information on migratory connectivity or pathways was found for this species.

Anthus cervinus RED-THROATED PIPIT

ORDER Passeriformes
FAMILY Motacillidae
BREEDING HABITAT Tundra, Arctic
NONBREEDING HABITAT Wetlands, Seasonally Wet Grasslands
GLOBAL POPULATION SIZE 2,000,000
POPULATION CHANGE SINCE 1970 NA
IUCN RED LIST CATEGORY Least Concern
R2R TIPPING POINT SPECIES No
US LISTING STATUS No
CANADA SPECIES AT RISK No
PIF CCS 10/20

DISTRIBUTION

MIGRATORY CONNECTIVITY

OVERALL Only a BirdLife distribution map is available; no eBird data or migratory connectivity information were found for this species.

Anthus rubescens AMERICAN PIPIT

DISTRIBUTION AND ABUNDANCE

In North America, it breeds in both Arctic tundra and high-elevation alpine tundra and meadows in Alaska, northwest and northeast Canada, and at high elevations above the timberline in the western US mountains, most densely in Colorado and western Wyoming (Fink et al. 2023). It is widespread in the nonbreeding season, often associated with shorelines and wetland edges, along the US Atlantic coast, the southeastern US, the western mountains, and central Mexico to Mexico City, most abundantly in the Central Valley of California, south Texas, and northeast Mexico, and the Gulf coast from New Orleans to Tamaulipas (Fink et al. 2023).

ORDER Passeriformes
FAMILY Motacillidae
BREEDING HABITAT Tundra, Alpine
NONBREEDING HABITAT Open Country, Developed/Disturbed
GLOBAL POPULATION SIZE 20,000,000
POPULATION CHANGE SINCE 1970 −10%
IUCN RED LIST CATEGORY Least Concern
R2R TIPPING POINT SPECIES No
US LISTING STATUS No
CANADA SPECIES AT RISK No
PIF CCS 9/20

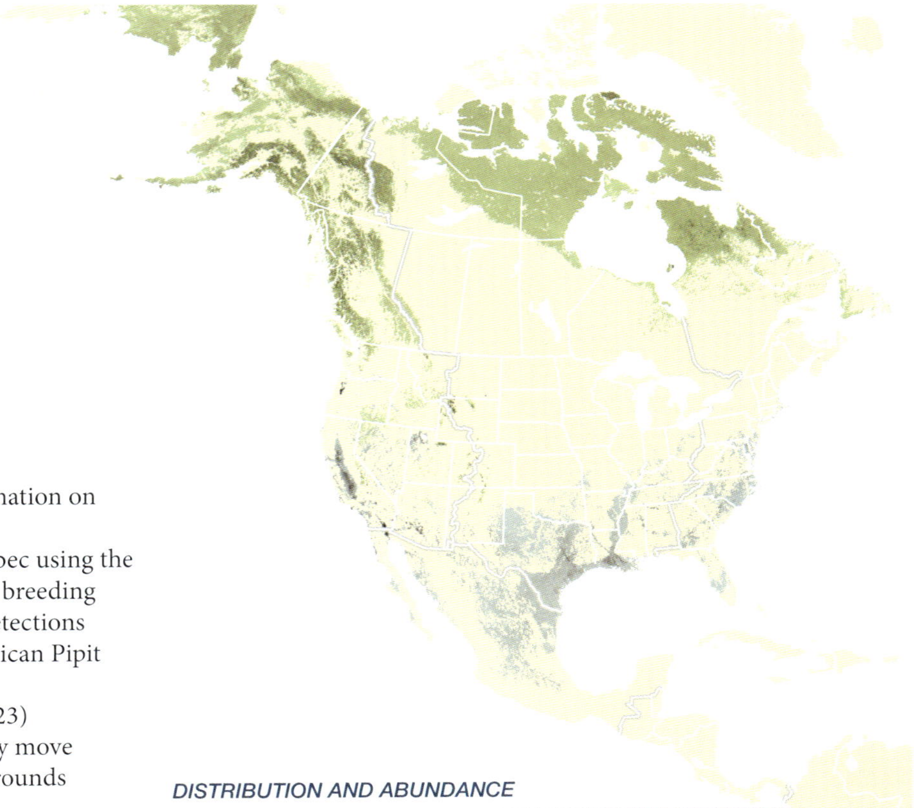

DISTRIBUTION AND ABUNDANCE

MIGRATORY CONNECTIVITY

OVERALL Migratory connectivity is unknown. No information on migratory connectivity was found for this species.

RADIO TRACKING Numerous tags were deployed in Quebec using the Motus Wildlife Tracking System. There are no individual breeding to nonbreeding locations, but there appear to be many detections during migration (Motus project, Fall migration of American Pipit and Horned Lark, unpublished data).

MIGRATORY PATHWAYS eBird animations (Fink et al. 2023) vaguely suggest that eastern and western populations may move independently across the US to and from nonbreeding grounds in the south.

SPRAGUE'S PIPIT *Anthus spragueii*

DISTRIBUTION AND ABUNDANCE

It breeds in large patches of prairie in the northern Great Plains in Alberta, Saskatchewan, Manitoba, Montana, and the Dakotas. It spends the nonbreeding season in grasslands, agricultural lands, and high desert in the south-central US and central Mexico south to Zacatecas. Thought to be most abundant along the Texas and Tamaulipas gulf coastal plains (Fink et al. 2023). Sprague's Pipit is one the most threatened of North America's grassland songbirds. Declines have been especially severe in the Prairie Pothole Region (Sauer et al. 2019).

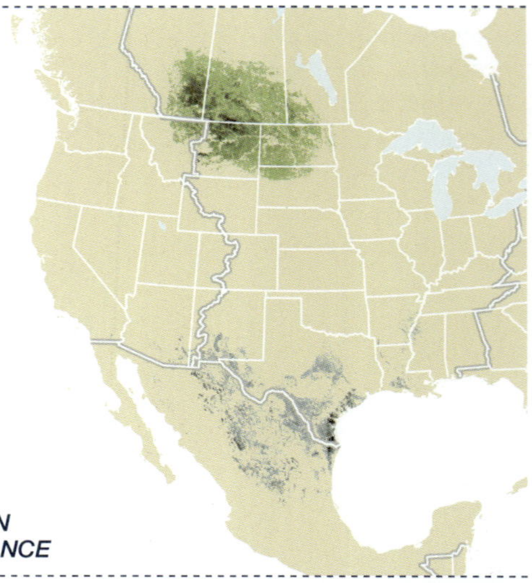

DISTRIBUTION AND ABUNDANCE

ORDER Passeriformes
FAMILY Motacillidae
BREEDING HABITAT Grasslands, Temperate
NONBREEDING HABITAT Grasslands, Chihuahuan
GLOBAL POPULATION SIZE 1,400,000
POPULATION CHANGE SINCE 1970 −77%
IUCN RED LIST CATEGORY Vulnerable
R2R TIPPING POINT SPECIES Orange-Alert
US LISTING STATUS No
CANADA SPECIES AT RISK Threatened
PIF CCS 14/20

INTERESTING NOTE

Motus data suggest individual birds move widely within the breeding range within and across years (A. Boyce, unpublished data).

MIGRATORY CONNECTIVITY

OVERALL Migratory connectivity is relatively unknown but likely to be low given that the breeding range is small and contiguous. There is one tracking study through the Motus Wildlife Tracking System; no other information on migratory connectivity was found for this species.

RADIO TRACKING Birds from Montana were tracked with Motus tags to nonbreeding areas on the gulf coast of Texas and northern Mexico (A. Boyce, unpublished data).

MIGRATORY PATHWAYS eBird animations (Fink et al. 2023) suggest a narrow migratory pathway through the central US, with an extended stopover in south-central Kansas during pre-breeding migration. The primary migratory route based on Motus data is shaped like a vertically elongated backward "C" from the northern Great Plains through the central plains of eastern Colorado, Kansas, and Nebraska to nonbreeding areas (A. Boyce, unpublished data).

FRINGILLIDAE | *FINCHES, EUPHONIAS, AND ALLIES*

EVENING GROSBEAK *Coccothraustes vespertinus*

MC = 0.59 ± 0.09*

DISTRIBUTION AND ABUNDANCE

Evening Grosbeaks have an alpine distribution on the western mountain ranges in British Columbia, Washington, Oregon, Idaho, western Montana, Colorado, and Northern California. East of the Rockies, they breed across the boreal region of Canada and the northern US. Birds are year-round residents on the breeding grounds, but some individuals from the extreme northern portion of the range abandon these areas post-breeding and move south in search of food. Post-breeding and nonbreeding irruptions of this species are common in the eastern US states and in Michigan, Pennsylvania, Virginia, Tennessee, and Indiana. In the west, nonbreeding birds are common in eastern Oregon, Nevada, southern Idaho, and Utah, central Arizona, and New Mexico (Gillihan and Byers 2020). This species is in steep decline and causes are not fully understood.

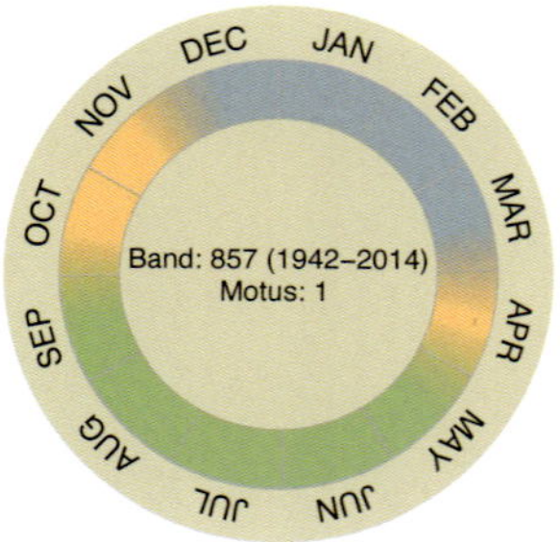

ORDER Passeriformes
FAMILY Fringillidae
BREEDING HABITAT Forests, Boreal
NONBREEDING HABITAT Forests, Temperate Generalist
GLOBAL POPULATION SIZE 3,800,000
POPULATION CHANGE SINCE 1970 −90%
IUCN RED LIST CATEGORY Vulnerable
R2R TIPPING POINT SPECIES Orange-Alert
US LISTING STATUS No
CANADA SPECIES AT RISK No
PIF CCS 14/20

MIGRATORY CONNECTIVITY

OVERALL The migratory connectivity of the Evening Grosbeak is known mostly from band reencounter data, though recent satellite tracking data may add to or clarify our knowledge of migratory connectivity for this species. Birds breeding in the Atlantic and Pacific flyways primarily spend the nonbreeding season within the breeding flyway (92% and 91%, respectively). The majority of Midcontinent Flyway breeding birds also spend the nonbreeding season within the Midcontinent Flyway (52%); however, others migrate to both the Atlantic (46%) and Pacific (3%) flyways. Despite this movement, and likely due to the high abundance of grosbeaks within the coastal flyways as compared to the midcontinent, migratory connectivity is strong (MC = 0.59 ± 0.09).

BANDING Band reencounter data show that populations that breed in the central US converge with eastern breeders in the midwestern US. Birds that breed in more northern parts of the range appear to migrate longer distances than those in the southern or year-round portions of the range.

RADIO TRACKING A single bird was detected, with the Motus Wildlife Tracking System, at its breeding location in Quebec and its nonbreeding location in Pennsylvania (L. DeGroote et al., unpublished data). There are no other breeding to nonbreeding detections.

SATELLITE TRACKING The Finch Research Network started the International Road to Recovery Evening Grosbeak Project, deploying 86 satellite tags (and 180 nanotags) from February 2022 to April 2025 across nonbreeding areas in Minnesota, Maine, Pennsylvania, and New York (Finch Research Network, unpublished data) with plans to expand to the western US and Canada. Tagged birds migrated to breeding areas in Quebec, Ontario, New Brunswick, and Manitoba (Finch Research Network, unpublished data).

MIGRATORY PATHWAYS There are three primary migration routes. Birds that breed west of the Rocky Mountains migrate south-southeast, following the orientation of inland mountain ranges. The central route also runs southeast through the boreal region to the Midwest. Eastern breeders migrate directly south or southeast, following inland mountain ranges (Brewer et al. 2000; Gillihan and Byers 2020)

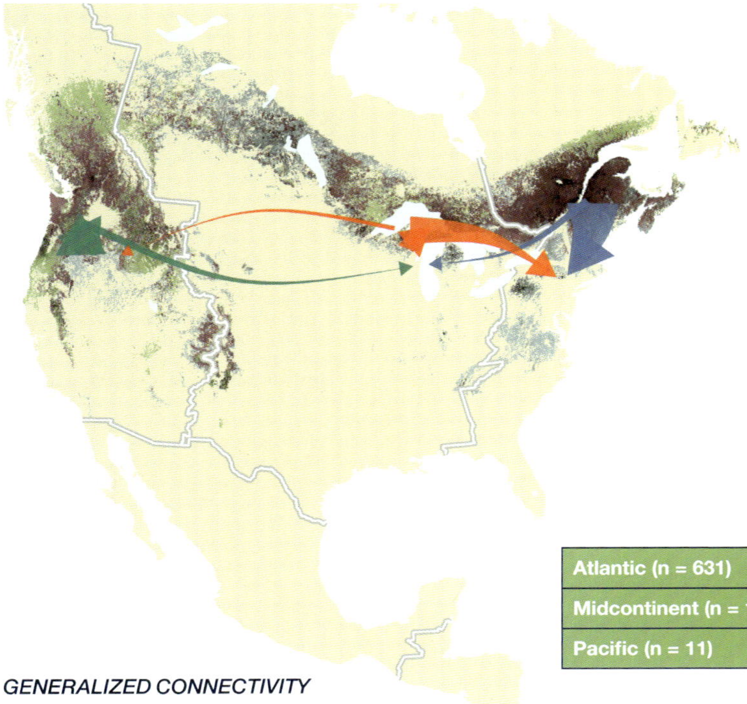

	Atlantic	Midcontinent	Pacific
Atlantic (n = 631)	92%	8%	–
Midcontinent (n = 114)	48%	52%	3%
Pacific (n = 11)	–	9%	91%

DATA CONTRIBUTORS
J. Cepek; L. DeGroote; J. Fowler; M. Gimpel; R. Jacob; L. Koitsch; T. Krynak; J. Rice; S. Sargent; M. Sarver; D. Yeany.

GENERALIZED CONNECTIVITY

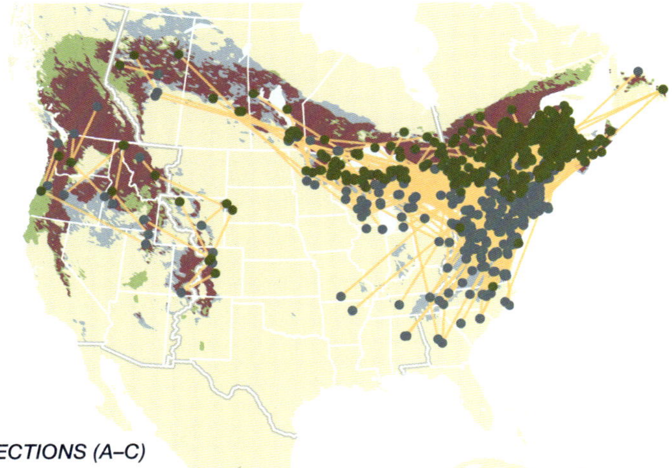

CONNECTIONS (A–C)

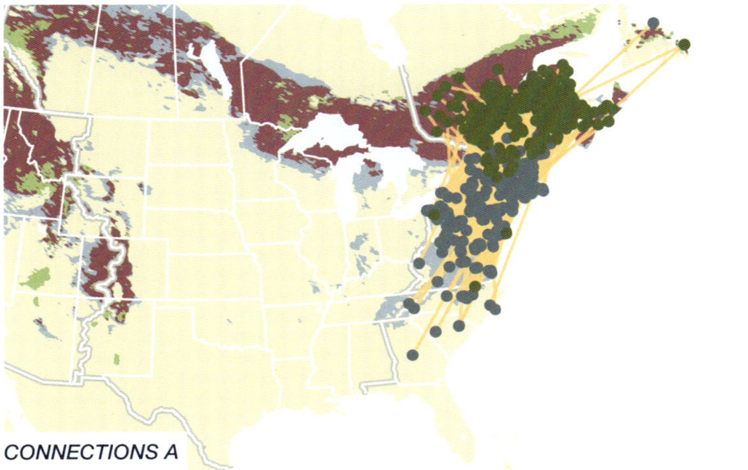

CONNECTIONS A

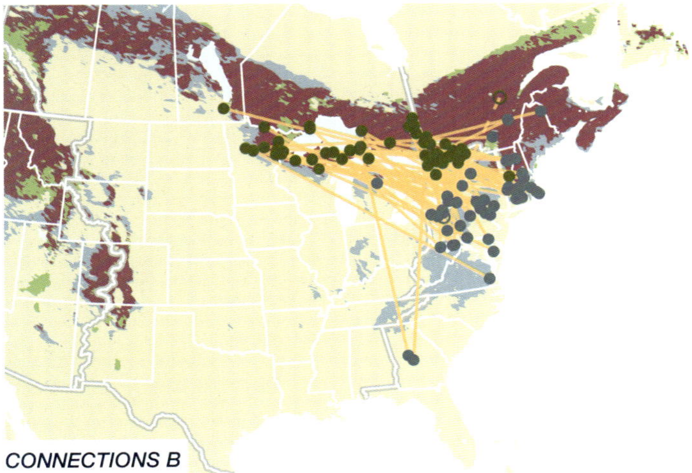

CONNECTIONS B

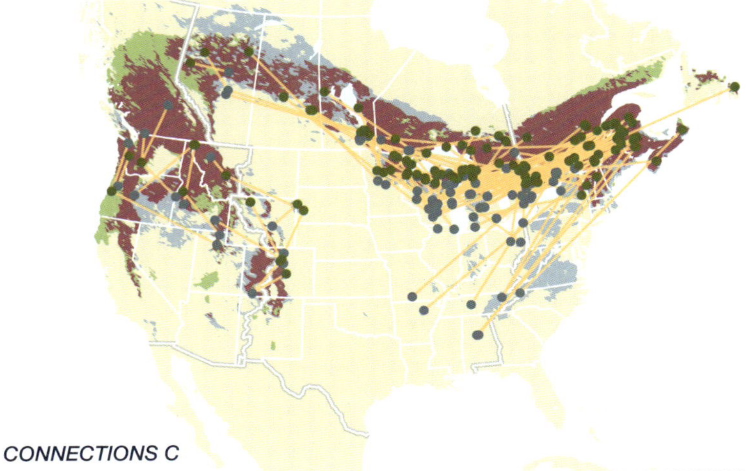

CONNECTIONS C

GRAY-CROWNED ROSY-FINCH *Leucostictecte tephrocotis*

DISTRIBUTION AND ABUNDANCE

The Gray-crowned Rosy-Finch breeds in low density throughout the alpine tundra regions of western Canada and Alaska. It spends the nonbreeding season in the western US, most densely in southern Idaho and western Wyoming, descending also to open-country habitats at lower elevations (Fink et al. 2023). Although the species is not well monitored, it is presumed to be potentially vulnerable to climate change due to its dependence on high-elevation snowfields (MacDougall-Shackleton et al. 2020).

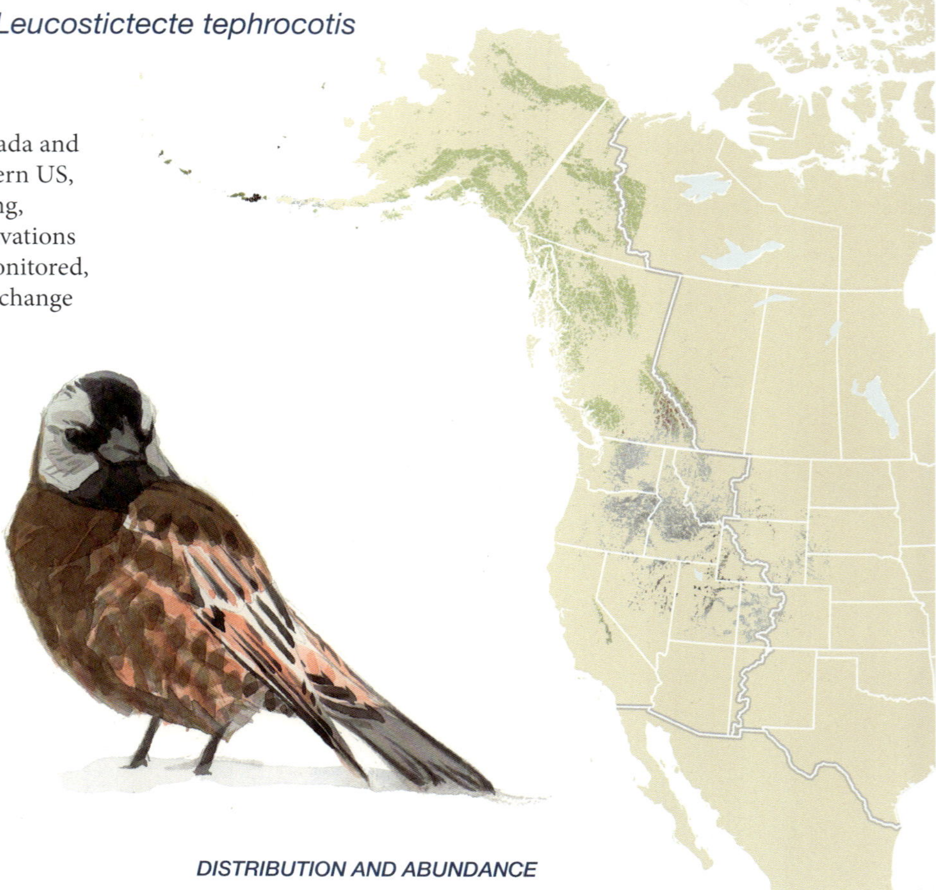

ORDER Passeriformes
FAMILY Fringillidae
BREEDING HABITAT Tundra, Alpine
NONBREEDING HABITAT Open Country, Habitat Mosaic
GLOBAL POPULATION SIZE 200,000
POPULATION CHANGE SINCE 1970 NA
IUCN RED LIST CATEGORY Least Concern
R2R TIPPING POINT SPECIES No
US LISTING STATUS No
CANADA SPECIES AT RISK No
PIF CCS 15/20

DISTRIBUTION AND ABUNDANCE

MIGRATORY CONNECTIVITY

OVERALL Migratory connectivity is unknown. No information on migratory connectivity or pathways was found for this species.

BLACK ROSY-FINCH *Leucosticte atrata*

DISTRIBUTION AND ABUNDANCE

The Black Rosy-Finch breeds on exposed rocky cliffs and alpine tundra of the northern US Rockies, most densely in southwest Montana and western Wyoming. It spends the nonbreeding season in the mountains, but depending on snow cover and storms, it tracks patchy snowfields or moves to a variety of thinly vegetated habitats at lower elevations (R.E. Johnson 2020).

ORDER Passeriformes
FAMILY Fringillidae
BREEDING HABITAT Tundra, Alpine
NONBREEDING HABITAT Tundra, Alpine
GLOBAL POPULATION SIZE 120,000
POPULATION CHANGE SINCE 1970 NA
IUCN RED LIST CATEGORY Endangered
R2R TIPPING POINT SPECIES Yellow-Alert
US LISTING STATUS No
CANADA SPECIES AT RISK No
PIF CCS 16/20

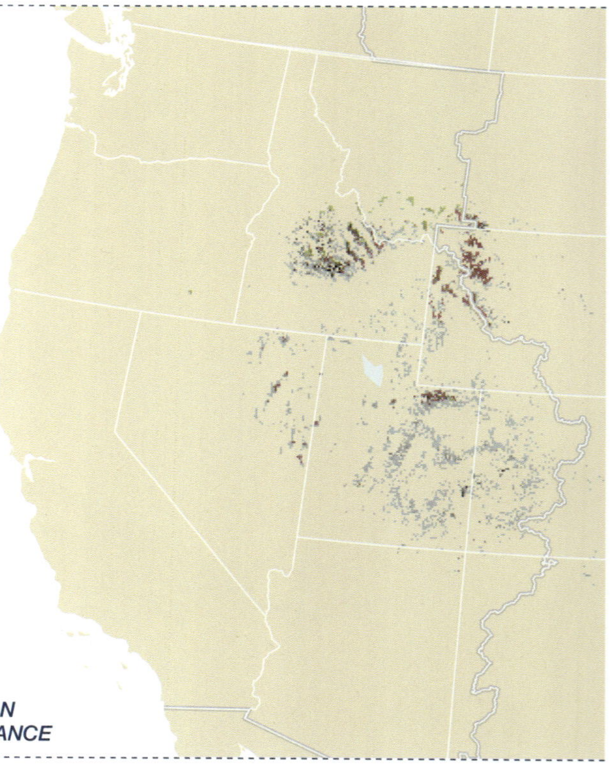

DISTRIBUTION AND ABUNDANCE

MIGRATORY CONNECTIVITY

OVERALL Migratory connectivity is unknown. No information on migratory connectivity or pathways was found for this species.

INTERESTING NOTE

Due to its dependence on high-elevation snowfields, this species is considered to be particularly vulnerable to climate change.

Leucosticte australis **BROWN-CAPPED ROSY-FINCH**

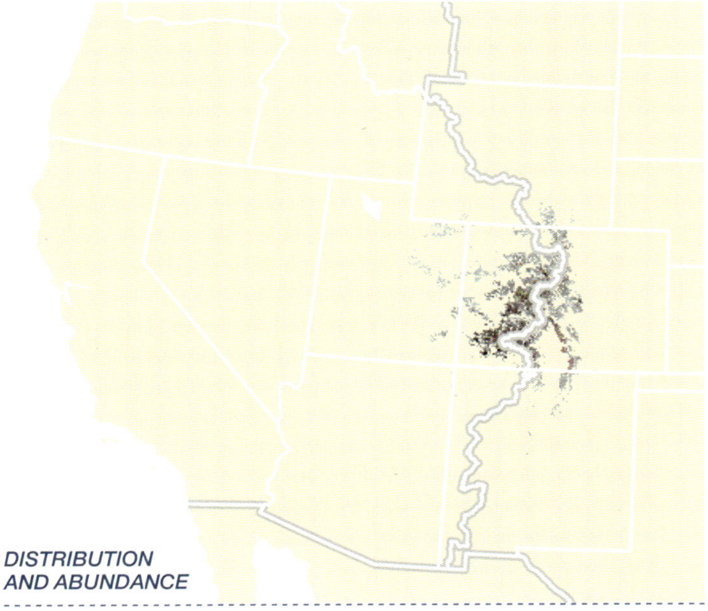

DISTRIBUTION AND ABUNDANCE

This highly specialized species has a very small range; it is found almost exclusively in Colorado, with small breeding populations in Wyoming and New Mexico, and some nonbreeding birds moving into Utah (R.E. Johnson et al. 2020; Fink et al. 2023).

ORDER Passeriformes
FAMILY Fringillidae
BREEDING HABITAT Tundra, Alpine
NONBREEDING HABITAT Tundra, Alpine
GLOBAL POPULATION SIZE 130,000
POPULATION CHANGE SINCE 1970 NA
IUCN RED LIST CATEGORY Endangered
R2R TIPPING POINT SPECIES Red-Alert
US LISTING STATUS No
CANADA SPECIES AT RISK No
PIF CCS 17/20

MIGRATORY CONNECTIVITY

OVERALL Regional movements of different populations are not well studied. This species mostly undergoes altitudinal movements seasonally and in response to harsh weather. MC is not able to be generated due to insufficient data.

GENETICS There are two genetically distinct populations identified within Colorado: the Southern Rockies and Pike's Peak. Genetic differentiation is weak, indicating potential ongoing gene flow among populations (DeSaix et al. 2022b).

OTHER APPROACHES No tracking, banding, or stable isotope studies of migratory connectivity were found for this species. A few tags were deployed through the Motus Wildlife Tracking System; there are no breeding to nonbreeding detections.

MIGRATORY PATHWAYS This species' altitudinal movements are largely in response to harsh weather in its alpine haunts. While movements are somewhat dictated by seasonality, birds will also undergo movements in response to weather at any time of the year, and will also commute to feeding areas during harsh weather to feed (R.E. Johnson et al. 2020).

Carpodacus purpureus **PURPLE FINCH**

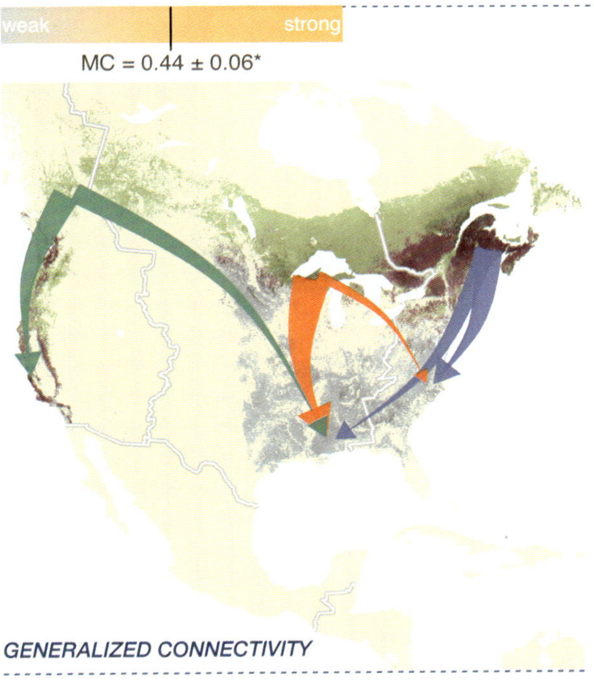

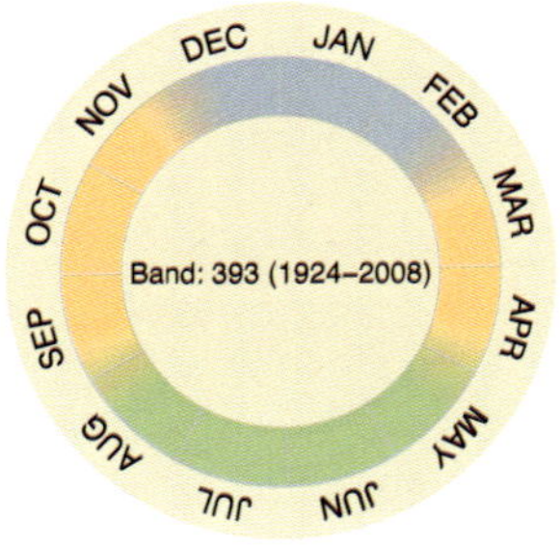

DISTRIBUTION AND ABUNDANCE

Eastern populations of the Purple Finch breed throughout the coniferous and mixed coniferous/deciduous forests of Canada and the northern US and south along the Appalachian Mountains, occurring most densely in southern Quebec and the northeastern US (Fink et al. 2023). Purple Finches occur sporadically throughout the East, from southern Canada to the Gulf Coast, during the nonbreeding season. Western populations in the coastal and Sierra Nevada ranges appear to be largely resident (Fink et al. 2023). Substantial range-wide declines since 1966 have leveled off over the last two decades, and populations have even increased since 2010, except in Maine and New Brunswick, where recent declines are evident (Sauer et al. 2019).

	Atlantic	Midcontinent	Pacific
Atlantic (n = 234)	61%	39%	–
Midcontinent (n = 105)	19%	81%	–
Pacific (n = 12)	–	42%	58%

ORDER Passeriformes
FAMILY Fringillidae
BREEDING HABITAT Forests, Boreal
NONBREEDING HABITAT Forests, Temperate Eastern
GLOBAL POPULATION SIZE 6,500,000
POPULATION CHANGE SINCE 1970 −45%
IUCN RED LIST CATEGORY Least Concern
R2R TIPPING POINT SPECIES No
US LISTING STATUS No
CANADA SPECIES AT RISK No
PIF CCS 8/20

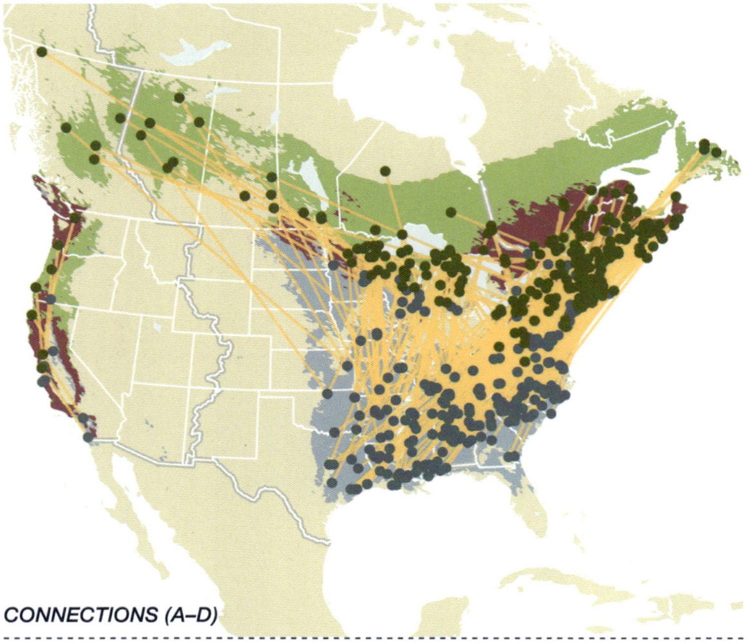

CONNECTIONS (A–D)

MIGRATORY CONNECTIVITY

OVERALL Migratory connectivity information available for Purple Finches comes from band reencounter data. Birds breeding in the Atlantic Flyway spend the nonbreeding season within the Atlantic (61%) and Midcontinent (39%) flyways. Birds breeding in the Midcontinent Flyway predominantly migrate within the flyway (81%), while birds breeding in the Pacific Flyway migrate within the flyway (58%) and the Midcontinent Flyway (42%). Despite the mixing of breeding populations on nonbreeding areas in the Midcontinent Flyway, the high transitions and abundance of finches within the coastal flyways as compared to the midcontinent results in strong migratory connectivity (MC = 0.44 ± 0.06).

BANDING Reencounters of banded birds indicate that individuals breeding in the US and Vancouver Pacific coastal mountains move

INTERESTING NOTE

Stedman (1989) banded 1,470 Purple Finches over five winters in Tennessee and found that 1.63% of the individuals returned to the same sites the following winter.

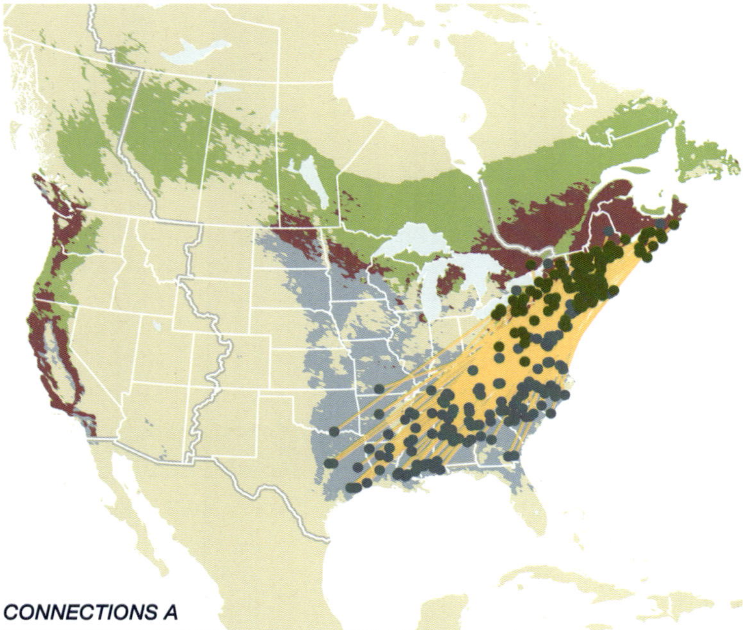

CONNECTIONS A

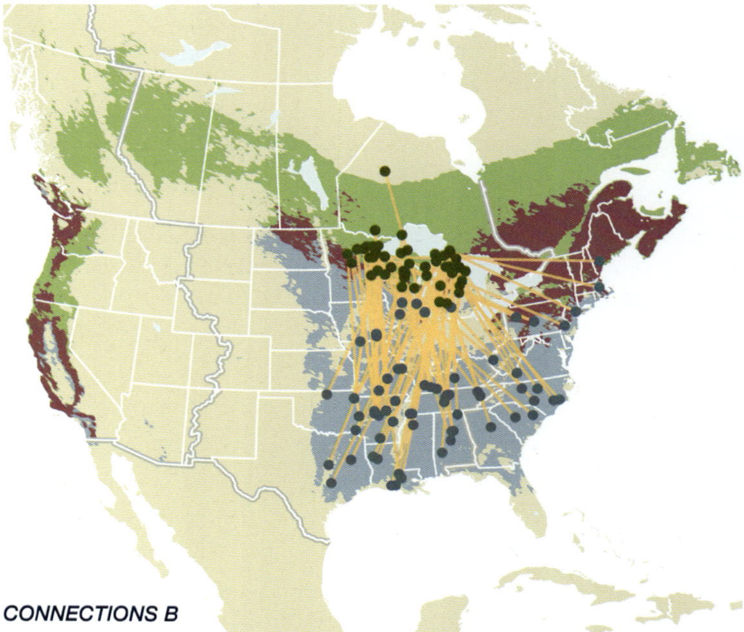

CONNECTIONS B

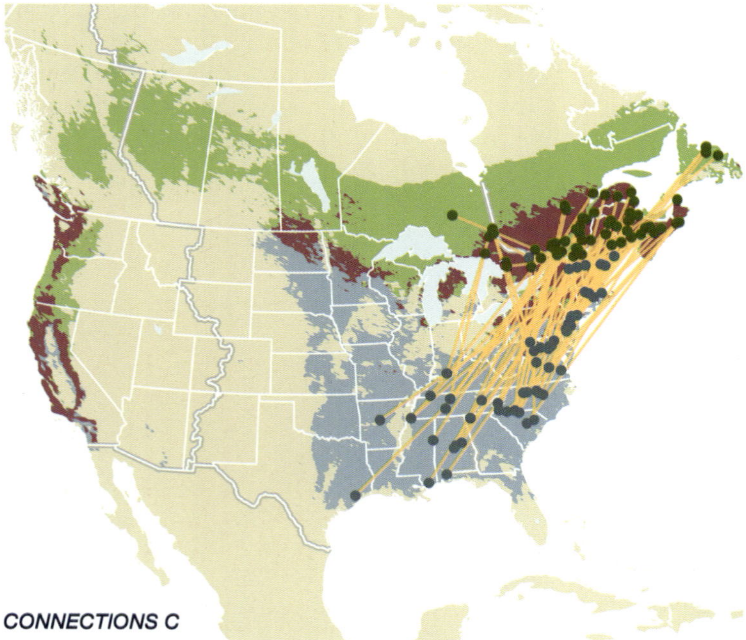

CONNECTIONS C

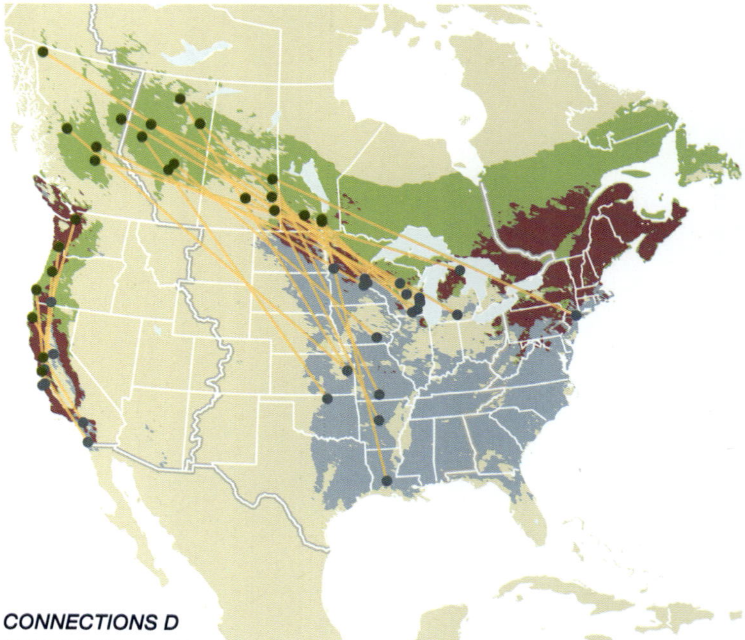

CONNECTIONS D

farther south into California in the nonbreeding season, while birds breeding in the interior coniferous forests of Canada, including those in British Columbia and Alberta, spend the nonbreeding season primarily in the eastern US.
OTHER APPROACHES No tracking, genetic, or stable isotope studies of migratory connectivity were found for this species.
MIGRATORY PATHWAYS It is generally considered an irruptive species, with nonbreeding movements dependent on annual variation of northern conifer cone production (Wootton 1996). Overall, populations of Purple Finches breeding in the Canadian Maritimes and New England appear to move south along the Atlantic coast to nonbreeding grounds in the southeastern US, while breeding individuals from the Midwest and western Canada move south to the US Gulf Coast. Purple Finches breeding in the western US migrate south to California and Arizona or to lower elevations (Wilson and Brown 2012).

Haemorhous cassinii **CASSIN'S FINCH**

DISTRIBUTION AND ABUNDANCE

Cassin's Finches breed at elevations of 1,000–3,000 m in western interior mountain ranges from southern British Columbia and extreme southwestern Alberta, running south through the eastern slope of the Rocky Mountains up to northern New Mexico, southern Utah, Nevada, and eastern California. Migration is not well understood, but there are year-round residents and elevational and latitudinal migrants throughout the range (Hahn 2020).

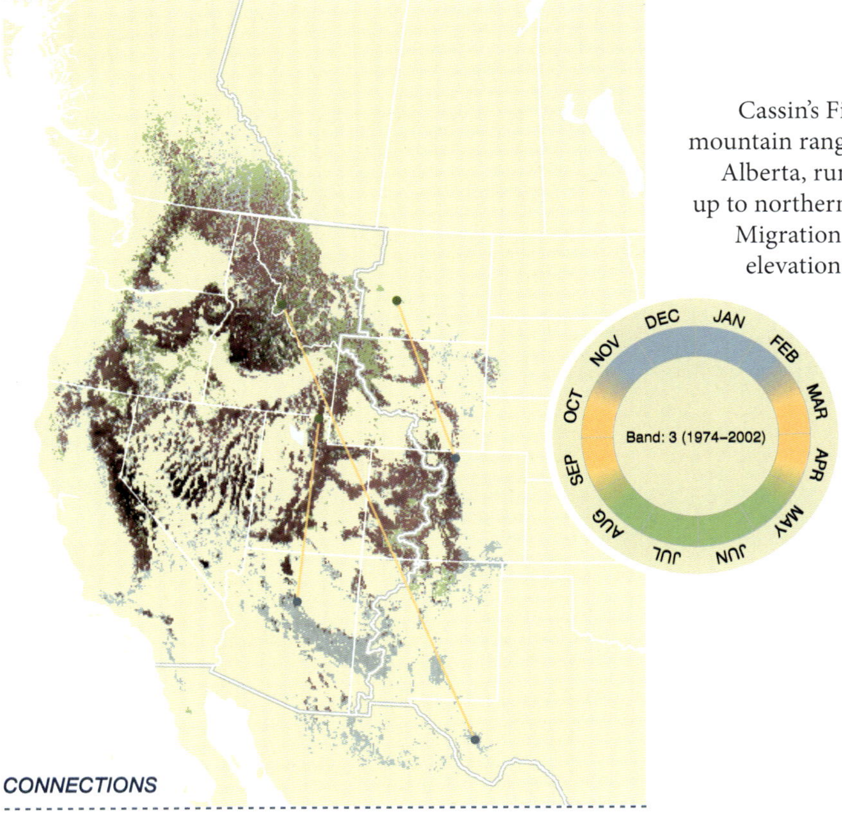

CONNECTIONS

MIGRATORY CONNECTIVITY

OVERALL Migratory connectivity is unknown. No information on migratory connectivity was found for this species.
BANDING A handful of band reencounters show that one bird banded in Montana spent the nonbreeding season in Colorado, and another banded in Idaho spent the nonbreeding season in northern Mexico (Chihuahua).
OTHER APPROACHES No stable isotope, genetic, or tracking studies of migratory connectivity were found for this species.
MIGRATORY PATHWAYS Parthways are not well understood, but latitudinal as well as altitudinal migration seem to follow the north-south orientation of the western mountain ranges (Hahn 2020).

ORDER Passeriformes
FAMILY Fringillidae
BREEDING HABITAT Forests, Temperate Western
NONBREEDING HABITAT Forests, Temperate Western
GLOBAL POPULATION SIZE 3,200,000
POPULATION CHANGE SINCE 1970 −61%
IUCN RED LIST CATEGORY Least Concern
R2R TIPPING POINT SPECIES No
US LISTING STATUS No
CANADA SPECIES AT RISK No
PIF CCS 12/20

REDPOLL *Acanthis flammea*

Until recently, the Redpoll was widely treated as three species: Common Redpoll (*Acanthis flammea*), Lesser Redpoll (*A. cabaret*), and Hoary Redpoll (*A. hornemanni*), but due to lack of genetic differentiation, the three were grouped into a single species (Knox et al. 2024). However, Redpoll taxonomy remains unsettled (Knox et al. 2024). Here we summarize migratory connectivity information for the Common Redpoll and the Hoary Redpoll.

— COMMON REDPOLL *A. f. flammea*

DISTRIBUTION AND ABUNDANCE

The Common Redpoll is a partial migrant that breeds in the boreal and taiga regions of both North America and Europe. Some populations in southern Alaska and Yukon and Northwest Territories appear to be year-round residents. The nonbreeding range covers the entire east-west gradient above approximately 46 degrees north latitude (Fink et al. 2023). The extent of southern distribution in the nonbreeding season is often correlated to a lack of seed production of birch trees in the northern portions of the range. When this happens, there are records of Redpoll irruptions into southern Canada and the US (Troy 1983; Knox et al. 2024). Distributional information is somewhat limited because of low eBird coverage.

ORDER Passeriformes
FAMILY Fringillidae
BREEDING HABITAT Tundra, Arctic
NONBREEDING HABITAT Open Country, Habitat Mosaic
GLOBAL POPULATION SIZE 250,000,000
POPULATION CHANGE SINCE 1970 −44%
IUCN RED LIST CATEGORY Least Concern
R2R TIPPING POINT SPECIES No
US LISTING STATUS No
CANADA SPECIES AT RISK No
PIF CCS 9/20

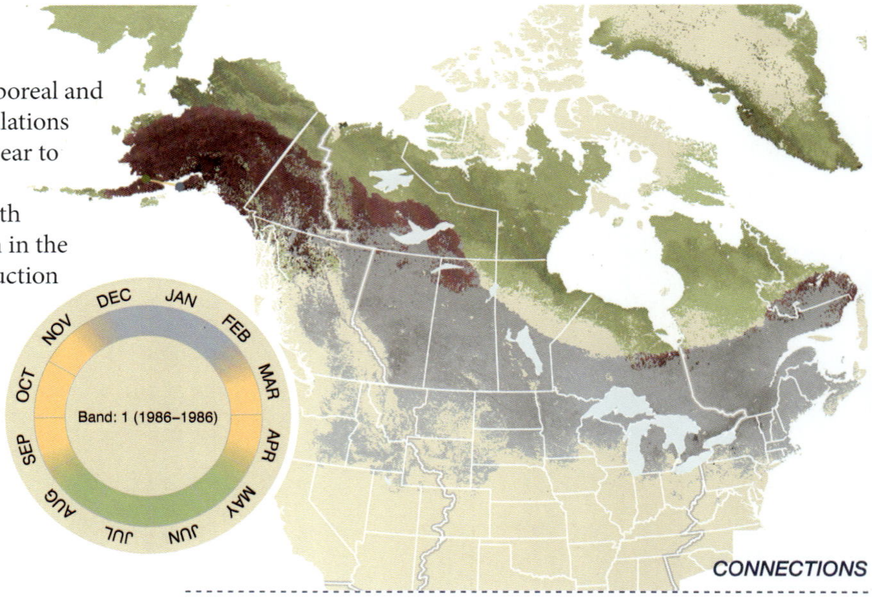

MIGRATORY CONNECTIVITY

OVERALL Migratory connectivity is unknown. There is one band reencounter, but no other information on migratory connectivity was found for this species.
BANDING There was one band reencounter in Alaska.
MIGRATORY PATHWAYS Specific migratory pathways are unknown; it appears to migrate across a broad front throughout the range (Fink et al. 2023).

— HOARY REDPOLL *A. f. hornemanni*

DISTRIBUTION AND ABUNDANCE

The Hoary Redpoll breeds across the high Arctic of North America, most densely in northern and western Alaska, in open forest and scrub and where shrubs extend into the tundra. It spends the nonbreeding season farther south in open boreal forests and shrubby field habitats of Alaska and Canada, just barely reaching into US northeastern states in some nonbreeding seasons (Knox et al. 2024).

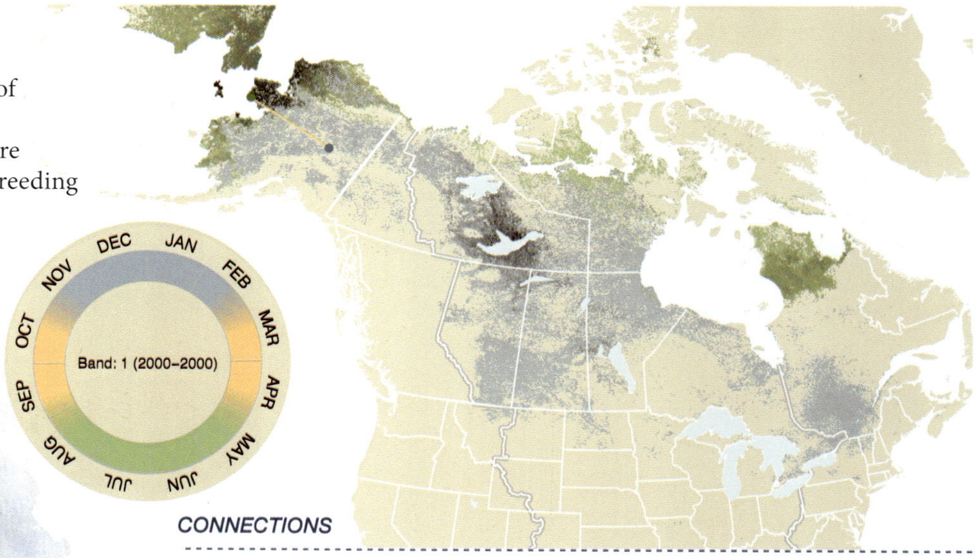

ORDER Passeriformes
FAMILY Fringillidae
BREEDING HABITAT Tundra, Arctic
NONBREEDING HABITAT Open Country, Habitat Mosaic
GLOBAL POPULATION SIZE 28,000,000
POPULATION CHANGE SINCE 1970 +10%
IUCN RED LIST CATEGORY Least Concern
R2R TIPPING POINT SPECIES No
US LISTING STATUS No
CANADA SPECIES AT RISK No
PIF CCS 9/20

MIGRATORY CONNECTIVITY

OVERALL Migratory connectivity is unknown. There is one band reencounter, but no other information on migratory connectivity was found for this species.

BANDING There is one breeding to nonbreeding reencounter from coastal Alaska in the breeding season to interior Alaska in the nonbreeding season.

MIGRATORY PATHWAYS Specific migratory pathways are unknown; it appears to migrate across a broad front throughout the range (Fink et al. 2023).

Loxia curvirostra RED CROSSBILL

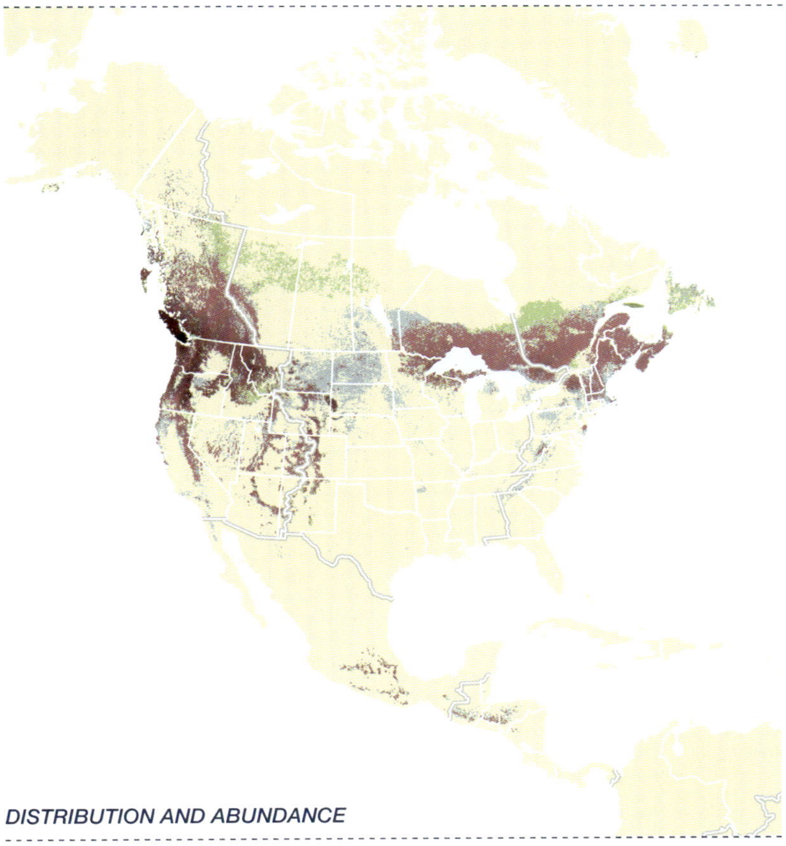

DISTRIBUTION AND ABUNDANCE

DISTRIBUTION AND ABUNDANCE

The Red Crossbill has a wide and complex distribution in both hemispheres, tied to the availability of northern and montane conifer cone seeds (Benkman and Young 2020). In North America, it occurs most densely in western Canada and the US Pacific Northwest. Since the species is nomadic and can breed opportunistically throughout the year where and when cone crops are abundant, its breeding and nonbreeding distributions are difficult to disentangle, although many populations periodically move farther south in the nonbreeding season (Benkman and Young 2020).

ORDER Passeriformes
FAMILY Fringillidae
BREEDING HABITAT Forests, Temperate Generalist
NONBREEDING HABITAT Forests, Temperate Generalist
GLOBAL POPULATION SIZE 26,000,000
POPULATION CHANGE SINCE 1970 +4%
IUCN RED LIST CATEGORY Least Concern
R2R TIPPING POINT SPECIES No
US LISTING STATUS No
CANADA SPECIES AT RISK Threatened *percna* subspecies
PIF CCS 9/20

MIGRATORY CONNECTIVITY

OVERALL Migratory connectivity is unknown. No information on migratory connectivity was found for North American populations.

MIGRATORY PATHWAYS There is no information on specific routes. It migrates or "wanders" throughout the range in search of seed crops (Benkman and Young 2020), with high abundance in the US Pacific Northwest and southwestern British Columbia, Canada (Fink et al. 2023).

INTERESTING NOTE

North American populations of the Red Crossbill are known to differentiate into at least ten subpopulations designated as "call types" or ecotypes, each with characteristic contact vocalizations, morphological variations, and preference for distinct conifer species. Each of the call types may migrate separately between distinct locations, irruptively or periodically, depending on cone crops of the conifer species on which each specializes. For example, 'Type 3' crossbills appear to move from the Pacific Northwest following western hemlock cone crop failures eastward across the continent to eastern hemlock forests on a two- to 5-year cycle (Benkman and Young 2020).

WHITE-WINGED CROSSBILL *Loxia leucoptera*

DISTRIBUTION AND ABUNDANCE

The White-winged Crossbill occurs throughout the boreal forests of Alaska, Canada, and the Northern Rockies, most densely in southern Quebec and southwest Alaska, including the spruce-fir-larch forests north of the distribution of the Red Crossbill. Like its congener, it potentially nests every month of the year, depending on the phenology and geography of conifer cone ripening (Benkman 2020). It is more likely to occur in the northern central and eastern US states during the nonbreeding season (Fink et al. 2023).

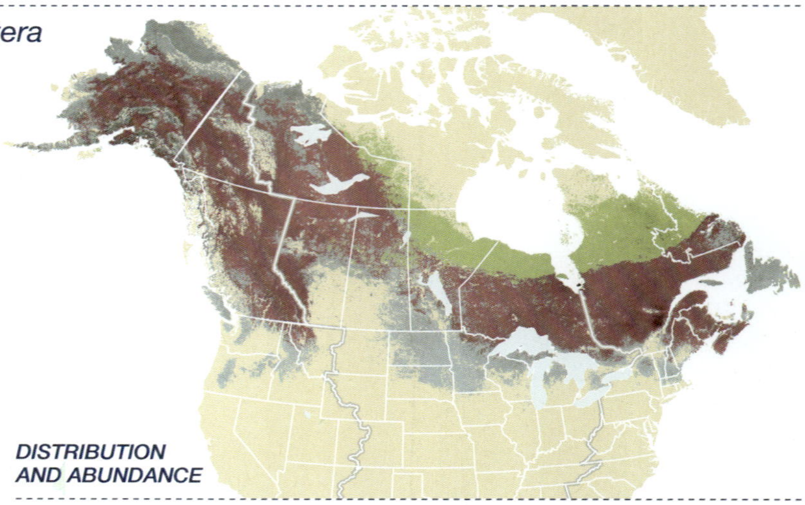

ORDER Passeriformes
FAMILY Fringillidae
BREEDING HABITAT Forests, Boreal
NONBREEDING HABITAT Forests, Boreal
GLOBAL POPULATION SIZE 79,000,000
POPULATION CHANGE SINCE 1970 +201%
IUCN RED LIST CATEGORY Least Concern
R2R TIPPING POINT SPECIES No
US LISTING STATUS No
CANADA SPECIES AT RISK No
PIF CCS 6/20

MIGRATORY CONNECTIVITY

OVERALL Migratory connectivity is unknown. No information on migratory connectivity was found for this species.
MIGRATORY PATHWAYS There is no information on specific routes. It migrates or "wanders" throughout the range in search of conifer cone crops (Benkman 2020). High abundance was observed in Quebec during pre-breeding migration and in both eastern and western portions of the range during post-breeding migration (Fink et al. 2023).

PINE SISKIN *Spinus spinus*

DISTRIBUTION AND ABUNDANCE

Pine Siskins breed across most of Canada from Newfoundland and Labrador west to the Yukon and into Alaska. In the continental US, they breed in the western United States, especially in the Rocky Mountain states, south into New Mexico, and in isolated areas of Mexico. They have an irruptive behavior, following abundance of cone crop and seed mast, which form the bulk of their diet. Because of the species' erratic movements, abundance varies greatly in a single location from year to year. They may breed throughout their range, but northernmost populations in Canada and the northwestern US appear to make regular migrations south, while southern populations appear to be more sedentary (Fink et al. 2023; Dawson 2020).

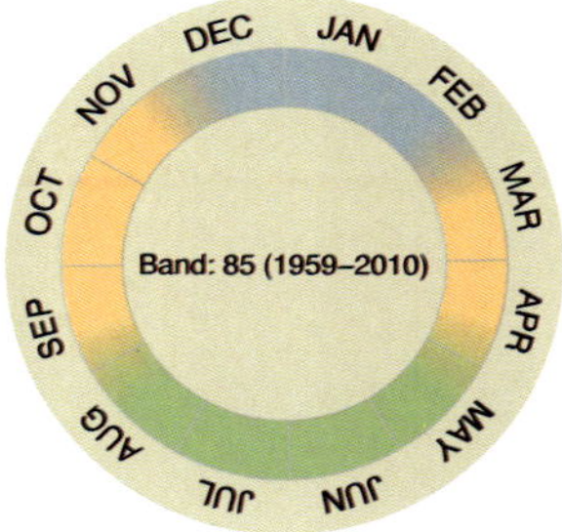

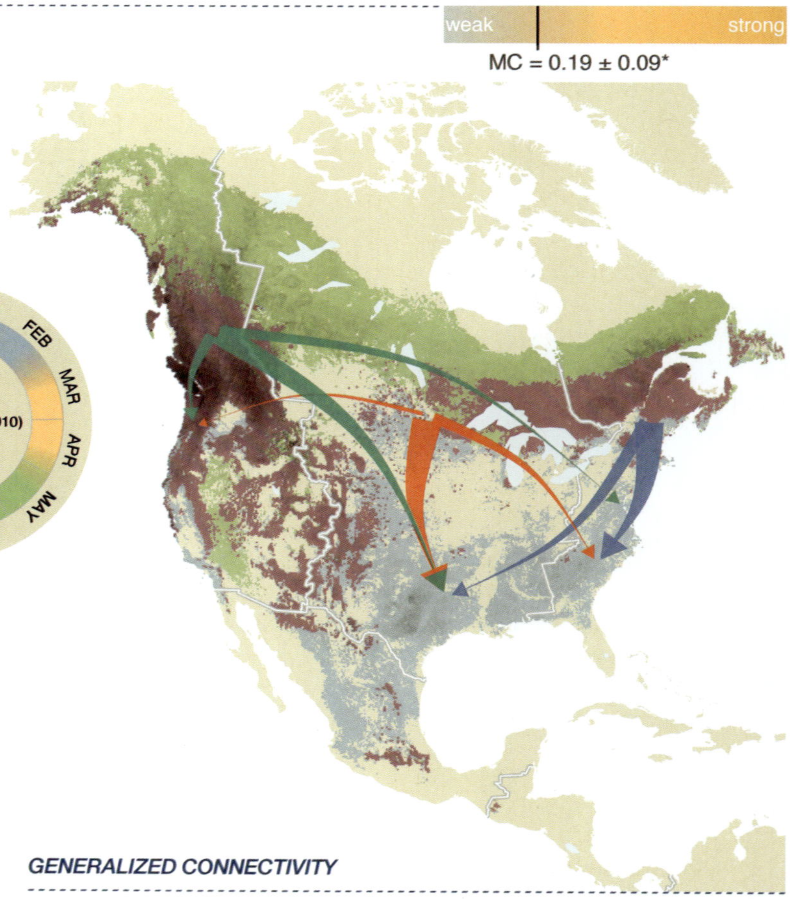

ORDER Passeriformes
FAMILY Fringillidae
BREEDING HABITAT Forests, Boreal
NONBREEDING HABITAT Forests, Temperate Generalist
GLOBAL POPULATION SIZE 46,000,000
POPULATION CHANGE SINCE 1970 −80%
IUCN RED LIST CATEGORY Least Concern
R2R TIPPING POINT SPECIES No
US LISTING STATUS No
CANADA SPECIES AT RISK No
PIF CCS 10/20

	Atlantic	Midcontinent	Pacific
Atlantic (n = 13)	13%	31%	–
Midcontinent (n = 47)	23%	70%	6%
Pacific (n = 23)	22%	52%	26%

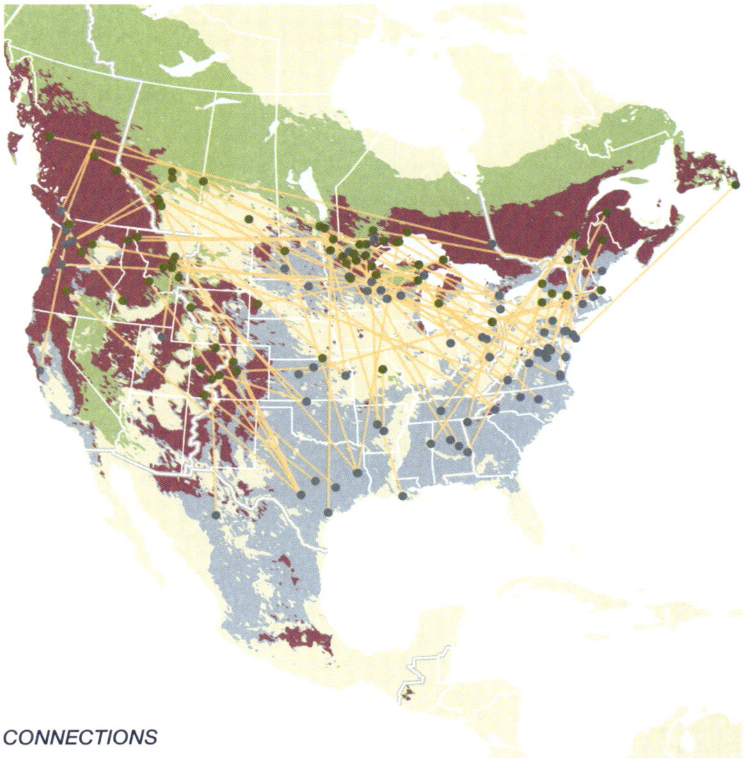

CONNECTIONS

MIGRATORY CONNECTIVITY

OVERALL The migratory connectivity of the Pine Siskin is known from band reencounter data, which suggest moderate migratory connectivity (MC = 0.19 ± 0.09). Birds breeding in the Atlantic Flyway migrate to nonbreeding areas within the flyway (69%) or to the Midcontinent Flyway (31%). Midcontinent Flyway breeding birds also primarily spend the nonbreeding period within the same flyway (70%), with some birds migrating to the Atlantic (23%) and Pacific (6%) flyways. Birds breeding in the Pacific Flyway largely migrate out of the flyway to the Midcontinent (52%) and Atlantic (22%) flyways.
BANDING Band reencounter data show a mixture of wandering movements, which can cover large distances, and regular migratory movements (Dawson 2020).
MIGRATORY PATHWAYS There are two known migration pathways. The western route mainly moves in a north-south direction; however, some western breeders also migrate southeast to Mexico or even east to the northeastern states of the US. The central route also runs north-south, and sometimes in a more southwest direction.

Spinus psaltria LESSER GOLDFINCH

DISTRIBUTION AND ABUNDANCE

Lesser Goldfinches have a broad distribution in the western United States, with population strongholds in California, Arizona, New Mexico, western Texas, and Colorado (Fink et al. 2023). Outside the US, the species breeds throughout Mexico, Central America, and northern South America from Venezuela to Peru. Birds are year-round residents on their breeding grounds, except on the northernmost (northwest US) and possibly the southernmost areas of their range (Peru and southern Ecuador; Fink et al. 2023).

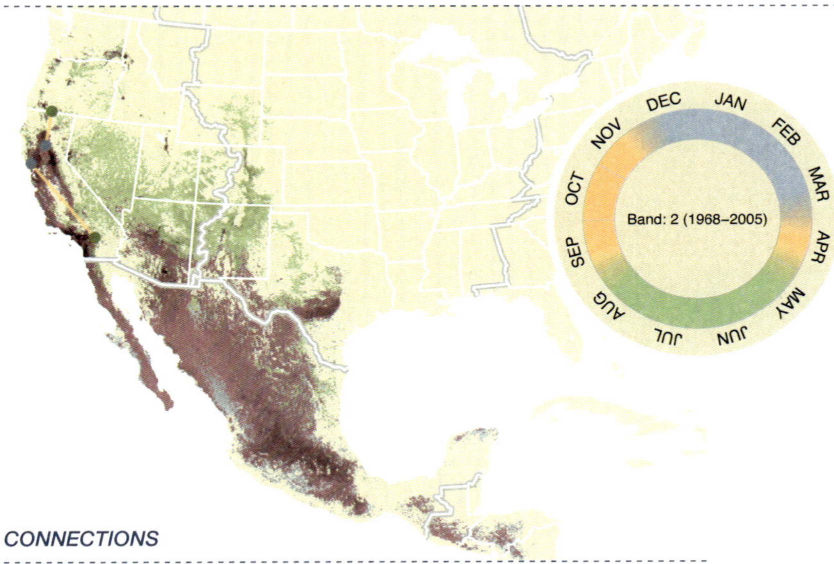

CONNECTIONS

MIGRATORY CONNECTIVITY

OVERALL The migratory connectivity of the Lesser Goldfinch is poorly known. Birds are known to leave the most northern extremes of the breeding range, but there is not enough data to understand where exactly they go. In addition, nomadic movements during the nonbreeding season make it difficult to understand their migratory connectivity patterns (Watt and Willoughby 2020).
BANDING One banded bird moved 260 km from Oregon to California (Watt and Willoughby 2020). Another migrated north from Southern to Northern California.
MIGRATORY PATHWAYS There is limited knowledge of migratory pathways. Birds are observed throughout the range during both pre- and post-breeding migration. In pre-breeding, there is high abundance north and south of the Central Valley in California into Baja California, Mexico, and through Arizona into the northern Sonoran Desert, and in post-breeding, in south-central Texas and Mexico (Fink et al. 2023).

ORDER Passeriformes
FAMILY Fringillidae
BREEDING HABITAT Forests, Temperate Western
NONBREEDING HABITAT Open Country, Habitat Mosaic
GLOBAL POPULATION SIZE 15,000,000
POPULATION CHANGE SINCE 1970 +57%
IUCN RED LIST CATEGORY Least Concern
R2R TIPPING POINT SPECIES No
US LISTING STATUS No
CANADA SPECIES AT RISK No
PIF CCS 8/20

LAWRENCE'S GOLDFINCH *Spinus lawrencei*

DISTRIBUTION AND ABUNDANCE

Lawrence's Goldfinch breeds in California and northern Baja California in arid and open woodlands near brush, annual weed fields, and water sources (Watt et al. 2020). It is largely resident within this range, although populations north of Kern County, California, generally migrate. Nonbreeding areas include Southern California, Arizona, southwestern New Mexico, Baja California, Sonora, and northwestern Chihuahua (Fink et al. 2023). Southern populations may be resident or irregular migrants depending on water and seed crops (Watt et al. 2020).

ORDER Passeriformes
FAMILY Fringillidae
BREEDING HABITAT Arid Lands, Chaparral
NONBREEDING HABITAT Arid Lands, Desert Scrub
GLOBAL POPULATION SIZE 380,000
POPULATION CHANGE SINCE 1970 −29%
IUCN RED LIST CATEGORY Least Concern
R2R TIPPING POINT SPECIES No
US LISTING STATUS No
CANADA SPECIES AT RISK No
PIF CCS 14/20

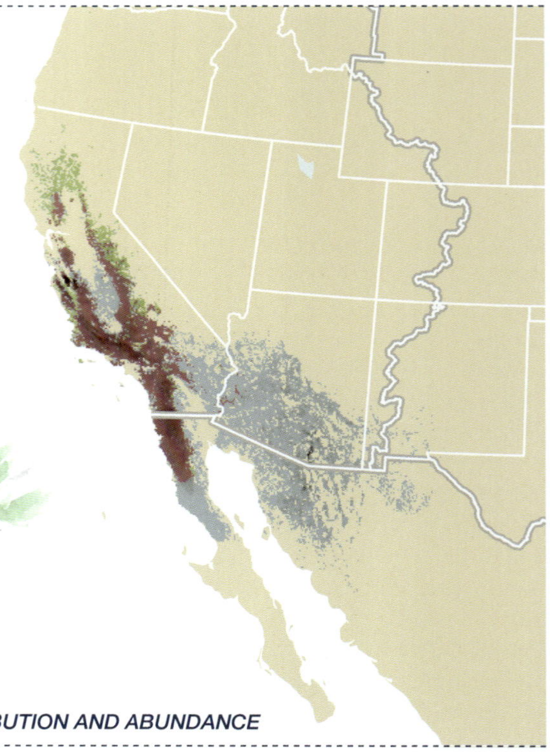

DISTRIBUTION AND ABUNDANCE

MIGRATORY CONNECTIVITY

OVERALL Migratory connectivity is unknown. No information on migratory connectivity was found for this species.
MIGRATORY PATHWAYS Seasonal movements are complex and vary annually. Movement may be nomadic or irruptive, and often is longitudinal rather than latitudinal. Irruptions were recorded throughout migration from Arizona and west to nonbreeding areas (Watt et al. 2020).

AMERICAN GOLDFINCH *Spinus tristis*

MC = 0.36 ± 0.07*

DISTRIBUTION AND ABUNDANCE

American Goldfinch distribution covers the entirety of the US east of the Rocky Mountains, as well as northern breeding strongholds in eastern Alberta, Saskatchewan, and Manitoba; the species is less abundant in Ontario and Quebec. West of the Rocky Mountains, there are breeding populations in Washington, Oregon, Idaho, Nevada, and California. Northern populations (in Canada and the upper midwestern US) are strictly migratory, and southernmost populations (North Carolina through Texas, northern Mexico, Arizona, and Southern California) are only nonbreeding populations. Populations in between are year-round residents (McGraw and Middleton 2020; Fink et al. 2023).

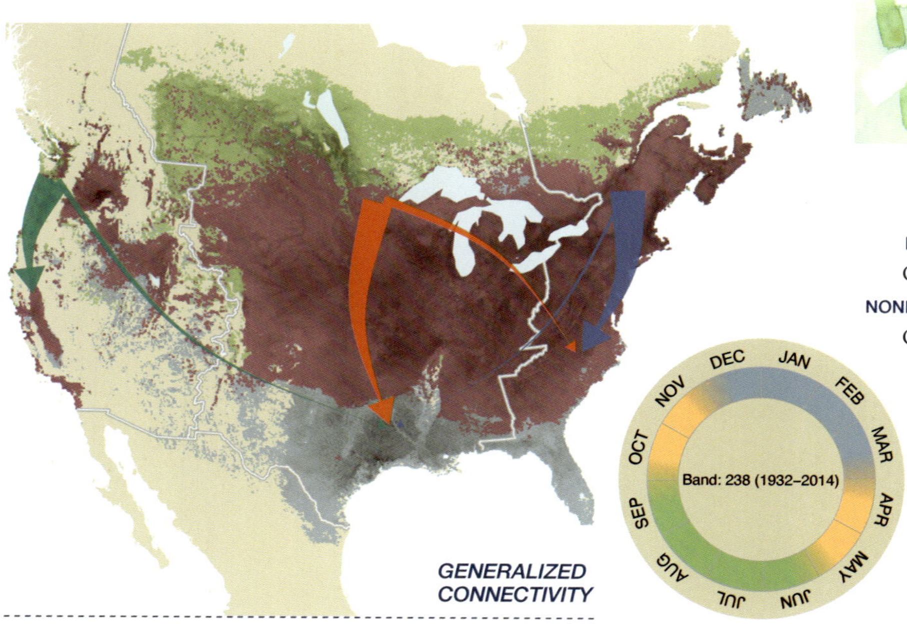

GENERALIZED CONNECTIVITY

ORDER Passeriformes
FAMILY Fringillidae
BREEDING HABITAT Open Country, Habitat Mosaic
NONBREEDING HABITAT Open Country, Habitat Mosaic
GLOBAL POPULATION SIZE 44,000,000
POPULATION CHANGE SINCE 1970 +3%
IUCN RED LIST CATEGORY Least Concern
R2R TIPPING POINT SPECIES No
US LISTING STATUS No
CANADA SPECIES AT RISK No
PIF CCS 8/20

	Atlantic	Midcontinent	Pacific
Atlantic (n = 99)	96%	4%	–
Midcontinent (n = 118)	19%	81%	–
Pacific (n = 16)	–	12%	88%

MIGRATORY CONNECTIVITY

OVERALL The migratory connectivity of the American Goldfinch is known from band reencounter data. Birds breeding in the Atlantic, Midcontinent, and Pacific flyways primarily spend the nonbreeding season within the same flyway as breeding; however, all three breeding populations overlap in high abundance in the Midcontinent Flyway, resulting in moderate migratory connectivity (MC = 0.36 ± 0.07).

BANDING Band reencounter data suggest overlap in nonbreeding areas, especially for the eastern and central populations, while the most northwestern breeders (Vancouver, Washington, and Oregon) spend the nonbreeding season in California.

OTHER APPROACHES No stable isotope, genetic, or tracking studies of migratory connectivity were found for this species.

MIGRATORY PATHWAYS The migration routes of American Goldfinches from eastern populations parallel the Atlantic coast and the Appalachian Mountains. Those from central populations move through the Mississippi River Valley in a north-south direction, while the westerly populations seem to orient southeast to reach their nonbreeding grounds. Migratory behavior varies by age, sex, and latitude (McGraw and Middleton 2020).

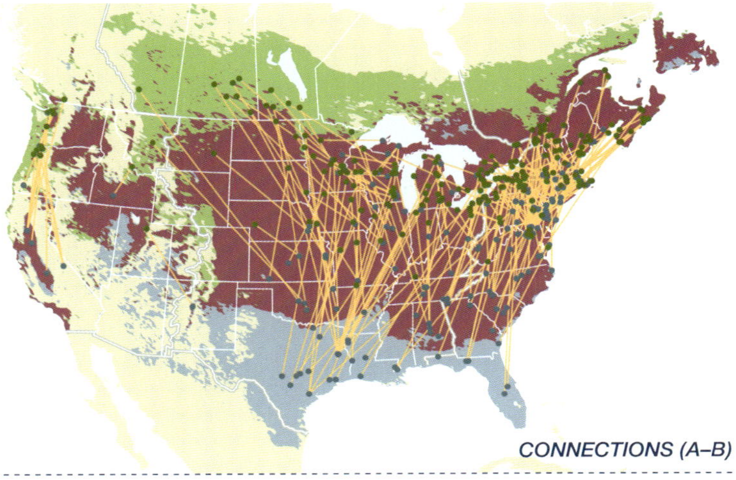

CONNECTIONS (A–B)

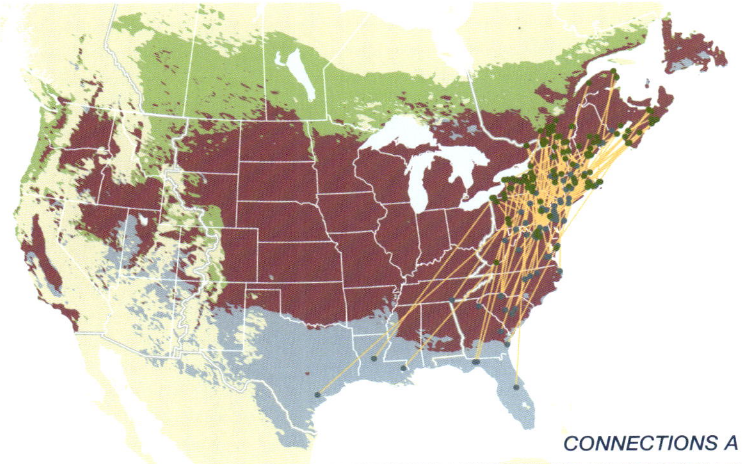

CONNECTIONS A

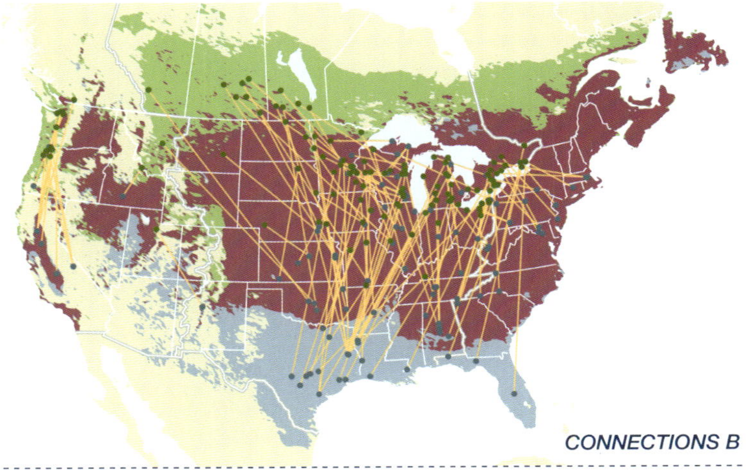

CONNECTIONS B

CALCARIIDAE | *LONGSPURS AND SNOW BUNTINGS*

Calcarius lapponicus LAPLAND LONGSPUR

DISTRIBUTION AND ABUNDANCE

This circumpolar Arctic breeder occurs in North America from western and northern Alaska across Arctic Canada to Greenland. Although it is broadly distributed across the US in the nonbreeding season, major centers of abundance are in the central Great Plains (Kansas, eastern Colorado, western Oklahoma) and in deforested agricultural regions of the upper midwestern states and the central Mississippi Valley (Hussell and Montgomerie 2020; Fink et al. 2023).

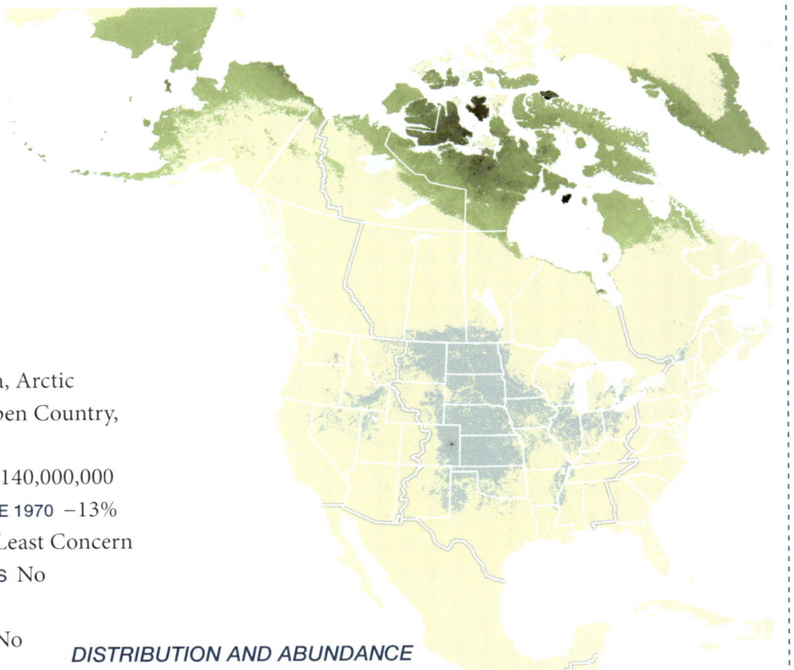

DISTRIBUTION AND ABUNDANCE

ORDER Passeriformes
FAMILY Calcariidae
BREEDING HABITAT Tundra, Arctic
NONBREEDING HABITAT Open Country, Developed/Disturbed
GLOBAL POPULATION SIZE 140,000,000
POPULATION CHANGE SINCE 1970 −13%
IUCN RED LIST CATEGORY Least Concern
R2R TIPPING POINT SPECIES No
US LISTING STATUS No
CANADA SPECIES AT RISK No
PIF CCS 10/20

MIGRATORY CONNECTIVITY

OVERALL Migratory connectivity is unknown. There is no information on migratory connectivity, other than broad separation of eastern and western subspecies within North America.

BANDING Banding data indicate that breeders from western and southeastern Greenland migrate to the eastern and central US (Hussell and Montgomerie 2020).

GEOLOCATOR TRACKING In 2024, a limited sample of barometric pressure geolocators were deployed on Lapland Longspurs in Utqiaġvik, Alaska (A. Scarpignato and A-L. Harrison, unpublished data). Recoveries have yet to be conducted.

OTHER APPROACHES Analysis of subspecies specimens indicates a broad separation of eastern and western breeders in North America, with birds from eastern Canada and Greenland migrating to the northeastern and north-central US, and birds from Alaska and northwestern Canada migrating to the western states; the two subspecies overlap in the core of the nonbreeding range in the Great Plains (American Ornithologists Union 1957; Hussell and Montgomerie 2020).

MIGRATORY PATHWAYS Migrations are highly concentrated in the Canadian Prairie provinces and the Dakotas, although smaller numbers migrate throughout the northern US and Canada. Detailed observations, including by Inuit residents, indicate that abundant breeders in Arctic Alaska arrive during pre-breeding migration through inland passes of the Brooks Range rather than from Arctic Yukon or from Asia (Irving 1961). Farther south, birds enter Alaska via three distinct routes: an early migration along the southeastern Alaska coast, a later route westward through the northern prairies, and a smaller passage through mountain passes of northern British Columbia and Yukon (West et al. 1968).

CHESTNUT-COLLARED LONGSPUR *Calcarius ornatus*

DISTRIBUTION AND ABUNDANCE

It breeds in the northern shortgrass prairies from southern Alberta, southern Saskatchewan, and southwestern Manitoba south to northeastern Colorado and western Nebraska. Nonbreeding distribution is in the southern shortgrass prairie from southeastern Colorado and southwestern Kansas south to desert grasslands of southeastern Arizona, southern New Mexico, western Texas, and northern Mexico (Fink et al. 2023). Declines are fairly uniform across the breeding range.

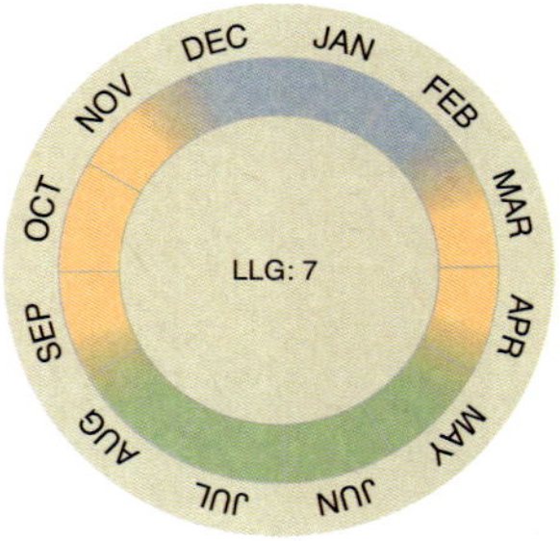

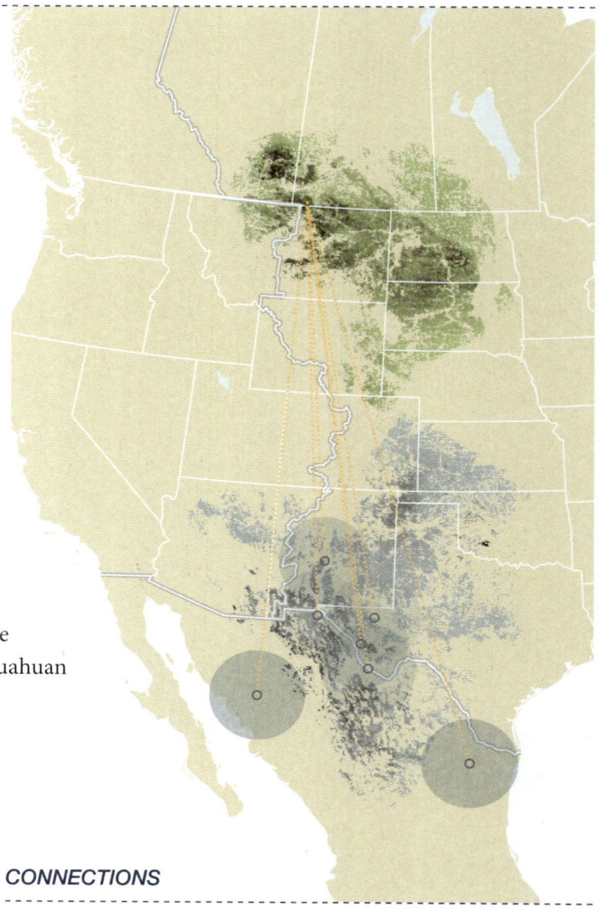

ORDER Passeriformes
FAMILY Calcariidae
BREEDING HABITAT Grasslands, Temperate
NONBREEDING HABITAT Grasslands, Chihuahuan
GLOBAL POPULATION SIZE 3,100,000
POPULATION CHANGE SINCE 1970 −88%
IUCN RED LIST CATEGORY Vulnerable
R2R TIPPING POINT SPECIES Red-Alert
US LISTING STATUS No
CANADA SPECIES AT RISK Endangered
PIF CCS 16/20

MIGRATORY CONNECTIVITY

OVERALL Migratory connectivity appears to be weak, based on a small sample of geolocator-tracked birds from a single site. MC is not calculated due to insufficient data.

RADIO TRACKING Several studies exist within the Motus Wildlife Tracking System. There were no detections between breeding and nonbreeding areas.

GEOLOCATOR TRACKING A small sample of birds tracked from southwestern Saskatchewan using light-level geolocators (n = 7) indicated low connectivity, with nonbreeding sites up to 1,200 km apart in southern Colorado, New Mexico, western Texas, and northern Mexico (Ellison et al. 2017).

OTHER APPROACHES No banding, stable isotope, or genetic studies of migratory connectivity were found for this species. The Bird Genoscape Project is currently working to build a genoscape for the Chestnut-collared Longspur (https://www.birdgenoscape.org/chestnutcollared-longspur/).

MIGRATORY PATHWAYS Based on geolocator-tracked individuals from southwestern Saskatchewan (n = 7), migration is eastward, then south through the western Great Plains in post-breeding; in pre-breeding migration, individuals retrace their path northward and then west to breeding sites (Ellison et al. 2017). Tracked birds used three to five stopover sites in post-breeding migration and two to four sites in pre-breeding migration. Flocks congregate around water sources during migration stopover (Bleho et al. 2020).

DISTRIBUTION AND ABUNDANCE

Calcarius pictus SMITH'S LONGSPUR

DISTRIBUTION AND ABUNDANCE

It breeds in a narrow zone near the tree line through the subarctic from Hudson Bay westward to north-central Alaska. Also there is a small disjunct population in the mountains of east-central Alaska and northwestern British Columbia, and possibly another in the mountains of central Yukon and western Northwest Territories. Nonbreeding range is in a small area of the southern Great Plains from southeastern Kansas through central Oklahoma to northeastern Texas and northwestern Louisiana. There are also small numbers in the central Mississippi Valley in eastern Arkansas and northwestern Mississippi (Fink et al. 2023).

ORDER Passeriformes
FAMILY Calcariidae
BREEDING HABITAT Tundra, Arctic
NONBREEDING HABITAT Grasslands, Temperate
GLOBAL POPULATION SIZE 75,000
POPULATION CHANGE SINCE 1970 +26%
IUCN RED LIST CATEGORY Least Concern
R2R TIPPING POINT SPECIES No
US LISTING STATUS No
CANADA SPECIES AT RISK No
PIF CCS 12/20

MIGRATORY CONNECTIVITY

OVERALL Migratory connectivity is unknown. No information on migratory connectivity was found for this species.

MIGRATORY PATHWAYS It exhibits an unusual counterclockwise, loop migration with little or no overlap in pre- and post-breeding routes. Post-breeding migration is through the central Great Plains in Manitoba, eastern North and South Dakota, eastern Nebraska, western Minnesota, western Iowa, and western Missouri. In pre-breeding, birds apparently migrate eastward to stage for several weeks in eastern Iowa, Illinois, Indiana, and western Ohio before moving rapidly north to breeding areas (Fink et al. 2023).

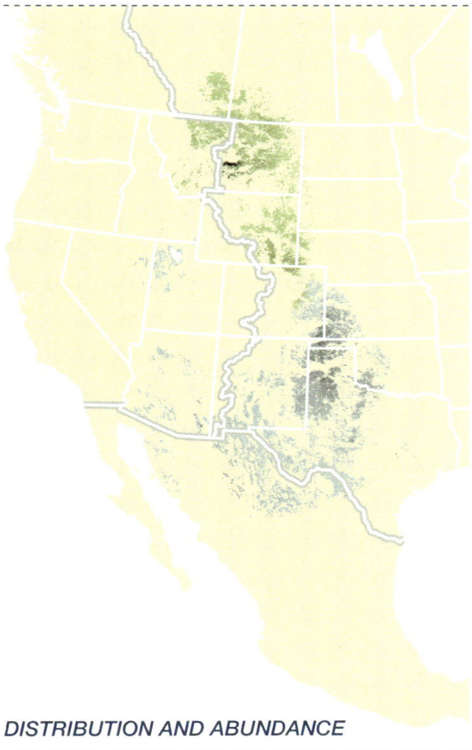

DISTRIBUTION AND ABUNDANCE

Rynchophanes mccownii THICK-BILLED LONGSPUR

DISTRIBUTION AND ABUNDANCE

Its breeding distribution is currently restricted to two disjunct areas in the northern shortgrass prairie region from southeastern Alberta and southwestern Saskatchewan south through eastern Montana and in southeastern Wyoming, western Nebraska, and northeastern Colorado; it was formerly more widespread. Its nonbreeding distribution is in the southern plains and arid grasslands from southeastern Colorado and southwestern Kansas through western Texas, southern and eastern New Mexico, and southeastern Arizona to the northern Central Mexican Plateau in Chihuahua and Coahuila. Steepest declines have been in the Canadian provinces, with recent increases noted in Montana and Wyoming (Fink et al. 2023).

ORDER Passeriformes
FAMILY Calcariidae
BREEDING HABITAT Grasslands, Temperate
NONBREEDING HABITAT Grasslands, Chihuahuan
GLOBAL POPULATION SIZE 840,000
POPULATION CHANGE SINCE 1970 −85%
IUCN RED LIST CATEGORY Least Concern
R2R TIPPING POINT SPECIES Red-Alert
US LISTING STATUS No
CANADA SPECIES AT RISK Threatened
PIF CCS 16/20

MIGRATORY CONNECTIVITY

OVERALL Migratory connectivity is unknown. No information on migratory connectivity was found for this species.

RADIO TRACKING In 2022, tags deployed on Thick-billed Longspurs at breeding grounds in northeastern Montana using the Motus Wildlife Tracking System resulted in no detections outside of the breeding areas (Motus project, Thick-billed Longspur migration, unpublished data).

MIGRATORY PATHWAYS It migrates through the western Great Plains in pre- and post-breeding seasons, with the largest numbers through eastern Colorado and western Nebraska; it regularly strays to other western states. It often migrates in flocks, especially in the pre-breeding season (With 2021).

SNOW BUNTING *Plectrophenax nivalis*

DISTRIBUTION AND ABUNDANCE

It breeds across the high Arctic tundra in Greenland, Canada, and Alaska, with smaller numbers in alpine tundra in interior Alaska and Yukon, and islands in the Bering Sea (where it is resident); it also breeds across the Palearctic in Iceland, northern Europe, and Siberia. Its nonbreeding range in North America is concentrated in the northern Great Plains and extends through open and agricultural habitats to Atlantic coastal Canada and the northeast US. It is also in the northern Great Basin and coastal western and southern Alaska, south to coastal Washington and Oregon (Fink et al. 2023).

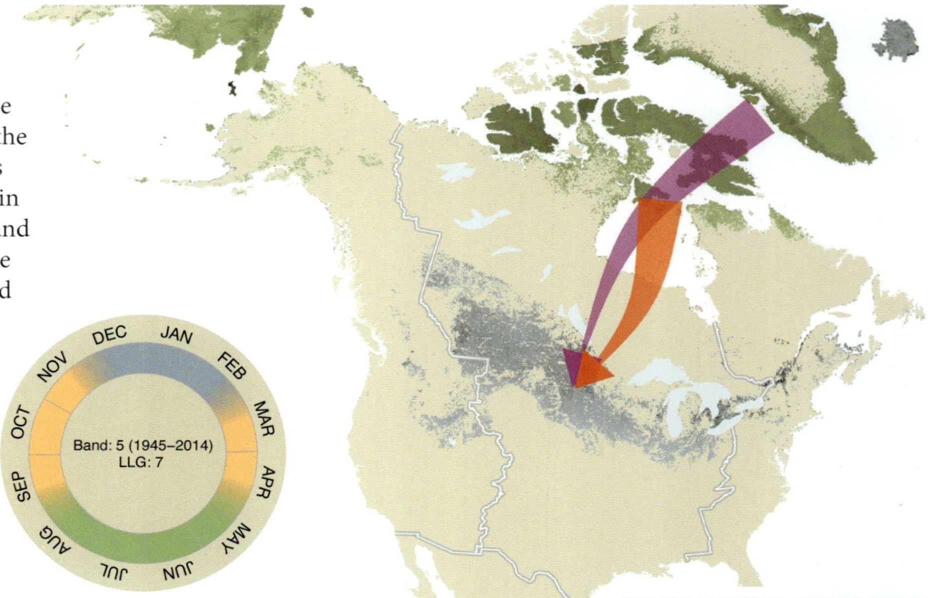

GENERALIZED CONNECTIVITY

	Midcontinent
Greenland (n = 5)	100%
Midcontinent (n = 7)	100%

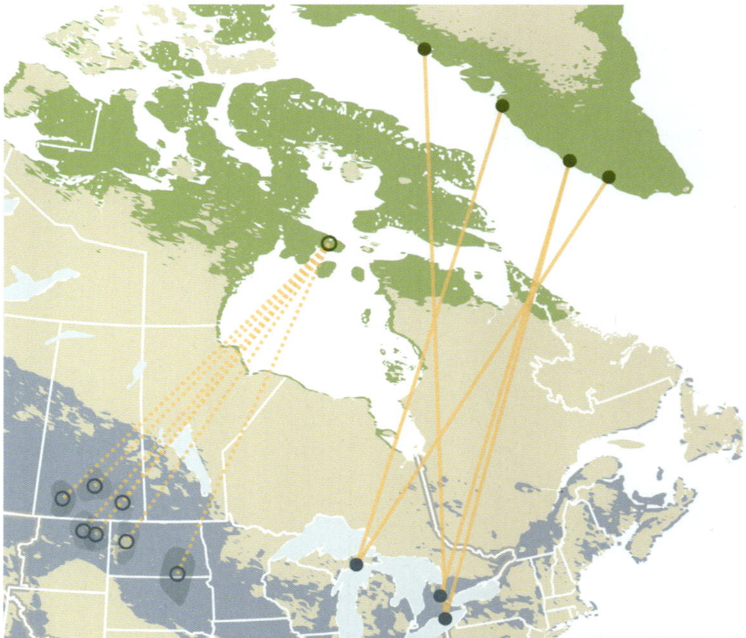

CONNECTIONS

ORDER Passeriformes
FAMILY Calcariidae
BREEDING HABITAT Tundra, Arctic
NONBREEDING HABITAT Open Country, Developed/Disturbed
GLOBAL POPULATION SIZE 29,000,000
POPULATION CHANGE SINCE 1970 −4%
IUCN RED LIST CATEGORY Least Concern
R2R TIPPING POINT SPECIES No
US LISTING STATUS No
CANADA SPECIES AT RISK No
PIF CCS 11/20

MIGRATORY CONNECTIVITY

OVERALL Migratory connectivity is moderately strong, and east-west parallel migration is suggested from band recoveries and geolocator data. MC is not calculated due to limited data. Birds breeding in both Greenland and the Midcontinent Flyway (Canadian Arctic) migrated to nonbreeding areas within the Midcontinent Flyway.

BANDING Breeding to nonbreeding band recoveries indicate that breeders from the west and southeast coast of Greenland migrated to northeastern North America (Macdonald et al. 2012).

GEOLOCATOR TRACKING A single study showed that breeding Snow Buntings on East Bay Island in northern Hudson Bay (n = 13) migrated to a relatively small area of the northern prairies in southern Manitoba, southwestern Saskatchewan, northeastern Montana, and North Dakota (Macdonald et al. 2012). In 2024 and 2025, barometric pressure geolocators (n = 54) were deployed on birds breeding in Utqiaġvik, Alaska. Study is ongoing but preliminary data show individuals migrating to Alberta for the nonbreeding season. (A. Scarpignato and A-L. Harrison, unpublished data).

MIGRATORY PATHWAYS Band recoveries indicate that Greenland breeders use a coastal route and stopovers around the Gulf of St. Lawrence in the pre-breeding season, presumably to avoid snow cover in interior regions during their early migration period (Macdonald et al. 2012). Flocks have been observed crossing the Davis Strait between Canada and Greenland in the pre-breeding season (Montgomerie and Lyon 2020b). East Bay Island breeders used pre-breeding stopovers north of the boreal forest in the western Hudson Bay region (Macdonald et al. 2012).

DATA CONTRIBUTORS
J.W. Fox; K.C. Fraser; H.G. Gilchrist; O.P. Love; C. Macdonald.

Plectrophenax hyperboreus MCKAY'S BUNTING

DISTRIBUTION AND ABUNDANCE

Little scientific data exists for this small Arctic songbird, which breeds only on St. Matthew and Hall islands in the Bering Sea, a total breeding range of only 326 square kilometers. In the nonbreeding season the species is found along the Bering Sea coast and a few nearshore islands, with the highest densities along the coast of the Yukon-Kuskokwim Delta (Montgomerie and Lyon 2020a; Fink et al. 2023).

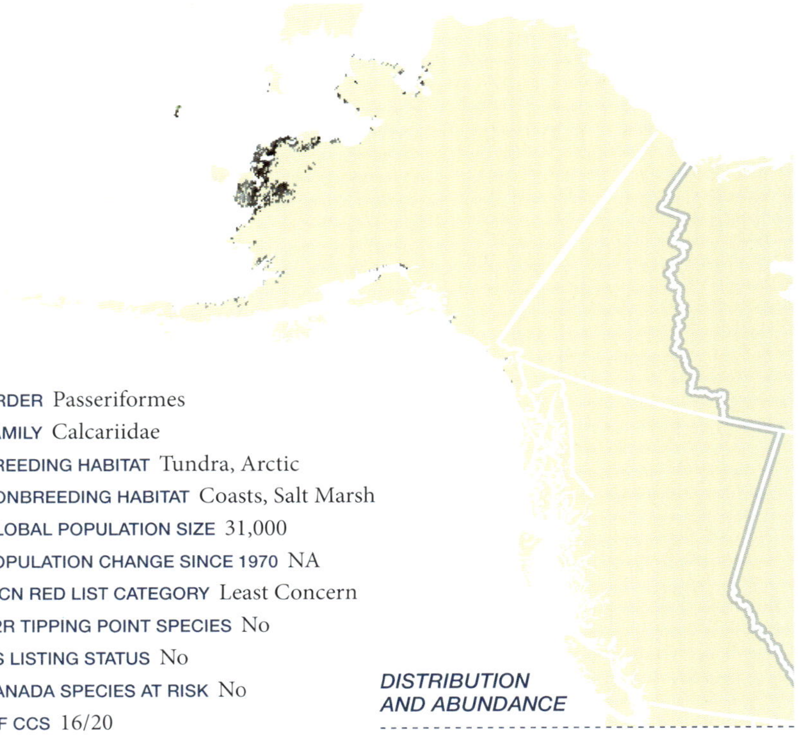

ORDER Passeriformes
FAMILY Calcariidae
BREEDING HABITAT Tundra, Arctic
NONBREEDING HABITAT Coasts, Salt Marsh
GLOBAL POPULATION SIZE 31,000
POPULATION CHANGE SINCE 1970 NA
IUCN RED LIST CATEGORY Least Concern
R2R TIPPING POINT SPECIES No
US LISTING STATUS No
CANADA SPECIES AT RISK No
PIF CCS 16/20

MIGRATORY CONNECTIVITY

OVERALL Migratory connectivity is unknown. No information on migratory connectivity was found for this species.
OTHER APPROACHES No tracking, banding, genetic, or stable isotope studies of migratory connectivity were found for this species.
MIGRATORY PATHWAYS The species moves annually on largely east-west migrations between breeding islands and the nonbreeding range. In the nonbreeding period, birds depart their island breeding areas and fly eastward toward the mainland, stopping on Nunivak Island before spreading out along the coast. Birds spend the nonbreeding season traveling with flocks of closely related Snow Buntings. Females appear to winter farther south than males (Montgomerie and Lyon 2020a).

PASSERELLIDAE | *NEW WORLD SPARROWS*

Peucaea botterii BOTTERI'S SPARROW

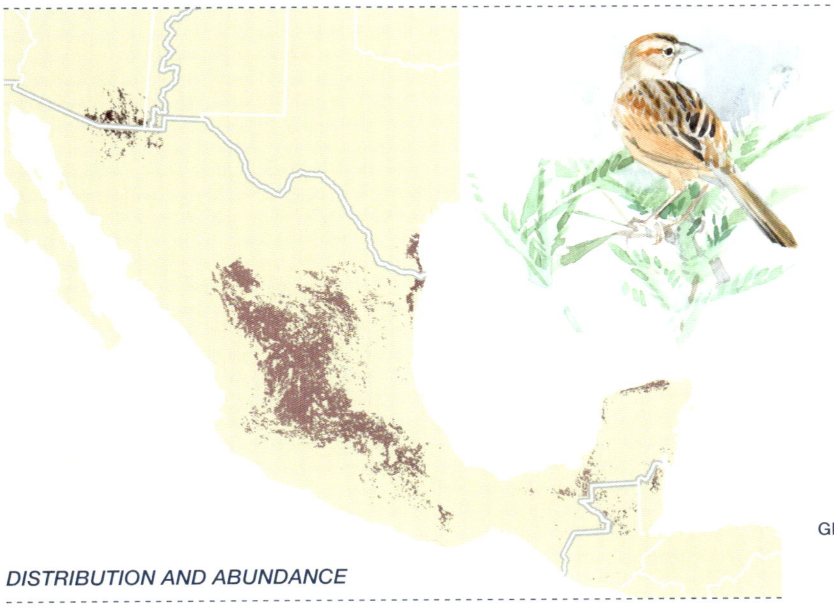

DISTRIBUTION AND ABUNDANCE

It is largely resident in semiarid grasslands in much of Mexico, with two populations barely reaching the US in southeastern Arizona and the coastal plain of southern Texas (Fink et al. 2023). US populations are believed to be migratory, but the whereabouts of northern migrants during the nonbreeding season in Mexico is largely unknown (Webb and Bock 2020).

ORDER Passeriformes
FAMILY Passerellidae
BREEDING HABITAT Grasslands, Chihuahuan
NONBREEDING HABITAT Grasslands, Chihuahuan
GLOBAL POPULATION SIZE 170,000
POPULATION CHANGE SINCE 1970 NA
IUCN RED LIST CATEGORY Least Concern
R2R TIPPING POINT SPECIES No
US LISTING STATUS No
CANADA SPECIES AT RISK No
PIF CCS 14/20

MIGRATORY CONNECTIVITY

OVERALL Migratory connectivity is unknown. No information on migratory connectivity or pathways was found for this species. A few tags were deployed through the Motus Wildlife Tracking System, with local detections only.

INTERESTING NOTE

The Arizona and Texas populations are part of two separate subspecies (Webb and Bock 2020).

CASSIN'S SPARROW *Peucaea cassini*

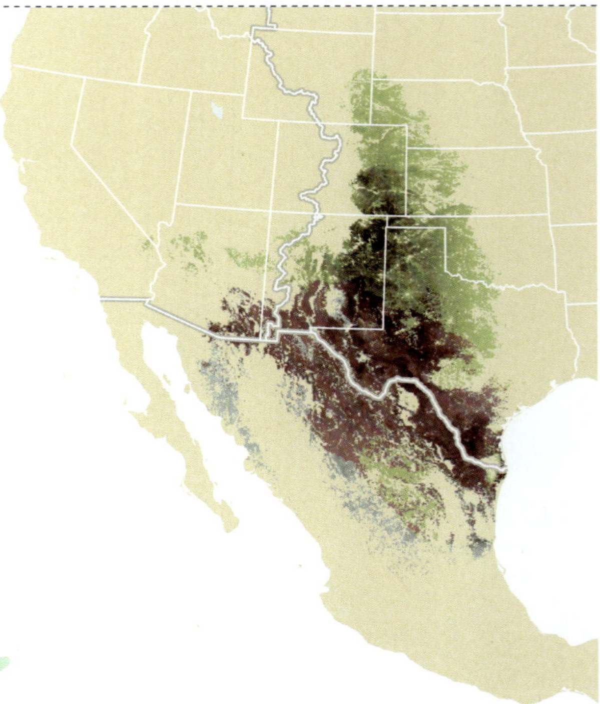

DISTRIBUTION AND ABUNDANCE

DISTRIBUTION AND ABUNDANCE

It breeds in dry grasslands from southeastern Wyoming and southwestern Nebraska south to southeastern Arizona, southern New Mexico, central and southern Texas, and northern Mexican states. Abundance, and even occurrence in northern and western parts of the range, is highly seasonal and dependent on fluctuating rainfall; July–August (monsoon) breeders in southeastern Arizona and northwestern Sonora appear to arrive from farther east (Dunning et al. 2020b). It withdraws from northern portions of the range in the nonbreeding season, when migrants also occur farther south in Mexico. Populations in the southern Great Plains are declining most steeply (Sauer et al. 2019).

ORDER Passeriformes
FAMILY Passerellidae
BREEDING HABITAT Grasslands, Temperate
NONBREEDING HABITAT Grasslands, Chihuahuan
GLOBAL POPULATION SIZE 14,000,000
POPULATION CHANGE SINCE 1970 −31%
IUCN RED LIST CATEGORY Least Concern
R2R TIPPING POINT SPECIES No
US LISTING STATUS No
CANADA SPECIES AT RISK No
PIF CCS 11/20

MIGRATORY CONNECTIVITY

OVERALL Migratory connectivity is unknown. No information on migratory connectivity was found for this species. However, movement patterns and song structure suggest that Great Plains populations are separate from those occurring seasonally in Arizona (Dunning et al. 2020b).

BANDING Banding studies in Arizona indicate low philopatry within and between years, suggesting that populations may be at least partially nomadic (Dunning et al. 2020b).

RADIO TRACKING A few tags were deployed through the Motus Wildlife Tracking System, with limited local detections.

BACHMAN'S SPARROW *Peucaea aestivalis*

DISTRIBUTION AND ABUNDANCE

It breeds in open pine forests across the southeastern US, from Florida, the Gulf states, and eastern Texas north to North Carolina and very patchily to Tennessee, Kentucky, southern Missouri, and eastern Oklahoma; historically, the species bred farther north. Northern populations appear migratory, withdrawing to southern portions of the range in the nonbreeding season (Fink et al. 2023). Historically, the species was more widespread, breeding north to Illinois, Ohio, Pennsylvania, and Maryland during a relatively brief expansion period (1890–1920) following forest clearing for agriculture (Dunning et al. 2020a). More recent declines have been steepest in peninsular Florida (Sauer et al. 2019).

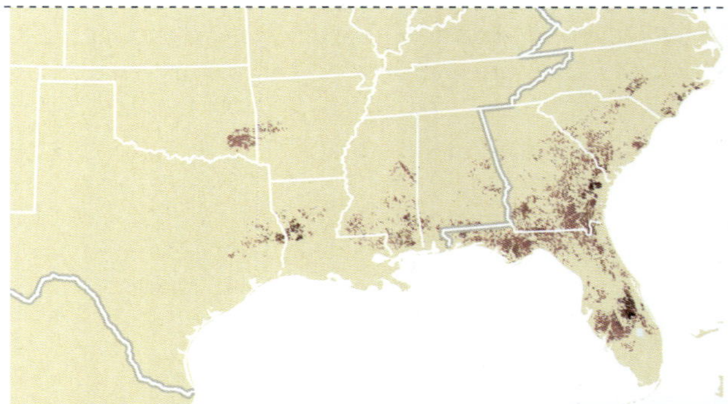

DISTRIBUTION AND ABUNDANCE

ORDER Passeriformes
FAMILY Passerellidae
BREEDING HABITAT Forests, Temperate Eastern
NONBREEDING HABITAT Forests, Temperate Eastern
GLOBAL POPULATION SIZE 170,000
POPULATION CHANGE SINCE 1970 −76%
IUCN RED LIST CATEGORY Near Threatened
R2R TIPPING POINT SPECIES Red-Alert
US LISTING STATUS No
CANADA SPECIES AT RISK No
PIF CCS 16/20

MIGRATORY CONNECTIVITY

OVERALL Migratory connectivity is relatively unknown. There is a separation of migratory populations east and west of the Mississippi River Valley, based on identification of specimens; finer-scale connectivity of breeding and nonbreeding populations is completely unknown.

OTHER APPROACHES Three subspecies have been recognized (American Ornithologists Union 1957; Dunning et al. 2020a), indicating that eastern and western migratory populations remain separate in the nonbreeding season, whereas migrants from eastern states overlap with the resident subspecies during the nonbreeding season in Florida.

MIGRATORY PATHWAYS This species is often considered a resident, but evidence for migration comes from nocturnal kills at communication towers, identification of specimens from northern subspecies in southern nonbreeding areas, and the appearance of individuals in unsuitable habitats during migration periods (Dunning et al. 2020a).

Ammodramus savannarum GRASSHOPPER SPARROW

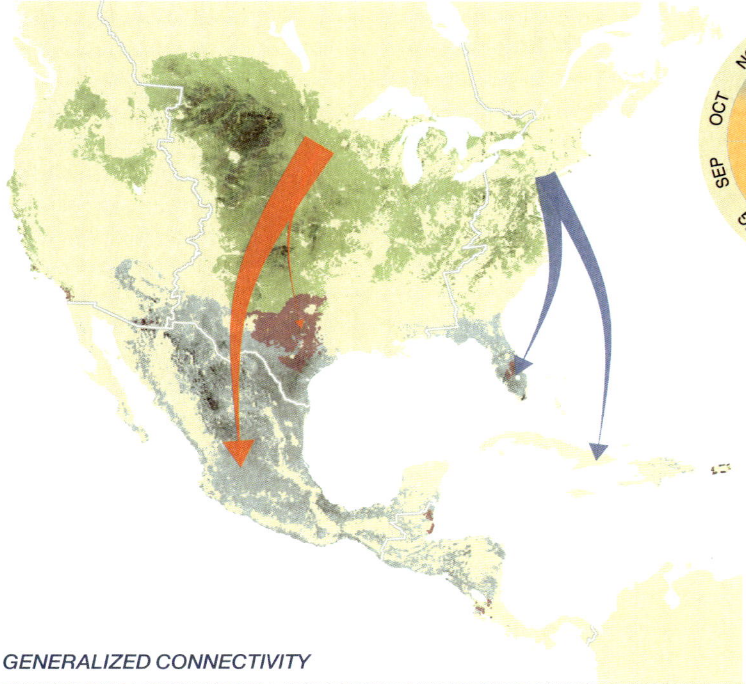

GENERALIZED CONNECTIVITY

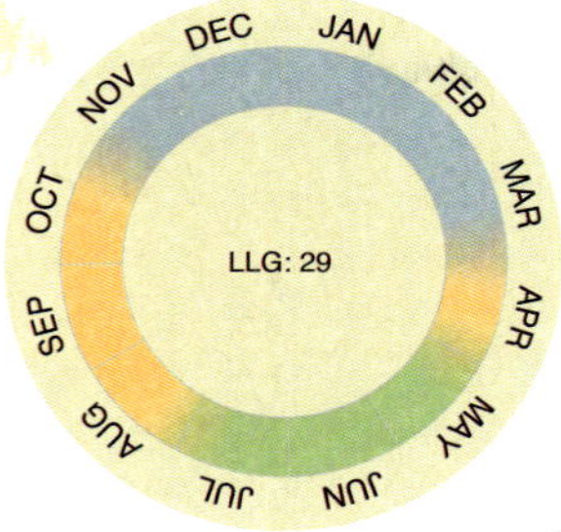

DISTRIBUTION AND ABUNDANCE

It breeds patchily in grasslands across most of the US and southern Canada, with by far the greatest abundance in the Great Plains from the Prairie provinces to central Texas; there are isolated resident populations in Florida, Arizona, Caribbean islands, and central and northwestern South America. The nonbreeding range is primarily in Mexico, southwestern US, and Florida, and sparsely south to Nicaragua. Eastern populations have declined more steeply than western populations (Sauer et al. 2019).

ORDER Passeriformes
FAMILY Passerellidae
BREEDING HABITAT Grasslands, Temperate
NONBREEDING HABITAT Grasslands, Chihuahuan
GLOBAL POPULATION SIZE 34,000,000
POPULATION CHANGE SINCE 1970 −70%
IUCN RED LIST CATEGORY Least Concern
R2R TIPPING POINT SPECIES No
US LISTING STATUS Endangered Florida subspecies
CANADA SPECIES AT RISK No
PIF CCS 12/20

	Atlantic	Caribbean	Mexico	Midcontinent
Atlantic (n = 19)	53%	47%	–	–
Midcontinent (n = 10)	–	–	90%	10%

MIGRATORY CONNECTIVITY

OVERALL Based on morphological variation and a single geolocator tracking study, migratory connectivity appears strong on a continental scale, with a migratory divide between Atlantic and Midcontinent flyway breeding populations. Birds breeding in the Atlantic Flyway migrated to nonbreeding areas either within the Atlantic Flyway (53%) or the Caribbean (47%), while birds breeding in the Midcontinent Flyway primarily migrated to Mexico (90%) or remained within the Midcontinent (10%) for the nonbreeding season. MC is not calculated due to limited sample size.

RADIO TRACKING There are several projects tracking Grasshopper Sparrow with the Motus Wildlife Tracking System across the range. There are no individual breeding to nonbreeding locations, but there appear to be many detections during migration.

GEOLOCATOR TRACKING A single light-level geolocator study found that individuals breeding in Kansas (n = 8), Wisconsin (n = 4), and North Dakota (n = 1) migrated to Texas and Mexico, whereas those breeding in Maryland (n = 10) and Massachusetts (n = 10) migrated to the Greater Antilles and Florida (Hill and Renfrew 2019). Migratory connectivity across the range was strong; however, it was weak among specific breeding sites within a region.

GENETICS The Bird Genoscape Project is currently working to build a genoscape for the Grasshopper Sparrow (https://www.birdgenoscape.org/grasshopper-sparrow/).

OTHER APPROACHES No banding, genetic, or stable isotope studies of migratory connectivity were found for this species. Twelve subspecies are recognized, and the two migratory subspecies in North America maintain separate migration and nonbreeding distributions (Vickery 2020). The paler *A. s. perpallidus* is the abundant bird of the Great Plains and westward to the Pacific coast, migrating to the southwestern US, Mexico, Guatemala, and El Salvador. The brighter *A. s. pratensis* breeds patchily in the eastern US and migrates to the southeastern US, Cuba, eastern and southern Mexico, and Belize. The Arizona subspecies is partially migratory to western Mexico.

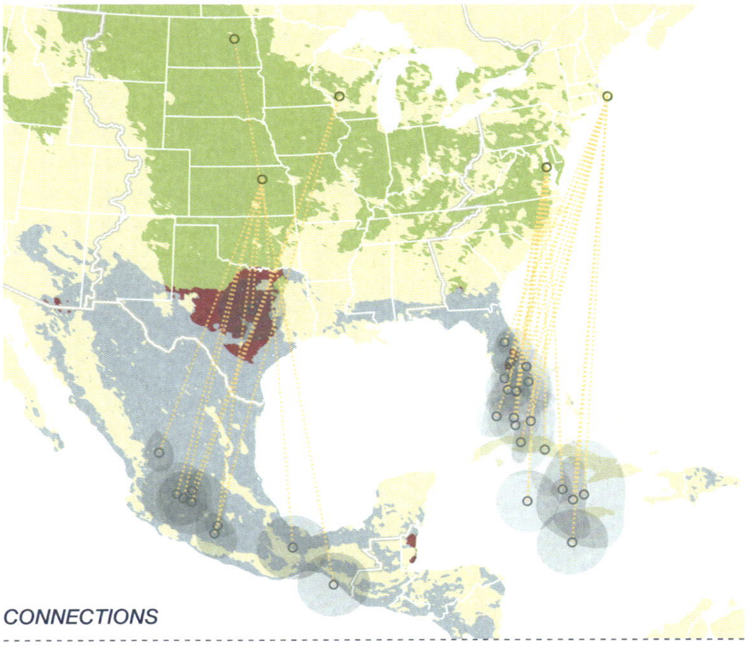

CONNECTIONS

MIGRATORY PATHWAYS Based on three geolocator-tracked individuals, migration follows a counterclockwise loop, with a pre-breeding migration path farther east than in post-breeding migration (Hill and Renfrew 2019). Regular occurrence on Bermuda indicates that long overwater flights are possible (Amos 1991).

DATA CONTRIBUTORS
J.M. Hill; R.B. Renfrew.

BLACK-THROATED SPARROW *Amphispiza bilineata*

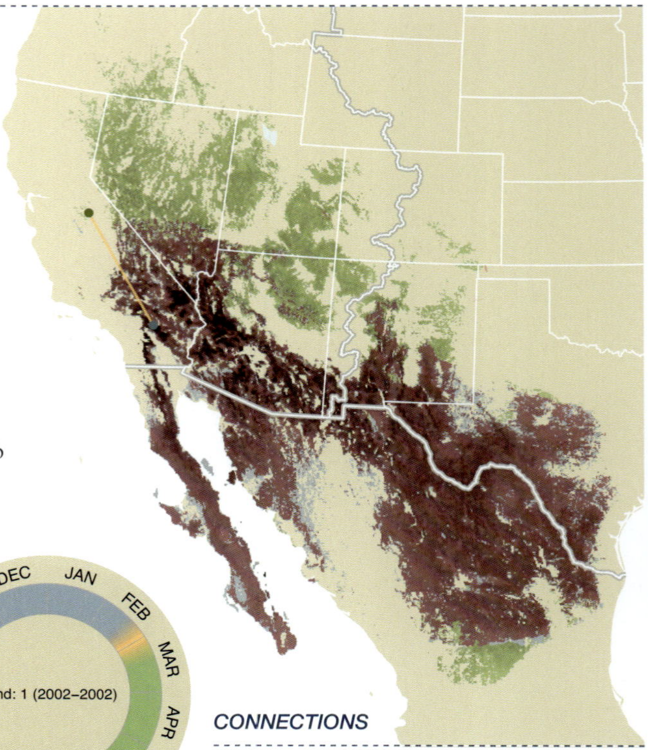

DISTRIBUTION AND ABUNDANCE

The species breeds throughout the major deserts of the southwestern US and northern Mexico, north to the southern Great Basin; it also breeds in dry grasslands of southeastern Colorado, northwestern Oklahoma, eastern New Mexico, and northern Texas. It is resident in most of the range, but northern breeders migrate to southern portions of the range in the nonbreeding season. Declines have been steepest in central and southern Texas, whereas populations in the Great Basin have increased and expanded northward (Sauer et al. 2019).

ORDER Passeriformes
FAMILY Passerellidae
BREEDING HABITAT Arid Lands, Desert Scrub
NONBREEDING HABITAT Arid Lands, Desert Scrub
GLOBAL POPULATION SIZE 62,000,000
POPULATION CHANGE SINCE 1970 −41%
IUCN RED LIST CATEGORY Least Concern
R2R TIPPING POINT SPECIES No
US LISTING STATUS No
CANADA SPECIES AT RISK No
PIF CCS 9/20

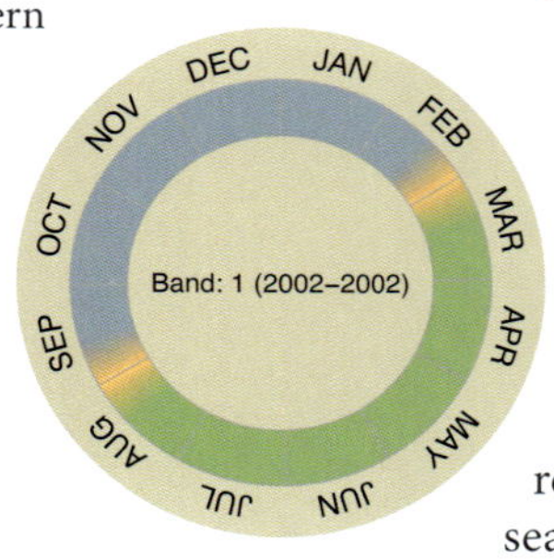

MIGRATORY CONNECTIVITY

OVERALL Migratory connectivity is unknown. Morphological specimen data and a single band return suggest that northern breeders are migratory and that the three subspecies populations remain geographically separate throughout the annual cycle.

BANDING A single band return links a breeder from central California to a nonbreeding site in the Mojave Desert of southeastern California. Three nonbreeding individuals in Arizona and New Mexico were recaptured at the same locality in the subsequent year (Knick et al. 2017).

OTHER APPROACHES No tracking, genetic, or stable isotope studies of migratory connectivity were found for this species. Analysis of the three migratory subspecies indicate that northern populations remain separate in the nonbreeding season, with breeders from northern Texas moving into southern Texas and northeastern Mexico, breeders from the Great Basin migrating to the southwestern US and northwestern Mexico, and breeders on the high plains of southeastern Colorado, northwestern Oklahoma, and eastern New Mexico moving south into western Texas and north-central Mexico (American Ornithologists Union 1957; M. Johnson et al. 2020).

MIGRATORY PATHWAYS Pathways are poorly known during migration; there is no information on migratory pathways or stopover sites.

LARK SPARROW *Chondestes grammacus*

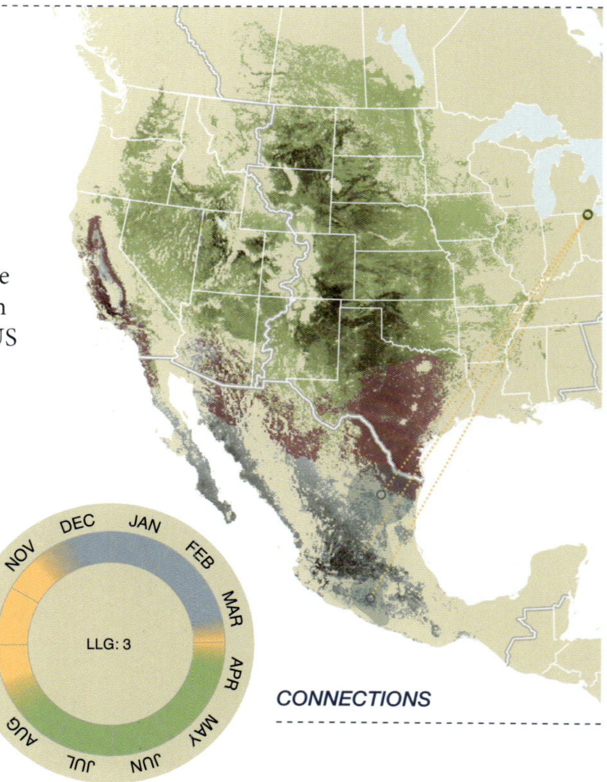

DISTRIBUTION AND ABUNDANCE

It breeds patchily in open habitats in much of the western and central United States (greatest abundance in western Great Plains, Great Basin, and Central Valley of California), sparsely east to Ohio, Tennessee, and Alabama, and south to northernmost Mexico. In the nonbreeding season, eastern and Great Plains breeders migrate primarily to southern Texas and central Mexico, whereas most Great Basin breeders migrate to southern Arizona, Baja California, and northwestern Mexico. The population in California appears to be largely resident. Declines have been steepest in California and the south-central US, whereas populations elsewhere in the western US appear to be stable but with recent declines (Sauer et al. 2019).

ORDER Passeriformes
FAMILY Passerellidae
BREEDING HABITAT Open Country, Habitat Mosaic
NONBREEDING HABITAT Open Country, Habitat Mosaic
GLOBAL POPULATION SIZE 11,000,000
POPULATION CHANGE SINCE 1970 −28%
IUCN RED LIST CATEGORY Least Concern
R2R TIPPING POINT SPECIES No
US LISTING STATUS No
CANADA SPECIES AT RISK No
PIF CCS 10/20

MIGRATORY CONNECTIVITY

OVERALL Limited tracking data from one geolocator study (Ross et al. 2014) shows connectivity between breeding populations in Ohio and a nonbreeding location in central Mexico. Genetic studies suggest some genetic differentiation between breeding populations and some mixing among distinct populations throughout the breeding range (Ross and Bouzat 2014), though it appears that they all share a similar nonbreeding range in the grasslands of central Mexico.

GEOLOCATOR TRACKING Two male and one female Lark Sparrows breeding in Ohio were tracked to their nonbreeding grounds in the central highlands of central Mexico (Ross et al. 2014). Nonbreeding areas appear to be within a narrow longitudinal range in the southern desert and semiarid highlands ecoregions in central Mexico, similar to core breeding habitats of the Great Plains (Ross et al. 2014).

GENETICS Analysis of mitochondrial DNA sequences across western, central, and eastern United States breeding zones distinguished western populations from the more similar central and eastern populations (Ross and Bouzat 2014).

OTHER APPROACHES There are significant morphological differences between populations, characterized by longer wings of adult birds from populations that are farthest from nonbreeding grounds and an east-to-west increase in beak depth (Ross and Bouzat 2014). Two weakly differentiated subspecies are recognized, separating eastern and western breeding populations (Martin and Parrish 2020). These populations maintain largely separate migration and nonbreeding distributions, possibly overlapping in central Mexico.

MIGRATORY PATHWAYS Tracked individuals from Ohio migrated south to the Gulf Coast and then westward through Texas to central Mexico; post-breeding migration was extremely early, prior to molt. (Ross et al. 2014). Pre-breeding migration was more direct and of shorter duration than post-breeding migration, as birds moved north-northeast from nonbreeding areas through northeast Texas and southwest Arkansas to breeding areas in Ohio (Ross et al. 2014).

INTERESTING NOTE

The return rate of geolocator-marked individuals was not significantly different than the return rate of color-banded adults in the same Ohio breeding population (Ross et al. 2014).

DATA CONTRIBUTORS
J.D. Ross.

LARK BUNTING

Calamospiza melanocorys

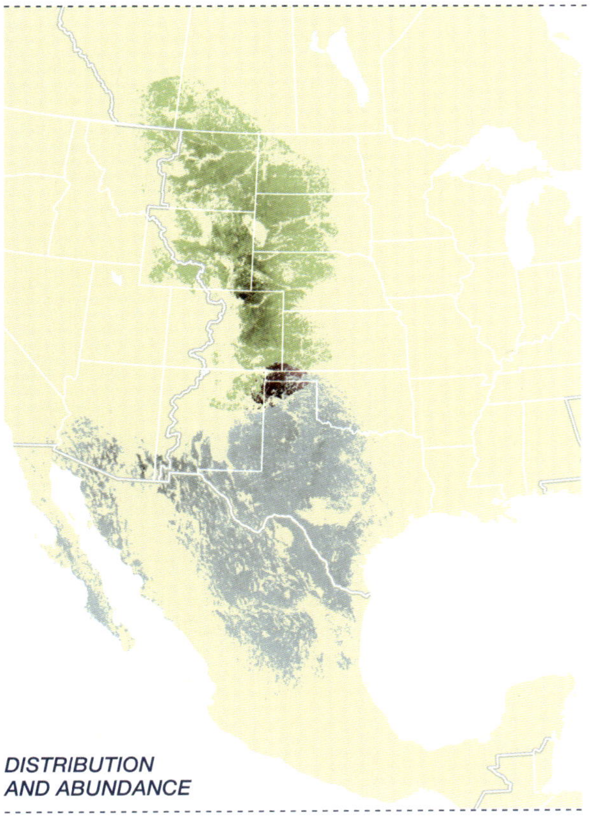

DISTRIBUTION AND ABUNDANCE

DISTRIBUTION AND ABUNDANCE

It breeds in shortgrass prairie from southeastern Alberta and southern Saskatchewan south to eastern Colorado and western Kansas. Core breeding abundance shifts from year to year with changing rainfall, and the species sporadically breeds in large numbers east, west, and south of its main breeding range. Nonbreeding distribution is in the Chihuahuan Desert and other arid grasslands in southern Arizona, southern New Mexico, western Texas, and northern Mexico. Declines are especially steep in the Canadian Prairies and North Dakota (Sauer et al. 2019).

ORDER Passeriformes
FAMILY Passerellidae
BREEDING HABITAT Grasslands, Temperate
NONBREEDING HABITAT Grasslands, Chihuahuan
GLOBAL POPULATION SIZE 12,000,000
POPULATION CHANGE SINCE 1970 −77%
IUCN RED LIST CATEGORY Least Concern
R2R TIPPING POINT SPECIES No
US LISTING STATUS No
CANADA SPECIES AT RISK Threatened
PIF CCS 12/20

MIGRATORY CONNECTIVITY

OVERALL Migratory connectivity is unknown. No information on migratory connectivity was found for this species. A few tags were deployed through the Motus Wildlife Tracking System, with local detections only.

MIGRATORY PATHWAYS It migrates diurnally in large flocks, primarily through the high plains east of the Rocky Mountains (Shane 2020). It departs early from many breeding areas (July) and arrives in northern Mexico by mid-August (Shane 2020). Flocks wander widely in search of food sources, and stray individuals have occurred in every US state and southern Canadian province (Fink et al. 2023).

INTERESTING NOTE

An individual banded at Tempe, Arizona, in March 1940 and recovered the following November 1,000 km to the east near San Angelo, Texas (Cooke 1945), suggests substantial east-west movement of populations from year to year.

CHIPPING SPARROW *Spizella passerina*

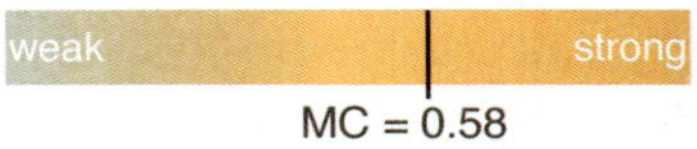

DISTRIBUTION AND ABUNDANCE

It breeds in much of forested North America, from eastern Alaska and northern Yukon across the boreal forest of Canada to southwestern Newfoundland, south through the eastern US to Georgia and the Gulf states, and in western mountains to Southern California, southeastern Arizona, southern New Mexico, and western Texas. There are also resident populations in highlands of Mexico and northern Central America. The nonbreeding range of migratory populations is primarily across the southern US (uncommon in California) and Mexico, south to central Florida and Oaxaca. The steepest long-term declines have been in the Pacific states and Appalachian Mountains region, although this species is experiencing steep recent declines virtually throughout the breeding range (Sauer et al. 2019).

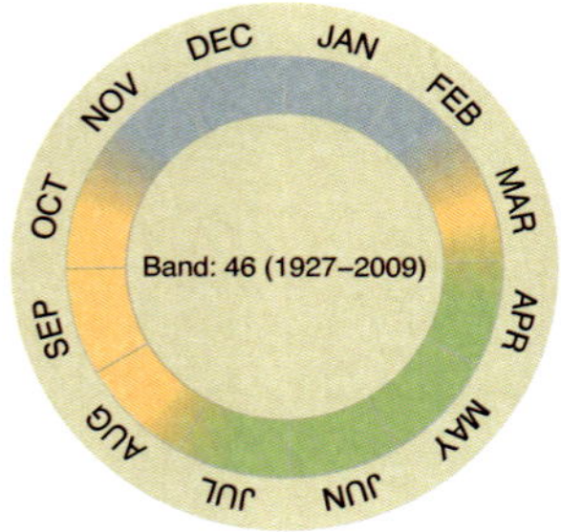

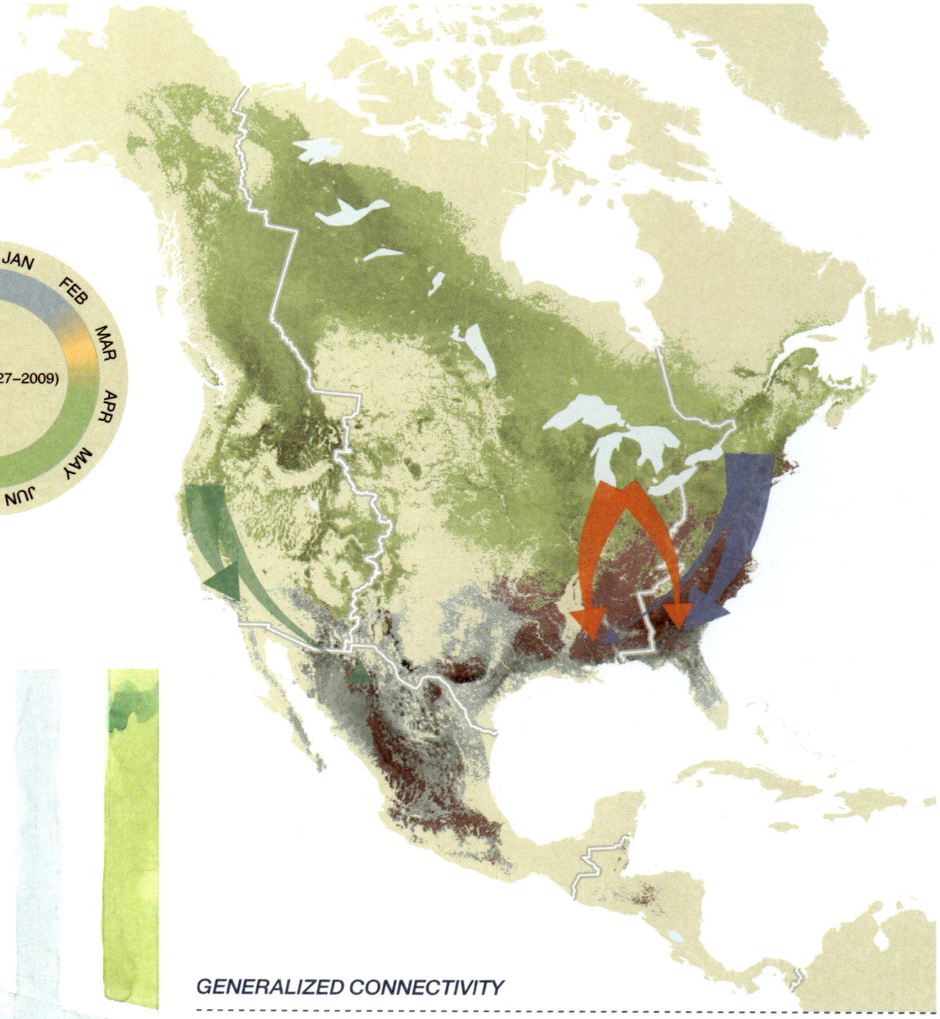

GENERALIZED CONNECTIVITY

ORDER Passeriformes
FAMILY Passerellidae
BREEDING HABITAT Forests, Temperate Generalist
NONBREEDING HABITAT Open Country, Habitat Mosaic
GLOBAL POPULATION SIZE 240,000,000
POPULATION CHANGE SINCE 1970 −24%
IUCN RED LIST CATEGORY Least Concern
R2R TIPPING POINT SPECIES No
US LISTING STATUS No
CANADA SPECIES AT RISK No
PIF CCS 9/20

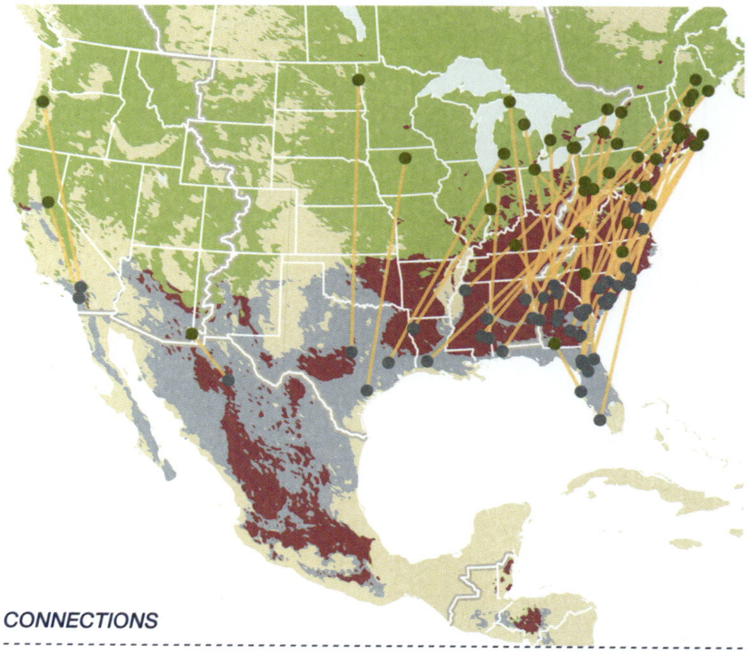

CONNECTIONS

	Atlantic	Mexico	Midcontinent	Pacific
Atlantic (n = 28)	82%	–	18%	–
Midcontinent (n = 12)	42%	–	58%	–
Pacific (n = 3)	–	33%	–	67%

MIGRATORY CONNECTIVITY

OVERALL Migratory connectivity appears strong (MC = 0.58) based on band reencounter data from eastern, central, and western North America, as well as on the distribution of morphologically variable subspecies. Atlantic (82%), Midcontinent (58%), and Pacific (67%) flyway breeding birds largely migrate to nonbreeding areas within the same flyway as breeding. Atlantic and Midcontinent flyway breeding populations appear to mix with each other on nonbreeding areas, while a distinct separation occurs from Pacific Flyway breeding populations, but sample size is limited.

BANDING Strong east-west parallel migration is indicated by breeding to nonbreeding band reencounters, with individuals from the east coast west to the Great Lakes region migrating south to the southern Atlantic coast, Florida, and the Gulf Coast, individuals in the north-central US and Canada migrating to Texas, and individuals from Pacific states migrating south to Southern California. Some mixing of eastern and central birds occurs along the Gulf Coast. An individual banded in Alberta, Canada, was recovered two months later in El Salvador, representing the only record of a migrant individual south of Oaxaca, Mexico (Thurber 1987).

OTHER APPROACHES No tracking, genetic, or stable isotope studies of migratory connectivity were found for this species. Analysis of specimens indicates that eastern and western migratory subspecies remain as separate populations throughout the year; eastern birds migrate to the southeastern US and northeastern Mexico in the nonbreeding season, whereas nonbreeders in most of Mexico originate in western North America (American Ornithologists Union 1957; Middleton 2020). Two resident subspecies also occur in Mexico and Central America.

MIGRATORY PATHWAYS Based on eBird animations, western boreal Chipping Sparrows move south through the western prairies and along the front range to northern Mexico, migrating much earlier than birds in the northeast part of the range or in the western US. By contrast, eastern birds migrate north much earlier in pre-breeding than western mountain or boreal populations (Fink et al. 2023).

Spizella pallida CLAY-COLORED SPARROW

ORDER Passeriformes
FAMILY Passerellidae
BREEDING HABITAT Grasslands, Temperate
NONBREEDING HABITAT Grasslands, Chihuahuan
GLOBAL POPULATION SIZE 60,000,000
POPULATION CHANGE SINCE 1970 −44%
IUCN RED LIST CATEGORY Least Concern
R2R TIPPING POINT SPECIES No
US LISTING STATUS No
CANADA SPECIES AT RISK No
PIF CCS 10/20

DISTRIBUTION AND ABUNDANCE

It breeds primarily in the northern Great Plains from Alberta to southern Manitoba and south to eastern Montana, South Dakota, and western Minnesota, very patchily westward to central British Columbia, northward into Northwest Territories, and eastward across southern Ontario, southern Quebec, and the Great Lakes states to northern New York. The nonbreeding range is primarily on the Central Mexican Plateau south to Oaxaca, and in smaller numbers in western Mexican lowlands and north to western and southern Texas and southeastern New Mexico. Declines are steepest in the center of the range, with some increases noted in the western and eastern portions (Sauer et al. 2019).

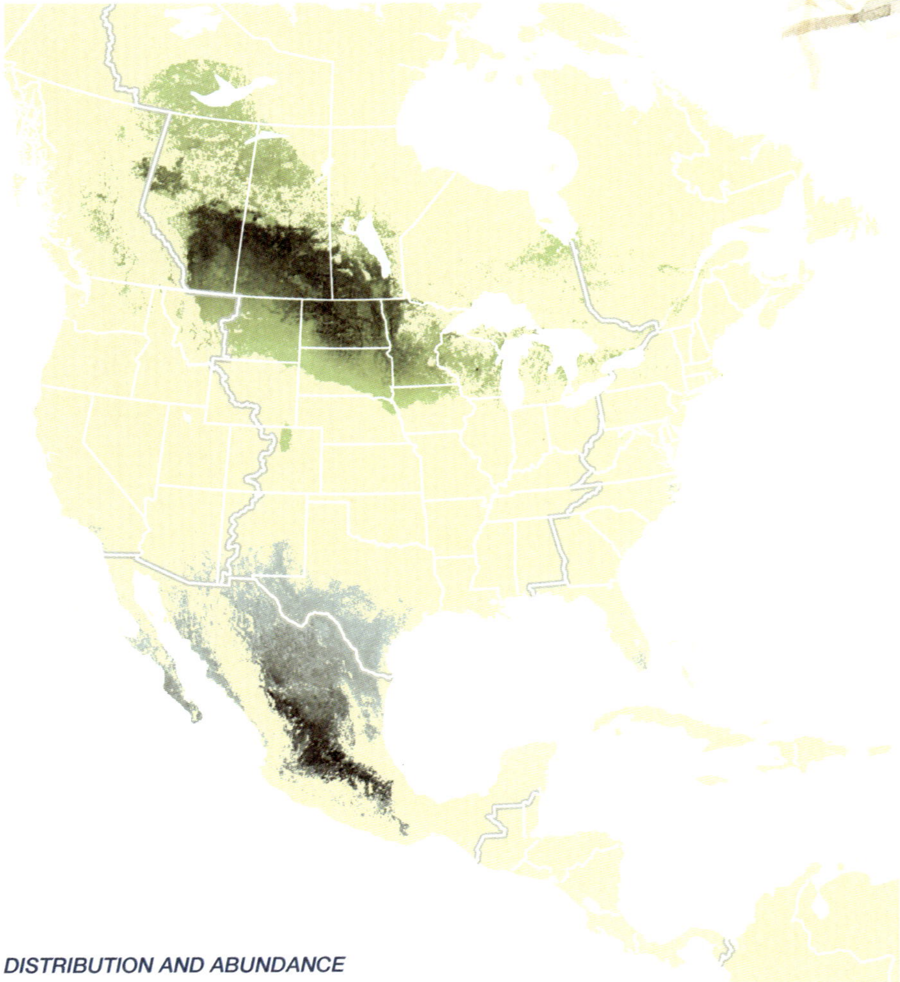

DISTRIBUTION AND ABUNDANCE

MIGRATORY CONNECTIVITY

OVERALL Migratory connectivity is unknown. No information on migratory connectivity was found for this species.

MIGRATORY PATHWAYS Migration is concentrated through the western Great Plains, from the Rocky Mountains eastward through Nebraska, Kansas, Oklahoma, and northern Texas. Small numbers occur regularly farther east and west, with strays reaching both coasts and even Bermuda. Pre-breeding of migration has been noted to occur in waves, with flocks of up to 3,000 birds reported (Grant and Knapton 2020).

BLACK-CHINNED SPARROW *Spizella atrogularis*

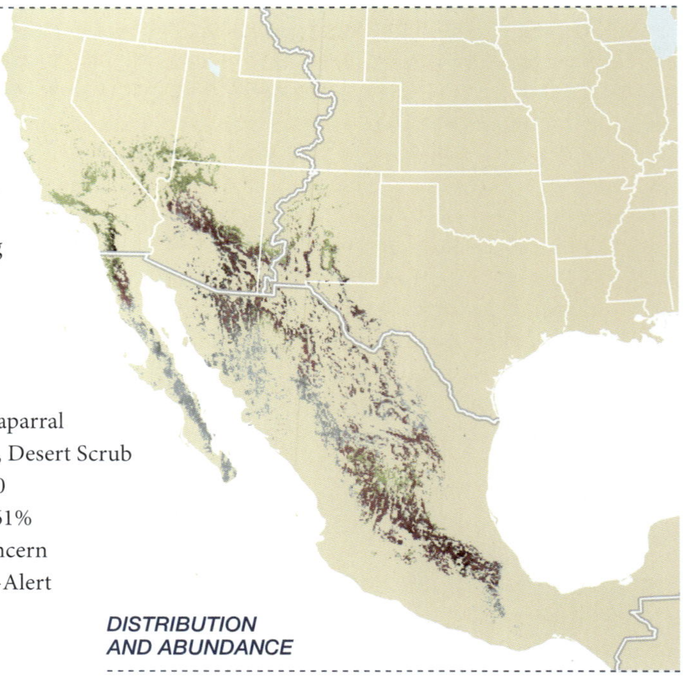

DISTRIBUTION AND ABUNDANCE

It breeds in chaparral and coastal sage habitats from central California, southern Nevada, southwestern Utah, and northern Arizona south to northwestern Baja California and in the central highlands to Oaxaca, Mexico. The nonbreeding range extends to southern Baja California and to the lower Pacific Slope of western Mexico. Western California populations are declining more steeply than those inland (Sauer et al. 2019).

ORDER Passeriformes
FAMILY Passerellidae
BREEDING HABITAT Arid Lands, Chaparral
NONBREEDING HABITAT Arid Lands, Desert Scrub
GLOBAL POPULATION SIZE 1,100,000
POPULATION CHANGE SINCE 1970 −61%
IUCN RED LIST CATEGORY Least Concern
R2R TIPPING POINT SPECIES Yellow-Alert
US LISTING STATUS No
CANADA SPECIES AT RISK No
PIF CCS 15/20

MIGRATORY CONNECTIVITY

OVERALL Migratory connectivity is largely unknown. Based on subspecies variation, coastal-slope populations remain separate from inland populations during the nonbreeding season; no further information on migratory connectivity was found for this species.
OTHER APPROACHES Analysis of specimens indicates that breeding populations in the coastal mountains of California (*S. a. cana*) migrate to Baja California in the nonbreeding season (rarely north to southwestern Arizona in wet nonbreeding seasons; Phillips et al. 1964). Inland-breeding *S. a. evura* populations presumably migrate into Mexico, but nonbreeding distribution is poorly defined (American Ornithologists Union 1957; Tenney 2020a).
MIGRATORY PATHWAYS Pathways during migration are poorly known; there is no information on migratory pathways or stopovers. eBird models suggest that breeders from the inland southwestern US may occupy the foothills of the Mexican Pacific Slope from Sonora to Durango during the nonbreeding season (Fink et al. 2023).

FIELD SPARROW *Spizella pusilla*

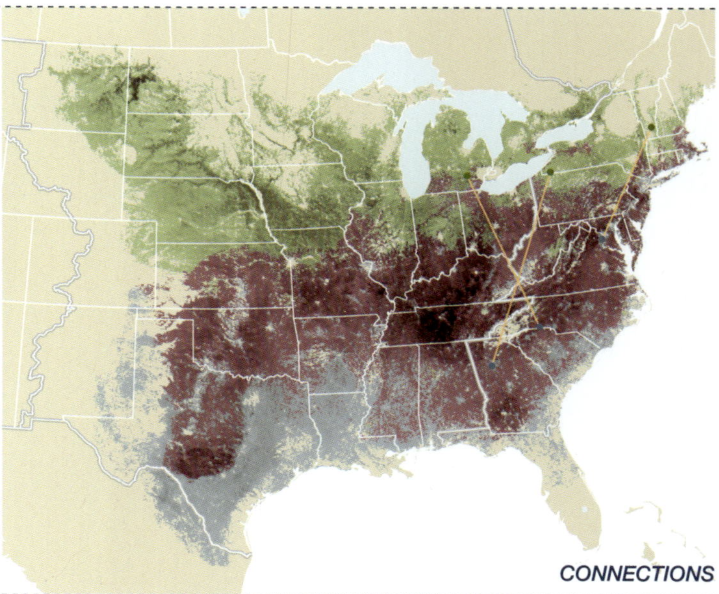

DISTRIBUTION AND ABUNDANCE

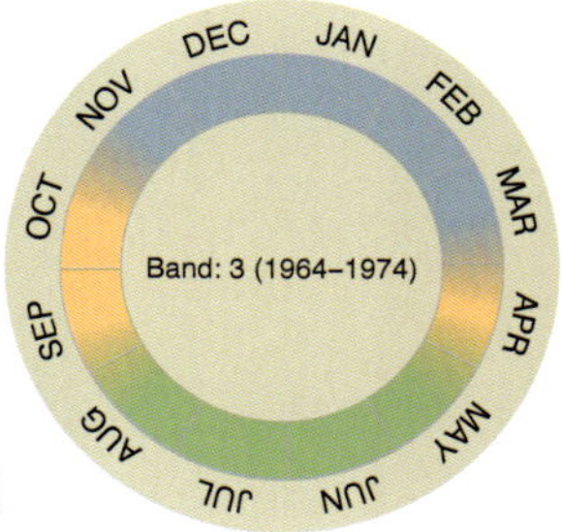

This species breeds throughout most of the eastern US but is absent from the southeastern and Gulf Coastal Plains and Florida; it also breeds westward through the Great Plains from eastern Montana south to central Texas. Its nonbreeding distribution overlaps with the southern half of the breeding range and extends south to the Gulf Coast, northern Florida, southern Texas, and northeastern Mexico; its greatest nonbreeding season abundance is in the Piedmont region, Kentucky, Tennessee, and Texas. Steepest population declines are in the Atlantic coastal states, whereas populations have increased in the northern Great Plains (Sauer et al. 2019).

ORDER Passeriformes
FAMILY Passerellidae
BREEDING HABITAT Open Country, Habitat Mosaic
NONBREEDING HABITAT Open Country, Habitat Mosaic
GLOBAL POPULATION SIZE 9,300,000
POPULATION CHANGE SINCE 1970 −64%
IUCN RED LIST CATEGORY Least Concern
R2R TIPPING POINT SPECIES Yellow-Alert
US LISTING STATUS No
CANADA SPECIES AT RISK No
PIF CCS 13/20

MIGRATORY CONNECTIVITY

OVERALL Migratory connectivity is essentially unknown, but the few band reencounters, analysis of specimens, and eBird animations suggest some east-west separation of populations.

BANDING A small number of breeding to nonbreeding band reencounters confirm that Field Sparrows migrate south for the nonbreeding season. One bird from Vermont was recovered in Virginia, and two from the Great Lakes region stayed west of the Appalachian Mountains. Combined with the larger sample of seasonal band recoveries from migration (Marra et al. 2014), a further suggestion of east-west parallel migration is apparent.

OTHER APPROACHES No tracking, genetic, or stable isotope studies of migratory connectivity were found for this species. There were a few detections, close to areas of deployment, from several tags deployed through the Motus Wildlife Tracking System. Although the two recognized subspecies may represent clinal variation and not be valid (Carey et al. 2020), identification of specimens from nonoverlap areas suggests an east-west separation of populations and at least some degree of migratory connectivity. The western subspecies, breeding primarily in the northern Great Plains, migrates to Oklahoma, central Texas, and northeastern Mexico, whereas eastern breeders are distributed throughout the southeastern US in the nonbreeding season (American Ornithologists Union 1957; Carey et al. 2020).

MIGRATORY PATHWAYS This species migrates over a broad front; however, high nonbreeding abundance occurs in three areas—Texas, Kentucky and Tennessee, and the Piedmont—that include birds originating in the northern Great Plains, upper Midwest, and northeastern US, respectively, indicating a north-south migration (Fink et al. 2023).

Spizella breweri BREWER'S SPARROW

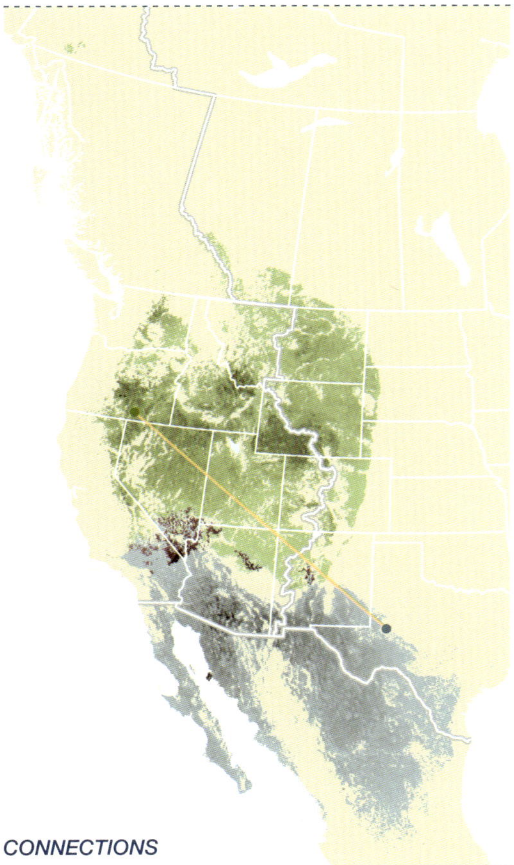

CONNECTIONS

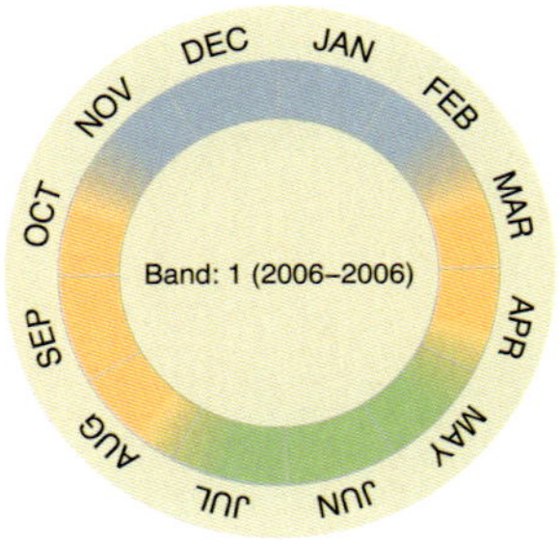

DISTRIBUTION AND ABUNDANCE

It breeds throughout the sagebrush ecosystem of the Great Basin, from southeastern British Columbia, southern Alberta, and southwestern Saskatchewan to eastern California, northern Arizona, and northern New Mexico, and sparsely in badlands east of the Rockies; it also breeds in subalpine shrublands of high mountains from southeastern Alaska and Yukon to northwestern Montana (timberline sparrow). Its nonbreeding range is in deserts from southeastern California to western Texas, south on the Central Mexican Plateau, and to lowlands of northwestern Mexico, including Baja California. Declines are steepest in the core of the range in eastern Oregon and Idaho (Sauer et al. 2019).

ORDER Passeriformes
FAMILY Passerellidae
BREEDING HABITAT Arid Lands, Sagebrush
NONBREEDING HABITAT Arid Lands, Desert Scrub
GLOBAL POPULATION SIZE 17,000,000
POPULATION CHANGE SINCE 1970 −40%
IUCN RED LIST CATEGORY Least Concern
R2R TIPPING POINT SPECIES No
US LISTING STATUS No
CANADA SPECIES AT RISK No
PIF CCS 12/20

MIGRATORY CONNECTIVITY

OVERALL Migratory connectivity data is unknown and comes primarily from banding reencounter and feather stable isotope data. Northern populations may move farthest south, and some southern populations probably only move short distances to nonbreeding grounds (Rotenberry et al. 2020).

BANDING There was one reencounter of a breeding individual from southeastern Oregon at a nonbreeding site in central Texas. Individuals banded at nonbreeding sites in Arizona and New Mexico were recaptured at the same site in up to four subsequent nonbreeding seasons (Knick et al. 2017).

RADIO TRACKING Over 100 tags were deployed using the Motus Wildlife Tracking System; all detections are within 100 km of deployment.

STABLE ISOTOPES Feather isotope analysis of deuterium and nitrogen found that individual nonbreeding locations were used by Brewer's Sparrows from potentially a wide portion of the breeding range, with sparrows captured in eastern locations coming from more northern breeding areas than individuals at western locations (Knick et al. 2014).

GENETICS The Bird Genoscape Project is currently in the process of building a genoscape for Brewer's Sparrow (https://www.birdgenoscape.org/brewers-sparrow/).

OTHER APPROACHES The high-elevation timberline sparrow (*S. b. taverneri*) has a disjunct breeding distribution but has been found during the nonbreeding season within the range of the nominate *S. b. breweri* (American Ornithologists Union 1957; Rotenberry et al. 2020; eBird).

MIGRATORY PATHWAYS Migration occurs along a broad front in pre- and post-breeding, with individuals straggling regularly to the Pacific coast and western Great Plains. Individual Timberline Sparrows may appear anywhere within the range of the nominate Brewer's Sparrow, but migrants move north into western Canada much later in pre-breeding (late May; Fink et al. 2023).

FOX SPARROW *Passerella iliaca*

DISTRIBUTION AND ABUNDANCE

It breeds across the boreal region, from Newfoundland to western Alaska and south in the mountains of western North America and along the Pacific coast from southern Alaska to southern British Columbia. Nonbreeding distribution is in two highly disjunct regions, with eastern-breeding Red Fox Sparrows (*P. i. iliaca* group) in southeastern and south-central US, and western populations in the Pacific coast and mountains from southwestern British Columbia to northwestern Baja California and sparsely eastward to southern Arizona. Pacific coast and Sierra Nevada breeding populations have declined steeply since the 1980s, whereas Rocky Mountain populations have been stable or increasing (Sauer et al. 2019).

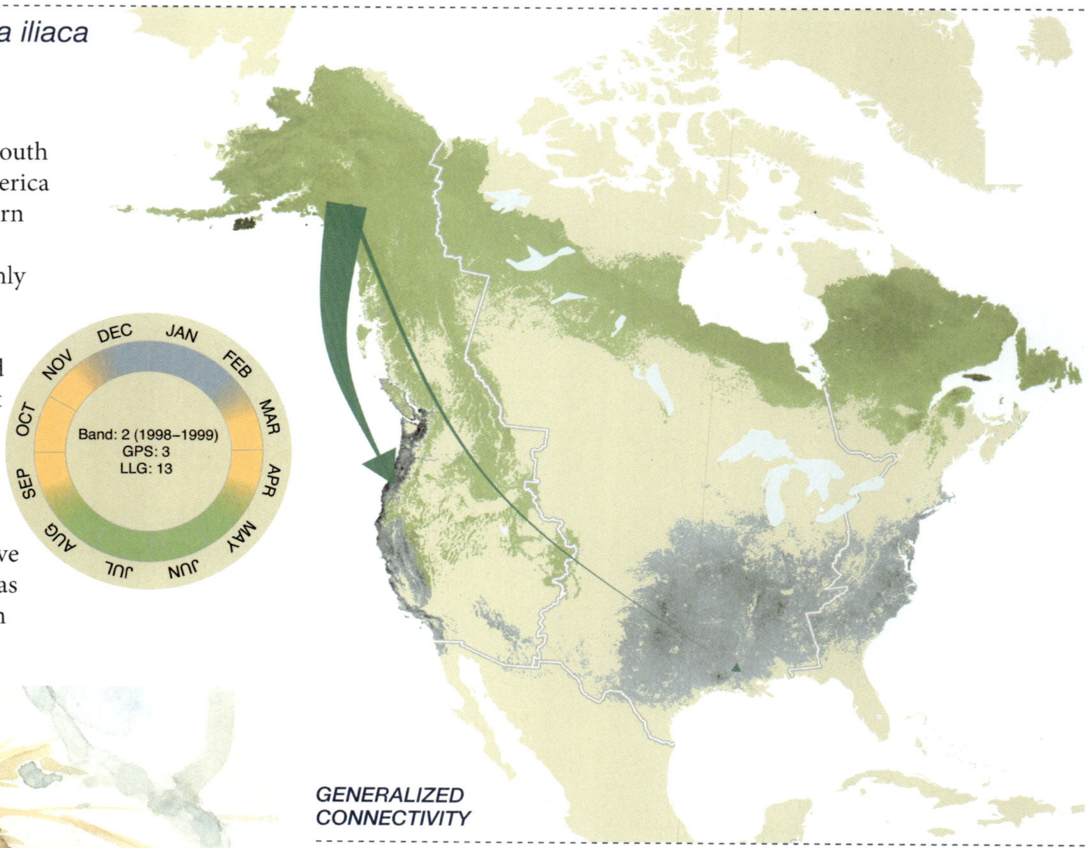

GENERALIZED CONNECTIVITY

ORDER Passeriformes
FAMILY Passerellidae
BREEDING HABITAT Forests, Boreal
NONBREEDING HABITAT Forests, Temperate Eastern
GLOBAL POPULATION SIZE 35,000,000
POPULATION CHANGE SINCE 1970 −43%
IUCN RED LIST CATEGORY Least Concern
R2R TIPPING POINT SPECIES No
US LISTING STATUS No
CANADA SPECIES AT RISK No
PIF CCS 9/20

	Midcontinent	Pacific
Pacific (n = 18)	6%	94%

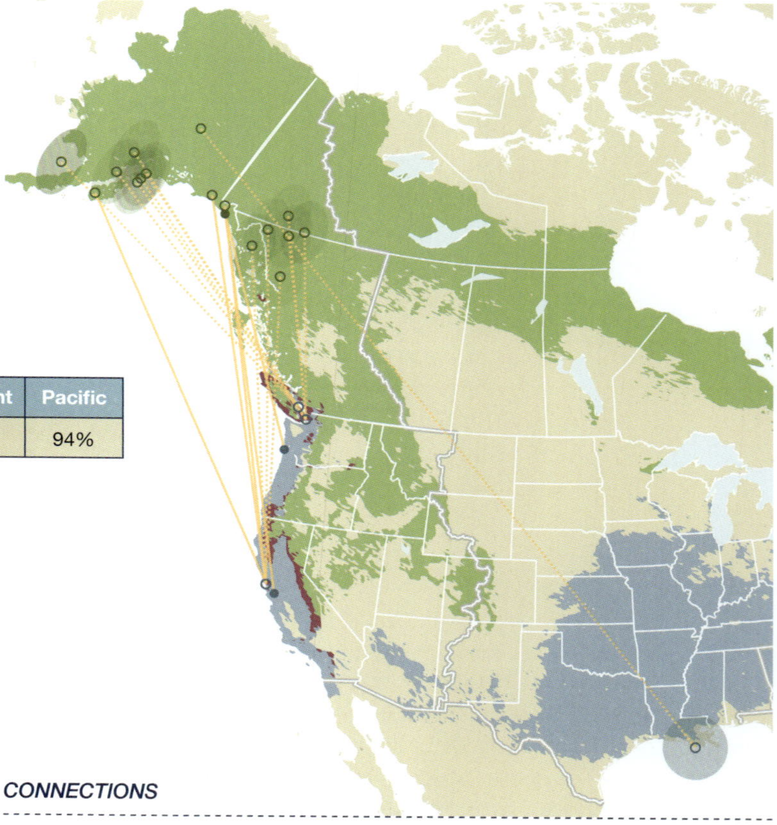

CONNECTIONS

MIGRATORY CONNECTIVITY

OVERALL Migratory connectivity information is available only for the sooty fox sparrow (*P. i. unalaschcensis* group); band reencounter and geolocator data show the majority of the population (94%) remain within the Pacific Flyway during both breeding and nonbreeding seasons, indicating strong connectivity across populations. However, within the Pacific Flyway population, connectivity is weak due to substantial mixing of breeding populations in Alaska and British Columbia at nonbreeding sites along the Pacific coast. East-west parallel migration is suggested within the broadly distributed red fox sparrow group. Morphological distinctions provide some insights into connectivity across the species; four distinct subspecies groups maintain separate breeding and largely separate migration and nonbreeding distributions, with all populations migrating either short or long distances (Weckstein et al. 2020).

BANDING Only two breeding to nonbreeding reencounters link southern coastal Alaska breeders to nonbreeding sites in coastal Washington and California. Breeders from the Canadian Maritimes migrated to the mid-Atlantic coast (Brewer et al. 2000, in Weckstein et al. 2020).

GEOLOCATOR TRACKING A small sample of sooty fox sparrows from nonbreeding sites in British Columbia (n = 9) and central California (n = 6) bred in the same general region of northwestern British

Columbia and southern coastal Alaska, suggesting weak connectivity (Fraser et al. 2018). These data do not support a previously suggested leapfrog migration described in this subspecies group (Swarth 1920; Bell 1997).

OTHER APPROACHES This species is extremely variable across its large range, with up to 18 subspecies described, falling into four recognized subspecies groups that may represent separate species (Weckstein et al. 2020). Each of the four recognized subspecies group maintains largely separate migration and nonbreeding distributions, but with substantial overlap among the three western groups in Southern California (Garrett et al. 2000). Within the subspecies groups, analysis of specimens suggests some finer-scale connectivity, including possible leapfrog migration of populations within the sooty fox sparrow group along the Pacific coast (Bell 1997) and fine-scale geographic separation (e.g., inland versus coastal) of nonbreeding sooty and slate-colored fox sparrow subspecies in California (Hampton 2016). Within the boreal red fox sparrow group, western-breeding *P. i. zaboria* migrate primarily to the south-central US, and eastern-breeding *P. i. iliaca* migrate to the southeastern US and mid-Atlantic coast (American Ornithologists Union 1957).

MIGRATORY PATHWAYS Tracking and banding data indicate a coastal migration route in pre- and post-breeding for Fox Sparrows breeding in the Pacific Northwest; although it has been suggested that some individuals may cross the northeastern Pacific Ocean while migrating between breeding areas in Alaska and California nonbreeding sites (Swarth 1920; Bell 1997). Both Pacific coast and Sierra Nevada Fox Sparrow populations migrate primarily westward to nonbreeding sites in central and Southern California.

DATA CONTRIBUTORS

E.M. Adams; J. Boelsma; B.F. Cousens; R.L. Cormier; K.C. Fraser; A. Gilbert; G. Gress; C. Grigsby; D.L. Humple; C. McIntyre; A. Nighingale; L.M. Phillips; A. Roberto-Charron; M. Simmons; I.J. Stenhouse; S. Weidensaul; E.J. Williams.

Spizelloides arborea AMERICAN TREE SPARROW

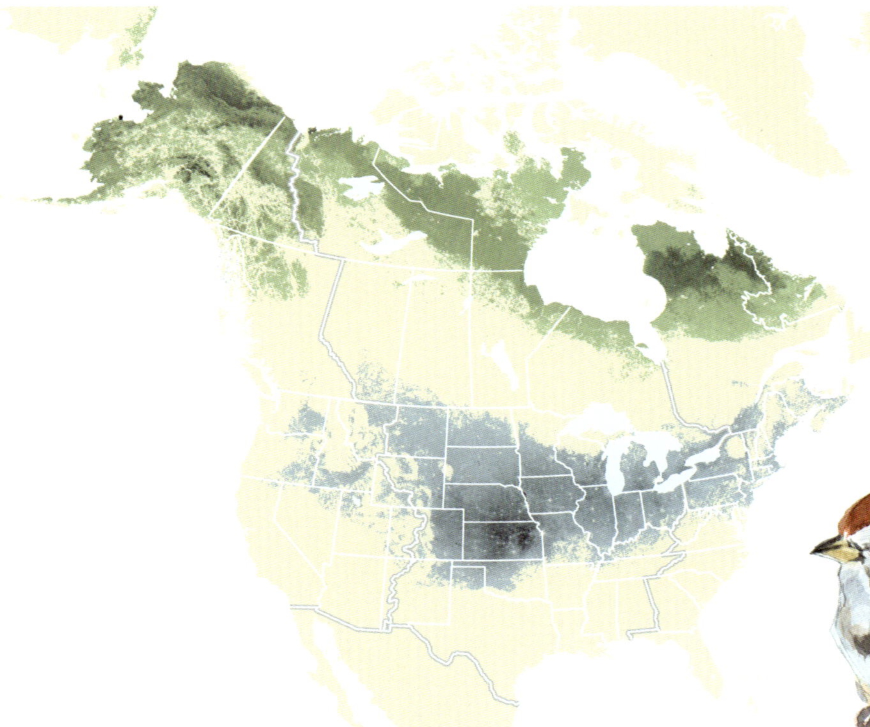

DISTRIBUTION AND ABUNDANCE

DISTRIBUTION AND ABUNDANCE: It breeds in shrubby Arctic tundra from western and northern Alaska, across northern Canada to Hudson Bay and Labrador; it also breeds near the tree line in the mountains of central Alaska, Yukon, and northern British Columbia. Its nonbreeding distribution is across most of the northern US, with highest abundance in the upper Midwest and central Great Plains states, and more patchily in the Great Basin (Fink et al. 2023). Abundance at the southern edge of the nonbreeding distribution is highly variable and dependent on temperatures during the nonbreeding season (Naugler et al. 2020).

ORDER Passeriformes
FAMILY Passerellidae
BREEDING HABITAT Tundra, Arctic
NONBREEDING HABITAT Open Country, Habitat Mosaic
GLOBAL POPULATION SIZE 26,000,000
POPULATION CHANGE SINCE 1970 −59%
IUCN RED LIST CATEGORY Least Concern
R2R TIPPING POINT SPECIES No
US LISTING STATUS No
CANADA SPECIES AT RISK No
PIF CCS 12/20

MIGRATORY CONNECTIVITY

OVERALL Migratory connectivity is relatively unknown. Based on morphological variation, eastern and western breeding populations maintain broadly separate nonbreeding distributions, but finer-scale connectivity is not known.

RADIO TRACKING There are several tracking projects with the Motus Wildlife Tracking System across the range. There are no individual breeding to nonbreeding locations, but there appear to be many detections through the Great Lakes region during migration.

OTHER APPROACHES Two subspecies are recognized, broadly separating eastern and western breeding populations, which migrate to nonbreeding areas primarily east and west of the Rocky Mountains, respectively (American Ornithologists Union 1957; Naugler et al. 2020).

MIGRATORY PATHWAYS There is some evidence from specimens and banding that western-breeding individuals migrate eastward before moving south, and that females are more prevalent in southern parts of the nonbreeding range and migrate northward later in the pre-breeding season (Baumgartner 1968, cited in Naugler et al. 2020).

DARK-EYED JUNCO *Junco hyemalis*

DISTRIBUTION AND ABUNDANCE

This species is among the most widespread and abundant North American birds. Migratory behavior varies across the range and includes migratory, partially migratory, and sedentary populations. It breeds widely throughout boreal and mixed forests from northern Alaska to, s Newfoundland outh along the Pacific coast to northern Baja California, in the western mountains to Arizona and New Mexico, and in the Appalachian Mountains to northern Georgia. Its nonbreeding range includes most of the US (although it is rare along the Gulf Coast and Florida), northern Mexico, and along the coasts north to British Columbia and Newfoundland. Declines are fairly uniform throughout the range and have accelerated in recent decades (Sauer et al. 2019).

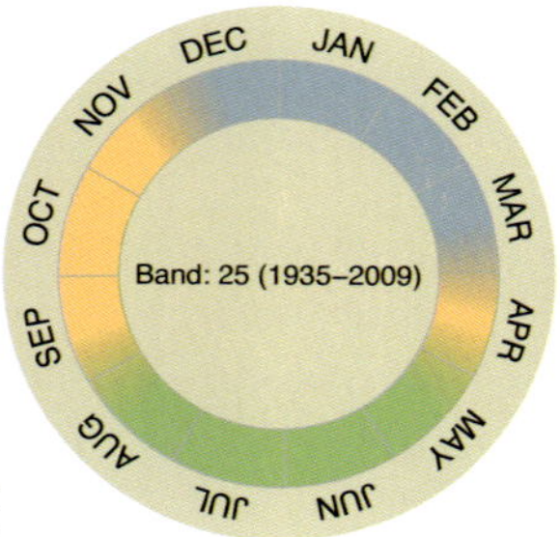

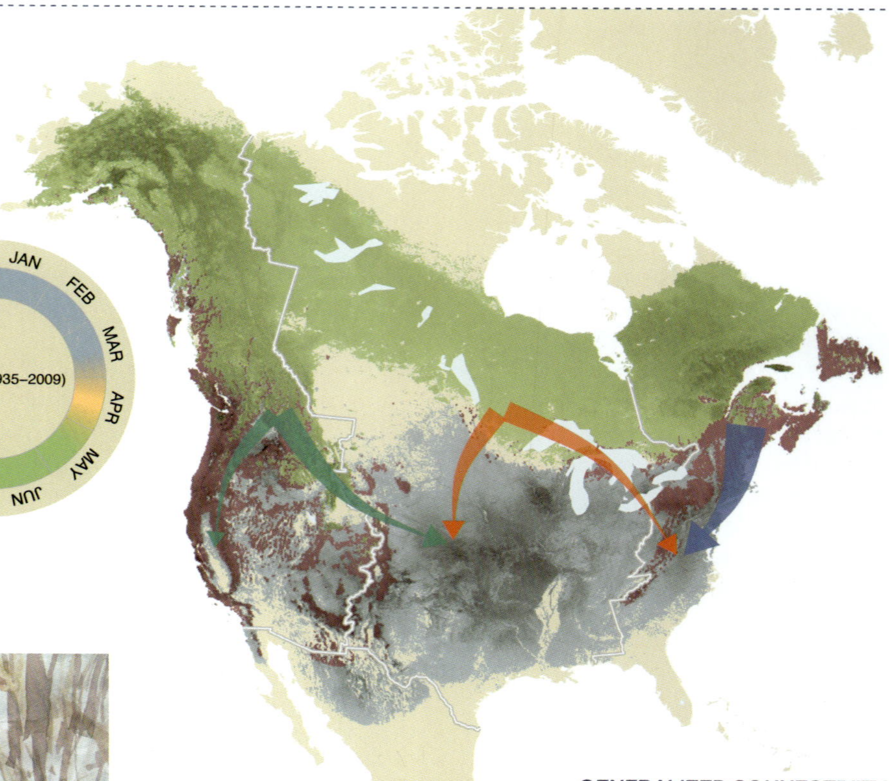

GENERALIZED CONNECTIVITY

ORDER Passeriformes
FAMILY Passerellidae
BREEDING HABITAT Forests, Temperate Generalist
NONBREEDING HABITAT Open Country, Habitat Mosaic
GLOBAL POPULATION SIZE 220,000,000
POPULATION CHANGE SINCE 1970 −46%
IUCN RED LIST CATEGORY Least Concern
R2R TIPPING POINT SPECIES No
US LISTING STATUS No
CANADA SPECIES AT RISK No
PIF CCS 9/20

	Atlantic	Midcontinent	Pacific
Atlantic (n = 8)	100%	–	–
Midcontinent (n = 6)	50%	50%	–
Pacific (n = 8)	–	62%	38%

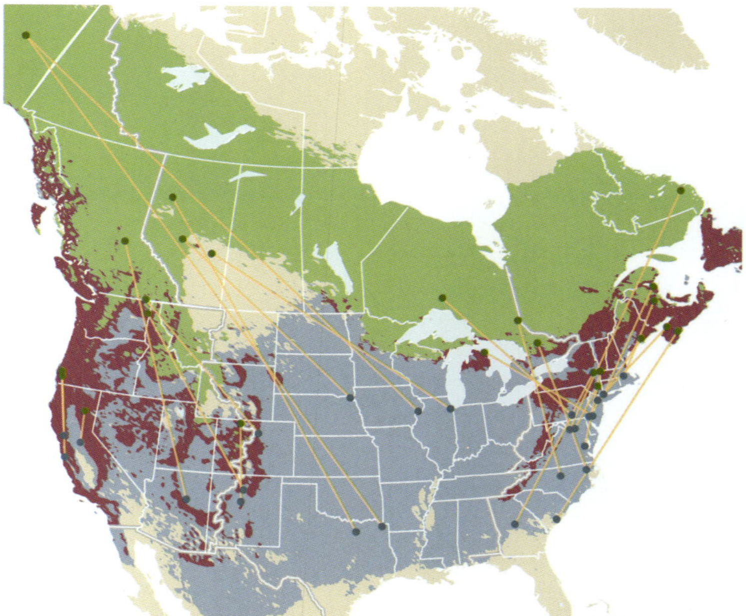

CONNECTIONS

MIGRATORY CONNECTIVITY

OVERALL Aside from broad separation of eastern/boreal and various western subspecies, migratory connectivity is poorly known; band reencounter data and analysis of geographically variable subspecies suggest moderate connectivity but with considerable mixing of regional populations in the nonbreeding season. MC is not calculated due to limited sample size. Atlantic Flyway breeding birds remain with the Atlantic Flyway during the nonbreeding season. Midcontinent Flyway birds migrate within the flyway (50%) and to the Atlantic Flyway (50%). Similarly, Pacific Flyway breeding birds migrate within the flyway (38%) and to the Midcontinent Flyway (62%), suggesting that an east-west divide may occur within the Midcontinent Flyway.

BANDING Band reencounter data suggest broad east-west separation and connectivity of populations, with eastern boreal breeders migrating to the eastern US, and Alaskan and western boreal breeders migrating to the midwestern US, northern Rocky Mountain breeders moving south in the western US, and Pacific coast breeders staying largely in Pacific states. More detailed analysis of banding

INTERESTING NOTE

Eastern populations exhibit differential migration, with female birds migrating farther than males (Nolan et al. 2020b).

data from the eastern US revealed differential migration among sex and age classes (Ketterson and Nolan 1976, 1983).

GENETICS The Bird Genoscape Project is constructing a genoscape to better understand the migratory connectivity of Dark-eyed Juncos (https://www.birdgenoscape.org/dark-eyed-junco/).

OTHER APPROACHES No tracking, genetic, or stable isotope studies of migratory connectivity were found for this species. The species is highly variable, with up to 15 subspecies recognized and at times considered as many as five full species (American Ornithologists Union 1957; Phillips et al. 1964; Nolan et al. 2020b). In general, major subspecies groups remain separated throughout the year, with boreal-breeding (Alaska to Newfoundland), Slate-colored Juncos (*J. h. hyemalis* group) migrating to the eastern and central US and western-breeding, Oregon Junco (*J. h. oreganus* group) remaining in the western US and northwestern Mexico. Across the Intermountain West, migration patterns are more complex, and although each subspecies (representing distinct breeding populations in isolated mountain ranges) maintains separate core nonbreeding ranges, considerable mixing occurs (e.g., in Arizona, New Mexico, and Colorado), and up to four subspecies may be present in a single nonbreeding flock. In addition, several montane breeding populations (e.g., Appalachian Mountains, California coastal ranges) undergo short-distance, largely altitudinal, migrations to remain close to breeding areas; others are largely sedentary (Nolan et al. 2020b).

MIGRATORY PATHWAYS Juncos migrate across broad fronts, with considerable mixing of subspecies in migratory flocks throughout the Intermountain West. Boreal breeding slate-colored juncos appear to separate more, however, with Alaskan and western boreal birds migrating in a southeasterly direction to the central US, and eastern boreal birds migrating southwesterly to the eastern US. The Appalachian Mountains, with its own largely resident population, may serve as a barrier to east-west mixing of northern breeding populations (Fink et al. 2023).

Zonotrichia leucophrys WHITE-CROWNED SPARROW

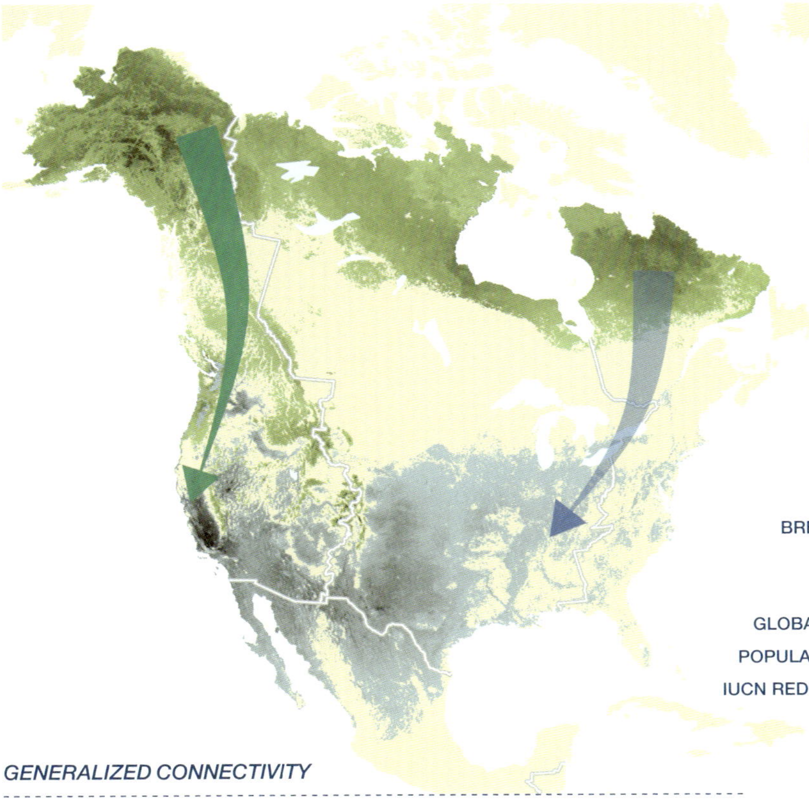

GENERALIZED CONNECTIVITY

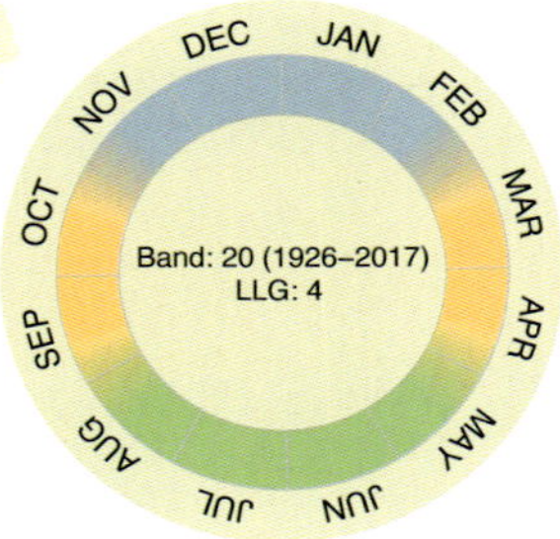

DISTRIBUTION AND ABUNDANCE

It breeds throughout Alaska and eastward across the Canadian Arctic tundra and taiga to Labrador, south through the high mountains of the western US, and along the Pacific coast to Southern California. Its nonbreeding range is large at low elevations throughout the southwestern US, Pacific states, and northern Mexico, and in decreasing abundance across the south-central and southeastern US. Breeding populations have declined in Alaska and the Pacific states, but have been stable elsewhere (Sauer et al. 2019).

ORDER Passeriformes
FAMILY Passerellidae
BREEDING HABITAT Tundra, Arctic
NONBREEDING HABITAT Open Country, Habitat Mosaic
GLOBAL POPULATION SIZE 79,000,000
POPULATION CHANGE SINCE 1970 −13%
IUCN RED LIST CATEGORY Least Concern
R2R TIPPING POINT SPECIES No
US LISTING STATUS No
CANADA SPECIES AT RISK No
PIF CCS 9/20

MIGRATORY CONNECTIVITY

OVERALL Migratory connectivity is relatively unknown due to limited data throughout the range. However Atlantic and Pacific flyway breeding populations remain within the same flyway during the nonbreeding season suggesting it may be moderate to strong.

BANDING Breeding to nonbreeding reencounters of northwestern *Z. l. pugetensis* in California were primarily in the San Francisco Bay area. Three more recent band recoveries link breeding *Z. l. gambelli* from Alaska to nonbreeding sites in Southern California and New Mexico (Lisovski et al. 2019). There is limited banding data from the eastern part of the range.

RADIO TRACKING There are several tracking projects through the Motus Wildlife Tracking System. There have been detections in both the eastern and western parts of the range but no individual breeding to nonbreeding detections.

	Midcontinent	Pacific
Atlantic (n = 1)	100%	–
Pacific (n = 22)	–	100%

GEOLOCATOR TRACKING Data from light-level geolocators showed that four tracked individuals from nonbreeding areas in Davis,

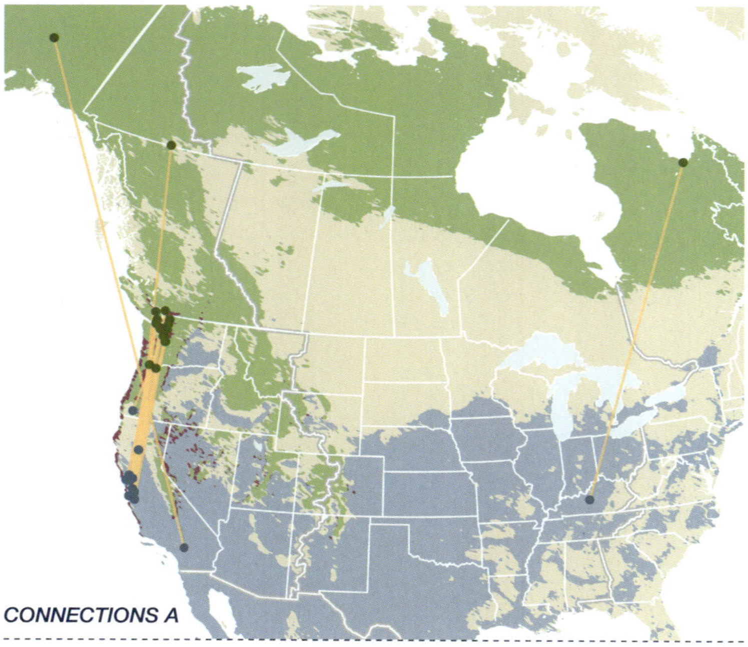

CONNECTIONS A

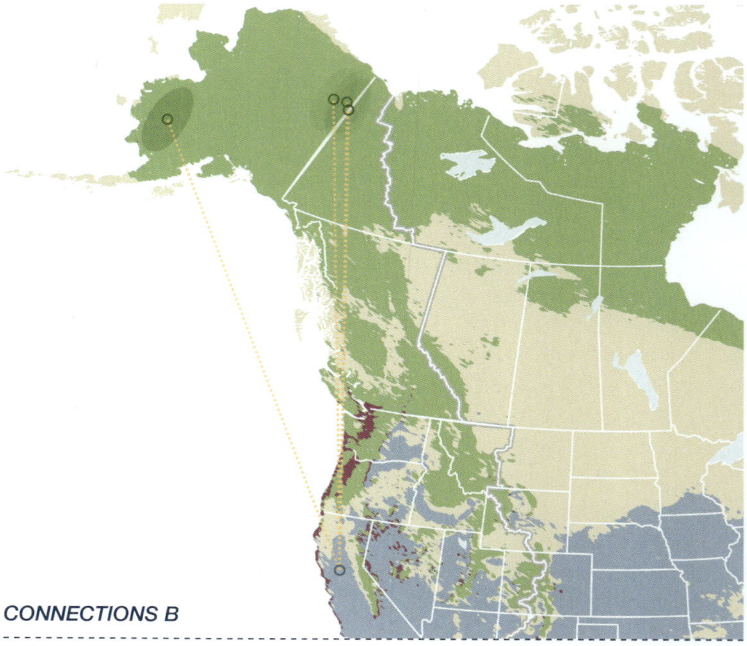

CONNECTIONS B

California, migrated to breeding sites in northern and western Alaska, separated by up to 1,500 km (Lisovski et al. 2019).

STABLE ISOTOPES Stable hydrogen isotopes in feathers collected along the Pacific Flyway (*Z. l. gambelli*) indicate that individuals at specific nonbreeding sites may come from throughout the large breeding range, suggesting weak connectivity (Losovski et al. 2019). Stable isotope data also suggest a leapfrog migration in western North America, with more northerly breeders migrating greater distances to more southerly nonbreeding areas.

GEOGRAPHIC SUBSPECIES Four migratory subspecies have been long known to occupy largely separate distributions throughout the annual cycle, indicating strong migratory connectivity at large regional scales: Alaskan and western boreal *Z. l. gambelli* migrates primarily to the southwestern US and northwestern Mexico; eastern boreal *Z. l. leucohrys* migrates to the southeastern and south-central US; Rocky Mountain breeding *Z. l. oriantha* migrates primarily to northern Mexico; and Pacific Northwestern *Z. l. pugetensis* remains along the South Pacific coast in the nonbreeding season, where they overlap partly with resident *Z. l. nutalli* and migratory *Z. l. gambelli* (Chilton et al. 2020).

MIGRATORY PATHWAYS Tracks of four *Z. l. gambelli* indicate a clockwise loop migration, with pre-breeding migration closer to the Pacific coast, and post-breeding migration beginning with an easterly movement from Alaska and then south along the eastern slope of the Rocky Mountains (Lisovski et al. 2019). More generally, the three migratory subspecies maintain separate migration routes, where *Z. l. gambelli* and *Z. l. oriantha* overlap during migration; timing of passage for *Z. l. oriantha* is earlier in post-breeding and later in pre-breeding (Phillips et al. 1964). Analysis of banding recoveries from 1920 to 1963 (n = 198) corroborated previous specimen data regarding migration routes for three of the four migratory subspecies of White-crowned Sparrows (Cortopassi and Mewaldt 1965).

DATA CONTRIBUTORS

S. Lisovski; Z. Németh; M. Ramenofsky; J.C. Wingfield.

GOLDEN-CROWNED SPARROW *Zonotrichia atricapilla*

DISTRIBUTION AND ABUNDANCE

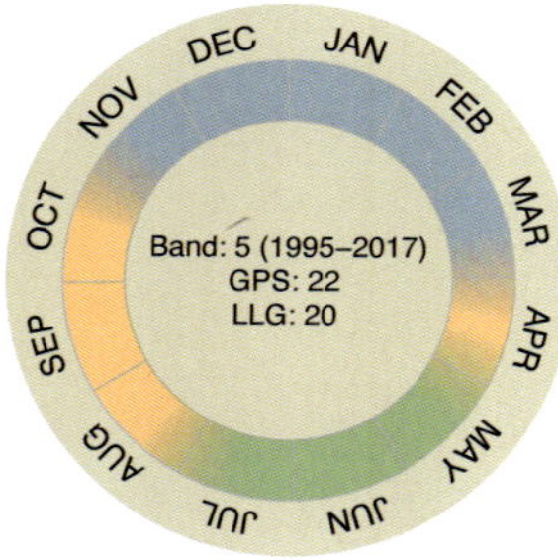

It breeds in shrubby coastal lowlands of southern and western Alaska and also near the tree line in mountains of interior Alaska, Yukon, and southwestern Northwest Territories, south to southern British Columbia and southwestern Alberta. Its nonbreeding range is primarily on the Pacific Slope of California, Oregon, and Washington, sparsely north to southwestern British Columbia, south to northwestern Baja California, and east of the coastal ranges and Sierra Nevada. Declines are steepest along the Alaska Coast, but the breeding population has increased in British Columbia (Sauer et al. 2019).

ORDER Passeriformes
FAMILY Passerellidae
BREEDING HABITAT Tundra, Alpine
NONBREEDING HABITAT Open Country, Habitat Mosaic
GLOBAL POPULATION SIZE 7,500,000
POPULATION CHANGE SINCE 1970 −28%
IUCN RED LIST CATEGORY Least Concern
R2R TIPPING POINT SPECIES No
US LISTING STATUS No
CANADA SPECIES AT RISK No
PIF CCS 10/20

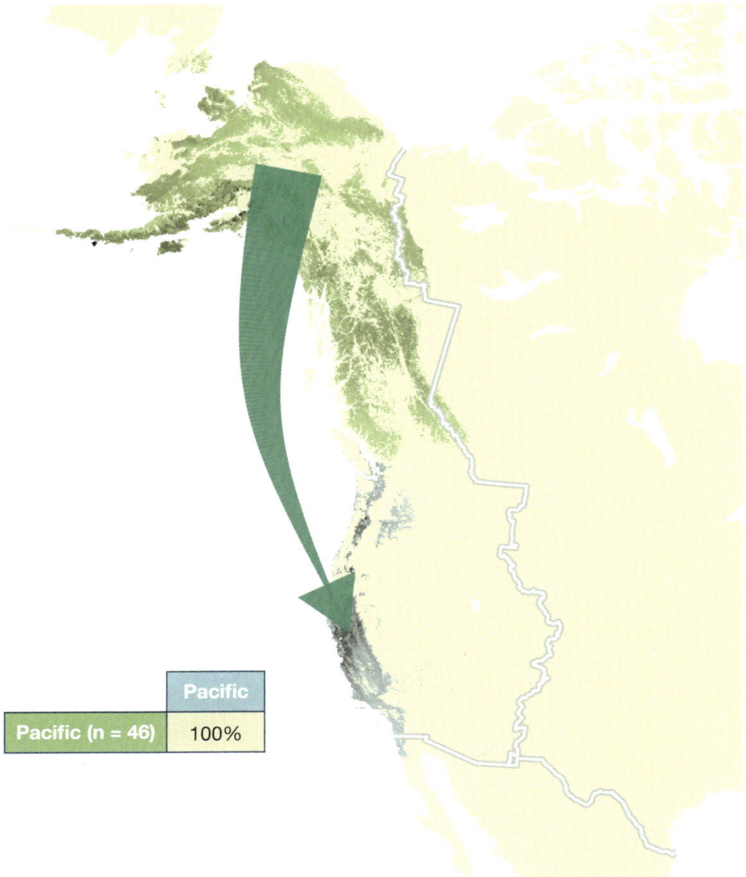

GENERALIZED CONNECTIVITY

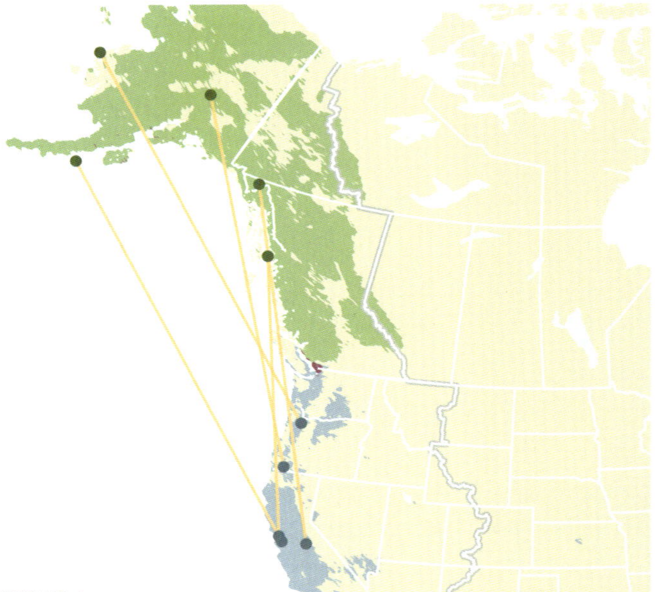

CONNECTIONS A

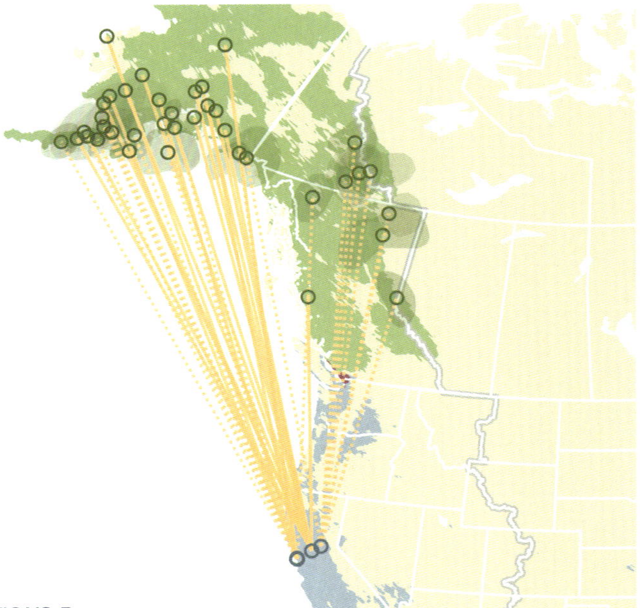

CONNECTIONS B

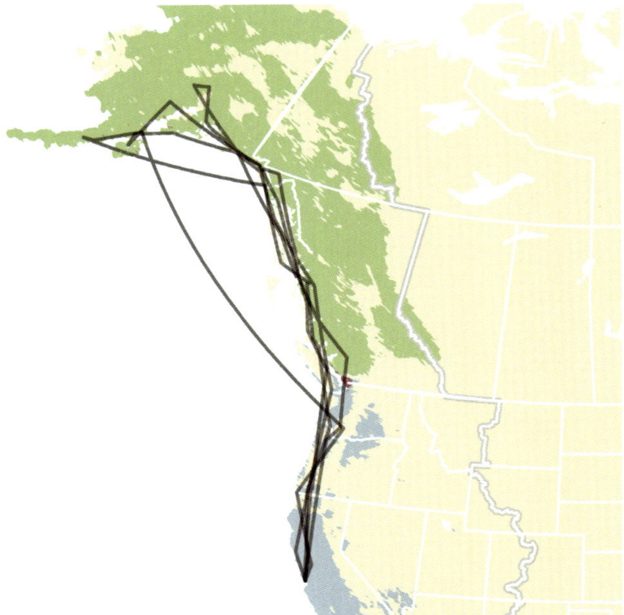

PATHWAYS

MIGRATORY CONNECTIVITY

OVERALL Based on several tracking studies (Seavy et al. 2012; Cormier et al. 2016; Iverson et al. 2023; Block et al. 2023) and band reencounter data, Golden-crowned Sparrows spend the breeding and nonbreeding season within the Pacific Flyway. A migratory divide may exist, however, between coastal and inland nonbreeding populations that follow separate migration routes and breed in disparate parts of the breeding range. MC was not calculated due to one breeding and nonbreeding region.

BANDING A small number of breeding to nonbreeding reencounters (n = 5) suggest that breeders in central and western Alaska may spend the nonbreeding season farther north than breeders along Alaska's southern coast.

RADIO TRACKING Detections from tags deployed through the Motus Wildlife Tracking System show connections between Vancouver, British Columbia, and southeastern Alaska.

GEOLOCATOR TRACKING Four individuals tracked with light-level geolocator tags from a nonbreeding site in Marin County, California, were found to breed along the Gulf of Alaska, at sites spanning 1,200 km of coastline west to the Aleutian Islands (Seavy et al. 2012). Another geolocator study found similar results with additional individuals tracked from Marin County (a coastal nonbreeding site) to breeding areas along the Gulf of Alaska's coast (Cormier et al. 2016). The same study also tracked individuals from an inland nonbreeding site in Placer County, California, to inland breeding sites in Yukon, Northwest Territories, and British Columbia—suggestive of a migratory divide on the nonbreeding grounds (Cormier et al. 2016).

SATELLITE TRACKING Similarly, nonbreeding individuals tracked with GPS archival tags migrated to breeding grounds in Alaska (n = 18) and British Columbia (n = 1) from nonbreeding areas in Marin and Yolo counties, California (Iverson et al. 2023). An additional four individuals tracked from the same nonbreeding site at Santa Cruz, California, to nonoverlapping breeding sites along the southern Alaska coast, up to 1,800 km apart indicates that individuals form separate social networks across seasons (Block et al. 2023).

OTHER APPROACHES No genetic or stable isotope studies of migratory connectivity were found for this species. Individuals in the same nonbreeding population in Santa Cruz, California, were found to not be strongly related to each other (Arnberg et al. 2015).

MIGRATORY PATHWAYS Coastal Alaska breeders follow a coastal migration route in pre- and post-breeding to nonbreeding areas along the Oregon and California coasts (Cormier et al. 2016; Iverson et al. 2023). Routes during pre-breeding migration were lower in elevation on average and shorter in duration, with fewer long stopovers than in post-breeding migration (Iverson et al. 2023). Tracking data also showed some individuals used the same stopover areas in both north- and southbound migration (Iverson et al. 2023). However, this was not found in individuals from the Santa Cruz population, but conclusions are limited due to low sample size (Block et al. 2023). Birds from inland nonbreeding sites in California followed a separate inland migration pathway to interior breeding sites in Yukon and British Columbia, providing evidence of a migratory divide between coastal and interior populations (Cormier et al. 2016).

DATA CONTRIBUTORS

T.A. Block; A.S. Chaine; R.L. Cormier; T. Gardali; J. Hull; D. Humple; A.R. Iverson; B.E. Lyon; N.E. Seavy; D. Shizuka.

HARRIS'S SPARROW *Zonotrichia querula*

DISTRIBUTION AND ABUNDANCE

It breeds in the transition zone between boreal forest and tundra in north-central Canada, from Northwest Territories and Nunavut south to northern Saskatchewan, northern Manitoba, and northwestern Ontario. Its nonbreeding distribution is in the south-central Great Plains, centered in Kansas, Oklahoma, and northern Texas (Fink et al. 2023).

ORDER Passeriformes
FAMILY Passerellidae
BREEDING HABITAT Tundra, Arctic
NONBREEDING HABITAT Forests, Temperate Eastern
GLOBAL POPULATION SIZE 2,000,000
POPULATION CHANGE SINCE 1970 −61%
IUCN RED LIST CATEGORY Near Threatened
R2R TIPPING POINT SPECIES Yellow-Alert
US LISTING STATUS No
CANADA SPECIES AT RISK No
PIF CCS 14/20

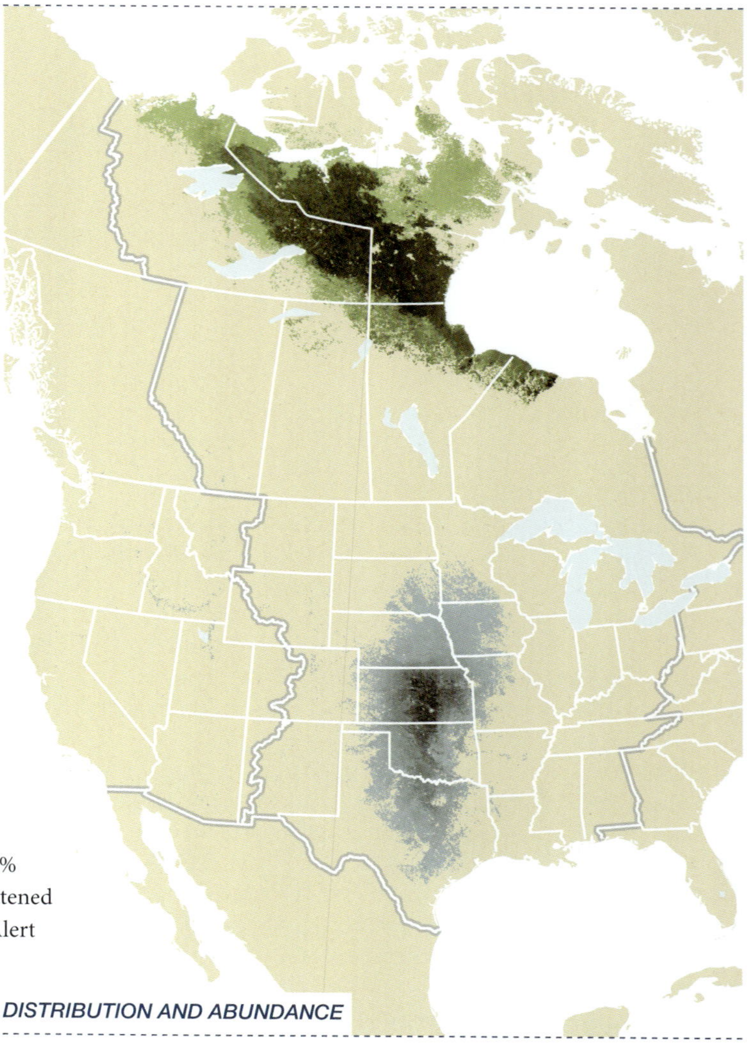

DISTRIBUTION AND ABUNDANCE

MIGRATORY CONNECTIVITY

OVERALL Migratory connectivity is unknown. No information on migratory connectivity was found for this species.

BANDING Banding studies during migration and the nonbreeding season indicate that birds migrate through a narrow north-south corridor in the central Great Plains, and that individuals exhibit moderate levels of site fidelity at nonbreeding sites (Brewer et al. 2000; Norment et al. 2020b).

RADIO TRACKING Several tags were deployed within the Motus Wildlife Tracking System. There are no detections between breeding and nonbreeding areas, but there are some long-distance detections within the Great Plains region.

INTERESTING NOTE

Harris's Sparrow is the only endemic breeding songbird in Canada and was among the last species in North America to have its nest discovered by western scientists (Norment et al. 2020b).

MIGRATORY PATHWAYS Although the core migration route is narrowly confined to the central Great Plains, this species regularly occurs in very small numbers throughout the western US, and less regularly in the east (Fink et al. 2023); it usually occurs in flocks with other sparrow species (Norment et al. 2020b).

Zonotrichia albicollis WHITE-THROATED SPARROW

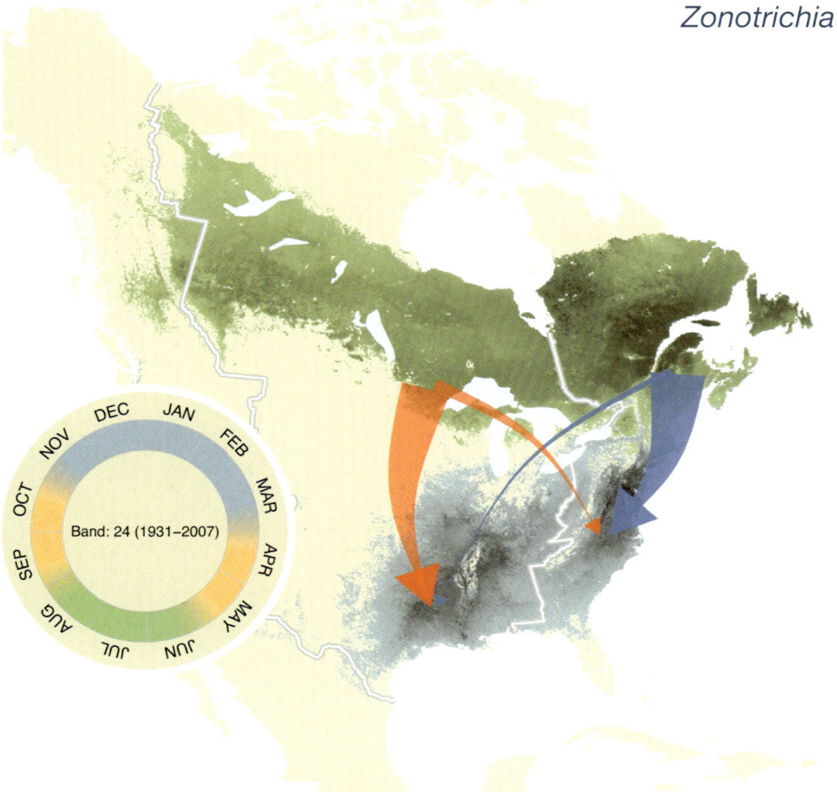

GENERALIZED CONNECTIVITY

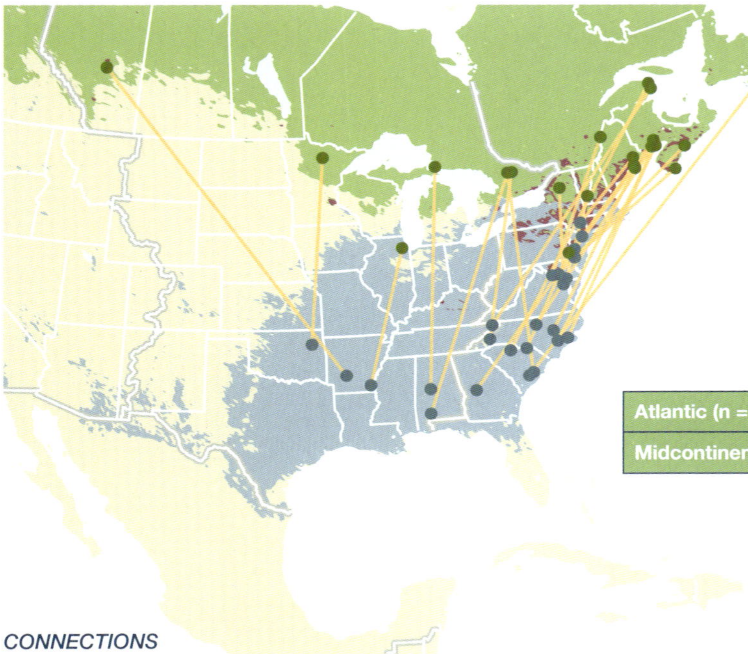

	Atlantic	Midcontinent
Atlantic (n = 16)	94%	6%
Midcontinent (n = 6)	17%	83%

CONNECTIONS

DISTRIBUTION AND ABUNDANCE

This is a widespread breeder across the boreal forest region from Newfoundland and the Canadian Maritimes to northeastern British Columbia and Northwest Territories and south to northern New England and the Great Lakes states; also patchily west of the Continental Divide in central British Columbia and Yukon. Its nonbreeding range is throughout the southeastern US except peninsular Florida, north to southern New England and west to Texas, Oklahoma, and Kansas; also very sparsely across the southwestern and Pacific states. A clear east-west pattern of population trends is apparent, with eastern boreal populations exhibiting steep long-term declines, northern and central boreal breeders showing steep decline in recent decades, western boreal breeders showing stable populations, and populations in the northern Rocky Mountains steeply increasing (Sauer et al. 2019).

ORDER Passeriformes
FAMILY Passerellidae
BREEDING HABITAT Forests, Boreal
NONBREEDING HABITAT Forests, Temperate Eastern
GLOBAL POPULATION SIZE 160,000,000
POPULATION CHANGE SINCE 1970 −37%
IUCN RED LIST CATEGORY Least Concern
R2R TIPPING POINT SPECIES No
US LISTING STATUS No
CANADA SPECIES AT RISK No
PIF CCS 9/20

INTERESTING NOTE

White-throated Sparrows are differential migrants, with females migrating longer distances than males and arriving later at migration stations (Jenkins and Cristol 2002; Mazerolle and Hobson 2007).

MIGRATORY CONNECTIVITY

OVERALL Migratory connectivity is probably at least moderate, based on broad-scale band reencounter and stable isotope data indicating east-west parallel migration throughout the range. MC is not calculated due to limited sample size.

BANDING Band reencounters show east-west parallel migration, with birds from the western breeding range migrating to the south-central US west of the Mississippi River, central breeders migrating mostly to the eastern Gulf states, and eastern breeders migrating to the Atlantic Coastal Plain from New England to Georgia (Mazerolle et al. 2005).

RADIO TRACKING No tracking or genetic studies of migratory connectivity were found for this species. There are numerous tracking projects through the Motus Wildlife Tracking System; there are no individual breeding to nonbreeding detections.

STABLE ISOTOPES Stable hydrogen isotope analysis of feathers from White-throated Sparrows captured in southern Alberta were found to likely originate from the westernmost breeding populations in British Columbia, Alberta, Northwest Territories, and Yukon (Hobson et al. 2015c).

MIGRATORY PATHWAYS Western boreal breeders move eastward across Canada before migrating south, whereas breeders west of the Continental Divide likely migrate through the western US (Fink et al. 2023).

BELL'S SPARROW *Artemesiospiza belli*

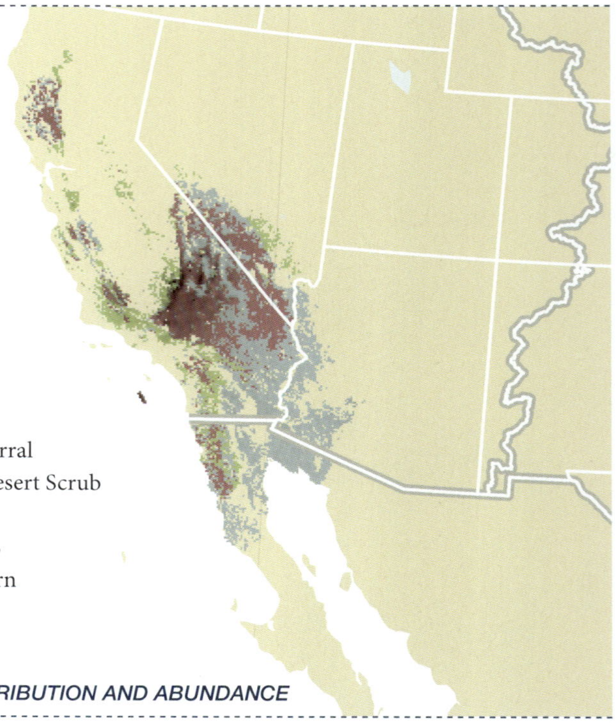

DISTRIBUTION AND ABUNDANCE

It breeds primarily in the coastal sage scrub and Mojave Desert of California, extending into southwestern Nevada and northern Baja California. Coastal sage populations are believed to be largely resident, whereas inland populations are at least partially migratory, moving to nonbreeding areas in the Colorado Desert of southeastern California and southwestern Arizona (Fink et al. 2023). Inland migratory populations are declining, whereas coastal populations appear to be stable (Sauer et al. 2019).

ORDER Passeriformes
FAMILY Passerellidae
BREEDING HABITAT Arid Lands, Chaparral
NONBREEDING HABITAT Arid Lands, Desert Scrub
GLOBAL POPULATION SIZE 270,000
POPULATION CHANGE SINCE 1970 −11%
IUCN RED LIST CATEGORY Least Concern
R2R TIPPING POINT SPECIES No
US LISTING STATUS No
CANADA SPECIES AT RISK No
PIF CCS 13/20

MIGRATORY CONNECTIVITY

OVERALL Migratory connectivity is unknown. No information on migratory connectivity was found for this species.

OTHER APPROACHES Of the four recognized subspecies, only the inland-breeding *A. b. canescens* is migratory, with nonbreeding season specimens identified from Southern California, northeastern Baja California, and western Arizona (American Ornithologists Union 1957; Martin and Carlson 2020a).

MIGRATORY PATHWAYS Some individuals move upslope during post-breeding dispersal, and the species withdraws from the northernmost and high-elevation sites in the nonbreeding season (Martin and Carlson 2020a).

SAGEBRUSH SPARROW *Artemisiospiza nevadensis*

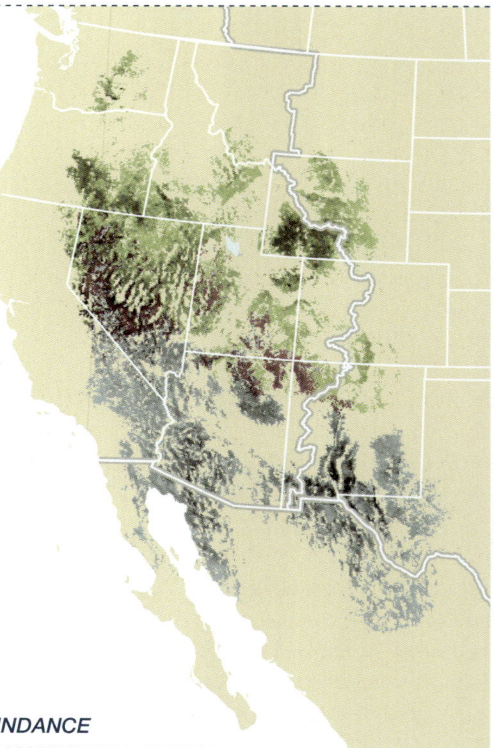

DISTRIBUTION AND ABUNDANCE

It breeds in sagebrush habitat throughout the Great Basin from eastern Washington, central Idaho, and southwestern Montana to eastern California, southern Nevada, northern Arizona, and northern New Mexico. Nonbreeding distribution is in the low deserts from southern Nevada, southeastern California, and southern Utah to northern Mexico, including Baja California. Declines are fairly uniform across the breeding range, but are steepest in eastern Oregon (Sauer et al. 2019).

ORDER Passeriformes
FAMILY Passerellidae
BREEDING HABITAT Arid Lands, Sagebrush
NONBREEDING HABITAT Arid Lands, Desert Scrub
GLOBAL POPULATION SIZE 5,400,000
POPULATION CHANGE SINCE 1970 −10%
IUCN RED LIST CATEGORY Least Concern
R2R TIPPING POINT SPECIES Yellow-Alert
US LISTING STATUS No
CANADA SPECIES AT RISK No
PIF CCS 13/20

MIGRATORY CONNECTIVITY

OVERALL Migration data on this species is limited, but stable isotopes suggest that there is weak migratory connectivity.

RADIO TRACKING A few tags were deployed through the Motus Wildlife Tracking System; one bird was detected in Wyoming and New Mexico (Motus project, University of Wyoming Coop, Sagebrush songbirds, unpublished data).

STABLE ISOTOPE Stable isotopes sampled from both nonbreeding and breeding Sagebrush Sparrows showed weak migratory connectivity in the species. Individuals from the same nonbreeding locations did not share predicted breeding locations (Knick et al. 2014).

MIGRATORY PATHWAYS Migration is through the Great Basin, primarily through Nevada and Utah; pre-breeding migration commences very early, with breeders arriving in eastern Washington by late February (Fink et al. 2023).

Pooecetes gramineus VESPER SPARROW

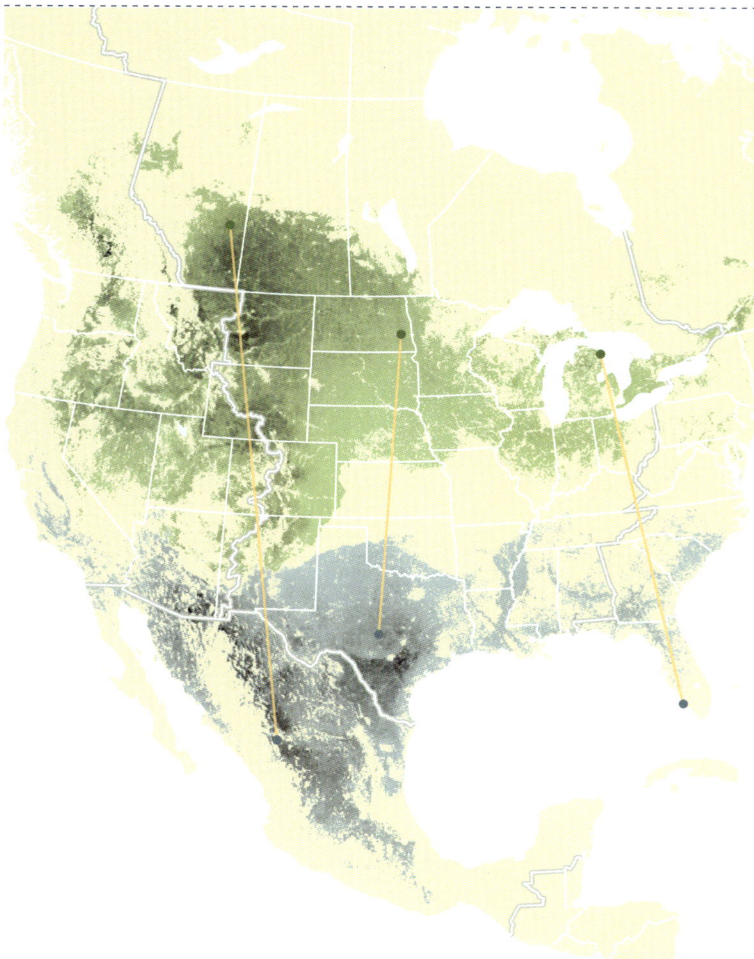

CONNECTIONS

DISTRIBUTION AND ABUNDANCE

It breeds in dry or sparse grasslands across the northern US and southern Canada, very patchily from Minnesota eastward to the Atlantic coast, and more abundantly in the western Great Plains and grasslands of the Great Basin and Intermountain West south to northern Arizona and central New Mexico. A small and threatened population breeds west of the Cascades in Washington and Oregon. Its nonbreeding range is across the southern US and northern Mexico, very patchily east of central Texas, and most abundantly in arid grasslands of the southwestern US and the Central Mexican Plateau. Declines are much steeper in eastern populations than in most of the west (Sauer et al. 2019).

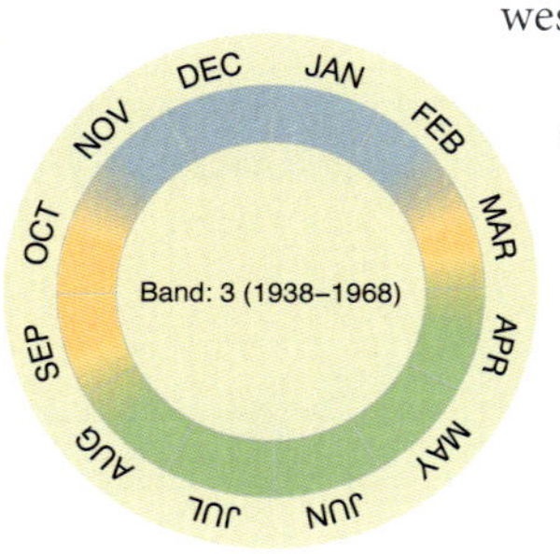

ORDER Passeriformes
FAMILY Passerellidae
BREEDING HABITAT Grasslands, Temperate
NONBREEDING HABITAT Grasslands, Chihuahuan
GLOBAL POPULATION SIZE 35,000,000
POPULATION CHANGE SINCE 1970 −34%
IUCN RED LIST CATEGORY Least Concern
R2R TIPPING POINT SPECIES No
US LISTING STATUS No
CANADA SPECIES AT RISK Endangered Coastal subspecies
PIF CCS 11/20

MIGRATORY CONNECTIVITY

OVERALL Very little is known about migratory connectivity. General east-west migratory connectivity is suggested, especially with the threatened Oregon population remaining on the Pacific Slope year-round.

BANDING Only three breeding to nonbreeding reencounters suggest an east-west parallel migration, with one bird from the eastern Great Lakes migrating to Florida, one from the northern Great Plains migrating to central Texas, and one from Alberta migrating to the Central Mexican Plateau.

RADIO TRACKING There are several tracking projects through the Motus Wildlife Tracking System. As of printing, there are no individual breeding to nonbreeding detections.

SATELLITE TRACKING Archival GPS tags were deployed in 2020–2022 on adult Oregon vesper sparrows (*P. g. affinis*) in the Klamath Mountains ecoregion (n = 26) and the Willamette Valley, Oregon (n = 14; 2022 only). Data from eight birds revealed substantial overlap in nonbreeding sites between the two breeding populations; nonbreeding areas in central California were all within 500 km of one another (S. Rockwell, unpublished data). Nonbreeding areas ranged from 39 to 35 degrees north latitude, in Colusa, Madera, Mariposa, and Tulare counties (S. Rockwell, unpublished data).

OTHER APPROACHES No genetic or stable isotope studies of migratory connectivity were found for this species. Three subspecies are usually recognized, with widespread eastern and western subspecies maintaining largely separate nonbreeding ranges (American Ornithologists Union 1957; Jones and Cornely 2020). A third distinct subspecies, the Oregon vesper sparrow, also remains separate from other populations throughout the annual cycle, therefore exhibiting strong migratory connectivity.

MIGRATORY PATHWAYS Migration is over a broad front, with breeding populations seeming to move due south to nonbreeding areas (Fink et al. 2023); post-breeding migration in the east is much later than in the west. Western birds are more abundant during post-breeding migration along the western Great Plains, compared with greater abundance through the Great Basin in pre-breeding migration, suggesting a clockwise loop migration pattern. Birds tracked from Oregon used post- and pre-breeding stopover sites in either the Central Valley or the western Sierra Nevada foothills (S. Rockwell, unpublished data).

LECONTE'S SPARROW *Ammospiza leconteii*

DISTRIBUTION AND ABUNDANCE

It breeds in wet grassland, sedge meadows, and bulrush marshes from northeastern British Columbia and Northwest Territories south through the Prairie Pothole Region of central Canada to North Dakota and Minnesota; also very patchily eastward to northeastern Ontario, western Quebec, Wisconsin, and Michigan. Its nonbreeding range is in the south-central US from southeastern Kansas, southwestern Missouri, Arkansas, and western Tennessee to the Gulf Coast, and in very small numbers east to the Gulf Coastal Plain and Florida. Declines are steepest in the core of the range in Manitoba, Saskatchewan, and North Dakota, with some increases noted in eastern parts of the range (Sauer et al. 2019).

ORDER Passeriformes
FAMILY Passerellidae
BREEDING HABITAT Wetlands, Seasonally Wet Grasslands
NONBREEDING HABITAT Grasslands, Temperate
GLOBAL POPULATION SIZE 5,100,000
POPULATION CHANGE SINCE 1970 −64%
IUCN RED LIST CATEGORY Least Concern
R2R TIPPING POINT SPECIES Orange-Alert
US LISTING STATUS No
CANADA SPECIES AT RISK No
PIF CCS 13/20

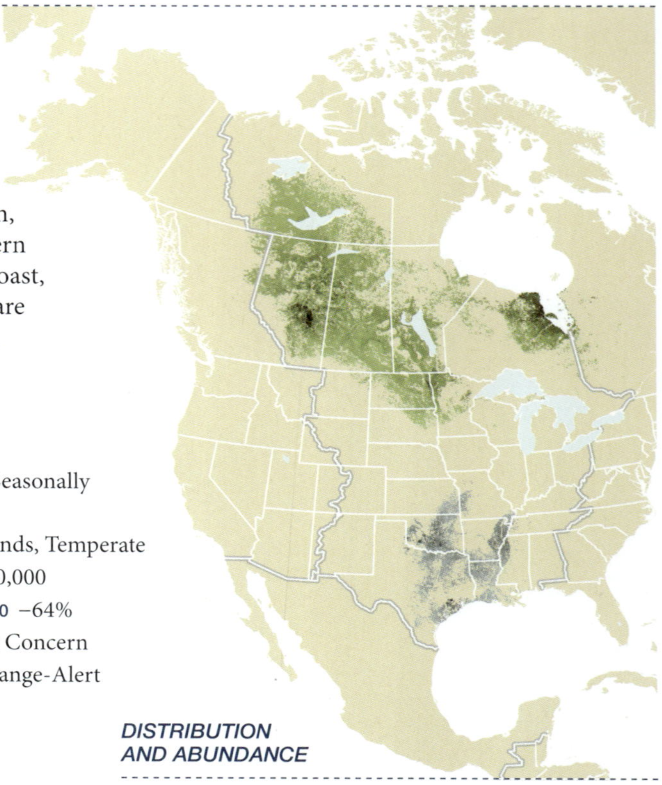

MIGRATORY CONNECTIVITY

OVERALL Migratory connectivity is unknown. No information on migratory connectivity was found for this species.

MIGRATORY PATHWAYS The species migrates primarily through a narrow region of the central prairies, with smaller numbers east to Illinois, Indiana, and the lower Mississippi Valley (Fink et al. 2023).

SEASIDE SPARROW *Ammospiza maritima*

DISTRIBUTION AND ABUNDANCE

It breeds in coastal salt marshes along the Atlantic and Gulf coasts, north to Massachusetts, and west to southern Texas. Only the northern Atlantic population (*A. m. maritima*) is migratory, spending the nonbreeding season from Virginia to northeastern Florida. The species is poorly monitored overall, but North Atlantic populations are believed to be declining most steeply, and Gulf Coast populations are showing a long-term increase. Several isolated resident populations are endangered or extinct.

ORDER Passeriformes
FAMILY Passerellidae
BREEDING HABITAT Coasts, Salt Marsh
NONBREEDING HABITAT Coasts, Salt Marsh
GLOBAL POPULATION SIZE 200,000
POPULATION CHANGE SINCE 1970 +104%
IUCN RED LIST CATEGORY Least Concern
R2R TIPPING POINT SPECIES No
US LISTING STATUS Endangered Cape Sable subspecies
CANADA SPECIES AT RISK No
PIF CCS 15/20

MIGRATORY CONNECTIVITY

OVERALL Very little is known about migratory connectivity, with the few data coming from banding and subspecific identification of specimens.

BANDING An individual banded in Connecticut was recaptured during three successive nonbreeding seasons in South Carolina (Greenlaw et al. 2022).

SATELLITE TRACKING Two individuals tagged with GPS data loggers at Kiawah Island, South Carolina, migrated to breeding sites on Delaware Bay, New Jersey (Greenlaw et al. 2022).
RADIO TRACKING Numerous tags were deployed through the Motus Wildlife Tracking System; as of printing, there are no breeding to nonbreeding detections.
OTHER APPROACHES Up to seven subspecies are usually recognized, most of isolated resident breeding populations, yet genetic analyses do not support subspecies boundaries based on phenotypic characters (American Ornithologists Union 1957; Greenlaw et al. 2022). Specimens of the migratory *A. m. maritima* have been identified during the nonbreeding season from Virginia to Florida, overlapping with the ranges of resident populations.
MIGRATORY PATHWAYS Migration of northern populations is believed to be strictly along the coast, with use of salt marshes as stopovers; however, there are several specimens collected from mortality due to collisions with communication towers while birds are migrating over inland areas in North Carolina (Greenlaw et al. 2022).

Ammospiza nelsoni **NELSON'S SPARROW**

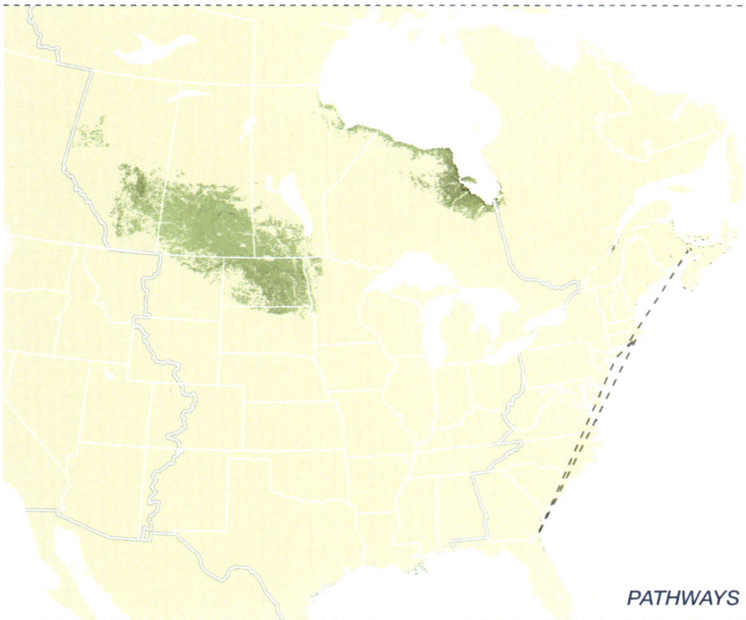

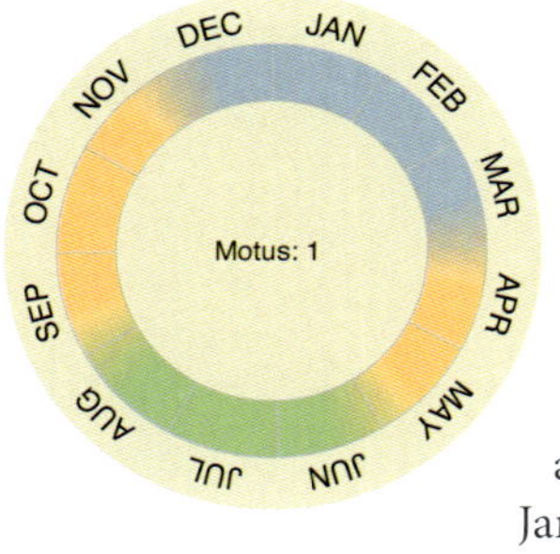

DISTRIBUTION AND ABUNDANCE
It breeds in three disjunct regions corresponding to three recognized subspecies: *A. n. nelsoni* in the prairie marshes from northeastern British Columbia south through the Canadian Prairie provinces to North Dakota and northwestern Minnesota; *A. n. alterus* along the southern shore of Hudson and James bays; and *A. n. subvirgatus* from the Gulf of St. Lawrence in Quebec east to Prince Edward Island, the Maritime provinces, and coastal northern New England (Shriver et al. 2020). Its nonbreeding range is strictly coastal, from New Jersey to northern Florida on the Atlantic coast and on the Gulf Coast from central Florida to central Texas. The prairie breeding population has increased since 1970, whereas Atlantic coastal breeders have declined (Sauer et al. 2019).

MIGRATORY CONNECTIVITY
OVERALL Migratory connectivity is unknown. There is little information on migratory connectivity other than the broadly overlapping distribution of subspecies populations during the nonbreeding season.
RADIO TRACKING There are no individual breeding to nonbreeding detections, but several detections through the Motus Wildlife Tracking System during migration connect a coastal route from Nova Scotia through the northeastern US to Georgia (A. Smith et al., unpublished data).
OTHER APPROACHES The two inland-breeding subspecies overlap extensively during the nonbreeding season along the South Atlantic coast, suggesting low connectivity across these populations, although nonbreeding birds along the Gulf Coast west of Florida are exclusively *A. n. nelsoni* from the prairies (Greenlaw and Wolfendon 2007). The Atlantic-breeding population remains along the Atlantic coast but also overlaps with inland populations during nonbreeding periods (American Ornithologists Union 1957; Greenlaw and Wolfendon 2007; Shriver et al. 2020). Individuals of all three subspecies may form mixed flocks in the nonbreeding season in coastal Virginia (Watts and Smith 2015).
MIGRATORY PATHWAYS Most prairie-breeding *A. n. nelsoni* migrate through the midwestern US to the Gulf Coast, whereas Atlantic *A. n. subvirgatus* migrate along the coast. Birds breeding in James Bay, *A. n. alterus*, and some prairie-nesters migrate across the northeast US or Appalachian Mountains to nonbreeding areas along the South Atlantic coast (Greenlaw and Wolfendon 2007). Pre-breeding migration for all populations is very late (late May to early June; Fink et al. 2023).

DATA CONTRIBUTORS
N. Ernst; K. O'Brien; N. Pau; A. Smith.

ORDER Passeriformes
FAMILY Passerellidae
BREEDING HABITAT Wetlands, Seasonally Wet Grasslands
NONBREEDING HABITAT Coasts, Salt Marsh
GLOBAL POPULATION SIZE 1,000,000
POPULATION CHANGE SINCE 1970 +149%
IUCN RED LIST CATEGORY Least Concern
R2R TIPPING POINT SPECIES No
US LISTING STATUS No
CANADA SPECIES AT RISK No
PIF CCS 14/20

SALTMARSH SPARROW *Ammospiza caudacuta*

DISTRIBUTION AND ABUNDANCE

It breeds in Atlantic coastal salt marshes from southern Maine to Virginia. Its nonbreeding range is along the Atlantic coast from New Jersey to Florida, and in smaller numbers on the Florida Gulf Coast (Fink et al. 2023).

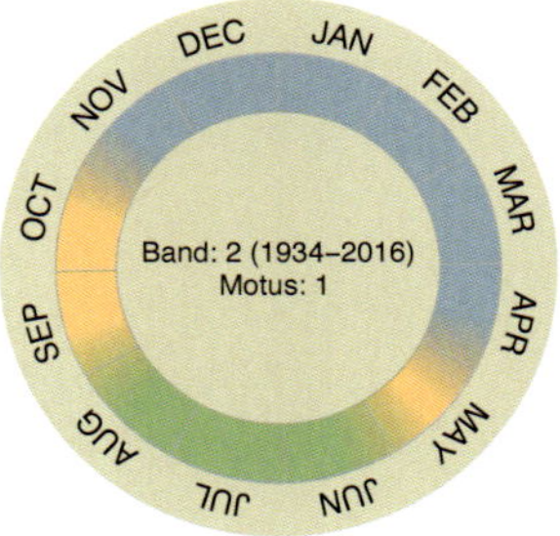

ORDER Passeriformes
FAMILY Passerellidae
BREEDING HABITAT Coasts, Salt Marsh
NONBREEDING HABITAT Coasts, Salt Marsh
GLOBAL POPULATION SIZE 60,000
POPULATION CHANGE SINCE 1970 NA
IUCN RED LIST CATEGORY Endangered
R2R TIPPING POINT SPECIES Red-Alert
US LISTING STATUS No
CANADA SPECIES AT RISK No
PIF CCS 19/20

CONNECTIONS AND PATHWAYS

MIGRATORY CONNECTIVITY

OVERALL Migratory connectivity is poorly known but appears to be weak based on limited band reencounter data and subspecific identification of specimens and trapped birds.
BANDING The few band reencounters away from breeding sites indicate that nonbreeding individuals in South Carolina originate from most of the breeding range (New Hampshire to New Jersey), whereas a recovery in Florida is from Maine. A bird banded on the Florida Gulf Coast was recovered at a nesting site in Rhode Island, having traversed a distance of 2,000 km, including across the north Florida peninsula (Greenlaw et al. 2020).
RADIO TRACKING There are no individual breeding to nonbreeding detections, but several detections during migration connect a coastal route from the northeastern US to Georgia (A. Smith et al., unpublished data).
OTHER APPROACHES Two weakly differentiated subspecies are recognized (*A. c. caudacuta* and *A. c. diversa*), representing breeding populations north and south of central New Jersey, and identification of specimens and banded individuals suggests that the nonbreeding distributions of the two subspecies overlap broadly, including on the Florida Gulf Coast (American Ornithologists Union 1957; Greenlaw and Wolfenden 2007); individuals of both subspecies, as well as all three subspecies of Nelson's Sparrow, form mixed flocks in coastal marshes of Virginia in the nonbreeding season (Watts and Smith 2015).
MIGRATORY PATHWAYS Migrations are strictly along the coast; however, Motus-tracked individuals will "cut inland" across coastal peninsulas, and there is a scattering of inland records of stray individuals during migration (Greenlaw et al. 2020). Individuals reaching the Gulf Coast presumably traverse the northern Florida peninsula.

DATA CONTRIBUTORS
N. Ernst; K. O'Brien; N. Pau; A. Smith.

BAIRD'S SPARROW *Centronyx bairdii*

DISTRIBUTION AND ABUNDANCE

It breeds in the northern Great Plains in southern Alberta, southern Saskatchewan, and southeastern Manitoba south to northeastern Wyoming and northwestern South Dakota; a few breed farther south to eastern Colorado. Its nonbreeding range is patchy in the arid grasslands from southeastern Arizona, southern New Mexico, and western Texas to southern Durango and northern Zacatecas, Mexico (Fink et al. 2023).

ORDER Passeriformes
FAMILY Passerellidae
BREEDING HABITAT Grasslands, Temperate
NONBREEDING HABITAT Grasslands, Chihuahuan
GLOBAL POPULATION SIZE 3,400,000
POPULATION CHANGE SINCE 1970 −68%
IUCN RED LIST CATEGORY Least Concern
R2R TIPPING POINT SPECIES Red-Alert
US LISTING STATUS No
CANADA SPECIES AT RISK No
PIF CCS 16/20

MIGRATORY CONNECTIVITY

OVERALL Migratory connectivity is unknown. No studies on migratory connectivity were found for this species.

RADIO TRACKING There are several tracking projects with the Motus Wildlife Tracking System across the range. As of printing, there are no individual breeding to nonbreeding locations, but there appear to be many detections through the Great Plains during migration.
GEOLOCATOR TRACKING One project is currently using light-level geolocators to track migration, but no data have been published (Bernath-Plaisted et al. 2019).
GENETICS The Bird Genoscape Project is currently working to build a genoscape for Baird's Sparrow (https://www.birdgenoscape.org/bairds-sparrow/).
MIGRATORY PATHWAYS Pathways are very poorly known; but known pathways occur in western Nebraska and western Kansas. There is some evidence for stopover in eastern Colorado and eastern Arizona (Green et al. 2020).

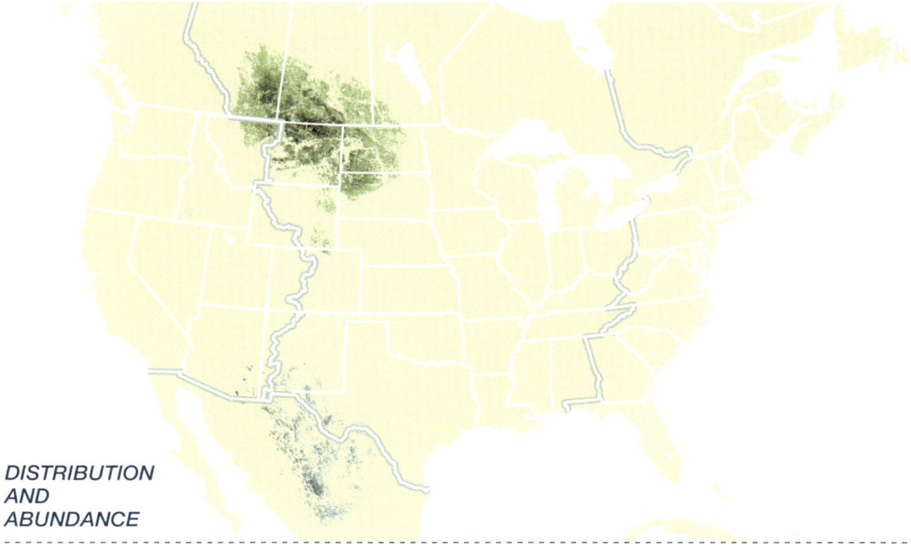

Centronyx henslowii **HENSLOW'S SPARROW**

ORDER Passeriformes
FAMILY Passerellidae
BREEDING HABITAT Grasslands, Temperate
NONBREEDING HABITAT Grasslands, Temperate
GLOBAL POPULATION SIZE 410,000
POPULATION CHANGE SINCE 1970 −23%
IUCN RED LIST CATEGORY Least Concern
R2R TIPPING POINT SPECIES Yellow-Alert
US LISTING STATUS No
CANADA SPECIES AT RISK Endangered
PIF CCS 15/20

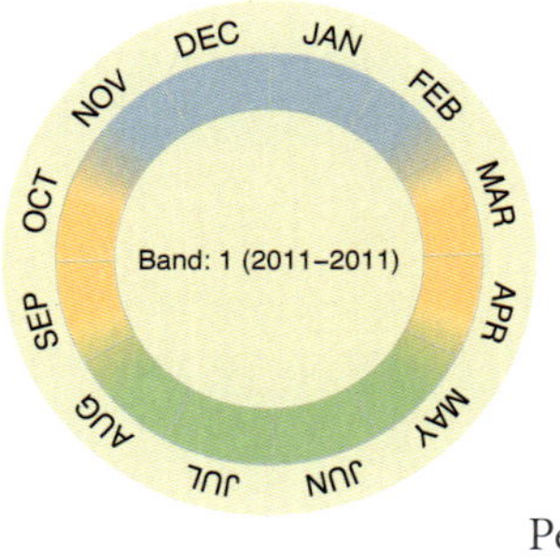

DISTRIBUTION AND ABUNDANCE
It breeds patchily in remnant tallgrass prairie and other grasslands in the upper midwestern US, from the southern Great Lakes states to northeastern Oklahoma, Kentucky, and southwestern Pennsylvania; there are isolated remnant breeding populations in northern New York and eastern North Carolina (extirpated from most of its former range in northeastern US). Its nonbreeding distribution is also patchy, from southern Arkansas, across the Gulf states to southeastern Georgia and northern Florida. Eastern populations continue to decline, whereas some increases have been noted in Iowa, Kansas, and Missouri (Sauer et al. 2019).

MIGRATORY CONNECTIVITY
OVERALL Migratory connectivity is relatively unknown. There is very little information on migratory connectivity for this species.
BANDING A single band reencounter connected a Missouri breeding site to a Louisiana nonbreeding site.
RADIO TRACKING Several tags were deployed through the Motus Wildlife Tracking System, with very few detections.
STABLE ISOTOPES A combination of stable isotope and abundance data suggested that Henslow's Sparrows at nonbreeding sites included individuals from various parts of the breeding range, but resolution for identifying populations was extremely low (Ibargüen 2004).
OTHER APPROACHES No tracking or genetic studies of migratory connectivity were found for this species. There are no detections from breeding to nonbreeding. Two subspecies are historically recognized (*C. h. henslowii* and *C. h. susurrans*), with breeding populations (largely extirpated) in eastern New York, southern New England, and the mid-Atlantic states remaining on the Atlantic Coastal Plain in the nonbreeding season, and populations west of the Appalachian Mountains migrating to the south-central US (American Ornithologists Union 1057; Herkert et al. 2020). It is unclear if isolated breeders in North Carolina represent a remnant of that subspecies.
MIGRATORY PATHWAYS This species is rarely detected on migration; there is no information on migration pathways.

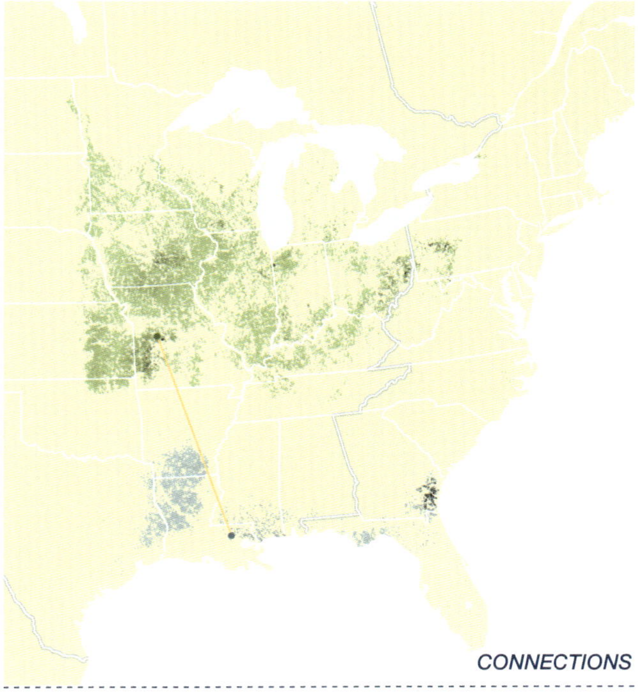

SAVANNAH SPARROW *Passerculus sandwichensis*

DISTRIBUTION AND ABUNDANCE

This widespread breeder has a complex distribution across Canada and the northern US; its centers of abundance are in Arctic Canada and Alaska, the northern Great Plains, the eastern Great Lakes and St. Lawrence Valley, and patchily throughout the Great Basin and northern Rocky Mountains. Its nonbreeding distribution is similarly patchy across the southern US and northern Mexico, rarely to Central America and the Caribbean, with centers of abundance in the Piedmont, lower Mississippi Valley, Texas, arid grasslands of the southwestern US and Central Mexican Plateau, and in the Central Valley of California. Its nonmigratory populations are along the southern Pacific coast and in Mexico. Population declines are steepest in the northeastern US, Great Lakes, and upper Midwest regions (Sauer et al. 2019).

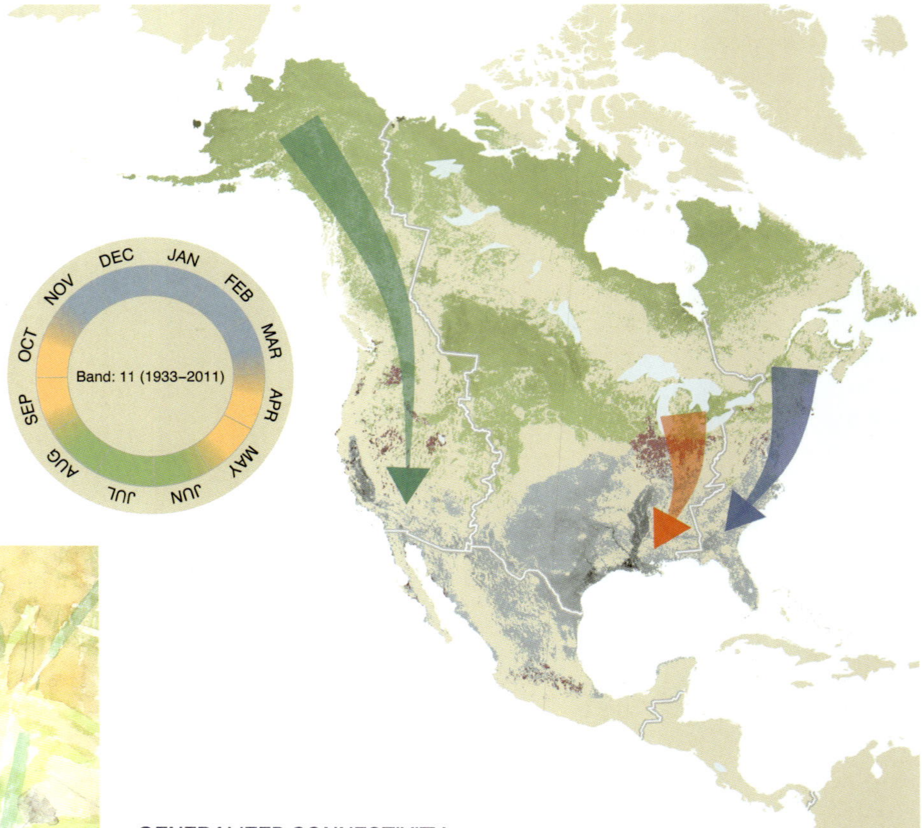

GENERALIZED CONNECTIVITY

ORDER Passeriformes

FAMILY Passerellidae

BREEDING HABITAT Grasslands, Temperate

NONBREEDING HABITAT Open Country, Developed/Disturbed

GLOBAL POPULATION SIZE 170,000,000

POPULATION CHANGE SINCE 1970 −46%

IUCN RED LIST CATEGORY Least Concern

R2R TIPPING POINT SPECIES No

US LISTING STATUS No

CANADA SPECIES AT RISK No

PIF CCS 10/20

	Atlantic	Midcontinent	Pacific
Atlantic (n = 6)	100%	–	–
Midcontinent (n = 2)	–	100%	–
Pacific (n = 2)	–	–	100%

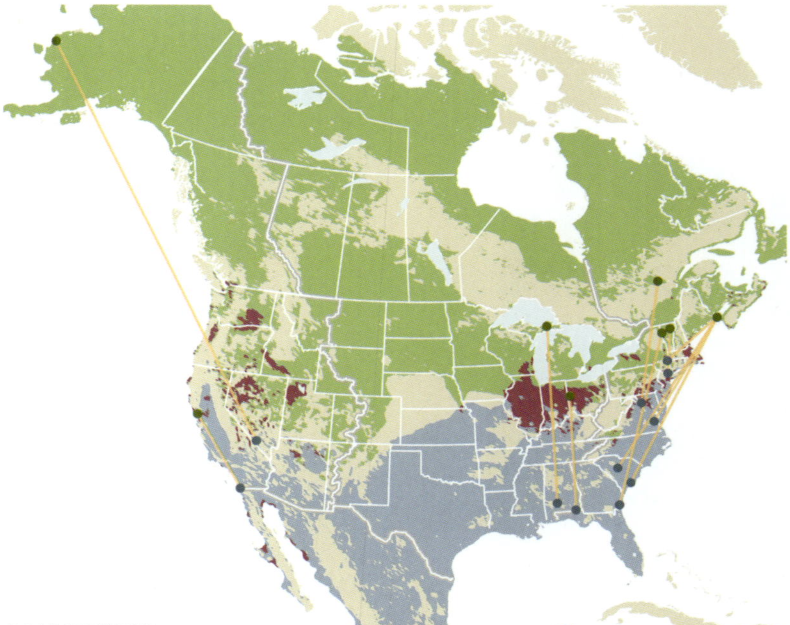

CONNECTIONS

MIGRATORY CONNECTIVITY

OVERALL Migratory connectivity, known from geolocator, band reencounter, and stable isotope data, appears strong, but MC is not calculated due to limited sample size. Atlantic, Midcontinent, and Pacific flyway breeding all remain separate and within the same flyway as breeding populations during the nonbreeding season.

BANDING Limited breeding to nonbreeding band reencounter data suggest east-west parallel migration, with movement of eastern populations in the Canadian Maritimes and New England to nonbreeding grounds from Chesapeake Bay south to Florida. Individuals breeding around the Great Lakes migrate to Gulf Coast states; an individual from Alaska was recovered in Nevada, and one from California moved south to Baja California.

RADIO TRACKING Over 500 tags were deployed through the Motus Wildlife Tracking System. As of printing, there are no individual breeding to nonbreeding detections.

GEOLOCATOR TRACKING Savannah Sparrows captured on Kent Island, New Brunswick, Canada, tracked with geolocators (n = 38), migrated mostly to the southern Atlantic coast from North Carolina to Florida, with males moving to nonbreeding sites on average

275 km farther north than females (Woodworth et al. 2016).
STABLE ISOTOPES Female Ipswich Sparrows (*P. s. princeps*) were found to migrate farther south than males, according to stable isotope signatures in claw samples (Dale and Leonard 2011).
OTHER APPROACHES This species has extensive geographic variation, with up to 28 named subspecies (American Ornithologists Union 1957; Wheelwright and Rising 2020). An analysis of diagnosable specimens indicates a high degree of separation and east-west migratory connectivity with: distinctive Ipswich sparrows migrating from Sable Island, Nova Scotia, to sites along the outer Atlantic coast from New England to Georgia; breeders from eastern Canada and northeastern US migrating to the southeastern US; breeders from the Great Plains migrating to the south-central US and Mexico; breeders from Alaska, interior Northwest, and Great Basin regions migrating primarily to the southwestern US and western Mexico; and breeders from the northern Pacific coast migrating to western California and Baja California (Wheelwright and Rising 2020).
MIGRATORY PATHWAYS Overall, this species migrates in a broad front across most of North America, with western populations migrating earlier during pre- and post-breeding than eastern populations (Fink et al. 2023). Juvenile Ipswich sparrows tracked using radio telemetry were found to leave breeding grounds on Sable Island approximately 24 days earlier than adults and followed a separate migration route from adults, spending less time over water, and making more frequent stopovers of longer duration (Crysler et al. 2016).

Melospiza melodia SONG SPARROW

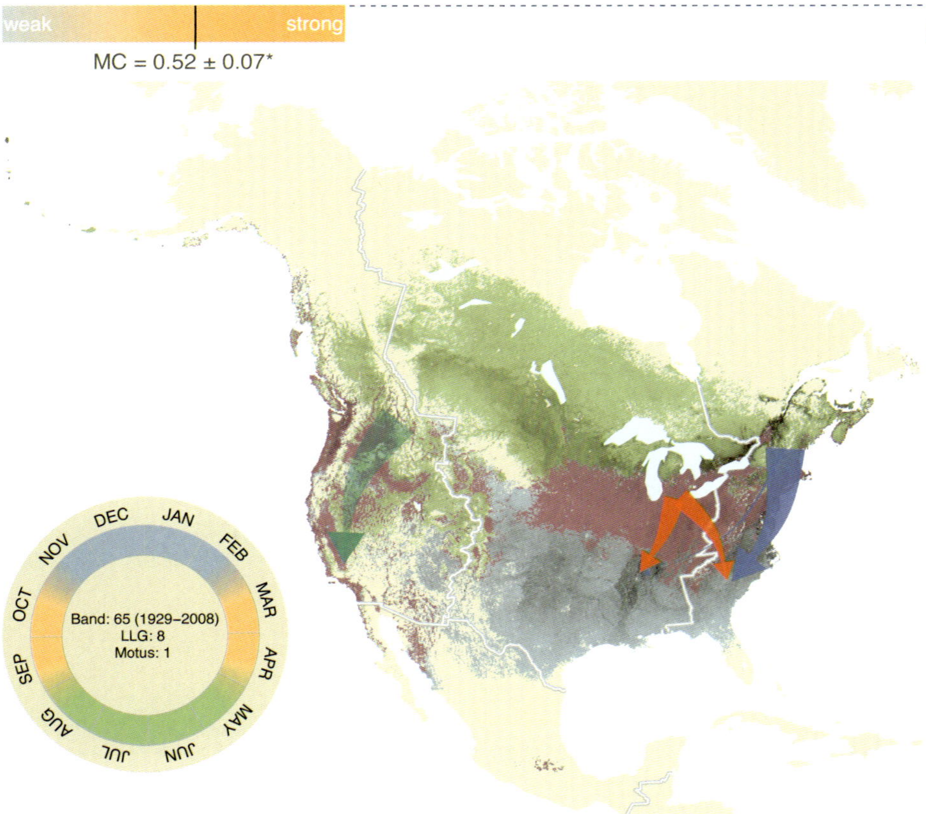

GENERALIZED CONNECTIVITY

	Atlantic	Midcontinent	Pacific
Atlantic (n = 46)	98%	2%	–
Midcontinent (n = 23)	48%	52%	–
Pacific (n = 1)	–	–	100%

DISTRIBUTION AND ABUNDANCE

The Song Sparrow has a widespread distribution across North America; the migratory eastern population breeds from Newfoundland to northeastern British Columbia and south to the upper Midwest, mid-Atlantic states, and the Appalachian Mountains to northern Georgia; its nonbreeding distribution is primarily across the southeastern and south-central US. It is patchily distributed and largely resident in the west from coastal Alaska and interior central British Columbia to Baja California, southern Arizona, northern Sonora, and northern New Mexico; migratory northern populations occur throughout the southwestern US and northern Mexico in the nonbreeding season. Long-term declines are apparent across Canada, in the Pacific Northwest, and northeastern US; populations in the Great Plains and southwestern US have been stable or increasing (Sauer et al. 2019).

ORDER Passeriformes
FAMILY Passerellidae
BREEDING HABITAT Open Country, Habitat Mosaic
NONBREEDING HABITAT Open Country, Habitat Mosaic
GLOBAL POPULATION SIZE 130,000,000
POPULATION CHANGE SINCE 1970 −27%
IUCN RED LIST CATEGORY Least Concern
R2R TIPPING POINT SPECIES No
US LISTING STATUS No
CANADA SPECIES AT RISK No
PIF CCS 9/20

MIGRATORY CONNECTIVITY

OVERALL Migratory connectivity is known from geolocator, band reencounter, and stable isotope data. Migratory populations maintain largely separate pathways and nonbreeding distributions, resulting in strong migratory connectivity (MC = 0.52 ± 0.07). As such, birds breeding in the Atlantic and Pacific flyways primarily remain within the same flyway as breeding during the nonbreeding season (98% and 100%, respectively), while birds breeding in the Midcontinent Flyway may remain within the flyway (52%) or migrate to the Atlantic Flyway (48%) for the nonbreeding period.
BANDING Breeding to nonbreeding band reencounter data indicate east-west parallel migration throughout the widespread eastern population, but with nonmigratory individuals as far north as New England and the Great Lakes. The proportion of birds migrating increases with latitude, and those breeding at higher latitudes migrate father

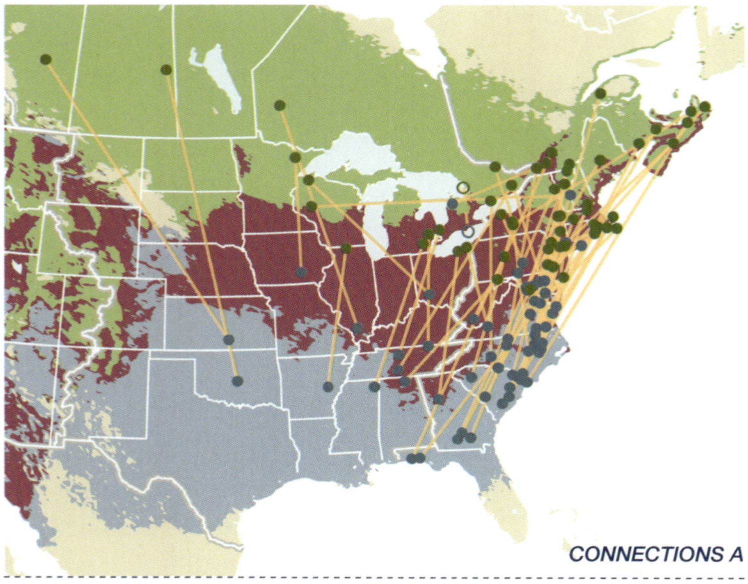

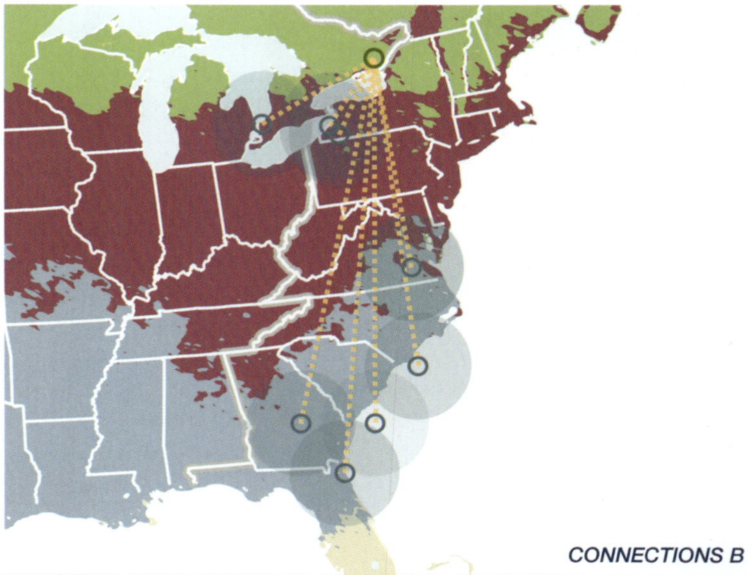

than those that breed at lower latitudes. Most banded individuals (90%) in western North America were nonmigratory, while most individuals from eastern and central North America migrate to the southeastern US (Davis and Arcese 1999).

RADIO TRACKING A few hundred tags were deployed through the Motus Wildlife Tracking System. There were numerous detections and one breeding to nonbreeding detection from Ontario to Ohio (A. Beauchamp et al., unpublished data).

GEOLOCATOR TRACKING Geolocator data from individuals breeding in western Ontario (n = 8) showed that birds primarily migrated to the southeastern US and males spend the nonbreeding season at more northerly latitudes than females (Kelly et al. 2019).

STABLE ISOTOPES Analysis of stable isotopes in claw tissue of breeders in western Ontario revealed that males migrated shorter distances, on average, than females (Kelly et al. 2019). Stable isotope assignment to nonbreeding locations also correlated with geolocator data from the same individuals.

OTHER APPROACHES Extensive variation in size and coloration results in up to 58 described subspecies, but with 24 diagnosable subspecies recognized in most modern assessments (Arcese et al. 2020). Analysis of the migratory subspecies provides evidence of strong broad-scale migratory connectivity, with the widespread eastern *M. m. melodia* migrating to the southeastern US; breeders in the Rocky Mountain and the Great Basin, *M. m. montana*, migrating to the southwestern US and northern Mexico; and partially migratory populations on the Pacific coast moving southward to Southern California and Baja California. The latter migratory birds overlap with local, resident populations in the nonbreeding season (Arcese et al. 2020)

MIGRATORY PATHWAYS Overall, migration is in a broad front and involves a complex pattern of long- and short-distance movements by different populations. Banding provides evidence of leapfrog migration in eastern North America, with northernmost breeding individuals migrating longer distances to more southerly nonbreeding areas (Arcese et al. 2020). Altitudinal migration is also documented in the west.

DATA CONTRIBUTORS

A. Beauchamp; C. Guglielmo; T.R. Kelly; Y. Morbey.

LINCOLN'S SPARROW *Melospiza lincolnii*

DISTRIBUTION AND ABUNDANCE

It breeds in wet, shrubby habitats across the entire boreal zone from western Alaska to Newfoundland, north to subarctic tundra, and south to the northern Great Lakes states and the mountains of northern New England; also along the Pacific coast to central British Columbia and in high mountains of the western US to Southern California, central Arizona, and northern New Mexico (Fink et al. 2023). Its nonbreeding distribution is primarily from central California and the southwestern US south through most of Mexico (including Baja California), sparsely to western Honduras and El Salvador; sparsely north to Oregon and Washington and east to Arkansas and Louisiana, but rare across the southeastern US. Eastern boreal populations have declined, whereas increases are noted throughout the western boreal and Rocky Mountain regions (Sauer et al. 2019).

ORDER Passeriformes
FAMILY Passerellidae
BREEDING HABITAT Wetlands, Boreal
NONBREEDING HABITAT Forests, Generalist
GLOBAL POPULATION SIZE 88,000,000
POPULATION CHANGE SINCE 1970 −3%
IUCN RED LIST CATEGORY Least Concern
R2R TIPPING POINT SPECIES No
US LISTING STATUS No
CANADA SPECIES AT RISK No
PIF CCS 7/20

MIGRATORY CONNECTIVITY

OVERALL Migratory connectivity is unknown. No information on migratory connectivity was found for this species.

RADIO TRACKING Tags were deployed through the Motus Wildlife Tracking System in the eastern part of the range. There were limited detections in the northeastern US and no detections of individual birds between breeding and nonbreeding.

STABLE ISOTOPES A single study with a small sample of migrants near Albany, New York, indicated a probable breeding origin in the eastern boreal forest of Canada (Kirchman et al. 2011).

OTHER APPROACHES Morphological variation is subtle through most of the range, and the two weakly differentiated subspecies seem to mix in the nonbreeding season (American Ornithologists Union 1957; Pandolfino et al. 2023). Some east-west separation is apparent, however, as specimens from the southeastern US are all identified as the northern boreal subspecies, and breeders from coastal Alaska and British Columbia migrate primarily to central California.

MIGRATORY PATHWAYS Migration is most concentrated in the southwestern US and through the central Great Plains, most likely representing breeding populations in the western mountains and the boreal forest, respectively (Fink et al. 2023). Birds from eastern Canada seem to migrate in a southwesterly direction, staying west of the Appalachian Mountains rather than down the Atlantic coast. Differential migration of eastern and western populations is also suggested by migration season banding data (Pandolfino et al. 2023).

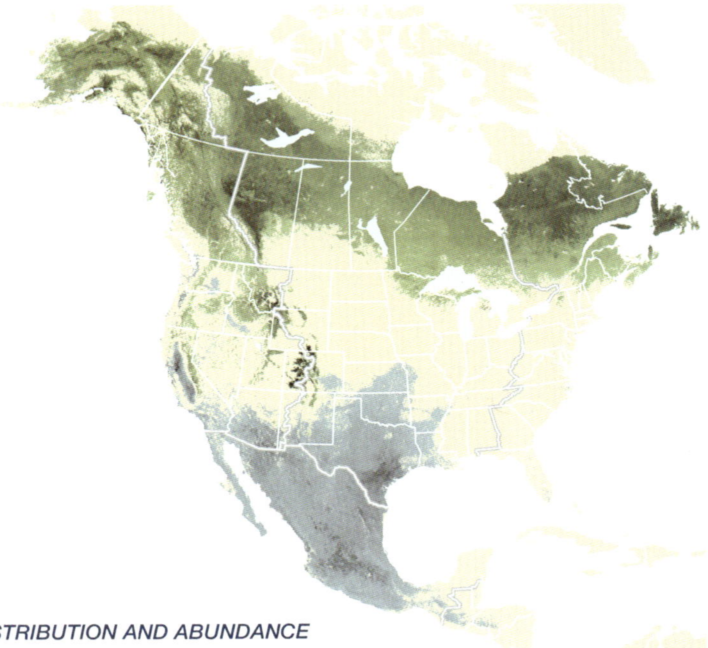

DISTRIBUTION AND ABUNDANCE

Melospiza georgiana **SWAMP SPARROW**

DISTRIBUTION AND ABUNDANCE

Breeds across the Canadian boreal forest from northeastern British Columbia and southern Yukon to Newfoundland; also south to the northeastern US from Minnesota to New England, sparsely farther south. Its nonbreeding range is at lower elevations across the southeastern US, north to the mid-Atlantic coast and the Ohio River Valley. Small numbers occur farther west during the nonbreeding season from Central Texas to Southern California. This species has increased in most of its breeding range, except along the mid-Atlantic coast where declines are apparent (Sauer et al. 2019).

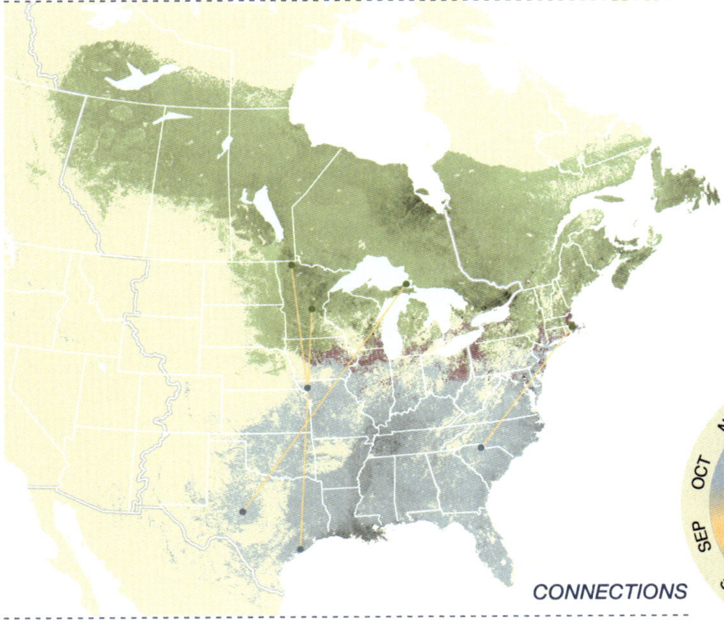

CONNECTIONS

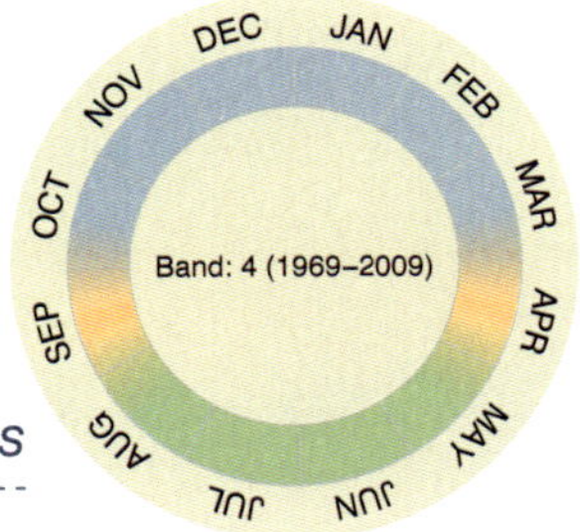

ORDER Passeriformes
FAMILY Passerellidae
BREEDING HABITAT Wetlands, Boreal
NONBREEDING HABITAT Wetlands, Generalist
GLOBAL POPULATION SIZE 23,000,000
POPULATION CHANGE SINCE 1970 +69%
IUCN RED LIST CATEGORY Least Concern
R2R TIPPING POINT SPECIES No
US LISTING STATUS No
CANADA SPECIES AT RISK No
PIF CCS 7/20

MIGRATORY CONNECTIVITY

OVERALL Migratory connectivity may be moderate, based on limited information from band reencounters and stable isotope analysis. MC is not calculated due to limited sample size.

BANDING Four breeding to nonbreeding reencounters show strong north-south latitudinal migrations.

RADIO TRACKING Detections from the Motus Wildlife Tracking System show migratory movements but not individual breeding to nonbreeding detections.

STABLE ISOTOPES Analysis and field confirmation of the coastal plain swamp sparrow (*M. g. nigrescens*) found that birds from New Jersey, Delaware, Maryland, and Virginia underwent pre-breeding molt at nonbreeding sites in South Carolina and Virginia (Greenberg et al. 2007).

OTHER APPROACHES No tracking or genetic studies of migratory connectivity were found for this species. Unlike most North American birds, subspecific variation is north-south, rather than east-west, with a northern *M. g. ericrypta* breeding across the boreal forest region and migrating as far south as central Mexico, the Gulf Coast, and Florida (American Ornithologists Union 1957). The more southern breeding *M. g. georgiana* is found farther north in the nonbreeding season, including in southern portions of the breeding range, suggesting that leapfrog migration may occur. The Coastal Plain subspecies

(*M. g. nigrescens*) remains isolated from other Swamp Sparrows in coastal salt marsh year-round (Herbert and Mowbray 2020).
MIGRATORY PATHWAYS Observational data suggest a clockwise loop migration, with greater abundance along the Atlantic coast in post-breeding than in pre-breeding (Fink et al. 2023). Banding returns, however, suggest that some individuals follow the same migration route each year, in both post- and pre-breeding (Herbert and Mowbray 2020).

GREEN-TAILED TOWHEE *Pipilo chlorurus*

DISTRIBUTION AND ABUNDANCE

It breeds in sagebrush steppe and chaparral throughout the Great Basin, from eastern Oregon, central Idaho, and southwestern Montana south patchily to New Mexico, northern Arizona, and Southern California. Its nonbreeding distribution is primarily in the Sonoran and Chihuahuan deserts from southern Arizona, southern New Mexico, and western Texas south through northwestern Mexico, including Baja California (Fink et al. 2023). This species has declined steadily since the 1980s, with steepest recent declines in Nevada, Colorado, and New Mexico (Sauer et al. 2019).

ORDER Passeriformes
FAMILY Passerellidae
BREEDING HABITAT Arid Lands, Sagebrush
NONBREEDING HABITAT Arid Lands, Desert Scrub
GLOBAL POPULATION SIZE 4,800,000
POPULATION CHANGE SINCE 1970 −18%
IUCN RED LIST CATEGORY Least Concern
R2R TIPPING POINT SPECIES No
US LISTING STATUS No
CANADA SPECIES AT RISK No
PIF CCS 12/20

DISTRIBUTION AND ABUNDANCE

MIGRATORY CONNECTIVITY

OVERALL There is no data on migratory connectivity and virtually no information on migration overall (Dobbs et al. 2020). Based on arrival and departure dates, and movements apparent in eBird animations, northernmost breeders may migrate the longest distance to nonbreeding areas in Mexico.
RADIO TRACKING A few tags were deployed through the Motus Wildlife Tracking System; detections during migration for one bird connect Colorado and southern Texas (Motus project, Climate Change and Migrating Songbirds, Ramos Lab, Colorado State University Pueblo, unpublished data).
MIGRATORY PATHWAYS There is no data from tracking or banding, but eBird animations show that the species appears to undertake a clockwise loop migration, with post-breeding migrants concentrated in Arizona and New Mexico, whereas pre-breeding migrants are more common through southeastern California and southern Nevada. Also, it is known to disperse to higher elevations post-breeding in some parts of the range, possibly part of a molt-migration system (Dobbs et al. 2020).

SPOTTED TOWHEE *Pipilo maculatus*

DISTRIBUTION AND ABUNDANCE

It breeds patchily throughout the western US and southwestern Canada, with greatest abundance in four or five distinct regions: in the northern Great Plains from Alberta to western Nebraska; in the Intermountain West from southeastern British Columbia to Arizona and New Mexico; along the Pacific Slope from southwestern British Columbia to Baja California; and in the mountains of Mexico. It is resident in the latter two regions, but migratory populations spend the nonbreeding season in the southwestern US from

ORDER Passeriformes
FAMILY Passerellidae
BREEDING HABITAT Forests, Temperate Western
NONBREEDING HABITAT Forests, Temperate Generalist
GLOBAL POPULATION SIZE 40,000,000
POPULATION CHANGE SINCE 1970 −8%
IUCN RED LIST CATEGORY Least Concern
R2R TIPPING POINT SPECIES No
US LISTING STATUS No
CANADA SPECIES AT RISK No
PIF CCS 9/20

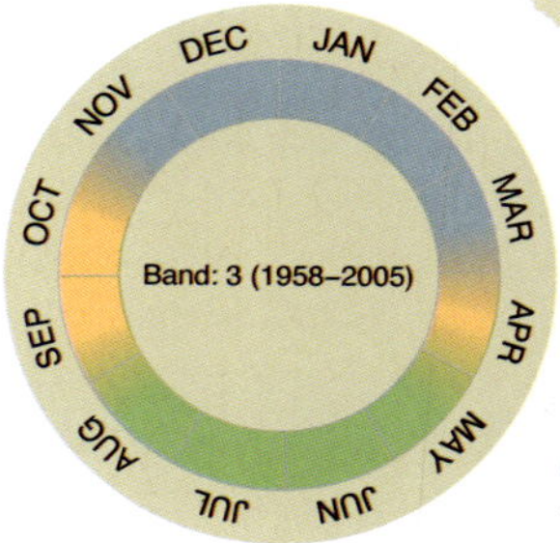

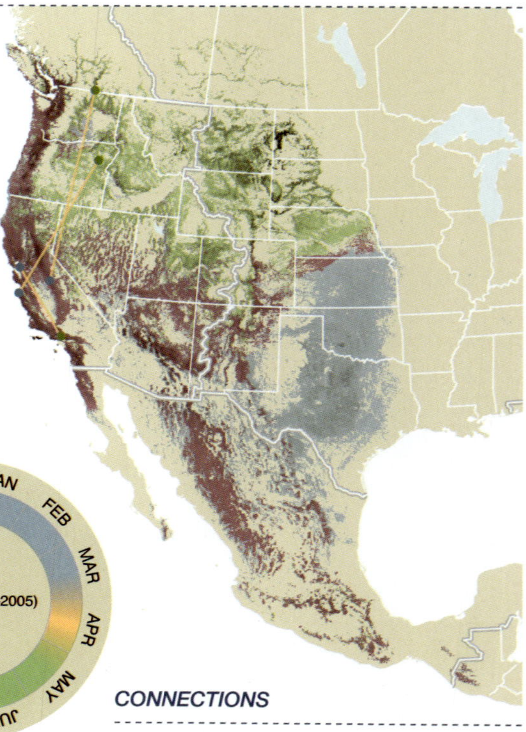

CONNECTIONS

southeastern California to central Texas and in adjacent northern Mexico. Most populations are stable or increasing, but declines are apparent in California, New Mexico, and Texas (Sauer et al. 2019).

MIGRATORY CONNECTIVITY

OVERALL Little is known about migratory connectivity, other than the distribution of various subspecies throughout their annual cycle. These indicate a complex pattern of long-distance and partial migration, altitudinal movements, and residency among populations.
BANDING Three breeding to nonbreeding band reencounters indicate movements from Washington and Oregon to central California, and one individual migrating between central and Southern California (Bartos Smith and Greenlaw 2020).
OTHER APPROACHES These are highly variable, with 21 described subspecies, most of which are resident along the Pacific coast or in Mexico (American Ornithologists Union 1957; Bartos Smith and Greenlaw 2020). Among the few migratory subspecies (identified primarily by female plumage; Phillips et al. 1964), a high degree of separation and migratory connectivity is apparent, with northern populations of *P. m. oregonus* from the Pacific Northwest migrating to central California; northern Great Basin breeders (*P. m. curtatus*) migrating mostly to Southern California and Arizona; southern Rocky Mountain breeders (*P. m. montanus*) migrating to the southwestern US and northern Mexico; and northern Great Plains breeders (*P. m. arcticus*) migrating east of the Rocky Mountains to the southern plains states and northeastern Mexico. Thus, migratory populations may overlap with resident birds in the nonbreeding season, and individuals of *P. m. curtatus* and *P. m. montanus* overlap substantially in Arizona (Phillips et al. 1964).
MIGRATORY PATHWAYS Separate migration pathways for the major subspecies groups (Bartos Smith and Greenlaw 2020) are visible in the eBird animations (Fink et al. 2023), with Great Plains breeders migrating southeastward and remaining east of the Rocky Mountains, northern Great Basin breeders migrating primarily southwestward to Southern California, and southern Rocky Mountain breeders moving southward into Arizona, New Mexico, and northern Mexico.

Pipilo erythrophthalmus

EASTERN TOWHEE

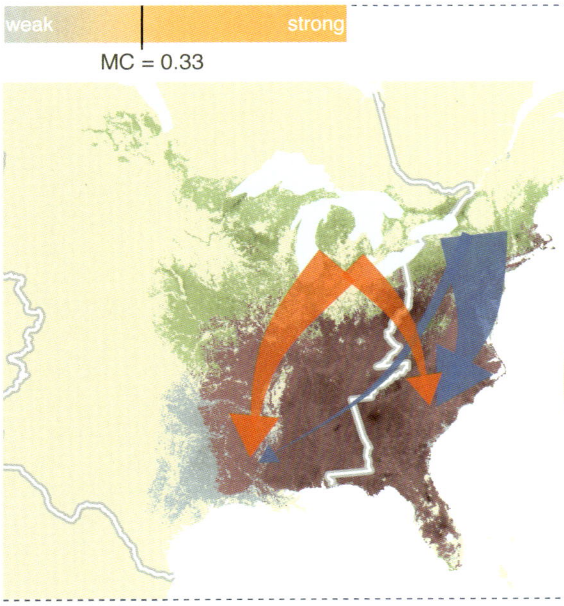

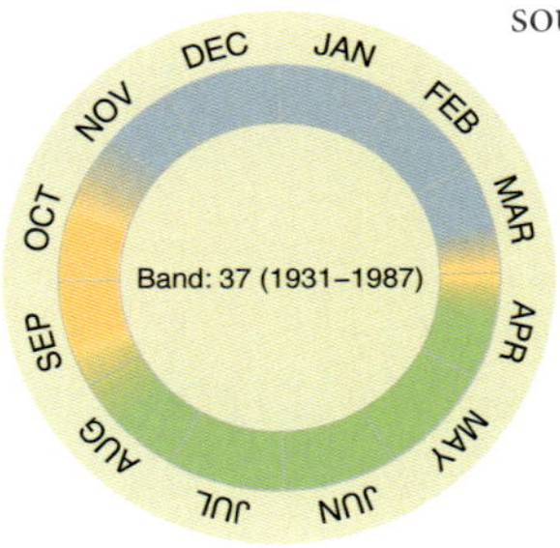

GENERALIZED CONNECTIVITY

DISTRIBUTION AND ABUNDANCE

Widely distributed as a breeder throughout the eastern US and southernmost Canada, this species extends west to the edge of the Great Plains from southeastern Saskatchewan to eastern Texas. Greatest abundance is throughout the Appalachian Mountains and across the southeastern US, where it is largely nonmigratory. Its nonbreeding distribution is in the southern half of the breeding range, with greatest abundance in the Piedmont, eastern Gulf Coastal Plain, and northern Florida. Declines are especially steep in New England, the mid-Atlantic states, and Florida, whereas slight increases are detected at the western edge of the range (Sauer et al. 2019). Northern populations are primarily migratory, with some individuals remaining on breeding areas for the nonbreeding season, while southern populations appear to be primarily sedentary (Greenlaw 2020).

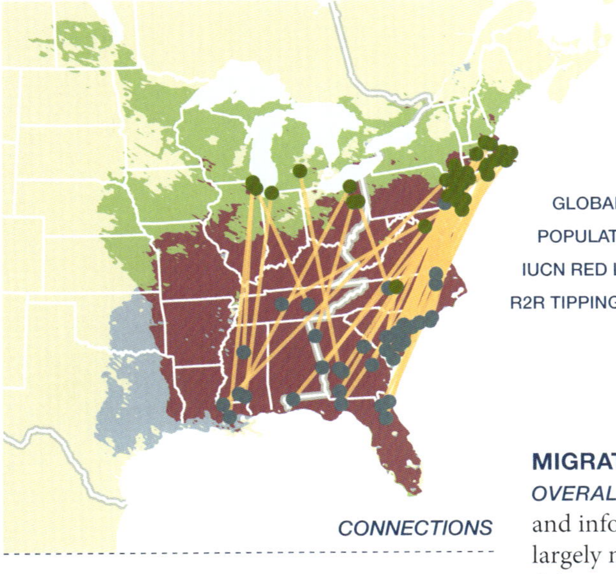

CONNECTIONS

ORDER Passeriformes
FAMILY Passerellidae
BREEDING HABITAT
Forests, Temperate Eastern
NONBREEDING HABITAT
Forests, Temperate Eastern
GLOBAL POPULATION SIZE 29,000,000
POPULATION CHANGE SINCE 1970 −44%
IUCN RED LIST CATEGORY Least Concern
R2R TIPPING POINT SPECIES Orange-Alert
US LISTING STATUS No
CANADA SPECIES AT RISK No
PIF CCS 13/20

	Atlantic	Midcontinent
Atlantic (n = 25)	88%	12%
Midcontinent (n = 8)	38%	62%

MIGRATORY CONNECTIVITY

OVERALL Migratory connectivity is moderate (MC = 0.33), based on band reencounter data and information on subspecies variation. Atlantic and Midcontinent flyway breeding birds largely migrate to nonbreeding areas within the breeding flyway (88% and 62%, respectively). A large percentage of Midcontinent Flyway breeders also migrate to the Atlantic Flyway (38%), though sample size is limited.
BANDING Band reencounters show an east-west parallel migration pattern, with northeastern populations migrating primarily to southern Atlantic coast states and midwestern populations migrating primarily to the Gulf Coast states.

OTHER APPROACHES No tracking, genetic, or stable isotope studies of migratory connectivity were found for this species. Four subspecies are recognized, with variable characters differing largely between northern and southern populations (American Ornithologists Union 1957; Greenlaw 2020). As such, analysis of specimens is useful for understanding migration behavior in northern breeding populations but sheds little light on east-west patterns of migratory connectivity shown by banded birds. Individuals from several northern migratory populations apparently mix with sedentary birds during the nonbreeding season in peninsular Florida (Greenlaw 2020).
MIGRATORY PATHWAYS There is no information on specific routes. Eastern populations migrate in a southeast-northwest direction, while central populations migrate north-south. The species is observed across a broad front in both pre- and post-breeding migration (Fink et al. 2023).

ICTERIIDAE | *YELLOW-BREASTED CHAT*

YELLOW-BREASTED CHAT *Icteria virens*

DISTRIBUTION AND ABUNDANCE
It breeds throughout the southeastern US, except peninsular Florida, sparsely north to the Great Lakes states and southern New England; also in lowland riparian corridors of western North America from extreme southern Canada to central Mexico. It spends the nonbreeding season mostly in Mexico (both Pacific and Atlantic lowlands) and northern Central America, sparsely to Costa Rica and western Panama. Populations have declined steeply in most of the eastern part of the range but have been stable or increasing in most of the west (Sauer et al. 2019).

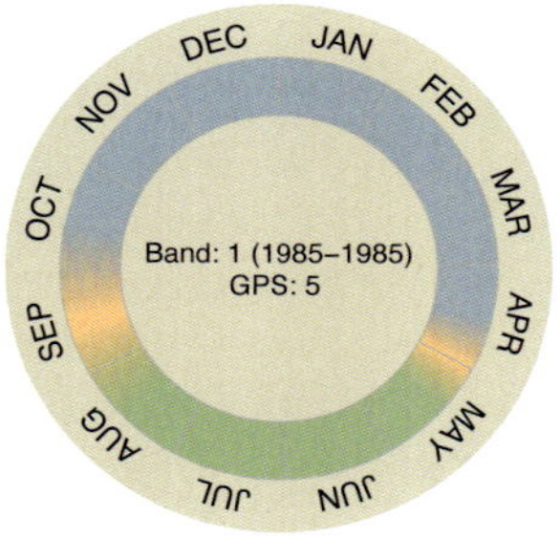

ORDER Passeriformes
FAMILY Icteriidae
BREEDING HABITAT Forests, Tropical Generalist
GLOBAL POPULATION SIZE 17,000,000
POPULATION CHANGE SINCE 1970 −15%
IUCN RED LIST CATEGORY Least Concern
R2R TIPPING POINT SPECIES No
US LISTING STATUS No
CANADA SPECIES AT RISK Endangered Southern Mountain (subspecies *I. v. auricollis*) and Ontario (subspecies *I. v. virens*) populations

MIGRATORY CONNECTIVITY
OVERALL Limited data prevents estimates of MC, but strong migratory connectivity is suggested between eastern and western subspecies based on multiple lines of evidence, including morphology, genetics, and stable isotopes.
BANDING The single breeding to nonbreeding band reencounter is a bird from Pennsylvania recovered in Veracruz, Mexico.
RADIO TRACKING Several projects tracking Yellow-breasted Chat through the Motus Wildlife Tracking System have resulted in several detections during migration.
GEOLOCATOR TRACKING Results from light-level geolocators (not shown on maps) were less precise, but similar GPS data showed individuals migrating to western Mexico (Mancuso et al. 2022).
SATELLITE TRACKING All five GPS tagged individuals from the South Okanagan Valley, British Columbia, spent the nonbreeding season within a relatively small area of Sinaloa and Nayarit, Mexico, indicating strong migratory connectivity for that endangered breeding population (Mancuso et al. 2022).
STABLE ISOTOPES A single study using both stable isotopes and genetics confirmed that individuals migrating through eastern Mexico originated from eastern breeding sites, whereas birds spending the nonbreeding season in Nayarit originated in the western breeding range—all corresponding with the two described subspecies (Mancuso et al. 2022).
GENETICS Analysis of mitochondrial DNA found differentiation between eastern and western breeding population haplotypes (Lovette et al. 2004). Nonbreeding locations along the Pacific coast of Mexico were found to have only western haplotypes, and sites in eastern Mexico and Central America were found to have only eastern haplotypes (Lovette et al. 2004). Mancuso et al. (2022) found similar results using SNP genotyping, with all individuals sampled in eastern (Veracruz, Chiapas) and western (Nayarit) Mexico assigned to eastern and western subspecies, respectively. The Bird Genoscape Project is currently working to construct a genoscape for the Yellow-breasted Chat (https://www.birdgenoscape.org/yellow-breasted-chat/).
OTHER APPROACHES Two recognized subspecies in the eastern and western parts of the breeding range also occupy separate nonbreeding

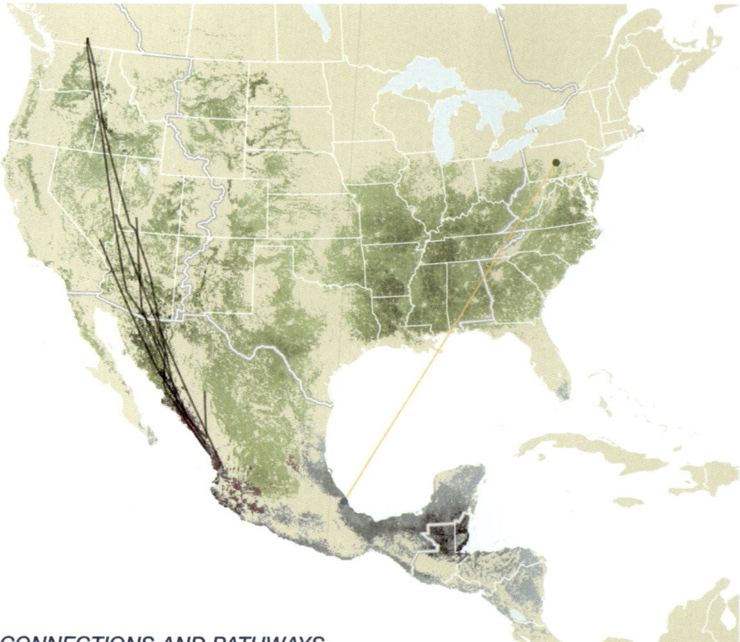

CONNECTIONS AND PATHWAYS

regions, indicating a migratory divide and strong east-west connectivity (Dunn and Garrett 1997).

MIGRATORY PATHWAYS Tracked individuals from British Columbia migrated through Washington, Oregon, Idaho, Utah, and Arizona, as well as through Sonora, Sinaloa, and Nayarit, Mexico. One individual stopped over in a riparian strip in Idaho for a minimum of eight days in September (Mancuso et al. 2022). Eastern and western chats maintain separate migration pathways, with western birds migrating through arid lowlands and river valleys of northwestern Mexico and southwestern US (rarely along the coast), and eastern birds migrating primarily around or across the western Gulf of Mexico (Thompson and Eckerle 2022; Dunn and Garrett 1997; Fink et al. 2023). Chats regularly appear on the Atlantic coast of New England and eastern Canada in Northern Hemisphere fall and early winter, suggesting a northeasterly post-breeding movement by some birds (Thompson and Eckerle 2022).

DATA CONTRIBUTORS

J. Alexander; C. Bishop; K. Hodges; K. Mancuso; S.M. Rockwell.

ICTERIDAE | *NEW WORLD BLACKBIRDS AND ALLIES*

weak strong

MC = 0.12 ± 0.07*

Xanthocephalus xanthocephalus

YELLOW-HEADED BLACKBIRD

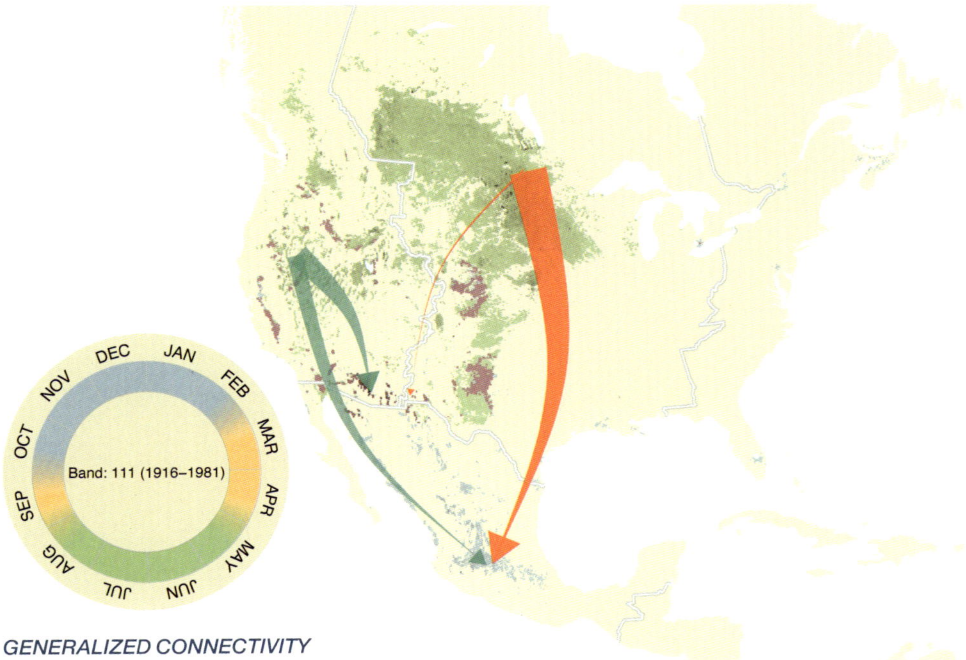

GENERALIZED CONNECTIVITY

DISTRIBUTION AND ABUNDANCE

Yellow-headed Blackbirds are very noticeable due to their unique plumage and harsh, grating calls; they are a conspicuous inhabitant of western North America, especially when breeding in freshwater cattail (*Typha* spp.) marshes (Twedt and Crawford 2020). They breed most abundantly in western Minnesota and the Dakotas north into southwestern Manitoba, Saskatchewan, and southeastern Alberta. Also, they are found more patchily breeding in appropriate habitats in many other western states. During the nonbreeding season, individuals can be found in several western states, including Texas and Louisiana but primarily in southern Arizona, New Mexico, and north and central Mexico (Fink et al. 2023).

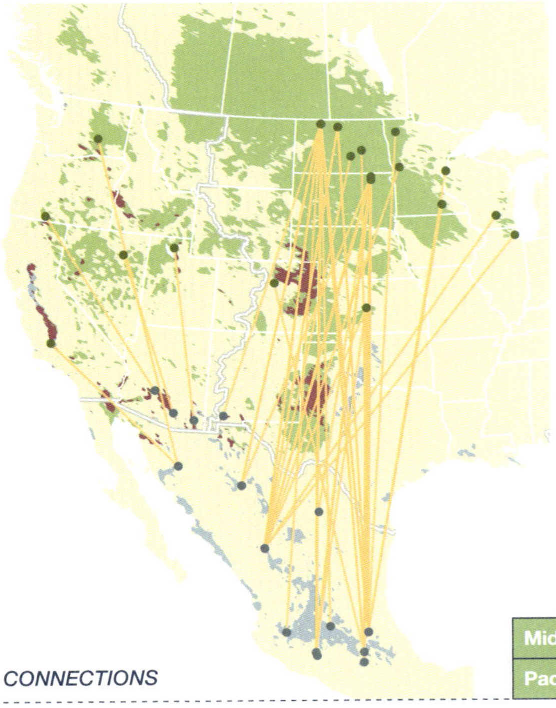

CONNECTIONS

ORDER Passeriformes
FAMILY Icteridae
BREEDING HABITAT Wetlands, Freshwater Marsh
NONBREEDING HABITAT Open Country, Habitat Mosaic
GLOBAL POPULATION SIZE 11,000,000
POPULATION CHANGE SINCE 1970 −10%
IUCN RED LIST CATEGORY Least Concern
R2R TIPPING POINT SPECIES No
US LISTING STATUS No
CANADA SPECIES AT RISK No
PIF CCS 9/20

	Mexico	Midcontinent	Pacific
Midcontinent (n = 105)	99%	1%	–
Pacific (n = 5)	40%	–	60%

INTERESTING NOTE

An individual with a Motus tag deployed in Nova Scotia was detected in Pennsylvania during migration (Motus project, BellVagrantsNS, unpublished data)

MIGRATORY CONNECTIVITY

OVERALL Migratory connectivity is known from band reencounter data. Nearly all (99%) Midcontinent Flyway breeders spend the nonbreeding season in Mexico, with very few (1%) remaining in the Midcontinent Flyway. Pacific Flyway breeding birds migrate to nonbreeding areas within the flyway (60%) and in Mexico (40%). A large portion of each breeding population, especially individuals in high abundant breeding areas, spend the nonbreeding season in Mexico, resulting in weak migratory connectivity (MC = 0.12 ± 0.07).

BANDING Band reencounter data show that birds breeding from the Great Lakes to Saskatchewan nearly all migrate to northern and central Mexico in the nonbreeding season. The limited banding data from birds breeding in the western states show migration to the US Southwest and northwest Mexico for the nonbreeding season. More data is needed from birds in the western part of the range.

OTHER APPROACHES No tracking, genetic, or stable isotope studies of migratory connectivity were found for this species.

MIGRATORY PATHWAYS Yellow-headed Blackbirds largely travel north-south during their annual migrations but seem to avoid major barriers, such as the Rockies or the Gulf of Mexico. Central US breeding birds undergo a longer migration to central Mexico, while more western birds spend the nonbreeding season in northern Mexico (Twedt and Crawford 2020; Fink et al. 2023).

BOBOLINK *Dolichonyx oryzivorus*

DISTRIBUTION AND ABUNDANCE

Bobolinks breed in southeastern Alberta, Saskatchewan, and Manitoba, and throughout the northern US from Montana to Maine south to Virginia. The nonbreeding range of this species is a relatively small region spanning southern Bolivia, Paraguay, northeastern Argentina, and southwestern Brazil (Fink et al. 2023).

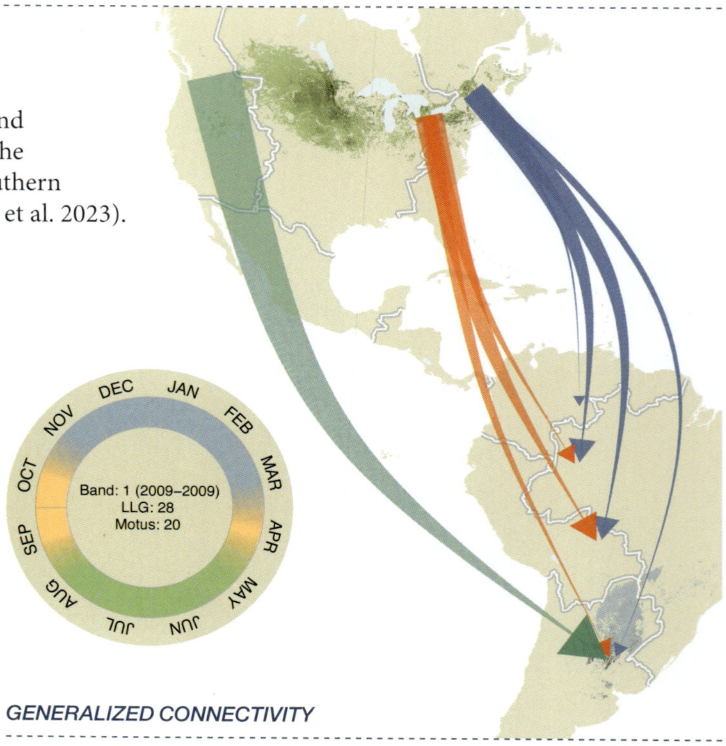

GENERALIZED CONNECTIVITY

ORDER Passeriformes
FAMILY Icteridae
BREEDING HABITAT Grasslands, Temperate
NONBREEDING HABITAT Grasslands, Pampas and Campos
GLOBAL POPULATION SIZE 10,000,000
POPULATION CHANGE SINCE 1970 −63%
IUCN RED LIST CATEGORY Least Concern
R2R TIPPING POINT SPECIES Orange-Alert
US LISTING STATUS No
CANADA SPECIES AT RISK Threatened
PIF CCS 14/20

	Andean West	Brazil	Northern SA	Southern Cone
Atlantic (n = 24)	42%	42%	4%	12%
Midcontinent (n = 4)	50%	25%	–	25%
Pacific (n = 1)	–	–	–	100%

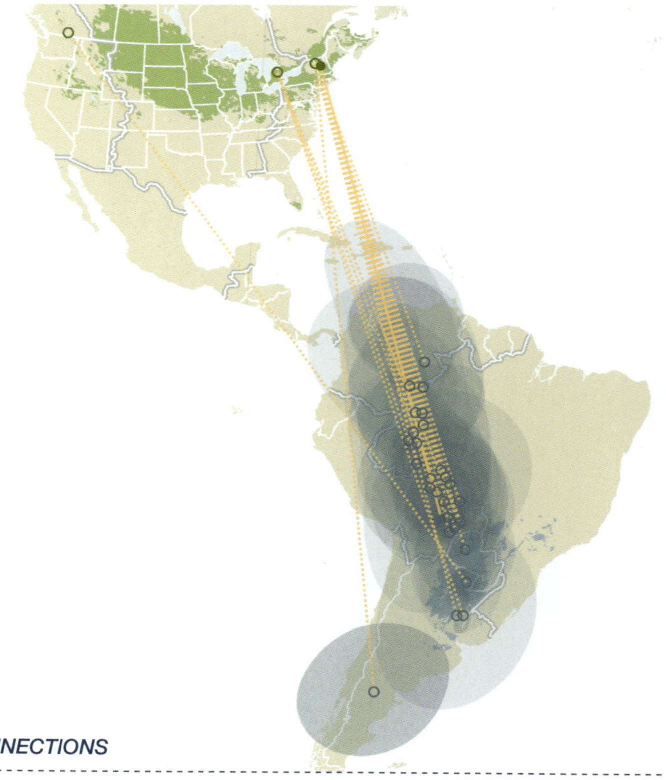

CONNECTIONS

INTERESTING NOTE

Southward movements through South America are synchronized to the southward greening of vegetation during the Southern Hemisphere's transition from spring into summer.

MIGRATORY CONNECTIVITY

OVERALL The migratory connectivity of the Bobolink is known from a mixture of band reencounter and geolocator data. Atlantic Flyway breeding populations primarily spend the nonbreeding season in the Andean West (42%) and Brazil (42%). Midcontinent Flyway breeding populations migrate to the Andean West (50%), Brazil (25%), and Southern Cone (25%) regions within South America. Pacific Flyway breeding birds migrate to nonbreeding areas in the Southern Cone region (100%). MC is not calculated due to limited sample size but appears weak, with breeding populations mixing at all nonbreeding regions.

BANDING There is one band reencounter connecting a breeding location in Vermont to a nonbreeding location in Bolivia.

RADIO TRACKING There are several projects and over 200 tag deployments through the Motus Wildlife Tracking System. As of printing, there are no individual breeding to nonbreeding detections, but there are many detections throughout migration (D. Brinker and M. Eshleman, unpublished data, J. Nocera unpublished data).

GEOLOCATOR TRACKING Tracking studies using both light-level and multisensor pressure geolocators show that Bobolink populations from throughout the range (British Columbia, Oregon, Nebraska, Vermont, Virginia, Ontario; Renfrew et al. 2013, 2019; A. Scarpignato and A. Johnson, unpublished data) arrive simultaneously at two separate nonbreeding areas, separated by thousands of kilometers, within one season. Birds first arrive in northern Venezuela and eventually move to Bolivia and Argentina.

OTHER APPROACHES No genetic or stable isotope studies of migratory connectivity were found for this species.

MIGRATORY PATHWAYS Bobolinks converge in a relatively narrow migration corridor along the east coast of the US (D. Brinker and M. Eshleman, unpublished data, J. Nocera, unpublished data); from there, they initiate transatlantic flights or fly across the Caribbean to South America (Renfrew et al. 2013, 2019; Perlut 2018). Bobolinks tagged in Vermont initiated transoceanic flights from Nova Scotia, Canada, and North Carolina, US (Perlut 2018). After critical stopovers on the northern coast of South America and in the Orinoco plains (*llanos*) of Colombia and Venezuela, Bobolinks make long-distance flights to Bolivia and then on to their southern nonbreeding grounds in Paraguay and Argentina (Renfrew et al. 2019; Perlut 2018; A. Scarpignato and A. Johnson, unpublished data). In pre-breeding migration, Bobolinks make an important stopover in the Llanos Region of Colombia and Venezuela before heading straight for their breeding grounds (Perlut 2018; Renfrew et al. 2019).

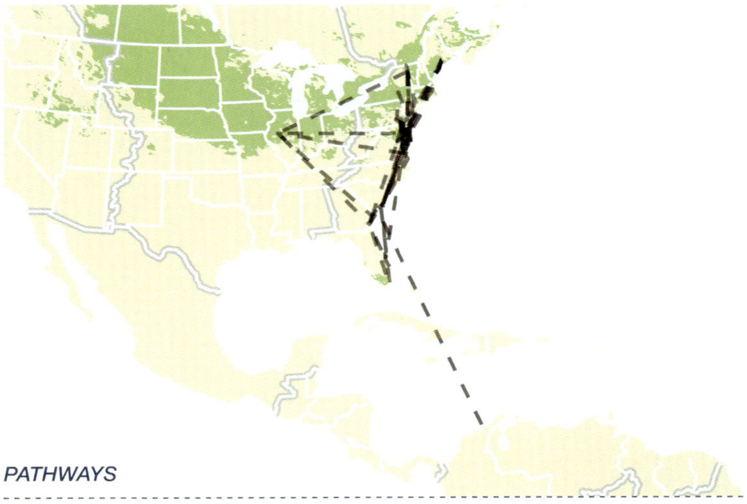

PATHWAYS

DATA CONTRIBUTORS
M. Cadman; D.H. Kim; J. Nocera; N.G. Perlut; R.B. Renfrew; A. Strong.

CHIHUAHUAN MEADOWLARK *Sturnella lilianae*

In 2022, the Chihuahuan Meadowlark was split from the Eastern Meadowlark (Beam et al. 2022). The Chihuahuan Meadowlark breeds throughout the desert grasslands of Arizona, New Mexico, West Texas, and into Mexico, with high abundances in the southern portions of the range during the nonbreeding season (Fink et al. 2023).

ORDER Passeriformes
FAMILY Icteridae
BREEDING HABITAT
Grasslands, Temperate
NONBREEDING HABITAT
Grasslands, Temperate
GLOBAL POPULATION SIZE 3,300,000
POPULATION CHANGE SINCE 1970 NA
IUCN RED LIST CATEGORY No
R2R TIPPING POINT SPECIES Yellow-Alert
US LISTING STATUS No
CANADA SPECIES AT RISK No
PIF CCS 15/20

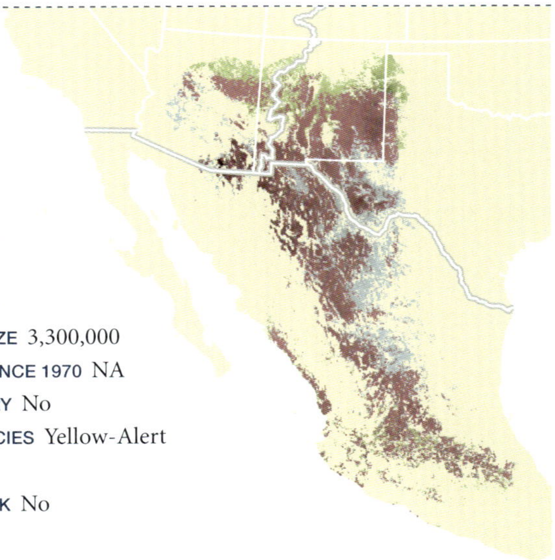

DISTRIBUTION AND ABUNDANCE

MIGRATORY CONNECTIVITY

OVERALL Migratory connectivity is unknown. No banding, genetic, or stable isotope studies of migratory connectivity were found for this species.

SATELLITE TRACKING In 2024, satellite tags (n = 7) were deployed on Chihuahuan Meadowlarks near Sonoita, Arizona; birds did not migrate and remained year-round (A. Scarpignato and A. Boyce, unpublished data).

GENETICS The Bird Genoscape Project is currently working to build a genoscape for Chihuahuan Meadowlark (https://www.birdgenoscape.org/eastern-western-meadowlarks/).

DATA CONTRIBUTORS
A.J. Boyce; A.L. Scarpignato; Migratory Connectivity Project.

EASTERN MEADOWLARK *Sturnella magna*

weak strong
MC = 0.69

DISTRIBUTION AND ABUNDANCE

Eastern Meadowlarks are common throughout native grasslands, agricultural fields, and pastures of eastern North America, Central America, and northern South America. Within the US, areas of high breeding abundance include the southern Great Plains, including Oklahoma, Kansas, Iowa, and Missouri, as well as the Texas Gulf Coast and Florida (Fink et al. 2023). Primary nonbreeding areas overlap with breeding areas but are also found throughout Texas into Mexico, in the Mississippi River Valley, and the southeastern coast of the US (Fink et al. 2023). Many individuals are year-round residents, especially southern populations. Northern individuals migrate south when snow and ice impede foraging, traveling variable distances (Jaster et al. 2022).

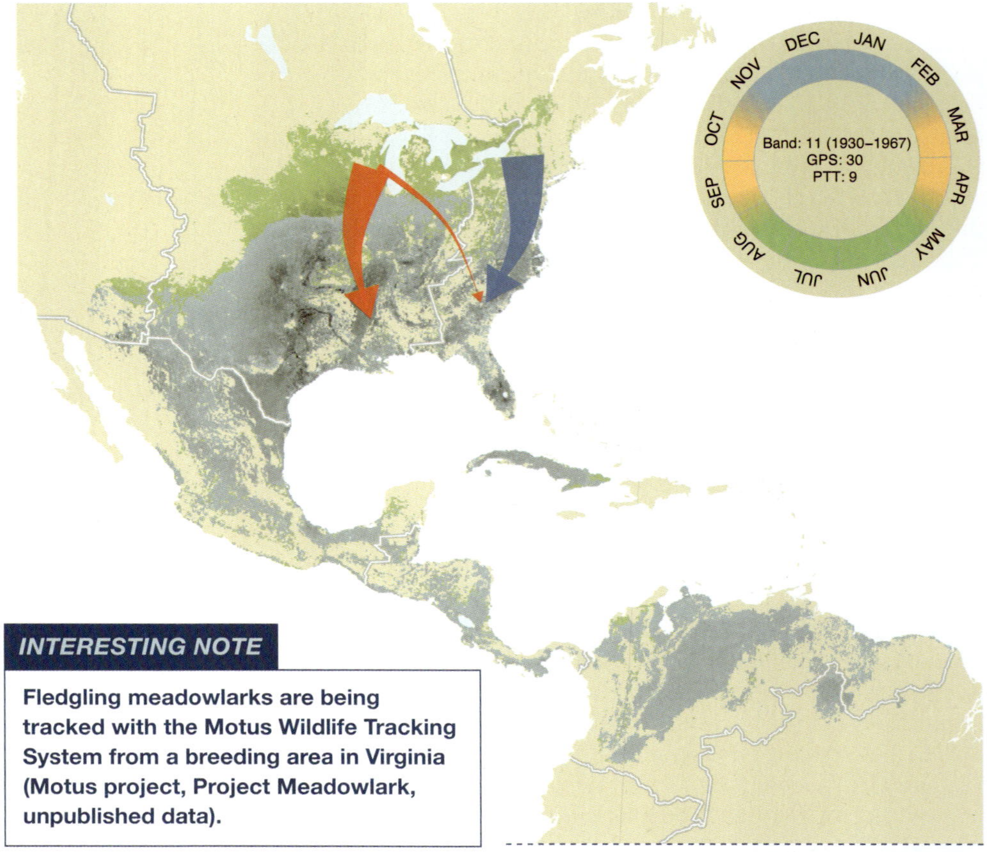

GENERALIZED CONNECTIVITY

INTERESTING NOTE

Fledgling meadowlarks are being tracked with the Motus Wildlife Tracking System from a breeding area in Virginia (Motus project, Project Meadowlark, unpublished data).

ORDER Passeriformes
FAMILY Icteridae
BREEDING HABITAT Grasslands, Temperate
NONBREEDING HABITAT Grasslands, Temperate
GLOBAL POPULATION SIZE 34,000,000
POPULATION CHANGE SINCE 1970 −79%
IUCN RED LIST CATEGORY Near Threatened
R2R TIPPING POINT SPECIES No
US LISTING STATUS No
CANADA SPECIES AT RISK Threatened
PIF CCS 11/20

MIGRATORY CONNECTIVITY

	Atlantic	Midcontinent
Atlantic (n = 15)	100%	–
Midcontinent (n = 16)	12%	88%

OVERALL Migratory connectivity is known from satellite, GPS, and band reencounter data. Birds breeding in the Atlantic Flyway migrate to nonbreeding areas within the Atlantic Flyway (100%). Similarly, birds breeding in the Midcontinent Flyway largely spend the nonbreeding season within the flyway (88%), as well as the Atlantic Flyway (12%). Migratory connectivity for this species is strong (MC = 0.69)

BANDING Band reencounter data support variable-length north-south movements, with some individuals migrating over 1,000 km from northern breeding grounds to nonbreeding areas. Southern birds exhibit shorter-distance migrations than northern birds; however, some northern birds also travel short distances, sometimes to coastal areas, in the nonbreeding period.

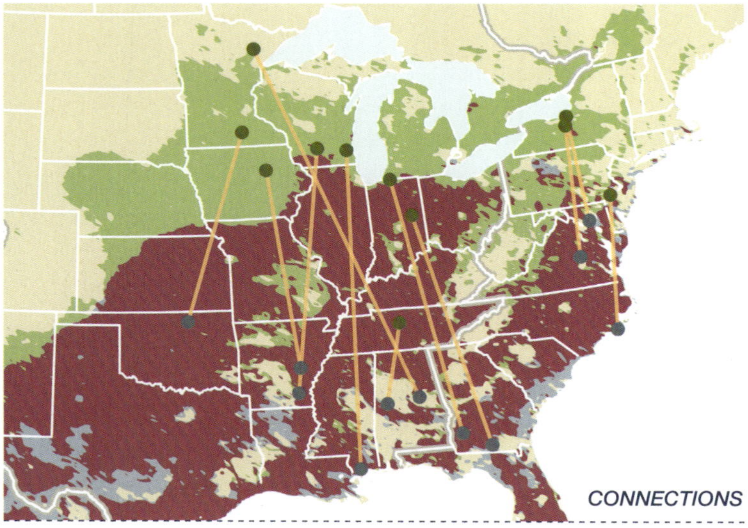

CONNECTIONS

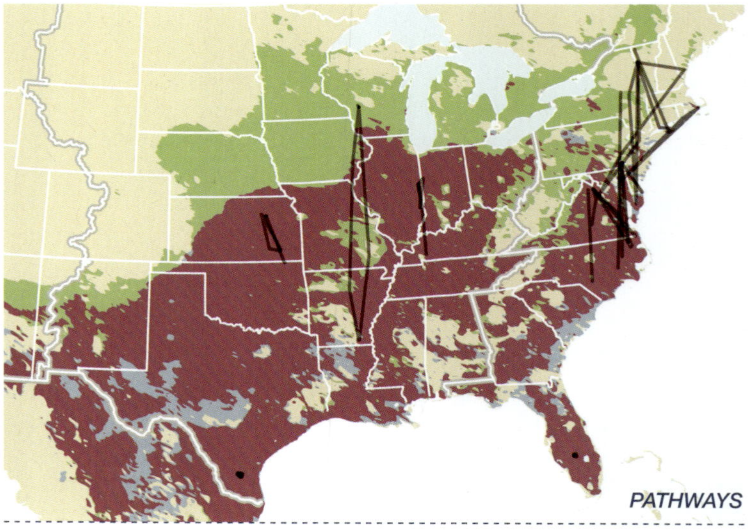

PATHWAYS

SATELLITE TRACKING Eastern Meadowlarks tracked by satellite and GPS tags throughout the range showed a diversity of migratory strategies, ranging from fully sedentary (Texas, Florida, Arkansas, Louisiana, Tennessee; A. Scarpignato, unpublished data; J. Giocomo, unpublished data) to partially migratory (Virginia, Kansas, Pennsylvania; Hill and Renfrew 2019; A. Scarpignato, unpublished data; B. Rigley and A.E.M. Johnson, unpublished data) to fully migratory (Vermont, Illinois, New York) with migrations up to 1,200 km (Hill and Renfrew 2019; A. Scarpignato, unpublished data). Within the range of migratory individuals, tagged birds exhibited a leapfrog migration, with the northernmost-breeding birds migrating farther south (Hill and Renfrew 2019; A. Scarpignato, unpublished data).
OTHER APPROACHES No genetic or stable isotope studies of migratory connectivity were found for this species.
MIGRATORY PATHWAYS Satellite and GPS data show birds migrated in a north-south direction, with some tagged birds making a loop migration, while others used the same pathways in both pre- and post-breeding (Hill and Renfrew 2019; A. Scarpignato, unpublished data; B. Rigley and A.E.M. Johnson, unpublished data). The Great Plains is an important area during migration (Fink et al. 2023).

DATA CONTRIBUTORS

T.J. Boves; W.A. Boyle; J. Bozzo; S. Barve; E.R. Donahue; M.A. Eshleman; J. Giocomo; M. Heather; J.M. Hill; A.E.M. Johnson; D.J. Newstead; P.P. Marra; N.G. Perlut; B. Rigley; A.L. Scarpignato; K. Tolan; Z.M. Warner; M.P. Ward; Migratory Connectivity Project.

Sturnella neglecta WESTERN MEADOWLARK

MC = 0.65

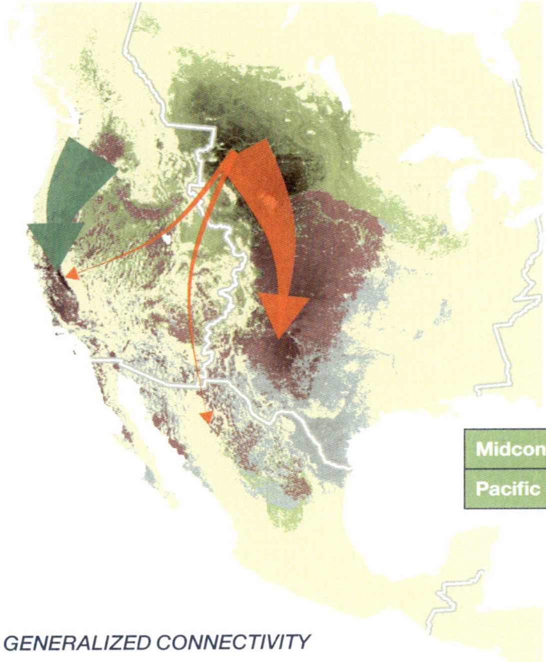

GENERALIZED CONNECTIVITY

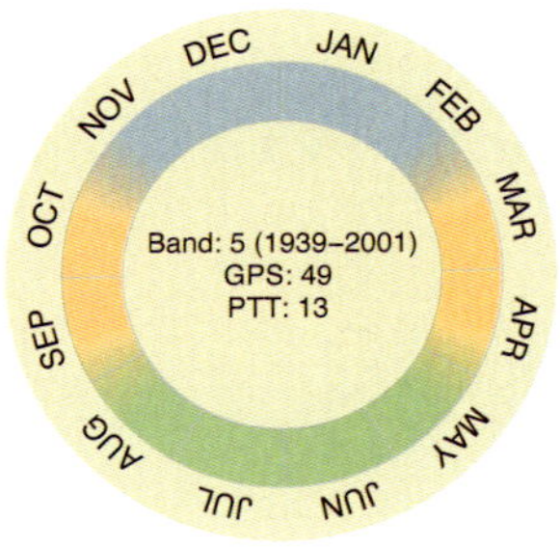

DISTRIBUTION AND ABUNDANCE

Western Meadowlarks are found throughout grasslands of western North America, with high year-round abundance throughout the Great Plains, Texas, and the Central Valley of California (Fink et al. 2023). Small populations in the Great Lakes region colonized there as agriculture dominated the landscape, and they are now in decline (Davis and Lanyon 2020). Although resident in the southern part of its range, northern birds undergo short- to medium-distance migrations and may vary their migratory distance from year to year, largely in response to weather conditions (Davis and Lanyon 2020).

	Mexico	Midcontinent	Pacific
Midcontinent (n = 32)	6%	88%	6%
Pacific (n = 16)	–	–	100%

ORDER Passeriformes
FAMILY Icteridae
BREEDING HABITAT Grasslands, Temperate
NONBREEDING HABITAT Grasslands, Chihuahuan
GLOBAL POPULATION SIZE 100,000,000
POPULATION CHANGE SINCE 1970 −42%
IUCN RED LIST CATEGORY Least Concern
R2R TIPPING POINT SPECIES No
US LISTING STATUS No
CANADA SPECIES AT RISK No
PIF CCS 10/20

MIGRATORY CONNECTIVITY

OVERALL Migratory connectivity is known from band reencounter, GPS, satellite, and Motus data. Individuals tend to remain within the same region during both breeding and nonbreeding periods, resulting in strong migratory connectivity (MC = 0.65). All Pacific Flyway breeding birds migrated to nonbreeding areas in the Pacific Flyway, and the majority of Midcontinent Flyway breeders (88%) migrated to areas within the Midcontinent Flyway for the nonbreeding season. The remainder of Midcontinent Flyway breeding birds migrated to nonbreeding areas in the Pacific Flyway (6%) or Mexico (6%).
BANDING Banding data are sparse and largely based on northern breeding individuals. Data demonstrate variable-length migrations, largely oriented north-south, with east or west movements away from montane areas during the nonbreeding period (Brewer et al. 2000).
RADIO TRACKING Similar to GPS and satellite data, data from tags using the Motus Wildlife Tracking System showed Western Meadowlarks migrated to nonbreeding areas in the Sacramento and San Joaquin valleys of California (Motus project, Intermountain West UM birds, unpublished data).
SATELLITE TRACKING Satellite and GPS data from Western Meadowlarks tagged throughout the range show a variety of migratory strategies. Birds breeding in Oklahoma remained resident; in Oregon, some birds remained resident, while others migrated to Northern California. Birds breeding in Idaho, Colorado, Montana, and North and South Dakota all migrated short-to-medium distances (A. Boyce, unpublished data; A. Scarpignato, unpublished data). Birds breeding in the northern Great Plains migrated to nonbreeding areas in Nebraska, Kansas, Oklahoma, Texas, and Mexico (A. Boyce, unpublished data; A. Scarpignato, unpublished data), while birds breeding in western Montana and Idaho migrated to valleys within California and Oregon (A. Scarpignato, unpublished data).
OTHER APPROACHES No genetic or stable isotope studies of migratory connectivity were found for this species.
MIGRATORY PATHWAYS Birds traveled in a loop migration pattern, with individuals breeding east of the Rocky Mountains migrating

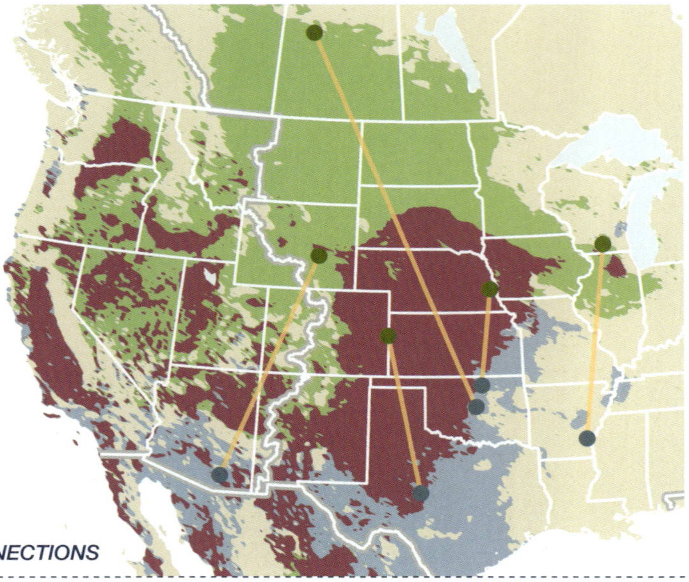

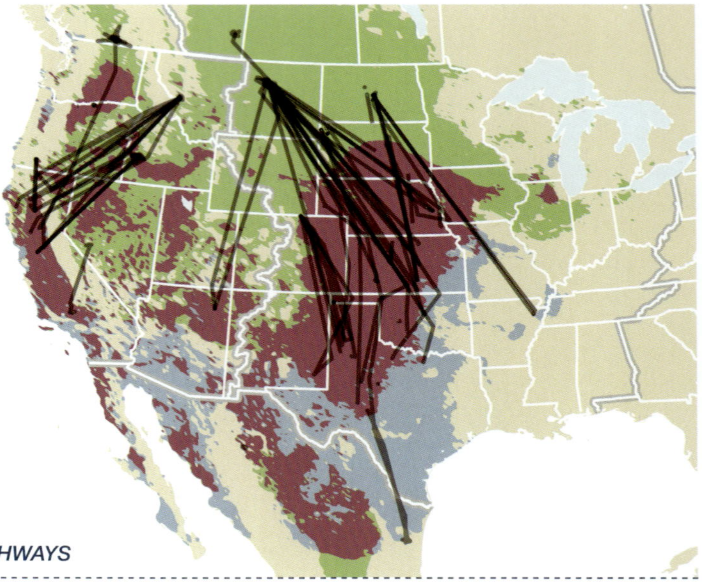

south or southeast through the Great Plains to nonbreeding areas and then traveling northwest back to breeding areas (A. Boyce, unpublished data; A. Scarpignato, unpublished data). Birds breeding west of the Rockies migrated in a southwest direction to nonbreeding areas and then traveled northeast back to breeding areas (A. Scarpignato, unpublished data; Motus project, Intermountain West UM birds, unpublished data).

DATA CONTRIBUTORS

M.R. Blake; A.J. Boyce; C.K. Boynton; J. Carlisle; K. Ellison; N. Mahony; B.G. Robinson; S.M. Rockwell; J.D. Ross; T.B. Ryder; A.L. Scarpignato; J.L. Stephens; Migratory Connectivity Project.

ORCHARD ORIOLE *Icterus spurius*

DISTRIBUTION AND ABUNDANCE

Orchard Orioles breed throughout most of the eastern US, in southern Saskatchewan and Manitoba in Canada, and from the Great Plains states to the Atlantic coast. The nonbreeding range extends from central Mexico, throughout Central America, and up to the northern Caribbean lowlands of Colombia in South America (Fink et al. 2023).

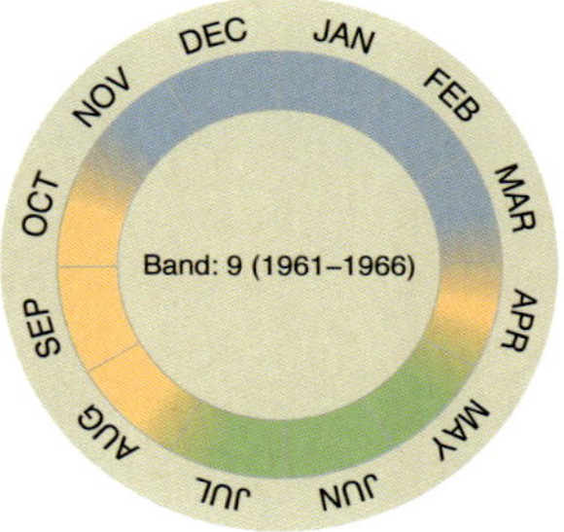

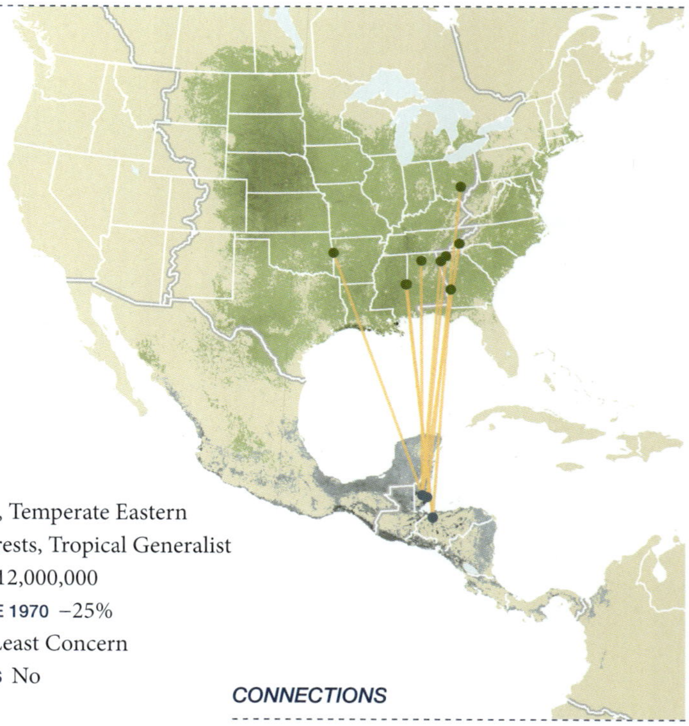

ORDER Passeriformes
FAMILY Icteridae
BREEDING HABITAT Forests, Temperate Eastern
NONBREEDING HABITAT Forests, Tropical Generalist
GLOBAL POPULATION SIZE 12,000,000
POPULATION CHANGE SINCE 1970 −25%
IUCN RED LIST CATEGORY Least Concern
R2R TIPPING POINT SPECIES No
US LISTING STATUS No
CANADA SPECIES AT RISK No
PIF CCS 10/20

MIGRATORY CONNECTIVITY

OVERALL The migratory connectivity of Orchard Orioles is unknown.
BANDING Band reencounter data link birds in Honduras and Belize on their nonbreeding grounds to breeding areas in the southeast US.
RADIO TRACKING While there are no breeding to nonbreeding detections, a Motus tag deployed in Nova Scotia was detected on the northwest coast of Florida 10 days later during post-breeding migration (Motus project, BellVagrantsNS, unpublished data).
OTHER APPROACHES No tracking, genetic or stable isotope studies of migratory connectivity were found for this species.
MIGRATORY PATHWAYS It is believed to migrate primarily overland through Mexico and Central America in the fall and to attempt Gulf of Mexico crossings in the spring (Scharf and Kren 2022). More study is needed.

Icterus cucullatus HOODED ORIOLE

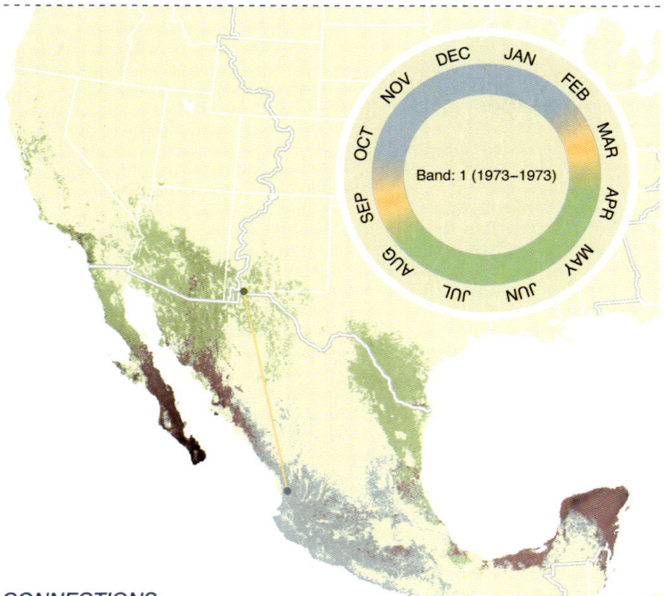

CONNECTIONS

DISTRIBUTION AND ABUNDANCE

The Hooded Oriole has a range restricted to the southwestern US and coastal Mexico. The main breeding strongholds are in San Francisco south along the coast into Baja California, southern Arizona, and southwestern Texas. There are year-round resident populations in southern Baja California, Sonora, and Quintana Roo (in the Yucatán Peninsula). Nonbreeding grounds are restricted to southern Baja California, the central Pacific coast, southwestern and Yucatán, Mexico (Pleasants and Albano 2020).

ORDER Passeriformes
FAMILY Icteridae
BREEDING HABITAT Forests, Generalist
NONBREEDING HABITAT Forests, Generalist
GLOBAL POPULATION SIZE 1,700,000
POPULATION CHANGE SINCE 1970 +85%
IUCN RED LIST CATEGORY Least Concern
R2R TIPPING POINT SPECIES No
US LISTING STATUS No
CANADA SPECIES AT RISK No
PIF CCS 9/20

MIGRATORY CONNECTIVITY

OVERALL The migratory connectivity of the Hooded Oriole is unknown.
BANDING Band reencounter data suggest western breeders primarily spend the nonbreeding period in western Mexico, while the south Texas breeding population is primarily on the Yucatán Peninsula (Pleasants and Albano 2020).
OTHER No tracking, genetic, or stable isotope studies of migratory connectivity were found for this species.

MIGRATORY PATHWAYS Pathways are not well studied, but banding records and eBird animations suggest a western pathway for westernmost breeding populations through Baja California and western Mexico, and a central pathway for central breeders, into the Yucatán Peninsula (Pleasants and Albano 2020; Fink et al. 2023).

Icterus bullockii BULLOCK'S ORIOLE

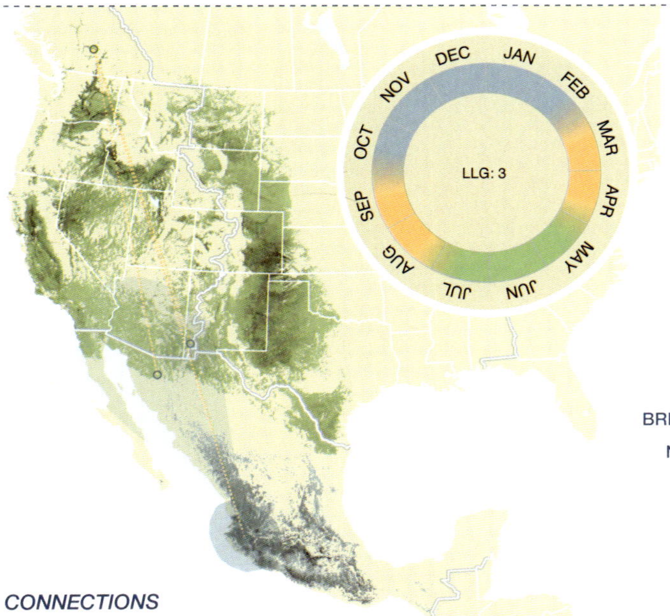

CONNECTIONS

DISTRIBUTION AND ABUNDANCE

Bullock's Orioles breed in western North America from southern British Columbia to northern Baja California, with an eastern limit along the Great Plains. They spend the nonbreeding season in a restricted area across southwestern Mexico, from Durango south to Oaxaca (Fink et al. 2023).

ORDER Passeriformes
FAMILY Icteridae
BREEDING HABITAT Forests, Temperate Western
NONBREEDING HABITAT Forests, Tropical Dry
GLOBAL POPULATION SIZE 7,400,000
POPULATION CHANGE SINCE 1970 −26%
IUCN RED LIST CATEGORY Least Concern
R2R TIPPING POINT SPECIES No
US LISTING STATUS No
CANADA SPECIES AT RISK No
PIF CCS 10/20

MIGRATORY CONNECTIVITY

OVERALL The migratory connectivity of Bullock's Oriole is partially known from a single geolocator tracking and stable isotope study.
RADIO TRACKING A couple of Motus tags were deployed in Idaho; one bird was detected near the Great Salt Lake, Utah (Motus project, Intermountain West Collaborative, unpublished data).
GEOLOCATOR TRACKING Bullock's Orioles carry out a molt migration. They take advantage of the fall monsoon in northern Mexico and go through an extended stopover to molt there, before heading farther south to their nonbreeding grounds (Pillar et al. 2016). Birds that bred in British Columbia were tracked to nonbreeding grounds in central Mexico (Pillar et al. 2016).
OTHER APPROACHES No banding or genetic studies of migratory connectivity were found for this species.
MIGRATORY PATHWAYS Although pathways are not well studied, survey and tracking data suggest all populations fly directly south and funnel into Mexico (Pillar et al. 2016; Fink et al. 2023).

DATA CONTRIBUTORS
N.J. Flood; P.P. Marra; A.G. Pillar; M.W. Reudink.

BALTIMORE ORIOLE *Icterus galbula*

DISTRIBUTION AND ABUNDANCE

Baltimore Orioles breed from Nebraska and Kansas east to Virginia and into the northeastern states and also into central and southern Alberta, Saskatchewan, and Manitoba. The nonbreeding range extends from central Mexico to central Colombia, but the species is particularly abundant in Central America from Guatemala to Costa Rica (Fink et al. 2023).

ORDER Passeriformes
FAMILY Icteridae
BREEDING HABITAT Forests, Temperate Eastern
NONBREEDING HABITAT Forests, Tropical Generalist
GLOBAL POPULATION SIZE 12,000,000
POPULATION CHANGE SINCE 1970 −45%
IUCN RED LIST CATEGORY Least Concern
R2R TIPPING POINT SPECIES No
US LISTING STATUS No
CANADA SPECIES AT RISK No
PIF CCS 11/20

MIGRATORY CONNECTIVITY

OVERALL The migratory connectivity of the Baltimore Oriole is known only from band reencounter data and appears moderate; however, MC is not calculated due to limited sample size. Atlantic Flyway breeding populations spend the nonbreeding period primarily within the Atlantic Flyway (71%), with smaller numbers in Mexico (14%) and Central America (14%). Midcontinent Flyway breeders spend the nonbreeding period exclusively in Central America (100%).
BANDING Band reencounter data suggest parallel migration of midwestern breeding populations to Central America.
RADIO TRACKING Several individuals were tracked through the Motus Wildlife Tracking System. There are no individual breeding to nonbreeding connections; however, there are many detections in the eastern portion of the range during migration.
OTHER APPROACHES No tracking, genetic, or stable isotope studies of migratory connectivity found for this species.
MIGRATORY PATHWAYS Baltimore Orioles migrate overland through Mexico and into Central America and northern South America (Rising and Flood 2020). Most eastern breeders migrate southwest overland, but some will island-hop through the Caribbean. Central breeders almost exclusively take a direct southward route. From what we know, pre-breeding migration pathways mirror the post-breeding movements (Rising and Flood 2020).

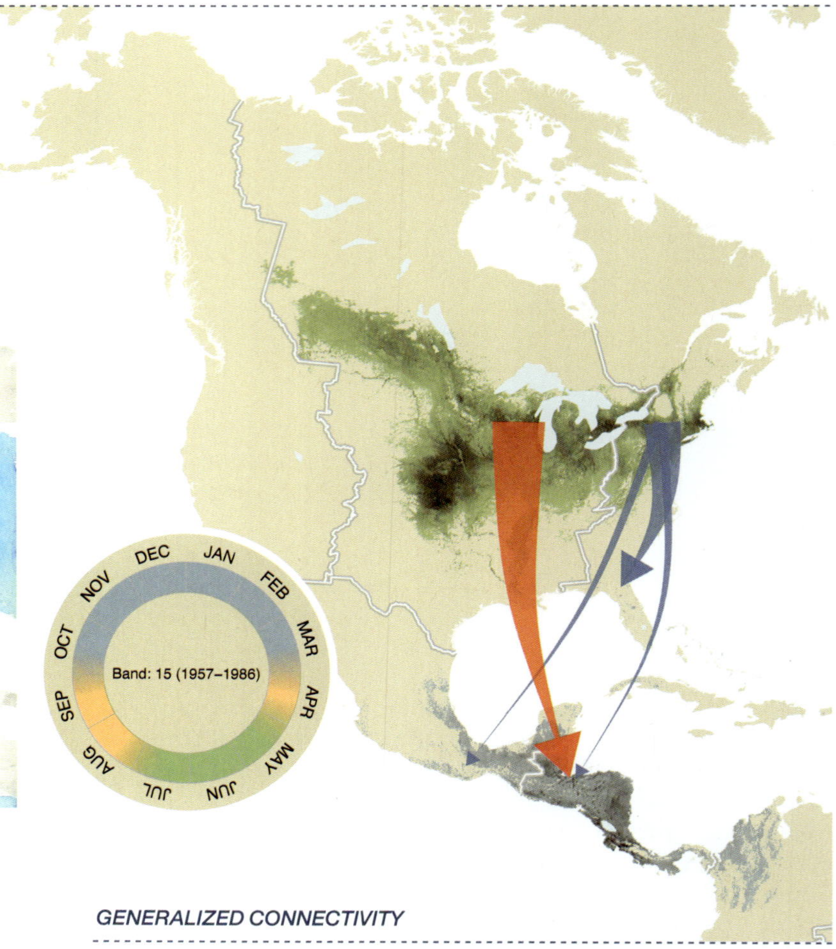

GENERALIZED CONNECTIVITY

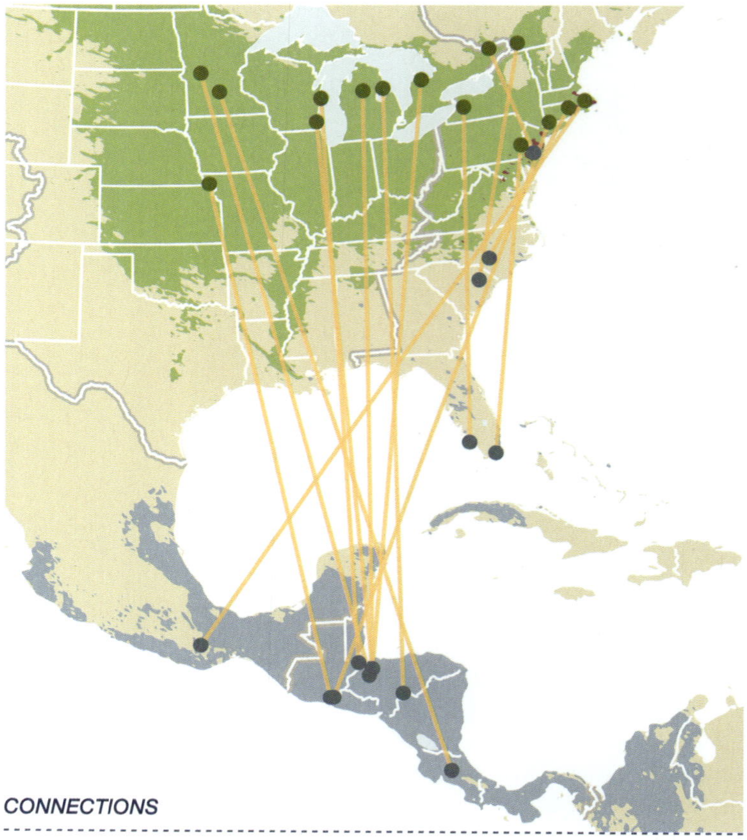

CONNECTIONS

	Atlantic	Central America	Mexico
Atlantic (n = 7)	71%	14%	14%
Midcontinent (n = 8)	–	100%	–

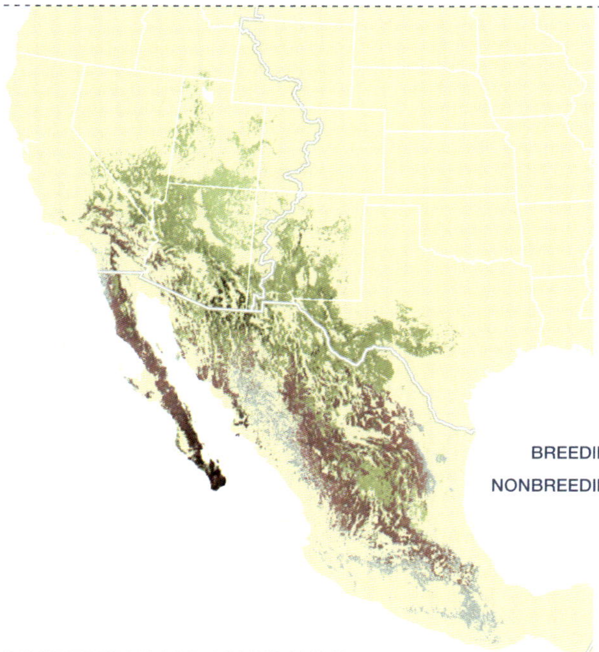

DISTRIBUTION AND ABUNDANCE

Icterus parisorum SCOTT'S ORIOLE

DISTRIBUTION AND ABUNDANCE

It breeds in desert-facing mountain slopes and semiarid plains (Flood 2020) in the southwestern US, Baja California peninsula, and central Mexico south to Mexico City, most densely in Baja California Sur. Except for a few localities in Southern California and Arizona, populations migrate completely from the US and most of northern Mexico during the nonbreeding season, presumably joining resident populations in south-central Mexico as far south as Oaxaca (Fink et al. 2023). The oriole's population in the US has decreased since 1970 (Rosenberg et al. 2019), with the steepest declines in California (Sauer et al. 2019).

ORDER Passeriformes
FAMILY Icteridae
BREEDING HABITAT Forests, Mesoamerican Pine-Oak
NONBREEDING HABITAT Forests, Mesoamerican Pine-Oak
GLOBAL POPULATION SIZE 4,900,000
POPULATION CHANGE SINCE 1970 −32%
IUCN RED LIST CATEGORY Least Concern
R2R TIPPING POINT SPECIES No
US LISTING STATUS No
CANADA SPECIES AT RISK No
PIF CCS 12/20

MIGRATORY CONNECTIVITY

OVERALL Migratory connectivity is unknown. No information on migratory connectivity was found for this species.

MIGRATORY PATHWAYS Departing from northern breeding grounds, this species moves south through the western US and northern Mexico utilizing its preferred semiarid habitats. Some lingering individuals will only depart their breeding grounds when the first winter weather conditions in the Northern Hemisphere arrive. Many Mexican birds may be year-round residents, but in the nonbreeding season, some individuals will move south from the core year-round range to areas of southern Mexico where the species does not breed (Fink et al. 2023; Flood 2020).

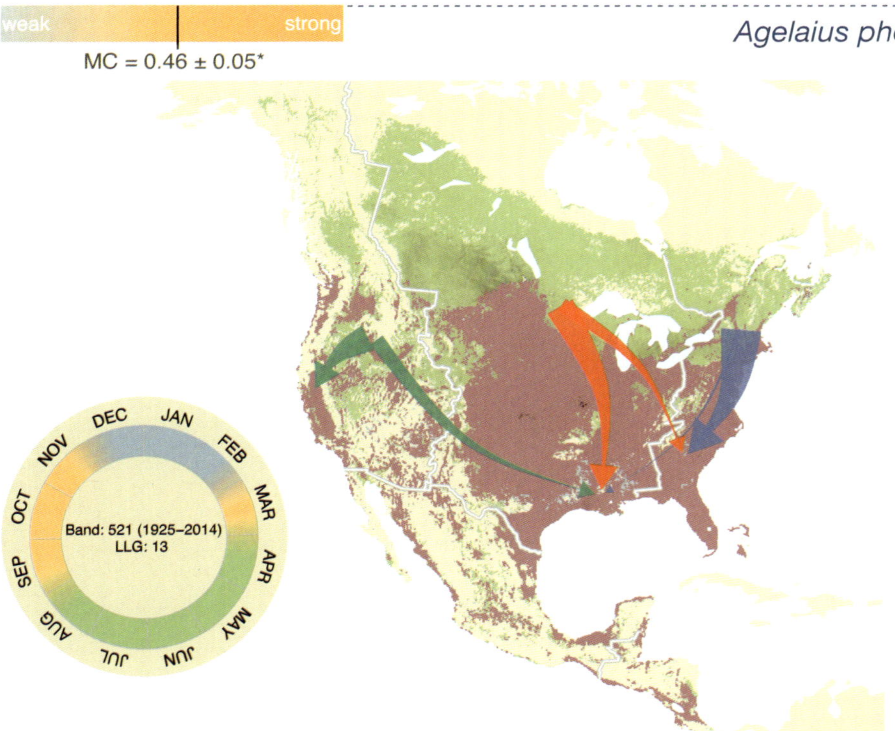

GENERALIZED CONNECTIVITY

Agelaius phoeniceus RED-WINGED BLACKBIRD

DISTRIBUTION AND ABUNDANCE

Red-winged Blackbirds have a broad breeding distribution throughout Canada, the US, Mexico, and Central America (to Costa Rica). Breeding strongholds, however, are located to the east of the Rocky Mountains, especially in Kansas north through the Dakotas, east to Massachusetts and eastern Alberta, Saskatchewan, and Manitoba (Fink et al. 2023). Northern populations are short-distance migrants and spend the nonbreeding period in the southeastern US, from North Carolina to Texas. Southern populations are year-round residents (Yasukawa and Searcy 2020).

ORDER Passeriformes
FAMILY Icteridae
BREEDING HABITAT Open Country, Habitat Mosaic
NONBREEDING HABITAT Open Country, Habitat Mosaic
GLOBAL POPULATION SIZE 180,000,000
POPULATION CHANGE SINCE 1970 −37%
IUCN RED LIST CATEGORY Least Concern
R2R TIPPING POINT SPECIES No
US LISTING STATUS No
CANADA SPECIES AT RISK No
PIF CCS 8/20

	Atlantic	Midcontinent	Pacific
Atlantic (n = 169)	98%	2%	–
Midcontinent (n = 329)	31%	69%	–
Pacific (n = 12)	–	33%	67%

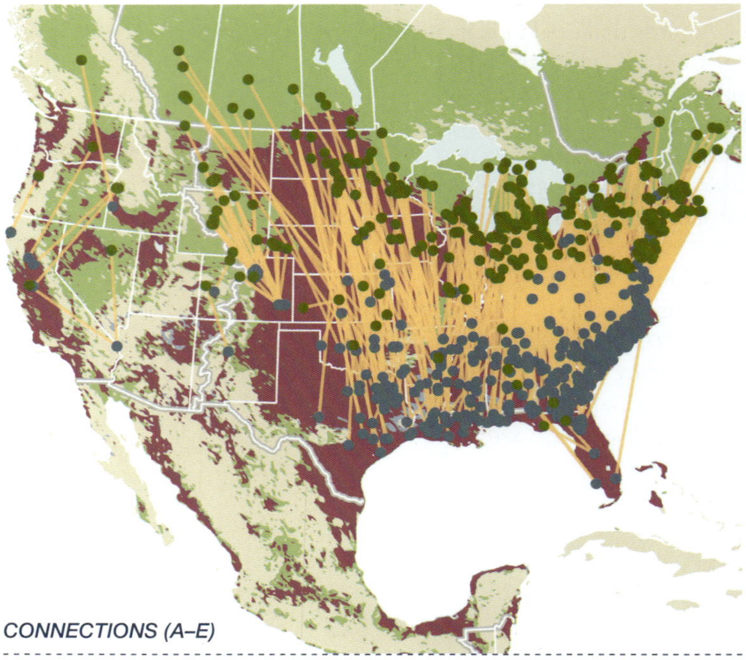
CONNECTIONS (A–E)

MIGRATORY CONNECTIVITY

OVERALL The migratory connectivity of Red-winged Blackbirds is known from a mixture of band reencounter and geolocator data. Atlantic Flyway breeding populations spend the nonbreeding period primarily within the Atlantic Flyway (98%) with a few individuals in the Midcontinent flyway (2%). Midcontinent Flyway breeders spend nonbreeding primarily within the Midcontinent (69%) and Atlantic (31%) flyways. Pacific Flyway breeding birds migrate to nonbreeding areas within the flyway (67%) or in the Midcontinent Flyway (33%). Migratory connectivity for this species is strong (MC = 0.46 ± 0.05)
BANDING Band reencounter data show that individuals that migrate use relatively narrow longitudinal corridors from their breeding grounds into southeastern US nonbreeding areas (Marra et al. 2014).
GEOLOCATOR TRACKING A tracking study found that Red-winged

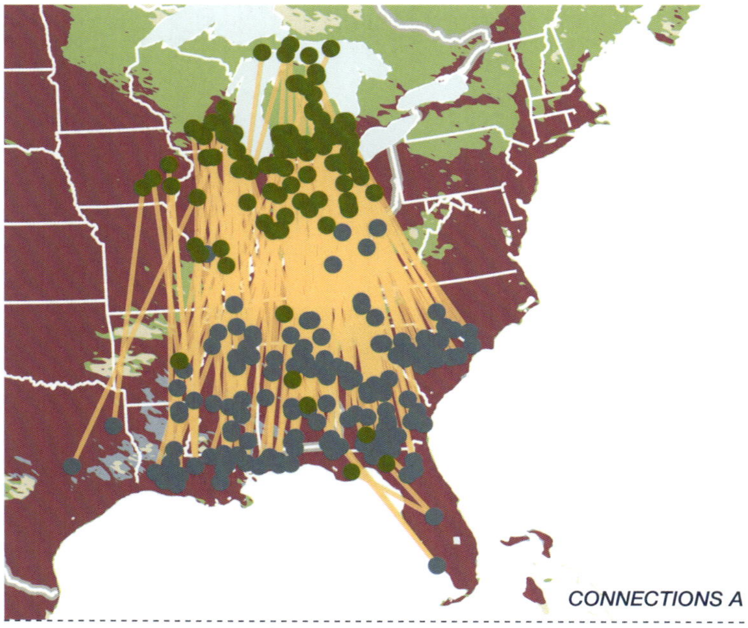
CONNECTIONS A

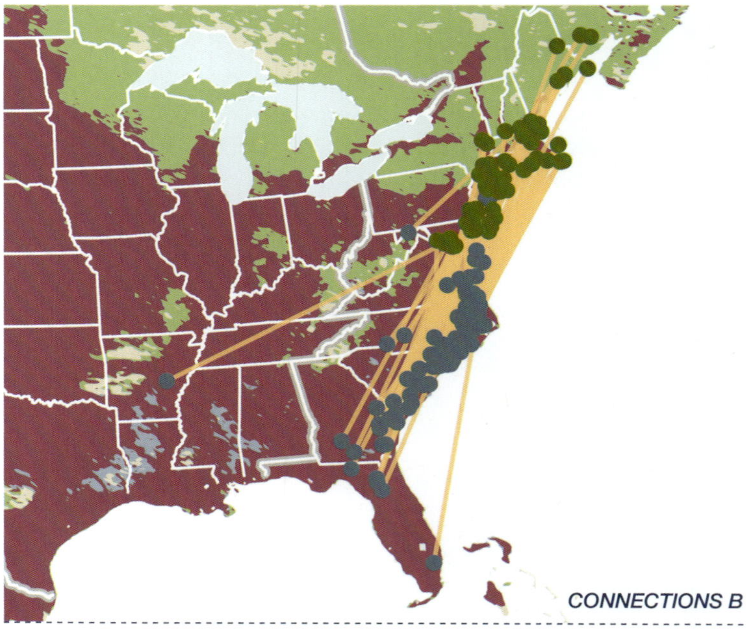
CONNECTIONS B

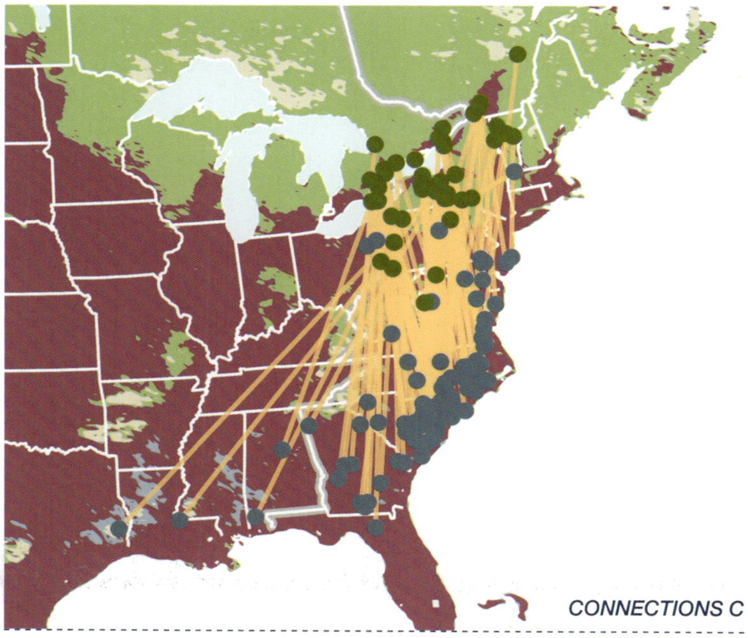
CONNECTIONS C

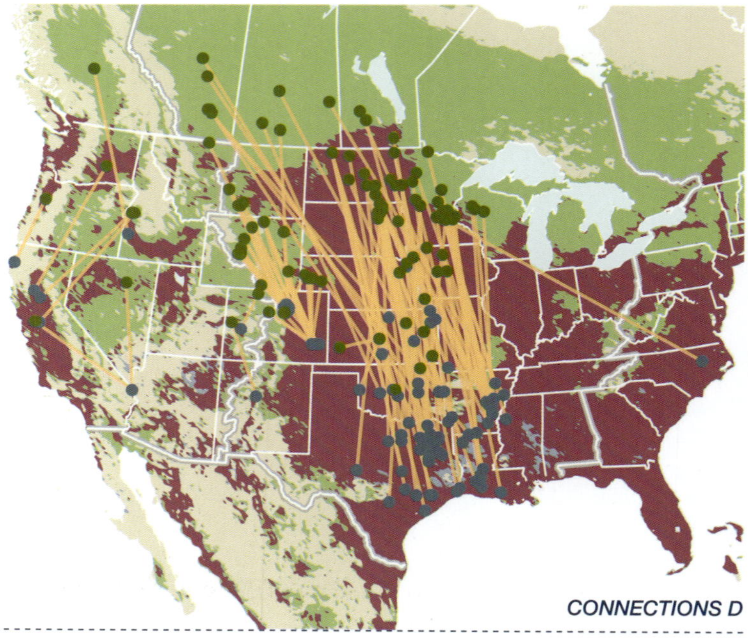
CONNECTIONS D

Blackbirds that bred in North Dakota migrated consistently through the Central Flyway and spent the nonbreeding period in the southeastern states (Stonefish et al. 2021).
STABLE ISOTOPES ^{2}H and ^{13}C isotopes measured from Red-winged Blackbirds were used to construct one of the initial isoscapes used for geographic assignment of breeding origins in North America (Wassenaar and Hobson 2000).
OTHER APPROACHES No genetic studies of migratory connectivity were found for this species. There were only detections near deployment areas for the few Motus tags deployed.
MIGRATORY PATHWAYS There is no information on specific routes, but there are observations of birds in high abundance through the Great Plains and the Mississippi River Valley during both pre- and post-breeding migration (Fink et al. 2023)

DATA CONTRIBUTORS
M.A. Eshleman; E.H. Gillam; T.J. Greives; H.J. Homan; P.E. Klug; G.M. Linz; D. Stonefish.

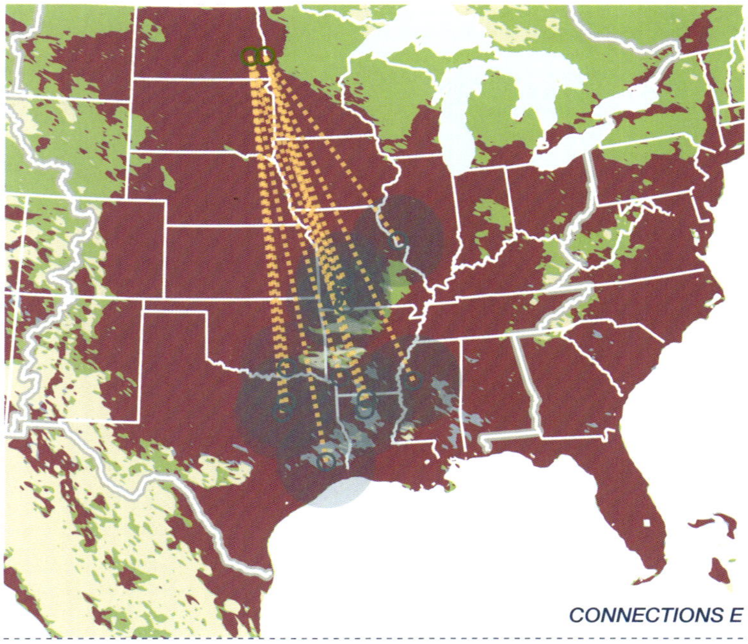
CONNECTIONS E

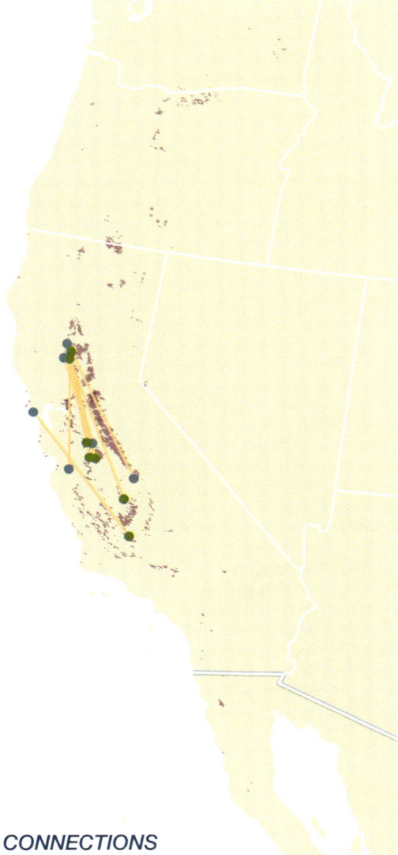
CONNECTIONS

Agelaius tricolor **TRICOLORED BLACKBIRD**

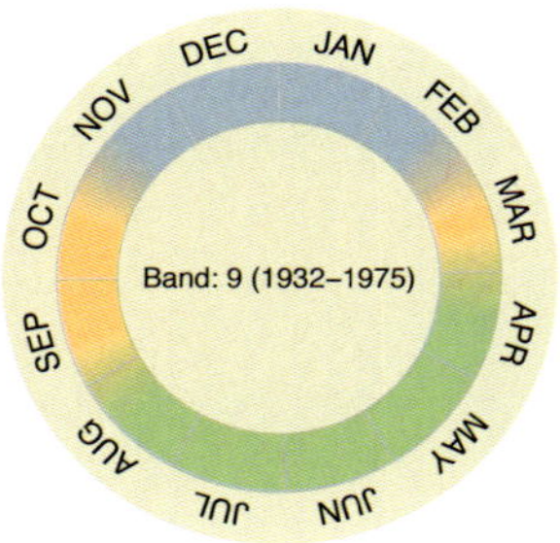

DISTRIBUTION AND ABUNDANCE

Although some birds range as far north as Washington, and as far south as Baja California, the vast majority of Tricolored Blackbirds are found in the Central Valley of California, where they form large breeding colonies in dense vegetation. Tricolored Blackbirds are found year-round within much of their range, but they likely do move between a breeding and nonbreeding location and may even move between and breed at multiple sites during the breeding season (DeHaven et al. 1975; C. Wilson et al. 2016; Beedy et al. 2023).

ORDER Passeriformes
FAMILY Icteridae
BREEDING HABITAT
Wetlands, Freshwater Marsh
NONBREEDING HABITAT
Wetlands, Freshwater Marsh
GLOBAL POPULATION SIZE 180,000
POPULATION CHANGE SINCE 1970 −17%
IUCN RED LIST CATEGORY Endangered
R2R TIPPING POINT SPECIES Red-Alert
US LISTING STATUS No
CANADA SPECIES AT RISK No
PIF CCS 17/20

MIGRATORY CONNECTIVITY

OVERALL Migratory connectivity is relatively unknown. All populations spend the breeding and nonbreeding seasons in the Pacific Flyway of the United States. MC is not calculated due to limited data.
BANDING Band reencounter data indicate that individuals move from breeding grounds in the California Central Valley to nonbreeding areas in the Sacramento Valley, San Joaquin Valley, and San Francisco Bay Area (DeHaven et al. 1975).
RADIO TRACKING A study of radio-tagged birds from several colonies showed some individuals moved between multiple active breeding colonies in the breeding season and documented the use of breeding areas and nearby habitats in the nonbreeding season (C. Wilson et al. 2016). As of printing, there are no detections from tags deployed with the Motus Wildlife Tracking System.
GENETICS Samples from across the breeding range identified a single genetically distinct population (Barr et al. 2020).
OTHER APPROACHES No tracking or stable isotope studies of migratory connectivity were found for this species.
MIGRATORY PATHWAYS There is no information on specific routes, likely because the establishment of breeding colonies and nonbreeding movements is largely tied to food availability (Beedy et al. 2023).

BRONZED COWBIRD *Molothrus aeneus*

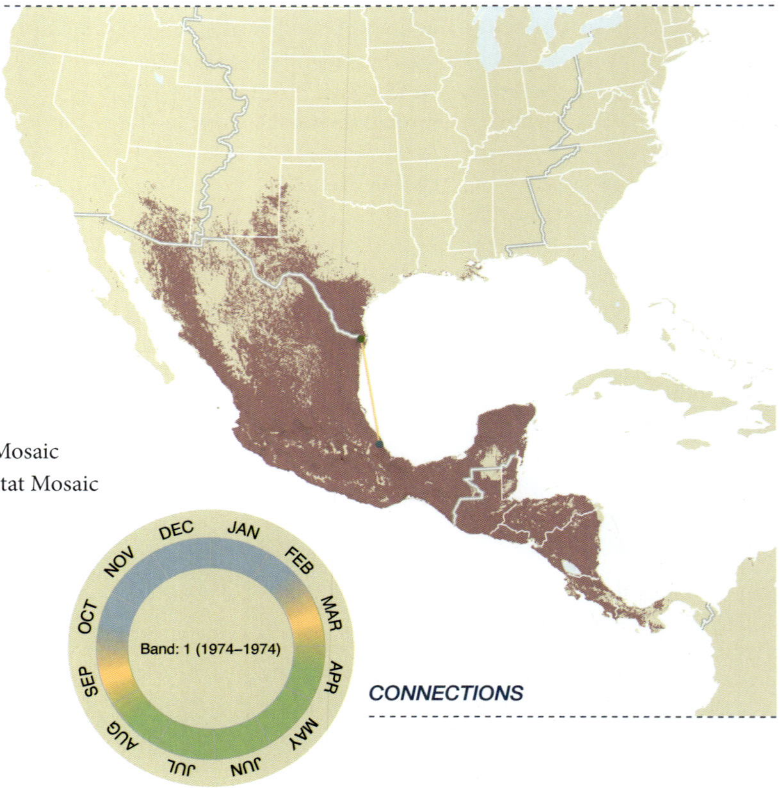

DISTRIBUTION AND ABUNDANCE

This species' range extends from the south-central United States, primarily from Arizona east through New Mexico, Texas, and into Louisiana, through Central America. Within the United States, it is most commonly found near the US–Mexico border, especially in southern Texas. In the nonbreeding season, individuals migrate south into Central America, with only a few birds remaining in the southern tip of Texas. Like other cowbirds, this species is a brood parasite (Ellison and Lowther 2020).

ORDER Passeriformes
FAMILY Icteridae
BREEDING HABITAT Open Country, Habitat Mosaic
NONBREEDING HABITAT Open Country, Habitat Mosaic
GLOBAL POPULATION SIZE 6,600,000
POPULATION CHANGE SINCE 1970 +21%
IUCN RED LIST CATEGORY Least Concern
R2R TIPPING POINT SPECIES No
US LISTING STATUS No
CANADA SPECIES AT RISK No
PIF CCS 7/20

MIGRATORY CONNECTIVITY

OVERALL Migratory connectivity is unknown.
BANDING There is one band reencounter from a breeding site near the US–Mexico border to central Mexico during the nonbreeding season.
OTHER APPROACHES No tracking, genetic, or stable isotope studies of migratory connectivity were found for this species.

MIGRATORY PATHWAYS Northernmost populations are migratory, mostly departing the American Southwest in the nonbreeding season. Some Central American birds may undertake altitudinal migrations, but little is known about the direction or extent of this species' movements (Ellison and Lowther 2020).

BROWN-HEADED COWBIRD *Molothrus ater*

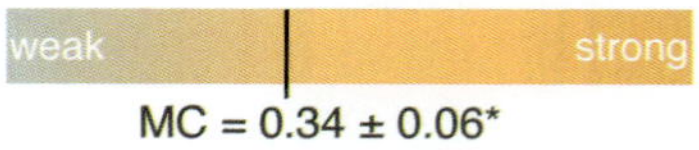

DISTRIBUTION AND ABUNDANCE

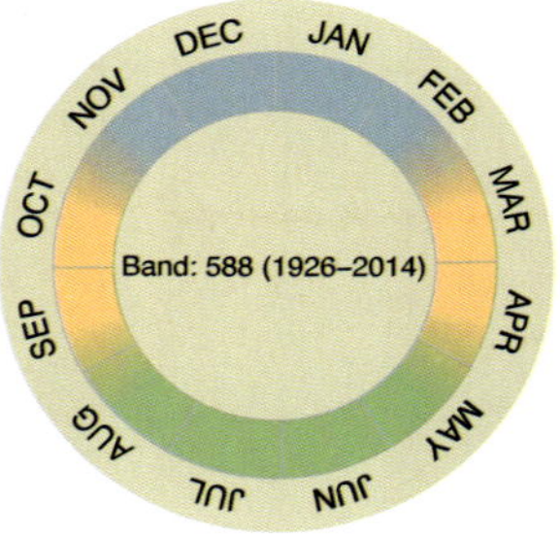

Brown-headed Cowbirds are the most common brood parasite in North America, with a range that has expanded since the 1900s as human development increased suitable habitat. This species is now found across the entire lower 48 states as well as much of northern Mexico and southern Canada (Lowther 2020b). High densities of Brown-headed Cowbirds breed in the Prairie Pothole Region, the Mississippi Delta, the eastern coastal plain, and riparian areas of the west. The species is found through much of its core range year-round, but retreats completely from the far north and montane west and moves into California and the southern Midwest into the southeastern US and into Mexico in the nonbreeding season (Fink et al. 2023).

ORDER Passeriformes
FAMILY Icteridae
BREEDING HABITAT Open Country, Habitat Mosaic
NONBREEDING HABITAT Open Country, Habitat Mosaic
GLOBAL POPULATION SIZE 130,000,000
POPULATION CHANGE SINCE 1970 −25%
IUCN RED LIST CATEGORY Least Concern
R2R TIPPING POINT SPECIES No
US LISTING STATUS No
CANADA SPECIES AT RISK No
PIF CCS 8/20

MIGRATORY CONNECTIVITY

OVERALL Migratory connectivity information comes from band reencounter data. The majority of Atlantic Flyway breeding birds (86%) remain within the Atlantic region in the nonbreeding season, with others (14%) moving into the Midcontinent Flyway. Likewise, birds breeding within the Midcontinent Flyway largely remain within the Midcontinent Flyway during the nonbreeding season (91%), as well as the Atlantic Flyway (6%) and Mexico (3%). Similarly, Pacific Flyway breeding birds migrate to nonbreeding areas within the flyway (89%) and to Mexico (11%). Although within-flyway transitions are high, mixing of breeding populations on other nonbreeding areas results in moderate migratory connectivity for this species (MC = 0.34 ± 0.06).

BANDING There are numerous band reencounter data from the eastern half of the US. There is limited data in the northern Great Plains and Pacific Northwest and no data in the Canadian Prairie provinces, and south through the Intermountain West, Southwest, and southern Great Plains.

	Atlantic	Mexico	Midcontinent	Pacific
Atlantic (n = 214)	86%	–	14%	–
Midcontinent (n = 293)	6%	3%	91%	–
Pacific (n = 18)	–	11%	–	89%

RADIO TRACKING There are no individual breeding to nonbreeding detections with the Motus Wildlife Tracking System, but several detections in the western part of the range from tags deployed in Illinois suggest birds in western portions of the range may migrate through Illinois (Motus project, Illinois Motus Network, unpublished data).

OTHER APPROACHES No tracking, genetic, or stable isotope studies of migratory connectivity were found for this species.

MIGRATORY PATHWAYS Birds move generally south to nonbreeding concentrations in the Atlantic, the central prairies, Mexico, and the US Southwest. (Lowther 2020b; Fink et al. 2023). Band reencounter data show individuals breeding in the Northeast migrate along the coastline to the southeastern US, while individuals breeding in the Midwest Great Lakes region migrate south along the Mississippi Flyway to the Gulf Coast.

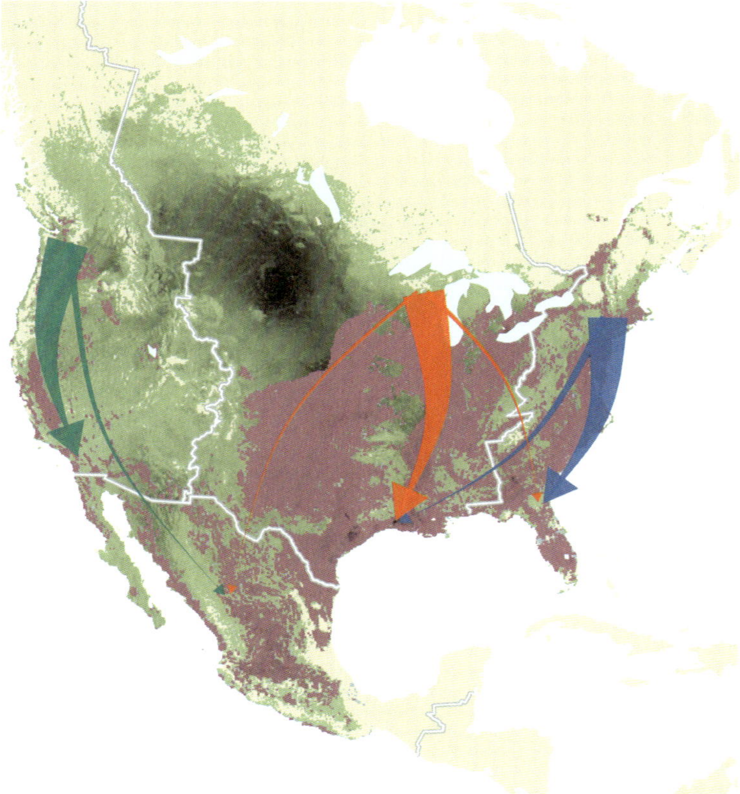

GENERALIZED CONNECTIVITY

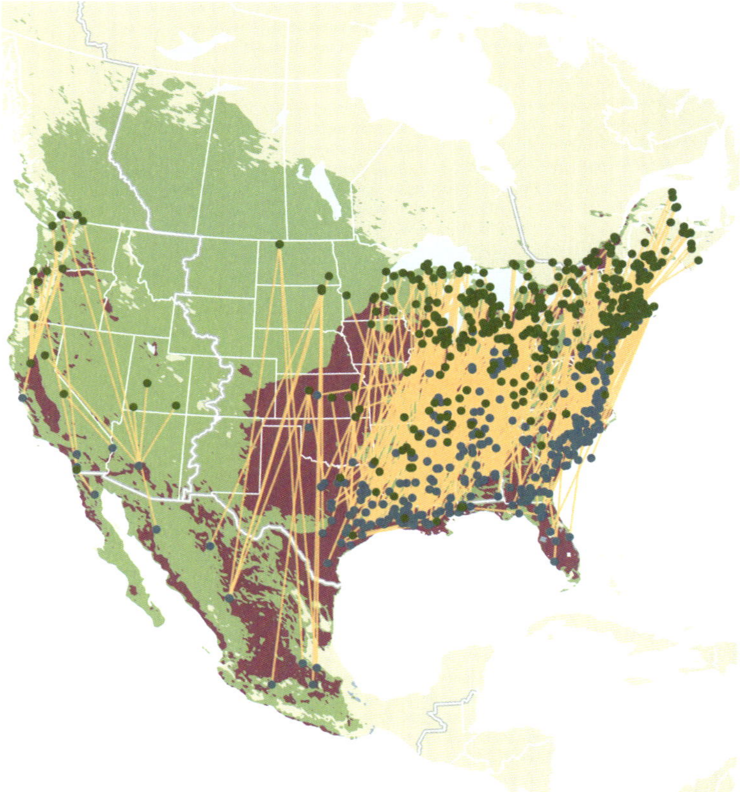

CONNECTIONS (A–B)

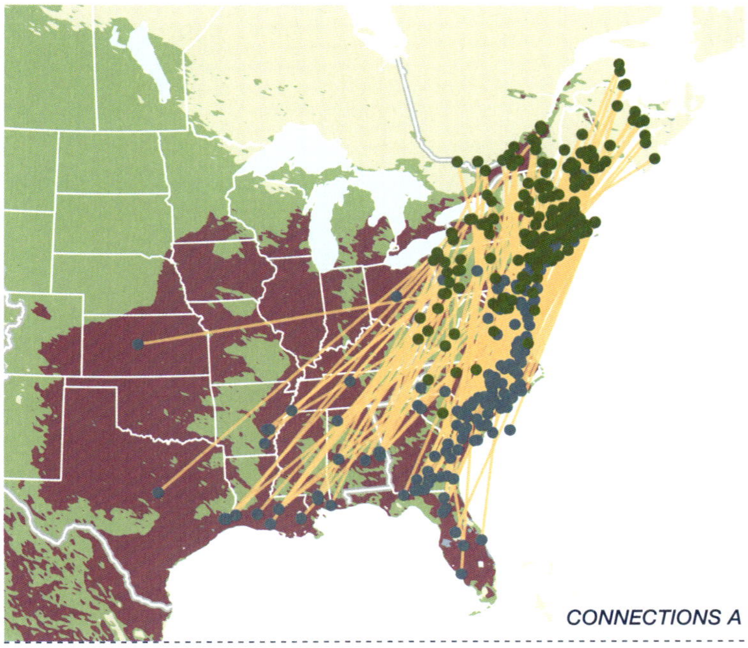

CONNECTIONS A

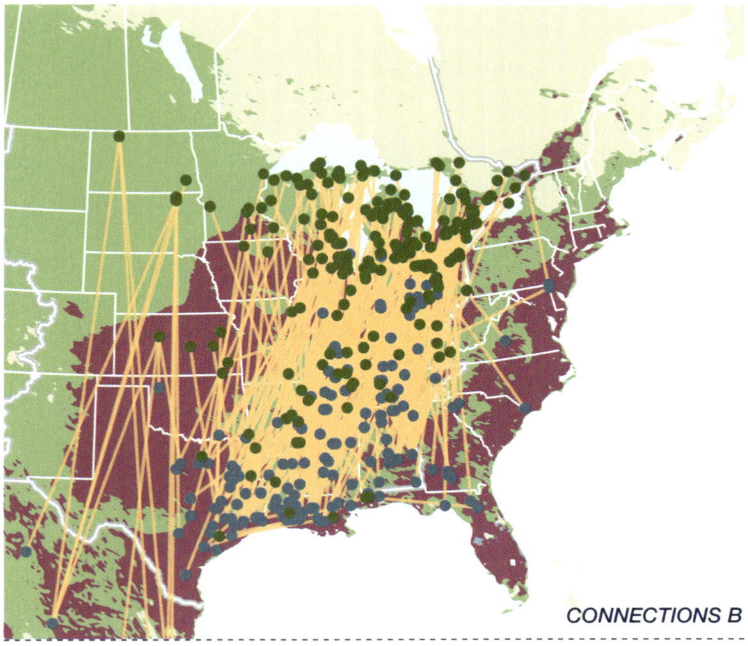

CONNECTIONS B

RUSTY BLACKBIRD *Euphagus carolinus*

DISTRIBUTION AND ABUNDANCE

The Rusty Blackbird is a boreal breeding species whose range extends from Alaska through the Canadian boreal region to the Atlantic provinces and into the northeastern US in Michigan, New York, Vermont, New Hampshire, and Maine. The nonbreeding range is concentrated in the eastern US, with strongholds in the southern Mississippi Alluvial Valley, the Black Belt region of Alabama and Mississippi, and the southeastern US Atlantic Coastal Plain region (Avery 2020; Fink et al. 2023).

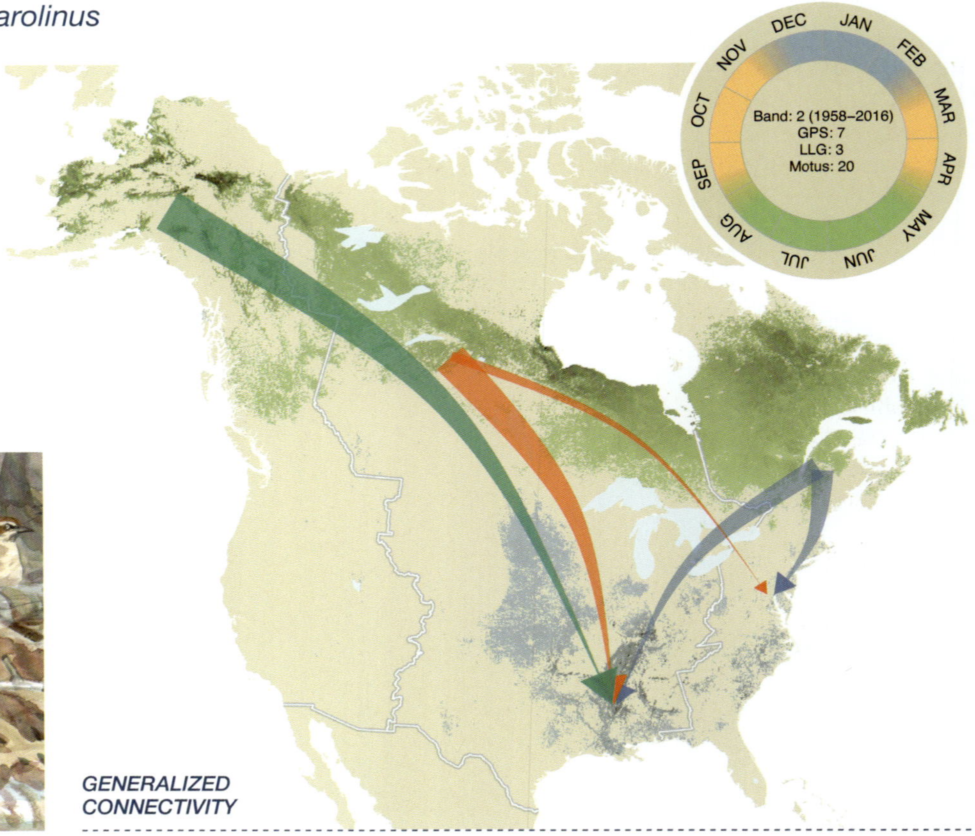

GENERALIZED CONNECTIVITY

ORDER Passeriformes
FAMILY Icteridae
BREEDING HABITAT Wetlands, Boreal
NONBREEDING HABITAT Wetlands, Forested
GLOBAL POPULATION SIZE 6,800,000
POPULATION CHANGE SINCE 1970 −79%
IUCN RED LIST CATEGORY Vulnerable
R2R TIPPING POINT SPECIES No
US LISTING STATUS No
CANADA SPECIES AT RISK No
PIF CCS 12/20

MIGRATORY CONNECTIVITY

	Atlantic	Midcontinent
Atlantic (n = 2)	50%	50%
Midcontinent (n = 5)	20%	80%
Pacific (n = 6)	–	100%

OVERALL The migratory connectivity of Rusty Blackbirds is known from a mixture of band reencounter data, stable isotope, and tracking studies. Atlantic Flyway breeding populations spend the nonbreeding season within the Atlantic (50%) or Midcontinent (50%) flyways. Midcontinent Flyway breeders primarily remain within the Midcontinent Flyway (80%) during the nonbreeding season, and other individuals migrate to the Atlantic Flyway (20%). Pacific Flyway breeders migrate to nonbreeding areas within the

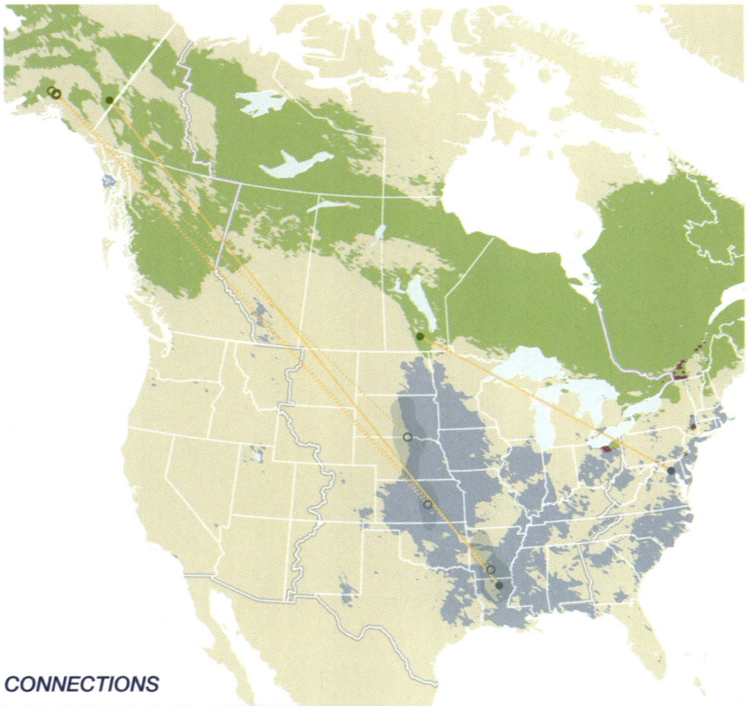

CONNECTIONS

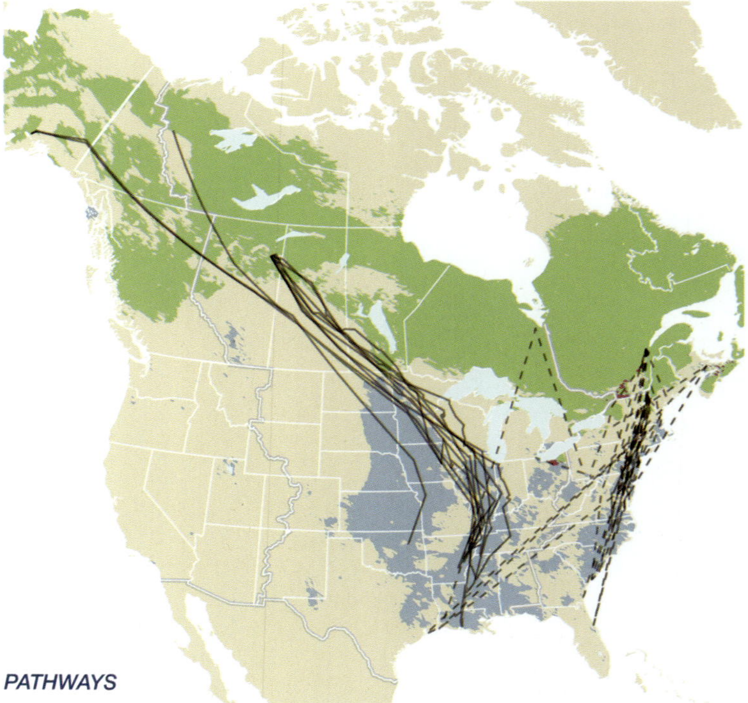

PATHWAYS

Midcontinent Flyway (100%). MC is not calculated due to limited sample size of breeding to nonbreeding connections.

BANDING There are breeding to nonbreeding band reencounters from Alaska to Louisiana and from Manitoba to the mid-Atlantic coast of the US.

RADIO TRACKING There was one breeding to nonbreeding detection and numerous migration detections by the Motus Wildlife Tracking System (C. Foss and L. Burford, unpublished data; C. Tonra, unpublished data; J. Tremblay and C. Foss, unpublished data). A bird breeding in the northeastern US was detected during the nonbreeding season on the Gulf Coast of the US (C. Tonra, unpublished data).

GEOLOCATOR TRACKING Rusty Blackbirds that bred in Alaska migrated to nonbreeding grounds in South Dakota and Louisiana (J.A. Johnson et al. 2012).

SATELLITE TRACKING Data from 7 of 30 individuals tagged with GPS in Alaska, Alberta, and New Hampshire found that Alberta individuals (n = 4) migrated to the Mississippi Alluvial Valley, that Alaska (n = 2) individuals spent the nonbreeding season in either the Mississippi Alluvial Valley or the Great Plains, and that the New Hampshire bird (n = 1) spent the nonbreeding season in the southeastern Coastal Plain (Wright et al. 2021).

STABLE ISOTOPES A study using stable isotopes found that western and central boreal breeding populations migrate through a Midcontinent Flyway, while eastern boreal breeders remain east and west of the Appalachians during the nonbreeding season (Hobson et al. 2010). They did not find significant temporal variation in migratory connectivity (Hobson et al. 2010).

MIGRATORY PATHWAYS Westernmost breeders carry out the longest migrations by flying east and then south through the Midcontinent Flyway. GPS data showed birds tracked from Alberta and one from Alaska follow a narrow corridor along the boreal transition / Prairie Pothole Region during both post- and pre-breeding migrations (Wright et al. 2021). Eastern breeders migrate straight south to their nonbreeding regions. A bird tracked from New Hampshire followed a migratory route between the Appalachians and the Atlantic coast (Wright et al. 2021). Interestingly, tracking studies using geolocators and the Motus network highlighted that the species often makes prolonged stopovers that are similar to the staging behavior of shorebirds, in areas such as the southwestern Lake Erie region (J.A. Johnson et al. 2012; Wright et al. 2018).

DATA CONTRIBUTORS

E.M. Bayne; L. Burford; C. Foss; J.A. Johnson; P.P. Marra; S. Matsuoka; L.L. Powell; D. Tessler; C. Tonra; J. Tremblay; J.R. Wright; Alaska Department of Fish and Game; Department of Defense; Migratory Connectivity Project; US Fish and Wildlife Service Migratory Bird Program, Alaska Region.

BREWER'S BLACKBIRD

Euphagus cyanocephalus

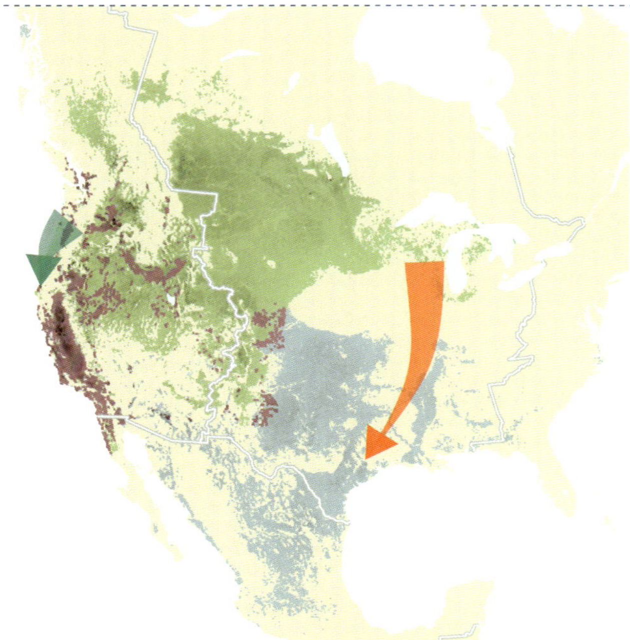

GENERALIZED CONNECTIVITY

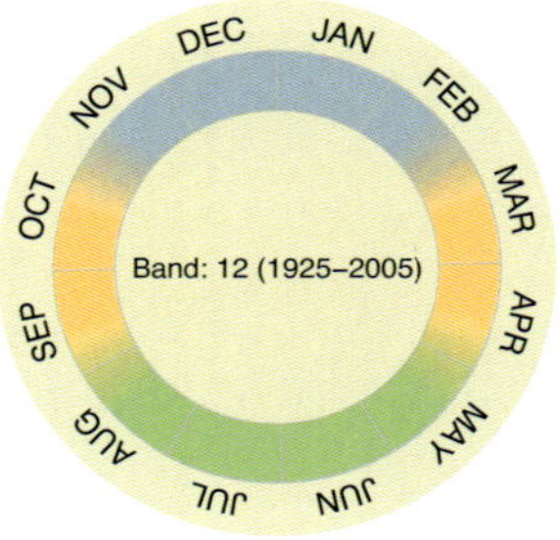

DISTRIBUTION AND ABUNDANCE

Brewer's Blackbirds breed throughout most of the western United States from Minnesota to Colorado west to the Pacific and north into Canada from Manitoba west to British Columbia. They spend the nonbreeding season in the southern midwestern states from Kansas south into Texas, with scattered populations west into California and into the northern states of Mexico. The species expanded its range eastward and westward in the twentieth century, likely due to land conversion (Martin 2020).

ORDER Passeriformes
FAMILY Icteridae
BREEDING HABITAT
Open Country, Habitat Mosaic
NONBREEDING HABITAT
Open Country, Habitat Mosaic
GLOBAL POPULATION SIZE 23,000,000
POPULATION CHANGE SINCE 1970 −62%
IUCN RED LIST CATEGORY Least Concern
R2R TIPPING POINT SPECIES No
US LISTING STATUS No
CANADA SPECIES AT RISK No
PIF CCS 11/20

	Midcontinent	Pacific
Midcontinent (n = 7)	100%	–
Pacific (n = 4)	–	100%

MIGRATORY CONNECTIVITY

OVERALL Migratory connectivity is relatively unknown but appears strong as both Midcontinent and Pacific flyway breeding birds migrate to nonbreeding areas within the same flyway as breeding. MC is not calculated due to limited data.

BANDING Breeding season banding data for this species comes from the Great Lakes region, Kansas, the Canadian Prairies, and along the west coast from British Columbia to Southern California. Nonbreeding data show that Great Lakes birds moved south to the southern Midwest and Mississippi. Birds from the Canadian Prairies and Kansas moved south to Texas. Birds from the northern west coast undertook shorter movements to central Washington and California. A Southern California breeding bird traveled north during the nonbreeding season to central California.

RADIO TRACKING A few Motus tags that were deployed yielded no breeding to nonbreeding detections thus far.

OTHER APPROACHES No tracking, genetic, or stable isotope studies of migratory connectivity were found for this species.
MIGRATORY PATHWAYS This species undergoes short- and medium-range migrations across much of the western two-thirds of North America. Overall, individuals breeding in central Canada and central US migrate longer distances, while individuals breeding in Pacific mountains and along the coast migrate shorter distances. Birds breeding east of the Rockies moved away from colder areas to spend the nonbreeding season in the central plains and the south. Western birds also tended to move north to south, but at least one bird underwent a short-distance northward migration. Some individuals that breed in the Canadian Prairies fly west to the Pacific coast rather than south in the nonbreeding season (Martin 2020).

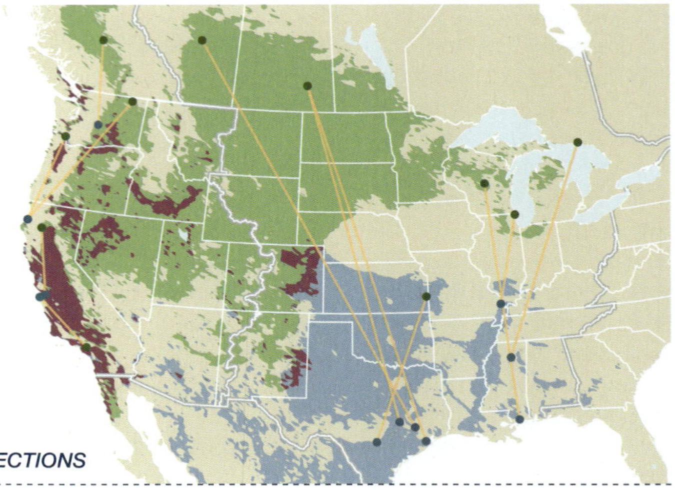

CONNECTIONS

COMMON GRACKLE *Quiscalus quiscula*

MC = 0.74 ± 0.03*

DISTRIBUTION AND ABUNDANCE

This large and raucous blackbird is a common breeder in the eastern and central United States and into southern Canada from Calgary, Alberta, and east to Nova Scotia. This species is found year-round in much of its eastern range south of the Canadian border, but there is a definite contraction in the nonbreeding season toward the American southeast (Peer and Bollinger 2020; Fink et al. 2023).

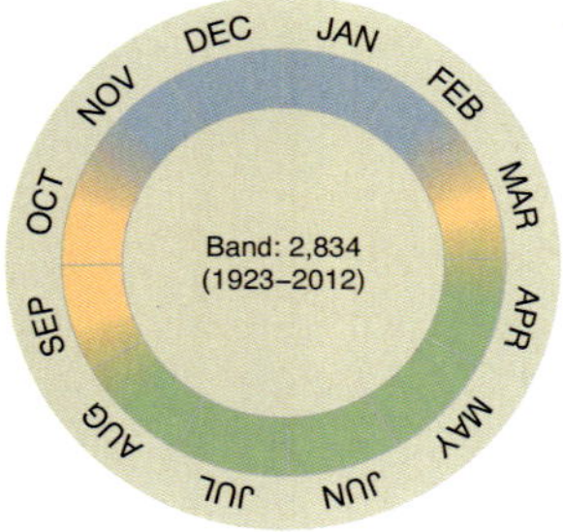

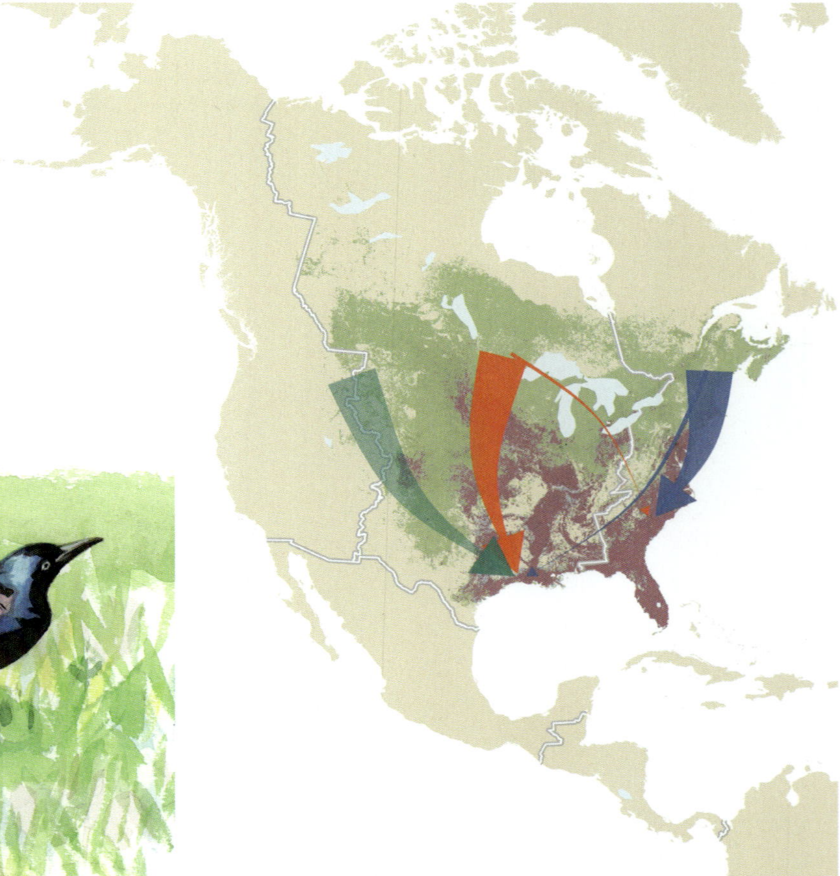

GENERALIZED CONNECTIVITY

ORDER Passeriformes
FAMILY Icteridae
BREEDING HABITAT Open Country, Habitat Mosaic
NONBREEDING HABITAT Open Country, Habitat Mosaic
GLOBAL POPULATION SIZE 67,000,000
POPULATION CHANGE SINCE 1970 –58%
IUCN RED LIST CATEGORY Near Threatened
R2R TIPPING POINT SPECIES No
US LISTING STATUS No
CANADA SPECIES AT RISK No
PIF CCS 10/20

	Atlantic	Midcontinent
Atlantic (n = 529)	94%	6%
Midcontinent (n = 2138)	3%	97%
Pacific (n = 1)	–	100%

MIGRATORY CONNECTIVITY

OVERALL Migratory connectivity is known from band reencounter data from across the range. The majority (94%) of Atlantic Flyway breeding birds migrated within their breeding flyway in the nonbreeding season, while Midcontinent (97%) and Pacific (100%) flyway birds migrated to nonbreeding areas in the Midcontinent Flyway, resulting in strong migratory connectivity for this species (MC = 0.74 ± 0.03).

BANDING There is extensive band reencounter data from across the range, except for breeding areas in Ontario and Quebec, north of the Great Lakes region. Breeding to nonbreeding band reencounter data from the US portion of the range show mostly southward movement from breeding sites in the northeast and central states to nonbreeding sites concentrated in the southern states and the mid-Atlantic coast.

OTHER APPROACHES No tracking, genetic, or stable isotope studies of migratory connectivity were found for this species.

MIGRATORY PATHWAYS Birds breeding in central Canada migrate southeast to the US states west of the Appalachians, where they join breeding birds from the central US in the nonbreeding season. Birds breeding in the Maritime provinces and New England spend the nonbreeding season mostly along the Atlantic coast. Birds in some southern areas, like the Gulf Coast and the Mississippi Delta, are present year-round and are largely nonmigratory (Peer and Bollinger 2020; Fink et al. 2023).

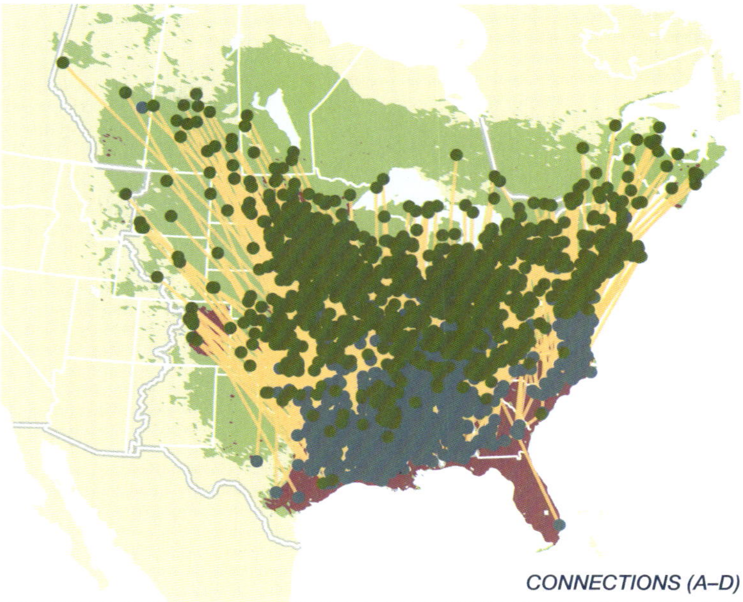

CONNECTIONS (A–D)

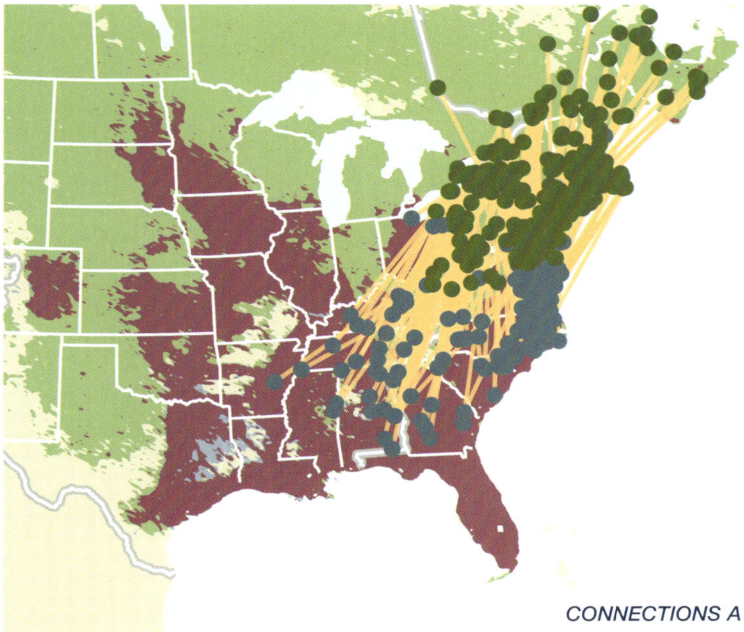

CONNECTIONS A

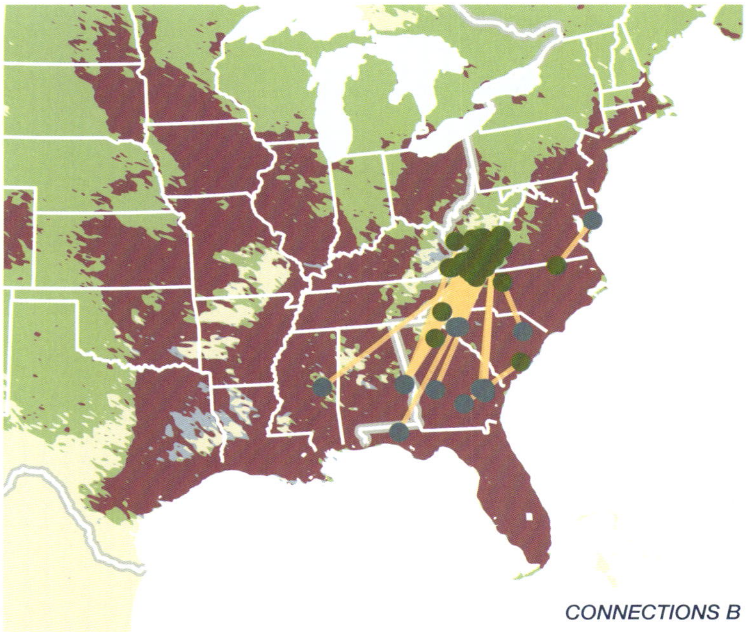

CONNECTIONS B

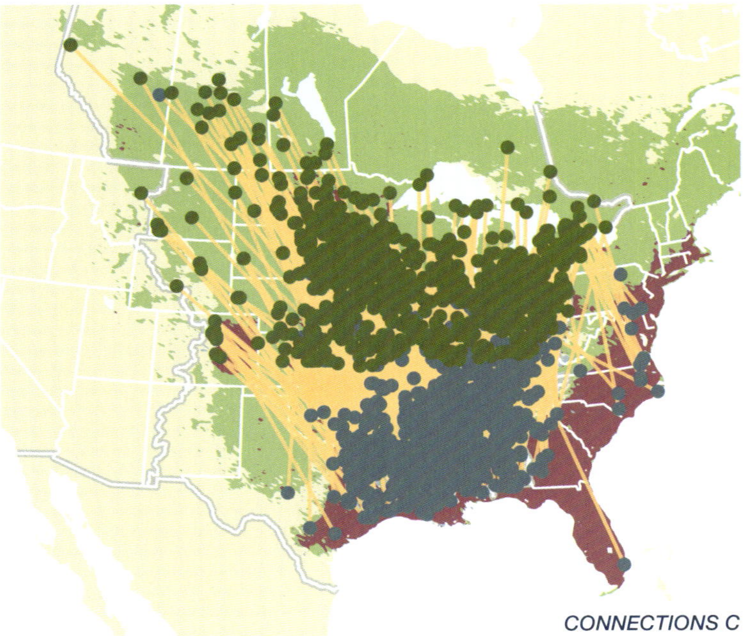

CONNECTIONS C

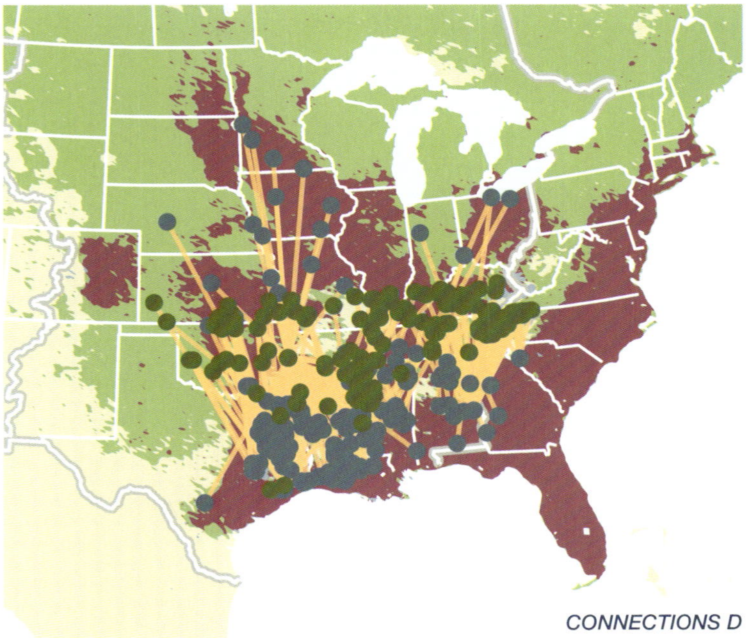

CONNECTIONS D

BOAT-TAILED GRACKLE *Quiscalus major*

DISTRIBUTION AND ABUNDANCE

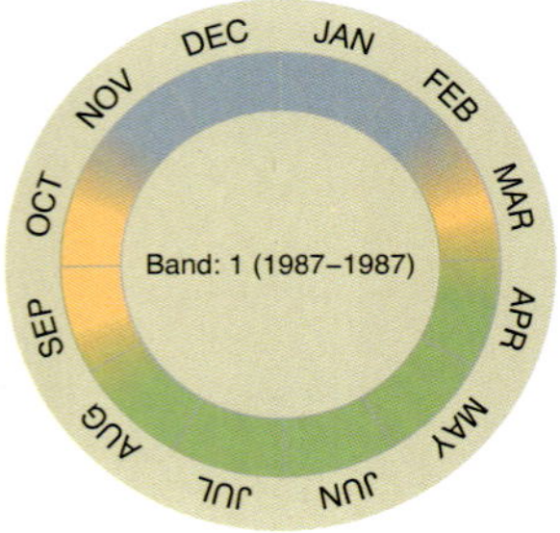

Closely associated with salt marsh habitats, this species occurs along the Atlantic and Gulf coasts of the United States year-round. Atlantic birds range as far north as Long Island, New York, while Gulf birds are found as far west as Galveston Bay, Texas. This species reaches its highest abundances in southern Florida, along the South Carolina and Georgia coasts, and along the Gulf Coast of Louisiana and far-eastern Texas (Fink et al. 2023). Most populations are nonmigratory, though some northern populations likely move south in the nonbreeding season (Post et al. 2020).

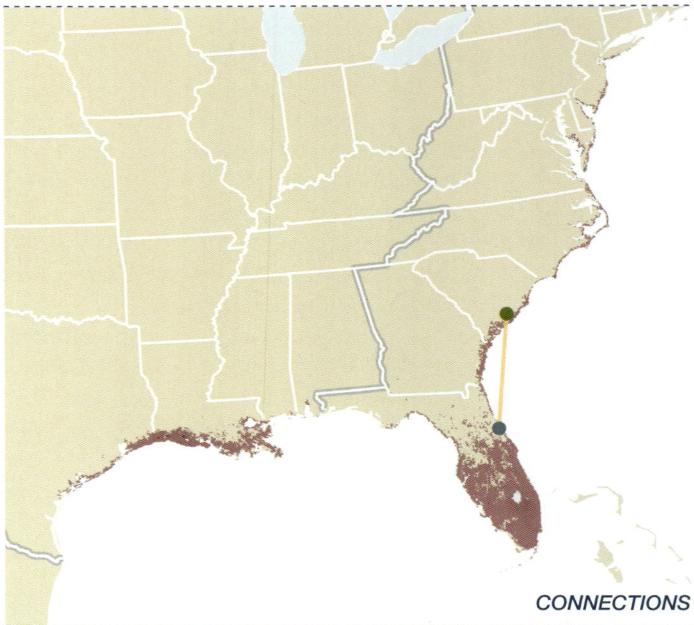

ORDER Passeriformes
FAMILY Icteridae
BREEDING HABITAT Coasts, Salt Marsh
NONBREEDING HABITAT Coasts, Salt Marsh
GLOBAL POPULATION SIZE 2,200,000
POPULATION CHANGE SINCE 1970 −34%
IUCN RED LIST CATEGORY Least Concern
R2R TIPPING POINT SPECIES No
US LISTING STATUS No
CANADA SPECIES AT RISK No
PIF CCS 13/20

MIGRATORY CONNECTIVITY

OVERALL Migratory connectivity is unknown.
BANDING One breeding to nonbreeding band reencounter showed southward movement in the nonbreeding season, connecting South Carolina to Florida (Post 1994).
OTHER APPROACHES No tracking, genetic, or stable isotope studies of migratory connectivity were found for this species.
MIGRATORY PATHWAYS Limited movements likely follow coastlines (Post et al. 2020).

GREAT-TAILED GRACKLE *Quiscalus mexicanus*

DISTRIBUTION AND ABUNDANCE

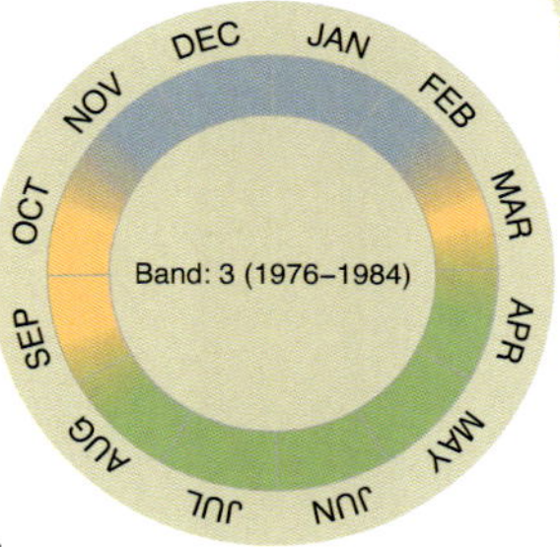

Distributed from the northwestern coasts of South America, throughout Central America, and into the central United States and along the coasts and some inland areas of Texas and Louisiana, this large and long-tailed blackbird is bold and conspicuous, with a unique and varied vocal repertoire. (Johnson and Peer 2020). Most individuals are thought to be residents, and migratory behavior is most common along the northern edge of its range expansion (Johnson and Peer 2020).

MIGRATORY CONNECTIVITY

OVERALL Migratory connectivity is unknown; little information about migratory connectivity and migratory behavior.
BANDING Limited band reencounter data exists for this species. Birds banded in Texas migrated south in the nonbreeding season and north in the breeding season (Arnold and Folse 1977).
OTHER APPROACHES No satellite, genetic, or stable isotope studies of migratory connectivity were found for this species.
MIGRATORY PATHWAYS There is no information on specific routes. Observational data show high abundance along the west coast of Mexico and the Gulf of Mexico during both pre- and post-breeding migration, suggesting this species follows coastlines and circumnavigates the Gulf of Mexico during migration (Fink et al. 2023).

ORDER Passeriformes
FAMILY Icteridae
BREEDING HABITAT Open Country, Developed/Disturbed
NONBREEDING HABITAT Open Country, Developed/Disturbed
GLOBAL POPULATION SIZE 30,000,000
POPULATION CHANGE SINCE 1970 +190%
IUCN RED LIST CATEGORY Least Concern
R2R TIPPING POINT SPECIES No
US LISTING STATUS No
CANADA SPECIES AT RISK No
PIF CCS 8/20

Seiurus aurocapilla OVENBIRD

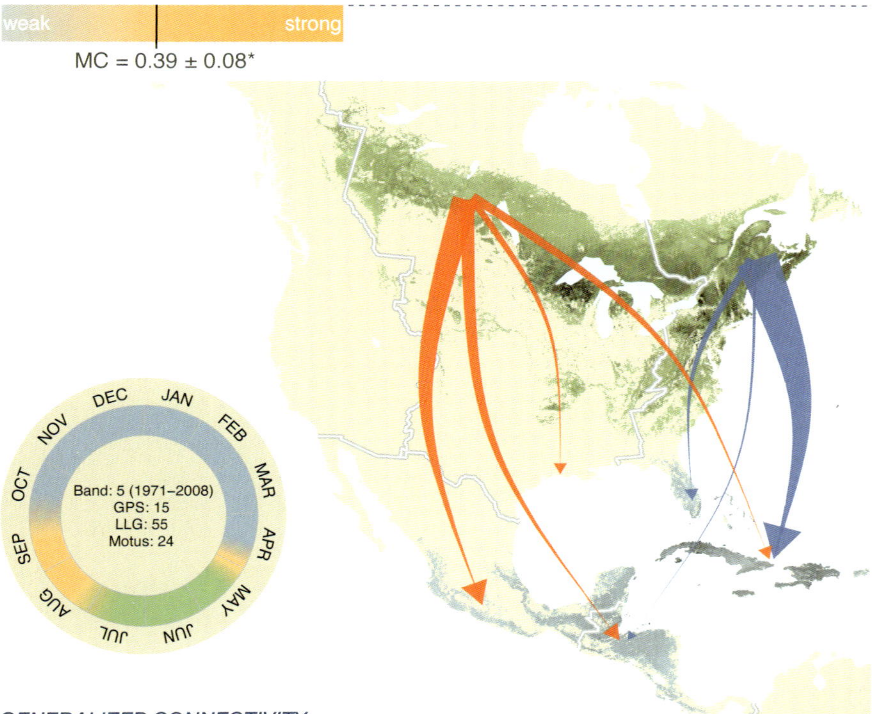

GENERALIZED CONNECTIVITY

DISTRIBUTION AND ABUNDANCE

This species is a widespread breeder of the eastern and central US and across the boreal region of Canada, decreasing in abundance from east to west; it is patchily distributed through the western Great Plains and the front range of the Rocky Mountains. Its nonbreeding range is in the lowlands of Mexico and northern Central America and throughout the Caribbean and southern Florida; a few individuals reach the Caribbean coast of Colombia and Venezuela. Populations have been stable or increasing in most of the range, with declines noted in southern New England, the mid-Atlantic coast, and the Great Plains region (Sauer et al. 2019).

	Atlantic	Caribbean	Central America	Mexico	Midcontinent
Atlantic (n = 54)	13%	85%	2%	–	–
Midcontinent (n = 20)	–	15%	25%	50%	10%

ORDER Passeriformes
FAMILY Parulidae
BREEDING HABITAT Forests, Temperate Eastern
NONBREEDING HABITAT Forests, Tropical Generalist
GLOBAL POPULATION SIZE 26,000,000
POPULATION CHANGE SINCE 1970 0%
IUCN RED LIST CATEGORY Least Concern
R2R TIPPING POINT SPECIES No
US LISTING STATUS No
CANADA SPECIES AT RISK No
PIF CCS 9/20

MIGRATORY CONNECTIVITY

OVERALL Migratory connectivity is strong (MC = 0.39 ± 0.08), with little overlap between breeding populations on the nonbreeding grounds, based on morphology, banding data, stable isotope analysis, and tracking studies. Individuals breeding in the Atlantic Flyway primarily spend the nonbreeding season in the Caribbean (85%), while individuals breeding in the Midcontinent Flyway spend the nonbreeding season in Mexico (50%), Central America (25%), the Caribbean (15%), and within the Midcontinent Flyway (10%).

GEOLOCATOR TRACKING Geolocator-tracked individuals breeding in the northeastern United States and New Brunswick, Canada, migrate along the Atlantic Flyway to nonbreeding grounds in the Caribbean; individuals breeding in the western boreal forest of Canada migrate along the Midcontinent Flyway to southern Mexican and Central American nonbreeding grounds; and individuals from Cypress Hills, Alberta, migrate to western Mexico (Hache et al. 2017; Hallworth et al. 2015).

BANDING Five nonbreeding band reencounters indicate an east-west separation, with two individuals from the western Great Lakes region recovered in eastern Mexico, two individuals from the eastern Great Lakes region recovered in Belize, and a single individual from the Atlantic coast recovered in the Bahamas.

RADIO TRACKING There are several projects tracking the Ovenbird through the Motus Wildlife Tracking System. There are no individual breeding to nonbreeding detections; however, there are numerous detections during migration.

GEOLOCATOR TRACKING Geolocator-tracked individuals breeding in the northeastern

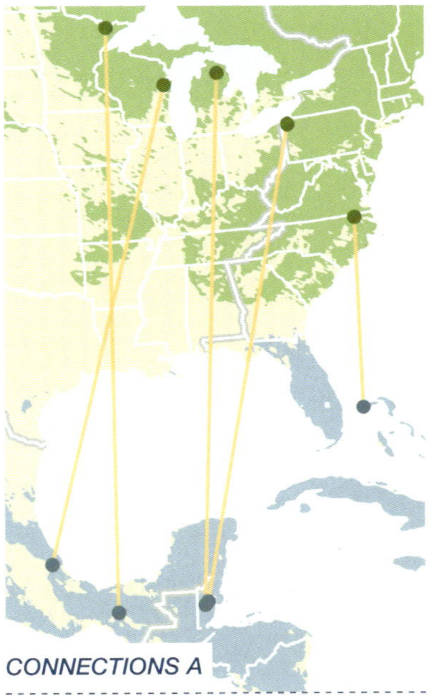

CONNECTIONS A

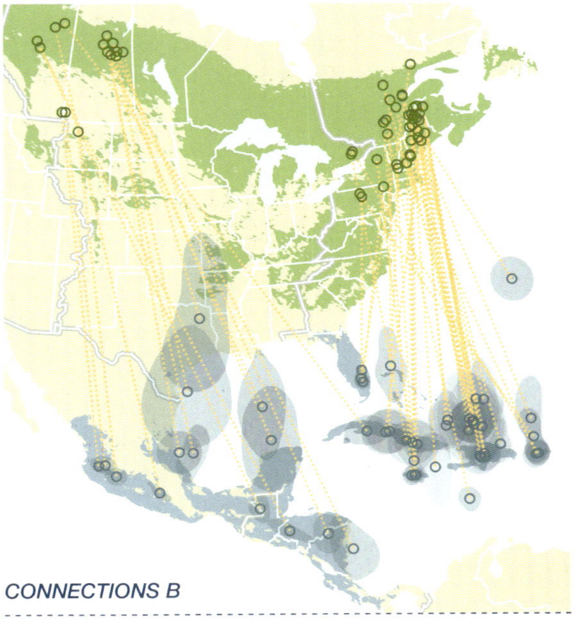

CONNECTIONS B

United States and New Brunswick, Canada, migrate along the Atlantic Flyway to nonbreeding grounds in the Caribbean; individuals breeding in the western boreal forest of Canada migrate along the Midcontinent Flyway to southern Mexican and Central American nonbreeding grounds; and individuals from Cypress Hills, Alberta, migrate to western Mexico (Hache et al. 2017; Hallworth et al. 2015).

SATELLITE TRACKING Individuals tracked with archival GPS tags from the breeding grounds in Maryland (n = 10) maintained nonbreeding territories in peninsular Florida and western Cuba, whereas individuals breeding in New Hampshire (n = 5) held nonbreeding territories in eastern Cuba and Hispaniola (Hallworth and Marra 2015). These studies also suggest a leapfrog migration pattern, with the southernmost breeding populations in both the east and west migrating much shorter distances than breeders from farther north.

STABLE ISOTOPES Stable isotope analysis for individuals sampled in Puerto Rico, were found to likely breed throughout the eastern United States (Dugger et al. 2004). Analysis of stable isotopes during pre-breeding migration through Louisiana suggests most individuals bred in more southerly portions of the range (Langin et al. 2009). A broader survey from three Caribbean Basin nonbreeding areas suggested at least moderate connectivity and a leapfrog migration pattern, with individuals sampled in Nicaragua and Belize likely originating in the western Great Lakes region, individuals from Florida, Bahamas, and western Cuba most likely originating in the eastern US, and individuals from Hispaniola and Puerto Rico most likely coming from the northeastern portion of the breeding range (Studds et al. 2021).

GENETICS Genetic analysis of nuclear DNA from feather samples confirms the distinctiveness of *S. a. cinereus* (sampled in Cypress Hills, Alberta), showing little gene flow with the widespread boreal and eastern Canadian populations, as well as strong connectivity to the nonbreeding range in western Mexico (Hache et al. 2017).

OTHER APPROACHES Examination of geographic subspecies suggests strong connectivity, with the most distinctive western population (*S. a. cinereus*) spending the nonbreeding season on the Pacific Slope from Mexico to Honduras; widespread *S. a. auricapilla* occupying the remainder of the nonbreeding range in eastern Mexico, Central America, and the Caribbean; and Newfoundland breeders (*S. a. furvior*) found mainly in Cuba, the Bahamas, and the Caribbean coast of Central America (Porneluzi et al. 2020; Dunn and Garrett 1997).

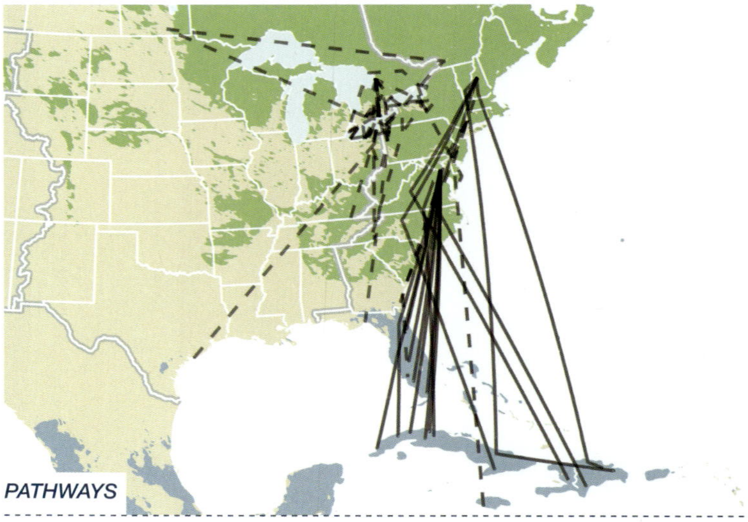

PATHWAYS

MIGRATORY PATHWAYS This species migrates broadly across the eastern US, with higher abundance in pre-breeding migration through the Atlantic coastal states and higher abundance in post-breeding migration through the Appalachian Mountains and Florida; hence this species appears to exhibit a counterclockwise loop migration (La Sorte et al. 2016). Evidence for migration both across and around the Gulf of Mexico and evidence of the species' common occurrence during post-breeding migration and the nonbreeding season on Bermuda indicate that Ovenbirds regularly cross the western Atlantic Ocean (Amos 1991).

DATA CONTRIBUTORS

A. Beauchamp; A.R. Brunner; J. Cepek; E. Cohen; L. DeGroote; B. Dossman; J. Fowler; M. Gimpel; C.C. Guglielmo; T. Guida; M.T. Hallworth; K.A. Hobson; R. Jacob; L. Koitsch; T. Krynak; P.P. Marra; Y. Morbey; C. Nemes; J. Rice; S. Sargent; M. Sarver; C. Seewagen; T.S. Sillett; S.L. Van Wilgenburg; D. Yeany.

WORM-EATING WARBLER *Helmitheros vermivorum*

DISTRIBUTION AND ABUNDANCE

It breeds patchily across the eastern and central US, with greatest abundance in the Appalachian Mountains and central hardwood forest regions, with smaller numbers on the mid-Atlantic and Gulf Coastal plains. Its greatest nonbreeding abundance is in Jamaica, Cuba, and the southern Yucatán Peninsula, with lower numbers elsewhere in the Greater Antilles, Bahamas, and in both Caribbean and Pacific lowlands from southern Mexico to western Panama. Populations have been stable or slightly increasing in most parts of the range (Sauer et al. 2019).

ORDER Passeriformes
FAMILY Parulidae
BREEDING HABITAT Forests, Temperate Eastern
NONBREEDING HABITAT Forests, Tropical Lowland Evergreen
GLOBAL POPULATION SIZE 780,000
POPULATION CHANGE SINCE 1970 +22%
IUCN RED LIST CATEGORY Least Concern
R2R TIPPING POINT SPECIES No
US LISTING STATUS No
CANADA SPECIES AT RISK No
PIF CCS 13/20

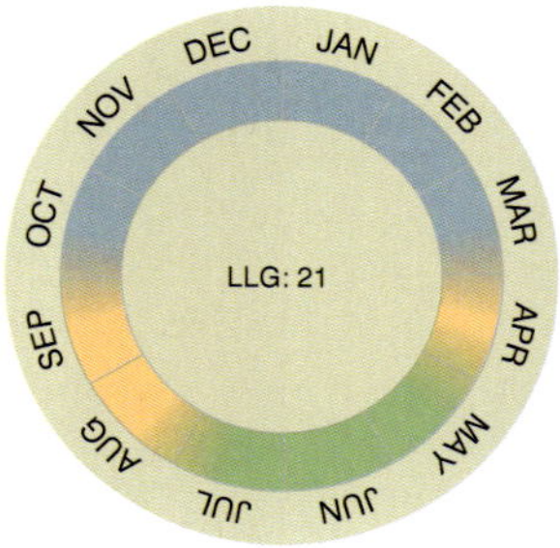

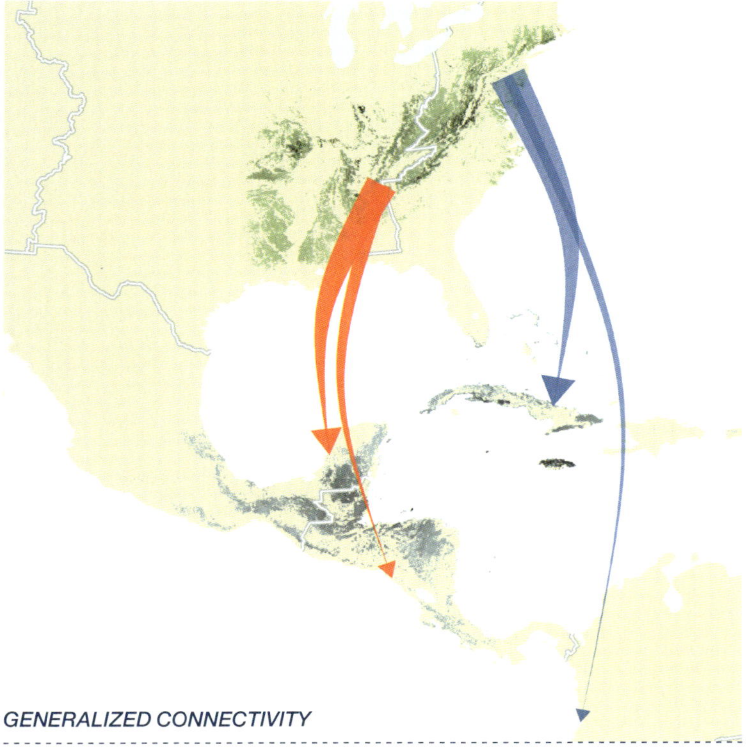

GENERALIZED CONNECTIVITY

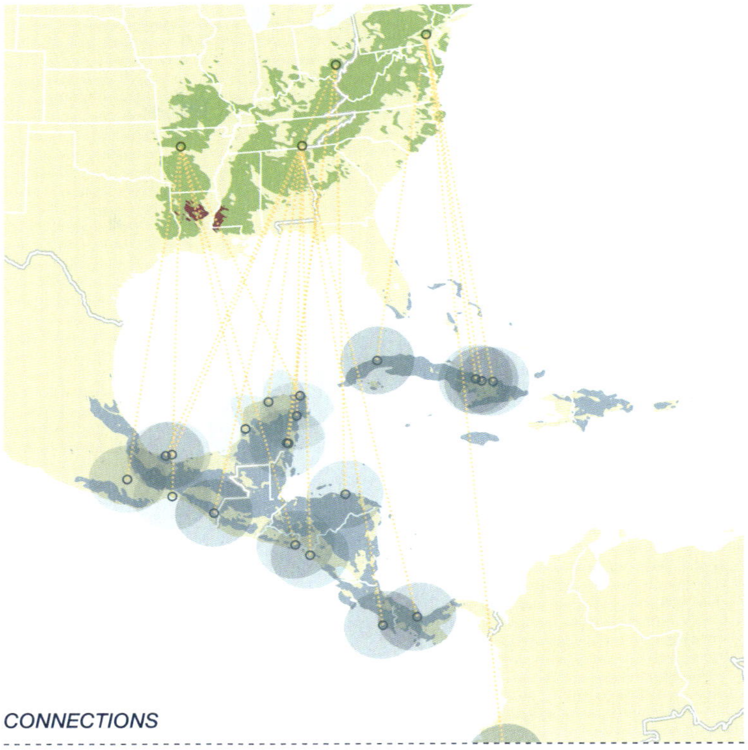

CONNECTIONS

	Caribbean	Central America	Mexico	Northern SA
Atlantic (n = 5)	80%	–	–	20%
Midcontinent (n = 16)	–	31%	69%	–

MIGRATORY CONNECTIVITY

OVERALL Information on migratory connectivity for this species comes mainly from a single study with light-level geolocators and appears strong (Kramer et al. 2025). Birds breeding in the Atlantic Flyway migrated to the Caribbean (80%) and Northern South America (20%) for the nonbreeding season, while birds breeding in the Midcontinent Flyway migrated to Mexico (69%) and Central America (31%).

RADIO TRACKING A few tags were deployed through the Motus Wildlife Tracking System. There are no breeding to nonbreeding detections, but there are a few detections during migration.

GEOLOCATOR TRACKING Kramer et al. (2025) attached 68 light-level geolocators to Worm-eating Warblers at four breeding sites (Arkansas, Tennessee, Ohio, and Pennsylvania). They recovered 21 geolocators and found moderately strong migratory connectivity between breeding and nonbreeding sites (MC = 0.46). Birds recovered in Arkansas (n = 5) spent the nonbreeding season primarily in Guatemala, Belize, and Yucatán; birds recovered in Tennessee (n = 7) spent the nonbreeding season primarily in Yucatán, Guatemala, Belize, and Honduras; birds recovered in Ohio (n = 4) spent the nonbreeding season primarily in Yucatán, Guatemala, Belize, and Honduras; and birds recovered in Pennsylvania (n = 5) spent the nonbreeding season primarily in Panama, Columbia, and Venezuela.

STABLE ISOTOPES Analysis of flight feathers indicated individuals captured on nonbreeding grounds in Puerto Rico likely originated from southern portions of the breeding range. (Hobson et al. 2014).

MIGRATORY PATHWAYS This species migrates through the southeastern US across a broad front from Texas to Florida and the Carolinas during both pre- and post-breeding, but the shifts in abundance suggest clockwise loop migration based on eBird data (La Sorte et al. 2016). Evidence for migration both across and around the Gulf of Mexico (Dunn and Garrett 1997; Vitz et al. 2020) and evidence of the species' frequent occurrence on Bermuda during nonbreeding and migration (up to 15 per day) suggest regular transoceanic crossings (Amos 1991).

Parksia motacilla LOUISIANA WATERTHRUSH

DISTRIBUTION AND ABUNDANCE

It breeds throughout most of the eastern US, west to the Edwards Plateau of Texas, avoiding flat coastal plains, the Mississippi Valley, and the Florida peninsula. It occurs at low densities in the nonbreeding season throughout most of Mexico (but not the Yucatán Peninsula), Central America, and the Greater Antilles, with small numbers south to the highlands of northern Colombia and adjacent Venezuela (Sierra de Perijá mountain range; Fink et al. 2023).

ORDER Passeriformes
FAMILY Parulidae
BREEDING HABITAT Wetlands, Lakes and Rivers
NONBREEDING HABITAT Wetlands, Lakes and Rivers
GLOBAL POPULATION SIZE 450,000

POPULATION CHANGE SINCE 1970 +47%
IUCN RED LIST CATEGORY Least Concern
R2R TIPPING POINT SPECIES No
US LISTING STATUS No
CANADA SPECIES AT RISK Threatened
PIF CCS 11/20

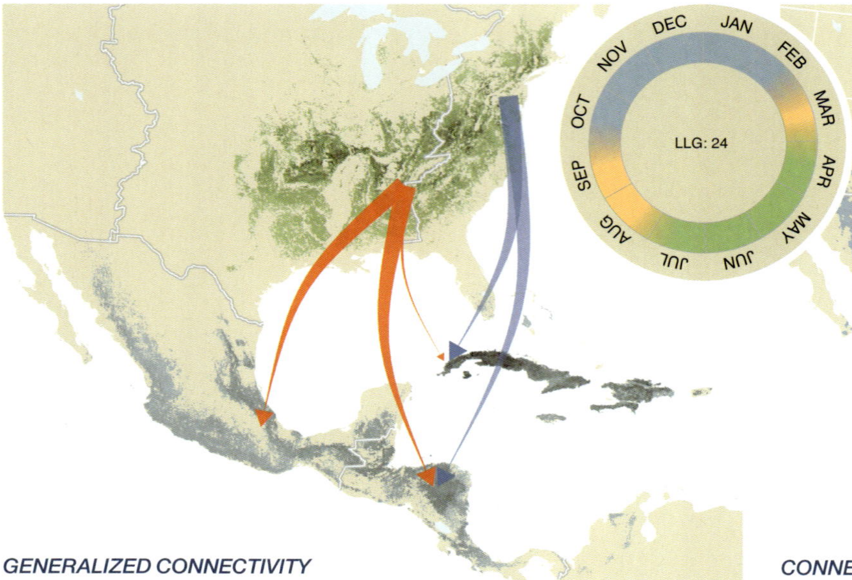

GENERALIZED CONNECTIVITY

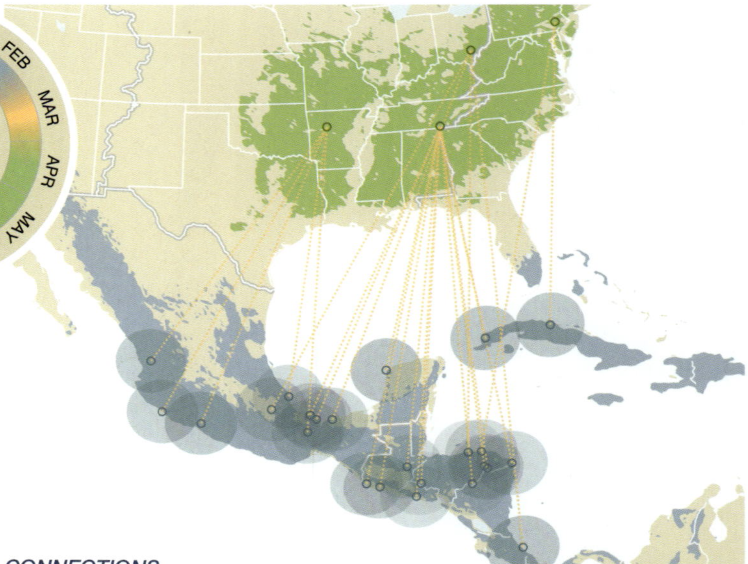

CONNECTIONS

MIGRATORY CONNECTIVITY

OVERALL Information on migratory connectivity for this species comes entirely from a single study with light-level geolocators (Kramer et al. 2025). Birds breeding in the Atlantic Flyway migrated to the Caribbean (50%) or Central America (50%) for the nonbreeding season, while birds breeding in the Midcontinent Flyway primarily migrated to Central America (50%) or Mexico (45%). MC is not calculated for this species, but Kramer et al. (2025) found strong migratory connectivity for the species.

RADIO TRACKING No breeding to nonbreeding detections from the Motus Wildlife Tracking System.

GEOLOCATOR TRACKING Kramer et al. (2025) attached 85 light-level geolocators to Louisiana Waterthrushes at four breeding sites (Arkansas, Tennessee, Ohio, and Pennsylvania). They recovered 24 and found strong migratory connectivity between breeding and nonbreeding sites (MC = 0.51). Birds recovered in Arkansas (n = 5) spent the nonbreeding season primarily in central and western Mexico; birds recovered in Tennessee (n = 13) spent the nonbreeding season primarily in Guatemala, Belize, and Honduras; birds recovered in Ohio (n = 4) spent the nonbreeding season primarily in Guatemala and Honduras; and birds recovered in Pennsylvania (n = 2) spent the nonbreeding season primarily in Nicaragua and Panama. Data are not presented in the Atlas.

	Caribbean	Central America	Mexico
Atlantic (n = 2)	50%	50%	–
Midcontinent (n = 22)	5%	50%	45%

MIGRATORY PATHWAYS Migrations are extremely early and across the Gulf of Mexico in both pre- and post-breeding migration (Mattsson et al. 2020), with many migrants passing through northern Central America and western Cuba as well. There is some evidence of loop migration; the species is more common on the Texas Coast in the pre-breeding season and through Florida and Bahamas in post-breeding fall (Dunn and Garrett 1997). The regular post-breeding occurrences in Bermuda (Amos 1991) also suggest transoceanic crossings.

NORTHERN WATERTHRUSH *Parkesia noveboracensis*

DISTRIBUTION AND ABUNDANCE

It breeds in forested wetlands and bogs across the entire boreal region from Alaska to Newfoundland southward into the northern Rocky Mountains and the northeastern US. The nonbreeding distribution is equally large and diverse, encompassing coastal mangroves and lowland regions from Mexico and Florida through the Caribbean and Central America to northern South America. Despite increases in much of the western

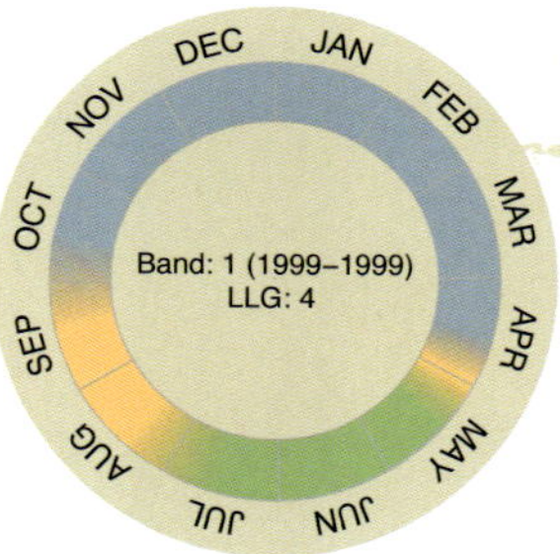

ORDER Passeriformes
FAMILY Parulidae
BREEDING HABITAT Forests, Boreal
NONBREEDING HABITAT Coasts, Mangroves
GLOBAL POPULATION SIZE 17,000,000
POPULATION CHANGE SINCE 1970 +51%
IUCN RED LIST CATEGORY Least Concern
R2R TIPPING POINT SPECIES No
US LISTING STATUS No
CANADA SPECIES AT RISK No
PIF CCS 8/20

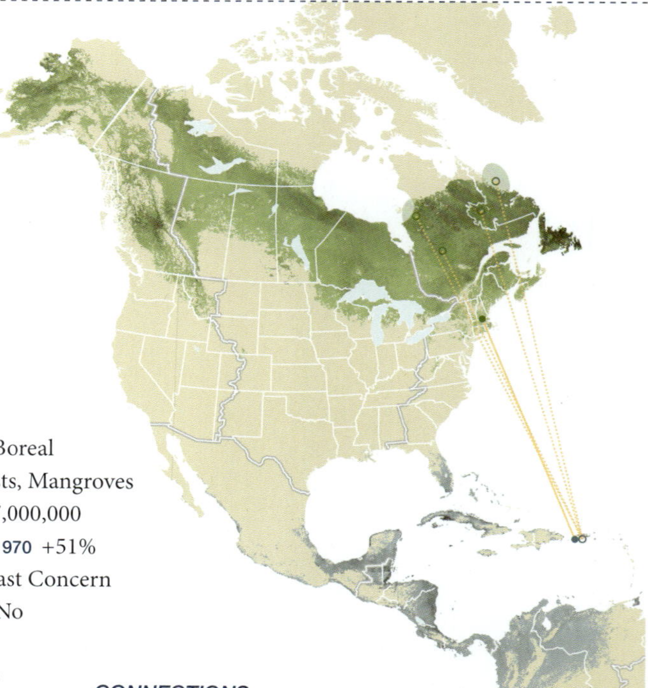

CONNECTIONS

part of the range, long-term declines have occurred in the eastern boreal forest and New England, as well as in Alaska (Fink et al. 2023).

MIGRATORY CONNECTIVITY

OVERALL Migratory connectivity is poorly known, with some suggestion of an east-west parallel migration, based on band reencounter, geolocator, and stable isotope data.
BANDING The single breeding to nonbreeding band recovery was a bird from New Hampshire recovered in Puerto Rico. An individual banded in Manitoba during post-breeding migration was recaptured 23 days later in Veracruz, Mexico (Schaffner 2013). Additional band returns from nonbreeding and migration sites suggest an east-west parallel migration pattern.
RADIO TRACKING Over 200 tags were deployed using the Motus Wildlife Tracking System. As of printing, there are no individual breeding to nonbreeding detections, yet there are many detections during migration along the eastern US.
GEOLOCATOR TRACKING Researchers deployed 40 light-level geolocators on Northern Waterthrushes in Puerto Rico in 2016–2017 (Schaffner et al. 2023). Of these, 11 geolocators were recovered the following year, but only 9 had usable data. However, only 4 of these tags had reliable location data, placing the breeding locations of these birds in northern Quebec.
MIGRATORY PATHWAYS Overall, migration is in a broad front through the eastern US, with birds crossing the Gulf of Mexico in both pre- and post-breeding. Common occurrence on Bermuda also suggests regular crossing of the western Atlantic Ocean (Amos 1991). Banding and stable isotope studies in Canada suggest that many Northern Waterthrushes breeding in Alaska and western Canada migrate eastward in post-breeding toward the Atlantic coast before heading south (Dunn et al. 2006). Individuals tagged in Puerto Rico migrated westward to Florida or the Bahamas in pre-breeding and then north along the Atlantic coast.

Vermivora chrysoptera GOLDEN-WINGED WARBLER

DISTRIBUTION AND ABUNDANCE

It breeds in the Great Lakes region of north-central US and southern Canada; remnant populations persist at high elevations in the Appalachian Mountains. Nonbreeding distribution is primarily in highland forests of Central America, northern Colombia, and adjacent Venezuela. Distribution has been highly dynamic over the past century, with severe declines and extirpation over most of the eastern and Appalachian range and expansion toward the northwest in northern Minnesota and Manitoba; the nonbreeding range has shifted accordingly, with birds becoming rare in South America and increasingly common in northern Central America (Rosenberg et al. 2016). Strong migratory connectivity has further been linked to differential population trajectories among breeding regions (Kramer et al. 2018).

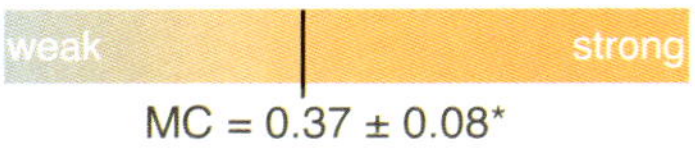

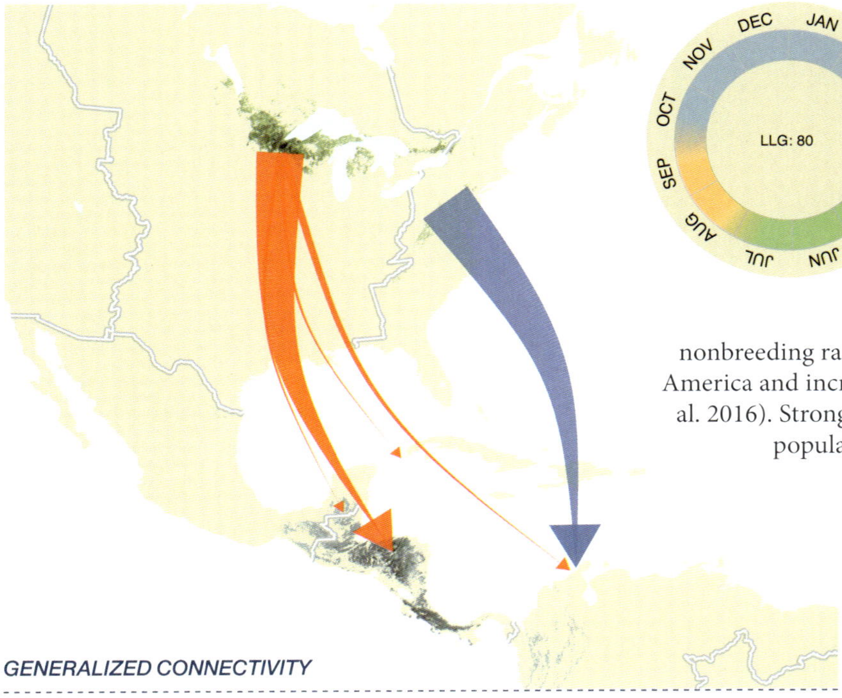

GENERALIZED CONNECTIVITY

	Caribbean	Central America	Mexico	Northern SA
Atlantic (n = 7)	–	–	–	100%
Midcontinent (n = 73)	3%	84%	4%	10%

MIGRATORY CONNECTIVITY

OVERALL Migratory connectivity is moderate (MC = 0.37 ± 0.08) based on light-level geolocator and stable isotope studies, with breeding populations in the Atlantic Flyway spending the nonbreeding season in northern South America (100%), and Midcontinent Flyway breeding populations primarily in Central America (84%). Within our broad geographic regions, the two breeding populations overlap in northern South America; however, at a finer scale, migratory connectivity may be strong.
RADIO TRACKING Birds were tracked through the Motus Wildlife Tracking System, with no individual breeding to nonbreeding detections.
GEOLOCATOR TRACKING Strong migratory connectivity confirmed through light-level geolocators deployed on males throughout the breeding range and the Central American nonbreeding range. Individuals spending the

ORDER Passeriformes
FAMILY Parulidae
BREEDING HABITAT Forests, Temperate Eastern
NONBREEDING HABITAT Forests, Tropical Montane Evergreen
GLOBAL POPULATION SIZE 390,000
POPULATION CHANGE SINCE 1970 −66%
IUCN RED LIST CATEGORY Near Threatened
R2R TIPPING POINT SPECIES Yellow-Alert
US LISTING STATUS No
CANADA SPECIES AT RISK Threatened
PIF CCS 15/20

INTERESTING NOTE

Based on geolocator results, hybrid birds (Golden-winged Warblers crossed with Blue-winged Warblers) from populations in the Great Lakes region (n = 4) spent nonbreeding seasons in northern Central America or Cuba, whereas hybrids from the Appalachian Mountains (n = 2) occurred in northern South America (Bennet et al. 2017; Kramer et al. 2018).

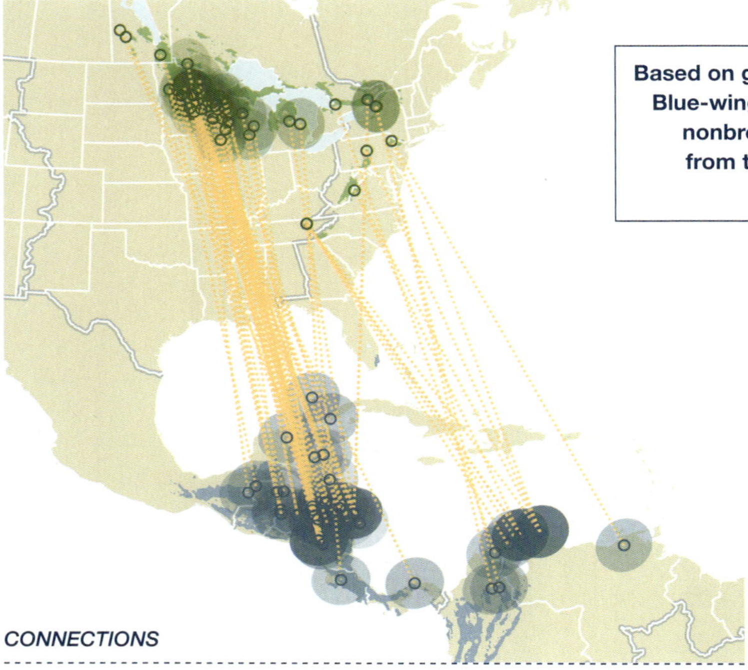

CONNECTIONS

nonbreeding season in Honduras occupied the same breeding region in the upper Midwest as birds spending the nonbreeding seasons in Guatemala, Nicaragua, and Costa Rica (n = 51 unique males; Larkin et al. 2017; Kramer et al. 2018; Bennett et al. 2019). Birds from the St. Lawrence Valley (n = 2) and from historically declining breeding populations in the Appalachian Mountains (n = 14) spent the nonbreeding season entirely in northern South America, primarily in northern Colombia and Venezuela (Bennett et al. 2017; Kramer et al. 2017, 2018).

STABLE ISOTOPES Stable hydrogen isotopes from feathers collected on nonbreeding grounds (Honduras, n = 68; Nicaragua, n = 19; Costa Rica, n = 65; Colombia, n = 16; Venezuela, n = 3) indicated individuals from Honduras originated in southern parts of the breeding distribution, while individuals from Costa Rica and Nicaragua were from northern areas, consistent with a divide between Appalachian and Great Lakes populations (Hobson et al. 2016).
OTHER APPROACHES No banding, telemetry, or genetic studies of migratory connectivity were found for this species.
MIGRATORY PATHWAYS Golden-winged Warblers exhibit a clockwise loop migration indicated from eBird data (La Sorte et al. 2016) and are trans-Gulf migrants in both pre- and post-breeding migration. Stopover ecology during pre-breeding migration was studied for the linked Central American–Great Lakes population (n = 42 males); 83% of those males conducted a prolonged stopover (approximately 7 days) in southern Mexico and northern Guatemala before crossing the Gulf of Mexico, with additional shorter stopovers in Texas, Louisiana, or Arkansas (Larkin et al. 2017; Bennett et al. 2019). Stopovers have not been described for post-breeding migration, but all populations appear to cross the Gulf of Mexico along more easterly routes in post-breeding than in pre-breeding, or pass through Florida and the western Caribbean to Central America (Kramer et al. 2017, 2023; Larkin et al. 2017; Bennett et al. 2019).

DATA CONTRIBUTORS

D.E. Andersen; R.E. Bennett; D.A. Buehler; R. Chandler; L. Chavarria-Duriaux; G. Duriaux; J.A. Gerwin; D.I. King; G.R. Kramer; J.L. Larkin; D. Raybuck; A.D. Rodewald; K.V. Rosenberg; A.M. Roth; M. Siles; C. Smalling; H.M. Streby; P.B. Wood.

BLUE-WINGED WARBLER *Vermivora cyanoptera*

DISTRIBUTION AND ABUNDANCE

It breeds patchily across the central and northeastern US. Little is known about this species on its nonbreeding grounds, which are restricted to the lower Gulf and Caribbean slope of southeastern Mexico, Yucatán Peninsula, and northern Central America; small numbers also spend the nonbreeding season in western Cuba and rarely to Panama and possibly northern Colombia. Despite this species expanding its range historically northward and eastward as humans altered northeastern forests, the relatively small breeding population has declined in recent decades, especially in the mid-Atlantic states and southern New England (F.B. Gill et al. 2020; Fink et al. 2023).

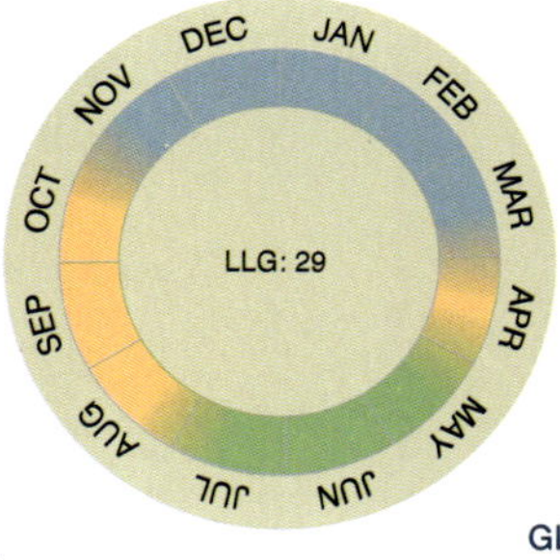

ORDER Passeriformes
FAMILY Parulidae
BREEDING HABITAT Forests, Temperate Eastern
NONBREEDING HABITAT Forests, Tropical Lowland Evergreen
GLOBAL POPULATION SIZE 680,000
POPULATION CHANGE SINCE 1970 −38%
IUCN RED LIST CATEGORY Least Concern
R2R TIPPING POINT SPECIES No
US LISTING STATUS No
CANADA SPECIES AT RISK No
PIF CCS 14/20

MIGRATORY CONNECTIVITY

OVERALL Migratory connectivity information is known from light-level geolocator data and appears weak, although possibly moderate at a finer scale. Over half of the birds breeding within the Atlantic Flyway (62%) overlap with the majority of the Midcontinent Flyway breeding birds (100%) in Mexico during the nonbreeding season. The other breeding populations within the Atlantic Flyway spend the nonbreeding season in Central America (23%), northern South America (8%), and the Caribbean (8%). Limited sample size prevents the calculation of MC.
GEOLOCATOR TRACKING Two studies using light-level geolocators indicated weak migratory connectivity. Eastern breeding populations

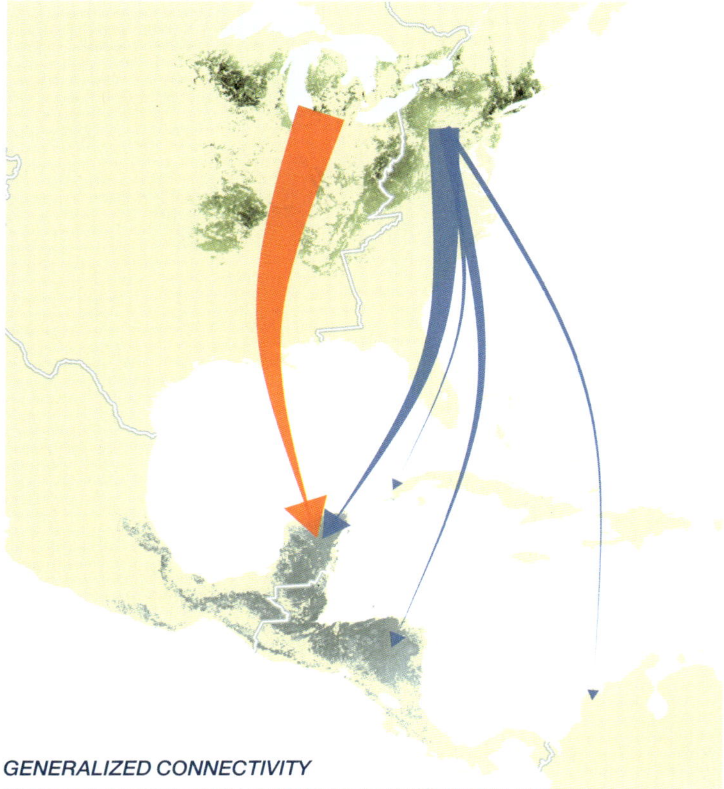

GENERALIZED CONNECTIVITY

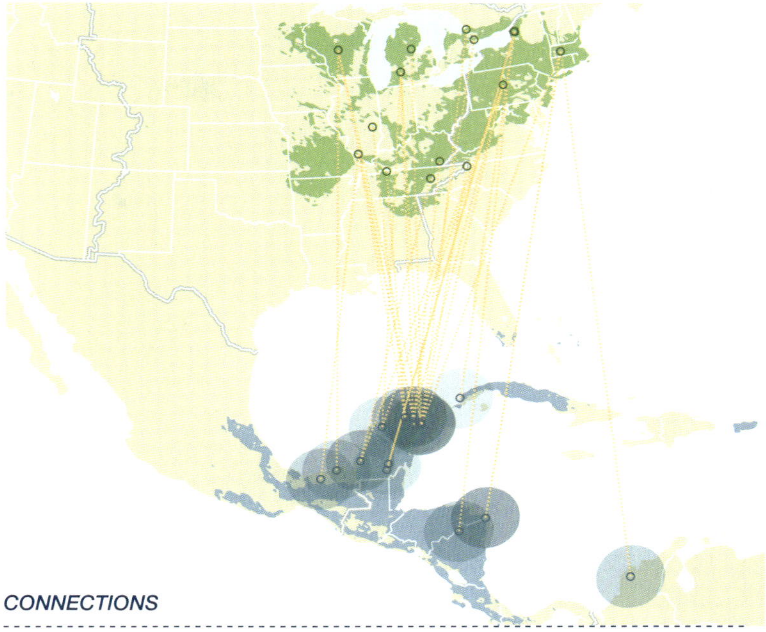

CONNECTIONS

	Caribbean	Central America	Mexico	Northern SA
Atlantic (n = 13)	8%	23%	62%	8%
Midcontinent (n = 16)	–	–	100%	–

INTERESTING NOTE

See Golden-winged Warbler for information on hybrid birds (Blue-winged Warblers crossed with Golden-winged Warblers).

tended to spend the nonbreeding season farther east in Central America and Cuba, and western populations spend the nonbreeding season almost exclusively in southeastern Mexico; however, considerable mixing of breeding populations occurred on the Yucatán Peninsula (Bennett et al. 2017; Kramer et al. 2018). It was estimated that one Blue-winged Warbler breeding in southern New England spent the nonbreeding period in northern Colombia.

OTHER APPROACHES No banding, radio or satellite tracking, stable isotope, or genetic studies of migratory connectivity were found for this species.

MIGRATORY PATHWAYS The species is a trans-Gulf migrant in pre- and post-breeding (F.B. Gill et al. 2020). It exhibits a clockwise loop migration, based on eBird data (La Sorte et al. 2016), with pre-breeding migrants concentrated on the Texas coast and post-breeding migration primarily through the eastern Gulf states and Florida (Kramer et al. 2023). Post-breeding concentrations on the mid-Atlantic coast (Massachusetts to New Jersey), as well as regular occurrence on Bermuda (Amos 1991) and the Bahamas, suggest crossings of the western Atlantic Ocean.

DATA CONTRIBUTORS
D.E. Andersen; R.E. Bennett; D.A. Buehler; G.R. Kramer; H.M. Streby; P.B. Wood.

Mniotilta varia BLACK-AND-WHITE WARBLER

DISTRIBUTION AND ABUNDANCE

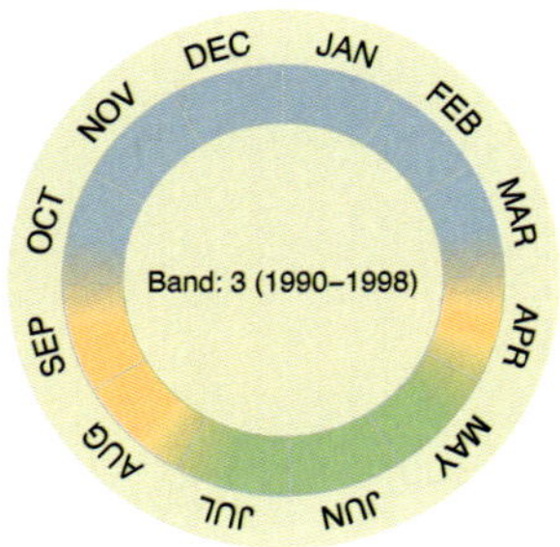

It is a widespread breeder in deciduous and mixed forests of eastern North America and westward, in decreasing abundance across the boreal forest of southern Canada. Nonbreeders occur widely in lowland and montane forests in the Caribbean, Mexico, and central and northwestern South America, with small numbers also in the southeastern US and Florida. There is regional variation in population trends, with declines steepest in New England and the mid-Atlantic coast and Appalachian regions. There are smaller declines across eastern Canada and increases in the western parts of the range (Sauer et al. 2019).

ORDER Passeriformes
FAMILY Parulidae
BREEDING HABITAT Forests, Temperate Eastern
NONBREEDING HABITAT Forests, Tropical Generalist
GLOBAL POPULATION SIZE 18,000,000
POPULATION CHANGE SINCE 1970 −30%
IUCN RED LIST CATEGORY Least Concern
R2R TIPPING POINT SPECIES No
US LISTING STATUS No
CANADA SPECIES AT RISK No
PIF CCS 9/20

MIGRATORY CONNECTIVITY

OVERALL Moderate migratory connectivity is suggested from small numbers of band reencounter and radio telemetry data and limited stable isotope work. More study is needed, especially in the western part of the range. MC is not calculated due to limited sample size.

BANDING Three breeding to nonbreeding band reencounters connect the northeastern US and eastern Canada to the Caribbean.

RADIO TRACKING Similar to band reencounter data, tracking studies using the Motus Wildlife Tracking System connect Jamaica to the northeastern US and Nova Scotia, Canada (Motus project, Ohio State University and Jamaica Long-Term Research, Font Hill Nature Preserve, unpublished data).

STABLE ISOTOPES Stable isotope analysis and North American Breeding Bird Survey abundance data indicated that nonbreeding birds in three areas likely originated from different parts of the breeding range, suggesting that moderate migratory connectivity exists in this species. Specifically, nonbreeding individuals sampled in Yucatán, Belize, and Nicaragua most likely bred in the northern part of the breeding range, from Manitoba and Minnesota to Newfoundland; nonbreeding individuals sampled in Jamaica, Cuba, Hispaniola, Bahamas, and Florida had high assignment probabilities in the Great Lakes region east to New England and the Maritime provinces of Canada; and nonbreeding individuals sampled in Puerto Rico and St. Martin likely bred in the southern portions of the breeding range from the mid-Atlantic states to Missouri and Arkansas (Studds et al. 2021). Note that large portions of the nonbreeding distribution were not sampled in this study.

MIGRATORY PATHWAYS Pathways exhibit a clockwise loop migration, based on eBird data, with average passage farther east in post-breeding than in pre-breeding migration (La Sorte et al. 2016). It is a trans-Gulf migrant in pre- and post-breeding, but post-breeding migration is mainly along the Atlantic coast. Common occurrence on Bermuda (up to 20 or more per day in post-breeding; Amos 1991) suggests regular transoceanic crossings to Caribbean nonbreeding grounds.

CONNECTIONS

INTERESTING NOTE

Additional studies using stable isotopes indicated that individuals that arrived earlier at a coastal Louisiana stopover site tended to be more southern breeders than those that arrived later (Langin et al. 2009; Paxton and Moore 2017).

PROTHONOTARY WARBLER *Protonotaria citrea*

DISTRIBUTION AND ABUNDANCE

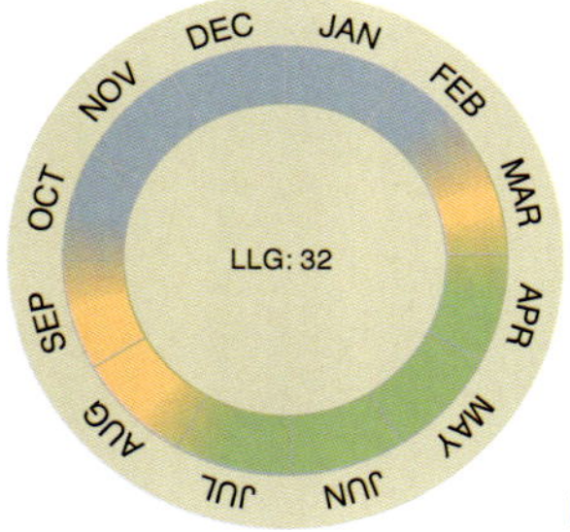

It breeds in the southeastern and central US, with greatest abundance on the southeastern and Gulf Coastal plains and in the Mississippi River drainage. Its nonbreeding distribution is primarily in northwestern Colombia and adjacent Venezuela and along the coasts of Central America north to the Yucatán Peninsula. Smaller numbers occur in the eastern Caribbean (Puerto Rico and Lesser Antilles) and along inland river valleys of northern Colombia. Population declines are steepest in the Mississippi Valley and Gulf Coastal Plain, with some increases in the northern and western parts of the range (Sauer et al. 2019).

ORDER Passeriformes
FAMILY Parulidae
BREEDING HABITAT Wetlands, Forested
NONBREEDING HABITAT Coasts, Mangroves
GLOBAL POPULATION SIZE 2,100,000
POPULATION CHANGE SINCE 1970 −41%
IUCN RED LIST CATEGORY Least Concern
R2R TIPPING POINT SPECIES No
US LISTING STATUS No
CANADA SPECIES AT RISK Endangered
PIF CCS 14/20

MIGRATORY CONNECTIVITY

OVERALL Migratory connectivity is known from two geolocator studies (Wolfe and Johnson 2015, Tonra et al. 2019a), as well as stable isotope (Reese et al. 2019) and genetic (DeSaix et al. 2019) studies. Birds breeding within the Atlantic Flyway spend the nonbreeding season in northern South America (71%) and Central America (29%), while Midcontinent Flyway breeding birds primarily spend the nonbreeding season in northern South America (92%). High overlap of breeding populations in northern South America results in weak migratory connectivity (MC = 0.04).

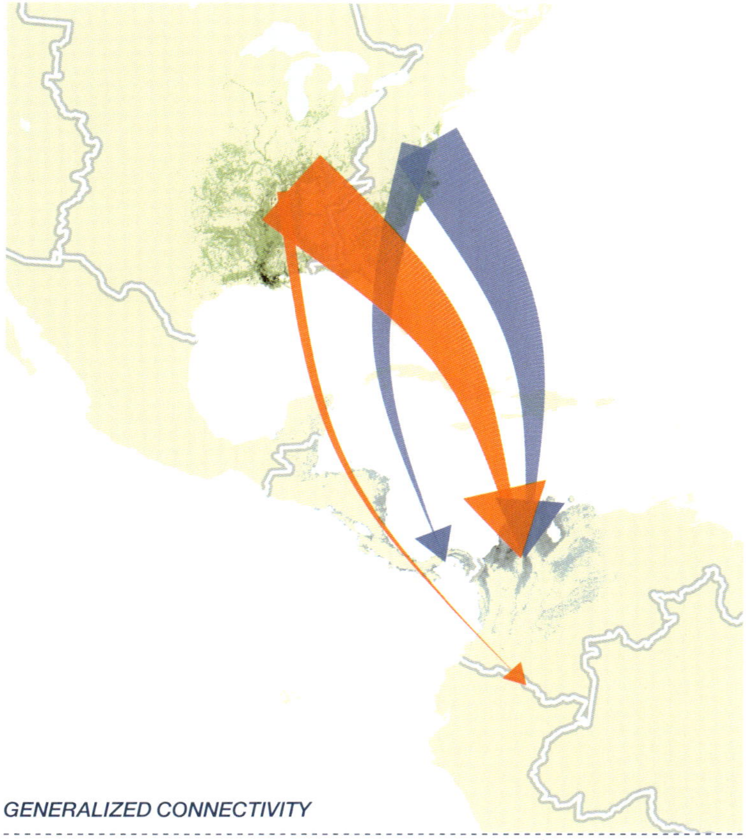

GENERALIZED CONNECTIVITY

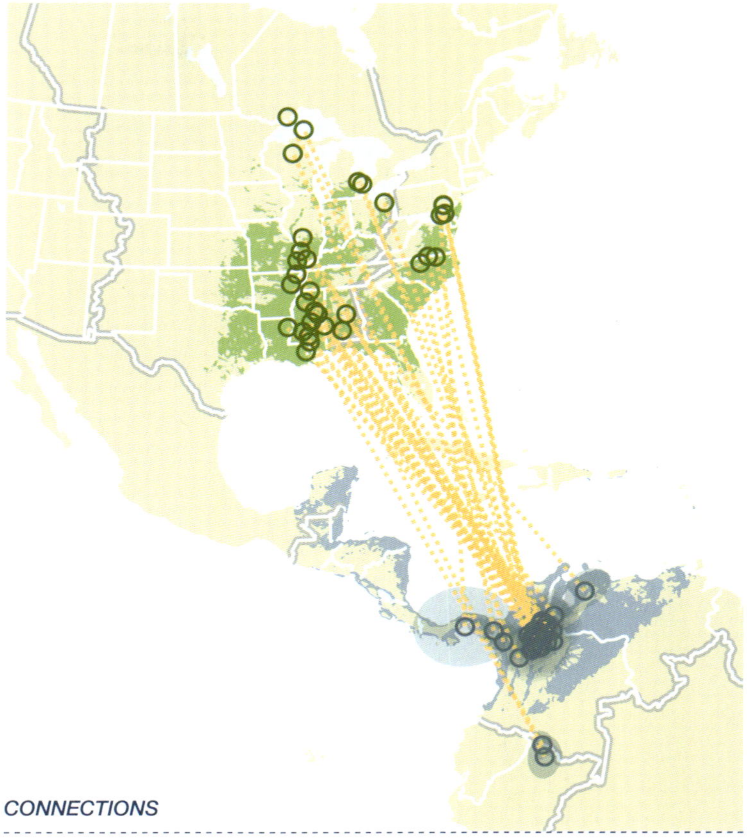

CONNECTIONS

	Andean West	Central America	Northern SA
Atlantic (n = 7)	–	29%	71%
Midcontinent (n = 25)	8%	–	92%

RADIO TRACKING There are several projects tracking Prothonotary Warblers with the Motus Wildlife Tracking System. There are no individual breeding to nonbreeding detections, but there are several detections during migration.

GEOLOCATOR TRACKING A single male was tracked using a light-level geolocator from its breeding territory in Louisiana, first to Central America; it then spent a month in the Greater Antilles, and finally flew south to nonbreeding grounds in Panama or Colombia (Wolfe and Johnson 2015). A second study recovered geolocators (n = 33) from six breeding locations (Arkansas, 7; Louisiana, 13; Ohio, 3; South Carolina, 3; Virginia, 4; Wisconsin, 3) and found that 91% of individuals from across the breeding grounds overlapped in their nonbreeding locations in northern Colombia, in an area 20% the size of their breeding range (Tonra et al. 2019a). However, the nonbreeding region described by geolocators is well to the south of known areas of abundance for this species, emphasizing the need for additional tracking studies to verify geolocator results.

STABLE ISOTOPES A stable isotope study from throughout the breeding and nonbreeding range found likely mixing of breeding populations at nonbreeding sites, suggesting weak connectivity, but the data for assigning breeding origins of nonbreeding samples were extremely weak (Reese et al. 2019).

GENETICS A genomic study using highly informative SNPs found that individuals assigned to both Atlantic coastal and Mississippi Valley populations mixed at nonbreeding sites in Colombia and Panama, consistent with weak migratory connectivity (DeSaix et al. 2019).

OTHER APPROACHES There are no band reencounters between breeding and nonbreeding sites.

MIGRATORY PATHWAYS A trans-Gulf migrant in pre and post-breeding, with peak post-breeding abundance farther east through Florida, the Bahamas, and western Cuba. It commonly occurs on Bermuda (up to 23 observed per day; Amos 1991), and Motus data also suggests regular crossings of the Atlantic Ocean in post-breeding. Geolocator tracks suggest that different breeding populations follow different migration routes in post-breeding, with midwestern breeders migrating through the Mississippi Valley and crossing the Gulf of Mexico to the Yucatán Peninsula, and Atlantic coastal breeders migrating through the Bahamas and/or Cuba before arriving in Central America (Tonra et al. 2019a). The mean migratory stopover duration in post-breeding was 21 days, and individuals from all breeding sites used stopover locations along the Honduras–Nicaragua border and in southern Costa Rica / northern Panama, suggesting these areas could be important during migration (Tonra et al. 2019a).

INTERESTING NOTE

Prothonotary Warblers have been documented to move from drier to wetter areas during the nonbreeding season, such as from the Pacific to the Caribbean coast in Panama and to higher elevations away from the coast in Colombia (Petit 2020; Bulluck et al. 2019).

DATA CONTRIBUTORS
C.M. Tonra.

SWAINSON'S WARBLER *Lymnothlypis swainsonii*

DISTRIBUTION AND ABUNDANCE

It breeds patchily on the South Atlantic and Gulf Coastal Plains from southeastern Virginia to northern Florida and west to eastern Texas; also north in the southern Appalachian Mountains and sparsely in the Mississippi drainage to southeastern Missouri and southern Illinois. Its nonbreeding distribution is poorly understood but is centered on the Greater Antilles and the Caribbean coast and offshore islands of the Yucatán Peninsula and northern Central America. It also occurs in the nonbreeding season in the Petén region of southeastern Mexico, Guatemala, and Belize, and in the Bahamas, Puerto Rico, southern Florida, and possibly elsewhere in the Caribbean Basin (e.g., Panama, Aruba, and northern Venezuela), where it is occasionally detected (Fink et al. 2023).

ORDER Passeriformes
FAMILY Parulidae
BREEDING HABITAT Forests, Temperate Eastern
NONBREEDING HABITAT Forests, Tropical Lowland Evergreen
GLOBAL POPULATION SIZE 160,000
POPULATION CHANGE SINCE 1970 +154%
IUCN RED LIST CATEGORY Least Concern
R2R TIPPING POINT SPECIES No
US LISTING STATUS No
CANADA SPECIES AT RISK No
PIF CCS 13/20

MIGRATORY CONNECTIVITY

OVERALL Evidence of parallel migration and strong east-west connectivity comes from a single light-level geolocator study from two populations (Brunner et al. 2022).
RADIO TRACKING There are no individual breeding to nonbreeding locations, but there are detections from Jamaica to the southeastern US coast during migration.
GEOLOCATOR TRACKING Geolocators recovered from nonbreeding birds in Jamaica (n = 3) indicate breeding locations in the southeastern US. Five birds with recovered geolocators from southern Louisiana spent the nonbreeding season in the northern Yucatán Peninsula to southern Guatemala and western Cuba (Brunner et al. 2022).
GENETICS A study of nuclear DNA showed subtle genetic structure among breeding populations but substantial mixing of genetic populations on nonbreeding sites in Mexico and Jamaica (Winker and Graves 2008).
OTHER APPROACHES No band reencounter data or stable isotope studies of migratory connectivity were found for this species. There are a few tracking studies with the Motus Wildlife Tracking System.
MIGRATORY PATHWAYS No precise data were available on migratory pathways from geolocator data, but they did reveal that birds migrated over the Gulf of Mexico and Caribbean Sea during both pre- and post-breeding migration. Pre-breeding arrival is primarily along the eastern Gulf Coast, and birds are more numerous along the Florida peninsula in post-breeding migration. In addition, both geolocators and Motus detections revealed differences in migration phenology, with eastern breeding populations migrating up to three weeks earlier in post-breeding and up to two weeks later in pre-breeding than western breeders (Brunner et al. 2022).

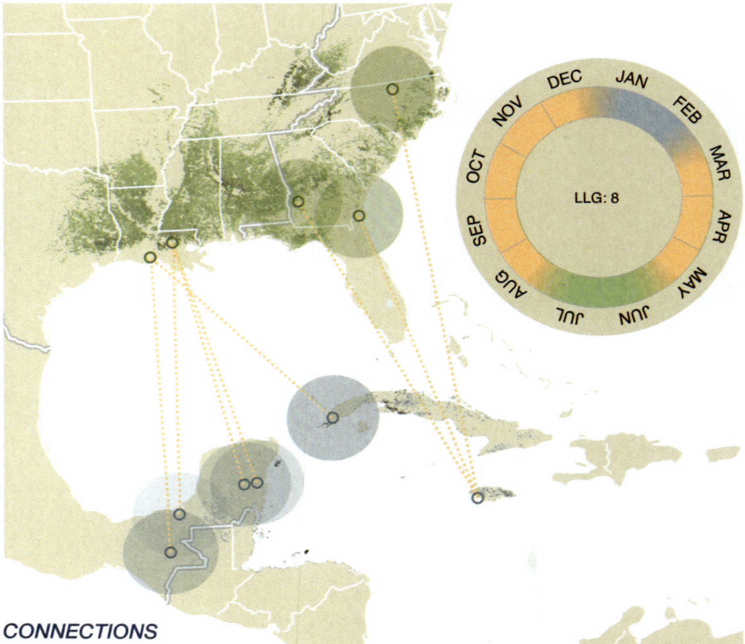

CONNECTIONS

DATA CONTRIBUTORS
A.R. Brunner; B.C. Dossman; V. Jirinec; E.I. Johnson; P.P. Marra; K. Percy.

TENNESSEE WARBLER *Leiothlypis peregrina*

DISTRIBUTION AND ABUNDANCE

This is a widespread breeder across the boreal region from Newfoundland to southeastern Alaska, with small numbers farther south to the northern Rocky Mountains and northeastern US. Its nonbreeding range is primarily in Central America with greatest abundance in the highlands and Pacific Slope from Chiapas, Mexico, to western Panama. Smaller numbers occur north to Veracruz and the Yucatán Peninsula, in western Cuba (rarely elsewhere in the Caribbean), and south to the Andes and other mountains of Colombia and northwestern Venezuela. Population declines have occurred over most of the eastern and southern parts of the range, and are steepest in the Atlantic northern forest region (Sauer et al. 2019).

ORDER Passeriformes
FAMILY Parulidae
BREEDING HABITAT Forests, Boreal
NONBREEDING HABITAT Forests, Tropical Generalist
GLOBAL POPULATION SIZE 110,000,000
POPULATION CHANGE SINCE 1970 −45%
IUCN RED LIST CATEGORY Least Concern
R2R TIPPING POINT SPECIES No
US LISTING STATUS No
CANADA SPECIES AT RISK No
PIF CCS 10/20

MIGRATORY CONNECTIVITY

OVERALL Migratory connectivity is unknown.

BANDING There is one breeding to nonbreeding band reencounter from Alberta to the Pacific Slope of El Salvador.

RADIO TRACKING There are several projects tracking the Tennessee Warbler with the Motus Wildlife Tracking System. There are many detections in the eastern part of the breeding range, but there are breeding to nonbreeding detections.

OTHER APPROACHES No tracking, stable isotope, or genetic studies of migratory connectivity were found for this species.

MIGRATORY PATHWAYS It exhibits clockwise loop migration (La Sorte et al. 2016). Western breeders move east across Canada in post-breeding and then south primarily through the upper Midwest and Appalachian Mountains and across the eastern Gulf of Mexico to the Yucatán and Central America. Pre-breeding migration is across the western Gulf of Mexico to Texas and Louisiana and northward through the midwestern US before spreading eastward and westward to breeding areas. Large numbers use montane forest in the Sierra Nevada de Santa Marta, Colombia, as a multiday pre-breeding stopover to refuel before crossing the Caribbean and Gulf of Mexico to the southern US (Bayly et al. 2019). Analysis of recaptures at banding sites indicate possible stopover site fidelity during post-breeding migration in the Appalachian Mountains (Vogt et al. 2012).

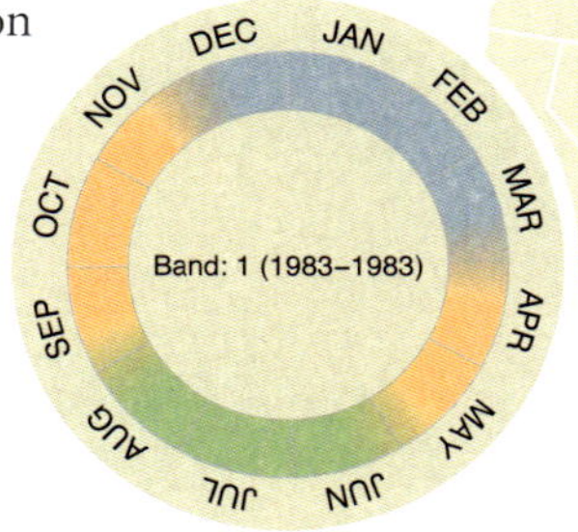

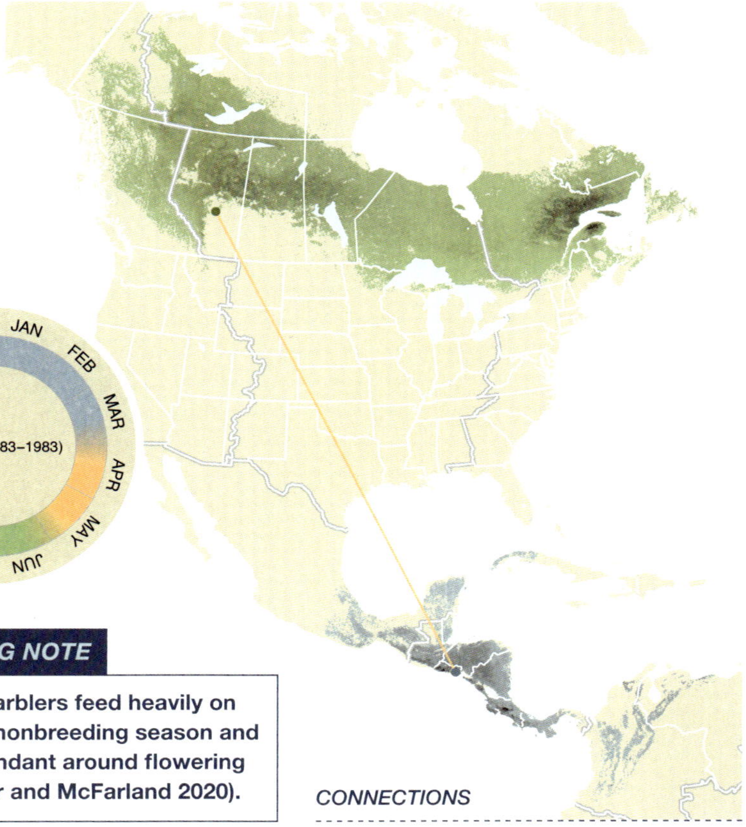

INTERESTING NOTE

Tennessee Warblers feed heavily on nectar in the nonbreeding season and are often abundant around flowering trees (Rimmer and McFarland 2020).

CONNECTIONS

Leiothlypis celata ORANGE-CROWNED WARBLER

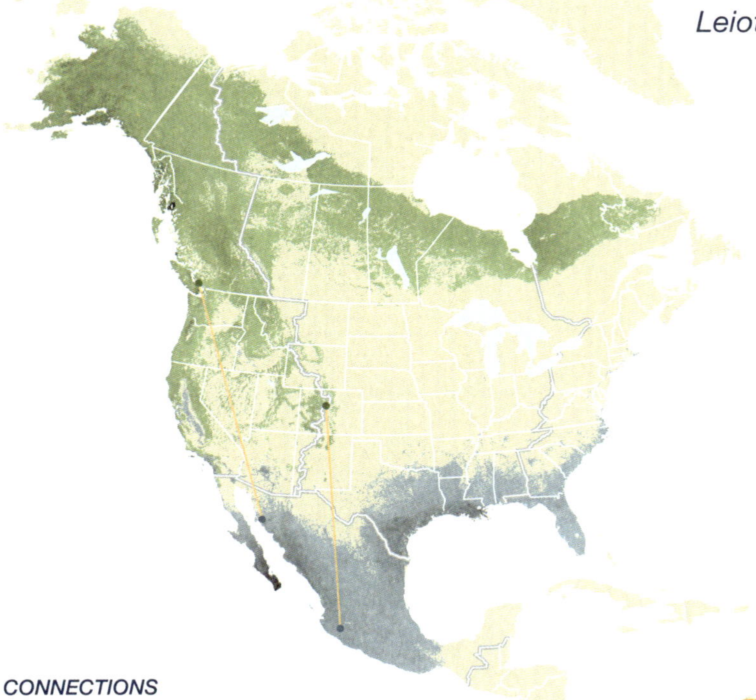

CONNECTIONS

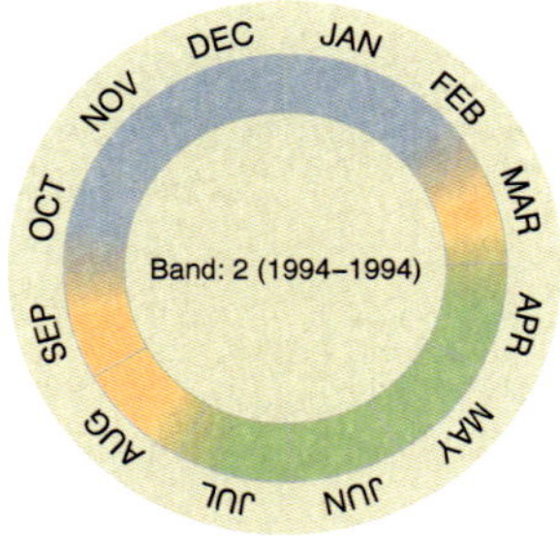

DISTRIBUTION AND ABUNDANCE

This is a widespread breeder across the taiga and boreal forest of Canada and Alaska and south through the Rocky Mountains and Pacific coastal forests to Southern California and New Mexico. The nonbreeding range is across the southern US and throughout Mexico north of the Isthmus of Tehuantepec, rarely to Guatemala. The two disjunct areas of highest nonbreeding abundance—along the Gulf slope from southern Louisiana to northeastern Mexico and on the Pacific Slope of Mexico and southern Baja California—reflect the two major geographic subspecies groups. Steepest declines have occurred on the Pacific coast from Alaska to California, whereas populations across most of Canada and the Rocky Mountains have been stable or slightly increasing (Sauer et al. 2019).

MIGRATORY CONNECTIVITY

OVERALL Migratory connectivity is mostly unknown. There is very limited data from band reencounter data and stable isotope analysis. However, examination of geographic subspecies suggests strong migratory connectivity among regional subspecific populations.

BANDING There are only two breeding to nonbreeding banding reencounters: an individual from British Columbia reencountered in Sonora, Mexico, and an individual from Colorado reencountered in Jalisco.

ORDER Passeriformes
FAMILY Parulidae
BREEDING HABITAT Forests, Temperate Western
NONBREEDING HABITAT Forests, Generalist
GLOBAL POPULATION SIZE 82,000,000
POPULATION CHANGE SINCE 1970 −35%
IUCN RED LIST CATEGORY Least Concern
R2R TIPPING POINT SPECIES No
US LISTING STATUS No
CANADA SPECIES AT RISK No
PIF CCS 8/20

RADIO TRACKING Few tags were deployed with the Motus Wildlife Tracking System, with no breeding to nonbreeding detections and few detections overall.

STABLE ISOTOPES Analysis of feathers found that Orange-crowned Warblers breeding at southern latitudes migrated through New Mexico earlier than those from northern latitudes, suggesting some separation of breeding populations during migration (Kelly 2006).

OTHER APPROACHES Examination of geographic subspecies from museum specimens suggests strong migratory connectivity, with the nonbreeding distribution of the nominate, boreal subspecies *L. c. celata* primarily in the southeastern US and eastern Mexico; the Rocky Mountain subspecies *L. c. orestera* in the southwestern US and western Mexico; and the Pacific subspecies *L. c. lutescens* from coastal Southern California and Baja California to southwestern Mexico. The Channel Islands subspecies disperses onto the California coastal mainland in the nonbreeding season (American Ornithologists Union 1957; Dunn and Garrett 1997).

MIGRATORY PATHWAYS Migration from the two primary nonbreeding areas are along separate pathways, with Pacific populations migrating earlier in pre- and post-breeding than eastern/boreal birds. Pacific birds exhibit a clockwise loop migration, with pre-breeding movement primarily through western deserts and along the Pacific coast, whereas post-breeding migration is more inland through mountainous regions. In contrast, the eastern/boreal population migrates primarily eastward from Alaska and western Canada in post-breeding and through the Great Plains states in both pre- and post-breeding. Pacific breeding birds have also been shown to be altitudinal molt-migrants prior to moving south to nonbreeding grounds (Wiegardt et al. 2017).

COLIMA WARBLER *Leiothlypis crissalis*

DISTRIBUTION AND ABUNDANCE

The tiny breeding distribution is in the mountains of western Texas (Chisos Mountains) and south through the northern Sierra Madre Oriental from Coahuila to Nueva Leon, Mexico, where the species occupies high-elevation pine–oak forest. The nonbreeding distribution is only slightly larger, occupying a belt of humid oak and pine forest at 1,500–3,500 m elevation on the Pacific Slope of Mexico, from southern Sinaloa to Colima and east to Morelos; also in coastal mountains of Guerrero and Oaxaca (Beason and Wauer 2020).

DISTRIBUTION AND ABUNDANCE

ORDER Passeriformes
FAMILY Parulidae
BREEDING HABITAT Forests, Mesoamerican Pine-Oak
NONBREEDING HABITAT Forests, Mesoamerican Pine-Oak
GLOBAL POPULATION SIZE 25,000
POPULATION CHANGE SINCE 1970 NA
IUCN RED LIST CATEGORY Least Concern
R2R TIPPING POINT SPECIES No
US LISTING STATUS No
CANADA SPECIES AT RISK No
PIF CCS 16/20

MIGRATORY CONNECTIVITY

OVERALL Migratory connectivity is unknown. No information on migratory connectivity was found for this species.

MIGRATORY PATHWAYS The complete lack of observational or banding records of transient birds suggests that individuals migrate directly between breeding and nonbreeding areas without stopovers.

LUCY'S WARBLER *Leiothlypis luciae*

DISTRIBUTION AND ABUNDANCE

It breeds patchily in desert riparian (especially mesquite) woodlands of the southwestern US and northwestern Mexico, with highest abundance in the Sonoran Desert region of southern Arizona and Sonora, Mexico (Fink et al. 2023). It occurs in the nonbreeding season in dry washes, riparian woodland, and thorn-scrub in a narrow band on the Pacific Slope of Mexico from southern Sonora and southern Baja California to Guerrero (R.R. Johnson et al. 2020b).

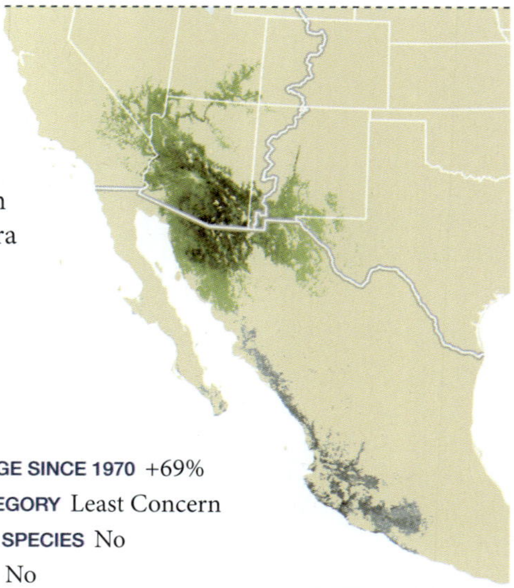

DISTRIBUTION AND ABUNDANCE

ORDER Passeriformes
FAMILY Parulidae
BREEDING HABITAT
Arid Lands, Desert Riparian
NONBREEDING HABITAT
Forests, Tropical Dry
GLOBAL POPULATION SIZE 3,000,000
POPULATION CHANGE SINCE 1970 +69%
IUCN RED LIST CATEGORY Least Concern
R2R TIPPING POINT SPECIES No
US LISTING STATUS No
CANADA SPECIES AT RISK No
PIF CCS 12/20

MIGRATORY CONNECTIVITY

OVERALL Migratory connectivity is unknown. No information on migratory connectivity was found for this species.
MIGRATORY PATHWAYS This species is rarely detected during migration, although rapid, en masse arrival on breeding grounds in Arizona in March coincides with early leaf-out of mesquite trees (Rosenberg et al. 1991). This species is a molt-migrant in early post-breeding, with individuals departing dry breeding areas by July to complete their molt during the monsoons in northwestern Mexico (Rohwer et al. 2007; Pyle et al. 2009).

Leiothlypis ruficapilla NASHVILLE WARBLER

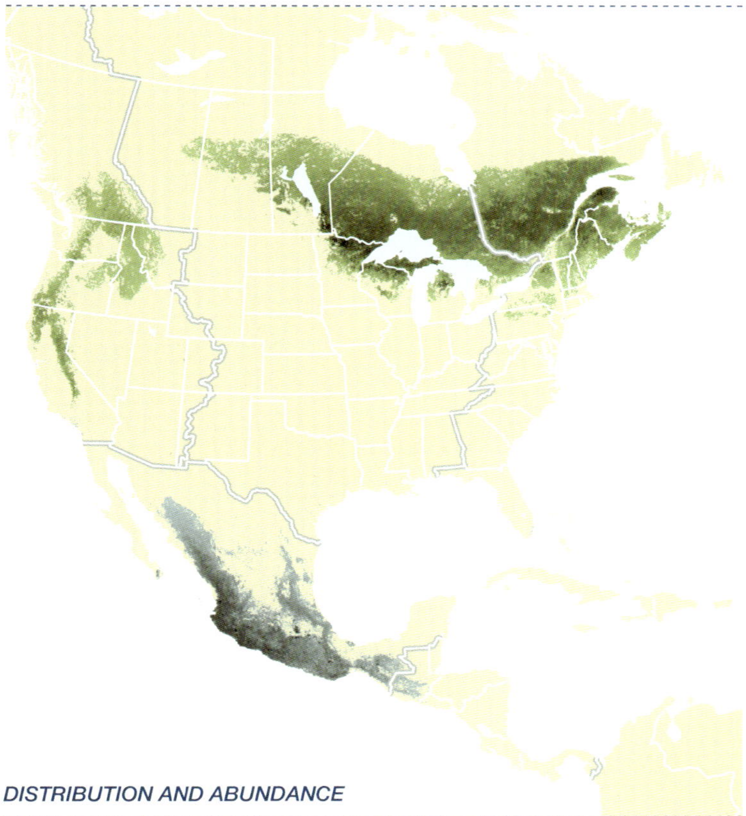

DISTRIBUTION AND ABUNDANCE

ORDER Passeriformes
FAMILY Parulidae
BREEDING HABITAT Forests, Boreal
NONBREEDING HABITAT Forests, Tropical Generalist
GLOBAL POPULATION SIZE 40,000,000
POPULATION CHANGE SINCE 1970 +5%
IUCN RED LIST CATEGORY Least Concern
R2R TIPPING POINT SPECIES No
US LISTING STATUS No
CANADA SPECIES AT RISK No
PIF CCS 11/20

DISTRIBUTION AND ABUNDANCE

There are two highly disjunct breeding distributions corresponding with an eastern and a western subspecies. Eastern birds breed in boreal forests from the Canadian Maritimes (not Newfoundland) and northern New England west only to eastern Alberta. The western population breeds in pine and pine–oak forest in the northern Rocky Mountains from southern British Columbia to Idaho and eastern Oregon, and in coastal ranges and the Sierra Nevada to central California. Its nonbreeding range is almost entirely within Mexico, most abundantly in dry forests of the Pacific Slope (apparently both eastern and western populations), with small numbers reaching Guatemala and El Salvador (Lowther and Williams 2020; Fink et al. 2023). Although long-term trends for both eastern and western populations have been stable, steep recent declines are apparent throughout most of the east and in the Sierra Nevada mountains in the west (Sauer et al. 2019).

MIGRATORY CONNECTIVITY

OVERALL Migratory connectivity is mostly unknown except for subspecies delineations. Although nonbreeding distributions of the two subspecies are largely separate, genetic analysis indicates that eastern and western populations show considerable overlap on the Pacific Slope of Mexico during the nonbreeding season.
RADIO TRACKING A few tags were deployed through the Motus Wildlife Tracking System, with local detections only.
GENETICS Analysis of mitochondrial DNA markers indicated relatively high genetic divergence between eastern and western subspecies, but also found considerable overlap between the two populations during nonbreeding in western Mexico (Lovette et al. 2004).
OTHER APPROACHES Based on diagnosable specimens, the western subspecies spends the nonbreeding season only on the Pacific Slope from southern Sonora to Oaxaca, whereas eastern birds spend the nonbreeding season widely throughout Mexico to northern Central America, including on the Pacific Slope from Jalisco south (American Ornithologists Union 1957; Dunn and Garrett 1997).
MIGRATORY PATHWAYS Differing routes and timing suggest that the eastern and western subspecies use largely separate migration systems, despite broad overlap in the nonbreeding season (Kelly and Hutto 2005; Fink et al. 2023). Eastern birds migrate primarily through the midwestern US and around the Gulf through Texas and northeastern Mexico in both pre- and post-breeding (Dunn and Garrett 1997). Based on observational data, the western subspecies migrates through the low deserts and along the Pacific coast in pre-breeding, whereas post-breeding migrants are more common through the Great Basin and mountains of Arizona and Sonora, suggesting a loop migration (Fink et al. 2023). Some individuals in the western population undergo a molt-migration during the July–August monsoons in Arizona and northwestern Mexico (Pyle et al. 2009).

VIRGINIA'S WARBLER *Leiothlypis virginiae*

DISTRIBUTION AND ABUNDANCE

It breeds at mid-elevations from the southern Rocky Mountains and Great Basin south to northern Mexico. It has recently expanded to the northeast in Wyoming and the Black Hills of South Dakota. Its nonbreeding range is in western and central Mexico, from Sinaloa, Durango, and southern Nuevo Leon to Oaxaca. Long-term declines are apparent in Colorado and New Mexico and, more recently, in Utah and Arizona; increases have been noted in Wyoming and eastern California (Sauer et al. 2019).

ORDER Passeriformes
FAMILY Parulidae
BREEDING HABITAT Forests, Temperate Western
NONBREEDING HABITAT Forests, Tropical Dry
GLOBAL POPULATION SIZE 900,000
POPULATION CHANGE SINCE 1970 −59%
IUCN RED LIST CATEGORY Least Concern
R2R TIPPING POINT SPECIES No
US LISTING STATUS No
CANADA SPECIES AT RISK No
PIF CCS 14/20

DISTRIBUTION AND ABUNDANCE

MIGRATORY CONNECTIVITY

OVERALL Migratory connectivity is unknown. No information on migratory connectivity was found for this species.
GENETICS Extensive gene flow found across the breeding range, including in a recently expanding population in South Dakota (Bubac and Spellman 2016).

MIGRATORY PATHWAYS It migrates primarily through mountains of southern Arizona, New Mexico, and western Mexico in both pre- and post-breeding; it rarely occurs in lowland deserts or along the Pacific coast (Dunn and Garrett 1997). It descends to lower elevations post-breeding and may depend on riparian corridors for stopover during migration (Olson and Martin 2020).

CONNECTICUT WARBLER *Oporornis agilis*

DISTRIBUTION AND ABUNDANCE

It breeds across the central and western boreal forest of Canada from central Quebec to eastern British Columbia, barely entering the northern US in Minnesota and Michigan (Fink et al. 2023). This species is known on nonbreeding grounds from only a handful of records including in Bolivia and southwestern Brazil; and the Gran Chaco ecoregion of northern Paraguay was also confirmed using geolocators (Hallworth et al. 2021; Fink et al. 2023). Population declines are steepest in the southeastern portion of the breeding range (Sauer et al. 2019).

ORDER Passeriformes
FAMILY Parulidae
BREEDING HABITAT Forests, Boreal
NONBREEDING HABITAT Forests, Tropical Lowland Evergreen
GLOBAL POPULATION SIZE 1,800,000
POPULATION CHANGE SINCE 1970 −59%
IUCN RED LIST CATEGORY Least Concern
R2R TIPPING POINT SPECIES No
US LISTING STATUS No
CANADA SPECIES AT RISK No
PIF CCS 13/20

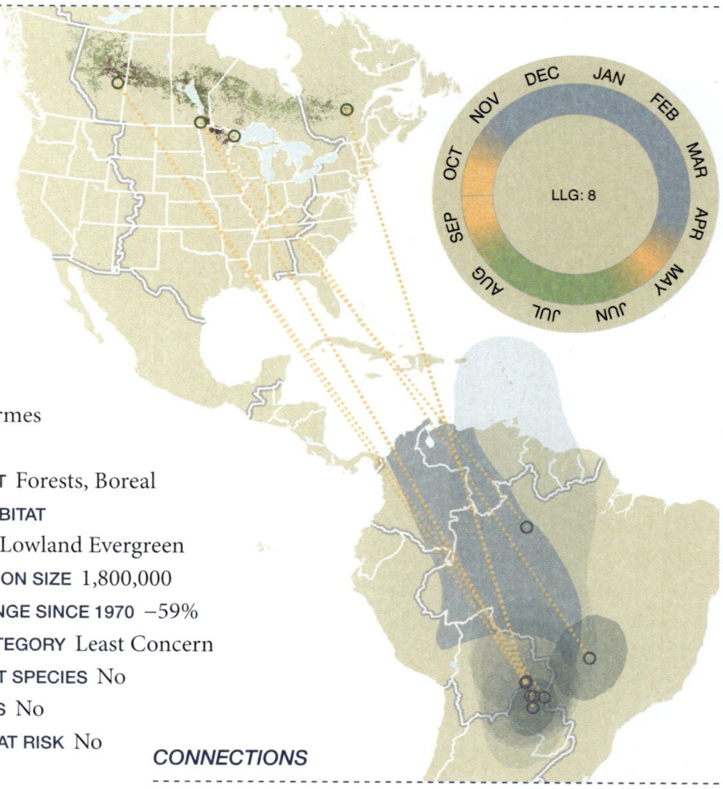

CONNECTIONS

MIGRATORY CONNECTIVITY

OVERALL Migratory connectivity is believed to be weak based on limited evidence from a single recent geolocator study (Hallworth et al. 2021), and because the nonbreeding distribution remains largely undefined. Due to small sample size, MC is not estimated (but see Hallworth et al. 2021).

RADIO TRACKING Very few tags were deployed with the Motus Wildlife Tracking System, and there are no individual breeding to nonbreeding detections.
GEOLOCATOR TRACKING Recovery of four light-level geolocators from a single site in Manitoba documented post-breeding migration routes and multiday transatlantic crossings but did not reveal

migratory connectivity (McKinnon et al. 2017). A larger study (n = 9 geolocator recoveries) from four breeding locations (Alberta, Manitoba, Minnesota, Quebec) found that Connecticut Warblers from the four breeding sites converged on a single nonbreeding region in South America. Migratory connectivity was stronger between breeding and post-breeding migration routes (MC = 0.24 ± 0.23) than between breeding and nonbreeding locations (MC = −0.19 ± 0.14; Hallworth et al. 2021).

OTHER APPROACHES No banding, genetic, or stable isotope studies of migratory connectivity were found for this species.

MIGRATORY PATHWAYS Strong clockwise loop migration in North America is shown by eBird data (La Sorte et al. 2016; Fink et al. 2023) and supported by geolocators (Hallworth et al. 2021). Birds make multiday, transoceanic flights in post-breeding from the mid-Atlantic coast of the US to the Greater Antilles or the northern coast of South America (McKinnon et al. 2017; Hallworth et al. 2021); large numbers (up to 60 per day) are also noted in post-breeding in Bermuda (Amos 1991). Pre-breeding migration is through Florida and the central US, west of the Appalachians. In South America, post-breeding migration appears to be through northern Colombia and western Amazonia, whereas pre-breeding migration records are farther east in Brazil and Venezuela. Individuals tracked with geolocators showed variation in migration route and stopover locations, which might be population specific. Individuals from Manitoba and Quebec departed from the Atlantic coast and made transatlantic crossings to Caribbean islands or to South America, whereas individuals from Minnesota and Alberta appeared to depart from the south-central US and make similar long flights directly to the Caribbean or South America.

DATA CONTRIBUTORS

K.C. Fraser; M.T. Hallworth; O.P. Love; P.P. Marra; E.A. McKinnon; Migratory Connectivity Project.

Geothlypis tolmiei MACGILLIVRAY'S WARBLER

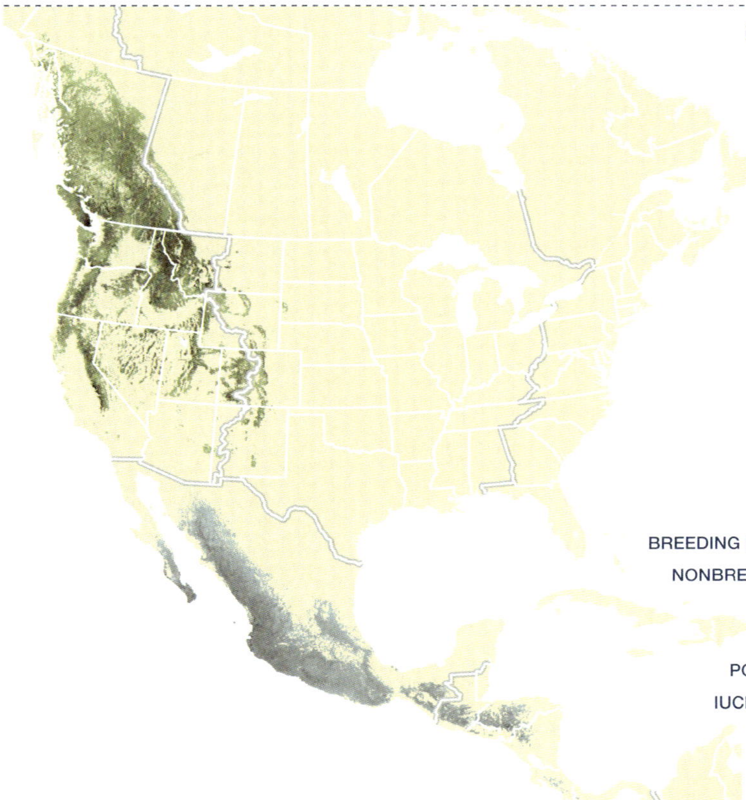

DISTRIBUTION AND ABUNDANCE

DISTRIBUTION AND ABUNDANCE

It breeds in forested areas of western North America from southeastern Alaska and southern Yukon to the Sierra Nevada and southern Rocky Mountains. Its nonbreeding range is on the Pacific Slope and in highlands of Mexico and northern Central America, sparsely to Costa Rica and western Panama (Fink et al. 2023). Populations in the Pacific Northwest and northern Rocky Mountains have declined most steeply, whereas California populations have increased slightly (Sauer et al. 2019).

ORDER Passeriformes
FAMILY Parulidae
BREEDING HABITAT Forests, Temperate Western
NONBREEDING HABITAT Forests, Tropical Dry
GLOBAL POPULATION SIZE 11,000,000
POPULATION CHANGE SINCE 1970 −36%
IUCN RED LIST CATEGORY Least Concern
R2R TIPPING POINT SPECIES No
US LISTING STATUS No
CANADA SPECIES AT RISK No
PIF CCS 12/20

MIGRATORY CONNECTIVITY

OVERALL Migratory connectivity is poorly known, but limited isotope data suggest a leapfrog migration pattern, and analysis of morphological variation indicates largely separate nonbreeding ranges for Pacific coast versus interior breeding populations.

STABLE ISOTOPES Stable hydrogen isotope analysis of tail feathers recovered during pre-breeding migration at two stopover sites along the lower Colorado River indicated that a majority of individuals originated from coastal regions of California north to British Columbia or the Sierra Nevada mountains. Stable isotope values also indicated that individuals that bred the previous year at more southern latitudes migrated earlier in pre-breeding than those that bred at more northern latitudes the previous year, suggesting a leapfrog migration pattern (Paxton and Van Riper 2006).

OTHER APPROACHES Subspecific variation indicates some separation of populations in the nonbreeding season, with the Pacific coastal subspecies being detected from western Mexico to Panama, and the interior subspecies found primarily in the highlands of central and southern Mexico (American Ornithologists Union 1957; Dunn and Garrett 1997).

MIGRATORY PATHWAYS It migrates broadly across most of western North America, including through low deserts and mountains, and especially riparian corridors, but rarely along the immediate California coast. Migration routes are completely over land, and individuals appear to take the shortest direct route possible (Kelly and Hutto 2005; Skagen et al. 2005). The two subspecies differ in timing and routes of pre-breeding migration. Pre-breeding migration through western Arizona and Southern California consists entirely of Pacific and Sierra Nevada breeders and occurs nearly a month earlier than the more easterly migration of interior-breeding populations (Dunn and Garrett 1997; Fink et al. 2023).

MOURNING WARBLER *Geothlypis philadelphia*

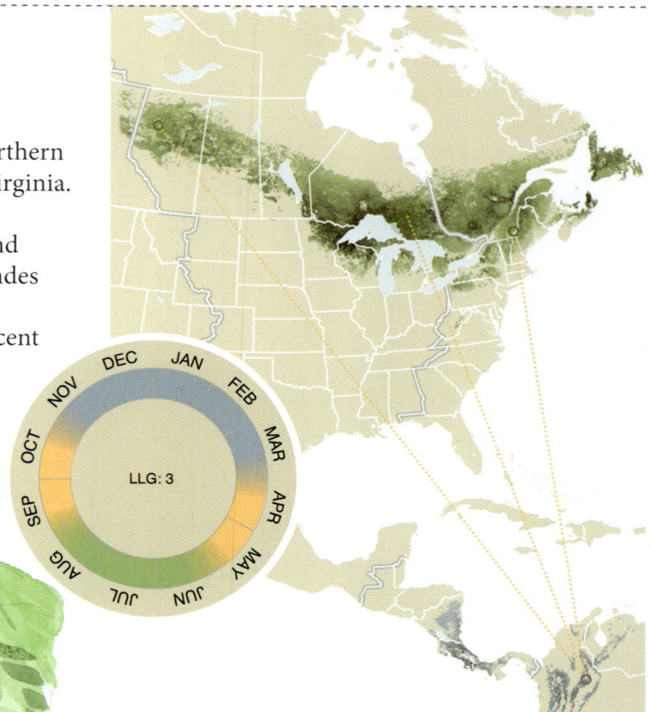

CONNECTIONS

DISTRIBUTION AND ABUNDANCE

It breeds across the boreal forest of Canada from Newfoundland to northeastern British Columbia and in the northeastern US from northern New England to northern Minnesota; it also breeds patchily in the northern Appalachians south to West Virginia. Its highest abundance is in the western Great Lakes region (Fink et al. 2023). Its nonbreeding distribution is primarily in disturbed habitats in humid lowlands and foothills of Central America, from Nicaragua to Panama, and in the northern Andes (mostly the western slopes) and other highlands of Colombia and northwestern Venezuela, rarely to Ecuador (Pitocchelli 2020b). Declines have accelerated in recent decades, especially in Maine and the Maritime provinces (Sauer et al. 2019).

ORDER Passeriformes
FAMILY Parulidae
BREEDING HABITAT Forests, Temperate Western
NONBREEDING HABITAT Forests, Tropical Dry
GLOBAL POPULATION SIZE 11,000,000
POPULATION CHANGE SINCE 1970 −36%
IUCN RED LIST CATEGORY Least Concern
R2R TIPPING POINT SPECIES No
US LISTING STATUS No
CANADA SPECIES AT RISK No
PIF CCS 12/20

MIGRATORY CONNECTIVITY

OVERALL Migratory connectivity is largely unknown. Very limited information on migratory connectivity was found for this species.

STABLE ISOTOPES A stable isotope study found that most Mourning Warblers sampled in northwestern Venezuela likely originated in the eastern portion of the breeding range (Hobson et al. 2014).

MIGRATORY PATHWAYS Migrations are west of the Appalachians and along the western Gulf of Mexico and through Central America in both pre- and post-breeding (Fink et al. 2023). The few band returns indicate that breeders from Minnesota to New York pass through Texas in pre-breeding migration.

RADIO TRACKING There are no Motus detections between breeding and nonbreeding areas, but there are several detections throughout migration.

GEOLOCATOR TRACKING Data from barometric pressure geolocators (n = 3) deployed on Mourning Warblers in coffee landscapes in Colombia showed birds migrated to breeding areas across the range suggesting weak connectivity (R. Bennett, N. Bayly, C. Gomez, N. Cooper, S. Szarmach, unpublished data).

DATA CONTRIBUTORS
N. Bayly, R. Bennett, N. Cooper, C. Gómez, S. Szarmach.

KENTUCKY WARBLER *Geothlypis formosa*

DISTRIBUTION AND ABUNDANCE

It breeds in deciduous forests across the southeastern US, except peninsular Florida, with greatest abundance west of the Appalachian Mountains in the central hardwood forests and Gulf Coastal Plain regions (Fink et al. 2023). Its nonbreeding range is primarily in the Gulf and Caribbean lowlands, and the foothills of southeastern Mexico, the Yucatán Peninsula, and northern Central America, with smaller numbers in the Pacific lowlands and south through Panama (Fink et al. 2023). Overall, it shows a moderate long-term decline, with steepest declines in the Appalachian Mountains, mid-Atlantic coast, and western Gulf Coastal Plain, and stable populations elsewhere (Sauer et al. 2019).

ORDER Passeriformes
FAMILY Parulidae
BREEDING HABITAT Forests, Temperate Eastern
NONBREEDING HABITAT Forests, Tropical Lowland Evergreen
GLOBAL POPULATION SIZE 2,600,000
POPULATION CHANGE SINCE 1970 −33%
IUCN RED LIST CATEGORY Least Concern
R2R TIPPING POINT SPECIES No
US LISTING STATUS No
CANADA SPECIES AT RISK No
PIF CCS 14/20

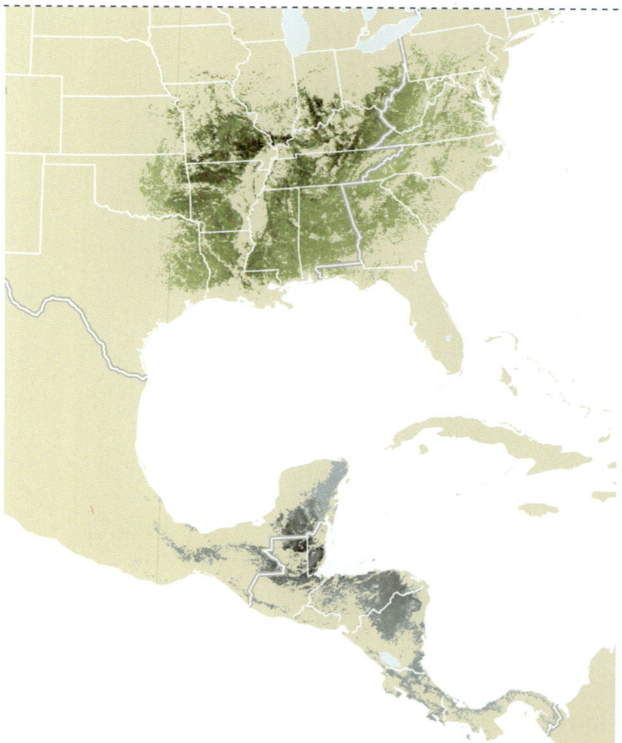
DISTRIBUTION AND ABUNDANCE

MIGRATORY CONNECTIVITY

OVERALL Migratory connectivity is unknown. No information on migratory connectivity was found for this species. A few tags were deployed through the Motus Wildlife Tracking System, with local detections only.

MIGRATORY PATHWAYS A trans-Gulf migrant passed through Florida and western Cuba in post-breeding and appeared along the entire Gulf Coast in pre-breeding (McDonald 2020); thus, there is some evidence that post-breeding migration is more easterly than pre-breeding migration (Dunn and Garrett 1997).

Geothlypis trichas

COMMON YELLOWTHROAT

DISTRIBUTION AND ABUNDANCE

A widespread breeder throughout North America, it is more continuously distributed across the eastern US from the Gulf states north to the Atlantic provinces and west across Canada, and more patchily distributed in the west from eastern Alaska south to central Mexico. Its nonbreeding range includes the southern US and low densities north to Massachusetts, Mexico, northern Central America, and the Caribbean, but distribution is complex due to nonmigratory populations in coastal and southern US and Mexico. Moderate declines are consistent across the entire boreal and eastern part of the range and are steepest in Florida and the southeastern and Gulf Coastal plains. In contrast, populations in the west, from Alaska to Southern California and Arizona, have increased. The exception is a resident population around San Francisco Bay, which is highly threatened (Sauer et al. 2019).

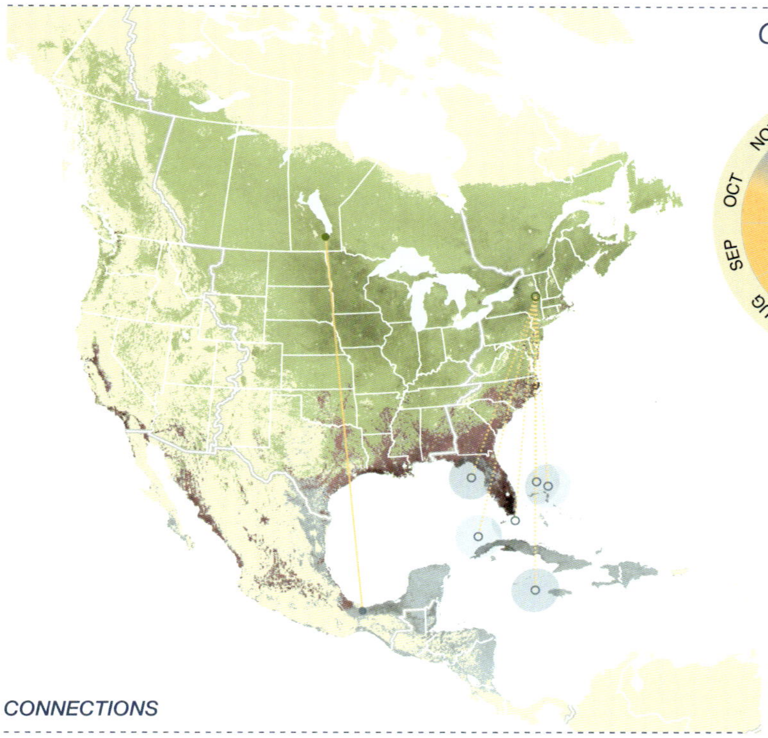

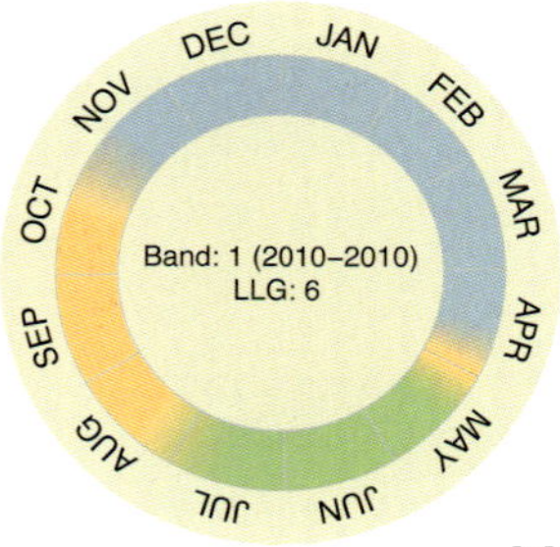

CONNECTIONS

ORDER Passeriformes
FAMILY Parulidae
BREEDING HABITAT Open Country, Habitat Mosaic
NONBREEDING HABITAT Open Country, Habitat Mosaic
GLOBAL POPULATION SIZE 77,000,000
POPULATION CHANGE SINCE 1970 −34%
IUCN RED LIST CATEGORY Least Concern
R2R TIPPING POINT SPECIES No
US LISTING STATUS No
CANADA SPECIES AT RISK No
PIF CCS 9/20

MIGRATORY CONNECTIVITY

INTERESTING NOTE

A study of malarial parasite lineages concluded that this technique was not useful for determining migratory connectivity in the Common Yellowthroat (Pagenkopp et al. 2008).

OVERALL Strong migratory connectivity and east-west parallel migration is documented by range-wide genetic analysis and geographic subspecies variation. However, there is limited information from banding and light-level geolocator data. Further study of the movement of individuals is necessary to understand finer-scale connectivity.

BANDING There is only one breeding to nonbreeding band recovery (Manitoba to southern Mexico).

RADIO TRACKING A few tags were deployed through the Motus Wildlife Tracking System, with no individual breeding to nonbreeding detections.

GEOLOCATOR TRACKING One study found that individuals breeding in New York had spent the nonbreeding season in Florida, Cuba, the Bahamas, or Jamaica (C.C. Taff et al., unpublished data).

GENETICS Early analysis of genetic markers in nonbreeding individuals was only useful for continent-wide, broad-scale identification of eastern versus western populations, suggesting a longitudinal migratory divide (Lovette et al. 2004; Mila et al. 2005). More recently, a complete genoscape based on SNP-based genetic markers, from 660 individuals and 88 breeding and nonbreeding locations, found that migratory connectivity is weak but with a general east versus west trend, and identified five genetic breeding regions connected to distinct but overlapping nonbreeding regions (Ruegg et al. 2020): Atlantic states and provinces breeders connected to southeast US and Caribbean nonbreeding areas; midwestern states and provinces connected primarily to southern Mexico; northwestern states and provinces connected to the southwest US and northern Mexico; southwestern breeders connected to Southern California and southern Mexico; and California breeders connected to Southern California and the Pacific coast of Mexico, including Baja California.

OTHER APPROACHES Based on 13 recognizable subspecies (American Ornithologists Union 1957; Dunn and Garrett 1997), similar, strong migratory connectivity is apparent among four regional breeding and nonbreeding populations: eastern populations migrate to the southeastern US, Caribbean, eastern Mexico, and Central America; South Atlantic coastal breeders migrate south to the Gulf Coast and northeastern Mexico; interior western breeders spend the nonbreeding season in the southwestern US and western Mexico to Guatemala; and Pacific coastal populations spend the nonbreeding season along the southern Pacific coast and northwestern Mexico.

Strong genetic divergence between eastern and western subspecies (Ball and Avise 1992) also is concordant with a migratory divide.
MIGRATORY PATHWAYS Migration behavior is complex, as populations may be long- or short-distance migrants, partial migrants, or sedentary. It is known to be a trans-Gulf migrant (Guzy et al. 2020) and to occur regularly in Bermuda (up to 64 per day in the post-breeding season; Amos 1991), which also suggests transoceanic crossings. Band reencounters during migration suggest a strong pattern of parallel migration. Birds banded in the northeastern US were tied to recoveries along the Atlantic coast, Florida, and the Caribbean; birds banded in the midwestern US were connected to recoveries along the Gulf Coast and eastern Mexico; and birds banded in the western US were connected to the Pacific coast and western Mexico.

DATA CONTRIBUTORS
C.R. Freeman-Gallant; G.R. Kramer; H.M. Streby; C.C. Taff.

HOODED WARBLER *Setophaga citrina*

DISTRIBUTION AND ABUNDANCE

It breeds in upland deciduous forests through most of the eastern US, with greatest abundance in the Appalachian Mountains and Gulf states (Fink et al. 2023). The nonbreeding range is primarily in the Gulf and Caribbean lowlands of southern Mexico, the Yucatán Peninsula, and northern Central America, with much smaller numbers in Pacific lowlands and throughout the Greater Antilles and Bahamas (Fink et al. 2023).

ORDER Passeriformes
FAMILY Parulidae
BREEDING HABITAT
Forests, Temperate Eastern
NONBREEDING HABITAT
Forests, Tropical Lowland Evergreen
GLOBAL POPULATION SIZE 5,200,000
POPULATION CHANGE SINCE 1970 +104%
IUCN RED LIST CATEGORY Least Concern
R2R TIPPING POINT SPECIES No
US LISTING STATUS No
CANADA SPECIES AT RISK No
PIF CCS 10/20

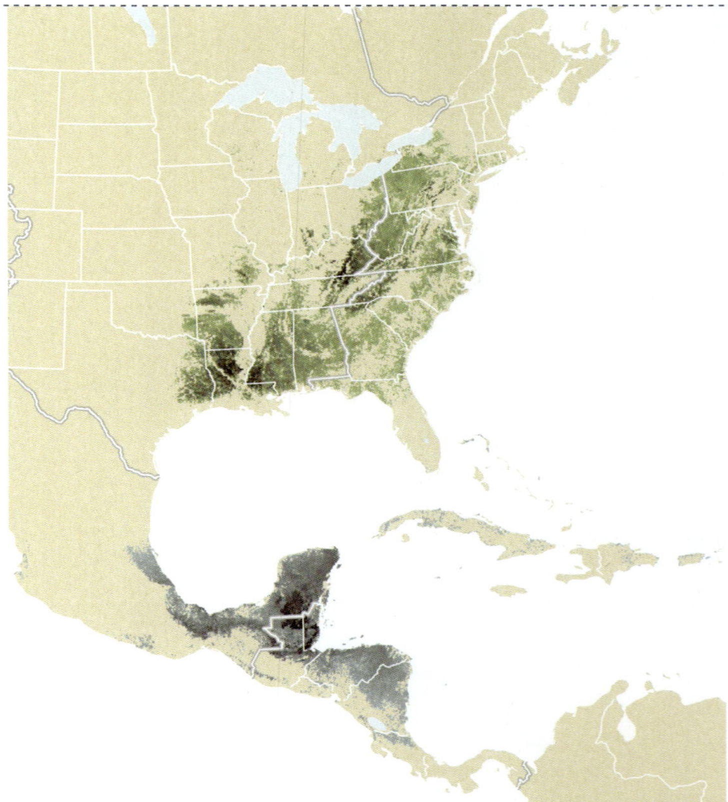

DISTRIBUTION AND ABUNDANCE

MIGRATORY CONNECTIVITY

OVERALL Migratory connectivity is relatively unknown. Besides a single stable isotope analysis, no information on migratory connectivity was found for this species.
STABLE ISOTOPES Stable isotope samples from Hooded Warblers spending the nonbreeding season in Puerto Rico and Guatemala were assigned to breeding regions at similar latitudes throughout the eastern United States and southern Canada, but with no resolution of longitudinal separation (Hobson et al. 2014).
RADIO TRACKING A few tags were deployed through the Motus Wildlife Tracking System; there are no breeding to nonbreeding detections, but there is one connection during migration from Cuba to the northwestern coast of Florida (K. Kardynal, unpublished data).
MIGRATORY PATHWAYS It is a trans-Gulf migrant in pre- and post-breeding, with pre-breeding migration slightly more westerly than post-breeding (Mumme et al. 2023; Dunn and Garrett 1997). This species also migrates along the Gulf Coast of Mexico, primarily in pre-breeding, and is more numerous in Florida and western Cuba in post-breeding. Regular occurrence on Bermuda suggests some transatlantic crossings (Amos 1991).

Setophaga ruticilla **AMERICAN REDSTART**

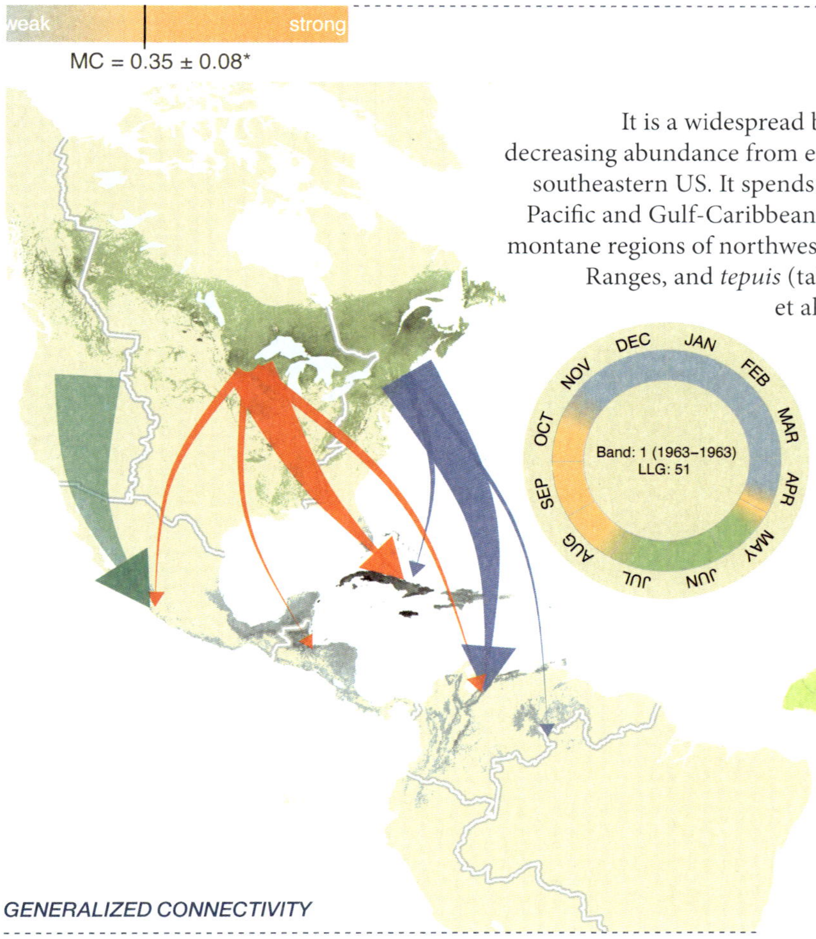

GENERALIZED CONNECTIVITY

DISTRIBUTION AND ABUNDANCE

It is a widespread breeder in the northeastern US and across southern Canada, with decreasing abundance from east to west; this species is a more local breeder in the western and southeastern US. It spends the nonbreeding season widely throughout the Caribbean, in the Pacific and Gulf-Caribbean lowlands of Mexico and northern Central America, and in lower montane regions of northwestern South America, including the northern Andes, Pacific Coast Ranges, and *tepuis* (tabletop mountains) of Venezuela; also in southern Florida (Sherry et al. 2020; Fink et al. 2023). Breeding Bird Survey trends are variable across the range, with steepest declines in New England, the Canadian Maritimes, and northern Rocky Mountain regions; there are stable or increasing populations in much of the remaining range (Sauer et al. 2019).

ORDER Passeriformes
FAMILY Parulidae
BREEDING HABITAT Forests, Temperate Generalist
NONBREEDING HABITAT Forests, Tropical Generalist
GLOBAL POPULATION SIZE 42,000,000

POPULATION CHANGE SINCE 1970 −11%
IUCN RED LIST CATEGORY Least Concern
R2R TIPPING POINT SPECIES No
US LISTING STATUS No
CANADA SPECIES AT RISK No
PIF CCS 9/20

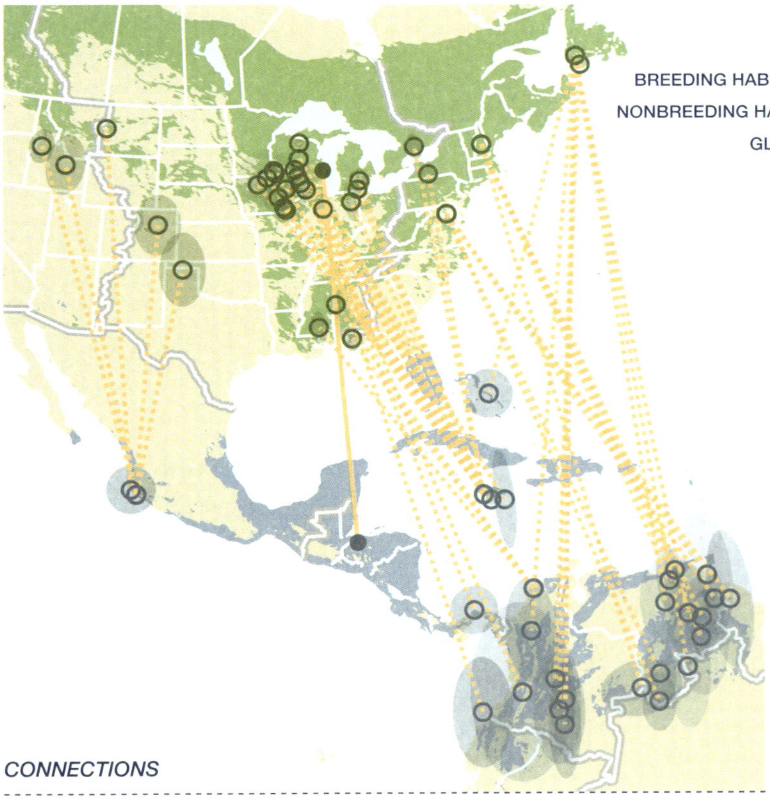

CONNECTIONS

	Brazil	Caribbean	Central America	Mexico	Northern SA
Atlantic (n = 21)	5%	10%	–	–	86%
Midcontinent (n = 29)	–	69%	7%	10%	14%
Pacific (n = 2)	–	–	–	100%	–

MIGRATORY CONNECTIVITY

OVERALL Moderate migratory connectivity (MC = 0.35 ± 0.08) is suggested in both latitude and longitude, based on multiple lines of evidence; however, considerable overlap among populations is apparent. Overall, birds breeding within the Atlantic Flyway primarily spend the nonbreeding season in northern South America (86%), while Midcontinent Flyway breeding birds primarily spend the nonbreeding season in the Caribbean (69%) and Pacific flyways breeding birds in Mexico (100%).

BANDING Data from band reencounters suggest east-west parallel migration, with birds breeding west of the Great Lakes spending the nonbreeding season in Central America, while individuals breeding east of the Great Lakes spend the nonbreeding season in the Caribbean (Norris et al. 2006).

RADIO TRACKING Over 35 redstarts fitted with Motus tags in Jamaica have been detected on migration, and these data support general geolocator results, with most birds migrating toward the upper Midwest. One female redstart wearing a Motus tag deployed in Jamaica was found nesting in northern Wisconsin (Dossman et al. 2023).

GEOLOCATOR TRACKING Multiple redstarts on nonbreeding and breeding grounds have been recaptured with geolocator data. Of 20 light-level geolocators recovered from redstarts on nonbreeding grounds at Font Hill, Jamaica (Dossman et al. 2023; Stevens et al. 2026), 14 birds migrated north across Cuba and through Florida, breeding in the upper Midwest in Iowa,

Wisconsin, Minnesota, and Illinois; three bred in Michigan, one near Toronto, one in northwestern Indiana, and one in western New York. Furthermore, geolocators recovered at a nonbreeding site in Cuyutlan, Mexico (n = 4), revealed that redstarts migrated to either the Rockies along the Idaho–Montana border (n = 2) or the Colorado Rockies (n = 2). One geolocator recovered near Villavicencio, Colombia, found that this redstart migrated along the Lesser Antilles and then the east coast of the US up to western Newfoundland. Geolocators recovered in the Santa Marta mountains of Colombia (n = 3) showed that these redstarts breed in the southeastern US, concentrated in eastern Louisiana and Mississippi, and that they arrive there by crossing the Gulf of Mexico or migrating up through Central America. On the breeding grounds, geolocators recovered from redstarts in Maryland (n = 5) and New Hampshire (n = 5) revealed that one redstart wintered in the southern Lesser Antilles and all others in eastern Venezuela. In Wisconsin (n = 6), three spent the nonbreeding season in Jamaica, two in western Colombia, and one along the Panama–Colombia border. One geolocator recovered in Montana found this redstart spent the nonbreeding season on the Pacific Slope of Mexico in the Michoacán region (Stevens et al. 2026).

STABLE ISOTOPES Bayesian assignments to breeding origin by combining δ^2H stable isotope values and North American Breeding Bird Survey abundance data found a broad pattern of east-west connectivity (Studds et al. 2021). More specifically, nonbreeding redstarts in Mexico likely bred in the northwestern portion of the range in British Columbia and Alberta; nonbreeding birds in Cuba, Bahamas, and Florida bred along the United States–Canada border from Saskatchewan to Newfoundland; those in Jamaica, Hispaniola, and Puerto Rico overlapped considerably with the previous group; and birds in St. Martin and Dominica bred in the eastern part of the range from Maine to North Carolina. Additional data show that birds in northwestern Colombia, Panama, and southern Costa Rica have the isotopic signature of birds in the southeastern US (Stevens et al. 2026).

DATA CONTRIBUTORS

B.C. Dossman; P.P. Marra; H. Stevens.

GENETICS Colbeck et al. (2008) used two different genetic markers to show that continental populations of redstarts were not differentiated across most of their range. They did find that redstarts in Newfoundland were genetically distinct from continental populations. A recent study using mitochondrial DNA found that some Newfoundland redstarts spent the nonbreeding season in Puerto Rico and the Dominican Republic (DeSaix et al. 2022a). A more complete genoscape map (DeSaix et al. 2023) using low-coverage whole genome sequence data identified fine-scale population structure. Principal components analysis with breeding samples revealed five genetic clusters grouped by geography: Maritime provinces (New Brunswick and Newfoundland sites), Northern Temperate (from Minnesota east to Quebec, south to Pennsylvania), Southern Temperate (from Missouri, east to Maryland, south to Louisiana), Western Boreal (British Columbia and Saskatchewan), and Basin Rockies (South Dakota and Montana). Mapping the nonbreeding assignment back to breeding areas reveals patterns of strong migratory connectivity across the breeding range. The Maritime provinces breeding population has strong connectivity with Colombia; the Northern Temperate breeding population with the Greater Antilles; the Southern Temperate breeding population with the Lesser Antilles; the Western Boreal breeding population with Central America and Mexico; and, although a limited sample (n=2), the Basin Rockies breeding population with Cuba.

MIGRATORY PATHWAYS Clockwise loop migration is evident from eBird data, with post-breeding migration on average farther east than pre-breeding migration (La Sorte et al. 2016). This is a trans-Gulf migrant in pre- and post-breeding (Sherry et al. 2020); up to 60–80 birds are seen on single days in Bermuda, which also suggests a regular crossing of the western Atlantic Ocean in post-breeding (Amos 1991).

INTERESTING NOTE

Two older males color banded at the Font Hill Nature Preserve in Jamaica returned to their mangrove non breeding territories for 12 years, accounting for 24 migrations to and from breeding areas.

KIRTLAND'S WARBLER *Setophaga kirtlandii*

DISTRIBUTION AND ABUNDANCE

Kirtland's Warblers are one of the rarest songbirds in North America, with a tiny breeding range centered on the northern Lower Peninsula of Michigan and a small nonbreeding range in the Bahamas, the Turks and Caicos Islands, and Cuba. In 2019, Kirtland's Warbler was delisted under the US Endangered Species Act, but the species will be reliant on conservation efforts for the foreseeable future (Cooper et al. 2019b).

ORDER Passeriformes
FAMILY Parulidae
BREEDING HABITAT Forests, Temperate Eastern
NONBREEDING HABITAT Forests, Tropical Dry
GLOBAL POPULATION SIZE 4,800
POPULATION CHANGE SINCE 1970 NA
IUCN RED LIST CATEGORY Near Threatened
R2R TIPPING POINT SPECIES Orange-Alert
US LISTING STATUS No
CANADA SPECIES AT RISK Endangered
PIF CCS 16/20

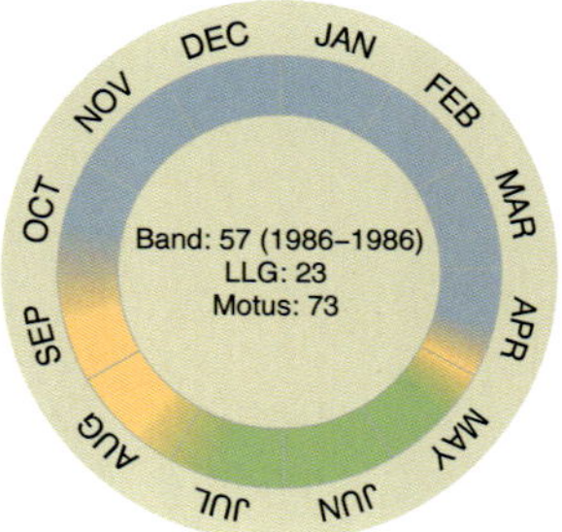

MIGRATORY CONNECTIVITY

OVERALL Migratory connectivity of Kirtland's Warblers has been studied using banding, geolocator, and automated radio telemetry data. Although the entire population spends the breeding and nonbreeding periods in a geographically restricted area, individuals within different parts of the nonbreeding range appear to completely mix on the breeding range (Cooper et al. 2018). Its restricted geographic distribution and the single breeding and nonbreeding areas prevent MC from being estimated in our analysis, but Cooper et al. (2018) found weak nonbreeding to breeding migratory connectivity (MC = 0.012).

BANDING Kirtland's Warblers banded on Eleuthera Island, the Bahamas (n = 220), during the nonbreeding season were later resighted at sites across the entire breeding range, including the small

GENERALIZED CONNECTIVITY

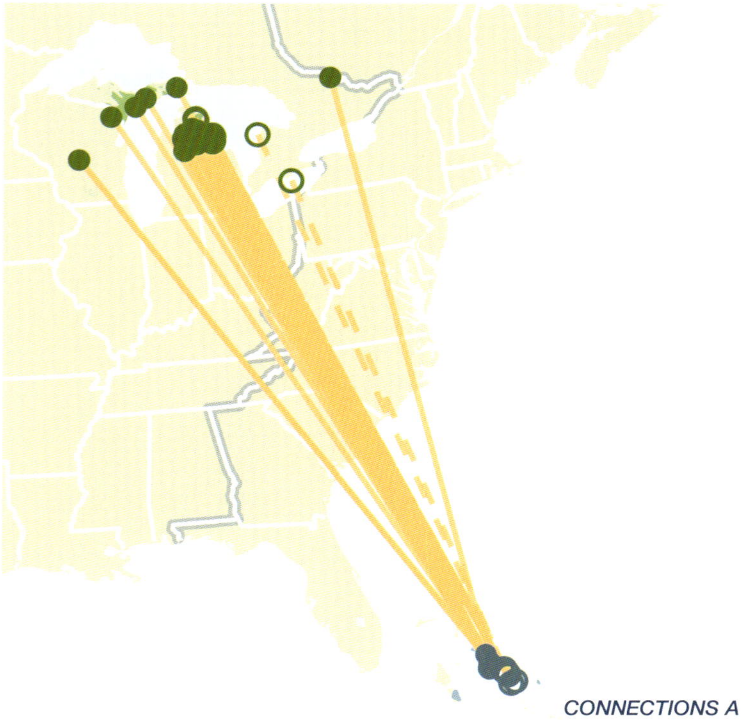

CONNECTIONS A

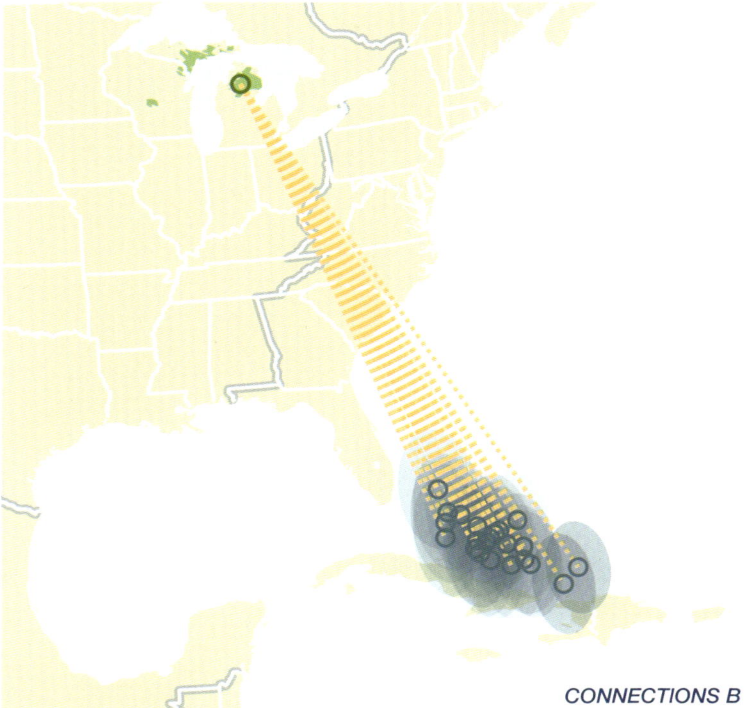

CONNECTIONS B

	Caribbean
Midcontinent (n = 153)	100%

subpopulations in Wisconsin, Ontario, and the Upper Peninsula of Michigan (Cooper et al. 2018).

RADIO TRACKING Similar to banded birds, Kirtland's Warblers radio-tagged on Cat Island, the Bahamas (n = 63), during the nonbreeding season were also resighted at sites across the entire breeding range, including the small subpopulations in Wisconsin, Ontario, and the Upper Peninsula of Michigan (Cooper et al. 2018).

GEOLOCATOR TRACKING Male Kirtland's (n = 27) marked with light-level geolocators on the breeding grounds in Michigan were found to spend the nonbreeding season across the Bahamian archipelago and Cuba (Cooper et al. 2017), though the central Bahamian islands of Eleuthera, Cat Island, San Salvador, and Long Island appear to host much of the population (Cooper et al. 2019a).

MIGRATORY PATHWAYS Geolocator data suggest that the species completes a loop migration, with post-breeding stopover in the eastern mid-Atlantic states and on the North and South Carolina coasts before an overwater flight directly to the Bahamas. In pre-breeding, migrants move westward to stopover sites in Florida, then northward to additional stopovers in Georgia, Ohio, and southwestern Ontario (Cooper et al. 2017). A more recent study of post-breeding migration using coded radio tags and the Motus Wildlife Tracking System has found very few detections in Ontario during post-breeding migration, suggesting that Kirtland's do not depart Michigan heading eastward into Canada as the geolocator data suggested; instead, they likely head south through Ohio and then east toward the Carolinas (Cooper et al. 2023a, 2023b).

DATA CONTRIBUTORS

N.W. Cooper; H. Haradon; P.P. Marra.

INTERESTING NOTE

As the population of Kirtland's Warbler has grown to over 2,500 males, their breeding range has expanded into Wisconsin and southern Ontario. Equally interesting is that their nonbreeding range may also be expanding, with recent confirmed sightings in southern Florida, Cuba, and southwestern Jamaica in the nonbreeding season (Cooper et al. 2019a; Brunner et al. 2021).

CAPE MAY WARBLER *Setophaga tigrina*

DISTRIBUTION AND ABUNDANCE

It breeds in boreal spruce forests across Canada and the northernmost US, with highest abundance in eastern Quebec. Its nonbreeding range is almost exclusively in the West Indies (especially Cuba and Hispaniola), with small numbers in Florida and along the Caribbean coast and offshore islands in Mexico, Central America, and Venezuela (Fink et al. 2023). Populations are notoriously cyclic in response to outbreaks of spruce budworm but have steadily increased in the western boreal region of Canada, while showing recent increases after a period of steep declines in the eastern parts of the range (Sauer et al. 2019; Baltz and Latta 2020).

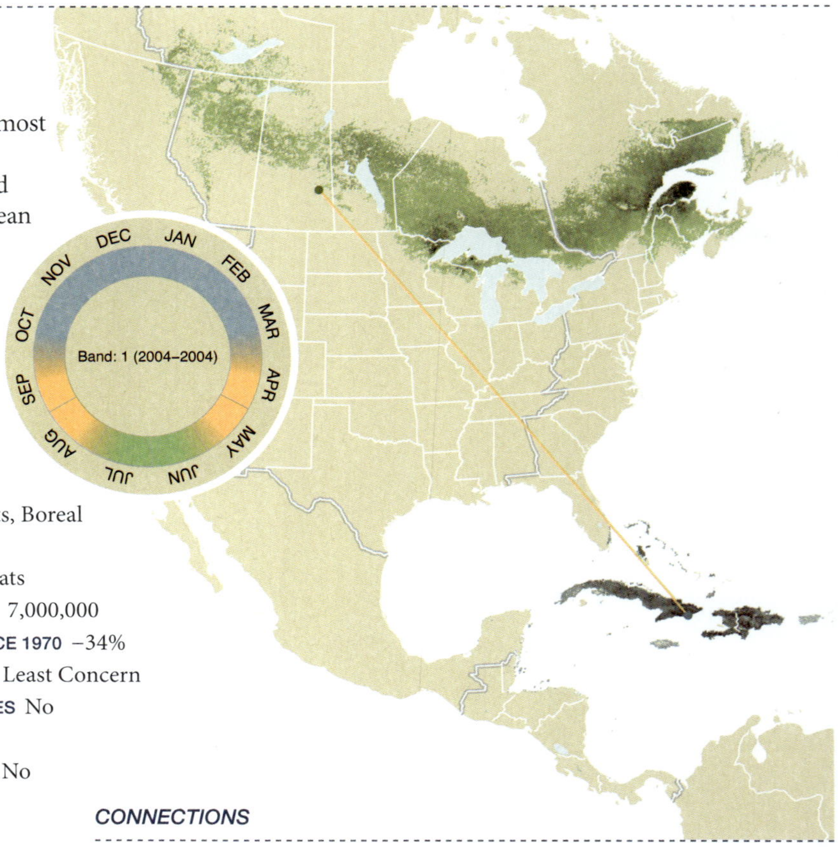

ORDER Passeriformes
FAMILY Parulidae
BREEDING HABITAT Forests, Boreal
NONBREEDING HABITAT Islands, Terrestrial Habitats
GLOBAL POPULATION SIZE 7,000,000
POPULATION CHANGE SINCE 1970 −34%
IUCN RED LIST CATEGORY Least Concern
R2R TIPPING POINT SPECIES No
US LISTING STATUS No
CANADA SPECIES AT RISK No
PIF CCS 11/20

MIGRATORY CONNECTIVITY

OVERALL Migratory connectivity is unknown, as data are only available from a few band recoveries and largely during migration. The nonbreeding biology of this species is virtually unstudied.
BANDING There is only one breeding to nonbreeding band return, a bird from Saskatchewan that was recovered in eastern Cuba.
OTHER APPROACHES No tracking, stable isotope, or genetic studies of migratory connectivity were found for this species.
MIGRATORY PATHWAYS This species exhibits a clockwise loop migration, with post-breeding migration farther east than in pre-breeding (La Sorte et al. 2016; Baltz and Latta 2020). Pre-breeding arrival in the US is almost exclusively in Florida, with northward movement through the Appalachian Mountains to northern breeding grounds. Post-breeding migrants move eastward to concentrate in the mid-Atlantic states and subsequently appear in southern Florida and the Bahamas (Fink et al. 2023). Regular post-breeding occurrence in Bermuda (up to 50 per day; Amos 1991) also suggests that many individuals cross the western Atlantic Ocean to reach Caribbean nonbreeding grounds.

INTERESTING NOTE

The Cape May Warbler is unique among warblers in having a semi-tubular tongue that allows it to feed heavily on nectar, especially during nonbreeding periods and migration; they often appear at hummingbird feeders and will follow Yellow-bellied Sapsuckers to feed at fresh drillings (Baltz and Latta 2020).

CERULEAN WARBLER *Setophaga cerulea*

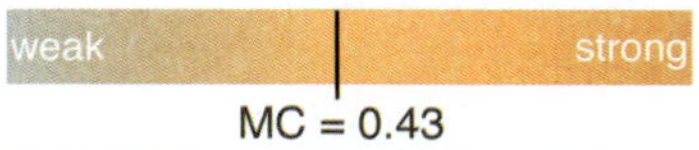

DISTRIBUTION AND ABUNDANCE

It breeds patchily in the eastern US and southeastern Canada, with the largest population in the Appalachian Mountains region. It spends the nonbreeding season at mid-elevations on the eastern slope of the South American Andes from Venezuela and Colombia to central Peru, and in smaller numbers south to northernmost Bolivia.

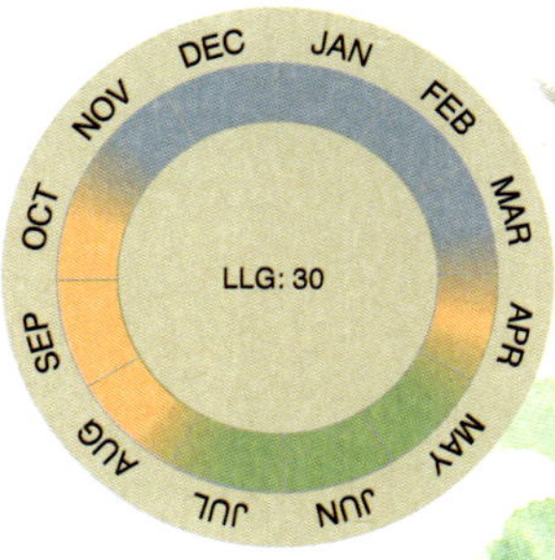

MIGRATORY CONNECTIVITY

OVERALL Migratory connectivity is strong (MC=0.43), with an east-west parallel pattern of breeding to nonbreeding connections, as documented by geolocator studies (n = 26, Raybuck et al. 2022; n = 4, Delancey et al. 2020) and inferred from stable isotopes (Jones et al. 2008).

ORDER Passeriformes
FAMILY Parulidae
BREEDING HABITAT Forests, Temperate Eastern
NONBREEDING HABITAT Forests, Tropical Montane Evergreen
GLOBAL POPULATION SIZE 530,000
POPULATION CHANGE SINCE 1970 −73%
IUCN RED LIST CATEGORY Near Threatened
R2R TIPPING POINT SPECIES Yellow-Alert
US LISTING STATUS No
CANADA SPECIES AT RISK Endangered
PIF CCS 14/20

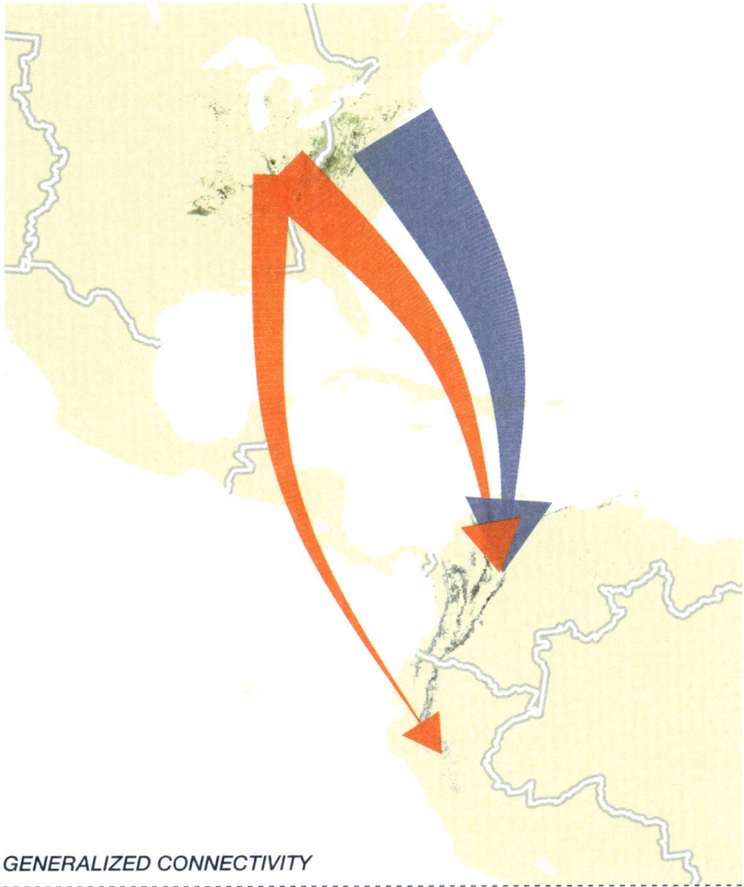

GENERALIZED CONNECTIVITY

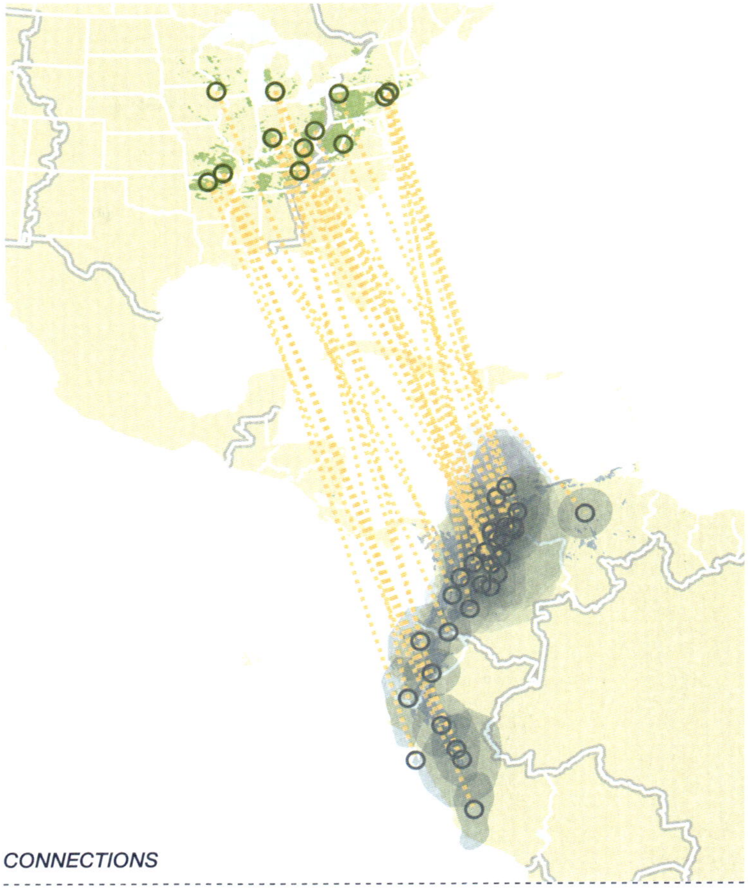

CONNECTIONS

	Andean West	Northern SA
Atlantic (n = 9)	–	100%
Midcontinent (n = 21)	38%	62%

In general, birds breeding in the Atlantic Flyway spend the nonbreeding season in northern South America, whereas birds breeding in the Midcontinent Flyway spend the nonbreeding season in northern South America (62%) and also farther southwest in the Andean West region of South America (38%).

RADIO TRACKING There are several tacking studies with the Motus Wildlife Tracking System. As of printing, there are no detections between breeding and nonbreeding areas.

GEOLOCATOR TRACKING A light-level geolocator study with data from 26 of 257 Cerulean Warbler males tagged at 11 study sites spread throughout the breeding range (Raybuck et al. 2022) revealed moderate east-west connectivity (MC=0.36 ± 0.16). Birds from the southwestern breeding regions (Arkansas and Missouri Ozarks) were most likely to go to the southwestern portion of the nonbreeding region in Peru and Ecuador. Birds tagged in eastern Tennessee, Kentucky, and southeastern Ohio, New Jersey, Pennsylvania, and Virginia migrated to nonbreeding areas in the Colombian and Venezuelan Andes.

STABLE ISOTOPES Analysis of crown feathers (grown at nonbreeding sites) from across their breeding range similarly indicated that Cerulean Warblers exhibit parallel migratory connectivity, with western breeding individuals migrating to southwestern nonbreeding grounds, and eastern breeding individuals migrating to northeastern nonbreeding grounds. (Jones et al. 2008). These results could also suggest latitudinal separation, with southernmost breeders (Tennessee) spending the nonbreeding season farthest south, and all northern populations showing broad overlap.

OTHER APPROACHES No banding or genetic studies of migratory connectivity were found for this species.

MIGRATORY PATHWAYS Cerulean Warblers exhibit a clockwise loop migration (La Sorte et al. 2016), which was confirmed with geolocators (Raybuck et al. 2022; Delancey et al. 2020). Post-breeding migration is primarily through the southern Appalachian Mountains, Alabama, Florida, and the Caribbean slope of Central America, whereas pre-breeding migration is through the highlands of Central America, and across the western Gulf of Mexico to Texas and the lower Mississippi Valley. Birds from throughout the breeding range converged on similar routes and relied on multiple stopover sites in Central America and southern Mexico (Raybuck et al. 2022).

DATA CONTRIBUTORS

N. Bayly; T.J. Boves; D. Buehler; L. Bulluck; J. Cox; G. George; K. Islam; L. Kearns; J. Larkin; S. Petzinger; D. Raybuck; K. Slankard; S. Stoleson.

NORTHERN PARULA *Setophaga americana*

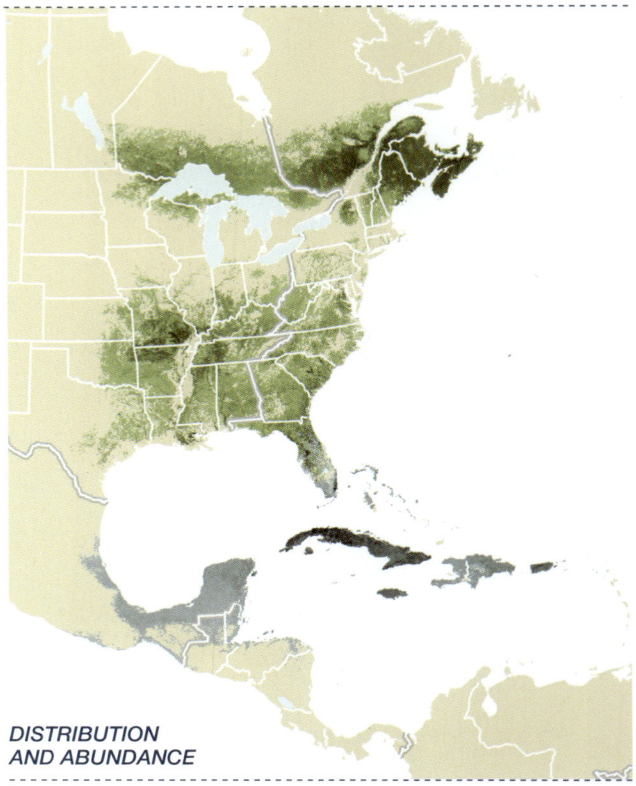

DISTRIBUTION AND ABUNDANCE

It breeds in two largely disjunct regions: across the southern boreal forest from the Canadian Maritimes and northern New England west to Minnesota and eastern Manitoba, and in a variety of forest types across the southeastern US west to Kansas, Oklahoma, and Texas. Its nonbreeding range is in Florida, throughout the Caribbean, and on the Gulf and Caribbean slope of Mexico and northern Central America, including the Yucatán Peninsula; rarely south to Panama and on the Pacific Slope from Oaxaca to El Salvador (Moldenhauer and Regelski 2020; Fink et al. 2023). Long-term trends have been stable or increasing in most of the range, except for steep recent declines noted in South Carolina, Louisiana, and Texas (Sauer et al. 2019).

ORDER Passeriformes
FAMILY Parulidae
BREEDING HABITAT Forests, Temperate Eastern
NONBREEDING HABITAT Forests, Tropical Generalist
GLOBAL POPULATION SIZE 18,000,000
POPULATION CHANGE SINCE 1970 +76%
IUCN RED LIST CATEGORY Least Concern
R2R TIPPING POINT SPECIES No
US LISTING STATUS No
CANADA SPECIES AT RISK No
PIF CCS 9/20

MIGRATORY CONNECTIVITY

OVERALL No banding, tracking, genetic, or stable isotope studies of migratory connectivity found for this species. Migratory connectivity is mostly unknown except for subspecies delineations. East-west or possibly latitudinal connectivity is suggested by morphological and song variation, as well as differential timing of migration.

RADIO TRACKING A few tags were deployed through the Motus Wildlife Tracking System; there were a few detections but no breeding to nonbreeding detections.

OTHER APPROACHES Analysis of weakly differentiated subspecies suggests that western populations migrate to Mexico and eastern populations migrate to the Caribbean (Dunn and Garrett 1997). This is also supported by eastern and western song types recorded during migration and the nonbreeding season (Moldenhauer and Regelski 2020).

MIGRATORY PATHWAYS Overall, this species exhibits a clockwise loop migration based on eBird data (La Sorte et al. 2016), with pre-breeding arrival primarily in Florida and along the Gulf Coast and post-breeding migration concentrated along the Atlantic coast. Common post-breeding occurrence on Bermuda suggests regular transoceanic crossings (Amos 1991). Earlier migration timing in the southeastern US corresponds to early arrival and departure from Mexico and Central America, whereas later migration along the mid-Atlantic coast and arrival of northeastern breeders corresponds with later departure from Caribbean nonbreeding areas (Fink et al. 2023).

TROPICAL PARULA *Setophaga pitiayumi*

DISTRIBUTION AND ABUNDANCE

This is a widespread tropical species with a very small breeding range in southern Texas, restricted to oak groves north of the Rio Grande Valley; the species has experienced recent expansion along rivers in the western Edwards Plateau region, where hybrids with Northern Parula have also occurred. Another migratory population breeds in northwestern Mexico and has occurred in southern Arizona (Fink et al. 2023).

ORDER Passeriformes
FAMILY Parulidae
BREEDING HABITAT
Forests, Tropical Generalist
NONBREEDING HABITAT
Forests, Tropical Generalist
GLOBAL POPULATION SIZE 20,000,000
POPULATION CHANGE SINCE 1970 NA
IUCN RED LIST CATEGORY Least Concern
R2R TIPPING POINT SPECIES No
US LISTING STATUS No
CANADA SPECIES AT RISK No
PIF CCS 10/20

MIGRATORY CONNECTIVITY

OVERALL Migratory connectivity is unknown. No information on migratory connectivity was found for US breeders.

MIGRATORY PATHWAYS Texas birds are migratory, but pathways and nonbreeding sites are not known; some birds disperse northward along the Gulf Coast in winter (Fink et al. 2023).

MAGNOLIA WARBLER *Setophaga magnolia*

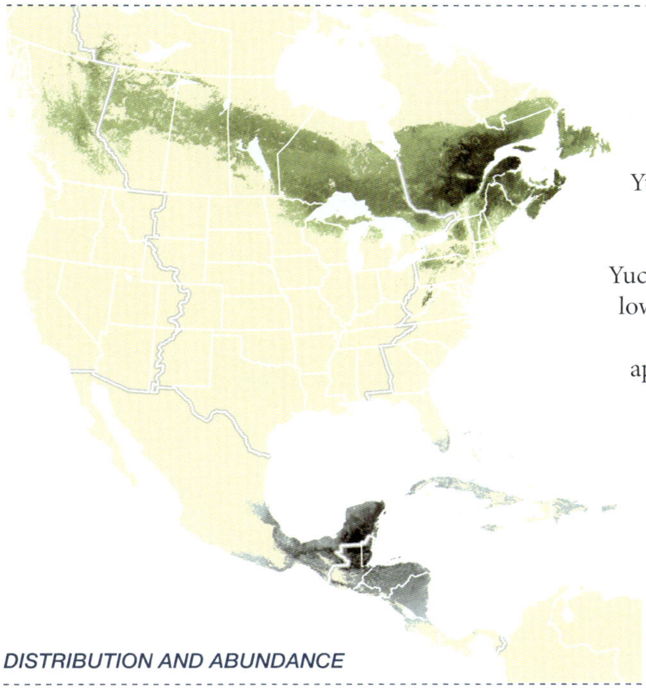

DISTRIBUTION AND ABUNDANCE

DISTRIBUTION AND ABUNDANCE

It breeds across the boreal forests of Canada, from Newfoundland and the Maritimes in decreasing abundance west to British Columbia and southeastern Yukon; it also occurs in the northeastern US in New England, northern Great Lakes states, and south in the northern Appalachians to West Virginia. Its nonbreeding range is primarily in the Gulf and Caribbean lowlands of southern Mexico, the Yucatán Peninsula, and northern Central America, with small numbers also in Pacific lowlands, south to Panama, in southern Florida, and throughout the Greater Antilles (Fink et al. 2023). Populations are stable overall, but with steep recent declines apparent in Ontario, New England, and the Canadian Maritimes (Sauer et al. 2019).

ORDER Passeriformes
FAMILY Parulidae
BREEDING HABITAT Forests, Boreal
NONBREEDING HABITAT Open Country, Habitat Mosaic
GLOBAL POPULATION SIZE 39,000,000
POPULATION CHANGE SINCE 1970 +34%
IUCN RED LIST CATEGORY Least Concern
R2R TIPPING POINT SPECIES No
US LISTING STATUS No
CANADA SPECIES AT RISK No
PIF CCS 9/20

MIGRATORY CONNECTIVITY

OVERALL Migratory connectivity is unknown. No information on migratory connectivity was found for this species.
BANDING There have been no breeding to nonbreeding recaptures, but the few captures at nonbreeding sites suggest possible east-west connectivity, with birds from Mexico and Guatemala captured in the Midwest and birds from Honduras and Nicaragua captured farther east.
RADIO TRACKING There are several projects tracking Magnolia Warblers with the Motus Wildlife Tracking System. As of printing, there are no individual breeding to nonbreeding locations, but there appear to be many detections during migration in the northeastern US.
MIGRATORY PATHWAYS Based on eBird data, this species exhibits a clockwise loop migration (La Sorte et al. 2016; Fink et al. 2023) and is a trans-Gulf migrant in both pre- and post-breeding (Dunn and Garrett 1997; Dunn and Hall 2020). Pre-breeding migration is primarily to Texas and southwestern Louisiana and northward west of the Appalachians, and post-breeding migration is primarily through the southern Appalachians and eastern Gulf states and is more numerous on the Atlantic coast, in Florida, the Bahamas, and Bermuda (up to 38 per day; Amos 1991), suggesting regular transoceanic crossings. It may exhibit stopover site fidelity, based on individual birds captured three years apart at the same site in both Tennessee and Maryland (Dunn and Hall 2020).

BAY-BREASTED WARBLER *Setophaga castanea*

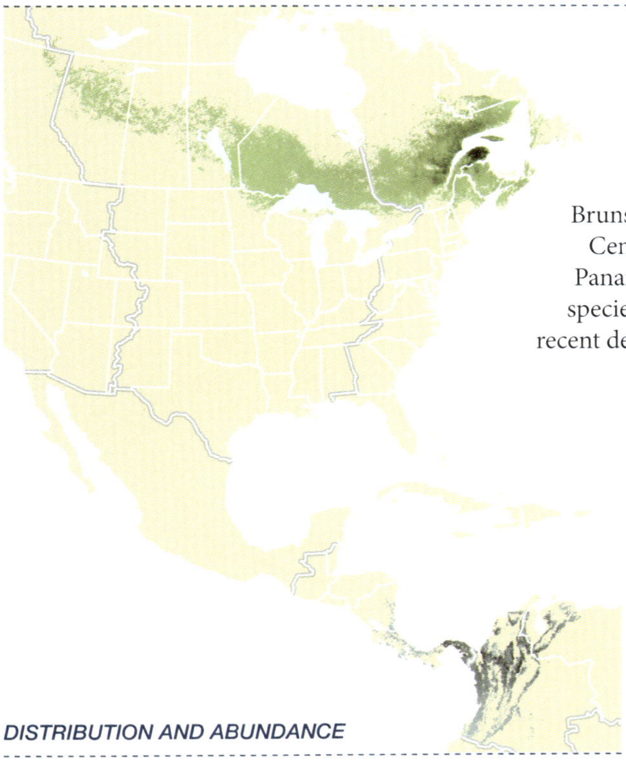

DISTRIBUTION AND ABUNDANCE

DISTRIBUTION AND ABUNDANCE

It breeds in the eastern boreal forest from western Newfoundland, the Canadian Maritimes, and northern New England west sparsely to northeastern British Columbia and southeastern Yukon; its highest abundance is in Quebec and New Brunswick. It has a relatively small nonbreeding range in the tropical lowlands of southern Central America and northwestern South America, with high abundance only in eastern Panama and west of the Andes in northwestern Colombia (Fink et al. 2023). Although the species has increased overall since 1970 (Rosenberg et al. 2019), regional populations show recent declines across the southern and eastern parts of the breeding range (Sauer et al. 2019).

ORDER Passeriformes
FAMILY Parulidae
BREEDING HABITAT Forests, Boreal
NONBREEDING HABITAT Forests, Tropical Lowland Evergreen
GLOBAL POPULATION SIZE 9,900,000
POPULATION CHANGE SINCE 1970 −18%
IUCN RED LIST CATEGORY Least Concern
R2R TIPPING POINT SPECIES No
US LISTING STATUS No
CANADA SPECIES AT RISK No
PIF CCS 10/20

MIGRATORY CONNECTIVITY

OVERALL Migratory connectivity for this species is unknown. No information on migratory connectivity was found for this species. One tag was deployed through the Motus Wildlife Tracking System; there are no breeding to nonbreeding detections.

MIGRATORY PATHWAYS It exhibits a clockwise loop migration (La Sorte et al. 2016; Fink et al. 2023), with pre-breeding migration primarily through Texas and west of the Appalachian Mountains and post-breeding migration primarily through the eastern Gulf states and Florida. It is rarely detected on the South Atlantic coast or in the Caribbean east of western Cuba (Venier et al. 2020). Farther south, migration is through the Yucatán Peninsula and the Caribbean slope of Central America in both pre- and post-breeding, with crossings of the Gulf of Mexico and the western Caribbean Sea.

BLACKBURNIAN WARBLER *Setophaga fusca*

DISTRIBUTION AND ABUNDANCE

An abundant breeder across the southern boreal and northern hardwood forests from the Canadian Maritimes and New England west sparsely to central Canada, and south at high elevations through the Appalachian Mountains. Its nonbreeding range is primarily in montane forest and adjacent disturbed areas in the northern Andes of South America, from western Venezuela through Colombia (all ranges) and Ecuador to central Peru, more sparsely south to northwestern Bolivia and in the highlands of Panama and Costa Rica (Fink et al. 2023). Breeding populations have been stable or slightly increasing since 1970 throughout most of the range, but recent declines are noted in the Appalachians (Rosenberg et al. 2019; Sauer et al. 2019).

ORDER Passeriformes
FAMILY Parulidae
BREEDING HABITAT Forests, Boreal
NONBREEDING HABITAT Forests, Tropical Montane Evergreen
GLOBAL POPULATION SIZE 13,000,000
POPULATION CHANGE SINCE 1970 +11%
IUCN RED LIST CATEGORY Least Concern
R2R TIPPING POINT SPECIES No
US LISTING STATUS No
CANADA SPECIES AT RISK No
PIF CCS 10/20

MIGRATORY CONNECTIVITY

OVERALL Migratory connectivity is unknown. No information on migratory connectivity was found for this species. A few tags were deployed through the Motus Wildlife Tracking System; there are no breeding to nonbreeding detections.

MIGRATORY PATHWAYS It exhibits a clockwise loop migration, based on eBird data (La Sorte et al. 2016; Fink et al. 2023). Migrations are through northern Central America and the eastern US, primarily crossing the western Gulf of Mexico to Texas and the midwestern US in pre-breeding, and from Florida and the southeastern US across the eastern Gulf and western Caribbean Sea in post-breeding (Fink et al. 2023). It is rare in the Caribbean east of Cuba, but regular occurrence on Bermuda (Amos 1991) suggests that some birds fly over the Atlantic Ocean in post-breeding migration.

YELLOW WARBLER *Setophaga petechia*

DISTRIBUTION AND ABUNDANCE

It breeds widely across North America to the northern limit of woody shrubs from western Alaska to Labrador and Newfoundland, south in the east to the mid-Atlantic and south-central US, and in the west to Southern California and northwestern Mexico. It is largely absent from purely coniferous regions of the eastern boreal forest. Its nonbreeding range extends from central Mexico (both Pacific and Gulf lowlands) through Central America and northern South America, sparsely south to Ecuador, northern Peru, and northern Brazil, and east to Trinidad and Tobago and the Guianas. Northern migrants are absent from the Caribbean, where resident populations occur. Steep declines, especially since the 1980s, are apparent in most of the eastern and western parts of the range, but populations have increased across the western Great Plains (Sauer et al. 2019).

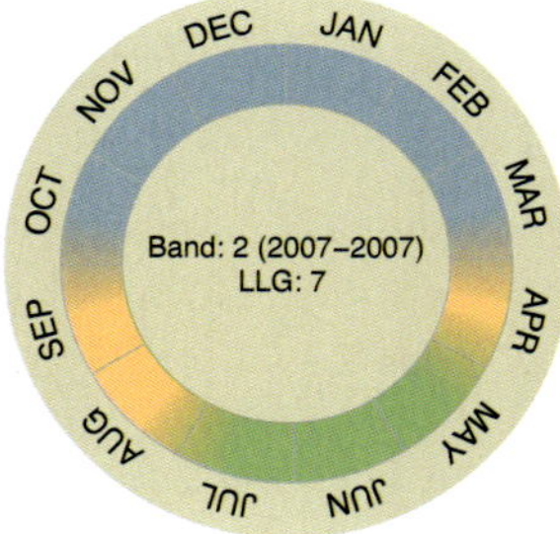

ORDER Passeriformes
FAMILY Parulidae
BREEDING HABITAT Forests, Temperate Generalist
NONBREEDING HABITAT Open Country, Habitat Mosaic
GLOBAL POPULATION SIZE 97,000,000
POPULATION CHANGE SINCE 1970 −26%
IUCN RED LIST CATEGORY Least Concern
R2R TIPPING POINT SPECIES No
US LISTING STATUS No
CANADA SPECIES AT RISK No
PIF CCS 8/20

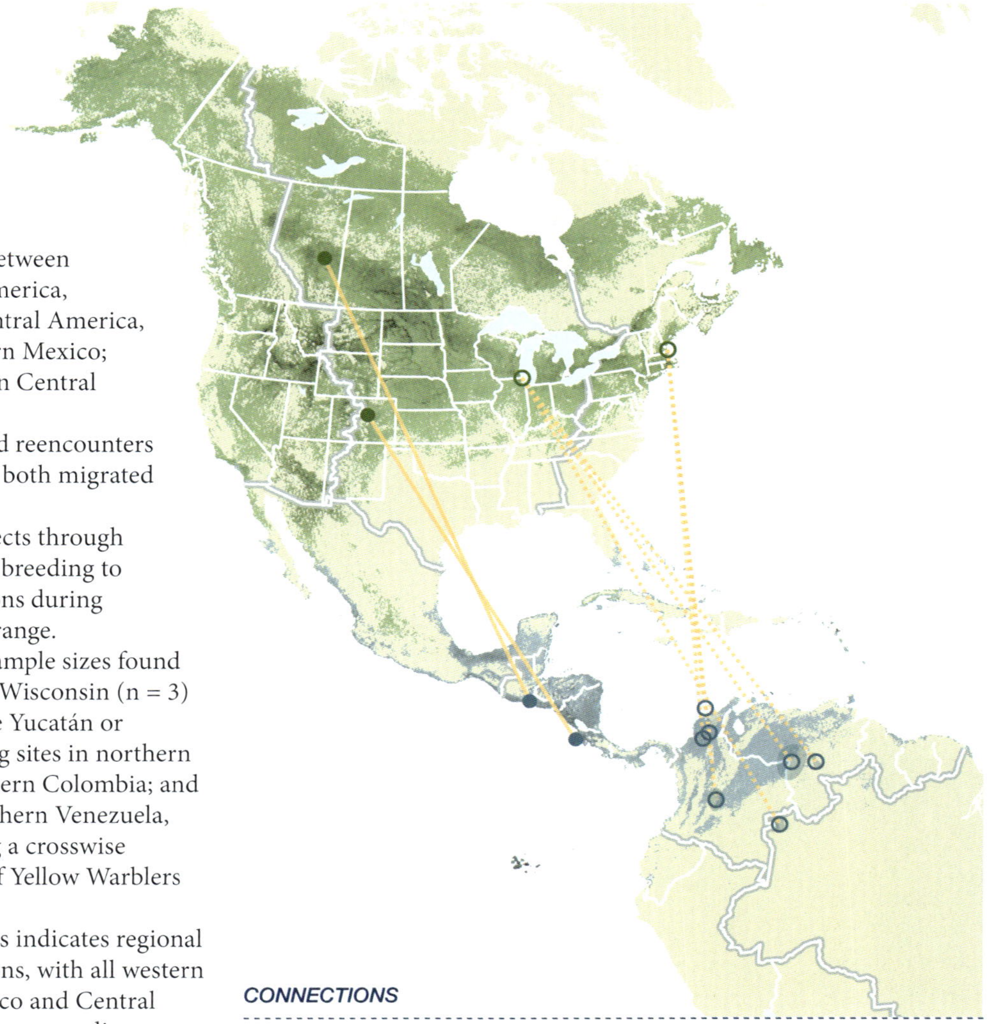

CONNECTIONS

MIGRATORY CONNECTIVITY

OVERALL Although there is only one small study of tracked individuals, moderate or strong migratory connectivity among regional populations is demonstrated by analysis of morphological subspecies, genetics, stable isotopes, and banding data. Multiple studies corroborate a clear divide between eastern populations migrating mostly to South America, western populations migrating to Mexico and Central America, and southwestern US breeders restricted to western Mexico; delineation of finer-scale connectivity, especially in Central America, requires additional study.

BANDING Only two breeding to nonbreeding band reencounters indicate that breeders from Alberta and Colorado both migrated to the Pacific Slope of Central America.

RADIO TRACKING There are several tracking projects through the Motus Wildlife Tracking System. There are no breeding to nonbreeding detections. There are several detections during migration in the eastern and western parts of the range.

GEOLOCATOR TRACKING One study with small sample sizes found that breeding individuals from Maine (n = 4) and Wisconsin (n = 3) crossed the Gulf of Mexico in post-breeding to the Yucatán or Central America before continuing to nonbreeding sites in northern South America; Maine breeders migrated to northern Colombia; and Wisconsin breeders migrated to the region of southern Venezuela, eastern Colombia, and northern Brazil, suggesting a crosswise migration pattern within the eastern population of Yellow Warblers (Witynski and Bonter 2018).

GENETICS A range-wide DNA analysis using SNPs indicates regional separation of breeding and nonbreeding populations, with all western populations migrating to the Pacific Slope of Mexico and Central America, but southwestern US breeding populations extending farther north within Mexico, central boreal and US breeders migrating primarily to eastern Mexico and Central America but also to South America, and easternmost breeders migrating exclusively to northern South America (Bay et al. 2021). Precipitation regimes were correlated across breeding and nonbreeding areas, suggesting that climate tracking and associated morphological adaptation may act as a selective force maintaining geographic variation and migratory connectivity (Bay et al. 2021).

OTHER APPROACHES Between six and nine subspecies are recognized within the migratory "Northern" Yellow Warbler group (Browning 1994; Garrett and Dunn 1997; Lowther et al. 2020b). Although variation is clinal, with overlap along distributional boundaries, analysis of distinct subspecies specimens reveals strong connectivity and complex migratory pathways among regional populations. Eastern and boreal-breeding populations spend the nonbreeding season from southeastern Mexico through Central America and northern South America, whereas western-breeding Yellow Warblers migrate to the Pacific Slope of Mexico and Central America to western Panama (possibly Ecuador).

MIGRATORY PATHWAYS Overall, individuals migrate over a broad area of North America, although different populations exhibit complex variation in migration pathways and behavior (Dunn and Garrett 1997). Combined analysis of mitochondrial DNA, stable isotopes, and banding data further indicate three migration patterns in northern Yellow Warblers, with breeders in the southwestern US migrating to the Pacific coast of Mexico, a western group breeding in western North America migrating to Mexico and Central America, and an eastern group breeding in central and eastern North America migrating to Central America and northern South America (Boulet et al. 2006). Eastern populations migrate extremely early (July–August), whereas migration in the west is later and more protracted (Lowther et al. 2020b; Dunn and Garrett 1997; eBird animations). Eastern Yellow Warblers are believed to migrate both across and around the Gulf in pre- and post-breeding, and frequent occurrence on Bermuda in post-breeding (up to 68 per day; Amos 1991) suggests regular transatlantic crossings. Boreal-breeding *S. p. amnicola* migrate through eastern North America later than more southerly-breeding *S. p. aestiva*; and in the west, earliest migrants are the southwestern *S. p. sonorana*, followed by Pacific coastal *S. p. brewsteri*, then Rocky Mountain *S. p. morcomi*, and finally Alaskan *S. p. banksi* (Garrett and Dunn 1997). Migration-season band recoveries suggest that birds from Alberta and Saskatchewan migrated eastward to the Great Lakes region before continuing south, whereas birds from Alaska, Yukon, and British Columbia migrated south through the western US.

INTERESTING NOTE

Resident populations in the Caribbean and south Florida (golden warbler), and along Mexican and Central American coasts (mangrove warbler) are considered separate subspecies and are not included in this account.

DATA CONTRIBUTORS

D.N. Bonter; M. Witynski.

CHESTNUT-SIDED WARBLER *Setophaga pennsylvatica*

DISTRIBUTION AND ABUNDANCE

This species breeds across the southern boreal and northern hardwood forests from the Canadian Maritimes and New England west through the Great Lakes region to central Canada, and south through the Appalachian Mountains. Its nonbreeding range is from southern Mexico through Central America; its highest nonbreeding abundance is in a relatively small area of lowland tropical forest from eastern Honduras to western Panama. Outside this core area, scattered individuals occur regularly throughout the Caribbean, on the Pacific Slope of Mexico, in northwestern South America, and across the southernmost US (Fink et al. 2023). Populations in the eastern part of the breeding range (Ontario, Quebec, New England) have declined most steeply, whereas populations have been stable or slightly increasing in the western part of the range (Sauer et al. 2019).

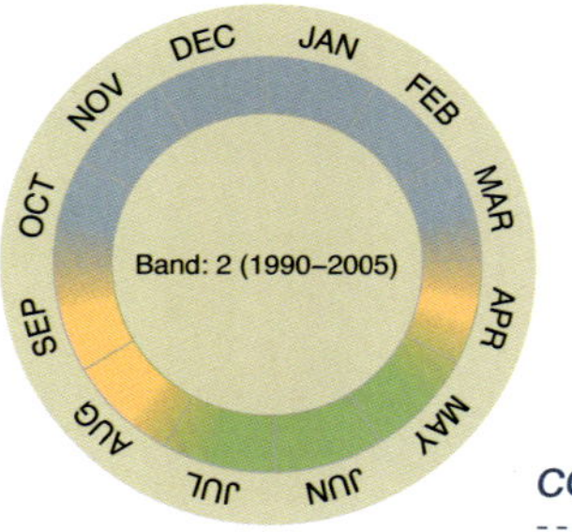

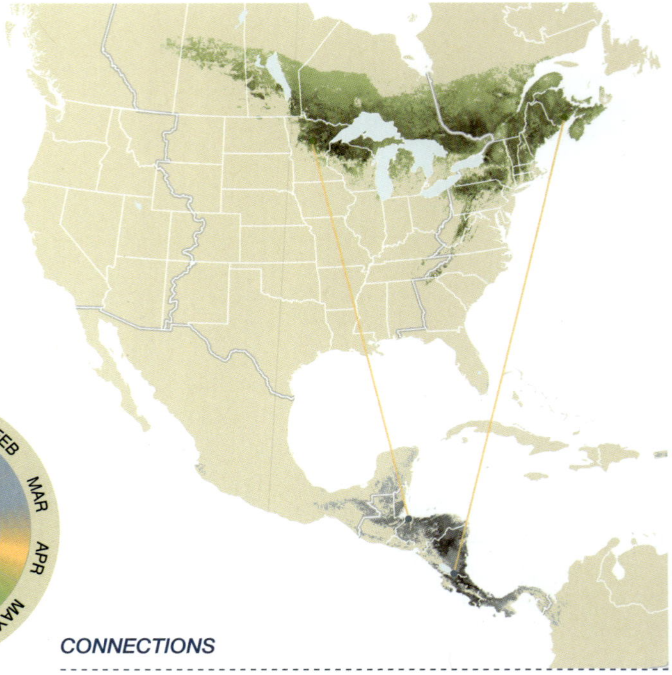

ORDER Passeriformes
FAMILY Parulidae
BREEDING HABITAT Forests, Boreal
NONBREEDING HABITAT Forests, Tropical Lowland Evergreen
GLOBAL POPULATION SIZE 18,000,000
POPULATION CHANGE SINCE 1970 −42%
IUCN RED LIST CATEGORY Least Concern
R2R TIPPING POINT SPECIES No
US LISTING STATUS No
CANADA SPECIES AT RISK No
PIF CCS 12/20

MIGRATORY CONNECTIVITY

OVERALL Almost no information on migratory connectivity was found for this species; the two breeding to nonbreeding band reencounters, separated by roughly 700 km, suggest that birds from throughout the large breeding range converge on a small area of high abundance in the nonbreeding season.

BANDING There are only two band reencounters, one from Minnesota to Honduras and one from New Brunswick to Nicaragua.

RADIO TRACKING There are no breeding to nonbreeding detections from tags deployed using the Motus Wildlife Tracking System.

OTHER APPROACHES No telemetry, genetic, or stable isotope studies of migratory connectivity were found for this species.

MIGRATORY PATHWAYS It exhibits a clockwise loop migration, based on observational data (La Sorte et al. 2016). It is a trans-Gulf migrant in pre- and post-breeding, with birds arriving primarily in Texas and moving north through the midwestern US in pre-breeding, and post-breeding migrants moving south through the Appalachian Mountains and departing through the eastern Gulf states. It is rare on the Atlantic coast, especially during pre-breeding migration (Fink et al. 2023).

BLACKPOLL WARBLER *Setophaga striata*

DISTRIBUTION AND ABUNDANCE

This species breeds from Newfoundland and mountainous areas of northern New England and New York to western Alaska and the northwestern Rocky Mountains. It is found during the nonbreeding season in northern South America, primarily in the Orinoco Basin and along the foothills of the northern Andes. Smaller numbers spend the nonbreeding season eastward to Guyana, throughout the northern Amazon basin, and south to Brazil's Atlantic Forest (DeLuca et al. 2020; Fink et al. 2023).

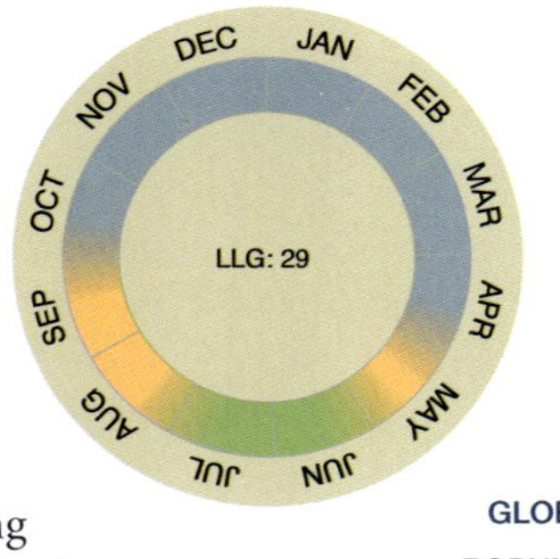

ORDER Passeriformes
FAMILY Parulidae
BREEDING HABITAT Forests, Boreal
NONBREEDING HABITAT Forests, Tropical Lowland Evergreen
GLOBAL POPULATION SIZE 60,000,000
POPULATION CHANGE SINCE 1970 −83%
IUCN RED LIST CATEGORY Near Threatened
R2R TIPPING POINT SPECIES No
US LISTING STATUS No
CANADA SPECIES AT RISK No
PIF CCS 11/20

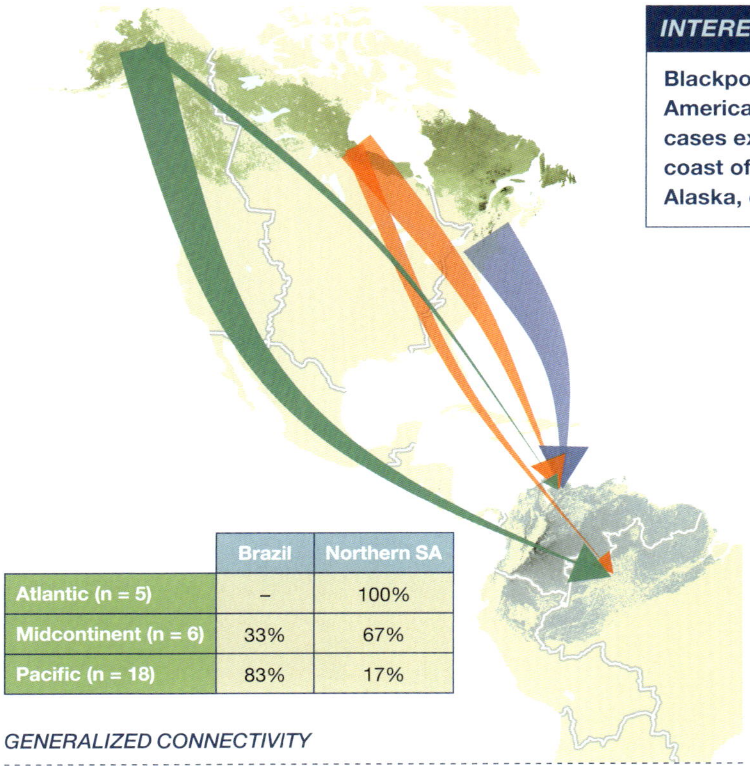

	Brazil	Northern SA
Atlantic (n = 5)	–	100%
Midcontinent (n = 6)	33%	67%
Pacific (n = 18)	83%	17%

GENERALIZED CONNECTIVITY

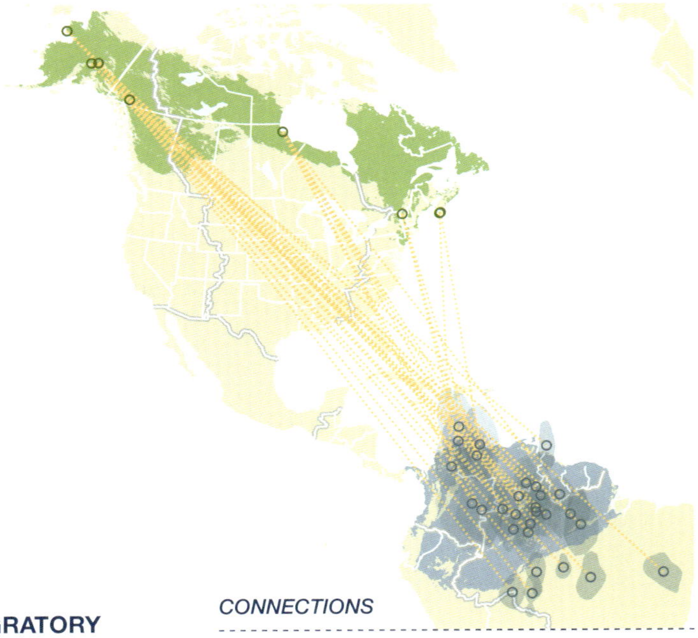

CONNECTIONS

INTERESTING NOTE

Blackpolls from throughout the breeding range departed eastern North America from points along the Atlantic coast and flew nonstop, in some cases exceeding 90 hours, over the ocean to the Greater Antilles and north coast of South America (DeLuca et al. 2015, 2019). Individuals from Nome, Alaska, completed a round-trip migration totaling almost 20,000 km.

MIGRATORY CONNECTIVITY

OVERALL Migratory connectivity appears weak due to mixing of breeding populations on nonbreeding areas. However, fine-scale migratory connectivity may be moderate, with distinct but overlapping nonbreeding areas. Atlantic Flyway breeding populations spend the nonbreeding season in northern South America (100%), while birds breeding in the Midcontinent Flyway spend the nonbreeding season in northern South America (67%) or Brazil (33%). Pacific Flyway breeders primarily migrate to Brazil (83%) with some individuals migrating to northern South America (17%).

RADIO TRACKING There are many tracking studies of the Blackpoll Warbler using the Motus Wildlife Tracking System. Detections occur throughout the range, yet as of printing, there are no individual breeding to nonbreeding detections.

GEOLOCATOR TRACKING Blackpoll Warblers tracked with geolocators originating from as far west as Nome, Alaska, and east to Nova Scotia suggest some connectivity between breeding and nonbreeding populations, with westernmost breeding individuals migrating to nonbreeding areas farther east in Amazonia, and northeastern breeders migrating to the base of the Andes in Colombia (DeLuca et al. 2015, 2019).

STABLE ISOTOPES A study of isotopes from modern-day and historical museum specimens from a site in the foothills of the Colombian Andes revealed a 600-km northward shift in the breeding origin of nonbreeding birds, suggesting that migratory connectivity can be dynamic (Gómez et al. 2021).

GENETICS Mitochondrial DNA analysis shows genetic differentiation between western, eastern, and Newfoundland populations (Ralston and Kirchman 2012), suggesting that additional connectivity research using samples from nonbreeding populations could further elucidate connectivity patterns.

OTHER APPROACHES No breeding to nonbreeding band reencounter data found for this species.

MIGRATORY PATHWAYS Individuals from throughout the large breeding range converge on similar migration routes and stopover sites, as indicated from limited band recoveries on migration, stable isotope analysis, and geolocators (DeLuca et al. 2015, 2019; Holberton et al. 2015). Geolocator data suggest a crossing rather than a parallel migration pattern (DeLuca et al. 2019) and a clockwise loop migration (DeLuca et al. 2020). Individuals from across the breeding population migrate east in post-breeding and undertake a transoceanic, multiple-day journey south over the Atlantic Ocean. A stable isotope study found that birds from throughout the broad breeding range converged in coastal New England before departing on this transatlantic crossing (Holberton et al. 2015). Sites in Wisconsin and Pennsylvania received birds from farther west and north, suggesting a possible east-west cline in breeding populations arriving north to south along the Atlantic coast. All individuals tracked on the southern migration used the northern coast of Colombia and Venezuela, highlighting the importance of this region for stopover habitat (DeLuca et al. 2015, 2019). A stable isotope study in northeastern Colombia found that birds arriving on the Guajira Peninsula originated from across the entire breeding range (Alaska to Newfoundland), also identifying that region as a major stopover (and bottleneck) during post-breeding migration (Bayly et al. 2020). Once birds arrive in northeastern Colombia, they replenish fat and muscle during a multiday stopover before continuing to nonbreeding grounds farther south in South America (Bayly et al. 2021). During pre-breeding, individuals migrated north along the Greater Antilles through Florida, with western breeders continuing largely through the midwestern United States, and eastern breeders moving north along the Atlantic coast. Individuals at a nonbreeding site in Colombia stop over in Cuba or the southeastern US before continuing to boreal breeding grounds (Bayly et al. 2020).

DATA CONTRIBUTORS

E. Adams; J. Boelsma; H.A. Cooke; W.V. DeLuca; A. Gilbert; G. Gress; C. Grigsby; N.E. Freeman; S.A. Mackenzie; P.P. Marra; K.P. McFarland; C. McIntyre; A.E.M. Newman; D.R. Norris; L.M. Phillips; C.C. Rimmer; I.J. Stenhouse; A.O. Sutton; L. Tauzer; P.D. Taylor; S. Weidensaul; E.J. Williams; B.K. Woodworth.

BLACK-THROATED BLUE WARBLER *Setophaga caerulescens*

DISTRIBUTION AND ABUNDANCE

This species breeds in northern New England, the Appalachian Mountains, and southeastern Canada, west only to the western Great Lakes region. It spends the nonbreeding season almost exclusively in the Greater Antilles, with smaller numbers elsewhere in the Caribbean, southern Florida, and the Caribbean coast of the Yucatán Peninsula and northern Central America. Populations throughout the range show a long-term increase, although steep declines in some populations recently (Rosenberg et al. 2019; Sauer et al. 2019).

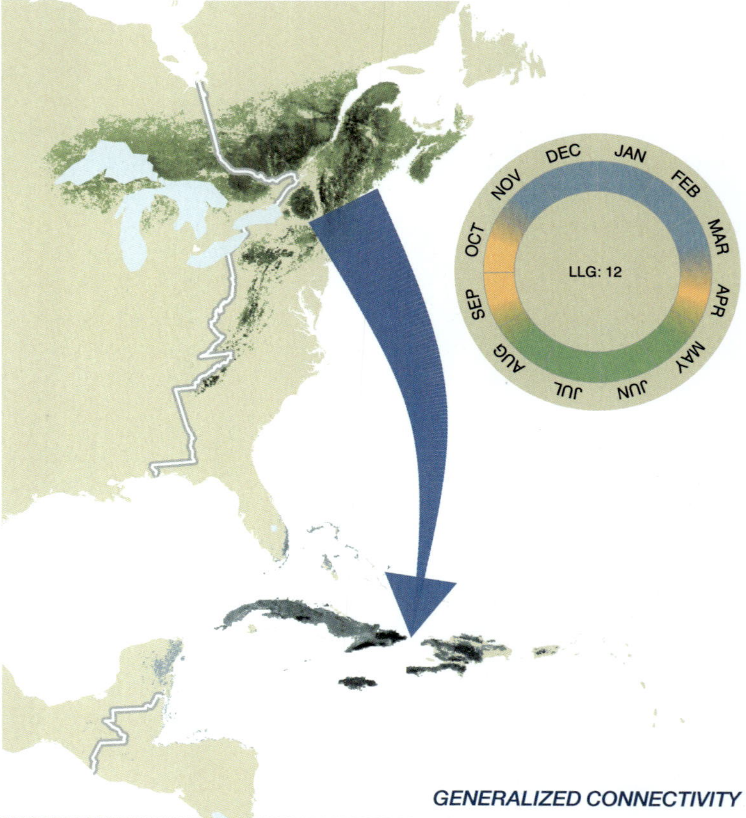

GENERALIZED CONNECTIVITY

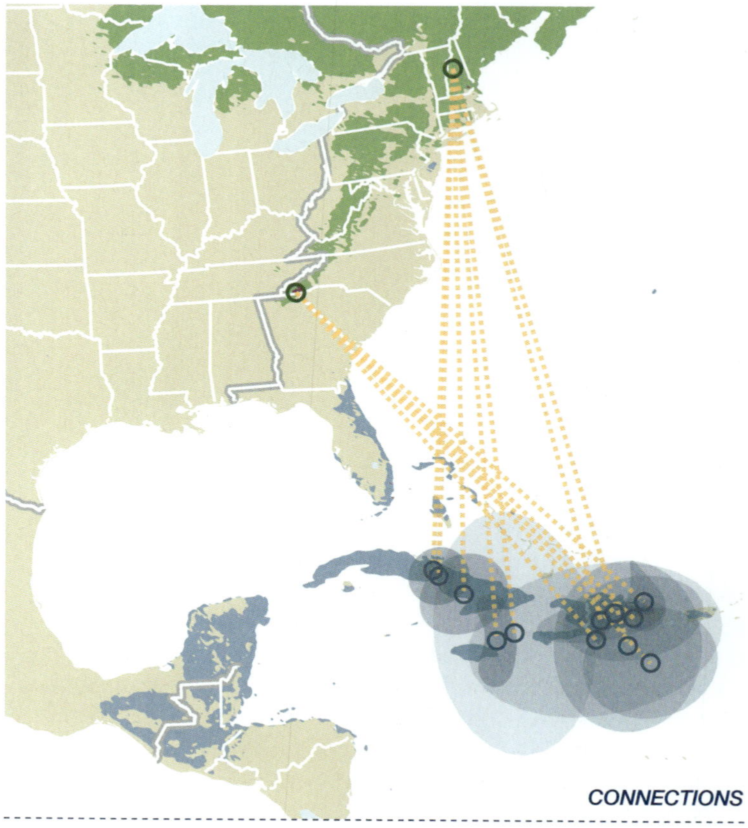

CONNECTIONS

ORDER Passeriformes
FAMILY Parulidae
BREEDING HABITAT Forests, Boreal
NONBREEDING HABITAT Islands, Terrestrial Habitats
GLOBAL POPULATION SIZE 2,400,000
POPULATION CHANGE SINCE 1970 +120%
IUCN RED LIST CATEGORY Least Concern
R2R TIPPING POINT SPECIES No
US LISTING STATUS No
CANADA SPECIES AT RISK No
PIF CCS 10/20

	Caribbean
Atlantic (n = 13)	100%

MIGRATORY CONNECTIVITY

OVERALL For such a well-studied species, there is surprisingly little data on migratory connectivity; based mostly on stable isotope studies (Chamberlain et al. 1997; Royle and Rubenstein 2004) and data from a small number of geolocator tags (13; Lewis et al. 2023), migratory connectivity appears weak, with all birds breeding in the Atlantic Flyway spending the nonbreeding season in the Caribbean. MC is not calculated due to limited sample size.

GEOLOCATOR TRACKING At a finer scale, migratory connectivity may be stronger, with more birds from northern breeding populations migrating to the western Greater Antilles (Cuba, Jamaica), and more birds from the Appalachian Mountains migrating farther east to Hispaniola and Puerto Rico. Geolocators were recovered from breeding individuals in New Hampshire (n = 8) and North Carolina (n = 5). Nonbreeding areas for birds from New Hampshire included islands across the Greater Antilles, ranging from Cuba to the Dominican Republic. However, the highest probability of occurrence was in central Cuba. Specifically, three individuals were located in central to eastern Cuba, two (including one female) were found in either Jamaica or southern Cuba, and three were on the island of Hispaniola.

In contrast to the multi-island spread of birds from New Hampshire, all five birds from North Carolina spent the nonbreeding season on the island of Hispaniola, largely in the Dominican Republic (Lewis et al. 2023).

RADIO TRACKING Nearly 400 tags were deployed with the Motus Wildlife Tracking System. As of printing, there are no individual breeding to nonbreeding connections; however, there have been many detections during migration.

STABLE ISOTOPES In one of the first applications of stable isotopes to identify geographic origin markers in a migratory bird, Chamberlain et al. (1997) found that Black-throated Blue Warblers sampled on nonbreeding grounds in the Greater Antilles likely originated in northern portions of the breeding grounds. Additionally, higher variation in stable isotope ratios in nonbreeding samples suggested that individuals from different parts of the breeding range mix on the nonbreeding grounds. Using a combination of abundance and stable isotopes, Royle and Rubenstein (2004) provided some additional resolution, with birds spending the nonbreeding season in Cuba and Puerto Rico assigned primarily to origins in the northern portion of the breeding range; birds in the Dominican Republic were assigned equally to the northwest and the south; and birds in Jamaica were assigned primarily to the northeast portion of the breeding range.

GENETICS The Bird Genoscape Project is currently in the process of constructing a genoscape from samples collected across the breeding range (https://www.birdgenoscape.org/black-throated-blue-warbler/).

OTHER APPROACHES There is no breeding to nonbreeding band reencounter data.

DATA CONTRIBUTORS

A.R. Brunner; R.J. Cooper; M.T. Hallworth; S.A. Kaiser; W.B. Lewis; T.S. Sillett; M. Webster.

MIGRATORY PATHWAYS Migration is through Florida and Atlantic-coastal and Great Lakes states in both pre- and post-breeding. Its greatest abundance is through the Appalachian Mountains, and it is rare along the Gulf Coast. Motus data show that individuals from the Great Lakes area migrate first east and then south along the Atlantic Flyway. Common occurrence (up to 32 per day) on Bermuda suggests regular crossing of the North Atlantic Ocean on post-breeding migration (Amos 1991).

INTERESTING NOTE

Individuals breeding in New Hampshire and spending the nonbreeding season in Jamaica experience much higher levels of mortality during migration than during stationary periods of the annual cycle. Males had higher annual survival than females in New Hampshire, but annual survival did not differentiate between sexes in Jamaica. Monthly survival was high for both sexes during the stationary breeding and nonbreeding seasons (Sillett and Holmes 2002).

Setophaga palmarum PALM WARBLER

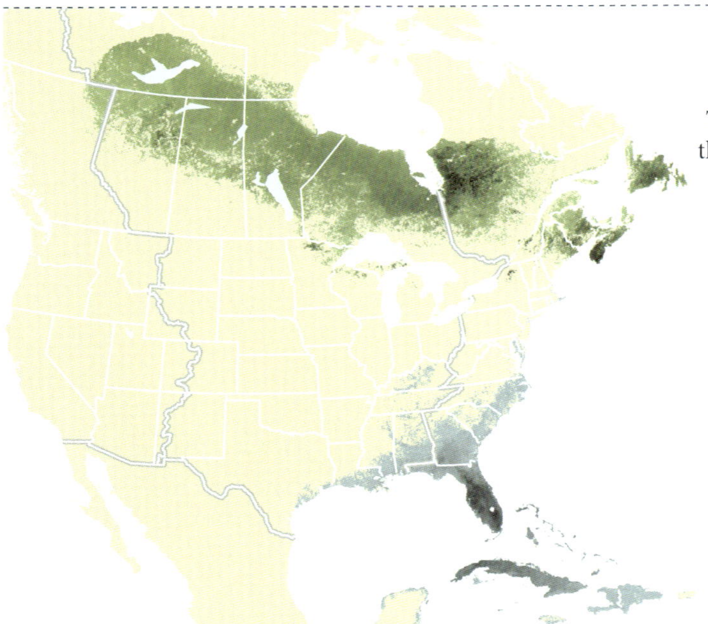

DISTRIBUTION AND ABUNDANCE

DISTRIBUTION AND ABUNDANCE

This species breeds across the boreal forest and taiga from Newfoundland to the Northwest Territories and British Columbia, south primarily in tamarack bogs to New England, the Adirondack Mountains, and Great Lakes states. Two centers of abundance in Atlantic Canada and from central Quebec westward correspond with the two geographic subspecies (Fink et al. 2023). Its greatest nonbreeding abundance is in Florida, the Bahamas, and Cuba, but it is also common across the southeastern and Gulf Coastal Plains, elsewhere in the Greater Antilles, and on the Gulf/Caribbean coast of the Yucatán Peninsula and Honduras; limited numbers spend the nonbreeding season north along the Atlantic coast to Massachusetts (Fink et al. 2023).

ORDER Passeriformes
FAMILY Parulidae
BREEDING HABITAT Wetlands, Boreal
NONBREEDING HABITAT Open Country, Habitat Mosaic
GLOBAL POPULATION SIZE 13,000,000
POPULATION CHANGE SINCE 1970 +1%
IUCN RED LIST CATEGORY Least Concern
R2R TIPPING POINT SPECIES No
US LISTING STATUS No
CANADA SPECIES AT RISK No
PIF CCS 10/20

MIGRATORY CONNECTIVITY

OVERALL Migratory connectivity is mostly unknown except for subspecies delineations. The eastern and western subspecies occupy largely separate nonbreeding distributions.

RADIO TRACKING A limited number of tags were deployed through the Motus Wildlife Tracking System; as of printing, there are no breeding to nonbreeding detections. No other studies of migratory connectivity found for this species.

OTHER APPROACHES Two distinct subspecies differ in their migration timing and routes (Wilson 2020) and in nonbreeding distribution and habitat use (American Ornithologists Union 1957; Dunn and Garrett 1997). The eastern subspecies, *S. p. hypochrysea*, occurs during the nonbreeding season mainly away from the coast in open pine habitats, from North Carolina (a few to Maryland) to northern Florida and southeastern Louisiana, whereas western *S. p. palmarum* occur primarily along the South Atlantic and Gulf coasts and throughout Florida and the remaining nonbreeding range.

MIGRATORY PATHWAYS Based on observational data, the much larger western population exhibits a clockwise loop migration, moving eastward to the mid-Atlantic coast in post-breeding and northward through Florida and west of the Appalachians to the Great Lakes region in pre-breeding. Common post-breeding and nonbreeding season sightings in Bermuda (Amos 1991) and observed landfall along the South Atlantic coast and the Bahamas indicate regular transoceanic crossings. In contrast, the eastern population migrates along a southwesterly-northeasterly axis in post- and pre-breeding, primarily east of the Appalachians but rarely along the coast (Dunn and Garrett 1997; Fink et al. 2023).

INTERESTING NOTE

A study in Fort McMurray in northern Alberta deployed approximately 20 geolocators in June 2015 on breeding Palm Warblers. Unfortunately, an extensive fire the following May destroyed the entire study area (P.P. Marra, unpublished data) and no returning birds were resighted.

PINE WARBLER *Setophaga pinus*

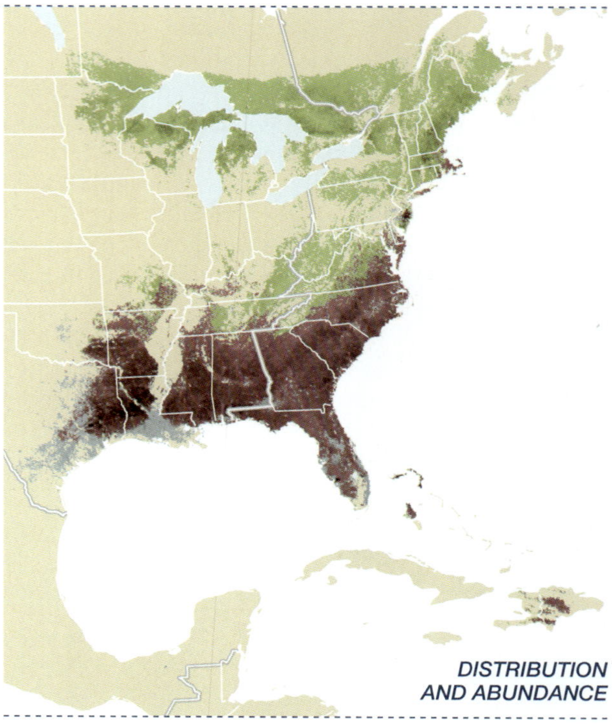

DISTRIBUTION AND ABUNDANCE

This species breeds in pine forests in two largely disjunct regions: across the southeastern US west to eastern Oklahoma and Texas, and more patchily from New England and the Maritimes across southern Canada and the northern US to Minnesota and Manitoba. Nonbreeding individuals occur throughout the southeastern breeding range, and also farther south along the Gulf Coast (rarely to Mexico) and along the Atlantic Coastal Plain north to New England (Fink et al. 2023).

ORDER Passeriformes
FAMILY Parulidae
BREEDING HABITAT
Forests, Temperate Eastern
NONBREEDING HABITAT
Forests, Temperate Eastern
GLOBAL POPULATION SIZE 13,000,000
POPULATION CHANGE SINCE 1970 +70%
IUCN RED LIST CATEGORY Least Concern
R2R TIPPING POINT SPECIES No
US LISTING STATUS No
CANADA SPECIES AT RISK No
PIF CCS 8/20

MIGRATORY CONNECTIVITY

OVERALL Migratory connectivity is unknown. No information on migratory connectivity was found for this species.
MIGRATORY PATHWAYS Migrations are difficult to detect because of year-round occurrence in most of the range. Based on eBird data, northern breeders appear earlier than most other warblers in pre-breeding, and post-breeding migrants appear south of breeding areas in Florida and Texas in October and November (Fink et al. 2023). Nocturnal migrants have been detected (tower kills) in Florida from mid-September to mid-December, but rarely during pre-breeding migration (Rodewald et al. 2020). Regular occurrence on Bermuda in post-breeding and during the nonbreeding season indicates that some individuals make transoceanic crossings (Amos 1991).

YELLOW-RUMPED WARBLER *Setophaga coronata*

DISTRIBUTION AND ABUNDANCE

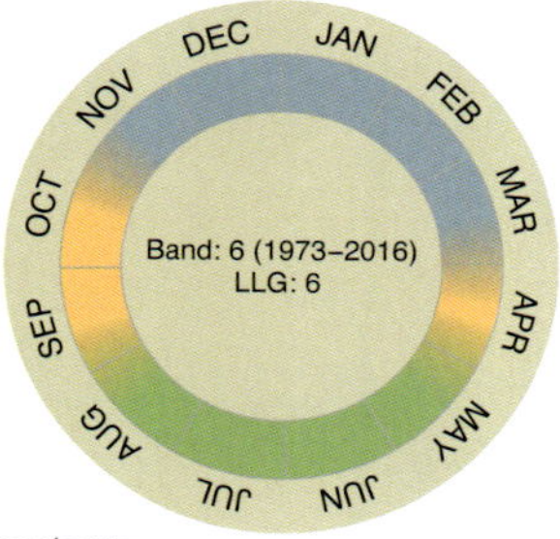

This species is among the most widespread North American warbler, breeding across the boreal and subarctic regions from Newfoundland to western Alaska, south to the northeastern US, and in the western mountains south to the Mexican border. Two distinct subspecies groups are recognized, with the "Myrtle" group (*S. c. coronata*) breeding across the entire northern and eastern parts of the range, and the "Audubon's" group (*S. c. auduboni*) breeding in mountains from central British Columbia and southern Alberta through the western US. Its nonbreeding distributions also distinct, with Myrtles migrating primarily to the southeastern US and irregularly south to the Caribbean, eastern Mexico, and Central America to Panama. Audubon's Warblers migrate primarily to western and central Mexico and Guatemala, with large numbers also in lowlands of the southwestern US and coastal California. Populations have been generally stable throughout the range but show steep recent declines in the northeastern US and Canadian Maritimes, as well as the Rocky Mountains region (Sauer et al. 2019; Hunt and Flaspohler 2020).

ORDER Passeriformes
FAMILY Parulidae
BREEDING HABITAT Forests, Boreal
NONBREEDING HABITAT Forests, Generalist
GLOBAL POPULATION SIZE 170,000,000
POPULATION CHANGE SINCE 1970 −5%
IUCN RED LIST CATEGORY Least Concern
R2R TIPPING POINT SPECIES No
US LISTING STATUS No
CANADA SPECIES AT RISK No
PIF CCS 8/20

MIGRATORY CONNECTIVITY

OVERALL Migratory connectivity is known from one geolocator tracking study (Szarmach et al. 2026), morphology and stable isotopes and is strong across the major subspecies groups, which maintain largely separate migration systems, and is also strong between high-latitude, northwestern-breeding myrtle warblers and those from the rest of the breeding range. Separate nonbreeding distributions between Pacific- and interior-breeding Audubon's warblers is also suggested by morphological analysis of specimens. Birds breeding in the Pacific Flyway primarily spent the nonbreeding season in the Midcontinent (55%) or Pacific (27%) Flyway. MC not calculated due to limited data.
BANDING The few band reencounters indicate that Audubon's warblers migrate to nonbreeding grounds along the Pacific coast of California and Mexico, while myrtle warblers primarily migrate to the southeastern United States during the nonbreeding season.
GEOLOCATOR TRACKING Of 30 multi-sensor geolocators deployed in Anchorage, Alaska, data from six geolocators recovered showed Myrtle warblers migrated to

GENERALIZED CONNECTIVITY

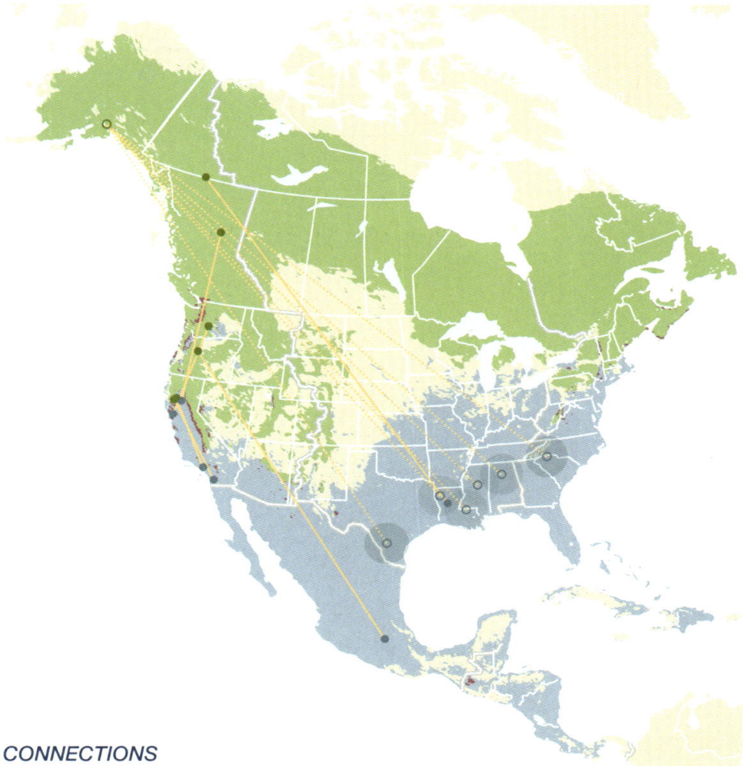

CONNECTIONS

nonbreeding areas in the southeastern United States, ranging from Texas to the Carolinas (Szarmach et al. 2026).

RADIO TRACKING Several hundred tags were deployed through the Motus Wildlife Tracking System. As of printing, there are no individual breeding to nonbreeding detections.

STABLE ISOTOPES Stable isotope analysis of individuals breeding near the hybrid zone (Myrtles crossed with Audubon's) in the Rocky Mountains of British Columbia and Alberta found that allopatric-breeding Audubon's and Myrtle warblers inhabit separate nonbreeding grounds. Most hybrid individuals spent the nonbreeding season in the southeastern US in the range of Myrtle warblers (Toews et al. 2014). Stable isotope analysis of Myrtle warblers migrating along the coast of British Columbia found that they were likely breeders in Alaska and the Yukon (Toews et al. 2017).

	Atlantic	Mexico	Midcontinent	Pacific
Pacific (n = 11)	9%	9%	55%	27%

Hydrogen isotope analysis showed that Myrtle warblers (n = 167) breeding in Alaska, British Columbia, and Alberta migrate partially through the Pacific to the Gulf Coast, with only a few reaching the Pacific Coast and no clear east–west divide (Szarmach et al. 2026).

MIGRATORY PATHWAYS This species migrates along a broad front across the continent, but complex movements and timing among subspecies are suggested from eBird animations. There is a large migration of Myrtle warblers along the Atlantic coast in post-breeding, whereas pre-breeding migration is more concentrated inland.

YELLOW-THROATED WARBLER

Setophaga dominica

DISTRIBUTION AND ABUNDANCE

This species breeds across the southeastern US north to New Jersey, southwestern Pennsylvania, southern Ohio, and southern Indiana, and west to eastern Kansas, Oklahoma, and Texas. Its nonbreeding range is primarily in Florida, the Bahamas, and the Greater Antilles (most abundant in Cuba), and also on the Gulf and Caribbean coasts from Mexico to northern Nicaragua, including the Yucatán Peninsula. Small numbers occur regularly during nonbreeding on the Pacific coast of Mexico, south to Costa Rica and western Panama, and along the US Gulf and South Atlantic coasts (Fink et al. 2023). Populations show long-term declines in Louisiana and Florida, but increases in the northern parts of the range (Sauer et al. 2019).

ORDER Passeriformes
FAMILY Parulidae
BREEDING HABITAT
Forests, Temperate Eastern
NONBREEDING HABITAT
Forests, Tropical Generalist
GLOBAL POPULATION SIZE 2,000,000

POPULATION CHANGE SINCE 1970 +53%
IUCN RED LIST CATEGORY Least Concern
R2R TIPPING POINT SPECIES No
US LISTING STATUS No
CANADA SPECIES AT RISK No
PIF CCS 10/20

MIGRATORY CONNECTIVITY

OVERALL Migratory connectivity is essentially unknown. Distribution of two historically recognized subspecies indicates strong east-west connectivity and a migratory divide. No band reencounter, tracking, or stable isotope data was found for this species.

GENETICS A mitochondrial DNA study from across the breeding range found no genetic differentiation (McKay 2009).

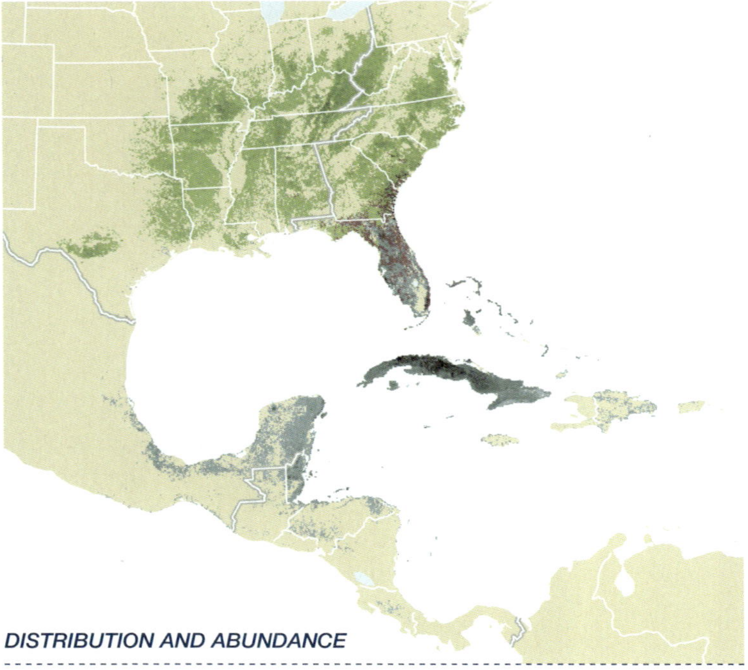

DISTRIBUTION AND ABUNDANCE

OTHER APPROACHES The historically recognized yellow-lored subspecies *S. d. dominica* breeds east of the Appalachians to the eastern Gulf states and Florida and migrates to the eastern part of the nonbreeding distribution in Florida and the Caribbean, whereas the white-lored *S. d. albilora* breeds to the west and migrates primarily to Mexico and Central America (American Ornithologists Union 1957; Dunn and Garrett 1997). Birds from both populations are found in western Cuba and the Yucatán Peninsula (Fink et al. 2023). Recent morphological and genetic analysis shows that variation is clinal, with little genetic differentiation (McKay 2008, 2009); however, those studies (based on small samples) do not negate the strong general pattern of east-west connectivity between breeding and nonbreeding populations (McKay and Hall 2020).
MIGRATORY PATHWAYS Migrations are very early, with evidence of trans-Gulf crossing in pre- and post-breeding (McKay and Hall 2020). Based on specimens (white- versus yellow-lored), there appears to be a migratory divide, with eastern populations migrating south along the Atlantic coast to Florida and the Caribbean, and western birds migrating through the Gulf states and Texas to Mexico (American Ornithologists Union 1957; Dunn and Garrett 1997).

PRAIRIE WARBLER *Setophaga discolor*

DISTRIBUTION AND ABUNDANCE

This species breeds across the southeastern US, north in the Appalachian Mountains and along the Atlantic Coastal Plain to southern New England; isolated breeding sites are north to Michigan and southern Ontario, and resident populations occur in mangroves of southern Florida. Its nonbreeding distribution is throughout the Caribbean and southern Florida, with small numbers along the Caribbean coast from Yucatán to Honduras (mostly on near-shore islands) and a few along the Gulf and South Atlantic coast to North Carolina (Fink et al. 2023).

ORDER Passeriformes
FAMILY Parulidae
BREEDING HABITAT
Forests, Temperate Eastern
NONBREEDING HABITAT
Islands, Terrestrial Habitats
GLOBAL POPULATION SIZE 3,600,000
POPULATION CHANGE SINCE 1970 −52%
IUCN RED LIST CATEGORY Least Concern
R2R TIPPING POINT SPECIES Yellow-Alert
US LISTING STATUS No
CANADA SPECIES AT RISK No
PIF CCS 14/20

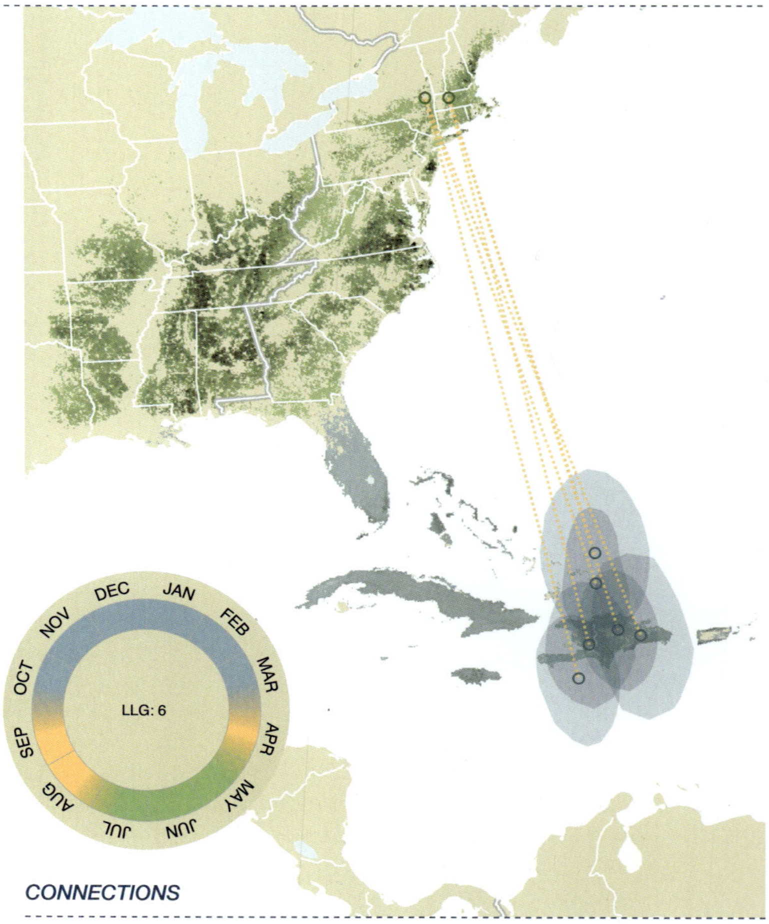

CONNECTIONS

MIGRATORY CONNECTIVITY

OVERALL There is limited data on migratory connectivity from geolocators and stable isotopes. MC is not calculated due to limited sample size, but migratory connectivity appears weak within our broad geographic regions. At a finer scale, the strength of migratory connectivity may increase as data show that individuals breeding in northern areas of the range migrated to Hispaniola, and nonbreeding individuals in Puerto Rico and the Bahamas bred in the southern portion of the range.
RADIO TRACKING There are a few projects tracking the Prairie Warbler through the Motus Wildlife Tracking System. There are no individual breeding to nonbreeding detections and a few detections during migration.
GEOLOCATOR TRACKING Individuals tracked with geolocators from breeding sites in western Massachusetts (n = 8) and eastern New York (n = 9) migrated to nonbreeding sites in the region of Hispaniola, though the location estimates were imprecise, with some birds likely spending the nonbreeding season in Haiti or the Dominican Republic

(subset of 6 individuals presented on the map; Akresh and King 2017; Campbell et al. 2021; Meehan et al. 2022; S.P. Campbell et al., unpublished data).

STABLE ISOTOPES Prairie Warblers sampled during the nonbreeding season in Puerto Rico were assigned to origins in an isotopic region encompassing the Gulf states and Florida (Schaffner 2014). Hobson et al. (2014) also found Puerto Rico birds likely originated from the southern US. Studds et al. (2021) found that most Prairie Warblers sampled throughout the Caribbean during the nonbreeding season were also assigned to the southern portion of the range. Additional Prairie Warblers captured in San Salvador Island, the Bahamas, were assigned to the southern US (M.E. Akresh and D.I. King, unpublished data).

OTHER APPROACHES The supposedly resident subspecies *D. d. paludicola*, which breeds in mangroves in southern Florida (Dunn and Garrett 1997; Nolan et al. 2020a), has been detected in Cuba, St. Croix (American Ornithologists Union 1957), and Puerto Rico (Schaffner 2014) in the nonbreeding season, suggesting at least some migration and possible connectivity in this population.

DATA CONTRIBUTORS

M.E. Akresh; S.P. Campbell; N.A. Gifford; D.I. King.

INTERESTING NOTE

Older males, spending the nonbreeding season in drier habitats, arrived later on the breeding grounds in Massachusetts, but experienced no reduction in breeding performance (Akresh et al. 2019). Juvenile males from nests that hatched later on the breeding grounds in Massachusetts occupied drier habitats on the nonbreeding grounds (Akresh et al. 2021).

MIGRATORY PATHWAYS This species migrates primarily through Florida and the southeastern US in pre- and post-breeding. It is thought to not be a frequent trans-Gulf migrant (Dunn and Garrett 1997), but higher abundance in the eastern Gulf states in pre-breeding and along the Atlantic coast in post-breeding suggests a clockwise loop migration (Fink et al. 2023). Regular occurrence on Bermuda in post- and pre-breeding also indicates transoceanic crossing by some individuals (Amos 1991).

Setophaga graciae GRACE'S WARBLER

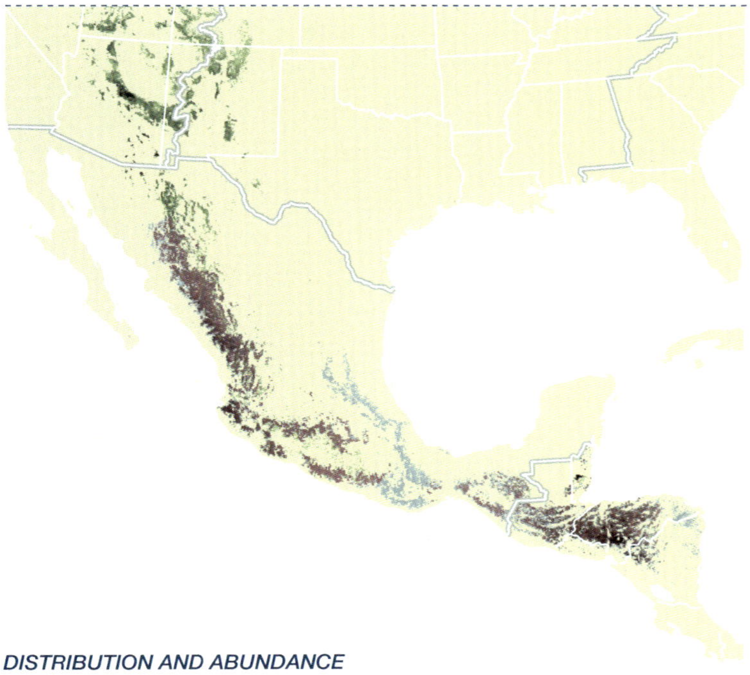

DISTRIBUTION AND ABUNDANCE

DISTRIBUTION AND ABUNDANCE

This species breeds in pine forests of the southern Rocky Mountains in southern Utah and southwestern Colorado and on the Mogollon Rim and the Sky Island mountains of southeastern Arizona and New Mexico; also through the mountains of western and southern Mexico and in the pine–oak forests and pine savannah of northern Central America. The US and northern Mexico populations are migratory, known in the nonbreeding season in the highlands of western Mexico from Sinaloa and Durango south to Oaxaca (Fink et al. 2023).

MIGRATORY CONNECTIVITY

OVERALL Migratory connectivity is unknown. No information on migratory connectivity was found for this species.

OTHER APPROACHES The migratory northern subspecies (*S. g. graciae*) breeds as far south as Sinaloa, but is known in the nonbreeding season from specimens collected in Nayarit, Morelos, and south to Oaxaca (Dunn and Garrett 1997; Staicer and Guzy 2020).

MIGRATORY PATHWAYS It is rarely detected on migration, but records in lowlands and foothills away from the breeding range in Sonora, Arizona, and New Mexico from late March to mid-May and from mid-August to early October suggest migration through those areas (Dunn and Garrett 1997; Staicer and Guzy 2020). Additional eBird records also suggest regular occurrence during migration along the Central Mexican Plateau and southern Sierra Madre Oriental from Queretaro to northern Veracruz (Fink et al. 2023).

ORDER Passeriformes
FAMILY Parulidae
BREEDING HABITAT Forests, Mesoamerican Pine-Oak
NONBREEDING HABITAT Forests, Mesoamerican Pine-Oak
GLOBAL POPULATION SIZE 3,000,000
POPULATION CHANGE SINCE 1970 −71%
IUCN RED LIST CATEGORY Least Concern
R2R TIPPING POINT SPECIES Yellow-Alert
US LISTING STATUS No
CANADA SPECIES AT RISK No
PIF CCS 14/20

BLACK-THROATED GRAY WARBLER *Setophaga nigrescens*

DISTRIBUTION AND ABUNDANCE

It breeds in oak–conifer forests and piñon–juniper woodlands from British Columbia to California and in the Great Basin and Rocky Mountains. Its nonbreeding range is primarily on the Pacific Slope and Sierra Madre Occidental of Mexico, with smaller numbers in Mexico's central highlands and a few north to the southwestern US (Fink et al. 2023). Pacific Northwest and Sierra Nevada populations show long-term declines becoming steeper in recent decades, whereas Great Basin and Rocky Mountain populations have been mostly stable (Sauer et al. 2019).

ORDER Passeriformes
FAMILY Parulidae
BREEDING HABITAT Forests, Temperate Western
NONBREEDING HABITAT Forests, Tropical Dry
GLOBAL POPULATION SIZE 3,200,000
POPULATION CHANGE SINCE 1970 −44%
IUCN RED LIST CATEGORY Least Concern
R2R TIPPING POINT SPECIES No
US LISTING STATUS No
CANADA SPECIES AT RISK No
PIF CCS 13/20

DISTRIBUTION AND ABUNDANCE

MIGRATORY CONNECTIVITY

OVERALL Migratory connectivity is unknown. No information on migratory connectivity was found for this species.
GENETICS Historically, two subspecies were recognized, and breeders in Washington differed genetically from those in Arizona (Bermingham et al. 1992); birds east and west of the Cascades also differ vocally (Morrison 1990). Phillips et al. (1964) assigned most migrants and nonbreeding specimens from southern and western Arizona to the Pacific Northwest population, *S. n. nigrescens*; but no assessment has been done elsewhere in the nonbreeding range, so connectivity has not been established.
MIGRATORY PATHWAYS It migrates broadly through the southwestern US and is one of the few western warblers to be equally common in lowland desert regions in both pre- and post-breeding (Dunn and Garrett 1997).

TOWNSEND'S WARBLER *Setophaga townsendi*

DISTRIBUTION AND ABUNDANCE

This species breeds in the Pacific Northwest from central Washington to coastal Alaska and in the northern Rocky Mountains from Idaho to central Alaska. Its nonbreeding range is in two disjunct regions; the majority of the species occurs in Mexico and northern Central America, and sparsely to Costa Rica (rarely western Panama). A second population occurs along the Pacific coast, centered in central California (Fink et al. 2023).

ORDER Passeriformes
FAMILY Parulidae
BREEDING HABITAT Forests, Temperate Western
NONBREEDING HABITAT Forests, Mesoamerican Pine-Oak
GLOBAL POPULATION SIZE 21,000,000
POPULATION CHANGE SINCE 1970 −30%
IUCN RED LIST CATEGORY Least Concern
R2R TIPPING POINT SPECIES No
US LISTING STATUS No
CANADA SPECIES AT RISK No
PIF CCS 11/20

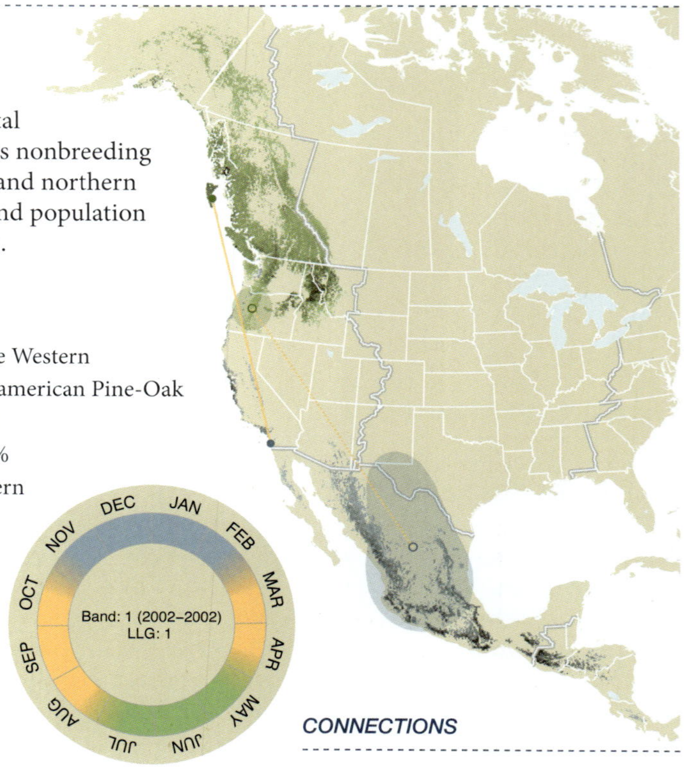

CONNECTIONS

MIGRATORY CONNECTIVITY

OVERALL Migratory connectivity is unknown. Limited information on connectivity is based on a few band returns, geolocators from a single site (H. Kim, unpublished data), and a morphological study (Morrison 1983). They reveal that the coastal nonbreeding population originates in the Queen Charlotte Islands, whereas the remainder of the species migrates to Mexico and Central America.
BANDING A single breeding to nonbreeding band reencounter, from a breeding site on Queen Charlotte Islands to a nonbreeding site in coastal Southern California, is consistent with the morphological analysis.
GEOLOCATOR TRACKING Data from a single recovered geolocator deployed on the breeding grounds in Oregon found the bird migrating to central Mexico in the nonbreeding season (H. Kim, unpublished data).

OTHER APPROACHES An analysis of museum specimens found that nonbreeding birds on the Pacific coast had significantly shorter wings than birds in Mexico and Central America. The shorter-winged individuals were assigned to breeders on Queen Charlotte Islands, British Columbia, and possibly other coastal archipelagos of British Columbia and southeastern Alaska, suggesting a migratory divide and connectivity between coastal and inland populations (Morrison 1983).

MIGRATORY PATHWAYS Aside from birds spending the nonbreeding season on the coast, the Townsend's Warbler exhibits a clockwise loop migration, with pre-breeding migration primarily across southern Arizona and lowland California regions and post-breeding migration primarily in the mountains east to Wyoming, Colorado, New Mexico, and western Texas (Dunn and Garrett 1997; Fink et al. 2023).

DATA CONTRIBUTORS
M. Betts; H. Kim; B. McComb.

HERMIT WARBLER

Setophaga occidentalis

DISTRIBUTION AND ABUNDANCE

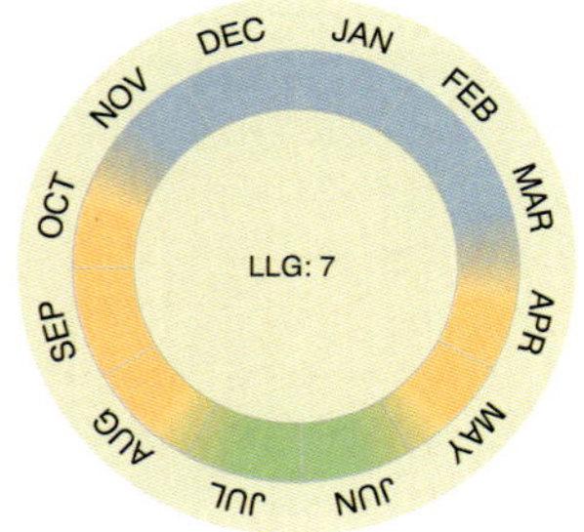

The species breeds in conifer forests of the Pacific Coast Ranges, and the Cascade and Sierra Nevada mountains from central Washington to central California. It spends the nonbreeding season in pine and pine–oak forests in mountains from Sinaloa and Neuvo Leon, Mexico, to northern Nicaragua and sparsely in the highlands of Costa Rica (Fink et al. 2023). Long-term populations have been stable in most of the range, but declines have occurred in the Sierra Nevada, especially since the 1980s.

ORDER Passeriformes
FAMILY Parulidae
BREEDING HABITAT Forests, Temperate Western
NONBREEDING HABITAT Forests, Mesoamerican Pine-Oak
GLOBAL POPULATION SIZE 2,500,000
POPULATION CHANGE SINCE 1970 −12%
IUCN RED LIST CATEGORY Least Concern
R2R TIPPING POINT SPECIES No
US LISTING STATUS No
CANADA SPECIES AT RISK No
PIF CCS 13/20

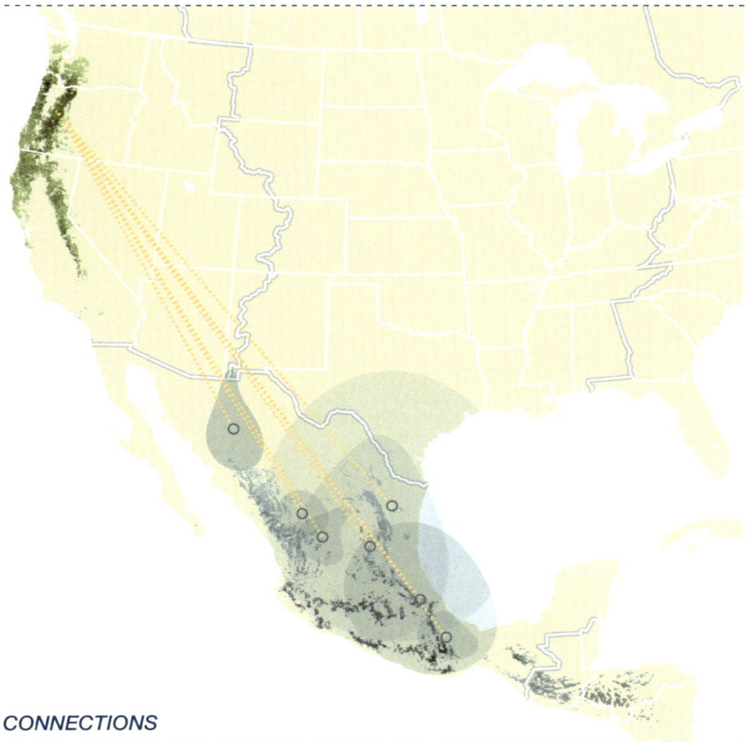

CONNECTIONS

MIGRATORY CONNECTIVITY

OVERVIEW Migratory connectivity is largely unknown and based on a single light-level geolocator study from a small part of the breeding range (Kim et al. 2024).

GEOLOCATOR TRACKING A single study (data from 2018–2019 only shown on map) from eastern Oregon and northeastern California (2018–2022; n = 22 full tracks) found that all individuals spend the nonbreeding season in Mexico, suggesting weak connectivity, but this study alone does not provide information on the strength of migratory connectivity (Kim et al. 2024). However, birds breeding at the southernmost breeding location migrated to nonbreeding areas farther east and south than birds from more northern breeding locations, providing weak evidence for potential chain migration (Kim et al. 2024).

OTHER APPROACHES No banding, radio or satellite tracking, genetic, or stable isotope studies of migratory connectivity were found for this species.

MIGRATORY PATHWAYS This species appears to exhibit a clockwise loop migration (La Sorte et al. 2016), with pre-breeding migration primarily through the Sonoran Desert, northern Baja California, and California lowlands, and post-breeding migration more through higher elevations to the east in southwestern Utah, Arizona, and New Mexico (Dunn and Garrett 1997; Fink et al. 2023). Birds showed shorter and faster migratory movements during pre-breeding migration than post-breeding migration (Kim et al. 2024).

DATA CONTRIBUTORS
M. Betts; H. Kim; B. McComb.

GOLDEN-CHEEKED WARBLER *Setophaga chrysoparia*

DISTRIBUTION AND ABUNDANCE

The very small breeding distribution is restricted to the ash juniper–oak woodlands on and around the Edwards Plateau in central Texas. The nonbreeding range is in humid pine–oak forest in the highlands from southern Mexico (Chiapas) through Guatemala, El Salvador, Honduras, and northern Nicaragua (Ladd and Gass 2020; Fink et al. 2023).

ORDER Passeriformes
FAMILY Parulidae
BREEDING HABITAT Forests, Temperate Western
NONBREEDING HABITAT Forests, Mesoamerican Pine-Oak
GLOBAL POPULATION SIZE 110,000
POPULATION CHANGE SINCE 1970 NA
IUCN RED LIST CATEGORY Endangered
R2R TIPPING POINT SPECIES Red-Alert
US LISTING STATUS No
CANADA SPECIES AT RISK No
PIF CCS 18/20

DISTRIBUTION AND ABUNDANCE

MIGRATORY CONNECTIVITY

OVERALL Migratory connectivity is unknown. No information on migratory connectivity was found for this species. Numerous tags were deployed through the Motus Wildlife Tracking System. As of printing, there are no detections between breeding and nonbreeding areas.

MIGRATION PATHWAYS Migration is through the mountains of eastern Mexico and is much earlier in both pre- and post-breeding (March; July–August) than for most other warblers (Ladd and Gass 2020).

BLACK-THROATED GREEN WARBLER *Setophaga virens*

DISTRIBUTION AND ABUNDANCE

This species breeds in boreal and northern hardwood forests in the northeastern US and eastern Canada west in decreasing abundance to northeastern British Columbia and south through the Appalachian Mountains. Its nonbreeding range is mostly from eastern and southern Mexico (including the Yucatán Peninsula) south through the Central American highlands to western Panama. Smaller numbers occur in the Greater Antilles, the Bahamas, and southern Florida (Fink et al. 2023). Long-term population trends have been mostly stable (Rosenberg et al. 2019), but with steep recent declines apparent in many eastern populations (Sauer et al. 2019).

ORDER Passeriformes
FAMILY Parulidae
BREEDING HABITAT Forests, Boreal
NONBREEDING HABITAT Forests, Tropical Montane Evergreen
GLOBAL POPULATION SIZE 9,200,000
POPULATION CHANGE SINCE 1970 +25%
IUCN RED LIST CATEGORY Least Concern
R2R TIPPING POINT SPECIES No
US LISTING STATUS No
CANADA SPECIES AT RISK No
PIF CCS 11/20

DISTRIBUTION AND ABUNDANCE

MIGRATORY CONNECTIVITY

OVERALL Migratory connectivity is unknown. No information on migratory connectivity was found for this species. A few tags were deployed through the Motus Wildlife Tracking System, but there are no breeding to nonbreeding detections.

MIGRATORY PATHWAYS It is a trans-Gulf migrant in pre- and post-breeding, although it is also believed to migrate over land through eastern Mexico (Morse and Poole 2020). It exhibits a clockwise loop migration based on eBird data (La Sorte et al. 2016), with migrants more common through Texas and the Great Plains in pre-breeding, and through the eastern Gulf states and Florida in post-breeding (Dunn and Garrett 1997). The bulk of migration is through and west of the Appalachians; it is relatively rare along the South Atlantic coast, although regular post-breeding occurrence in Bermuda (Amos 1991) suggests some transoceanic crossing.

INTERESTING NOTE

The weakly differentiated subspecies *S. v. waynei* arrives and breeds earlier in its disjunct distribution along the mid-Atlantic coast; it is known in the nonbreeding season only from Cuba (American Ornithologists Union 1957; Dunn and Garrett 1997).

Cardellina canadensis CANADA WARBLER

DISTRIBUTION AND ABUNDANCE

This species breeds in boreal and northern hardwood forests in the northeastern US, the Appalachian Mountains, and across southern Canada in decreasing abundance from the Maritime provinces to Alberta. It spends the nonbreeding season at lower and mid-elevations of the Andes from Colombia and northwestern Venezuela to northern Bolivia; it occurs in smaller numbers in the nonbreeding season in montane areas in southern Central America (Fink et al. 2023). Declines are apparent throughout the range but are steepest in the northern Atlantic Forest and New England / mid-Atlantic coast regions (Sauer et al. 2019).

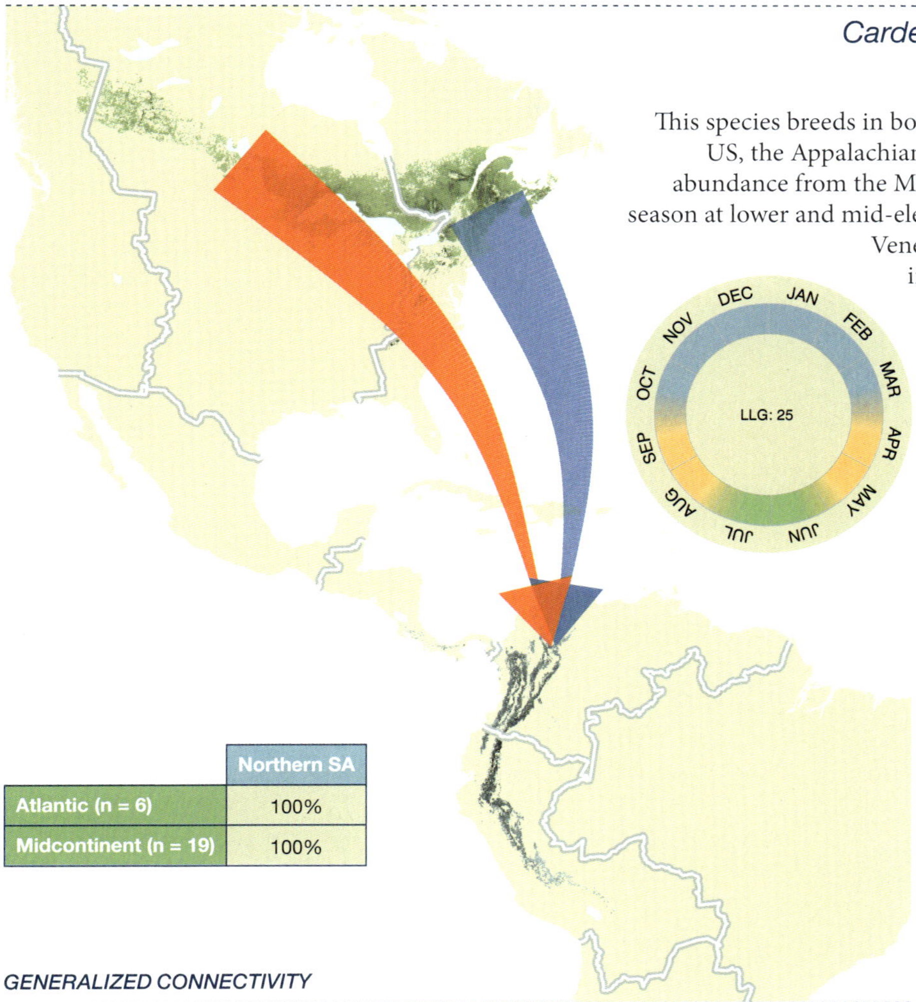

	Northern SA
Atlantic (n = 6)	100%
Midcontinent (n = 19)	100%

GENERALIZED CONNECTIVITY

ORDER Passeriformes
FAMILY Parulidae
BREEDING HABITAT Forests, Boreal
NONBREEDING HABITAT Forests, Tropical Montane Evergreen
GLOBAL POPULATION SIZE 2,600,000
POPULATION CHANGE SINCE 1970 −65%
IUCN RED LIST CATEGORY Least Concern
R2R TIPPING POINT SPECIES No
US LISTING STATUS No
CANADA SPECIES AT RISK Threatened
PIF CCS 13/20

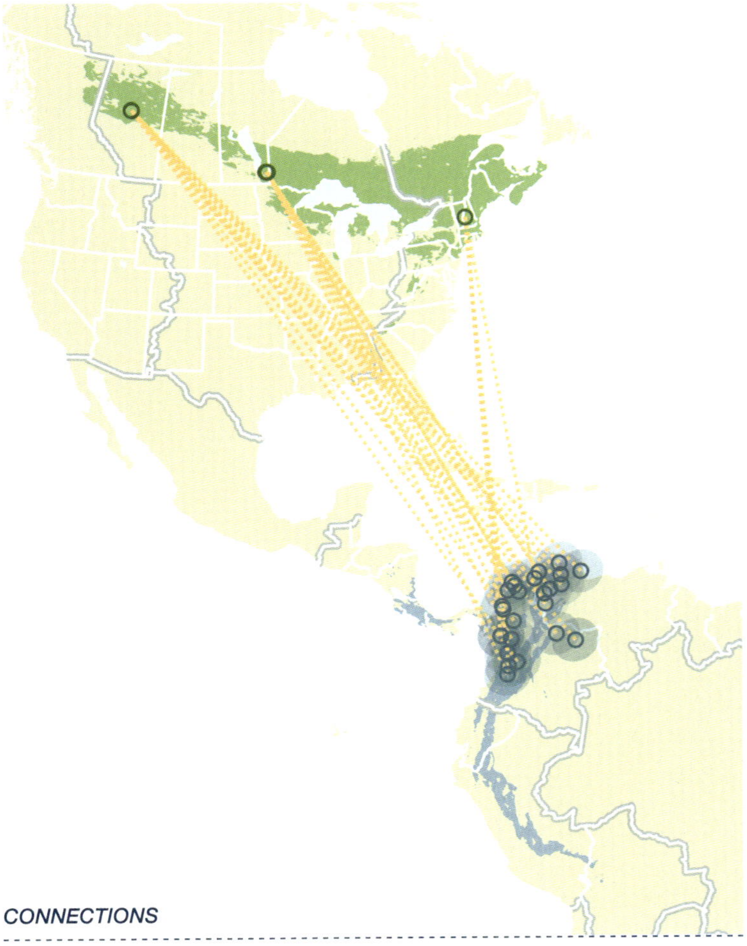

CONNECTIONS

MIGRATORY CONNECTIVITY

OVERALL At the broad regional scale, migratory connectivity is weak, with birds from Atlantic and Midcontinent flyways breeding populations converging on the Andes of northern South America. MC is not calculated due to limited sample size.

RADIO TRACKING Many Motus tags were deployed in the eastern part of the range, as well as in Colombia. There are limited detections of individual birds between breeding and nonbreeding areas.

GEOLOCATOR TRACKING Recovery of 25 light-level geolocators from across the northern breeding range (Alberta, Manitoba, New Hampshire) indicate considerable overlap in nonbreeding sites. Birds from all three sites spend the nonbreeding season in both Venezuela and northern Colombia, and birds from New Hampshire in Colombia (Roberto-Charron 2018; Roberto-Charron et al. 2020). Alberta breeders spent the nonbreeding season farther east than New Hampshire breeders, suggesting a crossing rather than parallel migration pattern, and contradicting results from stable isotopes in tail feathers in Colombia.

STABLE ISOTOPES Analysis of stable hydrogen isotopes from feathers indicated moderately strong east-west migratory connectivity between breeding and nonbreeding grounds for individuals in the three parallel ranges of the Colombian Andes (Gonzalez-Prieto et al. 2017). In contrast to the light-level geolocator results, stable isotopes suggested that birds that spent the nonbreeding season on the east slope of the Eastern Andes originated in the northeastern part of the breeding range, whereas individuals in the Western Andes (Cauca Valley) were from central Canada. In another stable isotope study, individuals migrating

through northern Colombia were found to originate in central and northeastern regions of the breeding range (Cardenas-Ortiz et al. 2017).
GENETICS There are no genetic studies of migratory connectivity, but three genetically distinct breeding populations have been identified—eastern, northwestern, and southern (Miller et al. 2023). Further sampling from the nonbreeding grounds will help better understand the migratory connectivity of this species (https://www.birdgenoscape.org/canada-warbler-2/).
OTHER APPROACHES No breeding to nonbreeding band reencounter data were found for this species.
MIGRATORY PATHWAYS Unlike many other boreal migrants, Canada Warblers do not follow a loop migration route (La Sorte et al. 2016) but rather travel primarily along the western Gulf of Mexico and the Central American mainland in both post- and pre-breeding. Populations from across the breeding range (Alberta to New Hampshire) converge over most of the migration route and especially in the narrow Darién region of Panama (Roberto-Charron et al. 2020; Cardenas-Ortiz et al. 2017). Based on geolocator tracks, New Hampshire birds migrated through the eastern US, and western Canadian breeders migrated through the central US, but all populations converged along the western Gulf of Mexico and in Central America. One-third of individuals made limited crossings of the western Gulf to the Yucatán Peninsula during post-breeding migration (Roberto-Charron et al. 2020).

DATA CONTRIBUTORS
K.C. Fraser; A. Roberto-Charron.

WILSON'S WARBLER *Cardellina pusilla*

DISTRIBUTION AND ABUNDANCE

This species is a widespread breeder across the boreal region from Newfoundland to western Alaska, and southward along the Pacific coast to Southern California and the Sierra Nevada, and in the Rocky Mountains to Colorado and northern New Mexico. It spends the nonbreeding season in high densities throughout most of Mexico (except on the Yucatán Peninsula) and northern Central America, in the mountains of Costa Rica and western Panama, and in low numbers in the southern US. Populations have declined in the eastern boreal and northern Atlantic Forest, as well as in the entire western range from Alaska to California, and especially in the Rocky Mountains (Sauer et al. 2019).

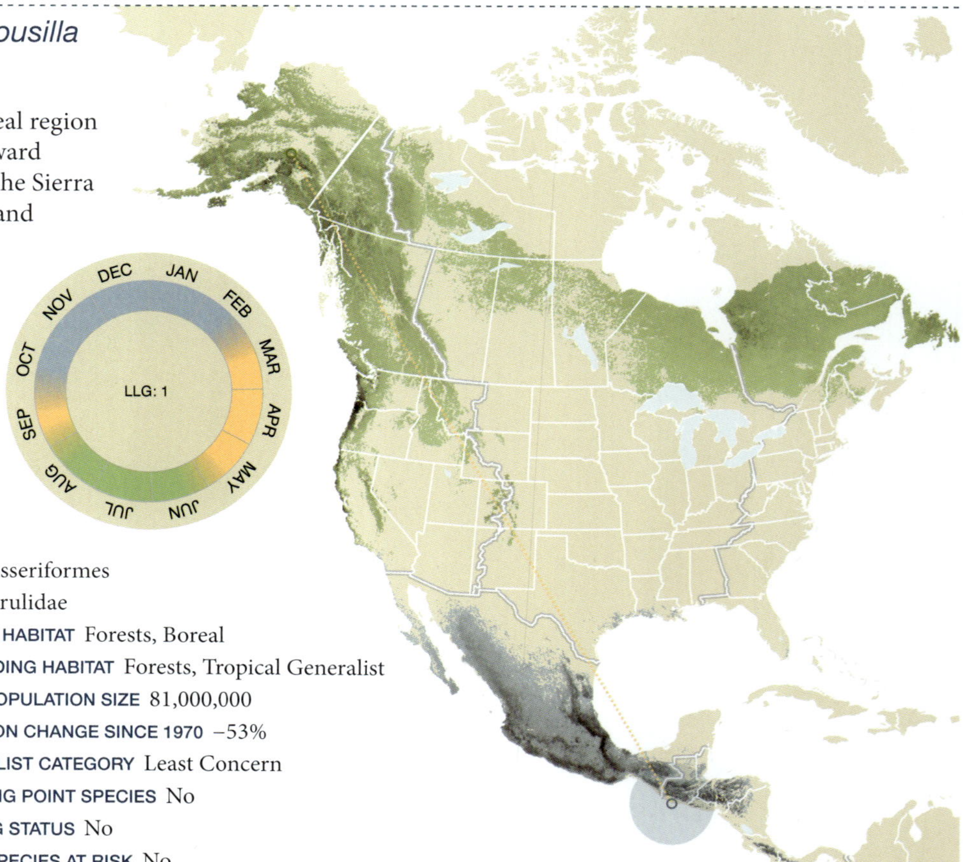

ORDER Passeriformes
FAMILY Parulidae
BREEDING HABITAT Forests, Boreal
NONBREEDING HABITAT Forests, Tropical Generalist
GLOBAL POPULATION SIZE 81,000,000
POPULATION CHANGE SINCE 1970 –53%
IUCN RED LIST CATEGORY Least Concern
R2R TIPPING POINT SPECIES No
US LISTING STATUS No
CANADA SPECIES AT RISK No
PIF CCS 9/20

CONNECTIONS

MIGRATORY CONNECTIVITY

OVERALL Limited data prevent estimates of MC, but moderate to strong connectivity is suggested from geolocator data and studies of genetic markers, stable isotopes, and subspecific variation in museum specimens. Overall, Pacific coast and Sierra Nevada breeders migrate to the Pacific Slope of Mexico and Central America; Alaska and Rocky Mountain breeders migrate to the highlands of Mexico and Central America; and boreal-Canadian breeders occupy only a small portion of the nonbreeding range in southeastern Mexico and northern Central America.
GEOLOCATOR TRACKING One bird was tracked from its breeding grounds in Denali National Park to its nonbreeding grounds near the border of Mexico and Guatemala (National Park Service, Critical Connections Program, unpublished data).
STABLE ISOTOPES Western individuals undergo a leapfrog migration pattern, with the northernmost breeders migrating earlier in post-breeding and spending the nonbreeding season at the southern end of the distribution in Central America (Kelly et al. 2002; Clegg et al. 2003). Pacific coastal birds spent the nonbreeding season in western Mexico, while inland and high-elevation individuals went farther east and south (Clegg et al. 2003).
GENETICS Genetic analysis of DNA variation has revealed strong differentiation between western and eastern breeding populations (Kimura et al. 2002; Clegg et al. 2003; Irwin et al. 2011) and has found that eastern breeders occupy a relatively small portion of the nonbreeding distribution, primarily in southeastern Mexico and northern Central America (Ruegg 2014). Further geographic differentiation among western groups indicates five distinct populations exhibiting moderately strong connectivity with nonbreeding sites (Ruegg et al. 2014). Nonbreeding individuals in western Mexico, including Baja California, originated from Pacific

coastal and Sierra Nevada populations; birds from the southern Rocky Mountains were found only at the southern end of the nonbreeding distribution in Central America; and birds from Alaska and the Northern Rockies were found throughout most of the nonbreeding distribution from southern Mexico to Costa Rica.

OTHER APPROACHES Integrated modeling of stable isotopes and genetic markers corroborated and strengthened information from single markers (Rundel et al. 2013). No radio or satellite tracking studies of migratory connectivity were found for this species. Analysis of the three traditionally recognized subspecies suggests largely separate nonbreeding distributions (Dunn and Garrett 1997), with the boreal-Canadian breeding *C. p. pusilla* migrating primarily to eastern Mexico and the Caribbean slope of Central America, *C. p. pileolata* breeding from Alaska through the Rocky Mountains and migrating mainly to the highlands from central Mexico to western Panama, and the Pacific-breeding *C. p. chryseola* migrating mainly to dry forests of western Mexico and northern Central America. The limits of the nonbreeding distribution and the degree of overlap are difficult to discern from these analyses; for example, specimens of all three subspecies have been identified from the Chiriqui Highlands of western Panama, although *C. p. pileolata* is the most frequent nonbreeding bird (Wetmore et al. 1984).

MIGRATORY PATHWAYS Different subspecies and populations have been shown to follow different migratory pathways and to differ in timing of migration (Ammon and Gilbert 2020; Dunn and Garrett 1997; Paxton et al. 2007, 2013; Ruegg et al. 2014; Fink et al. 2023). Pacific breeding populations (*C. p. chryseola*) migrate the earliest and follow the Pacific coast and arid lowlands of the southwestern US and northwestern Mexico in pre- and post-breeding. Alaskan and Rocky Mountain breeders (*C. p. pileolata*) migrate later and occur throughout the interior western US, including the western Great Plains. Eastern boreal breeding birds migrate (*C. p. pusilla*) even later than western birds, follow a pathway around the Gulf, and stay mainly west of the Appalachian Mountains in pre- and post-breeding.

DATA CONTRIBUTORS

E.M. Adams; J. Boelsma; A. Gilbert; G. Gress; C. Grigsby; C. McIntyre; L.M. Phillips; I.J. Stenhouse; S. Weidensaul; E.J. Williams.

Cardellina rubrifrons RED-FACED WARBLER

DISTRIBUTION AND ABUNDANCE

DISTRIBUTION AND ABUNDANCE

This species breeds from central Arizona and New Mexico south to northern Sinaloa and Durango in the Sierra Madre Occidental of Mexico. The nonbreeding distribution is from the southern limit of the breeding range south and east through the mountains of Mexico and patchily in highlands to western Honduras and El Salvador (Martin and Barber 2020; Fink et al. 2023).

ORDER Passeriformes
FAMILY Parulidae
BREEDING HABITAT Forests, Mesoamerican Pine-Oak
NONBREEDING HABITAT Forests, Mesoamerican Pine-Oak
GLOBAL POPULATION SIZE 350,000
POPULATION CHANGE SINCE 1970 −38%
IUCN RED LIST CATEGORY Least Concern
R2R TIPPING POINT SPECIES No
US LISTING STATUS No
CANADA SPECIES AT RISK No
PIF CCS 15/20

MIGRATORY CONNECTIVITY

OVERALL Migratory connectivity is unknown. No information on migratory connectivity was found for this species.

MIGRATORY PATHWAYS This species is almost entirely migratory, with migrations occurring primarily within breeding and nonbreeding ranges;
it is very rarely detected away from mountains. Migrations are also early; birds arrive in most of the breeding range by mid-April and fill in most of the nonbreeding distribution by mid-August (Martin and Barber 2020; Fink et al. 2023).

PAINTED REDSTART *Myoborus pictus*

DISTRIBUTION AND ABUNDANCE

This species breeds in the US in pine–oak forests and canyons of the Mogollon Rim and Sky Island mountains of Arizona and New Mexico, with small numbers in mountains of western Texas. It is mostly migratory within the US, with only a few remaining in southern Arizona in the nonbreeding season. Also, resident populations occur in mountains throughout Mexico and in Central America south to northern Nicaragua (Fink et al. 2023).

ORDER Passeriformes
FAMILY Parulidae
BREEDING HABITAT Forests, Mesoamerican Pine-Oak
NONBREEDING HABITAT Forests, Mesoamerican Pine-Oak
GLOBAL POPULATION SIZE 600,000
POPULATION CHANGE SINCE 1970 NA
IUCN RED LIST CATEGORY Least Concern
R2R TIPPING POINT SPECIES No
US LISTING STATUS No
CANADA SPECIES AT RISK No
PIF CCS 13/20

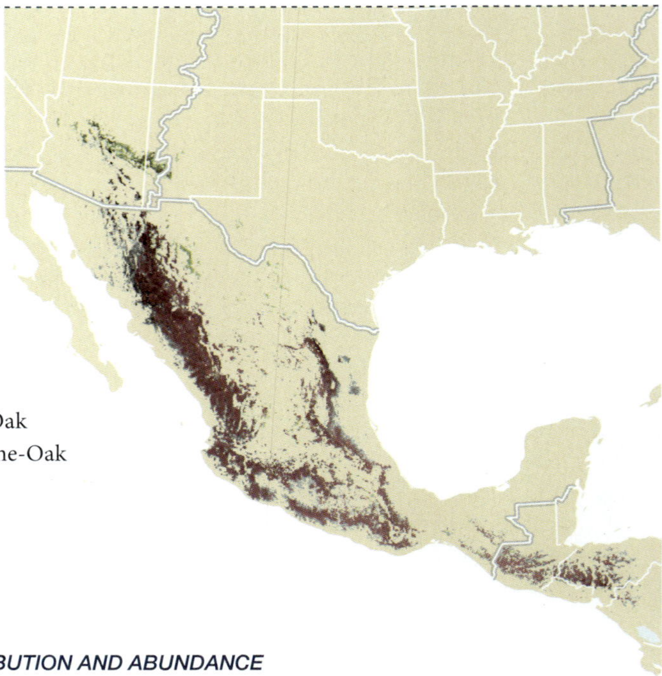

DISTRIBUTION AND ABUNDANCE

MIGRATORY CONNECTIVITY

OVERALL Migratory connectivity is unknown. No information on migratory connectivity was found for this species.

MIGRATORY PATHWAYS Although this species is rarely recorded away from breeding sites during migration, arrival on US breeding grounds occurs primarily in late March, and most lowland migrants and strays have been noted in March, April, and September (Dunn and Garrett 1997; Barber et al. 2020). eBird data indicate a withdrawal from breeding areas in central Arizona, New Mexico, and Texas by late September, coinciding with the regular occurrence of migrants in Southern California, southern Texas, and across the Central Mexican Plateau and Pacific lowlands of Mexico (Fink et al. 2023).

CARDINALIDAE | *TANAGERS, GROSBEAKS, AND BUNTINGS*

HEPATIC TANAGER *Piranga flava*

DISTRIBUTION AND ABUNDANCE

The Hepatic Tanager is the most widely distributed species in its genus, breeding from southwestern US mountains south to Uruguay and northern Argentina (Eddleman 2020). It is a permanent resident over most of its southern range, but most US birds migrate south out of the country during the nonbreeding season (Fink et al. 2023).

ORDER Passeriformes
FAMILY Cardinalidae
BREEDING HABITAT
Forests, Mesoamerican Pine-Oak
NONBREEDING HABITAT
Forests, Mesoamerican Pine-Oak
GLOBAL POPULATION SIZE 7,600,000
POPULATION CHANGE SINCE 1970 +254%
IUCN RED LIST CATEGORY Least Concern
R2R TIPPING POINT SPECIES No
US LISTING STATUS No
CANADA SPECIES AT RISK No
PIF CCS 7/20

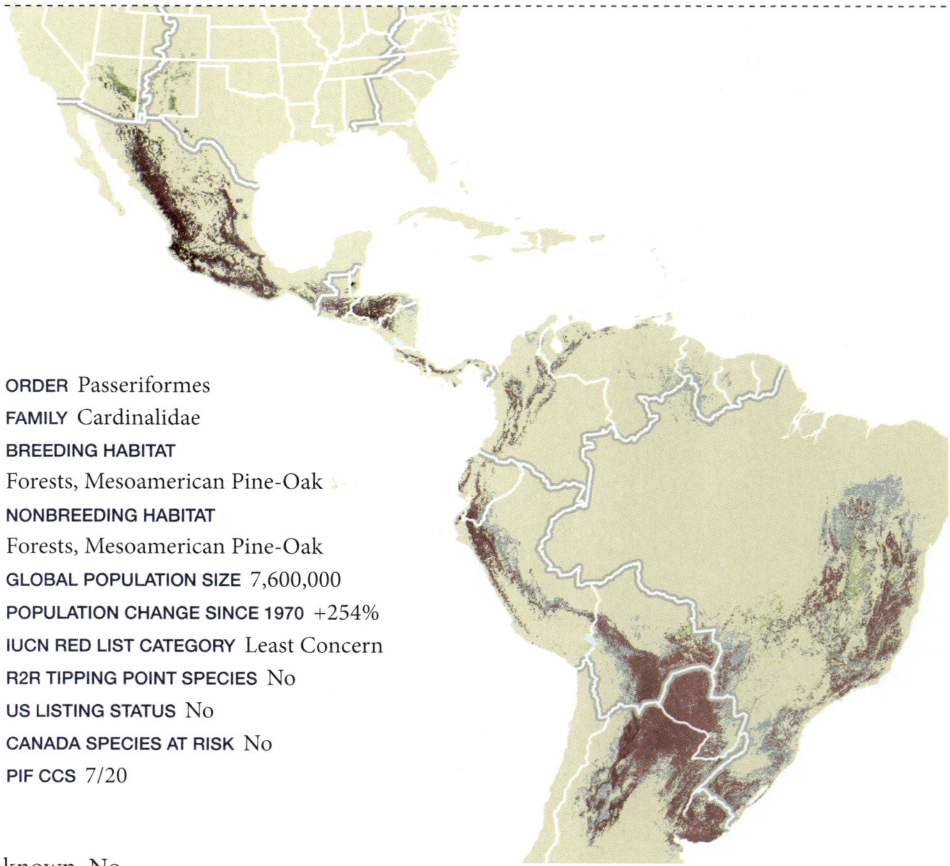

DISTRIBUTION AND ABUNDANCE

MIGRATORY CONNECTIVITY

OVERALL Migratory connectivity for the species is unknown. No information on migratory connectivity was found for this species.

Piranga rubra SUMMER TANAGER

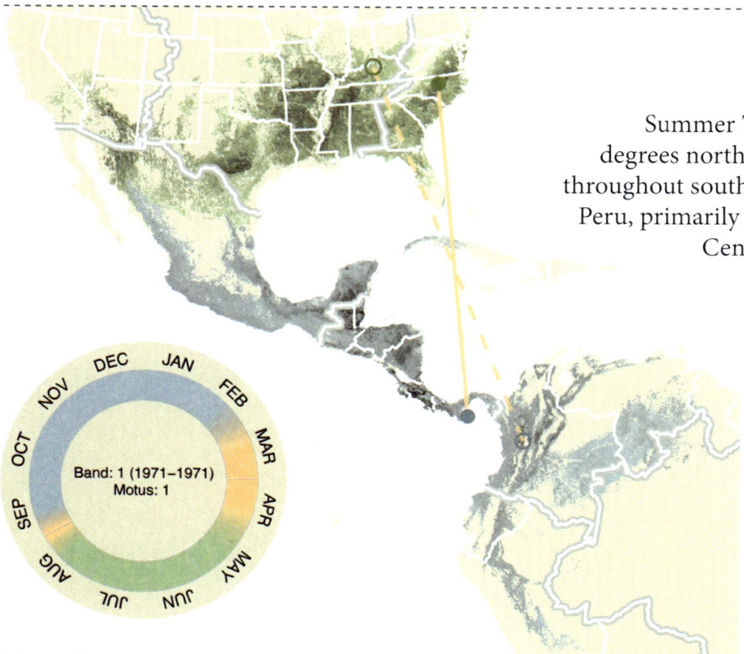

CONNECTIONS

DISTRIBUTION AND ABUNDANCE

Summer Tanagers breed throughout the southern United States below latitude 44 degrees north, and into northern Mexico. The species spends the nonbreeding season throughout southern Mexico, Central America, and in South America from Colombia to Peru, primarily associated with habitats at low and mid-elevations of the Andes and the Central American mountain ranges (W.D. Robinson 2020; Fink et al. 2023).

ORDER Passeriformes
FAMILY Cardinalidae
BREEDING HABITAT
Forests, Temperate Eastern
NONBREEDING HABITAT
Forests, Tropical Generalist
GLOBAL POPULATION SIZE 12,000,000
POPULATION CHANGE SINCE 1970 +15%
IUCN RED LIST CATEGORY Least Concern
R2R TIPPING POINT SPECIES No
US LISTING STATUS No
CANADA SPECIES AT RISK No
PIF CCS 8/20

MIGRATORY CONNECTIVITY

OVERALL Migratory connectivity is unknown. Little information is available on the migratory connectivity of the Summer Tanager.
BANDING There is one record connecting a bird breeding in North Carolina to a nonbreeding location in Panama.
RADIO TRACKING The Summer Tanager is currently being tracked with the Motus Wildlife Tracking System from coffee landscapes in Colombia. One bird was detected and resighted on its breeding grounds in Kentucky (R. Bennett, unpublished data).
STABLE ISOTOPES Stable isotope data suggest connectivity may be low, given that individuals from throughout the breeding range converged during post-breeding migration in the Darién province of Colombia (n = 34; Cardenas-Ortiz et al. 2020), but more study is needed.
OTHER APPROACHES No geolocator tracking or genetic studies of migratory connectivity were found for this species.
MIGRATORY PATHWAYS Eastern and central breeding populations converge during post-breeding migration on the Gulf Coast between southern Texas and northern Florida, before they cross the Gulf of Mexico to Central America and northern South America. Western populations seem to migrate directly south through Mexico and Central America (W.D. Robinson 2020).

DATA CONTRIBUTORS
N. Bayly; R.E. Bennett; C. Gómez; N.W. Cooper.

Piranga olivacea SCARLET TANAGER

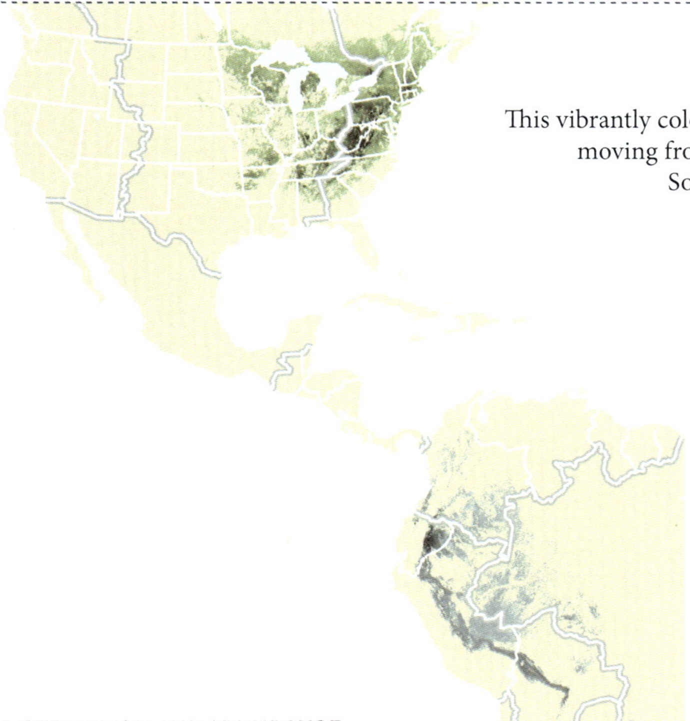
DISTRIBUTION AND ABUNDANCE

DISTRIBUTION AND ABUNDANCE

This vibrantly colored species undertakes the longest migration of any of the *Piranga* tanagers, moving from its eastern US breeding grounds to nonbreeding grounds in northwestern South America, mostly along the eastern slope of the Andes (Mowbray 2020c).

ORDER Passeriformes
FAMILY Cardinalidae
BREEDING HABITAT
Forests, Temperate Eastern
NONBREEDING HABITAT
Forests, Tropical Lowland Evergreen
GLOBAL POPULATION SIZE 2,600,000
POPULATION CHANGE SINCE 1970 −11%
IUCN RED LIST CATEGORY Least Concern
R2R TIPPING POINT SPECIES No
US LISTING STATUS No
CANADA SPECIES AT RISK No
PIF CCS 12/20

MIGRATORY CONNECTIVITY

OVERALL Migratory connectivity is unknown. There is one study of migratory connectivity with limited data; most of our knowledge of this species' movements is based on sighting records.

RADIO TRACKING The Scarlet Tanager is currently being tracked with the Motus Wildlife Tracking System. As of printing, there are no breeding to nonbreeding detections.

GEOLOCATOR TRACKING In 2014, 15 light-level geolocators were deployed on male Scarlet Tanagers in Pennsylvania. Five of the 15 tagged individuals were resighted, but only one bird was recaptured. The bird used two disjunct nonbreeding areas in South America, one near the border of Peru, Bolivia, and Brazil, and one in northern Peru near the border of Colombia and Ecuador (Raybuck et al. 2024). Data are not presented in the Atlas.

OTHER APPROACHES No banding, genetic, or stable isotope studies of migratory connectivity were found for this species.

MIGRATORY PATHWAYS This species is a trans-Gulf migrant; in pre-breeding, birds depart their nonbreeding range and travel to northern Central America and Mexico, especially the Yucatán Peninsula, before making direct flights to the United States. In post-breeding, the migration route is skewed farther east, and birds may use islands of the Caribbean as stopovers en route to the South American nonbreeding range (Mowbray 2020c; Fink et al. 2023). One tracked bird exhibited a loop migration that included a likely flight over the Pacific Ocean from South to Central America in April. The bird made at least six multiday stopovers during both pre- and post-breeding migration (Raybuck et al. 2024).

WESTERN TANAGER *Piranga ludoviciana*

DISTRIBUTION AND ABUNDANCE

Western Tanagers breed throughout the western United States and Canada, from the US–Mexico border north through British Columbia and Saskatchewan, Canada. The nonbreeding range extends throughout Baja California, western Mexico, and south through the Pacific side of Central America to Costa Rica. There are small pockets of a year-round range in California, southern Arizona, and northern Mexico (Hudon 2020).

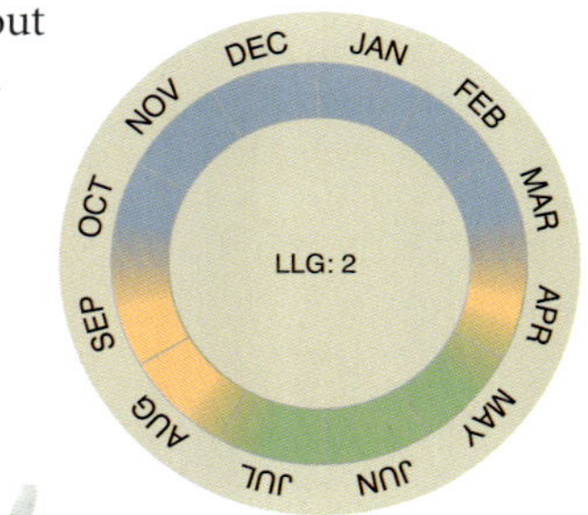

ORDER Passeriformes
FAMILY Cardinalidae
BREEDING HABITAT
Forests, Temperate Western
NONBREEDING HABITAT
Forests, Tropical Dry
GLOBAL POPULATION SIZE 15,000,000
POPULATION CHANGE SINCE 1970 +64%
IUCN RED LIST CATEGORY Least Concern
R2R TIPPING POINT SPECIES No
US LISTING STATUS No
CANADA SPECIES AT RISK No
PIF CCS 9/20

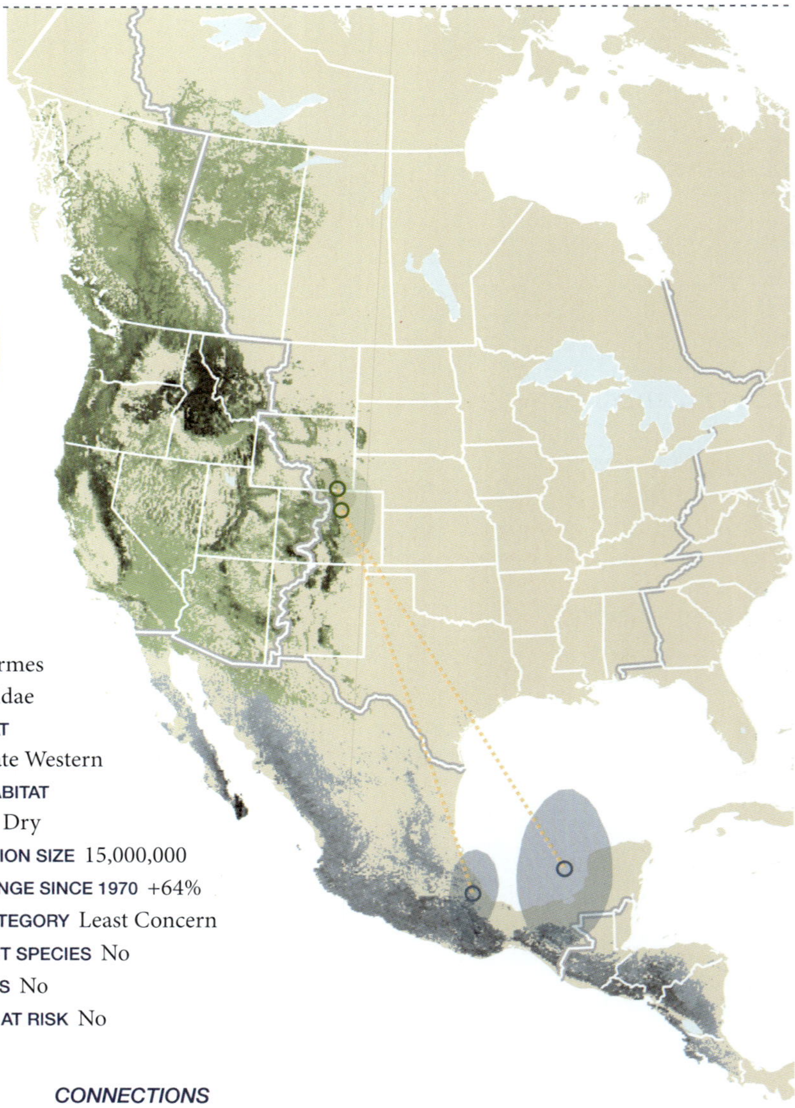

CONNECTIONS

MIGRATORY CONNECTIVITY

OVERALL Migratory connectivity is unknown. There is little information on the migratory connectivity of the Western Tanager.

GEOLOCATOR TRACKING Two individuals were tracked with geolocators from northern Colorado; one spent the nonbreeding season in Oaxaca, and the other in Chiapas or Guatemala (Bird Conservancy of the Rockies, unpublished data).

OTHER APPROACHES No banding, genetic, or stable isotope studies of migratory connectivity were found for this species.

MIGRATORY PATHWAYS All breeding populations seem to migrate directly south and concentrate on the western coast of the United States and Mexico before moving on to nonbreeding sites farther east in Central America (Hudon 2020). The two individuals tracked from Colorado underwent a molt migration, stopping for about a month in either West Texas or southeastern Mexico to molt (Bird Conservancy of the Rockies, unpublished data).

DATA CONTRIBUTORS
Bird Conservancy of the Rockies.

Pheucticus ludovicianus **ROSE-BREASTED GROSBEAK**

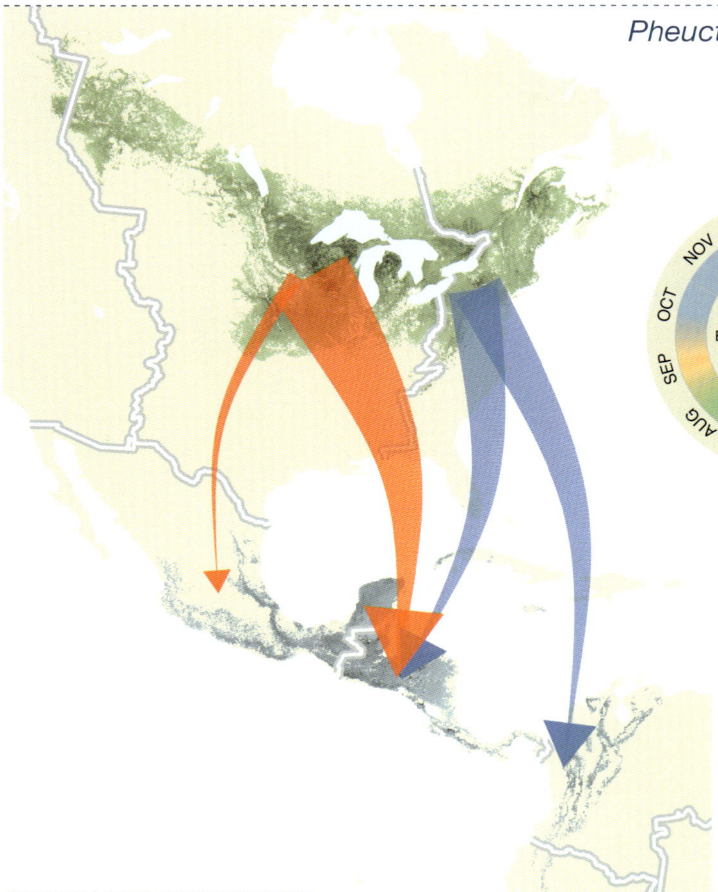

GENERALIZED CONNECTIVITY

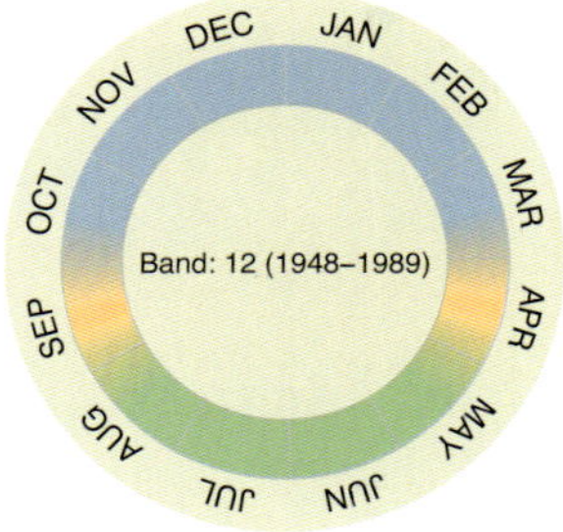

DISTRIBUTION AND ABUNDANCE

Rose-breasted Grosbeaks breed just south of the boreal region and east of the Rocky Mountains, in an extensive region from northeast British Columbia and eastward through Alberta, central Saskatchewan, and southern Manitoba, Ontario, and Quebec in Canada, with a southeastern limit in eastern Nebraska around 40 degrees north latitude all the way to the east coast of the United States. The species spends the nonbreeding season at mid-elevations from southern Mexico through Central America and in the northern Andean region of Venezuela, Colombia, and Ecuador (Wyatt and Francis 2020; Fink et al. 2023).

ORDER Passeriformes
FAMILY Cardinalidae
BREEDING HABITAT Forests, Temperate Eastern
NONBREEDING HABITAT Forests, Tropical Generalist
GLOBAL POPULATION SIZE 4,700,000
POPULATION CHANGE SINCE 1970 −34%
IUCN RED LIST CATEGORY Least Concern
R2R TIPPING POINT SPECIES No
US LISTING STATUS No
CANADA SPECIES AT RISK No
PIF CCS 10/20

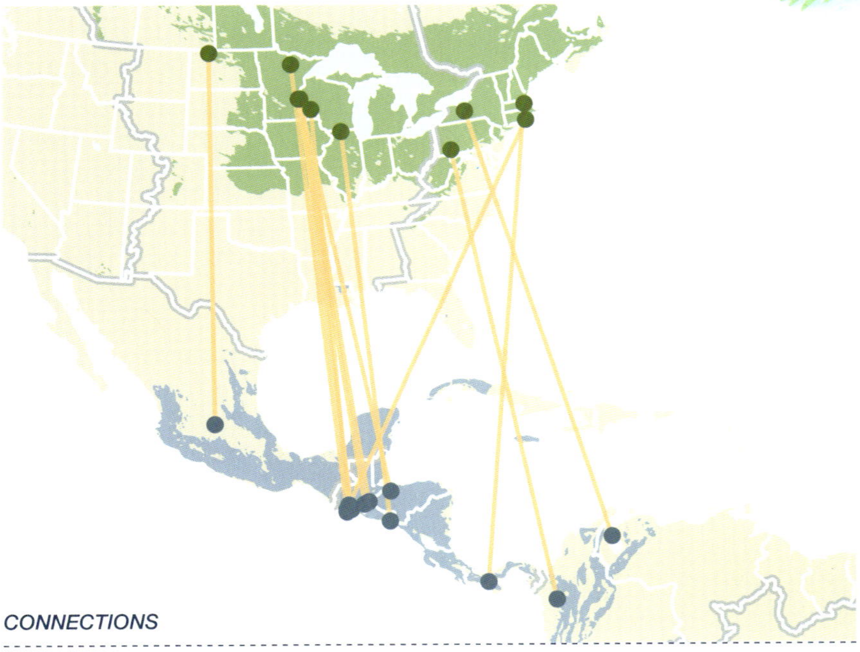

CONNECTIONS

	Central America	Mexico	Northern SA
Atlantic (n = 4)	50%	–	50%
Midcontinent (n = 8)	88%	12%	–

MIGRATORY CONNECTIVITY

OVERALL The migratory connectivity of the Rose-breasted Grosbeak is known superficially from band reencounter data. MC is not calculated due to low sample size. Breeding populations overlap in Central America, but more study is needed to understand the degree of mixing. Atlantic Flyway breeding populations spend the nonbreeding season either in northern South America (50%) or Central America (50%). Midcontinent Flyway breeders spend the nonbreeding season mostly in Central America (88%) and Mexico (12%).

BANDING Bands were reencountered from both eastern and midwestern breeding populations in Central America.

RADIO TRACKING Rose-breasted Grosbeaks are being tracked through the Motus Wildlife Tracking System from coffee landscapes in Colombia. As of printing, there are no detections between breeding and nonbreeding areas (R. Bennett, unpublished data).

OTHER APPROACHES No genetic, geolocator or satellite tracking or stable isotope studies of migratory connectivity were found.

MIGRATORY PATHWAYS All breeding populations converge during post-breeding migration on the Gulf Coast between southern Texas and northern Florida, before they cross the Gulf of Mexico to Central America and northern South America (Wyatt and Francis 2020).

BLACK-HEADED GROSBEAK *Pheucticus melanocephalus*

DISTRIBUTION AND ABUNDANCE

This species breeds throughout western North America from southwestern Canada to southern Mexico, most densely along the US west coast, northern Great Basin, and western mountains (Hill 2022). The nonbreeding range encompasses Mexico south to Oaxaca and spottily to Chiapas (Hill 2022). Populations have increased overall (Rosenberg et al. 2019), but there are some declines in Southern California and New Mexico (Sauer et al. 2019).

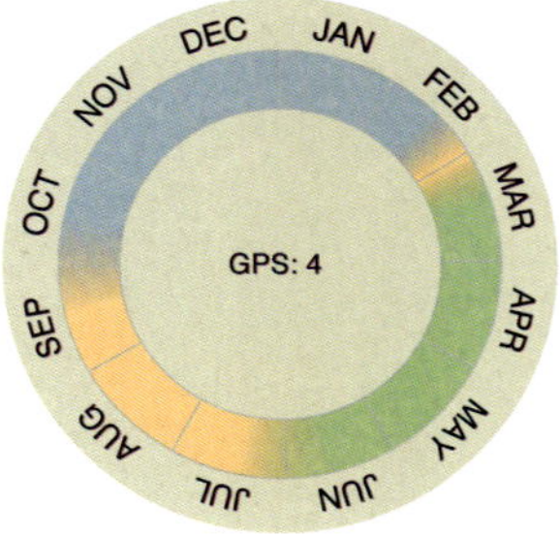

ORDER Passeriformes
FAMILY Cardinalidae
BREEDING HABITAT
Forests, Temperate Western
NONBREEDING HABITAT
Forests, Tropical Dry
GLOBAL POPULATION SIZE 15,000,000
POPULATION CHANGE SINCE 1970 +31%
IUCN RED LIST CATEGORY Least Concern
R2R TIPPING POINT SPECIES No
US LISTING STATUS No
CANADA SPECIES AT RISK No
PIF CCS 9/20

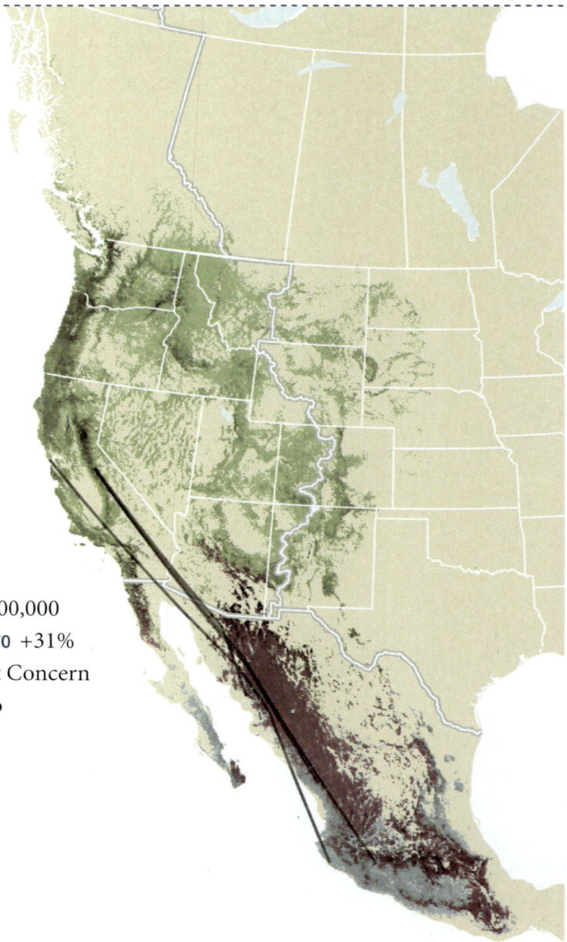

PATHWAYS

MIGRATORY CONNECTIVITY

OVERALL Migratory connectivity is unknown and data for this species is limited to GPS tracks from four individuals that were tagged at breeding sites in the Pacific Flyway and spent the nonbreeding season in Mexico. MC is not calculated due to insufficient data.

RADIO TRACKING Black-headed Grosbeaks are being tracked through the Motus Wildlife Tracking System. As of printing, there are no breeding to nonbreeding detections.

SATELLITE TRACKING Eighteen individuals were tagged with GPS archival tags at breeding sites in Yosemite National Park, and 15 individuals were tagged at sites in coastal California (Siegel et al. 2016; Saracco et al. 2022). Tags from three of the Yosemite birds were recovered (one tagged had stopped working in early October), and one bird was recovered that was tagged at a coastal California site in Golden Gate National Recreation Area. All of the recovered birds were males. Of the three birds with complete full annual-cycle data, secondary long-distance movements were made to nonbreeding sites farther south in Durango, Jalisco, and Michoacán (Saracco et al. 2022).

OTHER APPROACHES No banding, radio telemetry, genetic, or stable isotope studies of migratory connectivity were found for this species.

MIGRATORY PATHWAYS All tagged birds from California migrated to apparent molting areas in northwestern Mexico (Sonora/Sinaloa) during August to October, before continuing to nonbreeding locations (Siegel et al. 2016; Saracco et al. 2022).

DATA CONTRIBUTORS
L. Helton; J.F. Saracco; R.B. Siegel; S. Stock; R. Taylor.

Passerina caerulea BLUE GROSBEAK

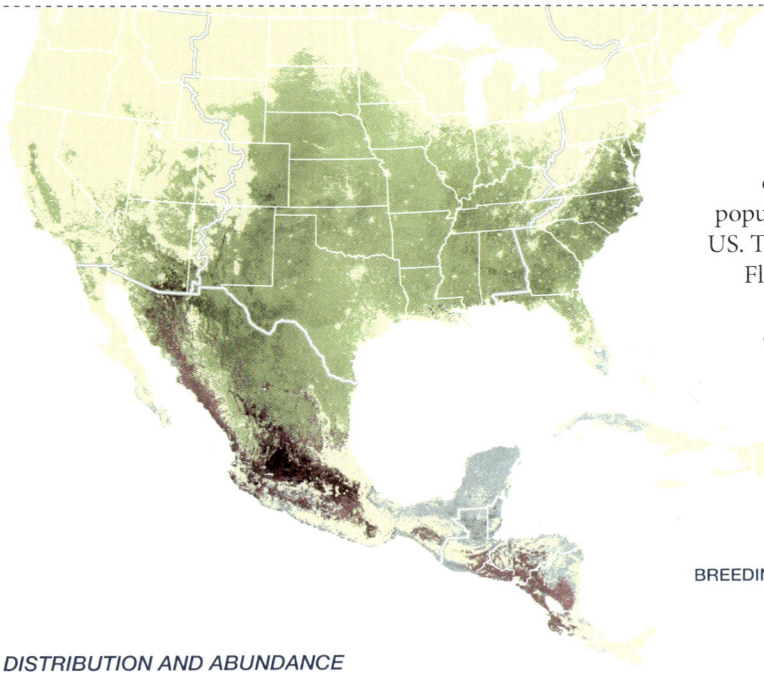

DISTRIBUTION AND ABUNDANCE

DISTRIBUTION AND ABUNDANCE

The Blue Grosbeak is widely distributed as a breeding species across the central and southern US, Mexico, and the *cordilleras* (mountain ranges) of Central America south through Costa Rica, with the densest breeding populations in central Mexico and the southeast, west-central, and southwest US. The species spends the nonbreeding period in small numbers in southern Florida and the Louisiana coast and more densely throughout Mexico and Central America to central Panama—including the Caribbean lowlands of the Yucatán and Central America (where it is not known as a breeder; Lowther and Ingold 2020; Fink et al. 2023). Over most of its range, Blue Grosbeak populations appear to be stable or increasing; the most severe declines are in the Great Basin where the species is thinly distributed (Sauer 2019).

ORDER Passeriformes
FAMILY Cardinalidae
BREEDING HABITAT Forests, Temperate Eastern
NONBREEDING HABITAT Open Country, Habitat Mosaic
GLOBAL POPULATION SIZE 35,000,000
POPULATION CHANGE SINCE 1970 +46%
IUCN RED LIST CATEGORY Least Concern
R2R TIPPING POINT SPECIES No
US LISTING STATUS No
CANADA SPECIES AT RISK No
PIF CCS 7/20

MIGRATORY CONNECTIVITY

OVERALL Migratory connectivity is unknown. No banding, tracking, stable isotope, or genetic studies of migratory connectivity were found for this species.

RADIO TRACKING Blue Grosbeaks are tracked through the Motus Wildlife Tracking System. As of printing, there are no breeding to nonbreeding detections.

OTHER APPROACHES Subspecies information and eBird animations suggest that eastern populations spend the nonbreeding season in the West Indies, Yucatán, and Caribbean lowlands of Central America, while western populations spend the nonbreeding season in western Mexico and western Central America south to Honduras (Lowther and Ingold 2020; Fink et al. 2023).

MIGRATORY PATHWAYS There is high relative abundance of Blue Grosbeaks along the Louisiana coast from September through October and virtually no detections along the Texas and Mexican Gulf Coast after mid-October, which strongly suggests a trans-Gulf migration in post-breeding for eastern populations (Fink et al. 2023).

Passerina amoena LAZULI BUNTING

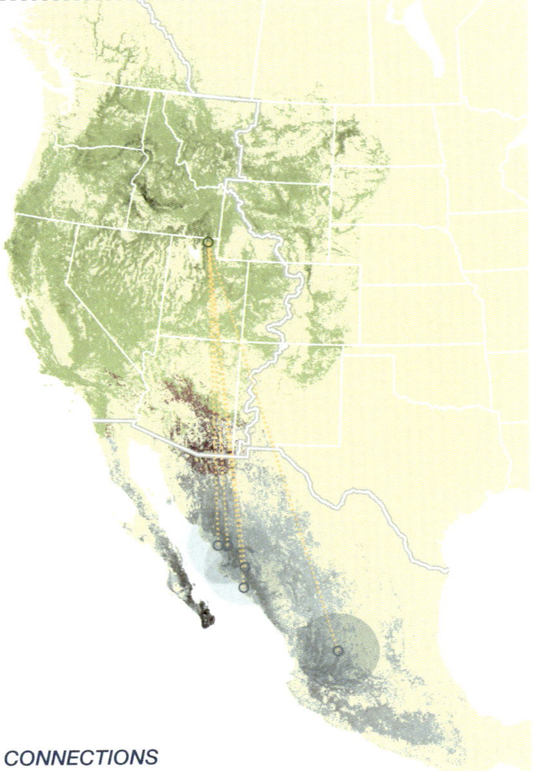

CONNECTIONS

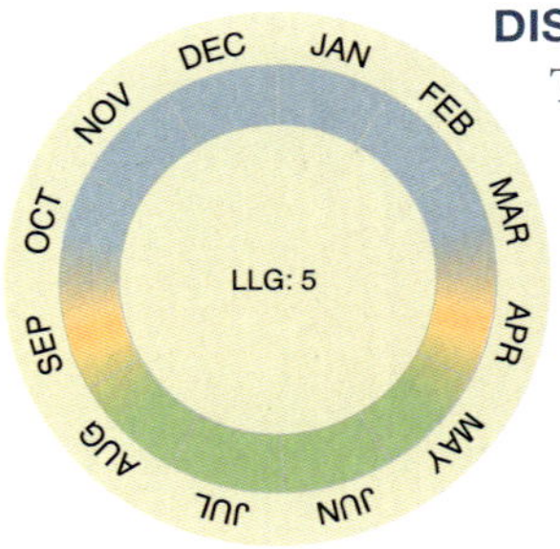

DISTRIBUTION AND ABUNDANCE

This species breeds in southern British Columbia, southwestern Alberta, throughout the western US west of the central plains, and in northwestern Sonora, Mexico. Lazuli Buntings migrate to western and central Mexico, most densely in Baja California Sur and the Sierra Madre Occidental for the nonbreeding season (Fink et al. 2023).

ORDER Passeriformes
FAMILY Cardinalidae
BREEDING HABITAT Forests, Temperate Western
NONBREEDING HABITAT Open Country, Habitat Mosaic
GLOBAL POPULATION SIZE 6,500,000
POPULATION CHANGE SINCE 1970 +20%
IUCN RED LIST CATEGORY Least Concern
R2R TIPPING POINT SPECIES No
US LISTING STATUS No
CANADA SPECIES AT RISK No
PIF CCS 8/20

MIGRATORY CONNECTIVITY

OVERALL Migratory connectivity is unknown. There is very limited data on migratory connectivity from one tracking study (Savides 2022).

GEOLOCATOR TRACKING Light-level geolocator data (n = 5) from tags deployed on breeding areas in northern Utah showed all birds spent the nonbreeding season in western Mexico: four birds along the Sonora–Sinaloa border and one in Durango (Savides 2022).

MIGRATORY PATHWAYS During post-breeding migration, migratory paths and timings differed between individuals, although each bird took a fairly direct southerly route toward the nonbreeding region (Savides 2022). All birds made multiday stopovers, with location and duration varying by individual. Stopover duration of one individual was adequate to molt. Other individuals may have undergone some molt while migrating or on the nonbreeding grounds (Savides 2022). Three birds took direct routes from nonbreeding to breeding areas, while two birds followed a clockwise, loop pathway through California and Nevada to Utah. Despite differences in distance between the direct and loop pathways, breeding ground arrival dates were similar (Savides 2022).

INDIGO BUNTING *Passerina cyanea*

DISTRIBUTION AND ABUNDANCE

The Indigo bunting is a widespread eastern-breeding species, with the core breeding range lying between a large area from Oklahoma to Iowa and from Georgia to Vermont. It spends the nonbreeding season primarily in southern Mexico and the Central American lowlands, with a few birds spending the nonbreeding season in southern Florida and the Caribbean, and a few individuals reaching northern Colombia (Payne 2020; Fink et al. 2023).

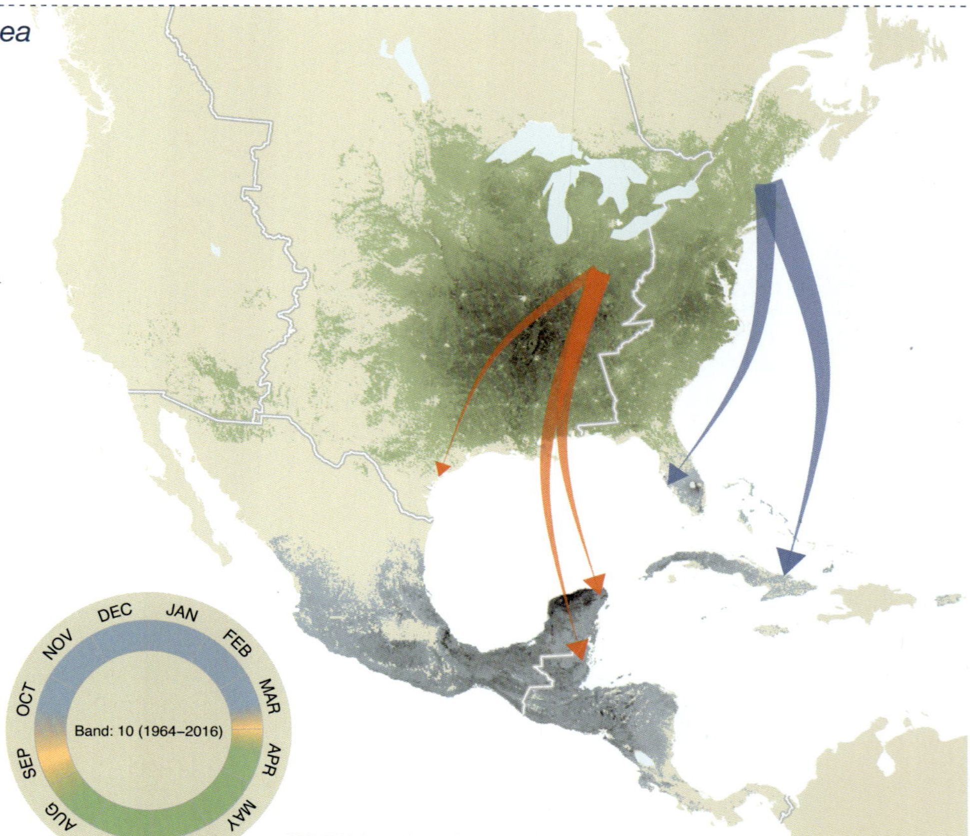

GENERALIZED CONNECTIVITY

ORDER Passeriformes
FAMILY Cardinalidae
BREEDING HABITAT Forests, Temperate Eastern
NONBREEDING HABITAT Open Country, Habitat Mosaic
GLOBAL POPULATION SIZE 77,000,000
POPULATION CHANGE SINCE 1970 −29%
IUCN RED LIST CATEGORY Least Concern
R2R TIPPING POINT SPECIES No
US LISTING STATUS No
CANADA SPECIES AT RISK No
PIF CCS 10/20

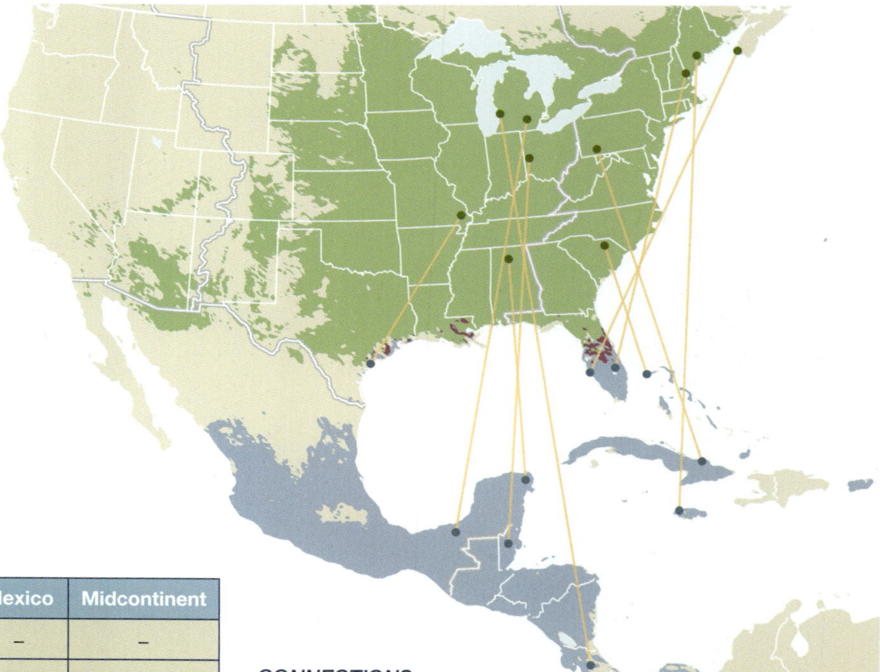

CONNECTIONS

	Atlantic	Caribbean	Central America	Mexico	Midcontinent
Atlantic (n = 5)	40%	60%	–	–	–
Midcontinent (n = 5)	–	–	40%	40%	20%

MIGRATORY CONNECTIVITY

OVERALL The migratory connectivity of Indigo Buntings is known primarily from band reencounter data. MC is not calculated due to limited sample size but appears strong, as Atlantic and Midcontinent flyway breeding populations maintain an east-west divide, with no overlap on the nonbreeding grounds. Birds breeding in the Atlantic Flyway primarily spend the nonbreeding season in the Caribbean (60%) and also within the Atlantic Flyway (40%). Birds breeding in the Midcontinent Flyway spend the nonbreeding season in Central America (40%), Mexico (40%), and within the Midcontinent Flyway (20%).

BANDING Individuals that breed in eastern and midwestern regions of North America remain geographically separate into the nonbreeding period.

OTHER APPROACHES No genetic, tracking, or stable isotope studies of migratory connectivity were found for this species.

MIGRATORY PATHWAYS There are two primary migration paths from breeding areas. The westerly pathway follows a straight line roughly from Iowa to Louisiana, then through the Gulf Coast to the Yucatán Peninsula and Central America. The more easterly route follows the eastern coastal states to Florida and the Gulf Coast on to Florida and Caribbean nonbreeding areas (Payne 2020; Fink et al. 2023).

VARIED BUNTING

Passerina versicolor

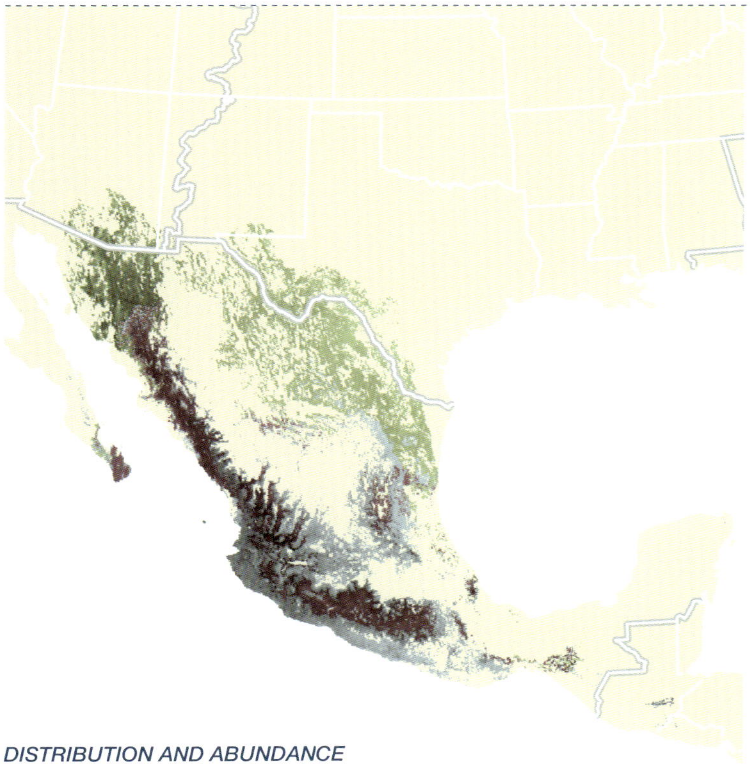

DISTRIBUTION AND ABUNDANCE

DISTRIBUTION AND ABUNDANCE

This species breeds throughout northern Mexico, disjunctly in central Chiapas, and locally in southwest Texas and southern Arizona and New Mexico in arid thorn brush, along desert washes, and in desert scrub and thorn forest. US populations and those in north-central Mexico withdraw south to spend the nonbreeding season in similar habitat within the same range as southern resident populations (Groschupf and Thompson 2020; Fink et al. 2023).

ORDER Passeriformes
FAMILY Cardinalidae
BREEDING HABITAT Forests, Tropical Dry
NONBREEDING HABITAT Forests, Tropical Dry
GLOBAL POPULATION SIZE 360,000
POPULATION CHANGE SINCE 1970 +55%
IUCN RED LIST CATEGORY Least Concern
R2R TIPPING POINT SPECIES No
US LISTING STATUS No
CANADA SPECIES AT RISK No
PIF CCS 14/20

MIGRATORY CONNECTIVITY

OVERALL Migratory connectivity is unknown. No information on migratory connectivity or migratory pathways was found for this species.

PAINTED BUNTING

Passerina ciris

MC = 0.34 ± 0.07*

ORDER Passeriformes
FAMILY Cardinalidae
BREEDING HABITAT Forests, Temperate Eastern
NONBREEDING HABITAT Open Country, Habitat Mosaic
GLOBAL POPULATION SIZE 15,000,000
POPULATION CHANGE SINCE 1970 −7%
IUCN RED LIST CATEGORY Least Concern
R2R TIPPING POINT SPECIES No
US LISTING STATUS No
CANADA SPECIES AT RISK No
PIF CCS 10/20

DISTRIBUTION AND ABUNDANCE

Painted Bunting breeding populations are separated into two completely disjunct ranges. A larger southern midwestern (interior) population breeds primarily in the several states of northwestern Mexico, throughout Texas, Oklahoma, southeastern Kansas, Arkansas, Louisiana, and at lower densities in western Mississippi. The smaller eastern population breeds along the southeastern United States coast from coastal southeastern North Carolina and South Carolina to Georgia and extreme northern Florida. The nonbreeding range of the midwestern population is from Mexico to northern Panama, and the eastern population's is from Florida to northern Cuba and most of the Bahamas (Lowther et al. 2020f; Fink et al. 2023).

	Atlantic	Caribbean	Central America	Mexico
Atlantic (n = 91)	49%	51%	–	–
Midcontinent (n = 20)	–	–	45%	55%

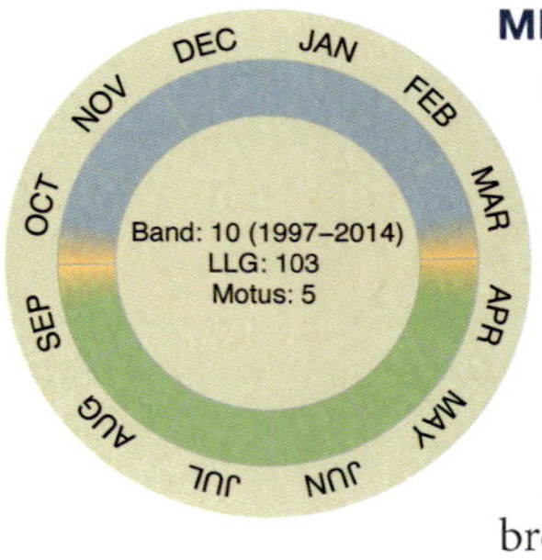

MIGRATORY CONNECTIVITY

OVERALL The migratory connectivity of Painted Buntings is known from a mixture of band reencounter data, geolocator, genetic, morphology, and stable isotope data, suggesting an east-west divide as well as a leapfrog migration pattern where northern breeding populations migrate to nonbreeding areas farther south than southern breeding populations (Contina et al. 2019; Battey et al. 2018; Rueda-Hernández et al. 2023). Atlantic and Midcontinent flyways breeding populations migrate to multiple nonbreeding areas but maintain an east-west divide on the nonbreeding grounds, resulting in moderate migratory connectivity (MC = 0.34 ± 0.07). Roughly half the birds breeding in the Atlantic Flyway spend the nonbreeding season in the Caribbean (51%) and within the Atlantic Flyway (49%). Birds breeding in the Midcontinent Flyway migrate either to nonbreeding areas in Mexico (55%) or to Central America (45%).

BANDING There are several band reencounters from the eastern breeding population in South Carolina and Georgia to nonbreeding areas in Florida and Cuba. One reencounter of a midwestern breeding bird was made from Louisiana to Veracruz, Mexico.

RADIO TRACKING Tags deployed through the Motus Wildlife Tracking System include both individual breeding to nonbreeding detections and detections during migration in the eastern portion of the range (A. Smith and D. Marain, unpublished data).

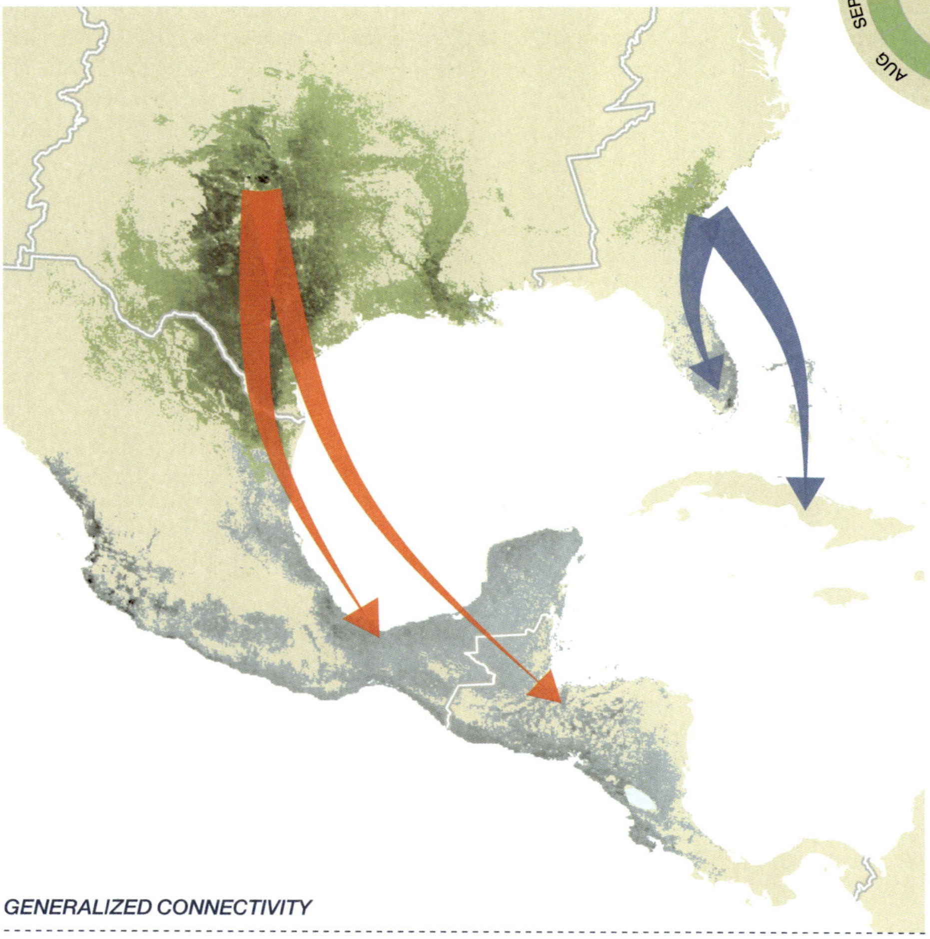

GENERALIZED CONNECTIVITY

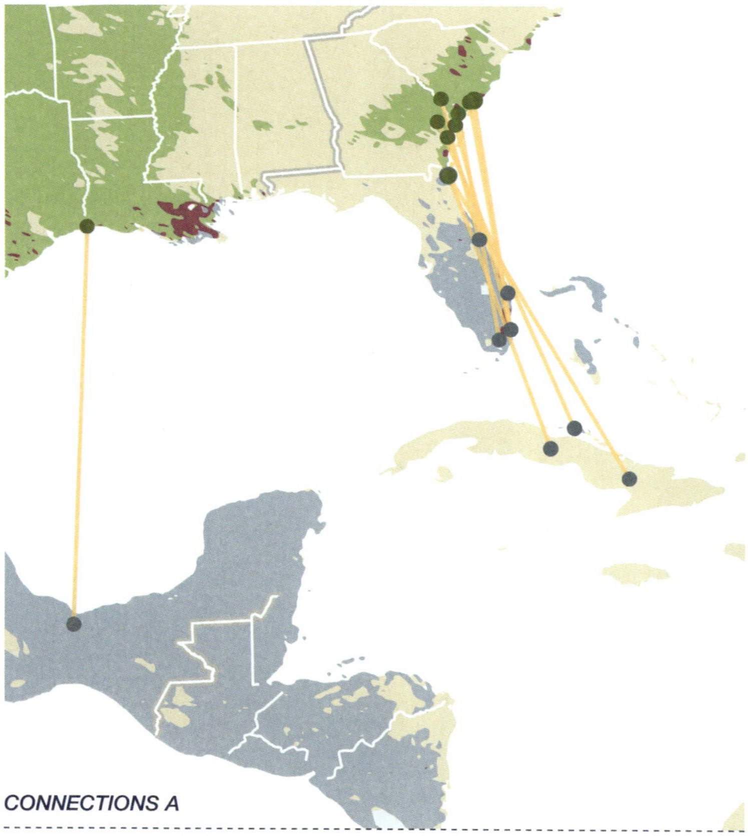

CONNECTIONS A

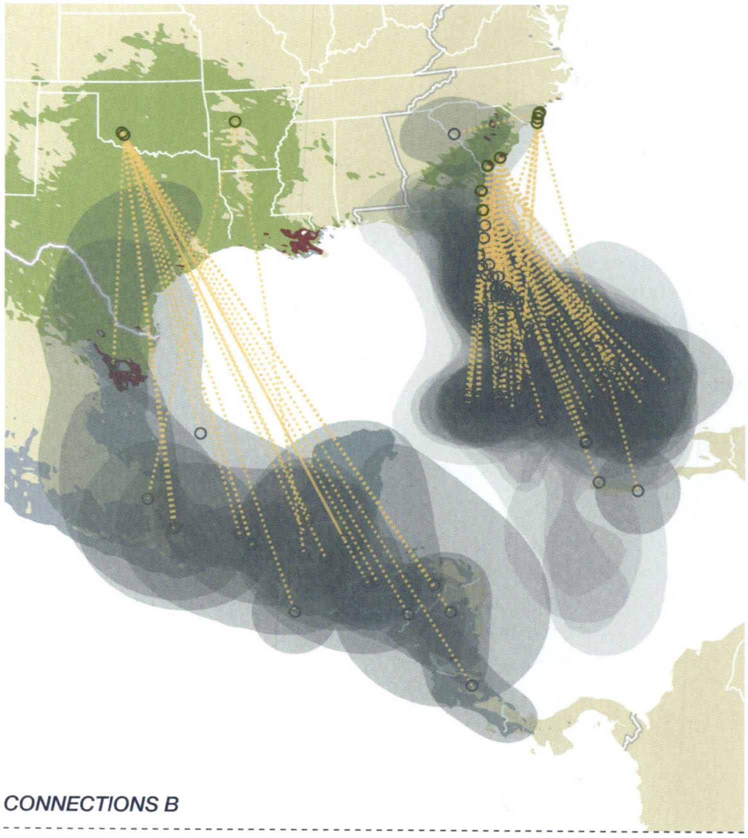

CONNECTIONS B

GEOLOCATOR TRACKING Based on light-level geolocator data from 112 individuals at 11 breeding sites, migratory connectivity between breeding and nonbreeding grounds was found to be strong at the range-wide scale, with midwestern (interior) and eastern Painted Buntings remaining separated throughout the annual cycle (Sharp et al. 2023). The interior breeding population (Oklahoma) migrated to nonbreeding areas in southern Mexico and Central America, while eastern breeding populations migrated to nonbreeding areas in Florida and Cuba (Contina et al. 2013; Bridge et al. 2016; Sharp et al. 2023). Conversely, within each of the interior and eastern populations, migratory connectivity was weak, with individuals from different breeding areas within the population mixing on nonbreeding areas (Sharp et al. 2023).

STABLE ISOTOPES A stable carbon isotopic study from Painted Bunting feathers provided the first evidence that a proportion of individuals performed a molt migration in northwestern Mexico, whereas others molted directly on breeding areas prior to commencing a post-breeding migration (Contina et al. 2013; Bridge et al. 2016).

GENETICS Using nuclear DNA, Contina et al. (2019) identified three breeding ground clusters across the east-west population divide: a western population covering Oklahoma and central Texas; a central cluster covering eastern and western Texas, Louisiana, Arkansas, and Mississippi; and an eastern group along the Atlantic coast. Many but not all buntings sampled on nonbreeding grounds could be assigned back to breeding areas (Contina et al. 2019; Battey et al. 2018). Within the central cluster, individuals primarily spend the nonbreeding season on the Yucatán Peninsula (Contina et al. 2019; Rueda-Hernández et al. 2023), while individuals in the western cluster spend the nonbreeding season in central and southern Mexico and Central America (Rueda-Hernández et al. 2023).

OTHER APPROACHES Abundance-index values from georeferenced natural history museum collections data for the midwestern US breeding population suggests a quasi-circular movement around its nonbreeding range. From July through September, an east-west split occurs, forming two clusters in northern Mexico: an eastern cluster in Nuevo Leon and Tamaulipas, and a western cluster in Sinaloa and Durango. Abundance values from October to December show southward movement to both coasts of Mexico, forming a western cluster in Guerrero, Michoacán, Oaxaca, Jalisco, and Colima, and an eastern cluster in the Veracruz, Tabasco, Campeche, and Yucatán. This period is followed by a cluster in northeast Mexico (Coahuila, Nuevo Leon, Tamaulipas) and an absence of abundance values elsewhere in the months of March and April. South of Mexico, specimen records indicate the presence of Painted Buntings at extremely low densities, mostly restricted to the nonbreeding months of November to March (Linck et al. 2016).

MIGRATORY PATHWAYS Approximately 60% of western Painted Buntings are molt-migrants in post-breeding, carrying out a molt migration in the monsoonal region of northwestern Mexico before heading to their final nonbreeding areas (Thompson 1991; Contina et al. 2013; Bridge et al. 2016; Rushing et al. 2021; Sharp et al. 2023). Eastern populations appear to molt on nonbreeding areas (Thompson 1991; Bridge et al. 2016). This difference likely accounts for differences in a direct (eastern populations) versus indirect (western populations) migratory pathway to nonbreeding areas taken by each population (E. Bridge, personal communication). Western-breeding populations first move in a southwesterly direction toward northwestern Mexico and then move south through Mexico and into Central America, displaying strong migratory connectivity as birds move from the breeding grounds to the molting grounds (Sharp et al. 2023) and back to their breeding grounds via the coastal plains of the Gulf of Mexico (Rueda-Hernández et al. 2023). One individual breeding in Louisiana flew directly south to the Yucatán Peninsula and into Central America. The eastern pathway follows the coast south into Florida, and from there to Cuba and the Bahamas.

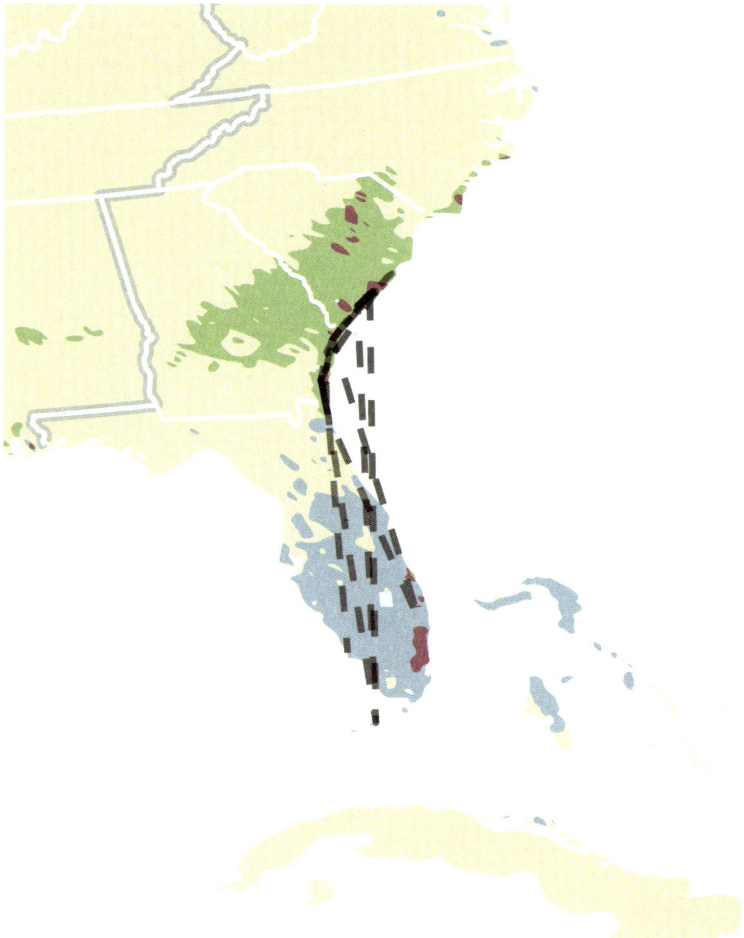

PATHWAYS

DATA CONTRIBUTORS

E.S. Bridge; A. Contina; E. Khazan; D. Marain; C.S. Rushing; A.J. Sharp; A. Smith.

INTERESTING NOTE

Data from geolocators deployed along the US east coast from North Carolina to Florida showed that buntings from South Carolina and Georgia were most likely to spend the nonbreeding season in south Florida, whereas those in North Carolina went to Cuba. Male Painted Buntings that migrated to Cuba were found to be 20% less likely to return to the breeding grounds than birds that migrated to nonbreeding areas elsewhere in their range, likely due to the illegal pet trade (Rushing et al. 2021).

DICKCISSEL *Spiza americana*

DISTRIBUTION AND ABUNDANCE

Dickcissels breed in open grassland habitats of all types throughout the midwestern US. Their abundance and distribution can vary dramatically from year to year, with seasonal movements both within their core range and sporadically to peripheral grassland areas (Sousa et al. 2022). Most individuals spend the nonbreeding season in large flocks in the seasonally flooded *llanos* grasslands of central Venezuela but may also be found sporadically in other parts of Central and South America during the nonbreeding season. The species declined substantially between 1966 and 1978 but has since stabilized at a lower population size throughout most of its range (Sauer et al. 2019).

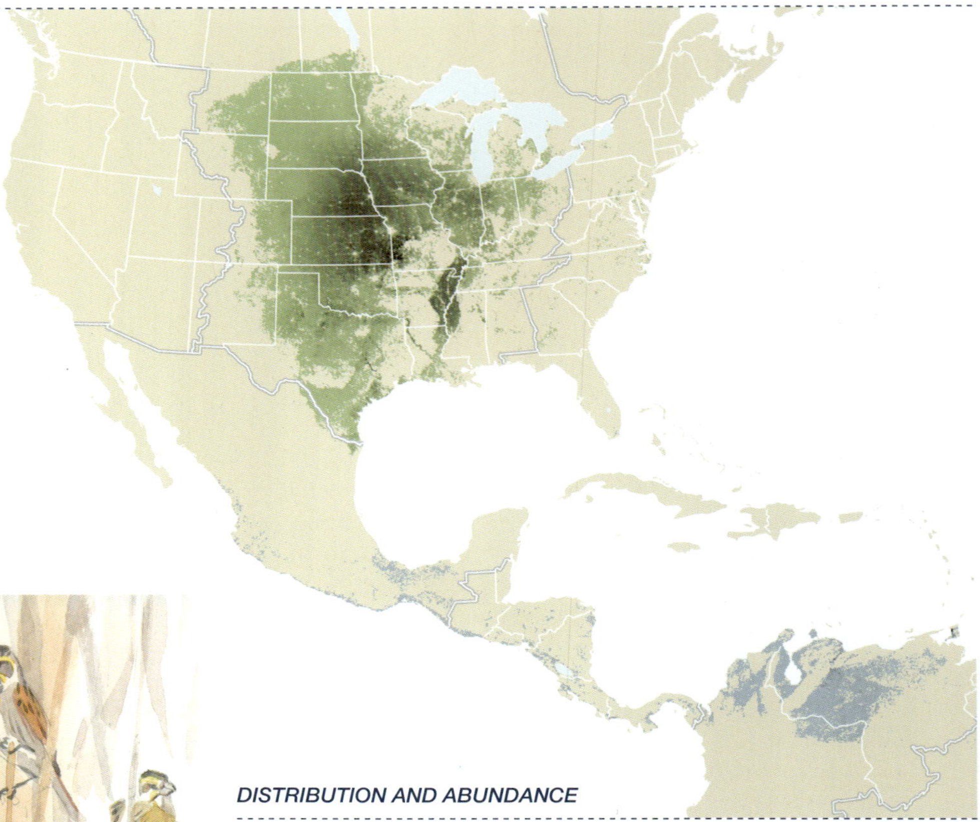

DISTRIBUTION AND ABUNDANCE

ORDER Passeriformes
FAMILY Cardinalidae
BREEDING HABITAT Grasslands, Temperate
NONBREEDING HABITAT Grasslands, Tropical
GLOBAL POPULATION SIZE 28,000,000
POPULATION CHANGE SINCE 1970 −9%
IUCN RED LIST CATEGORY Least Concern
R2R TIPPING POINT SPECIES No
US LISTING STATUS No
CANADA SPECIES AT RISK No
PIF CCS 10/20

INTERESTING NOTE

Dickcissels are persecuted for crop damage in Venezuela; their roosts have been targeted for toxic agrochemical poisoning (Sousa et al. 2022).

MIGRATORY CONNECTIVITY

OVERALL Although most Dickcissels are known to spend the nonbreeding season in central Venezuela, detailed migratory connectivity for the species is largely unknown.

RADIO TRACKING A few tags were deployed through the Motus Wildlife Tracking System, but with local detections only.

MIGRATORY PATHWAYS Dickcissels migrate along both coasts of Central America to nonbreeding areas in northern South America (Fink et al. 2023) and gather in flocks numbering into the thousands at stopover locations during migration (Sousa et al. 2022).

For all species for which the strength of migratory connectivity could be estimated (28%; 159/564), we present summaries by species and by order. Migratory connectivity ranges from −1 to 1 (± SE) along a weak-to-strong continuum, with 1 indicating that individuals from each population stay together between seasons, 0 indicating complete mixing of populations between seasons, and −1 indicating that individuals close to each other in one season are farther apart during the other season. We estimated migratory connectivity using the MC metric for landbirds (orange) and the Mantel metric for seabirds (blue). Understanding how sensitive these values are to increasing sample sizes of tracked individuals and the inclusion of additional sites will be important as more data become available.

LEFT: *Strength of migratory connectivity across species. We classified migratory connectivity as weak (0–33rd percentile), moderate (33rd–66th percentile), or strong (>66th percentile). For landbirds, the numerical ranges were −0.11 to 0.18 for weak, 0.19 to 0.38 for moderate, and 0.39 to 1 for strong connectivity. For seabirds, the numerical ranges were −0.08 to 0.05 for weak, 0.05 to 0.40 for moderate, and 0.41 to 0.77 for strong connectivity. Of the species, 54 exhibited strong migratory connectivity, 53 moderate connectivity, and 52 weak connectivity. Dashed lines indicate the median MC estimate for landbirds (M = 0.31) and the median Mantel estimate for seabirds (M = 0.12) across all species, whereas solid lines denote the percentile-based thresholds separating weak, moderate, and strong connectivity categories.*

BELOW: *Strength of migratory connectivity by taxonomic order. The smaller, closed circles each represent a different species and indicate the strength of migratory connectivity using the MC metric for landbirds (orange) and the Mantel metric for seabirds (blue). The larger, open circles in gray indicate the mean for each taxonomic order.*

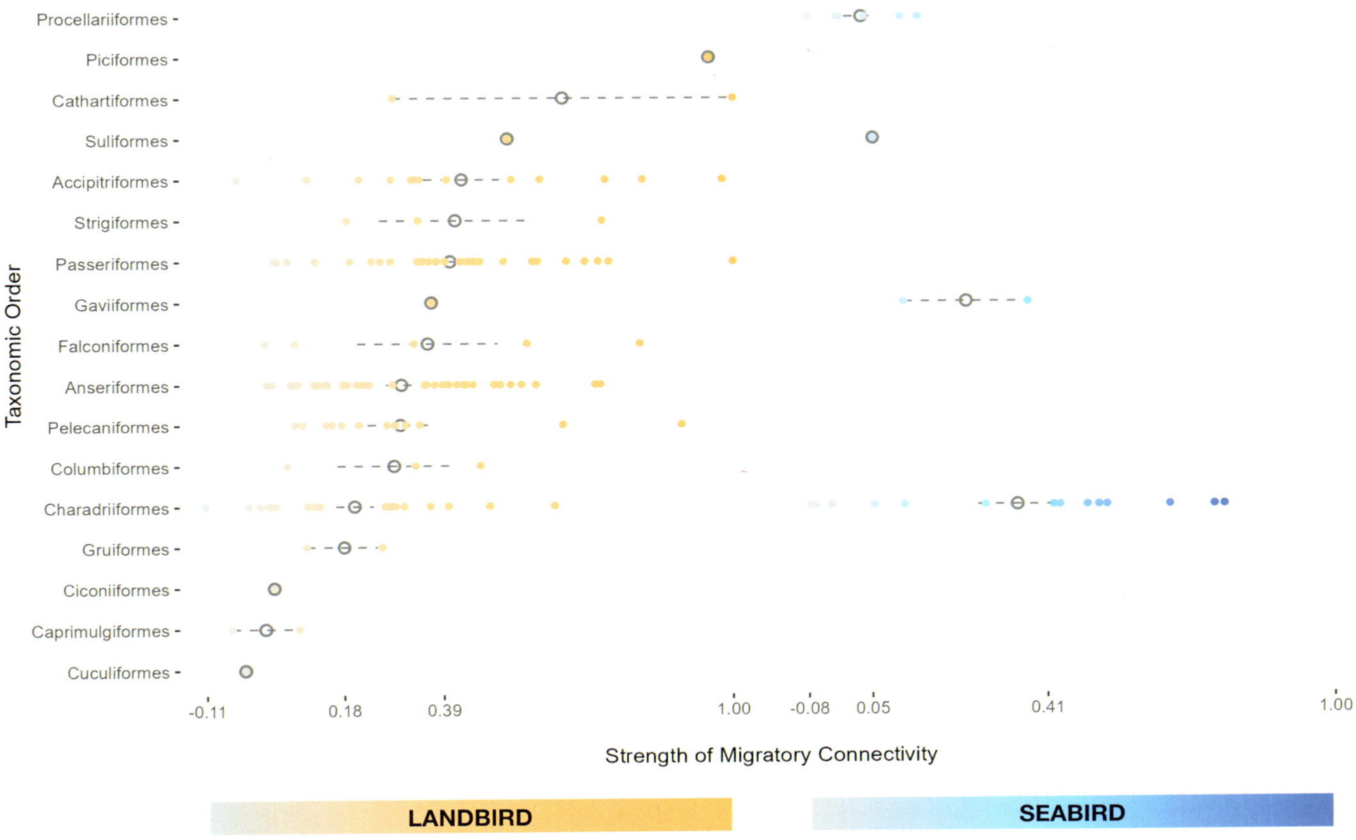

REFERENCES

Abbott C.L., R.L. Millikin, J.M. Hipfner, S. Hatch, M. Ito, Y. Watanuki, et al. 2014. Genetic structure of Rhinoceros Auklets, *Cerorhinca monocerata*, breeding in British Columbia, Alaska, and Japan. *Marine Biology* 161:275–283. https://doi.org/10.1007/s00227-013-2333-2.

Abraham, K.F., J.O. Leafloor, and D.H. Rusch. 1999. Molt migrant Canada Geese in northern Ontario and western James Bay. *Journal of Wildlife Management* 63(2):649–655.

Ackerman, J.T., C.A. Hartman, M.P. Herzog, J.Y. Takekawa, J.A. Robinson, L.W. Oring, J.P. Skorupa, and R. Boettcher. 2020. American Avocet (*Recurvirostra americana*). Version 1.0. In *Birds of the World*, ed. A.F. Poole. Cornell Lab of Ornithology, Ithaca, NY. https://doi.org/10.2173/bow.ameavo.01.

Adams, E.M., I.J. Stenhouse, A.T. Gilbert, J. Boelsma, G. Gress, C.S. Weidensaul, C. Grigsby, E.J. Williams, L. Phillips, and C.L. McIntyre. 2022. The first documentation of the Nearctic–Paleotropical migratory route of the Arctic Warbler. *Ecology and Evolution* 12(9):e9223.

Adams, J., D.G. Ainley, J.F. Penniman, C. Bailey, J. Tamayose, F. Duvall, and H. Freifeld. 2012a. Anti-cyclonic circulation and the long-range foraging movements of Hawaiian Petrels (*Pterdroma sandwichensis*) in the North Pacific. Pacific Seabird Group Annual Meeting. Honolulu, HI.

Adams, J., C. MacLeod, R.M. Suryan, K.D. Hyrenbach, and J.T. Harvey. 2012b. Summer-time use of west coast US National Marine Sanctuaries by migrating Sooty Shearwaters (*Puffinus griseus*). *Biological Conservation* 156:105–116.

Adams, J., and J.Y. Takekawa. 2008. At-sea distribution of radio-marked Ashy Storm-Petrels *Oceanodroma homochroa* captured on the California Channel Islands. *Marine Ornithology* 36:9–17.

Adams, J., J.Y. Takekawa, and H.R. Carter. 2004. Foraging distance and home range of Cassin's Auklets nesting at two colonies in the California Channel Islands. *Condor* 106(3):618–637.

Addison, L.M., S.E. Cameron, A.M. Dwyer, W. Golder, S. Maddock, and S.H. Schweitzer. 2017. Abundance, distribution, and geographic origin of non-breeding American Oystercatchers (*Haematopus palliatus*) in North Carolina. *Waterbirds* 40(sp1):86–94.

Afonso, R.O., F. Pina-Martins, V. Friesen, Z. Sun, L. Campioni, J. Madeiros, and M.C. Silva. 2023. No evidence of inbreeding depression despite a historical severe bottleneck in the endangered Bermuda Petrel (*Pterodroma cahow*). *Journal of Heredity*, esad030. https://doi.org/10.1093/jhered/esad030.

Aguillon, S.M., L. Campagna, R.G. Harrison, and I.J. Lovette. 2018. A flicker of hope: Genomic data distinguish Northern Flicker taxa despite low levels of divergence. *Auk* 135(3):748–766.

Ainley, D.G., D.A. Manuwal, J. Adams, and A.C. Thoresen. 2020a. Cassin's Auklet (*Ptychoramphus aleuticus*). Version 1.0. In *Birds of the World*, ed. A.F. Poole. Cornell Lab of Ornithology, Ithaca, NY. https://doi.org/10.2173/bow.casauk.01.

Ainley, D.G., W. McIver, J. Adams, and M. Parker. 2021a. Ashy Storm-Petrel (*Hydrobates homochroa*). Version 1.1. In *Birds of the World*, ed. P.G. Rodewald. Cornell Lab of Ornithology, Ithaca, NY. https://doi.org/10.2173/bow.asspet.01.1.

Ainley, D.G., D.N. Nettleship, and A.E. Storey. 2021b. Common Murre (*Uria aalge*). Version 2.0. In *Birds of the World*, ed. S.M. Billerman, P.G. Rodewald, and B.K. Keeney. Cornell Lab of Ornithology, Ithaca, NY. https://doi.org/10.2173/bow.commur.02.

Ainley, D.G., T.C. Telfer, and M.H. Reynolds. 2020b. Townsend's Shearwater (*Puffinus auricularis*). Version 1.0. In *Birds of the World*, ed. A.F. Poole. Cornell Lab of Ornithology, Ithaca, NY. https://doi.org/10.2173/bow.towshe1.01.

Ainley, D.G., T.C. Telfer, M.H. Reynolds, and A.F. Raine. 2020c. Newell's Shearwater (*Puffinus newelli*). Version 1.0. In *Birds of the World*, ed P.G. Rodewald. Cornell Lab of Ornithology, Ithaca, NY. https://doi.org/10.2173/bow.towshe2.01.

Airola, D., J. Estep, D. Krolick, R. Anderson, and J. Peters. 2019. Wintering areas and migration characteristics of Swainson's Hawks that breed in the Central Valley of California. *Journal of Raptor Research* 53(3):237–252.

Åkesson, S., R. Klaassen J. Holmgren, J.W. Fox, and A. Hedenström. 2012. Migration routes and strategies in a highly aerial migrant, the Common Swift *Apus apus*, revealed by light-level geolocators. *PLOS ONE* 7(7):e41195. https://doi.org/10.1371/journal.pone.0041195.

Akresh, M.E., and D.I. King. 2017. Examining migration and seasonal interaction of Prairie Warblers using geolocators and stable isotope analysis. Report for the Nuttall Ornithological Club, Cambridge, MA. https://www.nuttallclub.org/wp-content/uploads/2017/12/Examining-Migration-and-Seasonal-Interactions-of-Prairie-Warblers-Using-Geolocators-and-Stable-Isotope-Analysis.pdf.

Akresh, M.E., D.I. King, and P.P. Marra. 2019. Examining carry-over effects of winter habitat on breeding phenology and reproductive success in Prairie Warblers *Setophaga discolor*. *Journal of Avian Biology* 50(4);e02025.

———. 2021. Hatching date influences winter habitat occupancy: Examining seasonal interactions across the full life cycle in a migratory songbird. *Ecology and Evolution* 11:9241–9253.

Alaska Shorebird Group. 2019. Alaska Shorebird Conservation Plan. Version 3. Alaska Shorebird Group, Anchorage, AK.

Alberico, J.A.R., J.M. Reed, and L.W. Oring. 1992. Non-random philopatry of sibling Spotted Sandpipers. *Ornis Scandinavica* 23:504–508.

Albert C., H. Strøm, H.H. Helgason, V.S. Bråthen, F.T. Gudmundsson, P. Bustamante, and J. Fort. 2022. Spatial variations in winter Hg contamination affect egg volume in an Arctic seabird, the Great Skua (*Stercorarius skua*). *Environmental Pollution* 314:120322. https://doi.org/10.1016/j.envpol.2022.120322.

Albert, S.K., and R.B. Siegel. 2024. Improving the language of migratory bird science in North America. *Ornithological Applications* 126(1):duad059. https://doi.org/10.1093/ornithapp/duad059.

Aldrich, J.W., and A.J. Duvall. 1958. Distribution and migration of races of the Mourning Dove. *Condor* 60:108–128.

Alerstam T., J. Bäckman, J. Grönroos, P. Olofsson, and R. Strandberg. 2019. Hypotheses and tracking results about the longest migration: The case of the Arctic Tern. *Ecology and Evolution* 9:9511–9531

Alerstam T., M. Rosén, J. Bäckman, P.G.P. Ericson, and O. Hellgren. 2007. Flight speeds among bird species: Allometric and phylogenetic effects. *PLOS Biology* 5(8):e197. https://doi.org/10.1371/journal.pbio.0050197.

Alfrey, P., D. Monticelli, V. Legrand, and Corvo Birders. 2018. Nearctic vagrants on Corvo, Azores, in 2005–17. *Dutch Birding* 40:297–317.

Alisauskas, R.T., and K.A. Hobson. 1993. Determination of Lesser Snow Goose Diets and Winter Distribution Using Stable Isotope Analysis. *The Journal of Wildlife Management* 57(1):49–54. https://doi.org/10.2307/3808999.

Alisauskas, R.T., K.L. Drake, J.S. Caswell, and D.K. Kellett. 2012. Movement and persistence by Ross's Geese (*Chen rosii*) in Canada's Arctic. *Journal of Ornithology* 152(Suppl 2):S573–S584.

Alisauskas, R.T., K.L. Drake, S.M. Slattery, and D.K. Kellett. 2006. Neckbands, harvest, and survival of Ross's Geese from Canada's Central Arctic. *Journal of Wildlife Management* 70(1):89–100.

Alisauskas, R.T., R.F. Rockwell, K.W. Dufour, E.G. Cooch, G. Zimmerman, K.L. Drake, J.O. Leafloor, T.J. Moser, and E.T. Reed. 2011. Harvest, survival, and abundance of midcontinent Lesser Snow Geese relative to population reduction efforts. *Wildlife Monographs* 179:1–42.

Allen, P.E., L.J. Goodrich, and K.L. Bildstein. 1996. Within- and among-year effects of cold fronts on migrating raptors at Hawk Mountain, Pennsylvania, 1934–1991. *Auk* 113(2):329–338.

Allendorf, F.W., G.H. Luikart, and S.N. Aitken. 2012. *Conservation and the Genetics of Populations*. John Wiley & Sons, New York.

Allison, T.D., S. Perkins, K.F. Stryjewski, and M.D. Sorenson. 2013. Determining nocturnal locations, breeding ground locations, and genetic structure of Long-tailed Ducks wintering in Nantucket Sound. US Department of the Interior, Bureau of Ocean Energy Management, Final Report 2013-01141, https://marinecadastre.gov/espis/#/search/study/23163.

Almeida, J.B. 2009. Wintering ecology of Buff-breasted Sandpipers (*Tryngites subruficollis*) in southern Brazil. PhD diss., University of Nevada, Reno.

Altman, B., and R. Sallabanks. 2020. Olive-sided Flycatcher (*Contopus cooperi*). Version 1.0. In *Birds of the World*, ed. A.F. Poole. Cornell Lab of Ornithology, Ithaca, NY. https://doi.org/10.2173/bow.olsfly.01.

Alvarado, A.H., C.M. Bossu, R.J. Harrington, R.A. Bay, A.R.P. Nelson, T.B. Smith, and K.C. Ruegg. 2022. Genotype-environment associations across spatial scales reveal the importance of putative adaptive genetic variation in divergence. *Evolutionary Applications*. https://doi.org/10.1111/eva.13444.

Alvarado, A.H., T.L. Fuller, and T.B. Smith. 2014. Integrative tracking methods elucidate the evolutionary dynamics of a migratory divide. *Ecology and Evolution* 4:3456–3469.

Ambrosini, R., A.P. Møller, and N. Saino. 2009. A quantitative measure of migratory connectivity. *Journal of Theoretical Biology* 257(2):203–211.

Amélineau, F., B. Merkel, A. Tarroux, S. Descamps, T. Anker-Nilssen, O. Bjørnstad, … and H. Strøm. 2021. Six pelagic seabird species of the North Atlantic engage in a fly-and-forage strategy during their migratory movements. *Marine Ecology Progress Series* 676:127–144. https://doi.org/10.3354/meps13872.

American Ornithologists Union. 1957. *Checklist of North American Birds*. American Ornithologists Union.

Ammon, E.M., and W.M. Gilbert. 2020. Wilson's Warbler (*Cardellina pusilla*). Version 1.0. In *Birds of the World*, ed. P.G. Rodewald. Cornell Lab of Ornithology, Ithaca, NY. https://doi.org/10.2173/bow.wlswar.01.

Amos, E.J.R. 1991. *A Guide to the Birds of Bermuda*. Corncrake, Warwick, Bermuda.

Amundson, C.L., P.L. Flint, R.A. Stehn, R.M. Platte, H.M. Wilson, W.W. Larned, and J.B. Fischer. 2019. Spatio-temporal population change of Arctic-breeding waterbirds on the Arctic Coastal Plain of Alaska. *Avian Conservation & Ecology* 14(1).

Anderson, A.M., S. Duijns, P.A. Smith, C. Friis, and E. Nol. 2019. Migration distance and body condition influence shorebird migration strategies and stopover decisions during southbound migration. *Frontiers in Ecology and Evolution* 7:251.

Anderson, C., H. Gilchrist, R. Ronconi, K. Shlepr, D. Clark, D. Fifield, G. Robertson, and M. Mallory. 2020. Both short and long distance migrants use energy-minimizing migration strategies in North American Herring Gulls. *Movement Ecology* 8. https://doi.org/10.1186/s40462-020-00207-9.

Anderson, C., H. Gilchrist, R. Ronconi, K. Shlepr, D. Clark, D. Weseloh, G. Robertson, and M. Mallory. 2019. Winter home range and habitat selection differs among breeding populations of Herring Gulls in eastern North America. *Movement Ecology* 7. 10.1186/s40462-019-0152-x.

Anderson, D. Black-footed Albatross from Tern Island 1998–99. BirdLife International Seabird Tracking Database Dataset ID 452. https://data.seabirdtracking.org/dataset/452.

———. Laysan Albatross from Tern Island 1998–99. BirdLife International Seabird Tracking Database Dataset ID 453. https://data.seabirdtracking.org/dataset/453.

Anderson, E.M., R.D. Dickson, E.K. Lok, E.C. Palm, J.L. Savard, D. Bordage, and A. Reed. 2020. Surf Scoter (*Melanitta perspicillata*). Version 1.0. In *Birds of the World*, ed. P.G. Rodewald. Cornell Lab of Ornithology, Ithaca, NY. https://doi.org/10.2173/bow.sursco.01.

Anderson, J.G.T., and K.B. Anderson. 2005. An analysis of band returns of the American White Pelican 1922–1981. *Waterbirds* 28(special pub 1):55–60.

Anderson, K. 2013. Migration routes and chronology of American Black Ducks. Master's thesis, University of Delaware.

Andres, B.A., and G.A. Falxa. 2020. Black Oystercatcher (*Haematopus bachmani*). Version 1.0. In *Birds of the World*, ed. A.F. Poole and F.B. Gill. Cornell Lab of Ornithology, Ithaca, NY. https://doi.org/10.2173/bow.blkoys.01.

Andres, B.A., J.A. Johnson, J. Valenzuela, R.G. Morrison, L.A. Espinosa, and R.K. Ross. 2009. Estimating eastern Pacific coast populations of Whimbrels and Hudsonian Godwits, with an emphasis on Chiloé Island, Chile. *Waterbirds*, 32(2):216–224.

Andres, B.A., P.A. Smith, R.G. Morrison, C.L. Gratto-Trevor, S.C. Brown, and C.A. Friis. 2012. Population estimates of North American shorebirds. *Wader Study Group Bulletin* 119:178–194.

Antas, P.T.Z. 1983. Migration of Nearctic shorebirds (Charadriidae and Scolopacidae) in Brazil: Flyways and their different seasonal use. *Wader Study Group Bulletin* 39:52–56.

Arata, J.A., P.R. Sievert, and M.B. Naughton. 2009. Status assessment of Laysan and Black-footed Albatrosses, North Pacific Ocean, 1923–2005. *Scientific Investigations Report* 2009–5131. https://doi.org/10.3133/sir20095131.

Arcese, P., M.K. Sogge, A.B. Marr, and M.A. Patten. 2020. Song Sparrow (*Melospiza melodia*). Version 1.0. In *Birds of the World*, ed. A.F. Poole and F.B. Gill. Cornell Lab of Ornithology, Ithaca, NY. https://doi.org/10.2173/bow.sonspa.01.

Areta, J.I., S.A. Salvador, F.A. Gandoy, E.S. Bridge, F.C. Gorleri, T.M. Pegan, E.R. Gulson-Castillo, K.A. Hobson, and D.W. Winkler. 2021. Rapid adjustments of migration and life history in hemisphere-switching Cliff Swallows. *Current Biology* 31(13):2914–2919.e2. https://doi.org/10.1016/j.cub.2021.04.019.

Arizaga, J., D. Alonso, J.A. Cortés, T. Eggenhuizen, J. Foucher, D. Franz, J. García, F. Koning, M. Leconte, H. Rguibi, T. Valdenburg, P. Vera, and K.A. Hobson. 2015. Migratory connectivity in European bird populations: Feather stable isotope values correlate with biometrics of breeding and wintering Bluethroats *Luscinia svecica*. *Ardeola* 62(2):255–267.

Arizaga, J., and I. Tamayo. 2013. Connectivity patterns and key non-breeding areas of White-throated Bluethroat (*Luscinia svecica*) European populations. *Animal Biodiversity and Conservation* 36(1):69–78.

Arizaga, J., S.L. Van Wilgenburg, D. Alonso, J.A. Cortés, M. Leconte, H. Rguibi, T. Valkenburg, P. Vera, and K.A. Hobson. 2016. Breeding origins and pattern of migration of Bluethroats *Luscinia svecica* wintering from Iberia to Senegal as revealed by stable isotopes. *Bird Study* 63(2):196–202.

Arkhipov, V.Y. 2009. The first record of Wilson's Snipe (*Gallinago (g.) delicata*) in Russia and comments on its taxonomic status. *Zoologichesky zhurnal* 88(9):1146–1149.

Arnberg, N.N., D. Shizuka, A.S. Chaine, and B.E. Lyon. 2015. Social network structure in wintering Golden-crowned Sparrows is not correlated with kinship. *Molecular Ecology* 24:5034–5044. https://doi.org/10.1111/mec.13366.

Arnold, J.M., S.A. Oswald, I.C.T. Nisbet, P. Pyle, and M.A. Patten. 2020. Common Tern (*Sterna hirundo*). Version 1.0. In *Birds of the World*, ed. S.M. Billerman. Cornell Lab of Ornithology, Ithaca, NY. https://doi.org/10.2173/bow.comter.01.

Arnold, K.A., and L.J. Folse. 1977. Movements of the Great-tailed Grackle in Texas. *Wilson Bulletin* 89:602–608.

Asante, C.K., T.D. Jardine, S.L. Van Wilgenburg, and K.A. Hobson. 2017. Tracing origins of waterfowl using the Saskatchewan River Delta: Incorporating stable isotope approaches in continent-wide waterfowl management and conservation. *Condor* 119(2):261–274.

Ashley, P., K.A. Hobson, S.L. Van Wilgenburg, N. North, and S. A. Petrie. 2010. Linking Canadian harvested juvenile American Black Ducks to their natal areas using stable isotope (delta D, delta 13C, and delta 15N) methods. *Avian Conservation and Ecology* 5:13.

Askren, R.J., E.R. Massey, J.D., James, and D.C. Osborne. 2022. Migration chronology and multi-scale habitat selection of wintering midcontinent Greater White-fronted Geese. *Global Ecology and Conservation* 39:e02290. https://doi.org/10.1016/j.gecco.2022.e02290.

Atkinson, P.W., A.J. Baker, R.M. Bevan, N.A. Clark, K.B. Cole, P.M. Gonzalez, … and R.A. Robinson. 2005. Unravelling the migration and moult strategies of a long-distance migrant using stable isotopes: Red Knot *Calidris canutus* movements in the Americas. *Ibis* 147(4):738–749.

Atlantic Flyway Council. 2008. *Management Plan for the North Atlantic Population of Canada Geese*.

Austin, J.J., R.W.G. White, and J.R. Ovenden. 1994. Population-genetic structure of a philopatric, colonially nesting seabird, the Short-Tailed Shearwater (*Puffinus Tenuirostris*). *Auk* 111(1):70–79. https://doi.org/10.2307/4088506.

Avery, M.L. 1979. Review of avian mortality due to collisions with manmade structures. https://digitalcommons.unl.edu/icwdmbirdcontrol/2/.

Avery, M.L. 2020. Rusty Blackbird (*Euphagus carolinus*). Version 1.0. In *Birds of the World*, ed. A.F. Poole. Cornell Lab of Ornithology, Ithaca, NY. https://doi.org/10.2173/bow.rusbla.01

Avise, J.C., R.T. Alisauskas, W.S. Nelson, and C.D. Ankney. 1992. Matriarchal population genetic structure in an avian species with female natal philopatry. *Evolution* 46(4):1084–1096.

Avise J.C., W.S. Nelson, B.W. Bowen, and D. Walker. 2000. Phylogeography of colonially nesting seabirds, with special reference to global matrilineal patterns in the Sooty Tern (*Sterna fuscata*) *Molecular Ecology* 9:1783–1792. https://doi.org/10.1046/j.1365-294x.2000.01068.x.

Awkerman, J.A., D.J. Anderson, and G.C. Whittow. 2020a. Black-footed Albatross (*Phoebastria nigripes*). Version 1.0. In *Birds of the World*, ed. A.F. Poole. Cornell Lab of Ornithology, Ithaca, NY. https://doi.org/10.2173/bow.bkfalb.01.

———. 2020b. Laysan Albatross (*Phoebastria immutabilis*). Version 1.0. In *Birds of the World*, ed. A.F. Poole. Cornell Lab of Ornithology, Ithaca, NY. https://doi.org/10.2173/bow.layalb.01.

Baak, J.E., M.L. Leonard, C. Gjerdrum, M.D. Dodds, and R.A. Ronconi. 2021a. Non-breeding movements and foraging ecology of the Black Guillemot *Cepphus grylle* in Atlantic Canada. *Marine Ornithology* 49:57–70.

Baak, J.E., A. Patterson, H.G. Gilchrist, and K.H. Elliott. 2021b. First evidence of diverging migration and overwintering strategies in Glaucous Gulls (*Larus hyperboreus*) from the Canadian Arctic. *Animal Migration* 8(1):98–109. https://doi.org/10.1515/ami-2020-0107.

Badyaev, A.V., D.D. Gibson, B. Kessel, P. Pyle, and M.A. Patten. 2020b. White Wagtail (*Motacilla alba*). Version 1.0. In *Birds of the World*, ed. S.M. Billerman. Cornell Lab of Ornithology, Ithaca, NY. https://doi.org/10.2173/bow.whiwag.01.

Badyaev, A.V., B. Kessel, D. D. Gibson, J. del Hoyo, and N. Collar. 2020a. Eastern Yellow Wagtail (*Motacilla tschutschensis*). Version 1.0. In *Birds of the World*, ed. S.M. Billerman. Cornell Lab of Ornithology, Ithaca, NY. https://doi.org/10.2173/bow.eaywag.01

Baetscher, D.S., J. Beck, E.C. Anderson, K. Ruegg, A.M. Ramey, S. Hatch, H. Nevins, S.M. Fitzgerald, J.C. Garza. 2022. Genetic assignment of fisheries bycatch reveals disproportionate mortality among Alaska Northern Fulmar breeding colonies. *Evolutionary Applications* 15(3):447–458. https://doi.org/10.1111/eva.13357.

Bailey, A.M. 1925. A report on the birds of northwestern Alaska and regions adjacent to Bering Strait, part 4. *Condor* 27:164–171.

Bailey, I.E., G. Segelbacher, S.D. Healy, A. Hurly, and J.M. Pemberton. 2013. Microsatellite variation in Rufous Hummingbirds (*Selasphorus rufus*) and evidence for a weakly structured population. *Journal of Ornithology* 154:1029–1037.

Bailey, T.N., M.F. Porter, E.E. Bangs, W.W. Larned, R.A. Richey, and R.L. Delaney. 1990. Summer and migratory movements of Trumpeter Swans using the Kenai National Wildlife Refuge, Alaska. In *Proceedings and Papers of the 11th Trumpeter Swan Society Conference*, ed. D. Compton, 72–91. Trumpeter Swan Society, Maple Plain, MN.

Bairlein, F. Results of bird ringing in the study of migration routes and behaviour. 2001. *Ardea* 89(special issue):7–19.

Bairlein, F., D.R. Norris, R. Nagel, M. Bulte, C.C. Voigt, J.W. Fox, D.J.T. Hussell, and H. Schmaljohann. 2012. Cross-hemisphere migration of a 25 g songbird. *Biology Letters* 8:505–507.

Baker, A., P. Gonzalez, R.I.G. Morrison, and B.A. Harrington. 2020. Red Knot (*Calidris canutus*). Version 1.0. In *Birds of the World*, ed. S.M. Billerman. Cornell Lab of Ornithology, Ithaca, NY. https://doi.org/10.2173/bow.redkno.01.

Baker, A.J., and J.G. Strauch Jr. 1988. Genetic variation and differentiation in shorebirds. *Acta XIX Congressus Internationalis Ornithologici* 1639–1645.

Baker, J.D., C.L. Littnan, and D.W. Johnston. 2006. Potential effects of sea level rise on the terrestrial habitats of endangered and endemic megafauna in the Northwestern Hawaiian Islands. *Endangered Species Research* 4:1–10.

Baldassarre, D.T., L. Campagna, H.A. Thomassen, J.W. Atwell, M. Chu, L.H. Crampton, R.C. Fleischer, and C. Riehl. 2019. GPS tracking and population genomics suggest itinerant breeding across drastically different habitats in the Phainopepla. *Auk* 136:1–12.

Baldassarre, G. 2014. *Ducks, Geese, and Swans of North America*. 2 vols. John Hopkins University Press, Baltimore, MD.

Ball Jr., R.M., and J.C. Avise. 1992. Mitochondrial DNA phylogeographic differentiation among avian populations and the evolutionary significance of subspecies. *Auk* 109(3):626–636.

Baltz, M.E., and S.C. Latta. 2020. Cape May Warbler (*Setophaga tigrina*). Version 1.0. In *Birds of the World*, ed. P.G. Rodewald. Cornell Lab of Ornithology, Ithaca, NY. https://doi.org/10.2173/bow.camwar.01.

Banks, R.C. 1985. American Black Duck record from Korea. *Journal of Field Ornithology* 56:277–277.

Bannor, B.K., and E. Kiviat. 2020. Common Gallinule (*Gallinula galeata*). Version 1.0. In *Birds of the World*, ed. A.F. Poole and F.B. Gill. Cornell Lab of Ornithology, Ithaca, NY. https://doi.org/10.2173/bow.comgal1.01.

Baran, M.A., S.W. Kress, P. Shannon, D.E. Lyons, H.L. Major, and A.W. Diamond. 2022. Overwinter movement of Atlantic Puffins (*Fratercula arctica*) breeding in the Gulf of Maine: Inter- and intra-colony effects. *Waterbirds* 45(1):1–16.

Barbaree, B.A., M.E. Reiter, C.M. Hickey, N.K. Elliott, D. Schaffer-Smith, M.D. Reynolds, and G.W. Page. 2018. Dynamic surface water distributions influence wetland connectivity within a highly modified interior landscape. *Landscape Ecology* 33:829–844.

REFERENCES

Barbaree, B.A., M.E. Reiter, C.M. Hickey, and G.W. Page. 2016. Molt migration and migratory connectivity of the Long-billed Dowitcher. *Journal of Wildlife Management* 80:256–265.

Barber, D.R., P.M. Barber, and P.G. Jablonski. 2020. Painted Redstart (*Myioborus pictus*). Version 1.0. In *Birds of the World*, ed. A.F. Poole and F.B. Gill. Cornell Lab of Ornithology, Ithaca, NY. https://doi.org/10.2173/bow.paired.01.

Barisas, D.A., J. Amouret, G.T. Hallgrímsson, R.W. Summers, and S. Pálsson. 2015. A review of the subspecies status of the Icelandic Purple Sandpiper *Calidris maritima littoralis*. *Zoological Journal of the Linnean Society* 175(1):211–221.

Barlow, J.C., S.N. Leckie, and C.T. Baril. 2020. Gray Vireo (*Vireo vicinior*). Version 1.0. In *Birds of the World*, ed. A.F. Poole and F.B. Gill. Cornell Lab of Ornithology, Ithaca, NY. https://doi.org/10.2173/bow.gryvir.01.

Barr, K., A. Beichman, P. Kalhori, J. Rajbhandary, R. Bay, K. Ruegg, and T. Smith. 2020. Persistent panmixia despite extreme habitat loss and population decline in the threatened Tricolored Blackbird (*Agelaius tricolor*). *Evolutionary Applications* 14:674–684.

Barrett, V.A., and E.R. Vyse. 1982. Comparative genetics of three Trumpeter Swan populations. *Auk* 99:103–108.

Bart, J., et al. 2012. North Slope of Alaska. In *Arctic Shorebirds in North America: A Decade of Monitoring*, ed. J. Bart and V. Johnston, 37–96. *Studies in Avian Biology*, no. 44. University of California Press.

Bartolomé, C., X. Maside, C.J. Camphuysen, M. Heubeck, and R. Bao. 2011. Multilocus population analysis of *Gavia immer* (Aves: Gaviidae) mtDNA reveals low genetic diversity and lack of differentiation across the species breeding range. *Organisms Diversity & Evolution* 11:307–316.

Bartos Smith, S., and J.S. Greenlaw. 2020. Spotted Towhee (*Pipilo maculatus*). Version 1.0. In *Birds of the World*, ed. P.G. Rodewald. Cornell Lab of Ornithology, Ithaca, NY. https://doi.org/10.2173/bow.spotow.01.

Bartsch, P. 1904. Notes on the herons of the District of Columbia. Smithsonian Miscellaneous Collections. Pub. No. 1419. 45:104–111.

Bartzen, B.A., D.L. Dickson, and T.D. Bowman. 2016. Migration characteristics of Long-tailed Ducks (*Clangula hyemalis*) from the western Canadian Arctic. *Polar Biology* 40:1085–1099.

Basille, M., J. Watling, S. Romañach, and R. Borkhataria. 2020. Joint seasonality in geographic and ecological spaces, illustrated with a partially migratory bird. *Ecosphere* 11(5):e03110. https://doi.org/10.1002/ecs2.3110.

Bateman, M.C., and R.W. Daury. 1994. Update on Canada Goose band recoveries as they relate to the Atlantic Region. Canadian Wildlife Service, unpublished report. Sackville, NB.

Battey, C.J., E.B. Linck, K.L. Epperly, C. French, D.L. Slager, P.W. Sykes Jr, and J. Klicka. 2018. A migratory divide in the Painted Bunting (*Passerina ciris*). *American Naturalist* 191(2):259–268.

Battley, P.F., N. Warnock, T.L. Tibbitts, R.E. Gill Jr., T. Piersma, C.J. Hassell, D.C. Douglas, D.M. Mulcahy, B.D. Gartrell, R. Schuckard, D.S. Melville, and A.C. Riegen. 2012. Contrasting extreme long-distance migration patterns in Bar-tailed Godwits *Limosa lapponica*. *Journal of Avian Biology* 43:21–32.

Bauer, S., S. Lisovski, and S. Hahn. 2016. Timing is crucial for consequences of migratory connectivity. *Oikos* 125(5):605–612.

Baumann, M.J., S.V. Brant, S.M. Bauernfeind, C.R. Gerhart, J.L. Williamson, A.B. Johnson, G.M. Spellman, S.R. Uhrig, S. West, and C.C. Witt. 2023. Freshwater parasites as potential barriers to seabird dispersal: Evidence from vagrant booby specimens in western North America. *Wilson Journal of Ornithology* 135(3):327–344.

Bay, R.A., R.J. Harrigan, V.L. Underwood, H.L. Gibbs, T.B. Smith, and K. Ruegg. 2018. Genomic signals of selection predict climate-driven population declines in a migratory bird. *Science* 359(6371):83–86.

Bay, R.A., D.S. Karp, J.F. Saracco, W.R. Anderegg, L.O. Frishkoff, D. Wiedenfeld, T.B. Smith, and K. Ruegg. 2021. Genetic variation reveals individual-level climate tracking across the annual cycle of a migratory bird. *Ecology Letters* 24(4):819–828.

Bayly, N.J., L. Cárdenas-Ortiz, M. Rubio, and C. Gómez. 2014. Migration of raptors, swallows, and other diurnal migratory birds through the Darien of Colombia. *Ornitología Neotropical* 25:63–71.

Bayly, N.J., C. Gómez, K.A. Hobson, A.M. González, and K.V. Rosenberg. 2012. Fall migration of the Veery (*Catharus fuscescens*) in northern Colombia: Determining the energetic importance of a stopover site. *Auk* 129(3):449–459. https://doi.org/10.1525/auk.2012.11.188.

Bayly, N.J., C. Gómez, K.V. Rosenberg, and K.A. Hobson. 2019. Habitat choice shapes the stopover strategy of a Nearctic-Neotropical migratory songbird. *Journal of Ornithology* 160:377–388. https://doi.org/10.1007/s10336-019-01624-3.

Bayly, N.J., D.R. Norris, P.D. Taylor, K.A. Hobson, and A. Morales-Rozo. 2020. There's no place like home: Tropical overwintering sites may have a fundamental role in shaping migratory strategies. *Animal Behaviour* 162:95–104. https://doi.org/10.1016/j.anbehav.2020.02.003.

Bayly, N.J., K.V. Rosenberg, D.R. Norris, P.D. Taylor, and K.A. Hobson. 2021. Rapid recovery by fat- and muscle-depleted Blackpoll Warblers following trans-oceanic migration is driven by time-minimization. *Ornithology* 138:1–15. https://doi.org/10.1093/ornithology/ukab055.

Beam, J.K., L.A. Jaster, W.E. Jensen, and W.E. Lanyon. 2022. Chihuahuan Meadowlark (*Sturnella lilianae*). Version 1.0. In *Birds of the World*, ed. N.D. Sly. Cornell Lab of Ornithology, Ithaca, NY. https://doi.org/10.2173/bow.lilmea2.01.

Beason, J.P., C. Gunn, K.M. Potter, R.A. Sparks, and J.W. Fox. 2012. The Northern Black Swift: Migration path and wintering area revealed. *Wilson Journal of Ornithology* 124(1):1–8.

Beason, R.C. 2020. Horned Lark (*Eremophila alpestris*). Version 1.0. In *Birds of the World*, ed. S.M. Billerman. Cornell Lab of Ornithology, Ithaca, NY. https://doi.org/10.2173/bow.horlar.01.

Beason, R.C., J.S. Humphrey, N.E. Myers, and M.L. Avery. 2010. Synchronous monitoring of vulture movements with satellite telemetry and avian radar. *Journal of Zoology* 282:157–162. https://doi.org/10.1111/j.1469-7998.2010.00723.x.

Beason, R.C., and R.H. Wauer. 2020. Colima Warbler (*Leiothlypis crissalis*). Version 1.0. In *Birds of the World*, ed. A.F. Poole. Cornell Lab of Ornithology, Ithaca, NY. https://doi.org/10.2173/bow.colwar.01.

Beauchamp, W.D., F. Cooke, C. Lougheed, L.W. Lougheed, C.J. Ralph, and S. Courtney. 1999. Seasonal movements of Marbled Murrelets: Evidence from banded birds. *Condor* 101:671–674.

Bechard, M.J., C.S. Houston, J.H. Sarasola, and A.S. England. 2020. Swainson's Hawk (*Buteo swainsoni*). Version 1.0. In *Birds of the World*, ed. A.F. Poole. Cornell Lab of Ornithology, Ithaca, NY. https://doi.org/10.2173/bow.swahaw.01.

Becker, B.H., M.Z. Peery, and S.R. Beissinger. 2007. Ocean climate and prey availability affect the trophic level and reproductive success of the Marbled Murrelet, an endangered seabird. *Marine Ecology Progress Series* 329:267–279.

Beckett, S.R., and G.A. Proudfoot. 2011. Large-scale movement and migration of Northern Saw-whet Owls in eastern North America. *Wilson Journal of Ornithology* 123(3):521–535.

Bednarz, J.C., D. Klem Jr., L.J. Goodrich, and S.E. Senner. 1990. Migration counts of raptors at Hawk Mountain, Pennsylvania, as indicators of population trends, 1934–1986. *Auk* 107(1):96–109.

Bedrosian, B.E., S.L. Cain, S. Wolff, and D.J. Craighead. 2015. Migratory pathways, timing, and home ranges of Southern Greater Yellowstone Osprey. *Journal of Raptor Research* 49:325–332.

Bedrosian, B.E., R. Domenech, A. Shreading, M.M. Hayes, T.L. Booms, and C.R. Barger. 2018. Migration corridors of adult Golden Eagles originating in northwestern North America. *PLOS ONE* 13(11):e0205204.

Beedy, E.C., W.J. Hamilton III, R.J. Meese, D.A. Airola, W.S. Schackwitz, and P. Pyle. 2023. Tricolored Blackbird (*Agelaius tricolor*). Version 2.0. In *Birds of the World*, ed. P.G. Rodewald and B.K. Keeney. Cornell Lab of Ornithology, Ithaca, NY. https://doi.org/10.2173/bow.tribla.02.

Bégin-Marchand, C., A. Desrochers, P.D. Taylor, et al. 2021. Spatial structure in migration routes maintained despite regional convergence among eastern populations of Swainson's Thrushes. *Movement Ecology* 9:23. https://doi.org/10.1186/s40462-021-00263-9.

Behle, W.H., and W.A. Goates. 1957. Breeding biology of the California Gull. *Condor* 59:235–246.

Behle, W.H., E.D. Sorenson, and C.M. White. 1985. *Utah Birds: A Revised Checklist*. Utah Museum of Natural History, Salt Lake City, UT.

Belant, J.L., R.K. Anderson, and J.M. Wilson. 1991. Winter recoveries and territorial affinity of Common Loons banded in Wisconsin. *Wilson Bulletin* 103(1):141–142.

Belant, J.L., and R.A. Dolbeer. 1993. Migration and dispersal of Laughing Gulls in the United States (*Migración y patrones de dispersión de Individuos de Larus Atricilla en Los Estados Unidos*). *Journal of Field Ornithology* 64(4):557–565. www.jstor.org/stable/4513869.

Bell, C.P. 1997. Leap-frog migration in the Fox Sparrow: Minimizing the cost of spring migration. *Condor* 99:470–477.

Bell, D.A., 1996. Genetic differentiation, geographic variation, and hybridization in gulls of the *Larus glaucescens-occidentalis* complex. *Condor* 98(3):527–546.

Bellrose, F.C., and R.D. Crompton. 1970. Migrational behavior of Mallards and Black Ducks as determined from banding. *Illinois Natural History Survey Bulletin* 30:167–234.

Bellrose, F.C., and D.J. Holm. 1994. *Ecology and Management of the Wood Duck*. Stackpole Books, Harrisburg, PA.

Bemis, C., and J.D. Rising. 2020. Western Wood-Pewee (*Contopus sordidulus*). Version 1.0. In *Birds of the World*, ed. A.F. Poole and F.B. Gill. Cornell Lab of Ornithology, Ithaca, NY. https://doi.org/10.2173/bow.wewpew.01.

Benedict, L., N.B. Warning, N.A. Najar, S.G. Pitt, P.E. Lowther, D.E. Kroodsma, and G.H. Farley. 2021. Rock Wren (*Salpinctes obsoletus*). Version 2.0. In *Birds of the World*, ed. P.G. Rodewald and B.K. Keeney. Cornell Lab of Ornithology, Ithaca, NY. https://doi.org/10.2173/bow.rocwre.02.

Benkman, C.W. 2020. White-winged Crossbill (*Loxia leucoptera*). Version 1.0. In *Birds of the World*, ed. S.M. Billerman. Cornell Lab of Ornithology, Ithaca, NY. https://doi.org/10.2173/bow.whwcro.01.

Benkman, C.W., and M.A. Young. 2020. Red Crossbill (*Loxia curvirostra*). Version 1.0. In *Birds of the World*, ed. S.M. Billerman, B.K. Keeney, P.G. Rodewald, and T.S. Schulenberg. Cornell Lab of Ornithology, Ithaca, NY. https://doi.org/10.2173/bow.redcro.01.

Bennett, R.E., A.D. Rodewald, K.V. Rosenberg, R. Chandler, L. Chavarria-Duriaux, J.A. Gerwin, D.I. King, and J.L. Larkin. 2019. Drivers of variation in migration behavior for a linked population of long-distance migratory passerine. *Auk* 136(4):ukz051. doi.org/10.1093/auk/ukz051.

Bennett, R.E., S.B. Swarthout, J. Bolsinger, A.D. Rodewald, K.V. Rosenberg, and R. Rohrbaugh. 2017. Extreme genetic similarity does not predict non-breeding strategy of two closely related migratory warblers. *Journal of Field Ornithology* 88(2):156–168.

Bent, A.C. 1927. Life histories of North American shorebirds. Part 1. *US National Museum Bulletin* 142.

———. 1937. Life histories of North American birds of prey. Smithsonian Institution. *US National Museum Bulletin* 167:112–123.

Bentzen, R.L., and A.N. Powell. 2015. Dispersal, movements and site fidelity of post-fledging King Eiders *Somateria spectabilis* and their attendant females. *Ibis* 157:133–146.

Bermingham, E., S. Rohwer, S. Freeman, and C. Wood. 1992. Vicariance biogeography in the Pleistocene and speciation in North American Wood Warblers: A test of Mengel's model. *Proceedings of the National Academy of Sciences* 89:6624–6628.

Bernard, A., A.S. Rodrigues, V. Cazalis, and D. Grémillet. 2021. Toward a global strategy for seabird tracking. *Conservation Letters* 14(3):e12804.

Bernath-Plaisted, J., M.D. Correll, N.A. Guido, and A.O. Panjabi. 2019. *Demographic Monitoring of Breeding Grassland Songbirds in the Northern Great Plains*. 2018 Annual Report. Bird Conservancy of the Rockies, Brighton, Colorado.

Bertram, D.F., C.A. MacDonald, P.D. O'Hara, J.L. Cragg, R. Corcoran, R. Greene, P. Vincent, and K.J. Woo. 2023. Summer movements of Marbled Murrelets from Canada to Alaska. *Endangered Species Research* 51:215–225.

Bertram, D.F., C.A. MacDonald, P.D. O'Hara, J.L. Cragg, M.H. Janssen, M. McAdie, and W.S. Boyd. 2016. Marbled Murrelet *Brachyramphus marmoratus* movements and marine habitat use near proposed tanker routes to Kitimat, BC. *Marine Ornithology* 44:3–9.

Beske, A.E. 1982. Local and migratory movements of radio-tagged juvenile harriers. *Raptor Research* 16:39–53.

Bêty, J., J-F. Giroux, and G. Gauthier. 2004. Individual variation in timing of migration: Causes and reproductive consequences in Greater Snow Geese (*Anser caerulescens atlanticus*). *Behavioral Ecology and Sociobiology* 57:1–8.

Beuth, J.M. 2013. Body composition, movement phenology, and habitat use of Common Eider along the southern New England coast. Master's thesis, University of Rhode Island.

Bianchini, K., D.J. Newstead, and C.A. Morrissey. 2020. Differences in migration timing along the Midcontinental Flyway in Sanderling (*Calidris alba*) from three Gulf of Mexico staging areas. *Waterbirds* 43(3–4):225–356.

Bibles, B.D., R.L. Glinski, and R.R. Johnson. 2020. Gray Hawk (*Buteo plagiatus*). Version 1.0. In *Birds of the World*, ed. A.F. Poole and F.B. Gill. Cornell Lab of Ornithology, Ithaca, NY. https://doi.org/10.2173/bow.gryhaw2.01.

Bierregaard, R.O., K.B. Livezey, P. Pyle, K.M. Mazur, and P.C. James. 2024. Barred Owl (*Strix varia*). Version 2.0. In *Birds of the World*, ed. P.G. Rodewald, B.K. Keeney, and M.G. Smith. Cornell Lab of Ornithology, Ithaca, NY. https://doi.org/10.2173/bow.brdowl.02.

Bierregaard, R.O., A.F. Poole, M.S. Martell, P. Pyle, and M.A. Patten. 2020. Osprey (*Pandion haliaetus*). Version 2.0. In *Birds of the World*, ed. P.G. Rodewald. Cornell Lab of Ornithology, Ithaca, NY. https://doi.org/10.2173/bow.osprey.01.

Bildstein, K.L. 2004. Raptor migration in the Neotropics: Patterns, processes, and consequences. *Ornitologia Neotropical* 15(suppl.):83–99.

———. 2006. *Migrating Raptors of the World*. Cornell University Press, Ithaca, NY.

Bildstein, K.L., F.R. Santana, L.O. Melian, and M. Martell. 2002. A new migration route for Swallow-tailed Kite (*Elanoides forficatus*) in east Cuba. *Cotinga* 18:93–95.

Biles, K.S., J.C. Bednarz, S.E. Schulwitz, and J.A. Johnson. 2023. Tracking device attachment methods for American Kestrels: Backpack versus leg-loop harnesses. *Journal of Raptor Research* 57:304–313.

Birchard, K. 2019. Using population genomics to disentangle the evolutionary history and aid conservation of the Leach's Storm-petrel species complex (*Hydrobates* spp.). Master's thesis, Queen's University Kingston, ON.

Bird Conservancy of the Rockies. 2016. A tale of two Osprey continues. https://birdconservancy.org/a-tale-of-two-osprey/.

BirdLife International and *Handbook of the Birds of the World*. 2024. Bird species distribution maps of the world. Version 2024.2.

Birds of the World (website). 2026. Ed. S.M. Billerman, B.K. Keeney, P.G. Rodewald, and T.S. Schulenberg. Cornell Laboratory of Ornithology, Ithaca, NY. https://birdsoftheworld.org/.

Birt, T.P., H.R. Carter, D.L. Whitworth, A. McDonald, S.H. Newman, F. Gress, E. Palacios, J.S. Koepke, and V.L. Friesen. 2012. Rangewide population genetic structure of Xantus's Murrelet (*Synthliboramphus hypoleucus*). *Auk* 129(1):44–55. https://doi.org/10.1525/auk.2011.11011.

Birt, T.P., D. Mackinnon, J.F. Piatt, and V.L. Friesen, V. L. 2011. Genetic differentiation of the Kittlitz's Murrelet *Brachyramphus Brevirostris* in the Aleutian Islands and Gulf of Alaska. *Marine Ornithology* 39(1):article 6.

Birt-Friesen, V.L., W.A. Montevecchi, A.J. Gaston, and W.S. Davidson. 1992. Genetic structure of Thick-billed Murre (*Uria lomvia*) populations examined using direct sequence analysis of amplified DNA. *Evolution* 46:267–272.

Bishop, M.A., and N. Warnock. 1998. Migration of Western Sandpipers: Links between their Alaskan stopover areas and breeding grounds. *Wilson Bulletin* 110(4):457–462.

Blanco, D.E., and B. López-Lanús. 2008. Non-breeding distribution and conservation of the Upland Sandpiper (*Bartramia longicauda*) in South America. *Ornitologia Neotropical* 19:613–621.

Bleho, B., K. Ellison, D.P. Hill, and L.K. Gould. 2020. Chestnut-collared Longspur (*Calcarius ornatus*). Version 1.0. In *Birds of the World*, ed. P.G. Rodewald. Cornell Lab of Ornithology, Ithaca, NY. https://doi.org/10.2173/bow.chclon.01.

Blight, L.K., K.A. Hobson, T.K. Kyser, and P. Arcese. 2015. Changing gull diet in a changing world: A 150-year stable isotope (δ13C, δ15N) record from feathers collected in the Pacific Northwest of North America. *Global Change Biology* 21:1497–1507. https://doi.org/10.1111/gcb.12796.

Block, T.A., B.E. Lyon, Z. Mikalonis, A.S. Chaine, and D. Shizuka. 2023. Social connections across migration: Do Golden-crowned Sparrows that socialize in winter also breed together? *Wilson Journal of Ornithology* 135(4):569–575. https://doi.org/10.1676/23-00014.

Blohm, R.J. 1978. Migrational homing of male Gadwalls to breeding grounds. *Auk* 95(4):763–766.

Bloom, P.H., M.D. McCrary, J.M. Papp, and S.E. Thomas. 2017. Banding reveals potential northward migration of Cooper's Hawks from Southern California. *Journal of Raptor Research* 51(4):409–416.

Bloom, P.H., M.D. McCrary, J.M. Scott, J.M. Papp, K.J. Sernka, S.E. Thomas, J.W. Kidd, E.H. Henckel, J.L. Henckel, and M.J. Gibson. 2015. Northward summer migration of Red-Tailed Hawks fledged from southern latitudes. J. Raptor Res. 49(1):1–17.

Bloom, P.H., P.M. Scott, J.M. Papp, S.E. Thomas, and J.W. Kidd. 2011. Vagrant western Red-shouldered Hawks: Origins, natal dispersal patterns, and survival. *Condor* 113(3):538–546.

Blouin, F., J-F. Giroux, J. Ferron, G. Gauthier, and G.J. Doucet. 1999. The use of satellite telemetry to track Greater Snow Geese. *Journal of Field Ornithology* 70(2):187–199.

Bodey, T.W., I.R. Cleasby, F. Bell, N. Parr, A. Schultz, S.C. Votier, and S. Bearhop. 2018. A phylogenetically controlled meta-analysis of biologging device effects on birds: Deleterious effects and a call for more standardized reporting of study data. *Methods in Ecology and Evolution* 9(4):946–955.

Boersma, P.D., and M.C. Silva. 2021. Fork-tailed Storm-Petrel (*Hydrobates furcatus*). Version 1.1. In *Birds of the World*, ed. A.F. Poole and F.B. Gill. Cornell Lab of Ornithology, Ithaca, NY. https://doi.org/10.2173/bow.ftspet.01.1.

Bogdanova, Maria I., et al. 2011. Seasonal interactions in the Black-Legged Kittiwake, *Rissa tridactyla*: Links between breeding performance and winter distribution. *Proceedings of the Royal Society: Biological Sciences* 278(1717):2412–2418. www.jstor.org/stable/41314947.

Boland, J.M. 1990. Leapfrog migration in North American shorebirds: Intra-and interspecific examples. *Condor* 92(2):284–290.

Bond, A.L., K.A. Hobson, and B.A. Branfireun. 2015. Rapidly increasing methyl mercury in endangered Ivory Gull (*Pagophila eburnea*) feathers over a 130 year record. *Proceedings of the Royal Society: Biological Sciences* 282(1805):20150032. https://doi.org/10.1098/rspb.2015.0032. PMID: 25788594; PMCID: PMC4389618.

Bonerbo, Z., and L. Goodrich. 2020. Linking hawk watchsites to breeding and wintering catchment areas in Red-tailed Hawk. *Journal of Hawk Migration Studies* 46(2):9–16.

Bonnet-Lebrun, A.S., P. Catry, T.J. Clark, L. Campioni, et al. 2020. Habitat preferences, foraging behaviour and bycatch risk among breeding Sooty Shearwaters *Ardenna grisea* in the southwest Atlantic. *Marine Ecology Progress Series* 651:163–181. https://doi.org/10.3354/meps13439.

Booms, T.L., T.J. Cade, and N.J. Clum. 2020. Gyrfalcon (*Falco rusticolus*). Version 1.0. In *Birds of the World*, ed. S.M. Billerman. Cornell Lab of Ornithology, Ithaca, NY. https://doi.org/10.2173/bow.gyrfal.01.

Bordage, D., and J.-P.L. Savard. 2020. Black Scoter (*Melanitta americana*). Version 1.0. In *Birds of the World*, ed. A.F. Poole. Cornell Lab of Ornithology, Ithaca, NY. https://doi.org/10.2173/bow.blksco2.01.

Boulet, M., H.L. Gibbs, and K.A. Hobson. 2006. Integrated analysis of genetic, stable isotope, and banding data reveal migratory connectivity and flyways in the northern Yellow Warbler (*Dendroica petechia*; *aestiva* group). *Ornithological Monographs* 61:29–78.

Boutilier, S.T., S.A. Taylor, J.A. Morris-Pocock, R.A. Lavoie, and V.L. Friesen. 2014. Evidence for genetic differentiation among Caspian Tern (*Hydroprogne caspia*) populations in North America. *Conservation Genetics* 15:275–281.

Bowen, G.J., L.I. Wassenaar, and K.A. Hobson. 2005. Global application of stable hydrogen and oxygen isotopes to wildlife forensics. *Oecologia* 143:337–348.

Bowen, R.V. 2020. Townsend's Solitaire (*Myadestes townsendi*) Version 1.0. In *Birds of the World*, A.F. Poole and F.B. Gill. Cornell Lab of Ornithology, Ithaca, NY. https://doi.org/10.2173/bow.towsol.01.

Bowers Jr., R.K., and J.B. Dunning. 2020. Buff-collared Nightjar (*Antrostomus ridgwayi*) Version 1.0. In *Birds of the World*, ed. S.M. Billerman. Cornell Lab of Ornithology, Ithaca, NY. https://doi.org/10.2173/bow.bucnig.01.

Bowman, T., J. Schamber, D. Esler, D. Rosenberg, P. Flint, B. Stehn, and J. Pearce. 2007. Assessment of the Pacific Black Scoter population: Population size, distribution, and links among populations: An integrated approach. Sea Duck Joint Venture Project 38. http://seaduckjv.org/pdf/studies/pr38.pdf.

Boyce, A.J. 2024. Smithsonian Northern Great Plains (Project 386). Data accessed from the Motus Wildlife Tracking System. Birds Canada. http://www.motus.org/.

Boyd, S., and E. Esler. 2012. Migration and habitat use of Pacific Barrow's Goldeneye. Simon Frasier University. http://www.sfu.ca/biology/wildberg/CWESeaducksfolder/BAGOwebpage/BAGOMigrationHome.html.

Boyd, W.S., S.D. Schneider, and S.A. Cullen. 2000. Using radio telemetry to describe the fall migration of Eared Grebes. *Journal of Field Ornithology* 71(4):702–707.

REFERENCES

Boyd, W.S., D.H. Ward, D.K. Kraege, and A.A. Gerick. 2013. Migration patterns of western high Arctic (Grey-belly) Brant *Branta bernicla*. *Wildfowl* 3(special iss.):3–25.

Boyle, D.B., B.K. Goyen, D.G. Cameron, and R.G. Cameron. 2018. Movement of Short-tailed Shearwaters *Ardenna tenuirostris* banded at Phillip Island, Victoria from 1956 to 2010. *Corella* 42:80–85.

Boyle, D.B., D. Cameron, B. Goyen, and G. Cameron. 2012. A not so short tale: 100 years of muttonbird banding. *Australian Birdlife* 1(1):24–27.

Bracey, A., S. Lisovski, D. Moore, A. McKellar, E. Craig, S. Matteson, F. Strand, J. Costa, C. Pekarik, P. Curtis, G. Niemi, and F. Cuthbert. 2018. Migratory routes and wintering locations of declining inland North American Common Terns. *Auk* 135:385–399.

Brakhage, G.K. 1953. Migration and mortality of ducks hand-reared and wild-trapped at Delta, Manitoba. *Journal of Wildlife Management* 17:465–477.

Brandt, H. 1951. *Arizona and Its Bird Life: A Naturalist's Adventures with the Nesting Birds on the Deserts, Grasslands, Foothills, and Mountains of Southeastern Arizona*. Bird Research Foundation, Cleveland, OH.

Brewer, D., A.W. Diamond, E.J. Woodsworth, B.T. Collins, and E.H. Dunn. 2000. *Canadian Atlas of Bird Banding*. Vol. 1. *Doves, Cuckoos, and Hummingbirds through Passerines, 1921–1995*. Canadian Wildlife Services, Ottawa, ON.

Bridge, E.S., W.T. Honeycutt, A.J. Chen, R. Miller, and J.F. Kelly. 2024. Social behavior among nocturnally migrating birds revealed by automated moonwatching. *Ornithology* 141(1):ukad055.

Bridge, E.S., J.D. Ross, A. Contina, and J.F. Kelly. 2016. Do molt-migrant songbirds optimize migration routes based on primary productivity? *Behavioral Ecology* 27(3):784–792.

Bridge, E.S., K. Thorup, M.S. Bowlin, P.B. Chilson, R.H. Diehl, R.W. Fléron, P. Hartl, R. Kays, J.F. Kelly, W.D. Robinson, and M. Wikelski. 2011. Technology on the move: Recent and forthcoming innovations for tracking migratory birds. *BioScience* 61(9):689–698.

Briggs, C., A. Hull, J. Hull, J. Harley, P. Bloom, R. Rosenfield, and A. Fish. 2020. Natal dispersal distance and population origin of migrant Red-tailed Hawks and Cooper's Hawks. *Journal of Raptor Research* 54(1):47–56.

Brinkley, E.S., and K.E. Sutherland. 2020. Bermuda Petrel (*Pterodroma cahow*). Version 2.0. In *Birds of the World*, ed. T.S. Schulenberg, B.K. Keeney, and S.M. Billerman. Cornell Lab of Ornithology, Ithaca, NY. https://doi.org/10.2173/bow.berpet.02.

Brisbin Jr., I.L., and T.B. Mowbray. 2020. American Coot (*Fulica americana*). Version 1.0. In *Birds of the World*, ed. A.F. Poole and F.B. Gill. Cornell Lab of Ornithology, Ithaca, NY. https://doi.org/10.2173/bow.y00475.01.

Briskie, J.V. 2020. Smith's Longspur (*Calcarius pictus*). Version 1.0. In *Birds of the World*, ed. A.F. Poole. Cornell Lab of Ornithology, Ithaca, NY. https://doi.org/10.2173/bow.smilon.01.

Brodeur, S., R. Décarie, D.M. Bird, and M. Fuller. 1996. Complete migration cycle of Golden Eagles breeding in northern Quebec. *Condor* 98(2):293–299.

Brodeur, S., J.-P.L. Savard, M. Robert, P. Laporte, P. Lamothe, R.D. Titman, S. Marchand, S. Gilliland, and G. Fitzgerald. 2002. Harlequin duck *Histrionicus histrionicus* population structure in eastern Nearctic. *Journal of Avian Biology* 33:127–137.

Brown, C.E. 2012. Exploring connectivity in calliope hummingbird through stable isotope analysis of tail feathers. Master's thesis, Evergreen State College.

Brown, C.R., D.A. Airola, and S. Tarof. 2021. Purple Martin (*Progne Subis*). Version 2.0. In *Birds of the World*, ed. P.G. Rodewald. Cornell Lab of Ornithology, Ithaca, NY. 2021. https://doi.org/10.2173/bow.purmar.02.

Brown, C.R., A.M. Knott, and E.J. Damrose. 2020. Violet-green Swallow (*Tachycineta thalassina*). Version 1.0. In *Birds of the World*, ed. A.F. Poole. Cornell Lab of Ornithology, Ithaca, NY. https://doi.org/10.2173/bow.vigswa.01.

Brown, D., and G. Miller. 2016. Band recoveries reveal alternative migration strategies in American Robins. *Animal Migration* 3(1):35–47.

Brown, J.I., P. Lavretsky, R.E. Wilson, C.L. Haughey, W.S. Boyd, D. Esler, S.L. Talbot, and S.A. Sonsthagen. 2020. High site fidelity does not equate to population genetic structure for Common Goldeneye and Barrow's Goldeneye in North America. *Journal of Avian Biology* 51(12). https://doi.org/10.1111/jav.02600.

Brown, J.L., B. Bedrosian, D.A. Bell, M.A. Braham, J. Cooper, R.H. Crandall, J. DiDonato, R. Domenech, A.E. Duerr, T.E. Katzner, M.J. Lanzone, D.W. LaPlante, C.L. McIntyre, T.A. Miller, R.K. Murphy, A. Shreading, S.J. Slater, J.P. Smith, B.W. Smith, J.W. Watson, and B. Woodbridge. 2017. Patterns of spatial distribution of Golden Eagles across North America: How do they fit into existing landscape scale mapping systems? *Journal of Raptor Research* 51(3):197–215.

Brown, M.B., and C.R. Brown. 2020. Barn Swallow (*Hirundo rustica*). Version 1.0. In *Birds of the World*, ed. P.G. Rodewald. Cornell Lab of Ornithology, Ithaca, NY. https://doi.org/10.2173/bow.barswa.01.

Brown, R.G.B. 1973. Transatlantic migration of dark-phase fulmars from the European Arctic. *Canadian Field-Naturalist* 87:312–313.

———. 1986. *Revised Atlas of Eastern Canadian Seabirds: I. Shipboard Surveys*. Environment Canada, Canadian Wildlife Service, Ottawa, ON.

Brown, R.M., R.A. Nichols, C.G. Faulkes, C.G. Jones, L. Bugoni, V. Tatayah, D. Gottelli, and W.C. Jordan. 2010. Range expansion and hybridization in Round Island petrels (*Pterodroma* spp.): Evidence from microsatellite genotypes. *Molecular Ecology* 19(15):3157–3170. https://doi.org/10.1111/j.1365-294X.2010.04719.x.

Brown, S., C. Gratto-Trevor, R. Porter, E.L. Weiser, D. Mizrahi, R. Bentzen, ... and R.B. Lanctot. 2017. Migratory connectivity of Semipalmated Sandpipers and implications for conservation. *Condor: Ornithological Applications* 119(2):207–224.

Browning, R.M. 1994. A taxonomic review of *Dendroica petechia* (Yellow Warbler) (Aves: Parulinae). *Proceedings of the Biological Society of Washington* 107:27–51.

Brua, R.B. 2020. Ruddy Duck (*Oxyura jamaicensis*). Version 1.0. In *Birds of the World*, ed. P.G. Rodewald. Cornell Lab of Ornithology, Ithaca, NY. https://doi.org/10.2173/bow.rudduc.01.

Brunner, A.R., N.W. Cooper, and P.P. Marra. 2021. First known record of Kirtland's Warbler (*Setophaga kirtlandii*) observed and captured in Jamaica. *Wilson Journal of Ornithology* 132:746–749.

Brunner, A.R., B. Dossman, V. Jirinec, K. Percy, C. Tonra, E. Johnson, and P.P. Marra. 2022. Migratory behavior and connectivity revealed in a secretive Neotropical migratory songbird, the Swainson's Warbler. *Journal of Field Ornithology* 93(3):5. https://doi.org/10.5751/JFO-00134-930305.

Brush, T. 2020. Couch's Kingbird (*Tyrannus couchii*). Version 1.0. In *Birds of the World*, ed. A.F. Poole and F.B. Gill. Cornell Lab of Ornithology, Ithaca, NY. https://doi.org/10.2173/bow.coukin.01.

Bryan Jr., A.L., W. Brooks, J. Taylor, D. Richardson, C. Jeske, and I.L. Brisbin Jr. 2008. Satellite tracking large-scale movements of Wood Storks captured in the Gulf Coast region. *Waterbirds* 31(special pub. 1):35–41.

Brzorad, J., M.C. Allen, S. Elbin, R. Kays, A. Maccarone, S. Schweitzer, and N. Tsipoura. 2020. Movebank Study Egrets & Herons, ID 17469219.

Brzorad, J.N., M.C. Allen, S. Jennings, E. Condeso, S. Elbin, R. Kays, D. Lumpkin, S. Schweitzer, N. Tsipoura, and A.D. Maccarone. 2021. Seasonal patterns in daily flight distance and space use by Great Egrets (*Ardea alba*). *Waterbirds* 44:343–355.

Bubac, C.M, and G.M. Spellman. 2016. How connectivity shapes genetic structure during range expansion: Insights from the Virginia's Warbler. *Auk* 133:213–230. https://doi.org/10.1642/AUK-15-124.1.

Buckley, N.J., B.M. Kluever, R. Driver, and S.A. Rush. 2022. Black Vulture (*Coragyps atratus*). Version 2.0. In *Birds of the World*, ed. P.G. Rodewald and B.K. Keeney. Cornell Lab of Ornithology, Ithaca, NY. https://doi.org/10.2173/bow.blkvul.02.

Buckley, P.A., and F.G. Buckley. 2002. Royal Tern (*Thalasseus maximus*). Version 2.0. In *The Birds of North America*, ed. A.F. Poole and F.B. Gill. Cornell Lab of Ornithology, Ithaca, NY. https://doi.org/10.2173/bna.700.

Buehler, D.A. 2022. Bald Eagle (*Haliaeetus leucocephalus*). Version 2.0. In *Birds of the World*, ed. P.G. Rodewald and S.G. Mlodinow. Cornell Lab of Ornithology, Ithaca, NY. https://doi.org/10.2173/bow.baleag.02.

Buehler, D.A., J.D. Fraser, M.R. Fuller, L.S. Mcallister, and J.K.D. Seegar. 1995. Captive and field-tested radio transmitter attachments for Bald Eagles. *Journal of Field Ornithology* 66:173–180.

Buehler, D.M., and T. Piersma. 2008. Travelling on a budget: Predictions and ecological evidence for bottlenecks in the annual cycle of long-distance migrants. *Royal Society Philosophical Transactions Biological Sciences* 363(1490):247–266.

Buhnerkempe, M.G., C.T. Webb, A.A. Merton, J.E. Buhnerkempe, G.H. Givens, R.S. Miller, and J.A. Hoeting. 2016. Identification of migratory bird flyways in North America using community detection on biological networks. *Ecological Applications* 26(3):740–751.

Bulluck, L., E. Ames, N. Bayly, J. Reese, C. Viverette, J. Wright, A. Caguazango, and C. Tonra. 2019. Habitat-dependent occupancy and movement in a migrant songbird highlights the importance of mangroves and forested lagoons in Panama and Colombia. *Ecology and Evolution* 9(19):11064–11077. https://doi.org/10.1002/ece3.5610.

Bunnefeld, N., L. Börger, B. van Moorter, C.M. Rolandsen, H. Dettki, E.J. Solberg, and G. Ericsson. 2011. A model-driven approach to quantify migration patterns: Individual, regional and yearly differences. *Journal of Animal Ecology* 80:466–476. https://doi.org/10.1111/j.1365-2656.2010.01776.x.

Burger, J. 2015. Laughing Gull (*Leucophaeus atricilla*). Version 2.0. In *The Birds of North America*, ed. P.G. Rodewald. Cornell Lab of Ornithology, Ithaca, NY. https://doi.org/10.2173/bna.225.

Burger, J., and M. Gochfeld. 2020. Franklin's Gull (*Leucophaeus pipixcan*) Version 1.0. In *Birds of the World*, ed. A.F. Poole. Cornell Lab of Ornithology, Ithaca, NY. https://doi.org/10.2173/bow.fragul.01.

Burger, J., L.J. Niles, N.N. Porter, A.D. Dey, S. Koch, and C. Gordon. 2012. Migration and overwintering of Red Knots (*Calidris canutus rufa*) along the Atlantic Coast of the United States. *Condor* 114:302–313.

Burger, J., R. Porter, and L. Niles. 2024. The migration strategies of the Red Knots using the Western Atlantic Flyway as revealed by geolocators. *Wader Study* 131(2). https://doi.org/10.18194/ws.00341.

Burness, G. P., K. L. Lefevre, and C. T. Collins. 1999. Elegant Tern (*Thalasseus elegans*). Version 2.0. In *The Birds of North America*, ed A.F. Poole and F.B. Gill. Cornell Lab of Ornithology, Ithaca, NY. https://doi.org/10.2173/bna.404

Burnham, K.K., J.L. Burnham, J.A. Johnson, and A. Huffman. 2021. Migratory movements of Atlantic Puffins *Fratercula arctica naumanni* from high Arctic Greenland. *PLOS ONE* 16(5):e0252055. https://doi.org/10.1371/journal.pone.0252055. PMID: 34048451; PMCID: PMC8162707.

Burnham K.K., W.A. Burnham, I. Newton, J.A. Johnson, and A.G. Gosle. 2012. The history and range expansion of Peregrine Falcons in the Thule area, northwest Greenland. *Bioscience* 60; *Monographs on Greenland* 353. Copenhagen: Museum Tusculanum Press, University of Copenhagen.

Burnham, K.K., and I. Newton. 2011. Seasonal movements of Gyrfalcons, *Falco rusticolus*, include extensive periods at sea. *Ibis* 153:468–484.

Burnside, F.L. 1987. Long-distance movements by Loggerhead Shrikes. *Journal of Field Ornithology* 58(1):62–65.

Bustnes, J.O., A. Mosbech, C. Sonne, and G.H. Systad. 2010. Migration patterns, breeding and moulting locations of King Eiders wintering in north-eastern Norway. *Polar Biology* 33:1379–1385.

Butler, C.J., J.K. Wilson, S.R. Frazee, and J.F. Kelly. 2016. A comparison of the origins of Yellow Rails (*Coturnicops noveboracensis*) wintering in Oklahoma and Texas. *Waterbirds* 39(2):156–164.

Butler, R.G., D.E. Buckley, D.N. Nettleship, P.F.D. Boesman, and E.F.J. Garcia. 2020. Black Guillemot (*Cepphus grylle*). Version 1.0. In *Birds of the World*, ed. S.M. Billerman. Cornell Lab of Ornithology, Ithaca, NY. https://doi.org/10.2173/bow.blkgui.01.

Butler, R.W., et al. 1996. Migration routes of the Western Sandpiper. *Wilson Bulletin* 108(4):662–672.

Byerly, P.A., R.T. Chesser, R.C. Fleischer, et al. 2023. Conservation genomics reveals low connectivity among populations of threatened Roseate Terns (*Sterna dougallii*) in the Atlantic Basin. *Conservation Genetics* 24:331–345. https://doi.org/10.1007/s10592-023-01505-6.

Byers, B.E., M. Richardson, and D.W. Brauning. 2020. Chestnut-sided Warbler (*Setophaga pensylvanica*). Version 1.0. In *Birds of the World*, ed. A.F. Poole. Cornell Lab of Ornithology, Ithaca, NY. https://doi.org/10.2173/bow.chswar.01.

Byrd, G.V., and J.C. Williams. 1993. Red-legged Kittiwake (*Rissa brevirostris*). Version 2.0. In *The Birds of North America*, ed. A.F. Poole and F.B. Gill. Cornell Lab of Ornithology, Ithaca, NY. https://doi.org/10.2173/bna.60.

Caccamise, D.F., L.M. Reed, P.M. Castelli, S. Wainright, and T.C. Nichols. 2000. Distinguishing migratory and resident Canada Geese using stable isotope analysis. *Journal of Wildlife Management* 64(4):1084–1091.

Cade, D.E., W.T. Gough, M.F. Czapanskiy, J.A. Fahlbusch, S.R. Kahane-Rapport, J.M. Linsky, R.C. Nichols, W.K. Oestreich, D.M. Wisniewska, A.S. Friedlaender, and J.A. Goldbogen. 2021. Tools for integrating inertial sensor data with video bio-loggers, including estimation of animal orientation, motion, and position. *Animal Biotelemetry* 9(1):34.

Calder, W.A., and L.L. Calder. 2020. Calliope Hummingbird (*Selasphorus calliope*). Version 1.0. In *Birds of the World*, ed. A.F. Poole and F.B. Gill. Cornell Lab of Ornithology, Ithaca, NY. https://doi.org/10.2173/bow.calhum.01.

Calenge, C., and S. Fortmann-Roe. 2023. adehabitatHR: Home Range Estimation. R package version 0.4.21. https://CRAN.R-project.org/package=adehabitatHR.

Callo, P.A., E.S. Morton, and B.J.M. Stutchbury. 2013. Prolonged spring migration in the Red-eyed Vireo (*Vireo olivaceus*). *Auk* 130:240–246.

Cambrone, C. 2020. Movebank Study White-crowned Pigeon, *Patagioenas leucocephala*, within the Caribbean region [ID_PROG 1034]. https://www.movebank.org/cms/webapp?gwt_fragment=page=studies,path=study894865693.

Campbell, R.W., N.K. Dawe, I. McTaggart-Cowan, J.M. Cooper, G.W. Kaiser, and M.C.E. McNall. 1990. *The Birds of British Columbia.* Vol. 1. *Introduction and Loons through Waterfowl*. Royal British Columbia Museum, Victoria, BC.

Campbell, S.P., M.E. Akresh, D.I. King, and N. Gifford. 2021. Effects of light-level geolocators on return rates and inter-annual territory fidelity of Prairie Warblers *Setophaga discolor*. *Bird Study* 68:396–407.

Campos, L.F.A.S., A.B. Andrade, S. Bertrand, S., and M.A. Efe. 2018. Foraging behavior and at-sea distribution of White-Tailed Tropicbirds in tropical ocean. *Brazilian Journal of Biology* 78(3):556–563. https://doi.org/10.1590/1519-6984.173578.

Carboneras, C., F. Jutglar, and G.M. Kirwan. 2020a. Great Shearwater (*Ardenna gravis*). Version 1.0. In *Birds of the World*, ed. J. del Hoyo, A. Elliott, J. Sargatal, D.A. Christie, and E. de Juana. Cornell Lab of Ornithology, Ithaca, NY. https://doi.org/10.2173/bow.greshe.01.

———. 2020b. Murphy's Petrel (*Pterodroma ultima*). Version 1.0. In *Birds of the World*, ed. J. del Hoyo, A. Elliott, J. Sargatal, D.A. Christie, and E. de Juana. Cornell Lab of Ornithology, Ithaca, NY. https://doi.org/10.2173/bow.murpet.01.

———. 2020c. Short-tailed Albatross (*Phoebastria albatrus*). Version 1.0. In *Birds of the World*, ed. J. del Hoyo, A. Elliott, J. Sargatal, D.A. Christie, and E. de Juana. Cornell Lab of Ornithology, Ithaca, NY. https://doi.org/10.2173/bow.shtalb.01.

———. 2020d. Short-tailed Shearwater (*Ardenna tenuirostris*). Version 1.0. In *Birds of the World*, ed. J. del Hoyo, A. Elliott, J. Sargatal, D.A. Christie, and E. de Juana. Cornell Lab of Ornithology, Ithaca, NY. https://doi.org/10.2173/bow.shtshe.01.

———. 2020e. Sooty Shearwater (*Ardenna grisea*). Version 1.0. In *Birds of the World*, ed. J. del Hoyo, A. Elliott, J. Sargatal, D.A. Christie, and E. de Juana. Cornell Lab of Ornithology, Ithaca, NY. https://doi.org/10.2173/bow.sooshe.01.

Cardenas-Ortiz, L., N.J. Bayly, G.J. Colorado, and K.A. Hobson. 2017. Fall migration and breeding origins of Canada Warblers moving through northern Colombia. *Journal of Field Ornithology* 88(1):53–64.

Cardenas-Ortiz, L., N.J. Bayly, K.J. Kardynal, and K.A. Hobson. 2020. Defining catchment origins of a geographical bottleneck: Implications of population mixing and phenological overlap for the conservation of neotropical migratory birds. *Condor* 122(2):duaa004.

Cardiff, S.W., and D.L. Dittmann. 2020a. Ash-throated Flycatcher (*Myiarchus cinerascens*). Version 1.0. In *Birds of the World*, ed. A.F. Poole and F.B. Gill. Cornell Lab of Ornithology, Ithaca, NY. https://doi.org/10.2173/bow.astfly.01.

———. 2020b. Brown-crested Flycatcher (*Myiarchus tyrannulus*). Version 1.0. In *Birds of the World*, ed. A.F. Poole and F.B. Gill. Cornell Lab of Ornithology, Ithaca, NY. https://doi.org/10.2173/bow.bncfly.01.

Carey, M., D.E. Burhans, and D.A. Nelson. 2020. Field Sparrow (*Spizella pusilla*). Version 1.0. In *Birds of the World*, ed. A.F. Poole. Cornell Lab of Ornithology, Ithaca, NY. https://doi.org/10.2173/bow.fiespa.01.

Carey, M.J., R.A. Phillips, J.R.D. Silk, and S.A. Shaffer. 2014. Trans-equatorial migration of Short-tailed Shearwaters revealed by geolocators. *Emu* 114:352–359.

Carle, R.D., J.N. Beck, D.M. Calleri, and M.M. Hester. 2015. Temporal and sex-specific variability in Rhinoceros Auklet diet in the central California Current system. *Journal of Marine Systems* 146:99–108.

Carle, R.D., V. Colodro, J. Felis, J. Adams, and P. J. Hodum. 2022. Pink-footed Shearwater (*Ardenna creatopus*). Version 2.0. In *Birds of the World*, ed. P.G. Rodewald and B.K. Keeney. Cornell Lab of Ornithology, Ithaca, NY. https://doi.org/10.2173/bow.pifshe.02.

Carle, R.D., J.J. Felis, R. Vega, J. Beck, J. Adams, V. López, P.J. Hodum, A. González, V. Colodro, and T. Varela. 2019. Overlap of Pink-footed Shearwaters and central Chilean purse-seine fisheries: Implications for bycatch risk. Condor 121(3):duz026. https://doi.org/10.1093/condor/duz026.

Carleton, S., and C. Martinez del Rio. 2010. A genetic and isotopic characterization of eastern and western White-winged Dove breeding populations to determine wintering ground distribution and population genetic structure. Department of Zoology and Physiology, University of Wyoming. https://tpwd.texas.gov/huntwild/wild/research/highlights/taxa/publications/Carleton_MartinezdelRio_2010_WWDO.pdf.

Carleton, S.A., C. Martinez del Rio, and T.J. Robinson. 2015. Feather isotope analysis reveals differential patterns of habitat and resource use in populations of White-winged Doves. *Journal of Wildlife Management* 79:948–956. https://doi.org/10.1002/jwmg.916.

Carmona, R., N. Arce, V. Ayala-Perez, A. Hernández-Alvarez, J.B. Buchanan, L.J. Salzer, … and D. Newstead. 2013. Red Knot *Calidris canutus roselaari* migration connectivity, abundance and non-breeding distribution along the Pacific coast of the Americas. *Wader Study Group Bulletin* 120(3):168–180.

Carneiro, A.P.B., M.P. Dias, B.L. Clark, et al. 2024. The BirdLife Seabird Tracking Database: 20 years of Collaboration for Marine Conservation. *Biological Conservation* 299:110813. https://doi.org/10.1016/j.biocon.2024.110813.

Carpenter, A.M., B.A. Graham, G.M. Spellman, J. Klicka, and T.M. Burg. 2022. Genetic, bioacoustic and morphological analyses reveal cryptic speciation in the warbling vireo complex (*Vireo gilvus*: Vireonidae: Passeriformes). *Zoological Journal of the Linnean Society* 195:45–64. https://doi.org/10.1093/zoolinnean/zlab036.

Carpenter, T.W., A.L. Carpenter, and W.A. Lamb. 1990. Analysis of banding and recovery data for Sharp-shinned Hawks at Whitefish Point, Michigan, 1984–1987. *North American Bird Bander* 15(4):125–129.

Carter, H.R., D.G. Ainley, S.G. Wolf, and A.M. Weinstein. 2016. Range-wide conservation and science of the Ashy Storm-Petrel *Oceanodroma homochroa*. *Marine Ornithology* 44:53–62.

Carvalho, P.C., R.A. Ronconi, L. Bugoni, and G.K. Davoren. 2022. Moult chronology and strategies of sympatric Great (*Ardenna gravis*) and Sooty (*A. grisea*) Shearwaters based on stable isotope analysis. *Ibis* 164:998–1012. https://doi.org/10.1111/ibi.13060.

Carvey, Q.B., S.A. Pavey, A.W. Diamond, et al. 2024. Genetic structure of Atlantic Puffins (*Fratercula arctica*) breeding in Atlantic Canada. *Conservation Genetics* 25:1159–1174. https://doi.org/10.1007/s10592-024-01629-3.

Castelli, P.M., K.M. Dickson, and D.M. Cramer. 2010. Spatial and temporal distribution of Atlantic Brant. Report to the Atlantic Flyway Council.

Cavitt, J.F., and C.A. Haas. 2020. Brown Thrasher (*Toxostoma rufum*). Version 1.0. In *Birds of the World*, ed. A.F. Poole. Cornell Lab of Ornithology, Ithaca, NY. https://doi.org/10.2173/bow.brnthr.01.

Central Flyway Council, Mississippi Flyway Council, Pacific Flyway Council, Canadian Wildlife Service, and US Fish and Wildlife Service. 2015. Management Plan for Mid-continent Greater White-fronted Geese.

Chabot, A.A., K.A. Hobson, S.L. Van Wilgenburg, G.E. Pérez, and S.C. Lougheed. 2018. Migratory connectivity in the Loggerhead Shrike (*Lanius ludovicianus*). *Ecology and Evolution* 8(22):10662–10672. https://doi.org/10.1002/ece3.4415.

Chace, J.F., and R.C. Tweit. 2020. Greater Pewee (*Contopus pertinax*). Version 1.0. In *Birds of the World*, ed. A.F. Poole and F.B. Gill. Cornell Lab of Ornithology, Ithaca, NY. https://doi.org/10.2173/bow.grepew.01.

Chace, J.F., B.L. Woodworth, and A. Cruz. 2020. Black-whiskered Vireo (*Vireo altiloquus*). Version 1.0. In *Birds of the World*, ed. A.F. Poole and F.B. Gill. Cornell Lab of Ornithology, Ithaca, NY. https://doi.org/10.2173/bow.bkwvir.01.

Chamberlain, C.P., J.D. Blum, R.T. Holmes, X. Feng, T.W. Sherry, and G.R. Graves. 1997. The use of isotope tracers for identifying populations of migratory birds. *Oecologia* 109(1):132–141.

Champoux, L., J.-F. Rail, M. Houde, M. Giraudo, É. Lacaze, C.D. Franci, G.D. Fairhurst, K.A. Hobson, P. Brousseau, M. Guillemette, D. Pelletier, W.A. Montevecchi, S. Lair, J. Verreault, and C. Soos. 2020. An investigation of physiological effects of the Deepwater Horizon oil spill on a long-distance migratory seabird, the Northern Gannet. *Marine Pollution Bulletin* 153:110953. https://doi.org/10.1016/j.marpolbul.2020.110953.

Chan, Y.-C., M. Brugge, T.L. Tibbitts, A. Dekinga, R. Porter, R.H.G. Klaassen, and T. Piersma. 2016. Testing an attachment method for solar-powered tracking devices on a long-distance migrating shorebird. *Journal of Ornithology* 157(1):277–287. https://doi.org/10.1007/s10336-015-1276-4.

Chapdelaine, G. 1997. Pattern of recoveries of Banded Razorbills (*Alca Torda*) in the western Atlantic and survival rates of adults and immatures. *Colonial Waterbirds* 20(1):47–54.

Chardine, J.W., R.D. Morris, M. Gochfeld, J. Burger, G.M. Kirwan, and E.F.J. Garcia. 2020. Brown Noddy (*Anous stolidus*). Version 1.0. In *Birds of the World*, ed. S.M.

Billerman. Cornell Lab of Ornithology, Ithaca, NY. https://doi.org/10.2173/bow.brnnod.01.

Chardine, J.W., J.-F. Rail, and S. Wilhelm. 2013. Population dynamics of Northern Gannets in North America, 1984–2009. *Journal of Field Ornithology* 84(2):187–192. http://www.jstor.org/stable/24617968.

Chavez-Ramirez, F., and A. Moreno-Valdez. 2020. Buff-bellied Hummingbird (*Amazilia yucatanensis*). Version 1.0. In *Birds of the World*, ed. A.F. Poole. Cornell Lab of Ornithology, Ithaca, NY. https://doi.org/10.2173/bow.bubhum.01.

Chavez-Ramirez, F., G.P. Vose, and A. Tennant. 1994. Spring and fall migration of Peregrine Falcons from Padre Island, Texas. *Wilson Bulletin* 106(1):138–145.

Chesser, R.T., S.M. Billerman, K.J. Burns, C. Cicero, J.L. Dunn, B.E. Hernández-Baños, R.A. Jiménez, A.W. Kratter, N.A. Mason, P.C. Rasmussen, J.V. Remsen Jr., and K. Winker. 2023. Sixty-fourth Supplement to the American Ornithological Society's Check-list of North American Birds. *Ornithology* 140(3):1–11. https://doi.org/10.1093/ornithology/ukad023.

Chesser, R.T., S. M. Billerman, K. J. Burns, C. Cicero, J. L. Dunn, B. E. Hernández-Baños, R. A. Jiménez, O. Johnson, A.W. Kratter, N.A. Mason, P.C. Rasmussen, and J.V. Remsen Jr. 2024. Check-list of North American Birds (online). American Ornithological Society. https://checklist.americanornithology.org/taxa/.

Chesser, R.T., R.C. Burns, K.J. Cicero, J.L. Dunn, A.W. Kratter, I.J. Lovette, P.C. Rasmussen, J.V. Remsen Jr., J.D. Rising, D.F. Stotz, and K. Winker. 2017. Fifty-eighth supplement to the American Ornithological Society's Check-list of North American birds. *Auk* 134:751–773.

Chilton, G., M.C. Baker, C.D. Barrentine, and M.A. Cunningham. 2020. White-crowned Sparrow (*Zonotrichia leucophrys*). Version 1.0. In *Birds of the World*, ed. A.F. Poole and F.B. Gill. Cornell Lab of Ornithology, Ithaca, NY. https://doi.org/10.2173/bow.whcspa.01.

Christensen, T.C., and D.C. Ward. 2022. First description of Long-eared Owl (*Asio otus*) migration using GPS telemetry. *Wilson Journal of Ornithology* 134(4):708–715.

Christie, K., R.E. Wilson, J.A. Johnson, C. Friis, C.M. Harwood, L.A. McDuffie, E. Nol, and S.A. Sonsthagen. 2023. Movement and genomic methods reveal mechanisms promoting connectivity in a declining shorebird: The Lesser Yellowlegs. *Diversity* 15(5):595. https://doi.org/10.3390/d15050595.

Chu, M., and G. Walsberg. 2020. Phainopepla (*Phainopepla nitens*). Version 1.0. In *Birds of the World*, ed. A.F. Poole and F.B. Gill. Cornell Lab of Ornithology, Ithaca, NY. https://doi.org/10.2173/bow.phaino.01.

Chubbs, T.E., P.G. Trimper, G.W. Humphries, P.W. Thomas, L.T. Elson, and D.K. Laing. 2008. Tracking seasonal movements of adult male Harlequin Ducks from central Labrador using satellite telemetry. *Waterbirds* 31(sp2):173–182. https://doi.org/10.1675/1524-4695-31.sp2.173.

Cimprich, D.A., F.R. Moore, and M.P. Guilfoyle. 2020. Red-eyed Vireo (*Vireo olivaceus*). Version 1.0. In *Birds of the World*, ed. P.G. Rodewald. Cornell Lab of Ornithology, Ithaca, NY. https://doi.org/10.2173/bow.reevir1.01.

Clapp, R.B., and W.O. Wirtz II. 1975. The natural history of Lisianski Island, northwestern Hawaiian Islands. *Atoll Research Bulletin* 186.

Clark, C.J. 2017. eBird records show substantial growth of the Allen's Hummingbird (*Selasphorus sasin sedentarius*) population in urban southern California. *Condor* 119:122–130. https://doi.org/10.1650/CONDOR-16-153.1.

Clark, C.J., and D.E. Mitchell. 2020. Allen's Hummingbird (*Selasphorus sasin*). Version 1.0. In *Birds of the World*, ed. A.F. Poole. Cornell Lab of Ornithology, Ithaca, NY. https://doi.org/10.2173/bow.allhum.01.

Clark, C.J., and S.M. Russell. 2020. Anna's Hummingbird (*Calypte anna*). Version 1.0. In *Birds of the World*, ed. A.F. Poole. Cornell Lab of Ornithology, Ithaca, NY. https://doi.org/10.2173/bow.annhum.01.

Clark, D.E., K.K.G. Koenen, J.J. Whitney, K.G. MacKenzie, and S. DeStefano. 2016. Fidelity and persistence of Ring-Billed (*Larus delawarensis*) and Herring (*Larus argentatus*) Gulls to wintering sites. *Waterbirds* 39:220–234.

Clark, K.E., L.J. Niles, and J. Burger. 1993. Abundance and distribution of migrant shorebirds in Delaware Bay. *Condor* 95(3):694–705.

Clark, N.J. 2017. A "jaunt" across the pond: Investigating patterns of genetic differentiation and trans-Atlantic migration in a temperate seabird (*Morus bassanus*). Master's thesis, Queen's University.

Clark, T.J., J. Matthiopoulos, A.S. Bonnet-Lebrun, L. Campioni, et al. 2019. Integrating habitat and partial survey data to estimate the regional population of a globally declining seabird species, the Sooty Shearwater. *Global Ecology and Conservation* 17:e00554.

Clark, W.S. 1985. Migration of the Merlin along the coast of New Jersey. *Journal of Raptor Research* 19(2–3):85–93.

Clarke, T.C., A.W. Diamond, and J.W. Chardine. 2010. Origin of Canadian Razorbills (*Alca torda*) wintering in the outer Bay of Fundy confirmed by radio-tracking. *Waterbirds* 33(4):541–545.

Clarke, T.C.R. 2009. Offshore movements and behaviour of the Razorbill (*Alca torda*) in Atlantic Canada. Master's thesis, University of New Brunswick.

Clausen, P., M. Green, and T. Alerstam. 2003. Energy limitations for spring migration and breeding: The case of brent geese *Branta bernicla* tracked by satellite telemetry to Svalbard and Greenland. *Oikos* 103:426–445.

Clay, T.A., S. Oppel, J.L. Lavers, et al. 2019. Divergent foraging strategies during incubation of an unusually wide-ranging seabird, the Murphy's Petrel. *Marine Biology* 166:8. https://doi.org/10.1007/s00227-018-3451-7.

Clay, T.A., R.A. Phillips, A. Manica, H.A. Jackson, and M. de L. Brooke. 2017. Escaping the oligotrophic gyre? The year-round movements, foraging behavior, and habitat preferences of Murphy's Petrels. *Marine Ecology Progress Series* 579:139–155. https://doi.org/10.3354/meps12244.

Clegg, S.M., J.F. Kelly, M. Kimura, and T.B. Smith. 2003. Combining genetic markers and stable isotopes to reveal population connectivity and migration patterns in a Neotropical migrant, Wilson's Warbler (*Wilsonia pusilla*). *Molecular Ecology* 12(4):819–830.

Clements, S.J. 2022. Ecology of mid-continent migratory shorebirds, performance of tracking devices, and modelling animal social structure and demography. PhD diss., University of Missouri–Columbia.

Clements, S.J., J.P. Loghry, B.M. Ballard, and M.D. Weegman. 2022. Carry-over effects of weather and decision-making on nest success of a migratory shorebird. *Ecology and Evolution* 12:e9581. https://doi.org/10.1002/ece3.9581.

Cleminson, A., and S. Nebel. 2012. Bird banding. *Nature Education Knowledge* 3(1).

Cline, B.B., and S.M. Haig. 2011. Seasonal movement, residency, and migratory patterns of Wilson's snipe (*Gallinago delicata*). *Auk* 128(3):543–555.

Coates, S.E. 2018. Building the full annual cycle picture for Long-billed Curlews: Correlates of nest success in the breeding grounds and spatial distribution and site fidelity in the wintering grounds. Master's thesis, Boise State University.

Cochran, W.W., and R.D. Lord. 1963. A radio-tracking system for wild animals. *Journal of Wildlife Management* 27(1):9–24.

Cockle, K.L., O. Villalba, N. Fariña, A. Bodrati, L.G. Pagano, E.D. Stein, and A.R. Norris. 2023. Non-breeding ecology of a Neotropical-Nearctic migrant, the Common Nighthawk (*Chordeiles minor*): Habitat, activity patterns, molt, and migration. *Journal of Field Ornithology* 94(3):8. https://doi.org/10.5751/JFO-00293-940308.

Cohen, E.B., J.J. Buler, K.G. Horton, S.R. Loss, S.A. Cabrera-Cruz, J.A. Smolinsky, and P.P. Marra. 2022. Using weather radar to help minimize wind energy impacts on nocturnally migrating birds. *Conservation Letters* 15(4):e12887.

Cohen, E.B., K.G. Horton, P.P Marra, H.L. Clipp, A. Farnsworth, J.A. Smolinsky, D. Sheldon, and J.J. Buler. 2021. A place to land: Spatiotemporal drivers of stopover habitat use by migrating birds. *Ecology Letters* 24(1):38–49.

Cohen, E.B., J.A. Hostetler, M.T. Hallworth, C.S. Rushing, T.S. Sillett, and P.P. Marra. 2018. Quantifying the strength of migratory connectivity. *Methods in Ecology and Evolution* 9(3):513–524.

Cohen, E.B., J.A. Hostetler, J.A. Royle, and P.P. Marra. 2014. Estimating migratory connectivity of birds when re-encounter probabilities are heterogeneous. *Ecology and Evolution* 4:1659–1670.

Cohen, E.B., C.R. Rushing, F.R. Moore, M.T. Hallworth, J.A. Hostetler, M.G. Ramirez, and P.P. Marra. 2019. The strength of migratory connectivity for birds en route to breeding through the Gulf of Mexico. *Ecography* 42(4):658–669.

Colbeck, G.J., H.L. Gibbs, P.P. Marra, K. Hobson, and M.S. Webster. 2008. Phylogeography of a widespread North American migratory songbird (*Setophaga ruticilla*). *Journal of Heredity* 99:453–463.

Collins, D.P., C.M. Conring, B.A. Grisham, W.C. Conway, J.M. Knetter, S.A. Carleton, and M.A. Boggie. 2016. Evidence of new summer areas and mixing of two greater Sandhill Crane populations in the Intermountain West. *Journal of Fish and Wildlife Management* 7(1):141–152.

Colston-Nepali, L., J.F. Provencher, M.L. Mallory, R.P. Franckowiak, Z. Sun, G.J. Robertson, and V.L. Friesen. 2020. Using genomic tools to inform management of the Atlantic Northern Fulmar. *Conservation Genetics* 21:1037–1050.

Coluccy, J.M., K.A. Anderson, T. Yerkes, and J. L. Bowman. 2020. Migration routes and chronology of American Black Duck *Anas rubripes*. *Wildfowl* 70:148–166.

Colwell, M.A., J.D. Reynolds, C.L. Gratto, D. Schamel, and D. Tracy. 1988. Phalarope philopatry. Paper read at Proceedings of the XIXth International Ornithological Congress, Ottowa, ON.

———. 1988. Phalarope philopatry. In *Proceeding of the International Ornithological Congress* 19:585–593.

Comer, J., M. Coulter, and A. Bryan. 1987. Overwintering locations of Wood Storks captured in east-central Georgia. *Colonial Waterbirds* 10(2):162–166. https://doi.org/10.2307/1521255.

Confer, J.L., L.L. Kanda, and I. Li. 2014. Northern Saw-whet Owl: Regional patterns for fall migration and demographics revealed by banding data. *Wilson Journal of Ornithology* 126(2):305–320.

Conklin, J.R., P.F. Battley, M.A. Potter, and D.R. Ruthrauff. 2011. Geographic variation in morphology of Alaska-breeding Bar-tailed Godwits (*Limosa lapponica*) is not maintained on their nonbreeding grounds in New Zealand. *Auk* 128:363–373.

Conklin, J.R., T. Lok, D.S. Melville, A.C. Riegen, R. Schuckard, T. Piersma, and P.F. Battley. 2016a. Declining adult survival of New Zealand Bar-tailed Godwits during 2005–2012 despite apparent population stability. *Emu* 116:147–157.

Conklin, J.R., J. Reneerkens, Y.I. Verkuil, P.S. Tomkovich, P.J. Palsbøll, and T. Piersma. 2016b. Low genetic differentiation between Greenlandic and Siberian Sanderling populations implies a different phylogeographic history than found in Red Knots. *Journal of Ornithology* 157:325–332.

Conklin, J.R., Y.I. Verkuil, P.F. Battley, C.J. Hassell, J. ten Horn, J.A. Johnson, P.S. Tomkovich, A.J. Baker, T. Piersma, and M.C. Fontaine. 2022. Global flyway evolution in Red Knots *Calidris canutus* and genetic evidence for a Nearctic refugium. *Molecular Ecology* 31:2124–2139. https://doi.org/10.1111/mec.16379.

Conover, H.B. 1944a. The north Pacific allies of the Purple Sandpiper. *Fieldiana Zoology* 29(11).

———. 1944b. The races of the Solitary Sandpiper. *Auk* 61(4):537–544.

Conring, C.M., K. Brautigam, B.A. Grisham, D.P. Collins, and W.C. Conway. 2019. Identifying the migratory strategy of the Lower Colorado River Valley population of Greater Sandhill Cranes. *Avian Conservation and Ecology* 14(1):11.

Contina, A., J.L. Alcantara, E.S. Bridge, J.D. Ross, W.F. Oakley, J.F. Kelly, and K.C. Ruegg. 2019. Genetic structure of the Painted Bunting and its implications for conservation of migratory populations. *Ibis* 161(2):372–386.

Contina, A., E.S. Bridge, N.E. Seavy, J.M. Duckles, and J.F. Kelly. 2013. Using geologgers to investigate bimodal isotope patterns in Painted Buntings (*Passerina ciris*). *Auk* 130(2):265–272.

Contina, A., S.W. Yanco, A.K. Pierce, H.B. Vander Zanden, C.A. Stricker, G.J. Bowen, and M.B. Wunder. 2023. Data From: Dynamic environments generate geographic fluctuations in population structure of an inland shorebird. Version 1 [data set]. Zenodo. https://doi.org/10.5281/ZENODO.7991937.

Conway, C.J. 2020. Virginia Rail (*Rallus limicola*). Version 1.0. In *Birds of the World*, ed. A.F. Poole and F.B. Gill. Cornell Lab of Ornithology, Ithaca, NY. https://doi.org/10.2173/bow.virrai.01.

Cook, N.A., C.A. Nicolai, and K.T. Shoemaker. 2021. Inferring movement and staging locations for Canvasbacks using light-level geolocators. *Journal of Fish and Wildlife Management* 12(2):308–321.

Cooke, M.T. 1945. Returns from banded birds: Some interesting recoveries. *Bird Banding* 16:15–21.

———. 1946. Wanderings of the Mockingbird. *Bird Banding* 17:784.

Cooper, J.E., and M.E. Cooper. 2013. *Wildlife Forensic Investigation Principles and Practice*. Boca Raton: CRC Press.

Cooper, N.W., B.C. Dossman, L.E. Berrigan, J.M. Brown, A.R. Brunner, H.E. Chmura, D.A. Cormier, C. Bégin-Marchand, A.D. Rodewald, P.D. Taylor, C.M. Tonra, J.A. Tremblay, and P.P. Marra. 2023a. Songbirds initiate migratory flights synchronously relative to civil dusk. *Movement Ecology* 11:24.

Cooper, N.W., B.C. Dossman, L.E. Berrigan, J.M. Brown, D.A. Cormier, C. Bégin-Marchand, A.D. Rodewald, P.D. Taylor, J.A. Tremblay, and P.P. Marra. 2023b. Atmospheric pressure predicts probability of departure for migratory songbirds. *Movement Ecology* 11:23.

Cooper, N.W., D.N. Ewert, K.R. Hall, S.M. Rockwell, D. Currie, J.M. Wunderle Jr., J.D. White, and P.P. Marra. 2018. Re-sighting data reveal weak connectivity from wintering to breeding grounds in a range-restricted and endangered long-distance migratory passerine. *Avian Conservation and Ecology* 13:1–9.

Cooper, N.W., D.N. Ewert, J.M. Wunderle Jr., and P.P. Marra. 2019a. Revising the wintering distribution and habitat use of the Kirtland's Warbler using playback surveys, citizen-scientists, and geolocators. *Endangered Species Research* 38:79–89.

Cooper, N.W., M.T. Hallworth, and P.P. Marra. 2017. Light-level geolocation reveals wintering distribution, migration routes, and primary stopover locations of an endangered long-distance migratory songbird. *Journal of Avian Biology* 48(2):209–219.

Cooper, N.W., C.S. Rushing, and P.P. Marra. 2019b. Reducing the conservation-reliance of the endangered Kirtland's Warbler through adaptive management. *Journal of Wildlife Management* 83:1297–1305.

Cormier, R.L., D.L. Humple, T. Gardali, and N.E. Seavy. 2013. Light-level geolocators reveal strong migratory connectivity and within-winter movements for a coastal California Swainson's Thrush (*Catharus ustulatus*) population. *Auk* 130:283–290.

———. 2016. Migratory connectivity of Golden-crowned Sparrows from two wintering regions in California. *Animal Migration* 3:48–56. https://doi.org/10.1515/ami-2016-0005.

Corse, C., and R. Summers. 1999. The seasonal pattern of numbers, population structure, and migration of Purple Sandpiper *Calidris maritima* in Orkney. *Ringing & Migration* 19:275–282.

Cortopassi, A.J., and L.R. Mewaldt. 1965. The circumannual distribution of White-crowned Sparrows. *Bird Banding* 36:141–169.

Coulter, M.C. 1975. Post-breeding movements and mortality in the Western Gull, *Larus occidentalis*. *Condor* 77(3):243–249. www.jstor.org/stable/1366220.

Coulter, M.C., J.A. Rodgers Jr., J.C. Ogden, and F.C. Depkin. 2020. Wood Stork (*Mycteria americana*). Version 1.0. In *Birds of the World*, ed. A.F. Poole and F.B. Gill. Cornell Lab of Ornithology, Ithaca, NY. https://doi.org/10.2173/bow.woosto.01.

Coulter, M.W., and W.R. Miller. 1968. Nesting biology of Black Ducks and Mallards in northern New England. *Vermont Fish and Game Department Bulletin* 68.

Coulton, D.W., R.G. Clark, and C.E. Hebert. 2010. Determining natal origins of birds using stable isotopes (δ^{34}S, δD, δ^{15}N, δ^{13}C): Model validation and spatial resolution for mid-continent mallards. *Waterbirds* 33(1):10–21.

Courtot, K., D. Roby, J. Adkins, D. Lyons, D. King, and R. Larsen. 2012. Colony connectivity of Pacific coast Double-crested Cormorants based on post-breeding dispersal from the region's largest colony. *Journal of Wildlife Management* 76:1462–1471. https://doi.org/10.2307/23251445.

Cousineau, M.L., D. Bordage, J. Rodrigue, and P. Brousseau. 2014. Natal origin of Black Ducks harvested during the sport hunt in Quebec, Canada. *Journal of Wildlife Management* 78:1250–1260.

Couter, J.R. 2017. Changes in spring arrival dates of Rufous Hummingbirds (*Selasphorus rufus*) in western North America in the past century. *Wilson Journal of Ornithology* 129(3):535–544.

Craig, D.P., and K. Larson. 2019. Migratory connectivity of North American Caspian Tern (*Hydroprogne caspia*) populations. *Waterbirds* 40:58–62.

Craig, E.H., T.H. Craig, and L.R. Powers. 2016. Habitat use by wintering Golden Eagles and Rough-legged Hawks in southeastern Idaho. *Journal of Raptor Research* 20:69–71.

Craighead, D., R.H. Crandall, R.N. Smith, and S.L. Cain. 2016. Migration of Red-tailed Hawks (*Buteo jamaicensis*) from northwest Wyoming. *Wilson Journal of Ornithology* 128(1):150–158.

Craik, S., J. Pearce, and R.D. Titman. 2020. Red-breasted Merganser (*Mergus serrator*). Version 1.0. In *Birds of the World*, ed. S.M. Billerman. Cornell Lab of Ornithology, Ithaca, NY. https://doi.org/10.2173/bow.rebmer.01.

Cramp, S., and K.E.L. Simmons, eds. 1983. *The Birds of the Western Palearctic*, vol. 3. *Waders to Gulls*. Oxford University Press, Oxford.

Crandall, R.H., and D.J. Craighead. 2019. American Kestrel wintering locations identified using a global positioning system data logger. *Journal of Raptor Research* 53:223–226.

Craven, S.R., and D.H. Rusch. 1983. Winter distribution and affinities of Canada Geese marked on Hudson and James Bays. *Journal of Wildlife Management* 47(2):307–319.

Croll, D.A., A.J. Gaston, A.E. Burger, and D. Konnoff. 1992. Foraging behavior and physiological adaptation for diving in Thick-billed Murres. *Ecology* 73(1):344–356.

Crysler, Z.J., R.A. Ronconi, and P.D. Taylor. 2016. Differential fall migratory routes of adult and juvenile Ipswich Sparrows (*Passerculus sandwichensis princeps*). *Movement Ecology* 4:8.

Cullen, S.A., J.R. Jehl Jr., and G.L. Nuechterlein. 2020. Eared Grebe (*Podiceps nigricollis*). Version 1.0. In *Birds of the World*, ed. S.M. Billerman. Cornell Lab of Ornithology, Ithaca, NY. https://doi.org/10.2173/bow.eargre.01.

Cunningham, J.T., K.H. Elliott, K. Cottenie, S.A. Hatch, and S.R. Jacobs. 2018. Individual foraging location, but not dietary, specialization: Implications for Rhinoceros Auklets as samplers of forage fish. *Marine Ecology Progress Series* 605:225–240. https://doi.org/10.3354/meps12761.

Curk, T., I. Pokrovsky, N. Lecomte, T. Aarvak, D.F. Brinker, K. Burnham, et al. 2020. Arctic avian predators synchronise their spring migration with the northern progression of snowmelt. *Scientific Reports* 10:15450. https://doi.org/10.1038/s41598-020-72948-x.

Currier, H., J. Bowman, and K. Abraham. 2010. Migration and winter movements of Double-crested Cormorants breeding in Georgian Bay, Ontario. *Waterbirds* 33:451–460. https://doi.org/10.2307/40985864.

Cuthbert, F.J., and L.R. Wires. 2020. Caspian Tern (*Hydroprogne caspia*). Version 1.0. In *Birds of the World*, ed. S.M. Billerman. Cornell Lab of Ornithology, Ithaca, NY. https://doi.org/10.2173/bow.caster1.01.

Dale, C.A., and M.L. Leonard. 2011. Reproductive consequences of migration decisions by Ipswich Sparrows (*Passerculus sandwichensis princeps*). *Canadian Journal of Zoology* (*Revue Canadienne De Zoologie*) 89:100–108.

Dale, C.A., J.J. Nocera, S.E. Franks, T.K. Kyser, and L.M. Ratcliffe. 2019. Correlates of alternative migratory strategies in Western Bluebirds. *Journal of Avian Biology* e02031.

Dallin, N. 2004. Connecting students and swans. *Proceedings and Papers of the Nineteenth Trumpeter Swan Society Conference*, ed. D.K. Weaver, R.E. Shea, and M.H. Linck, 39. Trumpeter Swan Society, Maple Plain, MN.

Dau, C.P., and J.E. Sarvis. 2002. Tundra Swans of the lower Alaska Peninsula: Differences in migratory behavior and productivity. *Waterbirds* 25:241–249.

Dau, C.P., P.L. Flint, and M.R. Petersen. 2000. Distribution of recoveries of Steller's Eiders banded on the lower Alaska Peninsula, Alaska. *Journal of Field Ornithology* 71(3):541–548.

Davenport, L.C., K.S. Goodenough, and T. Haugaasen. 2016. Birds of two oceans? Trans-Andean and divergent migration of Black Skimmers (*Rynchops niger cinerascens*) from the Peruvian Amazon. *PLOS ONE* 11(1):e0144994. https://doi.org/10.1371/journal.pone.0144994.

Davidson, N.C., and R.E. Gill. 2008. How do Ruddy Turnstones *Arenaria interpres* prepare to cross the Pacific? *Wader Study Group Bulletin* 115(1):33–35.

Davies, T.E., A.P. Carneiro, B. Campos, C. Hazin, D.C. Dunn, K.M. Gjerde, D.E. Johnson, and M.P. Dias. 2021. Tracking data and the conservation of the high seas: Opportunities and challenges. *Journal of Applied Ecology* 58(12):2703–2710.

Davis, A., and P. Arcese. 1999. An examination of migration in Song Sparrows using banding recovery data. *North American Bird Bander* 24(4):122–128.

Davis, R.A., R.N. Jones, C.D. MacInnes, and A.J. Pakulak. 1985. Molt migration of large Canada Geese on the west coast of Hudson Bay. *Wilson Bulletin* 97(3):296–305.

Davis, S.E., M. Maftei, and M.L. Mallory. 2016. Migratory connectivity at high latitudes: Sabine's Gulls (*Xema sabini*) from a colony in the Canadian High Arctic migrate to different oceans. *PLOS ONE* 11(12):e0166043.

Davis, S.K., and W.E. Lanyon. 2020. Western Meadowlark (*Sturnella neglecta*). Version 1.0. In *Birds of the World*, ed. A.F. Poole. Cornell Lab of Ornithology, Ithaca, NY. https://doi.org/10.2173/bow.wesmea.01.

Davis Jr., W.E., and J.C. Kricher. 2020. Glossy Ibis (*Plegadis falcinellus*). Version 1.0. In *Birds of the World*, ed. S.M. Billerman. Cornell Lab of Ornithology, Ithaca, NY. https://doi.org/10.2173/bow.gloibi.01.

Davis Jr., W.E., and J.A. Kushlan. 2020. Green Heron (*Butorides virescens*). Version 1.0. In *Birds of the World*, ed. A.F. Poole and F.B. Gill. Cornell Lab of Ornithology, Ithaca, NY. https://doi.org/10.2173/bow.grnher.01.

Dawson, W.R. 2020. Pine Siskin (*Spinus Pinus*). Version 1.0. In *Birds of the World*, ed. A.F. Poole. Cornell Lab of Ornithology, Ithaca, NY. https://doi.org/10.2173/bow.pinsis.01.

Day, R.H., M.L. Kissling, K.J. Kuletz, D.A. Nigro, and P. Pyle. 2020a. Kittlitz's Murrelet (*Brachyramphus brevirostris*). Version 1.0. In *Birds of the World*, ed. P.G. Rodewald. Cornell Lab of Ornithology, Ithaca, NY. https://doi.org/10.2173/bow.kitmur.01

REFERENCES

Day, R.H., I.J. Stenhouse, and H.G. Gilchrist. 2020b. Sabine's Gull (*Xema sabini*). Version 1.0. In *Birds of the World*, ed. S.M. Billerman. Cornell Lab of Ornithology, Ithaca, NY. https://doi.org/10.2173/bow.sabgul.01.

Deguchi, T., J. Jacobs, T. Harada, L. Perriman, Y. Watanabe, F. Sato, N. Nakamura, K. Ozaki, and G. Balogh. 2012. Translocation and hand-rearing techniques for establishing a colony of threatened albatross. *Bird Conservation International* 22(1):66–81.

Deguchi, T., F. Sato, M. Eda, H. Izumi, H. Suzuki, R.M. Suryan, E.W. Lance, H. Hasegawa and K. Ozaki. 2017. Translocation and hand-rearing result in short-tailed albatrosses returning to breed in the Ogasawara Islands 80 years after extirpation. *Animal Conservation* 20:341–349. https://doi.org/10.1111/acv.12322.

DeHaven, R.W., F.T. Crase, and P.P. Woronecki. 1975. Movements of Tricolored Blackbirds banded in the Central Valley of California, 1965–1972. *Bird Banding* 46:220–229.

del Hoyo, J., C. Carboneras, F. Jutglar, N. Collar, and G.M. Kirwan. 2020a. Black-browed Albatross (*Thalassarche melanophris*). Version 1.0. In *Birds of the World*, ed. S.M. Billerman, B.K. Keeney, P.G. Rodewald, and T.S. Schulenberg. Cornell Lab of Ornithology, Ithaca, NY. https://doi.org/10.2173/bow.bkbalb.01.

———. 2020b. Cory's Shearwater (*Calonectris diomedea*). Version 1.0. In *Birds of the World*, ed. S.M. Billerman, B.K. Keeney, P.G. Rodewald, and T.S. Schulenberg. Cornell Lab of Ornithology, Ithaca, NY. https://doi.org/10.2173/bow.corshe.01.

De Jong, M.J. 2020. Northern Rough-winged Swallow (*Stelgidopteryx serripennis*). Version 1.0. In *Birds of the World*, ed. A.F. Poole and F.B. Gill. Cornell Lab of Ornithology, Ithaca, NY. https://doi.org/10.2173/bow.nrwswa.01.

De Juana, E., and E.F.J. Garcia. 2010. Vagrancy or migration: Why do American Teals cross the Atlantic? *Ardeola* 57(2):417–430.

De La Cruz, S.E.W., J.Y. Takekawa, M.T. Wilson, D.R. Nysewander, J.R. Evenson, D. Esler, W.S. Boyd, and D.H. Ward. 2009. Spring migration routes and chronology of Surf Scoters (*Melanitta perspicillata*): A synthesis of Pacific coast studies. *Canadian Journal of Zoology* 87:1069–1086.

Delancey, C.D., K.L. Islam, G.R. Kramer, G.J. MacDonald, A.R. Sharp, and B.M. Connare. 2020. Geolocators reveal migration routes, stopover sites, and non-breeding dispersion in a population of Cerulean Warblers. *Animal Migration* 7:19–26. https://doi.org/10.1515/ami-2020-0003.

Dellinger, R., P.B. Wood, P.W. Jones, and T.M. Donovan. 2020. Hermit Thrush (*Catharus guttatus*). Version 1.0. In *Birds of the World*, A.F. Poole. Cornell Lab of Ornithology, Ithaca, NY. https://doi.org/10.2173/bow.herthr.01.

Delmore, K.E., and D.E. Irwin. 2014. Hybrid songbirds employ intermediate routes in a migratory divide. *Ecology Letters* 17:1211–1218.

Delmore, K.E., J.W. Fox, and D.E. Irwin. 2012. Dramatic intraspecific differences in migratory routes, stopover sites and wintering areas, revealed using light-level geolocators. *Proceedings of the Royal Society: Biological Sciences* 279:4582–4589.

Delord, K., C. Barbraud, D. Pinaud, et al. 2020. Movements of three alcid species breeding sympatrically in Saint Pierre and Miquelon, northwestern Atlantic Ocean. *Journal of Ornithology* 161:359–371. https://doi.org/10.1007/s10336-019-01725-z.

Delta Waterfowl. 2020. [Ring-necked Ducks.] https://deltawaterfowl.org/ringneck/.

DeLuca, W., R. Holberton, P.D. Hunt, and B.C. Eliason. 2020. Blackpoll Warbler (*Setophaga striata*). Version 1.0. In *Birds of the World*, ed. A.F. Poole. Cornell Lab of Ornithology, Ithaca, NY. https://doi.org/10.2173/bow.bkpwar.01.

DeLuca, W.V., B.K. Woodworth, S.A. MacKenzie, A.E.M. Newman, H. Cooke, L.M. Phillips, N.E. Freeman, A.O. Sutton, L. Tauzer, C. McIntyre, I.J. Stenhouse, S. Weidensaul, P.D. Taylor, and D.R. Norris. 2019. A boreal songbird's 20,000 km migration across North America and the Atlantic Ocean. *Ecology*. https://doi.org/10.1002/ecy.2651.

DeLuca, W.V., B.K. Woodworth, C.C. Rimmer, P.P. Marra, P.D. Taylor, K.P. McFarland, S.A. Mackenzie, and D.R. Norris. 2015. Transoceanic migration by a 12 g songbird. *Biology Letters* 11:4.

d'Entremont, K.J.N., L.M. Zitske, A.J. Gladwell, N.K. Elliott, R.A. Mauck, and R.A. Ronconi. 2020. Breeding population decline and associations with nest site use of Leach's Storm-petrels on Kent Island, New Brunswick from 2001 to 2018. *Avian Conservation and Ecology* 15:1–11. https://doi.org/10.5751/ACE-01526-150111.

Derksen, D.V., K.S. Bollinger, D.H. Ward, J.S. Sedinger, and Y. Miyabayashi. 1996. Black Brant from Alaska staging and wintering in Japan. *Condor* 98:653–657.

De Ruyck, C., K.A. Hobson, N. Koper, K.W. Larson, and L.I. Wassenaar. 2013. An appraisal of the use of hydrogen-isotope methods to delineate origins of migratory Saw-whet Owls in North America. *Condor* 115(2):366–374.

DeSaix, M.G., E.C. Anderson, C.M. Bossu, C.E. Rayne, T.M. Schweizer, N.J. Bayly, D.S. Narang, J.C. Hagelin, H.L. Gibbs, J.F. Saracco, T.W. Sherry, M.S. Webster, T.B. Smith, P.P. Marra, and K.C. Ruegg. 2023. Low-coverage whole genome sequencing for highly accurate population assignment: Mapping migratory connectivity in the American Redstart (*Setophaga ruticilla*). *Molecular Ecology* 32:5528–5540. https://doi.org/10.1111/mec.17137.

DeSaix, M.G., L.P. Bulluck, A.J. Eckert, et al. 2019. Population assignment reveals low migratory connectivity in a weakly structured songbird. *Molecular Ecology* 28(9):2122–2135. https://doi.org/10.1111/mec.15083.

DeSaix, M.G., E.B. Connell, N. Cortes-Rodríguez, K.E. Omland, P.P. Marra, and C.E. Studds. 2022a. Migratory connectivity in a Newfoundland population of the American Redstart (*Setophaga ruticilla*). *Wilson Journal of Ornithology* 134(3):381–389. https://doi.org/10.1676/22-00004.

DeSaix, M.G., T.L. George, A.E. Seglund, G.M, Spellman, E.S. Zavaleta, and K.C. Ruegg. 2022b. Forecasting climate change response in an alpine specialist songbird reveals the importance of considering novel climate. *Diversity and Distributions* 28(10):2239–2254.

Desnoyers, M., G. Gauthier, and J. Lefebvre. 2012. Stable associations within flocks of Greater Snow Geese (*Chen caerulescens atlantica*): Do they exist beyond family bonds? *Auk* 129(4):611–622.

DeSorbo, C.R., C.P. Persico, A.T. Gilbert, M.E.H. Burton, and D. Meattey. 2022. Studying migrant raptors using the Atlantic Flyway. Block Island Raptor Research Station, Block Island, RI, 2022 season. BRI report no. 2023-11. Biodiversity Research Institute, Portland, ME.

Devereux, J., T. Carpenter, and K. Durham. 1985. Spring migration pattern of Sharp-shinned Hawks passing Whitefish Point, Michigan. *Journal of Field Ornithology* 56(4):346–355.

de Zwaan, D.R., S. Wilson, E.A. Gow, and K. Martin. 2019. Sex-specific spatiotemporal variation and carry-over effects in a migratory alpine bird. *Frontiers in Ecology and Evolution* 7(285).

Diamond, A.W., and E.A. Schreiber. 2020. Magnificent Frigatebird (*Fregata magnificens*). Version 1.0. In *Birds of the World*, ed. A.F. Poole and F.B. Gill. Cornell Lab of Ornithology, Ithaca, NY. https://doi.org/10.2173/bow.magfri.01.

Dias, M.P., J.P. Granadeiro, R.A. Phillips, H. Alonso, and P. Catry. 2011. Breaking the routine: Individual Cory's Shearwaters shift winter destinations between hemispheres and across ocean basins. *Proceedings of the Royal Society: Biological Sciences* 278(1713):1786–1793.

Dickson, D.L. 2012a. Movement of King Eiders from breeding grounds on Banks Island, NWT, to moulting and wintering areas. Technical Report Series no. 516. Canadian Wildlife Service, Edmonton, Alberta.

———. 2012b. Seasonal movement of King Eiders breeding in Western Arctic Canada and Northern Alaska. Canadian Wildlife Service Technical Report Series no. 520. Canadian Wildlife Service, Edmonton, Alberta.

———. 2012c. Seasonal movement of Pacific Common Eiders breeding in Arctic Canada. Canadian Wildlife Service Technical Report Series no. 521:i–v, 1–58.

Dickson, R.D., D. Ester, J.W. Hupp, E.M. Anderson, J.R. Evenson, and J. Barrett. 2012. Phenology and duration of remigial moult in Surf Scoters (*Melanitta perspicillata*) and White-winged Scoters (*Melanitta fusca*) on the Pacific coast of North America. *Canadian Journal of Zoology* (*Revue Canadienne De Zoologie*) 90:932–944.

Diebold, E.N., S. Matteson, F. Koontz, C. Koontz, and P.P. Manthey. 2008. Use of satellite telemetry to determine the migratory routes of a restored population of Trumpeter Swans (*Cygnus cygnus buccinators*). *Proceedings of the Association of Zoos and Aquariums Annual Conference*. https://www.aza.org/assets/2332/2008annual14.pdf.

Diefenbach, D.R., J.D. Nichols, and J.E. Hines. 1988. Distribution patterns of American Black Ducks and Mallard winter band recoveries. *Journal of Wildlife Management* 52:704–710.

Diem, K.L., and D.D. Condon. 1967. Banding studies of water birds on the Molly Islands, Yellowstone Lake, Wyoming. Yellowstone Library and Museum Association, Yellowstone National Park, WY.

Dieter, C.D., and B.J. Anderson. 2009. Molt migration by Giant Canada Geese in eastern South Dakota. *Human-Wildlife Conflict* 3(2):260–270.

Divoky, G.J. 1979. Sea ice as a factor in seabird distribution and ecology in the Beaufort, Chukchi, and Bering Seas. In *Conservation of Marine Birds of Northern North America*, ed. J.C. Bartonek and D.N. Nettleship, Wildlife Research Report 11:9–17. US Fish and Wildlife Service.

Divoky, G.J., D.C. Douglas, and I.J. Stenhouse. 2016. Arctic sea ice a major determinant in Mandt's Black Guillemot movement and distribution during non-breeding season. *Biology Letters* 12(9):20160275. https://doi.org/10.1098/rsbl.2016.0275.

Dobbs, R.C., P.R. Martin, and T.E. Martin. 2020. Green-tailed Towhee (*Pipilo chlorurus*). Version 1.0. In *Birds of the World*, ed. A.F. Poole. Cornell Lab of Ornithology, Ithaca, NY. https://doi.org/10.2173/bow.gnttow.01.

Dodds, M.D. 2024. Nonbreeding distribution and movement of adult Razorbills (*Alca torda*) breeding in Atlantic Canada and Québec. Master's thesis, University of New Brunswick. https://unbscholar.lib.unb.ca/items/e36151f0-5482-455b-b09b-419c26433404

Dodge, S., G. Bohrer, K. Bildstein, S.C. Davidson, R. Weinzierl, M.J. Bechard, et al. 2014. Environmental drivers of variability in the movement ecology of Turkey Vultures (*Cathartes aura*) in North and South America. *Philosophical Transactions of the Royal Society* 369:1–17.

Dokter, A.M., A. Farnsworth, D. Fink, V. Ruiz-Gutierrez, W.M. Hochachka, F.A. La Sorte, O.J. Robinson, K.V. Rosenberg, and S. Kelling. 2018. Seasonal abundance and survival of North America's migratory avifauna determined by weather radar. *Nature Ecology & Evolution* 2(10):1603–1609.

Dolbeer, R. 1991. Migration patterns of Double-crested Cormorants east of the Rocky Mountains. *Journal of Field Ornithology* 62.

Domalik, A.D., M. Maftei, K.G. Wright, S.A. Hudson (Trefry), and J.M. Hipfner. 2021. Migration and winter habitat use of Glaucous-Winged Gulls (*Larus glaucescens*) from Triangle Island, British Columbia. *Waterbirds* 44(4):438–448.

Domenech, R., T. Pitz, K. Gray, and M. Smith. 2015. Estimating natal origins of migratory juvenile Golden Eagles using stable hydrogen isotopes. *Journal of Raptor Research* 49(3):308–315.

Donohue, K.C. 2009. Exploring Red-tailed Hawk migration using stable isotope analysis and DNA sexing techniques. Master's thesis, Boise State University. https://scholarworks.boisestate.edu/td/80.

Donovan, T., J. Buzas, P. Jones, and H.L. Gibbs. 2006. Tracking dispersal in birds: Assessing the potential of elemental markers. *Auk* 123(2):500–511.

Dorr, B.S., J.J. Hatch, and D.V. Weseloh. 2021. Double-crested Cormorant (*Nannopterum auritum*). Version 1.1. In *Birds of the World*, ed. A.F. Poole. Cornell Lab of Ornithology, Ithaca, NY. https://doi.org/10.2173/bow.doccor.01.1.

Dorr, B., J. Taylor, S. Werner, D. King, J. Farquhar, I. Mazzocchi, and R. Mccullough. 2012. Summer and migrational movements of satellite-marked Double-Crested Cormorants from a breeding colony managed by egg-oiling in Lake Ontario. *Waterbirds* 35:114–123. https://doi.org/10.1675/063.035.sp112.

Dorr, B.S., J.J. Hatch, and D.V. Weseloh. 2021. Double-crested Cormorant (*Nannopterum auritum*). Version 1.1. In *Birds of the World*, ed. A.F. Poole. Cornell Lab of Ornithology, Ithaca, NY. https://doi.org/10.2173/bow.doccor.01.1.

Dossman, B.C., A.D. Rodewald, and P.P. Marra. 2024. Hidden space use behaviors of a nonbreeding migratory bird: The role of environment and social context. *Movement Ecology* 12(1):82.

Dossman, B.C., A.D. Rodewald, C.E. Studds, and P.P. Marra. 2023. Migratory birds with delayed spring departure migrate faster but pay the costs. *Ecology* 104(2):e3938.

Douglas, D.C., R. Weinzierl, S.C. Davidson, R. Kays, M. Wikelski, and G. Bohrer. 2012. Moderating Argos location errors in animal tracking data. *Methods in Ecology and Evolution* 3(6):999–1007. https://doi.org/10.1111/j.2041-210X.2012.00245.x.

Doyle, F.I., J.-F. Therrien, D.G. Reid, G. Gauthier, and C.J. Krebs. 2018. Seasonal movements of female Snowy Owls breeding in the western North American Arctic. *Journal of Raptor Research* 51(4):428–438.

Doyle, J.M., D.A. Bell, P.H. Bloom, G. Emmons, A. Fesnock, T.E. Katzner, L. LaPré, K. Leonard, P. SanMiguel, R. Westerman, and J.A. DeWoody. 2018. New insights into the phylogenetics and population structure of the Prairie Falcon (*Falco mexicanus*). *Genomics* 19:233–246.

Dragoo, D.E., H.M. Renner, and R.S.A. Kaler. 2020. Breeding status and population trends of seabirds in Alaska, 2019. Report no. AMNWR 2020/01. US Fish and Wildlife Service, Homer, AK.

Drewien, R.C., W.M. Brown, J.D. Varley, and D.C. Lockman. 1999. Seasonal movements of Sandhill Cranes radiomarked in Yellowstone National Park and Jackson Hole, Wyoming. *Journal of Wildlife Management* 63(1):126–136.

Drilling, N., R.D. Titman, and F. McKinney. 2020. Mallard (*Anas platyrhynchos*). Version 1.0. In *Birds of the World*, ed. S.M. Billerman. Cornell Lab of Ornithology, Ithaca, NY. https://doi.org/10.2173/bow.mallar3.01.

Drummond, B.A., R.A. Orben, A.M. Christ, A.B. Fleishman, H.M. Renner, N.A. Rojek, and M.D. Romano. 2021. Comparing non-breeding distribution and behavior of Red-legged Kittiwakes from two geographically distant colonies. *PLOS ONE* 16(7):e0254686. https://doi.org/10.1371/journal.pone.0254686. PMID: 34270622; PMCID: PMC8284784.

Drury Jr., W.H., and J.A. Keith. 1962. Radar studies of songbird migration in coastal New England. *Ibis* 104:449–489.

Dubowy, P.J., C. Carboneras, and G.M. Kirwan. 2020. Northern Shoveler (*Spatula clypeata*). Version 1.0. In *Birds of the World*, ed. S.M. Billerman. Cornell Lab of Ornithology, Ithaca, NY. https://doi.org/10.2173/bow.norsho.01.

Dugger, B.D., K.M. Dugger, and L.H. Fredrickson. 2020. Hooded Merganser (*Lophodytes cucullatus*). Version 1.0. In *Birds of the World*, ed. A.F. Poole. Cornell Lab of Ornithology, Ithaca, NY. https://doi.org/10.2173/bow.hoomer.01.

Dugger, K.M., J. Faaborg, W.J. Arendt, and K.A. Hobson. 2004. Understanding survival and abundance of overwintering warblers: Does rainfall matter? *Condor* 106:744–760.

Duijns, S., L.J. Niles, A. Dey, Y. Aubry, C. Friis, S. Koch, A.M. Anderson, and P.A. Smith. 2017. Body condition explains migratory performance of a long-distance migrant. *Proceedings of the Royal Society: Biological Sciences* 284(1866):20171374. https://doi.org/10.1098/rspb.2017.1374.

Dumas, J.V. 2020. Roseate Spoonbill (*Platalea ajaja*). Version 1.0. In *Birds of the World*, ed. A.F. Poole and F.B. Gill. Cornell Lab of Ornithology, Ithaca, NY. https://doi.org/10.2173/bow.rosspo1.01.

Duncan, B.W. 1982. Sharp-shinned Hawks banded at Hawk Cliff, Ontario, 1971–1981: Analysis of data. *Ontario Bird Banding* 15(2):24–38.

Dunn, E.H., D. Brewer, A.W. Diamond, E.J. Woodsworth, and B.T. Collins. 2009. *Canadian Atlas of Bird Banding*. Vol. 3. *Raptors and Waterbirds, 1921–1995*. Environment Canada.

Dunn, E.H., and G.A. Hall. 2020. Magnolia Warbler (*Setophaga magnolia*). Version 1.0. In *Birds of the World*, ed. P.G. Rodewald. Cornell Lab of Ornithology, Ithaca, NY. https://doi.org/10.2173/bow.magwar.01.

Dunn, E.H., D.J.T. Hussell, J. Kren, and A.C. Zoerb. 2022. Northern Wheatear (*Oenanthe oenanthe*). Version 2.1. In *Birds of the World*, ed. P.G. Rodewald and B.K. Keeney. Cornell Lab of Ornithology, Ithaca, NY. https://doi.org/10.2173/bow.norwhe.02.1.

Dunn, E.H., K.A. Hobson, L.I. Wassenaar, D.J.T. Hussell, and M.L. Allen. 2006. Identification of summer origins of songbirds migrating through southern Canada in autumn. *Avian Conservation and Ecology* (*Écologie et conservation des oiseaux*) 1(2):4. http://www.ace-eco.org/vol1/iss2/art4/.

Dunn, J.L., and K.L. Garrett. 1997. *A Field Guide to the Warblers*. Houghton Mifflin, New York.

Dunning Jr., J.B., P. Pyle, and M.A. Patten. 2020a. Bachman's Sparrow (*Peucaea aestivalis*). Version 1.0. In *Birds of the World*, ed. P.G. Rodewald. Cornell Lab of Ornithology, Ithaca, NY. https://doi.org/10.2173/bow.bacspa.01.

Dunning Jr., J.B., R.K. Bowers Jr., S.J. Suter, and C.E. Bock. 2020b. Cassin's Sparrow (*Peucaea cassinii*). Version 1.0. In *Birds of the World*, ed. P.G. Rodewald. Cornell Lab of Ornithology, Ithaca, NY. https://doi.org/10.2173/bow.casspa.01.

Dunstan, T.C. 1972. A harness for radio-tagging raptorial birds. *IBBA News* 44:4–8.

Durrant, K.L., P.P. Marra, S.M. Fallon, G.J. Colbeck, H.L. Gibbs, K.A. Hobson, D.R. Norris, B. Bernik, V.L. Lloyd, and R.C. Fleischer. 2008. Parasite assemblages distinguish populations of a migratory passerine on its breeding grounds. *Journal of Zoology* 274(4):318–326.

Duxbury, J. 2004. Stable isotope analysis and the investigation of the migrations and dispersal of Peregrine Falcons (*Falco peregrinus*) and Burrowing Owls (*Athene cunicularia hypugaea*). PhD diss., University of Alberta.

Dykstra, C.R., J.L. Hays, and S.T. Crocoll. 2020. Red-shouldered Hawk (*Buteo lineatus*). Version 1.0. In *Birds of the World*, ed. A.F. Poole. Cornell Lab of Ornithology, Ithaca, NY. https://doi.org/10.2173/bow.reshaw.01.

Dykstra, C.R., J.L. Hays, M.M. Simon, A.R. Wegman, K.A. Williams, and L.R. Dykstra. 2019. Dispersal and survival of Red-shouldered Hawks banded in suburban southern Ohio, 1996–2018. *Journal of Raptor Research* 53(3):276–292.

Eadie, J.M., M.L. Mallory, and H.G. Lumsden. 2020. Common Goldeneye (*Bucephala clangula*). Version 1.0. In *Birds of the World*, ed. S.M. Billerman. Cornell Lab of Ornithology, Ithaca, NY. https://doi.org/10.2173/bow.comgol.01.

Eda, M., H. Koike, M. Kuro-o, S. Mihara, H. Hasegawa, and H. Higuchi. 2012. Inferring the ancient population structure of the vulnerable albatross *Phoebastria albatrus*, combining ancient DNA, stable isotope, and morphometric analyses of archaeological samples. *Conservation Genetics* 13(1):143–151.

Eda, M., T. Shimada, T. Amano, K. Ushiyama, C. Mizota, and H. Koike. 2013. Phylogenetic relationship of the Greater White-fronted Goose *Anser albifrons* subspecies wintering in the Palaearctic region. *Ornithological Science* 12(1):35–42.

Eda, M., T. Yamasaki, H. Izumi, N. Tomita, S. Konno, M. Konno, H. Murakami, and F. Sato. 2020. Cryptic species in a vulnerable seabird: Short-tailed albatross consists of two species. *Endangered Species Research* 43:375–386.

Eddleman, W.R. 2020. Hepatic Tanager (*Piranga flava*). Version 1.0. In *Birds of the World*, ed. A.F. Poole and F.B. Gill. Cornell Lab of Ornithology, Ithaca, NY. https://doi.org/10.2173/bow.heptan.01.

Eddleman, W.R., R.E. Flores, and M. Legare. 2020. Black Rail (*Laterallus jamaicensis*). Version 1.0. In *Birds of the World*, ed. A.F. Poole and F.B. Gill. Cornell Lab of Ornithology, Ithaca, NY. https://doi.org/10.2173/bow.blkrai.01.

Edwards, A.E., S.M. Fitzgerald, J.K. Parrish, J.L. Klavitter, and M.D. Romano. 2015. Foraging strategies of Laysan Albatross inferred from stable isotopes: Implications for association with fisheries. *PLOS ONE* 10(7):e0133471. https://doi.org/10.1371/journal.pone.0133471.

Egerton, Z., K. Townsend, and D. Potvin. 2022. A review of the status and biology of an enigmatic pelagic species, the Red-tailed Tropicbird *Phaethon rubricauda*. *Corella* 46:80–88.

Egevang, C., I.J. Stenhouse, R.A. Phillips, A. Petersen, J.W. Fox, and J.R.D. Silk. 2010. Tracking of Arctic Terns *Sterna paradisaea* reveals longest animal migration. *Proceedings of the National Academy of Sciences* 107(5):2078–2081. https://doi.org/10.1073/pnas.0909493107.

Eichholz, M.W., and J.S. Sedinger. 2006. Staging, migration, and winter distribution of Canada and Cackling Geese staging in Interior Alaska. *Journal of Wildlife Management* 70:1308–1315.

Eichhorst, B.A. 1992. An analysis of Western Grebe banding and recovery data. *North American Bird Bander* 17(3):108–115.

Eisaguirre, J.M., T.L. Booms, P.F. Schempf, and S.B. Lewis. 2016. Gyrfalcon home ranges and movements on the Yukon-Kuskokwim Delta, Alaska. *Journal of Raptor Research* 50(1):109–114.

Eisenhauer, D.I., and C.M. Kirkpatrick. 1977. Ecology of the Emperor Goose in Alaska. *Wildlife Monographs* (57):3–62.

Elliott, J.E., M.C. Drever, K.R. Studholme, V. Silverthorn, A.A. Miller, K.H. Elliott, S.L. Lee, K.G. Drouillard, E. Porter, A.M. Idrissi, and G.T. Crossin. 2021. Exposure to persistent organic pollutants is linked to over-wintering latitude in a Pacific seabird, the Rhinoceros Auklet, *Cerorhinca monocerata*. *Environmental Pollution* 279:116928.

Elliott-Smith, E., and S.M. Haig. 2020. Piping Plover (*Charadrius melodus*). Version 1.0. In *Birds of the World*, ed. A.F. Poole. Cornell Lab of Ornithology, Ithaca, NY. https://doi.org/10.2173/bow.pipplo.01.

Ellison, K., and P. E. Lowther. 2020. Bronzed Cowbird (*Molothrus aeneus*). Version 1.0. In *Birds of the World*, ed. A.F. Poole. Cornell Lab of Ornithology, Ithaca, NY. https://doi.org/10.2173/bow.brocow.01.

Ellison, K., E. McKinnon, S. Zack, S. Olinb, R. Sparks, and E. Strasser. 2017. Migration and winter distribution of the Chestnut-collared Longspur. *Animal Migration* 4:37–50. https://doi.org/10.1515/ami-2917-000r.

Ellison, K., B.O. Wolf, and S.L. Jones. 2021. Vermilion Flycatcher (*Pyrocephalus rubinus*). Version 1.1. In *Birds of the World*, ed. A.F. Poole. Cornell Lab of Ornithology, Ithaca, NY. https://doi.org/10.2173/bow.verfly.01.1.

Elphick, C.S., and T. L. Tibbitts. 2020. Greater Yellowlegs (*Tringa melanoleuca*). Version 1.0. *In Birds of the World*, ed. A.F. Poole and F.B. Gill. Cornell Lab of Ornithology, Ithaca, NY. https://doi.org/10.2173/bow.greyel.01.

Ely, C.R., and B.W. Meixell. 2016. Demographic outcomes of diverse migration strategies assessed in a metapopulation of Tundra Swans. *Movement Ecology* 4:10.

Ely, C.R., K.S. Bollinger, J.W. Hupp, D.V. Derksen, J. Terenzi, J.Y. Takekawa, D.L. Orthmeyer, T.C. Rothe, M.J. Petrula, and D.R. Yparraguirre. 2006. Traversing a boreal forest landscape: Summer movements of Tule Greater White-fronted Geese. *Waterbirds* 29:43–55.

Ely, C.R., D.C. Douglas, A.C. Fowler, C.A. Babcock, D.V. Derksen, and J.Y. Takekawa. 1997. Migration behavior of Tundra Swans from the Yukon-Kuskokwim Delta, Alaska. *Wilson Bulletin* 109:679–692.

Ely, C.R., D.J. Nieman, R.T. Alisauskas, J.A. Schmutz, and J.E. Hines. 2013. Geographic variation in migration chronology and winter distribution of midcontinent Greater White-Fronted Geese. *Journal of Wildlife Management* 77:1182–1191.

Ely, C.R., W.L. Sladen, H.M. Wilson, S.E. Savage, K.M. Sowl, B. Henry, M. Schwitters, and J. Snowden. 2014. Delineation of Tundra Swan *Cygnus c. columbianus* populations in North America: Geographic boundaries and interchange. *Wildfowl* 64:132–147.

Ely, C.R., and J.Y. Takekawa. 1996. Geographic variation in migratory behavior of Greater White-fronted Geese (*Anser albifrons*). *Auk* 113:889–901.

Ely, C.R., R.E. Wilson, and S.L. Talbot. 2017. Genetic structure among Greater White-fronted Goose populations of the Pacific Flyway. *Ecology and Evolution* 7:2956–2968.

Enderson, J.H. 1964. A study of the Prairie Falcon in the Central Rocky Mountain region. *Auk* 81(3):332–352.

Engelmoer, M., and C.S. Roselaar. 1998. *Geographical Variation in Waders*. Kluwer Academic Publishers, Dordrecht, Netherlands.

England, A.S., and W.F. Laudenslayer Jr. 2020. Bendire's Thrasher (*Toxostoma bendirei*) Version 1.0. In *Birds of the World*, ed. A.F. Poole and F.B. Gill. Cornell Lab of Ornithology, Ithaca, NY. https://doi.org/10.2173/bow.benthr.01.

English, P.A., A.M. Mills, M.D. Cadman, A.E. Heagy, G.J. Rand, D.J. Green, and J.J. Nocera. 2017. Tracking the migration of a nocturnal aerial insectivore in the Americas. *BMC Zoology* 2, 5.

Enríquez-Paredes, L.M., C. Vilanova, and G. Fernández. 2012. Genetic diversity and population structure of wintering Western Sandpipers from the Sinaloa coast, Mexico. *Journal of Field Ornithology* 83:85–93.

Environment and Climate Change Canada. 2019. Status of Birds in Canada 2019. Gatineau, Quebec. https://wildlife-species.canada.ca/bird-status/index-eng.aspx?sY=2019&sL=e.

Evans, B.A., J.S. Humphrey, E.A. Tillman, M.L. Avery, and B.M. Kluever. 2024. Site-specific space use and resource selection by Black Vultures (*Coragyps atratus*) in the southeastern United States. *Ibis* 166:129–145. https://doi.org/10.1111/ibi.13244.

Evans, D., and C.R. Sindelar. 1974. First record of the Goshawk for Louisiana: A collected, banded bird. *Bird Banding* 45(3):270.

Evans, D.L., and R.N. Rosenfield. 1985. Migration and mortality of Sharp-shinned Hawks ringed at Duluth, Minnesota. *ICBP Technical Publication* No. 5:311–316.

Evenson, J.R., M.W. Wilson, and K.A. Spragens. Waterfowl telemetry studies from Washington state. Unpublished data.

Evers, D.C., J.A. Schmutz, N. Basu, C.R. DeSorbo, J. Fair, C.E. Gray, J.D. Paruk, M. Perkins, K. Regan, B.D. Uher-Koch, and K.G. Wright. 2014. Historic and contemporary mercury exposure and potential risk to Yellow-billed Loons (*Gaviaa adamsii*) breeding in Alaska and Canada. *Waterbirds* 37(sp1):147–159.

Ewins, P.J. 2020. Pigeon Guillemot (*Cepphus columba*). Version 1.0. In *Birds of the World*, ed. A.F. Poole and F.B. Gill. Cornell Lab of Ornithology, Ithaca, NY. https://doi.org/10.2173/bow.piggui.01.

Ezhov, A.V., M.V. Gavrilo, Y.V. Krasnov, V.S. Bråthen, B. Moe, A.V. Baranskaya, and H. Strøm. 2021. Transpolar and bi-directional migration strategies of Black-legged Kittiwakes *Rissa tridactyla* from a colony in Novaya Zemlya, Barents Sea, Russia. *Marine Ecology Progress Series* 676:189–203. https://doi.org/10.3354/meps13889.

Fallon, S.M., R.C. Fleischer, and G.R. Graves. 2006. Malarial parasites as geographical markers in migratory birds? *Biology Letters* 2(2):213–216.

Falls, J.B., and J.G. Kopachena. 2020. White-throated Sparrow (*Zonotrichia albicollis*). Version 1.0. In *Birds of the World*, ed. A.F. Poole. Cornell Lab of Ornithology, Ithaca, NY. https://doi.org/10.2173/bow.whtspa.01.

Farmer, A., M. Abril, M. Fernández, J. Torres, C. Kester, and C. Bern. 2004. Using stable isotopes to associate migratory shorebirds with their wintering locations in Argentina. *Ornitologia Neotropical* 15:377–384.

Farmer, A., R.T. Holmes, and F.A. Pitelka. 2020. Pectoral Sandpiper (*Calidris melanotos*). Version 1.0. In *Birds of the World*, ed. S.M. Billerman. Cornell Lab of Ornithology, Ithaca, NY. https://doi.org/10.2173/bow.pecsan.01.

Farmer, C.J., D.J.T. Hussell, and D. Mizrahi. 2007. Detecting population trends in migratory birds of prey. *Auk* 124(3):1047–1062.

Farmer, C.J., K. Safi, D.R. Barber, I. Newton, M. Martell, and K.L. Bildstein. 2010. Efficacy of migration counts for monitoring continental populations of raptors: An example using the Osprey (*Pandion haliaetus*). *Auk* 127(4):863–870. https://doi.org/10.1525/auk.2010.09152.

Farnsworth, A., K.G. Horton, and P.P. Marra. 2024. To mitigate bird collisions, enforce the Migratory Bird Treaty Act. *Proceedings of the National Academy of Sciences* 121(9):e2320411121.

Farnsworth, G., G.A. Londono, J.U. Martin, K.C. Derrickson, and R. Breitwisch. 2020. Northern Mockingbird (*Mimus polyglottos*). Version 1.0. In *Birds of the World*, ed. A.F. Poole. Cornell Lab of Ornithology, Ithaca, NY. https://doi.org/10.2173/bow.normoc.01.

Fayet, A.L., et al. 2016. Drivers and fitness consequences of dispersive migration in a pelagic seabird. *Behavioral Ecology* 27(4):1061–1072.

———. 2017. Ocean-wide drivers of migration strategies and their influence on population breeding performance in a declining seabird. *Current Biology* 27:3871–3878.

Fayet, A.L., P. Shannon, D.E. Lyons, and S.W. Kress. 2020. Manx Shearwaters *Puffinus puffinus* breeding in the western Atlantic follow a different migration route from their eastern Atlantic conspecifics. *Marine Ornithology* 48:179–183.

Felis, J.J., J. Adams, P.J. Hodum, R.D. Carle, and V. Colodro. 2019. Eastern Pacific migration strategies of Pink-footed Shearwaters *Ardenna creatopus*: Implications for fisheries interactions and international conservation. *Endangered Species Research* 39:269–282.

Fernandez, P., D.J. Anderson, P.R. Sievert, and K.P. Huyvaert. 2001. Foraging destinations of three low-latitude albatross (*Phoebastria*) species. *Journal of Zoology* 254:391–404.

Ferrer-Obiol, J., H.F. James, R.T. Chesser, V. Bretagnolle, J. González-Solís, J. Rozas, A.J. Welch, and M. Riutort. 2022. Paleoceanographic changes in the late Pliocene promoted rapid diversification in pelagic seabirds. *Journal of Biogeography* 49:171–188.

Ferris, R.W. 1940. Eight years of banding of Western Gulls. *Condor* 42(4):189–197.

Fifield, D.A., W.A. Montevecchi, S. Garthe, G.J. Robertson, U. Kubetzki, and J.-F. Rail. 2014. Migratory tactics and wintering areas of Northern Gannets (*Morus bassanus*) breeding in North America. *Ornithological Monographs* 79:i–63. https://www.researchgate.net/publication/260602079_Migratory_tactics_and_wintering_areas_of_Northern_Gannets_Morus_bassanus_breeding_in_North_America.

Fijn, R.C., D. Hiemstra, R.A. Phillips, and J. van der Winden. 2013. Arctic Terns *Sterna paradisaea* from the Netherlands migrate record distances across three oceans to Wilkes Land, East Antarctica. *Ardea* 101:3–12.

Finger, T.A., A.D. Afton, M.L. Schummer, S.A. Petrie, S.S. Badzinski, M.A. Johnson, M.L. Szymanski, K.J. Jacobs, G.H. Olsen, and M.A. Mitchell. 2016. Environmental factors influence Lesser Scaup migration chronology and population monitoring. *Journal of Wildlife Management* 80(8):1437–1449.

Fink, D., T. Auer, A. Johnston, M. Strimas-Mackey, S. Ligocki, O. Robinson, W. Hochachka, L. Jaromczyk, C. Crowley, K. Dunham, A. Stillman, I. Davies, A. Rodewald, V. Ruiz-Gutierrez, and C. Wood. 2023. eBird Status and Trends, Data Version 2022. Cornell Lab of Ornithology, Ithaca, NY. https://doi.org/10.2173/ebirdst.2022.

Finkelstein, M., B.S. Keitt, D.A. Croll, B. Tershy, W.M. Jarman, S. Rodriguez-Pastor, and D.J. Anderson. 2006. Albatross species demonstrate regional differences in North Pacific marine contamination. *Ecological Applications* 16(2):678–686.

Finkelstein, M.E., S. Wolf, M. Goldman, D.F. Doak, P.R. Sievert, G. Balogh, and H. Hasegawa. 2010. The anatomy of a (potential) disaster: Volcanoes, behavior, and population viability of the Short-tailed Albatross (*Phoebastria albatrus*). *Biological Conservation* 143:321–331.

Finnegan, L., S. Castillo, J. Hughes, K.F. Abraham, R.W. Brook, and C.J. Kyle. 2013. Fine-scale analysis reveals cryptic patterns of genetic structure in Canada Geese. *Condor* 115(4):738–749.

Fischer, K.N., R.M. Suryan, D.D. Roby, and G.R. Balogh. 2009. Post-breeding season distribution of Black-footed and Laysan Albatrosses satellite-tagged in Alaska: Inter-specific differences in spatial overlap with North Pacific fisheries. *Biological Conservation* 142(4):751–760.

Fischer, S.E. 2020. Post-fledging and migration ecology of Gray Vireos (*Vireo vicinior*) and using ArtScience to explore gender and identity. Master's thesis, University of Toledo. http://rave.ohiolink.edu/etdc/view?acc_num=toledo158895774132047.

Fischer, S.E., G.R. Kramer, M.J. Baumann, E.F. Fahy, K. Granillo, and H.M. Streby. 2025. Migratory connectivity and potential nonbreeding sexual segregation in Gray Vireos (Vireo vicinior). *The Wilson Journal of Ornithology* 137(3):441–462. https://doi.org/10.1080/15594491.2025.2497654.

Fisher, J., and R.T. Peterson. 1964. *The World of Birds*. Doubleday, Garden City, NY.

FitzGerald, A.M., D.M. Whitaker, J. Ralston, J.J. Kirchman, and I.G. Warkentin. 2017. Taxonomy and distribution of the imperilled Newfoundland Gray-cheeked Thrush, *Catharus minimus minimus*. *Avian Conservation and Ecology* 12(1):10. https://doi.org/10.5751/ace-00976-120110.

Fjeldså, J., C. Carboneras, G.M. Kirwan, F. Jutglar, C.J. Sharpe, and E.F.J. Garcia. 2020. Fea's Petrel (*Pterodroma feae*). Version 1.0. In *Birds of the World*, ed. S.M. Billerman, B.K. Keeney, P.G. Rodewald, and T.S. Schulenberg. Cornell Lab of Ornithology, Ithaca, NY. https://doi.org/10.2173/bow.feapet1.01.

Flanders Marine Institute. 2009. Longhurst Provinces. https://www.marineregions.org/.

Fleishman, A.B, R.A. Orben, N. Kokubun, A. Will, R. Paredes, J.T. Ackerman, A. Takahashi, A.S. Kitaysky, and S.A. Shaffer. 2019. Wintering in the western subarctic Pacific increases mercury contamination of Red-Legged Kittiwakes. *Environmental Science & Technology* 53(22):13398–13407. https://doi.org/10.1021/acs.est.9b03421.

Fleishman, E., J. Anderson, B.G. Dickson, D. Krolick, J.A. Estep, R.L. Anderson, C.S. Elphick, D.S. Dobkin, and D.A. Bell. 2016. Space use by Swainson's Hawk (*Buteo swainsoni*) that breed in the Natomas Basin, California. *Collabra* 2(1):5. https://doi.org/10.1525/collabra.35.

Fleming, C.H., and J.M. Calabrese. 2023. ctmm: Continuous-Time Movement Modeling. R package. Version 1.2.1, commit 565d2dbec88ed3ad2452e53e3bb6d4d8fbc3c6d3. https://github.com/ctmm-initiative/ctmm.

Fleskes, J.P., A.C. Fowler, M.L. Casazza, and J.M. Eadie. 2010. Population structure and relatedness among female Northern Pintails in three California wintering regions. *Waterbirds* 33(1):1–9.

Flickinger, E.L., K.A. King, and O. Heyland. 1973. Pen-raised Fulvous Tree Ducks used in movement studies of wild populations. *Journal of Wildlife Management* 37:171–175.

Flockhart, D.T., L.I. Wassenaar, T.G. Martin, K.A. Hobson, M.B. Wunder, and D.R. Norris. 2013. Tracking multi-generational colonization of the breeding grounds by Monarch butterflies in eastern North America. *Proceedings of the Royal Society: Biological Sciences* 280(1768):20131087.

Flood, N.J. 2020. Scott's Oriole (*Icterus parisorum*). Version 1.0. In *Birds of the World*, ed. A.F. Poole and F.B. Gill. Cornell Lab of Ornithology, Ithaca, NY. https://doi.org/10.2173/bow.scoori.01.

Forstner, T.M., W.S. Boyd, D. Esler, T.D. Bowman, J.L. Schamber, M. McAdie, J.E. Thompson, and D.J. Green. 2022. Migratory connectivity and variation in migration

phenology within the Pacific population of Barrow's Goldeneye (*Bucephala islandica*). *Avian Conservation and Ecology* 17(2):7. https://doi.org/10.5751/ACE-02195-170207.
Fort, J., E. Pettex, Y. Tremblay, S.H. Lorentsen, S. Garthe, S. Votier, J.B. Pons, F. Siorat, R.W. Furness, W.J. Grecian, and S. Bearhop. 2012. Meta-population evidence of oriented chain migration in Northern Gannets (*Morus bassanus*). *Frontiers in Ecology and the Environment* 10(5):237–242.
Fournier, A.M.V. 2021. Vagrancy occurrences in Sora (*Porzana carolina*). *Waterbirds* 44(4):539–543.
Fournier, A.M.V., K.L. Drake, and D.C. Tozer. 2017a. Using citizen science monitoring data in species distribution models to inform isotopic assignment of migratory connectivity in wetland birds. *Journal of Avian Biology* 48:1556–1562.
Fournier, A.M.V., A.R. Sullivan, J.K. Bump, M. Perkins, M.C. Shieldcastle, and S.L. King. 2017b. Combining citizen science species distribution models and stable isotopes reveals migratory connectivity in the secretive Virginia Rail. *Journal of Applied Ecology* 54:618–627.
Fowler, D.N., E.B. Webb, F.B. Baldwin, M.P Vrtiska, and K.A. Hobson. 2018. A multi-isotope ($\delta^{13}C$, $\delta^{15}N$, $\delta^{34}S$, $\delta^{2}H$) approach to establishing migratory connectivity in Lesser Snow Geese: Tracking an overabundant species. *PLOS ONE* 13(8):e0203077.
Fowler, M.A., J.B. Wong, and A.L. Harrison. 2023. Oxidative physiology of two small and highly migratory Arctic seabirds: Arctic Terns (*Sterna paradisaea*) and Long-tailed Jaegers (*Stercorarius longicaudus*). *Conservation Physiology* 11(1):coad060.
Francis, C.M., and F. Cooke. 1992. Migration routes and recovery rates of Lesser Snow Geese from southwestern Hudson Bay. *Journal of Wildlife Management* 56(2):279–286.
Franks, S., D.B. Lank, and W.H. Wilson Jr. 2020. Western Sandpiper (*Calidris mauri*). Version 2.0. In *The Birds of North America*, ed. A.F. Poole. Cornell Lab of Ornithology, Ithaca, NY.
Franks, S.E., et al. 2012. Range-wide patterns of migratory connectivity in the Western Sandpiper *Calidris mauri*. *Journal of Avian Biology* 43:155–167.
Fraser, K.C., A. Roberto-Charron, B. Cousens, M. Simmons, A. Nightingale, A.C. Shave, R.L. Cormier, and D.L. Humple. 2018. Classic pattern of leapfrog migration in Sooty Fox Sparrow (*Passerella iliaca unalaschensis*) is not supported by direct migration tracking of individual birds. *Auk* 135(3):572–582.
Fraser, K.C., A. Shave, A. Savage, A. Ritchie, K. Bell, J. Siegrist, J.D. Ray, K. Applegate, and M. Pearman. 2017. Determining fine-scale migratory connectivity and habitat selection for a migratory songbird by using new GPS technology. *Journal of Avian Biology* 48(3):339–45. https://doi.org/10.1111/jav.01091.
Fraser, K.C., B.J.M. Stutchbury, P. Kramer, C. Silveiro, J. Barrow, D. Newstead, N. Mickle, T. Shaheen, P. Mammenga, K. Applegate, E. Bridge, and J. Tautuin. 2013. Consistent range-wide pattern in fall migration strategy of Purple Martin (*Progne Subis*), despite different migration routes at the Gulf of Mexico. *Auk* 130(2):291–298.
Fraser, K.C., B.J.M. Stutchbury, C. Silverio, P.M. Kramer, J. Barrow, D. Newstead, N. Mickle, B.F. Cousens, J.C. Lee, D.M. Morisson, T. Shaheen, P. Mammenga, K. Applegate, E. Bridge, and J. Tautuin. 2012. Continent-wide tracking to determine migratory connectivity and tropical habitat associations of a declining aerial insectivore. *Proceedings of the Royal Society: Biological Sciences* 279:4901–4906.
Frederick, P.C. 2020. Tricolored Heron (*Egretta tricolor*). Version 1.0. In *Birds of the World*, ed. A.F. Poole. Cornell Lab of Ornithology, Ithaca, NY. https://doi.org/10.2173/bow.triher.01.
Frederick, P.C., and D. Siegel-Causey. 2020. Anhinga (*Anhinga anhinga*). Version 1.0. In *Birds of the World*, ed. A.F. Poole and F.B. Gill. Cornell Lab of Ornithology, Ithaca, NY. https://doi.org/10.2173/bow.anhing.01.
Frederiksen, M., S. Descamps, K.E. Erikstad, A.J. Gaston, H.G. Gilchrist, D. Grémillet, K.L. Johansen, Y. Kolbeinsson, J.F. Linnebjerg, M.L. Mallory, L.A.M. Tranquilla, F.R. Merkel, W.A. Montevecchi, A. Mosbech, T.K. Reiertsen, G.J. Robertson, H. Steen, H. Strøm, and T.L. Thórarinsson. 2016. Migration and wintering of a declining seabird, the Thick-billed Murre *Uria lomvia*, on an ocean basin scale: Conservation implications. *Biological Conservation* 200:26–35. https://doi.org/10.1016/j.biocon.2016.05.011.
Frederiksen, M., O. Gilg, and G. Yannic. 2021. Cross-icecap spring migration confirmed in a high-Arctic seabird, the Ivory Gull *Pagophila eburnea*. *Ibis* 163:706–714. https://doi.org/10.1111/ibi.12903.
Frederiksen, M., B. Moe, F. Daunt, R.A. Phillips, R.T. Barrett, M.I. Bogdanova, T. Boulinier, J.W. Chardine, O. Chastel, L.S. Chivers, S. Christensen-Dalsgaard, C. Clément-Chastel, K. Colhoun, R. Freeman, A.J. Gaston, J. González-Solís, A. Goutte, D. Grémillet, T. Guilford, G.H. Jensen, Y. Krasnov, S. Lorentsen, M.L. Mallory, M. Newell, B. Olsen, D. Shaw, H. Steen, H. Strøm, G.H. Systad, T.L. Thórarinsson, and T. Anker-Nilssen. 2012. Multicolony tracking reveals the winter distribution of a pelagic seabird on an ocean basin scale. *Diversity and Distributions* 18:530–542. https://doi.org/10.1111/j.1472-4642.2011.00864.x
Frei, B., K.G. Smith, J.H. Withgott, P.G. Rodewald, P. Pyle, and M.A. Patten. 2020. Red-headed Woodpecker (*Melanerpes erythrocephalus*). Version 1.0. In *Birds of the World*, ed. P.G. Rodewald. Cornell Lab of Ornithology, Ithaca, NY. https://doi.org/10.2173/bow.rehwoo.01.
Friesen, V.L., D.J. Anderson, T.E. Steeves, H. Jones, and E.A. Schreiber. 2002. Molecular support for species status of the Nazca Booby (*Sula granti*). *Auk* 119 (3):820–826.
Friesen V.L., T.P. Birt, J.F. Piatt, et al. 2005. Population genetic structure and conservation of Marbled Murrelets (*Brachyramphus marmoratus*). *Conservation Genetics* 6:607–614.
Friesen, V.L., R. Brunt, J.A. Morris-Pocock, D. Sauve, A.J. Baker, T.P. Birt, W.S. Davidson, K.H. Elliott, and W.A. Montevecchi. 2021. A test of mechanisms of population differentiation in Gannets (*Morus* spp.) using comparative phylogeography and morphometrics. *Marine Ornithology* 49:275–291.
Friesen, V.L., T.M. Burg, and K.D. McCoy. 2007. Mechanisms of population differentiation in seabirds. *Molecular Ecology* 16(9):1765–1785.
Friesen, V.L., W.A. Montevecchi, A.J. Gaston, R.T. Barret, and W.S. Davidson. 1996. Population differentiation and evolution in the Common Guillemot *Uria aalge*. *Molecular Ecology* 5:793–805.
Fronczak, D.L., D.E. Andersen, E.E. Hanna, and T.R. Cooper. 2017. Distribution and migration of eastern population Sandhill Cranes. *Journal of Wildlife Management* 81(6):1021–1032.
Frye, G.G. 2012. Autumn migration ecology of the Northern Saw-Whet Owl (*Aegolius acadius*) in northern Montana. *Journal of Raptor Research* 46(2):177–183.
Fuller, M., D. Holt, and L. Schueck. 2003. Snowy Eagle movements: Variation on the migration theme. In *Avian Migration*, ed. P. Berthold, E. Gwinner, and E. Sonnenschein, 359–366. Springer-Verlag.
Fuller, M.R., W.S. Seegar, and L.S. Schueck. 1998. Routes and travel rates of migrating Peregrine Falcons *Falco peregrinus* and Swainson's Hawks *Buteo swainsoni* in the Western Hemisphere. *Journal of Avian Biology* 29(4):433–440.
Furness, R.W., P.F.D. Boesman, and E.F.J. Garcia. 2020a. Great Skua (*Stercorarius skua*). Version 1.0. In *Birds of the World*, ed. J. del Hoyo, A. Elliott, J. Sargatal, D.A. Christie, and E. de Juana. Cornell Lab of Ornithology, Ithaca, NY. https://doi.org/10.2173/bow.gresku1.01.
———. 2020b. South Polar Skua (*Stercorarius maccormicki*). Version 1.0. In *Birds of the World*, ed. J. del Hoyo, A. Elliott, J. Sargatal, D.A. Christie, and E. de Juana. Cornell Lab of Ornithology, Ithaca, NY. https://doi.org/10.2173/bow.sopsku1.01.
Gabrey, S.W. 1996. Migration and dispersal in Great Lakes Ring-Billed and Herring Gulls. *Journal of Field Ornithology* 67:327–339.
Gale, R.S., E.O. Garton, and I.J. Ball. 1987. The history, ecology, and management of the Rocky Mountain population of Trumpeter Swans. Montana Cooperative Wildlife Research Unit, University of Montana. Missoula, MT.
Galushin, V.M. 1974. Synchronous fluctuations in populations of some raptors and their prey. *Ibis* 116(2):127–134.
Gammonley, J.H. 2020. Cinnamon Teal (*Spatula cyanoptera*). Version 1.0. In *Birds of the World*, ed. A.F. Poole. Cornell Lab of Ornithology, Ithaca, NY. https://doi.org/10.2173/bow.cintea.01.
Ganey, J.L., and W.M. Block. 2005. Winter movements and range use of radio-marked Mexican Spotted Owls: An evaluation of current management recommendations. Gen. Tech. Rep. RMRS-GTR-148-WWW. US Department of Agriculture, Forest Service, Rocky Mountain Research Station, Fort Collins, CO.
Garber, S.D., S. Chevalier, and J.R. Cohen. 1997. Twenty-eight-year study of Upland Sandpiper breeding population in New York. *North American Bird Bander* 22:109–113.
García-Pérez, B., and K.A. Hobson. 2014. A multi-isotope ($\Delta^{2}H$, $\Delta^{13}C$, $\Delta^{15}N$) approach to establishing migratory connectivity of Barn Swallow (*Hirundo rustica*). *Ecosphere* 5(2):1–12. https://doi.org/10.3161/000164514X682896.
Garcia-Walther, J.G. 2016. Abundance, distribution, and habitat use of shorebirds in the coast of Chile. Master's thesis, Georg-August-Universität Göttingen. https://doi.org/10.13140/RG.2.1.2129.1921.
Garcia and Associates (GANDA). 2016. Hatchet Ridge Wind Project, Bald Eagle Mitigation Plan. Report to Hatchet Ridge Wind, LLC. San Anselmo, CA.
Gardali, T., and G. Ballard. 2020. Warbling Vireo (*Vireo gilvus*). Version 1.0. In *Birds of the World*, ed. A.F. Poole and F.B. Gill. Cornell Lab of Ornithology, Ithaca, NY. https://doi.org/10.2173/bow.warvir.01.
Garrett, K.L., J.L. Dunn, and R. Righter. 2000. Call notes and winter distribution in the Fox Sparrow complex. *Birding* 32:412–417.
Garthe, S., G.T. Hallgrimsson, W.A. Montevecchi, D. Fifield, and R.W. Furness. 2016. East or west? Migration routes and wintering sites of Northern Gannets *Morus bassanus* from south-eastern Iceland. *Marine Biology* 163:1–9.
Garvon, J.M., A.M. Fedynich, M.J. Peterson, and D.B. Pence. 2011. Helminth community dynamics in populations of Blue-winged Teal (*Anas discors*) using two distinct migratory corridors. *Journal of Parasitology Research* 2011:1–9.
Gaston, A.J., and S.B. Dechesne. 2020. Rhinoceros Auklet (*Cerorhinca monocerata*). Version 1.0. In *Birds of the World*, ed. A.F. Poole and F.B. Gill. Cornell Lab of Ornithology, Ithaca, NY. https://doi.org/10.2173/bow.rhiauk.01.
Gaston, A.J., and J.M. Hipfner. 2020. Thick-billed Murre (*Uria lomvia*). Version 1.0. In *Birds of the World*, ed. S.M. Billerman. Cornell Lab of Ornithology, Ithaca, NY. https://doi.org/10.2173/bow.thbmur.01.
Gaston, A.J., and A. Shoji. 2020. Ancient Murrelet (*Synthliboramphus antiquus*). Version 1.0. In *Birds of the World*, ed. A.F. Poole. Cornell Lab of Ornithology, Ithaca, NY. https://doi.org/10.2173/bow.ancmur.01.
Gaston, A.J., and K. Woo. 2008. Razorbills (*Alca torda*) follow subarctic prey into the Canadian Arctic: Colonization results from climate change? *Auk* 125(4):939–942. https://doi.org/10.1525/auk.2008.07195.
Gaston, A.J., D. Brewer, A.W. Diamond, E.J. Woodsworth, and B.T. Collins. 2008. *Canadian Atlas of Bird Banding*. Vol. 2. *Seabirds, 1921–1995*. Canadian Wildlife Service, Ottawa, ON.
Gaston, A.J., Y. Hashimoto, and L. Wilson. 2015. First evidence of east-west migration across the North Pacific in a marine bird. *Ibis* 157:877–882.

———. 2017. Post-breeding movements of Ancient Murrelet *Synthliboramphus antiquus* family groups, subsequent migration of adults and implications for management. *PLOS ONE* 12(2):e0171726. https://doi.org/10.1371/journal.pone.017172.

Gaston, A.J., P.A. Smith, L.M. Tranquilla, et al. 2011. *Marine Biology* 158:1929–1941. https://doi.org/10.1007/s00227-011-1704-9.

Gauger, V.H. 2020. Black Noddy (*Anous minutus*). Version 1.0. In *Birds of the World*, ed. S.M. Billerman. Cornell Lab of Ornithology, Ithaca, NY. https://doi.org/10.2173/bow.blknod.01.

Gauthreaux Jr., S.A. 1970. Weather radar quantification of bird migration. *BioScience* 20(1):17–19.

Gauthreaux Jr., S.A., and C.G. Belser. 1998. Displays of bird movements on the WSR-88D: Patterns and quantification. *Weather and Forecasting* 13(2):453–464.

Geen, G.R., R.A. Robinson, and S.R. Baillie. 2019. Effects of tracking devices on individual birds: A review of the evidence. *Journal of Avian Biology* 50(2).

Geng, X., J. Summers, and N. Chen. 2024. Ecological niche contributes to the persistence of the western × Glaucous-winged Gull hybrid zone. *Ecology and Evolution* 14:e11678. https://doi.org/10.1002/ece3.11678.

George, T.L. 2020. Varied Thrush (*Ixoreus naevius*). Version 1.0. In *Birds of the World*, ed. A.F. Poole and F.B. Gill. Cornell Lab of Ornithology, Ithaca, NY. https://doi.org/10.2173/bow.varthr.01.

Gerber, B.D., J.F. Dwyer, S.A. Nesbitt, R.C. Drewien, C.D. Littlefield, T.C. Tacha, and P.A. Vohs. 2020. Sandhill Crane (*Antigone canadensis*). Version 1.0. In *Birds of the World*, ed. A.F. Poole. Cornell Lab of Ornithology, Ithaca, NY. https://doi.org/10.2173/bow.sancra.01.

Ghalambor, C.K., and T.E. Martin. 2020. Red-breasted Nuthatch (*Sitta canadensis*). Version 1.0. In *Birds of the World*, ed. A.F. Poole and F.B. Gill. Cornell Lab of Ornithology, Ithaca, NY. https://doi.org/10.2173/bow.rebnut.01.

Gibson, D.D., L.H. DeCicco, R.E. Gill Jr., S.C. Heinl, A.J. Lang, T.G. Tobish Jr., and J.J. Withrow. 2013. *Checklist of Alaska Birds*. 19th ed. http://www.universityofalaskamuseumbirds.org/products/checklist.pdf.

Gibson, D.D., S.C. Heinl, and T.G. Tobish Jr. 2008. Report of the Alaska Checklist Committee, 2003–2007. *Western Birds* 39(4):1.

Gilbert, W.M., M.K. Sogge, and C. van Riper. 2020. Orange-crowned Warbler (*Leiothlypis celata*). Version 1.0. In *Birds of the World*, ed. P.G. Rodewald. Cornell Lab of Ornithology, Ithaca, NY. https://doi.org/10.2173/bow.orcwar.01.

Gilchrist, H.G., and M.L. Mallory. 2005. Declines in abundance and distribution of the Ivory Gull (*Pagophila eburnea*) in Arctic Canada. *Biological Conservation* 121:303–309.

Giles, M.M., P.G.R. Jodice, R.F. Baldwin, J.D. Stanton, and M. Epstein. 2013. Spring migratory pathways and migration chronology of Canada Geese (*Branta canadensis interior*) wintering at the Santee National Wildlife Refuge, South Carolina. *Canadian Field-Naturalist* 127(1):17–25.

Gilg, O., A. Andreev, A. Aebischer, et al. 2016. Satellite tracking of Ross's Gull *Rhodostethia rosea* in the Arctic Ocean. *Journal of Ornithology* 157:249–253. https://doi.org/10.1007/s10336-015-1273-7.

Gilg, O., D. Boertmann, F. Merkel, A. Aebischer, and B. Sabard. 2009. Status of the endangered Ivory Gull, *Pagophila eburnea*, in Greenland. *Polar Biology* 32:1275–1286.

Gilg, O., B. Moe, S.A. Hanssen, N.M. Schmidt, B. Sittler, et al. 2013. Trans-equatorial migration routes, staging sites and wintering areas of a high-arctic avian predator: The Long-tailed Skua (*Stercorarius longicaudus*). *PLOS ONE* 8(5):e64614. https://doi.org/10.1371/journal.pone.0064614.

Gilg, O., H. Strøm, A. Aebischer, M.V. Gavrilo, A. Volkov, A.E., C. Miljeteig, and B. Sabard. 2010. Post-breeding movements of northeast Atlantic Ivory Gull *Pagophila eburnean* populations. *Journal of Avian Biology* 41:532–542. https://doi.org/10.1111/j.1600-048X.2010.05125.x.

Gill, F., D. Donsker, and P. Rasmussen, eds. 2023. Nicators, Bearded Reedling, larks. IOC World Bird List. Version 14.1. International Ornithologists' Union.

Gill, F.B., R.A. Canterbury, and J.L. Confer. 2020. Blue-winged Warbler (*Vermivora cyanoptera*). Version 1.0. In *Birds of the World*, ed. A.F. Poole and F.B. Gill. Cornell Lab of Ornithology, Ithaca, NY. https://doi.org/10.2173/bow.buwwar.01.

Gill, R.E., C.M. Handel, and D.R. Ruthrauff. 2013. Intercontinental migratory connectivity and population structuring of Dunlins from western Alaska. *Condor* 115:525–534.

Gill, R.E., B.J. McCaffery, and P.S. Tomkovich. 2020a. Wandering Tattler (*Tringa incana*). Version 1.0. In *Birds of the World*, ed. A.F. Poole and F.B. Gill. Cornell Lab of Ornithology, Ithaca, NY.

Gill, R.E., P.S. Tomkovich, R. David, and M. Dementyev. 2010. Longevity, movements, and site fidelity in Wandering Tattlers *Heteroscelus incanus*. *Wader Study Group Bulletin* 117(3):187–189.

Gill, R.E., P.S. Tomkovich, and B.J. McCaffery. 2020b. Rock Sandpiper (*Calidris ptilocnemis*). Version 1.0. In *Birds of the World*, ed. A.F. Poole and F.B. Gill. Cornell Lab of Ornithology, Ithaca, NY.

Gill Jr., R., and L.R. Mewaldt. 1979. Dispersal and migratory patterns of San Francisco Bay produced herons, egrets, and terns. *North American Bird Bander* 4:4–13.

Gill Jr., R.E., D.C. Douglas, C.M. Handel, T.L. Tibbitts, G. Hufford, and T. Piersma. 2014. Hemispheric-scale wind selection facilitates Bar-tailed Godwit circum-migration of the Pacific. *Animal Behaviour* 90:117–130.

Gill Jr., R.E., T. Piersma, G. Hufford, R. Servranckx, and A. Riegen. 2005. Crossing the ultimate ecological barrier: Evidence for an 11,000-km-long nonstop flight from Alaska to New Zealand and Eastern Australia by Bar-tailed Godwits. *Condor* 107:1–20.

Gill Jr., R.E., T.L. Tibbitts, D.C. Douglas, C.M. Handel, D.M. Mulcahy, J.C. Gottschalck, N. Warnock, B.J. McCaffery, P.F. Battley, and T. Piersma. 2009. Extreme endurance flight by landbirds crossing the Pacific Ocean: Ecological corridor rather than barrier? *Proceedings of the Royal Society*: Biological Sciences 276:447–457.

Gilligan, J., D. Rogers, M. Smith, and A. Contreras. 1994. *Birds of Oregon: Status and Distribution*. Cinclus Publications, McMinnville, OR.

Gillihan, S.W., and B.E. Byers. 2020. Evening Grosbeak (*Coccothraustes vespertinus*). Version 1.0. In *Birds of the World*, ed. A.F. Poole and F.B. Gill. Cornell Lab of Ornithology, Ithaca, NY. https://doi.org/10.2173/bow.evegro.01.

Gilmer, D.S., D.L. Evans, P.M. Konrad, and R.E. Stewart. 1985. Recoveries of Ferruginous Hawks banded in south-central North Dakota. *Journal of Field Ornithology* 56(2):184–187.

Girard, P., J.Y. Takekawa, and S.R. Beissinger. 2010. Uncloaking a cryptic, threatened rail with molecular markers: Origins, connectivity and demography of a recently-discovered population. *Conservation Genetics* 11:2409–2418.

Girvan, M.K., J. Jones, D.R. Norris, J.J. Barg, T.K. Kyser, and R.J. Robertson. 2007. Long-distance dispersal patterns of male Cerulean Warblers (*Dendroica cerulea*) measured by stable-hydrogen isotopes. *Avian Conservation and Ecology* 2(2):3.

Gochfeld, M., and J. Burger. 2020. Roseate Tern (*Sterna dougallii*). Version 1.0. In *Birds of the World*, ed. S.M. Billerman. Cornell Lab of Ornithology, Ithaca, NY. https://doi.org/10.2173/bow.roster.01.

Goguen, B., and D.R. Curson. 2020. Plumbeous Vireo (*Vireo plumbeus*). Version 1.0. In *Birds of the World*, ed. A.F. Poole. Cornell Lab of Ornithology, Ithaca, NY. https://doi.org/10.2173/bow.plsvir.01.

Goguen, B., D.R. Curson, S.G. Mlodinow, and P. Pyle. 2024. Cassin's Vireo (*Vireo cassinii*). Version 2.0. In *Birds of the World*, ed. N.C. García. Cornell Lab of Ornithology, Ithaca, NY. https://doi.org/10.2173/bow.casvir.02.

Goldstein, M.I., D.C. Duffy, S. Oehlers, N. Catterson, J. Frederick, and S. Payne. 2019. Interseasonal movements and non-breeding locations of Aleutian Terns *Onychoprion aleuticus*. *Marine Ornithology* 47:67–76.

Gómez, C., N.J. Bayly, D.R. Norris, S.A. Mackenzie, K.V. Rosenberg, P.D. Taylor, K.A. Hobson, and C.D. Cadena. 2017. Fuel loads acquired at a stopover site influence the pace of intercontinental migration in a boreal songbird. *Scientific Reports* 7:3405.

Gómez, C., N.J. Bayly, and K.V. Rosenberg. 2013. Seasonal variation in stopover site use: *Catharus* thrushes and vireos in northern Colombia. *Journal of Ornithology* 154:107–117. https://doi.org/10.1007/s10336-012-0876-5.

Gómez, C., S. Guerrero, A. Fitzgerald, N. Bayly, K. Hobson, and C. Cadena. 2019. Range-wide populations of a long-distance migratory songbird converge during stopover in the tropics. *Ecological Monographs* 89. 10.1002/ecm.1349.

Gómez, C., K.A. Hobson, N.J. Bayly, K.V. Rosenberg, A. Morales-Rozo, P. Cardozo, and C.D. Cadena. 2021. Migratory connectivity then and now: A northward shift in breeding origins of a long-distance migratory bird wintering in the tropics. Proceedings of the Royal Society: Biological Sciences 288(1948):2021.0188 http://doi.org/10.1098/rspb.2021.0188.

Gómez-Díaz, E., J. González-Solís, and M.A. Peinado. 2009. Population structure in a highly pelagic seabird, the Cory's Shearwater *Calonectris diomedea*: An examination of genetics, morphology and ecology. *Marine Ecology Progress Series* 382:197–209. https://doi.org/10.3354/meps07974.

González-Jaramillo, M., and A. Rocha-Olivares. 2011. Barriers acting on the genetic structure of a highly aerial seabird: The Magnificent Frigatebird. *Biochemical Systematics and Ecology* 39:392–396.

Gonzalez-Prieto, A.M., N.J. Bayly, G.J. Colorado, and K.A. Hobson. 2017. Topography of the Andes Mountains shapes the wintering distribution of a migratory bird. *Diversity and Distributions* 23:118–129.

Gonzalez-Prieto, A.M., K.A. Hobson, N.J. Bayly, and C. Gomez. 2011. Geographic origins and timing of fall migration of the Veery in northern Colombia. *Condor* 113:860–868.

González-Solís, J., J.P. Croxall, D. Oro, and X. Ruiz. 2007. Trans-equatorial migration and mixing in the wintering areas of a pelagic seabird. *Frontiers in Ecology and the Environment* 5(6):297–301. https://doi.org/10.1890/1540-9295(2007)5[297:TMAMIT]2.0.CO;2.

González-Solís, J., A. Felicísimo, J.W. Fox, V. Afanasyev, Y. Kolbeinsson, and J. Muñoz. 2009. Influence of sea surface winds on shearwater migration detours. *Marine Ecology Progress Series* 391:221–230. https://doi.org/10.3354/meps08128.

González-Solís, J., M. Smyrli, T. Militão, D. Gremillet, T. Tveraa, R.A. Phillips, and T. Boulinier. 2011. Combining stable isotope analyses and geolocation to reveal Kittiwake migration. *Marine Ecology Progress Series* 435:251–261.

Good, T.P. 2020. Great Black-backed Gull (*Larus marinus*). Version 1.0. In *Birds of the World*, ed. S.M. Billerman. Cornell Lab of Ornithology, Ithaca, NY. https://doi.org/10.2173/bow.gbbgul.01.

Goodenough, K.S. 2022. Ground-nesting birds in a changing climate: Case studies of proximate behavioral mechanisms used by skimmers and terns to address seasonal variation in resources. PhD diss., University of Oklahoma.

Goodenough, K.S., and R.T. Patton. 2019. Satellite telemetry reveals strong fidelity to migration routes and wintering grounds for the Gull-billed Tern (*Gelochelidon nilotica*). *Waterbirds* 42(4):400–410. https://doi.org/10.1675/063.042.0405.

Goodrich, L.J. 2010. Stopover ecology of autumn-migrating raptors in the Central Appalachians. PhD diss., Pennsylvania State University.

Goodrich, L., and J. Smith. 2008. Raptor migration in North America. In *State of North America's Birds of Prey*, 37–150, Series in Ornithology, no. 3, ed. K.L. Bildstein, J.P. Smith, E. Ruelas Inzunza, and R.R. Veit. Nuttall Ornithological Club, Cambridge, MA, and American Ornithologists Union, Washington, DC.

Goodrich, L.J., C.J. Farmer, D.R. Barber, and K.L. Bildstein. 2012. What banding tells us about movement ecology of raptors. *Journal of Raptor Research* 46(1):27–55.
Gosler, A. Birds in the hand. 2004. In *Bird Ecology and Conservation: A Handbook of Techniques*, 85–118, ed. W.J. Sutherland, I. Newton, and R.E. Green. Oxford University Press, New York.
Goudie, R.I., G.J. Robertson, and A. Reed. 2020. Common Eider (*Somateria mollissima*). Version 1.0. In *Birds of the World*, ed. S.M. Billerman Cornell Lab of Ornithology, Ithaca, NY. https://doi.org/10.2173/bow.comeid.01.
Gould, P.J. 1974. Sooty Tern (*Sterna fuscata*). In *Pelagic Studies of Seabirds in the Central and Eastern Pacific Ocean*, 6–52, ed. W.B. King. Smithsonian Contributions to Zoology 158, Smithsonian Institution Press, Washington, DC.
Gould, P.J., D.J. Forsell, and C.H. Lensink. 1982. Pelagic distribution and abundance of seabirds in the Gulf of Alaska and eastern Bering Sea. Migratory Bird Section, National Fisheries Research Center, Anchorage, AK.
Gould, P.J., P.H. Ostrom, and W. Walker. 1997. Trophic relationships of albatrosses associated with squid and large-mesh drift-net fisheries in the North Pacific Ocean. *Canadian Journal of Zoology* 75(4):549–562.
———. 2000. Foods, trophic relationships, and migration of Sooty and Short-tailed Shearwaters associated with squid and large-mesh driftnet fisheries in the North Pacific Ocean. *Waterbirds* 23(2):165–186. http://www.jstor.org/stable/4641138.
Gousy-Leblanc, M., J.F. Therrien, T. Broquet, D. Rioux, N. Curt-Grand-Gaudin, N. Tissot, S. Tissot, I. Szabo, L. Wilson, J. T. Evans, V. Bowes, G. Gauthier, K.L. Wiebe, G. Yannic, and N. Lecomte. 2023. Long-term population decline of a genetically homogenous continental-wide top Arctic predator. *Ibis* 165:1251–1266.
Gow, E.A., and K.L. Wiebe. 2014. Males migrate farther than females in a differential migrant: An examination of the fasting endurance hypothesis. *Royal Society Open Science* 1:140346.
Gowaty, P.A., and J.H. Plissner. 2020. Eastern Bluebird (*Sialia sialis*). Version 1.0. In *Birds of the World*, ed. A.F. Poole. Cornell Lab of Ornithology, Ithaca, NY. https://doi.org/10.2173/bow.easblu.01.
Gradev, G.G., V.G. Matarranz, E.H. Dobreva, D.V. Popov, I.I. Ivanov, E.V. Kmetova, and D.P. Vasilakis. 2011. First results of the tracking of an Eastern Imperial Eagle (*Aquila heliaca*) tagged with a GPS/GSM transmitter in Bulgaria. *Acta Zoologica Bulgarica* 3:15–20.
Grace, J., D.J. Anderson, C. Carboneras, D.A. Christie, F. Jutglar, E.F.J. Garcia, and G.M. Kirwan. 2020. Masked Booby (*Sula dactylatra*). Version 1.0. In *Birds of the World*, ed. S.M. Billerman. Cornell Lab of Ornithology, Ithaca, NY. https://doi.org/10.2173/bow.masboo.01.
Graham, B.A., J.M. Hipfner, N.A. Rojek, S.W. Stephensen, T.M. Burg. 2023. Tufted Puffins exhibit low levels of genetic differentiation among breeding colonies in North America. *Ornithological Applications* 125(3):duad023. https://doi.org/10.1093/ornithapp/duad023.
Graham, C.L., T. Steeves, and S.R. McWilliams. 2022. Cross-seasonal effects in the American Woodcock: Conditions prior to fall migration relate to migration strategy and implications for conservation. *Ornithological Applications* 124(2):duac011.
Grant, T.A., and R.W. Knapton. 2020. Clay-colored Sparrow (*Spizella pallida*). Version 1.0. In *Birds of the World*, ed. A.F. Poole. Cornell Lab of Ornithology, Ithaca, NY. https://doi.org/10.2173/bow.clcspa.01.
Gratto-Trevor, C.L. 2020. Marbled Godwit (*Limosa fedoa*). Version 1.0. In *Birds of the World*, ed. A.F. Poole and F.B. Gill. Cornell Lab of Ornithology, Ithaca, NY.
Gratto-Trevor, C., D. Amirault-Langlais, D. Catlin, F. Cuthbert, J. Fraser, S. Maddock, … and F. Shaffer. 2012b. Connectivity in Piping Plovers: Do breeding populations have distinct winter distributions? *Journal of Wildlife Management* 76(2):348–355.
Gratto-Trevor, C., R.G. Morrison, D. Mizrahi, D.B. Lank, P. Hicklin, and A.L. Spaans. 2012a. Migratory connectivity of Semipalmated Sandpipers: Winter distribution and migration routes of breeding populations. *Waterbirds* 35(1):83–95.
Gray, C.E., and B.J. Olsen. 2023. Migration routes, high-use areas, and network connectivity across the annual cycle of an arctic seabird. *Journal of Ornithology* 164:73–84. https://doi.org/10.1007/s10336-022-02010-2.
Gray, C.E., J.D. Paruk, C.R. DeSorbo, L.J. Savoy, D.E. Yates, M.D. Chickering, R.B. Gray, K.M. Taylor, D. Long IV, N. Schoch, W. Hanson, J. Cooley, and D.C. Evers. 2014. Body mass in Common Loons (*Gavia immer*) strongly associated with migration distance. *Waterbirds* 37(sp1):64–75.
Gray, J. 2010. Habitat use, movements, and spring migration chronology and corridors of female gadwalls that winter along the Louisiana Gulf Coast. Master's thesis, Louisiana State University.
Green, M.T., P.E. Lowther, S.L. Jones, S.K. Davis, and B.C. Dale. 2020. Baird's Sparrow (*Centronyx bairdii*). Version 1.0. In *Birds of the World*, ed. A.F. Poole and F.B. Gill. Cornell Lab of Ornithology, Ithaca, NY. https://doi.org/10.2173/bow.baispa.01.
Greenberg, R., and P.P. Marra, eds. 2005. *Birds of Two Worlds: The Ecology and Evolution of Migration*. Johns Hopkins University Press, Baltimore, MD.
Greenberg, R., P.P. Marra, and M.J. Wooller. 2007. Stable-isotope (C, N, H) analyses help locate the winter range of the Coastal Plain Swamp Sparrow (*Melospiza georgiana nigrescens*). *Auk* 124:1137–1148.
Greenlaw, J.S. 2015. Eastern Towhee (*Pipilo erythrophthalmus*). Version 2.0. In *The Birds of North America*, ed. P.G. Rodewald. Cornell Lab of Ornithology, Ithaca, NY. https://doi.org/10.2173/bna.262.
———. 2020. Rose-throated Becard (*Pachyramphus aglaiae*). Version 1.0. In *Birds of the World*, ed. T.S. Schulenberg. Cornell Lab of Ornithology, Ithaca, NY. https://doi.org/10.2173/bow.rotbec.01.
Greenlaw, J.S., and G.L. Wolfenden. 2007. Wintering distribution and migration of Saltmarsh and Nelson's Sharp-tailed Sparrows. *Wilson Journal of Ornithology* 119:361–377.
Greenlaw, J.S., C.S. Elphick, W. Post, and J.D. Rising. 2020. Saltmarsh Sparrow (*Ammospiza caudacuta*). Version 1.0. In *Birds of the World*, ed. P.G. Rodewald. Cornell Lab of Ornithology, Ithaca, NY. https://doi.org/10.2173/bow.sstspa.01.
Greenlaw, J.S., W.G. Shriver, and W. Post. 2022. Seaside Sparrow (*Ammospiza maritima*). Version 2.0. In *Birds of the World*, ed. P.G. Rodewald and B.K. Keeney. Cornell Lab of Ornithology, Ithaca, NY. https://doi.org/10.2173/bow.seaspa.02.
Grilli, M.G., S.A. Lambertucci, J.F. Therrien, and K.L. Bildstein. 2017. Wing size but not wing shape is related to migratory behavior in a soaring bird. *Journal of Avian Biology* 48(5):669–678. https://doi.org/10.1111/jav.01220.
Groschupf, K.D., and C.W. Thompson. 2020. Varied Bunting (*Passerina versicolor*). Version 1.0. *In Birds of the World*, ed. A.F. Poole and F.B. Gill. Cornell Lab of Ornithology, Ithaca, NY. https://doi.org/10.2173/bow.varbun.01.
Gross, D.A., and P.E. Lowther. 2020. Yellow-bellied Flycatcher (*Empidonax flaviventris*). Version 1.0. In *Birds of the World*, ed. A.F. Poole. Cornell Lab of Ornithology, Ithaca, NY. https://doi.org/10.2173/bow.yebfly.01.
Grosselet, M.O., G.J. Ruiz, and F.M. Zuniga. 2019. First migratory route of an Upland Sandpiper tracked with satellite transmitter during fall migration. *Wader Study* 126:125–128.
Gudmundsson, G.A., A. Lindstrom, and T. Alerstam. 1991. Optimal fat loads and long-distance flights by migrating knots *Calidris canutus*, sanderlings *C. alba* and turnstones *Arenaria interpres*. *Ibis* 133:140–152.
Guilford, T., J. Meade, J. Willis, R. Phillips, D. Boyle, S. Roberts, M. Collett, R. Freeman, and C. Perrins. 2009. Migration and stopover in a small pelagic seabird, the Manx Shearwater *Puffinus puffinus*: Insights from machine learning. *Proceedings of the Royal Society: Biological Sciences* 276(1660):1215–1223.
Guinan, J.A., P.A. Gowaty, and E.K. Eltzroth. 2020. Western Bluebird (*Sialia mexicana*). Version 1.0. In *Birds of the World*, ed. A.F. Poole. Cornell Lab of Ornithology, Ithaca, NY. https://doi.org/10.2173/bow.wesblu.01.
Guiry, E.J., M. James, C. Cheung, and T.C.A. Royle. 2022. Four millennia of long-term individual foraging site fidelity in a highly migratory marine predator. *Communications Biology* 5(1):368. https://doi.org/10.1038/s42003-022-03310-2.
Gulka, J., P.C. Carvalho, E. Jenkins, K. Johnson, L. Maynard, and G.K. Davoren. 2017. Dietary niche shifts of multiple marine predators under varying prey availability on the northeast Newfoundland coast. *Frontiers in Marine Science* 4:324.
Guo, F., J.J. Buler, J.A. Smolinsky, and D.S. Wilcove. 2023. Autumn stopover hotspots and multiscale habitat associations of migratory landbirds in the eastern United States. *Proceedings of the National Academy of Sciences* 120(3):e2203511120.
Gurney, K.E.B., C.J. Wood, R.T. Alisauskas, M. Wayland, J.M.A. DeVink, and S.M. Slattery. 2014. Identifying carry-over effects of wintering area on reproductive parameters in White-winged Scoters: An isotopic approach. *Condor* 116:251–264.
Gutiérrez, R.J., A.B. Franklin, and W.S. Lahaye. 2020. Spotted Owl (*Strix occidentalis*). Version 1.0. In *Birds of the World*, ed. A.F. Poole and F.B. Gill. Cornell Lab of Ornithology, Ithaca, NY. https://doi.org/10.2173/bow.spoowl.01.
Gutowsky, S.E., S.E. Davis, M. Maftei, and M.L. Mallory. 2021. Flexibility in migratory strategy contrasts with reliance on restricted staging and overwintering grounds for Sabine's Gulls from the Canadian High Arctic. *Animal Migration* 8(1):84–97. https://doi.org/10.1515/ami-2020-0106.
Gutowsky, S.E., L.F.G. Gutowsky, I.D. Jonsen, M.L. Leonard, M.B. Naughton, R.D. Romano, and S.A. Shaffer. 2014a. Daily activity budgets reveal a quasi-flightless stage during non-breeding in Hawaiian albatrosses. *Movement Ecology* 2:23. doi.org/10.1186/s40462-014-0023-4.
Gutowsky, S.E., J.M. Hipfner, M. Maftei, S. Boyd, M. Auger-Méthé, D.J. Yurkowski, and M.L. Mallory. 2020. First insights into Thayer's Gull *Larus glaucoides thayeri* migratory and overwinter patterns along the northeast Pacific coast. *Marine Ornithology* 48:9–16.
Gutowsky, S.E., Y. Tremblay, M.A. Kappes, E.N. Flint, J. Klavitter, L. Laniawe, B. Christenson, D.P. Costa, and S.A. Shaffer. 2014b. Divergent post-breeding distribution and habitat associations of fledgling and adult Black-footed Albatrosses *Phoebastria nigripes* in the North Pacific. *Ibis* 156:60–72.
Guzy, M.J., and G. Ritchison. 2020. Common Yellowthroat (*Geothlypis trichas*). Version 1.0. In *Birds of the World*, ed. P.G. Rodewald. Cornell Lab of Ornithology, Ithaca, NY. https://doi.org/10.2173/bow.comyel.01.
Guzy, M.J., B.J. McCaffery, and N. Collar. 2020. Bluethroat (*Luscinia svecica*). Version 1.0. In *Birds of the World*, ed. S.M. Billerman. Cornell Lab of Ornithology, Ithaca, NY. https://doi.org/10.2173/bow.blueth.01.
Guzzetti, B.M. 2008. Structure and dynamics of a colonizing population of Black Oystercatchers on an isolated island. PhD diss., University of Alaska Fairbanks.
Gyug, L.W., R.C. Dobbs, T.E. Martin, and C.J. Conway. 2023. Williamson's Sapsucker (*Sphyrapicus thyroideus*). Version 2.0. In *Birds of the World*, ed. S.M. Billerman and B.K. Keeney. Cornell Lab of Ornithology, Ithaca, NY. https://doi.org/10.2173/bow.wilsap.02.
Hache, S., E.M. Bayne, M.A. Villard, H. Proctor, C.S. Davis, D. Stralberg, J.K. Janes, M.T. Hallworth, K.R. Foster, E. Chidambara-vasi, A.A. Grossi, J.C. Gorrell, and R. Krikun. 2017. Phylogeography of a migratory songbird across its Canadian breeding range: Implications for conservation units. *Ecology and Evolution* 7:6078–6088.
Hagelin, J., M. Hallworth, C. Barger, J. Johnson, K. DuBour, G. Pendelton, L. Decicco, L. McDuffie, S. Matsuoka, M. Snively, P. Marra, and A. Taylor. 2021. Revealing migra-

tory path, important stopovers, and non-breeding areas of a boreal songbird in steep decline. *Animal Migration* 8:168–191. https://doi.org/10.1515/ami-2020-0116.

Hagelin, J.C., I.L. Jones, and L.E.L. Rasmussen. 2003. A tangerine-scented social odour in a monogamous birds. *Proceedings of the Royal Society: Biological Sciences* 270(1522). https://doi.org/10.1098/rspb.2003.2379.

Hahn, T.P. 2020. Cassin's Finch (*Haemorhous cassinii*). Version 1.0. In *Birds of the World*, ed. A.F. Poole and F.B. Gill. Cornell Lab of Ornithology, Ithaca, NY. https://doi.org/10.2173/bow.casfin.01.

Haig, S.M., and L.W. Oring. 1988. Distribution and dispersal in the Piping Plover. *Auk* 105:630–638.

Haig, S.M., and K. Winker. 2010. Avian subspecies: Summary and prospectus. In *Avian Subspecies*, 172–175, ed. K. Winker, and S.M. Haig. Ornithological Monographs 67. American Ornithologists Union, Washington, DC.

Haig, S.M., C.L. Gratto-Trevor, T.D. Mullins, and M.A. Colwell. 1997. Population identification of Western Hemisphere shorebirds throughout the annual cycle. *Molecular Ecology* 6(5):413–427.

Haig, S.M., L.W. Oring, P.M. Sanzenbacher, and O.W. Taft. 2002. Space use, migratory connectivity, and population segregation among Willets breeding in the western Great Basin. *Condor* 104(3):620–630.

Hailer, F., E.A. Schreiber, J.M. Miller, I.I. Levin, P.G. Parker, R.T Chesser, and R.C. Fleischer. 2010. Long-term isolation of a highly mobile seabird on the Galapagos. *Proceedings of the Royal Society: Biological Sciences* 278:817–825. https://doi.org/10.1098/rspb.2010.1342.

Haines, A.M., M.J. McGrady, M.S. Martell, B.J. Dayton, M.B. Henke, and W.S. Seegar. 2003. Migration routes and wintering locations of Broad-winged Hawks tracked by satellite telemetry. *Wilson Journal of Ornithology* 115(2):166–170.

Hall, H. 2022. The movement ecology and physiological health of Glaucous-winged Gulls wintering in the Salish Sea. Master's thesis, Simon Fraser University. https://summit.sfu.ca/item/35472.

Hall, L.A., and S.R. Beissinger. 2017. Inferring the timing of long-distance dispersal between Rail metapopulations using genetic and isotopic assignments. *Ecological Applications* 27(1):208–218.

Hall, L.A., P.J. Palsbøll, S.R. Beissinger, J.T. Harvey, M. Bérubé, M.G. Raphael, S.K. Nelson, R.T. Golightly, L. McFarlane-Tranquilla, S.H. Newman, and M.Z. Peery. 2009. Characterizing dispersal patterns in a threatened seabird with limited genetic structure. *Molecular Ecology* 18(24):5074–5085.

Hall, L.A., N.D. Van Schmidt, and S.R. Beissinger. 2018. Validating dispersal distances inferred from autoregressive occupancy models with genetic parentage assignments. *Journal of Animal Ecology* 87:691–702.

Hallworth, M.T., and P.P. Marra. 2015. Miniaturized GPS tags identify non-breeding territories of a small breeding migratory songbird. *Scientific Reports* 5:6.

Hallworth, M.T., E. Bayne, E. McKinnon, O. Love, J.A. Tremblay, B. Drolet, J. Ibarzabal, S. Van Wilgenburg, and P.P. Marra. 2021. Habitat loss on the breeding grounds is a major contributor to population declines in a long-distance migratory songbird. *Proceedings of the Royal Society: Biological Sciences* 288:20203164. http://doi.org/10.1098/rspb.2020.3164.

Hallworth, M.T., P.P. Marra, K.P. McFarland, S. Zahendra, and C.E. Studds. 2018. Tracking dragons: Stable isotopes reveal the annual cycle of a long-distance migratory insect. *Biology Letters* 14(12):20180741.

Hallworth, M.T., T.S. Sillett, S.L. Van Wilgenburg, K.A. Hobson, and P.P. Marra. 2015. Migratory connectivity of a Neotropical migratory songbird revealed by archival light-level geolocators. *Ecological Applications* 25:336–347.

Halpin, L.R., I.L. Pollet, C. Lee, K.H. Morgan, and H.R. Carter. 2018. Year-round movements of sympatric Fork-tailed (*Oceanodroma furcata*) and Leach's (*O. leucorhoa*) Storm-petrels. *Journal of Field Ornithology* 89:207–220.

Hampton, S. 2016. Status and identification of Fox Sparrow subspecies in the Central Valley of California. *CVBC Bulletin* 19:28–63.

Handel, C.M., and R.E. Gill Jr. 1992. Breeding distribution of the Black Turnstone. *Wilson Bulletin* 104:122–135.

———. 2000. Mate fidelity and breeding site tenacity in a monagous sandpiper, the Black Turnstone. *Animal Behaviour* 60:471–481.

———. 2020. Black Turnstone (*Arenaria melanocephala*). Version 1.0. In *Birds of the World*, ed. A.F. Poole and F.B. Gill. Cornell Lab of Ornithology, Ithaca, NY. https://doi.org/10.2173/bow.blktur.01.

Hanlon, N.J., R.S. van Bemmelen, K.R. Snell, G.J. Conway, C.B. Thaxter, H. Aiton, D. Aiton, D.E. Balmer, S.A. Hanssen, J.R. Calladine, and S. Hammer. 2024. Atlantic populations of a declining oceanic seabird have complex migrations and weak migratory connectivity to staging areas. *Marine Ecology Progress Series* 730:113–129.

Hanneman, M.J. 2021. Migration ecology, stopover and winter habitats, and mercury concentrations of Red-shouldered Hawks (*Buteo lineatus*) breeding in central and northeastern Wisconsin. Master's thesis, University of Wisconsin–Stevens Point.

Hansen, E.S., and A. Hedd. 2021. First year-round tracking of the globally threatened Leach's Storm-Petrel in Iceland. Poster presented at *Líffræðiráðstefnan*: The Icelandic Biology Conference. Abstract: http://biologia.is/files/agrip_2021/V18.html. Poster: https://nattsud.is/wp-content/uploads/2021/12/First-year-round-tracking-of-the-globally-threatened-Leachs-Storm-Petrel-in-Iceland-V2.pdf.

Haramis, G.M., and G.D. Kearns. 2007. Soras in tidal marsh: Banding and telemetry studies on the Patuxent River, Maryland. *Waterbirds* 30(spec. pub. 1):105–121.

Harfer, W.G. 1957. "Angels" on centimetric radars caused by birds. *Nature* 180(4591):847–849.

Harkness, B.A.S. 2017. An assessment of population genetic structure in guillemots (*Cepphus*). PhD diss., Queen's University.

Harkness, B.A.S., G. Ibarguchi, V.F. Poland, and V.L. Friesen. 2024. Historical fragmentation and stepping-stone gene flow led to population genetic differentiation in a coastal seabird. *Ecology and Evolution* 14:e11204. https://doi.org/10.1002/ece3.11204.

Harmata, A.R. 1982. Recoveries of Ferruginous Hawks banded in Colorado. *North American Bander* 6(4):144–147.

———. 1984. Bald eagles of the San Luis Valley, Colorado: Their winter ecology and spring migration. PhD diss., Montana State University, Bozeman.

Harmata, A.R., M. Restani, G.J. Montopoli, J.R. Zelenak, J.T. Ensign, and P.J. Harmata. 2001. Movements and mortality of Ferruginous Hawks banded in Montana. *Journal of Field Ornithology* 72(3):389–398.

Harrington, B.A., F.J. Leeuwenberg, S. Lara Resende, R. McNeil, B.T. Thomas, J.S. Grear, and E.F. Martinez. 1991. Migration and mass change of White-rumped Sandpipers in North and South America. *Wilson Bulletin* 103:621–636.

Harrington, B.A., R.W. Schreiber, and G.E. Woolfenden. 1972. The distribution of male and female Magnificent Frigatebirds, *Fregata magnificens*, along the Gulf Coast of Florida. *American Birds* 26:927–931.

Harris, S.W. 1974. Status, chronology and ecology of nesting storm petrels in northwestern California. *Condor* 76:249–261.

Harrison, A.-L., D.P. Costa, A.J. Winship, S.R. Benson, S.J. Bograd, M. Antolos, A.B. Carlisle, H. Dewar, P.H. Dutton, S.J. Jorgensen, and S. Kohin. 2018. The political biogeography of migratory marine predators. *Nature Ecology & Evolution* 2(10):1571–1578.

Harrison, A.-L., C. Stenzel, A. Anderson, J. Howell, R.B. Lanctot, M. Aikens, J. Aldabe, L.A. Berigan, J. Bêty, E. Blomberg, and J. Bosi de Almeida. 2024. The collective application of shorebird tracking data to conservation. *bioRxiv* 2024–01.

Harrison A.-L., B.D. Uher-Koch, J.A. Schmutz, and D.C. Douglas. 2020. Tracking data for Pacific Loons (*Gavia pacifica*). Version 1.0. US Geological Survey. https://doi.org/10.5066/P9NNN2XY.

Harrison, A.-L., P.F. Woodard, M.L. Mallory, and J. Rausch. 2022. Sympatrically breeding congeneric seabirds (*Stercorarius* spp.) from Arctic Canada migrate to four oceans. *Ecology and Evolution* 12(1):e8451.

Harrison, C.S., and D.L. Stone-Burner. 1981. Radiotelemetry of the Brown Noddy in Hawaii. *Journal of Wildlife Management* 45(4):1021–1025. www.jstor.org/stable/3808121.

Harrison, X.A., J.D. Blount, R. Inger, D.R. Norris, and S. Bearhop. 2011. Carry-over effects as drivers of fitness differences in animals. *Journal of Animal Ecology* 80(1):4–18.

Harten, T. 2009. Crested Auklets. PolarTREC (website). https://www.polartrec.com/expeditions/seabird-ecology-in-the-bering-sea/journals/2009-07-28.

Hatch, J.J., K.M. Brown, G.G. Hogan, R.D. Morris, J. Orta, E.F.J. Garcia, F. Jutglar, G.M. Kirwan, and P.F.D. Boesman. 2020. Great Cormorant (*Phalacrocorax carbo*). Version 1.0. In *Birds of the World*, ed. S.M. Billerman. Cornell Lab of Ornithology, Ithaca, NY. https://doi.org/10.2173/bow.grecor.01.

Hatch, J.J., M. Gochfeld, J. Burger, and E.F.J. Garcia. 2020. Arctic Tern (*Sterna paradisaea*). Version 1.0. In *Birds of the World*, ed. S.M. Billerman. Cornell Lab of Ornithology, Ithaca, NY. https://doi.org/10.2173/bow.arcter.01.

Hatch, N.R. 2011. Foraging ecology and reproductive energetics of the Kittlitz's Murrelet (*Brachyramphus brevirostris*) in Southeast Alaska. Master's thesis, Oregon State University.

Hatch, S.A., V.A. Gill, and D.M. Mulcahy. 2010. Individual and colony-specific wintering areas of Pacific Northern Fulmars (*Fulmarus glacialis*). *Canadian Journal of Fisheries and Aquatic Sciences* 67(2):386–400.

———. 2011a. Migration and wintering areas of Glaucous-Winged Gulls from south-central Alaska. *Condor* 113(2):340–351. www.jstor.org/stable/10.1525/cond.2011.090224.

———. 2011b. Migration and wintering sites of Pelagic Cormorants determined by satellite telemetry. *Journal of Field Ornithology* 82(3):269–278. https://doi.org/10.1111/j.1557-9263.2011.00330.x.

Hatch, S.A., P.M. Meyers, D.M. Mulcahy, and D.C. Douglas. 2000. Seasonal movements and pelagic habitat use of murres and puffins determined by satellite telemetry. *Condor* 102(1):145–154. https://doi.org/10.1093/condor/102.1.145.

———. 2020. Tracking data for Tufted Puffins (*Fratercula cirrhata*). Version 1.0. US Geological Survey. https://doi.org/10.5066/P9L9EIUI.

Hatch, S.A., G.J. Robertson, and P.H. Baird. 2020. Black-legged Kittiwake (*Rissa tridactyla*). Version 1.0. In *Birds of the World*, ed. S.M. Billerman. Cornell Lab of Ornithology, Ithaca, NY. https://doi.org/10.2173/bow.bklkit.01.

Haukos, D.A., M.R. Miller, D.L. Orthmeyer, J.Y. Takekawa, J.P. Fleskes, M.L. Casazza, W.M. Perry, and J.A. Moon. 2006. Spring migration of Northern Pintails from Texas and New Mexico. *Waterbirds* 29(2):127–136.

Haverland, A.A., and M.C. Green. 2022. Longevity and site fidelity of a banded Green Heron (*Butorides virescens*) in Texas. *Journal of Heron Biology and Conservation* 7:2. www.HeronConservation.org/JHBC/vol07/art02/.

Hayward, J.L., and N.A. Verbeek. 2020. Glaucous-winged Gull (*Larus glaucescens*). Version 1.0. In *Birds of the World*, ed. S.M. Billerman. Cornell Lab of Ornithology, Ithaca, NY. https://doi.org/10.2173/bow.glwgul.01.

Healy, S., and W.A. Calder. 2020. Rufous Hummingbird (*Selasphorus rufus*). Version 1.0. In *Birds of the World*, ed. A.F. Poole. Cornell Lab of Ornithology, Ithaca, NY. https://doi.org/10.2173/bow.rufhum.01.

Heath, J.A., P.C. Frederick, J.A. Kushlan, and K.L. Bildstein. 2020. White Ibis (*Eudocimus albus*). Version 1.0. In *Birds of the World*, ed. A.F. Poole. Cornell Lab of Ornithology, Ithaca, NY. https://doi.org/10.2173/bow.whiibi.01.

Heath, J.A., K. Steenhof, and M.A. Foster. 2012. Shorter migration distances associated with higher winter temperatures suggest a mechanism for advancing nesting phenology of American Kestrels *Falco sparverius*. *Journal of Avian Biology* 43:376–384.

Heath, S.A., J.K. Wilson, J. Smith, and N.W. Cooper. 2022. Migration and wintering locations of breeding Eastern Willet (*Tringa semipalmata semipalmata*) in the western Gulf of Mexico. *Waterbirds* 44(4):483–491.

Heath, S.R., E.H. Dunn, and D.J. Agro. 2009. Black Tern (*Chlidonias niger*). Version 2.0. In *The Birds of North America*, ed. A.F. Poole. Cornell Lab of Ornithology, Ithaca, NY. https://doi.org/10.2173/bna.147.

Heath-Acre, K.M., C.W. Boal, D.P. Collins, W.C. Conway, and W.P. Johnson. 2021. Using automated telemetry to identify population connectivity and migration phenology of Snowy Plovers breeding in the southern Great Plains. *Journal of Field Ornithology* 92(4):461–474.

Hebert, C.E., and L.I. Wassenaar. 2005a. Feather stable isotopes in western North American waterfowl: Spatial patterns, underlying factors, and management applications. *Wildlife Society Bulletin* 33(1):92–102.

———. 2005b. Stable isotopes provide evidence for poor Northern Pintail production on the Canadian Prairies. *Journal of Wildlife Management* 69(1):101–109.

Heckscher, C.M., L.R. Bevier, A.F. Poole, W. Moskoff, P. Pyle, and M.A. Patten. 2020. Veery (*Catharus fuscescens*). Version 1.0. In *Birds of the World*, ed. P.G. Rodewald. Cornell Lab of Ornithology, Ithaca, NY. https://doi.org/10.2173/bow.veery.01.

Heckscher, C.M., M. Gutierrez Ramirez, and A.H. Kneidel. 2017. Reproductive outcomes determine the timing of arrival and settlement of a single-brooded Nearctic–Neotropical migrant songbird (*Catharus fuscescens*) in South America. *Auk* 134(4):842–856. https://doi.org/10.1642/AUK-17-66.1.

Heckscher, C.M., M.R. Halley, and P.M. Stampul. 2015. Intratropical migration of a Nearctic-Neotropical migratory songbird (*Catharus fuscescens*) in South America with implications for migration theory. *Journal of Tropical Ecology* 31:285–289.

Heckscher, C.M., D. Mendez, K.A. Hobson, A.A. Aispuro, N. Kleponis, A.H. Kneidel, M.L. Mancuso, and K. Kardynal. 2021. Miniature GPS data loggers reveal habitat affiliations and movement of Veeries (*Catharus fuscescens*) during their first non-transient period in South America. *Ornitología Neotropical* 32(1):1–9. https://doi.org/10.58843/ornneo.v32i1.777.

Heckscher, C.M., S.M. Taylor, J.W. Fox, and V. Afanasyev. 2011. Veery (*Catharus fuscescens*) wintering locations, migratory connectivity, and a revision of its winter range using geolocator technology. *Auk* 128:531–542.

Hedd, A., D.A. Fifield, C.M. Burke, W.A. Montevecchi, L.A. McFarlane Tranquilla, P.M. Regular, A.D. Buren, and G.J. Robertson. 2010. Seasonal shift in the foraging niche of Atlantic Puffins *Fratercula arctica* revealed by stable isotope (δ^{15}N and δ^{13}C) analyses. *Aquatic Biology* 9:13–22.

Hedd, A., W.A. Montevecchi, H. Otley, R.A. Phillips, and D.A. Fifield. 2012. Trans-equatorial migration and habitat use by Sooty Shearwaters *Puffinus griseus* from the South Atlantic during the nonbreeding season. *Marine Ecology Progress Series* 449:277–290.

Hedd, A., I.L. Pollet, R.A. Mauck, C.M. Burke, M.L. Mallory, L.A. McFarlane Tranquilla LA, et al. 2018 Foraging areas, offshore habitat use, and colony overlap by incubating Leach's Storm-petrels *Oceanodroma leucorhoa* in the northwest Atlantic. *PLOS ONE* 13(5):e0194389. https://doi.org/10.1371/journal.pone.0194389

Hedley, R.W. 2019. Long-distance movements and evidence of post-breeding elevational movements by Cassin's Vireos. *Journal of Field Ornithology* 90(4):335–347.

Hegemann, A., P.P. Marra, and B.I. Tieleman. 2015. Causes and consequences of partial migration in a passerine bird. *American Naturalist* 186(4):531–546.

Hejl, S.J., J.A. Holmes, and D.E. Kroodsma. 2020. Winter Wren (*Troglodytes hiemalis*). Version 1.0. In *Birds of the World*, ed. S.M. Billerman. Cornell Lab of Ornithology, Ithaca, NY. https://doi.org/10.2173/bow.winwre3.01.

Hénaux, V., L.A. Powell, M.P. Vrtiska, and K.A. Hobson. 2012. Establishing winter origins of migrating Lesser Snow Geese using stable isotopes. *Avian Conservation and Ecology* 7(1).

Henny, C.J., and L.J. Blus. 1986. Radiotelemetry locates wintering grounds of DDE-contaminated Black-Crowned Night-Herons. *Wildlife Society Bulletin* (1973–2006) 14(3):236–241. http://www.jstor.org/stable/3782239.

Henny, C.J., and G.L. Brady. 1994. Partial migration and wintering localities of American Kestrels nesting in the Pacific Northwest. *Northwestern Naturalist* 75(2):37–43.

Henny, C.J., and N.E. Holgerson. 1974. Range expansion and population increase of the Gadwall in eastern North America. *Wildfowl* 25:95–101.

Henny, C.J., and M.S. Martell. 2017. Satellite-tagged Osprey nearly sets longevity record and productivity response to initial captures. *Journal of Raptor Research* 51(2):180–183. https://doi.org/10.3356/JRR-16-71.1.

Henry, B., and D. Costa. Laysan Albatross from Guadalupe 2002–2005. BirdLife International Seabird Tracking Database Dataset ID 521. https://data.seabirdtracking.org/dataset/521.

———. Laysan Albatross from Guadalupe 2005–2006. BirdLife International Seabird Tracking Database Dataset ID 524. https://data.seabirdtracking.org/dataset/524.

Henry III, R.W., S.A. Shaffer, M. Antolos, M. Félix-Lizárraga, D.G. Foley, E.L. Hazen, Y. Tremblay, D.P. Costa, B.R. Tershy, and D.A. Croll. 2021. Successful long-distance breeding range expansion of a top marine predator. *Frontiers in Ecology and Evolution* 9:620103.

Hepp, G.R., and F.C. Bellrose. 2020. Wood Duck (*Aix sponsa*). Version 1.0. In *Birds of the World*, ed. A.F. Poole. Cornell Lab of Ornithology, Ithaca, NY.

Herbert, J.A., and T.B. Mowbray. 2020. Swamp Sparrow (*Melospiza georgiana*). Version 1.0. In *Birds of the World*, ed. P.G. Rodewald. Cornell Lab of Ornithology, Ithaca, NY. https://doi.org/10.2173/bow.swaspa.01.

Herbert, J.A., D. Mizrahi, and C.M. Taylor. 2022. Migration tactics and connectivity of a Nearctic–Neotropical migratory shorebird. *Journal of Animal Ecology* 91(4):819–830.

Herkert, J.R., D.E. Kroodsma, and J.P. Gibbs. 2021. Sedge Wren (*Cistothorus stellaris*). Version 1.0. In *Birds of the World*, ed. A.F. Poole and F.B. Gill. Cornell Lab of Ornithology, Ithaca, NY. https://doi.org/10.2173/bow.sedwre1.01.

Herkert, J.R., P.D. Vickery, and D.E. Kroodsma. 2020. Henslow's Sparrow (*Centronyx henslowii*). Version 1.0. In *Birds of the World*, ed. P.G. Rodewald. Cornell Lab of Ornithology, Ithaca, NY. https://doi.org/10.2173/bow.henspa.01.

Herman, R.W., B.M. Winger, D.L. Dittmann, and M.G. Harvey. 2022. Fine-scale population genetic structure and barriers to gene flow in a widespread seabird (*Ardenna pacifica*). *Biological Journal of the Linnean Society* 137(1):125–136.

Hestbeck, J.B., and M.C. Bateman. 2000. Breeding, migration, and wintering affinities of Canada Geese marked in the Atlantic Provinces. In *Towards Conservation of the Diversity of Canada Geese (*Branta canadensis*)*, 73–83, ed. K.M. Dickson. Canadian Wildlife Service Occasional Paper no. 103.

Heveran, P., L. Goodrich, and D. Barber. 2022. Age-class difference in wintering distribution of Broad-winged Hawks. *Journal of Raptor Research* 56(4):471–478.

Hicklin, P., and C.L. Gratto-Trevor. 2020. Semipalmated Sandpiper (*Calidris pusilla*). Version 1.0. In *Birds of the World*, ed. A.F. Poole. Cornell Lab of Ornithology, Ithaca, NY.

Hill, G.E. 2022. Black-headed Grosbeak (*Pheucticus melanocephalus*). Version 2.0. In *Birds of the World*, ed. S.M. Billerman and B.K. Keeney. Cornell Lab of Ornithology, Ithaca, NY. https://doi.org/10.2173/bow.bkhgro.02.

Hill, J.M., and R.B. Renfrew. 2019. Migratory patterns and connectivity of two North American grassland bird species. *Ecology and Evolution* 9:680–692. https://doi.org/10.1002/ece3.4795.

Hill, J.M., B.K. Sandercock, and R.B. Renfrew. 2019. Migration patterns of Upland Sandpipers in the Western Hemisphere. *Frontiers in Ecology and Evolution* 7:426.

Hill, R.D., and M.J. Braun. 2001. Geolocation by light level: The next step: latitude. In *Electronic Tagging and Tracking in Marine Fisheries: Proceedings of the Symposium on Tagging and Tracking Marine Fish with Electronic Devices*, February 7–11, 2000, East-West Center, University of Hawaii, 315–330. Dordrecht: Springer Netherlands.

Hines, J.E., D.L. Dickson, B.C. Turner, M.O. Wiebe, S.J. Barry, T.A. Barry, R.H. Kerbes, D.J. Nieman, M.F. Kay, M.A. Fournier, and R.C. Cotter. 2000. Population status, distribution, and survival of shortgrass prairie Canada Geese from the Inuvialuit Settlement Region, western Canadian Arctic. In *Toward Conservation of the Diversity of Canada Geese (*Branta canadensis*)*, 27–58, ed. K.M. Dickson. Canadian Wildlife Service Occasional Paper no. 103.

Hipfner, J.M., M.R. Charette, and G.S. Blackburn. 2007. Subcolony variation in breeding success in the Tufted Puffin (*Fratercula cirrhata*): Association with foraging ecology and implications. *Auk* 124(4):1149–1157.

Hipfner, J.M., M.M. Prill, K.R. Studholme, A.D. Domalik, S. Tucker, C. Jardine, et al. 2020. Geolocator tagging links distributions in the nonbreeding season to population genetic structure in a sentinel North Pacific seabird. *PLOS ONE* 15(11):e0240056. https://doi.org/10.1371/journal.pone.0240056.

Hjertaas, D.G, D.H. Ellis, B.W. Johns, and S.L. Moon. 2001. Tracking Sandhill Crane migration from Saskatchewan to the Gulf Coast. *Proceedings North American Crane Workshop* 8:57–61.

Hobson, K.A. 1990. Stable isotope analysis of Marbled Murrelets: Evidence for freshwater feeding and determination of trophic level. *Condor* 92(4):897–903.

———. 2021. Pelagic Cormorant (*Urile pelagicus*). Version 1.1. In *Birds of the World*. Cornell Lab of Ornithology, Ithaca, NY. https://doi.org/10.2173/bow.pelcor.01.1.

Hobson, K.A., and K. J. Kardynal. 2015. Western Veeries use an eastern shortest-distance pathway: New insights to migration routes and phenology using light-level geolocators. *Auk* 132(3):540–550.

———. 2016. An isotope (^{34}S) filter and geolocator results constrain a dual feather isoscape (^{2}H, ^{13}C) to identify the wintering grounds of North American Barn Swallows. *Auk* 133(1):86–98.

Hobson, K.A., and L.I. Wassenaar. 1996. Linking breeding and wintering grounds of Neotropical–Nearctic migrant songbirds using stable hydrogen isotopic analysis of feathers. *Oecologia* 109:142–148. https://doi.org/10.1007/s004420050068.

Hobson, K.A., Y. Aubry, and L.I. Wassenaar. 2004. Migratory connectivity in Bicknell's Thrush: Locating missing populations with hydrogen isotopes. *Condor* 106:905–909.

Hobson, K.A. L.K. Blight, and P. Arcese. 2015a. Human-induced long-term shifts in gull diet from marine to terrestrial sources in North America's coastal Pacific: More evidence from more isotopes (δ^{2}H, δ^{34}S). *Environmental Science & Technology* 49(18):10834–10840. https://doi.org/10.1021/acs.est.5b02053.

Hobson, K.A., S.H. DeMent, S.L. Wilgenburg, and L.I. Wassenaar. 2009. Origins of American Kestrels wintering at two Southern US sites: An investigation using stable-isotope (δD, δ18O) methods. *Journal of Raptor Research* 43(4):325–337.

Hobson, K.A., R. Greenberg, S.L. Van Wilgenburg, and C. Mettke-Hofmann. 2010. Migratory connectivity in the Rusty Blackbird: Isotopic evidence from feathers of historical and contemporary specimens. *Condor* 112(4):778–788.

Hobson, K.A., K.J. Kardynal, S.L. Van Wilgenburg, G. Albrecht, A. Salvadori, M.D. Cadman, F. Liechti, and J.W. Fox. 2015b. A continent-wide migratory divide in North

American breeding Barn Swallows (*Hirundo Rustica*). *PLOS ONE* 10(6):e0129340. https://doi.org/10.1371/journal.pone.0129340.

Hobson, K.A., K.P. McFarland, L.I. Wassenaar, C.C. Rimmer, and J.E. Goetz. 2001. Linking breeding and wintering grounds of Bicknell's Thrushes using stable isotope analyses of feathers. *Auk* 118:16–23.

Hobson, K.A., S.L. Van Wilgenburg, E.H. Dunn, D.J.T. Hussell, P.D. Taylor, and D.M. Collister. 2015c. Predicting origins of passerines migrating through Canadian migration monitoring stations using stable-hydrogen isotope analyses of feathers: A new tool for bird conservation. *Avian Conservation and Ecology* 10:3.

Hobson, K.A., S.L. Van Wilgenburg, J. Faaborg, J.D. Toms, C. Rengifo, A. Llanes Sosa, Y. Aubry, and R. Brito Aguilar. 2014. Connecting breeding and wintering grounds of Neotropical migrant songbirds using stable hydrogen isotopes: A call for an isotopic atlas of migratory connectivity. *Journal of Field Ornithology* 85(3):237–257. https://doi.org/10.1111/jofo.12065.

Hobson, K.A., S.L. Van Wilgenburg, A.M. Roth, R.E. Bennett, N.J. Bayly, L. Chavarría-Duriaux, G.J. Colorado, P. Elizondo, C.G. Rengifo, and J.D. Ritterson. 2016. Golden-winged Warbler migratory connectivity derived from stable isotopes. In *Golden-winged Warbler Ecology, Conservation, and Habitat Management*, 193–203, ed. H.M. Streby, D.E. Andersen, and D.A. Buehler. Studies in Avian Biology 49, CRC Press, Boca Raton, FL.

Hobson, K.A., S Van Wilgenburg, L.I. Wassenaar, H. Hands, W.P. Johnson, M. O'Meilia, and P. Taylor. 2006. Using stable hydrogen isotope analysis of feathers to delineate origins of harvested Sandhill Cranes in the Central Flyway of North America. *Waterbirds* 29(2):137–147.

Hobson, K.A., S.L. Van Wilgenburg, L.I. Wassenaar, and K. Larson. 2012. Linking hydrogen (δ^2H) isotopes in feathers and precipitation: Sources of variance and consequences for assignment to isoscapes. *PLOS ONE* 7(4):e35137.

Hobson, K.A., S. Van Wilgenburg, L.I. Wassenaar, F. Moore, and J. Farrington. 2007. Estimating origins of three species of neotropical migrant songbirds at a Gulf Coast stopover site: Combining stable isotope and GIS tools. *Condor* 109:256–267.

Hobson, K.A., M.B. Wunder, S.L. Van Wilgenburg, R.G. Clark, and L.I. Wassenaar. 2011. A method for investigating population declines of migratory birds using stable isotopes: Origins of harvested Lesser Scaup in North America. *PLOS ONE* 4(11):e7915.

Hoffman, S.W., and J.P. Smith. 2003. Population trends of migratory raptors in western North America, 1977–2001. *Condor* 105(3):397–419.

Hoffman, S.W., J.P, Smith, and T.D. Meehan. 2002. Breeding grounds, winter ranges, and migratory routes of raptors in the mountain west. *Journal of Raptor Research* 36(2):97–110.

Hogan, D., J.E. Thompson, D. Esler, and W.S. Boyd. 2011. Discovery of important post-breeding sites for Barrow's Goldeneye in the boreal transition zone of Alberta. *Waterbirds* 34(3):261–388.

Hohman, W.L., and S.A. Lee. 2020. Fulvous Whistling-Duck (*Dendrocygna bicolor*). Version 1.0. In *Birds of the World*, ed. S.M. Billerman. Cornell Lab of Ornithology, Ithaca, NY. https://doi.org/10.2173/bow.fuwduc.01.

Hohman, W.L., and D.M. Richard. 1994. Timing of remigial molt in Fulvous Whistling Ducks nesting in Louisiana. *Southwestern Naturalist* 39:190–192.

Holberton, R.L., S.L. Van Wilgenburg, A.J. Leppold, and K.A. Hobson. 2015. Isotopic (δ^2H_f) evidence of "loop migration" and use of the Gulf of Maine Flyway by both western and eastern breeding populations of Blackpoll Warblers. *Journal of Field Ornithology* 86:213–228.

Holland A.E., M.E. Byrne, A.L. Bryan, T.L. DeVault, O.E. Rhodes, and J.C. Beasley. 2017. Fine-scale assessment of home ranges and activity patterns for resident Black Vultures (*Coragyps atratus*) and Turkey Vultures (*Cathartes aura*). *PLOS ONE* 12(7):e0179819. https://doi.org/10.1371/journal.pone.0179819.

Holland, A.E., M.E. Byrne, J. Hepinstall-Cymerman, A.L. Bryan, T.L. DeVault, O.E. Rhodes, Jr., and J.C. Beasley. 2019. Evidence of niche differentiation for two sympatric vulture species in the southeastern United States. *Movement Ecology* 7:31.

Holmes, R.T., S.A. Kaiser, N.L. Rodenhouse, T.S. Sillett, M.S. Webster, P. Pyle, and M.A. Patten. 2020. Black-throated Blue Warbler (*Setophaga caerulescens*). Version 1.0. In *Birds of the World*, ed. P.G. Rodewald. Cornell Lab of Ornithology, Ithaca, NY. https://doi.org/10.2173/bow.btbwar.01.

Holroyd, G.L., C.J. Conway, and H.E. Trefry. 2011. Breeding dispersal of a Burrowing Owl from Arizona to Saskatchewan. *Wilson Journal of Ornithology* 123(2):378–381.

Holroyd, G.L., and H.E. Trefry. 2011. Tracking movements of *Athene* owls: The application of North American experiences to Europe. Proceedings of the Third International Little Owl Conference, Vic, Spain. *Animal Biodiversity and Conservation* 34:379–387.

Holroyd, G.L., H.E. Trefry, and J. Duxbury. 2010. Winter destinations and habitats of Canadian Burrowing Owls. *Journal of Raptor Research* 44(4):294–299.

Holt, D.W., M.D. Larson, N. Smith, D.L. Evans, and D.F. Parmelee. 2020. Snowy Owl (*Bubo scandiacus*). Version 1.0. In *Birds of the World*, ed. S.M. Billerman. Cornell Lab of Ornithology, Ithaca, NY. https://doi.org/10.2173/bow.snoowl1.01.

Hong, P., D.N. Wiley, K.D. Powers, R.H. Michener, L. Kaufman, and Kent A. Hatch. 2019. Stable isotope analyses of multiple tissues of Great Shearwaters (*Ardenna gravis*) reveals long-term dietary stability, short-term changes in diet, and can be used as a tool to monitor food webs. *Diversity* 11(9):163. https://doi.org/10.3390/d11090163.

Hopkins III, J.B., K.A. Cutting, and J.M. Warren. 2013. Use of stable isotopes to investigate keratin deposition in the claw tips of ducks. *PLOS ONE* 8(11):e81026.

Hopkins Jr., M.N. 1972. Cattle Egret recoveries from south Georgia nesting colonies. *Bird Banding* 43(3):220–221.

Hopp, S.L. 2022. White-eyed Vireo (*Vireo griseus*). Version 2.0. In *Birds of the World*, ed. P.G. Rodewald and B.K. Keeney. Cornell Lab of Ornithology, Ithaca, NY. https://doi.org/10.2173/bow.whevir.02.

Horton, K.G., F.A. La Sorte, D. Sheldon, T.Y. Lin, K. Winner, G. Bernstein, S. Maji, W.M. Hochachka, and A. Farnsworth. 2020. Phenology of nocturnal avian migration has shifted at the continental scale. *Nature Climate Change* 10(1):63–68.

Horton, T.W., R.O. Bierregaard, P. Zawar-Reza, R.N. Holdaway, and P. Sagar. 2014. Juvenile Osprey navigation during trans-oceanic migration. *PLOS ONE* 9:32.

Hostetler J.A., and M. Hallworth. 2024. MigConnectivity: Estimate Migratory Connectivity for Migratory Animals. R package version 0.4.7, commit f8eddea3a4929d400db11914d40ddad5e61c3151, https://github.com/SMBC-NZP/MigConnectivity.

Hostetler, J.A., E.B. Cohen, C.M. Bossu, A.L. Scarpignato, K. Ruegg, A. Contina, C.S. Rushing, and M.T. Hallworth. 2025. Challenges and opportunities for data integration to improve estimation of migratory connectivity. *Methods in Ecology and Evolution* 16(2):362–376.

Hostetler, J.A., M. Hallworth, and C. Rushing. 2018. SMBC-NZP/MigConnectivity: v0.3.0. Zenodo. https://doi.org/10.5281/zenodo.1306198.

Houston, C.S. 1977. Movements of Saskatchewan banded California Gulls. *Bird Banding* 48(2):158–161.

———. 2005. Long-eared Owls, *Asio otus*: A review of North American banding. *Canadian Field Naturalist* 119(3):395–402.

Houston, C.S., and P.W. Brown. 1983. Recoveries of Saskatchewan-banded White-winged Scoters, *Melanitta fusca*. *Canadian Field-Naturalist* 97:454–455.

Houston, C.S., and J.K. Schmutz. 1995. Declining reproduction among Swainson's Hawks in prairie Canada. *Journal of Raptor Research* 29(3):198–201.

Houston, C.S., and B. Terry. 2003. Wing-tagging Turkey Vultures. *Blue Jay* 61:204–213.

Houston, C.S., S.C. Drury, and G. Holroyd. 1999. Upland Sandpiper banding in North America. *North American Bird Bander* 24:1–2.

Houston, C.S., C. Jackson, and D.E. Bowen Jr. 2020. Upland Sandpiper (*Bartramia longicauda*). Version 1.0. In *Birds of the World*, ed. A.F. Poole. Cornell Lab of Ornithology, Ithaca, NY.

Houston, C.S., and M. Martell. 2002. Speedy migration: Saskatchewan's first Osprey satellite transmitter. *Blue Jay* 60(2):74–79.

Howell, S.N.G., and S. Webb. 1995. *A Guide to the Birds of Mexico and Northern Central America*. Oxford University Press, New York.

Huang, A.C., J.E. Elliot, K.M. Cheng, K. Ritland, C.E. Ritland, S.K. Thomsen, S. Hindmarch, and K. Martin. 2016. Barn owls (*Tyto alba*) in western North America: Phylogeographic structure, connectivity, and genetic diversity. *Conservation Genetics* 17:357–367.

Huang, R.M., O.L. Bass Jr., and S.L. Pimm. 2017. Sooty Tern (*Onychoprion fuscatus*) survival, oil spills, shrimp fisheries, and hurricanes. *PeerJ* 5:e3287.

Hudon, J. 2020. Western Tanager (*Piranga ludoviciana*). Version 1.0. In *Birds of the World*, ed. A.F. Poole and F.B. Gill. Cornell Lab of Ornithology, Ithaca, NY. https://doi.org/10.2173/bow.westan.01.

Hughes, J.M. 2020a. Black-billed Cuckoo (*Coccyzus erythropthalmus*). Version 1.0. In *Birds of the World*, ed. P.G. Rodewald. Cornell Lab of Ornithology, Ithaca, NY. https://doi.org/10.2173/bow.bkbcuc.01.

———. 2020b. Yellow-billed Cuckoo (*Coccyzus americanus*). Version 1.0. In *Birds of the World*, ed. P.G. Rodewald. Cornell Lab of Ornithology, Ithaca, NY. https://doi.org/10.2173/bow.yebcuc.01.

Hull, J., H. Ernest, J. Harley, A. Fish and A.C. Hull. 2009. Differential migration between discrete populations of juvenile Red-tailed Hawks (*Buteo jamaicensis*). *Auk* 126(2):389–396.

Humeau, L., M. Le Corre, S.J. Reynolds, et al. 2020. Genetic structuring among colonies of a pantropical seabird: Implication for subspecies validation and conservation. *Ecology and Evolution* 10:11886–11905. https://doi.org/10.1002/ece3.6635.

Humphries, E.M., J.L. Peters, J.E. Jónsson, R. Stone, A.D. Afton, and K.E. Omland. 2009. Genetic differentiation between sympatric and allopatric wintering populations of Snow Geese. *Wilson Journal of Ornithology* 121(4):730–738.

Humple, D. 2009. Genetic structure and demographic impacts of oil spills in Western and Clark's Grebes. Master's thesis, Sonoma State University, Rohnert Park.

Humple, D.L., and J.B. Holcomb. 2014. Winter movements of Western Grebe's and Clark's Grebes: Insights from band recoveries. *North American Bird Bander* 39(1):21–28.

Humple, D.L., R.L. Cormier, T.W. Richardson, et al. 2020. Migration tracking reveals geographic variation in the vulnerability of a Nearctic-Neotropical migrant bird. *Scientific Reports* 10:5483. https://doi.org/10.1038/s41598-020-62132-6.

Hunt, A.R., J.L. Watson, J.M. Winiarski, R.R. Porter, and J. Heath. 2023. American Kestrel migration insights and challenges from tracking individuals across the annual cycle. *Journal of Raptor Research* 57:164–175.

Hunt, P.D., and D.J. Flaspohler. 2020. Yellow-rumped Warbler (*Setophaga coronata*). Version 1.0. In *Birds of the World*, ed. P.G. Rodewald. Cornell Lab of Ornithology, Ithaca, NY. https://doi.org/10.2173/bow.yerwar.01.

Hunt, W.G., D.E. Driscoll, R.I. Mesta, J.H. Barclay, and R.E. Jackman. 2009. Migration and survival of juvenile Bald Eagles from Arizona. *Journal of Raptor Research* 43(2):121–126.

Hunt, W.G., R.E. Jackman, J.M. Jenkins, C.G. Thelander, and R.N. Lehman. 1992. Northward post-fledging migration of California Bald Eagles. *Journal of Raptor Research* 26(1):19–23.

Hupp, J.W., J.I. Hodges, B.P. Conant, B.W. Meixell, and D.J. Groves. 2010. Winter distribution, movements, and annual survival of radiomarked Vancouver Canada Geese in southeast Alaska. *Journal of Wildlife Management* 74:274–284.

Hupp, J.W., J.A. Schmutz, and C.R. Ely. 2008. The annual migration cycle of Emperor Geese in western Alaska. *Arctic* 61:23–34.

Hupp, J.W., J.A. Schmutz, C.R. Ely, E.E. Syroechkovskiy Jr., A.V. Kondratyev, W.D. Eldridge, and E. Lappo. 2007. Moult migration of Emperor Geese *Chen canagica* between Alaska and Russia. *Journal of Avian Biology* 38:462–470.

Hupp, J.W., T.L. Tibbitts, and D.C. Douglas, D.C. 2019. Tracking data for Northern Pintails (*Anas acuta*). Version 1.0, September 2019. US Geological Survey data release. https://doi.org/10.5066/P90A6HW3.

Huschle, G., J. Toepfer, and D. Douglas. 2013. Migration and wintering areas of American Bitterns (*Botaurus lentiginosus*) that summer in central North America as determined by satellite and radio telemetry, 1998–2003. *Waterbirds* 36(3):300–309.

Hussell, D.J.T., and Ruelas Inzunza, E. 2008. Long-term monitoring: The raptor population index in principle. In *State of North Americas Birds of Prey*, ed. K.L. Bildstein. Series in Ornithology 3:151–164. Nuttall Ornithological Club, Cambridge, MA, and American Ornithologists Union, Washington, DC.

Hussell, D.J.T., and R. Montgomerie. 2020. Lapland Longspur (*Calcarius lapponicus*). Version 1.0. In *Birds of the World*, ed. S.M. Billerman, B.K. Keeney, P.G. Rodewald, and T.S. Schulenberg. Cornell Lab of Ornithology, Ithaca, NY. https://doi.org/10.2173/bow.laplon.01.

Hutcheson, C., L. Hendrix, and J.A. Moran. 2010. An isotopic analysis of migratory connectivity in Ruby-throated Hummingbirds. *North American Bird Bander* 35:5–11.

Huysman, A.E., N.W. Cooper, J.A. Smith, S.M. Haig, S.A. Heath, L. Johnson, … and P.P. Marra. 2022. Strong migratory connectivity indicates Willets need subspecies-specific conservation strategies. *Ornithological Applications* 124(3):duac015.

Hyrenbach, K.D., and R.C. Dotson. 2003. Assessing the susceptibility of female Black-footed Albatross (*Phoebastria nigripes*) to longline fisheries during their post-breeding dispersal: An integrated approach. *Biological Conservation* 112(3):391–404.

Hyrenbach, K.D., M.M. Hester, J. Adams, A.J. Titmus, P. Michael, T. Wahl, C.-W. Chang, A. Marie, and C. Vanderlip. 2017. Plastic ingestion by Black-footed Albatross *Phoebastria nigripes* from Kure Atoll, Hawai'i: Linking chick diet remains and parental at-sea foraging distributions. *Marine Ornithology* 45:225–236.

Ibargüen, S.B. 2004. Population connectivity: Combining methods for estimating avian dispersal and migratory linkages. PhD diss., Ohio State University.

Imber, M.J. 2004. Kermadec Petrels (*Pterodroma neglecta*) at Ilha da Trindade, South Atlantic Ocean and in the North Atlantic. *Notornis* 51(1):33–40.

Imlay, T.L., K.A. Hobson, A. Roberto-Charron, and M.L. Leonard. 2018. Wintering areas, migratory connectivity and habitat fidelity of three declining Nearctic–Neotropical migrant swallows. *Animal Migration* 5(1):1–16. https://doi.org/10.1515/ami-2018-0001.

Imlay, T.L., S. Saldanha, and P.D. Taylor. 2020. The fall migratory movements of Bank Swallows, *Riparia Riparia*: Fly-and-forage migration? *Avian Conservation and Ecology* 15(1):2.

Ineson, K.M., N.J. Hill, D.E. Clark, K.G. MacKenzie, J.J. Whitney, Y. Laskaris, R.A. Ronconi, J.C. Ellis, J.-F. Giroux, S. Lair, S. Stevens, W.B. Puryear, and J.A. Runstadler. 2022. Age and season predict influenza A virus dynamics in urban gulls: Consequences for natural hosts in unnatural landscapes. *Ecological Applications* 32(2):e02497. https://doi.org/10.1002/eap.2497.

Inoue, Y., S. Nakatsuka, Y. Niizuma, D. Ochi, N. Katsumata, K. Okamoto, Y. Ishihi, K. Oshima, and H. Minami. 2023. Stable isotope differences among species, sexes, and breeding stages of Laysan and Black-footed Albatrosses in the western North Pacific. *Ornithological Science* 22(1):3–14.

Irving, L. 1961. The migration of Lapland Longspurs to Alaska. *Auk* 78:327–342.

Irwin, D. E., J. H. Irwin, and T. B. Smith. 2011. Genetic variation and seasonal migratory connectivity in Wilson's Warblers (*Wilsonia pusilla*): Species-level differences in nuclear DNA between western and eastern populations. *Molecular Ecology* 20:3102–3115.

Islam, K., and E. Velarde. 2020. Heermann's Gull (*Larus heermanni*). Version 2.0. In *Birds of the World*, ed. P.G. Rodewald and B.K. Keeney. Cornell Lab of Ornithology, Ithaca, NY. https://doi.org/10.2173/bow.heegul.02.

Isleib, M.E. 1979. Migratory shorebird populations on the Copper River delta and eastern Prince William Sound, Alaska. In *Shorebirds in Marine Environments*, ed. F.A. Pitelka. *Studies in Avian Biology* 2:125–129.

Iverson, A.R., D.L. Humple, R.L. Cormier, and J. Hull. 2023. Land cover and NDVI are important predictors in habitat selection along migration for the Golden-crowned Sparrow, a temperate-zone migrating songbird. *Movement Ecology* 11. https://doi.org/10.1186/s40462-022-00353-2.

Iverson, G.C., et al. 1996. Spring migration of Western Sandpipers along the Pacific coast of North America: A telemetry study. *Condor* 98(1):10–21.

Iverson, S.A., D. Esler, and D.J. Rizzolo. 2004. Winter philopatry of Harlequin Ducks in Prince William Sound, Alaska. *Condor* 106:711–715.

Ivey, G.L., C.P. Herziger, and T.J. Hoffmann. 2005. Annual movements of Pacific coast Sandhill Cranes. *Proceedings North American Crane Workshop* 9:25–35.

Jackson, B.J., and J.A. Jackson. 2020. Killdeer (*Charadrius vociferus*). Version 1.0. In *Birds of the World*, ed. A.F. Poole and F.B. Gill. Cornell Lab of Ornithology, Ithaca, NY. https://doi.org/10.2173/bow.killde.01.

Jaeger, A., C.J. Feare, R.W. Summers, C. Lebarbenchon, C.S. Larose, and M. Le Corre. 2017. Geolocation reveals year-round at-sea distribution and activity of a superabundant tropical seabird, the Sooty Tern *Onychoprion fuscatus*. *Frontiers in Marine Science* 4:394.

Jahn, A.E., V.R. Cueto, J.W. Fox, M.S. Husak, D.H. Kim, D.V. Landoll, J.P. Ledezma, H.K. Lepage, D.J. Levey, M.T. Murphy, and R.B. Renfrew. 2013. Migration timing and wintering areas of three species of flycatchers (*Tyrannus*) breeding in the Great Plains of North America. *Auk* 130:247–257. https://doi.org/10.1525/auk.2013.13010.

Jahn, A.E., S.B. Lerman, L.M. Phillips, T.B. Ryder, and E.J. Williams. 2019. First tracking of individual American Robins (*Turdus migratorius*) across seasons. *Wilson Journal of Ornithology* 131(2):356–359.

Jakubas, D., K. Wojczulanis-Jakubas, and A. Petersen. 2022. A quiet extirpation of the breeding Little Auk *Alle alle* population in Iceland in the shadow of the famous cousin extermination. *Science of the Total Environment* 808:152167.

Jakubas, D., K. Wojczulanis-Jakubas, L.M. Iliszko, et al. 2020. Flexibility of Little Auks foraging in various oceanographic features in a changing Arctic. *Scientific Reports* 10:8283. https://doi.org/10.1038/s41598-020-65210-x.

James, J.D., and J.E. Thompson. 2020. Black-bellied Whistling-Duck *(Dendrocygna autumnalis*). Version 1.0. In *Birds of the World*, ed. A.F. Poole and F.B. Gill. Cornell Lab of Ornithology, Ithaca, NY. https://doi.org/10.2173/bow.bbwduc.01.

James, N.L., and G.S. Cumming. 2024. Climate change may impact habitat complementation and cause disassociation for mobile species. *Landscape Ecology* 39(8):139.

Jaramillo, A., R. Pittaway, and P. Burke. 1991. The identification and migration of breeding plumaged dowitchers in southern Ontario. *Birders Journal* 1:8–25.

Jaster, L.A., W.E. Jensen, W.E. Lanyon, and S.G. Mlodinow. 2022. Eastern Meadowlark (*Sturnella magna*). Version 1.1. In *Birds of the World*, ed. P. Pyle and N.D. Sly. Cornell Lab of Ornithology, Ithaca, NY. https://doi.org/10.2173/bow.easmea.01.1.

Jeglinski, J.W., J.V. Lane, S.C. Votier, R.W. Furness, K.C. Hamer, D.J. McCafferty, R.G. Nager, M. Sheddan, S. Wanless, and J. Matthiopoulos. 2024. HPAIV outbreak triggers short-term colony connectivity in a seabird metapopulation. *Scientific Reports* 14(1):3126.

Jehl Jr., J.R., J. Klima, and R.E. Harris. 2020. Short-billed Dowitcher (*Limnodromus griseus*). Version 1.0. In *Birds of the World*, ed. A.F. Poole and F.B. Gill. Cornell Lab of Ornithology, Ithaca, NY.

Jehl Jr., J.R., and P.K. Yochem. 1986. Movements of Eared Grebes indicated by banding recoveries. *Journal of Field Ornithology* 57(3):208–212.

Jenkins, K.D., and D.A. Cristol. 2002. Evidence of differential migration by sex in White-throated Sparrows (*Zonotrichia albicollis*). *Auk* 119:539–543.

Jodice, P.G.R, R.A. Ronconi, E. Rupp, G.E. Wallace, and Y. Satgé. 2015. First satellite tracks of the endangered Black-capped Petrel. *Endangered Species Research* 26:23–33.

Johns, M., and P. Warzybok. 2022. Northward migration, molting locations, and winter residency of California breeding Pigeon Guillemots *Cepphus columba*. *Marine Ecology Progress Series*. 701:133–143. https://doi.org/10.3354/meps14194.

Johnson, A.S., J. Perz, E. Nol, and N.R. Senner. 2016. Dichotomous strategies? The migration of Whimbrels breeding in the eastern Canadian sub-arctic. *Journal of Field Ornithology* 87:371–383.

Johnson, J.A., T.L. Booms, L.H. DeCicco, and D.C. Douglas. 2017. Seasonal movements of the Short-eared Owl (*Asio flammeus*) in western North America as revealed by satellite telemetry. *Journal of Raptor Research* 51:115–129.

Johnson, J.A., S.M. Matsouka, D.F. Tessler, R. Greenberg, and J.W. Fox. 2012. Identifying migratory pathways used by Rusty Blackbirds breeding in southcentral Alaska. *Wilson Journal of Ornithology* 124(4):698–703.

Johnson, J.A., S.L. Talbot, G.K. Sage, K.K. Burnham, J.W. Brown, T.L. Maechtle, W.S. Seegar, M.A. Yates, B. Anderson, and D.P. Mindell. 2010. The use of genetics for the management of a recovering population: Temporal assessment of migratory Peregrine Falcons in North America. *PLOS ONE* 5(11):e1402.

Johnson, K., and B.D. Peer. 2020. Great-tailed Grackle (*Quiscalus mexicanus*). Version 1.0. In *Birds of the World*, ed. A.F. Poole and F.B. Gill. Cornell Lab of Ornithology, Ithaca, NY. https://doi.org/10.2173/bow.grtgra.01.

Johnson, K., C. Carboneras, D.A. Christie, and G.M. Kirwan. 2020. Green-winged Teal (*Anas crecca*). Version 1.0. In *Birds of the World*, ed. S.M. Billerman. Cornell Lab of Ornithology, Ithaca, NY. https://doi.org/10.2173/bow.gnwtea.01.

Johnson, L.S. 2024. Northern House Wren (*Troglodytes aedon*). Version 1.1. In *Birds of the World*, ed. B.K. Keeney, A.F. Poole, M.G. Smith, and S.M. Billerman. Cornell Lab of Ornithology, Ithaca, NY. https://doi.org/10.2173/bow.houwre.01.1.

Johnson, L.S., and R.D. Dawson. 2020. Mountain Bluebird (*Sialia currucoides*). Version 1.0. In *Birds of the World*, ed. P.G. Rodewald. Cornell Lab of Ornithology, Ithaca, NY. https://doi.org/10.2173/bow.moublu.01.

Johnson, L.S., and J. H. Schnell. 2024. Common Black Hawk (*Buteogallus anthracinus*). Version 2.0. In *Birds of the World*, ed. G.M. Kirwan and B.K. Keeney. Cornell Lab of Ornithology, Ithaca, NY. https://doi.org/10.2173/bow.comblh1.02.

Johnson, L.S., and J. Wise. 2000. Wintering grounds of North American House Wrens as revealed by band recoveries. *Journal of Field Ornithology* 71:501–505.

Johnson, M., and J.R. Walters. 2008. Effects of mate and site fidelity on nest survival of Western Sandpipers (*Calidris mauri*). *Auk* 125(1):76–86.

Johnson, M., P. Clarkson, M.I. Goldstein, S.M. Haig, R.B. Lanctot, D.F. Tessler, and D. Zwiefelhofer. 2010. Seasonal movements, winter range use, and migratory connectivity of the Black Oystercatcher. *Condor* 112:731–743.

Johnson, M.J., C. van Riper, and K.M. Pearson. 2020. Black-throated Sparrow (*Amphispiza bilineata*). Version 1.0. In *Birds of the World*, ed. A.F. Poole and F.B. Gill. Cornell Lab of Ornithology, Ithaca, NY. https://doi.org/10.2173/bow.bktspa.01.

Johnson, O.W., P.G. Connors, and P. Pyle. 2021a. American Golden-Plover (*Pluvialis dominica*). Version 1.1. In *Birds of the World*, ed. P.G. Rodewald, B.K. Keeney, and

S.M. Billerman. Cornell Lab of Ornithology, Ithaca, NY. https://doi.org/10.2173/bow.amgplo.01.1.

———. 2021b. Pacific Golden-Plover (*Pluvialis fulva*). Version 1.1. In *Birds of the World*, ed. P.G. Rodewald, B.K. Keeney, and S.M. Billerman. Cornell Lab of Ornithology, Ithaca, NY. https://doi.org/10.2173/bow.pagplo.01.1.

Johnson, O.W., L. Fielding, J.P. Fisher, R.S. Gold, R.H. Goodwill, A.E. Bruner, J.F. Furey, P.A. Brusseau, N.H. Brusseau, P.M. Johnson, J. Jukema, L.L. Prince, M.J. Tenney, and J.W. Fox. 2012. New insight concerning transoceanic migratory pathways of Pacific Golden-Plovers (*Pluvialis fulva*): The Japan stopover and other linkages as revealed by geolocators. *Wader Study Group Bulletin* 119(1):1–8.

Johnson, O.W., L. Fielding, J.W. Fox, R.S. Gold, R.H. Goodwill, and P.M. Johnson. 2011. Tracking the migrations of Pacific Golden-Plovers (*Pluvialis fulva*) between Hawaii and Alaska: New insight on flight performance, breeding ground destinations, and nesting from birds carrying light level geolocators. *Wader Study Group Bulletin* 118:26–31.

Johnson, O.W., T.L. Tibbitts, M.F. Weber, D.R. Bybee, R.H. Goodwill, A.E. Bruner, E.J. Smith, E.L. Buss, T.Q.A. Waddell, D.C. Brooks, C.D. Smith, and J.-Y.K. Meyer. 2020. Tracking the migration of Pacific Golden-plovers from nonbreeding grounds at Moorea, French Polynesia, using Pinpoint GPS-Argos tags. *Wader Study* 127:53–57.

Johnson, O.W., N. Warnock, M.A. Bishop, A.J. Bennett, P.M. Johnson, and R.J. Kienholz. 1997. Migration by radio-tagged Pacific Golden-plovers from Hawaii to Alaska, and their subsequent survival. *Auk* 114:521–524. https://doi.org/10.2307/4089259.

Johnson, R.E. 2020. Black Rosy-Finch (*Leucosticte atrata*). Version 1.0. In *Birds of the World*, ed. A.F. Poole and F.B. Gill. Cornell Lab of Ornithology, Ithaca, NY. https://doi.org/10.2173/bow.bkrfin.01.

Johnson, R.E., P. Hendricks, D.L. Pattie, and K.B. Hunter. 2020. Brown-capped Rosy-Finch (*Leucosticte australis*). Version 1.0. In *Birds of the World*, ed. A.F. Poole and F.B. Gill. Cornell Lab of Ornithology, Ithaca, NY. https://doi.org/10.2173/bow.bcrfin.01.

Johnson, R.R., R.L. Glinski, and S.W. Matteson. 2020a. Zone-tailed Hawk (*Buteo albonotatus*). Version 1.0. In *Birds of the World*, ed. A.F. Poole and F.B. Gill. Cornell Lab of Ornithology, Ithaca, NY. https://doi.org/10.2173/bow.zothaw.01.

Johnson, R.R., H.K. Yard, and B.T. Brown. 2020b. Lucy's Warbler (*Leiothlypis luciae*). Version 1.0. In *Birds of the World*, ed. A.F. Poole. Cornell Lab of Ornithology, Ithaca, NY. https://doi.org/10.2173/bow.lucwar.01.

Johnson, S.R. 1996. Staging and wintering areas of Snow Geese nesting on Howe Island, Alaska. *Arctic* 49(1):86–93.

Joint Nature Conservation Committee (JNCC). 2021. Seabird population trends and causes of change: 1986–2019 Report, Manx Shearwater. https://jncc.gov.uk/our-work/smp-report-1986-2019.

Jones, I.L. 2020. Crested Auklet (*Aethia cristatella*). Version 1.0. In *Birds of the World*, ed. A.F. Poole and F.B. Gill. Cornell Lab of Ornithology, Ithaca, NY. https://doi.org/10.2173/bow.creauk.01.

Jones, I.L., N.B. Konyukhov, J.C. Williams, and G.V. Byrd. 2020. Parakeet Auklet (*Aethia psittacula*). Version 1.0. In *Birds of the World*, ed. A.F. Poole and F.B. Gill. Cornell Lab of Ornithology, Ithaca, NY. https://doi.org/10.2173/bow.parauk.01.

Jones, J., D.R. Norris, M.K. Girvan, J.J. Barg, T.K. Kyser, and R.J. Robertson. 2008. Migratory connectivity and rate of population decline in a vulnerable songbird. *Condor* 110:538–544.

Jones, K.L., G.L. Krapu, D.A. Brandt, and M.V. Ashley. 2005. Population genetic structure in migratory Sandhill Cranes and the role of Pleistocene glaciations. *Molecular Ecology* 14:2645–2657.

Jones, R.D. 1965. Returns from Steller's Eiders banded in Izembek Bay, Alaska. *Wildfowl Trust* 16:83–85.

Jones, S.L., and J.E. Cornely. 2020. Vesper Sparrow (*Pooecetes gramineus*). Version 1.0. In *Birds of the World*, ed. A.F. Poole and F.B. Gill. Cornell Lab of Ornithology, Ithaca, NY. https://doi.org/10.2173/bow.vesspa.01.

Jonsen, I., W.J. Grecian, and T. Patterson. 2022. aniMotum: Fit latent variable movement models to animal tracking data for location quality control and behavioural inference. Zenodo. https://zenodo.org/record/2628481#.YnUcwy_b1qs.

Jonsen, I., W.J. Grecian, L. Phillips, G. Carroll, C.R. McMahon, R.G. Harcourt, M.A. Hindell, and T.A. Patterson. 2023. aniMotum, an R package for animal movement data: Rapid quality control, behavioural estimation and simulation. *Methods in Ecology and Evolution*. https://doi.org/10.1111/2041-210X.14060.

Jonsen, I.D., T.A. Patterson, D.P. Costa, et al. 2020. A continuous-time state-space model for rapid quality control of Argos locations from animal-borne tags. *Movement Ecology* 8:31.

Jordan, F.D., S.A. Shaffer, M.G. Conners, J.E. Stepanuk, M.E. Gilmour, C.A. Clatterbuck, E.L. Hazen, D.M. Palacios, Y. Tremblay, M. Antolos, and D.G. Foley. 2022. Divergent post-breeding spatial habitat use of Laysan and Black-footed Albatross. *Frontiers in Ecology and Evolution* 10:1028317.

Juhant, M., and J. Areta. 2013. Distribution and migration of Mississippi Kite in South America. *Journal of Field Ornithology* 84(3):255–261.

Jukema, J., J.G. van Rhijn, and T. Piersma. 2015. Geographic variation in morphometrics, molt, and migration suggests ongoing subspeciation in Pacific Golden-Plovers (*Pluvialis fulva*). *Auk* 132:647–656.

Jung, J.F. 2014. Breeding ecology and modelling distribution of Bristle-thighed Curlews in the Southern Nulato Hills, Alaska. PhD diss., Tennessee Technological University, Cookeville, TN.

Käkelä, A., et al. 2006. Fatty acid signatures as indicators of diet in Great Skuas *Stercorarius skua*, Shetland. *Marine Ecology Progress Series* 319:297–310.

Kane, M.E, T.M. Gehrig, B.T. Shirkey, M.A. Picciuto, and J.W. Simpson. 2019. Migration chronology and wintering locations of King Rails (*Rallus elegans*) captured in the Upper Midwest. *Waterbirds* 42(2):210–216.

Kappes, M.A., S.A. Shaffer, Y. Tremblay, D.G. Foley, D.M. Palacios, S.J. Bograd, and D.P. Costa. 2015. Reproductive constraints influence habitat accessibility, segregation, and preference of sympatric albatross species. *Movement Ecology* 3:34. doi.org/10.1186/s40462-015-0063-4.

Kappes, M.A., S.A. Shaffer, Y. Tremblay, D.G. Foley, D.M. Palacios, P.W. Robinson, S.J. Bograd, and D.P. Costa. 2010. Hawaiian Albatrosses track interannual variability of marine habitats in the North Pacific. *Progress in Oceanography* 86:246–260. doi.org/10.1016/j.pocean.2010.04.012.

Kardynal, K.J., and K.A. Hobson. 2017. The pull of the Central Flyway? Veeries breeding in western Canada migrate using an ancestral eastern route. *Journal of Field Ornithology* 88:262–273.

Karnovsky, N.J., K.A. Hobson, Z.W. Brown, and G.L. Hunt. 2009. Distribution and diet of Ivory Gulls (*Pagophila eburnea*) in the North Water Polynya. *Arctic* 62(1):65–74.

Katzner, T.E., M.N. Kochert, K. Steenhof, C.L. McIntyre, E.H. Craig, and T.A. Miller. 2020. Golden Eagle (*Aquila chrysaetos*). Version 2.0. In *Birds of the World*, ed. P.G. Rodewald and B.K. Keeney. Cornell Lab of Ornithology, Ithaca, NY. https://doi.org/10.2173/bow.goleag.02.

Kays, R. 2020. Movebank Study Life Track—Great Egrets, ID 8849813. North Carolina Museum of Natural Sciences, North Carolina State University; Lenoir-Rhyne University; North Carolina Wildlife Resource Commission; and the Max Plank Institute for Animal Behavior.

Kays, R., M.C. Crofoot, W. Jetz, and M. Wikelski. 2015. Terrestrial animal tracking as an eye on life and planet. *Science* 348(6240):aaa2478.

Kays, R., S.C. Davidson, M. Berger, G. Bohrer, W. Fiedler, A. Flack, J. Hirt, C. Hahn, D. Gauggel, B. Russell, A. Kölzsch, A. Lohr, J. Partecke, M. Quetting, K. Safi, A. Scharf, G. Schneider, I., Lang, F. Schaeuffelhut, F., … and Wikelski, M. 2022. The Movebank system for studying global animal movement and demography. *Methods in Ecology and Evolution* 13(2):419–431. https://doi.org/10.1111/2041-210X.13767.

Keddy-Hector, D.P., P. Pyle, and M.A. Patten. 2020. Aplomado Falcon (*Falco femoralis*). Version 1.0. In *Birds of the World*, ed. P.G. Rodewald. Cornell Lab of Ornithology, Ithaca, NY. https://doi.org/10.2173/bow.aplfal.01.

Keitt, B.S., B.R. Tershy, and D.A. Croll. 2020. Black-vented Shearwater (*Puffinus opisthomelas*). Version 1.0. In *Birds of the World*, ed. A.F. Poole and F.B. Gill. Cornell Lab of Ornithology, Ithaca, NY. https://doi.org/10.2173/bow.bkvshe.01.

Kelly, J.F. 2006. Stable isotope evidence links breeding geography and migration timing in Wood Warblers (*Parulidae*). *Auk* 123:431–437.

Kelly, J.F., and R.L. Hutto. 2005. An east-west comparison of migration in North American Wood Warblers. *Condor* 107(2):197–211.

Kelly, J.F., V. Atudorei, Z.D. Sharp, and D.M. Finch. 2002. Insights into Wilson's Warbler migration from analyses of hydrogen stable-isotope ratios. *Oecologia* 130:216–221.

Kelly, J.F., E.S. Bridge, and M.J. Hamas. 2020. Belted Kingfisher (*Megaceryle alcyon*). Version 1.0. In *Birds of the World*, ed. A.F. Poole. Cornell Lab of Ornithology, Ithaca, NY. https://doi.org/10.2173/bow.belkin1.01.

Kelly, J.F., J.R. Shipley, P.B. Chilson, K.W. Howard, W.F. Frick, and T.H. Kunz. 2012. Quantifying animal phenology in the aerosphere at a continental scale using NEXRAD weather radars. *Ecosphere* 3(2):1–9.

Kelly, T.R., K.A. Hobson, G.W. Casbourn, E.A. MacDougall-Shackleton, and S.A. MacDougall-Shackleton. 2019. Long-term winter-site fidelity in Song Sparrows *Melospiza melodia*. *Auk* 136(2):1–12.

Kempenaers, B., and M. Valcu. 2017. Breeding site sampling across the Arctic by individual males of a polygynous shorebird. *Nature* 541(7638):528–531. https://doi.org/10.1038/nature20813.

Kennedy, E.D., and D.W. White. 2020. Bewick's Wren (*Thryomanes bewickii*). Version 1.0. In *Birds of the World*, ed. A.F. Poole. Cornell Lab of Ornithology, Ithaca, NY. https://doi.org/10.2173/bow.bewwre.01.

Kenow, K., L. Fara, S. Houdek, B. Gray, D. Heard, M. Meyer, T. Fox, R. Kratt, S. Ford, A. Gendron, and C. Henderson. 2021. Migration patterns and wintering distribution of Common Loons breeding in the upper Midwest. *Journal of Avian Biology* 52. https://doi.org/10.1111/jav.02609.

Kenow, K.P., D. Adams, N. Schoch, D.C. Evers, W. Hanson, D. Yates, L. Savoy, T.J. Fox, A. Major, R. Kratt, and J. Ozard. 2009. Migration patterns and wintering range of Common Loons breeding in the northeastern United States. *Waterbirds* 32(2):234–247.

Kenow, K.P., S.C. Houdek, L.J. Fara, B.R. Gray, B.R. Lubinski, D.J. Heard, M.W. Meyer, T.J. Fox, and R.J. Kratt. 2018. Distribution and foraging patterns of Common Loons on Lake Michigan with implication for exposure to type E avian botulism. *Journal of Great Lakes Research* 44:497–513.

Kenow, K.P., M.W. Meyer, D.C. Evers, D.C. Douglas, and J. Hines. 2002. Use of satellite telemetry to identify Common Loon migration routes, staging areas, and wintering range. *Waterbirds* 25(4):449–458.

Kent, G., K. Meyer, R. Miller, A. Martinez, P. Moore, and P. Watler. 2016. A snapshot of the range-wide movements of White-crowned Pigeons tracked by satellite. Avian Research and Conservation Institute. https://www.birdscaribbean.org/wp-content/uploads/2016/08/Kent-White-crowned-Pigeon-NAOC-2016.pdf.

Kerlinger, P. 1989. *Migration Strategies of Migrating Hawks*. University of Chicago Press.

Kerr, K.C., and C.J. Dove. 2013. Delimiting shades of gray: Phylogeography of the Northern Fulmar, *Fulmarus glacialis*. *Ecology and Evolution* 3(7):1915–1930.

Kershner, E.L., and W.G. Ellison. 2020. Blue-gray Gnatcatcher (*Polioptila caerulea*). Version 1.0. In *Birds of the World*, ed. A.F. Poole. Cornell Lab of Ornithology, Ithaca, NY. https://doi.org/10.2173/bow.buggna.01.

Kersten, O., B. Star, D.M. Leigh, T. Anker-Nilssen, H. Strøm, J. Danielsen, S. Descamps, K.E. Erikstad, M.G. Fitzsimmons, J. Fort, and E.S. Hansen. 2021. Complex population structure of the Atlantic Puffin revealed by whole genome analyses. *Communications Biology* 4(1), p.922.

Kersten, O., et al. 2023. Hybridization of Atlantic Puffins in the Arctic coincides with 20th-century climate change. *Science Advances* 9:eadh1407. https://doi.org/10.1126/sciadv.adh1407.

Kessel, B., D.A. Rocque, and J.S. Barclay. 2020. Greater Scaup (*Aythya marila*). Version 1.0. In *Birds of the World*, ed. S.M. Billerman. Cornell Lab of Ornithology, Ithaca, NY. https://doi.org/10.2173/bow.gresca.01.

Ketterson, E.D., and V. Nolan Jr. 1976. Geographic variation and its climatic correlates in the sex ratio of eastern-wintering Dark-eyed Juncos (*Junco hyemalis hyemalis*). *Ecology* 57:679–693.

———. 1982. The role of migration and winter mortality in the life history of a temperate-zone migrant, the Dark-eyed Junco, as determined from demographic analyses of winter populations. *Auk* 99(2):243–259.

Kidd, M.G., and V.L. Friesen. 1998. Analysis of mechanisms of microevolutionary change in *Cepphus* guillemots using patterns of control region variation. *Evolution* 52(4):1158–68. https://doi.org/10.2307/2411245.

Kidd-Weaver, A., J. Hepinstall-Cymerman, C.N. Welch, M.H. Murray, H.C. Adams, T.J. Ellison, M.J. Yabsley, and S.M. Hernandez. 2020. The movements of a recently urbanized wading bird reveal changes in season timing and length related to resource use. *PLOS ONE* 15(3):e0230158.

Killick, R., and I.A. Eckley. 2014. changepoint: An R package for changepoint analysis. *Journal of Statistical Software*, 58(3):1–19. https://www.jstatsoft.org/article/view/v058i03.

Killick, R., K. Haynes, and I.A. Eckley. 2022. changepoint: An R package for changepoint analysis. R package version 2.2.4. https://CRAN.R-project.org/package=changepoint.

Kim, D.H., L.J. Redmond, J.R. Fox, and M.T. Murphy. 2021. Inter- and intracontinental migration by the Eastern Kingbird (*Tyrannus Tyrannus*). *Wilson Journal of Ornithology* 133(2):202–214. https://doi.org/10.1676/19-00113.

Kim, H., R.B. Siegel, J.L. Stephens, J.C. Hagar, B.J. Furnas, M.-S. Jeong, B.C. McComb, and M.G. Betts. 2024. Annual migratory movement, apparent molt-migration, migration schedule, and diffuse migratory connectivity of Hermit Warblers. *Avian Conservation and Ecology* 19(2):6. https://doi.org/10.5751/ACE-02622-190206.

Kimmitt, A.A., T.M. Pegan, A.W. Jones, K. Winker, and B.M. Winger. 2023. How Veeries vary: Whole genome sequencing resolves genetic structure in a long-distance migratory bird. *Ornithology* 141:ukad061.

Kimura, M., S.M. Clegg, I.J. Lovette, K.R. Holder, D.J. Girman, B. Milá, P. Wade, and T.B. Smith. 2002. Phylogeographical approaches to assessing demographic connectivity between breeding and overwintering regions in a Nearctic–Neotropical warbler (*Wilsonia pusilla*). *Molecular Ecology* 11(9):1605–1616.

King, D.T., and A.H. Grewe. 2001. Movements and mortality of American White Pelicans banded at Marsh Lake, Minnesota. *North American Bird Bander* 26(2):57–60.

King, D.T., J. Fischer, B. Strickland, W.D. Walter, F.L. Cunningham, and G. Wang. 2016. Winter and summer home ranges of American White Pelicans (*Pelecanus erythrorhynchos*) captured at loafing sites in the southeastern United States. *Waterbirds* 39(3):287–294.

King, D.T., B.L. Goatcher, J.W. Fischer, J. Stanton, J.M. Lacour, S.C. Lemmons, and G. Wang. 2013. Home ranges and habitat use of Brown Pelicans (*Pelecanus Occidentalis*) in the northern Gulf of Mexico. *Waterbirds* 36(4):494–500.

King, D.T., B. Strickland, and A.A. Radomski. 2013. Migration patterns of Double-Crested Cormorants wintering in the southeastern United States. *Waterbirds* 35:132–137. https://doi.org/10.1675/063.035.sp114.

King, D.T., G. Wang, Z. Yang, and J.W. Fischer. 2017. Advances and environmental conditions of spring migration phenology of American White Pelicans. *Scientific Reports* 7:40339.

King, S.L., A.R. Pierce, K.R. Hersey, and N. Winstead. 2010. Migration patterns and movements of Sandhill Cranes wintering in central and southwestern Louisiana. *Proceedings of the North American Crane Workshop* 11:57–61.

Kinzelbach, R.K. 2005. *Das Buch vom Pfeilstorch* (*The Book of the Arrow Stork*). Basilisken-Presse.

Kirby R.E., and H.H. Obrecht. 1982. Recent changes in the North American distribution and abundance of wintering Atlantic Brant. *Journal of Field Ornithology* 53:333–341.

Kirchman, J.J., J. Ralston, and N.A. Gifford. 2011. Stable isotope analysis of fall migration stopover by six passerine species in an inland pitch pine–scrub oak barren. *Wilson Journal of Ornithology* 123:546–558.

Kirk, D.A., and C. Hyslop. 1998. Population status and recent trends in Canadian raptors: A review. *Biological Conservation* 83(1):91–118.

Kirk, D.A., and M.J. Mossman. 2020. Turkey Vulture (*Cathartes aura*). Version 1.0. In *Birds of the World*, ed. A.F. Poole and F.B. Gill. Cornell Lab of Ornithology, Ithaca, NY. https://doi.org/10.2173/bow.turvul.01.

Kirk, H. 2017. Annual and individual patterns in the behavioural cycle of the Manx Shearwater, *Puffinus puffinus*. PhD diss., University of Oxford.

Kirwan, G.M., C. Carboneras, F. Jutglar, C.J. Sharpe, and P. Pyle. 2022. Trindade Petrel (*Pterodroma arminjoniana*). Version 1.1. In *Birds of the World*, ed. N.D. Sly. Cornell Lab of Ornithology, Ithaca, NY. https://doi.org/10.2173/bow.tripet1.01.1.

Kirwan, G.M., J. Fjeldså, C. Carboneras, F. Jutglar, C.J. Sharpe, E. Garcia, T. Militão, and J. González-Solís. 2025. Fea's Petrel (*Pterodroma feae*). Version 1.1. In *Birds of the World*, ed. B.K. Keeney and S.M. Billerman. Cornell Lab of Ornithology, Ithaca, NY. https://doi.org/10.2173/bow.feapet1.01.1.

Kirwan, G.M., F. Medrano, J. Hernández-Montoya, A. Duarte, and J. González-Solís. 2023. Townsend's Storm-Petrel (*Hydrobates socorroensis*). Version 2.0. In *Birds of the World*, ed. B.K. Keeney. Cornell Lab of Ornithology, Ithaca, NY. https://doi.org/10.2173/bow.leastp5.02.

Kissling, M.L., S.M. Gende, S.B. Lewis, and P. M. Lukac. 2015. Reproductive performance of Kittlitz's Murrelet in a glaciated landscape, Icy Bay, Alaska. *Condor* 117(2):237–248. https://doi.org/10.1650/CONDOR-14-67.1

Kleinschmidt, B., C. Burger, P. Bustamante, et al. 2022. Annual movements of a migratory seabird—the NW European Red-throated Diver (*Gavia stellata*)—reveals high individual repeatability but low migratory connectivity. *Marine Biology* 169:114. https://doi.org/10.1007/s00227-022-04096-x.

Klicka, L.B., B.E. Kus, P.O. Title, and K.J. Burns. 2016. Conservation genomics reveals multiple evolutionary units within Bell's Vireo (*Vireo bellii*). *Conservation Genetics* 17:455–471.

Klima, J. and J.R. Jehl Jr. 2020. Stilt Sandpiper (*Calidris himantopus*). Version 1.0. In *Birds of the World*, ed. A.F. Poole. Cornell Lab of Ornithology, Ithaca, NY.

Knick, S.T., M. Leu, and S.E. Hanser. 2017. Subsequent-year recaptures at winter sites in three species of shrubland sparrows (*Emberizidae*). *Southwestern Naturalist* 62(2):165–169.

Knick, S.T., M. Leu, J.T. Rotenberry, S.E. Hanser, and K.A. Fesenmeyer. 2014. Diffuse migratory connectivity in two species of shrubland birds: Evidence from stable isotopes. *Oecologia* 174:595–608. https://doi.org/10.1007/s00442-013-2791-8.

Knight, E., J. Carlisle, A. Boyce, D. Bradley, P. Cimprich, S. Coates, S. Dinsmore, C. Gregory, J. Jorgensen, J. Kelly, D. Newstead, A. Olalla, L. Powell, A. Scarpignato, T.L. Tibbitts, N. Warnock, W. Wehtje, P. Marra, and A-L. Harrison. 2025. Delineating ecologically-distinct groups for annual cycle management of a declining shorebird. *Journal of Applied Ecology* 62(5):1152–1165. https://doi.org/10.1111/1365-2664.14885.

Knight, E.C., A.-L. Harrison, A.L. Scarpignato, S.L. Van Wilgenburg, E.M. Bayne, J.W. Ng, E. Angell, R. Bowman, R.M. Brigham, B. Drolet, W.E. Easton, T.R. Forrester, J.T. Foster, S. Haché, K.C. Hannah, K.G. Hick, J. Ibarzabal, T.L. Imlay, S.A. Mackenzie, A. Marsh, L.P. McGuire, G.N. Newberry, D. Newstead, A. Sidler, P.H. Sinclair, J.L. Stephens, D.L. Swanson, J.A. Tremblay, and P.P. Marra. 2021a. Comprehensive estimation of spatial and temporal migratory connectivity across the annual cycle to direct conservation efforts. *Ecography* 44:665–679. https://doi.org/10.1111/ecog.05111.

Knight, E.C., A.C. Smith, R. M. Brigham, and E.M. Bayne. 2021b. Combination of targeted monitoring and Breeding Bird Survey data improves population trend estimation and species distribution modeling for the Common Nighthawk. *Ornithological Applications* 123:1–14 https://doi.org/10.1093/ornithapp/duab005.

Knight, S.M., D.W. Bradley, R.G. Clark, E.A. Gow, M. Bélisle, L.L. Berzins, T. Blake, et al. 2018. Constructing and evaluating a continent-wide migratory songbird network across the annual cycle. *Ecological Monographs* 88(3):445–460. https://doi.org/10.1002/ecm.1298.

Knoche, M.J., A.N. Powell, L.T. Quakenbush, M.J. Wooller, and L.M. Phillips. 2007. Further evidence for site fidelity to wing molt locations by King Eiders: Integrating stable isotope analyses and satellite telemetry. *Waterbirds* 30(1):52–57.

Knopf, F.L., and R.M. Evans. 2020. American White Pelican (*Pelecanus erythrorhynchos*). Version 1.0. In *Birds of the World*, A.F. Poole. Cornell Lab of Ornithology, Ithaca, NY. https://doi.org/10.2173/bow.amwpel.01.

Knopf, F.L., and M.B. Wunder. 2020. Mountain Plover (*Charadrius montanus*). Version 1.0. In *Birds of the World*, ed. A.F. Poole. Cornell Lab of Ornithology, Ithaca, NY.

Knox, A.G., G. Sangster, P.E. Lowther, and P. Clement. 2024. Redpoll (*Acanthis flammea*). Version 1.0. In *Birds of the World*, ed. S.M. Billerman and M.G. Smith. Cornell Lab of Ornithology, Ithaca, NY. https://doi.org/10.2173/bow.redpol1.01.

Kochert, M.N., M.R. Fuller, L.S. Schueck, L. Bond, M.J. Bechard, B. Woodbridge, G.L. Holroyd, M.S. Martell, and U. Banasch. 2011. Migration patterns, use of stopover areas, and austral summer movements of Swainson's Hawks. *Condor* 113(1):89–106.

Kochert, M.N., and K. Steenhof. 2002. Golden Eagles in the US and Canada: Status, trends, and conservation challenges. *Journal of Raptor Research* 36(1 Supplement):32–40.

Koczur, L.M., M.C. Green, B.M. Ballard, P.E. Lowther, and R.T. Paul. 2020. Reddish Egret (*Egretta rufescens*). Version 1.0. In *Birds of the World*, ed. P.G. Rodewald. Cornell Lab of Ornithology, Ithaca, NY. https://doi.org/10.2173/bow.redegr.01.

Koczur, L.M., G.M. Kent, B.M. Ballard, K.D. Meyer, and M.C. Green. 2018. Space use and movements of adult Reddish Egrets (*Egretta rufescens*) during winter. *Waterbirds* 41(1):1–15.

Kohl, J.D., M.L. Casazza, C.T. Overton, M.P. Herzog, J.T. Ackerman, C.L. Feldheim, and J.M. Eadie. 2022. Post breeding movements and molting ecology of female gadwalls and mallards. *Journal of Wildlife Management* 86:e22314.

Kohno, H., A. Mizutani, K. Yoda, and T. Yamamoto. 2019. Movements and activity characteristics of the Brown Booby *Sula leucogaster* during the non-breeding period. *Marine Ornithology* 47:169–174.

Kolts, J.R., and S.B. McRae. 2017. Seasonal home range dynamics and sex differences in habitat use in a threatened, coastal marsh bird. *Ecology and Evolution* 7:1101–1111.

Komar, O. 2003. Notes on autumn bird migration in coastal El Salvador. *Ornitología Neotropical* 14:39–46.

Kopp, M., H.-U. Peter, O. Mustafa, S. Lisovski, M.S. Ritz, R.A. Phillips, and S. Hahn. 2011. South Polar Skuas from a single breeding population overwinter in different oceans though show similar migration patterns. *Marine Ecology Progress Series* 435:263–267.

Korner-Nievergelt, F., F. Liechti, and S. Hahn. 2012. Migratory connectivity derived from sparse ring reencounter data with unknown numbers of ringed birds. *Journal of Ornithology* 153:771–782. https://doi.org/10.1007/s10336-011-0793-z.

Korner-Nievergelt, F., C. Prévot, S. Hahn, L. Jenni, and F. Liechti. 2017. The integration of mark re-encounter and tracking data to quantify migratory connectivity. *Ecological Modelling* 344:87–94.

Korner-Nievergelt, F., A. Sauter, P.W. Atkinson, J. Guélat, W. Kania, M. Kéry, U. Köppen, R.A. Robinson, M. Schaub, K. Thorup, and H. Van Der Jeugd. 2010. Improving the analysis of movement data from marked individuals through explicit estimation of observer heterogeneity. *Journal of Avian Biology* 41(1):8–17.

Korpach, A.M., C.M. Davy, A. Mills, and K.C. Fraser. 2022. Migratory connectivity and timing for an at-risk Canadian landbird, Eastern Whip-poor-will (*Antrostomus vociferus*), from two geographically distant breeding areas. *Canadian Journal of Zoology* 100(10):660–668. https://doi.org/10.1139/cjz-2021-0175.

Korpach, A.M., A. Mills, C. Heidenreich, C.M. Davy, and K.C. Fraser. 2019. Blinded by the light? Circadian partitioning of migratory flights in a nightjar species. *Journal of Ornithology* 160:835–840.

Kramer, G.R., D.E. Andersen, D.A. Buehler, P.B. Wood, S.M. Peterson, J.A. Lehman, K.R. Aldinger, L.P. Bulluck, S. Harding, J.A. Jones, and J.P. Loegering. 2018. Population trends in *Vermivora* warblers are linked to strong migratory connectivity. *Proceedings of the National Academy of Sciences* 115(14):E3192–E3200.

Kramer, G.R., D.E. Andersen, D.A. Buehler, P.B. Wood, S.M. Peterson, J.A. Lehman, K.R. Aldinger, L.P. Bulluck, S. Harding, J.A. Jones, J.P. Loegering, C. Smalling, R. Vallender, and H.M. Streby. 2023. Exposure to risk factors experienced during migration is not associated with recent *Vermivora* warbler population trends. *Landscape Ecology* 38:2357–2380. https://doi.org/10.1007/s10980-023-01701-2.

Kramer, G.R., S.E. Fischer, P.J. Ruhl, E.S. Berz, R. Huffines, D.A. Aborn, and H.M. Streby. 2025. Spatial and temporal migratory connectivity of two sympatrically breeding wood-warblers with geographically discordant population trends. *Journal of Avian Biology* 2025(3):e03358.

Kramer, G.R., H.M. Streby, S.M. Peterson, J.A. Lehman, D.A. Buehler, J.L. Larkin, D.J. McNeil, P.D. Wood, and D.E. Andersen. 2017. Nonbreeding isolation and population-specific migration routes among three populations of Golden-winged Warblers. *Condor* 119:108–121. https://doi.org/10.1650/CONDOR-16-143.1.

Krapu, G.L., D.A Brandt, P.J. Kinzel, and A.T. Pearse. 2014. Spring migration ecology of the mid-continent Sandhill Crane population with an emphasis on use of the Central Platte River Valley, Nebraska. *Wildlife Monographs* 189:1–41.

Kraus, R.H.S., A. Zeddeman, P. van Hooft, D. Sartakov, S.A. Soloviev, R.C. Ydenberg, and H.H.T. Prins. 2011. Evolution and connectivity in the world-wide migration system of the mallard: Inferences from mitochondrial DNA. *BMC Genetics* 12:99.

Krementz, D.G., K. Asante, and L.W. Naylor. 2011. Spring migration of Mallards from Arkansas as determined by satellite telemetry. *Journal of Fish and Wildlife Management* 2(2):156–168.

Krementz, D.G., K. Asante, and L.W. Naylor. 2012. Autumn migration of Mississippi Flyway Mallards as determined by satellite telemetry. *Journal of Fish and Wildlife Management* 3(2):238–251.

Kricher, J.C. 2020. Black-and-white Warbler (*Mniotilta varia*). Version 2.0. In *The Birds of North America*, ed. P.G. Rodewald. Cornell Lab of Ornithology, Ithaca, NY. https://doi.org/10.2173/bna.158.

Krietsch, J., M. Valcu, and B. Kempenaers. 2020. Wind conditions influence breeding season movements in a nomadic polygynous shorebird. *Proceedings of the Royal Society: Biological Sciences* 287:20192789.

Kristiansen J.N., A.D. Fox, and N.S. Jarrett. 1999. Resightings and recoveries of Canada Geese *Branta canadensis* ringed in West Greenland. *Wildfowl* 50:199–203.

Krmpotich, A. 2012. Phylogeography and habitat associations of Franklin's Gulls in their United States breeding range. Master's thesis. 1254.https://commons.und.edu/theses/1254.

Kroodsma, D.E., and J. Verner. 2020. Marsh Wren (*Cistothorus palustris*). Version 1.0. In *Birds of the World*, ed. A.F. Poole. Cornell Lab of Ornithology, Ithaca, NY.

Krüger, L., V.H. Paiva, F.I. Colabuono, M.V. Petry, R.C. Montone, and J.A. Ramos. 2016. Year-round spatial movements and trophic ecology of Trindade Petrels (*Pterodroma arminjoniana*). *Journal of Field Ornithology* 87(4):404–416. http://www.jstor.org/stable/44994018.

Kruse, K.L., J.R. Lovvorn, J.Y. Takekawa, and J. Mackay. 2003. Winter distribution and survival of a high-desert breeding population of Canvasbacks. *Condor* 105:791–804.

Kucia, S.R., M.L. Schummer, J.W. Kusack, K.A. Hobson, and C.A. Nicolai. 2023. Natal origins of Mallards harvested in the Atlantic Flyway of North America: Implications for conservation and management. *Avian Conservation and Ecology* 18:10.

Kuletz, K.J., and E.A. Labunski. 2017. Seabird distribution and abundance in the offshore environment. Final Report. US Department of the Interior, Bureau of Ocean Energy Management, Anchorage, AK. Report No. BOEM 2017–004.

Kunzmann, M.R., L.S. Hall, R.R. Johnson, and N.R. Williams. 2020. Elegant Trogon (*Trogon elegans*). Version 1.0. In *Birds of the World*, ed. S.M. Billerman. Cornell Lab of Ornithology, Ithaca, NY. https://doi.org/10.2173/bow.eletro.01.

Kuro-o, M., H. Yonekawa, S. Saito, M. Eda, H. Higuchi, H. Koike, and H. Hasegawa. 2010. Unexpectedly high genetic diversity of mtDNA control region through severe bottleneck in vulnerable albatross *Phoebastria albatrus*. *Conservation Genetics* 11:127–137.

Kus, B., S.L. Hopp, R.R. Johnson, B.T. Brown, and B.M. Reiley. 2022. Bell's Vireo (*Vireo bellii*). Version 2.0. In *Birds of the World*, ed. P.G. Rodewald. Cornell Lab of Ornithology, Ithaca, NY. https://doi.org/10.2173/bow.belvir.02.

Kuyt, E. 1992. Aerial radio-tracking of Whooping Cranes migrating between Wood Buffalo National Park and Aransas National Wildlife Refuge, 1981–1984. Occasional Paper No. 74. Canadian Wildlife Service.

Kuznetsov, S.B., V.V. Baranyuk, and J.Y. Takekawa. 1998. Genetic differentiation between wintering populations of Lesser Snow Geese nesting on Wrangel Island. *Auk* 115(4):1053–1057.

Kwon, E., and B. Kempenaers. 2023. Lack of breeding site fidelity and mate fidelity in an enigmatic socially monogamous shorebird. *Animal Behaviour* 204:1–12. https://doi.org/10.1016/j.anbehav.2023.07.010.

Labarrère, C.R. 2016. Sanderling (*Calidris alba*) population structure and pollutant exposure at major winter and migratory stopover sites including Chaplin Lake, Saskatchewan. Master's thesis, University of Saskatchewan, Saskatoon.

Lack, D. 1968. Bird migration and natural selection. *Oikos* 19:1–9.

Ladd, C., and L. Gass. 2020. Golden-cheeked Warbler (*Setophaga chrysoparia*). Version 1.0. In *Birds of the World*, ed. P.G. Rodewald. Cornell Lab of Ornithology, Ithaca, NY. https://doi.org/10.2173/bow.gchwar.01.

Lagassé, B.J., R.B. Lanctot, M. Barter, S. Brown, C-Y. Chiang, C-Y. Choi, Y.N. Gerasimov, S. Kendall, J.R. Liebezeit, K.S. Maslovsky, A.I. Matsyna, E.L. Matsyna, D.C. Payer, S.T. Saalfeld, Y. Shigeta, I.M. Tiunov, P.S. Tomokovich, O.P. Valchuk, and M. Wunder. 2020. Dunlin subspecies exhibit regional segregation and high site fidelity along the East Asian–Australasian Flyway. *Condor* 122:1–15.

Lagassé, B.J., R.B. Lanctot, S. Brown, A.G. Dondua, S. Kendall, C.J. Latty, J.R. Liebezeit, E.Y. Loktionov, K.S. Maslovsky, A.I. Matsyna, E.L. Matsyna, R.L. McGuire, D.C. Payer, S.T. Saalfeld, J.C. Slaght, D.V. Solovyeva, P.S. Tomokovich, O.P. Valchuk, and M.B. Wunder. 2022. Migratory network reveals unique spatial-temporal migration dynamics of Dunlin subspecies along the East Asian–Australasian Flyway. *PLOS ONE* 17(8):e0270957. https://doi.org/10.1371/journal.pone.0270957.

Lamarre, J-F., G. Gauthier, R.B. Lanctot, S.T. Saalfeld, O.P. Love, E. Reed, O.W. Johnson, J. Liebezeit, R. McGuire, M. Russell, E. Nol, L. Kolosky, F. Sanders, L. McKinnon, P.A. Smith, S.A. Flemming, N. Lecomte, M-A. Giroux, S. Bauer, T. Emmenegger, and J. Bêty. 2021. Timing of breeding site availability across the North American Arctic partly determines spring migration schedule in a long-distance Neotropical migrant. *Frontiers in Ecology and Evolution* 9:710007.

Lamb, J.S. 2016. Ecological drivers of Brown Pelican movement patterns and reproductive success in the Gulf of Mexico. PhD diss., Clemson University.

Lamb, J.S., C.V. Fiorello, Y.G. Satgé, K. Mills, M. Ziccardi, and P.G.R. Jodice. 2018a. Movement patterns of California Brown Pelicans (*Pelecanus Occidentalis Californicus*) following oiling and rehabilitation. *Marine Pollution Bulletin* 131(Pt A):22–31.

Lamb, J.S., S.G. Gilliland, J.L. Savard, P.H. Loring, S.R. McWilliams, G.H. Olsen, J.E. Osenkowski, P.C. Paton, M.C. Perry, and T.D. Bowman. 2021. Annual-cycle movements and phenology of Black Scoters in eastern North America. *Journal of Wildlife Management* 85(8):1628–1645.

Lamb, J.S., D.J. Newstead, L.M. Koczur, B.M., Ballard, M.C. Green, and P.G. Jodice. 2018b. A bridge between oceans: Overland migration of marine birds in a wind energy corridor. *Journal of Avian Biology* 49(2):jav-01474.

Lamb, J.S., P.W.C. Paton, J.E. Osenkowski, S.S. Badzinski, A.M. Berlin, T. Bowman, C. Dwyer, L.J. Fara, S.G. Gilliland, K. Kenow, C. Lepage, M.L. Mallory, G.H. Olsen, M.C. Perry, S.A. Petrie, J.-P.L. Savard, L. Savoy, M. Schummer, C.S. Spiegel, and S.R. McWilliams. 2019. Spatially explicit network analysis reveals multi-species annual cycle movement patterns of sea ducks. *Ecological Applications* 29(5):e01919.

Lamb, J.S., Y.G. Satgé, and P.G.R. Jodice. 2017. Influence of density-dependent competition on foraging and migratory behavior of a subtropical colonial seabird. *Ecology and Evolution* 7(16):6469–6481.

Lance, E.W., and E.J. Mallek. 2004. One year of satellite telemetry data for four Alaskan Trumpeter Swans. In *Proceedings and Papers of the Nineteenth Trumpeter Swan Society Conference*, ed. D.K. Weaver, R.E. Shea, and M.H. Linck, 29–38. *Bulletin of the Trumpeter Swan Society* 32(1). Maple Plain, MN.

Lanctot, R.B., and P.J. Weatherhead. 1997. Ephemeral lekking behavior in the Buff-breasted Sandpiper, *Tryngites subruficollis*. *Behavioral Ecology* 8:268–278.

Lanctot, R., B. Goatcher, K. Scribner, S. Talbot, B. Pierson, D. Esler, and D. Zwiefelhofer. 1999. Harlequin Duck recovery from the Exxon Valdez oil spill: A population genetics perspective. *Auk* 116:781–791.

Lanctot, R.B., S. Yezerinac, J. Aldabe, J. Bosi de Almeida, G. Castresana, S. Brown, P. Rocca, S.T. Saalfeld, and J.W. Fox. 2016. Light-level geolocation reveals migration patterns of the Buff-breasted Sandpiper. *Wader Study* 123:29–43.

Langham, G.M., J.G. Schuetz, T. Distler, C.U. Soykan, and C.B. Wilsey. 2015. Conservation status of North American birds in the face of future climate change. *PLOS ONE* 10. https://doi.org/10.1371/journal.pone.0135350.

Langin, K.M., P.P. Marra, Z. Nemeth, F.R. Moore, T.K. Kyser, and L.M. Ratcliffe. 2009. Breeding latitude and timing of spring migration in songbirds crossing the Gulf of Mexico. *Journal of Avian Biology* 40:309–316.

LaPorte, N., R.W. Storer, and G.L. Nuechterlein. 2020. Western Grebe (*Aechmophorus occidentalis*). Version 1.0. In *Birds of the World*, ed. P.G. Rodewald. Cornell Lab of Ornithology, Ithaca, NY. https://doi.org/10.2173/bow.wesgre.01.

Larison, B., A.R. Lindsay, C. Bossu, M.D. Sorenson, J.D. Kaplan, D.C. Evers, J. Paruk, J.M. DaCosta, and T.B. Smith. 2020. Leveraging genomics to understand threats in a migratory waterfowl. *Evolutionary Applications* 14:1646–1658.

Larison, B., A.R. Lindsay, C. Bossu, M.D. Sorenson, J.D. Kaplan, D.C. Evers, J. Paruk, J.M. DaCosta, T.B. Smith, and K. Ruegg. 2021. Leveraging genomics to understand threats to migratory birds. *Evolutionary Applications* 14(6):1646–1658. https://doi.org/10.1111/eva.13231.

Larkin, J.L., D. Raybuck, A. Roth, L. Chavarria-Duriaux, G. Duriaux, M. Siles, and C. Smalling. 2017. Geolocators reveal migratory connectivity between wintering and breeding areas of Golden-winged Warblers. *Journal of Field Ornithology* 88:288–298.

L'Arrivee, L.P., and H. Blokpoel. 1990. Seasonal distribution and site-tenacity of Black-crowned Night-herons *Nycticorax nycticorax*, banded in Canada. *Canadian Field-Naturalist* 104:534–539.

Lascelles, B.G., P.R. Taylor, M.G.R. Miller, M.P. Dias, S. Oppel, L. Torres, A. Hedd, M. Le Corre, R.A. Phillips, S.A. Shaffer, and H. Weimerskirch. 2016. Applying global criteria to tracking data to define important areas for marine conservation. *Diversity and Distributions* 22(4):422–431.

La Sorte, F.A., and D. Fink. 2017. Projected changes in prevailing winds for transatlantic migratory birds under global warming. *Journal of Animal Ecology* 86:273–284.

La Sorte, F.A., D. Fink, W.M. Hochachka, and S. Kelling. 2016. Convergence of broad-scale migration strategies in terrestrial birds. *Proceedings of the Royal Society: Biological Sciences* 283:20152588. http://dx.doi.org/10.1098/rspb.2015.2588.

Latta, S.C., and M.E. Baltz. 2020. Lesser Nighthawk (*Chordeiles acutipennis*). Version 1.0. In *Birds of the World*, ed. A.F. Poole. Cornell Lab of Ornithology, Ithaca, NY. https://doi.org/10.2173/bow.lesnig.01.

Laughlin, A.J., C.M. Taylor, D.W. Bradley, D. LeClair, R.G. Clark, R.D. Dawson, P.O. Dunn, et al. 2013. Integrating information from geolocators, weather radar, and citizen science to uncover a key stopover area of an aerial insectivore. *Auk* 130(2):230–239.

Laurenson, K., K. Atkins, B. Clark, G. Morgan, L. Morgan, and S. Votier. 2021. Gannet *Morus Bassanus* ring recoveries suggest shorter migrations than geolocator tracking, but indicate no long-term change in migration distances, *Bird Study* 68(1):1–8. https://doi.org/10.1080/00063657.2021.1960270.

Lavers, J., J.M. Hipfner, and G. Chapdelaine. 2020. Razorbill (*Alca torda*). Version 1.0. In *Birds of the World*, ed. S.M. Billerman. Cornell Lab of Ornithology, Ithaca, NY. https://doi.org/10.2173/bow.razorb.01.

Lavoie, R., M. Tetreault, F. Bolduc, G. Bergeron, and D. Lieske. 2022. Predicting at-sea distribution of Razorbill in the St. Lawrence Gulf and Estuary, Québec, Canada, during the breeding period using GPS telemetry. *Avian Conservation and Ecology* 17(2):27. https://doi.org/10.5751/ACE-02188-170227.

Lavretsky, P., N.R. McInerney, J.E. Mohl, J.I. Brown, H.F. James, K.G. McCracken, and R.C. Fleischer. 2020. Assessing changes in genomic divergence following a century of human-mediated secondary contact among wild and captive-bred ducks. *Molecular Ecology* 29:578–595.

Lavretsky, P., J.H. Miller, V. Bahn, and J. L. Peters. 2014. Exploring fall migratory patterns of American Black Ducks using eight decades of band-recovery data. *Journal of Wildlife Management* 78:997–1004.

Lawson, S.L. 2006. Comparative reproductive strategies between Long-tailed Ducks and King Eiders at Karrak Lake, Nunavut: Use of energy resources during the nesting season. Master's thesis, University of Saskatchewan, Saskatoon. https://harvest.usask.ca/handle/10388/etd-09202006-233329.

Laymon, S.A. 1989. Altitudinal migration movements of Spotted Owls in the Sierra Nevada, California. *Condor* 91(4):837–841.

Leal, G.R., and L. Bugoni. 2021. Individual variability in habitat, migration routes and niche used by Trindade Petrels, *Pterodroma arminjoniana*. *Marine Biology* 168(134). https://doi.org/10.1007/s00227-021-03938-4.

Leal, G.R., R.W. Furness, R.A.R. McGill, et al. 2017. Feeding and foraging ecology of Trindade Petrels *Pterodroma arminjoniana* during the breeding period in the South Atlantic Ocean. *Marine Biology* 164(211). https://doi.org/10.1007/s00227-017-3240-8.

Léandri-Breton, D-J., J-F. Lamarre, and J. Bêty. 2019. Seasonal variation in migration strategies used to cross ecological barriers in a Nearctic migrant wintering in Africa. *Journal of Avian Biology* 50(6):e02101.

LeBlanc, N.M., D.T. Stewart, S. Pálsson, M.F. Elderkin, G. Mittelhauser, S. Mockford, … and M.L. Mallory. 2017. Population structure of Purple Sandpipers (*Calidris maritima*) as revealed by mitochondrial DNA and microsatellites. *Ecology and Evolution* 7(9):3225–3242.

Lecomte, N., G. Gauthier, J-F. Giroux, E. Milot, and L. Bernatchez. 2009. Tug of war between continental gene flow and rearing site philopatry in a migratory bird: The sex-biases dispersal paradigm reconsidered. *Molecular Ecology* 18:593–602.

Lee, D.E., J.M. Black, J.E. Moore, and J.S. Sedinger. 2007. Age-specific stopover ecology of Black Brant at Humboldt Bay, California. *Wilson Journal of Ornithology* 119(1):9–22.

Lee, D.S., and M. Walsh-McGee. 2020. White-tailed Tropicbird (*Phaethon lepturus*). Version 1.0. In *Birds of the World*, ed. S.M. Billerman. Cornell Lab of Ornithology, Ithaca, NY. https://doi.org/10.2173/bow.whttro.01.

Lee, D.S., J.C. Haney, C. Carboneras, F. Jutglar, and G.M. Kirwan. 2023. Manx Shearwater (*Puffinus puffinus*). Version 1.1. In *Birds of the World*, ed. N.D. Sly. Cornell Lab of Ornithology, Ithaca, NY. https://doi.org/10.2173/bow.manshe.01.1.

Lee, F.B., and H.K. Nelson. 1965. The role of artificial propagation in Wood Duck management. In *Wood Duck Management and Research: A Symposium*, 140–150. Wildlife Management Institute, Washington, DC.

Leigh, D.M., O. Kersten, B. Star, T. Anker-Nilssen, K. Burnham, J. Johnson, J. Provencher, and S. Boessenkool. 2023. Sympatry of genetically distinct Atlantic Puffins (*Fratercula arctica*) in the High Arctic. *Ibis* 165:1022–1030. https://doi.org/10.1111/ibi.13153.

Leschack, C.R., S.K. McKnight, and G.R. Hepp. 2020. Gadwall (*Mareca strepera*). Version 1.0. In *Birds of the World*, ed. S.M. Billerman. Cornell Lab of Ornithology, Ithaca, NY. https://doi.org/10.2173/bow.gadwal.01.

Lesku, J.A., N.C. Rattenborg, M. Valcu, A.L. Vyssotski, S. Kuhn, F. Kuemmeth, W. Heidrich, and B. Kempenaers. 2012. Adaptive sleep loss in polygynous Pectoral Sandpipers. *Science* 337(6102):1654–1658.

Lewis, H.F. 1937. Migrations of the American Brant (*Branta bernicla hrota*). *Auk* 54:73–95.

Lewis, T.L., D.H. Ward, J.S. Sedinger, A. Reed, D.V. Derksen, C. Carboneras, D.A. Christie, and G.M. Kirwan. 2020. Brant (*Branta bernicla*). Version 1.0. In *Birds of the World*, ed. S.M. Billerman. Cornell Lab of Ornithology, Ithaca, NY. https://doi.org/10.2173/bow.brant.01.

Lewis, W.B., R.J. Cooper, M.T. Hallworth, A.R. Brunner, and T.S. Sillett. 2023. Light-level geolocation reveals moderate levels of migratory connectivity for declining and stable populations of Black-throated Blue Warblers (*Setophaga caerulescens*). *Avian Conservation and Ecology* 18(2):12. https://doi.org/10.5751/ACE-02526-180212.

Lewison, R.L., and L.B. Crowder. 2003. Estimating fishery bycatch and effects on a vulnerable seabird population. *Ecological Applications* 13(3):743–753.

Liechti, F., S. Bauer, K.L. Dhanjal-Adams, et al. 2018. Miniaturized multi-sensor loggers provide new insight into year-round flight behaviour of small trans-Sahara avian migrants. *Movement Ecology* 6(19). https://doi.org/10.1186/s40462-018-0137-1.

Linck, E., E.S. Bridge J.M. Duckles, A.G. Navarro-Sigüenza, and S. Rohwer. 2016 Assessing migration patterns in *Passerina ciris* using the world's bird collections as an aggregated resource. *PeerJ* 4:e1871. https://doi.org/10.7717/peerj.1871.

Lincoln, F.C. 1921. The history and purposes of bird banding. *Auk* 38(2):217–228.

———. 1926. The migration of the Cackling Goose. *Condor* 28(4):153–157.

———. 1935. The waterfowl flyways of North America. US Department of Agriculture circular 342.

Lindberg, M.S., D.H. Ward, T.L. Tibbitts, and J. Roser. 2007. Winter movement dynamics of Black Brant. *Journal of Wildlife Management* 71(2):534–540.

Lindsey, K., B. Horwith and E.A. Schreiber. 2000. Status of the Magnificent Frigatebird in the Caribbean. In *Status and Conservation of West Indian Seabirds*, ed. E.A. Schreiber and D.S. Lee, 58–64. Society of Caribbean Ornithology, special publ. no. 1. Ruston, LA.

Linhart, R.C., D.J. Hamilton, J. Paquet, J.O.N. Monteiro, G.P. Ramires, and J.A. Mobley. 2022. Movement and habitat use of non-breeding Semipalmated Sandpiper (*Calidris Pusilla*) at the Banco Dos Cajuais in northeast Brazil. *Conservation Science and Practice* 4(6):e12683. https://doi.org/10.1111/csp2.12683.

Linkhart, B.D., and D.A. McCallum. 2020. Flammulated Owl (*Psiloscops flammeolus*). Version 1.0. In *Birds of the World*, ed. A.F. Poole. Cornell Lab of Ornithology, Ithaca, NY.

Linkhart, B.D., J.W. Fox, and S.W. Yanco. 2016. Migration timing and routes, and wintering areas of Flammulated Owls. *Journal of Field Ornithology* 87:42–54.

Linklater, E.L., S.A. Sonsthagen, G.J. Robertson, L. Colston-Nepali, F. Vigfúsdóttir, and V.L. Friesen. 2024. Reduced representation sequencing reveals weak genetic differentiation between Canadian and European *Larus hyperboreus* (Glaucous Gull). *Ornithological Applications* 126(4):duae037. https://doi.org/10.1093/ornithapp/duae037.

Linnebjerg, J.F., J. Fort, T. Guilford, A. Reuleaux, A. Mosbech, and M. Frederiksen. 2013. Sympatric breeding auks shift between dietary and spatial resource partitioning across the annual cycle. *PLOS ONE* 30(8):e72987. https://doi.org/10.1371/journal.pone.0072987. PMID: 24023663; PMCID: PMC3758292.

Linnebjerg, J.F., M. Frederiksen, Y. Kolbeinsson, et al. 2018. Non-breeding areas of three sympatric auk species breeding in three Icelandic colonies. *Polar Biology* 41:1951–1961. https://doi.org/10.1007/s00300-018-2334-1.

Linthicum, J., R.E. Jackman, B.C. Latta, J. Koshear, and M. Smith. 2007. Annual migrations of Bald Eagles to and from California. *Journal of Raptor Research* 41(2):106–112.

Lislevand, T., M. Briedis, and S.H. Oddvar Heggøy. 2016. Seasonal migration strategies of Common Ringed Plovers *Charadrius hiaticula*. *Ibis* 159:225–229.

Lisovski, S., K. Gosbell, M. Christie, B.J. Hoye, M. Klaassen, I.D. Stewart, A.J. Tayson, and C. Minton. 2016. Movement patterns of Sanderling (*Calidris alba*) in the East Asian-Australasian Flyway and a comparison of methods for identification of crucial areas for conservation. *Emu—Austral Ornithology* 116(2):168–177. https://doi.org/10.1071/MU15042.

Lisovski, S., C.M. Hewson, R.H. Klaassen, F. Korner-Nievergelt, M.W. Kristensen, and S. Hahn. 2012. Geolocation by light: Accuracy and precision affected by environmental factors. *Methods in Ecology and Evolution* 3(3):603–612.

Lisovski, S., Z. Nemeth, J. Wingfield, J. Krause, K. Hobson, N. Seavy, J. Gee, and M. Ramenofsky. 2019. Migration pattern of Gambel's White-crowned Sparrow along the Pacific Flyway. *Journal of Ornithology* 160:1097–1107. https://doi.org/10.1007/s10336-019-01685-4.

Littlefield, C.D. 1990. *Birds of Malheur National Wildlife Refuge, Oregon*. Oregon University Press, Corvalis.

Lobkov, E.G. 1986. *Breeding Birds of Kamchatka.*: DVNTs Academy of Science, Vladivostok.

REFERENCES

———. 2001. South-Kamchatkan Rock Sandpiper *Calidris ptilocnemis kurilensis* Yamashina 1929. In *The Red Data Book of Russian Federation (Animals)*, ed. V.I. Danilov-Danilian, 509–510. AST and Astrel, Moscow, Russia.

Longcore, J.R., D.G. McAuley, G.R. Hepp, and J.M. Rhymer. 2020. American Black Duck (*Anas rubripes*). Version 1.0. In *Birds of the World*, ed. A.F. Poole and F.B. Gill. Cornell Lab of Ornithology, Ithaca, NY. https://doi.org/10.2173/bow.ambduc.01.

Longhurst, A.R. 2006. *Ecological Geography of the Sea*. 2nd ed. Academic Press, San Diego, CA .

Loredo, S., R. Orben, R. Suryan, J. Adams, S. Stephensen, D. Lyons. 2020. Satellite tracking of Common Murres (*Uria aalge*) in the northern California Current System, 2015–2017. Research Workspace. 10.24431/rw1k47i, version: 10.24431_rw1k47i_20201211T201935Z.

Loredo, S.A., R.A. Orben, R.M. Suryan, D.E. Lyons, J. Adams, and S.W. Stephensen. 2019. Spatial and temporal diving behavior of non-breeding Common Murres during two summers of contrasting ocean conditions. *Journal of Experimental Marine Biology and Ecology* 517:13–24. http://doi.org/10.1016/j.jembe.2019.05.009.

Lorenz, T.J., M.G. Raphael, T.D. Bloxton, and P.G. Cunningham. 2017. Low breeding propensity and wide-ranging movements by Marbled Murrelets in Washington. *Journal of Wildlife Management* 81(2):306–321.

Loring, P.H., C.R. Griffin, P.R. Sievert, and C.S. Spiegel. 2017. Comparing satellite and digital radio telemetry to estimate space and habitat use of American Oystercatchers (*Haematopus palliatus*) in Massachusetts. *Waterbirds* 40:19–31.

Loring, P.H., J.D. McLaren, H.F. Goyert, and P.W. Paton. 2020. Supportive wind conditions influence offshore movements of Atlantic Coast Piping Plovers during fall migration. *Condor* 122(3):duaa028.

Loring, P.H., R.A. Ronconi, L.J. Welch, P.D. Taylor, and M.L. Mallory. 2017. Postbreeding dispersal and staging of Common and Arctic Terns throughout the western North Atlantic. *Avian Conservation Ecology* 12:20.

Lott, C.A., and J.P. Smith. 2006. A Geographic-Information-System approach to estimating the origin of migratory raptors in North America using stable hydrogen isotope ratios in feathers. *Auk* 123(3):822–835.

Lounsberry, Z.T., J.B. Almeida, T. Grace, R.B. Lanctot, J. Liebezeit, B.K. Sandercock, K.M. Strum, S. Zack, and S.M. Wisely. 2013. Range-wide conservation genetics of Buff-breasted Sandpipers (*Tryngites subruficollis*). *Auk* 130:429–439.

Lounsberry, Z.T., J.B. Almeida, R.B. Lanctot, J. Liebezeit, B.K. Sandercock, K.M. Strum, S. Zack, and S.M. Wisely. 2014. Museum collections reveal that Buff-breasted Sandpipers (*Calidris subruficollis*) maintained mtDNA variability despite large population declines during the past 135 years. *Conservation Genetics* 15:1197–1208.

Lovette, I.J., S.M. Clegg, and T.B. Smith. 2004. Limited utility of mtDNA markers for determining connectivity among breeding and overwintering locations in three neotropical migrant birds. *Conservation Biology* 18:156–166.

Lowery, G.H., and R.J. Newman. 1966. A continent wide view of bird migration on four nights in October. *Auk* 83(4):547–586.

Lowther, P.E. 2020a. Alder Flycatcher (*Empidonax alnorum*). Version 1.0. In *Birds of the World*, ed. A.F. Poole and F.B. Gill. Cornell Lab of Ornithology, Ithaca, NY. https://doi.org/10.2173/bow.aldfly.01.

———. 2020b. Brown-headed Cowbird (*Molothrus ater*). Version 1.0. In *Birds of the World*, ed. A.F. Poole and F.B. Gill. Cornell Lab of Ornithology, Ithaca, NY. https://doi.org/10.2173/bow.bnhcow.01.

———. 2020c. LeConte's Sparrow (*Ammospiza leconteii*). Version 1.0. In *Birds of the World*, ed. P.G. Rodewald. Cornell Lab of Ornithology, Ithaca, NY. https://doi.org/10.2173/bow.lecspa.01.

Lowther, P.E., and J.L. Ingold. 2020. Blue Grosbeak (*Passerina caerulea*). Version 1.0. In *Birds of the World*, ed. A.F. Poole. Cornell Lab of Ornithology, Ithaca, NY. https://doi.org/10.2173/bow.blugrb1.01.

Lowther, P.E., and J. Nocedal. 2020. Olive Warbler (*Peucedramus taeniatus*). Version 1.0. In *Birds of the World*, ed. A.F. Poole. Cornell Lab of Ornithology, Ithaca, NY. https://doi.org/10.2173/bow.oliwar.01.

Lowther, P.E., and S. Sharbaugh. 2020. Arctic Warbler (*Phylloscopus borealis*). Version 1.0. In *Birds of the World*, ed. S.M. Billerman. Cornell Lab of Ornithology, Ithaca, NY. https://doi.org/10.2173/bow.arcwar1.01.

Lowther, P.E., and D.F. Stotz. 2020. Sulphur-bellied Flycatcher (*Myiodynastes luteiventris*). Version 1.0. In *Birds of the World*, ed. A.F. Poole and F.B. Gill. Cornell Lab of Ornithology, Ithaca, NY. https://doi.org/10.2173/bow.subfly.01.

Lowther, P.E., and J.M. Williams. 2020. Nashville Warbler (*Leiothlypis ruficapilla*). Version 1.0. In *Birds of the World*, ed. A.F. Poole. Cornell Lab of Ornithology, Ithaca, NY. https://doi.org/10.2173/bow.naswar.01.

Lowther, P.E., J. Breeden, and J.C. Eitniear. 2020a. Red-billed Pigeon (*Patagioenas flavirostris*). Version 1.0. In *Birds of the World*, ed. P.G. Rodewald. Cornell Lab of Ornithology, Ithaca, NY. https://doi.org/10.2173/bow.rebpig1.01.

Lowther, P.E., C. Celada, N.K. Klein, C.C. Rimmer, and D.A. Spector. 2020b. Yellow Warbler (*Setophaga petechia*). Version 1.0. In *Birds of the World*, ed. A.F. Poole and F.B. Gill. Cornell Lab of Ornithology, Ithaca, NY. https://doi.org/10.2173/bow.yelwar.01.

Lowther, P.E., C.T. Collins, C. Gunn, J.P. Beason, K. Potter, and M. Webb. 2020c. Black Swift (*Cypseloides niger*). Version 1.0. In *Birds of the World*, ed. S.M. Billerman. Cornell Lab of Ornithology, Ithaca, NY. https://doi.org/10.2173/bow.blkswi.01.

Lowther, P.E., A.W. Diamond, S.W. Kress, G.J. Robertson, K. Russell, D.N. Nettleship, G.M. Kirwan, D.A. Christie, C.J. Sharpe, E.F.J. Garcia, and P.F.D. Boesman. 2020d. Atlantic Puffin (*Fratercula arctica*). Version 1.0. In *Birds of the World*, ed. S.M. Billerman. Cornell Lab of Ornithology, Ithaca, NY. https://doi.org/10.2173/bow.atlpuf.01.

Lowther, P.E., H.D. Douglas III, and C.L. Gratto-Trevor. 2020e. Willet (*Tringa semipalmata*). Version 1.0. In *Birds of the World*, ed. A.F. Poole and F.B. Gill. Cornell Lab of Ornithology, Ithaca, NY.

Lowther, P.E., S.M. Lanyon, C.W. Thompson, and T.S. Schulenberg. 2020f. Painted Bunting (*Passerina ciris*). Version 1.0. In *Birds of the World*, ed. S.M. Billerman. Cornell Lab of Ornithology, Ithaca, NY. 2020. https://doi.org/10.2173/bow.paibun.01.

Lowther, P.E., D.A. McCallum, P. Pyle, and W.A. Hopping. 2023. Western Flycatcher (*Empidonax difficilis*). Version 1.0. In *Birds of the World*, ed. B.K. Keeney and S.M. Billerman. Cornell Lab of Ornithology, Ithaca, NY. https://doi.org/10.2173/bow.wesfly.01.

Lowther, P.E., A.F. Poole, J.P. Gibbs, S.M. Melvin, and F.A. Reid. 2020g. American Bittern (*Botaurus lentiginosus*). Version 1.0. In *Birds of the World*, ed. A.F. Poole. Cornell Lab of Ornithology, Ithaca, NY. https://doi.org/10.2173/bow.amebit.01.

Lowther, P.E., P. Pyle, and M.A. Patten. 2020h. Thick-billed Kingbird (*Tyrannus crassirostris*). Version 1.0. In *Birds of the World*, ed. P.G. Rodewald. Cornell Lab of Ornithology, Ithaca, NY. https://doi.org/10.2173/bow.thbkin.01.

Lucia, M., H. Strøm, P. Bustamante, et al. 2017. Contamination of Ivory Gulls (*Pagophila eburnea*) at four colonies in Svalbard in relation to their trophic behaviour. *Polar Biology* 40:917–929. https://doi.org/10.1007/s00300-016-2018-7.

Lumpkin, D., S. Jennings, N. Warnock, and E. Condeso. 2023. Partial migration by Great Egrets *Ardea alba* in coastal California. *Waterbirds* 45:150–158.

Luttrell, S.A.M., and B. Lohr. 2017. Geographic variation in call structure, likelihood, and call-song associations across subspecies boundaries, migratory patterns, and habitat types in the Marsh Wren (*Cistothorus palustris*). *Auk* 135(1):127–151. https://doi.org/10.1642/AUK-17-110.1.

Luukkonen, D.R., H.H. Prince, and R.C. Mykut. 2008. Movements and survival of molt migrant Canada Geese from southern Michigan. *Journal of Wildlife Management* 72(2):449–462.

Lyons, D.E., A.G.L. Patterson, J. Tennyson, T.J. Lawes, and D.D. Roby. 2018. The Salton Sea: Critical migratory stopover habitat for Caspian Terns (*Hydroprogne caspia*) in the North American Pacific Flyway. *Waterbirds* 41:154–165.

Lyons, J.E., A.J. Baker, P.M. González, Y. Aubry, C. Buidin, and Y. Rochepault. 2017. Migration ecology and stopover population size of Red Knots *Calidris canutus rufa* at Mingan Archipelago after exiting the breeding grounds. *Wader Study* 124:197–205.

Lyons, J.E., B. Winn, T. Keyes, and K.S. Kalasz. 2018. Post-breeding migration and connectivity of Red Knots in the western Atlantic. *Journal of Wildlife Management* 82(2):383–396.

Macdonald, C.A., K.C. Fraser, H.G. Gilchrist, T.K. Kyser, J.W. Fox, and O.P. Love. 2012. Strong migratory connectivity in a declining Arctic passerine. *Animal Migration* 1(1):23–30. https://doi.org/10.2478/ami-2012-0003.

MacDonald, D.W. 1978. Radio-tracking: Some applications and limitations. In *Animal Marking*, 192–204. Palgrave, London.

MacDougall-Shackleton, S.A., R.E. Johnson, and T.P. Hahn. 2020. Gray-crowned Rosy-Finch (*Leucosticte tephrocotis*). Version 1.0. In *Birds of the World*, ed. A.F. Poole and F.B. Gill. Cornell Lab of Ornithology, Ithaca, NY. https://doi.org/10.2173/bow.gcrfin.01.

MacGregor-Fors, I., A.G. Navarro-Sigüenza, L. Morales-Pérez, and J.E. Schondube. 2012. Note on the distribution and migration of the Blackcapped Vireo (*Vireo atricapilla*). *Huitzil. Revista Mexicana de Ornitología* 13(1):17–21.

Macías-Duarte, A., C.J. Conway, M. Culver. 2020. Agriculture creates subtle genetic structure among migratory and nonmigratory populations of Burrowing Owls throughout North America. *Ecology and Evolution* 10(1):10697–10708. https://doi.org/10.1002/ece3.6725.

Mack, D.E., and W. Yong. 2020. Swainson's Thrush (*Catharus ustulatus*). Version 1.0. In *Birds of the World*, ed. A.F. Poole and F.B. Gill. Cornell Lab of Ornithology, Ithaca, NY. https://doi.org/10.2173/bow.swathr.01.

Mackay, R.H. 1957. Movements of Trumpeter Swans shown by band returns and observations. *Condor* 59:339.

Mackell, D.A., et al. 2021 Migration stopover ecology of Cinnamon Teal in western North America. *Ecology and Evolution* 11:14056–14069.

Macwhirter, R.B., P. Austin-Smith Jr., and D.E. Kroodsma. 2020. Sanderling (*Calidris alba*). Version 1.0. In *Birds of the World*, ed. S.M. Billerman. Cornell Lab of Ornithology, Ithaca, NY. https://doi.org/10.2173/bow.sander.01.

Madeiros, J. 2022. A record year for Bermuda's national bird: Cahow Recovery Program update. Nonsuch Expeditions. http://www.nonsuchisland.com/blog/2022-cahow-recovery-program-update-a-record-year-for-bermudas-national-bird

Madeiros, J., B. Flood, and K. Zufelt. 2014. Conservation and at-sea range of Bermuda Petrel (*Pterodroma cahow*). *North American Birds* 67:546–557.

Maftei, M., and R. Russ. 2014. A mass post-breeding movement of Crested Auklets *Aethia cristatella* in the Chukchi Sea. *Marine Ornithology* 42:31–34.

Maftei, M., S.E. Davis, and M.L. Mallory. 2015. Confirmation of a wintering ground of Ross's Gull *Rhodostethia rosea* in the northern Labrador Sea. *Ibis*. 157:642–647.

Magnusdottir, E., E.H.K. Leat, S. Bourgeon, H. Strøm, A. Petersen, R.A. Phillips, S.A. Hanssen, J.O. Bustnes, P. Hersteinsson, and R.W. Furness. 2012 Wintering areas of Great Skuas *Stercorarius skua* breeding in Scotland, Iceland, and Norway. *Bird Study* 59(1):1–9. https://doi.org/10.1080/00063657.2011.636798.

Malecki, R., S. Sheaffer, D. Howell, and T. Strange. 2006. Northern Pintails in eastern North America: Their seasonal distribution, movement patterns, and habitat affiliations. Atlantic Flyway Council. Technical Section, Final Report.

Malecki, R.A., B.D.J. Batt, and S.E. Sheaffer. 2001. Spatial and temporal distribution of Atlantic population Canada Geese. *Journal of Wildlife Management* 65:242–247.

Mallek, E.J., and D.J. Groves. 2011. Alaska-Yukon waterfowl breeding population survey. US Fish and Wildlife Service, Fairbanks and Juneau, Alaska.

Mallory, M.L., and C.D. Gilbert. 2008. Leg-loop harness design for attaching external transmitters to seabirds. *Marine Ornithology* 36:183–188. http://doi.org/10.5038/2074-1235.36.2.788.

Mallory, M.L., J.A. Akearok, D.B. Edwards, K. O'Donovan, and C.D. Gilbert. 2008. Autumn migration and wintering of Northern Fulmars (*Fulmarus glacialis*) from the Canadian high Arctic. *Polar Biology* 31:745–750.

Mallory, M.L., J. Akearok, N.R. North, D.V. Weseloh, and S. Lair. 2006. Movements of Long-tailed Ducks wintering on Lake Ontario to breeding areas in Nunavut, Canada. *Wilson Journal of Ornithology* 118(4):494–501.

Mallory, M.L., S.A. Hatch, and D.N. Nettleship. 2020a. Northern Fulmar (*Fulmarus glacialis*). Version 1.0. In *Birds of the World*, ed. S.M. Billerman. Cornell Lab of Ornithology, Ithaca, NY. https://doi.org/10.2173/bow.norful.01.

Mallory, M.L., R.A. Ronconi, R.B. Allen, C. Dwyer, S. Lair, C.D. Mallory, N.R. McLellan, G.R. Milton, G.J. Parsons, L. Savoy, and M.D. Tomlik. 2020b. Annual movement patterns of American Common Eiders (*Somateria mollissima dresseri*). *Wildlife Biology* 2020(2):1–10, wlb.00665.

Mallory, M.L., I.J. Stenhouse, H.G. Gilchrist, G.J. Robertson, J.C. Haney, and S.D. Macdonald. 2020c. Ivory Gull (*Pagophila eburnea*). Version 1.0. In *Birds of the World*, ed. S.M. Billerman. Cornell Lab of Ornithology, Ithaca, NY. https://doi.org/10.2173/bow.ivogul.01.

Mancilla-Morales, M.D., E. Velarde, A. Aguilar, A. Contreras-Rodríguez, E. Ezcurra, J.A. Rosas-Rodríguez, J.G. Soñanez-Organis, and E.A. Ruiz. 2022. Strong philopatry, isolation by distance, and local habitat have promoted genetic structure in Heermann's Gull. *Diversity* 14(2):108. https://doi.org/10.3390/d14020108.

Mancini, P.L., and L. Bugoni. 2014. Resources partitioning by seabirds and their relationship with other consumers at and around a small tropical archipelago. *ICES Journal of Marine Science* 71(9):2599–2607. https://doi.org/10.1093/icesjms/fsu105.

Mancuso, K.A., M.A. Fylling, C.A. Bishop, K.E. Hodges, M.B. Lancaster, and K.R. Stone. 2021. Migration ecology of western Gray Catbirds. *Movement Ecology* 9:10.

Mancuso, K.A., K.E. Hodges, J.D. Alexander, M. Grosselet, A.M. Bezener, L. Morales, S.C. Martinez, J. Castellanos-Labarcena, M.A. Russello, S.M. Rockwell, M.E. Bieber, and C.A. Bishop. 2022. Migration and non-breeding ecology of the Yellow-breasted Chat *Icteria virens*. *Journal of Ornithology* 163:37–50. https://doi.org/10.1007/s10336-021-01931-8.

Mandel, J.T., K.L. Bildstein, G. Bohrer, and D.W. Winkler. 2008. Movement ecology of migration in Turkey Vultures. *Proceedings of the National Academy of Sciences* 105:19102–19107.

Mandernack, B.A., M. Solensky, M. Martell, and R.T. Schmitz. 2012. Satellite tracking of Bald Eagles in the upper Midwest. *Journal of Raptor Research* 46(3):258–272.

Manly, B., B.S. Arbogast, D.S. Lee, and M. van Tuinen. 2013. Mitochondrial DNA analysis reveals substantial population structure within the endangered Black-capped Petrel (*Pterodroma hasitata*). *Waterbirds* 36(2), 228–233. http://www.jstor.org/stable/24641587.

Manthey, J.D., M. Geiger, and R.G. Moyle. 2017. Relationships of morphological groups in the Northern Flicker superspecies complex (*Colaptes auratus* & *C. chrysoides*). *Systematics and Biodiversity* 15(3):183–191.

Mariano-Jelicich, R., and E. Madrid. 2014. Microsatellite variability among Black Skimmer (*Rynchops niger intercedens*) populations in southern South America. *Waterbirds* 37:175–182.

Marks, J.S., and R.L. Redmond. 1996. Demography of Bristle-thighed Curlews *Numenius tahitiensis* wintering on Laysan Atoll. *Ibis* 138:438–447.

Marks, J.S., D.L. Evans, and D.W. Holt. 2020. Long-eared Owl (*Asio otus*).Version 1.0. In *Birds of the World*, ed. S.M. Billerman. Cornell Lab of Ornithology, Ithaca, NY. https://doi.org/10.2173/bow.loeowl.01.

Marks, J.S., T.L. Tibbitts, R.E. Gill, Jr., and B.J. McCaffery. 2020. Bristle-thighed Curlew (*Numenius tahitiensis*). Version 1.0 In *Birds of the World*, ed. A.F. Poole and F.B. Gill, Cornell Lab of Ornithology, Ithaca NY. https://doi.org/10.2173/bow.brtcur.01.

Marra, P.P., and R.T. Holmes. 2001. Consequences of dominance-mediated habitat segregation in American Redstarts during the nonbreeding season. *Auk* 118(1):92–104.

Marra, P.P., E.B. Cohen, S.R. Loss, J.E. Rutter, and C.M. Tonra. 2015. A call for full annual cycle research in animal ecology. *Biology Letters* 11(8):20150552.

Marra, P.P., L.A. Culp, A.L. Scarpignato, and E.B. Cohen. 2014. Full annual cycle climate change vulnerability assessment for migratory birds of the upper Midwest and Great Lakes region. Final Report to the Upper Midwest and Great Lakes Landscape Conservation Cooperative. Washington, DC. http://migratoryconnectivityproject.org/climate-change-vulnerability/.

Marra, P.P., K.A. Hobson, and R.T. Holmes. 1998. Linking winter and summer events in a migratory bird by using stable-carbon isotopes. *Science* 282(5395):1884–1886.

Martell, M.S., and D. Douglas. 2019. Osprey in North and South America 1995–2002 (Martell)-reference-data. Movebank Data Repository. https://doi.org/10.5441/001/1.sv6335t3/2.

Martell, M.S., R.O. Bierregaard, B.E. Washburn, J.E. Elliott, C.J. Henny, R.S. Kennedy, and I. MacLeod. 2014. The spring migration of adult North American Ospreys. *Journal of Raptor Research* 48:309–324.

Martell, M.S., C.J. Henny, P.E. Nye, and M.J. Solensky. 2001. Fall migration routes, timing, and wintering sites of North American Ospreys as determined by satellite telemetry. *Condor* 103:715–724.

Martell, M.S., M.J. Kennedy, C.J. Henny, and P.E. Nye. 1998. Highway to the tropics: Using satellite telemetry and the internet to track Ospreys in the Western Hemisphere. In *Migrating Birds Know No Boundaries: Scientific and Educational Aspects of Migrating Bird Conservation*; ed. Y. Leshem, E. Lachman, and P. Berhold, 163–172. Torgos, Tel Aviv, Israel.

Martell, M.S., M.A. McMillian, M.J. Solensky, and B.K. Mealy. 2004. Partial migration and wintering use of Florida by Ospreys. *Journal of Raptor Research* 38(1):55–61.

Marthinsen, G., L. Wennerberg, R. Solheim, and J.T. Lifjeld. 2009. No phylogeographic structure in the circumpolar Snowy Owl (*Bubo scandiacus*). *Conservation Genetics* 10(4):923–933.

Marti, C.D., A.F. Poole, L.R. Bevier, M.D. Bruce, D.A. Christie, G.M. Kirwan, J.S. Marks, and P. Pyle. 2024. American Barn Owl (*Tyto furcata*). Version 1.1. In *Birds of the World*, eds. S.M. Billerman, B.K. Keeney, and M.G. Smith. Cornell Lab of Ornithology, Ithaca, NY. https://doi.org/10.2173/bow.brnowl.01.1

Martin, J.W., and B.A. Carlson. 2020a. Bell's Sparrow (*Artemisiospiza belli*). Version 1.0. In *Birds of the World*, ed. A.F. Poole. Cornell Lab of Ornithology, Ithaca, NY. https://doi.org/10.2173/bow.belspa2.01.

———. 2020b. Sagebrush Sparrow (*Artemisiospiza nevadensis*). Version 1.0. In *Birds of the World*, ed. A.F. Poole. Cornell Lab of Ornithology, Ithaca, NY. https://doi.org/10.2173/bow.sagspa1.01.

Martin, J.W., and J.R. Parrish. 2020. Lark Sparrow (*Chondestes grammacus*). Version 1.0. In *Birds of the World*, ed. A.F. Poole and F.B. Gill. Cornell Lab of Ornithology, Ithaca, NY. https://doi.org/10.2173/bow.larspa.01.

Martin, P.D., D.C. Douglas, T. Obritschkewitsch, and S. Torrence 2015. Distribution and movements of Alaska-breeding Steller's Eiders in the nonbreeding period. *Condor* 117:341–353.

Martin, S.G. 2020. Brewer's Blackbird (*Euphagus cyanocephalus*). Version 1.0. In *Birds of the World*, ed. A.F. Poole and F.B. Gill. Cornell Lab of Ornithology, Ithaca, NY. https://doi.org/10.2173/bow.brebla.01.

Martin, T.E., and P.M. Barber. 2020. Red-faced Warbler (*Cardellina rubrifrons*). Version 1.0. In *Birds of the World*, ed. A.F. Poole and F.B. Gill. Cornell Lab of Ornithology, Ithaca, NY. https://doi.org/10.2173/bow.refwar.01.

Martínez-Gómez, J.E., N. Matías-Ferrer, R.N.M. Sehgal, et al. 2015. Phylogenetic placement of the critically endangered Townsend's Shearwater (*Puffinus auricularis auricularis*): Evidence for its conspecific status with Newell's Shearwater (*Puffinus a. newelli*) and a mismatch between genetic and phenotypic differentiation. *Journal of Ornithology* 156:1025–1034. https://doi.org/10.1007/s10336-015-1189-2.

Masto, N., O. Robinson, M. Brasher, A. Blake-Bradshaw, C. Highway, A. Keever, J. Feddersen, H.M. Hagy, D.C. Osborne, D. Combs, and B. Cohen. 2022. Citizen science reveals waterfowl responses to extreme winter weather. *Global Change Biology* 28:5469–5479. https://doi.org/10.1111/gcb.16288.

Mathiasson, S. 1963. Fulmars (*Fulmarus glacialis*) in Swedish waters, a biometrical-morphological study in order to establish their geographical origin. *Var Fagelvarld* 22:271–289.

Mattsson, B.J., T.L. Master, R.S. Mulvihill, and W.D. Robinson. 2020. Louisiana Waterthrush (*Parkesia motacilla*). Version 1.0. In *Birds of the World*, ed. A.F. Poole. Cornell Lab of Ornithology, Ithaca, NY. https://doi.org/10.2173/bow.louwat.01.

Mayfield, H.F. 1983. Kirtland's Warbler, victim of its own rarity? *Auk* 100(4):974–976.

Mazerolle, D.F., and K.A. Hobson. 2007. Patterns of differential migration in White-throated Sparrows evaluated with isotopic measurements of feathers. *Canadian Journal of Zoology* (*Revue Canadienne De Zoologie*) 85:413–420.

Mazerolle, D.F., K.A. Hobson, and L.I. Wassenaar. 2005. Stable isotope and band-encounter analyses delineate migratory patterns and catchment areas of White-throated Sparrows at a migration monitoring station. *Oecologia* 144:541–549.

Mazzochi, M.S., V. Muraro, N.J.R. Fagundes, and L. Bugoni. 2024. Absence of genetic structure among ecologically diverse populations indicate high plasticity in a pantropical seabird. *Conservation Genetics* 25:925–938.

McAuley, D.G., D.M. Keppie, and R.M. Whiting Jr. 2020. American Woodcock (*Scolopax minor*). Version 1.0. In *Birds of the World*, ed. A.F. Poole. Cornell Lab of Ornithology, Ithaca, NY. https://doi.org/10.2173/bow.amewoo.01.

McCabe, R.A., L.J. Goodrich, D.R. Barber, T.L. Master, J.L. Watson, E.M. Bayne, A.-L. Harrison, P.P. Marra, and K.L. Bildstein. 2020. Satellite tracking reveals age and origin differences in migration ecology of two populations of Broad-winged Hawks (*Buteo platypterus*). *Wilson Journal of Ornithology* 131(1):1–14.

McCaffery, B.J., and R.E. Gill. 2020. Bar-tailed Godwit (*Limosa lapponica*). Version 1.0. In *Birds of the World*, ed. S.M. Billerman. Cornell Lab of Ornithology, Ithaca, NY. https://doi.org/10.2173/bow.batgod.01.

McCarty, J.P., L.L. Wolfenbarger, C.D. Laredo, P. Pyle, and R.B. Lanctot. 2020. Buff-breasted Sandpiper (*Calidris subruficollis*). Version 2.0. In *The Birds of North America*, ed. P.G. Rodewald. Cornell Lab of Ornithology, Ithaca, NY. https://doi.org/10.2173/bna.bubsan.02.

McClelland, B.R., P.T. McClelland, R.E. Yates, E.L. Caton, and M.E. McFadzen. 1996. Fledging and migration of juvenile Bald Eagles from Glacier National Park, Montana. *Journal of Raptor Research* 30(2):79–89.

REFERENCES

McCloskey, S.E., B.D. Uher-Koch, J.A. Schmutz, and T.F. Fondell. 2018 International migration patterns of Red-throated Loons (*Gavia stellata*) from four breeding populations in Alaska. *PLOS ONE* 13(1):e0189954. https://doi.org/10.1371/journal.pone.0189954.

McCoy, K.D., T. Boulinier, and C. Tirard. 2005. Comparative host-parasite population structures: Disentangling prospecting and dispersal in the Black-legged Kittiwake *Rissa tridactyla*. *Molecular Ecology* 14(9):2825–2838. https://doi.org/10.1111/j.1365-294X.2005.02631.x. PMID: 16029481.

McCrimmon Jr., D.A., J.C. Ogden, G.T. Bancroft, A. Martínez-Vilalta, A. Motis, G.M. Kirwan, and P.F.D. Boesman. 2020. Great Egret (*Ardea alba*). Version 1.0. In *Birds of the World*, ed. S.M. Billerman. Cornell Lab of Ornithology, Ithaca, NY. https://doi.org/10.2173/bow.greegr.01.

McDonald, M.V. 2020. Kentucky Warbler (*Geothlypis formosa*). Version 1.0. In *Birds of the World*, ed. A.F. Poole. Cornell Lab of Ornithology, Ithaca, NY. https://doi.org/10.2173/bow.kenwar.01.

McDuffie, L.A., K.S. Christie, A.L. Harrison, A.R. Taylor, B.A. Andres, B. Laliberté, and J.A. Johnson. 2022a. Eastern-breeding Lesser Yellowlegs are more likely than western-breeding birds to visit areas with high shorebird hunting during southward migration. *Condor* 124(1):duab061.

McDuffie, L.A., K.S. Christie, A.R. Taylor, E. Nol, C. Friis, C.M. Harwood, ... and J.A. Johnson. 2022b. Flyway-scale GPS tracking reveals migratory routes and key stopover and non-breeding locations of Lesser Yellowlegs. *Ecology and Evolution* 12(11):e9495.

McDuie, F., and B.C. Congdon. 2016. Trans-equatorial migration and non-breeding habitat of tropical shearwaters: Implications for modelling pelagic Important Bird Areas. *Marine Ecology Progress Series* 550:219–234.

McFarlane Tranquilla, L.A., W.A. Montevecchi, D.A. Fifield, A. Hedd, A.J. Gaston, et al. 2014. Individual winter movement strategies in two species of murre (*Uria* spp.) in the northwest Atlantic. *PLOS ONE* 9(4): e90583. https://doi.org/10.1371/journal.pone.0090583.

McGowan, K.J. 2020. Fish Crow (*Corvus ossifragus*). Version 1.0. In *Birds of the World*, ed. A.F. Poole and F.B. Gill. Cornell Lab of Ornithology, Ithaca, NY. https://doi.org/10.2173/bow.fiscro.01.

McGrady, M.J., T.L. Maechtle, J.J. Vargas, W.S. Seegar, and M.C. Porras Pena. 2002. Migration and ranging of Peregrine Falcons wintering on the Gulf of Mexico Coast, Tamaulipas, Mexico. *Condor* 104(1):39–48.

McGraw, K.J., and A.L. Middleton. 2020. American Goldfinch (*Spinus tristis*). Version 1.0. In *Birds of the World*, ed. P.G. Rodewald. Cornell Lab of Ornithology, Ithaca, NY. https://doi.org/10.2173/bow.amegfi.01.

McIntyre, C., D.C. Douglas, L.G. Adams. 2009. Movements of juvenile Gyrfalcons from western and interior Alaska following departure from their natal areas. *Journal of Raptor Research* 43(2):99–109.

McIntyre, C.L., D.C. Douglas, and M.W. Collopy. 2008. Movements of Golden Eagles (*Aquila chrysaetos*) from interior Alaska during their first year of independence. *Auk* 125(1):214–224.

McKay, B., and G.A. Hall. 2020. Yellow-throated Warbler (*Setophaga dominica*). Version 1.0. In *Birds of the World*, ed. A.F. Poole. Cornell Lab of Ornithology, Ithaca, NY. https://doi.org/10.2173/bow.yetwar.01.McKay, B.D. 2008. Variation is clinal in the Yellow-throated Warbler. *Condor* 110:569–574.

———. 2009. Evolutionary history suggests rapid differentiation in the Yellow-throated Warbler *Dendroica dominica*. *Journal of Avian Biology* 40:181–190.

McKellar, A.E., P.P. Marra, S.J. Hannon, C.E. Studds, and L.M. Ratcliffe. 2013. Winter rainfall predicts phenology in widely separated populations of a migrant songbird. *Oecologia* (Berlin) 172:595–605.

McKinnon, E.A., C. Artuso, and O.P. Love. 2017. The mystery of the missing warbler. *Ecology* 98:1970–1972.

McKnight, A., A.J. Allyn, D.C. Duffy, and D.B. Irons. 2013. "Stepping stone" pattern in Pacific Arctic Tern migration reveals the importance of upwelling areas. *Marine Ecology Progress Series* 491:253–264.

McKnight, A., D.B. Irons, A.J. Allyn, K.M. Sullivan, and R.M. Suryan. 2011. Winter dispersal and activity patterns of post-breeding Black-legged Kittiwakes *Rissa tridactyla* from Prince William Sound, Alaska. *Marine Ecology Progress Series* 442:241–253.

McNeil, R. 1970. Hiverage et estivage d'oiseaux aquatiques Nord-Americains dans le nord-est du Venequela (Mue, accumulation de graisse, capacite de vol et routes de migration); Wintering and summering of North American waterfowl in northeastern Venezuela (Moulting, fat accumulation, flight capacity and migration routes). *L'oiseau et la Revue francaise d'ornithologie* 40:185–302.

———. 1982. Winter resident repeats and returns of austral and boreal migrant birds banded in Venezuela. *Journal of Field Ornithology* 53(2):125–132.

McNeil, R., and J. Burton. 1973. Dispersal of some southbound migrating North American shorebirds away from the Magdalen Is., Gulf of St. Lawrence, and Sable Is., Nova Scotia. *Caribbean Journal of Science* 13:257–267.

———. 1977. Southbound migration of shorebirds from the Gulf of St. Lawrence. *Wilson Bulletin* 89:167–171.

McNeil, S.E., D. Tracy, and C.D. Cappello. 2015. Loop migration by a western Yellow-billed Cuckoo wintering in the Gran Chaco. *Western Birds* 46:244–255.

McNicholl, M.K., P.E. Lowther, and J.A. Hall. 2001. Forster's Tern (*Sterna forsteri*). Version 2.0. In *The Birds of North America*, ed. A.F. Poole and F.B. Gill. Cornell Lab of Ornithology, Ithaca, NY. https://doi.org/10.2173/bna.595.

Meattey, D. 2017. Boston Harbor Common Eider Satellite Tracking Study. Biodiversity Research Institute. http://www.briloon.org/boston-harbor-common-eider-satellite-tracking-study.

Meattey, D.E., S.R. McWilliams, P.W.C. Paton, C. Lepage, S. Gilliland, L. Savoy, G. Olsen, and J. Osenkowski. 2018. Annual cycle of White-winged Scoters (*Melanitta fusca*) in eastern North America: Migration phenology, population delineation, and connectivity. *Canadian Journal of Zoology* 96:1353–1365.

Medrano, F., R. Barros, R. Peredo, and F. De Groote. 2018. Extension of the northward migratory route of the Upland Sandpiper (*Bartramia longicauda*) to the western slope of the Andes. *Wilson Journal of Ornithology* 130(3):805–809.

Medrano, F., J. Hernández-Montoya, S. Saldanha, Y. Bedolla-Guzmán, and J. González-Solís. 2024. Contrasting migratory ecology of two threatened and allochronic storm-petrels breeding in the Mexican Pacific. *Endangered Species Research* 54:331–339.

Medrano, F., J. Hernández-Montoya, S. Saldanha, A. Duarte, J. González-Solís, and G.M. Kirwan. 2023. Ainley's Storm-Petrel (*Hydrobates cheimomnestes*). Version 2.0. In *Birds of the World*, ed. S.M. Billerman. Cornell Lab of Ornithology, Ithaca, NY. https://doi.org/10.2173/bow.leastp2.02.

Medrano F., S. Saldanha, J. Hernández-Montoya, Y. Bedolla-Guzmán, and J. González-Solís. 2022. Foraging areas of nesting Ainley's Storm-petrel *Hydrobates cheimnomnestes*. *Marine Ornithology* 50(2):125–127.

Meehan, T.D., G.S. LeBaron, K. Dale, N.L. Michel, G.M. Verutes, and G.M. Langham. 2018. Abundance trends of birds wintering in the USA and Canada, from Audubon Christmas Bird Counts, 1966–2017. Version 2.1. National Audubon Society, New York.

Meehan, T.D., S.P. Saunders, W.V. DeLuca, N.L. Michel, J. Grand, ... and C.B. Wilsey. 2022. Data integration to estimate spatial patterns of avian migration across the Western Hemisphere. *Ecological Applications* 32:e2679.

Mehl, K.R., R.T. Alisauskas, K.A. Hobson, and D.K. Kellet. 2004. To winter east or west? Heterogeneity in winter philopatry in a Central-Arctic population of King Eiders. *Condor* 106:241–251.

Mehl, K.R., R.T. Alisauskas, K.A. Hobson, and F.R. Merkel. 2005. Linking breeding and wintering areas of King Eiders: Making use of polar isotopic gradients. *Journal of Wildlife Management* 69(3):1297–1304.

Mejías, M.A., Y.F. Wiersma, D.B. Wingate, and J.L. Madeiros, J.L. 2017. Distribution and at-sea behavior of Bermudan White-tailed Tropicbirds (*Phaethon lepturus catesbyi*) during the non-breeding season. *Journal of Field Ornithology* 88:184–197. https://doi.org/10.1111/jofo.12198.

Melvin, S.L., D.E. Gawlik, and T. Scharff. 1999. Long-term movement patterns for seven species of wading birds. *Waterbirds* 22(3):411–416.

Mendez, D.C. 2018. Migratory routes and stopover regions of eastern breeding Veeries (*Catharus fuscescens*) as revealed by archival light-level geolocators. Master's thesis, Delaware State University, Dover.

Menning, D.M., B.D. Uher-Koch, M.J. Flamme, T. Simmons, J.A. Schmutz, and S.L. Talbot. 2023. eDNA metabarcoding analyses of diet in Yellow-billed Loons of northern Alaska. *Waterbirds* 45(2):159–166. https://doi.org/10.1675/063.045.0206.

Mercer, D., S. Haig, and D. Roby. 2013. Phylogeography and population genetic structure of Double-crested Cormorants (*Phalacrocorax auritus*). *Conservation Genetics* 14. https://doi.org/10.1007/s10592-013-0477-8.

Merkel, B., S. Descamps, N.G. Yoccoz, D. Grémillet, F. Daunt, K.E. Erikstad, A.V. Ezhov, M.P. Harris, M. Gavrilo, S.H. Lorentsen, and T.K. Reiertsen. 2021. Individual migration strategy fidelity but no habitat specialization in two congeneric seabirds. *Journal of Biogeography* 48(2):263–275.

Merkel, F.R., K.L. Johansen, R.D. Nielsen, I.K. Petersen, J. Sterup, and A. Mosbech. 2019. Wintering seabirds in south-west Greenland, 2017. *Polar Research* 38. https://doi.org/10.33265/polar.v38.3462

Mestre, L.A.M., and R.O. Bierregaard. 2009. The role of Amazonian rivers for wintering Ospreys (*Pandion haliaetus*): Clues from North American band recoveries in Brazil between 1937 and 2006. *Studies on Neotropical Fauna and Environment* 44:141–147.

Meyer, K.D. 2020. Swallow-tailed Kite (*Elanoides forficatus*). Version 1.0. In *Birds of the World*, ed. A.F. Poole and F.B. Gill. Cornell Lab of Ornithology, Ithaca, NY. https://doi.org/10.2173/bow.swtkit.01.

Mezebish, T., G. Olsen, M. Goodman, F. Rohwer, and M. McConnell. 2022. Spring migration and breeding distribution of female Ring-necked Ducks wintering in the southern Atlantic Flyway. *Avian Conservation and Ecology* 17(2).

Middleton, A.L. 2020. Chipping Sparrow (*Spizella passerina*). Version 1.0. In *Birds of the World*, ed. A.F. Poole and F.B. Gill. Cornell Lab of Ornithology, Ithaca, NY. https://doi.org/10.2173/bow.chispa.01.

MigConnectivity R package v. 0.4.7. https://CRAN.R-project.org/package=MigConnectivity.

Mikkelsen, E.K., and J.T Weir. 2023. Phylogenomics reveals that mitochondrial capture and nuclear introgression characterize skua species proposed to be of hybrid origin. *Systematic Biology* 72(1):78–91. https://doi.org/10.1093/sysbio/syac078.

Mikuska, T., J.A. Kushlan, and S. Hartley. 1998. Key areas for wintering North American herons. *Colonial Waterbirds* 21(2):125–134.

Mila, B., S.M. Clegg, M. Kimura, K. Ruegg, I. Lovette, and T.B. Smith. 2005. Linking breeding and overwintering areas of five Nearctic–Neotropical migratory passerines using molecular genetic markers. US Forest Service General Technical Report PSW 191:889–892.

Miller, C., C. Bossu, J. Sarraco, D. Toews, C. Rushing, A. Roberto-Charron, J.A. Tremblay, R. Chandler, M. DeSaix, C. Fiss, J. Larkin, S. Haché, S. Nebel, and K. Ruegg. 2023. Genomics-informed conservation units reveal spatial variation in climate vulnerability in a migratory bird. *Molecular Ecology* 33(1):e17199.

Miller, K.E., and W. E. Lanyon. 2020. Great Crested Flycatcher (*Myiarchus crinitus*). Version 1.0. In *Birds of the World*, ed. A.F. Poole. Cornell Lab of Ornithology, Ithaca, NY. https://doi.org/10.2173/bow.grcfly.01.

Miller, K.E., and K.D. Meyer. 2020. Short-tailed Hawk (*Buteo brachyurus*). Version 1.0. In *Birds of the World*, ed. A.F. Poole and F.B. Gill. Cornell Lab of Ornithology, Ithaca, NY. https://doi.org/10.2173/bow.shthaw.01.

Miller, K.E., D.L. Leonard Jr., C.E. Shackelford, R.E. Brown, and R.N. Conner. 2020. Red-bellied Woodpecker (*Melanerpes carolinus*). Version 1.0. In *Birds of the World*, ed. P.G. Rodewald. Cornell Lab of Ornithology, Ithaca, NY. https://doi.org/10.2173/bow.rebwoo.01.

Miller, M.G.R., N. Carlile, J.S. Phillips, F. McDuie, and B.C. Congdon. 2018. Importance of tropical tuna for seabird foraging over a marine productivity gradient. *Marine Ecology Progress Series* 586:233–249.

Miller, M.P., C. Gratto-Trevor, S.M. Haig, D.S. Mizrahi, M.M. Mitchell, and T.D. Mullins. 2013. Population genetics and evaluation of genetic evidence for subspecies in the Semipalmated Sandpiper (*Calidris pusilla*). *Waterbirds* 36(2):166–178.

Miller, M.P., S.M. Haig, C.L. Gratto-Trevor, and T.D. Mullins. 2010. Subspecies status and population genetic structure in Piping Plover (*Charadrius melodus*). *Auk* 127(1):57–71.

Miller, M.P., R.J. Davis, E.D. Forsman, T.D. Mullins, and S.M. Haig. 2018. Isolation by distance versus landscape resistance: Understanding dominant patterns of genetic structure in Northern Spotted Owls (*Strix occidentalis caurina*). *PLOS ONE* 13(3):e0201720.

Miller, M.P., S.M. Haig, T.D. Mullins, L. Ruan, B. Casler, A. Dondua, H.R. Gates, J. M. Johnson, S. Kendall, P.S. Tomkovich, D. Tracy, O.P. Valchuk, and R.B. Lanctot. 2015. Intercontinental genetic structure and gene flow in Dunlin (*Calidris alpina*), a natural vector of avian influenza. *Evolutionary Applications* 8:149–171.

Miller, M.R., J.Y. Takekawa, J.P. Fleskes, D.L. Orthmeyer, M.L. Casazza, and W.M. Perry. 2005. Spring migration of Northern Pintails from California's Central Valley wintering area tracked with satellite telemetry: Routes, timing, and destinations. *Canadian Journal of Zoology* 83:1314–1332.

Miller, T.A., R.P. Brooks, M.J. Lanzone, D. Brandes, J. Cooper, J.A. Tremblay, J. Wilhelm, A. Duerr, and T.E. Katzner. 2016. Limitations and mechanisms influencing the migratory performance of soaring birds. *Ibis* 158:116–134.

Mills, K.L., J.K. Gaydos, C.V. Fiorello, E.R. Whitemer, S. De La Cruz, D.M. Mulcahy, I. Vilchis, and M.H. Ziccardi. 2016. Post-release survival and movement of Western Grebes (*Aechmophorus occidentalis*) implaced with intracoelomic satellite transmitters. *Waterbirds* 39(2):175–186.

Millspaugh, J.J., and J.M. Marzluff. 2001. Radio-tracking and animal populations: Past trends and future needs. In *Radio Tracking and Animal Populations*, 383–393. Academic Press, New York.

Milne, H. 1965. Seasonal movements and distribution of Eiders in northeast Scotland. *Bird Study* 12(3):170–180.

Minami, H., M. Minagawa, and H. Ogi. 1995. Changes in stable carbon and nitrogen isotope ratios in Sooty and Short-Tailed Shearwaters during their northward migration. *Condor* 97(2):565–574. https://doi.org/10.2307/1369042.

Mini, A.E., D.C. Bachman, J. Cocke, K.M. Griggs, K.A. Spragens, and J.M. Black. 2011. Recovery of the Aleutian Cackling Goose *Branta hutchinsii leucopareia*: 10-year review and future prospects. *Wildfowl* 61:3–29.

Mini, A.E., E.R. Harrington, E. Rucker, B.D. Dugger, and T.B. Mowbray. 2020. American Wigeon (*Mareca americana*). Version 1.0. In *Birds of the World*, ed. A.F. Poole. Cornell Lab of Ornithology, Ithaca, NY. https://doi.org/10.2173/bow.amewig.01.

Minton, C., K. Gosbell, P. Johns, M. Christie, J.W. Fox, and V. Afanasyev. 2010. Initial results from light level geolocator trials on Ruddy Turnstone *Arenaria interpres* reveal unexpected migration route. *Wader Study Group Bulletin* 117(1):9–14.

Minton, C., K. Gosbell, P. Johns, M. Christie, M., Klaassen, C. Hassell, A. Boyle, R. Jessop, and J. Fox. 2011. Geolocator studies on Ruddy Turnstones *Arenaria interpres* and Greater Sandplovers *Charadrius leschenaultii* in the East Asian–Australasia Flyway reveal widely different migration strategies. *Wader Study Group Bulletin* 118(2):87–96.

Mississippi Flyway Council. 2013. Management Plan for Midcontinent Cackling Geese in the Mississippi Flyway. US Fish and Wildlife Service, Bloomington, MN.

Mississippi Flyway Council. 2017. A Management Plan for Mississippi Flyway Canada Geese. US Fish and Wildlife Service, Bloomington, MN.

Mitchell, C.D., and M.W. Eichholz. 2020. Trumpeter Swan (*Cygnus buccinator*). Version 1.0. In *Birds of the World*, ed. P.G. Rodewald. Cornell Lab of Ornithology, Ithaca, NY.

Mlodinow, S.G. 2004. Manx Shearwaters in the North Pacific Ocean. *Birding* 36:608–615.

Mlodinow, S.G., J.W. Chardine, R.D. Morris, M. Gochfeld, J. Burger, G.M. Kirwan, and E. Garcia. 2025. Brown Noddy (*Anous stolidus*). Version 2.0. In *Birds of the World*, ed. S.M. Billerman and M.G. Smith. Cornell Lab of Ornithology, Ithaca, NY. https://doi.org/10.2173/bow.brnnod.02.

Mlodinow, S.G., P. Pyle, and M.J. Guzy. 2024. Antillean Nighthawk (*Chordeiles gundlachii*). Version 2.0. In *Birds of the World*, ed. B.K. Keeney. Cornell Lab of Ornithology, Ithaca, NY. https://doi.org/10.2173/bow.antnig.02.

Mlodinow, S.G., E.A. Schreiber, and R.L. Norton. 2024. Brown Booby (*Sula leucogaster*) Version 2.0. In *Birds of the World*, ed. B.K. Keeney and S.M. Billerman. Cornell Lab of Ornithology, Ithaca, NY. https://doi.org/10.2173/bow.brnboo.02.

Moermond, J.E., and M.A. Spindler. 1997. Migration route and wintering area of Tundra Swans *Cygnus columbianus* nesting in the Kobuk-Selawik lowlands, north-west Alaska. *Wildfowl* 48:16–25.

Moiteret, C.S., T.R. Martin, and P.D. Martin. 1996. Predevelopment surveys of nesting birds at two sites in the Kuparuk Oilfield, Alaska, 1988–1992. US Fish and Wildlife Service, Northern Alaska Ecological Services, Fairbanks, AK.

Mojica, E.K., J.M. Meyers, B.A. Millsap, and K.L. Haley. 2008. Migration of Florida subadult Bald Eagles. *Wilson Journal of Ornithology* 120(2):304–310.

Moldenhauer, R.R., and D.J. Regelski. 2020. Northern Parula (*Setophaga americana*). Version 1.0. In *Birds of the World*, ed. A.F. Poole. Cornell Lab of Ornithology, Ithaca, NY. https://doi.org/10.2173/bow.norpar.01.

Moller, H., D. Fletcher, P.N. Johnson, B.D. Bell, D. Flack, C. Bragg, D. Scott, J. Newman, S. McKechnie, and P.O. Lyver. 2009. Changes in Sooty Shearwater (*Puffinus griseus*) abundance and harvesting on the Rakiura Titi Islands. *New Zealand Journal of Zoology* 36:325e341.

Mong, T.W., and B.K. Sandercock. 2007. Optimizing radio retention and minimizing radio impacts in a field study of Upland Sandpipers. *Journal of Wildlife Management* 71:971–980.

Monroy-Ojeda, A., M. Grosselet, G. Ruiz, and E.D. Valle. 2013. Winter site fidelity and winter residency of six migratory Neotropical species in Mexico. *Wilson Journal of Ornithology* 125:192–196.

Montevecchi, W.A., and I.J. Stenhouse. 2020. Dovekie (*Alle alle*). Version 1.0. In *Birds of the World*, ed. S.M. Billerman. Cornell Lab of Ornithology, Ithaca, NY. https://doi.org/10.2173/bow.doveki.01.

Montevecchi, W.A., A. Hedd, L. McFarlane Tranquilla, D.A. Fifield, C.M. Burke, P.M. Regular, G.K. Davoren, S. Garthe, G.J. Robertson, R.A. Phillips. 2012. Tracking seabirds to identify ecologically important and high risk marine areas in the western North Atlantic. *Biological Conservation* 156:62–71. ISSN 0006-3207. https://doi.org/10.1016/j.biocon.2011.12.001.

Montgomerie, R., and B. Lyon. 2020a. McKay's Bunting (*Plectrophenax hyperboreus*). Version 1.0. In *Birds of the World*, ed. A.F. Poole. Cornell Lab of Ornithology, Ithaca, NY. https://doi.org/10.2173/bow.mckbun.01.

———. 2020b. Snow Bunting (*Plectrophenax nivalis*). Version 1.0. In *Birds of the World*, ed. S.M. Billerman, B.K. Keeney, P.G. Rodewald, and T.S. Schulenberg. Cornell Lab of Ornithology, Ithaca, NY. https://doi.org/10.2173/bow.snobun.01.

Monti, F., O. Duriez, V. Arnal, J.M. Dominici, A. Sforzi, L. Fusani, D. Gremillet, and C. Montgelard. 2015. Being cosmopolitan: Evolutionary history and phylogeography of a specialized raptor, the Osprey *Pandion haliaetus*. *BMC Evolutionary Biology* 15:15.

Moody, A., and K.A. Hobson. 2007. Alcid winter diet in the northwest Atlantic determined by stable isotope analysis. *Marine Ornithology* 35:39–46.

Moon, J. I., J.-G. Park, D.-H. Nam, and D.-H. Lee. 2019. Complete mitochondrial genome of the Pacific Loon, *Gavia pacifica* (Gaviiformes: Gaviidae). *Mitochondrial DNA Part B* 4(1):1200–1201. https://doi.org/10.1080/23802359.2019.1591229.

Moore, C.B., D.C. Osborne, L.G. Carlson, R.J. Askren, and M. Brasher. 2023. Distributional shifts of wintering midcontinent Greater White-fronted Geese. *Journal of Wildlife Management* e22401. https://doi.org/10.1002/jwmg.22401.

Moore, J.D., D.E. Anderson, T. Cooper, J.P. Duguay, S.L. Oldenburger, C.A. Steward, and D.G. Krementz. 2021. Migration phenology and patterns of American Woodcock in central North America derived using satellite telemetry. *Wildlife Biology* wlb.00816. https://doi.org/10.2981/wlb.00816.

Moore, J.D., and D.G. Krementz. 2017. Migratory connectivity of American Woodcock using band return data. *Journal of Wildlife Management* 81:1063–1072.

Moran, J.A., L.I. Wassenaar, J.C. Finlay, C. Hutcheson, L.A. Isaac, and S.M. Wethington. 2013. An exploration of migratory connectivity of the Rufous Hummingbird (*Selasphorus rufus*), using feather deuterium. *Journal of Ornithology* 154:423–430.

Morbey, Y.E., C.G. Guglielmo, P.D. Taylor, I. Maggini, J. Deakin, S.A. Mackenzie, J.M. Borwn, and L. Zhao. 2018. Evaluation of sex differences in the stopover behavior and postdeparture movements of wood-warblers. *Behavioral Ecology* 29(1):117–127.

Morera-Pujol, V., P. Catry, M. Magalhães, C.V. Péron, J.M. Reyes-González, J.P. Granadeiro, T. Militão, M.P. Dias, D. Oro, G. Dell'Omo, M. Muller, V.H. Paiva, B. Metzger, V. Neves, J. Navarro, G. Karris, S. Xirouchakis, J.G. Cecere, A. Zamora-López, M.G. Forero, F. De Felipe, Z. Zajková, M. Cruz-Flores, D. Grémillet, J. González-Solís, and R. Ramos. 2023. Methods to detect spatial biases in tracking studies caused by differential representativeness of individuals, populations, and time. *Diversity and Distributions* 29:19–38.

Morera-Pujol, V., P. Catry, M. Magalhães, C.V. Péron, J.M. Reyes-González, J.P. Granadeiro, T. Militão, M.P. Dias, D. Oro, G. Dell'Omo, M. Muller, V.H. Paiva, B. Metzger, V. Neves, J. Navarro, G. Karris, S. Xirouchakis, J.G. Cecere, A. Zamora-López, M.G. Forero, R. Ouni, M.S. Romdhane, F. De Felipe, Z. Zajková, M. Cruz-Flores, D. Grémillet, J. González-Solís, and R. Ramos. 2025. Migratory connectivity and non-breeding habitat segregation across biogeographical scales in three closely related seabird taxa. *Diversity and Distributions* 31(3):e70013. https://doi.org/10.1111/ddi.70013.

Morris-Pocock, J.A., D.J. Anderson, and V.L. Friesen. 2011. Mechanisms of global diversification in the Brown Booby (*Sula leucogaster*) revealed by uniting statistical phylogeographic and multilocus phylogenetic methods. *Molecular Ecology* 20(13):2835–2850.

Morris-Pocock, J.A., T.E. Steeves, F.A. Estela, D.J. Anderson, and V.L. Friesen. 2010. Comparative phylogeography of Brown (*Sula leucogaster*) and Red-footed Boobies (*S. sula*): The influence of physical barriers and habitat preference on gene flow in pelagic seabirds. *Molecular Phylogenetics and Evolution* 54(3):883–896.

Morris-Pocock, J.A., S.A. Taylor, T.P. Birt, M. Damus, J.F. Piatt, K.I. Warheit, and V.L. Friesen. 2008. Population genetic structure in Atlantic and Pacific Ocean Common Murres (*Uria aalge*): Natural replicate tests of post-Pleistocene evolution. *Molecular Ecology* 17(22):4859–4873.

Morrison, J.L., and J.M. Baird. 2016. Using banding and encounter data to investigate movements of Red-Tailed Hawks in the northeastern United States. *Journal of Raptor Research* 50(2):161–175.

Morrison, M.L. 1983. An analysis of geographic variation in the Townsend's Warbler. *Condor* 85:385–391.

———. 1990. Morphological and vocal variation in the Black-throated Gray Warbler in the Pacific Northwest. *Northwestern Naturalist* 71:53–58.

Morrison, R.I.G. 1978. Shorebird banding and color-marking studies in James Bay, 1977. *Wader Study Group Bulletin* 23:36–43.

Morrison, R.I.G., and J.P. Myers. 1987. Wader migration systems in the New World. *Wader Study Group Bulletin* 49(7):57–69.

Morse, D.H., and A.F. Poole. 2020. Black-throated Green Warbler (*Setophaga virens*). Version 1.0. In *Birds of the World*, ed. P.G. Rodewald. Cornell Lab of Ornithology, Ithaca, NY. https://doi.org/10.2173/bow.btnwar.01.

Morten, J.M., P. Buchanan, C. Egevang, I. Glissenaar, S. Maxwell, N. Parr, J. Screen, F. Vigfusdottir, N. Williams, M. Witt, and L. Hawkes. 2023. Global Warming and Arctic Terns: Estimating climate change impacts on the world's longest migration. *Global Change Biology* 29(19):5596–5614.

Morten J.M., J.M. Burgos, L. Collins, S.M. Maxwell, E.-J. Morin, N. Parr, W. Thurston, F. Vigfúsdóttir, M.J. Witt, and L.A. Hawkes. 2022. Foraging behaviours of breeding Arctic Terns *Sterna paradisaea* and the impact of local weather and fisheries. *Frontiers in Marine Science* 8. https://doi.org/10.3389/fmars.2021.760670.

Morten, J.M., A.P. Carneiro, M. Beal, A.S. Bonnet-Lebrun, M.P. Dias, M.M. Rouyer, A.L. Harrison, J. González-Solís, V.R. Jones, V.A. Garcia Alonso, and M. Antolos. 2025. Global marine flyways identified for long-distance migrating seabirds from tracking data. *Global Ecology and Biogeography* 34(2):e70004.

Morton, E.S., and R.D. James. 2020. Blue-headed Vireo (*Vireo solitarius*). Version 1.0. In *Birds of the World*, ed. A.F. Poole. Cornell Lab of Ornithology, Ithaca, NY. https://doi.org/10.2173/bow.buhvir.01.

Mosbech, A., and D. Boertmann 1999. Distribution, abundance, and reaction to aerial surveys of post-breeding King Eiders (*Somateria spectabilis*) in western Greenland. *Arctic* 52(2):188–203.

Mosbech, A., R.S. Danø, F. Merkel, C. Sonne, G. Gilchrist, and A. Flagstad. 2006. Use of satellite telemetry to locate key habitats for King Eiders *Somateria spectabilis* in West Greenland. In *Waterbirds around the World*, ed. G.C. Boere, C.A. Galbraith, and D.A. Stroud, 769–776. Stationery Office, Edinburgh.

Mosbech, A., G. Gilchrist, F. Merkel, C. Sonne, A. Flagstad, and H. Nyegaard. 2006. Year-round movements of Northern Common Eiders *Somateria mollissima borealis* breeding in Arctic Canada and West Greenland followed by satellite telemetry. *Ardea* 94:651–665.

Moskoff, W. 2020. Solitary Sandpiper (*Tringa solitaria*). Version 1.0. In *Birds of the World*, ed. A.F. Poole. Cornell Lab of Ornithology, Ithaca, NY.

Moskoff, W., and L.R. Bevier. 2021. Short-billed Gull (*Larus brachyrhynchus*). Version 1.1. In *Birds of the World*, ed. S.M. Billerman. Cornell Lab of Ornithology, Ithaca, NY. https://doi.org/10.2173/bow.mewgul2.01.1.

Moskoff, W., and R. Montgomerie. 2020. Baird's Sandpiper (*Calidris bairdii*). Version 1.0. In *Birds of the World*, ed. A.F. Poole and F.B. Gill. Cornell Lab of Ornithology, Ithaca, NY. https://doi.org/10.2173/bow.baisan.01.

Mostello, C., I. Nisbet, S. Oswald, and J. Fox. 2014. Non-breeding season movements of six North American Roseate Terns *Sterna dougallii* tracked with geolocators. *Seabird Journal* 1–20. https://doi.org/10.61350/sbj.27.1.

Mostello, C.S., N.A. Palaia, and R.B. Clapp. 2000. Gray-backed Tern (*Onychoprion lunatus*). Version 2.0. In *The Birds of North America*, ed. A.F. Poole and F.B. Gill. Cornell Lab of Ornithology, Ithaca, NY. https://doi.org/10.2173/bna.525.

Motus Wildlife Tracking System. 2023. Birds Canada, Port Rowan, Ontario. https://motus.org.

Moum, T., and E. Árnason. 2001. Genetic diversity and population history of two related seabird species based on mitochondrial DNA control region sequences. *Molecular Ecology* 10(10):2463–2478. https://doi.org/10.1046/j.0962-1083.2001.01375.x. PMID: 11703652.

Mowbray, T.B. 2020a. Canvasback (*Aythya valisineria*). Version 1.0. In *Birds of the World*, ed. A.F. Poole and F.B. Gill. Cornell Lab of Ornithology, Ithaca, NY. https://doi.org/10.2173/bow.canvas.01.

———. 2020b. Northern Gannet (*Morus bassanus*). Version 1.0. In *Birds of the World*, ed. S.M. Billerman. Cornell Lab of Ornithology, Ithaca, NY. https://doi.org/10.2173/bow.norgan.01.

———. 2020c. Scarlet Tanager (*Piranga olivacea*). Version 1.0. In *Birds of the World*, ed. A.F. Poole and F.B. Gill. Cornell Lab of Ornithology, Ithaca, NY. https://doi.org/10.2173/bow.scatan.01.

Mowbray, T.B., F. Cooke, and B. Ganter. 2020a. Snow Goose (*Anser caerulescens*). Version 1.0. In *Birds of the World*, ed. P.G. Rodewald. Cornell Lab of Ornithology, Ithaca, NY. https://doi.org/10.2173/bow.snogoo.01.

Mowbray, T.B., C.R. Ely, J.S. Sedinger, and R.E. Trost. 2020b. Cackling Goose (*Branta hutchinsii*). Version 1.0. In *Birds of the World*, ed. P.G. Rodewald and B.K. Keeney. Cornell Lab of Ornithology, Ithaca, NY. https://doi.org/10.2173/bow.cacgoo1.01.

———. 2020c. Canada Goose (*Branta canadensis*). Version 1.0. In *Birds of the World*, ed. P.G. Rodewald. Cornell Lab of Ornithology, Ithaca, NY. https://doi.org/10.2173/bow.cangoo.01.

Mueller, H. 2020. Wilson's Snipe (*Gallinago delicata*). Version 1.0. In *Birds of the World*, ed. A.F. Poole and F.B. Gill. Cornell Lab of Ornithology, Ithaca, NY.

Mueller, T., R.B. O'Hara, S.J. Converse, R.P. Urbanek, and W.F. Fagan. 2013. Social learning of migratory performance. *Science* 341:999–1002.

Muller, M.J., and R.W. Storer. 2020. Pied-billed Grebe (*Podilymbus podiceps*). Version 1.0. In *Birds of the World*, ed. A.F. Poole and F.B. Gill. Cornell Lab of Ornithology, Ithaca, NY. https://doi.org/10.2173/bow.pibgre.01.

Mumme, R.L., I. Chiver, L.J. Evans Ogden, and B.J. Stutchbury. 2023. Hooded Warbler (*Setophaga citrina*). Version 2.0. In *Birds of the World*, ed. P.G. Rodewald. Cornell Lab of Ornithology, Ithaca, NY. https://doi.org/10.2173/bow.hoowar.02.

Muraro, V., M.S. Mazzochi, A.M.C.R. Fregonezi, and L. Bugoni. 2024. Role of environmental factors in the genetic structure of a highly mobile seabird. *Journal of Biogeography* 51:1933–1943. https://doi.org/10.1111/jbi.14862.

Murphy, R.K., J.R. Dunk, B. Woodbridge, D.W. Stahlecker, D.W. LaPlante, B.A. Millsap, and K.V. Jacobson. 2017. First-year dispersal of Golden Eagles from natal areas in the southwestern United States and implications for second-year settling. *Journal of Raptor Research* 51(3):216–233.

Murphy, S.P., T. Virzi, and F. Sanders. 2017. Exploring differences in adult survival and site fidelity of migratory and non-migratory American Oystercatcher (*Haematopus palliatus*) populations. *Waterbirds* 40(sp1):32–43.

Murray, D.L., and M.R. Fuller. 2000. A critical review of the effects of marking on the biology of vertebrates. In *Research Techniques in Animal Ecology: Controversies and Consequences*, ed. M.C. Pearl, L. Boitani, and T.K. Fuller. Issues, Cases, and Methods in Biodiversity Conservation, 15–64. Columbia University Press, New York.

Myatt, N.A., and D.G. Krementz. 2007a. American Woodcock fall migration using Central Region band-recovery and wing-collection survey data. *Journal of Wildlife Management* 71:336–344.

———. 2007b. Fall migration and habitat use of American Woodcock in the central United States. *Journal of Wildlife Management* 71:1197–1205.

Myers, J.P., and L.P. Myers. 1979. Shorebirds of coastal Buenos Aires Province, Argentina. *Ibis* 121:186–200.

Myers, J.P., J.L. Maron, E. Ortiz, V. Gasto, M. Howe, R.I.G. Morrison, and B. Harrington. 1983. Rationale and suggestions for a hemispheric color-marking scheme for shorebirds: A way to avoid chaos. *Wader Study Group Bulletin* 38:30–32.

Myers, J.P., A.M. Sallaberry, E. Ortiz, G. Castro, L.M. Gordon, J.L. Maron, … and T. Below. 1990. Migration routes of new world sanderlings (*Calidris alba*). *Auk* 107(1):172–180.

Nakash, E., M. Malorodova, L. Howes, and A. Celis-Murillo. 2023. North American Bird Banding Program Dataset 1960–2023. US Geological Survey data release, https://doi.org/10.5066/P97LQNHY.

Naranjo, L.G., et al. 1994. Migration and wintering of Western Sandpipers on the Pacific coast of Colombia (*Migración e Invernada De Calidris Mauri En El Litoral Pacífico De Colombia*). *Journal of Field Ornithology* 65(2):194–200.

Naugler, C.T., P. Pyle, and M.A. Patten. 2020. American Tree Sparrow (*Spizelloides arborea*). Version 1.0. In *Birds of the World*, ed. P.G. Rodewald. Cornell Lab of Ornithology, Ithaca, NY. https://doi.org/10.2173/bow.amtspa.01.

Naves, L.C., J.M. Keating, T.L. Tibbitts, and D. Ruthrauff. 2019. Shorebird subsistence harvest and indigenous knowledge in Alaska: Informing harvest management and engaging users in shorebird conservation. *Condor* 121:1–19.

Nebel, S. 2006. Latitudinal clines in sex ratio, bill, and wing length in Least Sandpipers. *Journal of Field Ornithology* 77(1):39–45.

Nebel, S., and J.M. Cooper. 2020. Least Sandpiper (*Calidris minutilla*). Version 1.0. In *Birds of the World*, ed. A.F. Poole. Cornell Lab of Ornithology, Ithaca, NY. https://doi.org/10.2173/bow.leasan.01.

Nebel, S., D.B. Lank, et al. 2002. Western Sandpipers (*Calidris mauri*) during the non-breeding season: Spatial segregation on a hemispheric scale. *Auk* 119(4):922–928.

Neff, L. 1948. A Russian-banded Emperor Goose killed in California. *Condor* 50:271.

Nelson, A.R., R.L. Cormier, D.L. Humple, J.C. Scullen, R. Sehgal, and N.E. Seavy. 2016. Migration patterns of San Francisco Bay Area Hermit Thrushes differ across a fine spatial scale. *Animal Migration* 3. https://doi.org/10.1515/ami-2016-0001.

Nelson, A.R.P., W.T. Wiskes, and N.E. Seavy. 2021. Wintering destinations of Monterey Hermit Thrushes (*Catharus guttatus slevini*). *Journal of Field Ornithology* 92:43–53. https://doi.org/10.1111/jofo.12357.

Nelson, D.M., M. Braham, T.A. Miller, A.E. Duerr, J. Cooper, M. Lanzone, J. Lemaître, and T. Katzner. 2015. Stable hydrogen isotopes identify leapfrog migration, degree of connectivity, and summer distribution of Golden Eagles in eastern North America. *Condor* 117(3):414–429.

Nelson, S.K. 2020. Marbled Murrelet (*Brachyramphus marmoratus*). Version 1.0. In *Birds of the World*, ed. A.F. Poole and F.B. Gill. Cornell Lab of Ornithology, Ithaca, NY. https://doi.org/10.2173/bow.marmur.01.

Nesbitt, S.A., and L.E. Williams Jr. 1979. Summer range and migration routes of Florida wintering Greater Sandhill Cranes. *Wilson Bulletin* 91(1):137–141.

Nettleship, D.N. 2020. Ruddy Turnstone (*Arenaria interpres*). Version 1.0. In *Birds of the World*, ed. S.M. Billerman. Cornell Lab of Ornithology, Ithaca, NY. https:doi.org/10.2173/bow.rudtur.01.

Nettleship, D.N., and G.M. Kirwan. 2020. Guadalupe Murrelet (*Synthliboramphus hypoleucus*). Version 1.0. In *Birds of the World*, ed. J. del Hoyo, A. Elliott, J. Sargatal, D.A. Christie, and E. de Juana. Cornell Lab of Ornithology, Ithaca, NY. https://doi.org/10.2173/bow.xanmur1.01.

Newson, S.E., et al. 2008. Population decline of Leach's Storm-petrel *Oceanodroma leucorhoa* within the largest colony in Britain and Ireland. *Seabird* 21:77–84.

Newstead, D.J., L.J. Niles, R.R. Porter, A.D. Dey, J. Burger, and O.N. Fitzsimmons. 2013. Geolocation reveals mid-continent migratory routes and Texas wintering areas of Red Knots *Calidris canutus rufa*. *Waders Study Group Bulletin* 120(1):7.

Newton, I. 2023. *The Migration Ecology of Birds*. Elsevier.

Ng, J., M D. Giovanni, M.J. Bechard, J.K. Schmutz, and P. Pyle. 2020. Ferruginous Hawk (*Buteo regalis*). Version 1.0. In *Birds of the World*, ed. P.G. Rodewald. Cornell Lab of Ornithology, Ithaca, NY. https://doi.org/10.2173/bow.ferhaw.01.

Ng, J.W., E.C. Knight, A.L. Scarpignato, A.-L. Harrison, E.M. Bayne, and P.P. Marra. 2018. First full annual cycle tracking of a declining aerial insectivorous bird, the Common Nighthawk (*Chordeiles minor*), identifies migration routes, nonbreeding habitat, and breeding site fidelity. *Canadian Journal of Zoology* 96:869–875.

Nicolai, C.A., P.L. Flint, and M.L. Wege. 2005. Annual survival and site fidelity of Northern Pintails banded on the Yukon-Kuskokwim Delta, Alaska. *Journal of Wildlife Management* 69(3):1202–1210.

Nicoll, M., R.W. Summers, L.G. Underhill, K. Brockie, and R. Rae. 1988. Regional, seasonal, and annual variations in the structure of Purple Sandpiper *Calidris maritima* populations in Britain (United Kingdom). *Ibis* 130:221–233.

Nielsen, A., K.A. Bigelow, M.K. Musyl, and J.R. Sibert. 2006. Improving light-based geolocation by including sea surface temperature. *Fisheries Oceanography* 15(4):314–325.

Niles, L.J., J. Burger, R.R. Porter, A.D. Dey, S. Koch, B. Harrington, ... and M. Boarman. 2012. Migration pathways, migration speeds and non-breeding areas used by Northern Hemisphere wintering Red Knots *Calidris canutus* of the subspecies *rufa*. *Wader Study Group Bulletin* 119(2):195–203.

Niles, L.J., J. Burger, R.R. Porter, A.D. Dey, C.D. Minton, P.M. González, ... and C. Gordon. 2010. First results using light level geolocators to track Red Knots in the Western Hemisphere show rapid and long intercontinental flights and new details of migration pathways. *Wader Study Group Bulletin* 117(2):123–130.

Nisbet, I.C.T., C.S. Mostello, R.R. Veit, J.W. Fox, and V. Afanasyev. 2011. Migrations and winter quarters of five Common Terns tracked using geolocators. *Waterbirds* 34:32–39.

Nisbet, I.C.T., D.V. Weseloh, C.E. Hebert, M.L. Mallory, A.F. Poole, J.C. Ellis, P. Pyle, and M.A. Patten. 2017. Herring Gull (*Larus argentatus*). Version 3.0. In *The Birds of North America*, ed. P.G. Rodewald. Cornell Lab of Ornithology, Ithaca, NY. https://doi.org/10.2173/bna.hergul.03.

Noel, K., N. McLellan, S. Gilliland, K.A. Allard, B. Allen, S. Craik, A. Demagny, M.D. English, A. Diamond, J.-F. Giroux, A. Hanson, H.W. Heusmann, L.E. King, C. Lepage, H. Major, D. McAuley, D. Meattey, G.R. Milton, J. Osenkowski, A. Roberts, G.J. Robertson, M.-C. Roy, L. Savoy, K. Sullivan, and M.L. Mallory. 2021. Expert opinion on American Common Eiders in eastern North America: International information needs for future conservation. *Socio-Ecological Practice Research* 3:153–166.

Nol, E., and M.S. Blanken. 2020. Semipalmated Plover (*Charadrius semipalmatus*). Version 1.0. In *Birds of the World*, ed. A.F. Poole. Cornell Lab of Ornithology, Ithaca, NY.

Nolan Jr., V., E.D. Ketterson, and C.A. Buerkle. 2020a. Prairie Warbler (*Setophaga discolor*). Version 1.0. In *Birds of the World*, ed. A.F. Poole. Cornell Lab of Ornithology, Ithaca, NY. https://doi.org/10.2173/bow.prawar.01.

Nolan Jr., V., E.D. Ketterson, D.A. Cristol, C.M. Rogers, E.D. Clotfelter, R.C. Titus, S.J. Schoech, and E. Snajdr. 2020b. Dark-eyed Junco (*Junco hyemalis*). Version 1.0. In *Birds of the World*, ed. A.F. Poole and F.B. Gill. Cornell Lab of Ornithology, Ithaca, NY. https://doi.org/10.2173/bow.daejun.01.

Norment, C.J., P. Hendricks, and R. Santonocito. 2020a. Golden-crowned Sparrow (*Zonotrichia atricapilla*). Version 1.0. In *Birds of the World*, ed. A.F. Poole and F.B. Gill. Cornell Lab of Ornithology, Ithaca, NY. https://doi.org/10.2173/bow.gocspa.01.

Norment, C.J., S.A. MacDougall-Shackleton, D.J. Watt, P. Pyle, and M.A. Patten. 2020b. Harris's Sparrow (*Zonotrichia querula*). Version 1.0. In *Birds of the World*, ed. P.G. Rodewald. Cornell Lab of Ornithology, Ithaca, NY. https://doi.org/10.2173/bow.harspa.01.

Norris, D.R., P. Arcese, D. Preikshot, D.F. Bertram, and T.K. Kyser. 2007. Diet reconstruction and historic population dynamics in a threatened seabird. *Journal of Applied Ecology* 44(4):875–884.

Norris, D.R., P.P. Marra, G.J. Bowen, L.M. Ratcliffe, J.A. Royle, and T.K. Kyser. 2006. Migratory connectivity of a widely distributed songbird, the American Redstart (*Setophaga ruticilla*). *Ornithological Monographs* 61:14–28. https://doi.org/10.2307/40166836.

Norris, D.R., P.P. Marra, T.K. Kyser, and L.M. Ratcliffe. 2005. Tracking habitat use of a long-distance migratory bird, the American Redstart *Setophaga ruticilla*, using stable-carbon isotopes in cellular blood. *Journal of Avian Biology* 36:164–170.

North, M.R. 2020. Aleutian Tern (*Onychoprion aleuticus*). Version 1.0. In *Birds of the World*, ed. S.M. Billerman. Cornell Lab of Ornithology, Ithaca, NY. https://doi.org/10.2173/bow.aleter1.01.

Northrup, J.M., J.W. Rivers, S.K. Nelson, D.D. Roby, and M.G. Betts. 2018. Assessing the utility of satellite transmitters for identifying nest locations and foraging behavior of the threatened Marbled Murrelet *Brachyramphus marmoratus*. *Marine Ornithology* 46:47–55.

Norton, D.W. 1973. *Ecological Energetics of Calidridine Sandpipers Breeding in Northern Alaska*. University of Alaska Fairbanks.

Nuss A., C.J. Carlos, I.B. Moreno, N.J.R. Fagundes. 2016. Population genetic structure of the Magnificent Frigatebird *Fregata magnificens* (Aves, Suliformes) breeding colonies in the western Atlantic Ocean. *PLOS ONE* 11(2):e0149834. https://doi.org/10.1371/journal.pone.0149834.

Nussbaumer, R., M. Gravey, M. Briedis, and F. Liechti. 2023. Global positioning with animal-borne pressure sensors. *Methods in Ecology and Evolution* 14(4):1104–1117.

O'Connor, C.M., D.R. Norris, G.T. Crossin, and S.J. Cooke. 2014. Biological carryover effects: Linking common concepts and mechanisms in ecology and evolution. *Ecosphere* 5(3):1–11.

O'Hara, P.D., G. Fernández, B. Haase, H. de la Cueva, and D.B. Lank. 2006a. Differential migration in Western Sandpipers with respect to body size and wing length. *Condor* 108(1):225–232.

O'Hara, P.D., K.H. Morgan, and W.J. Sydeman. 2006b. Primary producer and seabird associations with AVHRR-derived sea surface temperatures and gradients in the southeastern Gulf of Alaska. *Deep Sea Research Part II: Topical Studies in Oceanography* 53(3–4):359–369.

Olalla-Kerstupp, A., G. Ruiz-Aymá, J.I. González-Rojas, and A. Guzmán-Velasco. 2015. High fidelity to wintering, stop-over and breeding sites shown by a Long-billed Curlew *Numenius americanus* tracked with satellite telemetry on migratory flights across North America. *Bird Study* 62:556–560.

Oleyar, D., and J. Watson. 2020. HawkWatch International 2019 Migration Network Report.

Oleyar, D., D. Ethier, L. Goodrich, D. Brandes, R. Smith, J. Brown, and J. Sodergren. 2021. *The Raptor Population Index: 2019 Analyses and Assessments*. rpi-project.org.

Oleyar, D., L.J. Goodrich, D. Ethier, D. Brandes, R. Smith, J. Brown, and J. Sodergren. 2023. Thirty years of migration and winter count data indicate regional differences in population trajectories for American Kestrels in North America. *Journal of Raptor Research* 57(2):146–153.

Oliver, R.Y., P.J. Mahoney, E. Gurarie, N. Krikun, B.C. Weeks, M. Hebblewhite, G. Liston, and N. Boelman. 2020. Behavioral responses to spring snow conditions contribute to long-term shift in migration phenology in American Robins. *Environmental Research Letters* 15 045003. https://doi.org/10.1088/1748-9326/ab71a0.

Olivo, C. 2004. Migration patterns of Mississippi Kites in the eastern lowlands of Bolivia. *Ornithologia Neotropical* 15(suppl.):261–266.

Olson, B.E., K.A. Sullivan, and A.H. Farmer. 2014. Marbled Godwit migration characterized with satellite telemetry. *Condor* 116(2):185–194.

Olson, C.R., and T.E. Martin. 2020. Virginia's Warbler (*Leiothlypis virginiae*). Version 1.0. In *Birds of the World*, ed. A.F. Poole and F.B. Gill. Cornell Lab of Ornithology, Ithaca, NY. https://doi.org/10.2173/bow.virwar.01.

Olson, D., J. Warren, and T. Reed. 2009. Satellite-tracking the seasonal locations of Trumpeter Swans *Cygnus buccinators* from Red Rock Lakes National Wildlife Refuge, Montana. *Wildfowl* 59:3–16.

Olson, S.L., and P.J. Hearty. 2003. Probable extirpation of a breeding colony of Short-tailed Albatross (*Phoebastria albatrus*) on Bermuda by Pleistocene sea-level rise. *Proceedings of the National Academies of Science* 100(22):12825–12829. https://doi.org/10.1073/pnas.1934576100.

Oomen, R.A., M.W. Reudink, J.J. Nocera, C.M. Somers, M.C. Green, and C.J. Kyle. 2011. Mitochondrial evidence for panmixia despite perceived barriers to gene flow in a widely distributed waterbird. *Journal of Heredity* 102(5):584–592.

Oppel, S., and A.N. Powell. 2008. Assigning King Eiders to wintering regions in the Bering Sea using stable isotopes of feathers and claws. *Marine Ecology Progress Series* 373:149–156.

Oppel, S., D.L. Dickson, and A.N. Powell. 2009. International importance of the eastern Chukchi Sea as a staging area for migrating King Eiders. *Polar Biology* 32:775–783.

Oppel, S., A.N. Powell, and D.L. Dickson. 2008. Timing and distance of King Eider migration and winter movements. *Condor* 110(2):296–305.

Orben, R.A., D.B. Irons, R. Paredes, D.D. Roby, R.A. Phillips, and S.A. Shaffer. 2015a. North or south? Niche separation of endemic Red-legged Kittiwakes and sympatric Blacklegged Kittiwakes during their non-breeding migrations. *Journal of Biogeography*. 42:401–412.

Orben, R.A., N. Kokubun, A.B. Fleishman, A.P. Will, et al. 2018. Persistent annual migration patterns of a specialist seabird. *Marine Ecology Progress Series* 593:231–245. https://doi.org/10.3354/meps12459.

Orben, R.A., A.J. O'Connor, R.M. Suryan, et al. 2018. Ontogenetic changes in at-sea distributions of immature short-tailed albatrosses *Phoebastria albatrus*. *Endangered Species Research* 35, 23–37.

Orben, R.A., R. Paredes, D.D. Roby, D.B. Irons, and S.A. Shaffer. 2015b. Body size affects individual winter foraging strategies of Thick-billed Murres in the Bering Sea. *Journal of Animal Ecology* 84(6), pp.1589–1599.

———. 2015c. Wintering North Pacific Black-legged Kittiwakes balance spatial flexibility and consistency. *Movement Ecology* 3, pp.1–14.

Oring, L.W., and W.M. Davis. 1966. Shorebird migration at Norman, Oklahoma: 1961–63. *Wilson Bulletin* 78:166–174.

Ornithological Society of Japan. 2000. *Check-list of Japanese Birds*. 6th rev. ed. Ornithological Society of Japan, Tokyo.

Ortega, B. 2010. Capture rates of shorebirds at managed and riverine freshwater wetlands near the central Texas coast. *Texas Ornithological Society* 43:25.

Oswald, J.A., M.G. Harvey, R.C. Remsen, D.U. Foxworth, S.W. Cardiff, D.L. Dittmann, L.C. Megna, M.D. Carling, and R.T. Brumfield. 2016. Willet be one species or two? A genomic view of the evolutionary history of *Tringa semipalmata*. *Auk* 133(4):593–614.

Oyler-McCance, S.J., F.A. Ransler, L.K. Berkman, and T.W. Quinn. 2007. A rangewide population genetic study of Trumpeter Swans. *Conservation Genetics* 8:1339–1353.

Oyler-McCance, S.J., J. St. John, F.L. Knopf, and T.W. Quinn. 2005. Population genetic analysis of Mountain Plover using mitochondrial DNA sequence data. *Condor* 107:354–362.

Oyler-McCance, S.J., J. St. John, R.F. Kysela, and F.L. Knopf. 2008. Population structure of Mountain Plover as determined using nuclear microsatellites. *Condor* 110:493–499.

Pacific Flyway Council. 1991. Pacific Flyway management plan for the Tule Greater White-fronted Goose. US Fish and Wildlife Service.

———. 2015. Pacific Flyway management plan for the Dusky Canada Goose. US Fish and Wildlife Service, Division of Migratory Bird Management, Vancouver, WA.

———. 2016. Management plan for the Cackling Canada Goose. US Fish and Wildlife Service, Vancouver, WA.

Page, G. 1974. Molt of wintering Least Sandpipers. *Bird Banding* 45:93–105.

Page, G.W., L.E. Stenzel, J.S. Warriner, J.C. Warriner, and P.W. Paton. 2020. Snowy Plover (*Charadrius nivosus*). Version 1.0. In *Birds of the World*, ed. A.F. Poole. Cornell Lab of Ornithology, Ithaca, NY.

Page, G.W., M.A. Stern, and P.W.C. Paton. 1995. Differences in wintering areas of Snowy Plovers from inland breeding sites in western North America. *Condor* 97:258–262.

Page, G.W., N. Warnock, T.L. Tibbitts, D. Jorgensen, C.A. Hartman, and L.E. Stenzel. 2014. Annual migratory patterns of Long-billed Curlews in the American West. *Condor* 116(1):50–61.

Pagenkopp, K.M., J. Klicka, K.L. Durrant, et al. 2008. Geographic variation in malarial parasite lineages in the Common Yellowthroat (*Geothlypis trichas*). *Conservation Genetics* 9:1577–1588. https://doi.org/10.1007/s10592-007-9497-6.

Palumbo, M.D., J.W. Kusack, D.C. Tozer, S.W. Meyer, C. Roy, and K.A. Hobson. 2020. Source areas of Blue-winged Teal harvested in Ontario and prairie Canada based on stable isotopes: Implications for sustainable management. *Journal of Field Ornithology* 91:64–76.

Pandolfino, E.R., E.M. Ammon, and K.W. Sockman. 2023. Lincoln's Sparrow (*Melospiza lincolnii*). Version 2.0. In *Birds of the World*, ed. P.G. Rodewald. Cornell Lab of Ornithology, Ithaca, NY. https://doi.org/10.2173/bow.linspa.02.

Panjabi, A.O., A.E. Shaw, P.J. Blancher, K.V. Rosenberg, M.A. Gahbauer, B. Bateman, A. Smith, D.W. Demarest, W.E. Easton, R. Dettmers, and T. Will. 2024. *Avian Conservation Assessment Database Handbook*. Version 2024. Partners in Flight Technical Series No. 8.3. http://pif.birdconservancy.org/acad.handbook.pdf.

Paprocki, N., J. Kidd, R. Warne, A. Macedo, and C.J. Conway. 2025. Causes of differential migration distance: Test of seven mechanistic hypotheses in an arctic raptor. *Behavioral Ecology* 36(4):araf072. https://doi-org.proxy.library.georgetown.edu/10.1093/beheco/araf072.

Paprocki, N., D. Oleyar, D. Brandes, L. Goodrich, T. Crewe, and S. W. Hoffman. 2017. Combining migration and wintering counts to enhance understanding of population change in a generalist raptor species, the North American Red-tailed Hawk. *Condor* 119(1):98–107.

Parker, J.W. 2020. Mississippi Kite (*Ictinia mississippiensis*). Version 1.0. In *Birds of the World*, ed. A.F. Poole and F.B. Gill. Cornell Lab of Ornithology, Ithaca, NY. https://doi.org/10.2173/bow.miskit.01.

Parmelee, D.F. 2020. White-rumped Sandpiper (*Calidris fuscicollis*). Version 1.0. In *Birds of the World*, ed. A.F. Poole, P.R. Stettenheim, and F.B. Gill. Cornell Lab of Ornithology, Ithaca, NY.

Parsons, K.C., and T.L. Master. 2020. Snowy Egret (*Egretta thula*). Version 1.0. In *Birds of the World*, ed. A.F. Poole and F.B. Gill. Cornell Lab of Ornithology, Ithaca, NY. https://doi.org/10.2173/bow.snoegr.01.

Paruk, J.D., T.J. Cade, E.C. Atkinson, P. Pyle, and M.A. Patten. 2020. Northern Shrike (*Lanius borealis*). Version 1.0. In *Birds of the World*, ed. S.M. Billerman. Cornell Lab of Ornithology, Ithaca, NY. https://doi.org/10.2173/bow.norshr4.01.

Paruk, J.D., M.D. Chickering, D. Long, H. Uher-Koch, A. East, D. Poleschook, V. Gumm, W. Hanson, E.M. Adams, K.A. Kovach, and D.C. Evers. 2015. Winter site fidelity and winter movements in Common Loons (*Gavia immer*) across North America. *Condor* 117(4):485–493.

Paruk, J.D., D.C. Evers, J.W. McIntyre, J.F. Barr, J. Mager, and W.H. Piper. 2021. Common Loon (*Gavia immer*). Version 2.0. In *Birds of the World*, ed. P.G. Rodewald and B.K. Keeney. Cornell Lab of Ornithology, Ithaca, NY. https://doi.org/10.2173/bow.comloo.02.

Paruk, J.D., D. Long, S.L. Ford, and D.C. Evers. 2014. Common Loons (*Gavia immer*) wintering off the Louisiana coast tracked to Saskatchewan during the breeding season. *Waterbirds* 37(sp1):47–52.

Paton, P.W.C., P. Loring, G. Cormons, K. Meyer, S. Williams, and L. Welch. 2021. Fate of Common (*Sterna hirundo*) and Roseate Terns (*S. dougallii*) with satellite transmitters attached with backpack harnesses. *Waterbirds* 43. https://doi.org/10.1675/063.043.0315.

Paton, P.W.C., C. Kneedy, and E. Sorensen. 1992. Chronology of shorebird and ibis use of selected marshes at Great Salt Lake. *Utah Birds* 8:1–19.

Patten, M.A. 2001. The roles of habitat and signaling in speciation: Evidence from a contact zone of two Song Sparrow subspecies. PhD diss., University of California, Riverside.

Patten, S.M. 1980. Interbreeding and evolution in the *Larus glaucescens–Larus argentatus* complex of the south coast of Alaska. PhD diss., Johns Hopkins University, Baltimore, MD.

Paullin, D.G., and E. Kridler. 1988. Spring and fall migration of Tundra Swans dyed at Malheur National Wildlife Refuge, Oregon. *Murrelet* 69:1–9.

Paulson, D.R. 1993. *Shorebirds of the Pacific Northwest*. University of Washington Press, Seattle.

Pavlovic, E.M. 2022. Estimating natal origins and migratory patterns of juvenile raptors banded during fall migration at Hawk Ridge Duluth, MN. PhD diss., University of Minnesota.

Paxton, E.H., P. Unitt, M.K. Sogge, M. Whitfield, and P. Keim. 2011. Winter distribution of Willow Flycatcher subspecies. *Condor* 113(3):608–618.

Paxton, K.L., and F.R. Moore. 2017. Connecting the dots: Stopover strategies of an intercontinental migratory songbird in the context of the annual cycle. *Ecology and Evolution* 7:6716–6728.

Paxton, K.L., and C. Van Riper III. 2006. Spatial and temporal migration patterns of neotropical migrants in the Southwest revealed by stable isotopes. Tucson, US Geological Survey, Southwest Biological Science Center, Sonoran Desert Research Station, University of Arizona.

Paxton, K.L., C. Van Riper, T.C. Theimer, and E.H. Paxton. 2007. Spatial and temporal migration patterns of Wilson's Warbler (*Wilsonia pusilla*) in the southwest as revealed by stable isotopes. *Auk* 124:162–175.

Paxton, K.L., M. Yau, F.R. Moore, and D.E. Irwin. 2013. Differential migratory timing of western populations of Wilson's Warbler (*Cardellina pusilla*) revealed by mitochondrial DNA and stable isotopes. *Auk* 130:689–698.

Payne, L.X., and E.P. Pierce. 2020. Purple Sandpiper (*Calidris maritima*). Version 1.0. In *Birds of the World*, ed. S.M. Billerman. Cornell Lab of Ornithology, Ithaca, NY.

Payne, R.B. 2020. Indigo Bunting (*Passerina cyanea*). Version 1.0. In *Birds of the World*, ed. A.F. Poole. Cornell Lab of Ornithology, Ithaca, NY. https://doi.org/10.2173/bow.indbun.01.

Pearce, J.M., and M. R. Petersen. 2009. Post-fledging movements of juvenile Common Mergansers (*Mergus merganser*) in Alaska as inferred by satellite telemetry. *Waterbirds* 32:133–137.

Pearce, J.M., P. Blums, and M.S. Lindberg. 2008. Site fidelity is an inconsistent determinant of population structure in the Hooded Merganser (*Lophodytes cucullatus*): Evidence from genetic, mark-recapture, and comparative data. *Auk* 125(3):711–722.

Pearce, J.M., J.M. Eadie, J-P.L. Savard, T.K. Christensen, J. Berdeen, E.J. Taylor, S. Boyd, A. Einarsson, and S.L. Talbot. 2014. Comparative population structure of cavity-nesting sea ducks. *Auk* 131(2):195–207.

Pearce, J.M., P.L. Flint, M.E. Whalen, S.A. Sonsthagen, J. Stiller, V.P. Patil, T. Bowman, S. Boyd, S.S. Badzinski, H.G. Gilchrist, S.G. Gilliland, C. Lepage, P. Loring, D. McAuley, N.R. McLellan, J. Osenkowski, E.T. Reed, A.J. Roberts, M.O. Robertson, T. Rothe, D.E. Safine, E.D. Silverman, and K. Spragens. 2019. Visualizing populations of North American sea ducks: Maps to guide research and management planning. US Geological Survey Open File Report 2019–1142. https://pubs.usgs.gov/publication/ofr20191142.

Pearce, J., M.L. Mallory, and K. Metz. 2020. Common Merganser (*Mergus merganser*). Version 1.0. In *Birds of the World*, ed. S.M. Billerman. Cornell Lab of Ornithology, Ithaca, NY. https://doi.org/10.2173/bow.commer.01.

Pearce, J.M., K.G. McCracken, T.K. Christensen, and Y.N. Zhuravlev. 2009a. Migratory patterns and population structure among breeding and wintering Red-breasted Mergansers (*Mergus serrator*) and Common Mergansers (*M. merganser*). *Auk* 126(4):784–798.

Pearce, J.M., J.A. Reed, and P.L. Flint. 2005a. Geographic variation in survival and migratory tendency among North American Common Mergansers. *Journal of Field Ornithology* 76:109–118.

Pearce, J.M., S.L. Talbot, M.R. Petersen, and J.R. Rearick. 2005b. Limited genetic differentiation among breeding, molting, and wintering groups of the threatened Steller's Eider: The role of historic and contemporary factors. *Conservation Genetics* 6:743–757.

Pearce, J.M., S.L. Talbot, B.J. Pierson, M.R. Petersen, K.T. Scribner, D. Lynne Dickson, and A. Mosbech. 2004. Lack of spatial genetic structure among nesting and wintering King Eiders. *Condor* 106:229–240.

Pearce, J.M., D. Zwiefelhofer, and N. Maryanski. 2009b. Mechanisms of population heterogeneity among molting Common Mergansers on Kodiak Island, Alaska: Implications for genetic assessments of migratory connectivity. *Condor* 111:283–293.

Pearce, R.L., J.J. Wood, Y. Artukhin, T.P. Birt, M. Damus, and V.L. Friesen. 2002. Mitochondrial DNA suggests high gene flow in Ancient Murrelets. *Condor* 104(1):84–91. https://doi.org/10.1093/condor/104.1.84.

Pearse, A.T., D.A. Brandt, D.M. Baasch, M.T. Bidwell, J.A. Conkin, M.J. Harner, W. Harrell, and K.L. Metzger. 2020. Location data for Whooping Cranes of the Aransas-Wood Buffalo Population, 2009–2018. US Geological Survey data release. https://doi.org/10.5066/P9Y8KZJ9.

Pearse, A.T., D.A. Brandt, W.C. Harrell, K.L. Metzger, D.M. Baasch, and T.J. Hefley. 2015. Whooping Crane stopover site use intensity within the Great Plains. US Geological Survey Open-File Report 2015–1166.

Pearse, A.T., D.A. Brandt, and G.L. Krapu. 2017. Sandhill Crane locations, autumn 2013 migration. US Geological Survey data release. https://doi.org/10.5066/F7F76BGR.

Pearse, A.T., G.L. Krapu, and D.A. Brandt. 2017. Sandhill Crane roost selection, human disturbance, and forage resources. *Journal of Wildlife Management* 81(3):477–486.

Pearse, A.T., M. Rabbe, L.M. Juliusson, M.T. Bidwell, L. Craig-Moore, D.A. Brandt, and W. Harrell. 2018. Delineating and identifying long-term changes in the Whooping Crane (*Grus americana*) migration corridor. *PLOS ONE* 13(2):e0192737.

Pearson, S.F. 2020. Hermit Warbler (*Setophaga occidentalis*). Version 1.0. In *Birds of the World*, ed. A.F. Poole. Cornell Lab of Ornithology, Ithaca, NY. https://doi.org/10.2173/bow.herwar.01.

Pearson, S.F., I. Keren, P.J. Hodum, et al. 2023. Range-wide changes in the North American Tufted Puffin *Fratercula cirrhata* breeding population over 115 years. *Bird Conservation International* 33:e24. https://doi.org/10.1017/S0959270922000193.

Peck, L.E., M.D. English, G.J. Robertson, S.R. Craik, and M.L. Mallory. 2022. Migration chronology and movements of adult American Black Ducks *Anas rubripes* wintering in Nova Scotia, Canada. *Wildlife Biology* 2022(1):e01000.

Peck-Richardson, A., B. Higgins, J. Soller, J. Lerczak, and R. Orben. 2024a. Cormorant Oceanography Project: Biologging Data from Pelagic Cormorants (*Urile pelagicus*), Middleton Island, AK, 2020. Research Workspace. https://doi.org/10.24431/rw1k8es. Version: https://doi.org/10.24431_rw1k8es_20240604T221827Z.

Peck-Richardson, A., A. Piggott, J. Lerczak, and R. Orben. 2024b. Cormorant Oceanography Project: Biologging Data from Brandt's and Pelagic Cormorants (*Urile penicillatus, U. pelagicus*), Columbia River, 2019–2022. Research Workspace. https://doi.org/10.24431/rw1k8ev. Version: https://doi.org/10.24431_rw1k8ev_20240605T002305Z.

Peer, B.D., and E.K. Bollinger. 2020. Common Grackle (*Quiscalus quiscula*). Version 1.0. In *Birds of the World*, ed. A.F. Poole and F.B. Gill. Cornell Lab of Ornithology, Ithaca, NY. https://doi.org/10.2173/bow.comgra.01.

Pennycuick, C.J., S. Åkesson, and A. Hedenström. 2013. Air speeds of migrating birds observed by ornithodolite and compared with predictions from flight theory. *Journal of the Royal Society Interface* 10(86):20130419. http://doi.org/10.1098/rsif.2013.0419.

Pennycuick, C.J., F.C. Schaffner, M.R. Fuller, H.H. Obrecht, and L. Sternberg. 1990. Foraging flights of the White-tailed Tropicbird *Phaethon lepturus*: Radiotracking and doubly-labelled water. *Colonial Waterbirds* 13:96–102.

Perez, G.E., and K.A. Hobson. 2007. Feather deuterium measurements reveal origins of migratory Western Loggerhead Shrikes (*Lanius ludovicianus excubitorides*) wintering in Mexico. *Diversity and Distributions* 13:166–171.

———. 2009. Winter habitat use by Loggerhead Shrikes (*Lanius Ludovicianus*) in Mexico: Separating migrants from residents using stable isotopes. *Journal of Ornithology* 150:459–467.

Perkins, M. 2007. The use of stable isotopes to determine the ratio of resident to migrant King Rails in southern Louisiana and Texas. Master's thesis, Louisiana State University.

Perkins, M.W., and W.J. Lindsey. 1983. Nesting studies of Ferruginous Hawks in the Ely BLM District, Nevada. *North American Bird Bander* 8(3):106–107.

Perlut, N., and A. Levesque. 2020. Light-level geolocation reveals the migration route and non-breeding location of an Antillean Nighthawk (*Chordeiles gundlachii*). *Journal of Caribbean Ornithology* 33:49–53.

Perlut, N.G. 2018. Prevalent transoceanic fall migration by a 30-gram songbird, the Bobolink. *Auk* 135(4):992–97. https://doi.org/10.1642/AUK-18-56.1.

Perold, V., R.A. Ronconi, C.L. Moloney, B.J. Dilley, M. Connan, and P.G. Ryan. 2024. Little change in plastic loads in South Atlantic seabirds since the 1980s. *Science of the Total Environment* 950:175343. https://doi.org/10.1016/j.scitotenv.2024.175343.

Peters, J.L., and K.E. Omland. 2007. Population structure and mitochondrial polyphyly in North American Gadwalls (*Anas strepera*). *Auk* 124:444–462.

Peters, J.L., K.A. Bolender, and J.M. Pearce. 2012. Behavioural vs. molecular sources of conflict between nuclear and mitochondrial DNA—The role of male-biased dispersal in a Holarctic sea duck. *Molecular Ecology* 21(14):3562–3575.

Peters, J.L., W. Gretes, and K.E. Omland. 2005. Late Pleistocene divergence between eastern and western populations of Wood Ducks (*Aix sponsa*) inferred by the "isolation with migration" coalescent method. *Molecular Ecology* 14:3407–3418.

Petersen, A., D.B. Irons, H.G. Gilchrist, G.J. Robertson, D. Boertmann, H. Strøm, M. Gavrilo, Y. Artukhin, D.S. Clausen, K.J. Kuletz, and M.L. Mallory. 2015. The status of Glaucous Gulls *Larus hyperboreus* in the circumpolar Arctic. *Arctic* 68(1):107–120.

Petersen, M.R., and P.L. Flint. 2002. Population structure of Pacific Common Eiders breeding in Alaska. *Condor* 104:780–787.

Petersen, M.R., J.O. Bustnes, and G.H. Systad. 2006. Breeding and moulting locations and migration patterns of the Atlantic population of Steller's Eiders *Polysticta stelleri* as determined from satellite telemetry. *Journal of Avian Biology* 37(1):58–68.

Petersen, M.R., D.C. Douglas, and D.M. Mulcahy. 1995. Use of implanted satellite transmitters to locate Spectacled Eiders at-sea. *Condor* 97(1):276–278.

Petersen, M.R., J.B. Grand, and C.P. Dau. 2020. Spectacled Eider (*Somateria fischeri*). Version 1.0. In *Birds of the World*, ed. A.F. Poole and F.B. Gill. Cornell Lab of Ornithology, Ithaca, NY. https://doi.org/10.2173/bow.speeid.01.

Petersen, M.R., W.W. Larned, and D.C. Douglas. 1999. At-sea distribution of Spectacled Eiders: A 120-year mystery resolved. *Auk* 116(4):1009–1020.

Petersen, M.R., B.J. McCaffery, and P.L. Flint. 2003. Post-breeding distribution of Long-tailed Ducks *Clangula hyemalis* from the Yukon-Kuskokwim Delta, Alaska. *Wildfowl* 54:103–113.

Petit, L.J. 2020. Prothonotary Warbler (*Protonotaria citrea*). Version 1.0. In *Birds of the World*, ed. A.F. Poole and F.B. Gill. Cornell Lab of Ornithology, Ithaca, NY. https://doi.org/10.2173/bow.prowar.01.

Petrie, S.A., and K.L. Wilcox. 2003. Migration chronology of eastern-population Tundra Swans. *Canadian Journal of Zoology* 81:861–870.

Petrula, M.J., and T.C. Rothe. 2005. Migration chronology, routes, and distribution of Pacific Flyway population Lesser Sandhill Cranes. *Proceedings North American Crane Workshop* 9:53–67.

Phillips, A.R. 1991. *The Known Birds of North and Middle America*, Part 2, *Bombycillidae; Sylviidae to Sturnidae; Vireonidae*. A.R. Phillips, Denver, CO.

Phillips, A.R., J.T. Marshall Jr., and G.B. Monson. 1964. *The Birds of Arizona*. University of Arizona Press, Tucson, AZ.

Phillips, L.M., A.N. Powell, E.J. Taylor, and E.A. Rexstad. 2007. Use of the Beaufort Sea by King Eiders breeding on the North Slope of Alaska. *Journal of Wildlife Management* 71(6):1892–1898.

Phillips, R.A., J.R.D. Silk, J.P. Croxall, V. Afanasyev, and D.R. Briggs. 2004. Accuracy of geolocation estimates for flying seabirds. *Marine Ecology Progress Series* 266:265–272.

Piatt, J.F., and A.S. Kitaysky. 2020. Tufted Puffin (*Fratercula cirrhata*). Version 1.0. In *Birds of the World*, ed. S.M. Billerman. Cornell Lab of Ornithology, Ithaca, NY. https://doi.org/10.2173/bow.tufpuf.01.

Piatt, J.F., D.C. Douglas, M.L. Arimitsu, M.L. Kissling, E.N. Madison, S.K. Schoen, … and G.S. Drew. 2021. Kittlitz's Murrelet seasonal distribution and post-breeding migration from the Gulf of Alaska to the Arctic Ocean. *Arctic* 74(4):482–495.

Piatt, J.F., J.K. Kuletz, A.E. Burger, et al. 2007 Status review of the Marbled Murrelet (*Brachyramphus marmoratus*) in Alaska and British Columbia. US Geological Survey Open-File Report 2006–1387. US Geological Survey, Reston, Virginia.

Picardi, S., R.R. Borkhataria, A.L. Bryan Jr, P.C. Frederick, and M. Basille. 2018. GPS telemetry reveals occasional dispersal of Wood Storks from the southeastern US to Mexico. *Caribbean Naturalist*, special issue 2:23–29.

Picardi, S., P. Frederick, R. Borkhataria, and M. Basille. 2020. Partial migration in a subtropical wading bird in the southeastern United States. *Ecosphere* 11(2):e03054. https://doi.org/10.1002/ecs2.3054.

Pickens, B.A., and B. Meanley. 2020. King Rail (*Rallus elegans*). Version 1.0. In *Birds of the World*, ed. P.G. Rodewald. Cornell Lab of Ornithology, Ithaca, NY. https://doi.org/10.2173/bow.kinrai4.01.

Pierce, A.K., S.J. Dinsmore, D. Jorgensen, and M.B. Wunder. 2017. Migration routes and timing of Mountain Plovers revealed by geolocators. *Journal of Field Ornithology* 88:30–38. https://doi.org/10.1111/jofo.12184.

Pierotti, R.J., C.A. Annett, S.A. Shaffer, and D.M. Devincenzi. 2024. Western Gull (*Larus occidentalis*). Version 2.0. In *Birds of the World*, ed. N.D. Sly and B.K. Keeney. Cornell Lab of Ornithology, Ithaca, NY. https://doi.org/10.2173/bow.wesgul.02.

Piersma, T., and N.C. Davidson. 1992. The migrations and annual cycles of five subspecies of knots in perspective. *Wader Study Group Bulletin* 64(Suppl.):187–197.

Pierson, B.J., J.M. Pearce, S.L. Talbot, G.F. Shields, and K.T. Scribner. 2000. Molecular genetic status of Aleutian Canada Geese from Buldir and the Semidi Islands, Alaska. *Condor* 102:172–180.

Pillar, A.G., P.P. Marra, N.J. Flood, and M.W. Reudink. 2016. Moult migration in Bullock's Orioles (*Icterus bullockii*) confirmed by geolocators and stable isotope analysis. *Journal of Ornithology* 157:265–275.

Pitocchelli, J. 2020a. MacGillivray's Warbler (*Geothlypis tolmiei*). Version 1.0. In *Birds of the World*, ed. A F. Poole. Cornell Lab of Ornithology, Ithaca, NY. https://doi.org/10.2173/bow.macwar.01.

———. 2020b. Mourning Warbler (*Geothlypis philadelphia*). Version 1.0. In *Birds of the World*, ed. P.G. Rodewald. Cornell Lab of Ornithology, Ithaca, NY. https://doi.org/10.2173/bow.mouwar.01.

Pitocchelli, J., J.L. Jones, and D. Jones. 2023. Connecticut Warbler (*Oporornis agilis*). Version 2.0. In *Birds of the World*, ed. P.G. Rodewald. Cornell Lab of Ornithology, Ithaca, NY. https://doi.org/10.2173/bow.conwar.02.

Pitocchelli, J., J.F. Piatt, and M.A. Cronin. 1995. Morphological and genetic divergence among Alaskan populations of *Brachyramphus* murrelets. *Wilson Bulletin* 107:235–250.

Pleasants, B.Y., and D.J. Albano. 2020. Hooded Oriole (*Icterus cucullatus*). Version 1.0. In *Birds of the World*, ed. A.F. Poole and F.B. Gill. Cornell Lab of Ornithology, Ithaca, NY. https://doi.org/10.2173/bow.hooori.01.

Plissner, J.H., S.M. Haig, and L.W. Oring. 2000. Postbreeding movements of American Avocets and implications for wetland connectivity in the western Great Basin. *Auk* 117:290–298.

Poessel, S.A., P.H. Bloom, M.A. Braham, and T.E. Katzner. 2016. Age- and season-specific variation in local and long-distance movement behavior of Golden Eagles. *European Journal of Wildlife Research* 62:377–393.

Poessel, S.A., B.D. Uher-Koch, J.M. Pearce, J.A. Schmutz, A.-L. Harrison, D.C. Douglas, … and T.E. Katzner. 2020. Movements and habitat use of loons for assessment of conservation buffer zones in the Arctic Coastal Plain of northern Alaska. *Global Ecology and Conservation* 22:e00980.

Pogson, T.H. 1987. Sandhill Crane banded in Alaska migrates to California and Oregon. *North American Bird Bander* 12(3):90–92.

Pogson, T.H., D.R. Herter, and T.W. Schlorff. 1988. Sightings of Lesser Sandhill Cranes color-marked in California. *Journal of Field Ornithology* 59(3):252–257.

Pokrovsky, I., T. Curk, A. Dietz, I. Fufachev, O. Kulikova, S. Rossler, and M. Wikelski. 2023. Conserving bird populations in the Anthropocene: The significance of non-breeding movements. *bioRxiv*. https://doi.org/10.1101/2023.03.13.532370.

Poli, C. 2015. Variability in movement patterns and habitat use of two species of Pelecaniformes. PhD diss., Clemson University.

Pollet, I.L., and D. Shutler. 2018. Leach's Storm Petrel *Oceanodroma leucorhoa* population trends on Bon Portage Island, Canada. *Seabird* 31:75–83.

Pollet, I.L., A.L. Bond, A. Hedd, C.E. Huntington, R.G. Butler, and R. Mauck. 2021. Leach's Storm-Petrel (*Hydrobates leucorhous*). Version 1.1. In *Birds of the World*. Cornell Lab of Ornithology, Ithaca, NY. https://doi.org/10.2173/bow.lcspet.01.1.

Pollet, I.L., S.E. Gutowsky, R.T. Alisauskas, et al. 2024. Trends in Arctic-nesting loon (*Gavia* spp.) populations in North America. *Polar Biology* 47:1001–1012. https://doi.org/10.1007/s00300-024-03277-2.

Pollet, I.L., R.A. Ronconi, I.D. Jonsen, M.L. Leonard, P.D. Taylor, and D. Shutler. 2014. Foraging movements of Leach's Storm-Petrels *Oceanodroma leucorhoa* during incubation. *Journal of Avian Biology* 45:305–314.

Pollet, I.L., R.A. Ronconi, M.L. Leonard, and D. Shutler. 2019. Migration routes and stop-over areas of Leach's Storm-Petrels. *Marine Ornithology* 47:55–65.

Pollet, I.L., D. Shutler, J.W. Chardine, and J.P. Ryder. 2020. Ring-billed Gull (*Larus delawarensis*). Version 1.0. In *Birds of the World*, ed. A.F. Poole. Cornell Lab of Ornithology, Ithaca, NY. https://doi.org/10.2173/bow.ribgul.01.

Poole, A.F., P.E. Lowther, J.P. Gibbs, F.A. Reid, and S.M. Melvin. 2020. Least Bittern (*Ixobrychus exilis*). Version 1.0. In *Birds of the World*, ed. A.F. Poole and F.B. Gill. Cornell Lab of Ornithology, Ithaca, NY. https://doi.org/10.2173/bow.comgal1.01.

Poole, A.F., P. Pyle, M.A. Patten, and D.R. Paulson. 2020. Black-bellied Plover (*Pluvialis squatarola*). Version 1.0. In *Birds of the World*, ed. S.M. Billerman. Cornell Lab of Ornithology, Ithaca, NY. https://doi.org/10.2173/bow.bkbplo.01.

Popovic, I., D.P.L. Toews, C.C. Keever, C.T. St. Clair, B.A. Barbaree, G. Fernández, and J. Rourke. 2019. Genotyping-by-sequencing reveals genomic homogeneity among overwintering Pacific Dunlin (*Calidris alpina pacifica*) aggregations along the Pacific coast of North America. *Condor* 121:1–12.

Porneluzi, P., M.A. Van Horn, and T.M. Donovan. 2020. Ovenbird (*Seiurus aurocapilla*). Version 1.0. In *Birds of the World*, ed. A.F. Poole. Cornell Lab of Ornithology, Ithaca, NY. https://doi.org/10.2173/bow.ovenbi1.01.

Porter, R., and P.A. Smith. 2013. Techniques to improve the accuracy of location estimation using light-level geolocation to track shorebirds. *Wader Study Group Bulletin* 120(3):147–158. Also see: https://www.waderstudygroup.org/projects/geolocator-project/maps-library-introduction/arenaria-interpres-rutu-ruddy-turnstone/.

Post, W. 1994. Banding confirmation that some middle Atlantic coast Boat-tailed Grackles visit Florida in the winter. *Florida Field Naturalist* 22:51–52.

Post, W., J.P. Poston, and G.T. Bancroft. 2020. Boat-tailed Grackle (*Quiscalus major*). Version 1.0. In *Birds of the World*, ed. A.F. Poole. Cornell Lab of Ornithology, Ithaca, NY. https://doi.org/10.2173/bow.botgra.01.

Poulin, R.G., L.D. Todd, E.A. Haug, B.A. Millsap, and M.S. Martell. 2020. Burrowing Owl (*Athene cunicularia*). Version 1.0. In *Birds of the World*, ed. A.F. Poole. Cornell Lab of Ornithology, Ithaca, NY.

Powell, A.N., and R.S. Suydam. 2020. King Eider (*Somateria spectabilis*). Version 1.0. In *Birds of the World*, ed. S.M. Billerman. Cornell Lab of Ornithology, Ithaca, NY. https://doi.org/10.2173/bow.kineid.01.

Powers, D.R., R. Partida-Lara, and P.L. Enríquez. 2020. Rivoli's Hummingbird (*Eugenes fulgens*). Version 1.0. In *Birds of the World*, ed. S.M. Billerman. Cornell Lab of Ornithology, Ithaca, NY. https://doi.org/10.2173/bow.maghum1.01.

Powers, K.D., I. Pratte, R.A. Ronconi, S.N. Wong, P.G. Ryan, L.J. Welch, T.L. Silva, K.A. Hatch, A.J. Westgate, and D.N. Wiley. 2022. Age-related interactions with wind during migration support the hypothesis of developmental learning in a migrating long-lived seabird. *Frontiers in Marine Science* 9:938033. https://doi.org/10.3389/fmars.2022.938033.

Powers, K.D., D.N. Wiley, A.J. Allyn, L.J. Welch, and R.A. Ronconi. 2017. Movements and foraging habitats of Great Shearwaters *Puffinus gravis* in the Gulf of Maine. *Marine Ecology Progress Series* 574:211–226. https://doi.org/10.3354/meps12168.

Pratte, I., G.J. Robertson, and M.L. Mallory. 2017 Four sympatrically nesting auks show clear resource segregation in their foraging environment. *Marine Ecology Progress Series* 572:243–254. https://doi.org/10.3354/meps12144.

Preston, C.R., and R.D. Beane. 2020. Red-tailed Hawk (*Buteo jamaicensis*). Version 1.0. In *Birds of the World*, ed. A.F. Poole. Cornell Lab of Ornithology, Ithaca, NY. https://doi.org/10.2173/bow.rethaw.01.

Preuss, N.O. 2001. Hans Christian Cornelius Mortensen: Aspects of his life and of the history of bird ringing. *Ardea* 89(1).

Priestley, L.T., C. Priestley, D.M. Collister, D. Zazelenchuk, and M. Hanneman. 2010. Encounters of Northern Saw-whet Owls (*Aegolius acadius*) from banding stations in Alberta and Saskatchewan, Canada. *Journal of Raptor Research* 44(4):300–310.

Probst, J.R. 1986. A review of factors limiting the Kirtland's Warbler on its breeding grounds. *American Midland Naturalist* 116:87–100.

Pruett, C.L., and K. Winker. 2005. Biological impacts of climatic change on a Beringian endemic: Cryptic refugia in the establishment and differentiation of the rock sandpiper (*Calidris ptilocnemis*). *Climatic Change* 68(1–2):219–240.

Pruitt, M.L., and K.G. Smith. 2016. History and fall migration of Northern Saw-whet Owls (*Aegolius acadicus*) in Arkansas. *Journal of the Arkansas Academy of Science* 70(32):190–198.

Pshenichnikova, O., A. Klenova, N. Konyukhov, V. Zubakin, Y. Artukhin, and C. Schacter. 2021. Signs of genetic and morphometric population differentiation in small planctivorous seabird, Parakeet Auklet *Aethia psittacula*, reinforces linkage between population structure and migratory strategy. *Acta Ornithologica* 56(1):109–124.

Pshenichnikova, O.S., A.V. Klenova, P.A. Sorokin, et al. 2015. The Crested Auklet, *Aethia cristatella* (Alcidae, Charadriiformes), does not vary geographically in genetics, morphology or vocalizations. *Marine Biology* 162:1329–1342. https://doi.org/10.1007/s00227-015-2672-2.

Pugesek, B.H., K.L. Diem, and C.L. Cordes. 1999. Seasonal movements, migration, and range sizes of subadult and adult Bamforth Lake California Gulls. *Waterbirds* 22(1):29–36.

Purcell, J., and A. Brodin. 2007. Factors influencing route choice by avian migrants: A dynamic programming model of Pacific Brant migration. *Journal of Theoretical Biology* 249:804–816.

Pyle, P. 2008. *Identification Guide to North American Birds*. Part 2, *Anatidae to Alcidae*. Slate Creek Press, Bolinas, CA.

Pyle, P., W.A. Leitner, L. Lozano-Angulo, F. Avilez-Teran, H. Swanson, E. Gómez Limón, and M.K. Chambers. 2009. Temporal, spatial, and annual variation in the occurrence of molt-migrant passerines in the Mexican monsoon region. *Condor* 111:583–590. https://doi.org/10.1525/cond.2009.090085.

Quinn, L.R., A.A. Meharg, J.A. van Franeker, I.M. Graham, and P.M. Thompson. 2016. Validating the use of intrinsic markers in body feathers to identify inter-individual differences in non-breeding areas of Northern Fulmars. *Marine Biology* 163:1–12.

Raffaele, H., J. Wiley, O. Garrido, A. Keith, and J. Raffaele. 1998. *A Guide to the Birds of the West Indies*. Princeton University Press, Princeton, NJ.

Raine, A.F., T. Anderson, M. Vynne, S. Driskill1, H. Raine, and J. Adams. 2020. Post-release survival of fallout Newell's Shearwater fledglings from a rescue and rehabilitation program on Kaua'i, Hawai'i. *Endangered Species Research* 43:39–50.

Raine, A.F., C. Gjerdrum, I. Pratte, J. Madeiros, J.J. Felis, and J. Adams. 2021. Marine distribution and foraging habitat highlight potential threats at sea for the endangered Bermuda Petrel *Pterodroma cahow*. *Endangered Species Research* 45:337–356. https://doi.org/10.3354/esr01139.

Rakhimberdiev, E., D.W. Winkler, E. Bridge, N.E. Seavy, D. Sheldon, T. Piersma, and A. Saveliev. 2015. A hidden Markov model for reconstructing animal paths from solar geolocation loggers using templates for light intensity. *Movement Ecology* 3(1):25. https://doi.org/10.1186/s40462-015-0062-5.

Ralston, J., and J.J. Kirchman. 2012. Continent-scale genetic structure in a boreal forest migrant, the Blackpoll Warbler (*Setophaga striata*). *Auk* 129:467–478.

Ramey, A.M., S.A. Hatch, C.A. Ahlstrom, M.L. van Toor, H. Woksepp, J.C. Chandler, J.A. Reed, A.B. Reeves, J. Waldenström, A.B. Franklin, J. Bonnedahl, V.A. Gill, D.M. Mulcahy, and D.C. Douglas. 2020. Tracking data for three large-bodied gull species and hybrids (*Larus* spp.). Version 2.0, March 2021. US Geological Survey data release, https://doi.org/10.5066/P9FZ4OJW.

Ramey, A.M., J.A. Reed, P. Walther, P. Link, J.A. Schmutz, D.C. Douglas, D.E. Stallknect, and C. Soos. 2016. Evidence for the exchange of blood parasites between North America and the Neotropics in Blue-winged Teal (*Anas discors*). *Parasitology Research* 115(10):3923–3939.

Ramey, A.M., C. Soos, P. Link, P. Walther, T.L. Tibbitts, and D.C. Douglas. 2019. Tracking data for Blue-winged Teal (*Anas discors*). Version 1.0, September 2019. US Geological Survey data release, https://doi.org/10.5066/P9Z9BA9F.

Ramírez, I., V.H. Paiva, I. Fagundes, D. Menezes, I. Silva, F.R. Ceia, R.A. Phillips, J.A. Ramos, and S. Garthe. 2016. Ecological consistency of a rare petrel species. *Animal Conservation* 19(2):139–152. https://doi.org/10.1111/acv.12227.

Ramírez, I., V.H. Paiva, D. Menezes, I. Silva, R.A. Phillips, J.A. Ramos, and S. Garthe. 2013. Year-round distribution and habitat preferences of the Bugio petrel. *Marine Ecology Progress Series* 476:269–284.

Ramos, R., N. Carlile, J. Madeiros, et al. 2017. It is the time for oceanic seabirds: Tracking year-round distribution of Gadfly Petrels across the Atlantic Ocean. *Diversity and Distributions* 23:794–805. https://doi.org/10.1111/ddi.12569.

Ramos, R., J. González-Solís, J.P. Croxall, D. Oro, and X. Ruiz. 2009. Understanding oceanic migrations with intrinsic biogeochemical markers. *PLOS ONE* 4(7):e6236. https://doi.org/10.1371/journal.pone.0006236.

Ramos, R., I. Ramírez, V. Paiva, et al. 2016. Global spatial ecology of three closely-related Gadfly Petrels. *Scientific Reports* 6:23447. https://doi.org/10.1038/srep23447.

Ranck, S.C., C.M. Garsvo, D.M. Schwartz, L.M. Reynard, M.J. Kohn, and J.A. Heath. 2023. Sex, body size, and winter weather explain migration strategies in a partial migrant population of American Kestrels. *Ornithology* 140(3):1–10. https://doi.org/10.1093/ornithology/ukad019.

Rappole, J.H., and A.R. Tipton. 1991. New harness design for attachment of radio transmitters to small passerines. *Journal of Field Ornithology* 62(3):335–337.

Rappole, J.H., M.A. Ramos, and K. Winker. 1989. Wintering Wood Thrush movements and mortality in southern Veracruz. *Auk* 106(3):402–410.

Rasmussen, J.L., S.G. Sealy, and R.J. Cannings. 2020. Northern Saw-whet Owl (*Aegolius acadicus*). Version 1.0. In *Birds of the World*, ed. A.F. Poole. Cornell Lab of Ornithology, Ithaca, NY. https://doi.org/10.2173/bow.nswowl.01.

Raybuck, D.W., T.H. Boves, S.H. Stoleson, J.L. Larkin, N.J. Bayly, L.P. Bullock, G.A. George, K.G. Slankard, L.J. Kearns, S. Petzinger, J.J. Cox, and D.A. Buehler. 2022. Cerulean Warblers exhibit parallel migration patterns and multiple migratory stopovers within the Central American isthmus. *Ornithological Applications* 124(4):duac031. https://doi.org/10.1093/ornithapp/duac031.

Raybuck, D.W., J.L. Larkin, S.H. Stoleson, and T.J. Boves. 2024. Use of multiple South American nonbreeding regions and seasonally distinct migration routes by an individual Scarlet Tanager (*Piranga olivacea*). *Wilson Journal of Ornithology* 136(4):464–471. https://doi.org/10.1676/23-00072.

Rayner, M.J., M.E. Hauber, T.E. Steeves, H.A. Lawrence, D.R. Thompson, P.M. Sagar, S. Bury, T.J. Landers, R.A. Phillips, L. Ranjard, and S.A. Shaffer. 2011. Contemporary and historic separation of transhemispheric migration between two genetically distinct seabird populations. *Nature Communications* 2:332.

Redfern, C.P.F. 2021. Pair bonds during the annual cycle of a long-distance migrant, the Arctic Tern (*Sterna paradisaea*). *Avian Research* 12:32. https://doi.org/10.1186/s40657-021-00268-3.

Redfern, C.P.F., and R.M. Bevan. 2020. Use of sea ice by Arctic Terns *Sterna paradisaea* in Antarctica and impacts of climate change. *Journal of Avian Biology* 51:e02318.

Reed, A. 1970. The breeding ecology of the Black Duck in the St. Lawrence estuary. DSc thesis. Universite Laval, Quebec.

Reed, A., M.A. Davison, and D.K. Kraege. 1989a. Segregation of Brent Geese *Branta bernicla* wintering and staging in Puget Sound and the Strait of Georgia. *Wildfowl* 40:22–31.

Reed, A., R. Stehn, and D. Ward. 1989b. Autumn use of Izembek Lagoon, Alaska, by Brant from different breeding areas. *Journal of Wildlife Management* 53(3):720–725.

Reed, E.T., J. Bêty, J. Mainguy, G. Gauthier, and J-F. Giroux. 2003. Molt migration in relation to breeding success in Greater Snow Geese. *Arctic* 56(1):76–81.

Reed, E.T., E.G. Cooch, R.I. Goudie, and F. Cooke. 1998. Site fidelity of Black Brant wintering and spring staging in the Strait of Georgia, British Columbia. *Condor* 100:426–437.

Reed, E.T., K.J. Kardynal, J.A. Horrocks, and K.A. Hobson. 2018. Shorebird hunting in Barbados: Using stable isotopes to link the harvest at a migratory stopover site with sources of production. *Condor* 120(2):357–370. https://doi.org/10.1650/CONDOR-17-127.1.

Reed, J.M., R.C. Fleischer, J. Eberhard, and L.W. Oring. 1996. Minisatellite DNA variability in two populations of Spotted Sandpipers *Actitis macularia* in Minnesota. *Wader Study Group Bulletin* 79:115–117.

Reed, J.M., L.W. Oring, and E.M. Gray. 2020. Spotted Sandpiper (*Actitis macularius*). Version 1.0. In *Birds of the World*, ed. A.F. Poole. Cornell Lab of Ornithology, Ithaca, NY. https://doi.org/10.2173/bow.sposan.01.

Reese, J., C. Viverette, C.M. Tonra, N.J. Bayly, T.J. Boves, E. Johnson, M. Johnson, P. Marra, E.M. Ames, A. Caguazango, M. DeSaix, A. Matthews, A. Molina, K. Percy, M.C. Slevin, and L. Bulluck. 2019. Using stable isotopes to estimate migratory connectivity for a patchily distributed, wetland-associated Neotropical migrant. *Condor* 121(4):1–15. https://doi.org/10.1093/condor/duz052.

Regehr, H.M. 2003. Survival and movement of postfledging juvenile Harlequin Ducks. *Wilson Bulletin* 115:423–430.

———. 2011. Movement rates and distances of wintering Harlequin Ducks: Implications for population structure. *Waterbirds* 34:19–31.

Regelski, D.J., and R.R. Moldenhauer. 2020. Tropical Parula (*Setophaga pitiayumi*). Version 1.0. In *Birds of the World*, ed. A.F. Poole. Cornell Lab of Ornithology, Ithaca, NY. https://doi.org/10.2173/bow.tropar.01.

Regosin, J.V. 2020. Scissor-tailed Flycatcher (*Tyrannus forficatus*). Version 1.0. In *Birds of the World*, ed. A.F. Poole. Cornell Lab of Ornithology, Ithaca, NY. https://doi.org/10.2173/bow.sctfly.01.

Reneerkens, J., A. Benhoussa, H. Boland, M. Collier, K. Grond, K. Günther, G.T. Hallgrímsson, J. Hansen, W. Meissner, B. de Meulenaer, Y. Ntiamoa-Baidu, T. Piersma, M. Poot, M. van Roomen, R.W. Summers, P.S. Tomkovich, and L.G. Underhill. 2009. Sanderlings using African-Eurasian flyways: A review of current knowledge. *Wader Study Group Bulletin* 116:2–20.

Renfrew, R., D. Kim, N.G. Perlut, and M.D. Cadman. 2019. Migration tactics of a long-distance migratory songbird from across a continental breeding range. *Wilson Journal of Ornithology* 131(4):735–49.

Renfrew, R.B., D. Kim, N. Perlut, J. Smith, J. Fox, and P.P. Marra. 2013. Phenological matching across hemispheres in a long-distance migratory bird. *Diversity and Distributions* 19(8):1008–1019.

Renner, H.M., M.D. Romano, M. Renner, S. Pyare, M.I. Goldstein, and Y. Arthukin. 2015. Assessing the breeding distribution and population trends of the Aleutian Tern *Onychoprion aleuticus*. *Marine Ornithology* 43(2):179–187.

Rense, W.A., 1946. Astronomy and Ornithology. *Popular Astronomy* 54:55.

Reppert, S.M., and J.C. de Roode. 2018. Demystifying Monarch Butterfly migration. *Current Biology* 28(17):R1009–R1022.

Restani, M., and E.M. Madden 2005. Movements of American White Pelicans breeding at Medicine Lake National Wildlife Refuge, Montana. Project Report to US Fish and Wildlife Service.

Reudink, M.W., C.J. Kyle, A.E. McKellar, C.M. Somers, R.L.F. Reudink, T.K. Kyser, S.E. Franks, and J.J. Nocera. 2016. Linking isotopes and panmixia: High within-colony variation in feather δ^2H, δ^{13}C, and δ^{15}N across the range of the American White Pelican. *PLOS ONE* 11(3):e0150810.

Reudink, M.W., C.J. Kyle, J.J. Nocera, R.A. Oomen, M.C. Green, and C.M. Somers. 2011. Panmixia on a continental scale in a widely-distributed waterbird. *Biological Journal of the Linnaean Society* 102:583–592

Reudink, M.W., S.L. Van Wilgenburg, L.S. Steele, A.G Pillar, P.P. Marra, and A.E. McKellar. 2015. Patterns of migratory connectivity in Vaux's Swifts at a northern migratory roost: A multi-isotope approach. *Condor* 117:670–682.

Reynolds, J.D., and F. Cooke. 1988. The influence of mating systems on philopatry: A test with polyandrous Red-necked Phalaropes. *Animal Behaviour* 36(6):1788–1795.

Reynolds, S.J., C.P. Wearn, B.J. Hughes, R.C. Dickey, L.J. Garrett, S. Walls, F.T. Hughes, N. Weber, S.B. Weber, E.H. Leat, and K. Andrews. 2021. Year-round movements of Sooty Terns (*Onychoprion fuscatus*) nesting within one of the Atlantic's largest marine protected Areas. *Frontiers in Marine Science* 8:744506.

Reynolds, T.D., T.D. Rich, and D.A. Stephens. 2020. Sage Thrasher (*Oreoscoptes montanus*). Version 1.0. In *Birds of the World*, ed. A.F. Poole and F.B. Gill. Cornell Lab of Ornithology, Ithaca, NY. https://doi.org/10.2173/bow.sagthr.01.

Rhodes, O.E., L.M. Smith, and R.K. Chesser. 1993. Temporal components of genetic variation in migrating and wintering American Wigeon. *Canadian Journal of Zoology* 71:2229–2235.

Rhymer, J., D. McAuley, and H.L. Ziel. 2005. Phylogeography of the American Woodcock (*Scolopax minor*): Are management units based on band recovery data reflected in genetically based management units? *Auk* 122(4):1149–1160.

Rich, S. 2019. Migration patterns of flammulated owls (*Psiloscops flammeolus*) using light-level geolocators. Master's thesis, California State University.

Ridgely, R.S., and J. Gwynne. 1989. *A Guide to the Birds of Panama, with Costa Rica, Nicaragua, and Honduras*. 2nd ed. Princeton University Press, Princeton, NJ.

Riley, T.Z., and B.A. Fistler. 1992. Necklace radio transmitter attachment for pheasants. *Journal of the Iowa Academy of Science* 99(2–3):65–66.

Rimmer, C.C., and C.H. Darmstadt. 1996. Non-breeding site fidelity in Northern Shrikes. *Journal of Field Ornithology* 67(3):360–366.

Rimmer, C.C., and K.P. McFarland. 2020. Tennessee Warbler (*Leiothlypis peregrina*). Version 1.0. In *Birds of the World*, ed. A.F. Poole. Cornell Lab of Ornithology, Ithaca, NY. https://doi.org/10.2173/bow.tenwar.01.

———. 2001. Known breeding and wintering sites of a Bicknell's Thrush. *Wilson Bulletin* 113:234–236.

Rising, J.D., and N.J. Flood. 2020. Baltimore Oriole (*Icterus galbula*). Version 1.0. In *Birds of the World*, ed. P.G. Rodewald. Cornell Lab of Ornithology, Ithaca, NY. https://doi.org/10.2173/bow.balori.01.

Ritz, M.S. 2009. Speciation and hybridisation in skuas (*Catharacta* spp.). PhD diss., Friedrich Schiller University, Jena, Thuringia, Germany. https://www.db-thueringen.de/receive/dbt_mods_00012901.

Ritz, M.S., C. Millar, G.D. Miller, R.A. Phillips, P. Ryan, V. Sternkopf, D. Liebers-Helbig, and H.U. Peter. 2008. Phylogeography of the southern skua complex: Rapid colonization of the Southern Hemisphere during a glacial period and reticulate evolution. *Molecular Phylogenetics and Evolution* 49(1):292–303.

Rizzolo, D.J. 2017. Contrasting diet, growth, and energy provisioning in loons breeding sympatrically in the Arctic. PhD diss., University of Alaska Fairbanks.

Rizzolo, D.J., C.E. Gray, J.A. Schmutz, J.F. Barr, C. Eberl, and J.W. McIntyre. 2020. Red-throated Loon (*Gavia stellata*). Version 2.0. In *Birds of the World*, ed. P.G. Rodewald and B.K. Keeney. Cornell Lab of Ornithology, Ithaca, NY. https://doi.org/10.2173/bow.retloo.02.

Robb, J.R. 2002. Band recovery and recapture rates of American Black Ducks and Mallards. *Journal of Wildlife Management* 66:153–161.

Robert, M. 2019. Barrow's Goldeneye. In *Second Atlas of the Breeding Birds of Southern Québec*, ed. M. Robert, M.-H. Hachey, D. Lepage, and A.R. Couturier, 136–137. Regroupement QuébecOiseaux, Canadian Wildlife Service (Environment and Climate Change Canada), and Bird Studies Canada, Montréal.

Robert, M., R. Benoit, and J.P.L. Savard. 2002. Relationship among breeding, molting, and wintering areas of male Barrow's Goldeneyes (*Bucephala islandica*) in eastern North America. *Auk* 119:676–684.

Robert, M., G.H. Mittelhauser, B. Jobin, G. Fitzgerald, and P. Lamothe. 2008. New insights on Harlequin Duck population structure in eastern North America as revealed by satellite telemetry. *Waterbirds* 31(sp2):159–172. https://doi.org/10.1675/1524-4695-31.sp2.159.

Roberto-Charron, A. 2018. Investigating methods of geolocator analysis in songbird migration research and their application to the study of a threatened, neotropical songbird. Master's thesis, University of Manitoba.

Roberto-Charron, A., J. Kennedy, L Reitsma, J.L. Tremblay, R. Krikun, K.A. Hobson, J. Ibazarbal, and K.C. Fraser. 2020. Widely distributed breeding populations of Canada Warbler (*Cardellina canadensis*) converge on migration through Central America. *BMC Zoology* 5:10.

Roberts, A., A.L. Scarpignato, A. Huysman, J.A. Hostetler, and E.B. Cohen. 2023. Migratory connectivity of North American waterfowl across administrative flyways. *Ecological Applications* 33(3):e2788.

Roberts, A.J., and M.R. Conover. 2015. Breeding origins of Northern Shovelers (*Anas clypeata*) wintering on the Great Salt Lake, Utah. *Wilson Journal of Ornithology* 127(2):233–238.

Robertson, G.J. 2002. Current status of the Manx Shearwater (*Puffinus puffinus*) colony on Middle Lawn Island, Newfoundland. *Northeastern Naturalist* 9(3):317–324.

———. 2016. Temporal comparisons of encounters of Herring Gulls (*Larus argentatus*) banded as chicks in eastern Newfoundland, Canada. *Waterbirds* 39(sp1):159–165.

Robertson, G.J., and R.I. Goudie. 2020. Harlequin Duck (*Histrionicus histrionicus*). Version 1.0. In *Birds of the World*, ed. S.M. Billerman. Cornell Lab of Ornithology, Ithaca, NY. https://doi.org/10.2173/bow.harduc.01.

Robertson G.J, M. Tomlik, G.R. Milton, G.J. Parsons, and M.L. Mallory. 2017. Increases in the number of American Black Ducks wintering in Nova Scotia, 1970–2015. *Journal of Fish and Wildlife Management* 8:669–675, e1944-687X. https://doi.org/10.3996/032017-JFWM-031.

REFERENCES

Robertson, G.J., S.N.P. Wong, M.D. Tomlik, G.R. Milton, G.J. Parsons, and M.L. Mallory. 2021. Common Eider wintering trends in Nova Scotia, 1970–2019. *Journal of Fish and Wildlife Management* 12:565–571.

Robertson, W.J. 1969. Transatlantic migration of juvenile Sooty Terns. *Nature* 222(5194):632–634.

Robillard, A., G. Gauthier, J-F. Therrien, and J. Bêty. 2018. Wintering space use and site fidelity in a nomadic species, the Snowy Owl. *Journal of Avian Biology* 49(5):jav-01707.

Robinson, J.A., and L.W. Oring. 1996. Long-distance movements by American Avocets and Black-necked Stilts. *Journal of Field Ornithology* 67:307–320.

Robinson, J.A., G.A. Gudmundsson, and P. Clausen. 2006. Flyways of the East Canadian high Arctic Light-bellied Brent Goose *Branta bernicla hrota*: Results of a satellite telemetry study. In *Waterbirds Around the World*, ed. G.C. Boere, C.A. Galbraith, and D.A. Stroud. Stationery Office, Edinburgh.

Robinson, J.A., J.M. Reed, J.P. Skorupa, and L.W. Oring. 2020. Black-necked Stilt (*Himantopus mexicanus*). Version 1.0. In *Birds of the World*, ed. A.F. Poole and F.B. Gill. Cornell Lab of Ornithology, Ithaca, NY. https://doi.org/10.2173/bow.bknsti.01.

Robinson, O.J., C.P. McGowan, and P.K. Devers. 2016a. Updating movement estimates for American Black Ducks (*Anas rubripes*). *PeerJ* 4:e1787.

Robinson, O.J., C.P. McGowan, P.K. Devers, R.W. Brook, M. Huang, M. Jones, D.G. McAuley, and G. Zimmerman. 2016b. A full annual cycle modeling framework for American Black Ducks. *Natural Resource Modeling* 29:159–174.

Robinson, W.D. 2020. Summer Tanager (*Piranga rubra*). Version 1.0. In *Birds of the World*, ed. A.F. Poole. Cornell Lab of Ornithology, Ithaca, NY. https://doi.org/10.2173/bow.sumtan.01.

Robuck, A.R., C.A. Hudak, L. Agvent, G. Emery, P.G. Ryan, V.A. Perold, K.D. Powers, J. Pedersen, M.A. Thompson, J.J. Suca, M.J. Moore, C. Harms, L. Bugoni, G. Shield, T. Glass, D.N. Wiley, and R. Lohmann. 2022. Birds of a feather eat plastic together: High levels of plastic ingestion in Great Shearwater adults and juveniles across their annual migratory cycle. *Frontiers in Marine Science* 8:719721. https://doi.org/10.3389/fmars.2021.719721.

Rocque, D.A., M. Ben-David, R.P. Barry, and K. Winker. 2006. Assigning birds to wintering and breeding grounds using stable isotopes: Lessons from two feather generations among three intercontinental migrants. *Journal of Ornithology* 147:395–404. https://doi.org/ 10.1007/s10336-006-0068-2.

Rodewald, P.G., and R.D. James. 2020. Yellow-throated Vireo (*Vireo flavifrons*). Version 1.0. In *Birds of the World*, ed. A.F. Poole. Cornell Lab of Ornithology, Ithaca, NY. https://doi.org/10.2173/bow.yetvir.01.

Rodewald, P.G., J.H. Withgott, and K.G. Smith. 2020. Pine Warbler (*Setophaga pinus*). Version 1.0. In *Birds of the World*, ed. A.F. Poole. Cornell Lab of Ornithology, Ithaca, NY. https://doi.org/10.2173/bow.pinwar.01.

Rodgers Jr., J.A., and H.T. Smith. 2020. Little Blue Heron (*Egretta caerulea*). Version 1.0. In *Birds of the World*, ed. A.F. Poole. Cornell Lab of Ornithology, Ithaca, NY. https://doi.org/10.2173/bow.libher.01.

Rodríguez, A., B. Rodríguez, T. Montelongo, J. Garcia-Porta, T. Pipa, M. Carty, J. Danielsen, J. Nunes, C. Silva, P. Geraldes, F.M. Medina, and J.C. Illera. 2020. Cryptic differentiation in the Manx Shearwater hinders the identification of a new endemic subspecies. *Journal of Avian Biology* 51(11). https://doi.org/10.1111/jav.02633.

Rohwer, F.C., W.P. Johnson, and E.R. Loos. 2020. Blue-winged Teal (*Spatula discors*). Version 1.0. In *Birds of the World*, ed. A.F. Poole and F.B. Gill. Cornell Lab of Ornithology, Ithaca, NY. https://doi.org/10.2173/bow.buwtea.01.

Rohwer, S., and G.S. Butcher. 1988. Winter versus summer explanations of delayed plumage maturation in temperate passerine birds. *American Naturalist* 131(4):556–572.

Rohwer, S., K.A. Hobson, and S. Yang. 2011. Stable isotopes (δD) reveal east-west differences in scheduling of molt and migration in northern Rough-Winged Swallows (*Stelgidopteryx serripennis*). *Auk* 128(3):522–530. https://doi.org/10.1525/auk.2011.10273.

Rohwer, S., A.G. Navarro, and G. Voelker. 2007. Rates versus counts: Fall molts of Lucy's Warblers (*Vermivora luciae*). *Auk* 124:806–814.

Romano, M.D., J.F. Piatt, and H.R. Carter. 2007. First successful radio-telemetry study of Kittlitz's Murrelet—Problems and potential. In *Proceedings of the Fourth Glacier Bay Science Symposium, October 26–28, 2004*, ed. J.F. Piatt, and S.M. Gende, 120–123. US Geological Survey Scientific Investigations Report 2007–5047.

Ronconi, R.A., H.N. Koopman, C.A.E. McKinstry, S.N.P. Wong, and A.J. Westgate. 2010. Inter-annual variability in diet of non-breeding pelagic seabirds *Puffinus* spp. at migratory staging areas: Evidence from stable isotopes and fatty acids. *Marine Ecology Progress Series* 419:267–282. https://doi.org/10.3354/meps08860.

Ronconi, R.A., D.J. Lieske, L.A. McFarlane Tranquilla, S. Abbott, K.A. Allard, B. Allen, A.L. Black, F. Bolduc, G.K. Davoren, A.W. Diamond, D.A. Fifield, S. Garthe, C. Gjerdrum, A. Hedd, M.L. Mallory, R.A. Mauck, J. McKnight, W.A. Montevecchi, I.L. Pollet, I. Pratte, J.-F. Rail, P.M. Regular, G.J. Robertson, J.C. Rock, L. Savoy, K.R. Shlepr, D. Shutler, S.C. Symons, P.D. Taylor, and S.I. Wilhelm. 2022. Predicting seabird foraging habitat for conservation planning in Atlantic Canada: Integrating telemetry and survey data across thousands of colonies. *Frontiers in Marine Science* 9. https://doi.org/10.3389/fmars.2022.816794.

Ronconi, R.A., S. Schoombie, A.J. Westgate, S.N.P. Wong, H.N. Koopman, and P.G. Ryan. 2018. Effects of age, sex, colony, and breeding phase on marine space use by Great Shearwaters *Ardenna gravis* in the South Atlantic. *Marine Biology* 165:58. https://doi.org/10.1007/s00227-018-3299-x.

Rosas-Hernández, M.P., Y.V. Albores-Barajas, C. Soldatini, J.C. Herguera, and G.D. 'Omo. 2022. The isotopic pattern variability in the breeding and non-breeding Black-Vented Shearwater. *Waterbirds* 45(3):300–311. https://doi.org/10.1675/063.045.0309.

Rosenberg, D.H., M.J. Petrula, J.L. Schamber, D. Zwiefelhofer, T.E. Hollmen, and D.D. Hill. 2014. Seasonal movements and distribution of Steller's Eiders (*Polysticta stelleri*) wintering at Kodiak Island Alaska. *Arctic* 67(3):347–359.

Rosenberg, K.V., A.M. Dokter, P.J. Blancher, J.R. Sauer, A.C. Smith, P.A. Smith, J.C. Stanton, A. Panjabi, L. Helft, M. Parr, and P.P. Marra. 2019. Decline of the North American avifauna. *Science* 366:120–124.

Rosenberg, K.V., R.D. Ohmart, W.C. Hunter, and B.W. Anderson. 1991. *Birds of the Lower Colorado River Valley*. University of Arizona Press, Tucson.

Rosenberg, K.V., T. Will, D.A. Buehler, S.B. Swarthout, W.E. Thogmartin, R.E. Bennett, and R. Chandler. 2016. Dynamic distributions and population declines of Golden-winged Warblers. In *Studies in Avian Biology* 49:3–28. CRC Press, Boca Raton, FL.

Ross, J.D., and J.L. Bouzat. 2014. Genetic and morphometric diversity in the Lark Sparrow (*Chondestes grammacus*) suggest discontinuous clinal variation across major breeding regions associated with previously characterized subspecies. *Auk* 131(3):298–313.

Ross, J.D., E.S. Bridge, M.J. Rozmarynowycz, and V.P. Bingman. 2014. Geolocator-tracked variation in migratory behavior within a highly site-faithful population of Lark Sparrows (*Chondestes grammacus*). *Animal Migration* 2(1):29–33. https://doi.org/10.2478/ami-2014-0003.

Rotenberry, J.T., M.A. Patten, and K.L. Preston. 2020. Brewer's Sparrow (*Spizella breweri*). Version 1.0. In *Birds of the World*, ed. A.F. Poole and F.B. Gill. Cornell Lab of Ornithology, Ithaca, NY. https://doi.org/10.2173/bow.brespa.01.

Roux, K.E., and P.P. Marra. 2007. The presence and impact of environmental lead in passerine birds along an urban to rural land use gradient. *Archives of Environmental Contamination and Toxicology* 53(2):261–268.

Rowlands, B.W., and T. Trueman. 1999. First Atlantic record of a Murphy's Petrel *Pterodroma ultima*, at St. Helena. *Bulletin of the African Bird Club* 6(1):25–28.

Roy, A., K. Delord, G.T. Nunes, C. Barbraud, L. Bugoni, and S. Lanco-Bertrand. 2021. Did the animal move? A cross-wavelet approach to geolocation data reveals year-round whereabouts of a resident seabird. *Marine Biology* 168(7):114.

Roy, C.L., C.M. Herwig, W.L. Hohman, and R.T. Eberhardt. 2020. Ring-necked Duck (*Aythya collaris*). Version 1.0. In *Birds of the World*, ed. A.F. Poole. Cornell Lab of Ornithology, Ithaca, NY. https://doi.org/10.2173/bow.rinduc.01.

Royle, J.A., and D.R. Rubenstein. 2004. The role of species abundance in determining breeding origins of migratory birds with stable isotopes. *Ecological Applications* 14(6):1780–1788.

Royston, S.R., and S.M. Carr. 2014. Conservation genetics of high-arctic gull species at risk. I. Diversity in the mtDNA control region of circumpolar populations of the Endangered Ivory Gull (*Pagophila eburnea*). *Mitochondrial DNA Part A* 27(6):3995–3999. https://doi.org/10.3109/19401736.2014.989520.

———. 2016. Conservation genetics of high-arctic gulls at risk. II. Diversity in the mtDNA control region of threatened Ross's Gull (*Rhodostethia rosea*). *Mitochondrial DNA Part A* 27(6):3941–3944.

Rubega, M.A., D. Schamel, and D.M. Tracy. 2020. Red-necked Phalarope (*Phalaropus lobatus*). Version 1.0. In *Birds of the World*, ed. S.M. Billerman. Cornell Lab of Ornithology, Ithaca, NY. https://doi.org/10.2173/bow.renpha.01.

Rubenstein, D.R., C.P. Chamberlain, R.T. Holmes, M.P. Ayres, J.R. Waldbauer, G.R. Graves, and N.C. Tuross. 2002. Linking breeding and wintering ranges of a migratory songbird using stable isotopes. *Science* 295(5557):1062–1065.

Rueda-Hernández, R., C.M. Bossu, T.B. Smith, A. Contina, R. Canales, K. Ruegg, and B.E. Hernández-Baños. 2023. Winter connectivity and leapfrog migration in a migratory passerine. *Ecology and Evolution* 13(2):e9769.

Ruegg, K. 2008. Genetic, morphological, and ecological characterization of a hybrid zone that spans a migratory divide. *Evolution* 62(2):452–466. https://doi.org/10.1111/j.1558-5646.2007.00263.x. PMID: 18039327.

Ruegg, K.C., and T.B. Smith. 2002. Not as the crow flies: A historical explanation for circuitous migration in Swainson's Thrush (*Catharus ustulatus*). *Proceedings of the Royal Society: Biological Sciences* 269(1498):1375–1381.

Ruegg, K.C., E.C. Anderson, R.J. Harrigan, K.L. Paxton, J.F. Kelly, F. Moore, and T.B. Smith. 2017. Genetic assignment with isotopes and habitat suitability (GAIAH), a migratory bird case study. *Methods in Ecology and Evolution* 8(10):1241–1252.

Ruegg, K.C., E.C. Anderson, K.L. Paxton, V. Apkenas, S. Lao, R.B. Siegel, D.F. Desante, F. Moore, and T.B. Smith. 2014. Mapping migration in a songbird using high-resolution genetic markers. *Molecular Ecology* 23:5726–5739.

Ruegg, K.C., E.C. Anderson, M. Somveille, R.A. Bay, M. Whitfield, E.H. Paxton, and T.B. Smith. 2021. Linking climate niches across seasons to assess population vulnerability in a migratory bird. *Global Change Biology* 27(15):3519–3531.

Ruegg, K.C., C.M. Bossu, J. Rajbhandary, T. Fuller, R. Harrigan, and T. Smith. 2020. A genoscape framework for assessing the population-level impacts of renewable energy development on migratory bird species in California. California Energy Commission, CEC-500-2020-036.

Ruegg, K.C., R.J. Hijmans, and C. Moritz. 2006. Climate change and the origin of migratory pathways in the Swainson's Thrush, *Catharus ustulatus*. *Journal of Biogeography* 33:1172–1182.

Ruelas Inzunza, E. 2005. Raptor and wading bird migration in Veracruz, Mexico: Spatial and temporal dynamics, flight performance, and monitoring applications. PhD diss., University of Missouri, Columbia, MO.

Ruelas Inzunza, E., L.J. Goodrich, and S.W. Hoffman. 2010. *Cambios en las poblaciones de aves rapaces migratorias en Veracruz, México, 1995–2005* (Changes in migratory raptor populations in Veracruz, Mexico, 1995–2005). *Acta Zoológica Mexicana* 26:3.

Ruelas Inzunza, E., L.J. Goodrich, S.W. Hoffman, E.M. Leyva, J.P. Smith, E.P. Rojas, R.R. Mesa, K.L. Scheuermann, S.L. M. Ortiz, Y.C. Carrasco, N. Ferriz, R. Straub, M.P. Perez, and J.G. Barrios. 2009. Long-term conservation of migratory birds in Mexico: The Veracruz River of Raptors Project. *Proceedings of the Fourth International Partners in Flight Conference: Tundra to Tropics*, 577–598.

Rundel, C.W., M.B. Wunder, A.H. Alvarado, K.C. Ruegg, R. Harrigan, A. Schuh, J.F. Kelly, R.B. Siegel, D.F. DeSante, T.B. Smith, and J. Novembre. 2013. Novel statistical methods for integrating genetic and stable isotope data to infer individual-level migratory connectivity. *Molecular Ecology* 22(16):4163–4176.

Rush, S.A., K.F. Gaines, W.R. Eddleman, and C.J. Conway. 2020. Clapper Rail (*Rallus crepitans*). Version 1.0. In *Birds of the World*, ed. P.G. Rodewald. Cornell Lab of Ornithology, Ithaca, NY. https://doi.org/10.2173/bow.clarai11.01.

Rushing, C.S., M.R. Dudash, C.E. Studds, and P.P. Marra. 2015. Annual variation in long-distance dispersal driven by breeding and non-breeding season climatic conditions in a migratory bird. *Ecography* 38:1006–1014.

Rushing, C.S., J.A. Hostetler, T.S. Sillett, P.P. Marra, J.A. Rotenberg, and T.B. Ryder. 2017. Spatial and temporal drivers of avian population dynamics across the annual cycle. *Ecology* 98(11):2837–2850.

Rushing, C.S., P.P. Marra, and C.E. Studds. 2017. Incorporating breeding abundance into spatial assignments on continuous surfaces. *Ecology and Evolution* 7(11):3847–3855. https://doi.org/10.1002/ece3.2605.

Rushing, C.S., T.B. Ryder, J.F. Saracco, and P.P. Marra. 2014. Assessing migratory connectivity for a long-distance migratory bird using multiple intrinsic markers. *Ecological Applications* 24(3):445–456. https://doi.org/10.1890/13-1091.1.

Rushing, C.S., T.B. Ryder, A.L. Scarpignato, J.F. Saracco, and P.P. Marra. 2016. Using demographic attributes from long-term monitoring data to delineate natural population structure. *Journal of Applied Ecology* 53(2):491–500.

Rushing, C.S., A.M. Van Tatenhove, A. Sharp, V. Ruiz-Gutierrez, M.C. Freeman, P.W. Sykes, A.M. Given, and T.S. Sillett. 2021. Integrating tracking and resight data enables unbiased inferences about migratory connectivity and winter range survival from archival tags. *Ornithological Applications* 123(1). https://doi.org/10.1093/ornithapp/duab010.

Russell, R.W. 2020a. Arctic Loon (*Gavia arctica*). Version 1.0. In *Birds of the World*, ed. S.M. Billerman. Cornell Lab of Ornithology, Ithaca, NY. https://doi.org/10.2173/bow.arcloo.01.

———. 2020b. Pacific Loon (*Gavia pacifica*). Version 1.0. In *Birds of the World*, ed. P.G. Rodewald. Cornell Lab of Ornithology, Ithaca, NY. https://doi.org/10.2173/bow.pacloo.01.

———. 2005. Interactions between migrating birds and offshore oil and gas platforms in the northern Gulf of Mexico, final report. US Department of the Interior, Minerals Management Service, New Orleans, LA. Contract 1435-01-99-CA-30951-16808. Report No.: OCS Study MMS 2005-009. https://espis.boem.gov/Technical%20Summaries/2966.pdf

Ruthrauff, D. 2014. On the frozen edge: Environmental and physiological constraints in the life history of a northerly-wintering shorebird. Doctoral diss., University of Groningen, Groningen, Netherlands.

Ruthrauff, D.R., C.M. Harwood, T.L. Tibbitts, N. Warnock, and R.E. Gill Jr. 2021. Diverse patterns of migratory timing, site use, and site fidelity by Alaska-breeding Whimbrels. *Journal of Field Ornithology* 92(2):156–172.

Ruthrauff, D.R., T.L. Tibbitts, and R.E. Gill Jr. 2019a. Flexible timing of annual movements across consistently used sites by Marbled Godwits breeding in Alaska. *Auk* 136(1):uky007.

Ruthrauff, D.R., T.L. Tibbitts, and V.P. Patil. 2019b. Survival of Bristle-thighed Curlews equipped with externally mounted transmitters. *Wader Study* 126(2):109–115.

Rutledge, L.Y., A. Coxon, and B.N. White. 2017. Genetic assessment of the San Clemente Island Loggerhead Shrike reveals evidence of historical gene flow with Santa Catalina Island. *Global Ecology and Conservation* 11:42–52.

Ryan, T. 2016. The use of light-level geolocators to determine off-site migratory routes and wintering locations of California Least Terns 2012–2016: 2015 Annual Report. San Diego Zoo, Global Beckman Center, Institute for Conservation Research. San Diego, CA.

Ryan, T.P., and C.T. Collins. 2020. White-throated Swift (*Aeronautes saxatalis*). Version 1.0. In *Birds of the World*, ed. A.F. Poole and F.B. Gill. Cornell Lab of Ornithology, Ithaca, NY. https://doi.org/10.2173/bow.whtswi.01.

Ryder, R.A. 1967. Distribution, migration, and mortality of the White-faced Ibis (*Plegadis chihi*) in North America. *Bird Banding* 38(4):257–277.

Ryder, R.A., and D.E. Manry. 2020. White-faced Ibis (*Plegadis chihi*). Version 1.0. In *Birds of the World*, ed. A.F. Poole and F.B. Gill. Cornell Lab of Ornithology, Ithaca, NY. https://doi.org/10.2173/bow.whfibi.01.

Ryder, T.B., J.W. Fox, and P.P. Marra. 2011. Estimating migratory connectivity of Gray Catbirds (*Dumatella carolinensis*) using geolocator and mark-recapture data. *Auk* 128(3):448–453.

Rypdal, E. 2018. Repeatability in isotopic signatures is linked to consistent individual migration strategies and individual specialization in a long-distance migratory seabird. Master's thesis, Norwegian University of Science and Technology.

Ryser Jr., R.A. 1985. *Birds of the Great Basin: A Natural History*. University of Nevada Press, Reno.

Saalfeld, S.T., and R.B. Lanctot. 2015. Conservative and opportunistic settlement strategies in Arctic-breeding shorebirds. *Auk* 132:212–234.

Saalfeld, S.T., M. Valcu, S. Brown, W. English, M-A. Giroux, A-L. Harrison, J. Krietsch, K. Kuletz, J-F. Lamarre, C. Latty, N. Lecomte, R. McGuire, M. Robards, A. Scarpignato, S. Schulte, P.A. Smith, B. Kempenaers, and R.B. Lanctot. 2024. From land to sea: The fall migration of the Red Phalarope through the Western Hemisphere. *Marine Ecology Progress Series* 729:1–29. https://doi.org/10.3354/meps14531.

Sakurayama, S., D. Nojima, M. Yoshizawa, T. Takeuchi, M. Ito, and T. Kitano. 2021. Genetic diversity of two populations of the Tufted Puffin *Fratercula cirrhata* (Pallas, 1769). *Genes & Genetic Systems* 96(3):119–128. https://doi.org/10.1266/ggs.20-00051. PMID: 34135205.

Sallaberry, M.A. 1990. Genetic and morphological differentiation of Sanderling (*Calidris alba*) populations on the wintering grounds. PhD diss., University of Pennsylvania.

Sallaberry, M., and M. Mann. 2007. Sanderlings (*Calidris alba*) banded in Canada recovered in Chile. *Ornitologia Neotropical* 18(4):623–626.

Salomonsen, F. 1944. The Atlantic Alcidae: The seasonal and geographical variation of the auks inhabiting the Atlantic Ocean and adjacent waters. Göteborgs Kungliga Vetenskaps, och Vitterhets-Samhälles Handlingar (*Proceedings of the Royal Swedish Academy of Sciences*), Series B 3:1–138.

———. 1955. *The Evolutionary Significance of Bird-Migration*. I kommission hos E. Munksgaard.

———. 1967. *Fuglene pa Gronland* (*The Birds of Greenland*). Copenhagen: Rhodos.

———. 1979. Fra Zoologisk Museum XXV. Trettende forelobige liste over genfundne gronlandske ringfugle (From the Zoological Museum XXV: Thirteenth preliminary list of recovered Greenlandic ringbirds). *Dansk Ornithologisk Forenings Tidsskrift* (*Journal of the Danish Ornithological Society*) 73:191–206.

Sanchez, G.H. 1993. The ecology of wintering Gyrfalcons (*Falco rusticolus*) in central South Dakota. Master's thesis, Boise State University. https://scholarworks.boisestate.edu/td/691.

Sanders, F., M. Spinks, and T. Magarian. 2013. American Oystercatcher winter roosting and foraging ecology at Cape Romain, South Carolina. *Wader Study Group Bulletin* 120(2):128–133.

Saracco, J.F., R.L. Cormier, D.L. Humple, S. Stock, R. Taylor, and R.B. Siegel. 2022. Demographic responses to climate-driven variation in habitat quality across the annual cycle of a migratory bird species. *Ecology and Evolution* 12:e8934. https://doi.org/10.1002/ece3.8934.

Sarasola, J.H., J.J. Negro, K.A. Hobson, G.R. Bortolotti, and K.L. Bildstein. 2008. Can a "wintering area effect" explain population status of Swainson's Hawks? A stable isotope approach. *Diversity and Distributions* 14:686–691.

Satgé, Y., A. Brown, J.A. Wheeler, and K.E. Sutherland. 2023a. Black-capped Petrel (*Pterodroma hasitata*). Version 2.0. In *Birds of the World*, ed. S.M. Billerman. Cornell Lab of Ornithology, Ithaca, NY.

Satgé, Y.G., B.S. Keitt, C.P. Gaskin, J.B. Patteson, and P.G.R. Jodice. 2023b. Spatial segregation between phenotypes of the diablotin Black-capped Petrel *Pterodroma hasitata* during the non-breeding period. *Endangered Species Research* 51:183–201. https://doi.org/10.3354/esr01254.

Satgé, Y.G., E. Rupp, and P.G.R. Jodice. 2019. A preliminary report of ongoing research of the ecology of Black-capped Petrel (*Pterodroma hasitata*) in Sierra de Bahoruco, Dominican Republic, I: GPS tracking of breeding adults. South Carolina Cooperative Research Unit, Clemson University.

Sato, T., Y. Yabuhara, J. Okado, Y. Watanuki, A. Yamauchi, and Y. Kawaguchi. 2022. At-sea habitat use of Rhinoceros Auklets breeding in the shelf region of Eastern Hokkaido. *Zoological Science* 39(3):261–269. https://doi.org/10.2108/zs210014.

Sauer, J.R. Breeding Bird Survey Counts, 1970–2017.

Sauer, J.R., J.E. Fallon, and R. Johnson. 2003. Use of North American Breeding Bird Survey data to estimate population change for bird conservation regions. *Journal of Wildlife Management* 67(2):372–389.

Sauer, J.R., D.K. Niven, J.E. Hines, D.J. Ziolkowski Jr., K.L. Pardieck, J.E. Fallon, and W.A. Link. 2019. The North American Breeding Bird Survey, Results and Analysis 1966–2019. Version 2.07.2019. https://www.mbr-pwrc.usgs.gov.

Sauve, D., A. Patirana, J.W. Chardine, and V.L. Friesen. 2018. Mitochondrial DNA reveals genetic structure within the Atlantic but not Pacific populations of a holarctic seabird, the Black-legged Kittiwake *Rissa tridactyla*. *Marine Ornithology* 47:199–208.

Savard, J.P.L., L. Lesage, S.G. Gilliland, H.G. Gilchrist, and J.F. Giroux. 2011. Molting, staging, and wintering locations of Common Eiders breeding in the Gyrfalcon Archipelago, Ungava Bay. *Arctic* 64:197–206.

Savard, J.P.L., and M. Robert. 2013. Relationships among breeding, molting and wintering areas of adult female Barrow's Goldeneyes (*Bucephala islandica*) in eastern North America. *Waterbirds* 36(1):34–42.

Savides, K. 2022. Characterizing the migratory phenology and routes of the Lazuli Bunting (*Passerina amoena*) in northern Utah. Master's thesis, Utah State University. https://digitalcommons.usu.edu/etd/8450.

Scarpignato, A.L., A-L. Harrison, D.J. Newstead, L.J. Niles, R.R. Porter, M. van den Tillaart, and P.P. Marra. 2016. Field-testing a new miniaturized GPS-Argos satellite transmitter (3.5 g) on migratory shorebirds. *Wader Study* 123(3):240–246. https://doi.org/10.18194/ws.00046.

Scarpignato, A.L., A.E. Huysman, M.F. Jimenez, C.J. Witko, A.L. Harrison, N.E. Seavy, M.A. Smith, J.L. Deppe, C.B. Wilsey, and P.P. Marra. 2023. Shortfalls in tracking data available to inform North American migratory bird conservation. *Biological Conservation* 286:110224.

Scarpignato, A.L., C.Q. Stanley, A.M. Anderson, N.W. Cooper, A.-L. Harrison, F.G. Lopez, M.E. Rebollo, K.M. Delaski, and P.P. Marra. 2026. Harnessing knowledge: A multi-taxa evaluation of device attachment elements, reporting gaps, and protocols for best practices in wild birds. *Ornithology* https://doi.org/10.1093/ornithology/ukag023

Scarpignato, A.L., K.A. Stein, E.B. Cohen, P.P. Marra, L.J. Kearns, S. Hallager, and C.M. Tonra. 2021. Full annual cycle tracking of Black-crowned Night-Herons suggests wintering areas do not explain differences in colony population trends. *Journal of Field Ornithology* 92:143–155. https://doi.org/10.1111/jofo.12369.

Schacter C. 2017. Migration dynamics: Testing ecological theory with tracking data for Aethia auklets in the North Pacific. PhD diss., Memorial University of Newfoundland, St. John's, Newfoundland & Labrador.

Schacter, C.R., and I.L. Jones. 2017. Effects of geolocation tracking devices on behavior, reproductive success, and return rate of Aethia Auklets: An evaluation of tag mass guidelines. *Wilson Journal of Ornithology* 129(3):459–468.

Schaefer, A.L., K.B. Gorman, and M.A. Bishop. 2022. Light-level geolocation reveals the short distance non-breeding movements and distribution of Tufted Puffins throughout the northeast Pacific Ocean. *Frontiers in Marine Science* 9:999461. https://doi.org/10.3389/fmars.2022.999461.

Schaffner, F.C. 2014. Interhabitat connectivity of North American wintering songbirds on the south coast of Puerto Rico. Report to the Nuttall Ornithological Club for 2013–2014. https://doi.org/10.13140/RG.2.2.30646.80964.

Schaffner, F.C., M.T. Hallworth, M.F. Jimenez, I. Rodríguez-Colón, and N.E. Seavy. 2023. Light-level tracking of Northern Waterthrush (*Parkesia noveboracensis*): Lessons learned from wintering ground deployments. *Journal of Field Ornithology* 94(4).

Schamber, J.L., P.L. Flint, and A.N. Powell. 2010. Patterns of use and distribution of King Eiders and Black Scoters during the annual cycle in northeastern Bristol Bay, Alaska. *Marine Biology* 157:2169–2176.

Schamber, J.L., J.S. Sedinger, D.H. Ward, and K.R. Hagmeier. 2007. Latitudinal variation in population structure of wintering Pacific Black Brant. *Journal of Field Ornithology* 78(1):74.82.

Schamel, D., and D. Tracy. 1977. Polyandry, replacement clutches, and site tenacity in the Red Phalarope (*Phalaropus fulicarius*) at Barrow, Alaska. *Bird Banding* 48:314–324.

———. 1991. Breeding site fidelity and natal philopatry in the sex role-reversed Red and Red-necked Phalaropes. *Journal of Field Ornithology* 62:390–398.

Scharf, W.C., and J. Kren. 2022. Orchard Oriole (*Icterus Spurius*). Version 2.0. In *Birds of the World*, ed. P.G. Rodewald. Cornell Lab of Ornithology, Ithaca, NY. https://doi.org/10.2173/bow.orcori.02.

Schimpf, D., L. Goodrich, A. Kocek, and D. La Puma. 2020. Northern Harrier has a geographically broad four-year cycle. *Journal of Raptor Research* 54(1):38.

Schlossberg, S., and J.C. Sterling. 2020. Gray Flycatcher (*Empidonax wrightii*). Version 1.0. In *Birds of the World*, ed. A.F. Poole. Cornell Lab of Ornithology, Ithaca, NY. https://doi.org/10.2173/bow.gryfly.01.

Schmaljohann, H., J.W. Fox, and F. Bairlein. 2012. Phenotypic response to environmental cues, orientation, and costs of migration in songbirds flying halfway around the world. *Animal Behaviour* 84:623–640.

Schmaljohann, H., S. Lisovski, and F. Bairlein. 2017. Flexible reaction norms to environmental variables along the migration route and the significance of stopover duration for total speed of migration in a songbird migrant. *Frontiers in Zoology* 14:16.

Schmidt, J.H., M.J. Flamme, and J. Walker. 2014. Habitat use and population status of Yellow-billed and Pacific Loons in western Alaska. *Condor* 116(3):483–492. https://doi.org/10.1650/CONDOR-14-28.1

Schmutz, J.A., and A.V. Kondratyev. 1995. Evidence of Emperor Geese breeding in Russia and staging in Alaska. *Auk* 112:1037–1038.

Schmutz, J.A., B. Uher-Koch., and D.C. Douglas. 2019. Tracking data for Yellow-billed Loons (*Gavia adamsii*). US Geological Survey data release, https://doi.org/10.5066/P9UKHJ22.

Schmutz, J.A., K.G. Wright, C.R. DeSorbo, J. Fair, D.C. Evers, B.D. Uher-Koch, and D.M. Mulcahy. 2014. Size and retention of breeding territories of Yellow-billed Loons (*Gavia adamsii*) in Alaska and Canada. *Waterbirds* 37(spec pub 1):53–63.

Schmutz, J.K., and R.W. Fyfe. 1987. Migration and mortality of Alberta Ferruginous Hawks. *Condor* 89(1):169–174.

Schmutz, J.K., R.W. Fyfe, U. Banasch, and H. Armbruster. 1991. Routes and timing of migration of falcons banded in Canada. *Wilson Bulletin* 103(1):44–58.

Schoombie, S., B.J. Dilley, D. Davies, et al. 2018. The foraging range of Great Shearwaters (*Ardenna gravis*) breeding on Gough Island. *Polar Biology* 41:2451–2458. https://doi.org/10.1007/s00300-018-2381-7.

Schreiber, B.A., and R.W. Schreiber. 2020. Red-tailed Tropicbird (*Phaethon rubricauda*). Version 1.0. In *Birds of the World*, ed. S.M. Billerman. Cornell Lab of Ornithology, Ithaca, NY. https://doi.org/10.2173/bow.rettro.01.

Schreiber, E.A., C.J. Feare, B.A. Harrington, B.G. Murray Jr., W.B. Robertson Jr., M.J. Robertson, and G.E. Woolfenden. 2020. Sooty Tern (*Onychoprion fuscatus*). Version 1.0. In *Birds of the World*, ed. S.M. Billerman. Cornell Lab of Ornithology, Ithaca, NY. https://doi.org/10.2173/bow.sooter1.01.

Schreiber, E.A., and R.L. Norton. 2020. Brown Booby (*Sula leucogaster*). Version 1.0. In *Birds of the World*, ed. S.M. Billerman. Cornell Lab of Ornithology, Ithaca, NY. https://doi.org/10.2173/bow.brnboo.01.

Schreiber, R.W. 1976. Movements of color marked Brown Pelicans. *Bird Banding* 47(2):101–111.

Schroeder, M.H., and C.A. Ely. 1972. Recoveries of Mourning Doves banded as nestlings in west-central Kansas. *Bird Banding* 43:257–260.

Schukman, J.M., and B.O. Wolf. 2020. Say's Phoebe (*Sayornis saya*). Version 1.0. In *Birds of the World*, ed. P.G. Rodewald. Cornell Lab of Ornithology, Ithaca, NY. https://doi.org/10.2173/bow.saypho.01.

Schummer, M.L., A.D. Afton, S.S. Badzinski, S.A. Petrie, G.H. Olsen, and M.A. Mitchell. 2018. Evaluating the waterfowl breeding population and habitat survey for scaup. *Journal of Wildlife Management* 82(6):1252–1262.

Schwertner, T.W., H.A. Mathewson, J.A. Roberson, and G.L. Waggerman. 2020. White-winged Dove (*Zenaida asiatica*). Version 1.0. In *Birds of the World*, ed. A.F. Poole and F.B. Gill. Cornell Lab of Ornithology, Ithaca, NY. https://doi.org/10.2173/bow.whwdov.01.

Schwitters, L., D. Schwitters, E.L. Bull, and C.T. Collins. 2020. Vaux's Swift (*Chaetura vauxi*). Version 1.0. In *Birds of the World*, ed. P.G. Rodewald. Cornell Lab of Ornithology, Ithaca, NY. https://doi.org/10.2173/bow.vauswi.01.

Scott, D., P. Scofield, C. Hunter, and D. Fletcher. 2008. Decline of Sooty Shearwaters, *Puffinus griseus*, on the Snares, New Zealand. *Papers and Proceedings of the Royal Society of Tasmania* 142:185–196.

Scribner, K.T., R.A. Malecki, B.D. Batt, R.L. Inman, S. Libants, and H.H. Prince. 2003. Identification of source population for Greenland Canada Geese: Genetic assessment of a recent colonization. *Condor* 105:771–782.

Scribner, K.T., M.R. Petersen, R.L. Fields, S.L. Talbot, J.M. Pearce, and R.K. Chesser. 2001. Sex-biased gene flow in Spectacled Eiders (Anatidae): Inferences from molecular markers with contrasting modes of inheritance. *Evolution* 55:2105–2115.

Sea Duck Joint Venture. 2014. Atlantic and Great Lakes Sea Duck Migration Study: Progress Report. http://seaduckjv.org/atlantic_migration_study.html.

Sea Duck Joint Venture. 2015. Atlantic and Great Lakes Sea Duck Migration Study: Progress Report. http://seaduckjv.org/science-resources/atlantic-and-great-lakes-sea-duck-migration-study/.

Sealy, S.G., and H.R. Carter. 2012. Rare inter-ocean vagrancy in Crested Auklet and Parakeet Auklet. *Waterbirds* 35(1):64–73. https://doi.org/10.1675/063.035.0107.

Seavy, N.E., D.L. Humple, R.L. Cormier, and T. Gardali. 2012. Establishing the breeding provenance of a temperate-wintering North American passerine, the Golden-Crowned Sparrow, using light-level geolocation. *PLOS ONE* 7:6.

Sechrist, J.D., E.H. Paxton, D.D. Ahlers, R.H. Doster, and V.M. Ryan. 2012. One year of migration data for a western Yellow-billed Cuckoo. *Western Birds* 43:2–11.

Sedgwick, J.A. 2020a. Hammond's Flycatcher (*Empidonax hammondii*). Version 1.0. In *Birds of the World*, ed. A.F. Poole and F.B. Gill. Cornell Lab of Ornithology, Ithaca, NY. https://doi.org/10.2173/bow.hamfly.01.

———. 2020b. Willow Flycatcher (*Empidonax traillii*). Version 1.0. In *Birds of the World*, ed. A.F. Poole and F.B. Gill. Cornell Lab of Ornithology, Ithaca, NY. https://doi.org/10.2173/bow.wilfly.01.

Sedinger, J.S., J.L. Schamber, D.H. Ward, C.A. Nicolai, and B. Conant. 2011. Carryover effects associated with winter location affect fitness, social status, and population dynamics in a long-distance migrant. *American Naturalist* 178(5):E110–E123.

Seegar, W.S., P.N. Cutchis, M.R. Fuller, J.H. Suter, V. Bhatnagar, and J.G. Wall. 1996. Fifteen years of satellite tracking, development and application to wildlife research and conservation. *Johns Hopkins APL Technical Digest* 17:401–411.

Senner, N.R. 2012. One species but two patterns: Populations of the Hudsonian Godwit (*Limosa haemastica*) differ in spring migration timing. *Auk* 129:670–682.

Senner, N.R., W.M. Hochachka, J.W. Fox, and V. Afanasyev. 2014. An exception to the rule: Carry-over effects do not accumulate in a long-distance migratory bird. *PLOS ONE* 9(2):e86588. https://doi.org/10.1371/journal.pone.0086588.

Senner, S.E., and B.J. McCaffery. 2020. Surfbird (*Calidris virgata*). Version 1.0. In *Birds of the World*, ed. A.F. Poole and F.B. Gill. Cornell Lab of Ornithology, Ithaca, NY.

Serie, J.R., D.L. Trauger, and D.E. Sharp. 1983. Migration and winter distributions of Canvasbacks staging on the upper Mississippi River. *Journal of Wildlife Management* 47:741–753.

Sesser, K.A. 2013. Habitat and space use by Long-billed Curlews (*Numenius americanus*) in California's Central Valley. Master's thesis, Humboldt State University.

Seto, N.W., and D.L. O'Daniel. 2020. Bonin Petrel (*Pterodroma hypoleuca*). Version 1.0. In *Birds of the World*, ed. A.F. Poole and F.B. Gill. Cornell Lab of Ornithology, Ithaca, NY. https://doi.org/10.2173/bow.bonpet.01.

Seward, A., R.C. Taylor, M.R. Perrow, R.J. Berridge, K.M. Bowgen, S. Dodd, I. Johnstone, and M. Bolton. 2021. Effect of GPS tagging on behaviour and marine distribution of breeding Arctic Terns *Sterna paradisaea*. *Ibis* 163:197–212. https://doi.org/10.1111/ibi.12849.

Sexon, M.G., J.M. Pearce, and M.R. Petersen. 2014. Spatiotemporal distribution and migratory patterns of Spectacled Eiders. BOEM 2014–665. Bureau of Ocean Energy Management, Alaska Outer Continental Shelf Region, Anchorage, Alaska.

Seyer, Y., G. Gauthier, J. Bêty, J.F. Therrien, and N. Lecomte. 2021. Seasonal variations in migration strategy of a long distance Arctic-breeding seabird. *Marine Ecology Progress Series* 677:1–16.

Seyer, Y., G. Gauthier, J. Bêty, J.F. Therrien, P. Legagneux, and N. Lecomte. 2023. Local food availability and nonbreeding carry-over effects affect breeding propensity and success of a tundra-nesting predator, the Long-tailed Jaeger. *Ornithology* 140(4):p.ukad032.

Seymour, N.R. 1991. Philopatry in male and female American Black Ducks. *Condor* 93:189–191.

Shaffer, S., and R. Orben. 2022. Laysan Albatross Midway Atoll GLS2008 12 postbreeding. BirdLife International Seabird Tracking Database, Dataset ID 1930. https://data.seabirdtracking.org/dataset/1930.

Shaffer, S., M. Antolos, M. Conners. 2022. Laysan albatross geolocators from Tern Island 2004–2006. BirdLife International Seabird Tracking Database, Dataset ID 529. https://data.seabirdtracking.org/dataset/529.

Shaffer S.A., S. Cockerham, P. Warzybok, R.W. Bradley, J. Jahncke, C.A. Clatterbuck, et al. 2017. Population-level plasticity in foraging behavior of Western Gulls (*Larus occidentalis*). *Movement Ecology* 5:27. https://doi.org/10.1186/s40462-017-0118-9.

Shaffer, S.A., Y. Tremblay, J. Awkerman, W.R. Henry, S.L.H. Teo, D.J. Anderson, D.A. Croll, B.A. Block, and D.P. Costa. 2005. Comparison of light- and SST-based geolocation with satellite telemetry in free-ranging albatrosses. *Marine Biology* 147:833–843. https://doi.org/doi.org/10.1007/s00227-005-1631-8.

Shaffer, S.A., Y. Tremblay, H. Weimerskirch, D. Scott, D.R. Thompson, P.M. Sagar, H. Moller, G.A. Taylor, D.G. Foley, B.A. Block, and D.P. Costa. 2006. Migratory shearwaters integrate oceanic resources across the Pacific Ocean in an endless summer. *Proceedings of the National Academy of Sciences* 103(34):12799–12802.

Shaffer, S.A., H. Weimerskirch, D. Scott, D. Pinaud, D.R. Thompson, P.M. Sagar, H. Moller, G.A. Taylor, D.G. Foley, Y. Tremblay, and D.P. Costa. 2009 Spatio-temporal habitat use by breeding Sooty Shearwaters (*Puffinus griseus*). *Marine Ecology Progress Series* 391:209–220. https://doi.org/doi.org/10.3354/meps07932.

Shahrokhi, G., D. Rodriguez, S. Collins, G. Kent, K. Meyer, E. Palacios-Castro, and M.C. Green. 2020. A re-evaluation of management units based on gene flow of a rare waterbird in the Americas. *Biotropica* 52(6):1107–1114.

Shane, T.G. 2020. Lark Bunting (*Calamospiza melanocorys*). Version 1.0. In *Birds of the World*, ed. A.F. Poole and F.B. Gill. Cornell Lab of Ornithology, Ithaca, NY. https://doi.org/10.2173/bow.larbun.01.

Sharp, A.J., A. Contina, V. Ruiz- Gutiérrez, T.S. Sillett, E.S. Bridge, E.M. Besozzi, J.A. Muller, J. Kelly, A.M. Given, and C.S. Rushing. 2023. The strength of migratory connectivity in Painted Buntings is spatial scale dependent and shaped by molting behavior. *Journal of Field Ornithology* 94(1):7.

Sharp, B. 1972. Eastward migration of blue-winged teal. *Journal of Wildlife Management* 36:1273–1277.

Sheaffer, S.E., R.A. Malecki, B.L. Swift, J. Dunn, and K. Scribner. 2007. Management implications of molt migration by the Atlantic Flyway resident population of Canada Geese, *Branta canadensis*. *Canadian Field Naturalist* 121:313–320.

Shealer, D., J.S. Liechty, A.R. Pierce, P. Pyle, and M.A. Patten. 2016. Sandwich Tern (*Thalasseus sandvicensis*). Version 3.0. In *The Birds of North America*, ed. P.G. Rodewald. Cornell Lab of Ornithology, Ithaca, NY.https://doi.org/10.2173/bna.santer1.03.

Shephard, N.G., P. Szczys, D.J. Moore, M.W. Reudink, J.N. Cosa, A.M. Bracey, S. Lisovski, and A.E. McKellar. 2023. Weak genetic structure, shared nonbreeding areas, and extensive movement in a declining waterbird. *Ornithological Applications* 125(1):duac053. https://doi.org/10.1093/ornithapp/duac053.

Sherry, T.W., R.T. Holmes, P. Pyle, and M.A. Patten. 2020. American Redstart (*Setophaga ruticilla*). Version 1.0. In *Birds of the World*, ed. P.G. Rodewald. Cornell Lab of Ornithology, Ithaca, NY. https://doi.org/10.2173/bow.amered.01.

Shields, L.J., and B.S. Mueller. 1982. A radio transmitter for quail. *National Quail Symposium Proceedings* 2: article 17. http://trace.tennessee.edu/nqsp/vol2/iss1/17.

Shields, M. 2020. Brown Pelican (*Pelecanus occidentalis*). Version 1.0. In *Birds of the World*, ed. A.F. Poole. Cornell Lab of Ornithology, Ithaca, NY. https://doi.org/10.2173/bow.brnpel.01.

Shirkey, B.T., R.J. Gates, and M.D. Ervin. 2018. Survival rates and harvest patterns of Ohio-banded Canada Geese. *Wildlife Society Bulletin* 42(3):394–402.

Shoji, A., K.W. Elliott, Y. Watanuki, N. Basu, S. Whelan, J. Cunningham, … and S. Aris-Brosou. 2021. Geolocators link marine mercury with levels in wild seabirds throughout their annual cycle: Consequences for trans-ecosystem biotransport. *Environmental Pollution* 284:117035.

Shoji, A., S. Whelan, J.T. Cunningham, S.A. Hatch, Y. Niizuma, C. Nakajima, and K.H. Elliott. 2023. Ecological niche partitioning in two Pacific puffins. *Marine Ecology Progress Series* 709:125–139. https://doi.org/10.3354/meps14282.

Shorey, R.I., K.T. Scribner, J. Kanefsky, M.D. Samuel, and S.V. Libants. 2011. Intercontinental gene flow among western Arctic populations of Lesser Snow Geese. *Condor* 113(4):735–746.

Shriver, W.G., T.P. Hodgman, and A.R. Hanson. 2020. Nelson's Sparrow (*Ammospiza nelsoni*). Version 1.0. In *Birds of the World*, ed. P.G. Rodewald. Cornell Lab of Ornithology, Ithaca, NY. https://doi.org/10.2173/bow.nstspa.01.

Shuford, W.D., and T. Gardali, eds. 2008. *California Bird Species of Special Concern: A Ranked Assessment of Species, Subspecies, and Distinct Populations of Birds of Immediate Conservation Concern in California*. Studies of Western Birds 1. Western Field Ornithologists, Camarillo, CA, and California Department of Fish and Game, Sacramento, CA.

Shuford, W.D., G.W. Page, J.G. Evans, and L.E. Stenzel. 1989. Seasonal abundance of waterbirds at Point Reyes: A coastal California perspective. *Western Birds* 20:137–265.

Siegel, R.B., R. Taylor, J.F. Saracco, L. Helton, and S. Stock. 2016. GPS-tracking reveals non-breeding locations and apparent molt migration of a Black-headed Grosbeak. *Journal of Field Ornithology* 57(2):196–203. https://doi.org/10.1111/jofo.12149.

Sillett, T.S., and R.T. Holmes. 2002. Variation in survivorship of a migratory songbird throughout its annual cycle. *Journal of Animal Ecology* 71:296–308.

Simons, T.R., and C.N. Bailey. 2020. Hawaiian Petrel (*Pterodroma sandwichensis*). Version 1.0. In *Birds of the World*, ed. A.F. Poole and F.B. Gill. Cornell Lab of Ornithology, Ithaca, NY. https://doi.org/10.2173/bow.hawpet1.01.

Sittler, B., A. Aebischer, and O. Gilg. 2011. Post-breeding migration of four Long-tailed Skuas (*Stercorarius longicaudus*) from north and east Greenland to West Africa. *Journal of Ornithology* 152:375–381.

Skagen, S.K., and F.L. Knopf. 1993. Towards conservation of midcontinental shorebird migrations. *Conservation Biology* 7:533–541.

Skagen, S.K., J.F. Kelly, C. Van Riper III, R.L. Hutto, D.M. Finch, D.J. Krueper, and C.P. Melcher. 2005. Geography of spring land bird migration through riparian habitats in southwestern North America. *Condor* 107(2):212–227.

Skagen, S.K., P.B. Sharpe, R.G. Waltermire, and M.B. Dillon. 1999. Biogeographical profiles of shorebird migration in midcontinental North America. Biological Science Report 2000-0003, US Geological Survey, Biological Resources Division. US Government Printing Office, Denver, CO.

Skeel, M.A., and E.P. Mallory. 2020. Whimbrel (*Numenius phaeopus*). Version 1.0. In *Birds of the World*, ed. S.M. Billerman. Cornell Lab of Ornithology, Ithaca, NY.

Skinner, A.A., M.P. Ward, I. Souza-Cole, J.R. Wright, F.R. Thompson, T.J. Benson, S.N. Matthews, and C.M. Tonra. 2022. High spatiotemporal overlap in the non-breeding season despite geographically dispersed breeding locations in the Eastern Whip-poor-will (*Antrostomus vociferus*). *Diversity and Distributions* 28:712–726. https://doi.org/10.1111/ddi.13477.

Skujina, I., A.J.E. Healey, S. de Becquevort, P.W. Shaw, R. McMahon, C. Morgan, C. Evans, R. Taylor, M. Hegarty, and N.J. McKeown. 2019. The complete mitochondrial genome of record-breaking migrant Arctic Tern (*Sterna paradisaea*). *Mitochondrial DNA Part B* 4(2):2738–2739. https://doi.org/10.1080/23802359.2019.1644225.

Sladen, W.J. 1973. A continental study of Whistling Swans using neck collars. *Wildfowl* 24:8–14.

Sladen, W.J., and J.C. Whissel. 2007. The winter distribution of Trumpeter Swans in relation to breeding areas: First neckband study, 1972–1981. *North American Swans* 33:117–128.

Slezak, C.R., E.J. Blomberg, A.M. Roth, L.A. Berigan, A.C. Fish, R. Darling, S.J. Clements, G. Balkcom, B. Carpenter, G. Costanzo, and J. Duguay. 2024. Unconventional life history in a migratory shorebird: Desegregating reproduction and migration. *Proceedings of the Royal Society: Biological Sciences* 291(2021):20240021.

Smith, A.D., P.W.C. Paton, and S.R. McWilliams. 2014. Using nocturnal flight calls to assess the fall migration of warblers and sparrows along a coastal ecological barrier. *PLOS ONE* 9:e92218.

Smith, G.A., and J.A. Jackson. 2020. Gray Kingbird (*Tyrannus dominicensis*). Version 1.0. In *Birds of the World*, ed. A.F. Poole and F.B. Gill. Cornell Lab of Ornithology, Ithaca, NY. https://doi.org/10.2173/bow.grykin.01.

Smith, J.A., K. Regan, N.W. Cooper, L. Johnson, E. Olson, A. Green, … and P.P. Marra. 2020. A green wave of saltmarsh productivity predicts the timing of the annual cycle in a long-distance migratory shorebird. *Scientific Reports* 10(1):1–13.

Smith, J.A., L.R. Reitsma, L.L. Rockwood, and P.P. Marra. 2008. Roosting behavior of a Neotropical migrant songbird, the Northern Waterthrush *Seiurus noveboracensis*, during the non-breeding season. *Journal of Avian Biology* 39(4):460–465.

Smith, J.P. 2010. Raptor migration in New Mexico. In *Raptors of New Mexico*, ed. J.-L.E. Cartron, 29–67. University of New Mexico Press, Albuquerque.

Smith, J.P., C.J. Farmer, S.W. Hoffman, C.A. Lott, L.J. Goodrich, J. Simon, C. Riley, and E. Ruelas Inzunza. 2008. Trends in autumn counts of migratory raptors around the Gulf of Mexico, 1995–2005. In *State of North America's Birds of Prey*, ed. K.L. Bildstein, J.P. Smith, E. Ruelas Inzunza, and R.R. Veit. Series in Ornithology 3:253–278. Nuttall Ornithological Club, Cambridge, MA, and American Ornithologists Union, Washington, DC.

Smith, K.G., K.A. Tarvin, and G.E. Woolfenden. 2020a. Blue Jay (*Cyanocitta cristata*). Version 1.0. In *Birds of the World*, ed. A.F. Poole. Cornell Lab of Ornithology, Ithaca, NY. 2020. https://doi.org/10.2173/bow.blujay.01.

Smith, K.G., S.R. Wittenberg, R.B. MacWhirter, and K.L. Bildstein. 2020b. Northern Harrier (*Circus hudsonius*). Version 1.0. In *Birds of the World*, ed. P.G. Rodewald. Cornell Lab of Ornithology, Ithaca, NY. https://doi.org/10.2173/bow.norhar2.01.

Smith, M., M. Bolton, D.J. Okill, P. Harris, G. Petrie, R. Riddington, et al. 2018. Further evidence of transatlantic migration routes and Pacific wintering grounds of Red-necked Phalaropes breeding in Shetland. *British Birds* 111:428–437.

Smith, M., M. Bolton, D.J. Okill, R.W. Summers, P. Ellis, F. Liechti, and J.D. Wilson. 2014. Geolocator tagging reveals Pacific migration of Red-necked Phalarope *Phalaropus lobatus* breeding in Scotland. *Ibis* 156(4):870–873.

Smith, N.R., A.K. Afton, and T.J. Hess Jr. 2017. Winter breeding and summer nonbreeding home ranges of Bald Eagles from Louisiana. *American Midland Naturalist* 178(2):203–214.

Smith, P.A., A.C. Smith, B. Andres, C.M. Francis, B. Harrington, C. Friis, R.G. Morrison, J. Paquet, B. Winn, and S. Brown. 2023. Accelerating declines of North America's shorebirds signal the need for urgent conservation action. *Ornithological Applications* 125(2):duad003.

Smith, R.B., T.D. Meehan, and B.O. Wolf. 2003. Assessing migration patterns of Sharp-Shinned Hawks *Accipiter striatus* using stable-isotope and band encounter analysis. *Journal of Avian Biology* 34(4):387–392.

Smith, R.J., M.I. Hatch, D.A. Cimprich, and F.R. Moore. 2020. Gray Catbird (*Dumetella carolinensis*). Version 1.0. In *Birds of the World*, ed. A.F. Poole. Cornell Lab of Ornithology, Ithaca, NY. https://doi.org/10.2173/bow.grycat.01.

Smith, S.M., and F.G. Stiles. 1979. Banding studies of migrating shorebirds in northwestern Costa Rica. In *Shorebirds in Marine Environments*, ed. F.A. Pitelka, 41–47. *Studies in Avian Biology* 2(1), article 7. Cooper Ornithological Society, Allen Press.

Snell, R.R., P. Pyle, and M.A. Patten. 2020. Iceland Gull (*Larus glaucoides*). Version 1.0. In *Birds of the World*, ed. P.G. Rodewald and B.K. Keeney. Cornell Lab of Ornithology, Ithaca, NY. https://doi.org/10.2173/bow.y00478.01.

Snyder, N.F.R., H.A. Snyder, J.L. Lincer, and R.T. Reynolds. 1973. Organochlorines, heavy metals, and the biology of North American Accipiters. *BioScience* 23(5):300–305.

Soanes, L.M., J.A. Bright, M. Bolton, J. Millett, F. Mukhida, and J.A. Green. 2015a. Foraging behaviour of Brown Boobies *Sula leucogaster* in Anguilla, Lesser Antilles: Preliminary identification of at-sea distribution using a time-in-area approach. *Bird Conservation International* 25(1):87–96.

Soanes, L.M., J.A. Bright, G. Brodin, F. Mukhida, and J.A. Green. 2015b. Tracking a small seabird: First records of foraging behaviour in the Sooty Tern *Onychoprion fuscatus*. *Marine Ornithology* 43:235–239.

Söhle, I.S., C.J.R. Robertson, D.G. Nicholls, H. Mouritsen, B. Frost, and H. Moller. 2007. Satellite tracking of Sooty Shearwaters (*Puffinus griseus*) during their pre-laying "exodus" and incubation. *Notornis* 54:180–188.

Soldatini, C., Y.V. Albores-Barajas, A. Ramos-Rodriguez, et al. 2019. Tracking reveals behavioural coordination driven by environmental constraints in the Black-vented Shearwater *Puffinus opisthomelas*. *Population Ecology* 61(2):227–239. https://doi.org/10.1002/1438-390X.1024.

Soldatini, C., M.P. Rosas Hernandez, Y.V. Albores-Barajas, C. Catoni, A. Ramos, G. Dell'Omo, et al. 2023. Individual variability in diving behavior of the Black-vented Shearwater in an ever-changing habitat. *Science of the Total Environment* 163286. https://doi.org/10.1016/j.scitotenv.2023.163286.

Sonsthagen, S.A., R.L. Rodriguez, and C.M. White. 2006. Satellite telemetry of Northern Goshawks breeding in Utah—I. Annual movements. *Studies in Avian Biology* 31:239–251.

Sonsthagen, S.A., R.N. Rosenfield, J. Bielefeldt, R.K. Murphy, A.C. Stewart, W.E. Stout, T.G. Driscoll, M.A. Bozek, B.L. Sloss, and S.L. Talbot. 2012. Genetic and morphological divergence among Cooper's Hawk (*Accipiter cooperii*) populations breeding in north-central and western North America. *Auk* 129(3):427–437.

Sonsthagen, S.A., S.L. Talbot, K.T. Scribner, and K.G. McCracken. 2011. Multilocus phylogeography and population structure of Common Eiders breeding in North America and Scandinavia. *Journal of Biogeography* 38:1368–1380.

Sonsthagen, S.A., T.L. Tibbitts, R.E. Gill Jr., I. Williams, and S.L. Talbot. 2014. Spatial genetic structure of Bristle-thighed Curlews (*Numenius tahitiensis*): Breeding area differentiation not reflected on the nonbreeding grounds. *Conservation Genetics* 16:223–233.

Sonsthagen, S.A., R.E. Wilson, P. Lavretsky, and S.L. Talbot. 2019. Coast to coast: High genomic connectivity in North American scoters. *Ecology and Evolution* 9:7246–7261.

Sorensen, M.C., J.M. Hipfner, T.K. Kyser, and D.R. Norris. 2010. Pre-breeding diet influences ornament size in the Rhinoceros Auklet *Cerorhinca monocerata*. *Ibis* 152(1):29–37.

Souchay, G., G. Gauthier, J. Lefebvre, and R. Pradel. 2015. Absence of difference in survival between two distant breeding sites of Greater Snow Geese. *Journal of Wildlife Management* 79(4):570–578.

Sousa, B.F., S.A. Temple, and G.D. Basili. 2022. Dickcissel (*Spiza americana*). Version 2.0. In *Birds of the World*, ed. T.S. Schulenberg and B.K. Keeney. Cornell Lab of Ornithology, Ithaca, NY. https://doi.org/10.2173/bow.dickci.02.

Southern, W.E. 1974. Florida distribution of Ring-Billed Gulls from the Great Lakes region. *Bird Banding* 45:341–352.

Sovada, M.A., P.J. Pietz, R.O. Woodward, A.J. Bartos, D.A. Buhl, and M.J. Assenmacher. 2013. American White Pelicans breeding in the northern plains: Productivity, behavior, movements, and migration. US Geological Survey Scientific Investigations Report 2013–5105.

Spear, L.B., and D.G. Ainley. 1999. Migration routes of Sooty Shearwaters in the Pacific Ocean. *Condor* 101(2):205–218. https://doi.org/10.2307/1369984.

———. 2005. At-sea distributions and abundance of tropicbirds in the eastern Pacific. *Ibis* 147(2):353–366.

———. 2007. Storm-Petrels of the eastern Pacific Ocean: Species assembly and diversity along marine habitat gradients. *Ornithological Monographs* 62:iii–77. https://doi.org/10.2307/40166847.

Spear, L.B., D.G. Ainley, N. Nur, and S. Howell. 1995. Population size and the behavioral and physical factors affecting at-sea distributions of four endangered procellariids in the tropical Pacific. *Condor* 97:613–638.

Species at Risk Act. Statues of Canada 2002, c 29. Last amended November 27, 2024; current to June 25, 2025. https://laws.justice.gc.ca/eng/acts/s-15.3/FullText.html.

Spencer, N.C., H.G. Gilchrist, and M.L. Mallory. 2014. Annual movement patterns of endangered Ivory Gulls: The importance of sea ice. *PLOS ONE* 9(12):e115231. https://doi.org/ 10.1371/journal.pone.0115231.

Spencer, N.C., H.G. Gilchrist, H. Strøm, K.A. Allard, and M.L. Mallory. 2016. Key winter habitat of the Ivory Gull *Pagophila eburnea* in the Canadian Arctic. *Endangered Species Research* 31:33–45. https://doi.org/10.3354/esr00747.

Spiegel, C.S., A.M. Berlin, A.T. Gilbert, C.O. Gray, W.A. Montevecchi, I.J. Stenhouse, S.L. Ford, G.H. Olsen, J.L. Fiely, L. Savoy, M.W. Goodale, and C.M. Burke. 2017. Determining fine- scale use and movement patterns of diving bird species in federal waters of the mid-Atlantic United States using satellite telemetry. OCS Study BOEM 2017-069. US Department of the Interior, Bureau of Ocean Energy Management. Sterling, VA.

Sprandel, G.L., H.A. Bolte, and K.T. Bowman. 1999. Wintering locations of Black Skimmers breeding in the Florida Panhandle. *Florida Field Naturalist* 27(3):109–111.

Squires, J.R., and R.T. Reynolds. 2023. American Goshawk (*Accipiter atricapillus*). Version 1.1. In *Birds of the World*, ed. N.D. Sly. Cornell Lab of Ornithology, Ithaca, NY. https://doi.org/10.2173/bow.norgos.01.1.

Staicer, C.A., and M.J. Guzy. 2020. Grace's Warbler (*Setophaga graciae*). Version 1.0. In *Birds of the World*, ed. A.F. Poole and F.B. Gill. Cornell Lab of Ornithology, Ithaca, NY. https://doi.org/10.2173/bow.grawar.01.

Stanley, C.Q., M.R. Dudash, T.B. Ryder, W.G. Shriver, and P.P. Marra. 2021a. Variable tropical moisture and food availability underlie mixed winter space-use strategies in a migratory songbird. *Proceedings of the Royal Society: Biological Sciences* 288(1955):20211220.

Stanley, C.Q., M.R. Dudash, T.B. Ryder, W.G. Shriver, K. Serno, S. Adalsteinsson, and P.P. Marra. 2021b. Seasonal variation in habitat selection for a Neotropical migratory songbird using high-resolution GPS tracking. *Ecosphere* 12(3). https://doi.org/10.1002/ecs2.3421.

Stanley, C.Q., E.A. McKinnon, K.C. Fraser, M.P. Macpherson, G. Casbourn, L. Friesen, P.P. Marra, C.E. Studds, T.B. Ryder, N.E. Diggs, and B.J.M. Stutchbury. 2015. Connectivity of Wood Thrush breeding, wintering, and migration sites based on range-wide tracking. *Conservation Biology* 29:164–174.

Stanley, C.Q., M. MacPherson, K.C. Fraser, E.A. McKinnon, and B.J. Stutchbury. 2012. Repeat tracking of individual songbirds reveals consistent migration timing but flexibility in route. *PLOS ONE* 7(7), e40688.

Stantial, M.L., and J.B. Cohen. 2020. Miniature GPS tags provide high resolution location data for a small-bodied threatened shorebird. *Waterbirds* 43(3–4):271–280. https://doi.org/10.1675/063.043.0305.

Stedman, S.J. 1989. Return and recovery rates for Purple Finches banded in Williamson County, Tennessee. *North American Bird Bander* (July–September):86–87.

———. 2020. Horned Grebe (*Podiceps auritus*). Version 1.0. In *Birds of the World*, ed. S.M. Billerman. Cornell Lab of Ornithology, Ithaca, NY. https://doi.org/10.2173/bow.horgre.01.

Steenhof, K. 2020. Prairie Falcon (*Falco mexicanus*). Version 1.0. In *Birds of the World*, ed. A.F. Poole. Cornell Lab of Ornithology, Ithaca, NY. https://doi.org/10.2173/bow.prafal.01.

Steenhof, K., M.R. Fuller, M.N. Kochert, and K.K. Bates. 2005. Long-range movements and breeding dispersal of Prairie Falcons from southwest Idaho. *Condor* 107(3):481–496.

Steenhof, K., M.N. Kochert, and M.Q. Moritsch. 1984. Dispersal and migration of southwestern Idaho raptors. *Journal of Field Ornithology* 55(3):357–368.

Steenweg, R.J., G.T. Crossin, T.K. Kyser, F.R. Merkel, H.G. Gilchrist, H.L. Hennin, G.J. Robertson, J.F. Provencher, J. Mills Flemming, and O.P. Love. 2017. Stable isotopes can be used to infer the overwintering locations of prebreeding marine birds in the Canadian Arctic. *Ecology and Evolution* 7:8742–8752.

Steeves, T.E., D.J. Anderson, H. McNally, M.H. Kim, and V.L. Friesen. 2003. Phylogeography of *Sula*: The role of physical barriers to gene flow in the diversification of tropical seabirds. *Journal of Avian Biology* 34(2):217–223.

Steeves, T.E., D.J. Anderson, and V.L. Friesen. 2005. A role for nonphysical barriers to gene flow in the diversification of a highly vagile seabird, the Masked Booby (*Sula dactylatra*). *Molecular Ecology* 14(12):3877–3887.

Steeves, T.K., S. B. Kearney-McGee, M.A. Rubega, C.L. Cink, and C.T. Collins. 2020. Chimney Swift (*Chaetura pelagica*). Version 1.0. In *Birds of the World*, ed. A.F. Poole. Cornell Lab of Ornithology, Ithaca, NY. https://doi.org/10.2173/bow.chiswi.01.

Stefan, S.J. 2008. An analysis of recovery data of Brown Pelicans (*Pelecanus occidentalis*) and Royal Terns (*Sterna maxima*) banded in South Carolina. Master's thesis, College of Charleston, Charleston, SC.

Stehn, T.V. 1992. Unusual movements and behaviors of color-banded Whooping Cranes during winter. *North American Crane Workshop Proceedings*. Paper 285.

Stenhouse, I.J., C. Egevang, and R.A. Phillips. 2012. Trans-equatorial migration, staging sites and wintering area of Sabine's Gulls *Larus sabini* in the Atlantic Ocean. *Ibis* 154:42–51. https://doi.org/10.1111/j.1474-919X.2011.01180.x.

Stenhouse, I.J., G.J. Robertson, and H.G. Gilchrist. 2004. Recoveries and survival rates of Ivory Gulls (*Pagophila eburnea*) banded in Nunavut, Canada, 1971–1999. *Waterbirds* 27:486–492.

Stephens, P.A., I.L. Boyd, J.M. McNamara, and A.I. Houston. 2009. Capital breeding and income breeding: Their meaning, measurement, and worth. *Ecology* 90(8):2057–2067.

Stevens, H.C., M. DeSaix, B.C. Re, J. Salgado-Ortiz, B.A. Razo Cárdenas, D. Ruíz Domínguez, Y. Caicedo-Ortiz, D. Garzón-Lozano, B.C. Dossman, K. Ruegg, and P. P. Marra. Integrated, range-wide migratory connectivity for a long-distance migratory bird. *Ecosphere*. In revision.

Stevenson, H.M., and B.H. Anderson. 1994. *The Birdlife of Florida*. University Press of Florida, Gainesville.

Stewart, K. 2020. Spatial visualization of migration of Mountain Blue Birds. https://storymaps.arcgis.com/stories/9a3864621c934f2f8e2a7e7491f8e995.

Stewart, M.T., W.S. Clark, B.A Millsap, B.D. Bibles, and T. Brush. 2023. Adult home range and juvenile movements of Gray Hawks in the lower Rio Grande Valley, Texas. *Journal of Raptor Research* 57:1–11.

Stewart, R.E. 1954. Migratory movements of the northern Clapper Rail. *Bird Banding* 25(1):1–5.

Stonefish, D., M.A. Eshleman, G.M. Linz, H.J. Homan, P.E. Klug, T.J. Greives, and E.H Gillan. 2021. Migration routes and wintering areas of male Red-winged Blackbirds as determined using geolocators. *Journal of Field Ornithology* 92(3):284–293.

Storer, R.W. 1950. Geographic variation in the Pigeon Guillemots of North America. *Condor* 52(1):28–31. www.jstor.org/stable/1364745.

Storey, A.E., and J. Lien. 1985. Development of the first North American colony of Manx Shearwaters. *Auk* 102:395–401.

Storm-Suke, A. 2012. The use of stable-hydrogen isotopes in connectivity studies: A test of assumptions and application with trace element analysis. PhD diss., Trent University, Peterborough, ON.

Stotz, D.F., R.O. Bierregaard, M. Cohn-Haft, P. Petermann, J. Smith, A. Whittaker, and S.V. Wilson. 1992. The status of North American migrants in central Amazonian Brazil. *Condor* 94(3):608–621.

Stouffer, P.C., R.T. Chesser, and A.E. Jahn. 2020. Tropical Kingbird (*Tyrannus melancholicus*). Version 1.0. In *Birds of the World*, ed. S.M. Billerman. Cornell Lab of Ornithology, Ithaca, NY. https://doi.org/10.2173/bow.trokin.01.

Stout, B.E., and F. Cooke. 2003. Timing and location of wing molt in Horned, Red-necked, and Western Grebes in North America. *Waterbirds* 26(1):88–93.

Stout, B.E., and G.L. Nuechterlein. 2020. Red-necked Grebe (*Podiceps grisegena*). Version 1.0. In *Birds of the World*, ed. S.M. Billerman. Cornell Lab of Ornithology, Ithaca, NY. https://doi.org/10.2173/bow.rengre.01.

Stout, W.E., V.L. Greene, and S. Postupalsky. 2009. Migration routes, reproduction, and lifespan of a translocated Osprey. *Wilson Journal of Ornithology* 121:203–206.

Straight, C.A., and R.J. Cooper. 2020. Chuck-will's-widow (*Antrostomus carolinensis*). Version 1.0. In *Birds of the World*, ed. A.F. Poole. Cornell Lab of Ornithology, Ithaca, NY. https://doi.org/10.2173/bow.chwwid.01.

Streby, H.M., T.L. McAllister, S.M. Peterson, G.R. Kramer, J.A Lehman, and D.E. Andersen. 2015. Minimizing marker mass and handling time when attaching radio transmitters and geolocators to small songbirds. *Condor* 115:249–255.

Strickler, S., and S. West. 2020. Cave Swallow (*Petrochelidon fulva*). Version 1.0. In *Birds of the World*, ed. A.F. Poole. Cornell Lab of Ornithology, Ithaca, NY. https://doi.org/10.2173/bow.cavswa.01.

Strikwerda, T.E., H.D. Black, N. Levanon, and P.W. Howey. 1985. The bird-borne transmitter. *Johns Hopkins APL Technical Digest* 6(1):60–67.

Strikwerda, T.E., M.R. Fuller, W.S. Seegar, P.W. Howey, and H.D. Black. 1986. Bird-borne satellite transmitter and location program. *Johns Hopkins APL Technical Digest* 7(2):203–208.

Stubbings, E.M., A. Ivanov, D. Dorofeev, and E. Khudyakova. 2020. To Russia with love: First record of Surfbird *Calidris virgata* in Eurasia. *Wader Study* 127(1):65–67.

Studds, C.E., B.E. Kendall, N.J. Murray, H.B. Wilson, D.I. Rogers, R.S. Clemens, K. Gosbell, C.J. Hassell, R. Jessop, D.S. Melville, D.A. Milton, C.D.T. Minton, H.P. Possingham, A.C. Riegen, P. Straw, E.J. Woehler, and R.A. Fuller. 2017. Rapid population decline in migratory shorebirds relying on Yellow Sea tidal mudflats as stopover sites. *Nature Communications* 8:14895.

Studds, C.E., T.K. Kyser, and P.P. Marra. 2008. Natal dispersal driven by environmental conditions interacting across the annual cycle of a migratory songbird. *Proceedings of the National Academy of Sciences* 105:2929–2933.

Studds, C.E., J.M. Wunderle, and P.P. Marra. 2021. Strong differences in migratory connectivity patterns among species of Neotropical-Nearctic migratory birds revealed by combining stable isotopes and abundance in a Bayesian assignment analysis. *Journal of Biogeography* 48:1746–1757. https:// doi.org/10.1111/jbi.14111.

Studholme, K.R., J.M. Hipfner, A.D. Domalik, S.J. Iverson, and G.T. Crossin. 2019. Year-round tracking reveals multiple migratory tactics in a sentinel North Pacific seabird, Cassin's Auklet. *Marine Ecology Progress Series* 619:169–185. https://doi.org/10.3354/meps12965.

Sugita, N., S. Matsui, N. Yoshikawa, and I. Nishiumi. 2021. Genetic characteristics of a small breeding population of Ancient Murrelets from Teuri Island, Japan, based on mitochondrial DNA sequences and newly-developed microsatellite markers. *Ecological Research* 36:637–647. https://doi.org/10.1111/1440-1703.12226.

Sullins, D.S., W.C. Conway, D.A. Haukos, K.A. Hobson, L.I. Wassenaar, C.E. Comer, and I. Hung. 2016. American Woodcock migratory connectivity as indicated by hydrogen isotopes. *Journal of Wildlife Management* 80:510–526.

Sullivan, A.R., J.K. Bump, L.A. Kruger, and R.O. Peterson. 2012. Bat-cave catchment areas: Using stable isotopes (δD) to determine the probable origins of hibernating bats. *Ecological Applications* 22(5):1428–1434.

Sullivan, B.L., T. Phillips, A.A. Dayer, C.L. Wood, A. Farnsworth, M.J. Iliff, I.J. Davies, A. Wiggins, D. Fink, W.M. Hochachka, and A.D. Rodewald. 2017. Using open access observational data for conservation action: A case study for birds. *Biological Conservation* 208:5–14.

Sullivan, B.L., C.L. Wood, M.J. Iliff, R.E. Bonney, D. Fink, and S. Kelling. 2009. eBird: A citizen-based bird observation network in the biological sciences. *Biological Conservation* 142(10):2282–2292.

Summers, R.W., H. Boland, K. Colhoun, N. Elkins, B. Etheridge, S. Foster, … and R.L. Swann. 2014. Contrasting trans-Atlantic migratory routes of Nearctic Purple Sandpipers *Calidris maritima* associated with low pressure systems in spring and winter. *Ardea* 102(2):139–152.

Surman, C.A., L.W. Nicholson, and R.A. Phillips. 2018. Distribution and patterns of migration of a tropical seabird community in the eastern Indian Ocean. *Journal of Ornithology* 159(3):867–877.

Suryan, R., and G.R. Balogh. Laysan albatross non-breeders 2005–2006. BirdLife International Seabird Tracking Database Dataset ID 516. https://data.seabirdtracking.org/dataset/516.

Suryan, R.M., and K.N. Fischer. 2010. Stable isotope analysis and satellite tracking reveal interspecific resource partitioning of nonbreeding albatrosses off Alaska. *Canadian Journal of Zoology* 88, 299–305.

Suryan, R.M., K.S. Dietrich, E.F. Melvin, G.R. Balogh, F. Sato, and K. Ozaki. 2007. Migratory routes of Short-tailed Albatrosses: Use of Exclusive Economic Zones of North Pacific Rim countries and spatial overlap with commercial fisheries in Alaska. *Biological Conservation* 137(3):450–460.

Swanson, D.L., J.L. Ingold, and R. Galati. 2020. Golden-crowned Kinglet (*Regulus satrapa*). Version 1.0. In *Birds of the World*, ed. A.F. Poole. Cornell Lab of Ornithology, Ithaca, NY. https://doi.org/10.2173/bow.gockin.01.

Swanson, D.L., J.L. Ingold, and G.E. Wallace. 2021. Ruby-Crowned Kinglet (*Corthylio Calendula*). Version 1.1. In *Birds of the World* (editor not available). Cornell Lab of Ornithology, Ithaca, NY. 2021. https://doi.org/10.2173/bow.ruckin.01.1.

Swarth, H.S. 1920. Revision of the avian genus *Passerella*, with special reference to the distribution and migration of the races in California. *University of California Publications in Zoology* 21(4):75–224.

Swoboda, C.J. 2007. Population delineation and wintering ground influence on vital rates of White-winged Scoters. Master's thesis, University of Saskatchewan, Saskatoon.

Symons, S., and T. Diamond. 2018. Ecological segregation between two closely related species: Exploring Atlantic Puffin and Razorbill foraging hotspots. Master's thesis, University of New Brunswick. 10.13140/RG.2.2.23114.00960.

Szarmach, S.J., J.K. Beam, M. Moore, B.M. Van Doren, A. Brelsford, and D.P.L. Toews. 2026. Unexpected migration patterns in a high-latitude breeding songbird: evidence from multi-sensor geolocators and isotopes. *Movement Ecology* 14:4. https://doi.org/10.1186/s40462-025-00618-6.

Tacha, M., A. Bishop, and J. Brei. 2010. Development of the Whooping Crane tracking project geographic information system. *Proceedings of the North American Crane Workshop* 11:9–104.

Tacha, T.C., P.A. Vohs, and G.C. Iverson. 1984. Migration routes of Sandhill Cranes from mid-continental North America. *Journal of Wildlife Management* 48(3):1028–1033.

Tacha, T.C., A. Woolf, W.D. Klimstra, and K.F. Abraham. 1991. Migration patterns of the Mississippi Valley population of Canada Geese. *Journal of Wildlife Management* 55:94–102.

Takahashi, A., M. Ito, Y. Suzuki, Y. Watanuki, J.-B. Thiebot, T. Yamamoto, et al. 2015. Migratory movements of Rhinoceros Auklets in the northwestern Pacific: Connecting seasonal productivities. *Marine Ecology Progress Series* 525:229–243. https://doi.org/10.3354/meps11179.

Takahashi, A., J.B. Thiebot, A. Will, S. Tsukamoto, B. Merkel, and A. Kitaysky. 2021. Breeding together, wintering an ocean apart: Foraging ecology of the northern Bering Sea Thick-billed and Common Murres in years of contrasting sea-ice conditions. *Polar Science* 27:100552.

Takekawa, J.Y., and N.D. Warnock. 2020. Long-billed Dowitcher (*Limnodromus scolopaceus*). Version 1.0. In *Birds of the World*, ed. A.F. Poole and F.B. Gill. Cornell Lab of Ornithology, Ithaca, NY. https://doi.org/10.2173/bow.lobdow.01.

Talbot, S.L., J.M. Pearce, B.J. Pierson, D.V. Derksen, and K.T. Scribner. 2003. Molecular status of the Dusky Canada Goose (*Branta canadensis occidentalis*): A genetic assessment of a translocation effort. *Conservation Genetics* 4:367–381.

Talbot, S.L., G.K. Sage, S.A. Sonsthagen, M.C. Gravley, T. Swem, J.C. Williams, J.L. Longmire, S. Ambrose, M.J. Flamme, S.B. Lewis, L. Phillips, C. Anderson, and C.M. White. 2017. Intraspecific evolutionary relationships among Peregrine Falcons in western North American high latitudes. *PLOS ONE* 12(11):e0188185.

Talbot, S.L., G.K. Sage, S.A. Sonsthagen, J.A. Schmutz, and M. Flamme. 2014. Preliminary results of the development of polymorphic microsatellite loci for the Yellow-billed Loon (*Gavia adamsii*). Interim report to the US Fish and Wildlife Service Region 7, Gates of the Arctic National Park and Preserve, and Yukon-Charley Rivers National Preserve, AK.

Tarof, S., and J.V. Briskie. 2020. Least Flycatcher (*Empidonax minimus*). Version 1.0. In *Birds of the World*, ed. A.F. Poole. Cornell Lab of Ornithology, Ithaca, NY. https://doi.org/10.2173/bow.leafly.01.

Tautin, J. 2008. A history of the Bird Banding Laboratory: 1920–2002. In *Bird Banding in North America: The First Hundred Years*, ed. J.A. Jacson, W.E. Davis Jr., and J. Tautin, 65–91. Nuttall Ornithological Club, Cambridge, MA.

Taylor, A., M.A. Bishop, A. Schaefer, R. Porter, and K. Sowl. 2022. Using geolocator data to address changes in migration patterns for Black Turnstone. *Animal Migration* 9:1–26.

Taylor, G.A., and M.J. Rayner. 2013 [updated 2022]. Cook's petrel | tītī. In New Zealand Birds Online (website), ed. C.M. Miskelly. https://www.nzbirdsonline.org.nz/species/cooks-petrel.

Taylor, P.D., T.L. Crewe, S.A. Mackenzie, D. Lepage, Y. Aubry, Z. Crysler, G. Finney, C.M. Francis, C.G. Guglielmo, D.J. Hamilton, R.L. Holberton, P.H. Loring, G.W. Mitchell, D. Norris, J. Paquet, R.A. Ronconi, J. Smetzer, P.A. Smith, L.J. Welch, and B.K. Woodworth. 2017. The Motus Wildlife Tracking System: A collaborative research network to enhance the understanding of wildlife movement. *Avian Conservation and Ecology* 12(1):8. https://doi.org/10.5751/ACE-00953-120108.

Teitelbaum, C.S., N.C. Bachner, and R.J. Hall. 2023. Post-migratory nonbreeding movements of birds: A review and case study. *Ecology and Evolution* 13(3):e9893.

Teitelbaum, C.S., S.J. Converse, W.F. Fagan, K. Bohning-Gaese, R.B. O'Hara, A.E. Lacy, and T. Mueller. 2016. Experience drives innovation of new migration patterns of Whooping Cranes in response to global change. *Nature Communications* 7:12793.

Telfair II, R.C. 2024. Western Cattle-Egret (*Ardea ibis*). Version 1.1. In *Birds of the World*, ed. P.G. Rodewald, B.K. Keeney, S.M. Billerman, and M.A. Bridwell. Cornell Lab of Ornithology, Ithaca, NY. https://doi.org/10.2173/bow.categr1.01.1.

Telfer, T.C., and J.K. Shisler. 1981. Record of movement of a Laughing Gull to Hawaii from New Jersey. *Journal of Field Ornithology* 52(4):340–341. www.jstor.org/stable/27639260.

Tenney, C.R. 2020a. Black-chinned Sparrow (*Spizella atrogularis*). Version 1.0. In *Birds of the World*, ed. A.F. Poole and F.B. Gill. Cornell Lab of Ornithology, Ithaca, NY. https://doi.org/10.2173/bow.bkcspa.01.

———. 2020b. Northern Beardless-Tyrannulet (*Camptostoma imberbe*). Version 1.0. In *Birds of the World*, ed. A.F. Poole and F.B. Gill. Cornell Lab of Ornithology, Ithaca, NY. https://doi.org/10.2173/bow.nobtyr.01.

Terrill, S.B., and R.D. Ohmart. 1984. Facultative extension of fall migration by Yellow-Rumped Warblers (*Dendroica coronata*). *Auk* 101(3):427–438.

Tessler, D.F., J.A. Johnson, B.A. Andres, S. Thomas, and R.B. Lanctot. 2014. A global assessment of the conservation status of the Black Oystercatcher *Haematopus bachmani*. *International Wader Studies* 20:83–96.

Therrien, J-F., G. Gauther, and J. Bêty. 2011. An avian terrestrial predator of the Arctic relies on the marine ecosystem during winter. *Journal of Avian Biology* 42:363–369.

———. 2012. Survival and reproduction of adult Snowy Owls tracked by satellite. *Journal of Wildlife Management* 76(8):1562–1567.

Therrien, J-F., G. Gauthier, D. Pinaud, and J. Bêty. 2014. Irruptive movements and breeding dispersal of Snowy Owls: A specialized predator exploring a pulsed resource. *Journal of Avian Biology* 45:536–544.

Thomas, P.W., G.H. Mittelhauser, T.E. Chubbs, P.G. Trimper, R.I. Goudie, G.J. Robertson, S. Brodeur, M. Robert, S. Gilliland, and J.-P.L. Savard. 2008. Movements of Harlequin Ducks in eastern North America. *Waterbirds* 31(sp2):188–193. https://doi.org/10.1675/1524-4695-31.sp2.188.

Thompson, B.C., J.A. Jackson, J. Burger, L.A. Hill, E.M. Kirsch, and J.L. Atwood. 2020. Least Tern (*Sternula antillarum*). Version 1.0. In *Birds of the World*, ed. A.F. Poole and F.B. Gill. Cornell Lab of Ornithology, Ithaca, NY. https://doi.org/10.2173/bow.leater1.01.

Thompson, C.F., and K.P. Eckerle. 2022. Yellow-breasted Chat (*Icteria virens*). Version 2.0. In *Birds of the World*, ed. P.G. Rodewald and B.K. Keeney. Cornell Lab of Ornithology, Ithaca, NY. https://doi.org/10.2173/bow.yebcha.02.

Thompson, C.W. 1991. The sequence of molts and plumages in Painted Buntings and implications for theories of delayed plumage maturation. *Condor* 93:209–235.

Thompson, D.R., L.G. Torres, G.A. Taylor, M.J. Rayner, et al. 2015. Stable isotope values delineate the non-breeding distributions of Sooty Shearwaters *Puffinus griseus* in the North Pacific Ocean. *Marine Ecology Progress Series* 521:277–282. https://doi.org/10.3354/meps11128.

Thompson, M.C. 1973. Migratory patterns of Ruddy Turnstones in the central Pacific region. *Living Bird* 12:5–23.

Thorup, K., T.E. Ortvad, and J. Rabøl. 2006. Do Nearctic Northern Wheatears (*Oenanthe oenanthe leucorhoa*) migrate nonstop to Africa? *Condor* 108:446–451.

Thul, J.E., D.J. Forrester, and C.L. Abercrombie. 1985. Ecology of parasitic helminths of Wood Ducks, *Aix sponsa*, in the Atlantic Flyway. *Proceeding of the Helminthological Society of Washington* 52:297–310.

Thurber, W.A., J.F. Serrano, A. Sermeno, and M. Benitez. 1987. Status of uncommon and previously unreported birds of El Salvador. *Proceedings of the Western Foundation of Vertebrate Zoology* 3:109–293.

Tibbitts, T.L., R.B. Lanctot, and D.C. Douglas. 2023. Tracking data for Buff-breasted Sandpipers (*Calidris subruficollis*). Version 1.0. US Geological Survey data release. https://doi.org/10.5066/P9TYMUID.

Tibbitts, T.L., and W. Moskoff. 2020. Lesser Yellowlegs (*Tringa flavipes*). Version 1.0. In *Birds of the World*, ed. A.F. Poole. Cornell Lab of Ornithology, Ithaca, NY.

Tigano, A., A.J. Shultz, S.V. Edwards, G.J. Robertson, and V.L. Friesen. 2017. Outlier analyses to test for local adaptation to breeding grounds in a migratory arctic seabird. *Ecology and Evolution* 7(7):2370–2381. https://doi.org/10.1002/ece3.2819.

Titus, K., and M.R. Fuller. 1990. Recent trends in counts of migrant hawks from northeastern North America. *Journal of Wildlife Management* 54(3):463–470.

Toews, D.P.L., and D.E. Irwin. 2020. Pacific Wren (*Troglodytes pacificus*). Version 1.0. In *Birds of the World*, ed. A.F. Poole. Cornell Lab of Ornithology, Ithaca, NY. https://doi.org/10.2173/bow.pacwre1.01.

Toews, D.P.L., A. Brelsford, and D.E. Irwin. 2014. Isotopic variation across the Audubon's–Myrtle Warbler hybrid zone. *Journal of Evolutionary Biology* 27:1179–1191.

Toews, D.P.L., J. Heavyside, and D.E. Irwin. 2017. Linking the wintering and breeding grounds of warblers along the Pacific Flyway. *Ecology and Evolution* 7:6649–6658.

Tomita, N., F. Sato, J.B. Thiebot, B. Nishizawa, M. Eda, H. Izumi, S. Konno, M. Konno, and Y. Watanuki. 2024. Incomplete isolation in the nonbreeding areas of two genetically separated but sympatric short-tailed albatross populations. *Endangered Species Research* 53:213–225. https://doi.org/10.3354/esr01302.

Tomkovich, P.S. 1990. On the post-fledging movements of Ivory Gulls (*Pagophila eburnea*) [in Russian]. *Modern Ornithology* 150–151. Nauka, Moscow.

Tomkovich, P.S., and M.Y. Soloviev. 1994. Site fidelity in high arctic breeding waders. *Ostrich* 65(2):174–180.

Tonra, C.M., M.T. Hallworth, T.J. Boves, J. Reese, L.P. Bulluck, M. Johnson, C. Viverette, K. Percy, E.M. Ames, A. Matthews, and M.C. Slevin. 2019a. Concentration of a widespread breeding population in a few critically important nonbreeding areas: Migratory connectivity in the Prothonotary Warbler. *Condor* 121(2):1–15. https://doi.org/10.1093/condor/duz019.

Tonra, C.M., J.R. Wright, and S.N. Matthews. 2019b. Remote estimation of overwintering home ranges in an elusive migratory nocturnal bird. *Ecology and Evolution* 9(22):12586–12599.

Torres-Dowdall, J., A.H. Farmer, M. Abril, E.H. Bucher, and I. Ridley. 2010. Trace elements have limited utility for studying migratory connectivity in shorebirds that winter in Argentina. *Condor* 112(3):490–498.

Torres-Dowdall, J., A.H. Farmer, E.H. Bucher, R.O. Rye, and G. Landis. 2009. Population variation in isotopic composition of shorebird feathers: Implications for determining molting grounds. *Waterbirds* 32:300–310.

Townsend, A.K., et al. 2018. Where do winter crows go? Characterizing partial migration of American Crows with satellite telemetry, stable isotopes, and molecular markers. *Auk* 135(4):964–974. https://doi.org/10.1642/AUK-18-23.1.

Townsend, J.M., and C.C. Rimmer. 2006. Known natal and wintering sites of a Bicknell's Thrush. *Journal of Field Ornithology* 77:452–454.

Townsend, J.M., K.P. McFarland, C.C. Rimmer, W.G. Ellison, and J.E. Goetz. 2020. Bicknell's Thrush (*Catharus bicknelli*). Version 1.0. In *Birds of the World*, ed. P.G. Rodewald. Cornell Lab of Ornithology, Ithaca, NY. https://doi.org/10.2173/bow.bicthr.01.

Trefry, S.A., and A.W. Diamond. 2017. Exploring hypotheses for sexual size dimorphism in frigatebirds. *Evolutionary Ecology Research* 18(1):225–252

Trefry, S.A., A.W. Diamond, and L.K. Jesson. 2013. Wing marker woes: A case study and meta-analysis of the impacts of wing and patagial tags. *Journal of Ornithology* 154:1–11.

Tremblay, J., and P. Cote. Fall migration of American Pipit and Horned Lark (Project 288). Motus Wildlife Tracking System. https://motus.org/.

Troy, D. 2007. From the top of the world to down under: Migration of a Pomarine Jaeger. *Tracker News* 8(1):2.

Troy, D.M. 1983. Recaptures of Redpolls: Movements of an irruptive species. *Journal of Field Ornithology* 54(2):146–151.

Trulio, L.A., D.A. Chromczak, P. Higgins, S. Menzel, C. Bossu, and K. Ruegg. 2023. Combining genomic and field analyses to reveal migratory status in a Burrowing Owl population. *Conservation Genetics* 25:427–437. https://doi.org/10.1007/s10592-023-01578-3.

Turbek, S.P., D.R. Schield, E.S. Scordato, A. Contina, X.W. Da, Y. Liu, Y. Liu, E. Pagani-Núñez, Q.M. Ren, C.C Smith, and C.A. Stricker. 2022. A migratory divide spanning two continents is associated with genomic and ecological divergence. *Evolution* 76(4):722–736.

Turcotte, Y., J.F. Lamarre, and J. Bêty. 2013. Staging ecology of Semipalmated Plover (*Charadrius semipalmatus*) and Semipalmated Sandpiper (*Calidris pusilla*) juveniles in the St. Lawrence River Estuary during fall migration. *Canadian Journal of Zoology* 91(11):802–809.

Twedt, D.J., and R.D. Crawford. 2020. Yellow-headed Blackbird (*Xanthocephalus xanthocephalus*). Version 1.0. In *Birds of the World*, ed. A.F. Poole and F.B. Gill. Cornell Lab of Ornithology, Ithaca, NY. https://doi.org/10.2173/bow.yehbla.01.

Tweit, R.C., and J.C. Tweit. 2020. Dusky-capped Flycatcher (*Myiarchus tuberculifer*). Version 1.0. In *Birds of the World*, ed. A.F. Poole and F.B. Gill. Cornell Lab of Ornithology, Ithaca, NY. https://doi.org/10.2173/bow.ducfly.01.

Uher-Koch, B.D., M.R. North, and J.A. Schmutz. 2020. Yellow-billed Loon (*Gavia adamsii*). Version 1.0. In *Birds of the World*, ed. S.M. Billerman. Cornell Lab of Ornithology, Ithaca, NY. https://doi.org/10.2173/bow.yebloo.01.

Uher-Koch, B.D., J.A. Schmutz, and D.C. Douglas. 2017 (revised 2020). Tracking data for Red-throated Loons (*Gavia stellata*). Version 2.0. US Geological Survey data release, https://doi.org/10.5066/F7TH8KVH.

Underwood, J., C.M. White, and R.L. Rodriguez. 2006. Winter movement and habitat use of Northern Goshawks breeding in Utah. *Studies in Avian Biology* 31:228–238.

Ungvari-Martin, J., C.M. Heckscher, and K.A. Hobson. 2016. Inter-annual site fidelity and breeding origins of Gray-cheeked Thrushes in white sand forests of the Peruvian Amazon. *Journal of Field Ornithology* 87(1):55–64.

Unitt, P., and R. Rodriguez Estrella. 1996. Winter distribution of Hermit Thrush subspecies in the Sierra de la Laguna, Baja California Sur. *Western Birds* 27:65–69.

Unitt, P., M.A. Faulkner, and C. Swanson. 2009. First record of Newell's Shearwater from the mainland of North America. *West Birds* 40:21–28.

Urbanek, R.P., and J.C. Lewis. 2020. Whooping Crane (*Grus americana*). Version 1.0. In *Birds of the World*, ed. A.F. Poole. Cornell Lab of Ornithology, Ithaca, NY. https://doi.org/10.2173/bow.whocra.01.

US Fish and Wildlife Service. Threatened & Endangered Species. Environmental Conservation Online System (ECOS). https://ecos.fws.gov/ecp/.

US Fish and Wildlife Service. 2008. Birds of Conservation Concern. US Department of Interior, Division of Migratory Bird Management, Arlington, VA.

US Fish and Wildlife Service. 2020. Short-tailed Albatross (*Phoebastria albatrus*) 5-year review: Summary and evaluation. https://www.fws.gov/node/5106431.

US Geological Survey, Patuxent Wildlife Research Center. 2014. North American Breeding Bird Survey 1966–2014 Analysis.

US Geological Survey, Alaska Science Center. 2019. Wildlife tracking data collection, data release. https://doi.org/10.5066/P9VYSWEH.

Ushakova, M.V. 2007. The Rhinoceros Auklet (*Cerorhinca monocerata*, Alcidae) colonies and number on the southern Kuril Islands (English abstract). *Zoologicheskiĭ Zhurnal* 86:955–965.

van Bemmelen, R., B. Moe, S.A. Hanssen, N.M. Schmidt, J. Hansen, J. Lang, B. Sittler, L. Bollache, I. Tulp, R. Klaassen, and O. Gilg. 2017. Flexibility in otherwise consistent non-breeding movements of a long-distance migratory seabird, the Long-tailed Skua. *Marine Ecology Progress Series* 578:197–211.

van Bemmelen, R.S., Y. Kolbeinsson, R. Ramos, O. Gilg, J.A. Alves, M. Smith, … and I. Tulp. 2019. A migratory divide among Red-necked Phalaropes in the Western Palearctic reveals contrasting migration and wintering movement strategies. *Frontiers in Ecology and Evolution* 86.

van Bemmelen, R.S., B. Moe, H. Schekkerman, S.A. Hansen, K.R. Snell, E.M. Humphreys, E. Mäntylä, G.T. Hallgrimsson, O. Gilg, D. Ehrich, and J. Calladine. 2024. Synchronous timing of return to breeding sites in a long-distance migratory seabird with ocean-scale variation in migration schedules. *Movement Ecology* 12(1):22.

van Bemmelen, R.S.A. 2019. Seabirds linking Arctic and ocean. PhD diss., Wageningen University & Research, Wageningen, Netherlands. https://doi.org/10.18174/499288.

van Brederode, N.E., H.J. Roersma, and C.J. de Graaf. 2009. Switch of flyway by a Ruddy Turnstone *Arenaria interpres*. *Wader Study Group Bulletin* 116(3):198–199.

Vanderhoff, N., P. Pyle, M.A. Patten, R. Sallabanks, and F.C. James. 2020. American Robin (*Turdus migratorius*). Version 1.0. In *Birds of the World*, ed. P.G. Rodewald. Cornell Lab of Ornithology, Ithaca, NY. https://doi.org/10.2173/bow.amerob.01.

Vansteelant, W., J. Shamoun-Baranes, J. McLaren, J. van Diermen, and W. Bouten. 2017. Soaring across continents: Decision-making of a soaring migrant under changing atmospheric conditions along an entire flyway. *Journal of Avian Biology* 48(6)887–896. https://doi.org/10.1111/jav.01298.

Varela, A.I., K. Brokordt, S.M.H. Ismar-Rebitz, C.P. Gaskin, N. Carlile, T. O'Dwyer, J. Adams, E.A. VanderWerf, and G. Luna-Jorquera. 2021. Genetic diversity, population structure, and historical demography of a highly vagile and human-impacted seabird in the Pacific Ocean: The Red-tailed Tropicbird, *Phaethon rubricauda*. *Aquatic Conservation: Marine and Freshwater Ecosystems* 31:367–377.

Varela, A.I., K. Brokordt, J.A. Vianna, M.J. Frugone, S.M. Ismar-Rebitz, C.P. Gaskin, N. Carlile, T. O'Dwyer, J. Adams, E.A. VanderWerf, and G. Luna-Jorquera. 2024. Are threatened seabird colonies of the Pacific Ocean genetically vulnerable? The case of the Red-tailed Tropicbird, *Phaethon rubricauda*, as a model species. *Biodiversity and Conservation* 33(3):1165–1184.

Vaught, R.W., and G.C. Arthur. 1965. Migration routes and mortality rates of Canada Geese banded in the Hudson Bay Lowlands. *Journal of Wildlife Management* 29(2):244–252.

Veit, R.R., and W. R. Petersen. 1993. *The Birds of Massachusetts*. Massachusetts Audubon Society, Lincoln, MA.

Velarde, E., E.A. Ruiz, A. Aguilar, and J.P. Gallo. 2015. Black-vented Shearwater *Puffinus opisthomelas* nesting in the Gulf of California: A major extension of breeding range. *Marine Ornithology* 43:249–254.

Venier, L., S. Holmes, and J.M. Williams. 2020. Bay-breasted Warbler (*Setophaga castanea*). Version 1.0. In *Birds of the World*, ed. A.F. Poole. Cornell Lab of Ornithology, Ithaca, NY. https://doi.org/10.2173/bow.babwar.01.

Ventura, F., J.P. Granadeiro, O. Padget, and P. Catry. 2020. Gadfly Petrels use knowledge of the windscape, not memorized foraging patches, to optimize foraging trips on ocean-wide scales. *Proceedings of the Royal Society: Biological Sciences* 287:20191775. http://dx.doi.org/10.1098/rspb.2019.1775.

Verbeek, N.A., C. Caffrey, A.B. Clark, K.J. McGowan, and P. Pyle. 2024. American Crow (*Corvus brachyrhynchos*). Version 1.2. In *Birds of the World*, ed. A.F. Poole, F.B. Gill, and M.G. Smith. Cornell Lab of Ornithology, Ithaca, NY. https://doi.org/10.2173/bow.amecro.01.2.

Vermeer, K. 1970. Breeding biology of California and Ring-Billed Gulls: A study of ecological adaptation to the inland habitat. Vol. 12. Queen's Printer for Canada.

Vickery, P.D. 2020. Grasshopper Sparrow (*Ammodramus savannarum*). Version 1.0. In *Birds of the World*, ed. A.F. Poole and F.B. Gill. Cornell Lab of Ornithology, Ithaca, NY. https://doi.org/10.2173/bow.graspa.01.

Vidaña, B., N. Busquets, S. Napp, E. Pérez-Ramírez, E., M.Á. Jiménez-Clavero, and N. Johnson. 2020. The role of birds of prey in West Nile virus epidemiology. *Vaccines* 8(3):550.

Vierling, K.T., V.A. Saab, and B.W. Tobalske. 2020. Lewis's Woodpecker (*Melanerpes lewis*). Version 1.0. In *Birds of the World*, ed. A.F. Poole. Cornell Lab of Ornithology, Ithaca, NY. https://doi.org/10.2173/bow.lewwoo.01.

Vitz, A.C., L.A. Hanners, and S.R. Patton. 2020. Worm-eating Warbler (*Helmitheros vermivorum*). Version 1.0. In *Birds of the World*, ed. A.F. Poole. Cornell Lab of Ornithology, Ithaca, NY. https://doi.org/10.2173/bow.woewar1.01.

Viverette, C.B., S. Struve, L.J. Goodrich, and K.L. Bildstein. 1996. Decreases in migrating Sharp-shinned Hawks (*Accipiter striatus*) at traditional raptor-migration watch sites in eastern North America. *Auk* 113(1):32–40.

Vogt, D.F., M.E. Hopey, G.R. Mayfield, E.C. Soehren, L.M. Lewis, J.A. Trent, and S.A. Rush. 2012. Stopover site fidelity by Tennessee Warblers at a southern Appalachian high-elevation Site. *Wilson Journal of Ornithology* 124:366–370.

Vokhshoori, N.L., M.D. McCarthy, P.W. Collins, M.A. Etnier, et al. 2019. Broader foraging range of ancient short-tailed albatross populations into California coastal waters based on bulk tissue and amino acid isotope analysis. *Marine Ecology Progress Series* 610:1–13. https://doi.org/10.3354/meps12839.

Volkov, A.E., and J.D. Korte. 1996. Distribution and numbers of breeding Ivory Gulls *Pagophila eburnea* in Severnaja Zemlja, Russian Arctic. *Polar Research* 15(1):11–21.

Von Rönn, J.A.C., C. Harrod, S. Bensch, and J.B.W. Wolf. 2015. Transcontinental migratory connectivity predicts parasite prevalence in breeding populations of the European barn swallow. *Journal of Evolutionary Biology* 28(3):535–546.

Vukovich, M., and J.C. Kilgo. 2013. Fall movements of Red-headed Woodpeckers in South Carolina. *Journal of Field Ornithology* 84(2):193–200.

Wahl, T.R., and B. Tweit. 2000. Where do Pigeon Guillomots from California go for the winter? *Western Birds* 31(3):203–206.

Wakefield, E.D., D.L. Miller, S.L. Bond, F. le Bouard, P.C. Carvalho, P. Catry, B.J. Dilley, D.A. Fifield, C. Cjerdrum, J. González-Solís, H. Hogan, V. Laptikhovsky, B. Merkel, J.A.O. Miller, P.I. Miller, S.J. Pinder, T. Pipa, P.M. Ryan, L.A. Thompson, and J. Matthiopoulos. 2021. The summer distribution, habitat associations and abundance of seabirds in the sub-polar frontal zone of the Northwest Atlantic. *Progress in Oceanography* 198:102657. https://doi.org/10.1016/j.pocean.2021.102657.

Walker, B.M., N.R. Senner, C.S. Elphick, and J. Klima. 2020. Hudsonian Godwit (*Limosa haemastica*). Version 2.0. In *The Birds of North America*, ed. A.F. Poole. Cornell Lab of Ornithology, Ithaca, NY. https://doi.org/10.2173/bow.hudgod.01.

Walker, J., J.-F. Therrien, C. Bégin-Marchand, P. Côté, A. Terrigeol, F. Gagnon, and J.A. Tremblay. 2025. Insights from a decade of using the Motus network to track boreal bird species from Observatoire d'oiseaux de Tadoussac, Québec to temperate and tropical wintering grounds. *Journal of Field Ornithology* 96:2.

Wall, J., D. Brinker, S. Weidensaul, D. Okines, P. Côté, and J.F. Therrien. 2020. Twenty-five year population trends in Northern Saw-whet Owl (*Aegolius acadicus*) in eastern North America. *Wilson Journal of Ornithology* 132(3):739–745.

Wallace, E.A., and G.E. Wallace. 2021. Brandt's Cormorant (*Urile penicillatus*). Version 1.1. In *Birds of the World*, ed. A.F. Poole and F.B. Gill. Cornell Lab of Ornithology, Ithaca, NY. https://doi.org/10.2173/bow.bracor.01.1.

Wallace, G.J. 1939. Bicknell's Thrush, its taxonomy, distribution, and life history. *Proceedings of the Boston Society of Natural History* 41:211–402.

Wallace, S.J., S.G. Wolf, R.W. Bradley, A. Laurie Harvey, and V.L. Friesen. 2015. The influence of biogeographical barriers on the population genetic structure and gene flow in a coastal Pacific seabird. *Journal of Biogeography* 42(2):390–400. https://doi.org/10.1111/jbi.12404.

Walsh, H.E., and S.V. Edwards. 2005. Conservation genetics and Pacific fisheries bycatch: Mitochondrial differentiation and population assignment in Black-footed Albatrosses (*Phoebastria nigripes*). *Conservation Genetics* 6(2):289–295.

Walsh, J., V. Elia, R. Kane, and T. Halliwell. 1999. *Birds of New Jersey*. New Jersey Audubon Society, Bernardsville.

Walters, E. L., E.H. Miller, and P.E. Lowther. 2020a. Red-naped Sapsucker (*Sphyrapicus nuchalis*). Version 1.0. In *Birds of the World*, ed. A.F. Poole. Cornell Lab of Ornithology, Ithaca, NY. https://doi.org/10.2173/bow.rensap.01.

———. 2020b. Yellow-bellied Sapsucker (*Sphyrapicus varius*). Version 1.0. In *Birds of the World*, ed. A.F. Poole and F.B. Gill. Cornell Lab of Ornithology, Ithaca, NY. https://doi.org/10.2173/bow.yebsap.01.

Ward, M.P., and A. Raim. 2011. The fly-and-social foraging hypothesis for diurnal migration: Why American Crows migrate during the day. *Behavioral Ecology and Sociobiology* 65:1411–1418.

Warkentin, I.G., N.S. Sodhi, R.H.M. Espie, A.F. Poole, L.W. Oliphant, and P.C. James. 2020. Merlin (*Falco columbarius*). Version 1.0. In *Birds of the World*, ed. S.M. Billerman. Cornell Lab of Ornithology, Ithaca, NY.

Warnock, N., and M.A. Bishop. 1998. Spring stopover ecology of migrant Western Sandpipers. *Condor* 100(3):456–467.

Warnock, N., J. Takekawa, and M.A. Bishop. 2004. Migration and stopover strategies of individual Dunlin along the Pacific Coast of North America. *Canadian Journal of Zoology* 82:1687–1697.

Warnock, N.D., and R.E. Gill. 2020. Dunlin (*Calidris alpina*). Version 1.0. In *Birds of the World*, ed. S.M. Billerman. Cornell Lab of Ornithology, Ithaca, NY. https://doi.org/10.2173/bow.dunlin.01.

Washburn, B.E., M.S. Martell, R.O. Bierregaard, C.J. Henny, B.S. Dorr, and T.J. Olexa. 2014. Wintering ecology of adult North American Ospreys. *Journal of Raptor Research* 48:325–333.

Wassenaar, L.I., and K.A. Hobson. 2000. Stable carbon and hydrogen isotope ratios reveal breeding origins of Red-Winged Blackbirds. *Ecological Applications* 10(3):911–916.

Watanabe, J., A. Koizumi, R. Nakagawa, K. Takahashi, T. Tanaka, and H. Matsuoka. 2019. Seabirds (Aves) from the Pleistocene Kazusa and Shimosa groups, central Japan. *Journal of Vertebrate Paleontology* 39(5). https://doi.org/10.1080/02724634.2019.1697277.

Watanuki, Y., T. Yamamoto, A. Yamashita, et al. 2015. Mercury concentrations in primary feathers reflect pollutant exposure in discrete non-breeding grounds used by Short-tailed Shearwaters. *Journal of Ornithology* 156:847–850. https://doi.org/10.1007/s10336-015-1205-6.

Watson, J.L. 2020. Ferruginous Hawk (*Buteo regalis*) home range and resource use on northern grasslands in Canada. Master's thesis, University of Alberta.

Watson, J.W., and I.N. Keren. 2019. Repeatability in migration of Ferruginous Hawks (*Buteo regalis*) and implications for nomadism. *Wilson Journal of Ornithology* 131(3):561–570. https://doi.org/10.1676/18-171.

Watson, J.W., U. Banasch, T. Byer, D.N. Svingen, R. McCready, M.A. Cruz, D. Hanni, A. Lafón, and R. Gerhardt. 2018. Migration patterns, timing, and seasonal destinations of adult Ferruginous Hawks (*Buteo regalis*). *Journal of Raptor Research* 52(3):267–281. https://doi.org/10.3356/JRR-17-30.1.

Watts, B., B. Truitt, F. Smith, E. Mojica, B. Paxton, A. Wilke, and A. Duerr. 2008. Whimbrel tracked with satellite transmitter on migratory flight across North America. *Wader Study Group Bulletin* 115:119–121.

Watts, B.D. 2020. Yellow-crowned Night-Heron (*Nyctanassa violacea*). Version 1.0. In *Birds of the World*, ed. A.F. Poole. Cornell Lab of Ornithology, Ithaca, NY. https://doi.org/10.2173/bow.ycnher.01.

Watts, B.D., and F.M. Smith. 2015. Winter composition of Nelson Sparrow (*Ammodramus nelsoni*) and Saltmarsh Sparrow (*Ammodramus caudacutus*) mixed flocks in coastal Virgina. *Wilson Journal of Ornithology* 127:387–394. http://dx.doi.org/10.1676/14-141.1.

Watts, B.D., F.M. Smith, C. Hines, L. Duval, D.J. Hamilton, T. Keyes, … and P. Woodard. 2021a. The annual cycle for whimbrel populations using the Western Atlantic Flyway. *PLOS ONE* 16(12):e0260339.

Watts, B.D., F.M. Smith, C. Hines, L. Duval, D.J. Hamilton, T. Keyes, … and P. Woodard. 2021b. Whimbrel populations differ in trans-Atlantic pathways and cyclone encounters. *Scientific Reports* 11:12919. https://doi.org/10.1038/s41598-021-92429-z.

Watt, D.J., and E.J. Willoughby. 2020. Lesser Goldfinch (*Spinus psaltria*). Version 1.0. In *Birds of the World*, ed. A.F. Poole. Cornell Lab of Ornithology, Ithaca, NY. https://doi.org/10.2173/bow.lesgol.01.

Watt, D.J., P. Pyle, M.A. Patten, and J.N. Davis. 2020. Lawrence's Goldfinch (*Spinus lawrencei*). Version 1.0. In *Birds of the World*, ed. P.G. Rodewald. Cornell Lab of Ornithology, Ithaca, NY. https://doi.org/10.2173/bow.lawgol.01.

Weaver, K.A. 2013. Tundra Swan (*Cygnus columbianus columbianus*) habitat selection during the nonbreeding period. Master's thesis, University of Western Ontario.

Webb, E.A., and C.E. Bock. 2020. Botteri's Sparrow (*Peucaea botterii*). Version 1.0. In *Birds of the World*, ed. A.F. Poole. Cornell Lab of Ornithology, Ithaca, NY. https://doi.org/10.2173/bow.botspa.01.

Webster, M.S., P.P. Marra, S.M. Haig, S. Bensch, and R.T. Holmes. 2002. Links between worlds: Unraveling migratory connectivity. *Trends in Ecology & Evolution* 17(2):76–83.

Weckstein, J.D., D.E. Kroodsma, and R.C. Faucett. 2020. Fox Sparrow (*Passerella iliaca*). Version 1.0. In *Birds of the World*, ed. A.F. Poole and F.B. Gill. Cornell Lab of Ornithology, Ithaca, NY. https://doi.org/10.2173/bow.foxspa.01.

Weeks Jr., H.P. 2020. Eastern Phoebe (*Sayornis phoebe*). Version 1.0. In *Birds of the World*, ed. A.F. Poole. Cornell Lab of Ornithology, Ithaca, NY. https://doi.org/10.2173/bow.easpho.01.

Wege, M.L., and D.G. Raveling. 1983. Factors influencing timing, distance, and path of migrations of Canada Geese. *Wilson Bulletin* 95(2):209–221.

Weimerskirch, H., and R.P. Wilson. 2000. Oceanic respite for wandering albatrosses. *Nature* 406(6799):955–956.

Weimerskirch, H., S. de Grissac, A. Ravache, A. Prudor, A. Corbeau, B.C. Congdon, F. McDuie, K. Bourgeois, S. Dromzée, J. Butscher, and C. Menkes. 2020. At-sea movements of Wedge-tailed Shearwaters during and outside the breeding season from four colonies in New Caledonia. *Marine Ecology Progress Series* 633:225–238.

Weimerskirch, H., S.A. Shaffer, Y. Tremblay, D.P. Costa, H. Gadenne, A. Kato, Y. Ropert-Coudert, K. Sato, and D. Aurioles. 2009. Species- and sex-specific differences in foraging behaviour and foraging zones in Blue-footed and Brown Boobies in the Gulf of California. *Marine Ecology Progress Series* 391:267–278.

Weimerskirch, H., A. Tarroux, O. Chastel, K. Delord, Y. Cherel, and S. Descamps. 2015. Population-specific wintering distributions of South Polar Skuas over the three oceans. *Marine Ecology Progress Series* 538:229–237.

Weir, D.N., A.C. Kitchener, and R.Y. McGowan. 2000. Hybridization and changes in the distribution of Iceland Gulls (*Larus glaucoides/kumlieni/thayeri*). *Journal of Zoology* 252(4):517–530. https://doi.org/10.1111/j.1469-7998.2000.tb01234.x.

Weir, R.D. 1976. Annotated bibliography of bird kills at man-made obstacles: A review of the state of the art and solutions. Department of Fisheries and the Environment, Canadian Wildlife Service, Ontario Region.

Weiser, E., and H.G. Gilchrist. 2020. Glaucous Gull (*Larus hyperboreus*). Version 1.0. In *Birds of the World*, ed. S.M. Billerman. Cornell Lab of Ornithology, Ithaca, NY. https://doi.org/10.2173/bow.glagul.01.

Weiser, E.L., R.B. Lanctot, S.C. Brown, J.A. Alves, P.F. Battley, R. Bentzen, J. Bêty, M.A. Bishop, M. Boldenow, L. Bollache, B. Casler, M. Christie, J.T. Coleman, J.R. Conklin, W.B. English, H.R. Gates, O. Gilg, M.-A. Giroux, K. Gosbell, C. Hassell, J. Helmericks, A. Johnson, B. Katrínardóttir, K. Koivula, E. Kwon, J.-F. Lamarre, J. Lang, D. B. Lank, N. Lecomte, J. Liebezeit, V. Loverti, L. McKinnon, C. Minton, D. Mizrahi, E. Nol, V.-M. Pakanen, J. Perz, R. Porter, J. Rausch, J. Reneerkens, N. Rönkä, S. Saalfeld, N. Senner, B. Sittler, P. A. Smith, K. Sowl, A. Taylor, D.H. Ward, S. Yezerinac, and B.K. Sandercock. 2016. Effects of geolocators on hatching success, return rates, breeding movements, and change in body mass in 16 species of Arctic-breeding shorebirds. *Movement Ecology* 4:12.

Welch, A.J., A.E. Wiley, H.F. James, P.H. Ostrom, T.W. Stafford Jr., and R.C. Fleischer. 2012. Ancient DNA reveals genetic stability despite demographic decline: 3,000 years of population history in the endemic Hawaiian Petrel. *Molecular Biology and Evolution* 29(12):3729–3740. https://doi.org/10.1093/molbev/mss185. PMID: 22844071.

Weller, M.W. 1964. Distribution and migration of the Redhead. *Journal of Wildlife Management* 28:64–103.

Wells, J.V., K.V. Rosenberg, D.L. Tessagalia and A.A. Dhondt. 1996. Population cycles in the Varied Thrush (*Ixoreus naevius*). *Canadian Journal of Zoology* 74:2062–2069.

Weseloh, D.V.C., D. Moore, and T. Knezevic. 2014. Wintering locations of Ontario-banded Great Egrets: New Jersey to the Caribbean. *Ontario Birds* 32(1): article 1. https://digitalcommons.usf.edu/ontario_birds/vol32/iss1/1.

West, G.C., L.J. Peyton, and L. Irving. 1968. Analysis of spring migration of Lapland Longspurs to Alaska. *Auk* 85:639–653.

West, R.L., and G.K. Hess. 2020. Purple Gallinule (*Porphyrio martinica*). Version 1.0. In *Birds of the World*, ed. A.F. Poole and F.B. Gill. Cornell Lab of Ornithology, Ithaca, NY. https://doi.org/10.2173/bow.purgal2.01.

Wethington, S.M. 2021. Violet-crowned Hummingbird (*Leucolia violiceps*). Version 1.1. In *Birds of the World* (editor not available). Cornell Lab of Ornithology, Ithaca, NY. https://doi.org/10.2173/bow.vichum.01.1.

Wetlands International. 2015. Waterbird Population Estimates. Wageningen, Netherlands.

Wetmore, A., R.F. Pasquier, and S.L. Olson. 1984. *The Birds of the Republic of Panama*. Smithsonian Institution Press.

Wheat, R.E., S.B. Lewis, Y. Wang, T. Levi, and C.C. Wilmers. 2017. To migrate, stay put, or wander? Varied movement strategies in Bald Eagles (*Haliaeetus leucocephalus*). *Movement Ecology* 5:9. https://doi.org/10.1186/s40462-017-0102-4.

Wheelwright, N.T., and J.D. Rising. 2020. Savannah Sparrow (*Passerculus sandwichensis*). Version 1.0. In *Birds of the World*, ed. A.F. Poole. Cornell Lab of Ornithology, Ithaca, NY. https://doi.org/10.2173/bow.savspa.01.

Whelan, S., S.A. Hatch, D.B. Irons, A. McKnight, and K.H. Elliott. 2020. Increased summer food supply decreases non-breeding movement in Black-legged Kittiwakes. *Biology Letters* 16(1):20190725.

Whitaker, D.M., and S.W. Eaton. 2020. Northern Waterthrush (*Parkesia noveboracensis*). Version 2.0. In *Birds of the World*, ed. P.G. Rodewald. Cornell Lab of Ornithology, Ithaca, NY https://doi.org/10.2173/bow.norwat.01.

Whitaker, D.M., I.G. Warkentin, J.P.B. McDermott, P.E. Lowther, C.C. Rimmer, B. Kessel, S.L. Johnson, and W.G. Ellison. 2020. Gray-cheeked Thrush (*Catharus minimus*). Version 1.0. In *Birds of the World*, ed. P.G. Rodewald. Cornell Lab of Ornithology, Ithaca, NY. https://doi.org/10.2173/bow.gycthr.01.

Whitaker, D.M, I.G. Warkentin, K.A Hobson, P. Thomas, and R. Boardman. 2018. Fall and winter movements of Newfoundland Gray-cheeked Thrushes (*Catharus minimus minimus*). *Animal Migration* 5:42–48.

White, C.M., N.J. Clum, T.J. Cade, and W.G. Hunt. 2020. Peregrine Falcon (*Falco peregrinus*). Version 1.0. In *Birds of the World*, ed. S.M. Billerman. Cornell Lab of Ornithology, Ithaca, NY. https://doi.org/10.2173/bow.perfal.01.

Whitfield, D.P. 1995. Behaviour and ecology of a polyandrous population of Grey Phalaropes *Phalaropus Fulicarius* in Iceland. *Journal of Avian Biology* 26(4):349–352. https://doi.org/10.2307/3677053.

Whittow, G.C. 2020. Wedge-tailed Shearwater (*Ardenna pacifica*). Version 1.0. In *Birds of the World*, ed. S.M. Billerman. Cornell Lab of Ornithology, Ithaca, NY. https://doi.org/10.2173/bow.wetshe.01.

Whitworth, D.L., S.K. Nelson, S.H. Newman, G.B. Van Vliet, and W.P. Smith. 2000. Foraging distances of radio-marked Marbled Murrelets from inland areas in southeast Alaska. *Condor* 102(2):452–456. https://doi.org/10.1093/condor/102.2.452.

Wiebe, K.L., and W.S. Moore. 2024. Northern Flicker (*Colaptes auratus*). Version 2.1. In *Birds of the World*, ed. P.G. Rodewald and B.K. Keeney. Cornell Lab of Ornithology, Ithaca, NY. https://doi.org/10.2173/bow.norfli.02.1.

Wiegardt, A., J. Wolfe, C.J. Ralph, J.L. Stephens, and J. Alexander. 2017. Postbreeding elevational movements of western songbirds in northern California and southern Oregon. *Ecology and Evolution* 7:7750–7764. https://doi.org/10.1002/ece3.3326.

Wiersma, P., G.M. Kirwan, and P.F.D. Boesman. 2020. Common Ringed Plover (*Charadrius hiaticula*). Version 1.0. In *Birds of the World*, ed. J. del Hoyo, A. Elliott, J. Sargatal, D.A. Christie, and E. de Juana. Cornell Lab of Ornithology, Ithaca, NY. https://doi.org/10.2173/bow.corplo.01.

Wiggins, D.A., D.W. Holt, and S.M. Leasure. 2020. Short-eared Owl (*Asio flammeus*). Version 1.0. In *Birds of the World*, ed. S.M. Billerman. Cornell Lab of Ornithology, Ithaca, NY. https://doi.org/10.2173/bow.sheowl.01.

Wikelski, M., R. W. Kays, N.J. Kasdin, K. Thorup, J.A. Smith, and G.W. Swenson. 2007. Going wild: What a global small-animal tracking system could do for experimental biologists. *Journal of Experimental Biology* 210:181–186. https://doi.org/10.1242/jeb.02629.

Wiley, A.E., S. Rossman, P.H. Ostrom, et al. 2019. From ecologically equivalent individuals to contrasting colonies: Quantifying isotopic niche and individual foraging specialization in an endangered oceanic seabird. *Marine Biology* 166(39). https://doi.org/10.1007/s00227-019-3483-7.

Wiley, R.H., and D.S. Lee. 2020a. Long-tailed Jaeger (*Stercorarius longicaudus*). Version 1.0. In *Birds of the World*, ed. S.M. Billerman. Cornell Lab of Ornithology, Ithaca, NY. https://doi.org/10.2173/bow.lotjae.01.

———. 2020b. Parasitic Jaeger (*Stercorarius parasiticus*). Version 1.0. In *Birds of the World*, ed. S.M. Billerman. Cornell Lab of Ornithology, Ithaca, NY. https://doi.org/10.2173/bow.parjae.01.

———. 2020c. Pomarine Jaeger (*Stercorarius pomarinus*). Version 1.0. In *Birds of the World*, ed. S.M. Billerman. Cornell Lab of Ornithology, Ithaca, NY. https://doi.org/10.2173/bow.pomjae.01.

Wilhelm, S.I., A. Hedd, G.J. Robertson, J. Mailhiot, P.M. Regular, P.C. Ryan, and R.D. Elliot. 2019. The world's largest breeding colony of Leach's Storm-petrel *Hydrobates leucorhous* has declined. *Bird Conservation International* 12. https://doi.org/10.1017/S0959270919000248.

Wilkins, K.A., R.A. Malecki, P.J. Sullivan, J.C. Fuller, J.P. Dunn, L.J. Hindman, G.R. Costanzo, and D. Luszcz. 2010. Migration routes and Bird Conservation Regions used by eastern population Tundra Swans (*Cygnus columbianus columbianus*) in North America. *Wildfowl* 60:20–37.

Wilkinson, B.P., and P.G.R. Jodice. 2023. Support for the fasting endurance hypothesis of partial migration in a nearshore seabird. *Ecosphere* 14(2):e4365. https://doi.org/10.1002/ecs2.4365.

Wilkinson, B.P., A.M. Haynes-Sutton, L. Meggs, and P.G. Jodice. 2020. High spatial fidelity among foraging trips of Masked Boobies from Pedro Cays, Jamaica. *PLOS ONE* 15(4):e0231654.

Wilkinson, B.P., J. Jahncke, P. Warzybok, R.W. Bradley, and S.A. Shaffer. 2018. Variable utilization of shelf break-associated habitats by chick-brooding Rhinoceros Auklets in the California Current System. *Marine Ecology Progress Series* 590:211–226. https://doi.org/10.3354/meps12500.

Will, A.P., and A.S. Kitaysky. 2018. Variability in trophic level and habitat use in response to environmental forcing: Isotopic niche dynamics of breeding seabirds in the southeastern Bering Sea. *Marine Ecology Progress Series* 593:247–260.

Williams, C.K., M.D. Samuel, V.V. Baranyuk, E.G. Cooch, and D. Kraege. 2008. Winter fidelity and apparent survival of Lesser Snow Goose populations in the Pacific Flyway. *Journal of Wildlife Management* 72(1):159–167.

Wilson, C., R.J. Meese, and A.C. Wyckoff. 2016. Breeding chronology, movements, and life history observations of Tricolored Blackbirds in the California Central Coast. *California Fish and Game* 102:162–174.

Wilson, H.M., W.W. Larned, and M.A. Swaim. 2018. Abundance and trends of waterbird breeding populations on the Arctic Coastal Plain, Alaska, 1986–2017. US Fish and Wildlife Service Report, Anchorage, AK.

Wilson, R.E., C.R. Ely, and S.L. Talbot. 2018. Flyway structure in the circumpolar Greater White-fronted Goose. *Ecology and Evolution* 8:8490–8507.

Wilson, R.E., J.R. Gust, M.R. Petersen, and S.L. Talbot. 2016. Genetic structure of Long-tailed Ducks (*Clangula hyemalis*) among Alaskan, Canadian, and Russian breeding populations. *Arctic* 69(1):65–78.

Wilson, S., K.A. Hobson, D.M. Collister, and A.G. Wilson. 2008. Breeding destinations and spring migration patterns of Swainson's Thrush (*Catharus ustulatus*) at a Costa Rican stopover site. *Auk* 125:95–104.

Wilson, S., S.L. LaDeau, A.P. Tottrup, and P.P. Marra. 2011. Range-wide effects of breeding- and nonbreeding-season climate on the abundance of a Neotropical migrant songbird. *Ecology* 92:1789–1798.

Wilson Jr., W.H. 2020. Palm Warbler (*Setophaga palmarum*). Version 1.0. In *Birds of the World*, ed. A.F. Poole. Cornell Lab of Ornithology, Ithaca, NY. https://doi.org/10.2173/bow.palwar.01.

Wilson, W.H., and B. Brown. 2012. Fidelity and interseasonal movements of Purple Finches (*Carpodacus purpureus* (Gmelin)): Analysis of band re-encounter data. *Open Ornithology Journal* 5:61–71.

Winker, K., and G.R. Graves. 2008. Genetic structure of breeding and wintering populations of Swainson's Warbler. *Wilson Journal of Ornithology* 120:433–445. https://doi.org/10.1676/07-073.1.

Winker, K., and S.M. Haig, eds. 2010. *Avian Subspecies*. Ornithological Monographs 67. American Ornithologists Union, Washington, DC.

Winkler, D.W., 2006. Roosts and migrations of swallows. Hornero, 21(2), pp.85–97.

———. 2020. California Gull (*Larus californicus*). Version 1.0. In *Birds of the World*, ed. A.F. Poole and F.B. Gill. Cornell Lab of Ornithology, Ithaca, NY. https://doi.org/10.2173/bow.calgul.01.

Winkler, D.W., F.A. Gandoy, J.I. Areta, M.J. Iliff, E. Rakhimberdiev, K.J. Kardynal, and K.A. Hobson. 2017. Long-distance range expansion and rapid adjustment of migration in a newly established population of Barn Swallows breeding in Argentina. *Current Biology* 27(7):1080–84. http://dx.doi.org/10.1016/j.cub.2017.03.006.

Winship, A.J., S.J. Jorgensen, S.A. Shaffer, I.D. Jonsen, P.W. Robinson, D.P. Costa, and B.A. Block. 2012. State-space framework for estimating measurement error from double-tagging telemetry experiments. *Methods in Ecology and Evolution* 3:291–302. https://doi.org/10.1111/j.2041-210X.2011.00161.x.

With, K.A. 2021. Thick-billed Longspur (*Rhynchophanes mccownii*). Version 1.1. In *Birds of the World*, ed. A.F. Poole. Cornell Lab of Ornithology, Ithaca, NY. https://doi.org/10.2173/bow.mcclon.01.1.

Witmer, M.C. 2024. Bohemian Waxwing (*Bombycilla garrulus*). Version 1.1. In *Birds of the World*, ed. S.M. Billerman. Cornell Lab of Ornithology, Ithaca, NY. https://doi.org/10.2173/bow.bohwax.01.1.

Witmer, M.C., D.J. Mountjoy, and L. Elliott. 2020. Cedar Waxwing (*Bombycilla cedrorum*). Version 1.0. In *Birds of the World*, ed. A.F. Poole. Cornell Lab of Ornithology, Ithaca, NY. https://doi.org/10.2173/bow.cedwax.01.

Wittenberg, S.R., and K.G. Smith. 2009. Lack of differential migration in juvenile Merlins and Northern Harriers during fall migration in the Florida Keys. *Wilson Journal of Ornithology* 121:841–843.

Witynski, M.L., and D.N. Bonter. 2018. Crosswise migration by Yellow Warblers, Nearctic-Neotropical passerine migrants. *Journal of Field Ornithology* 89:37–46. https://doi.org/10.1111/jofo.12237.

Wolf, B.O. 2020. Black Phoebe (*Sayornis nigricans*). Version 1.0. In *Birds of the World*, ed. A.F. Poole and F.B. Gill. Cornell Lab of Ornithology, Ithaca, NY. https://doi.org/10.2173/bow.blkpho.01.

Wolfe, J.D., and E.I. Johnson. 2015. Geolocator reveals migratory and winter movements of a Prothonotary Warbler. *Journal of Field Ornithology* 86:238–243.

Wolfson, D., J. Fieberg, J.S. Lawrence, T.R. Cooper, and D.E. Andersen. 2017. Range overlap between mid-continent and eastern Sandhill Cranes revealed by GPS-tracking. *Wildlife Society Bulletin* 41(3):489–498.

Wong, J.B., S. Lisovski, R.T. Alisauskas, W. English, M.-A. Giroux, A.L. Harrison, D. Kellett, N. Lecomte, M. Maftei, A. Nagy-MacArthur, R.A. Ronconi, P.A. Smith, M. Mallory, and M. Auger-Methe. 2021. Arctic Terns from circumpolar breeding colonies share common migratory routes. *Marine Ecology Progress Series* 671:191–206. https://doi.org/10.3354/meps13779.

Wong, J.B., S. Lisovski, R.T. Alisauskas, W. English, A.-L. Harrison, D. Kellett, M. Maftei, A. Nagy-MacArthur, R.A. Ronconi, P.A. Smith, M. Mallory, and M. Auger-Methe. 2022. Variation in migration behaviours used by Arctic Terns (*Sterna paradisaea*) breeding across a wide latitudinal gradient. *Polar Biology* 45:909–922.

Wood, H.B. 1945. The history of bird banding. *Auk* 256–265.

Woodbridge, B., K.K. Finley, and S.T. Seager. 1995. An investigation of the Swainson's Hawk in Argentina. *Journal of Raptor Research* 29(3):202–204.

Woodin, M.C., and T.C. Michot. 2020. Redhead (*Aythya americana*). Version 1.0. In *Birds of the World*, ed. A.F. Poole and F.B. Gill. Cornell Lab of Ornithology, Ithaca, NY. https://doi.org/10.2173/bow.redhea.01.

Woods, C.P., R.D. Csada, and R.M. Brigham. 2020. Common Poorwill (*Phalaenoptilus nuttallii*). Version 1.0. In *Birds of the World*, ed. A.F. Poole. Cornell Lab of Ornithology, Ithaca, NY. https://doi.org/10.2173/bow.compoo.01.

Woodward, P. 1972. Movement of a breeding White-Eyed Vireo from Maryland to British Honduras. *Bird Banding* 43(2):143.

Woodworth, B.K., A.E.M. Newman, S.P. Turbek, B.C. Dossman, K.A. Hobson, L.I. Wassenaar, G.W. Mitchell, N.T. Wheelwright, and D.R. Norris. 2016. Differential migration and the link between winter latitude, timing of migration, and breeding in a songbird. *Oecologia* 181:413–422.

Wootton, J. Timothy. 1996. Purple Finch (*Haemorhous purpureus*). Version 2.0. In *The Birds of North America*, ed. P.G. Rodewald. Cornell Lab of Ornithology, Ithaca, NY. https://doi.org/10.2173/bna.208

Working Group, American Oystercatcher, E. Nol, and R. C. Humphrey. 2020. American Oystercatcher (*Haematopus palliatus*). Version 1.0. In *Birds of the World*, ed. A.F. Poole. Cornell Lab of Ornithology, Ithaca, NY. https://doi.org/10.2173/bow.ameoys.01.

Worton, B.J. 1989. Kernel methods for estimating the utilization distribution in home-range studies. *Ecology* 70:164–168

———. 1995. Using Monte Carlo simulation to evaluate kernel-based home range estimators. *Journal of Wildlife Management* 59:794–800.

Wright, J.R., J.A. Johnson, E. Bayne, L.L. Powell, C.R. Foss, J.C. Kennedy, and P.P. Marra. 2021. Migratory connectivity and annual cycle phenology of Rusty Blackbirds (*Euphagus carolinus*) revealed through archival GPS tags. *Avian Conservation and Ecology* 16(1):20. https://doi. org/10.5751/ACE-01871-160120.

Wright, J.R., L.L. Powell, and C.M. Tonra. 2018. Automated telemetry reveals staging behavior in a declining migratory passerine. *Auk* 135(3):461–476. https://doi.org/10.1642/AUK-17-219.1.

Wright, L., E. Nol, R.B. Lanctot, and L. McKinnon. 2022. Rushing in the spring and relaxing in the fall: Seasonal and sex-specific migration profiles of Dunlin. *Journal of Field Ornithology* 93(2):10.

Wunder, M.B. 2007. Geographic structure and dynamics in Mountain Plover. PhD diss., Colorado State University.

Wunder, M.B., C.L. Kester, F.L. Knopf, and R.O. Rye. 2005. A test of geographic assignment using isotope tracers in feathers of known origin. *Oecologia* 144:607–617.

Wunderle Jr., J.M., D. Currie, E.H. Helmer, D.N. Ewert, J.D. White, T.S. Ruzycki, B. Parresol, and C. Kwit. 2010. Kirtland's Warblers in anthropogenically disturbed early-successional habitats on Eleuthera, the Bahamas. *Condor* 112(1):123–137.

Wunderle Jr., J.M., R.B. Waide, and J. Fernandez. 1989. Seasonal abundance of shorebirds in the Jobos Bay Estuary in southern Puerto Rico. *Journal of Field Ornithology* 60:329–339.

Wyatt, V.E., and C.M. Francis. 2020. Rose-Breasted Grosbeak (*Pheucticus ludovicianus*). Version 1.0. In *Birds of the World*, ed. A.F. Poole and F.B. Gill. Cornell Lab of Ornithology, Ithaca, NY. 2020. https://doi.org/10.2173/bow.robgro.01.

Yanco, S.W., B.D. Linkhart, P.P. Marra, M. Mika, M. Ciaglo, A. Carver, and M.B. Wunder. 2021. Niche dynamics suggest ecological factors influencing migration in an insectivorous owl. *Ecology* 103(3):e3617. https://doi.org/10.1002/ecy.3617.

Yannic, G., J.M. Yearsley, R. Sermier, et al. 2016. High connectivity in a long-lived high-Arctic seabird, the Ivory Gull *Pagophila eburnea*. *Polar Biology* 39:221–236. https://doi.org/10.1007/s00300-015-1775-z.

Yasukawa, K., and W.A. Searcy. 2020. Red-Winged Blackbird (*Agelaius phoeniceus*). Version 1.0. In *Birds of the World*, ed. P.G. Rodewald. Cornell Lab of Ornithology, Ithaca, NY. 2020. https://doi.org/10.2173/bow.rewbla.01.

Yates, M.A., M.R. Fuller, D. Evers, J.D. Paruk, and K.P. Kenow. 1999. Extant and causes of mercury contamination in the Walker River (NV) Basin, and implications of contaminants and the loss of fisheries for the western migratory loons. Unpublished report to the US Department of the Interior, Fish and Wildlife Service, Washington, DC.

Yates, M.A., M.R. Fuller, C.J. Henny, W.S. Seegar, and J. Garcia. 2009. Wintering area DDE source to migratory White-faced Ibis revealed by satellite telemetry and prey sampling. *Ecotoxicology* 19.

Yerkes, T., K.A. Hobson, L.J. Wassenaar, R. Macleod, and J.M. Coluccy. 2008. Stable isotopes (δD, $\delta^{13}C$, $\delta^{15}N$) reveal associations among geographic location and condition of Alaskan Northern Pintails. *Journal of Wildlife Management* 72(3):715–725.

Yosef, R. 2020. Loggerhead Shrike (*Lanius ludovicianus*). Version 1.0. In *Birds of the World*, ed. A.F. Poole and F.B. Gill. Cornell Lab of Ornithology, Ithaca, NY. https://doi.org/10.2173/bow.logshr.01.

Young, H.S., D.J. McCauley, R. Dirzo, R.B. Dunbar, and S.A. Shaffer. 2010. Niche partitioning among and within sympatric tropical seabirds revealed by stable isotope analysis. *Marine Ecology Progress Series* 416:285–294. https://doi.org/10.3354/meps08756.

REFERENCES

Young, L.C. 2010. Inferring colonization history and dispersal patterns of a long-lived seabird by combining genetic and empirical data. *Journal of Zoology* 281:232–240. https://doi.org/10.1111/j.1469-7998.2010.00698.x.

Young, L.C., C.R. Kohley, E.A. VanderWerf, L. Fowlke, D. Casillas, M. Dalton, M. Knight, A. Pesque, E.M. Dittmar, A.F. Raine, and M. Vynne. 2023. Successful translocation of Newell's Shearwaters and Hawaiian Petrels to create a new, predator free breeding colony. *Frontiers in Conservation Science* 4. doi.org/10.3389/fcosc.2023.1177789.

Young, L.C., C. Vanderlip, D.C. Duffy, V. Afanasyev, and S.A. Shaffer. 2009. Bringing home the trash: Do colony-based differences in foraging distribution lead to increased plastic ingestion in Laysan Albatrosses? *PLOS ONE* 4:11–13. https://doi.org/10.1371/journal.pone.0007623.

Young, L.C., B.J. Zaun, and E.A. Vanderwerf. 2008. Successful same-sex pairing in Laysan Albatross. *Biology Letters* 4(4):323–325. https://doi.org/10.1098/rsbl.2008.0191.

Young, R.C., R.A. Orben, A.P. Will, and A.S. Kitaysky. 2017 Relationship between telomere dynamics and movement and behavior during winter in the Thick-billed Murre. *Marine Ecology Progress Series* 578:253–261. https://doi.org/10.3354/meps12022.

Zarn, A.M., C.A. Valle, R. Brasso, W. Fetzner, and S.D. Emslie. 2020. Stable isotope and mercury analyses of the Galápagos Islands seabird community. *Marine Ornithology* 48:71–80.

Zdravkovic, M.G., C.A. Corbat, and P.W. Bergstrom. 2020. Wilson's Plover (*Charadrius wilsonia*). Version 1.0. In *Birds of the World*, ed. P.G. Rodewald. Cornell Lab of Ornithology, Ithaca, NY.

Zenzal, T.J., A.J. Contina, J.F. Kelly, and F.R. Moore. 2018. Temporal migration patterns between natal locations of Ruby-throated Hummingbirds (*Archilochus colubris*) and their Gulf Coast stopover site. *Movement Ecology* 6:7.

Zhang, G., B. Li, C. Li, M.T.P. Gilbert, E.D. Jarvis, and J. Wang. 2014. Genomic data of the Red throated Loon (*Gavia stellata*). Avian Genome Consortium. GigaScience Database. https://doi.org/10.5524/101026.

Zimmerman, G.M. 2004. Studies of the annual cycle of the Swallow-tailed Kite (*Elanoides forficatus*): Migration, habitat use, and parasites. Master's thesis, Georgia Southern University, Statesboro, GA.

Zimpfer, N.L., and M.J. Conroy. 2006. Modeling movement and fidelity of American Black Ducks. *Journal of Wildlife Management* 70:1770–1777.

Zwartjes, P.W. 2001. Genetic structuring among migratory populations of the Black-Whiskered Vireo, with a comparison to the Red-Eyed Vireo. *Condor* 103(3):439–448.

DUCKS, GEESE, AND WATERFOWL (ANSERIFORMES)

— FAMILY ANATIDAE—DUCKS, GEESE, AND WATERFOWL

Emperor Goose

Uher-Koch, B.D., J.A. Schmutz, J.W. Hupp, C.R. Ely, and D.C. Douglas. 2021. Tracking data for Emperor Geese (*Anser canagicus*). Version 1.0, April 2021. US Geological Survey data release. https://doi.org/10.5066/P9GJQ6LF.

Snow Goose

Evenson, J.R., M.W. Wilson, and K.A. Spragens. Waterfowl telemetry studies from Washington state. Unpublished data.

Legagneux, P., C.K. Williams, J. Lefebvre, P. Thomas, P. and J. Bêty. Greater Snow Goose migration. Unpublished data.

US Geological Survey and Canadian Wildlife Service. Unpublished data.

Ross's Goose

US Geological Survey. Unpublished data.

Greater White-fronted Goose

Askren, R.J., E.R. Massey, J.D., James, and D.C. Osborne. 2022. Migration chronology and multi-scale habitat selection of wintering midcontinent Greater White-fronted Geese. *Global Ecology and Conservation* 39:e02290. https://doi.org/10.1016/j.gecco.2022.e02290.

Ely, C.R., J.Y. Takekawa, M.L. Casazza, and D.C. Douglas. 2021. Tracking data for Greater White-fronted Geese (*Anser albifrons*). Version 1.0, April 2021. US Geological Survey data release. https://doi.org/10.5066/P9MBQ2GN.

Meixell, B.W. 2018. Satellite telemetry data for Greater White-fronted Geese (*Anser albifrons*) captured on the Arctic Coastal Plain of Alaska, 2013–2017 [Data set]. US Geological Survey. https://doi.org/10.5066/F7CV4GX2.

Moore, C.B., D.C. Osborne, L.G. Carlson, R.J. Askren, and M. Brasher. 2023. Distributional shifts of wintering midcontinent Greater White-fronted Geese. *Journal of Wildlife Management* e22401. https://doi.org/10.1002/jwmg.22401.

US Geological Survey. Unpublished data.

Brant

Boyd, W.S. Unpublished data.

Evenson, J.R., M.W. Wilson, and K.A. Spragens. Waterfowl telemetry studies from Washington state. Unpublished data.

Cackling Goose

Neufeld, L.R., F. Baldwin, K.C. and Fraser. 2021. Comparing migration ecology among geographically distinct populations of Canada Geese (*Branta canadensis*) and Cackling Geese (*Branta hutchinsii*). Master's thesis, University of Manitoba. Winnipeg, MB.

Canada Goose

Evenson, J.R., M.W. Wilson, and K.A. Spragens. Waterfowl telemetry studies from Washington state. Unpublished data.

Giles, M.M., P.G.R. Jodice, R.F. Baldwin, J.D. Stanton, and M. Epstein. 2013. Spring migratory pathways and migration chronology of Canada Geese (*Branta canadensis interior*) wintering at the Santee National Wildlife Refuge, South Carolina. *Canadian Field-Naturalist* 127(1):17–25.

Malecki R. A., B. D. J. Batt, and S. E. Sheaffer. 2001. Spatial and temporal distribution of Atlantic population Canada Geese. *Journal of Wildlife Management* 65:242–247.

Neufeld, L.R., F. Baldwin, and K.C. Fraser. 2021. Comparing migration ecology among geographically distinct populations of Canada Geese (*Branta canadensis*) and Cackling Geese (*Branta hutchinsii*). Master's thesis, University of Manitoba. Winnipeg, MB.

Trumpeter Swan

Caton, E.L., and G. Neudecker. 2021. Trumpeter Swan migrations from the Blackfoot Watershed, Montana, 2005–2020.

Diebold, E.N., S. Matteson, F. Koontz, C. Koontz, and P. Manthey. 2008. The use of satellite telemetry to determine the migratory routes of a restored population of Trumpeter Swans (*Cygnus cygnus buccinator*).

Tundra Swan

Ely, C.R., J. Terenzi, T.L. Tibbitts, and D.C. Douglas. 2020 Tracking data for Tundra Swans (*Cygnus columbianus*). Version 1.0, January 2020. US Geological Survey data release. https://doi.org/10.5066/P9KBR79C.

Wood Duck

Losito, M.P., R.J. Masse, E.B. Struening, and C. Slezak. Unpublished data.

Blue-winged Teal

Ramey, A.M., C. Soos, P. Link, P. Walther, T.L. Tibbitts, and D.C. Douglas. 2019. Tracking data for Blue-winged Teal (*Anas discors*). Version 1.0, September 2019). US Geological Survey data release. https://doi.org/10.5066/P9Z9BA9F.

US Geological Survey. Unpublished data.

Cinnamon Teal

US Geological Survey. Unpublished data.

Northern Shoveler

US Geological Survey. Unpublished data.

Gadwall

Gray, J. 2010. Habitat use, movements and spring migration chronology and corridors of female Gadwalls that winter along the Louisiana coast. Master's thesis, Louisiana State University.

Setash, C.M., J.H. Gammonley, A.C. Behney, M. Reddy, D.N. Koons. 2019. Breeding waterfowl productivity in a flood-irrigated landscape. Unpublished data.

US Geological Survey. Unpublished data.

American Wigeon

US Geological Survey. Unpublished data.

Mallard

Krementz, D.G., K. Asante, and L.W. Naylor. 2011. Spring migration of Mallards from Arkansas as determined by satellite telemetry. *Journal of Fish and Wildlife Management* 2(2):156–168.

Link, P. 2021. Unpublished data.

Masto, N., O. Robinson, M. Brasher, A. Blake-Bradshaw, C. Highway, A. Keever, J. Feddersen, H.M. Hagy, D.C. Osborne, D. Combs, and B. Cohen. 2022. Citizen science reveals waterfowl responses to extreme winter weather. *Global Change Biology* 28:5469–5479. https://doi.org/10.1111/gcb.16288.

Osborne, D. Unpublished data.

Setash, C.M., J.H. Gammonley, A.C. Behney, M. Reddy, and D.N. Koons. 2019. Breeding waterfowl productivity in a flood-irrigated landscape. Unpublished data.

US Geological Survey. Unpublished data.

American Black Duck

Anderson, K. 2013. Migration routes and chronology of American Black Ducks. Master's thesis, University of Delaware.

Coluccy, J.M., K.A. Anderson, T. Yerkes, and J.L. Bowman. 2020. Migration routes and chronology of American Black Duck *Anas rubripes*. *Wildfowl* 70:148–166.

Peck, L.E., M.D. English, G.J. Robertson, S.R. Craik, M.L. and Mallory. 2022. Migration chronology and movements of adult American Black Ducks *Anas rubripes* wintering in Nova Scotia, Canada. *Wildlife Biology* 2022(1):e01000. https://doi.org/10.1002/wlb3.01000.

Northern Pintail

Haukos, D.A., M.R. Miller, D.L. Orthmeyer, J.Y. Takekawa, J.P. Fleskes, M.L. Casazza, W.M. Perry, and J.A. Moon. 2006. Spring migration of Northern Pintails from Texas and New Mexico. *Waterbirds* 29:127–136.

Hayes, S.G., and M. Wilson. Unpublished data.

Hupp, J.W., T.L. Tibbitts, and D.C. Douglas, D.C. 2019. Tracking data for Northern Pintails (*Anas acuta*). Version 1.0, September 2019. US Geological Survey data release. https://doi.org/10.5066/P90A6HW3.

US Geological Survey. Unpublished data.

Green-winged Teal

US Geological Survey. Unpublished data.

Ring-necked Duck

Mezebish, T., G. Olsen, M. Goodman, F. Rohwer, and M. McConnell. 2022. Spring migration and breeding distribution of female Ring-necked Ducks wintering in the southern Atlantic Flyway. *Avian Conservation and Ecology* 17(2).

Lesser Scaup

Schummer, M.L., A.D. Afton, S.S. Badzinski, S.A. Petrie, G.H. Olsen, and M.A. Mitchell. 2018. Evaluating the waterfowl breeding population and habitat survey for scaup. *Journal of Wildlife Management* 82(6):1252–1262.

DATASET CITATIONS

Steller's Eider

Martin, P.D., D.C. Douglas, T. Obritschkewitsch, and S. Torrence 2015. Distribution and movements of Alaska-breeding Steller's Eiders in the nonbreeding period. *Condor* 117:341–353.

Martin, P.D., T. Obritschkewitsch, and D.C. Douglas. 2022. US Fish and Wildlife Service tracking data for Steller's Eiders (*Polysticta stelleri*). Version 1.0, November 2022. US Geological Survey data release. https://doi.org/10.5066/P9ECETBG.

Rosenberg, D.H., M.J. Petrula, J.L. Schamber, D. Zwiefelhofer, T.E. Hollmen, and D.D. Hill. 2014. Seasonal movements and distribution of Steller's Eiders (*Polysticta stelleri*) wintering at Kodiak Island Alaska. *Arctic* 67(3):347–359.

Spectacled Eider

Petersen, M.R., M.G. Sexson, and D.C. Douglas. 2020. Tracking data for Spectacled Eiders (*Somateria fischeri*). Version 1.1, September 2021. US Geological Survey data release. https://doi.org/10.5066/P9B091HG.

King Eider

Bentzen, R.L., and A.N. Powell. 2015. Dispersal, movements and site fidelity of post-fledging King Eiders *Somateria spectabilis* and their attendant females. *Ibis* 157:133–146.

Dickson, D.L. 2012b. Seasonal movement of King Eiders breeding in western Arctic Canada and northern Alaska. Canadian Wildlife Service Technical Report Series no. 520. Canadian Wildlife Service, Edmonton, AB.

Gilchrist, H.G. 2013. Common/King Eiders; East Bay Island, Nunavut.

Knoche, M.J., A.N. Powell, L.T. Quakenbush, M.J. Wooller, and L.M. Phillips. 2007. Further evidence for site fidelity to wing molt locations by King Eiders: Integrating stable isotope analyses and satellite telemetry. *Waterbirds* 30(1):52–57.

Mosbech, A., H.G. Gilchrist, F. Merkel, C. Sonne, A. Flagstad, and H. Nyegaard. 2006. Year-round movements of Northern Common Eiders *Somateria mollissima borealis* breeding in Arctic Canada and West Greenland followed by satellite telemetry. *Ardea* 94(3):651–665.

Mosbech, A., F. Merkel, and C. Sonne. 2021. Common Eider (*Somateria mollissima*) and King Eider (*Somateria spectabilis*) tracked with ARGOS PPTs from localities in West Greenland. Aarhus University and Greenland Institute for Natural Resources.

Oppel, S., D.L. Dickson, and A.N. Powell. 2009. International importance of the eastern Chukchi Sea as a staging area for migrating King Eiders. *Polar Biology* 32:775–783.

Oppel, S., A.N. Powell, and D.L. Dickson. 2008. Timing and distance of King Eider migration and winter movements. *Condor* 110:296–305.

Common Eider

Gilchrist, H.G. 2013. Common Eiders; East Bay Island, Nunavut.

Lamb, J.S. 2020. Data: Assessing year-round habitat use by migratory sea ducks in a multi-species context reveals seasonal variation in habitat selection and partitioning. http://doi.org/10.5281/zenodo.3971064.

Mallory, M.L., R.A. Ronconi, R.B. Allen, C. Dwyer, S. Lair, C.D. Mallory, N.R. McLellan, G.R. Milton, G.J. Parsons, L. Savoy, and M.D. Tomlik. 2020. Annual movement patterns of American Common Eiders (*Somateria molllissima dresseri*). *Wildlife Biology* wlb.00665. https://doi.org/10.2081/wlb.00665.

Mosbech, A., H.G. Gilchrist, F. Merkel, C. Sonne, A. Flagstad, and H. Nyegaard. 2006. Year-round movements of Northern Common Eiders *Somateria mollissima borealis* breeding in Arctic Canada and West Greenland followed by satellite telemetry. *Ardea* 94(3):651–665.

Mosbech, A., F. Merkel, and C. Sonne. 2021. Common Eider (*Somateria mollissima*) and King Eider (*Somateria spectabilis*) tracked with ARGOS PPTs from localities in West Greenland. Aarhus University and Greenland Institute for Natural Resources.

Petersen, M.R., G.V. Byrd, S.A. Sonsthagen, and M.G. Sexson. 2015. Re-colonization by Common Eiders *Somateria mollissima* in the Aleutian Archipelago following removal of introduced Arctic foxes (*Vulpes lagopus*). *Journal of Avian Biology* 46:538–549. https://doi.org/10.1111/jav.00626.

———. 2015 Data from: Re-colonization by Common Eiders (*Somateria mollissima*) in the Aleutian Archipelago following removal of introduced arctic foxes (*Vulpes lagopus*). Movebank Data Repository. https://doi.org/10.5441/001/1.s528h83q.

Petersen, M.R., P.L. Flint, D.M. Mulcahy, and D.C. Douglas. 2021. Tracking data for Common Eiders (*Somateria mollissima*). Version 1.0, September 2021. US Geological Survey data release. https://doi.org/10.5066/P9JH3EXV.

Harlequin Duck

Biodiversity Research Institute. Unpublished data.

Boyd, W.S. Unpublished data.

Evenson, J.R., M.W. Wilson, and K.A. Spragens. Waterfowl telemetry studies from Washington state. Unpublished data.

MacCallum, B. 2021. The Harlequin Duck Migration and Connectivity Project, Alberta. Unpublished ARGOS data May 2016 to June 2017.

Robert, M., G. Mittelhauser, B. Jobin, G. Fitzgérald, and P. Lamothe. 2008. New insights on Harlequin Duck population structure in eastern North America as revealed by satellite telemetry. *Waterbirds* 31(spec. iss. 2):159–172. https://doi.org/10.1675/1524-4695-31.sp2.159.

Surf Scoter

Boyd, W.S. Unpublished data.

De La Cruz, S.E.W., J.Y. Takekawa, M.T. Wilson, D.R. Nyeswander, J.R. Evenson, D. Esler, W.S. Boyd, and D.H. Ward. 2009. Surf Scoter (*Melanitta perspicillata*) spring migration routes and chronology: A synthesis of Pacific coast studies. *Canadian Journal of Zoology* 87:1069–1086. https://doi.org/10.1139/Z09-099.

Evenson, J.R., M.W. Wilson, and K.A. Spragens. Waterfowl telemetry studies from Washington state. Unpublished data.

Kirk, M., D. Esler, S.A. Iverson, and W.S. Boyd. 2008. Movements of wintering Surf Scoters: Predator responses to different prey landscapes. *Oecologia* 155:859–867. https://doi.org/10.1007/s00442-007-0947-0.

Lamb, J.S. 2020. Data: Assessing year-round habitat use by migratory sea ducks in a multi-species context reveals seasonal variation in habitat selection and partitioning. http://doi.org/10.5281/zenodo.3971064.

Lok, E.K., D. Esler, J.Y. Takekawa, S.W. De La Cruz, W.S. Boyd, D.R. Nyeswander, J.R. Evenson, and D.H. Ward. 2011. Stopover habitats of spring migrating Surf Scoters in southeast Alaska. *Journal of Wildlife Management* 75(1):92–100. https://doi.org/10.1002/jwmg.5.

———. 2012. Spatio-temporal associations between Pacific herring spawn and Surf Scoter spring migration: Evaluating a "silver wave" hypothesis. *Marine Ecology Progress Series* 457:139–150. https://doi.org/10.3354/meps09692.

Lok, E.K., M. Kirk, D. Esler, and W.S. Boyd. 2008. Movements of pre-migratory Surf and White-winged Scoters in response to Pacific herring spawn. *Waterbirds* 31:385–393. https://doi.org/10.1675/1524-4695-31.3.385.

Takekawa, J.Y., S.W. Del La Cruz, M.T. Wilson, E.C. Palm, J. Yee, D.R. Nysewander, J.R. Evenson, J.M. Eadie, D. Esler, W.S. Boyd, and D.H. Ward. 2011a. Breeding distribution and ecology of Pacific coast Surf Scoters. In *Boreal Birds of North America*, ed. J.V. Wells, 41–64. *Studies in Avian Biology*, no. 41, University of California Press, Berkeley.

Ward, D.H., D. Esler, W.S. Boyd, and D.C. Douglas. 2020. Tracking data for Surf Scoters (*Melanitta perspicillata*). Version 1.0, March 2020. US Geological Survey data release. https://doi.org/10.5066/P9LUFGEF.

White-winged Scoter

Evenson, J.R., M.W. Wilson, and K.A. Spragens. Waterfowl telemetry studies from Washington state. Unpublished data.

Lamb, J.S. 2020. Data: Assessing year-round habitat use by migratory sea ducks in a multi-species context reveals seasonal variation in habitat selection and partitioning. http://doi.org/10.5281/zenodo.3971064.

Meattey, D.E., S.R. McWilliams, P.W.C. Paton, C. Lepage, S. Gilliland, L. Savoy, G. Olsen, and J. Osenkowski. 2018. Annual cycle of White-winged Scoters (*Melanitta fusca*) in eastern North America: Migration phenology, population delineation, and connectivity. *Canadian Journal of Zoology* 96:1353–1365.

Black Scoter

Flint, P.L., J.L. Schamber, and D.C. Douglas. 2020. Tracking data for Black Scoters (*Melanitta americana*). Version 1.0, March 2020. US Geological Survey data release. https://doi.org/10.5066/P93EF3TH.

Lamb, J.S. 2020. Data: Assessing year-round habitat use by migratory sea ducks in a multi-species context reveals seasonal variation in habitat selection and partitioning. http://doi.org/10.5281/zenodo.3971064.

Long-tailed Duck

Bartzen, B.A., D.L. Dickson, and T.D. Bowman. 2017. Migration characteristics of Long-tailed Ducks (*Clangula hyemalis*) from the western Canadian Arctic. *Polar Biology* 40:1085–1099.

Lamb, J.S. 2020. Data: Assessing year-round habitat use by migratory sea ducks in a multi-species context reveals seasonal variation in habitat selection and partitioning. http://doi.org/10.5281/zenodo.3971064.

Mallory, M.L., J. Akearok, N.R. North, D. Vaughan Weseloh, and S. Lair. 2006. Movements of Long-tailed Ducks wintering on Lake Ontario to breeding areas in Nunavut, Canada. *Wilson Journal of Ornithology* 118:494–501.

Petersen, M.R., and D.C. Douglas. 2016. Data from: Movement of Long-tailed Ducks marked on the Yukon-Kuskokwim Delta, Alaska 1998–2000 (data from Petersen et al. 2003). Movebank Data Repository. https://www.doi.org/10.5441/001/1.1rc3hj8d.

Barrow's Goldeneye

Forstner, T.M., W.S. Boyd, D. Esler, T.D. Bowman, J.L. Schamber, M.McAdie, J.E. Thompson, and D.J. Green. 2022. Migratory connectivity and variation in migration phenology within the Pacific population of Barrow's Goldeneye (*Bucephala islandica*). *Avian Conservation and Ecology* 17(2):7. https://doi.org/10.5751/ACE-02195-170207.

Hogan, D., J.E. Thompson, D. Esler, and W.S. Boyd. 2011. Discovery of important post-breeding sites for Barrow's Goldeneye in the boreal transition zone of Alberta. *Waterbirds* 34(3):261–268. https://doi.org/10.1675/063.034.0301.

Robert, M., R. Benoit, and J.-P.L. Savard. 2002. Relationship among breeding, molting, and wintering areas of male Barrow's Goldeneye in eastern North America. *Auk* 119(3):676–684. https://doi.org/10.1093/auk/119.3.676.

Savard, J.-P.L., and M. Robert, 2013. Relationships among breeding, molting, and wintering areas of adult female Barrow's Goldeneyes in eastern North America. *Waterbirds* 36(1):34–42. https://doi.org/10.1675/063.036.0107.

Willie, M.C., D. Esler, W.S. Boyd, T. Bowman, J. Schamber, and J. Thompson. 2020. Annual winter site fidelity of Barrow's Goldeneye in the Pacific. *Journal of Wildlife Management* 84(1):161–171. https://doi.org/10.1002/jwmg.21767.

Common Merganser

Pearce, J.M., and M.R. Petersen. 2009. Post-fledging movements of juvenile Common Mergansers (*Mergus merganser*) in Alaska as inferred by satellite telemetry. *Waterbirds* 32:133–137. https://doi.org/10.1675/063.032.0116.

———. 2016. Data from: Post-fledging movements of juvenile Common Mergansers (*Mergus merganser*) in Alaska as inferred by satellite telemetry. Movebank Data Repository. https://doi.org/10.5441/001/1.3gc013f3.

PIGEONS AND DOVES (COLUMBIFORMES)

FAMILY COLUMBIDAE—PIGEONS AND DOVES

White-crowned Pigeon

Meyer, K.D., and G.M. Kent. Avian Research and Conservation Institute. Unpublished data.

Band-tailed Pigeon

Casazza, M. 2013. Band-tailed Pigeon satellite telemetry points. Data Basin Dataset. https://databasin.org/datasets/8a76106bf8954a74a02271c8a432f158/.

CUCKOOS (CUCULIFORMES)

FAMILY CUCULIDAE—CUCKOOS

Yellow-billed Cuckoo

McNeil, S.E., D. Tracy, and C.D. Capello. 2015. Loop migration by a western Yellow-billed Cuckoo wintering in the Gran Chaco. *Western Birds* 46:244–255.

Sechrist, J.D., E.H. Paxton, D.D. Ahlers, R.H. Doster, and V.M. Ryan. 2012. One year of migration data for a western Yellow-billed Cuckoo. *Western Birds* 43:2–11.

Stanley, C.Q. Unpublished data.

Black-billed Cuckoo

Bayly, N., and C. Gomez. 2015–2021. *Investigación para la Conservación en el Neotropico* (Research for Conservation in the Neotropic), Birds Canada, Environment and Climate Change Canada. SELVA Colombia, Project 57. Data accessed from Motus Wildlife Tracking System, Birds Canada. https://motus.org/.

NIGHTJARS AND ALLIES (CAPRIMULGIFORMES)

FAMILY CAPRIMULGIDAE—NIGHTJARS AND ALLIES

Common Nighthawk

Knight, E.C., A.-L. Harrison, A.L. Scarpignato, S.L. Van Wilgenburg, E.M. Bayne, J.W. Ng, E. Angell, R. Bowman, R.M. Brigham, B. Drolet, W.E. Easton, T.R. Forrester, J.T. Foster, S. Haché, K.C. Hannah, K.G. Hick, J. Ibarzabal, T.L. Imlay, S.A. Mackenzie, A. Marsh, L.P. McGuire, G.N. Newberry, D. Newstead, A. Sidler, P.H. Sinclair, J.L. Stephens, D.L. Swanson, J.A. Tremblay, and P.P. Marra. 2021. Comprehensive estimation of spatial and temporal migratory connectivity across the annual cycle to direct conservation efforts. *Ecography* 44:665–679. https://doi.org/10.1111/ecog.05111.

Ng, J.W., E.C. Knight, A.L. Scarpignato, A.-L. Harrison, E.M. Bayne, and P.P. Marra. 2018. First full annual cycle tracking of a declining aerial insectivorous bird, the Common Nighthawk (*Chordeiles minor*), identifies migration routes, nonbreeding habitat, and breeding site fidelity. *Canadian Journal of Zoology* 96:869–875. dx.doi.org/10.1139/cjz-2017-0098.

Eastern Whip-poor-will

Bakermans, M.H., and A.V. Vitz. 2021. Data from: Eastern whip-poor-will connectivity. Movebank Data Repository.

English, P.A., A.M. Mills, M.D. Cadman, A.E. Heagy, G.J. Rand, D.J. Green, and J.J. Nocera. 2017. Data from: Tracking the migration of a nocturnal aerial insectivore in the Americas. Movebank Data Repository. https://doi.org/10.5441/001/1.66jq0844.

Korpach, A.M., C.M. Davy, A. Mills, and K.C. Fraser. 2022. Migratory connectivity and timing for an at-risk Canadian landbird, Eastern Whip-poor-will (*Antrostomus vociferus*), from two geographically distant breeding areas. *Canadian Journal of Zoology* 100(10):1–9. https://doi.org/10.1139/cjz-2021-0175.

Skinner, A.A., M.P. Ward, I. Souza-Cole, J.R. Wright, F.R. Thompson, T.J. Benson, S.N. Matthews, and C.M. Tonra. 2023. Data from: GPS tracking of Eastern Whip-poor-will. Movebank Data Repository. https://www.doi.org/10.5441/001/1.k6s7916g.

SWIFTS AND HUMMINGBIRDS (APODIFORMES)

FAMILY APODIDAE—SWIFTS

Black Swift

Beason, J.P., C. Gunn, K.M. Potter, R.A. Sparks, and J.W. Fox. 2012. The northern Black Swift: Migration path and wintering area revealed. *Wilson Journal of Ornithology* 124(1):1–8.

Chimney Swift

Tixier, J., V. Lamarre, and J. Tremblay. Breeding habitat of Chimney Swift in Québec (Project 195). 2018–2020. Data accessed from Motus Wildlife Tracking System, Birds Canada. https://motus.org/.

RAILS, COOTS, AND CRANES (GRUIFORMES)

FAMILY RALLIDAE—RAILS, GALLINULES, AND COOTS

King Rail

Shirkey, B. Winous Point Marsh Conservancy King Rail Data.

Clapper Rail

Elizondo, E. Delaware Clapper Rail Research (Project 341). 2020–2020. Data accessed from Motus Wildlife Tracking System, Birds Canada. https://motus.org/.

Huang, M. Clapper Rail Survival in CT (Project 143). 2017–2018. Data accessed from Motus Wildlife Tracking System, Birds Canada. https://motus.org/.

Virginia Rail

Fournier, A. Forbes Bio Station Motus Tags (Project 314). 2020–2021. Data accessed from Motus Wildlife Tracking System, Birds Canada. https://motus.org/.

Kearns, G., and D. Brinker. Sora Migration (Project 172). 2017–2021. Data accessed from Motus Wildlife Tracking System, Birds Canada. https://motus.org/.

Patterson, A. New York Lake Plains Motus (Project 152). 2018–2021. Data accessed from Motus Wildlife Tracking System, Birds Canada. https://motus.org/.

Tonra, C. Ohio State University (Project 64). 2015–2021. Data accessed from Motus Wildlife Tracking System, Birds Canada. https://motus.org/.

Sora

Fournier, A. Forbes Bio Station Motus Tags (Project 314). 2020–2021. Data accessed from Motus Wildlife Tracking System, Birds Canada. https://motus.org/.

G.D. Kearns, K.A Dami, C. Miller, K. Harvey, M. Felperin, A. McDonough, A.G. Pellegrini, M. Janc, M. A. Lewis (Maryland—National Capital Park and Planning Commission) and D. Brinker (Maryland Department of Natural Resources). Sora tracking data.

Kearns, G., and D. Brinker. Sora Migration (Project 172). 2017–2021. Data accessed from Motus Wildlife Tracking System, Birds Canada. https://motus.org/.

FAMILY GRUIDAE—CRANES

Sandhill Crane

Casazza, M. 2013. Sandhill Cranes Satellite Telemetry Points. Data Basin Dataset. https://databasin.org/datasets/f2bd01bd303b4ce6a2f9cd9151e9a503/.

Collins, D.P., et al. Western Greater Sandhill Crane Movement Ecology Projects.

Hanna, E.E., and S.A. Petrie. 2017. Eastern Population Greater Sandhill Crane (*Antigone canadensis tabida*) migratory chronology habitat use.

Hjertaas, D.G, D.H. Ellis, B.W. Johns, and S.L. Moon. 2001. Tracking Sandhill Crane migration from Saskatchewan to the Gulf Coast. *Proceedings North American Crane Workshop* 8:57–61.

Ivey, G.L., C.P. Herziger, and T.J. Hoffmann. 2005. Annual movements of Pacific Coast Sandhill Cranes. *Proceedings North American Crane Workshop* 9:25–35.

Pearse, A.T., D.A. Brandt, and G.L. Krapu. 2017. Sandhill Crane locations, autumn 2013 migration. US Geological Survey data release. https://doi.org/10.5066/F7F76BGR.

Pogson, T.H. 1987. Sandhill Crane banded in Alaska migrates to California and Oregon. *North American Bird Bander* 12(3):90–92.

Whooping Crane

Kuyt, E. 1992. Aerial radio-tracking of Whooping Cranes migrating between Wood Buffalo National Park and Aransas National Wildlife Refuge, 1981–84. Occasional Paper Number 74. Canadian Wildlife Service.

Pearse, A.T., D.A. Brandt, D.M. Baasch, M.T. Bidwell, J.A. Conkin, M.J. Harner, W. Harrell, and K.L. Metzger. 2020. Location data for Whooping Cranes of the Aransas–Wood Buffalo Population, 2009–2018. US Geological Survey data release. https://doi.org/10.5066/P9Y8KZJ9.

Stehn, T.V. 1992. Unusual movements and behaviors of color-banded Whooping Cranes during winter. *North American Crane Workshop Proceedings*, Paper 285.

SHOREBIRDS, AUKS, GULLS, AND TERNS (CHARADRIIFORMES)

— FAMILY HAEMATOPODIDAE—OYSTERCATCHERS

American Oystercatcher

Loring, P. H., C.R. Griffin, P.R. Sievert, and C.S. Spiegel. 2017. Comparing satellite and digital radio telemetry to estimate space and habitat use of American Oystercatchers (*Haematopus palliatus*) in Massachusetts. *Waterbirds* 40(1):19–31.

Black Oystercatcher

Johnson, M., P. Clarkson, M.I. Goldstein, S.M. Haig, R.B. Lanctot, D.F. Tessler, and D. Zwiefelhofer. 2010. Seasonal movements, winter range use, and migratory connectivity of the Black Oystercatcher. *Condor* 112:731–743.

— FAMILY CHARADRIIDAE—PLOVERS AND LAPWINGS

Black-bellied Plover

Harrison, A.-L., et al. Movebank Study MCP Black-belled Plover Alaska, ID 77248725. Accessed on November 23, 2021.

Koch, S., P. Loring, A. MacDonald, K. O'Brien, N. Pau, P. Smith. Monomoy—Steph Koch (Project 88). 2014–2018. Data accessed from Motus Wildlife Tracking System, Birds Canada. https://motus.org/.

Rausch, J., and P. Woodard. 2021. Satellite tracking of Canadian Arctic breeding Black-bellied Plovers. Environment and Climate Change Canada, Canadian Wildlife Service, Northern Region, Yellowknife, NT.

American Golden-Plover

Lamarre, J.-F., et al. 2021. American Golden-Plover (*Pluvialis dominica*) migration data 2009–2016. Université du Québec à Rimouski, Québec.

Lanctot, R., et al. Arctic shorebird migration tracking study—American Golden-Plover. 2018–2019. US Fish and Wildlife Service.

Pacific Golden-Plover

Johnson, O.W., N. Warnock, M.A. Bishop, A.J. Bennett, P.M. Johnson, and R.J. Kienholz. 1997. Migration by radio-tagged Pacific Golden-Plovers from Hawaii to Alaska, and their subsequent survival. *Auk* 114:521–524. https://doi.org/10.2307/4089259.

Common Ringed Plover

Léandri-Breton, D.-J., J.-F. Lamarre, and J. Bêty. 2019. Seasonal variation in migration strategies used to cross ecological barriers in a Nearctic migrant wintering in Africa. *Journal of Avian Biology* e02101. https://doi.org/10.1111/jav.02101.

———. 2019. Data from: Seasonal variation in migration strategies used to cross ecological barriers in a Nearctic migrant wintering in Africa. Movebank Data Repository. https://doi.org/10.5441/001/1.f01v7r80.

Semipalmated Plover

Brown, G., K. Bennett, A. McKellar, and P. Smith. EC—Arctic Shorebirds (Project 40). 2016–2019. Data accessed from Motus Wildlife Tracking System, Birds Canada. https://motus.org/.

Nol, E., L. McKinnon, and P. Smith. Nol—Churchill Shorebirds (Project 67). 2014–2019. Data accessed from Motus Wildlife Tracking System, Birds Canada. https://motus.org/.

Smith, P., and A. MacDonald. Smith (Project 27). 2014–2019. Data accessed from Motus Wildlife Tracking System, Birds Canada. https://motus.org/.

Piping Plover

Stantial, M.L., and J.B. Cohen. 2020. Miniature GPS tags provide high resolution location data for a small-bodied threatened shorebird. *Waterbirds* 43(3–4):271–280. https://doi.org/10.1675/063.043.0305.

Mountain Plover

Pierce, A.K., and M.B. Wunder. Unpublished data.

Wilson's Plover

Williams, K. TU_WIPL_LA (Project 188). 2018–2019. Data accessed from Motus Wildlife Tracking System, Birds Canada. Available: https://motus.org/.

— FAMILY SCOLOPACIDAE—SANDPIPERS AND ALLIES

Upland Sandpiper

Gesmundo, C.F., Z.M. Pohlen, and J.A. Johnson. Movebank Study Upland Sandpiper migratory movements, ID 1608554329.

Hill, J. M., and R.B. Renfrew. 2019. Data from: Migration patterns of Upland Sandpipers in the Western Hemisphere. Movebank Data Repository. https://doi.org/10.5441/001/1.3pt25757.

Hudsonian Whimbrel

Brown, G., K. Bennett, A. McKellar, and P. Smith. EC—Arctic Shorebirds (Project 40). 2016–2019. Data accessed from Motus Wildlife Tracking System, Birds Canada. https://motus.org/.

Johnson, A.S., J. Perz, E. Nol, and N.R. Senner. 2016. Dichotomous strategies? The migration of Whimbrels breeding in the eastern Canadian sub-Arctic. *Journal of Field Ornithology* 87(4):371–383. https://doi.org/10.1111/jofo.12173.

Kneidel, A., and B. Winn. Unpublished data.

Nol, E. Unpublished data.

Tibbitts, T.L., D.R. Ruthrauff, R.E. Gill Jr., and D.C. Douglas. 2021. Tracking data for Whimbrels (*Numenius phaeopus*). Version 2.0, May 2021. US Geological Survey data release. https://doi.org/10.5066/P978PX2X.

Watts, B.D., F.M. Smith, D.J. Hamilton, T. Keyes, J. Paquet, J. Rausch, B. Truitt, B. Winn and P. Woodward. 2021. Tracking Whimbrel in North America. The Center for Conservation Biology, College of William and Mary, and Virginia Commonwealth University, Williamsburg, VA.

Long-billed Curlew

Birds Canada. British Columbia Long-billed Curlew study 2017–2021. Unpublished data.

Boyce, A.J., et al. Full annual cycle movement ecology of the Long-billed Curlew. Smithsonian Conservation Biology Institute. Unpublished data.

Carlisle, J., et al. Intermountain Bird Observatory, Boise State University. Unpublished data.

Coates, S.M.S. 2018. Building the full annual cycle picture for Long-billed Curlews: Correlates of nest success in the breeding grounds and spatial distribution and site fidelity in the wintering grounds. Master's thesis, Boise State University.

Migratory Connectivity Project. Movebank Study MCP Long-billed Curlews Texas Gulf Coast, ID 46195209. Accessed November 23, 2021.

Olalla-Kerstupp, A., J.I. González-Rojas, A. Guzmán-Velasco, and G. Ruiz-Aymá. Long-billed Curlew annual cycle satellite tracking. Unpublished data.

Ricketts Conservation Foundation. Unpublished data.

Hudsonian Godwit

Senner N.R., J.A. Linscott, J. Ruiz, M.D. Weegman, B.M. Ballard, and J.G. Navedo. Unpublished data.

Smith, F.M., B.D. Watts, and J. Rausch. 2021. Tracking Hudsonian Godwit in Canada. Center for Conservation Biology, College of William and Mary and Virginia Commonwealth University, Williamsburg, VA.

Marbled Godwit

Olson, B. 2011. Biogeography of Marbled Godwit (*Limosa fedoa*) populations in North America. Unpublished master's thesis, Utah State University, Logan.

Ruthrauff, D.R., T.L. Tibbitts, R.E. Gill Jr., and D.C. Douglas. 2019. Tracking data for Marbled Godwit (*Limosa fedoa*). US Geological Survey data release. https://doi.org/10.5066/P9NORY48.

Ruddy Turnstone

Niles, L., S. Koch, P. Loring, and P. Smith. Red Knot Staging and Migration Ecology (Project 47). 2014–2021. Data accessed from Motus Wildlife Tracking System, Birds Canada. https://motus.org/.

Sanders, F., and A. Smith. SCDNR Shorebirds (Project 140). 2017–2021. Data accessed from Motus Wildlife Tracking System, Birds Canada. https://motus.org/.

Smith, A., and D. Marain. Region 4 USFWS (Project 4). 2015–2020. Data accessed from Motus Wildlife Tracking System, Birds Canada. https://motus.org/.

Smith, P., S. Flemming, and A. MacDonald. EBM_Coats (#66) (Project 66). 2015–2017. Data accessed from Motus Wildlife Tracking System, Birds Canada. https://motus.org/.

Smith, P., and A. MacDonald. Smith (Project 27). 2014–2019. Data accessed from Motus Wildlife Tracking System, Birds Canada. https://motus.org/.

Black Turnstone

Taylor, A.R., and M.A. Bishop. Breeding and wintering locations of Black Turnstones, 2011–2015. Unpublished data.

Red Knot

Aubry, Y., A. Desrochers, A. MacDonald, and J. Tremblay. EC-Quebec-St.Laurent (Project 3). 2014–2019. Data accessed from Motus Wildlife Tracking System, Birds Canada. https://motus.org/.

Burger, J., L.J. Niles, R.R. Porter, A.D. Dey, S. Koch, and C. Gordon. 2012. Using a shore bird (Red Knot) fitted with geolocators to evaluate a conceptual risk model focusing on offshore wind. *Renewable Energy* 43:370–377. https://doi.org/10.1016/j.renene.2011.11.006.

Delaware Shorebird Project and US Fish and Wildlife Service. Movebank Study Red Knot Tracking Project, Mispillion Harbor, DE—Delaware Shorebird Project, ID 2783096023.

Friis, C., A. Anderson, K. Bennett, P. Loring, A. MacDonald, P. Smith, and R. Wood. James Bay Shorebirds (Project 38). 2014–2019. Data accessed from Motus Wildlife Tracking System, Birds Canada. https://motus.org/.

Gesmundo, C.F., and J.A. Johnson. Unpublished data.

Koch, S., P. Loring, A. MacDonald, K. O'Brien, N. Pau, and P. Smith. Monomoy—Steph Koch (Project 88). 2014–2018. Data accessed from Motus Wildlife Tracking System, Birds Canada. https://motus.org/.

Lefevre, K., and A. Smith. Lefevre—Florida Gulf Coast University (Project 96). 2017–2019. Data accessed from Motus Wildlife Tracking System, Birds Canada. https://motus.org/.

Mackenzie, S., and A. Smith. Bahia Lomas Shorebirds (Project 174). 2017–2019. Data accessed from Motus Wildlife Tracking System, Birds Canada. https://motus.org/.

Morrissey, C., A. McKellar, and P. Smith. Saskatchewan Migratory Shorebirds (Project 63). 2015–2020. Data accessed from Motus Wildlife Tracking System, Birds Canada. https://motus.org/.

Morrissey, C., D. Newstead, S. Heath, and A. McKellar. Texas Gulf Coast Migratory Shorebirds (Project 65). 2016–2016. Data accessed from Motus Wildlife Tracking System, Birds Canada. https://motus.org/.

Newstead, D.J., L.J. Niles, R.R. Porter, A.D. Dey, J. Burger, and O.N. Fitzsimmons. 2013. Geolocation reveals mid-continent migratory routes and Texas wintering areas of Red Knots *Calidris canutus rufa*. *Waders Study Group Bulletin* 120(1):7.

Niles, L., S. Koch, P. Loring, and P. Smith. Red Knot Staging and Migration Ecology (Project 47). 2014–2021. Data accessed from Motus Wildlife Tracking System, Birds Canada. https://motus.org/.

Niles, L.J., J. Burger, R.R. Porter, A.D. Dey, C.D.T. Minton, P.M. Gonzalez, A.J. Baker, J.W. Fox, and C. Gordon. 2010. First results using light level geolocators to track Red Knots in the Western Hemisphere show rapid and long intercontinental flights and new details of migration pathways. *Waders Study Group Bulletin* 117(2):123–130.

Sanders, F., and A. Smith. SCDNR Shorebirds (Project 140). 2017–2021. Data accessed from Motus Wildlife Tracking System, Birds Canada. https://motus.org/.

Scarpignato, A.L., A-L. Harrison, D.J. Newstead, L.J. Niles, R.R. Porter, M. van den Tillaart, and P.P. Marra. 2016. Field-testing a new miniaturized GPS-Argos satellite transmitter (3.5 g) on migratory shorebirds. *Wader Study* 123(3):240–246.

Smith, A., and D. Marain. Region 4 USFWS (Project 4). 2015–2020. Data accessed from Motus Wildlife Tracking System, Birds Canada. https://motus.org/.

South Carolina Department of Natural Resources. Unpublished data.

Stilt Sandpiper

Nol, E., L. McKinnon, and P. Smith. Nol—Churchill Shorebirds (Project 67). 2014–2019. Data accessed from Motus Wildlife Tracking System, Birds Canada. https://motus.org/.

Sanderling

Drever, M., W. Easton, S. Flemming, V. Loverti, and Y. Zharikov. Pacific Shorebirds (Project 191). 2018–2021. Data accessed from Motus Wildlife Tracking System, Birds Canada. https://motus.org/.

Morrissey, C., A. McKellar, and P. Smith. Saskatchewan Migratory Shorebirds (Project 63). 2015–2020. Data accessed from Motus Wildlife Tracking System, Birds Canada. https://motus.org/.

Niles, L., S. Koch, P. Loring, and P. Smith. Red Knot Staging and Migration Ecology (Project 47). 2014–2021. Data accessed from Motus Wildlife Tracking System, Birds Canada. https://motus.org/.

Woodard, P. Arctic Shorebirds—CWS Yellowknife (Project 68). 2015–2020. Data accessed from Motus Wildlife Tracking System, Birds Canada. https://motus.org/.

Dunlin

Drever, M., W. Easton, S. Flemming, V. Loverti, and Y. Zharikov. Pacific Shorebirds (Project 191). 2018–2021. Data accessed from Motus Wildlife Tracking System, Birds Canada. https://motus.org/.

Lagassé, B., R. Lanctot, et al. Unpublished data.

Lancot, R., and S. Saalfeld. Movebank Study Arctic shorebird migration tracking study—Dunlin, ID 565467203. Accessed November 23, 2021.

Nol, E., L. McKinnon, and P. Smith. Nol—Churchill Shorebirds (Project 67). 2014–2019. Data accessed from Motus Wildlife Tracking System, Birds Canada. https://motus.org/.

Wright, L.L., E. Nol, R. Lanctot, and L. McKinnon. 2022. Rushing in the spring and relaxing in the fall: Seasonal and sex-specific migration profiles of Dunlin. *Journal of Field Ornithology* 93(2):10. https://doi.org/10.5751/JFO-00102-930210.

Purple Sandpiper

Summers, R.W., H. Boland, K. Colhoun, N. Elkins, B. Etheridge, S. Foster, J.W. Fox, K. Mackie, L.R. Quinn, and R.L. Swann. 2014. Contrasting trans-Atlantic migratory routes of Nearctic Purple Sandpipers *Calidris maritima* associated with low pressure systems in spring and winter. *Ardea* 102:139–152.

White-rumped Sandpiper

Smith, P., and A. MacDonald. Smith (Project 27). 2014–2019. Data accessed from Motus Wildlife Tracking System, Birds Canada. https://motus.org/.

Buff-breasted Sandpiper

Tibbitts, T.L., R.B. Lanctot, and D.C. Douglas. 2023. Tracking data for Buff-breasted Sandpipers (*Calidris subruficollis*). Version 1.0, October 2023. US Geological Survey data release. https://doi.org/10.5066/P9TYMUID.

Pectoral Sandpiper

Kempenaers, B., and M. Valcu. Max Planck Institute for Biological Intelligence. Unpublished data.

Lanctot, R.B., et al. Arctic shorebird migration tracking study—Pectoral Sandpiper. 2018–2019. US Fish and Wildlife Service. Unpublished data.

Semipalmated Sandpiper

Aubry, Y., A. Desrochers, A. MacDonald, and J. Tremblay. EC-Quebec-St.Laurent (Project 3). 2014–2019. Data accessed from Motus Wildlife Tracking System, Birds Canada. https://motus.org/.

Brown, S. et al. 2017. Migratory connectivity of Semipalmated Sandpipers and implications for conservation. *Condor* 119(2):207–224. https://doi.org/10.1650/CONDOR-16-55.1.

English, W. Carleton University. Unpublished data.

Friis, C., A. Anderson, K. Bennett, P. Loring, A. MacDonald, P. Smith, and R. Wood. James Bay Shorebirds (Project 38). 2014–2019. Data accessed from Motus Wildlife Tracking System, Birds Canada. https://motus.org/.

Herbert, J., and A. MacDonald. Tulane_SESA_NGOM (Project 153). 2017–2019. Data accessed from Motus Wildlife Tracking System, Birds Canada. https://motus.org/.

Koch, S., P. Loring, A. MacDonald, K. O'Brien, N. Pau, and P. Smith. Monomoy—Steph Koch (Project 88). 2014–2018. Data accessed from Motus Wildlife Tracking System, Birds Canada. https://motus.org/.

Lanctot, R., and S. Saalfeld. Movebank Study Arctic shorebird migration tracking study—Semipalmated Sandpiper, ID 606508137.

Mizrahi, D., J. Herbert, and A. MacDonald. NJA—South America Shorebirds (Project 124). 2016–2018. Data accessed from Motus Wildlife Tracking System, Birds Canada. https://motus.org/.

Mizrahi, D., A. MacDonald, and P. Smith. NJ Audubon—Delaware Bay Shorebirds (Project 59). 2015–2017. Data accessed from Motus Wildlife Tracking System, Birds Canada. https://motus.org/.

Paquet, J., D. Hamilton, R. Linhart, and H. Mann. Atlantic Canada Shorebirds (Project 78). 2015–2020. Data accessed from Motus Wildlife Tracking System, Birds Canada. https://motus.org/.

Paquet, J., D. Hamilton, R. Linhart, and J. Mobley. Brazil Non-Breeding Shorebirds (Project 239). 2019–2020. Data accessed from Motus Wildlife Tracking System, Birds Canada. https://motus.org/.

Partridge, D., S. Elbin, D. Kriensky, K. Parkins, E. Tobon, J. Kolodzinski, and A. MacDonald. NYC Audubon (Project 103). 2016–2020. Data accessed from Motus Wildlife Tracking System, Birds Canada. https://motus.org/.

Smith, P., and A. MacDonald. Smith (Project 27). 2014–2019. Data accessed from Motus Wildlife Tracking System, Birds Canada. Available: https://motus.org/.

Smith, P., S. Flemming, and A. MacDonald. EBM_Coats (#66) (Project 66). 2015–2017. Data accessed from Motus Wildlife Tracking System, Birds Canada. https://motus.org/.

Western Sandpiper

Kwon, E. Unpublished data.

Short-billed Dowitcher

Maillet, O., and E. Nol. Unpublished data.

Migratory Connectivity Project. 2023. MCP Short-billed Dowitcher Texas. https://www.movebank.org/cms/webapp?gwt_fragment=page=studies,path=study2088568723.

Nol, E., L. McKinnon, and P. Smith. Nol—Churchill Shorebirds (Project 67). 2014–2019. Data accessed from Motus Wildlife Tracking System, Birds Canada. https://motus.org/.

US Fish and Wildlife Service Migratory Birds Alaska. 2023. Movebank Study Short-billed Dowitcher (*L. griseus caurinus*) migratory movements, ID 1803419380. https://www.movebank.org/cms/webapp?gwt_fragment=page=studies,path=study1803419380.

Long-billed Dowitcher

Kempenaers, B., and E. Kwon. 2023. Lack of breeding site fidelity and mate fidelity in an enigmatic socially monogamous shorebird. *Animal Behaviour* 204:1–12. https://doi.org/10.1016/j.anbehav.2023.07.010.

American Woodcock

Blomberg, E.J, A.C. Fish, L.A. Bergian, and A.M. Roth. 2022. Eastern Woodcock Migration Research Cooperative. University of Maine.

Brenner, S.J., J.G. Jorgensen, J. Laux, and W. Inselman. 2021. American Woodcock satellite telemetry project. Unpublished data. Nebraska Game and Parks Commission, Lincoln, NE.

Grinde, A., and A. Bracey. Unpublished data.

McAuley, D. Maine—American Woodcock (Project 82). 2015–2018. Data accessed from Motus Wildlife Tracking System, Birds Canada. https://motus.org/.

Moore, J.D., D.E. Andersen, T.R. Cooper, J.P. Duguay, S.L. Oldenburger, C.A. Stewart, and D.G. Krementz. 2021. Data from: Migration phenology and patterns of American Woodcock in central North America derived using satellite telemetry. Movebank Data Repository. https://doi.org/10.5441/001/1.8764q39q.

Solitary Sandpiper

Gesmundo, C.F., L.H. DeCicco, L.A. McDuffie, and J.A. Johnson. Unpublished data.

Wandering Tattler

Jiguet, F. Movebank Study Kivi Kuaka [ID_PROG 1099], ID 1381110575.

Lesser Yellowlegs

Christie, K., R.E. Wilson, J.A. Johnson, C. Friis, C.M. Harwood, L.A. McDuffie, E. Nol, and S.A. Sonsthagen. 2023. Movement and genomic methods reveal mechanisms promoting connectivity in a declining shorebird: The Lesser Yellowlegs. *Diversity* 15(5):595. https://doi.org/10.3390/d15050595.

McDuffie, L.A. et al. 2022a. Eastern-breeding Lesser Yellowlegs are more likely than western-breeding birds to visit areas with high shorebird harvest during southward migration. *Ornithological Applications* 124(1):duab061. https://doi.org/10.1093/ornithapp/duab061.

McDuffie, L.A., K.S. Christie, A.R. Taylor, E. Nol, C. Friis, C.M. Harwood, ... and J.A. Johnson. 2022b. Flyway-scale GPS tracking reveals migratory routes and key stopover and non-breeding locations of Lesser Yellowlegs. *Ecology and Evolution* 12(11):e9495.

Tibbitts, L., R. Lanctot, L. McDuffie, and J. Johnson. Unpublished data.

Willet

Haig, S.M., L.W. Oring, P.M. Sanzenbacher, and O.W. Taft. 2002. Space use, migratory connectivity, and population segregation among Willets breeding in the western Great Basin. *Condor* 104(3):620–630.

Heath, S.A., J.K. Wilson, J.A.M. Smith, and N.W. Cooper. 2021. Migration and wintering location of breeding Eastern Willet (*Tringa semipalmata semipalmata*) in the western Gulf of Mexico. *Waterbirds* 44(4):483–491. https://doi.org/10.1675/063.044.0409.

Regan, K.R. Migratory Connectivity of Eastern Willets Breeding in the Gulf of Maine. Biodiversity Research Institute, Portland, ME.

Smith, J.A.M., K. Regan, N.W. Cooper, L. Johnson, E. Olson, A. Green, J. Tash, D.C. Evers, and P.P. Marra. 2020. A green wave of saltmarsh productivity predicts the timing of the annual cycle in a long-distance migratory shorebird. *Scientific Reports* 10:20658. https://doi.org/10.1038/s41598-020-77784-7.

Red-necked Phalarope

Gilg, O. Migration routes and timing of one Red-necked Phalarope breeding in NE Greenland (GLS). Groupe de Recherche en Ecologie Arctique. Unpublished data.

Smith, M., M. Bolton, J.D. Okill, P. Harris, G. Petrie, R. Riddington, and J.D. Wilson. 2018. Further evidence of transatlantic migration routes and Pacific wintering grounds of Red-necked Phalaropes breeding in Shetland. *British Birds* 111:428–437.

van Bemmelen, R.S.A., Y. Kolbeinsson, R. Ramos, O. Gilg, M. Smith, J.A. Alves, H. Schekkerman, A. Lehikoinen, I.K. Petersen, B. Þórisson, A.A. Sokolov, K. Välimäki, T. van der Meer, J.D. Okill, M. Bolton, B. Moe, S.A. Hanssen, L. Bollache, A. Petersen, S. Thorstensen, J. González-Solís, R.H.G. Klaassen, and I. Tulp. 2019. A migratory divide among Red-necked Phalaropes in the western Palearctic reveals contrasting migration and wintering movement strategies. *Frontiers in Ecology and Evolution* 7:86. https://doi.org/10.3389/fevo.2019.00086.

Red Phalarope

Gilg, O. Migration routes and timing of Long-tailed Jaegers, Parasitic Jaeger, Ross's Gull and Red Phalarope. Groupe de Recherche en Ecologie Arctique. Unpublished data.

Lanctot, R.B., et al. Arctic shorebird migration tracking study—Red Phalarope. 2019–2020. US Fish and Wildlife Service. Unpublished data.

Ronconi, R., A. MacDonald, and P. Smith. EC—Phalaropes (Project 105). 2016–2017. Data accessed from Motus Wildlife Tracking System, Birds Canada. https://motus.org/.

— FAMILY STERCORARIIDAE—SKUAS AND JAEGERS

Great Skua

SEATRACK. Unpublished data.

Pomarine Jaeger

Harrison, A.-L., P.F. Woodard, M.L. Mallory, and J. Rausch. 2022. Sympatrically breeding congeneric seabirds (*Stercorarius* spp.) from Arctic Canada migrate to four oceans. *Ecology and Evolution* 12(1):e8451. https://doi.org/10.1002/ece3.8451.

Myles, J., and R.B. Lanctot. Unpublished data.

Parasitic Jaeger

Gilg, O. Movebank Study Migration routes and timing of parasitic jaegers (Groupe de Recherche en Ecologie Arctique), ID 260558178. https://www.movebank.org/cms/webapp?gwt_fragment=page%3Dstudies%2Cpath%3Dstudy1260558178.

Harrison, A.-L., P.F. Woodard, M.L. Mallory, and J. Rausch. 2022. Sympatrically breeding congeneric seabirds (*Stercorarius* spp.) from Arctic Canada migrate to four oceans. *Ecology and Evolution* 12(1):e8451. https://doi.org/10.1002/ece3.8451.

Long-tailed Jaeger

Gilg, O., L. Bollache, B. Sittler, J. Lang, and R. van Bemmelen. Migration routes and timing of Long-tailed Jaegers. Groupe de Recherche en Ecologie Arctique.

Harrison A.-L., P.F. Woodard, M.L. Mallory, and J. Rausch. 2022. Sympatrically breeding congeneric seabirds (*Stercorarius* spp.) from Arctic Canada migrate to four oceans. *Ecology and Evolution* 12(1):e8451. https://doi.org/10.1002/ece3.8451.

van Bemmelen, R., B. Moe, S.A. Hanssen, N.M. Schmidt, J. Hansen, J. Lang, B. Sittler, L. Bollache, I. Tulp, M. Klaasen, and O. Gilg. 2017 Flexibility in otherwise consistent non-breeding movements of a long-distance migratory seabird, the Long-tailed Skua. *Marine Ecology Progress Series* 578:197–211.

— FAMILY ALCIDAE—AUKS, MURRES, AND PUFFINS

Dovekie

Grémillet, D. and J. Fort. Unpublished data.

SEATRACK. Unpublished data.

Common Murre

Hatch, S.A., P.M. Meyers, D.M. Mulcahy, and D.C. Douglas. 2020. Tracking data for Common Murres (*Uria aalge*). Version 1.0, August 2020. US Geological Survey data release. https://doi.org/10.5066/P9SOW6V2.

Loredo, S., R. Orben, R. Suryan, J. Adams, S. Stephensen, and D. Lyons. 2020. Satellite tracking of Common Murres (*Uria aalge*) in the northern California Current System, 2015–2017. Research Workspace. 10.24431/rw1k47i, version 10.24431_rw1k47i_20201211T201935Z.

Loredo, S.A., R.A. Orben, R.M. Suryan, D.E. Lyons, J. Adams, and S.W. Stephensen. 2019. Spatial and temporal diving behavior of non-breeding Common Murres during two summers of contrasting ocean conditions. *Journal of Experimental Marine Biology and Ecology* 517:13–24. http://doi.org/10.1016/j.jembe.2019.05.009.

McFarlane Tranquilla L.A., W.A. Montevecchi, A. Hedd, D.A. Fifield, C.M. Burke, P.A. Smith, P.M. Regular, G.J. Robertson, A.J. Gaston, and R.A. Phillips. 2013. Multiple-colony winter habitat use by murres *Uria* spp. in the northwest Atlantic Ocean: Implications for marine risk assessment. *Marine Ecology Progress Series* 472:287–303. https:// doi.org/ 10.3354/ meps10 053.

SEATRACK.

Takahashi, A., J.B. Thiebot, A.P. Will, S. Tsukamoto, B. Merkel, and A.S. Kitaysky. 2021. Breeding together, wintering an ocean apart: Foraging ecology of the northern Bering Sea Thick-billed and Common Murres in years of contrasting sea-ice conditions. *Polar Science* 27:10052. https://doi.org/10.1016/j.polar.2020.100552.

Thick-billed Murre

Hatch, S.A., P.M. Meyers, D.M. Mulcahy, and D.C. Douglas. 2020. Tracking data for Thick-billed Murres (*Uria lomvia*). Version 1.0, August 2020. US Geological Survey data release. https://doi.org/10.5066/P983BSVQ.

McFarlane Tranquilla, L.A., W.A. Montevecchi, A. Hedd, D.A. Fifield, C.M. Burke, P.A. Smith, P.M. Regular, G.J. Robertson, A.J. Gaston, and R.A. Phillips. 2013. Multiple-colony winter habitat use by murres *Uria* spp. in the northwest Atlantic Ocean: Implications for marine risk assessment. *Marine Ecology Progress Series* 472:287–303. https:// doi.org/ 10.3354/ meps10 053.

SEATRACK.

Takahashi, A., J.B. Thiebot, A.P. Will, S. Tsukamoto, B. Merkel, and A.S. Kitaysky. 2021. Breeding together, wintering an ocean apart: Foraging ecology of the northern Bering Sea Thick-billed and Common Murres in years of contrasting sea-ice conditions. *Polar Science* 27:10052. https://doi.org/10.1016/j.polar.2020.100552.

Razorbill

Linnebjerg, J.F., and M. Frederiksen. 2009–2011. GLS tracking of Razorbills breeding at Kitsissut Avalliit, Greenland.

Black Guillemot

Baak, J.E., M.L. Leonard, C. Gjerdrum, M.D. Dodds, and R.A. Ronconi. 2021. Non-breeding movements and foraging ecology of the Black Guillemot *Cepphus grylle* in Atlantic Canada. *Marine Ornithology* 49:57–70.

Marbled Murrelet

Bertram, D.F., M.H. Janssen, J.L. Cragg, C.A. Macdonald, M. McAdie, A. Wilson, K. Woo, E. Gross, M. Maftei, P.D. O'Hara, S.E. Davis, R. Greene, and P. Vincent. 2022. Data from: Study Marbled Murrelet satellite tracking data from British Columbia, Canada, 2014–2016. Movebank Data Repository. https://www.doi.org/10.5441/001/1.j0d889f0.

Bertram, D.F., C.A. MacDonald, P.D. O'Hara, J.L. Cragg, R. Corcoran, R. Greene, P. Vincent, and K. Woo. 2023. Summer movements of Marbled Murrelets from Canada to Alaska. *Endangered Species Research* 51:215–225. https://doi.org/10.3354/esr01252.

Bertram, D.F., C.A. Macdonald, P.D. O'Hara, J.L. Cragg, M.H. Janssen, M. McAdie, and W.S. Boyd. 2016. Marbled Murrelet *Brachyramphus marmoratus* movements and marine habitat use near proposed tanker routes to Kitimat, BC, Canada. *Marine Ornithology* 44:3–9. http://www.marineornithology.org/content/get.cgi?rn=1148.

Kittlitz's Murrelet

Piatt, J.F., M.L. Kissling, M.L. Arimitsu, and D.C. Douglas. 2021 Tracking data for Kittlitz's Murrelets (*Brachyramphus brevirostris*). Version 1.0, June 2021. US Geological Survey data release. https://doi.org/10.5066/P9MMVP9I.

Ancient Murrelet

Gaston, A.J., Y. Hashimoto, and L. Wilson. 2015. First evidence of east–west migration across the North Pacific in a marine bird. *Ibis* 157:877–882.

———. 2017. Post-breeding movement of Ancient Murrelet *Synthliboramphus antiquus* family groups, subsequent migration of adults and implications for management. *PLOS ONE* 12(2):e0171726. https://doi.org/10.1371/journal.pone.0171726.

Cassin's Auklet

Studholme, K.R., J.M. Hipfner, A.D. Domalik, S.J. Iverson, and G.T. Crossin. 2019. Year-round tracking reveals multiple migratory tactics in a sentinel North Pacific seabird, Cassin's Auklet. *Marine Ecology Progress Series* 619:169–185. https://doi.org/10.3354/meps12965.

Rhinoceros Auklet

Hipfner, J.M., M.M. Prill, K.R. Studholme, A.D. Domalik, S. Tucker, C. Jardine, M. Maftei, K.G. Wright, J.N. Beck. R.W. Bradley, R.D. Carle, T.P. Good, S.A. Hatch, P.J. Hodum, M. Ito, S.F. Pearson, N.A. Rojek, L. Slater, Y. Watanuki, A.P. Will, A.D. Bindoff, G.T. Crossin, M.C. Drever, and T.M. Burg. 2020. Geolocator tagging links distributions in the non-breeding season to population genetic structure in a sentinel North Pacific seabird. *PLOS ONE* 15(11):e0240056. https://doi.org/10.1371/journal.pone.0240056.

Shoji, A., S. Whelan, J.T. Cunningham, S.A. Hatch, Y. Niizuma, C. Nakajima, K.H. Elliott. 2023. Ecological niche partitioning in two Pacific puffins. *Marine Ecology Progress Series* 709:125–139. https://doi.org/10.3354/meps14282.

Atlantic Puffin

Baran, M.A., S.W. Kress, P. Shannon, D.E. Lyons, H.L. Major, and A.W. Diamond. 2022. Overwinter movement of Atlantic Puffins (*Fratercula arctica*) breeding in the Gulf of Maine: Inter- and intra-colony effects. *Waterbirds* 45(1):1–15. https://doi.org/10.1675/063.045.0103.

Burnham, K.K., J.L. Burnham, J.A. Johnson, and A. Huffman. 2021. Migratory movements of Atlantic Puffins *Fratercula arctica naumanni* from High Arctic Greenland. *PLOS ONE* 16(5):e0252055. https://doi.org/10.1371/journal.pone.0252055.

———. Data from: Migratory movements of Atlantic Puffins *Fratercula arctica naumanni* from High Arctic Greenland. Movebank Data Repository. https://www.doi.org/10.5441/001/1.nr8qv005.

Fayet, A.L., R. Freeman, T. Anker-Nilssen, A. Diamond, K.E. Erikstad, D. Fifield, M.G. Fitzsimmons, E.S. Hansen, M.P. Harris, M. Jessopp, et al. 2017. Ocean-wide drivers of migration strategies and their influence on population breeding performance in a declining seabird. *Current Biology* 27(24). https://doi.org/10.1016/j.cub.2017.11.009.

Jessopp, M. Puffin overwintering. BirdLife International Seabird Tracking Database Dataset ID 1048. https://data.seabirdtracking.org/dataset/1048.

Jessopp, M.J., M. Cronin, T.K. Doyle, et al. 2013. Transatlantic migration by post-breeding puffins: A strategy to exploit a temporarily abundant food resource? *Marine Biology* 160:2755–2762. https://doi.org/10.1007/s00227-013-2268-7.

Petersen, A. Movebank Study Atlantic Puffin in Iceland (Petersen), ID 943122207. https://www.movebank.org/cms/webapp?gwt_fragment=page%3Dstudies%2Cpath%3Dstudy943122207.

SEATRACK. Unpublished data.

Tufted Puffin

Gorman, K.B., A.L. Schaefer, and M.A. Bishop. Resolving the annual pelagic distribution of Tufted Puffins. North Pacific Research Board Project 1812.

Hatch, S.A., P.M. Meyers, D.M. Mulcahy, and D.C. Douglas. 2000. Seasonal movements and pelagic habitat use of Murres and Puffins determined by satellite telemetry. *Condor* 102:145–154. https://doi.org/10.1093/condor/102.1.145.

———. 2020. Tracking data for Tufted Puffins (*Fratercula cirrhata*). Version 1.0, October 2020. US Geological Survey data release. https://doi.org/10.5066/P9L9EIUI.

Schaefer, A.L., K.B. Gorman, and M.A. Bishop. 2022. Light-level geolocation reveals the short-distance non-breeding movements and distribution of Tufted Puffins throughout the northeast Pacific Ocean. *Frontiers in Marine Science* 9:999461. https://doi.org/10.3389/fmars.2022.999461.

Shoji, A., S. Whelan, J.T. Cunningham, S.A. Hatch, Y. Niizuma, C. Nakajima, and K.H. Elliott. 2023. Ecological niche partitioning in two Pacific puffins. *Marine Ecology Progress Series* 709:125–139. https://doi.org/10.3354/meps14282.

— FAMILY LARIDAE—GULLS, TERNS, AND SKIMMERS

Black-legged Kittiwake

Frederiksen, M., A. Labansen, J.F. Linnebjerg, and F.R. Merkel. 2008–2020. GLS tracking of Black-legged Kittiwakes from four colonies in Greenland.

González-Solís, J., T. Boulinier, and D. Gremillet. Unpublished data.

Mallory, M., A.J. Gaston, and J. Akearok. Movebank Study Black-legged Kittiwakes; Mallory/Gaston/Akearok, Prince Leopold Island, Nunavut, Canada, ID 40008075. https://www.movebank.org/cms/webapp?gwt_fragment=page=studies,path=study40008075.

McKnight, A., D.B. Irons, A.J. Allyn, K.M. Sullivan, and R.M. Suryan. 2011. Winter dispersal and activity patterns of post-breeding Black-legged Kittiwakes *Rissa tridactyla* from Prince William Sound, Alaska. *Marine Ecology Progress Series* 442:241–253. https://doi.org/10.3354/meps09373.

Orben, R.A., R. Paredes, D.D. Roby, D.B. Irons, and S.A. Shaffer. 2015. Wintering North Pacific Black-legged Kittiwakes balance spatial flexibility and consistency. *Movement Ecology* 3(1):36. http://doi.org/10.1186/s40462-015-0059-0.

SEATRACK. Unpublished data.

Whelan, S., S.A. Hatch, Z.M. Benowitz-Fredericks, C. Parenteau, O. Chastel, and K.H. Elliott. 2021. The effects of food supply on reproductive hormones and timing of reproduction in an income-breeding seabird. *Hormones and Behavior* 127:104874. https://doi.org/10.1016/j.yhbeh.2020.104874.

Whelan, S., S.A. Hatch, D.B. Irons, A. McKnight, A. and K.H. Elliott., 2020. Increased summer food supply decreases non-breeding movement in Black-legged Kittiwakes. *Biology Letters* 16(1):20190725. https://doi.org/10.1098/rsbl.2019.0725.

Red-legged Kittiwake

Drummond, B.A., R.A. Orben, A.M. Christ, A.B. Fleishman, H.M. Renner, N.A. Rojek, and M.D. Romano. 2021. Comparing non-breeding distribution and behavior of Red-legged Kittiwakes from two geographically distant colonies. *PLOS ONE* 16(7):e0254686. https://doi.org/10.1371/journal.pone.0254686.

Orben, R.A., N. Kokubun, A.B. Fleishman, A.P. Will, T. Yamamoto, S.A. Shaffer, R. Paredes, A. Takahashi, and A.S. Kitaysky. 2018. Persistent annual migration patterns of a specialist seabird. *Marine Ecology Progress Series* 593:231–245. https://doi.org/10.3354/meps12459.

Ivory Gull

Frederiksen M., O. Gilg, and G. Yannic. 2021. Cross-icecap spring migration confirmed in a high-Arctic seabird, the Ivory Gull *Pagophila eburnea*. *Ibis* 163(2):706–714. https://doi.org/10.1111/ibi.12903.

Gilg, O., L. Istomina, G. Heygster, H. Strøm, M.V. Gavrilo, M.L. Mallory, G. Gilchrist, A. Aebischer, B. Sabard, M. Huntemann, A. Mosbech, and G. Yannic. 2016. Living on the edge of a shrinking habitat: The Ivory Gull, *Pagophila eburnea*, an endangered sea-ice specialist. *Biology Letters* 12:20160277. https://doi.org/10.1098/rsbl.2016.0277.

Spencer, N.C., H.G. Gilchrist, and M.L. Mallory. 2014. Annual movement patterns of endangered Ivory Gulls: The importance of sea ice. *PLOS ONE* 9(12):e115231. https://doi.org/10.1371/journal.pone.0115231.

Sabine's Gull

Davis, S.E., M. Maftei, and M.L. Mallory. 2016. Data from: Migratory connectivity at high latitudes: Sabine's Gulls (*Xema sabini*) from a colony in the Canadian High Arctic migrate to different oceans. Movebank Data Repository. https://www.doi.org/10.5441/001/1.c745vb70.

Maftei, M. Movebank Study Sabine's Gulls in the Juan de Fuca Eddy, ID 304527187. Accessed from movebank.org on December 9, 2021.

Stenhouse, I., C. Evevang, and R.A. Phillips. 2012. Trans-equatorial migration, staging sites and wintering area of Sabine's Gulls *Larus sabini* in the Atlantic Ocean. *Ibis* 154:42–51.

Ross's Gull

Maftei, M., S.E. Davis, and M.L. Mallory. 2015. Confirmation of a wintering ground of Ross's Gull *Rhodostethia rosea* in the northern Labrador Sea. *Ibis* 157(3):642–647. https://doi.org/10.1111/ibi.12261.

Ring-billed Gull

Clark, D.E. 2013. Roosting, site fidelity, and food sources of urban gulls in Massachusetts: Implications for protecting public water supplies. PhD diss., University of Massachusetts, Amherst.

Clark, D.E., S. DeStefano, K.G. MacKenzie, K.K.G. Koenen, and J.W. Whitney. 2016. Roost site selection by Ring-billed and Herring Gulls. *Journal of Wildlife Management* 80:708–719.

DeStefano. 2016. Fidelity and persistence of Ring-billed and Herring Gulls to wintering sites. *Waterbirds* 39:220–234.

Moore, D., and D. Costa. Unpublished data.

Herring Gull

Anderson, C.M., H.G. Gilchrist, R.A. Ronconi, K.R. Shlepr, D.E. Clark, D.A. Fifield, G.J. Robertson, and M.L. Mallory. 2020. Both short and long distance migrants use energy-minimizing strategies in North American Herring Gulls. *Movement Ecology* 8:26. https://doi.org/10.1186/s40462-020-00207-9.

Clark, D.E., S.A. Mackenzie, K. Koenen, J. Whitney, and S. DeStefano. 2020. Data from: Study Herring Gulls (*Larus argentatus*); Clark; Massachusetts, United States. Movebank Data Repository. https://www.doi.org/10.5441/001/1.3th8b5q3.

Fifield, D.A., R.A. Ronconi, and G.J. Robertson 2020. Data from: Study Herring Gulls (*Larus argentatus*); Fifield; Witless Bay, Canada. Movebank Data Repository. https://www.doi.org/10.5441/001/1.27244v55.

Gilchrist, H.G., C.A. Macdonald, M.H. Janssen, K.A. Allard, and C.M. Anderson. 2020. Data from: Study Herring Gulls (*Larus argentatus*); Gilchrist; East Bay Island, Canada. Movebank Data Repository. https://www.doi.org/10.5441/001/1.1r1s4v8d.

Massachusetts Department of Conservation and Recreation, Division of Water Supply Protection. Unpublished data.

Ramey, A.M., S.A. Hatch, C.A. Ahlstrom, M.L. van Toor, H. Woksepp, J.C. Chandler, J.A. Reed, A.B. Reeves, J. Waldenström, A.B. Franklin, J. Bonnedahl, V.A. Gill, D.M. Mulcahy, and D.C. Douglas. 2020. Tracking data for three large-bodied gull species and hybrids (*Larus* spp.). Version 2.0, March 2021. US Geological Survey data release. https://doi.org/10.5066/P9FZ4OJW.

Ronconi, R.A. and K.R. Shlepr. 2020. Data from: Study Herring Gulls (*Larus argentatus*); Ronconi; Brier Island, Canada. Movebank Data Repository. https://www.doi.org/10.5441/001/1.282vr7kd.

———. 2020. Data from: Study Herring Gulls (*Larus argentatus*); Ronconi; Kent Island, Canada. Movebank Data Repository. https://www.doi.org/10.5441/001/1.t5n4s456.

Ronconi, R.A., and P.D. Taylor. 2020. Data from: Study Herring Gulls (*Larus argentatus*); Ronconi; Sable Island, Canada. Movebank Data Repository. https://www.doi.org/10.5441/001/1.3264ss3v.

Iceland Gull

Davis, S., and M. Hipfner. Movebank Study Thayer's Gulls from the Canadian High Arctic, ID 176466071. https://www.movebank.org/cms/webapp?gwt_fragment=page%3Dstudies%2Cpath%3Dstudy176466071.

Gutowsky, S.E., J.M. Hipfner, M. Maftei, W.S. Boyd, M. Auger-Méthé, D.J. Yurkowski, and M.L. Mallory. 2024. First insights into Thayer's Gull Larus *Glaucoides thayeri* migratory and overwintering behavior along the Pacific coast of North America. *Marine Ornithology* 48:9–16.

Lesser Black-backed Gull

Barber, P. Unpublished data.

Glaucous-winged Gull

Domalik, A.D., M. Maftei, K.G. Wright, S.A. Hudson, and J.M. Hipfner. 2021. Migration and winter habitat use of Glaucous-winged Gulls (*Larus glaucescens*) from Triangle Island, British Columbia. *Waterbirds* 44:328–338. https://doi.org/10.1675/063.044.0405.

Elliott, K., and S. Whelan. Unpublished data.

Ramey, A.M., S.A. Hatch, C.A. Ahlstrom, M.L. van Toor, H. Woksepp, J.C. Chandler, J.A. Reed, A.B. Reeves, J. Waldenström, A.B. Franklin, J. Bonnedahl, V.A. Gill, D.M. Mulcahy, and D.C. Douglas. 2020. Tracking data for three large-bodied gull species and hybrids (*Larus* spp.). Version 2.0, March 2021. US Geological Survey data release. https://doi.org/10.5066/P9FZ4OJW.

Glaucous Gull

Baak, J.E., A. Patterson, H.G. Gilchrist, and K.H. Elliott. 2021. First evidence of diverging migration and overwintering strategies in Glaucous Gulls (*Larus hyperboreus*) from the Canadian Arctic. *Animal Migration* 8(1):98–109. https://doi.org/10.1515/ami-2020-0107.

Harrison, A-L. Movebank Study MCP Glaucous Gull Barrow Alaska, ID 116817431. https://www.movebank.org/cms/webapp?gwt_fragment=page=studies,path=-study116817431.

Ramey, A.M., S.A. Hatch, C.A. Ahlstrom, M.L. van Toor, H. Woksepp, J.C. Chandler, J.A. Reed, A.B. Reeves, J. Waldenström, A.B. Franklin, J. Bonnedahl, V.A. Gill, D.M. Mulcahy, and D.C. Douglas. 2020. Tracking data for three large-bodied gull species and hybrids (*Larus* spp.). Version 2.0, March 2021. US Geological Survey data release. https://doi.org/10.5066/P9FZ4OJW.

Great Black-backed Gull

Massachusetts Department of Conservation and Recreation, Division of Water Supply Protection. Unpublished data.

Maynard, L.D., and G.K. Davoren. 2017. Tracking data from migratory and wintering movement of Great Black-backed Gulls in Newfoundland. University of Manitoba. Unpublished data.

Sooty Tern

Huang, R.M., O.L. Bass Jr., and S.L. Pimm. 2017. Sooty Tern (*Onychoprion fuscatus*) survival, oil spills, shrimp fisheries, and hurricanes. *PeerJ* 5:e3287. https://doi.org/10.7717/peerj.3287.

Aleutian Tern

Goldstein, M.I., D.C. Duffy, S. Oehlers, N. Catterson, J. Frederick, and S. Pyare. 2019. Interseasonal movements and non-breeding locations of Aleutian Terns *Onychoprion aleuticus*. *Marine Ornithology* 47:67–76. http://www.marineornithology.org/PDF/47_1/47_1_67-76.pdf.

Gull-billed Tern

Goodenough, K.S., and R.T. Patton. 2019. Satellite telemetry reveals strong fidelity to migration routes and wintering grounds for the Gull-billed Tern (*Gelochelidon nilotica*). *Waterbirds* 42(4):400–410.

Caspian Tern

Lyons, D. and T. Lawes. Movebank Study Caspian Terns, Columbia Plateau, WA, 2016, ID 152496827. Accessed from movebank.org on November 2, 2020.

Black Tern

McKellar, A. Black Tern Dispersal—Saskatchewan (Project 359). 2020–2024. Data accessed from Motus Wildlife Tracking System, Birds Canada. https://motus.org/.

Moore, D., and J. Costa. Black Tern movements—breeding/migration (Project 156). 2017–2021. Data accessed from Motus Wildlife Tracking System, Birds Canada. https://motus.org/.

Shephard, N.G., P. Szczys, D.J. Moore, M.W. Reudink, J.N. Cosa, A.M. Bracey, S. Lisovski, and A.E. McKellar. 2023. Weak genetic structure, shared nonbreeding areas, and extensive movement in a declining waterbird. *Ornithological Applications* 125(1):duac053. https://doi.org/10.1093/ornithapp/duac053.

Roseate Tern

Mostello, C., I. Nisbet, S. Oswald, and J. Fox. 2014. Non-breeding season movements of six North American Roseate Terns *Sterna dougallii* tracked with geolocators. *Seabird Journal* 1–20. https://doi.org/10.61350/sbj.27.1.

Common Tern

Bracey, A., S. Lisovski, D. Moore, A. McKellar, E. Craig, S. Matteson, F. Strand, J. Costa, C. Pekarik, P. Curtis, G. Niemi, and F. Cuthbert. 2018. Migratory routes and wintering locations of declining inland North American Common Terns. *Auk* 135(2):385–399. https://doi.org/10.1642/AUK-17-210.1.

Nisbet, I.C.T., C.S. Mostello, R.R. Veit, J.W. Fox, and V. Afanasyev. 2011. Migrations and winter quarters of five Common Terns tracked using geolocators. *Waterbirds* 34:32–39.

Welch, L. Unpublished data.

Arctic Tern

Egevang, C., and I. Stenhouse. Greenland Institute of Natural Resources.

McKnight, A., A.J. Allyn, D.C. Duffy, and D.B. Irons. 2013. "Stepping stone" pattern in Pacific Arctic Tern migration reveals the importance of upwelling areas. *Marine Ecology Progress Series* 491:253–264. https://doi.org/10.3354/meps10469.

Welch, L., P. Loring, K. Myers, and P. Shannon. Unpublished data.

Wong, J.B., S. Lisovski, R.T. Alisauskas, W. English, M.-A. Giroux, A.-L. Harrison, D. Kellett, N. Lecomte, M. Maftei, A. Nagy-MacArthur, R.A. Ronconi, P.A. Smith, M.L. Mallory, and M. Auger-Méthé. 2021. Arctic Terns from circumpolar breeding colonies share common migratory routes. *Marine Ecology Progress Series* 671:191–206. https://doi.org/10.3354/meps13779.

Black Skimmer

Goodenough, K.S. 2022. Ground-nesting birds in a changing climate: Case studies of proximate behavioral mechanisms used by skimmers and terns to address seasonal variation in resources. PhD diss., University of Oklahoma.

Newstead, D. Unpublished data.

GREBES (PODICIPEDIFORMES)

FAMILY PODICIPEDIDAE—GREBES

Eared Grebe

Jehl Jr., J.R., and P.K. Yochem. 1986. Movements of Eared Grebes indicated by banding recoveries. *Journal of Field Ornithology* 57(3):208–212.

Western Grebe

Mills, K.L., J.K. Gaydos, C.V. Fiorello, E.R. Whitmer, S. De La Cruz, D.M. Mulcahy, L.I. Vilchis, and M.H. Ziccardi. 2016. Post-release survival and movement of Western Grebes (*Aechmophorus occidentalis*) implanted with intracoelomic satellite transmitters. *Waterbirds* 39(2):175–186. https://doi.org/10.1675/063.039.0208.

TROPICBIRDS (PHAETHONTIFORMES)

FAMILY PHAETHONTIDAE—TROPICBIRDS

White-tailed Tropicbird

Mejías, M.A., Y.F. Wiersma, D.B. Wingate, and J.L. Madeiros. 2017. Distribution and at-sea behavior of Bermudan White-tailed Tropicbirds (*Phaethon lepturus catesbyi*) during the non-breeding season. *Journal of Field Ornithology* 88:184–197.

LOONS (GAVIIFORMES)

FAMILY GAVIIDAE—LOONS

Red-throated Loon

Uher-Koch, B.D., J.A. Schmutz, and D.C. Douglas. 2017. Tracking data for Red-throated Loons (*Gavia stellata*). Version 2.0, February 2020. US Geological Survey data release. https://doi.org/10.5066/F7TH8KVH.

Pacific Loon

Harrison, A.-L., B.D. Uher-Koch, J.A. Schmutz, and D.C. Douglas DC. 2020. Tracking data for Pacific Loons (*Gavia pacifica*). Version 1.0, February 2020. US Geological Survey data release. https://doi.org/10.5066/P9NNN2XY.

Common Loon

Earthspan. Unpublished data.

Kenow, K.P., D. Adams, N. Schoch, D.C. Evers, W. Hanson, D. Yates, L. Savoy, T.J. Fox, A. Major, R. Kratt, and J. Ozard. 2009. Migration patterns and wintering range of Common Loons breeding in the northeastern United States. *Waterbirds* 32(2):234–247.

Kenow, K.P., M.W. Meyer, D.C. Evers, D.C. Douglas, and J. Hines. 2002. Use of satellite telemetry to identify Common Loon migration routes, staging areas, and wintering range. *Waterbirds* 25(4):449–458.

Paruk, J.D., D. Long, S.L. Ford, and D.C. Evers. 2014. Common Loons (*Gavia immer*) wintering off the Louisiana Coast tracked to Saskatchewan during the breeding season. *Waterbirds* 37(sp1):47.52.

Yellow-billed Loon

Schmutz, J.A., B. Uher-Koch., and D.C. Douglas. 2019. Tracking data for Yellow-billed Loons (*Gavia adamsii*). US Geological Survey data release. https://doi.org/10.5066/P9UKHJ22.

ALBATROSSES, PETRELS, AND SHEARWATERS (PROCELLARIIFORMES)

FAMILY DIOMEDEIDAE—ALBATROSSES

Laysan Albatross

Anderson, D. Laysan albatross from Tern Island 1998–99. BirdLife International Seabird Tracking Database Dataset, ID 453. https://data.seabirdtracking.org/dataset/453.

Henry, B., and D. Costa. Laysan albatross from Guadalupe 2002–05. BirdLife International Seabird Tracking Database Dataset, ID 521. https://data.seabirdtracking.org/dataset/521.

———. Laysan albatross from Guadalupe 2005–06. BirdLife International Seabird Tracking Database Dataset, ID 524. https://data.seabirdtracking.org/dataset/524.

Shaffer, S., and R. Orben. Laysan Albatross Midway Atoll GLS2008 12 postbreeding. BirdLife International Seabird Tracking Database Dataset, ID 1930. https://data.seabirdtracking.org/dataset/1930.

Shaffer, S., M. Antolos, and M. Conners. Laysan albatross geolocators from Tern Island 2004–06. BirdLife International Seabird Tracking Database Dataset, ID 529. https://data.seabirdtracking.org/dataset/529.

Suryan, R., and G.R. Balogh. Laysan albatross non-breeders 2005–06. BirdLife International Seabird Tracking Database Dataset, ID 516. https://data.seabirdtracking.org/dataset/516.

Black-footed Albatross

Anderson, D. Black-footed Albatross from Tern Island 1998–99. BirdLife International Seabird Tracking Database Dataset, ID 452. https://data.seabirdtracking.org/dataset/452.

Shaffer, S., and R. Orben. Black-footed Albatross Midway Atoll GLS2008 12 postbreeding. BirdLife International Seabird Tracking Database Dataset, ID 1927. https://data.seabirdtracking.org/dataset/1927.

Shaffer, S., M. Antolos, and M. Conners. Black-footed Albatross from Tern Island 2002–05. BirdLife International Seabird Tracking Database Dataset, ID 519. https://data.seabirdtracking.org/dataset/519.

———. Black-footed Albatross from Tern Island 2005–06. BirdLife International Seabird Tracking Database Dataset ID 522. https://data.seabirdtracking.org/dataset/522.

———. Black-footed Albatross geolocators from Tern Island 2004–2006. BirdLife International Seabird Tracking Database Dataset, ID 528. https://data.seabirdtracking.org/dataset/528.

Suryan, R., and G.R. Balogh. Black-footed Albatross non-breeders 2005–06. BirdLife International Seabird Tracking Database Dataset, ID 515. https://data.seabirdtracking.org/dataset/515.

Short-tailed Albatross

Suryan, R. Short-tailed albatross from Torishima 2002–07. BirdLife International Seabird Tracking Database Dataset, ID 512. https://data.seabirdtracking.org/dataset/512.

———. Short-tailed albatross from Torishima 2002–08. BirdLife International Seabird Tracking Database Dataset, ID 514. https://data.seabirdtracking.org/dataset/514.

———. Short-tailed albatross non-breeders 2003–06. BirdLife International Seabird Tracking Database Dataset, ID 513. https://data.seabirdtracking.org/dataset/513.

FAMILY HYDROBATIDAE—NORTHERN STORM-PETRELS

Fork-tailed Storm-Petrel

Halpin, L.R., I.L. Pollet, C. Lee, K.H. Morgan, and H.R. Carter. 2018. Year-round movements of sympatric Fork-tailed (*Oceanodroma furcata*) and Leach's (*O. leucorhoa*) storm-petrels. *Journal of Field Ornithology* 89(3):207–220. https://doi.org/10.1111/jofo.12255.

Leach's Storm-Petrel

Halpin, L.R., I.L. Pollet, C. Lee, K.H. Morgan, and H.R. Carter. 2018. Year-round movements of sympatric Fork-tailed (*Oceanodroma furcata*) and Leach's (*O. leucorhoa*) storm-petrels. *Journal of Field Ornithology* 89(3):270–220. https://doi.org/10.1111/jofo.12255.

Pollet, I.L. Leach's storm-petrel migration Bon Portage. BirdLife International Seabird Tracking Database Study, ID 1706. Accessed from seabirdtracking.org on September 21, 2021.

FAMILY PROCELLARIIDAE—SHEARWATERS AND PETRELS

Northern Fulmar

Hatch, S.A., V.A. Gill, D.M. Mulcahy, and D.C. Douglas. 2020 Tracking data for Northern Fulmars (*Fulmarus glacialis*). Version 1.0, October 2020. US Geological Survey data release. https://doi.org/10.5066/P992BR5E.

Mallory, M.L., J. Akearok, D.B. Edwards, K. O'Donovan, and C.D. Gilbert. 2008. Autumn migration and wintering of Northern Fulmars (*Fulmarus glacialis*) from the Canadian High Arctic. *Polar Biology* 31:745–750. https://doi.org/10.1007/s00300-008-0417-0.

SEATRACK. Unpublished data.

Murphy's Petrel

Clay, T.A., R.A. Phillips, A. Manica, H.A. Jackson, and M. de L. Brooke. 2017. Escaping the oligotrophic gyre? The year-round movements, foraging behaviour and habitat preferences of Murphy's Petrels. *Marine Ecology Progress Series* 579:139–155. https://doi.org/10.3354/meps12244.

Bermuda Petrel

Raine, A.F., C. Gjerdrum, I. Pratte, J. Madeiros, J.J. Felis, and J. Adams. 2021. Marine distribution and foraging habitat highlight potential threats at sea for the endangered Bermuda Petrel *Pterodroma cahow*. *Endangered Species Research* 45:337–356. https://doi.org/10.3354/esr01139.

Black-capped Petrel

Jodice, P.G., R.A. Ronconi, E. Rupp, G.E. Wallace, and Y. Satgé. 2015. First satellite tracks of the endangered Black-capped Petrel. *Endangered Species Research* 29(1):23–33. https://doi.org/10.3354/esr00697.

Satgé, Y.G., B.S. Keitt, C.P. Gaskin, J.B. Patteson, and P.G.R. Jodice. Spatial segregation between phenotypes of the diablotin Black-capped Petrel *Pterodroma hasitata* during the non-breeding period. *Endangered Species Research* 51:183–201. https://doi.org/10.3354/esr01254.

Satgé, Y.G., E. Rupp, E. and P.G.R. Jodice. 2019. A preliminary report of ongoing research of the ecology of Black-capped Petrel in Sierra de Bahoruco, Dominican Republic—I. GPS tracking of breeding adults. Unpublished report, South Carolina Cooperative Research Unit, Clemson, SC. https://doi.org/10.5066/P9UHASY4.

Hawaiian Petrel

Raine, A.F. Unpublished data.

Desertas Petrel

Paiva, V. and I. Ramirez. Pdeserta Bugio pd01 62 (Desertas Petrel, Bugio, GLS). BirdLife International Seabird Tracking Database Study, ID 1031. Accessed from seabirdtracking.org on December 6, 2021.

Cook's Petrel

Rayner, M.J. Cook Petrel Codfish Island. BirdLife International Seabird Tracking Database Study, ID 639. Accessed from seabirdtracking.org on October 26, 2021.

———. Cook Petrel, Little Barrier Island. BirdLife International Seabird Tracking Database Study, ID 888. Accessed from seabirdtracking.org on October 26, 2021.

Rayner, M.J., M.E. Hauber, T.E. Steeves, H.A. Lawrence, D.R. Thompson, P.M. Sagar, S. Bury, T.J. Landers, R.A. Phillips, L. Ranjard, and S.A. Shaffer 2011. Contemporary and historic separation of transhemispheric migration between two genetically distinct seabird populations. *Nature Communications* 2:332.

Cory's Shearwater

Dias, M.A., and P. Catry. Cory's Shearwater Selvagens 08. BirdLife International Seabird Tracking Database Study, ID 607. Accessed from seabirdtracking.org on November 30, 2021.

Dias, M.A., and J.P. Granadeiro. Cory's Shearwater Selvagens 06. BirdLife International Seabird Tracking Database Study, ID 610. Accessed from seabirdtracking.org on November 30, 2021.

González-Solís, J. Cory's Shearwater geolocators from Canarias 2002–04. BirdLife International Seabird Tracking Database Study, ID 509. Accessed from seabirdtracking.org on November 30, 2021.

———. Cory's Shearwater, GLS data, Veneguera (Canarias). BirdLife International Seabird Tracking Database Study, ID 974. Accessed from seabirdtracking.org on January 14, 2022.

———. Cory's Shearwater, GLS data, Vila (Azores). BirdLife International Seabird Tracking Database Study, ID 975. Accessed from seabirdtracking.org on January 25, 2022.

Granadeiro, J.P., and P. Catry. Cory's Shearwater Berlenga. BirdLife International Seabird Tracking Database Study, ID 609. Accessed from seabirdtracking.org on November 30, 2021.

Paiva, V., and J.A. Ramos. Cborealis Berlenga GLS 11 16 (Cory's Shearwater, Selvagens, GLS). BirdLife International Seabird Tracking Database Study, ID 1032. Accessed from seabirdtracking.org on November 30, 2021.

Wakefield, E.D., D.L. Miller, S.L. Bond, F. le Bouard, P.C. Carvalho, P. Catry, B.J. Dilley, D.A. Fifield, C. Cjerdrum, J. González-Solís, H. Hogan, V. Laptikhovsky, B. Merkel, J.A.O. Miller, P.I. Miller, S.J. Pinder, T. Pipa, P.M. Ryan, L.A. Thompson, and J. Matthiopoulos. 2021. The summer distribution, habitat associations, and abundance of seabirds in the sub-polar frontal zone of the northwest Atlantic. *Progress in Oceanography* 198:102657. https://doi.org/10.1016/j.pocean.2021.102657.

Sooty Shearwater

Adams, J., and J.J. Felis J. 2022. Satellite tracks of migratory Sooty Shearwater *Ardenna grisea* in the Pacific Ocean. US Geological Survey. Unpublished data.

Hedd, A., M. Montevecchi, and R. Phillips. Sooty Shearwaters Falkland Islands 2007–09. BirdLife International Seabird Tracking Database Study, ID 628. Accessed from seabirdtracking.org on January 18, 2022.

Ronconi, R.A. Sooty Shearwater, Bay of Fundy. BirdLife International Seabird Tracking Database Study, ID 656. Accessed from seabirdtracking.org on November 11, 2021.

Shaffer, S.A., Y. Tremblay, H. Weimerskirch, D. Scott, D.R. Thompson, P.M. Sagar, H. Moller, G.A. Taylor, D.G. Foley, B.A. Block, and D.P. Costa. 2006. Migratory shearwaters integrate oceanic resources across the Pacific Ocean in an endless summer. *Proceedings of the National Academy of Sciences* 103(34):12799–12802. https://doi.org/10.1073/pnas.0603715103.

Wakefield, E.D., D.L. Miller, S.L. Bond, F. le Bouard, P.C. Carvalho, P. Catry, B.J. Dilley, D.A. Fifield, C. Cjerdrum, J. González-Solís, H. Hogan, V. Laptikhovsky, B. Merkel, J.A.O. Miller, P.I. Miller, S.J. Pinder, T. Pipa, P.M. Ryan, L.A. Thompson, and J. Matthiopoulos. 2021. The summer distribution, habitat associations, and abundance of seabirds in the sub-polar frontal zone of the Northwest Atlantic. *Progress in Oceanography* 198:102657. https://doi.org/10.1016/j.pocean.2021.102657.

Great Shearwater

Hatch, J.M., D. Wiley, K.T. Murray, and L. Welch. 2016. Integrating satellite-tagged seabird and fishery-dependent data: A case study of Great Shearwaters (*Puffinus gravis*) and the US New England sink gillnet fishery. *Conservation Letters* 9(1):45–30. https://doi.org/10.1111/conl.12178.

Powers, K.D., I. Pratte, R.A. Ronconi, S.N.P. Wong, P.G. Ryan, L.J. Welch, T.L. Silva, K.A. Hatch, A.J. Westgate, and D.N. Wiley. 2022. Age-related interactions with wind during migration support the hypothesis of developmental learning in a migrating long-lived seabird. *Frontiers in Marine Science* 9:938033. https://doi.org/10.3389/fmars.2022.938033.

Powers, K.D., D.N. Wiley, A.J. Allyn, L.J. Welch, and R.A. Ronconi. 2017. Movements and foraging habitats of Great Shearwaters *Puffinus gravis* in the Gulf of Maine. *Marine Ecology Progress Series* 574:211–226.

Powers, K.D., D.N. Wiley, A.R. Robuck, Z.H Olson, L.J. Welch, M.A. Thompson, and L. Kaufman. 2020. Spatiotemporal characterization of non-breeding Great Shearwaters *Ardenna gravis* within their wintering range. *Marine Ornithology* 48:215–229.

Robuck, A. J. McCord, M. Strynar, M. Cantwell, D. Wiley, and R. Lohmann. 2021. Tissue-specific distribution of legacy and novel per- and polyfluoroalkyl substances in juvenile seabirds. *Environmental Science & Technology Letters* 8(6):457–462. https://doi.org/10.1021/acs.estlett.1c00222.

Ronconi, R.A., S. Schoombie, A.J. Westgate, S.N.P. Wong, H.N. Koopman, and P.G. Ryan. 2018. Effects of age, sex, colony and breeding phase on marine space use by Great Shearwaters *Ardenna gravis* in the South Atlantic. *Marine Biology* 165(58). https://doi.org/10.1007/s00227-018-3299-x.

Pink-footed Shearwater

Felis, J.J., J. Adams, P.J. Hodum, R.D. Carle, and V. Colodro. 2022. Satellite tracks of migratory Pink-footed Shearwater *Ardenna creatopus* in the eastern Pacific Ocean. US Geological Survey and Oikonos Ecosystem Knowledge. Unpublished data.

Manx Shearwater

Fayet, A.L., P. Shannon, D.E. Lyons, and S.W. Kress. 2020. Manx shearwaters *Puffinus puffinus* breeding in the western Atlantic follow a different migration route from their eastern-Atlantic conspecifics. *Marine Ornithology* 48(2):179–183.

Newell's Shearwater

Raine, A.F., T. Anderson, M. Vynne, S. Driskill, H. Raine, and J. Adams. 2020. Post-release survival of fallout Newell's Shearwater fledglings from a rescue and rehabilitation program on Kaua'i Hawai'i. *Endangered Species Research* 43:39–50. https://doi.org/10.3354/esr01051/.

STORKS (CICONIIFORMES)

— FAMILY CICONIIDAE—STORKS

Wood Stork

Basille, M., R.R. Borkhataria, A.L. Bryan Jr., D.N. Bucklin, S. Picardi and P.C. Frederick. 2021. GPS tracking of Wood Storks (*Mycteria americana*) in the southeastern United States from 2004 to 2019. Movebank Data Repository. https://doi.org/10.5441/001/1.r0h6725k.

Schweitzer, S., A.L. Bryan, J. Brzorad, and R. Kays. 2023. Data from: NC Wood Stork tracking. Movebank Data Repository. https://doi.org/10.5441/001/1.303.

FRIGATEBIRDS, BOOBIES, GANNETS, ANHINGAS, AND CORMORANTS (SULIFORMES)

— FAMILY FREGATIDAE—FRIGATEBIRDS

Magnificent Frigatebird

Meyer, K.D., and G.M. Kent. Avian Research and Conservation Institute. Unpublished data.

Voirin, B. 2011. Movebank Study Frigatebirds breeding at Iguana Island, Panama, ID 2760899. http://www.movebank.org/panel_embedded_movebank_webapp?gwt_fragment=page=studies,path=study2760899.

Wikelski, M. 2012. Movebank Study MPIAB PNIC hurricane frigate tracking, ID 6770990. http://www.movebank.org/panel_embedded_movebank_webapp?gwt_fragment=page=studies,path=study6770990.

— FAMILY SULIDAE—BOOBIES AND GANNETS

Northern Gannet

Fifield, D.A., W.A. Montevecchi, S. Garthe, G.J. Robertson, U. Kubetzki, and J.-F. Rail. 2014. Migratory tactics and wintering areas of Northern Gannets (*Morus bassanus*) breeding in North America. *Ornithological Monographs* 79:i–63. https://www.researchgate.net/publication/260602079_Migratory_tactics_and_wintering_areas_of_Northern_Gannets_Morus_bassanus_breeding_in_North_America.

Rail, J.-F. Canadian Wildlife Service. Unpublished data.

— FAMILY PHALACROCORACIDAE—CORMORANTS AND SHAGS

Brandt's Cormorant

Peck-Richardson, A., A. Piggott, J. Lerczak, and R. Orben. 2024. Cormorant Oceanography Project: Biologging Data from Brandt's and Pelagic Cormorants (*Urile penicillatus, U. pelagicus*), Columbia River, 2019–2022. Research Workspace. https://doi.org/10.24431/rw1k8ev. Version: 10.24431_rw1k8ev_20240605T002305Z.

Pelagic Cormorant

Hatch, S.A., V.A. Gill, D.M. Mulcahy, and D.C. Douglas. 2020. Tracking data for Pelagic Cormorants (*Phalacrocorax pelagicus*). Version 1.0, August 2020. US Geological Survey data release. https://doi.org/10.5066/P9Y5PQY3.

Peck-Richardson, A., B. Higgins, J. Soller, J. Lerczak, and R. Orben. 2024. Cormorant Oceanography Project: Biologging Data from Pelagic Cormorants (*Urile pelagicus*), Middleton Island, AK, 2020. Research Workspace. https://doi.org/10.24431/rw1k8es. Version: 10.24431_rw1k8es_20240604T221827Z.

Peck-Richardson, A., A. Piggott, J. Lerczak, and R. Orben. 2024. Cormorant Oceanography Project: Biologging Data from Brandt's and Pelagic Cormorants (*Urile penicillatus, U. pelagicus*), Columbia River, 2019–2022. Research Workspace. https://doi.org/10.24431/rw1k8ev. Version: 10.24431_rw1k8ev_20240605T002305Z.

Double-crested Cormorant

Courtot, K.N., D.D. Roby, J.Y. Adkins, D.E. Lyons, D.T. King, and R.S. Larsen. 2012. Colony connectivity of Pacific coast Double-crested Cormorants based on post-breeding dispersal from the region's largest colony. *Journal of Wildlife Management* 76:1462–1471. https://doi.org/10.1002/jwmg.403.

Currier, H., J. Bowman, and K. Abraham. 2010. Migration and winter movements of Double-crested Cormorants breeding in Georgian Bay, Ontario. *Waterbirds* 33:451–460. https://doi.org/10.2307/40985864.

Peck-Richardson, A.G. 2017. Double-crested Cormorants (*Phalacrocorax auritus*) and Brandt's Cormorants (*P. penicillatus*) breeding on East Sand Island in the Columbia River estuary: Foraging ecology, colony connectivity, and overwinter dispersal. Unpublished master's thesis, Oregon State University, Corvallis.

PELICANS, HERONS, EGRETS, BITTERNS, IBISES, AND SPOONBILLS (PELECANIFORMES)

FAMILY PELECANIDAE—PELICANS

American White Pelican

Audubon Minnesota. 2012–2015. American White Pelican tracking data. Unpublished data.

Earthspan. American White Pelican tracking data. www.earthspan.foundation. Unpublished data.

Idaho Department of Fish and Game. American White Pelican tracking data. Unpublished data.

Utah Division of Wildlife Resources. 2015–2022. American White Pelican telemetry and mark-resight data. Unpublished data.

Brown Pelican

Harrison, A.L., and D. Brinker. Data from: Brown Pelican Chesapeake Bay. Movebank Data Repository. https://www.movebank.org/cms/webapp?gwt_fragment=page=studies,path=study302686519.

Lamb, J.S., et al. 2017. Data from: A bridge between oceans: Overland migration of marine birds in a wind energy corridor. Dryad Dataset. https://doi.org/10.5061/dryad.1m12h.

Wilkinson, B.P., and P.G.R. Jodice. 2023. Support for the fasting endurance hypothesis of partial migration in a nearshore seabird. *Ecosphere* 14(2):e4365. https://doi.org/10.1002/ecs2.4365.

Wilkinson, B.P., Y.G. Satgé, J.S. Lamb, et al. 2019. Tropical cyclones alter short-term activity patterns of a coastal seabird. *Movement Ecology* 7:30. https://doi.org/10.1186/s40462-019-0178-0.

Wilkinson, B.P., Y.G. Satgé, J.S. Lamb, and P.G.R. Jodice. 2019. GPS tracking of Brown Pelican in the South Atlantic Bight during cyclonic activity (2017–2018). US Geological Survey data release. https://doi.org/10.5066/P9D5IP0G.

FAMILY ARDEIDAE—HERONS, EGRETS, AND BITTERNS

American Bittern

Huschle, G., J. Toepfer, and D. Douglas. 2013. Migration and wintering areas of American Bitterns (*Botaurus lentiginosus*) that summer in central North America as determined by satellite and radio telemetry, 1998–2003. *Waterbirds* 36(3):300–309.

Reddish Egret

Lamb, J.S., D.J. Newstead, L.M. Koczur, B.M. Ballard, M.C. Green, and P.G.R. Jodice. 2018. A bridge between oceans: Overland migration of marine birds in a wind energy corridor. *Journal of Avian Biology* 49:jav-01474.

———. 2017. Data from: A bridge between oceans: Overland migration of marine birds in a wind energy corridor. Dryad Dataset. https://doi.org/10.5061/dryad.1m12h.

Black-crowned Night Heron

Henny, C., and L. Blus. 1986. Radiotelemetry locates wintering grounds of DDE-contaminated Black-crowned Night-Herons. *Wildlife Society Bulletin* 14:236–241.

Scarpignato, A.L., K.A. Stein, E.B. Cohen, P.P. Marra, L.J. Kearns, S. Hallager, and C.M. Tonra. 2021. Full annual cycle tracking of Black-crowned Night-Herons suggests wintering areas do not explain differences in colony population trends. *Journal of Field Ornithology* 92:143–155.

Stein, K., et al. Movebank Study Black-crowned Night-Heron (*Nycticorax nycticorax*), Ohio, ID 230900284 (in preparation).

Tonra, C. Ohio State University (Project 64). 2015–2021. Data accessed from Motus Wildlife Tracking System, Birds Canada. https://motus.org/.

Great Egret

Brzorad, J., M.C. Allen, S. Elbin, R. Kays, A. Maccarone, S. Schweitzer, and N. Tsipoura. Movebank Study Egrets & Herons, ID 17469219.

Kays, R. Movebank Study Life Track—Great Egrets, ID 8849813. North Carolina Museum of Natural Sciences, North Carolina State University; Lenoir-Rhyne University; North Caroline Wildlife Resource Commission; and the Max Plank Institute for Animal Behavior.

Lumpkin, D., S. Jennings, N. Warnock, and E. Condeso. 2023. Partial migration by Great Egrets *Ardea alba* in coastal California. *Waterbirds* 45:150–158.

Weseloh, C. Unpublished data.

Weseloh, D.V.C., D. Moore, and T. Knezevic. 2014. Wintering locations of Ontario-banded Great Egrets: New Jersey to the Caribbean. *Ontario Birds* 32(1): article 1. https://digitalcommons.usf.edu/ontario_birds/vol32/iss1/1.

Great Blue Heron

Brzorad, J., M.C. Allen, S. Elbin, R. Kays, A. Maccarone, S. Schweitzer, and N. Tsipoura. Movebank Study Egrets & Herons, ID 17469219.

D'Auria, D.E. 2020. Heron Tracking Project. Maine Department of Inland Fisheries and Wildlife. Bangor, ME.

FAMILY THRESKIORNITHIDAE—IBISES AND SPOONBILLS

White Ibis

Kidd-Weaver, A., J. Hepinstall-Cymerman, C.C. Welch, M. Murray, H.C. Adams, T.J. Ellison, M.J. Yabsley, and S.M. Hernandez. 2020. The movements of a recently urbanized wading bird reveal changes in season timing and length related to resource use. *PLOS ONE*. https://doi.org/10.1371/journal.pone.0230158.

———. 2020 Data from: The movements of a recently urbanized wading bird reveal changes in season timing and length related to resource use. Movebank Data Repository. https://doi.org/10.5441/001/1.8ms50757.

White-faced Ibis

Yates, M. White-faced Ibis tracking data. Earthspan. www.earthspan.foundation.

Roseate Spoonbill

Lorenz, J. Roseate Spoonbill tracking data. Everglades Science Center, Audubon Florida.

NEW WORLD VULTURES (CATHARTIFORMES)

FAMILY CATHARTIDAE—NEW WORLD VULTURES

Black Vulture

Avery, M.L., J.S. Humphrey, E.A. Tillman, and B.M. Kluever. Turkey Vulture Tracking Data Set. US Department of Agriculture, Wildlife Services, National Wildlife Research Center, Florida Field Station.

Evans, B.A., J.S. Humphrey, E.A. Tillman, M.L. Avery, and B.M. Kluever. 2024. Site-specific space use and resource selection by Black Vultures (*Coragyps atratus*) in the southeastern United States. *Ibis* 166:129–145. https://doi.org/10.1111/ibi.13244.

Holland, A.E., M.E. Byrne, A.L. Bryan, T.L. DeVault, O. Rhodes Jr., and J.C. Beasley. 2017. Fine-scale assessment of monthly home ranges and activity patterns of resident Black (*Coragyps atratus*) and Turkey (*Cathartes aura*) Vultures. *PLOS ONE* 12(7):e0179819. https://doi.org/10.1371/journal.pone.0179819.

Holland, A.E., M.E. Byrne, J. Hepinstall-Cymerman, A.L. Bryan, T.L. DeVault, O.E. Rhodes Jr., and J.C. Beasley. 2019. Evidence of niche differentiation for two sympatric vulture species in the southeastern United States. *Movement Ecology* 7:31.

Vulture Movement Ecology Research, Hawk Mountain Sanctuary.

Turkey Vulture

Avery, M.L., J.S. Humphrey, E.A. Tillman, and B.M. Kluever. Turkey Vulture Tracking Data Set. US Department of Agriculture, Wildlife Services, National Wildlife Research Center, Florida Field Station.

Holland, A.E., M.E. Byrne, A.L. Bryan, T.L. DeVault, O. Rhodes Jr. and J.C. Beasley. 2017. Fine-scale assessment of monthly home ranges and activity patterns of resident Black (*Coragyps atratus*) and Turkey (*Cathartes aura*) Vultures. *PLOS ONE* 12(7):e0179819. https://doi.org/10.1371/journal.pone.0179819.

Holland, A.E., M.E. Byrne, J. Hepinstall-Cymerman, A.L. Bryan, T.L. DeVault, O.E. Rhodes Jr., and J.C. Beasley. 2019. Evidence of niche differentiation for two sympatric vulture species in the southeastern United States. *Movement Ecology* 7:31.

Vulture Movement Ecology Research, Hawk Mountain Sanctuary.

HAWKS, EAGLES, KITES, AND OSPREY (ACCIPITRIFORMES)

FAMILY PANDIONIDAE–OSPREY

Osprey

Bierregaard, Rob. Movebank Study Osprey Bierregaard North and South America, ID 8868155.

Bird Conservancy of the Rockies. https://birdconservancy.org/a-tale-of-two-osprey/.

Chubbs, T.E. Department of National Defence, Canada. Unpublished data.

Craighead Beringia South. Unpublished data.

Jensen, B. Movebank Study Pandion haliaetus Osprey—SouthEast Michigan, ID 10204361.
Martell, M.S., and D. Douglas. 2019. Data from: Osprey in North and South America 1995–2002 (Martell)-reference-data. Movebank Data Repository. https://doi.org/10.5441/001/1.sv6335t3/2.
Nye, P., G. Hewitt, T. Swenson, and R. Kays. 2018. Data from: New York State Bald Eagle report 2010. Movebank Data Repository. https://www.doi.org/10.5441/001/1.s65q50j0.
Stout, W.E., V.L. Greene, and S. Postupalsky. 2009. Migration routes, reproduction, and lifespan of a translocated Osprey. *Wilson Journal of Ornithology* 121:203–206.

— FAMILY ACCIPITRIDAE—HAWKS, EAGLES, AND KITES

Swallow-tailed Kite

Bildstein, K., F.R. Santana, L.O. Melian, and M. Martell. 2002. A new migration route for Swallow-tailed Kite (*Elanoides forficatus*) in east Cuba. *Cotinga* 18:93–99.
Meyer, K.D., and G.M. Kent. Avian Research and Conservation Institute. Unpublished data.
Zimmerman, G.M. 2004. Studies of the annual cycle of the Swallow-tailed Kite (*Elanoides forficatus*): Migration, habitat use, and parasites. Master's thesis, Georgia Southern University, Statesboro.

Golden Eagle

Barnes, J. Nevada Department of Wildlife. Unpublished data.
Booms, T. Alaska Department of Fish and Game. Unpublished data.
Brown, J.L., B. Bedrosian, D.A. Bell, M.A. Braham, ... J.P. Smith, ..., and B. Woodbridge (21 coauthors). 2017. Patterns of spatial distribution of Golden Eagles across North America: How do they fit into existing landscape-scale mapping systems? *Journal of Raptor Research* 51:197–215.
Craighead Beringia South. Movebank Study Beringia South Migrant Golden Eagle, ID 212420825.
Davidson, S.C., G. Bohrer, E. Gurarie, S. LaPoint, ..., J.P. Smith, ..., M. Hebblewhite (100+ coauthors). 2020. Ecological insights from three decades of animal movement tracking across a changing Arctic. *Science* 370:712–715.
Delaware-Otsego Audubon Society. Unpublished data.
Earthspan. Unpublished data.
Goodrich, L.J., and J.P. Smith. 2008. Raptor migration in North America. In *State of North America's Birds of Prey*, ed. K.L. Bildstein, J.P. Smith, E. Ruelas Inzunza, and R.R. Veit, Series in Ornithology 3, 37–149. Nuttall Ornithological Club, Cambridge, MA; and American Ornithologists Union, Washington, DC.
HawkWatch International. Unpublished data.
Lewis, S.B., and C.L. McIntyre. Unpublished data.
Martell, M., and S. Mehus. Unpublished data.
McIntyre, C.L., D.C. Douglas, and M.W. Collopy. 2008. Movements of Golden Eagles (*Aquila chrysaetos*) from interior Alaska during their first year of independence. *Auk* 125(1):214–224. https://doi.org/10.1525/auk.2008.125.1.214.
Nye, P., G. Hewitt, T. Swenson, and R. Kays. 2018. Data from: New York State Bald Eagle report 2010. Movebank Data Repository. https://www.doi.org/10.5441/001/1.s65q50j0.
Smith, B.W. USFWS Mountain-Prairie Region Study. Unpublished data.
Smith, J.P. 2010. Raptor migration in New Mexico. In *Raptors of New Mexico*, ed J.-L. E. Cartron, 29–67. University of New Mexico Press, Albuquerque.

Northern Harrier

DeSorbo, C.R. Biodiversity Research Institute. Unpublished data
Meyer, K.D., and G.M. Kent. Avian Research and Conservation Institute. Unpublished data.
New York State Department of Environmental Conservation. Unpublished data.
US Geological Survey. Unpublished data.

American Goshawk

Goodrich, L.J., and J.P. Smith. 2008. Raptor migration in North America. In *State of North America's Birds of Prey*, ed. K.L. Bildstein, J.P. Smith, E. Ruelas Inzunza, and R.R. Veit, Series in Ornithology 3, 37–149. Nuttall Ornithological Club, Cambridge, MA, and American Ornithologists Union, Washington, DC.
HawkWatch International. Unpublished data.
Jeffress, M., and J. Barnes. Nevada Department of Wildlife. Unpublished data.
Smith, J.P. 2010. Raptor migration in New Mexico. In *Raptors of New Mexico*, ed. J.-L. E. Cartron, 29–67. University of New Mexico Press, Albuquerque.
Sonsthagen, S.A., R.L. Rodriguez, and C.M. White. 2006. Satellite telemetry of Northern Goshawks breeding in Utah—I. Annual movements. *Studies in Avian Biology* 31:239–251.

Bald Eagle

Chubbs, T.E. Department of National Defence, Government of Canada. Unpublished data.
Clark, K.E. New Jersey Fish and Wildlife. Unpublished data.
Craighead Beringia South. Movebank Study Beringia South Bald Eagles, ID 1031368076. Accessed from movebank.org September 1, 2020.
DeSorbo, C.R. Biodiversity Research Institute. Unpublished data.
Garcia and Associates (GANDA). 2016. Hatchet Ridge Wind Project Bald Eagle Mitigation Plan—Final Report. Report to Hatchet Ridge Wind, LLC. San Anselmo, CA.
Harmata, A.R. 1984. Bald Eagles of the San Luis Valley, Colorado: Their winter ecology and spring migration. PhD diss., Montana State University, Bozeman.
Kays, R., and T. Simons. Movebank Study LifeTrack Bald Eagle, ID 20873986. Accessed from movebank.org on May 3, 2021.
Lamont, M., and D. Hancock. Hancock Wildlife Foundation.
Lewis, S.B. Unpublished data.
Linthicum, J., R.E. Jackman, B.C. Latta, J. Koshear, and M. Smith. 2007. Annual migrations of Bald Eagles to and from California. *Journal of Raptor Research* 41(2):106–112.
Mandernack, B.A., M. Solensky, M. Martell, and R.T. Schmitz. 2012. Satellite tracking of Bald Eagles in the Upper Midwest. *Journal of Raptor Research* 46(3):258–272.
———. 2021. Movebank Study satellite tracking of Bald Eagles in the Upper Midwest, ID 1428558871. Accessed from movebank.org on May 3, 2021.
McClelland, B. R., P.T. McClelland, R.E. Yates, E.L. Caton, and M.E. McFadzen. 1996. Fledging and migration of juvenile Bald Eagles from Glacier National Park, Montana. *Journal of Raptor Research* 30(2):79–89.
Merrill Creek Reservoir. Movebank Study Bald Eagle / *Haliaeetus leucocephalus* / J. Bullis, ID 45989874. Accessed from movebank.org on May 3, 2021.
Mojica, E.K., J.M. Meyers, B.A. Millsap, and K.L. Haley. 2008. Migration of Florida subadult Bald Eagles. *Wilson Journal of Ornithology* 120(2):304–310.
Nye, P., G. Hewitt, T. Swenson, and R. Kays. 2018. Data from: New York State Bald Eagle report 2010. Movebank Data Repository. https://www.doi.org/10.5441/001/1.s65q50j0.
Smith, N.R., A.D. Afton, and T.J. Hess Jr. 2017. Winter breeding and summer nonbreeding home ranges of Bald Eagles from Louisiana. *American Midland Naturalist* 178(2):203–214.
US Navy. 2021. Cooperative Ecosystems Studies Unit (CESU) Agreements between the US Navy and the Center for Conservation Biology at William & Mary to Conduct Bald Eagle Nest Surveys, Fledgling Banding, Fledgling Tracking, Nesting Habitat Delineations, and Communal Roost Location Delineations on Naval Air Station (NAS) Oceana Properties and within a 2,640 ft Buffer of the Properties. GIS Data from 2016 to 2021.
Watts et al. 2021. Bald Eagle tracking data from the Chesapeake Bay region as of July 8, 2021. Center for Conservation Biology, William and Mary, Williamsburg, VA.

Broad-winged Hawk

Haines, A.M., M.J. McGrady, M.S. Martell, B.J. Dayton, M.B. Henke, and W.S. Seegar. 2003. Migration routes and wintering locations of Broad-winged Hawks tracked by satellite telemetry. *Wilson Bulletin* 115(2):166–169. https://doi.org/10.1676/02-106.
McCabe, R.A., L.J. Goodrich, D.R. Barber, T.L. Master, J.L. Watson, E.M. Bayne, A.-L. Harrison, P.P. Marra, and K.L. Bildstein. 2020. Satellite tracking reveals age and origin differences in migration ecology of two populations of Broad-winged Hawks (*Buteo platypterus*). *Wilson Journal of Ornithology* 132(1):1–14. https://doi.org/10.1676/1559-4491-132.1.1.

Short-tailed Hawk

Meyer, K.D., and G.M. Kent. Avian Research and Conservation Institute. Unpublished data.

Swainson's Hawk

Fleishman, E., J. Anderson, B.G. Dickson, D. Krolick, J.A. Estep, R.L. Anderson, C.S. Elphick, D.S. Dobkin, and D.A. Bell. 2016. Space use by Swainson's Hawk (*Buteo swainsoni*) that breed in the Natomas Basin, California. *Collabra* 2(1):5. https://doi.org/10.1525/collabra.35.
Kochert, M.N., M.R. Fuller, L.S. Schueck, L. Bond, M.J. Bechard, B. Woodbridge, G.L. Holroyd, M.S. Martell, and U. Banasch. 2011. Migration patterns, use of stopover areas, and austral summer movements of Swainson's Hawks. *Condor* 113:89–106.
Salt Lake City Airport Wildlife Department. Unpublished data.
Watson, K.A., C.W. Boal, L.M. Groen, and J.R. Walker. 2017. Using GPS transmitters to explore movement ecology and to assess risk of the wind energy industry for Swainson's Hawks. Unpublished report to Consolidated Nuclear Security, LLC, Amarillo, TX.
Woodbridge, B., K.K. Finley, and S.T. Seager. 1995. An investigation of the Swainson's Hawk in Argentina. *Journal of Raptor Ressearch* 29(3):202–204.

Red-tailed Hawk

Alioto, N. 2021. Full cycle migration of Red-tailed Hawks through the Great Lakes. Mackinac Straits Raptor Watch and Michigan State University. Unpublished data.
Bickford, N. Colorado State University Pueblo. Unpublished data.
Bloom, P.H., and M.D. McCrary. 2020. Unpublished data.
Craighead Beringia South. Unpublished data.
Golden Gate Raptor Observatory. Golden Gate National Parks Conservancy, 2012–2015. Unpublished data.
Goodrich, L.J., and J.P. Smith. 2008. Raptor migration in North America. In *State of North America's Birds of Prey*, ed. K.L. Bildstein, J.P. Smith, E. Ruelas Inzunza, and R.R. Veit, Series in Ornithology 3, 37–149. Nuttall Ornithological Club, Cambridge, MA, and American Ornithologists Union, Washington, DC.

Hoffman, S.W., J.P, Smith, and T.D. Meehan. 2002. Breeding grounds, winter ranges, and migratory routes of raptors in the mountain west. *Journal of Raptor Research* 36(2):97–110.
Robinson, B.W., M.B. Robbins, L.H. DeCicco, and I.J. Lovette. The Red-tailed Hawk Project. Unpublished data.
Salt Lake City Wildlife Department. Unpublished data.
Smith, J.P. 2010. Raptor migration in New Mexico. In *Raptors of New Mexico*, ed. J.-L.E. Cartron, 29–67. University of New Mexico Press, Albuquerques.

Rough-legged Hawk

Craighead Beringia South. Movebank Study Grand Teton Rough-legged Hawks, ID 14857978. Accessed from movebank.org Sept. 1, 2020.
Pokrovsky, I., J.-F. Therrien, N. Lecomte, and M. Wikelski. Movebank Study LifeTrack Rough-legged buzzards, ID 9493874.

Ferruginous Hawk

Barnes, J.G. Nevada Department of Wildlife. Unpublished data.
Earthspan. Unpublished data.
Harmata, A.R. 1982. Recoveries of Ferruginous Hawks banded in Colorado. *North American Bander* 6(4):144–147.
Salt Lake City Airport Wildlife Department. Unpublished data.
Schmutz, J.K., and R.W. Fyfe. 1987. Migration and mortality of Alberta Ferruginous Hawks. *Condor* 89(1):169–174.
Watson, J.L. 2020. Ferruginous Hawk (*Buteo regalis*) home range and resource use on northern grasslands in Canada. Master's thesis, University of Alberta.
Watson, J.W., U. Banasch, T. Byer, D.N. Svingen, R. McCready, D. Hanni, A. Lafon, and R. Gerhard. 2018. Migration patterns, timing, and seasonal destinations of adult Ferruginous Hawks (*Buteo regalis*). *Journal of Raptor Research* 52:266–281. https://doi.org/10.3356/JRR-17-30.1.

OWLS (STRIGIFORMES)

— FAMILY STRIGIDAE—OWLS

Flammulated Owl

Linkhart, B.D., J.W. Fox, and S.W. Yanco. 2016. Migration timing and routes, and wintering areas of Flammulated Owls. *Journal of Field Ornithology* 87:42–54.
Rich, S. 2018. Migration patterns of Flammulated Owls (*Psilscops flammeolus*) using light-level geolocators. Unpublished master's thesis, California State University, Chico.
Yanco, S.W., B.D. Linkhart, P.P. Marra, M. Mika, M. Ciaglo, A. Carver, and M.B. Wunder. 2021. Niche dynamics suggest ecological factors influencing migration in an insectivorous owl. *Ecology* 103(3):e3617. https://doi.org/10.1002/ecy.3617.

Snowy Owl

Davidson, S.C., et al. 2020. Ecological insights from three decades of animal movement tracking across a changing Arctic. *Science* 370(6517):712–715. https://doi.org/10.1126/science.abb7080.
Doyle, F.I., J.F. Therrien, D.G. Reid, G. Gauthier, and C.J. Krebs. 2017. Seasonal movements of female Snowy Owls breeding in the western North American Arctic. *Journal of Raptor Research* 51(4):428–438. In partnership with International Polar Year project Arctic Wildlife Observatories Linking Vulnerable EcoSystems (ArcticWOLVES), administered by Université Laval (Gauthier and Berteaux 2011).
Holroyd, G. Beaverhill Bird Observatory and Canadian Wildlife Service. Unpublished data.
Project SNOWstorm. 2013–2021. Unpublished data.
Therrien, J.-F., D. Pinaud, G. Gauthier, N. Lecomte, K.L. Bildstein, and J. Bêty. 2015. Data from: Is pre-breeding prospecting behaviour affected by snow cover in the irruptive snowy owl? A test using state-space modelling and environmental data annotated via Movebank. Movebank Data Repository. https://www.doi.org/10.5441/001/1.32rf091b.
Wiebe, K. Unpublished data.

Burrowing Owl

Holroyd, G.L., and H.E. Trefry. 2010. Winter destinations and habitats of Canadian Burrowing Owls. *Journal of Raptor Research* 44(4):294–299.

Short-eared Owl

Holroyd, G. Beaverhill Bird Observatory and Canadian Wildlife Service. Unpublished data.
Johnson, J.A., T.L. Booms, L.H. DeCicco, and D.C. Douglas. 2017. Seasonal movements of the Short-eared Owl (*Asio flammeus*) in western North America as revealed by satellite telemetry. *Journal of Raptor Research* 51:115–129.
Kays, R., and T. Simons. Movebank Study NYSDEC Raptor Tracking, ID 2943485. Accessed from movebank.org on December 21, 2021.
New York State Department of Environmental Conservation. Unpublished data.

WOODPECKERS (PICIFORMES)

— FAMILY PICIDAE—WOODPECKERS

Northern Flicker

Gow, E.A., and K.L. Wiebe. 2014. Males migrate farther than females in a differential migrant: An examination of the fasting endurance hypothesis. *Royal Society Open Science* 1:140346.
———. 2014. Northern Flicker light-level geolocation data.

FALCONS AND CARACARAS (FALCONIFORMES)

— FAMILY FALCONIDAE—FALCONS AND CARACARAS

American Kestrel

Hall, K. AMKE research MN (Project 332). 2020–2021. Data accessed from Motus Wildlife Tracking System, Birds Canada. https://motus.org/.
Henny, C.J., and G.L. Brady. 1994. Partial migration and wintering localities of American Kestrels nesting in the Pacific Northwest. *Northwestern Naturalist* 75(2):37–43.
Hunt, A.R., J.L. Watson, J.M. Winiarksi, R.R. Porter, and J.A. Heath. 2023. American Kestrel migration: Insights and challenges from tracking individuals across the annual cycle. *Journal of Raptor Research* 57(2):165–175. https://doi.org/10.3356/JRR-22-05.

Merlin

DeSorbo, C.R. Unpublished data.
López, F.G., and N.W. Cooper. Unpublished data.

Gyrfalcon

Burnham, K.K. 2020. Data from: Seasonal movements of Gyrfalcons *Falco rusticolus* include extensive periods at sea [Central-West Greenland]. Movebank Data Repository. https://www.doi.org/10.5441/001/1.v81b20p9.
———. 2020. Data from: Seasonal movements of Gyrfalcons *Falco rusticolus* include extensive periods at sea [East Greenland]. Movebank Data Repository. https://www.doi.org/10.5441/001/1.n0m7gd47.
———. 2020. Data from: Seasonal movements of Gyrfalcons *Falco rusticolus* include extensive periods at sea [Maniitsoq]. Movebank Data Repository. https://www.doi.org/10.5441/001/1.qv4p2cg7.
———. 2020. Data from: Seasonal movements of Gyrfalcons *Falco rusticolus* include extensive periods at sea [Northwest Greenland]. Movebank Data Repository. https://www.doi.org/10.5441/001/1.3833b9n1.
Burnham, K.K., and I. Newton. 2011. Seasonal movements of Gyrfalcons *Falco rusticolus* include extensive periods at sea. *Ibis* 153(3):468–484. https://doi.org/10.1111/j.1474-919X.2011.01141.x.
Holroyd, G. Beaverhill Bird Observatory and Canadian Wildlife Service. Unpublished data.
McIntyre, C.L., D.C. Douglas, and L.G. Adams. 2009. Movements of juvenile Gyrfalcons from western and interior Alaska following departure from their natal areas. *Journal of Raptor Research* 43(2):99–109. https://doi.org/10.3356/JRR-08-43.1.

Peregrine Falcon

Burnham, K.K. 2020. Data from: The history and range expansion of Peregrine Falcons in the Thule area, northwest Greenland. Movebank Data Repository. https://www.doi.org/10.5441/001/1.b3b511d2.
Burnham, K.K., W.A. Burnham, I. Newton, J.A. Johnson, and A. Gosler. 2012. *The History and Range Expansion of Peregrine Falcons in the Thule Area, Northwest Greenland.* Monographs on Greenland 352, Bioscience 60. Museum Tusculanem Press.
Cochran, W.W., and M. Wikelski. Movebank Study Bill Cochran Peregrine Falcon migration 1973–75, ID 274355544.
DeSorbo, C. Unpublished data.
Earthspan. Unpublished data.
Holroyd, G. Beaverhill Bird Observatory and Canadian Wildlife Service.
McGrady, M.J., T.L. Maechtle, J.J. Vargas, W.S. Seegar, and M.C. Porras Pena. 2002. Migration and ranging of Peregrine Falcons wintering on the Gulf of Mexico Coast, Tamaulipas, Mexico. *Condor* 104(1):39–48.
New York State Department of Environmental Conservation. Unpublished data.
Nye, P., G. Hewitt, T. Swenson, and R. Kays. 2018. Data from: New York State Bald Eagle report 2010. Movebank Data Repository. https://www.doi.org/10.5441/001/1.s65q50j0.
Pokrovsky, I., J-F. Therrien, N. Lecomte. Movebank Study LifeTrack Peregrine falcon, ID 103426553.
Schmutz, J.K., R.W. Fyfe, U. Banasch, and H. Armbruster. 1991. Routes and timing of migration of falcons banded in Canada. *Wilson Bulletin* 103(1):44–58.

Prairie Falcon

Enderson, J.H. 1964. A study of the Prairie Falcon in the central Rocky Mountain region. *Auk* 81(3):332–352.

Holroyd, G. Beaverhill Bird Observatory and Canadian Wildlife Service. Unpublished data.

Steenhof, K., M. R. Fuller, M. N. Kochert, and K. K. Bates. 2005. Long-range movements and breeding dispersal of Prairie Falcons from southwest Idaho. *Condor* 107:481–496. https://doi.org/10.1093/condor/107.3.481.

SONGBIRDS (PASSERIFORMES)

— FAMILY TYRANNIDAE—TYRANT FLYCATCHERS

Western Kingbird

Jahn, A.E., V.R. Cueto, J.W. Fox, M.S. Husak, D.H. Kim, D.V. Landoll, J.P. Ledezma, H.K. Lepage, D.J. Levey, M.T. Murphy, and R.B. Renfrew. 2013. Migration timing and wintering areas of three species of flycatchers (*Tyrannus*) breeding in the Great Plains of North America. *Auk* 130:247–257.

Eastern Kingbird

Jahn, A.E., V.R. Cueto, J.W. Fox, M.S. Husak, D.H. Kim, D.V. Landoll, J.P. Ledezma, H.K. Lepage, D.J. Levey, M.T. Murphy, and R.B. Renfrew. 2013. Migration timing and wintering areas of three species of flycatchers (*Tyrannus*) breeding in the Great Plains of North America. *Auk* 130:247–257.

Kim, D.H., L.J. Redmond, J.R. Fox, and M.T. Murphy. 2021. Inter- and intracontinental migration by the Eastern Kingbird (*Tyrannus tyrannus*). *Wilson Journal of Ornithology* 133:202–214.

Scissor-tailed Flycatcher

Jahn, A.E., V.R. Cueto, J.W. Fox, M.S. Husak, D.H. Kim, D.V. Landoll, J.P. Ledezma, H.K. Lepage, D.J. Levey, M.T. Murphy, R.B. and Renfrew, R.B. 2013. Migration timing and wintering areas of three species of flycatchers (*Tyrannus*) breeding in the Great Plains of North America. *Auk* 130:247–257.

Olive-sided Flycatcher

Hagelin, J., M. Hallworth, C. Barger, J. Johnson, K. DuBour, G. Pendelton, L. Decicco, L. McDuffie, S. Matsuoka, M. Snively, P. Marra, and A. Taylor. 2021. Revealing migratory path, important stopovers and non-breeding areas of a boreal songbird in steep decline. *Animal Migration* 8:168–191. https://doi.org/10.1515/ami-2020-0116.

Willow Flycatcher

Bayly, N., and C. Gomez. *Investigación para la Conservación en el Neotropico* (Research for Conservation in the Neotropic), Birds Canada, Environment and Climate Change Canada. SELVA Colombia (Project 57). 2015–2021. Data accessed from Motus Wildlife Tracking System, Birds Canada. https://motus.org/.

— FAMILY VIREONIDAE—VIREOS

Cassin's Vireo

Hedley, R.W. 2019. Long-distance movements and evidence of post-breeding elevational movements by Cassin's Vireos. *Journal of Field Ornithology* 90(4):335–347.

Red-eyed Vireo

Callo, P.A., E.S. Morton, and B.J.M. Stutchbury. 2013. Prolonged spring migration in the Red-eyed Vireo (*Vireo olivaceus*). *Auk* 130:240–246.

Stutchbury, B. REVI—Timing of migration PA (Project 326). 2020–2020. Data accessed from Motus Wildlife Tracking System, Birds Canada. https://motus.org/.

— FAMILY CORVIDAE—CROWS, JAYS, AND MAGPIES

American Crow

Townsend, A.K., et al. 2018. Where do winter crows go? Characterizing partial migration of American Crows with satellite telemetry, stable isotopes, and molecular markers. *Auk* 135(4):964–974. https://doi.org/10.1642/AUK-18-23.1.

— FAMILY ALAUDIDAE—LARKS

Horned Lark

de Zwaan, D.R., S. Wilson, E.A. Gow, and K. Martin. 2019. Data from: Sex-specific spatiotemporal variation and carry-over effects in a migratory alpine songbird. *Frontiers in Ecology and Evolution* 7:285. 10.3389/fevo.2019.00285.

Tremblay, J. Fall migration of American Pipit and Horned Lark (Project 288). 2019–2021. Data accessed from Motus Wildlife Tracking System, Birds Canada. https://motus.org/.

— FAMILY HIRUNDINIDAE—SWALLOWS

Bank Swallow

Imlay, T., K.A. Hobson, A. Roberto-Charron, and M.L. Leonard. 2018. Wintering areas, migratory connectivity and habitat fidelity of three declining Nearctic-Neotropical migrant swallows. *Animal Migration* 5:1–16. https://doi.org/10.1515/ami-2018-0001.

Tree Swallow

Gow, E.A., D.R. Norris, and S.M. Knight. 2018.Tree swallow light-level geolocator dataset.

Knight, S.M., D.W. Bradley, R.G. Clark, E.A. Gow, M. Bélisle, L.L. Berzins, T. Blake, et al. 2018. Constructing and evaluating a continent-wide migratory songbird network across the annual cycle. *Ecological Monographs* 88(3):445–460. https://doi.org/10.1002/ecm.1298.

Purple Martin

Fraser, K.C., A. Shave, A. Savage, A. Ritchie, K. Bell, J. Siegrist, J.D. Ray, K. Applegate, and M. Pearman. 2017. Determining fine-scale migratory connectivity and habitat selection for a migratory songbird by using new GPS technology. *Journal of Avian Biology* 48(3):339–45. https://doi.org/10.1111/jav.01091.

———. 2016. Data from: Determining fine-scale migratory connectivity and habitat selection for a migratory songbird by using new GPS technology. Movebank Data Repository. https://www.doi.org/10.5441/001/1.5q5gn84d.

Fraser, K.C., B.J.M. Stutchbury, P. Kramer, C. Silveiro, J. Barrow, D. Newstead, N. Mickle, T. Shaheen, P. Mammenga, K. Applegate, E. Bridge, and J. Tautuin. 2013. Consistent range-wide pattern in fall migration strategy of Purple Martin (*Progne subis*), despite different migration routes at the Gulf of Mexico. *Auk* 130(2):291–298.

Fraser, K.C., B.J.M. Stutchbury, C. Silverio, P.M. Kramer, J. Barrow, D. Newstead, N. Mickle, B.F. Cousens, J.C. Lee, D.M. Morisson, T. Shaheen, P. Mammenga, K. Applegate, E. Bridge, and J. Tautuin. 2012. Continent-wide tracking to determine migratory connectivity and tropical habitat associations of a declining aerial insectivore. *Proceedings of the Royal Society: Biological Sciences* 279:4901–4906.

Barn Swallow

Kardynal, K. Swallow connectivity—Latin America to North America (Project 466). 2021–2024. Data accessed from Motus Wildlife Tracking System, Birds Canada. https://motus.org/.

Cliff Swallow

Imlay, T., K.A. Hobson, A. Roberto-Charron, and M.L. Leonard. 2018. Wintering areas, migratory connectivity and habitat fidelity of three declining Nearctic-Neotropical migrant swallows. *Animal Migration* 5:1–16. https://doi.org/10.1515/ami-2018-0001.

— FAMILY PTILIOGONATIDAE—SILKY-FLYCATCHERS

Phainopepla

Baldassarre, D.T., L. Campagna, H.A. Thomassen, J.W. Atwell, M. Chu, L.H. Crampton, R.C. Flesicher, and C. Riehl. 2019. GPS tracking and population genomics suggest itinerant breeding across drastically different habitats in the Phainopepla. *Auk* 136:1–12. https://doi.org/ 10.1093/auk/ukz058.

— FAMILY MIMIDAE—MOCKINGBIRDS AND THRASHERS

Gray Catbird

Adams, E.M., P.C. Keenan, K.A. Williams, and D.C. Evers. 2015. Veery and Gray Catbird breeding and wintering locations. Unpublished data.

Mancuso, K.A., M.A. Fylling, C.A. Bishop, K.E. Hodges, M.B. Lancaster, and K.R. Stone. 2021. Migration ecology of western gray catbirds. *Movement Ecology* 9:10. https://doi.org/10.1186/s40462-021-00249-7.

Ryder, T.B., B.S. Evans, D.L. Moseley, S. Lerman, and P.P. Marra. Unpublished data.

Ryder, T.B., J.W. Fox, and P.P. Marra. 2011. Estimating migratory connectivity of Gray Catbirds (*Dumatella carolinensis*) using geolocator and mark-recapture data. *Auk* 128:448–453.

— FAMILY TURDIDAE—THRUSHES

Mountain Bluebird

Roberto-Charron, A. 2018. Investigating methods of geolocator analysis in songbird migration research and their application to the study of a threatened, neotropical songbird. Master's thesis, University of Manitoba.

Veery

Adams, E.M. Unpublished data.

Heckscher, C.M., and D. Mendez. Delaware State University. Unpublished data.

Kardynal, K.J., and K.A. Hobson. 2017. The pull of the Central Flyway? Veeries breeding in western Canada migrate using an ancestral eastern route. *Journal of Field Ornithology* 88:262–273.

Gray-cheeked Thrush

Bayly, N., and C. Gomez. *Investigación para la Conservación en el Neotropico* (Research for Conservation in the Neotropic), Birds Canada, Environment and Climate Change Canada. SELVA Colombia (Project 57). 2015–2021. Data accessed from Motus Wildlife Tracking System, Birds Canada. https://motus.org/.

Gonzalez, A. Selva—GHTH—Colombia—Andes (Project 269). 2015–2017. Data accessed from Motus Wildlife Tracking System, Birds Canada. https://motus.org/.

Mitchell, G. Ontario Thrush Migration 2018 (Project 208). 2018–2019. Data accessed from Motus Wildlife Tracking System, Birds Canada. https://motus.org/.

National Park Service. Critical Connections Program. Unpublished data.

Whitaker, D.M, I.G. Warkentin, K.A Hobson, P. Thomas, and R. Boardman. 2018. Fall and winter movements of Newfoundland Gray-cheeked Thrushes (*Catharus minimus minimus*). *Animal Migration* 5:42–48. https://doi.org/10.1515/ami-2018-0004.

Bicknell's Thrush

McFarland, K.P. Movebank Study Annual Movements of Bicknell's Thrush (*Catharus bicknelli*) from Light-level Geolocation, ID 1416160002. Accessed from movebank.org on December 7, 2022.

Roth, A., and D. Buehler. UMaine Migration Ecology (Project 266). 2019–2021. Data accessed from Motus Wildlife Tracking System, Birds Canada. https://motus.org/.

Tremblay, J., C. Bégin Marchand, and A. Desrochers. Thrush post-breeding ecology in Quebec (Project 49). 2014–2021. Data accessed from Motus Wildlife Tracking System, Birds Canada. https://motus.org/.

Ward, C., D. Bell, and S. Mackenzie. Atlantic Canada BITH (Project 72). 2015–2015. Data accessed from Motus Wildlife Tracking System, Birds Canada. https://motus.org/.

Swainson's Thrush

Berrigan, L., and P. Taylor. Taylor—SWTH (Project 109). 2016–2017. Data accessed from Motus Wildlife Tracking System, Birds Canada. https://motus.org/.

Bird Conservancy of the Rockies. Unpublished data.

Blake, W. Intermountain West UM birds (Project 352). 2018–2020. Data accessed from Motus Wildlife Tracking System, Birds Canada. https://motus.org/.

Cohen, E., C. Nemes, and T. Guida. Clive Runnells Mad Island Marsh Preserve (Project 246). 2019–2021. Data accessed from Motus Wildlife Tracking System, Birds Canada. https://motus.org/.

Cormier, R.L., D.L. Humple, T. Gardali, and N.E. Seavy. 2013. Light-level geolocators reveal strong migratory connectivity and within-winter movements for a coastal California Swainson's Thrush (*Catharus ustulatus*) population. *Auk* 130:283–290. https://doi.org/10.1525/auk.2013.12228.

DeGroote, L., M. Sarver, J. Rice, D. Yeany, J. Fowler, R. Jacob, S. Sargent, L. Koitsch, M. Gimpel, J. Cepek, and T. Krynak. Powdermill Nature Reserve (Project 35). 2016–2021. Data accessed from Motus Wildlife Tracking System, Birds Canada. https://motus.org/.

Delmore, K., and W. Easton. BC Interior Thrushes (Project 280). 2019–2020. Data accessed from Motus Wildlife Tracking System, Birds Canada. https://motus.org/.

Delmore, K.E., and D.E. Irwin. 2014. Hybrid songbirds employ intermediate routes in a migratory divide. *Ecology Letters* 17:1211–1218.

Delmore, K.E., J.W. Fox, and D.E. Irwin. 2012. Dramatic intraspecific differences in migratory routes, stopover sites and wintering areas, revealed using light-level geolocators. *Proceedings of the Royal Society: Biological Sciences* 279:4582–4589.

Frei, B., A. Morales, V. Poirier, and K. Elliott. Montreal moult migrant research (Project 168). 2017–2019. Data accessed from Motus Wildlife Tracking System, Birds Canada. https://motus.org/.

Frei, B., V. Poirier, A. Morales, and K. Elliott. Montreal Connectivity (Project 373). 2021–2021. Data accessed from Motus Wildlife Tracking System, Birds Canada. https://motus.org/.

Gonzalez, A. Selva—Swainson's Thrush—Colombia (Project 270). 2015–2018. Data accessed from Motus Wildlife Tracking System, Birds Canada. https://motus.org/.

Guglielmo, C., and Y. Morbey. Gugli (Project 20). 2014–2021. Data accessed from Motus Wildlife Tracking System, Birds Canada. https://motus.org/.

Humple, D.L., R.L. Cormier, R.D. Burnett, T.W. Richardson, N.E. Seavy, K.E. Dybala, and T. Gardali. 2020. Migration tracking reveals geographic variation in the vulnerability of a Nearctic-Neotropical migrant bird. *Scientific Reports* 10:5483. https://doi.org/10.1038/s41598-020-62132-6.

Kardynal, K., and K.A. Hobson. Unpublished data.

Marty, J., M. Seymour, and P. Vasseur. Louisiana Chenier Stopover Habitat (Project 185). 2018–2019. Data accessed from Motus Wildlife Tracking System, Birds Canada. https://motus.org/.

Mitchell, G., and P. Taylor. Ontario Thrush Migration—2017 (Project 200). 2017–2019. Data accessed from Motus Wildlife Tracking System, Birds Canada. https://motus.org/.

National Park Service. Critical Connections Program. Unpublished data.

Sargent, S. Erie Bird Observatory—Presque Isle Migrants (Project 56). 2016–2017. Data accessed from Motus Wildlife Tracking System, Birds Canada. https://motus.org/.

Tremblay, J., C. Bégin Marchand, and A. Desrochers. Thrush post-breeding ecology in Quebec (Project 49). 2014–2021. Data accessed from Motus Wildlife Tracking System, Birds Canada. https://motus.org/.

Hermit Thrush

Alvarado, A.H., T.L. Fuller, and T.B. Smith. 2014. Integrative tracking methods elucidate the evolutionary dynamics of a migratory divide. *Ecology and Evolution* 4:3456–3469.

Guglielmo, C., and Y. Morbey. Gugli (Project 20). 2014–2021. Data accessed from Motus Wildlife Tracking System, Birds Canada. https://motus.org/.

National Park Service. Critical Connections Program. Unpublished data.

Nelson, A.R., R.L. Cormier, D.L. Humple, J.C. Scullen, R. Sehgal, and N.E. Seavy. 2016. Migration patterns of San Francisco Bay Area Hermit Thrushes differ across a fine spatial scale. *Animal Migration* 1:1–13. https://doi.org/10.1515/ami-2016-0001.

Wood Thrush

DeGroote, L., M. Sarver, J. Rice, D. Yeany, J. Fowler, R. Jacob, S. Sargent, L. Koitsch, M. Gimpel, J. Cepek, and T. Krynak. Powdermill Nature Reserve (Project 35). 2016–2021. Data accessed from Motus Wildlife Tracking System, Birds Canada. https://motus.org/.

Stanley, C.Q., M.R. Dudash, T.B. Ryder, W.G. Shriver, K. Serno, S. Adalsteinsson, and P.P. Marra. 2021. Seasonal variation in habitat selection for a Neotropical migratory songbird using high-resolution GPS tracking. *Ecosphere* 12(3):e03421. https://doi.org/10.1002/ecs2.3421.

Stanley, C.Q., E.A. McKinnon, K.C. Fraser, M. MacPherson, G. Casbourn, L. Friesen, P.P. Marra, C.E. Studds, T.B. Ryder, N.E. Diggs, and B.J.M Stutchbury. 2015. Connectivity of Wood Thrush breeding, wintering, and migration sites based on range-wide tracking. *Conservation Biology* 29:164–174.

Stutchbury, B. Stutchbury—Ontario Wood Thrush (Project 46). 2016–2019. Data accessed from Motus Wildlife Tracking System, Birds Canada. https://motus.org/.

American Robin

Jahn, A.E., S.B. Lerman, L.M. Phillips, T.B. Ryder, and E.J. Williams. 2019. First tracking of individual American Robins (*Turdus migratorius*) across seasons. *Wilson Journal of Ornithology* 131(2):356–359.

Oliver, R.Y., P.J. Mahoney, E. Gurarie, N. Krikun, B.C. Weeks, M. Hebblewhite, G. Liston, and N. Boelman. 2020. Behavioral responses to spring snow conditions contribute to long-term shift in migration phenology in American Robins. *Environmental Research Letters* 15(4):045003. https://doi.org/10.1088/1748-9326/ab71a0.

Williams, E.J., A. Jahn, and P.P. Marra. Unpublished data.

— FAMILY MUSCICAPIDAE—OLD WORLD FLYCATCHERS

Northern Wheatear

Bairlein, F., D.R. Norris, R. Nagel, M. Bulte, C.C. Voigt, J.W. Fox, D.J.T. Hussell, and H. Schmaljohann. 2012. Cross-hemisphere migration of a 25 g songbird. *Biology Letters* 8:505–507.

Schmaljohann, H., J.W. Fox, and F. Bairlein. 2012 Phenotypic response to environmental cues, orientation, and costs of migration in songbirds flying half around the world. *Animal Behaviour* 84:623–640.

Schmaljohann, H., S. Lisovski, and F. Bairlein. 2017 Flexible reaction norms to environmental variables along the migration route and the significance of stopover duration for total speed of migration in a songbird migrant. *Frontiers in Zoology* 14:17.

— FAMILY FRINGILLIDAE—FINCHES, EUPHONIAS, AND ALLIES

Evening Grosbeak

DeGroote, L., M. Sarver, J. Rice, D. Yeany, J. Fowler, R. Jacob, S. Sargent, L. Koitsch, M. Gimpel, J. Cepek, and T. Krynak. Powdermill Nature Reserve (Project 35). 2016–2021. Data accessed from Motus Wildlife Tracking System, Birds Canada. https://motus.org/.

— FAMILY CALCARIIDAE—LONGSPURS AND SNOW BUNTINGS

Chestnut-collared Longspur

Ellison, K., E. McKinnon, S. Zack, S. Olinb, R. Sparks, and E. Strasser. 2017. Migration and winter distribution of the Chestnut-collared Longspur. *Animal Migration* 4:37–50. https://doi.org/10.1515/ami-2917-000r.

Snow Bunting

Macdonald, C.A., K.C. Fraser, H.G. Gilchrist, T.K. Kyser, J.W. Fox, and O.P. Love. 2012. Strong migratory connectivity in a declining arctic passerine. *Animal Migration* 1:23–30. https://doi.org/10.2478/ami-2012-0003.

———. Movebank Study Snow Bunting Migration Phenology, ID 959128259. Accessed from movebank.org on September 27, 2022.

FAMILY PASSERELLIDAE—NEW WORLD SPARROWS

Grasshopper Sparrow

Hill, J.M., and R.B. Renfrew. 2018. Data from: Migratory patterns and connectivity of two North American grassland bird species [Grasshopper Sparrows]. Movebank Data Repository. https://doi.org/10.5441/001/1.c6b47s0r.

Lark Sparrow

Ross, J.D., E.S. Bridge, M.J. Rozmarynowycz, and V.P. Bingman. 2014. Individual variation in migratory path and behavior among Eastern Lark Sparrows. *Animal Migration* 2(1):29–33. https://doi.org/10.2478/ami-2014-0003.

Fox Sparrow

Fraser, K.C., A. Roberto-Charron, B. Cousens, M. Simmons, A. Nightingale, A.C. Shave, R.L. Cormier, and D.L. Humple. 2018. Classic pattern of leapfrog migration in Sooty Fox Sparrow (*Passerella iliaca unalaschcensis*) is not supported by direct migration tracking of individual birds. *Auk* 135:572–582. https://doi.org/10.1642/AUK-17-224.1.

National Park Service. Critical Connections Program. Unpublished data.

White-crowned Sparrow

Lisovski, S., Z. Németh, J.C. Wingfield, J.S. Krause, K.A. Hobson, N.E. Seavy, J. Gee, and M. Ramenofsky. 2019. Migration pattern of Gambel's White-crowned Sparrow along the Pacific Flyway. *Journal of Ornithology* 160:1097–1107. https://doi.org/10.1007/s10336-019-01685-4.

Golden-crowned Sparrow

Block, T.A., B.E. Lyon, Z. Mikalonis, A.S. Chaine, and D. Shizuka. 2023. Social connections across migration: Do Golden-crowned Sparrows that socialize in winter also breed together? *Wilson Journal of Ornithology* 135(4):569–575. https://doi.org/10.1676/23-00014.

Cormier, R.L., D.L. Humple, T. Gardali, and N.E. Seavy. 2016. Migratory connectivity of Golden-crowned Sparrows from two wintering regions in California. *Animal Migration* 3:48–56. https://doi.org/10.1515/ami-2016-0005.

Iverson, A.R., D.L. Humple, R.L. Cormier, and J. Hull. 2023. Land cover and NDVI are important predictors in habitat selection along migration for the Golden-crowned Sparrow, a temperate-zone migrating songbird. *Movement Ecology* 11. https://doi.org/10.1186/s40462-022-00353-2.

Seavy, N E., D.L. Humple, R.L. Cormier, and T.Gardali. 2012. Establishing the breeding provenance of a temperate-wintering North American passerine, the Golden-crowned Sparrow, using light-level geolocation. *PLOS ONE* 7(4):e34886. https://doi.org/10.1371/journal.pone.0034886.

Nelson's Sparrow

Smith, A., N. Ernst, K. O'Brien, N. Pau, and A. Smith. R5_SALS (Project 29). 2014–2017. Data accessed from Motus Wildlife Tracking System, Birds Canada. https://motus.org/.

Saltmarsh Sparrow

Smith, A., N. Ernst, K. O'Brien, N. Pau, and A. Smith. R5_SALS (Project 29). 2014–2017. Data accessed from Motus Wildlife Tracking System, Birds Canada. https://motus.org/.

Song Sparrow

Beauchamp, A., Y. Morbey, and C. Guglielmo. Georgian Bay songbirds (Project 417). 2021–2024. Data accessed from Motus Wildlife Tracking System, Birds Canada. https://motus.org/.

Kelly, T.R., K. Hobson, G. Casbourn, E.A. MacDougall-Shackleton, and S.A. MacDougall-Shackleton. 2019. Long-term winter-site fidelity in Song Sparrows (*Melospiza melodia*). *Auk* 136:1–12.

FAMILY ICTERIIDAE—YELLOW-BREASTED CHAT

Yellow-breasted Chat

Mancuso, K.A., K.E. Hodges, J.D. Alexander, M. Grosselet, A.M. Bezener, L. Morales, S.C. Martinez, J. Castellanos-Labarcena, M.A. Russello, S.M. Rockwell, M.E. Bieber, and C.A. Bishop. 2022. Migration and non-breeding ecology of the Yellow-breasted Chat *Icteria virens*. *Journal of Ornithology* 163:37–50. https://doi.org/10.1007/s10336-021-01931-8.

FAMILY ICTERIDAE— NEW WORLD BLACKBIRDS AND ALLIES

Bobolink

Brinker, D., M. Eshleman, Northeast Motus Collaboration Bobolink Project (Project 504). 2022–2026. Data accessed from Motus Wildlife Tracking System, Birds Canada. https://motus.org/.

Nocera, J. Northeast Motus Collaboration (Project 489). 2022–2022. Data accessed from Motus Wildlife Tracking System, Birds Canada. https://motus.org/.

Perlut, N. Unpublished data.

Renfrew, R.B., D. Kim, N. Perlut, J. Fox, J. Smith, and P.P. Marra. 2013. Phenological matching across hemispheres in a long-distance migratory bird. *Diversity and Distributions* 19:1008–1019. https://doi.org/10.1111/ddi.12080.

Renfrew, R., D. Kim, N. Perlut, and M. Cadman. 2019. Migration tactics of a long-distance migratory songbird from across a continental breeding range. *Wilson Journal of Ornithology* 131:735–749.

Chihuahuan Meadowlark

Scarpignato, A.L., and A.J. Boyce. Movebank Study MCP Meadowlarks North America Sunbirds, ID 2056253021.

Eastern Meadowlark

Giocomo, J. Movebank Study ABC Eastern Meadowlark, ID 3489134530.

Hill, J.M., and R.B. Renfrew. 2018. Data from: Migratory patterns and connectivity of two North American grassland bird species [Eastern Meadowlarks]. Movebank Data Repository. https://doi.org/10.5441/001/1.04qr5q47.

Rigley, B., and A.E.M. Johnson. Movebank Study VWL Eastern Meadowlarks, ID 3071650024.

Scarpignato, A.L. Movebank Study MCP Meadowlarks North America Pinpoint, ID 2327903822.

———. Movebank Study MCP Meadowlarks North America Sunbirds, ID 2056253021.

Western Meadowlark

Boyce, A.J. Movebank Study MCP Meadowlarks North America Pinpoint, ID 2327903822.

Scarpignato, A.L. Movebank Study MCP Meadowlarks North America Pinpoint, ID 2327903822.

———. Movebank Study MCP Meadowlarks North America Sunbirds, ID 2056253021.

Bullock's Oriole

Pillar, A.G., P.P. Marra, N.J. Flood, and M.W. Reudink. 2016. Moult-migration in Bullock's Orioles (*Icterus bullockii*) confirmed by geolocators and stable isotope analysis. *Journal of Ornithology* 157:265–275.

Red-winged Blackbird

Stonefish, D., M.A. Eshleman, G.M. Linz, H.J. Homan, P.E. Klug, T.J. Greives, and E.H. Gillam. 2021. Migration routes and wintering areas of male Red-winged Blackbirds as determined using geolocators. *Journal of Field Ornithology* 92:284–293. https://doi.org/10.1111/jofo.12373.

Rusty Blackbird

Foss, C., and L. Burford. New England Rusty Blackbird Project (Project 252). 2019–2019. Data accessed from Motus Wildlife Tracking System, Birds Canada. https://motus.org/.

Johnson, J., S. Matsuoka, D. Tessler, R. Greenburg, and J. Fox. 2012. Identifying migratory pathways of Rusty Blackbirds breeding in southcentral Alaska. *Wilson Journal of Ornithology* 124:698–703. Data available: Movebank Study Rusty Blackbird Migration (Alaska), ID 1456562300. https://www.movebank.org/cms/webapp?gwt_fragment=page=studies,path=study1456562300.

Tonra, C. Ohio State University (Project 64). 2015–2021. Data accessed from Motus Wildlife Tracking System, Birds Canada. https://motus.org/.

Tremblay, J., and C. Foss. Migration connectivity of the Rusty Blackbird (Project 161). 2017–2021. Data accessed from Motus Wildlife Tracking System, Birds Canada. https://motus.org/.

Wright, J.R., J.A. Johnson, E. Bayne, L.L. Powell, C.R. Foss, J.C. Kennedy, and P.P. Marra. 2021. Migratory connectivity and annual cycle phenology of Rusty Blackbirds (*Euphagus carolinus*) revealed through archival GPS tags. *Avian Conservation and Ecology* 16(1):20. https://doi. org/10.5751/ACE-01871-160120.

FAMILY PARULIDAE—NEW WORLD WARBLERS

Ovenbird

Beauchamp, A., Y. Morbey, and C. Guglielmo. Georgian Bay songbirds (Project 417). 2021–2024. Data accessed from Motus Wildlife Tracking System, Birds Canada. https://motus.org/.

Cohen, E., C. Nemes, and T. Guida. Clive Runnells Mad Island Marsh Preserve (Project 246). 2019–2021. Data accessed from Motus Wildlife Tracking System, Birds Canada. https://motus.org/.

DeGroote, L., M. Sarver, J. Rice, D. Yeany, J. Fowler, R. Jacob, S. Sargent, L. Koitsch, M. Gimpel, J. Cepek, and T. Krynak. Powdermill Nature Reserve (Project 35). 2016–2021. Data accessed from Motus Wildlife Tracking System, Birds Canada. https://motus.org/.

Dossman, B., and A. Brunner. Dossman_Jamaica (Project 19). 2016–2020. Data accessed from Motus Wildlife Tracking System, Birds Canada. https://motus.org/.

Haché, S., E.M. Bayne, M.-A. Villard, H. Proctor, C.S. Davis, D. Stralberg, E. Vasi, A. Grossi, J.C. Gorrell, J.K. Janes, M.T. Hallworth, K.R. Foster, and R. Krikun. 2017.

Phylogeography of a migratory songbird across its Canadian breeding range: Implications for conservation units. *Ecology and Evolution* 7(16):6078–6088. https://doi.org/10.1002/ece3.3170.

Hallworth, M.T., and P.P. Marra. 2015. Data from: Miniaturized GPS tags identify non-breeding territories of a small breeding migratory songbird. Movebank Data Repository. https://www.doi.org/10.5441/001/1.h43t9j6b.

Hallworth, M.T., T.S. Sillett, S.L. Van Wilgenburg, K.A. Hobson, and P.P. Marra. 2014. Migratory connectivity of a neotropical migratory songbird revealed by archival light-level geolocators. *Ecological Applications* 25(2):336–347. https://doi.org/10.1890/14-0195.1.

Seewagen, C. Effects of mercury on ovenbird migration (Project 142). 2017–2018. Data accessed from Motus Wildlife Tracking System, Birds Canada. https://motus.org/.

Golden-winged Warbler

Bennett, R.E., S. Barker Swarthout, J. Bolsinger, K.V. Rosenberg, R. Rourbaugh, and A.D. Rodewald. 2017. Extreme genetic similarity does not predict non-breeding strategy of two closely related migratory warblers. *Journal of Field Ornithology* 88(2):156–168.

Bennett, R.E., A.D. Rodewald, K.V. Rosenberg, R. Chandler, L. Chavarria-Duriaux, J.A. Gerwin, D.I. King, and J.L. Larkin. 2019. Drivers of variation in migration behavior for a linked population of long-distance migratory passerine. *Auk* https://doi.org/10.1093/auk/ukz051.

Kramer, G.R., D.E. Andersen, D.A. Buehler, P.B. Wood, S.M. Peterson, J.A. Lehman, K.R. Aldinger, L.P. Bulluck, S. Harding, J.A. Jones, J.P. Loegering, C. Smalling, R. Vallender, and H.M. Streby. 2018. Data from range-wide study of migratory connectivity of Vermivora warblers. Data Repository for the University of Minnesota. https://doi.org/10.13020/D6D97W.

———. 2018. Population trends in Vermivora warblers are linked to strong migratory connectivity. *Proceedings of the National Academy of Sciences* 115(14):E3192–E3200.

Blue-winged Warbler

Bennett, R.E. Unpublished data.

Kramer, G.R., D.E. Andersen, D.A. Buehler, P.B. Wood, S.M. Peterson, J.A. Lehman, K.R. Aldinger, L.P. Bulluck, S. Harding, J.A. Jones, J.P. Loegering, C. Smalling, R. Vallender, and H.M. Streby. 2018. Data from range-wide study of migratory connectivity of Vermivora warblers. Data Repository for the University of Minnesota. https://doi.org/10.13020/D6D97W.

———. 2018. Population trends in Vermivora warblers are linked to strong migratory connectivity. *Proceedings of the National Academy of Sciences* 115(14):E3192–E3200.

Prothonotary Warbler

Tonra, C.M., M.T. Hallworth, T.J. Boves, J. Reese, L.P. Bulluck, M. Johnson, C. Viverette, K. Percy, E.M. Ames, A. Matthews, M.C. Slevin, R.R. Wilson, and E.I. Johnson. 2019. Concentration of a widespread breeding population in few critically important nonbreeding areas: Migratory connectivity in the Prothonotary Warbler. *Condor* 121(2):duz019. https://doi.org/10.1093/condor/duz019.

Swainson's Warbler

Brunner, A., B. Dossman, V. Jirinec, K. Percy, C. Tonra, E. Johnson, and P.P. Marra. 2022. Migratory behavior and connectivity revealed in a secretive Neotropical migratory songbird, the Swainson's Warbler. *Journal of Field Ornithology* 93(3):5. https://doi.org/10.5751/JFO-00134-930305.

Connecticut Warbler

Hallworth, M.T., E. Bayne, E. McKinnon, O. Love, J.A. Tremblay, B. Drolet, J. Ibarzabal, S. Van Wilgenburg, and P.P. Marra. 2021. Habitat loss on the breeding grounds is a major contributor to population declines in a long-distance migratory songbird. *Proceedings of the Royal Society: Biological Sciences* 288(1949):20203164. https://doi.org/10.1098/rspb.2020.3164.

Common Yellowthroat

Taff, C.C., C.R. Freeman-Gallant, G.R. Kramer, and H.M. Streby. Unpublished data.

American Redstart

Dossman, B.C., A.D. Rodewald, C.E. Studds, and P.P. Marra. 2023. Migratory birds with delayed spring departure migrate faster but pay the costs. *Ecology* 104(2):e3938.

Stevens, H.C., M. DeSaix, B.C. Re, J. Salgado-Ortiz, B.A. Razo Cárdenas, D. Ruíz Domínguez, Y. Caicedo-Ortiz, D. Garzón-Lozano, B.C. Dossman, K. Ruegg, and P. P. Marra. Integrated, range-wide migratory connectivity for a long-distance migratory bird. *Ecosphere*. In revision.

Kirtland's Warbler

Cooper, N., and H. Haradon. Kirtland's Warbler Tracking, Bahamas—Michigan (Project 145). 2017–2019. Data accessed from Motus Wildlife Tracking System, Birds Canada. https://motus.org/.

Cooper, N.W., D.N. Ewert, K.R. Hall, S.M. Rockwell, D. Currie, J.M. Wunderle Jr., J.D. White, and P.P. Marra. 2018. Re-sighting data reveal weak connectivity from wintering to breeding grounds in a range-restricted and endangered long-distance migratory passerine. *Avian Conservation and Ecology* 13:1–9.

Cooper, N.W., M.T. Hallworth, and P.P. Marra. 2017. Light-level geolocation reveals wintering distribution, migration routes, and primary stopover locations of an endangered long-distance migratory songbird. *Journal of Avian Biology* 48:209–219.

Cerulean Warbler

Delancey, C., K. Islam, G. Kramer, G. MacDonald, A. Sharp, and B. Connare. 2020. Geolocators reveal migration routes, stopover sites, and nonbreeding dispersion in a population of Cerulean Warblers. *Animal Migration* 7(1):19–26. https://doi.org/10.1515/ami-2020-0003.

Raybuck, D.W., T.H. Boves, S.H. Stoleson, J.L. Larkin, N.J. Bayly, L.P. Bullock, G.A. George, K.G. Slankard, L.J. Kearns, S. Petzinger, J.J. Cox, and D.A. Buehler. 2022. Cerulean Warblers exhibit parallel migration patterns and multiple migratory stopovers within the Central American isthmus. *Ornithological Applications* 124(4):duac031. https://doi.org/10.1093/ornithapp/duac031.

Raybuck, D., et al. 2022. Cerulean Warbler light-level geolocator data. Dryad Dataset. https://doi.org/10.5061/dryad.crjdfn372.

Yellow Warbler

Witynski, M.L., and D.N. Bonter. 2018. Crosswise migration by Yellow Warblers, Nearctic-Neotropical passerine migrants. *Journal of Field Ornithology* 89(1):37–46.

Blackpoll Warbler

DeLuca, W.V., B.K. Woodworth, S.A. MacKenzie, A.E.M. Newman, H. Cooke, L.M. Phillips, N.E. Freeman, A.O. Sutton, L. Tauzer, C. McIntyre, I.J. Stenhouse, S. Weidensaul, P.D. Taylor, and D.R. Norris. 2019. A boreal songbird's 20,000 km migration across North America and the Atlantic Ocean. *Ecology*. https://doi.org/10.1002/ecy.2651.

DeLuca, W.V., B.K. Woodworth, C. Rimmer, P.P. Marra, P.D. Taylor, K. McFarland, S.A. Mackenzie, and D.R. Norris. 2015. Transoceanic migration of a 12 g songbird. *Biology Letters* 11:20141045. https://doi.org/10.1098/rsbl.2014.1045.

National Park Service. Critical Connections Program. Unpublished data.

Black-throated Blue Warbler

Lewis, W.B., R.J. Cooper, M.T. Hallworth, A.R. Brunner, and T.S. Sillett. 2023. Light-level geolocation reveals moderate levels of migratory connectivity for declining and stable populations of Black-throated Blue Warblers (*Setophaga caerulescens*). *Avian Conservation and Ecology* 18(2):12. https://doi.org/10.5751/ACE-02526-180212.

Prairie Warbler

Campbell, S.P., M.E. Akresh, N.A. Gifford, and D.I. King. 2020. Tracking the annual migration of Prairie Warblers breeding at the Albany Pine Bush Preserve and Montague Plains Wildlife Management Area using light-level geolocators. Unpublished data.

Campbell, S.P., M.E. Akresh, King, D.I., and N. Gifford. 2021. Effects of light-level geolocators on return rates and inter-annual territory fidelity of Prairie Warblers *Setophaga discolor*. *Bird Study* 68:396–407.

T.D. Meehan, S.P. Saunders, W.V. DeLuca, N.L. Michel, J. Grand, … M E. Akresh, … and C.B. Wilsey. 2022. Data integration to estimate spatial patterns of avian migration across the Western Hemisphere. *Ecological Applications* 32:e2679.

Townsend's Warbler

Kim, H. Unpublished data.

Hermit Warbler

Kim, H. Unpublished data.

Canada Warbler

Roberto-Charron, A., J. Kennedy, L. Reitsma, J.A. Tremblay, R. Krikun, K.A. Hobson, J. Ibarzabal, and K.C. Fraser. 2020. Widely distributed populations of Canada Warbler (*Cardellina canadensis*) converge on migration through Central America. *BMC Zoology* 5(1). https://doi.org/10.1186/s40850-020-00056-4.

Wilson's Warbler

National Park Service. Critical Connections Program. Unpublished data.

FAMILY CARDINALIDAE—TANAGERS, GROSBEAKS AND BUNTINGS

Summer Tanager

Cooper, N.W., N. Bayly, R.E. Bennett, and C. Gomez. Bird Friendly Coffee Full Lifecycle Ecology (Project 558). 2022–2024. Data accessed from Motus Wildlife Tracking System, Birds Canada. https://motus.org/.

Western Tanager

Bird Conservancy of the Rockies. Unpublished data.

Black-headed Grosbeak

Saracco, J.F., R.L. Cormier, D.L. Humple, S. Stock, R. Taylor, and R.B. Siegel. 2022. Demographic responses to climate-driven variation in habitat quality across the annual cycle of a migratory bird species. *Ecology and Evolution* 12:e8934. https://doi.org/10.1002/ece3.8934.

DATASET CITATIONS

Siegel, R.B., R. Taylor, J.F. Saracco, L. Helton, and S. Stock. 2016. GPS-tracking reveals non-breeding locations and apparent molt migration of a Black-headed Grosbeak. *Journal of Field Ornithology* 87:196–203.

Lazuli Bunting

Savides, K. 2022. Characterizing the migratory phenology and routes of the Lazuli Bunting (*Passerina amoena*) in northern Utah. Master's thesis, Utah State University. https://digitalcommons.usu.edu/etd/8450.

Painted Bunting

Contina, A., E.S. Bridge, N.E. Seavy, J.M. Duckles, and J.F. Kelly. 2013. Using geologgers to investigate bimodal isotope patterns in Painted Buntings (*Passerina ciris*). *Auk* 130(2):265–272.

Sharp, A.J., A. Contina, V. Ruiz-Gutiérrez, T. Scott Sillett, E. Bridge, E. Besozzi, J. Muller, J. Kelly, A. Given, and C. S. Rushing. 2023. The strength of migratory connectivity in Painted Buntings is spatial scale dependent and shaped by molting behavior. *Journal of Field Ornithology* 94(1):7. https://doi.org/10.5751/JFO-00233-940107.

Smith, A., and D. Marain. Region 4 USFWS (Project 4). 2015–2020. Data accessed from Motus Wildlife Tracking System, Birds Canada. https://motus.org/.

DATA CONTRIBUTORS

Aarhus University (King Eider)
Adalsteinsson, Solny—Washington University, St. Louis (Wood Thrush)
Adams, Evan M.—Biodiversity Research Institute (Blackpoll Warbler; Fox Sparrow; Gray-cheeked Thrush; Gray Catbird; Hermit Thrush; Swainson's Thrush; Veery; Wilson's Warbler)
Adams, Josh—US Geological Survey (Pink-footed Shearwater; Sooty Shearwater)
Adkins, Jessica Y.—US Geological Survey, Oregon Cooperative Fish and Wildlife Research Unit, Oregon State University (Double-crested Cormorant)
Aebischer, Adrian—Groupe de Recherche en Ecologie Arctique (Ivory Gull)
Ahlstrom, Christina A.—US Geological Survey, Alaska Science Center (Herring Gull)
Akearok, Jason A. (Black-legged Kittiwake)
Akresh, Michael E.—Antioch University of New England, and University of Massachusetts Amherst (Prairie Warbler)
Alaska Department of Fish and Game (Golden Eagle; Lesser Yellowlegs)
Aldabe, Joaquin—Universidad de la República de Uruguay (Buff-breasted Sandpiper)
Alexander, John—Klamath Bird Observatory (Yellow-breasted Chat)
Alioto, Nick—Michigan State University, and Mackinac Straits Raptor Watch (Red-tailed Hawk)
Allen, Brad—Maine Department of Inland Fisheries and Wildlife (Common Eider)
Allyn, Andrew J.—University of Massachusetts Amherst (Arctic Tern; Black-legged Kittiwake)
American Eagle Foundation (Bald Eagle)
Andersen, David E.—US Geological Survey, Minnesota Cooperative Fish and Wildlife Research Unit (Blue-winged Warbler; Golden-winged Warbler)
Anderson, Alexandra—Trent University (Red Knot; Semipalmated Sandpiper)
Anderson, Christine M.—Acadia University (Herring Gull)
Anderson, David—Wake Forest University (Black-footed Albatross; Laysan Albatross)
Anderson, Kurt—Ducks Unlimited (American Black Duck)
Anker-Nilssen, Tycho—SEATRACK (Black-legged Kittiwake)
Antolos, Michelle—Oregon State University (Black-footed Albatross); University of California, Santa Cruz (Laysan Albatross)
Applegate, Kelly—Mille Lacs Band of Ojibwe Department of Natural Resources (Purple Martin)
Atwell, Jonathan W.—Indiana University (Phainopepla)
Aubry, Yves—Canadian Wildlife Service (Bicknell's Thrush; Red Knot; Semipalmated Sandpiper)
Audubon Minnesota (American White Pelican)
Aurora Research Institute (Snowy Owl)
Avery, Michael L.—USDA/APHIS/WS National Wildlife Research Center (Black Vulture; Turkey Vulture)
Avian Research and Conservation Institute (Northern Harrier; Short-tailed Hawk; Swallow-tailed Kite; White-crowned Pigeon)
Baak, Julia E.—Dalhousie University (Black Guillemot)
Baffinland Iron Mines (Common Eider; King Eider)
Bairlein, Franz—Institute of Avian Research (Northern Wheatear)
Bakermans, Marja—Worcester Polytechnic Institute (Eastern Whip-poor-will)
Baldassarre, Daniel T.—SUNY Oswego (Phainopepla)
Baldwin, Frank—Environment and Climate Change Canada (Cackling Goose; Canada Goose)
Ballard, Bart M.—Texas A&M University (Brown Pelican; Hudsonian Godwit)
Balogh, Gregory R.—US Fish and Wildlife Service (Black-footed Albatross; Laysan Albatross; Short-tailed Albatross)
Banasch, Ursula—Canadian Wildlife Service (Ferruginous Hawk)
Baran, Mark A.—Atlantic Laboratory for Avian Research (Atlantic Puffin)
Barber, David R.—Hawk Mountain Sanctuary (Black Vulture; Turkey Vulture)
Barber, Patricia—Pennsylvania Game Commission (Lesser Black-backed Gull)
Barger, Chris—Alaska Department of Fish and Game (Golden Eagle)
Barnes, Joseph G.—Nevada Department of Wildlife (Golden Eagle); US Fish and Wildlife Service (American Goshawk; Ferruginous Hawk)
Barrow, John (Purple Martin)
Barter, Mark—Wetlands International (Dunlin)
Bartzen, Blake A.—Environment and Climate Change Canada (Long-tailed Duck)
Barve, Sahas—Archbold Biological Station (Eastern Meadowlark)
Basille, Mathieu—University of Florida (Wood Stork)
Bates, Kirk. K.—Boise State University, Raptor Research Center (Prairie Falcon)
Bathrick, R.—University of Massachusetts Amherst (Short-billed Dowitcher)
Bauer, Silke—Swiss Ornithological Institute (American Golden-Plover)
Bayly, Nicholas—SELVA (Black-billed Cuckoo; Cerulean Warbler; Gray-cheeked Thrush; Summer Tanager; Willow Flycatcher)
Bayne, Erin M.—University of Alberta (Broad-winged Hawk; Common Nighthawk; Ferruginous Hawk; Rusty Blackbird)
Beasley, James C.—University of Georgia, Savannah River Ecology Laboratory, Warnell School of Forestry and Natural Resources (Black Vulture; Turkey Vulture)
Beauchamp, Andrew—University of Western Ontario (Ovenbird; Song Sparrow)
Bégin Marchand, Camille—Environment and Climate Change Canada (Bicknell's Thrush; Swainson's Thrush)
Behney, Adam C.—Colorado Parks and Wildlife (Gadwall; Mallard)
Bell, David—Motus data contributor (Bicknell's Thrush)
Bell, Kelsey—Department of Biological Sciences, University of Manitoba (Purple Martin)
Bennett, Kim—Ontario Ministry of Natural Resources (Red Knot; Semipalmated Plover; Semipalmated Sandpiper; Whimbrel)
Bennett, Ruth E.—Smithsonian's National Zoo and Conservation Biology Institute, Migratory Bird Center (Blue-winged Warbler; Golden-winged Warbler; Summer Tanager)
Benoit, Réjean—Canadian Wildlife Service (Barrow's Goldeneye)
Bentzen, Rebecca—Wildlife Conservation Society (King Eider; Semipalmated Sandpiper)
Berigan, Liam A.—University of Maine (American Woodcock)
Berrigan, Lucas—Birds Canada (Swainson's Thrush)
Bertram, Douglas F.—Environment and Climate Change Canada (Marbled Murrelet)
Bertram, Mark—US Fish and Wildlife Service (Snowy Owl)
Betts, Matthew—Oregon State University (Hermit Warbler; Townsend's Warbler)
Bêty, Joel—University of Quebec at Rimouski, Quebec Centre for Northern Research (American Golden-Plover; Common Ringed Plover; Snow Goose)
Bickford, Nate—Colorado State University Pueblo (Red-tailed Hawk)
Bierregaard, Rob—Academy of Natural Sciences of Drexel University (Osprey)
Bildstein, Keith—Hawk Mountain Sanctuary (Black Vulture; Turkey Vulture)
Biodiversity Research Institute (Northern Harrier)
Bird Conservancy of the Rockies (Black Swift; Swainson's Thrush; Western Tanager)
Birds Canada (Sandhill Crane)
BirdsCaribbean (Short-tailed Hawk; Swallow-tailed Kite; White-crowned Pigeon)
Bishop, Christine—Environment and Climate Change Canada (Gray Catbird; Yellow-breasted Chat)
Bishop, Mary Anne—Prince William Sound Science Center (Black Turnstone; Tufted Puffin)
Bjornstad, Oskar—SEATRACK (Northern Fulmar)
Blake, Margaret R.—University of Montana (Western Meadowlark)
Blake, William—American Bird Conservancy (Swainson's Thrush)
Blanchford, A. (Snowy Owl)
Block, Theadora A.—University of California, Santa Cruz (Golden-crowned Sparrow)
Blomberg, Erik J.—University of Maine (American Woodcock)
Bloom, Peter H.—Bloom Research, Inc. (Red-tailed Hawk)
Boardman, Rinchen—Thunder Cape Bird Observatory (Gray-cheeked Thrush)
Boelman, Natalie—Columbia University (American Robin)
Boelsma, Jill—Denali Education Center (Blackpoll Warbler; Fox Sparrow; Gray-cheeked Thrush; Hermit Thrush; Swainson's Thrush; Wilson's Warbler)
Boise State University (Swainson's Hawk)
Boldenow, Megan—University of Alaska Fairbanks (Semipalmated Sandpiper)
Bollache, Loïc—Université de Bourgogne-Franche Comté (Long-tailed Jaeger; Parasitic Jaeger; Red Phalarope)
Bolton, Mark—Royal Society for the Protection of Birds (Red-necked Phalarope)
Bonter, David N.—Cornell Lab of Ornithology (Yellow Warbler)
Booms, Travis—Alaska Department of Fish and Game (Golden Eagle; Short-eared Owl)
Borkhataria, Rena R.—Everglades Research and Education Center (Wood Stork)
Bose de Almeida, Juliana—Manomet Conservation Sciences (Buff-breasted Sandpiper)
Boulinier, Thierry—SEATRACK (Black-legged Kittiwake)
Boves, Than J.—Arkansas State University (Cerulean Warbler; Eastern Meadowlark)
Bowman, Jeff—Ontario Ministry of Natural Resources (Double-crested Cormorant)
Bowman, Timothy D.—US Fish and Wildlife Service (Barrow's Goldeneye; Long-tailed Duck)
Boyce, Andy J.—Smithsonian's National Zoo and Conservation Biology Institute, Migratory Bird Center (Chihuahuan Meadowlark; Long-billed Curlew; Western Meadowlark)
Boyd, W. Sean—Environment and Climate Change Canada (Barrow's Goldeneye; Brant; Harlequin Duck; Surf Scoter)
Boyle, David (Short-tailed Shearwater)
Boyle, W. Alice—Kansas State University (Eastern Meadowlark)
Boynton, Chloe K.—Environment and Climate Change Canada (Western Meadowlark)
Bozzo, Julie—University of Illinois Urbana-Champaign (Eastern Meadowlark)
Bracey, Annie M.—Natural Resources Research Institute, University of Minnesota Duluth (American Woodcock; Common Tern)
Bradley, David—Birds Canada (Long-billed Curlew)
Brady, Caroline—California Waterfowl Association (Greater White-fronted Goose; Northern Pintail)
Bridge, Eli—Oklahoma Biological Survey, University of Oklahoma (Painted Bunting; Purple Martin)

DATA CONTRIBUTORS

Briggs, Chris W.—Golden Gate Raptor Observatory, Golden Gate National Parks Conservancy (Red-tailed Hawk)
Bringsvor, Ingar S.—SEATRACK (Black-legged Kittiwake)
Brinker, David—Maryland Department of Natural Resources (Brown Pelican; Snowy Owl; Sora; Virginia Rail)
Brooke, Michael de L.—University of Cambridge (Murphy's Petrel)
Brown, Glen—Ontario Ministry of Natural Resources (Semipalmated Plover; Whimbrel)
Brown, Stephen—Manomet Conservation Sciences (American Golden-Plover; Dunlin; Pectoral Sandpiper; Red Phalarope; Semipalmated Sandpiper)
Bruner, Phil—Brigham Young University–Hawaii (Black-bellied Plover)
Brunner, Alicia R.—Cornell University (Black-throated Blue Warbler; Ovenbird; Swainson's Warbler)
Bryan, A. Lawrence—Savannah River Ecology Laboratory, University of Georgia (Black Vulture; Turkey Vulture; Wood Stork)
Brzorad, John—Lenoir-Rhyne University (Great Blue Heron; Great Egret; Wood Stork)
Bucklin, David N.—University of Florida (Wood Stork)
Buehler, David A.—University of Tennessee, Knoxville (Bicknell's Thrush; Blue-winged Warbler; Cerulean Warbler; Golden-winged Warbler)
Bullis, Jane—Merrill Creek Reservoir (Bald Eagle)
Bulluck, Lesley—Virginia Commonwealth University (Cerulean Warbler)
Burch, J.—National Park Service (Snowy Owl)
Bureau of Land Management (Ferruginous Hawk; Golden Eagle)
Burford, Levi—New Hampshire Audubon's Rusty Blackbird Project (Rusty Blackbird)
Burger, Joanna (Red Knot)
Burnett, Ryan—Point Blue Conservation Science (Swainson's Thrush)
Burnham, Kurt K.—High Arctic Institute (Atlantic Puffin; Gyrfalcon; Peregrine Falcon)
Byer, Timothy—US Forest Service (Ferruginous Hawk)
Byrne, Michael E.—Savannah River Ecology Laboratory, University of Georgia (Black Vulture; Turkey Vulture)
Cadman, Michael D.—Environment and Climate Change Canada (Eastern Whip-poor-will)
Cadman, Mike—Environment Canada (Bobolink)
Campagna, Leonardo—Cornell Lab of Ornithology (Phainopepla)
Campbell, Greg—Bird Studies Canada (Bicknell's Thrush)
Campbell, Steven P.—Albany Pine Bush Preserve Commission (Prairie Warbler)
Canadian Wildlife Service (Bald Eagle; Sandhill Crane; Swainson's Hawk)
Carle, Ryan D.—Oikonos Ecosystem Knowledge (Pink-footed Shearwater)
Carlisle, Jay—Boise State University Intermountain Bird Observatory (Long-billed Curlew; Western Meadowlark)
Casazza, Michael—US Geological Survey, Western Ecological Research Center (Gadwall; Greater White-fronted Goose; Mallard; Northern Pintail; Northern Shoveler; Ross's Goose; Snow Goose; American Wigeon; Band-tailed Pigeon; Blue-winged Teal; Cinnamon Teal; Green-winged Teal; Northern Harrier; Sandhill Crane)
Casbourn, Garth—York University (Wood Thrush)
Caspari, Phil (Upland Sandpiper)
Castellanos, Gilbert—US Fish and Wildlife Service (Buff-breasted Sandpiper)
Castresana, Gabriel—Reserva Natural Bahía Samborombón (Buff-breasted Sandpiper)
Caton, Elaine—The Blackfoot Challenge (Trumpeter Swan)
Catry, Paulo—Marine and Environmental Sciences Centre (MARE), ISPA–Instituto Universitário (Cory's Shearwater; Sooty Shearwater)
Catterson, Nathaniel—US Department of Agriculture Forest Service (Aleutian Tern)
Center for Conservation Biology, College of William and Mary (Bald Eagle)
Cepek, Jonathon—Cleveland Metro Parks (Evening Grosbeak; Ovenbird; Swainson's Thrush; Wood Thrush)
Chaine, Alexis S.—Station d'Ecologie Théorique et Expérimentale du CNRS and Institute for Advanced Studies in Toulouse (Golden-crowned Sparrow)
Chandler, Richard—University of Georgia (Golden-winged Warbler)
Chastel, Olivier—SEATRACK (Black-legged Kittiwake)
Chavarria-Duriaux, Liliana—El Jaguar Reserve (Golden-winged Warbler)
Chiang, Chung-Yu—Tunghai University (Dunlin)
Choi, Chi-Yeung—Southern University of Science and Technology (Dunlin)
Christensen-Dalsgaard, Signe—SEATRACK (Black-legged Kittiwake)
Christie, Katie—Alaska Department of Fish and Game (Lesser Yellowlegs)
Chu, Miyoko—Cornell Lab of Ornithology (Phainopepla)
Chubbs, Tony E.—Department of National Defence, Government of Canada (Bald Eagle; Osprey)
Cimprich, Paula—University of Oklahoma (Long-billed Curlew)
Clark, Daniel E.—Massachusetts Department of Conservation and Recreation (Great Black-backed Gull; Herring Gull; Ring-billed Gull)
Clark, Kathleen E.—New Jersey Fish and Wildlife (Bald Eagle)
Clay, Rob—Manomet Conservation Sciences (Semipalmated Sandpiper)
Clay, Thomas A.—University of California, Santa Cruz (Murphy's Petrel)
Clemson University, College of Agriculture, Forestry, and Life Sciences (Canada Goose)
Coates, Stephanie—Boise State University Intermountain Bird Observatory; currently Point Blue Conservation Science (Long-billed Curlew)
Cohen, Emily—Smithsonian's National Zoo and Conservation Biology Institute (Ovenbird; Swainson's Thrush)
Cohen, Johnathan B.—State University of New York (SUNY) College of Environmental Science and Forestry (Piping Plover)
Collins, Dan P.—US Fish and Wildlife Service (Sandhill Crane)
Collis, Ken—Real Time Research (Double-crested Cormorant)
Colodro, Valentina—Oikonos Ecosystem Knowledge (Pink-footed Shearwater)
Coluccy, John—Ducks Unlimited (American Black Duck)
Conners, Melinda—Stony Brook University (Black-footed Albatross; Laysan Albatross)
Contina, Andrea—University of Texas Rio Grande Valley (Painted Bunting)
Cooke, Hilary A.—Wildlife Conservation Society Canada (Blackpoll Warbler)
Cooper, Nathan W.—Smithsonian's National Zoo and Conservation Biology Institute, Migratory Bird Center (Kirtland's Warbler; Merlin; Summer Tanager)
Cooper, Robert J.—University of Georgia, Warnell School of Forestry and Natural Resources (Black-throated Blue Warbler)
Cormier, Renée L.—Point Blue Conservation Science (Fox Sparrow; Golden-crowned Sparrow; Swainson's Thrush)
Costa, Daniel—University of California, Santa Cruz (Laysan Albatross)
Costa, Jeffrey—Environment and Climate Change Canada (Black Tern; Ring-billed Gull)
Cousens, Bruce F.—Georgia Basin Ecological Assessment and Restoration Society (Fox Sparrow; Purple Martin)
Cox, John—University of Kentucky (Cerulean Warbler)
Craig, Elizabeth—Cornell University (Common Tern)
Craighead Beringia South (Bald Eagle; Golden Eagle; Osprey; Red-tailed Hawk; Rough-legged Hawk)
Craik, Shawn—Université Sainte-Anne (American Black Duck)
Cramer, Bradley—Real Time Research (Double-crested Cormorant)
Crampton, Lisa H.—University of Hawai'i at Mānoa (Phainopepla)
Creed, Selena—Project SNOWstorm (Snowy Owl)
Cruz, Miguel Á.—Pronatura-Noreste (Ferruginous Hawk)
Currier (Scherr), Heidi—Simon Fraser University (Double-crested Cormorant)
Curtis, Paul—Cornell University (Common Tern)
D'Auria, Danielle—Maine Department of Inland Fisheries and Wildlife (Great Blue Heron)
Danielsen, Johannis—SEATRACK (Atlantic Puffin; Black-legged Kittiwake; Northern Fulmar)
Daunt, Francis—SEATRACK (Black-legged Kittiwake)
Davis, Shanti E.—High Arctic Gull Research Group (Iceland Gull; Ross's Gull; Sabine's Gull)
Davoren, Gail K.—University of Manitoba (Great Black-backed Gull)
Davy, Christina—Trent University and Government of Ontario (Eastern Whip-poor-will)
Dayton, B. James—University of Maryland, Baltimore County (Broad-winged Hawk)
De La Cruz, Susan—US Geological Survey, Western Ecological Research Center (Surf Scoter; Western Grebe)
de Zwaan, Devin R.—Environment and Climate Change Canada (Horned Lark)
Deavila, Candace—SLC Corporation/USDA (Ferruginous Hawk; Red-tailed Hawk; Swainson's Hawk)
DeCicco, Lucas H.—US Fish and Wildlife Service (Short-eared Owl); University of Kansas (Red-tailed Hawk; Solitary Sandpiper)
DeGroote, Luke—Powdermill Avian Research Center (Evening Grosbeak; Ovenbird; Swainson's Thrush; Wood Thrush)
Dehnhard, Nina—SEATRACK (Black-legged Kittiwake)
Delaware Shorebird Project (Red Knot)
Delmore, Kira—Texas A&M University (Swainson's Thrush)
DeLuca, William V.—National Audubon Society (Blackpoll Warbler)
Descamps, Sebastien—Norwegian Polar Institute, SEATRACK (Black-legged Kittiwake; Dovekie; Northern Fulmar)
DeSorbo, Christopher R.—Biodiversity Research Institute (Bald Eagle; Merlin; Northern Harrier; Peregrine Falcon; Snowy Owl)
Desrochers, André—Université Laval (Bicknell's Thrush; Red Knot; Semipalmated Sandpiper; Swainson's Thrush)
DeVault, Travis L.—USDA/APHIS/WS National Wildlife Research Center (Black Vulture; Turkey Vulture)
Dey, Amanda D. (Red Knot)
Diamond, Antony W.—Atlantic Laboratory for Avian Research (Atlantic Puffin)
Dias, Maria A.—Marine and Environmental Sciences Centre (MARE), ISPA–Instituto Universitário (Cory's Shearwater)
Dickson, D. Lynne—Environment and Climate Change Canada (King Eider; Long-tailed Duck)
Diebold, Ed—Riverbanks Zoo and Garden (Trumpeter Swan)
DiFolco, Donna—US National Park Service (Snowy Owl)
Diggs, Nora E.—Smithsonian's National Zoo and Conservation Biology Institute, Migratory Bird Center at time of study (Wood Thrush)
Donahue, Emily R.—Arkansas State University (Eastern Meadowlark)
Doolittle, T. (Snowy Owl)
Dossman, Bryant C.—Cornell and Georgetown University (American Redstart; Ovenbird; Swainson's Warbler)
Douglas, David C.—US Geological Survey, Alaska Science Center (Black Scoter; Blue-winged Teal; Common Eider; Common Merganser; Emperor Goose; Glaucous-winged Gull; Greater White-fronted Goose; Herring Gull; Kittlitz's Murrelet;

Northern Fulmar; Northern Pintail; Pacific Loon; Pelagic Cormorant; Red-throated Loon; Spectacled Eider; Steller's Eider; Surf Scoter; Thick-billed Murre; Tufted Puffin; Tundra Swan; Yellow-billed Loon)
Doyle, Frank—University of British Columbia Okanagan (Snowy Owl)
Drennan, Joseph E.—Garcia and Associates (Bald Eagle)
Drever, Mark—Environment and Climate Change Canada (Dunlin; Sanderling)
Drummond, Brie A.—US Fish and Wildlife Service (Red-legged Kittiwake)
DuBour, Kristin A.—US Fish and Wildlife Service (Olive-sided Flycatcher)
Dudash, Michele R.—South Dakota State University (Wood Thrush)
Duffy, David C.—University of Hawai'i at Mānoa (Aleutian Tern; Arctic Tern)
Duriaux, George—El Jaguar Reserve (Golden-winged Warbler)
Duxbury, Jason—University of Alberta at time of study (Peregrine Falcon; Gyrfalcon; Prairie Falcon; Snowy Owl)
Dwyer, Chris—US Fish and Wildlife Service, North Atlantic-Appalachian Region (Common Eider)
Dybala, Kristen E.—Point Blue Conservation Science (Swainson's Thrush)
Eadie, J. (Surf Scoter)
EA Engineering (Bald Eagle)
Earthspan (American White Pelican; Ferruginous Hawk; Golden Eagle)
Easton, Wendy—Environment and Climate Change Canada (Dunlin; Sanderling; Swainson's Thrush)
Egevang, Carsten—Greenland Institute of Natural Resources (Arctic Tern; Sabine's Gull)
Elbin, Susan—New York City Audubon (Semipalmated Sandpiper)
Elizondo, Elisa—Carleton University (Clapper Rail)
Elliott, Kyle H.—McGill University (American Golden-Plover; Black-legged Kittiwake; Glaucous-winged Gull; Glaucous Gull; Red Phalarope; Rhinoceros Auklet; Swainson's Thrush; Thick-billed Murre; Tufted Puffin)
Ellison, Kevin—American Bird Conservancy (Western Meadowlark)
Ely, Craig—US Geological Survey, Alaska Science Center (Greater White-fronted Goose; Tundra Swan)
Ely, Teresa E.—Golden Gate Raptor Observatory, Golden Gate National Parks Conservancy (Red-tailed Hawk)
Emmenegger, Tamara—Lund University (American Golden-Plover)
English, Matthew—Environment and Climate Change Canada (American Black Duck)
English, Philina—Fisheries and Oceans Canada (Eastern Whip-poor-will)
English, Willow—Carleton University (American Golden-Plover; Red Phalarope; Semipalmated Sandpiper)
Environment and Climate Change Canada, Science and Technology Branch, Wildlife Research Division (Marbled Murrelet)
Erikstad, Kjell E.—SEATRACK (Black-legged Kittiwake; Thick-billed Murre)
Ernst, Nick—Motus data contributor (Nelson's Sparrow; Saltmarsh Sparrow)
Eshleman, Michelle A.—North Dakota State University (Red-winged Blackbird); Willistown Conservation Trust (Eastern Meadowlark)
Esler, Daniel—US Geological Survey, Alaska Science Center (Barrow's Goldeneye; Surf Scoter)
Evans, Brian S.—Smithsonian's National Zoo and Conservation Biology Institute, Migratory Bird Center (Gray Catbird)
Evenson, Joe—Washington Department of Fish and Wildlife (Brant; Canada Goose; Harlequin Duck; Snow Goose; Surf Scoter; White-winged Scoter)
Evers, David C.—Biodiversity Research Institute (Gray Catbird; Veery)
Exelon Corporation (Bald Eagle)
Ezhov, Alexey—SEATRACK (Black-legged Kittiwake)
Farmer, Adrian—Wild Ecological Solutions (Marbled Godwit)
Fayet, Annette L.—Norwegian Institute for Nature Research (Manx Shearwater)
Feigin, Stephanie (Red Knot)
Feldheim, Cliff—Ducks Unlimited, Inc., Western Region (American Wigeon; Blue-winged Teal; Gadwall; Green-winged Teal; Mallard; Northern Shoveler)
Felis, Jonathan J.—US Geological Survey (Pink-footed Shearwater; Sooty Shearwater)
Fifield, David A.—Environment and Climate Change Canada (Atlantic Puffin; Herring Gull; Northern Gannet)
Fiorello, Christine V.—Oiled Wildlife Care Network (Western Grebe)
Fischer (Courtot), Karen.—US Geological Survey, Oregon Cooperative Fish and Wildlife Research Unit, Oregon State University (Double-crested Cormorant)
Fish, Alexander C.—University of Maine (American Woodcock)
Fish, Allen M.—Golden Gate Raptor Observatory, Golden Gate National Parks Conservancy (Red-tailed Hawk)
Fitzgérald, Guy—Union Québécoise de Réhabilitation des Oiseaux de Proie (UQROP) (Harlequin Duck)
Fleischer, Robert C.—Smithsonian Center for Conservation Genomics (Phainopepla)
Fleishman, Erica—Oregon State University (Swainson's Hawk)
Flemming, Scott A.—Canadian Wildlife Service, Environment and Climate Change Canada (American Golden-Plover; Dunlin; Ruddy Turnstone; Sanderling; Semipalmated Sandpiper)
Flint, Paul—US Geological Survey, Alaska Science Center (Black Scoter; Long-tailed Duck)
Flood, Nancy J.—Thompson Rivers University (Bullock's Oriole)
Florida Fish and Wildlife Conservation Commission (Bald Eagle; Short-tailed Hawk; Swallow-tailed Kite; White-crowned Pigeon)
Fort, Jérôme—French National Centre for Scientific Research, SEATRACK (Dovekie)
Foss, Carol—New Hampshire Audubon (Rusty Blackbird)
Fournier, Auriel—University of Illinois Urbana-Champaign (Sora; Virginia Rail)
Fowler, Jill—Motus data contributor (Evening Grosbeak; Ovenbird; Swainson's Thrush; Wood Thrush)
Fowler, Melinda—Northeastern University (Long-tailed Jaeger); Springfield College (Black-bellied Plover).
Fox, James W.—British Antarctic Survey (Bicknell's Thrush; Snow Bunting)
Fraser, Kevin C.—Department of Biological Sciences, University of Manitoba (Cackling Goose; Canada Goose; Canada Warbler; Connecticut Warbler; Eastern Whip-poor-will; Fox Sparrow; Purple Martin); York University (Snow Bunting; Wood Thrush)
Frederick, Jeffrey—University of Alaska Southeast (Aleutian Tern)
Frederick, Peter C.—University of Florida (Wood Stork)
Frederiksen, Morten—Aarhus University (Ivory Gull; Razorbill); SEATRACK (Black-legged Kittiwake; Thick-billed Murre)
Freeman-Gallant, Corey R.—Skidmore College (Common Yellowthroat)
Freeman, Nikole E.—University of Guelph (Blackpoll Warbler)
Freeman, Scott—Arctic National Wildlife Refuge (Semipalmated Sandpiper)
Frei, Barbara—Environment and Climate Change Canada (Swainson's Thrush)
French Polar Institute (Red-necked Phalarope)
Friesen, Lyle—Canadian Wildlife Service (Wood Thrush)
Friis, Christian—Environment Climate Change Canada (Lesser Yellowlegs; Red Knot; Semipalmated Sandpiper)
Fuller, Mark R.—Boise State University, Raptor Research Center (Prairie Falcon)
Fylling, Megan—University of Montana (Gray Catbird)
Gammonley, James H.—Colorado Parks and Wildlife (Gadwall; Mallard)
Gardali, Thomas—Point Blue Conservation Science (Golden-crowned Sparrow)
Garthe, Stefan—Christian-Albrechts-Universität zu Kiel (Northern Gannet)
Gaston, Anthony J.—Environment and Climate Change Canada (Ancient Murrelet; Black-legged Kittiwake)
Gates, H. River—National Audubon Society (Dunlin)
Gauthier, Gilles—Université Laval (American Golden-Plover; Snowy Owl)
Gavrilo, Maria V.—Arctic and Antarctic Research Institute (AARI), SEATRACK (Black-legged Kittiwake; Dovekie); National Park Russian Arctic (Ivory Gull)
Gaydos, Joseph K.—SeaDoc Society (Western Grebe)
George, Gregory—Delaware Valley University (Cerulean Warbler)
Georgia Department of Natural Resources (Bald Eagle)
Gerhardt, Rick—Northwest Wildlife Consultants, Inc. (Ferruginous Hawk)
Gerwin, John A.—North Carolina State Museum of Natural Sciences (Golden-winged Warbler)
Gesmundo, Callie F.—US Fish and Wildlife Service (Lesser Yellowlegs; Solitary Sandpiper; Upland Sandpiper)
Gifford, Neil A.—Albany Pine Bush Preserve Commission (Prairie Warbler)
Gilbert, Andrew—Biodiversity Research Institute (Blackpoll Warbler; Fox Sparrow; Gray-cheeked Thrush; Hermit Thrush; Swainson's Thrush; Wilson's Warbler)
Gilchrist, H. Grant—Environment and Climate Change Canada (Common Eider; Glaucous Gull; Herring Gull; Ivory Gull; King Eider; Snow Bunting; Thick-billed Murre)
Giles, Molly—Clemson University, College of Agriculture, Forestry, and Life Sciences (Canada Goose)
Gilg, Olivier—Groupe de Recherche en Ecologie Arctique (Ivory Gull; Long-tailed Jaeger; Parasitic Jaeger; Red-necked Phalarope; Red Phalarope)
Gillam, Erin H.—North Dakota State University (Red-winged Blackbird)
Gilliland, Scott—Canadian Wildlife Service (White-winged Scoter)
Gilpatrick, Lauren—Biodiversity Research Institute (Snowy Owl)
Gimpel, Maren—Washington College (Evening Grosbeak; Ovenbird; Swainson's Thrush; Wood Thrush)
Giocomo, James—American Bird Conservancy (Eastern Meadowlark)
Giroux, Marie-Andrée—Université de Moncton (American Golden-Plover; Red Phalarope; Semipalmated Sandpiper)
Gjerdrum, Carina—Canadian Wildlife Service (Bermuda Petrel)
Goldstein, Michael I.—US Department of Agriculture Forest Service (Aleutian Tern)
Gomez and Sullivan Engineers (Bald Eagle)
Gomez, Camila—SELVA (Black-billed Cuckoo; Gray-cheeked Thrush; Summer Tanager; Willow Flycatcher)
Gonzalez, Ana M.—Canadian Wildlife Service (Gray-cheeked Thrush; Swainson's Thrush)
González-Rojas, José Ignacio—Universidad Autónoma de Nuevo León (Long-billed Curlew)
González-Solís, Jacob—Universitat de Barcelona (Black-legged Kittiwake; Cory's Shearwater)
Goodenough, Katharine S.—University of Oklahoma (Black Skimmer; Gull-billed Tern; Long-billed Curlew)
Goodrich, Laurie—Hawk Mountain Sanctuary (Broad-winged Hawk)
Gorman, Kristen B.—University of Alaska Fairbanks (Tufted Puffin)
Gow, Elizabeth A.—Birds Canada (Tree Swallow); Environment and Climate Change Canada (Horned Lark); University of Saskatchewan (Northern Flicker)
Granadeiro, Jose P.—Universidade de Lisba (Cory's Shearwater)
Grand Manan Whale and Seabird Research Station (Sooty Shearwater)
Grand Teton National Park (Osprey; Rough-legged Hawk)

Gratto-Trevor, Cheri—Environment and Climate Change Canada (Marbled Godwit; Semipalmated Sandpiper)
Gray, Jacob—Louisiana State University (Gadwall)
Great Salt Lake Institute at Westminster College (American White Pelican)
Green, Clay—Texas State University (Brown Pelican)
Green, David J.—Simon Fraser University (Eastern Whip-poor-will)
Greives, Timothy J.—North Dakota State University (Red-winged Blackbird)
Grémillet, David—French National Centre for Scientific Research, SEATRACK (Black-legged Kittiwake; Dovekie)
Gress, George (Blackpoll Warbler; Fox Sparrow; Gray-cheeked Thrush; Hermit Thrush; Swainson's Thrush; Wilson's Warbler)
Grigsby, Charles—Denali National Park and Preserve (Blackpoll Warbler; Fox Sparrow; Gray-cheeked Thrush; Hermit Thrush; Swainson's Thrush; Wilson's Warbler)
Grinde, Alexis—Natural Resources Research Institute, University of Minnesota Duluth (American Woodcock)
Groupe de Recherche en Ecologie Arctique (Red-necked Phalarope)
Guglielmo, Christopher C.—University of Western Ontario, Advanced Facility for Avian Research (Hermit Thrush; Ovenbird; Song Sparrow; Swainson's Thrush)
Guida,Timothy—Smithsonian's National Zoo and Conservation Biology Institute, Migratory Bird Center (Long-tailed Jaeger); Motus data contributor (Ovenbird; Swainson's Thrush)
Guzmán-Velasco, Antonio—Universidad Autónoma de Nuevo León (Long-billed Curlew)
Hagelin, Julie C.—Alaska Department of Fish and Game (Golden Eagle; Olive-sided Flycatcher)
Haines, Aaron—Millersville University (Broad-winged Hawk)
Hall, Kristin—Minnesota Department of Natural Resources (American Kestrel; American White Pelican; Golden Eagle)
Hallworth, Michael T.—Smithsonian's National Zoo and Conservation Biology Institute, Migratory Bird Center at time of study (Black-throated Blue Warbler; Connecticut Warbler; Ovenbird)
Halpin, Luke R.—Halpin Wildlife Research (Fork-tailed Storm-Petrel; Leach's Storm-Petrel)
Hamilton, Diana—Mount Allison University (Semipalmated Sandpiper)
Hanna, Everett E.—Lethbridge College (Sandhill Crane)
Hanni, David—Bird Conservancy of the Rockies (Ferruginous Hawk)
Haradon, Haley—University of Illinois Urbana-Champaign (Kirtland's Warbler)
Harris, Phil—Shetland Seabird Tours (Red-necked Phalarope)
Harrison, Autumn-Lynn—Smithsonian's National Zoo and Conservation Biology Institute, Migratory Bird Center (American Golden-Plover; Arctic Tern; Black-bellied Plover; Brown Pelican; Common Nighthawk; Glaucous Gull; Long-billed Curlew; Long-tailed Jaeger; Pacific Loon; Parasitic Jaeger; Pomarine Jaeger; Red Phalarope; Short-billed Dowitcher)
Harwood, Chris M.—US Fish and Wildlife Service (Lesser Yellowlegs)
Hashimoto, Yuriko—Environment and Climate Change Canada (Ancient Murrelet)
Hatch, Kent A.—Long Island University (Great Shearwater)
Hatch, Scott A.—Institute for Seabird Research and Conservation, US Geological Survey, Alaska Science Center (Black-legged Kittiwake; Common Murre; Northern Fulmar; Pelagic Cormorant; Rhinoceros Auklet; Thick-billed Murre; Tufted Puffin)
Haukos, David—US Geological Survey, Kansas Cooperative Fish and Wildlife Research Unit (Northern Pintail)
Hawk Mountain Sanctuary (Broad-winged Hawk)
HawkWatch International (American Goshawk; Red-tailed Hawk)
Hayes, Stephen—Gonzaga University (Northern Pintail)
Heagy, Audrey E.—Bird Studies Canada (Eastern Whip-poor-will)
Heath, Julie A.—Boise State University (American Kestrel)
Heath, Susan—Gulf Coast Bird Observatory (Red Knot; Willet)
Heather, Meredith—Archbold Biological Station (Eastern Meadowlark)
Heckscher, Christopher M.—Delaware State University (Veery)
Hedd, April—Environment and Climate Change Canada (Atlantic Puffin; Sooty Shearwater); SEATRACK (Black-legged Kittiwake)
Hedley, Richard—University of Alberta (Cassin's Vireo)
Helberg, Morten—SEATRACK (Northern Fulmar)
Helgason, H.H.—SEATRACK (Northern Fulmar)
Helton, Lauren—Institute for Bird Populations (Black-headed Grosbeak)
Henke, M. Blake—University of Maryland, Baltimore County (Broad-winged Hawk)
Henry, R. William—University of California, Santa Cruz (Laysan Albatross)
Hepinstall-Cymerman, Jeffrey—University of Georgia (White Ibis)
Herbert, John—Mass Audubon (Semipalmated Sandpiper)
Heusmann, H.—Massachusetts Wildlife (Common Eider)
Hill, Jason M.—Vermont Center for Ecostudies (Bicknell's Thrush; Eastern Meadowlark; Grasshopper Sparrow; Upland Sandpiper)
Hinkes, Michael—US Fish and Wildlife Service (Snowy Owl)
Hipfner, J. Mark—Environment and Climate Change Canada (Cassin's Auklet; Glaucous-winged Gull; Iceland Gull; Rhinoceros Auklet)
Hobson, Keith A.—Environment and Climate Change Canada (Bank Swallow; Cliff Swallow; Gray-cheeked Thrush; Ovenbird; Swainson's Thrush; Veery)
Hodges, Karen—University of British Columbia Okanagan (Gray Catbird; Yellow-breasted Chat)
Hodum, Peter J.—Oikonos Ecosystem Knowledge (Pink-footed Shearwater)
Hogan, Danica—Environment and Climate Change Canada (Barrow's Goldeneye)
Holland, Amanda E.—University of Georgia (Black Vulture; Turkey Vulture)
Holroyd, Geoff L.—Environment and Climate Change Canada, Beaverhill Bird Observatory (Burrowing Owl; Gyrfalcon; Peregrine Falcon; Prairie Falcon; Short-eared Owl; Snowy Owl)
Homan, H. Jeffrey—USDA/APHIS/WS National Wildlife Research Center (Red-winged Blackbird)
Huang, Min—University of Connecticut (Clapper Rail)
Huang, Ryan—Duke University (Sooty Tern)
Hull, Joshua—University of California, Davis, US Fish and Wildlife Service (Golden-crowned Sparrow; Northern Harrier)
Humphrey, John S.—USDA/APHIS/WS National Wildlife Research Center (Black Vulture; Turkey Vulture)
Humple, Diana L.—Point Blue Conservation Science (Fox Sparrow; Golden-crowned Sparrow; Swainson's Thrush)
Hunt, Anjolene R.—Boise State University (American Kestrel)
Hupp, Jerry—US Geological Survey (Northern Pintail)
Husak, Michael S.—Cameron University (Eastern Kingbird; Scissor-tailed Flycatcher; Western Kingbird)
Huschle, Gary—US Fish and Wildlife Service (American Bittern)
Huy, Steve—Project SNOWstorm and Lambs Knoll Station (Snowy Owl)
Huysman, Allison—Smithsonian's National Zoo and Conservation Biology Institute, Migratory Bird Center (Long-tailed Jaeger)
Hycroft Gold Mine (Golden Eagle)
Idaho Department of Fish and Game (American White Pelican; Sandhill Crane)
Imlay, Tara L.—Dalhousie University, Acadia University (Bank Swallow; Cliff Swallow)
Intermountain West Joint Venture (Sandhill Crane)
Irons, David B.—US Fish and Wildlife Service (Arctic Tern; Black-legged Kittiwake)
Islam, Kamal—Ball State University (Cerulean Warbler)
Iverson, Autumn R.—University of California, Davis (Golden-crowned Sparrow)
Jackman, R.E.—Kleinfelder, Garcia and Associates (Bald Eagle)
Jacob, Ryan—Black Swamp Bird Observatory (Evening Grosbeak; Ovenbird; Swainson's Thrush; Wood Thrush)
Jacobs, Gene—Raptor Services, LLC (Snowy Owl)
Jahn, Alex E.—Indiana University (American Robin; Eastern Kingbird; Scissor-tailed Flycatcher; Western Kingbird)
James Bay Shorebird Project (Lesser Yellowlegs)
Jeffress, Mackenzie—Nevada Department of Wildlife (American Goshawk)
Jennings, Richard (Osprey)
Jennings, Scott—Audubon California (Great Egret)
Jensen, Barb—Michigan Osprey Association (Osprey)
Jerritt Canyon Gold, LLC (Ferruginous Hawk)
Jessopp, Mark—SEATRACK (Atlantic Puffin)
Jiguet, Frédéric—Muséum National d'Histoire Naturelle, Paris (Wandering Tattler)
Jirinec, Vitek—Louisiana State University at time of study, currently Integral Ecology Research Center (Swainson's Warbler)
Jobin, Benoit—Canadian Wildlife Service (Harlequin Duck)
Jodice, Patrick G.R.—US Geological Survey, South Carolina Cooperative Fish and Wildlife Research Unit, Clemson University (Black-capped Petrel; Brown Pelican)
Johnson, Amy E. M.—Smithsonian's National Zoo and Conservation Biology Institute, Virginia Working Landscapes (Eastern Meadowlark)
Johnson, Andrew S.—Cornell Lab of Ornithology (Whimbrel)
Johnson, Erik—Audubon Louisiana (Swainson's Warbler)
Johnson, James A.—US Fish and Wildlife Service (Lesser Yellowlegs; Olive-sided Flycatcher; Red Knot; Rusty Blackbird; Short-billed Dowitcher; Short-eared Owl; Solitary Sandpiper; Upland Sandpiper)
Johnson, Luanne—BiodiversityWorks (Willet)
Johnson, Oscar W.—University of Montana (American Golden-Plover)
Jorgensen, Joel G.—Nebraska Game and Parks Commission (American Woodcock)
Kaiser, Sara A.—Cornell Lab of Ornithology (Black-throated Blue Warbler)
Kardynal, Kevin J.—Environment and Climate Change Canada (Barn Swallow; Swainson's Thrush; Veery)
Kaufman, Les—Boston University (Great Shearwater)
Kays, Roland—North Carolina State University and North Carolina Museum of Natural Sciences (Bald Eagle; Golden Eagle; Great Egret; Osprey; Peregrine Falcon; Short-eared Owl; Wood Stork)
Kearns, Gregory D.—Maryland-National Capital Park and Planning Commission, Patuxent River Park (Sora; Virginia Rail)
Kearns, Laura—Ohio Division of Wildlife (Black-crowned Night Heron; Cerulean Warbler)
Keenan, Patrick—Biodiversity Research Institute (Gray Catbird; Veery)
Keitt, Brad—American Bird Conservancy (Black-capped Petrel)
Kelly, Jeff—University of Oklahoma (Long-billed Curlew)
Kelly, Tosha R.—Louisiana State University (Song Sparrow)
Kempenaers, Bart—Max Planck Institute for Biological Intelligence (Long-billed Dowitcher; Pectoral Sandpiper)
Kendall, Steve—US Fish and Wildlife Service (Dunlin)
Kenney, A. (Snowy Owl)

Kent, Gina M.—Avian Research and Conservation Institute (Magnificent Frigatebird; Northern Harrier; Short-tailed Hawk; Swallow-tailed Kite; White-crowned Pigeon)
Kentucky Department of Fish and Wildlife Research Units (Bald Eagle)
Kern-Lovick, Elsa—SEATRACK (Black-legged Kittiwake)
Ketterson, Ellen—Indiana University (American Robin)
Khazan, Emily—US Fish and Wildlife Service (Painted Bunting; Red Knot; Ruddy Turnstone)
Kim, Daniel H.—Platte River Whooping Crane Trust (Scissor-tailed Flycatcher; Western Kingbird); US Fish and Wildlife Service (Bobolink; Eastern Kingbird)
Kim, Hankyu—Oregon State University (Hermit Warbler; Townsend's Warbler)
King, D. Tommy—US Department of Agriculture, National Wildlife Research Center (Double-crested Cormorant)
King, David I.—US Forest Service Northern Research Station, University of Massachusetts Amherst (Prairie Warbler; Golden-winged Warbler)
Kitaysky, Alexander—University of Alaska Fairbanks (Common Murre; Thick-billed Murre)
Kleinschmidt Associates (Bald Eagle)
Kluever, Bryan M.—USDA/APHIS/WS National Wildlife Research Center (Black Vulture; Turkey Vulture)
Klug, Page E.—USDA/APHIS/WS National Wildlife Research Center (Red-winged Blackbird)
Kneidel, A.—Manomet Conservation Sciences (Whimbrel)
Knight, Elly C.—University of Alberta (Common Nighthawk)
Knight, Samantha M.—University of Guelph (Tree Swallow)
Koch, Stephanie—US Fish and Wildlife Service (Black-bellied Plover; Red Knot; Ruddy Turnstone; Sanderling; Semipalmated Sandpiper)
Kochert, Michael N.—US Geological Survey (Prairie Falcon; Swainson's Hawk)
Koczur, Lianne—Texas A&M University (Brown Pelican)
Koenen, Kiana K.G.—Massachusetts Department of Conservation and Recreation, retired (Great Black-backed Gull; Herring Gull; Ring-billed Gull)
Koitsch, Laura-Marie—Erie Bird Observatory (Evening Grosbeak; Ovenbird; Swainson's Thrush; Wood Thrush)
Kokubun, Nobuo—National Institute of Polar Research, Tachikawa, Tokyo, Japan (Red-legged Kittiwake)
Kolbeinsson, Yann—SEATRACK (Black-legged Kittiwake; Common Murre; Northern Fulmar; Thick-billed Murre)
Kolodzinski, Jeff—Port Authority of New York and New Jersey (Semipalmated Sandpiper)
Koloski, Laura—Trent University (American Golden-Plover)
Koons, David N.—Colorado State University (Gadwall; Mallard)
Koontz, Charles (Trumpeter Swan)
Koontz, Fred (Trumpeter Swan)
Koopman, Heather—Grand Manan Whale and Seabird Research Station (Great Shearwater; Sooty Shearwater)
Korpach, Alicia—University of Manitoba (Eastern Whip-poor-will)
Kramer, Gunnar R.—University of Toledo (Blue-winged Warbler; Common Yellowthroat; Golden-winged Warbler)
Kramer, Patrick M.—Department of Biology, York University (Purple Martin)
Krementz, David—US Geological Survey, Arkansas Cooperative Fish and Wildlife Research Unit (American Woodcock)
Kriensky, Debra—New York City Audubon (Semipalmated Sandpiper)
Krikun, Nicole—Lesser Slave Lake Bird Observatory (American Robin)
Krynak, Tim—Cleveland Metro Parks (Evening Grosbeak; Ovenbird; Swainson's Thrush; Wood Thrush)
Kubetzki, Ulrike—Christian-Albrechts-Universität zu Kiel (Northern Gannet)
Kwon, Eunbi—Max Planck Institute for Biological Intelligence (Long-billed Dowitcher; Western Sandpiper); Virginia Polytechnic Institute and State University (Semipalmated Sandpiper)
La Puma, David—Cellular Tracking Technologies (Snowy Owl)
Labansen, Aili—SEATRACK (Black-legged Kittiwake; Thick-billed Murre)
Lafón, Alberto—Universidad Autónoma de Chihuahua (Ferruginous Hawk)
Lagassé, Benjamin—University of Alaska Fairbanks (Dunlin)
Lamarre, Jean-François— Polar Knowledge Canada, Canadian High Arctic Research Station, Université du Québec à Rimouski (American Golden-Plover; Common Ringed Plover; Pectoral Sandpiper; Red Phalarope)
Lamarre, Vincent—Environment Canada (Chimney Swift)
Lamb, Juliet S.—Clemson University, South Carolina Cooperative Fish and Wildlife Research Unit at time of study (Brown Pelican; Reddish Egret); University of Rhode Island at time of study (Black Scoter; Common Eider; Long-tailed Duck; Surf Scoter; White-winged Scoter)
Lamont, Myles—Hancock Wildlife Foundation (Bald Eagle)
Lamothe, Pierre—Hydro-Québec (Harlequin Duck)
Lancaster, Michael B. (Gray Catbird)
Lanctot, Richard B.—US Fish and Wildlife Service (American Golden-Plover; Buff-breasted Sandpiper; Dunlin; Lesser Yellowlegs; Pectoral Sandpiper; Pomarine Jaeger; Red Phalarope; Semipalmated Sandpiper)
Landoll, Diane V.—Augustana University (Eastern Kingbird); University of Oklahoma (Scissor-tailed Flycatcher; Western Kingbird)
Lang, Johannes—Universität Gießen (Long-tailed Jaeger; Parasitic Jaeger; Red Phalarope)
Lank, David B.—Simon Fraser University (Semipalmated Sandpiper)
Lanzone, Michael—Cellular Tracking Technologies (Snowy Owl)
Larkin, Jeffery L.—Indiana University of Pennsylvania (Cerulean Warbler; Golden-winged Warbler)
Larsen, R. Scott—University of California, Davis (Double-crested Cormorant)
Latty, Christopher—US Fish and Wildlife Service (Pectoral Sandpiper; Red Phalarope)
Lawes, Timothy J.—Oregon State University (Caspian Tern)
Léandri-Breton, Don-Jean—University of Quebec at Rimouski, Quebec Centre for Northern Research (Common Ringed Plover)
Lecomte, Nicolas—Université de Moncton (American Golden-Plover; Peregrine Falcon; Red Phalarope; Rough-legged Hawk; Semipalmated Sandpiper)
Lee, J. Charlene—Georgia Basin Ecological Assessment and Restoration Society (Purple Martin)
Lefebvre, Josée—Environment and Climate Change Canada (Snow Goose)
Lefevre, Kara—Florida Gulf Coast University (Red Knot)
Legagneux, Pierre—Université Laval (Snow Goose)
Leonard, Marty L.—Dalhousie University (Bank Swallow; Cliff Swallow)
Lepage, Christine—Canadian Wildlife Service (White-winged Scoter)
Lerczak, James—Oregon State University (Brandt's Cormorant; Pelagic Cormorant)
Lerman, Susannah B.—University of Massachusetts Amherst (American Robin; Gray Catbird)
Lewis, M.A.—Maryland-National Capital Park and Planning Commission (Patuxent River Park) (Sora)
Lewis, Stephen B.—US Fish and Wildlife Service (Bald Eagle; Golden Eagle)
Lewis, William B.—Warnell School of Forestry and Natural Resources, University of Georgia (Black-throated Blue Warbler)
Liebezeit, Joe—Audubon Society of Portland (American Golden-Plover; Dunlin; Semipalmated Sandpiper)
Linhart, Rebeca—University of Rhode Island (Semipalmated Sandpiper)
Link, Paul (Mallard)
Linkhart, Brian—Colorado College (Flammulated Owl)
Linnebjerg, Jannie F.—Aarhus University (Black-legged Kittiwake; Razorbill)
Linscott, Jennifer A.—University of South Carolina (Hudsonian Godwit)
Linz, George M.—USDA/APHIS/WS National Wildlife Research Center (Red-winged Blackbird)
Lisovski, Simeon—University of California, Davis (White-crowned Sparrow)
Lloyd, J.D.—Vermont Center for Ecostudies (Bicknell's Thrush)
Lockhart, Mike—Wildlands Photography and Bio-Consulting (Golden Eagle)
Lok, E. (Surf Scoter)
Long Point Waterfowl (Sandhill Crane)
López, Fernando G.—Smithsonian's National Zoo and Conservation Biology Institute, Migratory Bird Center (Merlin)
Lorensten, Svein-Håkon—SEATRACK (Black-legged Kittiwake)
Lorenz, Jerome—Audubon Florida (Roseate Spoonbill)
Loring, Pamela H.—US Fish and Wildlife Service (American Oystercatcher; Arctic Tern; Black-bellied Plover; Red Knot; Ruddy Turnstone; Sanderling; Semipalmated Sandpiper)
Losito, Michael—State University of New York at Cobleskill (Wood Duck)
Louisiana Department of Wildlife and Fisheries (Bald Eagle)
Louisiana State University (Bald Eagle)
Love, Oliver P.—University of Windsor (American Golden-Plover; Connecticut Warbler; Snow Bunting)
Loverti, Vanessa—US Fish and Wildlife Service (Dunlin; Sanderling; Semipalmated Sandpiper)
Lovette, I.J.—Cornell University (Red-tailed Hawk)
Lyon, Bruce E.—University of California, Santa Cruz (Golden-crowned Sparrow)
Lyons, Donald E.—Oregon State University (Caspian Tern; Double-crested Cormorant); National Audubon Society (Manx Shearwater)
M. Kirk (Surf Scoter)
MacCallum, Beth—Bighorn Wildlife Technologies Ltd. (Harlequin Duck)
Maccarone, Alan—Friends University (Great Egret)
MacDonald, Amie—Birds Canada (Black-bellied Plover; Red Knot; Red Phalarope; Ruddy Turnstone; Semipalmated Plover; Semipalmated Sandpiper; White-rumped Sandpiper)
Macdonald, Christie—University of Windsor (Snow Bunting)
Mackell, Desmond—US Geological Survey, Western Ecological Research Center (Cinnamon Teal)
Mackentley, Nova—Project SNOWstorm (Snowy Owl)
MacKenzie, Kenneth G.—Massachusetts Department of Conservation and Recreation (Great Black-backed Gull; Herring Gull; Ring-billed Gull)
Mackenzie, Stuart A.—Birds Canada (Bicknell's Thrush; Blackpoll Warbler; Red Knot)
MacLeod, Iain (Osprey)
MacPherson, Maggie P.—York University (Wood Thrush)
Madeiros, Jeremy—Bermuda Department of Environment and Natural Resources (Bermuda Petrel)
Maftei, Mark S.—High Arctic Gull Research Group (Arctic Tern; Ross's Gull; Sabine's Gull)
Mahony, Nancy—Environment and Climate Change Canada (Western Meadowlark)
Maillet, Olivia—Trent University (Short-billed Dowitcher)
Major, Heather L.—Atlantic Laboratory for Avian Research (Atlantic Puffin)

Mallory, Mark L.—Acadia University (American Black Duck; Arctic Tern; Black-legged Kittiwake; Common Eider; Herring Gull; Ivory Gull; Long-tailed Duck; Northern Fulmar; Parasitic Jaeger; Ross's Gull; Sabine's Gull; Thick-billed Murre)
Mammenga, Paul—South Dakota Department of Game, Fish and Parks (Purple Martin)
Mance-Coniglio, Melissa—Project SNOWstorm (Snowy Owl)
Mancuso, Kristen—University of British Columbia Okanagan and Environment and Climate Change Canada (Gray Catbird; Yellow-breasted Chat)
Mandernack, B.A.—Eagle Valley Nature Preserve, Kohler Company and Kohler Trust for Preservation (Bald Eagle)
Mann, Hilary—Mount Allison University (Semipalmated Sandpiper)
Manthey, Pat—Wisconsin Department of Natural Resources (Trumpeter Swan)
Marain, Donna—Motus data contributor (Painted Bunting; Red Knot; Ruddy Turnstone)
Marra, Peter P.—Georgetown University (American Robin; Bicknell's Thrush; Flammulated Owl; Gray Catbird; Ovenbird; Swainson's Warbler; Yellow-billed Cuckoo); Smithsonian's National Zoo and Conservation Biology Institute, Migratory Bird Center at time of study (American Golden-Plover; American Redstart; Black-bellied Plover; Black-crowned Night Heron; Blackpoll Warbler; Broad-winged Hawk; Bullock's Oriole; Common Nighthawk; Connecticut Warbler; Eastern Meadowlark; Kirtland's Warbler; Lesser Yellowlegs; Long-billed Curlew; Rusty Blackbird; Willet; Wood Thrush)
Marsh, Brian—US Fish and Wildlife Service and Delaware Shorebird Project (Red Knot)
Martell, Mark—Audubon Minnesota (Osprey); The Raptor Center, University of Minnesota (Broad-winged Hawk; Golden Eagle)
Martin, Kathy—Environment and Climate Change Canada (Horned Lark)
Martin, Phillip—US Fish and Wildlife Service (Steller's Eider)
Marty, Joseph—Louisiana Department of Wildlife and Fisheries (Swainson's Thrush)
Masse, Roger—State University of New York at Cobleskill (Wood Duck)
Matsuoka, Steve—US Fish and Wildlife Service (Rusty Blackbird)
Matteson, Sumner—Wisconsin Department of Natural Resources (Trumpeter Swan)
Maynard, Laurie D.—Université de Moncton (Great Black-backed Gull)
McAdie, Malcolm—Marmot Recovery Foundation (Harlequin Duck)
McAuley, Dan—US Geological Survey (American Woodcock; Common Eider)
McCabe, Rebecca—Hawk Mountain Sanctuary at time of study (Broad-winged Hawk); McGill University (Snowy Owl)
McComb, Brenda—University of Oregon (Hermit Warbler; Townsend's Warbler)
McCrary, Michael D.—Bloom Research, Inc. (Red-tailed Hawk)
McCready, Robert—Nature Conservancy (Ferruginous Hawk)
McDonald, Tom—Project SNOWstorm (Snowy Owl)
McDuffie, Laura A.—US Fish and Wildlife Service at time of study (Lesser Yellowlegs; Short-billed Dowitcher; Solitary Sandpiper)
McFarland, Kent P.—Vermont Center for Ecostudies (Bicknell's Thrush; Blackpoll Warbler)
McGann, Andrew—Cellular Tracking Technologies (Snowy Owl)
McGrady, Mike—University of Maryland, Baltimore County (Broad-winged Hawk)
McGuire, Rebecca—ABR, Inc. (American Golden-Plover; Buff-breasted Sandpiper; Pectoral Sandpiper; Red Phalarope)
McIntyre, Carol—Denali National Park and Preserve (Blackpoll Warbler; Fox Sparrow; Golden Eagle; Gray-cheeked Thrush; Gyrfalcon; Hermit Thrush; Snowy Owl; Swainson's Thrush; Wilson's Warbler)
McKellar, Ann E.—Environment and Climate Change Canada (Black Tern; Common Tern; Red Knot; Sanderling; Semipalmated Plover; Whimbrel)
McKinnon, Emily A.—University of Manitoba (Wood Thrush); University of Windsor (Connecticut Warbler)
McKinnon, Laura—York University (American Golden-Plover; Dunlin; Semipalmated Plover; Short-billed Dowitcher; Stilt Sandpiper)
McKnight, Alyson—Unity College (Arctic Tern; Black-legged Kittiwake)
McNeil, Shannon E.—Southern Sierra Research Station (Yellow-billed Cuckoo)
McWilliams, Scott—University of Rhode Island (White-winged Scoter)
Meattey, Dustin—Biodiversity Research Institute (Common Eider; Harlequin Duck; White-winged Scoter)
Mehus, Scott—Golden Eagle Project (Golden Eagle)
Meixell, Brandt—US Geological Survey, Alaska Science Center (Greater White-fronted Goose)
Mejías, Miguel A.—Memorial University (White-tailed Tropicbird)
Mendez, Devin C.—Delaware State University (Veery)
Merkel, Flemming— Aarhus University (Common Eider; King Eider); SEATRACK (Black-legged Kittiwake; Thick-billed Murre)
Merrill Creek Reservoir (Bald Eagle)
Meyer, Kenneth D.—Avian Research and Conservation Institute (Magnificent Frigatebird; Northern Harrier; Short-tailed Hawk; Swallow-tailed Kite; White-crowned Pigeon)
Mickle, Nanette (Purple Martin)
Migratory Connectivity Project (Broad-winged Hawk; Chihuahuan Meadowlark; Common Nighthawk; Connecticut Warbler; Eastern Meadowlark; Flammulated Owl; Glaucous Gull; Lesser Yellowlegs; Long-billed Curlew; Long-tailed Jaeger; Mountain Plover; Pacific Loon; Parasitic Jaeger; Pomarine Jaeger; Red Knot; Rusty Blackbird; Western Meadowlark)
Miller, Trish—Conservation Science Global (Snowy Owl)
Mills, Alexander M.—York University (Eastern Whip-poor-will)
Mills, Kyra L.—Oiled Wildlife Care Network (Western Grebe)
Minnesota Department of Natural Resources (American White Pelican)
Mitchell, Greg—Environment Canada (Gray-cheeked Thrush; Swainson's Thrush)
Mittelhauser, Glen H.—Maine Natural History Observatory (Harlequin Duck)
Mizrahi, David—New Jersey Audubon Society (Semipalmated Sandpiper)
Mobley, Jason—Aquasis (Semipalmated Sandpiper)
Moe, Børge—Norwegian Institute for Nature Research, SEATRACK (Black-legged Kittiwake; Dovekie)
Montevecchi, William A.—Memorial University of Newfoundland (Northern Gannet; Sooty Shearwater); SEATRACK (Common Murre; Thick-billed Murre)
Moore, David—Environment and Climate Change Canada (Black Tern; Common Tern; Ring-billed Gull)
Moore, Joe—US Geological Survey (American Woodcock)
Morales, Ana—Environment and Climate Change Canada (Swainson's Thrush)
Morbey, Yolanda—University of Western Ontario, Advanced Facility for Avian Research (Hermit Thrush; Ovenbird; Song Sparrow; Swainson's Thrush)
Morrison, Danielle M.—Georgia Basin Ecological Assessment and Restoration Society (Purple Martin)
Morrissey, Christy—University of Saskatchewan (Red Knot; Sanderling)
Mosbech, Anders—Aarhus University (Common Eider; King Eider)
Moseley, Dana L.—Smithsonian's National Zoo and Conservation Biology Institute, Migratory Bird Center (Gray Catbird)
Mossop, David—Yukon Territorial Government (Snowy Owl)
Mott, Andrea—US Geological Survey, Western Ecological Research Center (Greater White-fronted Goose)
Moulton, Colleen—Idaho Department of Fish and Game (American White Pelican)
Mulcahy, Daniel M.—US Geological Survey, Alaska Science Center (Tufted Puffin; Western Grebe)
Murphy, Michael T.—Portland State University (Eastern Kingbird; Scissor-tailed Flycatcher; Western Kingbird)
Myers, Kenneth—Avian Research and Conservation Institute (Arctic Tern; Common Tern)
Myers, Paul M.—US Geological Survey, Alaska Science Center (Tufted Puffin)
Myles, John—Hamilton College (Pomarine Jaeger)
National Fish and Wildlife Foundation (Short-tailed Hawk; Swallow-tailed Kite; White-crowned Pigeon)
National Wildlife Refuges, US Fish and Wildlife Service (Short-tailed Hawk; Swallow-tailed Kite; White-crowned Pigeon)
Natural Resources Canada, Polar Continental Shelf Program (Snowy Owl)
Natural Resources, Ltd. (Bald Eagle)
Navedo, Juan G.—Universidad Austral de Chile (Hudsonian Godwit)
Nebraska Game and Parks Commission (American Woodcock)
Nelligan, M. (Snowy Owl)
Nelson, Allison R.—Gold Country Avian Studies (Hermit Thrush)
Nemes, Claire—University of Maryland Center for Environmental Science (Ovenbird; Swainson's Thrush)
Németh, Zoltan—University of California, Davis (White-crowned Sparrow)
Neri, Chris—Project SNOWstorm (Snowy Owl)
Neudecker, G.—US Fish and Wildlife Service (Trumpeter Swan)
Neufeld, Leanne—University of Manitoba (Cackling Goose; Canada Goose)
New Mexico Department of Game and Fish (Sandhill Crane)
New Mexico State University (Sandhill Crane)
New York State Department of Environmental Conservation (Northern Harrier; Peregrine Falcon; Short-eared Owl)
Newman, Amy E.M.—University of Guelph (Blackpoll Warbler)
Newstead, David J.—Coastal Bend Bays & Estuaries Program (Black Skimmer; Brown Pelican; Eastern Meadowlark; Long-billed Curlew; Purple Martin; Red Knot; Short-billed Dowitcher)
Ng, Janet W.—University of Alberta at time of study (Common Nighthawk)
Nicholai, Chris—Delta Waterfowl Association (Ross's Goose; Snow Goose)
Nightingale, Ann—Rocky Point Bird Observatory (Fox Sparrow)
Niles, Lawrence J.—Wildlife Restoration Partnerships (Red Knot; Ruddy Turnstone; Sanderling)
Nocera, Joseph J.—University of New Brunswick (Bobolink; Eastern Whip-poor-will)
Nol, Erica—Trent University (American Golden-Plover; Dunlin; Lesser Yellowlegs; Semipalmated Plover; Short-billed Dowitcher; Stilt Sandpiper; Whimbrel)
Norfolk Botanical Gardens (Bald Eagle)
Norris, D. Ryan—University of Guelph (Blackpoll Warbler; Tree Swallow)
North Dakota State University (American White Pelican)
Northwest Wildlife Consultants, Inc. (Golden Eagle)
Norvell, Russ—Utah Division of Wildlife Resources (American White Pelican)
Nye, Peter—New York State Department of Environmental Conservation (Bald Eagle; Golden Eagle; Osprey; Peregrine Falcon; Short-eared Owl)
Nyeswander, D. (Surf Scoter)
O'Brien, Kate—Motus data contributor (Nelson's Sparrow; Saltmarsh Sparrow)
O'Brien, Kathleen M.—New York State Department of Environmental Conservation (Black-bellied Plover; Northern Harrier; Peregrine Falcon; Red Knot; Semipalmated Sandpiper; Short-eared Owl)

Oehlers, Susan—US Forest Service (Aleutian Tern)
Okill, J. David—British Trust for Ornithology (Red-necked Phalarope)
Olalla-Kerstupp, Alina—Universidad Autónoma de Nuevo León (Long-billed Curlew)
Oliver, Ruth Y.—University of California, Santa Barbara (American Robin)
Olsen, Bergur—SEATRACK (Atlantic Puffin; Black-legged Kittiwake)
Olsen, Glenn—Patuxent Wildlife Research Center (White-winged Scoter); US Geological Survey, Eastern Ecological Research Center (Common Eider)
Olson, Bridget—US Fish and Wildlife Service (Marbled Godwit)
Olson, Dave—US Fish and Wildlife Service (Cinnamon Teal)
Olson, Liz—BiodiversityWorks (Willet)
Ontario Federation of Anglers and Hunters (Sandhill Crane)
Ontario Ministry of Natural Resources (Double-crested Cormorant)
Oppel, Steffen—Swiss Ornithological Institute (King Eider)
Orabona, Andrea—Wyoming Game and Fish Department (Harlequin Duck)
Orben, Rachael A.—Oregon State University (Black-footed Albatross; Black-legged Kittiwake; Brandt's Cormorant; Common Murre; Pelagic Cormorant; Red-legged Kittiwake)
Osborne, Douglas—University of Arkansas (Greater White-fronted Goose; Mallard)
Osenkowski, Jay—Rhode Island Department of Environmental Management (White-winged Scoter)
Overton, Cory—US Geological Survey, Western Ecological Research Center (American Wigeon; Band-tailed Pigeon; Blue-winged Teal; Cinnamon Teal; Gadwall; Greater White-fronted Goose; Green-winged Teal; Mallard; Northern Harrier; Northern Pintail; Northern Shoveler; Ross's Goose; Snow Goose)
Ozaki, Kiyoaki—Yamashina Institute for Ornithology (Short-tailed Albatross)
Paiva, Vitor—University of Coimbra, Marine and Environment Sciences Centre (MARE) (Cory's Shearwater; Desertas Petrel)
Palm, E. (Surf Scoter)
Papp, Joseph M.—Western Foundation of Vertebrate Zoology (Red-tailed Hawk)
Paquet, Julie—Environment and Climate Change Canada (Semipalmated Sandpiper; Whimbrel)
Paradis, Sebastien—Canadian Wildlife Service (Bicknell's Thrush)
Parkins, Kaitlyn—New York City Audubon (Semipalmated Sandpiper)
Partridge, Dustin—New York City Audubon (Semipalmated Sandpiper)
Patil, Vijay—US Geological Survey, Alaska Science Center (Ross's Goose; Snow Goose)
Patla, Susan—Wyoming Game and Fish Department (Harlequin Duck)
Paton, Peter—University of Rhode Island (White-winged Scoter)
Patterson, Allison—McGill University (Glaucous Gull)
Patterson, Andrea—Braddock Bay Bird Observatory (Virginia Rail)
Pau, Nancy—Motus data contributor (Black-bellied Plover; Nelson's Sparrow; Red Knot; Saltmarsh Sparrow; Semipalmated Sandpiper)
Paxton, Bart—Center for Conservation Biology, College of William and Mary (Bald Eagle)
Pearce, John—US Geological Survey, Alaska Science Center (Common Merganser; Surf Scoter)
Pearman, Myrna—Ellis Bird Farm (Purple Martin)
Pearse, Aaron T.—Northern Prairie Wildlife Research Center (Sandhill Crane; Whooping Crane)
Peck, Liam—Acadia University (American Black Duck)
Peck-Richardson, Adam—Oregon State University (Brandt's Cormorant; Double-crested Cormorant; Pelagic Cormorant)
Pellegrini, A.G.—Maryland-National Capital Park and Planning Commission, Patuxent River Park (Sora)
Percy, Katie—Audubon Louisiana (Swainson's Warbler)
Perkins, Genevive (Red Knot)
Perlut, Noah G.—University of New England (Bobolink; Eastern Meadowlark)
Petersen, Aevar (Atlantic Puffin)
Petersen, Margaret—US Geological Survey, Alaska Science Center (Common Eider; Common Merganser; Long-tailed Duck; Spectacled Eider)
Petrie, George (Red-necked Phalarope)
Petrie, Mark—Ducks Unlimited (Ross's Goose; Snow Goose)
Petrie, Scott A.—Delta Waterfowl (Sandhill Crane)
Petzinger, Sharon—New Jersey Fish and Wildlife (Cerulean Warbler)
Phillips, Laura M.—National Park Service, Denali National Park and Preserve (American Robin; Blackpoll Warbler; Fox Sparrow; Gray-cheeked Thrush; Hermit Thrush; King Eider; Long-tailed Jaeger; Swainson's Thrush; Wilson's Warbler)
Phillips, Richard—British Antarctic Survey (Sooty Shearwater)
Piatt, John F.—US Geological Survey, Alaska Science Center (Kittlitz's Murrelet)
Picardi, Simona—University of Florida (Wood Stork)
Pierce, Allison K.—University of Colorado Denver (Mountain Plover)
Pillar, Andrew G.—Thompson Rivers University (Bullock's Oriole)
Pitts, Marie—Center for Conservation Biology, College of William and Mary (Bald Eagle)
Pohlen, Zachary M.—US Fish and Wildlife Service (Upland Sandpiper)
Point Blue Conservation Science (Hermit Thrush)
Poirier, Vanessa—Motus data contributor (Swainson's Thrush)
Pokrovsky, Ivan—Max Planck Institute of Animal Behavior (Peregrine Falcon; Rough-legged Hawk)
Pollet, Ingrid L.—Dalhousie University (Leach's Storm-Petrel)
Porter, Ronald R.—Delaware Bay Shorebird Project (American Kestrel; Red Knot; Semipalmated Sandpiper)
Powell, Abby—US Geological Survey (King Eider)
Powell, Luke L.—Smithsonian's National Zoo and Conservation Biology Institute, Migratory Bird Center at time of study (Rusty Blackbird)
Powers, Kevin—US National Oceanic and Atmospheric Administration (Great Shearwater)
PPL Holtwood (Bald Eagle)
Pyare, Sanjay—University of Alaska Southeast (Aleutian Tern)
Rail, Jean-François—Canadian Wildlife Service, Environment and Climate Change Canada (Northern Gannet)
Raine, André F.—Archipelago Research and Conservation (Bermuda Petrel; Hawaiian Petrel; Newell's Shearwater)
Ramenofsky, Marilyn—University of California, Davis (White-crowned Sparrow)
Ramey, Andy—US Geological Survey, Alaska Science Center (Blue-winged Teal; Glaucous-winged Gull; Glaucous Gull; Herring Gull)
Ramirez, Ivan—BirdLife International (Desertas Petrel)
Ramos, Jaime A.—University of Coimbra (Cory's Shearwater)
Ramos, Raül—University of Barcelona (Cory's Shearwater)
Rand, Greg J.—Canadian Museum of Nature (Eastern Whip-poor-will)
Rausch, Jennie—Environment and Climate Change Canada (Black-bellied Plover; Hudsonian Godwit; Lesser Yellowlegs; Long-tailed Jaeger; Parasitic Jaeger; Pomarine Jaeger; Semipalmated Sandpiper; Whimbrel)
Ray, James D.—Consolidated Nuclear Security, LLC (Purple Martin)
Raybuck, Douglas—University of Tennessee, Knoxville, at time of study (Cerulean Warbler; Golden-winged Warbler)
Rayner, Matt J.—Auckland War Memorial Museum (Cook's Petrel)
Real Time Research (Double-crested Cormorant)
Reddy, Matt—Ducks Unlimited, Inc. (Gadwall; Mallard)
Reed, Eric—Environment and Climate Change Canada (American Golden-Plover; Ross's Goose; Snow Goose)
Regan, Kevin—Biodiversity Research Institute (Willet)
Reiertsen, Tone—SEATRACK (Thick-billed Murre)
Renfrew, Rosalind B.—Vermont Center for Ecostudies (Bobolink; Grasshopper Sparrow); Vermont Fish and Wildlife Department (Upland Sandpiper)
Reudink, Matthew W.—Thompson Rivers University (Bullock's Oriole)
Rhodes Jr., Olin E.—Savannah River Ecology Laboratory, University of Georgia (Black Vulture; Turkey Vulture)
Rice, Jonathan—Carnegie Museum of Natural History (Evening Grosbeak; Ovenbird; Swainson's Thrush; Wood Thrush)
Richardson, T. Will—Tahoe Institute for Natural Science (Swainson's Thrush)
Riddington, Roger (Red-necked Phalarope)
Riehl, Christina—Princeton University (Phainopepla)
Rigley, Bernadette—Smithsonian's National Zoo and Conservation Biology Institute, Virginia Working Landscapes (Eastern Meadowlark)
Rimmer, Christopher C.—Vermont Center for Ecostudies (Bicknell's Thrush; Blackpoll Warbler)
Ritchie, Alisha—Department of Biological Sciences, University of Manitoba (Purple Martin)
Robbins, Mark B.—University of Kansas Museum of Natural History (Red-tailed Hawk)
Robert, Michel—Canadian Wildlife Service (Barrow's Goldeneye; Harlequin Duck)
Roberto-Charron, Amélie—University of Manitoba (Bank Swallow; Canada Warbler; Cliff Swallow; Fox Sparrow)
Robertson, Gregory J.—Environment and Climate Change Canada (American Black Duck; Herring Gull; Northern Gannet)
Robillard, Audrey—Environment and Climate Change Canada (Snowy Owl)
Robinson, Barry G.—Environment and Climate Change Canada (Western Meadowlark)
Robinson, Bryce W.—Cornell University (Red-tailed Hawk)
Roby, Daniel D.—US Geological Survey, Oregon Cooperative Fish and Wildlife Research Unit, Oregon State University (Double-crested Cormorant)
Rockefeller Wildlife Refuge Trust (Bald Eagle)
Rockwell, Sarah M.—Klamath Bird Observatory (Western Meadowlark; Yellow-breasted Chat)
Rodewald, Amanda D.—Cornell Lab of Ornithology (Golden-winged Warbler)
Ronconi, Robert A.—Environment and Climate Change Canada (Black Guillemot; Great Shearwater; Herring Gull; Red Phalarope; Sooty Shearwater)
Rosenberg, Kenneth V.—Cornell Lab of Ornithology (Golden-winged Warbler)
Ross, Jeremy D.—University of Oklahoma (Lark Sparrow; Western Meadowlark)
Roth, Amber M.—University of Maine (American Woodcock; Bicknell's Thrush; Golden-winged Warbler)
Ruiz-Aymá, Gabriel—Universidad Autónoma de Nuevo León (Long-billed Curlew)
Ruiz, Jorge—Universidad Austral de Chile (Hudsonian Godwit)
Rushing, Clark S.—University of Georgia; Utah State University at time of study (American White Pelican; Painted Bunting)
Russell, Mike (American Golden-Plover)
Ruthrauff, Daniel R.—US Geological Survey, Alaska Science Center (Black-bellied Plover; Marbled Godwit; Pectoral Sandpiper; Short-billed Dowitcher; Whimbrel)
Ryan, Peter G.—University of Cape Town (Great Shearwater)

Ryder, T. Brandt—Bird Conservancy of the Rockies (American Robin; Gray Catbird; Western Meadowlark); Smithsonian's National Zoo and Conservation Biology Institute, Migratory Bird Center at time of study (Wood Thrush)
Saalfeld, Sarah T.—US Fish and Wildlife Service (American Golden-Plover; Dunlin; Pectoral Sandpiper; Red Phalarope; Semipalmated Sandpiper)
Saldanha, Sarah—Dalhousie University (Bank Swallow)
Salo, Thomas—Delaware-Otsego Audubon Society (Golden Eagle)
Salt Lake City International Airport (American White Pelican)
San Francisco Bay Bird Observatory (Hermit Thrush)
Sandercock, Brett K.—Kansas State University (Semipalmated Sandpiper); Max Planck Institute for Ornithology (Western Sandpiper)
Sanders, Felicia—South Carolina Department of Natural Resources (American Golden-Plover; Red Knot; Ruddy Turnstone)
Saracco, James F.— Institute for Bird Populations (Black-headed Grosbeak)
Sargent, Sarah—Erie Bird Observatory (Evening Grosbeak; Ovenbird; Swainson's Thrush; Wood Thrush)
Sarver, Mallory—Powdermill Avian Research Center (Evening Grosbeak; Ovenbird; Swainson's Thrush; Wood Thrush)
Satgé, Yvan—Clemson University, South Carolina Cooperative Fish and Wildlife Research Unit (Black-capped Petrel; Brown Pelican)
Sato, Fumio—Yamashina Institute for Ornithology (Short-tailed Albatross)
Savage, Anne—Disney's Animal Programs (Purple Martin)
Savard, Jean-Pierre L.—Canadian Wildlife Service (Barrow's Goldeneye)
Savoy, Lucas—Biodiversity Research Institute (Common Eider; Harlequin Duck; White-winged Scoter)
Scarpignato, Amy L.—Smithsonian's National Zoo and Conservation Biology Institute, Migratory Bird Center (Black-bellied Plover; Black-crowned Night Heron; Chihuahuan Meadowlark; Eastern Meadowlark; Long-billed Curlew; Long-tailed Jaeger; Parasitic Jaeger; Red Knot; Red Phalarope; Western Meadowlark)
Schaefer, Anne L.—Prince William Sound Science Center (Tufted Puffin)
Schamber, Jason—Alaska Department of Fish and Game (Barrow's Goldeneye; Black Scoter)
Schieldcastle, Mark—Black Swamp Bird Observatory (Snowy Owl)
Schmaljohann, Heiko—Carl von Ossietzky Universität & Institute of Avian Research (Northern Wheatear)
Schmitz, Ryan T.—Eagle Valley Nature Preserve, Kohler Company and Kohler Trust for Preservation (Bald Eagle)
Schmutz, Joel A.—US Geological Survey, Alaska Science Center (Pacific Loon; Red-throated Loon; Yellow-billed Loon)
Schulte, Shiloh—Manomet Conservation Sciences (American Golden-Plover; Pectoral Sandpiper; Red Phalarope; Semipalmated Sandpiper)
Schweitzer, Sara—North Carolina Wildlife Resources Commission (Wood Stork)
Sea Duck Joint Venture (White-winged Scoter)
SEATRACK (Great Skua)
Seavy, Nathaniel E.—National Audubon Society (Golden-crowned Sparrow; Swainson's Thrush)
Seegar, William—University of Maryland, Baltimore County (Broad-winged Hawk)
Seewagen, Chad—Great Hollow Nature Preserve and Ecological Research Center (Ovenbird)
Senner, Nathan R.—University of Massachusetts Amherst (Hudsonian Godwit; Short-billed Dowitcher; Whimbrel)
Serno, Kimberley—University of Delaware (Wood Thrush)
Setash, Casey—Colorado State University (Gadwall; Mallard)
Sexson, Matthew—US Geological Survey (Spectacled Eider)
Seymour, Michael—Louisiana Department of Wildlife and Fisheries (Swainson's Thrush)
Shaffer, Scott A.—University of California, Santa Cruz and San Jose State University (Black-footed Albatross; Laysan Albatross; Sooty Shearwater)
Shaheen, Tim—Purple Martin Conservation Association (Purple Martin)
Shannon, Paula—National Audubon Society (Arctic Tern; Common Tern)
Sharp, Andrew—Utah State University (Painted Bunting)
Shave, Amanda—Department of Biological Sciences, University of Manitoba (Purple Martin)
Shephard, Nicholas G.—University of Saskatchewan (Black Tern)
Shirkey, Brendan—Winous Point Marsh Conservancy (King Rail)
Shizuka, Dai—University of Nebraska Lincoln (Golden-crowned Sparrow)
Shlepr, Katherine R.—University of New Brunswick (Herring Gull)
Shoji, Akiko—University of Tsukuba (Rhinoceros Auklet; Tufted Puffin)
Shriver, W. Gregory—University of Delaware (Wood Thrush)
Siegel, Rodney B.— Institute for Bird Populations (Black-headed Grosbeak)
Siegrist, Joseph—Purple Martin Conservation Association (Purple Martin)
Siles, Moises—El Jaguar Reserve (Golden-winged Warbler)
Sillett, T. Scott—Smithsonian's National Zoo and Conservation Biology Institute, Migratory Bird Center (Black-throated Blue Warbler; Ovenbird)
Silva, Tammy L.—US National Oceanic and Atmospheric Administration (Great Shearwater)
Silverio, Cassandra—Department of Biology, York University (Purple Martin)
Simmons, Michael—Rocky Point Bird Observatory (Fox Sparrow)
Simon Fraser University (Bald Eagle)
Simons, Ted—North Carolina State University (Bald Eagle)
Sittler, Benoit—University of Freiburg (Long-tailed Jaeger; Parasitic Jaeger; Red Phalarope)
Skalos, Dan—California Department of Fish and Wildlife (Greater White-fronted Goose)
Skalos, Shannon—University of California, Davis, US Geological Survey, Western Ecological Research Center (Northern Harrier)
Skinner, Aaron A.— Ohio State University (Eastern Whip-poor-will)
Slankard, Kate—Kentucky Department of Fish and Wildlife Resources (Cerulean Warbler)
Slater, Steve—HawkWatch International (Golden Eagle)
Slezak, Casey—State University of New York at Cobleskill (Wood Duck)
Smalling, Curtis—Audubon North Carolina (Golden-winged Warbler)
Smith, Adam—American Bird Conservancy (Nelson's Sparrow; Saltmarsh Sparrow)
Smith, Adam—US Fish and Wildlife Service (Nelson's Sparrow; Painted Bunting; Red Knot; Ruddy Turnstone; Saltmarsh Sparrow)
Smith, Brian W.—US Fish and Wildlife Service (Golden Eagle)
Smith, Doug—National Park Service (Harlequin Duck)
Smith, Fletcher—Center for Conservation Biology, College of William and Mary (Hudsonian Godwit; Whimbrel)
Smith, Jeff P.—HawkWatch International and H.T. Harvey & Associates (American Goshawk; Golden Eagle; Red-tailed Hawk)
Smith, Joseph—Wildlife Restoration Partnerships (Willet)
Smith, Malcolm—Royal Society for the Protection of Birds (Red-necked Phalarope)
Smith, Nickolas—Louisiana State University (Bald Eagle)
Smith, Norman—Project SNOWstorm (Snowy Owl)
Smith, Paul A.—Environment and Climate Change Canada (American Golden-Plover; Black-bellied Plover; Dunlin; Red Knot; Red Phalarope; Ruddy Turnstone; Sanderling; Semipalmated Plover; Semipalmated Sandpiper; Short-billed Dowitcher; Stilt Sandpiper; Whimbrel; White-rumped Sandpiper)
Smithsonian's National Zoo and Conservation Biology Institute, Migratory Bird Center (Common Nighthawk)
Solensky, Matt—US Geological Survey, Northern Prairie Wildlife Research Center (Snowy Owl)
Sonne, Christian—Aarhus University (Common Eider; King Eider)
South Carolina Cooperative Fish and Wildlife Research Unit (Canada Goose)
South Carolina Department of Natural Resources (Canada Goose; Red Knot)
Sparks, Robert—Bird Conservancy of the Rockies (Black Swift)
Spragens, Kyle—Washington Department of Fish and Wildlife (Brant; Canada Goose; Harlequin Duck; Snow Goose; Surf Scoter; White-winged Scoter)
Stahlecker, Dale—Eagle Environmental, Inc (Golden Eagle)
Stanley, Calandra Q.—Georgetown University, York University, University of Maryland, and Smithsonian's National Zoo and Conservation Biology Institute, Migratory Bird Center at time of studies (Wood Thrush; Yellow-billed Cuckoo)
Stantial, Michelle—State University of New York (SUNY) College of Environmental Science and Forestry (Piping Plover)
Steenhof, Karen—US Geological Survey (Prairie Falcon)
Stein, Kristie—National Park Service, Sequoia National Park (Black-crowned Night Heron)
Stenhouse, Iain J.—Biodiversity Research Institute (Blackpoll Warbler; Fox Sparrow; Gray-cheeked Thrush; Hermit Thrush; Sabine's Gull; Swainson's Thrush; Wilson's Warbler); Greenland Institute of Natural Resources (Arctic Tern)
Stephens, Jaime L.—Klamath Bird Observatory (Western Meadowlark)
Stephenson, John—National Park Service (Harlequin Duck)
Stevens, Henry—Georgetown University (American Redstart)
Stewart, Rebecca L.M.—Bird Studies Canada (Bicknell's Thrush)
Stock, Sarah—Yosemite National Park (Black-headed Grosbeak)
Stoleson, Scott—US Forest Service, Northern Research Station (Cerulean Warbler)
Stone, Katherine—MPG Ranch (Gray Catbird)
Stonefish, Dereck—North Dakota State University (Red-winged Blackbird)
Streby, Henry M.—University of Toledo (Blue-winged Warbler; Common Yellowthroat; Golden-winged Warbler)
Strøm, Hallvard—Norwegian Polar Institute, SEATRACK (Black-legged Kittiwake; Common Murre; Dovekie; Great Skua; Ivory Gull; Northern Fulmar; Thick-billed Murre)
Strong, Allan—University of Vermont (Bobolink)
Struening, Eric—State University of New York at Cobleskill (Wood Duck)
Studds, Colin E.—Smithsonian's National Zoo and Conservation Biology Institute, Migratory Bird Center at time of study, currently University of Maryland, Baltimore County (Wood Thrush)
Studholme, Katharine R.—Dalhousie University (Cassin's Auklet)
Stutchbury, Bridget J.M.—Department of Biology, York University (Purple Martin; Red-eyed Vireo; Wood Thrush)
Sullivan, Kelsey M.—Maine Department of Inland Fisheries and Wildlife (Black-legged Kittiwake; Common Eider)
Sullivan, Kimberly—Utah State University (Marbled Godwit)
Summers, Ron—Highland Ringing Group (Purple Sandpiper)
Suryan, Robert M.—Oregon State University (Black-footed Albatross; Black-legged Kittiwake; Laysan Albatross; Short-tailed Albatross)
Sutton, Alex O.—University of Guelph (Blackpoll Warbler)
Suzuki, Yasuko—Oregon State University (Double-crested Cormorant)

Svingen, Daniel N.—US Forest Service (Ferruginous Hawk)
Swann, Robert—SEATRACK (Black-legged Kittiwake)
Systad, Geir H.—SEATRACK (Black-legged Kittiwake)
Taff, Conor C.—Cornell University (Common Yellowthroat)
Takahashi, Akinori—National Institute of Polar Research, Japan (Common Murre; Thick-billed Murre)
Takekawa, J. (Surf Scoter)
Tautin, John—Purple Martin Conservation Association (Purple Martin)
Tauzer, Lila—Wildlife Conservation Society Canada (Blackpoll Warbler)
Taylor, Audrey R.—University of Alaska Anchorage (Black Turnstone; Lesser Yellowlegs; Semipalmated Sandpiper)
Taylor, Phil D.—Acadia University (Bank Swallow; Blackpoll Warbler; Leach's Storm-Petrel; Swainson's Thrush)
Taylor, Ron—Institute for Bird Populations (Black-headed Grosbeak)
Terenzi, John—US Geological Survey, Alaska Science Center (Tundra Swan)
Tessler D.—Alaska Department of Fish and Game (Rusty Blackbird)
Texas Tech University (Sandhill Crane)
Therrien, Jean-François—Hawk Mountain Sanctuary (Black Vulture; Peregrine Falcon; Rough-legged Hawk; Snowy Owl; Turkey Vulture)
Thiebot, Jean-Baptiste—National Institute of Polar Research, Japan (Common Murre; Thick-billed Murre)
Thomas, Peter—Environment and Climate Change Canada (Gray-cheeked Thrush)
Thomas, Phil (Snow Goose)
Thomassen, Henri A.—University of Tübingen (Phainopepla)
Thompson, David R. (Sooty Shearwater)
Thompson, Jonathon—Alberta Environment and Parks (Barrow's Goldeneye)
Thompson, Michael—US National Oceanic and Atmospheric Administration (Great Shearwater)
Thompson, Paul—SEATRACK (Northern Fulmar)
Thorarinsson, Thorkell L.—SEATRACK (Black-legged Kittiwake; Common Murre; Northern Fulmar; Thick-billed Murre)
Tibbitts, T. Lee—US Geological Survey, Alaska Science Center (Black-bellied Plover; Buff-breasted Sandpiper; Lesser Yellowlegs; Marbled Godwit; Northern Pintail; Tundra Swan; Whimbrel)
Tillman, Eric A.—USDA/APHIS/WS National Wildlife Research Center (Black Vulture; Turkey Vulture)
Tixier, Jeremie—Université Laval (Chimney Swift)
Tobon, Emilio—New York City Audubon (Semipalmated Sandpiper)
Tolan, Kevin—Vermont Center for Ecostudies (Eastern Meadowlark)
Tonra, Christopher M.—Ohio State University (Black-crowned Night Heron; Eastern Whip-poor-will; Prothonotary Warbler; Rusty Blackbird; Virginia Rail)
Tope, Jay (Upland Sandpiper)
Townsend, Jason M.—State University of New York (SUNY) College of Environmental Science and Forestry (Bicknell's Thrush)
Tracey, Diane—Southern Sierra Research Station (Yellow-billed Cuckoo)
Tracy Aviary Conservation Science (American White Pelican)
Trefry, Helen E.—Canadian Wildlife Service, Environment and Climate Change Canada, Beaverhill Bird Observatory (Gyrfalcon; Peregrine Falcon; Prairie Falcon; Snowy Owl)
Tremblay, Junior—Environment and Climate Change Canada and Université Laval (Bicknell's Thrush; Chimney Swift; Horned Lark; Red Knot; Rusty Blackbird; Semipalmated Sandpiper; Swainson's Thrush)
US Army Corps of Engineers (Bald Eagle)
US Department of Defense (Bald Eagle; Rusty Blackbird)
US Department of Defense, Legacy Resource Management Program (American White Pelican; Ferruginous Hawk; Golden Eagle)
US Fish and Wildlife Service (Bald Eagle; Canada Goose; Golden Eagle; Red Knot; Swainson's Hawk)
US Fish and Wildlife Service Migratory Bird Program, Alaska Region (Rusty Blackbird)
US Forest Service (Ferruginous Hawk)
US Geological Survey (Swainson's Hawk)
US Geological Survey, Alaska Science Center (Lesser Yellowlegs; Long-tailed Duck; Pacific Loon)
US Geological Survey, Louisiana Cooperative Fish and Wildlife Research Unit (Bald Eagle)
US Geological Survey, Oregon Cooperative Fisheries and Wildlife Research Unit (Double-crested Cormorant)
US National Park Service (Bald Eagle)
Uher-Koch, Brian D.—US Geological Survey, Alaska Science Center (Emperor Goose; Pacific Loon; Red-throated Loon; Yellow-billed Loon)
Université de Bourgogne Franche Comté (Red-necked Phalarope)
University of Georgia (Bald Eagle)
University of Western Ontario (Sandhill Crane)
Utah Division of Wildlife Resources (American White Pelican)
Valcu, Mihai—Max Planck Institute for Biological Intelligence (Pectoral Sandpiper)
van Bemmelen, Rob—Bureau Waardenburg (Long-tailed Jaeger; Parasitic Jaeger; Red Phalarope)
Van Tatenhove, Aimee—Utah State University (American White Pelican)
Van Wilgenburg, Steven L.—Environment and Climate Change Canada (Common Nighthawk; Ovenbird)
Varland, Dan—Coastal Raptors (Black Vulture; Turkey Vulture)
Vasseur, Phillip L.—Louisiana Department of Wildlife and Fisheries (Swainson's Thrush)
Victorian Ornithological Research Group Inc. (Short-tailed Shearwater)
Vilchis, L. Ignacio—San Diego Zoo, Institute for Conservation Research (Western Grebe)
Virginia Commonwealth University (Bald Eagle)
Vitz, Andrew—Massachusetts Division of Fisheries and Wildlife (Eastern Whip-poor-will)
Voirin, Bryson—Max Planck Institute of Animal Behavior (Magnificent Frigatebird)
Wakefield, Ewan D.—University of Durham (Cory's Shearwater; Sooty Shearwater)
Wald, Eric—National Park Service (Snowy Owl)
Ward, Christopher—University of New Brunswick (Bicknell's Thrush)
Ward, David—US Geological Survey, Alaska Science Center (Surf Scoter)
Ward, Michael P.—University of Illinois Urbana-Champaign (Eastern Meadowlark)
Warkentin, Ian—Memorial University (Gray-cheeked Thrush)
Warner, M. Zoe—Willistown Conservation Trust (Eastern Meadowlark)
Warnock, Nils—Audubon Canyon Ranch (Long-billed Curlew)
Watson, James W.—Washington Department of Fish and Wildlife (Ferruginous Hawk)
Watson, Jesse L.—HawkWatch International (American Kestrel); University of Alberta at time of study (Broad-winged Hawk; Ferruginous Hawk)
Watts, Bryan—Center for Conservation Biology, College of William and Mary (Bald Eagle; Hudsonian Godwit; Whimbrel)
Weaver, Melanie—California Department of Fish and Wildlife (Greater White-fronted Goose)
Weber, Drew—Project SNOWstorm and Cornell Lab of Ornithology (Snowy Owl)
Webster, Michael—Cornell Lab of Ornithology (Black-throated Blue Warbler)
Weegman, Mitch D.—University of Missouri and University of Saskatchewan (Hudsonian Godwit)
Weeks, Brian—University of Michigan (American Robin)
Wehtje, Walter—Ricketts Conservation Foundation (Long-billed Curlew)
Weidensaul, Scott—Project SNOWstorm (Blackpoll Warbler; Fox Sparrow; Gray-cheeked Thrush; Hermit Thrush; Snowy Owl; Swainson's Thrush; Wilson's Warbler)
Weiser, Emily L.—Kansas State University (Semipalmated Sandpiper)
Welch, Linda—US Fish and Wildlife Service (Arctic Tern; Common Tern; Great Shearwater)
Wellicome, Troy I.—University of Alberta (Ferruginous Hawk)
Weseloh, Chip—Canadian Wildlife Service (Great Egret)
WEST, Inc. (Sandhill Crane)
Westgate, Andrew J.—Grand Manan Whale and Seabird Research Station (Great Shearwater; Sooty Shearwater)
Whelan, Shannon—McGill University (Black-legged Kittiwake; Glaucous-winged Gull; Rhinoceros Auklet; Tufted Puffin)
Whitaker, Darroch—Parks Canada (Gray-cheeked Thrush)
Whitmer, Emily R.—Oiled Wildlife Care Network (Western Grebe)
Whitney, Jillian J.—Massachusetts Department of Conservation and Recreation (Great Black-backed Gull; Herring Gull; Ring-billed Gull)
Whittam, Rebecca M.—Canadian Wildlife Service (Bicknell's Thrush)
Wiebe, Karen L.—University of Saskatchewan (Northern Flicker; Snowy Owl)
Wiersma, Yolanda F.—Memorial University of Newfoundland (White-tailed Tropicbird)
Wikelski, Martin—Max Planck Institute of Animal Behavior (Magnificent Frigatebird; Peregrine Falcon; Rough-legged Hawk)
Wildlife Habitat Canada (Sandhill Crane)
Wiley, David—US National Oceanic and Atmospheric Administration (Great Shearwater)
Wilkinson, Bradley—Clemson University, South Carolina Cooperative Fish and Wildlife Research Unit at time of study (Brown Pelican)
Will, Alexis—University of Alaska Fairbanks (Common Murre; Thick-billed Murre)
Williams, Chris—University of Delaware (Snow Goose)
Williams, Emily J.— Georgetown University (American Robin); Denali National Park and Preserve at time of study (Blackpoll Warbler; Fox Sparrow; Gray-cheeked Thrush; Hermit Thrush; Long-tailed Jaeger: Swainson's Thrush; Wilson's Warbler)
Williams, Kathryn A.—Biodiversity Research Institute (Gray Catbird; Veery)
Williams, Kiah—Tulane University (Wilson's Plover)
Williams, Michael—Yukon Territorial Government (Snowy Owl)
Wilson, Jeremy—Royal Society for the Protection of Birds (Red-necked Phalarope)
Wilson, Laurie—Environment and Climate Change Canada (Ancient Murrelet)
Wilson, Matthew—Washington Department of Fish and Wildlife (Brant; Canada Goose; Harlequin Duck; Northern Pintail; Snow Goose; Surf Scoter; White-winged Scoter)
Wilson, Scott—Environment and Climate Change Canada (Horned Lark)
Wingfield, John C.—University of California, Davis (White-crowned Sparrow)
Winiarski, Jason M.—Boise State University (American Kestrel)
Winn, Brad—Manomet Conservation Sciences (Semipalmated Sandpiper; Whimbrel)
Witynski, Max—Cornell University (Yellow Warbler)
Wojczulanis-Jakubas, Katarzyna—University of Gdansk, SEATRACK (Dovekie)

Wong, Sarah N. P.—Grand Manan Whale and Seabird Research Station (Great Shearwater; Sooty Shearwater)
Wood, Petra B.—US Geological Survey, West Virginia Cooperative Fish and Wildlife Research Unit (Blue-winged Warbler; Golden-winged Warbler)
Wood, Ross—Bird Studies Canada (Red Knot; Semipalmated Sandpiper)
Woodard, Paul—Environment and Climate Change Canada (Black-bellied Plover; Long-tailed Jaeger; Parasitic Jaeger; Pomarine Jaeger; Sanderling)
Woodworth, Bradley K.—University of Guelph (Blackpoll Warbler)
Wright, James R.—School of Environment and Natural Resources, Ohio State University (Lesser Yellowlegs; Rusty Blackbird)
Wright, Leah Lynn—York University (Dunlin)
Wunder, Michael B.—University of Colorado Denver (Flammulated Owl; Mountain Plover)
Yamamoto, Takahashi—National Institute of Polar Research, Tachikawa, Tokyo, Japan (Red-legged Kittiwake)
Yanco, Scott W.—University of Colorado Denver (Flammulated Owl)
Yannic, Glenn—Université Grenoble Alpes (Ivory Gull)
Yates, Michael—Earthspan (American White Pelican; Common Loon; Ferruginous Hawk; Golden Eagle; Peregrine Falcon; White-faced Ibis)
Yeany, David—Pennsylvania Natural Heritage Program, Western Pennsylvania Conservancy (Evening Grosbeak; Ovenbird; Swainson's Thrush; Wood Thrush)
Yee, J. (Surf Scoter)
Yezerinac, Stephen (Dunlin; Semipalmated Sandpiper)
Yukon's Qikiqtaruk (Herschel Island) Territorial Park (Snowy Owl)
Zharikov, Y.—Parks Canada (Dunlin; Sande rling)
Ziccardi, Michael H.—Oiled Wildlife Care Network (Western Grebe)

INDEX

A

B

C

D

E

F

G

H

I

J

K

L

M

N

O

P

R

S

T

U

V

W